Solar System Data

Body	Mass (kg)	Mean Radius (m)	Period (s)	Distance from the Sun (m)
Mercury	3.18×10^{23}	2.43×10^6	7.60×10^6	5.79×10^{10}
Venus	4.88×10^{24}	6.06×10^6	1.94×10^7	1.08×10^{11}
Earth	5.98×10^{24}	6.37×10^6	3.156×10^7	1.496×10^{11}
Mars	6.42×10^{23}	3.37×10^6	5.94×10^7	2.28×10^{11}
Jupiter	1.90×10^{27}	6.99×10^7	3.74×10^8	7.78×10^{11}
Saturn	5.68×10^{26}	5.85×10^7	9.35×10^8	1.43×10^{12}
Uranus	8.68×10^{25}	2.33×10^7	2.64×10^9	2.87×10^{12}
Neptune	1.03×10^{26}	2.21×10^7	5.22×10^9	4.50×10^{12}
Pluto	$\approx 1.4 \times 10^{22}$	$\approx 1.5 \times 10^6$	7.82×10^9	5.91×10^{12}
Moon	7.36×10^{22}	1.74×10^6	—	—
Sun	1.991×10^{30}	6.96×10^8	—	—

Physical Data Often Used[a]

Average Earth–Moon distance	3.84×10^8 m
Average Earth–Sun distance	1.496×10^{11} m
Average radius of the Earth	6.37×10^6 m
Density of air (20°C and 1 atm)	1.20 kg/m^3
Density of water (20°C and 1 atm)	1.00×10^3 kg/m^3
Free-fall acceleration	9.80 m/s^2
Mass of the Earth	5.98×10^{24} kg
Mass of the Moon	7.36×10^{22} kg
Mass of the Sun	1.99×10^{30} kg
Standard atmospheric pressure	1.013×10^5 Pa

[a] These are the values of the constants as used in the text.

Some Prefixes for Powers of Ten

Power	Prefix	Abbreviation	Power	Prefix	Abbreviation
10^{-24}	yocto	y	10^1	deka	da
10^{-21}	zepto	z	10^2	hecto	h
10^{-18}	atto	a	10^3	kilo	k
10^{-15}	femto	f	10^6	mega	M
10^{-12}	pico	p	10^9	giga	G
10^{-9}	nano	n	10^{12}	tera	T
10^{-6}	micro	μ	10^{15}	peta	P
10^{-3}	milli	m	10^{18}	exa	E
10^{-2}	centi	c	10^{21}	zetta	Z
10^{-1}	deci	d	10^{24}	yotta	Y

Pedagogical Color Chart

Part 1 (Chapters 1–15) : Mechanics

Displacement and
position vectors

Linear (**v**) and angular (ω)
velocity vectors

Velocity component vectors

Force vectors (**F**)
Force component vectors

Acceleration vectors (**a**)
Acceleration component vectors

Linear (**p**) and
angular (**L**)
momentum vectors

Torque vectors (τ)

Linear or rotational
motion directions

Springs

Pulleys

Part 4 (Chapters 23–34) : Electricity and Magnetism

Electric fields

Magnetic fields

Positive charges

Negative charges

Resistors

Batteries and other
DC power supplies

Switches

Capacitors

Inductors (coils)

Voltmeters

Ammeters

AC Generators

Ground symbol

Part 5 (Chapters 35–38) : Light and Optics

Light rays

Lenses and prisms

Mirrors

Objects

Images

Open the door to the fascinating world of
physics

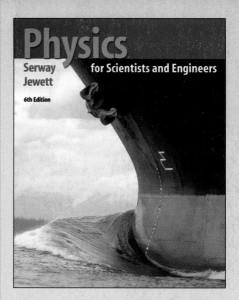

Physics for Scientists and Engineers
Serway
Jewett
6th Edition

Physics, the most fundamental of all natural sciences, will reveal to you the basic principles of the Universe. And while physics can seem challenging, its true beauty lies in the sheer simplicity of fundamental physical theories—theories and concepts that can alter and expand your view of the world around you. Other courses that follow will use the same principles, so it is important that you understand and are able to apply the various concepts and theories discussed in the text. **Physics for Scientists and Engineers, Sixth Edition** is your guide to this fascinating science.

Your quick start for studying smart

Achieve success in your physics course by making the most of what **Physics for Scientists and Engineers, Sixth Edition** has to offer you. From a host of in-text features to a range of Web resources, you'll have everything you need to understand the natural forces and principles of physics:

▶ **Dynamic built-in study aids.**
Throughout every chapter the authors have built in a wide range of examples, exercises, and illustrations that will help you understand and appreciate the laws of physics. *See pages 2 and 3 for more information.*

▶ **A powerful Web-based learning system**.
The text is fully integrated with **PhysicsNow**, an interactive learning system that tailors itself to your needs in the course. It's like having a personal tutor available whenever you need it! *See pages 4–7 to explore* **PhysicsNow**.

Your *Quick Start for Studying Smart* begins with this special tour through the book. On the following pages you'll discover how **Physics for Scientists and Engineers, Sixth Edition** and **PhysicsNow** not only enhance your experience in this course, but help you to succeed!

THOMSON ™ **BROOKS/COLE**

Quick Start for Studying Smart!

1

Everything you need to succeed in your course is available to you in **Physics for Scientists and Engineers, Sixth Edition.** Authors Serway and Jewett have filled their text with learning tools and study aids that will clarify concepts and help you build a solid base of knowledge. The end result: confidence in the classroom, in your study sessions, and in your exams.

THE RIGHT APPROACH

Start out right! Early on in the text the authors outline a general problem-solving strategy that will enable you to increase your accuracy in solving problems, enhance your understanding of physical concepts, eliminate initial worry or lack of direction in approaching a problem, and organize your work. The problem-solving strategy is integrated into the *Coached Problems* found on **PhysicsNow** to reinforce this key skill. (See pages 4–7 for more information about the **PhysicsNow** Web-based and student-centered learning system.)

PROBLEM-SOLVING HINTS

Problem-Solving Hints help you approach homework assignments with greater confidence. General strategies and suggestions are included for solving the types of problems featured in the worked examples, end-of-chapter problems, and **PhysicsNow**. This feature helps you identify the essential steps in solving problems and increases your skills as a problem solver.

WORKED EXAMPLES

Reinforce your understanding of essential problem-solving techniques using a large number of realistic *Worked Examples.* In many cases, these examples serve as models for solving the end-of-chapter problems. Numerous *Worked Examples* include specific references to the general problem-solving strategy to illustrate the underlying concepts and methodology used in arriving at a correct solution. This will help you understand the logic behind the solution and the advantage of using a particular approach to solve the problem. **PhysicsNow** also features a number of worked examples to further enhance your understanding of problem solving and to give you even more practice solving problems.

GENERAL PROBLEM-SOLVING STRATEGY

Conceptualize

- The first thing to do when approaching a problem is to *think about* and *understand* the situation. Study carefully any diagrams, graphs, tables, or photographs that accompany the problem. Imagine a movie, running in your mind, of what happens in the problem.
- If a diagram is not provided, you should almost always make a quick drawing of the situation. Indicate any known values, perhaps in a table or directly on your sketch.
- Now focus on what algebraic or numerical information is given in the problem. Carefully read the problem statement, looking for key phrases such as "starts from at rest" ($v_i = 0$), "stops" ($v_f = 0$), or "freely falls" ($a_y = -g = -9.80 \text{ m/s}^2$).
- Now focus on the expected result of solving the problem. Exactly what is the question asking? Will the final result be numerical or algebraic? Do you know what units to expect?
- Don't forget to incorporate information from your own experiences and common sense. What should a reasonable answer look like? You wouldn't expect to calculate the speed of an automobile to be

Analyze

- Now you must analyze the problem and strive for a mathematical solution. Because you have already categorized the problem, it should not be too difficult to select relevant equations that apply to the type of situation in the problem. For example, if the problem involves a particle moving under constant acceleration, Equations 2.9 to 2.13 are relevant.
- Use algebra (and calculus, if necessary) to solve symbolically for the unknown variable in terms of what is given. Substitute in the appropriate numbers, calculate the result, and round it to the proper number of significant figures.

Finalize

- This is the most important part. Examine your numerical answer. Does it have the correct units? Does it meet your expectations from your conceptualization of the problem? What about the algebraic form of the result — before you substituted numerical values? Does it make sense? Examine the variables in the problem to see whether the answer would change in a physically meaningful way if they were drastically increased or decreased or even became zero. Looking at limiting cases to see whether they yield expected values is a very useful way to make sure that you are obtaining reasonable results.

Think about how this problem compares with others you have done. How was it similar? In what critical ways did it differ? Why was this problem assigned? You should have learned something by doing it. Can you figure out what? If it is a new category of problem, be sure you understand it so that you can use it as a model for solving future problems in the same category.

When solving complex problems, you may need to identify a series of sub-problems and apply the problem-solv-

PROBLEM-SOLVING HINTS

Applying Newton's Laws

The following procedure is recommended when dealing with problems involving Newton's laws:

- Draw a simple, neat diagram of the system to help *conceptualize* the problem.
- *Categorize* the problem: if any acceleration component is zero, the particle is in equilibrium in this direction and $\Sigma F = 0$. If not, the particle is undergoing an acceleration, the problem is one of nonequilibrium in this direction, and $\Sigma F = ma$.
- *Analyze* the problem by isolating the object whose motion is being analyzed. Draw a free-body diagram for this object. For systems containing more than one object, draw *separate* free-body diagrams for each object. *Do not* include in the free-body diagram forces exerted by the object on its surroundings.
- Establish convenient coordinate axes for each object and find the components of the forces along these axes. Apply Newton's second law, $\Sigma F = ma$, in component form. Check your dimensions to make sure that all terms have units of force.
- Solve the comp[...] have as many i[...] complete soluti[...]
- *Finalize* by maki[...] Also check the [...] variables. By do[...]

Example 4.3 The Long Jump

A long-jumper (Fig. 4.12) leaves the ground at an angle of 20.0° above the horizontal and at a speed of 11.0 m/s.

(A) How far does he jump in the horizontal direction? (Assume his motion is equivalent to that of a particle.)

Solution We *conceptualize* the motion of the long-jumper as equivalent to that of a simple projectile such as the ball in Example 4.2, and *categorize* this problem as a projectile motion problem. Because the initial speed and launch angle are given, and because the final height is the same as the initial height, we further categorize this problem as satisfying the conditions for which Equations 4.13 and 4.14 can be used. This is the most direct way to *analyze* this problem, although generally we want to think about [...] describing [...] the gen[...]

provides a graphical representation of the flight of the long-jumper. As before, we set our origin of coordinates at the takeoff point and label the peak as Ⓐ and the landing point as Ⓑ. The horizontal motion is described by Equation 4.11:

$$x_f = x_B = (v_i \cos \theta_i) t_B = (11.0 \text{ m/s})(\cos 20.0°) t_B$$

The value of x_B can be found if the time of landing t_B is known. We can find t_B by remembering that $a_y = -g$ and by using the y part of Equation 4.8a. We also note that at the top of the jump the vertical component of velocity v_{yA} is zero:

$$v_{yf} = v_{yA} = v_i \sin \theta_i - g t_A$$

This is the time at which the long-jumper is at the *top* of the jump. Because of the symmetry of the vertical motion,

another 0.384 s passes before the jumper returns to the ground. Therefore, the time at which the jumper lands is $t_B = 2t_A = 0.768 \text{ s}$. Substituting this value into the above expression for x_f gives

$$x_f = x_B = (11.0 \text{ m/s})(\cos 20.0°)(0.768 \text{ s}) = \boxed{7.94 \text{ m}}$$

This is a reasonable distance for a world-class athlete.

(B) What is the maximum height reached?

Solution We find the maximum height reached by using Equation 4.12:

$$y_{max} = y_A = (v_i \sin \theta_i) t_A - \frac{1}{2} g t_A^2$$
$$= (11.0 \text{ m/s})(\sin 20.0°)(0.384 \text{ s})$$
$$- \frac{1}{2}(9.80 \text{ m/s}^2)(0.384 \text{ s})^2 = \boxed{0.722 \text{ m}}$$

To *finalize* this problem, find the answers to parts (a) and (b) using Equations 4.13 and 4.14. The results should agree. Treating the long-jumper as a particle is an oversimplification. Nevertheless, the values obtained are consistent with experience in sports. We learn that we can model a complicated system such as a long-jumper as a particle and still obtain results that are reasonable.

Figure 4.12 (Example 4.3) Mike Powell, current holder of the world long jump record of 8.95 m.

Mike Powell/Allsport/Getty Images

where k is a dimensionless constant of proportionality. Knowing the dimensions of a, r, and v, we see that the dimensional equation must be

$$\frac{L}{T^2} = L^n \left(\frac{L}{T}\right)^m = \frac{L^{n+m}}{T^m}$$

$n = -1$, and we can write the acceleration expression as

$$a = kr^{-1}v^2 = k\,\frac{v^2}{r}$$

When we discuss uniform circular motion later, we shall see that $k = 1$ if a consistent set of units is used. The constant k would not equal 1 if, for example, v were in km/h and you wanted a in m/s².

▲ **PITFALL PREVENTION**

1.5 Always Include Units

When performing calculations, include the units for every quantity and carry the units through the entire calculation. Avoid the temptation to drop the units early and then attach the expected units once you have an answer. By including the units in every step, you can detect errors if the units for the answer turn out to be incorrect.

1.5 Conversion of Units

Sometimes it is necessary to convert units from one measurement system to another, or to convert within a system, for example, from kilometers to meters. Equalities between SI and U.S. customary units of length are as follows:

$$1 \text{ mile} = 1\,609 \text{ m} = 1.609 \text{ km} \qquad 1 \text{ ft} = 0.304\,8 \text{ m} = 30.48 \text{ cm}$$

$$1 \text{ m} = 39.37 \text{ in.} = 3.281 \text{ ft} \qquad 1 \text{ in.} = 0.025\,4 \text{ m} = 2.54 \text{ cm (exactly)}$$

A more complete list of conversion factors can be found in Appendix A.

Units can be treated as algebraic quantities that can cancel each other. For example, suppose we wish to convert 15.0 in. to centimeters. Because 1 in. is defined as exactly 2.54 cm, we find that

---

Example 4.5 That's Quite an Arm!

A stone is thrown from the top of a building upward at an angle of 30.0° to the horizontal with an initial speed of 20.0 m/s, as shown in Figure 4.14. If the height of the building is 45.0 m,

(A) how long before the stone hits the ground?

Solution We *conceptualize* the problem by studying Figure 4.14, in which we have indicated the various parameters. By now, it should be natural to *categorize* this as a projectile motion problem.

To *analyze* the problem, let us once again separate motion into two components. The initial x and y components of the stone's velocity are

$$v_{xi} = v_i \cos \theta_i = (20.0 \text{ m/s})\cos 30.0° = 17.3 \text{ m/s}$$

$$v_{yi} = v_i \sin \theta_i = (20.0 \text{ m/s})\sin 30.0° = 10.0 \text{ m/s}$$

To find t, we can use $y_f = y_i + v_{yi}t + \frac{1}{2}a_y t^2$ (Eq. 4.9a) with $y_i = 0$, $y_f = -45.0$ m, $a_y = -g$, and $v_{yi} = 10.0$ m/s (there is a negative sign on the numerical value of y_f because we have chosen the top of the building as the origin):

$$-45.0 \text{ m} = (10.0 \text{ m/s})t - \frac{1}{2}(9.80 \text{ m/s}^2)t^2$$

Solving the quadratic equation for t gives, for the positive root, $t = 4.22$ s . To *finalize* this part, think: Does the negative root have any physical meaning?

(B) What is the speed of the stone just before it strikes the ground?

Solution We can use Equation 4.8a, $v_{yf} = v_{yi} + a_y t$, with $t = 4.22$ s to obtain the y component of the velocity just before the stone strikes the ground:

$$v_{yf} = 10.0 \text{ m/s} - (9.80 \text{ m/s}^2)(4.22 \text{ s}) = -31.4 \text{ m/s}$$

Because $v_{xf} = v_{xi} = 17.3$ m/s, the required speed is

$$v_f = \sqrt{v_{xf}^2 + v_{yf}^2} = \sqrt{(17.3)^2 + (-31.4)^2} \text{ m/s} = \boxed{35.9 \text{ m/s}}$$

What If? What if a horizontal wind is blowing in the same direction as the ball is thrown and it causes the ball to have a horizontal acceleration component $a_x = 0.500$ m/s². Which part of this example, (a) or (b), will have a different answer?

Answer Recall that the motions in the x and y directions are independent. Thus, the horizontal wind cannot affect the vertical motion. The vertical motion determines the time of the projectile in the air, so the answer to (a) does not change. The wind will cause the horizontal velocity component to increase with time, so that the final speed will change in part (b).

We can find the new final horizontal velocity component by using Equation 4.8a:

$$v_{xf} = v_{xi} + a_x t = 17.3 \text{ m/s} + (0.500 \text{ m/s}^2)(4.22 \text{ s})$$
$$= 19.4 \text{ m/s}$$

and the new final speed:

$$v_f = \sqrt{v_{xf}^2 + v_{yf}^2} = \sqrt{(19.4)^2 + (-31.4)^2} \text{ m/s} = 36.9 \text{ m/s}$$

Figure 4.14 (Example 4.5) A stone is thrown from the top of a building.

Interactive ... ample link at http://www.pse6.com.

Quick Quiz 5.2 An object experiences no acceleration. Which of the following *cannot* be true for the object? (a) A single force acts on the object. (b) No forces act on the object. (c) Forces act on the object, but the forces cancel.

Quick Quiz 5.3 An object experiences a net force and exhibits an acceleration in response. Which of the following statements is *always* true? (a) The object moves in the direction of the force. (b) The ... velocity. (c) The acceleration is in the same ... the object increases.

Quick Quiz 5.4 You push an object ... with a constant force for a time interval Δ... object. You repeat the experiment, but with ... terval is now required to reach the same fina... (e) Δt/4.

Answers to Quick Quizzes

5.1 (d). Choice (a) is true. Newton's first law tells us that motion requires no force: an object in motion continues to move at constant velocity in the absence of external forces. Choice (b) is also true. A stationary object can have several forces acting on it, but if the vector sum of all these external forces is zero, there is no net force and the object remains stationary.

5.2 (a). If a single force acts, this force constitutes the net force and there is an acceleration according to Newton's second law.

5.3 (c). Newton's second law relates only the force and the acceleration. Direction of motion is part of an object's *velocity*, and force determines the direction of acceleration, not that of velocity.

5.4 (d). With twice the force, the object will experience twice the acceleration. Because the force is constant, the acceler-

---

PITFALL PREVENTIONS

An easy way to make sure you avoid common mistakes—while studying *and* while taking exams! These helpful notes, found in the margins, allow you to correct misconceptions before applying them to new material.

WHAT IF?

What if you could get even more out of the text's worked examples? You can with this feature. The authors change some of the data or assumptions in about one-third of the *Worked Examples* and then explore the consequences. This allows you to apply concepts and problem-solving skills to new situations as well as test the final result to see if it's realistic.

QUICK QUIZZES

Test your understanding of the concepts before investigating the worked examples. Found throughout each chapter, these quizzes give you an opportunity to test your conceptual understanding while also making the material more interactive. Comprehensive answers are found at the end of each chapter. A supplementary set of *Quick Quizzes* is available on the text's Web site, accessible through **www.pse6.com**.

"You do not know anything until you have practiced."

R. P. Feynman, Nobel Laureate in Physics

Quick Start for Studying Smart!

INTERACTIVE LEARNING BEGINS WITH THIS BOOK

PhysicsNow directly links with **Physics for Scientists and Engineers, Sixth Edition. PhysicsNow** and this text were built in concert to enhance each other and provide you with a seamless, integrated learning system. As you work through the chapters you'll see notes that direct you to the media-enhanced activities on **PhysicsNow (www.pse6.com)**. This precise page-by-page integration means you'll spend less time flipping through pages and navigating Web sites for useful exercises. It is far easier to understand physics if you see it in action, and **PhysicsNow** enables you to *become* a part of the action!

Take a practice test for this chapter by clicking on the Practice Test link at http://www.pse6.com.

SUMMARY

Scalar quantities are those that have only magnitude and no associated direction. **Vector quantities** have both magnitude and direction and obey the laws of vector addition. The magnitude of a vector is *always* a positive number.

When two or more vectors are added together, all of them must have the same units and all of them must be the same type of quantity. We can add two vectors **A** and **B** graphically. In this method (Fig. 3.6), the resultant vector **R = A + B** runs from the tail of **A** to the tip of **B**.

A second method of adding vectors involves **components** of the vectors. The x component A_x of the vector **A** is equal to the projection of **A** along the x axis of a coordinate system, as shown in Figure 3.13, where $A_x = A\cos\theta$. The y component A_y of **A** is the projection of **A** along the y axis, where $A_y = A\sin\theta$. Be sure you can determine which trigonometric functions you should use in all situations, especially when θ is defined as something other than the counterclockwise angle from the positive x axis.

If a vector **A** has an x component A_x and a y component A_y, the vector can be expressed in unit–vector form as $\mathbf{A} = A_x\mathbf{i} + A_y\mathbf{j}$. In this notation, **i** is a unit vector pointing in the positive x direction, and **j** is a unit vector pointing in the positive y direction. Because **i** and **j** are unit vectors, $|\mathbf{i}| = |\mathbf{j}| = 1$.

We can find the resultant of two or more vectors by resolving all vectors into their x and y components, adding their resultant x and y components, and then using the Pythagorean theorem to find the magnitude of the resultant vector. We can find the angle that the resultant makes with respect to the x axis by using a suitable trigonometric function.

QUESTIONS

1. Two vectors have unequal magnitudes. Can their sum be zero? Explain.

2. Can the magnitude of a particle's displacement be greater

4. Which of the following are vectors and which are not: force, temperature, the volume of water in a can, the ratings of a TV show, the height of a building, the velocity of ... the Universe?

... y plane. For what orientations of **A** ... nts be negative? For what orienta- ... s have opposite signs?

Example 4.5 That's Quite an Arm!

A stone is thrown from the top of a building upward at an angle of 30.0° to the horizontal with an initial speed of 20.0 m/s, as shown in Figure 4.14. If the height of the building is 45.0 m,

(A) how long before the stone hits the ground?

Solution We *conceptualize* the problem by studying Figure 4.14, in which we have indicated the various parameters. By now, it should be natural to *categorize* this as a projectile motion problem.

To *analyze* the problem, let us once again separate motion into two components. The initial x and y components of the stone's velocity are

$$v_{xi} = v_i\cos\theta_i = (20.0 \text{ m/s})\cos 30.0° = 17.3 \text{ m/s}$$

$$v_{yi} = v_i\sin\theta_i = (20.0 \text{ m/s})\sin 30.0° = 10.0 \text{ m/s}$$

To find t, we can use $y_f = y_i + v_{yi}t + \frac{1}{2}a_yt^2$ (Eq. 4.9a) with $y_i = 0$, $y_f = -45.0$ m, $a_y = -g$, and $v_{yi} = 10.0$ m/s (there is a negative sign on the numerical value of y_f because we have chosen the top of the building as the origin):

$$-45.0 \text{ m} = (10.0 \text{ m/s})t - \frac{1}{2}(9.80 \text{ m/s}^2)t^2$$

Solving the quadratic equation for t gives, for the positive root, $t = \boxed{4.22 \text{ s}}$. To *finalize* this part, think: Does the negative root have any physical meaning?

(B) What is the speed of the stone just before it strikes the ground?

Solution We can use Equation 4.8a, $v_{yf} = v_{yi} + a_yt$, with $t = 4.22$ s to obtain the y component of the velocity just before the stone strikes the ground:

$$v_{yf} = 10.0 \text{ m/s} - (9.80 \text{ m/s}^2)(4.22 \text{ s}) = -31.4 \text{ m/s}$$

Because $v_{xf} = v_{xi} = 17.3$ m/s, the required speed is

$$v_f = \sqrt{v_{xf}^2 + v_{yf}^2} = \sqrt{(17.3)^2 + (-31.4)^2} \text{ m/s} = \boxed{35.9 \text{ m/s}}$$

To *finalize* this part, is it reasonable that the y component of the final velocity is negative? Is it reasonable that the final speed is larger than the initial speed of 20.0 m/s?

What If? What if a horizontal wind is blowing in the same direction as the ball is thrown and it causes the ball to have a horizontal acceleration component $a_x = 0.500$ m/s². Which part of this example, (a) or (b), will have a different answer?

Answer Recall that the motions in the x and y directions are independent. Thus, the horizontal wind cannot affect the vertical motion. The vertical motion determines the time of the projectile in the air, so the answer to (a) does not change. The wind will cause the horizontal velocity component to increase with time, so that the final speed will change in part (b).

We can find the new final horizontal velocity component by using Equation 4.8a:

$$v_{xf} = v_{xi} + a_xt = 17.3 \text{ m/s} + (0.500 \text{ m/s}^2)(4.22 \text{ s})$$
$$= 19.4 \text{ m/s}$$

and the new final speed:

$$v_f = \sqrt{v_{xf}^2 + v_{yf}^2} = \sqrt{(19.4)^2 + (-31.4)^2} \text{ m/s} = 36.9 \text{ m/s}$$

Figure 4.14 (Example 4.5) A stone is thrown from the top of a building.

Investigate this situation at the Interactive Worked Example link at http://www.pse6.com.

GO ONLINE AT www.pse6.com

Log on to **PhysicsNow** at **www.pse6.com** by using the free pincode packaged with this text.* You'll immediately notice the system's easy-to-use, browser-based format. Getting to where you need to go is as easy as a click of the mouse. The **PhysicsNow** system is made up of three interrelated parts:

- ► **How Much Do I Know?**
- ► **What Do I Need to Learn?**
- ► **What Have I Learned?**

These three interrelated elements work together, but are distinct enough to allow you the freedom to explore only those assets that meet your personal needs. You can use **PhysicsNow** like a traditional Web site, accessing all assets of a particular chapter and exploring on your own. The best way to maximize the system and *your* time is to start by taking the *Pre-Test*.

* Free PIN codes are only available with new copies of
  Physics for Scientists and Engineers, Sixth Edition.

HOW MUCH DO I KNOW?

The Pre-Test is the first step in creating your *Personalized Learning Plan*. Each *Pre-Test* is based on the end-of-chapter homework problems and includes approximately 15 questions.

Once you've completed the *Pre-Test* you'll be presented with a detailed *Learning Plan* that outlines the elements you need to review to master the chapter's most essential concepts.

At each stage, the text is referenced to reinforce its value as a learning tool.

Turn the page to view problems from a sample *Personalized Learning Plan.*

WHAT DO I NEED TO LEARN?

Once you've completed the *Pre-Test* you're ready to work the problems in your *Personalized Learning Plan*—problems that will help you master concepts essential to your success in this course.

More than 200 *Active Figures* are taken from the text and animated to help you visualize physics in action. Each figure is paired with a question to help you focus on physics at work, and a brief quiz ensures that you understand the concept played out in the animations.

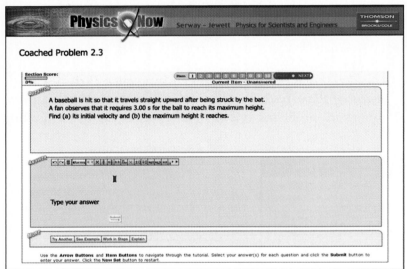

You'll continue to master the concepts though *Coached Problems*. These engaging problems reinforce the lessons in the text by taking a step-by-step approach to problem-solving methodology. Each *Coached Problem* gives you the option of working a question and receiving feedback, or seeing a solution worked for you. You'll find approximately five *Coached Problems* per chapter.

You'll strengthen your problem-solving and visualization skills by working through the *Interactive Examples*. Each step in the examples uses the authors' problem-solving methodology that is introduced in the text (see page 2 of this Visual Preface). You'll find *Interactive Examples* for each chapter of the text.

WHAT HAVE I LEARNED?

After working through the problems highlighted in your personal *Learning Plan* you'll move on to a *Chapter Quiz*. These multiple-choice quizzes present you with questions that are similar to those you might find in an exam. You can even e-mail your quiz results to your instructor.

Once you've completed the quiz you'll receive your results in the form of a percentage. If you need to improve your score, **PhysicsNow** will take you back through the system, beginning with *What Do I Know?*, and work with you as you continue to build your knowledge and skills and master concepts.

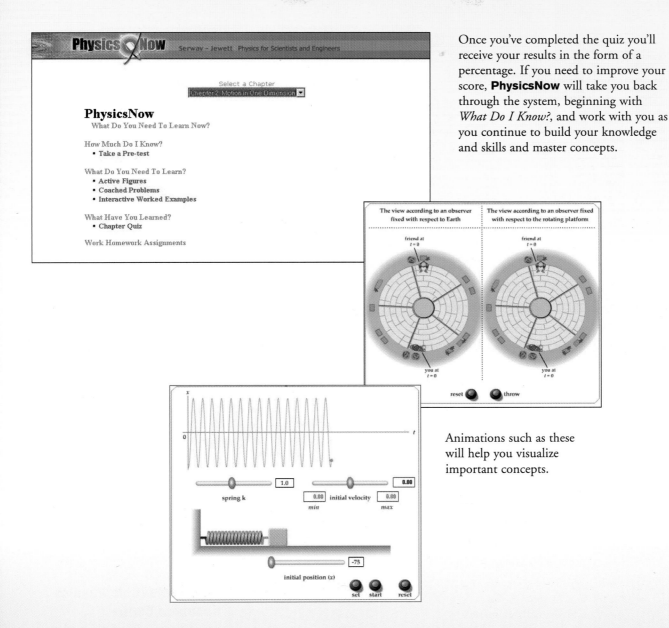

Animations such as these will help you visualize important concepts.

Chart your own course for success . . .

Log on to **www.pse6.com** to take advantage of **PhysicsNow!**

*We dedicate this book to the courageous astronauts
who died on the space shuttle* Columbia *on February 1, 2003.
The women and men of the international team lost their lives
not in a contest between countries or a struggle for
necessities but in advancing one of humankind's noblest
creations—science.*

6th Edition

PHYSICS

for Scientists and Engineers

Raymond A. Serway

John W. Jewett, Jr.

California State Polytechnic University–Pomona

THOMSON

BROOKS/COLE

Australia • Canada • Mexico • Singapore • Spain
United Kingdom • United States

THOMSON

BROOKS/COLE

Editor-in-Chief: Michelle Julet
Publisher: David Harris
Physics Editor: Chris Hall
Development Editor: Susan Dust Pashos
Assistant Editor: Rebecca Heider, Alyssa White
Editorial Assistant: Seth Dobrin
Technology Project Manager: Sam Subity
Marketing Manager: Kelley McAllister
Marketing Assistant: Sandra Perin
Advertising Project Manager: Stacey Purviance
Project Manager, Editorial Production: Teri Hyde
Print/Media Buyer: Barbara Britton
Permissions Editor: Joohee Lee

Production Service: Sparkpoint Communications, a division of J. B. Woolsey Associates
Text Designer: Lisa Devenish
Photo Researcher: Terri Wright
Copy Editor: Andrew Potter
Illustrator: Rolin Graphics
Cover Designer: Lisa Devenish
Cover Image: Water Displaced by Oil Tanker, © Stuart Westmorland/CORBIS
Compositor: Progressive Information Technologies
Cover Printer: Quebecor World, Versailles
Printer: Quebecor World, Versailles

Printed in the United States of America
3 4 5 6 7 07 06 05

For more information about our products, contact us at:
Thomson Learning Academic Resource Center
1-800-423-0563

For permission to use material from this text, contact us by:
Phone: 1-800-730-2214
Fax: 1-800-730-2215
Web: http://www.thomsonrights.com

Library of Congress Control Number: 2003104713

Student Edition: ISBN 0-534-40842-7

International Student Edition: ISBN 0-534-42398-1
(Not for sale in the United States)

Instructor's Edition: ISBN 0-534-40843-5

Brooks/Cole—Thomson Learning
10 Davis Drive
Belmont, CA 94002
USA

Asia
Thomson Learning
5 Shenton Way #01-01
UIC Building
Singapore 068808

Australia
Nelson Thomson Learning
102 Dodds Street
South Melbourne, Victoria 3205
Australia

Canada
Nelson Thomson Learning
1120 Birchmount Road
Toronto, Ontario M1K 5G4
Canada

Europe/Middle East/Africa
Thomson Learning
High Holborn House
50/51 Bedford Row
London WC1R 4LR
United Kingdom

Latin America
Thomson Learning
Seneca, 53
Colonia Polanco
11560 Mexico D.F.
Mexico

Spain
Paraninfo Thomson Learning
Calle/Magallanes, 25
28015 Madrid, Spain

Contents Overview

v

Table of Contents

elektraVision/Index Stock Imagery

Courtesy Tourism Malaysia

PART 2   **Oscillations and Mechanical Waves   451**

Don Bonsey/Getty Images

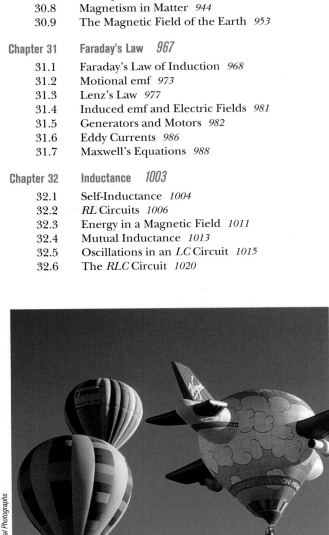

Richard Megna/Fundamental Photographs

About the Authors

Raymond A. Serway received his doctorate at Illinois Institute of Technology and is Professor Emeritus at James Madison University. Dr. Serway began his teaching career at Clarkson University, where he conducted research and taught from 1967 to 1980. His second academic appointment was at James Madison University as Professor of Physics and Head of the Physics Department from 1980 to 1986. He remained at James Madison University until his retirement in 1997. He was the recipient of the Madison Scholar Award at James Madison University in 1990, the Distinguished Teaching Award at Clarkson University in 1977, and the Alumni Achievement Award from Utica College in 1985. As Guest Scientist at the IBM Research Laboratory in Zurich, Switzerland, he worked with K. Alex Müller, 1987 Nobel Prize recipient. Dr. Serway also held research appointments at Rome Air Development Center from 1961 to 1963, at IIT Research Institute from 1963 to 1967, and as a visiting scientist at Argonne National Laboratory, where he collaborated with his mentor and friend, Sam Marshall. In addition to earlier editions of this textbook, Dr. Serway is the co-author of the high-school textbook *Physics* with Jerry Faughn, published by Holt, Rinehart, & Winston and co-author of the third edition of *Principles of Physics* with John Jewett, the sixth edition of *College Physics* with Jerry Faughn, and the second edition of *Modern Physics* with Clem Moses and Curt Moyer. In addition, Dr. Serway has published more than 40 research papers in the field of condensed matter physics and has given more than 70 presentations at professional meetings. Dr. Serway and his wife Elizabeth enjoy traveling, golfing, gardening, and spending quality time with their four children and five grandchildren.

John W. Jewett, Jr. earned his doctorate at Ohio State University, specializing in optical and magnetic properties of condensed matter. Dr. Jewett began his academic career at Richard Stockton College of New Jersey, where he taught from 1974 to 1984. He is currently Professor of Physics at California State Polytechnic University, Pomona. Throughout his teaching career, Dr. Jewett has been active in promoting science education. In addition to receiving four National Science Foundation grants, he helped found and direct the Southern California Area Modern Physics Institute (SCAMPI). He also directed Science IMPACT (Institute for Modern Pedagogy and Creative Teaching), which works with teachers and schools to develop effective science curricula. Dr. Jewett's honors include the Stockton Merit Award at Richard Stockton College, the Outstanding Professor Award at California State Polytechnic University for 1991–1992, and the Excellence in Undergraduate Physics Teaching Award from the American Association of Physics Teachers (AAPT) in 1998. He has given over 80 presentations at professional meetings, including presentations at international conferences in China and Japan. In addition to his work on this textbook, he is co-author of the third edition of *Principles of Physics* with Ray Serway and author of *The World of Physics . . . Mysteries, Magic, and Myth*. Dr. Jewett enjoys playing piano, traveling, and collecting antiques that can be used as demonstration apparatus in physics lectures, as well as spending time with his wife Lisa and their children and grandchildren.

Preface

In writing this sixth edition of *Physics for Scientists and Engineers,* we continue our ongoing efforts to improve the clarity of presentation and we again include new pedagogical features that help support the learning and teaching processes. Drawing on positive feedback from users of the fifth edition and reviewers' suggestions, we have refined the text in order to better meet the needs of students and teachers. We have for the first time integrated a powerful collection of media resources into many of the illustrations, examples, and end-of-chapter problems in the text. These resources compose the Web-based learning system *PhysicsNow* and are flagged by the media icon. Further details are described below.

This textbook is intended for a course in introductory physics for students majoring in science or engineering. The entire contents of the text in its extended version could be covered in a three-semester course, but it is possible to use the material in shorter sequences with the omission of selected chapters and sections. The mathematical background of the student taking this course should ideally include one semester of calculus. If that is not possible, the student should be enrolled in a concurrent course in introductory calculus.

Objectives

This introductory physics textbook has two main objectives: to provide the student with a clear and logical presentation of the basic concepts and principles of physics, and to strengthen an understanding of the concepts and principles through a broad range of interesting applications to the real world. To meet these objectives, we have placed emphasis on sound physical arguments and problem-solving methodology. At the same time, we have attempted to motivate the student through practical examples that demonstrate the role of physics in other disciplines, including engineering, chemistry, and medicine.

Changes in the Sixth Edition

A large number of changes and improvements have been made in preparing the sixth edition of this text. Some of the new features are based on our experiences and on current trends in science education. Other changes have been incorporated in response to comments and suggestions offered by users of the fifth edition and by reviewers of the manuscript. The following represent the major changes in the sixth edition:

Active Figures Many diagrams from the text have been animated to form **Active Figures,** part of the *PhysicsNow* integrated Web-based learning system. By visualizing phenomena and processes that cannot be fully represented on a static page, students greatly increase their conceptual understanding. **Active Figures** are identified with the media icon. An addition to the figure caption in blue type describes briefly the nature and contents of the animation.

Interactive Worked Examples Approximately 76 of the worked examples in the text have been identified as interactive, labeled with the media icon. As part of the *PhysicsNow* Web-based learning system, students can engage in an extension of the problem solved in the example. This often includes elements of both visualization and calculation, and may also involve prediction and intuition building. Often the interactivity is inspired by the **"What If?"** question we posed in the example text.

What If? Approximately one-third of the worked examples in the text contain this new feature. At the completion of the example solution, a **What If?** question offers a variation on the situation posed in the text of the example. For instance, this feature might explore the effects of changing the conditions of the situation, determine what happens when a quantity is taken to a particular limiting value, or question whether additional information can be determined about the problem situation. The answer to the question generally includes both a conceptual response and a mathematical response. This feature encourages students to think about the results of the example and assists in conceptual understanding of the principles. It also prepares students to encounter novel problems featured on exams. Some of the end-of-chapter problems also carry the **"What If?"** feature.

Quick Quizzes   The number of Quick Quiz questions in each chapter has been increased. Quick Quizzes provide students with opportunities to test their understanding of the physical concepts presented. The questions require students to make decisions on the basis of sound reasoning, and some of them have been written to help students overcome common misconceptions. Quick Quizzes have been cast in an objective format, including multiple choice, true–false, and ranking. Answers to all Quick Quiz questions are found at the end of each chapter. Additional Quick Quizzes that can be used in classroom teaching are available on the instructor's companion Web site. Many instructors choose to use such questions in a "peer instruction" teaching style, but they can be used in standard quiz format as well.

Pitfall Preventions   These new features are placed in the margins of the text and address common student misconceptions and situations in which students often follow unproductive paths. Over 200 Pitfall Preventions are provided to help students avoid common mistakes and misunderstandings.

General Problem-Solving Strategy   A general strategy to be followed by the student is outlined at the end of Chapter 2 and provides students with a structured process for solving problems. In Chapters 3 through 5, the strategy is employed explicitly in every example so that students learn how it is applied. In the remaining chapters, the strategy appears explicitly in one example per chapter so that students are encouraged throughout the course to follow the procedure.

Line-by-Line Revision   The entire text has been carefully edited to improve clarity of presentation and precision of language. We hope that the result is a book that is both accurate and enjoyable to read.

Problems   A substantial revision of the end-of-chapter problems was made in an effort to improve their variety and interest, while maintaining their clarity and quality. Approximately 17% of the problems (about 550) are new. All problems have been carefully edited. Solutions to approximately 20% of the end-of-chapter problems are included in the *Student Solutions Manual and Study Guide*. These problems are identified by boxes around their numbers. A smaller subset of solutions, identified by the media icon , are available on the World Wide Web (**http://www.pse6.com**) as coached solutions with hints. Targeted feedback is provided for students whose instructors adopt *Physics for Scientists and Engineers*, sixth edition. See the next section for a complete description of other features of the problem set.

Content Changes   The content and organization of the textbook is essentially the same as that of the fifth edition. An exception is that Chapter 13 (Oscillatory Motion) in the fifth edition has been moved to the Chapter 15 position in the sixth edition, in order to form a cohesive four-chapter Part 2 on oscillations and waves. Many sections in various chapters have been streamlined, deleted, or combined with other sections to allow for a more balanced presentation. The chapters on Modern Physics, Chapters 39–46, have been extensively rewritten to provide more up-to-date material as well as modern applications. A more detailed list of content changes can be found on the instructor's companion Web site.

Content

The material in this book covers fundamental topics in classical physics and provides an introduction to modern physics. The book is divided into six parts. Part 1 (Chapters 1 to 14) deals with the fundamentals of Newtonian mechanics and the physics of fluids, Part 2 (Chapters 15 to 18) covers oscillations, mechanical waves, and sound, Part 3 (Chapters 19 to 22) addresses heat and thermodynamics, Part 4 (Chapters 23 to 34) treats electricity and magnetism, Part 5 (Chapters 35 to 38) covers light and optics, and Part 6 (Chapters 39 to 46) deals with relativity and modern physics. Each part opener includes an overview of the subject matter covered in that part, as well as some historical perspectives.

Text Features

Most instructors would agree that the textbook selected for a course should be the student's primary guide for understanding and learning the subject matter. Furthermore, the textbook should be easily accessible and should be styled and written to facilitate instruction and learning. With these points in mind, we have included many pedagogical features in the textbook that are intended to enhance its usefulness to both students and instructors. These features are as follows:

Style   To facilitate rapid comprehension, we have attempted to write the book in a style that is clear, logical, and engaging. We have chosen a writing style that is somewhat informal and relaxed so that students will find the text appealing and enjoyable to read. New terms are carefully defined, and we have avoided the use of jargon.

Previews   All chapters begin with a brief preview that includes a discussion of the chapter's objectives and content.

Important Statements and Equations   Most important statements and definitions are set in **boldface** type or are highlighted with a background screen for added emphasis and ease of review. Similarly, important equations are highlighted with a background screen to facilitate location.

Bruce Ayers/Getty Images

Problem-Solving Hints   In several chapters, we have included general strategies for solving the types of problems featured both in the examples and in the end-of-chapter problems. This feature helps students to identify necessary steps in problem solving and to eliminate any uncertainty they might have. Problem-solving strategies are highlighted with a light red background screen for emphasis and ease of location.

Marginal Notes   Comments and notes appearing in blue type in the margin can be used to locate important statements, equations, and concepts in the text.

Pedagogical Use of Color   Readers should consult the **pedagogical color chart** (second page inside the front cover) for a listing of the color-coded symbols used in the text diagrams, Web-based **Active Figures,** and diagrams within **Interactive Worked Examples.** This system is followed consistently whenever possible, with slight variations made necessary by the complexity of physical situations depicted in Part 4.

Mathematical Level   We have introduced calculus gradually, keeping in mind that students often take introductory courses in calculus and physics concurrently. Most steps are shown when basic equations are developed, and reference is often made to mathematical appendices at the end of the textbook. Vector products are introduced later in the text, where they are needed in physical applications. The dot product is introduced in Chapter 7, which addresses energy and energy transfer; the cross product is introduced in Chapter 11, which deals with angular momentum.

Worked Examples   A large number of worked examples of varying difficulty are presented to promote students' understanding of concepts. In many cases, the examples serve as models for solving the end-of-chapter problems. Because of the increased emphasis on understanding physical concepts, many examples are conceptual in nature and are labeled as such. The examples are set off in boxes, and the answers to examples with numerical solutions are highlighted with a background screen. We have already mentioned that a number of examples are designated as interactive and are part of the *PhysicsNow* Web-based learning system.

Questions   Questions of a conceptual nature requiring verbal or written responses are provided at the end of each chapter. Over 1 000 questions are included in this edition. Some questions provide the student with a means of self-testing the concepts presented in the chapter. Others could serve as a basis for initiating classroom discussions. Answers to selected questions are included in the *Student Solutions Manual and Study Guide*, and answers to all questions are found in the *Instructor's Solutions Manual*.

Significant Figures   Significant figures in both worked examples and end-of-chapter problems have been handled with care. Most numerical examples are worked out to either two or three significant figures, depending on the precision of the data provided. End-of-chapter problems regularly state data and answers to three-digit precision.

Problems   An extensive set of problems is included at the end of each chapter; in all, over 3 000 problems are given throughout the text. Answers to odd-numbered problems are provided at the end of the book in a section whose pages have colored edges for ease of location. For the convenience of both the student and the instructor, about two thirds of the problems are keyed to specific sections of the chapter. The remaining problems, labeled "Additional Problems," are not keyed to specific sections.

Usually, the problems within a given section are presented so that the straightforward problems (those with black problem numbers) appear first. For ease of identification, the numbers of intermediate-level problems are printed in blue, and those of challenging problems are printed in magenta.

- **Review Problems**   Many chapters include review problems requiring the student to combine concepts covered in the chapter with those discussed in previous chapters. These problems reflect the cohesive nature of the principles in the text and verify that physics is not a scattered set of ideas. When facing real-world issues such as global warming or nuclear weapons, it may be necessary to call on ideas in physics from several parts of a textbook such as this one.

- **Paired Problems**   To allow focused practice in solving problems stated in symbolic terms, some end-of-chapter numerical problems are paired with the same problems in symbolic form. Paired problems are identified by a common light red background screen.

- **Computer- and Calculator-Based Problems**   Many chapters include one or more problems whose solution requires the use of a computer or graphing calculator. Computer modeling of physical phenomena enables students to obtain graphical representations of variables and to perform numerical analyses.

- **Coached Problems with Hints**   These have been described above as part of the *PhysicsNow* Web-based learning system. These problems are identified by the media icon and targeted feedback is provided to students of instructors adopting the sixth edition.

Units   The international system of units (SI) is used throughout the text. The U.S. customary system of units is used only to a limited extent in the chapters on mechanics, heat, and thermodynamics.

Summaries   Each chapter contains a summary that reviews the important concepts and equations discussed in that chapter. A marginal note in blue type next to each chapter summary directs students to a practice test (Post-Test) for the chapter.

Appendices and Endpapers   Several appendices are provided at the end of the textbook. Most of the appendix material represents a review of mathematical concepts and techniques used in the text, including scientific notation, algebra, geometry, trigonometry, differential calculus, and integral calculus. Reference to these appendices is made

Courtesy NASA

throughout the text. Most mathematical review sections in the appendices include worked examples and exercises with answers. In addition to the mathematical reviews, the appendices contain tables of physical data, conversion factors, atomic masses, and the SI units of physical quantities, as well as a periodic table of the elements. Other useful information, including fundamental constants and physical data, planetary data, a list of standard prefixes, mathematical symbols, the Greek alphabet, and standard abbreviations of units of measure, appears on the endpapers.

Student Ancillaries

Student Solutions Manual and Study Guide by John R. Gordon, Ralph McGrew, and Raymond Serway. This two-volume manual features detailed solutions to 20% of the end-of-chapter problems from the text. The manual also features a list of important equations, concepts, and notes from key sections of the text, in addition to answers to selected end-of-chapter questions. Volume 1 contains Chapters 1 through 22 and Volume 2 contains Chapters 23 through 46.

WebTutor™ on WebCT and Blackboard **WebTutor** offers students real-time access to a full array of study tools, including chapter outlines, summaries, learning objectives, glossary flashcards (with audio), practice quizzes, **InfoTrac® College Edition** exercises, and Web links.

InfoTrac® College Edition Adopters and their students automatically receive a four-month subscription to **InfoTrac® College Edition** with every new copy of this book. Newly improved, this extensive online library opens the door to the full text (not just abstracts) of countless articles from thousands of publications including *American Scientist, Physical Review, Science, Science Weekly,* and more! Available only to college and university students. Journals subject to change.

The Brooks/Cole Physics Resource Center You will find additional online quizzes, Web links and animations at **http://physics.brookscole.com.**

Ancillaries for Instructors

The first four ancillaries below are available to qualified adopters. Please consult your local sales representative for details.

Instructor's Solutions Manual by Ralph McGrew and James A. Currie. This two-volume manual contains complete worked solutions to all of the end-of-chapter problems in the textbook as well as answers to even-numbered problems. The solutions to problems new to the sixth edition are marked for easy identification by the instructor. New to this edition are complete answers to the conceptual questions in the main text. Volume 1 contains Chapters 1 through 22 and Volume 2 contains Chapters 23 through 46.

Printed Test Bank by Edward Adelson. This two-volume test bank contains approximately 2 300 multiple-choice questions. These questions are also available in electronic format with complete answers and solutions in the Brooks/Cole Assessment test program. Volume 1 contains Chapters 1 through 22 and Volume 2 contains Chapters 23 through 46.

Multimedia Manager This easy-to-use multimedia lecture tool allows you to quickly assemble art and database files with notes to create fluid lectures. The CD-ROM set (Volume 1, Chapters 1–22; Volume 2, Chapters 23–46) includes a database of animations, video clips, and digital art from the text as well as electronic files of the *Instructor's Solutions Manual and Test Bank.* The simple interface makes it easy for you to incorporate graphics, digital video, animations, and audio clips into your lectures.

Transparency Acetates Each volume contains approximately 100 acetates featuring art from the text. Volume 1 contains Chapters 1 through 22 and Volume 2 contains Chapters 23 through 46.

Brooks/Cole Assessment   With a balance of efficiency, high performance, simplicity and versatility, **Brooks/Cole Assessment (BCA)** gives you the power to transform the learning and teaching experience. **BCA** is fully integrated testing, tutorial, and course management software accessible by instructors and students anytime, anywhere. Delivered for FREE in a browser-based format without the need for any proprietary software or plug-ins, **BCA** uses correct scientific notation to provide the drill of basic skills that students need, enabling the instructor to focus more time in higher-level learning activities (i.e., concepts and applications). Students can have unlimited practice in questions and problems, building their own confidence and skills. Results flow automatically to a grade book for tracking so that instructors will be better able to assess student understanding of the material, even prior to class or an actual test.

WebTutor™ on WebCT and Blackboard   With **WebTutor's** text-specific, preformatted content and total flexibility, instructors can easily create and manage their own personal Web site. **WebTutor's** course management tool gives instructors the ability to provide virtual office hours, post syllabi, set up threaded discussions, track student progress with the quizzing material, and much more. **WebTutor** also provides robust communication tools, such as a course calendar, asynchronous discussion, real-time chat, a whiteboard, and an integrated e-mail system.

Additional Options for Online Homework   For detailed information and demonstrations, contact your Thomson•Brooks/Cole representative or visit the following:

George Sample

*   WebAssign: A Web-based Homework System
    http://www.webassign.net or contact WebAssign at *webassign@ncsu.edu*

*   Homework Service
    http://hw.ph.utexas.edu/hw.html or contact *moore@physics.utexas.edu*

*   CAPA: A Computer-Assisted Personalized Approach
    http://capa4.lite.msu.edu/homepage/

Instructor's Companion Web Site   Consult the instructor's site at *http://www.pse6.com* for additional Quick Quiz questions, a detailed list of content changes since the fifth edition, a problem correlation guide, images from the text, and sample PowerPoint lectures. Instructors adopting the sixth edition of *Physics for Scientists and Engineers* may download these materials after securing the appropriate password from their local Thomson•Brooks/Cole sales representative.

Teaching Options

The topics in this textbook are presented in the following sequence: classical mechanics, oscillations and mechanical waves, and heat and thermodynamics followed by electricity and magnetism, electromagnetic waves, optics, relativity, and modern physics. This presentation represents a traditional sequence, with the subject of mechanical waves being presented before electricity and magnetism. Some instructors may prefer to cover this material after completing electricity and magnetism (i.e., after Chapter 34). The chapter on relativity is placed near the end of the text because this topic often is treated as an introduction to the era of "modern physics." If time permits, instructors may choose to cover Chapter 39 after completing Chapter 13, as it concludes the material on Newtonian mechanics.

For those instructors teaching a two-semester sequence, some sections and chapters could be deleted without any loss of continuity. The following sections can be considered optional for this purpose:

2.7	Kinematic Equations Derived from Calculus	**6.4**	Motion in the Presence of Resistive Forces
4.6	Relative Velocity and Relative Acceleration	**6.5**	Numerical Modeling in Particle Dynamics
6.3	Motion in Accelerated Frames	**7.9**	Energy and the Automobile

Topham Picturepoint/The Image Works

Acknowledgments

The sixth edition of this textbook was prepared with the guidance and assistance of many professors who reviewed selections of the manuscript, the pre-revision text, or both. We wish to acknowledge the following scholars and express our sincere appreciation for their suggestions, criticisms, and encouragement:

Edward Adelson, *Ohio State University*

Michael R. Cohen, *Shippensburg University*

Jerry D. Cook, *Eastern Kentucky University*

J. William Dawicke, *Milwaukee School of Engineering*

N. John DiNardo, *Drexel University*

Andrew Duffy, *Boston University*

Robert J. Endorf, *University of Cincinnati*

F. Paul Esposito, *University of Cincinnati*

Joe L. Ferguson, *Mississippi State University*

Perry Ganas, *California State University, Los Angeles*

John C. Hardy, *Texas A&M University*

Michael Hayes, *University of Pretoria (South Africa)*

John T. Ho, *The State University of New York, Buffalo*

Joseph W. Howard, *Salisbury University*

Robert Hunt, *Johnson County Community College*

Walter S. Jaronski, *Radford University*

Sangyong Jeon, *McGill University, Quebec*

Stan Jones, *University of Alabama*

L. R. Jordan, *Palm Beach Community College*

Teruki Kamon, *Texas A & M University*

Louis E. Keiner, *Coastal Carolina University*

Mario Klarić, *Midlands Technical College*

Laird Kramer, *Florida International University*

Edwin H. Lo, *American University*

James G. McLean, *The State University of New York, Geneseo*

Richard E. Miers, *Indiana University–Purdue University, Fort Wayne*

Oscar Romulo Ochoa, *The College of New Jersey*

Paul S. Ormsby, *Moraine Valley Community College*

Didarul I. Qadir, *Central Michigan University*

Judith D. Redling, *New Jersey Institute of Technology*

Richard W. Robinett, *Pennsylvania State University*

Om P. Rustgi, *SUNY College at Buffalo*

Mesgun Sebhatu, *Winthrop University*

Natalia Semushkina, *Shippensburg University*

Daniel Stump, *Michigan State University*

Uwe C. Täuber, *Virginia Polytechnic Institute*

Perry A. Tompkins, *Samford University*

Doug Welch, *McMaster University, Ontario*

Augden Windelborn, *Northern Illinois University*

Jerzy M. Wrobel, *University of Missouri, Kansas City*

Jianshi Wu, *Fayetteville State University*

Michael Zincani, *University of Dallas*

Frank Oberle/Getty Images

This title was carefully checked for accuracy by Michael Kotlarchyk *(Rochester Institute of Technology)*, Chris Vuille *(Embry-Riddle Aeronautical University)*, Laurencin Dunbar *(St. Louis Community College)*, William Dawicke *(Milwaukee School of Engineering)*, Ioan Kosztin *(University of Missouri)*, Tom Barrett *(Ohio State University)*, Z. M. Stadnik *(University of Ottawa)*, Ronald E. Jodoin *(Rochester Institute of Technology)*, Brian A. Raue *(Florida International University)*, Peter Moeck *(Portland State University)*, and Grant Hart *(Brigham Young University)*. We thank them for their diligent efforts under schedule pressure!

We are grateful to Ralph McGrew for organizing the end-of-chapter problems, writing many new problems, and his excellent suggestions for improving the content of the textbook. Problems new to this edition were written by Edward Adelson, Ronald Bieniek, Michael Browne, Andrew Duffy, Robert Forsythe, Perry Ganas, Michael Hones, John Jewett, Boris Korsunsky, Edwin Lo, Ralph McGrew, Raymond Serway, and Jerzy Wrobel, with the help of Bennett Simpson and JoAnne Maniago. Students Alexander Coto, Karl Payne, and Eric Peterman made corrections to problems taken from previous editions, as did teachers David Aspnes, Robert Beichner, Joseph Biegen, Tom Devlin, Vasili Haralambous, Frank Hayes, Erika Hermon, Ken Menningen, Henry Nebel, and Charles Teague. We are grateful to authors John R. Gordon and Ralph McGrew and compositor Michael Rudmin for preparing the *Student Solutions Manual and Study Guide*. Authors Ralph McGrew and James Currie and compositor Mary Toscano have prepared an excellent *Instructor's Solutions Manual*, and we thank them. Edward Adelson has carefully edited and improved the Test Bank for the sixth edition. Kurt Vandervoort prepared extra Quick Quiz questions for the instructor's companion Web site.

Special thanks and recognition go to the professional staff at the Brooks/Cole Publishing Company—in particular Susan Pashos, Rebecca Heider and Alyssa White (who managed the ancillary program and so much more), Jessica Howard, Seth Dobrin, Peter McGahey, Teri Hyde, Michelle Julet, David Harris, and Chris Hall—for their fine work during the development and production of this textbook. We are most appreciative of Sam Subity's masterful management of the *PhysicsNow* media program. Kelley McAllister is our energetic Marketing Manager, and Stacey Purviance coordinates our marketing communications. We recognize the skilled production service provided by the staff at Sparkpoint Communications, the excellent artwork produced by Rolin Graphics, and the dedicated photo research efforts of Terri Wright.

Finally, we are deeply indebted to our wives and children for their love, support, and long-term sacrifices.

Raymond A. Serway
Leesburg, Virginia

John W. Jewett, Jr.
Pomona, California

To the Student

It is appropriate to offer some words of advice that should be of benefit to you, the student. Before doing so, we assume that you have read the Preface, which describes the various features of the text that will help you through the course.

How to Study

Very often instructors are asked, "How should I study physics and prepare for examinations?" There is no simple answer to this question, but we would like to offer some suggestions that are based on our own experiences in learning and teaching over the years.

First and foremost, maintain a positive attitude toward the subject matter, keeping in mind that physics is the most fundamental of all natural sciences. Other science courses that follow will use the same physical principles, so it is important that you understand and are able to apply the various concepts and theories discussed in the text.

Concepts and Principles

It is essential that you understand the basic concepts and principles before attempting to solve assigned problems. You can best accomplish this goal by carefully reading the textbook before you attend your lecture on the covered material. When reading the text, you should jot down those points that are not clear to you. We've purposely left wide margins in the text to give you space for making notes. Also be sure to make a diligent attempt at answering the questions in the Quick Quizzes as you come to them in your reading. We have worked hard to prepare questions that help you judge for yourself how well you understand the material. Study carefully the **What If?** features that appear with many of the worked examples. These will help you to extend your understanding beyond the simple act of arriving at a numerical result. The Pitfall Preventions will also help guide you away from common misunderstandings about physics. During class, take careful notes and ask questions about those ideas that are unclear to you. Keep in mind that few people are able to absorb the full meaning of scientific material after only one reading. Several readings of the text and your notes may be necessary. Your lectures and laboratory work supplement reading of the textbook and should clarify some of the more difficult material. You should minimize your memorization of material. Successful memorization of passages from the text, equations, and derivations does not necessarily indicate that you understand the material. Your understanding of the material will be enhanced through a combination of efficient study habits, discussions with other students and with instructors, and your ability to solve the problems presented in the textbook. Ask questions whenever you feel clarification of a concept is necessary.

Study Schedule

It is important that you set up a regular study schedule, preferably a daily one. Make sure that you read the syllabus for the course and adhere to the schedule set by your instructor. The lectures will make much more sense if you read the corresponding text material before attending them. As a general rule, you should devote about two hours of study time for every hour you are in class. If you are having trouble with the course, seek the advice of the instructor or other students who have taken the course. You may find it necessary to seek further instruction from experienced students. Very often, instructors offer review sessions in addition to regular class periods. It is important that you avoid the practice of delaying study until a day or two before an exam. More often than not, this approach has disastrous results. Rather than undertake an all-night study session, briefly review the basic concepts and equations, and get a good night's rest. If you feel you need additional help in understanding the concepts, in preparing for exams, or in problem solving, we suggest that you acquire a copy of the *Student Solutions Manual and Study Guide* that accompanies this textbook; this manual should be available at your college bookstore.

Use the Features

You should make full use of the various features of the text discussed in the Preface. For example, marginal notes are useful for locating and describing important equations and concepts, and **boldfaced** type indicates important statements and definitions. Many useful tables are contained in the Appendices, but most are incorporated in the text where they are most often referenced. Appendix B is a convenient review of mathematical techniques.

George Semple

Answers to odd-numbered problems are given at the end of the textbook, answers to Quick Quizzes are located at the end of each chapter, and answers to selected end-of-chapter questions are provided in the *Student Solutions Manual and Study Guide*. Problem-Solving Strategies and Hints are included in selected chapters throughout the text and give you additional information about how you should solve problems. The Table of Contents provides an overview of the entire text, while the Index enables you to locate specific material quickly. Footnotes sometimes are used to supplement the text or to cite other references on the subject discussed.

After reading a chapter, you should be able to define any new quantities introduced in that chapter and to discuss the principles and assumptions that were used to arrive at certain key relations. The chapter summaries and the review sections of the *Student Solutions Manual and Study Guide* should help you in this regard. In some cases, it may be necessary for you to refer to the index of the text to locate certain topics. You should be able to associate with each physical quantity the correct symbol used to represent that quantity and the unit in which the quantity is specified. Furthermore, you should be able to express each important equation in a concise and accurate prose statement.

Problem Solving

R. P. Feynman, Nobel laureate in physics, once said, "You do not know anything until you have practiced." In keeping with this statement, we strongly advise that you develop the skills necessary to solve a wide range of problems. Your ability to solve problems will be one of the main tests of your knowledge of physics, and therefore you should try to solve as many problems as possible. It is essential that you understand basic concepts and principles before attempting to solve problems. It is good practice to try to find alternate solutions to the same problem. For example, you can solve problems in mechanics using Newton's laws, but very often an alternative method that draws on energy considerations is more direct. You should not deceive yourself into thinking that you understand a problem merely because you have seen it solved in class. You must be able to solve the problem and similar problems on your own.

The approach to solving problems should be carefully planned. A systematic plan is especially important when a problem involves several concepts. First, read the problem several times until you are confident you understand what is being asked. Look for any key words that will help you interpret the problem and perhaps allow you to make certain assumptions. Your ability to interpret a question properly is an integral part of problem solving. Second, you should acquire the habit of writing down the information given in a problem and those quantities that need to be found; for example, you might construct a table listing both the quantities given and the quantities to be found. This procedure is sometimes used in the worked examples of the textbook. Finally, after you have decided on the method you feel is appropriate for a given problem, proceed with your solution. Specific problem-solving strategies (Hints) of this type are included in the text and are highlighted with a light red screen. We have also developed a General Problem-Solving Strategy to help guide you through complex problems. If you follow the steps of this procedure (*Conceptualize, Categorize, Analyze, Finalize*), you will not only find it easier to come up with a solution, but you will also gain more from your efforts. This Strategy is located at the end of Chapter 2 (page 47) and is used in all worked examples in Chapters 3 through 5 so that you can learn how to apply it. In the remaining chapters, the Strategy is used in one example per chapter as a reminder of its usefulness.

Often, students fail to recognize the limitations of certain equations or physical laws in a particular situation. It is very important that you understand and remember the assumptions that underlie a particular theory or formalism. For example, certain equations in kinematics apply only to a particle moving with constant acceleration. These equations are not valid for describing motion whose acceleration is not constant, such as the motion of an object connected to a spring or the motion of an object through a fluid.

Experiments

© Phil Degginger/Stone/Getty

Physics is a science based on experimental observations. In view of this fact, we recommend that you try to supplement the text by performing various types of "hands-on" experiments, either at home or in the laboratory. These can be used to test ideas and models discussed in class or in the textbook. For example, the common Slinky™ toy is excellent for studying traveling waves; a ball swinging on the end of a long string can be used to investigate pendulum motion; various masses attached to the end of a vertical spring or rubber band can be used to determine their elastic nature; an old pair of Polaroid sunglasses and some discarded lenses and a magnifying glass are the components of various experiments in optics; and an approximate measure of the free-fall acceleration can be determined simply by measuring with a stopwatch the time it takes for a ball to drop from a known height. The list of such experiments is endless. When physical models are not available, be imaginative and try to develop models of your own.

New Media

We strongly encourage you to use the ***PhysicsNow*** Web-based learning system that accompanies this textbook. It is far easier to understand physics if you see it in action, and these new materials will enable you to become a part of that action. ***PhysicsNow*** media described in the Preface are accessed at the URL *http://www.pse6.com,* and feature a three-step learning process consisting of a Pre-Test, a personalized learning plan, and a Post-Test.

In addition to other elements, ***PhysicsNow*** includes the following Active Figures and Interactive Worked Examples:

Chapter 2
Active Figures 2.1, 2.3, 2.9, 2.10, 2.11, and 2.13
Examples 2.8 and 2.12

Chapter 3
Active Figures 3.2, 3.3, 3.6, and 3.16
Example 3.5

Chapter 4
Active Figures 4.5, 4.7, and 4.11
Examples 4.4, 4.5, and 4.18

Chapter 5
Active Figure 5.16
Examples 5.9, 5.10, 5.12, and 5.14

Chapter 6
Active Figures 6.2, 6.8, 6.12, and 6.15
Examples 6.4, 6.5, and 6.7

Chapter 7
Active Figure 7.10
Examples 7.9 and 7.11

Chapter 8
Active Figures 8.3, 8.4, and 8.16
Examples 8.2 and 8.4

Chapter 9
Active Figures 9.8, 9.9, 9.13, 9.16, and 9.17
Examples 9.1, 9.5, and 9.8

Chapter 10
Active Figures 10.4, 10.14, and 10.30
Examples 10.12, 10.13, and 10.14

Chapter 11
Active Figures 11.1, 11.3, and 11.4
Examples 11.6 and 11.10

Chapter 12
Active Figures 12.14, 12.16, 12.17, and 12.18
Examples 12.3 and 12.4

Chapter 13
Active Figures 13.1, 13.5, and 13.7
Example 13.1

Chapter 14
Active Figures 14.9 and 14.10
Examples 14.2 and 14.10

Chapter 15
Active Figures 15.1, 15.2, 15.7, 15.9, 15.10,
15.11, 15.14, 15.17, and 15.22

Chapter 16
Active Figures 16.4, 16.7, 16.8, 16.10, 16.14,
16.15, and 16.17
Example 16.5

Chapter 17
Active Figures 17.2, 17.8, and 17.9
Examples 17.1 and 17.6

Chapter 18
Active Figures 18.1, 18.2, 18.4, 18.9, 18.10,
18.22, and 18.25
Examples 18.4 and 18.5

Chapter 19
Active Figures 19.8 and 19.12
Example 19.7

Chapter 20
Active Figures 20.4, 20.5, and 20.7
Example 20.10

Chapter 21
Active Figures 21.2, 21.4, 21.11, and 21.12
Example 21.4

Chapter 22
Active Figures 22.2, 22.5, 22.10, 22.11, 22.12,
22.13, and 22.19
Example 22.10

Chapter 23
Active Figures 23.7, 23.13, 23.24, and 23.26
Examples 23.3 and 23.11

Chapter 24
Active Figures 24.4 and 24.9
Examples 24.5 and 24.10

Chapter 25
Active Figures 25.10 and 25.27
Examples 25.2 and 25.3

Chapter 26
Active Figures 26.4, 26.9, and 26.10
Examples 26.4 and 26.5

Chapter 27
Active Figures 27.9 and 27.13
Examples 27.3 and 27.8

Chapter 28
Active Figures 28.2, 28.4, 28.6, 28.19, 28.21,
28.27, and 28.29
Examples 28.1, 28.5, and 28.9

Chapter 29
Active Figures 29.1, 29.14, 29.18, 29.19, 29.23,
and 29.24
Example 29.7

Chapter 30
Active Figures 30.8, 30.9, and 30.21
Examples 30.1, 30.3, and 30.8

Chapter 31
Active Figures 31.1, 31.2, 31.10, 31.21, 31.23,
and 31.26
Examples 31.4 and 31.5

Chapter 32
Active Figures 32.3, 32.4, 32.6, 32.7, 32.17,
32.18, 32.21, and 32.23
Examples 32.3 and 32.7

Chapter 33
Active Figures 33.2, 33.3, 33.6, 33.7, 33.9,
33.10, 33.13, 33.15, 33.19, 33.25, and 33.26
Examples 33.5 and 33.7

Chapter 34
Active Figure 34.3
Examples 34.1 and 34.4

Chapter 35
Active Figures 35.4, 35.6, 35.10, 35.11, 35.23,
and 35.26
Examples 35.2 and 35.6

Chapter 36
Active Figures 36.2, 36.15, 36.20, 36.28, 36.44,
and 36.45
Examples 36.4, 36.5, 36.9, 36.10, and 36.12

Chapter 37
Active Figures 37.2, 37.11, 37.13, 37.22
Examples 37.1 and 37.4

Chapter 38
Active Figures 38.4, 38.11, 38.17, 38.18, 38.30,
and 38.31
Examples 38.1 and 38.7

Chapter 39
Active Figures 39.4, 39.6, and 39.11
Examples 39.4 and 39.9

An Invitation to Physics

It is our sincere hope that you too will find physics an exciting and enjoyable experience and that you will profit from this experience, regardless of your chosen profession. Welcome to the exciting world of physics!

The scientist does not study nature because it is useful; he studies it because he delights in it, and he delights in it because it is beautiful. If nature were not beautiful, it would not be worth knowing, and if nature were not worth knowing, life would not be worth living.

—Henri Poincaré

Mechanics

Physics, the most fundamental physical science, is concerned with the basic principles of the Universe. It is the foundation upon which the other sciences—astronomy, biology, chemistry, and geology—are based. The beauty of physics lies in the simplicity of the fundamental physical theories and in the manner in which just a small number of fundamental concepts, equations, and assumptions can alter and expand our view of the world around us.

The study of physics can be divided into six main areas:

1. *classical mechanics,* which is concerned with the motion of objects that are large relative to atoms and move at speeds much slower than the speed of light;
2. *relativity,* which is a theory describing objects moving at any speed, even speeds approaching the speed of light;
3. *thermodynamics,* which deals with heat, work, temperature, and the statistical behavior of systems with large numbers of particles;
4. *electromagnetism,* which is concerned with electricity, magnetism, and electromagnetic fields;
5. *optics,* which is the study of the behavior of light and its interaction with materials;
6. *quantum mechanics,* a collection of theories connecting the behavior of matter at the submicroscopic level to macroscopic observations.

The disciplines of mechanics and electromagnetism are basic to all other branches of classical physics (developed before 1900) and modern physics (c. 1900–present). The first part of this textbook deals with classical mechanics, sometimes referred to as *Newtonian mechanics* or simply *mechanics.* This is an appropriate place to begin an introductory text because many of the basic principles used to understand mechanical systems can later be used to describe such natural phenomena as waves and the transfer of energy by heat. Furthermore, the laws of conservation of energy and momentum introduced in mechanics retain their importance in the fundamental theories of other areas of physics.

Today, classical mechanics is of vital importance to students from all disciplines. It is highly successful in describing the motions of different objects, such as planets, rockets, and baseballs. In the first part of the text, we shall describe the laws of classical mechanics and examine a wide range of phenomena that can be understood with these fundamental ideas. ■

◄ *Liftoff of the space shuttle* Columbia. *The tragic accident of February 1, 2003 that took the lives of all seven astronauts aboard happened just before Volume 1 of this book went to press. The launch and operation of a space shuttle involves many fundamental principles of classical mechanics, thermodynamics, and electromagnetism. We study the principles of classical mechanics in Part 1 of this text, and apply these principles to rocket propulsion in Chapter 9. (NASA)*

Chapter 1

Physics and Measurement

▲ The workings of a mechanical clock. Complicated timepieces have been built for centuries in an effort to measure time accurately. Time is one of the basic quantities that we use in studying the motion of objects. (elektraVision/Index Stock Imagery)

Like all other sciences, physics is based on experimental observations and quantitative measurements. The main objective of physics is to find the limited number of fundamental laws that govern natural phenomena and to use them to develop theories that can predict the results of future experiments. The fundamental laws used in developing theories are expressed in the language of mathematics, the tool that provides a bridge between theory and experiment.

When a discrepancy between theory and experiment arises, new theories must be formulated to remove the discrepancy. Many times a theory is satisfactory only under limited conditions; a more general theory might be satisfactory without such limitations. For example, the laws of motion discovered by Isaac Newton (1642–1727) in the 17th century accurately describe the motion of objects moving at normal speeds but do not apply to objects moving at speeds comparable with the speed of light. In contrast, the special theory of relativity developed by Albert Einstein (1879–1955) in the early 1900s gives the same results as Newton's laws at low speeds but also correctly describes motion at speeds approaching the speed of light. Hence, Einstein's special theory of relativity is a more general theory of motion.

Classical physics includes the theories, concepts, laws, and experiments in classical mechanics, thermodynamics, optics, and electromagnetism developed before 1900. Important contributions to classical physics were provided by Newton, who developed classical mechanics as a systematic theory and was one of the originators of calculus as a mathematical tool. Major developments in mechanics continued in the 18th century, but the fields of thermodynamics and electricity and magnetism were not developed until the latter part of the 19th century, principally because before that time the apparatus for controlled experiments was either too crude or unavailable.

A major revolution in physics, usually referred to as *modern physics,* began near the end of the 19th century. Modern physics developed mainly because of the discovery that many physical phenomena could not be explained by classical physics. The two most important developments in this modern era were the theories of relativity and quantum mechanics. Einstein's theory of relativity not only correctly described the motion of objects moving at speeds comparable to the speed of light but also completely revolutionized the traditional concepts of space, time, and energy. The theory of relativity also shows that the speed of light is the upper limit of the speed of an object and that mass and energy are related. Quantum mechanics was formulated by a number of distinguished scientists to provide descriptions of physical phenomena at the atomic level.

Scientists continually work at improving our understanding of fundamental laws, and new discoveries are made every day. In many research areas there is a great deal of overlap among physics, chemistry, and biology. Evidence for this overlap is seen in the names of some subspecialties in science—biophysics, biochemistry, chemical physics, biotechnology, and so on. Numerous technological advances in recent times are the result of the efforts of many scientists, engineers, and technicians. Some of the most notable developments in the latter half of the 20th century were (1) unmanned planetary explorations and manned moon landings, (2) microcircuitry and high-speed computers, (3) sophisticated imaging techniques used in scientific research and medicine, and

(4) several remarkable results in genetic engineering. The impacts of such developments and discoveries on our society have indeed been great, and it is very likely that future discoveries and developments will be exciting, challenging, and of great benefit to humanity.

1.1 Standards of Length, Mass, and Time

The laws of physics are expressed as mathematical relationships among physical quantities that we will introduce and discuss throughout the book. Most of these quantities are *derived quantities,* in that they can be expressed as combinations of a small number of *basic quantities.* In mechanics, the three basic quantities are length, mass, and time. All other quantities in mechanics can be expressed in terms of these three.

If we are to report the results of a measurement to someone who wishes to reproduce this measurement, a *standard* must be defined. It would be meaningless if a visitor from another planet were to talk to us about a length of 8 "glitches" if we do not know the meaning of the unit glitch. On the other hand, if someone familiar with our system of measurement reports that a wall is 2 meters high and our unit of length is defined to be 1 meter, we know that the height of the wall is twice our basic length unit. Likewise, if we are told that a person has a mass of 75 kilograms and our unit of mass is defined to be 1 kilogram, then that person is 75 times as massive as our basic unit.[1] Whatever is chosen as a standard must be readily accessible and possess some property that can be measured reliably. Measurements taken by different people in different places must yield the same result.

In 1960, an international committee established a set of standards for the fundamental quantities of science. It is called the **SI** (Système International), and its units of length, mass, and time are the *meter, kilogram,* and *second,* respectively. Other SI standards established by the committee are those for temperature (the *kelvin*), electric current (the *ampere*), luminous intensity (the *candela*), and the amount of substance (the *mole*).

Length

In A.D. 1120 the king of England decreed that the standard of length in his country would be named the *yard* and would be precisely equal to the distance from the tip of his nose to the end of his outstretched arm. Similarly, the original standard for the foot adopted by the French was the length of the royal foot of King Louis XIV. This standard prevailed until 1799, when the legal standard of length in France became the *meter,* defined as one ten-millionth the distance from the equator to the North Pole along one particular longitudinal line that passes through Paris.

Many other systems for measuring length have been developed over the years, but the advantages of the French system have caused it to prevail in almost all countries and in scientific circles everywhere. As recently as 1960, the length of the meter was defined as the distance between two lines on a specific platinum–iridium bar stored under controlled conditions in France. This standard was abandoned for several reasons, a principal one being that the limited accuracy with which the separation between the lines on the bar can be determined does not meet the current requirements of science and technology. In the 1960s and 1970s, the meter was defined as 1 650 763.73 wavelengths of orange-red light emitted from a krypton-86 lamp. However, in October 1983, **the meter (m) was redefined as the distance traveled by light in vacuum during a time of 1/299 792 458 second.** In effect, this

---

[1]   The need for assigning numerical values to various measured physical quantities was expressed by Lord Kelvin (William Thomson) as follows: "I often say that when you can measure what you are speaking about, and express it in numbers, you should know something about it, but when you cannot express it in numbers, your knowledge is of a meager and unsatisfactory kind. It may be the beginning of knowledge but you have scarcely in your thoughts advanced to the state of science."

Table 1.1

Approximate Values of Some Measured Lengths	
	Length (m)
Distance from the Earth to the most remote known quasar	1.4×10^{26}
Distance from the Earth to the most remote normal galaxies	9×10^{25}
Distance from the Earth to the nearest large galaxy (M 31, the Andromeda galaxy)	2×10^{22}
Distance from the Sun to the nearest star (Proxima Centauri)	4×10^{16}
One lightyear	9.46×10^{15}
Mean orbit radius of the Earth about the Sun	1.50×10^{11}
Mean distance from the Earth to the Moon	3.84×10^{8}
Distance from the equator to the North Pole	1.00×10^{7}
Mean radius of the Earth	6.37×10^{6}
Typical altitude (above the surface) of a satellite orbiting the Earth	2×10^{5}
Length of a football field	9.1×10^{1}
Length of a housefly	5×10^{-3}
Size of smallest dust particles	$\sim 10^{-4}$
Size of cells of most living organisms	$\sim 10^{-5}$
Diameter of a hydrogen atom	$\sim 10^{-10}$
Diameter of an atomic nucleus	$\sim 10^{-14}$
Diameter of a proton	$\sim 10^{-15}$

PITFALL PREVENTION

1.1 No Commas in Numbers with Many Digits

We will use the standard scientific notation for numbers with more than three digits, in which groups of three digits are separated by spaces rather than commas. Thus, 10 000 is the same as the common American notation of 10,000. Similarly, $\pi = 3.14159265$ is written as 3.141 592 65.

latest definition establishes that the speed of light in vacuum is precisely 299 792 458 meters per second.

Table 1.1 lists approximate values of some measured lengths. You should study this table as well as the next two tables and begin to generate an intuition for what is meant by a length of 20 centimeters, for example, or a mass of 100 kilograms or a time interval of 3.2×10^{7} seconds.

Mass

The SI unit of mass, **the kilogram (kg), is defined as the mass of a specific platinum–iridium alloy cylinder kept at the International Bureau of Weights and Measures at Sèvres, France.** This mass standard was established in 1887 and has not been changed since that time because platinum–iridium is an unusually stable alloy. A duplicate of the Sèvres cylinder is kept at the National Institute of Standards and Technology (NIST) in Gaithersburg, Maryland (Fig. 1.1a).

Table 1.2 lists approximate values of the masses of various objects.

Time

Before 1960, the standard of time was defined in terms of the *mean solar day* for the year 1900. (A solar day is the time interval between successive appearances of the Sun at the highest point it reaches in the sky each day.) The *second* was defined as $\left(\frac{1}{60}\right)\left(\frac{1}{60}\right)\left(\frac{1}{24}\right)$ of a mean solar day. The rotation of the Earth is now known to vary slightly with time, however, and therefore this motion is not a good one to use for defining a time standard.

In 1967, the second was redefined to take advantage of the high precision attainable in a device known as an *atomic clock* (Fig. 1.1b), which uses the characteristic frequency of the cesium-133 atom as the "reference clock." **The second (s) is now defined as 9 192 631 770 times the period of vibration of radiation from the cesium atom.**[2]

Table 1.2

Masses of Various Objects (Approximate Values)	
	Mass (kg)
Observable Universe	$\sim 10^{52}$
Milky Way galaxy	$\sim 10^{42}$
Sun	1.99×10^{30}
Earth	5.98×10^{24}
Moon	7.36×10^{22}
Shark	$\sim 10^{3}$
Human	$\sim 10^{2}$
Frog	$\sim 10^{-1}$
Mosquito	$\sim 10^{-5}$
Bacterium	$\sim 1 \times 10^{-15}$
Hydrogen atom	1.67×10^{-27}
Electron	9.11×10^{-31}

PITFALL PREVENTION

1.2 Reasonable Values

Generating intuition about typical values of quantities is important because when solving problems you must think about your end result and determine if it seems reasonable. If you are calculating the mass of a housefly and arrive at a value of 100 kg, this is *unreasonable*—there is an error somewhere.

---

[2] *Period* is defined as the time interval needed for one complete vibration.

(a)   (b)

(Courtesy of National Institute of Standards and Technology, U.S. Department of Commerce)

Figure 1.1 (a) The National Standard Kilogram No. 20, an accurate copy of the International Standard Kilogram kept at Sèvres, France, is housed under a double bell jar in a vault at the National Institute of Standards and Technology. (b) The nation's primary time standard is a cesium fountain atomic clock developed at the National Institute of Standards and Technology laboratories in Boulder, Colorado. The clock will neither gain nor lose a second in 20 million years.

To keep these atomic clocks—and therefore all common clocks and watches that are set to them—synchronized, it has sometimes been necessary to add leap seconds to our clocks.

Since Einstein's discovery of the linkage between space and time, precise measurement of time intervals requires that we know both the state of motion of the clock used to measure the interval and, in some cases, the location of the clock as well. Otherwise, for example, global positioning system satellites might be unable to pinpoint your location with sufficient accuracy, should you need to be rescued.

Approximate values of time intervals are presented in Table 1.3.

Table 1.3

Approximate Values of Some Time Intervals	Time Interval (s)
Age of the Universe	5×10^{17}
Age of the Earth	1.3×10^{17}
Average age of a college student	6.3×10^{8}
One year	3.2×10^{7}
One day (time interval for one revolution of the Earth about its axis)	8.6×10^{4}
One class period	3.0×10^{3}
Time interval between normal heartbeats	8×10^{-1}
Period of audible sound waves	$\sim 10^{-3}$
Period of typical radio waves	$\sim 10^{-6}$
Period of vibration of an atom in a solid	$\sim 10^{-13}$
Period of visible light waves	$\sim 10^{-15}$
Duration of a nuclear collision	$\sim 10^{-22}$
Time interval for light to cross a proton	$\sim 10^{-24}$

Table 1.4

Prefixes for Powers of Ten		
Power	**Prefix**	**Abbreviation**
10^{-24}	yocto	y
10^{-21}	zepto	z
10^{-18}	atto	a
10^{-15}	femto	f
10^{-12}	pico	p
10^{-9}	nano	n
10^{-6}	micro	μ
10^{-3}	milli	m
10^{-2}	centi	c
10^{-1}	deci	d
10^{3}	kilo	k
10^{6}	mega	M
10^{9}	giga	G
10^{12}	tera	T
10^{15}	peta	P
10^{18}	exa	E
10^{21}	zetta	Z
10^{24}	yotta	Y

In addition to SI, another system of units, the *U.S. customary system,* is still used in the United States despite acceptance of SI by the rest of the world. In this system, the units of length, mass, and time are the foot (ft), slug, and second, respectively. In this text we shall use SI units because they are almost universally accepted in science and industry. We shall make some limited use of U.S. customary units in the study of classical mechanics.

In addition to the basic SI units of meter, kilogram, and second, we can also use other units, such as millimeters and nanoseconds, where the prefixes *milli-* and *nano-* denote multipliers of the basic units based on various powers of ten. Prefixes for the various powers of ten and their abbreviations are listed in Table 1.4. For example, 10^{-3} m is equivalent to 1 millimeter (mm), and 10^{3} m corresponds to 1 kilometer (km). Likewise, 1 kilogram (kg) is 10^{3} grams (g), and 1 megavolt (MV) is 10^{6} volts (V).

1.2 Matter and Model Building

If physicists cannot interact with some phenomenon directly, they often imagine a **model** for a physical system that is related to the phenomenon. In this context, a model is a system of physical components, such as electrons and protons in an atom. Once we have identified the physical components, we make predictions about the behavior of the system, based on the interactions among the components of the system and/or the interaction between the system and the environment outside the system.

As an example, consider the behavior of *matter.* A 1-kg cube of solid gold, such as that at the left of Figure 1.2, has a length of 3.73 cm on a side. Is this cube nothing but wall-to-wall gold, with no empty space? If the cube is cut in half, the two pieces still retain their chemical identity as solid gold. But what if the pieces are cut again and again, indefinitely? Will the smaller and smaller pieces always be gold? Questions such as these can be traced back to early Greek philosophers. Two of them—Leucippus and his student Democritus—could not accept the idea that such cuttings could go on forever. They speculated that the process ultimately must end when it produces a particle

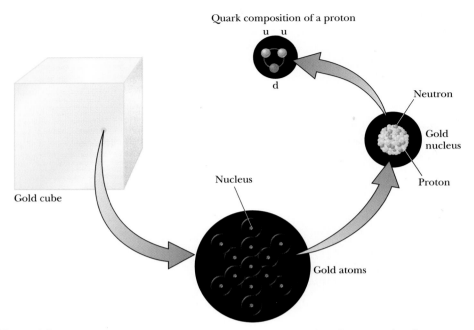

Figure 1.2 Levels of organization in matter. Ordinary matter consists of atoms, and at the center of each atom is a compact nucleus consisting of protons and neutrons. Protons and neutrons are composed of quarks. The quark composition of a proton is shown.

that can no longer be cut. In Greek, *atomos* means "not sliceable." From this comes our English word *atom*.

Let us review briefly a number of historical models of the structure of matter. The Greek model of the structure of matter was that all ordinary matter consists of atoms, as suggested to the lower right of the cube in Figure 1.2. Beyond that, no additional structure was specified in the model—atoms acted as small particles that interacted with each other, but internal structure of the atom was not a part of the model.

In 1897, J. J. Thomson identified the electron as a charged particle and as a constituent of the atom. This led to the first model of the atom that contained internal structure. We shall discuss this model in Chapter 42.

Following the discovery of the nucleus in 1911, a model was developed in which each atom is made up of electrons surrounding a central nucleus. A nucleus is shown in Figure 1.2. This model leads, however, to a new question—does the nucleus have structure? That is, is the nucleus a single particle or a collection of particles? The exact composition of the nucleus is not known completely even today, but by the early 1930s a model evolved that helped us understand how the nucleus behaves. Specifically, scientists determined that occupying the nucleus are two basic entities, protons and neutrons. The proton carries a positive electric charge, and a specific chemical element is identified by the number of protons in its nucleus. This number is called the **atomic number** of the element. For instance, the nucleus of a hydrogen atom contains one proton (and so the atomic number of hydrogen is 1), the nucleus of a helium atom contains two protons (atomic number 2), and the nucleus of a uranium atom contains 92 protons (atomic number 92). In addition to atomic number, there is a second number characterizing atoms—**mass number,** defined as the number of protons plus neutrons in a nucleus. The atomic number of an element never varies (i.e., the number of protons does not vary) but the mass number can vary (i.e., the number of neutrons varies).

The existence of neutrons was verified conclusively in 1932. A neutron has no charge and a mass that is about equal to that of a proton. One of its primary purposes is to act as a "glue" that holds the nucleus together. If neutrons were not present in the nucleus, the repulsive force between the positively charged particles would cause the nucleus to come apart.

But is this where the process of breaking down stops? Protons, neutrons, and a host of other exotic particles are now known to be composed of six different varieties of particles called **quarks,** which have been given the names of *up, down, strange, charmed, bottom,* and *top.* The up, charmed, and top quarks have electric charges of $+\frac{2}{3}$ that of the proton, whereas the down, strange, and bottom quarks have charges of $-\frac{1}{3}$ that of the proton. The proton consists of two up quarks and one down quark, as shown at the top in Figure 1.2. You can easily show that this structure predicts the correct charge for the proton. Likewise, the neutron consists of two down quarks and one up quark, giving a net charge of zero.

This process of building models is one that you should develop as you study physics. You will be challenged with many mathematical problems to solve in this study. One of the most important techniques is to build a model for the problem—identify a system of physical components for the problem, and make predictions of the behavior of the system based on the interactions among the components of the system and/or the interaction between the system and its surrounding environment.

1.3   Density and Atomic Mass

In Section 1.1, we explored three basic quantities in mechanics. Let us look now at an example of a derived quantity—**density.** The density ρ (Greek letter rho) of any substance is defined as its *mass per unit volume:*

$$\rho \equiv \frac{m}{V} \tag{1.1}$$

A table of the letters in the Greek alphabet is provided on the back endsheet of the textbook.

For example, aluminum has a density of 2.70 g/cm^3, and lead has a density of 11.3 g/cm^3. Therefore, a piece of aluminum of volume 10.0 cm^3 has a mass of 27.0 g, whereas an equivalent volume of lead has a mass of 113 g. A list of densities for various substances is given in Table 1.5.

The numbers of protons and neutrons in the nucleus of an atom of an element are related to the **atomic mass** of the element, which is defined as the mass of a single atom of the element measured in **atomic mass units** (u) where 1 u = 1.660 538 7 × 10^{-27} kg.

Table 1.5

Densities of Various Substances	
Substance	**Density ρ (10^3 kg/m^3)**
Platinum	21.45
Gold	19.3
Uranium	18.7
Lead	11.3
Copper	8.92
Iron	7.86
Aluminum	2.70
Magnesium	1.75
Water	1.00
Air at atmospheric pressure	0.0012

The atomic mass of lead is 207 u and that of aluminum is 27.0 u. However, the ratio of atomic masses, 207 u/27.0 u = 7.67, does not correspond to the ratio of densities, $(11.3 \times 10^3 \text{ kg/m}^3)/(2.70 \times 10^3 \text{ kg/m}^3) = 4.19$. This discrepancy is due to the difference in atomic spacings and atomic arrangements in the crystal structures of the two elements.

Quick Quiz 1.1 In a machine shop, two cams are produced, one of aluminum and one of iron. Both cams have the same mass. Which cam is larger? (a) the aluminum cam (b) the iron cam (c) Both cams have the same size.

Example 1.1   How Many Atoms in the Cube?

A solid cube of aluminum (density 2.70 g/cm³) has a volume of 0.200 cm³. It is known that 27.0 g of aluminum contains 6.02×10^{23} atoms. How many aluminum atoms are contained in the cube?

Solution Because density equals mass per unit volume, the mass of the cube is

$$m = \rho V = (2.70 \text{ g/cm}^3)(0.200 \text{ cm}^3) = 0.540 \text{ g}$$

To solve this problem, we will set up a ratio based on the fact that the mass of a sample of material is proportional to the number of atoms contained in the sample. This technique of solving by ratios is very powerful and should be studied and understood so that it can be applied in future problem solving. Let us express our proportionality as $m = kN$, where m is the mass of the sample, N is the number of atoms in the sample, and k is an unknown proportionality constant. We

write this relationship twice, once for the actual sample of aluminum in the problem and once for a 27.0-g sample, and then we divide the first equation by the second:

$$m_{\text{sample}} = kN_{\text{sample}}$$
$$m_{27.0\,g} = kN_{27.0\,g} \quad \rightarrow \quad \frac{m_{\text{sample}}}{m_{27.0\,g}} = \frac{N_{\text{sample}}}{N_{27.0\,g}}$$

Notice that the unknown proportionality constant k cancels, so we do not need to know its value. We now substitute the values:

$$\frac{0.540 \text{ g}}{27.0 \text{ g}} = \frac{N_{\text{sample}}}{6.02 \times 10^{23} \text{ atoms}}$$

$$N_{\text{sample}} = \frac{(0.540 \text{ g})(6.02 \times 10^{23} \text{ atoms})}{27.0 \text{ g}}$$

$$= 1.20 \times 10^{22} \text{ atoms}$$

 PITFALL PREVENTION

1.3 Setting Up Ratios

When using ratios to solve a problem, keep in mind that *ratios come from equations*. If you start from equations known to be correct and can divide one equation by the other as in Example 1.1 to obtain a useful ratio, you will avoid reasoning errors. So write the known equations first!

1.4 Dimensional Analysis

The word *dimension* has a special meaning in physics. It denotes the physical nature of a quantity. Whether a distance is measured in units of feet or meters or fathoms, it is still a distance. We say its dimension is *length*.

The symbols we use in this book to specify the dimensions of length, mass, and time are L, M, and T, respectively.[3] We shall often use brackets [ ] to denote the dimensions of a physical quantity. For example, the symbol we use for speed in this book is v, and in our notation the dimensions of speed are written $[v] = L/T$. As another example, the dimensions of area A are $[A] = L^2$. The dimensions and units of area, volume, speed, and acceleration are listed in Table 1.6. The dimensions of other quantities, such as force and energy, will be described as they are introduced in the text.

In many situations, you may have to derive or check a specific equation. A useful and powerful procedure called *dimensional analysis* can be used to assist in the derivation or to check your final expression. Dimensional analysis makes use of the fact that

---

[3] The *dimensions* of a quantity will be symbolized by a capitalized, non-italic letter, such as L. The *symbol* for the quantity itself will be italicized, such as L for the length of an object, or t for time.

Table 1.6

	Area (L^2)	Volume (L^3)	Speed (L/T)	Acceleration (L/T^2)
Units of Area, Volume, Velocity, Speed, and Acceleration				
System				
SI	m^2	m^3	m/s	m/s^2
U.S. customary	ft^2	ft^3	ft/s	ft/s^2

dimensions can be treated as algebraic quantities. For example, quantities can be added or subtracted only if they have the same dimensions. Furthermore, the terms on both sides of an equation must have the same dimensions. By following these simple rules, you can use dimensional analysis to help determine whether an expression has the correct form. The relationship can be correct only if the dimensions on both sides of the equation are the same.

To illustrate this procedure, suppose you wish to derive an equation for the position x of a car at a time t if the car starts from rest and moves with constant acceleration a. In Chapter 2, we shall find that the correct expression is $x = \frac{1}{2}at^2$. Let us use dimensional analysis to check the validity of this expression. The quantity x on the left side has the dimension of length. For the equation to be dimensionally correct, the quantity on the right side must also have the dimension of length. We can perform a dimensional check by substituting the dimensions for acceleration, L/T^2 (Table 1.6), and time, T, into the equation. That is, the dimensional form of the equation $x = \frac{1}{2}at^2$ is

$$L = \frac{L}{T^2} \cdot T^2 = L$$

The dimensions of time cancel as shown, leaving the dimension of length on the right-hand side.

A more general procedure using dimensional analysis is to set up an expression of the form

$$x \propto a^n t^m$$

where n and m are exponents that must be determined and the symbol $\propto$ indicates a proportionality. This relationship is correct only if the dimensions of both sides are the same. Because the dimension of the left side is length, the dimension of the right side must also be length. That is,

$$[a^n t^m] = L = L^1 T^0$$

Because the dimensions of acceleration are L/T^2 and the dimension of time is T, we have

$$(L/T^2)^n T^m = L^1 T^0$$

$$(L^n T^{m-2n}) = L^1 T^0$$

The exponents of L and T must be the same on both sides of the equation. From the exponents of L, we see immediately that $n = 1$. From the exponents of T, we see that $m - 2n = 0$, which, once we substitute for n, gives us $m = 2$. Returning to our original expression $x \propto a^n t^m$, we conclude that $x \propto at^2$. This result differs by a factor of $\frac{1}{2}$ from the correct expression, which is $x = \frac{1}{2}at^2$.

 PITFALL PREVENTION

1.4 Symbols for Quantities

Some quantities have a small number of symbols that represent them. For example, the symbol for time is almost always t. Others quantities might have various symbols depending on the usage. Length may be described with symbols such as x, y, and z (for position), r (for radius), a, b, and c (for the legs of a right triangle), ℓ (for the length of an object), d (for a distance), h (for a height), etc.

Quick Quiz 1.2 True or False: Dimensional analysis can give you the numerical value of constants of proportionality that may appear in an algebraic expression.

Example 1.2   Analysis of an Equation

Show that the expression $v = at$ is dimensionally correct, where v represents speed, a acceleration, and t an instant of time.

Solution For the speed term, we have from Table 1.6

$$[v] = \frac{L}{T}$$

The same table gives us L/T^2 for the dimensions of acceleration, and so the dimensions of at are

$$[at] = \frac{L}{T^2}\,T = \frac{L}{T}$$

Therefore, the expression is dimensionally correct. (If the expression were given as $v = at^2$ it would be dimensionally *incorrect*. Try it and see!)

Example 1.3   Analysis of a Power Law

Suppose we are told that the acceleration a of a particle moving with uniform speed v in a circle of radius r is proportional to some power of r, say r^n, and some power of v, say v^m. Determine the values of n and m and write the simplest form of an equation for the acceleration.

Solution Let us take a to be

$$a = kr^n v^m$$

where k is a dimensionless constant of proportionality. Knowing the dimensions of a, r, and v, we see that the dimensional equation must be

$$\frac{L}{T^2} = L^n\left(\frac{L}{T}\right)^m = \frac{L^{n+m}}{T^m}$$

This dimensional equation is balanced under the conditions

$$n + m = \boxed{1} \qquad \text{and} \qquad m = \boxed{2}$$

Therefore $n = -1$, and we can write the acceleration expression as

$$a = kr^{-1}v^2 = \boxed{k\frac{v^2}{r}}$$

When we discuss uniform circular motion later, we shall see that $k = 1$ if a consistent set of units is used. The constant k would not equal 1 if, for example, v were in km/h and you wanted a in m/s^2.

▲ PITFALL PREVENTION

1.5   Always Include Units

When performing calculations, include the units for every quantity and carry the units through the entire calculation. Avoid the temptation to drop the units early and then attach the expected units once you have an answer. By including the units in every step, you can detect errors if the units for the answer turn out to be incorrect.

1.5   Conversion of Units

Sometimes it is necessary to convert units from one measurement system to another, or to convert within a system, for example, from kilometers to meters. Equalities between SI and U.S. customary units of length are as follows:

$$1 \text{ mile} = 1\,609 \text{ m} = 1.609 \text{ km} \qquad 1 \text{ ft} = 0.304\,8 \text{ m} = 30.48 \text{ cm}$$

$$1 \text{ m} = 39.37 \text{ in.} = 3.281 \text{ ft} \qquad 1 \text{ in.} = 0.025\,4 \text{ m} = 2.54 \text{ cm (exactly)}$$

A more complete list of conversion factors can be found in Appendix A.

Units can be treated as algebraic quantities that can cancel each other. For example, suppose we wish to convert 15.0 in. to centimeters. Because 1 in. is defined as exactly 2.54 cm, we find that

$$15.0 \text{ in.} = (15.0 \text{ in.})\left(\frac{2.54 \text{ cm}}{1 \text{ in.}}\right) = 38.1 \text{ cm}$$

where the ratio in parentheses is equal to 1. Notice that we choose to put the unit of an inch in the denominator and it cancels with the unit in the original quantity. The remaining unit is the centimeter, which is our desired result.

Quick Quiz 1.3 The distance between two cities is 100 mi. The number of kilometers between the two cities is (a) smaller than 100 (b) larger than 100 (c) equal to 100.

Example 1.4   Is He Speeding?

On an interstate highway in a rural region of Wyoming, a car is traveling at a speed of 38.0 m/s. Is this car exceeding the speed limit of 75.0 mi/h?

Solution We first convert meters to miles:

$$(38.0 \text{ m/s}) \left(\frac{1 \text{ mi}}{1\,609 \text{ m}} \right) = 2.36 \times 10^{-2} \text{ mi/s}$$

Now we convert seconds to hours:

$$(2.36 \times 10^{-2} \text{ mi/s}) \left(\frac{60 \text{ s}}{1 \text{ min}} \right) \left(\frac{60 \text{ min}}{1 \text{ h}} \right) = 85.0 \text{ mi/h}$$

Thus, the car is exceeding the speed limit and should slow down.

What If? What if the driver is from outside the U.S. and is familiar with speeds measured in km/h? What is the speed of the car in km/h?

Answer We can convert our final answer to the appropriate units:

$$(85.0 \text{ mi/h}) \left(\frac{1.609 \text{ km}}{1 \text{ mi}} \right) = 137 \text{ km/h}$$

Figure 1.3 shows the speedometer of an automobile, with speeds in both mi/h and km/h. Can you check the conversion we just performed using this photograph?

Phil Boorman/Getty Images

Figure 1.3 The speedometer of a vehicle that shows speeds in both miles per hour and kilometers per hour.

1.6   Estimates and Order-of-Magnitude Calculations

It is often useful to compute an approximate answer to a given physical problem even when little information is available. This answer can then be used to determine whether or not a more precise calculation is necessary. Such an approximation is usually based on certain assumptions, which must be modified if greater precision is needed. We will sometimes refer to an *order of magnitude* of a certain quantity as the power of ten of the number that describes that quantity. Usually, when an order-of-magnitude calculation is made, the results are reliable to within about a factor of 10. If a quantity increases in value by three orders of magnitude, this means that its value increases by a factor of about $10^3 = 1\,000$. We use the symbol $\sim$ for "is on the order of." Thus,

$$0.008\,6 \sim 10^{-2} \qquad 0.002\,1 \sim 10^{-3} \qquad 720 \sim 10^3$$

The spirit of order-of-magnitude calculations, sometimes referred to as "guesstimates" or "ball-park figures," is given in the following quotation: "Make an estimate before every calculation, try a simple physical argument . . . before every derivation, guess the answer to every puzzle."[4] Inaccuracies caused by guessing too low for one number are often canceled out by other guesses that are too high. You will find that with practice your guesstimates become better and better. Estimation problems can be fun to work as you freely drop digits, venture reasonable approximations for

---

4   E. Taylor and J. A. Wheeler, *Spacetime Physics: Introduction to Special Relativity,* 2nd ed., San Francisco, W. H. Freeman & Company, Publishers, 1992, p. 20.

unknown numbers, make simplifying assumptions, and turn the question around into something you can answer in your head or with minimal mathematical manipulation on paper. Because of the simplicity of these types of calculations, they can be performed on a *small* piece of paper, so these estimates are often called "back-of-the-envelope calculations."

Example 1.5 Breaths in a Lifetime

Estimate the number of breaths taken during an average life span.

Solution We start by guessing that the typical life span is about 70 years. The only other estimate we must make in this example is the average number of breaths that a person takes in 1 min. This number varies, depending on whether the person is exercising, sleeping, angry, serene, and so forth. To the nearest order of magnitude, we shall choose 10 breaths per minute as our estimate of the average. (This is certainly closer to the true value than 1 breath per minute or 100 breaths per minute.) The number of minutes in a year is approximately

$$1 \text{ yr} \left(\frac{400 \text{ days}}{1 \text{ yr}} \right) \left(\frac{25 \text{ h}}{1 \text{ day}} \right) \left(\frac{60 \text{ min}}{1 \text{ h}} \right) = 6 \times 10^5 \text{ min}$$

Notice how much simpler it is in the expression above to multiply 400×25 than it is to work with the more accurate 365×24. These approximate values for the number of days

in a year and the number of hours in a day are close enough for our purposes. Thus, in 70 years there will be $(70 \text{ yr})(6 \times 10^5 \text{ min/yr}) = 4 \times 10^7 \text{ min}$. At a rate of 10 breaths/min, an individual would take $\boxed{4 \times 10^8 \text{ breaths}}$ in a lifetime, or on the order of 10^9 breaths.

What If? What if the average life span were estimated as 80 years instead of 70? Would this change our final estimate?

Answer We could claim that $(80 \text{ yr})(6 \times 10^5 \text{ min/yr}) = 5 \times 10^7 \text{ min}$, so that our final estimate should be 5×10^8 breaths. This is still on the order of 10^9 breaths, so an order-of-magnitude estimate would be unchanged. Furthermore, 80 years is 14% larger than 70 years, but we have overestimated the total time interval by using 400 days in a year instead of 365 and 25 hours in a day instead of 24. These two numbers together result in an overestimate of 14%, which cancels the effect of the increased life span!

Example 1.6 It's a Long Way to San Jose

Estimate the number of steps a person would take walking from New York to Los Angeles.

Solution Without looking up the distance between these two cities, you might remember from a geography class that they are about 3 000 mi apart. The next approximation we must make is the length of one step. Of course, this length depends on the person doing the walking, but we can estimate that each step covers about 2 ft. With our estimated step size, we can determine the number of steps in 1 mi. Because this is a rough calculation, we round 5 280 ft/mi to 5 000 ft/mi. (What percentage error does this introduce?) This conversion factor gives us

$$\frac{5\,000 \text{ ft/mi}}{2 \text{ ft/step}} = 2\,500 \text{ steps/mi}$$

Now we switch to scientific notation so that we can do the calculation mentally:

$$(3 \times 10^3 \text{ mi})(2.5 \times 10^3 \text{ steps/mi})$$

$$= \boxed{7.5 \times 10^6 \text{ steps} \sim 10^7 \text{ steps}}$$

So if we intend to walk across the United States, it will take us on the order of ten million steps. This estimate is almost certainly too small because we have not accounted for curving roads and going up and down hills and mountains. Nonetheless, it is probably within an order of magnitude of the correct answer.

Example 1.7 How Much Gas Do We Use?

Estimate the number of gallons of gasoline used each year by all the cars in the United States.

Solution Because there are about 280 million people in the United States, an estimate of the number of cars in the country is 100 million (guessing that there are between two and three people per car). We also estimate that the average

distance each car travels per year is 10 000 mi. If we assume a gasoline consumption of 20 mi/gal or 0.05 gal/mi, then each car uses about 500 gal/yr. Multiplying this by the total number of cars in the United States gives an estimated total consumption of $\boxed{5 \times 10^{10} \text{ gal} \sim 10^{11} \text{ gal.}}$

1.7 Significant Figures

When certain quantities are measured, the measured values are known only to within the limits of the experimental uncertainty. The value of this uncertainty can depend on various factors, such as the quality of the apparatus, the skill of the experimenter, and the number of measurements performed. The number of **significant figures** in a measurement can be used to express something about the uncertainty.

As an example of significant figures, suppose that we are asked in a laboratory experiment to measure the area of a computer disk label using a meter stick as a measuring instrument. Let us assume that the accuracy to which we can measure the length of the label is ± 0.1 cm. If the length is measured to be 5.5 cm, we can claim only that its length lies somewhere between 5.4 cm and 5.6 cm. In this case, we say that the measured value has two significant figures. Note that the significant figures include the first estimated digit. Likewise, if the label's width is measured to be 6.4 cm, the actual value lies between 6.3 cm and 6.5 cm. Thus we could write the measured values as (5.5 ± 0.1) cm and (6.4 ± 0.1) cm.

Now suppose we want to find the area of the label by multiplying the two measured values. If we were to claim the area is $(5.5 \text{ cm})(6.4 \text{ cm}) = 35.2 \text{ cm}^2$, our answer would be unjustifiable because it contains three significant figures, which is greater than the number of significant figures in either of the measured quantities. A good rule of thumb to use in determining the number of significant figures that can be claimed in a multiplication or a division is as follows:

> When multiplying several quantities, the number of significant figures in the final answer is the same as the number of significant figures in the quantity having the lowest number of significant figures. The same rule applies to division.

Applying this rule to the previous multiplication example, we see that the answer for the area can have only two significant figures because our measured quantities have only two significant figures. Thus, all we can claim is that the area is 35 cm^2, realizing that the value can range between $(5.4 \text{ cm})(6.3 \text{ cm}) = 34 \text{ cm}^2$ and $(5.6 \text{ cm})(6.5 \text{ cm}) = 36 \text{ cm}^2$.

Zeros may or may not be significant figures. Those used to position the decimal point in such numbers as 0.03 and 0.007 5 are not significant. Thus, there are one and two significant figures, respectively, in these two values. When the zeros come after other digits, however, there is the possibility of misinterpretation. For example, suppose the mass of an object is given as 1 500 g. This value is ambiguous because we do not know whether the last two zeros are being used to locate the decimal point or whether they represent significant figures in the measurement. To remove this ambiguity, it is common to use scientific notation to indicate the number of significant figures. In this case, we would express the mass as 1.5×10^3 g if there are two significant figures in the measured value, 1.50×10^3 g if there are three significant figures, and 1.500×10^3 g if there are four. The same rule holds for numbers less than 1, so that 2.3×10^{-4} has two significant figures (and so could be written 0.000 23) and 2.30×10^{-4} has three significant figures (also written 0.000 230). In general, **a significant figure in a measurement is a reliably known digit (other than a zero used to locate the decimal point) or the first estimated digit.**

For addition and subtraction, you must consider the number of decimal places when you are determining how many significant figures to report:

> When numbers are added or subtracted, the number of decimal places in the result should equal the smallest number of decimal places of any term in the sum.

 PITFALL PREVENTION

1.6 Read Carefully

Notice that the rule for addition and subtraction is different from that for multiplication and division. For addition and subtraction, the important consideration is the number of *decimal places,* not the number of *significant figures.*

For example, if we wish to compute $123 + 5.35$, the answer is 128 and not 128.35. If we compute the sum $1.000\ 1 + 0.000\ 3 = 1.000\ 4$, the result has five significant figures, even though one of the terms in the sum, 0.000 3, has only one significant figure. Likewise, if we perform the subtraction $1.002 - 0.998 = 0.004$, the result has only one significant figure even though one term has four significant figures and the other has three. In this book, **most of the numerical examples and end-of-chapter problems will yield answers having three significant figures.** When carrying out estimates we shall typically work with a single significant figure.

If the number of significant figures in the result of an addition or subtraction must be reduced, there is a general rule for rounding off numbers, which states that the last digit retained is to be increased by 1 if the last digit dropped is greater than 5. If the last digit dropped is less than 5, the last digit retained remains as it is. If the last digit dropped is equal to 5, the remaining digit should be rounded to the nearest even number. (This helps avoid accumulation of errors in long arithmetic processes.)

A technique for avoiding error accumulation is to delay rounding of numbers in a long calculation until you have the final result. Wait until you are ready to copy the final answer from your calculator before rounding to the correct number of significant figures.

Quick Quiz 1.4   Suppose you measure the position of a chair with a meter stick and record that the center of the seat is 1.043 860 564 2 m from a wall. What would a reader conclude from this recorded measurement?

Example 1.8   Installing a Carpet

A carpet is to be installed in a room whose length is measured to be 12.71 m and whose width is measured to be 3.46 m. Find the area of the room.

Solution If you multiply 12.71 m by 3.46 m on your calculator, you will see an answer of 43.976 6 m^2. How many of these numbers should you claim? Our rule of thumb for multiplication tells us that you can claim only the number of significant figures in your answer as are present in the measured quantity having the lowest number of significant figures. In this example, the lowest number of significant figures is three in 3.46 m, so we should express our final answer as 44.0 m^2.

SUMMARY

Take a practice test for this chapter by clicking on the Practice Test link at http://www.pse6.com.

The three fundamental physical quantities of mechanics are length, mass, and time, which in the SI system have the units meters (m), kilograms (kg), and seconds (s), respectively. Prefixes indicating various powers of ten are used with these three basic units.

The **density** of a substance is defined as its *mass per unit volume.* Different substances have different densities mainly because of differences in their atomic masses and atomic arrangements.

The method of **dimensional analysis** is very powerful in solving physics problems. Dimensions can be treated as algebraic quantities. By making estimates and performing order-of-magnitude calculations, you should be able to approximate the answer to a problem when there is not enough information available to completely specify an exact solution.

When you compute a result from several measured numbers, each of which has a certain accuracy, you should give the result with the correct number of **significant figures.** When multiplying several quantities, the number of significant figures in the final answer is the same as the number of significant figures in the quantity having the lowest number of significant figures. The same rule applies to division. When numbers are added or subtracted, the number of decimal places in the result should equal the smallest number of decimal places of any term in the sum.

QUESTIONS

1. What types of natural phenomena could serve as time standards?

2. Suppose that the three fundamental standards of the metric system were length, *density,* and time rather than length, *mass,* and time. The standard of density in this system is to be defined as that of water. What considerations about water would you need to address to make sure that the standard of density is as accurate as possible?

3. The height of a horse is sometimes given in units of "hands." Why is this a poor standard of length?

4. Express the following quantities using the prefixes given in Table 1.4: (a) 3×10^{-4} m (b) 5×10^{-5} s (c) 72×10^{2} g.

5. Suppose that two quantities A and B have different dimensions. Determine which of the following arithmetic operations *could* be physically meaningful: (a) $A + B$ (b) A/B (c) $B - A$ (d) AB.

6. If an equation is dimensionally correct, does this mean that the equation must be true? If an equation is not dimensionally correct, does this mean that the equation cannot be true?

7. Do an order-of-magnitude calculation for an everyday situation you encounter. For example, how far do you walk or drive each day?

8. Find the order of magnitude of your age in seconds.

9. What level of precision is implied in an order-of-magnitude calculation?

10. Estimate the mass of this textbook in kilograms. If a scale is available, check your estimate.

11. In reply to a student's question, a guard in a natural history museum says of the fossils near his station, "When I started work here twenty-four years ago, they were eighty million years old, so you can add it up." What should the student conclude about the age of the fossils?

PROBLEMS

1, 2, 3 = straightforward, intermediate, challenging ☐ = full solution available in the *Student Solutions Manual and Study Guide*

🌀 = coached solution with hints available at http://www.pse6.com 💻 = computer useful in solving problem

░ = paired numerical and symbolic problems

Section 1.2   Matter and Model Building

Note: Consult the endpapers, appendices, and tables in the text whenever necessary in solving problems. For this chapter, Appendix B.3 may be particularly useful. Answers to odd-numbered problems appear in the back of the book.

1. A crystalline solid consists of atoms stacked up in a repeating lattice structure. Consider a crystal as shown in Figure P1.1a. The atoms reside at the corners of cubes of side $L = 0.200$ nm. One piece of evidence for the regular arrangement of atoms comes from the flat surfaces along which a crystal separates, or cleaves, when it is broken. Suppose this crystal cleaves along a face diagonal, as shown in Figure P1.1b. Calculate the spacing d between two adjacent atomic planes that separate when the crystal cleaves.

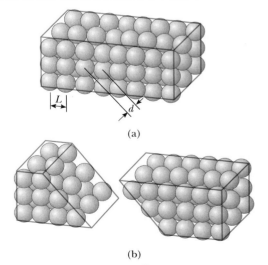

(a)

(b)

Figure P1.1

Section 1.3   Density and Atomic Mass

2. Use information on the endpapers of this book to calculate the average density of the Earth. Where does the value fit among those listed in Tables 1.5 and 14.1? Look up the density of a typical surface rock like granite in another source and compare the density of the Earth to it.

3. The standard kilogram is a platinum–iridium cylinder 39.0 mm in height and 39.0 mm in diameter. What is the density of the material?

4. A major motor company displays a die-cast model of its first automobile, made from 9.35 kg of iron. To celebrate its hundredth year in business, a worker will recast the model in gold from the original dies. What mass of gold is needed to make the new model?

5. What mass of a material with density ρ is required to make a hollow spherical shell having inner radius r_1 and outer radius r_2?

6. Two spheres are cut from a certain uniform rock. One has radius 4.50 cm. The mass of the other is five times greater. Find its radius.

7. Calculate the mass of an atom of (a) helium, (b) iron, and (c) lead. Give your answers in grams. The atomic masses of these atoms are 4.00 u, 55.9 u, and 207 u, respectively.

8. The paragraph preceding Example 1.1 in the text mentions that the atomic mass of aluminum is $27.0\,\text{u} = 27.0 \times 1.66 \times 10^{-27}$ kg. Example 1.1 says that 27.0 g of aluminum contains 6.02×10^{23} atoms. (a) Prove that each one of these two statements implies the other. (b) **What If?** What if it's not aluminum? Let M represent the numerical value of the mass of one atom of any chemical element in atomic mass units. Prove that M grams of the substance contains a particular number of atoms, the same number for all elements. Calculate this number precisely from the value for u quoted in the text. The number of atoms in M grams of an element is called *Avogadro's number* N_A. The idea can be extended: Avogadro's number of molecules of a chemical compound has a mass of M grams, where M atomic mass units is the mass of one molecule. Avogadro's number of atoms or molecules is called one *mole,* symbolized as 1 mol. A periodic table of the elements, as in Appendix C, and the chemical formula for a compound contain enough information to find the molar mass of the compound. (c) Calculate the mass of one mole of water, H_2O. (d) Find the molar mass of CO_2.

9. On your wedding day your lover gives you a gold ring of mass 3.80 g. Fifty years later its mass is 3.35 g. On the average, how many atoms were abraded from the ring during each second of your marriage? The atomic mass of gold is 197 u.

10. A small cube of iron is observed under a microscope. The edge of the cube is 5.00×10^{-6} cm long. Find (a) the mass of the cube and (b) the number of iron atoms in the cube. The atomic mass of iron is 55.9 u, and its density is 7.86 g/cm^3.

11. A structural I beam is made of steel. A view of its cross-section and its dimensions are shown in Figure P1.11. The density of the steel is 7.56×10^3 kg/m^3. (a) What is the mass of a section 1.50 m long? (b) Assume that the atoms are predominantly iron, with atomic mass 55.9 u. How many atoms are in this section?

Figure P1.11

12. A child at the beach digs a hole in the sand and uses a pail to fill it with water having a mass of 1.20 kg. The mass of one molecule of water is 18.0 u. (a) Find the number of water molecules in this pail of water. (b) Suppose the quantity of water on Earth is constant at 1.32×10^{21} kg. How many of the water molecules in this pail of water are likely to have been in an equal quantity of water that once filled one particular claw print left by a Tyrannosaur hunting on a similar beach?

Section 1.4   Dimensional Analysis

13. The position of a particle moving under uniform acceleration is some function of time and the acceleration. Suppose we write this position $s = ka^m t^n$, where k is a dimensionless constant. Show by dimensional analysis that this expression is satisfied if $m = 1$ and $n = 2$. Can this analysis give the value of k?

14. Figure P1.14 shows a *frustrum of a cone.* Of the following mensuration (geometrical) expressions, which describes (a) the total circumference of the flat circular faces (b) the volume (c) the area of the curved surface? (i) $\pi(r_1 + r_2)[h^2 + (r_1 - r_2)^2]^{1/2}$ (ii) $2\pi(r_1 + r_2)$ (iii) $\pi h(r_1^2 + r_1 r_2 + r_2^2)$.

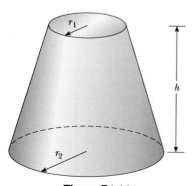

Figure P1.14

15. Which of the following equations are dimensionally correct?
(a) $v_f = v_i + ax$
(b) $y = (2\text{ m})\cos(kx)$, where $k = 2\text{ m}^{-1}$.

16. (a) A fundamental law of motion states that the acceleration of an object is directly proportional to the resultant force exerted on the object and inversely proportional to its mass. If the proportionality constant is defined to have no dimensions, determine the dimensions of force. (b) The newton is the SI unit of force. According to the results for (a), how can you express a force having units of newtons using the fundamental units of mass, length, and time?

17. Newton's law of universal gravitation is represented by

$$F = \frac{GMm}{r^2}$$

Here F is the magnitude of the gravitational force exerted by one small object on another, M and m are the masses of the objects, and r is a distance. Force has the SI units $kg \cdot m/s^2$. What are the SI units of the proportionality constant G?

Section 1.5 Conversion of Units

18. A worker is to paint the walls of a square room 8.00 ft high and 12.0 ft along each side. What surface area in square meters must she cover?

19. Suppose your hair grows at the rate 1/32 in. per day. Find the rate at which it grows in nanometers per second. Because the distance between atoms in a molecule is on the order of 0.1 nm, your answer suggests how rapidly layers of atoms are assembled in this protein synthesis.

20. The volume of a wallet is 8.50 in.^{3} Convert this value to m^3, using the definition 1 in. = 2.54 cm.

21. A rectangular building lot is 100 ft by 150 ft. Determine the area of this lot in m^2.

22. An auditorium measures 40.0 m $\times$ 20.0 m $\times$ 12.0 m. The density of air is 1.20 kg/m^3. What are (a) the volume of the room in cubic feet and (b) the weight of air in the room in pounds?

23. Assume that it takes 7.00 minutes to fill a 30.0-gal gasoline tank. (a) Calculate the rate at which the tank is filled in gallons per second. (b) Calculate the rate at which the tank is filled in cubic meters per second. (c) Determine the time interval, in hours, required to fill a 1-m^3 volume at the same rate. (1 U.S. gal = 231 in.^{3})

24. Find the height or length of these natural wonders in kilometers, meters and centimeters. (a) The longest cave system in the world is the Mammoth Cave system in central Kentucky. It has a mapped length of 348 mi. (b) In the United States, the waterfall with the greatest single drop is Ribbon Falls, which falls 1 612 ft. (c) Mount McKinley in Denali National Park, Alaska, is America's highest mountain at a height of 20 320 ft. (d) The deepest canyon in the United States is King's Canyon in California with a depth of 8 200 ft.

25. A solid piece of lead has a mass of 23.94 g and a volume of 2.10 cm^3. From these data, calculate the density of lead in SI units (kg/m^3).

26. A *section* of land has an area of 1 square mile and contains 640 acres. Determine the number of square meters in 1 acre.

27. An ore loader moves 1 200 tons/h from a mine to the surface. Convert this rate to lb/s, using 1 ton = 2 000 lb.

28. (a) Find a conversion factor to convert from miles per hour to kilometers per hour. (b) In the past, a federal law mandated that highway speed limits would be 55 mi/h. Use the conversion factor of part (a) to find this speed in kilometers per hour. (c) The maximum highway speed is now 65 mi/h in some places. In kilometers per hour, how much increase is this over the 55 mi/h limit?

29. At the time of this book's printing, the U.S. national debt is about $6 trillion. (a) If payments were made at the rate of $1 000 per second, how many years would it take to pay off the debt, assuming no interest were charged? (b) A dollar bill is about 15.5 cm long. If six trillion dollar bills were laid end to end around the Earth's equator, how many times would they encircle the planet? Take the radius of the Earth at the equator to be 6 378 km. (*Note:* Before doing any of these calculations, try to guess at the answers. You may be very surprised.)

30. The mass of the Sun is 1.99×10^{30} kg, and the mass of an atom of hydrogen, of which the Sun is mostly composed, is 1.67×10^{-27} kg. How many atoms are in the Sun?

31. One gallon of paint (volume = 3.78×10^{-3} m^3) covers an area of 25.0 m^2. What is the thickness of the paint on the wall?

32. A pyramid has a height of 481 ft and its base covers an area of 13.0 acres (Fig. P1.32). If the volume of a pyramid is given by the expression $V = \frac{1}{3} Bh$, where B is the area of the base and h is the height, find the volume of this pyramid in cubic meters. (1 acre = 43 560 ft^2)

Sylvain Grandadam/Photo Researchers, Inc.

Figure P1.32 Problems 32 and 33.

33. The pyramid described in Problem 32 contains approximately 2 million stone blocks that average 2.50 tons each. Find the weight of this pyramid in pounds.

34. Assuming that 70% of the Earth's surface is covered with water at an average depth of 2.3 mi, estimate the mass of the water on the Earth in kilograms.

35. A hydrogen atom has a diameter of approximately 1.06×10^{-10} m, as defined by the diameter of the spherical electron cloud around the nucleus. The hydrogen nucleus has a diameter of approximately 2.40×10^{-15} m. (a) For a scale model, represent the diameter of the hydrogen atom by the length of an American football field

(100 yd = 300 ft), and determine the diameter of the nucleus in millimeters. (b) The atom is how many times larger in volume than its nucleus?

36. The nearest stars to the Sun are in the Alpha Centauri multiple-star system, about 4.0×10^{13} km away. If the Sun, with a diameter of 1.4×10^9 m, and Alpha Centauri A are both represented by cherry pits 7.0 mm in diameter, how far apart should the pits be placed to represent the Sun and its neighbor to scale?

37. The diameter of our disk-shaped galaxy, the Milky Way, is about 1.0×10^5 lightyears (ly). The distance to Messier 31, which is Andromeda, the spiral galaxy nearest to the Milky Way, is about 2.0 million ly. If a scale model represents the Milky Way and Andromeda galaxies as dinner plates 25 cm in diameter, determine the distance between the two plates.

38. The mean radius of the Earth is 6.37×10^6 m, and that of the Moon is 1.74×10^8 cm. From these data calculate (a) the ratio of the Earth's surface area to that of the Moon and (b) the ratio of the Earth's volume to that of the Moon. Recall that the surface area of a sphere is $4\pi r^2$ and the volume of a sphere is $\frac{4}{3}\pi r^3$.

39. One cubic meter (1.00 m^3) of aluminum has a mass of 2.70×10^3 kg, and 1.00 m^3 of iron has a mass of 7.86×10^3 kg. Find the radius of a solid aluminum sphere that will balance a solid iron sphere of radius 2.00 cm on an equal-arm balance.

40. Let ρ_{Al} represent the density of aluminum and ρ_{Fe} that of iron. Find the radius of a solid aluminum sphere that balances a solid iron sphere of radius r_{Fe} on an equal-arm balance.

Section 1.6 Estimates and Order-of-Magnitude Calculations

41. Estimate the number of Ping-Pong balls that would fit into a typical-size room (without being crushed). In your solution state the quantities you measure or estimate and the values you take for them.

42. An automobile tire is rated to last for 50 000 miles. To an order of magnitude, through how many revolutions will it turn? In your solution state the quantities you measure or estimate and the values you take for them.

43. Grass grows densely everywhere on a quarter-acre plot of land. What is the order of magnitude of the number of blades of grass on this plot? Explain your reasoning. Note that 1 acre = 43 560 ft^2.

44. Approximately how many raindrops fall on a one-acre lot during a one-inch rainfall? Explain your reasoning.

45. Compute the order of magnitude of the mass of a bathtub half full of water. Compute the order of magnitude of the mass of a bathtub half full of pennies. In your solution list the quantities you take as data and the value you measure or estimate for each.

46. Soft drinks are commonly sold in aluminum containers. To an order of magnitude, how many such containers are thrown away or recycled each year by U.S. consumers?

How many tons of aluminum does this represent? In your solution state the quantities you measure or estimate and the values you take for them.

47. To an order of magnitude, how many piano tuners are in New York City? The physicist Enrico Fermi was famous for asking questions like this on oral Ph.D. qualifying examinations. His own facility in making order-of-magnitude calculations is exemplified in Problem 45.48.

Section 1.7 Significant Figures

> *Note:* Appendix B.8 on propagation of uncertainty may be useful in solving some problems in this section.

48. A rectangular plate has a length of (21.3 ± 0.2) cm and a width of (9.8 ± 0.1) cm. Calculate the area of the plate, including its uncertainty.

49. The radius of a circle is measured to be (10.5 ± 0.2) m. Calculate the (a) area and (b) circumference of the circle and give the uncertainty in each value.

50. How many significant figures are in the following numbers? (a) 78.9 ± 0.2 (b) 3.788×10^9 (c) 2.46×10^{-6} (d) 0.005 3.

51. The radius of a solid sphere is measured to be (6.50 ± 0.20) cm, and its mass is measured to be (1.85 ± 0.02) kg. Determine the density of the sphere in kilograms per cubic meter and the uncertainty in the density.

52. Carry out the following arithmetic operations: (a) the sum of the measured values 756, 37.2, 0.83, and 2.5; (b) the product $0.003\,2 \times 356.3$; (c) the product $5.620 \times \pi$.

53. The *tropical year*, the time from vernal equinox to the next vernal equinox, is the basis for our calendar. It contains 365.242 199 days. Find the number of seconds in a tropical year.

54. A farmer measures the distance around a rectangular field. The length of the long sides of the rectangle is found to be 38.44 m, and the length of the short sides is found to be 19.5 m. What is the total distance around the field?

55. A sidewalk is to be constructed around a swimming pool that measures (10.0 ± 0.1) m by (17.0 ± 0.1) m. If the sidewalk is to measure (1.00 ± 0.01) m wide by (9.0 ± 0.1) cm thick, what volume of concrete is needed, and what is the approximate uncertainty of this volume?

Additional Problems

56. In a situation where data are known to three significant digits, we write 6.379 m = 6.38 m and 6.374 m = 6.37 m. When a number ends in 5, we arbitrarily choose to write 6.375 m = 6.38 m. We could equally well write 6.375 m = 6.37 m, "rounding down" instead of "rounding up," because we would change the number 6.375 by equal increments in both cases. Now consider an order-of-magnitude

estimate, in which we consider factors rather than increments. We write 500 m $\sim 10^3$ m because 500 differs from 100 by a factor of 5 while it differs from 1 000 by only a factor of 2. We write 437 m $\sim 10^3$ m and 305 m $\sim 10^2$ m. What distance differs from 100 m and from 1 000 m by equal factors, so that we could equally well choose to represent its order of magnitude either as $\sim 10^2$ m or as $\sim 10^3$ m?

57. For many electronic applications, such as in computer chips, it is desirable to make components as small as possible to keep the temperature of the components low and to increase the speed of the device. Thin metallic coatings (films) can be used instead of wires to make electrical connections. Gold is especially useful because it does not oxidize readily. Its atomic mass is 197 u. A gold film can be no thinner than the size of a gold atom. Calculate the minimum coating thickness, assuming that a gold atom occupies a cubical volume in the film that is equal to the volume it occupies in a large piece of metal. This geometric model yields a result of the correct order of magnitude.

58. The basic function of the carburetor of an automobile is to "atomize" the gasoline and mix it with air to promote rapid combustion. As an example, assume that 30.0 cm^3 of gasoline is atomized into N spherical droplets, each with a radius of 2.00×10^{-5} m. What is the total surface area of these N spherical droplets?

59. The consumption of natural gas by a company satisfies the empirical equation $V = 1.50t + 0.008\ 00t^2$, where V is the volume in millions of cubic feet and t the time in months. Express this equation in units of cubic feet and seconds. Assign proper units to the coefficients. Assume a month is equal to 30.0 days.

60. In physics it is important to use mathematical approximations. Demonstrate that for small angles ($< 20°$)

$$\tan \alpha \approx \sin \alpha \approx \alpha = \pi\alpha'/180°$$

where α is in radians and α' is in degrees. Use a calculator to find the largest angle for which $\tan \alpha$ may be approximated by $\sin \alpha$ if the error is to be less than 10.0%.

61. A high fountain of water is located at the center of a circular pool as in Figure P1.61. Not wishing to get his feet wet,

Figure P1.61

a student walks around the pool and measures its circumference to be 15.0 m. Next, the student stands at the edge of the pool and uses a protractor to gauge the angle of elevation of the top of the fountain to be 55.0°. How high is the fountain?

62. Collectible coins are sometimes plated with gold to enhance their beauty and value. Consider a commemorative quarter-dollar advertised for sale at $4.98. It has a diameter of 24.1 mm, a thickness of 1.78 mm, and is completely covered with a layer of pure gold 0.180 μm thick. The volume of the plating is equal to the thickness of the layer times the area to which it is applied. The patterns on the faces of the coin and the grooves on its edge have a negligible effect on its area. Assume that the price of gold is $10.0 per gram. Find the cost of the gold added to the coin. Does the cost of the gold significantly enhance the value of the coin?

63. There are nearly $\pi \times 10^7$ s in one year. Find the percentage error in this approximation, where "percentage error" is defined as

$$\text{Percentage error} = \frac{|\text{assumed value} - \text{true value}|}{\text{true value}} \times 100\%$$

64. Assume that an object covers an area A and has a uniform height h. If its cross-sectional area is uniform over its height, then its volume is given by $V = Ah$. (a) Show that $V = Ah$ is dimensionally correct. (b) Show that the volumes of a cylinder and of a rectangular box can be written in the form $V = Ah$, identifying A in each case. (Note that A, sometimes called the "footprint" of the object, can have any shape and the height can be replaced by average thickness in general.)

65. A child loves to watch as you fill a transparent plastic bottle with shampoo. Every horizontal cross-section is a circle, but the diameters of the circles have different values, so that the bottle is much wider in some places than others. You pour in bright green shampoo with constant volume flow rate 16.5 cm^3/s. At what rate is its level in the bottle rising (a) at a point where the diameter of the bottle is 6.30 cm and (b) at a point where the diameter is 1.35 cm?

66. One cubic centimeter of water has a mass of 1.00×10^{-3} kg. (a) Determine the mass of 1.00 m^3 of water. (b) Biological substances are 98% water. Assume that they have the same density as water to estimate the masses of a cell that has a diameter of 1.0 μm, a human kidney, and a fly. Model the kidney as a sphere with a radius of 4.0 cm and the fly as a cylinder 4.0 mm long and 2.0 mm in diameter.

67. Assume there are 100 million passenger cars in the United States and that the average fuel consumption is 20 mi/gal of gasoline. If the average distance traveled by each car is 10 000 mi/yr, how much gasoline would be saved per year if average fuel consumption could be increased to 25 mi/gal?

68. A creature moves at a speed of 5.00 furlongs per fortnight (not a very common unit of speed). Given that 1 furlong = 220 yards and 1 fortnight = 14 days, determine the speed of the creature in m/s. What kind of creature do you think it might be?

69. The distance from the Sun to the nearest star is about 4×10^{16} m. The Milky Way galaxy is roughly a disk of diameter $\sim 10^{21}$ m and thickness $\sim 10^{19}$ m. Find the order of magnitude of the number of stars in the Milky Way. Assume the distance between the Sun and our nearest neighbor is typical.

70. The data in the following table represent measurements of the masses and dimensions of solid cylinders of aluminum, copper, brass, tin, and iron. Use these data to calculate the densities of these substances. Compare your results for aluminum, copper, and iron with those given in Table 1.5.

Substance	Mass (g)	Diameter (cm)	Length (cm)
Aluminum	51.5	2.52	3.75
Copper	56.3	1.23	5.06
Brass	94.4	1.54	5.69
Tin	69.1	1.75	3.74
Iron	216.1	1.89	9.77

71. (a) How many seconds are in a year? (b) If one micrometeorite (a sphere with a diameter of 1.00×10^{-6} m) strikes each square meter of the Moon each second, how many years will it take to cover the Moon to a depth of 1.00 m? To solve this problem, you can consider a cubic box on the Moon 1.00 m on each edge, and find how long it will take to fill the box.

Answers to Quick Quizzes

1.1 (a). Because the density of aluminum is smaller than that of iron, a larger volume of aluminum is required for a given mass than iron.

1.2 False. Dimensional analysis gives the units of the proportionality constant but provides no information about its numerical value. To determine its numerical value requires either experimental data or geometrical reasoning. For example, in the generation of the equation $x = \frac{1}{2}at^2$, because the factor $\frac{1}{2}$ is dimensionless, there is no way of determining it using dimensional analysis.

1.3 (b). Because kilometers are shorter than miles, a larger number of kilometers is required for a given distance than miles.

1.4 Reporting all these digits implies you have determined the location of the center of the chair's seat to the nearest $\pm 0.000\ 000\ 000\ 1$ m. This roughly corresponds to being able to count the atoms in your meter stick because each of them is about that size! It would be better to record the measurement as 1.044 m: this indicates that you know the position to the nearest millimeter, assuming the meter stick has millimeter markings on its scale.

Motion in One Dimension

▲ One of the physical quantities we will study in this chapter is the velocity of an object moving in a straight line. Downhill skiers can reach velocities with a magnitude greater than 100 km/h. (Jean Y. Ruszniewski/Getty Images)

As a first step in studying classical mechanics, we describe motion in terms of space and time while ignoring the agents that caused that motion. This portion of classical mechanics is called *kinematics*. (The word *kinematics* has the same root as *cinema*. Can you see why?) In this chapter we consider only motion in one dimension, that is, motion along a straight line. We first define position, displacement, velocity, and acceleration. Then, using these concepts, we study the motion of objects traveling in one dimension with a constant acceleration.

From everyday experience we recognize that motion represents a continuous change in the position of an object. In physics we can categorize motion into three types: translational, rotational, and vibrational. A car moving down a highway is an example of translational motion, the Earth's spin on its axis is an example of rotational motion, and the back-and-forth movement of a pendulum is an example of vibrational motion. In this and the next few chapters, we are concerned only with translational motion. (Later in the book we shall discuss rotational and vibrational motions.)

In our study of translational motion, we use what is called the **particle model**—we describe the moving object as a *particle* regardless of its size. In general, **a particle is a point-like object—that is, an object with mass but having infinitesimal size.** For example, if we wish to describe the motion of the Earth around the Sun, we can treat the Earth as a particle and obtain reasonably accurate data about its orbit. This approximation is justified because the radius of the Earth's orbit is large compared with the dimensions of the Earth and the Sun. As an example on a much smaller scale, it is possible to explain the pressure exerted by a gas on the walls of a container by treating the gas molecules as particles, without regard for the internal structure of the molecules.

2.1 Position, Velocity, and Speed

Position

The motion of a particle is completely known if the particle's position in space is known at all times. A particle's **position** is the location of the particle with respect to a chosen reference point that we can consider to be the origin of a coordinate system.

Consider a car moving back and forth along the x axis as in Figure 2.1a. When we begin collecting position data, the car is 30 m to the right of a road sign, which we will use to identify the reference position $x = 0$. (Let us assume that all data in this example are known to two significant figures. To convey this information, we should report the initial position as 3.0×10^1 m. We have written this value in the simpler form 30 m to make the discussion easier to follow.) We will use the particle model by identifying some point on the car, perhaps the front door handle, as a particle representing the entire car.

We start our clock and once every 10 s note the car's position relative to the sign at $x = 0$. As you can see from Table 2.1, the car moves to the right (which we have

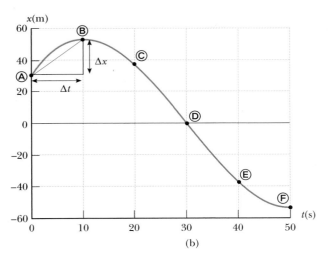

(b)

(a)

Table 2.1

Position of the Car at Various Times		
Position	**t(s)**	**x(m)**
Ⓐ	0	30
Ⓑ	10	52
Ⓒ	20	38
Ⓓ	30	0
Ⓔ	40	−37
Ⓕ	50	−53

Active Figure 2.1 (a) A car moves back and forth along a straight line taken to be the x axis. Because we are interested only in the car's translational motion, we can model it as a particle. (b) Position–time graph for the motion of the "particle."

At the Active Figures link at http://www.pse6.com, you can move each of the six points Ⓐ through Ⓕ and observe the motion of the car pictorially and graphically as it follows a smooth path through the six points.

defined as the positive direction) during the first 10 s of motion, from position Ⓐ to position Ⓑ. After Ⓑ, the position values begin to decrease, suggesting that the car is backing up from position Ⓑ through position Ⓕ. In fact, at Ⓓ, 30 s after we start measuring, the car is alongside the road sign (see Figure 2.1a) that we are using to mark our origin of coordinates. It continues moving to the left and is more than 50 m to the left of the sign when we stop recording information after our sixth data point. A graphical representation of this information is presented in Figure 2.1b. Such a plot is called a *position–time graph*.

Given the data in Table 2.1, we can easily determine the change in position of the car for various time intervals. The **displacement** of a particle is defined as its change in position in some time interval. As it moves from an initial position x_i to a final position x_f, the displacement of the particle is given by $x_f - x_i$. We use the Greek letter delta (Δ) to denote the *change* in a quantity. Therefore, we write the displacement, or change in position, of the particle as

Displacement

$$\Delta x \equiv x_f - x_i \qquad (2.1)$$

From this definition we see that Δx is positive if x_f is greater than x_i and negative if x_f is less than x_i.

It is very important to recognize the difference between displacement and distance traveled. **Distance** is the length of a path followed by a particle. Consider, for example, the basketball players in Figure 2.2. If a player runs from his own basket down the court to the other team's basket and then returns to his own basket, the *displacement* of the player during this time interval is zero, because he ended up at the same point as he started. During this time interval, however, he covered a *distance* of twice the length of the basketball court.

Displacement is an example of a vector quantity. Many other physical quantities, including position, velocity, and acceleration, also are vectors. In general, **a vector quantity requires the specification of both direction and magnitude.** By contrast, **a scalar quantity has a numerical value and no direction.** In this chapter, we use positive ($+$) and negative ($-$) signs to indicate vector direction. We can do this because the chapter deals with one-dimensional motion only; this means that any object we study can be moving only along a straight line. For example, for horizontal motion let us arbitrarily specify to the right as being the positive direction. It follows that any object always moving to the right undergoes a positive displacement $\Delta x > 0$, and any object moving to the left undergoes a negative displacement, so that $\Delta x < 0$. We shall treat vector quantities in greater detail in Chapter 3.

For our basketball player in Figure 2.2, if the trip from his own basket to the opposing basket is described by a displacement of $+28$ m, the trip in the reverse direction represents a displacement of -28 m. Each trip, however, represents a distance of 28 m, because distance is a scalar quantity. The total distance for the trip down the court and back is 56 m. Distance, therefore, is always represented as a positive number, while displacement can be either positive or negative.

There is one very important point that has not yet been mentioned. Note that the data in Table 2.1 results only in the six data points in the graph in Figure 2.1b. The smooth curve drawn through the six points in the graph is only a *possibility* of the actual motion of the car. We only have information about six instants of time—we have no idea what happened in between the data points. The smooth curve is a *guess* as to what happened, but keep in mind that it is *only* a guess.

If the smooth curve does represent the actual motion of the car, the graph contains information about the entire 50-s interval during which we watch the car move. It is much easier to see changes in position from the graph than from a verbal description or even a table of numbers. For example, it is clear that the car was covering more ground during the middle of the 50-s interval than at the end. Between positions Ⓒ and Ⓓ, the car traveled almost 40 m, but during the last 10 s, between positions Ⓔ and Ⓕ, it moved less than half that far. A common way of comparing these different motions is to divide the displacement Δx that occurs between two clock readings by the length of that particular time interval Δt. This turns out to be a very useful ratio, one that we shall use many times. This ratio has been given a special name—*average velocity*. **The average velocity $\bar{v}_x$ of a particle is defined as the**

Figure 2.2 On this basketball court, players run back and forth for the entire game. The distance that the players run over the duration of the game is nonzero. The displacement of the players over the duration of the game is approximately zero because they keep returning to the same point over and over again.

Ken White/Allsport/Getty Images

particle's displacement Δx divided by the time interval Δt during which that displacement occurs:

$$\bar{v}_x \equiv \frac{\Delta x}{\Delta t} \qquad (2.2)$$

◀ **Average velocity**

where the subscript x indicates motion along the x axis. From this definition we see that average velocity has dimensions of length divided by time (L/T)—meters per second in SI units.

The average velocity of a particle moving in one dimension can be positive or negative, depending on the sign of the displacement. (The time interval Δt is always positive.) If the coordinate of the particle increases in time (that is, if $x_f > x_i$), then Δx is positive and $\bar{v}_x = \Delta x / \Delta t$ is positive. This case corresponds to a particle moving in the positive x direction, that is, toward larger values of x. If the coordinate decreases in time (that is, if $x_f < x_i$) then Δx is negative and hence $\bar{v}_x$ is negative. This case corresponds to a particle moving in the negative x direction.

We can interpret average velocity geometrically by drawing a straight line between any two points on the position–time graph in Figure 2.1b. This line forms the hypotenuse of a right triangle of height Δx and base Δt. The slope of this line is the ratio $\Delta x / \Delta t$, which is what we have defined as average velocity in Equation 2.2. For example, the line between positions Ⓐ and Ⓑ in Figure 2.1b has a slope equal to the average velocity of the car between those two times, $(52\text{ m} - 30\text{ m})/(10\text{ s} - 0) = 2.2\text{ m/s}$.

In everyday usage, the terms *speed* and *velocity* are interchangeable. In physics, however, there is a clear distinction between these two quantities. Consider a marathon runner who runs more than 40 km, yet ends up at his starting point. His total displacement is zero, so his average velocity is zero! Nonetheless, we need to be able to quantify how fast he was running. A slightly different ratio accomplishes this for us. The **average speed** of a particle, a scalar quantity, is defined as **the total distance traveled divided by the total time interval required to travel that distance:**

$$\text{Average speed} = \frac{\text{total distance}}{\text{total time}} \qquad (2.3)$$

◀ **Average speed**

The SI unit of average speed is the same as the unit of average velocity: meters per second. However, unlike average velocity, average speed has no direction and hence carries no algebraic sign. Notice the distinction between average velocity and average speed—average velocity (Eq. 2.2) is the *displacement* divided by the time interval, while average speed (Eq. 2.3) is the *distance* divided by the time interval.

Knowledge of the average velocity or average speed of a particle does not provide information about the details of the trip. For example, suppose it takes you 45.0 s to travel 100 m down a long straight hallway toward your departure gate at an airport. At the 100-m mark, you realize you missed the rest room, and you return back 25.0 m along the same hallway, taking 10.0 s to make the return trip. The magnitude of the average *velocity* for your trip is $+ 75.0\text{ m}/55.0\text{ s} = + 1.36\text{ m/s}$. The average *speed* for your trip is $125\text{ m}/55.0\text{ s} = 2.27\text{ m/s}$. You may have traveled at various speeds during the walk. Neither average velocity nor average speed provides information about these details.

 PITFALL PREVENTION

2.1   Average Speed and Average Velocity

The magnitude of the average velocity is *not* the average speed. For example, consider the marathon runner discussed here. The magnitude of the average velocity is zero, but the average speed is clearly not zero.

Quick Quiz 2.1   Under which of the following conditions is the magnitude of the average velocity of a particle moving in one dimension smaller than the average speed over some time interval? (a) A particle moves in the $+ x$ direction without reversing. (b) A particle moves in the $- x$ direction without reversing. (c) A particle moves in the $+ x$ direction and then reverses the direction of its motion. (d) There are no conditions for which this is true.

Example 2.1   Calculating the Average Velocity and Speed

Find the displacement, average velocity, and average speed of the car in Figure 2.1a between positions Ⓐ and Ⓕ.

Solution From the position–time graph given in Figure 2.1b, note that $x_A = 30$ m at $t_A = 0$ s and that $x_F = -53$ m at $t_F = 50$ s. Using these values along with the definition of displacement, Equation 2.1, we find that

$$\Delta x = x_F - x_A = -53 \text{ m} - 30 \text{ m} = \boxed{-83 \text{ m}}$$

This result means that the car ends up 83 m in the negative direction (to the left, in this case) from where it started. This number has the correct units and is of the same order of magnitude as the supplied data. A quick look at Figure 2.1a indicates that this is the correct answer.

It is difficult to estimate the average velocity without completing the calculation, but we expect the units to be meters per second. Because the car ends up to the left of where we started taking data, we know the average velocity must be negative. From Equation 2.2,

$$\overline{v}_x = \frac{\Delta x}{\Delta t} = \frac{x_f - x_i}{t_f - t_i} = \frac{x_F - x_A}{t_F - t_A}$$

$$= \frac{-53 \text{ m} - 30 \text{ m}}{50 \text{ s} - 0 \text{ s}} = \frac{-83 \text{ m}}{50 \text{ s}}$$

$$= \boxed{-1.7 \text{ m/s}}$$

We cannot unambiguously find the average speed of the car from the data in Table 2.1, because we do not have information about the positions of the car between the data points. If we adopt the assumption that the details of the car's position are described by the curve in Figure 2.1b, then the distance traveled is 22 m (from Ⓐ to Ⓑ) plus 105 m (from Ⓑ to Ⓕ) for a total of 127 m. We find the car's average speed for this trip by dividing the distance by the total time (Eq. 2.3):

$$\text{Average speed} = \frac{127 \text{ m}}{50 \text{ s}} = \boxed{2.5 \text{ m/s}}$$

2.2   Instantaneous Velocity and Speed

Often we need to know the velocity of a particle at a particular instant in time, rather than the average velocity over a finite time interval. For example, even though you might want to calculate your average velocity during a long automobile trip, you would be especially interested in knowing your velocity at the *instant* you noticed the police car parked alongside the road ahead of you. In other words, you would like to be able to specify your velocity just as precisely as you can specify your position by noting what is happening at a specific clock reading—that is, at some specific instant. It may not be immediately obvious how to do this. What does it mean to talk about how fast something is moving if we "freeze time" and talk only about an individual instant? This is a subtle point not thoroughly understood until the late 1600s. At that time, with the invention of calculus, scientists began to understand how to describe an object's motion at any moment in time.

To see how this is done, consider Figure 2.3a, which is a reproduction of the graph in Figure 2.1b. We have already discussed the average velocity for the interval during which the car moved from position Ⓐ to position Ⓑ (given by the slope of the dark blue line) and for the interval during which it moved from Ⓐ to Ⓕ (represented by the slope of the light blue line and calculated in Example 2.1). Which of these two lines do you think is a closer approximation of the initial velocity of the car? The car starts out by moving to the right, which we defined to be the positive direction. Therefore, being positive, the value of the average velocity during the Ⓐ to Ⓑ interval is more representative of the initial value than is the value of the average velocity during the Ⓐ to Ⓕ interval, which we determined to be negative in Example 2.1. Now let us focus on the dark blue line and slide point Ⓑ to the left along the curve, toward point Ⓐ, as in Figure 2.3b. The line between the points becomes steeper and steeper, and as the two points become extremely close together, the line becomes a tangent line to the curve, indicated by the green line in Figure 2.3b. The slope of this tangent line

PITFALL PREVENTION

2.2   Slopes of Graphs

In any graph of physical data, the *slope* represents the ratio of the change in the quantity represented on the vertical axis to the change in the quantity represented on the horizontal axis. Remember that *a slope has units* (unless both axes have the same units). The units of slope in Figure 2.1b and Figure 2.3 are m/s, the units of velocity.

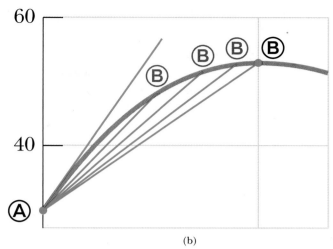

(a)

(b)

Active Figure 2.3 (a) Graph representing the motion of the car in Figure 2.1. (b) An enlargement of the upper-left-hand corner of the graph shows how the blue line between positions Ⓐ and Ⓑ approaches the green tangent line as point Ⓑ is moved closer to point Ⓐ.

 At the Active Figures link at http://www.pse6.com, *you can move point* Ⓑ *as suggested in (b) and observe the blue line approaching the green tangent line.*

represents the velocity of the car at the moment we started taking data, at point Ⓐ. What we have done is determine the *instantaneous velocity* at that moment. In other words, **the instantaneous velocity v_x equals the limiting value of the ratio $\Delta x/\Delta t$ as Δt approaches zero:**[1]

$$v_x \equiv \lim_{\Delta t \to 0} \frac{\Delta x}{\Delta t} \tag{2.4}$$

In calculus notation, this limit is called the *derivative* of x with respect to t, written dx/dt:

$$v_x \equiv \lim_{\Delta t \to 0} \frac{\Delta x}{\Delta t} = \frac{dx}{dt} \tag{2.5}$$

Instantaneous velocity

The instantaneous velocity can be positive, negative, or zero. When the slope of the position–time graph is positive, such as at any time during the first 10 s in Figure 2.3, v_x is positive—the car is moving toward larger values of x. After point Ⓑ, v_x is negative because the slope is negative—the car is moving toward smaller values of x. At point Ⓑ, the slope and the instantaneous velocity are zero—the car is momentarily at rest.

From here on, we use the word *velocity* to designate instantaneous velocity. When it is *average velocity* we are interested in, we shall always use the adjective *average*.

The **instantaneous speed** of a particle is defined as the magnitude of its instantaneous velocity. As with average speed, instantaneous speed has no direction associated with it and hence carries no algebraic sign. For example, if one particle has an instantaneous velocity of $+25$ m/s along a given line and another particle has an instantaneous velocity of -25 m/s along the same line, both have a speed[2] of 25 m/s.

▲ **PITFALL PREVENTION**

2.3 Instantaneous Speed and Instantaneous Velocity

In Pitfall Prevention 2.1, we argued that the magnitude of the average velocity is not the average speed. Notice the difference when discussing instantaneous values. The magnitude of the instantaneous velocity *is* the instantaneous speed. In an infinitesimal time interval, the magnitude of the displacement is equal to the distance traveled by the particle.

---

[1]   Note that the displacement Δx also approaches zero as Δt approaches zero, so that the ratio looks like 0/0. As Δx and Δt become smaller and smaller, the ratio $\Delta x/\Delta t$ approaches a value equal to the slope of the line tangent to the *x*-versus-*t* curve.

[2]   As with velocity, we drop the adjective for instantaneous speed: "Speed" means instantaneous speed.

Conceptual Example 2.2   The Velocity of Different Objects

Consider the following one-dimensional motions: **(A)** A ball thrown directly upward rises to a highest point and falls back into the thrower's hand. **(B)** A race car starts from rest and speeds up to 100 m/s. **(C)** A spacecraft drifts through space at constant velocity. Are there any points in the motion of these objects at which the instantaneous velocity has the same value as the average velocity over the entire motion? If so, identify the point(s).

Solution (A) The average velocity for the thrown ball is zero because the ball returns to the starting point; thus its displacement is zero. (Remember that average velocity is defined as $\Delta x/\Delta t$.) There is one point at which the instantaneous velocity is zero—at the top of the motion.

(B) The car's average velocity cannot be evaluated unambiguously with the information given, but it must be some value between 0 and 100 m/s. Because the car will have every instantaneous velocity between 0 and 100 m/s at some time during the interval, there must be some instant at which the instantaneous velocity is equal to the average velocity.

(C) Because the spacecraft's instantaneous velocity is constant, its instantaneous velocity at *any* time and its average velocity over *any* time interval are the same.

Example 2.3   Average and Instantaneous Velocity

A particle moves along the x axis. Its position varies with time according to the expression $x = -4t + 2t^2$ where x is in meters and t is in seconds.[3] The position–time graph for this motion is shown in Figure 2.4. Note that the particle moves in the negative x direction for the first second of motion, is momentarily at rest at the moment $t = 1$ s, and moves in the positive x direction at times $t > 1$ s.

(A) Determine the displacement of the particle in the time intervals $t = 0$ to $t = 1$ s and $t = 1$ s to $t = 3$ s.

Solution During the first time interval, the slope is negative and hence the average velocity is negative. Thus, we know that the displacement between Ⓐ and Ⓑ must be a negative number having units of meters. Similarly, we expect the displacement between Ⓑ and Ⓓ to be positive.

In the first time interval, we set $t_i = t_A = 0$ and $t_f = t_B = 1$ s. Using Equation 2.1, with $x = -4t + 2t^2$, we obtain for the displacement between $t = 0$ and $t = 1$ s,

$$\Delta x_{A \to B} = x_f - x_i = x_B - x_A$$
$$= [-4(1) + 2(1)^2] - [-4(0) + 2(0)^2]$$
$$= \boxed{-2 \text{ m}}$$

To calculate the displacement during the second time interval ($t = 1$ s to $t = 3$ s), we set $t_i = t_B = 1$ s and $t_f = t_D = 3$ s:

$$\Delta x_{B \to D} = x_f - x_i = x_D - x_B$$
$$= [-4(3) + 2(3)^2] - [-4(1) + 2(1)^2]$$
$$= \boxed{+8 \text{ m}}$$

These displacements can also be read directly from the position–time graph.

(B) Calculate the average velocity during these two time intervals.

Solution In the first time interval, $\Delta t = t_f - t_i = t_B - t_A = 1$ s. Therefore, using Equation 2.2 and the displacement calculated in (a), we find that

$$\bar{v}_{x(A \to B)} = \frac{\Delta x_{A \to B}}{\Delta t} = \frac{-2 \text{ m}}{1 \text{ s}} = \boxed{-2 \text{ m/s}}$$

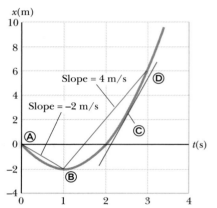

Figure 2.4 (Example 2.3) Position–time graph for a particle having an x coordinate that varies in time according to the expression $x = -4t + 2t^2$.

In the second time interval, $\Delta t = 2$ s; therefore,

$$\bar{v}_{x(B \to D)} = \frac{\Delta x_{B \to D}}{\Delta t} = \frac{8 \text{ m}}{2 \text{ s}} = \boxed{+4 \text{ m/s}}$$

These values are the same as the slopes of the lines joining these points in Figure 2.4.

(C) Find the instantaneous velocity of the particle at $t = 2.5$ s.

Solution We can guess that this instantaneous velocity must be of the same order of magnitude as our previous results, that is, a few meters per second. By measuring the slope of the green line at $t = 2.5$ s in Figure 2.4, we find that

$$v_x = \boxed{+6 \text{ m/s}}$$

---

[3] Simply to make it easier to read, we write the expression as $x = -4t + 2t^2$ rather than as $x = (-4.00 \text{ m/s})t + (2.00 \text{ m/s}^2)t^{2.00}$. When an equation summarizes measurements, consider its coefficients to have as many significant digits as other data quoted in a problem. Consider its coefficients to have the units required for dimensional consistency. When we start our clocks at $t = 0$, we usually do not mean to limit the precision to a single digit. Consider any zero value in this book to have as many significant figures as you need.

2.3 Acceleration

In the last example, we worked with a situation in which the velocity of a particle changes while the particle is moving. This is an extremely common occurrence. (How constant is your velocity as you ride a city bus or drive on city streets?) It is possible to quantify changes in velocity as a function of time similarly to the way in which we quantify changes in position as a function of time. When the velocity of a particle changes with time, the particle is said to be *accelerating*. For example, the magnitude of the velocity of a car increases when you step on the gas and decreases when you apply the brakes. Let us see how to quantify acceleration.

Suppose an object that can be modeled as a particle moving along the x axis has an initial velocity v_{xi} at time t_i and a final velocity v_{xf} at time t_f, as in Figure 2.5a.

> The average acceleration $\bar{a}_x$ of the particle is defined as the *change* in velocity Δv_x divided by the time interval Δt during which that change occurs:

$$\bar{a}_x \equiv \frac{\Delta v_x}{\Delta t} = \frac{v_{xf} - v_{xi}}{t_f - t_i} \qquad (2.6)$$

Average acceleration

As with velocity, when the motion being analyzed is one-dimensional, we can use positive and negative signs to indicate the direction of the acceleration. Because the dimensions of velocity are L/T and the dimension of time is T, acceleration has dimensions of length divided by time squared, or L/T^2. The SI unit of acceleration is meters per second squared (m/s^2). It might be easier to interpret these units if you think of them as meters per second per second. For example, suppose an object has an acceleration of $+2\ m/s^2$. You should form a mental image of the object having a velocity that is along a straight line and is increasing by 2 m/s during every interval of 1 s. If the object starts from rest, you should be able to picture it moving at a velocity of $+2$ m/s after 1 s, at $+4$ m/s after 2 s, and so on.

In some situations, the value of the average acceleration may be different over different time intervals. It is therefore useful to define the *instantaneous acceleration* as the limit of the average acceleration as Δt approaches zero. This concept is analogous to the definition of instantaneous velocity discussed in the previous section. If we imagine that point Ⓐ is brought closer and closer to point Ⓑ in Figure 2.5a and we take the limit of $\Delta v_x/\Delta t$ as Δt approaches zero, we obtain the instantaneous acceleration:

$$a_x \equiv \lim_{\Delta t \to 0} \frac{\Delta v_x}{\Delta t} = \frac{dv_x}{dt} \qquad (2.7)$$

Instantaneous acceleration

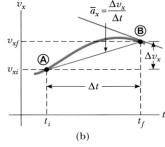

Figure 2.5 (a) A car, modeled as a particle, moving along the x axis from Ⓐ to Ⓑ has velocity v_{xi} at $t = t_i$ and velocity v_{xf} at $t = t_f$. (b) Velocity–time graph (rust) for the particle moving in a straight line. The slope of the blue straight line connecting Ⓐ and Ⓑ is the average acceleration in the time interval $\Delta t = t_f - t_i$.

PITFALL PREVENTION

2.4 Negative Acceleration

Keep in mind that *negative acceleration does not necessarily mean that an object is slowing down.* If the acceleration is negative, and the velocity is negative, the object is speeding up!

 PITFALL PREVENTION

2.5 Deceleration

The word *deceleration* has the common popular connotation of *slowing down.* We will not use this word in this text, because it further confuses the definition we have given for negative acceleration.

(a)

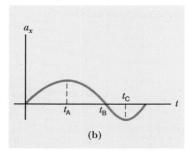

(b)

Figure 2.6 The instantaneous acceleration can be obtained from the velocity–time graph (a). At each instant, the acceleration in the a_x versus t graph (b) equals the slope of the line tangent to the v_x versus t curve (a).

That is, **the instantaneous acceleration equals the derivative of the velocity with respect to time,** which by definition is the slope of the velocity–time graph. The slope of the green line in Figure 2.5b is equal to the instantaneous acceleration at point Ⓑ. Thus, we see that just as the velocity of a moving particle is the slope at a point on the particle's x-t graph, the acceleration of a particle is the slope at a point on the particle's v_x-t graph. One can interpret the derivative of the velocity with respect to time as the time rate of change of velocity. If a_x is positive, the acceleration is in the positive x direction; if a_x is negative, the acceleration is in the negative x direction.

For the case of motion in a straight line, the direction of the velocity of an object and the direction of its acceleration are related as follows. **When the object's velocity and acceleration are in the same direction, the object is speeding up. On the other hand, when the object's velocity and acceleration are in opposite directions, the object is slowing down.**

To help with this discussion of the signs of velocity and acceleration, we can relate the acceleration of an object to the *force* exerted on the object. In Chapter 5 we formally establish that **force is proportional to acceleration:**

$$F \propto a$$

This proportionality indicates that acceleration is caused by force. Furthermore, force and acceleration are both vectors and the vectors act in the same direction. Thus, let us think about the signs of velocity and acceleration by imagining a force applied to an object and causing it to accelerate. Let us assume that the velocity and acceleration are in the same direction. This situation corresponds to an object moving in some direction that experiences a force acting in the same direction. In this case, the object speeds up! Now suppose the velocity and acceleration are in opposite directions. In this situation, the object moves in some direction and experiences a force acting in the opposite direction. Thus, the object slows down! It is very useful to equate the direction of the acceleration to the direction of a force, because it is easier from our everyday experience to think about what effect a force will have on an object than to think only in terms of the direction of the acceleration.

Quick Quiz 2.2 If a car is traveling eastward and slowing down, what is the direction of the force on the car that causes it to slow down? (a) eastward (b) westward (c) neither of these.

From now on we shall use the term *acceleration* to mean instantaneous acceleration. When we mean average acceleration, we shall always use the adjective *average.*

Because $v_x = dx/dt$, the acceleration can also be written

$$a_x = \frac{dv_x}{dt} = \frac{d}{dt}\left(\frac{dx}{dt}\right) = \frac{d^2x}{dt^2} \tag{2.8}$$

That is, in one-dimensional motion, the acceleration equals the *second derivative* of x with respect to time.

Figure 2.6 illustrates how an acceleration–time graph is related to a velocity–time graph. The acceleration at any time is the slope of the velocity–time graph at that time. Positive values of acceleration correspond to those points in Figure 2.6a where the velocity is increasing in the positive x direction. The acceleration reaches a maximum at time t_A, when the slope of the velocity–time graph is a maximum. The acceleration then goes to zero at time t_B, when the velocity is a maximum (that is, when the slope of the v_x-t graph is zero). The acceleration is negative when the velocity is decreasing in the positive x direction, and it reaches its most negative value at time t_C.

Quick Quiz 2.3 Make a velocity–time graph for the car in Figure 2.1a. The speed limit posted on the road sign is 30 km/h. True or false? The car exceeds the speed limit at some time within the interval.

Conceptual Example 2.4 Graphical Relationships between x, v_x, and a_x

The position of an object moving along the x axis varies with time as in Figure 2.7a. Graph the velocity versus time and the acceleration versus time for the object.

Solution The velocity at any instant is the slope of the tangent to the x-t graph at that instant. Between $t = 0$ and $t = t_A$, the slope of the x-t graph increases uniformly, and so the velocity increases linearly, as shown in Figure 2.7b. Between t_A and t_B, the slope of the x-t graph is constant, and so the velocity remains constant. At t_D, the slope of the x-t graph is zero, so the velocity is zero at that instant. Between t_D and t_E, the slope of the x-t graph and thus the velocity are negative and decrease uniformly in this interval. In the interval t_E to t_F, the slope of the x-t graph is still negative, and at t_F it goes to zero. Finally, after t_F, the slope of the x-t graph is zero, meaning that the object is at rest for $t > t_F$.

The acceleration at any instant is the slope of the tangent to the v_x-t graph at that instant. The graph of acceleration versus time for this object is shown in Figure 2.7c. The acceleration is constant and positive between 0 and t_A, where the slope of the v_x-t graph is positive. It is zero between t_A and t_B and for $t > t_F$ because the slope of the v_x-t graph is zero at these times. It is negative between t_B and t_E because the slope of the v_x-t graph is negative during this interval.

Note that the sudden changes in acceleration shown in Figure 2.7c are unphysical. Such instantaneous changes cannot occur in reality.

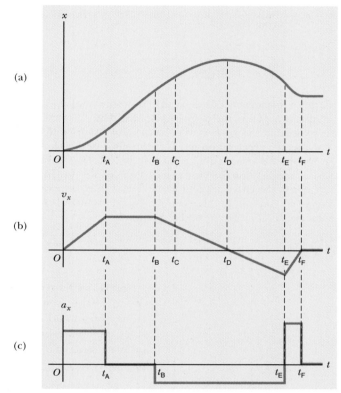

Figure 2.7 (Example 2.4) (a) Position–time graph for an object moving along the x axis. (b) The velocity–time graph for the object is obtained by measuring the slope of the position–time graph at each instant. (c) The acceleration–time graph for the object is obtained by measuring the slope of the velocity–time graph at each instant.

Example 2.5 Average and Instantaneous Acceleration

The velocity of a particle moving along the x axis varies in time according to the expression $v_x = (40 - 5t^2)$ m/s, where t is in seconds.

(A) Find the average acceleration in the time interval $t = 0$ to $t = 2.0$ s.

Solution Figure 2.8 is a v_x-t graph that was created from the velocity versus time expression given in the problem statement. Because the slope of the entire v_x-t curve is negative, we expect the acceleration to be negative.

We find the velocities at $t_i = t_A = 0$ and $t_f = t_B = 2.0$ s by substituting these values of t into the expression for the velocity:

$$v_{xA} = (40 - 5t_A{}^2)\ \text{m/s} = [40 - 5(0)^2]\ \text{m/s} = +40\ \text{m/s}$$

$$v_{xB} = (40 - 5t_B{}^2)\ \text{m/s} = [40 - 5(2.0)^2]\ \text{m/s} = +20\ \text{m/s}$$

Therefore, the average acceleration in the specified time interval $\Delta t = t_B - t_A = 2.0$ s is

$$\bar{a}_x = \frac{v_{xf} - v_{xi}}{t_f - t_i} = \frac{v_{xB} - v_{xA}}{t_B - t_A} = \frac{(20 - 40)\ \text{m/s}}{(2.0 - 0)\ \text{s}}$$

$$= -10\ \text{m/s}^2$$

The negative sign is consistent with our expectations—namely, that the average acceleration, which is represented by the slope of the line joining the initial and final points on the velocity–time graph, is negative.

(B) Determine the acceleration at $t = 2.0$ s.

Solution The velocity at any time t is $v_{xi} = (40 - 5t^2)$ m/s and the velocity at any later time $t + \Delta t$ is

$$v_{xf} = 40 - 5(t + \Delta t)^2 = 40 - 5t^2 - 10t\,\Delta t - 5(\Delta t)^2$$

Therefore, the change in velocity over the time interval Δt is

$$\Delta v_x = v_{xf} - v_{xi} = [-10t\,\Delta t - 5(\Delta t)^2]\ \text{m/s}$$

Dividing this expression by Δt and taking the limit of the result as Δt approaches zero gives the acceleration at *any* time t:

$$a_x = \lim_{\Delta t \to 0} \frac{\Delta v_x}{\Delta t} = \lim_{\Delta t \to 0}(-10t - 5\Delta t) = -10t\ \text{m/s}^2$$

Therefore, at $t = 2.0$ s,

$$a_x = (-10)(2.0)\ \text{m/s}^2 = \boxed{-20\ \text{m/s}^2}$$

Because the velocity of the particle is positive and the acceleration is negative, the particle is slowing down.

Note that the answers to parts (A) and (B) are different. The average acceleration in (A) is the slope of the blue line in Figure 2.8 connecting points Ⓐ and Ⓑ. The instantaneous acceleration in (B) is the slope of the green line tangent to the curve at point Ⓑ. Note also that the acceleration is *not* constant in this example. Situations involving constant acceleration are treated in Section 2.5.

Figure 2.8 (Example 2.5) The velocity–time graph for a particle moving along the x axis according to the expression $v_x = (40 - 5t^2)$ m/s. The acceleration at $t = 2$ s is equal to the slope of the green tangent line at that time.

So far we have evaluated the derivatives of a function by starting with the definition of the function and then taking the limit of a specific ratio. If you are familiar with calculus, you should recognize that there are specific rules for taking derivatives. These rules, which are listed in Appendix B.6, enable us to evaluate derivatives quickly. For instance, one rule tells us that the derivative of any constant is zero. As another example, suppose x is proportional to some power of t, such as in the expression

$$x = At^n$$

where A and n are constants. (This is a very common functional form.) The derivative of x with respect to t is

$$\frac{dx}{dt} = nAt^{n-1}$$

Applying this rule to Example 2.5, in which $v_x = 40 - 5t^2$, we find that the acceleration is $a_x = dv_x/dt = -10t$.

2.4 Motion Diagrams

The concepts of velocity and acceleration are often confused with each other, but in fact they are quite different quantities. It is instructive to use motion diagrams to describe the velocity and acceleration while an object is in motion.

A *stroboscopic* photograph of a moving object shows several images of the object, taken as the strobe light flashes at a constant rate. Figure 2.9 represents three sets of strobe photographs of cars moving along a straight roadway in a single direction, from left to right. The time intervals between flashes of the stroboscope are equal in each part of the diagram. In order not to confuse the two vector quantities, we use red for velocity vectors and violet for acceleration vectors in Figure 2.9. The vectors are

Active Figure 2.9 (a) Motion diagram for a car moving at constant velocity (zero acceleration). (b) Motion diagram for a car whose constant acceleration is in the direction of its velocity. The velocity vector at each instant is indicated by a red arrow, and the constant acceleration by a violet arrow. (c) Motion diagram for a car whose constant acceleration is in the direction *opposite* the velocity at each instant.

At the Active Figures link at http://www.pse6.com, *you can select the constant acceleration and initial velocity of the car and observe pictorial and graphical representations of its motion.*

sketched at several instants during the motion of the object. Let us describe the motion of the car in each diagram.

In Figure 2.9a, the images of the car are equally spaced, showing us that the car moves through the same displacement in each time interval. This is consistent with the car moving with *constant positive velocity* and *zero acceleration*. We could model the car as a particle and describe it as a particle moving with constant velocity.

In Figure 2.9b, the images become farther apart as time progresses. In this case, the velocity vector increases in time because the car's displacement between adjacent positions increases in time. This suggests that the car is moving with a *positive velocity* and a *positive acceleration*. The velocity and acceleration are in the same direction. In terms of our earlier force discussion, imagine a force pulling on the car in the same direction it is moving—it speeds up.

In Figure 2.9c, we can tell that the car slows as it moves to the right because its displacement between adjacent images decreases with time. In this case, this suggests that the car moves to the right with a constant negative acceleration. The velocity vector decreases in time and eventually reaches zero. From this diagram we see that the acceleration and velocity vectors are *not* in the same direction. The car is moving with a *positive velocity* but with a *negative acceleration*. (This type of motion is exhibited by a car that skids to a stop after applying its brakes.) The velocity and acceleration are in opposite directions. In terms of our earlier force discussion, imagine a force pulling on the car opposite to the direction it is moving—it slows down.

The violet acceleration vectors in Figures 2.9b and 2.9c are all of the same length. Thus, these diagrams represent motion with constant acceleration. This is an important type of motion that will be discussed in the next section.

Quick Quiz 2.4 Which of the following is true? (a) If a car is traveling eastward, its acceleration is eastward. (b) If a car is slowing down, its acceleration must be negative. (c) A particle with constant acceleration can never stop and stay stopped.

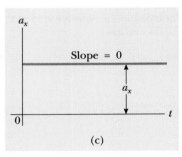

Active Figure 2.10 A particle moving along the *x* axis with constant acceleration a_x; (a) the position–time graph, (b) the velocity–time graph, and (c) the acceleration–time graph.

At the Active Figures link at http://www.pse6.com, you can adjust the constant acceleration and observe the effect on the position and velocity graphs.

Position as a function of velocity and time

Position as a function of time

2.5 One-Dimensional Motion with Constant Acceleration

If the acceleration of a particle varies in time, its motion can be complex and difficult to analyze. However, a very common and simple type of one-dimensional motion is that in which the acceleration is constant. When this is the case, the average acceleration $\bar{a}_x$ over any time interval is numerically equal to the instantaneous acceleration a_x at any instant within the interval, and the velocity changes at the same rate throughout the motion.

If we replace $\bar{a}_x$ by a_x in Equation 2.6 and take $t_i = 0$ and t_f to be any later time t, we find that

$$a_x = \frac{v_{xf} - v_{xi}}{t - 0}$$

or

$$v_{xf} = v_{xi} + a_x t \quad \text{(for constant } a_x) \tag{2.9}$$

This powerful expression enables us to determine an object's velocity at *any* time t if we know the object's initial velocity v_{xi} and its (constant) acceleration a_x. A velocity–time graph for this constant-acceleration motion is shown in Figure 2.10b. The graph is a straight line, the (constant) slope of which is the acceleration a_x; this is consistent with the fact that $a_x = dv_x/dt$ is a constant. Note that the slope is positive; this indicates a positive acceleration. If the acceleration were negative, then the slope of the line in Figure 2.10b would be negative.

When the acceleration is constant, the graph of acceleration versus time (Fig. 2.10c) is a straight line having a slope of zero.

Because velocity at constant acceleration varies linearly in time according to Equation 2.9, we can express the average velocity in any time interval as the arithmetic mean of the initial velocity v_{xi} and the final velocity v_{xf}:

$$\bar{v}_x = \frac{v_{xi} + v_{xf}}{2} \quad \text{(for constant } a_x) \tag{2.10}$$

Note that this expression for average velocity applies *only* in situations in which the acceleration is constant.

We can now use Equations 2.1, 2.2, and 2.10 to obtain the position of an object as a function of time. Recalling that Δx in Equation 2.2 represents $x_f - x_i$, and recognizing that $\Delta t = t_f - t_i = t - 0 = t$, we find

$$x_f - x_i = \bar{v}t = \tfrac{1}{2}(v_{xi} + v_{xf})t$$

$$x_f = x_i + \tfrac{1}{2}(v_{xi} + v_{xf})t \quad \text{(for constant } a_x) \tag{2.11}$$

This equation provides the final position of the particle at time t in terms of the initial and final velocities.

We can obtain another useful expression for the position of a particle moving with constant acceleration by substituting Equation 2.9 into Equation 2.11:

$$x_f = x_i + \tfrac{1}{2}[v_{xi} + (v_{xi} + a_x t)]t$$

$$x_f = x_i + v_{xi}t + \tfrac{1}{2}a_x t^2 \quad \text{(for constant } a_x) \tag{2.12}$$

This equation provides the final position of the particle at time t in terms of the initial velocity and the acceleration.

The position–time graph for motion at constant (positive) acceleration shown in Figure 2.10a is obtained from Equation 2.12. Note that the curve is a parabola.

The slope of the tangent line to this curve at $t = 0$ equals the initial velocity v_{xi}, and the slope of the tangent line at any later time t equals the velocity v_{xf} at that time.

Finally, we can obtain an expression for the final velocity that does not contain time as a variable by substituting the value of t from Equation 2.9 into Equation 2.11:

$$x_f = x_i + \tfrac{1}{2}(v_{xi} + v_{xf})\left(\frac{v_{xf} - v_{xi}}{a_x}\right) = \frac{v_{xf}^2 - v_{xi}^2}{2a_x}$$

$$v_{xf}^2 = v_{xi}^2 + 2a_x(x_f - x_i) \qquad \text{(for constant } a_x) \qquad (2.13)$$

Velocity as a function of position

This equation provides the final velocity in terms of the acceleration and the displacement of the particle.

For motion at *zero* acceleration, we see from Equations 2.9 and 2.12 that

$$\left.\begin{array}{r} v_{xf} = v_{xi} = v_x \\ x_f = x_i + v_x t \end{array}\right\} \quad \text{when } a_x = 0$$

That is, when the acceleration of a particle is zero, its velocity is constant and its position changes linearly with time.

Quick Quiz 2.5 In Figure 2.11, match each v_x-t graph on the left with the a_x-t graph on the right that best describes the motion.

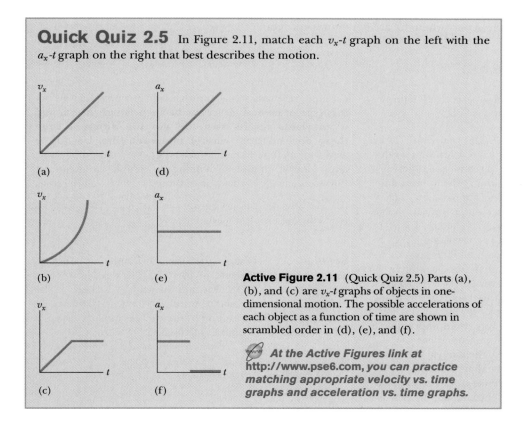

Active Figure 2.11 (Quick Quiz 2.5) Parts (a), (b), and (c) are v_x-t graphs of objects in one-dimensional motion. The possible accelerations of each object as a function of time are shown in scrambled order in (d), (e), and (f).

At the Active Figures link at http://www.pse6.com, you can practice matching appropriate velocity vs. time graphs and acceleration vs. time graphs.

Equations 2.9 through 2.13 are **kinematic equations that may be used to solve any problem involving one-dimensional motion at constant acceleration.** Keep in mind that these relationships were derived from the definitions of velocity and

Table 2.2

Kinematic Equations for Motion of a Particle Under Constant Acceleration	
Equation	**Information Given by Equation**
$v_{xf} = v_{xi} + a_x t$	Velocity as a function of time
$x_f = x_i + \frac{1}{2}(v_{xi} + v_{xf})t$	Position as a function of velocity and time
$x_f = x_i + v_{xi}t + \frac{1}{2}a_x t^2$	Position as a function of time
$v_{xf}{}^2 = v_{xi}{}^2 + 2a_x(x_f - x_i)$	Velocity as a function of position

Note: Motion is along the *x* axis.

acceleration, together with some simple algebraic manipulations and the requirement that the acceleration be constant.

The four kinematic equations used most often are listed in Table 2.2 for convenience. The choice of which equation you use in a given situation depends on what you know beforehand. Sometimes it is necessary to use two of these equations to solve for two unknowns. For example, suppose initial velocity v_{xi} and acceleration a_x are given. You can then find (1) the velocity at time t, using $v_{xf} = v_{xi} + a_x t$ and (2) the position at time t, using $x_f = x_i + v_{xi}t + \frac{1}{2}a_x t^2$. You should recognize that the quantities that vary during the motion are position, velocity, and time.

You will gain a great deal of experience in the use of these equations by solving a number of exercises and problems. Many times you will discover that more than one method can be used to obtain a solution. Remember that these equations of kinematics *cannot* be used in a situation in which the acceleration varies with time. They can be used only when the acceleration is constant.

Example 2.6   Entering the Traffic Flow

(A) Estimate your average acceleration as you drive up the entrance ramp to an interstate highway.

Solution This problem involves more than our usual amount of estimating! We are trying to come up with a value of a_x, but that value is hard to guess directly. The other variables involved in kinematics are position, velocity, and time. Velocity is probably the easiest one to approximate. Let us assume a final velocity of 100 km/h, so that you can merge with traffic. We multiply this value by (1 000 m/1 km) to convert kilometers to meters and then multiply by (1 h/3 600 s) to convert hours to seconds. These two calculations together are roughly equivalent to dividing by 3. In fact, let us just say that the final velocity is $v_{xf} \approx 30$ m/s. (Remember, this type of approximation and the dropping of digits when performing estimations is okay. If you were starting with U.S. customary units, you could approximate 1 mi/h as roughly 0.5 m/s and continue from there.)

Now we assume that you started up the ramp at about one third your final velocity, so that $v_{xi} \approx 10$ m/s. Finally, we assume that it takes about 10 s to accelerate from v_{xi} to v_{xf}, basing this guess on our previous experience in automobiles. We can then find the average acceleration, using Equation 2.6:

$$\bar{a}_x = \frac{v_{xf} - v_{xi}}{t} \approx \frac{30 \text{ m/s} - 10 \text{ m/s}}{10 \text{ s}}$$

$$= \boxed{2 \text{ m/s}^2}$$

Granted, we made many approximations along the way, but **this type of mental effort can be surprisingly useful and often yields results that are not too different from those derived from careful measurements.** Do not be afraid to attempt making educated guesses and doing some fairly drastic number rounding to simplify estimations. Physicists engage in this type of thought analysis all the time.

(B) How far did you go during the first half of the time interval during which you accelerated?

Solution Let us assume that the acceleration is constant, with the value calculated in part (A). Because the motion takes place in a straight line and the velocity is always in the same direction, the distance traveled from the starting point is equal to the final position of the car. We can calculate the final position at 5 s from Equation 2.12:

$$x_f = x_i + v_{xi}t + \frac{1}{2}a_x t^2$$

$$\approx 0 + (10 \text{ m/s})(5 \text{ s}) + \frac{1}{2}(2 \text{ m/s}^2)(5 \text{ s})^2 = 50 \text{ m} + 25 \text{ m}$$

$$= \boxed{75 \text{ m}}$$

This result indicates that if you had not accelerated, your initial velocity of 10 m/s would have resulted in a 50-m movement up the ramp during the first 5 s. The additional 25 m is the result of your increasing velocity during that interval.

Example 2.7   Carrier Landing

A jet lands on an aircraft carrier at 140 mi/h ($\approx$ 63 m/s).

(A) What is its acceleration (assumed constant) if it stops in 2.0 s due to an arresting cable that snags the airplane and brings it to a stop?

Solution We define our x axis as the direction of motion of the jet. A careful reading of the problem reveals that in addition to being given the initial speed of 63 m/s, we also know that the final speed is zero. We also note that we have no information about the change in position of the jet while it is slowing down. Equation 2.9 is the only equation in Table 2.2 that does not involve position, and so we use it to find the acceleration of the jet, modeled as a particle:

$$a_x = \frac{v_{xf} - v_{xi}}{t} \approx \frac{0 - 63 \text{ m/s}}{2.0 \text{ s}}$$

$$= -31 \text{ m/s}^2$$

(B) If the plane touches down at position $x_i = 0$, what is the final position of the plane?

Solution We can now use any of the other three equations in Table 2.2 to solve for the final position. Let us choose Equation 2.11:

$$x_f = x_i + \frac{1}{2}(v_{xi} + v_{xf})t = 0 + \frac{1}{2}(63 \text{ m/s} + 0)(2.0 \text{ s})$$

$$= 63 \text{ m}$$

If the plane travels much farther than this, it might fall into the ocean. The idea of using arresting cables to slow down landing aircraft and enable them to land safely on ships originated at about the time of the first World War. The cables are still a vital part of the operation of modern aircraft carriers.

What If? Suppose the plane lands on the deck of the aircraft carrier with a speed higher than 63 m/s but with the same acceleration as that calculated in part (A). How will that change the answer to part (B)?

Answer If the plane is traveling faster at the beginning, it will stop farther away from its starting point, so the answer to part (B) should be larger. Mathematically, we see in Equation 2.11 that if v_{xi} is larger, then x_f will be larger.

If the landing deck has a length of 75 m, we can find the maximum initial speed with which the plane can land and still come to rest on the deck at the given acceleration from Equation 2.13:

$$v_{xf}^2 = v_{xi}^2 + 2a_x(x_f - x_i)$$

$$\rightarrow v_{xi} = \sqrt{v_{xf}^2 - 2a_x(x_f - x_i)}$$

$$= \sqrt{0 - 2(-31 \text{ m/s}^2)(75 \text{ m} - 0)}$$

$$= 68 \text{ m/s}$$

Example 2.8   Watch Out for the Speed Limit!   `Interactive`

A car traveling at a constant speed of 45.0 m/s passes a trooper hidden behind a billboard. One second after the speeding car passes the billboard, the trooper sets out from the billboard to catch it, accelerating at a constant rate of 3.00 m/s². How long does it take her to overtake the car?

Solution Let us model the car and the trooper as particles. A sketch (Fig. 2.12) helps clarify the sequence of events.

First, we write expressions for the position of each vehicle as a function of time. It is convenient to choose the position of the billboard as the origin and to set $t_B = 0$ as the time the trooper begins moving. At that instant, the car has already traveled a distance of 45.0 m because it has traveled at a constant speed of $v_x = 45.0$ m/s for 1 s. Thus, the initial position of the speeding car is $x_B = 45.0$ m.

Because the car moves with constant speed, its acceleration is zero. Applying Equation 2.12 (with $a_x = 0$) gives for the car's position at any time t:

$$x_{car} = x_B + v_{x\,car}t = 45.0 \text{ m} + (45.0 \text{ m/s})t$$

A quick check shows that at $t = 0$, this expression gives the car's correct initial position when the trooper begins to move: $x_{car} = x_B = 45.0$ m.

The trooper starts from rest at $t_B = 0$ and accelerates at 3.00 m/s² away from the origin. Hence, her position at any

time t can be found from Equation 2.12:

$$x_f = x_i + v_{xi}t + \frac{1}{2}a_x t^2$$

$$x_{trooper} = 0 + (0)t + \frac{1}{2}a_x t^2 = \frac{1}{2}(3.00 \text{ m/s}^2)t^2$$

$v_{x\,car} = 45.0$ m/s

$a_{x\,car} = 0$

$a_{x\,trooper} = 3.00$ m/s²

$t_A = -1.00$ s $t_B = 0$ $t_C = ?$

Ⓐ   Ⓑ   Ⓒ

Figure 2.12 (Example 2.8) A speeding car passes a hidden trooper.

The trooper overtakes the car at the instant her position matches that of the car, which is position ©:

$$x_{\text{trooper}} = x_{\text{car}}$$

$$\tfrac{1}{2}(3.00 \text{ m/s}^2)t^2 = 45.0 \text{ m} + (45.0 \text{ m/s})t$$

This gives the quadratic equation

$$1.50t^2 - 45.0t - 45.0 = 0$$

The positive solution of this equation is $t =$ ▨ 31.0 s.

(For help in solving quadratic equations, see Appendix B.2.)

What If? What if the trooper had a more powerful motorcycle with a larger acceleration? How would that change the time at which the trooper catches the car?

Answer If the motorcycle has a larger acceleration, the trooper will catch up to the car sooner, so the answer for the time will be less than 31 s. Mathematically, let us cast the final quadratic equation above in terms of the parameters in the problem:

$$\tfrac{1}{2}a_x t^2 - v_{x\,\text{car}} t - x_{\text{B}} = 0$$

The solution to this quadratic equation is,

$$t = \frac{v_{x\,\text{car}} \pm \sqrt{v_{x\,\text{car}}^2 + 2a_x x_{\text{B}}}}{a_x}$$

$$= \frac{v_{x\,\text{car}}}{a_x} + \sqrt{\frac{v_{x\,\text{car}}^2}{a_x^2} + \frac{2x_{\text{B}}}{a_x}}$$

where we have chosen the positive sign because that is the only choice consistent with a time $t > 0$. Because all terms on the right side of the equation have the acceleration a_x in the denominator, increasing the acceleration will decrease the time at which the trooper catches the car.

 You can study the motion of the car and trooper for various velocities of the car at the Interactive Worked Example link at http://www.pse6.com.

Galileo Galilei
Italian physicist and astronomer (1564–1642)

Galileo formulated the laws that govern the motion of objects in free fall and made many other significant discoveries in physics and astronomy. Galileo publicly defended Nicholaus Copernicus's assertion that the Sun is at the center of the Universe (the heliocentric system). He published *Dialogue Concerning Two New World Systems* to support the Copernican model, a view which the Church declared to be heretical. *(North Wind)*

 PITFALL PREVENTION

2.6 *g* and g

Be sure not to confuse the italicized symbol *g* for free-fall acceleration with the nonitalicized symbol g used as the abbreviation for "gram."

2.6 Freely Falling Objects

It is well known that, in the absence of air resistance, all objects dropped near the Earth's surface fall toward the Earth with the same constant acceleration under the influence of the Earth's gravity. It was not until about 1600 that this conclusion was accepted. Before that time, the teachings of the great philosopher Aristotle (384–322 B.C.) had held that heavier objects fall faster than lighter ones.

The Italian Galileo Galilei (1564–1642) originated our present-day ideas concerning falling objects. There is a legend that he demonstrated the behavior of falling objects by observing that two different weights dropped simultaneously from the Leaning Tower of Pisa hit the ground at approximately the same time. Although there is some doubt that he carried out this particular experiment, it is well established that Galileo performed many experiments on objects moving on inclined planes. In his experiments he rolled balls down a slight incline and measured the distances they covered in successive time intervals. The purpose of the incline was to reduce the acceleration; with the acceleration reduced, Galileo was able to make accurate measurements of the time intervals. By gradually increasing the slope of the incline, he was finally able to draw conclusions about freely falling objects because a freely falling ball is equivalent to a ball moving down a vertical incline.

You might want to try the following experiment. Simultaneously drop a coin and a crumpled-up piece of paper from the same height. If the effects of air resistance are negligible, both will have the same motion and will hit the floor at the same time. In the idealized case, in which air resistance is absent, such motion is referred to as *free-fall*. If this same experiment could be conducted in a vacuum, in which air resistance is truly negligible, the paper and coin would fall with the same acceleration even when the paper is not crumpled. On August 2, 1971, such a demonstration was conducted on the Moon by astronaut David Scott. He simultaneously released a hammer and a feather, and they fell together to the lunar surface. This demonstration surely would have pleased Galileo!

When we use the expression *freely falling object*, we do not necessarily refer to an object dropped from rest. **A freely falling object is any object moving freely under the influence of gravity alone, regardless of its initial motion. Objects thrown upward or downward and those released from rest are all falling freely once they**

are released. Any freely falling object experiences an acceleration directed *downward*, regardless of its initial motion.

We shall denote the magnitude of the *free-fall acceleration* by the symbol g. The value of g near the Earth's surface decreases with increasing altitude. Furthermore, slight variations in g occur with changes in latitude. It is common to define "up" as the $+y$ direction and to use y as the position variable in the kinematic equations. At the Earth's surface, the value of g is approximately 9.80 m/s^2. Unless stated otherwise, we shall use this value for g when performing calculations. For making quick estimates, use $g = 10 \text{ m/s}^2$.

If we neglect air resistance and assume that the free-fall acceleration does not vary with altitude over short vertical distances, then the motion of a freely falling object moving vertically is equivalent to motion in one dimension under constant acceleration. Therefore, the equations developed in Section 2.5 for objects moving with constant acceleration can be applied. The only modification that we need to make in these equations for freely falling objects is to note that the motion is in the vertical direction (the y direction) rather than in the horizontal direction (x) and that the acceleration is downward and has a magnitude of 9.80 m/s^2. Thus, we always choose $a_y = -g = -9.80 \text{ m/s}^2$, where the negative sign means that the acceleration of a freely falling object is downward. In Chapter 13 we shall study how to deal with variations in g with altitude.

 PITFALL PREVENTION

2.7 The Sign of g

Keep in mind that g is a *positive number*—it is tempting to substitute -9.80 m/s^2 for g, but resist the temptation. Downward gravitational acceleration is indicated explicitly by stating the acceleration as $a_y = -g$.

 PITFALL PREVENTION

2.8 Acceleration at the Top of The Motion

It is a common misconception that the acceleration of a projectile at the top of its trajectory is zero. While the velocity at the top of the motion of an object thrown upward momentarily goes to zero, *the acceleration is still that due to gravity* at this point. If the velocity and acceleration were both zero, the projectile would stay at the top!

Quick Quiz 2.6 A ball is thrown upward. While the ball is in free fall, does its acceleration (a) increase (b) decrease (c) increase and then decrease (d) decrease and then increase (e) remain constant?

Quick Quiz 2.7 After a ball is thrown upward and is in the air, its speed (a) increases (b) decreases (c) increases and then decreases (d) decreases and then increases (e) remains the same.

Conceptual Example 2.9   The Daring Sky Divers

A sky diver jumps out of a hovering helicopter. A few seconds later, another sky diver jumps out, and they both fall along the same vertical line. Ignore air resistance, so that both sky divers fall with the same acceleration. Does the difference in their speeds stay the same throughout the fall? Does the vertical distance between them stay the same throughout the fall?

Solution At any given instant, the speeds of the divers are different because one had a head start. In any time interval

Δt after this instant, however, the two divers increase their speeds by the same amount because they have the same acceleration. Thus, the difference in their speeds remains the same throughout the fall.

The first jumper always has a greater speed than the second. Thus, in a given time interval, the first diver covers a greater distance than the second. Consequently, the separation distance between them increases.

Example 2.10   Describing the Motion of a Tossed Ball

A ball is tossed straight up at 25 m/s. Estimate its velocity at 1-s intervals.

Solution Let us choose the upward direction to be positive. Regardless of whether the ball is moving upward or downward, its vertical velocity changes by approximately -10 m/s for every second it remains in the air. It starts out at 25 m/s. After 1 s has elapsed, it is still moving upward but at 15 m/s because its acceleration is downward (downward acceleration causes its velocity to decrease). After another second, its upward velocity has dropped to 5 m/s. Now comes the tricky

part—after another half second, its velocity is zero. The ball has gone as high as it will go. After the last half of this 1-s interval, the ball is moving at -5 m/s. (The negative sign tells us that the ball is now moving in the negative direction, that is, *downward*. Its velocity has changed from $+5 \text{ m/s}$ to -5 m/s during that 1-s interval. The change in velocity is still $-5 \text{ m/s} - (+5 \text{ m/s}) = -10 \text{ m/s}$ in that second.) It continues downward, and after another 1 s has elapsed, it is falling at a velocity of -15 m/s. Finally, after another 1 s, it has reached its original starting point and is moving downward at -25 m/s.

Conceptual Example 2.11 Follow the Bouncing Ball

A tennis ball is dropped from shoulder height (about 1.5 m) and bounces three times before it is caught. Sketch graphs of its position, velocity, and acceleration as functions of time, with the $+y$ direction defined as upward.

Solution For our sketch let us stretch things out horizontally so that we can see what is going on. (Even if the ball were moving horizontally, this motion would not affect its vertical motion.)

From Figure 2.13a we see that the ball is in contact with the floor at points Ⓑ, Ⓓ, and Ⓕ. Because the velocity of the ball changes from negative to positive three times during these bounces (Fig. 2.13b), the slope of the position–time graph must change in the same way. Note that the time interval between bounces decreases. Why is that?

During the rest of the ball's motion, the slope of the velocity–time graph in Fig. 2.13b should be -9.80 m/s^2. The acceleration–time graph is a horizontal line at these times because the acceleration does not change when the ball is in free fall. When the ball is in contact with the floor, the velocity changes substantially during a very short time

interval, and so the acceleration must be quite large and positive. This corresponds to the very steep upward lines on the velocity–time graph and to the spikes on the acceleration–time graph.

(a)

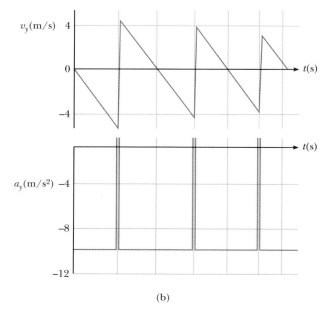

(b)

Active Figure 2.13 (Conceptual Example 2.11) (a) A ball is dropped from a height of 1.5 m and bounces from the floor. (The horizontal motion is not considered here because it does not affect the vertical motion.) (b) Graphs of position, velocity, and acceleration versus time.

At the Active Figures link at http://www.pse6.com, *you can adjust both the value for g and the amount of "bounce" of the ball, and observe the resulting motion of the ball both pictorially and graphically.*

Quick Quiz 2.8 Which values represent the ball's vertical velocity and acceleration at points Ⓐ, Ⓒ, and Ⓔ in Figure 2.13a?

(a) $v_y = 0$, $a_y = -9.80 \text{ m/s}^2$

(b) $v_y = 0$, $a_y = 9.80 \text{ m/s}^2$

(c) $v_y = 0$, $a_y = 0$

(d) $v_y = -9.80 \text{ m/s}$, $a_y = 0$

Example 2.12   Not a Bad Throw for a Rookie!   Interactive

A stone thrown from the top of a building is given an initial velocity of 20.0 m/s straight upward. The building is 50.0 m high, and the stone just misses the edge of the roof on its way down, as shown in Figure 2.14. Using $t_A = 0$ as the time the stone leaves the thrower's hand at position Ⓐ, determine **(A)** the time at which the stone reaches its maximum height, **(B)** the maximum height, **(C)** the time at which the stone returns to the height from which it was thrown, **(D)** the velocity of the stone at this instant, and **(E)** the velocity and position of the stone at $t = 5.00$ s.

Solution (A) As the stone travels from Ⓐ to Ⓑ, its velocity must change by 20 m/s because it stops at Ⓑ. Because gravity causes vertical velocities to change by about 10 m/s for every second of free fall, it should take the stone about 2 s to go from Ⓐ to Ⓑ in our drawing. To calculate the exact time t_B at which the stone reaches maximum height, we use Equation 2.9, $v_{yB} = v_{yA} + a_y t$, noting that $v_{yB} = 0$ and setting the start of our clock readings at $t_A = 0$:

$$0 = 20.0 \text{ m/s} + (-9.80 \text{ m/s}^2)t$$

$$t = t_B = \frac{20.0 \text{ m/s}}{9.80 \text{ m/s}^2} = \boxed{2.04 \text{ s}}$$

Our estimate was pretty close.

(B) Because the average velocity for this first interval is 10 m/s (the average of 20 m/s and 0 m/s) and because it travels for about 2 s, we expect the stone to travel about 20 m. By substituting our time into Equation 2.12, we can find the maximum height as measured from the position of the thrower, where we set $y_A = 0$:

$$y_{max} = y_B = y_A + v_{xA}t + \tfrac{1}{2}a_y t^2$$

$$y_B = 0 + (20.0 \text{ m/s})(2.04 \text{ s}) + \tfrac{1}{2}(-9.80 \text{ m/s}^2)(2.04 \text{ s})^2$$

$$= \boxed{20.4 \text{ m}}$$

Our free-fall estimates are very accurate.

(C) There is no reason to believe that the stone's motion from Ⓑ to Ⓒ is anything other than the reverse of its motion from Ⓐ to Ⓑ. The motion from Ⓐ to Ⓒ is symmetric. Thus, the time needed for it to go from Ⓐ to Ⓒ should be twice the time needed for it to go from Ⓐ to Ⓑ. When the stone is back at the height from which it was thrown (position Ⓒ), the y coordinate is again zero. Using Equation 2.12, with $y_C = 0$, we obtain

$$y_C = y_A + v_{yA}t + \tfrac{1}{2}a_y t^2$$
$$0 = 0 + 20.0t - 4.90t^2$$

This is a quadratic equation and so has two solutions for $t = t_C$. The equation can be factored to give

$$t(20.0 - 4.90t) = 0$$

One solution is $t = 0$, corresponding to the time the stone starts its motion. The other solution is $\boxed{t = 4.08 \text{ s}}$, which

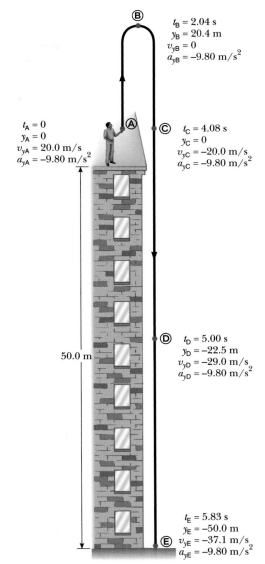

Figure 2.14 (Example 2.12) Position and velocity versus time for a freely falling stone thrown initially upward with a velocity $v_{yi} = 20.0$ m/s.

is the solution we are after. Notice that it is double the value we calculated for t_B.

(D) Again, we expect everything at Ⓒ to be the same as it is at Ⓐ, except that the velocity is now in the opposite direction. The value for t found in (c) can be inserted into Equation 2.9 to give

$$v_{yC} = v_{yA} + a_y t = 20.0 \text{ m/s} + (-9.80 \text{ m/s}^2)(4.08 \text{ s})$$

$$= \boxed{-20.0 \text{ m/s}}$$

The velocity of the stone when it arrives back at its original height is equal in magnitude to its initial velocity but opposite in direction.

(E) For this part we ignore the first part of the motion ($\textcircled{A} \rightarrow \textcircled{B}$) and consider what happens as the stone falls from position $\textcircled{B}$, where it has zero vertical velocity, to position $\textcircled{D}$. We define the initial time as $t_B = 0$. Because the given time for this part of the motion relative to our new zero of time is 5.00 s − 2.04 s = 2.96 s, we estimate that the acceleration due to gravity will have changed the speed by about 30 m/s. We can calculate this from Equation 2.9, where we take $t = 2.96$ s:

$$v_{yD} = v_{yB} + a_y t = 0 \text{ m/s} + (-9.80 \text{ m/s}^2)(2.96 \text{ s})$$

$$= \boxed{-29.0 \text{ m/s}}$$

We could just as easily have made our calculation between positions $\textcircled{A}$ (where we return to our original initial time $t_A = 0$) and $\textcircled{D}$:

$$v_{yD} = v_{yA} + a_y t = 20.0 \text{ m/s} + (-9.80 \text{ m/s}^2)(5.00 \text{ s})$$
$$= -29.0 \text{ m/s}$$

To further demonstrate that we can choose different initial instants of time, let us use Equation 2.12 to find the position of the stone at $t_D = 5.00$ s (with respect to $t_A = 0$) by defining a new initial instant, $t_C = 0$:

$$
\begin{aligned}
y_D &= y_C + v_{yC} t + \tfrac{1}{2} a_y t^2 \\
&= 0 + (-20.0 \text{ m/s})(5.00 \text{ s} - 4.08 \text{ s}) \\
&\quad + \tfrac{1}{2}(-9.80 \text{ m/s}^2)(5.00 \text{ s} - 4.08 \text{ s})^2 \\
&= \boxed{-22.5 \text{ m}}
\end{aligned}
$$

What If? What if the building were 30.0 m tall instead of 50.0 m tall? Which answers in parts (A) to (E) would change?

Answer None of the answers would change. All of the motion takes place in the air, and the stone does not interact with the ground during the first 5.00 s. (Notice that even for a 30.0-m tall building, the stone is above the ground at $t = 5.00$ s.) Thus, the height of the building is not an issue. Mathematically, if we look back over our calculations, we see that we never entered the height of the building into any equation.

 You can study the motion of the thrown ball at the Interactive Worked Example link at **http://www.pse6.com.**

2.7   Kinematic Equations Derived from Calculus

This section assumes the reader is familiar with the techniques of integral calculus. If you have not yet studied integration in your calculus course, you should skip this section or cover it after you become familiar with integration.

The velocity of a particle moving in a straight line can be obtained if its position as a function of time is known. Mathematically, the velocity equals the derivative of the position with respect to time. It is also possible to find the position of a particle if its velocity is known as a function of time. In calculus, the procedure used to perform this task is referred to either as *integration* or as finding the *antiderivative*. Graphically, it is equivalent to finding the area under a curve.

Suppose the v_x-t graph for a particle moving along the x axis is as shown in Figure 2.15. Let us divide the time interval $t_f - t_i$ into many small intervals, each of

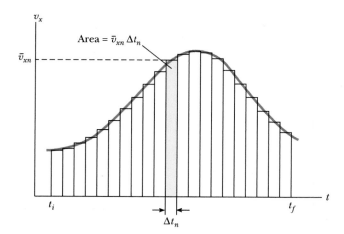

Figure 2.15 Velocity versus time for a particle moving along the x axis. The area of the shaded rectangle is equal to the displacement Δx in the time interval Δt_n, while the total area under the curve is the total displacement of the particle.

duration Δt_n. From the definition of average velocity we see that the displacement during any small interval, such as the one shaded in Figure 2.15, is given by $\Delta x_n = \bar{v}_{xn} \Delta t_n$ where $\bar{v}_{xn}$ is the average velocity in that interval. Therefore, the displacement during this small interval is simply the area of the shaded rectangle. The total displacement for the interval $t_f - t_i$ is the sum of the areas of all the rectangles:

$$\Delta x = \sum_n \bar{v}_{xn} \Delta t_n$$

where the symbol Σ (upper case Greek sigma) signifies a sum over all terms, that is, over all values of n. In this case, the sum is taken over all the rectangles from t_i to t_f. Now, as the intervals are made smaller and smaller, the number of terms in the sum increases and the sum approaches a value equal to the area under the velocity–time graph. Therefore, in the limit $n \to \infty$, or $\Delta t_n \to 0$, the displacement is

$$\Delta x = \lim_{\Delta t_n \to 0} \sum_n v_{xn} \Delta t_n \tag{2.14}$$

or

> Displacement = area under the v_x-t graph

Note that we have replaced the average velocity $\bar{v}_{xn}$ with the instantaneous velocity v_{xn} in the sum. As you can see from Figure 2.15, this approximation is valid in the limit of very small intervals. Therefore if we know the v_x-t graph for motion along a straight line, we can obtain the displacement during any time interval by measuring the area under the curve corresponding to that time interval.

The limit of the sum shown in Equation 2.14 is called a **definite integral** and is written

Definite integral

$$\lim_{\Delta t_n \to 0} \sum_n v_{xn} \Delta t_n = \int_{t_i}^{t_f} v_x(t)\, dt \tag{2.15}$$

where $v_x(t)$ denotes the velocity at any time t. If the explicit functional form of $v_x(t)$ is known and the limits are given, then the integral can be evaluated. Sometimes the v_x-t graph for a moving particle has a shape much simpler than that shown in Figure 2.15. For example, suppose a particle moves at a constant velocity v_{xi}. In this case, the v_x-t graph is a horizontal line, as in Figure 2.16, and the displacement of the particle during the time interval Δt is simply the area of the shaded rectangle:

$$\Delta x = v_{xi}\, \Delta t \qquad (\text{when } v_x = v_{xi} = \text{constant})$$

As another example, consider a particle moving with a velocity that is proportional to t, as in Figure 2.17. Taking $v_x = a_x t$, where a_x is the constant of proportionality (the

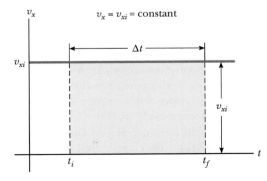

Figure 2.16 The velocity–time curve for a particle moving with constant velocity v_{xi}. The displacement of the particle during the time interval $t_f - t_i$ is equal to the area of the shaded rectangle.

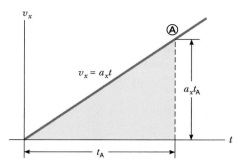

Figure 2.17 The velocity–time curve for a particle moving with a velocity that is proportional to the time.

acceleration), we find that the displacement of the particle during the time interval $t = 0$ to $t = t_A$ is equal to the area of the shaded triangle in Figure 2.17:

$$\Delta x = \tfrac{1}{2}(t_A)(a_x t_A) = \tfrac{1}{2} a_x t_A^2$$

Kinematic Equations

We now use the defining equations for acceleration and velocity to derive two of our kinematic equations, Equations 2.9 and 2.12.

The defining equation for acceleration (Eq. 2.7),

$$a_x = \frac{dv_x}{dt}$$

may be written as $dv_x = a_x dt$ or, in terms of an integral (or antiderivative), as

$$v_{xf} - v_{xi} = \int_0^t a_x dt$$

For the special case in which the acceleration is constant, a_x can be removed from the integral to give

$$v_{xf} - v_{xi} = a_x \int_0^t dt = a_x(t - 0) = a_x t \qquad (2.16)$$

which is Equation 2.9.

Now let us consider the defining equation for velocity (Eq. 2.5):

$$v_x = \frac{dx}{dt}$$

We can write this as $dx = v_x\, dt$, or in integral form as

$$x_f - x_i = \int_0^t v_x dt$$

Because $v_x = v_{xf} = v_{xi} + a_x t$, this expression becomes

$$x_f - x_i = \int_0^t (v_{xi} + a_x t)\, dt = \int_0^t v_{xi} dt + a_x \int_0^t t dt = v_{xi}(t - 0) + a_x\left(\frac{t^2}{2} - 0\right)$$
$$= v_{xi} t + \tfrac{1}{2} a_x t^2$$

which is Equation 2.12.

Besides what you might expect to learn about physics concepts, a very valuable skill you should hope to take away from your physics course is the ability to solve complicated problems. The way physicists approach complex situations and break them down into manageable pieces is extremely useful. On the next page is a general problem-solving strategy that will help guide you through the steps. To help you remember the steps of the strategy, they are called *Conceptualize, Categorize, Analyze, and Finalize.*

GENERAL PROBLEM-SOLVING STRATEGY

Conceptualize

- The first thing to do when approaching a problem is to *think about* and *understand* the situation. Study carefully any diagrams, graphs, tables, or photographs that accompany the problem. Imagine a movie, running in your mind, of what happens in the problem.

- If a diagram is not provided, you should almost always make a quick drawing of the situation. Indicate any known values, perhaps in a table or directly on your sketch.

- Now focus on what algebraic or numerical information is given in the problem. Carefully read the problem statement, looking for key phrases such as "starts from rest" ($v_i = 0$), "stops" ($v_f = 0$), or "freely falls" ($a_y = -g = -9.80 \text{ m/s}^2$).

- Now focus on the expected result of solving the problem. Exactly what is the question asking? Will the final result be numerical or algebraic? Do you know what units to expect?

- Don't forget to incorporate information from your own experiences and common sense. What should a reasonable answer look like? For example, you wouldn't expect to calculate the speed of an automobile to be 5×10^6 m/s.

Categorize

- Once you have a good idea of what the problem is about, you need to *simplify* the problem. Remove the details that are not important to the solution. For example, model a moving object as a particle. If appropriate, ignore air resistance or friction between a sliding object and a surface.

- Once the problem is simplified, it is important to *categorize* the problem. Is it a simple *plug-in problem,* such that numbers can be simply substituted into a definition? If so, the problem is likely to be finished when this substitution is done. If not, you face what we can call an *analysis problem*—the situation must be analyzed more deeply to reach a solution.

- If it is an analysis problem, it needs to be categorized further. Have you seen this type of problem before? Does it fall into the growing list of types of problems that you have solved previously? Being able to classify a problem can make it much easier to lay out a plan to solve it. For example, if your simplification shows that the problem can be treated as a particle moving under constant acceleration and you have already solved such a problem (such as the examples in Section 2.5), the solution to the present problem follows a similar pattern.

Analyze

- Now you must analyze the problem and strive for a mathematical solution. Because you have already categorized the problem, it should not be too difficult to select relevant equations that apply to the type of situation in the problem. For example, if the problem involves a particle moving under constant acceleration, Equations 2.9 to 2.13 are relevant.

- Use algebra (and calculus, if necessary) to solve symbolically for the unknown variable in terms of what is given. Substitute in the appropriate numbers, calculate the result, and round it to the proper number of significant figures.

Finalize

- This is the most important part. Examine your numerical answer. Does it have the correct units? Does it meet your expectations from your conceptualization of the problem? What about the algebraic form of the result—before you substituted numerical values? Does it make sense? Examine the variables in the problem to see whether the answer would change in a physically meaningful way if they were drastically increased or decreased or even became zero. Looking at limiting cases to see whether they yield expected values is a very useful way to make sure that you are obtaining reasonable results.

- Think about how this problem compares with others you have solved. How was it similar? In what critical ways did it differ? Why was this problem assigned? You should have learned something by doing it. Can you figure out what? If it is a new category of problem, be sure you understand it so that you can use it as a model for solving future problems in the same category.

When solving complex problems, you may need to identify a series of sub-problems and apply the problem-solving strategy to each. For very simple problems, you probably don't need this strategy at all. But when you are looking at a problem and you don't know what to do next, remember the steps in the strategy and use them as a guide.

For practice, it would be useful for you to go back over the examples in this chapter and identify the *Conceptualize, Categorize, Analyze,* and *Finalize* steps. In the next chapter, we will begin to show these steps explicitly in the examples.

SUMMARY

Take a practice test for this chapter by clicking the Practice Test link at http://www.pse6.com.

After a particle moves along the x axis from some initial position x_i to some final position x_f, its **displacement** is

$$\Delta x \equiv x_f - x_i \tag{2.1}$$

The **average velocity** of a particle during some time interval is the displacement Δx divided by the time interval Δt during which that displacement occurs:

$$\bar{v}_x \equiv \frac{\Delta x}{\Delta t} \tag{2.2}$$

The **average speed** of a particle is equal to the ratio of the total distance it travels to the total time interval during which it travels that distance:

$$\text{Average speed} = \frac{\text{total distance}}{\text{total time}} \tag{2.3}$$

The **instantaneous velocity** of a particle is defined as the limit of the ratio $\Delta x/\Delta t$ as Δt approaches zero. By definition, this limit equals the derivative of x with respect to t, or the time rate of change of the position:

$$v_x \equiv \lim_{\Delta t \to 0} \frac{\Delta x}{\Delta t} = \frac{dx}{dt} \tag{2.5}$$

The **instantaneous speed** of a particle is equal to the magnitude of its instantaneous velocity.

The **average acceleration** of a particle is defined as the ratio of the change in its velocity Δv_x divided by the time interval Δt during which that change occurs:

$$\bar{a}_x \equiv \frac{\Delta v_x}{\Delta t} = \frac{v_{xf} - v_{xi}}{t_f - t_i} \tag{2.6}$$

The **instantaneous acceleration** is equal to the limit of the ratio $\Delta v_x/\Delta t$ as Δt approaches 0. By definition, this limit equals the derivative of v_x with respect to t, or the time rate of change of the velocity:

$$a_x \equiv \lim_{\Delta t \to 0} \frac{\Delta v_x}{\Delta t} = \frac{dv_x}{dt} \tag{2.7}$$

When the object's velocity and acceleration are in the same direction, the object is speeding up. On the other hand, when the object's velocity and acceleration are in opposite directions, the object is slowing down. Remembering that $F \propto a$ is a useful way to identify the direction of the acceleration.

The **equations of kinematics** for a particle moving along the x axis with uniform acceleration a_x (constant in magnitude and direction) are

$$v_{xf} = v_{xi} + a_x t \tag{2.9}$$

$$x_f = x_i + \bar{v}_x t = x_i + \tfrac{1}{2}(v_{xi} + v_{xf})t \tag{2.11}$$

$$x_f = x_i + v_{xi}t + \tfrac{1}{2}a_x t^2 \tag{2.12}$$

$$v_{xf}{}^2 = v_{xi}{}^2 + 2a_x(x_f - x_i) \tag{2.13}$$

An object falling freely in the presence of the Earth's gravity experiences a free-fall acceleration directed toward the center of the Earth. If air resistance is neglected, if the motion occurs near the surface of the Earth, and if the range of the motion is small compared with the Earth's radius, then the free-fall acceleration g is constant over the range of motion, where g is equal to 9.80 m/s^2.

Complicated problems are best approached in an organized manner. You should be able to recall and apply the *Conceptualize, Categorize, Analyze,* and *Finalize* steps of the General Problem-Solving Strategy when you need them.

QUESTIONS

1. The speed of sound in air is 331 m/s. During the next thunderstorm, try to estimate your distance from a lightning bolt by measuring the time lag between the flash and the thunderclap. You can ignore the time it takes for the light flash to reach you. Why?

2. The average velocity of a particle moving in one dimension has a positive value. Is it possible for the instantaneous velocity to have been negative at any time in the interval? Suppose the particle started at the origin $x = 0$. If its average velocity is positive, could the particle ever have been in the $-x$ region of the axis?

3. If the average velocity of an object is zero in some time interval, what can you say about the displacement of the object for that interval?

4. Can the instantaneous velocity of an object at an instant of time ever be greater in magnitude than the average velocity over a time interval containing the instant? Can it ever be less?

5. If an object's average velocity is nonzero over some time interval, does this mean that its instantaneous velocity is never zero during the interval? Explain your answer.

6. If an object's average velocity is zero over some time interval, show that its instantaneous velocity must be zero at some time during the interval. It may be useful in your proof to sketch a graph of x versus t and to note that $v_x(t)$ is a continuous function.

7. If the velocity of a particle is nonzero, can its acceleration be zero? Explain.

8. If the velocity of a particle is zero, can its acceleration be nonzero? Explain.

9. Two cars are moving in the same direction in parallel lanes along a highway. At some instant, the velocity of car A exceeds the velocity of car B. Does this mean that the acceleration of A is greater than that of B? Explain.

10. Is it possible for the velocity and the acceleration of an object to have opposite signs? If not, state a proof. If so, give an example of such a situation and sketch a velocity–time graph to prove your point.

11. Consider the following combinations of signs and values for velocity and acceleration of a particle with respect to a one-dimensional x axis:

	Velocity	Acceleration
a.	Positive	Positive
b.	Positive	Negative
c.	Positive	Zero
d.	Negative	Positive
e.	Negative	Negative
f.	Negative	Zero
g.	Zero	Positive
h.	Zero	Negative

Describe what a particle is doing in each case, and give a real life example for an automobile on an east-west one-dimensional axis, with east considered the positive direction.

12. Can the equations of kinematics (Eqs. 2.9–2.13) be used in a situation where the acceleration varies in time? Can they be used when the acceleration is zero?

13. A stone is thrown vertically upward from the roof of a building. Does the position of the stone depend on the location chosen for the origin of the coordinate system? Does the stone's velocity depend on the choice of origin? Explain your answers.

14. A child throws a marble into the air with an initial speed v_i. Another child drops a ball at the same instant. Compare the accelerations of the two objects while they are in flight.

15. A student at the top of a building of height h throws one ball upward with a speed of v_i and then throws a second ball downward with the same initial speed $|v_i|$. How do the final velocities of the balls compare when they reach the ground?

16. An object falls freely from height h. It is released at time zero and strikes the ground at time t. (a) When the object is at height $0.5h$, is the time earlier than $0.5t$, equal to $0.5t$, or later than $0.5t$? (b) When the time is $0.5t$, is the height of the object greater than $0.5h$, equal to $0.5h$, or less than $0.5h$? Give reasons for your answers.

17. You drop a ball from a window on an upper floor of a building. It strikes the ground with speed v. You now repeat the drop, but you have a friend down on the street who throws another ball upward at speed v. Your friend throws the ball upward at exactly the same time that you drop yours from the window. At some location, the balls pass each other. Is this location *at* the halfway point between window and ground, *above* this point, or *below* this point?

PROBLEMS

1, 2, 3 = straightforward, intermediate, challenging ☐ = full solution available in the *Student Solutions Manual and Study Guide*

= coached solution with hints available at http://www.pse6.com 🖥 = computer useful in solving problem

= paired numerical and symbolic problems

Section 2.1 Position, Velocity, and Speed

1. The position of a pinewood derby car was observed at various times; the results are summarized in the following table. Find the average velocity of the car for (a) the first second, (b) the last 3 s, and (c) the entire period of observation.

t(s)	0	1.0	2.0	3.0	4.0	5.0
x(m)	0	2.3	9.2	20.7	36.8	57.5

2. (a) Sand dunes in a desert move over time as sand is swept up the windward side to settle in the lee side. Such "walking" dunes have been known to walk 20 feet in a year and can travel as much as 100 feet per year in particularly windy times. Calculate the average speed in each case in m/s. (b) Fingernails grow at the rate of drifting continents, on the order of 10 mm/yr. Approximately how long did it take for North America to separate from Europe, a distance of about 3 000 mi?

3. The position versus time for a certain particle moving along the x axis is shown in Figure P2.3. Find the average velocity in the time intervals (a) 0 to 2 s, (b) 0 to 4 s, (c) 2 s to 4 s, (d) 4 s to 7 s, (e) 0 to 8 s.

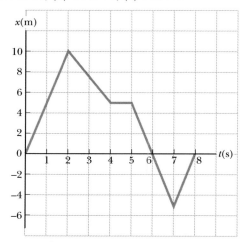

Figure P2.3 Problems 3 and 9

4. A particle moves according to the equation $x = 10t^2$ where x is in meters and t is in seconds. (a) Find the average velocity for the time interval from 2.00 s to 3.00 s. (b) Find the average velocity for the time interval from 2.00 to 2.10 s.

5. A person walks first at a constant speed of 5.00 m/s along a straight line from point A to point B and then back along the line from B to A at a constant speed of 3.00 m/s. What is (a) her average speed over the entire trip? (b) her average velocity over the entire trip?

Section 2.2 Instantaneous Velocity and Speed

6. The position of a particle moving along the x axis varies in time according to the expression $x = 3t^2$, where x is in meters and t is in seconds. Evaluate its position (a) at $t = 3.00$ s and (b) at $3.00 \text{ s} + \Delta t$. (c) Evaluate the limit of $\Delta x/\Delta t$ as Δt approaches zero, to find the velocity at $t = 3.00$ s.

7. A position-time graph for a particle moving along the x axis is shown in Figure P2.7. (a) Find the average velocity in the time interval $t = 1.50$ s to $t = 4.00$ s. (b) Determine the instantaneous velocity at $t = 2.00$ s by measuring the slope of the tangent line shown in the graph. (c) At what value of t is the velocity zero?

8. (a) Use the data in Problem 1 to construct a smooth graph of position versus time. (b) By constructing tangents to the $x(t)$ curve, find the instantaneous velocity of the car at several instants. (c) Plot the instantaneous velocity versus time and, from this, determine the average acceleration of the car. (d) What was the initial velocity of the car?

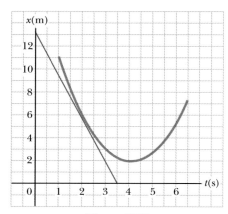

Figure P2.7

9. Find the instantaneous velocity of the particle described in Figure P2.3 at the following times: (a) $t = 1.0$ s, (b) $t = 3.0$ s, (c) $t = 4.5$ s, and (d) $t = 7.5$ s.

10. A hare and a tortoise compete in a race over a course 1.00 km long. The tortoise crawls straight and steadily at its maximum speed of 0.200 m/s toward the finish line. The hare runs at its maximum speed of 8.00 m/s toward the goal for 0.800 km and then stops to tease the tortoise. How close to the goal can the hare let the tortoise approach before resuming the race, which the tortoise wins in a photo finish? Assume that, when moving, both animals move steadily at their respective maximum speeds.

Section 2.3 Acceleration

11. A 50.0-g superball traveling at 25.0 m/s bounces off a brick wall and rebounds at 22.0 m/s. A high-speed camera records this event. If the ball is in contact with the wall for 3.50 ms, what is the magnitude of the average acceleration of the ball during this time interval? (*Note:* 1 ms = 10^{-3} s.)

12. A particle starts from rest and accelerates as shown in Figure P2.12. Determine (a) the particle's speed at $t = 10.0$ s and at $t = 20.0$ s, and (b) the distance traveled in the first 20.0 s.

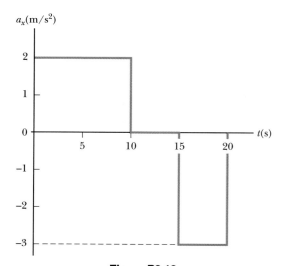

Figure P2.12

13. Secretariat won the Kentucky Derby with times for successive quarter-mile segments of 25.2 s, 24.0 s, 23.8 s, and 23.0 s. (a) Find his average speed during each quarter-mile segment. (b) Assuming that Secretariat's instantaneous speed at the finish line was the same as the average speed during the final quarter mile, find his average acceleration for the entire race. (Horses in the Derby start from rest.)

14. A velocity–time graph for an object moving along the x axis is shown in Figure P2.14. (a) Plot a graph of the acceleration versus time. (b) Determine the average acceleration of the object in the time intervals $t = 5.00$ s to $t = 15.0$ s and $t = 0$ to $t = 20.0$ s.

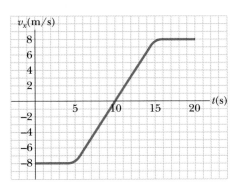

Figure P2.14

15. ⊘ A particle moves along the x axis according to the equation $x = 2.00 + 3.00t - 1.00t^2$, where x is in meters and t is in seconds. At $t = 3.00$ s, find (a) the position of the particle, (b) its velocity, and (c) its acceleration.

16. An object moves along the x axis according to the equation $x(t) = (3.00t^2 - 2.00t + 3.00)$ m. Determine (a) the average speed between $t = 2.00$ s and $t = 3.00$ s, (b) the instantaneous speed at $t = 2.00$ s and at $t = 3.00$ s, (c) the average acceleration between $t = 2.00$ s and $t = 3.00$ s, and (d) the instantaneous acceleration at $t = 2.00$ s and $t = 3.00$ s.

17. Figure P2.17 shows a graph of v_x versus t for the motion of a motorcyclist as he starts from rest and moves along the road in a straight line. (a) Find the average acceleration for the time interval $t = 0$ to $t = 6.00$ s. (b) Estimate the time at which the acceleration has its greatest positive value and the value of the acceleration at that instant. (c) When is the acceleration zero? (d) Estimate the maximum negative value of the acceleration and the time at which it occurs.

Figure P2.17

Section 2.4   Motion Diagrams

18. Draw motion diagrams for (a) an object moving to the right at constant speed, (b) an object moving to the right and speeding up at a constant rate, (c) an object moving to the right and slowing down at a constant rate, (d) an object moving to the left and speeding up at a constant rate, and (e) an object moving to the left and slowing down at a constant rate. (f) How would your drawings change if the changes in speed were not uniform; that is, if the speed were not changing at a constant rate?

Section 2.5   One-Dimensional Motion with Constant Acceleration

19. Jules Verne in 1865 suggested sending people to the Moon by firing a space capsule from a 220-m-long cannon with a launch speed of 10.97 km/s. What would have been the unrealistically large acceleration experienced by the space travelers during launch? Compare your answer with the free-fall acceleration 9.80 m/s^2.

20. A truck covers 40.0 m in 8.50 s while smoothly slowing down to a final speed of 2.80 m/s. (a) Find its original speed. (b) Find its acceleration.

21. ⊘ An object moving with uniform acceleration has a velocity of 12.0 cm/s in the positive x direction when its x coordinate is 3.00 cm. If its x coordinate 2.00 s later is -5.00 cm, what is its acceleration?

22. A 745i BMW car can brake to a stop in a distance of 121 ft. from a speed of 60.0 mi/h. To brake to a stop from a speed of 80.0 mi/h requires a stopping distance of 211 ft. What is the average braking acceleration for (a) 60 mi/h to rest, (b) 80 mi/h to rest, (c) 80 mi/h to 60 mi/h? Express the answers in mi/h/s and in m/s^2.

23. A speedboat moving at 30.0 m/s approaches a no-wake buoy marker 100 m ahead. The pilot slows the boat with a constant acceleration of -3.50 m/s^2 by reducing the throttle. (a) How long does it take the boat to reach the buoy? (b) What is the velocity of the boat when it reaches the buoy?

24. Figure P2.24 represents part of the performance data of a car owned by a proud physics student. (a) Calculate from the graph the total distance traveled. (b) What distance does the car travel between the times $t = 10$ s and $t = 40$ s? (c) Draw a graph of its acceleration versus time between $t = 0$ and $t = 50$ s. (d) Write an equation for x as a function of time for each phase of the motion, represented by (i) $0a$, (ii) ab, (iii) bc. (e) What is the average velocity of the car between $t = 0$ and $t = 50$ s?

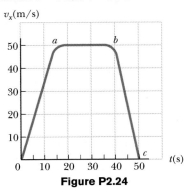

Figure P2.24

25. A particle moves along the x axis. Its position is given by the equation $x = 2 + 3t - 4t^2$ with x in meters and t in seconds. Determine (a) its position when it changes direction and (b) its velocity when it returns to the position it had at $t = 0$.

26. In the Daytona 500 auto race, a Ford Thunderbird and a Mercedes Benz are moving side by side down a straight-away at 71.5 m/s. The driver of the Thunderbird realizes he must make a pit stop, and he smoothly slows to a stop over a distance of 250 m. He spends 5.00 s in the pit and then accelerates out, reaching his previous speed of 71.5 m/s after a distance of 350 m. At this point, how far has the Thunderbird fallen behind the Mercedes Benz, which has continued at a constant speed?

27. A jet plane lands with a speed of 100 m/s and can accelerate at a maximum rate of -5.00 m/s^2 as it comes to rest. (a) From the instant the plane touches the runway, what is the minimum time interval needed before it can come to rest? (b) Can this plane land on a small tropical island airport where the runway is 0.800 km long?

28. A car is approaching a hill at 30.0 m/s when its engine suddenly fails just at the bottom of the hill. The car moves with a constant acceleration of -2.00 m/s^2 while coasting up the hill. (a) Write equations for the position along the slope and for the velocity as functions of time, taking $x = 0$ at the bottom of the hill, where $v_i = 30.0$ m/s. (b) Determine the maximum distance the car rolls up the hill.

29. The driver of a car slams on the brakes when he sees a tree blocking the road. The car slows uniformly with an acceleration of -5.60 m/s^2 for 4.20 s, making straight skid marks 62.4 m long ending at the tree. With what speed does the car then strike the tree?

30. *Help! One of our equations is missing!* We describe constant-acceleration motion with the variables and parameters v_{xi}, v_{xf}, a_x, t, and $x_f - x_i$. Of the equations in Table 2.2, the first does not involve $x_f - x_i$. The second does not contain a_x; the third omits v_{xf} and the last leaves out t. So to complete the set there should be an equation *not* involving v_{xi}. Derive it from the others. Use it to solve Problem 29 in one step.

31. For many years Colonel John P. Stapp, USAF, held the world's land speed record. On March 19, 1954, he rode a rocket-propelled sled that moved down a track at a speed of 632 mi/h. He and the sled were safely brought to rest in 1.40 s (Fig. P2.31). Determine (a) the negative acceleration he experienced and (b) the distance he traveled during this negative acceleration.

32. A truck on a straight road starts from rest, accelerating at 2.00 m/s^2 until it reaches a speed of 20.0 m/s. Then the truck travels for 20.0 s at constant speed until the brakes are applied, stopping the truck in a uniform manner in an additional 5.00 s. (a) How long is the truck in motion? (b) What is the average velocity of the truck for the motion described?

33. An electron in a cathode ray tube (CRT) accelerates from 2.00×10^4 m/s to 6.00×10^6 m/s over 1.50 cm. (a) How long does the electron take to travel this 1.50 cm? (b) What is its acceleration?

34. In a 100-m linear accelerator, an electron is accelerated to 1.00% of the speed of light in 40.0 m before it coasts for 60.0 m to a target. (a) What is the electron's acceleration during the first 40.0 m? (b) How long does the total flight take?

35. Within a complex machine such as a robotic assembly line, suppose that one particular part glides along a straight track. A control system measures the average velocity of the part during each successive interval of time $\Delta t_0 = t_0 - 0$, compares it with the value v_c it should be, and switches a servo motor on and off to give the part a correcting pulse of acceleration. The pulse consists of a constant acceleration a_m applied for time interval $\Delta t_m = t_m - 0$ within the next control time interval Δt_0. As shown in Fig. P2.35, the part may be modeled as having zero acceleration when the motor is off (between t_m and t_0). A computer in the control system chooses the size of the acceleration so that the final velocity of the part will have the correct value v_c. Assume the part is initially at rest and is to have instantaneous velocity v_c at time t_0. (a) Find the required value of a_m in terms of v_c and t_m. (b) Show that the displacement Δx of the part during the time interval Δt_0 is given by $\Delta x = v_c (t_0 - 0.5t_m)$. For specified values of v_c and t_0, (c) what is the minimum displacement of the part? (d) What is the maximum displacement of the part? (e) Are both the minimum and maximum displacements physically attainable?

Figure P2.35

Figure P2.31 *(Left)* Col. John Stapp on rocket sled. *(Right)* Col. Stapp's face is contorted by the stress of rapid negative acceleration.

36. A glider on an air track carries a flag of length ℓ through a stationary photogate, which measures the time interval Δt_d during which the flag blocks a beam of infrared light passing across the photogate. The ratio $v_d = \ell/\Delta t_d$ is the average velocity of the glider over this part of its motion. Suppose the glider moves with constant acceleration. (a) Argue for or against the idea that v_d is equal to the instantaneous velocity of the glider when it is halfway through the photogate in space. (b) Argue for or against the idea that v_d is equal to the instantaneous velocity of the glider when it is halfway through the photogate in time.

37. A ball starts from rest and accelerates at 0.500 m/s^2 while moving down an inclined plane 9.00 m long. When it reaches the bottom, the ball rolls up another plane, where, after moving 15.0 m, it comes to rest. (a) What is the speed of the ball at the bottom of the first plane? (b) How long does it take to roll down the first plane? (c) What is the acceleration along the second plane? (d) What is the ball's speed 8.00 m along the second plane?

38. Speedy Sue, driving at 30.0 m/s, enters a one-lane tunnel. She then observes a slow-moving van 155 m ahead traveling at 5.00 m/s. Sue applies her brakes but can accelerate only at -2.00 m/s^2 because the road is wet. Will there be a collision? If yes, determine how far into the tunnel and at what time the collision occurs. If no, determine the distance of closest approach between Sue's car and the van.

39. Solve Example 2.8, "Watch out for the Speed Limit!" by a graphical method. On the same graph plot position versus time for the car and the police officer. From the intersection of the two curves read the time at which the trooper overtakes the car.

Section 2.6 Freely Falling Objects

Note: In all problems in this section, ignore the effects of air resistance.

40. A golf ball is released from rest from the top of a very tall building. Neglecting air resistance, calculate (a) the position and (b) the velocity of the ball after 1.00, 2.00, and 3.00 s.

41. *Every morning at seven o'clock*
   There's twenty terriers drilling on the rock.
   The boss comes around and he says, "Keep still
   And bear down heavy on the cast-iron drill

   And drill, ye terriers, drill." And drill, ye terriers, drill.
   It's work all day for sugar in your tea
   Down beyond the railway. And drill, ye terriers, drill.

   The foreman's name was John McAnn.
   By God, he was a blamed mean man.
   One day a premature blast went off
   And a mile in the air went big Jim Goff. And drill ...

   Then when next payday came around
   Jim Goff a dollar short was found.
   When he asked what for, came this reply:
   "You were docked for the time you were up in the sky." And drill...
      —American folksong

   What was Goff's hourly wage? State the assumptions you make in computing it.

42. A ball is thrown directly downward, with an initial speed of 8.00 m/s, from a height of 30.0 m. After what time interval does the ball strike the ground?

43. A student throws a set of keys vertically upward to her sorority sister, who is in a window 4.00 m above. The keys are caught 1.50 s later by the sister's outstretched hand. (a) With what initial velocity were the keys thrown? (b) What was the velocity of the keys just before they were caught?

44. Emily challenges her friend David to catch a dollar bill as follows. She holds the bill vertically, as in Figure P2.44, with the center of the bill between David's index finger and thumb. David must catch the bill after Emily releases it without moving his hand downward. If his reaction time is 0.2 s, will he succeed? Explain your reasoning.

George Semple

Figure P2.44

45. In Mostar, Bosnia, the ultimate test of a young man's courage once was to jump off a 400-year-old bridge (now destroyed) into the River Neretva, 23.0 m below the bridge. (a) How long did the jump last? (b) How fast was the diver traveling upon impact with the water? (c) If the speed of sound in air is 340 m/s, how long after the diver took off did a spectator on the bridge hear the splash?

46. A ball is dropped from rest from a height h above the ground. Another ball is thrown vertically upwards from the ground at the instant the first ball is released. Determine the speed of the second ball if the two balls are to meet at a height $h/2$ above the ground.

47. A baseball is hit so that it travels straight upward after being struck by the bat. A fan observes that it takes 3.00 s for the ball to reach its maximum height. Find (a) its initial velocity and (b) the height it reaches.

48. It is possible to shoot an arrow at a speed as high as 100 m/s. (a) If friction is neglected, how high would an arrow launched at this speed rise if shot straight up? (b) How long would the arrow be in the air?

49. 🌐 A daring ranch hand sitting on a tree limb wishes to drop vertically onto a horse galloping under the tree. The constant speed of the horse is 10.0 m/s, and the distance from the limb to the level of the saddle is 3.00 m. (a) What must be the horizontal distance between the saddle and limb when the ranch hand makes his move? (b) How long is he in the air?

50. A woman is reported to have fallen 144 ft from the 17th floor of a building, landing on a metal ventilator box, which she crushed to a depth of 18.0 in. She suffered only minor injuries. Neglecting air resistance, calculate (a) the speed of the woman just before she collided with the ventilator, (b) her average acceleration while in contact with the box, and (c) the time it took to crush the box.

51. The height of a helicopter above the ground is given by $h = 3.00t^3$, where h is in meters and t is in seconds. After 2.00 s, the helicopter releases a small mailbag. How long after its release does the mailbag reach the ground?

52. A freely falling object requires 1.50 s to travel the last 30.0 m before it hits the ground. From what height above the ground did it fall?

Section 2.7   Kinematic Equations Derived from Calculus

53. Automotive engineers refer to the time rate of change of acceleration as the "jerk." If an object moves in one dimension such that its jerk J is constant, (a) determine expressions for its acceleration $a_x(t)$, velocity $v_x(t)$, and position $x(t)$, given that its initial acceleration, velocity, and position are a_{xi}, v_{xi}, and x_i, respectively. (b) Show that $a_x^2 = a_{xi}^2 + 2J(v_x - v_{xi})$.

54. A student drives a moped along a straight road as described by the velocity-versus-time graph in Figure P2.54. Sketch this graph in the middle of a sheet of graph paper. (a) Directly above your graph, sketch a graph of the position versus time, aligning the time coordinates of the two graphs. (b) Sketch a graph of the acceleration versus time directly below the v_x-t graph, again aligning the time coordinates. On each graph, show the numerical values of x and a_x for all points of inflection. (c) What is the acceleration at $t = 6$ s? (d) Find the position (relative to the starting point) at $t = 6$ s. (e) What is the moped's final position at $t = 9$ s?

55. The speed of a bullet as it travels down the barrel of a rifle toward the opening is given by $v = (-5.00 \times 10^7)t^2 + (3.00 \times 10^5)t$, where v is in meters per second and t is in seconds. The acceleration of the bullet just as it leaves the barrel is zero. (a) Determine the acceleration and position of the bullet as a function of time when the bullet is in the barrel. (b) Determine the length of time the bullet is accelerated. (c) Find the speed at which the bullet leaves the barrel. (d) What is the length of the barrel?

56. The acceleration of a marble in a certain fluid is proportional to the speed of the marble squared, and is given (in SI units) by $a = -3.00v^2$ for $v > 0$. If the marble enters this fluid with a speed of 1.50 m/s, how long will it take before the marble's speed is reduced to half of its initial value?

Additional Problems

57. A car has an initial velocity v_0 when the driver sees an obstacle in the road in front of him. His reaction time is Δt_r, and the braking acceleration of the car is a. Show that the total stopping distance is

$$s_{\text{stop}} = v_0 \Delta t_r - v_0^2/2a.$$

Remember that a is a negative number.

58. The yellow caution light on a traffic signal should stay on long enough to allow a driver to either pass through the intersection or safely stop before reaching the intersection. A car can stop if its distance from the intersection is greater than the stopping distance found in the previous problem. If the car is less than this stopping distance from the intersection, the yellow light should stay on long enough to allow the car to pass entirely through the intersection. (a) Show that the yellow light should stay on for a time interval

$$\Delta t_{\text{light}} = \Delta t_r - (v_0/2a) + (s_i/v_0)$$

where Δt_r is the driver's reaction time, v_0 is the velocity of the car approaching the light at the speed limit, a is the braking acceleration, and s_i is the width of the intersection. (b) As city traffic planner, you expect cars to approach an intersection 16.0 m wide with a speed of 60.0 km/h. Be cautious and assume a reaction time of 1.10 s to allow for a driver's indecision. Find the length of time the yellow light should remain on. Use a braking acceleration of -2.00 m/s^2.

59. The Acela is the Porsche of American trains. Shown in Figure P2.59a, the electric train whose name is pronounced ah-SELL-ah is in service on the Washington-New York-Boston run. With two power cars and six coaches, it can carry 304 passengers at 170 mi/h. The carriages tilt as much as 6° from the vertical to prevent passengers from feeling pushed to the side as they go around curves. Its braking mechanism uses electric generators to recover its energy of motion. A velocity-time graph for the Acela is shown in Figure P2.59b. (a) Describe the motion of the train in each successive time interval. (b) Find the peak positive acceleration of the train in the motion graphed. (c) Find the train's displacement in miles between $t = 0$ and $t = 200$ s.

Figure P2.54

(a)

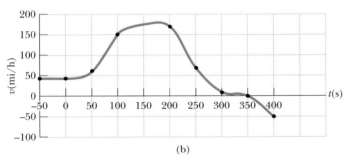

(b)

Figure P2.59 (a) The Acela—1 171 000 lb of cold steel thundering along at 150 mi/h. (b) Velocity-versus-time graph for the Acela.

60. Liz rushes down onto a subway platform to find her train already departing. She stops and watches the cars go by. Each car is 8.60 m long. The first moves past her in 1.50 s and the second in 1.10 s. Find the constant acceleration of the train.

61. A dog's hair has been cut and is now getting 1.04 mm longer each day. With winter coming on, this rate of hair growth is steadily increasing, by 0.132 mm/day every week. By how much will the dog's hair grow during 5 weeks?

62. A test rocket is fired vertically upward from a well. A catapult gives it an initial speed of 80.0 m/s at ground level. Its engines then fire and it accelerates upward at 4.00 m/s^2 until it reaches an altitude of 1 000 m. At that point its engines fail and the rocket goes into free fall, with an acceleration of -9.80 m/s^2. (a) How long is the rocket in motion above the ground? (b) What is its maximum altitude? (c) What is its velocity just before it collides with the Earth? (You will need to consider the motion while the engine is operating separate from the free-fall motion.)

63. A motorist drives along a straight road at a constant speed of 15.0 m/s. Just as she passes a parked motorcycle police officer, the officer starts to accelerate at 2.00 m/s^2 to overtake her. Assuming the officer maintains this acceleration, (a) determine the time it takes the police officer to reach the motorist. Find (b) the speed and (c) the total displacement of the officer as he overtakes the motorist.

64. In Figure 2.10b, the area under the velocity versus time curve and between the vertical axis and time t (vertical dashed line) represents the displacement. As shown, this area consists of a rectangle and a triangle. Compute their areas and compare the sum of the two areas with the expression on the right-hand side of Equation 2.12.

65. Setting a new world record in a 100-m race, Maggie and Judy cross the finish line in a dead heat, both taking 10.2 s. Accelerating uniformly, Maggie took 2.00 s and Judy 3.00 s to attain maximum speed, which they maintained for the rest of the race. (a) What was the acceleration of each sprinter? (b) What were their respective maximum speeds? (c) Which sprinter was ahead at the 6.00-s mark, and by how much?

66. A commuter train travels between two downtown stations. Because the stations are only 1.00 km apart, the train never reaches its maximum possible cruising speed. During rush hour the engineer minimizes the time interval Δt between two stations by accelerating for a time interval Δt_1 at a rate $a_1 = 0.100$ m/s^2 and then immediately braking with acceleration $a_2 = -0.500$ m/s^2 for a time interval Δt_2. Find the minimum time interval of travel Δt and the time interval Δt_1.

67. A hard rubber ball, released at chest height, falls to the pavement and bounces back to nearly the same height. When it is in contact with the pavement, the lower side of the ball is temporarily flattened. Suppose that the maximum depth of the dent is on the order of 1 cm. Compute an order-of-magnitude estimate for the maximum acceleration of the ball while it is in contact with the pavement. State your assumptions, the quantities you estimate, and the values you estimate for them.

68. At NASA's John H. Glenn research center in Cleveland, Ohio, free-fall research is performed by dropping experiment packages from the top of an evacuated shaft 145 m high. Free fall imitates the so-called microgravity environment of a satellite in orbit. (a) What is the maximum time interval for free fall if an experiment package were to fall the entire 145 m? (b) Actual NASA specifications allow for a 5.18-s drop time interval. How far do the packages drop and (c) what is their speed at 5.18 s? (d) What constant acceleration would be required to stop an experiment package in the distance remaining in the shaft after its 5.18-s fall?

69. An inquisitive physics student and mountain climber climbs a 50.0-m cliff that overhangs a calm pool of water. He throws two stones vertically downward, 1.00 s apart, and observes that they cause a single splash. The first stone has an initial speed of 2.00 m/s. (a) How long after release of the first stone do the two stones hit the water? (b) What initial velocity must the second stone have if they are to hit simultaneously? (c) What is the speed of each stone at the instant the two hit the water?

70. A rock is dropped from rest into a well. The well is not really 16 seconds deep, as in Figure P2.70. (a) The sound of the splash is actually heard 2.40 s after the rock is released from rest. How far below the top of the well is the surface of the water? The speed of sound in air (at the ambient temperature) is 336 m/s. (b) **What If?** If the travel time for the sound is neglected, what percentage error is introduced when the depth of the well is calculated?

71. To protect his food from hungry bears, a boy scout raises his food pack with a rope that is thrown over a tree limb at height h above his hands. He walks away from the vertical rope with constant velocity v_{boy}, holding the free end of the rope in his hands (Fig. P2.71). (a) Show that the speed v of the food pack is given by $x(x^2 + h^2)^{-1/2} v_{boy}$ where x

B.C.

By John Hart

By permission of John Hart and Creators Syndicate, Inc.

Figure P2.70

is the distance he has walked away from the vertical rope. (b) Show that the acceleration a of the food pack is $h^2(x^2 + h^2)^{-3/2} v_{boy}^2$. (c) What values do the acceleration a and velocity v have shortly after he leaves the point under the pack ($x = 0$)? (d) What values do the pack's velocity and acceleration approach as the distance x continues to increase?

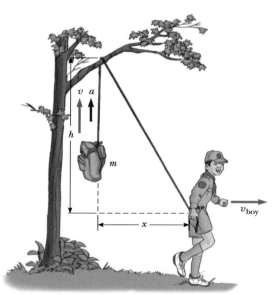

Figure P2.71 Problems 71 and 72.

72. In Problem 71, let the height h equal 6.00 m and the speed v_{boy} equal 2.00 m/s. Assume that the food pack starts from rest. (a) Tabulate and graph the speed–time graph. (b) Tabulate and graph the acceleration-time graph. Let the range of time be from 0 s to 5.00 s and the time intervals be 0.500 s.

73. Kathy Kool buys a sports car that can accelerate at the rate of 4.90 m/s². She decides to test the car by racing with another speedster, Stan Speedy. Both start from rest, but experienced Stan leaves the starting line 1.00 s before Kathy. If Stan moves with a constant acceleration of 3.50 m/s² and Kathy maintains an acceleration of 4.90 m/s², find (a) the time at which Kathy overtakes Stan, (b) the distance she travels before she catches him, and (c) the speeds of both cars at the instant she overtakes him.

74. Astronauts on a distant planet toss a rock into the air. With the aid of a camera that takes pictures at a steady rate, they record the height of the rock as a function of time as given in Table P2.74. (a) Find the average velocity of the rock in the time interval between each measurement and the next. (b) Using these average velocities to approximate instantaneous velocities at the midpoints of the time intervals, make a graph of velocity as a function of time. Does the rock move with constant acceleration? If so, plot a straight line of best fit on the graph and calculate its slope to find the acceleration.

Table P2.74

Height of a Rock versus Time			
Time (s)	**Height (m)**	**Time (s)**	**Height (m)**
0.00	5.00	2.75	7.62
0.25	5.75	3.00	7.25
0.50	6.40	3.25	6.77
0.75	6.94	3.50	6.20
1.00	7.38	3.75	5.52
1.25	7.72	4.00	4.73
1.50	7.96	4.25	3.85
1.75	8.10	4.50	2.86
2.00	8.13	4.75	1.77
2.25	8.07	5.00	0.58
2.50	7.90		

75. Two objects, A and B, are connected by a rigid rod that has a length L. The objects slide along perpendicular guide rails, as shown in Figure P2.75. If A slides to the left with a constant speed v, find the velocity of B when $\alpha = 60.0°$.

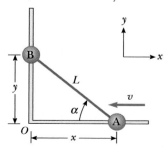

Figure P2.75

Answers to Quick Quizzes

2.1 (c). If the particle moves along a line without changing direction, the displacement and distance traveled over any time interval will be the same. As a result, the magnitude of the average velocity and the average speed will be the same. If the particle reverses direction, however, the displacement will be less than the distance traveled. In turn, the magnitude of the average velocity will be smaller than the average speed.

2.2 (b). If the car is slowing down, a force must be pulling in the direction opposite to its velocity.

2.3 False. Your graph should look something like the following. This v_x-t graph shows that the maximum speed is about 5.0 m/s, which is 18 km/h (= 11 mi/h), and so the driver was not speeding.

Answer to Quick Quiz 2.3

2.4 (c). If a particle with constant acceleration stops and its acceleration remains constant, it must begin to move again in the opposite direction. If it did not, the acceleration would change from its original constant value to zero. Choice (a) is not correct because the direction of acceleration is not specified by the direction of the velocity. Choice (b) is also not correct by counterexample—a car moving in the $-x$ direction and slowing down has a positive acceleration.

2.5 Graph (a) has a constant slope, indicating a constant acceleration; this is represented by graph (e).

Graph (b) represents a speed that is increasing constantly but not at a uniform rate. Thus, the acceleration must be increasing, and the graph that best indicates this is (d).

Graph (c) depicts a velocity that first increases at a constant rate, indicating constant acceleration. Then the velocity stops increasing and becomes constant, indicating zero acceleration. The best match to this situation is graph (f).

2.6 (e). For the entire time interval that the ball is in free fall, the acceleration is that due to gravity.

2.7 (d). While the ball is rising, it is slowing down. After reaching the highest point, the ball begins to fall and its speed increases.

2.8 (a). At the highest point, the ball is momentarily at rest, but still accelerating at $-g$.

Chapter 3

Vectors

▲ These controls in the cockpit of a commercial aircraft assist the pilot in maintaining control over the velocity of the aircraft—how fast it is traveling and in what direction it is traveling—allowing it to land safely. Quantities that are defined by both a magnitude and a direction, such as velocity, are called vector quantities. (Mark Wagner/Getty Images)

In our study of physics, we often need to work with physical quantities that have both numerical and directional properties. As noted in Section 2.1, quantities of this nature are vector quantities. This chapter is primarily concerned with vector algebra and with some general properties of vector quantities. We discuss the addition and subtraction of vector quantities, together with some common applications to physical situations.

Vector quantities are used throughout this text, and it is therefore imperative that you master both their graphical and their algebraic properties.

3.1 Coordinate Systems

Many aspects of physics involve a description of a location in space. In Chapter 2, for example, we saw that the mathematical description of an object's motion requires a method for describing the object's position at various times. This description is accomplished with the use of coordinates, and in Chapter 2 we used the Cartesian coordinate system, in which horizontal and vertical axes intersect at a point defined as the origin (Fig. 3.1). Cartesian coordinates are also called *rectangular coordinates.*

Sometimes it is more convenient to represent a point in a plane by its *plane polar coordinates* (r, θ), as shown in Figure 3.2a. In this *polar coordinate system, r* is the distance from the origin to the point having Cartesian coordinates (x, y), and θ is the angle between a line drawn from the origin to the point and a fixed axis. This fixed axis is usually the positive x axis, and θ is usually measured counterclockwise from it. From the right triangle in Figure 3.2b, we find that $\sin \theta = y/r$ and that $\cos \theta = x/r$. (A review of trigonometric functions is given in Appendix B.4.) Therefore, starting with the plane polar coordinates of any point, we can obtain the Cartesian coordinates by using the equations

$$x = r \cos \theta \tag{3.1}$$

$$y = r \sin \theta \tag{3.2}$$

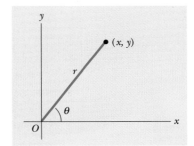

$$\sin \theta = \frac{y}{r}$$

$$\cos \theta = \frac{x}{r}$$

$$\tan \theta = \frac{y}{x}$$

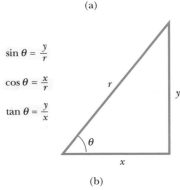

(b)

Active Figure 3.2 (a) The plane polar coordinates of a point are represented by the distance r and the angle θ, where θ is measured counterclockwise from the positive x axis. (b) The right triangle used to relate (x, y) to (r, θ).

At the Active Figures link at http://www.pse6.com, *you can move the point and see the changes to the rectangular and polar coordinates as well as to the sine, cosine, and tangent of angle θ.*

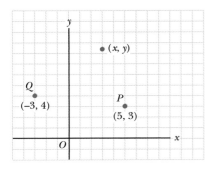

Figure 3.1 Designation of points in a Cartesian coordinate system. Every point is labeled with coordinates (x, y).

Furthermore, the definitions of trigonometry tell us that

$$\tan \theta = \frac{y}{x} \qquad (3.3)$$

$$r = \sqrt{x^2 + y^2} \qquad (3.4)$$

Equation 3.4 is the familiar Pythagorean theorem.

These four expressions relating the coordinates (x, y) to the coordinates (r, θ) apply only when θ is defined as shown in Figure 3.2a—in other words, when positive θ is an angle measured counterclockwise from the positive x axis. (Some scientific calculators perform conversions between Cartesian and polar coordinates based on these standard conventions.) If the reference axis for the polar angle θ is chosen to be one other than the positive x axis or if the sense of increasing θ is chosen differently, then the expressions relating the two sets of coordinates will change.

Example 3.1   Polar Coordinates

The Cartesian coordinates of a point in the xy plane are $(x, y) = (-3.50, -2.50)$ m, as shown in Figure 3.3. Find the polar coordinates of this point.

Solution For the examples in this and the next two chapters we will illustrate the use of the General Problem-Solving

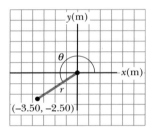

Active Figure 3.3 (Example 3.1) Finding polar coordinates when Cartesian coordinates are given.

At the Active Figures link at **http://www.pse6.com**, *you can move the point in the xy plane and see how its Cartesian and polar coordinates change.*

Strategy outlined at the end of Chapter 2. In subsequent chapters, we will make fewer explicit references to this strategy, as you will have become familiar with it and should be applying it on your own. The drawing in Figure 3.3 helps us to *conceptualize* the problem. We can *categorize* this as a plug-in problem. From Equation 3.4,

$$r = \sqrt{x^2 + y^2} = \sqrt{(-3.50 \text{ m})^2 + (-2.50 \text{ m})^2} = \boxed{4.30 \text{ m}}$$

and from Equation 3.3,

$$\tan \theta = \frac{y}{x} = \frac{-2.50 \text{ m}}{-3.50 \text{ m}} = 0.714$$

$$\theta = \boxed{216°}$$

Note that you must use the signs of x and y to find that the point lies in the third quadrant of the coordinate system. That is, $\theta = 216°$ and not $35.5°$.

3.2   Vector and Scalar Quantities

As noted in Chapter 2, some physical quantities are scalar quantities whereas others are vector quantities. When you want to know the temperature outside so that you will know how to dress, the only information you need is a number and the unit "degrees C" or "degrees F." Temperature is therefore an example of a *scalar quantity*:

A **scalar quantity** is completely specified by a single value with an appropriate unit and has no direction.

Other examples of scalar quantities are volume, mass, speed, and time intervals. The rules of ordinary arithmetic are used to manipulate scalar quantities.

If you are preparing to pilot a small plane and need to know the wind velocity, you must know both the speed of the wind and its direction. Because direction is important for its complete specification, velocity is a *vector quantity*:

A **vector quantity** is completely specified by a number and appropriate units plus a direction.

Another example of a vector quantity is displacement, as you know from Chapter 2. Suppose a particle moves from some point Ⓐ to some point Ⓑ along a straight path, as shown in Figure 3.4. We represent this displacement by drawing an arrow from Ⓐ to Ⓑ, with the tip of the arrow pointing away from the starting point. The direction of the arrowhead represents the direction of the displacement, and the length of the arrow represents the magnitude of the displacement. If the particle travels along some other path from Ⓐ to Ⓑ, such as the broken line in Figure 3.4, its displacement is still the arrow drawn from Ⓐ to Ⓑ. Displacement depends only on the initial and final positions, so the displacement vector is independent of the path taken between these two points.

In this text, we use a boldface letter, such as **A**, to represent a vector quantity. Another notation is useful when boldface notation is difficult, such as when writing on paper or on a chalkboard—an arrow is written over the symbol for the vector: $\vec{A}$. The magnitude of the vector **A** is written either A or $|\mathbf{A}|$. The magnitude of a vector has physical units, such as meters for displacement or meters per second for velocity. The magnitude of a vector is *always* a positive number.

> **Quick Quiz 3.1** Which of the following are vector quantities and which are scalar quantities? (a) your age (b) acceleration (c) velocity (d) speed (e) mass

3.3 Some Properties of Vectors

Equality of Two Vectors

For many purposes, two vectors **A** and **B** may be defined to be equal if they have the same magnitude and point in the same direction. That is, **A** = **B** only if $A = B$ and if **A** and **B** point in the same direction along parallel lines. For example, all the vectors in Figure 3.5 are equal even though they have different starting points. This property allows us to move a vector to a position parallel to itself in a diagram without affecting the vector.

Adding Vectors

The rules for adding vectors are conveniently described by graphical methods. To add vector **B** to vector **A**, first draw vector **A** on graph paper, with its magnitude represented by a convenient length scale, and then draw vector **B** to the same scale with its tail starting from the tip of **A**, as shown in Figure 3.6. The **resultant vector R =** **A** + **B** is the vector drawn from the tail of **A** to the tip of **B**.

For example, if you walked 3.0 m toward the east and then 4.0 m toward the north, as shown in Figure 3.7, you would find yourself 5.0 m from where you started, measured at an angle of 53° north of east. Your total displacement is the vector sum of the individual displacements.

A geometric construction can also be used to add more than two vectors. This is shown in Figure 3.8 for the case of four vectors. The resultant vector **R** = **A** + **B** + **C** + **D** is the vector that completes the polygon. In other words, **R is the vector drawn from the tail of the first vector to the tip of the last vector**.

When two vectors are added, the sum is independent of the order of the addition. (This fact may seem trivial, but as you will see in Chapter 11, the order is important

Figure 3.4 As a particle moves from Ⓐ to Ⓑ along an arbitrary path represented by the broken line, its displacement is a vector quantity shown by the arrow drawn from Ⓐ to Ⓑ.

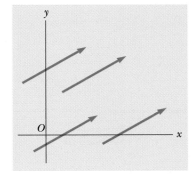

Figure 3.5 These four vectors are equal because they have equal lengths and point in the same direction.

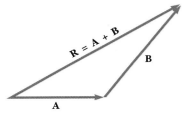

Active Figure 3.6 When vector **B** is added to vector **A**, the resultant **R** is the vector that runs from the tail of **A** to the tip of **B**.

 Go to the Active Figures link at **http://www.pse6.com.**

▲ **PITFALL PREVENTION**

3.1 Vector Addition versus Scalar Addition

Keep in mind that **A** + **B** = **C** is very different from $A + B = C$. The first is a vector sum, which must be handled carefully, such as with the graphical method described here. The second is a simple algebraic addition of numbers that is handled with the normal rules of arithmetic.

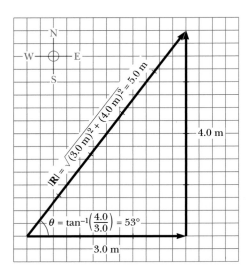

Figure 3.7 Vector addition. Walking first 3.0 m due east and then 4.0 m due north leaves you 5.0 m from your starting point.

when vectors are multiplied). This can be seen from the geometric construction in Figure 3.9 and is known as the **commutative law of addition:**

$$\mathbf{A} + \mathbf{B} = \mathbf{B} + \mathbf{A} \tag{3.5}$$

When three or more vectors are added, their sum is independent of the way in which the individual vectors are grouped together. A geometric proof of this rule for three vectors is given in Figure 3.10. This is called the **associative law of addition:**

$$\mathbf{A} + (\mathbf{B} + \mathbf{C}) = (\mathbf{A} + \mathbf{B}) + \mathbf{C} \tag{3.6}$$

In summary, **a vector quantity has both magnitude and direction and also obeys the laws of vector addition** as described in Figures 3.6 to 3.10. When two or more vectors are added together, all of them must have the same units and all of them must be the same type of quantity. It would be meaningless to add a velocity vector (for example, 60 km/h to the east) to a displacement vector (for example, 200 km to the north) because they represent different physical quantities. The same rule also applies to scalars. For example, it would be meaningless to add time intervals to temperatures.

Negative of a Vector

The negative of the vector **A** is defined as the vector that when added to **A** gives zero for the vector sum. That is, $\mathbf{A} + (-\mathbf{A}) = 0$. The vectors **A** and $-\mathbf{A}$ have the same magnitude but point in opposite directions.

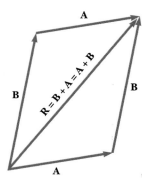

Figure 3.8 Geometric construction for summing four vectors. The resultant vector **R** is by definition the one that completes the polygon.

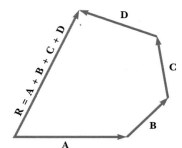

Figure 3.9 This construction shows that $\mathbf{A} + \mathbf{B} = \mathbf{B} + \mathbf{A}$—in other words, that vector addition is commutative.

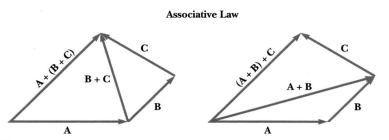

Figure 3.10 Geometric constructions for verifying the associative law of addition.

Vector Subtraction

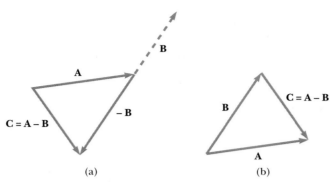

(a)   (b)

Figure 3.11 (a) This construction shows how to subtract vector **B** from vector **A**. The vector $-$ **B** is equal in magnitude to vector **B** and points in the opposite direction. To subtract **B** from **A**, apply the rule of vector addition to the combination of **A** and $-$ **B**: Draw **A** along some convenient axis, place the tail of $-$ **B** at the tip of **A**, and **C** is the difference **A** $-$ **B**. (b) A second way of looking at vector subtraction. The difference vector **C** = **A** $-$ **B** is the vector that we must add to **B** to obtain **A**.

Subtracting Vectors

The operation of vector subtraction makes use of the definition of the negative of a vector. We define the operation **A** $-$ **B** as vector $-$ **B** added to vector **A**:

$$\mathbf{A} - \mathbf{B} = \mathbf{A} + (-\mathbf{B}) \tag{3.7}$$

The geometric construction for subtracting two vectors in this way is illustrated in Figure 3.11a.

Another way of looking at vector subtraction is to note that the difference **A** $-$ **B** between two vectors **A** and **B** is what you have to add to the second vector to obtain the first. In this case, the vector **A** $-$ **B** points from the tip of the second vector to the tip of the first, as Figure 3.11b shows.

Quick Quiz 3.2 The magnitudes of two vectors **A** and **B** are $A = 12$ units and $B = 8$ units. Which of the following pairs of numbers represents the *largest* and *smallest* possible values for the magnitude of the resultant vector **R** = **A** + **B**? (a) 14.4 units, 4 units (b) 12 units, 8 units (c) 20 units, 4 units (d) none of these answers.

Quick Quiz 3.3 If vector **B** is added to vector **A**, under what condition does the resultant vector **A** + **B** have magnitude $A + B$? (a) **A** and **B** are parallel and in the same direction. (b) **A** and **B** are parallel and in opposite directions. (c) **A** and **B** are perpendicular.

Quick Quiz 3.4 If vector **B** is added to vector **A**, which *two* of the following choices must be true in order for the resultant vector to be equal to zero? (a) **A** and **B** are parallel and in the same direction. (b) **A** and **B** are parallel and in opposite directions. (c) **A** and **B** have the same magnitude. (d) **A** and **B** are perpendicular.

Example 3.2 A Vacation Trip

A car travels 20.0 km due north and then 35.0 km in a direction 60.0° west of north, as shown in Figure 3.12a. Find the magnitude and direction of the car's resultant displacement.

Solution The vectors **A** and **B** drawn in Figure 3.12a help us to *conceptualize* the problem. We can *categorize* this as a relatively simple analysis problem in vector addition. The displacement **R** is the resultant when the two individual displacements **A** and **B** are added. We can further categorize this as a problem about the analysis of triangles, so we appeal to our expertise in geometry and trigonometry.

In this example, we show two ways to *analyze* the problem of finding the resultant of two vectors. The first way is to solve the problem geometrically, using graph paper and a protractor to measure the magnitude of **R** and its direction in Figure 3.12a. (In fact, even when you know you are going to be carrying out a calculation, you should sketch the vectors to check your results.) With an ordinary ruler and protractor, a large diagram typically gives answers to two-digit but not to three-digit precision.

The second way to solve the problem is to analyze it algebraically. The magnitude of **R** can be obtained from the law of cosines as applied to the triangle (see Appendix B.4). With $\theta = 180° - 60° = 120°$ and $R^2 = A^2 + B^2 - 2AB \cos\theta$, we find that

$$R = \sqrt{A^2 + B^2 - 2AB\cos\theta}$$
$$= \sqrt{(20.0\,\text{km})^2 + (35.0\,\text{km})^2 - 2(20.0\,\text{km})(35.0\,\text{km})\cos 120°}$$
$$= \boxed{48.2\,\text{km}}$$

The direction of **R** measured from the northerly direction can be obtained from the law of sines (Appendix B.4):

$$\frac{\sin\beta}{B} = \frac{\sin\theta}{R}$$
$$\sin\beta = \frac{B}{R}\sin\theta = \frac{35.0\,\text{km}}{48.2\,\text{km}}\sin 120° = 0.629$$
$$\boxed{\beta = 39.0°}$$

The resultant displacement of the car is 48.2 km in a direction 39.0° west of north.

We now *finalize* the problem. Does the angle β that we calculated agree with an estimate made by looking at Figure 3.12a or with an actual angle measured from the diagram using the graphical method? Is it reasonable that the magnitude of **R** is larger than that of both **A** and **B**? Are the units of **R** correct?

While the graphical method of adding vectors works well, it suffers from two disadvantages. First, some individuals find using the laws of cosines and sines to be awkward. Second, a triangle only results if you are adding two vectors. If you are adding three or more vectors, the resulting geometric shape is not a triangle. In Section 3.4, we explore a new method of adding vectors that will address both of these disadvantages.

What If? Suppose the trip were taken with the two vectors in reverse order: 35.0 km at 60.0° west of north first, and then 20.0 km due north. How would the magnitude and the direction of the resultant vector change?

Answer They would not change. The commutative law for vector addition tells us that the order of vectors in an addition is irrelevant. Graphically, Figure 3.12b shows that the vectors added in the reverse order give us the same resultant vector.

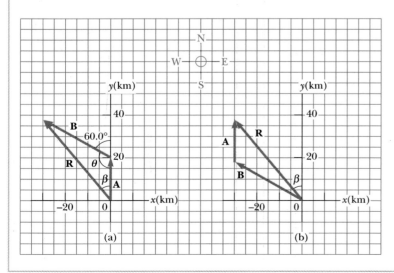

Figure 3.12 (Example 3.2) (a) Graphical method for finding the resultant displacement vector $\mathbf{R} = \mathbf{A} + \mathbf{B}$. (b) Adding the vectors in reverse order $(\mathbf{B} + \mathbf{A})$ gives the same result for **R**.

Multiplying a Vector by a Scalar

If vector **A** is multiplied by a positive scalar quantity m, then the product m**A** is a vector that has the same direction as **A** and magnitude mA. If vector **A** is multiplied by a negative scalar quantity $-m$, then the product $-m$**A** is directed opposite **A**. For example, the vector 5**A** is five times as long as **A** and points in the same direction as **A**; the vector $-\frac{1}{3}$**A** is one-third the length of **A** and points in the direction opposite **A**.

(a)

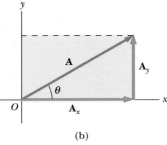

(b)

Figure 3.13 (a) A vector **A** lying in the xy plane can be represented by its component vectors **A**$_x$ and **A**$_y$. (b) The y component vector **A**$_y$ can be moved to the right so that it adds to **A**$_x$. The vector sum of the component vectors is **A**. These three vectors form a right triangle.

3.4 Components of a Vector and Unit Vectors

The graphical method of adding vectors is not recommended whenever high accuracy is required or in three-dimensional problems. In this section, we describe a method of adding vectors that makes use of the projections of vectors along coordinate axes. These projections are called the **components** of the vector. Any vector can be completely described by its components.

Consider a vector **A** lying in the xy plane and making an arbitrary angle θ with the positive x axis, as shown in Figure 3.13a. This vector can be expressed as the sum of two other vectors **A**$_x$ and **A**$_y$. From Figure 3.13b, we see that the three vectors form a right triangle and that **A** = **A**$_x$ + **A**$_y$. We shall often refer to the "components of a vector **A**," written A_x and A_y (without the boldface notation). The component A_x represents the projection of **A** along the x axis, and the component A_y represents the projection of **A** along the y axis. These components can be positive or negative. The component A_x is positive if **A**$_x$ points in the positive x direction and is negative if **A**$_x$ points in the negative x direction. The same is true for the component A_y.

From Figure 3.13 and the definition of sine and cosine, we see that $\cos\theta = A_x/A$ and that $\sin\theta = A_y/A$. Hence, the components of **A** are

$$A_x = A\cos\theta \tag{3.8}$$

$$A_y = A\sin\theta \tag{3.9}$$

Components of the vector A

These components form two sides of a right triangle with a hypotenuse of length A. Thus, it follows that the magnitude and direction of **A** are related to its components through the expressions

$$A = \sqrt{A_x^2 + A_y^2} \tag{3.10}$$

$$\theta = \tan^{-1}\left(\frac{A_y}{A_x}\right) \tag{3.11}$$

Note that **the signs of the components A_x and A_y depend on the angle θ.** For example, if $\theta = 120°$, then A_x is negative and A_y is positive. If $\theta = 225°$, then both A_x and A_y are negative. Figure 3.14 summarizes the signs of the components when **A** lies in the various quadrants.

When solving problems, you can specify a vector **A** either with its components A_x and A_y or with its magnitude and direction A and θ.

PITFALL PREVENTION

3.2 Component Vectors versus Components

The vectors **A**$_x$ and **A**$_y$ are the *component vectors* of **A**. These should not be confused with the scalars A_x and A_y, which we shall always refer to as the *components* of **A**.

> **Quick Quiz 3.5** Choose the correct response to make the sentence true: A component of a vector is (a) always, (b) never, or (c) sometimes larger than the magnitude of the vector.

Suppose you are working a physics problem that requires resolving a vector into its components. In many applications it is convenient to express the components in a coordinate system having axes that are not horizontal and vertical but are still perpendicular to each other. If you choose reference axes or an angle other than the axes and angle shown in Figure 3.13, the components must be modified accordingly. Suppose a

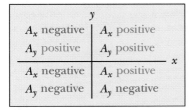

Figure 3.14 The signs of the components of a vector **A** depend on the quadrant in which the vector is located.

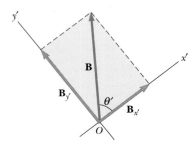

Figure 3.15 The component vectors of **B** in a coordinate system that is tilted.

(a)

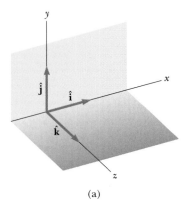

(b)

Active Figure 3.16 (a) The unit vectors $\hat{\mathbf{i}}$, $\hat{\mathbf{j}}$, and $\hat{\mathbf{k}}$ are directed along the x, y, and z axes, respectively. (b) Vector $\mathbf{A} = A_x\hat{\mathbf{i}} + A_y\hat{\mathbf{j}}$ lying in the xy plane has components A_x and A_y.

At the Active Figures link at http://www.pse6.com you can rotate the coordinate axes in 3-dimensional space and view a representation of vector A in three dimensions.

vector **B** makes an angle θ' with the x' axis defined in Figure 3.15. The components of **B** along the x' and y' axes are $B_{x'} = B\cos\theta'$ and $B_{y'} = B\sin\theta'$, as specified by Equations 3.8 and 3.9. The magnitude and direction of **B** are obtained from expressions equivalent to Equations 3.10 and 3.11. Thus, we can express the components of a vector in any coordinate system that is convenient for a particular situation.

Unit Vectors

Vector quantities often are expressed in terms of unit vectors. **A unit vector is a dimensionless vector having a magnitude of exactly 1.** Unit vectors are used to specify a given direction and have no other physical significance. They are used solely as a convenience in describing a direction in space. We shall use the symbols $\hat{\mathbf{i}}$, $\hat{\mathbf{j}}$, and $\hat{\mathbf{k}}$ to represent unit vectors pointing in the positive x, y, and z directions, respectively. (The "hats" on the symbols are a standard notation for unit vectors.) The unit vectors $\hat{\mathbf{i}}$, $\hat{\mathbf{j}}$, and $\hat{\mathbf{k}}$ form a set of mutually perpendicular vectors in a right-handed coordinate system, as shown in Figure 3.16a. The magnitude of each unit vector equals 1; that is, $|\hat{\mathbf{i}}| = |\hat{\mathbf{j}}| = |\hat{\mathbf{k}}| = 1$.

Consider a vector **A** lying in the xy plane, as shown in Figure 3.16b. The product of the component A_x and the unit vector $\hat{\mathbf{i}}$ is the vector $A_x\hat{\mathbf{i}}$, which lies on the x axis and has magnitude $|A_x|$. (The vector $A_x\hat{\mathbf{i}}$ is an alternative representation of vector $\mathbf{A}_x$.) Likewise, $A_y\hat{\mathbf{j}}$ is a vector of magnitude $|A_y|$ lying on the y axis. (Again, vector $A_y\hat{\mathbf{j}}$ is an alternative representation of vector $\mathbf{A}_y$.) Thus, the unit–vector notation for the vector **A** is

$$\mathbf{A} = A_x\hat{\mathbf{i}} + A_y\hat{\mathbf{j}} \tag{3.12}$$

For example, consider a point lying in the xy plane and having Cartesian coordinates (x, y), as in Figure 3.17. The point can be specified by the **position vector r**, which in unit–vector form is given by

$$\mathbf{r} = x\hat{\mathbf{i}} + y\hat{\mathbf{j}} \tag{3.13}$$

This notation tells us that the components of **r** are the lengths x and y.

Now let us see how to use components to add vectors when the graphical method is not sufficiently accurate. Suppose we wish to add vector **B** to vector **A** in Equation 3.12, where vector **B** has components B_x and B_y. All we do is add the x and y components separately. The resultant vector $\mathbf{R} = \mathbf{A} + \mathbf{B}$ is therefore

$$\mathbf{R} = (A_x\hat{\mathbf{i}} + A_y\hat{\mathbf{j}}) + (B_x\hat{\mathbf{i}} + B_y\hat{\mathbf{j}})$$

or

$$\mathbf{R} = (A_x + B_x)\hat{\mathbf{i}} + (A_y + B_y)\hat{\mathbf{j}} \tag{3.14}$$

Because $\mathbf{R} = R_x\hat{\mathbf{i}} + R_y\hat{\mathbf{j}}$, we see that the components of the resultant vector are

$$R_x = A_x + B_x$$
$$R_y = A_y + B_y \tag{3.15}$$

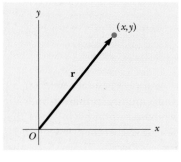

Figure 3.17 The point whose Cartesian coordinates are (x, y) can be represented by the position vector $\mathbf{r} = x\hat{\mathbf{i}} + y\hat{\mathbf{j}}$.

We obtain the magnitude of **R** and the angle it makes with the x axis from its components, using the relationships

$$R = \sqrt{R_x^2 + R_y^2} = \sqrt{(A_x + B_x)^2 + (A_y + B_y)^2} \tag{3.16}$$

$$\tan \theta = \frac{R_y}{R_x} = \frac{A_y + B_y}{A_x + B_x} \tag{3.17}$$

We can check this addition by components with a geometric construction, as shown in Figure 3.18. Remember that you must note the signs of the components when using either the algebraic or the graphical method.

At times, we need to consider situations involving motion in three component directions. The extension of our methods to three-dimensional vectors is straightforward. If **A** and **B** both have x, y, and z components, we express them in the form

$$\mathbf{A} = A_x \hat{\mathbf{i}} + A_y \hat{\mathbf{j}} + A_z \hat{\mathbf{k}} \tag{3.18}$$

$$\mathbf{B} = B_x \hat{\mathbf{i}} + B_y \hat{\mathbf{j}} + B_z \hat{\mathbf{k}} \tag{3.19}$$

The sum of **A** and **B** is

$$\mathbf{R} = (A_x + B_x)\hat{\mathbf{i}} + (A_y + B_y)\hat{\mathbf{j}} + (A_z + B_z)\hat{\mathbf{k}} \tag{3.20}$$

Note that Equation 3.20 differs from Equation 3.14: in Equation 3.20, the resultant vector also has a z component $R_z = A_z + B_z$. If a vector **R** has x, y, and z components, the magnitude of the vector is $R = \sqrt{R_x^2 + R_y^2 + R_z^2}$. The angle θ_x that **R** makes with the x axis is found from the expression $\cos \theta_x = R_x/R$, with similar expressions for the angles with respect to the y and z axes.

Quick Quiz 3.6 If at least one component of a vector is a positive number, the vector cannot (a) have any component that is negative (b) be zero (c) have three dimensions.

Quick Quiz 3.7 If **A** + **B** = 0, the corresponding components of the two vectors **A** and **B** must be (a) equal (b) positive (c) negative (d) of opposite sign.

Quick Quiz 3.8 For which of the following vectors is the magnitude of the vector equal to one of the components of the vector? (a) $\mathbf{A} = 2\hat{\mathbf{i}} + 5\hat{\mathbf{j}}$ (b) $\mathbf{B} = -3\hat{\mathbf{j}}$ (c) $\mathbf{C} = +5\hat{\mathbf{k}}$

PROBLEM-SOLVING HINTS

Adding Vectors

When you need to add two or more vectors, use this step-by-step procedure:

- Select a coordinate system that is convenient. (Try to reduce the number of components you need to calculate by choosing axes that line up with as many vectors as possible.)
- Draw a labeled sketch of the vectors described in the problem.
- Find the x and y components of all vectors and the resultant components (the algebraic sum of the components) in the x and y directions.
- If necessary, use the Pythagorean theorem to find the magnitude of the resultant vector and select a suitable trigonometric function to find the angle that the resultant vector makes with the x axis.

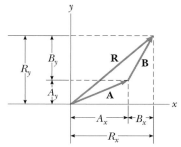

Figure 3.18 This geometric construction for the sum of two vectors shows the relationship between the components of the resultant **R** and the components of the individual vectors.

▲ **PITFALL PREVENTION**

3.3   *x* and *y* Components

Equations 3.8 and 3.9 associate the cosine of the angle with the x component and the sine of the angle with the y component. This is true *only* because we measured the angle θ with respect to the x axis, so don't memorize these equations. If θ is measured with respect to the y axis (as in some problems), these equations will be incorrect. Think about which side of the triangle containing the components is adjacent to the angle and which side is opposite, and assign the cosine and sine accordingly.

▲ **PITFALL PREVENTION**

3.4   Tangents on Calculators

Generally, the inverse tangent function on calculators provides an angle between $-90°$ and $+90°$. As a consequence, if the vector you are studying lies in the second or third quadrant, the angle measured from the positive x axis will be the angle your calculator returns plus $180°$.

Example 3.3 The Sum of Two Vectors

Find the sum of two vectors **A** and **B** lying in the xy plane and given by

$$\mathbf{A} = (2.0\hat{\mathbf{i}} + 2.0\hat{\mathbf{j}}) \text{ m} \quad \text{and} \quad \mathbf{B} = (2.0\hat{\mathbf{i}} - 4.0\hat{\mathbf{j}}) \text{ m}$$

Solution You may wish to draw the vectors to *conceptualize* the situation. We *categorize* this as a simple plug-in problem. Comparing this expression for **A** with the general expression $\mathbf{A} = A_x\hat{\mathbf{i}} + A_y\hat{\mathbf{j}}$, we see that $A_x = 2.0$ m and $A_y = 2.0$ m. Likewise, $B_x = 2.0$ m and $B_y = -4.0$ m. We obtain the resultant vector **R**, using Equation 3.14:

$$\mathbf{R} = \mathbf{A} + \mathbf{B} = (2.0 + 2.0)\hat{\mathbf{i}} \text{ m} + (2.0 - 4.0)\hat{\mathbf{j}} \text{ m}$$
$$= (4.0\hat{\mathbf{i}} - 2.0\hat{\mathbf{j}}) \text{ m}$$

or

$$R_x = 4.0 \text{ m} \qquad R_y = -2.0 \text{ m}$$

The magnitude of **R** is found using Equation 3.16:

$$R = \sqrt{R_x{}^2 + R_y{}^2} = \sqrt{(4.0 \text{ m})^2 + (-2.0 \text{ m})^2} = \sqrt{20} \text{ m}$$
$$= \boxed{4.5 \text{ m}}$$

We can find the direction of **R** from Equation 3.17:

$$\tan\theta = \frac{R_y}{R_x} = \frac{-2.0 \text{ m}}{4.0 \text{ m}} = -0.50$$

Your calculator likely gives the answer $-27°$ for $\theta = \tan^{-1}(-0.50)$. This answer is correct if we interpret it to mean 27° clockwise from the x axis. Our standard form has been to quote the angles measured counterclockwise from the $+x$ axis, and that angle for this vector is $\boxed{\theta = 333°}$.

Example 3.4 The Resultant Displacement

A particle undergoes three consecutive displacements: $\mathbf{d}_1 = (15\hat{\mathbf{i}} + 30\hat{\mathbf{j}} + 12\hat{\mathbf{k}})$ cm, $\mathbf{d}_2 = (23\hat{\mathbf{i}} - 14\hat{\mathbf{j}} - 5.0\hat{\mathbf{k}})$ cm and $\mathbf{d}_3 = (-13\hat{\mathbf{i}} + 15\hat{\mathbf{j}})$ cm. Find the components of the resultant displacement and its magnitude.

Solution Three-dimensional displacements are more difficult to imagine than those in two dimensions, because the latter can be drawn on paper. For this problem, let us *conceptualize* that you start with your pencil at the origin of a piece of graph paper on which you have drawn x and y axes. Move your pencil 15 cm to the right along the x axis, then 30 cm upward along the y axis, and then 12 cm *vertically away* from the graph paper. This provides the displacement described by $\mathbf{d}_1$. From this point, move your pencil 23 cm to the right parallel to the x axis, 14 cm parallel to the graph paper in the $-y$ direction, and then 5.0 cm vertically downward toward the graph paper. You are now at the displacement from the origin described by $\mathbf{d}_1 + \mathbf{d}_2$. From this point, move your pencil 13 cm to the left in the $-x$ direction, and (finally!) 15 cm parallel to the graph paper along the y axis.

Your final position is at a displacement $\mathbf{d}_1 + \mathbf{d}_2 + \mathbf{d}_3$ from the origin.

Despite the difficulty in conceptualizing in three dimensions, we can *categorize* this problem as a plug-in problem due to the careful bookkeeping methods that we have developed for vectors. The mathematical manipulation keeps track of this motion along the three perpendicular axes in an organized, compact way:

$$\mathbf{R} = \mathbf{d}_1 + \mathbf{d}_2 + \mathbf{d}_3$$
$$= (15 + 23 - 13)\hat{\mathbf{i}} \text{ cm} + (30 - 14 + 15)\hat{\mathbf{j}} \text{ cm} + (12 - 5.0 + 0)\hat{\mathbf{k}} \text{ cm}$$
$$= (25\hat{\mathbf{i}} + 31\hat{\mathbf{j}} + 7.0\hat{\mathbf{k}}) \text{ cm}$$

The resultant displacement has components $R_x = 25$ cm, $R_y = 31$ cm, and $R_z = 7.0$ cm. Its magnitude is

$$R = \sqrt{R_x{}^2 + R_y{}^2 + R_z{}^2}$$
$$= \sqrt{(25 \text{ cm})^2 + (31 \text{ cm})^2 + (7.0 \text{ cm})^2} = \boxed{40 \text{ cm}}$$

Example 3.5 Taking a Hike   `Interactive`

A hiker begins a trip by first walking 25.0 km southeast from her car. She stops and sets up her tent for the night. On the second day, she walks 40.0 km in a direction 60.0° north of east, at which point she discovers a forest ranger's tower.

(A) Determine the components of the hiker's displacement for each day.

Solution We *conceptualize* the problem by drawing a sketch as in Figure 3.19. If we denote the displacement vectors on the first and second days by **A** and **B**, respectively, and use the car as the origin of coordinates, we obtain the vectors shown in Figure 3.19. Drawing the resultant **R**, we can now *categorize* this as a problem we've solved before—an addition of two vectors. This should give you a hint of the power of categorization—many new problems are very similar to problems that we have already solved if we are careful to conceptualize them.

We will *analyze* this problem by using our new knowledge of vector components. Displacement **A** has a magnitude of 25.0 km and is directed 45.0° below the positive x axis. From Equations 3.8 and 3.9, its components are

$$A_x = A\cos(-45.0°) = (25.0 \text{ km})(0.707) = \boxed{17.7 \text{ km}}$$

$$A_y = A\sin(-45.0°) = (25.0 \text{ km})(-0.707) = \boxed{-17.7 \text{ km}}$$

The negative value of A_y indicates that the hiker walks in the negative y direction on the first day. The signs of A_x and A_y also are evident from Figure 3.19.

The second displacement **B** has a magnitude of 40.0 km and is 60.0° north of east. Its components are

$$B_x = B \cos 60.0° = (40.0 \text{ km})(0.500) = \boxed{20.0 \text{ km}}$$

$$B_y = B \sin 60.0° = (40.0 \text{ km})(0.866) = \boxed{34.6 \text{ km}}$$

(B) Determine the components of the hiker's resultant displacement **R** for the trip. Find an expression for **R** in terms of unit vectors.

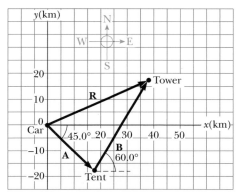

Figure 3.19 (Example 3.5) The total displacement of the hiker is the vector **R** = **A** + **B**.

Solution The resultant displacement for the trip **R** = **A** + **B** has components given by Equation 3.15:

$$R_x = A_x + B_x = 17.7 \text{ km} + 20.0 \text{ km} = \boxed{37.7 \text{ km}}$$

$$R_y = A_y + B_y = -17.7 \text{ km} + 34.6 \text{ km} = \boxed{16.9 \text{ km}}$$

In unit–vector form, we can write the total displacement as

$$\mathbf{R} = \boxed{(37.7\hat{\mathbf{i}} + 16.9\hat{\mathbf{j}}) \text{ km}}$$

Using Equations 3.16 and 3.17, we find that the vector **R** has a magnitude of 41.3 km and is directed 24.1° north of east.

Let us *finalize*. The units of **R** are km, which is reasonable for a displacement. Looking at the graphical representation in Figure 3.19, we estimate that the final position of the hiker is at about (38 km, 17 km) which is consistent with the components of **R** in our final result. Also, both components of **R** are positive, putting the final position in the first quadrant of the coordinate system, which is also consistent with Figure 3.19.

 Investigate this situation at the Interactive Worked Example link at **http://www.pse6.com.**

Example 3.6 Let's Fly Away!

A commuter airplane takes the route shown in Figure 3.20. First, it flies from the origin of the coordinate system shown to city A, located 175 km in a direction 30.0° north of east. Next, it flies 153 km 20.0° west of north to city B. Finally, it flies 195 km due west to city C. Find the location of city C relative to the origin.

Solution Once again, a drawing such as Figure 3.20 allows us to *conceptualize* the problem. It is convenient to choose the coordinate system shown in Figure 3.20, where the x axis points to the east and the y axis points to the north. Let us denote the three consecutive displacements by the vectors **a**, **b**, and **c**.

We can now *categorize* this problem as being similar to Example 3.5 that we have already solved. There are two primary differences. First, we are adding three vectors instead of two. Second, Example 3.5 guided us by first asking for the components in part (A). The current Example has no such guidance and simply asks for a result. We need to *analyze* the situation and choose a path. We will follow the same pattern that we did in Example 3.5, beginning with finding the components of the three vectors **a**, **b**, and **c**. Displacement **a** has a magnitude of 175 km and the components

$$a_x = a \cos(30.0°) = (175 \text{ km})(0.866) = 152 \text{ km}$$

$$a_y = a \sin(30.0°) = (175 \text{ km})(0.500) = 87.5 \text{ km}$$

Displacement **b**, whose magnitude is 153 km, has the components

$$b_x = b \cos(110°) = (153 \text{ km})(-0.342) = -52.3 \text{ km}$$

$$b_y = b \sin(110°) = (153 \text{ km})(0.940) = 144 \text{ km}$$

Finally, displacement **c,** whose magnitude is 195 km, has the components

$$c_x = c \cos(180°) = (195 \text{ km})(-1) = -195 \text{ km}$$

$$c_y = c \sin(180°) = 0$$

Therefore, the components of the position vector **R** from the starting point to city C are

$$R_x = a_x + b_x + c_x = 152 \text{ km} - 52.3 \text{ km} - 195 \text{ km}$$

$$= -95.3 \text{ km}$$

$$R_y = a_y + b_y + c_y = 87.5 \text{ km} + 144 \text{ km} + 0 = \boxed{232 \text{ km}}$$

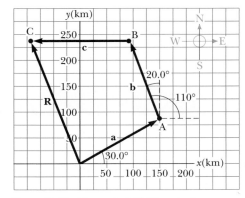

Figure 3.20 (Example 3.6) The airplane starts at the origin, flies first to city A, then to city B, and finally to city C.

In unit–vector notation, $\mathbf{R} = (-95.3\hat{\mathbf{i}} + 232\hat{\mathbf{j}})$ km . Using Equations 3.16 and 3.17, we find that the vector $\mathbf{R}$ has a magnitude of 251 km and is directed 22.3° west of north.

To *finalize* the problem, note that the airplane can reach city C from the starting point by first traveling 95.3 km due west and then by traveling 232 km due north. Or it could follow a straight-line path to C by flying a distance $R = 251$ km in a direction 22.3° west of north.

What If? After landing in city C, the pilot wishes to return to the origin along a single straight line. What are the components of the vector representing this displacement? What should the heading of the plane be?

Answer The desired vector $\mathbf{H}$ (for Home!) is simply the negative of vector $\mathbf{R}$:

$$\mathbf{H} = -\mathbf{R} = (+95.3\hat{\mathbf{i}} - 232\hat{\mathbf{j}})\ \text{km}$$

The heading is found by calculating the angle that the vector makes with the x axis:

$$\tan\theta = \frac{R_y}{R_x} = \frac{-232\ \text{m}}{95.3\ \text{m}} = -2.43$$

This gives a heading angle of $\theta = -67.7°$, or 67.7° south of east.

Take a practice test for this chapter by clicking on the Practice Test link at http://www.pse6.com.

SUMMARY

Scalar quantities are those that have only a numerical value and no associated direction. **Vector quantities** have both magnitude and direction and obey the laws of vector addition. The magnitude of a vector is *always* a positive number.

When two or more vectors are added together, all of them must have the same units and all of them must be the same type of quantity. We can add two vectors $\mathbf{A}$ and $\mathbf{B}$ graphically. In this method (Fig. 3.6), the resultant vector $\mathbf{R} = \mathbf{A} + \mathbf{B}$ runs from the tail of $\mathbf{A}$ to the tip of $\mathbf{B}$.

A second method of adding vectors involves **components** of the vectors. The x component A_x of the vector $\mathbf{A}$ is equal to the projection of $\mathbf{A}$ along the x axis of a coordinate system, as shown in Figure 3.13, where $A_x = A\cos\theta$. The y component A_y of $\mathbf{A}$ is the projection of $\mathbf{A}$ along the y axis, where $A_y = A\sin\theta$. Be sure you can determine which trigonometric functions you should use in all situations, especially when θ is defined as something other than the counterclockwise angle from the positive x axis.

If a vector $\mathbf{A}$ has an x component A_x and a y component A_y, the vector can be expressed in unit–vector form as $\mathbf{A} = A_x\hat{\mathbf{i}} + A_y\hat{\mathbf{j}}$. In this notation, $\hat{\mathbf{i}}$ is a unit vector pointing in the positive x direction, and $\hat{\mathbf{j}}$ is a unit vector pointing in the positive y direction. Because $\hat{\mathbf{i}}$ and $\hat{\mathbf{j}}$ are unit vectors, $|\hat{\mathbf{i}}| = |\hat{\mathbf{j}}| = 1$.

We can find the resultant of two or more vectors by resolving all vectors into their x and y components, adding their resultant x and y components, and then using the Pythagorean theorem to find the magnitude of the resultant vector. We can find the angle that the resultant vector makes with respect to the x axis by using a suitable trigonometric function.

QUESTIONS

1. Two vectors have unequal magnitudes. Can their sum be zero? Explain.

2. Can the magnitude of a particle's displacement be greater than the distance traveled? Explain.

3. The magnitudes of two vectors $\mathbf{A}$ and $\mathbf{B}$ are $A = 5$ units and $B = 2$ units. Find the largest and smallest values possible for the magnitude of the resultant vector $\mathbf{R} = \mathbf{A} + \mathbf{B}$.

4. Which of the following are vectors and which are not: force, temperature, the volume of water in a can, the ratings of a TV show, the height of a building, the velocity of a sports car, the age of the Universe?

5. A vector $\mathbf{A}$ lies in the xy plane. For what orientations of $\mathbf{A}$ will both of its components be negative? For what orientations will its components have opposite signs?

6. A book is moved once around the perimeter of a tabletop with the dimensions 1.0 m × 2.0 m. If the book ends up at its initial position, what is its displacement? What is the distance traveled?

7. While traveling along a straight interstate highway you notice that the mile marker reads 260. You travel until you reach mile marker 150 and then retrace your path to the mile marker 175. What is the magnitude of your resultant displacement from mile marker 260?

8. If the component of vector **A** along the direction of vector **B** is zero, what can you conclude about the two vectors?

9. Can the magnitude of a vector have a negative value? Explain.

10. Under what circumstances would a nonzero vector lying in the *xy* plane have components that are equal in magnitude?

11. If **A** = **B**, what can you conclude about the components of **A** and **B**?

12. Is it possible to add a vector quantity to a scalar quantity? Explain.

13. The resolution of vectors into components is equivalent to replacing the original vector with the sum of two vectors, whose sum is the same as the original vector. There are an infinite number of pairs of vectors that will satisfy this condition; we choose that pair with one vector parallel to the *x* axis and the second parallel to the *y* axis. What difficulties would be introduced by defining components relative to axes that are not perpendicular—for example, the *x* axis and a *y* axis oriented at 45° to the *x* axis?

14. In what circumstance is the *x* component of a vector given by the magnitude of the vector times the sine of its direction angle?

PROBLEMS

1, 2, 3 = straightforward, intermediate, challenging   ☐ = full solution available in the *Student Solutions Manual and Study Guide*

🌀 = coached solution with hints available at http://www.pse6.com   💻 = computer useful in solving problem

▨ = paired numerical and symbolic problems

Section 3.1   Coordinate Systems

1. 🌀 The polar coordinates of a point are $r = 5.50$ m and $\theta = 240°$. What are the Cartesian coordinates of this point?

2. Two points in a plane have polar coordinates (2.50 m, 30.0°) and (3.80 m, 120.0°). Determine (a) the Cartesian coordinates of these points and (b) the distance between them.

3. A fly lands on one wall of a room. The lower left-hand corner of the wall is selected as the origin of a two-dimensional Cartesian coordinate system. If the fly is located at the point having coordinates (2.00, 1.00) m, (a) how far is it from the corner of the room? (b) What is its location in polar coordinates?

4. Two points in the *xy* plane have Cartesian coordinates (2.00, − 4.00) m and (− 3.00, 3.00) m. Determine (a) the distance between these points and (b) their polar coordinates.

5. If the rectangular coordinates of a point are given by (2, *y*) and its polar coordinates are (*r*, 30°), determine *y* and *r*.

6. If the polar coordinates of the point (*x*, *y*) are (*r*, *θ*), determine the polar coordinates for the points: (a) (− *x*, *y*), (b) (− 2*x*, −2*y*), and (c) (3*x*, − 3*y*).

Section 3.2   Vector and Scalar Quantities
Section 3.3   Some Properties of Vectors

7. A surveyor measures the distance across a straight river by the following method: starting directly across from a tree on the opposite bank, she walks 100 m along the riverbank to establish a baseline. Then she sights across to the tree. The angle from her baseline to the tree is 35.0°. How wide is the river?

8. A pedestrian moves 6.00 km east and then 13.0 km north. Find the magnitude and direction of the resultant displacement vector using the graphical method.

9. A plane flies from base camp to lake A, 280 km away, in a direction of 20.0° north of east. After dropping off supplies it flies to lake B, which is 190 km at 30.0° west of north from lake A. Graphically determine the distance and direction from lake B to the base camp.

10. Vector **A** has a magnitude of 8.00 units and makes an angle of 45.0° with the positive *x* axis. Vector **B** also has a magnitude of 8.00 units and is directed along the negative *x* axis. Using graphical methods, find (a) the vector sum **A** + **B** and (b) the vector difference **A** − **B**.

11. 🌀 A skater glides along a circular path of radius 5.00 m. If he coasts around one half of the circle, find (a) the magnitude of the displacement vector and (b) how far the person skated. (c) What is the magnitude of the displacement if he skates all the way around the circle?

12. A force **F**₁ of magnitude 6.00 units acts at the origin in a direction 30.0° above the positive *x* axis. A second force **F**₂ of magnitude 5.00 units acts at the origin in the direction of the positive *y* axis. Find graphically the magnitude and direction of the resultant force **F**₁ + **F**₂.

13. Arbitrarily define the "instantaneous vector height" of a person as the displacement vector from the point halfway

between his or her feet to the top of the head. Make an order-of-magnitude estimate of the total vector height of all the people in a city of population 100 000 (a) at 10 o'clock on a Tuesday morning, and (b) at 5 o'clock on a Saturday morning. Explain your reasoning.

14. A dog searching for a bone walks 3.50 m south, then runs 8.20 m at an angle 30.0° north of east, and finally walks 15.0 m west. Find the dog's resultant displacement vector using graphical techniques.

15. Each of the displacement vectors **A** and **B** shown in Fig. P3.15 has a magnitude of 3.00 m. Find graphically (a) **A** + **B**, (b) **A** − **B**, (c) **B** − **A**, (d) **A** − 2**B**. Report all angles counterclockwise from the positive x axis.

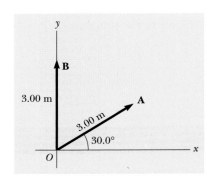

Figure P3.15 Problems 15 and 37.

16. Three displacements are **A** = 200 m, due south; **B** = 250 m, due west; **C** = 150 m, 30.0° east of north. Construct a separate diagram for each of the following possible ways of adding these vectors: **R**$_1$ = **A** + **B** + **C**; **R**$_2$ = **B** + **C** + **A**; **R**$_3$ = **C** + **B** + **A**.

17. A roller coaster car moves 200 ft horizontally, and then rises 135 ft at an angle of 30.0° above the horizontal. It then travels 135 ft at an angle of 40.0° downward. What is its displacement from its starting point? Use graphical techniques.

Section 3.4 Components of a Vector and Unit Vectors

18. Find the horizontal and vertical components of the 100-m displacement of a superhero who flies from the top of a tall building following the path shown in Fig. P3.18.

Figure P3.18

19. A vector has an x component of − 25.0 units and a y component of 40.0 units. Find the magnitude and direction of this vector.

20. A person walks 25.0° north of east for 3.10 km. How far would she have to walk due north and due east to arrive at the same location?

21. Obtain expressions in component form for the position vectors having the following polar coordinates: (a) 12.8 m, 150° (b) 3.30 cm, 60.0° (c) 22.0 in., 215°.

22. A displacement vector lying in the xy plane has a magnitude of 50.0 m and is directed at an angle of 120° to the positive x axis. What are the rectangular components of this vector?

23. A girl delivering newspapers covers her route by traveling 3.00 blocks west, 4.00 blocks north, and then 6.00 blocks east. (a) What is her resultant displacement? (b) What is the total distance she travels?

24. In 1992, Akira Matsushima, from Japan, rode a unicycle across the United States, covering about 4 800 km in six weeks. Suppose that, during that trip, he had to find his way through a city with plenty of one-way streets. In the city center, Matsushima had to travel in sequence 280 m north, 220 m east, 360 m north, 300 m west, 120 m south, 60.0 m east, 40.0 m south, 90.0 m west (road construction) and then 70.0 m north. At that point, he stopped to rest. Meanwhile, a curious crow decided to fly the distance from his starting point to the rest location directly ("as the crow flies"). It took the crow 40.0 s to cover that distance. Assuming the velocity of the crow was constant, find its magnitude and direction.

25. While exploring a cave, a spelunker starts at the entrance and moves the following distances. She goes 75.0 m north, 250 m east, 125 m at an angle 30.0° north of east, and 150 m south. Find the resultant displacement from the cave entrance.

26. A map suggests that Atlanta is 730 miles in a direction of 5.00° north of east from Dallas. The same map shows that Chicago is 560 miles in a direction of 21.0° west of north from Atlanta. Modeling the Earth as flat, use this information to find the displacement from Dallas to Chicago.

27. Given the vectors **A** = 2.00$\hat{\mathbf{i}}$ + 6.00$\hat{\mathbf{j}}$ and **B** = 3.00$\hat{\mathbf{i}}$ − 2.00$\hat{\mathbf{j}}$, (a) draw the vector sum **C** = **A** + **B** and the vector difference **D** = **A** − **B**. (b) Calculate **C** and **D**, first in terms of unit vectors and then in terms of polar coordinates, with angles measured with respect to the + x axis.

28. Find the magnitude and direction of the resultant of three displacements having rectangular components (3.00, 2.00) m, (− 5.00, 3.00) m, and (6.00, 1.00) m.

29. A man pushing a mop across a floor causes it to undergo two displacements. The first has a magnitude of 150 cm and makes an angle of 120° with the positive x axis. The resultant displacement has a magnitude of 140 cm and is directed at an angle of 35.0° to the positive x axis. Find the magnitude and direction of the second displacement.

30. Vector **A** has x and y components of − 8.70 cm and 15.0 cm, respectively; vector **B** has x and y components of 13.2 cm and − 6.60 cm, respectively. If **A** − **B** + 3**C** = 0, what are the components of **C**?

31. Consider the two vectors $\mathbf{A} = 3\hat{\mathbf{i}} - 2\hat{\mathbf{j}}$ and $\mathbf{B} = -\hat{\mathbf{i}} - 4\hat{\mathbf{j}}$. Calculate (a) $\mathbf{A} + \mathbf{B}$, (b) $\mathbf{A} - \mathbf{B}$, (c) $|\mathbf{A} + \mathbf{B}|$, (d) $|\mathbf{A} - \mathbf{B}|$, and (e) the directions of $\mathbf{A} + \mathbf{B}$ and $\mathbf{A} - \mathbf{B}$.

32. Consider the three displacement vectors $\mathbf{A} = (3\hat{\mathbf{i}} - 3\hat{\mathbf{j}})$ m, $\mathbf{B} = (\hat{\mathbf{i}} - 4\hat{\mathbf{j}})$ m, and $\mathbf{C} = (-2\hat{\mathbf{i}} + 5\hat{\mathbf{j}})$ m. Use the component method to determine (a) the magnitude and direction of the vector $\mathbf{D} = \mathbf{A} + \mathbf{B} + \mathbf{C}$, (b) the magnitude and direction of $\mathbf{E} = -\mathbf{A} - \mathbf{B} + \mathbf{C}$.

33. A particle undergoes the following consecutive displacements: 3.50 m south, 8.20 m northeast, and 15.0 m west. What is the resultant displacement?

34. In a game of American football, a quarterback takes the ball from the line of scrimmage, runs backward a distance of 10.0 yards, and then sideways parallel to the line of scrimmage for 15.0 yards. At this point, he throws a forward pass 50.0 yards straight downfield perpendicular to the line of scrimmage. What is the magnitude of the football's resultant displacement?

35. The helicopter view in Fig. P3.35 shows two people pulling on a stubborn mule. Find (a) the single force that is equivalent to the two forces shown, and (b) the force that a third person would have to exert on the mule to make the resultant force equal to zero. The forces are measured in units of newtons (abbreviated N).

Figure P3.35

36. A novice golfer on the green takes three strokes to sink the ball. The successive displacements are 4.00 m to the north, 2.00 m northeast, and 1.00 m at 30.0° west of south. Starting at the same initial point, an expert golfer could make the hole in what single displacement?

37. Use the component method to add the vectors $\mathbf{A}$ and $\mathbf{B}$ shown in Figure P3.15. Express the resultant $\mathbf{A} + \mathbf{B}$ in unit–vector notation.

38. In an assembly operation illustrated in Figure P3.38, a robot moves an object first straight upward and then also to the east, around an arc forming one quarter of a circle of radius 4.80 cm that lies in an east-west vertical plane. The robot then moves the object upward and to the north,

through a quarter of a circle of radius 3.70 cm that lies in a north-south vertical plane. Find (a) the magnitude of the total displacement of the object, and (b) the angle the total displacement makes with the vertical.

Figure P3.38

39. Vector $\mathbf{B}$ has x, y, and z components of 4.00, 6.00, and 3.00 units, respectively. Calculate the magnitude of $\mathbf{B}$ and the angles that $\mathbf{B}$ makes with the coordinate axes.

40. You are standing on the ground at the origin of a coordinate system. An airplane flies over you with constant velocity parallel to the x axis and at a fixed height of 7.60×10^3 m. At time $t = 0$ the airplane is directly above you, so that the vector leading from you to it is $\mathbf{P}_0 = (7.60 \times 10^3 \text{ m})\hat{\mathbf{j}}$. At $t = 30.0$ s the position vector leading from you to the airplane is $\mathbf{P}_{30} = (8.04 \times 10^3 \text{ m})\hat{\mathbf{i}} + (7.60 \times 10^3 \text{ m})\hat{\mathbf{j}}$. Determine the magnitude and orientation of the airplane's position vector at $t = 45.0$ s.

41. The vector $\mathbf{A}$ has x, y, and z components of 8.00, 12.0, and -4.00 units, respectively. (a) Write a vector expression for $\mathbf{A}$ in unit–vector notation. (b) Obtain a unit–vector expression for a vector $\mathbf{B}$ one fourth the length of $\mathbf{A}$ pointing in the same direction as $\mathbf{A}$. (c) Obtain a unit–vector expression for a vector $\mathbf{C}$ three times the length of $\mathbf{A}$ pointing in the direction opposite the direction of $\mathbf{A}$.

42. Instructions for finding a buried treasure include the following: Go 75.0 paces at 240°, turn to 135° and walk 125 paces, then travel 100 paces at 160°. The angles are measured counterclockwise from an axis pointing to the east, the $+x$ direction. Determine the resultant displacement from the starting point.

43. Given the displacement vectors $\mathbf{A} = (3\hat{\mathbf{i}} - 4\hat{\mathbf{j}} + 4\hat{\mathbf{k}})$ m and $\mathbf{B} = (2\hat{\mathbf{i}} + 3\hat{\mathbf{j}} - 7\hat{\mathbf{k}})$ m, find the magnitudes of the vectors (a) $\mathbf{C} = \mathbf{A} + \mathbf{B}$ and (b) $\mathbf{D} = 2\mathbf{A} - \mathbf{B}$, also expressing each in terms of its rectangular components.

44. A radar station locates a sinking ship at range 17.3 km and bearing 136° clockwise from north. From the same station a rescue plane is at horizontal range 19.6 km, 153° clockwise from north, with elevation 2.20 km. (a) Write the position vector for the ship relative to the plane, letting $\hat{\mathbf{i}}$ represent east, $\hat{\mathbf{j}}$ north, and $\hat{\mathbf{k}}$ up. (b) How far apart are the plane and ship?

45. As it passes over Grand Bahama Island, the eye of a hurricane is moving in a direction 60.0° north of west with a speed of 41.0 km/h. Three hours later, the course of the hurricane suddenly shifts due north, and its speed slows to 25.0 km/h. How far from Grand Bahama is the eye 4.50 h after it passes over the island?

46. (a) Vector **E** has magnitude 17.0 cm and is directed 27.0° counterclockwise from the +x axis. Express it in unit–vector notation. (b) Vector **F** has magnitude 17.0 cm and is directed 27.0° counterclockwise from the +y axis. Express it in unit–vector notation. (c) Vector **G** has magnitude 17.0 cm and is directed 27.0° clockwise from the −y axis. Express it in unit–vector notation.

47. Vector **A** has a negative x component 3.00 units in length and a positive y component 2.00 units in length. (a) Determine an expression for **A** in unit–vector notation. (b) Determine the magnitude and direction of **A**. (c) What vector **B** when added to **A** gives a resultant vector with no x component and a negative y component 4.00 units in length?

48. An airplane starting from airport A flies 300 km east, then 350 km at 30.0° west of north, and then 150 km north to arrive finally at airport B. (a) The next day, another plane flies directly from A to B in a straight line. In what direction should the pilot travel in this direct flight? (b) How far will the pilot travel in this direct flight? Assume there is no wind during these flights.

49. Three displacement vectors of a croquet ball are shown in Figure P3.49, where $|\mathbf{A}| = 20.0$ units, $|\mathbf{B}| = 40.0$ units, and $|\mathbf{C}| = 30.0$ units. Find (a) the resultant in unit–vector notation and (b) the magnitude and direction of the resultant displacement.

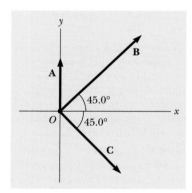

Figure P3.49

50. If $\mathbf{A} = (6.00\hat{\mathbf{i}} - 8.00\hat{\mathbf{j}})$ units, $\mathbf{B} = (-8.00\hat{\mathbf{i}} + 3.00\hat{\mathbf{j}})$ units, and $\mathbf{C} = (26.0\hat{\mathbf{i}} + 19.0\hat{\mathbf{j}})$ units, determine a and b such that $a\mathbf{A} + b\mathbf{B} + \mathbf{C} = 0$.

Additional Problems

51. Two vectors **A** and **B** have precisely equal magnitudes. In order for the magnitude of **A** + **B** to be one hundred times larger than the magnitude of **A** − **B**, what must be the angle between them?

52. Two vectors **A** and **B** have precisely equal magnitudes. In order for the magnitude of **A** + **B** to be larger than the magnitude of **A** − **B** by the factor n, what must be the angle between them?

53. A vector is given by $\mathbf{R} = 2\hat{\mathbf{i}} + \hat{\mathbf{j}} + 3\hat{\mathbf{k}}$. Find (a) the magnitudes of the x, y, and z components, (b) the magnitude of **R**, and (c) the angles between **R** and the x, y, and z axes.

54. The biggest stuffed animal in the world is a snake 420 m long, constructed by Norwegian children. Suppose the snake is laid out in a park as shown in Figure P3.54, forming two straight sides of a 105° angle, with one side 240 m long. Olaf and Inge run a race they invent. Inge runs directly from the tail of the snake to its head and Olaf starts from the same place at the same time but runs along the snake. If both children run steadily at 12.0 km/h, Inge reaches the head of the snake how much earlier than Olaf?

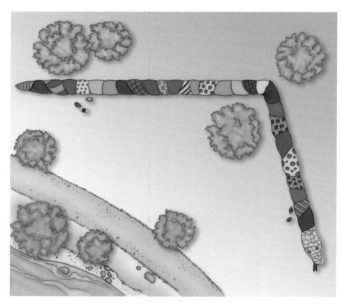

Figure P3.54

55. An air-traffic controller observes two aircraft on his radar screen. The first is at altitude 800 m, horizontal distance 19.2 km, and 25.0° south of west. The second aircraft is at altitude 1 100 m, horizontal distance 17.6 km, and 20.0° south of west. What is the distance between the two aircraft? (Place the x axis west, the y axis south, and the z axis vertical.)

56. A ferry boat transports tourists among three islands. It sails from the first island to the second island, 4.76 km away, in a direction 37.0° north of east. It then sails from the second island to the third island in a direction 69.0° west of north. Finally it returns to the first island, sailing in a direction 28.0° east of south. Calculate the distance between (a) the second and third islands (b) the first and third islands.

57. The rectangle shown in Figure P3.57 has sides parallel to the x and y axes. The position vectors of two corners are **A** = 10.0 m at 50.0° and **B** = 12.0 m at 30.0°. (a) Find the perimeter of the rectangle. (b) Find the magnitude and direction of the vector from the origin to the upper right corner of the rectangle.

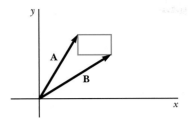

Figure P3.57

58. Find the sum of these four vector forces: 12.0 N to the right at 35.0° above the horizontal, 31.0 N to the left at 55.0° above the horizontal, 8.40 N to the left at 35.0° below the horizontal, and 24.0 N to the right at 55.0° below the horizontal. Follow these steps: Make a drawing of this situation and select the best axes for x and y so you have the least number of components. Then add the vectors by the component method.

59. A person going for a walk follows the path shown in Fig. P3.59. The total trip consists of four straight-line paths. At the end of the walk, what is the person's resultant displacement measured from the starting point?

Figure P3.59

60. The instantaneous position of an object is specified by its position vector $\mathbf{r}$ leading from a fixed origin to the location of the point object. Suppose that for a certain object the position vector is a function of time, given by $\mathbf{r} = 4\hat{\mathbf{i}} + 3\hat{\mathbf{j}} - 2t\hat{\mathbf{k}}$, where r is in meters and t is in seconds. Evaluate $d\mathbf{r}/dt$. What does it represent about the object?

61. A jet airliner, moving initially at 300 mi/h to the east, suddenly enters a region where the wind is blowing at 100 mi/h toward the direction 30.0° north of east. What are the new speed and direction of the aircraft relative to the ground?

62. Long John Silver, a pirate, has buried his treasure on an island with five trees, located at the following points:

(30.0 m, −20.0 m), (60.0 m, 80.0 m), (−10.0 m, −10.0 m), (40.0 m, −30.0 m), and (−70.0 m, 60.0 m), all measured relative to some origin, as in Figure P3.62. His ship's log instructs you to start at tree A and move toward tree B, but to cover only one half the distance between A and B. Then move toward tree C, covering one third the distance between your current location and C. Next move toward D, covering one fourth the distance between where you are and D. Finally move towards E, covering one fifth the distance between you and E, stop, and dig. (a) Assume that you have correctly determined the order in which the pirate labeled the trees as A, B, C, D, and E, as shown in the figure. What are the coordinates of the point where his treasure is buried? (b) **What if** you do not really know the way the pirate labeled the trees? Rearrange the order of the trees [for instance, B(30 m, −20 m), A(60 m, 80 m), E(−10 m, −10 m), C(40 m, −30 m), and D(−70 m, 60 m)] and repeat the calculation to show that the answer does not depend on the order in which the trees are labeled.

Figure P3.62

63. Consider a game in which N children position themselves at equal distances around the circumference of a circle. At the center of the circle is a rubber tire. Each child holds a rope attached to the tire and, at a signal, pulls on his rope. All children exert forces of the same magnitude F. In the case $N = 2$, it is easy to see that the net force on the tire will be zero, because the two oppositely directed force vectors add to zero. Similarly, if $N = 4$, 6, or any even integer, the resultant force on the tire must be zero, because the forces exerted by each pair of oppositely positioned children will cancel. When an odd number of children are around the circle, it is not so obvious whether the total force on the central tire will be zero. (a) Calculate the net force on the tire in the case $N = 3$, by adding the components of the three force vectors. Choose the x axis to lie along one of the ropes. (b) **What If?** Determine the net force for the general case where N is any integer, odd or even, greater than one. Proceed as follows: Assume that the total force is not zero. Then it must point in some particular direction. Let every child move one position clockwise. Give a reason that the total force must then have a direction turned clockwise by 360°/N. Argue that the total force must nevertheless be the same as before. Explain that the contradiction proves that the magnitude of the force is zero. This problem illustrates a widely useful technique of proving a result "by symmetry"—by using a bit of the mathematics of *group theory*. The particular situation

is actually encountered in physics and chemistry when an array of electric charges (ions) exerts electric forces on an atom at a central position in a molecule or in a crystal.

64. A rectangular parallelepiped has dimensions a, b, and c, as in Figure P3.64. (a) Obtain a vector expression for the face diagonal vector $\mathbf{R}_1$. What is the magnitude of this vector? (b) Obtain a vector expression for the body diagonal vector $\mathbf{R}_2$. Note that $\mathbf{R}_1$, $c\hat{\mathbf{k}}$, and $\mathbf{R}_2$ make a right triangle and prove that the magnitude of $\mathbf{R}_2$ is $\sqrt{a^2 + b^2 + c^2}$.

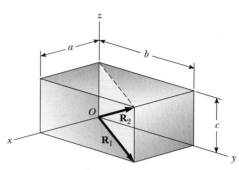

Figure P3.64

65. Vectors $\mathbf{A}$ and $\mathbf{B}$ have equal magnitudes of 5.00. If the sum of $\mathbf{A}$ and $\mathbf{B}$ is the vector $6.00\hat{\mathbf{j}}$, determine the angle between $\mathbf{A}$ and $\mathbf{B}$.

66. In Figure P3.66 a spider is resting after starting to spin its web. The gravitational force on the spider is 0.150 newton down. The spider is supported by different tension forces in the two strands above it, so that the resultant vector force on the spider is zero. The two strands are perpendicular to each other, so we have chosen the x and y directions to be along them. The tension T_x is 0.127 newton. Find (a) the tension T_y, (b) the angle the x axis makes with the horizontal, and (c) the angle the y axis makes with the horizontal.

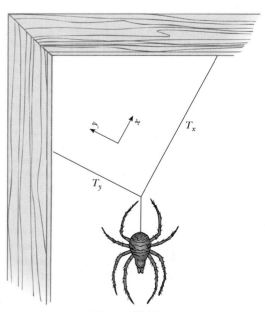

Figure P3.66

67. A point P is described by the coordinates (x, y) with respect to the normal Cartesian coordinate system shown in Fig. P3.67. Show that (x', y'), the coordinates of this point in the rotated coordinate system, are related to (x, y) and the rotation angle α by the expressions

$$x' = x \cos \alpha + y \sin \alpha$$

$$y' = -x \sin \alpha + y \cos \alpha$$

Figure P3.67

Answers to Quick Quizzes

3.1 Scalars: (a), (d), (e). None of these quantities has a direction. Vectors: (b), (c). For these quantities, the direction is necessary to specify the quantity completely.

3.2 (c). The resultant has its maximum magnitude $A + B = 12 + 8 = 20$ units when vector $\mathbf{A}$ is oriented in the same direction as vector $\mathbf{B}$. The resultant vector has its minimum magnitude $A - B = 12 - 8 = 4$ units when vector $\mathbf{A}$ is oriented in the direction opposite vector $\mathbf{B}$.

3.3 (a). The magnitudes will add numerically only if the vectors are in the same direction.

3.4 (b) and (c). In order to add to zero, the vectors must point in opposite directions and have the same magnitude.

3.5 (b). From the Pythagorean theorem, the magnitude of a vector is always larger than the absolute value of each component, unless there is only one nonzero component, in which case the magnitude of the vector is equal to the absolute value of that component.

3.6 (b). From the Pythagorean theorem, we see that the magnitude of a vector is nonzero if at least one component is nonzero.

3.7 (d). Each set of components, for example, the two x components A_x and B_x, must add to zero, so the components must be of opposite sign.

3.8 (c). The magnitude of $\mathbf{C}$ is 5 units, the same as the z component. Answer (b) is not correct because the magnitude of any vector is always a positive number while the y component of $\mathbf{B}$ is negative.

Chapter 4

Motion in Two Dimensions

▲ *Lava spews from a volcanic eruption. Notice the parabolic paths of embers projected into the air. We will find in this chapter that all projectiles follow a parabolic path in the absence of air resistance. (© Arndt/Premium Stock/PictureQuest)*

In this chapter we explore the kinematics of a particle moving in two dimensions. Knowing the basics of two-dimensional motion will allow us to examine—in future chapters—a wide variety of motions, ranging from the motion of satellites in orbit to the motion of electrons in a uniform electric field. We begin by studying in greater detail the vector nature of position, velocity, and acceleration. As in the case of one-dimensional motion, we derive the kinematic equations for two-dimensional motion from the fundamental definitions of these three quantities. We then treat projectile motion and uniform circular motion as special cases of motion in two dimensions. We also discuss the concept of relative motion, which shows why observers in different frames of reference may measure different positions, velocities, and accelerations for a given particle.

4.1 The Position, Velocity, and Acceleration Vectors

In Chapter 2 we found that the motion of a particle moving along a straight line is completely known if its position is known as a function of time. Now let us extend this idea to motion in the xy plane. We begin by describing the position of a particle by its **position vector r**, drawn from the origin of some coordinate system to the particle located in the xy plane, as in Figure 4.1. At time t_i the particle is at point Ⓐ, described by position vector $\mathbf{r}_i$. At some later time t_f it is at point Ⓑ, described by position vector $\mathbf{r}_f$. The path from Ⓐ to Ⓑ is not necessarily a straight line. As the particle moves from Ⓐ to Ⓑ in the time interval $\Delta t = t_f - t_i$, its position vector changes from $\mathbf{r}_i$ to $\mathbf{r}_f$. As we learned in Chapter 2, displacement is a vector, and the displacement of the particle is the difference between its final position and its initial position. We now define the **displacement vector Δr** for the particle of Figure 4.1 as being the difference between its final position vector and its initial position vector:

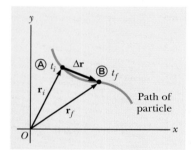

Figure 4.1 A particle moving in the xy plane is located with the position vector **r** drawn from the origin to the particle. The displacement of the particle as it moves from Ⓐ to Ⓑ in the time interval $\Delta t = t_f - t_i$ is equal to the vector $\Delta \mathbf{r} = \mathbf{r}_f - \mathbf{r}_i$.

Displacement vector

$$\Delta \mathbf{r} \equiv \mathbf{r}_f - \mathbf{r}_i \tag{4.1}$$

The direction of $\Delta \mathbf{r}$ is indicated in Figure 4.1. As we see from the figure, the magnitude of $\Delta \mathbf{r}$ is *less* than the distance traveled along the curved path followed by the particle.

As we saw in Chapter 2, it is often useful to quantify motion by looking at the ratio of a displacement divided by the time interval during which that displacement occurs, which gives the rate of change of position. In two-dimensional (or three-dimensional) kinematics, everything is the same as in one-dimensional kinematics except that we must now use full vector notation rather than positive and negative signs to indicate the direction of motion.

We define the **average velocity** of a particle during the time interval Δt as the displacement of the particle divided by the time interval:

Average velocity

$$\bar{\mathbf{v}} \equiv \frac{\Delta \mathbf{r}}{\Delta t} \tag{4.2}$$

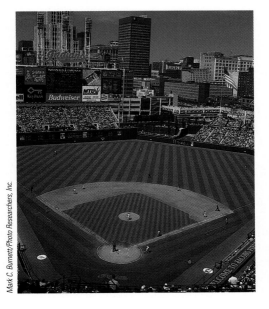

Figure 4.2 Bird's-eye view of a baseball diamond. A batter who hits a home run travels around the bases, ending up where he began. Thus, his average velocity for the entire trip is zero. His average speed, however, is not zero and is equal to the distance around the bases divided by the time interval during which he runs around the bases.

Mark C. Burnett/Photo Researchers, Inc.

Multiplying or dividing a vector quantity by a positive scalar quantity such as Δt changes only the magnitude of the vector, not its direction. Because displacement is a vector quantity and the time interval is a positive scalar quantity, we conclude that the average velocity is a vector quantity directed along $\Delta \mathbf{r}$.

Note that the average velocity between points is *independent of the path* taken. This is because average velocity is proportional to displacement, which depends only on the initial and final position vectors and not on the path taken. As with one-dimensional motion, we conclude that if a particle starts its motion at some point and returns to this point via any path, its average velocity is zero for this trip because its displacement is zero. Figure 4.2 suggests such a situation in a baseball park. When a batter hits a home run, he runs around the bases and returns to home plate. Thus, his average velocity is zero during this trip. His average speed, however, is not zero.

Consider again the motion of a particle between two points in the *xy* plane, as shown in Figure 4.3. As the time interval over which we observe the motion becomes smaller and smaller, the direction of the displacement approaches that of the line tangent to the path at Ⓐ. The **instantaneous velocity v** is defined as the limit of the average velocity $\Delta \mathbf{r}/\Delta t$ as Δt approaches zero:

$$\mathbf{v} \equiv \lim_{\Delta t \to 0} \frac{\Delta \mathbf{r}}{\Delta t} = \frac{d\mathbf{r}}{dt} \tag{4.3}$$

That is, the instantaneous velocity equals the derivative of the position vector with respect to time. The direction of the instantaneous velocity vector at any point in a particle's path is along a line tangent to the path at that point and in the direction of motion.

The magnitude of the instantaneous velocity vector $v = |\mathbf{v}|$ is called the *speed,* which is a scalar quantity.

As a particle moves from one point to another along some path, its instantaneous velocity vector changes from $\mathbf{v}_i$ at time t_i to $\mathbf{v}_f$ at time t_f. Knowing the velocity at these points allows us to determine the average acceleration of the particle—the **average acceleration $\bar{\mathbf{a}}$** of a particle as it moves is defined as the change in the instantaneous velocity vector $\Delta \mathbf{v}$ divided by the time interval Δt during which that change occurs:

$$\bar{\mathbf{a}} \equiv \frac{\mathbf{v}_f - \mathbf{v}_i}{t_f - t_i} = \frac{\Delta \mathbf{v}}{\Delta t} \tag{4.4}$$

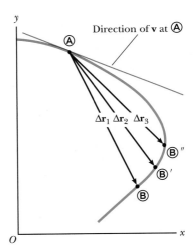

Figure 4.3 As a particle moves between two points, its average velocity is in the direction of the displacement vector $\Delta \mathbf{r}$. As the end point of the path is moved from Ⓑ to Ⓑ′ to Ⓑ″, the respective displacements and corresponding time intervals become smaller and smaller. In the limit that the end point approaches Ⓐ, Δt approaches zero, and the direction of $\Delta \mathbf{r}$ approaches that of the line tangent to the curve at Ⓐ. By definition, the instantaneous velocity at Ⓐ is directed along this tangent line.

PITFALL PREVENTION

4.1 Vector Addition

While the vector addition discussed in Chapter 3 involves *displacement* vectors, vector addition can be applied to *any* type of vector quantity. Figure 4.4, for example, shows the addition of *velocity* vectors using the graphical approach.

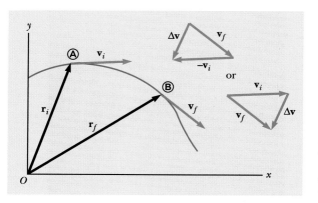

Figure 4.4 A particle moves from position Ⓐ to position Ⓑ. Its velocity vector changes from $\mathbf{v}_i$ to $\mathbf{v}_f$. The vector diagrams at the upper right show two ways of determining the vector $\Delta\mathbf{v}$ from the initial and final velocities.

Because $\bar{\mathbf{a}}$ is the ratio of a vector quantity $\Delta\mathbf{v}$ and a positive scalar quantity Δt, we conclude that average acceleration is a vector quantity directed along $\Delta\mathbf{v}$. As indicated in Figure 4.4, the direction of $\Delta\mathbf{v}$ is found by adding the vector $-\mathbf{v}_i$ (the negative of $\mathbf{v}_i$) to the vector $\mathbf{v}_f$, because by definition $\Delta\mathbf{v} = \mathbf{v}_f - \mathbf{v}_i$.

When the average acceleration of a particle changes during different time intervals, it is useful to define its instantaneous acceleration. The **instantaneous acceleration a** is defined as the limiting value of the ratio $\Delta\mathbf{v}/\Delta t$ as Δt approaches zero:

Instantaneous acceleration

$$\mathbf{a} \equiv \lim_{\Delta t \to 0} \frac{\Delta\mathbf{v}}{\Delta t} = \frac{d\mathbf{v}}{dt} \tag{4.5}$$

In other words, the instantaneous acceleration equals the derivative of the velocity vector with respect to time.

It is important to recognize that various changes can occur when a particle accelerates. First, the magnitude of the velocity vector (the speed) may change with time as in straight-line (one-dimensional) motion. Second, the direction of the velocity vector may change with time even if its magnitude (speed) remains constant, as in curved-path (two-dimensional) motion. Finally, both the magnitude and the direction of the velocity vector may change simultaneously.

Quick Quiz 4.1 Which of the following cannot *possibly* be accelerating? (a) An object moving with a constant speed (b) An object moving with a constant velocity (c) An object moving along a curve.

Quick Quiz 4.2 Consider the following controls in an automobile: gas pedal, brake, steering wheel. The controls in this list that cause an acceleration of the car are (a) all three controls (b) the gas pedal and the brake (c) only the brake (d) only the gas pedal.

4.2 Two-Dimensional Motion with Constant Acceleration

In Section 2.5, we investigated one-dimensional motion in which the acceleration is constant because this type of motion is common. Let us consider now two-dimensional motion during which the acceleration remains constant in both magnitude and direction. This will also be useful for analyzing some common types of motion.

The position vector for a particle moving in the *xy* plane can be written

$$\mathbf{r} = x\hat{\mathbf{i}} + y\hat{\mathbf{j}} \tag{4.6}$$

where x, y, and $\mathbf{r}$ change with time as the particle moves while the unit vectors $\hat{\mathbf{i}}$ and $\hat{\mathbf{j}}$ remain constant. If the position vector is known, the velocity of the particle can be obtained from Equations 4.3 and 4.6, which give

$$\mathbf{v} = \frac{d\mathbf{r}}{dt} = \frac{dx}{dt}\hat{\mathbf{i}} + \frac{dy}{dt}\hat{\mathbf{j}} = v_x\hat{\mathbf{i}} + v_y\hat{\mathbf{j}} \qquad (4.7)$$

Because $\mathbf{a}$ is assumed constant, its components a_x and a_y also are constants. Therefore, we can apply the equations of kinematics to the x and y components of the velocity vector. Substituting, from Equation 2.9, $v_{xf} = v_{xi} + a_xt$ and $v_{yf} = v_{yi} + a_yt$ into Equation 4.7 to determine the final velocity at any time t, we obtain

$$\mathbf{v}_f = (v_{xi} + a_xt)\hat{\mathbf{i}} + (v_{yi} + a_yt)\hat{\mathbf{j}}$$
$$= (v_{xi}\hat{\mathbf{i}} + v_{yi}\hat{\mathbf{j}}) + (a_x\hat{\mathbf{i}} + a_y\hat{\mathbf{j}})t$$
$$\mathbf{v}_f = \mathbf{v}_i + \mathbf{a}t \qquad (4.8)$$

Velocity vector as a function of time

This result states that the velocity of a particle at some time t equals the vector sum of its initial velocity $\mathbf{v}_i$ and the additional velocity $\mathbf{a}t$ acquired at time t as a result of constant acceleration. It is the vector version of Equation 2.9.

Similarly, from Equation 2.12 we know that the x and y coordinates of a particle moving with constant acceleration are

$$x_f = x_i + v_{xi}t + \tfrac{1}{2}a_xt^2 \qquad y_f = y_i + v_{yi}t + \tfrac{1}{2}a_yt^2$$

Substituting these expressions into Equation 4.6 (and labeling the final position vector $\mathbf{r}_f$) gives

$$\mathbf{r}_f = (x_i + v_{xi}t + \tfrac{1}{2}a_xt^2)\hat{\mathbf{i}} + (y_i + v_{yi}t + \tfrac{1}{2}a_yt^2)\hat{\mathbf{j}}$$
$$= (x_i\hat{\mathbf{i}} + y_i\hat{\mathbf{j}}) + (v_{xi}\hat{\mathbf{i}} + v_{yi}\hat{\mathbf{j}})t + \tfrac{1}{2}(a_x\hat{\mathbf{i}} + a_y\hat{\mathbf{j}})t^2$$
$$\mathbf{r}_f = \mathbf{r}_i + \mathbf{v}_it + \tfrac{1}{2}\mathbf{a}t^2 \qquad (4.9)$$

Position vector as a function of time

which is the vector version of Equation 2.12. This equation tells us that the position vector $\mathbf{r}_f$ is the vector sum of the original position $\mathbf{r}_i$, a displacement $\mathbf{v}_it$ arising from the initial velocity of the particle and a displacement $\tfrac{1}{2}\mathbf{a}t^2$ resulting from the constant acceleration of the particle.

Graphical representations of Equations 4.8 and 4.9 are shown in Figure 4.5. Note from Figure 4.5a that $\mathbf{v}_f$ is generally not along the direction of either $\mathbf{v}_i$ or $\mathbf{a}$ because the relationship between these quantities is a vector expression. For the same reason,

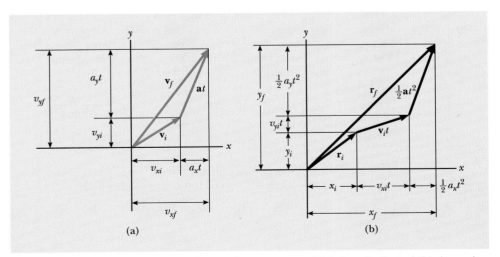

Active Figure 4.5 Vector representations and components of (a) the velocity and (b) the position of a particle moving with a constant acceleration $\mathbf{a}$.

At the Active Figures link at http://www.pse6.com, you can investigate the effect of different initial positions and velocities on the final position and velocity (for constant acceleration).

from Figure 4.5b we see that $\mathbf{r}_f$ is generally not along the direction of $\mathbf{v}_i$ or $\mathbf{a}$. Finally, note that $\mathbf{v}_f$ and $\mathbf{r}_f$ are generally not in the same direction.

Because Equations 4.8 and 4.9 are vector expressions, we may write them in component form:

$$\mathbf{v}_f = \mathbf{v}_i + \mathbf{a}t \qquad \begin{cases} v_{xf} = v_{xi} + a_x t \\ v_{yf} = v_{yi} + a_y t \end{cases} \qquad \text{(4.8a)}$$

$$\mathbf{r}_f = \mathbf{r}_i + \mathbf{v}_i t + \tfrac{1}{2}\mathbf{a}t^2 \qquad \begin{cases} x_f = x_i + v_{xi}t + \tfrac{1}{2}a_x t^2 \\ y_f = y_i + v_{yi}t + \tfrac{1}{2}a_y t^2 \end{cases} \qquad \text{(4.9a)}$$

These components are illustrated in Figure 4.5. The component form of the equations for $\mathbf{v}_f$ and $\mathbf{r}_f$ show us that two-dimensional motion at constant acceleration is equivalent to two *independent* motions—one in the x direction and one in the y direction—having constant accelerations a_x and a_y.

Example 4.1 Motion in a Plane

A particle starts from the origin at $t = 0$ with an initial velocity having an x component of 20 m/s and a y component of -15 m/s. The particle moves in the xy plane with an x component of acceleration only, given by $a_x = 4.0$ m/s^2.

(A) Determine the components of the velocity vector at any time and the total velocity vector at any time.

Solution After carefully reading the problem, we *conceptualize* what is happening to the particle. The components of the initial velocity tell us that the particle starts by moving toward the right and downward. The x component of velocity starts at 20 m/s and increases by 4.0 m/s every second. The y component of velocity never changes from its initial value of -15 m/s. We sketch a rough motion diagram of the situation in Figure 4.6. Because the particle is accelerating in the $+x$ direction, its velocity component in this direction will increase, so that the path will curve as shown in the diagram. Note that the spacing between successive images increases as time goes on because the speed is increasing. The placement of the acceleration and velocity vectors in Figure 4.6 helps us to further conceptualize the situation.

Because the acceleration is constant, we *categorize* this problem as one involving a particle moving in two dimensions with constant acceleration. To *analyze* such a problem, we use the equations developed in this section. To begin the mathematical analysis, we set $v_{xi} = 20$ m/s, $v_{yi} = -15$ m/s, $a_x = 4.0$ m/s^2, and $a_y = 0$.

Equations 4.8a give

$$(1) \qquad v_{xf} = v_{xi} + a_x t = (20 + 4.0t) \text{ m/s}$$

$$(2) \qquad v_{yf} = v_{yi} + a_y t = -15 \text{ m/s} + 0 = -15 \text{ m/s}$$

Therefore

$$\mathbf{v}_f = v_{xi}\hat{\mathbf{i}} + v_{yi}\hat{\mathbf{j}} = \; [(20 + 4.0t)\hat{\mathbf{i}} - 15\hat{\mathbf{j}}] \text{ m/s}$$

We could also obtain this result using Equation 4.8 directly, noting that $\mathbf{a} = 4.0\hat{\mathbf{i}}$ m/s^2 and $\mathbf{v}_i = [20\hat{\mathbf{i}} - 15\hat{\mathbf{j}}]$ m/s. To *finalize* this part, notice that the x component of velocity increases in time while the y component remains constant; this is consistent with what we predicted.

(B) Calculate the velocity and speed of the particle at $t = 5.0$ s.

Solution With $t = 5.0$ s, the result from part (A) gives

$$\mathbf{v}_f = [(20 + 4.0(5.0))\hat{\mathbf{i}} - 15\hat{\mathbf{j}}] \text{ m/s} = \; (40\hat{\mathbf{i}} - 15\hat{\mathbf{j}}) \text{ m/s}$$

This result tells us that at $t = 5.0$ s, $v_{xf} = 40$ m/s and $v_{yf} = -15$ m/s. Knowing these two components for this two-dimensional motion, we can find both the direction and the magnitude of the velocity vector. To determine the angle θ that $\mathbf{v}$ makes with the x axis at $t = 5.0$ s, we use the fact that $\tan\theta = v_{yf}/v_{xf}$:

$$(3) \qquad \theta = \tan^{-1}\left(\frac{v_{yf}}{v_{xf}}\right) = \tan^{-1}\left(\frac{-15 \text{ m/s}}{40 \text{ m/s}}\right)$$

$$= -21°$$

where the negative sign indicates an angle of 21° below the positive x axis. The speed is the magnitude of $\mathbf{v}_f$:

$$v_f = |\mathbf{v}_f| = \sqrt{v_{xf}{}^2 + v_{yf}{}^2} = \sqrt{(40)^2 + (-15)^2} \text{ m/s}$$

$$= \; 43 \text{ m/s}$$

To *finalize* this part, we notice that if we calculate v_i from the x and y components of $\mathbf{v}_i$, we find that $v_f > v_i$. Is this consistent with our prediction?

Figure 4.6 (Example 4.1) Motion diagram for the particle.

(C) Determine the x and y coordinates of the particle at any time t and the position vector at this time.

Solution Because $x_i = y_i = 0$ at $t = 0$, Equation 4.9a gives

$$x_f = v_{xi}t + \tfrac{1}{2}a_x t^2 = \boxed{(20t + 2.0t^2) \text{ m}}$$

$$y_f = v_{yi}t = \boxed{(-15t) \text{ m}}$$

Therefore, the position vector at any time t is

$$(4) \qquad \mathbf{r}_f = x_f\hat{\mathbf{i}} + y_f\hat{\mathbf{j}} = \boxed{[(20t + 2.0t^2)\hat{\mathbf{i}} - 15t\hat{\mathbf{j}}]\,\text{m}}$$

(Alternatively, we could obtain $\mathbf{r}_f$ by applying Equation 4.9 directly, with $\mathbf{v}_f = (20\hat{\mathbf{i}} - 15\hat{\mathbf{j}})$ m/s and $\mathbf{a} = 4.0\hat{\mathbf{i}}$ m/s^2. Try it!) Thus, for example, at $t = 5.0$ s, $x = 150$ m, $y = -75$ m, and $\mathbf{r}_f = (150\hat{\mathbf{i}} - 75\hat{\mathbf{j}})$ m. The magnitude of the displacement of the particle from the origin at $t = 5.0$ s is the magnitude of $\mathbf{r}_f$ at this time:

$$r_f = |\mathbf{r}_f| = \sqrt{(150)^2 + (-75)^2}\,\text{m} = 170\,\text{m}$$

Note that this is *not* the distance that the particle travels in this time! Can you determine this distance from the available data?

To *finalize* this problem, let us consider a limiting case for very large values of t in the following **What If?**

What If? What if we wait a very long time and then observe the motion of the particle? How would we describe the motion of the particle for large values of the time?

Answer Looking at Figure 4.6, we see the path of the particle curving toward the x axis. There is no reason to assume that this tendency will change, so this suggests that the path will become more and more parallel to the x axis as time grows large. Mathematically, let us consider Equations (1) and (2). These show that the y component of the velocity remains constant while the x component grows linearly with t. Thus, when t is very large, the x component of the velocity will be much larger than the y component, suggesting that the velocity vector becomes more and more parallel to the x axis.

Equation (3) gives the angle that the velocity vector makes with the x axis. Notice that $\theta \to 0$ as the denominator (v_{xf}) becomes much larger than the numerator (v_{yf}).

Despite the fact that the velocity vector becomes more and more parallel to the x axis, the particle does not approach a limiting value of y. Equation (4) shows that both x_f and y_f continue to grow with time, although x_f grows much faster.

4.3 Projectile Motion

Assumptions of projectile motion

Anyone who has observed a baseball in motion has observed projectile motion. The ball moves in a curved path, and its motion is simple to analyze if we make two assumptions: (1) the free-fall acceleration $\mathbf{g}$ is constant over the range of motion and is directed downward,[1] and (2) the effect of air resistance is negligible.[2] With these assumptions, we find that the path of a projectile, which we call its *trajectory*, is *always* a parabola. **We use these assumptions throughout this chapter.**

To show that the trajectory of a projectile is a parabola, let us choose our reference frame such that the y direction is vertical and positive is upward. Because air resistance is neglected, we know that $a_y = -g$ (as in one-dimensional free fall) and that $a_x = 0$. Furthermore, let us assume that at $t = 0$, the projectile leaves the origin ($x_i = y_i = 0$) with speed v_i, as shown in Figure 4.7. The vector $\mathbf{v}_i$ makes an angle θ_i with the horizontal. From the definitions of the cosine and sine functions we have

$$\cos\theta_i = v_{xi}/v_i \qquad \sin\theta_i = v_{yi}/v_i$$

Therefore, the initial x and y components of velocity are

$$v_{xi} = v_i\cos\theta_i \qquad v_{yi} = v_i\sin\theta_i \qquad (4.10)$$

Substituting the x component into Equation 4.9a with $x_i = 0$ and $a_x = 0$, we find that

$$x_f = v_{xi}t = (v_i\cos\theta_i)\,t \qquad (4.11)$$

---

[1] This assumption is reasonable as long as the range of motion is small compared with the radius of the Earth (6.4×10^6 m). In effect, this assumption is equivalent to assuming that the Earth is flat over the range of motion considered.

[2] This assumption is generally *not* justified, especially at high velocities. In addition, any spin imparted to a projectile, such as that applied when a pitcher throws a curve ball, can give rise to some very interesting effects associated with aerodynamic forces, which will be discussed in Chapter 14.

At the Active Figures link at http://www.pse6.com, you can change launch angle and initial speed. You can also observe the changing components of velocity along the trajectory of the projectile.

▲ **PITFALL PREVENTION**

4.2 Acceleration at the Highest Point

As discussed in Pitfall Prevention 2.8, many people claim that the acceleration of a projectile at the topmost point of its trajectory is zero. This mistake arises from confusion between zero vertical velocity and zero acceleration. If the projectile were to experience zero acceleration at the highest point, then its velocity at that point would not change—the projectile would move horizontally at constant speed from then on! This does not happen, because the acceleration is NOT zero anywhere along the trajectory.

A welder cuts holes through a heavy metal construction beam with a hot torch. The sparks generated in the process follow parabolic paths.

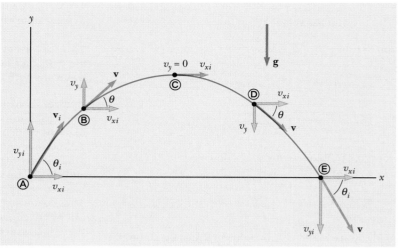

Active Figure 4.7 The parabolic path of a projectile that leaves the origin with a velocity $\mathbf{v}_i$. The velocity vector $\mathbf{v}$ changes with time in both magnitude and direction. This change is the result of acceleration in the negative y direction. The x component of velocity remains constant in time because there is no acceleration along the horizontal direction. The y component of velocity is zero at the peak of the path.

Repeating with the y component and using $y_i = 0$ and $a_y = -g$, we obtain

$$y_f = v_{yi}t + \tfrac{1}{2}a_y t^2 = (v_i \sin \theta_i)t - \tfrac{1}{2}g t^2 \tag{4.12}$$

Next, from Equation 4.11 we find $t = x_f/(v_i \cos \theta_i)$ and substitute this expression for t into Equation 4.12; this gives

$$y = (\tan \theta_i)x - \left(\frac{g}{2v_i{}^2 \cos^2 \theta_i}\right)x^2$$

This equation is valid for launch angles in the range $0 < \theta_i < \pi/2$. We have left the subscripts off the x and y because the equation is valid for any point (x, y) along the path of the projectile. The equation is of the form $y = ax - bx^2$, which is the equation of a parabola that passes through the origin. Thus, we have shown that the trajectory of a projectile is a parabola. Note that the trajectory is completely specified if both the initial speed v_i and the launch angle θ_i are known.

The vector expression for the position vector of the projectile as a function of time follows directly from Equation 4.9, with $\mathbf{a} = \mathbf{g}$:

$$\mathbf{r}_f = \mathbf{r}_i + \mathbf{v}_i t + \tfrac{1}{2}\mathbf{g}t^2$$

This expression is plotted in Figure 4.8, for a projectile launched from the origin, so that $\mathbf{r}_i = 0$.

The final position of a particle can be considered to be the superposition of the initial position $\mathbf{r}_i$, the term $\mathbf{v}_i t$, which is the displacement if no acceleration were present, and the term $\tfrac{1}{2}\mathbf{g}t^2$ that arises from the acceleration due to gravity. In other words, if there were no gravitational acceleration, the particle would continue to move along a straight path in the direction of $\mathbf{v}_i$. Therefore, the vertical distance $\tfrac{1}{2}\mathbf{g}t^2$ through which the particle "falls" off the straight-line path is the same distance that a freely falling object would fall during the same time interval.

In Section 4.2, we stated that two-dimensional motion with constant acceleration can be analyzed as a combination of two independent motions in the x and y directions, with accelerations a_x and a_y. Projectile motion is a special case of two-dimensional motion with constant acceleration and can be handled in this way, with zero acceleration in the x direction and $a_y = -g$ in the y direction. Thus, **when analyzing projectile motion, consider it to be the superposition of two motions:**

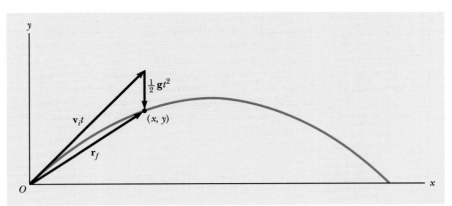

Figure 4.8 The position vector $\mathbf{r}_f$ of a projectile launched from the origin whose initial velocity at the origin is $\mathbf{v}_i$. The vector $\mathbf{v}_i t$ would be the displacement of the projectile if gravity were absent, and the vector $\frac{1}{2}\mathbf{g}t^2$ is its vertical displacement due to its downward gravitational acceleration.

(1) **constant-velocity motion in the horizontal direction** and (2) **free-fall motion in the vertical direction.** The horizontal and vertical components of a projectile's motion are completely independent of each other and can be handled separately, with time t as the common variable for both components.

Quick Quiz 4.3 Suppose you are running at constant velocity and you wish to throw a ball such that you will catch it as it comes back down. In what direction should you throw the ball relative to you? (a) straight up (b) at an angle to the ground that depends on your running speed (c) in the forward direction.

Quick Quiz 4.4 As a projectile thrown upward moves in its parabolic path (such as in Figure 4.8), at what point along its path are the velocity and acceleration vectors for the projectile perpendicular to each other? (a) nowhere (b) the highest point (c) the launch point.

Quick Quiz 4.5 As the projectile in Quick Quiz 4.4 moves along its path, at what point are the velocity and acceleration vectors for the projectile parallel to each other? (a) nowhere (b) the highest point (c) the launch point.

Example 4.2   Approximating Projectile Motion

A ball is thrown in such a way that its initial vertical and horizontal components of velocity are 40 m/s and 20 m/s, respectively. Estimate the total time of flight and the distance the ball is from its starting point when it lands.

Solution A motion diagram like Figure 4.9 helps us *conceptualize* the problem. The phrase "A ball is thrown" allows us to *categorize* this as a projectile motion problem, which we *analyze* by continuing to study Figure 4.9. The acceleration vectors are all the same, pointing downward with a magnitude of nearly 10 m/s². The velocity vectors change direction. Their horizontal components are all the same: 20 m/s.

Remember that the two velocity components are independent of each other. By considering the vertical motion

Figure 4.9 (Example 4.2) Motion diagram for a projectile.

first, we can determine how long the ball remains in the air. Because the vertical motion is free-fall, the vertical components of the velocity vectors change, second by second, from 40 m/s to roughly 30, 20, and 10 m/s in the upward direction, and then to 0 m/s. Subsequently, its velocity becomes 10, 20, 30, and 40 m/s in the downward direction. Thus it takes the ball about 4 s to go up and another 4 s to come back down, for a total time of flight of approximately 8 s.

Now we shift our analysis to the horizontal motion. Because the horizontal component of velocity is 20 m/s, and because the ball travels at this speed for 8 s, it ends up approximately 160 m from its starting point.

This is the first example that we have performed for projectile motion. In subsequent projectile motion problems, keep in mind the importance of separating the two components and of making approximations to give you rough expected results.

Horizontal Range and Maximum Height of a Projectile

Figure 4.10 A projectile launched from the origin at $t_i = 0$ with an initial velocity $\mathbf{v}_i$. The maximum height of the projectile is h, and the horizontal range is R. At Ⓐ, the peak of the trajectory, the particle has coordinates $(R/2, h)$.

Let us assume that a projectile is launched from the origin at $t_i = 0$ with a positive v_{yi} component, as shown in Figure 4.10. Two points are especially interesting to analyze: the peak point Ⓐ, which has Cartesian coordinates $(R/2, h)$, and the point Ⓑ, which has coordinates $(R, 0)$. The distance R is called the *horizontal range* of the projectile, and the distance h is its *maximum height*. Let us find h and R in terms of v_i, θ_i, and g.

We can determine h by noting that at the peak, $v_{yA} = 0$. Therefore, we can use Equation 4.8a to determine the time t_A at which the projectile reaches the peak:

$$v_{yf} = v_{yi} + a_y t$$

$$0 = v_i \sin \theta_i - g t_A$$

$$t_A = \frac{v_i \sin \theta_i}{g}$$

Substituting this expression for t_A into the y part of Equation 4.9a and replacing $y = y_A$ with h, we obtain an expression for h in terms of the magnitude and direction of the initial velocity vector:

$$h = (v_i \sin \theta_i)\frac{v_i \sin \theta_i}{g} - \frac{1}{2}g\left(\frac{v_i \sin \theta_i}{g}\right)^2$$

$$h = \frac{v_i^2 \sin^2 \theta_i}{2g} \tag{4.13}$$

The range R is the horizontal position of the projectile at a time that is twice the time at which it reaches its peak, that is, at time $t_B = 2t_A$. Using the x part of Equation 4.9a, noting that $v_{xi} = v_{xB} = v_i \cos \theta_i$ and setting $x_B = R$ at $t = 2t_A$, we find that

$$R = v_{xi} t_B = (v_i \cos \theta_i) 2t_A$$

$$= (v_i \cos \theta_i)\frac{2v_i \sin \theta_i}{g} = \frac{2v_i^2 \sin \theta_i \cos \theta_i}{g}$$

Using the identity $\sin 2\theta = 2\sin\theta\cos\theta$ (see Appendix B.4), we write R in the more compact form

$$R = \frac{v_i^2 \sin 2\theta_i}{g} \tag{4.14}$$

The maximum value of R from Equation 4.14 is $R_{\max} = v_i^2/g$. This result follows from the fact that the maximum value of $\sin 2\theta_i$ is 1, which occurs when $2\theta_i = 90°$. Therefore, R is a maximum when $\theta_i = 45°$.

Figure 4.11 illustrates various trajectories for a projectile having a given initial speed but launched at different angles. As you can see, the range is a maximum for $\theta_i = 45°$. In addition, for any θ_i other than $45°$, a point having Cartesian coordinates $(R, 0)$ can be reached by using either one of two complementary values of θ_i, such as $75°$ and $15°$. Of course, the maximum height and time of flight for one of these values of θ_i are different from the maximum height and time of flight for the complementary value.

▲ **PITFALL PREVENTION**

4.3   The Height and Range Equations

Equation 4.14 is useful for calculating R only for a symmetric path, as shown in Figure 4.10. If the path is not symmetric, *do not use this equation*. The general expressions given by Equations 4.8 and 4.9 are the *more important* results, because they give the position and velocity components of *any* particle moving in two dimensions at *any* time t.

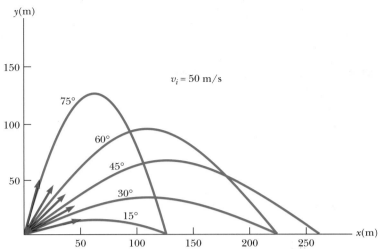

Active Figure 4.11 A projectile launched from the origin with an initial speed of 50 m/s at various angles of projection. Note that complementary values of θ_i result in the same value of R (range of the projectile).

At the Active Figures link at http://www.pse6.com, *you can vary the projection angle to observe the effect on the trajectory and measure the flight time.*

Quick Quiz 4.6 Rank the launch angles for the five paths in Figure 4.11 with respect to time of flight, from the shortest time of flight to the longest.

PROBLEM-SOLVING HINTS

Projectile Motion

We suggest that you use the following approach to solving projectile motion problems:

- Select a coordinate system and resolve the initial velocity vector into x and y components.

- Follow the techniques for solving constant-velocity problems to analyze the horizontal motion. Follow the techniques for solving constant-acceleration problems to analyze the vertical motion. The x and y motions share the same time t.

Example 4.3   The Long Jump

A long-jumper (Fig. 4.12) leaves the ground at an angle of 20.0° above the horizontal and at a speed of 11.0 m/s.

(A) How far does he jump in the horizontal direction? (Assume his motion is equivalent to that of a particle.)

Solution We *conceptualize* the motion of the long-jumper as equivalent to that of a simple projectile such as the ball in Example 4.2, and *categorize* this problem as a projectile motion problem. Because the initial speed and launch angle are given, and because the final height is the same as the initial height, we further categorize this problem as satisfying the conditions for which Equations 4.13 and 4.14 can be used. This is the most direct way to *analyze* this problem, although the general methods that we have been describing will always give the correct answer. We will take the general approach and use components. Figure 4.10

provides a graphical representation of the flight of the long-jumper. As before, we set our origin of coordinates at the takeoff point and label the peak as Ⓐ and the landing point as Ⓑ. The horizontal motion is described by Equation 4.11:

$$x_f = x_B = (v_i \cos \theta_i)\, t_B = (11.0 \text{ m/s})(\cos 20.0°)\, t_B$$

The value of x_B can be found if the time of landing t_B is known. We can find t_B by remembering that $a_y = -g$ and by using the y part of Equation 4.8a. We also note that at the top of the jump the vertical component of velocity v_{yA} is zero:

$$v_{yf} = v_{yA} = v_i \sin \theta_i - gt_A$$

$$0 = (11.0 \text{ m/s}) \sin 20.0° - (9.80 \text{ m/s}^2) t_A$$

$$t_A = 0.384 \text{ s}$$

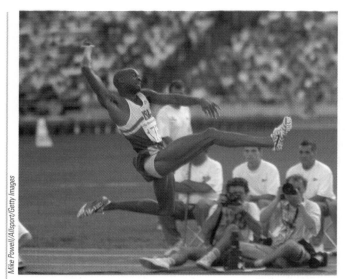

Mike Powell/Allsport/Getty Images

Figure 4.12 (Example 4.3) Mike Powell, current holder of the world long jump record of 8.95 m.

another 0.384 s passes before the jumper returns to the ground. Therefore, the time at which the jumper lands is $t_B = 2t_A = 0.768$ s. Substituting this value into the above expression for x_f gives

$$x_f = x_B = (11.0 \text{ m/s}) (\cos 20.0°) (0.768 \text{ s}) = \boxed{7.94 \text{ m}}$$

This is a reasonable distance for a world-class athlete.

(B) What is the maximum height reached?

Solution We find the maximum height reached by using Equation 4.12:

$$y_{max} = y_A = (v_i \sin \theta_i) t_A - \tfrac{1}{2} g t_A^2$$
$$= (11.0 \text{ m/s}) (\sin 20.0°) (0.384 \text{ s})$$
$$- \tfrac{1}{2}(9.80 \text{ m/s}^2)(0.384 \text{ s})^2 = \boxed{0.722 \text{ m}}$$

To *finalize* this problem, find the answers to parts (A) and (B) using Equations 4.13 and 4.14. The results should agree. Treating the long-jumper as a particle is an oversimplification. Nevertheless, the values obtained are consistent with experience in sports. We learn that we can model a complicated system such as a long-jumper as a particle and still obtain results that are reasonable.

This is the time at which the long-jumper is at the *top* of the jump. Because of the symmetry of the vertical motion,

---

Example 4.4   A Bull's-Eye Every Time   `Interactive`

In a popular lecture demonstration, a projectile is fired at a target T in such a way that the projectile leaves the gun at the same time the target is dropped from rest, as shown in Figure 4.13. Show that if the gun is initially aimed at the stationary target, the projectile hits the target.

Solution *Conceptualize* the problem by studying Figure 4.13. Notice that the problem asks for no numbers. The expected result must involve an algebraic argument. Because both objects are subject only to gravity, we *categorize* this problem as

one involving two objects in free-fall, one moving in one dimension and one moving in two. Let us now *analyze* the problem. A collision results under the conditions stated by noting that, as soon as they are released, the projectile and the target experience the same acceleration, $a_y = -g$. Figure 4.13b shows that the initial y coordinate of the target is $x_T \tan \theta_i$ and that it falls to a position $\tfrac{1}{2}gt^2$ below this coordinate at time t. Therefore, the y coordinate of the target at any moment after release is

$$y_T = x_T \tan \theta_i - \tfrac{1}{2} g t^2$$

Central Scientific Company

(a)   (b)

Figure 4.13 (Example 4.4) (a) Multiflash photograph of projectile–target demonstration. If the gun is aimed directly at the target and is fired at the same instant the target begins to fall, the projectile will hit the target. Note that the velocity of the projectile (red arrows) changes in direction and magnitude, while its downward acceleration (violet arrows) remains constant. (b) Schematic diagram of the projectile–target demonstration. Both projectile and target have fallen through the same vertical distance at time t, because both experience the same acceleration $a_y = -g$.

Now if we use Equation 4.9a to write an expression for the y coordinate of the projectile at any moment, we obtain

$$y_P = x_P \tan \theta_i - \tfrac{1}{2} g t^2$$

Thus, by comparing the two previous equations, we see that when the y coordinates of the projectile and target are the same, their x coordinates are the same and a collision

results. That is, when $y_P = y_T$, $x_P = x_T$. You can obtain the same result, using expressions for the position vectors for the projectile and target.

To *finalize* this problem, note that a collision can result only when $v_i \sin \theta_i \geq \sqrt{gd/2}$ where d is the initial elevation of the target above the floor. If $v_i \sin \theta_i$ is less than this value, the projectile will strike the floor before reaching the target.

 Investigate this situation at the Interactive Worked Example link at http://www.pse6.com.

Example 4.5   That's Quite an Arm!   `Interactive`

A stone is thrown from the top of a building upward at an angle of 30.0° to the horizontal with an initial speed of 20.0 m/s, as shown in Figure 4.14. If the height of the building is 45.0 m,

(A) how long does it take the stone to reach the ground?

Solution We *conceptualize* the problem by studying Figure 4.14, in which we have indicated the various parameters. By now, it should be natural to *categorize* this as a projectile motion problem.

To *analyze* the problem, let us once again separate motion into two components. The initial x and y components of the stone's velocity are

$$v_{xi} = v_i \cos \theta_i = (20.0 \text{ m/s}) \cos 30.0° = 17.3 \text{ m/s}$$

$$v_{yi} = v_i \sin \theta_i = (20.0 \text{ m/s}) \sin 30.0° = 10.0 \text{ m/s}$$

To find t, we can use $y_f = y_i + v_{yi}t + \tfrac{1}{2}a_y t^2$ (Eq. 4.9a) with $y_i = 0$, $y_f = -45.0$ m, $a_y = -g$, and $v_{yi} = 10.0$ m/s (there is a negative sign on the numerical value of y_f because we have chosen the top of the building as the origin):

$$-45.0 \text{ m} = (10.0 \text{ m/s}) t - \tfrac{1}{2} (9.80 \text{ m/s}^2) t^2$$

Solving the quadratic equation for t gives, for the positive root, $t = $ 4.22 s. To *finalize* this part, think: Does the negative root have any physical meaning?

(B) What is the speed of the stone just before it strikes the ground?

Solution We can use Equation 4.8a, $v_{yf} = v_{yi} + a_y t$, with $t = 4.22$ s to obtain the y component of the velocity just before the stone strikes the ground:

$$v_{yf} = 10.0 \text{ m/s} - (9.80 \text{ m/s}^2)(4.22 \text{ s}) = -31.4 \text{ m/s}$$

Because $v_{xf} = v_{xi} = 17.3$ m/s, the required speed is

$$v_f = \sqrt{v_{xf}^2 + v_{yf}^2} = \sqrt{(17.3)^2 + (-31.4)^2} \text{ m/s} = \boxed{35.9 \text{ m/s}}$$

To *finalize* this part, is it reasonable that the y component of the final velocity is negative? Is it reasonable that the final speed is larger than the initial speed of 20.0 m/s?

What If? What if a horizontal wind is blowing in the same direction as the ball is thrown and it causes the ball to have a horizontal acceleration component $a_x = 0.500$ m/s². Which part of this example, (A) or (B), will have a different answer?

Answer Recall that the motions in the x and y directions are independent. Thus, the horizontal wind cannot affect the vertical motion. The vertical motion determines the time of the projectile in the air, so the answer to (A) does not change. The wind will cause the horizontal velocity component to increase with time, so that the final speed will change in part (B).

We can find the new final horizontal velocity component by using Equation 4.8a:

$$v_{xf} = v_{xi} + a_x t = 17.3 \text{ m/s} + (0.500 \text{ m/s}^2)(4.22 \text{ s})$$
$$= 19.4 \text{ m/s}$$

and the new final speed:

$$v_f = \sqrt{v_{xf}^2 + v_{yf}^2} = \sqrt{(19.4)^2 + (-31.4)^2} \text{ m/s} = 36.9 \text{ m/s}$$

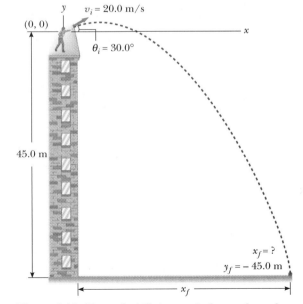

Figure 4.14 (Example 4.5) A stone is thrown from the top of a building.

 Investigate this situation at the Interactive Worked Example link at http://www.pse6.com.

Example 4.6   The Stranded Explorers

A plane drops a package of supplies to a party of explorers, as shown in Figure 4.15. If the plane is traveling horizontally at 40.0 m/s and is 100 m above the ground, where does the package strike the ground relative to the point at which it is released?

Solution *Conceptualize* what is happening with the assistance of Figure 4.15. The plane is traveling horizontally when it drops the package. Because the package is in free-fall while moving in the horizontal direction, we *categorize*

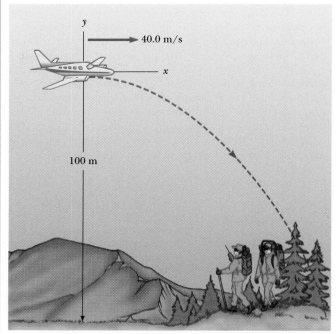

Figure 4.15 (Example 4.6) A package of emergency supplies is dropped from a plane to stranded explorers.

this as a projectile motion problem. To *analyze* the problem, we choose the coordinate system shown in Figure 4.15, in which the origin is at the point of release of the package. Consider first its horizontal motion. The only equation available for finding the position along the horizontal direction is $x_f = x_i + v_{xi}t$ (Eq. 4.9a). The initial x component of the package velocity is the same as that of the plane when the package is released: 40.0 m/s. Thus, we have

$$x_f = (40.0 \text{ m/s})t$$

If we know t, the time at which the package strikes the ground, then we can determine x_f, the distance the package travels in the horizontal direction. To find t, we use the equations that describe the vertical motion of the package. We know that, at the instant the package hits the ground, its y coordinate is $y_f = -100$ m. We also know that the initial vertical component of the package velocity v_{yi} is zero because at the moment of release, the package has only a horizontal component of velocity.

From Equation 4.9a, we have

$$y_f = -\tfrac{1}{2}gt^2$$
$$-100 \text{ m} = -\tfrac{1}{2}(9.80 \text{ m/s}^2)t^2$$
$$t = 4.52 \text{ s}$$

Substitution of this value for the time into the equation for the x coordinate gives

$$x_f = (40.0 \text{ m/s})(4.52 \text{ s}) = \boxed{181 \text{ m}}$$

The package hits the ground 181 m to the right of the drop point. To *finalize* this problem, we learn that an object dropped from a moving airplane does not fall straight down. It hits the ground at a point different from the one right below the plane where it was released. This was an important consideration for free-fall bombs such as those used in World War II.

Example 4.7   The End of the Ski Jump

A ski-jumper leaves the ski track moving in the horizontal direction with a speed of 25.0 m/s, as shown in Figure 4.16. The landing incline below him falls off with a slope of 35.0°. Where does he land on the incline?

Solution We can *conceptualize* this problem based on observations of winter Olympic ski competitions. We observe the skier to be airborne for perhaps 4 s and go a distance of about 100 m horizontally. We should expect the value of d, the distance traveled along the incline, to be of the same order of magnitude. We *categorize* the problem as that of a particle in projectile motion.

To *analyze* the problem, it is convenient to select the beginning of the jump as the origin. Because $v_{xi} = 25.0$ m/s and $v_{yi} = 0$, the x and y component forms of Equation 4.9a are

$$(1) \qquad x_f = v_{xi}t = (25.0 \text{ m/s})t$$
$$(2) \qquad y_f = v_{yi}t + \tfrac{1}{2}a_yt^2 = -\tfrac{1}{2}(9.80 \text{ m/s}^2)t^2$$

From the right triangle in Figure 4.16, we see that the jumper's x and y coordinates at the landing point are $x_f = d \cos 35.0°$ and $y_f = -d \sin 35.0°$. Substituting these relationships into (1) and (2), we obtain

$$(3) \qquad d \cos 35.0° = (25.0 \text{ m/s})t$$
$$(4) \qquad -d \sin 35.0° = -\tfrac{1}{2}(9.80 \text{ m/s}^2)t^2$$

Solving (3) for t and substituting the result into (4), we find that $d = 109$ m. Hence, the x and y coordinates of the point at which the skier lands are

$$x_f = d \cos 35.0° = (109 \text{ m})\cos 35.0° = \boxed{89.3 \text{ m}}$$

$$y_f = -d \sin 35.0° = -(109 \text{ m})\sin 35.0° = \boxed{-62.5 \text{ m}}$$

To *finalize* the problem, let us compare these results to our expectations. We expected the horizontal distance to be on the order of 100 m, and our result of 89.3 m is indeed on

Figure 4.16 (Example 4.7) A ski jumper leaves the track moving in a horizontal direction.

this order of magnitude. It might be useful to calculate the time interval that the jumper is in the air and compare it to our estimate of about 4 s.

What If? Suppose everything in this example is the same except that the ski jump is curved so that the jumper is projected upward at an angle from the end of the track. Is this a better design in terms of maximizing the length of the jump?

Answer If the initial velocity has an upward component, the skier will be in the air longer, and should therefore travel further. However, tilting the initial velocity vector upward will reduce the horizontal component of the initial velocity. Thus, angling the end of the ski track upward at a *large* angle may actually *reduce* the distance. Consider the extreme case. The skier is projected at 90° to the horizontal, and simply goes up and comes back down at the end of the ski track! This argument suggests that there must be an optimal angle between 0 and 90° that represents a balance between making the flight time longer and the horizontal velocity component smaller.

We can find this optimal angle mathematically. We modify equations (1) through (4) in the following way, assuming that the skier is projected at an angle θ with respect to the horizontal:

(1) and (3) $\rightarrow$ $x_f = (v_i \cos \theta)t = d \cos \phi$

(2) and (4) $\rightarrow$ $y_f = (v_i \sin \theta)t - \frac{1}{2}gt^2 = -d \sin \phi$

If we eliminate the time t between these equations and then use differentiation to maximize d in terms of θ, we arrive (after several steps—see Problem 72!) at the following equation for the angle θ that gives the maximum value of d:

$$\theta = 45° - \frac{\phi}{2}$$

For the slope angle in Figure 4.16, $\phi = 35.0°$; this equation results in an optimal launch angle of $\theta = 27.5°$. Notice that for a slope angle of $\phi = 0°$, which represents a horizontal plane, this equation gives an optimal launch angle of $\theta = 45°$, as we would expect (see Figure 4.11).

4.4   Uniform Circular Motion

Figure 4.17a shows a car moving in a circular path with *constant speed v*. Such motion is called **uniform circular motion,** and occurs in many situations. It is often surprising to students to find that **even though an object moves at a constant speed in a circular path, it still has an acceleration.** To see why, consider the defining equation for average acceleration, $\bar{\mathbf{a}} = \Delta\mathbf{v}/\Delta t$ (Eq. 4.4).

Note that the acceleration depends on *the change in the velocity vector*. Because velocity is a vector quantity, there are two ways in which an acceleration can occur, as mentioned in Section 4.1: by a change in the *magnitude* of the velocity and/or by a change in the *direction* of the velocity. The latter situation occurs for an object moving with constant speed in a circular path. The velocity vector is always tangent to the

▲ PITFALL PREVENTION

4.4   Acceleration of a Particle in Uniform Circular Motion

Remember that acceleration in physics is defined as a change in the *velocity,* not a change in the *speed* (contrary to the everyday interpretation). In circular motion, the velocity vector is changing in direction, so there is indeed an acceleration.

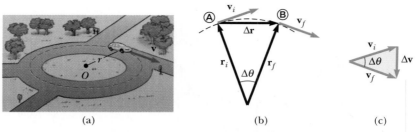

Figure 4.17 (a) A car moving along a circular path at constant speed experiences uniform circular motion. (b) As a particle moves from Ⓐ to Ⓑ, its velocity vector changes from $\mathbf{v}_i$ to $\mathbf{v}_f$. (c) The construction for determining the direction of the change in velocity $\Delta\mathbf{v}$, which is toward the center of the circle for small $\Delta\mathbf{r}$.

path of the object and perpendicular to the radius of the circular path. We now show that the acceleration vector in uniform circular motion is always perpendicular to the path and always points toward the center of the circle. An acceleration of this nature is called a **centripetal acceleration** (*centripetal* means *center-seeking*), and its magnitude is

Centripetal acceleration

$$a_c = \frac{v^2}{r} \tag{4.15}$$

where r is the radius of the circle. The subscript on the acceleration symbol reminds us that the acceleration is centripetal.

First note that the acceleration must be perpendicular to the path followed by the object, which we will model as a particle. If this were not true, there would be a component of the acceleration parallel to the path and, therefore, parallel to the velocity vector. Such an acceleration component would lead to a change in the speed of the particle along the path. But this is inconsistent with our setup of the situation—the particle moves with constant speed along the path. Thus, for *uniform* circular motion, the acceleration vector can only have a component perpendicular to the path, which is toward the center of the circle.

To derive Equation 4.15, consider the diagram of the position and velocity vectors in Figure 4.17b. In addition, the figure shows the vector representing the change in position $\Delta\mathbf{r}$. The particle follows a circular path, part of which is shown by the dotted curve. The particle is at Ⓐ at time t_i, and its velocity at that time is $\mathbf{v}_i$; it is at Ⓑ at some later time t_f, and its velocity at that time is $\mathbf{v}_f$. Let us also assume that $\mathbf{v}_i$ and $\mathbf{v}_f$ differ only in direction; their magnitudes are the same (that is, $v_i = v_f = v$, because it is *uniform* circular motion). In order to calculate the acceleration of the particle, let us begin with the defining equation for average acceleration (Eq. 4.4):

$$\overline{\mathbf{a}} \equiv \frac{\mathbf{v}_f - \mathbf{v}_i}{t_f - t_i} = \frac{\Delta\mathbf{v}}{\Delta t}$$

In Figure 4.17c, the velocity vectors in Figure 4.17b have been redrawn tail to tail. The vector $\Delta\mathbf{v}$ connects the tips of the vectors, representing the vector addition $\mathbf{v}_f = \mathbf{v}_i + \Delta\mathbf{v}$. In both Figures 4.17b and 4.17c, we can identify triangles that help us analyze the motion. The angle $\Delta\theta$ between the two position vectors in Figure 4.17b is the same as the angle between the velocity vectors in Figure 4.17c, because the velocity vector $\mathbf{v}$ is always perpendicular to the position vector $\mathbf{r}$. Thus, the two triangles are *similar*. (Two triangles are similar if the angle between any two sides is the same for both triangles and if the ratio of the lengths of these sides is the same.) This enables us to write a relationship between the lengths of the sides for the two triangles:

$$\frac{|\Delta\mathbf{v}|}{v} = \frac{|\Delta\mathbf{r}|}{r}$$

where $v = v_i = v_f$ and $r = r_i = r_f$. This equation can be solved for $|\Delta \mathbf{v}|$ and the expression so obtained can be substituted into $\overline{\mathbf{a}} = \Delta \mathbf{v}/\Delta t$ to give the magnitude of the average acceleration over the time interval for the particle to move from Ⓐ to Ⓑ:

$$|\overline{\mathbf{a}}| = \frac{|\Delta \mathbf{v}|}{\Delta t} = \frac{v}{r}\frac{|\Delta \mathbf{r}|}{\Delta t}$$

Now imagine that points Ⓐ and Ⓑ in Figure 4.17b become extremely close together. As Ⓐ and Ⓑ approach each other, Δt approaches zero, and the ratio $|\Delta \mathbf{r}|/\Delta t$ approaches the speed v. In addition, the average acceleration becomes the instantaneous acceleration at point Ⓐ. Hence, in the limit $\Delta t \rightarrow 0$, the magnitude of the acceleration is

$$a_c = \frac{v^2}{r}$$

Thus, in uniform circular motion the acceleration is directed inward toward the center of the circle and has magnitude v^2/r.

In many situations it is convenient to describe the motion of a particle moving with constant speed in a circle of radius r in terms of the **period** T, which is defined as the time required for one complete revolution. In the time interval T the particle moves a distance of $2\pi r$, which is equal to the circumference of the particle's circular path. Therefore, because its speed is equal to the circumference of the circular path divided by the period, or $v = 2\pi r/T$, it follows that

$$T \equiv \frac{2\pi r}{v} \qquad (4.16)$$

▲ **PITFALL PREVENTION**

4.5 Centripetal Acceleration is not Constant

We derived the magnitude of the centripetal acceleration vector and found it to be constant for uniform circular motion. But *the centripetal acceleration vector is not constant.* It always points toward the center of the circle, but continuously changes direction as the object moves around the circular path.

Period of circular motion

Quick Quiz 4.7 Which of the following correctly describes the centripetal acceleration vector for a particle moving in a circular path? (a) constant and always perpendicular to the velocity vector for the particle (b) constant and always parallel to the velocity vector for the particle (c) of constant magnitude and always perpendicular to the velocity vector for the particle (d) of constant magnitude and always parallel to the velocity vector for the particle.

Quick Quiz 4.8 A particle moves in a circular path of radius r with speed v. It then increases its speed to $2v$ while traveling along the same circular path. The centripetal acceleration of the particle has changed by a factor of (a) 0.25 (b) 0.5 (c) 2 (d) 4 (e) impossible to determine

---

Example 4.8 The Centripetal Acceleration of the Earth

What is the centripetal acceleration of the Earth as it moves in its orbit around the Sun?

Solution We *conceptualize* this problem by bringing forth our familiar mental image of the Earth in a circular orbit around the Sun. We will simplify the problem by modeling the Earth as a particle and approximating the Earth's orbit as circular (it's actually slightly elliptical). This allows us to *categorize* this problem as that of a particle in uniform circular motion. When we begin to *analyze* this problem, we realize that we do not know the orbital speed of the Earth in Equation 4.15. With the help of Equation 4.16, however, we can recast Equation 4.15 in terms of the period of the Earth's orbit, which we know is one year:

$$a_c = \frac{v^2}{r} = \frac{\left(\dfrac{2\pi r}{T}\right)^2}{r} = \frac{4\pi^2 r}{T^2}$$

$$= \frac{4\pi^2 (1.496 \times 10^{11}\ \text{m})}{(1\ \text{yr})^2}\left(\frac{1\ \text{yr}}{3.156 \times 10^7\ \text{s}}\right)^2$$

$$= \boxed{5.93 \times 10^{-3}\ \text{m/s}^2}$$

To *finalize* this problem, note that this acceleration is much smaller than the free-fall acceleration on the surface of the Earth. An important thing we learned here is the technique of replacing the speed v in terms of the period T of the motion.

4.5 Tangential and Radial Acceleration

Let us consider the motion of a particle along a smooth curved path where the velocity changes both in direction and in magnitude, as described in Figure 4.18. In this situation, the velocity vector is always tangent to the path; however, the acceleration vector **a** is at some angle to the path. At each of three points Ⓐ, Ⓑ, and Ⓒ in Figure 4.18, we draw dashed circles that represent a portion of the actual path at each point. The radius of the circles is equal to the radius of curvature of the path at each point.

As the particle moves along the curved path in Figure 4.18, the direction of the total acceleration vector **a** changes from point to point. This vector can be resolved into two components, based on an origin at the center of the dashed circle: a radial component a_r along the radius of the model circle, and a tangential component a_t perpendicular to this radius. The *total* acceleration vector **a** can be written as the vector sum of the component vectors:

Total acceleration

$$\mathbf{a} = \mathbf{a}_r + \mathbf{a}_t \qquad (4.17)$$

The tangential acceleration component causes the change in the speed of the particle. This component is parallel to the instantaneous velocity, and is given by

Tangential acceleration

$$a_t = \frac{d|\mathbf{v}|}{dt} \qquad (4.18)$$

The radial acceleration component arises from the change in direction of the velocity vector and is given by

Radial acceleration

$$a_r = -a_c = -\frac{v^2}{r} \qquad (4.19)$$

where r is the radius of curvature of the path at the point in question. We recognize the radial component of the acceleration as the centripetal acceleration discussed in Section 4.4. The negative sign indicates that the direction of the centripetal acceleration is toward the center of the circle representing the radius of curvature, which is opposite the direction of the radial unit vector $\hat{\mathbf{r}}$, which always points away from the center of the circle.

Because $\mathbf{a}_r$ and $\mathbf{a}_t$ are perpendicular component vectors of **a**, it follows that the magnitude of **a** is $a = \sqrt{a_r{}^2 + a_t{}^2}$. At a given speed, a_r is large when the radius of curvature is small (as at points Ⓐ and Ⓑ in Fig. 4.18) and small when r is large (such as at point Ⓒ). The direction of $\mathbf{a}_t$ is either in the same direction as **v** (if v is increasing) or opposite **v** (if v is decreasing).

In uniform circular motion, where v is constant, $a_t = 0$ and the acceleration is always completely radial, as we described in Section 4.4. In other words, uniform circular motion is a special case of motion along a general curved path. Furthermore, if the direction of **v** does not change, then there is no radial acceleration and the motion is one-dimensional (in this case, $a_r = 0$, but a_t may not be zero).

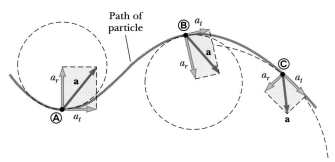

At the Active Figures link at http://www.pse6.com, you can study the acceleration components of a roller coaster car.

Active Figure 4.18 The motion of a particle along an arbitrary curved path lying in the *xy* plane. If the velocity vector **v** (always tangent to the path) changes in direction and magnitude, the components of the acceleration **a** are a tangential component a_t and a radial component a_r.

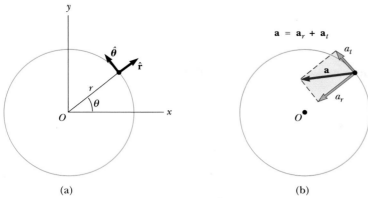

Figure 4.19 (a) Descriptions of the unit vectors $\hat{\mathbf{r}}$ and $\hat{\boldsymbol{\theta}}$. (b) The total acceleration **a** of a particle moving along a curved path (which at any instant is part of a circle of radius r) is the sum of radial and tangential component vectors. The radial component vector is directed toward the center of curvature. If the tangential component of acceleration becomes zero, the particle follows uniform circular motion.

It is convenient to write the acceleration of a particle moving in a circular path in terms of unit vectors. We do this by defining the unit vectors $\hat{\mathbf{r}}$ and $\hat{\boldsymbol{\theta}}$ shown in Figure 4.19a, where $\hat{\mathbf{r}}$ is a unit vector lying along the radius vector and directed radially outward from the center of the circle and $\hat{\boldsymbol{\theta}}$ is a unit vector tangent to the circle. The direction of $\hat{\boldsymbol{\theta}}$ is in the direction of increasing θ, where θ is measured counterclockwise from the positive x axis. Note that both $\hat{\mathbf{r}}$ and $\hat{\boldsymbol{\theta}}$ "move along with the particle" and so vary in time. Using this notation, we can express the total acceleration as

$$\mathbf{a} = \mathbf{a}_t + \mathbf{a}_r = \frac{d|\mathbf{v}|}{dt}\,\hat{\boldsymbol{\theta}} - \frac{v^2}{r}\,\hat{\mathbf{r}} \qquad (4.20)$$

These vectors are described in Figure 4.19b.

Quick Quiz 4.9 A particle moves along a path and its speed increases with time. In which of the following cases are its acceleration and velocity vectors parallel? (a) the path is circular (b) the path is straight (c) the path is a parabola (d) never.

Quick Quiz 4.10 A particle moves along a path and its speed increases with time. In which of the following cases are its acceleration and velocity vectors perpendicular everywhere along the path? (a) the path is circular (b) the path is straight (c) the path is a parabola (d) never.

Example 4.9   Over the Rise

A car exhibits a constant acceleration of 0.300 m/s^2 parallel to the roadway. The car passes over a rise in the roadway such that the top of the rise is shaped like a circle of radius 500 m. At the moment the car is at the top of the rise, its velocity vector is horizontal and has a magnitude of 6.00 m/s. What is the direction of the total acceleration vector for the car at this instant?

Solution *Conceptualize* the situation using Figure 4.20a. Because the car is moving along a curved path, we can *catego-*

rize this as a problem involving a particle experiencing both tangential and radial acceleration. Now we recognize that this is a relatively simple plug-in problem. The radial acceleration is given by Equation 4.19. With $v = 6.00$ m/s and $r = 500$ m, we find that

$$a_r = -\frac{v^2}{r} = -\frac{(6.00 \text{ m/s})^2}{500 \text{ m}} = \boxed{-0.0720 \text{ m/s}^2}$$

The radial acceleration vector is directed straight downward

$a_t = 0.300 \text{ m/s}^2$

$v = 6.00 \text{ m/s}$

(a)   (b)

Figure 4.20 (Example 4.9) (a) A car passes over a rise that is shaped like a circle. (b) The total acceleration vector **a** is the sum of the tangential and radial acceleration vectors **a**$_t$ and **a**$_r$.

while the tangential acceleration vector has magnitude 0.300 m/s² and is horizontal. Because $\mathbf{a} = \mathbf{a}_r + \mathbf{a}_t$, the magnitude of **a** is

$$a = \sqrt{a_r{}^2 + a_t{}^2} = \sqrt{(-0.0720)^2 + (0.300)^2} \text{ m/s}^2$$

$$= \boxed{0.309 \text{ m/s}^2}$$

If ϕ is the angle between **a** and the horizontal, then

$$\phi = \tan^{-1}\frac{a_r}{a_t} = \tan^{-1}\left(\frac{-0.0720 \text{ m/s}^2}{0.300 \text{ m/s}^2}\right) = \boxed{-13.5°}$$

This angle is measured downward from the horizontal. (See Figure 4.20b.)

4.6 Relative Velocity and Relative Acceleration

In this section, we describe how observations made by different observers in different frames of reference are related to each other. We find that observers in different frames of reference may measure different positions, velocities, and accelerations for a given particle. That is, two observers moving relative to each other generally do not agree on the outcome of a measurement.

As an example, consider two observers watching a man walking on a moving beltway at an airport in Figure 4.21. The woman standing on the moving beltway will see the man moving at a normal walking speed. The woman observing from the stationary floor will see the man moving with a higher speed, because the beltway speed combines with his walking speed. Both observers look at the same man and arrive at different values for his speed. Both are correct; the difference in their measurements is due to the relative velocity of their frames of reference.

Suppose a person riding on a skateboard (observer A) throws a ball in such a way that it appears in this person's frame of reference to move first straight upward and

Figure 4.21 Two observers measure the speed of a man walking on a moving beltway. The woman standing on the beltway sees the man moving with a slower speed than the woman observing from the stationary floor.

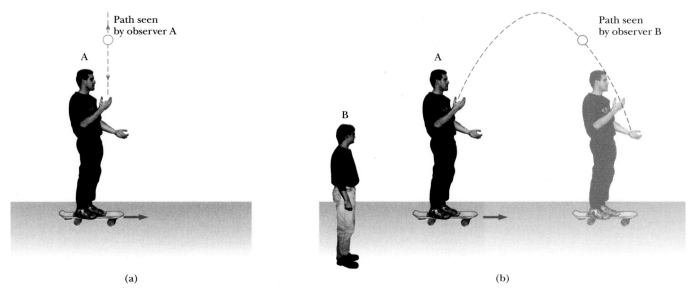

Figure 4.22 (a) Observer A on a moving skateboard throws a ball upward and sees it rise and fall in a straight-line path. (b) Stationary observer B sees a parabolic path for the same ball.

then straight downward along the same vertical line, as shown in Figure 4.22a. An observer B on the ground sees the path of the ball as a parabola, as illustrated in Figure 4.22b. Relative to observer B, the ball has a vertical component of velocity (resulting from the initial upward velocity and the downward acceleration due to gravity) *and* a horizontal component.

Another example of this concept is the motion of a package dropped from an airplane flying with a constant velocity—a situation we studied in Example 4.6. An observer on the airplane sees the motion of the package as a straight line downward toward Earth. The stranded explorer on the ground, however, sees the trajectory of the package as a parabola. Once the package is dropped, and the airplane continues to move horizontally with the same velocity, the package hits the ground directly beneath the airplane (if we assume that air resistance is neglected)!

In a more general situation, consider a particle located at point Ⓐ in Figure 4.23. Imagine that the motion of this particle is being described by two observers, one in reference frame S, fixed relative to Earth, and another in reference frame S', moving to the right relative to S (and therefore relative to Earth) with a constant velocity $\mathbf{v}_0$. (Relative to an observer in S', S moves to the left with a velocity $-\mathbf{v}_0$.) Where an observer stands in a reference frame is irrelevant in this discussion, but for purposes of this discussion let us place each observer at her or his respective origin.

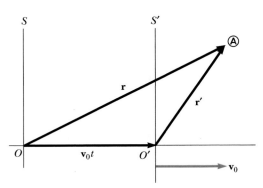

Figure 4.23 A particle located at Ⓐ is described by two observers, one in the fixed frame of reference S, and the other in the frame S', which moves to the right with a constant velocity $\mathbf{v}_0$. The vector $\mathbf{r}$ is the particle's position vector relative to S, and $\mathbf{r}'$ is its position vector relative to S'.

We define the time $t = 0$ as that instant at which the origins of the two reference frames coincide in space. Thus, at time t, the origins of the reference frames will be separated by a distance $v_0 t$. We label the position of the particle relative to the S frame with the position vector $\mathbf{r}$ and that relative to the S' frame with the position vector $\mathbf{r}'$, both at time t. The vectors $\mathbf{r}$ and $\mathbf{r}'$ are related to each other through the expression $\mathbf{r} = \mathbf{r}' + \mathbf{v}_0 t$, or

Galilean coordinate transformation

$$\mathbf{r}' = \mathbf{r} - \mathbf{v}_0 t \tag{4.21}$$

If we differentiate Equation 4.21 with respect to time and note that $\mathbf{v}_0$ is constant, we obtain

$$\frac{d\mathbf{r}'}{dt} = \frac{d\mathbf{r}}{dt} - \mathbf{v}_0$$

Galilean velocity transformation

$$\mathbf{v}' = \mathbf{v} - \mathbf{v}_0 \tag{4.22}$$

where $\mathbf{v}'$ is the velocity of the particle observed in the S' frame and $\mathbf{v}$ is its velocity observed in the S frame. Equations 4.21 and 4.22 are known as **Galilean transformation equations.** They relate the position and velocity of a particle as measured by observers in relative motion.

Although observers in two frames measure different velocities for the particle, they measure the *same acceleration* when $\mathbf{v}_0$ is constant. We can verify this by taking the time derivative of Equation 4.22:

$$\frac{d\mathbf{v}'}{dt} = \frac{d\mathbf{v}}{dt} - \frac{d\mathbf{v}_0}{dt}$$

Because $\mathbf{v}_0$ is constant, $d\mathbf{v}_0/dt = 0$. Therefore, we conclude that $\mathbf{a}' = \mathbf{a}$ because $\mathbf{a}' = d\mathbf{v}'/dt$ and $\mathbf{a} = d\mathbf{v}/dt$. That is, **the acceleration of the particle measured by an observer in one frame of reference is the same as that measured by any other observer moving with constant velocity relative to the first frame.**

Quick Quiz 4.11   A passenger, observer A, in a car traveling at a constant horizontal velocity of magnitude 60 mi/h pours a cup of coffee for the tired driver. Observer B stands on the side of the road and watches the pouring process through the window of the car as it passes. Which observer(s) sees a parabolic path for the coffee as it moves through the air? (a) A (b) B (c) both A and B (d) neither A nor B.

Example 4.10   A Boat Crossing a River

A boat heading due north crosses a wide river with a speed of 10.0 km/h relative to the water. The water in the river has a uniform speed of 5.00 km/h due east relative to the Earth. Determine the velocity of the boat relative to an observer standing on either bank.

Solution To *conceptualize* this problem, imagine moving across a river while the current pushes you along the river. You will not be able to move directly across the river, but will end up downstream, as suggested in Figure 4.24. Because of the separate velocities of you and the river, we can *categorize* this as a problem involving relative velocities. We will *analyze* this problem with the techniques discussed in this section. We know $\mathbf{v}_{br}$, the velocity of the *boat* relative to the *river*, and $\mathbf{v}_{rE}$, the velocity of the *river* relative to *Earth*. What we must find is $\mathbf{v}_{bE}$, the velocity of the *boat* relative to *Earth*. The relationship between these three quantities is

$$\mathbf{v}_{bE} = \mathbf{v}_{br} + \mathbf{v}_{rE}$$

Figure 4.24 (Example 4.10) A boat aims directly across a river and ends up downstream.

The terms in the equation must be manipulated as vector quantities; the vectors are shown in Figure 4.24. The quantity $\mathbf{v}_{br}$ is due north, $\mathbf{v}_{rE}$ is due east, and the vector sum of the two, $\mathbf{v}_{bE}$, is at an angle θ, as defined in Figure 4.24. Thus, we can find the speed v_{bE} of the boat relative to Earth by using the Pythagorean theorem:

$$v_{bE} = \sqrt{v_{br}{}^2 + v_{rE}{}^2} = \sqrt{(10.0)^2 + (5.00)^2}\ \text{km/h}$$

$$= \boxed{11.2\ \text{km/h}}$$

The direction of $\mathbf{v}_{bE}$ is

$$\theta = \tan^{-1}\left(\frac{v_{rE}}{v_{br}}\right) = \tan^{-1}\left(\frac{5.00}{10.0}\right) = \boxed{26.6°}$$

The boat is moving at a speed of 11.2 km/h in the direction 26.6° east of north relative to Earth. To *finalize* the problem, note that the speed of 11.2 km/h is faster than your boat speed of 10.0 km/h. The current velocity adds to yours to give you a larger speed. Notice in Figure 4.24 that your resultant velocity is at an angle to the direction straight across the river, so you will end up downstream, as we predicted.

Example 4.11   Which Way Should We Head?

If the boat of the preceding example travels with the same speed of 10.0 km/h relative to the river and is to travel due north, as shown in Figure 4.25, what should its heading be?

Solution This example is an extension of the previous one, so we have already *conceptualized* and *categorized* the problem. The *analysis* now involves the new triangle shown in Figure 4.25. As in the previous example, we know $\mathbf{v}_{rE}$ and the magnitude of the vector $\mathbf{v}_{br}$, and we want $\mathbf{v}_{bE}$ to be directed across the river. Note the difference between the triangle in Figure 4.24 and the one in Figure 4.25—the hypotenuse in Figure 4.25 is no longer $\mathbf{v}_{bE}$. Therefore, when we use the Pythagorean theorem to find $\mathbf{v}_{bE}$ in this situation, we obtain

$$v_{bE} = \sqrt{v_{br}{}^2 - v_{rE}{}^2} = \sqrt{(10.0)^2 - (5.00)^2}\ \text{km/h} = 8.66\ \text{km/h}$$

Now that we know the magnitude of $\mathbf{v}_{bE}$, we can find the direction in which the boat is heading:

$$\theta = \tan^{-1}\left(\frac{v_{rE}}{v_{bE}}\right) = \tan^{-1}\left(\frac{5.00}{8.66}\right) = \boxed{30.0°}$$

To *finalize* this problem, we learn that the boat must head upstream in order to travel directly northward across the river. For the given situation, the boat must steer a course 30.0° west of north.

What If? Imagine that the two boats in Examples 4.10 and 4.11 are racing across the river. Which boat arrives at the opposite bank first?

Answer In Example 4.10, the velocity of 10 km/h is aimed directly across the river. In Example 4.11, the velocity that is directed across the river has a magnitude of only 8.66 km/h. Thus, the boat in Example 4.10 has a larger velocity component directly across the river and will arrive first.

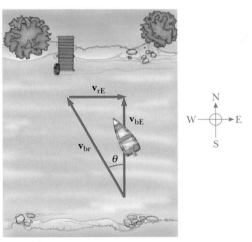

Figure 4.25 (Example 4.11) To move directly across the river, the boat must aim upstream.

SUMMARY

If a particle moves with *constant* acceleration $\mathbf{a}$ and has velocity $\mathbf{v}_i$ and position $\mathbf{r}_i$ at $t = 0$, its velocity and position vectors at some later time t are

$$\mathbf{v}_f = \mathbf{v}_i + \mathbf{a}t \tag{4.8}$$

$$\mathbf{r}_f = \mathbf{r}_i + \mathbf{v}_i t + \tfrac{1}{2}\mathbf{a}t^2 \tag{4.9}$$

Take a practice test for this chapter by clicking on the Practice Test link at http://www.pse6.com.

For two-dimensional motion in the xy plane under constant acceleration, each of these vector expressions is equivalent to two component expressions—one for the motion in the x direction and one for the motion in the y direction.

   Projectile motion is one type of two-dimensional motion under constant acceleration, where $a_x = 0$ and $a_y = -g$. It is useful to think of projectile motion as the superposition of two motions: (1) constant-velocity motion in the x direction and

(2) free-fall motion in the vertical direction subject to a constant downward acceleration of magnitude $g = 9.80$ m/s^2.

A particle moving in a circle of radius r with constant speed v is in **uniform circular motion.** It undergoes a radial acceleration $\mathbf{a}_r$ because the direction of $\mathbf{v}$ changes in time. The magnitude of $\mathbf{a}_r$ is the **centripetal acceleration** a_c:

$$a_c = \frac{v^2}{r} \qquad\qquad (4.19)$$

and its direction is always toward the center of the circle.

If a particle moves along a curved path in such a way that both the magnitude and the direction of $\mathbf{v}$ change in time, then the particle has an acceleration vector that can be described by two component vectors: (1) a radial component vector $\mathbf{a}_r$ that causes the change in direction of $\mathbf{v}$ and (2) a tangential component vector $\mathbf{a}_t$ that causes the change in magnitude of $\mathbf{v}$. The magnitude of $\mathbf{a}_r$ is v^2/r, and the magnitude of $\mathbf{a}_t$ is $d|\mathbf{v}|/dt$.

The velocity $\mathbf{v}$ of a particle measured in a fixed frame of reference S can be related to the velocity $\mathbf{v}'$ of the same particle measured in a moving frame of reference S' by

$$\mathbf{v}' = \mathbf{v} - \mathbf{v}_0 \qquad\qquad (4.22)$$

where $\mathbf{v}_0$ is the velocity of S' relative to S.

QUESTIONS

1. Can an object accelerate if its speed is constant? Can an object accelerate if its velocity is constant?

2. If you know the position vectors of a particle at two points along its path and also know the time it took to move from one point to the other, can you determine the particle's instantaneous velocity? Its average velocity? Explain.

3. Construct motion diagrams showing the velocity and acceleration of a projectile at several points along its path if (a) the projectile is launched horizontally and (b) the projectile is launched at an angle θ with the horizontal.

4. A baseball is thrown with an initial velocity of $(10\hat{\mathbf{i}} + 15\hat{\mathbf{j}})$ m/s. When it reaches the top of its trajectory, what are (a) its velocity and (b) its acceleration? Neglect the effect of air resistance.

5. A baseball is thrown such that its initial x and y components of velocity are known. Neglecting air resistance, describe how you would calculate, at the instant the ball reaches the top of its trajectory, (a) its position, (b) its velocity, and (c) its acceleration. How would these results change if air resistance were taken into account?

6. A spacecraft drifts through space at a constant velocity. Suddenly a gas leak in the side of the spacecraft gives it a constant acceleration in a direction perpendicular to the initial velocity. The orientation of the spacecraft does not change, so that the acceleration remains perpendicular to the original direction of the velocity. What is the shape of the path followed by the spacecraft in this situation?

7. A ball is projected horizontally from the top of a building. One second later another ball is projected horizontally from the same point with the same velocity. At what point in the motion will the balls be closest to each other? Will the first ball always be traveling faster than the second

ball? What will be the time interval between when the balls hit the ground? Can the horizontal projection velocity of the second ball be changed so that the balls arrive at the ground at the same time?

8. A rock is dropped at the same instant that a ball, at the same initial elevation, is thrown horizontally. Which will have the greater speed when it reaches ground level?

9. Determine which of the following moving objects obey the equations of projectile motion developed in this chapter. (a) A ball is thrown in an arbitrary direction. (b) A jet airplane crosses the sky with its engines thrusting the plane forward. (c) A rocket leaves the launch pad. (d) A rocket moving through the sky after its engines have failed. (e) A stone is thrown under water.

10. How can you throw a projectile so that it has zero speed at the top of its trajectory? So that it has nonzero speed at the top of its trajectory?

11. Two projectiles are thrown with the same magnitude of initial velocity, one at an angle θ with respect to the level ground and the other at angle $90° - \theta$. Both projectiles will strike the ground at the same distance from the projection point. Will both projectiles be in the air for the same time interval?

12. A projectile is launched at some angle to the horizontal with some initial speed v_i, and air resistance is negligible. Is the projectile a freely falling body? What is its acceleration in the vertical direction? What is its acceleration in the horizontal direction?

13. State which of the following quantities, if any, remain constant as a projectile moves through its parabolic trajectory: (a) speed, (b) acceleration, (c) horizontal component of velocity, (d) vertical component of velocity.

14. A projectile is fired at an angle of 30° from the horizontal with some initial speed. Firing the projectile at what other angle results in the same horizontal range if the initial speed is the same in both cases? Neglect air resistance.

15. The maximum range of a projectile occurs when it is launched at an angle of 45.0° with the horizontal, if air resistance is neglected. If air resistance is not neglected, will the optimum angle be greater or less than 45.0°? Explain.

16. A projectile is launched on the Earth with some initial velocity. Another projectile is launched on the Moon with the same initial velocity. Neglecting air resistance, which projectile has the greater range? Which reaches the greater altitude? (Note that the free-fall acceleration on the Moon is about 1.6 m/s².)

17. A coin on a table is given an initial horizontal velocity such that it ultimately leaves the end of the table and hits the floor. At the instant the coin leaves the end of the table, a ball is released from the same height and falls to the floor. Explain why the two objects hit the floor simultaneously, even though the coin has an initial velocity.

18. Explain whether or not the following particles have an acceleration: (a) a particle moving in a straight line with constant speed and (b) a particle moving around a curve with constant speed.

19. Correct the following statement: "The racing car rounds the turn at a constant velocity of 90 miles per hour."

20. At the end of a pendulum's arc, its velocity is zero. Is its acceleration also zero at that point?

21. An object moves in a circular path with constant speed v. (a) Is the velocity of the object constant? (b) Is its acceleration constant? Explain.

22. Describe how a driver can steer a car traveling at constant speed so that (a) the acceleration is zero or (b) the magnitude of the acceleration remains constant.

23. An ice skater is executing a figure eight, consisting of two equal, tangent circular paths. Throughout the first loop she increases her speed uniformly, and during the second loop she moves at a constant speed. Draw a motion diagram showing her velocity and acceleration vectors at several points along the path of motion.

24. Based on your observation and experience, draw a motion diagram showing the position, velocity, and acceleration vectors for a pendulum that swings in an arc carrying it from an initial position 45° to the right of the central vertical line to a final position 45° to the left of the central vertical line. The arc is a quadrant of a circle, and you should use the center of the circle as the origin for the position vectors.

25. What is the fundamental difference between the unit vectors $\hat{\mathbf{r}}$ and $\hat{\boldsymbol{\theta}}$ and the unit vectors $\hat{\mathbf{i}}$ and $\hat{\mathbf{j}}$?

26. A sailor drops a wrench from the top of a sailboat's mast while the boat is moving rapidly and steadily in a straight line. Where will the wrench hit the deck? (Galileo posed this question.)

27. A ball is thrown upward in the air by a passenger on a train that is moving with constant velocity. (a) Describe the path of the ball as seen by the passenger. Describe the path as seen by an observer standing by the tracks outside the train. (b) How would these observations change if the train were accelerating along the track?

28. A passenger on a train that is moving with constant velocity drops a spoon. What is the acceleration of the spoon relative to (a) the train and (b) the Earth?

PROBLEMS

1, 2, 3 = straightforward, intermediate, challenging   ☐ = full solution available in the *Student Solutions Manual and Study Guide*

🌐 = coached solution with hints available at http://www.pse6.com   🖥 = computer useful in solving problem

▨ = paired numerical and symbolic problems

Section 4.1   The Position, Velocity, and Acceleration Vectors

1. 🌐 A motorist drives south at 20.0 m/s for 3.00 min, then turns west and travels at 25.0 m/s for 2.00 min, and finally travels northwest at 30.0 m/s for 1.00 min. For this 6.00-min trip, find (a) the total vector displacement, (b) the average speed, and (c) the average velocity. Let the positive x axis point east.

2. A golf ball is hit off a tee at the edge of a cliff. Its x and y coordinates as functions of time are given by the following expressions:

$$x = (18.0 \text{ m/s})t$$

and $$y = (4.00 \text{ m/s})t - (4.90 \text{ m/s}^2)t^2$$

(a) Write a vector expression for the ball's position as a function of time, using the unit vectors $\hat{\mathbf{i}}$ and $\hat{\mathbf{j}}$. By taking derivatives, obtain expressions for (b) the velocity vector $\mathbf{v}$ as a function of time and (c) the acceleration vector $\mathbf{a}$ as a function of time. Next use unit-vector notation to write expressions for (d) the position, (e) the velocity, and (f) the acceleration of the golf ball, all at $t = 3.00$ s.

3. When the Sun is directly overhead, a hawk dives toward the ground with a constant velocity of 5.00 m/s at 60.0° below the horizontal. Calculate the speed of her shadow on the level ground.

4. The coordinates of an object moving in the xy plane vary with time according to the equations $x = -(5.00 \text{ m}) \sin(\omega t)$ and $y = (4.00 \text{ m}) - (5.00 \text{ m}) \cos(\omega t)$, where ω is a constant and t is in seconds. (a) Determine the components of velocity and components of acceleration at $t = 0$. (b) Write expressions for the position vector, the velocity vector, and the acceleration vector at any time $t > 0$. (c) Describe the path of the object in an xy plot.

Section 4.2 Two-Dimensional Motion with Constant Acceleration

5. At $t = 0$, a particle moving in the xy plane with constant acceleration has a velocity of $\mathbf{v}_i = (3.00\hat{\mathbf{i}} - 2.00\hat{\mathbf{j}})$ m/s and is at the origin. At $t = 3.00$ s, the particle's velocity is $\mathbf{v} = (9.00\hat{\mathbf{i}} + 7.00\hat{\mathbf{j}})$ m/s. Find (a) the acceleration of the particle and (b) its coordinates at any time t.

6. The vector position of a particle varies in time according to the expression $\mathbf{r} = (3.00\hat{\mathbf{i}} - 6.00t^2\hat{\mathbf{j}})$ m. (a) Find expressions for the velocity and acceleration as functions of time. (b) Determine the particle's position and velocity at $t = 1.00$ s.

7. A fish swimming in a horizontal plane has velocity $\mathbf{v}_i = (4.00\hat{\mathbf{i}} + 1.00\hat{\mathbf{j}})$ m/s at a point in the ocean where the position relative to a certain rock is $\mathbf{r}_i = (10.0\hat{\mathbf{i}} - 4.00\hat{\mathbf{j}})$ m. After the fish swims with constant acceleration for 20.0 s, its velocity is $\mathbf{v} = (20.0\hat{\mathbf{i}} - 5.00\hat{\mathbf{j}})$ m/s. (a) What are the components of the acceleration? (b) What is the direction of the acceleration with respect to unit vector $\hat{\mathbf{i}}$? (c) If the fish maintains constant acceleration, where is it at $t = 25.0$ s, and in what direction is it moving?

8. A particle initially located at the origin has an acceleration of $\mathbf{a} = 3.00\hat{\mathbf{j}}$ m/s^2 and an initial velocity of $\mathbf{v}_i = 500\hat{\mathbf{i}}$ m/s. Find (a) the vector position and velocity at any time t and (b) the coordinates and speed of the particle at $t = 2.00$ s.

9. It is not possible to see very small objects, such as viruses, using an ordinary light microscope. An electron microscope can view such objects using an electron beam instead of a light beam. Electron microscopy has proved invaluable for investigations of viruses, cell membranes and subcellular structures, bacterial surfaces, visual receptors, chloroplasts, and the contractile properties of muscles. The "lenses" of an electron microscope consist of electric and magnetic fields that control the electron beam. As an example of the manipulation of an electron beam, consider an electron traveling away from the origin along the x axis in the xy plane with initial velocity $\mathbf{v}_i = v_i\hat{\mathbf{i}}$. As it passes through the region $x = 0$ to $x = d$, the electron experiences acceleration $\mathbf{a} = a_x\hat{\mathbf{i}} + a_y\hat{\mathbf{j}}$, where a_x and a_y are constants. For the case $v_i = 1.80 \times 10^7$ m/s, $a_x = 8.00 \times 10^{14}$ m/s^2 and $a_y = 1.60 \times 10^{15}$ m/s^2, determine at $x = d = 0.0100$ m (a) the position of the electron, (b) the velocity of the electron, (c) the speed of the electron, and (d) the direction of travel of the electron (i.e., the angle between its velocity and the x axis).

Section 4.3 Projectile Motion

Note: Ignore air resistance in all problems and take $g = 9.80$ m/s^2 at the Earth's surface.

10. To start an avalanche on a mountain slope, an artillery shell is fired with an initial velocity of 300 m/s at 55.0° above the horizontal. It explodes on the mountainside 42.0 s after firing. What are the x and y coordinates of the shell where it explodes, relative to its firing point?

11. In a local bar, a customer slides an empty beer mug down the counter for a refill. The bartender is momentarily distracted and does not see the mug, which slides off the counter and strikes the floor 1.40 m from the base of the counter. If the height of the counter is 0.860 m, (a) with what velocity did the mug leave the counter, and (b) what was the direction of the mug's velocity just before it hit the floor?

12. In a local bar, a customer slides an empty beer mug down the counter for a refill. The bartender is momentarily distracted and does not see the mug, which slides off the counter and strikes the floor at distance d from the base of the counter. The height of the counter is h. (a) With what velocity did the mug leave the counter, and (b) what was the direction of the mug's velocity just before it hit the floor?

13. One strategy in a snowball fight is to throw a snowball at a high angle over level ground. While your opponent is watching the first one, a second snowball is thrown at a low angle timed to arrive before or at the same time as the first one. Assume both snowballs are thrown with a speed of 25.0 m/s. The first one is thrown at an angle of 70.0° with respect to the horizontal. (a) At what angle should the second snowball be thrown to arrive at the same point as the first? (b) How many seconds later should the second snowball be thrown after the first to arrive at the same time?

14. An astronaut on a strange planet finds that she can jump a maximum horizontal distance of 15.0 m if her initial speed is 3.00 m/s. What is the free-fall acceleration on the planet?

15. A projectile is fired in such a way that its horizontal range is equal to three times its maximum height. What is the angle of projection?

16. A rock is thrown upward from the level ground in such a way that the maximum height of its flight is equal to its horizontal range d. (a) At what angle θ is the rock thrown? (b) **What If?** Would your answer to part (a) be different on a different planet? (c) What is the range d_{max} the rock can attain if it is launched at the same speed but at the optimal angle for maximum range?

17. A ball is tossed from an upper-story window of a building. The ball is given an initial velocity of 8.00 m/s at an angle of 20.0° below the horizontal. It strikes the ground 3.00 s later. (a) How far horizontally from the base of the building does the ball strike the ground? (b) Find the height from which the ball was thrown. (c) How long does it take the ball to reach a point 10.0 m below the level of launching?

18. The small archerfish (length 20 to 25 cm) lives in brackish waters of southeast Asia from India to the Philippines. This aptly named creature captures its prey by shooting a stream of water drops at an insect, either flying or at rest. The bug falls into the water and the fish gobbles it up. The archerfish has high accuracy at distances of 1.2 m to 1.5 m, and it sometimes makes hits at distances up to 3.5 m. A groove in the roof of its mouth, along with a curled tongue, forms a tube that enables the fish to impart high velocity to the water in its mouth when it suddenly closes its gill flaps. Suppose the archerfish shoots at a target

2.00 m away, at an angle of 30.0º above the horizontal. With what velocity must the water stream be launched if it is not to drop more than 3.00 cm vertically on its path to the target?

19. [www] A place-kicker must kick a football from a point 36.0 m (about 40 yards) from the goal, and half the crowd hopes the ball will clear the crossbar, which is 3.05 m high. When kicked, the ball leaves the ground with a speed of 20.0 m/s at an angle of 53.0° to the horizontal. (a) By how much does the ball clear or fall short of clearing the crossbar? (b) Does the ball approach the crossbar while still rising or while falling?

20. A firefighter, a distance d from a burning building, directs a stream of water from a fire hose at angle θ_i above the horizontal as in Figure P4.20. If the initial speed of the stream is v_i, at what height h does the water strike the building?

21. A playground is on the flat roof of a city school, 6.00 m above the street below. The vertical wall of the building is 7.00 m high, to form a meter-high railing around the playground. A ball has fallen to the street below, and a passerby returns it by launching it at an angle of 53.0° above the horizontal at a point 24.0 meters from the base of the building wall. The ball takes 2.20 s to reach a point vertically above the wall. (a) Find the speed at which the ball was launched. (b) Find the vertical distance by which the ball clears the wall. (c) Find the distance from the wall to the point on the roof where the ball lands.

22. A dive bomber has a velocity of 280 m/s at an angle θ below the horizontal. When the altitude of the aircraft is 2.15 km, it releases a bomb, which subsequently hits a target on the ground. The magnitude of the displacement from the point of release of the bomb to the target is 3.25 km. Find the angle θ.

23. A soccer player kicks a rock horizontally off a 40.0-m high cliff into a pool of water. If the player hears the sound of the splash 3.00 s later, what was the initial speed given to the rock? Assume the speed of sound in air to be 343 m/s.

24. A basketball star covers 2.80 m horizontally in a jump to dunk the ball (Fig. P4.24). His motion through space can be modeled precisely as that of a particle at his *center of mass,* which we will define in Chapter 9. His center of mass is at elevation 1.02 m when he leaves the floor. It reaches a maximum height of 1.85 m above the floor, and is at elevation 0.900 m when he touches down again. Determine (a) his time of

Figure P4.20

Figure P4.24

flight (his "hang time"), (b) his horizontal and (c) vertical velocity components at the instant of takeoff, and (d) his takeoff angle. (e) For comparison, determine the hang time of a whitetail deer making a jump with center-of-mass elevations $y_i = 1.20$ m, $y_{max} = 2.50$ m, $y_f = 0.700$ m.

25. An archer shoots an arrow with a velocity of 45.0 m/s at an angle of 50.0° with the horizontal. An assistant standing on the level ground 150 m downrange from the launch point throws an apple straight up with the minimum initial speed necessary to meet the path of the arrow. (a) What is the initial speed of the apple? (b) At what time after the arrow launch should the apple be thrown so that the arrow hits the apple?

26. A fireworks rocket explodes at height h, the peak of its vertical trajectory. It throws out burning fragments in all directions, but all at the same speed v. Pellets of solidified metal fall to the ground without air resistance. Find the smallest angle that the final velocity of an impacting fragment makes with the horizontal.

Section 4.4   Uniform Circular Motion

Note: Problems 8, 10, 12, and 16 in Chapter 6 can also be assigned with this section.

27. The athlete shown in Figure P4.27 rotates a 1.00-kg discus along a circular path of radius 1.06 m. The maximum speed of the discus is 20.0 m/s. Determine the magnitude of the maximum radial acceleration of the discus.

Sam Sargent/Liaison International

Figure P4.27

28. From information on the endsheets of this book, compute the radial acceleration of a point on the surface of the Earth at the equator, due to the rotation of the Earth about its axis.

29. A tire 0.500 m in radius rotates at a constant rate of 200 rev/min. Find the speed and acceleration of a small stone lodged in the tread of the tire (on its outer edge).

30. As their booster rockets separate, Space Shuttle astronauts typically feel accelerations up to 3g, where $g = 9.80$ m/s². In their training, astronauts ride in a device where they experience such an acceleration as a centripetal acceleration. Specifically, the astronaut is fastened securely at the end of a mechanical arm that then turns at constant speed in a horizontal circle. Determine the rotation rate, in revolutions per second, required to give an astronaut a centripetal acceleration of 3.00g while in circular motion with radius 9.45 m.

31. Young David who slew Goliath experimented with slings before tackling the giant. He found that he could revolve a sling of length 0.600 m at the rate of 8.00 rev/s. If he increased the length to 0.900 m, he could revolve the sling only 6.00 times per second. (a) Which rate of rotation gives the greater speed for the stone at the end of the sling? (b) What is the centripetal acceleration of the stone at 8.00 rev/s? (c) What is the centripetal acceleration at 6.00 rev/s?

32. The astronaut orbiting the Earth in Figure P4.32 is preparing to dock with a Westar VI satellite. The satellite is in a circular orbit 600 km above the Earth's surface, where the free-fall acceleration is 8.21 m/s². Take the radius of the Earth as 6 400 km. Determine the speed of the satellite and the time interval required to complete one orbit around the Earth.

Courtesy of NASA

Figure P4.32

Section 4.5   Tangential and Radial Acceleration

33. A train slows down as it rounds a sharp horizontal turn, slowing from 90.0 km/h to 50.0 km/h in the 15.0 s that it takes to round the bend. The radius of the curve is 150 m. Compute the acceleration at the moment the train speed reaches 50.0 km/h. Assume it continues to slow down at this time at the same rate.

34. An automobile whose speed is increasing at a rate of 0.600 m/s² travels along a circular road of radius 20.0 m. When the instantaneous speed of the automobile is 4.00 m/s, find (a) the tangential acceleration component, (b) the centripetal acceleration component, and (c) the magnitude and direction of the total acceleration.

35. Figure P4.35 represents the total acceleration of a particle moving clockwise in a circle of radius 2.50 m at a certain instant of time. At this instant, find (a) the radial acceleration, (b) the speed of the particle, and (c) its tangential acceleration.

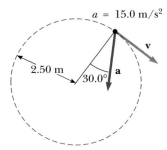

$a = 15.0 \text{ m/s}^2$

2.50 m

30.0°

v

a

Figure P4.35

36. A ball swings in a vertical circle at the end of a rope 1.50 m long. When the ball is 36.9° past the lowest point on its way up, its total acceleration is $(-22.5\hat{\mathbf{i}} + 20.2\hat{\mathbf{j}})$ m/s². At that instant, (a) sketch a vector diagram showing the components of its acceleration, (b) determine the magnitude of its radial acceleration, and (c) determine the speed and velocity of the ball.

37. A race car starts from rest on a circular track. The car increases its speed at a constant rate a_t as it goes once around the track. Find the angle that the total acceleration of the car makes—with the radius connecting the center of the track and the car—at the moment the car completes the circle.

Section 4.6 Relative Velocity and Relative Acceleration

38. Heather in her Corvette accelerates at the rate of $(3.00\hat{\mathbf{i}} - 2.00\hat{\mathbf{j}})$ m/s², while Jill in her Jaguar accelerates at $(1.00\hat{\mathbf{i}} + 3.00\hat{\mathbf{j}})$ m/s². They both start from rest at the origin of an xy coordinate system. After 5.00 s, (a) what is Heather's speed with respect to Jill, (b) how far apart are they, and (c) what is Heather's acceleration relative to Jill?

39. A car travels due east with a speed of 50.0 km/h. Raindrops are falling at a constant speed vertically with respect to the Earth. The traces of the rain on the side windows of the car make an angle of 60.0° with the vertical. Find the velocity of the rain with respect to (a) the car and (b) the Earth.

40. How long does it take an automobile traveling in the left lane at 60.0 km/h to pull alongside a car traveling in the same direction in the right lane at 40.0 km/h if the cars' front bumpers are initially 100 m apart?

41. A river has a steady speed of 0.500 m/s. A student swims upstream a distance of 1.00 km and swims back to the starting point. If the student can swim at a speed of 1.20 m/s in still water, how long does the trip take? Compare this with the time the trip would take if the water were still.

42. The pilot of an airplane notes that the compass indicates a heading due west. The airplane's speed relative to the air is 150 km/h. If there is a wind of 30.0 km/h toward the north, find the velocity of the airplane relative to the ground.

43. Two swimmers, Alan and Beth, start together at the same point on the bank of a wide stream that flows with a speed v. Both move at the same speed $c(c > v)$, relative to the water. Alan swims downstream a distance L and then upstream the same distance. Beth swims so that her motion relative to the Earth is perpendicular to the banks of the stream. She swims the distance L and then back the same distance, so that both swimmers return to the starting point. Which swimmer returns first? (*Note:* First guess the answer.)

44. A bolt drops from the ceiling of a train car that is accelerating northward at a rate of 2.50 m/s². What is the acceleration of the bolt relative to (a) the train car? (b) the Earth?

45. A science student is riding on a flatcar of a train traveling along a straight horizontal track at a constant speed of 10.0 m/s. The student throws a ball into the air along a path that he judges to make an initial angle of 60.0° with the horizontal and to be in line with the track. The student's professor, who is standing on the ground nearby, observes the ball to rise vertically. How high does she see the ball rise?

46. A Coast Guard cutter detects an unidentified ship at a distance of 20.0 km in the direction 15.0° east of north. The ship is traveling at 26.0 km/h on a course at 40.0° east of north. The Coast Guard wishes to send a speedboat to intercept the vessel and investigate it. If the speedboat travels 50.0 km/h, in what direction should it head? Express the direction as a compass bearing with respect to due north.

Additional Problems

47. *The "Vomit Comet."* In zero-gravity astronaut training and equipment testing, NASA flies a KC135A aircraft along a parabolic flight path. As shown in Figure P4.47, the aircraft climbs from 24 000 ft to 31 000 ft, where it enters the zero-g parabola with a velocity of 143 m/s nose-high at 45.0° and exits with velocity 143 m/s at 45.0° nose-low. During this portion of the flight the aircraft and objects inside its padded cabin are in free fall—they have gone ballistic. The aircraft then pulls out of the dive with an upward acceleration of 0.800g, moving in a vertical circle with radius 4.13 km. (During this portion of the flight, occupants of the plane perceive an acceleration of 1.8g.) What are the aircraft (a) speed and (b) altitude at the top of the maneuver? (c) What is the time spent in zero gravity? (d) What is the speed of the aircraft at the bottom of the flight path?

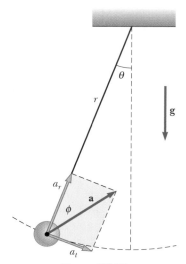

Figure P4.47

48. As some molten metal splashes, one droplet flies off to the east with initial velocity v_i at angle θ_i above the horizontal, and another droplet to the west with the same speed at the same angle above the horizontal, as in Figure P4.48. In terms of v_i and θ_i, find the distance between them as a function of time.

Figure P4.48

49. A ball on the end of a string is whirled around in a horizontal circle of radius 0.300 m. The plane of the circle is 1.20 m above the ground. The string breaks and the ball lands 2.00 m (horizontally) away from the point on the ground directly beneath the ball's location when the string breaks. Find the radial acceleration of the ball during its circular motion.

50. A projectile is fired up an incline (incline angle ϕ) with an initial speed v_i at an angle θ_i with respect to the horizontal ($\theta_i > \phi$), as shown in Figure P4.50. (a) Show that the projectile travels a distance d up the incline, where

$$d = \frac{2v_i^2 \cos\theta_i \sin(\theta_i - \phi)}{g \cos^2\phi}$$

(b) For what value of θ_i is d a maximum, and what is that maximum value?

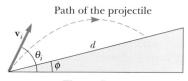

Figure P4.50

51. Barry Bonds hits a home run so that the baseball just clears the top row of bleachers, 21.0 m high, located 130 m from home plate. The ball is hit at an angle of 35.0° to the horizontal, and air resistance is negligible. Find (a) the initial speed of the ball, (b) the time at which the ball reaches the cheap seats, and (c) the velocity components and the speed of the ball when it passes over the top row. Assume the ball is hit at a height of 1.00 m above the ground.

52. An astronaut on the surface of the Moon fires a cannon to launch an experiment package, which leaves the barrel moving horizontally. (a) What must be the muzzle speed of the package so that it travels completely around the Moon and returns to its original location? (b) How long does this trip around the Moon take? Assume that the free-fall acceleration on the Moon is one-sixth that on the Earth.

53. A pendulum with a cord of length $r = 1.00$ m swings in a vertical plane (Fig. P4.53). When the pendulum is in the two horizontal positions $\theta = 90.0°$ and $\theta = 270°$, its speed is 5.00 m/s. (a) Find the magnitude of the radial acceleration and tangential acceleration for these positions. (b) Draw vector diagrams to determine the direc-

Figure P4.53

tion of the total acceleration for these two positions. (c) Calculate the magnitude and direction of the total acceleration.

54. A basketball player who is 2.00 m tall is standing on the floor 10.0 m from the basket, as in Figure P4.54. If he shoots the ball at a 40.0° angle with the horizontal, at what initial speed must he throw so that it goes through the hoop without striking the backboard? The basket height is 3.05 m.

Figure P4.54

55. When baseball players throw the ball in from the outfield, they usually allow it to take one bounce before it reaches the infield, on the theory that the ball arrives sooner that way. Suppose that the angle at which a bounced ball leaves the ground is the same as the angle at which the outfielder threw it, as in Figure P4.55, but that the ball's speed after the bounce is one half of what it was before the bounce. (a) Assuming the ball is always thrown with the same initial speed, at what angle θ should the fielder throw the ball to make it go the same distance D with one bounce (blue path) as a ball thrown upward at 45.0° with no bounce (green path)? (b) Determine the ratio of the times for the one-bounce and no-bounce throws.

Figure P4.55

56. A boy can throw a ball a maximum horizontal distance of R on a level field. How far can he throw the same ball vertically upward? Assume that his muscles give the ball the same speed in each case.

57. A stone at the end of a sling is whirled in a vertical circle of radius 1.20 m at a constant speed $v_0 = 1.50$ m/s as in Figure P4.57. The center of the sling is 1.50 m above the ground. What is the range of the stone if it is released when the sling is inclined at 30.0° with the horizontal (a) at Ⓐ ? (b) at Ⓑ ? What is the acceleration of the stone (c) just before it is released at Ⓐ ? (d) just after it is released at Ⓐ ?

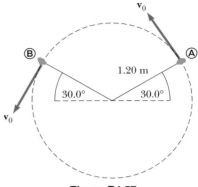

Figure P4.57

58. A quarterback throws a football straight toward a receiver with an initial speed of 20.0 m/s, at an angle of 30.0° above the horizontal. At that instant, the receiver is 20.0 m from the quarterback. In what direction and with what constant speed should the receiver run in order to catch the football at the level at which it was thrown?

59. Your grandfather is copilot of a bomber, flying horizontally over level terrain, with a speed of 275 m/s relative to the ground, at an altitude of 3 000 m. (a) The bombardier releases one bomb. How far will it travel horizontally between its release and its impact on the ground? Neglect the effects of air resistance. (b) Firing from the people on the ground suddenly incapacitates the bombardier before he can call, "Bombs away!" Consequently, the pilot maintains the plane's original course, altitude, and speed through a storm of flak. Where will the plane be when the bomb hits the ground? (c) The plane has a telescopic bomb sight set so that the bomb hits the target seen in the sight at the time of release. At what angle from the vertical was the bomb sight set?

60. A high-powered rifle fires a bullet with a muzzle speed of 1.00 km/s. The gun is pointed horizontally at a large bull's eye target—a set of concentric rings—200 m away. (a) How far below the extended axis of the rifle barrel does a bullet hit the target? The rifle is equipped with a telescopic sight. It is "sighted in" by adjusting the axis of the telescope so that it points precisely at the location where the bullet hits the target at 200 m. (b) Find the angle between the telescope axis and the rifle barrel axis. When shooting at a target at a distance other than 200 m, the marksman uses the telescopic sight, placing its crosshairs to "aim high" or "aim low" to compensate for the different range. Should she aim high or low, and approximately how far from the bull's eye, when the target is at a distance of (c) 50.0 m, (d) 150 m, or (e) 250 m? *Note:* The trajectory of the bullet is everywhere so nearly horizontal that it is a good approximation to model the bullet as fired horizontally in each case. **What if** the target is uphill or downhill? (f) Suppose the target is 200 m away, but the sight line to the target is above the horizontal by 30°. Should the marksman aim high, low, or right on? (g) Suppose the target is downhill by 30°. Should the marksman aim high, low, or right on? Explain your answers.

61. A hawk is flying horizontally at 10.0 m/s in a straight line, 200 m above the ground. A mouse it has been carrying struggles free from its grasp. The hawk continues on its path at the same speed for 2.00 seconds before attempting to retrieve its prey. To accomplish the retrieval, it dives in a straight line at constant speed and recaptures the mouse 3.00 m above the ground. (a) Assuming no air resistance, find the diving speed of the hawk. (b) What angle did the hawk make with the horizontal during its descent? (c) For how long did the mouse "enjoy" free fall?

62. A person standing at the top of a hemispherical rock of radius R kicks a ball (initially at rest on the top of the rock) to give it horizontal velocity $\mathbf{v}_i$ as in Figure P4.62. (a) What must be its minimum initial speed if the ball is never to hit the rock after it is kicked? (b) With this initial speed, how far from the base of the rock does the ball hit the ground?

Figure P4.62

63. A car is parked on a steep incline overlooking the ocean, where the incline makes an angle of 37.0° below the horizontal. The negligent driver leaves the car in neutral, and the parking brakes are defective. Starting from rest at $t = 0$, the car rolls down the incline with a constant acceleration of 4.00 m/s², traveling 50.0 m to the edge of a vertical cliff. The cliff is 30.0 m above the ocean. Find (a) the speed of the car when it reaches the edge of the cliff and the time at which it arrives there, (b) the velocity of the car when it lands in the ocean, (c) the total time interval that the car is in motion, and (d) the position of the car when it lands in the ocean, relative to the base of the cliff.

64. A truck loaded with cannonball watermelons stops suddenly to avoid running over the edge of a washed-out bridge (Fig. P4.64). The quick stop causes a number of melons to fly off the truck. One melon rolls over the edge with an initial speed $v_i = 10.0$ m/s in the horizontal direction. A cross-section of the bank has the shape of the bottom half of a parabola with its vertex at the edge of the road, and with the equation $y^2 = 16x$, where x and y are measured in meters. What are the x and y coordinates of the melon when it splatters on the bank?

65. The determined coyote is out once more in pursuit of the elusive roadrunner. The coyote wears a pair of Acme jet-powered roller skates, which provide a constant horizontal

Figure P4.64

acceleration of 15.0 m/s² (Fig. P4.65). The coyote starts at rest 70.0 m from the brink of a cliff at the instant the roadrunner zips past him in the direction of the cliff. (a) If the roadrunner moves with constant speed, determine the minimum speed he must have in order to reach the cliff before the coyote. At the edge of the cliff, the roadrunner escapes by making a sudden turn, while the coyote continues straight ahead. His skates remain horizontal and continue to operate while he is in flight, so that the coyote's acceleration while in the air is $(15.0\hat{\mathbf{i}} - 9.80\hat{\mathbf{j}})$ m/s². (b) If the cliff is 100 m above the flat floor of a canyon, determine where the coyote lands in the canyon. (c) Determine the components of the coyote's impact velocity.

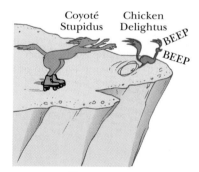

Figure P4.65

66. Do not hurt yourself; do not strike your hand against anything. Within these limitations, describe what you do to give your hand a large acceleration. Compute an order-of-magnitude estimate of this acceleration, stating the quantities you measure or estimate and their values.

67. A skier leaves the ramp of a ski jump with a velocity of 10.0 m/s, 15.0° above the horizontal, as in Figure P4.67. The slope is inclined at 50.0°, and air resistance is negligible. Find (a) the distance from the ramp to where the jumper lands and (b) the velocity components just before the landing. (How do you think the results might be affected if air resistance were included? Note that jumpers lean forward in the shape of an airfoil, with their hands at their sides, to increase their distance. Why does this work?)

Figure P4.67

68. In a television picture tube (a cathode ray tube) electrons are emitted with velocity $\mathbf{v}_i$ from a source at the origin of coordinates. The initial velocities of different electrons make different angles θ with the x axis. As they move a distance D along the x axis, the electrons are acted on by a constant electric field, giving each a constant acceleration $\mathbf{a}$ in the x direction. At $x = D$ the electrons pass through a circular aperture, oriented perpendicular to the x axis. At the aperture, the velocity imparted to the electrons by the electric field is much larger than $\mathbf{v}_i$ in magnitude. Show that velocities of the electrons going through the aperture radiate from a certain point on the x axis, which is not the origin. Determine the location of this point. This point is called a *virtual source*, and it is important in determining where the electron beam hits the screen of the tube.

69. A fisherman sets out upstream from Metaline Falls on the Pend Oreille River in northwestern Washington State. His small boat, powered by an outboard motor, travels at a constant speed v in still water. The water flows at a lower constant speed v_w. He has traveled upstream for 2.00 km when his ice chest falls out of the boat. He notices that the chest is missing only after he has gone upstream for another 15.0 minutes. At that point he turns around and heads back downstream, all the time traveling at the same speed relative to the water. He catches up with the floating ice chest just as it is about to go over the falls at his starting point. How fast is the river flowing? Solve this problem in two ways. (a) First, use the Earth as a reference frame. With respect to the Earth, the boat travels upstream at speed $v - v_w$ and downstream at $v + v_w$. (b) A second much simpler and more elegant solution is obtained by using the water as the reference frame. This approach has important applications in many more complicated problems; examples are calculating the motion of rockets and satellites and analyzing the scattering of subatomic particles from massive targets.

70. The water in a river flows uniformly at a constant speed of 2.50 m/s between parallel banks 80.0 m apart. You are to deliver a package directly across the river, but you can swim only at 1.50 m/s. (a) If you choose to minimize the time you spend in the water, in what direction should you head? (b) How far downstream will you be carried? (c) **What If?** If you choose to minimize the distance downstream that the river carries you, in what direction should you head? (d) How far downstream will you be carried?

71. An enemy ship is on the east side of a mountain island, as shown in Figure P4.71. The enemy ship has maneuvered to within 2 500 m of the 1 800-m-high mountain peak and can shoot projectiles with an initial speed of 250 m/s. If the western shoreline is horizontally 300 m from the peak, what are the distances from the western shore at which a ship can be safe from the bombardment of the enemy ship?

72. In the **What If?** section of Example 4.7, it was claimed that the maximum range of a ski-jumper occurs for a launch angle θ given by

$$\theta = 45° - \frac{\phi}{2}$$

where ϕ is the angle that the hill makes with the horizontal in Figure 4.16. Prove this claim by deriving the equation above.

Answers to Quick Quizzes

4.1 (b). An object moving with constant velocity has $\Delta\mathbf{v} = 0$, so, according to the definition of acceleration, $\mathbf{a} = \Delta\mathbf{v}/\Delta t = 0$. Choice (a) is not correct because a particle can move at a constant speed and change direction. This possibility also makes (c) an incorrect choice.

4.2 (a). Because acceleration occurs whenever the velocity changes in any way—with an increase or decrease in

Figure P4.71

speed, a change in direction, or both—all three controls are accelerators. The gas pedal causes the car to speed up; the brake pedal causes the car to slow down. The steering wheel changes the direction of the velocity vector.

4.3 (a). You should simply throw it straight up in the air. Because the ball is moving along with you, it will follow a parabolic trajectory with a horizontal component of velocity that is the same as yours.

4.4 (b). At only one point—the peak of the trajectory—are the velocity and acceleration vectors perpendicular to each other. The velocity vector is horizontal at that point and the acceleration vector is downward.

4.5 (a). The acceleration vector is always directed downward. The velocity vector is never vertical if the object follows a path such as that in Figure 4.8.

4.6 15°, 30°, 45°, 60°, 75°. The greater the maximum height, the longer it takes the projectile to reach that altitude and then fall back down from it. So, as the launch angle increases, the time of flight increases.

4.7 (c). We cannot choose (a) or (b) because the centripetal acceleration vector is not constant—it continuously changes in direction. Of the remaining choices, only (c) gives the correct perpendicular relationship between $\mathbf{a}_c$ and $\mathbf{v}$.

4.8 (d). Because the centripetal acceleration is proportional to the square of the speed, doubling the speed increases the acceleration by a factor of 4.

4.9 (b). The velocity vector is tangent to the path. If the acceleration vector is to be parallel to the velocity vector, it must also be tangent to the path. This requires that the acceleration vector have no component perpendicular to the path. If the path were to change direction, the acceleration vector would have a radial component, perpendicular to the path. Thus, the path must remain straight.

4.10 (d). The velocity vector is tangent to the path. If the acceleration vector is to be perpendicular to the velocity vector, it must have no component tangent to the path. On the other hand, if the speed is changing, there *must* be a component of the acceleration tangent to the path. Thus, the velocity and acceleration vectors are never perpendicular in this situation. They can only be perpendicular if there is no change in the speed.

4.11 (c). Passenger A sees the coffee pouring in a "normal" parabolic path, just as if she were standing on the ground pouring it. The stationary observer B sees the coffee moving in a parabolic path that is extended horizontally due to the constant horizontal velocity of 60 mi/h.

The Laws of Motion

▲ *A small tugboat exerts a force on a large ship, causing it to move. How can such a small boat move such a large object? (Steve Raymer/CORBIS)*

In Chapters 2 and 4, we described motion in terms of position, velocity, and acceleration without considering what might cause that motion. Now we consider the cause—what might cause one object to remain at rest and another object to accelerate? The two main factors we need to consider are the forces acting on an object and the mass of the object. We discuss the three basic laws of motion, which deal with forces and masses and were formulated more than three centuries ago by Isaac Newton. Once we understand these laws, we can answer such questions as "What mechanism changes motion?" and "Why do some objects accelerate more than others?"

5.1 The Concept of Force

Everyone has a basic understanding of the concept of force from everyday experience. When you push your empty dinner plate away, you exert a force on it. Similarly, you exert a force on a ball when you throw or kick it. In these examples, the word *force* is associated with muscular activity and some change in the velocity of an object. Forces do not always cause motion, however. For example, as you sit reading this book, a gravitational force acts on your body and yet you remain stationary. As a second example, you can push (in other words, exert a force) on a large boulder and not be able to move it.

What force (if any) causes the Moon to orbit the Earth? Newton answered this and related questions by stating that forces are what cause any change in the velocity of an object. The Moon's velocity is not constant because it moves in a nearly circular orbit around the Earth. We now know that this change in velocity is caused by the gravitational force exerted by the Earth on the Moon. Because only a force can cause a change in velocity, we can think of force as *that which causes an object to accelerate.* In this chapter, we are concerned with the relationship between the force exerted on an object and the acceleration of that object.

What happens when several forces act simultaneously on an object? In this case, the object accelerates only if the net force acting on it is not equal to zero. The **net force** acting on an object is defined as the vector sum of all forces acting on the object. (We sometimes refer to the net force as the *total force*, the *resultant force*, or the *unbalanced force*.) **If the net force exerted on an object is zero, the acceleration of the object is zero and its velocity remains constant.** That is, if the net force acting on the object is zero, the object either remains at rest or continues to move with constant velocity. When the velocity of an object is constant (including when the object is at rest), the object is said to be in **equilibrium.**

When a coiled spring is pulled, as in Figure 5.1a, the spring stretches. When a stationary cart is pulled sufficiently hard that friction is overcome, as in Figure 5.1b, the cart moves. When a football is kicked, as in Figure 5.1c, it is both deformed and set in motion. These situations are all examples of a class of forces called *contact forces.* That is, they involve physical contact between two objects. Other examples of contact forces are the force exerted by gas molecules on the walls of a container and the force exerted by your feet on the floor.

An object accelerates due to an external force

Definition of equilibrium

112

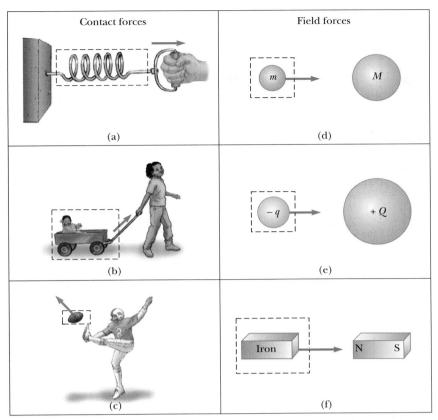

Contact forces	Field forces
(a)	(d)
(b)	(e)
(c)	(f)

Figure 5.1 Some examples of applied forces. In each case a force is exerted on the object within the boxed area. Some agent in the environment external to the boxed area exerts a force on the object.

Another class of forces, known as *field forces*, do not involve physical contact between two objects but instead act through empty space. The gravitational force of attraction between two objects, illustrated in Figure 5.1d, is an example of this class of force. This gravitational force keeps objects bound to the Earth and the planets in orbit around the Sun. Another common example of a field force is the electric force that one electric charge exerts on another (Fig. 5.1e). These charges might be those of the electron and proton that form a hydrogen atom. A third example of a field force is the force a bar magnet exerts on a piece of iron (Fig. 5.1f).

The distinction between contact forces and field forces is not as sharp as you may have been led to believe by the previous discussion. When examined at the atomic level, all the forces we classify as contact forces turn out to be caused by electric (field) forces of the type illustrated in Figure 5.1e. Nevertheless, in developing models for macroscopic phenomena, it is convenient to use both classifications of forces. The only known *fundamental* forces in nature are all field forces: (1) *gravitational forces* between objects, (2) *electromagnetic forces* between electric charges, (3) *nuclear forces* between subatomic particles, and (4) *weak forces* that arise in certain radioactive decay processes. In classical physics, we are concerned only with gravitational and electromagnetic forces.

Measuring the Strength of a Force

It is convenient to use the deformation of a spring to measure force. Suppose we apply a vertical force to a spring scale that has a fixed upper end, as shown in Figure 5.2a. The spring elongates when the force is applied, and a pointer on the scale reads the value of the applied force. We can calibrate the spring by defining a reference force $\mathbf{F}_1$ as the force that produces a pointer reading of 1.00 cm. (Because force is a vector

Isaac Newton,

English physicist and mathematician (1642–1727)

Isaac Newton was one of the most brilliant scientists in history. Before the age of 30, he formulated the basic concepts and laws of mechanics, discovered the law of universal gravitation, and invented the mathematical methods of calculus. As a consequence of his theories, Newton was able to explain the motions of the planets, the ebb and flow of the tides, and many special features of the motions of the Moon and the Earth. He also interpreted many fundamental observations concerning the nature of light. His contributions to physical theories dominated scientific thought for two centuries and remain important today. *(Giraudon/Art Resource)*

(a)   (b)   (c)   (d)

Figure 5.2 The vector nature of a force is tested with a spring scale. (a) A downward force $\mathbf{F}_1$ elongates the spring 1.00 cm. (b) A downward force $\mathbf{F}_2$ elongates the spring 2.00 cm. (c) When $\mathbf{F}_1$ and $\mathbf{F}_2$ are applied simultaneously, the spring elongates by 3.00 cm. (d) When $\mathbf{F}_1$ is downward and $\mathbf{F}_2$ is horizontal, the combination of the two forces elongates the spring $\sqrt{(1.00 \text{ cm})^2 + (2.00 \text{ cm})^2} = 2.24$ cm.

quantity, we use the bold-faced symbol $\mathbf{F}$.) If we now apply a different downward force $\mathbf{F}_2$ whose magnitude is twice that of the reference force $\mathbf{F}_1$, as seen in Figure 5.2b, the pointer moves to 2.00 cm. Figure 5.2c shows that the combined effect of the two collinear forces is the sum of the effects of the individual forces.

Now suppose the two forces are applied simultaneously with $\mathbf{F}_1$ downward and $\mathbf{F}_2$ horizontal, as illustrated in Figure 5.2d. In this case, the pointer reads $\sqrt{5.00 \text{ cm}^2} = 2.24$ cm. The single force $\mathbf{F}$ that would produce this same reading is the sum of the two vectors $\mathbf{F}_1$ and $\mathbf{F}_2$, as described in Figure 5.2d. That is, $|\mathbf{F}| = \sqrt{F_1{}^2 + F_2{}^2} = 2.24$ units, and its direction is $\theta = \tan^{-1}(-0.500) = -26.6°$. **Because forces have been experimentally verified to behave as vectors, you must use the rules of vector addition to obtain the net force on an object.**

5.2 Newton's First Law and Inertial Frames

We begin our study of forces by imagining some situations. Imagine placing a puck on a perfectly level air hockey table (Fig. 5.3). You expect that it will remain where it is placed. Now imagine your air hockey table is located on a train moving with constant velocity. If the puck is placed on the table, the puck again remains where it is placed. If the train were to accelerate, however, the puck would start moving along the table, just as a set of papers on your dashboard falls onto the front seat of your car when you step on the gas.

As we saw in Section 4.6, a moving object can be observed from any number of reference frames. **Newton's first law of motion,** sometimes called the *law of inertia,* defines a special set of reference frames called *inertial frames.* This law can be stated as follows:

Figure 5.3 On an air hockey table, air blown through holes in the surface allow the puck to move almost without friction. If the table is not accelerating, a puck placed on the table will remain at rest.

Newton's first law

If an object does not interact with other objects, it is possible to identify a reference frame in which the object has zero acceleration.

Such a reference frame is called an **inertial frame of reference**. When the puck is on the air hockey table located on the ground, you are observing it from an inertial reference frame—there are no horizontal interactions of the puck with any other objects and you observe it to have zero acceleration in that direction. When you are on the train moving at constant velocity, you are also observing the puck from an inertial reference frame. **Any reference frame that moves with constant velocity relative to an inertial frame is itself an inertial frame**. When the train accelerates, however, you are observing the puck from a **noninertial reference frame** because you and the train are accelerating relative to the inertial reference frame of the surface of the Earth. While the puck appears to be accelerating according to your observations, we can identify a reference frame in which the puck has zero acceleration. For example, an observer standing outside the train on the ground sees the puck moving with the same velocity as the train had before it started to accelerate (because there is almost no friction to "tie" the puck and the train together). Thus, Newton's first law is still satisfied even though your observations say otherwise.

A reference frame that moves with constant velocity relative to the distant stars is the best approximation of an inertial frame, and for our purposes we can consider the Earth as being such a frame. The Earth is not really an inertial frame because of its orbital motion around the Sun and its rotational motion about its own axis, both of which result in centripetal accelerations. However, these accelerations are small compared with g and can often be neglected. For this reason, we assume that the Earth is an inertial frame, as is any other frame attached to it.

Let us assume that we are observing an object from an inertial reference frame. (We will return to observations made in noninertial reference frames in Section 6.3.) Before about 1600, scientists believed that the natural state of matter was the state of rest. Observations showed that moving objects eventually stopped moving. Galileo was the first to take a different approach to motion and the natural state of matter. He devised thought experiments and concluded that it is not the nature of an object to stop once set in motion: rather, it is its nature to *resist changes in its motion*. In his words, "Any velocity once imparted to a moving body will be rigidly maintained as long as the external causes of retardation are removed." For example, a spacecraft drifting through empty space with its engine turned off will keep moving forever—it would *not* seek a "natural state" of rest.

Given our assumption of observations made from inertial reference frames, we can pose a more practical statement of Newton's first law of motion:

> In the absence of external forces, when viewed from an inertial reference frame, an object at rest remains at rest and an object in motion continues in motion with a constant velocity (that is, with a constant speed in a straight line).

In simpler terms, we can say that **when no force acts on an object, the acceleration of the object is zero.** If nothing acts to change the object's motion, then its velocity does not change. From the first law, we conclude that any *isolated object* (one that does not interact with its environment) is either at rest or moving with constant velocity. The tendency of an object to resist any attempt to change its velocity is called **inertia.**

Inertial frame of reference

 PITFALL PREVENTION

5.1 Newton's First Law

Newton's first law does *not* say what happens for an object with *zero net force,* that is, multiple forces that cancel; it says what happens *in the absence of a force.* This is a subtle but important difference that allows us to define force as that which causes a change in the motion. The description of an object under the effect of forces that balance is covered by Newton's second law.

Another statement of Newton's first law

Quick Quiz 5.1 Which of the following statements is most correct? (a) It is possible for an object to have motion in the absence of forces on the object. (b) It is possible to have forces on an object in the absence of motion of the object. (c) Neither (a) nor (b) is correct. (d) Both (a) and (b) are correct.

5.3 Mass

Imagine playing catch with either a basketball or a bowling ball. Which ball is more likely to keep moving when you try to catch it? Which ball has the greater tendency to remain motionless when you try to throw it? The bowling ball is more resistant to changes in its velocity than the basketball—how can we quantify this concept?

Definition of mass

Mass is that property of an object that specifies how much resistance an object exhibits to changes in its velocity, and as we learned in Section 1.1, the SI unit of mass is the kilogram. The greater the mass of an object, the less that object accelerates under the action of a given applied force.

To describe mass quantitatively, we begin by experimentally comparing the accelerations a given force produces on different objects. Suppose a force acting on an object of mass m_1 produces an acceleration $\mathbf{a}_1$, and the *same force* acting on an object of mass m_2 produces an acceleration $\mathbf{a}_2$. The ratio of the two masses is defined as the *inverse* ratio of the magnitudes of the accelerations produced by the force:

$$\frac{m_1}{m_2} \equiv \frac{a_2}{a_1} \tag{5.1}$$

For example, if a given force acting on a 3-kg object produces an acceleration of 4 m/s², the same force applied to a 6-kg object produces an acceleration of 2 m/s². If one object has a known mass, the mass of the other object can be obtained from acceleration measurements.

Mass is an inherent property of an object and is independent of the object's surroundings and of the method used to measure it. Also, **mass is a scalar quantity** and thus obeys the rules of ordinary arithmetic. That is, several masses can be combined in simple numerical fashion. For example, if you combine a 3-kg mass with a 5-kg mass, the total mass is 8 kg. We can verify this result experimentally by comparing the accelerations that a known force gives to several objects separately with the acceleration that the same force gives to the same objects combined as one unit.

Mass and weight are different quantities

Mass should not be confused with weight. **Mass and weight are two different quantities.** The weight of an object is equal to the magnitude of the gravitational force exerted on the object and varies with location (see Section 5.5). For example, a person who weighs 180 lb on the Earth weighs only about 30 lb on the Moon. On the other hand, the mass of an object is the same everywhere: an object having a mass of 2 kg on the Earth also has a mass of 2 kg on the Moon.

5.4 Newton's Second Law

Newton's first law explains what happens to an object when no forces act on it. It either remains at rest or moves in a straight line with constant speed. Newton's second law answers the question of what happens to an object that has a nonzero resultant force acting on it.

Imagine performing an experiment in which you push a block of ice across a frictionless horizontal surface. When you exert some horizontal force **F** on the block, it moves with some acceleration **a**. If you apply a force twice as great, you find that the acceleration of the block doubles. If you increase the applied force to 3**F**, the acceleration triples, and so on. From such observations, we conclude that **the acceleration of an object is directly proportional to the force acting on it.**

The acceleration of an object also depends on its mass, as stated in the preceding section. We can understand this by considering the following experiment. If you apply a force **F** to a block of ice on a frictionless surface, the block undergoes some acceleration **a**. If the mass of the block is doubled, the same applied force produces an acceleration **a**/2. If the mass is tripled, the same applied force produces an acceleration **a**/3, and so on. According to this observation, we conclude that **the magnitude of the acceleration of an object is inversely proportional to its mass.**

These observations are summarized in **Newton's second law:**

> When viewed from an inertial reference frame, the acceleration of an object is directly proportional to the net force acting on it and inversely proportional to its mass.

Thus, we can relate mass, acceleration, and force through the following mathematical statement of Newton's second law:[1]

$$\sum \mathbf{F} = m\mathbf{a} \qquad (5.2)$$

In both the textual and mathematical statements of Newton's second law above, we have indicated that the acceleration is due to the *net force* $\sum \mathbf{F}$ acting on an object. The **net force** on an object is the vector sum of all forces acting on the object. In solving a problem using Newton's second law, it is imperative to determine the correct net force on an object. There may be many forces acting on an object, but there is only one acceleration.

Note that Equation 5.2 is a vector expression and hence is equivalent to three component equations:

$$\sum F_x = ma_x \qquad \sum F_y = ma_y \qquad \sum F_z = ma_z \qquad (5.3)$$

Quick Quiz 5.2 An object experiences no acceleration. Which of the following *cannot* be true for the object? (a) A single force acts on the object. (b) No forces act on the object. (c) Forces act on the object, but the forces cancel.

Quick Quiz 5.3 An object experiences a net force and exhibits an acceleration in response. Which of the following statements is *always* true? (a) The object moves in the direction of the force. (b) The acceleration is in the same direction as the velocity. (c) The acceleration is in the same direction as the force. (d) The velocity of the object increases.

Quick Quiz 5.4 You push an object, initially at rest, across a frictionless floor with a constant force for a time interval Δt, resulting in a final speed of v for the object. You repeat the experiment, but with a force that is twice as large. What time interval is now required to reach the same final speed v? (a) $4\Delta t$ (b) $2\Delta t$ (c) Δt (d) $\Delta t/2$ (e) $\Delta t/4$.

Unit of Force

The SI unit of force is the **newton**, which is defined as the force that, when acting on an object of mass 1 kg, produces an acceleration of 1 m/s². From this definition and Newton's second law, we see that the newton can be expressed in terms of the following fundamental units of mass, length, and time:

$$1\,\text{N} \equiv 1\,\text{kg}\cdot\text{m/s}^2 \qquad (5.4)$$

---

[1] Equation 5.2 is valid only when the speed of the object is much less than the speed of light. We treat the relativistic situation in Chapter 39.

 PITFALL PREVENTION

5.2 Force is the Cause of Changes in Motion

Force *does not* cause motion. We can have motion in the absence of forces, as described in Newton's first law. Force is the cause of *changes* in motion, as measured by acceleration.

Newton's second law

Newton's second law—component form

 PITFALL PREVENTION

5.3 *m***a** is Not a Force

Equation 5.2 does *not* say that the product $m\mathbf{a}$ is a force. All forces on an object are added vectorially to generate the net force on the left side of the equation. This net force is then equated to the product of the mass of the object and the acceleration that results from the net force. Do *not* include an "$m\mathbf{a}$ force" in your analysis of the forces on an object.

Definition of the newton

Table 5.1

Units of Mass, Acceleration, and Force[a]			
System of Units	**Mass**	**Acceleration**	**Force**
SI	kg	m/s^2	$N = kg \cdot m/s^2$
U.S. customary	slug	ft/s^2	$lb = slug \cdot ft/s^2$

[a] $1 N = 0.225 \, lb$.

In the U.S. customary system, the unit of force is the **pound**, which is defined as the force that, when acting on a 1-slug mass,[2] produces an acceleration of $1 \, ft/s^2$:

$$1 \, lb \equiv 1 \, slug \cdot ft/s^2 \qquad (5.5)$$

A convenient approximation is that $1 \, N \approx \frac{1}{4} \, lb$.

The units of mass, acceleration, and force are summarized in Table 5.1.

Example 5.1   An Accelerating Hockey Puck

A hockey puck having a mass of 0.30 kg slides on the horizontal, frictionless surface of an ice rink. Two hockey sticks strike the puck simultaneously, exerting the forces on the puck shown in Figure 5.4. The force $\mathbf{F}_1$ has a magnitude of 5.0 N, and the force $\mathbf{F}_2$ has a magnitude of 8.0 N. Determine both the magnitude and the direction of the puck's acceleration.

Solution *Conceptualize* this problem by studying Figure 5.4. Because we can determine a net force and we want an acceleration, we *categorize* this problem as one that may be solved using Newton's second law. To *analyze* the problem, we resolve the force vectors into components. The net force acting on the puck in the *x* direction is

$$\sum F_x = F_{1x} + F_{2x} = F_1 \cos(-20°) + F_2 \cos 60°$$
$$= (5.0 \, N)(0.940) + (8.0 \, N)(0.500) = 8.7 \, N$$

The net force acting on the puck in the *y* direction is

$$\sum F_y = F_{1y} + F_{2y} = F_1 \sin(-20°) + F_2 \sin 60°$$
$$= (5.0 \, N)(-0.342) + (8.0 \, N)(0.866) = 5.2 \, N$$

Now we use Newton's second law in component form to find the *x* and *y* components of the puck's acceleration:

$$a_x = \frac{\sum F_x}{m} = \frac{8.7 \, N}{0.30 \, kg} = 29 \, m/s^2$$

$$a_y = \frac{\sum F_y}{m} = \frac{5.2 \, N}{0.30 \, kg} = 17 \, m/s^2$$

The acceleration has a magnitude of

$$a = \sqrt{(29)^2 + (17)^2} \, m/s^2 = \boxed{34 \, m/s^2}$$

and its direction relative to the positive *x* axis is

$$\theta = \tan^{-1}\left(\frac{a_y}{a_x}\right) = \tan^{-1}\left(\frac{17}{29}\right) = \boxed{30°}$$

To *finalize* the problem, we can graphically add the vectors in Figure 5.4 to check the reasonableness of our answer. Because the acceleration vector is along the direction of the resultant force, a drawing showing the resultant force helps us check the validity of the answer. (Try it!)

What If? Suppose three hockey sticks strike the puck simultaneously, with two of them exerting the forces shown in Figure 5.4. The result of the three forces is that the hockey puck shows *no* acceleration. What must be the components of the third force?

Answer If there is zero acceleration, the net force acting on the puck must be zero. Thus, the three forces must cancel. We have found the components of the combination of the first two forces. The components of the third force must be of equal magnitude and opposite sign in order that all of the components add to zero. Thus, $F_{3x} = -8.7 \, N$, $F_{3y} = -5.2 \, N$.

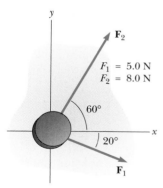

$F_1 = 5.0 \, N$
$F_2 = 8.0 \, N$

Figure 5.4 (Example 5.1) A hockey puck moving on a frictionless surface accelerates in the direction of the resultant force $\mathbf{F}_1 + \mathbf{F}_2$.

---

[2]   The *slug* is the unit of mass in the U.S. customary system and is that system's counterpart of the SI unit the *kilogram*. Because most of the calculations in our study of classical mechanics are in SI units, the slug is seldom used in this text.

5.5   The Gravitational Force and Weight

We are well aware that all objects are attracted to the Earth. The attractive force exerted by the Earth on an object is called the **gravitational force** $\mathbf{F}_g$. This force is directed toward the center of the Earth,[3] and its magnitude is called the **weight** of the object.

We saw in Section 2.6 that a freely falling object experiences an acceleration $\mathbf{g}$ acting toward the center of the Earth. Applying Newton's second law $\Sigma \mathbf{F} = m\mathbf{a}$ to a freely falling object of mass m, with $\mathbf{a} = \mathbf{g}$ and $\Sigma \mathbf{F} = \mathbf{F}_g$, we obtain

$$\mathbf{F}_g = m\mathbf{g} \tag{5.6}$$

Thus, the weight of an object, being defined as the magnitude of $\mathbf{F}_g$, is equal to mg.

Because it depends on g, weight varies with geographic location. Because g decreases with increasing distance from the center of the Earth, objects weigh less at higher altitudes than at sea level. For example, a 1 000-kg palette of bricks used in the construction of the Empire State Building in New York City weighed 9 800 N at street level, but weighed about 1 N less by the time it was lifted from sidewalk level to the top of the building. As another example, suppose a student has a mass of 70.0 kg. The student's weight in a location where $g = 9.80$ m/s^2 is $F_g = mg = 686$ N (about 150 lb). At the top of a mountain, however, where $g = 9.77$ m/s^2, the student's weight is only 684 N. Therefore, if you want to lose weight without going on a diet, climb a mountain or weigh yourself at 30 000 ft during an airplane flight!

Because weight is proportional to mass, we can compare the masses of two objects by measuring their weights on a spring scale. At a given location (at which two objects are subject to the same value of g), the ratio of the weights of two objects equals the ratio of their masses.

Equation 5.6 quantifies the gravitational force on the object, but notice that this equation does not require the object to be moving. Even for a stationary object, or an object on which several forces act, Equation 5.6 can be used to calculate the magnitude of the gravitational force. This results in a subtle shift in the interpretation of m in the equation. The mass m in Equation 5.6 is playing the role of determining the strength of the gravitational attraction between the object and the Earth. This is a completely different role from that previously described for mass, that of measuring the resistance to changes in motion in response to an external force. Thus, we call m in this type of equation the **gravitational mass.** Despite this quantity being different in behavior from inertial mass, it is one of the experimental conclusions in Newtonian dynamics that gravitational mass and inertial mass have the same value.

> **Quick Quiz 5.5** A baseball of mass m is thrown upward with some initial speed. A gravitational force is exerted on the ball (a) at all points in its motion (b) at all points in its motion except at the highest point (c) at no points in its motion.

> **Quick Quiz 5.6** Suppose you are talking by interplanetary telephone to your friend, who lives on the Moon. He tells you that he has just won a newton of gold in a contest. Excitedly, you tell him that you entered the Earth version of the same contest and also won a newton of gold! Who is richer? (a) You (b) Your friend (c) You are equally rich.

[3]   This statement ignores the fact that the mass distribution of the Earth is not perfectly spherical.

▲ **PITFALL PREVENTION**

5.4   "Weight of an Object"

We are familiar with the everyday phrase, the "weight of an object." However, weight is not an inherent property of an object, but rather a measure of the gravitational force between the object and the Earth. Thus, weight is a property of a *system* of items—the object and the Earth.

▲ **PITFALL PREVENTION**

5.5   Kilogram is Not a Unit of Weight

You may have seen the "conversion" 1 kg = 2.2 lb. Despite popular statements of weights expressed in kilograms, the kilogram is not a unit of *weight*, it is a unit of *mass*. The conversion statement is not an equality; it is an *equivalence* that is only valid on the surface of the Earth.

Courtesy of NASA

The life-support unit strapped to the back of astronaut Edwin Aldrin weighed 300 lb on the Earth. During his training, a 50-lb mock-up was used. Although this effectively simulated the reduced weight the unit would have on the Moon, it did not correctly mimic the unchanging mass. It was just as difficult to accelerate the unit (perhaps by jumping or twisting suddenly) on the Moon as on the Earth.

Conceptual Example 5.2   How Much Do You Weigh in an Elevator?

You have most likely had the experience of standing in an elevator that accelerates upward as it moves toward a higher floor. In this case, you feel heavier. In fact, if you are standing on a bathroom scale at the time, the scale measures a force having a magnitude that is greater than your weight. Thus, you have tactile and measured evidence that leads you to believe you are heavier in this situation. *Are* you heavier?

Solution No, your weight is unchanged. To provide the acceleration upward, the floor or scale must exert on your feet an upward force that is greater in magnitude than your weight. It is this greater force that you feel, which you interpret as feeling heavier. The scale reads this upward force, not your weight, and so its reading increases.

5.6   Newton's Third Law

If you press against a corner of this textbook with your fingertip, the book pushes back and makes a small dent in your skin. If you push harder, the book does the same and the dent in your skin is a little larger. This simple experiment illustrates a general principle of critical importance known as **Newton's third law:**

Newton's third law

> If two objects interact, the force $\mathbf{F}_{12}$ exerted by object 1 on object 2 is equal in magnitude and opposite in direction to the force $\mathbf{F}_{21}$ exerted by object 2 on object 1:

$$\mathbf{F}_{12} = -\mathbf{F}_{21} \tag{5.7}$$

When it is important to designate forces as interactions between two objects, we will use this subscript notation, where $\mathbf{F}_{ab}$ means "the force exerted *by* a *on* b." The third law, which is illustrated in Figure 5.5a, is equivalent to stating that **forces always occur in pairs,** or that **a single isolated force cannot exist.** The force that object 1 exerts on object 2 may be called the *action force* and the force of object 2 on object 1 the *reaction force.* In reality, either force can be labeled the action or reaction force. **The action force is equal in magnitude to the reaction force and opposite in direction. In all cases, the action and reaction forces act on different objects and must be of the same type.** For example, the force acting on a freely falling projectile is the gravitational force exerted by the Earth on the projectile $\mathbf{F}_g = \mathbf{F}_{Ep}$ (E = Earth, p = projectile), and the magnitude of this force is mg. The reaction to this force is the gravitational force exerted by the projectile on the Earth $\mathbf{F}_{pE} = -\mathbf{F}_{Ep}$. The reaction force $\mathbf{F}_{pE}$ must accelerate the Earth toward the projectile just as the action force $\mathbf{F}_{Ep}$ accelerates the projectile toward the Earth. However, because the Earth has such a large mass, its acceleration due to this reaction force is negligibly small.

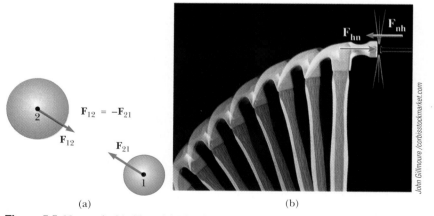

(a)   (b)

Figure 5.5 Newton's third law. (a) The force $\mathbf{F}_{12}$ exerted by object 1 on object 2 is equal in magnitude and opposite in direction to the force $\mathbf{F}_{21}$ exerted by object 2 on object 1. (b) The force $\mathbf{F}_{hn}$ exerted by the hammer on the nail is equal in magnitude and opposite to the force $\mathbf{F}_{nh}$ exerted by the nail on the hammer.

(a)   (b)

Figure 5.6 (a) When a computer monitor is at rest on a table, the forces acting on the monitor are the normal force **n** and the gravitational force $\mathbf{F}_g$. The reaction to **n** is the force $\mathbf{F}_{mt}$ exerted by the monitor on the table. The reaction to $\mathbf{F}_g$ is the force $\mathbf{F}_{mE}$ exerted by the monitor on the Earth. (b) The free-body diagram for the monitor.

Another example of Newton's third law is shown in Figure 5.5b. The force $\mathbf{F}_{hn}$ exerted by the hammer on the nail (the action) is equal in magnitude and opposite the force $\mathbf{F}_{nh}$ exerted by the nail on the hammer (the reaction). This latter force stops the forward motion of the hammer when it strikes the nail.

You experience the third law directly if you slam your fist against a wall or kick a football with your bare foot. You can feel the force back on your fist or your foot. You should be able to identify the action and reaction forces in these cases.

The Earth exerts a gravitational force $\mathbf{F}_g$ on any object. If the object is a computer monitor at rest on a table, as in Figure 5.6a, the reaction force to $\mathbf{F}_g = \mathbf{F}_{Em}$ is the force exerted by the monitor on the Earth $\mathbf{F}_{mE} = -\mathbf{F}_{Em}$. The monitor does not accelerate because it is held up by the table. The table exerts on the monitor an upward force $\mathbf{n} = \mathbf{F}_{tm}$, called the **normal force**.[4] This is the force that prevents the monitor from falling through the table; it can have any value needed, up to the point of breaking the table. From Newton's second law, we see that, because the monitor has zero acceleration, it follows that $\Sigma\mathbf{F} = \mathbf{n} - m\mathbf{g} = 0$, or $n = mg$. The normal force balances the gravitational force on the monitor, so that the net force on the monitor is zero. The reaction to **n** is the force exerted by the monitor downward on the table, $\mathbf{F}_{mt} = -\mathbf{F}_{tm} = -\mathbf{n}$.

Note that the forces acting on the monitor are $\mathbf{F}_g$ and **n**, as shown in Figure 5.6b. The two reaction forces $\mathbf{F}_{mE}$ and $\mathbf{F}_{mt}$ are exerted on objects other than the monitor. Remember, **the two forces in an action–reaction pair always act on two different objects.**

Figure 5.6 illustrates an extremely important step in solving problems involving forces. Figure 5.6a shows many of the forces in the situation—those acting on the monitor, one acting on the table, and one acting on the Earth. Figure 5.6b, by contrast, shows only the forces acting on *one object*, the monitor. This is a critical drawing called a **free-body diagram**. When analyzing an object subject to forces, we are interested in the net force acting on one object, which we will model as a particle. Thus, a free-body diagram helps us to isolate only those forces on the object and eliminate the other forces from our analysis. The free-body diagram can be simplified further by representing the object (such as the monitor) as a particle, by simply drawing a dot.

▲ PITFALL PREVENTION

5.6 *n* Does Not Always Equal *mg*

In the situation shown in Figure 5.6 and in many others, we find that $n = mg$ (the normal force has the same magnitude as the gravitational force). However, this is *not* generally true. If an object is on an incline, if there are applied forces with vertical components, or if there is a vertical acceleration of the system, then $n \neq mg$. *Always* apply Newton's second law to find the relationship between n and mg.

Definition of normal force

▲ PITFALL PREVENTION

5.7 Newton's Third Law

This is such an important and often misunderstood concept that it will be repeated here in a Pitfall Prevention. Newton's third law action and reaction forces act on *different* objects. Two forces acting on the same object, even if they are equal in magnitude and opposite in direction, *cannot* be an action–reaction pair.

---

[4] *Normal* in this context means *perpendicular*.

▲ **PITFALL PREVENTION**

5.8 Free-body Diagrams

The *most important* step in solving a problem using Newton's laws is to draw a proper sketch—the free-body diagram. Be sure to draw only those forces that act on the object that you are isolating. Be sure to draw *all* forces acting on the object, including any field forces, such as the gravitational force.

Quick Quiz 5.7 If a fly collides with the windshield of a fast-moving bus, which object experiences an impact force with a larger magnitude? (a) the fly (b) the bus (c) the same force is experienced by both.

Quick Quiz 5.8 If a fly collides with the windshield of a fast-moving bus, which object experiences the greater acceleration: (a) the fly (b) the bus (c) the same acceleration is experienced by both.

Quick Quiz 5.9 Which of the following is the reaction force to the gravitational force acting on your body as you sit in your desk chair? (a) The normal force exerted by the chair (b) The force you exert downward on the seat of the chair (c) Neither of these forces.

Quick Quiz 5.10 In a free-body diagram for a single object, you draw (a) the forces acting on the object and the forces the object exerts on other objects, or (b) only the forces acting on the object.

Conceptual Example 5.3   You Push Me and I'll Push You

A large man and a small boy stand facing each other on frictionless ice. They put their hands together and push against each other so that they move apart.

(A) Who moves away with the higher speed?

Solution This situation is similar to what we saw in Quick Quizzes 5.7 and 5.8. According to Newton's third law, the force exerted by the man on the boy and the force exerted by the boy on the man are an action–reaction pair, and so they must be equal in magnitude. (A bathroom scale placed between their hands would read the same, regard-

less of which way it faced.) Therefore, the boy, having the smaller mass, experiences the greater acceleration. Both individuals accelerate for the same amount of time, but the greater acceleration of the boy over this time interval results in his moving away from the interaction with the higher speed.

(B) Who moves farther while their hands are in contact?

Solution Because the boy has the greater acceleration and, therefore, the greater average velocity, he moves farther during the time interval in which the hands are in contact.

Rock climbers at rest are in equilibrium and depend on the tension forces in ropes for their safety.

5.7 Some Applications of Newton's Laws

In this section we apply Newton's laws to objects that are either in equilibrium ($\mathbf{a} = 0$) or accelerating along a straight line under the action of constant external forces. Remember that **when we apply Newton's laws to an object, we are interested only in external forces that act on the object.** We assume that the objects can be modeled as particles so that we need not worry about rotational motion. For now, we also neglect the effects of friction in those problems involving motion; this is equivalent to stating that the surfaces are *frictionless*. (We will incorporate the friction force in problems in Section 5.8.)

We usually neglect the mass of any ropes, strings, or cables involved. In this approximation, the magnitude of the force exerted at any point along a rope is the same at all points along the rope. In problem statements, the synonymous terms *light* and *of negligible mass* are used to indicate that a mass is to be ignored when you work the problems. When a rope attached to an object is pulling on the object, the rope exerts a force **T** on the object, and the magnitude T of that force is called the **tension** in the rope. Because it is the magnitude of a vector quantity, tension is a scalar quantity.

© John Elk III/Stock, Boston Inc./PictureQuest

Objects in Equilibrium

If the acceleration of an object that can be modeled as a particle is zero, the particle is in **equilibrium**. Consider a lamp suspended from a light chain fastened to the ceiling, as in Figure 5.7a. The free-body diagram for the lamp (Figure 5.7b) shows that the forces acting on the lamp are the downward gravitational force $\mathbf{F}_g$ and the upward force $\mathbf{T}$ exerted by the chain. If we apply the second law to the lamp, noting that $\mathbf{a} = 0$, we see that because there are no forces in the x direction, $\Sigma F_x = 0$ provides no helpful information. The condition $\Sigma F_y = ma_y = 0$ gives

$$\Sigma F_y = T - F_g = 0 \qquad \text{or} \qquad T = F_g$$

Again, note that $\mathbf{T}$ and $\mathbf{F}_g$ are *not* an action–reaction pair because they act on the same object—the lamp. The reaction force to $\mathbf{T}$ is $\mathbf{T}'$, the downward force exerted by the lamp on the chain, as shown in Figure 5.7c. The ceiling exerts on the chain a force $\mathbf{T}''$ that is equal in magnitude to the magnitude of $\mathbf{T}'$ and points in the opposite direction.

Figure 5.7 (a) A lamp suspended from a ceiling by a chain of negligible mass. (b) The forces acting on the lamp are the gravitational force $\mathbf{F}_g$ and the force $\mathbf{T}$ exerted by the chain. (c) The forces acting on the chain are the force $\mathbf{T}'$ exerted by the lamp and the force $\mathbf{T}''$ exerted by the ceiling.

Objects Experiencing a Net Force

If an object that can be modeled as a particle experiences an acceleration, then there must be a nonzero net force acting on the object. Consider a crate being pulled to the right on a frictionless, horizontal surface, as in Figure 5.8a. Suppose you are asked to find the acceleration of the crate and the force the floor exerts on it. First, note that the horizontal force $\mathbf{T}$ being applied to the crate acts through the rope. The magnitude of $\mathbf{T}$ is equal to the tension in the rope. The forces acting on the crate are illustrated in the free-body diagram in Figure 5.8b. In addition to the force $\mathbf{T}$, the free-body diagram for the crate includes the gravitational force $\mathbf{F}_g$ and the normal force $\mathbf{n}$ exerted by the floor on the crate.

We can now apply Newton's second law in component form to the crate. The only force acting in the x direction is $\mathbf{T}$. Applying $\Sigma F_x = ma_x$ to the horizontal motion gives

$$\Sigma F_x = T = ma_x \qquad \text{or} \qquad a_x = \frac{T}{m}$$

No acceleration occurs in the y direction. Applying $\Sigma F_y = ma_y$ with $a_y = 0$ yields

$$n + (-F_g) = 0 \qquad \text{or} \qquad n = F_g$$

That is, the normal force has the same magnitude as the gravitational force but acts in the opposite direction.

If $\mathbf{T}$ is a constant force, then the acceleration $a_x = T/m$ also is constant. Hence, the constant-acceleration equations of kinematics from Chapter 2 can be used to obtain the crate's position x and velocity v_x as functions of time. Because $a_x = T/m =$ constant, Equations 2.9 and 2.12 can be written as

$$v_{xf} = v_{xi} + \left(\frac{T}{m}\right) t$$

$$x_f = x_i + v_{xi}t + \frac{1}{2}\left(\frac{T}{m}\right) t^2$$

In the situation just described, the magnitude of the normal force $\mathbf{n}$ is equal to the magnitude of $\mathbf{F}_g$, but this is not always the case. For example, suppose a book is lying on a table and you push down on the book with a force $\mathbf{F}$, as in Figure 5.9. Because the book is at rest and therefore not accelerating, $\Sigma F_y = 0$, which gives $n - F_g - F = 0$, or $n = F_g + F$. In this situation, the normal force is *greater* than the force of gravity. Other examples in which $n \neq F_g$ are presented later.

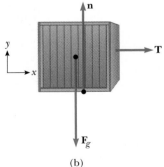

Figure 5.8 (a) A crate being pulled to the right on a frictionless surface. (b) The free-body diagram representing the external forces acting on the crate.

Figure 5.9 When one object pushes downward on another object with a force **F**, the normal force **n** is greater than the gravitational force: $n = F_g + F$.

Applying Newton's Laws

The following procedure is recommended when dealing with problems involving Newton's laws:

- Draw a simple, neat diagram of the system to help *conceptualize* the problem.

- *Categorize* the problem: if any acceleration component is zero, the particle is in equilibrium in this direction and $\Sigma F = 0$. If not, the particle is undergoing an acceleration, the problem is one of nonequilibrium in this direction, and $\Sigma F = ma$.

- *Analyze* the problem by isolating the object whose motion is being analyzed. Draw a free-body diagram for this object. For systems containing more than one object, draw *separate* free-body diagrams for each object. *Do not* include in the free-body diagram forces exerted by the object on its surroundings.

- Establish convenient coordinate axes for each object and find the components of the forces along these axes. Apply Newton's second law, $\Sigma \mathbf{F} = m\mathbf{a}$, in component form. Check your dimensions to make sure that all terms have units of force.

- Solve the component equations for the unknowns. Remember that you must have as many independent equations as you have unknowns to obtain a complete solution.

- *Finalize* by making sure your results are consistent with the free-body diagram. Also check the predictions of your solutions for extreme values of the variables. By doing so, you can often detect errors in your results.

Example 5.4 A Traffic Light at Rest

A traffic light weighing 122 N hangs from a cable tied to two other cables fastened to a support, as in Figure 5.10a. The upper cables make angles of 37.0° and 53.0° with the horizontal. These upper cables are not as strong as the vertical cable, and will break if the tension in them exceeds 100 N. Will the traffic light remain hanging in this situation, or will one of the cables break?

Solution We *conceptualize* the problem by inspecting the drawing in Figure 5.10a. Let us assume that the cables do not break so that there is no acceleration of any sort in this problem in any direction. This allows us to *categorize* the problem as one of equilibrium. Because the acceleration of the system is zero, we know that the net force on the light and the net force on the knot are both zero. To *analyze* the

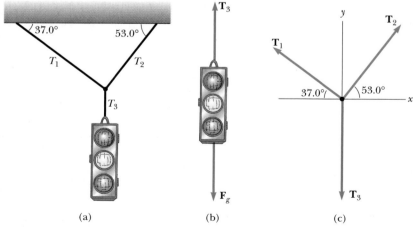

(a)   (b)   (c)

Figure 5.10 (Example 5.4) (a) A traffic light suspended by cables. (b) Free-body diagram for the traffic light. (c) Free-body diagram for the knot where the three cables are joined.

problem, we construct two free-body diagrams—one for the traffic light, shown in Figure 5.10b, and one for the knot that holds the three cables together, as in Figure 5.10c. This knot is a convenient object to choose because all the forces of interest act along lines passing through the knot.

With reference to Figure 5.10b, we apply the equilibrium condition in the y direction, $\sum F_y = 0 \rightarrow T_3 - F_g = 0$. This leads to $T_3 = F_g = 122$ N. Thus, the upward force T_3 exerted by the vertical cable on the light balances the gravitational force.

Next, we choose the coordinate axes shown in Figure 5.10c and resolve the forces acting on the knot into their components:

Force	x Component	y Component
T_1	$-T_1 \cos 37.0°$	$T_1 \sin 37.0°$
T_2	$T_2 \cos 53.0°$	$T_2 \sin 53.0°$
T_3	0	-122 N

Knowing that the knot is in equilibrium ($a = 0$) allows us to write

(1) $\sum F_x = -T_1 \cos 37.0° + T_2 \cos 53.0° = 0$

(2) $\sum F_y = T_1 \sin 37.0° + T_2 \sin 53.0° + (-122 \text{ N}) = 0$

From (1) we see that the horizontal components of T_1 and T_2 must be equal in magnitude, and from (2) we see that the sum of the vertical components of T_1 and T_2 must balance the downward force T_3, which is equal in magnitude to the weight of the light. We solve (1) for T_2 in terms of T_1 to obtain

(3) $T_2 = T_1 \left(\dfrac{\cos 37.0°}{\cos 53.0°} \right) = 1.33 T_1$

This value for T_2 is substituted into (2) to yield

$$T_1 \sin 37.0° + (1.33 T_1)(\sin 53.0°) - 122 \text{ N} = 0$$

$$T_1 = 73.4 \text{ N}$$

$$T_2 = 1.33 T_1 = 97.4 \text{ N}$$

Both of these values are less than 100 N (just barely for T_2), so the cables will not break. Let us *finalize* this problem by imagining a change in the system, as in the following **What If?**

What If? Suppose the two angles in Figure 5.10a are equal. What would be the relationship between T_1 and T_2?

Answer We can argue from the symmetry of the problem that the two tensions T_1 and T_2 would be equal to each other. Mathematically, if the equal angles are called θ, Equation (3) becomes

$$T_2 = T_1 \left(\dfrac{\cos \theta}{\cos \theta} \right) = T_1$$

which also tells us that the tensions are equal. Without knowing the specific value of θ, we cannot find the values of T_1 and T_2. However, the tensions will be equal to each other, regardless of the value of θ.

Conceptual Example 5.5   Forces Between Cars in a Train

Train cars are connected by *couplers*, which are under tension as the locomotive pulls the train. As you move through the train from the locomotive to the last car, does the tension in the couplers increase, decrease, or stay the same as the train speeds up? When the engineer applies the brakes, the couplers are under compression. How does this compression force vary from the locomotive to the last car? (Assume that only the brakes on the wheels of the engine are applied.)

Solution As the train speeds up, tension decreases from front to back. The coupler between the locomotive and the first car must apply enough force to accelerate the rest of the cars. As you move back along the train, each coupler is accelerating less mass behind it. The last coupler has to accelerate only the last car, and so it is under the least tension.

When the brakes are applied, the force again decreases from front to back. The coupler connecting the locomotive to the first car must apply a large force to slow down the rest of the cars, but the final coupler must apply a force large enough to slow down *only* the last car.

Example 5.6   The Runaway Car

A car of mass m is on an icy driveway inclined at an angle θ, as in Figure 5.11a.

(A) Find the acceleration of the car, assuming that the driveway is frictionless.

Solution *Conceptualize* the situation using Figure 5.11a. From everyday experience, we know that a car on an icy incline will accelerate down the incline. (It will do the same thing as a car on a hill with its brakes not set.) This allows us to *categorize* the situation as a nonequilibrium problem—that is, one in which an object accelerates. Figure 5.11b shows the free-body diagram for the car, which we can use to *analyze* the problem. The only forces acting on the car are the normal force **n** exerted by the inclined plane, which acts perpendicular to the plane, and the gravitational force $\mathbf{F}_g = m\mathbf{g}$, which acts vertically downward. For problems involving inclined planes, it is convenient to choose the coordinate axes with x along the incline and y perpendicular to it, as in Figure 5.11b. (It is possible to solve the problem with "standard" horizontal and vertical axes. You may want to try this, just for practice.) Then, we replace the gravitational force by a component of magnitude $mg \sin \theta$ along the positive x axis and one of magnitude $mg \cos \theta$ along the negative y axis.

Now we apply Newton's second law in component form, noting that $a_y = 0$:

(1) $\sum F_x = mg \sin \theta = ma_x$

(2) $\sum F_y = n - mg \cos \theta = 0$

(a)   (b)

Figure 5.11 (Example 5.6) (a) A car of mass m sliding down a frictionless incline. (b) The free-body diagram for the car. Note that its acceleration along the incline is $g \sin \theta$.

Solving (1) for a_x, we see that the acceleration along the incline is caused by the component of $\mathbf{F}_g$ directed down the incline:

$$(3) \qquad a_x = \boxed{g \sin \theta}$$

To *finalize* this part, note that this acceleration component is independent of the mass of the car! It depends only on the angle of inclination and on g.

From (2) we conclude that the component of $\mathbf{F}_g$ perpendicular to the incline is balanced by the normal force; that is, $n = mg \cos \theta$. This is another example of a situation in which the normal force is *not* equal in magnitude to the weight of the object.

(B) Suppose the car is released from rest at the top of the incline, and the distance from the car's front bumper to the bottom of the incline is d. How long does it take the front bumper to reach the bottom, and what is the car's speed as it arrives there?

Solution *Conceptualize* by imagining that the car is sliding down the hill and you are operating a stop watch to measure the entire time interval until it reaches the bottom. This part of the problem belongs to kinematics rather than to dynamics, and Equation (3) shows that the acceleration a_x is constant. Therefore you should *categorize* this problem as that of a particle undergoing constant acceleration. Apply Equation 2.12, $x_f = x_i + v_{xi}t + \frac{1}{2}a_x t^2$, to *analyze* the car's motion. Defining the initial position of the front bumper as $x_i = 0$ and its final position as $x_f = d$, and recognizing that $v_{xi} = 0$, we obtain

$$d = \tfrac{1}{2}a_x t^2$$

$$(4) \qquad t = \sqrt{\frac{2d}{a_x}} = \boxed{\sqrt{\frac{2d}{g \sin \theta}}}$$

Using Equation 2.13, with $v_{xi} = 0$, we find that

$$v_{xf}^2 = 2a_x d$$

$$(5) \qquad v_{xf} = \sqrt{2a_x d} = \boxed{\sqrt{2 gd \sin \theta}}$$

To *finalize* this part of the problem, we see from Equations (4) and (5) that the time t at which the car reaches the bottom and its final speed v_{xf} are independent of the car's mass, as was its acceleration. Note that we have combined techniques from Chapter 2 with new techniques from the present chapter in this example. As we learn more and more techniques in later chapters, this process of combining information from several parts of the book will occur more often. In these cases, use the General Problem-Solving Strategy to help you identify what techniques you will need.

What If? **(A)** What previously solved problem does this become if $\theta = 90°$? **(B)** What problem does this become if $\theta = 0$?

Answer (A) Imagine θ going to 90° in Figure 5.11. The inclined plane becomes vertical, and the car is an object in free-fall! Equation (3) becomes

$$a_x = g \sin \theta = g \sin 90° = g$$

which is indeed the free-fall acceleration. (We find $a_x = g$ rather than $a_x = -g$ because we have chosen positive x to be downward in Figure 5.11.) Notice also that the condition

$n = mg \cos \theta$ gives us $n = mg \cos 90° = 0$. This is consistent with the fact that the car is falling downward *next to* the vertical plane but there is no interaction force between the car and the plane. Equations (4) and (5) give us

$$t = \sqrt{\frac{2d}{g \sin 90°}} = \sqrt{\frac{2d}{g}} \quad \text{and} \quad v_{xf} = \sqrt{2gd \sin 90°} = \sqrt{2gd},$$

both of which are consistent with a falling object.

(B) Imagine θ going to 0 in Figure 5.11. In this case, the inclined plane becomes horizontal, and the car is on a horizontal surface. Equation (3) becomes

$$a_x = g \sin \theta = g \sin 0 = 0$$

which is consistent with the fact that the car is at rest in equilibrium. Notice also that the condition $n = mg \cos \theta$ gives us $n = mg \cos 0 = mg$, which is consistent with our expectation.

Equations (4) and (5) give us $t = \sqrt{\dfrac{2d}{g \sin 0}} \rightarrow \infty$ and $v_{xf} = \sqrt{2gd \sin 0} = 0$. These results agree with the fact that the car does not accelerate, so it will never achieve a non-zero final velocity, and it will take an infinite amount of time to reach the bottom of the "hill"!

Example 5.7 One Block Pushes Another

Two blocks of masses m_1 and m_2, with $m_1 > m_2$, are placed in contact with each other on a frictionless, horizontal surface, as in Figure 5.12a. A constant horizontal force **F** is applied to m_1 as shown. **(A)** Find the magnitude of the acceleration of the system.

Solution *Conceptualize* the situation using Figure 5.12a and realizing that both blocks must experience the *same* acceleration because they are in contact with each other and remain in contact throughout the motion. We *categorize* this as a Newton's second law problem because we have a force applied to a system and we are looking for an acceleration. To *analyze* the problem, we first address the combination of two blocks as a system. Because **F** is the only external horizontal force acting on the system, we have

$$\sum F_x(\text{system}) = F = (m_1 + m_2) a_x$$

$$(1) \qquad a_x = \boxed{\frac{F}{m_1 + m_2}}$$

To *finalize* this part, note that this would be the same acceleration as that of a single object of mass equal to the combined masses of the two blocks in Figure 5.12a and subject to the same force.

(B) Determine the magnitude of the contact force between the two blocks.

Solution *Conceptualize* by noting that the contact force is internal to the system of two blocks. Thus, we cannot find this force by modeling the whole system (the two blocks) as a single particle. We must now treat each of the two blocks individually by *categorizing* each as a particle subject to a net force. To *analyze* the situation, we first construct a free-body diagram for each block, as shown in Figures 5.12b and 5.12c, where the contact force is denoted by **P**. From Figure 5.12c we see that the only horizontal force acting on m_2 is the contact force $\mathbf{P}_{12}$ (the force exerted by m_1 on m_2), which is directed to the right. Applying Newton's second law to m_2 gives

$$(2) \qquad \sum F_x = P_{12} = m_2 a_x$$

Substituting the value of the acceleration a_x given by (1) into (2) gives

$$(3) \qquad P_{12} = m_2 a_x = \boxed{\left(\frac{m_2}{m_1 + m_2}\right) F}$$

To *finalize* the problem, we see from this result that the contact force P_{12} is *less* than the applied force F. This is consistent with the fact that the force required to accelerate block 2 alone must be less than the force required to produce the same acceleration for the two-block system.

To *finalize* further, it is instructive to check this expression for P_{12} by considering the forces acting on m_1, shown in Figure 5.12b. The horizontal forces acting on m_1 are the applied force **F** to the right and the contact force $\mathbf{P}_{21}$ to the left (the force exerted by m_2 on m_1). From Newton's third law, $\mathbf{P}_{21}$ is the reaction to $\mathbf{P}_{12}$, so $P_{21} = P_{12}$. Applying Newton's second law to m_1 gives

$$(4) \qquad \sum F_x = F - P_{21} = F - P_{12} = m_1 a_x$$

Substituting into (4) the value of a_x from (1), we obtain

$$P_{12} = F - m_1 a_x = F - m_1 \left(\frac{F}{m_1 + m_2}\right) = \left(\frac{m_2}{m_1 + m_2}\right) F$$

This agrees with (3), as it must.

Active Figure 5.12 (Example 5.7) A force is applied to a block of mass m_1, which pushes on a second block of mass m_2. (b) The free-body diagram for m_1. (c) The free-body diagram for m_2.

At the Active Figures link at http://www.pse6.com, you can study the forces involved in this two-block system.

What If? Imagine that the force F in Figure 5.12 is applied toward the left on the right-hand block of mass m_2. Is the magnitude of the force P_{12} the same as it was when the force was applied toward the right on m_1?

Answer With the force applied toward the left on m_2, the contact force must accelerate m_1. In the original situation, the contact force accelerates m_2. Because $m_1 > m_2$, this will require more force, so the magnitude of P_{12} is greater than in the original situation.

Example 5.8   Weighing a Fish in an Elevator

A person weighs a fish of mass m on a spring scale attached to the ceiling of an elevator, as illustrated in Figure 5.13. Show that if the elevator accelerates either upward or downward, the spring scale gives a reading that is different from the weight of the fish.

Solution *Conceptualize* by noting that the reading on the scale is related to the extension of the spring in the scale, which is related to the force on the end of the spring as in Figure 5.2. Imagine that a string is hanging from the end of the spring, so that the magnitude of the force exerted on the spring is equal to the tension T in the string. Thus, we are looking for T. The force T pulls down on the string and pulls up on the fish. Thus, we can *categorize* this problem as one of analyzing the forces and acceleration associated with the fish by means of Newton's second law. To *analyze* the problem, we inspect the free-body diagrams for the fish in Figure 5.13 and note that the external forces acting on the fish are the downward gravitational force

$F_g = m\mathbf{g}$ and the force T exerted by the scale. If the elevator is either at rest or moving at constant velocity, the fish does not accelerate, and so $\Sigma F_y = T - F_g = 0$ or $T = F_g = mg$. (Remember that the scalar mg is the weight of the fish.)

If the elevator moves with an acceleration a relative to an observer standing outside the elevator in an inertial frame (see Fig. 5.13), Newton's second law applied to the fish gives the net force on the fish:

$$(1) \qquad \sum F_y = T - mg = ma_y$$

where we have chosen upward as the positive y direction. Thus, we conclude from (1) that the scale reading T is greater than the fish's weight mg if a is upward, so that a_y is positive, and that the reading is less than mg if a is downward, so that a_y is negative.

For example, if the weight of the fish is 40.0 N and a is upward, so that $a_y = +2.00$ m/s^2, the scale reading from (1) is

(a)   (b)

Observer in
inertial frame

Figure 5.13 (Example 5.8) Apparent weight versus true weight. (a) When the elevator accelerates upward, the spring scale reads a value greater than the weight of the fish. (b) When the elevator accelerates downward, the spring scale reads a value less than the weight of the fish.

$$(2) \quad T = ma_y + mg = mg\left(\frac{a_y}{g} + 1\right)$$

$$= F_g\left(\frac{a_y}{g} + 1\right) = (40.0 \text{ N})\left(\frac{2.00 \text{ m/s}^2}{9.80 \text{ m/s}^2} + 1\right)$$

$$= \boxed{48.2 \text{ N}}$$

If **a** is downward so that $a_y = -2.00 \text{ m/s}^2$, then (2) gives us

$$T = F_g\left(\frac{a_y}{g} + 1\right) = (40.0 \text{ N})\left(\frac{-2.00 \text{ m/s}^2}{9.80 \text{ m/s}^2} + 1\right)$$

$$= \boxed{31.8 \text{ N}}$$

To *finalize* this problem, take this advice—if you buy a fish in an elevator, make sure the fish is weighed while the elevator is either at rest or accelerating downward! Furthermore, note that from the information given here, one cannot determine the direction of motion of the elevator.

What If? Suppose the elevator cable breaks, so that the elevator and its contents are in free-fall. What happens to the reading on the scale?

Answer If the elevator falls freely, its acceleration is $a_y = -g$. We see from (2) that the scale reading T is zero in this case; that is, the fish *appears* to be weightless.

Example 5.9 The Atwood Machine

When two objects of unequal mass are hung vertically over a frictionless pulley of negligible mass, as in Figure 5.14a, the arrangement is called an *Atwood machine*. The device is sometimes used in the laboratory to measure the free-fall acceleration. Determine the magnitude of the acceleration of the two objects and the tension in the lightweight cord.

Solution *Conceptualize* the situation pictured in Figure 5.14a—as one object moves upward, the other object moves

(a)

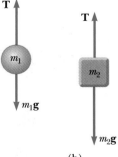

(b)

Active Figure 5.14 (Example 5.9) The Atwood machine. (a) Two objects ($m_2 > m_1$) connected by a massless inextensible cord over a frictionless pulley. (b) Free-body diagrams for the two objects.

At the Active Figures link at http://www.pse6.com, *you can adjust the masses of the objects on the Atwood machine and observe the motion.*

downward. Because the objects are connected by an inextensible string, their accelerations must be of equal magnitude. The objects in the Atwood machine are subject to the gravitational force as well as to the forces exerted by the strings connected to them—thus, we can *categorize* this as a Newton's second law problem. To *analyze* the situation, the free-body diagrams for the two objects are shown in Figure 5.14b. Two forces act on each object: the upward force **T** exerted by the string and the downward gravitational force. In problems such as this in which the pulley is modeled as massless and frictionless, the tension in the string on both sides of the pulley is the same. If the pulley has mass and/or is subject to friction, the tensions on either side are not the same and the situation requires techniques we will learn in Chapter 10.

We must be very careful with signs in problems such as this. In Figure 5.14a, notice that if object 1 accelerates upward, then object 2 accelerates downward. Thus, for consistency with signs, if we define the upward direction as positive for object 1, we must define the downward direction as positive for object 2. With this sign convention, both objects accelerate in the same direction as defined by the choice of sign. Furthermore, according to this sign convention, the y component of the net force exerted on object 1 is $T - m_1g$, and the y component of the net force exerted on object 2 is $m_2g - T$. Notice that we have chosen the signs of the forces to be consistent with the choices of signs for up and down for each object. If we assume that $m_2 > m_1$, then m_1 must accelerate upward, while m_2 must accelerate downward.

When Newton's second law is applied to object 1, we obtain

$$(1) \quad \sum F_y = T - m_1g = m_1a_y$$

Similarly, for object 2 we find

$$(2) \quad \sum F_y = m_2g - T = m_2a_y$$

When (2) is added to (1), T cancels and we have

$$-m_1g + m_2g = m_1a_y + m_2a_y$$

$$(3) \quad a_y = \boxed{\left(\frac{m_2 - m_1}{m_1 + m_2}\right)g}$$

The acceleration given by (3) can be interpreted as the ratio of the magnitude of the unbalanced force on the system $(m_2 - m_1)g$, to the total mass of the system $(m_1 + m_2)$, as expected from Newton's second law.

When (3) is substituted into (1), we obtain

$$(4) \qquad T = \left(\frac{2m_1 m_2}{m_1 + m_2} \right) g$$

Finalize this problem with the following **What If?**

What If? (A) Describe the motion of the system if the objects have equal masses, that is, $m_1 = m_2$.

(B) Describe the motion of the system if one of the masses is much larger than the other, $m_1 \gg m_2$.

Answer (A) If we have the same mass on both sides, the system is balanced and it should not accelerate. Mathematically, we see that if $m_1 = m_2$, Equation (3) gives us $a_y = 0$. (B) In the case in which one mass is infinitely larger than the other, we can ignore the effect of the smaller mass. Thus, the larger mass should simply fall as if the smaller mass were not there. We see that if $m_1 \gg m_2$, Equation (3) gives us $a_y = -g$.

 Investigate these limiting cases at the Interactive Worked Example link at http://www.pse6.com.

Example 5.10   Acceleration of Two Objects Connected by a Cord   `Interactive`

A ball of mass m_1 and a block of mass m_2 are attached by a lightweight cord that passes over a frictionless pulley of negligible mass, as in Figure 5.15a. The block lies on a frictionless incline of angle θ. Find the magnitude of the acceleration of the two objects and the tension in the cord.

Solution *Conceptualize* the motion in Figure 5.15. If m_2 moves down the incline, m_1 moves upward. Because the objects are connected by a cord (which we assume does not stretch), their accelerations have the same magnitude. We can identify forces on each of the two objects and we are looking for an acceleration, so we *categorize* this as a Newton's second-law problem. To *analyze* the problem, consider the free-body diagrams shown in Figures 5.15b and 5.15c. Applying Newton's second law in component form to the ball, choosing the upward direction as positive, yields

$$(1) \qquad \sum F_x = 0$$
$$(2) \qquad \sum F_y = T - m_1 g = m_1 a_y = m_1 a$$

Note that in order for the ball to accelerate upward, it is necessary that $T > m_1 g$. In (2), we replaced a_y with a because the acceleration has only a y component.

For the block it is convenient to choose the positive x' axis along the incline, as in Figure 5.15c. For consistency with our choice for the ball, we choose the positive direction to be down the incline. Applying Newton's second law in component form to the block gives

$$(3) \qquad \sum F_{x'} = m_2 g \sin \theta - T = m_2 a_{x'} = m_2 a$$
$$(4) \qquad \sum F_{y'} = n - m_2 g \cos \theta = 0$$

In (3) we replaced $a_{x'}$ with a because the two objects have accelerations of equal magnitude a. Equations (1) and (4) provide no information regarding the acceleration. However, if we solve (2) for T and then substitute this value for T into (3) and solve for a, we obtain

$$(5) \qquad a = \frac{m_2 g \sin \theta - m_1 g}{m_1 + m_2}$$

When this expression for a is substituted into (2), we find

$$(6) \qquad T = \frac{m_1 m_2 g (\sin \theta + 1)}{m_1 + m_2}$$

To *finalize* the problem, note that the block accelerates down the incline only if $m_2 \sin \theta > m_1$. If $m_1 > m_2 \sin \theta$,

(a)

(b)

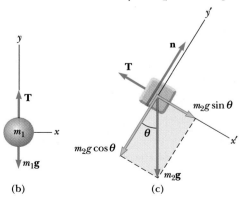

(c)

Figure 5.15 (Example 5.10) (a) Two objects connected by a lightweight cord strung over a frictionless pulley. (b) Free-body diagram for the ball. (c) Free-body diagram for the block. (The incline is frictionless.)

then the acceleration is up the incline for the block and downward for the ball. Also note that the result for the acceleration (5) can be interpreted as the magnitude of the net force acting on the system divided by the total mass of the system; this is consistent with Newton's second law.

What If? (A) What happens in this situation if the angle $\theta = 90°$?

(B) What happens if the mass $m_1 = 0$?

Answer (A) If $\theta = 90°$, the inclined plane becomes vertical and there is no interaction between its surface and m_2. Thus, this problem becomes the Atwood machine of Example 5.9. Letting $\theta \rightarrow 90°$ in Equations (5) and (6) causes them to reduce to Equations (3) and (4) of Example 5.9!
(B) If $m_1 = 0$, then m_2 is simply sliding down an inclined plane without interacting with m_1 through the string. Thus, this problem becomes the sliding car problem in Example 5.6. Letting $m_1 \rightarrow 0$ in Equation (5) causes it to reduce to Equation (3) of Example 5.6!

 Investigate these limiting cases at the Interactive Worked Example link at http://www.pse6.com.

5.8 Forces of Friction

When an object is in motion either on a surface or in a viscous medium such as air or water, there is resistance to the motion because the object interacts with its surroundings. We call such resistance a **force of friction.** Forces of friction are very important in our everyday lives. They allow us to walk or run and are necessary for the motion of wheeled vehicles.

Imagine that you are working in your garden and have filled a trash can with yard clippings. You then try to drag the trash can across the surface of your concrete patio, as in Figure 5.16a. This is a *real* surface, not an idealized, frictionless surface. If we apply an external horizontal force **F** to the trash can, acting to the right, the trash can remains stationary if **F** is small. The force that counteracts **F** and keeps the trash can from moving acts to the left and is called the **force of static friction** $\mathbf{f}_s$. As long as the trash can is not moving, $f_s = F$. Thus, if **F** is increased, $\mathbf{f}_s$ also increases. Likewise, if **F** decreases, $\mathbf{f}_s$ also

Force of static friction

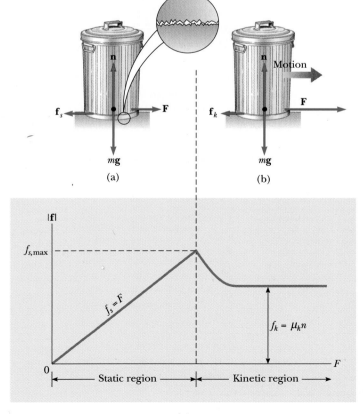

(a)

(b)

(c)

Active Figure 5.16 The direction of the force of friction **f** between a trash can and a rough surface is opposite the direction of the applied force **F**. Because both surfaces are rough, contact is made only at a few points, as illustrated in the "magnified" view. (a) For small applied forces, the magnitude of the force of static friction equals the magnitude of the applied force. (b) When the magnitude of the applied force exceeds the magnitude of the maximum force of static friction, the trash can breaks free. The applied force is now larger than the force of kinetic friction and the trash can accelerates to the right. (c) A graph of friction force versus applied force. Note that $f_{s,max} > f_k$.

 At the Active Figures link at http://www.pse6.com **you can vary the applied force on the trash can and practice sliding it on surfaces of varying roughness. Note the effect on the trash can's motion and the corresponding behavior of the graph in (c).**

decreases. Experiments show that the friction force arises from the nature of the two surfaces: because of their roughness, contact is made only at a few locations where peaks of the material touch, as shown in the magnified view of the surface in Figure 5.16a.

At these locations, the friction force arises in part because one peak physically blocks the motion of a peak from the opposing surface, and in part from chemical bonding ("spot welds") of opposing peaks as they come into contact. If the surfaces are rough, bouncing is likely to occur, further complicating the analysis. Although the details of friction are quite complex at the atomic level, this force ultimately involves an electrical interaction between atoms or molecules.

If we increase the magnitude of **F**, as in Figure 5.16b, the trash can eventually slips. When the trash can is on the verge of slipping, f_s has its maximum value $f_{s,max}$, as shown in Figure 5.16c. When F exceeds $f_{s,max}$, the trash can moves and accelerates to the right. When the trash can is in motion, the friction force is less than $f_{s,max}$ (Fig. 5.16c). We call the friction force for an object in motion the **force of kinetic friction** $\mathbf{f}_k$. The net force $F - f_k$ in the x direction produces an acceleration to the right, according to Newton's second law. If $F = f_k$, the acceleration is zero, and the trash can moves to the right with constant speed. If the applied force is removed, the friction force acting to the left provides an acceleration of the trash can in the $-x$ direction and eventually brings it to rest, again consistent with Newton's second law.

Experimentally, we find that, to a good approximation, both $f_{s,max}$ and f_k are proportional to the magnitude of the normal force. The following empirical laws of friction summarize the experimental observations:

- The magnitude of the force of static friction between any two surfaces in contact can have the values

$$f_s \leq \mu_s n \tag{5.8}$$

where the dimensionless constant μ_s is called the **coefficient of static friction** and n is the magnitude of the normal force exerted by one surface on the other. The equality in Equation 5.8 holds when the surfaces are on the verge of slipping, that is, when $f_s = f_{s,max} \equiv \mu_s n$. This situation is called *impending motion*. The inequality holds when the surfaces are not on the verge of slipping.

- The magnitude of the force of kinetic friction acting between two surfaces is

$$f_k = \mu_k n \tag{5.9}$$

where μ_k is the **coefficient of kinetic friction.** Although the coefficient of kinetic friction can vary with speed, we shall usually neglect any such variations in this text.

Force of kinetic friction

PITFALL PREVENTION

5.9   The Equal Sign is Used in Limited Situations

In Equation 5.8, the equal sign is used *only* in the case in which the surfaces are just about to break free and begin sliding. Do not fall into the common trap of using $f_s = \mu_s n$ in *any* static situation.

PITFALL PREVENTION

5.10   Friction Equations

Equations 5.8 and 5.9 are *not* vector equations. They are relationships between the *magnitudes* of the vectors representing the friction and normal forces. Because the friction and normal forces are perpendicular to each other, the vectors cannot be related by a multiplicative constant.

Table 5.2

Coefficients of Friction[a]		
	μ_s	μ_k
Steel on steel	0.74	0.57
Aluminum on steel	0.61	0.47
Copper on steel	0.53	0.36
Rubber on concrete	1.0	0.8
Wood on wood	0.25–0.5	0.2
Glass on glass	0.94	0.4
Waxed wood on wet snow	0.14	0.1
Waxed wood on dry snow	—	0.04
Metal on metal (lubricated)	0.15	0.06
Ice on ice	0.1	0.03
Teflon on Teflon	0.04	0.04
Synovial joints in humans	0.01	0.003

[a]   All values are approximate. In some cases, the coefficient of friction can exceed 1.0.

- The values of μ_k and μ_s depend on the nature of the surfaces, but μ_k is generally less than μ_s. Typical values range from around 0.03 to 1.0. Table 5.2 lists some reported values.

- The direction of the friction force on an object is parallel to the surface with which the object is in contact and opposite to the actual motion (kinetic friction) or the impending motion (static friction) of the object relative to the surface.

- The coefficients of friction are nearly independent of the area of contact between the surfaces. We might expect that placing an object on the side having the most area might increase the friction force. While this provides more points in contact, as in Figure 5.16a, the weight of the object is spread out over a larger area, so that the individual points are not pressed so tightly together. These effects approximately compensate for each other, so that the friction force is independent of the area.

▲ **PITFALL PREVENTION**

5.11 The Direction of the Friction Force

Sometimes, an incorrect statement about the friction force between an object and a surface is made—"the friction force on an object is opposite to its motion or impending motion"—rather than the correct phrasing, "the friction force on an object is opposite to its motion or impending motion *relative to the surface*." Think carefully about Quick Quiz 5.12.

Quick Quiz 5.11 You press your physics textbook flat against a vertical wall with your hand. What is the direction of the friction force exerted by the wall on the book? (a) downward (b) upward (c) out from the wall (d) into the wall.

Quick Quiz 5.12 A crate is located in the center of a flatbed truck. The truck accelerates to the east, and the crate moves with it, not sliding at all. What is the direction of the friction force exerted by the truck on the crate? (a) to the west (b) to the east (c) No friction force exists because the crate is not sliding.

Quick Quiz 5.13 You place your physics book on a wooden board. You raise one end of the board so that the angle of the incline increases. Eventually, the book starts sliding on the board. If you maintain the angle of the board at this value, the book (a) moves at constant speed (b) speeds up (c) slows down (d) none of these.

Quick Quiz 5.14 You are playing with your daughter in the snow. She sits on a sled and asks you to slide her across a flat, horizontal field. You have a choice of (a) pushing her from behind, by applying a force downward on her shoulders at 30° below the horizontal (Fig. 5.17a), or (b) attaching a rope to the front of the sled and pulling with a force at 30° above the horizontal (Fig 5.17b). Which would be easier for you and why?

(a)   (b)

Figure 5.17 (Quick Quiz 5.14) A father pushes his daughter on a sled either by (a) pushing down on her shoulders, or (b) pulling up on a rope.

Conceptual Example 5.11   Why Does the Sled Accelerate?

A horse pulls a sled along a level, snow-covered road, causing the sled to accelerate, as shown in Figure 5.18a. Newton's third law states that the sled exerts a force of equal magnitude and opposite direction on the horse. In view of this, how can the sled accelerate—don't the forces cancel? Under what condition does the system (horse plus sled) move with constant velocity?

Solution Remember that the forces described in Newton's third law act on *different* objects—the horse exerts a force on the sled, and the sled exerts an equal-magnitude and oppositely directed force on the horse. Because we are interested only in the motion of the sled, we do not consider the forces it exerts on the horse. When determining the motion

of an object, you must add only the forces on that object. (This is the principle behind drawing a free-body diagram.) The horizontal forces exerted on the sled are the forward force **T** exerted by the horse and the backward force of friction $\mathbf{f}_{sled}$ between sled and snow (see Fig. 5.18b). When the forward force on the sled exceeds the backward force, the sled accelerates to the right.

The horizontal forces exerted on the horse are the forward force $\mathbf{f}_{horse}$ exerted by the Earth and the backward tension force **T** exerted by the sled (Fig. 5.18c). The resultant of these two forces causes the horse to accelerate.

The force that accelerates the system (horse plus sled) is the net force $\mathbf{f}_{horse} - \mathbf{f}_{sled}$. When $\mathbf{f}_{horse}$ balances $\mathbf{f}_{sled}$, the system moves with constant velocity.

(a)   (b)   (c)

Figure 5.18 (Conceptual Example 5.11)

Example 5.12   Experimental Determination of μ_s and μ_k

The following is a simple method of measuring coefficients of friction: Suppose a block is placed on a rough surface inclined relative to the horizontal, as shown in Figure 5.19. The incline angle is increased until the block starts to move. Show that by measuring the critical angle θ_c at which this slipping just occurs, we can obtain μ_s.

Solution *Conceptualizing* from the free body diagram in Figure 5.19, we see that we can *categorize* this as a Newton's second law problem. To *analyze* the problem, note that the only forces acting on the block are the gravitational force $m\mathbf{g}$, the normal force **n**, and the force of static friction $\mathbf{f}_s$. These forces balance when the block is not moving. When we choose x to be parallel to the plane and y perpendicular to it, Newton's second law applied to the block for this balanced situation gives

$$(1) \qquad \sum F_x = mg \sin \theta - f_s = ma_x = 0$$

$$(2) \qquad \sum F_y = n - mg \cos \theta = ma_y = 0$$

We can eliminate mg by substituting $mg = n/\cos \theta$ from (2) into (1) to find

$$(3) \qquad f_s = mg \sin \theta = \left(\frac{n}{\cos \theta}\right) \sin \theta = n \tan \theta$$

When the incline angle is increased until the block is on the verge of slipping, the force of static friction has reached its maximum value $\mu_s n$. The angle θ in this situation is the critical angle θ_c, and (3) becomes

$$\mu_s n = n \tan \theta_c$$

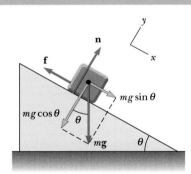

Figure 5.19 (Example 5.12) The external forces exerted on a block lying on a rough incline are the gravitational force $m\mathbf{g}$, the normal force **n**, and the force of friction **f**. For convenience, the gravitational force is resolved into a component along the incline $mg \sin \theta$ and a component perpendicular to the incline $mg \cos \theta$.

$$\mu_s = \tan \theta_c$$

For example, if the block just slips at $\theta_c = 20.0°$, then we find that $\mu_s = \tan 20.0° = 0.364$.

To *finalize* the problem, note that once the block starts to move at $\theta \geq \theta_c$, it accelerates down the incline and the force of friction is $f_k = \mu_k n$. However, if θ is reduced to a value less than θ_c, it may be possible to find an angle θ_c' such that the block moves down the incline with constant speed ($a_x = 0$). In this case, using (1) and (2) with f_s replaced by f_k gives

$$\mu_k = \tan \theta_c'$$

where $\theta_c' < \theta_c$.

Example 5.13 The Sliding Hockey Puck

A hockey puck on a frozen pond is given an initial speed of 20.0 m/s. If the puck always remains on the ice and slides 115 m before coming to rest, determine the coefficient of kinetic friction between the puck and ice.

Solution *Conceptualize* the problem by imagining that the puck in Figure 5.20 slides to the right and eventually comes to rest. To *categorize* the problem, note that we have forces identified in Figure 5.20, but that kinematic variables are provided in the text of the problem. Thus, we must combine the techniques of Chapter 2 with those of this chapter. (We assume that the friction force is constant, which will result in a constant horizontal acceleration.) To *analyze* the situation, note that the forces acting on the puck after it is in motion are shown in Figure 5.20. First, we find the acceleration algebraically in terms of the coefficient of kinetic friction, using Newton's second law. Knowing the acceleration of the puck and the distance it travels, we can then use the equations of kinematics to find the numerical value of the coefficient of kinetic friction.

Figure 5.20 (Example 5.13) *After* the puck is given an initial velocity to the right, the only external forces acting on it are the gravitational force $m\mathbf{g}$, the normal force $\mathbf{n}$, and the force of kinetic friction $\mathbf{f}_k$.

Defining rightward and upward as our positive directions, we apply Newton's second law in component form to the puck and obtain

$$(1) \qquad \sum F_x = -f_k = ma_x$$

$$(2) \qquad \sum F_y = n - mg = 0 \qquad (a_y = 0)$$

But $f_k = \mu_k n$, and from (2) we see that $n = mg$. Therefore, (1) becomes

$$-\mu_k n = -\mu_k mg = ma_x$$

$$a_x = -\mu_k g$$

The negative sign means the acceleration is to the left in Figure 5.20; because the velocity of the puck is to the right, this means that the puck is slowing down. The acceleration is independent of the mass of the puck and is constant because we assume that μ_k remains constant.

Because the acceleration is constant, we can use Equation 2.13, $v_{xf}^2 = v_{xi}^2 + 2a_x(x_f - x_i)$, with $x_i = 0$ and $v_f = 0$:

$$0 = v_{xi}^2 + 2a_x x_f = v_{xi}^2 - 2\mu_k g x_f$$

$$\mu_k = \frac{v_{xi}^2}{2gx_f}$$

$$\mu_k = \frac{(20.0 \text{ m/s})^2}{2(9.80 \text{ m/s}^2)(115 \text{ m})} = \boxed{0.117}$$

To *finalize* the problem, note that μ_k is dimensionless, as it should be, and that it has a low value, consistent with an object sliding on ice.

Example 5.14 Acceleration of Two Connected Objects When Friction Is Present

A block of mass m_1 on a rough, horizontal surface is connected to a ball of mass m_2 by a lightweight cord over a lightweight, frictionless pulley, as shown in Figure 5.21a. A force of magnitude F at an angle θ with the horizontal is applied to the block as shown. The coefficient of kinetic friction between the block and surface is μ_k. Determine the magnitude of the acceleration of the two objects.

Solution *Conceptualize* the problem by imagining what happens as $\mathbf{F}$ is applied to the block. Assuming that $\mathbf{F}$ is not large enough to lift the block, the block will slide to the right and the ball will rise. We can identify forces and we want an acceleration, so we *categorize* this as a Newton's second law problem, one that includes the friction force. To *analyze* the problem, we begin by drawing free-body diagrams for the two objects, as shown in Figures 5.21b and 5.21c. Next, we apply Newton's second law in component form to each object and use Equation 5.9, $f_k = \mu_k n$. Then we can solve for the acceleration in terms of the parameters given.

The applied force $\mathbf{F}$ has x and y components $F \cos \theta$ and $F \sin \theta$, respectively. Applying Newton's second law to both

objects and assuming the motion of the block is to the right, we obtain

Motion of block: $(1) \quad \sum F_x = F \cos \theta - f_k - T = m_1 a_x = m_1 a$

$(2) \quad \sum F_y = n + F \sin \theta - m_1 g = m_1 a_y = 0$

Motion of ball: $\quad \sum F_x = m_2 a_x = 0$

$(3) \quad \sum F_y = T - m_2 g = m_2 a_y = m_2 a$

Because the two objects are connected, we can equate the magnitudes of the x component of the acceleration of the block and the y component of the acceleration of the ball. From Equation 5.9 we know that $f_k = \mu_k n$, and from (2) we know that $n = m_1 g - F \sin \theta$ (in this case n is *not* equal to $m_1 g$); therefore,

$$(4) \qquad f_k = \mu_k (m_1 g - F \sin \theta)$$

That is, the friction force is reduced because of the positive y component of $\mathbf{F}$. Substituting (4) and the value of T from (3) into (1) gives

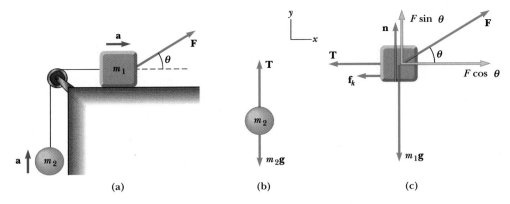

Figure 5.21 (Example 5.14) (a) The external force **F** applied as shown can cause the block to accelerate to the right. (b) and (c) The free-body diagrams assuming that the block accelerates to the right and the ball accelerates upward. The magnitude of the force of kinetic friction in this case is given by $f_k = \mu_k n = \mu_k(m_1 g - F\sin\theta)$.

$$F\cos\theta - \mu_k(m_1 g - F\sin\theta) - m_2(a + g) = m_1 a$$

Solving for a, we obtain

(5) $$a = \frac{F(\cos\theta + \mu_k\sin\theta) - g(m_2 + \mu_k m_1)}{m_1 + m_2}$$

To *finalize* the problem, note that the acceleration of the block can be either to the right or to the left, depending on the sign of the numerator in (5). If the motion is to the left,

then we must reverse the sign of f_k in (1) because the force of kinetic friction must oppose the motion of the block relative to the surface. In this case, the value of a is the same as in (5), with the two plus signs in the numerator changed to minus signs.

This is the final chapter in which we will explicitly show the steps of the General Problem-Solving Strategy in all worked examples. We will refer to them explicitly in occasional examples in future chapters, but you should use the steps in *all* of your problem solving.

Application   Automobile Antilock Braking Systems (ABS)

If an automobile tire is rolling and not slipping on a road surface, then the maximum friction force that the road can exert on the tire is the force of static friction $\mu_s n$. One must use static friction in this situation because at the point of contact between the tire and the road, no sliding of one surface over the other occurs if the tire is not skidding. However, if the tire starts to skid, the friction force exerted on it is reduced to the force of kinetic friction $\mu_k n$. Thus, to maximize the friction force and minimize stopping distance, the wheels must maintain pure rolling motion and not skid. An additional benefit of maintaining wheel rotation is that directional control is not lost as it is in skidding. Unfortunately, in emergency situations drivers typically press down as hard as they can on the brake pedal, "locking the brakes." This stops the wheels from rotating, ensuring a skid and reducing the friction force from the static to the kinetic value. To address this problem, automotive engineers have developed antilock braking systems (ABS). The purpose of the ABS is to help typical drivers (whose tendency is to lock the wheels in an emergency) to better maintain control of their automobiles and minimize stopping distance. The system briefly releases the brakes when a wheel is just about to stop turning. This

maintains rolling contact between the tire and the pavement. When the brakes are released momentarily, the stopping distance is greater than it would be if the brakes were being applied continuously. However, through the use of computer control, the "brake-off" time is kept to a minimum. As a result, the stopping distance is much less than what it would be if the wheels were to skid.

Let us model the stopping of a car by examining real data. In an issue of *AutoWeek*,[5] the braking performance for a Toyota Corolla was measured. These data correspond to the braking force acquired by a highly trained, professional driver. We begin by assuming constant acceleration. (Why do we need to make this assumption?) The magazine provided the initial speed and stopping distance in non-SI units, which we show in the left and middle sections of Table 5.3. After converting these values to SI, we use $v_f{}^2 = v_i{}^2 + 2ax$ to determine the acceleration at different speeds, shown in the right section. These do not vary greatly, and so our assumption of constant acceleration is reasonable.

[5] *AutoWeek* magazine, **48**:22–23, 1998.

Table 5.3

Data for a Toyota Corolla:				
Initial Speed		**Stopping Distance**		**Acceleration**
(mi/h)	**(m/s)**	**(ft)**	**(m)**	**(m/s²)**
30	13.4	34	10.4	− 8.63
60	26.8	143	43.6	− 8.24
80	35.8	251	76.5	− 8.38

Table 5.4

Stopping Distance With and Without Skidding		
Initial Speed	**Stopping Distance**	
(mi/h)	**no skid (m)**	**skidding (m)**
30	10.4	13.9
60	43.6	55.5
80	76.5	98.9

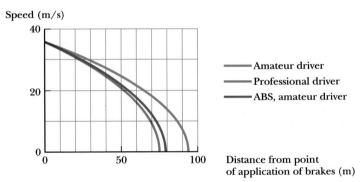

Figure 5.22 This plot of vehicle speed versus distance from the location at which the brakes were applied shows that an antilock braking system (ABS) approaches the performance of a trained professional driver.

We take an average value of acceleration of -8.4 m/s², which is approximately $0.86g$. We then calculate the coefficient of friction from $\Sigma F = \mu_s mg = ma$, which gives $\mu_s = 0.86$ for the Toyota. This is lower than the rubber-on-concrete value given in Table 5.2. Can you think of any reasons for this?

We now estimate the stopping distance of the car if the wheels were skidding. From Table 5.2, we see that the difference between the coefficients of static and kinetic friction for rubber against concrete is about 0.2. Let us assume that our coefficients differ by the same amount, so that $\mu_k \approx 0.66$. This allows us to estimate the stopping distances when the wheels are locked and the car skids across the pavement, as shown in the third column of Table 5.4. The results illustrate the advantage of not allowing the wheels to skid.

Because an ABS keeps the wheels rotating, the higher coefficient of static friction is maintained between the tires and road. This approximates the technique of a professional driver who is able to maintain the wheels at the point of maximum friction force. Let us estimate the ABS performance by assuming that the magnitude of the acceleration is not quite as good as that achieved by the professional driver but instead is reduced by 5%.

Figure 5.22 is a plot of vehicle speed versus distance from where the brakes were applied (at an initial speed of 80.0 mi/h = 35.8 m/s) for the three cases of amateur driver, professional driver, and estimated ABS performance (amateur driver). This shows that a markedly shorter distance is necessary for stopping without locking the wheels compared to skidding. In addition a satisfactory value of stopping distance is achieved when the ABS computer maintains tire rotation.

SUMMARY

Take a practice test for this chapter by clicking the Practice Test link at http://www.pse6.com.

An **inertial frame of reference** is one we can identify in which an object that does not interact with other objects experiences zero acceleration. Any frame moving with constant velocity relative to an inertial frame is also an inertial frame. **Newton's first law** states that it is possible to find such a frame, or, equivalently, in the absence of an external force, when viewed from an inertial frame, an object at rest remains at rest and an object in uniform motion in a straight line maintains that motion.

Newton's second law states that the acceleration of an object is directly proportional to the net force acting on it and inversely proportional to its mass. The net force acting on an object equals the product of its mass and its acceleration: $\Sigma\mathbf{F} = m\mathbf{a}$. If the object is either stationary or moving with constant velocity, then the object is in equilibrium and the force vectors must cancel each other.

The **gravitational force** exerted on an object is equal to the product of its mass (a scalar quantity) and the free-fall acceleration: $\mathbf{F}_g = m\mathbf{g}$. The **weight** of an object is the magnitude of the gravitational force acting on the object.

Newton's third law states that if two objects interact, the force exerted by object 1 on object 2 is equal in magnitude and opposite in direction to the force exerted by object 2 on object 1. Thus, an isolated force cannot exist in nature.

The **maximum force of static friction** $\mathbf{f}_{s,\max}$ between an object and a surface is proportional to the normal force acting on the object. In general, $f_s \leq \mu_s n$, where μ_s is the **coefficient of static friction** and n is the magnitude of the normal force. When an object slides over a surface, the direction of the **force of kinetic friction** $\mathbf{f}_k$ is opposite the direction of motion of the object relative to the surface and is also proportional to the magnitude of the normal force. The magnitude of this force is given by $f_k = \mu_k n$, where μ_k is the **coefficient of kinetic friction**.

QUESTIONS

1. A ball is held in a person's hand. (a) Identify all the external forces acting on the ball and the reaction to each. (b) If the ball is dropped, what force is exerted on it while it is falling? Identify the reaction force in this case. (Neglect air resistance.)

2. If a car is traveling westward with a constant speed of 20 m/s, what is the resultant force acting on it?

3. What is wrong with the statement "Because the car is at rest, there are no forces acting on it"? How would you correct this sentence?

4. In the motion picture *It Happened One Night* (Columbia Pictures, 1934), Clark Gable is standing inside a stationary bus in front of Claudette Colbert, who is seated. The bus suddenly starts moving forward and Clark falls into Claudette's lap. Why did this happen?

5. A passenger sitting in the rear of a bus claims that she was injured as the driver slammed on the brakes, causing a suitcase to come flying toward her from the front of the bus. If you were the judge in this case, what disposition would you make? Why?

6. A space explorer is moving through space far from any planet or star. She notices a large rock, taken as a specimen from an alien planet, floating around the cabin of the ship. Should she push it gently or kick it toward the storage compartment? Why?

7. A rubber ball is dropped onto the floor. What force causes the ball to bounce?

8. While a football is in flight, what forces act on it? What are the action–reaction pairs while the football is being kicked and while it is in flight?

9. The mayor of a city decides to fire some city employees because they will not remove the obvious sags from the cables that support the city traffic lights. If you were a lawyer, what defense would you give on behalf of the employees? Who do you think would win the case in court?

10. A weightlifter stands on a bathroom scale. He pumps a barbell up and down. What happens to the reading on the bathroom scale as this is done? **What if** he is strong enough to actually *throw* the barbell upward? How does the reading on the scale vary now?

11. Suppose a truck loaded with sand accelerates along a highway. If the driving force on the truck remains constant, what happens to the truck's acceleration if its trailer leaks sand at a constant rate through a hole in its bottom?

12. As a rocket is fired from a launching pad, its speed and acceleration increase with time as its engines continue to op-

erate. Explain why this occurs even though the thrust of the engines remains constant.

13. What force causes an automobile to move? A propeller-driven airplane? A rowboat?

14. Identify the action–reaction pairs in the following situations: a man takes a step; a snowball hits a girl in the back; a baseball player catches a ball; a gust of wind strikes a window.

15. In a contest of National Football League behemoths, teams from the Rams and the 49ers engage in a tug-of-war, pulling in opposite directions on a strong rope. The Rams exert a force of 9 200 N and they are winning, making the center of the rope move steadily toward themselves. Is it possible to know the tension in the rope from the information stated? Is it larger or smaller than 9 200 N? How hard are the 49ers pulling on the rope? Would it change your answer if the 49ers were winning or if the contest were even? The stronger team wins by exerting a larger force—on what? Explain your answers.

16. Twenty people participate in a tug-of-war. The two teams of ten people are so evenly matched that neither team wins. After the game they notice that a car is stuck in the mud. They attach the tug-of-war rope to the bumper of the car, and all the people pull on the rope. The heavy car has just moved a couple of decimeters when the rope breaks. Why did the rope break in this situation when it did not break when the same twenty people pulled on it in a tug-of-war?

17. "When the locomotive in Figure Q5.17 broke through the wall of the train station, the force exerted by the locomotive on the wall was greater than the force the wall could exert on the locomotive." Is this statement true or in need of correction? Explain your answer.

18. An athlete grips a light rope that passes over a low-friction pulley attached to the ceiling of a gym. A sack of sand precisely equal in weight to the athlete is tied to the other end of the rope. Both the sand and the athlete are initially at rest. The athlete climbs the rope, sometimes speeding up and slowing down as he does so. What happens to the sack of sand? Explain.

19. If the action and reaction forces are always equal in magnitude and opposite in direction to each other, then doesn't the net vector force on any object necessarily add up to zero? Explain your answer.

20. Can an object exert a force on itself? Argue for your answer.

21. If you push on a heavy box that is at rest, you must exert some force to start its motion. However, once the box is

Figure Q5.17

sliding, you can apply a smaller force to maintain that motion. Why?

22. The driver of a speeding empty truck slams on the brakes and skids to a stop through a distance d. (a) If the truck carried a load that doubled its mass, what would be the truck's "skidding distance"? (b) If the initial speed of the truck were halved, what would be the truck's skidding distance?

23. Suppose you are driving a classic car. Why should you avoid slamming on your brakes when you want to stop in the shortest possible distance? (Many cars have antilock brakes that avoid this problem.)

24. A book is given a brief push to make it slide up a rough incline. It comes to a stop and slides back down to the starting point. Does it take the same time to go up as to come down? **What if** the incline is frictionless?

25. A large crate is placed on the bed of a truck but not tied down. (a) As the truck accelerates forward, the crate remains at rest relative to the truck. What force causes the crate to accelerate forward? (b) If the driver slammed on the brakes, what could happen to the crate?

26. Describe a few examples in which the force of friction exerted on an object is in the direction of motion of the object.

PROBLEMS

1, 2, 3 = straightforward, intermediate, challenging ☐ = full solution available in the *Student Solutions Manual and Study Guide*

🪐 = coached solution with hints available at http://www.pse6.com   💻 = computer useful in solving problem

▨ = paired numerical and symbolic problems

Sections 5.1 through 5.6

1. A force **F** applied to an object of mass m_1 produces an acceleration of 3.00 m/s². The same force applied to a second object of mass m_2 produces an acceleration of 1.00 m/s². (a) What is the value of the ratio m_1/m_2? (b) If m_1 and m_2 are combined, find their acceleration under the action of the force **F**.

2. The largest-caliber antiaircraft gun operated by the German air force during World War II was the 12.8-cm Flak 40. This weapon fired a 25.8-kg shell with a muzzle speed of 880 m/s. What propulsive force was necessary to attain the muzzle speed within the 6.00-m barrel? (Assume the shell moves horizontally with constant acceleration and neglect friction.)

3. A 3.00-kg object undergoes an acceleration given by $\mathbf{a} = (2.00\hat{\mathbf{i}} + 5.00\hat{\mathbf{j}})$ m/s². Find the resultant force acting on it and the magnitude of the resultant force.

4. The gravitational force on a baseball is $-F_g\hat{\mathbf{j}}$. A pitcher throws the baseball with velocity $v\hat{\mathbf{i}}$ by uniformly accelerating it straight forward horizontally for a time interval $\Delta t = t - 0 = t$. If the ball starts from rest, (a) through what distance does it accelerate before its release? (b) What force does the pitcher exert on the ball?

5. 🪐 To model a spacecraft, a toy rocket engine is securely fastened to a large puck, which can glide with negligible friction over a horizontal surface, taken as the xy plane. The 4.00-kg puck has a velocity of $300\hat{\mathbf{i}}$ m/s at one instant. Eight seconds later, its velocity is to be $(800\hat{\mathbf{i}} + 10.0\hat{\mathbf{j}})$ m/s. Assuming the rocket engine exerts a constant horizontal force, find (a) the components of the force and (b) its magnitude.

6. The average speed of a nitrogen molecule in air is about 6.70×10^2 m/s, and its mass is 4.68×10^{-26} kg. (a) If it takes 3.00×10^{-13} s for a nitrogen molecule to hit a wall and rebound with the same speed but moving in the opposite direction, what is the average acceleration of the molecule during this time interval? (b) What average force does the molecule exert on the wall?

7. An electron of mass 9.11×10^{-31} kg has an initial speed of 3.00×10^5 m/s. It travels in a straight line, and its speed increases to 7.00×10^5 m/s in a distance of 5.00 cm. Assuming its acceleration is constant, (a) determine the force exerted on the electron and (b) compare this force with the weight of the electron, which we neglected.

8. A woman weighs 120 lb. Determine (a) her weight in newtons (N) and (b) her mass in kilograms (kg).

9. If a man weighs 900 N on the Earth, what would he weigh on Jupiter, where the acceleration due to gravity is 25.9 m/s²?

10. The distinction between mass and weight was discovered after Jean Richer transported pendulum clocks from Paris to French Guyana in 1671. He found that they ran slower there quite systematically. The effect was reversed when the clocks returned to Paris. How much weight would you personally lose in traveling from Paris, where $g = 9.809\ 5$ m/s², to Cayenne, where $g = 9.780\ 8$ m/s²? [We will consider how the free-fall acceleration influences the period of a pendulum in Section 15.5.]

11. Two forces $\mathbf{F}_1$ and $\mathbf{F}_2$ act on a 5.00-kg object. If $F_1 = 20.0$ N and $F_2 = 15.0$ N, find the accelerations in (a) and (b) of Figure P5.11.

Figure P5.11

12. Besides its weight, a 2.80-kg object is subjected to one other constant force. The object starts from rest and in 1.20 s experiences a displacement of $(4.20\hat{\mathbf{i}} - 3.30\hat{\mathbf{j}})$ m, where the direction of $\hat{\mathbf{j}}$ is the upward vertical direction. Determine the other force.

13. You stand on the seat of a chair and then hop off. (a) During the time you are in flight down to the floor, the Earth is lurching up toward you with an acceleration of what order of magnitude? In your solution explain your logic. Model the Earth as a perfectly solid object. (b) The Earth moves up through a distance of what order of magnitude?

14. Three forces acting on an object are given by $\mathbf{F}_1 = (-2.00\hat{\mathbf{i}} + 2.00\hat{\mathbf{j}})$ N, $\mathbf{F}_2 = (5.00\hat{\mathbf{i}} - 3.00\hat{\mathbf{j}})$ N, and $\mathbf{F}_3 = (-45.0\hat{\mathbf{i}})$ N. The object experiences an acceleration of magnitude 3.75 m/s². (a) What is the direction of the acceleration? (b) What is the mass of the object? (c) If the object is initially at rest, what is its speed after 10.0 s? (d) What are the velocity components of the object after 10.0 s?

15. A 15.0-lb block rests on the floor. (a) What force does the floor exert on the block? (b) If a rope is tied to the block and run vertically over a pulley, and the other end is attached to a free-hanging 10.0-lb weight, what is the force exerted by the floor on the 15.0-lb block? (c) If we replace the 10.0-lb weight in part (b) with a 20.0-lb weight, what is the force exerted by the floor on the 15.0-lb block?

Section 5.7   Some Applications of Newton's Laws

16. A 3.00-kg object is moving in a plane, with its x and y coordinates given by $x = 5t^2 - 1$ and $y = 3t^3 + 2$, where x and y are in meters and t is in seconds. Find the magnitude of the net force acting on this object at $t = 2.00$ s.

17. The distance between two telephone poles is 50.0 m. When a 1.00-kg bird lands on the telephone wire midway between the poles, the wire sags 0.200 m. Draw a free-body diagram of the bird. How much tension does the bird produce in the wire? Ignore the weight of the wire.

18. A bag of cement of weight 325 N hangs from three wires as suggested in Figure P5.18. Two of the wires make angles $\theta_1 = 60.0°$ and $\theta_2 = 25.0°$ with the horizontal. If the system is in equilibrium, find the tensions T_1, T_2, and T_3 in the wires.

Figure P5.18 Problems 18 and 19.

19. A bag of cement of weight F_g hangs from three wires as shown in Figure P5.18. Two of the wires make angles θ_1 and θ_2 with the horizontal. If the system is in equilibrium, show that the tension in the left-hand wire is

$$T_1 = F_g \cos \theta_2 / \sin (\theta_1 + \theta_2)$$

20. You are a judge in a children's kite-flying contest, and two children will win prizes for the kites that pull most strongly and least strongly on their strings. To measure string tensions, you borrow a weight hanger, some slotted weights, and a protractor from your physics teacher, and use the following protocol, illustrated in Figure P5.20: Wait for a child to get her kite well controlled, hook the hanger onto the kite string about 30 cm from her hand, pile on weight until that section of string is horizontal, record the mass required, and record the angle between the horizontal and the string running up to the kite. (a) Explain how this method works. As you construct your explanation, imagine that the children's parents ask you about your method, that they might make false assumptions about your ability without concrete evidence, and that your explanation is an opportunity to give them confidence in your evaluation

technique. (b) Find the string tension if the mass is 132 g and the angle of the kite string is 46.3°.

Figure P5.20

21. The systems shown in Figure P5.21 are in equilibrium. If the spring scales are calibrated in newtons, what do they read? (Neglect the masses of the pulleys and strings, and assume the incline in part (c) is frictionless.)

(a)

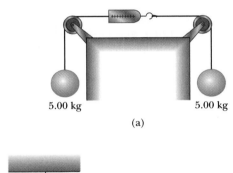

5.00 kg

30.0°

(c)

5.00 kg   5.00 kg

(b)

Figure P5.21

22. Draw a free-body diagram of a block which slides down a frictionless plane having an inclination of $\theta = 15.0°$ (Fig. P5.22). The block starts from rest at the top and the length of the incline is 2.00 m. Find (a) the acceleration of

the block and (b) its speed when it reaches the bottom of the incline.

Figure P5.22 Problems 22 and 25.

23. A 1.00-kg object is observed to have an acceleration of 10.0 m/s² in a direction 30.0° north of east (Fig. P5.23). The force $\mathbf{F}_2$ acting on the object has a magnitude of 5.00 N and is directed north. Determine the magnitude and direction of the force $\mathbf{F}_1$ acting on the object.

Figure P5.23

24. A 5.00-kg object placed on a frictionless, horizontal table is connected to a string that passes over a pulley and then is fastened to a hanging 9.00-kg object, as in Figure P5.24. Draw free-body diagrams of both objects. Find the acceleration of the two objects and the tension in the string.

Figure P5.24 Problems 24 and 43.

25. A block is given an initial velocity of 5.00 m/s up a frictionless 20.0° incline (Fig. P5.22). How far up the incline does the block slide before coming to rest?

26. Two objects are connected by a light string that passes over a frictionless pulley, as in Figure P5.26. Draw free-body diagrams of both objects. If the incline is frictionless and if $m_1 = 2.00$ kg, $m_2 = 6.00$ kg, and $\theta = 55.0°$, find (a) the accelerations of the objects, (b) the tension in the string, and (c) the speed of each object 2.00 s after being released from rest.

Figure P5.26

27. A tow truck pulls a car that is stuck in the mud, with a force of 2 500 N as in Fig. P5.27. The tow cable is under tension and therefore pulls downward and to the left on the pin at its upper end. The light pin is held in equilibrium by forces exerted by the two bars A and B. Each bar is a *strut*: that is, each is a bar whose weight is small compared to the forces it exerts, and which exerts forces only through hinge pins at its ends. Each strut exerts a force directed parallel to its length. Determine the force of tension or compression in each strut. Proceed as follows: Make a guess as to which way (pushing or pulling) each force acts on the top pin. Draw a free-body diagram of the pin. Use the condition for equilibrium of the pin to translate the free-body diagram into equations. From the equations calculate the forces exerted by struts A and B. If you obtain a positive answer, you correctly guessed the direction of the force. A negative answer means the direction should be reversed, but the absolute value correctly gives the magnitude of the force. If a strut pulls on a pin, it is in tension. If it pushes, the strut is in compression. Identify whether each strut is in tension or in compression.

Figure P5.27

28. Two objects with masses of 3.00 kg and 5.00 kg are connected by a light string that passes over a light frictionless pulley to form an Atwood machine, as in Figure 5.14a. Determine (a) the tension in the string, (b) the acceleration of each object, and (c) the distance each object will move in the first second of motion if they start from rest.

29. In Figure P5.29, the man and the platform together weigh 950 N. The pulley can be modeled as frictionless. Determine how hard the man has to pull on the rope to lift himself steadily upward above the ground. (Or is it impossible? If so, explain why.)

Figure P5.29

30. In the Atwood machine shown in Figure 5.14a, $m_1 = 2.00$ kg and $m_2 = 7.00$ kg. The masses of the pulley and string are negligible by comparison. The pulley turns without friction and the string does not stretch. The lighter object is released with a sharp push that sets it into motion at $v_i = 2.40$ m/s downward. (a) How far will m_1 descend below its initial level? (b) Find the velocity of m_1 after 1.80 seconds.

31. In the system shown in Figure P5.31, a horizontal force $\mathbf{F}_x$ acts on the 8.00-kg object. The horizontal surface is frictionless. (a) For what values of F_x does the 2.00-kg object accelerate upward? (b) For what values of F_x is the tension in the cord zero? (c) Plot the acceleration of the 8.00-kg object versus F_x. Include values of F_x from -100 N to $+100$ N.

Figure P5.31

32. A frictionless plane is 10.0 m long and inclined at 35.0°. A sled starts at the bottom with an initial speed of 5.00 m/s up the incline. When it reaches the point at which it momentarily stops, a second sled is released from the top of this incline with an initial speed v_i. Both sleds reach the bottom of the incline at the same moment. (a) Determine the distance that the first sled traveled up the incline. (b) Determine the initial speed of the second sled.

33. A 72.0-kg man stands on a spring scale in an elevator. Starting from rest, the elevator ascends, attaining its maximum speed of 1.20 m/s in 0.800 s. It travels with this constant speed for the next 5.00 s. The elevator then undergoes a uniform acceleration in the negative y direction for 1.50 s and comes to rest. What does the spring scale register (a) before the elevator starts to move? (b) during the first 0.800 s? (c) while the elevator is traveling at constant speed? (d) during the time it is slowing down?

34. An object of mass m_1 on a frictionless horizontal table is connected to an object of mass m_2 through a very light pulley P_1 and a light fixed pulley P_2 as shown in Figure P5.34. (a) If a_1 and a_2 are the accelerations of m_1 and m_2, respectively, what is the relation between these accelerations? Express (b) the tensions in the strings and (c) the accelerations a_1 and a_2 in terms of the masses m_1 and m_2, and g.

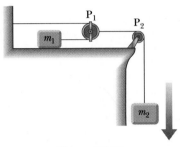

Figure P5.34

Section 5.8 Forces of Friction

35. The person in Figure P5.35 weighs 170 lb. As seen from the front, each light crutch makes an angle of 22.0° with the vertical. Half of the person's weight is supported by the crutches. The other half is supported by the vertical forces of the ground on his feet. Assuming the person is moving

Figure P5.35

with constant velocity and the force exerted by the ground on the crutches acts along the crutches, determine (a) the smallest possible coefficient of friction between crutches and ground and (b) the magnitude of the compression force in each crutch.

36. A 25.0-kg block is initially at rest on a horizontal surface. A horizontal force of 75.0 N is required to set the block in motion. After it is in motion, a horizontal force of 60.0 N is required to keep the block moving with constant speed. Find the coefficients of static and kinetic friction from this information.

37. A car is traveling at 50.0 mi/h on a horizontal highway. (a) If the coefficient of static friction between road and tires on a rainy day is 0.100, what is the minimum distance in which the car will stop? (b) What is the stopping distance when the surface is dry and $\mu_s = 0.600$?

38. Before 1960 it was believed that the maximum attainable coefficient of static friction for an automobile tire was less than 1. Then, about 1962, three companies independently developed racing tires with coefficients of 1.6. Since then, tires have improved, as illustrated in this problem. According to the 1990 Guinness Book of Records, the shortest time in which a piston-engine car initially at rest has covered a distance of one-quarter mile is 4.96 s. This record was set by Shirley Muldowney in September 1989. (a) Assume that, as in Figure P5.38, the rear wheels lifted the front wheels off the pavement. What minimum value of μ_s is necessary to achieve the record time? (b) Suppose Muldowney were able to double her engine power, keeping other things equal. How would this change affect the elapsed time?

Figure P5.38

39. To meet a U.S. Postal Service requirement, footwear must have a coefficient of static friction of 0.5 or more on a specified tile surface. A typical athletic shoe has a coefficient of 0.800. In an emergency, what is the minimum time interval in which a person starting from rest can move 3.00 m on a tile surface if she is wearing (a) footwear meeting the Postal Service minimum? (b) a typical athletic shoe?

40. A woman at an airport is towing her 20.0-kg suitcase at constant speed by pulling on a strap at an angle θ above the horizontal (Fig. P5.40). She pulls on the strap with a 35.0-N force, and the friction force on the suitcase is 20.0 N. Draw a free-body diagram of the suitcase. (a) What angle does the strap make with the horizontal? (b) What normal force does the ground exert on the suitcase?

Figure P5.40

41. A 3.00-kg block starts from rest at the top of a 30.0° incline and slides a distance of 2.00 m down the incline in 1.50 s. Find (a) the magnitude of the acceleration of the block, (b) the coefficient of kinetic friction between block and plane, (c) the friction force acting on the block, and (d) the speed of the block after it has slid 2.00 m.

42. A Chevrolet Corvette convertible can brake to a stop from a speed of 60.0 mi/h in a distance of 123 ft on a level roadway. What is its stopping distance on a roadway sloping downward at an angle of 10.0°?

43. A 9.00-kg hanging weight is connected by a string over a pulley to a 5.00-kg block that is sliding on a flat table (Fig. P5.24). If the coefficient of kinetic friction is 0.200, find the tension in the string.

44. Three objects are connected on the table as shown in Figure P5.44. The table is rough and has a coefficient of kinetic friction of 0.350. The objects have masses of 4.00 kg, 1.00 kg, and 2.00 kg, as shown, and the pulleys are frictionless. Draw free-body diagrams of each of the objects. (a) Determine the acceleration of each object and their directions. (b) Determine the tensions in the two cords.

Figure P5.44

45. Two blocks connected by a rope of negligible mass are being dragged by a horizontal force $\mathbf{F}$ (Fig. P5.45). Suppose that $F = 68.0$ N, $m_1 = 12.0$ kg, $m_2 = 18.0$ kg, and the coefficient of kinetic friction between each block and the surface is 0.100. (a) Draw a free-body diagram for each block.

(b) Determine the tension T and the magnitude of the acceleration of the system.

Figure P5.45

46. A block of mass 3.00 kg is pushed up against a wall by a force **P** that makes a 50.0° angle with the horizontal as shown in Figure P5.46. The coefficient of static friction between the block and the wall is 0.250. Determine the possible values for the magnitude of **P** that allow the block to remain stationary.

Figure P5.46

47. You and your friend go sledding. Out of curiosity, you measure the constant angle θ that the snow-covered slope makes with the horizontal. Next, you use the following method to determine the coefficient of friction μ_k between the snow and the sled. You give the sled a quick push up so that it will slide up the slope away from you. You wait for it to slide back down, timing the motion. It turns out that the sled takes twice as long to slide down as it does to reach the top point in the round trip. In terms of θ, what is the coefficient of friction?

48. The board sandwiched between two other boards in Figure P5.48 weighs 95.5 N. If the coefficient of friction between the boards is 0.663, what must be the magnitude of the compression forces (assume horizontal) acting on both sides of the center board to keep it from slipping?

Figure P5.48

49. A block weighing 75.0 N rests on a plane inclined at 25.0° to the horizontal. A force F is applied to the object at 40.0° to the horizontal, pushing it upward on the plane. The coefficients of static and kinetic friction between the block and the plane are, respectively, 0.363 and 0.156. (a) What is the minimum value of F that will prevent the block from slipping down the plane? (b) What is the minimum value of F that will start the block moving up the plane? (c) What

value of F will move the block up the plane with constant velocity?

50. **Review problem.** One side of the roof of a building slopes up at 37.0°. A student throws a Frisbee onto the roof. It strikes with a speed of 15.0 m/s and does not bounce, but slides straight up the incline. The coefficient of kinetic friction between the plastic and the roof is 0.400. The Frisbee slides 10.0 m up the roof to its peak, where it goes into free fall, following a parabolic trajectory with negligible air resistance. Determine the maximum height the Frisbee reaches above the point where it struck the roof.

Additional Problems

51. An inventive child named Pat wants to reach an apple in a tree without climbing the tree. Sitting in a chair connected to a rope that passes over a frictionless pulley (Fig. P5.51), Pat pulls on the loose end of the rope with such a force that the spring scale reads 250 N. Pat's true weight is 320 N, and the chair weighs 160 N. (a) Draw free-body diagrams for Pat and the chair considered as separate systems, and another diagram for Pat and the chair considered as one system. (b) Show that the acceleration of the system is *upward* and find its magnitude. (c) Find the force Pat exerts on the chair.

Figure P5.51

52. A time-dependent force, $\mathbf{F} = (8.00\hat{\mathbf{i}} - 4.00t\hat{\mathbf{j}})$ N, where t is in seconds, is exerted on a 2.00-kg object initially at rest. (a) At what time will the object be moving with a speed of 15.0 m/s? (b) How far is the object from its initial position when its speed is 15.0 m/s? (c) Through what total displacement has the object traveled at this time?

53. To prevent a box from sliding down an inclined plane, student A pushes on the box in the direction parallel to the incline, just hard enough to hold the box stationary. In an identical situation student B pushes on the box horizontally. Regard as known the mass m of the box, the coefficient of static friction μ_s between box and incline, and the inclination angle θ. (a) Determine the force A

has to exert. (b) Determine the force B has to exert. (c) If $m = 2.00$ kg, $\theta = 25.0°$, and $\mu_s = 0.160$, who has the easier job? (d) **What if** $\mu_s = 0.380$? Whose job is easier?

54. Three blocks are in contact with each other on a frictionless, horizontal surface, as in Figure P5.54. A horizontal force **F** is applied to m_1. Take $m_1 = 2.00$ kg, $m_2 = 3.00$ kg, $m_3 = 4.00$ kg, and $F = 18.0$ N. Draw a separate free-body diagram for each block and find (a) the acceleration of the blocks, (b) the *resultant* force on each block, and (c) the magnitudes of the contact forces between the blocks. (d) You are working on a construction project. A coworker is nailing up plasterboard on one side of a light partition, and you are on the opposite side, providing "backing" by leaning against the wall with your back pushing on it. Every blow makes your back sting. The supervisor helps you to put a heavy block of wood between the wall and your back. Using the situation analyzed in parts (a), (b), and (c) as a model, explain how this works to make your job more comfortable.

Figure P5.54

55. An object of mass M is held in place by an applied force **F** and a pulley system as shown in Figure P5.55. The pulleys are massless and frictionless. Find (a) the tension in each section of rope, T_1, T_2, T_3, T_4, and T_5 and (b) the magnitude of **F**. *Suggestion:* Draw a free-body diagram for each pulley.

Figure P5.55

56. A high diver of mass 70.0 kg jumps off a board 10.0 m above the water. If his downward motion is stopped 2.00 s after he enters the water, what average upward force did the water exert on him?

57. A crate of weight F_g is pushed by a force **P** on a horizontal floor. (a) If the coefficient of static friction is μ_s and **P** is directed at angle θ below the horizontal, show that the minimum value of P that will move the crate is given by

$$P = \frac{\mu_s F_g \sec \theta}{1 - \mu_s \tan \theta}$$

(b) Find the minimum value of P that can produce motion when $\mu_s = 0.400$, $F_g = 100$ N, and $\theta = 0°$, $15.0°$, $30.0°$, $45.0°$, and $60.0°$.

58. **Review problem**. A block of mass $m = 2.00$ kg is released from rest at $h = 0.500$ m above the surface of a table, at the top of a $\theta = 30.0°$ incline as shown in Figure P5.58. The frictionless incline is fixed on a table of height $H = 2.00$ m. (a) Determine the acceleration of the block as it slides down the incline. (b) What is the velocity of the block as it leaves the incline? (c) How far from the table will the block hit the floor? (d) How much time has elapsed between when the block is released and when it hits the floor? (e) Does the mass of the block affect any of the above calculations?

Figure P5.58 Problems 58 and 70.

59. A 1.30-kg toaster is not plugged in. The coefficient of static friction between the toaster and a horizontal countertop is 0.350. To make the toaster start moving, you carelessly pull on its electric cord. (a) For the cord tension to be as small as possible, you should pull at what angle above the horizontal? (b) With this angle, how large must the tension be?

60. Materials such as automobile tire rubber and shoe soles are tested for coefficients of static friction with an apparatus called a James tester. The pair of surfaces for which μ_s is to be measured are labeled B and C in Figure P5.60. Sample C is attached to a foot D at the lower end of a pivoting arm E, which makes angle θ with the vertical. The upper end of the arm is hinged at F to a vertical rod G, which slides freely in a guide H fixed to the frame of the apparatus and supports a load I of mass 36.4 kg. The hinge pin at F is also the axle of a wheel that can roll vertically on the frame. All of the moving parts have masses negligible in comparison to the 36.4-kg load. The pivots are nearly frictionless. The test surface B is attached to a

rolling platform A. The operator slowly moves the platform to the left in the picture until the sample C suddenly slips over surface B. At the critical point where sliding motion is ready to begin, the operator notes the angle θ_s of the pivoting arm. (a) Make a free-body diagram of the pin at F. It is in equilibrium under three forces. These forces are the gravitational force on the load I, a horizontal normal force exerted by the frame, and a force of compression directed upward along the arm E. (b) Draw a free-body diagram of the foot D and sample C, considered as one system. (c) Determine the normal force that the test surface B exerts on the sample for any angle θ. (d) Show that $\mu_s = \tan \theta_s$. (e) The protractor on the tester can record angles as large as 50.2°. What is the greatest coefficient of friction it can measure?

Figure P5.60

61. What horizontal force must be applied to the cart shown in Figure P5.61 in order that the blocks remain stationary relative to the cart? Assume all surfaces, wheels, and pulley are frictionless. (*Hint:* Note that the force exerted by the string accelerates m_1.)

Figure P5.61 Problems 61 and 63.

62. A student is asked to measure the acceleration of a cart on a "frictionless" inclined plane as in Figure 5.11, using an air track, a stopwatch, and a meter stick. The height of the incline is measured to be 1.774 cm, and the total length of the incline is measured to be $d = 127.1$ cm. Hence, the angle of inclination θ is determined from the relation

$\sin \theta = 1.774/127.1$. The cart is released from rest at the top of the incline, and its position x along the incline is measured as a function of time, where $x = 0$ refers to the initial position of the cart. For x values of 10.0 cm, 20.0 cm, 35.0 cm, 50.0 cm, 75.0 cm, and 100 cm, the measured times at which these positions are reached (averaged over five runs) are 1.02 s, 1.53 s, 2.01 s, 2.64 s, 3.30 s, and 3.75 s, respectively. Construct a graph of x versus t^2, and perform a linear least-squares fit to the data. Determine the acceleration of the cart from the slope of this graph, and compare it with the value you would get using $a' = g \sin \theta$, where $g = 9.80 \text{ m/s}^2$.

63. Initially the system of objects shown in Figure P5.61 is held motionless. All surfaces, pulley, and wheels are frictionless. Let the force **F** be zero and assume that m_2 can move only vertically. At the instant after the system of objects is released, find (a) the tension T in the string, (b) the acceleration of m_2, (c) the acceleration of M, and (d) the acceleration of m_1. (*Note:* The pulley accelerates along with the cart.)

64. One block of mass 5.00 kg sits on top of a second rectangular block of mass 15.0 kg, which in turn is on a horizontal table. The coefficients of friction between the two blocks are $\mu_s = 0.300$ and $\mu_k = 0.100$. The coefficients of friction between the lower block and the rough table are $\mu_s = 0.500$ and $\mu_k = 0.400$. You apply a constant horizontal force to the lower block, just large enough to make this block start sliding out from between the upper block and the table. (a) Draw a free-body diagram of each block, naming the forces on each. (b) Determine the magnitude of each force on each block at the instant when you have started pushing but motion has not yet started. In particular, what force must you apply? (c) Determine the acceleration you measure for each block.

65. A 1.00-kg glider on a horizontal air track is pulled by a string at an angle θ. The taut string runs over a pulley and is attached to a hanging object of mass 0.500 kg as in Fig. P5.65. (a) Show that the speed v_x of the glider and the

Figure P5.65

speed v_y of the hanging object are related by $v_x = uv_y$, where $u = z(z^2 - h_0^2)^{-1/2}$. (b) The glider is released from rest. Show that at that instant the acceleration a_x of the glider and the acceleration a_y of the hanging object are related by $a_x = ua_y$. (c) Find the tension in the string at the instant the glider is released for $h_0 = 80.0$ cm and $\theta = 30.0°$.

66. Cam mechanisms are used in many machines. For example, cams open and close the valves in your car engine to admit gasoline vapor to each cylinder and to allow the escape of exhaust. The principle is illustrated in Figure P5.66, showing a follower rod (also called a pushrod) of mass m resting on a wedge of mass M. The sliding wedge duplicates the function of a rotating eccentric disk on a camshaft in your car. Assume that there is no friction between the wedge and the base, between the pushrod and the wedge, or between the rod and the guide through which it slides. When the wedge is pushed to the left by the force F, the rod moves upward and does something, such as opening a valve. By varying the shape of the wedge, the motion of the follower rod could be made quite complex, but assume that the wedge makes a constant angle of $\theta = 15.0°$. Suppose you want the wedge and the rod to start from rest and move with constant acceleration, with the rod moving upward 1.00 mm in 8.00 ms. Take $m = 0.250$ kg and $M = 0.500$ kg. What force F must be applied to the wedge?

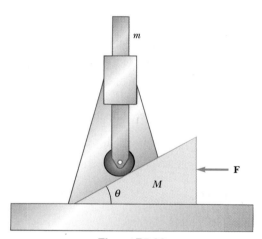

Figure P5.66

67. Any device that allows you to increase the force you exert is a kind of *machine*. Some machines, such as the pry-bar or the inclined plane, are very simple. Some machines do not even look like machines. An example is the following: Your car is stuck in the mud, and you can't pull hard enough to get it out. However, you have a long cable which you connect taut between your front bumper and the trunk of a stout tree. You now pull sideways on the cable at its midpoint, exerting a force f. Each half of the cable is displaced through a small angle θ from the straight line between the ends of the cable. (a) Deduce an expression for the force exerted on the car. (b) Evaluate the cable tension for the case where $\theta = 7.00°$ and $f = 100$ N.

68. Two blocks of mass 3.50 kg and 8.00 kg are connected by a massless string that passes over a frictionless pulley (Fig. P5.68). The inclines are frictionless. Find (a) the magnitude of the acceleration of each block and (b) the tension in the string.

Figure P5.68

69. A van accelerates down a hill (Fig. P5.69), going from rest to 30.0 m/s in 6.00 s. During the acceleration, a toy ($m = 0.100$ kg) hangs by a string from the van's ceiling. The acceleration is such that the string remains perpendicular to the ceiling. Determine (a) the angle θ and (b) the tension in the string.

Figure P5.69

70. In Figure P5.58 the incline has mass M and is fastened to the stationary horizontal tabletop. The block of mass m is placed near the bottom of the incline and is released with a quick push that sets it sliding upward. It stops near the top of the incline, as shown in the figure, and then slides down again, always without friction. Find the force that the tabletop exerts on the incline throughout this motion.

71. A magician pulls a tablecloth from under a 200-g mug located 30.0 cm from the edge of the cloth. The cloth exerts a friction force of 0.100 N on the mug, and the cloth is pulled with a constant acceleration of 3.00 m/s². How far does the mug move relative to the horizontal tabletop before the cloth is completely out from under it? Note that the cloth must move more than 30 cm relative to the tabletop during the process.

72. 🖥 An 8.40-kg object slides down a fixed, frictionless inclined plane. Use a computer to determine and tabulate the normal force exerted on the object and its acceleration for a series of incline angles (measured from the horizontal) ranging from 0° to 90° in 5° increments. Plot a graph of the normal force and the acceleration as functions of the incline angle. In the limiting cases of 0° and 90°, are your results consistent with the known behavior?

73. A mobile is formed by supporting four metal butterflies of equal mass m from a string of length L. The points of support are evenly spaced a distance ℓ apart as shown in Figure P5.73. The string forms an angle θ_1 with the ceiling at each end point. The center section of string is horizontal. (a) Find the tension in each section of string in terms of θ_1, m, and g. (b) Find the angle θ_2, in terms of θ_1, that the sections of string between the outside butterflies and the inside butterflies form with the horizontal. (c) Show that the distance D between the end points of the string is

$$D = \frac{L}{5} \left(2 \cos \theta_1 + 2 \cos \left[\tan^{-1}(\tfrac{1}{2} \tan \theta_1) \right] + 1 \right)$$

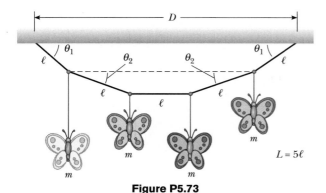

Figure P5.73

Answers to Quick Quizzes

5.1 (d). Choice (a) is true. Newton's first law tells us that motion requires no force: an object in motion continues to move at constant velocity in the absence of external forces. Choice (b) is also true. A stationary object can have several forces acting on it, but if the vector sum of all these external forces is zero, there is no net force and the object remains stationary.

5.2 (a). If a single force acts, this force constitutes the net force and there is an acceleration according to Newton's second law.

5.3 (c). Newton's second law relates only the force and the acceleration. Direction of motion is part of an object's *velocity*, and force determines the direction of acceleration, not that of velocity.

5.4 (d). With twice the force, the object will experience twice the acceleration. Because the force is constant, the acceleration is constant, and the speed of the object (starting from rest) is given by $v = at$. With twice the acceleration, the object will arrive at speed v at half the time.

5.5 (a). The gravitational force acts on the ball at *all* points in its trajectory.

5.6 (b). Because the value of g is smaller on the Moon than on the Earth, more mass of gold would be required to represent 1 newton of weight on the Moon. Thus, your friend on the Moon is richer, by about a factor of 6!

5.7 (c). In accordance with Newton's third law, the fly and bus experience forces that are equal in magnitude but opposite in direction.

5.8 (a). Because the fly has such a small mass, Newton's second law tells us that it undergoes a very large acceleration. The huge mass of the bus means that it more effectively resists any change in its motion and exhibits a small acceleration.

5.9 (c). The reaction force to your weight is an upward gravitational force on the Earth due to you.

5.10 (b). Remember the phrase "free-body." You draw *one* body (one object), free of all the others that may be interacting, and draw only the forces exerted on that object.

5.11 (b). The friction force acts opposite to the gravitational force on the book to keep the book in equilibrium. Because the gravitational force is downward, the friction force must be upward.

5.12 (b). The crate accelerates to the east. Because the only horizontal force acting on it is the force of static friction between its bottom surface and the truck bed, that force must also be directed to the east.

5.13 (b). At the angle at which the book breaks free, the component of the gravitational force parallel to the board is approximately equal to the maximum static friction force. Because the kinetic coefficient of friction is smaller than the static coefficient, at this angle, the component of the gravitational force parallel to the board is larger than the kinetic friction force. Thus, there is a net downhill force parallel to the board and the book speeds up.

5.14 (b). When pulling with the rope, there is a component of your applied force that is upward. This reduces the normal force between the sled and the snow. In turn, this reduces the friction force between the sled and the snow, making it easier to move. If you push from behind, with a force with a downward component, the normal force is larger, the friction force is larger, and the sled is harder to move.

Chapter 6

Circular Motion and Other Applications of Newton's Laws

▲ The London Eye, a ride on the River Thames in downtown London. Riders travel in a large vertical circle for a breathtaking view of the city. In this chapter, we will study the forces involved in circular motion. (© Paul Hardy/CORBIS)

In the preceding chapter we introduced Newton's laws of motion and applied them to situations involving linear motion. Now we discuss motion that is slightly more complicated. For example, we shall apply Newton's laws to objects traveling in circular paths. Also, we shall discuss motion observed from an accelerating frame of reference and motion of an object through a viscous medium. For the most part, this chapter consists of a series of examples selected to illustrate the application of Newton's laws to a wide variety of circumstances.

6.1 Newton's Second Law Applied to Uniform Circular Motion

In Section 4.4 we found that a particle moving with uniform speed v in a circular path of radius r experiences an acceleration that has a magnitude

$$a_c = \frac{v^2}{r}$$

The acceleration is called *centripetal acceleration* because $\mathbf{a}_c$ is directed toward the center of the circle. Furthermore, $\mathbf{a}_c$ is *always* perpendicular to $\mathbf{v}$. (If there were a component of acceleration parallel to $\mathbf{v}$, the particle's speed would be changing.)

Consider a ball of mass m that is tied to a string of length r and is being whirled at constant speed in a horizontal circular path, as illustrated in Figure 6.1. Its weight is supported by a frictionless table. Why does the ball move in a circle? According to Newton's first law, the ball tends to move in a straight line; however, the string prevents

An athlete in the process of throwing the hammer at the 1996 Olympic Games in Atlanta, Georgia. The force exerted by the chain causes the centripetal acceleration of the hammer. Only when the athlete releases the hammer will it move along a straight-line path tangent to the circle.

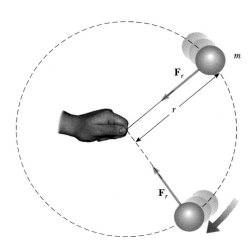

Figure 6.1 Overhead view of a ball moving in a circular path in a horizontal plane. A force $\mathbf{F}_r$ directed toward the center of the circle keeps the ball moving in its circular path.

motion along a straight line by exerting on the ball a radial force $\mathbf{F}_r$ that makes it follow the circular path. This force is directed along the string toward the center of the circle, as shown in Figure 6.1.

If we apply Newton's second law along the radial direction, we find that the net force causing the centripetal acceleration can be evaluated:

Force causing centripetal acceleration

$$\sum F = ma_c = m\frac{v^2}{r} \qquad (6.1)$$

A force causing a centripetal acceleration acts toward the center of the circular path and causes a change in the direction of the velocity vector. If that force should vanish, the object would no longer move in its circular path; instead, it would move along a straight-line path tangent to the circle. This idea is illustrated in Figure 6.2 for the ball whirling at the end of a string in a horizontal plane. If the string breaks at some instant, the ball moves along the straight-line path tangent to the circle at the point where the string breaks.

Figure 6.3 (Quick Quiz 6.1 and 6.2) A Ferris wheel located on the Navy Pier in Chicago, Illinois.

> **Quick Quiz 6.1** You are riding on a Ferris wheel (Fig. 6.3) that is rotating with constant speed. The car in which you are riding always maintains its correct upward orientation—it does not invert. What is the direction of your centripetal acceleration when you are at the *top* of the wheel? (a) upward (b) downward (c) impossible to determine. What is the direction of your centripetal acceleration when you are at the *bottom* of the wheel? (d) upward (e) downward (f) impossible to determine.

> **Quick Quiz 6.2** You are riding on the Ferris wheel of Quick Quiz 6.1. What is the direction of the normal force exerted by the seat on you when you are at the *top* of the wheel? (a) upward (b) downward (c) impossible to determine. What is the direction of the normal force exerted by the seat on you when you are at the *bottom* of the wheel? (d) upward (e) downward (f) impossible to determine.

▲ **PITFALL PREVENTION**

6.1 Direction of Travel When the String is Cut

Study Figure 6.2 very carefully. Many students (wrongly) think that the ball will move *radially* away from the center of the circle when the string is cut. The velocity of the ball is *tangent* to the circle. By Newton's first law, the ball continues to move in the direction that it is moving just as the force from the string disappears.

 At the Active Figures link at http://www.pse6.com, you can "break" the string yourself and observe the effect on the ball's motion.

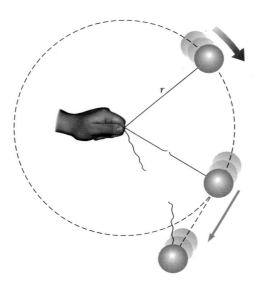

Active Figure 6.2 An overhead view of a ball moving in a circular path in a horizontal plane. When the string breaks, the ball moves in the direction tangent to the circle.

© Tom Carroll/Index Stock Imagery/PictureQuest

Conceptual Example 6.1   Forces That Cause Centripetal Acceleration

The force causing centripetal acceleration is sometimes called a *centripetal force*. We are familiar with a variety of forces in nature—friction, gravity, normal forces, tension, and so forth. Should we add *centripetal* force to this list?

Solution No; centripetal force *should not* be added to this list. This is a pitfall for many students. Giving the force causing circular motion a name—centripetal force—leads many students to consider it as a new *kind* of force rather than a new *role* for force. A common mistake in force diagrams is to draw all the usual forces and then to add another vector for the centripetal force. But it is not a separate force—it is simply one or more of our familiar forces *acting in the role of a force that causes a circular motion.*

Consider some examples. For the motion of the Earth around the Sun, the centripetal force is *gravity*. For an object sitting on a rotating turntable, the centripetal force is *friction*. For a rock whirled horizontally on the end of a string, the magnitude of the centripetal force is the *tension* in the string. For an amusement-park patron pressed against the inner wall of a rapidly rotating circular room, the centripetal force is the *normal force* exerted by the wall. Furthermore, the centripetal force could be a combination of two or more forces. For example, as you pass through the lowest point of the Ferris wheel in Quick Quiz 6.1, the centripetal force on you is the difference between the normal force exerted by the seat and the gravitational force. We will not use the term *centripetal force* in this book after this discussion.

Example 6.2   The Conical Pendulum

A small object of mass m is suspended from a string of length L. The object revolves with constant speed v in a horizontal circle of radius r, as shown in Figure 6.4. (Because the string sweeps out the surface of a cone, the system is known as a *conical pendulum.*) Find an expression for v.

Solution Conceptualize the problem with the help of Figure 6.4. We categorize this as a problem that combines equilibrium for the ball in the vertical direction with uniform circular motion in the horizontal direction. To analyze the problem, begin by letting θ represent the angle between the string and the vertical. In the free-body diagram shown, the force **T** exerted by the string is resolved into a vertical component $T \cos \theta$ and a horizontal component $T \sin \theta$ acting toward the center of revolution. Because the object does not accelerate in the vertical direction, $\Sigma F_y = ma_y = 0$ and the upward vertical component of **T** must balance the downward gravitational force. Therefore,

$$(1) \qquad T \cos \theta = mg$$

Because the force providing the centripetal acceleration in this example is the component $T \sin \theta$, we can use Equation 6.1 to obtain

$$(2) \qquad \sum F = T \sin \theta = ma_c = \frac{mv^2}{r}$$

Dividing (2) by (1) and using $\sin \theta / \cos \theta = \tan \theta$, we eliminate T and find that

$$\tan \theta = \frac{v^2}{rg}$$

$$v = \sqrt{rg \tan \theta}$$

From the geometry in Figure 6.4, we see that $r = L \sin \theta$; therefore,

$$v = \boxed{\sqrt{Lg \sin \theta \tan \theta}}$$

Note that the speed is independent of the mass of the object.

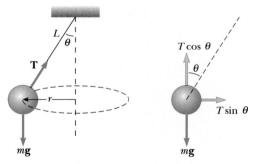

Figure 6.4 (Example 6.2) The conical pendulum and its free-body diagram.

Example 6.3   How Fast Can It Spin?

A ball of mass 0.500 kg is attached to the end of a cord 1.50 m long. The ball is whirled in a horizontal circle as shown in Figure 6.1. If the cord can withstand a maximum tension of 50.0 N, what is the maximum speed at which the ball can be whirled before the cord breaks? Assume that the string remains horizontal during the motion.

Solution It makes sense that the stronger the cord, the faster the ball can twirl before the cord breaks. Also, we expect a more massive ball to break the cord at a lower speed. (Imagine whirling a bowling ball on the cord!)

Because the force causing the centripetal acceleration in this case is the force **T** exerted by the cord on the ball,

Equation 6.1 yields

$$(1) \qquad T = m \frac{v^2}{r}$$

Solving for v, we have

$$v = \sqrt{\frac{Tr}{m}}$$

This shows that v increases with T and decreases with larger m, as we expect to see—for a given v, a large mass requires a large tension and a small mass needs only a small tension. The maximum speed the ball can have corresponds to the

maximum tension. Hence, we find

$$v_{max} = \sqrt{\frac{T_{max}r}{m}} = \sqrt{\frac{(50.0 \text{ N})\ (1.50 \text{ m})}{0.500 \text{ kg}}} = \boxed{12.2 \text{ m/s}}$$

What If? Suppose that the ball is whirled in a circle of larger radius at the same speed *v*. Is the cord more likely to break or less likely?

Answer The larger radius means that the change in the direction of the velocity vector will be smaller for a given time interval. Thus, the acceleration is smaller and the required force from the string is smaller. As a result, the string is less likely to break when the ball travels in a circle of larger radius. To understand this argument better, let us write

Equation (1) twice, once for each radius:

$$T_1 = \frac{mv^2}{r_1} \qquad T_2 = \frac{mv^2}{r_2}$$

Dividing the two equations gives us,

$$\frac{T_2}{T_1} = \frac{\left(\dfrac{mv^2}{r_2}\right)}{\left(\dfrac{mv^2}{r_1}\right)} = \frac{r_1}{r_2}$$

If we choose $r_2 > r_1$, we see that $T_2 < T_1$. Thus, less tension is required to whirl the ball in the larger circle and the string is less likely to break.

Example 6.4   What Is the Maximum Speed of the Car?   Interactive

A 1 500-kg car moving on a flat, horizontal road negotiates a curve, as shown in Figure 6.5. If the radius of the curve is 35.0 m and the coefficient of static friction between the tires and dry pavement is 0.500, find the maximum speed the car can have and still make the turn successfully.

Solution In this case, the force that enables the car to remain in its circular path is the force of static friction. (*Static* because no slipping occurs at the point of contact between road and tires. If this force of static friction were zero—for example, if the car were on an icy road—the car would continue in a straight line and slide off the road.) Hence, from Equation 6.1 we have

$$(1) \qquad f_s = m\frac{v^2}{r}$$

(a)

(b)

Figure 6.5 (Example 6.4) (a) The force of static friction directed toward the center of the curve keeps the car moving in a circular path. (b) The free-body diagram for the car.

The maximum speed the car can have around the curve is the speed at which it is on the verge of skidding outward. At this point, the friction force has its maximum value $f_{s,\ max} = \mu_s n$. Because the car shown in Figure 6.5b is in equilibrium in the vertical direction, the magnitude of the normal force equals the weight ($n = mg$) and thus $f_{s,\ max} = \mu_s mg$. Substituting this value for f_s into (1), we find that the maximum speed is

$$(2) \qquad v_{max} = \sqrt{\frac{f_{s,\ max}\ r}{m}} = \sqrt{\frac{\mu_s mg r}{m}} = \sqrt{\mu_s g r}$$
$$= \sqrt{(0.500)(9.80 \text{ m/s}^2)(35.0 \text{ m})}$$
$$= \boxed{13.1 \text{ m/s}}$$

Note that the maximum speed does not depend on the mass of the car. That is why curved highways do not need multiple speed limit signs to cover the various masses of vehicles using the road.

What If? Suppose that a car travels this curve on a wet day and begins to skid on the curve when its speed reaches only 8.00 m/s. What can we say about the coefficient of static friction in this case?

Answer The coefficient of friction between tires and a wet road should be smaller than that between tires and a dry road. This expectation is consistent with experience with driving, because a skid is more likely on a wet road than a dry road.

To check our suspicion, we can solve (2) for the coefficient of friction:

$$\mu_s = \frac{v_{max}^2}{g r}$$

Substituting the numerical values,

$$\mu_s = \frac{v_{max}^2}{g r} = \frac{(8.00 \text{ m/s})^2}{(9.80 \text{ m/s}^2)(35.0 \text{ m})} = 0.187$$

This is indeed smaller than the coefficient of 0.500 for the dry road.

  Study the relationship between the car's speed, radius of the turn, and the coefficient of static friction between road and tires at the Interactive Worked Example link at **http://www.pse6.com.**

Example 6.5   The Banked Exit Ramp

A civil engineer wishes to design a curved exit ramp for a highway in such a way that a car will not have to rely on friction to round the curve without skidding. In other words, a car moving at the designated speed can negotiate the curve even when the road is covered with ice. Such a ramp is usually *banked*; this means the roadway is tilted toward the inside of the curve. Suppose the designated speed for the ramp is to be 13.4 m/s (30.0 mi/h) and the radius of the curve is 50.0 m. At what angle should the curve be banked?

Solution On a level (unbanked) road, the force that causes the centripetal acceleration is the force of static friction between car and road, as we saw in the previous example. However, if the road is banked at an angle θ, as in Figure 6.6, the normal force **n** has a horizontal component $n \sin \theta$ pointing toward the center of the curve. Because the ramp is to be designed so that the force of static friction is zero, only the component $n_x = n \sin \theta$ causes the centripetal

Figure 6.6 (Example 6.5) A car rounding a curve on a road banked at an angle θ to the horizontal. When friction is neglected, the force that causes the centripetal acceleration and keeps the car moving in its circular path is the horizontal component of the normal force.

acceleration. Hence, Newton's second law for the radial direction gives

$$(1) \qquad \sum F_r = n \sin \theta = \frac{mv^2}{r}$$

The car is in equilibrium in the vertical direction. Thus, from $\sum F_y = 0$ we have

$$(2) \qquad n \cos \theta = mg$$

Dividing (1) by (2) gives

$$(3) \qquad \tan \theta = \frac{v^2}{rg}$$

$$\theta = \tan^{-1}\left(\frac{(13.4 \text{ m/s})^2}{(50.0 \text{ m})(9.80 \text{ m/s}^2)}\right) = \boxed{20.1°}$$

If a car rounds the curve at a speed less than 13.4 m/s, friction is needed to keep it from sliding down the bank (to the left in Fig. 6.6). A driver who attempts to negotiate the curve at a speed greater than 13.4 m/s has to depend on friction to keep from sliding up the bank (to the right in Fig. 6.6). The banking angle is independent of the mass of the vehicle negotiating the curve.

What If? What if this same roadway were built on Mars in the future to connect different colony centers; could it be traveled at the same speed?

Answer The reduced gravitational force on Mars would mean that the car is not pressed so tightly to the roadway. The reduced normal force results in a smaller component of the normal force toward the center of the circle. This smaller component will not be sufficient to provide the centripetal acceleration associated with the original speed. The centripetal acceleration must be reduced, which can be done by reducing the speed v.

Equation (3) shows that the speed v is proportional to the square root of g for a roadway of fixed radius r banked at a fixed angle θ. Thus, if g is smaller, as it is on Mars, the speed v with which the roadway can be safely traveled is also smaller.

 You can adjust the turn radius and banking angle at the Interactive Worked Example link at http://www.pse6.com.

Example 6.6   Let's Go Loop-the-Loop!

A pilot of mass m in a jet aircraft executes a loop-the-loop, as shown in Figure 6.7a. In this maneuver, the aircraft moves in a vertical circle of radius 2.70 km at a constant speed of 225 m/s. Determine the force exerted by the seat on the pilot **(A)** at the bottom of the loop and **(B)** at the top of the loop. Express your answers in terms of the weight of the pilot mg.

Solution To conceptualize this problem, look carefully at Figure 6.7. Based on experiences with driving over small

hills in a roadway, or riding over the top of a Ferris wheel, you would expect to feel lighter at the top of the path. Similarly, you would expect to feel heavier at the bottom of the path. By looking at Figure 6.7, we expect the answer for (A) to be greater than that for (B) because at the bottom of the loop the normal and gravitational forces act in *opposite* directions, whereas at the top of the loop these two forces act in the *same* direction. The vector sum of these two forces gives the force of constant magnitude that keeps the pilot moving in a circular path at a constant speed. To yield net

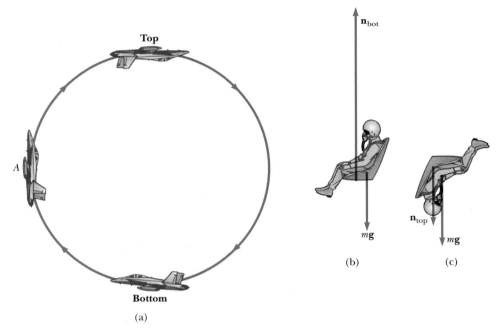

Figure 6.7 (Example 6.6) (a) An aircraft executes a loop-the-loop maneuver as it moves in a vertical circle at constant speed. (b) Free-body diagram for the pilot at the bottom of the loop. In this position the pilot experiences an apparent weight greater than his true weight. (c) Free-body diagram for the pilot at the top of the loop.

force vectors with the same magnitude, the normal force at the bottom must be greater than that at the top. Because the speed of the aircraft is constant (how likely is this?), we can categorize this as a uniform circular motion problem, complicated by the fact that the gravitational force acts at all times on the aircraft.

(A) Analyze the situation by drawing a free-body diagram for the pilot at the bottom of the loop, as shown in Figure 6.7b. The only forces acting on him are the downward gravitational force $\mathbf{F}_g = m\mathbf{g}$ and the upward force $\mathbf{n}_{\text{bot}}$ exerted by the seat. Because the net upward force that provides the centripetal acceleration has a magnitude $n_{\text{bot}} - mg$, Newton's second law for the radial direction gives

$$\sum F = n_{\text{bot}} - mg = m\frac{v^2}{r}$$

$$n_{\text{bot}} = mg + m\frac{v^2}{r} = mg\left(1 + \frac{v^2}{rg}\right)$$

Substituting the values given for the speed and radius gives

$$n_{\text{bot}} = mg\left(1 + \frac{(225 \text{ m/s})^2}{(2.70 \times 10^3 \text{ m})(9.80 \text{ m/s}^2)}\right) = \boxed{2.91\,mg}$$

Hence, the magnitude of the force $\mathbf{n}_{\text{bot}}$ exerted by the seat on the pilot is *greater* than the weight of the pilot by a factor of 2.91. This means that the pilot experiences an appar-

ent weight that is greater than his true weight by a factor of 2.91.

(B) The free-body diagram for the pilot at the top of the loop is shown in Figure 6.7c. As we noted earlier, both the gravitational force exerted by the Earth and the force $\mathbf{n}_{\text{top}}$ exerted by the seat on the pilot act downward, and so the net downward force that provides the centripetal acceleration has a magnitude $n_{\text{top}} + mg$. Applying Newton's second law yields

$$\sum F = n_{\text{top}} + mg = m\frac{v^2}{r}$$

$$n_{\text{top}} = m\frac{v^2}{r} - mg = mg\left(\frac{v^2}{rg} - 1\right)$$

$$n_{\text{top}} = mg\left(\frac{(225 \text{ m/s})^2}{(2.70 \times 10^3 \text{ m})(9.80 \text{ m/s}^2)} - 1\right)$$

$$= \boxed{0.913\,mg}$$

In this case, the magnitude of the force exerted by the seat on the pilot is *less* than his true weight by a factor of 0.913, and the pilot feels lighter. To finalize the problem, note that this is consistent with our prediction at the beginning of the solution.

6.2 Nonuniform Circular Motion

In Chapter 4 we found that if a particle moves with varying speed in a circular path, there is, in addition to the radial component of acceleration, a tangential component having magnitude dv/dt. Therefore, the force acting on the particle must also have a tangential and a radial component. Because the total acceleration is $\mathbf{a} = \mathbf{a}_r + \mathbf{a}_t$, the total force exerted on the particle is $\Sigma\mathbf{F} = \Sigma\mathbf{F}_r + \Sigma\mathbf{F}_t$, as shown in Figure 6.8. The vector $\Sigma\mathbf{F}_r$ is directed toward the center of the circle and is responsible for the centripetal acceleration. The vector $\Sigma\mathbf{F}_t$ tangent to the circle is responsible for the tangential acceleration, which represents a change in the speed of the particle with time.

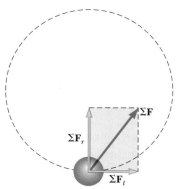

Active Figure 6.8 When the force acting on a particle moving in a circular path has a tangential component ΣF_t, the particle's speed changes. The total force exerted on the particle in this case is the vector sum of the radial force and the tangential force. That is, $\Sigma\mathbf{F} = \Sigma\mathbf{F}_r + \Sigma\mathbf{F}_t$.

 At the Active Figures link at http://www.pse6.com, you can adjust the initial position of the particle and compare the component forces acting on the particle to those for a child swinging on a swing set.

Quick Quiz 6.3 Which of the following is *impossible* for a car moving in a circular path? (a) the car has tangential acceleration but no centripetal acceleration. (b) the car has centripetal acceleration but no tangential acceleration. (c) the car has both centripetal acceleration and tangential acceleration.

Quick Quiz 6.4 A bead slides freely along a *horizontal*, curved wire at constant speed, as shown in Figure 6.9. Draw the vectors representing the force exerted by the wire on the bead at points Ⓐ, Ⓑ, and Ⓒ.

Figure 6.9 (Quick Quiz 6.4 and 6.5) A bead slides along a curved wire.

Quick Quiz 6.5 In Figure 6.9, the bead speeds up with constant tangential acceleration as it moves toward the right. Draw the vectors representing the force on the bead at points Ⓐ, Ⓑ, and Ⓒ.

Passengers on a "corkscrew" roller coaster experience a radial force toward the center of the circular track and a tangential force due to gravity.

Example 6.7   Keep Your Eye on the Ball

A small sphere of mass m is attached to the end of a cord of length R and set into motion in a *vertical* circle about a fixed point O, as illustrated in Figure 6.10a. Determine the tension in the cord at any instant when the speed of the sphere is v and the cord makes an angle θ with the vertical.

Solution Unlike the situation in Example 6.6, the speed is *not* uniform in this example because, at most points along the path, a tangential component of acceleration arises from the gravitational force exerted on the sphere. From the free-body diagram in Figure 6.10a, we see that the only forces acting on the sphere are the gravitational force $\mathbf{F}_g = m\mathbf{g}$ exerted by the Earth and the force $\mathbf{T}$ exerted by the cord. Now we resolve $\mathbf{F}_g$ into a tangential component $mg \sin \theta$ and a radial component $mg \cos \theta$. Applying Newton's second law to the forces acting on the sphere in the tangential direction yields

$$\sum F_t = mg \sin \theta = ma_t$$

$$a_t = g \sin \theta$$

This tangential component of the acceleration causes v to change in time because $a_t = dv/dt$.

Applying Newton's second law to the forces acting on the sphere in the radial direction and noting that both $\mathbf{T}$ and $\mathbf{a}_r$ are directed toward O, we obtain

$$\sum F_r = T - mg \cos \theta = \frac{mv^2}{R}$$

$$T = \boxed{m\left(\frac{v^2}{R} + g \cos \theta\right)}$$

What If? What if we set the ball in motion with a slower speed? **(A)** What speed would the ball have as it passes over the top of the circle if the tension in the cord goes to zero instantaneously at this point?

Answer At the top of the path (Fig. 6.10b), where $\theta = 180°$, we have $\cos 180° = -1$, and the tension equation becomes

$$T_{\text{top}} = m\left(\frac{v_{\text{top}}^2}{R} - g\right)$$

Let us set $T_{\text{top}} = 0$. Then,

$$0 = m\left(\frac{v_{\text{top}}^2}{R} - g\right)$$

$$v_{\text{top}} = \sqrt{gR}$$

(B) What if we set the ball in motion such that the speed at the top is less than this value? What happens?

Answer In this case, the ball never reaches the top of the circle. At some point on the way up, the tension in the string goes to zero and the ball becomes a projectile. It follows a segment of a parabolic path over the top of its motion, rejoining the circular path on the other side when the tension becomes nonzero again.

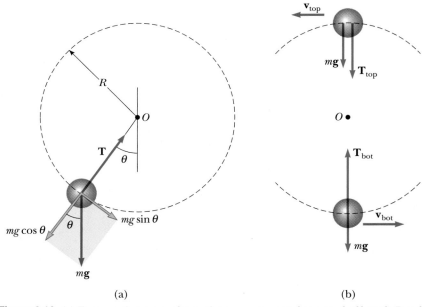

(a)   (b)

Figure 6.10 (a) Forces acting on a sphere of mass m connected to a cord of length R and rotating in a vertical circle centered at O. (b) Forces acting on the sphere at the top and bottom of the circle. The tension is a maximum at the bottom and a minimum at the top.

Investigate these alternatives at the Interactive Worked Example link at **http://www.pse6.com.**

6.3 Motion in Accelerated Frames

When Newton's laws of motion were introduced in Chapter 5, we emphasized that they are valid only when observations are made in an inertial frame of reference. In this section, we analyze how Newton's second law is applied by an observer in a noninertial frame of reference, that is, one that is accelerating. For example, recall the discussion of the air hockey table on a train in Section 5.2. The train moving at constant velocity represents an inertial frame. The puck at rest remains at rest, and Newton's first law is obeyed. The accelerating train is not an inertial frame. According to you as the observer on the train, there appears to be no visible force on the puck, yet it accelerates from rest toward the back of the train, violating Newton's first law.

As an observer on the accelerating train, if you apply Newton's second law to the puck as it accelerates toward the back of the train, you might conclude that a force has acted on the puck to cause it to accelerate. We call an apparent force such as this a **fictitious force,** because it is due to an accelerated reference frame. Remember that real forces are always due to interactions between two objects. A fictitious force appears to act on an object in the same way as a real force, but you cannot identify a second object for a fictitious force.

The train example above describes a fictitious force due to a change in the speed of the train. Another fictitious force is due to the change in the *direction* of the velocity vector. To understand the motion of a system that is noninertial because of a change in direction, consider a car traveling along a highway at a high speed and approaching a curved exit ramp, as shown in Figure 6.11a. As the car takes the sharp left turn onto the ramp, a person sitting in the passenger seat slides to the right and hits the door. At that point, the force exerted by the door on the passenger keeps her from being ejected from the car. What causes her to move toward the door? A popular but incorrect explanation is that a force acting toward the right in Figure 6.11b pushes her outward. This is often called the "centrifugal force," but it is a fictitious force due to the acceleration associated with the changing direction of the car's velocity vector. (The driver also experiences this effect but wisely holds on to the steering wheel to keep from sliding to the right.)

The phenomenon is correctly explained as follows. Before the car enters the ramp, the passenger is moving in a straight-line path. As the car enters the ramp and travels a curved path, the passenger tends to move along the original straight-line path. This is in accordance with Newton's first law: the natural tendency of an object is to continue moving in a straight line. However, if a sufficiently large force (toward the center of curvature) acts on the passenger, as in Figure 6.11c, she moves in a curved path along with the car. This force is the force of friction between her and the car seat. If this friction force is not large enough, she slides to the right as the seat turns to the left under her. Eventually, she encounters the door, which provides a force large enough to enable her to follow the same curved path as the car. She slides toward the door not because of an outward force but because **the force of friction is not sufficiently great to allow her to travel along the circular path followed by the car.**

Another interesting fictitious force is the "Coriolis force." This is an apparent force caused by changing the radial position of an object in a rotating coordinate system. For example, suppose you and a friend are on opposite sides of a rotating circular platform and you decide to throw a baseball to your friend. As Figure 6.12a shows, at $t = 0$ you throw the ball toward your friend, but by the time t_f when the ball has crossed the platform, your friend has moved to a new position.

Figure 6.12a represents what an observer would see if the ball is viewed while the observer is hovering at rest above the rotating platform. According to this observer, who is in an inertial frame, the ball follows a straight line, as it must according to Newton's first law. Now, however, consider the situation from your friend's viewpoint. Your friend is in a noninertial reference frame because he is undergoing a centripetal acceleration relative to the inertial frame of the Earth's surface. He starts off seeing the baseball coming toward him, but as it crosses the platform, it veers to one side, as shown in Figure 6.12b. Thus, your friend on the rotating platform claims that the ball

(a)

(b)

(c)

Figure 6.11 (a) A car approaching a curved exit ramp. What causes a front-seat passenger to move toward the right-hand door? (b) From the frame of reference of the passenger, a force appears to push her toward the right door, but this is a fictitious force. (c) Relative to the reference frame of the Earth, the car seat applies a leftward force to the passenger, causing her to change direction along with the rest of the car.

▲   **PITFALL PREVENTION**

6.2   Centrifugal Force

The commonly heard phrase "centrifugal force" is described as a force pulling *outward* on an object moving in a circular path. If you are feeling a "centrifugal force" on a rotating carnival ride, what is the other object with which you are interacting? You cannot identify another object because this is a fictitious force that occurs as a result of your being in a noninertial reference frame.

At the Active Figures link at http://www.pse6.com, you can observe the ball's path simultaneously from the reference frame of an inertial observer and from the reference frame of the rotating turntable.

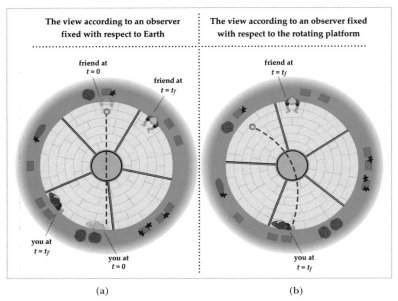

(a)   (b)

Active Figure 6.12 (a) You and your friend sit at the edge of a rotating turntable. In this overhead view observed by someone in an inertial reference frame attached to the Earth, you throw the ball at $t = 0$ in the direction of your friend. By the time t_f that the ball arrives at the other side of the turntable, your friend is no longer there to catch it. According to this observer, the ball followed a straight line path, consistent with Newton's laws. (b) From the point of view of your friend, the ball veers to one side during its flight. Your friend introduces a fictitious force to cause this deviation from the expected path. This fictitious force is called the "Coriolis force."

does not obey Newton's first law and claims that a force is causing the ball to follow a curved path. This fictitious force is called the Coriolis force.

Fictitious forces may not be real forces, but they can have real effects. An object on your dashboard *really* slides off if you press the accelerator of your car. As you ride on a merry-go-round, you feel pushed toward the outside as if due to the fictitious "centrifugal force." You are likely to fall over and injure yourself if you walk along a radial line while the merry-go-round rotates. The Coriolis force due to the rotation of the Earth is responsible for rotations of hurricanes and for large-scale ocean currents.

Quick Quiz 6.6 Consider the passenger in the car making a left turn in Figure 6.11. Which of the following is correct about forces in the horizontal direction if the person is making contact with the right-hand door? (a) The passenger is in equilibrium between real forces acting to the right and real forces acting to the left. (b) The passenger is subject only to real forces acting to the right. (c) The passenger is subject only to real forces acting to the left. (d) None of these is true.

Example 6.8   Fictitious Forces in Linear Motion

A small sphere of mass m is hung by a cord from the ceiling of a boxcar that is accelerating to the right, as shown in Figure 6.13. The noninertial observer in Figure 6.13b claims that a force, which we know to be fictitious, must act in order to cause the observed deviation of the cord from the vertical. How is the magnitude of this force related to the acceleration of the boxcar measured by the inertial observer in Figure 6.13a?

Solution According to the inertial observer at rest (Fig. 6.13a), the forces on the sphere are the force **T** exerted by the cord and the gravitational force. The inertial observer concludes that the acceleration of the sphere is the same as that of the boxcar and that this acceleration is provided by the horizontal component of **T**. Also, the vertical component of **T** balances the gravitational force because the sphere is in equilibrium in the vertical direction. Therefore,

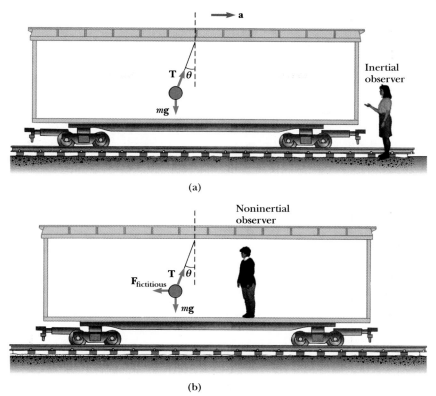

Figure 6.13 (Example 6.8) A small sphere suspended from the ceiling of a boxcar accelerating to the right is deflected as shown. (a) An inertial observer at rest outside the car claims that the acceleration of the sphere is provided by the horizontal component of **T**. (b) A noninertial observer riding in the car says that the net force on the sphere is zero and that the deflection of the cord from the vertical is due to a fictitious force **F**$_{\text{fictitious}}$ that balances the horizontal component of **T**.

she writes Newton's second law as $\Sigma \mathbf{F} = \mathbf{T} + m\mathbf{g} = m\mathbf{a}$, which in component form becomes

Inertial observer $\begin{cases} (1) & \sum F_x = T \sin \theta = ma \\ (2) & \sum F_y = T \cos \theta - mg = 0 \end{cases}$

According to the noninertial observer riding in the car (Fig. 6.13b), the cord also makes an angle θ with the vertical; however, to him the sphere is at rest and so its acceleration is zero. Therefore, he introduces a fictitious force in the horizontal direction to balance the horizontal component of **T** and claims that the net force on the sphere is *zero!* In this noninertial frame of reference, Newton's second law in component form yields

Noninertial observer $\begin{cases} \sum F'_x = T \sin \theta - F_{\text{fictitious}} = 0 \\ \sum F'_y = T \cos \theta - mg = 0 \end{cases}$

We see that these expressions are equivalent to (1) and (2) if $F_{\text{fictitious}} = ma$, where a is the acceleration according to the inertial observer. If we were to make this substitution in the equation for F'_x above, the noninertial observer obtains the same mathematical results as the inertial observer. However, the physical interpretation of the deflection of the cord differs in the two frames of reference.

What If? Suppose the inertial observer wants to measure the acceleration of the train by means of the pendulum (the sphere hanging from the cord). How could she do this?

Answer Our intuition tells us that the angle θ that the cord makes with the vertical should increase as the acceleration increases. By solving (1) and (2) simultaneously for a, the inertial observer can determine the magnitude of the car's acceleration by measuring the angle θ and using the relationship $a = g \tan \theta$. Because the deflection of the cord from the vertical serves as a measure of acceleration, *a simple pendulum can be used as an accelerometer.*

Example 6.9   Fictitious Force in a Rotating System

Suppose a block of mass m lying on a horizontal, frictionless turntable is connected to a string attached to the center of the turntable, as shown in Figure 6.14. How would each of the observers write Newton's second law for the block?

Solution According to an inertial observer (Fig. 6.14a), if the block rotates uniformly, it undergoes an acceleration of magnitude v^2/r, where v is its linear speed. The inertial observer concludes that this centripetal acceleration is provided by the force **T** exerted by the string and writes Newton's second law as $T = mv^2/r$.

According to a noninertial observer attached to the turntable (Fig 6.14b), the block is at rest and its acceleration is zero. Therefore, she must introduce a fictitious outward force of magnitude mv^2/r to balance the inward force exerted by the string. According to her, the net force on the block is zero, and she writes Newton's second law as $T - mv^2/r = 0$.

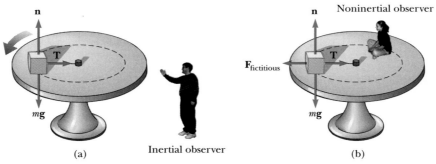

Figure 6.14 (Example 6.9) A block of mass m connected to a string tied to the center of a rotating turntable. (a) The inertial observer claims that the force causing the circular motion is provided by the force **T** exerted by the string on the block. (b) The noninertial observer claims that the block is not accelerating, and therefore she introduces a fictitious force of magnitude mv^2/r that acts outward and balances the force **T**.

6.4 Motion in the Presence of Resistive Forces

In the preceding chapter we described the force of kinetic friction exerted on an object moving on some surface. We completely ignored any interaction between the object and the medium through which it moves. Now let us consider the effect of that medium, which can be either a liquid or a gas. The medium exerts a **resistive force R** on the object moving through it. Some examples are the air resistance associated with moving vehicles (sometimes called *air drag*) and the viscous forces that act on objects moving through a liquid. The magnitude of **R** depends on factors such as the speed of the object, and the direction of **R** is always opposite the direction of motion of the object relative to the medium. Furthermore, the magnitude of **R** nearly always increases with increasing speed.

The magnitude of the resistive force can depend on speed in a complex way, and here we consider only two situations. In the first situation, we assume the resistive force is proportional to the speed of the moving object; this assumption is valid for objects falling slowly through a liquid and for very small objects, such as dust particles, moving through air. In the second situation, we assume a resistive force that is proportional to the square of the speed of the moving object; large objects, such as a skydiver moving through air in free fall, experience such a force.

Resistive Force Proportional to Object Speed

If we assume that the resistive force acting on an object moving through a liquid or gas is proportional to the object's speed, then the resistive force can be expressed as

$$\mathbf{R} = -b\mathbf{v} \tag{6.2}$$

where **v** is the velocity of the object and b is a constant whose value depends on the properties of the medium and on the shape and dimensions of the object. If the object is a sphere of radius r, then b is proportional to r. The negative sign indicates that **R** is in the opposite direction to **v**.

Consider a small sphere of mass m released from rest in a liquid, as in Figure 6.15a. Assuming that the only forces acting on the sphere are the resistive force $\mathbf{R} = -b\mathbf{v}$ and

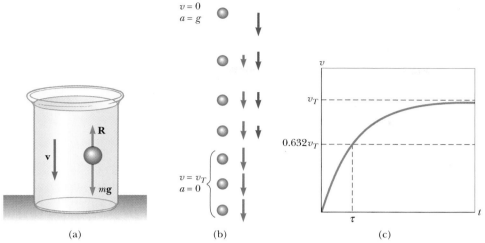

Active Figure 6.15 (a) A small sphere falling through a liquid. (b) Motion diagram of the sphere as it falls. (c) Speed–time graph for the sphere. The sphere reaches a maximum (or terminal) speed v_T, and the time constant τ is the time interval during which it reaches a speed of $0.632v_T$.

At the Active Figures link at http://www.pse6.com, you can vary the size and mass of the sphere and the viscosity (resistance to flow) of the surrounding medium, then observe the effects on the sphere's motion and its speed–time graph.

the gravitational force $\mathbf{F}_g$, let us describe its motion.[1] Applying Newton's second law to the vertical motion, choosing the downward direction to be positive, and noting that $\Sigma F_y = mg - bv$, we obtain

$$mg - bv = ma = m\frac{dv}{dt} \tag{6.3}$$

where the acceleration dv/dt is downward. Solving this expression for the acceleration gives

$$\frac{dv}{dt} = g - \frac{b}{m}v \tag{6.4}$$

This equation is called a *differential equation,* and the methods of solving it may not be familiar to you as yet. However, note that initially when $v = 0$, the magnitude of the resistive force bv is also zero, and the acceleration dv/dt is simply g. As t increases, the magnitude of the resistive force increases and the acceleration decreases. The acceleration approaches zero when the magnitude of the resistive force approaches the sphere's weight. In this situation, the speed of the sphere approaches its **terminal speed** v_T. In reality, the sphere only *approaches* terminal speed but never *reaches* terminal speed.

We can obtain the terminal speed from Equation 6.3 by setting $a = dv/dt = 0$. This gives

$$mg - bv_T = 0 \qquad \text{or} \qquad v_T = \frac{mg}{b}$$

The expression for v that satisfies Equation 6.4 with $v = 0$ at $t = 0$ is

$$v = \frac{mg}{b}(1 - e^{-bt/m}) = v_T(1 - e^{-t/\tau}) \tag{6.5}$$

This function is plotted in Figure 6.15c. The symbol e represents the base of the natural logarithm, and is also called *Euler's number:* $e = 2.718\ 28$. The **time constant** $\tau = m/b$ (Greek letter tau) is the time at which the sphere released from rest reaches 63.2% of its terminal speed. This can be seen by noting that when $t = \tau$, Equation 6.5 yields $v = 0.632v_T$.

Terminal speed

---

[1] There is also a *buoyant force* acting on the submerged object. This force is constant, and its magnitude is equal to the weight of the displaced liquid. This force changes the apparent weight of the sphere by a constant factor, so we will ignore the force here. We discuss buoyant forces in Chapter 14.

We can check that Equation 6.5 is a solution to Equation 6.4 by direct differentiation:

$$\frac{dv}{dt} = \frac{d}{dt}\left(\frac{mg}{b} - \frac{mg}{b}\,e^{-bt/m}\right) = -\frac{mg}{b}\frac{d}{dt}\,e^{-bt/m} = ge^{-bt/m}$$

(See Appendix Table B.4 for the derivative of e raised to some power.) Substituting into Equation 6.4 both this expression for dv/dt and the expression for v given by Equation 6.5 shows that our solution satisfies the differential equation.

Example 6.10   Sphere Falling in Oil

A small sphere of mass 2.00 g is released from rest in a large vessel filled with oil, where it experiences a resistive force proportional to its speed. The sphere reaches a terminal speed of 5.00 cm/s. Determine the time constant τ and the time at which the sphere reaches 90.0% of its terminal speed.

Solution Because the terminal speed is given by $v_T = mg/b$, the coefficient b is

$$b = \frac{mg}{v_T} = \frac{(2.00 \text{ g})(980 \text{ cm/s}^2)}{5.00 \text{ cm/s}} = 392 \text{ g/s}$$

Therefore, the time constant τ is

$$\tau = \frac{m}{b} = \frac{2.00 \text{ g}}{392 \text{ g/s}} = \boxed{5.10 \times 10^{-3} \text{ s}}$$

The speed of the sphere as a function of time is given by Equation 6.5. To find the time t at which the sphere reaches a speed of $0.900v_T$, we set $v = 0.900v_T$ in Equation 6.5 and solve for t:

$$0.900v_T = v_T(1 - e^{-t/\tau})$$

$$1 - e^{-t/\tau} = 0.900$$

$$e^{-t/\tau} = 0.100$$

$$-\frac{t}{\tau} = \ln(0.100) = -2.30$$

$$t = 2.30\tau = 2.30(5.10 \times 10^{-3} \text{ s})$$

$$= 11.7 \times 10^{-3} \text{ s} = \boxed{11.7 \text{ ms}}$$

Thus, the sphere reaches 90.0% of its terminal speed in a very short time interval.

Air Drag at High Speeds

For objects moving at high speeds through air, such as airplanes, sky divers, cars, and baseballs, the resistive force is approximately proportional to the square of the speed. In these situations, the magnitude of the resistive force can be expressed as

$$R = \tfrac{1}{2}D\rho Av^2 \tag{6.6}$$

where ρ is the density of air, A is the cross-sectional area of the moving object measured in a plane perpendicular to its velocity, and D is a dimensionless empirical quantity called the *drag coefficient*. The drag coefficient has a value of about 0.5 for spherical objects but can have a value as great as 2 for irregularly shaped objects.

Let us analyze the motion of an object in free-fall subject to an upward air resistive force of magnitude $R = \tfrac{1}{2}D\rho Av^2$. Suppose an object of mass m is released from rest. As Figure 6.16 shows, the object experiences two external forces:[2] the downward gravitational force $\mathbf{F}_g = m\mathbf{g}$ and the upward resistive force $\mathbf{R}$. Hence, the magnitude of the net force is

$$\sum F = mg - \tfrac{1}{2}D\rho Av^2 \tag{6.7}$$

where we have taken downward to be the positive vertical direction. Combining $\sum F = ma$ with Equation 6.7, we find that the object has a downward acceleration of magnitude

$$a = g - \left(\frac{D\rho A}{2m}\right)v^2 \tag{6.8}$$

We can calculate the terminal speed v_T by using the fact that when the gravitational force is balanced by the resistive force, the net force on the object is zero and therefore its acceleration is zero. Setting $a = 0$ in Equation 6.8 gives

$$g - \left(\frac{D\rho A}{2m}\right)v_T^2 = 0$$

Figure 6.16 An object falling through air experiences a resistive force $\mathbf{R}$ and a gravitational force $\mathbf{F}_g = m\mathbf{g}$. The object reaches terminal speed (on the right) when the net force acting on it is zero, that is, when $\mathbf{R} = -\mathbf{F}_g$ or $R = mg$. Before this occurs, the acceleration varies with speed according to Equation 6.8.

[2]   There is also an upward buoyant force that we neglect.

Table 6.1

Terminal Speed for Various Objects Falling Through Air			
Object	**Mass (kg)**	**Cross-Sectional Area (m^2)**	**v_T (m/s)**
Sky diver	75	0.70	60
Baseball (radius 3.7 cm)	0.145	4.2×10^{-3}	43
Golf ball (radius 2.1 cm)	0.046	1.4×10^{-3}	44
Hailstone (radius 0.50 cm)	4.8×10^{-4}	7.9×10^{-5}	14
Raindrop (radius 0.20 cm)	3.4×10^{-5}	1.3×10^{-5}	9.0

so that,

$$v_T = \sqrt{\frac{2mg}{D\rho A}} \qquad (6.9)$$

Using this expression, we can determine how the terminal speed depends on the dimensions of the object. Suppose the object is a sphere of radius r. In this case, $A \propto r^2$ (from $A = \pi r^2$) and $m \propto r^3$ (because the mass is proportional to the volume of the sphere, which is $V = \frac{4}{3}\pi r^3$). Therefore, $v_T \propto \sqrt{r}$.

Table 6.1 lists the terminal speeds for several objects falling through air.

Quick Quiz 6.7 A baseball and a basketball, having the same mass, are dropped through air from rest such that their bottoms are initially at the same height above the ground, on the order of 1 m or more. Which one strikes the ground first? (a) the baseball (b) the basketball (c) both strike the ground at the same time.

Conceptual Example 6.11   The Sky Surfer

Consider a sky surfer (Fig. 6.17) who jumps from a plane with her feet attached firmly to her surfboard, does some tricks, and then opens her parachute. Describe the forces acting on her during these maneuvers.

Solution When the surfer first steps out of the plane, she has no vertical velocity. The downward gravitational force causes her to accelerate toward the ground. As her downward speed increases, so does the upward resistive force exerted by the air on her body and the board. This upward force reduces their acceleration, and so their speed increases more slowly. Eventually, they are going so fast that the upward resistive force matches the downward gravitational force. Now the net force is zero and they no longer accelerate, but reach their terminal speed. At some point after reaching terminal speed, she opens her parachute, resulting in a drastic increase in the upward resistive force. The net force (and thus the acceleration) is now upward, in the direction opposite the direction of the velocity. This causes the downward velocity to decrease rapidly; this means the resistive force on the chute also decreases. Eventually the upward resistive force and the downward gravitational force balance each other and a much smaller terminal speed is reached, permitting a safe landing.

(Contrary to popular belief, the velocity vector of a sky diver never points upward. You may have seen a videotape in which a sky diver appears to "rocket" upward once the chute opens. In fact, what happens is that the diver slows down while the person holding the camera continues falling at high speed.)

Figure 6.17 (Conceptual Example 6.11) A sky surfer.

Jump Run Productions / Getty Images

Example 6.12 Falling Coffee Filters

The dependence of resistive force on speed is an empirical relationship. In other words, it is based on observation rather than on a theoretical model. Imagine an experiment in which we drop a series of stacked coffee filters, and measure their terminal speeds. Table 6.2 presents data for these coffee filters as they fall through the air. The time constant τ is small, so that a dropped filter quickly reaches terminal speed. Each filter has a mass of 1.64 g. When the filters are nested together, they stack in such a way that the front-facing surface area does not increase. Determine the relationship between the resistive force exerted by the air and the speed of the falling filters.

Solution At terminal speed, the upward resistive force balances the downward gravitational force. So, a single filter falling at its terminal speed experiences a resistive force of

$$R = mg = (1.64 \text{ g})\left(\frac{1 \text{ kg}}{100 \text{ 0 g}}\right)(9.80 \text{ m/s}^2) = 0.016 \text{ 1 N}$$

Two filters nested together experience 0.032 2 N of resistive force, and so forth. A graph of the resistive force on the filters as a function of terminal speed is shown in Figure 6.18a. A straight line would not be a good fit, indicating that the resistive force is *not* proportional to the speed. The behavior is more clearly seen in Figure 6.18b, in which the resistive force is plotted as a function of the square of the terminal speed. This indicates a proportionality of the resistive force to the *square* of the speed, as suggested by Equation 6.6.

Pleated coffee filters can be nested together so that the force of air resistance can be studied.

(a)

(b)

Figure 6.18 (Example 6.12) (a) Relationship between the resistive force acting on falling coffee filters and their terminal speed. The curved line is a second-order polynomial fit. (b) Graph relating the resistive force to the square of the terminal speed. The fit of the straight line to the data points indicates that the resistive force is proportional to the terminal speed squared. Can you find the proportionality constant?

Table 6.2

Terminal Speed for Stacked Coffee Filters	
Number of Filters	v_T **(m/s)**[a]
1	1.01
2	1.40
3	1.63
4	2.00
5	2.25
6	2.40
7	2.57
8	2.80
9	3.05
10	3.22

ᵃ All values of v_T are approximate.

Example 6.13 Resistive Force Exerted on a Baseball

A pitcher hurls a 0.145-kg baseball past a batter at 40.2 m/s (= 90 mi/h). Find the resistive force acting on the ball at this speed.

Solution We do not expect the air to exert a huge force on the ball, and so the resistive force we calculate from Equation 6.6 should not be more than a few newtons.

First, we must determine the drag coefficient D. We do this by imagining that we drop the baseball and allow it to reach terminal speed. We solve Equation 6.9 for D and substitute the appropriate values for m, v_T, and A from Table 6.1. Taking the density of air as 1.20 kg/m^3, we obtain

$$D = \frac{2mg}{v_T{}^2 \rho A} = \frac{2(0.145 \text{ kg})(9.80 \text{ m/s}^2)}{(43 \text{ m/s})^2(1.20 \text{ kg/m}^3)(4.2 \times 10^{-3} \text{ m}^2)}$$

$$= 0.305$$

This number has no dimensions. We have kept an extra digit beyond the two that are significant and will drop it at the end of our calculation.

We can now use this value for D in Equation 6.6 to find the magnitude of the resistive force:

$$R = \tfrac{1}{2} D\rho A v^2$$

$$= \tfrac{1}{2}(0.305)(1.20 \text{ kg/m}^3)(4.2 \times 10^{-3} \text{ m}^2)(40.2 \text{ m/s})^2$$

$$= \boxed{1.2 \text{ N}}$$

6.5 Numerical Modeling in Particle Dynamics[3]

As we have seen in this and the preceding chapter, the study of the dynamics of a particle focuses on describing the position, velocity, and acceleration as functions of time. Cause-and-effect relationships exist among these quantities: Velocity causes position to change, and acceleration causes velocity to change. Because acceleration is the direct result of applied forces, any analysis of the dynamics of a particle usually begins with an evaluation of the net force acting on the particle.

Until now, we have used what is called the *analytical method* to investigate the position, velocity, and acceleration of a moving particle. This method involves the identification of well-behaved functional expressions for the position of a particle (such as the kinematic equations of Chapter 2), generated from algebraic manipulations or the techniques of calculus. Let us review this method briefly before learning about a second way of approaching problems in dynamics. (Because we confine our discussion to one-dimensional motion in this section, boldface notation will not be used for vector quantities.)

If a particle of mass m moves under the influence of a net force ΣF, Newton's second law tells us that the acceleration of the particle is $a = \Sigma F/m$. In general, we apply the analytical method to a dynamics problem using the following procedure:

1. Sum all the forces acting on the particle to find the net force ΣF.
2. Use this net force to determine the acceleration from the relationship $a = \Sigma F/m$.
3. Use this acceleration to determine the velocity from the relationship $dv/dt = a$.
4. Use this velocity to determine the position from the relationship $dx/dt = v$.

The following straightforward example illustrates this method.

Example 6.14   An Object Falling in a Vacuum–Analytical Method

Consider a particle falling in a vacuum under the influence of the gravitational force, as shown in Figure 6.19. Use the analytical method to find the acceleration, velocity, and position of the particle.

Solution The only force acting on the particle is the downward gravitational force of magnitude F_g, which is also the net force. Applying Newton's second law, we set the net force acting on the particle equal to the mass of the particle times its acceleration (taking upward to be the positive y direction):

$$F_g = ma_y = -mg$$

Figure 6.19 (Example 6.14) An object falling in vacuum under the influence of gravity.

---

[3]   The authors are most grateful to Colonel James Head of the U.S. Air Force Academy for preparing this section.

Thus, $a_y = -g$, which means the acceleration is constant. Because $dv_y/dt = a_y$, we see that $dv_y/dt = -g$, which may be integrated to yield

$$v_y(t) = v_{yi} - gt$$

Then, because $v_y = dy/dt$, the position of the particle is obtained from another integration, which yields the well-known result

$$y(t) = y_i + v_{yi}t - \tfrac{1}{2}gt^2$$

In these expressions, y_i and v_{yi} represent the position and speed of the particle at $t_i = 0$.

The analytical method is straightforward for many physical situations. In the "real world," however, complications often arise that make analytical solutions difficult and perhaps beyond the mathematical abilities of most students taking introductory physics. For example, the net force acting on a particle may depend on the particle's position, as in cases where the gravitational acceleration varies with height. Or the force may vary with velocity, as in cases of resistive forces caused by motion through a liquid or gas.

Another complication arises because the expressions relating acceleration, velocity, position, and time are differential equations rather than algebraic ones. Differential equations are usually solved using integral calculus and other special techniques that introductory students may not have mastered.

When such situations arise, scientists often use a procedure called *numerical modeling* to study motion. The simplest numerical model is called the Euler method, after the Swiss mathematician Leonhard Euler (1707–1783).

The Euler Method

In the **Euler method** for solving differential equations, derivatives are approximated as ratios of finite differences. Considering a small increment of time Δt, we can approximate the relationship between a particle's speed and the magnitude of its acceleration as

$$a(t) \approx \frac{\Delta v}{\Delta t} = \frac{v(t + \Delta t) - v(t)}{\Delta t}$$

Then the speed $v(t + \Delta t)$ of the particle at the end of the time interval Δt is approximately equal to the speed $v(t)$ at the beginning of the time interval plus the magnitude of the acceleration during the interval multiplied by Δt:

$$v(t + \Delta t) \approx v(t) + a(t)\,\Delta t \tag{6.10}$$

Because the acceleration is a function of time, this estimate of $v(t + \Delta t)$ is accurate only if the time interval Δt is short enough such that the change in acceleration during the interval is very small (as is discussed later). Of course, Equation 6.10 is exact if the acceleration is constant.

The position $x(t + \Delta t)$ of the particle at the end of the interval Δt can be found in the same manner:

$$v(t) \approx \frac{\Delta x}{\Delta t} = \frac{x(t + \Delta t) - x(t)}{\Delta t}$$

$$x(t + \Delta t) \approx x(t) + v(t)\,\Delta t \tag{6.11}$$

You may be tempted to add the term $\tfrac{1}{2}a(\Delta t)^2$ to this result to make it look like the familiar kinematics equation, but this term is not included in the Euler method because Δt is assumed to be so small that $(\Delta t)^2$ is nearly zero.

If the acceleration at any instant t is known, the particle's velocity and position at a time $t + \Delta t$ can be calculated from Equations 6.10 and 6.11. The calculation then proceeds in a series of finite steps to determine the velocity and position at any later time.

Table 6.3

The Euler Method for Solving Dynamics Problems				
Step	**Time**	**Position**	**Velocity**	**Acceleration**
0	t_0	x_0	v_0	$a_0 = F(x_0, v_0, t_0)/m$
1	$t_1 = t_0 + \Delta t$	$x_1 = x_0 + v_0 \Delta t$	$v_1 = v_0 + a_0 \Delta t$	$a_1 = F(x_1, v_1, t_1)/m$
2	$t_2 = t_1 + \Delta t$	$x_2 = x_1 + v_1 \Delta t$	$v_2 = v_1 + a_1 \Delta t$	$a_2 = F(x_2, v_2, t_2)/m$
3	$t_3 = t_2 + \Delta t$	$x_3 = x_2 + v_2 \Delta t$	$v_3 = v_2 + a_2 \Delta t$	$a_3 = F(x_3, v_3, t_3)/m$
$\vdots$	$\vdots$	$\vdots$	$\vdots$	$\vdots$
n	t_n	x_n	v_n	a_n

The acceleration is determined from the net force acting on the particle, and this force may depend on position, velocity, or time:

$$a(x, v, t) = \frac{\sum F(x, v, t)}{m} \tag{6.12}$$

It is convenient to set up the numerical solution to this kind of problem by numbering the steps and entering the calculations in a table. Table 6.3 illustrates how to do this in an orderly way. Many small increments can be taken, and accurate results can usually be obtained with the help of a computer. The equations provided in the table can be entered into a spreadsheet and the calculations performed row by row to determine the velocity, position, and acceleration as functions of time. The calculations can also be carried out using a programming language, or with commercially available mathematics packages for personal computers. Graphs of velocity versus time or position versus time can be displayed to help you visualize the motion.

One advantage of the Euler method is that the dynamics is not obscured—the fundamental relationships between acceleration and force, velocity and acceleration, and position and velocity are clearly evident. Indeed, these relationships form the heart of the calculations. There is no need to use advanced mathematics, and the basic physics governs the dynamics.

The Euler method is completely reliable for infinitesimally small time increments, but for practical reasons a finite increment size must be chosen. For the finite difference approximation of Equation 6.10 to be valid, the time increment must be small enough that the acceleration can be approximated as being constant during the increment. We can determine an appropriate size for the time increment by examining the particular problem being investigated. The criterion for the size of the time increment may need to be changed during the course of the motion. In practice, however, we usually choose a time increment appropriate to the initial conditions and use the same value throughout the calculations.

The size of the time increment influences the accuracy of the result, but unfortunately it is not easy to determine the accuracy of an Euler-method solution without a knowledge of the correct analytical solution. One method of determining the accuracy of the numerical solution is to repeat the calculations with a smaller time increment and compare results. If the two calculations agree to a certain number of significant figures, you can assume that the results are correct to that precision.

Example 6.15   Euler and the Sphere in Oil Revisited

Consider the sphere falling in oil in Example 6.10. Using the Euler method, find the position and the acceleration of the sphere at the instant that the speed reaches 90.0% of terminal speed.

Solution The net force on the sphere is

$$\Sigma F = -mg + bv$$

Thus, the acceleration values in the last column of Table 6.3 are

$$a = \frac{\Sigma F(x, v, t)}{m} = \frac{-mg + bv}{m} = -g + \frac{bv}{m}$$

Choosing a time increment of 0.1 ms, the first few lines of the spreadsheet modeled after Table 6.3 look like Table 6.4. We see that the speed is increasing while the magnitude of the acceleration is decreasing due to the resistive force. We also see that the sphere does not fall very far in the first millisecond.

Further down the spreadsheet, as shown in Table 6.5, we find the instant at which the sphere reaches the speed

$0.900v_T$, which is 0.900×5.00 cm/s = 4.50 cm/s. This calculation shows that this occurs at $t = 11.6$ ms, which agrees within its uncertainty with the value obtained in Example 6.10. The 0.1-ms difference in the two values is due to the approximate nature of the Euler method. If a smaller time increment were used, the instant at which the speed reaches $0.900v_T$ approaches the value calculated in Example 6.10.

From Table 6.5, we see that the position and acceleration of the sphere when it reaches a speed of $0.900v_T$ are

$$y = \boxed{-0.035 \text{ cm}} \quad \text{and} \quad a = \boxed{-99 \text{ cm/s}^2}$$

Table 6.4

The Sphere Begins to Fall in Oil			
Step	Time (ms)	Position (cm)	Velocity (cm/s)	Acceleration (cm/s^2)
0	0.0	0.000 0	0.0	−980.0
1	0.1	0.000 0	−0.10	−960.8
2	0.2	0.000 0	−0.19	−942.0
3	0.3	0.000 0	−0.29	−923.5
4	0.4	−0.000 1	−0.38	−905.4
5	0.5	−0.000 1	−0.47	−887.7
6	0.6	−0.000 1	−0.56	−870.3
7	0.7	−0.000 2	−0.65	−853.2
8	0.8	−0.000 3	−0.73	−836.5
9	0.9	−0.000 3	−0.82	−820.1
10	1.0	−0.000 4	−0.90	−804.0

Table 6.5

The Sphere Reaches 0.900 v_T			
Step	Time (ms)	Position (cm)	Velocity (cm/s)	Acceleration (cm/s^2)
110	11.0	−0.032 4	−4.43	−111.1
111	11.1	−0.032 8	−4.44	−108.9
112	11.2	−0.033 3	−4.46	−106.8
113	11.3	−0.033 7	−4.47	−104.7
114	11.4	−0.034 2	−4.48	−102.6
115	11.5	−0.034 6	−4.49	−100.6
116	11.6	−0.035 1	−4.50	−98.6
117	11.7	−0.035 5	−4.51	−96.7
118	11.8	−0.036 0	−4.52	−94.8
119	11.9	−0.036 4	−4.53	−92.9
120	12.0	−0.036 9	−4.54	−91.1

SUMMARY

Take a practice test for this chapter by clicking the Practice Test link at http://www.pse6.com.

Newton's second law applied to a particle moving in uniform circular motion states that the net force causing the particle to undergo a centripetal acceleration is

$$\sum F = ma_c = \frac{mv^2}{r} \tag{6.1}$$

A particle moving in nonuniform circular motion has both a radial component of acceleration and a nonzero tangential component of acceleration. In the case of a particle rotating in a vertical circle, the gravitational force provides the tangential component of acceleration and part or all of the radial component of acceleration.

An observer in a noninertial (accelerating) frame of reference must introduce **fictitious forces** when applying Newton's second law in that frame. If these fictitious forces are properly defined, the description of motion in the noninertial frame is equivalent to that made by an observer in an inertial frame. However, the observers in the two frames do not agree on the causes of the motion.

An object moving through a liquid or gas experiences a speed-dependent **resistive force.** This resistive force, which opposes the motion relative to the medium, generally increases with speed. The magnitude of the resistive force depends on the size and shape of the object and on the properties of the medium through which the object is moving. In the limiting case for a falling object, when the magnitude of the resistive force equals the object's weight, the object reaches its **terminal speed.** **Euler's method** provides a means for analyzing the motion of a particle under the action of a force that is not simple.

QUESTIONS

1. Why does mud fly off a rapidly turning automobile tire?

2. Imagine that you attach a heavy object to one end of a spring, hold onto the other end of the spring, and then whirl the object in a horizontal circle. Does the spring stretch? If so, why? Discuss this in terms of the force causing the motion to be circular.

3. Describe a situation in which the driver of a car can have a centripetal acceleration but no tangential acceleration.

4. Describe the path of a moving body in the event that its acceleration is constant in magnitude at all times and (a) perpendicular to the velocity; (b) parallel to the velocity.

5. An object executes circular motion with constant speed whenever a net force of constant magnitude acts perpendicular to the velocity. What happens to the speed if the force is not perpendicular to the velocity?

6. Explain why the Earth is not spherical in shape and bulges at the equator.

7. Because the Earth rotates about its axis, it is a noninertial frame of reference. Assume the Earth is a uniform sphere. Why would the apparent weight of an object be greater at the poles than at the equator?

8. What causes a rotary lawn sprinkler to turn?

9. If someone told you that astronauts are weightless in orbit because they are beyond the pull of gravity, would you accept the statement? Explain.

10. It has been suggested that rotating cylinders about 10 mi in length and 5 mi in diameter be placed in space and used as colonies. The purpose of the rotation is to simulate gravity for the inhabitants. Explain this concept for producing an effective imitation of gravity.

11. Consider a rotating space station, spinning with just the right speed such that the centripetal acceleration on the inner surface is g. Thus, astronauts standing on this inner surface would feel pressed to the surface as if they were pressed into the floor because of the Earth's gravitational force. Suppose an astronaut in this station holds a ball above her head and "drops" it to the floor. Will the ball fall just like it would on the Earth?

12. A pail of water can be whirled in a vertical path such that none is spilled. Why does the water stay in the pail, even when the pail is above your head?

13. How would you explain the force that pushes a rider toward the side of a car as the car rounds a corner?

14. Why does a pilot tend to black out when pulling out of a steep dive?

15. The observer in the accelerating elevator of Example 5.8 would claim that the "weight" of the fish is T, the scale reading. This is obviously wrong. Why does this observation differ from that of a person outside the elevator, at rest with respect to the Earth?

16. If you have ever taken a ride in an express elevator of a high-rise building, you may have experienced a nauseating sensation of heaviness or lightness depending on the direction of the acceleration. Explain these sensations. Are we truly weightless in free-fall?

17. A falling sky diver reaches terminal speed with her parachute closed. After the parachute is opened, what parameters change to decrease this terminal speed?

18. Consider a small raindrop and a large raindrop falling through the atmosphere. Compare their terminal speeds. What are their accelerations when they reach terminal speed?

19. On long journeys, jet aircraft usually fly at high altitudes of about 30 000 ft. What is the main advantage of flying at these altitudes from an economic viewpoint?

20. Analyze the motion of a rock falling through water in terms of its speed and acceleration as it falls. Assume that the resistive force acting on the rock increases as the speed increases.

21. "If the current position and velocity of every particle in the Universe were known, together with the laws describing the forces that particles exert on one another, then the whole future of the Universe could be calculated. The future is determinate and preordained. Free will is an illusion." Do you agree with this thesis? Argue for or against it.

PROBLEMS

1, 2, 3 = straightforward, intermediate, challenging ☐ = full solution available in the *Student Solutions Manual and Study Guide*

🪐 = coached solution with hints available at http://www.pse6.com 🖥 = computer useful in solving problem

▨ = paired numerical and symbolic problems

Section 6.1 Newton's Second Law Applied to Uniform Circular Motion

1. A light string can support a stationary hanging load of 25.0 kg before breaking. A 3.00-kg object attached to the string rotates on a horizontal, frictionless table in a circle of radius 0.800 m, while the other end of the string is held fixed. What range of speeds can the object have before the string breaks?

2. A curve in a road forms part of a horizontal circle. As a car goes around it at constant speed 14.0 m/s, the total force on the driver has magnitude 130 N. What is the total vector force on the driver if the speed is 18.0 m/s instead?

3. In the Bohr model of the hydrogen atom, the speed of the electron is approximately 2.20×10^6 m/s. Find (a) the force acting on the electron as it revolves in a circular orbit of radius 0.530×10^{-10} m and (b) the centripetal acceleration of the electron.

4. In a cyclotron (one type of particle accelerator), a deuteron (of atomic mass 2.00 u) reaches a final speed of 10.0% of the speed of light while moving in a circular path of radius 0.480 m. The deuteron is maintained in the circular path by a magnetic force. What magnitude of force is required?

5. A coin placed 30.0 cm from the center of a rotating, horizontal turntable slips when its speed is 50.0 cm/s. (a) What force causes the centripetal acceleration when the coin is stationary relative to the turntable? (b) What is the coefficient of static friction between coin and turntable?

6. Whenever two *Apollo* astronauts were on the surface of the Moon, a third astronaut orbited the Moon. Assume the orbit to be circular and 100 km above the surface of the Moon, where the acceleration due to gravity is 1.52 m/s². The radius of the Moon is 1.70×10^6 m. Determine (a) the astronaut's orbital speed, and (b) the period of the orbit.

7. A crate of eggs is located in the middle of the flat bed of a pickup truck as the truck negotiates an unbanked curve in the road. The curve may be regarded as an arc of a circle of radius 35.0 m. If the coefficient of static friction between crate and truck is 0.600, how fast can the truck be moving without the crate sliding?

8. The cornering performance of an automobile is evaluated on a skidpad, where the maximum speed that a car can maintain around a circular path on a dry, flat surface is measured. Then the centripetal acceleration, also called the lateral acceleration, is calculated as a multiple of the free-fall acceleration *g*. The main factors affecting the performance are the tire characteristics and the suspension system of the car. A Dodge Viper GTS can negotiate a skidpad of radius 61.0 m at 86.5 km/h. Calculate its maximum lateral acceleration.

9. Consider a conical pendulum with an 80.0-kg bob on a 10.0-m wire making an angle of 5.00° with the vertical (Fig. P6.9). Determine (a) the horizontal and vertical components of the force exerted by the wire on the pendulum and (b) the radial acceleration of the bob.

Figure P6.9

10. A car initially traveling eastward turns north by traveling in a circular path at uniform speed as in Figure P6.10. The length of the arc *ABC* is 235 m, and the car completes the turn in 36.0 s. (a) What is the acceleration when the car is at *B* located at an angle of 35.0°? Express your answer in terms of the unit vectors $\hat{\mathbf{i}}$ and $\hat{\mathbf{j}}$. Determine (b) the car's average speed and (c) its average acceleration during the 36.0-s interval.

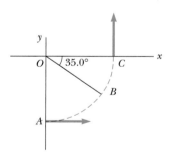

Figure P6.10

11. A 4.00-kg object is attached to a vertical rod by two strings, as in Figure P6.11. The object rotates in a horizontal circle at constant speed 6.00 m/s. Find the tension in (a) the upper string and (b) the lower string.

12. Casting of molten metal is important in many industrial processes. *Centrifugal casting* is used for manufacturing pipes, bearings and many other structures. A variety of sophisticated techniques have been invented, but the basic idea is as illustrated in Figure P6.12. A cylindrical enclosure is rotated rapidly and steadily about a horizontal axis. Molten metal is poured into the rotating cylinder and then cooled, forming the finished product. Turning the cylin-

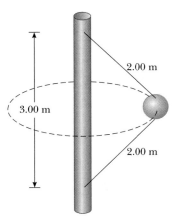

Figure P6.11

der at a high rotation rate forces the solidifying metal strongly to the outside. Any bubbles are displaced toward the axis, so unwanted voids will not be present in the casting. Sometimes it is desirable to form a composite casting, such as for a bearing. Here a strong steel outer surface is poured, followed by an inner lining of special low-friction metal. In some applications a very strong metal is given a coating of corrosion-resistant metal. Centrifugal casting results in strong bonding between the layers.

Suppose that a copper sleeve of inner radius 2.10 cm and outer radius 2.20 cm is to be cast. To eliminate bubbles and give high structural integrity, the centripetal acceleration of each bit of metal should be 100g. What rate of rotation is required? State the answer in revolutions per minute.

Section 6.2   Nonuniform Circular Motion

13. A 40.0-kg child swings in a swing supported by two chains, each 3.00 m long. If the tension in each chain at the lowest point is 350 N, find (a) the child's speed at the lowest point and (b) the force exerted by the seat on the child at the lowest point. (Neglect the mass of the seat.)

14. A child of mass m swings in a swing supported by two chains, each of length R. If the tension in each chain at the lowest point is T, find (a) the child's speed at the lowest point and (b) the force exerted by the seat on the child at the lowest point. (Neglect the mass of the seat.)

15. Tarzan ($m = 85.0$ kg) tries to cross a river by swinging from a vine. The vine is 10.0 m long, and his speed at the bottom of the swing (as he just clears the water) will be 8.00 m/s. Tarzan doesn't know that the vine has a breaking strength of 1 000 N. Does he make it safely across the river?

16. A hawk flies in a horizontal arc of radius 12.0 m at a constant speed of 4.00 m/s. (a) Find its centripetal acceleration. (b) It continues to fly along the same horizontal arc but increases its speed at the rate of 1.20 m/s². Find the acceleration (magnitude and direction) under these conditions.

17. A pail of water is rotated in a vertical circle of radius 1.00 m. What is the minimum speed of the pail at the top of the circle if no water is to spill out?

18. A 0.400-kg object is swung in a vertical circular path on a string 0.500 m long. If its speed is 4.00 m/s at the top of the circle, what is the tension in the string there?

19. A roller coaster car (Fig. P6.19) has a mass of 500 kg when fully loaded with passengers. (a) If the vehicle has a speed of 20.0 m/s at point Ⓐ, what is the force exerted by the track on the car at this point? (b) What is the maximum speed the vehicle can have at Ⓑ and still remain on the track?

Figure P6.19

20. A roller coaster at the Six Flags Great America amusement park in Gurnee, IL, incorporates some clever design technology and some basic physics. Each vertical loop, instead of being circular, is shaped like a teardrop (Fig. P6.20). The cars ride on the inside of the loop at the top, and the speeds are high enough to ensure that the cars remain on the track. The biggest loop is 40.0 m high, with a maximum speed of 31.0 m/s (nearly 70 mi/h) at the bottom. Suppose

Figure P6.12

the speed at the top is 13.0 m/s and the corresponding centripetal acceleration is 2g. (a) What is the radius of the arc of the teardrop at the top? (b) If the total mass of a car plus the riders is M, what force does the rail exert on the car at the top? (c) Suppose the roller coaster had a circular loop of radius 20.0 m. If the cars have the same speed, 13.0 m/s at the top, what is the centripetal acceleration at the top? Comment on the normal force at the top in this situation.

Figure P6.20

Section 6.3 Motion in Accelerated Frames

21. An object of mass 5.00 kg, attached to a spring scale, rests on a frictionless, horizontal surface as in Figure P6.21. The spring scale, attached to the front end of a boxcar, has a constant reading of 18.0 N when the car is in motion. (a) If the spring scale reads zero when the car is at rest, determine the acceleration of the car. (b) What constant reading will the spring scale show if the car moves with constant velocity? (c) Describe the forces on the object as observed by someone in the car and by someone at rest outside the car.

5.00 kg

Figure P6.21

22. If the coefficient of static friction between your coffee cup and the horizontal dashboard of your car is $\mu_s = 0.800$, how fast can you drive on a horizontal roadway around a right turn of radius 30.0 m before the cup starts to slide? If you go too fast, in what direction will the cup slide relative to the dashboard?

23. A 0.500-kg object is suspended from the ceiling of an accelerating boxcar as in Figure 6.13. If $a = 3.00 \text{ m/s}^2$, find

(a) the angle that the string makes with the vertical and (b) the tension in the string.

24. A small container of water is placed on a carousel inside a microwave oven, at a radius of 12.0 cm from the center. The turntable rotates steadily, turning through one revolution in each 7.25 s. What angle does the water surface make with the horizontal?

25. A person stands on a scale in an elevator. As the elevator starts, the scale has a constant reading of 591 N. As the elevator later stops, the scale reading is 391 N. Assume the magnitude of the acceleration is the same during starting and stopping, and determine (a) the weight of the person, (b) the person's mass, and (c) the acceleration of the elevator.

26. The Earth rotates about its axis with a period of 24.0 h. Imagine that the rotational speed can be increased. If an object at the equator is to have zero apparent weight, (a) what must the new period be? (b) By what factor would the speed of the object be increased when the planet is rotating at the higher speed? Note that the apparent weight of the object becomes zero when the normal force exerted on it is zero.

27. A small block is at rest on the floor at the front of a railroad boxcar that has length ℓ. The coefficient of kinetic friction between the floor of the car and the block is μ_k. The car, originally at rest, begins to move with acceleration a. The block slides back horizontally until it hits the back wall of the car. At that moment, what is its speed (a) relative to the car? (b) relative to Earth?

28. A student stands in an elevator that is continuously accelerating upward with acceleration a. Her backpack is sitting on the floor next to the wall. The width of the elevator car is L. The student gives her backpack a quick kick at $t = 0$, imparting to it speed v, and making it slide across the elevator floor. At time t, the backpack hits the opposite wall. Find the coefficient of kinetic friction μ_k between the backpack and the elevator floor.

29. A child on vacation wakes up. She is lying on her back. The tension in the muscles on both sides of her neck is 55.0 N as she raises her head to look past her toes and out the motel window. Finally it is not raining! Ten minutes later she is screaming feet first down a water slide at terminal speed 5.70 m/s, riding high on the outside wall of a horizontal curve of radius 2.40 m (Figure P6.29). She raises her head to look forward past her toes; find the tension in the muscles on both sides of her neck.

Figure P6.29

30. One popular design of a household juice machine is a conical, perforated stainless steel basket 3.30 cm high with a closed bottom of diameter 8.00 cm and open top of diameter 13.70 cm that spins at 20 000 revolutions per minute about a vertical axis (Figure P6.30). Solid pieces of fruit are chopped into granules by cutters at the bottom of the spinning cone. Then the fruit granules rapidly make their way to the sloping surface where the juice is extracted to the outside of the cone through the mesh perforations. The dry pulp spirals upward along the slope to be ejected from the top of the cone. The juice is collected in an enclosure immediately surrounding the sloped surface of the cone. (a) What centripetal acceleration does a bit of fruit experience when it is spinning with the basket at a point midway between the top and bottom? Express the answer as a multiple of g. (b) Observe that the weight of the fruit is a negligible force. What is the normal force on 2.00 g of fruit at that point? (c) If the effective coefficient of kinetic friction between the fruit and the cone is 0.600, with what acceleration relative to the cone will the bit of fruit start to slide up the wall of the cone at that point, after being temporarily stuck?

Figure P6.30

31. A plumb bob does not hang exactly along a line directed to the center of the Earth's rotation. How much does the plumb bob deviate from a radial line at 35.0° north latitude? Assume that the Earth is spherical.

Section 6.4 Motion in the Presence of Resistive Forces

32. A sky diver of mass 80.0 kg jumps from a slow-moving aircraft and reaches a terminal speed of 50.0 m/s. (a) What is the acceleration of the sky diver when her speed is 30.0 m/s? What is the drag force on the diver when her speed is (b) 50.0 m/s? (c) 30.0 m/s?

33. A small piece of Styrofoam packing material is dropped from a height of 2.00 m above the ground. Until it reaches terminal speed, the magnitude of its acceleration is given by $a = g - bv$. After falling 0.500 m, the Styrofoam effectively reaches terminal speed, and then takes 5.00 s

more to reach the ground. (a) What is the value of the constant b? (b) What is the acceleration at $t = 0$? (c) What is the acceleration when the speed is 0.150 m/s?

34. (a) Estimate the terminal speed of a wooden sphere (density 0.830 g/cm^3) falling through air if its radius is 8.00 cm and its drag coefficient is 0.500. (b) From what height would a freely falling object reach this speed in the absence of air resistance?

35. Calculate the force required to pull a copper ball of radius 2.00 cm upward through a fluid at the constant speed 9.00 cm/s. Take the drag force to be proportional to the speed, with proportionality constant 0.950 kg/s. Ignore the buoyant force.

36. A fire helicopter carries a 620-kg bucket at the end of a cable 20.0 m long as in Figure P6.36. As the helicopter flies to a fire at a constant speed of 40.0 m/s, the cable makes an angle of 40.0° with respect to the vertical. The bucket presents a cross-sectional area of 3.80 m^2 in a plane perpendicular to the air moving past it. Determine the drag coefficient assuming that the resistive force is proportional to the square of the bucket's speed.

Figure P6.36

37. A small, spherical bead of mass 3.00 g is released from rest at $t = 0$ in a bottle of liquid shampoo. The terminal speed is observed to be $v_T = 2.00$ cm/s. Find (a) the value of the constant b in Equation 6.2, (b) the time τ at which the bead reaches $0.632v_T$, and (c) the value of the resistive force when the bead reaches terminal speed.

38. The mass of a sports car is 1 200 kg. The shape of the body is such that the aerodynamic drag coefficient is 0.250 and the frontal area is 2.20 m^2. Neglecting all other sources of friction, calculate the initial acceleration of the car if it has been traveling at 100 km/h and is now shifted into neutral and allowed to coast.

39. A motorboat cuts its engine when its speed is 10.0 m/s and coasts to rest. The equation describing the motion of the motorboat during this period is $v = v_i e^{-ct}$, where v is the speed at time t, v_i is the initial speed, and c is a constant. At $t = 20.0$ s, the speed is 5.00 m/s. (a) Find the constant c. (b) What is the speed at $t = 40.0$ s? (c) Differentiate the expression for $v(t)$ and thus show that the acceleration of the boat is proportional to the speed at any time.

40. Consider an object on which the net force is a resistive force proportional to the square of its speed. For example, assume that the resistive force acting on a speed skater is $f = - kmv^2$, where k is a constant and m is the skater's mass. The skater crosses the finish line of a straight-line race with

speed v_0 and then slows down by coasting on his skates. Show that the skater's speed at any time t after crossing the finish line is $v(t) = v_0/(1 + ktv_0)$. This problem also provides the background for the two following problems.

41. (a) Use the result of Problem 40 to find the position x as a function of time for an object of mass m, located at $x = 0$ and moving with velocity $v_0\hat{\mathbf{i}}$ at time $t = 0$ and thereafter experiencing a net force $-kmv^2\hat{\mathbf{i}}$. (b) Find the object's velocity as a function of position.

42. At major league baseball games it is commonplace to flash on the scoreboard a speed for each pitch. This speed is determined with a radar gun aimed by an operator positioned behind home plate. The gun uses the Doppler shift of microwaves reflected from the baseball, as we will study in Chapter 39. The gun determines the speed at some particular point on the baseball's path, depending on when the operator pulls the trigger. Because the ball is subject to a drag force due to air, it slows as it travels 18.3 m toward the plate. Use the result of Problem 41(b) to find how much its speed decreases. Suppose the ball leaves the pitcher's hand at 90.0 mi/h = 40.2 m/s. Ignore its vertical motion. Use data on baseballs from Example 6.13 to determine the speed of the pitch when it crosses the plate.

43. You can feel a force of air drag on your hand if you stretch your arm out of the open window of a speeding car. [*Note:* Do not endanger yourself.] What is the order of magnitude of this force? In your solution state the quantities you measure or estimate and their values.

Section 6.5 Numerical Modeling in Particle Dynamics

44. A 3.00-g leaf is dropped from a height of 2.00 m above the ground. Assume the net downward force exerted on the leaf is $F = mg - bv$, where the drag factor is $b = 0.030\ 0$ kg/s. (a) Calculate the terminal speed of the leaf. (b) Use Euler's method of numerical analysis to find the speed and position of the leaf, as functions of time, from the instant it is released until 99% of terminal speed is reached. (*Suggestion:* Try $\Delta t = 0.005$ s.)

45. A hailstone of mass 4.80×10^{-4} kg falls through the air and experiences a net force given by

$$F = -mg + Cv^2$$

where $C = 2.50 \times 10^{-5}$ kg/m. (a) Calculate the terminal speed of the hailstone. (b) Use Euler's method of numerical analysis to find the speed and position of the hailstone at 0.2-s intervals, taking the initial speed to be zero. Continue the calculation until the hailstone reaches 99% of terminal speed.

46. A 0.142-kg baseball has a terminal speed of 42.5 m/s (95 mi/h). (a) If a baseball experiences a drag force of magnitude $R = Cv^2$, what is the value of the constant C? (b) What is the magnitude of the drag force when the speed of the baseball is 36.0 m/s? (c) Use a computer to determine the motion of a baseball thrown vertically upward at an initial speed of 36 m/s. What maximum height does the ball reach? How long is it in the air? What is its speed just before it hits the ground?

47. A 50.0-kg parachutist jumps from an airplane and falls to Earth with a drag force proportional to the square of the speed, $R = Cv^2$. Take $C = 0.200$ kg/m (with the parachute closed) and $C = 20.0$ kg/m (with the chute open). (a) Determine the terminal speed of the parachutist in both configurations, before and after the chute is opened. (b) Set up a numerical analysis of the motion and compute the speed and position as functions of time, assuming the jumper begins the descent at 1 000 m above the ground and is in free fall for 10.0 s before opening the parachute. (*Suggestion:* When the parachute opens, a sudden large acceleration takes place; a smaller time step may be necessary in this region.)

48. Consider a 10.0-kg projectile launched with an initial speed of 100 m/s, at an elevation angle of 35.0°. The resistive force is $\mathbf{R} = -b\mathbf{v}$, where $b = 10.0$ kg/s. (a) Use a numerical method to determine the horizontal and vertical coordinates of the projectile as functions of time. (b) What is the range of this projectile? (c) Determine the elevation angle that gives the maximum range for the projectile. (*Suggestion:* Adjust the elevation angle by trial and error to find the greatest range.)

49. A professional golfer hits her 5-iron 155 m (170 yd). A 46.0-g golf ball experiences a drag force of magnitude $R = Cv^2$, and has a terminal speed of 44.0 m/s. (a) Calculate the drag constant C for the golf ball. (b) Use a numerical method to calculate the trajectory of this shot. If the initial velocity of the ball makes an angle of 31.0° (the loft angle) with the horizontal, what initial speed must the ball have to reach the 155-m distance? (c) If this same golfer hits her 9-iron (47.0° loft) a distance of 119 m, what is the initial speed of the ball in this case? Discuss the differences in trajectories between the two shots.

Additional Problems

50. In a home laundry dryer, a cylindrical tub containing wet clothes is rotated steadily about a horizontal axis, as shown in Figure P6.50. So that the clothes will dry uniformly, they are made to tumble. The rate of rotation of the smooth-walled tub is chosen so that a small piece of cloth will lose contact with the tub when the cloth is at an angle of 68.0°

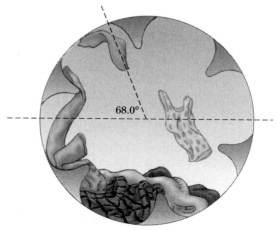

Figure P6.50

above the horizontal. If the radius of the tub is 0.330 m, what rate of revolution is needed?

51. We will study the most important work of Nobel laureate Arthur Compton in Chapter 40. Disturbed by speeding cars outside the physics building at Washington University in St. Louis, Compton designed a speed bump and had it installed. Suppose that a 1 800-kg car passes over a bump in a roadway that follows the arc of a circle of radius 20.4 m as in Figure P6.51. (a) What force does the road exert on the car as the car passes the highest point of the bump if the car travels at 30.0 km/h? (b) **What If?** What is the maximum speed the car can have as it passes this highest point without losing contact with the road?

Figure P6.51 Problems 51 and 52.

52. A car of mass m passes over a bump in a road that follows the arc of a circle of radius R as in Figure P6.51. (a) What force does the road exert on the car as the car passes the highest point of the bump if the car travels at a speed v? (b) **What If?** What is the maximum speed the car can have as it passes this highest point without losing contact with the road?

53. Interpret the graph in Figure 6.18(b). Proceed as follows: (a) Find the slope of the straight line, including its units. (b) From Equation 6.6, $R = \frac{1}{2}D\rho Av^2$, identify the theoretical slope of a graph of resistive force versus squared speed. (c) Set the experimental and theoretical slopes equal to each other and proceed to calculate the drag coefficient of the filters. Use the value for the density of air listed on the book's endpapers. Model the cross-sectional area of the filters as that of a circle of radius 10.5 cm. (d) Arbitrarily choose the eighth data point on the graph and find its vertical separation from the line of best fit. Express this scatter as a percentage. (e) In a short paragraph state what the graph demonstrates and compare it to the theoretical prediction. You will need to make reference to the quantities plotted on the axes, to the shape of the graph line, to the data points, and to the results of parts (c) and (d).

54. A student builds and calibrates an accelerometer, which she uses to determine the speed of her car around a certain unbanked highway curve. The accelerometer is a plumb bob with a protractor that she attaches to the roof of her car. A friend riding in the car with her observes that the plumb bob hangs at an angle of 15.0° from the vertical when the car has a speed of 23.0 m/s. (a) What is the centripetal acceleration of the car rounding the curve? (b) What is the radius of the curve? (c) What is the speed of the car if the plumb bob deflection is 9.00° while rounding the same curve?

55. Suppose the boxcar of Figure 6.13 is moving with constant acceleration a up a hill that makes an angle ϕ with the

horizontal. If the pendulum makes a constant angle θ with the perpendicular to the ceiling, what is a?

56. (a) A luggage carousel at an airport has the form of a section of a large cone, steadily rotating about its vertical axis. Its metallic surface slopes downward toward the outside, making an angle of 20.0° with the horizontal. A piece of luggage having mass 30.0 kg is placed on the carousel, 7.46 m from the axis of rotation. The travel bag goes around once in 38.0 s. Calculate the force of static friction between the bag and the carousel. (b) The drive motor is shifted to turn the carousel at a higher constant rate of rotation, and the piece of luggage is bumped to another position, 7.94 m from the axis of rotation. Now going around once in every 34.0 s, the bag is on the verge of slipping. Calculate the coefficient of static friction between the bag and the carousel.

57. Because the Earth rotates about its axis, a point on the equator experiences a centripetal acceleration of 0.033 7 m/s², while a point at the poles experiences no centripetal acceleration. (a) Show that at the equator the gravitational force on an object must exceed the normal force required to support the object. That is, show that the object's true weight exceeds its apparent weight. (b) What is the apparent weight at the equator and at the poles of a person having a mass of 75.0 kg? (Assume the Earth is a uniform sphere and take $g = 9.800$ m/s².)

58. An air puck of mass m_1 is tied to a string and allowed to revolve in a circle of radius R on a frictionless horizontal table. The other end of the string passes through a hole in the center of the table, and a counterweight of mass m_2 is tied to it (Fig. P6.58). The suspended object remains in equilibrium while the puck on the tabletop revolves. What is (a) the tension in the string? (b) the radial force acting on the puck? (c) the speed of the puck?

Figure P6.58

59. The pilot of an airplane executes a constant-speed loop-the-loop maneuver in a vertical circle. The speed of the airplane is 300 mi/h, and the radius of the circle is 1 200 ft. (a) What is the pilot's apparent weight at the lowest point if his true weight is 160 lb? (b) What is his apparent weight at the highest point? (c) **What If?** Describe how the pilot could experience weightlessness if both the radius and the speed can be varied. (*Note:* His apparent weight is equal to the magnitude of the force exerted by the seat on his body.)

60. A penny of mass 3.10 g rests on a small 20.0-g block supported by a spinning disk (Fig. P6.60). The coefficients of friction between block and disk are 0.750 (static) and

0.640 (kinetic) while those for the penny and block are 0.520 (static) and 0.450 (kinetic). What is the maximum rate of rotation in revolutions per minute that the disk can have, without the block or penny sliding on the disk?

Figure P6.60

61. Figure P6.61 shows a Ferris wheel that rotates four times each minute. It carries each car around a circle of diameter 18.0 m. (a) What is the centripetal acceleration of a rider? What force does the seat exert on a 40.0-kg rider (b) at the lowest point of the ride and (c) at the highest point of the ride? (d) What force (magnitude and direction) does the seat exert on a rider when the rider is halfway between top and bottom?

Figure P6.61

62. A space station, in the form of a wheel 120 m in diameter, rotates to provide an "artificial gravity" of 3.00 m/s² for persons who walk around on the inner wall of the outer rim. Find the rate of rotation of the wheel (in revolutions per minute) that will produce this effect.

63. An amusement park ride consists of a rotating circular platform 8.00 m in diameter from which 10.0-kg seats are suspended at the end of 2.50-m massless chains (Fig. P6.63). When the system rotates, the chains make an angle $\theta = 28.0°$ with the vertical. (a) What is the speed of each seat? (b) Draw a free-body diagram of a 40.0-kg child riding in a seat and find the tension in the chain.

64. A piece of putty is initially located at point A on the rim of a grinding wheel rotating about a horizontal axis. The putty is dislodged from point A when the diameter through A is horizontal. It then rises vertically and returns to A at the instant the wheel completes one revolution.

Figure P6.63

(a) Find the speed of a point on the rim of the wheel in terms of the acceleration due to gravity and the radius R of the wheel. (b) If the mass of the putty is m, what is the magnitude of the force that held it to the wheel?

65. An amusement park ride consists of a large vertical cylinder that spins about its axis fast enough such that any person inside is held up against the wall when the floor drops away (Fig. P6.65). The coefficient of static friction between person and wall is μ_s, and the radius of the cylinder is R. (a) Show that the maximum period of revolution necessary to keep the person from falling is $T = (4\pi^2 R\mu_s/g)^{1/2}$. (b) Obtain a numerical value for T if $R = 4.00$ m and $\mu_s = 0.400$. How many revolutions per minute does the cylinder make?

Figure P6.65

66. *An example of the Coriolis effect.* Suppose air resistance is negligible for a golf ball. A golfer tees off from a location precisely at $\phi_i = 35.0°$ north latitude. He hits the ball due south, with range 285 m. The ball's initial velocity is at 48.0° above the horizontal. (a) For how long is the ball in flight? The cup is due south of the golfer's location, and he would have a hole-in-one if the Earth were not rotating. The Earth's rotation makes the tee move in a circle of radius $R_E \cos \phi_i = (6.37 \times 10^6 \text{ m}) \cos 35.0°$, as shown in Figure P6.66. The tee completes one revolution each day. (b) Find the eastward speed of the tee, relative to the stars. The hole is also moving east, but it is 285 m farther south, and thus at a slightly lower latitude ϕ_f. Because the hole moves in a slightly larger circle, its speed must be greater than that of the tee. (c) By how much does the hole's speed exceed that of the tee? During the time the ball is in flight, it moves upward and downward as well as southward with the projectile motion you studied in Chapter 4, but it also moves eastward with the speed you found in part (b). The hole moves to the east at a faster speed, however, pulling ahead of the ball with the relative speed you found in part (c). (d) How far to the west of the hole does the ball land?

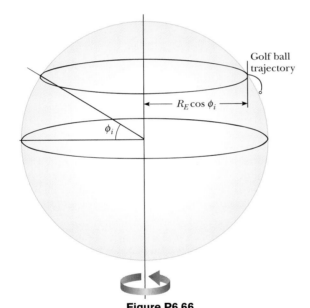

Golf ball trajectory

$R_E \cos \phi_i$

ϕ_i

Figure P6.66

67. A car rounds a banked curve as in Figure 6.6. The radius of curvature of the road is R, the banking angle is θ, and the coefficient of static friction is μ_s. (a) Determine the range of speeds the car can have without slipping up or down the road. (b) Find the minimum value for μ_s such that the minimum speed is zero. (c) What is the range of speeds possible if $R = 100$ m, $\theta = 10.0°$, and $\mu_s = 0.100$ (slippery conditions)?

68. A single bead can slide with negligible friction on a wire that is bent into a circular loop of radius 15.0 cm, as in Figure P6.68. The circle is always in a vertical plane and rotates steadily about its vertical diameter with (a) a period of 0.450 s. The position of the bead is described by the angle θ that the radial line, from the center of the loop to the bead, makes with the vertical. At what angle up from the bottom of the circle can the bead stay motionless relative

θ

Figure P6.68

to the turning circle? (b) **What If?** Repeat the problem if the period of the circle's rotation is 0.850 s.

69. The expression $F = arv + br^2v^2$ gives the magnitude of the resistive force (in newtons) exerted on a sphere of radius r (in meters) by a stream of air moving at speed v (in meters per second), where a and b are constants with appropriate SI units. Their numerical values are $a = 3.10 \times 10^{-4}$ and $b = 0.870$. Using this expression, find the terminal speed for water droplets falling under their own weight in air, taking the following values for the drop radii: (a) 10.0 μm, (b) 100 μm, (c) 1.00 mm. Note that for (a) and (c) you can obtain accurate answers without solving a quadratic equation, by considering which of the two contributions to the air resistance is dominant and ignoring the lesser contribution.

70. A 9.00-kg object starting from rest falls through a viscous medium and experiences a resistive force $\mathbf{R} = -b\mathbf{v}$, where $\mathbf{v}$ is the velocity of the object. If the object reaches one-half its terminal speed in 5.54 s, (a) determine the terminal speed. (b) At what time is the speed of the object three-fourths the terminal speed? (c) How far has the object traveled in the first 5.54 s of motion?

71. A model airplane of mass 0.750 kg flies in a horizontal circle at the end of a 60.0-m control wire, with a speed of 35.0 m/s. Compute the tension in the wire if it makes a constant angle of 20.0° with the horizontal. The forces exerted on the airplane are the pull of the control wire, the gravitational force, and aerodynamic lift, which acts at 20.0° inward from the vertical as shown in Figure P6.71.

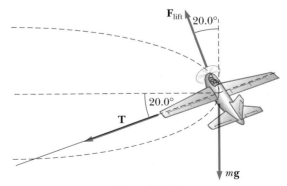

$\mathbf{F}_{\text{lift}}$ 20.0°

20.0°

T

$m\mathbf{g}$

Figure P6.71

72. 💻 Members of a skydiving club were given the following data to use in planning their jumps. In the table, d is the distance fallen from rest by a sky diver in a "free-fall stable spread position," versus the time of fall t. (a) Convert the distances in feet into meters. (b) Graph d (in meters) versus t. (c) Determine the value of the terminal speed v_T by finding the slope of the straight portion of the curve. Use a least-squares fit to determine this slope.

t (s)	d (ft)	t (s)	d (ft)
1	16	11	1 309
2	62	12	1 483
3	138	13	1 657
4	242	14	1 831
5	366	15	2 005
6	504	16	2 179
7	652	17	2 353
8	808	18	2 527
9	971	19	2 701
10	1 138	20	2 875

73. If a single constant force acts on an object that moves on a straight line, the object's velocity is a linear function of time. The equation $v = v_i + at$ gives its velocity v as a function of time, where a is its constant acceleration. **What if** velocity is instead a linear function of position? Assume that as a particular object moves through a resistive medium, its speed decreases as described by the equation $v = v_i - kx$, where k is a constant coefficient and x is the position of the object. Find the law describing the total force acting on this object.

Answers to Quick Quizzes

6.1 (b), (d). The centripetal acceleration is always toward the center of the circular path.

6.2 (a), (d). The normal force is always perpendicular to the surface that applies the force. Because your car maintains its orientation at all points on the ride, the normal force is always upward.

6.3 (a). If the car is moving in a circular path, it must have centripetal acceleration given by Equation 4.15.

6.4 Because the speed is constant, the only direction the force can have is that of the centripetal acceleration. The force is larger at Ⓒ than at Ⓐ because the radius at Ⓒ is smaller. There is no force at Ⓑ because the wire is straight.

6.5 In addition to the forces in the centripetal direction in Quick Quiz 6.4, there are now tangential forces to provide the tangential acceleration. The tangential force is the same at all three points because the tangential acceleration is constant.

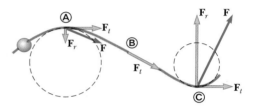

6.6 (c). The only forces acting on the passenger are the contact force with the door and the friction force from the seat. Both of these are real forces and both act to the left in Figure 6.11. Fictitious forces should never be drawn in a force diagram.

6.7 (a). The basketball, having a larger cross-sectional area, will have a larger force due to air resistance than the baseball. This will result in a smaller net force in the downward direction and a smaller downward acceleration.

Chapter 7

Energy and Energy Transfer

▲ On a wind farm, the moving air does work on the blades of the windmills, causing the blades and the rotor of an electrical generator to rotate. Energy is transferred out of the system of the windmill by means of electricity. (Billy Hustace/Getty Images)

181

The concept of energy is one of the most important topics in science and engineering. In everyday life, we think of energy in terms of fuel for transportation and heating, electricity for lights and appliances, and foods for consumption. However, these ideas do not really define energy. They merely tell us that fuels are needed to do a job and that those fuels provide us with something we call energy.

The definitions of quantities such as position, velocity, acceleration, and force and associated principles such as Newton's second law have allowed us to solve a variety of problems. Some problems that could theoretically be solved with Newton's laws, however, are very difficult in practice. These problems can be made much simpler with a different approach. In this and the following chapters, we will investigate this new approach, which will include definitions of quantities that may not be familiar to you. Other quantities may sound familiar, but they may have more specific meanings in physics than in everyday life. We begin this discussion by exploring the notion of energy.

Energy is present in the Universe in various forms. *Every* physical process that occurs in the Universe involves energy and energy transfers or transformations. Unfortunately, despite its extreme importance, energy cannot be easily defined. The variables in previous chapters were relatively concrete; we have everyday experience with velocities and forces, for example. The notion of energy is more abstract, although we do have *experiences* with energy, such as running out of gasoline, or losing our electrical service if we forget to pay the utility bill.

The concept of energy can be applied to the dynamics of a mechanical system without resorting to Newton's laws. This "energy approach" to describing motion is especially useful when the force acting on a particle is not constant; in such a case, the acceleration is not constant, and we cannot apply the constant acceleration equations that were developed in Chapter 2. Particles in nature are often subject to forces that vary with the particles' positions. These forces include gravitational forces and the force exerted on an object attached to a spring. We shall describe techniques for treating such situations with the help of an important concept called *conservation of energy*. This approach extends well beyond physics, and can be applied to biological organisms, technological systems, and engineering situations.

Our problem-solving techniques presented in earlier chapters were based on the motion of a particle or an object that could be modeled as a particle. This was called the *particle model*. We begin our new approach by focusing our attention on a *system* and developing techniques to be used in a *system model*.

7.1 Systems and Environments

In the system model mentioned above, we focus our attention on a small portion of the Universe—the **system**—and ignore details of the rest of the Universe outside of the system. A critical skill in applying the system model to problems is *identifying the system*.

A valid system may

- be a single object or particle
- be a collection of objects or particles
- be a region of space (such as the interior of an automobile engine combustion cylinder)
- vary in size and shape (such as a rubber ball, which deforms upon striking a wall)

Identifying the *need* for a system approach to solving a problem (as opposed to a particle approach) is part of the "categorize" step in the General Problem-Solving Strategy outlined in Chapter 2. Identifying the particular system and its nature is part of the "analyze" step.

No matter what the particular system is in a given problem, there is a **system boundary,** an imaginary surface (not necessarily coinciding with a physical surface) that divides the Universe into the system and the **environment** surrounding the system.

As an example, imagine a force applied to an object in empty space. We can define the object as the system. The force applied to it is an influence on the system from the environment that acts across the system boundary. We will see how to analyze this situation from a system approach in a subsequent section of this chapter.

Another example is seen in Example 5.10 (page 130). Here the system can be defined as the combination of the ball, the cube, and the string. The influence from the environment includes the gravitational forces on the ball and the cube, the normal and friction forces on the cube, and the force exerted by the pulley on the string. The forces exerted by the string on the ball and the cube are internal to the system and, therefore, are not included as an influence from the environment.

We shall find that there are a number of mechanisms by which a system can be influenced by its environment. The first of these that we shall investigate is *work*.

▲ **PITFALL PREVENTION**

7.1 Identify the System

The most important step to take in solving a problem using the energy approach is to identify the appropriate system of interest. Make sure this is the *first* step you take in solving a problem.

7.2 Work Done by a Constant Force

Almost all the terms we have used thus far—velocity, acceleration, force, and so on—convey a similar meaning in physics as they do in everyday life. Now, however, we encounter a term whose meaning in physics is distinctly different from its everyday meaning—*work*.

To understand what work means to the physicist, consider the situation illustrated in Figure 7.1. A force is applied to a chalkboard eraser, and the eraser slides

Charles D. Winters

(a)   (b)   (c)

Figure 7.1 An eraser being pushed along a chalkboard tray.

Figure 7.2 If an object undergoes a displacement $\Delta\mathbf{r}$ under the action of a constant force $\mathbf{F}$, the work done by the force is $F\Delta r\cos\theta$.

Work done by a constant force

 PITFALL PREVENTION

7.2 What is being Displaced?

The displacement in Equation 7.1 is that of *the point of application of the force.* If the force is applied to a particle or a non-deformable, non-rotating system, this displacement is the same as the displacement of the particle or system. For deformable systems, however, these two displacements are often not the same.

 PITFALL PREVENTION

7.3 Work is Done by . . . on . . .

Not only must you identify the system, you must also identify the interaction of the system with the environment. When discussing work, always use the phrase, "the work done by . . . on . . . ." After "by," insert the part of the environment that is interacting directly with the system. After "on," insert the system. For example, "the work done by the hammer on the nail" identifies the nail as the system and the force from the hammer represents the interaction with the environment. This is similar to our use in Chapter 5 of "the force exerted by . . . on . . . ."

along the tray. If we want to know how effective the force is in moving the eraser, we must consider not only the magnitude of the force but also its direction. Assuming that the magnitude of the applied force is the same in all three photographs, the push applied in Figure 7.1b does more to move the eraser than the push in Figure 7.1a. On the other hand, Figure 7.1c shows a situation in which the applied force does not move the eraser at all, regardless of how hard it is pushed. (Unless, of course, we apply a force so great that we break the chalkboard tray.) So, in analyzing forces to determine the work they do, we must consider the vector nature of forces. We must also know how far the eraser moves along the tray if we want to determine the work associated with that displacement. Moving the eraser 3 m requires more work than moving it 2 cm.

Let us examine the situation in Figure 7.2, where an object undergoes a displacement along a straight line while acted on by a constant force $\mathbf{F}$ that makes an angle θ with the direction of the displacement.

> The **work** W done on a system by an agent exerting a constant force on the system is the product of the magnitude F of the force, the magnitude Δr of the displacement of the point of application of the force, and $\cos\theta$, where θ is the angle between the force and displacement vectors:
>
> $$W \equiv F\,\Delta r\cos\theta \qquad (7.1)$$

As an example of the distinction between this definition of work and our everyday understanding of the word, consider holding a heavy chair at arm's length for 3 min. At the end of this time interval, your tired arms may lead you to think that you have done a considerable amount of work on the chair. According to our definition, however, you have done no work on it whatsoever.[1] You exert a force to support the chair, but you do not move it. A force does no work on an object if the force does not move through a displacement. This can be seen by noting that if $\Delta r = 0$, Equation 7.1 gives $W = 0$—the situation depicted in Figure 7.1c.

Also note from Equation 7.1 that the work done by a force on a moving object is zero when the force applied is perpendicular to the displacement of its point of

The weightlifter does no work on the weights as he holds them on his shoulders. (If he could rest the bar on his shoulders and lock his knees, he would be able to support the weights for quite some time.) Did he do any work when he raised the weights to this height?

---

[1]   Actually, you do work while holding the chair at arm's length because your muscles are continuously contracting and relaxing; this means that they are exerting internal forces on your arm. Thus, work is being done by your body—but internally on itself rather than on the chair.

application. That is, if $\theta = 90°$, then $W = 0$ because $\cos 90° = 0$. For example, in Figure 7.3, the work done by the normal force on the object and the work done by the gravitational force on the object are both zero because both forces are perpendicular to the displacement and have zero components along an axis in the direction of $\Delta \mathbf{r}$.

The sign of the work also depends on the direction of $\mathbf{F}$ relative to $\Delta \mathbf{r}$. The work done by the applied force is positive when the projection of $\mathbf{F}$ onto $\Delta \mathbf{r}$ is in the same direction as the displacement. For example, when an object is lifted, the work done by the applied force is positive because the direction of that force is upward, in the same direction as the displacement of its point of application. When the projection of $\mathbf{F}$ onto $\Delta \mathbf{r}$ is in the direction opposite the displacement, W is negative. For example, as an object is lifted, the work done by the gravitational force on the object is negative. The factor $\cos \theta$ in the definition of W (Eq. 7.1) automatically takes care of the sign.

If an applied force $\mathbf{F}$ is in the same direction as the displacement $\Delta \mathbf{r}$, then $\theta = 0$ and $\cos 0 = 1$. In this case, Equation 7.1 gives

$$W = F \Delta r$$

Work is a scalar quantity, and its units are force multiplied by length. Therefore, the SI unit of work is the **newton·meter** (N·m). This combination of units is used so frequently that it has been given a name of its own: the **joule** (J).

An important consideration for a system approach to problems is to note that **work is an energy transfer.** If W is the work done on a system and W is positive, energy is transferred *to* the system; if W is negative, energy is transferred *from* the system. Thus, if a system interacts with its environment, this interaction can be described as a transfer of energy across the system boundary. This will result in a change in the energy stored in the system. We will learn about the first type of energy storage in Section 7.5, after we investigate more aspects of work.

Figure 7.3 When an object is displaced on a frictionless, horizontal surface, the normal force **n** and the gravitational force $m\mathbf{g}$ do no work on the object. In the situation shown here, **F** is the only force doing work on the object.

⚠ PITFALL PREVENTION

7.4   Cause of the Displacement

We can calculate the work done by a force on an object, but that force is *not* necessarily the cause of the object's displacement. For example, if you lift an object, work is done by the gravitational force, although gravity is not the cause of the object moving upward!

Quick Quiz 7.1 The gravitational force exerted by the Sun on the Earth holds the Earth in an orbit around the Sun. Let us assume that the orbit is perfectly circular. The work done by this gravitational force during a short time interval in which the Earth moves through a displacement in its orbital path is (a) zero (b) positive (c) negative (d) impossible to determine.

Quick Quiz 7.2 Figure 7.4 shows four situations in which a force is applied to an object. In all four cases, the force has the same magnitude, and the displacement of the object is to the right and of the same magnitude. Rank the situations in order of the work done by the force on the object, from most positive to most negative.

Figure 7.4 (Quick Quiz 7.2)

Example 7.1 Mr. Clean

A man cleaning a floor pulls a vacuum cleaner with a force of magnitude $F = 50.0$ N at an angle of 30.0° with the horizontal (Fig. 7.5a). Calculate the work done by the force on the vacuum cleaner as the vacuum cleaner is displaced 3.00 m to the right.

Solution Figure 7.5a helps conceptualize the situation. We are given a force, a displacement, and the angle between the two vectors, so we can categorize this as a simple problem that will need minimal analysis. To analyze the situation, we identify the vacuum cleaner as the system and draw a free-body diagram as shown in Figure 7.5b. Using the definition of work (Eq. 7.1),

$$W = F \, \Delta r \cos \theta = (50.0 \text{ N})(3.00 \text{ m})(\cos 30.0°)$$

$$= 130 \text{ N} \cdot \text{m} = \boxed{130 \text{ J}}$$

To finalize this problem, notice in this situation that the normal force $\mathbf{n}$ and the gravitational $\mathbf{F}_g = m\mathbf{g}$ do no work on the vacuum cleaner because these forces are perpendicular to its displacement.

Figure 7.5 (Example 7.1) (a) A vacuum cleaner being pulled at an angle of 30.0° from the horizontal. (b) Free-body diagram of the forces acting on the vacuum cleaner.

7.3 The Scalar Product of Two Vectors

Because of the way the force and displacement vectors are combined in Equation 7.1, it is helpful to use a convenient mathematical tool called the **scalar product** of two vectors. We write this scalar product of vectors $\mathbf{A}$ and $\mathbf{B}$ as $\mathbf{A} \cdot \mathbf{B}$. (Because of the dot symbol, the scalar product is often called the **dot product.**)

In general, the scalar product of any two vectors $\mathbf{A}$ and $\mathbf{B}$ is a scalar quantity equal to the product of the magnitudes of the two vectors and the cosine of the angle θ between them:

Scalar product of any two vectors A and B

$$\mathbf{A} \cdot \mathbf{B} \equiv AB \cos \theta \tag{7.2}$$

Note that $\mathbf{A}$ and $\mathbf{B}$ need not have the same units, as is the case with any multiplication.

Comparing this definition to Equation 7.1, we see that we can express Equation 7.1 as a scalar product:

$$W = F \, \Delta r \cos \theta = \mathbf{F} \cdot \Delta \mathbf{r} \tag{7.3}$$

In other words, $\mathbf{F} \cdot \Delta \mathbf{r}$ (read "$\mathbf{F}$ dot $\Delta \mathbf{r}$") is a shorthand notation for $F \Delta r \cos \theta$.

Before continuing with our discussion of work, let us investigate some properties of the dot product. Figure 7.6 shows two vectors $\mathbf{A}$ and $\mathbf{B}$ and the angle θ between them that is used in the definition of the dot product. In Figure 7.6, $B \cos \theta$ is the projection of $\mathbf{B}$ onto $\mathbf{A}$. Therefore, Equation 7.2 means that $\mathbf{A} \cdot \mathbf{B}$ is the product of the magnitude of $\mathbf{A}$ and the projection of $\mathbf{B}$ onto $\mathbf{A}$.[2]

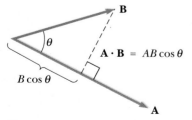

Figure 7.6 The scalar product $\mathbf{A} \cdot \mathbf{B}$ equals the magnitude of $\mathbf{A}$ multiplied by $B \cos \theta$, which is the projection of $\mathbf{B}$ onto $\mathbf{A}$.

---

[2] This is equivalent to stating that $\mathbf{A} \cdot \mathbf{B}$ equals the product of the magnitude of $\mathbf{B}$ and the projection of $\mathbf{A}$ onto $\mathbf{B}$.

From the right-hand side of Equation 7.2 we also see that the scalar product is **commutative**.[3] That is,

$$\mathbf{A} \cdot \mathbf{B} = \mathbf{B} \cdot \mathbf{A}$$

Finally, the scalar product obeys the **distributive law of multiplication,** so that

$$\mathbf{A} \cdot (\mathbf{B} + \mathbf{C}) = \mathbf{A} \cdot \mathbf{B} + \mathbf{A} \cdot \mathbf{C}$$

The dot product is simple to evaluate from Equation 7.2 when $\mathbf{A}$ is either perpendicular or parallel to $\mathbf{B}$. If $\mathbf{A}$ is perpendicular to $\mathbf{B}$ ($\theta = 90°$), then $\mathbf{A} \cdot \mathbf{B} = 0$. (The equality $\mathbf{A} \cdot \mathbf{B} = 0$ also holds in the more trivial case in which either $\mathbf{A}$ or $\mathbf{B}$ is zero.) If vector $\mathbf{A}$ is parallel to vector $\mathbf{B}$ and the two point in the same direction ($\theta = 0$), then $\mathbf{A} \cdot \mathbf{B} = AB$. If vector $\mathbf{A}$ is parallel to vector $\mathbf{B}$ but the two point in opposite directions ($\theta = 180°$), then $\mathbf{A} \cdot \mathbf{B} = -AB$. The scalar product is negative when $90° < \theta \le 180°$.

The unit vectors $\hat{\mathbf{i}}$, $\hat{\mathbf{j}}$, and $\hat{\mathbf{k}}$, which were defined in Chapter 3, lie in the positive x, y, and z directions, respectively, of a right-handed coordinate system. Therefore, it follows from the definition of $\mathbf{A} \cdot \mathbf{B}$ that the scalar products of these unit vectors are

$$\hat{\mathbf{i}} \cdot \hat{\mathbf{i}} = \hat{\mathbf{j}} \cdot \hat{\mathbf{j}} = \hat{\mathbf{k}} \cdot \hat{\mathbf{k}} = 1 \qquad (7.4)$$

$$\hat{\mathbf{i}} \cdot \hat{\mathbf{j}} = \hat{\mathbf{i}} \cdot \hat{\mathbf{k}} = \hat{\mathbf{j}} \cdot \hat{\mathbf{k}} = 0 \qquad (7.5)$$

Equations 3.18 and 3.19 state that two vectors $\mathbf{A}$ and $\mathbf{B}$ can be expressed in component vector form as

$$\mathbf{A} = A_x \hat{\mathbf{i}} + A_y \hat{\mathbf{j}} + A_z \hat{\mathbf{k}}$$

$$\mathbf{B} = B_x \hat{\mathbf{i}} + B_y \hat{\mathbf{j}} + B_z \hat{\mathbf{k}}$$

Using the information given in Equations 7.4 and 7.5 shows that the scalar product of $\mathbf{A}$ and $\mathbf{B}$ reduces to

$$\mathbf{A} \cdot \mathbf{B} = A_x B_x + A_y B_y + A_z B_z \qquad (7.6)$$

(Details of the derivation are left for you in Problem 6.) In the special case in which $\mathbf{A} = \mathbf{B}$, we see that

$$\mathbf{A} \cdot \mathbf{A} = A_x{}^2 + A_y{}^2 + A_z{}^2 = A^2$$

 PITFALL PREVENTION

7.5 Work is a Scalar

Although Equation 7.3 defines the work in terms of two vectors, *work is a scalar*—there is no direction associated with it. *All* types of energy and energy transfer are scalars. This is a major advantage of the energy approach—we don't need vector calculations!

Dot products of unit vectors

Quick Quiz 7.3 Which of the following statements is true about the relationship between $\mathbf{A} \cdot \mathbf{B}$ and $(-\mathbf{A}) \cdot (-\mathbf{B})$? (a) $\mathbf{A} \cdot \mathbf{B} = -[(-\mathbf{A}) \cdot (-\mathbf{B})]$; (b) If $\mathbf{A} \cdot \mathbf{B} = AB \cos \theta$, then $(-\mathbf{A}) \cdot (-\mathbf{B}) = AB \cos (\theta + 180°)$; (c) Both (a) and (b) are true. (d) Neither (a) nor (b) is true.

Quick Quiz 7.4 Which of the following statements is true about the relationship between the dot product of two vectors and the product of the magnitudes of the vectors? (a) $\mathbf{A} \cdot \mathbf{B}$ is larger than AB; (b) $\mathbf{A} \cdot \mathbf{B}$ is smaller than AB; (c) $\mathbf{A} \cdot \mathbf{B}$ could be larger or smaller than AB, depending on the angle between the vectors; (d) $\mathbf{A} \cdot \mathbf{B}$ could be equal to AB.

---

[3] This may seem obvious, but in Chapter 11 you will see another way of combining vectors that proves useful in physics and is not commutative.

Example 7.2 The Scalar Product

The vectors $\mathbf{A}$ and $\mathbf{B}$ are given by $\mathbf{A} = 2\hat{\mathbf{i}} + 3\hat{\mathbf{j}}$ and $\mathbf{B} = -\hat{\mathbf{i}} + 2\hat{\mathbf{j}}$.

(A) Determine the scalar product $\mathbf{A} \cdot \mathbf{B}$.

Solution Substituting the specific vector expressions for $\mathbf{A}$ and $\mathbf{B}$, we find,

$$\mathbf{A} \cdot \mathbf{B} = (2\hat{\mathbf{i}} + 3\hat{\mathbf{j}}) \cdot (-\hat{\mathbf{i}} + 2\hat{\mathbf{j}})$$
$$= -2\hat{\mathbf{i}} \cdot \hat{\mathbf{i}} + 2\hat{\mathbf{i}} \cdot 2\hat{\mathbf{j}} - 3\hat{\mathbf{j}} \cdot \hat{\mathbf{i}} + 3\hat{\mathbf{j}} \cdot 2\hat{\mathbf{j}}$$
$$= -2(1) + 4(0) - 3(0) + 6(1)$$
$$= -2 + 6 = \boxed{4}$$

where we have used the facts that $\hat{\mathbf{i}} \cdot \hat{\mathbf{i}} = \hat{\mathbf{j}} \cdot \hat{\mathbf{j}} = 1$ and $\hat{\mathbf{i}} \cdot \hat{\mathbf{j}} = \hat{\mathbf{j}} \cdot \hat{\mathbf{i}} = 0$. The same result is obtained when we use Equation 7.6 directly, where $A_x = 2$, $A_y = 3$, $B_x = -1$, and $B_y = 2$.

(B) Find the angle θ between $\mathbf{A}$ and $\mathbf{B}$.

Solution The magnitudes of $\mathbf{A}$ and $\mathbf{B}$ are

$$A = \sqrt{A_x^2 + A_y^2} = \sqrt{(2)^2 + (3)^2} = \sqrt{13}$$
$$B = \sqrt{B_x^2 + B_y^2} = \sqrt{(-1)^2 + (2)^2} = \sqrt{5}$$

Using Equation 7.2 and the result from part (a) we find that

$$\cos\theta = \frac{\mathbf{A} \cdot \mathbf{B}}{AB} = \frac{4}{\sqrt{13}\sqrt{5}} = \frac{4}{\sqrt{65}}$$

$$\theta = \cos^{-1}\frac{4}{8.06} = \boxed{60.2°}$$

Example 7.3 Work Done by a Constant Force

A particle moving in the xy plane undergoes a displacement $\Delta\mathbf{r} = (2.0\hat{\mathbf{i}} + 3.0\hat{\mathbf{j}})$ m as a constant force $\mathbf{F} = (5.0\hat{\mathbf{i}} + 2.0\hat{\mathbf{j}})$ N acts on the particle.

(A) Calculate the magnitudes of the displacement and the force.

Solution We use the Pythagorean theorem:

$$\Delta r = \sqrt{(\Delta x)^2 + (\Delta y)^2} = \sqrt{(2.0)^2 + (3.0)^2} = \boxed{3.6\,\text{m}}$$

$$F = \sqrt{F_x^2 + F_y^2} = \sqrt{(5.0)^2 + (2.0)^2} = \boxed{5.4\,\text{N}}$$

(B) Calculate the work done by $\mathbf{F}$.

Solution Substituting the expressions for $\mathbf{F}$ and $\Delta\mathbf{r}$ into Equation 7.3 and using Equations 7.4 and 7.5, we obtain

$$W = \mathbf{F} \cdot \Delta\mathbf{r} = [(5.0\hat{\mathbf{i}} + 2.0\hat{\mathbf{j}})\,\text{N}] \cdot [(2.0\hat{\mathbf{i}} + 3.0\hat{\mathbf{j}})\,\text{m}]$$
$$= (5.0\hat{\mathbf{i}} \cdot 2.0\hat{\mathbf{i}} + 5.0\hat{\mathbf{i}} \cdot 3.0\hat{\mathbf{j}} + 2.0\hat{\mathbf{j}} \cdot 2.0\hat{\mathbf{i}} + 2.0\hat{\mathbf{j}} \cdot 3.0\hat{\mathbf{j}})\,\text{N} \cdot \text{m}$$
$$= [10 + 0 + 0 + 6]\,\text{N} \cdot \text{m} = \boxed{16\,\text{J}}$$

7.4 Work Done by a Varying Force

Consider a particle being displaced along the x axis under the action of a force that varies with position. The particle is displaced in the direction of increasing x from $x = x_i$ to $x = x_f$. In such a situation, we cannot use $W = F\Delta r \cos\theta$ to calculate the work done by the force because this relationship applies only when $\mathbf{F}$ is constant in magnitude and direction. However, if we imagine that the particle undergoes a very small displacement Δx, shown in Figure 7.7a, the x component F_x of the force is approximately constant over this small interval; for this small displacement, we can approximate the work done by the force as

$$W \approx F_x \Delta x$$

This is just the area of the shaded rectangle in Figure 7.7a. If we imagine that the F_x versus x curve is divided into a large number of such intervals, the total work done for the displacement from x_i to x_f is approximately equal to the sum of a large number of such terms:

$$W \approx \sum_{x_i}^{x_f} F_x \Delta x$$

If the size of the displacements is allowed to approach zero, the number of terms in the sum increases without limit but the value of the sum approaches a definite value equal to the area bounded by the F_x curve and the x axis:

$$\lim_{\Delta x \to 0} \sum_{x_i}^{x_f} F_x \Delta x = \int_{x_i}^{x_f} F_x \, dx$$

Therefore, we can express the work done by F_x as the particle moves from x_i to x_f as

$$W = \int_{x_i}^{x_f} F_x \, dx \qquad (7.7)$$

This equation reduces to Equation 7.1 when the component $F_x = F \cos \theta$ is constant.

If more than one force acts on a system *and the system can be modeled as a particle*, the total work done on the system is just the work done by the net force. If we express the net force in the x direction as ΣF_x, then the total work, or *net work*, done as the particle moves from x_i to x_f is

$$\sum W = W_{\text{net}} = \int_{x_i}^{x_f} \left(\sum F_x \right) dx \qquad (7.8)$$

If the system cannot be modeled as a particle (for example, if the system consists of multiple particles that can move with respect to each other), we cannot use Equation 7.8. This is because different forces on the system may move through different displacements. In this case, we must evaluate the work done by each force separately and then add the works algebraically.

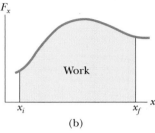

Figure 7.7 (a) The work done by the force component F_x for the small displacement Δx is $F_x \Delta x$, which equals the area of the shaded rectangle. The total work done for the displacement from x_i to x_f is approximately equal to the sum of the areas of all the rectangles. (b) The work done by the component F_x of the varying force as the particle moves from x_i to x_f is *exactly* equal to the area under this curve.

Example 7.4   Calculating Total Work Done from a Graph

A force acting on a particle varies with x, as shown in Figure 7.8. Calculate the work done by the force as the particle moves from $x = 0$ to $x = 6.0$ m.

Solution The work done by the force is equal to the area under the curve from $x_A = 0$ to $x_C = 6.0$ m. This area is equal to the area of the rectangular section from Ⓐ to Ⓑ plus the area of the triangular section from Ⓑ to Ⓒ. The area of the rectangle is $(5.0 \text{ N})(4.0 \text{ m}) = 20$ J, and the area of the triangle is $\frac{1}{2}(5.0 \text{ N})(2.0 \text{ m}) = 5.0$ J. Therefore, the total work done by the force on the particle is 25 J.

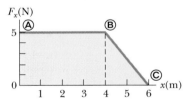

Figure 7.8 (Example 7.4) The force acting on a particle is constant for the first 4.0 m of motion and then decreases linearly with x from $x_B = 4.0$ m to $x_C = 6.0$ m. The net work done by this force is the area under the curve.

Example 7.5   Work Done by the Sun on a Probe

The interplanetary probe shown in Figure 7.9a is attracted to the Sun by a force given by

$$F = -\frac{1.3 \times 10^{22}}{x^2}$$

in SI units, where x is the Sun-probe separation distance. Graphically and analytically determine how much work is done by the Sun on the probe as the probe–Sun separation changes from 1.5×10^{11} m to 2.3×10^{11} m.

Graphical Solution The negative sign in the equation for the force indicates that the probe is attracted to the Sun. Because the probe is moving away from the Sun, we expect to obtain a negative value for the work done on it. A spreadsheet or other numerical means can be used to generate a graph like that in Figure 7.9b. Each small square of the grid corresponds to an area $(0.05 \text{ N})(0.1 \times 10^{11} \text{ m}) = 5 \times 10^8$ J. The work done is equal to the shaded area in Figure 7.9b. Because there are approximately 60 squares shaded, the total

(a)

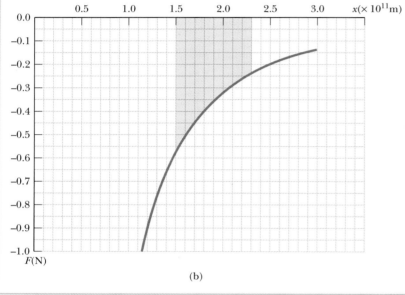

(b)

area (which is negative because the curve is below the x axis) is about -3×10^{10} J. This is the work done by the Sun on the probe.

Analytical Solution We can use Equation 7.7 to calculate a more precise value for the work done on the probe by the Sun. To solve this integral, we make use of the integral $\int x^n dx = x^{n+1}/(n+1)$ with $n = -2$:

$$W = \int_{1.5 \times 10^{11}}^{2.3 \times 10^{11}} \left(\frac{-1.3 \times 10^{22}}{x^2} \right) dx$$

$$= (-1.3 \times 10^{22}) \int_{1.5 \times 10^{11}}^{2.3 \times 10^{11}} x^{-2} dx$$

$$= (-1.3 \times 10^{22}) \left(\frac{x^{-1}}{-1} \right) \Big|_{1.5 \times 10^{11}}^{2.3 \times 10^{11}}$$

$$= (-1.3 \times 10^{22}) \left(\frac{-1}{2.3 \times 10^{11}} - \frac{-1}{1.5 \times 10^{11}} \right)$$

$$= \boxed{-3.0 \times 10^{10} \, \text{J}}$$

Figure 7.9 (Example 7.5) (a) An interplanetary probe moves from a position near the Earth's orbit radially outward from the Sun, ending up near the orbit of Mars. (b) Attractive force versus distance for the interplanetary probe.

Work Done by a Spring

A model of a common physical system for which the force varies with position is shown in Figure 7.10. A block on a horizontal, frictionless surface is connected to a spring. If the spring is either stretched or compressed a small distance from its unstretched (equilibrium) configuration, it exerts on the block a force that can be expressed as

Spring force

$$F_s = -kx \qquad (7.9)$$

where x is the position of the block relative to its equilibrium ($x = 0$) position and k is a positive constant called the **force constant** or the **spring constant** of the spring. In other words, the force required to stretch or compress a spring is proportional to the amount of stretch or compression x. This force law for springs is known as **Hooke's law**. The value of k is a measure of the *stiffness* of the spring. Stiff springs have large k values, and soft springs have small k values. As can be seen from Equation 7.9, the units of k are N/m.

The negative sign in Equation 7.9 signifies that the force exerted by the spring is always directed *opposite* to the displacement from equilibrium. When $x > 0$ as in Figure 7.10a, so that the block is to the right of the equilibrium position, the spring force is directed to the left, in the negative x direction. When $x < 0$ as in Figure 7.10c, the block is to the left of equilibrium and the spring force is directed to the right, in the positive x direction. When $x = 0$ as in Figure 7.10b, the spring is unstretched and $F_s = 0$.

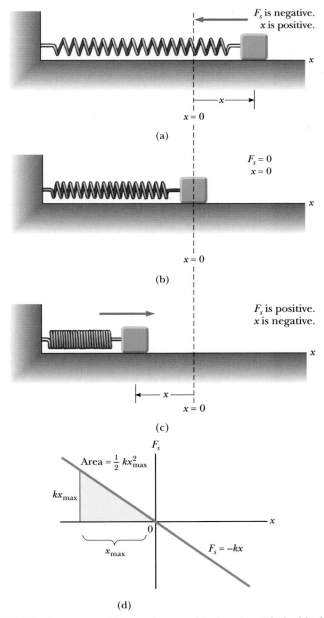

F_s is negative.
x is positive.

x

$x = 0$

(a)

$F_s = 0$
$x = 0$

x

$x = 0$

(b)

F_s is positive.
x is negative.

x

x

$x = 0$

(c)

F_s

Area $= \frac{1}{2} kx_{max}^2$

kx_{max}

x

0

x_{max}

$F_s = -kx$

(d)

Active Figure 7.10 The force exerted by a spring on a block varies with the block's position x relative to the equilibrium position $x = 0$. (a) When x is positive (stretched spring), the spring force is directed to the left. (b) When x is zero (natural length of the spring), the spring force is zero. (c) When x is negative (compressed spring), the spring force is directed to the right. (d) Graph of F_s versus x for the block–spring system. The work done by the spring force as the block moves from $-x_{max}$ to 0 is the area of the shaded triangle, $\frac{1}{2}kx_{max}^2$.

At the Active Figures link at http://www.pse6.com, *you can observe the block's motion for various maximum displacements and spring constants.*

Because the spring force always acts toward the equilibrium position $(x = 0)$, it is sometimes called a *restoring force*. If the spring is compressed until the block is at the point $-x_{max}$ and is then released, the block moves from $-x_{max}$ through zero to $+x_{max}$. If the spring is instead stretched until the block is at the point $+x_{max}$ and is then released, the block moves from $+x_{max}$ through zero to $-x_{max}$. It then reverses direction, returns to $+x_{max}$, and continues oscillating back and forth.

Suppose the block has been pushed to the left to a position $-x_{max}$ and is then released. Let us identify the block as our system and calculate the work W_s done by the spring force on the block as the block moves from $x_i = -x_{max}$ to $x_f = 0$. Applying

Equation 7.7 and assuming the block may be treated as a particle, we obtain

$$W_s = \int_{x_i}^{x_f} F_s dx = \int_{-x_{max}}^{0} (-kx)\, dx = \tfrac{1}{2} kx_{max}^2 \qquad (7.10)$$

where we have used the integral $\int x^n dx = x^{n+1}/(n+1)$ with $n = 1$. The work done by the spring force is positive because the force is in the same direction as the displacement of the block (both are to the right). Because the block arrives at $x = 0$ with some speed, it will continue moving, until it reaches a position $+x_{max}$. When we consider the work done by the spring force as the block moves from $x_i = 0$ to $x_f = x_{max}$, we find that $W_s = -\tfrac{1}{2} kx_{max}^2$ because for this part of the motion the displacement is to the right and the spring force is to the left. Therefore, the *net* work done by the spring force as the block moves from $x_i = -x_{max}$ to $x_f = x_{max}$ is *zero*.

Figure 7.10d is a plot of F_s versus x. The work calculated in Equation 7.10 is the area of the shaded triangle, corresponding to the displacement from $-x_{max}$ to 0. Because the triangle has base x_{max} and height kx_{max}, its area is $\tfrac{1}{2} kx_{max}^2$, the work done by the spring as given by Equation 7.10.

If the block undergoes an arbitrary displacement from $x = x_i$ to $x = x_f$, the work done by the spring force on the block is

$$W_s = \int_{x_i}^{x_f} (-kx)\, dx = \tfrac{1}{2} kx_i^2 - \tfrac{1}{2} kx_f^2 \qquad (7.11)$$

Work done by a spring

For example, if the spring has a force constant of 80 N/m and is compressed 3.0 cm from equilibrium, the work done by the spring force as the block moves from $x_i = -3.0$ cm to its unstretched position $x_f = 0$ is 3.6×10^{-2} J. From Equation 7.11 we also see that the work done by the spring force is zero for any motion that ends where it began ($x_i = x_f$). We shall make use of this important result in Chapter 8, in which we describe the motion of this system in greater detail.

Equations 7.10 and 7.11 describe the work done by the spring on the block. Now let us consider the work done on the spring by an *external agent* that stretches the spring very slowly from $x_i = 0$ to $x_f = x_{max}$, as in Figure 7.11. We can calculate this work by noting that at any value of the position, the *applied force* $\mathbf{F}_{app}$ is equal in magnitude and opposite in direction to the spring force $\mathbf{F}_s$, so that $F_{app} = -(-kx) = kx$. Therefore, the work done by this applied force (the external agent) on the block–spring system is

Figure 7.11 A block being pulled from $x_i = 0$ to $x_f = x_{max}$ on a frictionless surface by a force $\mathbf{F}_{app}$. If the process is carried out very slowly, the applied force is equal in magnitude and opposite in direction to the spring force at all times.

$$W_{F_{app}} = \int_{0}^{x_{max}} F_{app}\, dx = \int_{0}^{x_{max}} kx\, dx = \tfrac{1}{2} kx_{max}^2$$

This work is equal to the negative of the work done by the spring force for this displacement.

The work done by an applied force on a block–spring system between arbitrary positions of the block is

$$W_{F_{app}} = \int_{x_i}^{x_f} F_{app}\, dx = \int_{x_i}^{x_f} kx\, dx = \tfrac{1}{2} kx_f^2 - \tfrac{1}{2} kx_i^2 \qquad (7.12)$$

Notice that this is the negative of the work done by the spring as expressed by Equation 7.11. This is consistent with the fact that the spring force and the applied force are of equal magnitude but in opposite directions.

Quick Quiz 7.5 A dart is loaded into a spring-loaded toy dart gun by pushing the spring in by a distance d. For the next loading, the spring is compressed a distance $2d$. How much work is required to load the second dart compared to that required to load the first? (a) four times as much (b) two times as much (c) the same (d) half as much (e) one-fourth as much.

Example 7.6   Measuring *k* for a Spring

A common technique used to measure the force constant of a spring is demonstrated by the setup in Figure 7.12. The spring is hung vertically, and an object of mass *m* is attached to its lower end. Under the action of the "load" *m***g**, the spring stretches a distance *d* from its equilibrium position.

(A) If a spring is stretched 2.0 cm by a suspended object having a mass of 0.55 kg, what is the force constant of the spring?

Solution Because the object (the system) is at rest, the upward spring force balances the downward gravitational force *m***g**. In this case, we apply Hooke's law to give $|\mathbf{F}_s| = kd = mg$, or

$$k = \frac{mg}{d} = \frac{(0.55\ \text{kg})(9.80\ \text{m/s}^2)}{2.0 \times 10^{-2}\ \text{m}} = \boxed{2.7 \times 10^2\ \text{N/m}}$$

(B) How much work is done by the spring as it stretches through this distance?

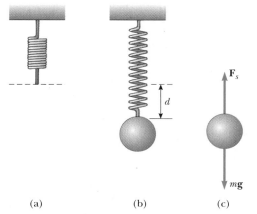

(a)   (b)   (c)

Figure 7.12 (Example 7.6) Determining the force constant *k* of a spring. The elongation *d* is caused by the attached object, which has a weight *m***g**. Because the spring force balances the gravitational force, it follows that $k = mg/d$.

Solution Using Equation 7.11,

$$W_s = 0 - \tfrac{1}{2}kd^2 = -\tfrac{1}{2}(2.7 \times 10^2\ \text{N/m})(2.0 \times 10^{-2}\ \text{m})^2$$

$$= \boxed{-5.4 \times 10^{-2}\,\text{J}}$$

What If? Suppose this measurement is made on an elevator with an upward vertical acceleration a. Will the unaware experimenter arrive at the same value of the spring constant?

Answer The force $\mathbf{F}_s$ in Figure 7.12 must be larger than *m***g** to produce an upward acceleration of the object. Because $\mathbf{F}_s$ must increase in magnitude, and $|\mathbf{F}_s| = kd$, the spring must extend farther. The experimenter sees a larger extension for the same hanging weight and therefore measures the spring constant to be smaller than the value found in part (A) for **a** = 0.

Newton's second law applied to the hanging object gives

$$\sum F_y = |\mathbf{F}_s| - mg = ma_y$$

$$kd - mg = ma_y$$

$$d = \frac{m(g + a_y)}{k}$$

where *k* is the *actual* spring constant. Now, the experimenter is unaware of the acceleration, so she claims that $|\mathbf{F}_s| = k'd = mg$ where *k'* is the spring constant as measured by the experimenter. Thus,

$$k' = \frac{mg}{d} = \frac{mg}{\left(\dfrac{m(g + a_y)}{k}\right)} = \frac{g}{g + a_y}k$$

If the acceleration of the elevator is upward so that a_y is positive, this result shows that the measured spring constant will be smaller, consistent with our conceptual argument.

7.5   Kinetic Energy and the Work–Kinetic Energy Theorem

We have investigated work and identified it as a mechanism for transferring energy into a system. One of the possible outcomes of doing work on a system is that the system changes its speed. In this section, we investigate this situation and introduce our first type of energy that a system can possess, called *kinetic energy*.

Consider a system consisting of a single object. Figure 7.13 shows a block of mass *m* moving through a displacement directed to the right under the action of a net force $\sum\mathbf{F}$, also directed to the right. We know from Newton's second law that the block moves with an acceleration **a**. If the block moves through a displacement $\Delta\mathbf{r} = \Delta x\hat{\mathbf{i}} = (x_f - x_i)\hat{\mathbf{i}}$, the work done by the net force $\sum\mathbf{F}$ is

$$\sum W = \int_{x_i}^{x_f} \sum F\,dx \tag{7.13}$$

Using Newton's second law, we can substitute for the magnitude of the net force $\sum F = ma$, and then perform the following chain-rule manipulations on the integrand:

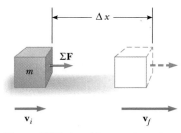

Figure 7.13 An object undergoing a displacement $\Delta\mathbf{r} = \Delta x\hat{\mathbf{i}}$ and a change in velocity under the action of a constant net force $\sum\mathbf{F}$.

$$\sum W = \int_{x_i}^{x_f} ma\,dx = \int_{x_i}^{x_f} m\frac{dv}{dt}\,dx = \int_{x_i}^{x_f} m\frac{dv}{dx}\frac{dx}{dt}\,dx = \int_{v_i}^{v_f} mv\,dv$$

$$\sum W = \tfrac{1}{2}mv_f^2 - \tfrac{1}{2}mv_i^2 \qquad (7.14)$$

where v_i is the speed of the block when it is at $x = x_i$ and v_f is its speed at x_f.

This equation was generated for the specific situation of one-dimensional motion, but it is a general result. It tells us that the work done by the net force on a particle of mass m is equal to the difference between the initial and final values of a quantity $\tfrac{1}{2}mv^2$. The quantity $\tfrac{1}{2}mv^2$ represents the energy associated with the motion of the particle. This quantity is so important that it has been given a special name—**kinetic energy.** Equation 7.14 states that the net work done on a particle by a net force $\Sigma\mathbf{F}$ acting on it equals the change in kinetic energy of the particle.

In general, the kinetic energy K of a particle of mass m moving with a speed v is defined as

Kinetic energy

$$K \equiv \tfrac{1}{2}mv^2 \qquad (7.15)$$

Kinetic energy is a scalar quantity and has the same units as work. For example, a 2.0 kg object moving with a speed of 4.0 m/s has a kinetic energy of 16 J. Table 7.1 lists the kinetic energies for various objects.

It is often convenient to write Equation 7.14 in the form

Work–kinetic energy theorem

$$\sum W = K_f - K_i = \Delta K \qquad (7.16)$$

Another way to write this is $K_f = K_i + \Sigma W$, which tells us that the final kinetic energy is equal to the initial kinetic energy plus the change due to the work done.

Equation 7.16 is an important result known as the **work–kinetic energy theorem:**

> In the case in which work is done on a system and the only change in the system is in its speed, the work done by the net force equals the change in kinetic energy of the system.

The work–kinetic energy theorem indicates that the speed of a particle will *increase* if the net work done on it is *positive*, because the final kinetic energy will be greater than the initial kinetic energy. The speed will *decrease* if the net work is *negative*, because the final kinetic energy will be less than the initial kinetic energy.

 PITFALL PREVENTION

7.6   Conditions for the Work–Kinetic Energy Theorem

The work–kinetic energy theorem is important, but limited in its application—it is not a general principle. There are many situations in which other changes in the system occur besides its speed, and there are other interactions with the environment besides work. A more general principle involving energy is conservation of energy in Section 7.6.

Table 7.1

Kinetic Energies for Various Objects			
Object	**Mass (kg)**	**Speed (m/s)**	**Kinetic Energy (J)**
Earth orbiting the Sun	5.98×10^{24}	2.98×10^4	2.66×10^{33}
Moon orbiting the Earth	7.35×10^{22}	1.02×10^3	3.82×10^{28}
Rocket moving at escape speed[a]	500	1.12×10^4	3.14×10^{10}
Automobile at 65 mi/h	2 000	29	8.4×10^5
Running athlete	70	10	3 500
Stone dropped from 10 m	1.0	14	98
Golf ball at terminal speed	0.046	44	45
Raindrop at terminal speed	3.5×10^{-5}	9.0	1.4×10^{-3}
Oxygen molecule in air	5.3×10^{-26}	500	6.6×10^{-21}

[a] Escape speed is the minimum speed an object must reach near the Earth's surface in order to move infinitely far away from the Earth.

Because we have only investigated translational motion through space so far, we arrived at the work–kinetic energy theorem by analyzing situations involving translational motion. Another type of motion is *rotational motion,* in which an object spins about an axis. We will study this type of motion in Chapter 10. The work–kinetic energy theorem is also valid for systems that undergo a change in the rotational speed due to work done on the system. The windmill in the chapter opening photograph is an example of work causing rotational motion.

The work–kinetic energy theorem will clarify a result that we have seen earlier in this chapter that may have seemed odd. In Section 7.4, we arrived at a result of zero net work done when we let a spring push a block from $x_i = -x_{max}$ to $x_f = x_{max}$. Notice that the speed of the block is continually changing during this process, so it may seem complicated to analyze this process. The quantity ΔK in the work–kinetic energy theorem, however, only refers to the initial and final points for the speeds—it does not depend on details of the path followed between these points. Thus, because the speed is zero at both the initial and final points of the motion, the net work done on the block is zero. We will see this concept of path independence often in similar approaches to problems.

Earlier, we indicated that work can be considered as a mechanism for transferring energy into a system. Equation 7.16 is a mathematical statement of this concept. We do work ΣW on a system and the result is a transfer of energy across the boundary of the system. The result on the system, in the case of Equation 7.16, is a change ΔK in kinetic energy. We will explore this idea more fully in the next section.

▲ **PITFALL PREVENTION**

7.7 The Work–Kinetic Energy Theorem: Speed, not Velocity

The work–kinetic energy theorem relates work to a change in the *speed* of an object, not a change in its velocity. For example, if an object is in uniform circular motion, the speed is constant. Even though the velocity is changing, no work is done by the force causing the circular motion.

Quick Quiz 7.6 A dart is loaded into a spring-loaded toy dart gun by pushing the spring in by a distance d. For the next loading, the spring is compressed a distance $2d$. How much faster does the second dart leave the gun compared to the first? (a) four times as fast (b) two times as fast (c) the same (d) half as fast (e) one-fourth as fast.

Example 7.7 A Block Pulled on a Frictionless Surface

A 6.0-kg block initially at rest is pulled to the right along a horizontal, frictionless surface by a constant horizontal force of 12 N. Find the speed of the block after it has moved 3.0 m.

Solution We have made a drawing of this situation in Figure 7.14. We could apply the equations of kinematics to determine the answer, but let us practice the energy approach. The block is the system, and there are three external forces acting on the system. The normal force balances the gravitational force on the block, and neither of these vertically acting forces does work on the block because their points of application are horizontally displaced. Thus, the net external force acting on the block is the 12-N force. The work done by this force is

$$W = F\Delta x = (12\,\text{N})(3.0\,\text{m}) = 36\,\text{J}$$

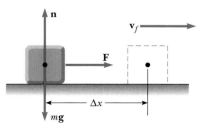

Figure 7.14 (Example 7.7) A block pulled to the right on a frictionless surface by a constant horizontal force.

Using the work–kinetic energy theorem and noting that the initial kinetic energy is zero, we obtain

$$W = K_f - K_i = \tfrac{1}{2}mv_f^2 - 0$$

$$v_f = \sqrt{\frac{2W}{m}} = \sqrt{\frac{2(36\,\text{J})}{6.0\,\text{kg}}} = \boxed{3.5\,\text{m/s}}$$

What If? Suppose the magnitude of the force in this example is doubled to $F' = 2F$. The 6.0-kg block accelerates to 3.5 m/s due to this applied force while moving through a displacement $\Delta x'$. **(A)** How does the displacement $\Delta x'$ compare to the original displacement Δx? **(B)** How does the time interval $\Delta t'$ for the block to accelerate from rest to 3.5 m/s compare to the original interval Δt?

Answer (A) If we pull harder, the block should accelerate to a given speed in a shorter distance, so we expect $\Delta x' < \Delta x$. Mathematically, from the work–kinetic energy theorem $W = \Delta K$, we find

$$F'\Delta x' = \Delta K = F\Delta x$$

$$\Delta x' = \frac{F}{F'}\Delta x = \frac{F}{2F}\Delta x = \tfrac{1}{2}\Delta x$$

and the distance is shorter as suggested by our conceptual argument.

(B) If we pull harder, the block should accelerate to a given speed in a shorter time interval, so we expect $\Delta t' < \Delta t$. Mathematically, from the definition of average velocity,

$$\overline{v} = \frac{\Delta x}{\Delta t} \quad \rightarrow \quad \Delta t = \frac{\Delta x}{\overline{v}}$$

Because both the original force and the doubled force cause the same change in velocity, the average velocity $\overline{v}$ is the same in both cases. Thus,

$$\Delta t' = \frac{\Delta x'}{\overline{v}} = \frac{\frac{1}{2}\Delta x}{\overline{v}} = \frac{1}{2}\Delta t$$

and the time interval is shorter, consistent with our conceptual argument.

Conceptual Example 7.8 Does the Ramp Lessen the Work Required?

A man wishes to load a refrigerator onto a truck using a ramp, as shown in Figure 7.15. He claims that less work would be required to load the truck if the length L of the ramp were increased. Is his statement valid?

Solution No. Suppose the refrigerator is wheeled on a dolly up the ramp at constant speed. Thus, $\Delta K = 0$. The normal force exerted by the ramp on the refrigerator is directed at 90° to the displacement and so does no work on the refrigerator. Because $\Delta K = 0$, the work–kinetic energy theorem gives

$$W_{\text{net}} = W_{\text{by man}} + W_{\text{by gravity}} = 0$$

The work done by the gravitational force equals the product of the weight mg of the refrigerator, the height h through which it is displaced, and cos 180°, or $W_{\text{by gravity}} = -mgh$. (The negative sign arises because the downward gravitational force is opposite the displacement.) Thus, the man must do the same amount of work mgh on the refrigerator, *regardless* of the length of the ramp. Although less force is required with a longer ramp, that force must act over a greater distance.

Figure 7.15 (Conceptual Example 7.8) A refrigerator attached to a frictionless wheeled dolly is moved up a ramp at constant speed.

7.6 The Nonisolated System–Conservation of Energy

Figure 7.16 A book sliding to the right on a horizontal surface slows down in the presence of a force of kinetic friction acting to the left. The initial velocity of the book is $\mathbf{v}_i$, and its final velocity is $\mathbf{v}_f$. The normal force and the gravitational force are not included in the diagram because they are perpendicular to the direction of motion and therefore do not influence the book's speed.

We have seen examples in which an object, modeled as a particle, is acted on by various forces, resulting in a change in its kinetic energy. This very simple situation is the first example of the **nonisolated system**—a common scenario in physics problems. Physical problems for which this scenario is appropriate involve systems that interact with or are influenced by their environment, causing some kind of change in the system. If a system does not interact with its environment it is an **isolated system**, which we will study in Chapter 8.

The work–kinetic energy theorem is our first example of an energy equation appropriate for a nonisolated system. In the case of the work–kinetic energy theorem, the interaction is the work done by the external force, and the quantity in the system that changes is the kinetic energy.

In addition to kinetic energy, we now introduce a second type of energy that a system can possess. Let us imagine the book in Figure 7.16 sliding to the right on the surface of a heavy table and slowing down due to the friction force. Suppose the *surface* is the system. Then the friction force from the sliding book does work on the surface. The force on the surface is to the right and the displacement of the point of application of the force is to the right—the work is positive. But the surface is not moving after the book has stopped. Positive work has been done on the surface, yet there is no increase in the surface's kinetic energy. Is this a violation of the work–kinetic energy theorem?

It is not really a violation, because this situation does not fit the description of the conditions given for the work–kinetic energy theorem. Work is done on the system of the surface, but the result of that work is *not* an increase in kinetic energy. From your everyday experience with sliding over surfaces with friction, you can probably guess that the surface will be *warmer* after the book slides over it. (Rub your hands together briskly to experience this!) Thus, the work that was done on the surface has gone into warming the surface rather than increasing its speed. We call the energy associated with an object's temperature its **internal energy,** symbolized E_{int}. (We will define internal energy more generally in Chapter 20.) In this case, the work done on the surface does indeed represent energy transferred into the system, but it appears in the system as internal energy rather than kinetic energy.

We have now seen two methods of storing energy in a system—kinetic energy, related to motion of the system, and internal energy, related to its temperature. A third method, which we cover in Chapter 8, is *potential energy*. This is energy related to the configuration of a system in which the components of the system interact by forces. For example, when a spring is stretched, *elastic potential energy* is stored in the spring due to the force of interaction between the spring coils. Other types of potential energy include gravitational and electric.

We have seen only one way to transfer energy into a system so far—work. We mention below a few other ways to transfer energy into or out of a system. The details of these processes will be studied in other sections of the book. We illustrate these in Figure 7.17 and summarize them as follows:

Work, as we have learned in this chapter, is a method of transferring energy to a system by applying a force to the system and causing a displacement of the point of application of the force (Fig. 7.17a).

Mechanical waves (Chapters 16–18) are a means of transferring energy by allowing a disturbance to propagate through air or another medium. This is the method by which energy (which you detect as sound) leaves your clock radio through the loudspeaker and enters your ears to stimulate the hearing process (Fig. 7.17b). Other examples of mechanical waves are seismic waves and ocean waves.

Heat (Chapter 20) is a mechanism of energy transfer that is driven by a temperature difference between two regions in space. One clear example is thermal conduction, a mechanism of transferring energy by microscopic collisions. For example, a metal spoon in a cup of coffee becomes hot because fast-moving electrons and atoms in the submerged portion of the spoon bump into slower ones in the nearby part of the handle (Fig. 7.17c). These particles move faster because of the collisions and bump into the next group of slow particles. Thus, the internal energy of the spoon handle rises from energy transfer due to this bumping process.[4]

Matter transfer (Chapter 20) involves situations in which matter physically crosses the boundary of a system, carrying energy with it. Examples include filling your automobile tank with gasoline (Fig. 7.17d), and carrying energy to the rooms of your home by circulating warm air from the furnace, a process called *convection*.

---

[4]   The process we call heat can also proceed by convection and radiation, as well as conduction. Convection and radiation, described in Chapter 20, overlap with other types of energy transfer in our list of six.

 PITFALL PREVENTION

7.8 Heat is not a Form of Energy

The word *heat* is one of the most misused words in our popular language. In this text, heat is a method of *transferring* energy, *not* a form of storing energy. Thus, phrases such as "heat content," "the heat of the summer," and "the heat escaped" all represent uses of this word that are inconsistent with our physics definition. See Chapter 20.

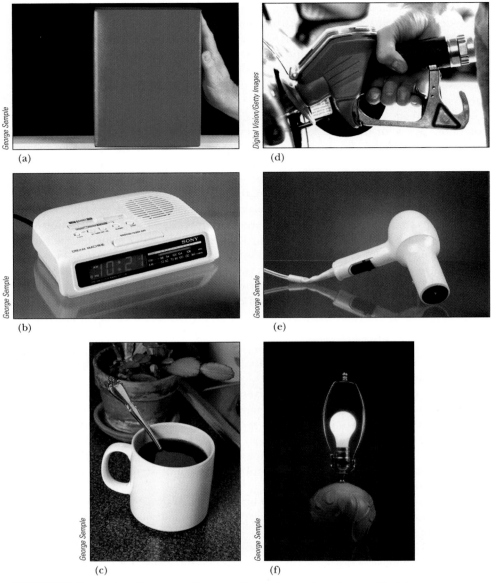

Figure 7.17 Energy transfer mechanisms. (a) Energy is transferred to the block by *work*; (b) energy leaves the radio from the speaker by *mechanical waves*; (c) energy transfers up the handle of the spoon by *heat*; (d) energy enters the automobile gas tank by *matter transfer*; (e) energy enters the hair dryer by *electrical transmission*; and (f) energy leaves the light bulb by *electromagnetic radiation*.

Electrical Transmission (Chapters 27–28) involves energy transfer by means of electric currents. This is how energy transfers into your hair dryer (Fig. 7.17e), stereo system, or any other electrical device.

Electromagnetic radiation (Chapter 34) refers to electromagnetic waves such as light, microwaves, radio waves, and so on (Fig. 7.17f). Examples of this method of transfer include cooking a baked potato in your microwave oven and light energy traveling from the Sun to the Earth through space.[5]

---

[5] Electromagnetic radiation and work done by field forces are the only energy transfer mechanisms that do not require molecules of the environment to be available at the system boundary. Thus, systems surrounded by a vacuum (such as planets) can only exchange energy with the environment by means of these two possibilities.

One of the central features of the energy approach is the notion that **we can neither create nor destroy energy—energy is always** *conserved*. Thus, **if the total amount of energy in a system changes, it can** *only* **be due to the fact that energy has crossed the boundary of the system by a transfer mechanism such as one of the methods listed above.** This is a general statement of the principle of **conservation of energy.** We can describe this idea mathematically as follows:

$$\Delta E_{\text{system}} = \sum T \qquad (7.17)$$

◀ Conservation of energy

where E_{system} is the total energy of the system, including all methods of energy storage (kinetic, internal, and potential, as discussed in Chapter 8) and T is the amount of energy transferred across the system boundary by some mechanism. Two of our transfer mechanisms have well-established symbolic notations. For work, $T_{\text{work}} = W$, as we have seen in the current chapter, and for heat, $T_{\text{heat}} = Q$, as defined in Chapter 20. The other four members of our list do not have established symbols.

This is no more complicated in theory than is balancing your checking account statement. If your account is the system, the change in the account balance for a given month is the sum of all the transfers—deposits, withdrawals, fees, interest, and checks written. It may be useful for you to think of energy as the *currency of nature*!

Suppose a force is applied to a nonisolated system and the point of application of the force moves through a displacement. Suppose further that the only effect on the system is to change its speed. Then the only transfer mechanism is work (so that ΣT in Equation 7.17 reduces to just W) and the only kind of energy in the system that changes is the kinetic energy (so that ΔE_{system} reduces to just ΔK). Equation 7.17 then becomes

$$\Delta K = W$$

which is the work–kinetic energy theorem. The work–kinetic energy theorem is a special case of the more general principle of conservation of energy. We shall see several more special cases in future chapters.

Quick Quiz 7.7 By what transfer mechanisms does energy enter and leave (a) your television set; (b) your gasoline-powered lawn mower; (c) your hand-cranked pencil sharpener?

Quick Quiz 7.8 Consider a block sliding over a horizontal surface with friction. Ignore any sound the sliding might make. If we consider the system to be the *block*, this system is (a) isolated (b) nonisolated (c) impossible to determine.

Quick Quiz 7.9 If we consider the system in Quick Quiz 7.8 to be the *surface*, this system is (a) isolated (b) nonisolated (c) impossible to determine.

Quick Quiz 7.10 If we consider the system in Quick Quiz 7.8 to be the *block and the surface*, this system is (a) isolated (b) nonisolated (c) impossible to determine.

7.7 Situations Involving Kinetic Friction

Consider again the book in Figure 7.16 sliding to the right on the surface of a heavy table and slowing down due to the friction force. Work is done by the friction force because there is a force and a displacement. Keep in mind, however, that our equations for work involve the displacement *of the point of application of the force*. The friction force is spread out over the entire contact area of an object sliding on a surface, so the force is not localized at a point. In addition, the magnitudes of the friction forces at various points are constantly changing as spot welds occur, the surface and the book deform locally, and so on. The points of application of the friction force on the book are jumping all over the face of the book in contact with the surface. This means that the displacement of the point of application of the friction force (assuming we could calculate it!) is not the same as the displacement of the book.

The work–kinetic energy theorem is valid for a particle or an object that can be modeled as a particle. When an object cannot be treated as a particle, however, things become more complicated. For these kinds of situations, Newton's second law is still valid for the system, even though the work–kinetic energy theorem is not. In the case of a nondeformable object like our book sliding on the surface,[6] we can handle this in a relatively straightforward way.

Starting from a situation in which a constant force is applied to the book, we can follow a similar procedure to that in developing Equation 7.14. We start by multiplying each side of Newton's second law (x component only) by a displacement Δx of the book:

$$\left(\sum F_x\right)\Delta x = (ma_x)\Delta x \tag{7.18}$$

For a particle under constant acceleration, we know that the following relationships (Eqs. 2.9 and 2.11) are valid:

$$a_x = \frac{v_f - v_i}{t} \qquad \Delta x = \tfrac{1}{2}(v_i + v_f)t$$

where v_i is the speed at $t = 0$ and v_f is the speed at time t. Substituting these expressions into Equation 7.18 gives

$$\left(\sum F_x\right)\Delta x = m\left(\frac{v_f - v_i}{t}\right)\tfrac{1}{2}(v_i + v_f)t$$

$$\left(\sum F_x\right)\Delta x = \tfrac{1}{2}mv_f{}^2 - \tfrac{1}{2}mv_i{}^2$$

This *looks* like the work–kinetic energy theorem, but *the left hand side has not been called work*. The quantity Δx is the displacement of the book—*not* the displacement of the point of application of the friction force.

Let us now apply this equation to a book that has been projected across a surface. We imagine that the book has an initial speed and slows down due to friction, the only force in the horizontal direction. The net force on the book is the kinetic friction force $\mathbf{f}_k$, which is directed opposite to the displacement Δx. Thus,

$$\left(\sum F_x\right)\Delta x = -f_k\Delta x = \tfrac{1}{2}mv_f{}^2 - \tfrac{1}{2}mv_i{}^2 = \Delta K$$

$$-f_k\Delta x = \Delta K \tag{7.19}$$

which mathematically describes the decrease in kinetic energy due to the friction force.

We have generated these results by assuming that a book is moving along a straight line. An object could also slide over a surface with friction and follow a curved path. In this case, Equation 7.19 must be generalized as follows:

Change in kinetic energy due to friction

$$-f_k d = \Delta K \tag{7.20}$$

where d is the length of the path followed by an object.

If there are other forces besides friction acting on an object, the change in kinetic energy is the sum of that due to the other forces from the work–kinetic energy theorem, and that due to friction:

---

[6]   The overall shape of the book remains the same, which is why we are saying it is nondeformable. On a microscopic level, however, there is deformation of the book's face as it slides over the surface.

$$\Delta K = -f_k d + \sum W_{\text{other forces}} \qquad (7.21\text{a})$$

or

$$K_f = K_i - f_k d + \sum W_{\text{other forces}} \qquad (7.21\text{b})$$

Now consider the larger system of the book *and* the surface as the book slows down under the influence of a friction force alone. There is no work done across the boundary of this system—the system does not interact with the environment. There are no other types of energy transfer occurring across the boundary of the system, assuming we ignore the inevitable sound the sliding book makes! In this case, Equation 7.17 becomes

$$\Delta E_{\text{system}} = \Delta K + \Delta E_{\text{int}} = 0$$

The change in kinetic energy of this book-plus-surface system is the same as the change in kinetic energy of the the book alone in Equation 7.20, because the book is the only part of the book-surface system that is moving. Thus,

$$-f_k d + \Delta E_{\text{int}} = 0$$

$$\Delta E_{\text{int}} = f_k d \qquad (7.22)$$

Change in internal energy due to friction

Thus, the increase in internal energy of the system is equal to the product of the friction force and the displacement of the book.

The conclusion of this discussion is that **the result of a friction force is to transform kinetic energy into internal energy, and the increase in internal energy is equal to the decrease in kinetic energy.**

Quick Quiz 7.11 You are traveling along a freeway at 65 mi/h. Your car has kinetic energy. You suddenly skid to a stop because of congestion in traffic. Where is the kinetic energy that your car once had? (a) All of it is in internal energy in the road. (b) All of it is in internal energy in the tires. (c) Some of it has transformed to internal energy and some of it transferred away by mechanical waves. (d) All of it is transferred away from your car by various mechanisms.

Example 7.9   A Block Pulled on a Rough Surface

`Interactive`

A 6.0-kg block initially at rest is pulled to the right along a horizontal surface by a constant horizontal force of 12 N.

(A) Find the speed of the block after it has moved 3.0 m if the surfaces in contact have a coefficient of kinetic friction of 0.15. (This is Example 7.7, modified so that the surface is no longer frictionless.)

Solution Conceptualize this problem by realizing that the rough surface is going to apply a friction force opposite to the applied force. As a result, we expect the speed to be lower than that found in Example 7.7. The surface is rough and we are given forces and a distance, so we categorize this as a situation involving kinetic friction that must be handled by means of Equation 7.21. To analyze the problem, we have made a drawing of this situation in Figure 7.18a. We identify the block as the system, and there are four external forces interacting with the system. The normal force balances the gravitational force on the

block, and neither of these vertically acting forces does work on the block because their points of application are displaced horizontally. The applied force does work just as in Example 7.7:

$$W = F \Delta x = (12 \text{ N})(3.0 \text{ m}) = 36 \text{ J}$$

In this case we must use Equation 7.21a to calculate the kinetic energy change due to friction, $\Delta K_{\text{friction}}$. Because the block is in equilibrium in the vertical direction, the normal force $\mathbf{n}$ counterbalances the gravitational force $m\mathbf{g}$, so we have $n = mg$. Hence, the magnitude of the friction force is

$$f_k = \mu_k n = \mu_k mg = (0.15)(6.0 \text{ kg})(9.80 \text{ m/s}^2) = 8.82 \text{ N}$$

The change in kinetic energy of the block due to friction is

$$\Delta K_{\text{friction}} = -f_k d = -(8.82 \text{ N})(3.0 \text{ m}) = -26.5 \text{ J}$$

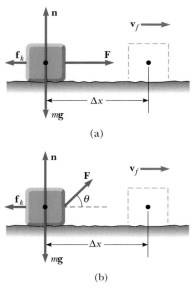

(a)

(b)

Figure 7.18 (Example 7.9) (a) A block pulled to the right on a rough surface by a constant horizontal force. (b) The applied force is at an angle θ to the horizontal.

The final speed of the block follows from Equation 7.21b:

$$\tfrac{1}{2}mv_f{}^2 = \tfrac{1}{2}mv_i{}^2 - f_kd + \sum W_{\text{other forces}}$$

$$v_f = \sqrt{v_i{}^2 + \frac{2}{m}\left(-f_kd + \sum W_{\text{other forces}}\right)}$$

$$= \sqrt{0 + \frac{2}{6.0\,\text{kg}}(-26.5\,\text{J} + 36\,\text{J})}$$

$$= \boxed{1.8\,\text{m/s}}$$

To finalize this problem note that, after covering the same distance on a frictionless surface (see Example 7.7), the speed of the block was 3.5 m/s.

(B) Suppose the force **F** is applied at an angle θ as shown in Figure 7.18b. At what angle should the force be applied to achieve the largest possible speed after the block has moved 3.0 m to the right?

Solution The work done by the applied force is now

$$W = F\,\Delta x \cos\theta = Fd\cos\theta$$

where $\Delta x = d$ because the path followed by the block is a straight line. The block is in equilibrium in the vertical direction, so

$$\sum F_y = n + F\sin\theta - mg = 0$$

and

$$n = mg - F\sin\theta$$

Because $K_i = 0$, Equation 7.21b can be written,

$$K_f = -f_kd + \sum W_{\text{other forces}}$$
$$= -\mu_knd + Fd\cos\theta$$
$$= -\mu_k(mg - F\sin\theta)d + Fd\cos\theta$$

Maximizing the speed is equivalent to maximizing the final kinetic energy. Consequently, we differentiate K_f with respect to θ and set the result equal to zero:

$$\frac{d(K_f)}{d\theta} = -\mu_k(0 - F\cos\theta)d - Fd\sin\theta = 0$$

$$\mu_k\cos\theta - \sin\theta = 0$$

$$\tan\theta = \mu_k$$

For $\mu_k = 0.15$, we have,

$$\theta = \tan^{-1}(\mu_k) = \tan^{-1}(0.15) = \boxed{8.5°}$$

 Try out the effects of pulling the block at various angles at the Interactive Worked Example link at **http://www.pse6.com.**

Conceptual Example 7.10   Useful Physics for Safer Driving

A car traveling at an initial speed v slides a distance d to a halt after its brakes lock. Assuming that the car's initial speed is instead $2v$ at the moment the brakes lock, estimate the distance it slides.

Solution Let us assume that the force of kinetic friction between the car and the road surface is constant and the same for both speeds. According to Equation 7.20, the friction force multiplied by the distance d is equal to the initial kinetic energy of the car (because $K_f = 0$). If the speed is doubled, as it is in this example, the kinetic energy is quadrupled. For a given friction force, the distance traveled is four times as great when the initial speed is doubled, and so the estimated distance that the car slides is $4d$.

Example 7.11   A Block–Spring System   `Interactive`

A block of mass 1.6 kg is attached to a horizontal spring that has a force constant of 1.0×10^3 N/m, as shown in Figure 7.10. The spring is compressed 2.0 cm and is then released from rest.

(A) Calculate the speed of the block as it passes through the equilibrium position $x = 0$ if the surface is frictionless.

Solution In this situation, the block starts with $v_i = 0$ at $x_i = -2.0$ cm, and we want to find v_f at $x_f = 0$. We use Equation 7.10 to find the work done by the spring with $x_{\max} = x_i = -2.0$ cm $= -2.0 \times 10^{-2}$ m:

$$W_s = \tfrac{1}{2}kx_{\max}^2 = \tfrac{1}{2}(1.0 \times 10^3\,\text{N/m})(-2.0 \times 10^{-2}\,\text{m})^2 = 0.20\,\text{J}$$

Using the work–kinetic energy theorem with $v_i = 0$, we set the change in kinetic energy of the block equal to the work done on it by the spring:

$$W_s = \tfrac{1}{2}mv_f{}^2 - \tfrac{1}{2}mv_i{}^2$$

$$v_f = \sqrt{v_i{}^2 + \frac{2}{m}W_s}$$

$$= \sqrt{0 + \frac{2}{1.6\text{ kg}}(0.20\text{ J})}$$

$$= \boxed{0.50\,\text{m/s}}$$

(B) Calculate the speed of the block as it passes through the equilibrium position if a constant friction force of 4.0 N retards its motion from the moment it is released.

Solution Certainly, the answer has to be less than what we found in part (A) because the friction force retards the motion. We use Equation 7.20 to calculate the kinetic energy lost because of friction and add this negative value to the kinetic energy we calculated in the absence of friction. The kinetic energy lost due to friction is

$$\Delta K = -f_k d = -(4.0\text{ N})(2.0 \times 10^{-2}\text{ m}) = -0.080\text{ J}$$

In part (A), the work done by the spring was found to be 0.20 J. Therefore, the final kinetic energy in the presence of friction is

$$K_f = 0.20\text{ J} - 0.080\text{ J} = 0.12\text{ J} = \tfrac{1}{2}mv_f{}^2$$

$$v_f = \sqrt{\frac{2K_f}{m}} = \sqrt{\frac{2(0.12\text{ J})}{1.6\text{ kg}}} = \boxed{0.39\,\text{m/s}}$$

As expected, this value is somewhat less than the 0.50 m/s we found in part (A). If the friction force were greater, then the value we obtained as our answer would have been even smaller.

What If? What if the friction force were increased to 10.0 N? What is the block's speed at $x = 0$?

Answer In this case, the loss of kinetic energy as the block moves to $x = 0$ is

$$\Delta K = -f_k d = -(10.0\text{ N})(2.0 \times 10^{-2}\text{ m}) = -0.20\text{ J}$$

which is equal in magnitude to the kinetic energy at $x = 0$ without the loss due to friction. Thus, all of the kinetic energy has been transformed by friction when the block arrives at $x = 0$ and its speed at this point is $v = 0$.

In this situation as well as that in part (B), the speed of the block reaches a maximum at some position other than $x = 0$. Problem 70 asks you to locate these positions.

 Investigate the role of the spring constant, amount of spring compression, and surface friction at the Interactive Worked Example link at **http://www.pse6.com.**

7.8 Power

Consider Conceptual Example 7.8 again, which involved rolling a refrigerator up a ramp into a truck. Suppose that the man is not convinced by our argument that the work is the same regardless of the length of the ramp and sets up a long ramp with a gentle rise. Although he will do the same amount of work as someone using a shorter ramp, he will take longer to do the work simply because he has to move the refrigerator over a greater distance. While the work done on both ramps is the same, there is *something* different about the tasks—the *time interval* during which the work is done.

The time rate of energy transfer is called **power.** We will focus on work as the energy transfer method in this discussion, but keep in mind that the notion of power is valid for *any* means of energy transfer. If an external force is applied to an object (which we assume acts as a particle), and if the work done by this force in the time interval Δt is W, then the **average power** during this interval is defined as

$$\overline{\mathcal{P}} \equiv \frac{W}{\Delta t}$$

Thus, while the same work is done in rolling the refrigerator up both ramps, less power is required for the longer ramp.

In a manner similar to the way we approached the definition of velocity and acceleration, we define the **instantaneous power** $\mathcal{P}$ as the limiting value of the average power as Δt approaches zero:

$$\mathcal{P} \equiv \lim_{\Delta t \to 0} \frac{W}{\Delta t} = \frac{dW}{dt}$$

where we have represented the infinitesimal value of the work done by dW. We find from Equation 7.3 that $dW = \mathbf{F} \cdot d\mathbf{r}$. Therefore, the instantaneous power can be written

Instantaneous power

$$\mathcal{P} = \frac{dW}{dt} = \mathbf{F} \cdot \frac{d\mathbf{r}}{dt} = \mathbf{F} \cdot \mathbf{v} \qquad (7.23)$$

where we use the fact that $\mathbf{v} = d\mathbf{r}/dt$.

In general, power is defined for any type of energy transfer. Therefore, the most general expression for power is

$$\mathcal{P} = \frac{dE}{dt} \qquad (7.24)$$

where dE/dt is the rate at which energy is crossing the boundary of the system by a given transfer mechanism.

The SI unit of power is joules per second (J/s), also called the **watt** (W) (after James Watt):

The watt

$$1 \text{ W} = 1 \text{ J/s} = 1 \text{ kg} \cdot \text{m}^2/\text{s}^3$$

A unit of power in the U.S. customary system is the **horsepower** (hp):

$$1 \text{ hp} = 746 \text{ W}$$

A unit of energy (or work) can now be defined in terms of the unit of power. One **kilowatt-hour** (kWh) is the energy transferred in 1 h at the constant rate of $1 \text{ kW} = 1\,000 \text{ J/s}$. The amount of energy represented by 1 kWh is

$$1 \text{ kWh} = (10^3 \text{ W})(3\,600 \text{ s}) = 3.60 \times 10^6 \text{ J}$$

Note that a kilowatt-hour is a unit of energy, not power. When you pay your electric bill, you are buying energy, and the amount of energy transferred by electrical transmission into a home during the period represented by the electric bill is usually expressed in kilowatt-hours. For example, your bill may state that you used 900 kWh of energy during a month, and you are being charged at the rate of 10¢ per kWh. Your obligation is then $90 for this amount of energy. As another example, suppose an electric bulb is rated at 100 W. In 1.00 hour of operation, it would have energy transferred to it by electrical transmission in the amount of $(0.100 \text{ kW})(1.00 \text{ h}) = 0.100 \text{ kWh} = 3.60 \times 10^5 \text{ J}$.

▲ **PITFALL PREVENTION**

7.9 W, *W*, and watts

Do not confuse the symbol W for the watt with the italic symbol *W* for work. Also, remember that the watt already represents a rate of energy transfer, so that "watts per second" does not make sense. The watt is *the same as* a joule per second.

Quick Quiz 7.12 An older model car accelerates from rest to speed v in 10 seconds. A newer, more powerful sports car accelerates from rest to $2v$ in the same time period. What is the ratio of the power of the newer car to that of the older car? (a) 0.25 (b) 0.5 (c) 1 (d) 2 (e) 4

Example 7.12 Power Delivered by an Elevator Motor

An elevator car has a mass of 1 600 kg and is carrying passengers having a combined mass of 200 kg. A constant friction force of 4 000 N retards its motion upward, as shown in Figure 7.19a.

(A) What power delivered by the motor is required to lift the elevator car at a constant speed of 3.00 m/s?

Solution The motor must supply the force of magnitude T that pulls the elevator car upward. The problem states that the speed is constant, which provides the hint that $a = 0$. Therefore we know from Newton's second law that

$\Sigma F_y = 0$. The free-body diagram in Figure 7.19b specifies the upward direction as positive. From Newton's second law we obtain

$$\sum F_y = T - f - Mg = 0$$

where M is the *total* mass of the system (car plus passengers), equal to 1 800 kg. Therefore,

$$\begin{aligned} T &= f + Mg \\ &= 4.00 \times 10^3 \text{ N} + (1.80 \times 10^3 \text{ kg})(9.80 \text{ m/s}^2) \\ &= 2.16 \times 10^4 \text{ N} \end{aligned}$$

Figure 7.19 (Example 7.12) (a) The motor exerts an upward force **T** on the elevator car. The magnitude of this force is the tension T in the cable connecting the car and motor. The downward forces acting on the car are a friction force **f** and the gravitational force $\mathbf{F}_g = M\mathbf{g}$. (b) The free-body diagram for the elevator car.

Using Equation 7.23 and the fact that **T** is in the same direction as **v**, we find that

$$\mathcal{P} = \mathbf{T} \cdot \mathbf{v} = Tv$$

$$= (2.16 \times 10^4 \, \text{N})(3.00 \, \text{m/s}) = \boxed{6.48 \times 10^4 \, \text{W}}$$

(B) What power must the motor deliver at the instant the speed of the elevator is v if the motor is designed to provide the elevator car with an upward acceleration of $1.00 \, \text{m/s}^2$?

Solution We expect to obtain a value greater than we did in part (A), where the speed was constant, because the motor must now perform the additional task of accelerating the car. The only change in the setup of the problem is that in this case, $a > 0$. Applying Newton's second law to the car gives

$$\sum F_y = T - f - Mg = Ma$$

$$T = M(a + g) + f$$

$$= (1.80 \times 10^3 \, \text{kg})(1.00 \, \text{m/s}^2 + 9.80 \, \text{m/s}^2)$$

$$+ 4.00 \times 10^3 \, \text{N}$$

$$= 2.34 \times 10^4 \, \text{N}$$

Therefore, using Equation 7.23, we obtain for the required power

$$\mathcal{P} = Tv = \boxed{(2.34 \times 10^4 \, \text{N})v}$$

where v is the instantaneous speed of the car in meters per second. To compare to part (A), let $v = 3.00 \, \text{m/s}$, giving a power of

$$\mathcal{P} = (2.34 \times 10^4 \, \text{N})(3.00 \, \text{m/s}) = 7.02 \times 10^4 \, \text{W}$$

This is larger than the power found in part (A), as we expect.

7.9   Energy and the Automobile

Automobiles powered by gasoline engines are very inefficient machines. Even under ideal conditions, less than 15% of the chemical energy in the fuel is used to power the vehicle. The situation is much worse than this under stop-and-go driving conditions in a city. In this section, we use the concepts of energy, power, and friction to analyze automobile fuel consumption.

Many mechanisms contribute to energy loss in an automobile. About 67% of the energy available from the fuel is lost in the engine. This energy ends up in the atmosphere, partly via the exhaust system and partly via the cooling system. (As explained in Chapter 22, energy loss from the exhaust and cooling systems is required by a fundamental law of thermodynamics.) Approximately 10% of the available energy is lost to friction in the transmission, drive shaft, wheel and axle bearings, and differential. Friction in other moving parts transforms approximately 6% of the energy to internal energy, and 4% of the energy is used to operate fuel and oil pumps and such accessories as power steering and air conditioning. This leaves a mere 13% of the available energy to propel the automobile! This energy is used mainly to balance the energy loss due to flexing of the tires and the friction caused by the air, which is more commonly referred to as *air resistance*.

Let us examine the power required to provide a force in the forward direction that balances the combination of the two friction forces. The coefficient of rolling friction μ between the tires and the road is about 0.016. For a 1 450-kg car, the weight is 14 200 N and on a horizontal roadway the force of rolling friction has a magnitude of $\mu n = \mu mg = 227 \, \text{N}$. As the car's speed increases, a small reduction in the normal force

Table 7.2

Friction Forces and Power Requirements for a Typical Car[a]						
v(mi/h)	v(m/s)	n(N)	f_r(N)	f_a(N)	f_t(N)	$\mathcal{P} = f_t v$(kW)
0	0	14 200	227	0	227	0
20	8.9	14 100	226	48	274	2.4
40	17.9	13 900	222	192	414	7.4
60	26.8	13 600	218	431	649	17.4
80	35.8	13 200	211	767	978	35.0
100	44.7	12 600	202	1 199	1 400	62.6

[a] In this table, n is the normal force, f_r is rolling friction, f_a is air friction, f_t is total friction, and $\mathcal{P}$ is the power delivered to the wheels.

occurs as a result of decreased pressure as air flows over the top of the car. (This phenomenon is discussed in Chapter 14.) This reduction in the normal force causes a reduction in the force of rolling friction f_r with increasing speed, as the data in Table 7.2 indicate.

Now let us consider the effect of the resistive force that results from the movement of air past the car. For large objects, the resistive force f_a associated with air friction is proportional to the square of the speed (see Section 6.4) and is given by Equation 6.6:

$$f_a = \tfrac{1}{2}D\rho A v^2$$

where D is the drag coefficient, ρ is the density of air, and A is the cross-sectional area of the moving object. We can use this expression to calculate the f_a values in Table 7.2, using $D = 0.50$, $\rho = 1.20 \text{ kg/m}^3$, and $A \approx 2 \text{ m}^2$.

The magnitude of the total friction force f_t is the sum of the rolling friction force and the air resistive force:

$$f_t = f_r + f_a$$

At low speeds, rolling friction is the predominant resistive force, but at high speeds air drag predominates, as shown in Table 7.2. Rolling friction can be decreased by a reduction in tire flexing (for example, by an increase in the air pressure slightly above recommended values) and by the use of radial tires. Air drag can be reduced through the use of a smaller cross-sectional area and by streamlining the car. Although driving a car with the windows open increases air drag and thus results in a 3% decrease in mileage, driving with the windows closed and the air conditioner running results in a 12% decrease in mileage.

The total power needed to maintain a constant speed v is $f_t v$, and this is the power that must be delivered to the wheels. For example, from Table 7.2 we see that at $v = 26.8$ m/s (60 mi/h) the required power is

$$\mathcal{P} = f_t v = (649 \text{ N})(26.8 \text{ m/s}) = 17.4 \text{ kW}$$

This power can be broken down into two parts: (1) the power $f_r v$ needed to compensate for rolling friction, and (2) the power $f_a v$ needed to compensate for air drag. At $v = 26.8$ m/s, we obtain the values

$$\mathcal{P}_r = f_r v = (218 \text{ N})(26.8 \text{ m/s}) = 5.84 \text{ kW}$$

$$\mathcal{P}_a = f_a v = (431 \text{ N})(26.8 \text{ m/s}) = 11.6 \text{ kW}$$

Note that $\mathcal{P} = \mathcal{P}_r + \mathcal{P}_a$ and 67% of the power is used to compensate for air drag.

On the other hand, at $v = 44.7$ m/s(100 mi/h), $\mathcal{P}_r = 9.03$ kW, $\mathcal{P}_a = 53.6$ kW, $\mathcal{P} = 62.6$ kW and 86% of the power is associated with air drag. This shows the importance of air drag at high speeds.

Example 7.13 Gas Consumed by a Compact Car

A compact car has a mass of 800 kg, and its efficiency is rated at 18%. (That is, 18% of the available fuel energy is delivered to the wheels.) Find the amount of gasoline used to accelerate the car from rest to 27 m/s (60 mi/h). Use the fact that the energy equivalent of 1 gal of gasoline is 1.3×10^8 J.

Solution The energy required to accelerate the car from rest to a speed v is equal to its final kinetic energy, $\frac{1}{2}mv^2$:

$$K = \tfrac{1}{2}mv^2 = \tfrac{1}{2}(800 \text{ kg})(27 \text{ m/s})^2 = 2.9 \times 10^5 \text{ J}$$

If the engine were 100% efficient, each gallon of gasoline would supply 1.3×10^8 J of energy. Because the engine is only 18% efficient, each gallon delivers an energy of only

$(0.18)(1.3 \times 10^8 \text{ J}) = 2.3 \times 10^7$ J. Hence, the number of gallons used to accelerate the car is

$$\text{Number of gal} = \frac{2.9 \times 10^5 \text{ J}}{2.3 \times 10^7 \text{ J/gal}} = \boxed{0.013 \text{ gal}}$$

Let us estimate that it takes 10 s to achieve the indicated speed. The distance traveled during this acceleration is

$$\Delta x = \bar{v}\Delta t = \frac{v_{xf} + v_{xi}}{2}(\Delta t) = \frac{27 \text{ m/s} + 0}{2}(10 \text{ s})$$

$$= 135 \text{ m} \approx 0.08 \text{ mi}$$

At a constant cruising speed, 0.013 gal of gasoline is sufficient to propel the car nearly 0.5 mi, over six times farther. This demonstrates the extreme energy requirements of stop-and-start driving.

Example 7.14 Power Delivered to the Wheels

Suppose the compact car in Example 7.13 has a gas mileage of 35 mi/gal at 60 mi/h. How much power is delivered to the wheels?

Solution We find the rate of gasoline consumption by dividing the speed by the gas mileage:

$$\frac{60 \text{ mi/h}}{35 \text{ mi/gal}} = 1.7 \text{ gal/h}$$

Using the fact that each gallon is equivalent to 1.3×10^8 J, we find that the total power used is

$$\mathcal{P} = (1.7 \text{ gal/h})(1.3 \times 10^8 \text{ J/gal})\left(\frac{1 \text{ h}}{3.6 \times 10^3 \text{ s}}\right)$$

$$= 62 \text{ kW}$$

Because 18% of the available power is used to propel the car, the power delivered to the wheels is $(0.18)(62 \text{ kW}) = $ $\boxed{11 \text{ kW.}}$ This is 37% less than the 17.4-kW value obtained for the 1 450-kg car discussed in the text. Vehicle mass is clearly an important factor in power-loss mechanisms.

Example 7.15 Car Accelerating Up a Hill

Consider a car of mass m that is accelerating up a hill, as shown in Figure 7.20. An automotive engineer measures the magnitude of the total resistive force to be

$$f_t = (218 + 0.70v^2) \text{ N}$$

where v is the speed in meters per second. Determine the power the engine must deliver to the wheels as a function of speed.

Solution The forces on the car are shown in Figure 7.20, in which **F** is the force of friction from the road that propels the car; the remaining forces have their usual meaning.

Figure 7.20 (Example 7.15) A car climbs a hill.

Applying Newton's second law to the motion along the road surface, we find that

$$\sum F_x = F - f_t - mg \sin \theta = ma$$

$$F = ma + mg \sin \theta + f_t$$

$$= ma + mg \sin \theta + (218 + 0.70v^2)$$

Therefore, the power required to move the car forward is

$$\mathcal{P} = Fv = mva + mvg \sin \theta + 218v + 0.70v^3$$

The term mva represents the power that the engine must deliver to accelerate the car. If the car moves at constant speed, this term is zero and the total power requirement is reduced. The term $mvg \sin \theta$ is the power required to provide a force to balance a component of the gravitational force as the car moves up the incline. This term would be zero for motion on a horizontal surface. The term $218v$ is the power required to provide a force to balance rolling friction, and the term $0.70v^3$ is the power needed against air drag.

If we take $m = 1\,450$ kg, $v = 27$ m/s ($= 60$ mi/h), $a = 1.0$ m/s^2, and $\theta = 10°$, then the various terms in $\mathcal{P}$ are calculated to be

$$mva = (1\,450 \text{ kg})(27 \text{ m/s})(1.0 \text{ m/s}^2)$$

$$= 39 \text{ kW} = 52 \text{ hp}$$

$$mvg \sin \theta = (1\,450 \text{ kg})(27 \text{ m/s})(9.80 \text{ m/s}^2)(\sin 10°)$$
$$= 67 \text{ kW} = 89 \text{ hp}$$

$$218v = 218(27 \text{ m/s}) = 5.9 \text{ kW} = 7.9 \text{ hp}$$

$$0.70v^3 = 0.70(27 \text{ m/s})^3 = 14 \text{ kW} = 18 \text{ hp}$$

Hence, the total power required is 126 kW or   167 hp.

Note that the power requirements for traveling at constant speed on a horizontal surface are only 20 kW, or 27 hp (the sum of the last two terms). Furthermore, if the mass were halved (as in the case of a compact car), then the power required also is reduced by almost the same factor.

SUMMARY

Take a practice test for this chapter by clicking on the Practice Test link at http://www.pse6.com.

A **system** is most often a single particle, a collection of particles or a region of space. A **system boundary** separates the system from the **environment**. Many physics problems can be solved by considering the interaction of a system with its environment.

The **work** W done on a system by an agent exerting a constant force $\mathbf{F}$ on the system is the product of the magnitude Δr of the displacement of the point of application of the force and the component $F \cos \theta$ of the force along the direction of the displacement $\Delta \mathbf{r}$:

$$W \equiv F \Delta r \cos \theta \qquad (7.1)$$

The **scalar product** (dot product) of two vectors $\mathbf{A}$ and $\mathbf{B}$ is defined by the relationship

$$\mathbf{A} \cdot \mathbf{B} \equiv AB \cos \theta \qquad (7.2)$$

where the result is a scalar quantity and θ is the angle between the two vectors. The scalar product obeys the commutative and distributive laws.

If a varying force does work on a particle as the particle moves along the x axis from x_i to x_f, the work done by the force on the particle is given by

$$W \equiv \int_{x_i}^{x_f} F_x \, dx \qquad (7.7)$$

where F_x is the component of force in the x direction.

The **kinetic energy** of a particle of mass m moving with a speed v is

$$K \equiv \tfrac{1}{2} mv^2 \qquad (7.15)$$

The **work–kinetic energy theorem** states that if work is done on a system by external forces and the only change in the system is in its speed, then

$$\sum W = K_f - K_i = \tfrac{1}{2} mv_f{}^2 - \tfrac{1}{2} mv_i{}^2 \qquad (7.14,\ 7.16)$$

For a nonisolated system, we can equate the change in the total energy stored in the system to the sum of all the transfers of energy across the system boundary. For an isolated system, the total energy is constant—this is a statement of **conservation of energy.**

If a friction force acts, the kinetic energy of the system is reduced and the appropriate equation to be applied is

$$\Delta K = -f_k d + \sum W_{\text{other forces}} \qquad (7.21a)$$

or

$$K_f = K_i - f_k d + \sum W_{\text{other forces}} \qquad (7.21b)$$

The **instantaneous power** $\mathcal{P}$ is defined as the time rate of energy transfer. If an agent applies a force $\mathbf{F}$ to an object moving with a velocity $\mathbf{v}$, the power delivered by that agent is

$$\mathcal{P} \equiv \frac{dW}{dt} = \mathbf{F} \cdot \mathbf{v} \qquad (7.23)$$

QUESTIONS

1. When a particle rotates in a circle, a force acts on it directed toward the center of rotation. Why is it that this force does no work on the particle?

2. Discuss whether any work is being done by each of the following agents and, if so, whether the work is positive or negative: (a) a chicken scratching the ground, (b) a person studying, (c) a crane lifting a bucket of concrete, (d) the gravitational force on the bucket in part (c), (e) the leg muscles of a person in the act of sitting down.

3. When a punter kicks a football, is he doing any work on the ball while his toe is in contact with it? Is he doing any work on the ball after it loses contact with his toe? Are any forces doing work on the ball while it is in flight?

4. Cite two examples in which a force is exerted on an object without doing any work on the object.

5. As a simple pendulum swings back and forth, the forces acting on the suspended object are the gravitational force, the tension in the supporting cord, and air resistance. (a) Which of these forces, if any, does no work on the pendulum? (b) Which of these forces does negative work at all times during its motion? (c) Describe the work done by the gravitational force while the pendulum is swinging.

6. If the dot product of two vectors is positive, does this imply that the vectors must have positive rectangular components?

7. For what values of θ is the scalar product (a) positive and (b) negative?

8. As the load on a vertically hanging spiral spring is increased, one would not expect the F_s-versus-x graph line to remain straight, as shown in Figure 7.10d. Explain qualitatively what you would expect for the shape of this graph as the load on the spring is increased.

9. A certain uniform spring has spring constant k. Now the spring is cut in half. What is the relationship between k and the spring constant k' of each resulting smaller spring? Explain your reasoning.

10. Can kinetic energy be negative? Explain.

11. Discuss the work done by a pitcher throwing a baseball. What is the approximate distance through which the force acts as the ball is thrown?

12. One bullet has twice the mass of a second bullet. If both are fired so that they have the same speed, which has more kinetic energy? What is the ratio of the kinetic energies of the two bullets?

13. Two sharpshooters fire 0.30-caliber rifles using identical shells. A force exerted by expanding gases in the barrels accelerates the bullets. The barrel of rifle A is 2.00 cm longer than the barrel of rifle B. Which rifle will have the higher muzzle speed?

14. (a) If the speed of a particle is doubled, what happens to its kinetic energy? (b) What can be said about the speed of a particle if the net work done on it is zero?

15. A car salesman claims that a souped-up 300-hp engine is a necessary option in a compact car, in place of the conventional 130-hp engine. Suppose you intend to drive the car within speed limits (≤ 65 mi/h) on flat terrain. How would you counter this sales pitch?

16. Can the average power over a time interval ever be equal to the instantaneous power at an instant within the interval? Explain.

17. In Example 7.15, does the required power increase or decrease as the force of friction is reduced?

18. The kinetic energy of an object depends on the frame of reference in which its motion is measured. Give an example to illustrate this point.

19. Words given precise definitions in physics are sometimes used in popular literature in interesting ways. For example, a rock falling from the top of a cliff is said to be "gathering force as it falls to the beach below." What does the phrase "gathering force" mean, and can you repair this phrase?

20. In most circumstances, the normal force acting on an object and the force of static friction do zero work on the object. However, the reason that the work is zero is different for the two cases. Explain why each does zero work.

21. "A level air track can do no work." Argue for or against this statement.

22. Who first stated the work–kinetic energy theorem? Who showed that it is useful for solving many practical problems? Do some research to answer these questions.

PROBLEMS

1, 2, 3 = straightforward, intermediate, challenging ☐ = full solution available in the *Student Solutions Manual and Study Guide*

🌎 = coached solution with hints available at http://www.pse6.com   💻 = computer useful in solving problem

▨ = paired numerical and symbolic problems

Section 7.2 Work Done by a Constant Force

1. A block of mass 2.50 kg is pushed 2.20 m along a frictionless horizontal table by a constant 16.0-N force directed 25.0° below the horizontal. Determine the work done on the block by (a) the applied force, (b) the normal force exerted by the table, and (c) the gravitational force. (d) Determine the total work done on the block.

2. A shopper in a supermarket pushes a cart with a force of 35.0 N directed at an angle of 25.0° downward from the horizontal. Find the work done by the shopper on the cart as he moves down an aisle 50.0 m long.

3. 🌎 Batman, whose mass is 80.0 kg, is dangling on the free end of a 12.0-m rope, the other end of which is fixed to a tree limb above. He is able to get the rope in motion

as only Batman knows how, eventually getting it to swing enough that he can reach a ledge when the rope makes a 60.0° angle with the vertical. How much work was done by the gravitational force on Batman in this maneuver?

4. A raindrop of mass 3.35×10^{-5} kg falls vertically at constant speed under the influence of gravity and air resistance. Model the drop as a particle. As it falls 100 m, what is the work done on the raindrop (a) by the gravitational force and (b) by air resistance?

Section 7.3 The Scalar Product of Two Vectors

5. Vector **A** has a magnitude of 5.00 units, and **B** has a magnitude of 9.00 units. The two vectors make an angle of 50.0° with each other. Find **A** · **B**.

6. For any two vectors **A** and **B**, show that $\mathbf{A} \cdot \mathbf{B} = A_x B_x + A_y B_y + A_z B_z$. (*Suggestion:* Write **A** and **B** in unit vector form and use Equations 7.4 and 7.5.)

Note: In Problems 7 through 10, calculate numerical answers to three significant figures as usual.

---

7. A force $\mathbf{F} = (6\hat{\mathbf{i}} - 2\hat{\mathbf{j}})$ N acts on a particle that undergoes a displacement $\Delta \mathbf{r} = (3\hat{\mathbf{i}} + \hat{\mathbf{j}})$ m. Find (a) the work done by the force on the particle and (b) the angle between **F** and $\Delta \mathbf{r}$.

8. Find the scalar product of the vectors in Figure P7.8.

Figure P7.8

9. Using the definition of the scalar product, find the angles between (a) $\mathbf{A} = 3\hat{\mathbf{i}} - 2\hat{\mathbf{j}}$ and $\mathbf{B} = 4\hat{\mathbf{i}} - 4\hat{\mathbf{j}}$; (b) $\mathbf{A} = -2\hat{\mathbf{i}} + 4\hat{\mathbf{j}}$ and $\mathbf{B} = 3\hat{\mathbf{i}} - 4\hat{\mathbf{j}} + 2\hat{\mathbf{k}}$; (c) $\mathbf{A} = \hat{\mathbf{i}} - 2\hat{\mathbf{j}} + 2\hat{\mathbf{k}}$ and $\mathbf{B} = 3\hat{\mathbf{j}} + 4\hat{\mathbf{k}}$.

10. For $\mathbf{A} = 3\hat{\mathbf{i}} + \hat{\mathbf{j}} - \hat{\mathbf{k}}$, $\mathbf{B} = -\hat{\mathbf{i}} + 2\hat{\mathbf{j}} + 5\hat{\mathbf{k}}$, and $\mathbf{C} = 2\hat{\mathbf{j}} - 3\hat{\mathbf{k}}$, find $\mathbf{C} \cdot (\mathbf{A} - \mathbf{B})$.

Section 7.4 Work Done by a Varying Force

11. The force acting on a particle varies as in Figure P7.11. Find the work done by the force on the particle as it moves (a) from $x = 0$ to $x = 8.00$ m, (b) from $x = 8.00$ m to $x = 10.0$ m, and (c) from $x = 0$ to $x = 10.0$ m.

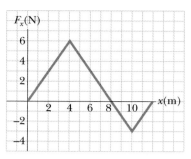

Figure P7.11

12. The force acting on a particle is $F_x = (8x - 16)$ N, where x is in meters. (a) Make a plot of this force versus x from $x = 0$ to $x = 3.00$ m. (b) From your graph, find the net work done by this force on the particle as it moves from $x = 0$ to $x = 3.00$ m.

13. A particle is subject to a force F_x that varies with position as in Figure P7.13. Find the work done by the force on the particle as it moves (a) from $x = 0$ to $x = 5.00$ m, (b) from $x = 5.00$ m to $x = 10.0$ m, and (c) from $x = 10.0$ m to $x = 15.0$ m. (d) What is the total work done by the force over the distance $x = 0$ to $x = 15.0$ m?

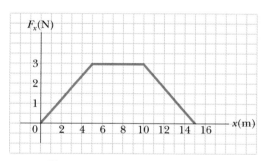

Figure P7.13 Problems 13 and 28.

14. A force $\mathbf{F} = (4x\hat{\mathbf{i}} + 3y\hat{\mathbf{j}})$ N acts on an object as the object moves in the x direction from the origin to $x = 5.00$ m. Find the work $W = \int \mathbf{F} \cdot d\mathbf{r}$ done on the object by the force.

15. When a 4.00-kg object is hung vertically on a certain light spring that obeys Hooke's law, the spring stretches 2.50 cm. If the 4.00-kg object is removed, (a) how far will the spring stretch if a 1.50-kg block is hung on it, and (b) how much work must an external agent do to stretch the same spring 4.00 cm from its unstretched position?

16. An archer pulls her bowstring back 0.400 m by exerting a force that increases uniformly from zero to 230 N. (a) What is the equivalent spring constant of the bow? (b) How much work does the archer do in pulling the bow?

17. Truck suspensions often have "helper springs" that engage at high loads. One such arrangement is a leaf spring with a helper coil spring mounted on the axle, as in Figure P7.17. The helper spring engages when the main leaf spring is compressed by distance y_0, and then helps to support any additional load. Consider a leaf spring constant of 5.25×10^5 N/m, helper spring constant of 3.60×10^5 N/m, and $y_0 = 0.500$ m. (a) What is the

Figure P7.17

compression of the leaf spring for a load of 5.00×10^5 N? (b) How much work is done in compressing the springs?

18. A 100-g bullet is fired from a rifle having a barrel 0.600 m long. Assuming the origin is placed where the bullet begins to move, the force (in newtons) exerted by the expanding gas on the bullet is $15\,000 + 10\,000x - 25\,000x^2$, where x is in meters. (a) Determine the work done by the gas on the bullet as the bullet travels the length of the barrel. (b) **What If?** If the barrel is 1.00 m long, how much work is done, and how does this value compare to the work calculated in (a)?

19. If it takes 4.00 J of work to stretch a Hooke's-law spring 10.0 cm from its unstressed length, determine the extra work required to stretch it an additional 10.0 cm.

20. A small particle of mass m is pulled to the top of a frictionless half-cylinder (of radius R) by a cord that passes over the top of the cylinder, as illustrated in Figure P7.20. (a) If the particle moves at a constant speed, show that $F = mg \cos \theta$. (*Note:* If the particle moves at constant speed, the component of its acceleration tangent to the cylinder must be zero at all times.) (b) By directly integrating $W = \int \mathbf{F} \cdot d\mathbf{r}$, find the work done in moving the particle at constant speed from the bottom to the top of the half-cylinder.

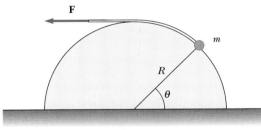

Figure P7.20

21. A light spring with spring constant 1 200 N/m is hung from an elevated support. From its lower end a second light spring is hung, which has spring constant 1 800 N/m. An object of mass 1.50 kg is hung at rest from the lower end of the second spring. (a) Find the total extension distance of the pair of springs. (b) Find the effective spring constant of the pair of springs as a system. We describe these springs as *in series*.

22. A light spring with spring constant k_1 is hung from an elevated support. From its lower end a second light spring is hung, which has spring constant k_2. An object of mass m is hung at rest from the lower end of the second spring. (a) Find the total extension distance of the pair of springs. (b) Find the effective spring constant of the pair of springs as a system. We describe these springs as *in series*.

23. Express the units of the force constant of a spring in SI base units.

Section 7.5   Kinetic Energy and the Work–Kinetic Energy Theorem

Section 7.6   The Nonisolated System–Conservation of Energy

24. A 0.600-kg particle has a speed of 2.00 m/s at point Ⓐ and kinetic energy of 7.50 J at point Ⓑ. What is (a) its kinetic energy at Ⓐ? (b) its speed at Ⓑ? (c) the total work done on the particle as it moves from Ⓐ to Ⓑ?

25. A 0.300-kg ball has a speed of 15.0 m/s. (a) What is its kinetic energy? (b) **What If?** If its speed were doubled, what would be its kinetic energy?

26. A 3.00-kg object has a velocity $(6.00\hat{\mathbf{i}} - 2.00\hat{\mathbf{j}})$ m/s. (a) What is its kinetic energy at this time? (b) Find the total work done on the object if its velocity changes to $(8.00\hat{\mathbf{i}} + 4.00\hat{\mathbf{j}})$ m/s. (*Note:* From the definition of the dot product, $v^2 = \mathbf{v} \cdot \mathbf{v}$.)

27. A 2 100-kg pile driver is used to drive a steel I-beam into the ground. The pile driver falls 5.00 m before coming into contact with the top of the beam, and it drives the beam 12.0 cm farther into the ground before coming to rest. Using energy considerations, calculate the average force the beam exerts on the pile driver while the pile driver is brought to rest.

28. A 4.00-kg particle is subject to a total force that varies with position as shown in Figure P7.13. The particle starts from rest at $x = 0$. What is its speed at (a) $x = 5.00$ m, (b) $x = 10.0$ m, (c) $x = 15.0$ m?

29. You can think of the work–kinetic energy theorem as a second theory of motion, parallel to Newton's laws in describing how outside influences affect the motion of an object. In this problem, solve parts (a) and (b) separately from parts (c) and (d) to compare the predictions of the two

theories. In a rifle barrel, a 15.0-g bullet is accelerated from rest to a speed of 780 m/s. (a) Find the work that is done on the bullet. (b) If the rifle barrel is 72.0 cm long, find the magnitude of the average total force that acted on it, as $F = W/(\Delta r \cos \theta)$. (c) Find the constant acceleration of a bullet that starts from rest and gains a speed of 780 m/s over a distance of 72.0 cm. (d) If the bullet has mass 15.0 g, find the total force that acted on it as $\Sigma F = ma$.

30. In the neck of the picture tube of a certain black-and-white television set, an electron gun contains two charged metallic plates 2.80 cm apart. An electric force accelerates each electron in the beam from rest to 9.60% of the speed of light over this distance. (a) Determine the kinetic energy of the electron as it leaves the electron gun. Electrons carry this energy to a phosphorescent material on the inner surface of the television screen, making it glow. For an electron passing between the plates in the electron gun, determine (b) the magnitude of the constant electric force acting on the electron, (c) the acceleration, and (d) the time of flight.

Section 7.7   Situations Involving Kinetic Friction

31. A 40.0-kg box initially at rest is pushed 5.00 m along a rough, horizontal floor with a constant applied horizontal force of 130 N. If the coefficient of friction between box and floor is 0.300, find (a) the work done by the applied force, (b) the increase in internal energy in the box-floor system due to friction, (c) the work done by the normal force, (d) the work done by the gravitational force, (e) the change in kinetic energy of the box, and (f) the final speed of the box.

32. A 2.00-kg block is attached to a spring of force constant 500 N/m as in Figure 7.10. The block is pulled 5.00 cm to the right of equilibrium and released from rest. Find the speed of the block as it passes through equilibrium if (a) the horizontal surface is frictionless and (b) the coefficient of friction between block and surface is 0.350.

33. A crate of mass 10.0 kg is pulled up a rough incline with an initial speed of 1.50 m/s. The pulling force is 100 N parallel to the incline, which makes an angle of 20.0° with the horizontal. The coefficient of kinetic friction is 0.400, and the crate is pulled 5.00 m. (a) How much work is done by the gravitational force on the crate? (b) Determine the increase in internal energy of the crate–incline system due to friction. (c) How much work is done by the 100-N force on the crate? (d) What is the change in kinetic energy of the crate? (e) What is the speed of the crate after being pulled 5.00 m?

34. A 15.0-kg block is dragged over a rough, horizontal surface by a 70.0-N force acting at 20.0° above the horizontal. The block is displaced 5.00 m, and the coefficient of kinetic friction is 0.300. Find the work done on the block by (a) the 70-N force, (b) the normal force, and (c) the gravitational force. (d) What is the increase in internal energy of the block-surface system due to friction? (e) Find the total change in the block's kinetic energy.

35. A sled of mass m is given a kick on a frozen pond. The kick imparts to it an initial speed of 2.00 m/s. The coefficient of kinetic friction between sled and ice is 0.100. Use energy considerations to find the distance the sled moves before it stops.

Section 7.8   Power

36. The electric motor of a model train accelerates the train from rest to 0.620 m/s in 21.0 ms. The total mass of the train is 875 g. Find the average power delivered to the train during the acceleration.

37. A 700-N Marine in basic training climbs a 10.0-m vertical rope at a constant speed in 8.00 s. What is his power output?

38. Make an order-of-magnitude estimate of the power a car engine contributes to speeding the car up to highway speed. For concreteness, consider your own car if you use one. In your solution state the physical quantities you take as data and the values you measure or estimate for them. The mass of the vehicle is given in the owner's manual. If you do not wish to estimate for a car, consider a bus or truck that you specify.

39. A skier of mass 70.0 kg is pulled up a slope by a motor-driven cable. (a) How much work is required to pull him a distance of 60.0 m up a 30.0° slope (assumed frictionless) at a constant speed of 2.00 m/s? (b) A motor of what power is required to perform this task?

40. A 650-kg elevator starts from rest. It moves upward for 3.00 s with constant acceleration until it reaches its cruising speed of 1.75 m/s. (a) What is the average power of the elevator motor during this period? (b) How does this power compare with the motor power when the elevator moves at its cruising speed?

41. An energy-efficient lightbulb, taking in 28.0 W of power, can produce the same level of brightness as a conventional bulb operating at power 100 W. The lifetime of the energy efficient bulb is 10 000 h and its purchase price is $17.0, whereas the conventional bulb has lifetime 750 h and costs $0.420 per bulb. Determine the total savings obtained by using one energy-efficient bulb over its lifetime, as opposed to using conventional bulbs over the same time period. Assume an energy cost of $0.080 0 per kilowatt-hour.

42. Energy is conventionally measured in Calories as well as in joules. One Calorie in nutrition is one kilocalorie, defined as 1 kcal = 4 186 J. Metabolizing one gram of fat can release 9.00 kcal. A student decides to try to lose weight by exercising. She plans to run up and down the stairs in a football stadium as fast as she can and as many times as necessary. Is this in itself a practical way to lose weight? To evaluate the program, suppose she runs up a flight of 80 steps, each 0.150 m high, in 65.0 s. For simplicity, ignore the energy she uses in coming down (which is small). Assume that a typical efficiency for human muscles is 20.0%. This means that when your body converts 100 J from metabolizing fat, 20 J goes into doing mechanical work (here, climbing stairs). The remainder goes into extra internal energy. Assume the student's mass is 50.0 kg. (a) How many times must she run the flight of stairs to lose one pound of fat? (b) What is her average power output, in watts and in horsepower, as she is running up the stairs?

43. For saving energy, bicycling and walking are far more efficient means of transportation than is travel by automobile. For example, when riding at 10.0 mi/h a cyclist uses food energy at a rate of about 400 kcal/h above what he would use if merely sitting still. (In exercise physiology, power is often measured in kcal/h rather than in watts. Here 1 kcal = 1 nutritionist's Calorie = 4186 J.) Walking at 3.00 mi/h requires about 220 kcal/h. It is interesting to compare these values with the energy consumption required for travel by car. Gasoline yields about 1.30×10^8 J/gal. Find the fuel economy in equivalent miles per gallon for a person (a) walking, and (b) bicycling.

Section 7.9   Energy and the Automobile

44. Suppose the empty car described in Table 7.2 has a fuel economy of 6.40 km/liter (15 mi/gal) when traveling at 26.8 m/s (60 mi/h). Assuming constant efficiency, determine the fuel economy of the car if the total mass of passengers plus driver is 350 kg.

45. A compact car of mass 900 kg has an overall motor efficiency of 15.0%. (That is, 15% of the energy supplied by the fuel is delivered to the wheels of the car.) (a) If burning one gallon of gasoline supplies 1.34×10^8 J of energy, find the amount of gasoline used in accelerating the car from rest to 55.0 mi/h. Here you may ignore the effects of air resistance and rolling friction. (b) How many such accelerations will one gallon provide? (c) The mileage claimed for the car is 38.0 mi/gal at 55 mi/h. What power is delivered to the wheels (to overcome frictional effects) when the car is driven at this speed?

Additional Problems

46. A baseball outfielder throws a 0.150-kg baseball at a speed of 40.0 m/s and an initial angle of 30.0°. What is the kinetic energy of the baseball at the highest point of its trajectory?

47. While running, a person dissipates about 0.600 J of mechanical energy per step per kilogram of body mass. If a 60.0-kg runner dissipates a power of 70.0 W during a race, how fast is the person running? Assume a running step is 1.50 m long.

48. The direction of any vector **A** in three-dimensional space can be specified by giving the angles α, β, and γ that the vector makes with the x, y, and z axes, respectively. If **A** = $A_x\hat{\mathbf{i}} + A_y\hat{\mathbf{j}} + A_z\hat{\mathbf{k}}$, (a) find expressions for cos α, cos β, and cos γ (these are known as *direction cosines*), and (b) show that these angles satisfy the relation $\cos^2\alpha + \cos^2\beta + \cos^2\gamma = 1$. (*Hint:* Take the scalar product of **A** with $\hat{\mathbf{i}}$, $\hat{\mathbf{j}}$, and **k** separately.)

49. A 4.00-kg particle moves along the x axis. Its position varies with time according to $x = t + 2.0t^3$, where x is in meters and t is in seconds. Find (a) the kinetic energy at any time t, (b) the acceleration of the particle and the force acting on it at time t, (c) the power being delivered to the particle at time t, and (d) the work done on the particle in the interval $t = 0$ to $t = 2.00$ s.

50. The spring constant of an automotive suspension spring increases with increasing load due to a spring coil that is widest at the bottom, smoothly tapering to a smaller diameter near the top. The result is a softer ride on normal road surfaces from the narrower coils, but the car does not bottom out on bumps because when the upper coils col-

lapse, they leave the stiffer coils near the bottom to absorb the load. For a tapered spiral spring that compresses 12.9 cm with a 1 000-N load and 31.5 cm with a 5 000-N load, (a) evaluate the constants a and b in the empirical equation $F = ax^b$ and (b) find the work needed to compress the spring 25.0 cm.

51. A bead at the bottom of a bowl is one example of an object in a stable equilibrium position. When a physical system is displaced by an amount x from stable equilibrium, a restoring force acts on it, tending to return the system to its equilibrium configuration. The magnitude of the restoring force can be a complicated function of x. For example, when an ion in a crystal is displaced from its lattice site, the restoring force may not be a simple function of x. In such cases we can generally imagine the function $F(x)$ to be expressed as a power series in x, as $F(x) = -(k_1x + k_2x^2 + k_3x^3 + \ldots)$. The first term here is just Hooke's law, which describes the force exerted by a simple spring for small displacements. For small excursions from equilibrium we generally neglect the higher order terms, but in some cases it may be desirable to keep the second term as well. If we model the restoring force as $F = -(k_1x + k_2x^2)$, how much work is done in displacing the system from $x = 0$ to $x = x_{max}$ by an applied force $-F$?

52. A traveler at an airport takes an escalator up one floor, as in Figure P7.52. The moving staircase would itself carry him upward with vertical velocity component v between entry and exit points separated by height h. However, while the escalator is moving, the hurried traveler climbs the steps of the escalator at a rate of n steps/s. Assume that the height of each step is h_s. (a) Determine the amount of chemical energy converted into mechanical energy by the traveler's leg muscles during his escalator ride, given that

Figure P7.52

Figure P7.56

his mass is m. (b) Determine the work the escalator motor does on this person.

53. A mechanic pushes a car of mass m, doing work W in making it accelerate from rest. Neglecting friction between car and road, (a) what is the final speed of the car? During this time, the car moves a distance d. (b) What constant horizontal force did the mechanic exert on the car?

54. A 5.00-kg steel ball is dropped onto a copper plate from a height of 10.0 m. If the ball leaves a dent 3.20 mm deep, what is the average force exerted by the plate on the ball during the impact?

55. A single constant force $\mathbf{F}$ acts on a particle of mass m. The particle starts at rest at $t = 0$. (a) Show that the instantaneous power delivered by the force at any time t is $\mathcal{P} = (F^2/m)\,t$. (b) If $F = 20.0$ N and $m = 5.00$ kg, what is the power delivered at $t = 3.00$ s?

56. Two springs with negligible masses, one with spring constant k_1 and the other with spring constant k_2, are attached to the endstops of a level air track as in Figure P7.56. A glider attached to both springs is located between them. When the glider is in equilibrium, spring 1 is stretched by extension x_{i1} to the right of its unstretched length and spring 2 is stretched by x_{i2} to the left. Now a horizontal force $\mathbf{F}_{app}$ is applied to the glider to move it a distance x_a to the right from its equilibrium position. Show that in this process (a) the work done on spring 1 is $\frac{1}{2}k_1(x_a^2 + 2x_a x_{i1})$, (b) the work done on spring 2 is $\frac{1}{2}k_2(x_a^2 - 2x_a x_{i2})$, (c) x_{i2} is related to x_{i1} by $x_{i2} = k_1 x_{i1}/k_2$, and (d) the total work done by the force F_{app} is $\frac{1}{2}(k_1 + k_2)x_a^2$.

57. As the driver steps on the gas pedal, a car of mass 1 160 kg accelerates from rest. During the first few seconds of motion, the car's acceleration increases with time according to the expression

$$a = (1.16 \text{ m/s}^3)\,t - (0.210 \text{ m/s}^4)\,t^2 + (0.240 \text{ m/s}^5)\,t^3$$

(a) What work is done by the wheels on the car during the interval from $t = 0$ to $t = 2.50$ s? (b) What is the output power of the wheels at the instant $t = 2.50$ s?

58. A particle is attached between two identical springs on a horizontal frictionless table. Both springs have spring constant k and are initially unstressed. (a) If the particle is pulled a distance x along a direction perpendicular to the

initial configuration of the springs, as in Figure P7.58, show that the force exerted by the springs on the particle is

$$\mathbf{F} = -2kx\left(1 - \frac{L}{\sqrt{x^2 + L^2}}\right)\hat{\mathbf{i}}$$

(b) Determine the amount of work done by this force in moving the particle from $x = A$ to $x = 0$.

Top view
Figure P7.58

59. A rocket body of mass M will fall out of the sky with terminal speed v_T after its fuel is used up. What power output must the rocket engine produce if the rocket is to fly (a) at its terminal speed straight up; (b) at three times the terminal speed straight down? In both cases assume that the mass of the fuel and oxidizer remaining in the rocket is negligible compared to M. Assume that the force of air resistance is proportional to the square of the rocket's speed.

60. **Review problem.** Two constant forces act on a 5.00-kg object moving in the xy plane, as shown in Figure P7.60. Force $\mathbf{F}_1$ is 25.0 N at 35.0°, while $\mathbf{F}_2$ is 42.0 N at 150°. At time $t = 0$, the object is at the origin and has velocity $(4.00\hat{\mathbf{i}} + 2.50\hat{\mathbf{j}})$ m/s. (a) Express the two forces in unit-vector notation. Use unit-vector notation for your other answers. (b) Find the total force on the object. (c) Find the object's acceleration. Now, considering the instant $t = 3.00$ s, (d) find the object's velocity, (e) its location, (f) its kinetic energy from $\frac{1}{2}mv_f^2$, and (g) its kinetic energy from $\frac{1}{2}mv_i^2 + \Sigma\mathbf{F}\cdot\Delta\mathbf{r}$.

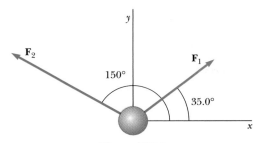

Figure P7.60

61. A 200-g block is pressed against a spring of force constant 1.40 kN/m until the block compresses the spring 10.0 cm. The spring rests at the bottom of a ramp inclined at 60.0° to the horizontal. Using energy considerations, determine how far up the incline the block moves before it stops (a) if there is no friction between the block and the ramp and (b) if the coefficient of kinetic friction is 0.400.

62. When different weights are hung on a spring, the spring stretches to different lengths as shown in the following table. (a) Make a graph of the applied force versus the extension of the spring. By least-squares fitting, determine the straight line that best fits the data. (You may not want to use all the data points.) (b) From the slope of the best-fit line, find the spring constant k. (c) If the spring is extended to 105 mm, what force does it exert on the suspended weight?

F (N)	2.0	4.0	6.0	8.0	10	12	14	16	18	20	22
L (mm)	15	32	49	64	79	98	112	126	149	175	190

63. The ball launcher in a pinball machine has a spring that has a force constant of 1.20 N/cm (Fig. P7.63). The surface on which the ball moves is inclined 10.0° with respect to the horizontal. If the spring is initially compressed 5.00 cm, find the launching speed of a 100-g ball when the plunger is released. Friction and the mass of the plunger are negligible.

Figure P7.63

64. A 0.400-kg particle slides around a horizontal track. The track has a smooth vertical outer wall forming a circle with a radius of 1.50 m. The particle is given an initial speed of 8.00 m/s. After one revolution, its speed has dropped to 6.00 m/s because of friction with the rough floor of the track. (a) Find the energy converted from mechanical to internal in the system due to friction in one revolution. (b) Calculate the coefficient of kinetic friction. (c) What is the total number of revolutions the particle makes before stopping?

65. In diatomic molecules, the constituent atoms exert attractive forces on each other at large distances and repulsive forces at short distances. For many molecules, the Lennard-Jones law is a good approximation to the magnitude of these forces:

$$F = F_0 \left[2\left(\frac{\sigma}{r}\right)^{13} - \left(\frac{\sigma}{r}\right)^{7} \right]$$

where r is the center-to-center distance between the atoms in the molecule, σ is a length parameter, and F_0 is the force when $r = \sigma$. For an oxygen molecule, we find that $F_0 = 9.60 \times 10^{-11}$ N and $\sigma = 3.50 \times 10^{-10}$ m. Determine the work done by this force if the atoms are pulled apart from $r = 4.00 \times 10^{-10}$ m to $r = 9.00 \times 10^{-10}$ m.

66. As it plows a parking lot, a snowplow pushes an ever-growing pile of snow in front of it. Suppose a car moving through the air is similarly modeled as a cylinder pushing a growing plug of air in front of it. The originally stationary air is set into motion at the constant speed v of the cylinder, as in Figure P7.66. In a time interval Δt, a new disk of air of mass Δm must be moved a distance $v\Delta t$ and hence must be given a kinetic energy $\frac{1}{2}(\Delta m)v^2$. Using this model, show that the automobile's power loss due to air resistance is $\frac{1}{2}\rho A v^3$ and that the resistive force acting on the car is $\frac{1}{2}\rho A v^2$, where ρ is the density of air. Compare this result with the empirical expression $\frac{1}{2}D\rho A v^2$ for the resistive force.

Figure P7.66

67. A particle moves along the x axis from $x = 12.8$ m to $x = 23.7$ m under the influence of a force

$$F = \frac{375}{x^3 + 3.75x}$$

where F is in newtons and x is in meters. Using numerical integration, determine the total work done by this force on the particle during this displacement. Your result should be accurate to within 2%.

68. A windmill, such as that in the opening photograph of this chapter, turns in response to a force of high-speed air resistance, $R = \frac{1}{2}D\rho A v^2$. The power available is $\mathcal{P} = Rv = \frac{1}{2}D\rho \pi r^2 v^3$, where v is the wind speed and we have assumed a circular face for the windmill, of radius r. Take the drag coefficient as $D = 1.00$ and the density of air from the front endpaper. For a home windmill with $r = 1.50$ m, calculate the power available if (a) $v = 8.00$ m/s and (b) $v = 24.0$ m/s. The power delivered to the generator is limited by the efficiency of the system, which is about 25%. For comparison, a typical home needs about 3 kW of electric power.

69. More than 2 300 years ago the Greek teacher Aristotle wrote the first book called *Physics*. Put into more precise terminology, this passage is from the end of its Section Eta:

Let $\mathscr{P}$ be the power of an agent causing motion; w, the thing moved; d, the distance covered; and Δt, the time interval required. Then (1) a power equal to $\mathscr{P}$ will in a period of time equal to Δt move $w/2$ a distance $2d$; or (2) it will move $w/2$ the given distance d in the time interval $\Delta t/2$. Also, if (3) the given power $\mathscr{P}$ moves the given object w a distance $d/2$ in time interval $\Delta t/2$, then (4) $\mathscr{P}/2$ will move $w/2$ the given distance d in the given time interval Δt.

(a) Show that Aristotle's proportions are included in the equation $\mathscr{P}\Delta t = bwd$ where b is a proportionality constant. (b) Show that our theory of motion includes this part of Aristotle's theory as one special case. In particular, describe a situation in which it is true, derive the equation representing Aristotle's proportions, and determine the proportionality constant.

70. Consider the block-spring-surface system in part (b) of Example 7.11. (a) At what position x of the block is its speed a maximum? (b) In the **What If?** section of this example, we explored the effects of an increased friction force of 10.0 N. At what position of the block does its maximum speed occur in this situation?

Answers to Quick Quizzes

7.1 (a). The force does no work on the Earth because the force is pointed toward the center of the circle and is therefore perpendicular to the direction of the displacement.

7.2 c, a, d, b. The work in (c) is positive and of the largest possible value because the angle between the force and the displacement is zero. The work done in (a) is zero because the force is perpendicular to the displacement. In (d) and (b), negative work is done by the applied force because in neither case is there a component of the force in the direction of the displacement. Situation (b) is the most negative value because the angle between the force and the displacement is 180°.

7.3 (d). Answer (a) is incorrect because the scalar product $(-\mathbf{A}) \cdot (-\mathbf{B})$ is equal to $\mathbf{A} \cdot \mathbf{B}$. Answer (b) is incorrect because $AB \cos(\theta + 180°)$ gives the negative of the correct value.

7.4 (d). Because of the range of values of the cosine function, $\mathbf{A} \cdot \mathbf{B}$ has values that range from AB to $-AB$.

7.5 (a). Because the work done in compressing a spring is proportional to the square of the compression distance x, doubling the value of x causes the work to increase fourfold.

7.6 (b). Because the work is proportional to the square of the compression distance x and the kinetic energy is proportional to the square of the speed v, doubling the compression distance doubles the speed.

7.7 (a) For the television set, energy enters by electrical transmission (through the power cord) and electromagnetic radiation (the television signal). Energy leaves by heat (from hot surfaces into the air), mechanical waves (sound from the speaker), and electromagnetic radiation (from the screen). (b) For the gasoline-powered lawn mower, energy enters by matter transfer (gasoline). Energy leaves by work (on the blades of grass), mechanical waves (sound), and heat (from hot surfaces into the air). (c) For the hand-cranked pencil sharpener, energy enters by work (from your hand turning the crank). Energy leaves by work (done on the pencil) and mechanical waves (sound).

7.8 (b). The friction force represents an interaction with the environment of the block.

7.9 (b). The friction force represents an interaction with the environment of the surface.

7.10 (a). The friction force is internal to the system, so there are no interactions with the environment.

7.11 (c). The brakes and the roadway are warmer, so their internal energy has increased. In addition, the sound of the skid represents transfer of energy away by mechanical waves.

7.12 (e). Because the speed is doubled, the kinetic energy is four times as large. This kinetic energy was attained for the newer car in the same time interval as the smaller kinetic energy for the older car, so the power is four times as large.

Chapter 8

Potential Energy

▲ A strobe photograph of a pole vaulter. During this process, several types of energy transformations occur. The two types of potential energy that we study in this chapter are evident in the photograph. *Gravitational potential energy* is associated with the change in vertical position of the vaulter relative to the Earth. *Elastic potential energy* is evident in the bending of the pole. (©Harold E. Edgerton/Courtesy of Palm Press, Inc.)

In Chapter 7 we introduced the concepts of kinetic energy associated with the motion of members of a system and internal energy associated with the temperature of a system. In this chapter we introduce *potential energy*, the energy associated with the configuration of a system of objects that exert forces on each other.

The potential energy concept can be used only when dealing with a special class of forces called *conservative forces*. When only conservative forces act within an isolated system, the kinetic energy gained (or lost) by the system as its members change their relative positions is balanced by an equal loss (or gain) in potential energy. This balancing of the two forms of energy is known as the *principle of conservation of mechanical energy*.

Potential energy is present in the Universe in various forms, including gravitational, electromagnetic, chemical, and nuclear. Furthermore, one form of energy in a system can be converted to another. For example, when a system consists of an electric motor connected to a battery, the chemical energy in the battery is converted to kinetic energy as the shaft of the motor turns. The transformation of energy from one form to another is an essential part of the study of physics, engineering, chemistry, biology, geology, and astronomy.

8.1 Potential Energy of a System

In Chapter 7, we defined a system in general, but focused our attention primarily on single particles or objects under the influence of an external force. In this chapter, we consider systems of two or more particles or objects interacting via a force that is *internal* to the system. The kinetic energy of such a system is the algebraic sum of the kinetic energies of all members of the system. There may be systems, however, in which one object is so massive that it can be modeled as stationary and its kinetic energy can be neglected. For example, if we consider a ball–Earth system as the ball falls to the ground, the kinetic energy of the system can be considered as just the kinetic energy of the ball. The Earth moves so slowly in this process that we can ignore its kinetic energy. On the other hand, the kinetic energy of a system of two electrons must include the kinetic energies of both particles.

Let us imagine a system consisting of a book and the Earth, interacting via the gravitational force. We do some work on the system by lifting the book slowly through a height $\Delta y = y_b - y_a$, as in Figure 8.1. According to our discussion of energy and energy transfer in Chapter 7, this work done on the system must appear as an increase in energy of the system. The book is at rest before we perform the work and is at rest after we perform the work. Thus, there is no change in the kinetic energy of the system. There is no reason why the temperature of the book or the Earth should change, so there is no increase in the internal energy of the system.

Because the energy change of the system is not in the form of kinetic energy or internal energy, it must appear as some other form of energy storage. After lifting the book, we could release it and let it fall back to the position y_a. Notice that the book (and, therefore, the system) will now have kinetic energy, and its source is in the work that was done

Figure 8.1 The work done by an external agent on the system of the book and the Earth as the book is lifted from a height y_a to a height y_b is equal to $mgy_b - mgy_a$.

in lifting the book. While the book was at the highest point, the energy of the system had the *potential* to become kinetic energy, but did not do so until the book was allowed to fall. Thus, we call the energy storage mechanism before we release the book **potential energy.** We will find that a potential energy can only be associated with specific types of forces. In this particular case, we are discussing **gravitational potential energy.**

Let us now derive an expression for the gravitational potential energy associated with an object at a given location above the surface of the Earth. Consider an external agent lifting an object of mass m from an initial height y_a above the ground to a final height y_b, as in Figure 8.1. We assume that the lifting is done slowly, with no acceleration, so that the lifting force can be modeled as being equal in magnitude to the weight of the object—the object is in equilibrium and moving at constant velocity. The work done by the external agent on the system (object and Earth) as the object undergoes this upward displacement is given by the product of the upward applied force $\mathbf{F}_{app}$ and the upward displacement $\Delta\mathbf{r} = \Delta y\hat{\mathbf{j}}$:

$$W = (\mathbf{F}_{app}) \cdot \Delta\mathbf{r} = (mg\hat{\mathbf{j}}) \cdot [(y_b - y_a)\hat{\mathbf{j}}] = mgy_b - mgy_a \qquad (8.1)$$

Notice how similar this equation is to Equation 7.14 in the preceding chapter. In each equation, the work done on a system equals a difference between the final and initial values of a quantity. In Equation 7.14, the work represents a transfer of energy into the system, and the increase in energy of the system is kinetic in form. In Equation 8.1, the work represents a transfer of energy into the system, and the system energy appears in a different form, which we have called gravitational potential energy.

Thus, we can identify the quantity mgy as the gravitational potential energy U_g:

$$U_g \equiv mgy \qquad (8.2)$$

The units of gravitational potential energy are joules, the same as those of work and kinetic energy. Potential energy, like work and kinetic energy, is a scalar quantity. Note that Equation 8.2 is valid only for objects near the surface of the Earth, where g is approximately constant.[1]

Using our definition of gravitational potential energy, Equation 8.1 can now be rewritten as

$$W = \Delta U_g \qquad (8.3)$$

which mathematically describes the fact that the work done on the system in this situation appears as a change in the gravitational potential energy of the system.

The gravitational potential energy depends only on the vertical height of the object above the surface of the Earth. The same amount of work must be done on an object–Earth system whether the object is lifted vertically from the Earth or is pushed starting from the same point up a frictionless incline, ending up at the same height. This can be shown by calculating the work with a displacement having both vertical and horizontal components:

$$W = (\mathbf{F}_{app}) \cdot \Delta\mathbf{r} = (mg\hat{\mathbf{j}}) \cdot [(x_b - x_a)\hat{\mathbf{i}} + (y_b - y_a)\hat{\mathbf{j}}] = mgy_b - mgy_a$$

where there is no term involving x in the final result because $\hat{\mathbf{j}} \cdot \hat{\mathbf{i}} = 0$.

In solving problems, you must choose a reference configuration for which the gravitational potential energy is set equal to some reference value, which is normally zero. The choice of reference configuration is completely arbitrary because the important quantity is the *difference* in potential energy and this difference is independent of the choice of reference configuration.

It is often convenient to choose as the reference configuration for zero potential energy the configuration in which an object is at the surface of the Earth, but this is not essential. Often, the statement of the problem suggests a convenient configuration to use.

[1]   The assumption that g is constant is valid as long as the vertical displacement is small compared with the Earth's radius.

 PITFALL PREVENTION

8.1   Potential Energy Belongs to a System

Potential energy is always associated with a *system* of two or more interacting objects. When a small object moves near the surface of the Earth under the influence of gravity, we may sometimes refer to the potential energy "associated with the object" rather than the more proper "associated with the system" because the Earth does not move significantly. We will not, however, refer to the potential energy "of the object" because this clearly ignores the role of the Earth.

Gravitational potential energy

Quick Quiz 8.1 Choose the correct answer. The gravitational potential energy of a system (a) is always positive (b) is always negative (c) can be negative or positive.

Quick Quiz 8.2 An object falls off a table to the floor. We wish to analyze the situation in terms of kinetic and potential energy. In discussing the kinetic energy of the system, we (a) must include the kinetic energy of both the object and the Earth (b) can ignore the kinetic energy of the Earth because it is not part of the system (c) can ignore the kinetic energy of the Earth because the Earth is so massive compared to the object.

Quick Quiz 8.3 An object falls off a table to the floor. We wish to analyze the situation in terms of kinetic and potential energy. In discussing the potential energy of the system, we identify the system as (a) both the object and the Earth (b) only the object (c) only the Earth.

Example 8.1 The Bowler and the Sore Toe

A bowling ball held by a careless bowler slips from the bowler's hands and drops on the bowler's toe. Choosing floor level as the $y = 0$ point of your coordinate system, estimate the change in gravitational potential energy of the ball–Earth system as the ball falls. Repeat the calculation, using the top of the bowler's head as the origin of coordinates.

Solution First, we need to estimate a few values. A bowling ball has a mass of approximately 7 kg, and the top of a person's toe is about 0.03 m above the floor. Also, we shall assume the ball falls from a height of 0.5 m. Keeping nonsignificant digits until we finish the problem, we calculate the gravitational potential energy of the ball–Earth system just before the ball is released to be $U_i = mgy_i = (7 \text{ kg})(9.80 \text{ m/s}^2)(0.5 \text{ m}) = 34.3 \text{ J}$. A similar calculation for

when the ball reaches his toe gives $U_f = mgy_f = (7 \text{ kg})(9.80 \text{ m/s}^2)(0.03 \text{ m}) = 2.06 \text{ J}$. So, the change in gravitational potential energy of the ball–Earth system is $\Delta U_g = U_f - U_i = -32.24 \text{ J}$. We should probably keep only one digit because of the roughness of our estimates; thus, we estimate that the change in gravitational potential energy is -30 J. The system had 30 J of gravitational potential energy relative to the top of the toe before the ball began its fall.

When we use the bowler's head (which we estimate to be 1.50 m above the floor) as our origin of coordinates, we find that $U_i = mgy_i = (7 \text{ kg})(9.80 \text{ m/s}^2)(-1 \text{ m}) = -68.6 \text{ J}$ and $U_f = mgy_f = (7 \text{ kg})(9.80 \text{ m/s}^2)(-1.47 \text{ m}) = -100.8 \text{ J}$. The change in gravitational potential energy of the ball–Earth system is $\Delta U_g = U_f - U_i = -32.24 \text{ J} \approx \boxed{-30 \text{ J}}$. This is the same value as before, as it must be.

8.2 The Isolated System–Conservation of Mechanical Energy

The introduction of potential energy allows us to generate a powerful and universally applicable principle for solving problems that are difficult to solve with Newton's laws. Let us develop this new principle by thinking about the book–Earth system in Figure 8.1 again. After we have lifted the book, there is gravitational potential energy stored in the system, which we can calculate from the work done by the external agent on the system, using $W = \Delta U_g$.

Let us now shift our focus to the work done on the book alone by the gravitational force (Fig. 8.2) as the book falls back to its original height. As the book falls from y_b to y_a, the work done by the gravitational force on the book is

$$W_{\text{on book}} = (m\mathbf{g}) \cdot \Delta \mathbf{r} = (-mg\hat{\mathbf{j}}) \cdot [(y_a - y_b)\hat{\mathbf{j}}] = mgy_b - mgy_a \quad (8.4)$$

From the work–kinetic energy theorem of Chapter 7, the work done on the book is equal to the change in the kinetic energy of the book:

$$W_{\text{on book}} = \Delta K_{\text{book}}$$

Figure 8.2 The work done by the gravitational force on the book as the book falls from y_b to a height y_a is equal to $mgy_b - mgy_a$.

Therefore, equating these two expressions for the work done on the book,

$$\Delta K_{\text{book}} = mgy_b - mgy_a \qquad (8.5)$$

Now, let us relate each side of this equation to the *system* of the book and the Earth. For the right-hand side,

$$mgy_b - mgy_a = -(mgy_a - mgy_b) = -(U_f - U_i) = -\Delta U_g$$

where U_g is the gravitational potential energy of the system. For the left-hand side of Equation 8.5, because the book is the only part of the system that is moving, we see that $\Delta K_{\text{book}} = \Delta K$, where K is the kinetic energy of the system. Thus, with each side of Equation 8.5 replaced with its system equivalent, the equation becomes

$$\Delta K = -\Delta U_g \qquad (8.6)$$

This equation can be manipulated to provide a very important general result for solving problems. First, we bring the change in potential energy to the left side of the equation:

$$\Delta K + \Delta U_g = 0 \qquad (8.7)$$

On the left, we have a sum of changes of the energy stored in the system. The right hand is zero because there are no transfers of energy across the boundary of the system—the book–Earth system is *isolated* from the environment.

We define the sum of kinetic and potential energies as **mechanical energy:**

$$E_{\text{mech}} = K + U_g$$

We will encounter other types of potential energy besides gravitational later in the text, so we can write the general form of the definition for mechanical energy without a subscript on U:

$$E_{\text{mech}} \equiv K + U \qquad (8.8)$$

where U represents the total of *all* types of potential energy.

Let us now write the changes in energy in Equation 8.7 explicitly:

$$(K_f - K_i) + (U_f - U_i) = 0$$

$$K_f + U_f = K_i + U_i \qquad (8.9)$$

For the gravitational situation that we have described, Equation 8.9 can be written as

$$\tfrac{1}{2}mv_f^2 + mgy_f = \tfrac{1}{2}mv_i^2 + mgy_i$$

As the book falls to the Earth, the book–Earth system loses potential energy and gains kinetic energy, such that the total of the two types of energy always remains constant.

Equation 8.9 is a statement of **conservation of mechanical energy** for an **isolated system.** An isolated system is one for which there are no energy transfers across the boundary. The energy in such a system is conserved—the sum of the kinetic and potential energies remains constant. (This statement assumes that no *nonconservative forces* act within the system; see Pitfall Prevention 8.2.)

 PITFALL PREVENTION

8.2 Conditions on Equation 8.6

Equation 8.6 is true for only one of two categories of forces. These forces are called *conservative forces,* as discussed in the next section.

Mechanical energy of a system

The mechanical energy of an isolated, friction-free system is conserved.

 PITFALL PREVENTION

8.3 Mechanical Energy in an Isolated System

Equation 8.9 is not the only statement we can make for an isolated system. This describes conservation of *mechanical energy only* for the isolated system. We will see shortly how to include internal energy. In later chapters, we will generate new conservation statements (and associated equations) related to other conserved quantities.

> **Quick Quiz 8.4** In an isolated system, which of the following is a correct statement of the quantity that is conserved? (a) kinetic energy (b) potential energy (c) kinetic energy plus potential energy (d) both kinetic energy and potential energy.

Active Figure 8.3 (Quick Quiz 8.6) Three identical balls are thrown with the same initial speed from the top of a building.

At the Active Figures link at http://www.pse6.com, you can throw balls at different angles from the top of the building and compare the trajectories and the speeds as the balls hit the ground.

Elastic Potential Energy

We are familiar now with gravitational potential energy; let us explore a second type of potential energy. Consider a system consisting of a block plus a spring, as shown in Figure 8.4. The force that the spring exerts on the block is given by $F_s = -kx$. In the previous chapter, we learned that the work done by an external applied force F_{app} on a system consisting of a block connected to the spring is given by Equation 7.12:

$$W_{F_{app}} = \tfrac{1}{2}kx_f{}^2 - \tfrac{1}{2}kx_i{}^2 \tag{8.10}$$

In this situation, the initial and final x coordinates of the block are measured from its equilibrium position, $x = 0$. Again (as in the gravitational case), we see that the work done on the system is equal to the difference between the initial and final values of an expression related to the configuration of the system. The **elastic potential energy** function associated with the block–spring system is defined by

Elastic potential energy stored in a spring

$$U_s \equiv \tfrac{1}{2}kx^2 \tag{8.11}$$

The elastic potential energy of the system can be thought of as the energy stored in the deformed spring (one that is either compressed or stretched from its equilibrium position). To visualize this, consider Figure 8.4, which shows a spring on a frictionless, horizontal surface. When a block is pushed against the spring (Fig. 8.4b) and the spring is compressed a distance x, the elastic potential energy stored in the spring is $\tfrac{1}{2}kx^2$.

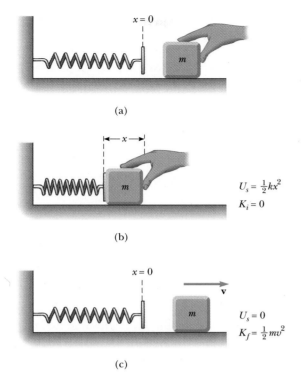

(a)

$U_s = \frac{1}{2}kx^2$
$K_i = 0$

(b)

$U_s = 0$
$K_f = \frac{1}{2}mv^2$

(c)

Active Figure 8.4 (a) An undeformed spring on a frictionless horizontal surface. (b) A block of mass m is pushed against the spring, compressing it a distance x. (c) When the block is released from rest, the elastic potential energy stored in the spring is transferred to the block in the form of kinetic energy.

At the Active Figures link at http://www.pse6.com, *you can compress the spring by varying amounts and observe the effect on the block's speed.*

When the block is released from rest, the spring exerts a force on the block and returns to its original length. The stored elastic potential energy is transformed into kinetic energy of the block (Fig. 8.4c).

The elastic potential energy stored in a spring is zero whenever the spring is undeformed ($x = 0$). Energy is stored in the spring only when the spring is either stretched or compressed. Furthermore, the elastic potential energy is a maximum when the spring has reached its maximum compression or extension (that is, when $|x|$ is a maximum). Finally, because the elastic potential energy is proportional to x^2, we see that U_s is always positive in a deformed spring.

Quick Quiz 8.7 A ball is connected to a light spring suspended vertically, as shown in Figure 8.5. When displaced downward from its equilibrium position and released, the ball oscillates up and down. In the system of *the ball, the spring, and the Earth,* what forms of energy are there during the motion? (a) kinetic and elastic potential (b) kinetic and gravitational potential (c) kinetic, elastic potential, and gravitational potential (d) elastic potential and gravitational potential.

Quick Quiz 8.8 Consider the situation in Quick Quiz 8.7 once again. In the system of *the ball and the spring,* what forms of energy are there during the motion? (a) kinetic and elastic potential (b) kinetic and gravitational potential (c) kinetic, elastic potential, and gravitational potential (d) elastic potential and gravitational potential.

Figure 8.5 (Quick Quizzes 8.7 and 8.8) A ball connected to a massless spring suspended vertically. What forms of potential energy are associated with the system when the ball is displaced downward?

PROBLEM-SOLVING HINTS

Isolated Systems–Conservation of Mechanical Energy

We can solve many problems in physics using the principle of conservation of mechanical energy. You should incorporate the following procedure when you apply this principle:

- Define your isolated system, which may include two or more interacting particles, as well as springs or other structures in which elastic potential energy can be stored. Be sure to include all components of the system that exert forces on each other. Identify the initial and final configurations of the system.

- Identify configurations for zero potential energy (both gravitational and spring). If there is more than one force acting within the system, write an expression for the potential energy associated with each force.

- If friction or air resistance is present, mechanical energy of the system is not conserved and the techniques of Section 8.4 must be employed.

- If mechanical energy of the system is conserved, you can write the total energy $E_i = K_i + U_i$ for the initial configuration. Then, write an expression for the total energy $E_f = K_f + U_f$ for the final configuration that is of interest. Because mechanical energy is conserved, you can equate the two total energies and solve for the quantity that is unknown.

Example 8.2   Ball in Free Fall

Interactive

A ball of mass m is dropped from a height h above the ground, as shown in Figure 8.6.

(A) Neglecting air resistance, determine the speed of the ball when it is at a height y above the ground.

Solution Figure 8.6 and our everyday experience with falling objects allow us to conceptualize the situation. While we can readily solve this problem with the techniques of Chapter 2,

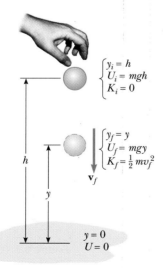

Figure 8.6 (Example 8.2) A ball is dropped from a height h above the ground. Initially, the total energy of the ball–Earth system is potential energy, equal to mgh relative to the ground. At the elevation y, the total energy is the sum of the kinetic and potential energies.

let us take an energy approach and categorize this as an energy problem for practice. To analyze the problem, we identify the system as the ball and the Earth. Because there is no air resistance and the system is isolated, we apply the principle of conservation of mechanical energy to the ball–Earth system.

At the instant the ball is released, its kinetic energy is $K_i = 0$ and the potential energy of the system is $U_i = mgh$. When the ball is at a distance y above the ground, its kinetic energy is $K_f = \frac{1}{2}mv_f{}^2$ and the potential energy relative to the ground is $U_f = mgy$. Applying Equation 8.9, we obtain

$$K_f + U_f = K_i + U_i$$

$$\tfrac{1}{2} mv_f{}^2 + mgy = 0 + mgh$$

$$v_f{}^2 = 2g(h - y)$$

$$v_f = \boxed{\sqrt{2g(h - y)}}$$

The speed is always positive. If we had been asked to find the ball's velocity, we would use the negative value of the square root as the y component to indicate the downward motion.

(B) Determine the speed of the ball at y if at the instant of release it already has an initial upward speed v_i at the initial altitude h.

Solution In this case, the initial energy includes kinetic energy equal to $\frac{1}{2}mv_i{}^2$ and Equation 8.9 gives

$$\tfrac{1}{2} mv_f{}^2 + mgy = \tfrac{1}{2} mv_i{}^2 + mgh$$

$$v_f{}^2 = v_i{}^2 + 2g(h - y)$$

$$v_f = \sqrt{v_i{}^2 + 2g(h - y)}$$

Note that this result is consistent with the expression $v_{yf}{}^2 = v_{yi}{}^2 - 2g(y_f - y_i)$ from kinematics, where $y_i = h$. Furthermore, this result is valid even if the initial velocity is at an angle to the horizontal (Quick Quiz 8.6) for two reasons: (1) energy is a scalar, and the kinetic energy depends only on the magnitude of the velocity; and (2) the change in the

gravitational potential energy depends only on the change in position in the vertical direction.

What If? What if the initial velocity v_i in part (B) were downward? How would this affect the speed of the ball at position y?

Answer We might be tempted to claim that throwing it downward would result in it having a higher speed at y than if we threw it upward. Conservation of mechanical energy, however, depends on kinetic and potential energies, which are scalars. Thus, the direction of the initial velocity vector has no bearing on the final speed.

 Compare the effect of upward, downward, and zero initial velocities at the Interactive Worked Example link at http://www.pse6.com.

Example 8.3   The Pendulum

A pendulum consists of a sphere of mass m attached to a light cord of length L, as shown in Figure 8.7. The sphere is released from rest at point Ⓐ when the cord makes an angle θ_A with the vertical, and the pivot at P is frictionless.

(A) Find the speed of the sphere when it is at the lowest point Ⓑ.

Solution The only force that does work on the sphere is the gravitational force. (The force applied by the cord is always perpendicular to each element of the displacement and hence does no work.) Because the pendulum–Earth system is isolated, the energy of the system is conserved. As the pendulum swings, continuous transformation between potential and kinetic energy occurs. At the instant the pendulum is released, the energy of the system is entirely potential energy. At point Ⓑ the pendulum has kinetic energy, but the system has lost some potential energy. At Ⓒ the system

has regained its initial potential energy, and the kinetic energy of the pendulum is again zero.

If we measure the y coordinates of the sphere from the center of rotation, then $y_A = -L \cos \theta_A$ and $y_B = -L$. Therefore, $U_A = -mgL \cos \theta_A$ and $U_B = -mgL$.

Applying the principle of conservation of mechanical energy to the system gives

$$K_B + U_B = K_A + U_A$$

$$\tfrac{1}{2}mv_B{}^2 - mgL = 0 - mgL \cos \theta_A$$

(1) $$v_B = \sqrt{2gL(1 - \cos \theta_A)}$$

(B) What is the tension T_B in the cord at Ⓑ?

Solution Because the tension force does no work, it does not enter into an energy equation, and we cannot determine the tension using the energy method. To find T_B, we can apply Newton's second law to the radial direction. First, recall that the centripetal acceleration of a particle moving in a circle is equal to v^2/r directed toward the center of rotation. Because $r = L$ in this example, Newton's second law gives

(2) $$\sum F_r = mg - T_B = ma_r = -m\frac{v_B{}^2}{L}$$

Substituting Equation (1) into Equation (2) gives the tension at point Ⓑ as a function of θ_A:

(3) $$T_B = mg + 2mg(1 - \cos \theta_A) = mg(3 - 2 \cos \theta_A)$$

From Equation (2) we see that the tension at Ⓑ is greater than the weight of the sphere. Furthermore, Equation (3) gives the expected result that $T_B = mg$ when the initial angle $\theta_A = 0$. Note also that part (A) of this example is categorized as an energy problem while part (B) is categorized as a Newton's second law problem.

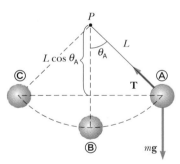

Figure 8.7 (Example 8.3) If the sphere is released from rest at the angle θ_A, it will never swing above this position during its motion. At the start of the motion, when the sphere is at position Ⓐ, the energy of the sphere–Earth system is entirely potential. This initial potential energy is transformed into kinetic energy when the sphere is at the lowest elevation Ⓑ. As the sphere continues to move along the arc, the energy again becomes entirely potential energy when the sphere is at Ⓒ.

Example 8.4   A Grand Entrance   **Interactive**

You are designing an apparatus to support an actor of mass 65 kg who is to "fly" down to the stage during the performance of a play. You attach the actor's harness to a 130-kg sandbag by means of a lightweight steel cable running

smoothly over two frictionless pulleys, as in Figure 8.8a. You need 3.0 m of cable between the harness and the nearest pulley so that the pulley can be hidden behind a curtain. For the apparatus to work successfully, the sandbag must

(a)

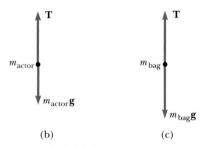

(b)   (c)

Figure 8.8 (Example 8.4) (a) An actor uses some clever staging to make his entrance. (b) Free-body diagram for the actor at the bottom of the circular path. (c) Free-body diagram for the sandbag.

never lift above the floor as the actor swings from above the stage to the floor. Let us call the initial angle that the actor's cable makes with the vertical θ. What is the maximum value θ can have before the sandbag lifts off the floor?

Solution We must use several concepts to solve this problem. To conceptualize, imagine what happens as the actor approaches the bottom of the swing. At the bottom, the cable is vertical and must support his weight as well as provide centripetal acceleration of his body in the upward direction. At this point, the tension in the cable is the highest and the sandbag is most likely to lift off the floor. Looking first at the swinging of the actor from the initial point to the lowest point, we categorize this as an energy problem involving an isolated system—the actor and the Earth. We use the principle of conservation of mechanical energy for the system to find the actor's speed as he arrives at the floor as a function of the initial angle θ and the radius R of the circular path through which he swings.

Applying conservation of mechanical energy to the actor–Earth system gives

$$K_f + U_f = K_i + U_i$$

$$(1) \qquad \tfrac{1}{2}m_{\text{actor}}v_f{}^2 + 0 = 0 + m_{\text{actor}}gy_i$$

where y_i is the initial height of the actor above the floor and v_f is the speed of the actor at the instant before he lands. (Note that $K_i = 0$ because he starts from rest and that $U_f = 0$ because we define the configuration of the actor at the floor as having a gravitational potential energy of zero.) From the geometry in Figure 8.8a, and noting that $y_f = 0$, we see that $y_i = R - R \cos \theta = R(1 - \cos \theta)$. Using this relationship in Equation (1), we obtain

$$(2) \qquad v_f{}^2 = 2gR(1 - \cos \theta)$$

Next, we focus on the instant the actor is at the lowest point. Because the tension in the cable is transferred as a force applied to the sandbag, we categorize the situation at this instant as a Newton's second law problem. We apply Newton's second law to the actor at the bottom of his path, using the free-body diagram in Figure 8.8b as a guide:

$$\sum F_y = T - m_{\text{actor}}g = m_{\text{actor}}\frac{v_f{}^2}{R}$$

$$(3) \qquad T = m_{\text{actor}}g + m_{\text{actor}}\frac{v_f{}^2}{R}$$

Finally, we note that the sandbag lifts off the floor when the upward force exerted on it by the cable exceeds the gravitational force acting on it; the normal force is zero when this happens. Thus, when we focus our attention on the sandbag, we categorize this part of the situation as another Newton's second law problem. A force T of the magnitude given by Equation (3) is transmitted by the cable to the sandbag. If the sandbag is to be just lifted off the floor, the normal force on it becomes zero and we require that $T = m_{\text{bag}}g$, as in Figure 8.8c. Using this condition together with Equations (2) and (3), we find that

$$m_{\text{bag}}g = m_{\text{actor}}g + m_{\text{actor}}\frac{2gR(1 - \cos \theta)}{R}$$

Solving for $\cos \theta$ and substituting in the given parameters, we obtain

$$\cos \theta = \frac{3m_{\text{actor}} - m_{\text{bag}}}{2m_{\text{actor}}} = \frac{3(65 \text{ kg}) - 130 \text{ kg}}{2(65 \text{ kg})} = 0.50$$

$$\theta = 60°$$

Note that we had to combine techniques from different areas of our study—energy and Newton's second law. Furthermore, we see that the length R of the cable from the actor's harness to the leftmost pulley did not appear in the final algebraic equation. Thus, the final answer is independent of R.

What If? What if a stagehand locates the sandbag so that the cable from the sandbag to the right-hand pulley in Figure 8.8a is not vertical but makes an angle ϕ with the vertical? If the actor swings from the angle found in the solution above, will the sandbag lift off the floor? Assume that the length R remains the same.

Answer In this situation, the gravitational force acting on the sandbag is no longer parallel to the cable. Thus, only a component of the force in the cable acts against the gravitational force, and the vertical resultant of this force component and the gravitational force should be downward. As a result, there should be a nonzero normal force to balance this resultant, and the sandbag should *not* lift off the floor.

If the sandbag is in equilibrium in the y direction and the normal force from the floor goes to zero, Newton's second law gives us $T\cos\phi = m_{bag}g$. In this case, Equation (3) gives

$$\frac{m_{bag}g}{\cos\phi} = m_{actor}g + m_{actor}\frac{v_f^2}{R}$$

Substituting for v_f from Equation (2) gives

$$\frac{m_{bag}g}{\cos\phi} = m_{actor}g + m_{actor}\frac{2gR(1-\cos\theta)}{R}$$

Solving for $\cos\theta$, we have

$$(4) \qquad \cos\theta = \frac{3m_{actor} - \dfrac{m_{bag}}{\cos\phi}}{2m_{actor}}$$

For $\phi = 0$, which is the situation in Figure 8.8a, $\cos\phi = 1$. For nonzero values of ϕ, the term $\cos\phi$ is smaller than 1.

This makes the numerator of the fraction in Equation (4) smaller, which makes the angle θ larger. Thus, the sandbag remains on the floor if the actor swings from a larger angle. If he swings from the original angle, the sandbag remains on the floor. For example, suppose $\phi = 10°$. Then, Equation (4) gives

$$\cos\theta = \frac{3(65\text{ kg}) - \dfrac{130\text{ kg}}{\cos 10°}}{2(65\text{ kg})} = 0.48 \longrightarrow \theta = 61°$$

Thus, if he swings from 60°, he is swinging from an angle below the new maximum allowed angle, and the sandbag remains on the floor.

One factor we have not addressed is the friction force between the sandbag and the floor. If this is not large enough, the sandbag may break free and start to slide horizontally as the actor reaches some point in his swing. This will cause the length R to increase, and the actor may have a frightening moment as he begins to drop in addition to swinging!

 Let the actor fly or crash without injury to people at the Interactive Worked Example link at http://www.pse6.com. You may choose to include the effect of friction between the sandbag and the floor.

Example 8.5 The Spring-Loaded Popgun

The launching mechanism of a toy gun consists of a spring of unknown spring constant (Fig. 8.9a). When the spring is compressed 0.120 m, the gun, when fired vertically, is able to launch a 35.0-g projectile to a maximum height of 20.0 m above the position of the projectile before firing.

(A) Neglecting all resistive forces, determine the spring constant.

Solution Because the projectile starts from rest, its initial kinetic energy is zero. If we take the zero configuration for the gravitational potential energy of the projectile–spring–Earth system to be when the projectile is at the lowest position x_A, then the initial gravitational potential energy of the system also is zero. The mechanical energy of this system is conserved because the system is isolated.

Initially, the only mechanical energy in the system is the elastic potential energy stored in the spring of the gun, $U_{sA} = \frac{1}{2}kx^2$, where the compression of the spring is $x = 0.120$ m. The projectile rises to a maximum height $x_C = h = 20.0$ m, and so the final gravitational potential energy of the system when the projectile reaches its peak is mgh. The final kinetic energy of the projectile is zero, and the final elastic potential energy stored in the spring is zero. Because the mechanical energy of the system is conserved, we find that

$$E_C = E_A$$

$$K_C + U_{gC} + U_{sC} = K_A + U_{gA} + U_{sA}$$

$$0 + mgh + 0 = 0 + 0 + \tfrac{1}{2}kx^2$$

$$k = \frac{2mgh}{x^2} = \frac{2(0.035\,0\text{ kg})(9.80\text{ m/s}^2)(20.0\text{ m})}{(0.120\text{ m})^2}$$

$$= \boxed{953\text{ N/m}}$$

© ○ $x_C = 20.0$ m

$x_B = 0.120$ m

$x_A = 0$

(a)　　　　(b)

Figure 8.9 (Example 8.5) A spring-loaded popgun.

(B) Find the speed of the projectile as it moves through the equilibrium position of the spring (where $x_B = 0.120\,\text{m}$) as shown in Figure 8.9b.

Solution As already noted, the only mechanical energy in the system at Ⓐ is the elastic potential energy $\frac{1}{2}kx^2$. The total energy of the system as the projectile moves through the equilibrium position of the spring includes the kinetic energy of the projectile $\frac{1}{2}mv_B{}^2$ and the gravitational potential energy mgx_B of the system. Hence, the principle of conservation of mechanical energy in this case gives

$$E_B = E_A$$

$$K_B + U_{gB} + U_{sB} = K_A + U_{gA} + U_{sA}$$

$$\tfrac{1}{2}mv_B{}^2 + mgx_B + 0 = 0 + 0 + \tfrac{1}{2}kx^2$$

Solving for v_B gives

$$v_B = \sqrt{\frac{kx^2}{m} - 2gx_B}$$

$$= \sqrt{\frac{(953\ \text{N/m})(0.120\ \text{m})^2}{(0.0350\ \text{kg})} - 2(9.80\ \text{m/s}^2)(0.120\ \text{m})}$$

$$= \boxed{19.7\ \text{m/s}}$$

8.3 Conservative and Nonconservative Forces

As an object moves downward near the surface of the Earth, the work done by the gravitational force on the object does not depend on whether it falls vertically or slides down a sloping incline. All that matters is the change in the object's elevation. However, the energy loss due to friction on that incline depends on the distance the object slides. In other words, the path makes no difference when we consider the work done by the gravitational force, but it does make a difference when we consider the energy loss due to friction forces. We can use this varying dependence on path to classify forces as either conservative or nonconservative.

Of the two forces just mentioned, the gravitational force is conservative and the friction force is nonconservative.

Conservative Forces

Conservative forces have these two equivalent properties:

Properties of a conservative force

1. The work done by a conservative force on a particle moving between any two points is independent of the path taken by the particle.

2. The work done by a conservative force on a particle moving through any closed path is zero. (A closed path is one in which the beginning and end points are identical.)

The gravitational force is one example of a conservative force, and the force that a spring exerts on any object attached to the spring is another. As we learned in the preceding section, the work done by the gravitational force on an object moving between any two points near the Earth's surface is $W_g = mgy_i - mgy_f$. From this equation, we see that W_g depends only on the initial and final y coordinates of the object and hence is independent of the path. Furthermore, W_g is zero when the object moves over any closed path (where $y_i = y_f$).

For the case of the object–spring system, the work W_s done by the spring force is given by $W_s = \frac{1}{2}kx_i{}^2 - \frac{1}{2}kx_f{}^2$ (Eq. 7.11). Again, we see that the spring force is conservative because W_s depends only on the initial and final x coordinates of the object and is zero for any closed path.

We can associate a potential energy for a system with any conservative force acting between members of the system and can do this only for conservative forces. In the previous section, the potential energy associated with the gravitational force was defined as $U_g \equiv mgy$. In general, the work W_c done by a conservative force on an object that is a member of a system as the object moves from one position to another is equal to the initial value of the potential energy of the system minus the final value:

$$W_c = U_i - U_f = -\Delta U \tag{8.12}$$

 PITFALL PREVENTION

8.4 Similar Equation Warning

Compare Equation 8.12 to Equation 8.3. These equations are similar except for the negative sign, which is a common source of confusion. Equation 8.3 tells us that the work done *by an outside agent* on a system causes an increase in the potential energy of the system (with no change in the kinetic or internal energy). Equation 8.12 states that work done *on a component of a system by a conservative force internal to an isolated system* causes a decrease in the potential energy of the system (with a corresponding increase in kinetic energy).

This equation should look familiar to you. It is the general form of the equation for work done by the gravitational force (Eq. 8.4) as an object moves relative to the Earth and that for the work done by the spring force (Eq. 7.11) as the extension of the spring changes.

Nonconservative Forces

A force is **nonconservative** if it does not satisfy properties 1 and 2 for conservative forces. Nonconservative forces acting within a system cause a *change* in the mechanical energy E_{mech} of the system. We have defined mechanical energy as the sum of the kinetic and all potential energies. For example, if a book is sent sliding on a horizontal surface that is not frictionless, the force of kinetic friction reduces the book's kinetic energy. As the book slows down, its kinetic energy decreases. As a result of the friction force, the temperatures of the book and surface increase. The type of energy associated with temperature is internal energy, which we introduced in Chapter 7. Only part of the book's kinetic energy is transformed to internal energy in the book. The rest appears as internal energy in the surface. (When you trip and fall while running across a gymnasium floor, not only does the skin on your knees warm up, so does the floor!) Because the force of kinetic friction transforms the mechanical energy of a system into internal energy, it is a nonconservative force.

As an example of the path dependence of the work, consider Figure 8.10. Suppose you displace a book between two points on a table. If the book is displaced in a straight line along the blue path between points Ⓐ and Ⓑ in Figure 8.10, you do a certain amount of work against the kinetic friction force to keep the book moving at a constant speed. Now, imagine that you push the book along the brown semicircular path in Figure 8.10. You perform more work against friction along this longer path than along the straight path. The work done depends on the path, so the friction force cannot be conservative.

Figure 8.10 The work done against the force of kinetic friction depends on the path taken as the book is moved from Ⓐ to Ⓑ. The work is greater along the red path than along the blue path.

8.4 Changes in Mechanical Energy for Nonconservative Forces

As we have seen, if the forces acting on objects within a system are conservative, then the mechanical energy of the system is conserved. However, if some of the forces acting on objects within the system are not conservative, then the mechanical energy of the system changes.

Consider the book sliding across the surface in the preceding section. As the book moves through a distance d, the only force that does work on it is the force of kinetic friction. This force causes a decrease in the kinetic energy of the book. This decrease was calculated in Chapter 7, leading to Equation 7.20, which we repeat here:

$$\Delta K = -f_k d \tag{8.13}$$

Suppose, however, that the book is part of a system that also exhibits a change in potential energy. In this case, $-f_k d$ is the amount by which the *mechanical* energy of the system changes because of the force of kinetic friction. For example, if the book moves on an incline that is not frictionless, there is a change in both the kinetic energy and the gravitational potential energy of the book–Earth system. Consequently,

$$\Delta E_{mech} = \Delta K + \Delta U_g = -f_k d$$

In general, if a friction force acts within a system,

$$\Delta E_{mech} = \Delta K + \Delta U = -f_k d \tag{8.14}$$

Change in mechanical energy of a system due to friction within the system

where ΔU is the change in *all* forms of potential energy. Notice that Equation 8.14 reduces to Equation 8.9 if the friction force is zero.

Quick Quiz 8.9 A block of mass m is projected across a horizontal surface with an initial speed v. It slides until it stops due to the friction force between the block and the surface. The same block is now projected across the horizontal surface with an initial speed $2v$. When the block has come to rest, how does the distance from the projection point compare to that in the first case? (a) It is the same. (b) It is twice as large. (c) It is four times as large. (d) The relationship cannot be determined.

Quick Quiz 8.10 A block of mass m is projected across a horizontal surface with an initial speed v. It slides until it stops due to the friction force between the block and the surface. The surface is now tilted at 30°, and the block is projected up the surface with the same initial speed v. Assume that the friction force remains the same as when the block was sliding on the horizontal surface. When the block comes to rest momentarily, how does the decrease in mechanical energy of the block–surface–Earth system compare to that when the block slid over the horizontal surface? (a) It is the same. (b) It is larger. (c) It is smaller. (d) The relationship cannot be determined.

PROBLEM-SOLVING HINTS

Isolated Systems–Nonconservative Forces

You should incorporate the following procedure when you apply energy methods to a system in which nonconservative forces are acting:

- Follow the procedure in the first three bullets of the Problem-Solving Hints in Section 8.2. If nonconservative forces act within the system, the third bullet should tell you to use the techniques of this section.

- Write expressions for the total initial and total final mechanical energies of the system. The difference between the total final mechanical energy and the total initial mechanical energy equals the change in mechanical energy of the system due to friction.

Example 8.6 Crate Sliding Down a Ramp

A 3.00-kg crate slides down a ramp. The ramp is 1.00 m in length and inclined at an angle of 30.0°, as shown in Figure 8.11. The crate starts from rest at the top, experiences a constant friction force of magnitude 5.00 N, and continues to move a short distance on the horizontal floor after it leaves the ramp. Use energy methods to determine the speed of the crate at the bottom of the ramp.

Solution Because $v_i = 0$, the initial kinetic energy of the crate–Earth system when the crate is at the top of the ramp is zero. If the y coordinate is measured from the bottom of the ramp (the final position of the crate, for which the gravitational potential energy of the system is zero) with the upward direction being positive, then $y_i = 0.500$ m. Therefore, the total mechanical energy of the system when the crate is at the top is all potential energy:

$$E_i = K_i + U_i = 0 + U_i = mgy_i$$
$$= (3.00 \text{ kg})(9.80 \text{ m/s}^2)(0.500 \text{ m}) = 14.7 \text{ J}$$

When the crate reaches the bottom of the ramp, the potential energy of the system is zero because the elevation of

Figure 8.11 (Example 8.6) A crate slides down a ramp under the influence of gravity. The potential energy decreases while the kinetic energy increases.

the crate is $y_f = 0$. Therefore, the total mechanical energy of the system when the crate reaches the bottom is all kinetic energy:

$$E_f = K_f + U_f = \tfrac{1}{2} m v_f{}^2 + 0$$

We cannot say that $E_i = E_f$ because a nonconservative force reduces the mechanical energy of the system. In this case, Equation 8.14 gives $\Delta E_{\text{mech}} = -f_k d$, where d is the distance the crate moves along the ramp. (Remember that the forces normal to the ramp do no work on the crate because they are perpendicular to the displacement.) With $f_k = 5.00\,\text{N}$ and $d = 1.00\,\text{m}$, we have

$$\text{(1)} \qquad -f_k d = (-5.00\,\text{N})(1.00\,\text{m}) = -5.00\,\text{J}$$

Applying Equation 8.14 gives

$$E_f - E_i = \tfrac{1}{2} m v_f{}^2 - m g y_i = -f_k d$$
$$\text{(2)} \qquad \tfrac{1}{2} m v_f{}^2 = 14.7\,\text{J} - 5.00\,\text{J} = 9.70\,\text{J}$$
$$v_f{}^2 = \frac{19.4\,\text{J}}{3.00\,\text{kg}} = 6.47\,\text{m}^2/\text{s}^2$$
$$v_f = \boxed{2.54\,\text{m/s}}$$

What If? A cautious worker decides that the speed of the crate when it arrives at the bottom of the ramp may be so large

that its contents may be damaged. Therefore, he replaces the ramp with a longer one such that the new ramp makes an angle of 25° with the ground. Does this new ramp reduce the speed of the crate as it reaches the ground?

Answer Because the ramp is longer, the friction force will act over a longer distance and transform more of the mechanical energy into internal energy. This reduces the kinetic energy of the crate, and we expect a lower speed as it reaches the ground.

We can find the length d of the new ramp as follows:

$$\sin 25° = \frac{0.500\,\text{m}}{d} \longrightarrow d = \frac{0.500\,\text{m}}{\sin 25°} = 1.18\,\text{m}$$

Now, Equation (1) becomes

$$-f_k d = (-5.00\,\text{N})(1.18\,\text{m}) = -5.90\,\text{J}$$

and Equation (2) becomes

$$\tfrac{1}{2} m v_f{}^2 = 14.7\,\text{J} - 5.90\,\text{J} = 8.80\,\text{J}$$

leading to

$$v_f = 2.42\,\text{m/s}$$

The final speed is indeed lower than in the higher-angle case.

Example 8.7   Motion on a Curved Track

A child of mass m rides on an irregularly curved slide of height $h = 2.00\,\text{m}$, as shown in Figure 8.12. The child starts from rest at the top.

(A) Determine his speed at the bottom, assuming no friction is present.

Figure 8.12 (Example 8.7) If the slide is frictionless, the speed of the child at the bottom depends only on the height of the slide.

Solution Although you have no experience on totally frictionless surfaces, you can conceptualize that your speed at the bottom of a frictionless ramp would be greater than in the situation in which friction acts. If we tried to solve this problem with Newton's laws, we would have a difficult time because the acceleration of the child continuously varies in direction due to the irregular shape of the slide. The child–Earth system is isolated and frictionless, however, so we can categorize this as a conservation of energy problem and search for a solution using the energy approach. (Note that the normal force **n** does no work on the child because this force is always perpendicular to each element of the displacement.) To analyze the situation, we measure the y coordinate in the upward direction from the bottom of the slide so that $y_i = h$, $y_f = 0$, and we obtain

$$K_f + U_f = K_i + U_i$$
$$\tfrac{1}{2} m v_f{}^2 + 0 = 0 + m g h$$
$$v_f = \sqrt{2gh}$$

Note that the result is the same as it would be had the child fallen vertically through a distance h! In this example, $h = 2.00\,\text{m}$, giving

$$v_f = \sqrt{2gh} = \sqrt{2(9.80\,\text{m/s}^2)(2.00\,\text{m})} = \boxed{6.26\,\text{m/s}}$$

(B) If a force of kinetic friction acts on the child, how much mechanical energy does the system lose? Assume that $v_f = 3.00$ m/s and $m = 20.0$ kg.

Solution We categorize this case, with friction, as a problem in which a nonconservative force acts. Hence, mechanical energy is not conserved, and we must use Equation 8.14 to find the loss of mechanical energy due to friction:

$$\Delta E_{mech} = (K_f + U_f) - (K_i + U_i)$$
$$= (\tfrac{1}{2} mv_f^2 + 0) - (0 + mgh) = \tfrac{1}{2} mv_f^2 - mgh$$
$$= \tfrac{1}{2}(20.0 \text{ kg})(3.00 \text{ m/s})^2$$
$$\quad - (20.0 \text{ kg})(9.80 \text{ m/s}^2)(2.00 \text{ m})$$
$$= \boxed{-302 \text{ J}}$$

Again, ΔE_{mech} is negative because friction is reducing the mechanical energy of the system. (The final mechanical energy is less than the initial mechanical energy.)

What If? Suppose you were asked to find the coefficient of friction μ_k for the child on the slide. Could you do this?

Answer We can argue that the same final speed could be obtained by having the child travel down a short slide with large friction or a long slide with less friction. Thus, there does not seem to be enough information in the problem to determine the coefficient of friction.

The energy loss of 302 J must be equal to the product of the friction force and the length of the slide:

$$-f_k d = -302 \text{ J}$$

We can also argue that the friction force can be expressed as $\mu_k n$, where n is the magnitude of the normal force. Thus,

$$\mu_k n d = 302 \text{ J}$$

If we try to evaluate the coefficient of friction from this relationship, we run into two problems. First, there is no single value of the normal force n unless the angle of the slide relative to the horizontal remains fixed. Even if the angle were fixed, we do not know its value. The second problem is that we do not have information about the length d of the slide. Thus, we cannot find the coefficient of friction from the information given.

Example 8.8   Let's Go Skiing!

A skier starts from rest at the top of a frictionless incline of height 20.0 m, as shown in Figure 8.13. At the bottom of the incline, she encounters a horizontal surface where the coefficient of kinetic friction between the skis and the snow is 0.210. How far does she travel on the horizontal surface before coming to rest, if she simply coasts to a stop?

Solution The system is the skier plus the Earth, and we choose as our configuration of zero potential energy that in

which the skier is at the bottom of the incline. While the skier is on the frictionless incline, the mechanical energy of the system remains constant, and we find, as we did in Example 8.7, that

$$v_B = \sqrt{2gh} = \sqrt{2(9.80 \text{ m/s}^2)(20.0 \text{ m})} = 19.8 \text{ m/s}$$

Now we apply Equation 8.14 as the skier moves along the rough horizontal surface from Ⓑ to Ⓒ. The change in mechanical energy along the horizontal surface is

Figure 8.13 (Example 8.8) The skier slides down the slope and onto a level surface, stopping after a distance d from the bottom of the hill.

$\Delta E_{mech} = -f_k d$, where d is the horizontal distance traveled by the skier.

To find the distance the skier travels before coming to rest, we take $K_C = 0$. With $v_B = 19.8$ m/s and the friction force given by $f_k = \mu_k n = \mu_k mg$, we obtain

$$\Delta E_{mech} = E_C - E_B = -\mu_k mgd$$

$$(K_C + U_C) - (K_B + U_B) = (0 + 0) - (\tfrac{1}{2}mv_B^2 + 0)$$

$$= -\mu_k mgd$$

$$d = \frac{v_B^2}{2\mu_k g} = \frac{(19.8 \text{ m/s})^2}{2(0.210)(9.80 \text{ m/s}^2)} = \boxed{95.2 \text{ m}}$$

Example 8.9   Block–Spring Collision

A block having a mass of 0.80 kg is given an initial velocity $v_A = 1.2$ m/s to the right and collides with a spring of negligible mass and force constant $k = 50$ N/m, as shown in Figure 8.14.

(A) Assuming the surface to be frictionless, calculate the maximum compression of the spring after the collision.

Solution Our system in this example consists of the block and spring. All motion takes place in a horizontal plane, so we do not need to consider changes in gravitational potential energy. Before the collision, when the block is at Ⓐ, it has kinetic energy and the spring is uncompressed, so the elastic potential energy stored in the spring is zero. Thus, the total mechanical energy of the system before the collision is just $\tfrac{1}{2}mv_A^2$. After the collision, when the block is at Ⓒ, the spring is fully compressed; now the block is at rest and so has zero kinetic energy, while the energy stored in the spring has its maximum value

$\tfrac{1}{2}kx^2 = \tfrac{1}{2}kx_{max}^2$, where the origin of coordinates $x = 0$ is chosen to be the equilibrium position of the spring and x_{max} is the maximum compression of the spring, which in this case happens to be x_C. The total mechanical energy of the system is conserved because no nonconservative forces act on objects within the system.

Because the mechanical energy of the system is conserved, the kinetic energy of the block before the collision equals the maximum potential energy stored in the fully compressed spring:

$$E_C = E_A$$

$$K_C + U_{sC} = K_A + U_{sA}$$

$$0 + \tfrac{1}{2}kx_{max}^2 = \tfrac{1}{2}mv_A^2 + 0$$

$$x_{max} = \sqrt{\frac{m}{k}}\, v_A = \sqrt{\frac{0.80 \text{ kg}}{50 \text{ N/m}}}\,(1.2 \text{ m/s})$$

$$= \boxed{0.15 \text{ m}}$$

(B) Suppose a constant force of kinetic friction acts between the block and the surface, with $\mu_k = 0.50$. If the speed of the block at the moment it collides with the spring is $v_A = 1.2$ m/s, what is the maximum compression x_C in the spring?

Solution In this case, the mechanical energy of the system is *not* conserved because a friction force acts on the block. The magnitude of the friction force is

$$f_k = \mu_k n = \mu_k mg = 0.50(0.80 \text{ kg})(9.80 \text{ m/s}^2) = 3.92 \text{ N}$$

Therefore, the change in the mechanical energy of the system due to friction as the block is displaced from the equilibrium position of the spring (where we have set our origin) to x_C is

$$\Delta E_{mech} = -f_k x_C = (-3.92 x_C)$$

Substituting this into Equation 8.14 gives

$$\Delta E_{mech} = E_f - E_i = (0 + \tfrac{1}{2}kx_C^2) - (\tfrac{1}{2}mv_A^2 + 0) = -f_k x_C$$

$$\tfrac{1}{2}(50)x_C^2 - \tfrac{1}{2}(0.80)(1.2)^2 = -3.92 x_C$$

$$25x_C^2 + 3.92 x_C - 0.576 = 0$$

Solving the quadratic equation for x_C gives $x_C = 0.092$ m and $x_C = -0.25$ m. The physically meaningful root is $x_C = 0.092$ m. The negative root does not apply to this situation because the block must be to the right of the origin (positive value of x) when it comes to rest. Note that the value of 0.092 m is less than the distance obtained in the frictionless case of part (A). This result is what we expect because friction retards the motion of the system.

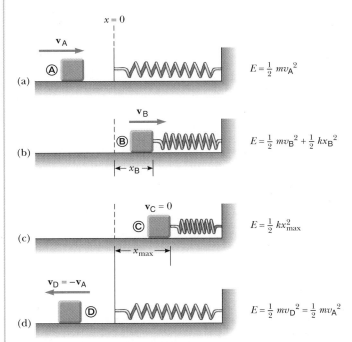

$x = 0$

v$_A$

(a)   Ⓐ $E = \tfrac{1}{2}mv_A^2$

v$_B$

(b)   Ⓑ $E = \tfrac{1}{2}mv_B^2 + \tfrac{1}{2}kx_B^2$

$\leftarrow x_B \rightarrow$

v$_C$ = 0

(c)   Ⓒ $E = \tfrac{1}{2}kx_{max}^2$

$\leftarrow x_{max} \rightarrow$

v$_D$ = -**v**$_A$

(d)   Ⓓ $E = \tfrac{1}{2}mv_D^2 = \tfrac{1}{2}mv_A^2$

Figure 8.14 (Example 8.9) A block sliding on a smooth, horizontal surface collides with a light spring. (a) Initially the mechanical energy is all kinetic energy. (b) The mechanical energy is the sum of the kinetic energy of the block and the elastic potential energy in the spring. (c) The energy is entirely potential energy. (d) The energy is transformed back to the kinetic energy of the block. The total energy of the system remains constant throughout the motion.

Example 8.10 Connected Blocks in Motion

Two blocks are connected by a light string that passes over a frictionless pulley, as shown in Figure 8.15. The block of mass m_1 lies on a horizontal surface and is connected to a spring of force constant k. The system is released from rest when the spring is unstretched. If the hanging block of mass m_2 falls a distance h before coming to rest, calculate the coefficient of kinetic friction between the block of mass m_1 and the surface.

Solution The key word *rest* appears twice in the problem statement. This suggests that the configurations associated with rest are good candidates for the initial and final configurations because the kinetic energy of the system is zero for these configurations. (Also note that because we are concerned only with the beginning and ending points of the motion, we do not need to label events with circled letters as we did in the previous two examples. Simply using i and f is sufficient to keep track of the situation.) In this situation, the system consists of the two blocks, the spring, and the Earth. We need to consider two forms of potential energy: gravitational and elastic. Because the initial and final kinetic energies of the system are zero, $\Delta K = 0$, and we can write

$$(1) \qquad \Delta E_{\text{mech}} = \Delta U_g + \Delta U_s$$

where $\Delta U_g = U_{gf} - U_{gi}$ is the change in the system's gravitational potential energy and $\Delta U_s = U_{sf} - U_{si}$ is the change in the system's elastic potential energy. As the hanging block falls a distance h, the horizontally moving block moves the same distance h to the right. Therefore, using Equation 8.14, we find that the loss in mechanical energy in the system due to friction between the horizontally sliding block and the surface is

$$(2) \qquad \Delta E_{\text{mech}} = -f_k h = -\mu_k m_1 g h$$

The change in the gravitational potential energy of the system is associated with only the falling block because the vertical coordinate of the horizontally sliding block does not change. Therefore, we obtain

$$(3) \qquad \Delta U_g = U_{gf} - U_{gi} = 0 - m_2 g h$$

where the coordinates have been measured from the lowest position of the falling block.

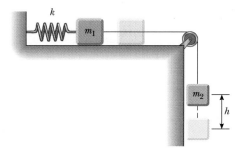

Figure 8.15 (Example 8.10) As the hanging block moves from its highest elevation to its lowest, the system loses gravitational potential energy but gains elastic potential energy in the spring. Some mechanical energy is lost because of friction between the sliding block and the surface.

The change in the elastic potential energy of the system is that stored in the spring:

$$(4) \qquad \Delta U_s = U_{sf} - U_{si} = \tfrac{1}{2} k h^2 - 0$$

Substituting Equations (2), (3), and (4) into Equation (1) gives

$$-\mu_k m_1 g h = -m_2 g h + \tfrac{1}{2} k h^2$$

$$\mu_k = \frac{m_2 g - \tfrac{1}{2} k h}{m_1 g}$$

This setup represents a way of measuring the coefficient of kinetic friction between an object and some surface. As you can see from the problem, sometimes it is easier to work with the changes in the various types of energy rather than the actual values. For example, if we wanted to calculate the numerical value of the gravitational potential energy associated with the horizontally sliding block, we would need to specify the height of the horizontal surface relative to the lowest position of the falling block. Fortunately, this is not necessary because the gravitational potential energy associated with the first block does not change.

8.5 Relationship Between Conservative Forces and Potential Energy

In an earlier section we found that the work done on a member of a system by a conservative force between the members does not depend on the path taken by the moving member. The work depends only on the initial and final coordinates. As a consequence, we can define a **potential energy function U** such that the work done by a conservative force equals the decrease in the potential energy of the system. Let us imagine a system of particles in which the configuration changes due to the motion of one particle along the x axis. The work done by a conservative force $\mathbf{F}$ as a particle moves along the x axis is[2]

---

[2] For a general displacement, the work done in two or three dimensions also equals $-\Delta U$, where $U = U(x, y, z)$. We write this formally as $W = \int_i^f \mathbf{F} \cdot d\mathbf{r} = U_i - U_f$.

$$W_c = \int_{x_i}^{x_f} F_x \, dx = -\Delta U \qquad (8.15)$$

where F_x is the component of $\mathbf{F}$ in the direction of the displacement. That is, the work done by a conservative force acting between members of a system equals the negative of the change in the potential energy associated with that force when the configuration of the system changes, where the change in the potential energy is defined as $\Delta U = U_f - U_i$. We can also express Equation 8.15 as

$$\Delta U = U_f - U_i = -\int_{x_i}^{x_f} F_x \, dx \qquad (8.16)$$

Therefore, ΔU is negative when F_x and dx are in the same direction, as when an object is lowered in a gravitational field or when a spring pushes an object toward equilibrium.

The term *potential energy* implies that the system has the potential, or capability, of either gaining kinetic energy or doing work when it is released under the influence of a conservative force exerted on an object by some other member of the system. It is often convenient to establish some particular location x_i of one member of a system as representing a reference configuration and measure all potential energy differences with respect to it. We can then define the potential energy function as

$$U_f(x) = -\int_{x_i}^{x_f} F_x \, dx + U_i \qquad (8.17)$$

The value of U_i is often taken to be zero for the reference configuration. It really does not matter what value we assign to U_i because any nonzero value merely shifts $U_f(x)$ by a constant amount and only the *change* in potential energy is physically meaningful.

If the conservative force is known as a function of position, we can use Equation 8.17 to calculate the change in potential energy of a system as an object within the system moves from x_i to x_f.

If the point of application of the force undergoes an infinitesimal displacement dx, we can express the infinitesimal change in the potential energy of the system dU as

$$dU = -F_x \, dx$$

Therefore, the conservative force is related to the potential energy function through the relationship[3]

$$F_x = -\frac{dU}{dx} \qquad (8.18)$$

Relation of force between members of a system to the potential energy of the system

That is, **the x component of a conservative force acting on an object within a system equals the negative derivative of the potential energy of the system with respect to x.**

We can easily check this relationship for the two examples already discussed. In the case of the deformed spring, $U_s = \frac{1}{2}kx^2$, and therefore

$$F_s = -\frac{dU_s}{dx} = -\frac{d}{dx}\left(\tfrac{1}{2}kx^2\right) = -kx$$

which corresponds to the restoring force in the spring (Hooke's law). Because the gravitational potential energy function is $U_g = mgy$, it follows from Equation 8.18 that $F_g = -mg$ when we differentiate U_g with respect to y instead of x.

We now see that U is an important function because a conservative force can be derived from it. Furthermore, Equation 8.18 should clarify the fact that adding a constant to the potential energy is unimportant because the derivative of a constant is zero.

---

[3]   In three dimensions, the expression is, $\mathbf{F} = -\frac{\partial U}{\partial x}\,\hat{\mathbf{i}} - \frac{\partial U}{\partial y}\,\hat{\mathbf{j}} - \frac{\partial U}{\partial z}\,\hat{\mathbf{k}}$ where $\frac{\partial U}{\partial x}$ etc. are partial derivatives. In the language of vector calculus, $\mathbf{F}$ equals the negative of the *gradient* of the scalar quantity $U(x, y, z)$.

Quick Quiz 8.11 What does the slope of a graph of $U(x)$ versus x represent? (a) the magnitude of the force on the object (b) the negative of the magnitude of the force on the object (c) the x component of the force on the object (d) the negative of the x component of the force on the object.

8.6 Energy Diagrams and Equilibrium of a System

The motion of a system can often be understood qualitatively through a graph of its potential energy versus the position of a member of the system. Consider the potential energy function for a block–spring system, given by $U_s = \frac{1}{2}kx^2$. This function is plotted versus x in Figure 8.16a. (A common mistake is to think that potential energy on the graph represents height. This is clearly not the case here, where the block is only moving horizontally.) The force F_s exerted by the spring on the block is related to U_s through Equation 8.18:

$$F_s = -\frac{dU_s}{dx} = -kx$$

As we saw in Quick Quiz 8.11, the x component of the force is equal to the negative of the slope of the U-versus-x curve. When the block is placed at rest at the equilibrium position of the spring ($x = 0$), where $F_s = 0$, it will remain there unless some external force F_{ext} acts on it. If this external force stretches the spring from equilibrium, x is positive and the slope dU/dx is positive; therefore, the force F_s exerted by the spring is negative and the block accelerates back toward $x = 0$ when released. If the external force compresses the spring, then x is negative and the slope is negative; therefore, F_s is positive and again the mass accelerates toward $x = 0$ upon release.

From this analysis, we conclude that the $x = 0$ position for a block–spring system is one of **stable equilibrium.** That is, any movement away from this position results in a force directed back toward $x = 0$. In general, **configurations of stable equilibrium correspond to those for which $U(x)$ is a minimum.**

From Figure 8.16 we see that if the block is given an initial displacement x_{max} and is released from rest, its total energy initially is the potential energy $\frac{1}{2}kx_{max}^2$ stored in the

Stable equilibrium

At the Active Figures link at http://www.pse6.com, you can observe the block oscillate between its turning points and trace the corresponding points on the potential energy curve for varying values of k.

(a)

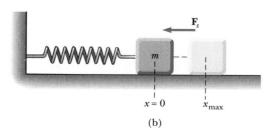

(b)

Active Figure 8.16 (a) Potential energy as a function of x for the frictionless block–spring system shown in (b). The block oscillates between the turning points, which have the coordinates $x = \pm x_{max}$. Note that the restoring force exerted by the spring always acts toward $x = 0$, the position of stable equilibrium.

spring. As the block starts to move, the system acquires kinetic energy and loses an equal amount of potential energy. Because the total energy of the system must remain constant, the block oscillates (moves back and forth) between the two points $x = -x_{max}$ and $x = +x_{max}$, called the *turning points*. In fact, because no energy is lost (no friction), the block will oscillate between $-x_{max}$ and $+x_{max}$ forever. (We discuss these oscillations further in Chapter 15.) From an energy viewpoint, the energy of the system cannot exceed $\frac{1}{2}kx_{max}^2$; therefore, the block must stop at these points and, because of the spring force, must accelerate toward $x = 0$.

Another simple mechanical system that has a configuration of stable equilibrium is a ball rolling about in the bottom of a bowl. Anytime the ball is displaced from its lowest position, it tends to return to that position when released.

Now consider a particle moving along the x axis under the influence of a conservative force F_x, where the U-versus-x curve is as shown in Figure 8.17. Once again, $F_x = 0$ at $x = 0$, and so the particle is in equilibrium at this point. However, this is a position of **unstable equilibrium** for the following reason: Suppose that the particle is displaced to the right ($x > 0$). Because the slope is negative for $x > 0$, $F_x = -dU/dx$ is positive, and the particle accelerates away from $x = 0$. If instead the particle is at $x = 0$ and is displaced to the left ($x < 0$), the force is negative because the slope is positive for $x < 0$, and the particle again accelerates away from the equilibrium position. The position $x = 0$ in this situation is one of unstable equilibrium because for any displacement from this point, the force pushes the particle farther away from equilibrium. The force pushes the particle toward a position of lower potential energy. A pencil balanced on its point is in a position of unstable equilibrium. If the pencil is displaced slightly from its absolutely vertical position and is then released, it will surely fall over. In general, **configurations of unstable equilibrium correspond to those for which $U(x)$ is a maximum.**

Finally, a situation may arise where U is constant over some region. This is called a configuration of **neutral equilibrium.** Small displacements from a position in this region produce neither restoring nor disrupting forces. A ball lying on a flat horizontal surface is an example of an object in neutral equilibrium.

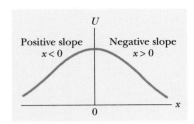

Figure 8.17 A plot of U versus x for a particle that has a position of unstable equilibrium located at $x = 0$. For any finite displacement of the particle, the force on the particle is directed away from $x = 0$.

Unstable equilibrium

Neutral equilibrium

Example 8.11 Force and Energy on an Atomic Scale

The potential energy associated with the force between two neutral atoms in a molecule can be modeled by the Lennard–Jones potential energy function:

$$U(x) = 4\epsilon \left[\left(\frac{\sigma}{x} \right)^{12} - \left(\frac{\sigma}{x} \right)^{6} \right]$$

where x is the separation of the atoms. The function $U(x)$ contains two parameters σ and ϵ that are determined from experiments. Sample values for the interaction between two atoms in a molecule are $\sigma = 0.263$ nm and $\epsilon = 1.51 \times 10^{-22}$ J.

(A) Using a spreadsheet or similar tool, graph this function and find the most likely distance between the two atoms.

Solution We expect to find stable equilibrium when the two atoms are separated by some equilibrium distance and the potential energy of the system of two atoms (the molecule) is a minimum. One can minimize the function $U(x)$ by taking its derivative and setting it equal to zero:

$$\frac{dU(x)}{dx} = 4\epsilon \frac{d}{dx} \left[\left(\frac{\sigma}{x} \right)^{12} - \left(\frac{\sigma}{x} \right)^{6} \right] = 0$$

$$= 4\epsilon \left[\frac{-12\sigma^{12}}{x^{13}} - \frac{-6\sigma^{6}}{x^{7}} \right] = 0$$

Solving for x—the equilibrium separation of the two atoms in the molecule—and inserting the given information yields

$$x = \quad 2.95 \times 10^{-10} \text{ m}.$$

We graph the Lennard–Jones function on both sides of this critical value to create our energy diagram, as shown in Figure 8.18a. Notice that $U(x)$ is extremely large when the atoms are very close together, is a minimum when the atoms are at their critical separation, and then increases again as the atoms move apart. When $U(x)$ is a minimum, the atoms are in stable equilibrium; this indicates that this is the most likely separation between them.

(B) Determine $F_x(x)$—the force that one atom exerts on the other in the molecule as a function of separation—and argue that the way this force behaves is physically plausible when the atoms are close together and far apart.

Solution Because the atoms combine to form a molecule, the force must be attractive when the atoms are far apart. On the other hand, the force must be repulsive when the two atoms are very close together. Otherwise, the molecule would collapse in on itself. Thus, the force must change sign at the critical separation, similar to the way spring forces switch sign in the change from extension to compression. Applying Equation 8.18 to the Lennard–Jones potential energy function gives

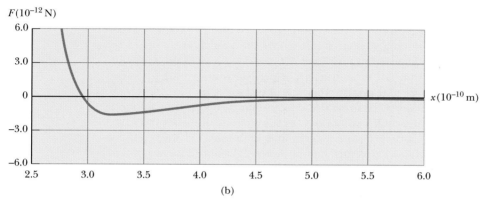

Figure 8.18 (Example 8.11) (a) Potential energy curve associated with a molecule. The distance x is the separation between the two atoms making up the molecule. (b) Force exerted on one atom by the other.

$$F_x = -\frac{dU(x)}{dx} = -4\epsilon \frac{d}{dx}\left[\left(\frac{\sigma}{x}\right)^{12} - \left(\frac{\sigma}{x}\right)^6\right]$$

$$= 4\epsilon\left[\frac{12\sigma^{12}}{x^{13}} - \frac{6\sigma^6}{x^7}\right]$$

This result is graphed in Figure 8.18b. As expected, the force is positive (repulsive) at small atomic separations, zero when the atoms are at the position of stable equilibrium [recall how we found the minimum of $U(x)$], and negative (attractive) at greater separations. Note that the force approaches zero as the separation between the atoms becomes very great.

SUMMARY

Take a Practice Test for this chapter by clicking on the Practice Test link at http://www.pse6.com.

If a particle of mass m is at a distance y above the Earth's surface, the **gravitational potential energy** of the particle–Earth system is

$$U_g \equiv mgy \tag{8.2}$$

The **elastic potential energy** stored in a spring of force constant k is

$$U_s \equiv \tfrac{1}{2}kx^2 \tag{8.11}$$

A reference configuration of the system should be chosen, and this configuration is often assigned a potential energy of zero.

A force is **conservative** if the work it does on a particle moving between two points is independent of the path the particle takes between the two points. Furthermore, a force is conservative if the work it does on a particle is zero when the particle moves through an arbitrary closed path and returns to its initial position. A force that does not meet these criteria is said to be **nonconservative.**

The **total mechanical energy of a system** is defined as the sum of the kinetic energy and the potential energy:

$$E_{\text{mech}} \equiv K + U \tag{8.8}$$

If a system is isolated and if no nonconservative forces are acting on objects inside the system, then the total mechanical energy of the system is constant:

$$K_f + U_f = K_i + U_i \tag{8.9}$$

If nonconservative forces (such as friction) act on objects inside a system, then mechanical energy is not conserved. In these situations, the difference between the total final mechanical energy and the total initial mechanical energy of the system equals the energy transformed to internal energy by the nonconservative forces.

A **potential energy function** U can be associated only with a conservative force. If a conservative force $\mathbf{F}$ acts between members of a system while one member moves along the x axis from x_i to x_f, then the change in the potential energy of the system equals the negative of the work done by that force:

$$U_f - U_i = - \int_{x_i}^{x_f} F_x \, dx \tag{8.16}$$

Systems can be in three types of equilibrium configurations when the net force on a member of the system is zero. Configurations of **stable equilibrium** correspond to those for which $U(x)$ is a minimum. Configurations of **unstable equilibrium** correspond to those for which $U(x)$ is a maximum. **Neutral equilibrium** arises where U is constant as a member of the system moves over some region.

QUESTIONS

1. If the height of a playground slide is kept constant, will the length of the slide or the presence of bumps make any difference in the final speed of children playing on it? Assume the slide is slick enough to be considered frictionless. Repeat this question assuming friction is present.

2. Explain why the total energy of a system can be either positive or negative, whereas the kinetic energy is always positive.

3. One person drops a ball from the top of a building while another person at the bottom observes its motion. Will these two people agree on the value of the gravitational potential energy of the ball–Earth system? On the change in potential energy? On the kinetic energy?

4. Discuss the changes in mechanical energy of an object–Earth system in (a) lifting the object, (b) holding the object at a fixed position, and (c) lowering the object slowly. Include the muscles in your discussion.

5. In Chapter 7, the work–kinetic energy theorem, $W = \Delta K$, was introduced. This equation states that work done on a system appears as a change in kinetic energy. This is a special-case equation, valid if there are no changes in any other type of energy such as potential or internal. Give some examples in which work is done on a system, but the change in energy of the system is not that of kinetic energy.

6. If three conservative forces and one nonconservative force act within a system, how many potential-energy terms appear in the equation that describes the system?

7. If only one external force acts on a particle, does it necessarily change the particle's (a) kinetic energy? (b) velocity?

8. A driver brings an automobile to a stop. If the brakes lock so that the car skids, where is the original kinetic energy of the car, and in what form is it after the car stops? Answer the same question for the case in which the brakes do not lock, but the wheels continue to turn.

9. You ride a bicycle. In what sense is your bicycle solar-powered?

10. In an earthquake, a large amount of energy is "released" and spreads outward, potentially causing severe damage. In what form does this energy exist before the earthquake, and by what energy transfer mechanism does it travel?

11. A bowling ball is suspended from the ceiling of a lecture hall by a strong cord. The ball is drawn away from its equilibrium position and released from rest at the tip of the demonstrator's nose as in Figure Q8.11. If the demonstrator remains stationary, explain why she is not struck by the ball on its return swing. Would this demonstrator be safe if the ball were given a push from its starting position at her nose?

12. Roads going up mountains are formed into switchbacks, with the road weaving back and forth along the face of the slope such that there is only a gentle rise on any portion of the roadway. Does this require any less work to be done by an automobile climbing the mountain compared to driving on a roadway that is straight up the slope? Why are switchbacks used?

13. As a sled moves across a flat snow-covered field at constant velocity, is any work done? How does air resistance enter into the picture?

14. You are working in a library, reshelving books. You lift a book from the floor to the top shelf. The kinetic energy of the book on the floor was zero, and the kinetic energy of the book on the top shelf is zero, so there is no change

Figure Q8.11

in kinetic energy. Yet you did some work in lifting the book. Is the work–kinetic energy theorem violated?

15. A ball is thrown straight up into the air. At what position is its kinetic energy a maximum? At what position is the gravitational potential energy of the ball–Earth system a maximum?

16. A pile driver is a device used to drive objects into the Earth by repeatedly dropping a heavy weight on them. By how much does the energy of the pile driver–Earth system increase when the weight it drops is doubled? Assume the weight is dropped from the same height each time.

17. Our body muscles exert forces when we lift, push, run, jump, and so forth. Are these forces conservative?

18. A block is connected to a spring that is suspended from the ceiling. If the block is set in motion and air resistance is neglected, describe the energy transformations that occur within the system consisting of the block, Earth, and spring.

19. Describe the energy transformations that occur during (a) the pole vault (b) the shot put (c) the high jump. What is the source of energy in each case?

20. Discuss the energy transformations that occur during the operation of an automobile.

21. What would the curve of U versus x look like if a particle were in a region of neutral equilibrium?

22. A ball rolls on a horizontal surface. Is the ball in stable, unstable, or neutral equilibrium?

23. Consider a ball fixed to one end of a rigid rod whose other end pivots on a horizontal axis so that the rod can rotate in a vertical plane. What are the positions of stable and unstable equilibrium?

PROBLEMS

1, 2, 3 = straightforward, intermediate, challenging   ☐ = full solution available in the *Student Solutions Manual and Study Guide*

🌀 = coached solution with hints available at http://www.pse6.com   💻 = computer useful in solving problem

▬ = paired numerical and symbolic problems

Section 8.1   Potential Energy of a System

1. A 1 000-kg roller coaster train is initially at the top of a rise, at point Ⓐ. It then moves 135 ft, at an angle of 40.0° below the horizontal, to a lower point Ⓑ. (a) Choose point Ⓑ to be the zero level for gravitational potential energy. Find the potential energy of the roller coaster–Earth system at points Ⓐ and Ⓑ, and the change in potential energy as the coaster moves. (b) Repeat part (a), setting the zero reference level at point Ⓐ.

2. A 400-N child is in a swing that is attached to ropes 2.00 m long. Find the gravitational potential energy of the child–Earth system relative to the child's lowest position when (a) the ropes are horizontal, (b) the ropes make a 30.0° angle with the vertical, and (c) the child is at the bottom of the circular arc.

3. A person with a remote mountain cabin plans to install her own hydroelectric plant. A nearby stream is 3.00 m wide and 0.500 m deep. Water flows at 1.20 m/s over the brink of a waterfall 5.00 m high. The manufacturer promises only 25.0% efficiency in converting the potential energy of the water–Earth system into electric energy. Find the power she can generate. (Large-scale hydroelectric plants, with a much larger drop, are more efficient.)

Section 8.2   The Isolated System–Conservation of Mechanical Energy

4. At 11:00 A.M. on September 7, 2001, more than 1 million British school children jumped up and down for one minute. The curriculum focus of the "Giant Jump" was on earthquakes, but it was integrated with many other topics, such as exercise, geography, cooperation, testing hypotheses, and setting world records. Children built their own seismographs, which registered local effects. (a) Find the mechanical energy released in the experiment. Assume that 1 050 000 children of average mass 36.0 kg jump twelve times each, raising their centers of mass by 25.0 cm each time and briefly resting between one jump and the next. The free-fall acceleration in Britain is 9.81 m/s². (b) Most of the energy is converted very rapidly into internal energy within the bodies of the children and the floors of the school buildings. Of the energy that propagates into the ground, most produces high-frequency "microtremor" vibrations that are rapidly damped and cannot travel far. Assume that 0.01% of the energy is carried away by a long-range seismic wave. The magnitude of an earthquake on the Richter scale is given by

$$M = \frac{\log E - 4.8}{1.5}$$

where E is the seismic wave energy in joules. According to this model, what is the magnitude of the demonstration quake? (It did not register above background noise overseas or on the seismograph of the Wolverton Seismic Vault, Hampshire.)

5. A bead slides without friction around a loop-the-loop (Fig. P8.5). The bead is released from a height $h = 3.50R$. (a) What is its speed at point Ⓐ? (b) How large is the normal force on it if its mass is 5.00 g?

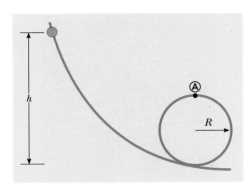

Figure P8.5

6. Dave Johnson, the bronze medalist at the 1992 Olympic decathlon in Barcelona, leaves the ground at the high jump with vertical velocity component 6.00 m/s. How far does his center of mass move up as he makes the jump?

7. A glider of mass 0.150 kg moves on a horizontal frictionless air track. It is permanently attached to one end of a massless horizontal spring, which has a force constant of 10.0 N/m both for extension and for compression. The other end of the spring is fixed. The glider is moved to compress the spring by 0.180 m and then released from rest. Calculate the speed of the glider (a) at the point where it has moved 0.180 m from its starting point, so that the spring is momentarily exerting no force and (b) at the point where it has moved 0.250 m from its starting point.

8. A loaded ore car has a mass of 950 kg and rolls on rails with negligible friction. It starts from rest and is pulled up a mine shaft by a cable connected to a winch. The shaft is inclined at 30.0° above the horizontal. The car accelerates uniformly to a speed of 2.20 m/s in 12.0 s and then continues at constant speed. (a) What power must the winch motor provide when the car is moving at constant speed? (b) What maximum power must the winch motor provide? (c) What total energy transfers out of the motor by work by the time the car moves off the end of the track, which is of length 1 250 m?

9. A simple pendulum, which we will consider in detail in Chapter 15, consists of an object suspended by a string. The object is assumed to be a particle. The string, with its top end fixed, has negligible mass and does not stretch. In the absence of air friction, the system oscillates by swinging back and forth in a vertical plane. If the string is 2.00 m long and makes an initial angle of 30.0° with the

vertical, calculate the speed of the particle (a) at the lowest point in its trajectory and (b) when the angle is 15.0°.

10. An object of mass m starts from rest and slides a distance d down a frictionless incline of angle θ. While sliding, it contacts an unstressed spring of negligible mass as shown in Figure P8.10. The object slides an additional distance x as it is brought momentarily to rest by compression of the spring (of force constant k). Find the initial separation d between object and spring.

Figure P8.10

11. A block of mass 0.250 kg is placed on top of a light vertical spring of force constant 5 000 N/m and pushed downward so that the spring is compressed by 0.100 m. After the block is released from rest, it travels upward and then leaves the spring. To what maximum height above the point of release does it rise?

12. A circus trapeze consists of a bar suspended by two parallel ropes, each of length ℓ, allowing performers to swing in a vertical circular arc (Figure P8.12). Suppose a performer with mass m holds the bar and steps off an elevated platform, starting from rest with the ropes at an angle θ_i with respect to the vertical. Suppose the size of the performer's body is small compared to the length ℓ, that she does not pump the trapeze to swing higher, and that air resistance is negligible. (a) Show that when the ropes make an angle θ with the vertical, the performer must exert a force

$$mg(3\cos\theta - 2\cos\theta_i)$$

in order to hang on. (b) Determine the angle θ_i for which

Figure P8.12

the force needed to hang on at the bottom of the swing is twice the performer's weight.

13. Two objects are connected by a light string passing over a light frictionless pulley as shown in Figure P8.13. The object of mass 5.00 kg is released from rest. Using the principle of conservation of energy, (a) determine the speed of the 3.00-kg object just as the 5.00-kg object hits the ground. (b) Find the maximum height to which the 3.00-kg object rises.

$m_1 = 5.00 \text{ kg}$

$m_2 = 3.00 \text{ kg}$ $h = 4.00 \text{ m}$

Figure P8.13 Problems 13 and 14.

14. Two objects are connected by a light string passing over a light frictionless pulley as in Figure P8.13. The object of mass m_1 is released from rest at height h. Using the principle of conservation of energy, (a) determine the speed of m_2 just as m_1 hits the ground. (b) Find the maximum height to which m_2 rises.

15. A light rigid rod is 77.0 cm long. Its top end is pivoted on a low-friction horizontal axle. The rod hangs straight down at rest with a small massive ball attached to its bottom end. You strike the ball, suddenly giving it a horizontal velocity so that it swings around in a full circle. What minimum speed at the bottom is required to make the ball go over the top of the circle?

16. Air moving at 11.0 m/s in a steady wind encounters a windmill of diameter 2.30 m and having an efficiency of 27.5%. The energy generated by the windmill is used to pump water from a well 35.0 m deep into a tank 2.30 m above the ground. At what rate in liters per minute can water be pumped into the tank?

17. A 20.0-kg cannon ball is fired from a cannon with muzzle speed of 1 000 m/s at an angle of 37.0° with the horizontal. A second ball is fired at an angle of 90.0°. Use the conservation of energy principle to find (a) the maximum height reached by each ball and (b) the total mechanical energy at the maximum height for each ball. Let $y = 0$ at the cannon.

18. A 2.00-kg ball is attached to the bottom end of a length of fishline with a breaking strength of 10 lb (44.5 N). The top end of the fishline is held stationary. The ball is released from rest with the line taut and horizontal ($\theta = 90.0°$). At what angle θ (measured from the vertical) will the fishline break?

19. A daredevil plans to bungee-jump from a balloon 65.0 m above a carnival midway (Figure P8.19). He will use a uniform elastic cord, tied to a harness around his body, to stop his fall at a point 10.0 m above the ground. Model his body as a particle and the cord as having negligible mass and obeying Hooke's force law. In a preliminary test, hanging at rest from a 5.00-m length of the cord, he finds that his body weight stretches it by 1.50 m. He will drop from rest at the point where the top end of a longer section of the cord is attached to the stationary balloon. (a) What length of cord should he use? (b) What maximum acceleration will he experience?

Figure P8.19

20. Review problem. The system shown in Figure P8.20 consists of a light inextensible cord, light frictionless pulleys, and blocks of equal mass. It is initially held at rest so that the blocks are at the same height above the ground. The blocks are then released. Find the speed of block A at the moment when the vertical separation of the blocks is h.

Figure P8.20

Section 8.3 Conservative and Nonconservative Forces

21. A 4.00-kg particle moves from the origin to position C, having coordinates $x = 5.00$ m and $y = 5.00$ m. One force on the particle is the gravitational force acting in the negative y direction (Fig. P8.21). Using Equation 7.3, calculate the work done by the gravitational force in going from O to C along (a) OAC. (b) OBC. (c) OC. Your results should all be identical. Why?

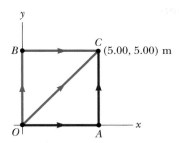

Figure P8.21 Problems 21, 22 and 23.

22. (a) Suppose that a constant force acts on an object. The force does not vary with time, nor with the position or the velocity of the object. Start with the general definition for work done by a force

$$W = \int_i^f \mathbf{F} \cdot d\mathbf{r}$$

and show that the force is conservative. (b) As a special case, suppose that the force $\mathbf{F} = (3\hat{\mathbf{i}} + 4\hat{\mathbf{j}})$ N acts on a particle that moves from O to C in Figure P8.21. Calculate the work done by $\mathbf{F}$ if the particle moves along each one of the three paths OAC, OBC, and OC. (Your three answers should be identical.)

23. A force acting on a particle moving in the xy plane is given by $\mathbf{F} = (2y\hat{\mathbf{i}} + x^2\hat{\mathbf{j}})$ N, where x and y are in meters. The particle moves from the origin to a final position having coordinates $x = 5.00$ m and $y = 5.00$ m, as in Figure P8.21. Calculate the work done by $\mathbf{F}$ along (a) OAC, (b) OBC, (c) OC. (d) Is $\mathbf{F}$ conservative or nonconservative? Explain.

24. A particle of mass $m = 5.00$ kg is released from point Ⓐ and slides on the frictionless track shown in Figure P8.24. Determine (a) the particle's speed at points Ⓑ and Ⓒ and (b) the net work done by the gravitational force in moving the particle from Ⓐ to Ⓒ.

Figure P8.24

25. A single constant force $\mathbf{F} = (3\hat{\mathbf{i}} + 5\hat{\mathbf{j}})$ N acts on a 4.00-kg particle. (a) Calculate the work done by this force if the particle moves from the origin to the point having the vector position $\mathbf{r} = (2\hat{\mathbf{i}} - 3\hat{\mathbf{j}})$ m. Does this result depend on the path? Explain. (b) What is the speed of the particle at $\mathbf{r}$ if its speed at the origin is 4.00 m/s? (c) What is the change in the potential energy?

Section 8.4 Changes in Mechanical Energy for Nonconservative Forces

26. At time t_i, the kinetic energy of a particle is 30.0 J and the potential energy of the system to which it belongs is 10.0 J. At some later time t_f, the kinetic energy of the particle is 18.0 J. (a) If only conservative forces act on the particle, what are the potential energy and the total energy at time t_f? (b) If the potential energy of the system at time t_f is 5.00 J, are there any nonconservative forces acting on the particle? Explain.

27. In her hand a softball pitcher swings a ball of mass 0.250 kg around a vertical circular path of radius 60.0 cm before releasing it from her hand. The pitcher maintains a component of force on the ball of constant magnitude 30.0 N in the direction of motion around the complete path. The speed of the ball at the top of the circle is 15.0 m/s. If she releases the ball at the bottom of the circle, what is its speed upon release?

28. An electric scooter has a battery capable of supplying 120 Wh of energy. If friction forces and other losses account for 60.0% of the energy usage, what altitude change can a rider achieve when driving in hilly terrain, if the rider and scooter have a combined weight of 890 N?

29. The world's biggest locomotive is the MK5000C, a behemoth of mass 160 metric tons driven by the most powerful engine ever used for rail transportation, a Caterpillar diesel capable of 5 000 hp. Such a huge machine can provide a gain in efficiency, but its large mass presents challenges as well. The engineer finds that the locomotive handles differently from conventional units, notably in braking and climbing hills. Consider the locomotive pulling no train, but traveling at 27.0 m/s on a level track while operating with output power 1 000 hp. It comes to a 5.00% grade (a slope that rises 5.00 m for every 100 m along the track). If the throttle is not advanced, so that the power level is held steady, to what value will the speed drop? Assume that friction forces do not depend on the speed.

30. A 70.0-kg diver steps off a 10.0-m tower and drops straight down into the water. If he comes to rest 5.00 m beneath the surface of the water, determine the average resistance force exerted by the water on the diver.

31. The coefficient of friction between the 3.00-kg block and the surface in Figure P8.31 is 0.400. The system starts from rest. What is the speed of the 5.00-kg ball when it has fallen 1.50 m?

Figure P8.31

32. A boy in a wheelchair (total mass 47.0 kg) wins a race with a skateboarder. The boy has speed 1.40 m/s at the crest of a slope 2.60 m high and 12.4 m long. At the bottom of the slope his speed is 6.20 m/s. If air resistance and rolling resistance can be modeled as a constant friction force of 41.0 N, find the work he did in pushing forward on his wheels during the downhill ride.

33. A 5.00-kg block is set into motion up an inclined plane with an initial speed of 8.00 m/s (Fig. P8.33). The block comes to rest after traveling 3.00 m along the plane, which is inclined at an angle of 30.0° to the horizontal. For this motion determine (a) the change in the block's kinetic energy, (b) the change in the potential energy of the block–Earth system, and (c) the friction force exerted on the block (assumed to be constant). (d) What is the coefficient of kinetic friction?

$v_i = 8.00$ m/s

3.00 m

30.0°

Figure P8.33

34. An 80.0-kg skydiver jumps out of a balloon at an altitude of 1 000 m and opens the parachute at an altitude of 200 m. (a) Assuming that the total retarding force on the diver is constant at 50.0 N with the parachute closed and constant at 3 600 N with the parachute open, what is the speed of the diver when he lands on the ground? (b) Do you think the skydiver will be injured? Explain. (c) At what height should the parachute be opened so that the final speed of the skydiver when he hits the ground is 5.00 m/s? (d) How realistic is the assumption that the total retarding force is constant? Explain.

35. A toy cannon uses a spring to project a 5.30-g soft rubber ball. The spring is originally compressed by 5.00 cm and has a force constant of 8.00 N/m. When the cannon is fired, the ball moves 15.0 cm through the horizontal barrel of the cannon, and there is a constant friction force of 0.032 0 N between the barrel and the ball. (a) With what speed does the projectile leave the barrel of the cannon? (b) At what point does the ball have maximum speed? (c) What is this maximum speed?

36. A 50.0-kg block and a 100-kg block are connected by a string as in Figure P8.36. The pulley is frictionless and of

50.0 kg

100 kg

Ⓑ

Ⓐ

v

37.0°

Figure P8.36

negligible mass. The coefficient of kinetic friction between the 50.0 kg block and incline is 0.250. Determine the change in the kinetic energy of the 50.0-kg block as it moves from Ⓐ to Ⓑ, a distance of 20.0 m.

37. A 1.50-kg object is held 1.20 m above a relaxed massless vertical spring with a force constant of 320 N/m. The object is dropped onto the spring. (a) How far does it compress the spring? (b) **What If?** How far does it compress the spring if the same experiment is performed on the Moon, where $g = 1.63$ m/s²? (c) **What If?** Repeat part (a), but this time assume a constant air-resistance force of 0.700 N acts on the object during its motion.

38. A 75.0-kg skysurfer is falling straight down with terminal speed 60.0 m/s. Determine the rate at which the skysurfer–Earth system is losing mechanical energy.

39. A uniform board of length L is sliding along a smooth (frictionless) horizontal plane as in Figure P8.39a. The board then slides across the boundary with a rough horizontal surface. The coefficient of kinetic friction between the board and the second surface is μ_k. (a) Find the acceleration of the board at the moment its front end has traveled a distance x beyond the boundary. (b) The board stops at the moment its back end reaches the boundary, as in Figure P8.39b. Find the initial speed v of the board.

v

Boundary

(a)

(b)

Figure P8.39

Section 8.5 Relationship Between Conservative Forces and Potential Energy

40. A single conservative force acting on a particle varies as $\mathbf{F} = (-Ax + Bx^2)\hat{\mathbf{i}}$ N, where A and B are constants and x is in meters. (a) Calculate the potential-energy function $U(x)$ associated with this force, taking $U = 0$ at $x = 0$. (b) Find the change in potential energy and the change in kinetic energy as the particle moves from $x = 2.00$ m to $x = 3.00$ m.

41. A single conservative force acts on a 5.00-kg particle. The equation $F_x = (2x + 4)$ N describes the force, where x is in meters. As the particle moves along the x axis from $x = 1.00$ m to $x = 5.00$ m, calculate (a) the work done by this force, (b) the change in the potential energy of the system, and (c) the kinetic energy of the particle at $x = 5.00$ m if its speed is 3.00 m/s at $x = 1.00$ m.

42. A potential-energy function for a two-dimensional force is of the form $U = 3x^3y - 7x$. Find the force that acts at the point (x, y).

43. The potential energy of a system of two particles separated by a distance r is given by $U(r) = A/r$, where A is a constant. Find the radial force $\mathbf{F}_r$ that each particle exerts on the other.

Section 8.6 Energy Diagrams and Equilibrium of a System

44. A right circular cone can be balanced on a horizontal surface in three different ways. Sketch these three equilibrium configurations, and identify them as positions of stable, unstable, or neutral equilibrium.

45. For the potential energy curve shown in Figure P8.45, (a) determine whether the force F_x is positive, negative, or zero at the five points indicated. (b) Indicate points of stable, unstable, and neutral equilibrium. (c) Sketch the curve for F_x versus x from $x = 0$ to $x = 9.5$ m.

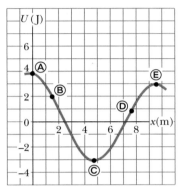

Figure P8.45

46. A particle moves along a line where the potential energy of its system depends on its position r as graphed in Figure P8.46. In the limit as r increases without bound, $U(r)$ approaches $+1$ J. (a) Identify each equilibrium position for this particle. Indicate whether each is a point of stable, unstable, or neutral equilibrium. (b) The particle will be bound if the total energy of the system is in what range? Now suppose that the system has energy -3 J. Determine (c) the range of positions where the particle can be found, (d) its maximum kinetic energy, (e) the location where it has maximum kinetic energy, and (f) the *binding energy* of the system—that is, the additional energy that it would have to be given in order for the particle to move out to $r \to \infty$.

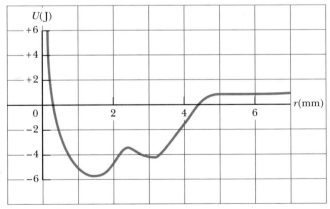

Figure P8.46

47. A particle of mass 1.18 kg is attached between two identical springs on a horizontal frictionless tabletop. The springs have force constant k and each is initially unstressed. (a) If the particle is pulled a distance x along a direction perpendicular to the initial configuration of the springs, as in Figure P8.47, show that the potential energy of the system is

$$U(x) = kx^2 + 2kL\left(L - \sqrt{x^2 + L^2}\right)$$

(*Hint*: See Problem 58 in Chapter 7.) (b) Make a plot of $U(x)$ versus x and identify all equilibrium points. Assume that $L = 1.20$ m and $k = 40.0$ N/m. (c) If the particle is pulled 0.500 m to the right and then released, what is its speed when it reaches the equilibrium point $x = 0$?

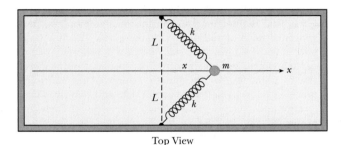

Top View

Figure P8.47

Additional Problems

48. A block slides down a curved frictionless track and then up an inclined plane as in Figure P8.48. The coefficient of kinetic friction between block and incline is μ_k. Use energy methods to show that the maximum height reached by the block is

$$y_{max} = \frac{h}{1 + \mu_k \cot \theta}$$

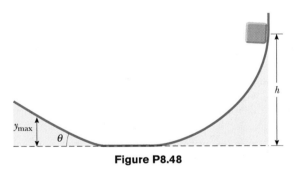

Figure P8.48

49. Make an order-of-magnitude estimate of your power output as you climb stairs. In your solution, state the physical quantities you take as data and the values you measure or estimate for them. Do you consider your peak power or your sustainable power?

50. **Review problem.** The mass of a car is 1 500 kg. The shape of the body is such that its aerodynamic drag coefficient is $D = 0.330$ and the frontal area is 2.50 m^2. Assuming that the drag force is proportional to v^2 and neglecting other sources of friction, calculate the power required to maintain a speed of 100 km/h as the car climbs a long hill sloping at 3.20°.

51. Assume that you attend a state university that started out as an agricultural college. Close to the center of the campus is a tall silo topped with a hemispherical cap. The cap is frictionless when wet. Someone has somehow balanced a pumpkin at the highest point. The line from the center of curvature of the cap to the pumpkin makes an angle $\theta_i = 0°$ with the vertical. While you happen to be standing nearby in the middle of a rainy night, a breath of wind makes the pumpkin start sliding downward from rest. It loses contact with the cap when the line from the center of the hemisphere to the pumpkin makes a certain angle with the vertical. What is this angle?

52. A 200-g particle is released from rest at point Ⓐ along the horizontal diameter on the inside of a frictionless, hemispherical bowl of radius $R = 30.0$ cm (Fig. P8.52). Calculate (a) the gravitational potential energy of the particle–Earth system when the particle is at point Ⓐ relative to point Ⓑ, (b) the kinetic energy of the particle at point Ⓑ, (c) its speed at point Ⓑ, and (d) its kinetic energy and the potential energy when the particle is at point Ⓒ.

Figure P8.52 Problems 52 and 53.

53. **What If?** The particle described in Problem 52 (Fig. P8.52) is released from rest at Ⓐ, and the surface of the bowl is rough. The speed of the particle at Ⓑ is 1.50 m/s. (a) What is its kinetic energy at Ⓑ? (b) How much mechanical energy is transformed into internal energy as the particle moves from Ⓐ to Ⓑ? (c) Is it possible to determine the coefficient of friction from these results in any simple manner? Explain.

54. A 2.00-kg block situated on a rough incline is connected to a spring of negligible mass having a spring constant of 100 N/m (Fig. P8.54). The pulley is frictionless. The block is released from rest when the spring is unstretched. The

Figure P8.54 Problems 54 and 55.

block moves 20.0 cm down the incline before coming to rest. Find the coefficient of kinetic friction between block and incline.

55. **Review problem.** Suppose the incline is frictionless for the system described in Problem 54 (Fig. P8.54). The block is released from rest with the spring initially unstretched. (a) How far does it move down the incline before coming to rest? (b) What is its acceleration at its lowest point? Is the acceleration constant? (c) Describe the energy transformations that occur during the descent.

56. A child's pogo stick (Fig. P8.56) stores energy in a spring with a force constant of 2.50×10^4 N/m. At position Ⓐ ($x_A = -0.100$ m), the spring compression is a maximum and the child is momentarily at rest. At position Ⓑ ($x_B = 0$), the spring is relaxed and the child is moving upward. At position Ⓒ, the child is again momentarily at rest at the top of the jump. The combined mass of child and pogo stick is 25.0 kg. (a) Calculate the total energy of the child–stick–Earth system if both gravitational and elastic potential energies are zero for $x = 0$. (b) Determine x_C. (c) Calculate the speed of the child at $x = 0$. (d) Determine the value of x for which the kinetic energy of the system is a maximum. (e) Calculate the child's maximum upward speed.

Figure P8.56

57. A 10.0-kg block is released from point Ⓐ in Figure P8.57. The track is frictionless except for the portion between points Ⓑ and Ⓒ, which has a length of 6.00 m. The block travels down the track, hits a spring of force constant 2 250 N/m, and compresses the spring 0.300 m from its equilibrium position before coming to rest momentarily. Determine the coefficient of kinetic friction between the block and the rough surface between Ⓑ and Ⓒ.

Figure P8.57

58. The potential energy function for a system is given by $U(x) = -x^3 + 2x^2 + 3x$. (a) Determine the force F_x as a function of x. (b) For what values of x is the force equal to zero? (c) Plot $U(x)$ versus x and F_x versus x, and indicate points of stable and unstable equilibrium.

59. A 20.0-kg block is connected to a 30.0-kg block by a string that passes over a light frictionless pulley. The 30.0-kg block is connected to a spring that has negligible mass and a force constant of 250 N/m, as shown in Figure P8.59. The spring is unstretched when the system is as shown in the figure, and the incline is frictionless. The 20.0-kg block is pulled 20.0 cm down the incline (so that the 30.0-kg block is 40.0 cm above the floor) and released from rest. Find the speed of each block when the 30.0-kg block is 20.0 cm above the floor (that is, when the spring is unstretched).

Figure P8.59

60. A 1.00-kg object slides to the right on a surface having a coefficient of kinetic friction 0.250 (Fig. P8.60). The object has a speed of $v_i = 3.00$ m/s when it makes contact with a light spring that has a force constant of 50.0 N/m. The object comes to rest after the spring has been compressed a distance d. The object is then forced toward the left by the spring and continues to move in that direction beyond the spring's unstretched position. Finally, the object comes to rest a distance D to the left of the unstretched spring. Find (a) the distance of compression d, (b) the speed v at the unstretched position when the object is moving to the left, and (c) the distance D where the object comes to rest.

61. A block of mass 0.500 kg is pushed against a horizontal spring of negligible mass until the spring is compressed a distance x (Fig. P8.61). The force constant of the spring is 450 N/m. When it is released, the block travels along a frictionless, horizontal surface to point B, the bottom of a vertical circular track of radius $R = 1.00$ m, and continues to move up the track. The speed of the block at the bottom of the track is $v_B = 12.0$ m/s, and the block experi-

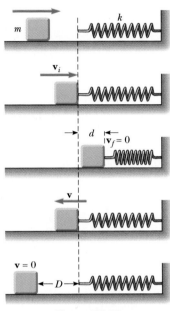

Figure P8.60

ences an average friction force of 7.00 N while sliding up the track. (a) What is x? (b) What speed do you predict for the block at the top of the track? (c) Does the block actually reach the top of the track, or does it fall off before reaching the top?

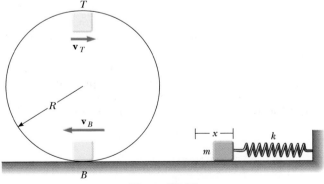

Figure P8.61

62. A uniform chain of length 8.00 m initially lies stretched out on a horizontal table. (a) If the coefficient of static friction between chain and table is 0.600, show that the chain will begin to slide off the table if at least 3.00 m of it hangs over the edge of the table. (b) Determine the speed of the chain

as all of it leaves the table, given that the coefficient of kinetic friction between the chain and the table is 0.400.

63. A child slides without friction from a height h along a curved water slide (Fig. P8.63). She is launched from a height $h/5$ into the pool. Determine her maximum airborne height y in terms of h and θ.

Figure P8.63

64. Refer to the situation described in Chapter 5, Problem 65. A 1.00-kg glider on a horizontal air track is pulled by a string at angle θ. The taut string runs over a light pulley at height $h_0 = 40.0$ cm above the line of motion of the glider. The other end of the string is attached to a hanging mass of 0.500 kg as in Fig. P5.65. (a) Show that the speed of the glider v_x and the speed of the hanging mass v_y are related by $v_y = v_x \cos \theta$. The glider is released from rest when $\theta = 30.0°$. Find (b) v_x and (c) v_y when $\theta = 45.0°$. (d) Explain why the answers to parts (b) and (c) to Chapter 5, Problem 65 do not help to solve parts (b) and (c) of this problem.

65. Jane, whose mass is 50.0 kg, needs to swing across a river (having width D) filled with man-eating crocodiles to save Tarzan from danger. She must swing into a wind exerting constant horizontal force $\mathbf{F}$, on a vine having length L and initially making an angle θ with the vertical (Fig. P8.65). Taking $D = 50.0$ m, $F = 110$ N, $L = 40.0$ m, and $\theta = 50.0°$, (a) with what minimum speed must Jane begin her swing

in order to just make it to the other side? (b) Once the rescue is complete, Tarzan and Jane must swing back across the river. With what minimum speed must they begin their swing? Assume that Tarzan has a mass of 80.0 kg.

66. A 5.00-kg block free to move on a horizontal, frictionless surface is attached to one end of a light horizontal spring. The other end of the spring is held fixed. The spring is compressed 0.100 m from equilibrium and released. The speed of the block is 1.20 m/s when it passes the equilibrium position of the spring. The same experiment is now repeated with the frictionless surface replaced by a surface for which the coefficient of kinetic friction is 0.300. Determine the speed of the block at the equilibrium position of the spring.

67. A skateboarder with his board can be modeled as a particle of mass 76.0 kg, located at his center of mass (which we will study in Chapter 9). As in Figure P8.67, the skateboarder starts from rest in a crouching position at one lip of a half-pipe (point Ⓐ). The half-pipe is a dry water channel, forming one half of a cylinder of radius 6.80 m with its axis horizontal. On his descent, the skateboarder moves without friction so that his center of mass moves through one quarter of a circle of radius 6.30 m. (a) Find his speed at the bottom of the half-pipe (point Ⓑ). (b) Find his centripetal acceleration. (c) Find the normal force n_B acting on the skateboarder at point Ⓑ. Immediately after passing point Ⓑ, he stands up and raises his arms, lifting his center of mass from 0.500 m to 0.950 m above the concrete (point Ⓒ). To account for the conversion of chemical into mechanical energy, model his legs as doing work by pushing him vertically up, with a constant force equal to the normal force n_B, over a distance of 0.450 m. (You will be able to solve this problem with a more accurate model in Chapter 11.) (d) What is the work done on the skateboarder's body in this process? Next, the skateboarder glides upward with his center of mass moving in a quarter circle of radius 5.85 m. His body is horizontal when he passes point Ⓓ, the far lip of the half-pipe. (e) Find his speed at this location. At last he goes ballistic, twisting around while his center of mass moves vertically. (f) How high above point Ⓓ does he rise? (g) Over what time interval is he airborne before he touches down, 2.34 m below the level of point Ⓓ? [*Caution:* Do not try this yourself without the required skill and protective equipment, or in a drainage channel to which you do not have legal access.]

Figure P8.65

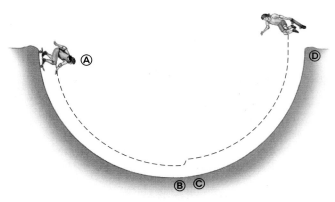

Figure P8.67

68. A block of mass M rests on a table. It is fastened to the lower end of a light vertical spring. The upper end of the spring is fastened to a block of mass m. The upper block is pushed down by an additional force $3mg$, so the spring compression is $4mg/k$. In this configuration the upper block is released from rest. The spring lifts the lower block off the table. In terms of m, what is the greatest possible value for M?

69. A ball having mass m is connected by a strong string of length L to a pivot point and held in place in a vertical position. A wind exerting constant force of magnitude F is blowing from left to right as in Figure P8.69a. (a) If the ball is released from rest, show that the maximum height H reached by the ball, as measured from its initial height, is

$$H = \frac{2L}{1 + (mg/F)^2}$$

Check that the above result is valid both for cases when $0 \le H \le L$ and for $L \le H \le 2L$. (b) Compute the value of H using the values $m = 2.00$ kg, $L = 2.00$ m, and $F = 14.7$ N. (c) Using these same values, determine the *equilibrium* height of the ball. (d) Could the equilibrium height ever be larger than L? Explain.

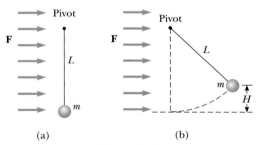

(a) (b)

Figure P8.69

70. A ball is tied to one end of a string. The other end of the string is held fixed. The ball is set moving around a vertical circle without friction, and with speed $v_i = \sqrt{Rg}$ at the top of the circle, as in Figure P8.70. At what angle θ should the string be cut so that the ball will then travel through the center of the circle?

Figure P8.70

71. A ball whirls around in a vertical circle at the end of a string. If the total energy of the ball–Earth system remains constant, show that the tension in the string at the bottom is greater than the tension at the top by six times the weight of the ball.

72. A pendulum, comprising a light string of length L and a small sphere, swings in the vertical plane. The string hits a peg located a distance d below the point of suspension (Fig. P8.72). (a) Show that if the sphere is released from a height below that of the peg, it will return to this height after the string strikes the peg. (b) Show that if the pendulum is released from the horizontal position ($\theta = 90°$) and is to swing in a complete circle centered on the peg, then the minimum value of d must be $3L/5$.

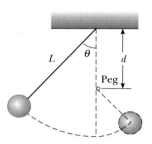

Figure P8.72

73. A roller-coaster car is released from rest at the top of the first rise and then moves freely with negligible friction. The roller coaster shown in Figure P8.73 has a circular loop of radius R in a vertical plane. (a) Suppose first that the car barely makes it around the loop: at the top of the loop the riders are upside down and feel weightless. Find the required height of the release point above the bottom of the loop in terms of R. (b) Now assume that the release point is at or above the minimum required height. Show that the normal force on the car at the bottom of the loop exceeds the normal force at the top of the loop by six times the weight of the car. The normal force on each rider follows the same rule. Such a large normal force is dangerous and very uncomfortable for the riders. Roller coasters are therefore not built with circular loops in vertical planes. Figure P6.20 and the photograph on page 157 show two actual designs.

Figure P8.73

74. **Review problem.** In 1887 in Bridgeport, Connecticut, C. J. Belknap built the water slide shown in Figure P8.74. A rider on a small sled, of total mass 80.0 kg, pushed off to start at the top of the slide (point Ⓐ) with a speed of 2.50 m/s. The chute was 9.76 m high at the top, 54.3 m long, and 0.51 m wide. Along its length, 725 wheels made

friction negligible. Upon leaving the chute horizontally at its bottom end (point Ⓒ), the rider skimmed across the water of Long Island Sound for as much as 50 m, "skipping along like a flat pebble," before at last coming to rest and swimming ashore, pulling his sled after him. According to *Scientific American*, "The facial expression of novices taking their first adventurous slide is quite remarkable, and the sensations felt are correspondingly novel and peculiar." (a) Find the speed of the sled and rider at point Ⓒ. (b) Model the force of water friction as a constant retarding force acting on a particle. Find the work done by water friction in stopping the sled and rider. (c) Find the magnitude of the force the water exerts on the sled. (d) Find the magnitude of the force the chute exerts on the sled at point Ⓑ. (e) At point Ⓒ the chute is horizontal but curving in the vertical plane. Assume its radius of curvature is 20.0 m. Find the force the chute exerts on the sled at point Ⓒ.

Engraving from Scientific American, July 1888

Figure P8.74

Answers to Quick Quizzes

8.1 (c). The sign of the gravitational potential energy depends on your choice of zero configuration. If the two objects in the system are closer together than in the zero configuration, the potential energy is negative. If they are farther apart, the potential energy is positive.

8.2 (c). The reason that we can ignore the kinetic energy of the massive Earth is that this kinetic energy is so small as to be essentially zero.

8.3 (a). We must include the Earth if we are going to work with gravitational potential energy.

8.4 (c). The total mechanical energy, kinetic plus potential, is conserved.

8.5 (a). The more massive rock has twice as much gravitational potential energy associated with it compared to the lighter rock. Because mechanical energy of an isolated system is conserved, the more massive rock will arrive at the ground with twice as much kinetic energy as the lighter rock.

8.6 $v_1 = v_2 = v_3$. The first and third balls speed up after they are thrown, while the second ball initially slows down but then speeds up after reaching its peak. The paths of all three balls are parabolas, and the balls take different times to reach the ground because they have different initial velocities. However, all three balls have the same speed at the moment they hit the ground because all start with the same kinetic energy and the ball–Earth system undergoes the same change in gravitational potential energy in all three cases.

8.7 (c). This system exhibits changes in kinetic energy as well as in both types of potential energy.

8.8 (a). Because the Earth is not included in the system, there is no gravitational potential energy associated with the system.

8.9 (c). The friction force must transform four times as much mechanical energy into internal energy if the speed is doubled, because kinetic energy depends on the square of the speed. Thus, the force must act over four times the distance.

8.10 (c). The decrease in mechanical energy of the system is $f_k d$, where d is the distance the block moves along the incline. While the force of kinetic friction remains the same, the distance d is smaller because a component of the gravitational force is pulling on the block in the direction opposite to its velocity.

8.11 (d). The slope of a $U(x)$-versus-x graph is by definition $dU(x)/dx$. From Equation 8.18, we see that this expression is equal to the negative of the x component of the conservative force acting on an object that is part of the system.

Chapter 9

Linear Momentum and Collisions

▲ *A moving bowling ball carries momentum, the topic of this chapter. In the collision between the ball and the pins, momentum is transferred to the pins. (Mark Cooper/Corbis Stock Market)*

Consider what happens when a bowling ball strikes a pin, as in the opening photograph. The pin is given a large velocity as a result of the collision; consequently, it flies away and hits other pins or is projected toward the backstop. Because the average force exerted on the pin during the collision is large (resulting in a large acceleration), the pin achieves the large velocity very rapidly and experiences the force for a very short time interval. According to Newton's third law, the pin exerts a reaction force on the ball that is equal in magnitude and opposite in direction to the force exerted by the ball on the pin. This reaction force causes the ball to accelerate, but because the ball is so much more massive than the pin, the ball's acceleration is much less than the pin's acceleration.

Although F and a are large for the pin, they vary in time—a complicated situation! One of the main objectives of this chapter is to enable you to understand and analyze such events in a simple way. First, we introduce the concept of *momentum*, which is useful for describing objects in motion. Imagine that you have intercepted a football and see two players from the opposing team approaching you as you run with the ball. One of the players is the 180-lb quarterback who threw the ball; the other is a 300-lb lineman. Both of the players are running toward you at 5 m/s. However, because the two players have different masses, intuitively you know that you would rather collide with the quarterback than with the lineman. The momentum of an object is related to both its mass and its velocity. The concept of momentum leads us to a second conservation law, that of conservation of momentum. This law is especially useful for treating problems that involve collisions between objects and for analyzing rocket propulsion. In this chapter we also introduce the concept of the center of mass of a system of particles. We find that the motion of a system of particles can be described by the motion of one representative particle located at the center of mass.

9.1 Linear Momentum and Its Conservation

In the preceding two chapters we studied situations that are complex to analyze with Newton's laws. We were able to solve problems involving these situations by applying a conservation principle—conservation of energy. Consider another situation—a 60-kg archer stands on frictionless ice and fires a 0.50-kg arrow horizontally at 50 m/s. From Newton's third law, we know that the force that the bow exerts on the arrow will be matched by a force in the opposite direction on the bow (and the archer). This will cause the archer to begin to slide backward on the ice. But with what speed? We cannot answer this question directly using *either* Newton's second law or an energy approach—there is not enough information.

Despite our inability to solve the archer problem using our techniques learned so far, this is a very simple problem to solve if we introduce a new quantity that describes motion, *linear momentum*. Let us apply the General Problem-Solving Strategy and *conceptualize* an isolated system of two particles (Fig. 9.1) with masses m_1 and m_2 and moving with velocities $\mathbf{v}_1$ and $\mathbf{v}_2$ at an instant of time. Because the system is isolated, the only force on one particle is that from the other particle and we can *categorize* this as a situation in which Newton's laws will be useful. If a force from particle 1 (for example, a gravitational force) acts on particle 2, then there must be a second force—equal in magnitude but opposite in direction—that particle 2 exerts on particle 1. That is, they form a Newton's third law action–reaction pair, so that $\mathbf{F}_{12} = -\mathbf{F}_{21}$. We can express this condition as

$$\mathbf{F}_{21} + \mathbf{F}_{12} = 0$$

Let us further *analyze* this situation by incorporating Newton's second law. Over some time interval, the interacting particles in the system will accelerate. Thus, replacing each force with $m\mathbf{a}$ gives

$$m_1\mathbf{a}_1 + m_2\mathbf{a}_2 = 0$$

Now we replace the acceleration with its definition from Equation 4.5:

$$m_1\frac{d\mathbf{v}_1}{dt} + m_2\frac{d\mathbf{v}_2}{dt} = 0$$

If the masses m_1 and m_2 are constant, we can bring them into the derivatives, which gives

$$\frac{d(m_1\mathbf{v}_1)}{dt} + \frac{d(m_2\mathbf{v}_2)}{dt} = 0$$

$$\frac{d}{dt}(m_1\mathbf{v}_1 + m_2\mathbf{v}_2) = 0 \tag{9.1}$$

To *finalize* this discussion, note that the derivative of the sum $m_1\mathbf{v}_1 + m_2\mathbf{v}_2$ with respect to time is zero. Consequently, this sum must be constant. We learn from this discussion that the quantity $m\mathbf{v}$ for a particle is important, in that the sum of these quantities for an isolated system is conserved. We call this quantity *linear momentum*:

The **linear momentum** of a particle or an object that can be modeled as a particle of mass m moving with a velocity $\mathbf{v}$ is defined to be the product of the mass and velocity:

$$\mathbf{p} \equiv m\mathbf{v} \tag{9.2}$$

Linear momentum is a vector quantity because it equals the product of a scalar quantity m and a vector quantity $\mathbf{v}$. Its direction is along $\mathbf{v}$, it has dimensions ML/T, and its SI unit is kg·m/s.

If a particle is moving in an arbitrary direction, $\mathbf{p}$ must have three components, and Equation 9.2 is equivalent to the component equations

$$p_x = mv_x \quad p_y = mv_y \quad p_z = mv_z$$

As you can see from its definition, the concept of momentum[1] provides a quantitative distinction between heavy and light particles moving at the same velocity. For example, the momentum of a bowling ball moving at 10 m/s is much greater than that of a tennis ball moving at the same speed. Newton called the product $m\mathbf{v}$ *quantity of motion*; this is perhaps a more graphic description than our present-day word *momentum*, which comes from the Latin word for movement.

Using Newton's second law of motion, we can relate the linear momentum of a particle to the resultant force acting on the particle. We start with Newton's second law and substitute the definition of acceleration:

$$\sum\mathbf{F} = m\mathbf{a} = m\frac{d\mathbf{v}}{dt}$$

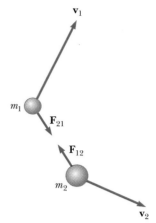

Figure 9.1 Two particles interact with each other. According to Newton's third law, we must have $\mathbf{F}_{12} = -\mathbf{F}_{21}$.

Definition of linear momentum of a particle

---

[1] In this chapter, the terms *momentum* and *linear momentum* have the same meaning. Later, in Chapter 11, we shall use the term *angular momentum* when dealing with rotational motion.

In Newton's second law, the mass m is assumed to be constant. Thus, we can bring m inside the derivative notation to give us

Newton's second law for a particle

$$\sum \mathbf{F} = \frac{d(m\mathbf{v})}{dt} = \frac{d\mathbf{p}}{dt} \qquad (9.3)$$

This shows that **the time rate of change of the linear momentum of a particle is equal to the net force acting on the particle.**

This alternative form of Newton's second law is the form in which Newton presented the law and is actually more general than the form we introduced in Chapter 5. In addition to situations in which the velocity vector varies with time, we can use Equation 9.3 to study phenomena in which the mass changes. For example, the mass of a rocket changes as fuel is burned and ejected from the rocket. We cannot use $\sum \mathbf{F} = m\mathbf{a}$ to analyze rocket propulsion; we must use Equation 9.3, as we will show in Section 9.7.

The real value of Equation 9.3 as a tool for analysis, however, arises if we apply it to a *system* of two or more particles. As we have seen, this leads to a law of conservation of momentum for an isolated system. Just as the law of conservation of energy is useful in solving complex motion problems, the law of conservation of momentum can greatly simplify the analysis of other types of complicated motion.

Quick Quiz 9.1 Two objects have equal kinetic energies. How do the magnitudes of their momenta compare? (a) $p_1 < p_2$ (b) $p_1 = p_2$ (c) $p_1 > p_2$ (d) not enough information to tell.

Quick Quiz 9.2 Your physical education teacher throws a baseball to you at a certain speed, and you catch it. The teacher is next going to throw you a medicine ball whose mass is ten times the mass of the baseball. You are given the following choices: You can have the medicine ball thrown with (a) the same speed as the baseball (b) the same momentum (c) the same kinetic energy. Rank these choices from easiest to hardest to catch.

Using the definition of momentum, Equation 9.1 can be written

$$\frac{d}{dt}(\mathbf{p}_1 + \mathbf{p}_2) = 0$$

Because the time derivative of the total momentum $\mathbf{p}_{\text{tot}} = \mathbf{p}_1 + \mathbf{p}_2$ is *zero*, we conclude that the *total* momentum of the system must remain constant:

$$\mathbf{p}_{\text{tot}} = \mathbf{p}_1 + \mathbf{p}_2 = \text{constant} \qquad (9.4)$$

or, equivalently,

$$\mathbf{p}_{1i} + \mathbf{p}_{2i} = \mathbf{p}_{1f} + \mathbf{p}_{2f} \qquad (9.5)$$

where $\mathbf{p}_{1i}$ and $\mathbf{p}_{2i}$ are the initial values and $\mathbf{p}_{1f}$ and $\mathbf{p}_{2f}$ the final values of the momenta for the two particles for the time interval during which the particles interact. Equation 9.5 in component form demonstrates that the total momenta in the x, y, and z directions are all independently conserved:

$$p_{ix} = p_{fx} \qquad p_{iy} = p_{fy} \qquad p_{iz} = p_{fz} \qquad (9.6)$$

This result, known as the **law of conservation of linear momentum**, can be extended to any number of particles in an isolated system. It is considered one of the most important laws of mechanics. We can state it as follows:

▲ **PITFALL PREVENTION**

9.1 Momentum of a *System* is Conserved

Remember that the momentum of an isolated *system* is conserved. The momentum of one particle within an isolated system is not necessarily conserved, because other particles in the system may be interacting with it. Always apply conservation of momentum to an isolated *system*.

Whenever two or more particles in an isolated system interact, the total momentum of the system remains constant.

Conservation of momentum

This law tells us that **the total momentum of an isolated system at all times equals its initial momentum.**

Notice that we have made no statement concerning the nature of the forces acting on the particles of the system. The only requirement is that the forces must be *internal* to the system.

Quick Quiz 9.3 A ball is released and falls toward the ground with no air resistance. The isolated system for which momentum is conserved is (a) the ball (b) the Earth (c) the ball and the Earth (d) impossible to determine.

Quick Quiz 9.4 A car and a large truck traveling at the same speed make a head-on collision and stick together. Which vehicle experiences the larger change in the magnitude of momentum? (a) the car (b) the truck (c) The change in the magnitude of momentum is the same for both. (d) impossible to determine.

Example 9.1   The Archer   **Interactive**

Let us consider the situation proposed at the beginning of this section. A 60-kg archer stands at rest on frictionless ice and fires a 0.50-kg arrow horizontally at 50 m/s (Fig. 9.2). With what velocity does the archer move across the ice after firing the arrow?

Solution We *cannot* solve this problem using Newton's second law, $\Sigma \mathbf{F} = m\mathbf{a}$, because we have no information about the force on the arrow or its acceleration. We *cannot* solve this problem using an energy approach because we do not know how much work is done in pulling the bow back or how much potential energy is stored in the bow. However, we *can* solve this problem very easily with conservation of momentum.

Let us take the system to consist of the archer (including the bow) and the arrow. The system is not isolated because the gravitational force and the normal force act on the system. However, these forces are vertical and perpendicular to the motion of the system. Therefore, there are no external forces in the horizontal direction, and we can consider the system to be isolated in terms of momentum components in this direction.

The total horizontal momentum of the system before the arrow is fired is zero ($m_1\mathbf{v}_{1i} + m_2\mathbf{v}_{2i} = 0$), where the archer is particle 1 and the arrow is particle 2. Therefore, the total horizontal momentum after the arrow is fired must be zero; that is,

$$m_1\mathbf{v}_{1f} + m_2\mathbf{v}_{2f} = 0$$

We choose the direction of firing of the arrow as the positive x direction. With $m_1 = 60$ kg, $m_2 = 0.50$ kg, and $\mathbf{v}_{2f} = 50\hat{\mathbf{i}}$ m/s, solving for $\mathbf{v}_{1f}$, we find the recoil velocity of the archer to be

$$\mathbf{v}_{1f} = -\frac{m_2}{m_1}\mathbf{v}_{2f} = -\left(\frac{0.50 \text{ kg}}{60 \text{ kg}}\right)(50\hat{\mathbf{i}} \text{ m/s}) = \boxed{-0.42\hat{\mathbf{i}} \text{ m/s}}$$

The negative sign for $\mathbf{v}_{1f}$ indicates that the archer is moving to the left after the arrow is fired, in the direction opposite the direction of motion of the arrow, in accordance with Newton's third law. Because the archer is much more massive than the arrow, his acceleration and consequent velocity are much smaller than the acceleration and velocity of the arrow.

What If? What if the arrow were shot in a direction that makes an angle θ with the horizontal? How will this change the recoil velocity of the archer?

Answer The recoil velocity should decrease in magnitude because only a component of the velocity is in the x direction.

Figure 9.2 (Example 9.1) An archer fires an arrow horizontally to the right. Because he is standing on frictionless ice, he will begin to slide to the left across the ice.

If the arrow were shot straight up, for example, there would be no recoil at all—the archer would just be pressed down into the ice because of the firing of the arrow.

Only the *x* component of the momentum of the arrow should be used in a conservation of momentum statement, because momentum is only conserved in the *x* direction. In the *y* direction, the normal force from the ice and the gravitational force are external influences on the system. Conservation of momentum in the *x* direction gives us

$$m_1 v_{1f} + m_2 v_{2f} \cos \theta = 0$$

leading to

$$v_{1f} = -\frac{m_2}{m_1} v_{2f} \cos \theta$$

For $\theta = 0$, $\cos \theta = 1$ and this reduces to the value when the arrow is fired horizontally. For nonzero values of θ, the cosine function is less than 1 and the recoil velocity is less than the value calculated for $\theta = 0$. If $\theta = 90°$, $\cos \theta = 0$, and there is no recoil velocity v_{1f}, as we argued conceptually.

 At the Interactive Worked Example link at **http://www.pse6.com,** *you can change the mass of the archer and the mass and speed of the arrow.*

Example 9.2 Breakup of a Kaon at Rest

One type of nuclear particle, called the *neutral kaon* (K^0), breaks up into a pair of other particles called *pions* (π^+ and π^-) that are oppositely charged but equal in mass, as illustrated in Figure 9.3. Assuming the kaon is initially at rest, prove that the two pions must have momenta that are equal in magnitude and opposite in direction.

Solution The breakup of the kaon can be written

$$K^0 \longrightarrow \pi^+ + \pi^-$$

If we let $\mathbf{p}^+$ be the final momentum of the positive pion and $\mathbf{p}^-$ the final momentum of the negative pion, the final momentum of the system consisting of the two pions can be written

$$\mathbf{p}_f = \mathbf{p}^+ + \mathbf{p}^-$$

Because the kaon is at rest before the breakup, we know that $\mathbf{p}_i = 0$. Because the momentum of the isolated system (the kaon before the breakup, the two pions afterward) is conserved, $\mathbf{p}_i = \mathbf{p}_f = 0$, so that $\mathbf{p}^+ + \mathbf{p}^- = 0$, or

$$\mathbf{p}^+ = -\mathbf{p}^-$$

An important point to learn from this problem is that even though it deals with objects that are very different from those in the preceding example, the physics is identical: *linear momentum is conserved in an isolated system.*

Figure 9.3 (Example 9.2) A kaon at rest breaks up spontaneously into a pair of oppositely charged pions. The pions move apart with momenta that are equal in magnitude but opposite in direction.

9.2 Impulse and Momentum

According to Equation 9.3, the momentum of a particle changes if a net force acts on the particle. Knowing the change in momentum caused by a force is useful in solving some types of problems. To build a better understanding of this important concept, let us assume that a single force **F** acts on a particle and that this force may vary with time. According to Newton's second law, $\mathbf{F} = d\mathbf{p}/dt$, or

$$d\mathbf{p} = \mathbf{F}\, dt \tag{9.7}$$

We can integrate[2] this expression to find the change in the momentum of a particle when the force acts over some time interval. If the momentum of the particle changes from $\mathbf{p}_i$ at time t_i to $\mathbf{p}_f$ at time t_f, integrating Equation 9.7 gives

---

[2] Note that here we are integrating force with respect to time. Compare this with our efforts in Chapter 7, where we integrated force with respect to position to find the work done by the force.

$$\Delta\mathbf{p} = \mathbf{p}_f - \mathbf{p}_i = \int_{t_i}^{t_f} \mathbf{F}\,dt \qquad (9.8)$$

To evaluate the integral, we need to know how the force varies with time. The quantity on the right side of this equation is called the **impulse** of the force $\mathbf{F}$ acting on a particle over the time interval $\Delta t = t_f - t_i$. Impulse is a vector defined by

$$\mathbf{I} \equiv \int_{t_i}^{t_f} \mathbf{F}\,dt \qquad (9.9)$$

◄ Impulse of a force

Equation 9.8 is an important statement known as the **impulse–momentum theorem:**[3]

> The impulse of the force $\mathbf{F}$ acting on a particle equals the change in the momentum of the particle.

◄ Impulse–momentum theorem

This statement is equivalent to Newton's second law. From this definition, we see that impulse is a vector quantity having a magnitude equal to the area under the force–time curve, as described in Figure 9.4a. In this figure, it is assumed that the force varies in time in the general manner shown and is nonzero in the time interval $\Delta t = t_f - t_i$. The direction of the impulse vector is the same as the direction of the change in momentum. Impulse has the dimensions of momentum—that is, ML/T. Note that impulse is *not* a property of a particle; rather, it is a measure of the degree to which an external force changes the momentum of the particle. Therefore, when we say that an impulse is given to a particle, we mean that momentum is transferred from an external agent to that particle.

Because the force imparting an impulse can generally vary in time, it is convenient to define a time-averaged force

$$\overline{\mathbf{F}} \equiv \frac{1}{\Delta t}\int_{t_i}^{t_f} \mathbf{F}\,dt \qquad (9.10)$$

where $\Delta t = t_f - t_i$. (This is an application of the mean value theorem of calculus.) Therefore, we can express Equation 9.9 as

$$\mathbf{I} \equiv \overline{\mathbf{F}}\,\Delta t \qquad (9.11)$$

(a)

(b)

Airbags in automobiles have saved countless lives in accidents. The airbag increases the time interval during which the passenger is brought to rest, thereby decreasing the force on (and resultant injury to) the passenger.

Courtesy of Saab

Figure 9.4 (a) A force acting on a particle may vary in time. The impulse imparted to the particle by the force is the area under the force-versus-time curve. (b) In the time interval Δt, the time-averaged force (horizontal dashed line) gives the same impulse to a particle as does the time-varying force described in part (a).

Area $= \overline{F}\Delta t$

---

[3]   Although we assumed that only a single force acts on the particle, the impulse–momentum theorem is valid when several forces act; in this case, we replace $\mathbf{F}$ in Equation 9.8 with $\Sigma\mathbf{F}$.

This time-averaged force, shown in Figure 9.4b, can be interpreted as the constant force that would give to the particle in the time interval Δt the same impulse that the time-varying force gives over this same interval.

In principle, if **F** is known as a function of time, the impulse can be calculated from Equation 9.9. The calculation becomes especially simple if the force acting on the particle is constant. In this case, $\overline{\mathbf{F}} = \mathbf{F}$ and Equation 9.11 becomes

$$\mathbf{I} = \mathbf{F}\Delta t \qquad (9.12)$$

In many physical situations, we shall use what is called the **impulse approximation, in which we assume that one of the forces exerted on a particle acts for a short time but is much greater than any other force present.** This approximation is especially useful in treating collisions in which the duration of the collision is very short. When this approximation is made, we refer to the force as an *impulsive force.* For example, when a baseball is struck with a bat, the time of the collision is about 0.01 s and the average force that the bat exerts on the ball in this time is typically several thousand newtons. Because this contact force is much greater than the magnitude of the gravitational force, the impulse approximation justifies our ignoring the gravitational forces exerted on the ball and bat. When we use this approximation, it is important to remember that $\mathbf{p}_i$ and $\mathbf{p}_f$ represent the momenta *immediately* before and after the collision, respectively. Therefore, in any situation in which it is proper to use the impulse approximation, the particle moves very little during the collision.

Quick Quiz 9.5 Two objects are at rest on a frictionless surface. Object 1 has a greater mass than object 2. When a constant force is applied to object 1, it accelerates through a distance d. The force is removed from object 1 and is applied to object 2. At the moment when object 2 has accelerated through the same distance d, which statements are true? (a) $p_1 < p_2$ (b) $p_1 = p_2$ (c) $p_1 > p_2$ (d) $K_1 < K_2$ (e) $K_1 = K_2$ (f) $K_1 > K_2$.

Quick Quiz 9.6 Two objects are at rest on a frictionless surface. Object 1 has a greater mass than object 2. When a force is applied to object 1, it accelerates for a time interval Δt. The force is removed from object 1 and is applied to object 2. After object 2 has accelerated for the same time interval Δt, which statements are true? (a) $p_1 < p_2$ (b) $p_1 = p_2$ (c) $p_1 > p_2$ (d) $K_1 < K_2$ (e) $K_1 = K_2$ (f) $K_1 > K_2$.

Quick Quiz 9.7 Rank an automobile dashboard, seatbelt, and airbag in terms of (a) the impulse and (b) the average force they deliver to a front-seat passenger during a collision, from greatest to least.

Example 9.3 Teeing Off

A golf ball of mass 50 g is struck with a club (Fig. 9.5). The force exerted by the club on the ball varies from zero, at the instant before contact, up to some maximum value and then back to zero when the ball leaves the club. Thus, the force–time curve is qualitatively described by Figure 9.4. Assuming that the ball travels 200 m, estimate the magnitude of the impulse caused by the collision.

Solution Let us use Ⓐ to denote the position of the ball when the club first contacts it, Ⓑ to denote the position of the ball when the club loses contact with the ball, and Ⓒ to denote the position of the ball upon landing. Neglecting air resistance, we can use Equation 4.14 for the range of a projectile:

$$R = x_C = \frac{v_B^2}{g} \sin 2\theta_B$$

Let us assume that the launch angle θ_B is 45°, the angle that provides the maximum range for any given launch velocity. This assumption gives $\sin 2\theta_B = 1$, and the launch velocity of the ball is

$$v_B = \sqrt{Rg} \approx \sqrt{(200 \text{ m})(9.80 \text{ m/s}^2)} = 44 \text{ m/s}$$

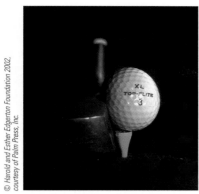

Figure 9.5 (Example 9.3) A golf ball being struck by a club. Note the deformation of the ball due to the large force from the club.

Considering initial and final values of the ball's velocity for the time interval for the collision, $v_i = v_A = 0$ and $v_f = v_B$. Hence, the magnitude of the impulse imparted to the ball is

$$I = \Delta p = mv_B - mv_A = (50 \times 10^{-3} \text{ kg})(44 \text{ m/s}) - 0$$

$$= \boxed{2.2 \text{ kg} \cdot \text{m/s}}$$

What If? What if you were asked to find the average force on the ball during the collision with the club? Can you determine this value?

Answer With the information given in the problem, we cannot find the average force. Considering Equation 9.11, we would need to know the time interval of the collision in order to calculate the average force. If we *assume* that the time interval is 0.01 s as it was for the baseball in the discussion after Equation 9.12, we can estimate the magnitude of the average force:

$$\bar{F} = \frac{I}{\Delta t} = \frac{2.2 \text{ kg} \cdot \text{m/s}}{0.01 \text{ s}} = 2 \times 10^2 \text{ N}$$

where we have kept only one significant figure due to our rough estimate of the time interval.

Example 9.4   How Good Are the Bumpers?

In a particular crash test, a car of mass 1 500 kg collides with a wall, as shown in Figure 9.6. The initial and final velocities of the car are $\mathbf{v}_i = -15.0\hat{\mathbf{i}}$ m/s and $\mathbf{v}_f = 2.60\hat{\mathbf{i}}$ m/s, respectively. If the collision lasts for 0.150 s, find the impulse caused by the collision and the average force exerted on the car.

Solution Let us assume that the force exerted by the wall on the car is large compared with other forces on the car so that we can apply the impulse approximation. Furthermore, we note that the gravitational force and the normal force

exerted by the road on the car are perpendicular to the motion and therefore do not affect the horizontal momentum.

The initial and final momenta of the car are

$$\mathbf{p}_i = m\mathbf{v}_i = (1\,500 \text{ kg})(-15.0\hat{\mathbf{i}} \text{ m/s})$$

$$= -2.25 \times 10^4 \hat{\mathbf{i}} \text{ kg} \cdot \text{m/s}$$

$$\mathbf{p}_f = m\mathbf{v}_f = (1\,500 \text{ kg})(2.60\hat{\mathbf{i}} \text{ m/s})$$

$$= 0.39 \times 10^4 \hat{\mathbf{i}} \text{ kg} \cdot \text{m/s}$$

Hence, the impulse is equal to

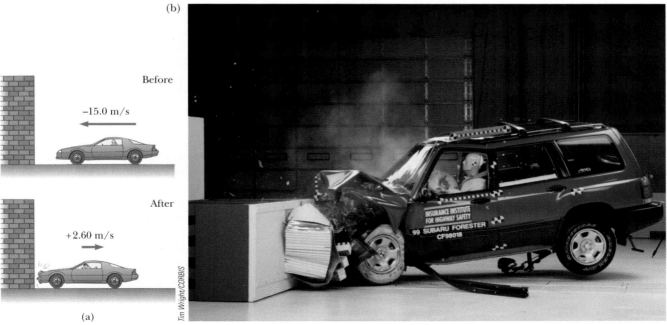

(b)

Before

−15.0 m/s

After

+2.60 m/s

(a)

Figure 9.6 (Example 9.4) (a) This car's momentum changes as a result of its collision with the wall. (b) In a crash test, much of the car's initial kinetic energy is transformed into energy associated with the damage to the car.

$$\mathbf{I} = \Delta\mathbf{p} = \mathbf{p}_f - \mathbf{p}_i = 0.39 \times 10^4 \hat{\mathbf{i}} \text{ kg·m/s}$$
$$- (-2.25 \times 10^4 \hat{\mathbf{i}} \text{ kg·m/s})$$
$$\mathbf{I} = \boxed{2.64 \times 10^4 \hat{\mathbf{i}} \text{ kg·m/s}}$$

The average force exerted by the wall on the car is

$$\overline{\mathbf{F}} = \frac{\Delta\mathbf{p}}{\Delta t} = \frac{2.64 \times 10^4 \hat{\mathbf{i}} \text{ kg·m/s}}{0.150 \text{ s}} = \boxed{1.76 \times 10^5 \hat{\mathbf{i}} \text{ N}}$$

In this problem, note that the signs of the velocities indicate the reversal of directions. What would the mathematics be describing if both the initial and final velocities had the same sign?

What If? What if the car did not rebound from the wall? Suppose the final velocity of the car is zero and the time interval of the collision remains at 0.150 s. Would this represent a larger or a smaller force by the wall on the car?

Answer In the original situation in which the car rebounds, the force by the wall on the car does two things in the time interval—it (1) stops the car and (2) causes it to move away from the wall at 2.60 m/s after the collision. If the car does not rebound, the force is only doing the first of these, stopping the car. This will require a *smaller* force.

Mathematically, in the case of the car that does not rebound, the impulse is

$$\mathbf{I} = \Delta\mathbf{p} = \mathbf{p}_f - \mathbf{p}_i = 0 - (-2.25 \times 10^4 \hat{\mathbf{i}} \text{ kg·m/s})$$
$$= 2.25 \times 10^4 \hat{\mathbf{i}} \text{ kg·m/s}$$

The average force exerted by the wall on the car is

$$\overline{\mathbf{F}} = \frac{\Delta\mathbf{p}}{\Delta t} = \frac{2.25 \times 10^4 \hat{\mathbf{i}} \text{ kg·m/s}}{0.150 \text{ s}} = 1.50 \times 10^5 \hat{\mathbf{i}} \text{ N}$$

which is indeed smaller than the previously calculated value, as we argued conceptually.

9.3 Collisions in One Dimension

Figure 9.7 (a) The collision between two objects as the result of direct contact. (b) The "collision" between two charged particles.

In this section we use the law of conservation of linear momentum to describe what happens when two particles collide. We use the term **collision** to represent an event during which two particles come close to each other and interact by means of forces. The time interval during which the velocities of the particles change from initial to final values is assumed to be short. The interaction forces are assumed to be much greater than any external forces present, so we can use the impulse approximation.

A collision may involve physical contact between two macroscopic objects, as described in Figure 9.7a, but the notion of what we mean by collision must be generalized because "physical contact" on a submicroscopic scale is ill-defined and hence meaningless. To understand this, consider a collision on an atomic scale (Fig. 9.7b), such as the collision of a proton with an alpha particle (the nucleus of a helium atom). Because the particles are both positively charged, they repel each other due to the strong electrostatic force between them at close separations and never come into "physical contact."

When two particles of masses m_1 and m_2 collide as shown in Figure 9.7, the impulsive forces may vary in time in complicated ways, such as that shown in Figure 9.4. Regardless of the complexity of the time behavior of the force of interaction, however, this force is internal to the system of two particles. Thus, the two particles form an isolated system, and the momentum of the system must be conserved. Therefore, the total momentum of an isolated system just before a collision equals the total momentum of the system just after the collision.

In contrast, the total kinetic energy of the system of particles may or may not be conserved, depending on the type of collision. In fact, whether or not kinetic energy is conserved is used to classify collisions as either *elastic* or *inelastic*.

An **elastic collision** between two objects is one in which **the total kinetic energy (as well as total momentum) of the system is the same before and after the collision.** Collisions between certain objects in the macroscopic world, such as billiard balls, are only *approximately* elastic because some deformation and loss of kinetic energy take place. For example, you can hear a billiard ball collision, so you know that some of the energy is being transferred away from the system by sound. An elastic collision must be perfectly silent! *Truly* elastic collisions occur between atomic and subatomic particles.

An **inelastic collision** is one in which **the total kinetic energy of the system is not the same before and after the collision (even though the momentum of the system is conserved).** Inelastic collisions are of two types. When the colliding objects stick together after the collision, as happens when a meteorite collides with the Earth,

Elastic collision

Inelastic collision

the collision is called **perfectly inelastic.** When the colliding objects do not stick together, but some kinetic energy is lost, as in the case of a rubber ball colliding with a hard surface, the collision is called **inelastic** (with no modifying adverb). When the rubber ball collides with the hard surface, some of the kinetic energy of the ball is lost when the ball is deformed while it is in contact with the surface.

In most collisions, the kinetic energy of the system is *not* conserved because some of the energy is converted to internal energy and some of it is transferred away by means of sound. Elastic and perfectly inelastic collisions are limiting cases; most collisions fall somewhere between them.

In the remainder of this section, we treat collisions in one dimension and consider the two extreme cases—perfectly inelastic and elastic collisions. The important distinction between these two types of collisions is that **momentum of the system is conserved in all collisions, but kinetic energy of the system is conserved only in elastic collisions.**

Perfectly Inelastic Collisions

Consider two particles of masses m_1 and m_2 moving with initial velocities $\mathbf{v}_{1i}$ and $\mathbf{v}_{2i}$ along the same straight line, as shown in Figure 9.8. The two particles collide head-on, stick together, and then move with some common velocity $\mathbf{v}_f$ after the collision. Because the momentum of an isolated system is conserved in *any* collision, we can say that the total momentum before the collision equals the total momentum of the composite system after the collision:

$$m_1\mathbf{v}_{1i} + m_2\mathbf{v}_{2i} = (m_1 + m_2)\mathbf{v}_f \tag{9.13}$$

Solving for the final velocity gives

$$\mathbf{v}_f = \frac{m_1\mathbf{v}_{1i} + m_2\mathbf{v}_{2i}}{m_1 + m_2} \tag{9.14}$$

Elastic Collisions

Consider two particles of masses m_1 and m_2 moving with initial velocities $\mathbf{v}_{1i}$ and $\mathbf{v}_{2i}$ along the same straight line, as shown in Figure 9.9. The two particles collide head-on and then leave the collision site with different velocities, $\mathbf{v}_{1f}$ and $\mathbf{v}_{2f}$. If the collision is elastic, both the momentum and kinetic energy of the system are conserved. Therefore, considering velocities along the horizontal direction in Figure 9.9, we have

$$m_1v_{1i} + m_2v_{2i} = m_1v_{1f} + m_2v_{2f} \tag{9.15}$$

$$\tfrac{1}{2}m_1v_{1i}{}^2 + \tfrac{1}{2}m_2v_{2i}{}^2 = \tfrac{1}{2}m_1v_{1f}{}^2 + \tfrac{1}{2}m_2v_{2f}{}^2 \tag{9.16}$$

Because all velocities in Figure 9.9 are either to the left or the right, they can be represented by the corresponding speeds along with algebraic signs indicating direction. We shall indicate v as positive if a particle moves to the right and negative if it moves to the left.

In a typical problem involving elastic collisions, there are two unknown quantities, and Equations 9.15 and 9.16 can be solved simultaneously to find these. An alternative approach, however—one that involves a little mathematical manipulation of Equation 9.16—often simplifies this process. To see how, let us cancel the factor $\tfrac{1}{2}$ in Equation 9.16 and rewrite it as

$$m_1(v_{1i}{}^2 - v_{1f}{}^2) = m_2(v_{2f}{}^2 - v_{2i}{}^2)$$

and then factor both sides:

$$m_1(v_{1i} - v_{1f})(v_{1i} + v_{1f}) = m_2(v_{2f} - v_{2i})(v_{2f} + v_{2i}) \tag{9.17}$$

Next, let us separate the terms containing m_1 and m_2 in Equation 9.15 to obtain

$$m_1(v_{1i} - v_{1f}) = m_2(v_{2f} - v_{2i}) \tag{9.18}$$

▲ **PITFALL PREVENTION**

9.2 Inelastic Collisions

Generally, inelastic collisions are hard to analyze unless additional information is provided. This appears in the mathematical representation as having more unknowns than equations.

Before collision

(a)

After collision

$m_1 + m_2$

(b)

Active Figure 9.8 Schematic representation of a perfectly inelastic head-on collision between two particles: (a) before collision and (b) after collision.

At the Active Figures link at http://www.pse6.com, *you can adjust the masses and velocities of the colliding objects to see the effect on the final velocity.*

Before collision

(a)

After collision

(b)

Active Figure 9.9 Schematic representation of an elastic head-on collision between two particles: (a) before collision and (b) after collision.

At the Active Figures link at http://www.pse6.com, *you can adjust the masses and velocities of the colliding objects to see the effect on the final velocities.*

PITFALL PREVENTION

9.3 Not a General Equation

We have spent some effort on deriving Equation 9.19, but remember that it can only be used in a very *specific* situation—a one-dimensional, elastic collision between two objects. The *general* concept is conservation of momentum (and conservation of kinetic energy if the collision is elastic) for an isolated system.

Elastic collision: particle 2 initially at rest

 PITFALL PREVENTION

9.4 Momentum and Kinetic Energy in Collisions

Momentum of an isolated system is conserved in *all* collisions. Kinetic energy of an isolated system is conserved *only* in elastic collisions. Why? Because there are several types of energy into which kinetic energy can transform, or be transferred out of the system (so that the system may *not* be isolated in terms of energy during the collision). However, there is only one type of momentum.

To obtain our final result, we divide Equation 9.17 by Equation 9.18 and obtain

$$v_{1i} + v_{1f} = v_{2f} + v_{2i}$$
$$v_{1i} - v_{2i} = -(v_{1f} - v_{2f}) \tag{9.19}$$

This equation, in combination with Equation 9.15, can be used to solve problems dealing with elastic collisions. According to Equation 9.19, the *relative* velocity of the two particles before the collision, $v_{1i} - v_{2i}$, equals the negative of their relative velocity after the collision, $-(v_{1f} - v_{2f})$.

Suppose that the masses and initial velocities of both particles are known. Equations 9.15 and 9.19 can be solved for the final velocities in terms of the initial velocities because there are two equations and two unknowns:

$$v_{1f} = \left(\frac{m_1 - m_2}{m_1 + m_2}\right)v_{1i} + \left(\frac{2m_2}{m_1 + m_2}\right)v_{2i} \tag{9.20}$$

$$v_{2f} = \left(\frac{2m_1}{m_1 + m_2}\right)v_{1i} + \left(\frac{m_2 - m_1}{m_1 + m_2}\right)v_{2i} \tag{9.21}$$

It is important to use the appropriate signs for v_{1i} and v_{2i} in Equations 9.20 and 9.21. For example, if particle 2 is moving to the left initially, then v_{2i} is negative.

Let us consider some special cases. If $m_1 = m_2$, then Equations 9.20 and 9.21 show us that $v_{1f} = v_{2i}$ and $v_{2f} = v_{1i}$. That is, the particles exchange velocities if they have equal masses. This is approximately what one observes in head-on billiard ball collisions—the cue ball stops, and the struck ball moves away from the collision with the same velocity that the cue ball had.

If particle 2 is initially at rest, then $v_{2i} = 0$, and Equations 9.20 and 9.21 become

$$v_{1f} = \left(\frac{m_1 - m_2}{m_1 + m_2}\right)v_{1i} \tag{9.22}$$

$$v_{2f} = \left(\frac{2m_1}{m_1 + m_2}\right)v_{1i} \tag{9.23}$$

If m_1 is much greater than m_2 and $v_{2i} = 0$, we see from Equations 9.22 and 9.23 that $v_{1f} \approx v_{1i}$ and $v_{2f} \approx 2v_{1i}$. That is, when a very heavy particle collides head-on with a very light one that is initially at rest, the heavy particle continues its motion unaltered after the collision and the light particle rebounds with a speed equal to about twice the initial speed of the heavy particle. An example of such a collision would be that of a moving heavy atom, such as uranium, striking a light atom, such as hydrogen.

If m_2 is much greater than m_1 and particle 2 is initially at rest, then $v_{1f} \approx -v_{1i}$ and $v_{2f} \approx 0$. That is, when a very light particle collides head-on with a very heavy particle that is initially at rest, the light particle has its velocity reversed and the heavy one remains approximately at rest.

Quick Quiz 9.8 In a perfectly inelastic one-dimensional collision between two objects, what condition alone is necessary so that *all* of the original kinetic energy of the system is gone after the collision? (a) The objects must have momenta with the same magnitude but opposite directions. (b) The objects must have the same mass. (c) The objects must have the same velocity. (d) The objects must have the same speed, with velocity vectors in opposite directions.

Quick Quiz 9.9 A table-tennis ball is thrown at a stationary bowling ball. The table-tennis ball makes a one-dimensional elastic collision and bounces back along the same line. After the collision, compared to the bowling ball, the table-tennis ball has (a) a larger magnitude of momentum and more kinetic energy (b) a smaller magnitude of momentum and more kinetic energy (c) a larger magnitude of momentum and less kinetic energy (d) a smaller magnitude of momentum and less kinetic energy (e) the same magnitude of momentum and the same kinetic energy.

Example 9.5   The Executive Stress Reliever

Interactive

An ingenious device that illustrates conservation of momentum and kinetic energy is shown in Figure 9.10. It consists of five identical hard balls supported by strings of equal lengths. When ball 1 is pulled out and released, after the almost-elastic collision between it and ball 2, ball 5 moves out, as shown in Figure 9.10b. If balls 1 and 2 are pulled out and released, balls 4 and 5 swing out, and so forth. Is it ever possible that when ball 1 is released, balls 4 and 5 will swing out on the opposite side and travel with half the speed of ball 1, as in Figure 9.10c?

Solution No, such movement can never occur if we assume the collisions are elastic. The momentum of the system before the collision is mv, where m is the mass of ball 1 and v is its speed just before the collision. After the collision, we would have two balls, each of mass m moving with speed $v/2$. The total momentum of the system after the collision would be $m(v/2) + m(v/2) = mv$. Thus, momentum of the system is conserved. However, the kinetic energy just before the collision is $K_i = \frac{1}{2}mv^2$ and that after the collision is $K_f = \frac{1}{2}m(v/2)^2 + \frac{1}{2}m(v/2)^2 = \frac{1}{4}mv^2$. Thus, kinetic energy of the system is *not* conserved. The only way to have both momentum and kinetic energy conserved is for one ball to move out when one ball is released, two balls to move out when two are released, and so on.

What If? Consider what would happen if balls 4 and 5 are glued together so that they must move together. Now what happens when ball 1 is pulled out and released?

Answer We are now forcing balls 4 and 5 to come out together. We have argued that we cannot conserve both momentum and energy in this case. However, we assumed that ball 1 stopped after striking ball 2. What if we do not make this assumption? Consider the conservation equations with the assumption that ball 1 moves after the collision. For conservation of momentum,

$$p_i = p_f$$

$$mv_{1i} = mv_{1f} + 2mv_{4,5f}$$

where $v_{4,5f}$ refers to the final speed of the ball 4–ball 5 combination. Conservation of kinetic energy gives us

$$K_i = K_f$$

$$\tfrac{1}{2}mv_{1i}^2 = \tfrac{1}{2}mv_{1f}^2 + \tfrac{1}{2}(2m)v_{4,5f}^2$$

Combining these equations, we find

$$v_{4,5f} = \tfrac{2}{3}v_{1i} \qquad v_{1f} = -\tfrac{1}{3}v_{1i}$$

Thus, balls 4 and 5 come out together and ball 1 bounces back from the collision with one third of its original speed.

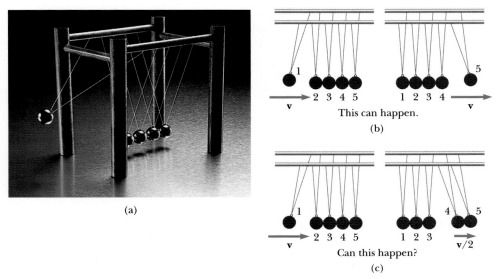

(a)

This can happen.
(b)

Can this happen?
(c)

Figure 9.10 (Example 9.5) An executive stress reliever.

 At the Interactive Worked Example link at **http://www.pse6.com,** you can "glue" balls 4 and 5 together to see the situation discussed above.

Example 9.6 Carry Collision Insurance!

An 1 800-kg car stopped at a traffic light is struck from the rear by a 900-kg car, and the two become entangled, moving along the same path as that of the originally moving car. If the smaller car were moving at 20.0 m/s before the collision, what is the velocity of the entangled cars after the collision?

Solution The phrase "become entangled" tells us that this is a perfectly inelastic collision. We can guess that the final speed is less than 20.0 m/s, the initial speed of the smaller car. The total momentum of the system (the two cars) before the collision must equal the total momentum immediately after the collision because momentum of an isolated system is conserved in any type of collision. The magnitude of the total momentum of the system before the collision is equal to that of the smaller car because the larger car is initially at rest:

$$p_i = m_1 v_i = (900 \text{ kg})(20.0 \text{ m/s}) = 1.80 \times 10^4 \text{ kg·m/s}$$

After the collision, the magnitude of the momentum of the entangled cars is

$$p_f = (m_1 + m_2)v_f = (2\ 700 \text{ kg})v_f$$

Equating the initial and final momenta of the system and solving for v_f, the final velocity of the entangled cars, we have

$$v_f = \frac{p_i}{m_1 + m_2} = \frac{1.80 \times 10^4 \text{ kg·m/s}}{2\ 700 \text{ kg}} = \boxed{6.67 \text{ m/s}}$$

Because the final velocity is positive, the direction of the final velocity is the same as the velocity of the initially moving car.

What If? Suppose we reverse the masses of the cars–a stationary 900-kg car is struck by a moving 1 800-kg car. Is the final speed the same as before?

Answer Intuitively, we can guess that the final speed will be higher, based on common experiences in driving. Mathematically, this should be the case because the system has a larger momentum if the initially moving car is the more massive one. Solving for the new final velocity, we find

$$v_f = \frac{p_i}{m_1 + m_2} = \frac{(1\ 800 \text{ kg})(20.0 \text{ m/s})}{2\ 700 \text{ kg}} = 13.3 \text{ m/s}$$

which is indeed higher than the previous final velocity.

Example 9.7 The Ballistic Pendulum

The ballistic pendulum (Fig. 9.11) is an apparatus used to measure the speed of a fast-moving projectile, such as a bullet. A bullet of mass m_1 is fired into a large block of wood of mass m_2 suspended from some light wires. The bullet embeds in the block, and the entire system swings through a height h. How can we determine the speed of the bullet from a measurement of h?

Solution Figure 9.11a helps to conceptualize the situation. Let configuration Ⓐ be the bullet and block before the collision, and configuration Ⓑ be the bullet and block immediately after colliding. The bullet and the block form an isolated system, so we can categorize the collision between them as a conservation of momentum problem. The collision is perfectly inelastic. To analyze the collision, we note that Equation 9.14 gives the speed of the system right after the collision when we assume the impulse approximation. Noting that $v_{2A} = 0$, Equation 9.14 becomes

$$(1) \qquad v_B = \frac{m_1 v_{1A}}{m_1 + m_2}$$

For the process during which the bullet–block combination swings upward to height h (ending at configuration Ⓒ), we focus on a *different* system—the bullet, the block, and the Earth. This is an isolated system for energy, so we categorize this part of the motion as a conservation of mechanical energy problem:

$$K_B + U_B = K_C + U_C$$

We begin to analyze the problem by finding the total kinetic energy of the system right after the collision:

$$(2) \qquad K_B = \tfrac{1}{2}(m_1 + m_2)v_B^2$$

(a)

(b)

Courtesy of Central Scientific Company

Figure 9.11 (Example 9.7) (a) Diagram of a ballistic pendulum. Note that $\mathbf{v}_{1A}$ is the velocity of the bullet just before the collision and $\mathbf{v}_B$ is the velocity of the bullet-block system just after the perfectly inelastic collision. (b) Multiflash photograph of a ballistic pendulum used in the laboratory.

Substituting the value of v_B from Equation (1) into Equation (2) gives

$$K_B = \frac{m_1^2 v_{1A}^2}{2(m_1 + m_2)}$$

This kinetic energy immediately after the collision is *less* than the initial kinetic energy of the bullet, as expected in an inelastic collision.

We define the gravitational potential energy of the system for configuration Ⓑ to be zero. Thus, $U_B = 0$ while $U_C = (m_1 + m_2)gh$. Conservation of energy now leads to

$$\frac{m_1^2 v_{1A}^2}{2(m_1 + m_2)} + 0 = 0 + (m_1 + m_2)gh$$

Solving for v_{1A}, we obtain

$$v_{1A} = \left(\frac{m_1 + m_2}{m_1}\right)\sqrt{2gh}$$

To finalize this problem, note that we had to solve this problem in two steps. Each step involved a different system and a different conservation principle. Because the collision was assumed to be perfectly inelastic, some mechanical energy was converted to internal energy. It would have been *incorrect* to equate the initial kinetic energy of the incoming bullet to the final gravitational potential energy of the bullet–block–Earth combination.

Example 9.8 A Two-Body Collision with a Spring

A block of mass $m_1 = 1.60$ kg initially moving to the right with a speed of 4.00 m/s on a frictionless horizontal track collides with a spring attached to a second block of mass $m_2 = 2.10$ kg initially moving to the left with a speed of 2.50 m/s, as shown in Figure 9.12a. The spring constant is 600 N/m.

(A) Find the velocities of the two blocks after the collision.

Solution Because the spring force is conservative, no kinetic energy is converted to internal energy during the compression of the spring. Ignoring any sound made when the block hits the spring, we can model the collision as being elastic. Equation 9.15 gives us

$$m_1 v_{1i} + m_2 v_{2i} = m_1 v_{1f} + m_2 v_{2f}$$

$$(1.60 \text{ kg})(4.00 \text{ m/s}) + (2.10 \text{ kg})(-2.50 \text{ m/s})$$

$$= (1.60 \text{ kg})v_{1f} + (2.10 \text{ kg})v_{2f}$$

$$(1) \qquad 1.15 \text{ kg·m/s} = (1.60 \text{ kg})v_{1f} + (2.10 \text{ kg})v_{2f}$$

Equation 9.19 gives us

$$v_{1i} - v_{2i} = -(v_{1f} - v_{2f})$$

$$4.00 \text{ m/s} - (-2.50 \text{ m/s}) = -v_{1f} + v_{2f}$$

$$(2) \qquad 6.50 \text{ m/s} = -v_{1f} + v_{2f}$$

Multiplying Equation (2) by 1.60 kg gives us

$$(3) \qquad 10.4 \text{ kg·m/s} = -(1.60 \text{ kg})v_{1f} + (1.60 \text{ kg})v_{2f}$$

Adding Equations (1) and (3) allows us to find v_{2f}:

$$11.55 \text{ kg·m/s} = (3.70 \text{ kg})v_{2f}$$

$$v_{2f} = \frac{11.55 \text{ kg·m/s}}{3.70 \text{ kg}} = \boxed{3.12 \text{ m/s}}$$

Now, Equation (2) allows us to find v_{1f}:

$$6.50 \text{ m/s} = -v_{1f} + 3.12 \text{ m/s}$$

$$v_{1f} = \boxed{-3.38 \text{ m/s}}$$

(B) During the collision, at the instant block 1 is moving to the right with a velocity of $+3.00$ m/s, as in Figure 9.12b, determine the velocity of block 2.

Solution Because the momentum of the system of two blocks is conserved *throughout* the collision for the system of two blocks, we have, for *any* instant during the collision,

$$m_1 v_{1i} + m_2 v_{2i} = m_1 v_{1f} + m_2 v_{2f}$$

We choose the final instant to be that at which block 1 is moving with a velocity of $+3.00$ m/s:

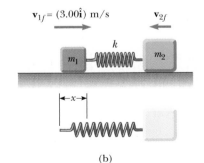

(a)

(b)

Figure 9.12 (Example 9.8) A moving block approaches a second moving block that is attached to a spring.

$$(1.60 \text{ kg})(4.00 \text{ m/s}) + (2.10 \text{ kg})(-2.50 \text{ m/s})$$

$$= (1.60 \text{ kg})(3.00 \text{ m/s}) + (2.10 \text{ kg}) v_{2f}$$

$$v_{2f} = \boxed{-1.74 \text{ m/s}}$$

The negative value for v_{2f} means that block 2 is still moving to the left at the instant we are considering.

(C) Determine the distance the spring is compressed at that instant.

Solution To determine the distance that the spring is compressed, shown as x in Figure 9.12b, we can use the principle of conservation of mechanical energy for the system of the spring and two blocks because no friction or other nonconservative forces are acting within the system. We choose the initial configuration of the system to be that existing just before block 1 strikes the spring and the final configuration to be that when block 1 is moving to the right at 3.00 m/s. Thus, we have

$$K_i + U_i = K_f + U_f$$

$$\tfrac{1}{2}m_1 v_{1i}^2 + \tfrac{1}{2}m_2 v_{2i}^2 + 0 = \tfrac{1}{2}m_1 v_{1f}^2 + \tfrac{1}{2}m_2 v_{2f}^2 + \tfrac{1}{2}kx^2$$

Substituting the given values and the result to part (B) into this expression gives

$$x = \boxed{0.173 \text{ m}}$$

(D) What is the *maximum* compression of the spring during the collision?

Solution The maximum compression would occur when the two blocks are moving with the same velocity. The conservation of momentum equation for the system can be written

$$m_1 v_{1i} + m_2 v_{2i} = (m_1 + m_2) v_f$$

where the initial instant is just before the collision and the final instant is when the blocks are moving with the same velocity v_f. Solving for v_f,

$$v_f = \frac{m_1 v_{1i} + m_2 v_{2i}}{m_1 + m_2}$$

$$= \frac{(1.60 \text{ kg})(4.00 \text{ m/s}) + (2.10 \text{ kg})(-2.50 \text{ m/s})}{1.60 \text{ kg} + 2.10 \text{ kg}}$$

$$= 0.311 \text{ m/s}$$

Now, we apply conservation of mechanical energy between these two instants as in part (C):

$$K_i + U_i = K_f + U_f$$

$$\tfrac{1}{2}m_1 v_{1i}^2 + \tfrac{1}{2}m_2 v_{2i}^2 + 0 = \tfrac{1}{2}(m_1 + m_2) v_f^2 + \tfrac{1}{2}kx^2$$

Substituting the given values into this expression gives

$$x = \boxed{0.253 \text{ m}}$$

 At the Interactive Worked Example link at **http://www.pse6.com,** *you can change the masses and speeds of the blocks and freeze the motion at the maximum compression of the spring.*

Example 9.9 Slowing Down Neutrons by Collisions

In a nuclear reactor, neutrons are produced when an atom splits in a process called *fission*. These neutrons are moving at about 10^7 m/s and must be slowed down to about 10^3 m/s before they take part in another fission event. They are slowed down by passing them through a solid or liquid material called a *moderator*. The slowing-down process involves elastic collisions. Show that a neutron can lose most of its kinetic energy if it collides elastically with a moderator containing light nuclei, such as deuterium (in "heavy water," D_2O) or carbon (in graphite).

Solution Let us assume that the moderator nucleus of mass m_m is at rest initially and that a neutron of mass m_n and initial speed v_{ni} collides with it head-on. Because these are elastic collisions, both momentum and kinetic energy of the neutron–nucleus system are conserved. Therefore, Equations 9.22 and 9.23 can be applied to the head-on collision of a neutron with a moderator nucleus. We can represent this process by a drawing such as Figure 9.9 with $\mathbf{v}_{2i} = 0$.

The initial kinetic energy of the neutron is

$$K_{ni} = \tfrac{1}{2}m_n v_{ni}^2$$

After the collision, the neutron has kinetic energy $\tfrac{1}{2}m_n v_{nf}^2$, and we can substitute into this the value for v_{nf} given by

Equation 9.22:

$$K_{nf} = \tfrac{1}{2}m_n v_{nf}^2 = \tfrac{1}{2}m_n \left(\frac{m_n - m_m}{m_n + m_m}\right)^2 v_{ni}^2$$

Therefore, the fraction f_n of the initial kinetic energy possessed by the neutron after the collision is

$$(1) \qquad f_n = \frac{K_{nf}}{K_{ni}} = \left(\frac{m_n - m_m}{m_n + m_m}\right)^2$$

From this result, we see that the final kinetic energy of the neutron is small when m_m is close to m_n and zero when $m_n = m_m$.

We can use Equation 9.23, which gives the final speed of the particle that was initially at rest, to calculate the kinetic energy of the moderator nucleus after the collision:

$$K_{mf} = \tfrac{1}{2}m_m v_{mf}^2 = \frac{2m_n^2 m_m}{(m_n + m_m)^2} v_{ni}^2$$

Hence, the fraction f_m of the initial kinetic energy transferred to the moderator nucleus is

$$(2) \qquad f_m = \frac{K_{mf}}{K_{ni}} = \frac{4m_n m_m}{(m_n + m_m)^2}$$

Because the total kinetic energy of the system is conserved, Equation (2) can also be obtained from Equation (1) with the condition that $f_n + f_m = 1$, so that $f_m = 1 - f_n$.

Suppose that heavy water is used for the moderator. For collisions of the neutrons with deuterium nuclei in D_2O ($m_m = 2m_n$), $f_n = 1/9$ and $f_m = 8/9$. That is, 89% of the neutron's kinetic energy is transferred to the deuterium nucleus. In practice, the moderator efficiency is reduced because head-on collisions are very unlikely.

How do the results differ when graphite (^{12}C, as found in pencil lead) is used as the moderator?

9.4 Two-Dimensional Collisions

In Section 9.1, we showed that the momentum of a system of two particles is conserved when the system is isolated. For any collision of two particles, this result implies that the momentum in each of the directions x, y, and z is conserved. An important subset of collisions takes place in a plane. The game of billiards is a familiar example involving multiple collisions of objects moving on a two-dimensional surface. For such two-dimensional collisions, we obtain two component equations for conservation of momentum:

$$m_1 v_{1ix} + m_2 v_{2ix} = m_1 v_{1fx} + m_2 v_{2fx}$$

$$m_1 v_{1iy} + m_2 v_{2iy} = m_1 v_{1fy} + m_2 v_{2fy}$$

where we use three subscripts in these equations to represent, respectively, (1) the identification of the object, (2) initial and final values, and (3) the velocity component.

Let us consider a two-dimensional problem in which particle 1 of mass m_1 collides with particle 2 of mass m_2, where particle 2 is initially at rest, as in Figure 9.13. After the collision, particle 1 moves at an angle θ with respect to the horizontal and particle 2 moves at an angle ϕ with respect to the horizontal. This is called a *glancing* collision. Applying the law of conservation of momentum in component form and noting that the initial y component of the momentum of the two-particle system is zero, we obtain

$$m_1 v_{1i} = m_1 v_{1f} \cos \theta + m_2 v_{2f} \cos \phi \qquad (9.24)$$

$$0 = m_1 v_{1f} \sin \theta - m_2 v_{2f} \sin \phi \qquad (9.25)$$

where the minus sign in Equation 9.25 comes from the fact that after the collision, particle 2 has a y component of velocity that is downward. We now have two independent equations. As long as no more than two of the seven quantities in Equations 9.24 and 9.25 are unknown, we can solve the problem.

If the collision is elastic, we can also use Equation 9.16 (conservation of kinetic energy) with $v_{2i} = 0$ to give

$$\tfrac{1}{2} m_1 v_{1i}^2 = \tfrac{1}{2} m_1 v_{1f}^2 + \tfrac{1}{2} m_2 v_{2f}^2 \qquad (9.26)$$

▲ **PITFALL PREVENTION**

9.5 Don't Use Equation 9.19

Equation 9.19, relating the initial and final relative velocities of two colliding objects, is only valid for one-dimensional elastic collisions. Do not use this equation when analyzing two-dimensional collisions.

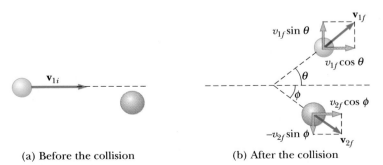

(a) Before the collision   (b) After the collision

Active Figure 9.13 An elastic glancing collision between two particles.

At the Active Figures link at http://www.pse6.com, *you can adjust the speed and position of the blue particle and the masses of both particles to see the effects.*

Knowing the initial speed of particle 1 and both masses, we are left with four unknowns (v_{1f}, v_{2f}, θ, and ϕ). Because we have only three equations, one of the four remaining quantities must be given if we are to determine the motion after the collision from conservation principles alone.

If the collision is inelastic, kinetic energy is *not* conserved and Equation 9.26 does *not* apply.

PROBLEM-SOLVING HINTS

Two-Dimensional Collisions

The following procedure is recommended when dealing with problems involving two-dimensional collisions between two objects:

- Set up a coordinate system and define your velocities with respect to that system. It is usually convenient to have the x axis coincide with one of the initial velocities.

- In your sketch of the coordinate system, draw and label all velocity vectors and include all the given information.

- Write expressions for the x and y components of the momentum of each object before and after the collision. Remember to include the appropriate signs for the components of the velocity vectors.

- Write expressions for the total momentum of the system in the x direction before and after the collision and equate the two. Repeat this procedure for the total momentum of the system in the y direction.

- If the collision is inelastic, kinetic energy of the system is *not* conserved, and additional information is probably required. If the collision is *perfectly* inelastic, the final velocities of the two objects are equal. Solve the momentum equations for the unknown quantities.

- If the collision is *elastic*, kinetic energy of the system is conserved, and you can equate the total kinetic energy before the collision to the total kinetic energy after the collision to obtain an additional relationship between the velocities.

Example 9.10   Collision at an Intersection

A 1 500-kg car traveling east with a speed of 25.0 m/s collides at an intersection with a 2 500-kg van traveling north at a speed of 20.0 m/s, as shown in Figure 9.14. Find the direction and magnitude of the velocity of the wreckage after the collision, assuming that the vehicles undergo a perfectly inelastic collision (that is, they stick together).

Solution Let us choose east to be along the positive x direction and north to be along the positive y direction. Before the collision, the only object having momentum in the x direction is the car. Thus, the magnitude of the total initial momentum of the system (car plus van) in the x direction is

$$\sum p_{xi} = (1\,500\text{ kg})(25.0\text{ m/s}) = 3.75 \times 10^4 \text{ kg·m/s}$$

Let us assume that the wreckage moves at an angle θ and speed v_f after the collision. The magnitude of the total momentum in the x direction after the collision is

Figure 9.14 (Example 9.10) An eastbound car colliding with a northbound van.

$$\sum p_{xf} = (4\ 000\ \text{kg})\,v_f \cos\theta$$

Because the total momentum in the x direction is conserved, we can equate these two equations to obtain

(1) $3.75 \times 10^4\ \text{kg}\cdot\text{m/s} = (4\ 000\ \text{kg})\,v_f \cos\theta$

Similarly, the total initial momentum of the system in the y direction is that of the van, and the magnitude of this momentum is $(2\ 500\ \text{kg})(20.0\ \text{m/s}) = 5.00 \times 10^4\ \text{kg}\cdot\text{m/s}$. Applying conservation of momentum to the y direction, we have

$$\sum p_{yi} = \sum p_{yf}$$

(2) $5.00 \times 10^4\ \text{kg}\cdot\text{m/s} = (4\ 000\ \text{kg})\,v_f \sin\theta$

If we divide Equation (2) by Equation (1), we obtain

$$\frac{\sin\theta}{\cos\theta} = \tan\theta = \frac{5.00 \times 10^4}{3.75 \times 10^4} = 1.33$$

$$\theta = \boxed{53.1°}$$

When this angle is substituted into Equation (2), the value of v_f is

$$v_f = \frac{5.00 \times 10^4\ \text{kg}\cdot\text{m/s}}{(4\ 000\ \text{kg})\sin 53.1°} = \boxed{15.6\ \text{m/s}}$$

It might be instructive for you to draw the momentum vectors of each vehicle before the collision and the two vehicles together after the collision.

Example 9.11 Proton–Proton Collision

A proton collides elastically with another proton that is initially at rest. The incoming proton has an initial speed of 3.50×10^5 m/s and makes a glancing collision with the second proton, as in Figure 9.13. (At close separations, the protons exert a repulsive electrostatic force on each other.) After the collision, one proton moves off at an angle of $37.0°$ to the original direction of motion, and the second deflects at an angle of ϕ to the same axis. Find the final speeds of the two protons and the angle ϕ.

Solution The pair of protons is an isolated system. Both momentum and kinetic energy of the system are conserved in this glancing elastic collision. Because $m_1 = m_2$, $\theta = 37.0°$, and we are given that $v_{1i} = 3.50 \times 10^5$ m/s, Equations 9.24, 9.25, and 9.26 become

(1) $v_{1f} \cos 37° + v_{2f} \cos\phi = 3.50 \times 10^5\ \text{m/s}$

(2) $v_{1f} \sin 37.0° - v_{2f} \sin\phi = 0$

(3) $v_{1f}^2 + v_{2f}^2 = (3.50 \times 10^5\ \text{m/s})^2$

$$= 1.23 \times 10^{11}\ \text{m}^2/\text{s}^2$$

We rewrite Equations (1) and (2) as follows:

$$v_{2f} \cos\phi = 3.50 \times 10^5\ \text{m/s} - v_{1f} \cos 37.0°$$

$$v_{2f} \sin\phi = v_{1f} \sin 37.0°$$

Now we square these two equations and add them:

$$v_{2f}^2 \cos^2\phi + v_{2f}^2 \sin^2\phi$$

$$= 1.23 \times 10^{11}\ \text{m}^2/\text{s}^2 - (7.00 \times 10^5\ \text{m/s})v_{1f} \cos 37.0°$$

$$+ v_{1f}^2 \cos^2 37.0° + v_{1f}^2 \sin^2 37.0°$$

$$v_{2f}^2 = 1.23 \times 10^{11} - (5.59 \times 10^5)v_{1f} + v_{1f}^2$$

Substituting into Equation (3) gives

$$v_{1f}^2 + [1.23 \times 10^{11} - (5.59 \times 10^5)v_{1f} + v_{1f}^2]$$

$$= 1.23 \times 10^{11}$$

$$2v_{1f}^2 - (5.59 \times 10^5)v_{1f} = (2v_{1f} - 5.59 \times 10^5)v_{1f} = 0$$

One possibility for the solution of this equation is $v_{1f} = 0$, which corresponds to a head-on collision—the first proton stops and the second continues with the same speed in the same direction. This is not what we want. The other possibility is

$$2v_{1f} - 5.59 \times 10^5 = 0 \quad\longrightarrow\quad v_{1f} = \boxed{2.80 \times 10^5\ \text{m/s}}$$

From Equation (3),

$$v_{2f} = \sqrt{1.23 \times 10^{11} - v_{1f}^2} = \sqrt{1.23 \times 10^{11} - (2.80 \times 10^5)^2}$$

$$= 2.12 \times 10^5\ \text{m/s}$$

and from Equation (2),

$$\phi = \sin^{-1}\left(\frac{v_{1f} \sin 37.0°}{v_{2f}}\right) = \sin^{-1}\left(\frac{(2.80 \times 10^5) \sin 37.0°}{2.12 \times 10^5}\right)$$

$$= \boxed{53.0°}$$

It is interesting to note that $\theta + \phi = 90°$. This result is *not* accidental. Whenever two objects of equal mass collide elastically in a glancing collision and one of them is initially at rest, their final velocities are perpendicular to each other. The next example illustrates this point in more detail.

Example 9.12 Billiard Ball Collision

In a game of billiards, a player wishes to sink a target ball in the corner pocket, as shown in Figure 9.15. If the angle to the corner pocket is $35°$, at what angle θ is the cue ball deflected? Assume that friction and rotational motion are unimportant and that the collision is elastic. Also assume that all billiard balls have the same mass m.

Solution Let ball 1 be the cue ball and ball 2 be the target ball. Because the target ball is initially at rest, conservation of kinetic energy (Eq. 9.16) for the two-ball system gives

$$\tfrac{1}{2}m_1 v_{1i}^2 = \tfrac{1}{2}m_1 v_{1f}^2 + \tfrac{1}{2}m_2 v_{2f}^2$$

But $m_1 = m_2 = m$, so that

(1) $v_{1i}{}^2 = v_{1f}{}^2 + v_{2f}{}^2$

Applying conservation of momentum to the two-dimensional collision gives

(2) $m_1\mathbf{v}_{1i} = m_1\mathbf{v}_{1f} + m_2\mathbf{v}_{2f}$

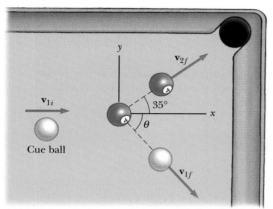

Figure 9.15 (Example 9.12) The cue ball (white) strikes the number 4 ball (blue) and sends it toward the corner pocket.

Note that because $m_1 = m_2 = m$, the masses also cancel in Equation (2). If we square both sides of Equation (2) and use the definition of the dot product of two vectors from Section 7.3, we obtain

$$v_{1i}{}^2 = (\mathbf{v}_{1f} + \mathbf{v}_{2f}) \cdot (\mathbf{v}_{1f} + \mathbf{v}_{2f}) = v_{1f}{}^2 + v_{2f}{}^2 + 2\mathbf{v}_{1f} \cdot \mathbf{v}_{2f}$$

Because the angle between $\mathbf{v}_{1f}$ and $\mathbf{v}_{2f}$ is $\theta + 35°$, $\mathbf{v}_{1f} \cdot \mathbf{v}_{2f} = v_{1f}v_{2f}\cos(\theta + 35°)$, and so

(3) $v_{1i}{}^2 = v_{1f}{}^2 + v_{2f}{}^2 + 2v_{1f}v_{2f}\cos(\theta + 35°)$

Subtracting Equation (1) from Equation (3) gives

$$0 = 2v_{1f}v_{2f}\cos(\theta + 35°)$$

$$0 = \cos(\theta + 35°)$$

$$\theta + 35° = 90° \text{ or } \boxed{\theta = 55°}$$

This result shows that whenever two equal masses undergo a glancing elastic collision and one of them is initially at rest, they move in perpendicular directions after the collision. The same physics describes two very different situations, protons in Example 9.11 and billiard balls in this example.

9.5   The Center of Mass

In this section we describe the overall motion of a mechanical system in terms of a special point called the **center of mass** of the system. The mechanical system can be either a group of particles, such as a collection of atoms in a container, or an extended object, such as a gymnast leaping through the air. We shall see that the center of mass of the system moves as if all the mass of the system were concentrated at that point. Furthermore, if the resultant external force on the system is $\Sigma\mathbf{F}_{ext}$ and the total mass of the system is M, the center of mass moves with an acceleration given by $\mathbf{a} = \Sigma\mathbf{F}_{ext}/M$. That is, the system moves as if the resultant external force were applied to a single particle of mass M located at the center of mass. This behavior is independent of other motion, such as rotation or vibration of the system. This is the *particle model* that was introduced in Chapter 2.

Consider a mechanical system consisting of a pair of particles that have different masses and are connected by a light, rigid rod (Fig. 9.16). The position of the center of mass of a system can be described as being the *average position* of the system's mass. The center of mass of the system is located somewhere on the line joining the two particles and is closer to the particle having the larger mass. If a single force is applied at a point on the rod somewhere between the center of mass and the less massive particle, the system rotates clockwise (see Fig. 9.16a). If the force is applied at a point on the rod somewhere between the center of mass and the more massive particle, the system rotates counterclockwise (see Fig. 9.16b). If the force is applied at the center of mass, the system moves in the direction of $\mathbf{F}$ without rotating (see Fig. 9.16c). Thus, the center of mass can be located with this procedure.

The center of mass of the pair of particles described in Figure 9.17 is located on the x axis and lies somewhere between the particles. Its x coordinate is given by

$$x_{CM} \equiv \frac{m_1 x_1 + m_2 x_2}{m_1 + m_2} \tag{9.27}$$

For example, if $x_1 = 0$, $x_2 = d$, and $m_2 = 2m_1$, we find that $x_{CM} = \frac{2}{3}d$. That is, the center of mass lies closer to the more massive particle. If the two masses are equal, the center of mass lies midway between the particles.

We can extend this concept to a system of many particles with masses m_i in three dimensions. The x coordinate of the center of mass of n particles is defined to be

$$x_{CM} \equiv \frac{m_1 x_1 + m_2 x_2 + m_3 x_3 + \cdots + m_n x_n}{m_1 + m_2 + m_3 + \cdots + m_n} = \frac{\sum_i m_i x_i}{\sum_i m_i} = \frac{\sum_i m_i x_i}{M} \qquad (9.28)$$

where x_i is the x coordinate of the ith particle. For convenience, we express the total mass as $M \equiv \sum_i m_i$ where the sum runs over all n particles. The y and z coordinates of the center of mass are similarly defined by the equations

$$y_{CM} \equiv \frac{\sum_i m_i y_i}{M} \quad \text{and} \quad z_{CM} \equiv \frac{\sum_i m_i z_i}{M} \qquad (9.29)$$

The center of mass can also be located by its position vector $\mathbf{r}_{CM}$. The Cartesian coordinates of this vector are x_{CM}, y_{CM}, and z_{CM}, defined in Equations 9.28 and 9.29. Therefore,

$$\mathbf{r}_{CM} = x_{CM}\hat{\mathbf{i}} + y_{CM}\hat{\mathbf{j}} + z_{CM}\hat{\mathbf{k}} = \frac{\sum_i m_i x_i \hat{\mathbf{i}} + \sum_i m_i y_i \hat{\mathbf{j}} + \sum_i m_i z_i \hat{\mathbf{k}}}{M}$$

$$\mathbf{r}_{CM} \equiv \frac{\sum_i m_i \mathbf{r}_i}{M} \qquad (9.30)$$

where $\mathbf{r}_i$ is the position vector of the ith particle, defined by

$$\mathbf{r}_i \equiv x_i \hat{\mathbf{i}} + y_i \hat{\mathbf{j}} + z_i \hat{\mathbf{k}}$$

Although locating the center of mass for an extended object is somewhat more cumbersome than locating the center of mass of a system of particles, the basic ideas we have discussed still apply. We can think of an extended object as a system containing a large number of particles (Fig. 9.18). The particle separation is very small, and so the object can be considered to have a continuous mass distribution. By dividing the object into elements of mass Δm_i with coordinates x_i, y_i, z_i, we see that the x coordinate of the center of mass is approximately

$$x_{CM} \approx \frac{\sum_i x_i \Delta m_i}{M}$$

with similar expressions for y_{CM} and z_{CM}. If we let the number of elements n approach infinity, then x_{CM} is given precisely. In this limit, we replace the sum by an integral and Δm_i by the differential element dm:

$$x_{CM} = \lim_{\Delta m_i \to 0} \frac{\sum_i x_i \Delta m_i}{M} = \frac{1}{M} \int x \, dm \qquad (9.31)$$

Likewise, for y_{CM} and z_{CM} we obtain

$$y_{CM} = \frac{1}{M} \int y \, dm \quad \text{and} \quad z_{CM} = \frac{1}{M} \int z \, dm \qquad (9.32)$$

We can express the vector position of the center of mass of an extended object in the form

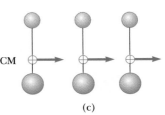

(a)

(b)

(c)

Active Figure 9.16 Two particles of unequal mass are connected by a light, rigid rod. (a) The system rotates clockwise when a force is applied between the less massive particle and the center of mass. (b) The system rotates counterclockwise when a force is applied between the more massive particle and the center of mass. (c) The system moves in the direction of the force without rotating when a force is applied at the center of mass.

At the Active Figures link at http://www.pse6.com, you can choose the point at which to apply the force.

Active Figure 9.17 The center of mass of two particles of unequal mass on the x axis is located at x_{CM}, a point between the particles, closer to the one having the larger mass.

At the Active Figures link at http://www.pse6.com, you can adjust the masses and positions of the particles to see the effect on the location of the center of mass.

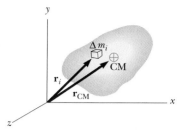

Figure 9.18 An extended object can be considered to be a distribution of small elements of mass Δm_i. The center of mass is located at the vector position $\mathbf{r}_{CM}$, which has coordinates x_{CM}, y_{CM}, and z_{CM}.

Figure 9.19 An experimental technique for determining the center of mass of a wrench. The wrench is hung freely first from point A and then from point C. The intersection of the two lines AB and CD locates the center of mass.

$$\mathbf{r}_{CM} = \frac{1}{M} \int \mathbf{r}\, dm \qquad (9.33)$$

which is equivalent to the three expressions given by Equations 9.31 and 9.32.

The center of mass of any symmetric object lies on an axis of symmetry and on any plane of symmetry.[4] For example, the center of mass of a uniform rod lies in the rod, midway between its ends. The center of mass of a sphere or a cube lies at its geometric center.

The center of mass of an irregularly shaped object such as a wrench can be determined by suspending the object first from one point and then from another. In Figure 9.19, a wrench is hung from point A, and a vertical line AB (which can be established with a plumb bob) is drawn when the wrench has stopped swinging. The wrench is then hung from point C, and a second vertical line CD is drawn. The center of mass is halfway through the thickness of the wrench, under the intersection of these two lines. In general, if the wrench is hung freely from any point, the vertical line through this point must pass through the center of mass.

Because an extended object is a continuous distribution of mass, each small mass element is acted upon by the gravitational force. The net effect of all these forces is equivalent to the effect of a single force $M\mathbf{g}$ acting through a special point, called the **center of gravity.** If $\mathbf{g}$ is constant over the mass distribution, then the center of gravity coincides with the center of mass. If an extended object is pivoted at its center of gravity, it balances in any orientation.

Quick Quiz 9.10 A baseball bat is cut at the location of its center of mass as shown in Figure 9.20. The piece with the smaller mass is (a) the piece on the right (b) the piece on the left (c) Both pieces have the same mass. (d) impossible to determine.

Figure 9.20 (Quick Quiz 9.10) A baseball bat cut at the location of its center of mass.

Example 9.13   The Center of Mass of Three Particles

A system consists of three particles located as shown in Figure 9.21a. Find the center of mass of the system.

Solution We set up the problem by labeling the masses of the particles as shown in the figure, with $m_1 = m_2 = 1.0$ kg and $m_3 = 2.0$ kg. Using the defining equations for the coordinates of the center of mass and noting that $z_{CM} = 0$, we obtain

$$
\begin{aligned}
x_{CM} &= \frac{\sum_i m_i x_i}{M} = \frac{m_1 x_1 + m_2 x_2 + m_3 x_3}{m_1 + m_2 + m_3} \\
&= \frac{(1.0\ \text{kg})(1.0\ \text{m}) + (1.0\ \text{kg})(2.0\ \text{m}) + (2.0\ \text{kg})(0)}{1.0\ \text{kg} + 1.0\ \text{kg} + 2.0\ \text{kg}} \\
&= \frac{3.0\ \text{kg} \cdot \text{m}}{4.0\ \text{kg}} = 0.75\ \text{m}
\end{aligned}
$$

---

[4]   This statement is valid only for objects that have a uniform mass per unit volume.

(a)

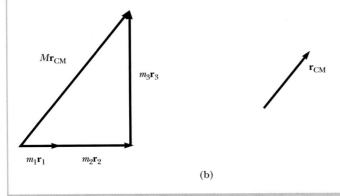

(b)

$$y_{CM} = \frac{\sum\limits_{i} m_i y_i}{M} = \frac{m_1 y_1 + m_2 y_2 + m_3 y_3}{m_1 + m_2 + m_3}$$

$$= \frac{(1.0 \text{ kg})(0) + (1.0 \text{ kg})(0) + (2.0 \text{ kg})(2.0 \text{ m})}{4.0 \text{ kg}}$$

$$= \frac{4.0 \text{ kg} \cdot \text{m}}{4.0 \text{ kg}} = 1.0 \text{ m}$$

The position vector to the center of mass measured from the origin is therefore

$$\mathbf{r}_{CM} \equiv x_{CM}\hat{\mathbf{i}} + y_{CM}\hat{\mathbf{j}} = \boxed{(0.75\hat{\mathbf{i}} + 1.0\hat{\mathbf{j}}) \text{ m}}$$

We can verify this result graphically by adding together $m_1\mathbf{r}_1 + m_2\mathbf{r}_2 + m_3\mathbf{r}_3$ and dividing the vector sum by M, the total mass. This is shown in Figure 9.21b.

Figure 9.21 (Example 9.13) (a) Two 1.0-kg particles are located on the x axis and a single 2.0-kg particle is located on the y axis as shown. The vector indicates the location of the system's center of mass. (b) The vector sum of $m_i \mathbf{r}_i$ and the resulting vector for $\mathbf{r}_{CM}$.

Example 9.14   The Center of Mass of a Rod

(A) Show that the center of mass of a rod of mass M and length L lies midway between its ends, assuming the rod has a uniform mass per unit length.

Solution The rod is shown aligned along the x axis in Figure 9.22, so that $y_{CM} = z_{CM} = 0$. Furthermore, if we call the mass per unit length λ (this quantity is called the *linear mass density*), then $\lambda = M/L$ for the uniform rod we assume here. If we divide the rod into elements of length dx, then the mass of each element is $dm = \lambda \, dx$. Equation 9.31 gives

$$x_{CM} = \frac{1}{M} \int x \, dm = \frac{1}{M} \int_0^L x\lambda \, dx = \frac{\lambda}{M} \frac{x^2}{2} \Big|_0^L = \frac{\lambda L^2}{2M}$$

Because $\lambda = M/L$, this reduces to

$$x_{CM} = \frac{L^2}{2M}\left(\frac{M}{L}\right) = \boxed{\frac{L}{2}}$$

One can also use symmetry arguments to obtain the same result.

(B) Suppose a rod is *nonuniform* such that its mass per unit length varies linearly with x according to the expression $\lambda = \alpha x$, where α is a constant. Find the x coordinate of the center of mass as a fraction of L.

Solution In this case, we replace dm by $\lambda \, dx$, where λ is not constant. Therefore, x_{CM} is

$$x_{CM} = \frac{1}{M} \int x \, dm = \frac{1}{M} \int_0^L x\lambda \, dx = \frac{1}{M} \int_0^L x\alpha x \, dx$$

$$= \frac{\alpha}{M} \int_0^L x^2 \, dx = \frac{\alpha L^3}{3M}$$

Figure 9.22 (Example 9.14) The geometry used to find the center of mass of a uniform rod.

We can eliminate α by noting that the total mass of the rod is related to α through the relationship

$$M = \int dm = \int_0^L \lambda \, dx = \int_0^L \alpha x \, dx = \frac{\alpha L^2}{2}$$

Substituting this into the expression for x_{CM} gives

$$x_{CM} = \frac{\alpha L^3}{3\alpha L^2/2} = \boxed{\tfrac{2}{3}L}$$

Example 9.15   The Center of Mass of a Right Triangle

You have been asked to hang a metal sign from a single vertical wire. The sign has the triangular shape shown in Figure 9.23a. The bottom of the sign is to be parallel to the ground. At what distance from the left end of the sign should you attach the support wire?

Solution The wire must be attached at a point directly above the center of gravity of the sign, which is the same as its center of mass because it is in a uniform gravitational field. We assume that the triangular sign has a uniform density and total mass M. Because the sign is a continuous distribution of mass, we must use the integral expression in Equation 9.31 to find the x coordinate of the center of mass.

We divide the triangle into narrow strips of width dx and height y as shown in Figure 9.23b, where y is the height of the hypotenuse of the triangle above the x axis for a given value of x. The mass of each strip is the product of the volume of the strip and the density ρ of the material from which the sign is made: $dm = \rho y t \, dx$, where t is the thickness of the metal sign. The density of the material is the total mass of the sign divided by its total volume (area of the triangle times thickness), so

$$dm = \rho y t \, dx = \left(\frac{M}{\frac{1}{2}abt}\right) y t \, dx = \frac{2My}{ab} \, dx$$

Using Equation 9.31 to find the x coordinate of the center of mass gives

$$x_{CM} = \frac{1}{M}\int x \, dm = \frac{1}{M}\int_0^a x \, \frac{2My}{ab} \, dx = \frac{2}{ab}\int_0^a xy \, dx$$

To proceed further and evaluate the integral, we must express y in terms of x. The line representing the hypotenuse of the triangle in Figure 9.23b has a slope of b/a and passes through the origin, so the equation of this line is $y = (b/a)x$. With this substitution for y in the integral, we have

$$x_{CM} = \frac{2}{ab}\int_0^a x\left(\frac{b}{a}x\right) dx = \frac{2}{a^2}\int_0^a x^2 \, dx = \frac{2}{a^2}\left[\frac{x^3}{3}\right]_0^a$$

$$= \boxed{\tfrac{2}{3}a}$$

Thus, the wire must be attached to the sign at a distance two thirds of the length of the bottom edge from the left end. We could also find the y coordinate of the center of mass of the sign, but this is not needed in order to determine where the wire should be attached.

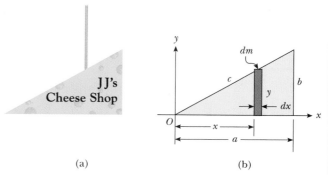

Figure 9.23 (Example 9.15) (a) A triangular sign to be hung from a single wire. (b) Geometric construction for locating the center of mass.

9.6   Motion of a System of Particles

We can begin to understand the physical significance and utility of the center of mass concept by taking the time derivative of the position vector given by Equation 9.30. From Section 4.1 we know that the time derivative of a position vector is by definition a velocity. Assuming M remains constant for a system of particles, that is, no particles enter or leave the system, we obtain the following expression for the **velocity of the center of mass** of the system:

Velocity of the center of mass

$$\mathbf{v}_{CM} = \frac{d\mathbf{r}_{CM}}{dt} = \frac{1}{M}\sum_i m_i \frac{d\mathbf{r}_i}{dt} = \frac{\sum_i m_i \mathbf{v}_i}{M} \qquad (9.34)$$

where $\mathbf{v}_i$ is the velocity of the ith particle. Rearranging Equation 9.34 gives

Total momentum of a system of particles

$$M\mathbf{v}_{CM} = \sum_i m_i \mathbf{v}_i = \sum_i \mathbf{p}_i = \mathbf{p}_{tot} \qquad (9.35)$$

Therefore, we conclude that the **total linear momentum of the system equals the total mass multiplied by the velocity of the center of mass.** In other words, the total linear momentum of the system is equal to that of a single particle of mass M moving with a velocity $\mathbf{v}_{CM}$.

If we now differentiate Equation 9.34 with respect to time, we obtain the **acceleration of the center of mass** of the system:

$$\mathbf{a}_{CM} = \frac{d\mathbf{v}_{CM}}{dt} = \frac{1}{M} \sum_i m_i \frac{d\mathbf{v}_i}{dt} = \frac{1}{M} \sum_i m_i \mathbf{a}_i \qquad (9.36)$$

Acceleration of the center of mass

Rearranging this expression and using Newton's second law, we obtain

$$M\mathbf{a}_{CM} = \sum_i m_i \mathbf{a}_i = \sum_i \mathbf{F}_i \qquad (9.37)$$

where $\mathbf{F}_i$ is the net force on particle i.

The forces on any particle in the system may include both external forces (from outside the system) and internal forces (from within the system). However, by Newton's third law, the internal force exerted by particle 1 on particle 2, for example, is equal in magnitude and opposite in direction to the internal force exerted by particle 2 on particle 1. Thus, when we sum over all internal forces in Equation 9.37, they cancel in pairs and we find that the net force on the system is caused *only* by external forces. Thus, we can write Equation 9.37 in the form

$$\sum \mathbf{F}_{ext} = M\mathbf{a}_{CM} \qquad (9.38)$$

Newton's second law for a system of particles

That is, **the net external force on a system of particles equals the total mass of the system multiplied by the acceleration of the center of mass.** If we compare this with Newton's second law for a single particle, we see that the particle model that we have used for several chapters can be described in terms of the center of mass:

The center of mass of a system of particles of combined mass M moves like an equivalent particle of mass M would move under the influence of the net external force on the system.

Finally, we see that if the net external force is zero, then from Equation 9.38 it follows that

$$M\mathbf{a}_{CM} = M \frac{d\mathbf{v}_{CM}}{dt} = 0$$

so that

$$M\mathbf{v}_{CM} = \mathbf{p}_{tot} = \text{constant} \qquad (\text{when } \sum \mathbf{F}_{ext} = 0) \qquad (9.39)$$

That is, the total linear momentum of a system of particles is conserved if no net external force is acting on the system. It follows that for an isolated system of particles, both the total momentum and the velocity of the center of mass are constant in time, as shown in Figure 9.24. This is a generalization to a many-particle system of the law of conservation of momentum discussed in Section 9.1 for a two-particle system.

Suppose an isolated system consisting of two or more members is at rest. The center of mass of such a system remains at rest unless acted upon by an external force. For example, consider a system made up of a swimmer standing on a raft, with the system initially at rest. When the swimmer dives horizontally off the raft, the raft moves in the direction opposite to that of the swimmer and the center of mass of the system remains at rest (if we neglect friction between raft and water). Furthermore, the linear momentum of the diver is equal in magnitude to that of the raft, but opposite in direction.

As another example, suppose an unstable atom initially at rest suddenly breaks up into two fragments of masses M_1 and M_2, with velocities $\mathbf{v}_1$ and $\mathbf{v}_2$, respectively. Because the total momentum of the system before the breakup is zero, the total momentum of the

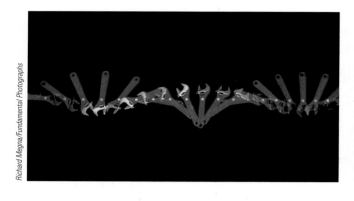

Figure 9.24 Multiflash photograph showing an overhead view of a wrench moving on a horizontal surface. The white dots are located at the center of mass of the wrench and show that the center of mass moves in a straight line as the wrench rotates.

Richard Megna/Fundamental Photographs

system after the breakup must also be zero. Therefore, $M_1\mathbf{v}_1 + M_2\mathbf{v}_2 = 0$. If the velocity of one of the fragments is known, the recoil velocity of the other fragment can be calculated.

Quick Quiz 9.11 The vacationers on a cruise ship are eager to arrive at their next destination. They decide to try to speed up the cruise ship by gathering at the bow (the front) and running all at once toward the stern (the back) of the ship. While they are running toward the stern, the speed of the ship is (a) higher than it was before (b) unchanged (c) lower than it was before (d) impossible to determine.

Quick Quiz 9.12 The vacationers in Quick Quiz 9.11 stop running when they reach the stern of the ship. After they have all stopped running, the speed of the ship is (a) higher than it was before they started running (b) unchanged from what it was before they started running (c) lower than it was before they started running (d) impossible to determine.

Conceptual Example 9.16 The Sliding Bear

Suppose you tranquilize a polar bear on a smooth glacier as part of a research effort. How might you estimate the bear's mass using a measuring tape, a rope, and knowledge of your own mass?

Solution Tie one end of the rope around the bear, and then lay out the tape measure on the ice with one end at the bear's original position, as shown in Figure 9.25. Grab hold of the free end of the rope and position yourself as

Figure 9.25 (Conceptual Example 9.16) The center of mass of an isolated system remains at rest unless acted on by an external force. How can you determine the mass of the polar bear?

shown, noting your location. Take off your spiked shoes, and pull on the rope hand over hand. Both you and the bear will slide over the ice until you meet. From the tape, observe how far you slide, x_p, and how far the bear slides, x_b. The point where you meet the bear is the fixed location of the center of mass of the system (bear plus you), and so you can determine the mass of the bear from $m_b x_b = m_p x_p$. (Unfortunately, you cannot return to your spiked shoes and so you are in big trouble if the bear wakes up!)

Conceptual Example 9.17   Exploding Projectile

A projectile fired into the air suddenly explodes into several fragments (Fig. 9.26). What can be said about the motion of the center of mass of the system made up of all the fragments after the explosion?

Solution Neglecting air resistance, the only external force on the projectile is the gravitational force. Thus, if the projectile did not explode, it would continue to move along the parabolic path indicated by the dashed line in Figure 9.26. Because the forces caused by the explosion are internal, they do not affect the motion of the center of mass of the system (the fragments). Thus, after the explosion, the center of mass of the fragments follows the same parabolic path the projectile would have followed if there had been no explosion.

Figure 9.26 (Conceptual Example 9.17) When a projectile explodes into several fragments, the center of mass of the system made up of all the fragments follows the same parabolic path the projectile would have taken had there been no explosion.

Example 9.18   The Exploding Rocket

A rocket is fired vertically upward. At the instant it reaches an altitude of 1 000 m and a speed of 300 m/s, it explodes into three fragments having equal mass. One fragment continues to move upward with a speed of 450 m/s following the explosion. The second fragment has a speed of 240 m/s and is moving east right after the explosion. What is the velocity of the third fragment right after the explosion?

Solution Let us call the total mass of the rocket M; hence, the mass of each fragment is $M/3$. Because the forces of the explosion are internal to the system and cannot affect its total momentum, the total momentum $\mathbf{p}_i$ of the rocket just before the explosion must equal the total momentum $\mathbf{p}_f$ of the fragments right after the explosion.

Before the explosion,

$$\mathbf{p}_i = M\mathbf{v}_i = M(300\hat{\mathbf{j}} \text{ m/s})$$

After the explosion,

$$\mathbf{p}_f = \frac{M}{3}(240\hat{\mathbf{i}} \text{ m/s}) + \frac{M}{3}(450\hat{\mathbf{j}} \text{ m/s}) + \frac{M}{3}\mathbf{v}_f$$

where $\mathbf{v}_f$ is the unknown velocity of the third fragment. Equating these two expressions (because $\mathbf{p}_i = \mathbf{p}_f$) gives

$$\frac{M}{3}\mathbf{v}_f + \frac{M}{3}(240\hat{\mathbf{i}} \text{ m/s}) + \frac{M}{3}(450\hat{\mathbf{j}} \text{ m/s})$$

$$= M(300\hat{\mathbf{j}} \text{ m/s})$$

$$\mathbf{v}_f = \boxed{(-240\hat{\mathbf{i}} + 450\hat{\mathbf{j}}) \text{ m/s}}$$

What does the sum of the momentum vectors for all the fragments look like?

9.7 Rocket Propulsion

When ordinary vehicles such as cars and locomotives are propelled, the driving force for the motion is friction. In the case of the car, the driving force is the force exerted by the road on the car. A locomotive "pushes" against the tracks; hence, the driving force is the force exerted by the tracks on the locomotive. However, a rocket moving in space has no road or tracks to push against. Therefore, the source of the propulsion of a rocket must be something other than friction. Figure 9.27 is a dramatic photograph of a spacecraft at liftoff. **The operation of a rocket depends upon the law of conservation of linear momentum as applied to a system of particles, where the system is the rocket plus its ejected fuel.**

Figure 9.27 At liftoff, enormous thrust is generated by the space shuttle's liquid-fuel engines, aided by the two solid-fuel boosters. This photograph shows the liftoff of the space shuttle *Columbia,* which was lost in a tragic accident during its landing attempt on February 1, 2003 (shortly before this volume went to press).

Rocket propulsion can be understood by first considering a mechanical system consisting of a machine gun mounted on a cart on wheels. As the gun is fired, each bullet receives a momentum $m\mathbf{v}$ in some direction, where $\mathbf{v}$ is measured with respect to a stationary Earth frame. The momentum of the system made up of cart, gun, and bullets must be conserved. Hence, for each bullet fired, the gun and cart must receive a compensating momentum in the opposite direction. That is, the reaction force exerted by the bullet on the gun accelerates the cart and gun, and the cart moves in the direction opposite that of the bullets. If n is the number of bullets fired each second, then the average force exerted on the gun is $\overline{\mathbf{F}} = nm\mathbf{v}$.

In a similar manner, as a rocket moves in free space, its linear momentum changes when some of its mass is released in the form of ejected gases. **Because the gases are given momentum when they are ejected out of the engine, the rocket receives a compensating momentum in the opposite direction.** Therefore, the rocket is accelerated as a result of the "push," or thrust, from the exhaust gases. In free space, the center of mass of the system (rocket plus expelled gases) moves uniformly, independent of the propulsion process.[5]

Suppose that at some time t, the magnitude of the momentum of a rocket plus its fuel is $(M + \Delta m)v$, where v is the speed of the rocket relative to the Earth (Fig. 9.28a). Over a short time interval Δt, the rocket ejects fuel of mass Δm, and so at the end of this interval the rocket's speed is $v + \Delta v$, where Δv is the change in speed of the rocket (Fig. 9.28b). If the fuel is ejected with a speed v_e relative to the rocket (the subscript "e" stands for *exhaust*, and v_e is usually called the *exhaust speed*), the velocity of the fuel relative to a stationary frame of reference is $v - v_e$. Thus, if we equate the total initial mo-

---

[5] It is interesting to note that the rocket and machine gun represent cases of the reverse of a perfectly inelastic collision: momentum is conserved, but the kinetic energy of the system increases (at the expense of chemical potential energy in the fuel).

$$\mathbf{v}$$
$$M + \Delta m$$

$$\mathbf{p}_i = (M + \Delta m)\mathbf{v}$$

(a)

$$\Delta m$$
$$M$$

$$\mathbf{v} + \Delta\mathbf{v}$$

(b)

Figure 9.28 Rocket propulsion. (a) The initial mass of the rocket plus all its fuel is $M + \Delta m$ at a time t, and its speed is v. (b) At a time $t + \Delta t$, the rocket's mass has been reduced to M and an amount of fuel Δm has been ejected. The rocket's speed increases by an amount Δv.

mentum of the system to the total final momentum, we obtain

$$(M + \Delta m)v = M(v + \Delta v) + \Delta m(v - v_e)$$

where M represents the mass of the rocket and its remaining fuel after an amount of fuel having mass Δm has been ejected. Simplifying this expression gives

$$M \, \Delta v = v_e \Delta m$$

We also could have arrived at this result by considering the system in the center-of-mass frame of reference, which is a frame having the same velocity as the center of mass of the system. In this frame, the total momentum of the system is zero; therefore, if the rocket gains a momentum $M \, \Delta v$ by ejecting some fuel, the exhausted fuel obtains a momentum $v_e \Delta m$ in the *opposite* direction, so that $M\Delta v - v_e \Delta m = 0$. If we now take the limit as Δt goes to zero, we let $\Delta v \rightarrow dv$ and $\Delta m \rightarrow dm$. Furthermore, the increase in the exhaust mass dm corresponds to an equal decrease in the rocket mass, so that $dm = -dM$. Note that dM is negative because it represents a decrease in mass, so $-dM$ is a positive number. Using this fact, we obtain

$$M \, dv = v_e \, dm = -v_e \, dM \tag{9.40}$$

We divide the equation by M and integrate, taking the initial mass of the rocket plus fuel to be M_i and the final mass of the rocket plus its remaining fuel to be M_f. This gives

$$\int_{v_i}^{v_f} dv = -v_e \int_{M_i}^{M_f} \frac{dM}{M}$$

$$v_f - v_i = v_e \ln\left(\frac{M_i}{M_f}\right) \tag{9.41}$$

Courtesy of NASA

The force from a nitrogen-propelled hand-controlled device allows an astronaut to move about freely in space without restrictive tethers, using the thrust force from the expelled nitrogen.

Expression for rocket propulsion

This is the basic expression for rocket propulsion. First, it tells us that the increase in rocket speed is proportional to the exhaust speed v_e of the ejected gases. Therefore, the exhaust speed should be very high. Second, the increase in rocket speed is proportional to the natural logarithm of the ratio M_i/M_f. Therefore, this ratio should be as large as possible, which means that the mass of the rocket without its fuel should be as small as possible and the rocket should carry as much fuel as possible.

The **thrust** on the rocket is the force exerted on it by the ejected exhaust gases. We can obtain an expression for the thrust from Equation 9.40:

$$\text{Thrust} = M \frac{dv}{dt} = \left| v_e \frac{dM}{dt} \right| \tag{9.42}$$

This expression shows us that the thrust increases as the exhaust speed increases and as the rate of change of mass (called the *burn rate*) increases.

Example 9.19 A Rocket in Space

A rocket moving in space, far from all other objects, has a speed of 3.0×10^3 m/s relative to the Earth. Its engines are turned on, and fuel is ejected in a direction opposite the rocket's motion at a speed of 5.0×10^3 m/s relative to the rocket.

(A) What is the speed of the rocket relative to the Earth once the rocket's mass is reduced to half its mass before ignition?

Solution We can guess that the speed we are looking for must be greater than the original speed because the rocket is accelerating. Applying Equation 9.41, we obtain

$$v_f = v_i + v_e \ln\left(\frac{M_i}{M_f}\right)$$

$$= 3.0 \times 10^3 \text{ m/s} + (5.0 \times 10^3 \text{ m/s})\ln\left(\frac{M_i}{0.5 \, M_i}\right)$$

$$= \boxed{6.5 \times 10^3 \text{ m/s}}$$

(B) What is the thrust on the rocket if it burns fuel at the rate of 50 kg/s?

Solution Using Equation 9.42,

$$\text{Thrust} = \left| v_e \frac{dM}{dt} \right| = (5.0 \times 10^3 \text{ m/s})(50 \text{ kg/s})$$

$$= \boxed{2.5 \times 10^5 \text{ N}}$$

Example 9.20 Fighting a Fire

Two firefighters must apply a total force of 600 N to steady a hose that is discharging water at the rate of 3 600 L/min. Estimate the speed of the water as it exits the nozzle.

Solution The water is exiting at 3 600 L/min, which is 60 L/s. Knowing that 1 L of water has a mass of 1 kg, we estimate that about 60 kg of water leaves the nozzle every second. As the water leaves the hose, it exerts on the hose a thrust that must be counteracted by the 600-N force exerted by the firefighters. So, applying Equation 9.42 gives

$$\text{Thrust} = \left| v_e \frac{dM}{dt} \right|$$

$$600 \text{ N} = \left| v_e (60 \text{ kg/s}) \right|$$

$$v_e = \boxed{10 \text{ m/s}}$$

Firefighting is dangerous work. If the nozzle should slip from their hands, the movement of the hose due to the thrust it receives from the rapidly exiting water could injure the firefighters.

Take a practice test for this chapter by clicking on the Practice Test link at http://www.pse6.com.

SUMMARY

The **linear momentum p** of a particle of mass m moving with a velocity $\mathbf{v}$ is

$$\mathbf{p} \equiv m\mathbf{v} \tag{9.2}$$

The law of **conservation of linear momentum** indicates that the total momentum of an isolated system is conserved. If two particles form an isolated system, the momentum of the system is conserved regardless of the nature of the force between them. Therefore, the total momentum of the system at all times equals its initial total momentum, or

$$\mathbf{p}_{1i} + \mathbf{p}_{2i} = \mathbf{p}_{1f} + \mathbf{p}_{2f} \tag{9.5}$$

The **impulse** imparted to a particle by a force $\mathbf{F}$ is equal to the change in the momentum of the particle:

$$\mathbf{I} \equiv \int_{t_i}^{t_f} \mathbf{F} \, dt = \Delta\mathbf{p} \tag{9.8, 9.9}$$

This is known as the **impulse–momentum theorem.**

Impulsive forces are often very strong compared with other forces on the system and usually act for a very short time, as in the case of collisions.

When two particles collide, the total momentum of the isolated system before the collision always equals the total momentum after the collision, regardless of the nature of the collision. An **inelastic collision** is one for which the total kinetic energy of the system is not conserved. A **perfectly inelastic collision** is one in which the colliding bodies stick together after the collision. An **elastic collision** is one in which the kinetic energy of the system is conserved.

In a two- or three-dimensional collision, the components of momentum of an isolated system in each of the directions (x, y, and z) are conserved independently.

The position vector of the center of mass of a system of particles is defined as

$$\mathbf{r}_{\text{CM}} \equiv \frac{\sum_i m_i \mathbf{r}_i}{M} \tag{9.30}$$

where $M = \sum_i m_i$ is the total mass of the system and $\mathbf{r}_i$ is the position vector of the ith particle.

The position vector of the center of mass of an extended object can be obtained from the integral expression

$$\mathbf{r}_{CM} = \frac{1}{M} \int \mathbf{r}\, dm \qquad (9.33)$$

The velocity of the center of mass for a system of particles is

$$\mathbf{v}_{CM} = \frac{\sum_i m_i \mathbf{v}_i}{M} \qquad (9.34)$$

The total momentum of a system of particles equals the total mass multiplied by the velocity of the center of mass.

Newton's second law applied to a system of particles is

$$\sum \mathbf{F}_{ext} = M \mathbf{a}_{CM} \qquad (9.38)$$

where $\mathbf{a}_{CM}$ is the acceleration of the center of mass and the sum is over all external forces. The center of mass moves like an imaginary particle of mass M under the influence of the resultant external force on the system.

QUESTIONS

1. Does a large force always produce a larger impulse on an object than a smaller force does? Explain.

2. If the speed of a particle is doubled, by what factor is its momentum changed? By what factor is its kinetic energy changed?

3. If two particles have equal kinetic energies, are their momenta necessarily equal? Explain.

4. While in motion, a pitched baseball carries kinetic energy and momentum. (a) Can we say that it carries a force that it can exert on any object it strikes? (b) Can the baseball deliver more kinetic energy to the object it strikes than the ball carries initially? (c) Can the baseball deliver to the object it strikes more momentum than the ball carries initially? Explain your answers.

5. An isolated system is initially at rest. Is it possible for parts of the system to be in motion at some later time? If so, explain how this might occur.

6. If two objects collide and one is initially at rest, is it possible for both to be at rest after the collision? Is it possible for one to be at rest after the collision? Explain.

7. Explain how linear momentum is conserved when a ball bounces from a floor.

8. A bomb, initially at rest, explodes into several pieces. (a) Is linear momentum of the system conserved? (b) Is kinetic energy of the system conserved? Explain.

9. A ball of clay is thrown against a brick wall. The clay stops and sticks to the wall. Is the principle of conservation of momentum violated in this example?

10. You are standing perfectly still, and then you take a step forward. Before the step your momentum was zero, but afterward you have some momentum. Is the principle of conservation of momentum violated in this case?

11. When a ball rolls down an incline, its linear momentum increases. Is the principle of conservation of momentum violated in this process?

12. Consider a perfectly inelastic collision between a car and a large truck. Which vehicle experiences a larger change in kinetic energy as a result of the collision?

13. A sharpshooter fires a rifle while standing with the butt of the gun against his shoulder. If the forward momentum of a bullet is the same as the backward momentum of the gun, why isn't it as dangerous to be hit by the gun as by the bullet?

14. A pole-vaulter falls from a height of 6.0 m onto a foam rubber pad. Can you calculate his speed just before he reaches the pad? Can you calculate the force exerted on him by the pad? Explain.

15. Firefighters must apply large forces to hold a fire hose steady (Fig. Q9.15). What factors related to the projection of the water determine the magnitude of the force needed to keep the end of the fire hose stationary?

Figure Q9.15

16. A large bed sheet is held vertically by two students. A third student, who happens to be the star pitcher on the baseball team, throws a raw egg at the sheet. Explain why the egg does not break when it hits the sheet, regardless of its initial speed. (If you try this demonstration, make sure the pitcher hits the sheet near its center, and do not allow the egg to fall on the floor after being caught.)

17. A skater is standing still on a frictionless ice rink. Her friend throws a Frisbee straight at her. In which of the following cases is the largest momentum transferred to the skater? (a) The skater catches the Frisbee and holds onto it. (b) The skater catches the Frisbee momentarily, but then drops it vertically downward. (c) The skater catches the Frisbee, holds it momentarily, and throws it back to her friend.

18. In an elastic collision between two particles, does the kinetic energy of each particle change as a result of the collision?

19. Three balls are thrown into the air simultaneously. What is the acceleration of their center of mass while they are in motion?

20. A person balances a meter stick in a horizontal position on the extended index fingers of her right and left hands. She slowly brings the two fingers together. The stick remains balanced and the two fingers always meet at the 50-cm mark regardless of their original positions. (Try it!) Explain.

21. NASA often uses the gravity of a planet to "slingshot" a probe on its way to a more distant planet. The interaction of the planet and the spacecraft is a collision in which the objects do not touch. How can the probe have its speed increased in this manner?

22. The Moon revolves around the Earth. Model its orbit as circular. Is the Moon's linear momentum conserved? Is its kinetic energy conserved?

23. A raw egg dropped to the floor breaks upon impact. However, a raw egg dropped onto a thick foam rubber cushion from a height of about 1 m rebounds without breaking. Why is this possible? If you try this experiment, be sure to catch the egg after its first bounce.

24. Can the center of mass of an object be located at a position at which there is no mass? If so, give examples.

25. A juggler juggles three balls in a continuous cycle. Any one ball is in contact with his hands for one fifth of the time. Describe the motion of the center of mass of the three balls. What average force does the juggler exert on one ball while he is touching it?

26. Does the center of mass of a rocket in free space accelerate? Explain. Can the speed of a rocket exceed the exhaust speed of the fuel? Explain.

27. Early in the twentieth century, Robert Goddard proposed sending a rocket to the moon. Critics objected that in a vacuum, such as exists between the Earth and the Moon, the gases emitted by the rocket would have nothing to push against to propel the rocket. According to *Scientific American* (January 1975), Goddard placed a gun in a vacuum and fired a blank cartridge from it. (A blank cartridge contains no bullet and fires only the wadding and the hot gases produced by the burning gunpowder.) What happened when the gun was fired?

28. Explain how you could use a balloon to demonstrate the mechanism responsible for rocket propulsion.

29. On the subject of the following positions, state your own view and argue to support it. (a) The best theory of motion is that force causes acceleration. (b) The true measure of a force's effectiveness is the work it does, and the best theory of motion is that work done on an object changes its energy. (c) The true measure of a force's effect is impulse, and the best theory of motion is that impulse injected into an object changes its momentum.

PROBLEMS

1, 2, 3 = straightforward, intermediate, challenging   ☐ = full solution available in the *Student Solutions Manual and Study Guide*

🌐 = coached solution with hints available at http://www.pse6.com   💻 = computer useful in solving problem

▨ = paired numerical and symbolic problems

Section 9.1 Linear Momentum and its Conservation

1. A 3.00-kg particle has a velocity of $(3.00\hat{i} - 4.00\hat{j})$ m/s. (a) Find its x and y components of momentum. (b) Find the magnitude and direction of its momentum.

2. A 0.100-kg ball is thrown straight up into the air with an initial speed of 15.0 m/s. Find the momentum of the ball (a) at its maximum height and (b) halfway up to its maximum height.

3. How fast can you set the Earth moving? In particular, when you jump straight up as high as you can, what is the order of magnitude of the maximum recoil speed that you give to the Earth? Model the Earth as a perfectly solid object. In your solution, state the physical quantities you take as data and the values you measure or estimate for them.

4. Two blocks of masses M and $3M$ are placed on a horizontal, frictionless surface. A light spring is attached to one

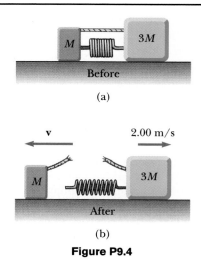

Figure P9.4

of them, and the blocks are pushed together with the spring between them (Fig. P9.4). A cord initially holding the blocks together is burned; after this, the block of mass $3M$ moves to the right with a speed of 2.00 m/s. (a) What is the speed of the block of mass M? (b) Find the original elastic potential energy in the spring if $M = 0.350$ kg.

5. (a) A particle of mass m moves with momentum p. Show that the kinetic energy of the particle is $K = p^2/2m$. (b) Express the magnitude of the particle's momentum in terms of its kinetic energy and mass.

Section 9.2 Impulse and Momentum

6. A friend claims that, as long as he has his seatbelt on, he can hold on to a 12.0-kg child in a 60.0 mi/h head-on collision with a brick wall in which the car passenger compartment comes to a stop in 0.050 0 s. Show that the violent force during the collision will tear the child from his arms. A child should always be in a toddler seat secured with a seat belt in the back seat of a car.

7. An estimated force–time curve for a baseball struck by a bat is shown in Figure P9.7. From this curve, determine (a) the impulse delivered to the ball, (b) the average force exerted on the ball, and (c) the peak force exerted on the ball.

Figure P9.7

8. A ball of mass 0.150 kg is dropped from rest from a height of 1.25 m. It rebounds from the floor to reach a height of 0.960 m. What impulse was given to the ball by the floor?

9. A 3.00-kg steel ball strikes a wall with a speed of 10.0 m/s at an angle of 60.0° with the surface. It bounces off with the same speed and angle (Fig. P9.9). If the ball is in contact with the wall for 0.200 s, what is the average force exerted by the wall on the ball ?

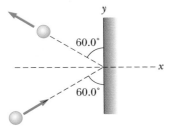

Figure P9.9

10. A tennis player receives a shot with the ball (0.060 0 kg) traveling horizontally at 50.0 m/s and returns the shot with the ball traveling horizontally at 40.0 m/s in the opposite direction. (a) What is the impulse delivered to the ball by the racquet? (b) What work does the racquet do on the ball?

11. In a slow-pitch softball game, a 0.200-kg softball crosses the plate at 15.0 m/s at an angle of 45.0° below the horizontal. The batter hits the ball toward center field, giving it a velocity of 40.0 m/s at 30.0° above the horizontal. (a) Determine the impulse delivered to the ball. (b) If the force on the ball increases linearly for 4.00 ms, holds constant for 20.0 ms, and then decreases to zero linearly in another 4.00 ms, what is the maximum force on the ball?

12. A professional diver performs a dive from a platform 10 m above the water surface. Estimate the order of magnitude of the average impact force she experiences in her collision with the water. State the quantities you take as data and their values.

13. A garden hose is held as shown in Figure P9.13. The hose is originally full of motionless water. What additional force is necessary to hold the nozzle stationary after the water flow is turned on, if the discharge rate is 0.600 kg/s with a speed of 25.0 m/s?

Figure P9.13

14. A glider of mass m is free to slide along a horizontal air track. It is pushed against a launcher at one end of the track. Model the launcher as a light spring of force constant k compressed by a distance x. The glider is released from rest. (a) Show that the glider attains a speed of $v = x(k/m)^{1/2}$. (b) Does a glider of large or of small mass attain a greater speed? (c) Show that the impulse imparted to the glider is given by the expression $x(km)^{1/2}$. (d) Is a greater impulse injected into a large or a small mass? (e) Is more work done on a large or a small mass?

Section 9.3 Collisions in One Dimension

15. High-speed stroboscopic photographs show that the head of a golf club of mass 200 g is traveling at 55.0 m/s just before it strikes a 46.0-g golf ball at rest on a tee. After the collision, the club head travels (in the same direction) at 40.0 m/s. Find the speed of the golf ball just after impact.

16. An archer shoots an arrow toward a target that is sliding toward her with a speed of 2.50 m/s on a smooth, slippery surface. The 22.5-g arrow is shot with a speed of 35.0 m/s and passes through the 300-g target, which is stopped by the impact. What is the speed of the arrow after passing through the target?

17. A 10.0-g bullet is fired into a stationary block of wood ($m = 5.00$ kg). The relative motion of the bullet stops inside the block. The speed of the bullet-plus-wood combination immediately after the collision is 0.600 m/s. What was the original speed of the bullet?

18. A railroad car of mass 2.50×10^4 kg is moving with a speed of 4.00 m/s. It collides and couples with three other coupled railroad cars, each of the same mass as the single car and moving in the same direction with an initial speed of 2.00 m/s. (a) What is the speed of the four cars after the collision? (b) How much mechanical energy is lost in the collision?

19. Four railroad cars, each of mass 2.50×10^4 kg, are coupled together and coasting along horizontal tracks at speed v_i toward the south. A very strong but foolish movie actor, riding on the second car, uncouples the front car and gives it a big push, increasing its speed to 4.00 m/s southward. The remaining three cars continue moving south, now at 2.00 m/s. (a) Find the initial speed of the cars. (b) How much work did the actor do? (c) State the relationship between the process described here and the process in Problem 18.

20. Two blocks are free to slide along the frictionless wooden track *ABC* shown in Figure P9.20. The block of mass $m_1 = 5.00$ kg is released from *A*. Protruding from its front end is the north pole of a strong magnet, repelling the north pole of an identical magnet embedded in the back end of the block of mass $m_2 = 10.0$ kg, initially at rest. The two blocks never touch. Calculate the maximum height to which m_1 rises after the elastic collision.

Figure P9.20

21. A 45.0-kg girl is standing on a plank that has a mass of 150 kg. The plank, originally at rest, is free to slide on a frozen lake, which is a flat, frictionless supporting surface. The girl begins to walk along the plank at a constant speed of 1.50 m/s relative to the plank. (a) What is her speed relative to the ice surface? (b) What is the speed of the plank relative to the ice surface?

22. Most of us know intuitively that in a head-on collision between a large dump truck and a subcompact car, you are better off being in the truck than in the car. Why is this? Many people imagine that the collision force exerted on the car is much greater than that experienced by the truck. To substantiate this view, they point out that the car is crushed, whereas the truck is only dented. This idea of unequal

forces, of course, is false. Newton's third law tells us that both objects experience forces of the same magnitude. The truck suffers less damage because it is made of stronger metal. But what about the two drivers? Do they experience the same forces? To answer this question, suppose that each vehicle is initially moving at 8.00 m/s and that they undergo a perfectly inelastic head-on collision. Each driver has mass 80.0 kg. Including the drivers, the total vehicle masses are 800 kg for the car and 4 000 kg for the truck. If the collision time is 0.120 s, what force does the seatbelt exert on each driver?

23. A neutron in a nuclear reactor makes an elastic head-on collision with the nucleus of a carbon atom initially at rest. (a) What fraction of the neutron's kinetic energy is transferred to the carbon nucleus? (b) If the initial kinetic energy of the neutron is 1.60×10^{-13} J, find its final kinetic energy and the kinetic energy of the carbon nucleus after the collision. (The mass of the carbon nucleus is nearly 12.0 times the mass of the neutron.)

24. As shown in Figure P9.24, a bullet of mass m and speed v passes completely through a pendulum bob of mass M. The bullet emerges with a speed of $v/2$. The pendulum bob is suspended by a stiff rod of length ℓ and negligible mass. What is the minimum value of v such that the pendulum bob will barely swing through a complete vertical circle?

Figure P9.24

25. A 12.0-g wad of sticky clay is hurled horizontally at a 100-g wooden block initially at rest on a horizontal surface. The clay sticks to the block. After impact, the block slides 7.50 m before coming to rest. If the coefficient of friction between the block and the surface is 0.650, what was the speed of the clay immediately before impact?

26. A 7.00-g bullet, when fired from a gun into a 1.00-kg block of wood held in a vise, penetrates the block to a depth of 8.00 cm. **What If?** This block of wood is placed on a frictionless horizontal surface, and a second 7.00-g bullet is fired from the gun into the block. To what depth will the bullet penetrate the block in this case?

27. (a) Three carts of masses 4.00 kg, 10.0 kg, and 3.00 kg move on a frictionless horizontal track with speeds of 5.00 m/s, 3.00 m/s, and 4.00 m/s, as shown in Figure P9.27. Velcro couplers make the carts stick together after colliding. Find the final velocity of the train of three carts. (b) **What If?** Does your answer require that all the carts collide and stick together at the same time? What if they collide in a different order?

Figure P9.27

Section 9.4 Two-Dimensional Collisions

28. A 90.0-kg fullback running east with a speed of 5.00 m/s is tackled by a 95.0-kg opponent running north with a speed of 3.00 m/s. If the collision is perfectly inelastic, (a) calculate the speed and direction of the players just after the tackle and (b) determine the mechanical energy lost as a result of the collision. Account for the missing energy.

29. Two shuffleboard disks of equal mass, one orange and the other yellow, are involved in an elastic, glancing collision. The yellow disk is initially at rest and is struck by the orange disk moving with a speed of 5.00 m/s. After the collision, the orange disk moves along a direction that makes an angle of 37.0° with its initial direction of motion. The velocities of the two disks are perpendicular after the collision. Determine the final speed of each disk.

30. Two shuffleboard disks of equal mass, one orange and the other yellow, are involved in an elastic, glancing collision. The yellow disk is initially at rest and is struck by the orange disk moving with a speed v_i. After the collision, the orange disk moves along a direction that makes an angle θ with its initial direction of motion. The velocities of the two disks are perpendicular after the collision. Determine the final speed of each disk.

31. The mass of the blue puck in Figure P9.31 is 20.0% greater than the mass of the green one. Before colliding, the pucks approach each other with momenta of equal magnitudes and opposite directions, and the green puck has an initial speed of 10.0 m/s. Find the speeds of the pucks after the collision if half the kinetic energy is lost during the collision.

Figure P9.31

32. Two automobiles of equal mass approach an intersection. One vehicle is traveling with velocity 13.0 m/s toward the east, and the other is traveling north with speed v_{2i}. Neither driver sees the other. The vehicles collide in the intersection and stick together, leaving parallel skid marks at an angle of 55.0° north of east. The speed limit for both roads is 35 mi/h, and the driver of the northward-moving vehicle claims he was within the speed limit when the collision occurred. Is he telling the truth?

33. A billiard ball moving at 5.00 m/s strikes a stationary ball of the same mass. After the collision, the first ball moves, at 4.33 m/s, at an angle of 30.0° with respect to the original line of motion. Assuming an elastic collision (and ignoring friction and rotational motion), find the struck ball's velocity after the collision.

34. A proton, moving with a velocity of $v_i\hat{\mathbf{i}}$, collides elastically with another proton that is initially at rest. If the two protons have equal speeds after the collision, find (a) the speed of each proton after the collision in terms of v_i and (b) the direction of the velocity vectors after the collision.

35. An object of mass 3.00 kg, moving with an initial velocity of $5.00\hat{\mathbf{i}}$ m/s, collides with and sticks to an object of mass 2.00 kg with an initial velocity of $-3.00\hat{\mathbf{j}}$ m/s. Find the final velocity of the composite object.

36. Two particles with masses m and $3m$ are moving toward each other along the x axis with the same initial speeds v_i. Particle m is traveling to the left, while particle $3m$ is traveling to the right. They undergo an elastic glancing collision such that particle m is moving downward after the collision at right angles from its initial direction. (a) Find the final speeds of the two particles. (b) What is the angle θ at which the particle $3m$ is scattered?

37. An unstable atomic nucleus of mass 17.0×10^{-27} kg initially at rest disintegrates into three particles. One of the particles, of mass 5.00×10^{-27} kg, moves along the y axis with a speed of 6.00×10^6 m/s. Another particle, of mass 8.40×10^{-27} kg, moves along the x axis with a speed of 4.00×10^6 m/s. Find (a) the velocity of the third particle and (b) the total kinetic energy increase in the process.

Section 9.5 The Center of Mass

38. Four objects are situated along the y axis as follows: a 2.00 kg object is at $+3.00$ m, a 3.00-kg object is at $+2.50$ m, a 2.50-kg object is at the origin, and a 4.00-kg object is at -0.500 m. Where is the center of mass of these objects?

39. A water molecule consists of an oxygen atom with two hydrogen atoms bound to it (Fig. P9.39). The angle between the two bonds is 106°. If the bonds are 0.100 nm long, where is the center of mass of the molecule?

Figure P9.39

40. The mass of the Earth is 5.98×10^{24} kg, and the mass of the Moon is 7.36×10^{22} kg. The distance of separation, measured between their centers, is 3.84×10^8 m. Locate the center of mass of the Earth–Moon system as measured from the center of the Earth.

41. A uniform piece of sheet steel is shaped as in Figure P9.41. Compute the x and y coordinates of the center of mass of the piece.

Figure P9.41

42. (a) Consider an extended object whose different portions have different elevations. Assume the free-fall acceleration is uniform over the object. Prove that the gravitational potential energy of the object–Earth system is given by $U_g = Mgy_{CM}$ where M is the total mass of the object and y_{CM} is the elevation of its center of mass above the chosen reference level. (b) Calculate the gravitational potential energy associated with a ramp constructed on level ground with stone with density 3 800 kg/m^3 and everywhere 3.60 m wide. In a side view, the ramp appears as a right triangle with height 15.7 m at the top end and base 64.8 m (Figure P9.42).

Figure P9.42

43. A rod of length 30.0 cm has linear density (mass-per-length) given by

$$\lambda = 50.0 \,\text{g/m} + 20.0x \,\text{g/m}^2,$$

where x is the distance from one end, measured in meters. (a) What is the mass of the rod? (b) How far from the $x = 0$ end is its center of mass?

44. In the 1968 Olympic Games, University of Oregon jumper Dick Fosbury introduced a new technique of high jumping called the "Fosbury flop." It contributed to raising the world record by about 30 cm and is presently used by nearly every world-class jumper. In this technique, the jumper goes over the bar face up while arching his back as much as possible, as in Figure P9.44a. This action places his center of mass outside his body, below his back. As his body goes over the bar, his center of mass passes below the bar. Because a given energy input implies a certain elevation for his center of mass, the action of arching his back means his body is higher than if his back were straight. As a model, consider the jumper as a thin uniform rod of length L. When the rod is straight, its center of mass is at its center. Now bend the rod in a circular arc so that it subtends an angle of $90.0°$ at the center of the arc, as shown in Figure P9.44b. In this configuration, how far outside the rod is the center of mass?

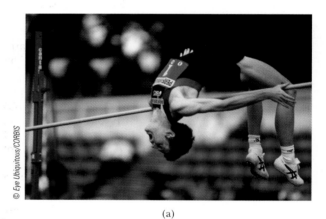

(a)

(b)

Figure P9.44

Section 9.6 Motion of a System of Particles

45. A 2.00-kg particle has a velocity $(2.00\hat{\mathbf{i}} - 3.00\hat{\mathbf{j}})$ m/s, and a 3.00-kg particle has a velocity $(1.00\hat{\mathbf{i}} + 6.00\hat{\mathbf{j}})$ m/s. Find (a) the velocity of the center of mass and (b) the total momentum of the system.

46. Consider a system of two particles in the xy plane: $m_1 = 2.00$ kg is at the location $\mathbf{r}_1 = (1.00\hat{\mathbf{i}} + 2.00\hat{\mathbf{j}})$ m and has a velocity of $(3.00\hat{\mathbf{i}} + 0.500\hat{\mathbf{j}})$ m/s; $m_2 = 3.00$ kg is at $\mathbf{r}_2 = (-4.00\hat{\mathbf{i}} - 3.00\hat{\mathbf{j}})$ m and has velocity $(3.00\hat{\mathbf{i}} - 2.00\hat{\mathbf{j}})$ m/s.

(a) Plot these particles on a grid or graph paper. Draw their position vectors and show their velocities. (b) Find the position of the center of mass of the system and mark it on the grid. (c) Determine the velocity of the center of mass and also show it on the diagram. (d) What is the total linear momentum of the system?

47. Romeo (77.0 kg) entertains Juliet (55.0 kg) by playing his guitar from the rear of their boat at rest in still water, 2.70 m away from Juliet, who is in the front of the boat. After the serenade, Juliet carefully moves to the rear of the boat (away from shore) to plant a kiss on Romeo's cheek. How far does the 80.0-kg boat move toward the shore it is facing?

48. A ball of mass 0.200 kg has a velocity of $150\hat{\mathbf{i}}$ m/s; a ball of mass 0.300 kg has a velocity of $-0.400\hat{\mathbf{i}}$ m/s. They meet in a head-on elastic collision. (a) Find their velocities after the collision. (b) Find the velocity of their center of mass before and after the collision.

Section 9.7 Rocket Propulsion

49. The first stage of a Saturn V space vehicle consumed fuel and oxidizer at the rate of 1.50×10^4 kg/s, with an exhaust speed of 2.60×10^3 m/s. (a) Calculate the thrust produced by these engines. (b) Find the acceleration of the vehicle just as it lifted off the launch pad on the Earth if the vehicle's initial mass was 3.00×10^6 kg. *Note:* You must include the gravitational force to solve part (b).

50. Model rocket engines are sized by thrust, thrust duration, and total impulse, among other characteristics. A size C5 model rocket engine has an average thrust of 5.26 N, a fuel mass of 12.7 g, and an initial mass of 25.5 g. The duration of its burn is 1.90 s. (a) What is the average exhaust speed of the engine? (b) If this engine is placed in a rocket body of mass 53.5 g, what is the final velocity of the rocket if it is fired in outer space? Assume the fuel burns at a constant rate.

51. A rocket for use in deep space is to be capable of boosting a total load (payload plus rocket frame and engine) of 3.00 metric tons to a speed of 10 000 m/s. (a) It has an engine and fuel designed to produce an exhaust speed of 2 000 m/s. How much fuel plus oxidizer is required? (b) If a different fuel and engine design could give an exhaust speed of 5 000 m/s, what amount of fuel and oxidizer would be required for the same task?

52. *Rocket Science.* A rocket has total mass $M_i = 360$ kg, including 330 kg of fuel and oxidizer. In interstellar space it starts from rest, turns on its engine at time $t = 0$, and puts out exhaust with relative speed $v_e = 1\ 500$ m/s at the constant rate $k = 2.50$ kg/s. The fuel will last for an actual burn time of 330 kg/$(2.5$ kg/s$) = 132$ s, but define a "projected depletion time" as $T_p = M_i/k = 144$ s. (This would be the burn time if the rocket could use its payload and fuel tanks as fuel, and even the walls of the combustion chamber.) (a) Show that during the burn the velocity of the rocket is given as a function of time by

$$v(t) = -v_e \ln[1 - (t/T_p)]$$

(b) Make a graph of the velocity of the rocket as a function of time for times running from 0 to 132 s. (c) Show that

the acceleration of the rocket is

$$a(t) = v_e/(T_p - t)$$

(d) Graph the acceleration as a function of time. (e) Show that the position of the rocket is

$$x(t) = v_e(T_p - t)\ln[1 - (t/T_p)] + v_e t$$

(f) Graph the position during the burn.

53. An orbiting spacecraft is described not as a "zero-g," but rather as a "microgravity" environment for its occupants and for on-board experiments. Astronauts experience slight lurches due to the motions of equipment and other astronauts, and due to venting of materials from the craft. Assume that a 3 500-kg spacecraft undergoes an acceleration of 2.50 $\mu g = 2.45 \times 10^{-5}$ m/s² due to a leak from one of its hydraulic control systems. The fluid is known to escape with a speed of 70.0 m/s into the vacuum of space. How much fluid will be lost in 1 h if the leak is not stopped?

Additional Problems

54. Two gliders are set in motion on an air track. A spring of force constant k is attached to the near side of one glider. The first glider, of mass m_1, has velocity $\mathbf{v}_1$, and the second glider, of mass m_2, moves more slowly, with velocity $\mathbf{v}_2$, as in Figure P9.54. When m_1 collides with the spring attached to m_2 and compresses the spring to its maximum compression $x_{\max}$, the velocity of the gliders is $\mathbf{v}$. In terms of $\mathbf{v}_1$, $\mathbf{v}_2$, m_1, m_2, and k, find (a) the velocity $\mathbf{v}$ at maximum compression, (b) the maximum compression $x_{\max}$, and (c) the velocity of each glider after m_1 has lost contact with the spring.

Figure P9.54

55. **Review problem**. A 60.0-kg person running at an initial speed of 4.00 m/s jumps onto a 120-kg cart initially at rest (Figure P9.55). The person slides on the cart's top surface and finally comes to rest relative to the cart. The coefficient of kinetic friction between the person and the cart is 0.400. Friction between the cart and ground can be neglected. (a) Find the final velocity of the person and cart relative to the ground. (b) Find the friction force acting on the person while he is sliding across the top surface of the cart. (c) How long does the friction force act on the person? (d) Find the change in momentum of the person and the change in momentum of the cart. (e) Determine the displacement of the person relative to the ground while he is sliding on the cart. (f) Determine the displacement of the cart relative to the ground while the person is sliding. (g) Find the change in

kinetic energy of the person. (h) Find the change in kinetic energy of the cart. (i) Explain why the answers to (g) and (h) differ. (What kind of collision is this, and what accounts for the loss of mechanical energy?)

Figure P9.55

56. A golf ball ($m = 46.0$ g) is struck with a force that makes an angle of 45.0° with the horizontal. The ball lands 200 m away on a flat fairway. If the golf club and ball are in contact for 7.00 ms, what is the average force of impact? (Neglect air resistance.)

57. An 80.0-kg astronaut is working on the engines of his ship, which is drifting through space with a constant velocity. The astronaut, wishing to get a better view of the Universe, pushes against the ship and much later finds himself 30.0 m behind the ship. Without a thruster, the only way to return to the ship is to throw his 0.500-kg wrench directly away from the ship. If he throws the wrench with a speed of 20.0 m/s relative to the ship, how long does it take the astronaut to reach the ship?

58. A bullet of mass m is fired into a block of mass M initially at rest at the edge of a frictionless table of height h (Fig. P9.58). The bullet remains in the block, and after impact the block lands a distance d from the bottom of the table. Determine the initial speed of the bullet.

Figure P9.58

59. A 0.500-kg sphere moving with a velocity $(2.00\hat{i} - 3.00\hat{j} + 1.00\hat{k})$ m/s strikes another sphere of mass 1.50 kg moving with a velocity $(-1.00\hat{i} + 2.00\hat{j} - 3.00\hat{k})$ m/s. (a) If the velocity of the 0.500-kg sphere after the collision is $(-1.00\hat{i} + 3.00\hat{j} - 8.00\hat{k})$ m/s, find the final velocity of the 1.50-kg sphere and identify the kind of collision (elastic, inelastic, or perfectly inelastic). (b) If the velocity of the 0.500-kg sphere after the collision is $(-0.250\hat{i} + 0.750\hat{j} -$

$2.00\hat{k})$ m/s, find the final velocity of the 1.50-kg sphere and identify the kind of collision. (c) **What If?** If the velocity of the 0.500-kg sphere after the collision is $(-1.00\hat{i} + 3.00\hat{j} + a\hat{k})$ m/s, find the value of a and the velocity of the 1.50-kg sphere after an elastic collision.

60. A small block of mass $m_1 = 0.500$ kg is released from rest at the top of a curve-shaped frictionless wedge of mass $m_2 = 3.00$ kg, which sits on a frictionless horizontal surface as in Figure P9.60a. When the block leaves the wedge, its velocity is measured to be 4.00 m/s to the right, as in Figure P9.60b. (a) What is the velocity of the wedge after the block reaches the horizontal surface? (b) What is the height h of the wedge?

(a)   (b)

Figure P9.60

61. A bucket of mass m and volume V is attached to a light cart, completely covering its top surface. The cart is given a quick push along a straight, horizontal, smooth road. It is raining, so as the cart cruises along without friction, the bucket gradually fills with water. By the time the bucket is full, its speed is v. (a) What was the initial speed v_i of the cart? Let ρ represent the density of water. (b) **What If?** Assume that when the bucket is half full, it develops a slow leak at the bottom, so that the level of the water remains constant thereafter. Describe qualitatively what happens to the speed of the cart after the leak develops.

62. A 75.0-kg firefighter slides down a pole while a constant friction force of 300 N retards her motion. A horizontal 20.0-kg platform is supported by a spring at the bottom of the pole to cushion the fall. The firefighter starts from rest 4.00 m above the platform, and the spring constant is 4 000 N/m. Find (a) the firefighter's speed just before she collides with the platform and (b) the maximum distance the spring is compressed. (Assume the friction force acts during the entire motion.)

63. George of the Jungle, with mass m, swings on a light vine hanging from a stationary tree branch. A second vine of equal length hangs from the same point, and a gorilla of larger mass M swings in the opposite direction on it. Both vines are horizontal when the primates start from rest at the same moment. George and the gorilla meet at the lowest point of their swings. Each is afraid that one vine will break, so they grab each other and hang on. They swing upward together, reaching a point where the vines make an angle of 35.0° with the vertical. (a) Find the value of the ratio m/M. (b) **What If?** Try this at home. Tie a small magnet and a steel screw to opposite ends of a string. Hold the cen-

ter of the string fixed to represent the tree branch, and reproduce a model of the motions of George and the gorilla. What changes in your analysis will make it apply to this situation? **What If?** Assume the magnet is strong, so that it noticeably attracts the screw over a distance of a few centimeters. Then the screw will be moving faster just before it sticks to the magnet. Does this make a difference?

64. A cannon is rigidly attached to a carriage, which can move along horizontal rails but is connected to a post by a large spring, initially unstretched and with force constant $k = 2.00 \times 10^4$ N/m, as in Figure P9.64. The cannon fires a 200-kg projectile at a velocity of 125 m/s directed 45.0° above the horizontal. (a) If the mass of the cannon and its carriage is 5 000 kg, find the recoil speed of the cannon. (b) Determine the maximum extension of the spring. (c) Find the maximum force the spring exerts on the carriage. (d) Consider the system consisting of the cannon, carriage, and shell. Is the momentum of this system conserved during the firing? Why or why not?

45.0°

Figure P9.64

65. A student performs a ballistic pendulum experiment using an apparatus similar to that shown in Figure 9.11b. She obtains the following average data: $h = 8.68$ cm, $m_1 = 68.8$ g, and $m_2 = 263$ g. The symbols refer to the quantities in Figure 9.11a. (a) Determine the initial speed v_{1A} of the projectile. (b) The second part of her experiment is to obtain v_{1A} by firing the same projectile horizontally (with the pendulum removed from the path), by measuring its final horizontal position x and distance of fall y (Fig. P9.65). Show that the initial speed of the projectile is related to x and y

$\mathbf{v}_{1A}$

y

x

Figure P9.65

through the relation

$$v_{1A} = \frac{x}{\sqrt{2y/g}}$$

What numerical value does she obtain for v_{1A} based on her measured values of $x = 257$ cm and $y = 85.3$ cm? What factors might account for the difference in this value compared to that obtained in part (a)?

66. Small ice cubes, each of mass 5.00 g, slide down a frictionless track in a steady stream, as shown in Figure P9.66. Starting from rest, each cube moves down through a net vertical distance of 1.50 m and leaves the bottom end of the track at an angle of 40.0° above the horizontal. At the highest point of its subsequent trajectory, the cube strikes a vertical wall and rebounds with half the speed it had upon impact. If 10.0 cubes strike the wall per second, what average force is exerted on the wall?

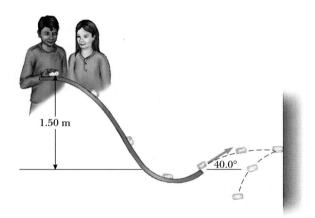

1.50 m

40.0°

Figure P9.66

67. A 5.00-g bullet moving with an initial speed of 400 m/s is fired into and passes through a 1.00-kg block, as in Figure P9.67. The block, initially at rest on a frictionless, horizontal surface, is connected to a spring with force constant 900 N/m. If the block moves 5.00 cm to the right after impact, find (a) the speed at which the bullet emerges from the block and (b) the mechanical energy converted into internal energy in the collision.

400 m/s

5.00 cm

$\mathbf{v}$

Figure P9.67

68. Consider as a system the Sun with the Earth in a circular orbit around it. Find the magnitude of the change in the velocity of the Sun relative to the center of mass of the

system over a period of 6 months. Neglect the influence of other celestial objects. You may obtain the necessary astronomical data from the endpapers of the book.

69. **Review problem.** There are (one can say) three coequal theories of motion: Newton's second law, stating that the total force on an object causes its acceleration; the work–kinetic energy theorem, stating that the total work on an object causes its change in kinetic energy; and the impulse–momentum theorem, stating that the total impulse on an object causes its change in momentum. In this problem, you compare predictions of the three theories in one particular case. A 3.00-kg object has velocity $7.00\hat{\mathbf{j}}$ m/s. Then, a total force $12.0\hat{\mathbf{i}}$ N acts on the object for 5.00 s. (a) Calculate the object's final velocity, using the impulse–momentum theorem. (b) Calculate its acceleration from $\mathbf{a} = (\mathbf{v}_f - \mathbf{v}_i)/\Delta t$. (c) Calculate its acceleration from $\mathbf{a} = \Sigma\mathbf{F}/m$. (d) Find the object's vector displacement from $\Delta\mathbf{r} = \mathbf{v}_i t + \frac{1}{2}\mathbf{a}t^2$. (e) Find the work done on the object from $W = \mathbf{F}\cdot\Delta\mathbf{r}$. (f) Find the final kinetic energy from $\frac{1}{2}mv_f^2 = \frac{1}{2}m\mathbf{v}_f\cdot\mathbf{v}_f$. (g) Find the final kinetic energy from $\frac{1}{2}mv_i^2 + W$.

70. 🖥 A rocket has total mass $M_i = 360$ kg, including 330 kg of fuel and oxidizer. In interstellar space it starts from rest. Its engine is turned on at time $t = 0$, and it puts out exhaust with relative speed $v_e = 1\,500$ m/s at the constant rate 2.50 kg/s. The burn lasts until the fuel runs out, at time 330 kg$/(2.5$ kg/s$) = 132$ s. Set up and carry out a computer analysis of the motion according to Euler's method. Find (a) the final velocity of the rocket and (b) the distance it travels during the burn.

71. A chain of length L and total mass M is released from rest with its lower end just touching the top of a table, as in Figure P9.71a. Find the force exerted by the table on the chain after the chain has fallen through a distance x, as in Figure P9.71b. (Assume each link comes to rest the instant it reaches the table.)

(a)   (b)

Figure P9.71

72. Sand from a stationary hopper falls onto a moving conveyor belt at the rate of 5.00 kg/s as in Figure P9.72. The conveyor belt is supported by frictionless rollers and moves at a constant speed of 0.750 m/s under the action of a constant horizontal external force $\mathbf{F}_{ext}$ supplied by the motor that drives the belt. Find (a) the sand's rate of change of momentum in the horizontal direction, (b) the force of friction exerted by the belt on the sand, (c) the external force $\mathbf{F}_{ext}$, (d) the work done by $\mathbf{F}_{ext}$ in 1 s, and (e) the kinetic energy acquired by the falling sand each second due to the change in its horizontal motion. (f) Why are the answers to (d) and (e) different?

73. A golf club consists of a shaft connected to a club head. The golf club can be modeled as a uniform rod of length ℓ and mass m_1 extending radially from the surface of a sphere of radius R and mass m_2. Find the location of the club's center of mass, measured from the center of the club head.

Answers to Quick Quizzes

9.1 (d). Two identical objects ($m_1 = m_2$) traveling at the same speed ($v_1 = v_2$) have the same kinetic energies and the same magnitudes of momentum. It also is possible, however, for particular combinations of masses and velocities to satisfy $K_1 = K_2$ but not $p_1 = p_2$. For example, a 1-kg object moving at 2 m/s has the same kinetic energy as a 4-kg object moving at 1 m/s, but the two clearly do not have the same momenta. Because we have no information about masses and speeds, we cannot choose among (a), (b), or (c).

9.2 (b), (c), (a). The slower the ball, the easier it is to catch. If the momentum of the medicine ball is the same as the momentum of the baseball, the speed of the medicine ball must be $1/10$ the speed of the baseball because the medicine ball has 10 times the mass. If the kinetic energies are the same, the speed of the medicine ball must be $1/\sqrt{10}$ the speed of the baseball because of the squared speed term in the equation for K. The medicine ball is hardest to catch when it has the same speed as the baseball.

9.3 (c). The ball and the Earth exert forces on each other, so neither is an isolated system. We must include both in the system so that the interaction force is internal to the system.

9.4 (c). From Equation 9.4, if $\mathbf{p}_1 + \mathbf{p}_2 = $ constant, then it follows that $\Delta\mathbf{p}_1 + \Delta\mathbf{p}_2 = 0$ and $\Delta\mathbf{p}_1 = -\Delta\mathbf{p}_2$. While the change in momentum is the same, the change in the *velocity* is a lot larger for the car!

9.5 (c) and (e). Object 2 has a greater acceleration because of its smaller mass. Therefore, it takes less time to travel the distance d. Even though the force applied to objects 1 and 2 is the same, the change in momentum is less for object 2 because Δt is smaller. The work $W = Fd$ done on

0.750 m/s

$\mathbf{F}_{ext}$

Figure P9.72

both objects is the same because both F and d are the same in the two cases. Therefore, $K_1 = K_2$.

9.6 (b) and (d). The same impulse is applied to both objects, so they experience the same change in momentum. Object 2 has a larger acceleration due to its smaller mass. Thus, the distance that object 2 covers in the time interval Δt is larger than that for object 1. As a result, more work is done on object 2 and $K_2 > K_1$.

9.7 (a) All three are the same. Because the passenger is brought from the car's initial speed to a full stop, the change in momentum (equal to the impulse) is the same regardless of what stops the passenger. (b) Dashboard, seatbelt, airbag. The dashboard stops the passenger very quickly in a front-end collision, resulting in a very large force. The seatbelt takes somewhat more time, so the force is smaller. Used along with the seatbelt, the airbag can extend the passenger's stopping time further, notably for his head, which would otherwise snap forward.

9.8 (a). If all of the initial kinetic energy is transformed, then nothing is moving after the collision. Consequently, the final momentum of the system is necessarily zero and, therefore, the initial momentum of the system must be zero. While (b) and (d) *together* would satisfy the conditions, neither one *alone* does.

9.9 (b). Because momentum of the two-ball system is conserved, $\mathbf{p}_{Ti} + 0 = \mathbf{p}_{Tf} + \mathbf{p}_B$. Because the table-tennis ball bounces back from the much more massive bowling ball with approximately the same speed, $\mathbf{p}_{Tf} = -\mathbf{p}_{Ti}$. As a consequence, $\mathbf{p}_B = 2\mathbf{p}_{Ti}$. Kinetic energy can be expressed as $K = p^2/2m$. Because of the much larger mass of the

bowling ball, its kinetic energy is much smaller than that of the table-tennis ball.

9.10 (b). The piece with the handle will have less mass than the piece made up of the end of the bat. To see why this is so, take the origin of coordinates as the center of mass before the bat was cut. Replace each cut piece by a small sphere located at the center of mass for each piece. The sphere representing the handle piece is farther from the origin, but the product of less mass and greater distance balances the product of greater mass and less distance for the end piece:

9.11 (a). This is the same effect as the swimmer diving off the raft that we just discussed. The vessel–passengers system is isolated. If the passengers all start running one way, the speed of the vessel increases (a *small* amount!) the other way.

9.12 (b). Once they stop running, the momentum of the system is the same as it was before they started running—you cannot change the momentum of an isolated system by means of internal forces. In case you are thinking that the passengers could do this over and over to take advantage of the speed increase *while* they are running, remember that they will slow the ship down every time they return to the bow!

Chapter 10

Rotation of a Rigid Object About a Fixed Axis

▲ *The Malaysian pastime of gasing involves the spinning of tops that can have masses up to 20 kg. Professional spinners can spin their tops so that they might rotate for hours before stopping. We will study the rotational motion of objects such as these tops in this chapter. (Courtesy Tourism Malaysia)*

When an extended object such as a wheel rotates about its axis, the motion cannot be analyzed by treating the object as a particle because at any given time different parts of the object have different linear velocities and linear accelerations. We can, however, analyze the motion by considering an extended object to be composed of a collection of particles, each of which has its own linear velocity and linear acceleration.

In dealing with a rotating object, analysis is greatly simplified by assuming that the object is rigid. A **rigid object** is one that is nondeformable—that is, the relative locations of all particles of which the object is composed remain constant. All real objects are deformable to some extent; however, our rigid-object model is useful in many situations in which deformation is negligible.

Rigid object

10.1 Angular Position, Velocity, and Acceleration

Figure 10.1 illustrates an overhead view of a rotating compact disc. The disc is rotating about a fixed axis through O. The axis is perpendicular to the plane of the figure. Let us investigate the motion of only one of the millions of "particles" making up the disc. A particle at P is at a fixed distance r from the origin and rotates about it in a circle of radius r. (In fact, *every* particle on the disc undergoes circular motion about O.) It is convenient to represent the position of P with its polar coordinates (r, θ), where r is the distance from the origin to P and θ is measured *counterclockwise* from some reference line as shown in Figure 10.1a. In this representation, the only coordinate for the particle that changes in time is the angle θ; r remains constant. As the particle moves along the circle from the reference line ($\theta = 0$), it moves through an arc of length s, as in Figure 10.1b. The arc length s is related to the angle θ through the relationship

(a)

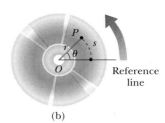

(b)

Figure 10.1 A compact disc rotating about a fixed axis through O perpendicular to the plane of the figure. (a) In order to define angular position for the disc, a fixed reference line is chosen. A particle at P is located at a distance r from the rotation axis at O. (b) As the disc rotates, point P moves through an arc length s on a circular path of radius r.

$$s = r\theta \qquad (10.1a)$$

$$\theta = \frac{s}{r} \qquad (10.1b)$$

Note the dimensions of θ in Equation 10.1b. Because θ is the ratio of an arc length and the radius of the circle, it is a pure number. However, we commonly give θ the artificial unit **radian** (rad), where

one radian is the angle subtended by an arc length equal to the radius of the arc.

Because the circumference of a circle is $2\pi r$, it follows from Equation 10.1b that 360° corresponds to an angle of $(2\pi r/r)$ rad $= 2\pi$ rad. (Also note that 2π rad corresponds

PITFALL PREVENTION

10.1 Remember the Radian

In rotational equations, we must use angles expressed in *radians*. Don't fall into the trap of using angles measured in degrees in rotational equations.

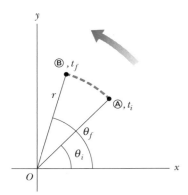

Figure 10.2 A particle on a rotating rigid object moves from Ⓐ to Ⓑ along the arc of a circle. In the time interval $\Delta t = t_f - t_i$, the radius vector moves through an angular displacement $\Delta\theta = \theta_f - \theta_i$.

Average angular speed

Instantaneous angular speed

to one complete revolution.) Hence, 1 rad = $360°/2\pi \approx 57.3°$. To convert an angle in degrees to an angle in radians, we use the fact that π rad = $180°$, or

$$\theta\,(\text{rad}) = \frac{\pi}{180°}\,\theta\,(\text{deg})$$

For example, $60°$ equals $\pi/3$ rad and $45°$ equals $\pi/4$ rad.

Because the disc in Figure 10.1 is a rigid object, as the particle moves along the circle from the reference line, every other particle on the object rotates through the same angle θ. Thus, **we can associate the angle θ with the entire rigid object as well as with an individual particle.** This allows us to define the *angular position* of a rigid object in its rotational motion. We choose a reference line on the object, such as a line connecting O and a chosen particle on the object. The **angular position** of the rigid object is the angle θ between this reference line on the object and the fixed reference line in space, which is often chosen as the x axis. This is similar to the way we identify the position of an object in translational motion—the distance x between the object and the reference position, which is the origin, $x = 0$.

As the particle in question on our rigid object travels from position Ⓐ to position Ⓑ in a time interval Δt as in Figure 10.2, the reference line of length r sweeps out an angle $\Delta\theta = \theta_f - \theta_i$. This quantity $\Delta\theta$ is defined as the **angular displacement** of the rigid object:

$$\Delta\theta \equiv \theta_f - \theta_i$$

The rate at which this angular displacement occurs can vary. If the rigid object spins rapidly, this displacement can occur in a short time interval. If it rotates slowly, this displacement occurs in a longer time interval. These different rotation rates can be quantified by introducing *angular speed*. We define the **average angular speed** $\overline{\omega}$ (Greek omega) as the ratio of the angular displacement of a rigid object to the time interval Δt during which the displacement occurs:

$$\overline{\omega} \equiv \frac{\theta_f - \theta_i}{t_f - t_i} = \frac{\Delta\theta}{\Delta t} \tag{10.2}$$

In analogy to linear speed, the **instantaneous angular speed** ω is defined as the limit of the ratio $\Delta\theta/\Delta t$ as Δt approaches zero:

$$\omega \equiv \lim_{\Delta t \to 0} \frac{\Delta\theta}{\Delta t} = \frac{d\theta}{dt} \tag{10.3}$$

Angular speed has units of radians per second (rad/s), which can be written as second^{-1} (s^{-1}) because radians are not dimensional. We take ω to be positive when θ is increasing (counterclockwise motion in Figure 10.2) and negative when θ is decreasing (clockwise motion in Figure 10.2).

Quick Quiz 10.1 A rigid object is rotating in a counterclockwise sense around a fixed axis. Each of the following pairs of quantities represents an initial angular position and a final angular position of the rigid object. Which of the sets can *only* occur if the rigid object rotates through more than 180°? (a) 3 rad, 6 rad (b) −1 rad, 1 rad (c) 1 rad, 5 rad.

Quick Quiz 10.2 Suppose that the change in angular position for each of the pairs of values in Quick Quiz 10.1 occurs in 1 s. Which choice represents the lowest average angular speed?

Figure 10.3 The right-hand rule for determining the direction of the angular velocity vector.

If the instantaneous angular speed of an object changes from ω_i to ω_f in the time interval Δt, the object has an angular acceleration. The **average angular acceleration** $\overline{\alpha}$ (Greek alpha) of a rotating rigid object is defined as the ratio of the change in the angular speed to the time interval Δt during which the change in the angular speed occurs:

$$\overline{\alpha} \equiv \frac{\omega_f - \omega_i}{t_f - t_i} = \frac{\Delta\omega}{\Delta t} \qquad (10.4)$$

◀ **Average angular acceleration**

In analogy to linear acceleration, the **instantaneous angular acceleration** is defined as the limit of the ratio $\Delta\omega/\Delta t$ as Δt approaches zero:

$$\alpha \equiv \lim_{\Delta t \to 0} \frac{\Delta\omega}{\Delta t} = \frac{d\omega}{dt} \qquad (10.5)$$

◀ **Instantaneous angular acceleration**

Angular acceleration has units of radians per second squared (rad/s^2), or just second^{-2} (s^{-2}). Note that α is positive when a rigid object rotating counterclockwise is speeding up or when a rigid object rotating clockwise is slowing down during some time interval.

When a rigid object is rotating about a *fixed* axis, **every particle on the object rotates through the same angle in a given time interval and has the same angular speed and the same angular acceleration.** That is, the quantities θ, ω, and α characterize the rotational motion of the entire rigid object as well as individual particles in the object. Using these quantities, we can greatly simplify the analysis of rigid-object rotation.

Angular position (θ), angular speed (ω), and angular acceleration (α) are analogous to linear position (x), linear speed (v), and linear acceleration (a). The variables θ, ω, and α differ dimensionally from the variables x, v, and a only by a factor having the unit of length. (See Section 10.3.)

We have not specified any direction for angular speed and angular acceleration. Strictly speaking, ω and α are the magnitudes of the angular velocity and the angular acceleration vectors[1] $\boldsymbol{\omega}$ and $\boldsymbol{\alpha}$, respectively, and they should always be positive. Because we are considering rotation about a fixed axis, however, we can use nonvector notation and indicate the directions of the vectors by assigning a positive or negative sign to ω and α, as discussed earlier with regard to Equations 10.3 and 10.5. For rotation about a fixed axis, the only direction that uniquely specifies the rotational motion is the direction along the axis of rotation. Therefore, the directions of $\boldsymbol{\omega}$ and $\boldsymbol{\alpha}$ are along this axis. If an object rotates in the xy plane as in Figure 10.1, the direction of $\boldsymbol{\omega}$ is out of the plane of the diagram when the rotation is counterclockwise and into the plane of the diagram when the rotation is clockwise. To illustrate this convention, it is convenient to use the *right-hand rule* demonstrated in Figure 10.3. When the four fingers of the right

  PITFALL PREVENTION

10.2 Specify Your Axis

In solving rotation problems, you must specify an axis of rotation. This is a new feature not found in our study of translational motion. The choice is arbitrary, but once you make it, you must maintain that choice consistently throughout the problem. In some problems, the physical situation suggests a natural axis, such as the center of an automobile wheel. In other problems, there may not be an obvious choice, and you must exercise judgement.

---

[1]   Although we do not verify it here, the instantaneous angular velocity and instantaneous angular acceleration are vector quantities, but the corresponding average values are not. This is because angular displacements do not add as vector quantities for finite rotations.

hand are wrapped in the direction of rotation, the extended right thumb points in the direction of $\boldsymbol{\omega}$. The direction of $\boldsymbol{\alpha}$ follows from its definition $\boldsymbol{\alpha} \equiv d\boldsymbol{\omega}/dt$. It is in the same direction as $\boldsymbol{\omega}$ if the angular speed is increasing in time, and it is antiparallel to $\boldsymbol{\omega}$ if the angular speed is decreasing in time.

Quick Quiz 10.3 A rigid object is rotating with an angular speed $\omega < 0$. The angular velocity vector $\boldsymbol{\omega}$ and the angular acceleration vector $\boldsymbol{\alpha}$ are antiparallel. The angular speed of the rigid object is (a) clockwise and increasing (b) clockwise and decreasing (c) counterclockwise and increasing (d) counterclockwise and decreasing.

10.2 Rotational Kinematics: Rotational Motion with Constant Angular Acceleration

In our study of linear motion, we found that the simplest form of accelerated motion to analyze is motion under constant linear acceleration. Likewise, for rotational motion about a fixed axis, the simplest accelerated motion to analyze is motion under constant angular acceleration. Therefore, we next develop kinematic relationships for this type of motion. If we write Equation 10.5 in the form $d\omega = \alpha\, dt$, and let $t_i = 0$ and $t_f = t$, integrating this expression directly gives

Rotational kinematic equations

$$\omega_f = \omega_i + \alpha t \qquad \text{(for constant } \alpha\text{)} \tag{10.6}$$

where ω_i is the angular speed of the rigid object at time $t = 0$. Equation 10.6 allows us to find the angular speed ω_f of the object at any later time t. Substituting Equation 10.6 into Equation 10.3 and integrating once more, we obtain

$$\theta_f = \theta_i + \omega_i t + \tfrac{1}{2}\alpha t^2 \qquad \text{(for constant } \alpha\text{)} \tag{10.7}$$

where θ_i is the angular position of the rigid object at time $t = 0$. Equation 10.7 allows us to find the angular position θ_f of the object at any later time t. If we eliminate t from Equations 10.6 and 10.7, we obtain

$$\omega_f{}^2 = \omega_i{}^2 + 2\alpha(\theta_f - \theta_i) \qquad \text{(for constant } \alpha\text{)} \tag{10.8}$$

This equation allows us to find the angular speed ω_f of the rigid object for any value of its angular position θ_f. If we eliminate α between Equations 10.6 and 10.7, we obtain

$$\theta_f = \theta_i + \tfrac{1}{2}(\omega_i + \omega_f)t \qquad \text{(for constant } \alpha\text{)} \tag{10.9}$$

Notice that these kinematic expressions for rotational motion under constant angular acceleration are of the same mathematical form as those for linear motion under constant linear acceleration. They can be generated from the equations for linear motion by making the substitutions $x \rightarrow \theta$, $v \rightarrow \omega$, and $a \rightarrow \alpha$. Table 10.1 compares the kinematic equations for rotational and linear motion.

 PITFALL PREVENTION

10.3 Just Like Translation?

Equations 10.6 to 10.9 and Table 10.1 suggest that rotational kinematics is just like translational kinematics. That is almost true, with two key differences: (1) in rotational kinematics, you must specify a rotation axis (per Pitfall Prevention 10.2); (2) in rotational motion, the object keeps returning to its original orientation—thus, you may be asked for the number of revolutions made by a rigid object. This concept has no meaning in translational motion, but is specified by $\Delta\theta$, which is analogous to Δx.

Table 10.1

Kinematic Equations for Rotational and Linear Motion Under Constant Acceleration	
Rotational Motion About Fixed Axis	**Linear Motion**
$\omega_f = \omega_i + \alpha t$	$v_f = v_i + at$
$\theta_f = \theta_i + \omega_i t + \frac{1}{2}\alpha t^2$	$x_f = x_i + v_i t + \frac{1}{2}at^2$
$\omega_f^2 = \omega_i^2 + 2\alpha(\theta_f - \theta_i)$	$v_f^2 = v_i^2 + 2a(x_f - x_i)$
$\theta_f = \theta_i + \frac{1}{2}(\omega_i + \omega_f)t$	$x_f = x_i + \frac{1}{2}(v_i + v_f)t$

Quick Quiz 10.4 Consider again the pairs of angular positions for the rigid object in Quick Quiz 10.1. If the object starts from rest at the initial angular position, moves counterclockwise with constant angular acceleration, and arrives at the final angular position with the same angular speed in all three cases, for which choice is the angular acceleration the highest?

Example 10.1   Rotating Wheel

A wheel rotates with a constant angular acceleration of 3.50 rad/s^2.

(A) If the angular speed of the wheel is 2.00 rad/s at $t_i = 0$, through what angular displacement does the wheel rotate in 2.00 s?

Solution We can use Figure 10.2 to represent the wheel. We arrange Equation 10.7 so that it gives us angular displacement:

$$\Delta\theta = \theta_f - \theta_i = \omega_i t + \frac{1}{2}\alpha t^2$$

$$= (2.00 \text{ rad/s})(2.00 \text{ s}) + \frac{1}{2}(3.50 \text{ rad/s}^2)(2.00 \text{ s})^2$$

$$= \boxed{11.0 \text{ rad}} = (11.0 \text{ rad})(57.3°/\text{rad}) = \boxed{630°}$$

(B) Through how many revolutions has the wheel turned during this time interval?

Solution We multiply the angular displacement found in part (A) by a conversion factor to find the number of revolutions:

$$\Delta\theta = 630°\left(\frac{1 \text{ rev}}{360°}\right) = \boxed{1.75 \text{ rev}}$$

(C) What is the angular speed of the wheel at $t = 2.00$ s?

Solution Because the angular acceleration and the angular speed are both positive, our answer must be greater than 2.00 rad/s. Using Equation 10.6, we find

$$\omega_f = \omega_i + \alpha t = 2.00 \text{ rad/s} + (3.50 \text{ rad/s}^2)(2.00 \text{ s})$$

$$= \boxed{9.00 \text{ rad/s}}$$

We could also obtain this result using Equation 10.8 and the results of part (A). Try it!

What If? Suppose a particle moves along a straight line with a constant acceleration of 3.50 m/s^2. If the velocity of the particle is 2.00 m/s at $t_i = 0$, through what displacement does the particle move in 2.00 s? What is the velocity of the particle at $t = 2.00$ s?

Answer Notice that these questions are translational analogs to parts (A) and (C) of the original problem. The mathematical solution follows exactly the same form. For the displacement,

$$\Delta x = x_f - x_i = v_i t + \frac{1}{2}at^2$$

$$= (2.00 \text{ m/s})(2.00 \text{ s}) + \frac{1}{2}(3.50 \text{ m/s}^2)(2.00 \text{ s})^2$$

$$= 11.0 \text{ m}$$

and for the velocity,

$$v_f = v_i + at = 2.00 \text{ m/s} + (3.50 \text{ m/s}^2)(2.00 \text{ s}) = 9.00 \text{ m/s}$$

Note that there is no translational analog to part (B) because translational motion is not repetitive like rotational motion.

10.3 Angular and Linear Quantities

In this section we derive some useful relationships between the angular speed and acceleration of a rotating rigid object and the linear speed and acceleration of a point in the object. To do so, we must keep in mind that when a rigid object rotates about a fixed axis, as in Figure 10.4, **every particle of the object moves in a circle whose center is the axis of rotation.**

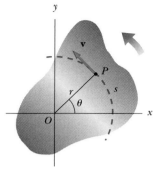

Active Figure 10.4 As a rigid object rotates about the fixed axis through O, the point P has a tangential velocity $\mathbf{v}$ that is always tangent to the circular path of radius r.

At the Active Figures link at http://www.pse6.com, you can move point P and observe the tangential velocity as the object rotates.

Because point P in Figure 10.4 moves in a circle, the linear velocity vector $\mathbf{v}$ is always tangent to the circular path and hence is called *tangential velocity*. The magnitude of the tangential velocity of the point P is by definition the tangential speed $v = ds/dt$, where s is the distance traveled by this point measured along the circular path. Recalling that $s = r\theta$ (Eq. 10.1a) and noting that r is constant, we obtain

$$v = \frac{ds}{dt} = r \frac{d\theta}{dt}$$

Because $d\theta/dt = \omega$ (see Eq. 10.3), we see that

$$v = r\omega \tag{10.10}$$

That is, the tangential speed of a point on a rotating rigid object equals the perpendicular distance of that point from the axis of rotation multiplied by the angular speed. Therefore, although every point on the rigid object has the same *angular* speed, not every point has the same *tangential* speed because r is not the same for all points on the object. Equation 10.10 shows that the tangential speed of a point on the rotating object increases as one moves outward from the center of rotation, as we would intuitively expect. The outer end of a swinging baseball bat moves much faster than the handle.

We can relate the angular acceleration of the rotating rigid object to the tangential acceleration of the point P by taking the time derivative of v:

$$a_t = \frac{dv}{dt} = r \frac{d\omega}{dt}$$

Relation between tangential and angular acceleration

$$a_t = r\alpha \tag{10.11}$$

That is, the tangential component of the linear acceleration of a point on a rotating rigid object equals the point's distance from the axis of rotation multiplied by the angular acceleration.

In Section 4.4 we found that a point moving in a circular path undergoes a radial acceleration a_r of magnitude v^2/r directed toward the center of rotation (Fig. 10.5). Because $v = r\omega$ for a point P on a rotating object, we can express the centripetal acceleration at that point in terms of angular speed as

$$a_c = \frac{v^2}{r} = r\omega^2 \tag{10.12}$$

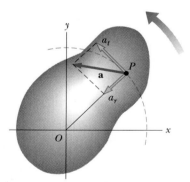

Figure 10.5 As a rigid object rotates about a fixed axis through O, the point P experiences a tangential component of linear acceleration a_t and a radial component of linear acceleration a_r. The total linear acceleration of this point is $\mathbf{a} = \mathbf{a}_t + \mathbf{a}_r$.

The total linear acceleration vector at the point is $\mathbf{a} = \mathbf{a}_t + \mathbf{a}_r$, where the magnitude of $\mathbf{a}_r$ is the centripetal acceleration a_c. Because $\mathbf{a}$ is a vector having a radial and a tangential component, the magnitude of $\mathbf{a}$ at the point P on the rotating rigid object is

$$a = \sqrt{a_t^2 + a_r^2} = \sqrt{r^2\alpha^2 + r^2\omega^4} = r\sqrt{\alpha^2 + \omega^4} \tag{10.13}$$

Quick Quiz 10.5 Andy and Charlie are riding on a merry-go-round. Andy rides on a horse at the outer rim of the circular platform, twice as far from the center of the circular platform as Charlie, who rides on an inner horse. When the merry-go-round is rotating at a constant angular speed, Andy's angular speed is (a) twice Charlie's (b) the same as Charlie's (c) half of Charlie's (d) impossible to determine.

Quick Quiz 10.6 Consider again the merry-go-round situation in Quick Quiz 10.5. When the merry-go-round is rotating at a constant angular speed, Andy's tangential speed is (a) twice Charlie's (b) the same as Charlie's (c) half of Charlie's (d) impossible to determine.

Example 10.2   CD Player

On a compact disc (Fig. 10.6), audio information is stored in a series of pits and flat areas on the surface of the disc. The information is stored digitally, and the alternations between pits and flat areas on the surface represent binary ones and zeroes to be read by the compact disc player and converted back to sound waves. The pits and flat areas are detected by a system consisting of a laser and lenses. The length of a string of ones and zeroes representing one piece of information is the same everywhere on the disc, whether the information is near the center of the disc or near its outer edge. In order that this length of ones and zeroes always passes by the laser–lens system in the same time period, the tangential speed of the disc surface at the location of the lens must be constant. This requires, according to Equation 10.10, that the angular speed vary as the laser–lens system moves radially along the disc. In a typical compact disc player, the constant speed of the surface at the point of the laser–lens system is 1.3 m/s.

(A) Find the angular speed of the disc in revolutions per minute when information is being read from the innermost first track ($r = 23$ mm) and the outermost final track ($r = 58$ mm).

Solution Using Equation 10.10, we can find the angular speed that will give us the required tangential speed at the position of the inner track,

$$\omega_i = \frac{v}{r_i} = \frac{1.3 \text{ m/s}}{2.3 \times 10^{-2} \text{ m}} = 57 \text{ rad/s}$$

$$= (57 \text{ rad/s})\left(\frac{1 \text{ rev}}{2\pi \text{ rad}}\right)\left(\frac{60 \text{ s}}{1 \text{ min}}\right)$$

$$= \boxed{5.4 \times 10^2 \text{ rev/min}}$$

For the outer track,

$$\omega_f = \frac{v}{r_f} = \frac{1.3 \text{ m/s}}{5.8 \times 10^{-2} \text{ m}} = 22 \text{ rad/s}$$

$$= \boxed{2.1 \times 10^2 \text{ rev/min}}$$

The player adjusts the angular speed ω of the disc within this range so that information moves past the objective lens at a constant rate.

(B) The maximum playing time of a standard music CD is 74 min and 33 s. How many revolutions does the disc make during that time?

Solution We know that the angular speed is always decreasing, and we assume that it is decreasing steadily, with α constant. If $t = 0$ is the instant that the disc begins, with angular speed of 57 rad/s, then the final value of the time t is (74 min)(60 s/min) + 33 s = 4 473 s. We are looking for the angular displacement $\Delta\theta$ during this time interval. We use Equation 10.9:

Figure 10.6 (Example 10.2) A compact disc.

$$\Delta\theta = \theta_f - \theta_i = \tfrac{1}{2}(\omega_i + \omega_f)t$$

$$= \tfrac{1}{2}(57 \text{ rad/s} + 22 \text{ rad/s})(4 \text{ 473 s})$$

$$= 1.8 \times 10^5 \text{ rad}$$

We convert this angular displacement to revolutions:

$$\Delta\theta = 1.8 \times 10^5 \text{ rad}\left(\frac{1 \text{ rev}}{2\pi \text{ rad}}\right) = \boxed{2.8 \times 10^4 \text{ rev}}$$

(C) What total length of track moves past the objective lens during this time?

Solution Because we know the (constant) linear velocity and the time interval, this is a straightforward calculation:

$$x_f = v_i t = (1.3 \text{ m/s})(4 \text{ 473 s}) = \boxed{5.8 \times 10^3 \text{ m}}$$

More than 5.8 km of track spins past the objective lens!

(D) What is the angular acceleration of the CD over the 4 473-s time interval? Assume that α is constant.

Solution The most direct approach to solving this problem is to use Equation 10.6 and the results to part (A). We should obtain a negative number for the angular acceleration because the disc spins more and more slowly in the positive direction as time goes on. Our answer should also be relatively small because it takes such a long time—more than an hour—for the change in angular speed to be accomplished:

$$\alpha = \frac{\omega_f - \omega_i}{t} = \frac{22 \text{ rad/s} - 57 \text{ rad/s}}{4 \text{ 473 s}}$$

$$= \boxed{-7.8 \times 10^{-3} \text{ rad/s}^2}$$

The disc experiences a very gradual decrease in its rotation rate, as expected.

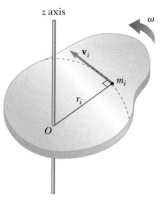

Figure 10.7 A rigid object rotating about the z axis with angular speed ω. The kinetic energy of the particle of mass m_i is $\frac{1}{2}m_i v_i{}^2$. The total kinetic energy of the object is called its *rotational* kinetic energy.

10.4 Rotational Kinetic Energy

In Chapter 7, we defined the kinetic energy of an object as the energy associated with its motion through space. An object rotating about a fixed axis remains stationary in space, so there is no kinetic energy associated with translational motion. The individual particles making up the rotating object, however, are moving through space—they follow circular paths. Consequently, there should be kinetic energy associated with rotational motion.

Let us consider an object as a collection of particles and assume that it rotates about a fixed z axis with an angular speed ω. Figure 10.7 shows the rotating object and identifies one particle on the object located at a distance r_i from the rotation axis. Each such particle has kinetic energy determined by its mass and tangential speed. If the mass of the ith particle is m_i and its tangential speed is v_i, its kinetic energy is

$$K_i = \tfrac{1}{2}m_i v_i{}^2$$

To proceed further, recall that although every particle in the rigid object has the same angular speed ω, the individual tangential speeds depend on the distance r_i from the axis of rotation according to the expression $v_i = r_i\omega$ (see Eq. 10.10). The *total* kinetic energy of the rotating rigid object is the sum of the kinetic energies of the individual particles:

$$K_R = \sum_i K_i = \sum_i \tfrac{1}{2}m_i v_i{}^2 = \tfrac{1}{2}\sum_i m_i r_i{}^2 \omega^2$$

We can write this expression in the form

$$K_R = \tfrac{1}{2}\left(\sum_i m_i r_i{}^2\right)\omega^2 \tag{10.14}$$

where we have factored ω^2 from the sum because it is common to every particle. We simplify this expression by defining the quantity in parentheses as the **moment of inertia I**:

Moment of inertia

$$I \equiv \sum_i m_i r_i{}^2 \tag{10.15}$$

From the definition of moment of inertia, we see that it has dimensions of ML^2 ($\mathrm{kg \cdot m^2}$ in SI units).[2] With this notation, Equation 10.14 becomes

Rotational kinetic energy

$$K_R = \tfrac{1}{2}I\omega^2 \tag{10.16}$$

 PITFALL PREVENTION

10.4 No Single Moment of Inertia

There is one major difference between mass and moment of inertia. Mass is an inherent property of an object. The moment of inertia of an object depends on your choice of rotation axis. Thus, there is no single value of the moment of inertia for an object. There is a minimum value of the moment of inertia, which is that calculated about an axis passing through the center of mass of the object.

Although we commonly refer to the quantity $\frac{1}{2}I\omega^2$ as **rotational kinetic energy**, it is not a new form of energy. It is ordinary kinetic energy because it is derived from a sum over individual kinetic energies of the particles contained in the rigid object. However, the mathematical form of the kinetic energy given by Equation 10.16 is convenient when we are dealing with rotational motion, provided we know how to calculate I.

It is important that you recognize the analogy between kinetic energy associated with linear motion $\frac{1}{2}mv^2$ and rotational kinetic energy $\frac{1}{2}I\omega^2$. The quantities I and ω in rotational motion are analogous to m and v in linear motion, respectively. (In fact, I takes the place of m and ω takes the place of v every time we compare a linear-motion equation with its rotational counterpart.) The moment of inertia is a measure of the resistance of an object to changes in its rotational motion, just as mass is a measure of the tendency of an object to resist changes in its linear motion.

---

[2]   Civil engineers use moment of inertia to characterize the elastic properties (rigidity) of such structures as loaded beams. Hence, it is often useful even in a nonrotational context.

Quick Quiz 10.7 A section of hollow pipe and a solid cylinder have the same radius, mass, and length. They both rotate about their long central axes with the same angular speed. Which object has the higher rotational kinetic energy? (a) the hollow pipe (b) the solid cylinder (c) they have the same rotational kinetic energy (d) impossible to determine.

Example 10.3   The Oxygen Molecule

Consider an oxygen molecule (O_2) rotating in the xy plane about the z axis. The rotation axis passes through the center of the molecule, perpendicular to its length. The mass of each oxygen atom is 2.66×10^{-26} kg, and at room temperature the average separation between the two atoms is $d = 1.21 \times 10^{-10}$ m. (The atoms are modeled as particles.)

(A) Calculate the moment of inertia of the molecule about the z axis.

Solution This is a straightforward application of the definition of I. Because each atom is a distance $d/2$ from the z axis, the moment of inertia about the axis is

$$I = \sum_i m_i r_i^2 = m\left(\frac{d}{2}\right)^2 + m\left(\frac{d}{2}\right)^2 = \frac{md^2}{2}$$

$$= \frac{(2.66 \times 10^{-26}\ \text{kg})(1.21 \times 10^{-10}\ \text{m})^2}{2}$$

$$= \boxed{1.95 \times 10^{-46}\ \text{kg} \cdot \text{m}^2}$$

This is a very small number, consistent with the minuscule masses and distances involved.

(B) If the angular speed of the molecule about the z axis is 4.60×10^{12} rad/s, what is its rotational kinetic energy?

Solution We apply the result we just calculated for the moment of inertia in the equation for K_R:

$$K_R = \tfrac{1}{2} I \omega^2$$
$$= \tfrac{1}{2}(1.95 \times 10^{-46}\ \text{kg} \cdot \text{m}^2)(4.60 \times 10^{12}\ \text{rad/s})^2$$

$$= \boxed{2.06 \times 10^{-21}\ \text{J}}$$

Example 10.4   Four Rotating Objects

Four tiny spheres are fastened to the ends of two rods of negligible mass lying in the xy plane (Fig. 10.8). We shall assume that the radii of the spheres are small compared with the dimensions of the rods.

(A) If the system rotates about the y axis (Fig. 10.8a) with an angular speed ω, find the moment of inertia and the rotational kinetic energy about this axis.

Solution First, note that the two spheres of mass m, which lie on the y axis, do not contribute to I_y (that is, $r_i = 0$ for these spheres about this axis). Applying Equation 10.15, we obtain

$$I_y = \sum_i m_i r_i^2 = Ma^2 + Ma^2 = \boxed{2Ma^2}$$

Therefore, the rotational kinetic energy about the y axis is

$$K_R = \tfrac{1}{2} I_y \omega^2 = \tfrac{1}{2}(2Ma^2)\omega^2 = \boxed{Ma^2\omega^2}$$

The fact that the two spheres of mass m do not enter into this result makes sense because they have no motion about the axis of rotation; hence, they have no rotational kinetic energy. By similar logic, we expect the moment of inertia about the x axis to be $I_x = 2mb^2$ with a rotational kinetic energy about that axis of $K_R = mb^2\omega^2$.

(B) Suppose the system rotates in the xy plane about an axis (the z axis) through O (Fig. 10.8b). Calculate the moment of inertia and rotational kinetic energy about this axis.

Solution Because r_i in Equation 10.15 is the distance between a sphere and the axis of rotation, we obtain

$$I_z = \sum_i m_i r_i^2 = Ma^2 + Ma^2 + mb^2 + mb^2 = \boxed{2Ma^2 + 2mb^2}$$

$$K_R = \tfrac{1}{2} I_z \omega^2 = \tfrac{1}{2}(2Ma^2 + 2mb^2)\omega^2 = \boxed{(Ma^2 + mb^2)\omega^2}$$

Comparing the results for parts (A) and (B), we conclude that the moment of inertia and therefore the rotational kinetic energy associated with a given angular speed depend on the axis of rotation. In part (B), we expect the result to include all four spheres and distances because all four spheres are rotating in the xy plane. Furthermore, the fact that the rotational kinetic energy in part (A) is smaller than that in part (B) indicates, based on the work–kinetic energy theorem, that it would require less work to set the system into rotation about the y axis than about the z axis.

What If? What if the mass *M* is much larger than *m*? How do the answers to parts (A) and (B) compare?

Answer If $M \gg m$, then *m* can be neglected and the moment of inertia and rotational kinetic energy in part (B) become

$$I_z = 2Ma^2 \quad \text{and} \quad K_R = Ma^2\omega^2$$

which are the same as the answers in part (A). If the masses *m* of the two red spheres in Figure 10.8 are negligible, then these spheres can be removed from the figure and rotations about the *y* and *z* axes are equivalent.

Figure 10.8 (Example 10.4) Four spheres form an unusual baton. (a) The baton is rotated about the *y* axis. (b) The baton is rotated about the *z* axis.

10.5 Calculation of Moments of Inertia

We can evaluate the moment of inertia of an extended rigid object by imagining the object to be divided into many small volume elements, each of which has mass Δm_i. We use the definition $I = \sum_i r_i^2 \Delta m_i$ and take the limit of this sum as $\Delta m_i \to 0$. In this limit, the sum becomes an integral over the volume of the object:

Moment of inertia of a rigid object

$$I = \lim_{\Delta m_i \to 0} \sum_i r_i^2 \Delta m_i = \int r^2 \, dm \qquad (10.17)$$

It is usually easier to calculate moments of inertia in terms of the volume of the elements rather than their mass, and we can easily make that change by using Equation 1.1, $\rho = m/V$, where ρ is the density of the object and *V* is its volume. From this equation, the mass of a small element is $dm = \rho \, dV$. Substituting this result into Equation 10.17 gives

$$I = \int \rho r^2 \, dV$$

If the object is homogeneous, then ρ is constant and the integral can be evaluated for a known geometry. If ρ is not constant, then its variation with position must be known to complete the integration.

The density given by $\rho = m/V$ sometimes is referred to as *volumetric mass density* because it represents mass per unit volume. Often we use other ways of expressing density. For instance, when dealing with a sheet of uniform thickness *t*, we can define a *surface mass density* $\sigma = \rho t$, which represents *mass per unit area*. Finally, when mass is distributed along a rod of uniform cross-sectional area *A*, we sometimes use *linear mass density* $\lambda = M/L = \rho A$, which is the *mass per unit length*.

Example 10.5 Uniform Thin Hoop

Find the moment of inertia of a uniform thin hoop of mass M and radius R about an axis perpendicular to the plane of the hoop and passing through its center (Fig. 10.9).

Solution Because the hoop is thin, all mass elements dm are the same distance $r = R$ from the axis, and so, applying Equation 10.17, we obtain for the moment of inertia about the z axis through O:

$$I_z = \int r^2 \, dm = R^2 \int dm = \boxed{MR^2}$$

Note that this moment of inertia is the same as that of a single particle of mass M located a distance R from the axis of rotation.

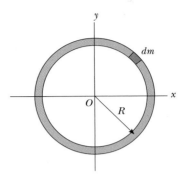

Figure 10.9 (Example 10.5) The mass elements dm of a uniform hoop are all the same distance from O.

Example 10.6 Uniform Rigid Rod

Calculate the moment of inertia of a uniform rigid rod of length L and mass M (Fig. 10.10) about an axis perpendicular to the rod (the y axis) and passing through its center of mass.

Solution The shaded length element dx in Figure 10.10 has a mass dm equal to the mass per unit length λ multiplied by dx:

$$dm = \lambda \, dx = \frac{M}{L} \, dx$$

Substituting this expression for dm into Equation 10.17, with $r^2 = x^2$, we obtain

$$I_y = \int r^2 \, dm = \int_{-L/2}^{L/2} x^2 \frac{M}{L} \, dx = \frac{M}{L} \int_{-L/2}^{L/2} x^2 \, dx$$

$$= \frac{M}{L} \left[\frac{x^3}{3} \right]_{-L/2}^{L/2} = \boxed{\tfrac{1}{12} ML^2}$$

Figure 10.10 (Example 10.6) A uniform rigid rod of length L. The moment of inertia about the y axis is less than that about the y' axis. The latter axis is examined in Example 10.8.

Example 10.7 Uniform Solid Cylinder

A uniform solid cylinder has a radius R, mass M, and length L. Calculate its moment of inertia about its central axis (the z axis in Fig. 10.11).

Solution It is convenient to divide the cylinder into many cylindrical shells, each of which has radius r, thickness dr, and length L, as shown in Figure 10.11. The volume dV of each shell is its cross-sectional area multiplied by its length: $dV = L \, dA = L(2\pi r) \, dr$. If the mass per unit volume is ρ, then the mass of this differential volume element is $dm = \rho \, dV = 2\pi \rho L r \, dr$. Substituting this expression for dm into Equation 10.17, we obtain

$$I_z = \int r^2 \, dm = \int r^2 (2\pi \rho L r \, dr) = 2\pi \rho L \int_0^R r^3 \, dr = \tfrac{1}{2} \pi \rho L R^4$$

Because the total volume of the cylinder is $\pi R^2 L$, we see that $\rho = M/V = M/\pi R^2 L$. Substituting this value for ρ into the above result gives

$$I_z = \boxed{\tfrac{1}{2} MR^2}$$

Figure 10.11 (Example 10.7) Calculating I about the z axis for a uniform solid cylinder.

What If? What if the length of the cylinder in Figure 10.11 is increased to **2L**, while the mass **M** and radius **R** are held fixed? How does this change the moment of inertia of the cylinder?

Answer Note that the result for the moment of inertia of a cylinder does not depend on L, the length of the cylinder. In other words, it applies equally well to a long cylinder and a flat disk having the same mass M and radius R. Thus, the moment of inertia of the cylinder would not be affected by changing its length.

Table 10.2 gives the moments of inertia for a number of objects about specific axes. The moments of inertia of rigid objects with simple geometry (high symmetry) are relatively easy to calculate provided the rotation axis coincides with an axis of symmetry. The calculation of moments of inertia about an arbitrary axis can be cumbersome, however, even for a highly symmetric object. Fortunately, use of an important theorem, called the **parallel-axis theorem,** often simplifies the calculation. Suppose the moment of inertia about an axis through the center of mass of an object is I_{CM}. The parallel-axis theorem states that the moment of inertia about any axis parallel to and a distance D away from this axis is

Parallel-axis theorem

$$I = I_{\text{CM}} + MD^2 \tag{10.18}$$

To prove the parallel-axis theorem, suppose that an object rotates in the xy plane about the z axis, as shown in Figure 10.12, and that the coordinates of the center of mass are x_{CM}, y_{CM}. Let the mass element dm have coordinates x, y. Because this

Table 10.2

Moments of Inertia of Homogeneous Rigid Objects with Different Geometries

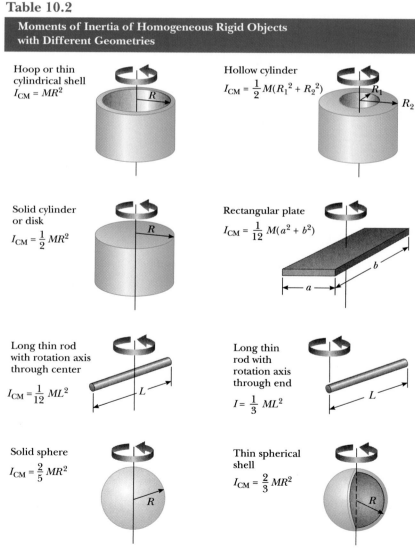

Hoop or thin cylindrical shell
$I_{\text{CM}} = MR^2$

Hollow cylinder
$I_{\text{CM}} = \frac{1}{2} M(R_1{}^2 + R_2{}^2)$

Solid cylinder or disk
$I_{\text{CM}} = \frac{1}{2} MR^2$

Rectangular plate
$I_{\text{CM}} = \frac{1}{12} M(a^2 + b^2)$

Long thin rod with rotation axis through center
$I_{\text{CM}} = \frac{1}{12} ML^2$

Long thin rod with rotation axis through end
$I = \frac{1}{3} ML^2$

Solid sphere
$I_{\text{CM}} = \frac{2}{5} MR^2$

Thin spherical shell
$I_{\text{CM}} = \frac{2}{3} MR^2$

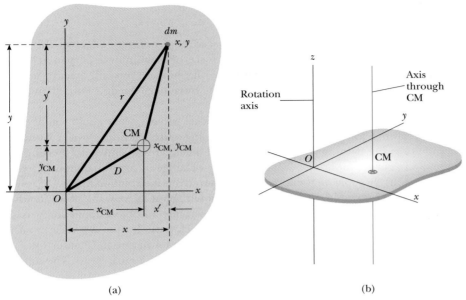

Figure 10.12 (a) The parallel-axis theorem: if the moment of inertia about an axis perpendicular to the figure through the center of mass is I_{CM}, then the moment of inertia about the z axis is $I_z = I_{CM} + MD^2$. (b) Perspective drawing showing the z axis (the axis of rotation) and the parallel axis through the CM.

element is a distance $r = \sqrt{x^2 + y^2}$ from the z axis, the moment of inertia about the z axis is

$$I = \int r^2 \, dm = \int (x^2 + y^2) \, dm$$

However, we can relate the coordinates x, y of the mass element dm to the coordinates of this same element located in a coordinate system having the object's center of mass as its origin. If the coordinates of the center of mass are x_{CM}, y_{CM} in the original coordinate system centered on O, then from Figure 10.12a we see that the relationships between the unprimed and primed coordinates are $x = x' + x_{CM}$ and $y = y' + y_{CM}$. Therefore,

$$I = \int [(x' + x_{CM})^2 + (y' + y_{CM})^2] \, dm$$

$$= \int [(x')^2 + (y')^2] \, dm + 2x_{CM} \int x' \, dm + 2y_{CM} \int y' \, dm + (x_{CM}^2 + y_{CM}^2) \int dm$$

The first integral is, by definition, the moment of inertia about an axis that is parallel to the z axis and passes through the center of mass. The second two integrals are zero because, by definition of the center of mass, $\int x' \, dm = \int y' \, dm = 0$. The last integral is simply MD^2 because $\int dm = M$ and $D^2 = x_{CM}^2 + y_{CM}^2$. Therefore, we conclude that

$$I = I_{CM} + MD^2$$

Example 10.8   Applying the Parallel-Axis Theorem

Consider once again the uniform rigid rod of mass M and length L shown in Figure 10.10. Find the moment of inertia of the rod about an axis perpendicular to the rod through one end (the y' axis in Fig. 10.10).

Solution Intuitively, we expect the moment of inertia to be greater than $I_{CM} = \frac{1}{12}ML^2$ because there is mass up to a distance of L away from the rotation axis, while the farthest distance in Example 10.6 was only $L/2$. Because the distance

between the center-of-mass axis and the y' axis is $D = L/2$, the parallel-axis theorem gives

$$I = I_{CM} + MD^2 = \tfrac{1}{12}ML^2 + M\left(\frac{L}{2}\right)^2 = \boxed{\tfrac{1}{3}ML^2}$$

So, it is four times more difficult to change the rotation of a rod spinning about its end than it is to change the motion of one spinning about its center.

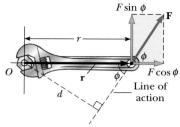

Figure 10.13 The force **F** has a greater rotating tendency about O as F increases and as the moment arm d increases. The component $F \sin \phi$ tends to rotate the wrench about O.

 PITFALL PREVENTION

10.5 Torque Depends on Your Choice of Axis

Like moment of inertia, there is no unique value of the torque—its value depends on your choice of rotation axis.

Moment arm

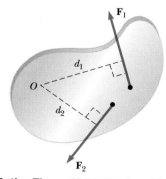

Active Figure 10.14 The force **F**$_1$ tends to rotate the object counterclockwise about O, and **F**$_2$ tends to rotate it clockwise.

At the Active Figures link at http://www.pse6.com, you can change the magnitudes, directions, and points of application of forces F_1 and F_2 to see how the object accelerates under the action of the two forces.

10.6 Torque

Why are a door's hinges and its doorknob placed near opposite edges of the door? Imagine trying to rotate a door by applying a force of magnitude F perpendicular to the door surface but at various distances from the hinges. You will achieve a more rapid rate of rotation for the door by applying the force near the doorknob than by applying it near the hinges.

If you cannot loosen a stubborn bolt with a socket wrench, what would you do in an effort to loosen the bolt? You may intuitively try using a wrench with a longer handle or slip a pipe over the existing wrench to make it longer. This is similar to the situation with the door. You are more successful at causing a change in rotational motion (of the door or the bolt) by applying the force farther away from the rotation axis.

When a force is exerted on a rigid object pivoted about an axis, the object tends to rotate about that axis. The tendency of a force to rotate an object about some axis is measured by a vector quantity called **torque** τ (Greek tau). Torque is a vector, but we will consider only its magnitude here and explore its vector nature in Chapter 11.

Consider the wrench pivoted on the axis through O in Figure 10.13. The applied force **F** acts at an angle ϕ to the horizontal. We define the magnitude of the torque associated with the force **F** by the expression

$$\tau \equiv rF \sin \phi = Fd \tag{10.19}$$

where r is the distance between the pivot point and the point of application of **F** and d is the perpendicular distance from the pivot point to the line of action of **F**. (The *line of action* of a force is an imaginary line extending out both ends of the vector representing the force. The dashed line extending from the tail of **F** in Figure 10.13 is part of the line of action of **F**.) From the right triangle in Figure 10.13 that has the wrench as its hypotenuse, we see that $d = r \sin \phi$. The quantity d is called the **moment arm** (or *lever arm*) of **F**.

In Figure 10.13, the only component of **F** that tends to cause rotation is $F \sin \phi$, the component perpendicular to a line drawn from the rotation axis to the point of application of the force. The horizontal component $F \cos \phi$, because its line of action passes through O, has no tendency to produce rotation about an axis passing through O. From the definition of torque, we see that the rotating tendency increases as F increases and as d increases. This explains the observation that it is easier to rotate a door if we push at the doorknob rather than at a point close to the hinge. We also want to apply our push as closely perpendicular to the door as we can. Pushing sideways on the doorknob will not cause the door to rotate.

If two or more forces are acting on a rigid object, as in Figure 10.14, each tends to produce rotation about the axis at O. In this example, **F**$_2$ tends to rotate the object clockwise and **F**$_1$ tends to rotate it counterclockwise. We use the convention that the sign of the torque resulting from a force is positive if the turning tendency of the force is counterclockwise and is negative if the turning tendency is clockwise. For example, in Figure 10.14, the torque resulting from **F**$_1$, which has a moment arm d_1, is positive and equal to $+F_1 d_1$; the torque from **F**$_2$ is negative and equal to $-F_2 d_2$. Hence, the *net* torque about O is

$$\sum \tau = \tau_1 + \tau_2 = F_1 d_1 - F_2 d_2$$

Torque should not be confused with force. Forces can cause a change in linear motion, as described by Newton's second law. Forces can also cause a change in rotational motion, but the effectiveness of the forces in causing this change depends on both the forces and the moment arms of the forces, in the combination that we call *torque*. Torque has units of force times length—newton · meters in SI units—and should be reported in these units. Do not confuse torque and work, which have the same units but are very different concepts.

Quick Quiz 10.8 If you are trying to loosen a stubborn screw from a piece of wood with a screwdriver and fail, should you find a screwdriver for which the handle is (a) longer or (b) fatter?

Quick Quiz 10.9 If you are trying to loosen a stubborn bolt from a piece of metal with a wrench and fail, should you find a wrench for which the handle is (a) longer (b) fatter?

Example 10.9 **The Net Torque on a Cylinder**

A one-piece cylinder is shaped as shown in Figure 10.15, with a core section protruding from the larger drum. The cylinder is free to rotate about the central axis shown in the drawing. A rope wrapped around the drum, which has radius R_1, exerts a force $\mathbf{T}_1$ to the right on the cylinder. A rope wrapped around the core, which has radius R_2, exerts a force $\mathbf{T}_2$ downward on the cylinder.

(A) What is the net torque acting on the cylinder about the rotation axis (which is the z axis in Figure 10.15)?

Solution The torque due to $\mathbf{T}_1$ is $-R_1T_1$. (The sign is negative because the torque tends to produce clockwise rotation.) The torque due to $\mathbf{T}_2$ is $+R_2T_2$. (The sign is positive because the torque tends to produce counterclockwise rotation.) Therefore, the net torque about the rotation axis is

$$\sum \tau = \tau_1 + \tau_2 = \boxed{R_2T_2 - R_1T_1}$$

We can make a quick check by noting that if the two forces are of equal magnitude, the net torque is negative because $R_1 > R_2$. Starting from rest with both forces of equal magnitude acting on it, the cylinder would rotate clockwise because $\mathbf{T}_1$ would be more effective at turning it than would $\mathbf{T}_2$.

(B) Suppose $T_1 = 5.0$ N, $R_1 = 1.0$ m, $T_2 = 15.0$ N, and $R_2 = 0.50$ m. What is the net torque about the rotation axis, and which way does the cylinder rotate starting from rest?

Solution Evaluating the net torque,

$$\sum \tau = (15\text{ N})(0.50\text{ m}) - (5.0\text{ N})(1.0\text{ m}) = \boxed{2.5\text{ N} \cdot \text{m}}$$

Because this torque is positive, the cylinder will begin to rotate in the counterclockwise direction.

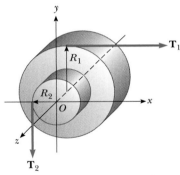

Figure 10.15 (Example 10.9) A solid cylinder pivoted about the z axis through O. The moment arm of $\mathbf{T}_1$ is R_1, and the moment arm of $\mathbf{T}_2$ is R_2.

10.7 Relationship Between Torque and Angular Acceleration

In Chapter 4, we learned that a net force on an object causes an acceleration of the object and that the acceleration is proportional to the net force (Newton's second law). In this section we show the rotational analog of Newton's second law—the angular acceleration of a rigid object rotating about a fixed axis is proportional to the net torque acting about that axis. Before discussing the more complex case of rigid-object rotation, however, it is instructive first to discuss the case of a particle moving in a circular path about some fixed point under the influence of an external force.

Figure 10.16 A particle rotating in a circle under the influence of a tangential force $\mathbf{F}_t$. A force $\mathbf{F}_r$ in the radial direction also must be present to maintain the circular motion.

Consider a particle of mass m rotating in a circle of radius r under the influence of a tangential force $\mathbf{F}_t$ and a radial force $\mathbf{F}_r$, as shown in Figure 10.16. The tangential force provides a tangential acceleration $\mathbf{a}_t$, and

$$F_t = ma_t$$

The magnitude of the torque about the center of the circle due to $\mathbf{F}_t$ is

$$\tau = F_t r = (ma_t) r$$

Because the tangential acceleration is related to the angular acceleration through the relationship $a_t = r\alpha$ (see Eq. 10.11), the torque can be expressed as

$$\tau = (mr\alpha) r = (mr^2) \alpha$$

Recall from Equation 10.15 that mr^2 is the moment of inertia of the particle about the z axis passing through the origin, so that

$$\tau = I\alpha \tag{10.20}$$

That is, **the torque acting on the particle is proportional to its angular acceleration,** and the proportionality constant is the moment of inertia. Note that $\tau = I\alpha$ is the rotational analog of Newton's second law of motion, $F = ma$.

Now let us extend this discussion to a rigid object of arbitrary shape rotating about a fixed axis, as in Figure 10.17. The object can be regarded as an infinite number of mass elements dm of infinitesimal size. If we impose a Cartesian coordinate system on the object, then each mass element rotates in a circle about the origin, and each has a tangential acceleration $\mathbf{a}_t$ produced by an external tangential force $d\mathbf{F}_t$. For any given element, we know from Newton's second law that

$$dF_t = (dm) a_t$$

The torque $d\tau$ associated with the force $d\mathbf{F}_t$ acts about the origin and is given by

$$d\tau = r\, dF_t = a_t r\, dm$$

Because $a_t = r\alpha$, the expression for $d\tau$ becomes

$$d\tau = \alpha r^2\, dm$$

Although each mass element of the rigid object may have a different linear acceleration $\mathbf{a}_t$, they all have the *same* angular acceleration α. With this in mind, we can integrate the above expression to obtain the net torque $\Sigma\tau$ about O due to the external forces:

$$\sum \tau = \int \alpha r^2\, dm = \alpha \int r^2\, dm$$

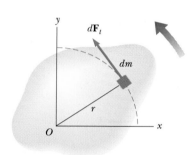

Figure 10.17 A rigid object rotating about an axis through O. Each mass element dm rotates about O with the same angular acceleration α, and the net torque on the object is proportional to α.

where α can be taken outside the integral because it is common to all mass elements. From Equation 10.17, we know that $\int r^2\, dm$ is the moment of inertia of the object about the rotation axis through O, and so the expression for $\Sigma\tau$ becomes

$$\sum \tau = I\alpha \tag{10.21}$$

Torque is proportional to angular acceleration

Note that this is the same relationship we found for a particle moving in a circular path (see Eq. 10.20). So, again we see that the net torque about the rotation axis is proportional to the angular acceleration of the object, with the proportionality factor being I, a quantity that depends upon the axis of rotation and upon the size and shape of the object. In view of the complex nature of the system, the relationship $\Sigma\tau = I\alpha$ is strikingly simple and in complete agreement with experimental observations.

Finally, note that the result $\Sigma\tau = I\alpha$ also applies when the forces acting on the mass elements have radial components as well as tangential components. This is because the line of action of all radial components must pass through the axis of rotation, and hence all radial components produce zero torque about that axis.

Quick Quiz 10.10 You turn off your electric drill and find that the time interval for the rotating bit to come to rest due to frictional torque in the drill is Δt. You replace the bit with a larger one that results in a doubling of the moment of inertia of the entire rotating mechanism of the drill. When this larger bit is rotated at the same angular speed as the first and the drill is turned off, the frictional torque remains the same as that for the previous situation. The time for this second bit to come to rest is (a) $4\Delta t$ (b) $2\Delta t$ (c) Δt (d) $0.5\Delta t$ (e) $0.25\Delta t$ (f) impossible to determine.

Example 10.10 Rotating Rod

A uniform rod of length L and mass M is attached at one end to a frictionless pivot and is free to rotate about the pivot in the vertical plane, as in Figure 10.18. The rod is released from rest in the horizontal position. What is the initial angular acceleration of the rod and the initial linear acceleration of its right end?

Solution We cannot use our kinematic equations to find α or a because the torque exerted on the rod varies with its angular position and so neither acceleration is constant. We have enough information to find the torque, however, which we can then use in Equation 10.21 to find the initial α and then the initial a.

The only force contributing to the torque about an axis through the pivot is the gravitational force $M\mathbf{g}$ exerted on the rod. (The force exerted by the pivot on the rod has zero torque about the pivot because its moment arm is zero.) To compute the torque on the rod, we assume that the gravitational force acts at the center of mass of the rod, as shown in Figure 10.18. The magnitude of the torque due to this force about an axis through the pivot is

$$\tau = Mg\left(\frac{L}{2}\right)$$

With $\Sigma\tau = I\alpha$, and $I = \frac{1}{3}ML^2$ for this axis of rotation (see Table 10.2), we obtain

$$(1) \qquad \alpha = \frac{\tau}{I} = \frac{Mg(L/2)}{\frac{1}{3}ML^2} = \boxed{\frac{3g}{2L}}$$

All points on the rod have this initial angular acceleration.

To find the initial linear acceleration of the right end of the rod, we use the relationship $a_t = r\alpha$ (Eq. 10.11), with $r = L$:

$$a_t = L\alpha = \boxed{\frac{3}{2}g}$$

Figure 10.18 (Example 10.10) A rod is free to rotate around a pivot at the left end.

What If? What if we were to place a penny on the end of the rod and release the rod? Would the penny stay in contact with the rod?

Answer The result for the initial acceleration of a point on the end of the rod shows that $a_t > g$. A penny will fall at acceleration g. This means that if we place a penny at the end of the rod and then release the rod, the end of the rod falls faster than the penny does! The penny does not stay in contact with the rod. (Try this with a penny and a meter stick!)

This raises the question as to the location on the rod at which we can place a penny that *will* stay in contact as both begin to fall. To find the linear acceleration of an arbitrary point on the rod at a distance $r < L$ from the pivot point, we combine (1) with Equation 10.11:

$$a_t = r\alpha = \frac{3g}{2L}r$$

For the penny to stay in contact with the rod, the limiting case is that the linear acceleration must be equal to that due to gravity:

$$a_t = g = \frac{3g}{2L}r$$

$$r = \frac{2}{3}L$$

Thus, a penny placed closer to the pivot than two thirds of the length of the rod will stay in contact with the falling rod while a penny farther out than this point will lose contact.

Conceptual Example 10.11   Falling Smokestacks and Tumbling Blocks

When a tall smokestack falls over, it often breaks somewhere along its length before it hits the ground, as shown in Figure 10.19. The same thing happens with a tall tower of children's toy blocks. Why does this happen?

Solution As the smokestack rotates around its base, each higher portion of the smokestack falls with a larger tangential acceleration than the portion below it. (The tangential acceleration of a given point on the smokestack is proportional to the distance of that portion from the base.) As the angular acceleration increases as the smokestack tips farther, higher portions of the smokestack experience an acceleration greater than that which could result from gravity alone; this is similar to the situation described in Example 10.10. This can happen only if these portions are being pulled downward by a force in addition to the gravitational force. The force that causes this to occur is the shear force from lower portions of the smokestack. Eventually the shear force that provides this acceleration is greater than the smokestack can withstand, and the smokestack breaks.

Figure 10.19 (Conceptual Example 10.11) A falling smokestack breaks at some point along its length.

Example 10.12   Angular Acceleration of a Wheel

Interactive

A wheel of radius R, mass M, and moment of inertia I is mounted on a frictionless horizontal axle, as in Figure 10.20. A light cord wrapped around the wheel supports an object of mass m. Calculate the angular acceleration of the wheel, the linear acceleration of the object, and the tension in the cord.

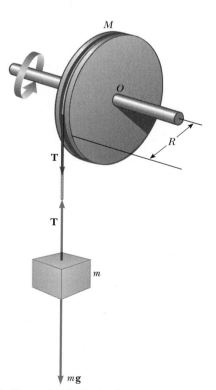

Figure 10.20 (Example 10.12) An object hangs from a cord wrapped around a wheel.

Solution The magnitude of the torque acting on the wheel about its axis of rotation is $\tau = TR$, where T is the force exerted by the cord on the rim of the wheel. (The gravitational force exerted by the Earth on the wheel and the normal force exerted by the axle on the wheel both pass through the axis of rotation and thus produce no torque.) Because $\sum \tau = I\alpha$, we obtain

$$\sum \tau = I\alpha = TR$$

$$(1) \qquad \alpha = \frac{TR}{I}$$

Now let us apply Newton's second law to the motion of the object, taking the downward direction to be positive:

$$\sum F_y = mg - T = ma$$

$$(2) \qquad a = \frac{mg - T}{m}$$

Equations (1) and (2) have three unknowns: α, a, and T. Because the object and wheel are connected by a cord that does not slip, the linear acceleration of the suspended object is equal to the tangential acceleration of a point on the rim of the wheel. Therefore, the angular acceleration α of the wheel and the linear acceleration of the object are related by $a = R\alpha$. Using this fact together with Equations (1) and (2), we obtain

$$(3) \qquad a = R\alpha = \frac{TR^2}{I} = \frac{mg - T}{m}$$

$$(4) \qquad T = \frac{mg}{1 + (mR^2/I)}$$

Substituting Equation (4) into Equation (2) and solving for a and α, we find that

$$(5) \qquad a = \frac{g}{1 + (I/mR^2)}$$

$$\alpha = \frac{a}{R} = \frac{g}{R + (I/mR)}$$

What If? **What if the wheel were to become very massive so that I becomes very large? What happens to the acceleration a of the object and the tension T?**

Answer If the wheel becomes infinitely massive, we can imagine that the object of mass m will simply hang from the cord without causing the wheel to rotate.

We can show this mathematically by taking the limit $I \to \infty$, so that Equation (5) becomes

$$a = \frac{g}{1 + (I/mR^2)} \qquad \longrightarrow \qquad 0$$

This agrees with our conceptual conclusion that the object will hang at rest. We also find that Equation (4) becomes

$$T = \frac{mg}{1 + (mR^2/I)} \qquad \longrightarrow \qquad \frac{mg}{1 + 0} = mg$$

This is consistent with the fact that the object simply hangs at rest in equilibrium between the gravitational force and the tension in the string.

At the Interactive Worked Example link at **http://www.pse6.com**, you can change the masses of the object and the wheel as well as the radius of the wheel to see the effect on how the system moves.

Example 10.13 Atwood's Machine Revisited `Interactive`

Two blocks having masses m_1 and m_2 are connected to each other by a light cord that passes over two identical frictionless pulleys, each having a moment of inertia I and radius R, as shown in Figure 10.21a. Find the acceleration of each block and the tensions T_1, T_2, and T_3 in the cord. (Assume no slipping between cord and pulleys.)

Solution Compare this situation with the Atwood machine of Example 5.9 (p. 129). The motion of m_1 and m_2 is similar to the motion of the two blocks in that example. The primary differences are that in the present example we have two pulleys and each of the pulleys has mass. Despite these differences, the apparatus in the present example is indeed an Atwood machine.

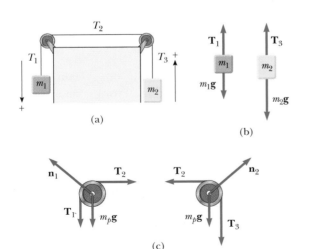

Figure 10.21 (Example 10.13) (a) Another look at Atwood's machine. (b) Free-body diagrams for the blocks. (c) Free-body diagrams for the pulleys, where $m_p\mathbf{g}$ represents the gravitational force acting on each pulley.

We shall define the downward direction as positive for m_1 and upward as the positive direction for m_2. This allows us to represent the acceleration of both masses by a single variable a and also enables us to relate a positive a to a positive (counterclockwise) angular acceleration α of the pulleys. Let us write Newton's second law of motion for each block, using the free-body diagrams for the two blocks as shown in Figure 10.21b:

$$(1) \qquad m_1g - T_1 = m_1a$$

$$(2) \qquad T_3 - m_2g = m_2a$$

Next, we must include the effect of the pulleys on the motion. Free-body diagrams for the pulleys are shown in Figure 10.21c. The net torque about the axle for the pulley on the left is $(T_1 - T_2)R$, while the net torque for the pulley on the right is $(T_2 - T_3)R$. Using the relation $\Sigma\tau = I\alpha$ for each pulley and noting that each pulley has the same angular acceleration α, we obtain

$$(3) \qquad (T_1 - T_2)R = I\alpha$$

$$(4) \qquad (T_2 - T_3)R = I\alpha$$

We now have four equations with five unknowns: α, a, T_1, T_2, and T_3. We also have a fifth equation that relates the accelerations, $a = R\alpha$. These equations can be solved simultaneously. Adding Equations (3) and (4) gives

$$(5) \qquad (T_1 - T_3)R = 2I\alpha$$

Adding Equations (1) and (2) gives

$$T_3 - T_1 + m_1g - m_2g = (m_1 + m_2)a$$

$$(6) \qquad T_1 - T_3 = (m_1 - m_2)g - (m_1 + m_2)a$$

Substituting Equation (6) into Equation (5), we have

$$[(m_1 - m_2)g - (m_1 + m_2)a]R = 2I\alpha$$

Because $\alpha = a/R$, this expression can be simplified to

$$(m_1 - m_2)g - (m_1 + m_2)a = 2I\frac{a}{R^2}$$

$$(7) \qquad a = \frac{(m_1 - m_2)g}{m_1 + m_2 + 2(I/R^2)}$$

Note that if $m_1 > m_2$, the acceleration is positive; this means that the left block accelerates downward, the right block accelerates upward, and both pulleys accelerate counterclockwise. If $m_1 < m_2$, the acceleration is negative and the motions are reversed. If $m_1 = m_2$, no acceleration occurs at all. You should compare these results with those found in Example 5.9.

The expression for a can be substituted into Equations (1) and (2) to give T_1 and T_3. From Equation (1),

$$T_1 = m_1g - m_1a = m_1(g - a)$$

$$= m_1\left(g - \frac{(m_1 - m_2)g}{m_1 + m_2 + 2(I/R^2)}\right)$$

$$= 2m_1g\left(\frac{m_2 + (I/R^2)}{m_1 + m_2 + 2(I/R^2)}\right)$$

Similarly, from Equation (2),

$$T_3 = m_2g + m_2a = 2m_2g\left(\frac{m_1 + (I/R^2)}{m_1 + m_2 + 2(I/R^2)}\right)$$

Finally, T_2 can be found from Equation (3):

$$T_2 = T_1 - \frac{I\alpha}{R} = T_1 - \frac{Ia}{R^2}$$

$$= 2m_1g\left(\frac{m_2 + (I/R^2)}{m_1 + m_2 + 2(I/R^2)}\right)$$

$$- \frac{I}{R^2}\left(\frac{(m_1 - m_2)g}{m_1 + m_2 + 2(I/R^2)}\right)$$

$$= \frac{2m_1m_2 + (m_1 + m_2)(I/R^2)}{m_1 + m_2 + 2(I/R^2)}g$$

What If? What if the pulleys become massless? Does this reduce to a previously solved problem?

Answer If the pulleys become massless, the system should behave in the same way as the massless-pulley Atwood machine that we investigated in Example 5.9. The only difference is the existence of two pulleys instead of one.

Mathematically, if $I \to 0$, Equation (7) becomes

$$a = \frac{(m_1 - m_2)g}{m_1 + m_2 + 2(I/R^2)} \longrightarrow a = \left(\frac{m_1 - m_2}{m_1 + m_2}\right)g$$

which is the same result as Equation (3) in Example 5.9. Although the expressions for the three tensions in the present example are different from each other, all three expressions become, in the limit $I \to 0$,

$$T = \left(\frac{2m_1m_2}{m_1 + m_2}\right)g$$

which is the same as Equation (4) in Example 5.9.

 At the Interactive Worked Example link at **http://www.pse6.com,** you can change the masses of the blocks and the pulleys to see the effect on the motion of the system.

10.8 Work, Power, and Energy in Rotational Motion

Up to this point in our discussion of rotational motion in this chapter, we focused on an approach involving force, leading to a description of torque on a rigid object. We now see how an energy approach can be useful to us in solving rotational problems.

We begin by considering the relationship between the torque acting on a rigid object and its resulting rotational motion in order to generate expressions for power and a rotational analog to the work–kinetic energy theorem. Consider the rigid object pivoted at O in Figure 10.22. Suppose a single external force $\mathbf{F}$ is applied at P, where $\mathbf{F}$ lies in the plane of the page. The work done by $\mathbf{F}$ on the object as it rotates through an infinitesimal distance $ds = r\,d\theta$ is

$$dW = \mathbf{F} \cdot d\mathbf{s} = (F\sin\phi)r\,d\theta$$

where $F\sin\phi$ is the tangential component of $\mathbf{F}$, or, in other words, the component of the force along the displacement. Note that *the radial component of* $\mathbf{F}$ *does no work because it is perpendicular to the displacement.*

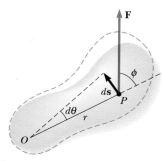

Figure 10.22 A rigid object rotates about an axis through O under the action of an external force $\mathbf{F}$ applied at P.

Because the magnitude of the torque due to $\mathbf{F}$ about O is defined as $rF \sin \phi$ by Equation 10.19, we can write the work done for the infinitesimal rotation as

$$dW = \tau \, d\theta \tag{10.22}$$

The rate at which work is being done by $\mathbf{F}$ as the object rotates about the fixed axis through the angle $d\theta$ in a time interval dt is

$$\frac{dW}{dt} = \tau \, \frac{d\theta}{dt}$$

Because dW/dt is the instantaneous power $\mathscr{P}$ (see Section 7.8) delivered by the force and $d\theta/dt = \omega$, this expression reduces to

$$\mathscr{P} = \frac{dW}{dt} = \tau\omega \tag{10.23}$$

Power delivered to a rotating rigid object

This expression is analogous to $\mathscr{P} = Fv$ in the case of linear motion, and the expression $dW = \tau \, d\theta$ is analogous to $dW = F_x \, dx$.

In studying linear motion, we found the energy approach extremely useful in describing the motion of a system. From what we learned of linear motion, we expect that when a symmetric object rotates about a fixed axis, the work done by external forces equals the change in the rotational energy.

To show that this is in fact the case, let us begin with $\Sigma\tau = I\alpha$. Using the chain rule from calculus, we can express the resultant torque as

$$\sum \tau = I\alpha = I \, \frac{d\omega}{dt} = I \, \frac{d\omega}{d\theta} \, \frac{d\theta}{dt} = I \, \frac{d\omega}{d\theta} \, \omega$$

Rearranging this expression and noting that $\Sigma\tau \, d\theta = dW$, we obtain

$$\sum \tau \, d\theta = dW = I\omega \, d\omega$$

Integrating this expression, we obtain for the total work done by the net external force acting on a rotating system

$$\sum W = \int_{\omega_i}^{\omega_f} I\omega \, d\omega = \tfrac{1}{2}I\omega_f{}^2 - \tfrac{1}{2}I\omega_i{}^2 \tag{10.24}$$

Work–kinetic energy theorem for rotational motion

where the angular speed changes from ω_i to ω_f. That is, the **work–kinetic energy theorem for rotational motion** states that

> the net work done by external forces in rotating a symmetric rigid object about a fixed axis equals the change in the object's rotational energy.

In general, then, combining this with the translational form of the work–kinetic energy theorem from Chapter 7, the net work done by external forces on an object is the change in its *total* kinetic energy, which is the sum of the translational and rotational kinetic energies. For example, when a pitcher throws a baseball, the work done by the pitcher's hands appears as kinetic energy associated with the ball moving through space as well as rotational kinetic energy associated with the spinning of the ball.

In addition to the work–kinetic energy theorem, other energy principles can also be applied to rotational situations. For example, if a system involving rotating objects is isolated, the principle of conservation of energy can be used to analyze the system, as in Example 10.14 below.

Table 10.3 lists the various equations we have discussed pertaining to rotational motion, together with the analogous expressions for linear motion. The last two equations in Table 10.3, involving angular momentum L, are discussed in Chapter 11 and are included here only for the sake of completeness.

Table 10.3

Useful Equations in Rotational and Linear Motion	
Rotational Motion About a Fixed Axis	**Linear Motion**
Angular speed $\omega = d\theta/dt$	Linear speed $v = dx/dt$
Angular acceleration $\alpha = d\omega/dt$	Linear acceleration $a = dv/dt$
Net torque $\Sigma\tau = I\alpha$	Net force $\Sigma F = ma$
If $\alpha = $ constant $\begin{cases} \omega_f = \omega_i + \alpha t \\ \theta_f = \theta_i + \omega_i t + \frac{1}{2}\alpha t^2 \\ \omega_f^2 = \omega_i^2 + 2\alpha(\theta_f - \theta_i) \end{cases}$	If $a = $ constant $\begin{cases} v_f = v_i + at \\ x_f = x_i + v_i t + \frac{1}{2}at^2 \\ v_f^2 = v_i^2 + 2a(x_f - x_i) \end{cases}$
Work $W = \displaystyle\int_{\theta_i}^{\theta_f} \tau\, d\theta$	Work $W = \displaystyle\int_{x_i}^{x_f} F_x\, dx$
Rotational kinetic energy $K_R = \frac{1}{2}I\omega^2$	Kinetic energy $K = \frac{1}{2}mv^2$
Power $\mathcal{P} = \tau\omega$	Power $\mathcal{P} = Fv$
Angular momentum $L = I\omega$	Linear momentum $p = mv$
Net torque $\Sigma\tau = dL/dt$	Net force $\Sigma F = dp/dt$

Quick Quiz 10.11 A rod is attached to the shaft of a motor at the center of the rod so that the rod is perpendicular to the shaft, as in Figure 10.23a. The motor is turned on and performs work W on the rod, accelerating it to an angular speed ω. The system is brought to rest, and the rod is attached to the shaft of the motor at one end of the rod as in Figure 10.23b. The motor is turned on and performs work W on the rod. The angular speed of the rod in the second situation is (a) 4ω (b) 2ω (c) ω (d) 0.5ω (e) 0.25ω (f) impossible to determine.

(a)   (b)

Figure 10.23 (Quick Quiz 10.11) (a) A rod is rotated about its midpoint by a motor. (b) The rod is rotated about one of its ends.

Example 10.14   Rotating Rod Revisited   **Interactive**

A uniform rod of length L and mass M is free to rotate on a frictionless pin passing through one end (Fig 10.24). The rod is released from rest in the horizontal position.

(A) What is its angular speed when it reaches its lowest position?

Solution To conceptualize this problem, consider Figure 10.24 and imagine the rod rotating downward through a

quarter turn about the pivot at the left end. In this situation, the angular acceleration of the rod is not constant. Thus, the kinematic equations for rotation (Section 10.2) cannot be used to solve this problem. As we found with translational motion, however, an energy approach can make such a seemingly insoluble problem relatively easy. We categorize this as a conservation of energy problem.

$$\frac{1}{2}I\omega^2 + 0 = \frac{1}{2}(\frac{1}{3}ML^2)\omega^2 = 0 + \frac{1}{2}MgL$$

$$\omega = \sqrt{\frac{3g}{L}}$$

Figure 10.24 (Example 10.14) A uniform rigid rod pivoted at O rotates in a vertical plane under the action of the gravitational force.

(In the figure) $E_i = U = MgL/2$

$L/2$

O'

$E_f = K_R = \frac{1}{2}I\omega^2$

(B) Determine the tangential speed of the center of mass and the tangential speed of the lowest point on the rod when it is in the vertical position.

Solution These two values can be determined from the relationship between tangential and angular speeds. We know ω from part (A), and so the tangential speed of the center of mass is

$$v_{CM} = r\omega = \frac{L}{2}\omega = \frac{1}{2}\sqrt{3gL}$$

Because r for the lowest point on the rod is twice what it is for the center of mass, the lowest point has a tangential speed v equal to

$$v = 2v_{CM} = \sqrt{3gL}$$

To finalize this problem, note that the initial configuration in this example is the same as that in Example 10.10. In Example 10.10, however, we could only find the initial angular acceleration of the rod. We cannot use this and the kinematic equations to find the angular speed of the rod at its lowest point because the angular acceleration is not constant. Applying an energy approach in the current example allows us to find something that we cannot in Example 10.10.

To analyze the problem, we consider the mechanical energy of the system of the rod and the Earth. We choose the configuration in which the rod is hanging straight down as the reference configuration for gravitational potential energy and assign a value of zero for this configuration. When the rod is in the horizontal position, it has no rotational kinetic energy. The potential energy of the system in this configuration relative to the reference configuration is $MgL/2$ because the center of mass of the rod is at a height $L/2$ higher than its position in the reference configuration. When the rod reaches its lowest position, the energy is entirely rotational energy $\frac{1}{2}I\omega^2$, where I is the moment of inertia about the pivot, and the potential energy of the system is zero. Because $I = \frac{1}{3}ML^2$ (see Table 10.2) and because the system is isolated with no nonconservative forces acting, we apply conservation of mechanical energy for the system:

$$K_f + U_f = K_i + U_i$$

At the Interactive Worked Example link at **http://www.pse6.com,** you can alter the mass and length of the rod and see the effect on the velocity at the lowest point.

Example 10.15   Energy and the Atwood Machine

Consider two cylinders having different masses m_1 and m_2, connected by a string passing over a pulley, as shown in Figure 10.25. The pulley has a radius R and moment of inertia I about its axis of rotation. The string does not slip on the pulley, and the system is released from rest. Find the linear speeds of the cylinders after cylinder 2 descends through a distance h, and the angular speed of the pulley at this time.

Solution We will solve this problem by applying energy methods to an Atwood machine with a massive pulley. Because the string does not slip, the pulley rotates about the axle. We can neglect friction in the axle because the axle's radius is small relative to that of the pulley, so the frictional torque is much smaller than the torque applied by the two cylinders, provided that their masses are quite different. Consequently, the system consisting of the two cylinders, the pulley, and the Earth is isolated with no nonconservative forces acting; thus, the mechanical energy of the system is conserved.

We define the zero configuration for gravitational potential energy as that which exists when the system is re-

leased. From Figure 10.25, we see that the descent of cylinder 2 is associated with a decrease in system potential energy and the rise of cylinder 1 represents an increase in

Figure 10.25 (Example 10.15) An Atwood machine.

potential energy. Because $K_i = 0$ (the system is initially at rest), we have

$$K_f + U_f = K_i + U_i$$

$$(\tfrac{1}{2}m_1v_f^2 + \tfrac{1}{2}m_2v_f^2 + \tfrac{1}{2}I\omega_f^2) + (m_1gh - m_2gh) = 0 + 0$$

where v_f is the same for both blocks. Because $v_f = R\omega_f$, this expression becomes

$$\left(\tfrac{1}{2}m_1v_f^2 + \tfrac{1}{2}m_2v_f^2 + \tfrac{1}{2}\frac{I}{R^2}v_f^2\right) = (m_2gh - m_1gh)$$

$$\tfrac{1}{2}\left(m_1 + m_2 + \frac{I}{R^2}\right)v_f^2 = (m_2gh - m_1gh)$$

Solving for v_f, we find

$$v_f = \left[\frac{2(m_2 - m_1)gh}{[m_1 + m_2 + (I/R^2)]}\right]^{1/2}$$

The angular speed of the pulley at this instant is

$$\omega_f = \frac{v_f}{R} = \frac{1}{R}\left[\frac{2(m_2 - m_1)gh}{(m_1 + m_2 + (I/R^2))}\right]^{1/2}$$

10.9 Rolling Motion of a Rigid Object

In this section we treat the motion of a rigid object rolling along a flat surface. In general, such motion is very complex. Suppose, for example, that a cylinder is rolling on a straight path such that the axis of rotation remains parallel to its initial orientation in space. As Figure 10.26 shows, a point on the rim of the cylinder moves in a complex path called a *cycloid*. However, we can simplify matters by focusing on the center of mass rather than on a point on the rim of the rolling object. As we see in Figure 10.26, the center of mass moves in a straight line. If an object such as a cylinder rolls without slipping on the surface (we call this *pure rolling motion*), we can show that a simple relationship exists between its rotational and translational motions.

Consider a uniform cylinder of radius R rolling without slipping on a horizontal surface (Fig. 10.27). As the cylinder rotates through an angle θ, its center of mass moves a linear distance $s = R\theta$ (see Eq. 10.1a). Therefore, the linear speed of the center of mass for pure rolling motion is given by

$$v_{\text{CM}} = \frac{ds}{dt} = R\frac{d\theta}{dt} = R\omega \tag{10.25}$$

where ω is the angular speed of the cylinder. Equation 10.25 holds whenever a cylinder or sphere rolls without slipping and is the **condition for pure rolling motion.**

▲ **PITFALL PREVENTION**

10.6 Equation 10.25 Looks Familiar

Equation 10.25 looks very similar to Equation 10.10, so be sure that you are clear on the difference. Equation 10.10 gives the *tangential* speed of a point on a *rotating* object located a distance r from the rotation axis if the object is rotating with angular speed ω. Equation 10.25 gives the *translational* speed of the center of mass of a *rolling* object of radius R rotating with angular speed ω.

Henry Leap and Jim Lehman

Figure 10.26 One light source at the center of a rolling cylinder and another at one point on the rim illustrate the different paths these two points take. The center moves in a straight line (green line), while the point on the rim moves in the path called a cycloid (red curve).

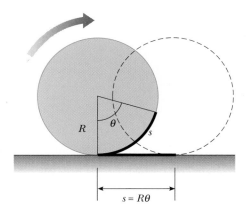

Figure 10.27 For pure rolling motion, as the cylinder rotates through an angle θ, its center moves a linear distance $s = R\theta$.

The magnitude of the linear acceleration of the center of mass for pure rolling motion is

$$a_{CM} = \frac{dv_{CM}}{dt} = R\,\frac{d\omega}{dt} = R\alpha \tag{10.26}$$

where α is the angular acceleration of the cylinder.

The linear velocities of the center of mass and of various points on and within the cylinder are illustrated in Figure 10.28. A short time after the moment shown in the drawing, the rim point labeled P might rotate from the six o'clock position to, say, the seven o'clock position, while the point Q would rotate from the ten o'clock position to the eleven o'clock position, and so on. Note that the linear velocity of any point is in a direction perpendicular to the line from that point to the contact point P. At any instant, the part of the rim that is at point P is at rest relative to the surface because slipping does not occur.

All points on the cylinder have the same angular speed. Therefore, because the distance from P' to P is twice the distance from P to the center of mass, P' has a speed $2v_{CM} = 2R\omega$. To see why this is so, let us model the rolling motion of the cylinder in Figure 10.29 as a combination of translational (linear) motion and rotational motion. For the pure translational motion shown in Figure 10.29a, imagine that the cylinder does not rotate, so that each point on it moves to the right with speed v_{CM}. For the pure rotational motion shown in Figure 10.29b, imagine that a rotation axis through the center of mass is stationary, so that each point on the cylinder has the same angular speed ω. The combination of these two motions represents the rolling motion shown in Figure 10.29c. Note in Figure 10.29c that the top of the cylinder has linear speed $v_{CM} + R\omega = v_{CM} + v_{CM} = 2v_{CM}$, which is greater than the linear speed of any other point on the cylinder. As mentioned earlier, the center of mass moves with linear speed v_{CM} while the contact point between the surface and cylinder has a linear speed of zero.

We can express the total kinetic energy of the rolling cylinder as

$$K = \tfrac{1}{2} I_P \omega^2 \tag{10.27}$$

where I_P is the moment of inertia about a rotation axis through P. Applying the parallel-axis theorem, we can substitute $I_P = I_{CM} + MR^2$ into Equation 10.27 to obtain

$$K = \tfrac{1}{2} I_{CM}\omega^2 + \tfrac{1}{2} MR^2\omega^2$$

or, because $v_{CM} = R\omega$,

$$K = \tfrac{1}{2} I_{CM}\omega^2 + \tfrac{1}{2} M v_{CM}^2 \tag{10.28}$$

Total kinetic energy of a rolling object

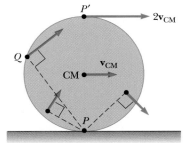

Figure 10.28 All points on a rolling object move in a direction perpendicular to an axis through the instantaneous point of contact P. In other words, all points rotate about P. The center of mass of the object moves with a velocity $\mathbf{v}_{CM}$, and the point P' moves with a velocity $2\mathbf{v}_{CM}$.

(a) Pure translation   (b) Pure rotation

(c) Combination of translation and rotation

Figure 10.29 The motion of a rolling object can be modeled as a combination of pure translation and pure rotation.

The term $\frac{1}{2}I_{CM}\omega^2$ represents the rotational kinetic energy of the cylinder about its center of mass, and the term $\frac{1}{2}Mv_{CM}^2$ represents the kinetic energy the cylinder would have if it were just translating through space without rotating. Thus, we can say that the **total kinetic energy of a rolling object is the sum of the rotational kinetic energy about the center of mass and the translational kinetic energy of the center of mass.**

We can use energy methods to treat a class of problems concerning the rolling motion of an object down a rough incline. For example, consider Figure 10.30, which shows a sphere rolling without slipping after being released from rest at the top of the incline. Note that accelerated rolling motion is possible only if a friction force is present between the sphere and the incline to produce a net torque about the center of mass. Despite the presence of friction, no loss of mechanical energy occurs because the contact point is at rest relative to the surface at any instant. (On the other hand, if the sphere were to slip, mechanical energy of the sphere–incline–Earth system would be lost due to the nonconservative force of kinetic friction.)

Using the fact that $v_{CM} = R\omega$ for pure rolling motion, we can express Equation 10.28 as

$$K = \frac{1}{2}I_{CM}\left(\frac{v_{CM}}{R}\right)^2 + \frac{1}{2}Mv_{CM}^2$$

$$K = \frac{1}{2}\left(\frac{I_{CM}}{R^2} + M\right)v_{CM}^2 \qquad (10.29)$$

For the system of the sphere and the Earth, we define the zero configuration of gravitational potential energy to be when the sphere is at the bottom of the incline. Thus, conservation of mechanical energy gives us

$$K_f + U_f = K_i + U_i$$

$$\frac{1}{2}\left(\frac{I_{CM}}{R^2} + M\right)v_{CM}^2 + 0 = 0 + Mgh$$

$$v_{CM} = \left(\frac{2gh}{1 + (I_{CM}/MR^2)}\right)^{1/2} \qquad (10.30)$$

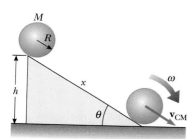

Active Figure 10.30 A sphere rolling down an incline. Mechanical energy of the sphere–incline–Earth system is conserved if no slipping occurs.

At the Active Figures link at http://www.pse6.com, you can roll several objects down the hill and see how the final speed depends on the type of object.

Quick Quiz 10.12 A ball rolls without slipping down incline A, starting from rest. At the same time, a box starts from rest and slides down incline B, which is identical to incline A except that it is frictionless. Which arrives at the bottom first? (a) the ball (b) the box (c) Both arrive at the same time. (d) impossible to determine

Quick Quiz 10.13 Two solid spheres roll down an incline, starting from rest. Sphere A has twice the mass and twice the radius of sphere B. Which arrives at the bottom first? (a) sphere A (b) sphere B (c) Both arrive at the same time. (d) impossible to determine

Quick Quiz 10.14 Two spheres roll down an incline, starting from rest. Sphere A has the same mass and radius as sphere B, but sphere A is solid while sphere B is hollow. Which arrives at the bottom first? (a) sphere A (b) sphere B (c) Both arrive at the same time. (d) impossible to determine

---

Example 10.16   Sphere Rolling Down an Incline

For the solid sphere shown in Figure 10.30, calculate the linear speed of the center of mass at the bottom of the incline and the magnitude of the linear acceleration of the center of mass.

Solution For a uniform solid sphere, $I_{CM} = \frac{2}{5}MR^2$ (see Table 10.2), and therefore Equation 10.30 gives

$$v_{CM} = \left(\frac{2gh}{1 + (\frac{2}{5}MR^2/MR^2)} \right)^{1/2} = \left(\tfrac{10}{7} gh \right)^{1/2}$$

Notice that this is less than $\sqrt{2gh}$, which is the speed an object would have if it simply slid down the incline without rotating (see Example 8.7).

To calculate the linear acceleration of the center of mass, we note that the vertical displacement is related to the displacement x along the incline through the relationship $h = x \sin \theta$. Hence, after squaring both sides, we can express the equation above as

$$v_{CM}{}^2 = \tfrac{10}{7} gx \sin \theta$$

Comparing this with the expression from kinematics, $v_{CM}{}^2 = 2a_{CM}x$ (see Eq. 2.13), we see that the acceleration of the center of mass is

$$a_{CM} = \tfrac{5}{7} g \sin \theta$$

These results are interesting because both the speed and the acceleration of the center of mass are *independent* of the mass and the radius of the sphere! That is, **all homogeneous solid spheres experience the same speed and acceleration on a given incline,** as we argued in the answer to Quick Quiz 10.13.

If we were to repeat the acceleration calculation for a hollow sphere, a solid cylinder, or a hoop, we would obtain similar results in which only the factor in front of $g \sin \theta$ would differ. The constant factors that appear in the expressions for v_{CM} and a_{CM} depend only on the moment of inertia about the center of mass for the specific object. In all cases, the acceleration of the center of mass is *less* than $g \sin \theta$, the value the acceleration would have if the incline were frictionless and no rolling occurred.

SUMMARY

If a particle moves in a circular path of radius r through an angle θ (measured in radians), the arc length it moves through is $s = r\theta$.

The **angular position** of a rigid object is defined as the angle θ between a reference line attached to the object and a reference line fixed in space. The **angular displacement** of a particle moving in a circular path or a rigid object rotating about a fixed axis is $\Delta\theta \equiv \theta_f - \theta_i$.

Take a practice test for this chapter by clicking on the Practice Test link at http://www.pse6.com.

The **instantaneous angular speed** of a particle moving in a circular path or of a rigid object rotating about a fixed axis is

$$\omega \equiv \frac{d\theta}{dt} \tag{10.3}$$

The **instantaneous angular acceleration** of a particle moving in a circular path or a rotating rigid object is

$$\alpha \equiv \frac{d\omega}{dt} \tag{10.5}$$

When a rigid object rotates about a fixed axis, every part of the object has the same angular speed and the same angular acceleration.

If an object rotates about a fixed axis under constant angular acceleration, one can apply equations of kinematics that are analogous to those for linear motion under constant linear acceleration:

$$\omega_f = \omega_i + \alpha t \tag{10.6}$$

$$\theta_f = \theta_i + \omega_i t + \tfrac{1}{2}\alpha t^2 \tag{10.7}$$

$$\omega_f^2 = \omega_i^2 + 2\alpha(\theta_f - \theta_i) \tag{10.8}$$

$$\theta_f = \theta_i + \tfrac{1}{2}(\omega_i + \omega_f)t \tag{10.9}$$

A useful technique in solving problems dealing with rotation is to visualize a linear version of the same problem.

When a rigid object rotates about a fixed axis, the angular position, angular speed, and angular acceleration are related to the linear position, linear speed, and linear acceleration through the relationships

$$s = r\theta \tag{10.1a}$$

$$v = r\omega \tag{10.10}$$

$$a_t = r\alpha \tag{10.11}$$

The **moment of inertia of a system of particles** is defined as

$$I \equiv \sum_i m_i r_i^2 \tag{10.15}$$

If a rigid object rotates about a fixed axis with angular speed ω, its **rotational kinetic energy** can be written

$$K_R = \tfrac{1}{2}I\omega^2 \tag{10.16}$$

where I is the moment of inertia about the axis of rotation.

The **moment of inertia of a rigid object** is

$$I = \int r^2\,dm \tag{10.17}$$

where r is the distance from the mass element dm to the axis of rotation.

The magnitude of the **torque** associated with a force $\mathbf{F}$ acting on an object is

$$\tau = Fd \tag{10.19}$$

where d is the moment arm of the force, which is the perpendicular distance from the rotation axis to the line of action of the force. Torque is a measure of the tendency of the force to change the rotation of the object about some axis.

If a rigid object free to rotate about a fixed axis has a **net external torque** acting on it, the object undergoes an angular acceleration α, where

$$\sum \tau = I\alpha \tag{10.21}$$

The rate at which work is done by an external force in rotating a rigid object about a fixed axis, or the **power** delivered, is

$$\mathcal{P} = \tau\omega \qquad (10.23)$$

If work is done on a rigid object and the only result of the work is rotation about a fixed axis, the net work done by external forces in rotating the object equals the change in the rotational kinetic energy of the object:

$$\sum W = \tfrac{1}{2} I\omega_f^2 - \tfrac{1}{2} I\omega_i^2 \qquad (10.24)$$

The **total kinetic energy** of a rigid object rolling on a rough surface without slipping equals the rotational kinetic energy about its center of mass, $\tfrac{1}{2} I_{CM}\omega^2$, plus the translational kinetic energy of the center of mass, $\tfrac{1}{2} Mv_{CM}^2$:

$$K = \tfrac{1}{2} I_{CM}\omega^2 + \tfrac{1}{2} Mv_{CM}^2 \qquad (10.28)$$

QUESTIONS

1. What is the angular speed of the second hand of a clock? What is the direction of $\boldsymbol{\omega}$ as you view a clock hanging on a vertical wall? What is the magnitude of the angular acceleration vector $\boldsymbol{\alpha}$ of the second hand?

2. One blade of a pair of scissors rotates counterclockwise in the xy plane. What is the direction of $\boldsymbol{\omega}$? What is the direction of $\boldsymbol{\alpha}$ if the magnitude of the angular velocity is decreasing in time?

3. Are the kinematic expressions for θ, ω, and α valid when the angular position is measured in degrees instead of in radians?

4. If a car's standard tires are replaced with tires of larger outside diameter, will the reading of the speedometer change? Explain.

5. Suppose $a = b$ and $M > m$ for the system of particles described in Figure 10.8. About which axis (x, y, or z) does the moment of inertia have the smallest value? the largest value?

6. Suppose that the rod in Figure 10.10 has a nonuniform mass distribution. In general, would the moment of inertia about the y axis still be equal to $ML^2/12$? If not, could the moment of inertia be calculated without knowledge of the manner in which the mass is distributed?

7. Suppose that just two external forces act on a stationary rigid object and the two forces are equal in magnitude and opposite in direction. Under what condition does the object start to rotate?

8. Suppose a pencil is balanced on a perfectly frictionless table. If it falls over, what is the path followed by the center of mass of the pencil?

9. Explain how you might use the apparatus described in Example 10.12 to determine the moment of inertia of the wheel. (If the wheel does not have a uniform mass density, the moment of inertia is not necessarily equal to $\tfrac{1}{2} MR^2$.)

10. Using the results from Example 10.12, how would you calculate the angular speed of the wheel and the linear speed of the suspended counterweight at $t = 2$ s, if the system is released from rest at $t = 0$? Is the expression $v = R\omega$ valid in this situation?

11. If a small sphere of mass M were placed at the end of the rod in Figure 10.24, would the result for ω be greater than, less than, or equal to the value obtained in Example 10.14?

12. Explain why changing the axis of rotation of an object changes its moment of inertia.

13. The moment of inertia of an object depends on the choice of rotation axis, as suggested by the parallel-axis theorem. Argue that an axis passing through the center of mass of an object must be the axis with the smallest moment of inertia.

14. Suppose you remove two eggs from the refrigerator, one hard-boiled and the other uncooked. You wish to determine which is the hard-boiled egg without breaking the eggs. This can be done by spinning the two eggs on the floor and comparing the rotational motions. Which egg spins faster? Which rotates more uniformly? Explain.

15. Which of the entries in Table 10.2 applies to finding the moment of inertia of a long straight sewer pipe rotating about its axis of symmetry? Of an embroidery hoop rotating about an axis through its center and perpendicular to its plane? Of a uniform door turning on its hinges? Of a coin turning about an axis through its center and perpendicular to its faces?

16. Is it possible to change the translational kinetic energy of an object without changing its rotational energy?

17. Must an object be rotating to have a nonzero moment of inertia?

18. If you see an object rotating, is there necessarily a net torque acting on it?

19. Can a (momentarily) stationary object have a nonzero angular acceleration?

20. In a tape recorder, the tape is pulled past the read-and-write heads at a constant speed by the drive mechanism. Consider the reel from which the tape is pulled. As the tape is pulled from it, the radius of the roll of remaining tape decreases. How does the torque on the reel change with time? How does the angular speed of the reel change in time? If the drive mechanism is switched on so that the

tape is suddenly jerked with a large force, is the tape more likely to break when it is being pulled from a nearly full reel or from a nearly empty reel?

21. The polar diameter of the Earth is slightly less than the equatorial diameter. How would the moment of inertia of the Earth about its axis of rotation change if some mass from near the equator were removed and transferred to the polar regions to make the Earth a perfect sphere?

22. Suppose you set your textbook sliding across a gymnasium floor with a certain initial speed. It quickly stops moving because of a friction force exerted on it by the floor. Next, you start a basketball rolling with the same initial speed. It keeps rolling from one end of the gym to the other. Why does the basketball roll so far? Does friction significantly affect its motion?

23. When a cylinder rolls on a horizontal surface as in Figure 10.28, do any points on the cylinder have only a vertical component of velocity at some instant? If so, where are they?

24. Three objects of uniform density—a solid sphere, a solid cylinder, and a hollow cylinder—are placed at the top of

an incline (Fig. Q10.24). They are all released from rest at the same elevation and roll without slipping. Which object reaches the bottom first? Which reaches it last? Try this at home and note that the result is independent of the masses and the radii of the objects.

Figure Q10.24 Which object wins the race?

25. In a soap-box derby race, the cars have no engines; they simply coast down a hill to race with one another. Suppose you are designing a car for a coasting race. Do you want to use large wheels or small wheels? Do you want to use solid disk-like wheels or hoop-like wheels? Should the wheels be heavy or light?

PROBLEMS

1, 2, 3 = straightforward, intermediate, challenging ☐ = full solution available in the *Student Solutions Manual and Study Guide*

= coached solution with hints available at http://www.pse6.com 🖳 = computer useful in solving problem

▨ = paired numerical and symbolic problems

Section 10.1 Angular Position, Velocity, and Acceleration

1. During a certain period of time, the angular position of a swinging door is described by $\theta = 5.00 + 10.0t + 2.00t^2$, where θ is in radians and t is in seconds. Determine the angular position, angular speed, and angular acceleration of the door (a) at $t = 0$ (b) at $t = 3.00$ s.

Section 10.2 Rotational Kinematics: Rotational Motion with Constant Angular Acceleration

2. A dentist's drill starts from rest. After 3.20 s of constant angular acceleration, it turns at a rate of 2.51×10^4 rev/min. (a) Find the drill's angular acceleration. (b) Determine the angle (in radians) through which the drill rotates during this period.

3. A wheel starts from rest and rotates with constant angular acceleration to reach an angular speed of 12.0 rad/s in 3.00 s. Find (a) the magnitude of the angular acceleration of the wheel and (b) the angle in radians through which it rotates in this time.

4. An airliner arrives at the terminal, and the engines are shut off. The rotor of one of the engines has an initial clockwise angular speed of 2 000 rad/s. The engine's rotation slows with an angular acceleration of magnitude 80.0 rad/s². (a) Determine the angular speed after 10.0 s. (b) How long does it take the rotor to come to rest?

5. An electric motor rotating a grinding wheel at 100 rev/min is switched off. With constant negative angular acceleration of magnitude 2.00 rad/s², (a) how long does it take the wheel to stop? (b) Through how many radians does it turn while it is slowing down?

6. A centrifuge in a medical laboratory rotates at an angular speed of 3 600 rev/min. When switched off, it rotates 50.0 times before coming to rest. Find the constant angular acceleration of the centrifuge.

7. The tub of a washer goes into its spin cycle, starting from rest and gaining angular speed steadily for 8.00 s, at which time it is turning at 5.00 rev/s. At this point the person doing the laundry opens the lid, and a safety switch turns off the washer. The tub smoothly slows to rest in 12.0 s. Through how many revolutions does the tub turn while it is in motion?

8. A rotating wheel requires 3.00 s to rotate through 37.0 revolutions. Its angular speed at the end of the 3.00-s interval is 98.0 rad/s. What is the constant angular acceleration of the wheel?

9. (a) Find the angular speed of the Earth's rotation on its axis. As the Earth turns toward the east, we see the sky turning toward the west at this same rate.
(b) *The rainy Pleiads wester*
    And seek beyond the sea
    The head that I shall dream of
    That shall not dream of me.

    –A. E. Housman (© Robert E. Symons)

Cambridge, England, is at longitude 0°, and Saskatoon, Saskatchewan, is at longitude 107° west. How much time elapses after the Pleiades set in Cambridge until these stars fall below the western horizon in Saskatoon?

10. A merry-go-round is stationary. A dog is running on the ground just outside its circumference, moving with a constant angular speed of 0.750 rad/s. The dog does not change his pace when he sees what he has been looking for: a bone resting on the edge of the merry-go-round one third of a revolution in front of him. At the instant the dog sees the bone ($t = 0$), the merry-go-round begins to move in the direction the dog is running, with a constant angular acceleration of 0.015 0 rad/s². (a) At what time will the dog reach the bone? (b) The confused dog keeps running and passes the bone. How long after the merry-go-round starts to turn do the dog and the bone draw even with each other for the second time?

Section 10.3 Angular and Linear Quantities

11. Make an order-of-magnitude estimate of the number of revolutions through which a typical automobile tire turns in 1 yr. State the quantities you measure or estimate and their values.

12. A racing car travels on a circular track of radius 250 m. If the car moves with a constant linear speed of 45.0 m/s, find (a) its angular speed and (b) the magnitude and direction of its acceleration.

13. A wheel 2.00 m in diameter lies in a vertical plane and rotates with a constant angular acceleration of 4.00 rad/s². The wheel starts at rest at $t = 0$, and the radius vector of a certain point P on the rim makes an angle of 57.3° with the horizontal at this time. At $t = 2.00$ s, find (a) the angular speed of the wheel, (b) the tangential speed and the total acceleration of the point P, and (c) the angular position of the point P.

14. Figure P10.14 shows the drive train of a bicycle that has wheels 67.3 cm in diameter and pedal cranks 17.5 cm long. The cyclist pedals at a steady angular rate of 76.0 rev/min. The chain engages with a front sprocket 15.2 cm in diameter and a rear sprocket 7.00 cm in diameter. (a) Calculate the speed of a link of the chain relative to the bicycle frame. (b) Calculate the angular speed of the bicycle wheels. (c) Calculate the speed of the bicycle relative to the road. (d) What pieces of data, if any, are not necessary for the calculations?

15. A discus thrower (Fig. P10.15) accelerates a discus from rest to a speed of 25.0 m/s by whirling it through 1.25 rev. Assume the discus moves on the arc of a circle 1.00 m in radius. (a) Calculate the final angular speed of the discus. (b) Determine the magnitude of the angular acceleration of the discus, assuming it to be constant. (c) Calculate the time interval required for the discus to accelerate from rest to 25.0 m/s.

Bruce Ayers/Stone/Getty

Figure P10.15

16. A car accelerates uniformly from rest and reaches a speed of 22.0 m/s in 9.00 s. If the diameter of a tire is 58.0 cm, find (a) the number of revolutions the tire makes during this motion, assuming that no slipping occurs. (b) What is the final angular speed of a tire in revolutions per second?

17. A disk 8.00 cm in radius rotates at a constant rate of 1 200 rev/min about its central axis. Determine (a) its angular speed, (b) the tangential speed at a point 3.00 cm from its center, (c) the radial acceleration of a point on the rim, and (d) the total distance a point on the rim moves in 2.00 s.

18. A car traveling on a flat (unbanked) circular track accelerates uniformly from rest with a tangential acceleration of 1.70 m/s². The car makes it one quarter of the way around the circle before it skids off the track. Determine the coefficient of static friction between the car and track from these data.

19. Consider a tall building located on the Earth's equator. As the Earth rotates, a person on the top floor of the building moves faster than someone on the ground with respect to an inertial reference frame, because the latter person is closer to the Earth's axis. Consequently, if an object is dropped from the top floor to the ground a distance h below, it lands east of the point vertically below where it was dropped. (a) How far to the east will the object land? Express your answer in terms of h, g, and the angular speed ω of the Earth. Neglect air resistance, and assume that the free-fall acceleration is constant over this range of heights. (b) Evaluate the eastward displacement for $h = 50.0$ m. (c) In your judgment, were we justified in ignoring this aspect of the *Coriolis effect* in our previous study of free fall?

Sprocket   Crank
Chain

Figure P10.14

Section 10.4 Rotational Kinetic Energy

20. Rigid rods of negligible mass lying along the y axis connect three particles (Fig. P10.20). If the system rotates about the x axis with an angular speed of 2.00 rad/s, find (a) the moment of inertia about the x axis and the total rotational kinetic energy evaluated from $\frac{1}{2}I\omega^2$ and (b) the tangential speed of each particle and the total kinetic energy evaluated from $\sum \frac{1}{2}m_i v_i^2$.

4.00 kg $y = 3.00$ m

O x

2.00 kg $y = -2.00$ m

3.00 kg $y = -4.00$ m

Figure P10.20

21. The four particles in Figure P10.21 are connected by rigid rods of negligible mass. The origin is at the center of the rectangle. If the system rotates in the xy plane about the z axis with an angular speed of 6.00 rad/s, calculate (a) the moment of inertia of the system about the z axis and (b) the rotational kinetic energy of the system.

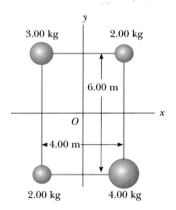

3.00 kg   2.00 kg

6.00 m

O x

◄4.00 m►

2.00 kg   4.00 kg

Figure P10.21

22. Two balls with masses M and m are connected by a rigid rod of length L and negligible mass as in Figure P10.22. For an axis perpendicular to the rod, show that the system has the minimum moment of inertia when the axis passes

M L m

x $L - x$

Figure P10.22

through the center of mass. Show that this moment of inertia is $I = \mu L^2$, where $\mu = mM/(m + M)$.

Section 10.5 Calculation of Moments of Inertia

23. Three identical thin rods, each of length L and mass m, are welded perpendicular to one another as shown in Figure P10.23. The assembly is rotated about an axis that passes through the end of one rod and is parallel to another. Determine the moment of inertia of this structure.

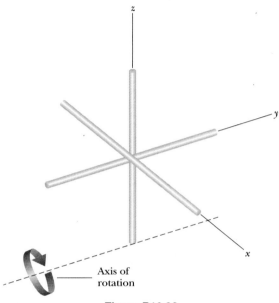

Axis of rotation

Figure P10.23

24. Figure P10.24 shows a side view of a car tire. Model it as having two sidewalls of uniform thickness 0.635 cm and a tread wall of uniform thickness 2.50 cm and width 20.0 cm. Assume the rubber has uniform density 1.10×10^3 kg/m³. Find its moment of inertia about an axis through its center.

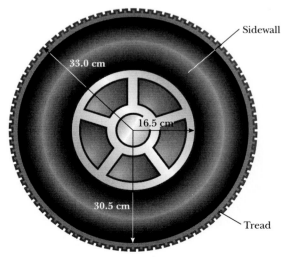

Sidewall

33.0 cm

16.5 cm

30.5 cm

Tread

Figure P10.24

25. A uniform thin solid door has height 2.20 m, width 0.870 m, and mass 23.0 kg. Find its moment of inertia for rotation on its hinges. Is any piece of data unnecessary?

26. *Attention! About face!* Compute an order-of-magnitude estimate for the moment of inertia of your body as you stand tall and turn about a vertical axis through the top of your head and the point halfway between your ankles. In your solution state the quantities you measure or estimate and their values.

27. The density of the Earth, at any distance r from its center, is approximately

$$\rho = [14.2 - 11.6(r/R)] \times 10^3 \text{ kg/m}^3$$

where R is the radius of the Earth. Show that this density leads to a moment of inertia $I = 0.330MR^2$ about an axis through the center, where M is the mass of the Earth.

28. Calculate the moment of inertia of a thin plate, in the shape of a right triangle, about an axis that passes through one end of the hypotenuse and is parallel to the opposite leg of the triangle, as in Figure P10.28a. Let M represent the mass of the triangle and L the length of the base of the triangle perpendicular to the axis of rotation. Let h represent the height of the triangle and w the thickness of the plate, much smaller than L or h. Do the calculation in either or both of the following ways, as your instructor assigns:

(a) Use Equation 10.17. Let an element of mass consist of a vertical ribbon within the triangle, of width dx, height y, and thickness w. With x representing the location of the ribbon, show that $y = hx/L$. Show that the density of the material is given by $\rho = 2M/Lwh$. Show that the mass of the ribbon is $dm = \rho yw\, dx = 2Mx\, dx/L^2$. Proceed to use Equation 10.17 to calculate the moment of inertia.

(b) Let I represent the unknown moment of inertia about an axis through the corner of the triangle. Note that Example 9.15 demonstrates that the center of mass of the triangle is two thirds of the way along the length L, from the corner toward the side of height h. Let I_{CM} represent the moment of inertia of the triangle about an axis through the center of mass and parallel to side h. Demonstrate that $I = I_{CM} + 4ML^2/9$. Figure P10.28b shows the same object in a different orientation.

Demonstrate that the moment of inertia of the triangular plate, about the y axis is $I_h = I_{CM} + ML^2/9$. Demonstrate that the sum of the moments of inertia of the triangles shown in parts (a) and (b) of the figure must be the moment of inertia of a rectangular sheet of mass $2M$ and length L, rotating like a door about an axis along its edge of height h. Use information in Table 10.2 to write down the moment of inertia of the rectangle, and set it equal to the sum of the moments of inertia of the two triangles. Solve the equation to find the moment of inertia of a triangle about an axis through its center of mass, in terms of M and L. Proceed to find the original unknown I.

29. Many machines employ cams for various purposes, such as opening and closing valves. In Figure P10.29, the cam is a circular disk rotating on a shaft that does not pass through the center of the disk. In the manufacture of the cam, a uniform solid cylinder of radius R is first machined. Then an off-center hole of radius $R/2$ is drilled, parallel to the axis of the cylinder, and centered at a point a distance $R/2$ from the center of the cylinder. The cam, of mass M, is then slipped onto the circular shaft and welded into place. What is the kinetic energy of the cam when it is rotating with angular speed ω about the axis of the shaft?

Figure P10.29

Section 10.6 Torque

30. The fishing pole in Figure P10.30 makes an angle of 20.0° with the horizontal. What is the torque exerted by the fish about an axis perpendicular to the page and passing through the fisher's hand?

(a)   (b)

Figure P10.28

Figure P10.30

31. 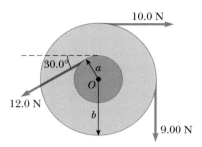 Find the net torque on the wheel in Figure P10.31 about the axle through O if $a = 10.0$ cm and $b = 25.0$ cm.

Figure P10.31

32. The tires of a 1 500-kg car are 0.600 m in diameter, and the coefficients of friction with the road surface are $\mu_s = 0.800$ and $\mu_k = 0.600$. Assuming that the weight is evenly distributed on the four wheels, calculate the maximum torque that can be exerted by the engine on a driving wheel without spinning the wheel. If you wish, you may assume the car is at rest.

33. Suppose the car in Problem 32 has a disk brake system. Each wheel is slowed by the friction force between a single brake pad and the disk-shaped rotor. On this particular car, the brake pad contacts the rotor at an average distance of 22.0 cm from the axis. The coefficients of friction between the brake pad and the disk are $\mu_s = 0.600$ and $\mu_k = 0.500$. Calculate the normal force that the pad must apply to the rotor in order to slow the car as quickly as possible.

Section 10.7 Relationship between Torque and Angular Acceleration

34. A grinding wheel is in the form of a uniform solid disk of radius 7.00 cm and mass 2.00 kg. It starts from rest and accelerates uniformly under the action of the constant torque of 0.600 N·m that the motor exerts on the wheel. (a) How long does the wheel take to reach its final operating speed of 1 200 rev/min? (b) Through how many revolutions does it turn while accelerating?

35. 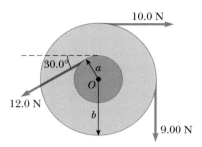 A model airplane with mass 0.750 kg is tethered by a wire so that it flies in a circle 30.0 m in radius. The airplane engine provides a net thrust of 0.800 N perpendicular to the tethering wire. (a) Find the torque the net thrust produces about the center of the circle. (b) Find the angular acceleration of the airplane when it is in level flight. (c) Find the linear acceleration of the airplane tangent to its flight path.

36. The combination of an applied force and a friction force produces a constant total torque of 36.0 N·m on a wheel rotating about a fixed axis. The applied force acts for 6.00 s. During this time the angular speed of the wheel increases from 0 to 10.0 rad/s. The applied force is then removed, and the wheel comes to rest in 60.0 s. Find (a) the moment of inertia of the wheel, (b) the magnitude of the frictional torque, and (c) the total number of revolutions of the wheel.

37. A block of mass $m_1 = 2.00$ kg and a block of mass $m_2 = 6.00$ kg are connected by a massless string over a pulley in the shape of a solid disk having radius $R = 0.250$ m and mass $M = 10.0$ kg. These blocks are allowed to move on a fixed block-wedge of angle $\theta = 30.0°$ as in Figure P10.37. The coefficient of kinetic friction is 0.360 for both blocks. Draw free-body diagrams of both blocks and of the pulley. Determine (a) the acceleration of the two blocks and (b) the tensions in the string on both sides of the pulley.

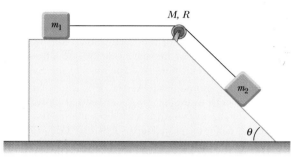

Figure P10.37

38. A potter's wheel—a thick stone disk of radius 0.500 m and mass 100 kg—is freely rotating at 50.0 rev/min. The potter can stop the wheel in 6.00 s by pressing a wet rag against the rim and exerting a radially inward force of 70.0 N. Find the effective coefficient of kinetic friction between wheel and rag.

39. An electric motor turns a flywheel through a drive belt that joins a pulley on the motor and a pulley that is rigidly attached to the flywheel, as shown in Figure P10.39. The flywheel is a solid disk with a mass of 80.0 kg and a diameter of 1.25 m. It turns on a frictionless axle. Its pulley has much smaller mass and a radius of 0.230 m. If the tension in the upper (taut) segment of the belt is 135 N and the flywheel has a clockwise angular acceleration of 1.67 rad/s², find the tension in the lower (slack) segment of the belt.

Figure P10.39

Section 10.8 Work, Power, and Energy in Rotational Motion

40. Big Ben, the Parliament tower clock in London, has an hour hand 2.70 m long with a mass of 60.0 kg, and a minute hand 4.50 m long with a mass of 100 kg (Fig. P10.40). Calculate the total rotational kinetic energy of the two hands about the axis of rotation. (You may model the hands as long, thin rods.)

John Lawrence /Getty

Figure P10.40 Problems 40 and 74.

41. In a city with an air-pollution problem, a bus has no combustion engine. It runs on energy drawn from a large, rapidly rotating flywheel under the floor of the bus. The flywheel is spun up to its maximum rotation rate of 4 000 rev/min by an electric motor at the bus terminal. Every time the bus speeds up, the flywheel slows down slightly. The bus is equipped with regenerative braking so that the flywheel can speed up when the bus slows down. The flywheel is a uniform solid cylinder with mass 1 600 kg and radius 0.650 m. The bus body does work against air resistance and rolling resistance at the average rate of 18.0 hp as it travels with an average speed of 40.0 km/h. How far can the bus travel before the flywheel has to be spun up to speed again?

42. The top in Figure P10.42 has a moment of inertia of 4.00×10^{-4} kg·m^2 and is initially at rest. It is free to rotate about the stationary axis AA'. A string, wrapped around a peg along the axis of the top, is pulled in such a manner as to maintain a constant tension of 5.57 N. If the string does not slip while it is unwound from the peg, what is the angular speed of the top after 80.0 cm of string has been pulled off the peg?

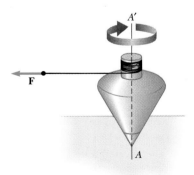

Figure P10.42

43. In Figure P10.43 the sliding block has a mass of 0.850 kg, the counterweight has a mass of 0.420 kg, and the pulley is a hollow cylinder with a mass of 0.350 kg, an inner radius of 0.020 0 m, and an outer radius of 0.030 0 m. The coefficient of kinetic friction between the block and the horizontal surface is 0.250. The pulley turns without friction on its axle. The light cord does not stretch and does not slip on the pulley. The block has a velocity of 0.820 m/s toward the pulley when it passes through a photogate. (a) Use energy methods to predict its speed after it has moved to a second photogate, 0.700 m away. (b) Find the angular speed of the pulley at the same moment.

Figure P10.43

44. A cylindrical rod 24.0 cm long with mass 1.20 kg and radius 1.50 cm has a ball of diameter 8.00 cm and mass 2.00 kg attached to one end. The arrangement is originally vertical and stationary, with the ball at the top. The system is free to pivot about the bottom end of the rod after being given a slight nudge. (a) After the rod rotates through ninety degrees, what is its rotational kinetic energy? (b) What is the angular speed of the rod and ball? (c) What is the linear speed of the ball? (d) How does this compare to the speed if the ball had fallen freely through the same distance of 28 cm?

45. An object with a weight of 50.0 N is attached to the free end of a light string wrapped around a reel of radius 0.250 m and mass 3.00 kg. The reel is a solid disk, free to rotate in a vertical plane about the horizontal axis passing through its center. The suspended object is released 6.00 m above the floor. (a) Determine the tension in the string, the acceleration of the object, and the speed with which the object hits the floor. (b) Verify your last answer by using the principle of conservation of energy to find the speed with which the object hits the floor.

46. A 15.0-kg object and a 10.0-kg object are suspended, joined by a cord that passes over a pulley with a radius of 10.0 cm and a mass of 3.00 kg (Fig. P10.46). The cord has a negligible mass and does not slip on the pulley. The pulley rotates on its axis without friction. The objects start from rest 3.00 m apart. Treat the pulley as a uniform disk, and determine the speeds of the two objects as they pass each other.

$M = 3.00$ kg
$R = 10.0$ cm
$m_1 = 15.0$ kg
$m_2 = 10.0$ kg

3.00 m

Figure P10.46

47. This problem describes one experimental method for determining the moment of inertia of an irregularly shaped object such as the payload for a satellite. Figure P10.47 shows a counterweight of mass m suspended by a cord wound around a spool of radius r, forming part of a turntable supporting the object. The turntable can rotate without friction. When the counterweight is released from rest, it descends through a distance h, acquiring a speed v. Show that the moment of inertia I of the rotating apparatus (including the turntable) is $mr^2(2gh/v^2 - 1)$.

Figure P10.47

48. A horizontal 800-N merry-go-round is a solid disk of radius 1.50 m, started from rest by a constant horizontal force of 50.0 N applied tangentially to the edge of the disk. Find the kinetic energy of the disk after 3.00 s.

49. (a) A uniform solid disk of radius R and mass M is free to rotate on a frictionless pivot through a point on its rim (Fig. P10.49). If the disk is released from rest in the position shown by the blue circle, what is the speed of its center of mass when the disk reaches the position indicated by the dashed circle? (b) What is the speed of the lowest point on the disk in the dashed position? (c) **What If?** Repeat part (a) using a uniform hoop.

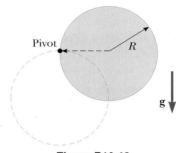

Pivot

R

g

Figure P10.49

50. The head of a grass string trimmer has 100 g of cord wound in a light cylindrical spool with inside diameter 3.00 cm and outside diameter 18.0 cm, as in Figure P10.50. The cord has a linear density of 10.0 g/m. A single strand of the cord extends 16.0 cm from the outer edge of the spool. (a) When switched on, the trimmer speeds up from 0 to 2 500 rev/min in 0.215 s. (a) What average power is delivered to the head by the trimmer motor while it is accelerating? (b) When the trimmer is cutting grass, it spins at 2 000 rev/min and the grass exerts an average tangential force of 7.65 N on the outer end of the cord, which is still at a radial distance of 16.0 cm from the outer edge of the spool. What is the power delivered to the head under load?

16.0 cm

3.0 cm

18.0 cm

Figure P10.50

Section 10.9 Rolling Motion of a Rigid Object

51. A cylinder of mass 10.0 kg rolls without slipping on a horizontal surface. At the instant its center of mass has a speed of 10.0 m/s, determine (a) the translational kinetic energy of its center of mass, (b) the rotational kinetic energy about its center of mass, and (c) its total energy.

52. A bowling ball has mass M, radius R, and a moment of inertia of $\frac{2}{5}MR^2$. If it starts from rest, how much work must be done on it to set it rolling without slipping at a linear speed v? Express the work in terms of M and v.

53. (a) Determine the acceleration of the center of mass of a uniform solid disk rolling down an incline making angle θ with the horizontal. Compare this acceleration with that of a uniform hoop. (b) What is the minimum coefficient of friction required to maintain pure rolling motion for the disk?

54. A uniform solid disk and a uniform hoop are placed side by side at the top of an incline of height h. If they are released from rest and roll without slipping, which object reaches the bottom first? Verify your answer by calculating their speeds when they reach the bottom in terms of h.

55. A metal can containing condensed mushroom soup has mass 215 g, height 10.8 cm, and diameter 6.38 cm. It is placed at rest on its side at the top of a 3.00-m-long incline that is at 25.0° to the horizontal, and it is then released to roll straight down. Assuming mechanical energy conservation, calculate the moment of inertia of the can if it takes 1.50 s to reach the bottom of the incline. Which pieces of data, if any, are unnecessary for calculating the solution?

56. A tennis ball is a hollow sphere with a thin wall. It is set rolling without slipping at 4.03 m/s on a horizontal section of a track, as shown in Figure P10.56. It rolls around the inside of a vertical circular loop 90.0 cm in diameter and finally leaves the track at a point 20.0 cm below the horizontal section. (a) Find the speed of the ball at the top of the loop. Demonstrate that it will fall from the track. (b) Find its speed as it leaves the track. **What If?** (c) Suppose that static friction between ball and track were negligible, so that the ball slid instead of rolling. Would its speed then be higher, lower, or the same at the top of the loop? Explain.

Figure P10.56

Additional Problems

57. As in Figure P10.57, toppling chimneys often break apart in mid-fall because the mortar between the bricks cannot withstand much shear stress. As the chimney begins to fall, shear forces must act on the topmost sections to accelerate them tangentially so that they can keep up with the rotation of the lower part of the stack. For simplicity, let us model the chimney as a uniform rod of length ℓ pivoted at the lower end. The rod starts at rest in a vertical position (with the frictionless pivot at the bottom) and falls over under the influence of gravity. What fraction of the length of the rod has a tangential acceleration greater than $g \sin \theta$, where θ is the angle the chimney makes with the vertical axis?

Figure P10.57 A building demolition site in Baltimore, MD. At the left is a chimney, mostly concealed by the building, that has broken apart on its way down. Compare with Figure 10.19.

58. **Review problem.** A mixing beater consists of three thin rods, each 10.0 cm long. The rods diverge from a central hub, separated from each other by 120°, and all turn in the same plane. A ball is attached to the end of each rod. Each ball has cross-sectional area 4.00 cm^2 and is so shaped that it has a drag coefficient of 0.600. Calculate the power input required to spin the beater at 1 000 rev/min (a) in air and (b) in water.

59. A 4.00-m length of light nylon cord is wound around a uniform cylindrical spool of radius 0.500 m and mass 1.00 kg. The spool is mounted on a frictionless axle and is initially at rest. The cord is pulled from the spool with a constant acceleration of magnitude 2.50 m/s^2. (a) How much work has been done on the spool when it reaches an angular speed of 8.00 rad/s? (b) Assuming there is enough cord on the spool, how long does it take the spool to reach this angular speed? (c) Is there enough cord on the spool?

60. A videotape cassette contains two spools, each of radius r_s, on which the tape is wound. As the tape unwinds from the first spool, it winds around the second spool. The tape moves at constant linear speed v past the heads between the spools. When all the tape is on the first spool, the tape has an outer radius r_t. Let r represent the outer radius of the tape on the first spool at any instant while the tape is being played. (a) Show that at any instant the angular speeds of the two spools are

$$\omega_1 = v/r \quad \text{and} \quad \omega_2 = v/(r_s{}^2 + r_t{}^2 - r^2)^{1/2}$$

(b) Show that these expressions predict the correct maximum and minimum values for the angular speeds of the two spools.

61. A long uniform rod of length L and mass M is pivoted about a horizontal, frictionless pin through one end. The rod is released from rest in a vertical position, as shown in Figure P10.61. At the instant the rod is horizontal, find (a) its angular speed, (b) the magnitude of its angular acceleration, (c) the x and y components of the acceleration of its center of mass, and (d) the components of the reaction force at the pivot.

Figure P10.61

62. A shaft is turning at 65.0 rad/s at time $t = 0$. Thereafter, its angular acceleration is given by

$$\alpha = -10.0 \text{ rad/s}^2 - 5.00t \text{ rad/s}^3,$$

where t is the elapsed time. (a) Find its angular speed at $t = 3.00$ s. (b) How far does it turn in these 3 s?

63. A bicycle is turned upside down while its owner repairs a flat tire. A friend spins the other wheel, of radius 0.381 m, and observes that drops of water fly off tangentially. She measures the height reached by drops moving vertically (Fig. P10.63). A drop that breaks loose from the tire on one turn rises $h = 54.0$ cm above the tangent point. A drop that breaks loose on the next turn rises 51.0 cm above the tangent point. The height to which the drops rise decreases because the angular speed of the wheel decreases. From this information, determine the magnitude of the average angular acceleration of the wheel.

Figure P10.63 Problems 63 and 64.

64. A bicycle is turned upside down while its owner repairs a flat tire. A friend spins the other wheel, of radius R, and observes that drops of water fly off tangentially. She measures the height reached by drops moving vertically (Fig. P10.63). A drop that breaks loose from the tire on one turn rises a distance h_1 above the tangent point. A drop that breaks loose on the next turn rises a distance $h_2 < h_1$ above the tangent point. The height to which the drops rise decreases because the angular speed of the wheel decreases. From this information, determine the magnitude of the average angular acceleration of the wheel.

65. A cord is wrapped around a pulley of mass m and radius r. The free end of the cord is connected to a block of mass M. The block starts from rest and then slides down an incline that makes an angle θ with the horizontal. The coefficient of kinetic friction between block and incline is μ. (a) Use energy methods to show that the block's speed as a function of position d down the incline is

$$v = \sqrt{\frac{4gdM(\sin \theta - \mu \cos \theta)}{m + 2M}}$$

(b) Find the magnitude of the acceleration of the block in terms of μ, m, M, g, and θ.

66. (a) What is the rotational kinetic energy of the Earth about its spin axis? Model the Earth as a uniform sphere and use data from the endpapers. (b) The rotational kinetic energy of the Earth is decreasing steadily because of tidal friction. Find the change in one day, assuming that the rotational period decreases by 10.0 μs each year.

67. Due to a gravitational torque exerted by the Moon on the Earth, our planet's rotation period slows at a rate on the order of 1 ms/century. (a) Determine the order of magnitude of the Earth's angular acceleration. (b) Find the order of magnitude of the torque. (c) Find the order of magnitude of the size of the wrench an ordinary person would need to exert such a torque, as in Figure P10.67. Assume the person can brace his feet against a solid firmament.

Figure P10.67

68. The speed of a moving bullet can be determined by allowing the bullet to pass through two rotating paper disks mounted a distance d apart on the same axle (Fig. P10.68). From the angular displacement $\Delta\theta$ of the two bullet holes in the disks and the rotational speed of the disks, we can determine the speed v of the bullet. Find the bullet speed for the following data: $d = 80$ cm, $\omega = 900$ rev/min, and $\Delta\theta = 31.0°$.

Figure P10.68

69. A uniform, hollow, cylindrical spool has inside radius $R/2$, outside radius R, and mass M (Fig. P10.69). It is mounted so that it rotates on a fixed horizontal axle. A counterweight of mass m is connected to the end of a string wound around the spool. The counterweight falls from rest at $t = 0$ to a position y at time t. Show that the torque due to the friction forces between spool and axle is

$$\tau_f = R\left[m\left(g - \frac{2y}{t^2}\right) - M\frac{5y}{4t^2}\right]$$

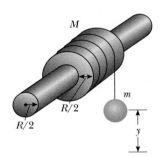

Figure P10.69

70. The reel shown in Figure P10.70 has radius R and moment of inertia I. One end of the block of mass m is connected to a spring of force constant k, and the other end

Figure P10.70

is fastened to a cord wrapped around the reel. The reel axle and the incline are frictionless. The reel is wound counterclockwise so that the spring stretches a distance d from its unstretched position and is then released from rest. (a) Find the angular speed of the reel when the spring is again unstretched. (b) Evaluate the angular speed numerically at this point if $I = 1.00$ kg·m², $R = 0.300$ m, $k = 50.0$ N/m, $m = 0.500$ kg, $d = 0.200$ m, and $\theta = 37.0°$.

71. Two blocks, as shown in Figure P10.71, are connected by a string of negligible mass passing over a pulley of radius 0.250 m and moment of inertia I. The block on the frictionless incline is moving up with a constant acceleration of 2.00 m/s². (a) Determine T_1 and T_2, the tensions in the two parts of the string. (b) Find the moment of inertia of the pulley.

Figure P10.71

72. A common demonstration, illustrated in Figure P10.72, consists of a ball resting at one end of a uniform board of length ℓ, hinged at the other end, and elevated at an angle θ. A light cup is attached to the board at r_c so that it will catch the ball when the support stick is suddenly removed. (a) Show that the ball will lag behind the falling board when θ is less than 35.3°. (b) If the board is 1.00 m long and is supported at this limiting angle, show that the cup must be 18.4 cm from the moving end.

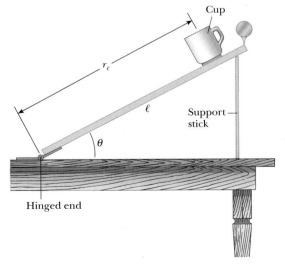

Figure P10.72

73. As a result of friction, the angular speed of a wheel changes with time according to

$$\frac{d\theta}{dt} = \omega_0 e^{-\sigma t}$$

where ω_0 and σ are constants. The angular speed changes from 3.50 rad/s at $t = 0$ to 2.00 rad/s at $t = 9.30$ s. Use this information to determine σ and ω_0. Then determine (a) the magnitude of the angular acceleration at $t = 3.00$ s, (b) the number of revolutions the wheel makes in the first 2.50 s, and (c) the number of revolutions it makes before coming to rest.

74. 🖥 The hour hand and the minute hand of Big Ben, the Parliament tower clock in London, are 2.70 m and 4.50 m long and have masses of 60.0 kg and 100 kg, respectively (see Figure P10.40). (a) Determine the total torque due to the weight of these hands about the axis of rotation when the time reads (i) 3:00 (ii) 5:15 (iii) 6:00 (iv) 8:20 (v) 9:45. (You may model the hands as long, thin uniform rods.) (b) Determine all times when the total torque about the axis of rotation is zero. Determine the times to the nearest second, solving a transcendental equation numerically.

75. (a) Without the wheels, a bicycle frame has a mass of 8.44 kg. Each of the wheels can be roughly modeled as a uniform solid disk with a mass of 0.820 kg and a radius of 0.343 m. Find the kinetic energy of the whole bicycle when it is moving forward at 3.35 m/s. (b) Before the invention of a wheel turning on an axle, ancient people moved heavy loads by placing rollers under them. (Modern people use rollers too. Any hardware store will sell you a roller bearing for a lazy susan.) A stone block of mass 844 kg moves forward at 0.335 m/s, supported by two uniform cylindrical tree trunks, each of mass 82.0 kg and radius 0.343 m. No slipping occurs between the block and the rollers or between the rollers and the ground. Find the total kinetic energy of the moving objects.

76. A uniform solid sphere of radius r is placed on the inside surface of a hemispherical bowl with much larger radius R. The sphere is released from rest at an angle θ to the vertical and rolls without slipping (Fig. P10.76). Determine the angular speed of the sphere when it reaches the bottom of the bowl.

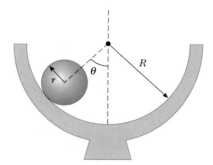

Figure P10.76

77. A string is wound around a uniform disk of radius R and mass M. The disk is released from rest with the string vertical and its top end tied to a fixed bar (Fig. P10.77). Show that (a) the tension in the string is one third of the weight of the disk, (b) the magnitude of the acceleration of the center of mass is $2g/3$, and (c) the speed of the center of mass is $(4gh/3)^{1/2}$ after the disk has descended through distance h. Verify your answer to (c) using the energy approach.

Figure P10.77

78. A constant horizontal force $\mathbf{F}$ is applied to a lawn roller in the form of a uniform solid cylinder of radius R and mass M (Fig. P10.78). If the roller rolls without slipping on the horizontal surface, show that (a) the acceleration of the center of mass is $2\mathbf{F}/3M$ and (b) the minimum coefficient of friction necessary to prevent slipping is $F/3Mg$. (*Hint:* Take the torque with respect to the center of mass.)

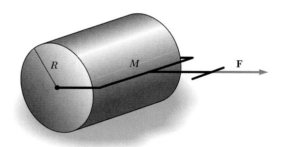

Figure P10.78

79. A solid sphere of mass m and radius r rolls without slipping along the track shown in Figure P10.79. It starts from rest with the lowest point of the sphere at height h above the bottom of the loop of radius R, much larger than r. (a) What is the minimum value of h (in terms of R) such that the sphere completes the loop? (b) What are the force components on the sphere at the point P if $h = 3R$?

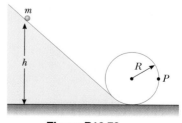

Figure P10.79

80. A thin rod of mass 0.630 kg and length 1.24 m is at rest, hanging vertically from a strong fixed hinge at its top end. Suddenly a horizontal impulsive force $(14.7\hat{\mathbf{i}})$ N is applied to it. (a) Suppose the force acts at the bottom end of the rod. Find the acceleration of its center of mass and the horizontal force the hinge exerts. (b) Suppose the force acts at the midpoint of the rod. Find the acceleration of this point and the horizontal hinge reaction. (c) Where can the impulse be applied so that the hinge will exert no horizontal force? This point is called the *center of percussion*.

81. A bowler releases a bowling ball with no spin, sending it sliding straight down the alley toward the pins. The ball continues to slide for a distance of what order of magnitude, before its motion becomes rolling without slipping? State the quantities you take as data, the values you measure or estimate for them, and your reasoning.

82. Following Thanksgiving dinner your uncle falls into a deep sleep, sitting straight up facing the television set. A naughty grandchild balances a small spherical grape at the top of his bald head, which itself has the shape of a sphere. After all the children have had time to giggle, the grape starts from rest and rolls down without slipping. It will leave contact with your uncle's scalp when the radial line joining it to the center of curvature makes what angle with the vertical?

83. (a) A thin rod of length h and mass M is held vertically with its lower end resting on a frictionless horizontal surface. The rod is then released to fall freely. Determine the speed of its center of mass just before it hits the horizontal surface. (b) **What If?** Now suppose the rod has a fixed pivot at its lower end. Determine the speed of the rod's center of mass just before it hits the surface.

84. A large, cylindrical roll of tissue paper of initial radius R lies on a long, horizontal surface with the outside end of the paper nailed to the surface. The roll is given a slight shove $(v_i \approx 0)$ and commences to unroll. Assume the roll has a uniform density and that mechanical energy is conserved in the process. (a) Determine the speed of the center of mass of the roll when its radius has diminished to r. (b) Calculate a numerical value for this speed at $r = 1.00$ mm, assuming $R = 6.00$ m. (c) **What If?** What happens to the energy of the system when the paper is completely unrolled?

85. A spool of wire of mass M and radius R is unwound under a constant force $\mathbf{F}$ (Fig. P10.85). Assuming the spool is a

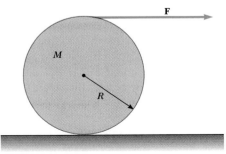

Figure P10.85

uniform solid cylinder that doesn't slip, show that (a) the acceleration of the center of mass is $4\mathbf{F}/3M$ and (b) the force of friction is to the *right* and equal in magnitude to $F/3$. (c) If the cylinder starts from rest and rolls without slipping, what is the speed of its center of mass after it has rolled through a distance d?

86. A plank with a mass $M = 6.00$ kg rides on top of two identical solid cylindrical rollers that have $R = 5.00$ cm and $m = 2.00$ kg (Fig. P10.86). The plank is pulled by a constant horizontal force $\mathbf{F}$ of magnitude 6.00 N applied to the end of the plank and perpendicular to the axes of the cylinders (which are parallel). The cylinders roll without slipping on a flat surface. There is also no slipping between the cylinders and the plank. (a) Find the acceleration of the plank and of the rollers. (b) What friction forces are acting?

Figure P10.86

87. A spool of wire rests on a horizontal surface as in Figure P10.87. As the wire is pulled, the spool does not slip at the contact point P. On separate trials, each one of the forces $\mathbf{F}_1$, $\mathbf{F}_2$, $\mathbf{F}_3$, and $\mathbf{F}_4$ is applied to the spool. For each one of these forces, determine the direction the spool will roll. Note that the line of action of $\mathbf{F}_2$ passes through P.

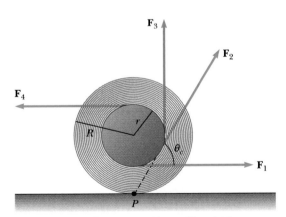

Figure P10.87 Problems 87 and 88.

88. Refer to Problem 87 and Figure P10.87. The spool of wire has an inner radius r and an outer radius R. The angle θ between the applied force and the horizontal can be varied. Show that the critical angle for which the spool does not roll is given by

$$\cos \theta_c = \frac{r}{R}$$

If the wire is held at this angle and the force increased, the spool will remain stationary until it slips along the floor.

89. In a demonstration known as the ballistics cart, a ball is projected vertically upward from a cart moving with constant velocity along the horizontal direction. The ball lands in the catching cup of the cart because both the cart a ball have the same horizontal component of velocity. **What If?** Now consider a ballistics cart on an incline making an angle θ with the horizontal as in Figure P10.89. The cart (including wheels) has a mass M and the moment of inertia of each of the two wheels is $mR^2/2$. (a) Using conservation of energy (assuming no friction between cart and axles) and assuming pure rolling motion (no slipping), show that the acceleration of the cart along the incline is

$$a_x = \left(\frac{M}{M + 2m}\right) g \sin \theta$$

(b) Note that the x component of acceleration of the ball released by the cart is $g \sin \theta$. Thus, the x component of the cart's acceleration is *smaller* than that of the ball by the factor $M/(M + 2m)$. Use this fact and kinematic equations to show that the ball overshoots the cart by an amount Δx, where

$$\Delta x = \left(\frac{4m}{M + 2m}\right) \left(\frac{\sin \theta}{\cos^2 \theta}\right) \frac{v_{yi}^2}{g}$$

and v_{yi} is the initial speed of the ball imparted to it by the spring in the cart. (c) Show that the distance d that the ball travels measured along the incline is

$$d = \frac{2v_{yi}^2}{g} \frac{\sin \theta}{\cos^2 \theta}$$

Figure P10.89

90. A spool of thread consists of a cylinder of radius R_1 with end caps of radius R_2 as in the end view shown in Figure P10.90. The mass of the spool, including the thread, is m and its moment of inertia about an axis through its center is I. The spool is placed on a rough horizontal surface so that it rolls without slipping when a force **T** acting to the right is applied to the free end of the thread. Show that the magnitude of the friction force exerted by the surface on the spool is given by

$$f = \left(\frac{I + mR_1R_2}{I + mR_2^2}\right) T$$

Determine the direction of the force of friction.

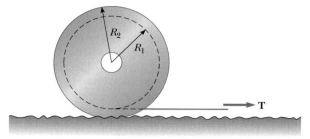

Figure P10.90

Answers to Quick Quizzes

10.1 (c). For a rotation of more than 180°, the angular displacement must be larger than $\pi = 3.14$ rad. The angular displacements in the three choices are (a) 6 rad − 3 rad = 3 rad (b) 1 rad − (− 1) rad = 2 rad (c) 5 rad − 1 rad = 4 rad.

10.2 (b). Because all angular displacements occur in the same time interval, the displacement with the lowest value will be associated with the lowest average angular speed.

10.3 (b). The fact that ω is negative indicates that we are dealing with an object that is rotating in the clockwise direction. We also know that when $\boldsymbol{\omega}$ and $\boldsymbol{\alpha}$ are antiparallel, ω must be decreasing—the object is slowing down. Therefore, the object is spinning more and more slowly (with less and less angular speed) in the clockwise, or negative, direction.

10.4 (b). In Equation 10.8, both the initial and final angular speeds are the same in all three cases. As a result, the angular acceleration is inversely proportional to the angular displacement. Thus, the highest angular acceleration is associated with the lowest angular displacement.

10.5 (b). The system of the platform, Andy, and Charlie is a rigid object, so all points on the rigid object have the same angular speed.

10.6 (a). The tangential speed is proportional to the radial distance from the rotation axis.

10.7 (a). Almost all of the mass of the pipe is at the same distance from the rotation axis, so it has a larger moment of inertia than the solid cylinder.

10.8 (b). The fatter handle of the screwdriver gives you a larger moment arm and increases the torque that you can apply with a given force from your hand.

10.9 (a). The longer handle of the wrench gives you a larger moment arm and increases the torque that you can apply with a given force from your hand.

10.10 (b). With twice the moment of inertia and the same frictional torque, there is half the angular acceleration. With half the angular acceleration, it will require twice as long to change the speed to zero.

10.11 (d). When the rod is attached at its end, it offers four times as much moment of inertia as when attached in the center (see Table 10.2). Because the rotational kinetic energy of the rod depends on the square of the angular speed, the same work will result in half of the angular speed.

10.12 (b). All of the gravitational potential energy of the box–Earth system is transformed to kinetic energy of translation. For the ball, some of the gravitational potential energy of the ball–Earth system is transformed to rotational kinetic energy, leaving less for translational kinetic energy, so the ball moves downhill more slowly than the box does.

10.13 (c). In Equation 10.30, I_{CM} for a sphere is $\frac{2}{5}MR^2$. Thus, MR^2 will cancel and the remaining expression on the right-hand side of the equation is independent of mass and radius.

10.14 (a). The moment of inertia of the hollow sphere B is larger than that of sphere A. As a result, Equation 10.30 tells us that the center of mass of sphere B will have a smaller speed, so sphere A should arrive first.

Chapter 11

Angular Momentum

▲ Mark Ruiz undergoes a rotation during a dive at the U.S. Olympic trials in June 2000. He spins at a higher rate when he curls up and grabs his ankles due to the principle of conservation of angular momentum, as discussed in this chapter. (Otto Greule/Allsport/Getty)

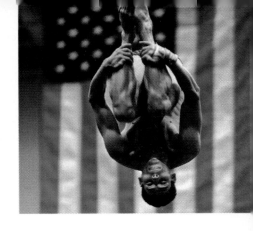

The central topic of this chapter is angular momentum, a quantity that plays a key role in rotational dynamics. In analogy to the principle of conservation of linear momentum, we find that the angular momentum of a system is conserved if no external torques act on the system. Like the law of conservation of linear momentum, the law of conservation of angular momentum is a fundamental law of physics, equally valid for relativistic and quantum systems.

11.1 The Vector Product and Torque

An important consideration in defining angular momentum is the process of multiplying two vectors by means of the operation called the *vector product*. We will introduce the vector product by considering torque as introduced in the preceding chapter.

Consider a force **F** acting on a rigid object at the vector position **r** (Fig. 11.1). As we saw in Section 10.6, the *magnitude* of the torque due to this force relative to the origin is $rF \sin \phi$, where ϕ is the angle between **r** and **F**. The axis about which **F** tends to produce rotation is perpendicular to the plane formed by **r** and **F**.

The torque vector $\boldsymbol{\tau}$ is related to the two vectors **r** and **F**. We can establish a mathematical relationship between $\boldsymbol{\tau}$, **r**, and **F** using a mathematical operation called the **vector product,** or **cross product:**

$$\boldsymbol{\tau} \equiv \mathbf{r} \times \mathbf{F} \tag{11.1}$$

We now give a formal definition of the vector product. Given any two vectors **A** and **B**, the **vector product A × B** is defined as a third vector **C**, which has a magnitude of $AB \sin \theta$, where θ is the angle between **A** and **B**. That is, if **C** is given by

$$\mathbf{C} = \mathbf{A} \times \mathbf{B} \tag{11.2}$$

then its magnitude is

$$C \equiv AB \sin \theta \tag{11.3}$$

The quantity $AB \sin \theta$ is equal to the area of the parallelogram formed by **A** and **B**, as shown in Figure 11.2. The *direction* of **C** is perpendicular to the plane formed by **A** and **B**, and the best way to determine this direction is to use the right-hand rule illustrated in Figure 11.2. The four fingers of the right hand are pointed along **A** and then "wrapped" into **B** through the angle θ. The direction of the upright thumb is the direction of **A × B = C**. Because of the notation, **A × B** is often read "**A** cross **B**"; hence, the term *cross product*.

Some properties of the vector product that follow from its definition are as follows:

1. Unlike the scalar product, the vector product is *not* commutative. Instead, the order in which the two vectors are multiplied in a cross product is important:

$$\mathbf{A} \times \mathbf{B} = -\mathbf{B} \times \mathbf{A} \tag{11.4}$$

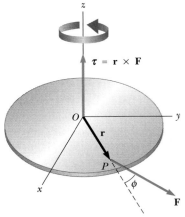

Active Figure 11.1 The torque vector $\boldsymbol{\tau}$ lies in a direction perpendicular to the plane formed by the position vector **r** and the applied force vector **F**.

At the Active Figures link at http://www.pse6.com, *you can move point P and change the force vector F to see the effect on the torque vector.*

▲ PITFALL PREVENTION

11.1 The Cross Product is a Vector

Remember that the result of taking a cross product between two vectors is *a third vector*. Equation 11.3 gives only the magnitude of this vector.

Right-hand rule

Figure 11.2 The vector product $\mathbf{A} \times \mathbf{B}$ is a third vector $\mathbf{C}$ having a magnitude $AB \sin \theta$ equal to the area of the parallelogram shown. The direction of $\mathbf{C}$ is perpendicular to the plane formed by $\mathbf{A}$ and $\mathbf{B}$, and this direction is determined by the right-hand rule.

Therefore, if you change the order of the vectors in a cross product, you must change the sign. You can easily verify this relationship with the right-hand rule.

2. If $\mathbf{A}$ is parallel to $\mathbf{B}$ ($\theta = 0°$ or $180°$), then $\mathbf{A} \times \mathbf{B} = 0$; therefore, it follows that $\mathbf{A} \times \mathbf{A} = 0$.

3. If $\mathbf{A}$ is perpendicular to $\mathbf{B}$, then $|\mathbf{A} \times \mathbf{B}| = AB$.

4. The vector product obeys the distributive law:

Properties of the vector product

$$\mathbf{A} \times (\mathbf{B} + \mathbf{C}) = \mathbf{A} \times \mathbf{B} + \mathbf{A} \times \mathbf{C} \tag{11.5}$$

5. The derivative of the cross product with respect to some variable such as t is

$$\frac{d}{dt}(\mathbf{A} \times \mathbf{B}) = \frac{d\mathbf{A}}{dt} \times \mathbf{B} + \mathbf{A} \times \frac{d\mathbf{B}}{dt} \tag{11.6}$$

where it is important to preserve the multiplicative order of $\mathbf{A}$ and $\mathbf{B}$, in view of Equation 11.4.

It is left as an exercise (Problem 10) to show from Equations 11.3 and 11.4 and from the definition of unit vectors that the cross products of the rectangular unit vectors $\hat{\mathbf{i}}$, $\hat{\mathbf{j}}$, and $\hat{\mathbf{k}}$ obey the following rules:

Cross products of unit vectors

$$\hat{\mathbf{i}} \times \hat{\mathbf{i}} = \hat{\mathbf{j}} \times \hat{\mathbf{j}} = \hat{\mathbf{k}} \times \hat{\mathbf{k}} = 0 \tag{11.7a}$$

$$\hat{\mathbf{i}} \times \hat{\mathbf{j}} = -\hat{\mathbf{j}} \times \hat{\mathbf{i}} = \hat{\mathbf{k}} \tag{11.7b}$$

$$\hat{\mathbf{j}} \times \hat{\mathbf{k}} = -\hat{\mathbf{k}} \times \hat{\mathbf{j}} = \hat{\mathbf{i}} \tag{11.7c}$$

$$\hat{\mathbf{k}} \times \hat{\mathbf{i}} = -\hat{\mathbf{i}} \times \hat{\mathbf{k}} = \hat{\mathbf{j}} \tag{11.7d}$$

Signs are interchangeable in cross products. For example, $\mathbf{A} \times (-\mathbf{B}) = -\mathbf{A} \times \mathbf{B}$ and $\hat{\mathbf{i}} \times (-\hat{\mathbf{j}}) = -\hat{\mathbf{i}} \times \hat{\mathbf{j}}$.

The cross product of any two vectors $\mathbf{A}$ and $\mathbf{B}$ can be expressed in the following determinant form:

$$\mathbf{A} \times \mathbf{B} = \begin{vmatrix} \hat{\mathbf{i}} & \hat{\mathbf{j}} & \hat{\mathbf{k}} \\ A_x & A_y & A_z \\ B_x & B_y & B_z \end{vmatrix} = \begin{vmatrix} A_y & A_z \\ B_y & B_z \end{vmatrix} \hat{\mathbf{i}} - \begin{vmatrix} A_x & A_z \\ B_x & B_z \end{vmatrix} \hat{\mathbf{j}} + \begin{vmatrix} A_x & A_y \\ B_x & B_y \end{vmatrix} \hat{\mathbf{k}}$$

Expanding these determinants gives the result

$$\mathbf{A} \times \mathbf{B} = (A_y B_z - A_z B_y)\hat{\mathbf{i}} - (A_x B_z - A_z B_x)\hat{\mathbf{j}} + (A_x B_y - A_y B_x)\hat{\mathbf{k}} \tag{11.8}$$

Given the definition of the cross product, we can now assign a direction to the torque vector. If the force lies in the xy plane, as in Figure 11.1, the torque $\boldsymbol{\tau}$ is represented by a vector parallel to the z axis. The force in Figure 11.1 creates a torque that tends to rotate the object counterclockwise about the z axis; thus the direction of $\boldsymbol{\tau}$ is toward increasing z, and $\boldsymbol{\tau}$ is therefore in the positive z direction. If we reversed the direction of $\mathbf{F}$ in Figure 11.1, then $\boldsymbol{\tau}$ would be in the negative z direction.

Quick Quiz 11.1 Which of the following is equivalent to the following *scalar* product: $(\mathbf{A} \times \mathbf{B}) \cdot (\mathbf{B} \times \mathbf{A})$? (a) $\mathbf{A} \cdot \mathbf{B} + \mathbf{B} \cdot \mathbf{A}$ (b) $(\mathbf{A} \times \mathbf{A}) \cdot (\mathbf{B} \times \mathbf{B})$ (c) $(\mathbf{A} \times \mathbf{B}) \cdot (\mathbf{A} \times \mathbf{B})$ (d) $-(\mathbf{A} \times \mathbf{B}) \cdot (\mathbf{A} \times \mathbf{B})$

Quick Quiz 11.2 Which of the following statements is true about the relationship between the magnitude of the cross product of two vectors and the product of the magnitudes of the vectors? (a) $|\mathbf{A} \times \mathbf{B}|$ is larger than AB; (b) $|\mathbf{A} \times \mathbf{B}|$ is smaller than AB; (c) $|\mathbf{A} \times \mathbf{B}|$ could be larger or smaller than AB, depending on the angle between the vectors; (d) $|\mathbf{A} \times \mathbf{B}|$ could be equal to AB.

Example 11.1   The Vector Product

Two vectors lying in the xy plane are given by the equations $\mathbf{A} = 2\hat{\mathbf{i}} + 3\hat{\mathbf{j}}$ and $\mathbf{B} = -\hat{\mathbf{i}} + 2\hat{\mathbf{j}}$. Find $\mathbf{A} \times \mathbf{B}$ and verify that $\mathbf{A} \times \mathbf{B} = -\mathbf{B} \times \mathbf{A}$.

Solution Using Equations 11.7a through 11.7d, we obtain

$$\mathbf{A} \times \mathbf{B} = (2\hat{\mathbf{i}} + 3\hat{\mathbf{j}}) \times (-\hat{\mathbf{i}} + 2\hat{\mathbf{j}})$$
$$= 2\hat{\mathbf{i}} \times 2\hat{\mathbf{j}} + 3\hat{\mathbf{j}} \times (-\hat{\mathbf{i}}) = 4\hat{\mathbf{k}} + 3\hat{\mathbf{k}} = \boxed{7\hat{\mathbf{k}}}$$

(We have omitted the terms containing $\hat{\mathbf{i}} \times \hat{\mathbf{i}}$ and $\hat{\mathbf{j}} \times \hat{\mathbf{j}}$ because, as Equation 11.7a shows, they are equal to zero.)

We can show that $\mathbf{A} \times \mathbf{B} = -\mathbf{B} \times \mathbf{A}$, because

$$\mathbf{B} \times \mathbf{A} = (-\hat{\mathbf{i}} + 2\hat{\mathbf{j}}) \times (2\hat{\mathbf{i}} + 3\hat{\mathbf{j}})$$
$$= -\hat{\mathbf{i}} \times 3\hat{\mathbf{j}} + 2\hat{\mathbf{j}} \times 2\hat{\mathbf{i}} = -3\hat{\mathbf{k}} - 4\hat{\mathbf{k}} = \boxed{-7\hat{\mathbf{k}}}$$

Therefore, $\mathbf{A} \times \mathbf{B} = -\mathbf{B} \times \mathbf{A}$.

As an alternative method for finding $\mathbf{A} \times \mathbf{B}$, we could use Equation 11.8, with $A_x = 2$, $A_y = 3$, $A_z = 0$ and $B_x = -1$, $B_y = 2$, $B_z = 0$:

$$\mathbf{A} \times \mathbf{B} = (0)\hat{\mathbf{i}} - (0)\hat{\mathbf{j}} + [(2)(2) - (3)(-1)]\hat{\mathbf{k}} = 7\hat{\mathbf{k}}$$

Example 11.2   The Torque Vector

A force of $\mathbf{F} = (2.00\hat{\mathbf{i}} + 3.00\hat{\mathbf{j}})$ N is applied to an object that is pivoted about a fixed axis aligned along the z coordinate axis. If the force is applied at a point located at $\mathbf{r} = (4.00\hat{\mathbf{i}} + 5.00\hat{\mathbf{j}})$ m, find the torque vector $\boldsymbol{\tau}$.

Solution The torque vector is defined by means of a cross product in Equation 11.1:

$$\boldsymbol{\tau} = \mathbf{r} \times \mathbf{F} = [(4.00\hat{\mathbf{i}} + 5.00\hat{\mathbf{j}}) \text{ m}]$$
$$\times [(2.00\hat{\mathbf{i}} + 3.00\hat{\mathbf{j}}) \text{ N}]$$
$$= [(4.00)(2.00)\hat{\mathbf{i}} \times \hat{\mathbf{i}} + (4.00)(3.00)\hat{\mathbf{i}} \times \hat{\mathbf{j}}$$
$$+ (5.00)(2.00)\hat{\mathbf{j}} \times \hat{\mathbf{i}}$$
$$+ (5.00)(3.00)\hat{\mathbf{j}} \times \hat{\mathbf{j}}] \text{ N} \cdot \text{m}$$

$$= [12.0\hat{\mathbf{i}} \times \hat{\mathbf{j}} + 10.0\hat{\mathbf{j}} \times \hat{\mathbf{i}}] \text{ N} \cdot \text{m}$$
$$= [12.0\hat{\mathbf{k}} - 10.0\hat{\mathbf{k}}] \text{ N} \cdot \text{m}$$
$$= \boxed{2.0\hat{\mathbf{k}} \text{ N} \cdot \text{m}}$$

Notice that both $\mathbf{r}$ and $\mathbf{F}$ are in the xy plane. As expected, the torque vector is perpendicular to this plane, having only a z component.

11.2   Angular Momentum

Imagine a rigid pole sticking up through the ice on a frozen pond (Fig. 11.3). A skater glides rapidly toward the pole, aiming a little to the side so that she does not hit it. As she approaches the pole, she reaches out and grabs it, an action that causes her to move in a circular path around the pole. Just as the idea of linear momentum helps us analyze translational motion, a rotational analog—*angular momentum*—helps us analyze the motion of this skater and other objects undergoing rotational motion.

Active Figure 11.3 As the skater passes the pole, she grabs hold of it. This causes her to swing around the pole rapidly in a circular path.

At the Active Figures link at http://www.pse6.com, you can change the speed of the skater and her distance to the pole and watch her spin when she grabs the pole.

Angular momentum of a particle

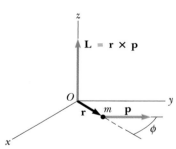

Active Figure 11.4 The angular momentum $\mathbf{L}$ of a particle of mass m and linear momentum $\mathbf{p}$ located at the vector position $\mathbf{r}$ is a vector given by $\mathbf{L} = \mathbf{r} \times \mathbf{p}$. The value of $\mathbf{L}$ depends on the origin about which it is measured and is a vector perpendicular to both $\mathbf{r}$ and $\mathbf{p}$.

At the Active Figures link at http://www.pse6.com, you can change the position vector r and the momentum vector p to see the effect on the angular momentum vector.

In Chapter 9, we began by developing the mathematical form of linear momentum and then proceeded to show how this new quantity was valuable in problem-solving. We will follow a similar procedure for angular momentum.

Consider a particle of mass m located at the vector position $\mathbf{r}$ and moving with linear momentum $\mathbf{p}$ as in Figure 11.4. In describing linear motion, we found that the net force on the particle equals the time rate of change of its linear momentum, $\Sigma \mathbf{F} = d\mathbf{p}/dt$ (see Eq. 9.3). Let us take the cross product of each side of Equation 9.3 with $\mathbf{r}$, which gives us the net torque on the particle on the left side of the equation:

$$\mathbf{r} \times \sum \mathbf{F} = \sum \boldsymbol{\tau} = \mathbf{r} \times \frac{d\mathbf{p}}{dt}$$

Now let us add to the right-hand side the term $\dfrac{d\mathbf{r}}{dt} \times \mathbf{p}$, which is zero because $d\mathbf{r}/dt = \mathbf{v}$ and $\mathbf{v}$ and $\mathbf{p}$ are parallel. Thus,

$$\sum \boldsymbol{\tau} = \mathbf{r} \times \frac{d\mathbf{p}}{dt} + \frac{d\mathbf{r}}{dt} \times \mathbf{p}$$

We recognize the right-hand side of this equation as the derivative of $\mathbf{r} \times \mathbf{p}$ (see Equation 11.6). Therefore,

$$\sum \boldsymbol{\tau} = \frac{d(\mathbf{r} \times \mathbf{p})}{dt} \tag{11.9}$$

This looks very similar in form to Equation 9.3, $\Sigma \mathbf{F} = d\mathbf{p}/dt$. This suggests that the combination $\mathbf{r} \times \mathbf{p}$ should play the same role in rotational motion that $\mathbf{p}$ plays in translational motion. We call this combination the *angular momentum* of the particle:

The instantaneous **angular momentum L** of a particle relative to the origin O is defined by the cross product of the particle's instantaneous position vector $\mathbf{r}$ and its instantaneous linear momentum $\mathbf{p}$:

$$\mathbf{L} \equiv \mathbf{r} \times \mathbf{p} \tag{11.10}$$

This allows us to write Equation 11.9 as

$$\sum \boldsymbol{\tau} = \frac{d\mathbf{L}}{dt} \tag{11.11}$$

which is the rotational analog of Newton's second law, $\Sigma \mathbf{F} = d\mathbf{p}/dt$. Note that torque causes the angular momentum $\mathbf{L}$ to change just as force causes linear momentum $\mathbf{p}$ to change. Equation 11.11 states that **the torque acting on a particle is equal to the time rate of change of the particle's angular momentum.**

Note that Equation 11.11 is valid only if $\Sigma\boldsymbol{\tau}$ and $\mathbf{L}$ are measured about the same origin. (Of course, the same origin must be used in calculating all of the torques.) Furthermore, **the expression is valid for any origin fixed in an inertial frame.**

The SI unit of angular momentum is $\text{kg} \cdot \text{m}^2/\text{s}$. Note also that both the magnitude and the direction of $\mathbf{L}$ depend on the choice of origin. Following the right-hand rule, we see that the direction of $\mathbf{L}$ is perpendicular to the plane formed by $\mathbf{r}$ and $\mathbf{p}$. In Figure 11.4, $\mathbf{r}$ and $\mathbf{p}$ are in the xy plane, and so $\mathbf{L}$ points in the z direction. Because $\mathbf{p} = m\mathbf{v}$, the magnitude of $\mathbf{L}$ is

$$L = mvr \sin \phi \tag{11.12}$$

where ϕ is the angle between $\mathbf{r}$ and $\mathbf{p}$. It follows that L is zero when $\mathbf{r}$ is parallel to $\mathbf{p}$ ($\phi = 0$ or $180°$). In other words, when the linear velocity of the particle is along a line that passes through the origin, the particle has zero angular momentum with respect to the origin. On the other hand, if $\mathbf{r}$ is perpendicular to $\mathbf{p}$ ($\phi = 90°$), then $L = mvr$. At that instant, the particle moves exactly as if it were on the rim of a wheel rotating about the origin in a plane defined by $\mathbf{r}$ and $\mathbf{p}$.

Quick Quiz 11.3 Recall the skater described at the beginning of this section. Let her mass be m. What would be her angular momentum relative to the pole at the instant she is a distance d from the pole if she were skating directly toward it at speed v? (a) zero (b) mvd (c) impossible to determine

Quick Quiz 11.4 Consider again the skater in Quick Quiz 11.3. What would be her angular momentum relative to the pole at the instant she is a distance d from the pole if she were skating at speed v along a straight line that would pass within a distance a from the pole? (a) zero (b) mvd (c) mva (d) impossible to determine

 PITFALL PREVENTION

11.2 Is Rotation Necessary for Angular Momentum?

Notice that we can define angular momentum even if the particle is not moving in a circular path. Even a particle moving in a straight line has angular momentum about any axis displaced from the path of the particle.

Example 11.3 Angular Momentum of a Particle in Circular Motion

A particle moves in the xy plane in a circular path of radius r, as shown in Figure 11.5. Find the magnitude and direction of its angular momentum relative to O when its linear velocity is **v**.

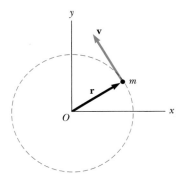

Figure 11.5 (Example 11.3) A particle moving in a circle of radius r has an angular momentum about O that has magnitude mvr. The vector $\mathbf{L} = \mathbf{r} \times \mathbf{p}$ points *out* of the diagram.

Solution The linear momentum of the particle is always changing (in direction, not magnitude). You might be tempted, therefore, to conclude that the angular momentum of the particle is always changing. In this situation, however, this is not the case—let us see why. From Equation 11.12, the magnitude of **L** is given by

$$L = mvr\sin 90° = \boxed{mvr}$$

where we have used $\phi = 90°$ because **v** is perpendicular to **r**. This value of L is constant because all three factors on the right are constant.

The direction of **L** also is constant, even though the direction of $\mathbf{p} = m\mathbf{v}$ keeps changing. You can visualize this by applying the right-hand rule to find the direction of $\mathbf{L} = \mathbf{r} \times \mathbf{p} = m\mathbf{r} \times \mathbf{v}$ in Figure 11.5. Your thumb points upward and away from the page; this is the direction of **L** Hence, we can write the vector expression $\mathbf{L} = (mvr)\hat{\mathbf{k}}$. If the particle were to move clockwise, **L** would point downward and into the page. **A particle in uniform circular motion has a constant angular momentum about an axis through the center of its path.**

Angular Momentum of a System of Particles

In Section 9.6, we showed that Newton's second law for a particle could be extended to a system of particles, resulting in:

$$\sum \mathbf{F}_{\text{ext}} = \frac{d\mathbf{p}_{\text{tot}}}{dt}$$

This equation states that the net external force on a system of particles is equal to the time rate of change of the total linear momentum of the system. Let us see if there is a similar statement that can be made in rotational motion. The total angular momentum of a system of particles about some point is defined as the vector sum of the angular momenta of the individual particles:

$$\mathbf{L}_{\text{tot}} = \mathbf{L}_1 + \mathbf{L}_2 + \cdots + \mathbf{L}_n = \sum_i \mathbf{L}_i$$

where the vector sum is over all n particles in the system.

Let us differentiate this equation with respect to time:

$$\frac{d\mathbf{L}_{\text{tot}}}{dt} = \sum_i \frac{d\mathbf{L}_i}{dt} = \sum_i \boldsymbol{\tau}_i$$

where we have used Equation 11.11 to replace the time rate of change of the angular momentum of each particle with the net torque on the particle.

The torques acting on the particles of the system are those associated with internal forces between particles and those associated with external forces. However, the net torque associated with all internal forces is zero. To understand this, recall that Newton's third law tells us that internal forces between particles of the system are equal in magnitude and opposite in direction. If we assume that these forces lie along the line of separation of each pair of particles, then the total torque around some axis passing through an origin O due to each action–reaction force pair is zero. That is, the moment arm d from O to the line of action of the forces is equal for both particles and the forces are in opposite directions. In the summation, therefore, we see that the net internal torque vanishes. We conclude that the total angular momentum of a system can vary with time only if a net external torque is acting on the system, so that we have

The net external torque on a system equals the time rate of change of angular momentum of the system

$$\sum \boldsymbol{\tau}_{\text{ext}} = \frac{d\mathbf{L}_{\text{tot}}}{dt} \qquad (11.13)$$

That is

> the net external torque acting on a system about some axis passing through an origin in an inertial frame equals the time rate of change of the total angular momentum of the system about that origin.

Note that Equation 11.13 is indeed the rotational analog of $\Sigma \mathbf{F}_{\text{ext}} = d\mathbf{p}_{\text{tot}}/dt$, for a system of particles.

Although we do not prove it here, the following statement is an important theorem concerning the angular momentum of a system relative to the system's center of mass:

> The resultant torque acting on a system about an axis through the center of mass equals the time rate of change of angular momentum of the system regardless of the motion of the center of mass.

This theorem applies even if the center of mass is accelerating, provided $\boldsymbol{\tau}$ and $\mathbf{L}$ are evaluated relative to the center of mass.

Example 11.4 Two Connected Objects

A sphere of mass m_1 and a block of mass m_2 are connected by a light cord that passes over a pulley, as shown in Figure 11.6. The radius of the pulley is R, and the mass of the rim is M. The spokes of the pulley have negligible mass. The block slides on a frictionless, horizontal surface. Find an expression for the linear acceleration of the two objects, using the concepts of angular momentum and torque.

Solution We need to determine the angular momentum of the system, which consists of the two objects and the pulley.

Figure 11.6 (Example 11.4) When the system is released, the sphere moves downward and the block moves to the left.

Let us calculate the angular momentum about an axis that coincides with the axle of the pulley. The angular momentum of the system includes that of two objects moving translationally (the sphere and the block) and one object undergoing pure rotation (the pulley).

At any instant of time, the sphere and the block have a common speed v, so the angular momentum of the sphere is m_1vR, and that of the block is m_2vR. At the same instant, all points on the rim of the pulley also move with speed v, so the angular momentum of the pulley is MvR. Hence, the total angular momentum of the system is

(1) $L = m_1vR + m_2vR + MvR = (m_1 + m_2 + M)vR$

Now let us evaluate the total external torque acting on the system about the pulley axle. Because it has a moment arm of zero, the force exerted by the axle on the pulley does not contribute to the torque. Furthermore, the normal force acting on the block is balanced by the gravitational force $m_2\mathbf{g}$, and so these forces do not contribute to the torque. The gravitational force $m_1\mathbf{g}$ acting on the sphere produces a torque about the axle equal in magnitude to m_1gR, where R is the moment arm of the force about the axle. This is the total external torque about the pulley axle;

that is, $\Sigma\tau_{\text{ext}} = m_1gR$. Using this result, together with Equation (1) and Equation 11.13, we find

$$\sum \tau_{\text{ext}} = \frac{dL}{dt}$$

$$m_1gR = \frac{d}{dt}[(m_1 + m_2 + M)vR]$$

(2) $$m_1gR = (m_1 + m_2 + M)R\frac{dv}{dt}$$

Because $dv/dt = a$, we can solve this for a to obtain

$$a = \frac{m_1g}{m_1 + m_2 + M}$$

You may wonder why we did not include the forces that the cord exerts on the objects in evaluating the net torque about the axle. The reason is that these forces are internal to the system under consideration, and we analyzed the system as a whole. Only *external* torques contribute to the change in the system's angular momentum.

11.3 Angular Momentum of a Rotating Rigid Object

In Example 11.4, we considered the angular momentum of a deformable system. Let us now restrict our attention to a nondeformable system—a rigid object. Consider a rigid object rotating about a fixed axis that coincides with the z axis of a coordinate system, as shown in Figure 11.7. Let us determine the angular momentum of this object. Each *particle* of the object rotates in the xy plane about the z axis with an angular speed ω. The magnitude of the angular momentum of a particle of mass m_i about the z axis is $m_iv_ir_i$. Because $v_i = r_i\omega$, we can express the magnitude of the angular momentum of this particle as

$$L_i = m_ir_i^2\omega$$

The vector $\mathbf{L}_i$ is directed along the z axis, as is the vector $\boldsymbol{\omega}$.

We can now find the angular momentum (which in this situation has only a z component) of the whole object by taking the sum of L_i over all particles:

$$L_z = \sum_i L_i = \sum_i m_ir_i^2\omega = \left(\sum_i m_ir_i^2\right)\omega$$

$$L_z = I\omega \tag{11.14}$$

where we have recognized $\sum_i m_ir_i^2$ as the moment of inertia I of the object about the z axis (Equation 10.15).

Now let us differentiate Equation 11.14 with respect to time, noting that I is constant for a rigid object:

$$\frac{dL_z}{dt} = I\frac{d\omega}{dt} = I\alpha \tag{11.15}$$

where α is the angular acceleration relative to the axis of rotation. Because dL_z/dt is equal to the net external torque (see Eq. 11.13), we can express Equation 11.15 as

$$\sum \tau_{\text{ext}} = I\alpha \tag{11.16}$$

Rotational form of Newton's second law

That is, the net external torque acting on a rigid object rotating about a fixed axis equals the moment of inertia about the rotation axis multiplied by the object's angular acceleration relative to that axis. This result is the same as Equation 10.21, which was derived using a force approach, but we derived Equation 11.16 using the concept of angular momentum. This equation is also valid for a rigid object rotating about a moving axis provided the moving axis (1) passes through the center of mass and (2) is a symmetry axis.

If a symmetrical object rotates about a fixed axis passing through its center of mass, you can write Equation 11.14 in vector form as $\mathbf{L} = I\boldsymbol{\omega}$, where $\mathbf{L}$ is the total angular

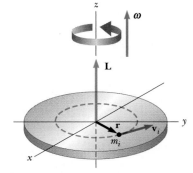

Figure 11.7 When a rigid object rotates about an axis, the angular momentum $\mathbf{L}$ is in the same direction as the angular velocity $\boldsymbol{\omega}$, according to the expression $\mathbf{L} = I\boldsymbol{\omega}$.

momentum of the object measured with respect to the axis of rotation. Furthermore, the expression is valid for any object, regardless of its symmetry, if **L** stands for the component of angular momentum along the axis of rotation.[1]

> **Quick Quiz 11.5** A solid sphere and a hollow sphere have the same mass and radius. They are rotating with the same angular speed. The one with the higher angular momentum is (a) the solid sphere (b) the hollow sphere (c) they both have the same angular momentum (d) impossible to determine.

Example 11.5   Bowling Ball

Estimate the magnitude of the angular momentum of a bowling ball spinning at 10 rev/s, as shown in Figure 11.8.

Solution We start by making some estimates of the relevant physical parameters and model the ball as a uniform solid sphere. A typical bowling ball might have a mass of 6.0 kg and a radius of 12 cm. The moment of inertia of a solid sphere about an axis through its center is, from Table 10.2,

$$I = \tfrac{2}{5}MR^2 = \tfrac{2}{5}(6.0 \text{ kg})(0.12 \text{ m})^2 = 0.035 \text{ kg} \cdot \text{m}^2$$

Therefore, the magnitude of the angular momentum is

$$L_z = I\omega = (0.035 \text{ kg} \cdot \text{m}^2)(10 \text{ rev/s})(2\pi \text{ rad/rev})$$
$$= 2.2 \text{ kg} \cdot \text{m}^2/\text{s}$$

Because of the roughness of our estimates, we probably want to keep only one significant figure, and so $L_z \approx \boxed{2 \text{ kg} \cdot \text{m}^2/\text{s}}$.

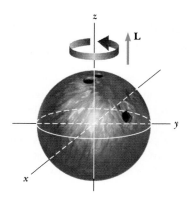

Figure 11.8 (Example 11.5) A bowling ball that rotates about the z axis in the direction shown has an angular momentum **L** in the positive z direction. If the direction of rotation is reversed, **L** points in the negative z direction.

Example 11.6   The Seesaw   `Interactive`

A father of mass m_f and his daughter of mass m_d sit on opposite ends of a seesaw at equal distances from the pivot at the center (Fig. 11.9). The seesaw is modeled as a rigid rod of mass M and length ℓ and is pivoted without friction. At a given moment, the combination rotates in a vertical plane with an angular speed ω.

(A) Find an expression for the magnitude of the system's angular momentum.

Solution The moment of inertia of the system equals the sum of the moments of inertia of the three components: the seesaw and the two individuals, whom we will model as particles. Referring to Table 10.2 to obtain the expression for the moment of inertia of the rod, and using the expression $I = mr^2$ for each person, we find that the total moment of inertia about the z axis through O is

$$I = \tfrac{1}{12}M\ell^2 + m_f\left(\frac{\ell}{2}\right)^2 + m_d\left(\frac{\ell}{2}\right)^2 = \frac{\ell^2}{4}\left(\frac{M}{3} + m_f + m_d\right)$$

Therefore, the magnitude of the angular momentum is

$$L = I\omega = \boxed{\frac{\ell^2}{4}\left(\frac{M}{3} + m_f + m_d\right)\omega}$$

(B) Find an expression for the magnitude of the angular acceleration of the system when the seesaw makes an angle θ with the horizontal.

Solution To find the angular acceleration of the system at any angle θ, we first calculate the net torque on the system and then use $\Sigma\tau_{\text{ext}} = I\alpha$ to obtain an expression for α.

The torque due to the force $m_f g$ about the pivot is

$$\tau_f = m_f g\,\frac{\ell}{2}\cos\theta \qquad (\tau_f \text{ out of page})$$

The torque due to the force $m_d g$ about the pivot is

$$\tau_d = -m_d g\,\frac{\ell}{2}\cos\theta \qquad (\tau_d \text{ into page})$$

---

[1]   In general, the expression $\mathbf{L} = I\boldsymbol{\omega}$ is not always valid. If a rigid object rotates about an *arbitrary* axis, **L** and **ω** may point in different directions. In this case, the moment of inertia cannot be treated as a scalar. Strictly speaking, $\mathbf{L} = I\boldsymbol{\omega}$ applies only to rigid objects of any shape that rotate about one of three mutually perpendicular axes (called *principal axes*) through the center of mass. This is discussed in more advanced texts on mechanics.

Hence, the net torque exerted on the system about O is

$$\sum \tau_{\text{ext}} = \tau_f + \tau_d = \tfrac{1}{2}(m_f - m_d)g\ell \cos \theta$$

To find α, we use $\sum \tau_{\text{ext}} = I\alpha$, where I was obtained in part (A):

$$\alpha = \frac{\sum \tau_{\text{ext}}}{I} = \frac{2(m_f - m_d)g\cos \theta}{\ell\left(\dfrac{M}{3} + m_f + m_d\right)}$$

Generally, fathers are more massive than daughters, so the angular acceleration is positive. If the seesaw begins in a horizontal orientation ($\theta = 0$) and is released, the rotation will be counterclockwise in Figure 11.9 and the father's end of the seesaw drops. This is consistent with everyday experience.

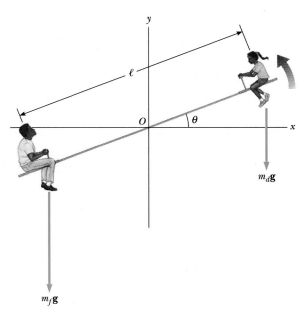

Figure 11.9 (Example 11.6) A father and daughter demonstrate angular momentum on a seesaw.

What If? After several complaints from the daughter that she simply rises into the air rather than moving up and down as planned, the father moves inward on the seesaw to try to balance the two sides. He moves in to a position that is a distance d from the pivot. What is the angular acceleration of the system in this case when it is released from an arbitrary angle θ?

Answer The angular acceleration of the system should decrease if the system is more balanced. As the father continues to slide inward, he should reach a point at which the seesaw is balanced and there is no angular acceleration of the system when released.

The total moment of inertia about the z axis through O for the modified system is

$$I = \tfrac{1}{12}M\ell^2 + m_f d^2 + m_d\left(\frac{\ell}{2}\right)^2$$

$$= \frac{\ell^2}{4}\left(\frac{M}{3} + m_d\right) + m_f d^2$$

The net torque exerted on the system about O is

$$\tau_{\text{net}} = \tau_f + \tau_d = m_f g d \cos \theta - \tfrac{1}{2}m_d g\ell \cos \theta$$

Now, the angular acceleration of the system is

$$\alpha = \frac{\tau_{\text{net}}}{I} = \frac{m_f g d \cos \theta - \tfrac{1}{2}m_d g\ell \cos \theta}{\dfrac{\ell^2}{4}\left[(M/3) + m_d\right] + m_f d^2}$$

The seesaw will be balanced when the angular acceleration is zero. In this situation, both father and daughter can push off the ground and rise to the highest possible point. We find the required position of the father by setting $\alpha = 0$:

$$\alpha = \frac{m_f g d \cos \theta - \tfrac{1}{2}m_d g\ell \cos \theta}{(\ell^2/4)\left[(M/3) + m_d\right] + m_f d^2} = 0$$

$$m_f g d \cos \theta - \tfrac{1}{2}m_d g\ell \cos \theta = 0$$

$$d = \left(\frac{m_d}{m_f}\right)\tfrac{1}{2}\ell$$

In the rare case that the father and daughter have the same mass, the father is located at the end of the seesaw, $d = \ell/2$.

 At the Interactive Worked Example link at http://www.pse6.com, *you can move the father and daughter to see the effect on the motion of the system.*

11.4 Conservation of Angular Momentum

In Chapter 9 we found that the total linear momentum of a system of particles remains constant if the system is isolated, that is, if the resultant external force acting on the system is zero. We have an analogous conservation law in rotational motion:

> The total angular momentum of a system is constant in both magnitude and direction if the resultant external torque acting on the system is zero, that is, if the system is isolated.

Conservation of angular momentum

©1998 David Madison

Figure 11.10 Figure skater Todd Eldridge is demonstrating angular momentum conservation. When he pulls his arms toward his body, he spins faster.

This follows directly from Equation 11.13, which indicates that if

$$\sum \boldsymbol{\tau}_{\text{ext}} = \frac{d\mathbf{L}_{\text{tot}}}{dt} = 0 \qquad (11.17)$$

then

$$\mathbf{L}_{\text{tot}} = \text{constant} \qquad \text{or} \qquad \mathbf{L}_i = \mathbf{L}_f \qquad (11.18)$$

For an isolated system consisting of a number of particles, we write this conservation law as $\mathbf{L}_{\text{tot}} = \Sigma \mathbf{L}_n = \text{constant}$, where the index n denotes the nth particle in the system.

If the mass of an isolated rotating system undergoes redistribution in some way, the system's moment of inertia changes. Because the magnitude of the angular momentum of the system is $L = I\omega$ (Eq. 11.14), conservation of angular momentum requires that the product of I and ω must remain constant. Thus, a change in I for an isolated system requires a change in ω. In this case, we can express the principle of conservation of angular momentum as

$$I_i \omega_i = I_f \omega_f = \text{constant} \qquad (11.19)$$

This expression is valid both for rotation about a fixed axis and for rotation about an axis through the center of mass of a moving system as long as that axis remains fixed in direction. We require only that the net external torque be zero.

There are many examples that demonstrate conservation of angular momentum for a deformable system. You may have observed a figure skater spinning in the finale of a program (Fig. 11.10). The angular speed of the skater increases when the skater pulls his hands and feet close to his body, thereby decreasing I. Neglecting friction between skates and ice, no external torques act on the skater. Because the angular momentum of the skater is conserved, the product $I\omega$ remains constant, and a decrease in the moment of inertia of the skater causes an increase in the angular speed. Similarly, when divers or acrobats wish to make several somersaults, they pull their hands and feet close to their bodies to rotate at a higher rate, as in the opening photograph of this chapter. In these cases, the external force due to gravity acts through the center of mass and hence exerts no torque about this point. Therefore, the angular momentum about the center of mass must be conserved—that is, $I_i \omega_i = I_f \omega_f$. For example, when divers wish to double their angular speed, they must reduce their moment of inertia to half its initial value.

In Equation 11.18 we have a third conservation law to add to our list. We can now state that the energy, linear momentum, and angular momentum of an isolated system all remain constant:

$$\left. \begin{array}{l} E_i = E_f \\ \mathbf{p}_i = \mathbf{p}_f \\ \mathbf{L}_i = \mathbf{L}_f \end{array} \right\} \qquad \text{For an isolated system}$$

Quick Quiz 11.6 A competitive diver leaves the diving board and falls toward the water with her body straight and rotating slowly. She pulls her arms and legs into a tight tuck position. Her angular speed (a) increases (b) decreases (c) stays the same (d) is impossible to determine.

Quick Quiz 11.7 Consider the competitive diver in Quick Quiz 11.6 again. When she goes into the tuck position, the rotational kinetic energy of her body (a) increases (b) decreases (c) stays the same (d) is impossible to determine.

Example 11.7 Formation of a Neutron Star

A star rotates with a period of 30 days about an axis through its center. After the star undergoes a supernova explosion, the stellar core, which had a radius of 1.0×10^4 km, collapses into a neutron star of radius 3.0 km. Determine the period of rotation of the neutron star.

Solution The same physics that makes a skater spin faster with his arms pulled in describes the motion of the neutron star. Let us assume that during the collapse of the stellar core, (1) no external torque acts on it, (2) it remains spherical with the same relative mass distribution, and (3) its mass remains constant. Also, let us use the symbol T for the period, with T_i being the initial period of the star and T_f being the period of the neutron star. The period is the length of time a point on the star's equator takes to make one complete circle around the axis of rotation. The angular speed of the star is given by $\omega = 2\pi/T$. Therefore, Equation 11.19 gives

$$I_i \omega_i = I_f \omega_f$$

$$I_i \left(\frac{2\pi}{T_i} \right) = I_f \left(\frac{2\pi}{T_f} \right)$$

We don't know the mass distribution of the star, but we have assumed that the distribution is symmetric, so that the moment of inertia can be expressed as kMR^2, where k is some numerical constant. (From Table 10.2, for example, we see that $k = 2/5$ for a solid sphere and $k = 2/3$ for a spherical shell.) Thus, we can rewrite the preceding equation as

$$kMR_i^2 \left(\frac{2\pi}{T_i} \right) = kMR_f^2 \left(\frac{2\pi}{T_f} \right)$$

$$T_f = \left(\frac{R_f^2}{R_i^2} \right) T_i$$

Substituting numerical values gives

$$T_f = (30 \text{ days}) \left(\frac{3.0 \text{ km}}{1.0 \times 10^4 \text{ km}} \right)^2 = 2.7 \times 10^{-6} \text{ days}$$

$$= \boxed{0.23 \text{ s}}$$

Thus, the neutron star rotates about four times each second.

Example 11.8 The Merry-Go-Round

A horizontal platform in the shape of a circular disk rotates freely in a horizontal plane about a frictionless vertical axle (Fig. 11.11). The platform has a mass $M = 100$ kg and a radius $R = 2.0$ m. A student whose mass is $m = 60$ kg walks slowly from the rim of the disk toward its center. If the angular speed of the system is 2.0 rad/s when the student is at the rim, what is the angular speed when he reaches a point $r = 0.50$ m from the center?

Solution The speed change here is similar to the increase in angular speed of the spinning skater when he pulls his arms inward. Let us denote the moment of inertia of the

Figure 11.11 (Example 11.8) As the student walks toward the center of the rotating platform, the angular speed of the system increases because the angular momentum of the system remains constant.

platform as I_p and that of the student as I_s. Modeling the student as a particle, we can write the initial moment of inertia I_i of the system (student plus platform) about the axis of rotation:

$$I_i = I_{pi} + I_{si} = \tfrac{1}{2}MR^2 + mR^2$$

When the student walks to the position $r < R$, the moment of inertia of the system reduces to

$$I_f = I_{pf} + I_{sf} = \tfrac{1}{2}MR^2 + mr^2$$

Note that we still use the greater radius R when calculating I_{pf} because the radius of the platform does not change. Because no external torques act on the system about the axis of rotation, we can apply the law of conservation of angular momentum:

$$I_i \omega_i = I_f \omega_f$$

$$\left(\tfrac{1}{2}MR^2 + mR^2 \right) \omega_i = \left(\tfrac{1}{2}MR^2 + mr^2 \right) \omega_f$$

$$\omega_f = \left(\frac{\tfrac{1}{2}MR^2 + mR^2}{\tfrac{1}{2}MR^2 + mr^2} \right) \omega_i$$

$$= \left(\frac{\tfrac{1}{2}(100 \text{ kg})(2.0\,\text{m})^2 + (60 \text{ kg})(2.0\,\text{m})^2}{\tfrac{1}{2}(100 \text{ kg})(2.0\,\text{m})^2 + (60 \text{ kg})(0.50\,\text{m})^2} \right) (2.0 \text{ rad/s})$$

$$= \left(\frac{440 \text{ kg} \cdot \text{m}^2}{215 \text{ kg} \cdot \text{m}^2} \right) (2.0 \text{ rad/s}) = \boxed{4.1 \text{ rad/s}}$$

As expected, the angular speed increases.

What If? What if we were to measure the kinetic energy of the system before and after the student walks inward? Are they the same?

Answer You may be tempted to say yes because the system is isolated. But remember that energy comes in several forms, so we have to handle an energy question carefully. The initial kinetic energy is

$$K_i = \tfrac{1}{2}I_i\omega_i^2 = \tfrac{1}{2}(440 \text{ kg}\cdot\text{m}^2)(2.0 \text{ rad/s})^2 = 880 \text{ J}$$

The final kinetic energy is

$$K_f = \tfrac{1}{2}I_f\omega_f^2 = \tfrac{1}{2}(215 \text{ kg}\cdot\text{m}^2)(4.1 \text{ rad/s})^2 = 1.81 \times 10^3 \text{ J}$$

Thus, the kinetic energy of the system *increases*. The student must do work to move himself closer to the center of rotation, so this extra kinetic energy comes from chemical potential energy in the body of the student.

Example 11.9 The Spinning Bicycle Wheel

In a favorite classroom demonstration, a student holds the axle of a spinning bicycle wheel while seated on a stool that is free to rotate (Fig. 11.12). The student and stool are initially at rest while the wheel is spinning in a horizontal plane with an initial angular momentum $\mathbf{L}_i$ that points upward. When the wheel is inverted about its cen-

Figure 11.12 (Example 11.9) The wheel is initially spinning when the student is at rest. What happens when the wheel is inverted?

ter by 180°, the student and stool start rotating. In terms of $\mathbf{L}_i$, what are the magnitude and the direction of $\mathbf{L}$ for the student plus stool?

Solution The system consists of the student, the wheel, and the stool. Initially, the total angular momentum of the system $\mathbf{L}_i$ comes entirely from the spinning wheel. As the wheel is inverted, the student applies a torque to the wheel, but this torque is internal to the system. No external torque is acting on the system about the vertical axis. Therefore, the angular momentum of the system is conserved. Initially, we have

$$\mathbf{L}_{\text{system}} = \mathbf{L}_i = \mathbf{L}_{\text{wheel}} \qquad \text{(upward)}$$

After the wheel is inverted, we have $\mathbf{L}_{\text{inverted wheel}} = -\mathbf{L}_i$. For angular momentum to be conserved, some other part of the system has to start rotating so that the total final angular momentum equals the initial angular momentum $\mathbf{L}_i$. That other part of the system is the student plus the stool she is sitting on. So, we can now state that

$$\mathbf{L}_f = \mathbf{L}_i = \mathbf{L}_{\text{student}+\text{stool}} - \mathbf{L}_i$$

$$\boxed{\mathbf{L}_{\text{student}+\text{stool}} = 2\mathbf{L}_i}$$

Example 11.10 Disk and Stick   **Interactive**

A 2.0-kg disk traveling at 3.0 m/s strikes a 1.0-kg stick of length 4.0 m that is lying flat on nearly frictionless ice, as shown in Figure 11.13. Assume that the collision is elastic and that the disk does not deviate from its original line of motion. Find the translational speed of the disk, the translational speed of the stick, and the angular speed of the stick after the collision. The moment of inertia of the stick about its center of mass is 1.33 kg·m².

Solution Conceptualize the situation by considering Figure 11.13 and imagining what happens after the disk hits the

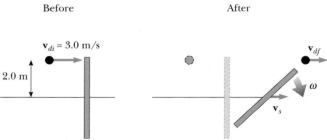

Figure 11.13 (Example 11.10) Overhead view of a disk striking a stick in an elastic collision, which causes the stick to rotate and move to the right.

stick. Because the disk and stick form an isolated system, we can assume that total energy, linear momentum, and angular momentum are all conserved. Thus, we can categorize this as a problem in which all three conservation laws might play a part. To analyze the problem, first note that we have three unknowns, and so we need three equations to solve simultaneously. The first comes from the law of the conservation of linear momentum:

$$p_i = p_f$$

$$m_d v_{di} = m_d v_{df} + m_s v_s$$

$$(2.0 \text{ kg})(3.0 \text{ m/s}) = (2.0 \text{ kg})v_{df} + (1.0 \text{ kg})v_s$$

(1) $$6.0 \text{ kg}\cdot\text{m/s} - (2.0 \text{ kg})v_{df} = (1.0 \text{ kg})v_s$$

Now we apply the law of conservation of angular momentum, using the initial position of the center of the stick as our reference point. We know that the component of angular momentum of the disk along the axis perpendicular to the plane of the ice is negative. (The right-hand rule shows that $\mathbf{L}_d$ points into the ice.) Applying conservation of angular momentum to the system gives

$$L_i = L_f$$

$$-rm_d v_{di} = -rm_d v_{df} + I\omega$$

$$- (2.0 \text{ m})(2.0 \text{ kg})(3.0 \text{ m/s}) = - (2.0 \text{ m})(2.0 \text{ kg})v_{df}$$
$$+ (1.33 \text{ kg} \cdot \text{m}^2)\omega$$
$$- 12 \text{ kg} \cdot \text{m}^2/\text{s} = - (4.0 \text{ kg} \cdot \text{m})v_{df}$$
$$+ (1.33 \text{ kg} \cdot \text{m}^2)\omega$$

$$(2) \quad - 9.0 \text{ rad/s} + (3.0 \text{ rad/m})v_{df} = \omega$$

We use the fact that radians are dimensionless to ensure consistent units for each term.

Finally, the elastic nature of the collision tells us that kinetic energy is conserved; in this case, the kinetic energy consists of translational and rotational forms:

$$K_i = K_f$$

$$\tfrac{1}{2}m_d v_{di}^2 = \tfrac{1}{2}m_d v_{df}^2 + \tfrac{1}{2}m_s v_s^2 + \tfrac{1}{2}I\omega^2$$

$$\tfrac{1}{2}(2.0 \text{ kg})(3.0 \text{ m/s})^2 = \tfrac{1}{2}(2.0 \text{ kg})v_{df}^2 + \tfrac{1}{2}(1.0 \text{ kg})v_s^2$$
$$+ \tfrac{1}{2}(1.33 \text{ kg} \cdot \text{m}^2)\omega^2$$

$$(3) \quad 18 \text{ m}^2/\text{s}^2 = 2.0\, v_{df}^2 + v_s^2 + (1.33 \text{ m}^2)\omega^2$$

In solving Equations (1), (2), and (3) simultaneously, we find that $v_{df} = $ 2.3 m/s, $v_s = $ 1.3 m/s, and $\omega = $ -2.0 rad/s.

To finalize the problem, note that these values seem reasonable. The disk is moving more slowly after the collision than it was before the collision, and the stick has a small translational speed. Table 11.1 summarizes the initial and final values of variables for the disk and the stick, and verifies the conservation of linear momentum, angular momentum, and kinetic energy.

What If? What if the collision between the disk and the stick is perfectly inelastic? How does this change the analysis?

Answer In this case, the disk adheres to the end of the stick upon collision. The conservation of linear momentum principle leading to Equation (1) would be altered to

$$p_i = p_f$$
$$m_d v_{di} = (m_d + m_s)v_{CM}$$
$$(2.0 \text{ kg})(3.0 \text{ m/s}) = (2.0 \text{ kg} + 1.0 \text{ kg})v_{CM}$$
$$v_{CM} = 2.0 \text{ m/s}$$

For the rotational part of this question, we need to find the center of mass of the system of the disk and the stick. Choosing the center of the stick as the origin, the y position of the center of mass along the vertical stick is

$$y_{CM} = \frac{(2.0 \text{ kg})(2.0 \text{ m}) + (1.0 \text{ kg})(0)}{(2.0 \text{ kg} + 1.0 \text{ kg})} = 1.33 \text{ m}$$

Thus, the center of mass of the system is 2.0 m − 1.33 m = 0.67 m from the upper end of the stick.

The conservation of angular momentum principle leading to Equation (2) would be altered to the following, evaluating angular momenta around the center of mass of the system:

$$L_i = L_f$$
$$- rm_d v_{di} = I_d\omega + I_s\omega$$
$$(4) \quad - (0.67 \text{ m})m_d v_{di} = [m_d(0.67 \text{ m})^2]\omega + I_s\omega$$

The moment of inertia of the stick around the center of mass *of the system* is found from the parallel-axis theorem:

$$I_s = I_{CM} + MD^2$$
$$= 1.33 \text{ kg} \cdot \text{m}^2 + (1.0 \text{ kg})(1.33 \text{ m})^2 = 3.1 \text{ kg} \cdot \text{m}^2$$

Thus, Equation (4) becomes

$$- (0.67 \text{ m})(2.0 \text{ kg})(3.0 \text{ m/s}) = [(2.0 \text{ kg})(0.67 \text{ m})^2]\omega$$
$$+ (3.1 \text{ kg} \cdot \text{m}^2)\omega$$

$$- 4.0 \text{ kg} \cdot \text{m}^2/\text{s} = (4.0 \text{ kg} \cdot \text{m}^2)\omega$$

$$\omega = \frac{- 4.0 \text{ kg} \cdot \text{m}^2/\text{s}}{4.0 \text{ kg} \cdot \text{m}^2} = - 1.0 \text{ rad/s}$$

Evaluating the total kinetic energy of the system after the collision shows that it is less than that before the collision because kinetic energy is not conserved in an inelastic collision.

Table 11.1

Comparison of Values in Example 11.10 Before and After the Collision[a]

	v (m/s)	ω (rad/s)	p (kg·m/s)	L (kg·m²/s)	K_{trans} (J)	K_{rot} (J)
Before						
Disk	3.0	—	6.0	− 12	9.0	—
Stick	0	0	0	0	0	0
Total for System	—	—	6.0	− 12	9.0	0
After						
Disk	2.3	—	4.7	− 9.3	5.4	—
Stick	1.3	− 2.0	1.3	− 2.7	0.9	2.7
Total for System	—	—	6.0	− 12	6.3	2.7

[a] Notice that linear momentum, angular momentum, and total kinetic energy are conserved.

 At the Interactive Worked Example link at **http://www.pse6.com,** *you can adjust the speed and position of the disk and observe the collision.*

11.5 The Motion of Gyroscopes and Tops

A very unusual and fascinating type of motion you probably have observed is that of a top spinning about its axis of symmetry, as shown in Figure 11.14a. If the top spins very rapidly, the symmetry axis rotates about the z axis, sweeping out a cone (see Fig. 11.14b). The motion of the symmetry axis about the vertical—known as **precessional motion**—is usually slow relative to the spinning motion of the top.

It is quite natural to wonder why the top does not fall over. Because the center of mass is not directly above the pivot point O, a net torque is clearly acting on the top about O—a torque resulting from the gravitational force $M\mathbf{g}$. The top would certainly fall over if it were not spinning. Because it is spinning, however, it has an angular momentum $\mathbf{L}$ directed along its symmetry axis. We shall show that this symmetry axis moves about the z axis (precessional motion occurs) because the torque produces a change in the *direction* of the symmetry axis. This is an excellent example of the importance of the directional nature of angular momentum.

The essential features of precessional motion can be illustrated by considering the simple gyroscope shown in Figure 11.15a. The two forces acting on the top are the downward gravitational force $M\mathbf{g}$ and the normal force $\mathbf{n}$ acting upward at the pivot point O. The normal force produces no torque about the pivot because its moment arm through that point is zero. However, the gravitational force produces a torque $\boldsymbol{\tau} = \mathbf{r} \times M\mathbf{g}$ about O, where the direction of $\boldsymbol{\tau}$ is perpendicular to the plane formed by $\mathbf{r}$ and $M\mathbf{g}$. By necessity, the vector $\boldsymbol{\tau}$ lies in a horizontal xy plane perpendicular to the angular momentum vector. The net torque and angular momentum of the gyroscope are related through Equation 11.13:

$$\boldsymbol{\tau} = \frac{d\mathbf{L}}{dt}$$

From this expression, we see that the nonzero torque produces a change in angular momentum $d\mathbf{L}$—a change that is in the same direction as $\boldsymbol{\tau}$. Therefore, like the torque vector, $d\mathbf{L}$ must also be perpendicular to $\mathbf{L}$. Figure 11.15b illustrates the resulting precessional motion of the symmetry axis of the gyroscope. In a time interval dt, the change in angular momentum is $d\mathbf{L} = \mathbf{L}_f - \mathbf{L}_i = \boldsymbol{\tau}\, dt$. Because $d\mathbf{L}$ is perpendicular to $\mathbf{L}$, the magnitude of $\mathbf{L}$ does not change ($|\mathbf{L}_i| = |\mathbf{L}_f|$). Rather, what is changing is the *direction* of $\mathbf{L}$. Because the change in angular momentum $d\mathbf{L}$ is in the direction of $\boldsymbol{\tau}$, which lies in the xy plane, the gyroscope undergoes precessional motion.

Precessional motion

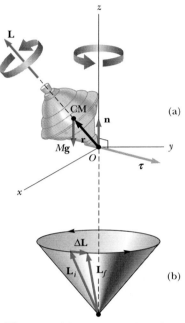

Figure 11.14 Precessional motion of a top spinning about its symmetry axis. (a) The only external forces acting on the top are the normal force $\mathbf{n}$ and the gravitational force $M\mathbf{g}$. The direction of the angular momentum $\mathbf{L}$ is along the axis of symmetry. The right-hand rule indicates that $\boldsymbol{\tau} = \mathbf{r} \times \mathbf{F} = \mathbf{r} \times M\mathbf{g}$ is in the xy plane. (b) The direction of $\Delta\mathbf{L}$ is parallel to that of $\boldsymbol{\tau}$ in part (a). The fact that $\mathbf{L}_f = \Delta\mathbf{L} + \mathbf{L}_i$ indicates that the top precesses about the z axis.

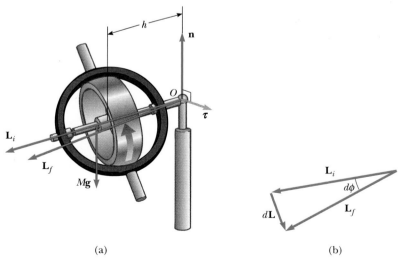

Figure 11.15 (a) The motion of a simple gyroscope pivoted a distance h from its center of mass. The gravitational force $M\mathbf{g}$ produces a torque about the pivot, and this torque is perpendicular to the axle. (b) Overhead view of the initial and final angular momentum vectors. The torque results in a change in angular momentum $d\mathbf{L}$ in a direction perpendicular to the axle. The axle sweeps out an angle $d\phi$ in a time interval dt.

To simplify the description of the system, we must make an assumption: The total angular momentum of the precessing wheel is the sum of the angular momentum $I\omega$ due to the spinning and the angular momentum due to the motion of the center of mass about the pivot. In our treatment, we shall neglect the contribution from the center-of-mass motion and take the total angular momentum to be just $I\omega$. In practice, this is a good approximation if ω is made very large.

The vector diagram in Figure 11.15b shows that in the time interval dt, the angular momentum vector rotates through an angle $d\phi$, which is also the angle through which the axle rotates. From the vector triangle formed by the vectors $\mathbf{L}_i$, $\mathbf{L}_f$, and $d\mathbf{L}$, we see that

$$\sin(d\phi) \approx d\phi = \frac{dL}{L} = \frac{\tau\, dt}{L} = \frac{(Mgh)\, dt}{L}$$

where we have used the fact that, for small values of any angle θ, $\sin\theta \approx \theta$. Dividing through by dt and using the relationship $L = I\omega$, we find that the rate at which the axle rotates about the vertical axis is

$$\omega_p = \frac{d\phi}{dt} = \frac{Mgh}{I\omega} \tag{11.20}$$

The angular speed ω_p is called the **precessional frequency.** This result is valid only when $\omega_p \ll \omega$. Otherwise, a much more complicated motion is involved. As you can see from Equation 11.20, the condition $\omega_p \ll \omega$ is met when ω is large, that is, the wheel spins rapidly. Furthermore, note that the precessional frequency decreases as ω increases—that is, as the wheel spins faster about its axis of symmetry.

As an example of the usefulness of gyroscopes, suppose you are in a spacecraft in deep space and you need to alter your trajectory. You need to turn the spacecraft around in order to fire the engines in the correct direction. But how do you turn a spacecraft around in empty space? One way is to have small rocket engines that fire perpendicularly out the side of the spacecraft, providing a torque around its center of mass. This is desirable, and many spacecraft have such rockets.

Let us consider another method, however, that is related to angular momentum and does not require the consumption of rocket fuel. Suppose that the spacecraft carries a gyroscope that is not rotating, as in Figure 11.16a. In this case, the angular momentum of the spacecraft about its center of mass is zero. Suppose the gyroscope is set into rotation, giving the gyroscope a nonzero angular momentum. There is no external torque on the isolated system (spacecraft + gyroscope), so the angular momentum of this system must remain zero according to the principle of conservation of angular momentum. This principle can be satisfied if the spacecraft rotates in the direction opposite to that of the gyroscope, so that the angular momentum vectors of the gyroscope and the spacecraft cancel, resulting in no angular momentum of the system. The result of rotating the gyroscope, as in Figure 11.16b, is that the spacecraft turns around! By including three gyroscopes with mutually perpendicular axles, any desired rotation in space can be achieved.

This effect created an undesirable situation with the *Voyager 2* spacecraft during its flight. The spacecraft carried a tape recorder whose reels rotated at high speeds. Each time the tape recorder was turned on, the reels acted as gyroscopes, and the spacecraft started an undesirable rotation in the opposite direction. This had to be counteracted by Mission Control by using the sideward-firing jets to stop the rotation!

(a)

Spacecraft rotates clockwise

Gyroscope rotates counterclockwise

(b)

Figure 11.16 (a) A spacecraft carries a gyroscope that is not spinning. (b) When the gyroscope is set into rotation, the spacecraft turns the other way so that the angular momentum of the system is conserved.

11.6 Angular Momentum as a Fundamental Quantity

We have seen that the concept of angular momentum is very useful for describing the motion of macroscopic systems. However, the concept also is valid on a submicroscopic scale and has been used extensively in the development of modern theories of atomic, molecular, and nuclear physics. In these developments, it has been found that the angular momentum of a system is a fundamental quantity. The word *fundamental* in this context implies that angular momentum is an intrinsic property of atoms, molecules, and their constituents, a property that is a part of their very nature.

To explain the results of a variety of experiments on atomic and molecular systems, we rely on the fact that the angular momentum has discrete values. These discrete values are multiples of the fundamental unit of angular momentum $\hbar = h/2\pi$, where h is called Planck's constant:

$$\text{Fundamental unit of angular momentum} = \hbar = 1.054 \times 10^{-34} \text{ kg} \cdot \text{m}^2/\text{s}$$

Let us accept this postulate without proof for the time being and show how it can be used to estimate the angular speed of a diatomic molecule. Consider the O_2 molecule as a rigid rotor, that is, two atoms separated by a fixed distance d and rotating about the center of mass (Fig. 11.17). Equating the angular momentum to the fundamental unit $\hbar$, we can find the order of magnitude of the lowest angular speed:

$$I_{CM}\omega \approx \hbar \qquad \text{or} \qquad \omega \approx \frac{\hbar}{I_{CM}}$$

In Example 10.3, we found that the moment of inertia of the O_2 molecule about this axis of rotation is 1.95×10^{-46} kg $\cdot$ m^2. Therefore,

$$\omega \approx \frac{\hbar}{I_{CM}} = \frac{1.054 \times 10^{-34} \text{ kg} \cdot \text{m}^2/\text{s}}{1.95 \times 10^{-46} \text{ kg} \cdot \text{m}^2} \sim 10^{12} \text{ rad/s}$$

Actual angular speeds are found to be multiples of a number with this order of magnitude.

This simple example shows that certain classical concepts and models, when properly modified, are useful in describing some features of atomic and molecular systems. A wide variety of phenomena on the submicroscopic scale can be explained only if we assume discrete values of the angular momentum associated with a particular type of motion.

The Danish physicist Niels Bohr (1885–1962) accepted and adopted this radical idea of discrete angular momentum values in developing his theory of the hydrogen atom. Strictly classical models were unsuccessful in describing many of the hydrogen atom's properties. Bohr postulated that the electron could occupy only those circular orbits about the proton for which the orbital angular momentum was equal to $n\hbar$, where n is an integer. That is, he made the bold claim that orbital angular momentum is quantized. One can use this simple model to estimate the rotational frequencies of the electron in the various orbits (see Problem 42).

Figure 11.17 The rigid-rotor model of a diatomic molecule. The rotation occurs about the center of mass in the plane of the page.

Take a practice test for this chapter by clicking on the Practice Test link at http://www.pse6.com.

SUMMARY

The **torque** $\boldsymbol{\tau}$ due to a force $\mathbf{F}$ about an origin in an inertial frame is defined to be

$$\boldsymbol{\tau} \equiv \mathbf{r} \times \mathbf{F} \tag{11.1}$$

Given two vectors $\mathbf{A}$ and $\mathbf{B}$, the **cross product** $\mathbf{A} \times \mathbf{B}$ is a vector $\mathbf{C}$ having a magnitude

$$C \equiv AB \sin \theta \tag{11.3}$$

where θ is the angle between $\mathbf{A}$ and $\mathbf{B}$. The direction of the vector $\mathbf{C} = \mathbf{A} \times \mathbf{B}$ is perpendicular to the plane formed by $\mathbf{A}$ and $\mathbf{B}$, and this direction is determined by the right-hand rule.

The **angular momentum** $\mathbf{L}$ of a particle having linear momentum $\mathbf{p} = m\mathbf{v}$ is

$$\mathbf{L} \equiv \mathbf{r} \times \mathbf{p} \tag{11.10}$$

where $\mathbf{r}$ is the vector position of the particle relative to an origin in an inertial frame.

The **net external torque** acting on a system is equal to the time rate of change of its angular momentum:

$$\sum \tau_{\text{ext}} = \frac{d\mathbf{L}_{\text{tot}}}{dt} \tag{11.13}$$

The z component of **angular momentum** of a rigid object rotating about a fixed z axis is

$$L_z = I\omega \tag{11.14}$$

where I is the moment of inertia of the object about the axis of rotation and ω is its angular speed.

The **net external torque** acting on a rigid object equals the product of its moment of inertia about the axis of rotation and its angular acceleration:

$$\sum \tau_{\text{ext}} = I\alpha \tag{11.16}$$

If the net external torque acting on a system is zero, then the total angular momentum of the system is constant:

$$\mathbf{L}_i = \mathbf{L}_f \tag{11.18}$$

Applying this **law of conservation of angular momentum** to a system whose moment of inertia changes gives

$$I_i\omega_i = I_f\omega_f = \text{constant} \tag{11.19}$$

QUESTIONS

1. Is it possible to calculate the torque acting on a rigid object without specifying an axis of rotation? Is the torque independent of the location of the axis of rotation?

2. Is the triple product defined by $\mathbf{A} \cdot (\mathbf{B} \times \mathbf{C})$ a scalar or a vector quantity? Explain why the operation $(\mathbf{A} \cdot \mathbf{B}) \times \mathbf{C}$ has no meaning.

3. Vector $\mathbf{A}$ is in the negative y direction, and vector $\mathbf{B}$ is in the negative x direction. What are the directions of (a) $\mathbf{A} \times \mathbf{B}$ (b) $\mathbf{B} \times \mathbf{A}$?

4. If a single force acts on an object and the torque caused by the force is nonzero about some point, is there any other point about which the torque is zero?

5. Suppose that the vector velocity of a particle is completely specified. What can you conclude about the direction of its angular momentum vector with respect to the direction of motion?

6. If a system of particles is in motion, is it possible for the total angular momentum to be zero about some origin? Explain.

7. If the torque acting on a particle about a certain origin is zero, what can you say about its angular momentum about that origin?

8. A ball is thrown in such a way that it does not spin about its own axis. Does this mean that the angular momentum is zero about an arbitrary origin? Explain.

9. For a helicopter to be stable as it flies, it must have at least two propellers. Why?

10. A particle is moving in a circle with constant speed. Locate one point about which the particle's angular momentum is constant and another point about which it changes in time.

11. Why does a long pole help a tightrope walker stay balanced?

12. Often when a high diver wants to turn a flip in midair, she draws her legs up against her chest. Why does this make her rotate faster? What should she do when she wants to come out of her flip?

13. In some motorcycle races, the riders drive over small hills, and the motorcycle becomes airborne for a short time. If the motorcycle racer keeps the throttle open while leaving the hill and going into the air, the motorcycle tends to nose upward. Why does this happen?

14. Stars originate as large bodies of slowly rotating gas. Because of gravitation, these clumps of gas slowly decrease in size. What happens to the angular speed of a star as it shrinks? Explain.

15. If global warming occurs over the next century, it is likely that some polar ice will melt and the water will be distributed closer to the Equator. How would this change the moment of inertia of the Earth? Would the length of the day (one revolution) increase or decrease?

16. A mouse is initially at rest on a horizontal turntable mounted on a frictionless vertical axle. If the mouse

Gerard Lacz/NHPA

Figure Q11.17

begins to walk clockwise around the perimeter, what happens to the turntable? Explain.

17. A cat usually lands on its feet regardless of the position from which it is dropped. A slow-motion film of a cat falling shows that the upper half of its body twists in one direction while the lower half twists in the opposite direction. (See Figure Q11.17.) Why does this type of rotation occur?

18. As the cord holding a tether ball winds around a thin pole, what happens to the angular speed of the ball? Explain.

19. If you toss a textbook into the air, rotating it each time about one of the three axes perpendicular to the textbook, you will find that it will not rotate smoothly about one of these axis. (Try placing a strong rubber band around the book before the toss so it will stay closed.) Its rotation is stable about those axes having the largest and smallest moment of inertia but unstable about the axis of intermediate moment. Try this on your own to find the axis that has this intermediate moment.

20. A scientist arriving at a hotel asks a bellhop to carry a heavy suitcase. When the bellhop rounds a corner, the suitcase suddenly swings away from him for some unknown reason. The alarmed bellhop drops the suitcase and runs away. What might be in the suitcase?

PROBLEMS

1, 2, 3 = straightforward, intermediate, challenging ☐ = full solution available in the *Student Solutions Manual and Study Guide*

= coached solution with hints available at http://www.pse6.com   = computer useful in solving problem

= paired numerical and symbolic problems

Section 11.1   The Vector Product and Torque

1. Given $\mathbf{M} = 6\hat{\mathbf{i}} + 2\hat{\mathbf{j}} - \hat{\mathbf{k}}$ and $\mathbf{N} = 2\hat{\mathbf{i}} - \hat{\mathbf{j}} - 3\hat{\mathbf{k}}$, calculate the vector product $\mathbf{M} \times \mathbf{N}$.

2. The vectors 42.0 cm at 15.0° and 23.0 cm at 65.0° both start from the origin. Both angles are measured counterclockwise from the *x* axis. The vectors form two sides of a parallelogram. (a) Find the area of the parallelogram. (b) Find the length of its longer diagonal.

3. Two vectors are given by $\mathbf{A} = -3\hat{\mathbf{i}} + 4\hat{\mathbf{j}}$ and $\mathbf{B} = 2\hat{\mathbf{i}} + 3\hat{\mathbf{j}}$. Find (a) $\mathbf{A} \times \mathbf{B}$ and (b) the angle between $\mathbf{A}$ and $\mathbf{B}$.

4. Two vectors are given by $\mathbf{A} = -3\hat{\mathbf{i}} + 7\hat{\mathbf{j}} - 4\hat{\mathbf{k}}$ and $\mathbf{B} = 6\hat{\mathbf{i}} - 10\hat{\mathbf{j}} + 9\hat{\mathbf{k}}$. Evaluate the quantities (a) $\cos^{-1}[\mathbf{A}\cdot\mathbf{B}/AB]$ and (b) $\sin^{-1}[|\mathbf{A} \times \mathbf{B}|/AB]$. (c) Which give(s) the angle between the vectors?

5. The wind exerts on a flower the force 0.785 N horizontally to the east. The stem of the flower is 0.450 m long and tilts toward the east, making an angle of 14.0° with the vertical. Find the vector torque of the wind force about the base of the stem.

6. A student claims that she has found a vector $\mathbf{A}$ such that $(2\hat{\mathbf{i}} - 3\hat{\mathbf{j}} + 4\hat{\mathbf{k}}) \times \mathbf{A} = (4\hat{\mathbf{i}} + 3\hat{\mathbf{j}} - \hat{\mathbf{k}})$. Do you believe this claim? Explain.

7. If $|\mathbf{A} \times \mathbf{B}| = \mathbf{A}\cdot\mathbf{B}$, what is the angle between $\mathbf{A}$ and $\mathbf{B}$?

8. A particle is located at the vector position $\mathbf{r} = (\hat{\mathbf{i}} + 3\hat{\mathbf{j}})$ m, and the force acting on it is $\mathbf{F} = (3\hat{\mathbf{i}} + 2\hat{\mathbf{j}})$ N. What is the torque about (a) the origin and (b) the point having coordinates (0, 6) m?

9. Two forces $\mathbf{F}_1$ and $\mathbf{F}_2$ act along the two sides of an equilateral triangle as shown in Figure P11.9. Point *O* is the intersection of the altitudes of the triangle. Find a third force $\mathbf{F}_3$ to be applied at *B* and along *BC* that will make the total torque

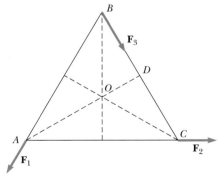

Figure P11.9

zero about the point *O*. **What If?** Will the total torque change if $\mathbf{F}_3$ is applied not at *B* but at any other point along *BC*?

10. Use the definition of the vector product and the definitions of the unit vectors $\hat{\mathbf{i}}$, $\hat{\mathbf{j}}$, and $\hat{\mathbf{k}}$ to prove Equations 11.7. You may assume that the *x* axis points to the right, the *y* axis up, and the *z* axis toward you (not away from you). This choice is said to make the coordinate system *right-handed*.

Section 11.2   Angular Momentum

11. A light rigid rod 1.00 m in length joins two particles, with masses 4.00 kg and 3.00 kg, at its ends. The combination rotates in the *xy* plane about a pivot through the center of the rod (Fig. P11.11). Determine the angular momentum of the system about the origin when the speed of each particle is 5.00 m/s.

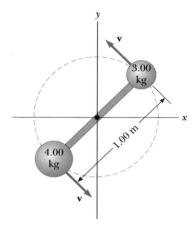

Figure P11.11

12. A 1.50-kg particle moves in the *xy* plane with a velocity of $\mathbf{v} = (4.20\hat{\mathbf{i}} - 3.60\hat{\mathbf{j}})$ m/s. Determine the angular momentum of the particle when its position vector is $\mathbf{r} = (1.50\hat{\mathbf{i}} + 2.20\hat{\mathbf{j}})$ m.

13. The position vector of a particle of mass 2.00 kg is given as a function of time by $\mathbf{r} = (6.00\hat{\mathbf{i}} + 5.00t\hat{\mathbf{j}})$ m. Determine the angular momentum of the particle about the origin, as a function of time.

14. A conical pendulum consists of a bob of mass *m* in motion in a circular path in a horizontal plane as shown in Figure

Figure P11.14

P11.14. During the motion, the supporting wire of length ℓ maintains the constant angle θ with the vertical. Show that the magnitude of the angular momentum of the bob about the center of the circle is

$$L = \left(\frac{m^2 g \, \ell^3 \sin^4 \theta}{\cos \theta} \right)^{1/2}$$

15. A particle of mass *m* moves in a circle of radius *R* at a constant speed *v*, as shown in Figure P11.15. If the motion begins at point *Q* at time $t = 0$, determine the angular momentum of the particle about point *P* as a function of time.

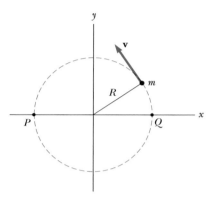

Figure P11.15

16. A 4.00-kg counterweight is attached to a light cord, which is wound around a spool (refer to Fig. 10.20). The spool is a uniform solid cylinder of radius 8.00 cm and mass 2.00 kg. (a) What is the net torque on the system about the point *O*? (b) When the counterweight has a speed *v*, the pulley has an angular speed $\omega = v/R$. Determine the total angular momentum of the system about *O*. (c) Using the fact that $\tau = d\mathbf{L}/dt$ and your result from (b), calculate the acceleration of the counterweight.

17. A particle of mass *m* is shot with an initial velocity $\mathbf{v}_i$ making an angle θ with the horizontal as shown in Figure P11.17. The particle moves in the gravitational field of the Earth. Find the angular momentum of the particle about the origin when the particle is (a) at the origin, (b) at the highest point of its trajectory, and (c) just before it hits the ground. (d) What torque causes its angular momentum to change?

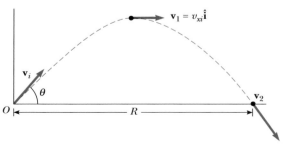

Figure P11.17

18. Heading straight toward the summit of Pike's Peak, an airplane of mass 12 000 kg flies over the plains of Kansas at nearly constant altitude 4.30 km, with a constant velocity of 175 m/s west. (a) What is the airplane's vector angular momentum relative to a wheat farmer on the ground directly below the airplane? (b) Does this value change as the airplane continues its motion along a straight line? (c) **What If?** What is its angular momentum relative to the summit of Pike's Peak?

19. A ball having mass m is fastened at the end of a flagpole that is connected to the side of a tall building at point P shown in Figure P11.19. The length of the flagpole is ℓ and it makes an angle θ with the horizontal. If the ball becomes loose and starts to fall, determine the angular momentum (as a function of time) of the ball about point P. Neglect air resistance.

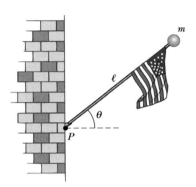

Figure P11.19

20. A fireman clings to a vertical ladder and directs the nozzle of a hose horizontally toward a burning building. The rate of water flow is 6.31 kg/s, and the nozzle speed is 12.5 m/s. The hose passes vertically between the fireman's feet, which are 1.30 m below the nozzle. Choose the origin to be inside the hose between the fireman's feet. What torque must the fireman exert on the hose? That is, what is the rate of change of the angular momentum of the water?

Section 11.3 Angular Momentum of a Rotating Rigid Object

21. Show that the kinetic energy of an object rotating about a fixed axis with angular momentum $L = I\omega$ can be written as $K = L^2/2I$.

22. A uniform solid sphere of radius 0.500 m and mass 15.0 kg turns counterclockwise about a vertical axis through its center. Find its vector angular momentum when its angular speed is 3.00 rad/s.

23. A uniform solid disk of mass 3.00 kg and radius 0.200 m rotates about a fixed axis perpendicular to its face. If the angular frequency of rotation is 6.00 rad/s, calculate the angular momentum of the disk when the axis of rotation (a) passes through its center of mass and (b) passes through a point midway between the center and the rim.

24. Big Ben (Figure P10.40), the Parliament Building tower clock in London, has hour and minute hands with lengths of 2.70 m and 4.50 m and masses of 60.0 kg and 100 kg, respectively. Calculate the total angular momentum of these hands about the center point. Treat the hands as long, thin uniform rods.

25. A particle of mass 0.400 kg is attached to the 100-cm mark of a meter stick of mass 0.100 kg. The meter stick rotates on a horizontal, frictionless table with an angular speed of 4.00 rad/s. Calculate the angular momentum of the system when the stick is pivoted about an axis (a) perpendicular to the table through the 50.0-cm mark and (b) perpendicular to the table through the 0-cm mark.

26. The distance between the centers of the wheels of a motorcycle is 155 cm. The center of mass of the motorcycle, including the biker, is 88.0 cm above the ground and halfway between the wheels. Assume the mass of each wheel is small compared to the body of the motorcycle. The engine drives the rear wheel only. What horizontal acceleration of the motorcycle will make the front wheel rise off the ground?

27. A space station is constructed in the shape of a hollow ring of mass 5.00×10^4 kg. Members of the crew walk on a deck formed by the inner surface of the outer cylindrical wall of the ring, with radius 100 m. At rest when constructed, the ring is set rotating about its axis so that the people inside experience an effective free-fall acceleration equal to g. (Figure P11.27 shows the ring together with some other parts that make a negligible contribution to the total moment of inertia.) The rotation is achieved by firing two small rockets attached tangentially to opposite points on the outside of the ring. (a) What angular momentum does the space station acquire? (b) How long must the rockets be fired if each exerts a thrust of 125 N? (c) Prove that the total torque on the ring, multiplied by the time interval found in part (b), is equal to the change in angular momentum, found in part (a). This equality represents the *angular impulse–angular momentum theorem*.

Figure P11.27 Problems 27 and 36.

Section 11.4 Conservation of Angular Momentum

28. A cylinder with moment of inertia I_1 rotates about a vertical, frictionless axle with angular speed ω_i. A second cylinder, this one having moment of inertia I_2 and initially not rotating, drops onto the first cylinder (Fig. P11.28). Because of friction between the surfaces, the two eventually reach the same angular speed ω_f. (a) Calculate ω_f. (b) Show that the kinetic energy of the system decreases in this interaction, and calculate the ratio of the final to the initial rotational energy.

Before   After

Figure P11.28

29. A playground merry-go-round of radius $R = 2.00$ m has a moment of inertia $I = 250$ kg·m² and is rotating at 10.0 rev/min about a frictionless vertical axle. Facing the axle, a 25.0-kg child hops onto the merry-go-round and manages to sit down on the edge. What is the new angular speed of the merry-go-round?

30. A student sits on a freely rotating stool holding two weights, each of mass 3.00 kg (Figure P11.30). When his arms are extended horizontally, the weights are 1.00 m from the axis of rotation and he rotates with an angular speed of 0.750 rad/s. The moment of inertia of the student plus stool is 3.00 kg·m² and is assumed to be constant. The student pulls the weights inward horizontally to a position 0.300 m from the rotation axis. (a) Find the new angular speed of the student. (b) Find the kinetic energy of the rotating system before and after he pulls the weights inward.

31. A uniform rod of mass 100 g and length 50.0 cm rotates in a horizontal plane about a fixed, vertical, frictionless pin through its center. Two small beads, each of mass 30.0 g, are mounted on the rod so that they are able to slide without friction along its length. Initially the beads are held by catches at positions 10.0 cm on each side of center, at which time the system rotates at an angular speed of 20.0 rad/s. Suddenly, the catches are released and the small beads slide outward along the rod. (a) Find the angular speed of the system at the instant the beads reach the ends of the rod. (b) **What if** the beads fly off the ends? What is the angular speed of the rod after this occurs?

32. An umbrella consists of a circle of cloth, a thin rod with the handle at one end and the center of the cloth at the other end, and several straight uniform ribs hinged to the top end of the rod and holding the cloth taut. With the ribs perpendicular to the rod, the umbrella is set rotating about the rod with an angular speed of 1.25 rad/s. The cloth is so light and the rod is so thin that they make negligible contributions to the moment of inertia, in comparison with the ribs. The spinning umbrella is balanced on its handle and keeps rotating without friction. Suddenly its latch breaks and the umbrella partly folds up, until each rib makes an angle of 22.5° with the rod. What is the final angular speed of the umbrella?

33. A 60.0-kg woman stands at the rim of a horizontal turntable having a moment of inertia of 500 kg·m² and a radius of 2.00 m. The turntable is initially at rest and is free to rotate about a frictionless, vertical axle through its center. The woman then starts walking around the rim clockwise (as viewed from above the system) at a constant speed of 1.50 m/s relative to the Earth. (a) In what direction and with what angular speed does the turntable rotate? (b) How much work does the woman do to set herself and the turntable into motion?

34. A puck of mass 80.0 g and radius 4.00 cm slides along an air table at a speed of 1.50 m/s as shown in Figure P11.34a. It makes a glancing collision with a second puck of radius 6.00 cm and mass 120 g (initially at rest) such that their rims just touch. Because their rims are coated with instant-acting glue, the pucks stick together and spin after the collision (Fig. P11.34b). (a) What is the angular momentum of the system relative to the center of mass? (b) What is the angular speed about the center of mass?

(a)   (b)

Figure P11.30

1.50 m/s

(a)   (b)

Figure P11.34

35. A wooden block of mass M resting on a frictionless horizontal surface is attached to a rigid rod of length ℓ and of negligible mass (Fig. P11.35). The rod is pivoted at the other end. A bullet of mass m traveling parallel to the horizontal surface and perpendicular to the rod with speed v hits the block and becomes embedded in it. (a) What is the angular momentum of the bullet–block system? (b) What fraction of the original kinetic energy is lost in the collision?

Figure P11.35

36. A space station shaped like a giant wheel has a radius of 100 m and a moment of inertia of 5.00×10^8 kg·m². A crew of 150 is living on the rim, and the station's rotation causes the crew to experience an apparent free-fall acceleration of g (Fig. P11.27). When 100 people move to the center of the station for a union meeting, the angular speed changes. What apparent free-fall acceleration is experienced by the managers remaining at the rim? Assume that the average mass for each inhabitant is 65.0 kg.

37. A wad of sticky clay with mass m and velocity $\mathbf{v}_i$ is fired at a solid cylinder of mass M and radius R (Figure P11.37). The cylinder is initially at rest and is mounted on a fixed horizontal axle that runs through its center of mass. The line of motion of the projectile is perpendicular to the axle and at a distance $d < R$ from the center. (a) Find the angular speed of the system just after the clay strikes and sticks to the surface of the cylinder. (b) Is mechanical energy of the clay–cylinder system conserved in this process? Explain your answer.

Figure P11.37

38. A thin uniform rectangular sign hangs vertically above the door of a shop. The sign is hinged to a stationary horizontal rod along its top edge. The mass of the sign is 2.40 kg and its vertical dimension is 50.0 cm. The sign is swinging without friction, becoming a tempting target for children armed with snowballs. The maximum angular displacement of the sign is 25.0° on both sides of the vertical. At a

moment when the sign is vertical and moving to the left, a snowball of mass 400 g, traveling horizontally with a velocity of 160 cm/s to the right, strikes perpendicularly the lower edge of the sign and sticks there. (a) Calculate the angular speed of the sign immediately before the impact. (b) Calculate its angular speed immediately after the impact. (c) The spattered sign will swing up through what maximum angle?

39. Suppose a meteor of mass 3.00×10^{13} kg, moving at 30.0 km/s relative to the center of the Earth, strikes the Earth. What is the order of magnitude of the maximum possible decrease in the angular speed of the Earth due to this collision? Explain your answer.

Section 11.5 The Motion of Gyroscopes and Tops

40. A spacecraft is in empty space. It carries on board a gyroscope with a moment of inertia of $I_g = 20.0$ kg·m² about the axis of the gyroscope. The moment of inertia of the spacecraft around the same axis is $I_s = 5.00 \times 10^5$ kg·m². Neither the spacecraft nor the gyroscope is originally rotating. The gyroscope can be powered up in a negligible period of time to an angular speed of 100 s⁻¹. If the orientation of the spacecraft is to be changed by 30.0°, for how long should the gyroscope be operated?

41. The angular momentum vector of a precessing gyroscope sweeps out a cone, as in Figure 11.14b. Its angular speed, called its precessional frequency, is given by $\omega_p = \tau/L$, where τ is the magnitude of the torque on the gyroscope and L is the magnitude of its angular momentum. In the motion called *precession of the equinoxes*, the Earth's axis of rotation precesses about the perpendicular to its orbital plane with a period of 2.58×10^4 yr. Model the Earth as a uniform sphere and calculate the torque on the Earth that is causing this precession.

Section 11.6 Angular Momentum as a Fundamental Quantity

42. In the Bohr model of the hydrogen atom, the electron moves in a circular orbit of radius 0.529×10^{-10} m around the proton. Assuming the orbital angular momentum of the electron is equal to $h/2\pi$, calculate (a) the orbital speed of the electron, (b) the kinetic energy of the electron, and (c) the angular frequency of the electron's motion.

Additional Problems

43. We have all complained that there aren't enough hours in a day. In an attempt to change that, suppose that all the people in the world line up at the equator, and all start running east at 2.50 m/s relative to the surface of the Earth. By how much does the length of a day increase? Assume that the world population is 5.50×10^9 people with an average mass of 70.0 kg each, and that the Earth is a solid homogeneous sphere. In addition, you may use the approximation $1/(1 - x) \approx 1 + x$ for small x.

44. A skateboarder with his board can be modeled as a particle of mass 76.0 kg, located at his center of mass. As shown in Figure P8.67 on page 248, the skateboarder starts from rest in a crouching position at one lip of a half-pipe (point Ⓐ). The half-pipe forms one half of a cylinder of radius 6.80 m

with its axis horizontal. On his descent, the skateboarder moves without friction and maintains his crouch, so that his center of mass moves through one quarter of a circle of radius 6.30 m. (a) Find his speed at the bottom of the half-pipe (point Ⓑ). (b) Find his angular momentum about the center of curvature. (c) Immediately after passing point Ⓑ, he stands up and raises his arms, lifting his center of gravity from 0.500 m to 0.950 m above the concrete (point Ⓒ). Explain why his angular momentum is constant in this maneuver, while his linear momentum and his mechanical energy are not constant. (d) Find his speed immediately after he stands up, when his center of mass is moving in a quarter circle of radius 5.85 m. (e) What work did the skateboarder's legs do on his body as he stood up? Next, the skateboarder glides upward with his center of mass moving in a quarter circle of radius 5.85 m. His body is horizontal when he passes point Ⓓ, the far lip of the half-pipe. (f) Find his speed at this location. At last he goes ballistic, twisting around while his center of mass moves vertically. (g) How high above point Ⓓ does he rise? (h) Over what time interval is he airborne before he touches down, facing downward and again in a crouch, 2.34 m below the level of point Ⓓ? (i) Compare the solution to this problem with the solution to Problem 8.67. Which is more accurate? Why? (*Caution*: Do not try this yourself without the required skill and protective equipment, or in a drainage channel to which you do not have legal access.)

45. A rigid, massless rod has three particles with equal masses attached to it as shown in Figure P11.45. The rod is free to rotate in a vertical plane about a frictionless axle perpendicular to the rod through the point *P*, and is released from rest in the horizontal position at $t = 0$. Assuming *m* and *d* are known, find (a) the moment of inertia of the system (rod plus particles) about the pivot, (b) the torque acting on the system at $t = 0$, (c) the angular acceleration of the system at $t = 0$, (d) the linear acceleration of the particle labeled 3 at $t = 0$, (e) the maximum kinetic energy of the system, (f) the maximum angular speed reached by the rod, (g) the maximum angular momentum of the system, and (h) the maximum speed reached by the particle labeled 2.

Figure P11.45

46. A 100-kg uniform horizontal disk of radius 5.50 m turns without friction at 2.50 rev/s on a vertical axis through its center, as in Figure P11.46. A feedback mechanism senses the angular speed of the disk, and a drive motor at *A* maintains the angular speed constant while a 1.20 kg block on top of the disk slides outward in a radial slot. The 1.20-kg block starts at the center of the disk at time $t = 0$ and moves outward with constant speed 1.25 cm/s relative to the disk until it reaches the edge at $t = 440$ s. The sliding block feels no friction. Its motion is constrained to have constant radial speed by a brake at *B*, producing tension in a light string tied to the block. (a) Find the torque that the drive motor must provide as a function of time, while the block is sliding. (b) Find the value of this torque at $t = 440$ s, just before the sliding block finishes its motion. (c) Find the power that the drive motor must deliver as a function of time. (d) Find the value of the power when the sliding block is just reaching the end of the slot. (e) Find the string tension as a function of time. (f) Find the work done by the drive motor during the 440-s motion. (g) Find the work done by the string brake on the sliding block. (h) Find the total work on the system consisting of the disk and the sliding block.

Figure P11.46

47. Comet Halley moves about the Sun in an elliptical orbit, with its closest approach to the Sun being about 0.590 AU and its greatest distance 35.0 AU (1 AU = the Earth–Sun distance). If the comet's speed at closest approach is 54.0 km/s, what is its speed when it is farthest from the Sun? The angular momentum of the comet about the Sun is conserved, because no torque acts on the comet. The gravitational force exerted by the Sun has zero moment arm.

48. A light rope passes over a light, frictionless pulley. One end is fastened to a bunch of bananas of mass *M*, and a monkey of mass *M* clings to the other end (Fig. P11.48). The mon-

Figure P11.48

key climbs the rope in an attempt to reach the bananas. (a) Treating the system as consisting of the monkey, bananas, rope, and pulley, evaluate the net torque about the pulley axis. (b) Using the results of (a), determine the total angular momentum about the pulley axis and describe the motion of the system. Will the monkey reach the bananas?

49. A puck of mass m is attached to a cord passing through a small hole in a frictionless, horizontal surface (Fig. P11.49). The puck is initially orbiting with speed v_i in a circle of radius r_i. The cord is then slowly pulled from below, decreasing the radius of the circle to r. (a) What is the speed of the puck when the radius is r? (b) Find the tension in the cord as a function of r. (c) How much work W is done in moving m from r_i to r? (*Note:* The tension depends on r.) (d) Obtain numerical values for v, T, and W when $r = 0.100$ m, $m = 50.0$ g, $r_i = 0.300$ m, and $v_i = 1.50$ m/s.

Figure P11.49

50. A projectile of mass m moves to the right with a speed v_i (Fig. P11.50a). The projectile strikes and sticks to the end of a stationary rod of mass M, length d, pivoted about a frictionless axle through its center (Fig. P11.50b). (a) Find the angular speed of the system right after the collision. (b) Determine the fractional loss in mechanical energy due to the collision.

Figure P11.50

51. Two astronauts (Fig. P11.51), each having a mass of 75.0 kg, are connected by a 10.0-m rope of negligible mass.

They are isolated in space, orbiting their center of mass at speeds of 5.00 m/s. Treating the astronauts as particles, calculate (a) the magnitude of the angular momentum of the system and (b) the rotational energy of the system. By pulling on the rope, one of the astronauts shortens the distance between them to 5.00 m. (c) What is the new angular momentum of the system? (d) What are the astronauts' new speeds? (e) What is the new rotational energy of the system? (f) How much work does the astronaut do in shortening the rope?

Figure P11.51 Problems 51 and 52.

52. Two astronauts (Fig. P11.51), each having a mass M, are connected by a rope of length d having negligible mass. They are isolated in space, orbiting their center of mass at speeds v. Treating the astronauts as particles, calculate (a) the magnitude of the angular momentum of the system and (b) the rotational energy of the system. By pulling on the rope, one of the astronauts shortens the distance between them to $d/2$. (c) What is the new angular momentum of the system? (d) What are the astronauts' new speeds? (e) What is the new rotational energy of the system? (f) How much work does the astronaut do in shortening the rope?

53. Global warming is a cause for concern because even small changes in the Earth's temperature can have significant consequences. For example, if the Earth's polar ice caps were to melt entirely, the resulting additional water in the oceans would flood many coastal cities. Would it appreciably change the length of a day? Calculate the resulting change in the duration of one day. Model the polar ice as having mass 2.30×10^{19} kg and forming two flat disks of radius 6.00×10^5 m. Assume the water spreads into an unbroken thin spherical shell after it melts.

54. A solid cube of wood of side $2a$ and mass M is resting on a horizontal surface. The cube is constrained to rotate about an axis AB (Fig. P11.54). A bullet of mass m and speed v is shot at the face opposite $ABCD$ at a height of $4a/3$. The bullet becomes embedded in the cube. Find the minimum value of v required to tip the cube so that it falls on face $ABCD$. Assume $m \ll M$.

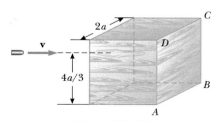

Figure P11.54

55. A solid cube of side $2a$ and mass M is sliding on a frictionless surface with uniform velocity $\mathbf{v}$ as in Figure P11.55a. It hits a small obstacle at the end of the table, which causes the cube to tilt as in Figure P11.55b. Find the minimum value of $\mathbf{v}$ such that the cube falls off the table. Note that the moment of inertia of the cube about an axis along one of its edges is $8Ma^2/3$. (*Hint:* The cube undergoes an inelastic collision at the edge.)

(a)

Figure P11.55

56. A uniform solid disk is set into rotation with an angular speed ω_i about an axis through its center. While still

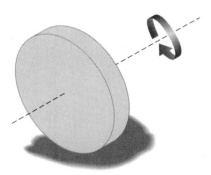

Figure P11.56 Problems 56 and 57.

rotating at this speed, the disk is placed into contact with a horizontal surface and released as in Figure P11.56. (a) What is the angular speed of the disk once pure rolling takes place? (b) Find the fractional loss in kinetic energy from the time the disk is released until pure rolling occurs. (*Hint:* Consider torques about the center of mass.)

57. Suppose a solid disk of radius R is given an angular speed ω_i about an axis through its center and then lowered to a horizontal surface and released, as in Problem 56 (Fig. P11.56). Furthermore, assume that the coefficient of friction between disk and surface is μ. (a) Show that the time interval before pure rolling motion occurs is $R\omega_i/3\mu g$. (b) Show that the distance the disk travels before pure rolling occurs is $R^2\omega_i^2/18\mu g$.

Answers to Quick Quizzes

11.1 (d). This result can be obtained by replacing $\mathbf{B} \times \mathbf{A}$ with $-(\mathbf{A} \times \mathbf{B})$, according to Equation 11.4.

11.2 (d). Because of the $\sin\theta$ function, $|\mathbf{A} \times \mathbf{B}|$ is either equal to or smaller than AB, depending on the angle θ.

11.3 (a). If $\mathbf{p}$ and $\mathbf{r}$ are parallel or antiparallel, the angular momentum is zero. For a nonzero angular momentum, the linear momentum vector must be offset from the rotation axis.

11.4 (c). The angular momentum is the product of the linear momentum and the perpendicular distance from the rotation axis to the line along which the linear momentum vector lies.

11.5 (b). The hollow sphere has a larger moment of inertia than the solid sphere.

11.6 (a). The diver is an isolated system, so the product $I\omega$ remains constant. Because her moment of inertia decreases, her angular speed increases.

11.7 (a). As the moment of inertia of the diver decreases, the angular speed increases by the same factor. For example, if I goes down by a factor of 2, ω goes up by a factor of 2. The rotational kinetic energy varies as the square of ω. If I is halved, ω^2 increases by a factor of 4 and the energy increases by a factor of 2.

Chapter 12

Static Equilibrium and Elasticity

▲ Balanced Rock in Arches National Park, Utah, is a 3 000 000-kg boulder that has been in stable equilibrium for several millennia. It had a smaller companion nearby, called "Chip Off the Old Block," which fell during the winter of 1975. Balanced Rock appeared in an early scene of the movie Indiana Jones and the Last Crusade. We will study the conditions under which an object is in equilibrium in this chapter. (John W. Jewett, Jr.)

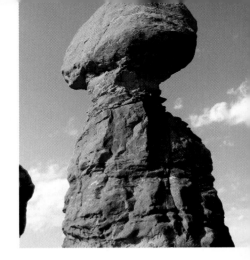

In Chapters 10 and 11 we studied the dynamics of rigid objects. Part of this current chapter addresses the conditions under which a rigid object is in equilibrium. The term *equilibrium* implies either that the object is at rest or that its center of mass moves with constant velocity relative to the observer. We deal here only with the former case, in which the object is in *static equilibrium*. Static equilibrium represents a common situation in engineering practice, and the principles it involves are of special interest to civil engineers, architects, and mechanical engineers. If you are an engineering student, you will undoubtedly take an advanced course in statics in the future.

The last section of this chapter deals with how objects deform under load conditions. An *elastic* object returns to its original shape when the deforming forces are removed. Several elastic constants are defined, each corresponding to a different type of deformation.

12.1   The Conditions for Equilibrium

In Chapter 5 we found that one necessary condition for equilibrium is that the net force acting on an object must be zero. If the object is modeled as a particle, then this is the only condition that must be satisfied for equilibrium. The situation with real (extended) objects is more complex, however, because these objects often cannot be modeled as particles. For an extended object to be in static equilibrium, a second condition must be satisfied. This second condition involves the net torque acting on the extended object.

Consider a single force **F** acting on a rigid object, as shown in Figure 12.1. The effect of the force depends on the location of its point of application *P*. If **r** is the position vector of this point relative to *O*, the torque associated with the force **F** about *O* is given by Equation 11.1:

$$\boldsymbol{\tau} = \mathbf{r} \times \mathbf{F}$$

Recall from the discussion of the vector product in Section 11.1 that the vector $\boldsymbol{\tau}$ is perpendicular to the plane formed by **r** and **F**. You can use the right-hand rule to determine the direction of $\boldsymbol{\tau}$ as shown in Figure 11.2. Hence, in Figure 12.1 $\boldsymbol{\tau}$ is directed toward you out of the page.

As you can see from Figure 12.1, the tendency of **F** to rotate the object about an axis through *O* depends on the moment arm *d*, as well as on the magnitude of **F**. Recall that the magnitude of $\boldsymbol{\tau}$ is *Fd* (see Eq. 10.19). According to Equation 10.21, the net torque on a rigid object will cause it to undergo an angular acceleration.

In the current discussion, we want to look at those rotational situations in which the angular acceleration of a rigid object is zero. Such an object is in **rotational equilibrium**. Because $\Sigma\boldsymbol{\tau} = I\boldsymbol{\alpha}$ for rotation about a fixed axis, the necessary condition for rotational equilibrium is that **the net torque about any axis must be zero**. We now

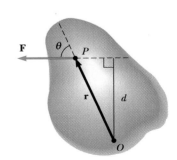

Figure 12.1 A single force **F** acts on a rigid object at the point *P*.

have two necessary conditions for equilibrium of an object:

1. The resultant external force must equal zero:

$$\sum \mathbf{F} = 0 \tag{12.1}$$

2. The resultant external torque about *any* axis must be zero:

$$\sum \boldsymbol{\tau} = 0 \tag{12.2}$$

The first condition is a statement of translational equilibrium; it tells us that the linear acceleration of the center of mass of the object must be zero when viewed from an inertial reference frame. The second condition is a statement of rotational equilibrium and tells us that the angular acceleration about any axis must be zero. In the special case of **static equilibrium**, which is the main subject of this chapter, the object is at rest relative to the observer and so has no linear or angular speed (that is, $v_{\text{CM}} = 0$ and $\omega = 0$).

> ### Quick Quiz 12.1
> Consider the object subject to the two forces in Figure 12.2. Choose the correct statement with regard to this situation. (a) The object is in force equilibrium but not torque equilibrium. (b) The object is in torque equilibrium but not force equilibrium. (c) The object is in both force and torque equilibrium. (d) The object is in neither force nor torque equilibrium.

> ### Quick Quiz 12.2
> Consider the object subject to the three forces in Figure 12.3. Choose the correct statement with regard to this situation. (a) The object is in force equilibrium but not torque equilibrium. (b) The object is in torque equilibrium but not force equilibrium. (c) The object is in both force and torque equilibrium. (d) The object is in neither force nor torque equilibrium.

The two vector expressions given by Equations 12.1 and 12.2 are equivalent, in general, to six scalar equations: three from the first condition for equilibrium, and three from the second (corresponding to *x*, *y*, and *z* components). Hence, in a complex system involving several forces acting in various directions, you could be faced with solving a set of equations with many unknowns. Here, we restrict our discussion to situations in which all the forces lie in the *xy* plane. (Forces whose vector representations are in the same plane are said to be *coplanar*.) With this restriction, we must deal with only three scalar equations. Two of these come from balancing the forces in the *x* and *y* directions. The third comes from the torque equation—namely, that the net torque about a perpendicular axis through *any* point in the *xy* plane must be zero. Hence, the two conditions of equilibrium provide the equations

$$\sum F_x = 0 \qquad \sum F_y = 0 \qquad \sum \tau_z = 0 \tag{12.3}$$

where the location of the axis of the torque equation is arbitrary, as we now show.

Regardless of the number of forces that are acting, if an object is in translational equilibrium and if the net torque is zero about one axis, then the net torque must also be zero about any other axis. The axis can pass through a point that is inside or outside the boundaries of the object. Consider an object being acted on by several forces such that the resultant force $\sum \mathbf{F} = \mathbf{F}_1 + \mathbf{F}_2 + \mathbf{F}_3 + \cdots = 0$. Figure 12.4 describes this situation (for clarity, only four forces are shown). The point of application of $\mathbf{F}_1$ relative to O is specified by the position vector $\mathbf{r}_1$. Similarly, the points of application of $\mathbf{F}_2$, $\mathbf{F}_3$, . . . are specified by $\mathbf{r}_2$, $\mathbf{r}_3$, . . . (not shown). The net torque about an axis through O is

$$\sum \boldsymbol{\tau}_O = \mathbf{r}_1 \times \mathbf{F}_1 + \mathbf{r}_2 \times \mathbf{F}_2 + \mathbf{r}_3 \times \mathbf{F}_3 + \cdots$$

Now consider another arbitrary point O' having a position vector $\mathbf{r}'$ relative to O. The point of application of $\mathbf{F}_1$ relative to O' is identified by the vector $\mathbf{r}_1 - \mathbf{r}'$. Like-

▲ **PITFALL PREVENTION**

12.1 Zero Torque

Zero net torque does not mean an absence of rotational motion. An object which is rotating at a constant angular speed can be under the influence of a net torque of zero. This is analogous to the translational situation—zero net force does not mean an absence of translational motion.

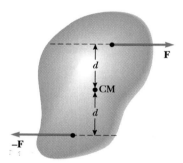

Figure 12.2 (Quick Quiz 12.1) Two forces of equal magnitude are applied at equal distances from the center of mass of a rigid object.

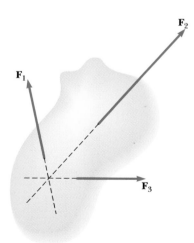

Figure 12.3 (Quick Quiz 12.2) Three forces act on an object. Notice that the lines of action of all three forces pass through a common point.

wise, the point of application of $\mathbf{F}_2$ relative to O' is $\mathbf{r}_2 - \mathbf{r}'$, and so forth. Therefore, the torque about an axis through O' is

$$\sum \boldsymbol{\tau}_{O'} = (\mathbf{r}_1 - \mathbf{r}') \times \mathbf{F}_1 + (\mathbf{r}_2 - \mathbf{r}') \times \mathbf{F}_2 + (\mathbf{r}_3 - \mathbf{r}') \times (\mathbf{F}_3 + \cdots$$
$$= \mathbf{r}_1 \times \mathbf{F}_1 + \mathbf{r}_2 \times \mathbf{F}_2 + \mathbf{r}_3 \times \mathbf{F}_3 + \cdots - \mathbf{r}' \times (\mathbf{F}_1 + \mathbf{F}_2 + \mathbf{F}_3 + \cdots)$$

Because the net force is assumed to be zero (given that the object is in translational equilibrium), the last term vanishes, and we see that the torque about an axis through O' is equal to the torque about an axis through O. Hence, **if an object is in translational equilibrium and the net torque is zero about one axis, then the net torque must be zero about any other axis.**

12.2 More on the Center of Gravity

We have seen that the point at which a force is applied can be critical in determining how an object responds to that force. For example, two equal-magnitude but oppositely directed forces result in equilibrium if they are applied at the same point on an object. However, if the point of application of one of the forces is moved, so that the two forces no longer act along the same line of action, then the object undergoes an angular acceleration.

Whenever we deal with a rigid object, one of the forces we must consider is the gravitational force acting on it, and we must know the point of application of this force. As we learned in Section 9.5, associated with every object is a special point called its center of gravity. All the various gravitational forces acting on all the various mass elements of the object are equivalent to a single gravitational force acting through this point. Thus, to compute the torque due to the gravitational force on an object of mass M, we need only consider the force $M\mathbf{g}$ acting at the center of gravity of the object.

How do we find this special point? As we mentioned in Section 9.5, if we assume that $\mathbf{g}$ is uniform over the object, then the center of gravity of the object coincides with its center of mass. To see that this is so, consider an object of arbitrary shape lying in the xy plane, as illustrated in Figure 12.5. Suppose the object is divided into a large number of particles of masses m_1, m_2, m_3, . . . having coordinates (x_1, y_1), (x_2, y_2), (x_3, y_3), ... In Equation 9.28 we defined the x coordinate of the center of mass of such an object to be

$$x_{\text{CM}} = \frac{m_1 x_1 + m_2 x_2 + m_3 x_3 + \cdots}{m_1 + m_2 + m_3 + \cdots} = \frac{\sum_i m_i x_i}{\sum_i m_i}$$

We use a similar equation to define the y coordinate of the center of mass, replacing each x with its y counterpart.

Let us now examine the situation from another point of view by considering the gravitational force exerted on each particle, as shown in Figure 12.6. Each particle contributes a torque about the origin equal in magnitude to the particle's weight mg multiplied by its moment arm. For example, the magnitude of the torque due to the force $m_1 \mathbf{g}_1$ is $m_1 g_1 x_1$, where g_1 is the value of the gravitational acceleration at the position of the particle of mass m_1. We wish to locate the center of gravity, the point at which application of the single gravitational force $M\mathbf{g}$ (where $M = m_1 + m_2 + m_3 + \cdots$ is the total mass of the object) has the same effect on rotation as does the combined effect of all the individual gravitational forces $m_i \mathbf{g}_i$. Equating the torque resulting from $M\mathbf{g}$ acting at the center of gravity to the sum of the torques acting on the individual particles gives

$$(m_1 g_1 + m_2 g_2 + m_3 g_3 + \cdots) x_{\text{CG}} = m_1 g_1 x_1 + m_2 g_2 x_2 + m_3 g_3 x_3 + \cdots$$

This expression accounts for the fact that the value of g can in general vary over the object. If we assume uniform g over the object (as is usually the case), then the

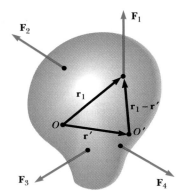

Figure 12.4 Construction showing that if the net torque is zero about origin O, it is also zero about any other origin, such as O'.

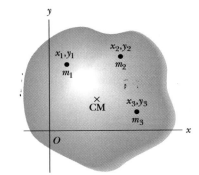

Figure 12.5 An object can be divided into many small particles each having a specific mass and specific coordinates. These particles can be used to locate the center of mass.

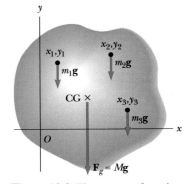

Figure 12.6 The center of gravity of an object is located at the center of mass if $\mathbf{g}$ is constant over the object.

g terms cancel and we obtain

$$x_{CG} = \frac{m_1 x_1 + m_2 x_2 + m_3 x_3 + \cdots}{m_1 + m_2 + m_3 + \cdots} \qquad (12.4)$$

Comparing this result with Equation 9.28, we see that **the center of gravity is located at the center of mass as long as g is uniform over the entire object**. In several examples presented in the next section, we will deal with homogeneous, symmetric objects. The center of gravity for any such object coincides with its geometric center.

Quick Quiz 12.3   A meter stick is supported on a fulcrum at the 25-cm mark. A 0.50-kg object is hung from the zero end of the meter stick, and the stick is balanced horizontally. The mass of the meter stick is (a) 0.25 kg (b) 0.50 kg (c) 0.75 kg (d) 1.0 kg (e) 2.0 kg (f) impossible to determine.

Charles D. Winters

Figure 12.7 This one-bottle wine holder is a surprising display of static equilibrium. The center of gravity of the system (bottle plus holder) is directly over the support point.

12.3   Examples of Rigid Objects in Static Equilibrium

The photograph of the one-bottle wine holder in Figure 12.7 shows one example of a balanced mechanical system that seems to defy gravity. For the system (wine holder plus bottle) to be in equilibrium, the net external force must be zero (see Eq. 12.1) and the net external torque must be zero (see Eq. 12.2). The second condition can be satisfied only when the center of gravity of the system is directly over the support point.

When working static equilibrium problems, you must recognize all the external forces acting on the object. Failure to do so results in an incorrect analysis. When analyzing an object in equilibrium under the action of several external forces, use the following procedure.

PROBLEM-SOLVING STRATEGY

OBJECTS IN STATIC EQUILIBRIUM

● Draw a simple, neat diagram of the system.

● Isolate the object being analyzed. Draw a free-body diagram. Then show and label all external forces acting on the object, indicating where those forces are applied. Do not include forces exerted by the object on its surroundings. (For systems that contain more than one object, draw a *separate* free-body diagram for each one.) Try to guess the correct direction for each force.

● Establish a convenient coordinate system and find the components of the forces on the object along the two axes. Then apply the first condition for equilibrium. Remember to keep track of the signs of the various force components.

● Choose a convenient axis for calculating the net torque on the object. Remember that the choice of origin for the torque equation is arbitrary; therefore choose an origin that simplifies your calculation as much as possible. Note that a force that acts along a line passing through the point chosen as the origin gives zero contribution to the torque and so can be ignored.

● The first and second conditions for equilibrium give a set of linear equations containing several unknowns, and these equations can be solved simultaneously. If the direction you selected for a force leads to a negative value, do not be alarmed; this merely means that the direction of the force is the opposite of what you guessed.

Example 12.1   The Seesaw Revisited

A seesaw consisting of a uniform board of mass M and length ℓ supports a father and daughter with masses m_f and m_d, respectively, as shown in Figure 12.8. The support (called the *fulcrum*) is under the center of gravity of the board, the father is a distance d from the center, and the daughter is a distance $\ell/2$ from the center.

(A) Determine the magnitude of the upward force **n** exerted by the support on the board.

Solution First note that, in addition to **n**, the external forces acting on the board are the downward forces exerted by each person and the gravitational force acting on the board. We know that the board's center of gravity is at its geometric center because we are told that the board is uniform. Because the system is in static equilibrium, the net force on the board is zero. Thus, the upward force **n** must balance all the downward forces. From $\Sigma F_y = 0$, and defining upward as the positive y direction, we have

$$n - m_f g - m_d g - Mg = 0$$

$$n = \boxed{m_f g + m_d g + Mg}$$

(The equation $\Sigma F_x = 0$ also applies, but we do not need to consider it because no forces act horizontally on the board.)

(B) Determine where the father should sit to balance the system.

Solution To find this position, we must invoke the second condition for equilibrium. If we take an axis perpendicular

to the page through the center of gravity of the board as the axis for our torque equation, the torques produced by **n** and the gravitational force acting on the board are zero. We see from $\Sigma \tau = 0$ that

$$(m_f g)(d) - (m_d g)\frac{\ell}{2} = 0$$

$$d = \boxed{\left(\frac{m_d}{m_f}\right)\frac{1}{2}\ell}$$

This is the same result that we obtained in Example 11.6 by evaluating the angular acceleration of the system and setting the angular acceleration equal to zero.

What If? Suppose we had chosen another point through which the rotation axis were to pass. For example, suppose the axis is perpendicular to the page and passes through the location of the father. Does this change the results to parts (A) and (B)?

Answer Part (A) is unaffected because the calculation of the net force does not involve a rotation axis. In part (B), we would conceptually expect there to be no change if a different rotation axis is chosen because the second condition of equilibrium claims that the torque is zero about any rotation axis.

   Let us verify this mathematically. Recall that the sign of the torque associated with a force is positive if that force tends to rotate the system counterclockwise, while the sign of the torque is negative if the force tends to rotate the system clockwise. In the case of a rotation axis passing through the location of the father, $\Sigma \tau = 0$ yields

$$n(d) - (Mg)(d) - (m_d g)(d + \ell/2) = 0$$

From part (A) we know that $n = m_f g + m_d g + Mg$. Thus, we can substitute this expression for n and solve for d:

$$(m_f g + m_d g + Mg)(d) - (Mg)(d) - (m_d g)\left(d + \frac{\ell}{2}\right) = 0$$

$$(m_f g)(d) - (m_d g)\left(\frac{\ell}{2}\right) = 0$$

$$d = \left(\frac{m_d}{m_f}\right)\frac{1}{2}\ell$$

This result is in agreement with the one we obtained in part (B).

Figure 12.8 (Example 12.1) A balanced system.

Example 12.2   A Weighted Hand

A person holds a 50.0-N sphere in his hand. The forearm is horizontal, as shown in Figure 12.9a. The biceps muscle is attached 3.00 cm from the joint, and the sphere is 35.0 cm from the joint. Find the upward force exerted by the biceps on the forearm and the downward force exerted by the upper arm on the forearm and acting at the joint. Neglect the weight of the forearm.

Solution We simplify the situation by modeling the forearm as a bar as shown in Figure 12.9b, where **F** is the upward force exerted by the biceps and **R** is the downward force exerted by the upper arm at the joint. From the first condition for equilibrium, we have, with upward as the positive y direction,

$$(1) \qquad \Sigma F_y = F - R - 50.0\,\text{N} = 0$$

Figure 12.9 (Example 12.2) (a) The biceps muscle pulls upward with a force **F** that is essentially at a right angle to the forearm. (b) The mechanical model for the system described in part (a).

From the second condition for equilibrium, we know that the sum of the torques about any point must be zero. With the joint O as the axis, we have

$$\sum \tau = Fd - mg\ell = 0$$

$$F(3.00 \text{ cm}) - (50.0 \text{ N})(35.0 \text{ cm}) = 0$$

$$F = \boxed{583 \text{ N}}$$

This value for F can be substituted into Equation (1) to give $R = 533$ N. As this example shows, the forces at joints and in muscles can be extremely large.

Example 12.3   Standing on a Horizontal Beam

Interactive

A uniform horizontal beam with a length of 8.00 m and a weight of 200 N is attached to a wall by a pin connection. Its far end is supported by a cable that makes an angle of 53.0° with the beam (Fig. 12.10a). If a 600-N person stands 2.00 m from the wall, find the tension in the cable as well as the magnitude and direction of the force exerted by the wall on the beam.

Solution Conceptualize this problem by imagining that the person in Figure 12.10 moves outward on the beam. It seems reasonable that the farther he moves outward, the larger the torque that he applies about the pivot and the larger the tension in the cable must be to balance this torque. Because the system is at rest, we categorize this as a static equilibrium problem. We begin to analyze the problem by identifying all the external forces acting on the beam: the 200-N gravitational force, the force **T** exerted by the cable, the force **R** exerted by the wall at the pivot, and the 600-N force that the person exerts on the beam. These forces are all indicated in the free-body diagram for the beam shown in Figure 12.10b. When we assign directions for forces, it is sometimes helpful to imagine what would happen if a force were suddenly removed. For example, if the wall were to vanish suddenly, the left end of the beam would move to the left as it begins to fall. This tells us that the wall is not only holding the beam up but is also pressing outward against it. Thus, we draw the vector **R** as shown in Figure 12.10b. If we resolve **T** and **R** into horizontal and vertical components, as shown in Figure 12.10c, and apply the first condition for equilibrium, we obtain

(1) $$\sum F_x = R \cos \theta - T \cos 53.0° = 0$$

(2) $$\sum F_y = R \sin \theta + T \sin 53.0° - 600 \text{ N} - 200 \text{ N} = 0$$

where we have chosen rightward and upward as our positive directions. Because R, T, and θ are all unknown, we cannot obtain a solution from these expressions alone. (The number of simultaneous equations must equal the number of unknowns for us to be able to solve for the unknowns.)

Now let us invoke the condition for rotational equilibrium. A convenient axis to choose for our torque equation is the one that passes through the pin connection. The feature that makes this point so convenient is that the force **R** and the horizontal component of **T** both have a moment arm of zero; hence, these forces provide no torque about this point. Recalling our counterclockwise-equals-positive convention for the sign of the torque about an axis and noting that the moment arms of the 600-N, 200-N, and $T \sin 53.0°$ forces are 2.00 m, 4.00 m, and 8.00 m, respectively, we obtain

(3) $$\sum \tau = (T \sin 53.0°)(8.00 \text{ m}) - (600 \text{ N})(2.00 \text{ m})$$
$$- (200 \text{ N})(4.00 \text{ m}) = 0$$

$$T = \boxed{313 \text{ N}}$$

Thus, the torque equation with this axis gives us one of the unknowns directly! We now substitute this value into Equations (1) and (2) and find that

$$R \cos \theta = 188 \text{ N}$$

$$R \sin \theta = 550 \text{ N}$$

We divide the second equation by the first and, recalling the trigonometric identity $\sin \theta / \cos \theta = \tan \theta$, we obtain

$$\tan \theta = \frac{550 \text{ N}}{188 \text{ N}} = 2.93$$

Figure 12.10 (Example 12.3) (a) A uniform beam supported by a cable. A person walks outward on the beam. (b) The free-body diagram for the beam. (c) The free-body diagram for the beam showing the components of **R** and **T**.

$$\theta = \boxed{71.1°}$$

This positive value indicates that our estimate of the direction of **R** was accurate.

Finally,

$$R = \frac{188 \text{ N}}{\cos \theta} = \frac{188 \text{ N}}{\cos 71.1°} = \boxed{580 \text{ N}}$$

To finalize this problem, note that if we had selected some other axis for the torque equation, the solution might differ in the details, but the answers would be the same. For example, if we had chosen an axis through the center of gravity of the beam, the torque equation would involve both T and R. However, this equation, coupled with Equations (1) and (2), could still be solved for the unknowns. Try it!

When many forces are involved in a problem of this nature, it is convenient in your analysis to set up a table. For instance, for the example just given, we could construct the following table. Setting the sum of the terms in the last column equal to zero represents the condition of rotational equilibrium.

Force component	Moment arm relative to O (m)	Torque about O (N·m)
$T \sin 53.0°$	8.00	$(8.00)T \sin 53.0°$
$T \cos 53.0°$	0	0
200 N	4.00	$-(4.00)(200)$
600 N	2.00	$-(2.00)(600)$
$R \sin \theta$	0	0
$R \cos \theta$	0	0

What If? What if the person walks farther out on the beam? Does T change? Does R change? Does θ change?

Answer T must increase because the weight of the person exerts a larger torque about the pin connection, which must be countered by a larger torque in the opposite direction due to an increased value of T. If T increases, the vertical component of **R** decreases to maintain force equilibrium in the vertical direction. But force equilibrium in the horizontal direction requires an increased horizontal component of **R** to balance the horizontal component of the increased **T**. This suggests that θ will become smaller, but it is hard to predict what will happen to R. Problem 26 allows you to explore the behavior of R.

 At the Interactive Worked Example link at **http://www.pse6.com,** *you can adjust the position of the person and observe the effect on the forces.*

Example 12.4   The Leaning Ladder   `Interactive`

A uniform ladder of length ℓ rests against a smooth, vertical wall (Fig. 12.11a). If the mass of the ladder is m and the coefficient of static friction between the ladder and the ground is $\mu_s = 0.40$, find the minimum angle θ_{min} at which the ladder does not slip.

Solution The free-body diagram showing all the external forces acting on the ladder is illustrated in Figure 12.11b. The force exerted by the ground on the ladder is the vector sum of a normal force **n** and the force of static friction $\mathbf{f}_s$.

The reaction force **P** exerted by the wall on the ladder is horizontal because the wall is frictionless. Notice how we have included only forces that act on the ladder. For example, the forces exerted by the ladder on the ground and on the wall are not part of the problem and thus do not appear in the free-body diagram. Applying the first condition for equilibrium to the ladder, we have

$$(1) \qquad \sum F_x = f_s - P = 0$$
$$(2) \qquad \sum F_y = n - mg = 0$$

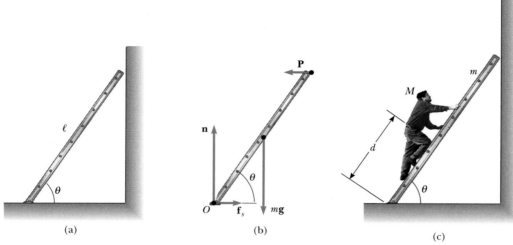

Figure 12.11 (Example 12.4) (a) A uniform ladder at rest, leaning against a smooth wall. The ground is rough. (b) The free-body diagram for the ladder. (c) A person of mass M begins to climb the ladder when it is at the minimum angle found in part (a) of the example. Will the ladder slip?

The first equation tells us that $P = f_s$. From the second equation we see that $n = mg$. Furthermore, when the ladder is on the verge of slipping, the force of friction must be a maximum, which is given by $f_{s,\,max} = \mu_s n$. (Recall Eq. 5.8: $f_s \le \mu_s n$.) Thus, we must have $P = f_s = \mu_s n = \mu_s mg$.

To find θ_{min}, we must use the second condition for equilibrium. When we take the torques about an axis through the origin O at the bottom of the ladder, we have

$$(3) \qquad \sum \tau_O = P\ell \sin \theta - mg\,\frac{\ell}{2} \cos \theta = 0$$

This expression gives

$$\tan \theta_{min} = \frac{mg}{2P} = \frac{mg}{2\mu_s mg} = \frac{1}{2\mu_s} = 1.25$$

$$\theta_{min} = \boxed{51°}$$

What If? What if a person begins to climb the ladder when the angle is 51°? Will the presence of a person on the ladder make it more or less likely to slip?

Answer The presence of the additional weight of a person on the ladder will increase the clockwise torque about its base in Figure 12.11b. To maintain static equilibrium, the counterclockwise torque must increase, which can occur if P increases. Because equilibrium in the horizontal direction tells us that $P = f_s$, this would suggest that the friction force rises above the maximum value $f_{s,\,max}$ and the ladder slips. However, the increased weight of the person also causes n to increase, which increases the maximum friction force $f_{s,\,max}$! Thus, it is not clear conceptually whether the ladder is more or less likely to slip.

Imagine that the person of mass M is at a position d that is measured along the ladder from its base (Fig. 12.11c). Equations (1) and (2) can be rewritten

$$(4) \qquad \sum F_x = f_s - P = 0$$
$$(5) \qquad \sum F_y = n - (m + M)g = 0$$

Equation (3) can be rewritten

$$\sum \tau_O = P\ell \sin \theta - mg\,\frac{\ell}{2} \cos \theta - Mgd \cos \theta = 0$$

Solving this equation for $\tan \theta$, we find

$$\tan \theta = \frac{mg(\ell/2) + Mgd}{P\ell}$$

Incorporating Equations (4) and (5), and imposing the condition that the ladder is about to slip, this becomes

$$(6) \qquad \tan \theta_{min} = \frac{m(\ell/2) + Md}{\mu_s \ell (m + M)}$$

When the person is at the bottom of the ladder, $d = 0$. In this case, there is no additional torque about the bottom of the ladder and the increased normal force causes the maximum static friction force to increase. Thus, the ladder is less likely to slip than in the absence of the person. As the person climbs and d becomes larger, however, the numerator in Equation (6) becomes larger. Thus, the minimum angle at which the ladder does not slip increases. Eventually, as the person climbs higher, the minimum angle becomes larger than 51° and the ladder slips. The particular value of d at which the ladder slips depends on the coefficient of friction and the masses of the person and the ladder.

At the Interactive Worked Example link at **http://www.pse6.com,** *you can adjust the angle of the ladder and watch what happens when it is released.*

Example 12.5 Negotiating a Curb

(A) Estimate the magnitude of the force **F** a person must apply to a wheelchair's main wheel to roll up over a sidewalk curb (Fig. 12.12a). This main wheel that comes in contact with the curb has a radius r, and the height of the curb is h.

Solution Normally, the person's hands supply the required force to a slightly smaller wheel that is concentric with the main wheel. For simplicity, we assume that the radius of the smaller wheel is the same as the radius of the main wheel. Let us estimate a combined weight of $mg = 1\ 400$ N for the person and the wheelchair and choose a wheel radius of $r = 30$ cm. We also pick a curb height of $h = 10$ cm. We assume that the wheelchair and occupant are symmetric, and that each wheel supports a weight of 700 N. We then proceed to analyze only one of the wheels. Figure 12.12b shows the geometry for a single wheel.

When the wheel is just about to be raised from the street, the reaction force exerted by the ground on the wheel at point B goes to zero. Hence, at this time only three forces act on the wheel, as shown in the free-body diagram in Figure 12.12c. However, the force **R**, which is the force exerted by the curb on the wheel, acts at point A, and so if we choose to have our axis of rotation pass through point A, we do not need to include **R** in our torque equation. From the triangle OAC shown in Figure 12.12b, we see that the moment arm d of the gravitational force $m\mathbf{g}$ acting on the wheel relative to point A is

$$d = \sqrt{r^2 - (r-h)^2} = \sqrt{2rh - h^2}$$

The moment arm of **F** relative to point A is $2r - h$ (see Figure 12.12c). Therefore, the net torque acting on the wheel about point A is

$$mgd - F(2r - h) = 0$$

$$mg\sqrt{2rh - h^2} - F(2r - h) = 0$$

$$F = \frac{mg\sqrt{2rh - h^2}}{2r - h}$$

$$= \frac{(700\ \text{N})\sqrt{2(0.3\ \text{m})(0.1\ \text{m}) - (0.1\ \text{m})^2}}{2(0.3\ \text{m}) - 0.1\ \text{m}}$$

$$= \boxed{3 \times 10^2\ \text{N}}$$

(Notice that we have kept only one digit as significant.) This result indicates that the force that must be applied to each wheel is substantial. You may want to estimate the force required to roll a wheelchair up a typical sidewalk accessibility ramp for comparison.

(B) Determine the magnitude and direction of **R**.

Solution We use the first condition for equilibrium to determine the direction:

$$\sum F_x = F - R\cos\theta = 0$$

$$\sum F_y = R\sin\theta - mg = 0$$

Dividing the second equation by the first gives

$$\tan\theta = \frac{mg}{F} = \frac{700\ \text{N}}{300\ \text{N}}$$

(a)

(b)

(c)

(d)

Figure 12.12 (Example 12.5) (a) A person in a wheelchair attempts to roll up over a curb. (b) Details of the wheel and curb. The person applies a force **F** to the top of the wheel. (c) The free-body diagram for the wheel when it is just about to be raised. Three forces act on the wheel at this instant: **F**, which is exerted by the hand; **R**, which is exerted by the curb; and the gravitational force $m\mathbf{g}$. (d) The vector sum of the three external forces acting on the wheel is zero.

$$\theta = 70°$$

We can use the right triangle shown in Figure 12.12d to obtain R:

$$R = \sqrt{(mg)^2 + F^2} = \sqrt{(700\ \text{N})^2 + (300\ \text{N})^2} = \boxed{800\ \text{N}}$$

Roofs, bridges, and other structures that must be both strong and lightweight often are made of trusses similar to the one shown in Figure 12.13a. Imagine that this truss structure represents part of a bridge. To approach this problem, we assume that the structural components are connected by pin joints. We also assume that the entire structure is free to slide horizontally because it rests on "rockers" on each end, which allow it to move back and forth as it undergoes thermal expansion and contraction. We assume the mass of the bridge structure is negligible compared with the load. In this situation, the force exerted by each of the bars (struts) on the hinge pins is a force of tension or of compression and must be along the length of the bar. Let us calculate the force in each strut when the bridge is supporting a 7 200-N load at its center. We will do this by determining the forces that act at the pins.

The force notation that we use here is not of our usual format. Until now, we have used the notation F_{AB} to mean "the force exerted by A on B." For this application, however, the first letter in a double-letter subscript on F indicates the location of the pin on which the force is exerted. The combination of two letters identifies the strut exerting the force on the pin. For example, in Figure 12.13b, F_{AB} is the force exerted by strut AB on the pin at A. The subscripts are symmetric in that strut AB is the same as strut BA and $F_{AB} = F_{BA}$.

First, we apply Newton's second law to the truss as a whole in the vertical direction. Internal forces do not enter into this accounting. We balance the weight of the load with the normal forces exerted at the two ends by the supports on which the bridge rests:

$$\sum F_y = n_A + n_E - F_g = 0$$

$$n_A + n_E = 7\ 200\ \text{N}$$

Next, we calculate the torque about A, noting that the overall length of the bridge structure is $L = 50$ m:

$$\sum \tau = L n_E - (L/2) F_g = 0$$

$$n_E = F_g / 2 = 3\ 600\ \text{N}$$

Although we could repeat the torque calculation for the right end (point E), it should be clear from symmetry arguments that $n_A = 3\ 600$ N.

Now let us balance the vertical forces acting on the pin at point A. If we assume that strut AB is in compression, then the force F_{AB} that the strut exerts on the pin at point A has a negative y component. (If the strut is actually in tension, our calculations will result in a negative value for the magnitude of the force, still of the correct size):

$$\sum F_y = n_A - F_{AB} \sin 30° = 0$$

$$F_{AB} = 7\ 200\ \text{N}$$

The positive result shows that our assumption of compression was correct.

We can now find the force F_{AC} by considering the horizontal forces acting on the pin at point A. Because point A is not accelerating, we can safely assume that F_{AC} must point toward the right (Fig. 12.13b); this indicates that the bar between points A and C is under tension:

$$\sum F_x = F_{AC} - F_{AB} \cos 30° = 0$$

$$F_{AC} = (7\ 200\ \text{N}) \cos 30° = 6\ 200\ \text{N}$$

Now consider the vertical forces acting on the pin at point C. We shall assume that strut CB is in tension. (Imagine the subsequent motion of the pin at point C if strut CB were to

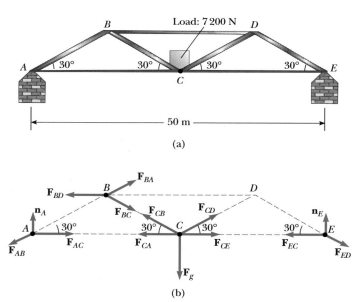

Figure 12.13 (a) Truss structure for a bridge. (b) The forces acting on the pins at points A, B, C, and E. Force vectors are not to scale.

break suddenly.) On the basis of symmetry, we assert that $F_{CB} = F_{CD}$ and $F_{CA} = F_{CE}$:

$$\sum F_y = 2F_{CB} \sin 30° - 7\,200 \text{ N} = 0$$

$$F_{CB} = 7\,200 \text{ N}$$

Finally, we balance the horizontal forces on B, assuming that strut BD is in compression:

$$\sum F_x = F_{BA} \cos 30° + F_{BC} \cos 30° - F_{BD} = 0$$

$$(7\,200 \text{ N}) \cos 30° + (7\,200 \text{ N}) \cos 30° - F_{BD} = 0$$

$$F_{BD} = 12\,000 \text{ N}$$

Thus, the top bar in a bridge of this design must be very strong.

12.4 Elastic Properties of Solids

Except for our discussion about springs in earlier chapters, we have assumed that objects remain rigid when external forces act on them. In reality, all objects are deformable. That is, it is possible to change the shape or the size (or both) of an object by applying external forces. As these changes take place, however, internal forces in the object resist the deformation.

We shall discuss the deformation of solids in terms of the concepts of *stress* and *strain*. **Stress** is a quantity that is proportional to the force causing a deformation; more specifically, stress is the external force acting on an object per unit cross-sectional area. The result of a stress is **strain**, which is a measure of the degree of deformation. It is found that, for sufficiently small stresses, **strain is proportional to stress**; the constant of proportionality depends on the material being deformed and on the nature of the deformation. We call this proportionality constant the **elastic modulus**. The elastic modulus is therefore defined as the ratio of the stress to the resulting strain:

$$\text{Elastic modulus} \equiv \frac{\text{stress}}{\text{strain}} \tag{12.5}$$

The elastic modulus in general relates what is done to a solid object (a force is applied) to how that object responds (it deforms to some extent).

We consider three types of deformation and define an elastic modulus for each:

1. **Young's modulus**, which measures the resistance of a solid to a change in its length
2. **Shear modulus**, which measures the resistance to motion of the planes within a solid parallel to each other
3. **Bulk modulus**, which measures the resistance of solids or liquids to changes in their volume

Young's Modulus: Elasticity in Length

Consider a long bar of cross-sectional area A and initial length L_i that is clamped at one end, as in Figure 12.14. When an external force is applied perpendicular to the cross section, internal forces in the bar resist distortion ("stretching"), but the bar reaches an equilibrium situation in which its final length L_f is greater than L_i and in which the external force is exactly balanced by internal forces. In such a situation, the bar is said to be stressed. We define the **tensile stress** as the ratio of the magnitude of the external force F to the cross-sectional area A. The **tensile strain** in this case is defined as the ratio of the change in length ΔL to the original length L_i. We define **Young's modulus** by a combination of these two ratios:

$$Y \equiv \frac{\text{tensile stress}}{\text{tensile strain}} = \frac{F/A}{\Delta L/L_i} \tag{12.6}$$

Active Figure 12.14 A long bar clamped at one end is stretched by an amount ΔL under the action of a force $\mathbf{F}$.

At the Active Figures link at http://www.pse6.com, *you can adjust the values of the applied force and Young's modulus to observe the change in length of the bar.*

Young's modulus

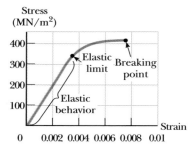

Stress
(MN/m^2)

Figure 12.15 Stress-versus-strain curve for an elastic solid.

(a)

(b)

Active Figure 12.16 (a) A shear deformation in which a rectangular block is distorted by two forces of equal magnitude but opposite directions applied to two parallel faces. (b) A book under shear stress.

At the Active Figures link at http://www.pse6.com, you can adjust the values of the applied force and the shear modulus to observe the change in shape of the block in part (a).

Shear modulus

Table 12.1

Typical Values for Elastic Moduli			
Substance	**Young's Modulus** (N/m^2)	**Shear Modulus** (N/m^2)	**Bulk Modulus** (N/m^2)
Tungsten	35×10^{10}	14×10^{10}	20×10^{10}
Steel	20×10^{10}	8.4×10^{10}	6×10^{10}
Copper	11×10^{10}	4.2×10^{10}	14×10^{10}
Brass	9.1×10^{10}	3.5×10^{10}	6.1×10^{10}
Aluminum	7.0×10^{10}	2.5×10^{10}	7.0×10^{10}
Glass	$6.5–7.8 \times 10^{10}$	$2.6–3.2 \times 10^{10}$	$5.0–5.5 \times 10^{10}$
Quartz	5.6×10^{10}	2.6×10^{10}	2.7×10^{10}
Water	—	—	0.21×10^{10}
Mercury	—	—	2.8×10^{10}

Young's modulus is typically used to characterize a rod or wire stressed under either tension or compression. Note that because strain is a dimensionless quantity, Y has units of force per unit area. Typical values are given in Table 12.1. Experiments show (a) that for a fixed applied force, the change in length is proportional to the original length and (b) that the force necessary to produce a given strain is proportional to the cross-sectional area. Both of these observations are in accord with Equation 12.6.

For relatively small stresses, the bar will return to its initial length when the force is removed. The **elastic limit** of a substance is defined as the maximum stress that can be applied to the substance before it becomes permanently deformed and does not return to its initial length. It is possible to exceed the elastic limit of a substance by applying a sufficiently large stress, as seen in Figure 12.15. Initially, a stress-versus-strain curve is a straight line. As the stress increases, however, the curve is no longer a straight line. When the stress exceeds the elastic limit, the object is permanently distorted and does not return to its original shape after the stress is removed. As the stress is increased even further, the material ultimately breaks.

Shear Modulus: Elasticity of Shape

Another type of deformation occurs when an object is subjected to a force parallel to one of its faces while the opposite face is held fixed by another force (Fig. 12.16a). The stress in this case is called a shear stress. If the object is originally a rectangular block, a shear stress results in a shape whose cross section is a parallelogram. A book pushed sideways, as shown in Figure 12.16b, is an example of an object subjected to a shear stress. To a first approximation (for small distortions), no change in volume occurs with this deformation.

We define the **shear stress** as F/A, the ratio of the tangential force to the area A of the face being sheared. The **shear strain** is defined as the ratio $\Delta x/h$, where Δx is the horizontal distance that the sheared face moves and h is the height of the object. In terms of these quantities, the **shear modulus** is

$$S \equiv \frac{\text{shear stress}}{\text{shear strain}} = \frac{F/A}{\Delta x/h} \qquad (12.7)$$

Values of the shear modulus for some representative materials are given in Table 12.1. Like Young's modulus, the unit of shear modulus is the ratio of that for force to that for area.

Bulk Modulus: Volume Elasticity

Bulk modulus characterizes the response of an object to changes in a force of uniform magnitude applied perpendicularly over the entire surface of the object, as shown in Figure 12.17. (We assume here that the object is made of a single substance.) As we shall see in Chapter 14, such a uniform distribution of forces occurs when an object is immersed in a fluid. An object subject to this type of deformation undergoes a change in volume but no change in shape. The **volume stress** is defined as the ratio of the magnitude of the total force F exerted on a surface to the area A of the surface. The quantity $P = F/A$ is called **pressure**, which we will study in more detail in Chapter 14. If the pressure on an object changes by an amount $\Delta P = \Delta F/A$, then the object will experience a volume change ΔV. The **volume strain** is equal to the change in volume ΔV divided by the initial volume V_i. Thus, from Equation 12.5, we can characterize a volume ("bulk") compression in terms of the **bulk modulus**, which is defined as

$$B \equiv \frac{\text{volume stress}}{\text{volume strain}} = -\frac{\Delta F/A}{\Delta V/V_i} = -\frac{\Delta P}{\Delta V/V_i} \qquad (12.8)$$

Bulk modulus

A negative sign is inserted in this defining equation so that B is a positive number. This maneuver is necessary because an increase in pressure (positive ΔP) causes a decrease in volume (negative ΔV) and vice versa.

Table 12.1 lists bulk moduli for some materials. If you look up such values in a different source, you often find that the reciprocal of the bulk modulus is listed. The reciprocal of the bulk modulus is called the **compressibility** of the material.

Note from Table 12.1 that both solids and liquids have a bulk modulus. However, no shear modulus and no Young's modulus are given for liquids because a liquid does not sustain a shearing stress or a tensile stress. If a shearing force or a tensile force is applied to a liquid, the liquid simply flows in response.

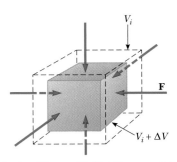

Active Figure 12.17 When a solid is under uniform pressure, it undergoes a change in volume but no change in shape. This cube is compressed on all sides by forces normal to its six faces.

At the Active Figures link at http://www.pse6.com, *you can adjust the values of the applied force and the bulk modulus to observe the change in volume of the cube.*

Quick Quiz 12.4 A block of iron is sliding across a horizontal floor. The friction force between the block and the floor causes the block to deform. To describe the relationship between stress and strain for the block, you would use (a) Young's modulus (b) shear modulus (c) bulk modulus (d) none of these.

Quick Quiz 12.5 A trapeze artist swings through a circular arc. At the bottom of the swing, the wires supporting the trapeze are longer than when the trapeze artist simply hangs from the trapeze, due to the increased tension in them. To describe the relationship between stress and strain for the wires, you would use (a) Young's modulus (b) shear modulus (c) bulk modulus (d) none of these.

Quick Quiz 12.6 A spacecraft carries a steel sphere to a planet on which atmospheric pressure is much higher than on the Earth. The higher pressure causes the radius of the sphere to decrease. To describe the relationship between stress and strain for the sphere, you would use (a) Young's modulus (b) shear modulus (c) bulk modulus (d) none of these.

Prestressed Concrete

If the stress on a solid object exceeds a certain value, the object fractures. The maximum stress that can be applied before fracture occurs depends on the nature of the material and on the type of applied stress. For example, concrete has a tensile strength of about 2×10^6 N/m^2, a compressive strength of 20×10^6 N/m^2, and a shear strength of 2×10^6 N/m^2. If the applied stress exceeds these values, the concrete fractures. It is common practice to use large safety factors to prevent failure in concrete structures.

Active Figure 12.18 (a) A concrete slab with no reinforcement tends to crack under a heavy load. (b) The strength of the concrete is increased by using steel reinforcement rods. (c) The concrete is further strengthened by prestressing it with steel rods under tension.

Concrete is normally very brittle when it is cast in thin sections. Thus, concrete slabs tend to sag and crack at unsupported areas, as shown in Figure 12.18a. The slab can be strengthened by the use of steel rods to reinforce the concrete, as illustrated in Figure 12.18b. Because concrete is much stronger under compression (squeezing) than under tension (stretching) or shear, vertical columns of concrete can support very heavy loads, whereas horizontal beams of concrete tend to sag and crack. However, a significant increase in shear strength is achieved if the reinforced concrete is prestressed, as shown in Figure 12.18c. As the concrete is being poured, the steel rods are held under tension by external forces. The external forces are released after the concrete cures; this results in a permanent tension in the steel and hence a compressive stress on the concrete. This enables the concrete slab to support a much heavier load.

Example 12.6 Stage Design

Recall Example 8.4, in which we analyzed a cable used to support an actor as he swung onto the stage. Suppose that the tension in the cable is 940 N as the actor reaches the lowest point. What diameter should a 10-m-long steel wire have if we do not want it to stretch more than 0.5 cm under these conditions?

Solution From the definition of Young's modulus, we can solve for the required cross-sectional area. Assuming that the cross section is circular, we can determine the diameter of the wire. From Equation 12.6, we have

$$Y = \frac{F/A}{\Delta L/L_i}$$

$$A = \frac{FL_i}{Y\Delta L} = \frac{(940 \text{ N})(10 \text{ m})}{(20 \times 10^{10} \text{ N/m}^2)(0.005 \text{ m})}$$

$$= 9.4 \times 10^{-6} \text{ m}^2$$

Because $A = \pi r^2$, the radius of the wire can be found from

$$r = \sqrt{\frac{A}{\pi}} = \sqrt{\frac{9.4 \times 10^{-6} \text{ m}^2}{\pi}} = 1.7 \times 10^{-3} \text{ m} = 1.7 \text{ mm}$$

$$d = 2r = 2(1.7 \text{ mm}) = \boxed{3.4 \text{ mm}}$$

To provide a large margin of safety, we would probably use a flexible cable made up of many smaller wires having a total cross-sectional area substantially greater than our calculated value.

Example 12.7 Squeezing a Brass Sphere

A solid brass sphere is initially surrounded by air, and the air pressure exerted on it is $1.0 \times 10^5 \text{ N/m}^2$ (normal atmospheric pressure). The sphere is lowered into the ocean to a depth where the pressure is $2.0 \times 10^7 \text{ N/m}^2$. The volume of the sphere in air is 0.50 m³. By how much does this volume change once the sphere is submerged?

Solution From the definition of bulk modulus, we have

$$B = -\frac{\Delta P}{\Delta V/V_i}$$

$$\Delta V = -\frac{V_i \Delta P}{B}$$

Substituting the numerical values, we obtain

$$\Delta V = -\frac{(0.50 \text{ m}^3)(2.0 \times 10^7 \text{ N/m}^2 - 1.0 \times 10^5 \text{ N/m}^2)}{6.1 \times 10^{10} \text{ N/m}^2}$$

$$= \boxed{-1.6 \times 10^{-4} \text{ m}^3}$$

The negative sign indicates that the volume of the sphere decreases.

SUMMARY

Take a practice test for this chapter by clicking on the Practice Test link at http://www.pse6.com.

A rigid object is in **equilibrium** if and only if **the resultant external force acting on it is zero and the resultant external torque on it is zero about any axis:**

$$\Sigma \mathbf{F} = 0 \tag{12.1}$$

$$\Sigma \boldsymbol{\tau} = 0 \tag{12.2}$$

The first condition is the condition for translational equilibrium, and the second is the condition for rotational equilibrium. These two equations allow you to analyze a great variety of problems. Make sure you can identify forces unambiguously, create a free-body diagram, and then apply Equations 12.1 and 12.2 and solve for the unknowns.

The gravitational force exerted on an object can be considered as acting at a single point called the **center of gravity**. The center of gravity of an object coincides with its center of mass if the object is in a uniform gravitational field.

We can describe the elastic properties of a substance using the concepts of stress and strain. **Stress** is a quantity proportional to the force producing a deformation; **strain** is a measure of the degree of deformation. Strain is proportional to stress, and the constant of proportionality is the **elastic modulus:**

$$\text{Elastic modulus} \equiv \frac{\text{stress}}{\text{strain}} \tag{12.5}$$

Three common types of deformation are represented by (1) the resistance of a solid to elongation under a load, characterized by **Young's modulus** Y; (2) the resistance of a solid to the motion of internal planes sliding past each other, characterized by the **shear modulus** S; and (3) the resistance of a solid or fluid to a volume change, characterized by the **bulk modulus** B.

QUESTIONS

1. Stand with your back against a wall. Why can't you put your heels firmly against the wall and then bend forward without falling?

2. Can an object be in equilibrium if it is in motion? Explain.

3. Can an object be in equilibrium when only one force acts upon it? If you believe the answer is yes, give an example to support your conclusion.

4. (a) Give an example in which the net force acting on an object is zero and yet the net torque is nonzero. (b) Give an example in which the net torque acting on an object is zero and yet the net force is nonzero.

5. Can an object be in equilibrium if the only torques acting on it produce clockwise rotation?

6. If you measure the net force and the net torque on a system to be zero, (a) could the system still be rotating with respect to you? (b) Could it be translating with respect to you?

7. The center of gravity of an object may be located outside the object. Give a few examples for which this is the case.

8. Assume you are given an arbitrarily shaped piece of plywood, together with a hammer, nail, and plumb bob. How could you use these items to locate the center of gravity of the plywood? *Suggestion:* Use the nail to suspend the plywood.

9. For a chair to be balanced on one leg, where must the center of gravity of the chair be located?

10. A girl has a large, docile dog she wishes to weigh on a small bathroom scale. She reasons that she can determine her dog's weight with the following method: First she puts the dog's two front feet on the scale and records the scale reading. Then she places the dog's two back feet on the scale and records the reading. She thinks that the sum of the readings will be the dog's weight. Is she correct? Explain your answer.

11. A tall crate and a short crate of equal mass are placed side by side on an incline, without touching each other. As the incline angle is increased, which crate will topple first? Explain.

12. A ladder stands on the ground, leaning against a wall. Would you feel safer climbing up the ladder if you were told that the ground is frictionless but the wall is rough, or that the wall is frictionless but the ground is rough? Justify your answer.

13. When you are lifting a heavy object, it is recommended that you keep your back as nearly vertical as possible, lifting from your knees. Why is this better than bending over and lifting from your waist?

14. What kind of deformation does a cube of Jell-O exhibit when it jiggles?

15. Ruins of ancient Greek temples often have intact vertical columns, but few horizontal slabs of stone are still in place. Can you think of a reason why this is so?

PROBLEMS

1, 2, 3 = straightforward, intermediate, challenging ☐ = full solution available in the *Student Solutions Manual and Study Guide*

🌀 = coached solution with hints available at http://www.pse6.com 💻 = computer useful in solving problem

▨ = paired numerical and symbolic problems

Section 12.1 The Conditions for Equilibrium of a Rigid Body

1. A baseball player holds a 36-oz bat (weight = 10.0 N) with one hand at the point O (Fig. P12.1). The bat is in equilibrium. The weight of the bat acts along a line 60.0 cm to the right of O. Determine the force and the torque exerted by the player on the bat around an axis through O.

— 60.0 cm —

$m\mathbf{g}$

Figure P12.1

2. Write the necessary conditions for equilibrium of the object shown in Figure P12.2. Take the origin of the torque equation at the point O.

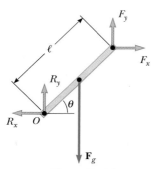

F_y

ℓ

F_x

R_y

θ

R_x O

$\mathbf{F}_g$

Figure P12.2

3. 🌀 A uniform beam of mass m_b and length ℓ supports blocks with masses m_1 and m_2 at two positions, as in Figure P12.3. The beam rests on two knife edges. For what value of x will the beam be balanced at P such that the normal force at O is zero?

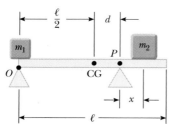

$\frac{\ell}{2}$ d

m_1 m_2

P

O CG

x

ℓ

Figure P12.3

Section 12.2 More on the Center of Gravity

Problems 38, 39, 41, 43, and 44 in Chapter 9 can also be assigned with this section.

4. A circular pizza of radius R has a circular piece of radius $R/2$ removed from one side as shown in Figure P12.4. The center of gravity has moved from C to C' along the x axis. Show that the distance from C to C' is $R/6$. Assume the thickness and density of the pizza are uniform throughout.

C C'

Figure P12.4

5. A carpenter's square has the shape of an L, as in Figure P12.5. Locate its center of gravity.

4.0 cm

18.0 cm

4.0 cm

— 12.0 cm —

Figure P12.5

6. Pat builds a track for his model car out of wood, as in Figure P12.6. The track is 5.00 cm wide, 1.00 m high, and 3.00 m

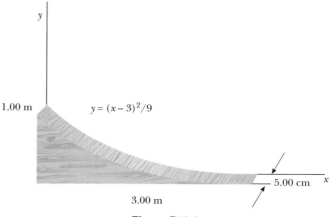

y

1.00 m

$y = (x-3)^2/9$

5.00 cm x

3.00 m

Figure P12.6

long, and is solid. The runway is cut such that it forms a parabola with the equation $y = (x - 3)^2/9$. Locate the horizontal coordinate of the center of gravity of this track.

7. Consider the following mass distribution: 5.00 kg at (0, 0) m, 3.00 kg at (0, 4.00) m, and 4.00 kg at (3.00, 0) m. Where should a fourth object of mass 8.00 kg be placed so that the center of gravity of the four-object arrangement will be at (0, 0)?

8. Figure P12.8 shows three uniform objects: a rod, a right triangle, and a square. Their masses and their coordinates in meters are given. Determine the center of gravity for the three-object system.

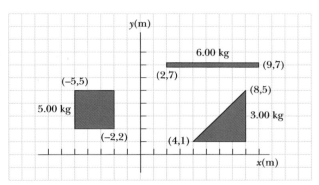

Figure P12.8

Section 12.3 Examples of Rigid Objects in Static Equilibrium

Problems 17, 18, 19, 20, 21, 27, 40, 46, 57, 59, and 73 in Chapter 5 can also be assigned with this section.

9. Find the mass m of the counterweight needed to balance the 1 500-kg truck on the incline shown in Figure P12.9. Assume all pulleys are frictionless and massless.

Figure P12.9

10. A mobile is constructed of light rods, light strings, and beach souvenirs, as shown in Figure P12.10. Determine the masses of the objects (a) m_1, (b) m_2, and (c) m_3.

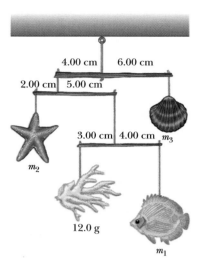

Figure P12.10

11. Two pans of a balance are 50.0 cm apart. The fulcrum of the balance has been shifted 1.00 cm away from the center by a dishonest shopkeeper. By what percentage is the true weight of the goods being marked up by the shopkeeper? (Assume the balance has negligible mass.)

12. A 20.0-kg floodlight in a park is supported at the end of a horizontal beam of negligible mass that is hinged to a pole, as shown in Figure P12.12. A cable at an angle of 30.0° with the beam helps to support the light. Find (a) the tension in the cable and (b) the horizontal and vertical forces exerted on the beam by the pole.

Figure P12.12

13. A 15.0-m uniform ladder weighing 500 N rests against a frictionless wall. The ladder makes a 60.0° angle with the horizontal. (a) Find the horizontal and vertical forces the ground exerts on the base of the ladder when an 800-N firefighter is 4.00 m from the bottom. (b) If the ladder is just on the verge of slipping when the firefighter is 9.00 m up, what is the coefficient of static friction between ladder and ground?

14. A uniform ladder of length L and mass m_1 rests against a frictionless wall. The ladder makes an angle θ with the horizontal. (a) Find the horizontal and vertical forces the ground exerts on the base of the ladder when a firefighter of mass m_2 is a distance x from the bottom. (b) If the ladder is just on the verge of slipping when the firefighter is a distance d from the bottom, what is the coefficient of static friction between ladder and ground?

15. Figure P12.15 shows a claw hammer as it is being used to pull a nail out of a horizontal board. If a force of 150 N is exerted horizontally as shown, find (a) the force exerted by the hammer claws on the nail and (b) the force exerted by the surface on the point of contact with the hammer head. Assume that the force the hammer exerts on the nail is parallel to the nail.

F

30.0 cm

30.0°

Single point of contact

5.00 cm

Figure P12.15

16. A uniform plank of length 6.00 m and mass 30.0 kg rests horizontally across two horizontal bars of a scaffold. The bars are 4.50 m apart, and 1.50 m of the plank hangs over one side of the scaffold. Draw a free-body diagram of the plank. How far can a painter of mass 70.0 kg walk on the overhanging part of the plank before it tips?

17. A 1 500-kg automobile has a wheel base (the distance between the axles) of 3.00 m. The center of mass of the automobile is on the center line at a point 1.20 m behind the front axle. Find the force exerted by the ground on each wheel.

18. A vertical post with a square cross section is 10.0 m tall. Its bottom end is encased in a base 1.50 m tall, which is precisely square but slightly loose. A force 5.50 N to the right acts on the top of the post. The base maintains the post in equilibrium. Find the force that the top of the right side wall of the base exerts on the post. Find the force that the bottom of the left side wall of the base exerts on the post.

19. A flexible chain weighing 40.0 N hangs between two hooks located at the same height (Fig. P12.19). At each hook, the tangent to the chain makes an angle $\theta = 42.0°$ with

θ

Figure P12.19

the horizontal. Find (a) the magnitude of the force each hook exerts on the chain and (b) the tension in the chain at its midpoint. (*Suggestion:* for part (b), make a free-body diagram for half of the chain.)

20. Sir Lost-a-Lot dons his armor and sets out from the castle on his trusty steed in his quest to improve communication between damsels and dragons (Fig. P12.20). Unfortunately his squire lowered the drawbridge too far and finally stopped it 20.0° below the horizontal. Lost-a-Lot and his horse stop when their combined center of mass is 1.00 m from the end of the bridge. The uniform bridge is 8.00 m long and has mass 2 000 kg. The lift cable is attached to the bridge 5.00 m from the hinge at the castle end, and to a point on the castle wall 12.0 m above the bridge. Lost-a-Lot's mass combined with his armor and steed is 1 000 kg. Determine (a) the tension in the cable and the (b) horizontal and (c) vertical force components acting on the bridge at the hinge.

Figure P12.20 Problems 20 and 21.

21. Review problem. In the situation described in Problem 20 and illustrated in Figure P12.20, the lift cable suddenly breaks! The hinge between the castle wall and the bridge is frictionless, and the bridge swings freely until it is vertical. (a) Find the angular acceleration of the bridge once it starts to move. (b) Find the angular speed of the bridge when it strikes the vertical castle wall below the hinge. (c) Find the force exerted by the hinge on the bridge immediately after the cable breaks. (d) Find the force exerted by the hinge on the bridge immediately before it strikes the castle wall.

22. Stephen is pushing his sister Joyce in a wheelbarrow when it is stopped by a brick 8.00 cm high (Fig. P12.22).

The handles make an angle of 15.0° below the horizontal. A downward force of 400 N is exerted on the wheel, which has a radius of 20.0 cm. (a) What force must Stephen apply along the handles in order to just start the wheel over the brick? (b) What is the force (magnitude and direction) that the brick exerts on the wheel just as the wheel begins to lift over the brick? Assume in both parts that the brick remains fixed and does not slide along the ground.

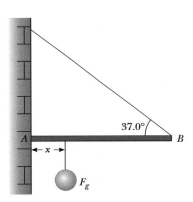

Figure P12.22

23. One end of a uniform 4.00-m-long rod of weight F_g is supported by a cable. The other end rests against the wall, where it is held by friction, as in Figure P12.23. The coefficient of static friction between the wall and the rod is $\mu_s = 0.500$. Determine the minimum distance x from point A at which an additional weight F_g (the same as the weight of the rod) can be hung without causing the rod to slip at point A.

Figure P12.23

24. Two identical uniform bricks of length L are placed in a stack over the edge of a horizontal surface with the maxi-

Figure P12.24

mum overhang possible without falling, as in Figure P12.24. Find the distance x.

25. A vaulter holds a 29.4-N pole in equilibrium by exerting an upward force $\mathbf{U}$ with her leading hand and a downward force $\mathbf{D}$ with her trailing hand, as shown in Figure P12.25. Point C is the center of gravity of the pole. What are the magnitudes of $\mathbf{U}$ and $\mathbf{D}$?

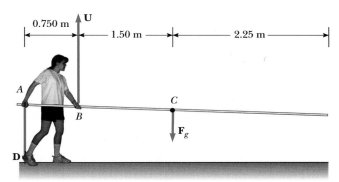

Figure P12.25

26. In the **What If?** section of Example 12.3, let x represent the distance in meters between the person and the hinge at the left end of the beam. (a) Show that the cable tension in newtons is given by $T = 93.9x + 125$. Argue that T increases as x increases. (b) Show that the direction angle θ of the hinge force is described by

$$\tan \theta = \left(\frac{32}{3x + 4} - 1 \right) \tan 53.0°$$

How does θ change as x increases? (c) Show that the magnitude of the hinge force is given by

$$R = \sqrt{8.82 \times 10^3 x^2 - 9.65 \times 10^4 x + 4.96 \times 10^5}$$

How does R change as x increases?

Section 12.4 Elastic Properties of Solids

27. A 200-kg load is hung on a wire having a length of 4.00 m, cross-sectional area 0.200×10^{-4} m², and Young's modulus 8.00×10^{10} N/m². What is its increase in length?

28. Assume that Young's modulus is 1.50×10^{10} N/m² for bone and that the bone will fracture if stress greater than 1.50×10^8 N/m² is imposed on it. (a) What is the maximum force that can be exerted on the femur bone in the leg if it has a minimum effective diameter of 2.50 cm? (b) If this much force is applied compressively, by how much does the 25.0-cm-long bone shorten?

29. Evaluate Young's modulus for the material whose stress-versus-strain curve is shown in Figure 12.15.

30. A steel wire of diameter 1 mm can support a tension of 0.2 kN. A cable to support a tension of 20 kN should have diameter of what order of magnitude?

31. A child slides across a floor in a pair of rubber-soled shoes. The friction force acting on each foot is 20.0 N.

The footprint area of each shoe sole is 14.0 cm², and the thickness of each sole is 5.00 mm. Find the horizontal distance by which the upper and lower surfaces of each sole are offset. The shear modulus of the rubber is 3.00 MN/m².

32. Review problem. A 30.0-kg hammer strikes a steel spike 2.30 cm in diameter while moving with speed 20.0 m/s. The hammer rebounds with speed 10.0 m/s after 0.110 s. What is the average strain in the spike during the impact?

33. If the shear stress in steel exceeds 4.00×10^8 N/m², the steel ruptures. Determine the shearing force necessary to (a) shear a steel bolt 1.00 cm in diameter and (b) punch a 1.00-cm-diameter hole in a steel plate 0.500 cm thick.

34. Review problem. A 2.00-m-long cylindrical steel wire with a cross-sectional diameter of 4.00 mm is placed over a light frictionless pulley, with one end of the wire connected to a 5.00-kg object and the other end connected to a 3.00-kg object. By how much does the wire stretch while the objects are in motion?

35. When water freezes, it expands by about 9.00%. What pressure increase would occur inside your automobile engine block if the water in it froze? (The bulk modulus of ice is 2.00×10^9 N/m².)

36. The deepest point in the ocean is in the Mariana Trench, about 11 km deep. The pressure at this depth is huge, about 1.13×10^8 N/m². (a) Calculate the change in volume of 1.00 m³ of seawater carried from the surface to this deepest point in the Pacific ocean. (b) The density of seawater at the surface is 1.03×10^3 kg/m³. Find its density at the bottom. (c) Is it a good approximation to think of water as incompressible?

37. A walkway suspended across a hotel lobby is supported at numerous points along its edges by a vertical cable above each point and a vertical column underneath. The steel cable is 1.27 cm in diameter and is 5.75 m long before loading. The aluminum column is a hollow cylinder with an inside diameter of 16.14 cm, an outside diameter of 16.24 cm, and unloaded length of 3.25 m. When the walkway exerts a load force of 8 500 N on one of the support points, how much does the point move down?

Additional Problems

38. A lightweight, rigid beam 10.0 m long is supported by a cable attached to a spring of force constant $k = 8.25$ kN/m as shown in Figure P12.38. When no load is hung on the beam ($F_g = 0$), the length L is equal to 5.00 m. (a) Find the angle θ in this situation. (b) Now a load of $F_g = 250$ N is hung on the end of the beam. Temporarily ignore the extension of the spring and the change in the angle θ. Calculate the tension in the cable with this approximation. (c) Use the answer to part (b) to calculate the spring elongation and a new value for the angle θ. (d) With the value of θ from part (c), find a second approximation for the tension in the cable. (e) Use the answer to part (d) to calculate more precise values for the spring elongation and

the angle θ. (f) To three-digit precision, what is the actual value of θ under load?

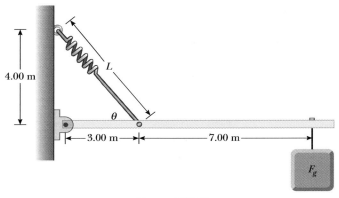

Figure P12.38

39. A bridge of length 50.0 m and mass 8.00×10^4 kg is supported on a smooth pier at each end as in Figure P12.39. A truck of mass 3.00×10^4 kg is located 15.0 m from one end. What are the forces on the bridge at the points of support?

Figure P12.39

40. Refer to Figure 12.18(c). A lintel of prestressed reinforced concrete is 1.50 m long. The cross-sectional area of the concrete is 50.0 cm². The concrete encloses one steel reinforcing rod with cross-sectional area 1.50 cm². The rod joins two strong end plates. Young's modulus for the concrete is 30.0×10^9 N/m². After the concrete cures and the original tension T_1 in the rod is released, the concrete is to be under compressive stress 8.00×10^6 N/m². (a) By what distance will the rod compress the concrete when the original tension in the rod is released? (b) The rod will still be under what tension T_2? (c) The rod will then be how much longer than its unstressed length? (d) When the concrete was poured, the rod should have been stretched by what extension distance from its unstressed length? (e) Find the required original tension T_1 in the rod.

41. A uniform pole is propped between the floor and the ceiling of a room. The height of the room is 7.80 ft, and the coefficient of static friction between the pole and the ceiling is 0.576. The coefficient of static friction between the pole and the floor is greater than that. What is the length of the longest pole that can be propped between the floor and the ceiling?

42. A solid sphere of radius R and mass M is placed in a trough as shown in Figure P12.42. The inner surfaces of the trough are frictionless. Determine the forces exerted by the trough on the sphere at the two contact points.

Figure P12.42

43. A hungry bear weighing 700 N walks out on a beam in an attempt to retrieve a basket of food hanging at the end of the beam (Fig. P12.43). The beam is uniform, weighs 200 N, and is 6.00 m long; the basket weighs 80.0 N. (a) Draw a free-body diagram for the beam. (b) When the bear is at $x = 1.00$ m, find the tension in the wire and the components of the force exerted by the wall on the left end of the beam. (c) **What If?** If the wire can withstand a maximum tension of 900 N, what is the maximum distance the bear can walk before the wire breaks?

Figure P12.43

44. A farm gate (Fig. P12.44) is 3.00 m wide and 1.80 m high, with hinges attached to the top and bottom. The guy wire makes an angle of 30.0° with the top of the gate and is tightened by a turnbuckle to a tension of 200 N. The mass of the gate is 40.0 kg. (a) Determine the horizontal force exerted by the bottom hinge on the gate. (b) Find the horizontal force exerted by the upper hinge. (c) Determine the combined vertical force exerted by both hinges. (d) **What If?** What must be the tension in the guy wire so that the horizontal force exerted by the upper hinge is zero?

Figure P12.44

45. A uniform sign of weight F_g and width $2L$ hangs from a light, horizontal beam, hinged at the wall and supported by a cable (Fig. P12.45). Determine (a) the tension in the cable and (b) the components of the reaction force exerted by the wall on the beam, in terms of F_g, d, L, and θ.

Figure P12.45

46. A 1 200-N uniform boom is supported by a cable as in Figure P12.46. The boom is pivoted at the bottom, and a 2 000-N object hangs from its top. Find the tension in the cable and the components of the reaction force exerted by the floor on the boom.

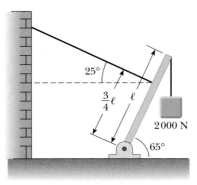

Figure P12.46

47. A crane of mass 3 000 kg supports a load of 10 000 kg as in Figure P12.47. The crane is pivoted with a frictionless pin at *A* and rests against a smooth support at *B*. Find the reaction forces at *A* and *B*.

Figure P12.47

48. A ladder of uniform density and mass *m* rests against a frictionless vertical wall, making an angle of 60.0° with the horizontal. The lower end rests on a flat surface where the coefficient of static friction is $\mu_s = 0.400$. A window cleaner with mass $M = 2m$ attempts to climb the ladder. What fraction of the length *L* of the ladder will the worker have reached when the ladder begins to slip?

49. A 10 000-N shark is supported by a cable attached to a 4.00-m rod that can pivot at the base. Calculate the tension in the tie-rope between the rod and the wall if it is holding the system in the position shown in Figure P12.49. Find the horizontal and vertical forces exerted on the base of the rod. (Neglect the weight of the rod.)

Figure P12.49

50. When a person stands on tiptoe (a strenuous position), the position of the foot is as shown in Figure P12.50a. The gravitational force on the body $\mathbf{F}_g$ is supported by the force **n** exerted by the floor on the toe. A mechanical model for the situation is shown in Figure P12.50b, where **T** is the force exerted by the Achilles tendon on the foot

and **R** is the force exerted by the tibia on the foot. Find the values of *T*, *R*, and θ when $F_g = 700$ N.

Figure P12.50

51. A person bending forward to lift a load "with his back" (Fig. P12.51a) rather than "with his knees" can be injured by large forces exerted on the muscles and vertebrae. The spine pivots mainly at the fifth lumbar vertebra, with the principal supporting force provided by the erector spinalis muscle in the back. To see the magnitude of the forces involved, and to understand why back problems are common among humans, consider the model shown in Figure P12.51b for a person bending forward to lift a 200-N object. The spine and upper body are represented as a uniform horizontal rod of weight 350 N, pivoted at the base of the spine. The erector spinalis muscle, attached at a

Figure P12.51

point two thirds of the way up the spine, maintains the position of the back. The angle between the spine and this muscle is 12.0°. Find the tension in the back muscle and the compressional force in the spine.

52. A uniform rod of weight F_g and length L is supported at its ends by a frictionless trough as shown in Figure P12.52. (a) Show that the center of gravity of the rod must be vertically over point O when the rod is in equilibrium. (b) Determine the equilibrium value of the angle θ.

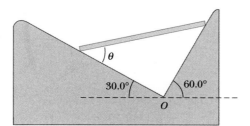

Figure P12.52

53. A force acts on a rectangular cabinet weighing 400 N, as in Figure P12.53. (a) If the cabinet slides with constant speed when $F = 200$ N and $h = 0.400$ m, find the coefficient of kinetic friction and the position of the resultant normal force. (b) If $F = 300$ N, find the value of h for which the cabinet just begins to tip.

Figure P12.53 Problems 53 and 54.

54. Consider the rectangular cabinet of Problem 53, but with a force **F** applied horizontally at the upper edge. (a) What is the minimum force required to start to tip the cabinet? (b) What is the minimum coefficient of static friction required for the cabinet not to slide with the application of a force of this magnitude? (c) Find the magnitude and direction of the minimum force required to tip the cabinet if the point of application can be chosen anywhere on the cabinet.

55. A uniform beam of mass m is inclined at an angle θ to the horizontal. Its upper end produces a ninety-degree bend in a very rough rope tied to a wall, and its lower end rests on a rough floor (Fig. P12.55). (a) If the coefficient of static friction between beam and floor is μ_s, determine an expression for the maximum mass M that can be suspended from the top before the beam slips. (b) Determine the magnitude of the reaction force at the floor and the magnitude of the force exerted by the beam on the rope at P in terms of m, M, and μ_s.

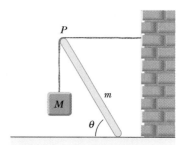

Figure P12.55

56. Figure P12.56 shows a truss that supports a downward force of 1 000 N applied at the point B. The truss has negligible weight. The piers at A and C are smooth. (a) Apply the conditions of equilibrium to prove that $n_A = 366$ N and $n_C = 634$ N. (b) Show that, because forces act on the light truss only at the hinge joints, each bar of the truss must exert on each hinge pin only a force along the length of that bar—a force of tension or compression. (c) Find the force of tension or of compression in each of the three bars.

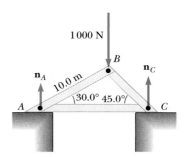

Figure P12.56

57. A stepladder of negligible weight is constructed as shown in Figure P12.57. A painter of mass 70.0 kg stands on the ladder 3.00 m from the bottom. Assuming the floor is frictionless, find (a) the tension in the horizontal bar connecting the two halves of the ladder, (b) the normal forces at A

Figure P12.57

and *B*, and (c) the components of the reaction force at the single hinge *C* that the left half of the ladder exerts on the right half. (*Suggestion:* Treat the ladder as a single object, but also each half of the ladder separately.)

58. A flat dance floor of dimensions 20.0 m by 20.0 m has a mass of 1 000 kg. Three dance couples, each of mass 125 kg, start in the top left, top right, and bottom left corners. (a) Where is the initial center of gravity? (b) The couple in the bottom left corner moves 10.0 m to the right. Where is the new center of gravity? (c) What was the average velocity of the center of gravity if it took that couple 8.00 s to change positions?

59. A shelf bracket is mounted on a vertical wall by a single screw, as shown in Figure P12.59. Neglecting the weight of the bracket, find the horizontal component of the force that the screw exerts on the bracket when an 80.0 N vertical force is applied as shown. (*Hint:* Imagine that the bracket is slightly loose.)

80.0 N 5.00 cm

3.00 cm

6.00 cm

Figure P12.59

60. Figure P12.60 shows a vertical force applied tangentially to a uniform cylinder of weight F_g. The coefficient of static friction between the cylinder and all surfaces is 0.500. In terms of F_g, find the maximum force **P** that can be applied that does not cause the cylinder to rotate. (*Hint:* When the cylinder is on the verge of slipping, both friction forces are at their maximum values. Why?)

P

Figure P12.60

61. **Review problem.** A wire of length *L*, Young's modulus *Y*, and cross-sectional area *A* is stretched elastically by an amount ΔL. By Hooke's law (Section 7.4), the restoring force is $- k \Delta L$. (a) Show that $k = YA/L$. (b) Show that the

work done in stretching the wire by an amount ΔL is

$$W = \tfrac{1}{2} YA(\Delta L)^2/L$$

62. Two racquetballs are placed in a glass jar, as shown in Figure P12.62. Their centers and the point *A* lie on a straight line. (a) Assume that the walls are frictionless, and determine P_1, P_2, and P_3. (b) Determine the magnitude of the force exerted by the left ball on the right ball. Assume each ball has a mass of 170 g.

Figure P12.62

63. In exercise physiology studies it is sometimes important to determine the location of a person's center of mass. This can be done with the arrangement shown in Figure P12.63. A light plank rests on two scales, which give readings of $F_{g1} = 380$ N and $F_{g2} = 320$ N. The scales are separated by a distance of 2.00 m. How far from the woman's feet is her center of mass?

2.00 m

F_{g1} F_{g2}

Figure P12.63

64. A steel cable 3.00 cm² in cross-sectional area has a mass of 2.40 kg per meter of length. If 500 m of the cable is hung over a vertical cliff, how much does the cable stretch under its own weight? $Y_{steel} = 2.00 \times 10^{11}$ N/m².

65. (a) Estimate the force with which a karate master strikes a board if the hand's speed at time of impact is 10.0 m/s, decreasing to 1.00 m/s during a 0.002 00-s time-of-contact with the board. The mass of his hand and arm is 1.00 kg. (b) Estimate the shear stress if this force is exerted on a 1.00-cm-thick pine board that is 10.0 cm wide. (c) If the maximum shear stress a pine board can support before breaking is 3.60×10^6 N/m², will the board break?

66. A bucket is made from thin sheet metal. The bottom and top of the bucket have radii of 25.0 cm and 35.0 cm, respectively. The bucket is 30.0 cm high and filled with water. Where is the center of gravity? (Ignore the weight of the bucket itself.)

67. **Review problem.** An aluminum wire is 0.850 m long and has a circular cross section of diameter 0.780 mm. Fixed at the top end, the wire supports a 1.20-kg object that swings in a horizontal circle. Determine the angular velocity required to produce a strain of 1.00×10^{-3}.

68. A bridge truss extends 200 m across a river (Fig. P12.68). The structure is free to slide horizontally to permit thermal expansion. The structural components are connected by pin joints, and the masses of the bars are small compared with the mass of a 1 360-kg car at the center. Calculate the force of tension or compression in each structural component.

Figure P12.68

69. A bridge truss extends 100 m across a river (Fig. P12.69). The structure is free to slide horizontally to permit thermal expansion. The structural components are connected by pin joints, and the masses of the bars are small compared with the mass of a 1 500-kg car halfway between points A and C. Show that the weight of the car is in effect equally distributed between points A and C. Specify

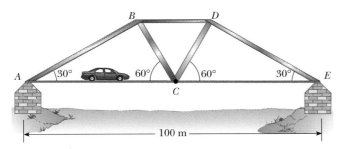

Figure P12.69

whether each structural component is under tension or compression and find the force in each.

70. **Review problem.** A cue strikes a cue ball and delivers a horizontal impulse in such a way that the ball rolls without slipping as it starts to move. At what height above the ball's center (in terms of the radius of the ball) was the blow struck?

71. **Review problem.** A trailer with loaded weight F_g is being pulled by a vehicle with a force **P**, as in Figure P12.71. The trailer is loaded such that its center of mass is located as shown. Neglect the force of rolling friction and let a represent the x component of the acceleration of the trailer. (a) Find the vertical component of **P** in terms of the given parameters. (b) If $a = 2.00 \text{ m/s}^2$ and $h = 1.50$ m, what must be the value of d in order that $P_y = 0$ (no vertical load on the vehicle)? (c) Find the values of P_x and P_y given that $F_g = 1\,500$ N, $d = 0.800$ m, $L = 3.00$ m, $h = 1.50$ m, and $a = -2.00 \text{ m/s}^2$.

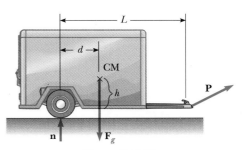

Figure P12.71

72. **Review problem.** A bicycle is traveling downhill at a high speed. Suddenly, the cyclist sees that a bridge ahead has collapsed, so she has to stop. What is the maximum magnitude of acceleration the bicycle can have if it is not to flip over its front wheel—in particular, if its rear wheel is not to leave the ground? The slope makes an angle of 20.0° with the horizontal. On level ground, the center of mass of the woman–bicycle system is at a point 1.05 m above the ground, 65.0 cm horizontally behind the axle of the front wheel, and 35.0 cm in front of the rear axle. Assume that the tires do not skid.

73. **Review problem.** A car moves with speed v on a horizontal circular track of radius R. A head-on view of the car is shown in Figure P12.73. The height of the car's center of mass above the ground is h, and the separation between its inner and outer wheels is d. The road is dry, and the car does not skid. Show that the maximum speed the car can

Figure P12.73

have without overturning is given by

$$v_{max} = \sqrt{\frac{gRd}{2h}}$$

To reduce the risk of rollover, should one increase or decrease h? Should one increase or decrease the width d of the wheel base?

Answers to Quick Quizzes

12.1 (a). The unbalanced torques due to the forces in Figure 12.2 cause an angular acceleration even though the linear acceleration is zero.

12.2 (b). Notice that the lines of action of all the forces in Figure 12.3 intersect at a common point. Thus, the net torque about this point is zero. This zero value of the net torque is independent of the values of the forces. Because no force has a downward component, there is a net force and the object is not in force equilibrium.

12.3 (b). Both the object and the center of gravity of the meter stick are 25 cm from the pivot point. Thus, the meter stick and the object must have the same mass if the system is balanced.

12.4 (b). The friction force on the block as it slides along the surface is parallel to the lower surface and will cause the block to undergo a shear deformation.

12.5 (a). The stretching of the wire due to the increased tension is described by Young's modulus.

12.6 (c). The pressure of the atmosphere results in a force of uniform magnitude perpendicular at all points on the surface of the sphere.

Chapter 13

Universal Gravitation

▲ An understanding of the law of universal gravitation has allowed scientists to send spacecraft on impressively accurate journeys to other parts of our solar system. This photo of a volcano on Io, a moon of Jupiter, was taken by the Galileo spacecraft, which has been orbiting Jupiter since 1995. The red material has been vented from below the surface. (Univ. of Arizona/JPL/NASA)

389

Before 1687, a large amount of data had been collected on the motions of the Moon and the planets, but a clear understanding of the forces related to these motions was not available. In that year, Isaac Newton provided the key that unlocked the secrets of the heavens. He knew, from his first law, that a net force had to be acting on the Moon because without such a force the Moon would move in a straight-line path rather than in its almost circular orbit. Newton reasoned that this force was the gravitational attraction exerted by the Earth on the Moon. He realized that the forces involved in the Earth–Moon attraction and in the Sun–planet attraction were not something special to those systems, but rather were particular cases of a general and universal attraction between objects. In other words, Newton saw that the same force of attraction that causes the Moon to follow its path around the Earth also causes an apple to fall from a tree. As he put it, "I deduced that the forces which keep the planets in their orbs must be reciprocally as the squares of their distances from the centers about which they revolve; and thereby compared the force requisite to keep the Moon in her orb with the force of gravity at the surface of the Earth; and found them answer pretty nearly."

In this chapter we study the law of universal gravitation. We emphasize a description of planetary motion because astronomical data provide an important test of this law's validity. We then show that the laws of planetary motion developed by Johannes Kepler follow from the law of universal gravitation and the concept of conservation of angular momentum. We conclude by deriving a general expression for gravitational potential energy and examining the energetics of planetary and satellite motion.

13.1   Newton's Law of Universal Gravitation

You may have heard the legend that Newton was struck on the head by a falling apple while napping under a tree. This alleged accident supposedly prompted him to imagine that perhaps all objects in the Universe were attracted to each other in the same way the apple was attracted to the Earth. Newton analyzed astronomical data on the motion of the Moon around the Earth. From that analysis, he made the bold assertion that the force law governing the motion of planets was the *same* as the force law that attracted a falling apple to the Earth. This was the first time that "earthly" and "heavenly" motions were unified. We shall look at the mathematical details of Newton's analysis in this section.

In 1687 Newton published his work on the law of gravity in his treatise *Mathematical Principles of Natural Philosophy*. **Newton's law of universal gravitation** states that

The law of universal gravitation

every particle in the Universe attracts every other particle with a force that is directly proportional to the product of their masses and inversely proportional to the square of the distance between them.

If the particles have masses m_1 and m_2 and are separated by a distance r, the magnitude of this gravitational force is

$$F_g = G \frac{m_1 m_2}{r^2} \tag{13.1}$$

where G is a constant, called the *universal gravitational constant,* that has been measured experimentally. Its value in SI units is

$$G = 6.673 \times 10^{-11} \ \text{N} \cdot \text{m}^2/\text{kg}^2 \tag{13.2}$$

The form of the force law given by Equation 13.1 is often referred to as an **inverse-square law** because the magnitude of the force varies as the inverse square of the separation of the particles.[1] We shall see other examples of this type of force law in subsequent chapters. We can express this force in vector form by defining a unit vector $\hat{\mathbf{r}}_{12}$ (Fig. 13.1). Because this unit vector is directed from particle 1 toward particle 2, the force exerted by particle 1 on particle 2 is

$$\mathbf{F}_{12} = -G \frac{m_1 m_2}{r^2} \hat{\mathbf{r}}_{12} \tag{13.3}$$

where the negative sign indicates that particle 2 is attracted to particle 1, and hence the force on particle 2 must be directed toward particle 1. By Newton's third law, the force exerted by particle 2 on particle 1, designated $\mathbf{F}_{21}$, is equal in magnitude to $\mathbf{F}_{12}$ and in the opposite direction. That is, these forces form an action–reaction pair, and $\mathbf{F}_{21} = -\mathbf{F}_{12}$.

Several features of Equation 13.3 deserve mention. The gravitational force is a field force that always exists between two particles, regardless of the medium that separates them. Because the force varies as the inverse square of the distance between the particles, it decreases rapidly with increasing separation.

Another important point that we can show from Equation 13.3 is that **the gravitational force exerted by a finite-size, spherically symmetric mass distribution on a particle outside the distribution is the same as if the entire mass of the distribution were concentrated at the center.** For example, the magnitude of the force exerted by the Earth on a particle of mass m near the Earth's surface is

$$F_g = G \frac{M_E m}{R_E^2} \tag{13.4}$$

where M_E is the Earth's mass and R_E its radius. This force is directed toward the center of the Earth.

In formulating his law of universal gravitation, Newton used the following reasoning, which supports the assumption that the gravitational force is proportional to the inverse square of the separation between the two interacting objects. He compared the acceleration of the Moon in its orbit with the acceleration of an object falling near the Earth's surface, such as the legendary apple (Fig. 13.2). Assuming that both accelerations had the same cause—namely, the gravitational attraction of the Earth—Newton used the inverse-square law to reason that the acceleration of the Moon toward the Earth (centripetal acceleration) should be proportional to $1/r_M^2$, where r_M is the distance between the centers of the Earth and the Moon. Furthermore, the acceleration of the apple toward the Earth should be proportional to $1/R_a^2$, where R_a is the distance between the centers of the Earth and the apple. Because the apple is located at the surface of the earth, $R_a = R_E$, the radius of the Earth. Using the values $r_M = 3.84 \times 10^8$ m and $R_E = 6.37 \times 10^6$ m, Newton predicted that the ratio of the Moon's acceleration a_M to the apple's acceleration g would be

$$\frac{a_M}{g} = \frac{(1/r_M)^2}{(1/R_E)^2} = \left(\frac{R_E}{r_M}\right)^2 = \left(\frac{6.37 \times 10^6 \ \text{m}}{3.84 \times 10^8 \ \text{m}}\right)^2 = 2.75 \times 10^{-4}$$

[1] An *inverse* proportionality between two quantities x and y is one in which $y = k/x$, where k is a constant. A *direct* proportion between x and y exists when $y = kx$.

 PITFALL PREVENTION

13.1 Be Clear on *g* and *G*

The symbol g represents the magnitude of the free-fall acceleration near a planet. At the surface of the Earth, g has the value 9.80 m/s². On the other hand, G is a universal constant that has the same value everywhere in the Universe.

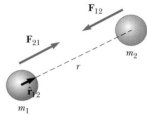

Active Figure 13.1 The gravitational force between two particles is attractive. The unit vector $\hat{\mathbf{r}}_{12}$ is directed from particle 1 toward particle 2. Note that $\mathbf{F}_{21} = -\mathbf{F}_{12}$.

At the Active Figures link at http://www.pse6.com, *you can change the masses of the particles and the separation distance between the particles to see the effect on the gravitational force.*

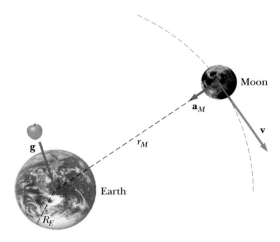

Figure 13.2 As it revolves around the Earth, the Moon experiences a centripetal acceleration $\mathbf{a}_M$ directed toward the Earth. An object near the Earth's surface, such as the apple shown here, experiences an acceleration $\mathbf{g}$. (Dimensions are not to scale.)

Therefore, the centripetal acceleration of the Moon is

$$a_M = (2.75 \times 10^{-4})(9.80 \text{ m/s}^2) = 2.70 \times 10^{-3} \text{ m/s}^2$$

Newton also calculated the centripetal acceleration of the Moon from a knowledge of its mean distance from the Earth and the known value of its orbital period, $T = 27.32$ days $= 2.36 \times 10^6$ s. In a time interval T, the Moon travels a distance $2\pi r_M$, which equals the circumference of its orbit. Therefore, its orbital speed is $2\pi r_M / T$ and its centripetal acceleration is

$$a_M = \frac{v^2}{r_M} = \frac{(2\pi r_M / T)^2}{r_M} = \frac{4\pi^2 r_M}{T^2} = \frac{4\pi^2 (3.84 \times 10^8 \text{ m})}{(2.36 \times 10^6 \text{ s})^2}$$

$$= 2.72 \times 10^{-3} \text{ m/s}^2$$

The nearly perfect agreement between this value and the value Newton obtained using g provides strong evidence of the inverse-square nature of the gravitational force law.

Although these results must have been very encouraging to Newton, he was deeply troubled by an assumption he made in the analysis. To evaluate the acceleration of an object at the Earth's surface, Newton treated the Earth as if its mass were all concentrated at its center. That is, he assumed that the Earth acted as a particle as far as its influence on an exterior object was concerned. Several years later, in 1687, on the basis of his pioneering work in the development of calculus, Newton proved that this assumption was valid and was a natural consequence of the law of universal gravitation.

We have evidence that the gravitational force acting on an object is directly proportional to its mass from our observations of falling objects, discussed in Chapter 2. All objects, regardless of mass, fall in the absence of air resistance at the same acceleration g near the surface of the Earth. According to Newton's second law, this acceleration is given by $g = F_g/m$, where m is the mass of the falling object. If this ratio is to be the same for all falling objects, then F_g must be directly proportional to m, so that the mass cancels in the ratio. If we consider the more general situation of a gravitational force between any two objects with mass, such as two planets, this same argument can be applied to show that the gravitational force is proportional to one of the masses. We can choose *either* of the masses in the argument, however; thus, the gravitational force must be directly proportional to *both* masses, as can be seen in Equation 13.3.

Quick Quiz 13.1 The Moon remains in its orbit around the Earth rather than falling to the Earth because (a) it is outside of the gravitational influence of the Earth (b) it is in balance with the gravitational forces from the Sun and other planets (c) the net force on the Moon is zero (d) none of these (e) all of these.

> **Quick Quiz 13.2** A planet has two moons of equal mass. Moon 1 is in a circular orbit of radius r. Moon 2 is in a circular orbit of radius $2r$. The magnitude of the gravitational force exerted by the planet on moon 2 is (a) four times as large as that on moon 1 (b) twice as large as that on moon 1 (c) equal to that on moon 1 (d) half as large as that on moon 1 (e) one fourth as large as that on moon 1.

Example 13.1   Billiards, Anyone?   **Interactive**

Three 0.300-kg billiard balls are placed on a table at the corners of a right triangle, as shown in Figure 13.3. Calculate the gravitational force on the cue ball (designated m_1) resulting from the other two balls.

Solution First we calculate separately the individual forces on the cue ball due to the other two balls, and then we find the vector sum to obtain the resultant force. We can see graphically that this force should point upward and toward the right. We locate our coordinate axes as shown in Figure 13.3, placing our origin at the position of the cue ball.

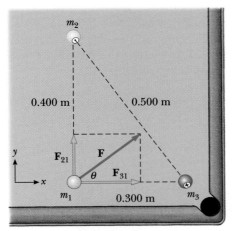

Figure 13.3 (Example 13.1) The resultant gravitational force acting on the cue ball is the vector sum $\mathbf{F}_{21} + \mathbf{F}_{31}$.

The force exerted by m_2 on the cue ball is directed upward and is given by

$$\mathbf{F}_{21} = G\,\frac{m_2 m_1}{r_{21}^{2}}\,\hat{\mathbf{j}}$$

$$= (6.67 \times 10^{-11}\ \text{N}\cdot\text{m}^2/\text{kg}^2)\,\frac{(0.300\ \text{kg})(0.300\ \text{kg})}{(0.400\ \text{m})^2}\,\hat{\mathbf{j}}$$

$$= 3.75 \times 10^{-11}\,\hat{\mathbf{j}}\ \text{N}$$

This result shows that the gravitational forces between everyday objects have extremely small magnitudes. The force exerted by m_3 on the cue ball is directed to the right:

$$\mathbf{F}_{31} = G\,\frac{m_3 m_1}{r_{31}^{2}}\,\hat{\mathbf{i}}$$

$$= (6.67 \times 10^{-11}\ \text{N}\cdot\text{m}^2/\text{kg}^2)\,\frac{(0.300\ \text{kg})(0.300\ \text{kg})}{(0.300\ \text{m})^2}\,\hat{\mathbf{i}}$$

$$= 6.67 \times 10^{-11}\,\hat{\mathbf{i}}\ \text{N}$$

Therefore, the net gravitational force on the cue ball is

$$\mathbf{F} = \mathbf{F}_{21} + \mathbf{F}_{31} = \boxed{(6.67\hat{\mathbf{i}} + 3.75\hat{\mathbf{j}}) \times 10^{-11}\ \text{N}}$$

and the magnitude of this force is

$$F = \sqrt{F_{21}^{2} + F_{31}^{2}} = \sqrt{(3.75)^2 + (6.67)^2} \times 10^{-11}\ \text{N}$$

$$= 7.65 \times 10^{-11}\ \text{N}$$

From $\tan\theta = 3.75/6.67 = 0.562$, the direction of the net gravitational force is $\theta = 29.3°$ counterclockwise from the x axis.

 At the Interactive Worked Example link at **http://www.pse6.com**, *you can move balls 2 and 3 to see the effect on the net gravitational force on ball 1.*

13.2   Measuring the Gravitational Constant

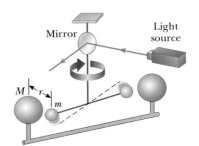

Figure 13.4 Cavendish apparatus for measuring G. The dashed line represents the original position of the rod.

The universal gravitational constant G was measured in an important experiment by Henry Cavendish (1731–1810) in 1798. The Cavendish apparatus consists of two small spheres, each of mass m, fixed to the ends of a light horizontal rod suspended by a fine fiber or thin metal wire, as illustrated in Figure 13.4. When two large spheres, each of mass M, are placed near the smaller ones, the attractive force between smaller and larger spheres causes the rod to rotate and twist the wire suspension to a new equilibrium orientation. The angle of rotation is measured by the deflection of a light beam reflected from a mirror attached to the vertical suspension. The deflection of the light beam is an effective technique for amplifying the motion. The experiment is carefully repeated with different masses at various separations. In addition to providing a value for G, the results show experimentally that the force is attractive, proportional to the product mM, and inversely proportional to the square of the distance r.

13.3 Free-Fall Acceleration and the Gravitational Force

In Chapter 5, when defining mg as the weight of an object of mass m, we referred to g as the magnitude of the free-fall acceleration. Now we are in a position to obtain a more fundamental description of g. Because the magnitude of the force acting on a freely falling object of mass m near the Earth's surface is given by Equation 13.4, we can equate mg to this force to obtain

$$mg = G\frac{M_E m}{R_E{}^2}$$

$$g = G\frac{M_E}{R_E{}^2} \tag{13.5}$$

Now consider an object of mass m located a distance h above the Earth's surface or a distance r from the Earth's center, where $r = R_E + h$. The magnitude of the gravitational force acting on this object is

$$F_g = G\frac{M_E m}{r^2} = G\frac{M_E m}{(R_E + h)^2}$$

The magnitude of the gravitational force acting on the object at this position is also $F_g = mg$, where g is the value of the free-fall acceleration at the altitude h. Substituting this expression for F_g into the last equation shows that g is

Variation of g with altitude

$$g = \frac{GM_E}{r^2} = \frac{GM_E}{(R_E + h)^2} \tag{13.6}$$

Thus, it follows that g *decreases* with *increasing altitude*. Because the weight of an object is mg, we see that as $r \rightarrow \infty$, its weight approaches zero.

Courtesy NASA

Astronauts F. Story Musgrave and Jeffrey A. Hoffman, along with the Hubble Space Telescope and the space shuttle *Endeavor*, are all in free fall while orbiting the Earth.

Quick Quiz 13.3 Superman stands on top of a very tall mountain and throws a baseball horizontally with a speed such that the baseball goes into a circular orbit around the Earth. While the baseball is in orbit, the acceleration of the ball (a) depends on how fast the baseball is thrown (b) is zero because the ball does not fall to the ground (c) is slightly less than 9.80 m/s² (d) is equal to 9.80 m/s².

Example 13.2   Variation of g with Altitude h

The International Space Station operates at an altitude of 350 km. When final construction is completed, it will have a weight (measured at the Earth's surface) of 4.22×10^6 N. What is its weight when in orbit?

Solution We first find the mass of the space station from its weight at the surface of the Earth:

$$m = \frac{F_g}{g} = \frac{4.22 \times 10^6 \text{ N}}{9.80 \text{ m/s}^2} = 4.31 \times 10^5 \text{ kg}$$

This mass is fixed—it is independent of the location of the space station. Because the station is above the surface of the Earth, however, we expect its weight in orbit to be less than its weight on the Earth. Using Equation 13.6 with $h = 350$ km, we obtain

$$g = \frac{GM_E}{(R_E + h)^2}$$

$$= \frac{(6.67 \times 10^{-11} \text{ N} \cdot \text{m}^2/\text{kg}^2)(5.98 \times 10^{24} \text{ kg})}{(6.37 \times 10^6 \text{ m} + 0.350 \times 10^6 \text{ m})^2}$$

$$= 8.83 \text{ m/s}^2$$

Because this value is about 90% of the value of g at the Earth surface, we expect that the weight of the station at an altitude of 350 km is 90% of the value at the Earth's surface.

Using the value of g at the location of the station, the station's weight in orbit is

$$mg = (4.31 \times 10^5 \text{ kg})(8.83 \text{ m/s}^2) = \boxed{3.80 \times 10^6 \text{ N}}$$

Values of g at other altitudes are listed in Table 13.1.

Table 13.1

Free-Fall Acceleration g at Various Altitudes Above the Earth's Surface	
Altitude h (km)	g (m/s²)
1 000	7.33
2 000	5.68
3 000	4.53
4 000	3.70
5 000	3.08
6 000	2.60
7 000	2.23
8 000	1.93
9 000	1.69
10 000	1.49
50 000	0.13
∞	0

Example 13.3   The Density of the Earth

Using the known radius of the Earth and the fact that $g = 9.80$ m/s² at the Earth's surface, find the average density of the Earth.

Solution From Eq. 1.1, we know that the average density is

$$\rho = \frac{M_E}{V_E}$$

where M_E is the mass of the Earth and V_E is its volume.
From Equation 13.5, we can relate the mass of the Earth to the value of g:

$$g = G\frac{M_E}{R_E^2} \longrightarrow M_E = \frac{gR_E^2}{G}$$

Substituting this into the definition of density, we obtain

$$\rho_E = \frac{M_E}{V_E} = \frac{(gR_E^2/G)}{\frac{4}{3}\pi R_E^3} = \frac{3}{4}\frac{g}{\pi G R_E}$$

$$= \frac{3}{4}\frac{9.80 \text{ m/s}^2}{\pi(6.67 \times 10^{-11} \text{ N} \cdot \text{m}^2/\text{kg}^2)(6.37 \times 10^6 \text{ m})}$$

$$= \boxed{5.51 \times 10^3 \text{ kg/m}^3}$$

What If? What if you were told that a typical density of granite at the Earth's surface were 2.75×10^3 kg/m³—what would you conclude about the density of the material in the Earth's interior?

Answer Because this value is about half the density that we calculated as an average for the entire Earth, we conclude that the inner core of the Earth has a density much higher than the average value. It is most amazing that the Cavendish experiment, which determines G and can be done on a table-top, combined with simple free-fall measurements of g provides information about the core of the Earth!

Johannes Kepler

German astronomer
(1571–1630)

The German astronomer Kepler is best known for developing the laws of planetary motion based on the careful observations of Tycho Brahe. *(Art Resource)*

13.4 Kepler's Laws and the Motion of Planets

People have observed the movements of the planets, stars, and other celestial objects for thousands of years. In early history, scientists regarded the Earth as the center of the Universe. This so-called *geocentric model* was elaborated and formalized by the Greek astronomer Claudius Ptolemy (c. 100–c. 170) in the second century A.D. and was accepted for the next 1 400 years. In 1543 the Polish astronomer Nicolaus Copernicus (1473–1543) suggested that the Earth and the other planets revolved in circular orbits around the Sun (the *heliocentric model*).

The Danish astronomer Tycho Brahe (1546–1601) wanted to determine how the heavens were constructed, and thus he developed a program to determine the positions of both stars and planets. It is interesting to note that those observations of the planets and 777 stars visible to the naked eye were carried out with only a large sextant and a compass. (The telescope had not yet been invented.)

The German astronomer Johannes Kepler was Brahe's assistant for a short while before Brahe's death, whereupon he acquired his mentor's astronomical data and spent 16 years trying to deduce a mathematical model for the motion of the planets. Such data are difficult to sort out because the Earth is also in motion around the Sun. After many laborious calculations, Kepler found that Brahe's data on the revolution of Mars around the Sun provided the answer.

Kepler's complete analysis of planetary motion is summarized in three statements known as **Kepler's laws:**

Kepler's laws

1. All planets move in elliptical orbits with the Sun at one focus.
2. The radius vector drawn from the Sun to a planet sweeps out equal areas in equal time intervals.
3. The square of the orbital period of any planet is proportional to the cube of the semimajor axis of the elliptical orbit.

We discuss each of these laws below.

Kepler's First Law

We are familiar with circular orbits of objects around gravitational force centers from our discussions in this chapter. Kepler's first law indicates that the circular orbit is a very special case and elliptical orbits are the general situation. This was a difficult notion for scientists of the time to accept, because they felt that perfect circular orbits of the planets reflected the perfection of heaven.

Figure 13.5 shows the geometry of an ellipse, which serves as our model for the elliptical orbit of a planet. An ellipse is mathematically defined by choosing two points F_1 and F_2, each of which is a called a **focus**, and then drawing a curve through points for which the sum of the distances r_1 and r_2 from F_1 and F_2, respectively, is a constant. The longest distance through the center between points on the ellipse (and passing through both foci) is called the **major axis**, and this distance is $2a$. In Figure 13.5, the major axis is drawn along the x direction. The distance a is called the **semimajor axis.** Similarly, the shortest distance through the center between points on the ellipse is called the **minor axis** of length $2b$, where the distance b is the **semiminor axis.** Either focus of the ellipse is located at a distance c from the center of the ellipse, where $a^2 = b^2 + c^2$. In the elliptical orbit of a planet around the Sun, the Sun is at one focus of the ellipse. There is nothing at the other focus.

The **eccentricity** of an ellipse is defined as $e = c/a$ and describes the general shape of the ellipse. For a circle, $c = 0$, and the eccentricity is therefore zero. The smaller b is than a, the shorter the ellipse is along the y direction compared to its extent in the x direction in Figure 13.5. As b decreases, c increases, and the eccentricity e increases.

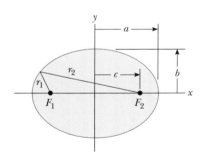

Active Figure 13.5 Plot of an ellipse. The semimajor axis has length a, and the semiminor axis has length b. Each focus is located at a distance c from the center on each side of the center.

At the Active Figures link at http://www.pse6.com, you can move the focal points or enter values for a, b, c, and e to see the resulting elliptical shape.

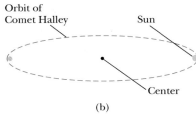

(a)

(b)

Figure 13.6 (a) The shape of the orbit of Pluto, which has the highest eccentricity ($e = 0.25$) among the planets in the solar system. The Sun is located at the large yellow dot, which is a focus of the ellipse. There is nothing physical located at the center (the small dot) or the other focus (the blue dot). (b) The shape of the orbit of Comet Halley.

Thus, higher values of eccentricity correspond to longer and thinner ellipses. The range of values of the eccentricity for an ellipse is $0 < e < 1$.

Eccentricities for planetary orbits vary widely in the solar system. The eccentricity of the Earth's orbit is 0.017, which makes it nearly circular. On the other hand, the eccentricity of Pluto's orbit is 0.25, the highest of all the nine planets. Figure 13.6a shows an ellipse with the eccentricity of that of Pluto's orbit. Notice that even this highest-eccentricity orbit is difficult to distinguish from a circle. This is why Kepler's first law is an admirable accomplishment. The eccentricity of the orbit of Comet Halley is 0.97, describing an orbit whose major axis is much longer than its minor axis, as shown in Figure 13.6b. As a result, Comet Halley spends much of its 76-year period far from the Sun and invisible from the Earth. It is only visible to the naked eye during a small part of its orbit when it is near the Sun.

Now imagine a planet in an elliptical orbit such as that shown in Figure 13.5, with the Sun at focus F_2. When the planet is at the far left in the diagram, the distance between the planet and the Sun is $a + c$. This point is called the *aphelion*, where the planet is the farthest away from the Sun that it can be in the orbit. (For an object in orbit around the Earth, this point is called the *apogee*). Conversely, when the planet is at the right end of the ellipse, the point is called the *perihelion* (for an Earth orbit, the *perigee*), and the distance between the planet and the Sun is $a - c$.

Kepler's first law is a direct result of the inverse square nature of the gravitational force. We have discussed circular and elliptical orbits. These are the allowed shapes of orbits for objects that are *bound* to the gravitational force center. These objects include planets, asteroids, and comets that move repeatedly around the Sun, as well as moons orbiting a planet. There could also be *unbound* objects, such as a meteoroid from deep space that might pass by the Sun once and then never return. The gravitational force between the Sun and these objects also varies as the inverse square of the separation distance, and the allowed paths for these objects include parabolas ($e = 1$) and hyperbolas ($e > 1$).

Kepler's Second Law

Kepler's second law can be shown to be a consequence of angular momentum conservation as follows. Consider a planet of mass M_P moving about the Sun in an elliptical orbit (Fig. 13.7a). Let us consider the planet as a system. We will model the Sun to be

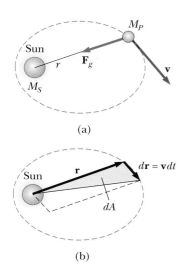

(a)

(b)

Active Figure 13.7 (a) The gravitational force acting on a planet is directed toward the Sun. (b) As a planet orbits the Sun, the area swept out by the radius vector in a time interval dt is equal to half the area of the parallelogram formed by the vectors $\mathbf{r}$ and $d\mathbf{r} = \mathbf{v}\,dt$.

🌀 **At the Active Figures link at http://www.pse6.com, you can assign a value of the eccentricity and see the resulting motion of the planet around the Sun.**

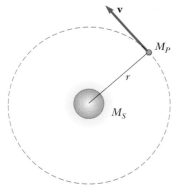

Figure 13.8 A planet of mass M_P moving in a circular orbit around the Sun. The orbits of all planets except Mercury and Pluto are nearly circular.

so much more massive than the planet that the Sun does not move. The gravitational force acting on the planet is a central force, always along the radius vector, directed toward the Sun (Fig. 13.7a). The torque on the planet due to this central force is clearly zero, because $\mathbf{F}$ is parallel to $\mathbf{r}$. That is

$$\boldsymbol{\tau} = \mathbf{r} \times \mathbf{F} = \mathbf{r} \times F(r)\hat{\mathbf{r}} = 0$$

Recall that the external net torque on a system equals the time rate of change of angular momentum of the system; that is, $\boldsymbol{\tau} = d\mathbf{L}/dt$. Therefore, because $\boldsymbol{\tau} = 0$, **the angular momentum L of the planet is a constant of the motion:**

$$\mathbf{L} = \mathbf{r} \times \mathbf{p} = M_P \mathbf{r} \times \mathbf{v} = \text{constant}$$

We can relate this result to the following geometric consideration. In a time interval dt, the radius vector $\mathbf{r}$ in Figure 13.7b sweeps out the area dA, which equals half the area $|\mathbf{r} \times d\mathbf{r}|$ of the parallelogram formed by the vectors $\mathbf{r}$ and $d\mathbf{r}$. Because the displacement of the planet in the time interval dt is given by $d\mathbf{r} = \mathbf{v}\,dt$, we have

$$dA = \tfrac{1}{2}|\mathbf{r} \times d\mathbf{r}| = \tfrac{1}{2}|\mathbf{r} \times \mathbf{v}\,dt| = \frac{L}{2M_P}\,dt$$

$$\frac{dA}{dt} = \frac{L}{2M_P} = \text{constant} \tag{13.7}$$

where L and M_P are both constants. Thus, we conclude that **the radius vector from the Sun to any planet sweeps out equal areas in equal times.**

It is important to recognize that this result is a consequence of the fact that the gravitational force is a central force, which in turn implies that angular momentum of the planet is constant. Therefore, the law applies to *any* situation that involves a central force, whether inverse-square or not.

Kepler's Third Law

It is informative to show that Kepler's third law can be predicted from the inverse-square law for circular orbits.[2] Consider a planet of mass M_P that is assumed to be moving about the Sun (mass M_S) in a circular orbit, as in Figure 13.8. Because the gravitational force provides the centripetal acceleration of the planet as it moves in a circle, we use Newton's second law for a particle in uniform circular motion,

$$\frac{GM_S M_P}{r^2} = \frac{M_P v^2}{r}$$

The orbital speed of the planet is $2\pi r/T$, where T is the period; therefore, the preceding expression becomes

$$\frac{GM_S}{r^2} = \frac{(2\pi r/T)^2}{r}$$

$$T^2 = \left(\frac{4\pi^2}{GM_S}\right) r^3 = K_S r^3$$

where K_S is a constant given by

$$K_S = \frac{4\pi^2}{GM_S} = 2.97 \times 10^{-19} \text{ s}^2/\text{m}^3$$

---

[2] The orbits of all planets except Mercury and Pluto are very close to being circular; hence, we do not introduce much error with this assumption. For example, the ratio of the semiminor axis to the semimajor axis for the Earth's orbit is $b/a = 0.999\,86$.

Table 13.2

Useful Planetary Data					
Body	Mass (kg)	Mean Radius (m)	Period of Revolution (s)	Mean Distance from Sun (m)	$\dfrac{T^2}{r^3}$ (s^2/m^3)
Mercury	3.18×10^{23}	2.43×10^6	7.60×10^6	5.79×10^{10}	2.97×10^{-19}
Venus	4.88×10^{24}	6.06×10^6	1.94×10^7	1.08×10^{11}	2.99×10^{-19}
Earth	5.98×10^{24}	6.37×10^6	3.156×10^7	1.496×10^{11}	2.97×10^{-19}
Mars	6.42×10^{23}	3.37×10^6	5.94×10^7	2.28×10^{11}	2.98×10^{-19}
Jupiter	1.90×10^{27}	6.99×10^7	3.74×10^8	7.78×10^{11}	2.97×10^{-19}
Saturn	5.68×10^{26}	5.85×10^7	9.35×10^8	1.43×10^{12}	2.99×10^{-19}
Uranus	8.68×10^{25}	2.33×10^7	2.64×10^9	2.87×10^{12}	2.95×10^{-19}
Neptune	1.03×10^{26}	2.21×10^7	5.22×10^9	4.50×10^{12}	2.99×10^{-19}
Pluto	$\approx 1.4 \times 10^{22}$	$\approx 1.5 \times 10^6$	7.82×10^9	5.91×10^{12}	2.96×10^{-19}
Moon	7.36×10^{22}	1.74×10^6	—	—	—
Sun	1.991×10^{30}	6.96×10^8	—	—	—

This equation is also valid for elliptical orbits if we replace r with the length a of the semimajor axis (Fig. 13.5):

$$T^2 = \left(\frac{4\pi^2}{GM_S}\right) a^3 = K_S a^3 \qquad (13.8) \qquad \textbf{Kepler's third law}$$

Equation 13.8 is Kepler's third law. Because the semimajor axis of a circular orbit is its radius, Equation 13.8 is valid for both circular and elliptical orbits. Note that the constant of proportionality K_S is independent of the mass of the planet. Equation 13.8 is therefore valid for *any* planet.[3] If we were to consider the orbit of a satellite such as the Moon about the Earth, then the constant would have a different value, with the Sun's mass replaced by the Earth's mass, that is, $K_E = 4\pi^2/GM_E$.

Table 13.2 is a collection of useful planetary data. The last column verifies that the ratio T^2/r^3 is constant. The small variations in the values in this column are due to uncertainties in the data measured for the periods and semimajor axes of the planets.

Recent astronomical work has revealed the existence of a large number of solar system objects beyond the orbit of Neptune. In general, these lie in the *Kuiper belt,* a region that extends from about 30 AU (the orbital radius of Neptune) to 50 AU. (An AU is an *astronomical unit*—the radius of the Earth's orbit.) Current estimates identify at least 70 000 objects in this region with diameters larger than 100 km. The first KBO (Kuiper Belt Object) was discovered in 1992. Since then, many more have been detected and some have been given names, such as Varuna (diameter about 900–1 000 km, discovered in 2000), Ixion (diameter about 900–1 000 km, discovered in 2001), and Quaoar (diameter about 800 km, discovered in 2002).

A subset of about 1 400 KBOs are called "Plutinos" because, like Pluto, they exhibit a resonance phenomenon, orbiting the Sun two times in the same time interval as Neptune revolves three times. Some astronomers even claim that Pluto should not be considered a planet but should be identified as a KBO. The contemporary application of Kepler's laws and such exotic proposals as planetary angular momentum exchange and migrating planets[4] suggest the excitement of this active area of current research.

Quick Quiz 13.4 Pluto, the farthest planet from the Sun, has an orbital period that is (a) greater than a year (b) less than a year (c) equal to a year.

---

[3]   Equation 13.8 is indeed a proportion because the ratio of the two quantities T^2 and a^3 is a constant. The variables in a proportion are not required to be limited to the first power only.

[4]   Malhotra, R., "Migrating Planets," *Scientific American,* September 1999, volume 281, number 3.

Quick Quiz 13.5 An asteroid is in a highly eccentric elliptical orbit around the Sun. The period of the asteroid's orbit is 90 days. Which of the following statements is true about the possibility of a collision between this asteroid and the Earth? (a) There is no possible danger of a collision. (b) There is a possibility of a collision. (c) There is not enough information to determine whether there is danger of a collision.

Quick Quiz 13.6 A satellite moves in an elliptical orbit about the Earth such that, at perigee and apogee positions, its distances from the Earth's center are respectively D and $4D$. The relationship between the speeds at these two positions is (a) $v_p = v_a$ (b) $v_p = 4v_a$ (c) $v_a = 4v_p$ (d) $v_p = 2v_a$ (e) $v_a = 2v_p$.

Example 13.4 The Mass of the Sun

Calculate the mass of the Sun using the fact that the period of the Earth's orbit around the Sun is 3.156×10^7 s and its distance from the Sun is 1.496×10^{11} m.

Solution Using Equation 13.8, we find that

$$M_S = \frac{4\pi^2 r^3}{GT^2} = \frac{4\pi^2 (1.496 \times 10^{11}\ \text{m})^3}{(6.67 \times 10^{-11}\ \text{N} \cdot \text{m}^2/\text{kg}^2)(3.156 \times 10^7\ \text{s})^2}$$

$$= 1.99 \times 10^{30}\ \text{kg}$$

In Example 13.3, an understanding of gravitational forces enabled us to find out something about the density of the Earth's core, and now we have used this understanding to determine the mass of the Sun!

What If? Suppose you were asked for the mass of Mars. How could you determine this value?

Answer Kepler's third law is valid for any system of objects in orbit around an object with a large mass. Mars has two moons, Phobos and Deimos. If we rewrite Equation 13.8 for these moons of Mars, we have

$$T^2 = \left(\frac{4\pi^2}{GM_M} \right) a^3$$

where M_M is the mass of Mars. Solving for this mass,

$$M_M = \left(\frac{4\pi^2}{G} \right) \frac{a^3}{T^2} = \left(\frac{4\pi^2}{6.67 \times 10^{-11}\ \text{N} \cdot \text{m}^2/\text{kg}^2} \right) \frac{a^3}{T^2}$$

$$= (5.92 \times 10^{11}\ \text{kg} \cdot \text{s}^2/\text{m}^3) \frac{a^3}{T^2}$$

Phobos has an orbital period of 0.32 days and an almost circular orbit of radius 9 380 km. The orbit of Deimos is even more circular, with a radius of 23 460 km and an orbital period of 1.26 days. Let us calculate the mass of Mars using each of these sets of data:

Phobos:

$$M_M = (5.92 \times 10^{11}\ \text{kg} \cdot \text{s}^2/\text{m}^3)$$
$$\times \frac{(9.380 \times 10^6\ \text{m})^3}{(0.32\ \text{d})^2} \left(\frac{1\ \text{d}}{86\ 400\ \text{s}} \right)^2 = 6.39 \times 10^{23}\ \text{kg}$$

Deimos:

$$M_M = (5.92 \times 10^{11}\ \text{kg} \cdot \text{s}^2/\text{m}^3)$$
$$\times \frac{(2.346 \times 10^7\ \text{m})^3}{(1.26\ \text{d})^2} \left(\frac{1\ \text{d}}{86\ 400\ \text{s}} \right)^2 = 6.45 \times 10^{23}\ \text{kg}$$

These two calculations are within 1% of each other and both are within 0.5% of the value of the mass of Mars given in Table 13.2.

Example 13.5 A Geosynchronous Satellite | Interactive

Consider a satellite of mass m moving in a circular orbit around the Earth at a constant speed v and at an altitude h above the Earth's surface, as illustrated in Figure 13.9.

(A) Determine the speed of the satellite in terms of G, h, R_E (the radius of the Earth), and M_E (the mass of the Earth).

Solution Conceptualize by imagining the satellite moving around the Earth in a circular orbit under the influence of the gravitational force. The satellite must have a centripetal acceleration. Thus, we categorize this problem as one involving Newton's second law, the law of universal gravitation, and circular motion. To analyze the problem,

note that the only external force acting on the satellite is the gravitational force, which acts toward the center of the Earth and keeps the satellite in its circular orbit. Therefore, the net force on the satellite is the gravitational force

$$F_r = F_g = G \frac{M_E m}{r^2}$$

From Newton's second law and the fact that the acceleration of the satellite is centripetal, we obtain

$$G \frac{M_E m}{r^2} = m \frac{v^2}{r}$$

Solving for v and remembering that the distance r from the center of the Earth to the satellite is $r = R_E + h$, we obtain

$$(1) \qquad v = \sqrt{\frac{GM_E}{r}} = \sqrt{\frac{GM_E}{R_E + h}}$$

(B) If the satellite is to be *geosynchronous* (that is, appearing to remain over a fixed position on the Earth), how fast is it moving through space?

Solution In order to appear to remain over a fixed position on the Earth, the period of the satellite must be 24 h and the satellite must be in orbit directly over the equator. From Kepler's third law (Equation 13.8) with $a = r$ and $M_S \rightarrow M_E$, we find the radius of the orbit:

$$T^2 = \left(\frac{4\pi^2}{GM_E} \right) r^3$$

$$r = \sqrt[3]{\frac{GM_E T^2}{4\pi^2}}$$

Substituting numerical values and noting that the period is $T = 24$ h $= 86\ 400$ s, we find

$$r = \sqrt[3]{\frac{(6.67 \times 10^{-11}\ \text{N} \cdot \text{m}^2/\text{kg}^2)(5.98 \times 10^{24}\ \text{kg})(86\ 400\ \text{s})^2}{4\pi^2}}$$

$$= 4.23 \times 10^7\ \text{m}$$

To find the speed of the satellite, we use Equation (1):

$$v = \sqrt{\frac{GM_E}{r}}$$

$$= \sqrt{\frac{(6.67 \times 10^{-11}\ \text{N} \cdot \text{m}^2/\text{kg}^2)(5.98 \times 10^{24}\ \text{kg})}{4.23 \times 10^7\ \text{m}}}$$

$$= 3.07 \times 10^3\ \text{m/s}$$

To finalize this problem, it is interesting to note that the value of r calculated here translates to a height of the satel-

Figure 13.9 (Example 13.5) A satellite of mass m moving around the Earth in a circular orbit of radius r with constant speed v. The only force acting on the satellite is the gravitational force $\mathbf{F}_g$. (Not drawn to scale.)

lite above the surface of the Earth of almost 36 000 km. Thus, geosynchronous satellites have the advantage of allowing an earthbound antenna to be aimed in a fixed direction, but there is a disadvantage in that the signals between Earth and the satellite must travel a long distance. It is difficult to use geosynchronous satellites for optical observation of the Earth's surface because of their high altitude.

What If? What if the satellite motion in part (A) were taking place at height h above the surface of another planet more massive than the Earth but of the same radius? Would the satellite be moving at a higher or a lower speed than it does around the Earth?

Answer If the planet pulls downward on the satellite with more gravitational force due to its larger mass, the satellite would have to move with a higher speed to avoid moving toward the surface. This is consistent with the predictions of Equation (1), which shows that because the speed v is proportional to the square root of the mass of the planet, as the mass increases, the speed also increases.

 You can adjust the altitude of the satellite at the Interactive Worked Example link at **http://www.pse6.com.**

13.5 The Gravitational Field

When Newton published his theory of universal gravitation, it was considered a success because it satisfactorily explained the motion of the planets. Since 1687 the same theory has been used to account for the motions of comets, the deflection of a Cavendish balance, the orbits of binary stars, and the rotation of galaxies. Nevertheless, both Newton's contemporaries and his successors found it difficult to accept the concept of a force that acts at a distance, as mentioned in Section 5.1. They asked how it was possible for two objects to interact when they were not in contact with each other. Newton himself could not answer that question.

An approach to describing interactions between objects that are not in contact came well after Newton's death, and it enables us to look at the gravitational interaction in a different way, using the concept of a **gravitational field** that exists at every point in space. When a particle of mass m is placed at a point where the gravitational

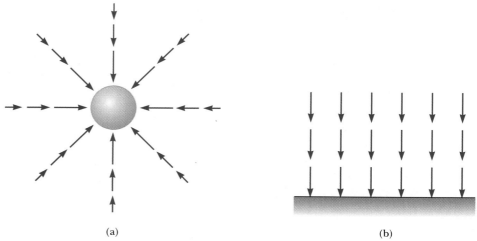

(a)   (b)

Figure 13.10 (a) The gravitational field vectors in the vicinity of a uniform spherical mass such as the Earth vary in both direction and magnitude. The vectors point in the direction of the acceleration a particle would experience if it were placed in the field. The magnitude of the field vector at any location is the magnitude of the free-fall acceleration at that location. (b) The gravitational field vectors in a small region near the Earth's surface are uniform in both direction and magnitude.

field is **g**, the particle experiences a force $\mathbf{F}_g = m\mathbf{g}$. In other words, the field exerts a force on the particle. The gravitational field **g** is defined as

Gravitational field

$$\mathbf{g} \equiv \frac{\mathbf{F}_g}{m} \tag{13.9}$$

That is, the gravitational field at a point in space equals the gravitational force experienced by a *test particle* placed at that point divided by the mass of the test particle. Notice that the presence of the test particle is not necessary for the field to exist—the Earth creates the gravitational field. We call the object creating the field the *source particle*. (Although the Earth is clearly not a particle, it is possible to show that we can approximate the Earth as a particle for the purpose of finding the gravitational field that it creates.) We can detect the presence of the field and measure its strength by placing a test particle in the field and noting the force exerted on it.

Although the gravitational force is inherently an interaction between two objects, the concept of a gravitational field allows us to "factor out" the mass of one of the objects. In essence, we are describing the "effect" that any object (in this case, the Earth) has on the empty space around itself in terms of the force that *would* be present *if* a second object were somewhere in that space.[5]

As an example of how the field concept works, consider an object of mass m near the Earth's surface. Because the gravitational force acting on the object has a magnitude GM_Em/r^2 (see Eq. 13.4), the field **g** at a distance r from the center of the Earth is

$$\mathbf{g} = \frac{\mathbf{F}_g}{m} = -\frac{GM_E}{r^2}\,\hat{\mathbf{r}} \tag{13.10}$$

where $\hat{\mathbf{r}}$ is a unit vector pointing radially outward from the Earth and the negative sign indicates that the field points toward the center of the Earth, as illustrated in Figure 13.10a. Note that the field vectors at different points surrounding the Earth vary in both direction and magnitude. In a small region near the Earth's surface, the downward field **g** is approximately constant and uniform, as indicated in Figure 13.10b. Equation 13.10 is valid at all points *outside* the Earth's surface, assuming that the Earth is spherical. At the Earth's surface, where $r = R_E$, **g** has a magnitude of 9.80 N/kg. (The unit N/kg is the same as m/s².)

[5]   We shall return to this idea of mass affecting the space around it when we discuss Einstein's theory of gravitation in Chapter 39.

13.6 Gravitational Potential Energy

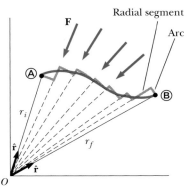

In Chapter 8 we introduced the concept of gravitational potential energy, which is the energy associated with the configuration of a system of objects interacting via the gravitational force. We emphasized that the gravitational potential-energy function *mgy* for a particle–Earth system is valid only when the particle is near the Earth's surface, where the gravitational force is constant. Because the gravitational force between two particles varies as $1/r^2$, we expect that a more general potential-energy function—one that is valid without the restriction of having to be near the Earth's surface—will be significantly different from $U = mgy$.

Before we calculate this general form for the gravitational potential energy function, let us first verify that *the gravitational force is conservative.* (Recall from Section 8.3 that a force is conservative if the work it does on an object moving between any two points is independent of the path taken by the object.) To do this, we first note that the gravitational force is a central force. By definition, a central force is any force that is directed along a radial line to a fixed center and has a magnitude that depends only on the radial coordinate r. Hence, a central force can be represented by $F(r)\hat{\mathbf{r}}$ where $\hat{\mathbf{r}}$ is a unit vector directed from the origin toward the particle, as shown in Figure 13.11.

Consider a central force acting on a particle moving along the general path Ⓐ to Ⓑ in Figure 13.11. The path from Ⓐ to Ⓑ can be approximated by a series of steps according to the following procedure. In Figure 13.11, we draw several thin wedges, which are shown as dashed lines. The outer boundary of our set of wedges is a path consisting of short radial line segments and arcs (gray in the figure). We select the length of the radial dimension of each wedge such that the short arc at the wedge's wide end intersects the actual path of the particle. Then we can approximate the actual path with a series of zigzag movements that alternate between moving along an arc and moving along a radial line.

By definition, a central force is always directed along one of the radial segments; therefore, the work done by $\mathbf{F}$ along any radial segment is

$$dW = \mathbf{F} \cdot d\mathbf{r} = F(r)\ dr$$

By definition, the work done by a force that is perpendicular to the displacement is zero. Hence, the work done in moving along any arc is zero because $\mathbf{F}$ is perpendicular to the displacement along these segments. Therefore, the total work done by $\mathbf{F}$ is the sum of the contributions along the radial segments:

$$W = \int_{r_i}^{r_f} F(r)\ dr$$

where the subscripts i and f refer to the initial and final positions. Because the integrand is a function only of the radial position, this integral depends only on the initial and final values of r. Thus, the work done is the same over *any* path from Ⓐ to Ⓑ. Because the work done is independent of the path and depends only on the end points, we conclude that *any central force is conservative*. We are now assured that a potential energy function can be obtained once the form of the central force is specified.

Recall from Equation 8.15 that the change in the gravitational potential energy of a system associated with a given displacement of a member of the system is defined as the negative of the work done by the gravitational force on that member during the displacement:

$$\Delta U = U_f - U_i = -\int_{r_i}^{r_f} F(r)\ dr \tag{13.11}$$

We can use this result to evaluate the gravitational potential energy function. Consider a particle of mass m moving between two points Ⓐ and Ⓑ above the Earth's surface (Fig. 13.12). The particle is subject to the gravitational force given by Equation 13.1. We can express this force as

$$F(r) = -\frac{GM_E m}{r^2}$$

Figure 13.11 A particle moves from Ⓐ to Ⓑ while acted on by a central force $\mathbf{F}$, which is directed radially. The path is broken into a series of radial segments and arcs. Because the work done along the arcs is zero, the work done is independent of the path and depends only on r_f and r_i.

Work done by a central force

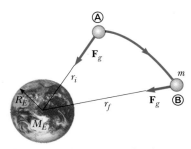

Figure 13.12 As a particle of mass m moves from Ⓐ to Ⓑ above the Earth's surface, the gravitational potential energy changes according to Equation 13.11.

where the negative sign indicates that the force is attractive. Substituting this expression for $F(r)$ into Equation 13.11, we can compute the change in the gravitational potential energy function:

$$U_f - U_i = GM_E m \int_{r_i}^{r_f} \frac{dr}{r^2} = GM_E m \left[-\frac{1}{r} \right]_{r_i}^{r_f}$$

$$U_f - U_i = -GM_E m \left(\frac{1}{r_f} - \frac{1}{r_i} \right) \tag{13.12}$$

As always, the choice of a reference configuration for the potential energy is completely arbitrary. It is customary to choose the reference configuration for zero potential energy to be the same as that for which the force is zero. Taking $U_i = 0$ at $r_i = \infty$, we obtain the important result

$$U(r) = -\frac{GM_E m}{r} \tag{13.13}$$

This expression applies to the Earth–particle system where the particle is separated from the center of the Earth by a distance r, provided that $r \geq R_E$. The result is not valid for particles inside the Earth, where $r < R_E$. Because of our choice of U_i, the function U is always negative (Fig. 13.13).

Although Equation 13.13 was derived for the particle–Earth system, it can be applied to any two particles. That is, the gravitational potential energy associated with any pair of particles of masses m_1 and m_2 separated by a distance r is

$$U = -\frac{Gm_1 m_2}{r} \tag{13.14}$$

This expression shows that the gravitational potential energy for any pair of particles varies as $1/r$, whereas the force between them varies as $1/r^2$. Furthermore, the potential energy is negative because the force is attractive and we have taken the potential energy as zero when the particle separation is infinite. Because the force between the particles is attractive, we know that an external agent must do positive work to increase the separation between them. The work done by the external agent produces an increase in the potential energy as the two particles are separated. That is, U becomes less negative as r increases.

When two particles are at rest and separated by a distance r, an external agent has to supply an energy at least equal to $+Gm_1 m_2/r$ in order to separate the particles to an infinite distance. It is therefore convenient to think of the absolute value of the potential energy as the *binding energy* of the system. If the external agent supplies an energy greater than the binding energy, the excess energy of the system will be in the form of kinetic energy of the particles when the particles are at an infinite separation.

We can extend this concept to three or more particles. In this case, the total potential energy of the system is the sum over all pairs of particles.[6] Each pair contributes a term of the form given by Equation 13.14. For example, if the system contains three particles, as in Figure 13.14, we find that

$$U_{\text{total}} = U_{12} + U_{13} + U_{23} = -G \left(\frac{m_1 m_2}{r_{12}} + \frac{m_1 m_3}{r_{13}} + \frac{m_2 m_3}{r_{23}} \right) \tag{13.15}$$

The absolute value of U_{total} represents the work needed to separate the particles by an infinite distance.

Gravitational potential energy of the Earth–particle system for $r > R_E$

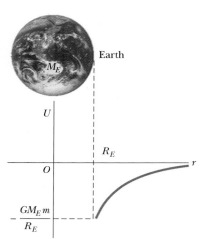

Figure 13.13 Graph of the gravitational potential energy U versus r for an object above the Earth's surface. The potential energy goes to zero as r approaches infinity.

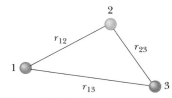

Figure 13.14 Three interacting particles.

---

[6]   The fact that potential energy terms can be added for all pairs of particles stems from the experimental fact that gravitational forces obey the superposition principle.

Example 13.6   The Change in Potential Energy

A particle of mass m is displaced through a small vertical distance Δy near the Earth's surface. Show that in this situation the general expression for the change in gravitational potential energy given by Equation 13.12 reduces to the familiar relationship $\Delta U = mg\,\Delta y$.

Solution We can express Equation 13.12 in the form

$$\Delta U = -GM_E m \left(\frac{1}{r_f} - \frac{1}{r_i} \right) = GM_E m \left(\frac{r_f - r_i}{r_i r_f} \right)$$

If both the initial and final positions of the particle are close to the Earth's surface, then $r_f - r_i = \Delta y$ and $r_i r_f \approx R_E{}^2$. (Recall that r is measured from the center of the Earth.) Therefore, the change in potential energy becomes

$$\Delta U \approx \frac{GM_E m}{R_E{}^2}\,\Delta y = mg\,\Delta y$$

where we have used the fact that $g = GM_E/R_E{}^2$ (Eq. 13.5). Keep in mind that the reference configuration is arbitrary because it is the *change* in potential energy that is meaningful.

What If? Suppose you are performing upper-atmosphere studies and are asked by your supervisor to find the height in the Earth's atmosphere at which the "surface equation" $\Delta U = mg\,\Delta y$ gives a 1.0% error in the change in the potential energy. What is this height?

Answer Because the surface equation assumes a constant value for g, it will give a ΔU value that is larger than the value given by the general equation, Equation 13.12. Thus, a 1.0% error would be described by the ratio

$$\frac{\Delta U_{\text{surface}}}{\Delta U_{\text{general}}} = 1.010$$

Substituting the expressions for each of these changes ΔU, we have

$$\frac{mg\,\Delta y}{GM_E m(\Delta y/r_i r_f)} = \frac{g r_i r_f}{GM_E} = 1.010$$

where $r_i = R_E$ and $r_f = R_E + \Delta y$. Substituting for g from Equation 13.5, we find

$$\frac{(GM_E/R_E{}^2)R_E(R_E + \Delta y)}{GM_E} = \frac{R_E + \Delta y}{R_E} = 1 + \frac{\Delta y}{R_E} = 1.010$$

Thus,

$$\Delta y = 0.010 R_E = 0.010(6.37 \times 10^6 \text{ m})$$
$$= 6.37 \times 10^4 \text{ m} = 63.7 \text{ km}$$

13.7 Energy Considerations in Planetary and Satellite Motion

Consider an object of mass m moving with a speed v in the vicinity of a massive object of mass M, where $M \gg m$. The system might be a planet moving around the Sun, a satellite in orbit around the Earth, or a comet making a one-time flyby of the Sun. If we assume that the object of mass M is at rest in an inertial reference frame, then the total mechanical energy E of the two-object system when the objects are separated by a distance r is the sum of the kinetic energy of the object of mass m and the potential energy of the system, given by Equation 13.14:[7]

$$E = K + U$$

$$E = \tfrac{1}{2}mv^2 - \frac{GMm}{r} \qquad (13.16)$$

---

[7] You might recognize that we have ignored the kinetic energy of the larger body. To see that this simplification is reasonable, consider an object of mass m falling toward the Earth. Because the center of mass of the object–Earth system is effectively stationary, it follows from conservation of momentum that $mv = M_E v_E$. Thus, the Earth acquires a kinetic energy equal to

$$\tfrac{1}{2}M_E v_E{}^2 = \tfrac{1}{2}\frac{m^2}{M_E}v^2 = \frac{m}{M_E}K$$

where K is the kinetic energy of the object. Because $M_E \gg m$, this result shows that the kinetic energy of the Earth is negligible.

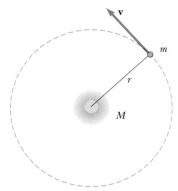

Figure 13.15 An object of mass m moving in a circular orbit about a much larger object of mass M.

This equation shows that E may be positive, negative, or zero, depending on the value of v. However, for a bound system,[8] such as the Earth–Sun system, E is necessarily *less than zero* because we have chosen the convention that $U \to 0$ as $r \to \infty$.

We can easily establish that $E < 0$ for the system consisting of an object of mass m moving in a circular orbit about an object of mass $M \gg m$ (Fig. 13.15). Newton's second law applied to the object of mass m gives

$$\frac{GMm}{r^2} = ma = \frac{mv^2}{r}$$

Multiplying both sides by r and dividing by 2 gives

$$\tfrac{1}{2}mv^2 = \frac{GMm}{2r} \qquad (13.17)$$

Substituting this into Equation 13.16, we obtain

$$E = \frac{GMm}{2r} - \frac{GMm}{r}$$

Total energy for circular orbits

$$E = -\frac{GMm}{2r} \qquad \text{(circular orbits)} \qquad (13.18)$$

This result clearly shows that **the total mechanical energy is negative in the case of circular orbits.** Note that **the kinetic energy is positive and equal to half the absolute value of the potential energy.** The absolute value of E is also equal to the binding energy of the system, because this amount of energy must be provided to the system to move the two objects infinitely far apart.

The total mechanical energy is also negative in the case of elliptical orbits. The expression for E for elliptical orbits is the same as Equation 13.18 with r replaced by the semimajor axis length a:

Total energy for elliptical orbits

$$E = -\frac{GMm}{2a} \qquad \text{(elliptical orbits)} \qquad (13.19)$$

Furthermore, the total energy is constant if we assume that the system is isolated. Therefore, as the object of mass m moves from Ⓐ to Ⓑ in Figure 13.12, the total energy remains constant and Equation 13.16 gives

$$E = \tfrac{1}{2}mv_i^2 - \frac{GMm}{r_i} = \tfrac{1}{2}mv_f^2 - \frac{GMm}{r_f} \qquad (13.20)$$

Combining this statement of energy conservation with our earlier discussion of conservation of angular momentum, we see that **both the total energy and the total angular momentum of a gravitationally bound, two-object system are constants of the motion.**

Quick Quiz 13.7 A comet moves in an elliptical orbit around the Sun. Which point in its orbit (perihelion or aphelion) represents the highest value of (a) the speed of the comet (b) the potential energy of the comet–Sun system (c) the kinetic energy of the comet (d) the total energy of the comet–Sun system?

[8] Of the three examples provided at the beginning of this section, the planet moving around the Sun and a satellite in orbit around the Earth are bound systems—the Earth will always stay near the Sun, and the satellite will always stay near the Earth. The one-time comet flyby represents an unbound system—the comet interacts once with the Sun but is not bound to it. Thus, in theory the comet can move infinitely far away from the Sun.

Example 13.7 Changing the Orbit of a Satellite

The space shuttle releases a 470-kg communications satellite while in an orbit 280 km above the surface of the Earth. A rocket engine on the satellite boosts it into a geosynchronous orbit, which is an orbit in which the satellite stays directly over a single location on the Earth. How much energy does the engine have to provide?

Solution We first determine the initial radius (not the altitude above the Earth's surface) of the satellite's orbit when it is still in the shuttle's cargo bay. This is simply

$$R_E + 280 \text{ km} = 6.65 \times 10^6 \text{ m} = r_i$$

In Example 13.5, we found that the radius of the orbit of a geosynchronous satellite is $r_f = 4.23 \times 10^7$ m. Applying Equation 13.18, we obtain, for the total initial and final energies,

$$E_i = -\frac{GM_E m}{2r_i} \qquad E_f = -\frac{GM_E m}{2r_f}$$

The energy required from the engine to boost the satellite is

$$\Delta E = E_f - E_i = -\frac{GM_E m}{2}\left(\frac{1}{r_f} - \frac{1}{r_i}\right)$$

$$= -\frac{(6.67 \times 10^{-11} \text{ N} \cdot \text{m}^2/\text{kg}^2)(5.98 \times 10^{24} \text{kg})(470 \text{ kg})}{2}$$

$$\times \left(\frac{1}{4.23 \times 10^7 \text{ m}} - \frac{1}{6.65 \times 10^6 \text{ m}}\right)$$

$$= \boxed{1.19 \times 10^{10} \text{ J}}$$

This is the energy equivalent of 89 gal of gasoline. NASA engineers must account for the changing mass of the spacecraft as it ejects burned fuel, something we have not done here. Would you expect the calculation that includes the effect of this changing mass to yield a greater or lesser amount of energy required from the engine?

If we wish to determine how the energy is distributed after the engine is fired, we find from Equation 13.17 that the change in kinetic energy is $\Delta K = (GM_E m/2)(1/r_f - 1/r_i) = -1.19 \times 10^{10}$ J (a decrease), and the corresponding change in potential energy is $\Delta U = -GM_E m(1/r_f - 1/r_i) = 2.38 \times 10^{10}$ J (an increase). Thus, the change in orbital energy of the system is $\Delta E = \Delta K + \Delta U = 1.19 \times 10^{10}$ J, as we already calculated. The firing of the engine results in a transformation of chemical potential energy in the fuel to orbital energy of the system. Because an increase in gravitational potential energy is accompanied by a decrease in kinetic energy, we conclude that the speed of an orbiting satellite decreases as its altitude increases.

Escape Speed

Suppose an object of mass m is projected vertically upward from the Earth's surface with an initial speed v_i, as illustrated in Figure 13.16. We can use energy considerations to find the minimum value of the initial speed needed to allow the object to move infinitely far away from the Earth. Equation 13.16 gives the total energy of the system at any point. At the surface of the Earth, $v = v_i$ and $r = r_i = R_E$. When the object reaches its maximum altitude, $v = v_f = 0$ and $r = r_f = r_{max}$. Because the total energy of the system is constant, substituting these conditions into Equation 13.20 gives

$$\tfrac{1}{2}mv_i^2 - \frac{GM_E m}{R_E} = -\frac{GM_E m}{r_{max}}$$

Solving for v_i^2 gives

$$v_i^2 = 2GM_E\left(\frac{1}{R_E} - \frac{1}{r_{max}}\right) \tag{13.21}$$

Therefore, if the initial speed is known, this expression can be used to calculate the maximum altitude h because we know that

$$h = r_{max} - R_E$$

We are now in a position to calculate **escape speed**, which is the minimum speed the object must have at the Earth's surface in order to approach an infinite separation distance from the Earth. Traveling at this minimum speed, the object continues to

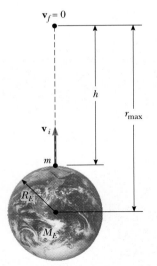

Figure 13.16 An object of mass m projected upward from the Earth's surface with an initial speed v_i reaches a maximum altitude h.

PITFALL PREVENTION

13.3 You Can't Really Escape

Although Equation 13.22 provides the "escape speed" from the Earth, *complete* escape from the Earth's gravitational influence is impossible because the gravitational force is of infinite range. No matter how far away you are, you will always feel some gravitational force due to the Earth.

move farther and farther away from the Earth as its speed asymptotically approaches zero. Letting $r_{max} \rightarrow \infty$ in Equation 13.21 and taking $v_i = v_{esc}$, we obtain

$$v_{esc} = \sqrt{\frac{2GM_E}{R_E}} \qquad (13.22)$$

Note that this expression for v_{esc} is independent of the mass of the object. In other words, a spacecraft has the same escape speed as a molecule. Furthermore, the result is independent of the direction of the velocity and ignores air resistance.

If the object is given an initial speed equal to v_{esc}, the total energy of the system is equal to zero. This can be seen by noting that when $r \rightarrow \infty$, the object's kinetic energy and the potential energy of the system are both zero. If v_i is greater than v_{esc}, the total energy of the system is greater than zero and the object has some residual kinetic energy as $r \rightarrow \infty$.

Example 13.8 Escape Speed of a Rocket

Calculate the escape speed from the Earth for a 5 000-kg spacecraft, and determine the kinetic energy it must have at the Earth's surface in order to move infinitely far away from the Earth.

Solution Using Equation 13.22 gives

$$v_{esc} = \sqrt{\frac{2GM_E}{R_E}}$$

$$= \sqrt{\frac{2(6.67 \times 10^{-11}\ \text{N} \cdot \text{m}^2/\text{kg}^2)(5.98 \times 10^{24}\ \text{kg})}{6.37 \times 10^6\ \text{m}}}$$

$$= \boxed{1.12 \times 10^4\ \text{m/s}}$$

This corresponds to about 25 000 mi/h.

The kinetic energy of the spacecraft is

$$K = \tfrac{1}{2}mv_{esc}^2 = \tfrac{1}{2}(5.00 \times 10^3\ \text{kg})(1.12 \times 10^4\ \text{m/s})^2$$

$$= \boxed{3.14 \times 10^{11}\ \text{J}}$$

This is equivalent to about 2 300 gal of gasoline.

What If? What if we wish to launch a 1 000-kg spacecraft at the escape speed? How much energy does this require?

Answer In Equation 13.22, the mass of the object moving with the escape speed does not appear. Thus, the escape speed for the 1 000-kg spacecraft is the same as that for the 5 000-kg spacecraft. The only change in the kinetic energy is due to the mass, so the 1 000-kg spacecraft will require one fifth of the energy of the 5 000-kg spacecraft:

$$K = \tfrac{1}{5}(3.14 \times 10^{11}\ \text{J}) = 6.28 \times 10^{10}\ \text{J}$$

Table 13.3

Escape Speeds from the Surfaces of the Planets, Moon, and Sun	
Planet	v_{esc} **(km/s)**
Mercury	4.3
Venus	10.3
Earth	11.2
Mars	5.0
Jupiter	60
Saturn	36
Uranus	22
Neptune	24
Pluto	1.1
Moon	2.3
Sun	618

Equations 13.21 and 13.22 can be applied to objects projected from any planet. That is, in general, the escape speed from the surface of any planet of mass M and radius R is

$$v_{esc} = \sqrt{\frac{2GM}{R}} \qquad (13.23)$$

Escape speeds for the planets, the Moon, and the Sun are provided in Table 13.3. Note that the values vary from 1.1 km/s for Pluto to about 618 km/s for the Sun. These results, together with some ideas from the kinetic theory of gases (see Chapter 21), explain why some planets have atmospheres and others do not. As we shall see later, at a given temperature the average kinetic energy of a gas molecule depends only on the mass of the molecule. Lighter molecules, such as hydrogen and helium, have a higher average speed than heavier molecules at the same temperature. When the average speed of the lighter molecules is not much less than the escape speed of a planet, a significant fraction of them have a chance to escape.

This mechanism also explains why the Earth does not retain hydrogen molecules and helium atoms in its atmosphere but does retain heavier molecules, such as oxygen and nitrogen. On the other hand, the very large escape speed for Jupiter enables that planet to retain hydrogen, the primary constituent of its atmosphere.

Black Holes

In Example 11.7 we briefly described a rare event called a supernova—the catastrophic explosion of a very massive star. The material that remains in the central core of such an object continues to collapse, and the core's ultimate fate depends on its mass. If the core has a mass less than 1.4 times the mass of our Sun, it gradually cools down and ends its life as a white dwarf star. However, if the core's mass is greater than this, it may collapse further due to gravitational forces. What remains is a neutron star, discussed in Example 11.7, in which the mass of a star is compressed to a radius of about 10 km. (On Earth, a teaspoon of this material would weigh about 5 billion tons!)

An even more unusual star death may occur when the core has a mass greater than about three solar masses. The collapse may continue until the star becomes a very small object in space, commonly referred to as a **black hole.** In effect, black holes are remains of stars that have collapsed under their own gravitational force. If an object such as a spacecraft comes close to a black hole, it experiences an extremely strong gravitational force and is trapped forever.

The escape speed for a black hole is very high, due to the concentration of the mass of the star into a sphere of very small radius (see Eq. 13.23). If the escape speed exceeds the speed of light c, radiation from the object (such as visible light) cannot escape, and the object appears to be black; hence the origin of the terminology "black hole." The critical radius R_S at which the escape speed is c is called the **Schwarzschild radius** (Fig. 13.17). The imaginary surface of a sphere of this radius surrounding the black hole is called the **event horizon.** This is the limit of how close you can approach the black hole and hope to escape.

Although light from a black hole cannot escape, light from events taking place near the black hole should be visible. For example, it is possible for a binary star system to consist of one normal star and one black hole. Material surrounding the ordinary star can be pulled into the black hole, forming an **accretion disk** around the black hole, as suggested in Figure 13.18. Friction among particles in the accretion disk results in transformation of mechanical energy into internal energy. As a result, the orbital height of the material above the event horizon decreases and the temperature rises. This high-temperature material emits a large amount of radiation, extending well into the x-ray region of the electromagnetic spectrum. These x-rays are characteristic of a black hole. Several possible candidates for black holes have been identified by observation of these x-rays.

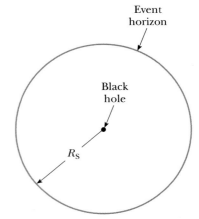

Figure 13.17 A black hole. The distance R_S equals the Schwarzschild radius. Any event occurring within the boundary of radius R_S, called the event horizon, is invisible to an outside observer.

Figure 13.18 A binary star system consisting of an ordinary star on the left and a black hole on the right. Matter pulled from the ordinary star forms an accretion disk around the black hole, in which matter is raised to very high temperatures, resulting in the emission of x-rays.

H. Ford et al. & NASA

Figure 13.19 Hubble Space Telescope images of the galaxy M87. The inset shows the center of the galaxy. The wider view shows a jet of material moving away from the center of the galaxy toward the upper right of the figure at about one tenth of the speed of light. Such jets are believed to be evidence of a supermassive black hole at the galaxy center.

There is also evidence that supermassive black holes exist at the centers of galaxies, with masses very much larger than the Sun. (There is strong evidence of a supermassive black hole of mass 2–3 million solar masses at the center of our galaxy.) Theoretical models for these bizarre objects predict that jets of material should be evident along the rotation axis of the hole. Figure 13.19 shows a Hubble Space Telescope photograph of galaxy M87. The jet of material coming from this galaxy is believed to be evidence for a supermassive black hole at the center of the galaxy.

SUMMARY

Take a practice test for this chapter by clicking on the Practice Test link at http://www.pse6.com.

Newton's law of universal gravitation states that the gravitational force of attraction between any two particles of masses m_1 and m_2 separated by a distance r has the magnitude

$$F_g = G \frac{m_1 m_2}{r^2} \tag{13.1}$$

where $G = 6.673 \times 10^{-11}$ N·m²/kg² is the **universal gravitational constant.** This equation enables us to calculate the force of attraction between masses under a wide variety of circumstances.

An object at a distance h above the Earth's surface experiences a gravitational force of magnitude mg, where g is the free-fall acceleration at that elevation:

$$g = \frac{GM_E}{r^2} = \frac{GM_E}{(R_E + h)^2} \tag{13.6}$$

In this expression, M_E is the mass of the Earth and R_E is its radius. Thus, the weight of an object decreases as the object moves away from the Earth's surface.

Kepler's laws of planetary motion state that

1. All planets move in elliptical orbits with the Sun at one focus.

2. The radius vector drawn from the Sun to a planet sweeps out equal areas in equal time intervals.

3. The square of the orbital period of any planet is proportional to the cube of the semimajor axis of the elliptical orbit.

Kepler's third law can be expressed as

$$T^2 = \left(\frac{4\pi^2}{GM_S} \right) a^3 \tag{13.8}$$

where M_S is the mass of the Sun and a is the semimajor axis. For a circular orbit, a can be replaced in Equation 13.8 by the radius r. Most planets have nearly circular orbits around the Sun.

The **gravitational field** at a point in space is defined as the gravitational force experienced by any test particle located at that point divided by the mass of the test particle:

$$\mathbf{g} \equiv \frac{\mathbf{F}_g}{m} \tag{13.9}$$

The gravitational force is conservative, and therefore a potential energy function can be defined for a system of two objects interacting gravitationally. The **gravitational potential energy** associated with two particles separated by a distance r is

$$U = -\frac{Gm_1 m_2}{r} \tag{13.14}$$

where U is taken to be zero as $r \rightarrow \infty$. The total potential energy for a system of particles is the sum of energies for all pairs of particles, with each pair represented by a term of the form given by Equation 13.14.

If an isolated system consists of an object of mass m moving with a speed v in the vicinity of a massive object of mass M, the total energy E of the system is the sum of the kinetic and potential energies:

$$E = \tfrac{1}{2} mv^2 - \frac{GMm}{r} \tag{13.16}$$

The total energy is a constant of the motion. If the object moves in an elliptical orbit of semimajor axis a around the massive object and if $M \gg m$, the total energy of the system is

$$E = -\frac{GMm}{2a} \tag{13.19}$$

For a circular orbit, this same equation applies with $a = r$. The total energy is negative for any bound system.

The **escape speed** for an object projected from the surface of a planet of mass M and radius R is

$$v_{\text{esc}} = \sqrt{\frac{2GM}{R}} \tag{13.23}$$

QUESTIONS

1. If the gravitational force on an object is directly proportional to its mass, why don't objects with large masses fall with greater acceleration than small ones?

2. The gravitational force exerted by the Sun on you is downward into the Earth at night, and upward into the sky during the day. If you had a sensitive enough bathroom scale,

would you expect to weigh more at night than during the day? Note also that you are farther away from the Sun at night than during the day. Would you expect to weigh less?

3. Use Kepler's second law to convince yourself that the Earth must move faster in its orbit during December, when it is closest to the Sun, than during June, when it is farthest from the Sun.

4. The gravitational force that the Sun exerts on the Moon is about twice as great as the gravitational force that the Earth exerts on the Moon. Why doesn't the Sun pull the Moon away from the Earth during a total eclipse of the Sun?

5. A satellite in orbit is not truly traveling through a vacuum. It is moving through very, very thin air. Does the resulting air friction cause the satellite to slow down?

6. How would you explain the fact that Jupiter and Saturn have periods much greater than one year?

7. If a system consists of five particles, how many terms appear in the expression for the total potential energy? How many terms appear if the system consists of N particles?

8. Does the escape speed of a rocket depend on its mass? Explain.

9. Compare the energies required to reach the Moon for a 10^5-kg spacecraft and a 10^3-kg satellite.

10. Explain why it takes more fuel for a spacecraft to travel from the Earth to the Moon than for the return trip. Estimate the difference.

11. A particular set of directions forms the *celestial equator*. If you live at 40° north latitude, these directions lie in an arc across your southern sky, including horizontally east, horizontally west, and south at 50° above the horizontal. In order to enjoy satellite TV, you need to install a dish with an unobstructed view to a particular point on the celestial equator. Why is this requirement so specific?

12. Why don't we put a geosynchronous weather satellite in orbit around the 45th parallel? Wouldn't this be more useful in the United States than one in orbit around the equator?

13. Is the absolute value of the potential energy associated with the Earth–Moon system greater than, less than, or equal to the kinetic energy of the Moon relative to the Earth?

14. Explain why no work is done on a planet as it moves in a circular orbit around the Sun, even though a gravitational force is acting on the planet. What is the net work done on a planet during each revolution as it moves around the Sun in an elliptical orbit?

15. Explain why the force exerted on a particle by a uniform sphere must be directed toward the center of the sphere. Would this be the case if the mass distribution of the sphere were not spherically symmetric?

16. At what position in its elliptical orbit is the speed of a planet a maximum? At what position is the speed a minimum?

17. If you are given the mass and radius of planet X, how would you calculate the free-fall acceleration on the surface of this planet?

18. If a hole could be dug to the center of the Earth, would the force on an object of mass m still obey Equation 13.1 there? What do you think the force on m would be at the center of the Earth?

19. In his 1798 experiment, Cavendish was said to have "weighed the Earth." Explain this statement.

20. The *Voyager* spacecraft was accelerated toward escape speed from the Sun by Jupiter's gravitational force exerted on the spacecraft. How is this possible?

21. How would you find the mass of the Moon?

22. The *Apollo 13* spacecraft developed trouble in the oxygen system about halfway to the Moon. Why did the mission continue on around the Moon, and then return home, rather than immediately turn back to Earth?

PROBLEMS

1, 2, 3 = straightforward, intermediate, challenging ☐ = full solution available in the *Student Solutions Manual and Study Guide*

🌐 = coached solution with hints available at http://www.pse6.com 💻 = computer useful in solving problem

▭ = paired numerical and symbolic problems

Section 13.1 Newton's Law of Universal Gravitation

Problem 17 in Chapter 1 can also be assigned with this section.

1. Determine the order of magnitude of the gravitational force that you exert on another person 2 m away. In your solution state the quantities you measure or estimate and their values.

2. Two ocean liners, each with a mass of 40 000 metric tons, are moving on parallel courses, 100 m apart. What is the magnitude of the acceleration of one of the liners toward the other due to their mutual gravitational attraction? Treat the ships as particles.

3. A 200-kg object and a 500-kg object are separated by 0.400 m. (a) Find the net gravitational force exerted by these objects on a 50.0-kg object placed midway between them. (b) At what position (other than an infinitely remote one) can the 50.0-kg object be placed so as to experience a net force of zero?

4. Two objects attract each other with a gravitational force of magnitude 1.00×10^{-8} N when separated by 20.0 cm. If the total mass of the two objects is 5.00 kg, what is the mass of each?

5. Three uniform spheres of mass 2.00 kg, 4.00 kg, and 6.00 kg are placed at the corners of a right triangle as in Figure P13.5. Calculate the resultant gravitational force on the 4.00-kg object, assuming the spheres are isolated from the rest of the Universe.

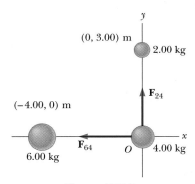

Figure P13.5

6. During a solar eclipse, the Moon, Earth, and Sun all lie on the same line, with the Moon between the Earth and the Sun. (a) What force is exerted by the Sun on the Moon? (b) What force is exerted by the Earth on the Moon? (c) What force is exerted by the Sun on the Earth?

Section 13.2 Measuring the Gravitational Constant

7. In introductory physics laboratories, a typical Cavendish balance for measuring the gravitational constant G uses lead spheres with masses of 1.50 kg and 15.0 g whose centers are separated by about 4.50 cm. Calculate the gravitational force between these spheres, treating each as a particle located at the center of the sphere.

8. A student proposes to measure the gravitational constant G by suspending two spherical objects from the ceiling of a tall cathedral and measuring the deflection of the cables from the vertical. Draw a free-body diagram of one of the objects. If two 100.0-kg objects are suspended at the lower ends of cables 45.00 m long and the cables are attached to the ceiling 1.000 m apart, what is the separation of the objects?

Section 13.3 Free-Fall Acceleration and the Gravitational Force

9. When a falling meteoroid is at a distance above the Earth's surface of 3.00 times the Earth's radius, what is its acceleration due to the Earth's gravitation?

10. The free-fall acceleration on the surface of the Moon is about one sixth of that on the surface of the Earth. If the radius of the Moon is about $0.250R_E$, find the ratio of their average densities, $\rho_{\text{Moon}}/\rho_{\text{Earth}}$.

11. On the way to the Moon the *Apollo* astronauts reached a point where the Moon's gravitational pull became stronger than the Earth's. (a) Determine the distance of this point from the center of the Earth. (b) What is the acceleration due to the Earth's gravitation at this point?

Section 13.4 Kepler's Laws and the Motion of Planets

12. The center-to-center distance between Earth and Moon is 384 400 km. The Moon completes an orbit in 27.3 days. (a) Determine the Moon's orbital speed. (b) If gravity were switched off, the Moon would move along a straight line tangent to its orbit, as described by Newton's first law. In its actual orbit in 1.00 s, how far does the Moon fall below the tangent line and toward the Earth?

13. Plaskett's binary system consists of two stars that revolve in a circular orbit about a center of mass midway between them. This means that the masses of the two stars are equal (Fig. P13.13). Assume the orbital speed of each star is 220 km/s and the orbital period of each is 14.4 days. Find the mass M of each star. (For comparison, the mass of our Sun is 1.99×10^{30} kg.)

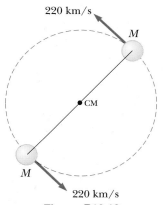

Figure P13.13

14. A particle of mass m moves along a straight line with constant speed in the x direction, a distance b from the x axis (Fig. P13.14). Show that Kepler's second law is satisfied by showing that the two shaded triangles in the figure have the same area when $t_4 - t_3 = t_2 - t_1$.

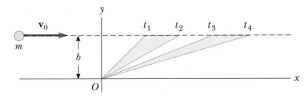

Figure P13.14

15. Io, a moon of Jupiter, has an orbital period of 1.77 days and an orbital radius of 4.22×10^5 km. From these data, determine the mass of Jupiter.

16. The *Explorer VIII* satellite, placed into orbit November 3, 1960, to investigate the ionosphere, had the following orbit parameters: perigee, 459 km; apogee, 2 289 km (both distances above the Earth's surface); period, 112.7 min. Find the ratio v_p/v_a of the speed at perigee to that at apogee.

17. Comet Halley (Figure P13.17) approaches the Sun to within 0.570 AU, and its orbital period is 75.6 years. (AU is the symbol for astronomical unit, where 1 AU = 1.50×10^{11} m is the mean Earth–Sun distance.) How far from the Sun will Halley's comet travel before it starts its return journey?

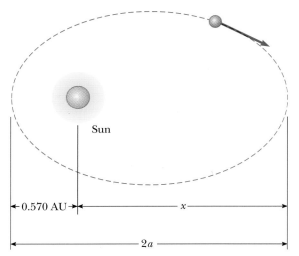

Figure P13.17

18. Two planets X and Y travel counterclockwise in circular orbits about a star as in Figure P13.18. The radii of their orbits are in the ratio 3:1. At some time, they are aligned as in Figure P13.18a, making a straight line with the star. During the next five years, the angular displacement of planet X is 90.0°, as in Figure P13.18b. Where is planet Y at this time?

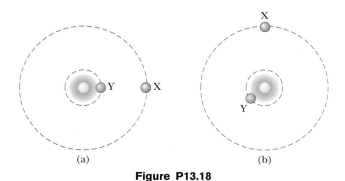

(a)   (b)

Figure P13.18

19. A synchronous satellite, which always remains above the same point on a planet's equator, is put in orbit around Jupiter to study the famous red spot. Jupiter rotates about its axis once every 9.84 h. Use the data of Table 13.2 to find the altitude of the satellite.

20. Neutron stars are extremely dense objects that are formed from the remnants of supernova explosions. Many rotate very rapidly. Suppose that the mass of a certain spherical neutron star is twice the mass of the Sun and its radius is 10.0 km. Determine the greatest possible angular speed it can have so that the matter at the surface of the star on its equator is just held in orbit by the gravitational force.

21. Suppose the Sun's gravity were switched off. The planets would leave their nearly circular orbits and fly away in straight lines, as described by Newton's first law. Would Mercury ever be farther from the Sun than Pluto? If so, find how long it would take for Mercury to achieve this passage. If not, give a convincing argument that Pluto is always farther from the Sun.

22. As thermonuclear fusion proceeds in its core, the Sun loses mass at a rate of 3.64×10^9 kg/s. During the 5 000-yr period of recorded history, by how much has the length of the year changed due to the loss of mass from the Sun? *Suggestions:* Assume the Earth's orbit is circular. No external torque acts on the Earth–Sun system, so its angular momentum is conserved. If x is small compared to 1, then $(1 + x)^n$ is nearly equal to $1 + nx$.

Section 13.5   The Gravitational Field

23. Three objects of equal mass are located at three corners of a square of edge length ℓ as in Figure P13.23. Find the gravitational field at the fourth corner due to these objects.

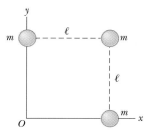

Figure P13.23

24. A spacecraft in the shape of a long cylinder has a length of 100 m, and its mass with occupants is 1 000 kg. It has strayed too close to a black hole having a mass 100 times that of the Sun (Fig. P13.24). The nose of the spacecraft points toward the black hole, and the distance between the nose and the center of the black hole is 10.0 km. (a) Determine the total force on the spacecraft. (b) What is the difference in the gravitational fields acting on the occupants in the nose of the ship and on those in the rear of the ship, farthest from the black hole? This difference in accelerations grows rapidly as the ship approaches the black hole. It puts the body of the ship under extreme tension and eventually tears it apart.

Figure P13.24

25. Compute the magnitude and direction of the gravitational field at a point P on the perpendicular bisector of the line joining two objects of equal mass separated by a distance $2a$ as shown in Figure P13.25.

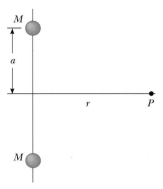

Figure P13.25

Section 13.6   Gravitational Potential Energy

> Assume $U = 0$ at $r = \infty$.

26. A satellite of the Earth has a mass of 100 kg and is at an altitude of 2.00×10^6 m. (a) What is the potential energy of the satellite–Earth system? (b) What is the magnitude of the gravitational force exerted by the Earth on the satellite? (c) **What If?** What force does the satellite exert on the Earth?

27. How much energy is required to move a 1 000-kg object from the Earth's surface to an altitude twice the Earth's radius?

28. At the Earth's surface a projectile is launched straight up at a speed of 10.0 km/s. To what height will it rise? Ignore air resistance and the rotation of the Earth.

29. After our Sun exhausts its nuclear fuel, its ultimate fate may be to collapse to a *white dwarf* state, in which it has approximately the same mass as it has now, but a radius equal to the radius of the Earth. Calculate (a) the average density of the white dwarf, (b) the free-fall acceleration, and (c) the gravitational potential energy of a 1.00-kg object at its surface.

30. How much work is done by the Moon's gravitational field as a 1 000-kg meteor comes in from outer space and impacts on the Moon's surface?

31. A system consists of three particles, each of mass 5.00 g, located at the corners of an equilateral triangle with sides of 30.0 cm. (a) Calculate the potential energy of the system. (b) If the particles are released simultaneously, where will they collide?

32. 🖥 An object is released from rest at an altitude h above the surface of the Earth. (a) Show that its speed at a distance r from the Earth's center, where $R_E \leq r \leq R_E + h$, is given by

$$v = \sqrt{2GM_E\left(\frac{1}{r} - \frac{1}{R_E + h}\right)}$$

(b) Assume the release altitude is 500 km. Perform the integral

$$\Delta t = \int_i^f dt = -\int_i^f \frac{dr}{v}$$

to find the time of fall as the object moves from the release point to the Earth's surface. The negative sign appears because the object is moving opposite to the radial direction, so its speed is $v = -dr/dt$. Perform the integral numerically.

Section 13.7   Energy Considerations in Planetary and Satellite Motion

33. 🪐 A space probe is fired as a projectile from the Earth's surface with an initial speed of 2.00×10^4 m/s. What will its speed be when it is very far from the Earth? Ignore friction and the rotation of the Earth.

By permission of John Hart and Creators Syndicate, Inc.

Figure P13.35

34. (a) What is the minimum speed, relative to the Sun, necessary for a spacecraft to escape the solar system if it starts at the Earth's orbit? (b) *Voyager 1* achieved a maximum speed of 125 000 km/h on its way to photograph Jupiter. Beyond what distance from the Sun is this speed sufficient to escape the solar system?

35. A "treetop satellite" (Fig. P13.35) moves in a circular orbit just above the surface of a planet, assumed to offer no air resistance. Show that its orbital speed v and the escape speed from the planet are related by the expression

$$v_{esc} = \sqrt{2}\,v.$$

36. A 500-kg satellite is in a circular orbit at an altitude of 500 km above the Earth's surface. Because of air friction, the satellite eventually falls to the Earth's surface, where it hits the ground with a speed of 2.00 km/s. How much energy was transformed into internal energy by means of friction?

37. A satellite of mass 200 kg is placed in Earth orbit at a height of 200 km above the surface. (a) With a circular orbit, how long does the satellite take to complete one orbit? (b) What is the satellite's speed? (c) What is the minimum energy input necessary to place this satellite in orbit? Ignore air resistance but include the effect of the planet's daily rotation.

38. A satellite of mass m, originally on the surface of the Earth, is placed into Earth orbit at an altitude h. (a) With a circular orbit, how long does the satellite take to complete one orbit? (b) What is the satellite's speed? (c) What is the minimum energy input necessary to place this satellite in orbit? Ignore air resistance but include the effect of the planet's daily rotation. At what location on the Earth's surface and in what direction should the satellite be launched to minimize the required energy investment? Represent the mass and radius of the Earth as M_E and R_E.

39. A 1 000-kg satellite orbits the Earth at a constant altitude of 100 km. How much energy must be added to the system to move the satellite into a circular orbit with altitude 200 km?

40. The planet Uranus has a mass about 14 times the Earth's mass, and its radius is equal to about 3.7 Earth radii. (a) By setting up ratios with the corresponding Earth values, find the free-fall acceleration at the cloud tops of Uranus. (b) Ignoring the rotation of the planet, find the minimum escape speed from Uranus.

41. Determine the escape speed for a rocket on the far side of Ganymede, the largest of Jupiter's moons (Figure P13.41). The radius of Ganymede is 2.64×10^6 m, and its mass is

1.495 × 10^{23} kg. The mass of Jupiter is 1.90×10^{27} kg, and the distance between Jupiter and Ganymede is 1.071×10^9 m. Be sure to include the gravitational effect due to Jupiter, but you may ignore the motion of Jupiter and Ganymede as they revolve about their center of mass.

42. In Robert Heinlein's "The Moon is a Harsh Mistress," the colonial inhabitants of the Moon threaten to launch rocks down onto the Earth if they are not given independence (or at least representation). Assuming that a rail gun could launch a rock of mass m at twice the lunar escape speed, calculate the speed of the rock as it enters the Earth's atmosphere. (By *lunar escape speed* we mean the speed required to move infinitely far away from a stationary Moon alone in the Universe. Problem 61 in Chapter 30 describes a rail gun.)

43. An object is fired vertically upward from the surface of the Earth (of radius R_E) with an initial speed v_i that is comparable to but less than the escape speed v_{esc}. (a) Show that the object attains a maximum height h given by

$$h = \frac{R_E v_i^2}{v_{esc}^2 - v_i^2}$$

(b) A space vehicle is launched vertically upward from the Earth's surface with an initial speed of 8.76 km/s, which is less than the escape speed of 11.2 km/s. What maximum height does it attain? (c) A meteorite falls toward the Earth. It is essentially at rest with respect to the Earth when it is at a height of 2.51×10^7 m. With what speed does the meteorite strike the Earth? (d) **What If?** Assume that a baseball is tossed up with an initial speed that is very small compared to the escape speed. Show that the equation from part (a) is consistent with Equation 4.13.

44. Derive an expression for the work required to move an Earth satellite of mass m from a circular orbit of radius $2R_E$ to one of radius $3R_E$.

45. A comet of mass 1.20×10^{10} kg moves in an elliptical orbit around the Sun. Its distance from the Sun ranges between 0.500 AU and 50.0 AU. (a) What is the eccentricity of its orbit? (b) What is its period? (c) At aphelion what is the potential energy of the comet–Sun system? *Note:* 1 AU = one astronomical unit = the average distance from Sun to Earth = 1.496×10^{11} m.

46. A satellite moves around the Earth in a circular orbit of radius r. (a) What is the speed v_0 of the satellite? Suddenly, an explosion breaks the satellite into two pieces, with masses m and $4m$. Immediately after the explosion the smaller piece of mass m is stationary with respect to the Earth and falls directly toward the Earth. (b) What is the speed v_i of the larger piece immediately after the explosion? (c) Because of the increase in its speed, this larger piece now moves in a new elliptical orbit. Find its distance away from the center of the Earth when it reaches the other end of the ellipse.

Additional Problems

47. The Solar and Heliospheric Observatory (SOHO) spacecraft has a special orbit, chosen so that its view of the Sun is never eclipsed and it is always close enough to the Earth to

Ganymede

Jupiter

Figure P13.41

transmit data easily. It moves in a near-circle around the Sun that is smaller than the Earth's circular orbit. Its period, however, is just equal to 1 yr. It is always located between the Earth and the Sun along the line joining them. Both objects exert gravitational forces on the observatory. Show that its distance from the Earth must be between 1.47×10^9 m and 1.48×10^9 m. In 1772 Joseph Louis Lagrange determined theoretically the special location allowing this orbit. The SOHO spacecraft took this position on February 14, 1996. *Suggestion:* Use data that are precise to four digits. The mass of the Earth is 5.983×10^{24} kg.

48. Let Δg_M represent the difference in the gravitational fields produced by the Moon at the points on the Earth's surface nearest to and farthest from the Moon. Find the fraction $\Delta g_M/g$, where g is the Earth's gravitational field. (This difference is responsible for the occurrence of the *lunar tides* on the Earth.)

49. **Review problem.** Two identical hard spheres, each of mass m and radius r, are released from rest in otherwise empty space with their centers separated by the distance R. They are allowed to collide under the influence of their gravitational attraction. (a) Show that the magnitude of the impulse received by each sphere before they make contact is given by $[Gm^3(1/2r - 1/R)]^{1/2}$. (b) **What If?** Find the magnitude of the impulse each receives if they collide elastically.

50. Two spheres having masses M and $2M$ and radii R and $3R$, respectively, are released from rest when the distance between their centers is $12R$. How fast will each sphere be moving when they collide? Assume that the two spheres interact only with each other.

51. In Larry Niven's science-fiction novel *Ringworld,* a rigid ring of material rotates about a star (Fig. P13.51). The tangential speed of the ring is 1.25×10^6 m/s, and its radius is 1.53×10^{11} m. (a) Show that the centripetal acceleration of the inhabitants is 10.2 m/s². (b) The inhabitants of this ring world live on the starlit inner surface of the ring. Each person experiences a normal contact force **n**. Acting alone, this normal force would produce an inward acceleration of 9.90 m/s². Additionally, the star at the center of the ring exerts a gravitational force on the ring and its inhabitants. The difference between the total acceleration and the acceleration provided by the normal force is due to the gravitational attraction of the central star. Show that the mass of the star is approximately 10^{32} kg.

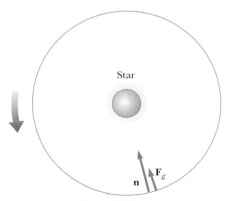

Figure P13.51

52. (a) Show that the rate of change of the free-fall acceleration with distance above the Earth's surface is

$$\frac{dg}{dr} = -\frac{2GM_E}{R_E^{\,3}}$$

This rate of change over distance is called a *gradient*. (b) If h is small in comparison to the radius of the Earth, show that the difference in free-fall acceleration between two points separated by vertical distance h is

$$|\Delta g| = \frac{2GM_E h}{R_E^{\,3}}$$

(c) Evaluate this difference for $h = 6.00$ m, a typical height for a two-story building.

53. A ring of matter is a familiar structure in planetary and stellar astronomy. Examples include Saturn's rings and a ring nebula. Consider a uniform ring of mass 2.36×10^{20} kg and radius 1.00×10^8 m. An object of mass 1 000 kg is placed at a point A on the axis of the ring, 2.00×10^8 m from the center of the ring (Figure P13.53). When the object is released, the attraction of the ring makes the object move along the axis toward the center of the ring (point B). (a) Calculate the gravitational potential energy of the object–ring system when the object is at A. (b) Calculate the gravitational potential energy of the system when the object is at B. (c) Calculate the speed of the object as it passes through B.

Figure P13.53

54. *Voyagers 1* and *2* surveyed the surface of Jupiter's moon Io and photographed active volcanoes spewing liquid sulfur to heights of 70 km above the surface of this moon. Find

the speed with which the liquid sulfur left the volcano. Io's mass is 8.9×10^{22} kg, and its radius is 1 820 km.

55. As an astronaut, you observe a small planet to be spherical. After landing on the planet, you set off, walking always straight ahead, and find yourself returning to your spacecraft from the opposite side after completing a lap of 25.0 km. You hold a hammer and a falcon feather at a height of 1.40 m, release them, and observe that they fall together to the surface in 29.2 s. Determine the mass of the planet.

56. A certain quaternary star system consists of three stars, each of mass m, moving in the same circular orbit of radius r about a central star of mass M. The stars orbit in the same sense, and are positioned one third of a revolution apart from each other. Show that the period of each of the three stars is given by

$$T = 2\pi\sqrt{\frac{r^3}{g\left(M + m/\sqrt{3}\right)}}$$

57. **Review problem.** A cylindrical habitat in space 6.00 km in diameter and 30 km long has been proposed (by G. K. O'Neill, 1974). Such a habitat would have cities, land, and lakes on the inside surface and air and clouds in the center. This would all be held in place by rotation of the cylinder about its long axis. How fast would the cylinder have to rotate to imitate the Earth's gravitational field at the walls of the cylinder?

58. Newton's law of universal gravitation is valid for distances covering an enormous range, but it is thought to fail for very small distances, where the structure of space itself is uncertain. Far smaller than an atomic nucleus, this crossover distance is called the Planck length. It is determined by a combination of the constants G, c, and h, where c is the speed of light in vacuum and h is Planck's constant (introduced in Chapter 11) with units of angular momentum. (a) Use dimensional analysis to find a combination of these three universal constants that has units of length. (b) Determine the order of magnitude of the Planck length. You will need to consider noninteger powers of the constants.

59. Show that the escape speed from the surface of a planet of uniform density is directly proportional to the radius of the planet.

60. Many people assume that air resistance acting on a moving object will always make the object slow down. It can actually be responsible for making the object speed up. Consider a 100-kg Earth satellite in a circular orbit at an altitude of 200 km. A small force of air resistance makes the satellite drop into a circular orbit with an altitude of 100 km. (a) Calculate its initial speed. (b) Calculate its final speed in this process. (c) Calculate the initial energy of the satellite–Earth system. (d) Calculate the final energy of the system. (e) Show that the system has lost mechanical energy and find the amount of the loss due to friction. (f) What force makes the satellite's speed increase? You will find a free-body diagram useful in explaining your answer.

61. ![www icon] Two hypothetical planets of masses m_1 and m_2 and radii r_1 and r_2, respectively, are nearly at rest when they are an infinite distance apart. Because of their gravitational attraction, they head toward each other on a collision course. (a) When their center-to-center separation is d, find expressions for the speed of each planet and for their relative speed. (b) Find the kinetic energy of each planet just before they collide, if $m_1 = 2.00 \times 10^{24}$ kg, $m_2 = 8.00 \times 10^{24}$ kg, $r_1 = 3.00 \times 10^6$ m, and $r_2 = 5.00 \times 10^6$ m. (*Note:* Both energy and momentum of the system are conserved.)

62. The maximum distance from the Earth to the Sun (at our aphelion) is 1.521×10^{11} m, and the distance of closest approach (at perihelion) is 1.471×10^{11} m. If the Earth's orbital speed at perihelion is 3.027×10^4 m/s, determine (a) the Earth's orbital speed at aphelion, (b) the kinetic and potential energies of the Earth–Sun system at perihelion, and (c) the kinetic and potential energies at aphelion. Is the total energy constant? (Ignore the effect of the Moon and other planets.)

63. (a) Determine the amount of work (in joules) that must be done on a 100-kg payload to elevate it to a height of 1 000 km above the Earth's surface. (b) Determine the amount of additional work that is required to put the payload into circular orbit at this elevation.

64. X-ray pulses from Cygnus X-1, a celestial x-ray source, have been recorded during high-altitude rocket flights. The signals can be interpreted as originating when a blob of ionized matter orbits a black hole with a period of 5.0 ms. If the blob were in a circular orbit about a black hole whose mass is $20M_{\text{Sun}}$, what is the orbit radius?

65. Studies of the relationship of the Sun to its galaxy—the Milky Way—have revealed that the Sun is located near the outer edge of the galactic disk, about 30 000 lightyears from the center. The Sun has an orbital speed of approximately 250 km/s around the galactic center. (a) What is the period of the Sun's galactic motion? (b) What is the order of magnitude of the mass of the Milky Way galaxy? Suppose that the galaxy is made mostly of stars of which the Sun is typical. What is the order of magnitude of the number of stars in the Milky Way?

66. The oldest artificial satellite in orbit is *Vanguard I*, launched March 3, 1958. Its mass is 1.60 kg. In its initial orbit, its minimum distance from the center of the Earth was 7.02 Mm, and its speed at this perigee point was 8.23 km/s. (a) Find the total energy of the satellite–Earth system. (b) Find the magnitude of the angular momentum of the satellite. (c) Find its speed at apogee and its maximum (apogee) distance from the center of the Earth. (d) Find the semimajor axis of its orbit. (e) Determine its period.

67. Astronomers detect a distant meteoroid moving along a straight line that, if extended, would pass at a distance $3R_E$ from the center of the Earth, where R_E is the radius of the Earth. What minimum speed must the meteoroid have if the Earth's gravitation is not to deflect the meteoroid to make it strike the Earth?

68. A spherical planet has uniform density ρ. Show that the minimum period for a satellite in orbit around it is

$$T_{\text{min}} = \sqrt{\frac{3\pi}{G\rho}}$$

independent of the radius of the planet.

69. Two stars of masses M and m, separated by a distance d, revolve in circular orbits about their center of mass (Fig. P13.69). Show that each star has a period given by

$$T^2 = \frac{4\pi^2 d^3}{G(M + m)}$$

Proceed as follows: Apply Newton's second law to each star. Note that the center-of-mass condition requires that $Mr_2 = mr_1$, where $r_1 + r_2 = d$.

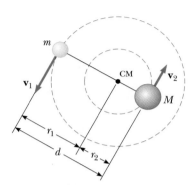

Figure P13.69

70. (a) A 5.00-kg object is released 1.20×10^7 m from the center of the Earth. It moves with what acceleration relative to the Earth? (b) **What If?** A 2.00×10^{24} kg object is released 1.20×10^7 m from the center of the Earth. It moves with what acceleration relative to the Earth? Assume that the objects behave as pairs of particles, isolated from the rest of the Universe.

71. 🖥 The acceleration of an object moving in the gravitational field of the Earth is

$$\mathbf{a} = -\frac{GM_E\mathbf{r}}{r^3}$$

where $\mathbf{r}$ is the position vector directed from the center of the Earth toward the object. Choosing the origin at the center of the Earth and assuming that the small object is moving in the xy plane, we find that the rectangular (Cartesian) components of its acceleration are

$$a_x = -\frac{GM_E x}{(x^2 + y^2)^{3/2}} \qquad a_y = -\frac{GM_E y}{(x^2 + y^2)^{3/2}}$$

Use a computer to set up and carry out a numerical prediction of the motion of the object, according to Euler's method. Assume the initial position of the object is $x = 0$ and $y = 2R_E$, where R_E is the radius of the Earth. Give the object an initial velocity of 5 000 m/s in the x direction. The time increment should be made as small as practical. Try 5 s. Plot the x and y coordinates of the object as time goes on. Does the object hit the Earth? Vary the initial velocity until you find a circular orbit.

Answers to Quick Quizzes

13.1 (d). The gravitational force exerted by the Earth on the Moon provides a net force that causes the Moon's centripetal acceleration.

13.2 (e). The gravitational force follows an inverse-square behavior, so doubling the distance causes the force to be one fourth as large.

13.3 (c). An object in orbit is simply falling while it moves around the Earth. The acceleration of the object is that due to gravity. Because the object was launched from a very tall mountain, the value for g is slightly less than that at the surface.

13.4 (a). Kepler's third law (Eq. 13.8), which applies to all the planets, tells us that the period of a planet is proportional to $a^{3/2}$. Because Pluto is farther from the Sun than the Earth, it has a longer period. The Sun's gravitational field is much weaker at Pluto than it is at the Earth. Thus, this planet experiences much less centripetal acceleration than the Earth does, and it has a correspondingly longer period.

13.5 (a). From Kepler's third law and the given period, the major axis of the asteroid can be calculated. It is found to be 1.2×10^{11} m. Because this is smaller than the Earth–Sun distance, the asteroid cannot possibly collide with the Earth.

13.6 (b). From conservation of angular momentum, $mv_p r_p = mv_a r_a$, so that $v_p = (r_a/r_p)v_a = (4D/D)v_a = 4v_a$.

13.7 (a) Perihelion. Because of conservation of angular momentum, the speed of the comet is highest at its closest position to the Sun. (b) Aphelion. The potential energy of the comet–Sun system is highest when the comet is at its farthest distance from the Sun. (c) Perihelion. The kinetic energy is highest at the point at which the speed of the comet is highest. (d) All points. The total energy of the system is the same regardless of where the comet is in its orbit.

Chapter 14

Fluid Mechanics

▲ These hot-air balloons float because they are filled with air at high temperature and are surrounded by denser air at a lower temperature. In this chapter, we will explore the buoyant force that supports these balloons and other floating objects. (Richard Megna/Fundamental Photographs)

Matter is normally classified as being in one of three states: solid, liquid, or gas. From everyday experience, we know that a solid has a definite volume and shape. A brick maintains its familiar shape and size day in and day out. We also know that a liquid has a definite volume but no definite shape. Finally, we know that an unconfined gas has neither a definite volume nor a definite shape. These descriptions help us picture the states of matter, but they are somewhat artificial. For example, asphalt and plastics are normally considered solids, but over long periods of time they tend to flow like liquids. Likewise, most substances can be a solid, a liquid, or a gas (or a combination of any of these), depending on the temperature and pressure. In general, the time it takes a particular substance to change its shape in response to an external force determines whether we treat the substance as a solid, a liquid, or a gas.

A **fluid** is a collection of molecules that are randomly arranged and held together by weak cohesive forces and by forces exerted by the walls of a container. Both liquids and gases are fluids.

In our treatment of the mechanics of fluids, we do not need to learn any new physical principles to explain such effects as the buoyant force acting on a submerged object and the dynamic lift acting on an airplane wing. First, we consider the mechanics of a fluid at rest—that is, *fluid statics*. We then treat the mechanics of fluids in motion—that is, *fluid dynamics*. We can describe a fluid in motion by using a model that is based upon certain simplifying assumptions.

Figure 14.1 At any point on the surface of a submerged object, the force exerted by the fluid is perpendicular to the surface of the object. The force exerted by the fluid on the walls of the container is perpendicular to the walls at all points.

14.1 Pressure

Fluids do not sustain shearing stresses or tensile stresses; thus, the only stress that can be exerted on an object submerged in a static fluid is one that tends to compress the object from all sides. In other words, the force exerted by a static fluid on an object is always perpendicular to the surfaces of the object, as shown in Figure 14.1.

The pressure in a fluid can be measured with the device pictured in Figure 14.2. The device consists of an evacuated cylinder that encloses a light piston connected to a spring. As the device is submerged in a fluid, the fluid presses on the top of the piston and compresses the spring until the inward force exerted by the fluid is balanced by the outward force exerted by the spring. The fluid pressure can be measured directly if the spring is calibrated in advance. If F is the magnitude of the force exerted on the piston and A is the surface area of the piston, then the **pressure** P of the fluid at the level to which the device has been submerged is defined as the ratio F/A:

Figure 14.2 A simple device for measuring the pressure exerted by a fluid.

$$P \equiv \frac{F}{A} \tag{14.1}$$

Definition of pressure

Note that pressure is a scalar quantity because it is proportional to the magnitude of the force on the piston.

If the pressure varies over an area, we can evaluate the infinitesimal force dF on an infinitesimal surface element of area dA as

Earl Young/Getty Images

Snowshoes keep you from sinking into soft snow because they spread the downward force you exert on the snow over a large area, reducing the pressure on the snow surface.

 PITFALL PREVENTION

14.1 Force and Pressure

Equations 14.1 and 14.2 make a clear distinction between force and pressure. Another important distinction is that *force is a vector* and *pressure is a scalar*. There is no direction associated with pressure, but the direction of the force associated with the pressure is perpendicular to the surface of interest.

$$dF = P\, dA \qquad (14.2)$$

where P is the pressure at the location of the area dA. The pressure exerted by a fluid varies with depth. Therefore, to calculate the total force exerted on a flat vertical wall of a container, we must integrate Equation 14.2 over the surface area of the wall.

Because pressure is force per unit area, it has units of newtons per square meter (N/m^2) in the SI system. Another name for the SI unit of pressure is **pascal** (Pa):

$$1\ Pa \equiv 1\ N/m^2 \qquad (14.3)$$

> **Quick Quiz 14.1** Suppose you are standing directly behind someone who steps back and accidentally stomps on your foot with the heel of one shoe. Would you be better off if that person were (a) a large professional basketball player wearing sneakers (b) a petite woman wearing spike-heeled shoes?

Example 14.1 The Water Bed

The mattress of a water bed is 2.00 m long by 2.00 m wide and 30.0 cm deep.

(A) Find the weight of the water in the mattress.

Solution The density of fresh water is 1 000 kg/m³ (see Table 14.1 on page 423), and the volume of the water filling the mattress is $V = (2.00\ m)(2.00\ m)(0.300\ m) = 1.20\ m^3$. Hence, using Equation 1.1, the mass of the water in the bed is

$$M = \rho V = (1\ 000\ kg/m^3)(1.20\ m^3) = 1.20 \times 10^3\ kg$$

and its weight is

$$Mg = (1.20 \times 10^3\ kg)(9.80\ m/s^2) = \boxed{1.18 \times 10^4\ N}$$

This is approximately 2 650 lb. (A regular bed weighs approximately 300 lb.) Because this load is so great, such a water bed is best placed in the basement or on a sturdy, well-supported floor.

(B) Find the pressure exerted by the water on the floor when the bed rests in its normal position. Assume that the entire lower surface of the bed makes contact with the floor.

Solution When the bed is in its normal position, the area in contact with the floor is 4.00 m²; thus, from Equation

14.1, we find that

$$P = \frac{1.18 \times 10^4\ N}{4.00\ m^2} = \boxed{2.95 \times 10^3\ Pa}$$

What If? What if the water bed is replaced by a 300-lb ordinary bed that is supported by four legs? Each leg has a circular cross section of radius 2.00 cm. What pressure does this bed exert on the floor?

Answer The weight of the bed is distributed over four circular cross sections at the bottom of the legs. Thus, the pressure is

$$P = \frac{F}{A} = \frac{mg}{4(\pi r^2)} = \frac{300\ lb}{4\pi(0.0200\ m)^2}\left(\frac{1\ N}{0.225\ lb}\right)$$
$$= 2.65 \times 10^5\ Pa$$

Note that this is almost 100 times larger than the pressure due to the water bed! This is because the weight of the ordinary bed, even though it is much less than the weight of the water bed, is applied over the very small area of the four legs. The high pressure on the floor at the feet of an ordinary bed could cause denting of wood floors or permanently crush carpet pile. In contrast, a water bed requires a sturdy floor to support the very large weight.

Table 14.1

Densities of Some Common Substances at Standard Temperature (0°C) and Pressure (Atmospheric)			
Substance	$\rho\,(kg/m^3)$	**Substance**	$\rho\,(kg/m^3)$
Air	1.29	Ice	0.917×10^3
Aluminum	2.70×10^3	Iron	7.86×10^3
Benzene	0.879×10^3	Lead	11.3×10^3
Copper	8.92×10^3	Mercury	13.6×10^3
Ethyl alcohol	0.806×10^3	Oak	0.710×10^3
Fresh water	1.00×10^3	Oxygen gas	1.43
Glycerin	1.26×10^3	Pine	0.373×10^3
Gold	19.3×10^3	Platinum	21.4×10^3
Helium gas	1.79×10^{-1}	Seawater	1.03×10^3
Hydrogen gas	8.99×10^{-2}	Silver	10.5×10^3

14.2 Variation of Pressure with Depth

As divers well know, water pressure increases with depth. Likewise, atmospheric pressure decreases with increasing altitude; for this reason, aircraft flying at high altitudes must have pressurized cabins.

We now show how the pressure in a liquid increases with depth. As Equation 1.1 describes, the *density* of a substance is defined as its mass per unit volume; Table 14.1 lists the densities of various substances. These values vary slightly with temperature because the volume of a substance is temperature-dependent (as shown in Chapter 19). Under standard conditions (at 0°C and at atmospheric pressure) the densities of gases are about 1/1 000 the densities of solids and liquids. This difference in densities implies that the average molecular spacing in a gas under these conditions is about ten times greater than that in a solid or liquid.

Now consider a liquid of density ρ at rest as shown in Figure 14.3. We assume that ρ is uniform throughout the liquid; this means that the liquid is incompressible. Let us select a sample of the liquid contained within an imaginary cylinder of cross-sectional area A extending from depth d to depth $d + h$. The liquid external to our sample exerts forces at all points on the surface of the sample, perpendicular to the surface. The pressure exerted by the liquid on the bottom face of the sample is P, and the pressure on the top face is P_0. Therefore, the upward force exerted by the outside fluid on the bottom of the cylinder has a magnitude PA, and the downward force exerted on the top has a magnitude P_0A. The mass of liquid in the cylinder is $M = \rho V = \rho Ah$; therefore, the weight of the liquid in the cylinder is $Mg = \rho Ahg$. Because the cylinder is in equilibrium, the net force acting on it must be zero. Choosing upward to be the positive y direction, we see that

$$\sum \mathbf{F} = PA\hat{\mathbf{j}} - P_0A\hat{\mathbf{j}} - Mg\hat{\mathbf{j}} = 0$$

or

$$PA - P_0A - \rho Ahg = 0$$
$$PA - P_0A = \rho Ahg$$

$$P = P_0 + \rho gh \tag{14.4}$$

Figure 14.3 A parcel of fluid (darker region) in a larger volume of fluid is singled out. The net force exerted on the parcel of fluid must be zero because it is in equilibrium.

◀ Variation of pressure with depth

That is, **the pressure P at a depth h below a point in the liquid at which the pressure is P_0 is greater by an amount ρgh.** If the liquid is open to the atmosphere and P_0

(a)   (b)

Figure 14.4 (a) Diagram of a hydraulic press. Because the increase in pressure is the same on the two sides, a small force $\mathbf{F}_1$ at the left produces a much greater force $\mathbf{F}_2$ at the right. (b) A vehicle undergoing repair is supported by a hydraulic lift in a garage.

is the pressure at the surface of the liquid, then P_0 is atmospheric pressure. In our calculations and working of end-of-chapter problems, we usually take atmospheric pressure to be

$$P_0 = 1.00 \text{ atm} = 1.013 \times 10^5 \text{ Pa}$$

Equation 14.4 implies that the pressure is the same at all points having the same depth, independent of the shape of the container.

In view of the fact that the pressure in a fluid depends on depth and on the value of P_0, any increase in pressure at the surface must be transmitted to every other point in the fluid. This concept was first recognized by the French scientist Blaise Pascal (1623–1662) and is called **Pascal's law: a change in the pressure applied to a fluid is transmitted undiminished to every point of the fluid and to the walls of the container.**

An important application of Pascal's law is the hydraulic press illustrated in Figure 14.4a. A force of magnitude F_1 is applied to a small piston of surface area A_1. The pressure is transmitted through an incompressible liquid to a larger piston of surface area A_2. Because the pressure must be the same on both sides, $P = F_1/A_1 = F_2/A_2$. Therefore, the force F_2 is greater than the force F_1 by a factor A_2/A_1. By designing a hydraulic press with appropriate areas A_1 and A_2, a large output force can be applied by means of a small input force. Hydraulic brakes, car lifts, hydraulic jacks, and forklifts all make use of this principle (Fig. 14.4b).

Because liquid is neither added nor removed from the system, the volume of liquid pushed down on the left in Figure 14.4a as the piston moves downward through a displacement Δx_1 equals the volume of liquid pushed up on the right as the right piston moves upward through a displacement Δx_2. That is, $A_1 \Delta x_1 = A_2 \Delta x_2$; thus, $A_2/A_1 = \Delta x_1/\Delta x_2$. We have already shown that $A_2/A_1 = F_2/F_1$. Thus, $F_2/F_1 = \Delta x_1/\Delta x_2$, so $F_1 \Delta x_1 = F_2 \Delta x_2$. Each side of this equation is the work done by the force. Thus, the work done by $\mathbf{F}_1$ on the input piston equals the work done by $\mathbf{F}_2$ on the output piston, as it must in order to conserve energy.

Pascal's law

Quick Quiz 14.2 The pressure at the bottom of a filled glass of water ($\rho =$ 1 000 kg/m³) is P. The water is poured out and the glass is filled with ethyl alcohol ($\rho = 806$ kg/m³). The pressure at the bottom of the glass is (a) smaller than P (b) equal to P (c) larger than P (d) indeterminate.

Example 14.2   The Car Lift

Interactive

In a car lift used in a service station, compressed air exerts a force on a small piston that has a circular cross section and a radius of 5.00 cm. This pressure is transmitted by a liquid to a piston that has a radius of 15.0 cm. What force must the compressed air exert to lift a car weighing 13 300 N? What air pressure produces this force?

Solution Because the pressure exerted by the compressed air is transmitted undiminished throughout the liquid, we have

$$F_1 = \left(\frac{A_1}{A_2}\right)F_2 = \frac{\pi(5.00 \times 10^{-2}\text{ m})^2}{\pi(15.0 \times 10^{-2}\text{ m})^2}(1.33 \times 10^4\text{ N})$$

$$= \boxed{1.48 \times 10^3\text{ N}}$$

The air pressure that produces this force is

$$P = \frac{F_1}{A_1} = \frac{1.48 \times 10^3\text{ N}}{\pi(5.00 \times 10^{-2}\text{ m})^2}$$

$$= \boxed{1.88 \times 10^5\text{ Pa}}$$

This pressure is approximately twice atmospheric pressure.

 You can adjust the weight of the truck in Figure 14.4a at the Interactive Worked Example link at **http://www.pse6.com.**

Example 14.3   A Pain in Your Ear

Estimate the force exerted on your eardrum due to the water above when you are swimming at the bottom of a pool that is 5.0 m deep.

Solution First, we must find the unbalanced pressure on the eardrum; then, after estimating the eardrum's surface area, we can determine the force that the water exerts on it.

The air inside the middle ear is normally at atmospheric pressure P_0. Therefore, to find the net force on the eardrum, we must consider the difference between the total pressure at the bottom of the pool and atmospheric pressure:

$$P_{\text{bot}} - P_0 = \rho g h$$
$$= (1.00 \times 10^3\text{ kg/m}^3)(9.80\text{ m/s}^2)(5.0\text{ m})$$
$$= 4.9 \times 10^4\text{ Pa}$$

We estimate the surface area of the eardrum to be approximately 1 cm² = 1 × 10⁻⁴ m². This means that the force on it is $F = (P_{\text{bot}} - P_0)A \approx 5$ N. Because a force on the eardrum of this magnitude is extremely uncomfortable, swimmers often "pop their ears" while under water, an action that pushes air from the lungs into the middle ear. Using this technique equalizes the pressure on the two sides of the eardrum and relieves the discomfort.

Example 14.4   The Force on a Dam

Water is filled to a height H behind a dam of width w (Fig. 14.5). Determine the resultant force exerted by the water on the dam.

Solution Because pressure varies with depth, we cannot calculate the force simply by multiplying the area by the pressure. We can solve the problem by using Equation 14.2 to find the force dF exerted on a narrow horizontal strip at depth h and then integrating the expression to find the total force. Let us imagine a vertical y axis, with $y = 0$ at the bottom of the dam and our strip a distance y above the bottom.

We can use Equation 14.4 to calculate the pressure at the depth h; we omit atmospheric pressure because it acts on both sides of the dam:

$$P = \rho g h = \rho g(H - y)$$

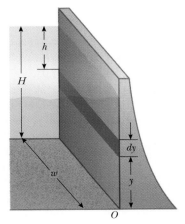

Figure 14.5 (Example 14.4)

Using Equation 14.2, we find that the force exerted on the shaded strip of area $dA = w\,dy$ is

$$dF = P\,dA = \rho g(H - y)\,w\,dy$$

Therefore, the total force on the dam is

$$F = \int P\,dA = \int_0^H \rho g(H - y)\,w\,dy = \boxed{\tfrac{1}{2}\rho g w H^2}$$

Note that the thickness of the dam shown in Figure 14.5 increases with depth. This design accounts for the greater and greater pressure that the water exerts on the dam at greater depths.

What If? What if you were asked to find this force without using calculus? How could you determine its value?

Answer We know from Equation 14.4 that the pressure varies linearly with depth. Thus, the average pressure due to the water over the face of the dam is the average of the pressure at the top and the pressure at the bottom:

$$P_{av} = \frac{P_{top} + P_{bottom}}{2} = \frac{0 + \rho g H}{2} = \tfrac{1}{2}\rho g H$$

Now, the total force is equal to the average pressure times the area of the face of the dam:

$$F = P_{av}A = (\tfrac{1}{2}\rho g H)(Hw) = \tfrac{1}{2}\rho g w H^2$$

which is the same result we obtained using calculus.

14.3 Pressure Measurements

During the weather report on a television news program, the *barometric pressure* is often provided. This is the current pressure of the atmosphere, which varies over a small range from the standard value provided earlier. How is this pressure measured?

One instrument used to measure atmospheric pressure is the common barometer, invented by Evangelista Torricelli (1608–1647). A long tube closed at one end is filled with mercury and then inverted into a dish of mercury (Fig. 14.6a). The closed end of the tube is nearly a vacuum, so the pressure at the top of the mercury column can be taken as zero. In Figure 14.6a, the pressure at point A, due to the column of mercury, must equal the pressure at point B, due to the atmosphere. If this were not the case, there would be a net force that would move mercury from one point to the other until equilibrium is established. Therefore, it follows that $P_0 = \rho_{Hg}gh$, where ρ_{Hg} is the density of the mercury and h is the height of the mercury column. As atmospheric pressure varies, the height of the mercury column varies, so the height can be calibrated to measure atmospheric pressure. Let us determine the height of a mercury column for one atmosphere of pressure, $P_0 = 1\text{ atm} = 1.013 \times 10^5\text{ Pa}$:

(a)

$$P_0 = \rho_{Hg}gh \longrightarrow h = \frac{P_0}{\rho_{Hg}g} = \frac{1.013 \times 10^5\text{ Pa}}{(13.6 \times 10^3\text{ kg/m}^3)(9.80\text{ m/s}^2)} = 0.760\text{ m}$$

Based on a calculation such as this, one atmosphere of pressure is defined to be the pressure equivalent of a column of mercury that is exactly 0.760 0 m in height at 0°C.

A device for measuring the pressure of a gas contained in a vessel is the open-tube manometer illustrated in Figure 14.6b. One end of a U-shaped tube containing a liquid is open to the atmosphere, and the other end is connected to a system of unknown pressure P. The pressures at points A and B must be the same (otherwise, the curved portion of the liquid would experience a net force and would accelerate), and the pressure at A is the unknown pressure of the gas. Therefore, equating the unknown pressure P to the pressure at point B, we see that $P = P_0 + \rho g h$. The difference in pressure $P - P_0$ is equal to $\rho g h$. The pressure P is called the **absolute pressure,** while the difference $P - P_0$ is called the **gauge pressure.** For example, the pressure you measure in your bicycle tire is gauge pressure.

(b)

Figure 14.6 Two devices for measuring pressure: (a) a mercury barometer and (b) an open-tube manometer.

Quick Quiz 14.3 Several common barometers are built, with a variety of fluids. For which fluid will the column of fluid in the barometer be the highest? (a) mercury (b) water (c) ethyl alcohol (d) benzene

Quick Quiz 14.4 You have invented a spacesuit with a straw passing through the faceplate so that you can drink from a glass while on the surface of a planet. Out on the surface of the Moon, you attempt to drink through the straw from an open glass of water. The value of *g* on the Moon is about one sixth of that on Earth. Compared to the difficulty in drinking through a straw on Earth, you find drinking through a straw on the Moon to be (a) easier (b) equally difficult (c) harder (d) impossible.

14.4 Buoyant Forces and Archimedes's Principle

Have you ever tried to push a beach ball under water (Fig. 14.7a)? This is extremely difficult to do because of the large upward force exerted by the water on the ball. The upward force exerted by a fluid on any immersed object is called a **buoyant force.** We can determine the magnitude of a buoyant force by applying some logic. Imagine a beach ball–sized parcel of water beneath the water surface, as in Figure 14.7b. Because this parcel is in equilibrium, there must be an upward force that balances the downward gravitational force on the parcel. This upward force is the buoyant force, and its magnitude is equal to the weight of the water in the parcel. The buoyant force is the resultant force due to all forces applied by the fluid surrounding the parcel.

Now imagine replacing the beach ball–sized parcel of water with a beach ball of the same size. The resultant force applied by the fluid surrounding the beach ball is the same, regardless of whether it is applied to a beach ball or to a parcel of water. Consequently, we can claim that **the magnitude of the buoyant force always equals the weight of the fluid displaced by the object.** This statement is known as **Archimedes's principle.**

With the beach ball under water, the buoyant force, equal to the weight of a beach ball-sized parcel of water, is much larger than the weight of the beach ball. Thus, there is a net upward force of large magnitude—this is why it is so hard to hold the beach ball under the water. Note that Archimedes's principle does not refer to the makeup of the object experiencing the buoyant force. The object's composition is not a factor in the buoyant force because the buoyant force is exerted by the fluid.

Archimedes
(287–212 B.C.)

Archimedes, a Greek mathematician, physicist, and engineer, was perhaps the greatest scientist of antiquity. He was the first to compute accurately the ratio of a circle's circumference to its diameter, and he also showed how to calculate the volume and surface area of spheres, cylinders, and other geometric shapes. He is well known for discovering the nature of the buoyant force and was also a gifted inventor. One of his practical inventions, still in use today, is Archimedes's screw, an inclined, rotating, coiled tube used originally to lift water from the holds of ships. He also invented the catapult and devised systems of levers, pulleys, and weights for raising heavy loads. Such inventions were successfully used to defend his native city, Syracuse, during a two-year siege by Romans.

(a)   (b)

Figure 14.7. (a) A swimmer attempts to push a beach ball underwater. (b) The forces on a beach ball–sized parcel of water. The buoyant force **B** on a beach ball that replaces this parcel is exactly the same as the buoyant force on the parcel.

Figure 14.8 The external forces acting on the cube of liquid are the gravitational force $\mathbf{F}_g$ and the buoyant force $\mathbf{B}$. Under equilibrium conditions, $B = F_g$.

 PITFALL PREVENTION

14.2 Buoyant Force is Exerted by the Fluid

Remember that **the buoyant force is exerted by the fluid.** It is not determined by properties of the object, except for the amount of fluid displaced by the object. Thus, if several objects of different densities but the same volume are immersed in a fluid, they will all experience the same buoyant force. Whether they sink or float will be determined by the relationship between the buoyant force and the weight.

To understand the origin of the buoyant force, consider a cube immersed in a liquid as in Figure 14.8. The pressure P_b at the bottom of the cube is greater than the pressure P_t at the top by an amount $\rho_{\text{fluid}} gh$, where h is the height of the cube and ρ_{fluid} is the density of the fluid. The pressure at the bottom of the cube causes an *upward* force equal to $P_b A$, where A is the area of the bottom face. The pressure at the top of the cube causes a *downward* force equal to $P_t A$. The resultant of these two forces is the buoyant force $\mathbf{B}$:

$$B = (P_b - P_t)A = (\rho_{\text{fluid}} gh)A = \rho_{\text{fluid}} gV \qquad (14.5)$$

where V is the volume of the fluid displaced by the cube. Because the product $\rho_{\text{fluid}} V$ is equal to the mass of fluid displaced by the object, we see that

$$B = Mg$$

where Mg is the weight of the fluid displaced by the cube. This is consistent with our initial statement about Archimedes's principle above, based on the discussion of the beach ball.

Before we proceed with a few examples, it is instructive for us to discuss two common situations—a totally submerged object and a floating (partly submerged) object.

Case 1: Totally Submerged Object When an object is totally submerged in a fluid of density ρ_{fluid}, the magnitude of the upward buoyant force is $B = \rho_{\text{fluid}} gV = \rho_{\text{fluid}} gV_{\text{obj}}$, where V_{obj} is the volume of the object. If the object has a mass M and density ρ_{obj}, its weight is equal to $F_g = Mg = \rho_{\text{obj}} gV_{\text{obj}}$, and the net force on it is $B - F_g = (\rho_{\text{fluid}} - \rho_{\text{obj}})gV_{\text{obj}}$. Hence, if the density of the object is less than the density of the fluid, then the downward gravitational force is less than the buoyant force, and the unsupported object accelerates upward (Fig. 14.9a). If the density of the object is greater than the density of the fluid, then the upward buoyant force is less than the downward gravitational force, and the unsupported object sinks (Fig. 14.9b). If the density of the submerged object equals the density of the fluid, the net force on the object is zero and it remains in equilibrium. Thus, **the motion of an object submerged in a fluid is determined only by the densities of the object and the fluid.**

Case 2: Floating Object Now consider an object of volume V_{obj} and density $\rho_{\text{obj}} < \rho_{\text{fluid}}$ in static equilibrium floating on the surface of a fluid—that is, an object that is only *partially* submerged (Fig. 14.10). In this case, the upward buoyant force is balanced by the downward gravitational force acting on the object. If V_{fluid} is the volume of the fluid displaced by the object (this volume is the same as the volume of that part of the object that is beneath the surface of the fluid), the buoyant force has a magnitude $B = \rho_{\text{fluid}} gV_{\text{fluid}}$. Because the weight of the object is $F_g = Mg = \rho_{\text{obj}} gV_{\text{obj}}$, and because $F_g = B$, we see that $\rho_{\text{fluid}} gV_{\text{fluid}} = \rho_{\text{obj}} gV_{\text{obj}}$, or

$$\frac{V_{\text{fluid}}}{V_{\text{obj}}} = \frac{\rho_{\text{obj}}}{\rho_{\text{fluid}}} \qquad (14.6)$$

At the Active Figures link at http://www.pse6.com, you can move the object to new positions as well as change the density of the object to see the results.

(a)　　　　　　　(b)

Active Figure 14.9 (a) A totally submerged object that is less dense than the fluid in which it is submerged experiences a net upward force. (b) A totally submerged object that is denser than the fluid experiences a net downward force and sinks.

This equation tells us that **the fraction of the volume of a floating object that is below the fluid surface is equal to the ratio of the density of the object to that of the fluid.**

Under normal conditions, the weight of a fish is slightly greater than the buoyant force due to water. It follows that the fish would sink if it did not have some mechanism for adjusting the buoyant force. The fish accomplishes this by internally regulating the size of its air-filled swim bladder to increase its volume and the magnitude of the buoyant force acting on it. In this manner, fish are able to swim to various depths.

Active Figure 14.10 An object floating on the surface of a fluid experiences two forces, the gravitational force $\mathbf{F}_g$ and the buoyant force **B**. Because the object floats in equilibrium, $B = F_g$.

At the Active Figures link at **http://www.pse6.com,** *you can change the densities of the object and the fluid.*

Quick Quiz 14.5 An apple is held completely submerged just below the surface of a container of water. The apple is then moved to a deeper point in the water. Compared to the force needed to hold the apple just below the surface, the force needed to hold it at a deeper point is (a) larger (b) the same (c) smaller (d) impossible to determine.

Quick Quiz 14.6 A glass of water contains a single floating ice cube (Fig. 14.11). When the ice melts, does the water level (a) go up (b) go down (c) remain the same?

Figure 14.11 (Quick Quiz 14.6) An ice cube floats on the surface of water. What happens to the water level as the ice cube melts?

Quick Quiz 14.7 You are shipwrecked and floating in the middle of the ocean on a raft. Your cargo on the raft includes a treasure chest full of gold that you found before your ship sank, and the raft is just barely afloat. To keep you floating as high as possible in the water, should you (a) leave the treasure chest on top of the raft (b) secure the treasure chest to the underside of the raft (c) hang the treasure chest in the water with a rope attached to the raft? (Assume that throwing the treasure chest overboard is not an option you wish to consider!)

Example 14.5   Eureka!

Archimedes supposedly was asked to determine whether a crown made for the king consisted of pure gold. Legend has it that he solved this problem by weighing the crown first in air and then in water, as shown in Figure 14.12. Suppose the scale read 7.84 N in air and 6.84 N in water. What should Archimedes have told the king?

Solution Figure 14.12 helps us to conceptualize the problem. Because of our understanding of the buoyant force, we realize that the scale reading will be smaller in Figure 14.12b than in Figure 14.12a. The scale reading is a measure of one of the forces on the crown, and we recognize that the crown is stationary. Thus, we can categorize this as a force equilibrium problem. To analyze the problem, note that when the crown is

suspended in air, the scale reads the true weight $T_1 = F_g$ (neglecting the buoyancy of air). When it is immersed in water, the buoyant force **B** reduces the scale reading to an *apparent* weight of $T_2 = F_g - B$. Because the crown is in equilibrium, the net force on it is zero. When the crown is in water,

$$\sum F = B + T_2 - F_g = 0$$

so that

$$B = F_g - T_2 = 7.84 \text{ N} - 6.84 \text{ N} = 1.00 \text{ N}$$

Because this buoyant force is equal in magnitude to the weight of the displaced water, we have $\rho_w g V_w = 1.00$ N, where V_w is the volume of the displaced water and ρ_w is its

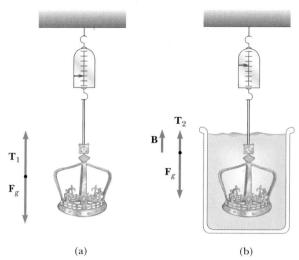

(a)   (b)

Figure 14.12 (Example 14.5) (a) When the crown is suspended in air, the scale reads its true weight because $T_1 = F_g$ (the buoyancy of air is negligible). (b) When the crown is immersed in water, the buoyant force **B** changes the scale reading to a lower value $T_2 = F_g - B$.

density. Also, the volume of the crown V_c is equal to the volume of the displaced water because the crown is completely submerged. Therefore,

$$V_c = V_w = \frac{1.00 \text{ N}}{\rho_w g} = \frac{1.00 \text{ N}}{(1\,000 \text{ kg/m}^3)(9.80 \text{ m/s}^2)}$$

$$= 1.02 \times 10^{-4} \text{ m}^3$$

Finally, the density of the crown is

$$\rho_c = \frac{m_c}{V_c} = \frac{m_c g}{V_c g} = \frac{7.84 \text{ N}}{(1.02 \times 10^{-4} \text{ m}^3)(9.80 \text{ m/s}^2)}$$

$$= 7.84 \times 10^3 \text{ kg/m}^3$$

To finalize the problem, from Table 14.1 we see that the density of gold is 19.3×10^3 kg/m³. Thus, Archimedes should have told the king that he had been cheated. Either the crown was hollow, or it was not made of pure gold.

What If? Suppose the crown has the same weight but were indeed pure gold and not hollow. What would the scale reading be when the crown is immersed in water?

Answer We first find the volume of the solid gold crown:

$$V_c = \frac{m_c}{\rho_c} = \frac{m_c g}{\rho_c g} = \frac{7.84 \text{ N}}{(19.3 \times 10^3 \text{ kg/m}^3)(9.80 \text{ m/s}^2)}$$

$$= 4.15 \times 10^{-5} \text{ m}^3$$

Now, the buoyant force on the crown will be

$$B = \rho_w g V_w = \rho_w g V_c$$

$$= (1.00 \times 10^3 \text{ kg/m}^3)(9.80 \text{ m/s}^2)(4.15 \times 10^{-5} \text{ m}^3)$$

$$= 0.406 \text{ N}$$

and the tension in the string hanging from the scale is

$$T_2 = F_g - B = 7.84 \text{ N} - 0.406 \text{ N} = 7.43 \text{ N}$$

Example 14.6 A Titanic Surprise

An iceberg floating in seawater, as shown in Figure 14.13a, is extremely dangerous because most of the ice is below the surface. This hidden ice can damage a ship that is still a considerable distance from the visible ice. What fraction of the iceberg lies below the water level?

Solution This problem corresponds to Case 2. The weight of the iceberg is $F_g = \rho_i V_i g$, where $\rho_i = 917$ kg/m³ and V_i is the volume of the whole iceberg. The magnitude of the up-

ward buoyant force equals the weight of the displaced water: $B = \rho_w V_w g$, where V_w, the volume of the displaced water, is equal to the volume of the ice beneath the water (the shaded region in Fig. 14.13b) and ρ_w is the density of seawater, $\rho_w = 1\,030$ kg/m³. Because $\rho_i V_i g = \rho_w V_w g$, the fraction of ice beneath the water's surface is

$$f = \frac{V_w}{V_i} = \frac{\rho_i}{\rho_w} = \frac{917 \text{ kg/m}^3}{1\,030 \text{ kg/m}^3} = \boxed{0.890} \text{ or } \boxed{89.0\%}$$

Geraldine Prentice/Getty

(a)   (b)

Figure 14.13 (Example 14.6) (a) Much of the volume of this iceberg is beneath the water. (b) A ship can be damaged even when it is not near the visible ice.

Figure 14.14 Laminar flow around an automobile in a test wind tunnel.

14.5 Fluid Dynamics

Thus far, our study of fluids has been restricted to fluids at rest. We now turn our attention to fluids in motion. When fluid is in motion, its flow can be characterized as being one of two main types. The flow is said to be **steady,** or **laminar,** if each particle of the fluid follows a smooth path, such that the paths of different particles never cross each other, as shown in Figure 14.14. In steady flow, the velocity of fluid particles passing any point remains constant in time.

Above a certain critical speed, fluid flow becomes **turbulent;** turbulent flow is irregular flow characterized by small whirlpool-like regions, as shown in Figure 14.15.

The term **viscosity** is commonly used in the description of fluid flow to characterize the degree of internal friction in the fluid. This internal friction, or *viscous force,* is associated with the resistance that two adjacent layers of fluid have to moving relative to each other. Viscosity causes part of the kinetic energy of a fluid to be converted to internal energy. This mechanism is similar to the one by which an object sliding on a rough horizontal surface loses kinetic energy.

Because the motion of real fluids is very complex and not fully understood, we make some simplifying assumptions in our approach. In our model of **ideal fluid flow,** we make the following four assumptions:

1. **The fluid is nonviscous.** In a nonviscous fluid, internal friction is neglected. An object moving through the fluid experiences no viscous force.

2. **The flow is steady.** In steady (laminar) flow, the velocity of the fluid at each point remains constant.

3. **The fluid is incompressible.** The density of an incompressible fluid is constant.

4. **The flow is irrotational.** In irrotational flow, the fluid has no angular momentum about any point. If a small paddle wheel placed anywhere in the fluid does not rotate about the wheel's center of mass, then the flow is irrotational.

The path taken by a fluid particle under steady flow is called a **streamline.** The velocity of the particle is always tangent to the streamline, as shown in Figure 14.16. A set of streamlines like the ones shown in Figure 14.16 form a *tube of flow.* Note that fluid

Figure 14.15 Hot gases from a cigarette made visible by smoke particles. The smoke first moves in laminar flow at the bottom and then in turbulent flow above.

Figure 14.16 A particle in laminar flow follows a streamline, and at each point along its path the particle's velocity is tangent to the streamline.

Figure 14.17 A fluid moving with steady flow through a pipe of varying cross-sectional area. The volume of fluid flowing through area A_1 in a time interval Δt must equal the volume flowing through area A_2 in the same time interval. Therefore, $A_1 v_1 = A_2 v_2$.

particles cannot flow into or out of the sides of this tube; if they could, then the streamlines would cross each other.

Consider an ideal fluid flowing through a pipe of nonuniform size, as illustrated in Figure 14.17. The particles in the fluid move along streamlines in steady flow. In a time interval Δt, the fluid at the bottom end of the pipe moves a distance $\Delta x_1 = v_1 \Delta t$. If A_1 is the cross-sectional area in this region, then the mass of fluid contained in the left shaded region in Figure 14.17 is $m_1 = \rho A_1 \Delta x_1 = \rho A_1 v_1 \Delta t$, where ρ is the (unchanging) density of the ideal fluid. Similarly, the fluid that moves through the upper end of the pipe in the time interval Δt has a mass $m_2 = \rho A_2 v_2 \Delta t$. However, because the fluid is incompressible and because the flow is steady, the mass that crosses A_1 in a time interval Δt must equal the mass that crosses A_2 in the same time interval. That is, $m_1 = m_2$, or $\rho A_1 v_1 = \rho A_2 v_2$; this means that

$$A_1 v_1 = A_2 v_2 = \text{constant} \qquad (14.7)$$

This expression is called the **equation of continuity for fluids.** It states that

> the product of the area and the fluid speed at all points along a pipe is constant for an incompressible fluid.

Equation 14.7 tells us that the speed is high where the tube is constricted (small A) and low where the tube is wide (large A). The product Av, which has the dimensions of volume per unit time, is called either the *volume flux* or the *flow rate*. The condition $Av =$ constant is equivalent to the statement that the volume of fluid that enters one end of a tube in a given time interval equals the volume leaving the other end of the tube in the same time interval if no leaks are present.

You demonstrate the equation of continuity each time you water your garden with your thumb over the end of a garden hose as in Figure 14.18. By partially blocking the opening with your thumb, you reduce the cross-sectional area through which the water passes. As a result, the speed of the water increases as it exits the hose, and it can be sprayed over a long distance.

Quick Quiz 14.8 You tape two different soda straws together end-to-end to make a longer straw with no leaks. The two straws have radii of 3 mm and 5 mm. You drink a soda through your combination straw. In which straw is the speed of the liquid the highest? (a) whichever one is nearest your mouth (b) the one of radius 3 mm (c) the one of radius 5 mm (d) Neither—the speed is the same in both straws.

George Semple

Figure 14.18 The speed of water spraying from the end of a garden hose increases as the size of the opening is decreased with the thumb.

Example 14.7   Niagara Falls

Each second, 5 525 m^3 of water flows over the 670-m-wide cliff of the Horseshoe Falls portion of Niagara Falls. The water is approximately 2 m deep as it reaches the cliff. What is its speed at that instant?

Solution The cross-sectional area of the water as it reaches the edge of the cliff is $A = (670 \text{ m})(2 \text{ m}) = 1\,340 \text{ m}^2$.

The flow rate of 5 525 m^3/s is equal to Av. This gives

$$v = \frac{5\,525 \text{ m}^3/\text{s}}{A} = \frac{5\,525 \text{ m}^3/\text{s}}{1\,340 \text{ m}^2} = \boxed{4 \text{ m/s}}$$

Note that we have kept only one significant figure because our value for the depth has only one significant figure.

Example 14.8   Watering a Garden

A water hose 2.50 cm in diameter is used by a gardener to fill a 30.0-L bucket. The gardener notes that it takes 1.00 min to fill the bucket. A nozzle with an opening of cross-sectional area 0.500 cm^2 is then attached to the hose. The nozzle is held so that water is projected horizontally from a point 1.00 m above the ground. Over what horizontal distance can the water be projected?

Solution We identify point 1 within the hose and point 2 at the exit of the nozzle. We first find the speed of the water in the hose from the bucket-filling information. The cross-sectional area of the hose is

$$A_1 = \pi r^2 = \pi \frac{d^2}{4} = \pi \left(\frac{(2.50 \text{ cm})^2}{4} \right) = 4.91 \text{ cm}^2$$

According to the data given, the volume flow rate is equal to 30.0 L/min:

$$A_1 v_1 = 30.0 \text{ L/min} = \frac{30.0 \times 10^3 \text{ cm}^3}{60.0 \text{ s}} = 500 \text{ cm}^3/\text{s}$$

$$v_1 = \frac{500 \text{ cm}^3/\text{s}}{A_1} = \frac{500 \text{ cm}^3/\text{s}}{4.91 \text{ cm}^2} = 102 \text{ cm/s} = 1.02 \text{ m/s}$$

Now we use the continuity equation for fluids to find the speed $v_2 = v_{xi}$ with which the water exits the nozzle. The subscript i anticipates that this will be the *initial* velocity

component of the water projected from the hose, and the subscript x recognizes that the initial velocity vector of the projected water is in the horizontal direction.

$$A_1 v_1 = A_2 v_2 = A_2 v_{xi} \quad \longrightarrow \quad v_{xi} = \frac{A_1}{A_2} v_1$$

$$v_{xi} = \frac{4.91 \text{ cm}^2}{0.500 \text{ cm}^2} (1.02 \text{ m/s})$$

$$= 10.0 \text{ m/s}$$

We now shift our thinking away from fluids and to projectile motion because the water is in free fall once it exits the nozzle. A particle of the water falls through a vertical distance of 1.00 m starting from rest, and lands on the ground at a time that we find from Equation 2.12:

$$y_f = y_i + v_{yi}t - \tfrac{1}{2}gt^2$$

$$-1.00 \text{ m} = 0 + 0 - \tfrac{1}{2}(9.80 \text{ m/s}^2)t^2$$

$$t = \sqrt{\frac{2(1.00 \text{ m})}{9.80 \text{ m/s}^2}} = 0.452 \text{ s}$$

In the horizontal direction, we apply Equation 2.12 with $a_x = 0$ to a particle of water to find the horizontal distance:

$$x_f = x_i + v_{xi}t = 0 + (10.0 \text{ m/s})(0.452 \text{ s}) = \boxed{4.52 \text{ m}}$$

14.6   Bernoulli's Equation

You have probably had the experience of driving on a highway and having a large truck pass you at high speed. In this situation, you may have had the frightening feeling that your car was being pulled in toward the truck as it passed. We will investigate the origin of this effect in this section.

As a fluid moves through a region where its speed and/or elevation above the Earth's surface changes, the pressure in the fluid varies with these changes. The relationship between fluid speed, pressure, and elevation was first derived in 1738 by the Swiss physicist Daniel Bernoulli. Consider the flow of a segment of an ideal fluid through a nonuniform pipe in a time interval Δt, as illustrated in Figure 14.19. At the beginning of the time interval, the segment of fluid consists of the blue shaded portion (portion 1) at the left and the unshaded portion. During the time interval, the left end of the segment moves to the right by a distance Δx_1, which is the length of the blue shaded portion at the left. Meanwhile, the right end of the segment moves to the right through a distance Δx_2, which is the length of the blue shaded portion (portion 2) at the upper right of Figure 14.19.

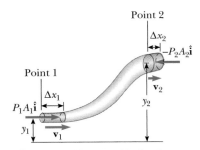

Figure 14.19 A fluid in laminar flow through a constricted pipe. The volume of the shaded portion on the left is equal to the volume of the shaded portion on the right.

Daniel Bernoulli

Swiss physicist
(1700–1782)

Daniel Bernoulli made important discoveries in fluid dynamics. Born into a family of mathematicians, he was the only member of the family to make a mark in physics.

Bernoulli's most famous work, *Hydrodynamica*, was published in 1738; it is both a theoretical and a practical study of equilibrium, pressure, and speed in fluids. He showed that as the speed of a fluid increases, its pressure decreases. Referred to as "Bernoulli's principle," his work is used to produce a partial vacuum in chemical laboratories by connecting a vessel to a tube through which water is running rapidly.

In *Hydrodynamica*, Bernoulli also attempted the first explanation of the behavior of gases with changing pressure and temperature; this was the beginning of the kinetic theory of gases, a topic we study in Chapter 21. *(Corbis–Bettmann)*

Thus, at the end of the time interval, the segment of fluid consists of the unshaded portion and the blue shaded portion at the upper right.

Now consider forces exerted on this segment by fluid to the left and the right of the segment. The force exerted by the fluid on the left end has a magnitude $P_1 A_1$. The work done by this force on the segment in a time interval Δt is $W_1 = F_1 \Delta x_1 = P_1 A_1 \Delta x_1 = P_1 V$, where V is the volume of portion 1. In a similar manner, the work done by the fluid to the right of the segment in the same time interval Δt is $W_2 = -P_2 A_2 \Delta x_2 = -P_2 V$. (The volume of portion 1 equals the volume of portion 2.) This work is negative because the force on the segment of fluid is to the left and the displacement is to the right. Thus, the net work done on the segment by these forces in the time interval Δt is

$$W = (P_1 - P_2) V$$

Part of this work goes into changing the kinetic energy of the segment of fluid, and part goes into changing the gravitational potential energy of the segment–Earth system. Because we are assuming streamline flow, the kinetic energy of the unshaded portion of the segment in Figure 14.19 is unchanged during the time interval. The only change is as follows: before the time interval we have portion 1 traveling at v_1, whereas after the time interval, we have portion 2 traveling at v_2. Thus, the change in the kinetic energy of the segment of fluid is

$$\Delta K = \tfrac{1}{2}mv_2^2 - \tfrac{1}{2}mv_1^2$$

where m is the mass of both portion 1 and portion 2. (Because the volumes of both portions are the same, they also have the same mass.)

Considering the gravitational potential energy of the segment–Earth system, once again there is no change during the time interval for the unshaded portion of the fluid. The net change is that the mass of the fluid in portion 1 has effectively been moved to the location of portion 2. Consequently, the change in gravitational potential energy is

$$\Delta U = mgy_2 - mgy_1$$

The total work done on the system by the fluid outside the segment is equal to the change in mechanical energy of the system: $W = \Delta K + \Delta U$. Substituting for each of these terms, we obtain

$$(P_1 - P_2) V = \tfrac{1}{2}mv_2^2 - \tfrac{1}{2}mv_1^2 + mgy_2 - mgy_1$$

If we divide each term by the portion volume V and recall that $\rho = m/V$, this expression reduces to

$$P_1 - P_2 = \tfrac{1}{2}\rho v_2^2 - \tfrac{1}{2}\rho v_1^2 + \rho gy_2 - \rho gy_1$$

Rearranging terms, we obtain

$$P_1 + \tfrac{1}{2}\rho v_1^2 + \rho gy_1 = P_2 + \tfrac{1}{2}\rho v_2^2 + \rho gy_2 \qquad (14.8)$$

This is **Bernoulli's equation** as applied to an ideal fluid. It is often expressed as

$$P + \tfrac{1}{2}\rho v^2 + \rho gy = \text{constant} \qquad (14.9)$$

Bernoulli's equation

This expression shows that the pressure of a fluid decreases as the speed of the fluid increases. In addition, the pressure decreases as the elevation increases. This explains why water pressure from faucets on the upper floors of a tall building is weak unless measures are taken to provide higher pressure for these upper floors.

When the fluid is at rest, $v_1 = v_2 = 0$ and Equation 14.8 becomes

$$P_1 - P_2 = \rho g(y_2 - y_1) = \rho gh$$

This is in agreement with Equation 14.4.

While Equation 14.9 was derived for an incompressible fluid, the general behavior of pressure with speed is true even for gases—as the speed increases, the pressure decreases. This *Bernoulli effect* explains the experience with the truck on the highway at the opening of this section. As air passes between you and the truck, it must pass through a relatively narrow channel. According to the continuity equation, the speed of the air is higher. According to the Bernoulli effect, this higher speed air exerts less pressure on your car than the slower moving air on the other side of your car. Thus, there is a net force pushing you toward the truck!

Quick Quiz 14.9 You observe two helium balloons floating next to each other at the ends of strings secured to a table. The facing surfaces of the balloons are separated by 1–2 cm. You blow through the small space between the balloons. What happens to the balloons? (a) They move toward each other. (b) They move away from each other. (c) They are unaffected.

Example 14.9   The Venturi Tube

The horizontal constricted pipe illustrated in Figure 14.20, known as a *Venturi tube,* can be used to measure the flow speed of an incompressible fluid. Determine the flow speed at point 2 if the pressure difference $P_1 - P_2$ is known.

A_1

(a)   (b)

Figure 14.20 (Example 14.9) (a) Pressure P_1 is greater than pressure P_2 because $v_1 < v_2$. This device can be used to measure the speed of fluid flow. (b) A Venturi tube, located at the top of the photograph. The higher level of fluid in the middle column shows that the pressure at the top of the column, which is in the constricted region of the Venturi tube, is lower.

Courtesy of Central Scientific Company

Solution Because the pipe is horizontal, $y_1 = y_2$, and applying Equation 14.8 to points 1 and 2 gives

$$(1) \qquad P_1 + \tfrac{1}{2}\rho v_1^2 = P_2 + \tfrac{1}{2}\rho v_2^2$$

From the equation of continuity, $A_1 v_1 = A_2 v_2$, we find that

$$(2) \qquad v_1 = \frac{A_2}{A_1}\, v_2$$

Substituting this expression into Equation (1) gives

$$P_1 + \tfrac{1}{2}\rho\left(\frac{A_2}{A_1}\right)^2 v_2^2 = P_2 + \tfrac{1}{2}\rho v_2^2$$

$$v_2 = A_1 \sqrt{\frac{2(P_1 - P_2)}{\rho(A_1^2 - A_2^2)}}$$

We can use this result and the continuity equation to obtain an expression for v_1. Because $A_2 < A_1$, Equation (2) shows us that $v_2 > v_1$. This result, together with Equation (1), indicates that $P_1 > P_2$. In other words, the pressure is reduced in the constricted part of the pipe.

Example 14.10   Torricelli's Law   Interactive

An enclosed tank containing a liquid of density ρ has a hole in its side at a distance y_1 from the tank's bottom (Fig. 14.21). The hole is open to the atmosphere, and its diameter is much smaller than the diameter of the tank. The air above the liquid is maintained at a pressure P. Determine the speed of the liquid as it leaves the hole when the liquid's level is a distance h above the hole.

Solution Because $A_2 \gg A_1$, the liquid is approximately at rest at the top of the tank, where the pressure is P. Applying Bernoulli's equation to points 1 and 2 and noting that at the hole P_1 is equal to atmospheric pressure P_0, we find that

$$P_0 + \tfrac{1}{2}\rho v_1^2 + \rho g y_1 = P + \rho g y_2$$

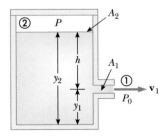

Figure 14.21 (Example 14.10) A liquid leaves a hole in a tank at speed v_1.

But $y_2 - y_1 = h$; thus, this expression reduces to

$$v_1 = \sqrt{\frac{2(P - P_0)}{\rho} + 2gh}$$

When P is much greater than P_0 (so that the term $2gh$ can be neglected), the exit speed of the water is mainly a function of P. If the tank is open to the atmosphere, then $P = P_0$ and $v_1 = \sqrt{2gh}$. In other words, for an open tank, the speed of liquid coming out through a hole a distance h below the surface is equal to that acquired by an object falling freely through a vertical distance h. This phenomenon is known as **Torricelli's law.**

What If? What if the position of the hole in Figure 14.21 could be adjusted vertically? If the tank is open to the atmosphere and sitting on a table, what position of the hole would cause the water to land on the table at the farthest distance from the tank?

Answer We model a parcel of water exiting the hole as a projectile. We find the time at which the parcel strikes the table from a hole at an arbitrary position:

$$y_f = y_i + v_{yi}t - \tfrac{1}{2}gt^2$$
$$0 = y_1 + 0 - \tfrac{1}{2}gt^2$$
$$t = \sqrt{\frac{2y_1}{g}}$$

Thus, the horizontal position of the parcel at the time it strikes the table is

$$x_f = x_i + v_{xi}t = 0 + \sqrt{2g(y_2 - y_1)}\sqrt{\frac{2y_1}{g}}$$
$$= 2\sqrt{(y_2 y_1 - y_1^2)}$$

Now we maximize the horizontal position by taking the derivative of x_f with respect to y_1 (because y_1, the height of the hole, is the variable that can be adjusted) and setting it equal to zero:

$$\frac{dx_f}{dy_1} = \tfrac{1}{2}(2)(y_2 y_1 - y_1^2)^{-1/2}(y_2 - 2y_1) = 0$$

This is satisfied if

$$y_1 = \tfrac{1}{2}y_2$$

Thus, the hole should be halfway between the bottom of the tank and the upper surface of the water to maximize the horizontal distance. Below this location, the water is projected at a higher speed, but falls for a short time interval, reducing the horizontal range. Above this point, the water is in the air for a longer time interval, but is projected with a smaller horizontal speed.

 At the Interactive Worked Example link at http://www.pse6.com, you can move the hole vertically to see where the water lands.

14.7 Other Applications of Fluid Dynamics

Consider the streamlines that flow around an airplane wing as shown in Figure 14.22. Let us assume that the airstream approaches the wing horizontally from the right with a velocity $\mathbf{v}_1$. The tilt of the wing causes the airstream to be deflected downward with a velocity $\mathbf{v}_2$. Because the airstream is deflected by the wing, the wing must exert a force on the airstream. According to Newton's third law, the airstream exerts a force $\mathbf{F}$ on the wing that is equal in magnitude and opposite in direction. This force has a vertical component called the **lift** (or aerodynamic lift) and a horizontal component called **drag.** The lift depends on several factors, such as the speed of the airplane, the area of the wing, its curvature, and the angle between the wing and the horizontal. The curvature of the wing surfaces causes the pressure above the wing to be lower than that below the wing, due to the Bernoulli effect. This assists with the lift on the wing. As the angle between the wing and the horizontal increases, turbulent flow can set in above the wing to reduce the lift.

In general, an object moving through a fluid experiences lift as the result of any effect that causes the fluid to change its direction as it flows past the object. Some factors that influence lift are the shape of the object, its orientation with respect to the fluid flow, any spinning motion it might have, and the texture of its surface. For example, a golf ball struck with a club is given a rapid backspin due to the slant of the club. The dimples on the ball increase the friction force between the ball and the air so that air adheres to the ball's surface. This effect is most pronounced on the top half of the ball, where the ball's surface is moving in the same direction as the air flow. Figure 14.23 shows air adhering to the ball and being deflected downward as a result. Because the ball pushes the air down, the air must push up on the ball. Without the dimples, the friction force is lower, and the golf ball does not travel as far. It may seem counterintuitive to increase the range by increasing the friction force, but the lift gained by spinning the ball more than compensates for the loss of range due to the effect of friction

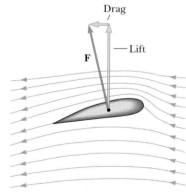

Figure 14.22 Streamline flow around a moving airplane wing. The air approaching from the right is deflected downward by the wing. By Newton's third law, this must coincide with an upward force on the wing from the air—*lift*. Because of air resistance, there is also a force opposite the velocity of the wing—*drag.*

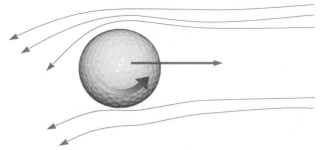

Figure 14.23 Because of the deflection of air, a spinning golf ball experiences a lifting force that allows it to travel much farther than it would if it were not spinning.

on the translational motion of the ball! For the same reason, a baseball's cover helps the spinning ball "grab" the air rushing by and helps to deflect it when a "curve ball" is thrown.

A number of devices operate by means of the pressure differentials that result from differences in a fluid's speed. For example, a stream of air passing over one end of an open tube, the other end of which is immersed in a liquid, reduces the pressure above the tube, as illustrated in Figure 14.24. This reduction in pressure causes the liquid to rise into the air stream. The liquid is then dispersed into a fine spray of droplets. You might recognize that this so-called atomizer is used in perfume bottles and paint sprayers.

Figure 14.24 A stream of air passing over a tube dipped into a liquid causes the liquid to rise in the tube.

SUMMARY

The **pressure** P in a fluid is the force per unit area exerted by the fluid on a surface:

$$P \equiv \frac{F}{A} \tag{14.1}$$

In the SI system, pressure has units of newtons per square meter (N/m^2), and 1 N/m^2 = 1 **pascal** (Pa).

The pressure in a fluid at rest varies with depth h in the fluid according to the expression

$$P = P_0 + \rho g h \tag{14.4}$$

where P_0 is the pressure at $h = 0$ and ρ is the density of the fluid, assumed uniform.

Pascal's law states that when pressure is applied to an enclosed fluid, the pressure is transmitted undiminished to every point in the fluid and to every point on the walls of the container.

When an object is partially or fully submerged in a fluid, the fluid exerts on the object an upward force called the **buoyant force.** According to **Archimedes's principle,** the magnitude of the buoyant force is equal to the weight of the fluid displaced by the object:

$$B = \rho_{\text{fluid}} g V \tag{14.5}$$

You can understand various aspects of a fluid's dynamics by assuming that the fluid is nonviscous and incompressible, and that the fluid's motion is a steady flow with no rotation.

Two important concepts regarding ideal fluid flow through a pipe of nonuniform size are as follows:

1. The flow rate (volume flux) through the pipe is constant; this is equivalent to stating that the product of the cross-sectional area A and the speed v at any point is a constant. This result is expressed in the **equation of continuity for fluids:**

$$A_1 v_1 = A_2 v_2 = \text{constant} \tag{14.7}$$

Take a practice test for this chapter by clicking on the Practice Test link at http://www.pse6.com.

2. The sum of the pressure, kinetic energy per unit volume, and gravitational potential energy per unit volume has the same value at all points along a streamline. This result is summarized in **Bernoulli's equation:**

$$P + \tfrac{1}{2}\rho v^2 + \rho gy = \text{constant} \qquad (14.9)$$

QUESTIONS

1. Two drinking glasses having equal weights but different shapes and different cross-sectional areas are filled to the same level with water. According to the expression $P = P_0 + \rho gh$, the pressure is the same at the bottom of both glasses. In view of this, why does one weigh more than the other?

2. Figure Q14.2 shows aerial views from directly above two dams. Both dams are equally wide (the vertical dimension in the diagram) and equally high (into the page in the diagram). The dam on the left holds back a very large lake, while the dam on the right holds back a narrow river. Which dam has to be built stronger?

Dam   Dam

Figure Q14.2

3. Some physics students attach a long tube to the opening of a hot water bottle made of strong rubber. Leaving the hot water bottle on the ground, they hoist the other end of the tube to the roof of a multistory campus building. Students at the top of the building pour water into the tube. The students on the ground watch the bottle fill with water. On the roof, the students are surprised to see that the tube never seems to fill up—they can continue to pour more and more water down the tube. On the ground, the hot water bottle swells up like a balloon and bursts, drenching the students. Explain these observations.

4. If the top of your head has a surface area of 100 cm², what is the weight of the air above your head?

5. A helium-filled balloon rises until its density becomes the same as that of the surrounding air. If a sealed submarine begins to sink, will it go all the way to the bottom of the ocean or will it stop when its density becomes the same as that of the surrounding water?

6. A fish rests on the bottom of a bucket of water while the bucket is being weighed on a scale. When the fish begins to swim around, does the scale reading change?

7. Will a ship ride higher in the water of an inland lake or in the ocean? Why?

8. Suppose a damaged ship can just barely keep afloat in the ocean. It is towed toward shore and into a river, heading toward a dry dock for repair. As it is pulled up the river, it sinks. Why?

9. Lead has a greater density than iron, and both are denser than water. Is the buoyant force on a lead object greater than, less than, or equal to the buoyant force on an iron object of the same volume?

10. The water supply for a city is often provided from reservoirs built on high ground. Water flows from the reservoir, through pipes, and into your home when you turn the tap on your faucet. Why is the water flow more rapid out of a faucet on the first floor of a building than in an apartment on a higher floor?

11. Smoke rises in a chimney faster when a breeze is blowing. Use the Bernoulli effect to explain this phenomenon.

12. If the air stream from a hair dryer is directed over a Ping-Pong ball, the ball can be levitated. Explain.

13. When ski jumpers are airborne (Fig. Q14.13), why do they bend their bodies forward and keep their hands at their sides?

Galen Powell/Peter Arnold, Inc.

Figure Q14.13

14. When an object is immersed in a liquid at rest, why is the net force on the object in the horizontal direction equal to zero?

15. Explain why a sealed bottle partially filled with a liquid can float in a basin of the same liquid.

16. When is the buoyant force on a swimmer greater—after exhaling or after inhaling?

17. A barge is carrying a load of gravel along a river. It approaches a low bridge and the captain realizes that the top of the pile of gravel is not going to make it under the bridge. The captain orders the crew to quickly shovel gravel from the pile into the water. Is this a good decision?

18. A person in a boat floating in a small pond throws an anchor overboard. Does the level of the pond rise, fall, or remain the same?

19. An empty metal soap dish barely floats in water. A bar of Ivory soap floats in water. When the soap is stuck in the soap dish, the combination sinks. Explain why.

20. A piece of unpainted porous wood barely floats in a container partly filled with water. If the container is sealed and pressurized above atmospheric pressure, does the wood rise, fall, or remain at the same level?

21. A flat plate is immersed in a liquid at rest. For what orientation of the plate is the pressure on its flat surface uniform?

22. Because atmospheric pressure is about 10^5 N/m^2 and the area of a person's chest is about 0.13 m^2, the force of the atmosphere on one's chest is around 13 000 N. In view of this enormous force, why don't our bodies collapse?

23. How would you determine the density of an irregularly shaped rock?

24. Why do airplane pilots prefer to take off into the wind?

25. If you release a ball while inside a freely falling elevator, the ball remains in front of you rather than falling to the floor, because the ball, the elevator, and you all experience the same downward acceleration **g**. What happens if you repeat this experiment with a helium-filled balloon? (This one is tricky.)

26. Two identical ships set out to sea. One is loaded with a cargo of Styrofoam, and the other is empty. Which ship is more submerged?

27. A small piece of steel is tied to a block of wood. When the wood is placed in a tub of water with the steel on top, half of the block is submerged. If the block is inverted so that the steel is under water, does the amount of the block submerged increase, decrease, or remain the same? What happens to the water level in the tub when the block is inverted?

28. Prairie dogs (Fig. Q14.28) ventilate their burrows by building a mound around one entrance, which is open to a stream of air when wind blows from any direction. A second entrance at ground level is open to almost stagnant air. How does this construction create an air flow through the burrow?

29. An unopened can of diet cola floats when placed in a tank of water, whereas a can of regular cola of the same brand sinks in the tank. What do you suppose could explain this behavior?

30. Figure Q14.30 shows a glass cylinder containing four liquids of different densities. From top to bottom, the liquids are oil (orange), water (yellow), salt water (green), and mercury (silver). The cylinder also contains, from top to bottom, a Ping-Pong ball, a piece of wood, an egg, and a steel ball. (a) Which of these liquids has the lowest density, and which has the greatest? (b) What can you conclude about the density of each object?

Figure Q14.30

31. In Figure Q14.31, an air stream moves from right to left through a tube that is constricted at the middle. Three Ping-Pong balls are levitated in equilibrium above the vertical columns through which the air escapes. (a) Why is the ball at the right higher than the one in the middle? (b) Why is the ball at the left lower than the ball at the right even though the horizontal tube has the same dimensions at these two points?

Figure Q14.31

Figure Q14.28

32. You are a passenger on a spacecraft. For your survival and comfort, the interior contains air just like that at the surface of the Earth. The craft is coasting through a very empty region of space. That is, a nearly perfect vacuum exists just outside the wall. Suddenly, a meteoroid pokes a hole, about the size of a large coin, right through the wall next to your seat. What will happen? Is there anything you can or should do about it?

PROBLEMS

1, 2, 3 = straightforward, intermediate, challenging ☐ = full solution available in the *Student Solutions Manual and Study Guide*

🌀 = coached solution with hints available at http://www.pse6.com 💻 = computer useful in solving problem

▨ = paired numerical and symbolic problems

Section 14.1   Pressure

1. Calculate the mass of a solid iron sphere that has a diameter of 3.00 cm.

2. Find the order of magnitude of the density of the nucleus of an atom. What does this result suggest concerning the structure of matter? Model a nucleus as protons and neutrons closely packed together. Each has mass 1.67×10^{-27} kg and radius on the order of 10^{-15} m.

3. A 50.0-kg woman balances on one heel of a pair of high-heeled shoes. If the heel is circular and has a radius of 0.500 cm, what pressure does she exert on the floor?

4. The four tires of an automobile are inflated to a gauge pressure of 200 kPa. Each tire has an area of 0.024 0 m^2 in contact with the ground. Determine the weight of the automobile.

5. What is the total mass of the Earth's atmosphere? (The radius of the Earth is 6.37×10^6 m, and atmospheric pressure at the surface is 1.013×10^5 N/m^2.)

Section 14.2   Variation of Pressure with Depth

6. (a) Calculate the absolute pressure at an ocean depth of 1 000 m. Assume the density of seawater is 1 024 kg/m^3 and that the air above exerts a pressure of 101.3 kPa. (b) At this depth, what force must the frame around a circular submarine porthole having a diameter of 30.0 cm exert to counterbalance the force exerted by the water?

7. The spring of the pressure gauge shown in Figure 14.2 has a force constant of 1 000 N/m, and the piston has a diameter of 2.00 cm. As the gauge is lowered into water, what change in depth causes the piston to move in by 0.500 cm?

8. The small piston of a hydraulic lift has a cross-sectional area of 3.00 cm^2, and its large piston has a cross-sectional area of 200 cm^2 (Figure 14.4). What force must be applied to the small piston for the lift to raise a load of 15.0 kN? (In service stations, this force is usually exerted by compressed air.)

9. 🌀 What must be the contact area between a suction cup (completely exhausted) and a ceiling if the cup is to support the weight of an 80.0-kg student?

10. (a) A very powerful vacuum cleaner has a hose 2.86 cm in diameter. With no nozzle on the hose, what is the weight of the heaviest brick that the cleaner can lift? (Fig. P14.10a) (b) **What If?** A very powerful octopus uses one sucker of diameter 2.86 cm on each of the two shells of a clam in an attempt to pull the shells apart (Fig. P14.10b). Find the greatest force the octopus can exert in salt water 32.3 m deep. *Caution*: Experimental verification can be interesting, but do not drop a brick on your foot. Do not overheat the motor of a vacuum cleaner. Do not get an octopus mad at you.

11. For the cellar of a new house, a hole is dug in the ground, with vertical sides going down 2.40 m. A concrete foundation wall is built all the way across the 9.60-m width of the

(a)

(b)

Figure P14.10

excavation. This foundation wall is 0.183 m away from the front of the cellar hole. During a rainstorm, drainage from the street fills up the space in front of the concrete wall, but not the cellar behind the wall. The water does not soak into the clay soil. Find the force the water causes on the foundation wall. For comparison, the weight of the water is given by 2.40 m × 9.60 m × 0.183 m × 1 000 kg/m^3 × 9.80 m/s^2 = 41.3 kN.

12. A swimming pool has dimensions 30.0 m × 10.0 m and a flat bottom. When the pool is filled to a depth of 2.00 m with fresh water, what is the force caused by the water on the bottom? On each end? On each side?

13. A sealed spherical shell of diameter d is rigidly attached to a cart, which is moving horizontally with an acceleration a as in Figure P14.13. The sphere is nearly filled with a fluid having density ρ and also contains one small bubble of air at atmospheric pressure. Determine the pressure P at the center of the sphere.

Figure P14.13

14. The tank in Figure P14.14 is filled with water 2.00 m deep. At the bottom of one side wall is a rectangular hatch 1.00 m high and 2.00 m wide, which is hinged at the top of the hatch. (a) Determine the force the water causes on the hatch. (b) Find the torque caused by the water about the hinges.

Figure P14.14

15. Review problem. The Abbott of Aberbrothock paid to have a bell moored to the Inchcape Rock to warn seamen of the hazard. Assume the bell was 3.00 m in diameter, cast from brass with a bulk modulus of 14.0 × 10^{10} N/m^2. The pirate Ralph the Rover cut loose the warning bell and threw it into the ocean. By how much did the diameter of the bell decrease as it sank to a depth of 10.0 km? Years later, Ralph drowned when his ship collided with the rock. *Note*: The brass is compressed uniformly, so you may model the bell as a sphere of diameter 3.00 m.

Section 14.3 Pressure Measurements

16. Figure P14.16 shows Superman attempting to drink water through a very long straw. With his great strength he achieves maximum possible suction. The walls of the tubular straw do not collapse. (a) Find the maximum height through which he can lift the water. (b) **What If?** Still thirsty, the Man of Steel repeats his attempt on the Moon, which has no atmosphere. Find the difference between the water levels inside and outside the straw.

Figure P14.16

17. 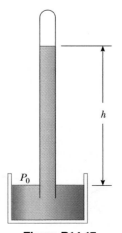 Blaise Pascal duplicated Torricelli's barometer using a red Bordeaux wine, of density 984 kg/m^3, as the working liquid (Fig. P14.17). What was the height h of the wine

Figure P14.17

column for normal atmospheric pressure? Would you expect the vacuum above the column to be as good as for mercury?

18. Mercury is poured into a U-tube as in Figure P14.18a. The left arm of the tube has cross-sectional area A_1 of 10.0 cm^2, and the right arm has a cross-sectional area A_2 of 5.00 cm^2. One hundred grams of water are then poured into the right arm as in Figure P14.18b. (a) Determine the length of the water column in the right arm of the U-tube. (b) Given that the density of mercury is 13.6 g/cm^3, what distance h does the mercury rise in the left arm?

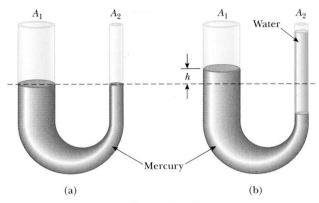

(a)   (b)

Figure P14.18

19. Normal atmospheric pressure is 1.013×10^5 Pa. The approach of a storm causes the height of a mercury barometer to drop by 20.0 mm from the normal height. What is the atmospheric pressure? (The density of mercury is 13.59 g/cm^3.)

20. A U-tube of uniform cross-sectional area, open to the atmosphere, is partially filled with mercury. Water is then poured into both arms. If the equilibrium configuration of the tube is as shown in Figure P14.20, with $h_2 = 1.00$ cm, determine the value of h_1.

Water
Mercury

Figure P14.20

21. The human brain and spinal cord are immersed in the cerebrospinal fluid. The fluid is normally continuous between the cranial and spinal cavities. It normally exerts a

pressure of 100 to 200 mm of H$_2$O above the prevailing atmospheric pressure. In medical work pressures are often measured in units of millimeters of H$_2$O because body fluids, including the cerebrospinal fluid, typically have the same density as water. The pressure of the cerebrospinal fluid can be measured by means of a *spinal tap*, as illustrated in Figure P14.21. A hollow tube is inserted into the spinal column, and the height to which the fluid rises is observed. If the fluid rises to a height of 160 mm, we write its gauge pressure as 160 mm H$_2$O. (a) Express this pressure in pascals, in atmospheres, and in millimeters of mercury. (b) Sometimes it is necessary to determine if an accident victim has suffered a crushed vertebra that is blocking flow of the cerebrospinal fluid in the spinal column. In other cases a physician may suspect a tumor or other growth is blocking the spinal column and inhibiting flow of cerebrospinal fluid. Such conditions can be investigated by means of the *Queckensted test*. In this procedure, the veins in the patient's neck are compressed, to make the blood pressure rise in the brain. The increase in pressure in the blood vessels is transmitted to the cerebrospinal fluid. What should be the normal effect on the height of the fluid in the spinal tap? (c) Suppose that compressing the veins had no effect on the fluid level. What might account for this?

Figure P14.21

Section 14.4 Buoyant Forces and Archimedes's Principle

22. (a) A light balloon is filled with 400 m^3 of helium. At 0°C, the balloon can lift a payload of what mass? (b) **What If?** In Table 14.1, observe that the density of hydrogen is nearly half the density of helium. What load can the balloon lift if filled with hydrogen?

23. A Ping-Pong ball has a diameter of 3.80 cm and average density of 0.084 0 g/cm^3. What force is required to hold it completely submerged under water?

24. A Styrofoam slab has thickness h and density ρ_s. When a swimmer of mass m is resting on it, the slab floats in fresh water with its top at the same level as the water surface. Find the area of the slab.

25. A piece of aluminum with mass 1.00 kg and density 2 700 kg/m^3 is suspended from a string and then completely immersed in a container of water (Figure P14.25). Calculate the tension in the string (a) before and (b) after the metal is immersed.

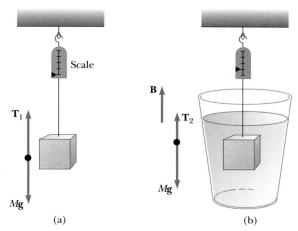

Figure P14.25 Problems 25 and 27

26. The weight of a rectangular block of low-density material is 15.0 N. With a thin string, the center of the horizontal bottom face of the block is tied to the bottom of a beaker partly filled with water. When 25.0% of the block's volume is submerged, the tension in the string is 10.0 N. (a) Sketch a free-body diagram for the block, showing all forces acting on it. (b) Find the buoyant force on the block. (c) Oil of density 800 kg/m³ is now steadily added to the beaker, forming a layer above the water and surrounding the block. The oil exerts forces on each of the four side walls of the block that the oil touches. What are the directions of these forces? (d) What happens to the string tension as the oil is added? Explain how the oil has this effect on the string tension. (e) The string breaks when its tension reaches 60.0 N. At this moment, 25.0% of the block's volume is still below the water line; what additional fraction of the block's volume is below the top surface of the oil? (f) After the string breaks, the block comes to a new equilibrium position in the beaker. It is now in contact only with the oil. What fraction of the block's volume is submerged?

27. A 10.0-kg block of metal measuring 12.0 cm × 10.0 cm × 10.0 cm is suspended from a scale and immersed in water as in Figure P14.25b. The 12.0-cm dimension is vertical, and the top of the block is 5.00 cm below the surface of the water. (a) What are the forces acting on the top and on the bottom of the block? (Take $P_0 = 1.013\ 0 \times 10^5$ N/m².) (b) What is the reading of the spring scale? (c) Show that the buoyant force equals the difference between the forces at the top and bottom of the block.

28. To an order of magnitude, how many helium-filled toy balloons would be required to lift you? Because helium is an irreplaceable resource, develop a theoretical answer rather than an experimental answer. In your solution state what physical quantities you take as data and the values you measure or estimate for them.

29. 🪐 A cube of wood having an edge dimension of 20.0 cm and a density of 650 kg/m³ floats on water. (a) What is the distance from the horizontal top surface of the cube to the water level? (b) How much lead weight has to be placed on top of the cube so that its top is just level with the water?

30. A spherical aluminum ball of mass 1.26 kg contains an empty spherical cavity that is concentric with the ball. The ball just barely floats in water. Calculate (a) the outer radius of the ball and (b) the radius of the cavity.

31. Determination of the density of a fluid has many important applications. A car battery contains sulfuric acid, for which density is a measure of concentration. For the battery to function properly, the density must be inside a range specified by the manufacturer. Similarly, the effectiveness of antifreeze in your car's engine coolant depends on the density of the mixture (usually ethylene glycol and water). When you donate blood to a blood bank, its screening includes determination of the density of the blood, since higher density correlates with higher hemoglobin content. A *hydrometer* is an instrument used to determine liquid density. A simple one is sketched in Figure P14.31. The bulb of a syringe is squeezed and released to let the atmosphere lift a sample of the liquid of interest into a tube containing a calibrated rod of known density. The rod, of length L and average density ρ_0, floats partially immersed in the liquid of density ρ. A length h of the rod protrudes above the surface of the liquid. Show that the density of the liquid is given by

$$\rho = \frac{\rho_0 L}{L - h}$$

Figure P14.31 Problems 31 and 32

32. Refer to Problem 31 and Figure P14.31. A hydrometer is to be constructed with a cylindrical floating rod. Nine fiduciary marks are to be placed along the rod to indicate densities of 0.98 g/cm³, 1.00 g/cm³, 1.02 g/cm³, 1.04 g/cm³, . . . 1.14 g/cm³. The row of marks is to start 0.200 cm from the top end of the rod and end 1.80 cm from the top end. (a) What is the required length of the rod? (b) What must be its average density? (c) Should the marks be equally spaced? Explain your answer.

33. How many cubic meters of helium are required to lift a balloon with a 400-kg payload to a height of 8 000 m? (Take $\rho_{He} = 0.180$ kg/m³.) Assume that the balloon maintains a

constant volume and that the density of air decreases with the altitude z according to the expression $\rho_{air} = \rho_0 e^{-z/8\,000}$, where z is in meters and $\rho_0 = 1.25$ kg/m^3 is the density of air at sea level.

34. A frog in a hemispherical pod (Fig. P14.34) just floats without sinking into a sea of blue-green ooze with density 1.35 g/cm^3. If the pod has radius 6.00 cm and negligible mass, what is the mass of the frog?

Figure P14.34

35. A plastic sphere floats in water with 50.0 percent of its volume submerged. This same sphere floats in glycerin with 40.0 percent of its volume submerged. Determine the densities of the glycerin and the sphere.

36. A bathysphere used for deep-sea exploration has a radius of 1.50 m and a mass of 1.20×10^4 kg. To dive, this submarine takes on mass in the form of seawater. Determine the amount of mass the submarine must take on if it is to descend at a constant speed of 1.20 m/s, when the resistive force on it is 1 100 N in the upward direction. The density of seawater is 1.03×10^3 kg/m^3.

37. The United States possesses the eight largest warships in the world—aircraft carriers of the *Nimitz* class—and is building two more. Suppose one of the ships bobs up to float 11.0 cm higher in the water when 50 fighters take off from it in 25 min, at a location where the free-fall acceleration is 9.78 m/s^2. Bristling with bombs and missiles, the planes have average mass 29 000 kg. Find the horizontal area enclosed by the waterline of the \$4-billion ship. By comparison, its flight deck has area 18 000 m^2. Below decks are passageways hundreds of meters long, so narrow that two large men cannot pass each other.

Section 14.5 Fluid Dynamics
Section 14.6 Bernoulli's Equation

38. A horizontal pipe 10.0 cm in diameter has a smooth reduction to a pipe 5.00 cm in diameter. If the pressure of the water in the larger pipe is 8.00×10^4 Pa and the pressure in the smaller pipe is 6.00×10^4 Pa, at what rate does water flow through the pipes?

39. A large storage tank, open at the top and filled with water, develops a small hole in its side at a point 16.0 m below the water level. If the rate of flow from the leak is equal to 2.50×10^{-3} m^3/min, determine (a) the speed at which the water leaves the hole and (b) the diameter of the hole.

40. A village maintains a large tank with an open top, containing water for emergencies. The water can drain from the tank through a hose of diameter 6.60 cm. The hose ends with a nozzle of diameter 2.20 cm. A rubber stopper is inserted into the nozzle. The water level in the tank is kept 7.50 m above the nozzle. (a) Calculate the friction force exerted by the nozzle on the stopper. (b) The stopper is removed. What mass of water flows from the nozzle in 2.00 h? (c) Calculate the gauge pressure of the flowing water in the hose just behind the nozzle.

41. Water flows through a fire hose of diameter 6.35 cm at a rate of 0.0120 m^3/s. The fire hose ends in a nozzle of inner diameter 2.20 cm. What is the speed with which the water exits the nozzle?

42. Water falls over a dam of height h with a mass flow rate of R, in units of kg/s. (a) Show that the power available from the water is

$$\mathcal{P} = Rgh$$

where g is the free-fall acceleration. (b) Each hydroelectric unit at the Grand Coulee Dam takes in water at a rate of 8.50×10^5 kg/s from a height of 87.0 m. The power developed by the falling water is converted to electric power with an efficiency of 85.0%. How much electric power is produced by each hydroelectric unit?

43. Figure P14.43 shows a stream of water in steady flow from a kitchen faucet. At the faucet the diameter of the stream is 0.960 cm. The stream fills a 125-cm^3 container in 16.3 s. Find the diameter of the stream 13.0 cm below the opening of the faucet.

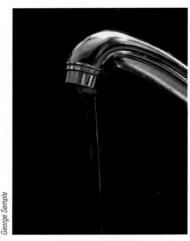

Figure P14.43

44. A legendary Dutch boy saved Holland by plugging a hole in a dike with his finger, which is 1.20 cm in diameter. If the hole was 2.00 m below the surface of the North Sea (density 1 030 kg/m^3), (a) what was the force on his finger? (b) If he pulled his finger out of the hole, how long would it take the released water to fill 1 acre of land to a depth of 1 ft, assuming the hole remained constant in size? (A typical U.S. family of four uses 1 acre-foot of water, 1 234 m^3, in 1 year.)

45. Through a pipe 15.0 cm in diameter, water is pumped from the Colorado River up to Grand Canyon Village, located on the rim of the canyon. The river is at an elevation of 564 m, and the village is at an elevation of 2 096 m.

(a) What is the minimum pressure at which the water must be pumped if it is to arrive at the village? (b) If 4 500 m³ are pumped per day, what is the speed of the water in the pipe? (c) What additional pressure is necessary to deliver this flow? *Note*: Assume that the free-fall acceleration and the density of air are constant over this range of elevations.

46. Old Faithful Geyser in Yellowstone Park (Fig. P14.46) erupts at approximately 1-h intervals, and the height of the water column reaches 40.0 m. (a) Model the rising stream as a series of separate drops. Analyze the free-fall motion of one of the drops to determine the speed at which the water leaves the ground. (b) **What If?** Model the rising stream as an ideal fluid in streamline flow. Use Bernoulli's equation to determine the speed of the water as it leaves ground level. (c) What is the pressure (above atmospheric) in the heated underground chamber if its depth is 175 m? You may assume that the chamber is large compared with the geyser's vent.

Figure P14.46

47. A Venturi tube may be used as a fluid flow meter (see Fig. 14.20). If the difference in pressure is $P_1 - P_2 = 21.0$ kPa, find the fluid flow rate in cubic meters per second, given that the radius of the outlet tube is 1.00 cm, the radius of the inlet tube is 2.00 cm, and the fluid is gasoline ($\rho = 700$ kg/m³).

Section 14.7 Other Applications of Fluid Dynamics

48. An airplane has a mass of 1.60×10^4 kg, and each wing has an area of 40.0 m². During level flight, the pressure on the lower wing surface is 7.00×10^4 Pa. Determine the pressure on the upper wing surface.

49. A Pitot tube can be used to determine the velocity of air flow by measuring the difference between the total pressure and the static pressure (Fig. P14.49). If the fluid in the tube is mercury, density $\rho_{Hg} = 13\ 600$ kg/m³, and $\Delta h = 5.00$ cm, find the speed of air flow. (Assume that the air is stagnant at point A, and take $\rho_{air} = 1.25$ kg/m³.)

Figure P14.49

50. An airplane is cruising at an altitude of 10 km. The pressure outside the craft is 0.287 atm; within the passenger compartment the pressure is 1.00 atm and the temperature is 20°C. A small leak occurs in one of the window seals in the passenger compartment. Model the air as an ideal fluid to find the speed of the stream of air flowing through the leak.

51. A siphon is used to drain water from a tank, as illustrated in Figure P14.51. The siphon has a uniform diameter. Assume steady flow without friction. (a) If the distance $h = 1.00$ m, find the speed of outflow at the end of the siphon. (b) **What If?** What is the limitation on the height of the top of the siphon above the water surface? (For the flow of the liquid to be continuous, the pressure must not drop below the vapor pressure of the liquid.)

Figure P14.51

52. The Bernoulli effect can have important consequences for the design of buildings. For example, wind can blow around a skyscraper at remarkably high speed, creating low pressure. The higher atmospheric pressure in the still air inside the buildings can cause windows to pop out. As originally constructed, the John Hancock building in Boston popped window panes, which fell many stories to the sidewalk below. (a) Suppose that a horizontal wind blows in streamline flow with a speed of 11.2 m/s outside a large pane of plate glass with dimensions 4.00 m × 1.50 m. Assume the density of the air to be uniform at 1.30 kg/m³. The air inside the building is at atmospheric pressure. What is the total force exerted by air on the window pane? (b) **What If?** If a second skyscraper is built nearby, the air speed can be especially high where wind passes through the narrow separation between the buildings. Solve part (a) again if the wind speed is 22.4 m/s, twice as high.

53. A hypodermic syringe contains a medicine with the density of water (Figure P14.53). The barrel of the syringe has a cross-sectional area $A = 2.50 \times 10^{-5}$ m², and the needle has a cross-sectional area $a = 1.00 \times 10^{-8}$ m². In the absence of a force on the plunger, the pressure everywhere is 1 atm. A force **F** of magnitude 2.00 N acts on the plunger, making medicine squirt horizontally from the needle. Determine the speed of the medicine as it leaves the needle's tip.

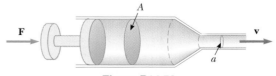

Figure P14.53

Additional Problems

54. Figure P14.54 shows a water tank with a valve at the bottom. If this valve is opened, what is the maximum height attained by the water stream coming out of the right side of the tank? Assume that $h = 10.0$ m, $L = 2.00$ m, and $\theta = 30.0°$, and that the cross-sectional area at A is very large compared with that at B.

Figure P14.54

55. A helium-filled balloon is tied to a 2.00-m-long, 0.050 0-kg uniform string. The balloon is spherical with a radius of 0.400 m. When released, it lifts a length h of string and then remains in equilibrium, as in Figure P14.55. Determine the value of h. The envelope of the balloon has mass 0.250 kg.

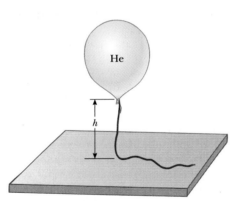

Figure P14.55

56. Water is forced out of a fire extinguisher by air pressure, as shown in Figure P14.56. How much gauge air pressure in the tank (above atmospheric) is required for the water jet to have a speed of 30.0 m/s when the water level in the tank is 0.500 m below the nozzle?

Figure P14.56

57. The true weight of an object can be measured in a vacuum, where buoyant forces are absent. An object of volume V is weighed in air on a balance with the use of weights of density ρ. If the density of air is ρ_{air} and the balance reads F_g', show that the true weight F_g is

$$F_g = F_g' + \left(V - \frac{F_g'}{\rho g}\right)\rho_{air}\,g$$

58. A wooden dowel has a diameter of 1.20 cm. It floats in water with 0.400 cm of its diameter above water (Fig. P14.58). Determine the density of the dowel.

Figure P14.58

59. A light spring of constant $k = 90.0$ N/m is attached vertically to a table (Fig. P14.59a). A 2.00-g balloon is filled with helium (density $= 0.180$ kg/m^3) to a volume of 5.00 m^3 and is then connected to the spring, causing it to stretch as in Figure P14.59b. Determine the extension distance L when the balloon is in equilibrium.

(a)   (b)

Figure P14.59

60. Evangelista Torricelli was the first person to realize that we live at the bottom of an ocean of air. He correctly surmised that the pressure of our atmosphere is attributable to the weight of the air. The density of air at 0°C at the Earth's surface is 1.29 kg/m^3. The density decreases with increasing altitude (as the atmosphere thins). On the other hand, if we assume that the density is constant at 1.29 kg/m^3 up to some altitude h, and zero above that altitude, then h would represent the depth of the ocean of air. Use this model to determine the value of h that gives a pressure of 1.00 atm at the surface of the Earth. Would the peak of

Mount Everest rise above the surface of such an atmosphere?

61. **Review problem**. With reference to Figure 14.5, show that the total torque exerted by the water behind the dam about a horizontal axis through O is $\frac{1}{6}\rho g w H^3$. Show that the effective line of action of the total force exerted by the water is at a distance $\frac{1}{3}H$ above O.

62. In about 1657 Otto von Guericke, inventor of the air pump, evacuated a sphere made of two brass hemispheres. Two teams of eight horses each could pull the hemispheres apart only on some trials, and then "with greatest difficulty," with the resulting sound likened to a cannon firing (Fig. P14.62). (a) Show that the force F required to pull the evacuated hemispheres apart is $\pi R^2 (P_0 - P)$, where R is the radius of the hemispheres and P is the pressure inside the hemispheres, which is much less than P_0. (b) Determine the force if $P = 0.100 P_0$ and $R = 0.300$ m.

Figure P14.62 The colored engraving, dated 1672, illustrates Otto von Guericke's demonstration of the force due to air pressure as performed before Emperor Ferdinand III in 1657.

63. A 1.00-kg beaker containing 2.00 kg of oil (density = 916.0 kg/m³) rests on a scale. A 2.00-kg block of iron is suspended from a spring scale and completely submerged in the oil as in Figure P14.63. Determine the equilibrium readings of both scales.

64. A beaker of mass m_{beaker} containing oil of mass m_{oil} (density = ρ_{oil}) rests on a scale. A block of iron of mass

Figure P14.63 Problems 63 and 64

m_{iron} is suspended from a spring scale and completely submerged in the oil as in Figure P14.63. Determine the equilibrium readings of both scales.

65. In 1983, the United States began coining the cent piece out of copper-clad zinc rather than pure copper. The mass of the old copper penny is 3.083 g, while that of the new cent is 2.517 g. Calculate the percentage of zinc (by volume) in the new cent. The density of copper is 8.960 g/cm³ and that of zinc is 7.133 g/cm³. The new and old coins have the same volume.

66. A thin spherical shell of mass 4.00 kg and diameter 0.200 m is filled with helium (density = 0.180 kg/m³). It is then released from rest on the bottom of a pool of water that is 4.00 m deep. (a) Neglecting frictional effects, show that the shell rises with constant acceleration and determine the value of that acceleration. (b) How long will it take for the top of the shell to reach the water surface?

67. **Review problem**. A uniform disk of mass 10.0 kg and radius 0.250 m spins at 300 rev/min on a low-friction axle. It must be brought to a stop in 1.00 min by a brake pad that makes contact with the disk at average distance 0.220 m from the axis. The coefficient of friction between pad and disk is 0.500. A piston in a cylinder of diameter 5.00 cm presses the brake pad against the disk. Find the pressure required for the brake fluid in the cylinder.

68. Show that the variation of atmospheric pressure with altitude is given by $P = P_0 e^{-\alpha y}$, where $\alpha = \rho_0 g / P_0$, P_0 is atmospheric pressure at some reference level $y = 0$, and ρ_0 is the atmospheric density at this level. Assume that the decrease in atmospheric pressure over an infinitesimal change in altitude (so that the density is approximately uniform) is given by $dP = -\rho g\, dy$, and that the density of air is proportional to the pressure.

69. An incompressible, nonviscous fluid is initially at rest in the vertical portion of the pipe shown in Figure P14.69a, where $L = 2.00$ m. When the valve is opened, the fluid flows into the horizontal section of the pipe. What is the speed of the fluid when all of it is in the horizontal section, as in Figure P14.69b? Assume the cross-sectional area of the entire pipe is constant.

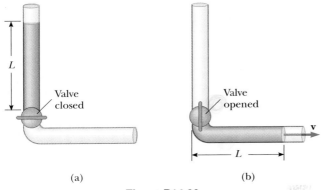

(a)   (b)

Figure P14.69

70. A cube of ice whose edges measure 20.0 mm is floating in a glass of ice-cold water with one of its faces parallel to the water's surface. (a) How far below the water surface is the bottom face of the block? (b) Ice-cold ethyl alcohol is gently poured onto the water surface to form a layer 5.00 mm thick above the water. The alcohol does not mix with the water. When the ice cube again attains hydrostatic equilibrium, what will be the distance from the top of the water to the bottom face of the block? (c) Additional cold ethyl alcohol is poured onto the water's surface until the top surface of the alcohol coincides with the top surface of the ice cube (in hydrostatic equilibrium). How thick is the required layer of ethyl alcohol?

71. A U-tube open at both ends is partially filled with water (Fig. P14.71a). Oil having a density of 750 kg/m^3 is then poured into the right arm and forms a column $L = 5.00$ cm high (Fig. P14.71b). (a) Determine the difference h in the heights of the two liquid surfaces. (b) The right arm is then shielded from any air motion while air is blown across the top of the left arm until the surfaces of the two liquids are at the same height (Fig. P14.71c). Determine the speed of the air being blown across the left arm. Take the density of air as 1.29 kg/m^3.

(a)   (b)   (c)

Figure P14.71

72. The water supply of a building is fed through a main pipe 6.00 cm in diameter. A 2.00-cm-diameter faucet tap, located 2.00 m above the main pipe, is observed to fill a 25.0-L container in 30.0 s. (a) What is the speed at which the water leaves the faucet? (b) What is the gauge pressure in the 6-cm main pipe? (Assume the faucet is the only "leak" in the building.)

73. The *spirit-in-glass thermometer*, invented in Florence, Italy, around 1654, consists of a tube of liquid (the spirit) containing a number of submerged glass spheres with slightly differ-

ent masses (Fig. P14.73). At sufficiently low temperatures all the spheres float, but as the temperature rises, the spheres sink one after another. The device is a crude but interesting tool for measuring temperature. Suppose that the tube is filled with ethyl alcohol, whose density is 0.789 45 g/cm^3 at 20.0°C and decreases to 0.780 97 g/cm^3 at 30.0°C. (a) If one of the spheres has a radius of 1.000 cm and is in equilibrium halfway up the tube at 20.0°C, determine its mass. (b) When the temperature increases to 30.0°C, what mass must a second sphere of the same radius have in order to be in equilibrium at the halfway point? (c) At 30.0°C the first sphere has fallen to the bottom of the tube. What upward force does the bottom of the tube exert on this sphere?

Figure P14.73

74. A woman is draining her fish tank by siphoning the water into an outdoor drain, as shown in Figure P14.74. The rectangular tank has footprint area A and depth h. The drain is located a distance d below the surface of the water in the tank, where $d \gg h$. The cross-sectional area of the siphon tube is A'. Model the water as flowing without friction. (a) Show that the time interval required to empty the tank is given by

$$\Delta t = \frac{Ah}{A'\sqrt{2gd}}$$

(b) Evaluate the time interval required to empty the tank if it is a cube 0.500 m on each edge, if $A' = 2.00$ cm^2, and $d = 10.0$ m.

Figure P14.74

75. The hull of an experimental boat is to be lifted above the water by a hydrofoil mounted below its keel, as shown in Figure P14.75. The hydrofoil has a shape like that of an airplane wing. Its area projected onto a horizontal surface is A. When the boat is towed at sufficiently high speed, water of density ρ moves in streamline flow so that its average speed at the top of the hydrofoil is n times larger than its speed v_b below the hydrofoil. (a) Neglecting the buoyant force, show that the upward lift force exerted by the water on the hydrofoil has a magnitude given by

$$F \approx \tfrac{1}{2}(n^2 - 1)\rho v_b^2 A$$

(b) The boat has mass M. Show that the liftoff speed is given by

$$v \approx \sqrt{\frac{2Mg}{(n^2 - 1)A\rho}}$$

(c) Assume that an 800-kg boat is to lift off at 9.50 m/s. Evaluate the area A required for the hydrofoil if its design yields $n = 1.05$.

Figure P14.75

Answer to Quick Quizzes

14.1 (a). Because the basketball player's weight is distributed over the larger surface area of the shoe, the pressure (F/A) that he applies is relatively small. The woman's lesser weight is distributed over the very small cross-sectional area of the spiked heel, so the pressure is high.

14.2 (a). Because both fluids have the same depth, the one with the smaller density (alcohol) will exert the smaller pressure.

14.3 (c). All barometers will have the same pressure at the bottom of the column of fluid—atmospheric pressure. Thus, the barometer with the highest column will be the one with the fluid of lowest density.

14.4 (d). Because there is no atmosphere on the Moon, there is no atmospheric pressure to provide a force to push the water up the straw.

14.5 (b). For a totally submerged object, the buoyant force does not depend on the depth in an incompressible fluid.

14.6 (c). The ice cube displaces a volume of water that has a weight equal to that of the ice cube. When the ice cube melts, it becomes a parcel of water with the same weight and exactly the volume that was displaced by the ice cube before.

14.7 (b) or (c). In all three cases, the weight of the treasure chest causes a downward force on the raft that makes it sink into the water. In (b) and (c), however, the treasure chest also displaces water, which provides a buoyant force in the upward direction, reducing the effect of the chest's weight.

14.8 (b). The liquid moves at the highest speed in the straw with the smaller cross sectional area.

14.9 (a). The high-speed air between the balloons results in low pressure in this region. The higher pressure on the outer surfaces of the balloons pushes them toward each other.

PART 2

Oscillations and Mechanical Waves

W e begin this new part of the text by studying a special type of motion called *periodic* motion. This is a *repeating* motion of an object in which the object continues to return to a given position after a fixed time interval. Familiar objects that exhibit periodic motion include a pendulum and a beach ball floating on the waves at a beach. The back and forth movements of such an object are called *oscillations*. We will focus our attention on a special case of periodic motion called *simple harmonic motion*. We shall find that all periodic motions can be modeled as combinations of simple harmonic motions. Thus, simple harmonic motion forms a basic building block for more complicated periodic motion.

Simple harmonic motion also forms the basis for our understanding of *mechanical waves*. Sound waves, seismic waves, waves on stretched strings, and water waves are all produced by some source of oscillation. As a sound wave travels through the air, elements of the air oscillate back and forth; as a water wave travels across a pond, elements of the water oscillate up and down and backward and forward. In general, as waves travel through any medium, the elements of the medium move in repetitive cycles. Therefore, the motion of the elements of the medium bears a strong resemblance to the periodic motion of an oscillating pendulum or an object attached to a spring.

To explain many other phenomena in nature, we must understand the concepts of oscillations and waves. For instance, although skyscrapers and bridges appear to be rigid, they actually oscillate, a fact that the architects and engineers who design and build them must take into account. To understand how radio and television work, we must understand the origin and nature of electromagnetic waves and how they propagate through space. Finally, much of what scientists have learned about atomic structure has come from information carried by waves. Therefore, we must first study oscillations and waves if we are to understand the concepts and theories of atomic physics. ■

◄ *Drops of water fall from a leaf into a pond. The disturbance caused by the falling water causes the water surface to oscillate. These oscillations are associated with waves moving away from the point at which the water fell. In Part 2 of the text, we will explore the principles related to oscillations and waves. (Don Bonsey/Getty Images)*

Chapter 15

Oscillatory Motion

▲ In the Bay of Fundy, Nova Scotia, the tides undergo oscillations with very large amplitudes, such that boats often end up sitting on dry ground for part of the day. In this chapter, we will investigate the physics of oscillatory motion. (www.comstock.com)

P*eriodic motion* is motion of an object that regularly repeats—the object returns to a given position after a fixed time interval. With a little thought, we can identify several types of periodic motion in everyday life. Your car returns to the driveway each afternoon. You return to the dinner table each night to eat. A bumped chandelier swings back and forth, returning to the same position at a regular rate. The Earth returns to the same position in its orbit around the Sun each year, resulting in the variation among the four seasons. The Moon returns to the same relationship with the Earth and the Sun, resulting in a full Moon approximately once a month.

In addition to these everyday examples, numerous other systems exhibit periodic motion. For example, the molecules in a solid oscillate about their equilibrium positions; electromagnetic waves, such as light waves, radar, and radio waves, are characterized by oscillating electric and magnetic field vectors; and in alternating-current electrical circuits, voltage, current, and electric charge vary periodically with time.

A special kind of periodic motion occurs in mechanical systems when the force acting on an object is proportional to the position of the object relative to some equilibrium position. If this force is always directed toward the equilibrium position, the motion is called *simple harmonic motion,* which is the primary focus of this chapter.

Active Figure 15.1 A block attached to a spring moving on a frictionless surface. (a) When the block is displaced to the right of equilibrium ($x > 0$), the force exerted by the spring acts to the left. (b) When the block is at its equilibrium position ($x = 0$), the force exerted by the spring is zero. (c) When the block is displaced to the left of equilibrium ($x < 0$), the force exerted by the spring acts to the right.

At the Active Figures link, at http://www.pse6.com, you can choose the spring constant and the initial position and velocities of the block to see the resulting simple harmonic motion.

15.1 Motion of an Object Attached to a Spring

As a model for simple harmonic motion, consider a block of mass m attached to the end of a spring, with the block free to move on a horizontal, frictionless surface (Fig. 15.1). When the spring is neither stretched nor compressed, the block is at the position called the **equilibrium position** of the system, which we identify as $x = 0$. We know from experience that such a system oscillates back and forth if disturbed from its equilibrium position.

We can understand the motion in Figure 15.1 qualitatively by first recalling that when the block is displaced to a position x, the spring exerts on the block a force that is proportional to the position and given by **Hooke's law** (see Section 7.4):

$$F_s = -kx \tag{15.1}$$

Hooke's law

We call this a **restoring force** because it is always directed toward the equilibrium position and therefore *opposite* the displacement from equilibrium. That is, when the block is displaced to the right of $x = 0$ in Figure 15.1, then the position is positive and the restoring force is directed to the left. When the block is displaced to the left of $x = 0$, then the position is negative and the restoring force is directed to the right.

Applying Newton's second law $\Sigma F_x = ma_x$ to the motion of the block, with Equation 15.1 providing the net force in the x direction, we obtain

$$-kx = ma_x$$

$$a_x = -\frac{k}{m}x \tag{15.2}$$

15.1 The Orientation of the Spring

Figure 15.1 shows a *horizontal* spring, with an attached block sliding on a frictionless surface. Another possibility is a block hanging from a *vertical* spring. All of the results that we discuss for the horizontal spring will be the same for the vertical spring, except that when the block is placed on the vertical spring, its weight will cause the spring to extend. If the resting position of the block is defined as $x = 0$, the results of this chapter will apply to this vertical system also.

That is, the acceleration is proportional to the position of the block, and its direction is opposite the direction of the displacement from equilibrium. Systems that behave in this way are said to exhibit **simple harmonic motion. An object moves with simple harmonic motion whenever its acceleration is proportional to its position and is oppositely directed to the displacement from equilibrium.**

If the block in Figure 15.1 is displaced to a position $x = A$ and released from rest, its *initial* acceleration is $-kA/m$. When the block passes through the equilibrium position $x = 0$, its acceleration is zero. At this instant, its speed is a maximum because the acceleration changes sign. The block then continues to travel to the left of equilibrium with a positive acceleration and finally reaches $x = -A$, at which time its acceleration is $+kA/m$ and its speed is again zero, as discussed in Sections 7.4 and 8.6. The block completes a full cycle of its motion by returning to the original position, again passing through $x = 0$ with maximum speed. Thus, we see that the block oscillates between the turning points $x = \pm A$. In the absence of friction, because the force exerted by the spring is conservative, this idealized motion will continue forever. Real systems are generally subject to friction, so they do not oscillate forever. We explore the details of the situation with friction in Section 15.6.

As Pitfall Prevention 15.1 points out, the principles that we develop in this chapter are also valid for an object hanging from a vertical spring, as long as we recognize that the weight of the object will stretch the spring to a new equilibrium position $x = 0$. To prove this statement, let x_s represent the total extension of the spring from its equilibrium position *without* the hanging object. Then, $x_s = -(mg/k) + x$, where $-(mg/k)$ is the extension of the spring due to the weight of the hanging object and x is the instantaneous extension of the spring due to the simple harmonic motion. The magnitude of the net force on the object is then $F_s - F_g = -k(-(mg/k) + x) - mg = -kx$. The net force on the object is the same as that on a block connected to a horizontal spring as in Equation 15.1, so the same simple harmonic motion results.

> **Quick Quiz 15.1** A block on the end of a spring is pulled to position $x = A$ and released. In one full cycle of its motion, through what total distance does it travel? (a) $A/2$ (b) A (c) $2A$ (d) $4A$

15.2 Mathematical Representation of Simple Harmonic Motion

Let us now develop a mathematical representation of the motion we described in the preceding section. We model the block as a particle subject to the force in Equation 15.1. We will generally choose x as the axis along which the oscillation occurs; hence, we will drop the subscript-x notation in this discussion. Recall that, by definition, $a = dv/dt = d^2x/dt^2$, and so we can express Equation 15.2 as

15.2 A Nonconstant Acceleration

Notice that the acceleration of the particle in simple harmonic motion is not constant. Equation 15.3 shows that it varies with position x. Thus, we *cannot* apply the kinematic equations of Chapter 2 in this situation.

$$\frac{d^2x}{dt^2} = -\frac{k}{m}x \tag{15.3}$$

If we denote the ratio k/m with the symbol ω^2 (we choose ω^2 rather than ω in order to make the solution that we develop below simpler in form), then

$$\omega^2 = \frac{k}{m} \tag{15.4}$$

and Equation 15.3 can be written in the form

$$\frac{d^2x}{dt^2} = -\omega^2x \tag{15.5}$$

What we now require is a mathematical solution to Equation 15.5—that is, a function $x(t)$ that satisfies this second-order differential equation. This is a mathematical representation of the position of the particle as a function of time. We seek a function $x(t)$ whose second derivative is the same as the original function with a negative sign and multiplied by ω^2. The trigonometric functions sine and cosine exhibit this behavior, so we can build a solution around one or both of these. The following cosine function is a solution to the differential equation:

$$x(t) = A\cos(\omega t + \phi) \tag{15.6}$$

where A, ω, and ϕ are constants. To see explicitly that this equation satisfies Equation 15.5, note that

$$\frac{dx}{dt} = A\frac{d}{dt}\cos(\omega t + \phi) = -\omega A\sin(\omega t + \phi) \tag{15.7}$$

$$\frac{d^2x}{dt^2} = -\omega A\frac{d}{dt}\sin(\omega t + \phi) = -\omega^2 A\cos(\omega t + \phi) \tag{15.8}$$

Comparing Equations 15.6 and 15.8, we see that $d^2x/dt^2 = -\omega^2 x$ and Equation 15.5 is satisfied.

The parameters A, ω, and ϕ are constants of the motion. In order to give physical significance to these constants, it is convenient to form a graphical representation of the motion by plotting x as a function of t, as in Figure 15.2a. First, note that A, called the **amplitude** of the motion, is simply **the maximum value of the position of the particle in either the positive or negative x direction.** The constant ω is called the **angular frequency,** and has units of rad/s.[1] It is a measure of how rapidly the oscillations are occurring—the more oscillations per unit time, the higher is the value of ω. From Equation 15.4, the angular frequency is

$$\omega = \sqrt{\frac{k}{m}} \tag{15.9}$$

The constant angle ϕ is called the **phase constant** (or initial phase angle) and, along with the amplitude A, is determined uniquely by the position and velocity of the particle at $t = 0$. If the particle is at its maximum position $x = A$ at $t = 0$, the phase constant is $\phi = 0$ and the graphical representation of the motion is shown in Figure 15.2b. The quantity $(\omega t + \phi)$ is called the **phase** of the motion. Note that the function $x(t)$ is periodic and its value is the same each time ωt increases by 2π radians.

Equations 15.1, 15.5, and 15.6 form the basis of the mathematical representation of simple harmonic motion. If we are analyzing a situation and find that the force on a particle is of the mathematical form of Equation 15.1, we know that the motion will be that of a simple harmonic oscillator and that the position of the particle is described by Equation 15.6. If we analyze a system and find that it is described by a differential equation of the form of Equation 15.5, the motion will be that of a simple harmonic oscillator. If we analyze a situation and find that the position of a particle is described by Equation 15.6, we know the particle is undergoing simple harmonic motion.

Position versus time for an object in simple harmonic motion

 PITFALL PREVENTION

15.3 Where's the Triangle?

Equation 15.6 includes a trigonometric function, a *mathematical function* that can be used whether it refers to a triangle or not. In this case, the cosine function happens to have the correct behavior for representing the position of a particle in simple harmonic motion.

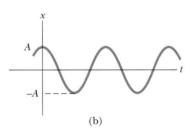

Active Figure 15.2 (a) An x-vs.-t graph for an object undergoing simple harmonic motion. The amplitude of the motion is A, the period (page 456) is T, and the phase constant is ϕ. (b) The x-vs.-t graph in the special case in which $x = A$ at $t = 0$ and hence $\phi = 0$.

At the Active Figures link at http://www.pse6.com, you can adjust the graphical representation and see the resulting simple harmonic motion of the block in Figure 15.1.

---

[1] We have seen many examples in earlier chapters in which we evaluate a trigonometric function of an angle. The argument of a trigonometric function, such as sine or cosine, *must* be a pure number. The radian is a pure number because it is a ratio of lengths. Angles in degrees are pure numbers simply because the degree is a completely artificial "unit"—it is not related to measurements of lengths. The notion of requiring a pure number for a trigonometric function is important in Equation 15.6, where the angle is expressed in terms of other measurements. Thus, ω *must* be expressed in rad/s (and not, for example, in revolutions per second) if t is expressed in seconds. Furthermore, other types of functions such as logarithms and exponential functions require arguments that are pure numbers.

Figure 15.3 An experimental apparatus for demonstrating simple harmonic motion. A pen attached to the oscillating object traces out a sinusoidal pattern on the moving chart paper.

An experimental arrangement that exhibits simple harmonic motion is illustrated in Figure 15.3. An object oscillating vertically on a spring has a pen attached to it. While the object is oscillating, a sheet of paper is moved perpendicular to the direction of motion of the spring, and the pen traces out the cosine curve in Equation 15.6.

Quick Quiz 15.2 Consider a graphical representation (Fig. 15.4) of simple harmonic motion, as described mathematically in Equation 15.6. When the object is at point Ⓐ on the graph, its (a) position and velocity are both positive (b) position and velocity are both negative (c) position is positive and its velocity is zero (d) position is negative and its velocity is zero (e) position is positive and its velocity is negative (f) position is negative and its velocity is positive.

Figure 15.4 (Quick Quiz 15.2) An *x-t* graph for an object undergoing simple harmonic motion. At a particular time, the object's position is indicated by Ⓐ in the graph.

Quick Quiz 15.3 Figure 15.5 shows two curves representing objects undergoing simple harmonic motion. The correct description of these two motions is that the simple harmonic motion of object B is (a) of larger angular frequency and larger amplitude than that of object A (b) of larger angular frequency and smaller amplitude than that of object A (c) of smaller angular frequency and larger amplitude than that of object A (d) of smaller angular frequency and smaller amplitude than that of object A.

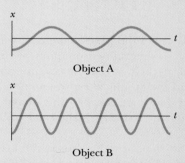

Figure 15.5 (Quick Quiz 15.3) Two *x-t* graphs for objects undergoing simple harmonic motion. The amplitudes and frequencies are different for the two objects.

Let us investigate further the mathematical description of simple harmonic motion. The **period** T of the motion is the time interval required for the particle to go through one full cycle of its motion (Fig. 15.2a). That is, the values of x and v for the particle at time t equal the values of x and v at time $t + T$. We can relate the period to the angular frequency by using the fact that the phase increases by 2π radians in a time interval of T:

$$[\omega(t + T) + \phi] - (\omega t + \phi) = 2\pi$$

Simplifying this expression, we see that $\omega T = 2\pi$, or

$$T = \frac{2\pi}{\omega}$$

(15.10)

The inverse of the period is called the **frequency** f of the motion. Whereas the period is the time interval per oscillation, the frequency represents the **number of oscillations that the particle undergoes per unit time interval:**

$$f = \frac{1}{T} = \frac{\omega}{2\pi} \qquad (15.11)$$

The units of f are cycles per second, or **hertz** (Hz). Rearranging Equation 15.11 gives

$$\omega = 2\pi f = \frac{2\pi}{T} \qquad (15.12)$$

We can use Equations 15.9, 15.10, and 15.11 to express the period and frequency of the motion for the particle–spring system in terms of the characteristics m and k of the system as

$$T = \frac{2\pi}{\omega} = 2\pi \sqrt{\frac{m}{k}} \qquad (15.13)$$

$$f = \frac{1}{T} = \frac{1}{2\pi} \sqrt{\frac{k}{m}} \qquad (15.14)$$

That is, the period and frequency depend *only* on the mass of the particle and the force constant of the spring, and *not* on the parameters of the motion, such as A or ϕ. As we might expect, the frequency is larger for a stiffer spring (larger value of k) and decreases with increasing mass of the particle.

We can obtain the velocity and acceleration[2] of a particle undergoing simple harmonic motion from Equations 15.7 and 15.8:

$$v = \frac{dx}{dt} = -\omega A \sin(\omega t + \phi) \qquad (15.15)$$

$$a = \frac{d^2x}{dt^2} = -\omega^2 A \cos(\omega t + \phi) \qquad (15.16)$$

From Equation 15.15 we see that, because the sine and cosine functions oscillate between ± 1, the extreme values of the velocity v are $\pm \omega A$. Likewise, Equation 15.16 tells us that the extreme values of the acceleration a are $\pm \omega^2 A$. Therefore, the *maximum* values of the magnitudes of the velocity and acceleration are

$$v_{\text{max}} = \omega A = \sqrt{\frac{k}{m}} \, A \qquad (15.17)$$

$$a_{\text{max}} = \omega^2 A = \frac{k}{m} \, A \qquad (15.18)$$

Figure 15.6a plots position versus time for an arbitrary value of the phase constant. The associated velocity–time and acceleration–time curves are illustrated in Figures 15.6b and 15.6c. They show that the phase of the velocity differs from the phase of the position by $\pi/2$ rad, or 90°. That is, when x is a maximum or a minimum, the velocity is zero. Likewise, when x is zero, the speed is a maximum. Furthermore, note that the

▲ **PITFALL PREVENTION**

15.4 Two Kinds of Frequency

We identify two kinds of frequency for a simple harmonic oscillator—f, called simply the *frequency*, is measured in hertz, and ω, the *angular frequency*, is measured in radians per second. Be sure that you are clear about which frequency is being discussed or requested in a given problem. Equations 15.11 and 15.12 show the relationship between the two frequencies.

Period

Frequency

Velocity of an object in simple harmonic motion

Acceleration of an object in simple harmonic motion

Maximum magnitudes of speed and acceleration in simple harmonic motion

---

[2] Because the motion of a simple harmonic oscillator takes place in one dimension, we will denote velocity as v and acceleration as a, with the direction indicated by a positive or negative sign, as in Chapter 2.

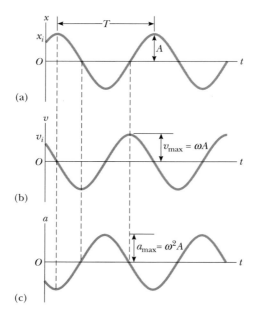

(a)

(b)

(c)

Figure 15.6 Graphical representation of simple harmonic motion. (a) Position versus time. (b) Velocity versus time. (c) Acceleration versus time. Note that at any specified time the velocity is 90° out of phase with the position and the acceleration is 180° out of phase with the position.

phase of the acceleration differs from the phase of the position by π radians, or 180°. For example, when x is a maximum, a has a maximum magnitude in the opposite direction.

Quick Quiz 15.4 Consider a graphical representation (Fig. 15.4) of simple harmonic motion, as described mathematically in Equation 15.6. When the object is at position Ⓐ on the graph, its (a) velocity and acceleration are both positive (b) velocity and acceleration are both negative (c) velocity is positive and its acceleration is zero (d) velocity is negative and its acceleration is zero (e) velocity is positive and its acceleration is negative (f) velocity is negative and its acceleration is positive.

Quick Quiz 15.5 An object of mass m is hung from a spring and set into oscillation. The period of the oscillation is measured and recorded as T. The object of mass m is removed and replaced with an object of mass $2m$. When this object is set into oscillation, the period of the motion is (a) $2T$ (b) $\sqrt{2}T$ (c) T (d) $T/\sqrt{2}$ (e) $T/2$.

Equation 15.6 describes simple harmonic motion of a particle in general. Let us now see how to evaluate the constants of the motion. The angular frequency ω is evaluated using Equation 15.9. The constants A and ϕ are evaluated from the initial conditions, that is, the state of the oscillator at $t = 0$.

Suppose we initiate the motion by pulling the particle from equilibrium by a distance A and releasing it from rest at $t = 0$, as in Figure 15.7. We must then require that

Active Figure 15.7 A block–spring system that begins its motion from rest with the block at $x = A$ at $t = 0$. In this case, $\phi = 0$ and thus $x = A \cos \omega t$.

At the Active Figures link at http://www.pse6.com, you can compare the oscillations of two blocks starting from different initial positions to see that the frequency is independent of the amplitude.

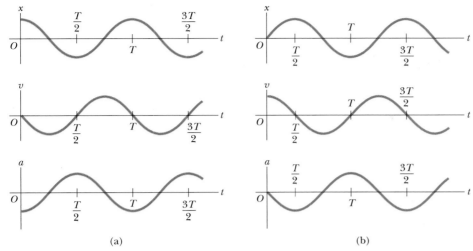

(a)   (b)

Figure 15.8 (a) Position, velocity, and acceleration versus time for a block undergoing simple harmonic motion under the initial conditions that at $t = 0$, $x(0) = A$ and $v(0) = 0$. (b) Position, velocity, and acceleration versus time for a block undergoing simple harmonic motion under the initial conditions that at $t = 0$, $x(0) = 0$ and $v(0) = v_i$.

our solutions for $x(t)$ and $v(t)$ (Eqs. 15.6 and 15.15) obey the initial conditions that $x(0) = A$ and $v(0) = 0$:

$$x(0) = A \cos \phi = A$$

$$v(0) = -\omega A \sin \phi = 0$$

These conditions are met if we choose $\phi = 0$, giving $x = A \cos \omega t$ as our solution. To check this solution, note that it satisfies the condition that $x(0) = A$, because $\cos 0 = 1$.

The position, velocity, and acceleration versus time are plotted in Figure 15.8a for this special case. The acceleration reaches extreme values of $\mp \omega^2 A$ when the position has extreme values of $\pm A$. Furthermore, the velocity has extreme values of $\pm \omega A$, which both occur at $x = 0$. Hence, the quantitative solution agrees with our qualitative description of this system.

Let us consider another possibility. Suppose that the system is oscillating and we define $t = 0$ as the instant that the particle passes through the unstretched position of the spring while moving to the right (Fig. 15.9). In this case we must require that our solutions for $x(t)$ and $v(t)$ obey the initial conditions that $x(0) = 0$ and $v(0) = v_i$:

$$x(0) = A \cos \phi = 0$$

$$v(0) = -\omega A \sin \phi = v_i$$

The first of these conditions tells us that $\phi = \pm \pi/2$. With these choices for ϕ, the second condition tells us that $A = \mp v_i/\omega$. Because the initial velocity is positive and the amplitude must be positive, we must have $\phi = -\pi/2$. Hence, the solution is given by

$$x = \frac{v_i}{\omega} \cos \left(\omega t - \frac{\pi}{2} \right)$$

The graphs of position, velocity, and acceleration versus time for this choice of $t = 0$ are shown in Figure 15.8b. Note that these curves are the same as those in Figure 15.8a, but shifted to the right by one fourth of a cycle. This is described mathematically by the phase constant $\phi = -\pi/2$, which is one fourth of a full cycle of 2π.

Active Figure 15.9 The block–spring system is undergoing oscillation, and $t = 0$ is defined at an instant when the block passes through the equilibrium position $x = 0$ and is moving to the right with speed v_i.

At the *Active Figures link* at **http://www.pse6.com**, *you can compare the oscillations of two blocks with different velocities at t = 0 to see that the frequency is independent of the amplitude.*

Example 15.1   An Oscillating Object

An object oscillates with simple harmonic motion along the x axis. Its position varies with time according to the equation

$$x = (4.00 \text{ m}) \cos \left(\pi t + \frac{\pi}{4} \right)$$

where t is in seconds and the angles in the parentheses are in radians.

(A) Determine the amplitude, frequency, and period of the motion.

Solution By comparing this equation with Equation 15.6, $x = A \cos (\omega t + \phi)$, we see that $A = \boxed{4.00 \text{ m}}$ and $\omega = \pi \text{ rad/s}$. Therefore, $f = \omega/2\pi = \pi/2\pi = \boxed{0.500 \text{ Hz}}$ and $T = 1/f = \boxed{2.00 \text{ s}}$.

(B) Calculate the velocity and acceleration of the object at any time t.

Solution Differentiating x to find v, and v to find a, we obtain

$$v = \frac{dx}{dt} = -(4.00 \text{ m/s}) \sin \left(\pi t + \frac{\pi}{4} \right) \frac{d}{dt} (\pi t)$$

$$= \boxed{-(4.00\pi \text{ m/s}) \sin \left(\pi t + \frac{\pi}{4} \right)}$$

$$a = \frac{dv}{dt} = -(4.00\pi \text{ m/s}) \cos \left(\pi t + \frac{\pi}{4} \right) \frac{d}{dt} (\pi t)$$

$$= \boxed{-(4.00\pi^2 \text{ m/s}^2) \cos \left(\pi t + \frac{\pi}{4} \right)}$$

(C) Using the results of part (B), determine the position, velocity, and acceleration of the object at $t = 1.00$ s.

Solution Noting that the angles in the trigonometric functions are in radians, we obtain, at $t = 1.00$ s,

$$x = (4.00 \text{ m}) \cos \left(\pi + \frac{\pi}{4} \right) = (4.00 \text{ m}) \cos \left(\frac{5\pi}{4} \right)$$

$$= (4.00 \text{ m})(-0.707) = \boxed{-2.83 \text{ m}}$$

$$v = -(4.00\pi \text{ m/s}) \sin \left(\frac{5\pi}{4} \right)$$

$$= -(4.00\pi \text{ m/s})(-0.707) = \boxed{8.89 \text{ m/s}}$$

$$a = -(4.00\pi^2 \text{ m/s}^2) \cos \left(\frac{5\pi}{4} \right)$$

$$= -(4.00\pi^2 \text{ m/s}^2)(-0.707) = \boxed{27.9 \text{ m/s}^2}$$

(D) Determine the maximum speed and maximum acceleration of the object.

Solution In the general expressions for v and a found in part (B), we use the fact that the maximum values of the sine and cosine functions are unity. Therefore, v varies between $\pm 4.00\pi$ m/s, and a varies between $\pm 4.00\pi^2$ m/s^2. Thus,

$$v_{max} = 4.00\pi \text{ m/s} = \boxed{12.6 \text{ m/s}}$$

$$a_{max} = 4.00\pi^2 \text{ m/s}^2 = \boxed{39.5 \text{ m/s}^2}$$

We obtain the same results using the relations $v_{max} = \omega A$ and $a_{max} = \omega^2 A$, where $A = 4.00$ m and $\omega = \pi$ rad/s.

(E) Find the displacement of the object between $t = 0$ and $t = 1.00$ s.

Solution The position at $t = 0$ is

$$x_i = (4.00 \text{ m}) \cos \left(0 + \frac{\pi}{4} \right) = (4.00 \text{ m})(0.707) = 2.83 \text{ m}$$

In part (C), we found that the position at $t = 1.00$ s is -2.83 m; therefore, the displacement between $t = 0$ and $t = 1.00$ s is

$$\Delta x = x_f - x_i = -2.83 \text{ m} - 2.83 \text{ m} = \boxed{-5.66 \text{ m}}$$

Because the object's velocity changes sign during the first second, the magnitude of Δx is not the same as the distance traveled in the first second. (By the time the first second is over, the object has been through the point $x = -2.83$ m once, traveled to $x = -4.00$ m, and come back to $x = -2.83$ m.)

Example 15.2   Watch Out for Potholes!

A car with a mass of 1 300 kg is constructed so that its frame is supported by four springs. Each spring has a force constant of 20 000 N/m. If two people riding in the car have a combined mass of 160 kg, find the frequency of vibration of the car after it is driven over a pothole in the road.

Solution To conceptualize this problem, think about your experiences with automobiles. When you sit in a car, it moves downward a small distance because your weight is compressing the springs further. If you push down on the front bumper and release, the front of the car oscillates a

couple of times. We can model the car as being supported by a single spring and categorize this as an oscillation problem based on our simple spring model. To analyze the problem, we first need to consider the effective spring constant of the four springs combined. For a given extension x of the springs, the combined force on the car is the sum of the forces from the individual springs:

$$F_{total} = \sum (-kx) = -\left(\sum k \right) x$$

where x has been factored from the sum because it is the

same for all four springs. We see that the effective spring constant for the combined springs is the sum of the individual spring constants:

$$k_{eff} = \sum k = 4 \times 20\ 000\ \text{N/m} = 80\ 000\ \text{N/m}$$

Hence, the frequency of vibration is, from Equation 15.14,

$$f = \frac{1}{2\pi} \sqrt{\frac{k_{eff}}{m}} = \frac{1}{2\pi} \sqrt{\frac{80\ 000\ \text{N/m}}{1\ 460\ \text{kg}}} = \boxed{1.18\ \text{Hz}}$$

To finalize the problem, note that the mass we used here is that of the car plus the people, because this is the total mass that is oscillating. Also note that we have explored only up-and-down motion of the car. If an oscillation is established in which the car rocks back and forth such that the front end goes up when the back end goes down, the frequency will be different.

What If? Suppose the two people exit the car on the side of the road. One of them pushes downward on the car and releases it so that it oscillates vertically. Is the frequency of the oscillation the same as the value we just calculated?

Answer The suspension system of the car is the same, but the mass that is oscillating is smaller—it no longer includes the mass of the two people. Thus, the frequency should be higher. Let us calculate the new frequency:

$$f = \frac{1}{2\pi} \sqrt{\frac{k_{eff}}{m}} = \frac{1}{2\pi} \sqrt{\frac{80\ 000\ \text{N/m}}{1\ 300\ \text{kg}}} = 1.25\ \text{Hz}$$

As we predicted conceptually, the frequency is a bit higher.

Example 15.3 A Block–Spring System

A 200-g block connected to a light spring for which the force constant is 5.00 N/m is free to oscillate on a horizontal, frictionless surface. The block is displaced 5.00 cm from equilibrium and released from rest, as in Figure 15.7.

(A) Find the period of its motion.

Solution From Equations 15.9 and 15.10, we know that the angular frequency of a block–spring system is

$$\omega = \sqrt{\frac{k}{m}} = \sqrt{\frac{5.00\ \text{N/m}}{200 \times 10^{-3}\ \text{kg}}} = 5.00\ \text{rad/s}$$

and the period is

$$T = \frac{2\pi}{\omega} = \frac{2\pi}{5.00\ \text{rad/s}} = \boxed{1.26\ \text{s}}$$

(B) Determine the maximum speed of the block.

Solution We use Equation 15.17:

$$v_{max} = \omega A = (5.00\ \text{rad/s})(5.00 \times 10^{-2}\ \text{m}) = \boxed{0.250\ \text{m/s}}$$

(C) What is the maximum acceleration of the block?

Solution We use Equation 15.18:

$$a_{max} = \omega^2 A = (5.00\ \text{rad/s})^2(5.00 \times 10^{-2}\ \text{m}) = \boxed{1.25\ \text{m/s}^2}$$

(D) Express the position, speed, and acceleration as functions of time.

Solution We find the phase constant from the initial condition that $x = A$ at $t = 0$:

$$x(0) = A \cos \phi = A$$

which tells us that $\phi = 0$. Thus, our solution is $x = A \cos \omega t$. Using this expression and the results from (A), (B), and (C), we find that

$$x = A \cos \omega t = \boxed{(0.050\ 0\ \text{m}) \cos 5.00t}$$

$$v = \omega A \sin \omega t = \boxed{-(0.250\ \text{m/s}) \sin 5.00t}$$

$$a = -\omega^2 A \cos \omega t = \boxed{-(1.25\ \text{m/s}^2) \cos 5.00t}$$

What If? What if the block is released from the same initial position, $x_i = 5.00$ cm, but with an initial velocity of $v_i = -0.100$ m/s? Which parts of the solution change and what are the new answers for those that do change?

Answers Part (A) does not change—the period is independent of how the oscillator is set into motion. Parts (B), (C), and (D) will change. We begin by considering position and velocity expressions for the initial conditions:

$$\text{(1)} \qquad x(0) = A \cos \phi = x_i$$

$$\text{(2)} \qquad v(0) = -\omega A \sin \phi = v_i$$

Dividing Equation (2) by Equation (1) gives us the phase constant:

$$\frac{-\omega A \sin \phi}{A \cos \phi} = \frac{v_i}{x_i}$$

$$\tan \phi = -\frac{v_i}{\omega x_i} = -\frac{-0.100\ \text{m}}{(5.00\ \text{rad/s})(0.050\ 0\ \text{m})} = 0.400$$

$$\phi = 0.12\pi$$

Now, Equation (1) allows us to find A:

$$A = \frac{x_i}{\cos \phi} = \frac{0.050\ 0\ \text{m}}{\cos(0.12\pi)} = 0.053\ 9\ \text{m}$$

The new maximum speed is

$$v_{max} = \omega A = (5.00\ \text{rad/s})(5.39 \times 10^{-2}\ \text{m}) = 0.269\ \text{m/s}$$

The new magnitude of the maximum acceleration is

$$a_{max} = \omega^2 A = (5.00\ \text{rad/s})^2(5.39 \times 10^{-2}\ \text{m}) = 1.35\ \text{m/s}^2$$

The new expressions for position, velocity, and acceleration are

$$x = (0.053\ 9\ \text{m}) \cos(5.00t + 0.12\pi)$$

$$v = -(0.269\ \text{m/s}) \sin(5.00t + 0.12\pi)$$

$$a = -(1.35\ \text{m/s}^2) \cos(5.00t + 0.12\pi)$$

As we saw in Chapters 7 and 8, many problems are easier to solve with an energy approach rather than one based on variables of motion. This particular **What If?** is easier to solve from an energy approach. Therefore, in the next section we shall investigate the energy of the simple harmonic oscillator.

15.3 Energy of the Simple Harmonic Oscillator

Let us examine the mechanical energy of the block–spring system illustrated in Figure 15.1. Because the surface is frictionless, we expect the total mechanical energy of the system to be constant, as was shown in Chapter 8. We assume a massless spring, so the kinetic energy of the system corresponds only to that of the block. We can use Equation 15.15 to express the kinetic energy of the block as

$$K = \tfrac{1}{2}mv^2 = \tfrac{1}{2}m\omega^2 A^2 \sin^2(\omega t + \phi) \qquad (15.19)$$

The elastic potential energy stored in the spring for any elongation x is given by $\tfrac{1}{2}kx^2$ (see Eq. 8.11). Using Equation 15.6, we obtain

$$U = \tfrac{1}{2}kx^2 = \tfrac{1}{2}kA^2 \cos^2(\omega t + \phi) \qquad (15.20)$$

We see that K and U are *always* positive quantities. Because $\omega^2 = k/m$, we can express the total mechanical energy of the simple harmonic oscillator as

$$E = K + U = \tfrac{1}{2}kA^2[\sin^2(\omega t + \phi) + \cos^2(\omega t + \phi)]$$

From the identity $\sin^2\theta + \cos^2\theta = 1$, we see that the quantity in square brackets is unity. Therefore, this equation reduces to

$$E = \tfrac{1}{2}kA^2 \qquad (15.21)$$

That is, **the total mechanical energy of a simple harmonic oscillator is a constant of the motion and is proportional to the square of the amplitude.** Note that U is small when K is large, and vice versa, because the sum must be constant. In fact, the total mechanical energy is equal to the maximum potential energy stored in the spring when $x = \pm A$ because $v = 0$ at these points and thus there is no kinetic energy. At the equilibrium position, where $U = 0$ because $x = 0$, the total energy, all in the form of kinetic energy, is again $\tfrac{1}{2}kA^2$. That is,

$$E = \tfrac{1}{2}mv_{\text{max}}^2 = \tfrac{1}{2}m\omega^2 A^2 = \tfrac{1}{2}m\frac{k}{m}A^2 = \tfrac{1}{2}kA^2 \qquad (\text{at } x = 0)$$

Plots of the kinetic and potential energies versus time appear in Figure 15.10a, where we have taken $\phi = 0$. As already mentioned, both K and U are always positive, and at all times their sum is a constant equal to $\tfrac{1}{2}kA^2$, the total energy of the system. The variations of K and U with the position x of the block are plotted in Figure 15.10b.

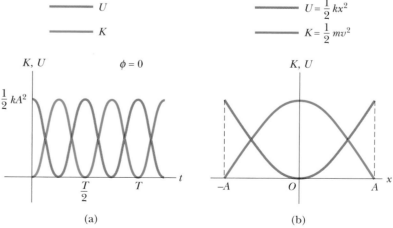

Active Figure 15.10 (a) Kinetic energy and potential energy versus time for a simple harmonic oscillator with $\phi = 0$. (b) Kinetic energy and potential energy versus position for a simple harmonic oscillator. In either plot, note that $K + U = $ constant.

At the Active Figures link at **http://www.pse6.com,** *you can compare the physical oscillation of a block with energy graphs in this figure as well as with energy bar graphs.*

Energy is continuously being transformed between potential energy stored in the spring and kinetic energy of the block.

Figure 15.11 illustrates the position, velocity, acceleration, kinetic energy, and potential energy of the block–spring system for one full period of the motion. Most of the ideas discussed so far are incorporated in this important figure. Study it carefully.

Finally, we can use the principle of conservation of energy to obtain the velocity for an arbitrary position by expressing the total energy at some arbitrary position x as

$$E = K + U = \tfrac{1}{2}mv^2 + \tfrac{1}{2}kx^2 = \tfrac{1}{2}kA^2$$

$$v = \pm\sqrt{\frac{k}{m}(A^2 - x^2)} = \pm\omega\sqrt{A^2 - x^2} \qquad (15.22)$$

When we check Equation 15.22 to see whether it agrees with known cases, we find that it verifies the fact that the speed is a maximum at $x = 0$ and is zero at the turning points $x = \pm A$.

You may wonder why we are spending so much time studying simple harmonic oscillators. We do so because they are good models of a wide variety of physical phenomena. For example, recall the Lennard–Jones potential discussed in Example 8.11. This complicated function describes the forces holding atoms together. Figure 15.12a shows that, for small displacements from the equilibrium position, the potential energy curve

Velocity as a function of position for a simple harmonic oscillator

t	x	v	a	K	U
0	A	0	$-\omega^2 A$	0	$\tfrac{1}{2}kA^2$
$T/4$	0	$-\omega A$	0	$\tfrac{1}{2}kA^2$	0
$T/2$	$-A$	0	$\omega^2 A$	0	$\tfrac{1}{2}kA^2$
$3T/4$	0	ωA	0	$\tfrac{1}{2}kA^2$	0
T	A	0	$-\omega^2 A$	0	$\tfrac{1}{2}kA^2$

Active Figure 15.11 Simple harmonic motion for a block–spring system and its analogy to the motion of a simple pendulum (Section 15.5). The parameters in the table at the right refer to the block–spring system, assuming that at $t = 0$, $x = A$ so that $x = A \cos \omega t$.

At the Active Figures link at http://www.pse6.com, you can set the initial position of the block and see the block–spring system and the analogous pendulum in motion.

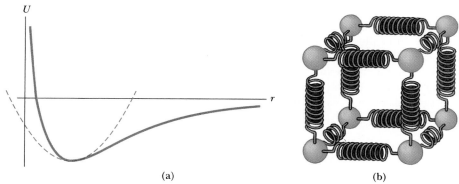

Figure 15.12 (a) If the atoms in a molecule do not move too far from their equilibrium positions, a graph of potential energy versus separation distance between atoms is similar to the graph of potential energy versus position for a simple harmonic oscillator (blue curve). (b) The forces between atoms in a solid can be modeled by imagining springs between neighboring atoms.

for this function approximates a parabola, which represents the potential energy function for a simple harmonic oscillator. Thus, we can model the complex atomic binding forces as being due to tiny springs, as depicted in Figure 15.12b.

The ideas presented in this chapter apply not only to block–spring systems and atoms, but also to a wide range of situations that include bungee jumping, tuning in a television station, and viewing the light emitted by a laser. You will see more examples of simple harmonic oscillators as you work through this book.

Example 15.4 Oscillations on a Horizontal Surface

A 0.500-kg cart connected to a light spring for which the force constant is 20.0 N/m oscillates on a horizontal, frictionless air track.

(A) Calculate the total energy of the system and the maximum speed of the cart if the amplitude of the motion is 3.00 cm.

Solution Using Equation 15.21, we obtain

$$E = \tfrac{1}{2} kA^2 = \tfrac{1}{2} (20.0 \text{ N/m})(3.00 \times 10^{-2} \text{ m})^2$$

$$= \boxed{9.00 \times 10^{-3} \text{ J}}$$

When the cart is located at $x = 0$, we know that $U = 0$ and $E = \tfrac{1}{2} mv_{\max}^2$; therefore,

$$\tfrac{1}{2} mv_{\max}^2 = 9.00 \times 10^{-3} \text{ J}$$

$$v_{\max} = \sqrt{\frac{2(9.00 \times 10^{-3} \text{ J})}{0.500 \text{ kg}}} = \boxed{0.190 \text{ m/s}}$$

(B) What is the velocity of the cart when the position is 2.00 cm?

Solution We can apply Equation 15.22 directly:

$$v = \pm \sqrt{\frac{k}{m}(A^2 - x^2)}$$

$$= \pm \sqrt{\frac{20.0 \text{ N/m}}{0.500 \text{ kg}}[(0.0300 \text{ m})^2 - (0.0200 \text{ m})^2]}$$

$$= \boxed{\pm 0.141 \text{ m/s}}$$

The positive and negative signs indicate that the cart could be moving to either the right or the left at this instant.

(C) Compute the kinetic and potential energies of the system when the position is 2.00 cm.

Solution Using the result of (B), we find that

$$K = \tfrac{1}{2} mv^2 = \tfrac{1}{2}(0.500 \text{ kg})(0.141 \text{ m/s})^2 = \boxed{5.00 \times 10^{-3} \text{ J}}$$

$$U = \tfrac{1}{2} kx^2 = \tfrac{1}{2}(20.0 \text{ N/m})(0.0200 \text{ m})^2 = \boxed{4.00 \times 10^{-3} \text{ J}}$$

Note that $K + U = E$.

What If? The motion of the cart in this example could have been initiated by releasing the cart from rest at $x = 3.00$ cm. What if the cart were released from the same position, but with an initial velocity of $v = -0.100$ m/s? What are the new amplitude and maximum speed of the cart?

Answer This is the same type of question as we asked at the end of Example 15.3, but here we apply an energy approach. First let us calculate the total energy of the system at $t = 0$, which consists of both kinetic energy and potential energy:

$$E = \tfrac{1}{2} mv^2 + \tfrac{1}{2} kx^2$$

$$= \tfrac{1}{2}(0.500 \text{ kg})(-0.100 \text{ m/s})^2 + \tfrac{1}{2}(20.0 \text{ N/m})(0.030\,0 \text{ m})^2$$

$$= 1.15 \times 10^{-2} \text{ J}$$

To find the new amplitude, we equate this total energy to the potential energy when the cart is at the end point of the motion:

$$E = \tfrac{1}{2} k A^2$$

$$A = \sqrt{\frac{2E}{k}} = \sqrt{\frac{2(1.15 \times 10^{-2}\,\text{J})}{20.0\,\text{N/m}}} = 0.033\,9\,\text{m}$$

Note that this is larger than the previous amplitude of 0.030 0 m. To find the new maximum speed, we equate this

total energy to the kinetic energy when the cart is at the equilibrium position:

$$E = \tfrac{1}{2} m v_{\text{max}}^2$$

$$v_{\text{max}} = \sqrt{\frac{2E}{m}} = \sqrt{\frac{2(1.15 \times 10^{-2}\,\text{J})}{0.500\,\text{kg}}} = 0.214\,\text{m/s}$$

This is larger than the value found in part (a) as expected because the cart has an initial velocity at $t = 0$.

15.4 Comparing Simple Harmonic Motion with Uniform Circular Motion

Some common devices in our everyday life exhibit a relationship between oscillatory motion and circular motion. For example, the pistons in an automobile engine (Figure 15.13a) go up and down—oscillatory motion—yet the net result of this motion is circular motion of the wheels. In an old-fashioned locomotive (Figure 15.13b), the drive shaft goes back and forth in oscillatory motion, causing a circular motion of the wheels. In this section, we explore this interesting relationship between these two types of motion. We shall use this relationship again when we study electromagnetism and when we explore optics.

Figure 15.14 is an overhead view of an experimental arrangement that shows this relationship. A ball is attached to the rim of a turntable of radius A, which is illuminated from the side by a lamp. The ball casts a shadow on a screen. We find that **as the turntable rotates with constant angular speed, the shadow of the ball moves back and forth in simple harmonic motion.**

Active Figure 15.14 An experimental setup for demonstrating the connection between simple harmonic motion and uniform circular motion. As the ball rotates on the turntable with constant angular speed, its shadow on the screen moves back and forth in simple harmonic motion.

At the Active Figures link at http://www.pse6.com, *you can adjust the frequency and radial position of the ball and see the resulting simple harmonic motion of the shadow.*

Courtesy of Ford Motor Company

(a)

© Link / Visuals Unlimited

(b)

Figure 15.13 (a) The pistons of an automobile engine move in periodic motion along a single dimension. This photograph shows a cutaway view of two of these pistons. This motion is converted to circular motion of the crankshaft, at the lower right, and ultimately of the wheels of the automobile. (b) The back-and-forth motion of pistons (in the curved housing at the left) in an old-fashioned locomotive is converted to circular motion of the wheels.

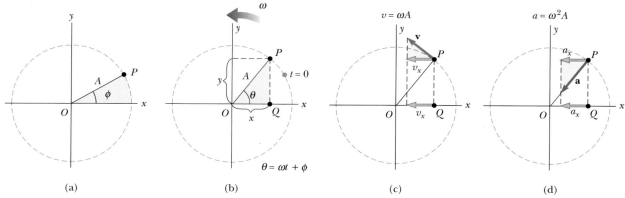

Figure 15.15 Relationship between the uniform circular motion of a point P and the simple harmonic motion of a point Q. A particle at P moves in a circle of radius A with constant angular speed ω. (a) A reference circle showing the position of P at $t = 0$. (b) The x coordinates of points P and Q are equal and vary in time according to the expression $x = A\cos(\omega t + \phi)$. (c) The x component of the velocity of P equals the velocity of Q. (d) The x component of the acceleration of P equals the acceleration of Q.

Consider a particle located at point P on the circumference of a circle of radius A, as in Figure 15.15a, with the line OP making an angle ϕ with the x axis at $t = 0$. We call this circle a *reference circle* for comparing simple harmonic motion with uniform circular motion, and we take the position of P at $t = 0$ as our reference position. If the particle moves along the circle with constant angular speed ω until OP makes an angle θ with the x axis, as in Figure 15.15b, then at some time $t > 0$, the angle between OP and the x axis is $\theta = \omega t + \phi$. As the particle moves along the circle, the projection of P on the x axis, labeled point Q, moves back and forth along the x axis between the limits $x = \pm A$.

Note that points P and Q always have the same x coordinate. From the right triangle OPQ, we see that this x coordinate is

$$x(t) = A\cos(\omega t + \phi) \tag{15.23}$$

This expression is the same as Equation 15.6 and shows that the point Q moves with simple harmonic motion along the x axis. Therefore, we conclude that

> simple harmonic motion along a straight line can be represented by the projection of uniform circular motion along a diameter of a reference circle.

We can make a similar argument by noting from Figure 15.15b that the projection of P along the y axis also exhibits simple harmonic motion. Therefore, **uniform circular motion can be considered a combination of two simple harmonic motions, one along the x axis and one along the y axis, with the two differing in phase by 90°.**

This geometric interpretation shows that the time interval for one complete revolution of the point P on the reference circle is equal to the period of motion T for simple harmonic motion between $x = \pm A$. That is, the angular speed ω of P is the same as the angular frequency ω of simple harmonic motion along the x axis. (This is why we use the same symbol.) The phase constant ϕ for simple harmonic motion corresponds to the initial angle that OP makes with the x axis. The radius A of the reference circle equals the amplitude of the simple harmonic motion.

Because the relationship between linear and angular speed for circular motion is $v = r\omega$ (see Eq. 10.10), the particle moving on the reference circle of radius A has a velocity of magnitude ωA. From the geometry in Figure 15.15c, we see that the x component of this velocity is $-\omega A\sin(\omega t + \phi)$. By definition, point Q has a velocity given by dx/dt. Differentiating Equation 15.23 with respect to time, we find that the velocity of Q is the same as the x component of the velocity of P.

The acceleration of P on the reference circle is directed radially inward toward O and has a magnitude $v^2/A = \omega^2 A$. From the geometry in Figure 15.15d, we see that the x component of this acceleration is $-\omega^2 A \cos(\omega t + \phi)$. This value is also the acceleration of the projected point Q along the x axis, as you can verify by taking the second derivative of Equation 15.23.

Quick Quiz 15.6 Figure 15.16 shows the position of an object in uniform circular motion at $t = 0$. A light shines from above and projects a shadow of the object on a screen below the circular motion. The correct values for the *amplitude* and *phase constant* (relative to an x axis to the right) of the simple harmonic motion of the shadow are (a) 0.50 m and 0 (b) 1.00 m and 0 (c) 0.50 m and π (d) 1.00 m and π.

Figure 15.16 (Quick Quiz 15.6) An object moves in circular motion, casting a shadow on the screen below. Its position at an instant of time is shown.

Example 15.5 Circular Motion with Constant Angular Speed

A particle rotates counterclockwise in a circle of radius 3.00 m with a constant angular speed of 8.00 rad/s. At $t = 0$, the particle has an x coordinate of 2.00 m and is moving to the right.

(A) Determine the x coordinate as a function of time.

Solution Because the amplitude of the particle's motion equals the radius of the circle and $\omega = 8.00$ rad/s, we have

$$x = A \cos(\omega t + \phi) = (3.00 \text{ m})\cos(8.00t + \phi)$$

We can evaluate ϕ by using the initial condition that $x = 2.00$ m at $t = 0$:

$$2.00 \text{ m} = (3.00 \text{ m})\cos(0 + \phi)$$

$$\phi = \cos^{-1}\left(\frac{2.00 \text{ m}}{3.00 \text{ m}}\right)$$

If we were to take our answer as $\phi = 48.2° = 0.841$ rad, then the coordinate $x = (3.00 \text{ m})\cos(8.00t + 0.841)$ would be decreasing at time $t = 0$ (that is, moving to the left). Because our particle is first moving to the right, we must choose $\phi = -0.841$ rad. The x coordinate as a function of time is then

$$x = \boxed{(3.00 \text{ m})\cos(8.00t - 0.841)}$$

Note that the angle ϕ in the cosine function must be in radians.

(B) Find the x components of the particle's velocity and acceleration at any time t.

Solution

$$v_x = \frac{dx}{dt} = (-3.00 \text{ m})(8.00 \text{ rad/s})\sin(8.00t - 0.841)$$

$$= \boxed{-(24.0 \text{ m/s})\sin(8.00t - 0.841)}$$

$$a_x = \frac{dv}{dt} = (-24.0 \text{ m/s})(8.00 \text{ rad/s})\cos(8.00t - 0.841)$$

$$= \boxed{-(192 \text{ m/s}^2)\cos(8.00t - 0.841)}$$

From these results, we conclude that $v_{max} = 24.0$ m/s and that $a_{max} = 192$ m/s^2.

Active Figure 15.17 When θ is small, a simple pendulum oscillates in simple harmonic motion about the equilibrium position $\theta = 0$. The restoring force is $- mg \sin \theta$, the component of the gravitational force tangent to the arc.

 At the Active Figures link at http://www.pse6.com, you can adjust the mass of the bob, the length of the string, and the initial angle and see the resulting oscillation of the pendulum.

▲ PITFALL PREVENTION

15.5   Not True Simple Harmonic Motion

Remember that the pendulum *does not* exhibit true simple harmonic motion for *any* angle. If the angle is less than about 10°, the motion is close to and can be *modeled* as simple harmonic.

Angular frequency for a simple pendulum

Period of a simple pendulum

15.5   The Pendulum

The **simple pendulum** is another mechanical system that exhibits periodic motion. It consists of a particle-like bob of mass m suspended by a light string of length L that is fixed at the upper end, as shown in Figure 15.17. The motion occurs in the vertical plane and is driven by the gravitational force. We shall show that, provided the angle θ is small (less than about 10°), the motion is very close to that of a simple harmonic oscillator.

The forces acting on the bob are the force **T** exerted by the string and the gravitational force $m\mathbf{g}$. The tangential component $mg \sin \theta$ of the gravitational force always acts toward $\theta = 0$, opposite the displacement of the bob from the lowest position. Therefore, the tangential component is a restoring force, and we can apply Newton's second law for motion in the tangential direction:

$$F_t = - mg \sin \theta = m \frac{d^2 s}{dt^2}$$

where s is the bob's position measured along the arc and the negative sign indicates that the tangential force acts toward the equilibrium (vertical) position. Because $s = L\theta$ (Eq. 10.1a) and L is constant, this equation reduces to

$$\frac{d^2 \theta}{dt^2} = - \frac{g}{L} \sin \theta$$

Considering θ as the position, let us compare this equation to Equation 15.3—does it have the same mathematical form? The right side is proportional to $\sin \theta$ rather than to θ; hence, we would not expect simple harmonic motion because this expression is not of the form of Equation 15.3. However, if we assume that θ is *small*, we can use the approximation $\sin \theta \approx \theta$; thus, in this approximation, the equation of motion for the simple pendulum becomes

$$\frac{d^2 \theta}{dt^2} = - \frac{g}{L} \theta \qquad \text{(for small values of } \theta) \tag{15.24}$$

Now we have an expression that has the same form as Equation 15.3, and we conclude that the motion for small amplitudes of oscillation is simple harmonic motion. Therefore, the function θ can be written as $\theta = \theta_{max} \cos(\omega t + \phi)$, where θ_{max} is the *maximum angular position* and the angular frequency ω is

$$\omega = \sqrt{\frac{g}{L}} \tag{15.25}$$

The period of the motion is

$$T = \frac{2\pi}{\omega} = 2\pi \sqrt{\frac{L}{g}} \tag{15.26}$$

In other words, **the period and frequency of a simple pendulum depend only on the length of the string and the acceleration due to gravity.** Because the period is independent of the mass, we conclude that all simple pendula that are of equal length and are at the same location (so that g is constant) oscillate with the same period. The analogy between the motion of a simple pendulum and that of a block–spring system is illustrated in Figure 15.11.

The simple pendulum can be used as a timekeeper because its period depends only on its length and the local value of g. It is also a convenient device for making precise measurements of the free-fall acceleration. Such measurements are important because variations in local values of g can provide information on the location of oil and of other valuable underground resources.

Quick Quiz 15.7 A grandfather clock depends on the period of a pendulum to keep correct time. Suppose a grandfather clock is calibrated correctly and then a mischievous child slides the bob of the pendulum downward on the oscillating rod. Does the grandfather clock run (a) slow (b) fast (c) correctly?

Quick Quiz 15.8 Suppose a grandfather clock is calibrated correctly at sea level and is then taken to the top of a very tall mountain. Does the grandfather clock run (a) slow (b) fast (c) correctly?

Example 15.6   A Connection Between Length and Time

Christian Huygens (1629–1695), the greatest clockmaker in history, suggested that an international unit of length could be defined as the length of a simple pendulum having a period of exactly 1 s. How much shorter would our length unit be had his suggestion been followed?

Solution Solving Equation 15.26 for the length gives

$$L = \frac{T^2 g}{4\pi^2} = \frac{(1.00 \text{ s})^2 (9.80 \text{ m/s}^2)}{4\pi^2} = \boxed{0.248 \text{ m}}$$

Thus, the meter's length would be slightly less than one fourth of its current length. Note that the number of significant digits depends only on how precisely we know g because the time has been defined to be exactly 1 s.

What If? What if Huygens had been born on another planet? What would the value for g have to be on that planet such that the meter based on Huygens's pendulum would have the same value as our meter?

Answer We solve Equation 15.26 for g:

$$g = \frac{4\pi^2 L}{T^2} = \frac{4\pi^2 (1.00 \text{ m})}{(1.00 \text{ s})^2} = 4\pi^2 \text{ m/s}^2 = 39.5 \text{ m/s}^2$$

No planet in our solar system has an acceleration due to gravity that is this large.

Physical Pendulum

Suppose you balance a wire coat hanger so that the hook is supported by your extended index finger. When you give the hanger a small angular displacement (with your other hand) and then release it, it oscillates. If a hanging object oscillates about a fixed axis that does not pass through its center of mass and the object cannot be approximated as a point mass, we cannot treat the system as a simple pendulum. In this case the system is called a **physical pendulum.**

Consider a rigid object pivoted at a point O that is a distance d from the center of mass (Fig. 15.18). The gravitational force provides a torque about an axis through O, and the magnitude of that torque is $mgd \sin \theta$, where θ is as shown in Figure 15.18. Using the rotational form of Newton's second law, $\Sigma \tau = I\alpha$, where I is the moment of inertia about the axis through O, we obtain

$$-mgd \sin \theta = I \frac{d^2 \theta}{dt^2}$$

The negative sign indicates that the torque about O tends to decrease θ. That is, the gravitational force produces a restoring torque. If we again assume that θ is small, the approximation $\sin \theta \approx \theta$ is valid, and the equation of motion reduces to

$$\frac{d^2 \theta}{dt^2} = -\left(\frac{mgd}{I}\right)\theta = -\omega^2 \theta \qquad (15.27)$$

Because this equation is of the same form as Equation 15.3, the motion is simple harmonic motion. That is, the solution of Equation 15.27 is $\theta = \theta_{max} \cos(\omega t + \phi)$, where θ_{max} is the maximum angular position and

$$\omega = \sqrt{\frac{mgd}{I}}$$

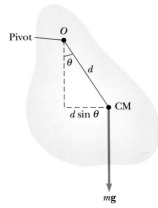

Figure 15.18 A physical pendulum pivoted at O.

The period is

Period of a physical pendulum

$$T = \frac{2\pi}{\omega} = 2\pi \sqrt{\frac{I}{mgd}}$$ (15.28)

One can use this result to measure the moment of inertia of a flat rigid object. If the location of the center of mass—and hence the value of d—is known, the moment of inertia can be obtained by measuring the period. Finally, note that Equation 15.28 reduces to the period of a simple pendulum (Eq. 15.26) when $I = md^2$—that is, when all the mass is concentrated at the center of mass.

Example 15.7   A Swinging Rod

A uniform rod of mass M and length L is pivoted about one end and oscillates in a vertical plane (Fig. 15.19). Find the period of oscillation if the amplitude of the motion is small.

Solution In Chapter 10 we found that the moment of inertia of a uniform rod about an axis through one end is $\frac{1}{3}ML^2$. The distance d from the pivot to the center of mass is $L/2$. Substituting these quantities into Equation 15.28 gives

$$T = 2\pi \sqrt{\frac{\frac{1}{3}ML^2}{Mg(L/2)}} = 2\pi \sqrt{\frac{2L}{3g}}$$

Comment In one of the Moon landings, an astronaut walking on the Moon's surface had a belt hanging from his space suit, and the belt oscillated as a physical pendulum. A scientist on the Earth observed this motion on television

and used it to estimate the free-fall acceleration on the Moon. How did the scientist make this calculation?

Figure 15.19 A rigid rod oscillating about a pivot through one end is a physical pendulum with $d = L/2$ and, from Table 10.2, $I = \frac{1}{3}ML^2$.

Figure 15.20 A torsional pendulum consists of a rigid object suspended by a wire attached to a rigid support. The object oscillates about the line OP with an amplitude θ_{max}.

Period of a torsional pendulum

Torsional Pendulum

Figure 15.20 shows a rigid object suspended by a wire attached at the top to a fixed support. When the object is twisted through some angle θ, the twisted wire exerts on the object a restoring torque that is proportional to the angular position. That is,

$$\tau = -\kappa\theta$$

where κ (kappa) is called the *torsion constant* of the support wire. The value of κ can be obtained by applying a known torque to twist the wire through a measurable angle θ. Applying Newton's second law for rotational motion, we find

$$\tau = -\kappa\theta = I\frac{d^2\theta}{dt^2}$$

$$\frac{d^2\theta}{dt^2} = -\frac{\kappa}{I}\theta$$ (15.29)

Again, this is the equation of motion for a simple harmonic oscillator, with $\omega = \sqrt{\kappa/I}$ and a period

$$T = 2\pi \sqrt{\frac{I}{\kappa}}$$ (15.30)

This system is called a *torsional pendulum*. There is no small-angle restriction in this situation as long as the elastic limit of the wire is not exceeded.

15.6 Damped Oscillations

The oscillatory motions we have considered so far have been for ideal systems—that is, systems that oscillate indefinitely under the action of only one force—a linear restoring force. In many real systems, nonconservative forces, such as friction, retard the motion. Consequently, the mechanical energy of the system diminishes in time, and the motion is said to be *damped*. Figure 15.21 depicts one such system: an object attached to a spring and submersed in a viscous liquid.

One common type of retarding force is the one discussed in Section 6.4, where the force is proportional to the speed of the moving object and acts in the direction opposite the motion. This retarding force is often observed when an object moves through air, for instance. Because the retarding force can be expressed as $\mathbf{R} = -b\mathbf{v}$ (where b is a constant called the *damping coefficient*) and the restoring force of the system is $-kx$, we can write Newton's second law as

$$\sum F_x = -kx - bv_x = ma_x$$

$$-kx - b\frac{dx}{dt} = m\frac{d^2x}{dt^2} \tag{15.31}$$

The solution of this equation requires mathematics that may not be familiar to you; we simply state it here without proof. When the retarding force is small compared with the maximum restoring force—that is, when b is small—the solution to Equation 15.31 is

$$x = Ae^{-\frac{b}{2m}t}\cos(\omega t + \phi) \tag{15.32}$$

where the angular frequency of oscillation is

$$\omega = \sqrt{\frac{k}{m} - \left(\frac{b}{2m}\right)^2} \tag{15.33}$$

This result can be verified by substituting Equation 15.32 into Equation 15.31.

Figure 15.22 shows the position as a function of time for an object oscillating in the presence of a retarding force. We see that **when the retarding force is small, the oscillatory character of the motion is preserved but the amplitude decreases in time, with the result that the motion ultimately ceases.** Any system that behaves in this way is known as a **damped oscillator.** The dashed blue lines in Figure 15.22, which define the *envelope* of the oscillatory curve, represent the exponential factor in Equation 15.32. This envelope shows that **the amplitude decays exponentially with time.** For motion with a given spring constant and object mass, the oscillations dampen more rapidly as the maximum value of the retarding force approaches the maximum value of the restoring force.

It is convenient to express the angular frequency (Eq. 15.33) of a damped oscillator in the form

$$\omega = \sqrt{\omega_0^2 - \left(\frac{b}{2m}\right)^2}$$

where $\omega_0 = \sqrt{k/m}$ represents the angular frequency in the absence of a retarding force (the undamped oscillator) and is called the **natural frequency** of the system.

When the magnitude of the maximum retarding force $R_{max} = bv_{max} < kA$, the system is said to be **underdamped.** The resulting motion is represented by the blue curve in Figure 15.23. As the value of b increases, the amplitude of the oscillations decreases more and more rapidly. When b reaches a critical value b_c such that $b_c/2m = \omega_0$, the system does not oscillate and is said to be **critically damped.** In this case the system, once released from rest at some nonequilibrium position, approaches but does not pass through the equilibrium position. The graph of position versus time for this case is the red curve in Figure 15.23.

Figure 15.21 One example of a damped oscillator is an object attached to a spring and submersed in a viscous liquid.

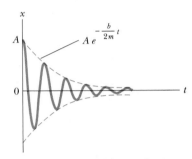

Active Figure 15.22 Graph of position versus time for a damped oscillator. Note the decrease in amplitude with time.

At the Active Figures link at http://www.pse6.com, you can adjust the spring constant, the mass of the object, and the damping constant and see the resulting damped oscillation of the object.

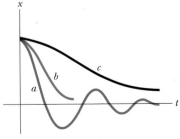

Figure 15.23 Graphs of position versus time for (a) an underdamped oscillator, (b) a critically damped oscillator, and (c) an overdamped oscillator.

If the medium is so viscous that the retarding force is greater than the restoring force—that is, if $R_{max} = bv_{max} > kA$ and $b/2m > \omega_0$—the system is **overdamped.** Again, the displaced system, when free to move, does not oscillate but simply returns to its equilibrium position. As the damping increases, the time interval required for the system to approach equilibrium also increases, as indicated by the black curve in Figure 15.23. For critically damped and overdamped systems, there is no angular frequency ω and the solution in Equation 15.32 is not valid.

Whenever friction is present in a system, whether the system is overdamped or underdamped, the energy of the oscillator eventually falls to zero. The lost mechanical energy is transformed into internal energy in the object and the retarding medium.

Quick Quiz 15.9 An automotive suspension system consists of a combination of springs and shock absorbers, as shown in Figure 15.24. If you were an automotive engineer, would you design a suspension system that was (a) underdamped (b) critically damped (c) overdamped?

(a)   (b)

Figure 15.24 (a) A shock absorber consists of a piston oscillating in a chamber filled with oil. As the piston oscillates, the oil is squeezed through holes between the piston and the chamber, causing a damping of the piston's oscillations. (b) One type of automotive suspension system, in which a shock absorber is placed inside a coil spring at each wheel.

15.7 Forced Oscillations

We have seen that the mechanical energy of a damped oscillator decreases in time as a result of the resistive force. It is possible to compensate for this energy decrease by applying an external force that does positive work on the system. At any instant, energy can be transferred into the system by an applied force that acts in the direction of motion of the oscillator. For example, a child on a swing can be kept in motion by appropriately timed "pushes." The amplitude of motion remains constant if the energy input per cycle of motion exactly equals the decrease in mechanical energy in each cycle that results from resistive forces.

A common example of a forced oscillator is a damped oscillator driven by an external force that varies periodically, such as $F(t) = F_0 \sin \omega t$, where ω is the angular frequency of the driving force and F_0 is a constant. In general, the frequency ω of the driving force is variable while the natural frequency ω_0 of the oscillator is fixed by the values of k and m. Newton's second law in this situation gives

$$\sum F = ma \longrightarrow F_0 \sin \omega t - b \frac{dx}{dt} - kx = m \frac{d^2x}{dt^2} \qquad (15.34)$$

Again, the solution of this equation is rather lengthy and will not be presented. After the driving force on an initially stationary object begins to act, the amplitude of the oscillation will increase. After a sufficiently long period of time, when the energy input per cycle from the driving force equals the amount of mechanical energy transformed to internal energy for each cycle, a steady-state condition is reached in which the oscillations proceed with constant amplitude. In this situation, Equation 15.34 has the solution

$$x = A \cos(\omega t + \phi) \qquad (15.35)$$

where

$$A = \frac{F_0/m}{\sqrt{(\omega^2 - \omega_0{}^2)^2 + \left(\dfrac{b\omega}{m}\right)^2}} \qquad (15.36)$$

◀ **Amplitude of a driven oscillator**

and where $\omega_0 = \sqrt{k/m}$ is the natural frequency of the undamped oscillator ($b = 0$).

Equations 15.35 and 15.36 show that the forced oscillator vibrates at the frequency of the driving force and that the amplitude of the oscillator is constant for a given driving force because it is being driven in steady-state by an external force. For small damping, the amplitude is large when the frequency of the driving force is near the natural frequency of oscillation, or when $\omega \approx \omega_0$. The dramatic increase in amplitude near the natural frequency is called **resonance,** and the natural frequency ω_0 is also called the **resonance frequency** of the system.

The reason for large-amplitude oscillations at the resonance frequency is that energy is being transferred to the system under the most favorable conditions. We can better understand this by taking the first time derivative of x in Equation 15.35, which gives an expression for the velocity of the oscillator. We find that v is proportional to $\sin(\omega t + \phi)$, which is the same trigonometric function as that describing the driving force. Thus, the applied force $\mathbf{F}$ is in phase with the velocity. The rate at which work is done on the oscillator by $\mathbf{F}$ equals the dot product $\mathbf{F} \cdot \mathbf{v}$; this rate is the power delivered to the oscillator. Because the product $\mathbf{F} \cdot \mathbf{v}$ is a maximum when $\mathbf{F}$ and $\mathbf{v}$ are in phase, we conclude that **at resonance the applied force is in phase with the velocity and the power transferred to the oscillator is a maximum.**

Figure 15.25 is a graph of amplitude as a function of frequency for a forced oscillator with and without damping. Note that the amplitude increases with decreasing damping ($b \rightarrow 0$) and that the resonance curve broadens as the damping increases. Under steady-state conditions and at any driving frequency, the energy transferred into the system equals the energy lost because of the damping force; hence, the average total energy of the oscillator remains constant. In the absence of a damping force ($b = 0$), we see from Equation 15.36 that the steady-state amplitude approaches infinity as ω approaches ω_0. In other words, if there are no losses in the system and if we continue to drive an initially motionless oscillator with a periodic force that is in phase with the velocity, the amplitude of motion builds without limit (see the brown curve in Fig. 15.25). This limitless building does not occur in practice because some damping is always present in reality.

Later in this book we shall see that resonance appears in other areas of physics. For example, certain electric circuits have natural frequencies. A bridge has natural frequencies that can be set into resonance by an appropriate driving force. A dramatic example of such resonance occurred in 1940, when the Tacoma Narrows Bridge in the state of Washington was destroyed by resonant vibrations. Although the winds were not particularly strong on that occasion, the "flapping" of the wind across the roadway (think of the "flapping" of a flag in a strong wind) provided a periodic driving force whose frequency matched that of the bridge. The resulting oscillations of the bridge caused it to ultimately collapse (Fig. 15.26) because the bridge design had inadequate built-in safety features.

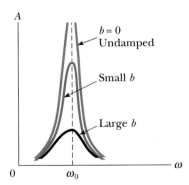

Figure 15.25 Graph of amplitude versus frequency for a damped oscillator when a periodic driving force is present. When the frequency ω of the driving force equals the natural frequency ω_0 of the oscillator, resonance occurs. Note that the shape of the resonance curve depends on the size of the damping coefficient b.

(a)   (b)

Figure 15.26 (a) In 1940 turbulent winds set up torsional vibrations in the Tacoma Narrows Bridge, causing it to oscillate at a frequency near one of the natural frequencies of the bridge structure. (b) Once established, this resonance condition led to the bridge's collapse.

Many other examples of resonant vibrations can be cited. A resonant vibration that you may have experienced is the "singing" of telephone wires in the wind. Machines often break if one vibrating part is in resonance with some other moving part. Soldiers marching in cadence across a bridge have been known to set up resonant vibrations in the structure and thereby cause it to collapse. Whenever any real physical system is driven near its resonance frequency, you can expect oscillations of very large amplitudes.

SUMMARY

Take a practice test for this chapter by clicking on the Practice Test link at http://www.pse6.com.

When the acceleration of an object is proportional to its position and is in the direction opposite the displacement from equilibrium, the object moves with simple harmonic motion. The position x of a simple harmonic oscillator varies periodically in time according to the expression

$$x(t) = A\cos(\omega t + \phi) \tag{15.6}$$

where A is the **amplitude** of the motion, ω is the **angular frequency,** and ϕ is the **phase constant.** The value of ϕ depends on the initial position and initial velocity of the oscillator.

The time interval T needed for one complete oscillation is defined as the **period** of the motion:

$$T = \frac{2\pi}{\omega} \tag{15.10}$$

A block–spring system moves in simple harmonic motion on a frictionless surface with a period

$$T = \frac{2\pi}{\omega} = 2\pi\sqrt{\frac{m}{k}} \tag{15.13}$$

The inverse of the period is the **frequency** of the motion, which equals the number of oscillations per second.

The velocity and acceleration of a simple harmonic oscillator are

$$v = \frac{dx}{dt} = -\omega A\sin(\omega t + \phi) \tag{15.15}$$

$$a = \frac{d^2x}{dt^2} = -\omega^2 A\cos(\omega t + \phi) \tag{15.16}$$

$$v = \pm\omega\sqrt{A^2 - x^2} \tag{15.22}$$

Thus, the maximum speed is ωA, and the maximum acceleration is $\omega^2 A$. The speed is zero when the oscillator is at its turning points $x = \pm A$ and is a maximum when the oscillator is at the equilibrium position $x = 0$. The magnitude of the acceleration is a maximum at the turning points and zero at the equilibrium position.

The kinetic energy and potential energy for a simple harmonic oscillator vary with time and are given by

$$K = \tfrac{1}{2} mv^2 = \tfrac{1}{2} m\omega^2 A^2 \sin^2 (\omega t + \phi) \qquad (15.19)$$

$$U = \tfrac{1}{2} kx^2 = \tfrac{1}{2} kA^2 \cos^2 (\omega t + \phi) \qquad (15.20)$$

The total energy of a simple harmonic oscillator is a constant of the motion and is given by

$$E = \tfrac{1}{2} kA^2 \qquad (15.21)$$

The potential energy of the oscillator is a maximum when the oscillator is at its turning points and is zero when the oscillator is at the equilibrium position. The kinetic energy is zero at the turning points and a maximum at the equilibrium position.

A **simple pendulum** of length L moves in simple harmonic motion for small angular displacements from the vertical. Its period is

$$T = 2\pi \sqrt{\frac{L}{g}} \qquad (15.26)$$

For small angular displacements from the vertical, a **physical pendulum** moves in simple harmonic motion about a pivot that does not go through the center of mass. The period of this motion is

$$T = 2\pi \sqrt{\frac{I}{mgd}} \qquad (15.28)$$

where I is the moment of inertia about an axis through the pivot and d is the distance from the pivot to the center of mass.

If an oscillator experiences a damping force $\mathbf{R} = -b\mathbf{v}$, its position for small damping is described by

$$x = Ae^{-\frac{b}{2m}t} \cos(\omega t + \phi) \qquad (15.32)$$

where

$$\omega = \sqrt{\frac{k}{m} - \left(\frac{b}{2m}\right)^2} \qquad (15.33)$$

If an oscillator is subject to a sinusoidal driving force $F(t) = F_0 \sin \omega t$, it exhibits **resonance,** in which the amplitude is largest when the driving frequency matches the natural frequency of the oscillator.

QUESTIONS

1. Is a bouncing ball an example of simple harmonic motion? Is the daily movement of a student from home to school and back simple harmonic motion? Why or why not?

2. If the coordinate of a particle varies as $x = -A \cos \omega t$, what is the phase constant in Equation 15.6? At what position is the particle at $t = 0$?

3. Does the displacement of an oscillating particle between $t = 0$ and a later time t necessarily equal the position of the particle at time t? Explain.

4. Determine whether or not the following quantities can be in the same direction for a simple harmonic oscillator: (a) position and velocity, (b) velocity and acceleration, (c) position and acceleration.

5. Can the amplitude A and phase constant ϕ be determined for an oscillator if only the position is specified at $t = 0$? Explain.

6. Describe qualitatively the motion of a block–spring system when the mass of the spring is not neglected.

7. A block is hung on a spring, and the frequency f of the oscillation of the system is measured. The block, a second identical block, and the spring are carried in the Space Shuttle to space. The two blocks are attached to the ends of the spring, and the system is taken out into space on a space walk. The spring is extended, and the system is released to oscillate while floating in space. What is the frequency of oscillation for this system, in terms of f?

8. A block–spring system undergoes simple harmonic motion with amplitude A. Does the total energy change if the mass is doubled but the amplitude is not changed? Do the kinetic and potential energies depend on the mass? Explain.

9. The equations listed in Table 2.2 give position as a function of time, velocity as a function of time, and velocity as function of position for an object moving in a straight line with constant acceleration. The quantity v_{xi} appears in every equation. Do any of these equations apply to an object moving in a straight line with simple harmonic motion? Using a similar format, make a table of equations describing simple harmonic motion. Include equations giving acceleration as a function of time and acceleration as a function of position. State the equations in such a form that they apply equally to a block–spring system, to a pendulum, and to other vibrating systems. What quantity appears in every equation?

10. What happens to the period of a simple pendulum if the pendulum's length is doubled? What happens to the period if the mass of the suspended bob is doubled?

11. A simple pendulum is suspended from the ceiling of a stationary elevator, and the period is determined. Describe the changes, if any, in the period when the elevator (a) accelerates upward, (b) accelerates downward, and (c) moves with constant velocity.

12. Imagine that a pendulum is hanging from the ceiling of a car. As the car coasts freely down a hill, is the equilibrium position of the pendulum vertical? Does the period of oscillation differ from that in a stationary car?

13. A simple pendulum undergoes simple harmonic motion when θ is small. Is the motion periodic when θ is large? How does the period of motion change as θ increases?

14. If a grandfather clock were running slow, how could we adjust the length of the pendulum to correct the time?

15. Will damped oscillations occur for any values of b and k? Explain.

16. Is it possible to have damped oscillations when a system is at resonance? Explain.

17. At resonance, what does the phase constant ϕ equal in Equation 15.35? (*Suggestion:* Compare this equation with the expression for the driving force, which must be in phase with the velocity at resonance.)

18. You stand on the end of a diving board and bounce to set it into oscillation. You find a maximum response, in terms of the amplitude of oscillation of the end of the board, when you bounce at frequency f. You now move to the middle of the board and repeat the experiment. Is the resonance frequency for forced oscillations at this point higher, lower, or the same as f? Why?

19. Some parachutes have holes in them to allow air to move smoothly through the chute. Without the holes, the air gathered under the chute as the parachutist falls is sometimes released from under the edges of the chute alternately and periodically from one side and then the other. Why might this periodic release of air cause a problem?

20. You are looking at a small tree. You do not notice any breeze, and most of the leaves on the tree are motionless. However, one leaf is fluttering back and forth wildly. After you wait for a while, that leaf stops moving and you notice a different leaf moving much more than all the others. Explain what could cause the large motion of one particular leaf.

21. A pendulum bob is made with a sphere filled with water. What would happen to the frequency of vibration of this pendulum if there were a hole in the sphere that allowed the water to leak out slowly?

PROBLEMS

1, 2, 3 = straightforward, intermediate, challenging ☐ = full solution available in the *Student Solutions Manual and Study Guide*

🌐 = coached solution with hints available at http://www.pse6.com 💻 = computer useful in solving problem

▥ = paired numerical and symbolic problems

Note: Neglect the mass of every spring, except in problems 66 and 68.

Section 15.1 Motion of an Object Attached to a Spring

Problems 15, 16, 19, 23, 56, and 62 in Chapter 7 can also be assigned with this section.

1. A ball dropped from a height of 4.00 m makes a perfectly elastic collision with the ground. Assuming no mechanical energy is lost due to air resistance, (a) show that the ensuing motion is periodic and (b) determine the period of the motion. (c) Is the motion simple harmonic? Explain.

Section 15.2 Mathematical Representation of Simple Harmonic Motion

2. In an engine, a piston oscillates with simple harmonic motion so that its position varies according to the expression

$$x = (5.00 \text{ cm})\cos(2t + \pi/6)$$

where x is in centimeters and t is in seconds. At $t = 0$, find (a) the position of the piston, (b) its velocity, and (c) its acceleration. (d) Find the period and amplitude of the motion.

3. The position of a particle is given by the expression $x = (4.00 \text{ m})\cos(3.00 \pi t + \pi)$, where x is in meters and t is in seconds. Determine (a) the frequency and period of the motion, (b) the amplitude of the motion, (c) the phase constant, and (d) the position of the particle at $t = 0.250$ s.

4. (a) A hanging spring stretches by 35.0 cm when an object of mass 450 g is hung on it at rest. In this situation, we define its position as $x = 0$. The object is pulled down an additional 18.0 cm and released from rest to oscillate without friction. What is its position x at a time 84.4 s later? (b) **What If?** A hanging spring stretches by 35.5 cm when an object of mass 440 g is hung on it at rest. We define this new position as $x = 0$. This object is also pulled down an additional 18.0 cm and released from rest to oscillate without friction. Find its position 84.4 s later. (c) Why are the answers to (a) and (b) different by such a large percentage when the data are so similar? Does this circumstance reveal a fundamental difficulty in calculating the future? (d) Find the distance traveled by the vibrating object in part (a). (e) Find the distance traveled by the object in part (b).

5. A particle moving along the x axis in simple harmonic motion starts from its equilibrium position, the origin, at $t = 0$ and moves to the right. The amplitude of its motion is 2.00 cm, and the frequency is 1.50 Hz. (a) Show that the position of the particle is given by

$$x = (2.00 \text{ cm})\sin(3.00\pi t)$$

Determine (b) the maximum speed and the earliest time $(t > 0)$ at which the particle has this speed, (c) the maximum acceleration and the earliest time $(t > 0)$ at which the particle has this acceleration, and (d) the total distance traveled between $t = 0$ and $t = 1.00$ s.

6. The initial position, velocity, and acceleration of an object moving in simple harmonic motion are x_i, v_i, and a_i; the angular frequency of oscillation is ω. (a) Show that the position and velocity of the object for all time can be written as

$$x(t) = x_i \cos \omega t + \left(\frac{v_i}{\omega}\right)\sin \omega t$$

$$v(t) = -x_i\omega \sin \omega t + v_i \cos \omega t$$

(b) If the amplitude of the motion is A, show that

$$v^2 - ax = v_i{}^2 - a_ix_i = \omega^2 A^2$$

7. A simple harmonic oscillator takes 12.0 s to undergo five complete vibrations. Find (a) the period of its motion, (b) the frequency in hertz, and (c) the angular frequency in radians per second.

8. A vibration sensor, used in testing a washing machine, consists of a cube of aluminum 1.50 cm on edge mounted on one end of a strip of spring steel (like a hacksaw blade) that lies in a vertical plane. The mass of the strip is small compared to that of the cube, but the length of the strip is large compared to the size of the cube. The other end of the strip is clamped to the frame of the washing machine, which is not operating. A horizontal force of 1.43 N applied to the cube is required to hold it 2.75 cm away from its equilibrium position. If the cube is released, what is its frequency of vibration?

9. A 7.00-kg object is hung from the bottom end of a vertical spring fastened to an overhead beam. The object is set into vertical oscillations having a period of 2.60 s. Find the force constant of the spring.

10. A piston in a gasoline engine is in simple harmonic motion. If the extremes of its position relative to its center point are ± 5.00 cm, find the maximum velocity and acceleration of the piston when the engine is running at the rate of 3 600 rev/min.

11. A 0.500-kg object attached to a spring with a force constant of 8.00 N/m vibrates in simple harmonic motion with an amplitude of 10.0 cm. Calculate (a) the maximum value of its speed and acceleration, (b) the speed and acceleration when the object is 6.00 cm from the equilibrium position, and (c) the time interval required for the object to move from $x = 0$ to $x = 8.00$ cm.

12. A 1.00-kg glider attached to a spring with a force constant of 25.0 N/m oscillates on a horizontal, frictionless air track. At $t = 0$ the glider is released from rest at $x = -3.00$ cm. (That is, the spring is compressed by 3.00 cm.) Find (a) the period of its motion, (b) the maximum values of its speed and acceleration, and (c) the position, velocity, and acceleration as functions of time.

13. A 1.00-kg object is attached to a horizontal spring. The spring is initially stretched by 0.100 m, and the object is released from rest there. It proceeds to move without friction. The next time the speed of the object is zero is 0.500 s later. What is the maximum speed of the object?

14. A particle that hangs from a spring oscillates with an angular frequency ω. The spring is suspended from the ceiling of an elevator car and hangs motionless (relative to the elevator car) as the car descends at a constant speed v. The car then stops suddenly. (a) With what amplitude does the particle oscillate? (b) What is the equation of motion for the particle? (Choose the upward direction to be positive.)

Section 15.3   Energy of the Simple Harmonic Oscillator

15. A block of unknown mass is attached to a spring with a spring constant of 6.50 N/m and undergoes simple harmonic motion with an amplitude of 10.0 cm. When the block is halfway between its equilibrium position and the end point, its speed is measured to be 30.0 cm/s. Calculate (a) the mass of the block, (b) the period of the motion, and (c) the maximum acceleration of the block.

16. A 200-g block is attached to a horizontal spring and executes simple harmonic motion with a period of 0.250 s. If the total energy of the system is 2.00 J, find (a) the force constant of the spring and (b) the amplitude of the motion.

17. An automobile having a mass of 1 000 kg is driven into a brick wall in a safety test. The bumper behaves like a spring of force constant 5.00×10^6 N/m and compresses 3.16 cm as the car is brought to rest. What was the speed of the car before impact, assuming that no mechanical energy is lost during impact with the wall?

18. A block–spring system oscillates with an amplitude of 3.50 cm. If the spring constant is 250 N/m and the mass of the block is 0.500 kg, determine (a) the mechanical energy of the system, (b) the maximum speed of the block, and (c) the maximum acceleration.

19. A 50.0-g object connected to a spring with a force constant of 35.0 N/m oscillates on a horizontal, frictionless surface with an amplitude of 4.00 cm. Find (a) the total energy of the system and (b) the speed of the object when the position is 1.00 cm. Find (c) the kinetic energy and (d) the potential energy when the position is 3.00 cm.

20. A 2.00-kg object is attached to a spring and placed on a horizontal, smooth surface. A horizontal force of 20.0 N is required to hold the object at rest when it is pulled 0.200 m from its equilibrium position (the origin of the x axis). The object is now released from rest with an initial position of $x_i = 0.200$ m, and it subsequently undergoes simple harmonic oscillations. Find (a) the force constant of the spring, (b) the frequency of the oscillations, and (c) the maximum speed of the object. Where does this maximum speed occur? (d) Find the maximum acceleration of the object. Where does it occur? (e) Find the total energy of the oscillating system. Find (f) the speed and (g) the acceleration of the object when its position is equal to one third of the maximum value.

21. The amplitude of a system moving in simple harmonic motion is doubled. Determine the change in (a) the total energy, (b) the maximum speed, (c) the maximum acceleration, and (d) the period.

22. A 65.0-kg bungee jumper steps off a bridge with a light bungee cord tied to herself and to the bridge (Figure P15.22). The unstretched length of the cord is 11.0 m. She reaches the bottom of her motion 36.0 m below the bridge before bouncing back. Her motion can be separated into an 11.0-m free fall and a 25.0-m section of simple harmonic oscillation. (a) For what time interval is she in free fall? (b) Use the principle of conservation of energy to find the spring constant of the bungee cord. (c) What is the location of the equilibrium point where the spring force balances the gravitational force acting on the jumper? Note that this point is taken as the origin in our mathematical description of simple harmonic oscillation. (d) What is the angular frequency of the oscillation? (e) What time interval is required for the cord to stretch by 25.0 m? (f) What is the total time interval for the entire 36.0-m drop?

23. A particle executes simple harmonic motion with an amplitude of 3.00 cm. At what position does its speed equal half its maximum speed?

24. A cart attached to a spring with constant 3.24 N/m vibrates with position given by $x = (5.00 \text{ cm}) \cos(3.60t \text{ rad/s})$. (a) During the first cycle, for $0 < t < 1.75$ s, just when is the system's potential energy changing most rapidly into kinetic energy? (b) What is the maximum rate of energy transformation?

Section 15.4 Comparing Simple Harmonic Motion with Uniform Circular Motion

25. While riding behind a car traveling at 3.00 m/s, you notice that one of the car's tires has a small hemispherical bump on its rim, as in Figure P15.25. (a) Explain why the bump, from your viewpoint behind the car, executes simple harmonic motion. (b) If the radii of the car's tires are 0.300 m, what is the bump's period of oscillation?

Figure P15.25

26. Consider the simplified single-piston engine in Figure P15.26. If the wheel rotates with constant angular speed, explain why the piston rod oscillates in simple harmonic motion.

Figure P15.26

Section 15.5 The Pendulum

Problem 60 in Chapter 1 can also be assigned with this section.

Figure P15.22 Problems 22 and 58.

27. A man enters a tall tower, needing to know its height. He notes that a long pendulum extends from the ceiling almost to the floor and that its period is 12.0 s. (a) How tall is the tower? (b) **What If?** If this pendulum is taken to the Moon, where the free-fall acceleration is 1.67 m/s^2, what is its period there?

28. A "seconds pendulum" is one that moves through its equilibrium position once each second. (The period of the pendulum is precisely 2 s.) The length of a seconds pendulum is 0.992 7 m at Tokyo, Japan and 0.994 2 m at Cambridge, England. What is the ratio of the free-fall accelerations at these two locations?

29. A rigid steel frame above a street intersection supports standard traffic lights, each of which is hinged to hang immediately below the frame. A gust of wind sets a light swinging in a vertical plane. Find the order of magnitude of its period. State the quantities you take as data and their values.

30. The angular position of a pendulum is represented by the equation $\theta = (0.320 \text{ rad})\cos \omega t$, where θ is in radians and $\omega = 4.43$ rad/s. Determine the period and length of the pendulum.

31. A simple pendulum has a mass of 0.250 kg and a length of 1.00 m. It is displaced through an angle of 15.0° and then released. What are (a) the maximum speed, (b) the maximum angular acceleration, and (c) the maximum restoring force? **What If?** Solve this problem by using the simple harmonic motion model for the motion of the pendulum, and then solve the problem more precisely by using more general principles.

32. **Review problem.** A simple pendulum is 5.00 m long. (a) What is the period of small oscillations for this pendulum if it is located in an elevator accelerating upward at 5.00 m/s^2? (b) What is its period if the elevator is accelerating downward at 5.00 m/s^2? (c) What is the period of this pendulum if it is placed in a truck that is accelerating horizontally at 5.00 m/s^2?

33. A particle of mass m slides without friction inside a hemispherical bowl of radius R. Show that, if it starts from rest with a small displacement from equilibrium, the particle moves in simple harmonic motion with an angular frequency equal to that of a simple pendulum of length R. That is, $\omega = \sqrt{g/R}$.

34. A small object is attached to the end of a string to form a simple pendulum. The period of its harmonic motion is measured for small angular displacements and three lengths, each time clocking the motion with a stopwatch for 50 oscillations. For lengths of 1.000 m, 0.750 m, and 0.500 m, total times of 99.8 s, 86.6 s, and 71.1 s are measured for 50 oscillations. (a) Determine the period of motion for each length. (b) Determine the mean value of g obtained from these three independent measurements, and compare it with the accepted value. (c) Plot T^2 versus L, and obtain a value for g from the slope of your best-fit straight-line graph. Compare this value with that obtained in part (b).

35. A physical pendulum in the form of a planar body moves in simple harmonic motion with a frequency of 0.450 Hz. If the pendulum has a mass of 2.20 kg and the pivot is located 0.350 m from the center of mass, determine the moment of inertia of the pendulum about the pivot point.

36. A very light rigid rod with a length of 0.500 m extends straight out from one end of a meter stick. The stick is suspended from a pivot at the far end of the rod and is set into oscillation. (a) Determine the period of oscillation. *Suggestion*: Use the parallel-axis theorem from Section 10.5. (b) By what percentage does the period differ from the period of a simple pendulum 1.00 m long?

37. Consider the physical pendulum of Figure 15.18. (a) If its moment of inertia about an axis passing through its center of mass and parallel to the axis passing through its pivot point is I_{CM}, show that its period is

$$T = 2\pi \sqrt{\frac{I_{CM} + md^2}{mgd}}$$

where d is the distance between the pivot point and center of mass. (b) Show that the period has a minimum value when d satisfies $md^2 = I_{CM}$.

38. A torsional pendulum is formed by taking a meter stick of mass 2.00 kg, and attaching to its center a wire. With its upper end clamped, the vertical wire supports the stick as the stick turns in a horizontal plane. If the resulting period is 3.00 minutes, what is the torsion constant for the wire?

39. A clock balance wheel (Fig. P15.39) has a period of oscillation of 0.250 s. The wheel is constructed so that its mass of 20.0 g is concentrated around a rim of radius 0.500 cm. What are (a) the wheel's moment of inertia and (b) the torsion constant of the attached spring?

Figure P15.39

George Semple

Section 15.6 Damped Oscillations

40. Show that the time rate of change of mechanical energy for a damped, undriven oscillator is given by $dE/dt = -bv^2$ and hence is always negative. Proceed as follows: Differentiate the expression for the mechanical energy of an oscillator, $E = \frac{1}{2}mv^2 + \frac{1}{2}kx^2$, and use Equation 15.31.

41. A pendulum with a length of 1.00 m is released from an initial angle of 15.0°. After 1 000 s, its amplitude has been reduced by friction to 5.50°. What is the value of $b/2m$?

42. Show that Equation 15.32 is a solution of Equation 15.31 provided that $b^2 < 4mk$.

43. A 10.6-kg object oscillates at the end of a vertical spring that has a spring constant of 2.05×10^4 N/m. The effect of air resistance is represented by the damping coefficient $b = 3.00$ N·s/m. (a) Calculate the frequency of the damped oscillation. (b) By what percentage does the amplitude of the oscillation decrease in each cycle? (c) Find the time interval that elapses while the energy of the system drops to 5.00% of its initial value.

Section 15.7 Forced Oscillations

44. The front of her sleeper wet from teething, a baby rejoices in the day by crowing and bouncing up and down in her crib. Her mass is 12.5 kg, and the crib mattress can be modeled as a light spring with force constant 4.30 kN/m. (a) The baby soon learns to bounce with maximum amplitude and minimum effort by bending her knees at what frequency? (b) She learns to use the mattress as a trampoline—losing contact with it for part of each cycle—when her amplitude exceeds what value?

45. A 2.00-kg object attached to a spring moves without friction and is driven by an external force given by $F = (3.00 \text{ N})\sin(2\pi t)$. If the force constant of the spring is 20.0 N/m, determine (a) the period and (b) the amplitude of the motion.

46. Considering an undamped, forced oscillator ($b = 0$), show that Equation 15.35 is a solution of Equation 15.34, with an amplitude given by Equation 15.36.

47. A weight of 40.0 N is suspended from a spring that has a force constant of 200 N/m. The system is undamped and is subjected to a harmonic driving force of frequency 10.0 Hz, resulting in a forced-motion amplitude of 2.00 cm. Determine the maximum value of the driving force.

48. Damping is negligible for a 0.150-kg object hanging from a light 6.30-N/m spring. A sinusoidal force with an amplitude of 1.70 N drives the system. At what frequency will the force make the object vibrate with an amplitude of 0.440 m?

49. You are a research biologist. You take your emergency pager along to a fine restaurant. You switch the small pager to vibrate instead of beep, and you put it into a side pocket of your suit coat. The arm of your chair presses the light cloth against your body at one spot. Fabric with a length of 8.21 cm hangs freely below that spot, with the pager at the bottom. A coworker urgently needs instructions and calls you from your laboratory. The motion of the pager makes the hanging part of your coat swing back and forth with remarkably large amplitude. The waiter and nearby diners notice immediately and fall silent. Your daughter pipes up and says, "Daddy, look! Your cockroaches must have gotten out again!" Find the frequency at which your pager vibrates.

50. Four people, each with a mass of 72.4 kg, are in a car with a mass of 1 130 kg. An earthquake strikes. The driver manages to pull off the road and stop, as the vertical oscillations of the ground surface make the car bounce up and down on its suspension springs. When the frequency of the shaking is 1.80 Hz, the car exhibits a maximum amplitude of vibration. The earthquake ends, and the four people leave the car as fast as they can. By what distance does the car's undamaged suspension lift the car body as the people get out?

Additional Problems

51. A small ball of mass M is attached to the end of a uniform rod of equal mass M and length L that is pivoted at the top (Fig. P15.51). (a) Determine the tensions in the rod at the pivot and at the point P when the system is stationary. (b) Calculate the period of oscillation for small displacements from equilibrium, and determine this period for $L = 2.00$ m. (*Suggestions:* Model the object at the end of the rod as a particle and use Eq. 15.28.)

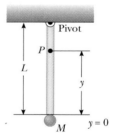

Figure P15.51

52. An object of mass $m_1 = 9.00$ kg is in equilibrium while connected to a light spring of constant $k = 100$ N/m that is fastened to a wall as shown in Figure P15.52a. A second object, $m_2 = 7.00$ kg, is slowly pushed up against m_1, compressing the spring by the amount $A = 0.200$ m, (see Figure P15.52b). The system is then released, and both objects start moving to the right on the frictionless surface. (a) When m_1 reaches the equilibrium point, m_2 loses contact with m_1 (see Fig. P15.5c) and moves to the right with speed v. Determine the value of v. (b) How far apart are the objects when the spring is fully stretched for the first time (D in Fig. P15.52d)? (*Suggestion:* First determine the period of oscillation and the amplitude of the m_1–spring system after m_2 loses contact with m_1.)

(a)

(b)

$\leftarrow A \rightarrow$

(c)

v

(d)

$\leftarrow D \rightarrow$

Figure P15.52

53. A large block P executes horizontal simple harmonic motion as it slides across a frictionless surface with a frequency $f = 1.50$ Hz. Block B rests on it, as shown in Figure P15.53, and the coefficient of static friction between the two is $\mu_s = 0.600$. What maximum amplitude of oscillation can the system have if block B is not to slip?

Figure P15.53 Problems 53 and 54.

54. A large block P executes horizontal simple harmonic motion as it slides across a frictionless surface with a frequency f. Block B rests on it, as shown in Figure P15.53, and the coefficient of static friction between the two is μ_s. What maximum amplitude of oscillation can the system have if the upper block is not to slip?

55. The mass of the deuterium molecule (D_2) is twice that of the hydrogen molecule (H_2). If the vibrational frequency of H_2 is 1.30×10^{14} Hz, what is the vibrational frequency of D_2? Assume that the "spring constant" of attracting forces is the same for the two molecules.

56. A solid sphere (radius $= R$) rolls without slipping in a cylindrical trough (radius $= 5R$) as shown in Figure P15.56. Show that, for small displacements from equilibrium perpendicular to the length of the trough, the sphere executes simple harmonic motion with a period $T = 2\pi\sqrt{28R/5g}$.

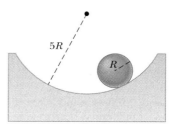

$5R$

R

Figure P15.56

57. A light, cubical container of volume a^3 is initially filled with a liquid of mass density ρ. The cube is initially supported by a light string to form a simple pendulum of length L_i, measured from the center of mass of the filled container, where $L_i \gg a$. The liquid is allowed to flow from the bottom of the container at a constant rate (dM/dt). At any time t, the level of the fluid in the container is h and the length of the pendulum is L (measured relative to the instantaneous center of mass). (a) Sketch the apparatus and label the dimensions a, h, L_i, and L. (b) Find the time rate of change of the period as a function of time t. (c) Find the period as a function of time.

58. After a thrilling plunge, bungee-jumpers bounce freely on the bungee cord through many cycles (Fig. P15.22). After the first few cycles, the cord does not go slack. Your little brother can make a pest of himself by figuring out the mass of each person, using a proportion which you set up by solving this problem: An object of mass m is oscillating freely on a vertical spring with a period T. An object of unknown mass m' on the same spring oscillates with a period T'. Determine (a) the spring constant and (b) the unknown mass.

59. A pendulum of length L and mass M has a spring of force constant k connected to it at a distance h below its point of suspension (Fig. P15.59). Find the frequency of vibration

h

θ

L

k

M

Figure P15.59

of the system for small values of the amplitude (small θ). Assume the vertical suspension of length L is rigid, but ignore its mass.

60. A particle with a mass of 0.500 kg is attached to a spring with a force constant of 50.0 N/m. At time $t = 0$ the particle has its maximum speed of 20.0 m/s and is moving to the left. (a) Determine the particle's equation of motion, specifying its position as a function of time. (b) Where in the motion is the potential energy three times the kinetic energy? (c) Find the length of a simple pendulum with the same period. (d) Find the minimum time interval required for the particle to move from $x = 0$ to $x = 1.00$ m.

61. A horizontal plank of mass m and length L is pivoted at one end. The plank's other end is supported by a spring of force constant k (Fig P15.61). The moment of inertia of the plank about the pivot is $\frac{1}{3} mL^2$. The plank is displaced by a small angle θ from its horizontal equilibrium position and released. (a) Show that it moves with simple harmonic motion with an angular frequency $\omega = \sqrt{3k/m}$. (b) Evaluate the frequency if the mass is 5.00 kg and the spring has a force constant of 100 N/m.

Figure P15.61

62. **Review problem.** A particle of mass 4.00 kg is attached to a spring with a force constant of 100 N/m. It is oscillating on a horizontal frictionless surface with an amplitude of 2.00 m. A 6.00-kg object is dropped vertically on top of the 4.00-kg object as it passes through its equilibrium point. The two objects stick together. (a) By how much does the amplitude of the vibrating system change as a result of the collision? (b) By how much does the period change? (c) By how much does the energy change? (d) Account for the change in energy.

63. A simple pendulum with a length of 2.23 m and a mass of 6.74 kg is given an initial speed of 2.06 m/s at its equilibrium position. Assume it undergoes simple harmonic motion, and determine its (a) period, (b) total energy, and (c) maximum angular displacement.

64. **Review problem.** One end of a light spring with force constant 100 N/m is attached to a vertical wall. A light string is tied to the other end of the horizontal spring. The string changes from horizontal to vertical as it passes over a solid pulley of diameter 4.00 cm. The pulley is free to turn on a fixed smooth axle. The vertical section of the string supports a 200-g object. The string does not slip at its contact with the pulley. Find the frequency of oscillation of the object if the mass of the pulley is (a) negligible, (b) 250 g, and (c) 750 g.

65. People who ride motorcycles and bicycles learn to look out for bumps in the road, and especially for *washboarding*, a condition in which many equally spaced ridges are worn into the road. What is so bad about washboarding? A motorcycle has several springs and shock absorbers in its suspension, but you can model it as a single spring supporting a block. You can estimate the force constant by thinking about how far the spring compresses when a big biker sits down on the seat. A motorcyclist traveling at highway speed must be particularly careful of washboard bumps that are a certain distance apart. What is the order of magnitude of their separation distance? State the quantities you take as data and the values you measure or estimate for them.

66. A block of mass M is connected to a spring of mass m and oscillates in simple harmonic motion on a horizontal, frictionless track (Fig. P15.66). The force constant of the spring is k and the equilibrium length is ℓ. Assume that all portions of the spring oscillate in phase and that the velocity of a segment dx is proportional to the distance x from the fixed end; that is, $v_x = (x/\ell)v$. Also, note that the mass of a segment of the spring is $dm = (m/\ell)\,dx$. Find (a) the kinetic energy of the system when the block has a speed v and (b) the period of oscillation.

Figure P15.66

67. A ball of mass m is connected to two rubber bands of length L, each under tension T, as in Figure P15.67. The ball is displaced by a small distance y perpendicular to the length of the rubber bands. Assuming that the tension does not change, show that (a) the restoring force is $-(2T/L)y$ and (b) the system exhibits simple harmonic motion with an angular frequency $\omega = \sqrt{2T/mL}$.

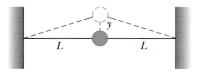

Figure P15.67

68. When a block of mass M, connected to the end of a spring of mass $m_s = 7.40$ g and force constant k, is set into simple harmonic motion, the period of its motion is

$$T = 2\pi\sqrt{\frac{M + (m_s/3)}{k}}$$

A two-part experiment is conducted with the use of blocks of various masses suspended vertically from the spring, as shown in Figure P15.68. (a) Static extensions of 17.0, 29.3, 35.3, 41.3, 47.1, and 49.3 cm are measured for M values of 20.0, 40.0, 50.0, 60.0, 70.0, and 80.0 g, respectively. Construct a graph of Mg versus x, and perform a linear least-squares fit to the data. From the slope of your graph, determine a value for k for this spring. (b) The system is now set into simple harmonic motion, and periods are measured with a stopwatch. With $M = 80.0$ g, the total time for 10 oscillations is measured to be 13.41 s. The experiment is repeated with M values of 70.0, 60.0, 50.0, 40.0, and 20.0 g, with corresponding times for 10 oscillations of 12.52, 11.67, 10.67, 9.62, and 7.03 s. Compute the experimental value for T from each of these measurements. Plot a graph of T^2 versus M, and determine a value for k from the slope of the linear least-squares fit through the data points. Compare this value of k with that obtained in part (a). (c) Obtain a value for m_s from your graph and compare it with the given value of 7.40 g.

Figure P15.68

69. A smaller disk of radius r and mass m is attached rigidly to the face of a second larger disk of radius R and mass M as shown in Figure P15.69. The center of the small disk is located at the edge of the large disk. The large disk is mounted at its center on a frictionless axle. The assembly is rotated through a small angle θ from its equilibrium position and released. (a) Show that the speed of the center of the small disk as it passes through the equilibrium position is

$$v = 2\left[\frac{Rg(1 - \cos\theta)}{(M/m) + (r/R)^2 + 2}\right]^{1/2}$$

(b) Show that the period of the motion is

$$T = 2\pi\left[\frac{(M + 2m)R^2 + mr^2}{2mgR}\right]^{1/2}$$

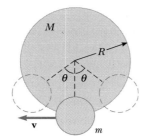

Figure P15.69

70. Consider a damped oscillator as illustrated in Figures 15.21 and 15.22. Assume the mass is 375 g, the spring constant is 100 N/m, and $b = 0.100$ N·s/m. (a) How long does it takes for the amplitude to drop to half its initial value? (b) **What If?** How long does it take for the mechanical energy to drop to half its initial value? (c) Show that, in general, the fractional rate at which the amplitude decreases in a damped harmonic oscillator is half the fractional rate at which the mechanical energy decreases.

71. A block of mass m is connected to two springs of force constants k_1 and k_2 as shown in Figures P15.71a and P15.71b. In each case, the block moves on a frictionless table after it is displaced from equilibrium and released. Show that in the two cases the block exhibits simple harmonic motion with periods

(a)　　$T = 2\pi\sqrt{\dfrac{m(k_1 + k_2)}{k_1k_2}}$

(b)　　$T = 2\pi\sqrt{\dfrac{m}{k_1 + k_2}}$

(a)

(b)

Figure P15.71

72. A lobsterman's buoy is a solid wooden cylinder of radius r and mass M. It is weighted at one end so that it floats upright in calm sea water, having density ρ. A passing shark tugs on the slack rope mooring the buoy to a lobster trap, pulling the buoy down a distance x from its equilibrium position and releasing it. Show that the buoy will execute simple harmonic motion if the resistive effects of the water are neglected, and determine the period of the oscillations.

73. Consider a bob on a light stiff rod, forming a simple pendulum of length $L = 1.20$ m. It is displaced from the vertical by an angle θ_{max} and then released. Predict the subsequent angular positions if θ_{max} is small or if it is large. Proceed as follows: Set up and carry out a numerical method to integrate the equation of motion for the simple pendulum:

$$\frac{d^2\theta}{dt^2} = -\frac{g}{L}\sin\theta$$

Take the initial conditions to be $\theta = \theta_{max}$ and $d\theta/dt = 0$ at $t = 0$. On one trial choose $\theta_{max} = 5.00°$, and on another trial take $\theta_{max} = 100°$. In each case find the position θ as a function of time. Using the same values of θ_{max}, compare your results for θ with those obtained from $\theta(t) = \theta_{max} \cos \omega t$. How does the period for the large value of θ_{max} compare with that for the small value of θ_{max}? *Note:* Using the Euler method to solve this differential equation, you may find that the amplitude tends to increase with time. The fourth-order Runge–Kutta method would be a better choice to solve the differential equation. However, if you choose Δt small enough, the solution using Euler's method can still be good.

74. Your thumb squeaks on a plate you have just washed. Your sneakers often squeak on the gym floor. Car tires squeal when you start or stop abruptly. You can make a goblet sing by wiping your moistened finger around its rim. As you slide it across the table, a Styrofoam cup may not make much sound, but it makes the surface of some water inside it dance in a complicated resonance vibration. When chalk squeaks on a blackboard, you can see that it makes a row of regularly spaced dashes. As these examples suggest, vibration commonly results when friction acts on a moving elastic object. The oscillation is not simple harmonic motion, but is called *stick-and-slip*. This problem models stick-and-slip motion.

A block of mass m is attached to a fixed support by a horizontal spring with force constant k and negligible mass (Fig. P15.74). Hooke's law describes the spring both in extension and in compression. The block sits on a long horizontal board, with which it has coefficient of static friction μ_s and a smaller coefficient of kinetic friction μ_k. The board moves to the right at constant speed v. Assume that the block spends most of its time sticking to the board and moving to the right, so that the speed v is small in comparison to the average speed the block has as it slips back toward the left. (a) Show that the maximum extension of the spring from its unstressed position is very nearly given by $\mu_s mg/k$. (b) Show that the block oscillates around an equilibrium position at which the spring is stretched by $\mu_k mg/k$. (c) Graph the block's position versus time. (d) Show that the amplitude of the block's motion is

$$A = \frac{(\mu_s - \mu_k)\, mg}{k}$$

(e) Show that the period of the block's motion is

$$T = \frac{2(\mu_s - \mu_k)\, mg}{vk} + \pi\sqrt{\frac{m}{k}}$$

(f) Evaluate the frequency of the motion if $\mu_s = 0.400$, $\mu_k = 0.250$, $m = 0.300$ kg, $k = 12.0$ N/m, and $v = 2.40$ cm/s. (g) **What If?** What happens to the frequency if the mass increases? (h) If the spring constant increases? (i) If the speed of the board increases? (j) If the coefficient of static friction increases relative to the coefficient of kinetic friction? Note that it is the excess of static over kinetic friction that is important for the vibration. "The squeaky wheel gets the grease" because even a viscous fluid cannot exert a force of static friction.

Figure P15.74

75. **Review problem.** Imagine that a hole is drilled through the center of the Earth to the other side. An object of mass m at a distance r from the center of the Earth is pulled toward the center of the Earth only by the mass within the sphere of radius r (the reddish region in Fig. P15.75). (a) Write Newton's law of gravitation for an object at the distance r from the center of the Earth, and show that the force on it is of Hooke's law form, $F = -kr$, where the effective force constant is $k = (4/3)\pi\rho Gm$. Here ρ is the density of the Earth, assumed uniform, and G is the gravitational constant. (b) Show that a sack of mail dropped into the hole will execute simple harmonic motion if it moves without friction. When will it arrive at the other side of the Earth?

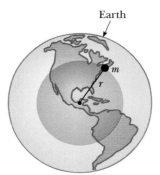

Figure P15.75

Answers to Quick Quizzes

15.1 (d). From its maximum positive position to the equilibrium position, the block travels a distance A. It then goes an equal distance past the equilibrium position to its maximum negative position. It then repeats these two motions in the reverse direction to return to its original position and complete one cycle.

15.2 (f). The object is in the region $x < 0$, so the position is negative. Because the object is moving back toward the origin in this region, the velocity is positive.

15.3 (a). The amplitude is larger because the curve for Object B shows that the displacement from the origin (the vertical axis on the graph) is larger. The frequency is larger for Object B because there are more oscillations per unit time interval.

15.4 (a). The velocity is positive, as in Quick Quiz 15.2. Because the spring is pulling the object toward equilibrium from the negative x region, the acceleration is also positive.

15.5 (b). According to Equation 15.13, the period is proportional to the square root of the mass.

15.6 (c). The amplitude of the simple harmonic motion is the same as the radius of the circular motion. The initial position of the object in its circular motion is π radians from the positive x axis.

15.7 (a). With a longer length, the period of the pendulum will increase. Thus, it will take longer to execute each swing, so that each second according to the clock will take longer than an actual second—the clock will run *slow*.

15.8 (a). At the top of the mountain, the value of g is less than that at sea level. As a result, the period of the pendulum will increase and the clock will run slow.

15.9 (a). If your goal is simply to stop the bounce from an absorbed shock as rapidly as possible, you should critically damp the suspension. Unfortunately, the stiffness of this design makes for an uncomfortable ride. If you underdamp the suspension, the ride is more comfortable but the car bounces. If you overdamp the suspension, the wheel is displaced from its equilibrium position longer than it should be. (For example, after hitting a bump, the spring stays compressed for a short time and the wheel does not quickly drop back down into contact with the road after the wheel is past the bump—a dangerous situation.) Because of all these considerations, automotive engineers usually design suspensions to be slightly underdamped. This allows the suspension to absorb a shock rapidly (minimizing the roughness of the ride) and then return to equilibrium after only one or two noticeable oscillations.

Chapter 16

Wave Motion

▲ The rich sound of a piano is due to waves on strings that are under tension. Many such strings can be seen in this photograph. Waves also travel on the soundboard, which is visible below the strings. In this chapter, we study the fundamental principles of wave phenomena. (Kathy Ferguson Johnson/PhotoEdit/PictureQuest)

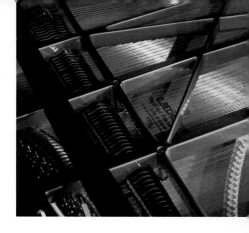

Most of us experienced waves as children when we dropped a pebble into a pond. At the point where the pebble hits the water's surface, waves are created. These waves move outward from the creation point in expanding circles until they reach the shore. If you were to examine carefully the motion of a beach ball floating on the disturbed water, you would see that the ball moves vertically and horizontally about its original position but does not undergo any net displacement away from or toward the point where the pebble hit the water. The small elements of water in contact with the beach ball, as well as all the other water elements on the pond's surface, behave in the same way. That is, the water *wave* moves from the point of origin to the shore, but the water is not carried with it.

The world is full of waves, the two main types being *mechanical* waves and *electromagnetic* waves. In the case of mechanical waves, some physical medium is being disturbed—in our pebble and beach ball example, elements of water are disturbed. Electromagnetic waves do not require a medium to propagate; some examples of electromagnetic waves are visible light, radio waves, television signals, and x-rays. Here, in this part of the book, we study only mechanical waves.

The wave concept is abstract. When we observe what we call a water wave, what we see is a rearrangement of the water's surface. Without the water, there would be no wave. A wave traveling on a string would not exist without the string. Sound waves could not travel from one point to another if there were no air molecules between the two points. With mechanical waves, what we interpret as a wave corresponds to the propagation of a disturbance through a medium.

Considering further the beach ball floating on the water, note that we have caused the ball to move at one point in the water by dropping a pebble at another location. The ball has gained kinetic energy from our action, so energy must have transferred from the point at which we drop the pebble to the position of the ball. This is a central feature of wave motion—*energy* is transferred over a distance, but *matter* is not.

All waves carry energy, but the amount of energy transmitted through a medium and the mechanism responsible for that transport of energy differ from case to case. For instance, the power of ocean waves during a storm is much greater than the power of sound waves generated by a single human voice.

16.1   Propagation of a Disturbance

In the introduction, we alluded to the essence of wave motion—the transfer of energy through space without the accompanying transfer of matter. In the list of energy transfer mechanisms in Chapter 7, two mechanisms depend on waves—mechanical waves and electromagnetic radiation. By contrast, in another mechanism—matter transfer—the energy transfer is accompanied by a movement of matter through space.

All mechanical waves require (1) some source of disturbance, (2) a medium that can be disturbed, and (3) some physical mechanism through which elements of the medium can influence each other. One way to demonstrate wave

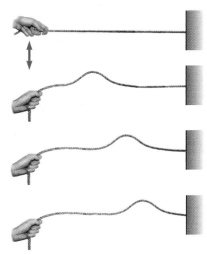

Figure 16.1 A pulse traveling down a stretched rope. The shape of the pulse is approximately unchanged as it travels along the rope.

motion is to flick one end of a long rope that is under tension and has its opposite end fixed, as shown in Figure 16.1. In this manner, a single bump (called a *pulse*) is formed and travels along the rope with a definite speed. Figure 16.1 represents four consecutive "snapshots" of the creation and propagation of the traveling pulse. The rope is the medium through which the pulse travels. The pulse has a definite height and a definite speed of propagation along the medium (the rope). As we shall see later, the properties of this particular medium that determine the speed of the disturbance are the tension in the rope and its mass per unit length. The shape of the pulse changes very little as it travels along the rope.[1]

We shall first focus our attention on a pulse traveling through a medium. Once we have explored the behavior of a pulse, we will then turn our attention to a *wave*, which is a *periodic* disturbance traveling through a medium. We created a pulse on our rope by flicking the end of the rope once, as in Figure 16.1. If we were to move the end of the rope up and down repeatedly, we would create a traveling wave, which has characteristics that a pulse does not have. We shall explore these characteristics in Section 16.2.

As the pulse in Figure 16.1 travels, each disturbed element of the rope moves in a direction *perpendicular* to the direction of propagation. Figure 16.2 illustrates this point for one particular element, labeled *P*. Note that no part of the rope ever moves in the direction of the propagation.

> A traveling wave or pulse that causes the elements of the disturbed medium to move perpendicular to the direction of propagation is called a **transverse wave.**

Compare this with another type of pulse—one moving down a long, stretched spring, as shown in Figure 16.3. The left end of the spring is pushed briefly to the right and then pulled briefly to the left. This movement creates a sudden compression of a region of the coils. The compressed region travels along the spring (to the right in Figure 16.3). The compressed region is followed by a region where the coils are extended. Notice that the direction of the displacement of the coils is *parallel* to the direction of propagation of the compressed region.

> A traveling wave or pulse that causes the elements of the medium to move parallel to the direction of propagation is called a **longitudinal wave.**

Figure 16.2 A transverse pulse traveling on a stretched rope. The direction of motion of any element *P* of the rope (blue arrows) is perpendicular to the direction of propagation (red arrows).

Sound waves, which we shall discuss in Chapter 17, are another example of longitudinal waves. The disturbance in a sound wave is a series of high-pressure and low-pressure regions that travel through air.

Some waves in nature exhibit a combination of transverse and longitudinal displacements. Surface water waves are a good example. When a water wave travels on the surface of deep water, elements of water at the surface move in nearly circular paths, as shown in Figure 16.4. Note that the disturbance has both transverse and longitudinal

Figure 16.3 A longitudinal pulse along a stretched spring. The displacement of the coils is parallel to the direction of the propagation.

---

[1]   In reality, the pulse changes shape and gradually spreads out during the motion. This effect is called *dispersion* and is common to many mechanical waves as well as to electromagnetic waves. We do not consider dispersion in this chapter.

Active Figure 16.4 The motion of water elements on the surface of deep water in which a wave is propagating is a combination of transverse and longitudinal displacements, with the result that elements at the surface move in nearly circular paths. Each element is displaced both horizontally and vertically from its equilibrium position.

At the Active Figures link at http://www.pse6.com, *you can observe the displacement of water elements at the surface of the moving waves.*

components. The transverse displacements seen in Figure 16.4 represent the variations in vertical position of the water elements. The longitudinal displacement can be explained as follows: as the wave passes over the water's surface, water elements at the highest points move in the direction of propagation of the wave, whereas elements at the lowest points move in the direction opposite the propagation.

The three-dimensional waves that travel out from points under the Earth's surface along a fault at which an earthquake occurs are of both types—transverse and longitudinal. The longitudinal waves are the faster of the two, traveling at speeds in the range of 7 to 8 km/s near the surface. These are called **P waves** (with "P" standing for *primary*) because they travel faster than the transverse waves and arrive at a seismograph (a device used to detect waves due to earthquakes) first. The slower transverse waves, called **S waves** (with "S" standing for *secondary*), travel through the Earth at 4 to 5 km/s near the surface. By recording the time interval between the arrivals of these two types of waves at a seismograph, the distance from the seismograph to the point of origin of the waves can be determined. A single measurement establishes an imaginary sphere centered on the seismograph, with the radius of the sphere determined by the difference in arrival times of the P and S waves. The origin of the waves is located somewhere on that sphere. The imaginary spheres from three or more monitoring stations located far apart from each other intersect at one region of the Earth, and this region is where the earthquake occurred.

Consider a pulse traveling to the right on a long string, as shown in Figure 16.5. Figure 16.5a represents the shape and position of the pulse at time $t = 0$. At this time, the shape of the pulse, whatever it may be, can be represented by some mathematical function which we will write as $y(x, 0) = f(x)$. This function describes the transverse position y of the element of the string located at each value of x at time $t = 0$. Because the speed of the pulse is v, the pulse has traveled to the right a distance vt at the time t (Fig. 16.5b). We assume that the shape of the pulse does not change with time. Thus, at time t, the shape of the pulse is the same as it was at time $t = 0$, as in Figure 16.5a.

(a) Pulse at $t = 0$ (b) Pulse at time t

Figure 16.5 A one-dimensional pulse traveling to the right with a speed v. (a) At $t = 0$, the shape of the pulse is given by $y = f(x)$. (b) At some later time t, the shape remains unchanged and the vertical position of an element of the medium any point P is given by $y = f(x - vt)$.

Consequently, an element of the string at x at this time has the same y position as an element located at $x - vt$ had at time $t = 0$:

$$y(x, t) = y(x - vt, 0)$$

In general, then, we can represent the transverse position y for all positions and times, measured in a stationary frame with the origin at O, as

Pulse traveling to the right

$$y(x, t) = f(x - vt) \qquad (16.1)$$

Similarly, if the pulse travels to the left, the transverse positions of elements of the string are described by

Pulse traveling to the left

$$y(x, t) = f(x + vt) \qquad (16.2)$$

The function y, sometimes called the **wave function,** depends on the two variables x and t. For this reason, it is often written $y(x, t)$, which is read "y as a function of x and t."

It is important to understand the meaning of y. Consider an element of the string at point P, identified by a particular value of its x coordinate. As the pulse passes through P, the y coordinate of this element increases, reaches a maximum, and then decreases to zero. **The wave function $y(x, t)$ represents the y coordinate—the transverse position—of any element located at position x at any time t.** Furthermore, if t is fixed (as, for example, in the case of taking a snapshot of the pulse), then the wave function $y(x)$, sometimes called the **waveform,** defines a curve representing the actual geometric shape of the pulse at that time.

Quick Quiz 16.1 In a long line of people waiting to buy tickets, the first person leaves and a pulse of motion occurs as people step forward to fill the gap. As each person steps forward, the gap moves through the line. Is the propagation of this gap (a) transverse (b) longitudinal?

Quick Quiz 16.2 Consider the "wave" at a baseball game: people stand up and shout as the wave arrives at their location, and the resultant pulse moves around the stadium. Is this wave (a) transverse (b) longitudinal?

Example 16.1 A Pulse Moving to the Right

A pulse moving to the right along the x axis is represented by the wave function

$$y(x, t) = \frac{2}{(x - 3.0t)^2 + 1}$$

where x and y are measured in centimeters and t is measured in seconds. Plot the wave function at $t = 0$, $t = 1.0$ s, and $t = 2.0$ s.

Solution First, note that this function is of the form $y = f(x - vt)$. By inspection, we see that the wave speed is $v = 3.0$ cm/s. Furthermore, the maximum value of y is given by $A = 2.0$ cm. (We find the maximum value of the function representing y by letting $x - 3.0t = 0$.) The wave function expressions are

$$y(x, 0) = \frac{2}{x^2 + 1} \qquad \text{at } t = 0$$

$$y(x, 1.0) = \frac{2}{(x - 3.0)^2 + 1} \qquad \text{at } t = 1.0 \text{ s}$$

$$y(x, 2.0) = \frac{2}{(x - 6.0)^2 + 1} \qquad \text{at } t = 2.0 \text{ s}$$

We now use these expressions to plot the wave function versus x at these times. For example, let us evaluate $y(x, 0)$ at $x = 0.50$ cm:

$$y(0.50, 0) = \frac{2}{(0.50)^2 + 1} = 1.6 \text{ cm}$$

Likewise, at $x = 1.0$ cm, $y(1.0, 0) = 1.0$ cm, and at $y = 2.0$ cm, $y(2.0, 0) = 0.40$ cm. Continuing this procedure for other values of x yields the wave function shown in Figure 16.6a. In a similar manner, we obtain the graphs of $y(x, 1.0)$ and $y(x, 2.0)$, shown in Figure 16.6b and c, respectively. These snapshots show that the pulse moves to the right without changing its shape and that it has a constant speed of 3.0 cm/s.

What If? (A) What if the wave function were

$$y(x, t) = \frac{2}{(x + 3.0t)^2 + 1}$$

How would this change the situation?

(B) What if the wave function were

$$y(x, t) = \frac{4}{(x - 3.0t)^2 + 1}$$

How would this change the situation?

Answer (A) The new feature in this expression is the plus sign in the denominator rather than the minus sign. This results in a pulse with the same shape as that in Figure 16.6, but moving to the left as time progresses.

(B) The new feature here is the numerator of 4 rather than 2. This results in a pulse moving to the right, but with twice the height of that in Figure 16.6.

(a)

(b)

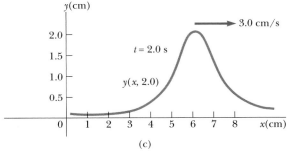

(c)

Figure 16.6 (Example 16.1) Graphs of the function $y(x, t) = 2/[(x - 3.0t)^2 + 1]$ at (a) $t = 0$, (b) $t = 1.0$ s, and (c) $t = 2.0$ s.

16.2 Sinusoidal Waves

In this section, we introduce an important wave function whose shape is shown in Figure 16.7. The wave represented by this curve is called a **sinusoidal wave** because the curve is the same as that of the function $\sin \theta$ plotted against θ. On a rope, a sinusoidal wave could be established by shaking the end of the rope up and down in simple harmonic motion.

The sinusoidal wave is the simplest example of a periodic continuous wave and can be used to build more complex waves (see Section 18.8). The brown curve in Figure 16.7 represents a snapshot of a traveling sinusoidal wave at $t = 0$, and the blue curve represents a snapshot of the wave at some later time t. Notice two types of motion that can be seen in your mind. First, the entire waveform in Figure 16.7 moves to the right, so that the brown curve moves toward the right and eventually reaches the position of the blue curve. This is the motion of the *wave*. If we focus on one element of the medium, such as the element at $x = 0$, we see that each element moves up and down along the y axis in simple harmonic motion. This is the motion of the *elements of the medium*. It is important to differentiate between the motion of the wave and the motion of the elements of the medium.

Figure 16.8a shows a snapshot of a wave moving through a medium. Figure 16.8b shows a graph of the position of one element of the medium as a function of time. The

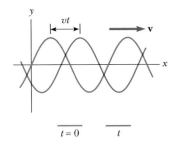

Active Figure 16.7 A one-dimensional sinusoidal wave traveling to the right with a speed v. The brown curve represents a snapshot of the wave at $t = 0$, and the blue curve represents a snapshot at some later time t.

At the Active Figures link at **http://www.pse6.com,** *you can watch the wave move and take snapshots of it at various times.*

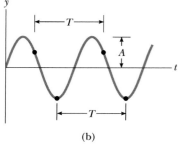

Active Figure 16.8 (a) The wavelength λ of a wave is the distance between adjacent crests or adjacent troughs. (b) The period T of a wave is the time interval required for the wave to travel one wavelength.

At the Active Figures link at http://www.pse6.com, you can change the parameters to see the effect on the wave function.

⚠️ **PITFALL PREVENTION**

16.1 What's the Difference Between Figure 16.8a and 16.8b?

Notice the visual similarity between Figures 16.8a and 16.8b. The shapes are the same, but (a) is a graph of vertical position versus horizontal position while (b) is vertical position versus time. Figure 16.8a is a pictorial representation of the wave *for a series of particles of the medium*—this is what you would see at an instant of time. Figure 16.8b is a graphical representation of the position of *one element of the medium* as a function of time. The fact that both figures have the identical shape represents Equation 16.1—a wave is the *same* function of both *x* and *t*.

point at which the displacement of the element from its normal position is highest is called the **crest** of the wave. The distance from one crest to the next is called the **wavelength** λ (Greek lambda). More generally, **the wavelength is the minimum distance between any two identical points (such as the crests) on adjacent waves,** as shown in Figure 16.8a.

If you count the number of seconds between the arrivals of two adjacent crests at a given point in space, you are measuring the **period** T of the waves. In general, **the period is the time interval required for two identical points (such as the crests) of adjacent waves to pass by a point.** The period of the wave is the same as the period of the simple harmonic oscillation of one element of the medium.

The same information is more often given by the inverse of the period, which is called the **frequency** f. In general, **the frequency of a periodic wave is the number of crests (or troughs, or any other point on the wave) that pass a given point in a unit time interval.** The frequency of a sinusoidal wave is related to the period by the expression

$$f = \frac{1}{T} \tag{16.3}$$

The frequency of the wave is the same as the frequency of the simple harmonic oscillation of one element of the medium. The most common unit for frequency, as we learned in Chapter 15, is second^{-1}, or **hertz** (Hz). The corresponding unit for T is seconds.

The maximum displacement from equilibrium of an element of the medium is called the **amplitude** A of the wave.

Waves travel with a specific speed, and this speed depends on the properties of the medium being disturbed. For instance, sound waves travel through room-temperature air with a speed of about 343 m/s (781 mi/h), whereas they travel through most solids with a speed greater than 343 m/s.

Consider the sinusoidal wave in Figure 16.8a, which shows the position of the wave at $t = 0$. Because the wave is sinusoidal, we expect the wave function at this instant to be expressed as $y(x, 0) = A \sin ax$, where A is the amplitude and a is a constant to be determined. At $x = 0$, we see that $y(0, 0) = A \sin a(0) = 0$, consistent with Figure 16.8a. The next value of x for which y is zero is $x = \lambda/2$. Thus,

$$y\left(\frac{\lambda}{2}, 0\right) = A \sin a\left(\frac{\lambda}{2}\right) = 0$$

For this to be true, we must have $a(\lambda/2) = \pi$, or $a = 2\pi/\lambda$. Thus, the function describing the positions of the elements of the medium through which the sinusoidal wave is traveling can be written

$$y(x, 0) = A \sin\left(\frac{2\pi}{\lambda} x\right) \tag{16.4}$$

where the constant A represents the wave amplitude and the constant λ is the wavelength. We see that the vertical position of an element of the medium is the same whenever x is increased by an integral multiple of λ. If the wave moves to the right with a speed v, then the wave function at some later time t is

$$y(x, t) = A \sin\left[\frac{2\pi}{\lambda}(x - vt)\right] \tag{16.5}$$

That is, the traveling sinusoidal wave moves to the right a distance vt in the time t, as shown in Figure 16.7. Note that the wave function has the form $f(x - vt)$ (Eq. 16.1). If the wave were traveling to the left, the quantity $x - vt$ would be replaced by $x + vt$, as we learned when we developed Equations 16.1 and 16.2.

By definition, the wave travels a distance of one wavelength in one period T. Therefore, the wave speed, wavelength, and period are related by the expression

$$v = \frac{\lambda}{T} \tag{16.6}$$

Substituting this expression for v into Equation 16.5, we find that

$$y = A \sin\left[2\pi\left(\frac{x}{\lambda} - \frac{t}{T}\right)\right] \tag{16.7}$$

This form of the wave function shows the *periodic* nature of y. (We will often use y rather than $y(x, t)$ as a shorthand notation.) At any given time t, y has the *same* value at the positions x, $x + \lambda$, $x + 2\lambda$, and so on. Furthermore, at any given position x, the value of y is the same at times t, $t + T$, $t + 2T$, and so on.

We can express the wave function in a convenient form by defining two other quantities, the **angular wave number** k (usually called simply the **wave number**) and the **angular frequency** ω:

$$k \equiv \frac{2\pi}{\lambda} \tag{16.8}$$

◀ Angular wave number

$$\omega \equiv \frac{2\pi}{T} \tag{16.9}$$

◀ Angular frequency

Using these definitions, we see that Equation 16.7 can be written in the more compact form

$$y = A \sin(kx - \omega t) \tag{16.10}$$

◀ Wave function for a sinusoidal wave

Using Equations 16.3, 16.8, and 16.9, we can express the wave speed v originally given in Equation 16.6 in the alternative forms

$$v = \frac{\omega}{k} \tag{16.11}$$

$$v = \lambda f \tag{16.12}$$

◀ Speed of a sinusoidal wave

The wave function given by Equation 16.10 assumes that the vertical position y of an element of the medium is zero at $x = 0$ and $t = 0$. This need not be the case. If it is not, we generally express the wave function in the form

$$y = A \sin(kx - \omega t + \phi) \tag{16.13}$$

◀ General expression for a sinusoidal wave

where ϕ is the **phase constant,** just as we learned in our study of periodic motion in Chapter 15. This constant can be determined from the initial conditions.

Quick Quiz 16.3 A sinusoidal wave of frequency f is traveling along a stretched string. The string is brought to rest, and a second traveling wave of frequency $2f$ is established on the string. The wave speed of the second wave is (a) twice that of the first wave (b) half that of the first wave (c) the same as that of the first wave (d) impossible to determine.

Quick Quiz 16.4 Consider the waves in Quick Quiz 16.3 again. The wavelength of the second wave is (a) twice that of the first wave (b) half that of the first wave (c) the same as that of the first wave (d) impossible to determine.

Quick Quiz 16.5 Consider the waves in Quick Quiz 16.3 again. The amplitude of the second wave is (a) twice that of the first wave (b) half that of the first wave (c) the same as that of the first wave (d) impossible to determine.

Example 16.2 A Traveling Sinusoidal Wave

A sinusoidal wave traveling in the positive x direction has an amplitude of 15.0 cm, a wavelength of 40.0 cm, and a frequency of 8.00 Hz. The vertical position of an element of the medium at $t = 0$ and $x = 0$ is also 15.0 cm, as shown in Figure 16.9.

(A) Find the wave number k, period T, angular frequency ω, and speed v of the wave.

Solution Using Equations 16.8, 16.3, 16.9, and 16.12, we find the following:

$$k = \frac{2\pi}{\lambda} = \frac{2\pi \text{ rad}}{40.0 \text{ cm}} = \boxed{0.157 \text{ rad/cm}}$$

$$T = \frac{1}{f} = \frac{1}{8.00 \text{ s}^{-1}} = \boxed{0.125 \text{ s}}$$

$$\omega = 2\pi f = 2\pi(8.00 \text{ s}^{-1}) = \boxed{50.3 \text{ rad/s}}$$

$$v = \lambda f = (40.0 \text{ cm})(8.00 \text{ s}^{-1}) = \boxed{320 \text{ cm/s}}$$

(B) Determine the phase constant ϕ, and write a general expression for the wave function.

Solution Because $A = 15.0$ cm and because $y = 15.0$ cm at $x = 0$ and $t = 0$, substitution into Equation 16.13 gives

$$15.0 = (15.0) \sin \phi \qquad \text{or} \qquad \sin \phi = 1$$

We may take the principal value $\phi = \pi/2$ rad (or 90°). Hence, the wave function is of the form

$$y = A \sin\left(kx - \omega t + \frac{\pi}{2}\right) = A \cos(kx - \omega t)$$

By inspection, we can see that the wave function must have this form, noting that the cosine function has the same shape as the sine function displaced by 90°. Substituting the values for A, k, and ω into this expression, we obtain

$$y = \boxed{(15.0 \text{ cm}) \cos(0.157x - 50.3t)}$$

Figure 16.9 (Example 16.2) A sinusoidal wave of wavelength $\lambda = 40.0$ cm and amplitude $A = 15.0$ cm. The wave function can be written in the form $y = A \cos(kx - \omega t)$.

Sinusoidal Waves on Strings

In Figure 16.1, we demonstrated how to create a pulse by jerking a taut string up and down once. To create a series of such pulses—a wave—we can replace the hand with an oscillating blade. If the wave consists of a series of identical waveforms, whatever their shape, the relationships $f = 1/T$ and $v = f\lambda$ among speed, frequency, period, and wavelength hold true. We can make more definite statements about the wave function if the source of the waves vibrates in simple harmonic motion. Figure 16.10 represents snapshots of the wave created in this way at intervals of $T/4$. Because the end of the blade oscillates in simple harmonic motion, **each element of the string, such as that at P, also oscillates vertically with simple harmonic motion.** This must be the case because each element follows the simple harmonic motion of the blade. Therefore, every element of the string can be treated as a simple harmonic oscillator vibrating with a frequency equal to the frequency of oscillation of the blade.[2] Note that although each element oscillates in the y direction, the wave travels in the x direction with a speed v. Of course, this is the definition of a transverse wave.

If the wave at $t = 0$ is as described in Figure 16.10b, then the wave function can be written as

$$y = A \sin(kx - \omega t)$$

---

[2] In this arrangement, we are assuming that a string element always oscillates in a vertical line. The tension in the string would vary if an element were allowed to move sideways. Such motion would make the analysis very complex.

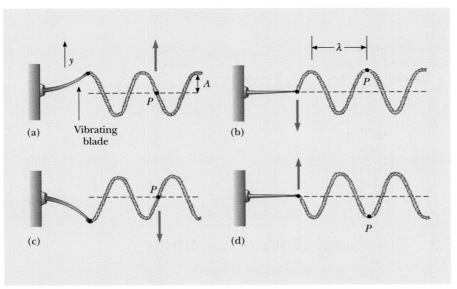

At the Active Figures link at http://www.pse6.com, you can adjust the frequency of the blade.

Active Figure 16.10 One method for producing a sinusoidal wave on a string. The left end of the string is connected to a blade that is set into oscillation. Every element of the string, such as that at point P, oscillates with simple harmonic motion in the vertical direction.

We can use this expression to describe the motion of any element of the string. An element at point P (or any other element of the string) moves only vertically, and so its x coordinate remains constant. Therefore, the **transverse speed** v_y (not to be confused with the wave speed v) and the **transverse acceleration** a_y of elements of the string are

$$v_y = \frac{dy}{dt}\bigg]_{x=\text{constant}} = \frac{\partial y}{\partial t} = -\omega A \cos(kx - \omega t) \tag{16.14}$$

$$a_y = \frac{dv_y}{dt}\bigg]_{x=\text{constant}} = \frac{\partial v_y}{\partial t} = -\omega^2 A \sin(kx - \omega t) \tag{16.15}$$

In these expressions, we must use partial derivatives (see Section 8.5) because y depends on both x and t. In the operation $\partial y/\partial t$, for example, we take a derivative with respect to t while holding x constant. The maximum values of the transverse speed and transverse acceleration are simply the absolute values of the coefficients of the cosine and sine functions:

$$v_{y,\,\text{max}} = \omega A \tag{16.16}$$

$$a_{y,\,\text{max}} = \omega^2 A \tag{16.17}$$

The transverse speed and transverse acceleration of elements of the string do not reach their maximum values simultaneously. The transverse speed reaches its maximum value (ωA) when $y = 0$, whereas the magnitude of the transverse acceleration reaches its maximum value ($\omega^2 A$) when $y = \pm A$. Finally, Equations 16.16 and 16.17 are identical in mathematical form to the corresponding equations for simple harmonic motion, Equations 15.17 and 15.18.

 PITFALL PREVENTION

16.2 Two Kinds of Speed/Velocity

Do not confuse v, the speed of the wave as it propagates along the string, with v_y, the transverse velocity of a point on the string. The speed v is constant while v_y varies sinusoidally.

Quick Quiz 16.6 The amplitude of a wave is doubled, with no other changes made to the wave. As a result of this doubling, which of the following statements is correct? (a) The speed of the wave changes. (b) The frequency of the wave changes. (c) The maximum transverse speed of an element of the medium changes. (d) All of these are true. (e) None of these is true.

Example 16.3   A Sinusoidally Driven String

The string shown in Figure 16.10 is driven at a frequency of 5.00 Hz. The amplitude of the motion is 12.0 cm, and the wave speed is 20.0 m/s. Determine the angular frequency ω and wave number k for this wave, and write an expression for the wave function.

Solution   Using Equations 16.3, 16.9, and 16.11, we find that

$$\omega = \frac{2\pi}{T} = 2\pi f = 2\pi(5.00 \text{ Hz}) = \boxed{31.4 \text{ rad/s}}$$

$$k = \frac{\omega}{v} = \frac{31.4 \text{ rad/s}}{20.0 \text{ m/s}} = \boxed{1.57 \text{ rad/m}}$$

Because $A = 12.0$ cm $= 0.120$ m, we have

$$y = A \sin(kx - \omega t)$$

$$= \boxed{(0.120 \text{ m}) \sin(1.57x - 31.4t)}$$

16.3   The Speed of Waves on Strings

In this section, we focus on determining the speed of a transverse pulse traveling on a taut string. Let us first conceptually predict the parameters that determine the speed. If a string under tension is pulled sideways and then released, the tension is responsible for accelerating a particular element of the string back toward its equilibrium position. According to Newton's second law, the acceleration of the element increases with increasing tension. If the element returns to equilibrium more rapidly due to this increased acceleration, we would intuitively argue that the wave speed is greater. Thus, we expect the wave speed to increase with increasing tension.

Likewise, the wave speed should decrease as the mass per unit length of the string increases. This is because it is more difficult to accelerate a massive element of the string than a light element. If the tension in the string is T and its mass per unit length is μ (Greek mu), then as we shall show, the wave speed is

Speed of a wave on a stretched string

$$v = \sqrt{\frac{T}{\mu}} \qquad (16.18)$$

First, let us verify that this expression is dimensionally correct. The dimensions of T are ML/T^2, and the dimensions of μ are M/L. Therefore, the dimensions of T/μ are L^2/T^2; hence, the dimensions of $\sqrt{T/\mu}$ are L/T, the dimensions of speed. No other combination of T and μ is dimensionally correct, and if we assume that these are the only variables relevant to the situation, the speed must be proportional to $\sqrt{T/\mu}$.

Now let us use a mechanical analysis to derive Equation 16.18. Consider a pulse moving on a taut string to the right with a uniform speed v measured relative to a stationary frame of reference. Instead of staying in this reference frame, it is more convenient to choose as our reference frame one that moves along with the pulse with the same speed as the pulse, so that the pulse is at rest within the frame. This change of reference frame is permitted because Newton's laws are valid in either a stationary frame or one that moves with constant velocity. In our new reference frame, all elements of the string move to the left—a given element of the string initially to the right of the pulse moves to the left, rises up and follows the shape of the pulse, and then continues to move to the left. Figure 16.11a shows such an element at the instant it is located at the top of the pulse.

The small element of the string of length Δs shown in Figure 16.11a, and magnified in Figure 16.11b, forms an approximate arc of a circle of radius R. In our moving frame of reference (which is moving to the right at a speed v along with the pulse), the shaded element is moving to the left with a speed v. This element has a centripetal acceleration equal to v^2/R, which is supplied by components of the force **T** whose magnitude is the tension in the string. The force **T** acts on both sides of the element and is tangent to the arc, as shown in Figure 16.11b. The horizontal components of **T** cancel, and each vertical component $T \sin \theta$ acts radially toward the center of the arc. Hence, the total

 PITFALL PREVENTION

16.3   Multiple T's

Do not confuse the T in Equation 16.18 for the tension with the symbol T used in this chapter for the period of a wave. The context of the equation should help you to identify which quantity is meant. There simply aren't enough letters in the alphabet to assign a unique letter to each variable!

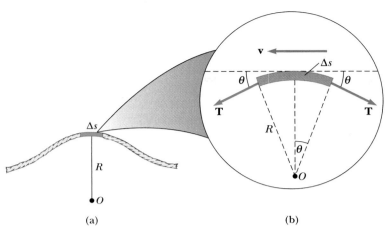

Figure 16.11 (a) To obtain the speed v of a wave on a stretched string, it is convenient to describe the motion of a small element of the string in a moving frame of reference. (b) In the moving frame of reference, the small element of length Δs moves to the left with speed v. The net force on the element is in the radial direction because the horizontal components of the tension force cancel.

radial force on the element is $2T \sin \theta$. Because the element is small, θ is small, and we can use the small-angle approximation $\sin \theta \approx \theta$. Therefore, the total radial force is

$$F_r = 2T \sin \theta \approx 2T\theta$$

The element has a mass $m = \mu \, \Delta s$. Because the element forms part of a circle and subtends an angle 2θ at the center, $\Delta s = R(2\theta)$, we find that

$$m = \mu \, \Delta s = 2\mu R\theta$$

If we apply Newton's second law to this element in the radial direction, we have

$$F_r = ma = \frac{mv^2}{R}$$

$$2T\theta = \frac{2\mu R\theta v^2}{R} \quad \longrightarrow \quad v = \sqrt{\frac{T}{\mu}}$$

This expression for v is Equation 16.18.

Notice that this derivation is based on the assumption that the pulse height is small relative to the length of the string. Using this assumption, we were able to use the approximation $\sin \theta \approx \theta$. Furthermore, the model assumes that the tension T is not affected by the presence of the pulse; thus, T is the same at all points on the string. Finally, this proof does *not* assume any particular shape for the pulse. Therefore, we conclude that a pulse of *any shape* travels along the string with speed $v = \sqrt{T/\mu}$ without any change in pulse shape.

Quick Quiz 16.7 Suppose you create a pulse by moving the free end of a taut string up and down once with your hand beginning at $t = 0$. The string is attached at its other end to a distant wall. The pulse reaches the wall at time t. Which of the following actions, taken by itself, decreases the time interval that it takes for the pulse to reach the wall? More than one choice may be correct. (a) moving your hand more quickly, but still only up and down once by the same amount (b) moving your hand more slowly, but still only up and down once by the same amount (c) moving your hand a greater distance up and down in the same amount of time (d) moving your hand a lesser distance up and down in the same amount of time (e) using a heavier string of the same length and under the same tension (f) using a lighter string of the same length and under the same tension (g) using a string of the same linear mass density but under decreased tension (h) using a string of the same linear mass density but under increased tension

Example 16.4 The Speed of a Pulse on a Cord

A uniform cord has a mass of 0.300 kg and a length of 6.00 m (Fig. 16.12). The cord passes over a pulley and supports a 2.00-kg object. Find the speed of a pulse traveling along this cord.

Solution The tension T in the cord is equal to the weight of the suspended 2.00-kg object:

$$T = mg = (2.00 \text{ kg})(9.80 \text{ m/s}^2) = 19.6 \text{ N}$$

(This calculation of the tension neglects the small mass of the cord. Strictly speaking, the cord can never be exactly horizontal, and therefore the tension is not uniform.) The mass per unit length μ of the cord is

$$\mu = \frac{m}{\ell} = \frac{0.300 \text{ kg}}{6.00 \text{ m}} = 0.050 \text{ 0 kg/m}$$

Therefore, the wave speed is

$$v = \sqrt{\frac{T}{\mu}} = \sqrt{\frac{19.6 \text{ N}}{0.050 \text{ 0 kg/m}}} = \boxed{19.8 \text{ m/s}}$$

What If? What if the block were swinging back and forth between maximum angles of $\pm 20°$ with respect to the vertical? What range of wave speeds would this create on the horizontal cord?

Answer Figure 16.13 shows the swinging block at three positions—its highest position, its lowest position, and an arbitrary position. Summing the forces on the block in the radial direction when the block is at an arbitrary position, Newton's second law gives

$$(1) \qquad \sum F = T - mg \cos\theta = m\frac{v_{\text{block}}^2}{L}$$

where the acceleration of the block is centripetal, L is the length of the vertical piece of string, and v_{block} is the instantaneous speed of the block at the arbitrary position. Now consider conservation of mechanical energy for the block–Earth system. We define the zero of gravitational potential energy for the system when the block is at its lowest point, point Ⓒ in Figure 16.13. Equating the mechanical energy of the system when the block is at Ⓐ to the mechanical energy when the block is at an arbitrary position Ⓑ, we have,

Figure 16.12 (Example 16.4) The tension T in the cord is maintained by the suspended object. The speed of any wave traveling along the cord is given by $v = \sqrt{T/\mu}$.

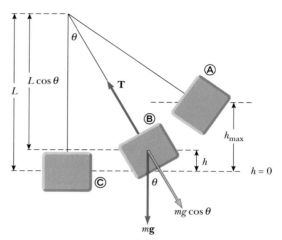

Figure 16.13 (Example 16.4) If the block swings back and forth, the tension in the cord changes, which causes a variation in the wave speed on the horizontal section of cord in Figure 16.12. The forces on the block when it is at arbitrary position Ⓑ are shown. Position Ⓐ is the highest position and Ⓒ is the lowest. (The maximum angle is exaggerated for clarity.)

$$E_A = E_B$$
$$mgh_{\max} = mgh + \tfrac{1}{2}mv_{\text{block}}^2$$
$$mv_{\text{block}}^2 = 2mg(h_{\max} - h)$$

Substituting this into Equation (1), we find an expression for T as a function of angle θ and height h:

$$T - mg\cos\theta = \frac{2mg(h_{\max} - h)}{L}$$
$$T = mg\left[\cos\theta + \frac{2}{L}(h_{\max} - h)\right]$$

The maximum value of T occurs when $\theta = 0$ and $h = 0$:

$$T_{\max} = mg\left[\cos 0 + \frac{2}{L}(h_{\max} - 0)\right] = mg\left(1 + \frac{2h_{\max}}{L}\right)$$

The minimum value of T occurs when $h = h_{\max}$ and $\theta = \theta_{\max}$:

$$T_{\min} = mg\left[\cos\theta_{\max} + \frac{2}{L}(h_{\max} - h_{\max})\right] = mg\cos\theta_{\max}$$

Now we find the maximum and minimum values of the wave speed v, using the fact that, as we see from Figure 16.13, h and θ are related by $h = L - L\cos\theta$:

$$v_{\max} = \sqrt{\frac{T_{\max}}{\mu}} = \sqrt{\frac{mg[1 + (2h_{\max}/L)]}{\mu}}$$
$$= \sqrt{\frac{mg\{1 + [2(L - L\cos\theta_{\max})/L]\}}{\mu}}$$

$$= \sqrt{\frac{mg(3 - 2\cos\theta_{max})}{\mu}}$$

$$v_{min} = \sqrt{\frac{T_{min}}{\mu}} = \sqrt{\frac{mg\cos\theta_{max}}{\mu}}$$

$$= \sqrt{\frac{(2.00\ \text{kg})(9.80\ \text{m/s}^2)(3 - 2\cos 20°)}{0.050\ 0\ \text{kg/m}}} = 21.0\ \text{m/s}$$

$$= \sqrt{\frac{(2.00\ \text{kg})(9.80\ \text{m/s}^2)(\cos 20°)}{0.050\ 0\ \text{kg/m}}} = 19.2\ \text{m/s}$$

Example 16.5   Rescuing the Hiker   `Interactive`

An 80.0-kg hiker is trapped on a mountain ledge following a storm. A helicopter rescues the hiker by hovering above him and lowering a cable to him. The mass of the cable is 8.00 kg, and its length is 15.0 m. A chair of mass 70.0 kg is attached to the end of the cable. The hiker attaches himself to the chair, and the helicopter then accelerates upward. Terrified by hanging from the cable in midair, the hiker tries to signal the pilot by sending transverse pulses up the cable. A pulse takes 0.250 s to travel the length of the cable. What is the acceleration of the helicopter?

Solution To conceptualize this problem, imagine the effect of the acceleration of the helicopter on the cable. The higher the upward acceleration, the larger is the tension in the cable. In turn, the larger the tension, the higher is the speed of pulses on the cable. Thus, we categorize this problem as a combination of one involving Newton's laws and one involving the speed of pulses on a string. To analyze the problem, we use the time interval for the pulse to travel from the hiker to the helicopter to find the speed of the pulses on the cable:

$$v = \frac{\Delta x}{\Delta t} = \frac{15.0\ \text{m}}{0.250\ \text{s}} = 60.0\ \text{m/s}$$

The speed of pulses on the cable is given by Equation 16.18, which allows us to find the tension in the cable:

$$v = \sqrt{\frac{T}{\mu}} \longrightarrow T = \mu v^2 = \left(\frac{8.00\ \text{kg}}{15.0\ \text{m}}\right)(60.0\ \text{m/s})^2$$

$$T = 1.92 \times 10^3\ \text{N}$$

Newton's second law relates the tension in the cable to the acceleration of the hiker and the chair, which is the same as the acceleration of the helicopter:

$$\sum F = ma \longrightarrow T - mg = ma$$

$$a = \frac{T}{m} - g = \frac{1.92 \times 10^3\ \text{N}}{150.0\ \text{kg}} - 9.80\ \text{m/s}^2$$

$$= 3.00\ \text{m/s}^2$$

To finalize this problem, note that a real cable has stiffness in addition to tension. Stiffness tends to return a wire to its original straight-line shape even when it is not under tension. For example, a piano wire straightens if released from a curved shape; package wrapping string does not.

Stiffness represents a restoring force in addition to tension, and increases the wave speed. Consequently, for a real cable, the speed of 60.0 m/s that we determined is most likely associated with a tension lower than 1.92×10^3 N and a correspondingly smaller acceleration of the helicopter.

 Investigate this situation at the Interactive Worked Example link at **http://www.pse6.com.**

16.4 Reflection and Transmission

We have discussed waves traveling through a uniform medium. We now consider how a traveling wave is affected when it encounters a change in the medium. For example, consider a pulse traveling on a string that is rigidly attached to a support at one end as in Figure 16.14. When the pulse reaches the support, a severe change in the medium occurs—the string ends. The result of this change is that the pulse undergoes **reflection**—that is, the pulse moves back along the string in the opposite direction.

Note that the reflected pulse is *inverted*. This inversion can be explained as follows. When the pulse reaches the fixed end of the string, the string produces an upward force on the support. By Newton's third law, the support must exert an equal-magnitude and oppositely directed (downward) reaction force on the string. This downward force causes the pulse to invert upon reflection.

Now consider another case: this time, the pulse arrives at the end of a string that is free to move vertically, as in Figure 16.15. The tension at the free end is maintained because the string is tied to a ring of negligible mass that is free to slide vertically on a smooth post without friction. Again, the pulse is reflected, but this time it is not inverted. When it reaches the post, the pulse exerts a force on the free end of the string, causing the ring to accelerate upward. The ring rises as high as the incoming pulse,

 At the Active Figures link at **http://www.pse6.com,** *you can adjust the linear mass density of the string and the transverse direction of the initial pulse.*

Incident pulse

(a)

(b)

(c)

(d)

(e)   Reflected pulse

Active Figure 16.14 The reflection of a traveling pulse at the fixed end of a stretched string. The reflected pulse is inverted, but its shape is otherwise unchanged.

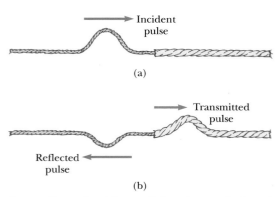

Active Figure 16.15 The reflection of a traveling pulse at the free end of a stretched string. The reflected pulse is not inverted.

At the Active Figures link at http://www.pse6.com, you can adjust the linear mass density of the string and the transverse direction of the initial pulse.

Figure 16.16 (a) A pulse traveling to the right on a light string attached to a heavier string. (b) Part of the incident pulse is reflected (and inverted), and part is transmitted to the heavier string. See Figure 16.17 for an animation available for both figures at the Active Figures link.

and then the downward component of the tension force pulls the ring back down. This movement of the ring produces a reflected pulse that is not inverted and that has the same amplitude as the incoming pulse.

Finally, we may have a situation in which the boundary is intermediate between these two extremes. In this case, part of the energy in the incident pulse is reflected and part undergoes **transmission**—that is, some of the energy passes through the boundary. For instance, suppose a light string is attached to a heavier string, as in Figure 16.16. When a pulse traveling on the light string reaches the boundary between the two, part of the pulse is reflected and inverted and part is transmitted to the heavier string. The reflected pulse is inverted for the same reasons described earlier in the case of the string rigidly attached to a support.

Note that the reflected pulse has a smaller amplitude than the incident pulse. In Section 16.5, we show that the energy carried by a wave is related to its amplitude. According to the principle of the conservation of energy, when the pulse breaks up into a reflected pulse and a transmitted pulse at the boundary, the sum of the energies of these two pulses must equal the energy of the incident pulse. Because the reflected pulse contains only part of the energy of the incident pulse, its amplitude must be smaller.

When a pulse traveling on a heavy string strikes the boundary between the heavy string and a lighter one, as in Figure 16.17, again part is reflected and part is transmitted. In this case, the reflected pulse is not inverted.

In either case, the relative heights of the reflected and transmitted pulses depend on the relative densities of the two strings. If the strings are identical, there is no discontinuity at the boundary and no reflection takes place.

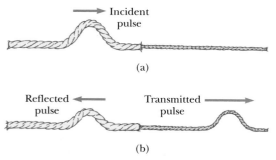

At the Active Figures link at http://www.pse6.com, you can adjust the linear mass densities of the strings and the transverse direction of the initial pulse.

Active Figure 16.17 (a) A pulse traveling to the right on a heavy string attached to a lighter string. (b) The incident pulse is partially reflected and partially transmitted, and the reflected pulse is not inverted.

According to Equation 16.18, the speed of a wave on a string increases as the mass per unit length of the string decreases. In other words, a wave travels more slowly on a heavy string than on a light string if both are under the same tension. The following general rules apply to reflected waves: **when a wave or pulse travels from medium A to medium B and $v_A > v_B$ (that is, when B is denser than A), it is inverted upon reflection. When a wave or pulse travels from medium A to medium B and $v_A < v_B$ (that is, when A is denser than B), it is not inverted upon reflection.**

16.5 Rate of Energy Transfer by Sinusoidal Waves on Strings

Waves transport energy when they propagate through a medium. We can easily demonstrate this by hanging an object on a stretched string and then sending a pulse down the string, as in Figure 16.18a. When the pulse meets the suspended object, the object is momentarily displaced upward, as in Figure 16.18b. In the process, energy is transferred to the object and appears as an increase in the gravitational potential energy of the object–Earth system. This section examines the rate at which energy is transported along a string. We shall assume a one-dimensional sinusoidal wave in the calculation of the energy transferred.

Consider a sinusoidal wave traveling on a string (Fig. 16.19). The source of the energy is some external agent at the left end of the string, which does work in producing the oscillations. We can consider the string to be a nonisolated system. As the external agent performs work on the end of the string, moving it up and down, energy enters the system of the string and propagates along its length. Let us focus our attention on an element of the string of length Δx and mass Δm. Each such element moves vertically with simple harmonic motion. Thus, we can model each element of the string as a simple harmonic oscillator, with the oscillation in the y direction. All elements have the same angular frequency ω and the same amplitude A. The kinetic energy K associated with a moving particle is $K = \frac{1}{2}mv^2$. If we apply this equation to an element of length Δx and mass Δm, we see that the kinetic energy ΔK of this element is

$$\Delta K = \tfrac{1}{2}(\Delta m)\, v_y^{\,2}$$

where v_y is the transverse speed of the element. If μ is the mass per unit length of the string, then the mass Δm of the element of length Δx is equal to $\mu\, \Delta x$. Hence, we can express the kinetic energy of an element of the string as

$$\Delta K = \tfrac{1}{2}(\mu\, \Delta x)\, v_y^{\,2} \qquad (16.19)$$

As the length of the element of the string shrinks to zero, this becomes a differential relationship:

$$dK = \tfrac{1}{2}(\mu\, dx)\, v_y^{\,2}$$

We substitute for the general transverse speed of a simple harmonic oscillator using Equation 16.14:

$$dK = \tfrac{1}{2}\mu[\omega A \cos(kx - \omega t)]^2\, dx$$
$$= \tfrac{1}{2}\mu\omega^2 A^2 \cos^2(kx - \omega t)\, dx$$

(a)

(b)

Figure 16.18 (a) A pulse traveling to the right on a stretched string that has an object suspended from it. (b) Energy is transmitted to the suspended object when the pulse arrives.

Figure 16.19 A sinusoidal wave traveling along the x axis on a stretched string. Every element moves vertically, and every element has the same total energy.

If we take a snapshot of the wave at time $t = 0$, then the kinetic energy of a given element is

$$dK = \tfrac{1}{2}\mu\omega^2 A^2 \cos^2 kx \, dx$$

Let us integrate this expression over all the string elements in a wavelength of the wave, which will give us the total kinetic energy K_λ in one wavelength:

$$K_\lambda = \int dK = \int_0^\lambda \tfrac{1}{2}\mu\omega^2 A^2 \cos^2 kx \, dx = \tfrac{1}{2}\mu\omega^2 A^2 \int_0^\lambda \cos^2 kx \, dx$$

$$= \tfrac{1}{2}\mu\omega^2 A^2 \left[\tfrac{1}{2}x + \frac{1}{4k} \sin 2kx \right]_0^\lambda = \tfrac{1}{2}\mu\omega^2 A^2 [\tfrac{1}{2}\lambda] = \tfrac{1}{4}\mu\omega^2 A^2 \lambda$$

In addition to kinetic energy, each element of the string has potential energy associated with it due to its displacement from the equilibrium position and the restoring forces from neighboring elements. A similar analysis to that above for the total potential energy U_λ in one wavelength will give exactly the same result:

$$U_\lambda = \tfrac{1}{4}\mu\omega^2 A^2 \lambda$$

The total energy in one wavelength of the wave is the sum of the potential and kinetic energies:

$$E_\lambda = U_\lambda + K_\lambda = \tfrac{1}{2}\mu\omega^2 A^2 \lambda \qquad (16.20)$$

As the wave moves along the string, this amount of energy passes by a given point on the string during a time interval of one period of the oscillation. Thus, the power, or rate of energy transfer, associated with the wave is

$$\mathcal{P} = \frac{\Delta E}{\Delta t} = \frac{E_\lambda}{T} = \frac{\tfrac{1}{2}\mu\omega^2 A^2 \lambda}{T} = \tfrac{1}{2}\mu\omega^2 A^2 \left(\frac{\lambda}{T}\right)$$

Power of a wave

$$\mathcal{P} = \tfrac{1}{2}\mu\omega^2 A^2 v \qquad (16.21)$$

This expression shows that the rate of energy transfer by a sinusoidal wave on a string is proportional to (a) the square of the frequency, (b) the square of the amplitude, and (c) the wave speed. In fact: **the rate of energy transfer in any sinusoidal wave is proportional to the square of the angular frequency and to the square of the amplitude.**

Quick Quiz 16.8 Which of the following, taken by itself, would be most effective in increasing the rate at which energy is transferred by a wave traveling along a string? (a) reducing the linear mass density of the string by one half (b) doubling the wavelength of the wave (c) doubling the tension in the string (d) doubling the amplitude of the wave

---

Example 16.6 Power Supplied to a Vibrating String

A taut string for which $\mu = 5.00 \times 10^{-2}$ kg/m is under a tension of 80.0 N. How much power must be supplied to the string to generate sinusoidal waves at a frequency of 60.0 Hz and an amplitude of 6.00 cm?

Solution The wave speed on the string is, from Equation 16.18,

$$v = \sqrt{\frac{T}{\mu}} = \sqrt{\frac{80.0 \text{ N}}{5.00 \times 10^{-2} \text{ kg/m}}} = 40.0 \text{ m/s}$$

Because $f = 60.0$ Hz, the angular frequency ω of the sinusoidal waves on the string has the value

$$\omega = 2\pi f = 2\pi(60.0 \text{ Hz}) = 377 \text{ s}^{-1}$$

Using these values in Equation 16.21 for the power, with $A = 6.00 \times 10^{-2}$ m, we obtain

$$\mathcal{P} = \tfrac{1}{2}\mu\omega^2 A^2 v$$

$$= \tfrac{1}{2}(5.00 \times 10^{-2} \text{ kg/m})(377 \text{ s}^{-1})^2$$

$$\times (6.00 \times 10^{-2} \text{ m})^2 (40.0 \text{ m/s})$$

$$= \boxed{512 \text{ W}}$$

What If? What if the string is to transfer energy at a rate of 1 000 W? What must be the required amplitude if all other parameters remain the same?

Answer We set up a ratio of the new and old power, reflecting only a change in the amplitude:

$$\frac{\mathcal{P}_{\text{new}}}{\mathcal{P}_{\text{old}}} = \frac{\tfrac{1}{2}\mu\omega^2 A_{\text{new}}^2 v}{\tfrac{1}{2}\mu\omega^2 A_{\text{old}}^2 v} = \frac{A_{\text{new}}^2}{A_{\text{old}}^2}$$

Solving for the new amplitude,

$$A_{\text{new}} = A_{\text{old}}\sqrt{\frac{\mathcal{P}_{\text{new}}}{\mathcal{P}_{\text{old}}}} = (6.00 \text{ cm})\sqrt{\frac{1\,000 \text{ W}}{512 \text{ W}}}$$

$$= 8.39 \text{ cm}$$

16.6 The Linear Wave Equation

In Section 16.1 we introduced the concept of the wave function to represent waves traveling on a string. All wave functions $y(x, t)$ represent solutions of an equation called the *linear wave equation*. This equation gives a complete description of the wave motion, and from it one can derive an expression for the wave speed. Furthermore, the linear wave equation is basic to many forms of wave motion. In this section, we derive this equation as applied to waves on strings.

Suppose a traveling wave is propagating along a string that is under a tension T. Let us consider one small string element of length Δx (Fig. 16.20). The ends of the element make small angles θ_A and θ_B with the x axis. The net force acting on the element in the vertical direction is

$$\sum F_y = T\sin\theta_B - T\sin\theta_A = T(\sin\theta_B - \sin\theta_A)$$

Because the angles are small, we can use the small-angle approximation $\sin\theta \approx \tan\theta$ to express the net force as

$$\sum F_y \approx T(\tan\theta_B - \tan\theta_A) \tag{16.22}$$

Figure 16.20 An element of a string under tension T.

Imagine undergoing an infinitesimal displacement outward from the end of the rope element in Figure 16.20 along the blue line representing the force **T**. This displacement has infinitesimal x and y components and can be represented by the vector $dx\hat{\mathbf{i}} + dy\hat{\mathbf{j}}$. The tangent of the angle with respect to the x axis for this displacement is dy/dx. Because we are evaluating this tangent at a particular instant of time, we need to express this in partial form as $\partial y/\partial x$. Substituting for the tangents in Equation 16.22 gives

$$\sum F_y \approx T\left[\left(\frac{\partial y}{\partial x}\right)_B - \left(\frac{\partial y}{\partial x}\right)_A\right] \tag{16.23}$$

We now apply Newton's second law to the element, with the mass of the element given by $m = \mu\,\Delta x$:

$$\sum F_y = ma_y = \mu\,\Delta x\left(\frac{\partial^2 y}{\partial t^2}\right) \tag{16.24}$$

Combining Equation 16.23 with Equation 16.24, we obtain

$$\mu\,\Delta x\left(\frac{\partial^2 y}{\partial t^2}\right) = T\left[\left(\frac{\partial y}{\partial x}\right)_B - \left(\frac{\partial y}{\partial x}\right)_A\right]$$

$$\frac{\mu}{T}\frac{\partial^2 y}{\partial t^2} = \frac{(\partial y/\partial x)_B - (\partial y/dx)_A}{\Delta x} \tag{16.25}$$

The right side of this equation can be expressed in a different form if we note that the partial derivative of any function is defined as

$$\frac{\partial f}{\partial x} \equiv \lim_{\Delta x \to 0} \frac{f(x + \Delta x) - f(x)}{\Delta x}$$

If we associate $f(x + \Delta x)$ with $(\partial y/\partial x)_B$ and $f(x)$ with $(\partial y/\partial x)_A$, we see that, in the limit $\Delta x \to 0$, Equation 16.25 becomes

Linear wave equation for a string

$$\frac{\mu}{T} \frac{\partial^2 y}{\partial t^2} = \frac{\partial^2 y}{\partial x^2} \qquad (16.26)$$

This is the linear wave equation as it applies to waves on a string.

We now show that the sinusoidal wave function (Eq. 16.10) represents a solution of the linear wave equation. If we take the sinusoidal wave function to be of the form $y(x, t) = A \sin(kx - \omega t)$, then the appropriate derivatives are

$$\frac{\partial^2 y}{\partial t^2} = -\omega^2 A \sin(kx - \omega t)$$

$$\frac{\partial^2 y}{\partial x^2} = -k^2 A \sin(kx - \omega t)$$

Substituting these expressions into Equation 16.26, we obtain

$$-\frac{\mu \omega^2}{T} \sin(kx - \omega t) = -k^2 \sin(kx - \omega t)$$

This equation must be true for all values of the variables x and t in order for the sinusoidal wave function to be a solution of the wave equation. Both sides of the equation depend on x and t through the same function $\sin(kx - \omega t)$. Because this function divides out, we do indeed have an identity, provided that

$$k^2 = \frac{\mu}{T} \omega^2$$

Using the relationship $v = \omega/k$ (Eq. 16.11) in this expression, we see that

$$v^2 = \frac{\omega^2}{k^2} = \frac{T}{\mu}$$

$$v = \sqrt{\frac{T}{\mu}}$$

which is Equation 16.18. This derivation represents another proof of the expression for the wave speed on a taut string.

The linear wave equation (Eq. 16.26) is often written in the form

Linear wave equation in general

$$\frac{\partial^2 y}{\partial x^2} = \frac{1}{v^2} \frac{\partial^2 y}{\partial t^2} \qquad (16.27)$$

This expression applies in general to various types of traveling waves. For waves on strings, y represents the vertical position of elements of the string. For sound waves, y corresponds to longitudinal position of elements of air from equilibrium or variations in either the pressure or the density of the gas through which the sound waves are propagating. In the case of electromagnetic waves, y corresponds to electric or magnetic field components.

We have shown that the sinusoidal wave function (Eq. 16.10) is one solution of the linear wave equation (Eq. 16.27). Although we do not prove it here, the linear wave equation is satisfied by *any* wave function having the form $y = f(x \pm vt)$. Furthermore, we have seen that the linear wave equation is a direct consequence of Newton's second law applied to any element of a string carrying a traveling wave.

SUMMARY

A **transverse wave** is one in which the elements of the medium move in a direction *perpendicular* to the direction of propagation. An example is a wave on a taut string. A **longitudinal wave** is one in which the elements of the medium move in a direction *parallel* to the direction of propagation. Sound waves in fluids are longitudinal.

Any one-dimensional wave traveling with a speed v in the x direction can be represented by a wave function of the form

$$y(x, t) = f(x \pm vt) \tag{16.1, 16.2}$$

where the positive sign applies to a wave traveling in the negative x direction and the negative sign applies to a wave traveling in the positive x direction. The shape of the wave at any instant in time (a snapshot of the wave) is obtained by holding t constant.

The **wave function** for a one-dimensional sinusoidal wave traveling to the right can be expressed as

$$y = A \sin \left[\frac{2\pi}{\lambda} (x - vt) \right] = A \sin(kx - \omega t) \tag{16.5, 16.10}$$

where A is the **amplitude,** λ is the **wavelength,** k is the **angular wave number,** and ω is the **angular frequency.** If T is the **period** and f the **frequency,** v, k, and ω can be written

$$v = \frac{\lambda}{T} = \lambda f \tag{16.6, 16.12}$$

$$k \equiv \frac{2\pi}{\lambda} \tag{16.8}$$

$$\omega \equiv \frac{2\pi}{T} = 2\pi f \tag{16.3, 16.9}$$

The speed of a wave traveling on a taut string of mass per unit length μ and tension T is

$$v = \sqrt{\frac{T}{\mu}} \tag{16.18}$$

A wave is totally or partially reflected when it reaches the end of the medium in which it propagates or when it reaches a boundary where its speed changes discontinuously. If a wave traveling on a string meets a fixed end, the wave is reflected and inverted. If the wave reaches a free end, it is reflected but not inverted.

The **power** transmitted by a sinusoidal wave on a stretched string is

$$\mathcal{P} = \tfrac{1}{2}\mu\omega^2 A^2 v \tag{16.21}$$

Wave functions are solutions to a differential equation called the **linear wave equation:**

$$\frac{\partial^2 y}{\partial x^2} = \frac{1}{v^2} \frac{\partial^2 y}{\partial t^2} \tag{16.27}$$

QUESTIONS

1. Why is a pulse on a string considered to be transverse?

2. How would you create a longitudinal wave in a stretched spring? Would it be possible to create a transverse wave in a spring?

3. By what factor would you have to multiply the tension in a stretched string in order to double the wave speed?

4. When traveling on a taut string, does a pulse always invert upon reflection? Explain.

5. Does the vertical speed of a segment of a horizontal taut string, through which a wave is traveling, depend on the wave speed?

6. If you shake one end of a taut rope steadily three times each second, what would be the period of the sinusoidal wave set up in the rope?

7. A vibrating source generates a sinusoidal wave on a string under constant tension. If the power delivered to the

string is doubled, by what factor does the amplitude change? Does the wave speed change under these circumstances?

8. Consider a wave traveling on a taut rope. What is the difference, if any, between the speed of the wave and the speed of a small segment of the rope?

9. If a long rope is hung from a ceiling and waves are sent up the rope from its lower end, they do not ascend with constant speed. Explain.

10. How do transverse waves differ from longitudinal waves?

11. When all the strings on a guitar are stretched to the same tension, will the speed of a wave along the most massive bass string be faster, slower, or the same as the speed of a wave on the lighter strings?

12. If one end of a heavy rope is attached to one end of a light rope, the speed of a wave will change as the wave goes from the heavy rope to the light one. Will it increase or decrease? What happens to the frequency? To the wavelength?

13. If you stretch a rubber hose and pluck it, you can observe a pulse traveling up and down the hose. What happens to the speed of the pulse if you stretch the hose more tightly? What happens to the speed if you fill the hose with water?

14. In a longitudinal wave in a spring, the coils move back and forth in the direction of wave motion. Does the speed of the wave depend on the maximum speed of each coil?

15. Both longitudinal and transverse waves can propagate through a solid. A wave on the surface of a liquid can involve both longitudinal and transverse motion of elements of the medium. On the other hand, a wave propagating through the volume of a fluid must be purely longitudinal, not transverse. Why?

16. In an earthquake both S (transverse) and P (longitudinal) waves propagate from the focus of the earthquake. The focus is in the ground below the epicenter on the surface. The S waves travel through the Earth more slowly than the P waves (at about 5 km/s versus 8 km/s). By detecting the time of arrival of the waves, how can one determine the distance to the focus of the quake? How many detection stations are necessary to locate the focus unambiguously?

17. In mechanics, massless strings are often assumed. Why is this not a good assumption when discussing waves on strings?

PROBLEMS

1, 2, 3 = straightforward, intermediate, challenging ☐ = full solution available in the *Student Solutions Manual and Study Guide*

🌐 = coached solution with hints available at http://www.pse6.com 🖥 = computer useful in solving problem

▨ = paired numerical and symbolic problems

Section 16.1 Propagation of a Disturbance

1. At $t = 0$, a transverse pulse in a wire is described by the function

$$y = \frac{6}{x^2 + 3}$$

where x and y are in meters. Write the function $y(x, t)$ that describes this pulse if it is traveling in the positive x direction with a speed of 4.50 m/s.

2. Ocean waves with a crest-to-crest distance of 10.0 m can be described by the wave function

$$y(x, t) = (0.800 \text{ m}) \sin[0.628(x - vt)]$$

where $v = 1.20$ m/s. (a) Sketch $y(x, t)$ at $t = 0$. (b) Sketch $y(x, t)$ at $t = 2.00$ s. Note that the entire wave form has shifted 2.40 m in the positive x direction in this time interval.

3. A pulse moving along the x axis is described by

$$y(x, t) = 5.00e^{-(x + 5.00t)^2}$$

where x is in meters and t is in seconds. Determine (a) the direction of the wave motion, and (b) the speed of the pulse.

4. Two points A and B on the surface of the Earth are at the same longitude and 60.0° apart in latitude. Suppose that an earthquake at point A creates a P wave that reaches point B by traveling straight through the body of the Earth at a constant speed of 7.80 km/s. The earthquake also radiates a *Rayleigh wave*, which travels across the surface of the Earth in an analogous way to a surface wave on water, at 4.50 km/s.

(a) Which of these two seismic waves arrives at B first? (b) What is the time difference between the arrivals of the two waves at B? Take the radius of the Earth to be 6 370 km.

5. S and P waves, simultaneously radiated from the hypocenter of an earthquake, are received at a seismographic station 17.3 s apart. Assume the waves have traveled over the same path at speeds of 4.50 km/s and 7.80 km/s. Find the distance from the seismograph to the hypocenter of the quake.

Section 16.2 Sinusoidal Waves

6. For a certain transverse wave, the distance between two successive crests is 1.20 m, and eight crests pass a given point along the direction of travel every 12.0 s. Calculate the wave speed.

7. A sinusoidal wave is traveling along a rope. The oscillator that generates the wave completes 40.0 vibrations in 30.0 s. Also, a given maximum travels 425 cm along the rope in 10.0 s. What is the wavelength?

8. When a particular wire is vibrating with a frequency of 4.00 Hz, a transverse wave of wavelength 60.0 cm is produced. Determine the speed of waves along the wire.

9. A wave is described by $y = (2.00 \text{ cm}) \sin(kx - \omega t)$, where $k = 2.11$ rad/m, $\omega = 3.62$ rad/s, x is in meters, and t is in seconds. Determine the amplitude, wavelength, frequency, and speed of the wave.

10. A sinusoidal wave on a string is described by

$$y = (0.51 \text{ cm}) \sin(kx - \omega t)$$

where $k = 3.10$ rad/cm and $\omega = 9.30$ rad/s. How far does a wave crest move in 10.0 s? Does it move in the positive or negative x direction?

11. Consider further the string shown in Figure 16.10 and treated in Example 16.3. Calculate (a) the maximum transverse speed and (b) the maximum transverse acceleration of a point on the string.

12. Consider the sinusoidal wave of Example 16.2, with the wave function

$$y = (15.0 \text{ cm}) \cos(0.157x - 50.3t).$$

At a certain instant, let point A be at the origin and point B be the first point along the x axis where the wave is 60.0° out of phase with point A. What is the coordinate of point B?

13. A sinusoidal wave is described by

$$y = (0.25 \text{ m}) \sin(0.30x - 40t)$$

where x and y are in meters and t is in seconds. Determine for this wave the (a) amplitude, (b) angular frequency, (c) angular wave number, (d) wavelength, (e) wave speed, and (f) direction of motion.

14. (a) Plot y versus t at $x = 0$ for a sinusoidal wave of the form $y = (15.0 \text{ cm}) \cos(0.157x - 50.3t)$, where x and y are in centimeters and t is in seconds. (b) Determine the period of vibration from this plot and compare your result with the value found in Example 16.2.

15. (a) Write the expression for y as a function of x and t for a sinusoidal wave traveling along a rope in the *negative* x direction with the following characteristics: $A = 8.00$ cm, $\lambda = 80.0$ cm, $f = 3.00$ Hz, and $y(0, t) = 0$ at $t = 0$. (b) **What If?** Write the expression for y as a function of x and t for the wave in part (a) assuming that $y(x, 0) = 0$ at the point $x = 10.0$ cm.

16. A sinusoidal wave traveling in the $-x$ direction (to the left) has an amplitude of 20.0 cm, a wavelength of 35.0 cm, and a frequency of 12.0 Hz. The transverse position of an element of the medium at $t = 0$, $x = 0$ is $y = -3.00$ cm, and the element has a positive velocity here. (a) Sketch the wave at $t = 0$. (b) Find the angular wave number, period, angular frequency, and wave speed of the wave. (c) Write an expression for the wave function $y(x, t)$.

17. A transverse wave on a string is described by the wave function

$$y = (0.120 \text{ m}) \sin[(\pi x/8) + 4\pi t]$$

(a) Determine the transverse speed and acceleration at $t = 0.200$ s for the point on the string located at $x = 1.60$ m. (b) What are the wavelength, period, and speed of propagation of this wave?

18. A transverse sinusoidal wave on a string has a period $T = 25.0$ ms and travels in the negative x direction with a speed of 30.0 m/s. At $t = 0$, a particle on the string at $x = 0$ has a transverse position of 2.00 cm and is traveling downward with a speed of 2.00 m/s. (a) What is the amplitude of the wave? (b) What is the initial phase angle? (c) What is the maximum transverse speed of the string? (d) Write the wave function for the wave.

19. A sinusoidal wave of wavelength 2.00 m and amplitude 0.100 m travels on a string with a speed of 1.00 m/s to the right. Initially, the left end of the string is at the origin. Find (a) the frequency and angular frequency, (b) the angular wave number, and (c) the wave function for this wave. Determine the equation of motion for (d) the left end of the string and (e) the point on the string at $x = 1.50$ m to the right of the left end. (f) What is the maximum speed of any point on the string?

20. A wave on a string is described by the wave function $y = (0.100 \text{ m}) \sin(0.50x - 20t)$. (a) Show that a particle in the string at $x = 2.00$ m executes simple harmonic motion. (b) Determine the frequency of oscillation of this particular point.

Section 16.3 The Speed of Waves on Strings

21. A telephone cord is 4.00 m long. The cord has a mass of 0.200 kg. A transverse pulse is produced by plucking one end of the taut cord. The pulse makes four trips down and back along the cord in 0.800 s. What is the tension in the cord?

22. Transverse waves with a speed of 50.0 m/s are to be produced in a taut string. A 5.00-m length of string with a total mass of 0.060 0 kg is used. What is the required tension?

23. A piano string having a mass per unit length equal to 5.00×10^{-3} kg/m is under a tension of 1 350 N. Find the speed of a wave traveling on this string.

24. A transverse traveling wave on a taut wire has an amplitude of 0.200 mm and a frequency of 500 Hz. It travels with a speed of 196 m/s. (a) Write an equation in SI units of the form $y = A \sin(kx - \omega t)$ for this wave. (b) The mass per unit length of this wire is 4.10 g/m. Find the tension in the wire.

25. An astronaut on the Moon wishes to measure the local value of the free-fall acceleration by timing pulses traveling down a wire that has an object of large mass suspended from it. Assume a wire has a mass of 4.00 g and a length of 1.60 m, and that a 3.00-kg object is suspended from it. A pulse requires 36.1 ms to traverse the length of the wire. Calculate g_{Moon} from these data. (You may ignore the mass of the wire when calculating the tension in it.)

26. Transverse pulses travel with a speed of 200 m/s along a taut copper wire whose diameter is 1.50 mm. What is the tension in the wire? (The density of copper is 8.92 g/cm³.)

27. Transverse waves travel with a speed of 20.0 m/s in a string under a tension of 6.00 N. What tension is required for a wave speed of 30.0 m/s in the same string?

28. A simple pendulum consists of a ball of mass M hanging from a uniform string of mass m and length L, with $m \ll M$. If the period of oscillations for the pendulum is T, determine the speed of a transverse wave in the string when the pendulum hangs at rest.

29. The elastic limit of the steel forming a piece of wire is equal to 2.70×10^8 Pa. What is the maximum speed at which transverse wave pulses can propagate along this wire without exceeding this stress? (The density of steel is 7.86×10^3 km/m³.)

30. **Review problem.** A light string with a mass per unit length of 8.00 g/m has its ends tied to two walls separated by a

distance equal to three fourths of the length of the string (Fig. P16.30). An object of mass m is suspended from the center of the string, putting a tension in the string. (a) Find an expression for the transverse wave speed in the string as a function of the mass of the hanging object. (b) What should be the mass of the object suspended from the string in order to produce a wave speed of 60.0 m/s?

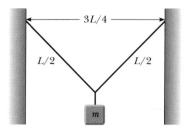

Figure P16.30

31. A 30.0-m steel wire and a 20.0-m copper wire, both with 1.00-mm diameters, are connected end to end and stretched to a tension of 150 N. How long does it take a transverse wave to travel the entire length of the two wires?

32. **Review problem.** A light string of mass m and length L has its ends tied to two walls that are separated by the distance D. Two objects, each of mass M, are suspended from the string as in Figure P16.32. If a wave pulse is sent from point A, how long does it take to travel to point B?

Figure P16.32

33. A student taking a quiz finds on a reference sheet the two equations

$$f = 1/T \quad \text{and} \quad v = \sqrt{T/\mu}$$

She has forgotten what T represents in each equation. (a) Use dimensional analysis to determine the units required for T in each equation. (b) Identify the physical quantity each T represents.

Section 16.5   Rate of Energy Transfer by Sinusoidal Waves on Strings

34. A taut rope has a mass of 0.180 kg and a length of 3.60 m. What power must be supplied to the rope in order to generate sinusoidal waves having an amplitude of 0.100 m

and a wavelength of 0.500 m and traveling with a speed of 30.0 m/s?

35. A two-dimensional water wave spreads in circular ripples. Show that the amplitude A at a distance r from the initial disturbance is proportional to $1/\sqrt{r}$. (*Suggestion:* Consider the energy carried by one outward-moving ripple.)

36. Transverse waves are being generated on a rope under constant tension. By what factor is the required power increased or decreased if (a) the length of the rope is doubled and the angular frequency remains constant, (b) the amplitude is doubled and the angular frequency is halved, (c) both the wavelength and the amplitude are doubled, and (d) both the length of the rope and the wavelength are halved?

37. Sinusoidal waves 5.00 cm in amplitude are to be transmitted along a string that has a linear mass density of 4.00×10^{-2} kg/m. If the source can deliver a maximum power of 300 W and the string is under a tension of 100 N, what is the highest frequency at which the source can operate?

38. It is found that a 6.00-m segment of a long string contains four complete waves and has a mass of 180 g. The string is vibrating sinusoidally with a frequency of 50.0 Hz and a peak-to-valley distance of 15.0 cm. (The "peak-to-valley" distance is the vertical distance from the farthest positive position to the farthest negative position.) (a) Write the function that describes this wave traveling in the positive x direction. (b) Determine the power being supplied to the string.

39. A sinusoidal wave on a string is described by the equation

$$y = (0.15 \text{ m}) \sin(0.80x - 50t)$$

where x and y are in meters and t is in seconds. If the mass per unit length of this string is 12.0 g/m, determine (a) the speed of the wave, (b) the wavelength, (c) the frequency, and (d) the power transmitted to the wave.

40. The wave function for a wave on a taut string is

$$y(x, t) = (0.350 \text{ m})\sin(10\pi t - 3\pi x + \pi/4)$$

where x is in meters and t in seconds. (a) What is the average rate at which energy is transmitted along the string if the linear mass density is 75.0 g/m? (b) What is the energy contained in each cycle of the wave?

41. A horizontal string can transmit a maximum power $\mathcal{P}_0$ (without breaking) if a wave with amplitude A and angular frequency ω is traveling along it. In order to increase this maximum power, a student folds the string and uses this "double string" as a medium. Determine the maximum power that can be transmitted along the "double string," assuming that the tension is constant.

42. In a region far from the epicenter of an earthquake, a seismic wave can be modeled as transporting energy in a single direction without absorption, just as a string wave does. Suppose the seismic wave moves from granite into mudfill with similar density but with a much lower bulk modulus. Assume the speed of the wave gradually drops by a factor of 25.0, with negligible reflection of the wave. Will the amplitude of the ground shaking increase or decrease? By

what factor? This phenomenon led to the collapse of part of the Nimitz Freeway in Oakland, California, during the Loma Prieta earthquake of 1989.

Section 16.6   The Linear Wave Equation

43. (a) Evaluate A in the scalar equality $(7 + 3)4 = A$. (b) Evaluate A, B, and C in the vector equality $7.00\hat{\mathbf{i}} + 3.00\hat{\mathbf{k}} = A\hat{\mathbf{i}} + B\hat{\mathbf{j}} + C\hat{\mathbf{k}}$. Explain how you arrive at the answers to convince a student who thinks that you cannot solve a single equation for three different unknowns. (c) **What If?** The functional equality or identity

$$A + B\cos(Cx + Dt + E) = (7.00 \text{ mm}) \cos(3x + 4t + 2)$$

is true for all values of the variables x and t, which are measured in meters and in seconds, respectively. Evaluate the constants A, B, C, D, and E. Explain how you arrive at the answers.

44. Show that the wave function $y = e^{b(x - vt)}$ is a solution of the linear wave equation (Eq. 16.27), where b is a constant.

45. Show that the wave function $y = \ln[b(x - vt)]$ is a solution to Equation 16.27, where b is a constant.

46. (a) Show that the function $y(x, t) = x^2 + v^2t^2$ is a solution to the wave equation. (b) Show that the function in part (a) can be written as $f(x + vt) + g(x - vt)$, and determine the functional forms for f and g. (c) **What If?** Repeat parts (a) and (b) for the function $y(x, t) = \sin(x)\cos(vt)$.

Additional Problems

47. "The wave" is a particular type of pulse that can propagate through a large crowd gathered at a sports arena to watch a soccer or American football match (Figure P16.47). The elements of the medium are the spectators, with zero posi-

Figure P16.47

tion corresponding to their being seated and maximum position corresponding to their standing and raising their arms. When a large fraction of the spectators participate in the wave motion, a somewhat stable pulse shape can develop. The wave speed depends on people's reaction time, which is typically on the order of 0.1 s. Estimate the order of magnitude, in minutes, of the time required for such a pulse to make one circuit around a large sports stadium. State the quantities you measure or estimate and their values.

48. A traveling wave propagates according to the expression $y = (4.0 \text{ cm}) \sin(2.0x - 3.0t)$, where x is in centimeters and t is in seconds. Determine (a) the amplitude, (b) the wavelength, (c) the frequency, (d) the period, and (e) the direction of travel of the wave.

49. The wave function for a traveling wave on a taut string is (in SI units)

$$y(x, t) = (0.350 \text{ m}) \sin(10\pi t - 3\pi x + \pi/4)$$

(a) What are the speed and direction of travel of the wave? (b) What is the vertical position of an element of the string at $t = 0$, $x = 0.100$ m? (c) What are the wavelength and frequency of the wave? (d) What is the maximum magnitude of the transverse speed of the string?

50. A transverse wave on a string is described by the equation

$$y(x, t) = (0.350 \text{ m}) \sin[(1.25 \text{ rad/m})x + (99.6 \text{ rad/s})t]$$

Consider the element of the string at $x = 0$. (a) What is the time interval between the first two instants when this element has a position of $y = 0.175$ m? (b) What distance does the wave travel during this time interval?

51. Motion picture film is projected at 24.0 frames per second. Each frame is a photograph 19.0 mm high. At what constant speed does the film pass into the projector?

52. **Review problem.** A block of mass M, supported by a string, rests on a frictionless incline making an angle θ with the horizontal (Fig. P16.52). The length of the string is L, and its mass is $m \ll M$. Derive an expression for the time interval required for a transverse wave to travel from one end of the string to the other.

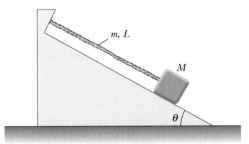

Figure P16.52

53. **Review problem.** A 2.00-kg block hangs from a rubber cord, being supported so that the cord is not stretched. The unstretched length of the cord is 0.500 m, and its mass is 5.00 g. The "spring constant" for the cord is 100 N/m. The block is released and stops at the lowest point. (a) Determine the tension in the cord when the block is at this lowest point. (b) What is the length of the cord in this "stretched" position? (c) Find the speed of a transverse wave in the cord if the block is held in this lowest position.

54. **Review problem.** A block of mass M hangs from a rubber cord. The block is supported so that the cord is not stretched. The unstretched length of the cord is L_0 and its mass is m, much less than M. The "spring constant" for the cord is k. The block is released and stops at the lowest point. (a) Determine the tension in the string when the block is at this lowest point. (b) What is the length of the cord in this "stretched" position? (c) Find the speed of a transverse wave in the cord if the block is held in this lowest position.

55. (a) Determine the speed of transverse waves on a string under a tension of 80.0 N if the string has a length of 2.00 m and a mass of 5.00 g. (b) Calculate the power required to generate these waves if they have a wavelength of 16.0 cm and an amplitude of 4.00 cm.

56. A sinusoidal wave in a rope is described by the wave function

$$y = (0.20 \text{ m}) \sin(0.75\pi x + 18\pi t)$$

where x and y are in meters and t is in seconds. The rope has a linear mass density of 0.250 kg/m. If the tension in the rope is provided by an arrangement like the one illustrated in Figure 16.12, what is the value of the suspended mass?

57. A block of mass 0.450 kg is attached to one end of a cord of mass 0.003 20 kg; the other end of the cord is attached to a fixed point. The block rotates with constant angular speed in a circle on a horizontal frictionless table. Through what angle does the block rotate in the time that a transverse wave takes to travel along the string from the center of the circle to the block?

58. A wire of density ρ is tapered so that its cross-sectional area varies with x according to

$$A = (1.0 \times 10^{-3} x + 0.010) \text{ cm}^2$$

(a) If the wire is subject to a tension T, derive a relationship for the speed of a wave as a function of position. (b) **What If?** If the wire is aluminum and is subject to a tension of 24.0 N, determine the speed at the origin and at $x = 10.0$ m.

59. A rope of total mass m and length L is suspended vertically. Show that a transverse pulse travels the length of the rope in a time interval $\Delta t = 2\sqrt{L/g}$. (*Suggestion:* First find an expression for the wave speed at any point a distance x from the lower end by considering the tension in the rope as resulting from the weight of the segment below that point.)

60. If an object of mass M is suspended from the bottom of the rope in Problem 59, (a) show that the time interval for a transverse pulse to travel the length of the rope is

$$\Delta t = 2\sqrt{\frac{L}{mg}}\left(\sqrt{M + m} - \sqrt{M}\right)$$

What If? (b) Show that this reduces to the result of Problem 59 when $M = 0$. (c) Show that for $m \ll M$, the

expression in part (a) reduces to

$$\Delta t = \sqrt{\frac{mL}{Mg}}$$

61. It is stated in Problem 59 that a pulse travels from the bottom to the top of a hanging rope of length L in a time interval $\Delta t = 2\sqrt{L/g}$. Use this result to answer the following questions. (It is not necessary to set up any new integrations.) (a) How long does it take for a pulse to travel halfway up the rope? Give your answer as a fraction of the quantity $2\sqrt{L/g}$. (b) A pulse starts traveling up the rope. How far has it traveled after a time interval $\sqrt{L/g}$?

62. Determine the speed and direction of propagation of each of the following sinusoidal waves, assuming that x and y are measured in meters and t in seconds.

(a) $y = 0.60 \cos(3.0x - 15t + 2)$

(b) $y = 0.40 \cos(3.0x + 15t - 2)$

(c) $y = 1.2 \sin(15t + 2.0x)$

(d) $y = 0.20 \sin[12t - (x/2) + \pi]$

63. An aluminum wire is clamped at each end under zero tension at room temperature. The tension in the wire is increased by reducing the temperature, which results in a decrease in the wire's equilibrium length. What strain $(\Delta L/L)$ results in a transverse wave speed of 100 m/s? Take the cross-sectional area of the wire to be $5.00 \times 10^{-6} \text{ m}^2$, the density to be $2.70 \times 10^3 \text{ kg/m}^3$, and Young's modulus to be $7.00 \times 10^{10} \text{ N/m}^2$.

64. If a loop of chain is spun at high speed, it can roll along the ground like a circular hoop without slipping or collapsing. Consider a chain of uniform linear mass density μ whose center of mass travels to the right at a high speed v_0. (a) Determine the tension in the chain in terms of μ and v_0. (b) If the loop rolls over a bump, the resulting deformation of the chain causes two transverse pulses to propagate along the chain, one moving clockwise and one moving counterclockwise. What is the speed of the pulses traveling along the chain? (c) Through what angle does each pulse travel during the time it takes the loop to make one revolution?

65. (a) Show that the speed of longitudinal waves along a spring of force constant k is $v = \sqrt{kL/\mu}$, where L is the unstretched length of the spring and μ is the mass per unit length. (b) A spring with a mass of 0.400 kg has an unstretched length of 2.00 m and a force constant of 100 N/m. Using the result you obtained in (a), determine the speed of longitudinal waves along this spring.

66. A string of length L consists of two sections. The left half has mass per unit length $\mu = \mu_0/2$, while the right has a mass per unit length $\mu' = 3\mu = 3\mu_0/2$. Tension in the string is T_0. Notice from the data given that this string has the same total mass as a uniform string of length L and mass per unit length μ_0. (a) Find the speeds v and v' at which transverse pulses travel in the two sections. Express the speeds in terms of T_0 and μ_0, and also as multiples of the speed $v_0 = (T_0/\mu_0)^{1/2}$. (b) Find the time interval required for a pulse to travel from one end

of the string to the other. Give your result as a multiple of $\Delta t_0 = L/v_0$.

67. A pulse traveling along a string of linear mass density μ is described by the wave function

$$y = [A_0 e^{-bx}] \sin(kx - \omega t)$$

where the factor in brackets before the sine function is said to be the amplitude. (a) What is the power $\mathcal{P}(x)$ carried by this wave at a point x? (b) What is the power carried by this wave at the origin? (c) Compute the ratio $\mathcal{P}(x)/\mathcal{P}(0)$.

68. An earthquake on the ocean floor in the Gulf of Alaska produces a *tsunami* (sometimes incorrectly called a "tidal wave") that reaches Hilo, Hawaii, 4 450 km away, in a time interval of 9 h 30 min. Tsunamis have enormous wavelengths (100 to 200 km), and the propagation speed for these waves is $v \approx \sqrt{gd}$, where d is the average depth of the water. From the information given, find the average wave speed and the average ocean depth between Alaska and Hawaii. (This method was used in 1856 to estimate the average depth of the Pacific Ocean long before soundings were made to give a direct determination.)

69. A string on a musical instrument is held under tension T and extends from the point $x = 0$ to the point $x = L$. The string is overwound with wire in such a way that its mass per unit length $\mu(x)$ increases uniformly from μ_0 at $x = 0$ to μ_L at $x = L$. (a) Find an expression for $\mu(x)$ as a function of x over the range $0 \le x \le L$. (b) Show that the time interval required for a transverse pulse to travel the length of the string is given by

$$\Delta t = \frac{2L\left(\mu_L + \mu_0 + \sqrt{\mu_L \mu_0}\right)}{3\sqrt{T}\left(\sqrt{\mu_L} + \sqrt{\mu_0}\right)}$$

Answers to Quick Quizzes

16.1 (b). It is longitudinal because the disturbance (the shift of position of the people) is parallel to the direction in which the wave travels.

16.2 (a). It is transverse because the people stand up and sit down (vertical motion), whereas the wave moves either to the left or to the right.

16.3 (c). The wave speed is determined by the medium, so it is unaffected by changing the frequency.

16.4 (b). Because the wave speed remains the same, the result of doubling the frequency is that the wavelength is half as large.

16.5 (d). The amplitude of a wave is unrelated to the wave speed, so we cannot determine the new amplitude without further information.

16.6 (c). With a larger amplitude, an element of the string has more energy associated with its simple harmonic motion, so the element passes through the equilibrium position with a higher maximum transverse speed.

16.7 Only answers (f) and (h) are correct. (a) and (b) affect the transverse speed of a particle of the string, but not the wave speed along the string. (c) and (d) change the amplitude. (e) and (g) increase the time interval by decreasing the wave speed.

16.8 (d). Doubling the amplitude of the wave causes the power to be larger by a factor of 4. In (a), halving the linear mass density of the string causes the power to change by a factor of 0.71—the rate decreases. In (b), doubling the wavelength of the wave halves the frequency and causes the power to change by a factor of 0.25—the rate decreases. In (c), doubling the tension in the string changes the wave speed and causes the power to change by a factor of 1.4—not as large as in part (d).

Chapter 17

Sound Waves

▲ Human ears have evolved to detect sound waves and interpret them as music or speech. Some animals, such as this young bat-eared fox, have ears adapted for the detection of very weak sounds. (Getty Images)

Sound waves are the most common example of longitudinal waves. They travel through any material medium with a speed that depends on the properties of the medium. As the waves travel through air, the elements of air vibrate to produce changes in density and pressure along the direction of motion of the wave. If the source of the sound waves vibrates sinusoidally, the pressure variations are also sinusoidal. The mathematical description of sinusoidal sound waves is very similar to that of sinusoidal string waves, which were discussed in the previous chapter.

Sound waves are divided into three categories that cover different frequency ranges. (1) *Audible waves* lie within the range of sensitivity of the human ear. They can be generated in a variety of ways, such as by musical instruments, human voices, or loudspeakers. (2) *Infrasonic waves* have frequencies below the audible range. Elephants can use infrasonic waves to communicate with each other, even when separated by many kilometers. (3) *Ultrasonic waves* have frequencies above the audible range. You may have used a "silent" whistle to retrieve your dog. The ultrasonic sound it emits is easily heard by dogs, although humans cannot detect it at all. Ultrasonic waves are also used in medical imaging.

We begin this chapter by discussing the speed of sound waves and then wave intensity, which is a function of wave amplitude. We then provide an alternative description of the intensity of sound waves that compresses the wide range of intensities to which the ear is sensitive into a smaller range for convenience. We investigate the effects of the motion of sources and/or listeners on the frequency of a sound. Finally, we explore digital reproduction of sound, focusing in particular on sound systems used in modern motion pictures.

17.1 Speed of Sound Waves

Let us describe pictorially the motion of a one-dimensional longitudinal pulse moving through a long tube containing a compressible gas (Fig. 17.1). A piston at the left end can be moved to the right to compress the gas and create the pulse. Before the piston is moved, the gas is undisturbed and of uniform density, as represented by the uniformly shaded region in Figure 17.1a. When the piston is suddenly pushed to the right (Fig. 17.1b), the gas just in front of it is compressed (as represented by the more heavily shaded region); the pressure and density in this region are now higher than they were before the piston moved. When the piston comes to rest (Fig. 17.1c), the compressed region of the gas continues to move to the right, corresponding to a longitudinal pulse traveling through the tube with speed v. Note that the piston speed does *not* equal v. Furthermore, the compressed region does not "stay with" the piston as the piston moves, because the speed of the wave is usually greater than the speed of the piston.

The speed of sound waves in a medium depends on the compressibility and density of the medium. If the medium is a liquid or a gas and has a bulk modulus B (see

Undisturbed gas

(a)

Compressed region

(b)

(c)

(d)

Figure 17.1 Motion of a longitudinal pulse through a compressible gas. The compression (darker region) is produced by the moving piston.

Table 17.1

Speed of Sound in Various Media	
Medium	**v (m/s)**
Gases	
Hydrogen (0°C)	1 286
Helium (0°C)	972
Air (20°C)	343
Air (0°C)	331
Oxygen (0°C)	317
Liquids at 25°C	
Glycerol	1 904
Seawater	1 533
Water	1 493
Mercury	1 450
Kerosene	1 324
Methyl alcohol	1 143
Carbon tetrachloride	926
Solids[a]	
Pyrex glass	5 640
Iron	5 950
Aluminum	6 420
Brass	4 700
Copper	5 010
Gold	3 240
Lucite	2 680
Lead	1 960
Rubber	1 600

[a] Values given are for propagation of longitudinal waves in bulk media. Speeds for longitudinal waves in thin rods are smaller, and speeds of transverse waves in bulk are smaller yet.

Section 12.4) and density ρ, the speed of sound waves in that medium is

$$v = \sqrt{\frac{B}{\rho}} \tag{17.1}$$

It is interesting to compare this expression with Equation 16.18 for the speed of transverse waves on a string, $v = \sqrt{T/\mu}$. In both cases, the wave speed depends on an elastic property of the medium—bulk modulus B or string tension T—and on an inertial property of the medium—ρ or μ. In fact, the *speed of all mechanical waves* follows an expression of the general form

$$v = \sqrt{\frac{\text{elastic property}}{\text{inertial property}}}$$

For longitudinal sound waves in a solid rod of material, for example, the speed of sound depends on Young's modulus Y and the density ρ. Table 17.1 provides the speed of sound in several different materials.

The speed of sound also depends on the temperature of the medium. For sound traveling through air, the relationship between wave speed and medium temperature is

$$v = (331 \text{ m/s})\sqrt{1 + \frac{T_C}{273°C}}$$

where 331 m/s is the speed of sound in air at 0°C, and T_C is the air temperature in degrees Celsius. Using this equation, one finds that at 20°C the speed of sound in air is approximately 343 m/s.

This information provides a convenient way to estimate the distance to a thunderstorm. You count the number of seconds between seeing the flash of lightning and hearing the thunder. Dividing this time by 3 gives the approximate distance to the lightning in kilometers, because 343 m/s is approximately $\frac{1}{3}$ km/s. Dividing the time in seconds by 5 gives the approximate distance to the lightning in miles, because the speed of sound in ft/s (1 125 ft/s) is approximately $\frac{1}{5}$ mi/s.

> **Quick Quiz 17.1** The speed of sound in air is a function of (a) wavelength (b) frequency (c) temperature (d) amplitude.

Example 17.1   Speed of Sound in a Liquid   Interactive

(A) Find the speed of sound in water, which has a bulk modulus of 2.1×10^9 N/m² at a temperature of 0°C and a density of 1.00×10^3 kg/m³.

Solution Using Equation 17.1, we find that

$$v_{\text{water}} = \sqrt{\frac{B}{\rho}} = \sqrt{\frac{2.1 \times 10^9 \text{ N/m}^2}{1.00 \times 10^3 \text{ kg/m}^3}} = \boxed{1.4 \text{ km/s}}$$

In general, sound waves travel more slowly in liquids than in solids because liquids are more compressible than solids. Note that the speed of sound in water is lower at 0°C than at 25°C (Table 17.1).

(B) Dolphins use sound waves to locate food. Experiments have shown that a dolphin can detect a 7.5-cm target 110 m away, even in murky water. For a bit of "dinner" at that distance, how much time passes between the moment the dolphin emits a sound pulse and the moment the dolphin hears its reflection and thereby detects the distant target?

Solution The total distance covered by the sound wave as it travels from dolphin to target and back is 2×110 m = 220 m. From Equation 2.2, we have, for 25°C water

$$\Delta t = \frac{\Delta x}{v_x} = \frac{220 \text{ m}}{1\,533 \text{ m/s}} = \boxed{0.14 \text{ s}}$$

 *At the Interactive Worked Example link at **http://www.pse6.com**, you can compare the speed of sound through the various media found in Table 17.1.*

17.2 Periodic Sound Waves

This section will help you better comprehend the nature of sound waves. An important fact for understanding how our ears work is that *pressure variations control what we hear.*

One can produce a one-dimensional periodic sound wave in a long, narrow tube containing a gas by means of an oscillating piston at one end, as shown in Figure 17.2. The darker parts of the colored areas in this figure represent regions where the gas is compressed and thus the density and pressure are above their equilibrium values. A compressed region is formed whenever the piston is pushed into the tube. This compressed region, called a **compression,** moves through the tube as a pulse, continuously compressing the region just in front of itself. When the piston is pulled back, the gas in front of it expands, and the pressure and density in this region fall below their equilibrium values (represented by the lighter parts of the colored areas in Fig. 17.2). These low-pressure regions, called **rarefactions,** also propagate along the tube, following the compressions. Both regions move with a speed equal to the speed of sound in the medium.

As the piston oscillates sinusoidally, regions of compression and rarefaction are continuously set up. The distance between two successive compressions (or two successive rarefactions) equals the wavelength λ. As these regions travel through the tube, any small element of the medium moves with simple harmonic motion parallel to the direction of the wave. If $s(x, t)$ is the position of a small element relative to its equilibrium position,[1] we can express this harmonic position function as

$$s(x, t) = s_{\max} \cos(kx - \omega t) \tag{17.2}$$

where $s_{\max}$ **is the maximum position of the element relative to equilibrium.** This is often called the **displacement amplitude** of the wave. The parameter k is the wave number and ω is the angular frequency of the piston. Note that the displacement of the element is along x, in the direction of propagation of the sound wave, which means we are describing a longitudinal wave.

The variation in the gas pressure ΔP measured from the equilibrium value is also periodic. For the position function in Equation 17.2, ΔP is given by

$$\Delta P = \Delta P_{\max} \sin(kx - \omega t) \tag{17.3}$$

where **the pressure amplitude** $\Delta P_{\max}$—which is the **maximum change in pressure from the equilibrium value**—is given by

$$\Delta P_{\max} = \rho v \omega s_{\max} \tag{17.4}$$

Thus, we see that a sound wave may be considered as either a displacement wave or a pressure wave. A comparison of Equations 17.2 and 17.3 shows that **the pressure wave is 90° out of phase with the displacement wave.** Graphs of these functions are shown in Figure 17.3. Note that the pressure variation is a maximum when the displacement from equilibrium is zero, and the displacement from equilibrium is a maximum when the pressure variation is zero.

Active Figure 17.2 A longitudinal wave propagating through a gas-filled tube. The source of the wave is an oscillating piston at the left.

At the Active Figures link at http://www.pse6.com, you can adjust the frequency of the piston.

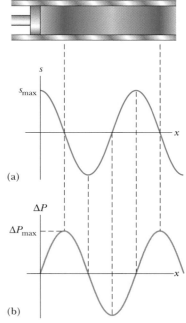

Figure 17.3 (a) Displacement amplitude and (b) pressure amplitude versus position for a sinusoidal longitudinal wave.

> **Quick Quiz 17.2** If you blow across the top of an empty soft-drink bottle, a pulse of sound travels down through the air in the bottle. At the moment the pulse reaches the bottom of the bottle, the correct descriptions of the displacement of elements of air from their equilibrium positions and the pressure of the air at this point are (a) the displacement and pressure are both at a maximum (b) the displacement and pressure are both at a minimum (c) the displacement is zero and the pressure is a maximum (d) the displacement is zero and the pressure is a minimum.

---

[1] We use $s(x, t)$ here instead of $y(x, t)$ because the displacement of elements of the medium is not perpendicular to the x direction.

Derivation of Equation 17.3

Consider a thin disk-shaped element of gas whose circular cross section is parallel to the piston in Figure 17.2. This element will undergo changes in position, pressure, and density as a sound wave propagates through the gas. From the definition of bulk modulus (see Eq. 12.8), the pressure variation in the gas is

$$\Delta P = -B \frac{\Delta V}{V_i}$$

The element has a thickness Δx in the horizontal direction and a cross-sectional area A, so its volume is $V_i = A\Delta x$. The change in volume ΔV accompanying the pressure change is equal to $A\Delta s$, where Δs is the difference between the value of s at $x + \Delta x$ and the value of s at x. Hence, we can express ΔP as

$$\Delta P = -B \frac{\Delta V}{V_i} = -B \frac{A\Delta s}{A\Delta x} = -B \frac{\Delta s}{\Delta x}$$

As Δx approaches zero, the ratio $\Delta s/\Delta x$ becomes $\partial s/\partial x$. (The partial derivative indicates that we are interested in the variation of s with position at a *fixed* time.) Therefore,

$$\Delta P = -B \frac{\partial s}{\partial x}$$

If the position function is the simple sinusoidal function given by Equation 17.2, we find that

$$\Delta P = -B \frac{\partial}{\partial x} [s_{max} \cos(kx - \omega t)] = Bs_{max}k \sin(kx - \omega t)$$

Because the bulk modulus is given by $B = \rho v^2$ (see Eq. 17.1), the pressure variation reduces to

$$\Delta P = \rho v^2 s_{max}k \sin(kx - \omega t)$$

From Equation 16.11, we can write $k = \omega/v$; hence, ΔP can be expressed as

$$\Delta P = \rho v \omega s_{max} \sin(kx - \omega t)$$

Because the sine function has a maximum value of 1, we see that the maximum value of the pressure variation is $\Delta P_{max} = \rho v \omega s_{max}$ (see Eq. 17.4), and we arrive at Equation 17.3:

$$\Delta P = \Delta P_{max} \sin(kx - \omega t)$$

17.3 Intensity of Periodic Sound Waves

In the preceding chapter, we showed that a wave traveling on a taut string transports energy. The same concept applies to sound waves. Consider an element of air of mass Δm and width Δx in front of a piston oscillating with a frequency ω, as shown in Figure 17.4.

Figure 17.4 An oscillating piston transfers energy to the air in the tube, causing the element of air of width Δx and mass Δm to oscillate with an amplitude s_{max}.

The piston transmits energy to this element of air in the tube, and the energy is propagated away from the piston by the sound wave. To evaluate the rate of energy transfer for the sound wave, we shall evaluate the kinetic energy of this element of air, which is undergoing simple harmonic motion. We shall follow a procedure similar to that in Section 16.5, in which we evaluated the rate of energy transfer for a wave on a string.

As the sound wave propagates away from the piston, the position of any element of air in front of the piston is given by Equation 17.2. To evaluate the kinetic energy of this element of air, we need to know its speed. We find the speed by taking the time derivative of Equation 17.2:

$$v(x, t) = \frac{\partial}{\partial t} s(x, t) = \frac{\partial}{\partial t} [s_{max} \cos(kx - \omega t)] = -\omega s_{max} \sin(kx - \omega t)$$

Imagine that we take a "snapshot" of the wave at $t = 0$. The kinetic energy of a given element of air at this time is

$$\Delta K = \tfrac{1}{2}\Delta m(v)^2 = \tfrac{1}{2}\Delta m(-\omega s_{max} \sin kx)^2 = \tfrac{1}{2}\rho A\,\Delta x(-\omega s_{max} \sin kx)^2$$

$$= \tfrac{1}{2}\rho A\,\Delta x(\omega s_{max})^2 \sin^2 kx$$

where A is the cross-sectional area of the element and $A\,\Delta x$ is its volume. Now, as in Section 16.5, we integrate this expression over a full wavelength to find the total kinetic energy in one wavelength. Letting the element of air shrink to infinitesimal thickness, so that $\Delta x \rightarrow dx$, we have

$$K_\lambda = \int dK = \int_0^\lambda \tfrac{1}{2}\rho A(\omega s_{max})^2 \sin^2 kx\,dx = \tfrac{1}{2}\rho A(\omega s_{max})^2 \int_0^\lambda \sin^2 kx\,dx$$

$$= \tfrac{1}{2}\rho A(\omega s_{max})^2 (\tfrac{1}{2}\lambda) = \tfrac{1}{4}\rho A(\omega s_{max})^2\lambda$$

As in the case of the string wave in Section 16.5, the total potential energy for one wavelength has the same value as the total kinetic energy; thus, the total mechanical energy for one wavelength is

$$E_\lambda = K_\lambda + U_\lambda = \tfrac{1}{2}\rho A(\omega s_{max})^2\lambda$$

As the sound wave moves through the air, this amount of energy passes by a given point during one period of oscillation. Hence, the rate of energy transfer is

$$\mathcal{P} = \frac{\Delta E}{\Delta t} = \frac{E_\lambda}{T} = \frac{\tfrac{1}{2}\rho A(\omega s_{max})^2\lambda}{T} = \tfrac{1}{2}\rho A(\omega s_{max})^2 \left(\frac{\lambda}{T}\right) = \tfrac{1}{2}\rho A v(\omega s_{max})^2$$

where v is the speed of sound in air.

We define the **intensity** I of a wave, or the power per unit area, to be the rate at which the energy being transported by the wave transfers through a unit area A perpendicular to the direction of travel of the wave:

$$I \equiv \frac{\mathcal{P}}{A} \tag{17.5}$$

In the present case, therefore, the intensity is

$$I = \frac{\mathcal{P}}{A} = \tfrac{1}{2}\rho v(\omega s_{max})^2$$

Intensity of a sound wave

Thus, we see that the intensity of a periodic sound wave is proportional to the square of the displacement amplitude and to the square of the angular frequency (as in the case of a periodic string wave). This can also be written in terms of the pressure amplitude $\Delta P_{\max}$; in this case, we use Equation 17.4 to obtain

$$I = \frac{\Delta P_{\max}^2}{2\rho v} \qquad (17.6)$$

Now consider a point source emitting sound waves equally in all directions. From everyday experience, we know that the intensity of sound decreases as we move farther from the source. We identify an imaginary sphere of radius r centered on the source. When a source emits sound equally in all directions, we describe the result as a **spherical wave.** The average power $\mathcal{P}_{\text{av}}$ emitted by the source must be distributed uniformly over this spherical surface of area $4\pi r^2$. Hence, the wave intensity at a distance r from the source is

Inverse-square behavior of intensity for a point source

$$I = \frac{\mathcal{P}_{\text{av}}}{A} = \frac{\mathcal{P}_{\text{av}}}{4\pi r^2} \qquad (17.7)$$

This inverse-square law, which is reminiscent of the behavior of gravity in Chapter 13, states that the intensity decreases in proportion to the square of the distance from the source.

Quick Quiz 17.3 An *ear trumpet* is a cone-shaped shell, like a megaphone, that was used before hearing aids were developed to help persons who were hard of hearing. The small end of the cone was held in the ear, and the large end was aimed toward the source of sound as in Figure 17.5. The ear trumpet increases the intensity of sound because (a) it increases the speed of sound (b) it reflects sound back toward the source (c) it gathers sound that would normally miss the ear and concentrates it into a smaller area (d) it increases the density of the air.

Courtesy Kenneth Burger Museum Archives/Kent State University

Figure 17.5 (Quick Quiz 17.3) An ear trumpet, used before hearing aids to make sounds intense enough for people who were hard of hearing. You can simulate the effect of an ear trumpet by cupping your hands behind your ears.

Quick Quiz 17.4 A vibrating guitar string makes very little sound if it is not mounted on the guitar. But if this vibrating string is attached to the guitar body, so that the body of the guitar vibrates, the sound is higher in intensity. This is because (a) the power of the vibration is spread out over a larger area (b) the energy leaves the guitar at a higher rate (c) the speed of sound is higher in the material of the guitar body (d) none of these.

Example 17.2 Hearing Limits

The faintest sounds the human ear can detect at a frequency of 1 000 Hz correspond to an intensity of about 1.00×10^{-12} W/m²—the so-called *threshold of hearing*. The loudest sounds the ear can tolerate at this frequency correspond to an intensity of about 1.00 W/m²—the *threshold of pain*. Determine the pressure amplitude and displacement amplitude associated with these two limits.

Solution First, consider the faintest sounds. Using Equation 17.6 and taking $v = 343$ m/s as the speed of sound waves in air and $\rho = 1.20$ kg/m³ as the density of air, we obtain

$$\Delta P_{max} = \sqrt{2\rho v I}$$

$$= \sqrt{2(1.20 \text{ kg/m}^3)(343 \text{ m/s})(1.00 \times 10^{-12} \text{ W/m}^2)}$$

$$= \boxed{2.87 \times 10^{-5} \text{ N/m}^2}$$

Because atmospheric pressure is about 10^5 N/m², this result tells us that the ear is sensitive to pressure fluctuations as small as 3 parts in 10^{10}!

We can calculate the corresponding displacement amplitude by using Equation 17.4, recalling that $\omega = 2\pi f$ (see Eqs. 16.3 and 16.9):

$$s_{max} = \frac{\Delta P_{max}}{\rho v \omega} = \frac{2.87 \times 10^{-5} \text{ N/m}^2}{(1.20 \text{ kg/m}^3)(343 \text{ m/s})(2\pi \times 1\,000 \text{ Hz})}$$

$$= \boxed{1.11 \times 10^{-11} \text{ m}}$$

This is a remarkably small number! If we compare this result for s_{max} with the size of an atom (about 10^{-10} m), we see that the ear is an extremely sensitive detector of sound waves.

In a similar manner, one finds that the loudest sounds the human ear can tolerate correspond to a pressure amplitude of 28.7 N/m² and a displacement amplitude equal to 1.11×10^{-5} m.

Example 17.3 Intensity Variations of a Point Source

A point source emits sound waves with an average power output of 80.0 W.

(A) Find the intensity 3.00 m from the source.

Solution A point source emits energy in the form of spherical waves. Using Equation 17.7, we have

$$I = \frac{\mathcal{P}_{av}}{4\pi r^2} = \frac{80.0 \text{ W}}{4\pi(3.00 \text{ m})^2} = \boxed{0.707 \text{ W/m}^2}$$

an intensity that is close to the threshold of pain.

(B) Find the distance at which the intensity of the sound is 1.00×10^{-8} W/m².

Solution Using this value for I in Equation 17.7 and solving for r, we obtain

$$r = \sqrt{\frac{\mathcal{P}_{av}}{4\pi I}} = \sqrt{\frac{80.0 \text{ W}}{4\pi(1.00 \times 10^{-8} \text{ W/m}^2)}}$$

$$= \boxed{2.52 \times 10^4 \text{ m}}$$

which equals about 16 miles!

Sound Level in Decibels

Example 17.2 illustrates the wide range of intensities the human ear can detect. Because this range is so wide, it is convenient to use a logarithmic scale, where the **sound level** β (Greek beta) is defined by the equation

$$\beta \equiv 10 \log\left(\frac{I}{I_0}\right) \tag{17.8}$$

Sound level in decibels

The constant I_0 is the *reference intensity*, taken to be at the threshold of hearing ($I_0 = 1.00 \times 10^{-12}$ W/m²), and I is the intensity in watts per square meter to which the sound level β corresponds, where β is measured[2] in **decibels** (dB). On this scale,

---

[2] The unit *bel* is named after the inventor of the telephone, Alexander Graham Bell (1847–1922). The prefix *deci-* is the SI prefix that stands for 10^{-1}.

Table 17.2

Sound Levels	
Source of Sound	**β (dB)**
Nearby jet airplane	150
Jackhammer; machine gun	130
Siren; rock concert	120
Subway; power mower	100
Busy traffic	80
Vacuum cleaner	70
Normal conversation	50
Mosquito buzzing	40
Whisper	30
Rustling leaves	10
Threshold of hearing	0

the threshold of pain ($I = 1.00 \text{ W/m}^2$) corresponds to a sound level of $\beta = 10 \log[(1 \text{ W/m}^2)/(10^{-12} \text{ W/m}^2)] = 10 \log(10^{12}) = 120$ dB, and the threshold of hearing corresponds to $\beta = 10 \log[(10^{-12} \text{ W/m}^2)/(10^{-12} \text{ W/m}^2)] = 0$ dB.

Prolonged exposure to high sound levels may seriously damage the ear. Ear plugs are recommended whenever sound levels exceed 90 dB. Recent evidence suggests that "noise pollution" may be a contributing factor to high blood pressure, anxiety, and nervousness. Table 17.2 gives some typical sound-level values.

Quick Quiz 17.5 A violin plays a melody line and is then joined by a second violin, playing at the same intensity as the first violin, in a repeat of the same melody. With both violins playing, what physical parameter has doubled compared to the situation with only one violin playing? (a) wavelength (b) frequency (c) intensity (d) sound level in dB (e) none of these.

Quick Quiz 17.6 Increasing the intensity of a sound by a factor of 100 causes the sound level to increase by (a) 100 dB (b) 20 dB (c) 10 dB (d) 2 dB.

Example 17.4 Sound Levels

Two identical machines are positioned the same distance from a worker. The intensity of sound delivered by each machine at the location of the worker is $2.0 \times 10^{-7} \text{ W/m}^2$. Find the sound level heard by the worker

(A) when one machine is operating

(B) when both machines are operating.

Solution

(A) The sound level at the location of the worker with one machine operating is calculated from Equation 17.8:

$$\beta_1 = 10 \log\left(\frac{2.0 \times 10^{-7} \text{ W/m}^2}{1.00 \times 10^{-12} \text{ W/m}^2}\right) = 10 \log(2.0 \times 10^5)$$

$$= \boxed{53 \text{ dB}}$$

(B) When both machines are operating, the intensity is doubled to $4.0 \times 10^{-7} \text{ W/m}^2$; therefore, the sound level now is

$$\beta_2 = 10 \log\left(\frac{4.0 \times 10^{-7} \text{ W/m}^2}{1.00 \times 10^{-12} \text{ W/m}^2}\right) = 10 \log(4.0 \times 10^5)$$

$$= \boxed{56 \text{ dB}}$$

From these results, we see that when the intensity is doubled, the sound level increases by only 3 dB.

What If? Loudness is a psychological response to a sound and depends on both the intensity and the frequency of the sound. As a rule of thumb, a doubling in loudness is approximately associated with an increase in sound level of 10 dB. (Note that this rule of thumb is relatively inaccurate at very low or very high frequencies.) If the loudness of the machines in this example is to be doubled, how many machines must be running?

Answer Using the rule of thumb, a doubling of loudness corresponds to a sound level increase of 10 dB. Thus,

$$\beta_2 - \beta_1 = 10 \text{ dB} = 10 \log \left(\frac{I_2}{I_0} \right) - 10 \log \left(\frac{I_1}{I_0} \right) = 10 \log \left(\frac{I_2}{I_1} \right)$$

$$\log \left(\frac{I_2}{I_1} \right) = 1$$

$$I_2 = 10 I_1$$

Thus, ten machines must be operating to double the loudness.

Loudness and Frequency

The discussion of sound level in decibels relates to a *physical* measurement of the strength of a sound. Let us now consider how we describe the *psychological* "measurement" of the strength of a sound.

Of course, we don't have meters in our bodies that can read out numerical values of our reactions to stimuli. We have to "calibrate" our reactions somehow by comparing different sounds to a reference sound. However, this is not easy to accomplish. For example, earlier we mentioned that the threshold intensity is 10^{-12} W/m², corresponding to an intensity level of 0 dB. In reality, this value is the threshold only for a sound of frequency 1 000 Hz, which is a standard reference frequency in acoustics. If we perform an experiment to measure the threshold intensity at other frequencies, we find a distinct variation of this threshold as a function of frequency. For example, at 100 Hz, a sound must have an intensity level of about 30 dB in order to be just barely audible! Unfortunately, there is no simple relationship between physical measurements and psychological "measurements." The 100-Hz, 30-dB sound is psychologically "equal" to the 1 000-Hz, 0-dB sound (both are just barely audible) but they are not physically equal (30 dB ≠ 0 dB).

By using test subjects, the human response to sound has been studied, and the results are shown in Figure 17.6 (the white area), along with the approximate frequency and sound-level ranges of other sound sources. The lower curve of the white area corresponds to the threshold of hearing. Its variation with frequency is clear from this diagram. Note that humans are sensitive to frequencies ranging from about 20 Hz to about 20 000 Hz. The upper bound of the white area is the threshold of pain. Here the

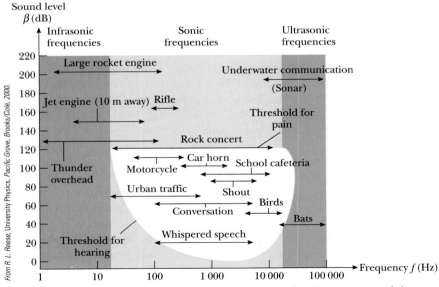

Figure 17.6 Approximate frequency and sound level ranges of various sources and that of normal human hearing, shown by the white area.

boundary of the white area is straight, because the psychological response is relatively independent of frequency at this high sound level.

The most dramatic change with frequency is in the lower left region of the white area, for low frequencies and low intensity levels. Our ears are particularly insensitive in this region. If you are listening to your stereo and the bass (low frequencies) and treble (high frequencies) sound balanced at a high volume, try turning the volume down and listening again. You will probably notice that the bass seems weak, which is due to the insensitivity of the ear to low frequencies at low sound levels, as shown in Figure 17.6.

17.4 The Doppler Effect

Perhaps you have noticed how the sound of a vehicle's horn changes as the vehicle moves past you. The frequency of the sound you hear as the vehicle approaches you is higher than the frequency you hear as it moves away from you. This is one example of the **Doppler effect.**[3]

To see what causes this apparent frequency change, imagine you are in a boat that is lying at anchor on a gentle sea where the waves have a period of $T = 3.0$ s. This means that every 3.0 s a crest hits your boat. Figure 17.7a shows this situation, with the water waves moving toward the left. If you set your watch to $t = 0$ just as one crest hits, the watch reads 3.0 s when the next crest hits, 6.0 s when the third crest hits, and so on. From these observations you conclude that the wave frequency is $f = 1/T = 1/(3.0$ s$) = 0.33$ Hz. Now suppose you start your motor and head directly into the oncoming waves, as in Figure 17.7b. Again you set your watch to $t = 0$ as a crest hits the front of your boat. Now, however, because you are moving toward the next wave crest as it moves toward you, it hits you less than 3.0 s after the first hit. In other words, the period you observe is shorter than the 3.0-s period you observed when you were stationary. Because $f = 1/T$, you observe a higher wave frequency than when you were at rest.

If you turn around and move in the same direction as the waves (see Fig. 17.7c), you observe the opposite effect. You set your watch to $t = 0$ as a crest hits the back of the boat. Because you are now moving away from the next crest, more than 3.0 s has elapsed on your watch by the time that crest catches you. Thus, you observe a lower frequency than when you were at rest.

These effects occur because the *relative* speed between your boat and the waves depends on the direction of travel and on the speed of your boat. When you are moving toward the right in Figure 17.7b, this relative speed is higher than that of the wave speed, which leads to the observation of an increased frequency. When you turn around and move to the left, the relative speed is lower, as is the observed frequency of the water waves.

Let us now examine an analogous situation with sound waves, in which the water waves become sound waves, the water becomes the air, and the person on the boat becomes an observer listening to the sound. In this case, an observer O is moving and a sound source S is stationary. For simplicity, we assume that the air is also stationary and that the observer moves directly toward the source (Fig. 17.8). The observer moves with a speed v_O toward a stationary point source ($v_S = 0$), where *stationary* means at rest with respect to the medium, air.

If a point source emits sound waves and the medium is uniform, the waves move at the same speed in all directions radially away from the source; this is a spherical wave, as was mentioned in Section 17.3. It is useful to represent these waves with a series of circular arcs concentric with the source, as in Figure 17.8. Each arc represents a surface over which the phase of the wave is constant. For example, the surface could pass through the crests of all waves. We call such a surface of constant phase a **wave front.** The distance between adjacent wave fronts equals the wavelength λ. In Figure 17.8, the

[3] Named after the Austrian physicist Christian Johann Doppler (1803–1853), who in 1842 predicted the effect for both sound waves and light waves.

(a)

(b)

(c)

Figure 17.7 (a) Waves moving toward a stationary boat. The waves travel to the left, and their source is far to the right of the boat, out of the frame of the photograph. (b) The boat moving toward the wave source. (c) The boat moving away from the wave source.

circles are the intersections of these three-dimensional wave fronts with the two-dimensional paper.

We take the frequency of the source in Figure 17.8 to be f, the wavelength to be λ, and the speed of sound to be v. If the observer were also stationary, he or she would detect wave fronts at a rate f. (That is, when $v_O = 0$ and $v_S = 0$, the observed frequency equals the source frequency.) When the observer moves toward the source, the speed of the waves relative to the observer is $v' = v + v_O$, as in the case of the boat, but the

Active Figure 17.8 An observer O (the cyclist) moves with a speed v_O toward a stationary point source S, the horn of a parked truck. The observer hears a frequency f' that is greater than the source frequency.

At the Active Figures link at **http://www.pse6.com,** *you can adjust the speed of the observer.*

wavelength λ is unchanged. Hence, using Equation 16.12, $v = \lambda f$, we can say that the frequency f' heard by the observer is *increased* and is given by

$$f' = \frac{v'}{\lambda} = \frac{v + v_O}{\lambda}$$

Because $\lambda = v/f$, we can express f' as

$$f' = \left(\frac{v + v_O}{v}\right)f \qquad \text{(observer moving toward source)} \qquad (17.9)$$

If the observer is moving away from the source, the speed of the wave relative to the observer is $v' = v - v_O$. The frequency heard by the observer in this case is *decreased* and is given by

$$f' = \left(\frac{v - v_O}{v}\right)f \qquad \text{(observer moving away from source)} \qquad (17.10)$$

In general, whenever an observer moves with a speed v_O relative to a stationary source, the frequency heard by the observer is given by Equation 17.9, with a sign convention: a positive value is substituted for v_O when the observer moves toward the source and a negative value is substituted when the observer moves away from the source.

Now consider the situation in which the source is in motion and the observer is at rest. If the source moves directly toward observer A in Figure 17.9a, the wave fronts heard by the observer are closer together than they would be if the source were not moving. As a result, the wavelength λ' measured by observer A is shorter than the wavelength λ of the source. During each vibration, which lasts for a time interval T (the period), the source moves a distance $v_S T = v_S/f$ and the wavelength is *shortened* by this amount. Therefore, the observed wavelength λ' is

$$\lambda' = \lambda - \Delta\lambda = \lambda - \frac{v_S}{f}$$

Because $\lambda = v/f$, the frequency f' heard by observer A is

$$f' = \frac{v}{\lambda'} = \frac{v}{\lambda - (v_S/f)} = \frac{v}{(v/f) - (v_S/f)}$$

(b)

(a)

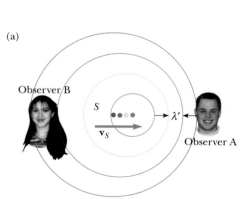

🌐 *At the Active Figures link*
at http://www.pse6.com, you
can adjust the speed of the
source.

Active Figure 17.9 (a) A source S moving with a speed v_S toward a stationary observer A and away from a stationary observer B. Observer A hears an increased frequency, and observer B hears a decreased frequency. (b) The Doppler effect in water, observed in a ripple tank. A point source is moving to the right with speed v_S. Letters shown in the photo refer to Quick Quiz 17.7.

Courtesy of the Educational Development Center, Newton, MA

$$f' = \left(\frac{v}{v - v_S}\right)f \qquad \text{(source moving toward observer)} \qquad (17.11)$$

That is, the observed frequency is *increased* whenever the source is moving toward the observer.

When the source moves away from a stationary observer, as is the case for observer B in Figure 17.9a, the observer measures a wavelength λ' that is *greater* than λ and hears a *decreased* frequency:

$$f' = \left(\frac{v}{v + v_S}\right)f \qquad \text{(source moving away from observer)} \qquad (17.12)$$

We can express the general relationship for the observed frequency when a source is moving and an observer is at rest as Equation 17.11, with the same sign convention applied to v_S as was applied to v_O: a positive value is substituted for v_S when the source moves toward the observer and a negative value is substituted when the source moves away from the observer.

Finally, we find the following general relationship for the observed frequency:

$$f' = \left(\frac{v + v_O}{v - v_S}\right)f \qquad (17.13)$$

In this expression, the signs for the values substituted for v_O and v_S depend on the direction of the velocity. A positive value is used for motion of the observer or the source *toward* the other, and a negative sign for motion of one *away from* the other.

A convenient rule concerning signs for you to remember when working with all Doppler-effect problems is as follows:

> The word *toward* is associated with an *increase* in observed frequency. The words *away from* are associated with a *decrease* in observed frequency.

Although the Doppler effect is most typically experienced with sound waves, it is a phenomenon that is common to all waves. For example, the relative motion of source and observer produces a frequency shift in light waves. The Doppler effect is used in police radar systems to measure the speeds of motor vehicles. Likewise, astronomers use the effect to determine the speeds of stars, galaxies, and other celestial objects relative to the Earth.

General Doppler-shift expression

 PITFALL PREVENTION

17.1 Doppler Effect Does Not Depend on Distance

Many people think that the Doppler effect depends on the distance between the source and the observer. While the intensity of a sound varies as the distance changes, the apparent frequency depends only on the relative speed of source and observer. As you listen to an approaching source, you will detect increasing intensity but constant frequency. As the source passes, you will hear the frequency suddenly drop to a new constant value and the intensity begin to decrease.

Quick Quiz 17.7 Consider detectors of water waves at three locations A, B, and C in Figure 17.9b. Which of the following statements is true? (a) The wave speed is highest at location A. (b) The wave speed is highest at location C. (c) The detected wavelength is largest at location B. (c) The detected wavelength is largest at location C. (e) The detected frequency is highest at location C. (f) The detected frequency is highest at location A.

Quick Quiz 17.8 You stand on a platform at a train station and listen to a train approaching the station at a constant velocity. While the train approaches, but before it arrives, you hear (a) the intensity and the frequency of the sound both increasing (b) the intensity and the frequency of the sound both decreasing (c) the intensity increasing and the frequency decreasing (d) the intensity decreasing and the frequency increasing (e) the intensity increasing and the frequency remaining the same (f) the intensity decreasing and the frequency remaining the same.

Example 17.5 The Broken Clock Radio

Your clock radio awakens you with a steady and irritating sound of frequency 600 Hz. One morning, it malfunctions and cannot be turned off. In frustration, you drop the clock radio out of your fourth-story dorm window, 15.0 m from the ground. Assume the speed of sound is 343 m/s.

(A) As you listen to the falling clock radio, what frequency do you hear just before you hear the radio striking the ground?

(B) At what rate does the frequency that you hear change with time just before you hear the radio striking the ground?

Solution

(A) In conceptualizing the problem, note that the speed of the radio increases as it falls. Thus, it is a source of sound moving away from you with an increasing speed. We categorize this problem as one in which we must combine our understanding of falling objects with that of the frequency shift due to the Doppler effect. To analyze the problem, we identify the clock radio as a moving source of sound for which the Doppler-shifted frequency is given by

$$f' = \left(\frac{v}{v - v_S}\right)f$$

The speed of the source of sound is given by Equation 2.9 for a falling object:

$$v_S = v_{yi} + a_y t = 0 - gt = -gt$$

Thus, the Doppler-shifted frequency of the falling clock radio is

(1) $$f' = \left(\frac{v}{v - (-gt)}\right)f = \left(\frac{v}{v + gt}\right)f$$

The time at which the radio strikes the ground is found from Equation 2.12:

$$y_f = y_i + v_{yi}t - \tfrac{1}{2}gt^2$$

$$-15.0 \text{ m} = 0 + 0 - \tfrac{1}{2}(9.80 \text{ m/s}^2)t^2$$

$$t = 1.75 \text{ s}$$

Thus, the Doppler-shifted frequency just as the radio strikes the ground is

$$f' = \left(\frac{v}{v + gt}\right)f$$

$$= \left(\frac{343 \text{ m/s}}{343 \text{ m/s} + (9.80 \text{ m/s}^2)(1.75 \text{ s})}\right)(600 \text{ Hz})$$

$$= \boxed{571 \text{ Hz}}$$

(B) The rate at which the frequency changes is found by differentiating Equation (1) with respect to t:

$$\frac{df'}{dt} = \frac{d}{dt}\left(\frac{vf}{v + gt}\right) = \frac{-vg}{(v + gt)^2}f$$

$$= \frac{-(343 \text{ m/s})(9.80 \text{ m/s}^2)}{[343 \text{ m/s} + (9.80 \text{ m/s}^2)(1.75 \text{ s})]^2}(600 \text{ Hz})$$

$$= \boxed{-15.5 \text{ Hz/s}}$$

To finalize this problem, consider the following **What If?**

What If? Suppose you live on the eighth floor instead of the fourth floor. If you repeat the radio-dropping activity, does the frequency shift in part (A) and the rate of change of frequency in part (B) of this example double?

Answer The doubled height does not give a time at which the radio lands that is twice the time found in part (A). From Equation 2.12:

$$y_f = y_i + v_{yi}t - \tfrac{1}{2}gt^2$$

$$-30.0 \text{ m} = 0 + 0 - \tfrac{1}{2}(9.80 \text{ m/s}^2)t^2$$

$$t = 2.47 \text{ s}$$

The new frequency heard just before you hear the radio strike the ground is

$$f' = \left(\frac{v}{v + gt}\right)f$$

$$= \left(\frac{343 \text{ m/s}}{343 \text{ m/s} + (9.80 \text{ m/s}^2)(2.47 \text{ s})}\right)(600 \text{ Hz})$$

$$= \boxed{560 \text{ Hz}}$$

The frequency shift heard on the fourth floor is 600 Hz − 571 Hz = 29 Hz, while the frequency shift heard from the eighth floor is 600 Hz − 560 Hz = 40 Hz, which is not twice as large.

The new rate of change of frequency is

$$\frac{df'}{dt} = \frac{-vg}{(v + gt)^2}f$$

$$= \frac{-(343 \text{ m/s})(9.80 \text{ m/s}^2)}{[343 \text{ m/s} + (9.80 \text{ m/s}^2)(2.47 \text{ s})]^2}(600 \text{ Hz})$$

$$= -15.0 \text{ Hz/s}$$

Note that this value is actually *smaller* in magnitude than the previous value of − 15.5 Hz/s!

Example 17.6 Doppler Submarines

Interactive

A submarine (sub A) travels through water at a speed of 8.00 m/s, emitting a sonar wave at a frequency of 1 400 Hz. The speed of sound in the water is 1 533 m/s. A second submarine (sub B) is located such that both submarines are traveling directly toward one another. The second submarine is moving at 9.00 m/s.

(A) What frequency is detected by an observer riding on sub B as the subs approach each other?

(B) The subs barely miss each other and pass. What frequency is detected by an observer riding on sub B as the subs recede from each other?

Solution

(A) We use Equation 17.13 to find the Doppler-shifted frequency. As the two submarines approach each other, the observer in sub B hears the frequency

$$f' = \left(\frac{v + v_O}{v - v_S}\right)f$$

$$= \left(\frac{1\,533\,\text{m/s} + (+9.00\,\text{m/s})}{1\,533\,\text{m/s} - (+8.00\,\text{m/s})}\right)(1\,400\,\text{Hz}) = \boxed{1\,416\,\text{Hz}}$$

(B) As the two submarines recede from each other, the observer in sub B hears the frequency

$$f' = \left(\frac{v + v_O}{v - v_S}\right)f$$

$$= \left(\frac{1\,533\,\text{m/s} + (-9.00\,\text{m/s})}{1\,533\,\text{m/s} - (-8.00\,\text{m/s})}\right)(1\,400\,\text{Hz}) = \boxed{1\,385\,\text{Hz}}$$

What If? While the subs are approaching each other, some of the sound from sub A will reflect from sub B and return to sub A. If this sound were to be detected by an observer on sub A, what is its frequency?

Answer The sound of apparent frequency 1 416 Hz found in part (A) will be reflected from a moving source (sub B) and then detected by a moving observer (sub A). Thus, the frequency detected by sub A is

$$f'' = \left(\frac{v + v_O}{v - v_S}\right)f'$$

$$= \left(\frac{1\,533\,\text{m/s} + (+8.00\,\text{m/s})}{1\,533\,\text{m/s} - (+9.00\,\text{m/s})}\right)(1\,416\,\text{Hz}) = 1\,432\,\text{Hz}$$

This technique is used by police officers to measure the speed of a moving car. Microwaves are emitted from the police car and reflected by the moving car. By detecting the Doppler-shifted frequency of the reflected microwaves, the police officer can determine the speed of the moving car.

 At the Interactive Worked Example link at http://www.pse6.com, you can alter the relative speeds of the submarines and observe the Doppler-shifted frequency.

Shock Waves

Now consider what happens when the speed v_S of a source *exceeds* the wave speed v. This situation is depicted graphically in Figure 17.10a. The circles represent spherical wave fronts emitted by the source at various times during its motion. At $t = 0$, the source is at S_0, and at a later time t, the source is at S_n. At the time t, the wave front

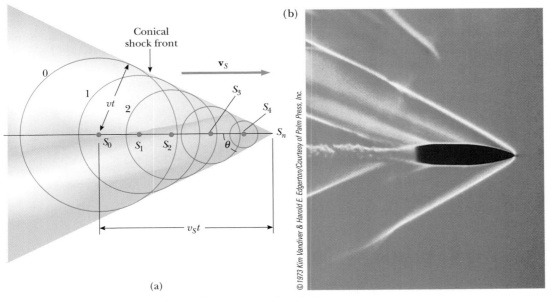

Figure 17.10 (a) A representation of a shock wave produced when a source moves from S_0 to S_n with a speed v_S, which is greater than the wave speed v in the medium. The envelope of the wave fronts forms a cone whose apex half-angle is given by $\sin \theta = v/v_S$. (b) A stroboscopic photograph of a bullet moving at supersonic speed through the hot air above a candle. Note the shock wave in the vicinity of the bullet.

www.comstock.com

Figure 17.11 The **V**-shaped bow wave of a boat is formed because the boat speed is greater than the speed of the water waves it generates. A bow wave is analogous to a shock wave formed by an airplane traveling faster than sound.

centered at S_0 reaches a radius of vt. In this same time interval, the source travels a distance $v_S t$ to S_n. At the instant the source is at S_n, waves are just beginning to be generated at this location, and hence the wave front has zero radius at this point. The tangent line drawn from S_n to the wave front centered on S_0 is tangent to all other wave fronts generated at intermediate times. Thus, we see that the envelope of these wave fronts is a cone whose apex half-angle θ (the "Mach angle") is given by

$$\sin \theta = \frac{vt}{v_S t} = \frac{v}{v_S}$$

The ratio v_S/v is referred to as the *Mach number*, and the conical wave front produced when $v_S > v$ (supersonic speeds) is known as a *shock wave*. An interesting analogy to shock waves is the **V**-shaped wave fronts produced by a boat (the bow wave) when the boat's speed exceeds the speed of the surface-water waves (Fig. 17.11).

Jet airplanes traveling at supersonic speeds produce shock waves, which are responsible for the loud "sonic boom" one hears. The shock wave carries a great deal of energy concentrated on the surface of the cone, with correspondingly great pressure variations. Such shock waves are unpleasant to hear and can cause damage to buildings when aircraft fly supersonically at low altitudes. In fact, an airplane flying at supersonic speeds produces a double boom because two shock waves are formed, one from the nose of the plane and one from the tail. People near the path of the space shuttle as it glides toward its landing point often report hearing what sounds like two very closely spaced cracks of thunder.

Quick Quiz 17.9 An airplane flying with a constant velocity moves from a cold air mass into a warm air mass. Does the Mach number (a) increase (b) decrease (c) stay the same?

17.5 Digital Sound Recording

The first sound recording device, the phonograph, was invented by Thomas Edison in the nineteenth century. Sound waves were recorded in early phonographs by encoding the sound waveforms as variations in the depth of a continuous groove cut in tin foil wrapped around a cylinder. During playback, as a needle followed along the groove of the rotating cylinder, the needle was pushed back and forth according to the sound

Figure 17.12 An Edison phonograph. Sound information is recorded in a groove on a rotating cylinder of wax. A needle follows the groove and vibrates according to the sound information. A diaphragm and a horn make the sound intense enough to hear.

waves encoded on the record. The needle was attached to a diaphragm and a horn (Fig. 17.12), which made the sound loud enough to be heard.

As the development of the phonograph continued, sound was recorded on cardboard cylinders coated with wax. During the last decade of the nineteenth century and the first half of the twentieth century, sound was recorded on disks made of shellac and clay. In 1948, the plastic phonograph disk was introduced and dominated the recording industry market until the advent of compact discs in the 1980s.

There are a number of problems with phonograph records. As the needle follows along the groove of the rotating phonograph record, the needle is pushed back and forth according to the sound waves encoded on the record. By Newton's third law, the needle also pushes on the plastic. As a result, the recording quality diminishes with each playing as small pieces of plastic break off and the record wears away.

Another problem occurs at high frequencies. The wavelength of the sound on the record is so small that natural bumps and graininess in the plastic create signals as loud as the sound signal, resulting in noise. The noise is especially noticeable during quiet passages in which high frequencies are being played. This is handled electronically by a process known as *pre-emphasis*. In this process, the high frequencies are recorded with more intensity than they actually have, which increases the amplitude of the vibrations and overshadows the sources of noise. Then, an *equalization circuit* in the playback system is used to reduce the intensity of the high-frequency sounds, which also reduces the intensity of the noise.

Example 17.7   Wavelengths on a Phonograph Record

Consider a 10 000-Hz sound recorded on a phonograph record which rotates at $33\frac{1}{3}$ rev/min. How far apart are the crests of the wave for this sound on the record

(A) at the outer edge of the record, 6.0 inches from the center?

(B) at the inner edge, 1.0 inch from the center?

Solution

(A) The linear speed v of a point at the outer edge of the record is $2\pi r/T$ where T is the period of the rotation and r

is the distance from the center. We first find T:

$$T = \frac{1}{f} = \frac{1}{33.33 \text{ rev/min}} = 0.030 \text{ min} \left(\frac{60 \text{ s}}{1 \text{ min}}\right) = 1.8 \text{ s}$$

Now, the linear speed at the outer edge is

$$v = \frac{2\pi r}{T} = \frac{2\pi(6.0 \text{ in.})}{1.8 \text{ s}} = 21 \text{ in./s} \left(\frac{2.54 \text{ cm}}{1 \text{ in.}}\right)$$
$$= 53 \text{ cm/s}$$

Thus, the wave on the record is moving past the needle at this speed. The wavelength is

$$\lambda = \frac{v}{f} = \frac{53 \text{ cm/s}}{10\,000 \text{ Hz}} = 5.3 \times 10^{-5} \text{ m}$$
$$= \boxed{53 \ \mu\text{m}}$$

(B) The linear speed at the inner edge is

$$v = \frac{2\pi r}{T} = \frac{2\pi(1.0 \text{ in.})}{1.8 \text{ s}} = 3.5 \text{ in./s} \left(\frac{2.54 \text{ cm}}{1 \text{ in.}}\right)$$
$$= 8.9 \text{ cm/s}$$

The wavelength is

$$\lambda = \frac{v}{f} = \frac{8.9 \text{ cm/s}}{10\,000 \text{ Hz}} = 8.9 \times 10^{-6} \text{ m}$$
$$= \boxed{8.9 \ \mu\text{m}}$$

Thus, the problem with noise interfering with the recorded sound is more severe at the inner edge of the disk than at the outer edge.

Digital Recording

In digital recording, information is converted to binary code (ones and zeroes), similar to the dots and dashes of Morse code. First, the waveform of the sound is *sampled*, typically at the rate of 44 100 times per second. Figure 17.13 illustrates this process. The sampling frequency is much higher than the upper range of hearing, about 20 000 Hz, so all frequencies of sound are sampled at this rate. During each sampling, the pressure of the wave is measured and converted to a voltage. Thus, there are 44 100 numbers associated with each second of the sound being sampled.

These measurements are then converted to *binary numbers*, which are numbers expressed using base 2 rather than base 10. Table 17.3 shows some sample binary numbers. Generally, voltage measurements are recorded in 16-bit "words," where each bit is a one or a zero. Thus, the number of different voltage levels that can be assigned codes is $2^{16} = 65\,536$. The number of bits in one second of sound is $16 \times 44\,100 = 705\,600$. It is these strings of ones and zeroes, in 16-bit words, that are recorded on the surface of a compact disc.

Figure 17.14 shows a magnification of the surface of a compact disc. There are two types of areas that are detected by the laser playback system—*lands* and *pits*. The lands are untouched regions of the disc surface that are highly reflective. The pits, which are areas burned into the surface, scatter light rather than reflecting it back to the detection system. The playback system samples the reflected light 705 600 times per second. When the laser moves from a pit to a flat or from a flat to a pit, the reflected light changes during the sampling and the bit is recorded as a one. If there is no change during the sampling, the bit is recorded as a zero.

Figure 17.13 Sound is digitized by electronically sampling the sound waveform at periodic intervals. During each time interval between the blue lines, a number is recorded for the average voltage during the interval. The sampling rate shown here is much slower than the actual sampling rate of 44 100 samples per second.

Table 17.3

Sample Binary Numbers		
Number in Base 10	**Number in Binary**	**Sum**
1	0000000000000001	1
2	0000000000000010	2 + 0
3	0000000000000011	2 + 1
10	0000000000001010	8 + 0 + 2 + 0
37	0000000000100101	32 + 0 + 0 + 4 + 0 + 1
275	0000000100010011	256 + 0 + 0 + 0 + 16 + 0 + 0 + 2 + 1

The binary numbers read from the CD are converted back to voltages, and the waveform is reconstructed, as shown in Figure 17.15. Because the sampling rate is so high—44 100 voltage readings each second—the fact that the waveform is constructed from step-wise discrete voltages is not evident in the sound.

The advantage of digital recording is in the high fidelity of the sound. With analog recording, any small imperfection in the record surface or the recording equipment can cause a distortion of the waveform. If all peaks of a maximum in a waveform are clipped off so as to be only 90% as high, for example, this will have a major effect on the spectrum of the sound in an analog recording. With digital recording, however, it takes a major imperfection to turn a one into a zero. If an imperfection causes the magnitude of a one to be 90% of the original value, it still registers as a one, and there is no distortion. Another advantage of digital recording is that the information is extracted optically, so that there is no mechanical wear on the disc.

Figure 17.14 The surface of a compact disc, showing the pits. Transitions between pits and lands correspond to ones. Regions without transitions correspond to zeroes.

Figure 17.15 The reconstruction of the sound wave sampled in Figure 17.13. Notice that the reconstruction is step-wise, rather than the continuous waveform in Figure 17.13.

Example 17.8 How Big Are the Pits?

In Example 10.2, we mentioned that the speed with which the CD surface passes the laser is 1.3 m/s. What is the average length of the audio track on a CD associated with each bit of the audio information?

Solution In one second, a 1.3-m length of audio track passes by the laser. This length includes 705 600 bits of audio information. Thus, the average length per bit is

$$\frac{1.3 \text{ m}}{705\ 600 \text{ bits}} = 1.8 \times 10^{-6} \text{ m/bit}$$

$$= \boxed{1.8 \ \mu\text{m/bit}}$$

The average length per bit of *total* information on the CD is smaller than this because there is additional information on the disc besides the audio information. This information includes error correction codes, song numbers, timing codes, etc. As a result, the shortest length per bit is actually about 0.8 μm.

Example 17.9 What's the Number?

Consider the photograph of the compact disc surface in Figure 17.14. Audio data undergoes complicated processing in order to reduce a variety of errors in reading the data. Thus, an audio "word" is not laid out linearly on the disc. Suppose that data has been read from the disc, the error encoding has been removed, and the resulting audio word is

$$1\ 0\ 1\ 1\ 1\ 0\ 1\ 1\ 1\ 0\ 1\ 1\ 1\ 0\ 1\ 1$$

What is the decimal number represented by this 16-bit word?

Solution We convert each of these bits to a power of 2 and add the results:

$1 \times 2^{15} = 32\ 768$	$1 \times 2^9 = 512$	$1 \times 2^3 = 8$
$0 \times 2^{14} = 0$	$1 \times 2^8 = 256$	$0 \times 2^2 = 0$
$1 \times 2^{13} = 8\ 192$	$1 \times 2^7 = 128$	$1 \times 2^1 = 2$
$1 \times 2^{12} = 4\ 096$	$0 \times 2^6 = 0$	$1 \times 2^0 = 1$
$1 \times 2^{11} = 2\ 048$	$1 \times 2^5 = 32$	
$0 \times 2^{10} = 0$	$1 \times 2^4 = 16$	sum $= \boxed{48\ 059}$

This number is converted by the CD player into a voltage, representing one of the 44 100 values that will be used to build one second of the electronic waveform that represents the recorded sound.

17.6 Motion Picture Sound

Another interesting application of digital sound is the soundtrack in a motion picture. Early twentieth-century movies recorded sound on phonograph records, which were synchronized with the action on the screen. Beginning with early newsreel films, the *variable-area optical soundtrack* process was introduced, in which sound was recorded on an optical track on the film. The width of the transparent portion of the track varied according to the sound wave that was recorded. A photocell detecting light passing through the track converted the varying light intensity to a sound wave. As with phonograph recording, there are a number of difficulties with this recording system. For example, dirt or fingerprints on the film cause fluctuations in intensity and loss of fidelity.

Digital recording on film first appeared with *Dick Tracy* (1990), using the Cinema Digital Sound (CDS) system. This system suffered from lack of an analog backup system in case of equipment failure and is no longer used in the film industry. It did, however, introduce the use of 5.1 channels of sound—Left, Center, Right, Right Surround, Left Surround, and Low Frequency Effects (LFE). The LFE channel, which is the "0.1

channel" of 5.1, carries very low frequencies for dramatic sound from explosions, earthquakes, and the like.

Current motion pictures are produced with three systems of digital sound recording:

Dolby Digital; In this format, 5.1 channels of digital sound are optically stored between the sprocket holes of the film. There is an analog optical backup in case the digital system fails. The first film to use this technique was *Batman Returns* (1992).

DTS (Digital Theater Sound); 5.1 channels of sound are stored on a separate CD-ROM which is synchronized to the film print by time codes on the film. There is an analog optical backup in case the digital system fails. The first film to use this technique was *Jurassic Park* (1993).

SDDS (Sony Dynamic Digital Sound); Eight full channels of digital sound are optically stored outside the sprocket holes on both sides of film. There is an analog optical backup in case the digital system fails. The first film to use this technique was *Last Action Hero* (1993). The existence of information on both sides of the tape is a system of redundancy—in case one side is damaged, the system will still operate. SDDS employs a full-spectrum LFE channel and two additional channels (left center and right center behind the screen). In Figure 17.16, showing a section of SDDS film, both the analog optical soundtrack and the dual digital soundtracks can be seen.

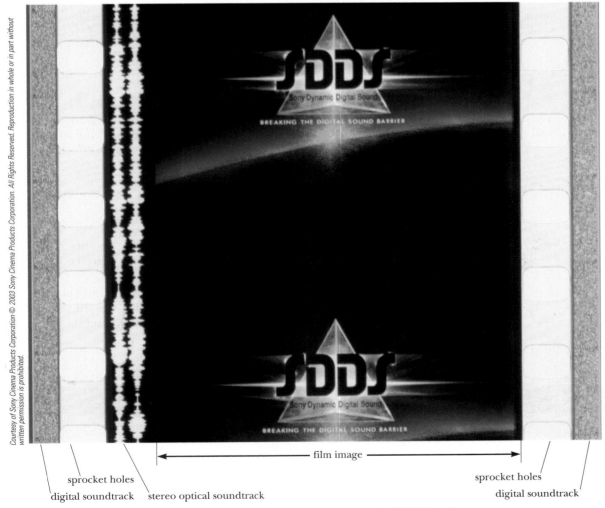

Figure 17.16 The layout of information on motion picture film using the SDDS digital sound system.

SUMMARY

Take a practice test for this chapter by clicking on the Practice Test link at http://www.pse6.com.

Sound waves are longitudinal and travel through a compressible medium with a speed that depends on the elastic and inertial properties of that medium. The speed of sound in a liquid or gas having a bulk modulus B and density ρ is

$$v = \sqrt{\frac{B}{\rho}} \tag{17.1}$$

For sinusoidal sound waves, the variation in the position of an element of the medium is given by

$$s(x, t) = s_{\max} \cos(kx - \omega t) \tag{17.2}$$

and the variation in pressure from the equilibrium value is

$$\Delta P = \Delta P_{\max} \sin(kx - \omega t) \tag{17.3}$$

where $\Delta P_{\max}$ is the **pressure amplitude.** The pressure wave is 90° out of phase with the displacement wave. The relationship between $s_{\max}$ and $\Delta P_{\max}$ is given by

$$\Delta P_{\max} = \rho v \omega s_{\max} \tag{17.4}$$

The intensity of a periodic sound wave, which is the power per unit area, is

$$I \equiv \frac{\mathcal{P}}{A} = \frac{\Delta P_{\max}^2}{2\rho v} \tag{17.5, 17.6}$$

The sound level of a sound wave, in decibels, is given by

$$\beta \equiv 10 \log\left(\frac{I}{I_0}\right) \tag{17.8}$$

The constant I_0 is a reference intensity, usually taken to be at the threshold of hearing $(1.00 \times 10^{-12} \text{ W/m}^2)$, and I is the intensity of the sound wave in watts per square meter.

The change in frequency heard by an observer whenever there is relative motion between a source of sound waves and the observer is called the **Doppler effect.** The observed frequency is

$$f' = \left(\frac{v + v_O}{v - v_S}\right)f \tag{17.13}$$

In this expression, the signs for the values substituted for v_O and v_S depend on the direction of the velocity. A positive value for the velocity of the observer or source is substituted if the velocity of one is toward the other, while a negative value represents a velocity of one away from the other.

In digital recording of sound, the sound waveform is sampled 44 100 times per second. The pressure of the wave for each sampling is measured and converted to a binary number. In playback, these binary numbers are read and used to build the original waveform.

QUESTIONS

1. Why are sound waves characterized as longitudinal?

2. If an alarm clock is placed in a good vacuum and then activated, no sound is heard. Explain.

3. A sonic ranger is a device that determines the distance to an object by sending out an ultrasonic sound pulse and measuring how long it takes for the wave to return after it reflects from the object. Typically these devices cannot

reliably detect an object that is less than half a meter from the sensor. Why is that?

4. A friend sitting in her car far down the road waves to you and beeps her horn at the same time. How far away must she be for you to calculate the speed of sound to two significant figures by measuring the time it takes for the sound to reach you?

5. If the wavelength of sound is reduced by a factor of 2, what happens to its frequency? Its speed?

6. By listening to a band or orchestra, how can you determine that the speed of sound is the same for all frequencies?

7. In Example 17.3 we found that a point source with a power output of 80 W produces sound with an intensity of 1.00×10^{-8} W/m², which corresponds to 40 dB, at a distance of about 16 miles. Why do you suppose you cannot normally hear a rock concert that is going on 16 miles away? (See Table 17.2.)

8. If the distance from a point source is tripled, by what factor does the intensity decrease?

9. *The Tunguska Event.* On June 30, 1908, a meteor burned up and exploded in the atmosphere above the Tunguska River valley in Siberia. It knocked down trees over thousands of square kilometers and started a forest fire, but apparently caused no human casualties. A witness sitting on his doorstep outside the zone of falling trees recalled events in the following sequence: He saw a moving light in the sky, brighter than the sun and descending at a low angle to the horizon. He felt his face become warm. He felt the ground shake. An invisible agent picked him up and immediately dropped him about a meter farther away from where the light had been. He heard a very loud protracted rumbling. Suggest an explanation for these observations and for the order in which they happened.

10. Explain how the Doppler effect with microwaves is used to determine the speed of an automobile.

11. Explain what happens to the frequency of the echo of your car horn as you move in a vehicle toward the wall of a canyon. What happens to the frequency as you move away from the wall?

12. Of the following sounds, which is most likely to have a sound level of 60 dB: a rock concert, the turning of a page in this textbook, normal conversation, or a cheering crowd at a football game?

13. Estimate the decibel level of each of the sounds in the previous question.

14. A binary star system consists of two stars revolving about their common center of mass. If we observe the light reaching us from one of these stars as it makes one complete revolution, what does the Doppler effect predict will happen to this light?

15. How can an object move with respect to an observer so that the sound from it is not shifted in frequency?

16. Suppose the wind blows. Does this cause a Doppler effect for sound propagating through the air? Is it like a moving source or a moving observer?

17. Why is it not possible to use sonar (sound waves) to determine the speed of an object traveling faster than the speed of sound?

18. Why is it so quiet after a snowfall?

19. Why is the intensity of an echo less than that of the original sound?

20. A loudspeaker built into the exterior wall of an airplane produces a large-amplitude burst of vibration at 200 Hz, then a burst at 300 Hz, and then a burst at 400 Hz (Boop . . . baap . . . beep), all while the plane is flying faster than the speed of sound. Describe qualitatively what an observer hears if she is in front of the plane, close to its flight path. **What If?** What will the observer hear if the pilot uses the loudspeaker to say, "How are you?"

21. In several cases, a nearby star has been found to have a large planet orbiting about it, although the planet could not be seen. Using the ideas of a system rotating about its center of mass and of the Doppler shift for light (which is in several ways similar to the Doppler effect for sound), explain how an astronomer could determine the presence of the invisible planet.

PROBLEMS

1, 2, 3 = straightforward, intermediate, challenging   ☐ = full solution available in the *Student Solutions Manual and Study Guide*

 = coached solution with hints available at http://www.pse6.com   💻 = computer useful in solving problem

▨ = paired numerical and symbolic problems

Section 17.1   Speed of Sound Waves

1. Suppose that you hear a clap of thunder 16.2 s after seeing the associated lightning stroke. The speed of sound waves in air is 343 m/s, and the speed of light is 3.00×10^8 m/s. How far are you from the lightning stroke?

2. Find the speed of sound in mercury, which has a bulk modulus of approximately 2.80×10^{10} N/m² and a density of 13 600 kg/m³.

3. A flowerpot is knocked off a balcony 20.0 m above the sidewalk and falls toward an unsuspecting 1.75-m-tall man who is standing below. How close to the sidewalk can the flower pot fall before it is too late for a warning shouted from the balcony to reach the man in time? Assume that the man below requires 0.300 s to respond to the warning.

4. The speed of sound in air (in m/s) depends on temperature according to the approximate expression

$$v = 331.5 + 0.607 T_C$$

where T_C is the Celsius temperature. In dry air the temperature decreases about 1°C for every 150 m rise in altitude. (a) Assuming this change is constant up to an altitude of 9 000 m, how long will it take the sound from an airplane flying at 9 000 m to reach the ground on a day when the ground temperature is 30°C? (b) **What If?** Compare this to the time interval required if the air were a constant 30°C. Which time interval is longer?

5. A cowboy stands on horizontal ground between two parallel vertical cliffs. He is not midway between the cliffs. He fires a shot and hears its echoes. The second echo arrives 1.92 s after the first and 1.47 s before the third. Consider only the sound traveling parallel to the ground and reflecting from the cliffs. Take the speed of sound as 340 m/s. (a) What is the distance between the cliffs? (b) **What If?** If he can hear a fourth echo, how long after the third echo does it arrive?

6. A rescue plane flies horizontally at a constant speed searching for a disabled boat. When the plane is directly above the boat, the boat's crew blows a loud horn. By the time the plane's sound detector perceives the horn's sound, the plane has traveled a distance equal to half its altitude above the ocean. If it takes the sound 2.00 s to reach the plane, determine (a) the speed of the plane and (b) its altitude. Take the speed of sound to be 343 m/s.

Section 17.2 Periodic Sound Waves

Note: Use the following values as needed unless otherwise specified: the equilibrium density of air at 20°C is $\rho = 1.20$ kg/m^3. The speed of sound in air is $v = 343$ m/s. Pressure variations ΔP are measured relative to atmospheric pressure, 1.013×10^5 N/m^2. Problem 70 in Chapter 2 can also be assigned with this section.

7. A bat (Fig. P17.7) can detect very small objects, such as an insect whose length is approximately equal to one wavelength of the sound the bat makes. If a bat emits chirps at a frequency of 60.0 kHz, and if the speed of sound in air is 340 m/s, what is the smallest insect the bat can detect?

Figure P17.7 Problems 7 and 60.

8. An ultrasonic tape measure uses frequencies above 20 MHz to determine dimensions of structures such as buildings. It does this by emitting a pulse of ultrasound into air and then measuring the time for an echo to return from a reflecting surface whose distance away is to be measured. The distance is displayed as a digital read-out. For a tape measure that emits a pulse of ultrasound with a frequency of 22.0 MHz, (a) What is the distance to an object from which the echo pulse returns after 24.0 ms when the air temperature is 26°C? (b) What should be the duration of the emitted pulse if it is to include 10 cycles of the ultrasonic wave? (c) What is the spatial length of such a pulse?

9. Ultrasound is used in medicine both for diagnostic imaging and for therapy. For diagnosis, short pulses of ultrasound are passed through the patient's body. An echo reflected from a structure of interest is recorded, and from the time delay for the return of the echo the distance to the structure can be determined. A single transducer emits and detects the ultrasound. An image of the structure is obtained by reducing the data with a computer. With sound of low intensity, this technique is noninvasive and harmless. It is used to examine fetuses, tumors, aneurysms, gallstones, and many other structures. A Doppler ultrasound unit is used to study blood flow and functioning of the heart. To reveal detail, the wavelength of the reflected ultrasound must be small compared to the size of the object reflecting the wave. For this reason, frequencies in the range 1.00 to 20.0 MHz are used. What is the range of wavelengths corresponding to this range of frequencies? The speed of ultrasound in human tissue is about 1 500 m/s (nearly the same as the speed of sound in water).

10. A sound wave in air has a pressure amplitude equal to 4.00×10^{-3} N/m^2. Calculate the displacement amplitude of the wave at a frequency of 10.0 kHz.

11. A sinusoidal sound wave is described by the displacement wave function

$$s(x, t) = (2.00 \ \mu m) \cos[(15.7 \ m^{-1})x - (858 \ s^{-1})t]$$

(a) Find the amplitude, wavelength, and speed of this wave. (b) Determine the instantaneous displacement from equilibrium of the elements of air at the position $x = 0.050 \ 0$ m at $t = 3.00$ ms. (c) Determine the maximum speed of the element's oscillatory motion.

12. As a certain sound wave travels through the air, it produces pressure variations (above and below atmospheric pressure) given by $\Delta P = 1.27 \sin(\pi x - 340\pi t)$ in SI units. Find (a) the amplitude of the pressure variations, (b) the frequency, (c) the wavelength in air, and (d) the speed of the sound wave.

13. Write an expression that describes the pressure variation as a function of position and time for a sinusoidal sound wave in air, if $\lambda = 0.100$ m and $\Delta P_{max} = 0.200$ N/m^2.

14. Write the function that describes the displacement wave corresponding to the pressure wave in Problem 13.

15. An experimenter wishes to generate in air a sound wave that has a displacement amplitude of 5.50×10^{-6} m. The pressure amplitude is to be limited to 0.840 N/m^2. What is the minimum wavelength the sound wave can have?

16. The tensile stress in a thick copper bar is 99.5% of its elastic breaking point of 13.0×10^{10} N/m^2. If a 500-Hz sound wave is transmitted through the material, (a) what displacement amplitude will cause the bar to break? (b) What is the maximum speed of the elements of copper at this moment? (c) What is the sound intensity in the bar?

17. Prove that sound waves propagate with a speed given by Equation 17.1. Proceed as follows. In Figure 17.3, consider a thin cylindrical layer of air in the cylinder, with face area A and thickness Δx. Draw a free-body diagram of this thin layer. Show that $\Sigma F_x = ma_x$ implies that $-[\partial(\Delta P)/\partial x]A\,\Delta x = \rho A\,\Delta x(\partial^2 s/\partial t^2)$. By substituting $\Delta P = -B(\partial s/\partial x)$, obtain the wave equation for sound, $(B/\rho)(\partial^2 s/\partial x^2) = (\partial^2 s/\partial t^2)$. To a mathematical physicist, this equation demonstrates the existence of sound waves and determines their speed. As a physics student, you must take another step or two. Substitute into the wave equation the trial solution $s(x, t) = s_{max} \cos(kx - \omega t)$. Show that this function satisfies the wave equation provided that $\omega/k = \sqrt{B/\rho}$. This result reveals that sound waves exist provided that they move with the speed $v = f\lambda = (2\pi f)(\lambda/2\pi) = \omega/k = \sqrt{B/\rho}$.

Section 17.3 Intensity of Periodic Sound Waves

18. The area of a typical eardrum is about 5.00×10^{-5} m^2. Calculate the sound power incident on an eardrum at (a) the threshold of hearing and (b) the threshold of pain.

19. Calculate the sound level in decibels of a sound wave that has an intensity of 4.00 μW/m^2.

20. A vacuum cleaner produces sound with a measured sound level of 70.0 dB. (a) What is the intensity of this sound in W/m^2? (b) What is the pressure amplitude of the sound?

21. The intensity of a sound wave at a fixed distance from a speaker vibrating at 1.00 kHz is 0.600 W/m^2. (a) Determine the intensity if the frequency is increased to 2.50 kHz while a constant displacement amplitude is maintained. (b) Calculate the intensity if the frequency is reduced to 0.500 kHz and the displacement amplitude is doubled.

22. The intensity of a sound wave at a fixed distance from a speaker vibrating at a frequency f is I. (a) Determine the intensity if the frequency is increased to f' while a constant displacement amplitude is maintained. (b) Calculate the intensity if the frequency is reduced to $f/2$ and the displacement amplitude is doubled.

23. The most soaring vocal melody is in Johann Sebastian Bach's *Mass in B minor*. A portion of the score for the Credo section, number 9, bars 25 to 33, appears in Figure P17.23. The repeating syllable O in the phrase "resurrectionem mortuorum" (the resurrection of the dead) is seamlessly passed from basses to tenors to altos to first sopranos, like a baton in a relay. Each voice carries the melody up in a run of an octave or more. Together they carry it from D below middle C to A above a tenor's high C. In concert pitch, these notes are now assigned frequencies of 146.8 Hz and 880.0 Hz. (a) Find the wavelengths of the initial and final notes. (b) Assume that the choir sings the melody with a uniform sound level of 75.0 dB. Find the pressure amplitudes of the initial and final notes. (c) Find the displacement amplitudes of the initial and final notes. (d) **What If?** In Bach's time, before the invention of the tuning fork, frequencies were assigned to notes as a matter of immediate local convenience. Assume that the rising melody was sung starting from 134.3 Hz and ending at 804.9 Hz. How would the answers to parts (a) through (c) change?

24. The tube depicted in Figure 17.2 is filled with air at 20°C and equilibrium pressure 1 atm. The diameter of the tube is 8.00 cm. The piston is driven at a frequency of 600 Hz with an amplitude of 0.120 cm. What power must be supplied to maintain the oscillation of the piston?

25. A family ice show is held at an enclosed arena. The skaters perform to music with level 80.0 dB. This is too loud for your baby, who yells at 75.0 dB. (a) What total sound intensity engulfs you? (b) What is the combined sound level?

26. Consider sinusoidal sound waves propagating in these three different media: air at 0°C, water, and iron. Use densities and speeds from Tables 14.1 and 17.1. Each wave has the same intensity I_0 and the same angular frequency ω_0. (a) Compare the values of the wavelength in the three media. (b) Compare the values of the displacement amplitude in the three media. (c) Compare the values of the pressure amplitude in the three media. (d) For values of $\omega_0 = 2\,000\,\pi$ rad/s and $I_0 = 1.00 \times 10^{-6}$ W/m^2, evaluate the wavelength, displacement amplitude, and pressure amplitude in each of the three media.

27. The power output of a certain public address speaker is 6.00 W. Suppose it broadcasts equally in all directions. (a) Within what distance from the speaker would the sound be painful to the ear? (b) At what distance from the speaker would the sound be barely audible?

Figure P17.23 Bass (blue), tenor (green), alto (brown), and first soprano (red) parts for a portion of Bach's *Mass in B minor*. For emphasis, the line we choose to call the melody is printed in black. Parts for the second soprano, violins, viola, flutes, oboes, and continuo are omitted. The tenor part is written as it is sung.

28. Show that the difference between decibel levels β_1 and β_2 of a sound is related to the ratio of the distances r_1 and r_2 from the sound source by

$$\beta_2 - \beta_1 = 20\log\left(\frac{r_1}{r_2}\right)$$

29. A firework charge is detonated many meters above the ground. At a distance of 400 m from the explosion, the acoustic pressure reaches a maximum of 10.0 N/m². Assume that the speed of sound is constant at 343 m/s throughout the atmosphere over the region considered, that the ground absorbs all the sound falling on it, and that the air absorbs sound energy as described by the rate 7.00 dB/km. What is the sound level (in dB) at 4.00 km from the explosion?

30. A loudspeaker is placed between two observers who are 110 m apart, along the line connecting them. If one observer records a sound level of 60.0 dB and the other records a sound level of 80.0 dB, how far is the speaker from each observer?

31. Two small speakers emit sound waves of different frequencies. Speaker A has an output of 1.00 mW, and speaker B has an output of 1.50 mW. Determine the sound level (in dB) at point C (Fig. P17.31) if (a) only speaker A emits sound, (b) only speaker B emits sound, and (c) both speakers emit sound.

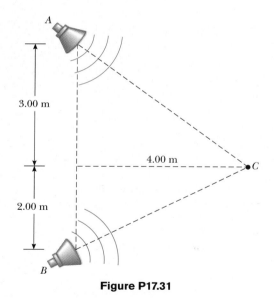

Figure P17.31

32. A jackhammer, operated continuously at a construction site, behaves as a point source of spherical sound waves. A construction supervisor stands 50.0 m due north of this sound source and begins to walk due west. How far does she have to walk in order for the amplitude of the wave function to drop by a factor of 2.00?

33. The sound level at a distance of 3.00 m from a source is 120 dB. At what distance will the sound level be (a) 100 dB and (b) 10.0 dB?

34. A fireworks rocket explodes at a height of 100 m above the ground. An observer on the ground directly under the explosion experiences an average sound intensity of 7.00×10^{-2} W/m² for 0.200 s. (a) What is the total sound energy of the explosion? (b) What is the sound level in decibels heard by the observer?

35. As the people sing in church, the sound level everywhere inside is 101 dB. No sound is transmitted through the massive walls, but all the windows and doors are open on a summer morning. Their total area is 22.0 m². (a) How much sound energy is radiated in 20.0 min? (b) Suppose the ground is a good reflector and sound radiates uniformly in all horizontal and upward directions. Find the sound level 1 km away.

36. The smallest change in sound level that a person can distinguish is approximately 1 dB. When you are standing next to your power lawnmower as it is running, can you hear the steady roar of your neighbor's lawnmower? Perform an order-of-magnitude calculation to substantiate your answer, stating the data you measure or estimate.

Section 17.4   The Doppler Effect

37. A train is moving parallel to a highway with a constant speed of 20.0 m/s. A car is traveling in the same direction as the train with a speed of 40.0 m/s. The car horn sounds at a frequency of 510 Hz, and the train whistle sounds at a frequency of 320 Hz. (a) When the car is behind the train, what frequency does an occupant of the car observe for the train whistle? (b) After the car passes and is in front of the train, what frequency does a train passenger observe for the car horn?

38. Expectant parents are thrilled to hear their unborn baby's heartbeat, revealed by an ultrasonic motion detector. Suppose the fetus's ventricular wall moves in simple harmonic motion with an amplitude of 1.80 mm and a frequency of 115 per minute. (a) Find the maximum linear speed of the heart wall. Suppose the motion detector in contact with the mother's abdomen produces sound at 2 000 000.0 Hz, which travels through tissue at 1.50 km/s. (b) Find the maximum frequency at which sound arrives at the wall of the baby's heart. (c) Find the maximum frequency at which reflected sound is received by the motion detector. By electronically "listening" for echoes at a frequency different from the broadcast frequency, the motion detector can produce beeps of audible sound in synchronization with the fetal heartbeat.

39. Standing at a crosswalk, you hear a frequency of 560 Hz from the siren of an approaching ambulance. After the ambulance passes, the observed frequency of the siren is 480 Hz. Determine the ambulance's speed from these observations.

40. A block with a speaker bolted to it is connected to a spring having spring constant $k = 20.0$ N/m as in Figure P17.40. The total mass of the block and speaker is 5.00 kg, and the amplitude of this unit's motion is 0.500 m. (a) If the speaker emits sound waves of frequency 440 Hz, determine the highest and lowest frequencies heard by the person to the right of the speaker. (b) If the maximum sound level heard by the person is 60.0 dB when he is closest to the

speaker, 1.00 m away, what is the minimum sound level heard by the observer? Assume that the speed of sound is 343 m/s.

Figure P17.40

41. A tuning fork vibrating at 512 Hz falls from rest and accelerates at 9.80 m/s². How far below the point of release is the tuning fork when waves of frequency 485 Hz reach the release point? Take the speed of sound in air to be 340 m/s.

42. At the Winter Olympics, an athlete rides her luge down the track while a bell just above the wall of the chute rings continuously. When her sled passes the bell, she hears the frequency of the bell fall by the musical interval called a minor third. That is, the frequency she hears drops to five sixths of its original value. (a) Find the speed of sound in air at the ambient temperature −10.0°C. (b) Find the speed of the athlete.

43. A siren mounted on the roof of a firehouse emits sound at a frequency of 900 Hz. A steady wind is blowing with a speed of 15.0 m/s. Taking the speed of sound in calm air to be 343 m/s, find the wavelength of the sound (a) upwind of the siren and (b) downwind of the siren. Firefighters are approaching the siren from various directions at 15.0 m/s. What frequency does a firefighter hear (c) if he or she is approaching from an upwind position, so that he or she is moving in the direction in which the wind is blowing? (d) if he or she is approaching from a downwind position and moving against the wind?

44. The Concorde can fly at Mach 1.50, which means the speed of the plane is 1.50 times the speed of sound in air. What is the angle between the direction of propagation of the shock wave and the direction of the plane's velocity?

45. When high-energy charged particles move through a transparent medium with a speed greater than the speed of light in that medium, a shock wave, or bow wave, of light is produced. This phenomenon is called the *Cerenkov effect*. When a nuclear reactor is shielded by a large pool of water, Cerenkov radiation can be seen as a blue glow in the vicinity of the reactor core, due to high-speed electrons moving through the water. In a particular case, the Cerenkov radiation produces a wave front with an apex half-angle of 53.0°. Calculate the speed of the electrons in the water. (The speed of light in water is 2.25×10^8 m/s.)

46. The loop of a circus ringmaster's whip travels at Mach 1.38 (that is, $v_S/v = 1.38$). What angle does the shock wave make with the direction of the whip's motion?

47. A supersonic jet traveling at Mach 3.00 at an altitude of 20 000 m is directly over a person at time $t = 0$ as in Figure P17.47. (a) How long will it be before the person encounters the shock wave? (b) Where will the plane be when it is finally heard? (Assume the speed of sound in air is 335 m/s.)

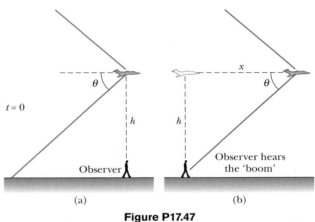

Figure P17.47

Section 17.5   Digital Sound Recording
Section 17.6   Motion Picture Sound

48. This problem represents a possible (but not recommended) way to code instantaneous pressures in a sound wave into 16-bit digital words. Example 17.2 mentions that the pressure amplitude of a 120-dB sound is 28.7 N/m². Let this pressure variation be represented by the digital code 65 536. Let zero pressure variation be represented on the recording by the digital word 0. Let other intermediate pressures be represented by digital words of intermediate size, in direct proportion to the pressure. (a) What digital word would represent the maximum pressure in a 40 dB sound? (b) Explain why this scheme works poorly for soft sounds. (c) Explain how this coding scheme would clip off half of the waveform of any sound, ignoring the actual shape of the wave and turning it into a string of zeros. By introducing sharp corners into every recorded waveform, this coding scheme would make everything sound like a buzzer or a kazoo.

49. Only two recording channels are required to give the illusion of sound coming from any point located between two speakers of a stereophonic sound system. If the same signal is recorded in both channels, a listener will hear it coming from a single direction halfway between the two speakers. This "phantom orchestra" illusion can be heard in the two-channel original Broadway cast recording of the song "Do-Re-Mi" from *The Sound of Music* (Columbia Records KOS 2020). Each of the eight singers can be heard at a different location between the loudspeakers. All listeners with normal hearing will agree on their locations. The brain can sense the direction of sound by noting how much earlier a sound is heard in one ear than in the other. Model your ears as two sensors 19.0 cm apart in a flat screen. If a click from a distant source is heard 210 μs earlier in the left ear than in the right, from what direction does it appear to originate?

50. Assume that a loudspeaker broadcasts sound equally in all directions and produces sound with a level of 103 dB at a distance of 1.60 m from its center. (a) Find its sound power output. (b) If the salesperson claims to be giving you 150 W per channel, he is referring to the electrical power input to the speaker. Find the efficiency of the speaker—that is, the fraction of input power that is converted into useful output power.

Additional Problems

51. A large set of unoccupied football bleachers has solid seats and risers. You stand on the field in front of the bleachers and fire a starter's pistol or sharply clap two wooden boards together once. The sound pulse you produce has no definite frequency and no wavelength. The sound you hear reflected from the bleachers has an identifiable frequency and may remind you of a short toot on a trumpet, or of a buzzer or kazoo. Account for this sound. Compute order-of-magnitude estimates for its frequency, wavelength, and duration, on the basis of data you specify.

52. Many artists sing very high notes in *ad lib* ornaments and cadenzas. The highest note written for a singer in a published score was F-sharp above high C, 1.480 kHz, for Zerbinetta in the original version of Richard Strauss's opera *Ariadne auf Naxos*. (a) Find the wavelength of this sound in air. (b) Suppose people in the fourth row of seats hear this note with level 81.0 dB. Find the displacement amplitude of the sound. (c) **What If?** Because of complaints, Strauss later transposed the note down to F above high C, 1.397 kHz. By what increment did the wavelength change?

53. A sound wave in a cylinder is described by Equations 17.2 through 17.4. Show that $\Delta P = \pm \rho v \omega \sqrt{s_{max}^2 - s^2}$.

54. On a Saturday morning, pickup trucks and sport utility vehicles carrying garbage to the town dump form a nearly steady procession on a country road, all traveling at 19.7 m/s. From one direction, two trucks arrive at the dump every 3 min. A bicyclist is also traveling toward the dump, at 4.47 m/s. (a) With what frequency do the trucks pass him? (b) **What If?** A hill does not slow down the trucks, but makes the out-of-shape cyclist's speed drop to 1.56 m/s. How often do noisy, smelly, inefficient, garbage-dripping, roadhogging trucks whiz past him now?

55. The ocean floor is underlain by a layer of basalt that constitutes the crust, or uppermost layer, of the Earth in that region. Below this crust is found denser periodotite rock, which forms the Earth's mantle. The boundary between these two layers is called the Mohorovicic discontinuity ("Moho" for short). If an explosive charge is set off at the surface of the basalt, it generates a seismic wave that is reflected back out at the Moho. If the speed of this wave in basalt is 6.50 km/s and the two-way travel time is 1.85 s, what is the thickness of this oceanic crust?

56. For a certain type of steel, stress is always proportional to strain with Young's modulus as shown in Table 12.1. The steel has the density listed for iron in Table 14.1. It will fail by bending permanently if subjected to compressive stress greater than its yield strength $\sigma_y = 400$ MPa. A rod 80.0 cm long, made of this steel, is fired at 12.0 m/s straight at a very hard wall, or at another identical rod moving in the opposite direction. (a) The speed of a one-dimensional compressional wave moving along the rod is given by $\sqrt{Y/\rho}$, where ρ is the density and Y is Young's modulus for the rod. Calculate this speed. (b) After the front end of the rod hits the wall and stops, the back end of the rod keeps moving, as described by Newton's first law, until it is stopped by excess pressure in a sound wave moving back through the rod. How much time elapses before the back end of the rod receives the message that it should stop? (c) How far has the back end of the rod moved in this time? Find (d) the strain in the rod and (e) the stress. (f) If it is not to fail, show that the maximum impact speed a rod can have is given by the expression $\sigma_y/\sqrt{\rho Y}$.

57. To permit measurement of her speed, a skydiver carries a buzzer emitting a steady tone at 1 800 Hz. A friend on the ground at the landing site directly below listens to the amplified sound he receives. Assume that the air is calm and that the sound speed is 343 m/s, independent of altitude. While the skydiver is falling at terminal speed, her friend on the ground receives waves of frequency 2 150 Hz. (a) What is the skydiver's speed of descent? (b) **What If?** Suppose the skydiver can hear the sound of the buzzer reflected from the ground. What frequency does she receive?

58. A train whistle ($f = 400$ Hz) sounds higher or lower in frequency depending on whether it approaches or recedes. (a) Prove that the difference in frequency between the approaching and receding train whistle is

$$\Delta f = \frac{2u/v}{1 - u^2/v^2} f$$

where u is the speed of the train and v is the speed of sound. (b) Calculate this difference for a train moving at a speed of 130 km/h. Take the speed of sound in air to be 340 m/s.

59. Two ships are moving along a line due east. The trailing vessel has a speed relative to a land-based observation point of 64.0 km/h, and the leading ship has a speed of 45.0 km/h relative to that point. The two ships are in a region of the ocean where the current is moving uniformly due west at 10.0 km/h. The trailing ship transmits a sonar signal at a frequency of 1 200.0 Hz. What frequency is monitored by the leading ship? (Use 1 520 m/s as the speed of sound in ocean water.)

60. A bat, moving at 5.00 m/s, is chasing a flying insect (Fig. P17.7). If the bat emits a 40.0 kHz chirp and receives back an echo at 40.4 kHz, at what speed is the insect moving toward or away from the bat? (Take the speed of sound in air to be $v = 340$ m/s.)

61. A supersonic aircraft is flying parallel to the ground. When the aircraft is directly overhead, an observer sees a rocket fired from the aircraft. Ten seconds later the observer

hears the sonic boom, followed 2.80 s later by the sound of the rocket engine. What is the Mach number of the aircraft?

62. A police car is traveling east at 40.0 m/s along a straight road, overtaking a car ahead of it moving east at 30.0 m/s. The police car has a malfunctioning siren that is stuck at 1 000 Hz. (a) Sketch the appearance of the wave fronts of the sound produced by the siren. Show the wave fronts both to the east and to the west of the police car. (b) What would be the wavelength in air of the siren sound if the police car were at rest? (c) What is the wavelength in front of the police car? (d) What is it behind the police car? (e) What is the frequency heard by the driver being chased?

63. The speed of a one-dimensional compressional wave traveling along a thin copper rod is 3.56 km/s. A copper bar is given a sharp compressional blow at one end. The sound of the blow, traveling through air at 0°C, reaches the opposite end of the bar 6.40 later than the sound transmitted through the metal of the bar. What is the length of the bar?

64. A jet flies toward higher altitude at a constant speed of 1 963 m/s in a direction making an angle θ with the horizontal (Fig. P17.64). An observer on the ground hears the jet for the first time when it is directly overhead. Determine the value of θ if the speed of sound in air is 340 m/s.

Figure P17.64

65. A meteoroid the size of a truck enters the earth's atmosphere at a speed of 20.0 km/s and is not significantly slowed before entering the ocean. (a) What is the Mach angle of the shock wave from the meteoroid in the atmosphere? (Use 331 m/s as the sound speed.) (b) Assuming that the meteoroid survives the impact with the ocean surface, what is the (initial) Mach angle of the shock wave that the meteoroid produces in the water? (Use the wave speed for seawater given in Table 17.1.)

66. An interstate highway has been built through a poor neighborhood in a city. In the afternoon, the sound level in a rented room is 80.0 dB, as 100 cars pass outside the window every minute. Late at night, when the tenant is working in a factory, the traffic flow is only five cars per minute. What is the average late-night sound level?

67. With particular experimental methods, it is possible to produce and observe in a long thin rod both a longitudinal wave and a transverse wave whose speed depends primarily on tension in the rod. The speed of the longitudinal wave is determined by the Young's modulus and the density of the material as $\sqrt{Y/\rho}$. The transverse wave can be modeled as a wave in a stretched string. A particular metal rod is 150 cm long and has a radius of 0.200 cm and a mass of 50.9 g. Young's modulus for the material is 6.80×10^{10} N/m². What must the tension in the rod be if the ratio of the speed of longitudinal waves to the speed of transverse waves is 8.00?

68. A siren creates sound with a level β at a distance d from the speaker. The siren is powered by a battery that delivers a total energy E. Let e represent the efficiency of the siren. (That is, e is equal to the output sound energy divided by the supplied energy). Determine the total time the siren can sound.

69. The Doppler equation presented in the text is valid when the motion between the observer and the source occurs on a straight line, so that the source and observer are moving either directly toward or directly away from each other. If this restriction is relaxed, one must use the more general Doppler equation

$$f' = \left(\frac{v + v_O \cos\theta_O}{v - v_S \cos\theta_S} \right) f$$

where θ_O and θ_S are defined in Figure P17.69a. (a) Show that if the observer and source are moving away from each other, the preceding equation reduces to Equation 17.13 with negative values for both v_O and v_S. (b) Use the preceding equation to solve the following problem. A train moves at a constant speed of 25.0 m/s toward the intersection shown in Figure P17.69b. A car is stopped near the intersection, 30.0 m from the tracks. If the train's horn emits a frequency of 500 Hz, what is the frequency heard by the passengers in the car when the train is 40.0 m from the intersection? Take the speed of sound to be 343 m/s.

(a)   (b)

Figure P17.69

70. Equation 17.7 states that, at distance r away from a point source with power $\mathcal{P}_{av}$, the wave intensity is

$$I = \frac{\mathcal{P}_{av}}{4\pi r^2}$$

Study Figure 17.9 and prove that, at distance r straight in front of a point source with power $\mathcal{P}_{av}$ moving with constant speed v_S, the wave intensity is

$$I = \frac{\mathcal{P}_{av}}{4\pi r^2}\left(\frac{v - v_S}{v}\right)$$

71. Three metal rods are located relative to each other as shown in Figure P17.71, where $L_1 + L_2 = L_3$. The speed of sound in a rod is given by $v = \sqrt{Y/\rho}$, where ρ is the density and Y is Young's modulus for the rod. Values of density and Young's modulus for the three materials are $\rho_1 = 2.70 \times 10^3$ kg/m^3, $Y_1 = 7.00 \times 10^{10}$ N/m^2, $\rho_2 = 11.3 \times 10^3$ kg/m^3, $Y_2 = 1.60 \times 10^{10}$ N/m^2, $\rho_3 = 8.80 \times 10^3$ kg/m^3, $Y_3 = 11.0 \times 10^{10}$ N/m^2. (a) If $L_3 = 1.50$ m, what must the ratio L_1/L_2 be if a sound wave is to travel the length of rods 1 and 2 in the same time as it takes for the wave to travel the length of rod 3? (b) If the frequency of the source is 4.00 kHz, determine the phase difference between the wave traveling along rods 1 and 2 and the one traveling along rod 3.

Figure P17.71

72. The smallest wavelength possible for a sound wave in air is on the order of the separation distance between air molecules. Find the order of magnitude of the highest-frequency sound wave possible in air, assuming a wave speed of 343 m/s, density 1.20 kg/m^3, and an average molecular mass of 4.82×10^{-26} kg.

Answers to Quick Quizzes

17.1 (c). Although the speed of a wave is given by the product of its wavelength (a) and frequency (b), it is not affected by changes in either one. The amplitude (d) of a sound wave determines the size of the oscillations of elements of air but does not affect the speed of the wave through the air.

17.2 (c). Because the bottom of the bottle is a rigid barrier, the displacement of elements of air at the bottom is zero. Because the pressure variation is a minimum or a maximum when the displacement is zero, and the pulse is moving downward, the pressure variation at the bottom is a maximum.

17.3 (c). The ear trumpet collects sound waves from the large area of its opening and directs it toward the ear. Most of the sound in this large area would miss the ear in the absence of the trumpet.

17.4 (b). The large area of the guitar body sets many elements of air into oscillation and allows the energy to leave the system by mechanical waves at a much larger rate than from the thin vibrating string.

17.5 (c). The only parameter that adds directly is intensity. Because of the logarithm function in the definition of sound level, sound levels cannot be added directly.

17.6 (b). The factor of 100 is two powers of ten. Thus, the logarithm of 100 is 2, which multiplied by 10 gives 20 dB.

17.7 (e). The wave speed cannot be changed by moving the source, so (a) and (b) are incorrect. The detected wavelength is largest at A, so (c) and (d) are incorrect. Choice (f) is incorrect because the detected frequency is lowest at location A.

17.8 (e). The intensity of the sound increases because the train is moving closer to you. Because the train moves at a constant velocity, the Doppler-shifted frequency remains fixed.

17.9 (b). The Mach number is the ratio of the plane's speed (which does not change) to the speed of sound, which is greater in the warm air than in the cold. The denominator of this ratio increases while the numerator stays constant. Therefore, the ratio as a whole—the Mach number—decreases.

Superposition and Standing Waves

▲ Guitarist Carlos Santana takes advantage of standing waves on strings. He changes to a higher note on the guitar by pushing the strings against the frets on the fingerboard, shortening the lengths of the portions of the strings that vibrate. (Bettmann/Corbis)

In the previous two chapters, we introduced the wave model. We have seen that waves are very different from particles. A particle is of zero size, while a wave has a characteristic size—the wavelength. Another important difference between waves and particles is that we can explore the possibility of two or more waves combining at one point in the same medium. We can combine particles to form extended objects, but the particles must be at *different* locations. In contrast, two waves can both be present at the same location, and the ramifications of this possibility are explored in this chapter.

When waves are combined, only certain allowed frequencies can exist on systems with boundary conditions—the frequencies are *quantized*. Quantization is a notion that is at the heart of quantum mechanics, a subject that we introduce formally in Chapter 40. There we show that waves under boundary conditions explain many of the quantum phenomena. For our present purposes in this chapter, quantization enables us to understand the behavior of the wide array of musical instruments that are based on strings and air columns.

We also consider the combination of waves having different frequencies and wavelengths. When two sound waves having nearly the same frequency interfere, we hear variations in the loudness called *beats*. The beat frequency corresponds to the rate of alternation between constructive and destructive interference. Finally, we discuss how any nonsinusoidal periodic wave can be described as a sum of sine and cosine functions.

18.1   Superposition and Interference

Many interesting wave phenomena in nature cannot be described by a single traveling wave. Instead, one must analyze complex waves in terms of a combination of traveling waves. To analyze such wave combinations, one can make use of the **superposition principle:**

Superposition principle

> If two or more traveling waves are moving through a medium, the resultant value of the wave function at any point is the algebraic sum of the values of the wave functions of the individual waves.

Waves that obey this principle are called *linear waves*. In the case of mechanical waves, linear waves are generally characterized by having amplitudes much smaller than their wavelengths. Waves that violate the superposition principle are called *nonlinear waves* and are often characterized by large amplitudes. In this book, we deal only with linear waves.

One consequence of the superposition principle is that **two traveling waves can pass through each other without being destroyed or even altered.** For instance,

Active Figure 18.1 (a–d) Two pulses traveling on a stretched string in opposite directions pass through each other. When the pulses overlap, as shown in (b) and (c), the net displacement of the string equals the sum of the displacements produced by each pulse. Because each pulse produces positive displacements of the string, we refer to their superposition as *constructive interference*. (e) Photograph of the superposition of two equal, symmetric pulses traveling in opposite directions on a stretched spring.

At the Active Figures link at http://www.pse6.com, *you can choose the amplitude and orientation of each of the pulses and study the interference between them as they pass each other.*

when two pebbles are thrown into a pond and hit the surface at different places, the expanding circular surface waves do not destroy each other but rather pass through each other. The complex pattern that is observed can be viewed as two independent sets of expanding circles. Likewise, when sound waves from two sources move through air, they pass through each other.

Figure 18.1 is a pictorial representation of the superposition of two pulses. The wave function for the pulse moving to the right is y_1, and the wave function for the pulse moving to the left is y_2. The pulses have the same speed but different shapes, and the displacement of the elements of the medium is in the positive y direction for both pulses. When the waves begin to overlap (Fig. 18.1b), the wave function for the resulting complex wave is given by $y_1 + y_2$. When the crests of the pulses coincide (Fig. 18.1c), the resulting wave given by $y_1 + y_2$ has a larger amplitude than that of the individual pulses. The two pulses finally separate and continue moving in their original directions (Fig. 18.1d). Note that the pulse shapes remain unchanged after the interaction, as if the two pulses had never met!

The combination of separate waves in the same region of space to produce a resultant wave is called **interference.** For the two pulses shown in Figure 18.1, the displacement of the elements of the medium is in the positive y direction for both pulses, and the resultant pulse (created when the individual pulses overlap) exhibits an amplitude greater than that of either individual pulse. Because the displacements caused by the two pulses are in the same direction, we refer to their superposition as **constructive interference.**

Now consider two pulses traveling in opposite directions on a taut string where one pulse is inverted relative to the other, as illustrated in Figure 18.2. In this case, when the pulses begin to overlap, the resultant pulse is given by $y_1 + y_2$, but the values of the function y_2 are negative. Again, the two pulses pass through each other; however, because the displacements caused by the two pulses are in opposite directions, we refer to their superposition as **destructive interference.**

▲ **PITFALL PREVENTION**

18.1 Do Waves Really *Interfere*?

In popular usage, the term *interfere* implies that an agent affects a situation in some way so as to preclude something from happening. For example, in American football, *pass interference* means that a defending player has affected the receiver so that he is unable to catch the ball. This is very different from its use in physics, where waves pass through each other and interfere, but do not affect each other in any way. In physics, interference is similar to the notion of *combination* as described in this chapter.

Constructive interference

Destructive interference

Education Development Center, Newton, MA

Education Development Center, Newton, MA

At the Active Figures link at http://www.pse6.com, you can choose the amplitude and orientation of each of the pulses and watch the interference as they pass each other.

Active Figure 18.2 (a–e) Two pulses traveling in opposite directions and having displacements that are inverted relative to each other. When the two overlap in (c), their displacements partially cancel each other. (f) Photograph of the superposition of two symmetric pulses traveling in opposite directions, where one is inverted relative to the other.

Quick Quiz 18.1 Two pulses are traveling toward each other, each at 10 cm/s on a long string, as shown in Figure 18.3. Sketch the shape of the string at $t = 0.6$ s.

Figure 18.3 (Quick Quiz 18.1) The pulses on this string are traveling at 10 cm/s.

Quick Quiz 18.2 Two pulses move in opposite directions on a string and are identical in shape except that one has positive displacements of the elements of the string and the other has negative displacements. At the moment that the two pulses completely overlap on the string, (a) the energy associated with the pulses has disappeared (b) the string is not moving (c) the string forms a straight line (d) the pulses have vanished and will not reappear.

Superposition of Sinusoidal Waves

Let us now apply the principle of superposition to two sinusoidal waves traveling in the same direction in a linear medium. If the two waves are traveling to the right and have the same frequency, wavelength, and amplitude but differ in phase, we can express their individual wave functions as

$$y_1 = A \sin(kx - \omega t) \qquad y_2 = A \sin(kx - \omega t + \phi)$$

where, as usual, $k = 2\pi/\lambda$, $\omega = 2\pi f$, and ϕ is the phase constant, which we discussed in Section 16.2. Hence, the resultant wave function y is

$$y = y_1 + y_2 = A[\sin(kx - \omega t) + \sin(kx - \omega t + \phi)]$$

To simplify this expression, we use the trigonometric identity

$$\sin a + \sin b = 2 \cos\left(\frac{a - b}{2}\right) \sin\left(\frac{a + b}{2}\right)$$

If we let $a = kx - \omega t$ and $b = kx - \omega t + \phi$, we find that the resultant wave function y reduces to

$$y = 2A \cos\left(\frac{\phi}{2}\right) \sin\left(kx - \omega t + \frac{\phi}{2}\right)$$

Resultant of two traveling sinusoidal waves

This result has several important features. The resultant wave function y also is sinusoidal and has the same frequency and wavelength as the individual waves because the sine function incorporates the same values of k and ω that appear in the original wave functions. The amplitude of the resultant wave is $2A \cos(\phi/2)$, and its phase is $\phi/2$. If the phase constant ϕ equals 0, then $\cos(\phi/2) = \cos 0 = 1$, and the amplitude of the resultant wave is $2A$—twice the amplitude of either individual wave. In this case the waves are said to be everywhere *in phase* and thus interfere constructively. That is, the crests and troughs of the individual waves y_1 and y_2 occur at the same positions and combine to form the red curve y of amplitude $2A$ shown in Figure 18.4a. Because the

(a)

(b)

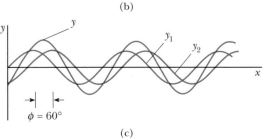

(c)

Active Figure 18.4 The superposition of two identical waves y_1 and y_2 (blue and green) to yield a resultant wave (red). (a) When y_1 and y_2 are in phase, the result is constructive interference. (b) When y_1 and y_2 are π rad out of phase, the result is destructive interference. (c) When the phase angle has a value other than 0 or π rad, the resultant wave y falls somewhere between the extremes shown in (a) and (b).

At the Active Figures link at http://www.pse6.com, you can change the phase relationship between the waves and observe the wave representing the superposition.

individual waves are in phase, they are indistinguishable in Figure 18.4a, in which they appear as a single blue curve. In general, constructive interference occurs when $\cos(\phi/2) = \pm 1$. This is true, for example, when $\phi = 0, 2\pi, 4\pi, \ldots$ rad—that is, when ϕ is an *even* multiple of π.

When ϕ is equal to π rad or to any *odd* multiple of π, then $\cos(\phi/2) = \cos(\pi/2) = 0$, and the crests of one wave occur at the same positions as the troughs of the second wave (Fig. 18.4b). Thus, the resultant wave has *zero* amplitude everywhere, as a consequence of destructive interference. Finally, when the phase constant has an arbitrary value other than 0 or an integer multiple of π rad (Fig. 18.4c), the resultant wave has an amplitude whose value is somewhere between 0 and $2A$.

Interference of Sound Waves

One simple device for demonstrating interference of sound waves is illustrated in Figure 18.5. Sound from a loudspeaker S is sent into a tube at point P, where there is a T-shaped junction. Half of the sound energy travels in one direction, and half travels in the opposite direction. Thus, the sound waves that reach the receiver R can travel along either of the two paths. The distance along any path from speaker to receiver is called the **path length** r. The lower path length r_1 is fixed, but the upper path length r_2 can be varied by sliding the U-shaped tube, which is similar to that on a slide trombone. When the difference in the path lengths $\Delta r = |r_2 - r_1|$ is either zero or some integer multiple of the wavelength λ (that is $\Delta r = n\lambda$, where $n = 0, 1, 2, 3, \ldots$), the two waves reaching the receiver at any instant are in phase and interfere constructively, as shown in Figure 18.4a. For this case, a maximum in the sound intensity is detected at the receiver. If the path length r_2 is adjusted such that the path difference $\Delta r = \lambda/2$, $3\lambda/2, \ldots, n\lambda/2$ (for n odd), the two waves are exactly π rad, or 180°, out of phase at the receiver and hence cancel each other. In this case of destructive interference, no sound is detected at the receiver. This simple experiment demonstrates that a phase difference may arise between two waves generated by the same source when they travel along paths of unequal lengths. This important phenomenon will be indispensable in our investigation of the interference of light waves in Chapter 37.

It is often useful to express the path difference in terms of the phase angle ϕ between the two waves. Because a path difference of one wavelength corresponds to a phase angle of 2π rad, we obtain the ratio $\phi/2\pi = \Delta r/\lambda$ or

Relationship between path difference and phase angle

$$\Delta r = \frac{\phi}{2\pi}\lambda \tag{18.1}$$

Using the notion of path difference, we can express our conditions for constructive and destructive interference in a different way. If the path difference is any even multiple of $\lambda/2$, then the phase angle $\phi = 2n\pi$, where $n = 0, 1, 2, 3, \ldots$, and the interference is constructive. For path differences of odd multiples of $\lambda/2$, $\phi = (2n + 1)\pi$, where $n = 0, 1, 2, 3, \ldots$, and the interference is destructive. Thus, we have the conditions

Figure 18.5 An acoustical system for demonstrating interference of sound waves. A sound wave from the speaker (S) propagates into the tube and splits into two parts at point P. The two waves, which combine at the opposite side, are detected at the receiver (R). The upper path length r_2 can be varied by sliding the upper section.

$$\Delta r = (2n)\,\frac{\lambda}{2} \qquad \text{for constructive interference}$$

and   (18.2)

$$\Delta r = (2n + 1)\,\frac{\lambda}{2} \qquad \text{for destructive interference}$$

This discussion enables us to understand why the speaker wires in a stereo system should be connected properly. When connected the wrong way—that is, when the positive (or red) wire is connected to the negative (or black) terminal on one of the speakers and the other is correctly wired—the speakers are said to be "out of phase"—one speaker cone moves outward while the other moves inward. As a consequence, the sound wave coming from one speaker destructively interferes with the wave coming from the other—along a line midway between the two, a rarefaction region due to one speaker is superposed on a compression region from the other speaker. Although the two sounds probably do not completely cancel each other (because the left and right stereo signals are usually not identical), a substantial loss of sound quality occurs at points along this line.

Example 18.1   Two Speakers Driven by the Same Source

A pair of speakers placed 3.00 m apart are driven by the same oscillator (Fig. 18.6). A listener is originally at point O, which is located 8.00 m from the center of the line connecting the two speakers. The listener then walks to point P, which is a perpendicular distance 0.350 m from O, before reaching the *first minimum* in sound intensity. What is the frequency of the oscillator?

Solution To find the frequency, we must know the wavelength of the sound coming from the speakers. With this information, combined with our knowledge of the speed of sound, we can calculate the frequency. The wavelength can be determined from the interference information given. The first minimum occurs when the two waves reaching the listener at point P are 180° out of phase—in other words, when their path difference Δr equals $\lambda/2$. To calculate the path difference, we must first find the path lengths r_1 and r_2.

Figure 18.6 shows the physical arrangement of the speakers, along with two shaded right triangles that can be drawn on the basis of the lengths described in the problem. From these triangles, we find that the path lengths are

$$r_1 = \sqrt{(8.00 \text{ m})^2 + (1.15 \text{ m})^2} = 8.08 \text{ m}$$

and

$$r_2 = \sqrt{(8.00 \text{ m})^2 + (1.85 \text{ m})^2} = 8.21 \text{ m}$$

Hence, the path difference is $r_2 - r_1 = 0.13$ m. Because we require that this path difference be equal to $\lambda/2$ for the first minimum, we find that $\lambda = 0.26$ m.

To obtain the oscillator frequency, we use Equation 16.12, $v = \lambda f$, where v is the speed of sound in air, 343 m/s:

$$f = \frac{v}{\lambda} = \frac{343 \text{ m/s}}{0.26 \text{ m}} = \boxed{1.3 \text{ kHz}}$$

What If? What if the speakers were connected out of phase? What happens at point P in Figure 18.6?

Answer In this situation, the path difference of $\lambda/2$ combines with a phase difference of $\lambda/2$ due to the incorrect wiring to give a full phase difference of λ. As a result, the waves are in phase and there is a *maximum* intensity at point P.

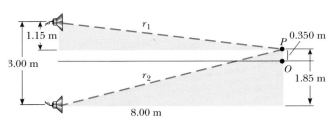

Figure 18.6 (Example 18.1) Two speakers emit sound waves to a listener at P.

18.2   Standing Waves

The sound waves from the speakers in Example 18.1 leave the speakers in the forward direction, and we considered interference at a point in front of the speakers. Suppose that we turn the speakers so that they face each other and then have them emit sound of the same frequency and amplitude. In this situation, two identical waves travel in

Figure 18.7 Two speakers emit sound waves toward each other. When they overlap, identical waves traveling in opposite directions will combine to form standing waves.

opposite directions in the same medium, as in Figure 18.7. These waves combine in accordance with the superposition principle.

We can analyze such a situation by considering wave functions for two transverse sinusoidal waves having the same amplitude, frequency, and wavelength but traveling in opposite directions in the same medium:

$$y_1 = A \sin(kx - \omega t) \qquad y_2 = A \sin(kx + \omega t)$$

where y_1 represents a wave traveling in the $+x$ direction and y_2 represents one traveling in the $-x$ direction. Adding these two functions gives the resultant wave function y:

$$y = y_1 + y_2 = A \sin(kx - \omega t) + A \sin(kx + \omega t)$$

When we use the trigonometric identity $\sin(a \pm b) = \sin(a)\cos(b) \pm \cos(a)\sin(b)$, this expression reduces to

$$y = (2A \sin kx) \cos \omega t \qquad \text{(18.3)}$$

Equation 18.3 represents the wave function of a **standing wave.** A standing wave, such as the one shown in Figure 18.8, is an oscillation pattern *with a stationary outline* that results from the superposition of two identical waves traveling in opposite directions.

Notice that Equation 18.3 does not contain a function of $kx - \omega t$. Thus, it is not an expression for a traveling wave. If we observe a standing wave, we have no sense of motion in the direction of propagation of either of the original waves. If we compare this equation with Equation 15.6, we see that Equation 18.3 describes a special kind of simple harmonic motion. Every element of the medium oscillates in simple harmonic motion with the same frequency ω (according to the $\cos \omega t$ factor in the equation). However, the amplitude of the simple harmonic motion of a given element (given by the factor $2A \sin kx$, the coefficient of the cosine function) depends on the location x of the element in the medium.

The maximum amplitude of an element of the medium has a minimum value of zero when x satisfies the condition $\sin kx = 0$, that is, when

$$kx = \pi, 2\pi, 3\pi, \ldots$$

Because $k = 2\pi/\lambda$, these values for kx give

$$x = \frac{\lambda}{2}, \lambda, \frac{3\lambda}{2}, \ldots = \frac{n\lambda}{2} \qquad n = 0, 1, 2, 3, \ldots \qquad \text{(18.4)}$$

These points of zero amplitude are called **nodes.**

▲ **PITFALL PREVENTION**

18.2 Three Types of Amplitude

We need to distinguish carefully here between the **amplitude of the individual waves,** which is A, and the **amplitude of the simple harmonic motion of the elements of the medium,** which is $2A \sin kx$. A given element in a standing wave vibrates within the constraints of the *envelope* function $2A \sin kx$, where x is that element's position in the medium. This is in contrast to traveling sinusoidal waves, in which all elements oscillate with the same amplitude and the same frequency, and the amplitude A of the wave is the same as the amplitude A of the simple harmonic motion of the elements. Furthermore, we can identify the **amplitude of the standing wave** as $2A$.

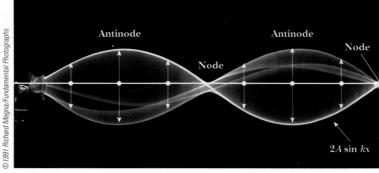

Figure 18.8 Multiflash photograph of a standing wave on a string. The time behavior of the vertical displacement from equilibrium of an individual element of the string is given by $\cos \omega t$. That is, each element vibrates at an angular frequency ω. The amplitude of the vertical oscillation of any elements of the string depends on the horizontal position of the element. Each element vibrates within the confines of the envelope function $2A \sin kx$.

The element with the *greatest* possible displacement from equilibrium has an amplitude of 2A, and we define this as the amplitude of the standing wave. The positions in the medium at which this maximum displacement occurs are called **antinodes.** The antinodes are located at positions for which the coordinate x satisfies the condition $\sin kx = \pm 1$, that is, when

$$kx = \frac{\pi}{2}, \frac{3\pi}{2}, \frac{5\pi}{2}, \ldots$$

Thus, the positions of the antinodes are given by

$$x = \frac{\lambda}{4}, \frac{3\lambda}{4}, \frac{5\lambda}{4}, \ldots = \frac{n\lambda}{4} \qquad n = 1, 3, 5, \ldots \qquad (18.5)$$

Position of antinodes

In examining Equations 18.4 and 18.5, we note the following important features of the locations of nodes and antinodes:

The distance between adjacent antinodes is equal to $\lambda/2$.

The distance between adjacent nodes is equal to $\lambda/2$.

The distance between a node and an adjacent antinode is $\lambda/4$.

Wave patterns of the elements of the medium produced at various times by two waves traveling in opposite directions are shown in Figure 18.9. The blue and green curves are the wave patterns for the individual traveling waves, and the red curves are the wave patterns for the resultant standing wave. At $t = 0$ (Fig. 18.9a), the two traveling waves are in phase, giving a wave pattern in which each element of the medium is experiencing its maximum displacement from equilibrium. One quarter of a period later, at $t = T/4$ (Fig. 18.9b), the traveling waves have moved one quarter of a wavelength (one to the right and the other to the left). At this time, the traveling waves are out of phase, and each element of the medium is passing through the equilibrium position in its simple harmonic motion. The result is zero displacement for elements at all values of x—that is, the wave pattern is a straight line. At $t = T/2$ (Fig. 18.9c), the traveling waves are again in phase, producing a wave pattern that is inverted relative to the $t = 0$ pattern. In the standing wave, the elements of the medium alternate in time between the extremes shown in Figure 18.9a and c.

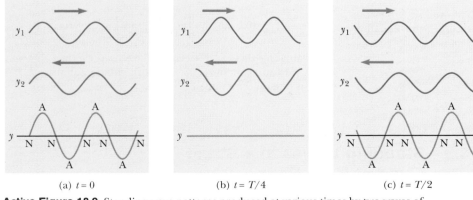

(a) $t = 0$ (b) $t = T/4$ (c) $t = T/2$

Active Figure 18.9 Standing-wave patterns produced at various times by two waves of equal amplitude traveling in opposite directions. For the resultant wave y, the nodes (N) are points of zero displacement, and the antinodes (A) are points of maximum displacement.

At the Active Figures link at http://www.pse6.com, *you can choose the wavelength of the waves and see the standing wave that results.*

Quick Quiz 18.3 Consider a standing wave on a string as shown in Figure 18.9. Define the velocity of elements of the string as positive if they are moving upward in the figure. At the moment the string has the shape shown by the red curve in Figure 18.9a, the instantaneous velocity of elements along the string (a) is zero for all elements (b) is positive for all elements (c) is negative for all elements (d) varies with the position of the element.

Quick Quiz 18.4 Continuing with the scenario in Quick Quiz 18.3, at the moment the string has the shape shown by the red curve in Figure 18.9b, the instantaneous velocity of elements along the string (a) is zero for all elements (b) is positive for all elements (c) is negative for all elements (d) varies with the position of the element.

Example 18.2 Formation of a Standing Wave

Two waves traveling in opposite directions produce a standing wave. The individual wave functions are

$$y_1 = (4.0 \text{ cm}) \sin(3.0x - 2.0t)$$
$$y_2 = (4.0 \text{ cm}) \sin(3.0x + 2.0t)$$

where x and y are measured in centimeters.

(A) Find the amplitude of the simple harmonic motion of the element of the medium located at $x = 2.3$ cm.

Solution The standing wave is described by Equation 18.3; in this problem, we have $A = 4.0$ cm, $k = 3.0$ rad/cm, and $\omega = 2.0$ rad/s. Thus,

$$y = (2A \sin kx) \cos \omega t = [(8.0 \text{ cm}) \sin 3.0x] \cos 2.0t$$

Thus, we obtain the amplitude of the simple harmonic motion of the element at the position $x = 2.3$ cm by evaluating the coefficient of the cosine function at this position:

$$y_{max} = (8.0 \text{ cm}) \sin 3.0x|_{x=2.3}$$

$$= (8.0 \text{ cm}) \sin (6.9 \text{ rad}) = \boxed{4.6 \text{ cm}}$$

(B) Find the positions of the nodes and antinodes if one end of the string is at $x = 0$.

Solution With $k = 2\pi/\lambda = 3.0$ rad/cm, we see that the wavelength is $\lambda = (2\pi/3.0)$ cm. Therefore, from Equation 18.4 we find that the nodes are located at

$$x = n\frac{\lambda}{2} = n\left(\frac{\pi}{3}\right) \text{cm} \qquad n = \boxed{0, 1, 2, 3, \ldots}$$

and from Equation 18.5 we find that the antinodes are located at

$$x = n\frac{\lambda}{4} = n\left(\frac{\pi}{6}\right) \text{cm} \qquad n = \boxed{1, 3, 5, \ldots}$$

(C) What is the maximum value of the position in the simple harmonic motion of an element located at an antinode?

Solution According to Equation 18.3, the maximum position of an element at an antinode is the amplitude of the standing wave, which is twice the amplitude of the individual traveling waves:

$$y_{max} = 2A(\sin kx)_{max} = 2(4.0 \text{ cm})(\pm 1) = \boxed{\pm 8.0 \text{ cm}}$$

where we have used the fact that the maximum value of $\sin kx$ is ± 1. Let us check this result by evaluating the coefficient of our standing-wave function at the positions we found for the antinodes:

$$y_{max} = (8.0 \text{ cm}) \sin 3.0x|_{x=n(\pi/6)}$$

$$= (8.0 \text{ cm}) \sin \left[3.0n\left(\frac{\pi}{6}\right) \text{rad}\right]$$

$$= (8.0 \text{ cm}) \sin \left[n\left(\frac{\pi}{2}\right) \text{rad}\right] = \pm 8.0 \text{ cm}$$

In evaluating this expression, we have used the fact that n is an odd integer; thus, the sine function is equal to ± 1, depending on the value of n.

18.3 Standing Waves in a String Fixed at Both Ends

Consider a string of length L fixed at both ends, as shown in Figure 18.10. Standing waves are set up in the string by a continuous superposition of waves incident on and reflected from the ends. Note that there is a boundary condition for the waves on the

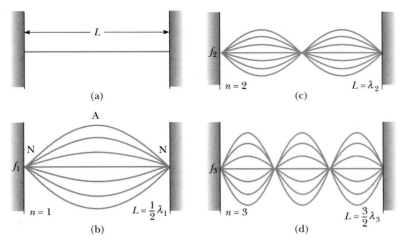

Active Figure 18.10 (a) A string of length L fixed at both ends. The normal modes of vibration form a harmonic series: (b) the fundamental, or first harmonic; (c) the second harmonic; (d) the third harmonic.

At the Active Figures link at http://www.pse6.com, you can choose the mode number and see the corresponding standing wave.

string. The ends of the string, because they are fixed, must necessarily have zero displacement and are, therefore, nodes by definition. The boundary condition results in the string having a number of natural patterns of oscillation, called **normal modes,** each of which has a characteristic frequency that is easily calculated. This situation in which only certain frequencies of oscillation are allowed is called **quantization.** Quantization is a common occurrence when waves are subject to boundary conditions and will be a central feature in our discussions of quantum physics in the extended version of this text.

Figure 18.11 shows one of the normal modes of oscillation of a string fixed at both ends. Except for the nodes, which are always stationary, all elements of the string oscillate vertically with the same frequency but with different amplitudes of simple harmonic motion. Figure 18.11 represents snapshots of the standing wave at various times over one half of a period. The red arrows show the velocities of various elements of the string at various times. As we found in Quick Quizzes 18.3 and 18.4,

Figure 18.11 A standing-wave pattern in a taut string. The five "snapshots" were taken at intervals of one eighth of the period. (a) At $t = 0$, the string is momentarily at rest. (b) At $t = T/8$, the string is in motion, as indicated by the red arrows, and different parts of the string move in different directions with different speeds. (c) At $t = T/4$, the string is moving but horizontal (undeformed). (d) The motion continues as indicated. (e) At $t = T/2$, the string is again momentarily at rest, but the crests and troughs of (a) are reversed. The cycle continues until ultimately, when a time interval equal to T has passed, the configuration shown in (a) is repeated.

all elements of the string have zero velocity at the extreme positions (Figs. 18.11a and 18.11e) and elements have varying velocities at other positions (Figs. 18.11b through 18.11d).

The normal modes of oscillation for the string can be described by imposing the requirements that the ends be nodes and that the nodes and antinodes be separated by one fourth of a wavelength. The first normal mode that is consistent with the boundary conditions, shown in Figure 18.10b, has nodes at its ends and one antinode in the middle. This is the longest-wavelength mode that is consistent with our requirements. This first normal mode occurs when the length of the string is half the wavelength λ_1, as indicated in Figure 18.10b, or $\lambda_1 = 2L$. The next normal mode (see Fig. 18.10c) of wavelength λ_2 occurs when the wavelength equals the length of the string, that is, when $\lambda_2 = L$. The third normal mode (see Fig. 18.10d) corresponds to the case in which $\lambda_3 = 2L/3$. In general, the wavelengths of the various normal modes for a string of length L fixed at both ends are

Wavelengths of normal modes

$$\lambda_n = \frac{2L}{n} \qquad n = 1, 2, 3, \ldots \qquad (18.6)$$

where the index n refers to the nth normal mode of oscillation. These are the *possible* modes of oscillation for the string. The *actual* modes that are excited on a string are discussed shortly.

The natural frequencies associated with these modes are obtained from the relationship $f = v/\lambda$, where the wave speed v is the same for all frequencies. Using Equation 18.6, we find that the natural frequencies f_n of the normal modes are

Frequencies of normal modes as functions of wave speed and length of string

$$f_n = \frac{v}{\lambda_n} = n \frac{v}{2L} \qquad n = 1, 2, 3, \ldots \qquad (18.7)$$

These natural frequencies are also called the *quantized frequencies* associated with the vibrating string fixed at both ends.

Because $v = \sqrt{T/\mu}$ (see Eq. 16.18), where T is the tension in the string and μ is its linear mass density, we can also express the natural frequencies of a taut string as

Frequencies of normal modes as functions of string tension and linear mass density

$$f_n = \frac{n}{2L} \sqrt{\frac{T}{\mu}} \qquad n = 1, 2, 3, \ldots \qquad (18.8)$$

The lowest frequency f_1, which corresponds to $n = 1$, is called either the **fundamental** or the **fundamental frequency** and is given by

Fundamental frequency of a taut string

$$f_1 = \frac{1}{2L} \sqrt{\frac{T}{\mu}} \qquad (18.9)$$

The frequencies of the remaining normal modes are integer multiples of the fundamental frequency. Frequencies of normal modes that exhibit an integer-multiple relationship such as this form a **harmonic series,** and the normal modes are called

Multiflash photographs of standing-wave patterns in a cord driven by a vibrator at its left end. The single-loop pattern represents the first normal mode (the fundamental), $n = 1$). The double-loop pattern represents the second normal mode ($n = 2$), and the triple-loop pattern represents the third normal mode ($n = 3$).

harmonics. The fundamental frequency f_1 is the frequency of the first harmonic; the frequency $f_2 = 2f_1$ is the frequency of the second harmonic; and the frequency $f_n = nf_1$ is the frequency of the nth harmonic. Other oscillating systems, such as a drumhead, exhibit normal modes, but the frequencies are not related as integer multiples of a fundamental. Thus, we do not use the term *harmonic* in association with these types of systems.

In obtaining Equation 18.6, we used a technique based on the separation distance between nodes and antinodes. We can obtain this equation in an alternative manner. Because we require that the string be fixed at $x = 0$ and $x = L$, the wave function $y(x, t)$ given by Equation 18.3 must be zero at these points for all times. That is, the *boundary conditions* require that $y(0, t) = 0$ and $y(L, t) = 0$ for all values of t. Because the standing wave is described by $y = 2A(\sin kx) \cos \omega t$, the first boundary condition, $y(0, t) = 0$, is automatically satisfied because $\sin kx = 0$ at $x = 0$. To meet the second boundary condition, $y(L, t) = 0$, we require that $\sin kL = 0$. This condition is satisfied when the angle kL equals an integer multiple of π rad. Therefore, the allowed values of k are given by[1]

$$k_n L = n\pi \qquad n = 1, 2, 3, \ldots \tag{18.10}$$

Because $k_n = 2\pi/\lambda_n$, we find that

$$\left(\frac{2\pi}{\lambda_n}\right) L = n\pi \qquad \text{or} \qquad \lambda_n = \frac{2L}{n}$$

which is identical to Equation 18.6.

Let us examine further how these various harmonics are created in a string. If we wish to excite just a single harmonic, we must distort the string in such a way that its distorted shape corresponds to that of the desired harmonic. After being released, the string vibrates at the frequency of that harmonic. This maneuver is difficult to perform, however, and it is not how we excite a string of a musical instrument. If the string is distorted such that its distorted shape is not that of just one harmonic, the resulting vibration includes various harmonics. Such a distortion occurs in musical instruments when the string is plucked (as in a guitar), bowed (as in a cello), or struck (as in a piano). When the string is distorted into a nonsinusoidal shape, only waves that satisfy the boundary conditions can persist on the string. These are the harmonics.

The frequency of a string that defines the musical note that it plays is that of the fundamental. The frequency of the string can be varied by changing either the tension or the string's length. For example, the tension in guitar and violin strings is varied by a screw adjustment mechanism or by tuning pegs located on the neck of the instrument. As the tension is increased, the frequency of the normal modes increases in accordance with Equation 18.8. Once the instrument is "tuned," players vary the frequency by moving their fingers along the neck, thereby changing the length of the oscillating portion of the string. As the length is shortened, the frequency increases because, as Equation 18.8 specifies, the normal-mode frequencies are inversely proportional to string length.

Quick Quiz 18.5 When a standing wave is set up on a string fixed at both ends, (a) the number of nodes is equal to the number of antinodes (b) the wavelength is equal to the length of the string divided by an integer (c) the frequency is equal to the number of nodes times the fundamental frequency (d) the shape of the string at any time is symmetric about the midpoint of the string.

---

[1] We exclude $n = 0$ because this value corresponds to the trivial case in which no wave exists ($k = 0$).

Example 18.3 Give Me a C Note!

Middle C on a piano has a fundamental frequency of 262 Hz, and the first A above middle C has a fundamental frequency of 440 Hz.

(A) Calculate the frequencies of the next two harmonics of the C string.

Solution Knowing that the frequencies of higher harmonics are integer multiples of the fundamental frequency $f_1 = 262$ Hz, we find that

$$f_2 = 2f_1 = \boxed{524 \text{ Hz}}$$

$$f_3 = 3f_1 = \boxed{786 \text{ Hz}}$$

(B) If the A and C strings have the same linear mass density μ and length L, determine the ratio of tensions in the two strings.

Solution Using Equation 18.9 for the two strings vibrating at their fundamental frequencies gives

$$f_{1A} = \frac{1}{2L}\sqrt{\frac{T_A}{\mu}} \quad \text{and} \quad f_{1C} = \frac{1}{2L}\sqrt{\frac{T_C}{\mu}}$$

Setting up the ratio of these frequencies, we find that

$$\frac{f_{1A}}{f_{1C}} = \sqrt{\frac{T_A}{T_C}}$$

$$\frac{T_A}{T_C} = \left(\frac{f_{1A}}{f_{1C}}\right)^2 = \left(\frac{440}{262}\right)^2 = \boxed{2.82}$$

What If? What if we look inside a real piano? In this case, the assumption we made in part (B) is only partially true. The string densities are equal, but the length of the A string is only 64 percent of the length of the C string. What is the ratio of their tensions?

Answer Using Equation 18.8 again, we set up the ratio of frequencies:

$$\frac{f_{1A}}{f_{1C}} = \frac{L_C}{L_A}\sqrt{\frac{T_A}{T_C}} = \left(\frac{100}{64}\right)\sqrt{\frac{T_A}{T_C}}$$

$$\frac{T_A}{T_C} = (0.64)^2\left(\frac{440}{262}\right)^2 = 1.16$$

Example 18.4 Guitar Basics Interactive

The high E string on a guitar measures 64.0 cm in length and has a fundamental frequency of 330 Hz. By pressing down so that the string is in contact with the first fret (Fig. 18.12), the string is shortened so that it plays an F note that has a frequency of 350 Hz. How far is the fret from the neck end of the string?

Solution Equation 18.7 relates the string's length to the fundamental frequency. With $n = 1$, we can solve for the speed of the wave on the string,

$$v = \frac{2L}{n}f_n = \frac{2(0.640 \text{ m})}{1}(330 \text{ Hz}) = 422 \text{ m/s}$$

Because we have not adjusted the tuning peg, the tension in the string, and hence the wave speed, remain constant. We can again use Equation 18.7, this time solving for L and

Figure 18.12 (Example 18.4) Playing an F note on a guitar.

substituting the new frequency to find the shortened string length:

$$L = n\frac{v}{2f_n} = (1)\frac{422 \text{ m/s}}{2(350 \text{ Hz})} = 0.603 \text{ m} = 60.3 \text{ cm}$$

The difference between this length and the measured length of 64.0 cm is the distance from the fret to the neck end of the string, or $\boxed{3.7 \text{ cm.}}$

What If? What if we wish to play an F sharp, which we do by pressing down on the second fret from the neck in Figure 18.12? The frequency of F sharp is 370 Hz. Is this fret another 3.7 cm from the neck?

Answer If you inspect a guitar fingerboard, you will find that the frets are *not* equally spaced. They are far apart near the neck and close together near the opposite end. Consequently, from this observation, we would not expect the F sharp fret to be another 3.7 cm from the end.

Let us repeat the calculation of the string length, this time for the frequency of F sharp:

$$L = n\frac{v}{2f_n} = (1)\frac{422 \text{ m/s}}{2(370 \text{ Hz})} = 0.571 \text{ m}$$

This gives a distance of $0.640 \text{ m} - 0.571 \text{ m} = 0.069 \text{ m} = 6.9$ cm from the neck. Subtracting the distance from the neck to the first fret, the separation distance between the first and second frets is $6.9 \text{ cm} - 3.7 \text{ cm} = 3.2 \text{ cm.}$

 Explore this situation at the Interactive Worked Example link at **http://www.pse6.com.**

Example 18.5   Changing String Vibration with Water

One end of a horizontal string is attached to a vibrating blade and the other end passes over a pulley as in Figure 18.13a. A sphere of mass 2.00 kg hangs on the end of the string. The string is vibrating in its second harmonic. A container of water is raised under the sphere so that the sphere is completely submerged. After this is done, the string vibrates in its fifth harmonic, as shown in Figure 18.13b. What is the radius of the sphere?

Solution To conceptualize the problem, imagine what happens when the sphere is immersed in the water. The buoyant force acts upward on the sphere, reducing the tension in the string. The change in tension causes a change in the speed of waves on the string, which in turn causes a change in the wavelength. This altered wavelength results in the string vibrating in its fifth normal mode rather than the second. We categorize the problem as one in which we will need to combine our understanding of Newton's second law, buoyant forces, and standing waves on strings. We begin to analyze the problem by studying Figure 18.13a. Newton's second law applied to the sphere tells us that the tension in the string is equal to the weight of the sphere:

$$\Sigma F = T_1 - mg = 0$$
$$T_1 = mg = (2.00 \text{ kg})(9.80 \text{ m/s}^2) = 19.6 \text{ N}$$

where the subscript 1 is used to indicate initial variables before we immerse the sphere in water. Once the sphere is immersed in water, the tension in the string decreases to T_2. Applying Newton's second law to the sphere again in this situation, we have

$$T_2 + B - mg = 0$$
$$(1) \qquad B = mg - T_2$$

The desired quantity, the radius of the sphere, will appear in the expression for the buoyant force B. Before proceeding in this direction, however, we must evaluate T_2. We do this from the standing wave information. We write the equation for the frequency of a standing wave on a string (Equation 18.8) twice, once before we immerse the sphere and once after, and divide the equations:

$$f = \frac{n_1}{2L}\sqrt{\frac{T_1}{\mu}}$$
$$f = \frac{n_2}{2L}\sqrt{\frac{T_2}{\mu}} \qquad \longrightarrow \qquad 1 = \frac{n_1}{n_2}\sqrt{\frac{T_1}{T_2}}$$

where the frequency f is the same in both cases, because it is determined by the vibrating blade. In addition, the linear mass density μ and the length L of the vibrating portion of the string are the same in both cases. Solving for T_2, we have

$$T_2 = \left(\frac{n_1}{n_2}\right)^2 T_1 = \left(\frac{2}{5}\right)^2 (19.6 \text{ N}) = 3.14 \text{ N}$$

Substituting this into Equation (1), we can evaluate the buoyant force on the sphere:

$$B = mg - T_2 = 19.6 \text{ N} - 3.14 \text{ N} = 16.5 \text{ N}$$

Finally, expressing the buoyant force (Eq. 14.5) in terms of the radius of the sphere, we solve for the radius:

$$B = \rho_{\text{water}} g V_{\text{sphere}} = \rho_{\text{water}} g \left(\tfrac{4}{3}\pi r^3\right)$$

$$r = \sqrt[3]{\frac{3B}{4\pi\rho_{\text{water}} g}} = \sqrt[3]{\frac{3(16.5 \text{ N})}{4\pi(1\,000 \text{ kg/m}^3)(9.80 \text{ m/s}^2)}}$$

$$= 7.38 \times 10^{-2} \text{ m} = \boxed{7.38 \text{ cm}}$$

To finalize this problem, note that only certain radii of the sphere will result in the string vibrating in a normal mode. This is because the speed of waves on the string must be changed to a value such that the length of the string is an integer multiple of half wavelengths. This is a feature of the *quantization* that we introduced earlier in this chapter—the sphere radii that cause the string to vibrate in a normal mode are *quantized*.

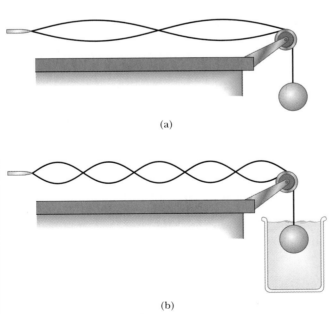

(a)

(b)

Figure 18.13 (Example 18.5) When the sphere hangs in air, the string vibrates in its second harmonic. When the sphere is immersed in water, the string vibrates in its fifth harmonic.

You can adjust the mass at the Interactive Worked Example link at **http://www.pse6.com.**

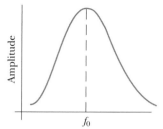

Figure 18.14 Graph of the amplitude (response) versus driving frequency for an oscillating system. The amplitude is a maximum at the resonance frequency f_0.

Figure 18.15 An example of resonance. If pendulum A is set into oscillation, only pendulum C, whose length matches that of A, eventually oscillates with large amplitude, or resonates. The arrows indicate motion in a plane perpendicular to the page.

Vibrating blade

Figure 18.16 Standing waves are set up in a string when one end is connected to a vibrating blade. When the blade vibrates at one of the natural frequencies of the string, large-amplitude standing waves are created.

18.4 Resonance

We have seen that a system such as a taut string is capable of oscillating in one or more normal modes of oscillation. **If a periodic force is applied to such a system, the amplitude of the resulting motion is greatest when the frequency of the applied force is equal to one of the natural frequencies of the system.** We discussed this phenomenon, known as *resonance*, briefly in Section 15.7. Although a block–spring system or a simple pendulum has only one natural frequency, standing-wave systems have a whole set of natural frequencies, such as that given by Equation 18.7 for a string. Because an oscillating system exhibits a large amplitude when driven at any of its natural frequencies, these frequencies are often referred to as **resonance frequencies.**

Figure 18.14 shows the response of an oscillating system to various driving frequencies, where one of the resonance frequencies of the system is denoted by f_0. Note that the amplitude of oscillation of the system is greatest when the frequency of the driving force equals the resonance frequency. The maximum amplitude is limited by friction in the system. If a driving force does work on an oscillating system that is initially at rest, the input energy is used both to increase the amplitude of the oscillation and to overcome the friction force. Once maximum amplitude is reached, the work done by the driving force is used only to compensate for mechanical energy loss due to friction.

Examples of Resonance

A playground swing is a pendulum having a natural frequency that depends on its length. Whenever we use a series of regular impulses to push a child in a swing, the swing goes higher if the frequency of the periodic force equals the natural frequency of the swing. We can demonstrate a similar effect by suspending pendulums of different lengths from a horizontal support, as shown in Figure 18.15. If pendulum A is set into oscillation, the other pendulums begin to oscillate as a result of waves transmitted along the beam. However, pendulum C, the length of which is close to the length of A, oscillates with a much greater amplitude than pendulums B and D, the lengths of which are much different from that of pendulum A. Pendulum C moves the way it does because its natural frequency is nearly the same as the driving frequency associated with pendulum A.

Next, consider a taut string fixed at one end and connected at the opposite end to an oscillating blade, as illustrated in Figure 18.16. The fixed end is a node, and the end connected to the blade is very nearly a node because the amplitude of the blade's motion is small compared with that of the elements of the string. As the blade oscillates, transverse waves sent down the string are reflected from the fixed end. As we learned in Section 18.3, the string has natural frequencies that are determined by its length, tension, and linear mass density (see Eq. 18.8). When the frequency of the blade equals one of the natural frequencies of the string, standing waves are produced and the string oscillates with a large amplitude. In this resonance case, the wave generated by the oscillating blade is in phase with the reflected wave, and the string absorbs energy from the blade. If the string is driven at a frequency that is not one of its natural frequencies, then the oscillations are of low amplitude and exhibit no stable pattern.

Once the amplitude of the standing-wave oscillations is a maximum, the mechanical energy delivered by the blade and absorbed by the system is transformed to internal energy because of the damping forces caused by friction in the system. If the applied frequency differs from one of the natural frequencies, energy is transferred to the string at first, but later the phase of the wave becomes such that it forces the blade to receive energy from the string, thereby reducing the energy in the string.

Resonance is very important in the excitation of musical instruments based on air columns. We shall discuss this application of resonance in Section 18.5.

Quick Quiz 18.6 A wine glass can be shattered through resonance by maintaining a certain frequency of a high-intensity sound wave. Figure 18.17a shows a side view of a wine glass vibrating in response to such a sound wave. Sketch the standing-wave pattern in the rim of the glass as seen from above. If an integral number of waves "fit" around the circumference of the vibrating rim, how many wavelengths fit around the rim in Figure 18.17a?

(a)   (b)

Figure 18.17 (Quick Quiz 18.6) (a) Standing-wave pattern in a vibrating wine glass. The glass shatters if the amplitude of vibration becomes too great. (b) A wine glass shattered by the amplified sound of a human voice.

18.5 Standing Waves in Air Columns

Standing waves can be set up in a tube of air, such as that inside an organ pipe, as the result of interference between longitudinal sound waves traveling in opposite directions. The phase relationship between the incident wave and the wave reflected from one end of the pipe depends on whether that end is open or closed. This relationship is analogous to the phase relationships between incident and reflected transverse waves at the end of a string when the end is either fixed or free to move (see Figs. 16.14 and 16.15).

In a pipe closed at one end, **the closed end is a displacement node because the wall at this end does not allow longitudinal motion of the air.** As a result, at a closed end of a pipe, the reflected sound wave is 180° out of phase with the incident wave. Furthermore, because the pressure wave is 90° out of phase with the displacement wave (see Section 17.2), **the closed end of an air column corresponds to a pressure antinode** (that is, a point of maximum pressure variation).

The open end of an air column is approximately a displacement antinode[2] and a pressure node. We can understand why no pressure variation occurs at an open end by noting that the end of the air column is open to the atmosphere; thus, the pressure at this end must remain constant at atmospheric pressure.

You may wonder how a sound wave can reflect from an open end, as there may not appear to be a change in the medium at this point. It is indeed true that the medium

---

[2] Strictly speaking, the open end of an air column is not exactly a displacement antinode. A compression reaching an open end does not reflect until it passes beyond the end. For a tube of circular cross section, an end correction equal to approximately $0.6R$, where R is the tube's radius, must be added to the length of the air column. Hence, the effective length of the air column is longer than the true length L. We ignore this end correction in this discussion.

through which the sound wave moves is air both inside and outside the pipe. However, sound is a pressure wave, and a compression region of the sound wave is constrained by the sides of the pipe as long as the region is inside the pipe. As the compression region exits at the open end of the pipe, the constraint of the pipe is removed and the compressed air is free to expand into the atmosphere. Thus, there is a change in the *character* of the medium between the inside of the pipe and the outside even though there is no change in the *material* of the medium. This change in character is sufficient to allow some reflection.

With the boundary conditions of nodes or antinodes at the ends of the air column, we have a set of normal modes of oscillation, as we do for the string fixed at both ends. Thus, the air column has quantized frequencies.

The first three normal modes of oscillation of a pipe open at both ends are shown in Figure 18.18a. Note that both ends are displacement antinodes (approximately). In the first normal mode, the standing wave extends between two adjacent antinodes, which is a distance of half a wavelength. Thus, the wavelength is twice the length of the pipe, and the fundamental frequency is $f_1 = v/2L$. As Figure 18.18a shows, the frequencies of the higher harmonics are $2f_1$, $3f_1$, ... Thus, we can say that

> In a pipe open at both ends, the natural frequencies of oscillation form a harmonic series that includes all integral multiples of the fundamental frequency.

▲ **PITFALL PREVENTION**

18.3 Sound Waves in Air Are Longitudinal, not Transverse

Note that the standing longitudinal waves are drawn as transverse waves in Figure 18.18. This is because it is difficult to draw longitudinal displacements—they are in the same direction as the propagation. Thus, it is best to interpret the curves in Figure 18.18 as a graphical representation of the waves (our diagrams of string waves are pictorial representations), with the vertical axis representing horizontal displacement of the elements of the medium.

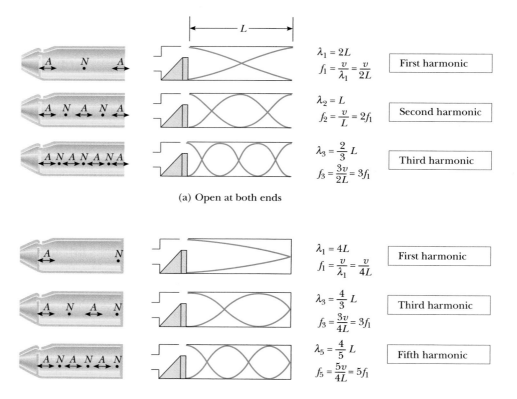

(a) Open at both ends

(b) Closed at one end, open at the other

Figure 18.18 Motion of elements of air in standing longitudinal waves in a pipe, along with schematic representations of the waves. In the schematic representations, the structure at the left end has the purpose of exciting the air column into a normal mode. The hole in the upper edge of the column assures that the left end acts as an open end. The graphs represent the displacement amplitudes, not the pressure amplitudes. (a) In a pipe open at both ends, the harmonic series created consists of all integer multiples of the fundamental frequency: f_1, $2f_1$, $3f_1$, ... (b) In a pipe closed at one end and open at the other, the harmonic series created consists of only odd-integer multiples of the fundamental frequency: f_1, $3f_1$, $5f_1$, ...

Because all harmonics are present, and because the fundamental frequency is given by the same expression as that for a string (see Eq. 18.7), we can express the natural frequencies of oscillation as

$$f_n = n \frac{v}{2L} \qquad n = 1, 2, 3, \ldots \qquad (18.11)$$

Natural frequencies of a pipe open at both ends

Despite the similarity between Equations 18.7 and 18.11, you must remember that v in Equation 18.7 is the speed of waves on the string, whereas v in Equation 18.11 is the speed of sound in air.

If a pipe is closed at one end and open at the other, the closed end is a displacement node (see Fig. 18.18b). In this case, the standing wave for the fundamental mode extends from an antinode to the adjacent node, which is one fourth of a wavelength. Hence, the wavelength for the first normal mode is $4L$, and the fundamental frequency is $f_1 = v/4L$. As Figure 18.18b shows, the higher-frequency waves that satisfy our conditions are those that have a node at the closed end and an antinode at the open end; this means that the higher harmonics have frequencies $3f_1$, $5f_1$, ...

In a pipe closed at one end, the natural frequencies of oscillation form a harmonic series that includes only odd integral multiples of the fundamental frequency.

We express this result mathematically as

$$f_n = n \frac{v}{4L} \qquad n = 1, 3, 5, \ldots \qquad (18.12)$$

Natural frequencies of a pipe closed at one end and open at the other

It is interesting to investigate what happens to the frequencies of instruments based on air columns and strings during a concert as the temperature rises. The sound emitted by a flute, for example, becomes sharp (increases in frequency) as it warms up because the speed of sound increases in the increasingly warmer air inside the flute (consider Eq. 18.11). The sound produced by a violin becomes flat (decreases in frequency) as the strings thermally expand because the expansion causes their tension to decrease (see Eq. 18.8).

Musical instruments based on air columns are generally excited by resonance. The air column is presented with a sound wave that is rich in many frequencies. The air column then responds with a large-amplitude oscillation to the frequencies that match the quantized frequencies in its set of harmonics. In many woodwind instruments, the initial rich sound is provided by a vibrating reed. In the brasses, this excitation is provided by the sound coming from the vibration of the player's lips. In a flute, the initial excitation comes from blowing over an edge at the mouthpiece of the instrument. This is similar to blowing across the opening of a bottle with a narrow neck. The sound of the air rushing across the edge has many frequencies, including one that sets the air cavity in the bottle into resonance.

Quick Quiz 18.7 A pipe open at both ends resonates at a fundamental frequency f_{open}. When one end is covered and the pipe is again made to resonate, the fundamental frequency is f_{closed}. Which of the following expressions describes how these two resonant frequencies compare? (a) $f_{closed} = f_{open}$ (b) $f_{closed} = \frac{1}{2} f_{open}$ (c) $f_{closed} = 2 f_{open}$ (d) $f_{closed} = \frac{3}{2} f_{open}$

Quick Quiz 18.8 Balboa Park in San Diego has an outdoor organ. When the air temperature increases, the fundamental frequency of one of the organ pipes (a) stays the same (b) goes down (c) goes up (d) is impossible to determine.

Example 18.6 Wind in a Culvert

A section of drainage culvert 1.23 m in length makes a howling noise when the wind blows.

(A) Determine the frequencies of the first three harmonics of the culvert if it is cylindrical in shape and open at both ends. Take $v = 343$ m/s as the speed of sound in air.

Solution The frequency of the first harmonic of a pipe open at both ends is

$$f_1 = \frac{v}{2L} = \frac{343 \text{ m/s}}{2(1.23 \text{ m})} = \boxed{139 \text{ Hz}}$$

Because both ends are open, all harmonics are present; thus,

$$f_2 = 2f_1 = \boxed{278 \text{ Hz}} \quad \text{and} \quad f_3 = 3f_1 = \boxed{417 \text{ Hz}}$$

(B) What are the three lowest natural frequencies of the culvert if it is blocked at one end?

Solution The fundamental frequency of a pipe closed at one end is

$$f_1 = \frac{v}{4L} = \frac{343 \text{ m/s}}{4(1.23 \text{ m})} = \boxed{69.7 \text{ Hz}}$$

In this case, only odd harmonics are present; hence, the next two harmonics have frequencies $f_3 = 3f_1 = \boxed{209 \text{ Hz}}$ and $f_5 = 5f_1 = \boxed{349 \text{ Hz}}$.

(C) For the culvert open at both ends, how many of the harmonics present fall within the normal human hearing range (20 to 20 000 Hz)?

Solution Because all harmonics are present for a pipe open at both ends, we can express the frequency of the highest harmonic heard as $f_n = nf_1$ where n is the number of harmonics that we can hear. For $f_n = 20 000$ Hz, we find that the number of harmonics present in the audible range is

$$n = \frac{20\ 000 \text{ Hz}}{139 \text{ Hz}} = \boxed{143}$$

Only the first few harmonics are of sufficient amplitude to be heard.

Example 18.7 Measuring the Frequency of a Tuning Fork

A simple apparatus for demonstrating resonance in an air column is depicted in Figure 18.19. A vertical pipe open at both ends is partially submerged in water, and a tuning fork vibrating at an unknown frequency is placed near the top of the pipe. The length L of the air column can be adjusted by moving the pipe vertically. The sound waves generated by the fork are reinforced when L corresponds to one of the resonance frequencies of the pipe.

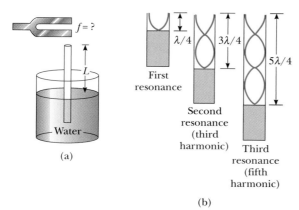

Figure 18.19 (Example 18.7) (a) Apparatus for demonstrating the resonance of sound waves in a pipe closed at one end. The length L of the air column is varied by moving the pipe vertically while it is partially submerged in water. (b) The first three normal modes of the system shown in part (a).

For a certain pipe, the smallest value of L for which a peak occurs in the sound intensity is 9.00 cm. What are

(A) the frequency of the tuning fork

(B) the values of L for the next two resonance frequencies?

Solution

(A) Although the pipe is open at its lower end to allow the water to enter, the water's surface acts like a wall at one end. Therefore, this setup can be modeled as an air column closed at one end, and so the fundamental frequency is given by $f_1 = v/4L$. Taking $v = 343$ m/s for the speed of sound in air and $L = 0.090\ 0$ m, we obtain

$$f_1 = \frac{v}{4L} = \frac{343 \text{ m/s}}{4(0.090\ 0 \text{ m})} = \boxed{953 \text{ Hz}}$$

Because the tuning fork causes the air column to resonate at this frequency, this must also be the frequency of the tuning fork.

(B) Because the pipe is closed at one end, we know from Figure 18.18b that the wavelength of the fundamental mode is $\lambda = 4L = 4(0.090\ 0 \text{ m}) = 0.360$ m. Because the frequency of the tuning fork is constant, the next two normal modes (see Fig. 18.19b) correspond to lengths of

$$L = 3\lambda/4 = \boxed{0.270 \text{ m}} \quad \text{and} \quad L = 5\lambda/4 = \boxed{0.450 \text{ m}}.$$

18.6 Standing Waves in Rods and Membranes

Standing waves can also be set up in rods and membranes. A rod clamped in the middle and stroked parallel to the rod at one end oscillates, as depicted in Figure 18.20a. The oscillations of the elements of the rod are longitudinal, and so the broken lines in Figure 18.20 represent *longitudinal* displacements of various parts of the rod. For clarity, we have drawn them in the transverse direction, just as we did for air columns. The midpoint is a displacement node because it is fixed by the clamp, whereas the ends are displacement antinodes because they are free to oscillate. The oscillations in this setup are analogous to those in a pipe open at both ends. The broken lines in Figure 18.20a represent the first normal mode, for which the wavelength is $2L$ and the frequency is $f = v/2L$, where v is the speed of longitudinal waves in the rod. Other normal modes may be excited by clamping the rod at different points. For example, the second normal mode (Fig. 18.20b) is excited by clamping the rod a distance $L/4$ away from one end.

Musical instruments that depend on standing waves in rods include triangles, marimbas, xylophones, glockenspiels, chimes, and vibraphones. Other devices that make sounds from bars include music boxes and wind chimes.

Two-dimensional oscillations can be set up in a flexible membrane stretched over a circular hoop, such as that in a drumhead. As the membrane is struck at some point, waves that arrive at the fixed boundary are reflected many times. The resulting sound is not harmonic because the standing waves have frequencies that are *not* related by integer multiples. Without this relationship, the sound may be more correctly described as *noise* than as music. This is in contrast to the situation in wind and stringed instruments, which produce sounds that we describe as musical.

Some possible normal modes of oscillation for a two-dimensional circular membrane are shown in Figure 18.21. While nodes are *points* in one-dimensional standing

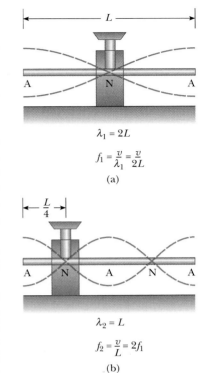

$$\lambda_1 = 2L$$

$$f_1 = \frac{v}{\lambda_1} = \frac{v}{2L}$$

(a)

$$\lambda_2 = L$$

$$f_2 = \frac{v}{L} = 2f_1$$

(b)

Figure 18.20 Normal-mode longitudinal vibrations of a rod of length L (a) clamped at the middle to produce the first normal mode and (b) clamped at a distance $L/4$ from one end to produce the second normal mode. Note that the broken lines represent oscillations parallel to the rod (longitudinal waves).

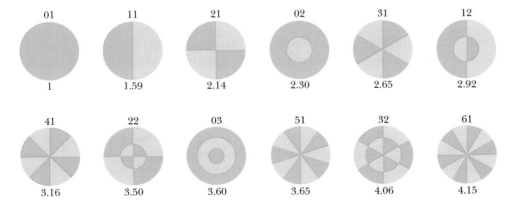

Elements of the medium moving out of the page at an instant of time.

Elements of the medium moving into the page at an instant of time.

Figure 18.21 Representation of some of the normal modes possible in a circular membrane fixed at its perimeter. The pair of numbers above each pattern corresponds to the number of radial nodes and the number of circular nodes. Below each pattern is a factor by which the frequency of the mode is larger than that of the 01 mode. The frequencies of oscillation do not form a harmonic series because these factors are not integers. In each diagram, elements of the membrane on either side of a nodal line move in opposite directions, as indicated by the colors. (*Adapted from T. D. Rossing*, The Science of Sound, 2nd ed, *Reading, Massachusetts, Addison-Wesley Publishing Co., 1990*) waves on strings and in air columns, a two-dimensional oscillator has *curves* along which there is no displacement of the elements of the medium. The lowest normal mode, which has a frequency f_1, contains only one nodal curve; this curve runs around the outer edge of the membrane. The other possible normal modes show additional nodal curves that are circles and straight lines across the diameter of the membrane.

18.7 Beats: Interference in Time

The interference phenomena with which we have been dealing so far involve the superposition of two or more waves having the same frequency. Because the amplitude of the oscillation of elements of the medium varies with the position in space of the element, we refer to the phenomenon as *spatial interference*. Standing waves in strings and pipes are common examples of spatial interference.

We now consider another type of interference, one that results from the superposition of two waves having slightly *different* frequencies. In this case, when the two waves are observed at the point of superposition, they are periodically in and out of phase. That is, there is a *temporal* (time) alternation between constructive and destructive interference. As a consequence, we refer to this phenomenon as *interference in time* or *temporal interference*. For example, if two tuning forks of slightly different frequencies are struck, one hears a sound of periodically varying amplitude. This phenomenon is called **beating:**

Definition of beating

> Beating is the periodic variation in amplitude at a given point due to the superposition of two waves having slightly different frequencies.

The number of amplitude maxima one hears per second, or the *beat frequency*, equals the difference in frequency between the two sources, as we shall show below. The maximum beat frequency that the human ear can detect is about 20 beats/s. When the beat frequency exceeds this value, the beats blend indistinguishably with the sounds producing them.

A piano tuner can use beats to tune a stringed instrument by "beating" a note against a reference tone of known frequency. The tuner can then adjust the string tension until the frequency of the sound it emits equals the frequency of the reference tone. The tuner does this by tightening or loosening the string until the beats produced by it and the reference source become too infrequent to notice.

Consider two sound waves of equal amplitude traveling through a medium with slightly different frequencies f_1 and f_2. We use equations similar to Equation 16.10 to represent the wave functions for these two waves at a point that we choose as $x = 0$:

$$y_1 = A \cos \omega_1 t = A \cos 2\pi f_1 t$$
$$y_2 = A \cos \omega_2 t = A \cos 2\pi f_2 t$$

Using the superposition principle, we find that the resultant wave function at this point is

$$y = y_1 + y_2 = A(\cos 2\pi f_1 t + \cos 2\pi f_2 t)$$

The trigonometric identity

$$\cos a + \cos b = 2 \cos\left(\frac{a - b}{2}\right) \cos\left(\frac{a + b}{2}\right)$$

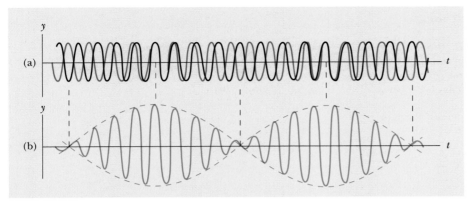

Active Figure 18.22 Beats are formed by the combination of two waves of slightly different frequencies. (a) The individual waves. (b) The combined wave has an amplitude (broken line) that oscillates in time.

At the Active Figures link at http://www.pse6.com, *you can choose the two frequencies and see the corresponding beats.*

allows us to write the expression for y as

$$y = \left[2A \cos 2\pi \left(\frac{f_1 - f_2}{2} \right) t \right] \cos 2\pi \left(\frac{f_1 + f_2}{2} \right) t \qquad (18.13)$$

Resultant of two waves of different frequencies but equal amplitude

Graphs of the individual waves and the resultant wave are shown in Figure 18.22. From the factors in Equation 18.13, we see that the resultant sound for a listener standing at any given point has an effective frequency equal to the average frequency $(f_1 + f_2)/2$ and an amplitude given by the expression in the square brackets:

$$A_{\text{resultant}} = 2A \cos 2\pi \left(\frac{f_1 - f_2}{2} \right) t \qquad (18.14)$$

That is, the **amplitude and therefore the intensity of the resultant sound vary in time.** The broken blue line in Figure 18.22b is a graphical representation of Equation 18.14 and is a sine wave varying with frequency $(f_1 - f_2)/2$.

Note that a maximum in the amplitude of the resultant sound wave is detected whenever

$$\cos 2\pi \left(\frac{f_1 - f_2}{2} \right) t = \pm 1$$

This means there are *two* maxima in each period of the resultant wave. Because the amplitude varies with frequency as $(f_1 - f_2)/2$, the number of beats per second, or the beat frequency f_{beat}, is twice this value. That is,

$$f_{\text{beat}} = |f_1 - f_2| \qquad (18.15)$$

Beat frequency

For instance, if one tuning fork vibrates at 438 Hz and a second one vibrates at 442 Hz, the resultant sound wave of the combination has a frequency of 440 Hz (the musical note A) and a beat frequency of 4 Hz. A listener would hear a 440-Hz sound wave go through an intensity maximum four times every second.

Quick Quiz 18.9 You are tuning a guitar by comparing the sound of the string with that of a standard tuning fork. You notice a beat frequency of 5 Hz when both sounds are present. You tighten the guitar string and the beat frequency rises to 8 Hz. In order to tune the string exactly to the tuning fork, you should (a) continue to tighten the string (b) loosen the string (c) impossible to determine.

Example 18.8 **The Mistuned Piano Strings**

Two identical piano strings of length 0.750 m are each tuned exactly to 440 Hz. The tension in one of the strings is then increased by 1.0%. If they are now struck, what is the beat frequency between the fundamentals of the two strings?

Solution We find the ratio of frequencies if the tension in one string is 1.0% larger than the other:

$$\frac{f_2}{f_1} = \frac{(v_2/2L)}{(v_1/2L)} = \frac{v_2}{v_1} = \frac{\sqrt{T_2/\mu}}{\sqrt{T_1/\mu}} = \sqrt{\frac{T_2}{T_1}} = \sqrt{\frac{1.010 T_1}{T_1}}$$

$$= 1.005$$

Thus, the frequency of the tightened string is

$$f_2 = 1.005 f_1 = 1.005(440 \text{ Hz}) = 442 \text{ Hz}$$

and the beat frequency is

$$f_{\text{beat}} = 442 \text{ Hz} - 440 \text{ Hz} = \boxed{2 \text{ Hz.}}$$

18.8 Nonsinusoidal Wave Patterns

The sound wave patterns produced by the majority of musical instruments are nonsinusoidal. Characteristic patterns produced by a tuning fork, a flute, and a clarinet, each playing the same note, are shown in Figure 18.23. Each instrument has its own characteristic pattern. Note, however, that despite the differences in the patterns, each pattern is periodic. This point is important for our analysis of these waves.

It is relatively easy to distinguish the sounds coming from a violin and a saxophone even when they are both playing the same note. On the other hand, an individual untrained in music may have difficulty distinguishing a note played on a clarinet from the same note played on an oboe. We can use the pattern of the sound waves from various sources to explain these effects.

This is in contrast to a musical instrument that makes a noise, such as the drum, in which the combination of frequencies do not form a harmonic series. When frequencies that are integer multiples of a fundamental frequency are combined, the result is a *musical* sound. A listener can assign a pitch to the sound, based on the fundamental frequency. Pitch is a psychological reaction to a sound that allows the listener to place the sound on a scale of low to high (bass to treble). Combinations of frequencies that are not integer multiples of a fundamental result in a *noise*, rather than a musical sound. It is much harder for a listener to assign a pitch to a noise than to a musical sound.

The wave patterns produced by a musical instrument are the result of the superposition of various harmonics. This superposition results in the corresponding richness of musical tones. The human perceptive response associated with various mixtures of harmonics is the *quality* or *timbre* of the sound. For instance, the sound of the trumpet is perceived to have a "brassy" quality (that is, we have learned to associate the adjective *brassy* with that sound); this quality enables us to distinguish the sound of the trumpet from that of the saxophone, whose quality is perceived as "reedy." The clarinet and oboe, however, both contain air columns excited by reeds; because of this similarity, it is more difficult for the ear to distinguish them on the basis of their sound quality.

The problem of analyzing nonsinusoidal wave patterns appears at first sight to be a formidable task. However, if the wave pattern is periodic, it can be represented as closely as desired by the combination of a sufficiently large number of sinusoidal waves that form a harmonic series. In fact, we can represent any periodic function as a series of sine and cosine terms by using a mathematical technique based on **Fourier's theorem.**[3] The corresponding sum of terms that represents the periodic wave pattern

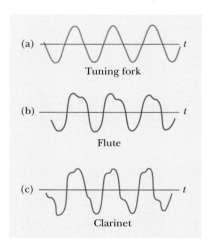

(a) Tuning fork

(b) Flute

(c) Clarinet

Figure 18.23 Sound wave patterns produced by (a) a tuning fork, (b) a flute, and (c) a clarinet, each at approximately the same frequency. (*Adapted from C. A. Culver, Musical Acoustics, 4th ed., New York, McGraw-Hill Book Company, 1956, p. 128.*)

---

[3] Developed by Jean Baptiste Joseph Fourier (1786–1830).

Figure 18.24 Harmonics of the wave patterns shown in Figure 18.23. Note the variations in intensity of the various harmonics. (*Adapted from C. A. Culver*, Musical Acoustics, 4th ed., *New York, McGraw-Hill Book Company, 1956.*)

is called a **Fourier series.** Let $y(t)$ be any function that is periodic in time with period T, such that $y(t + T) = y(t)$. Fourier's theorem states that this function can be written as

$$y(t) = \sum_n (A_n \sin 2\pi f_n t + B_n \cos 2\pi f_n t) \qquad (18.16)$$

Fourier's theorem

where the lowest frequency is $f_1 = 1/T$. The higher frequencies are integer multiples of the fundamental, $f_n = nf_1$, and the coefficients A_n and B_n represent the amplitudes of the various waves. Figure 18.24 represents a harmonic analysis of the wave patterns shown in Figure 18.23. Note that a struck tuning fork produces only one harmonic (the first), whereas the flute and clarinet produce the first harmonic and many higher ones.

Note the variation in relative intensity of the various harmonics for the flute and the clarinet. In general, any musical sound consists of a fundamental frequency f plus other frequencies that are integer multiples of f, all having different intensities.

⚠ **PITFALL PREVENTION**

18.4 Pitch vs. Frequency

Do not confuse the term *pitch* with *frequency*. Frequency is the physical measurement of the number of oscillations per second. Pitch is a psychological reaction to sound that enables a person to place the sound on a scale from high to low, or from treble to bass. Thus, frequency is the stimulus and pitch is the response. Although pitch is related mostly (but not completely) to frequency, they are not the same. A phrase such as "the pitch of the sound" is incorrect because pitch is not a physical property of the sound.

Photographs courtesy of (a) © 1989 Gary Buss/FPG; (b) and (c) © 1989 Richard Laird/FPG

Each musical instrument has its own characteristic sound and mixture of harmonics. Instruments shown are (a) the violin, (b) the saxophone, and (c) the trumpet.

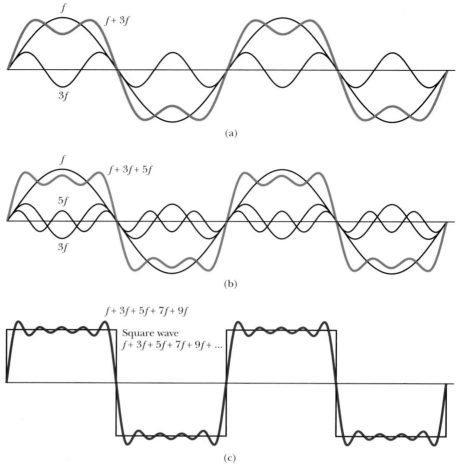

Active Figure 18.25 Fourier synthesis of a square wave, which is represented by the sum of odd multiples of the first harmonic, which has frequency f. (a) Waves of frequency f and $3f$ are added. (b) One more odd harmonic of frequency $5f$ is added. (c) The synthesis curve approaches closer to the square wave when odd frequencies up to $9f$ are added.

We have discussed the *analysis* of a wave pattern using Fourier's theorem. The analysis involves determining the coefficients of the harmonics in Equation 18.16 from a knowledge of the wave pattern. The reverse process, called *Fourier synthesis*, can also be performed. In this process, the various harmonics are added together to form a resultant wave pattern. As an example of Fourier synthesis, consider the building of a square wave, as shown in Figure 18.25. The symmetry of the square wave results in only odd multiples of the fundamental frequency combining in its synthesis. In Figure 18.25a, the orange curve shows the combination of f and $3f$. In Figure 18.25b, we have added $5f$ to the combination and obtained the green curve. Notice how the general shape of the square wave is approximated, even though the upper and lower portions are not flat as they should be.

Figure 18.25c shows the result of adding odd frequencies up to $9f$. This approximation (purple curve) to the square wave is better than the approximations in parts a and b. To approximate the square wave as closely as possible, we would need to add all odd multiples of the fundamental frequency, up to infinite frequency.

Using modern technology, we can generate musical sounds electronically by mixing different amplitudes of any number of harmonics. These widely used electronic music synthesizers are capable of producing an infinite variety of musical tones.

SUMMARY

The **superposition principle** specifies that when two or more waves move through a medium, the value of the resultant wave function equals the algebraic sum of the values of the individual wave functions.

When two traveling waves having equal amplitudes and frequencies superimpose, the resultant wave has an amplitude that depends on the phase angle ϕ between the two waves. **Constructive interference** occurs when the two waves are in phase, corresponding to $\phi = 0, 2\pi, 4\pi, \ldots$ rad. **Destructive interference** occurs when the two waves are 180° out of phase, corresponding to $\phi = \pi, 3\pi, 5\pi, \ldots$ rad.

Standing waves are formed from the superposition of two sinusoidal waves having the same frequency, amplitude, and wavelength but traveling in opposite directions. The resultant standing wave is described by the wave function

$$y = (2A \sin kx) \cos \omega t \tag{18.3}$$

Hence, the amplitude of the standing wave is $2A$, and the amplitude of the simple harmonic motion of any particle of the medium varies according to its position as $2A \sin kx$. The points of zero amplitude (called **nodes**) occur at $x = n\lambda/2$ ($n = 0, 1, 2, 3, \ldots$). The maximum amplitude points (called **antinodes**) occur at $x = n\lambda/4$ ($n = 1, 3, 5, \ldots$). Adjacent antinodes are separated by a distance $\lambda/2$. Adjacent nodes also are separated by a distance $\lambda/2$.

The natural frequencies of vibration of a taut string of length L and fixed at both ends are quantized and are given by

$$f_n = \frac{n}{2L}\sqrt{\frac{T}{\mu}} \qquad n = 1, 2, 3, \ldots \tag{18.8}$$

where T is the tension in the string and μ is its linear mass density. The natural frequencies of vibration $f_1, 2f_1, 3f_1, \ldots$ form a **harmonic series.**

An oscillating system is in **resonance** with some driving force whenever the frequency of the driving force matches one of the natural frequencies of the system. When the system is resonating, it responds by oscillating with a relatively large amplitude.

Standing waves can be produced in a column of air inside a pipe. If the pipe is open at both ends, all harmonics are present and the natural frequencies of oscillation are

$$f_n = n\frac{v}{2L} \qquad n = 1, 2, 3, \ldots \tag{18.11}$$

If the pipe is open at one end and closed at the other, only the odd harmonics are present, and the natural frequencies of oscillation are

$$f_n = n\frac{v}{4L} \qquad n = 1, 3, 5, \ldots \tag{18.12}$$

The phenomenon of **beating** is the periodic variation in intensity at a given point due to the superposition of two waves having slightly different frequencies.

Take a practice test for this chapter by clicking on the Practice Test link at http://www.pse6.com.

QUESTIONS

1. Does the phenomenon of wave interference apply only to sinusoidal waves?

2. As oppositely moving pulses of the same shape (one upward, one downward) on a string pass through each other, there is one instant at which the string shows no displacement from the equilibrium position at any point. Has the energy carried by the pulses disappeared at this instant of time? If not, where is it?

3. Can two pulses traveling in opposite directions on the same string reflect from each other? Explain.

4. When two waves interfere, can the amplitude of the resultant wave be greater than either of the two original waves? Under what conditions?

5. For certain positions of the movable section shown in Figure 18.5, no sound is detected at the receiver—a situation

corresponding to destructive interference. This suggests that energy is somehow lost. What happens to the energy transmitted by the speaker?

6. When two waves interfere constructively or destructively, is there any gain or loss in energy? Explain.

7. A standing wave is set up on a string, as shown in Figure 18.10. Explain why no energy is transmitted along the string.

8. What limits the amplitude of motion of a real vibrating system that is driven at one of its resonant frequencies?

9. Explain why your voice seems to sound better than usual when you sing in the shower.

10. What is the purpose of the slide on a trombone or of the valves on a trumpet?

11. Explain why all harmonics are present in an organ pipe open at both ends, but only odd harmonics are present in a pipe closed at one end.

12. Explain how a musical instrument such as a piano may be tuned by using the phenomenon of beats.

13. To keep animals away from their cars, some people mount short, thin pipes on the fenders. The pipes give out a high-pitched wail when the cars are moving. How do they create the sound?

14. When a bell is rung, standing waves are set up around the bell's circumference. What boundary conditions must be satisfied by the resonant wavelengths? How does a crack in the bell, such as in the Liberty Bell, affect the satisfying of the boundary conditions and the sound emanating from the bell?

15. An archer shoots an arrow from a bow. Does the string of the bow exhibit standing waves after the arrow leaves? If so, and if the bow is perfectly symmetric so that the arrow leaves from the center of the string, what harmonics are excited?

16. Despite a reasonably steady hand, a person often spills his coffee when carrying it to his seat. Discuss resonance as a possible cause of this difficulty, and devise a means for solving the problem.

17. An airplane mechanic notices that the sound from a twin-engine aircraft rapidly varies in loudness when both engines are running. What could be causing this variation from loud to soft?

18. When the base of a vibrating tuning fork is placed against a chalkboard, the sound that it emits becomes louder. This is because the vibrations of the tuning fork are transmitted to the chalkboard. Because it has a larger area than the tuning fork, the vibrating chalkboard sets more air into vibration. Thus, the chalkboard is a better radiator of sound than the tuning fork. How does this affect the length of time during which the fork vibrates? Does this agree with the principle of conservation of energy?

19. If you wet your finger and lightly run it around the rim of a fine wineglass, a high-frequency sound is heard. Why? How could you produce various musical notes with a set of wineglasses, each of which contains a different amount of water?

20. If you inhale helium from a balloon and do your best to speak normally, your voice will have a comical quacky quality. Explain why this "Donald Duck effect" happens. *Caution*: Helium is an asphyxiating gas and asphyxiation can cause panic. Helium can contain poisonous contaminants.

21. You have a standard tuning fork whose frequency is 262 Hz and a second tuning fork with an unknown frequency. When you tap both of them on the heel of one of your sneakers, you hear beats with a frequency of 4 per second. Thoughtfully chewing your gum, you wonder whether the unknown frequency is 258 Hz or 266 Hz. How can you decide?

PROBLEMS

1, 2, 3 = straightforward, intermediate, challenging ☐ = full solution available in the *Student Solutions Manual and Study Guide*

 = coached solution with hints available at http://www.pse6.com 🖳 = computer useful in solving problem

= paired numerical and symbolic problems

Section 18.1 Superposition and Interference

1. Two waves in one string are described by the wave functions

$$y_1 = 3.0 \cos(4.0x - 1.6t)$$

and

$$y_2 = 4.0 \sin(5.0x - 2.0t)$$

where y and x are in centimeters and t is in seconds. Find the superposition of the waves $y_1 + y_2$ at the points (a) $x = 1.00$, $t = 1.00$, (b) $x = 1.00$, $t = 0.500$, and (c) $x = 0.500$, $t = 0$. (Remember that the arguments of the trigonometric functions are in radians.)

2. Two pulses A and B are moving in opposite directions along a taut string with a speed of 2.00 cm/s. The ampli-

tude of A is twice the amplitude of B. The pulses are shown in Figure P18.2 at $t = 0$. Sketch the shape of the string at $t = 1, 1.5, 2, 2.5,$ and 3 s.

Figure P18.2

3. Two pulses traveling on the same string are described by

$$y_1 = \frac{5}{(3x - 4t)^2 + 2} \quad \text{and} \quad y_2 = \frac{-5}{(3x + 4t - 6)^2 + 2}$$

(a) In which direction does each pulse travel? (b) At what time do the two cancel everywhere? (c) At what point do the two pulses always cancel?

4. Two waves are traveling in the same direction along a stretched string. The waves are 90.0° out of phase. Each wave has an amplitude of 4.00 cm. Find the amplitude of the resultant wave.

5. Two traveling sinusoidal waves are described by the wave functions

$$y_1 = (5.00 \text{ m}) \sin[\pi(4.00x - 1\,200t)]$$

and

$$y_2 = (5.00 \text{ m}) \sin[\pi(4.00x - 1\,200t - 0.250)]$$

where x, y_1, and y_2 are in meters and t is in seconds. (a) What is the amplitude of the resultant wave? (b) What is the frequency of the resultant wave?

6. Two identical sinusoidal waves with wavelengths of 3.00 m travel in the same direction at a speed of 2.00 m/s. The second wave originates from the same point as the first, but at a later time. Determine the minimum possible time interval between the starting moments of the two waves if the amplitude of the resultant wave is the same as that of each of the two initial waves.

7. **Review problem.** A series of pulses, each of amplitude 0.150 m, is sent down a string that is attached to a post at one end. The pulses are reflected at the post and travel back along the string without loss of amplitude. What is the net displacement at a point on the string where two pulses are crossing, (a) if the string is rigidly attached to the post? (b) if the end at which reflection occurs is free to slide up and down?

8. Two loudspeakers are placed on a wall 2.00 m apart. A listener stands 3.00 m from the wall directly in front of one of the speakers. A single oscillator is driving the speakers at a frequency of 300 Hz. (a) What is the phase difference between the two waves when they reach the observer? (b) **What If?** What is the frequency closest to 300 Hz to which the oscillator may be adjusted such that the observer hears minimal sound?

9. Two speakers are driven by the same oscillator whose frequency is 200 Hz. They are located on a vertical pole a distance of 4.00 m from each other. A man walks straight toward the lower speaker in a direction perpendicular to the pole as shown in Figure P18.9. (a) How many times will he hear a minimum in sound intensity, and (b) how far is he from the pole at these moments? Take the speed of sound to be 330 m/s and ignore any sound reflections coming off the ground.

10. Two speakers are driven by the same oscillator whose frequency is f. They are located a distance d from each other on a vertical pole. A man walks straight toward the lower

Figure P18.9 Problems 9 and 10.

speaker in a direction perpendicular to the pole, as shown in Figure P18.9. (a) How many times will he hear a minimum in sound intensity? (b) How far is he from the pole at these moments? Let v represent the speed of sound, and assume that the ground does not reflect sound.

11. Two sinusoidal waves in a string are defined by the functions

$$y_1 = (2.00 \text{ cm}) \sin(20.0x - 32.0t)$$

and

$$y_2 = (2.00 \text{ cm}) \sin(25.0x - 40.0t)$$

where y_1, y_2, and x are in centimeters and t is in seconds. (a) What is the phase difference between these two waves at the point $x = 5.00$ cm at $t = 2.00$ s? (b) What is the positive x value closest to the origin for which the two phases differ by $\pm \pi$ at $t = 2.00$ s? (This is where the two waves add to zero.)

12. Two identical speakers 10.0 m apart are driven by the same oscillator with a frequency of $f = 21.5$ Hz (Fig. P18.12). (a) Explain why a receiver at point A records a minimum in sound intensity from the two speakers. (b) If the receiver is moved in the plane of the speakers, what path should it take so that the intensity remains at a minimum? That is, deter-

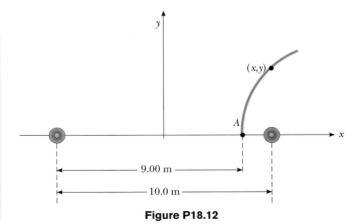

Figure P18.12

mine the relationship between x and y (the coordinates of the receiver) that causes the receiver to record a minimum in sound intensity. Take the speed of sound to be 344 m/s.

Section 18.2 Standing Waves

13. Two sinusoidal waves traveling in opposite directions interfere to produce a standing wave with the wave function

$$y = (1.50 \text{ m}) \sin(0.400x) \cos(200t)$$

where x is in meters and t is in seconds. Determine the wavelength, frequency, and speed of the interfering waves.

14. Two waves in a long string have wave functions given by

$$y_1 = (0.015\ 0 \text{ m}) \cos\left(\frac{x}{2} - 40t\right)$$

and

$$y_2 = (0.015\ 0 \text{ m}) \cos\left(\frac{x}{2} + 40t\right)$$

where y_1, y_2, and x are in meters and t is in seconds. (a) Determine the positions of the nodes of the resulting standing wave. (b) What is the maximum transverse position of an element of the string at the position $x = 0.400$ m?

15. ✏️ Two speakers are driven in phase by a common oscillator at 800 Hz and face each other at a distance of 1.25 m. Locate the points along a line joining the two speakers where relative minima of sound pressure amplitude would be expected. (Use $v = 343$ m/s.)

16. Verify by direct substitution that the wave function for a standing wave given in Equation 18.3,

$$y = 2A \sin kx \cos \omega t$$

is a solution of the general linear wave equation, Equation 16.27:

$$\frac{\partial^2 y}{\partial x^2} = \frac{1}{v^2} \frac{\partial^2 y}{\partial t^2}$$

17. Two sinusoidal waves combining in a medium are described by the wave functions

$$y_1 = (3.0 \text{ cm}) \sin \pi(x + 0.60t)$$

and

$$y_2 = (3.0 \text{ cm}) \sin \pi(x - 0.60t)$$

where x is in centimeters and t is in seconds. Determine the *maximum* transverse position of an element of the medium at (a) $x = 0.250$ cm, (b) $x = 0.500$ cm, and (c) $x = 1.50$ cm. (d) Find the three smallest values of x corresponding to antinodes.

18. Two waves that set up a standing wave in a long string are given by the wave functions

$$y_1 = A \sin(kx - \omega t + \phi) \qquad \text{and} \qquad y_2 = A \sin(kx + \omega t)$$

Show (a) that the addition of the arbitrary phase constant ϕ changes only the position of the nodes and, in particular, (b) that the distance between nodes is still one half the wavelength.

Section 18.3 Standing Waves in a String Fixed at Both Ends

19. Find the fundamental frequency and the next three frequencies that could cause standing-wave patterns on a string that is 30.0 m long, has a mass per length of 9.00×10^{-3} kg/m, and is stretched to a tension of 20.0 N.

20. A string with a mass of 8.00 g and a length of 5.00 m has one end attached to a wall; the other end is draped over a pulley and attached to a hanging object with a mass of 4.00 kg. If the string is plucked, what is the fundamental frequency of vibration?

21. In the arrangement shown in Figure P18.21, an object can be hung from a string (with linear mass density $\mu = 0.002\ 00$ kg/m) that passes over a light pulley. The string is connected to a vibrator (of constant frequency f), and the length of the string between point P and the pulley is $L = 2.00$ m. When the mass m of the object is either 16.0 kg or 25.0 kg, standing waves are observed; however, no standing waves are observed with any mass between these values. (a) What is the frequency of the vibrator? (*Note:* The greater the tension in the string, the smaller the number of nodes in the standing wave.) (b) What is the largest object mass for which standing waves could be observed?

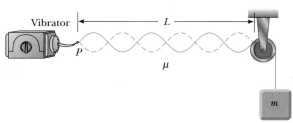

Figure P18.21 Problems 21 and 22.

22. A vibrator, pulley, and hanging object are arranged as in Figure P18.21, with a compound string, consisting of two strings of different masses and lengths fastened together end-to-end. The first string, which has a mass of 1.56 g and a length of 65.8 cm, runs from the vibrator to the junction of the two strings. The second string runs from the junction over the pulley to the suspended 6.93-kg object. The mass and length of the string from the junction to the pulley are, respectively, 6.75 g and 95.0 cm. (a) Find the lowest frequency for which standing waves are observed in both strings, with a node at the junction. The standing wave patterns in the two strings may have different numbers of nodes. (b) What is the total number of nodes observed along the compound string at this frequency, excluding the nodes at the vibrator and the pulley?

23. Example 18.4 tells you that the adjacent notes E, F, and F-sharp can be assigned frequencies of 330 Hz, 350 Hz, and 370 Hz. You might not guess how the pattern continues. The next notes, G, G-sharp, and A, have frequencies of 392 Hz, 416 Hz, and 440 Hz. On the equally tempered or chromatic scale used in Western music, the frequency of each higher note is obtained by multiplying the previous frequency by $^{12}\sqrt{2}$. A standard guitar has strings 64.0 cm long and nineteen frets. In Example 18.4, we found the

spacings of the first two frets. Calculate the distance between the last two frets.

24. The top string of a guitar has a fundamental frequency of 330 Hz when it is allowed to vibrate as a whole, along all of its 64.0-cm length from the neck to the bridge. A fret is provided for limiting vibration to just the lower two-thirds of the string. (a) If the string is pressed down at this fret and plucked, what is the new fundamental frequency? (b) **What If?** The guitarist can play a "natural harmonic" by gently touching the string at the location of this fret and plucking the string at about one sixth of the way along its length from the bridge. What frequency will be heard then?

25. A string of length L, mass per unit length μ, and tension T is vibrating at its fundamental frequency. What effect will the following have on the fundamental frequency? (a) The length of the string is doubled, with all other factors held constant. (b) The mass per unit length is doubled, with all other factors held constant. (c) The tension is doubled, with all other factors held constant.

26. A 60.000-cm guitar string under a tension of 50.000 N has a mass per unit length of 0.100 00 g/cm. What is the highest resonant frequency that can be heard by a person capable of hearing frequencies up to 20 000 Hz?

27. A cello A-string vibrates in its first normal mode with a frequency of 220 Hz. The vibrating segment is 70.0 cm long and has a mass of 1.20 g. (a) Find the tension in the string. (b) Determine the frequency of vibration when the string vibrates in three segments.

28. A violin string has a length of 0.350 m and is tuned to concert G, with $f_G = 392$ Hz. Where must the violinist place her finger to play concert A, with $f_A = 440$ Hz? If this position is to remain correct to half the width of a finger (that is, to within 0.600 cm), what is the maximum allowable percentage change in the string tension?

29. **Review problem.** A sphere of mass M is supported by a string that passes over a light horizontal rod of length L (Fig. P18.29). Given that the angle is θ and that f represents the fundamental frequency of standing waves in the portion of the string above the rod, determine the mass of this portion of the string.

30. **Review problem.** A copper cylinder hangs at the bottom of a steel wire of negligible mass. The top end of the wire is fixed. When the wire is struck, it emits sound with a fundamental frequency of 300 Hz. If the copper cylinder is then submerged in water so that half its volume is below the water line, determine the new fundamental frequency.

31. A standing-wave pattern is observed in a thin wire with a length of 3.00 m. The equation of the wave is

$$y = (0.002 \text{ m}) \sin(\pi x) \cos(100\pi t)$$

where x is in meters and t is in seconds. (a) How many loops does this pattern exhibit? (b) What is the fundamental frequency of vibration of the wire? (c) **What If?** If the original frequency is held constant and the tension in the wire is increased by a factor of 9, how many loops are present in the new pattern?

Section 18.4 Resonance

32. The chains suspending a child's swing are 2.00 m long. At what frequency should a big brother push to make the child swing with largest amplitude?

33. An earthquake can produce a *seiche* in a lake, in which the water sloshes back and forth from end to end with remarkably large amplitude and long period. Consider a seiche produced in a rectangular farm pond, as in the cross-sectional view of Figure P18.33. (The figure is not drawn to scale.) Suppose that the pond is 9.15 m long and of uniform width and depth. You measure that a pulse produced at one end reaches the other end in 2.50 s. (a) What is the wave speed? (b) To produce the seiche, several people stand on the bank at one end and paddle together with snow shovels, moving them in simple harmonic motion. What should be the frequency of this motion?

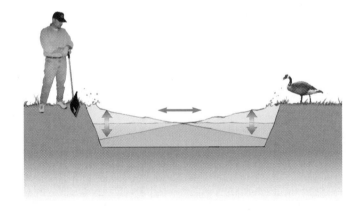

Figure P18.33

34. The Bay of Fundy, Nova Scotia, has the highest tides in the world, as suggested in the photographs on page 452. Assume that in mid-ocean and at the mouth of the bay, the Moon's gravity gradient and the Earth's rotation make the water surface oscillate with an amplitude of a few centimeters and a period of 12 h 24 min. At the head of the bay, the amplitude is several meters. Argue for or against the

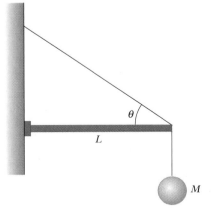

Figure P18.29

proposition that the tide is amplified by standing-wave resonance. Assume the bay has a length of 210 km and a uniform depth of 36.1 m. The speed of long-wavelength water waves is given by $\sqrt{gd}$, where d is the water's depth.

35. Standing-wave vibrations are set up in a crystal goblet with four nodes and four antinodes equally spaced around the 20.0-cm circumference of its rim. If transverse waves move around the glass at 900 m/s, an opera singer would have to produce a high harmonic with what frequency to shatter the glass with a resonant vibration?

Section 18.5 Standing Waves in Air Columns

Note: Unless otherwise specified, assume that the speed of sound in air is 343 m/s at 20°C, and is described by

$$v = (331 \text{ m/s}) \sqrt{1 + \frac{T_C}{273°}}$$

at any Celsius temperature T_C.

36. The overall length of a piccolo is 32.0 cm. The resonating air column vibrates as in a pipe open at both ends. (a) Find the frequency of the lowest note that a piccolo can play, assuming that the speed of sound in air is 340 m/s. (b) Opening holes in the side effectively shortens the length of the resonant column. If the highest note a piccolo can sound is 4 000 Hz, find the distance between adjacent antinodes for this mode of vibration.

37. Calculate the length of a pipe that has a fundamental frequency of 240 Hz if the pipe is (a) closed at one end and (b) open at both ends.

38. The fundamental frequency of an open organ pipe corresponds to middle C (261.6 Hz on the chromatic musical scale). The third resonance of a closed organ pipe has the same frequency. What are the lengths of the two pipes?

39. The windpipe of one typical whooping crane is 5.00 ft long. What is the fundamental resonant frequency of the bird's trachea, modeled as a narrow pipe closed at one end? Assume a temperature of 37°C.

40. Do not stick anything into your ear! Estimate the length of your ear canal, from its opening at the external ear to the eardrum. If you regard the canal as a narrow tube that is open at one end and closed at the other, at approximately what fundamental frequency would you expect your hearing to be most sensitive? Explain why you can hear especially soft sounds just around this frequency.

41. A shower stall measures 86.0 cm × 86.0 cm × 210 cm. If you were singing in this shower, which frequencies would sound the richest (because of resonance)? Assume that the stall acts as a pipe closed at both ends, with nodes at opposite sides. Assume that the voices of various singers range from 130 Hz to 2 000 Hz. Let the speed of sound in the hot shower stall be 355 m/s.

42. As shown in Figure P18.42, water is pumped into a tall vertical cylinder at a volume flow rate R. The radius of the cylinder is r, and at the open top of the cylinder a tuning fork is vibrating with a frequency f. As the water rises, how much time elapses between successive resonances?

Figure P18.42

43. If two adjacent natural frequencies of an organ pipe are determined to be 550 Hz and 650 Hz, calculate the fundamental frequency and length of this pipe. (Use $v = 340$ m/s.)

44. A glass tube (open at both ends) of length L is positioned near an audio speaker of frequency $f = 680$ Hz. For what values of L will the tube resonate with the speaker?

45. An air column in a glass tube is open at one end and closed at the other by a movable piston. The air in the tube is warmed above room temperature, and a 384-Hz tuning fork is held at the open end. Resonance is heard when the piston is 22.8 cm from the open end and again when it is 68.3 cm from the open end. (a) What speed of sound is implied by these data? (b) How far from the open end will the piston be when the next resonance is heard?

46. A tuning fork with a frequency of 512 Hz is placed near the top of the pipe shown in Figure 18.19a. The water level is lowered so that the length L slowly increases from an initial value of 20.0 cm. Determine the next two values of L that correspond to resonant modes.

47. When an open metal pipe is cut into two pieces, the lowest resonance frequency for the air column in one piece is 256 Hz and that for the other is 440 Hz. (a) What resonant frequency would have been produced by the original length of pipe? (b) How long was the original pipe?

48. With a particular fingering, a flute plays a note with frequency 880 Hz at 20.0°C. The flute is open at both ends. (a) Find the air column length. (b) Find the frequency it produces at the beginning of the half-time performance at a late-season American football game, when the ambient temperature is −5.00°C and the musician has not had a chance to warm up the flute.

Section 18.6 Standing Waves in Rods and Membranes

49. An aluminum rod 1.60 m long is held at its center. It is stroked with a rosin-coated cloth to set up a longitudinal vibration. The speed of sound in a thin rod of aluminum is 5 100 m/s. (a) What is the fundamental frequency of the waves established in the rod? (b) What harmonics are set up in the rod held in this manner? (c) **What If?** What would be the fundamental frequency if the rod were made of copper, in which the speed of sound is 3 560 m/s?

50. An aluminum rod is clamped one quarter of the way along its length and set into longitudinal vibration by a variable-frequency driving source. The lowest frequency that produces resonance is 4 400 Hz. The speed of sound in an aluminum rod is 5 100 m/s. Find the length of the rod.

Section 18.7 Beats: Interference in Time

51. In certain ranges of a piano keyboard, more than one string is tuned to the same note to provide extra loudness. For example, the note at 110 Hz has two strings at this frequency. If one string slips from its normal tension of 600 N to 540 N, what beat frequency is heard when the hammer strikes the two strings simultaneously?

52. While attempting to tune the note C at 523 Hz, a piano tuner hears 2 beats/s between a reference oscillator and the string. (a) What are the possible frequencies of the string? (b) When she tightens the string slightly, she hears 3 beats/s. What is the frequency of the string now? (c) By what percentage should the piano tuner now change the tension in the string to bring it into tune?

53. A student holds a tuning fork oscillating at 256 Hz. He walks toward a wall at a constant speed of 1.33 m/s. (a) What beat frequency does he observe between the tuning fork and its echo? (b) How fast must he walk away from the wall to observe a beat frequency of 5.00 Hz?

54. When beats occur at a rate higher than about 20 per second, they are not heard individually but rather as a steady hum, called a *combination tone*. The player of a typical pipe organ can press a single key and make the organ produce sound with different fundamental frequencies. She can select and pull out different stops to make the same key for the note C produce sound at the following frequencies: 65.4 Hz from a so-called eight-foot pipe; $2 \times 65.4 = 131$ Hz from a four-foot pipe; $3 \times 65.4 = 196$ Hz from a two-and-two-thirds-foot pipe; $4 \times 65.4 = 262$ Hz from a two-foot pipe; or any combination of these. With notes at low frequencies, she obtains sound with the richest quality by pulling out all the stops. When an air leak develops in one of the pipes, that pipe cannot be used. If a leak occurs in an eight-foot pipe, playing a combination of other pipes can create the sensation of sound at the frequency that the eight-foot pipe would produce. Which sets of stops, among those listed, could be pulled out to do this?

Section 18.8 Nonsinusoidal Wave Patterns

55. An A-major chord consists of the notes called A, C#, and E. It can be played on a piano by simultaneously striking strings with fundamental frequencies of 440.00 Hz, 554.37 Hz, and 659.26 Hz. The rich consonance of the chord is associated with near equality of the frequencies of some of the higher harmonics of the three tones. Consider the first five harmonics of each string and determine which harmonics show near equality.

56. Suppose that a flutist plays a 523-Hz C note with first harmonic displacement amplitude $A_1 = 100$ nm. From Figure 18.24b read, by proportion, the displacement amplitudes of harmonics 2 through 7. Take these as the values A_2 through A_7 in the Fourier analysis of the sound, and assume that $B_1 = B_2 = \cdots = B_7 = 0$. Construct a graph of the waveform of the sound. Your waveform will not look exactly like the flute waveform in Figure 18.23b because you simplify by ignoring cosine terms; nevertheless, it produces the same sensation to human hearing.

Additional Problems

57. On a marimba (Fig. P18.57), the wooden bar that sounds a tone when struck vibrates in a transverse standing wave having three antinodes and two nodes. The lowest frequency note is 87.0 Hz, produced by a bar 40.0 cm long. (a) Find the speed of transverse waves on the bar. (b) A resonant pipe suspended vertically below the center of the bar enhances the loudness of the emitted sound. If the pipe is open at the top end only and the speed of sound in air is 340 m/s, what is the length of the pipe required to resonate with the bar in part (a)?

Murray Greenberg

Figure P18.57 Marimba players in Mexico City.

58. A loudspeaker at the front of a room and an identical loudspeaker at the rear of the room are being driven by the same oscillator at 456 Hz. A student walks at a uniform rate of 1.50 m/s along the length of the room. She hears a single tone, repeatedly becoming louder and softer. (a) Model these variations as beats between the Doppler-shifted sounds the student receives. Calculate the number of beats the student hears each second. (b) **What If?** Model the two speakers as producing a standing wave in the room and the student as walking between antinodes. Calculate the number of intensity maxima the student hears each second.

59. Two train whistles have identical frequencies of 180 Hz. When one train is at rest in the station and the other is moving nearby, a commuter standing on the station platform hears beats with a frequency of 2.00 beats/s when the whistles sound at the same time. What are the two possible speeds and directions that the moving train can have?

60. A string fixed at both ends and having a mass of 4.80 g, a length of 2.00 m, and a tension of 48.0 N vibrates in its second ($n = 2$) normal mode. What is the wavelength in air of the sound emitted by this vibrating string?

61. A student uses an audio oscillator of adjustable frequency to measure the depth of a water well. The student hears two successive resonances at 51.5 Hz and 60.0 Hz. How deep is the well?

62. A string has a mass per unit length of 9.00×10^{-3} kg/m and a length of 0.400 m. What must be the tension in the string if its second harmonic has the same frequency as the second resonance mode of a 1.75-m-long pipe open at one end?

63. Two wires are welded together end to end. The wires are made of the same material, but the diameter of one is twice that of the other. They are subjected to a tension of 4.60 N. The thin wire has a length of 40.0 cm and a linear mass density of 2.00 g/m. The combination is fixed at both ends and vibrated in such a way that two antinodes are present, with the node between them being right at the weld. (a) What is the frequency of vibration? (b) How long is the thick wire?

64. **Review problem.** For the arrangement shown in Figure P18.64, $\theta = 30.0°$, the inclined plane and the small pulley are frictionless, the string supports the object of mass M at the bottom of the plane, and the string has mass m that is small compared to M. The system is in equilibrium and the vertical part of the string has a length h. Standing waves are set up in the vertical section of the string. (a) Find the tension in the string. (b) Model the shape of the string as one leg and the hypotenuse of a right triangle. Find the whole length of the string. (c) Find the mass per unit length of the string. (d) Find the speed of waves on the string. (e) Find the lowest frequency for a standing wave. (f) Find the period of the standing wave having three nodes. (g) Find the wavelength of the standing wave having three nodes. (h) Find the frequency of the beats resulting from the interference of the sound wave of lowest frequency generated by the string with another sound wave having a frequency that is 2.00% greater.

the string are fixed. When the vibrator has a frequency f, in a string of length L and under tension T, n antinodes are set up in the string. (a) If the length of the string is doubled, by what factor should the frequency be changed so that the same number of antinodes is produced? (b) If the frequency and length are held constant, what tension will produce $n + 1$ antinodes? (c) If the frequency is tripled and the length of the string is halved, by what factor should the tension be changed so that twice as many antinodes are produced?

66. A 0.010 0-kg wire, 2.00 m long, is fixed at both ends and vibrates in its simplest mode under a tension of 200 N. When a vibrating tuning fork is placed near the wire, a beat frequency of 5.00 Hz is heard. (a) What could be the frequency of the tuning fork? (b) What should be the tension in the wire be if the beats are to disappear?

67. Two waves are described by the wave functions

$$y_1(x, t) = 5.0 \sin(2.0x - 10t)$$

and

$$y_2(x, t) = 10 \cos(2.0x - 10t)$$

where y_1, y_2, and x are in meters and t is in seconds. Show that the wave resulting from their superposition is also sinusoidal. Determine the amplitude and phase of this sinusoidal wave.

68. The wave function for a standing wave is given in Equation 18.3 as $y = 2A \sin kx \cos \omega t$. (a) Rewrite this wave function in terms of the wavelength λ and the wave speed v of the wave. (b) Write the wave function of the simplest standing-wave vibration of a stretched string of length L. (c) Write the wave function for the second harmonic. (d) Generalize these results and write the wave function for the nth resonance vibration.

69. **Review problem.** A 12.0-kg object hangs in equilibrium from a string with a total length of $L = 5.00$ m and a linear mass density of $\mu = 0.001\,00$ kg/m. The string is wrapped around two light, frictionless pulleys that are separated by a distance of $d = 2.00$ m (Fig. P18.69a). (a) Determine the tension in the string. (b) At what frequency must the string between the pulleys vibrate in order to form the standing wave pattern shown in Figure P18.69b?

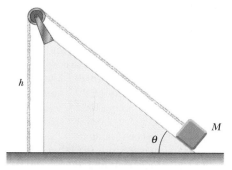

Figure P18.64

65. A standing wave is set up in a string of variable length and tension by a vibrator of variable frequency. Both ends of

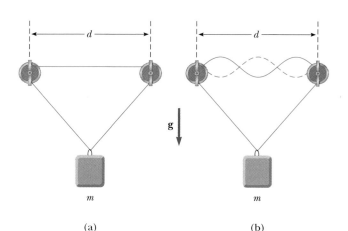

(a)   (b)

Figure P18.69

70. A quartz watch contains a crystal oscillator in the form of a block of quartz that vibrates by contracting and expanding. Two opposite faces of the block, 7.05 mm apart, are antinodes, moving alternately toward each other and away from each other. The plane halfway between these two faces is a node of the vibration. The speed of sound in quartz is 3.70 km/s. Find the frequency of the vibration. An oscillating electric voltage accompanies the mechanical oscillation—the quartz is described as *piezoelectric*. An electric circuit feeds in energy to maintain the oscillation and also counts the voltage pulses to keep time.

Answers to Quick Quizzes

18.1 The shape of the string at $t = 0.6$ s is shown below.

1 cm

18.2 (c). The pulses completely cancel each other in terms of displacement of elements of the string from equilibrium, but the string is still moving. A short time later, the string will be displaced again and the pulses will have passed each other.

18.3 (a). The pattern shown at the bottom of Figure 18.9a corresponds to the extreme position of the string. All elements of the string have momentarily come to rest.

18.4 (d). Near a nodal point, elements on one side of the point are moving upward at this instant and elements on the other side are moving downward.

18.5 (d). Choice (a) is incorrect because the number of nodes is one greater than the number of antinodes. Choice (b) is only true for half of the modes; it is not true for any odd-numbered mode. Choice (c) would be correct if we replace the word *nodes* with *antinodes*.

18.6 For each natural frequency of the glass, the standing wave must "fit" exactly around the rim. In Figure 18.17a we see three antinodes on the near side of the glass, and thus there must be another three on the far side. This corresponds to three complete waves. In a top view, the wave pattern looks like this (although we have greatly exaggerated the amplitude):

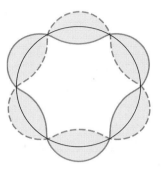

18.7 (b). With both ends open, the pipe has a fundamental frequency given by Equation 18.11: $f_{open} = v/2L$. With one end closed, the pipe has a fundamental frequency given by Equation 18.12:

$$f_{closed} = \frac{v}{4L} = \frac{1}{2}\frac{v}{2L} = \frac{1}{2}f_{open}$$

18.8 (c). The increase in temperature causes the speed of sound to go up. According to Equation 18.11, this will result in an increase in the fundamental frequency of a given organ pipe.

18.9 (b). Tightening the string has caused the frequencies to be farther apart, based on the increase in the beat frequency.

Thermodynamics

W e now direct our attention to the study of thermodynamics, which involves situations in which the temperature or state (solid, liquid, gas) of a system changes due to energy transfers. As we shall see, thermodynamics is very successful in explaining the bulk properties of matter and the correlation between these properties and the mechanics of atoms and molecules.

Historically, the development of thermodynamics paralleled the development of the atomic theory of matter. By the 1820s, chemical experiments had provided solid evidence for the existence of atoms. At that time, scientists recognized that a connection between thermodynamics and the structure of matter must exist. In 1827, the botanist Robert Brown reported that grains of pollen suspended in a liquid move erratically from one place to another, as if under constant agitation. In 1905, Albert Einstein used kinetic theory to explain the cause of this erratic motion, which today is known as *Brownian motion*. Einstein explained this phenomenon by assuming that the grains are under constant bombardment by "invisible" molecules in the liquid, which themselves move erratically. This explanation gave scientists insight into the concept of molecular motion and gave credence to the idea that matter is made up of atoms. A connection was thus forged between the everyday world and the tiny, invisible building blocks that make up this world.

Thermodynamics also addresses more practical questions. Have you ever wondered how a refrigerator is able to cool its contents, what types of transformations occur in a power plant or in the engine of your automobile, or what happens to the kinetic energy of a moving object when the object comes to rest? The laws of thermodynamics can be used to provide explanations for these and other phenomena. ■

◄ *The Alyeska oil pipeline near the Tazlina River in Alaska. The oil in the pipeline is warm, and energy transferring from the pipeline could melt environmentally sensitive permafrost in the ground. The finned structures on top of the support posts are thermal radiators that allow the energy to be transferred into the air in order to protect the permafrost. (Topham Picturepoint/The Image Works)*

Chapter 19

Temperature

▲ Why would someone designing a pipeline include these strange loops? Pipelines carrying liquids often contain loops such as these to allow for expansion and contraction as the temperature changes. We will study thermal expansion in this chapter. (Lowell Georgia/CORBIS)

In our study of mechanics, we carefully defined such concepts as *mass, force,* and *kinetic energy* to facilitate our quantitative approach. Likewise, a quantitative description of thermal phenomena requires careful definitions of such important terms as *temperature, heat,* and *internal energy.* This chapter begins with a discussion of temperature and with a description of one of the laws of thermodynamics (the so-called "zeroth law").

Next, we consider why an important factor when we are dealing with thermal phenomena is the particular substance we are investigating. For example, gases expand appreciably when heated, whereas liquids and solids expand only slightly.

This chapter concludes with a study of ideal gases on the macroscopic scale. Here, we are concerned with the relationships among such quantities as pressure, volume, and temperature. In Chapter 21, we shall examine gases on a microscopic scale, using a model that represents the components of a gas as small particles.

19.1 Temperature and the Zeroth Law of Thermodynamics

We often associate the concept of temperature with how hot or cold an object feels when we touch it. Thus, our senses provide us with a qualitative indication of temperature. However, our senses are unreliable and often mislead us. For example, if we remove a metal ice tray and a cardboard box of frozen vegetables from the freezer, the ice tray feels colder than the box *even though both are at the same temperature.* The two objects feel different because metal transfers energy by heat at a higher rate than cardboard does. What we need is a reliable and reproducible method for measuring the relative hotness or coldness of objects rather than the rate of energy transfer. Scientists have developed a variety of thermometers for making such quantitative measurements.

We are all familiar with the fact that two objects at different initial temperatures eventually reach some intermediate temperature when placed in contact with each other. For example, when hot water and cold water are mixed in a bathtub, the final temperature of the mixture is somewhere between the initial hot and cold temperatures. Likewise, when an ice cube is dropped into a cup of hot coffee, it melts and the coffee's temperature decreases.

To understand the concept of temperature, it is useful to define two often-used phrases: *thermal contact* and *thermal equilibrium.* To grasp the meaning of thermal contact, imagine that two objects are placed in an insulated container such that they interact with each other but not with the environment. If the objects are at different temperatures, energy is exchanged between them, even if they are initially not in physical contact with each other. The energy transfer mechanisms from Chapter 7 that we will focus on are heat and electromagnetic radiation. For purposes of the current discussion, we assume that two objects are in **thermal contact** with each other if energy can be exchanged between them by these processes due to a temperature difference.

Figure 19.1 The zeroth law of thermodynamics. (a) and (b) If the temperatures of A and B are measured to be the same by placing them in thermal contact with a thermometer (object C), no energy will be exchanged between them when they are placed in thermal contact with each other (c).

Thermal equilibrium is a situation in which two objects would not exchange energy by heat or electromagnetic radiation if they were placed in thermal contact.

Let us consider two objects A and B, which are not in thermal contact, and a third object C, which is our thermometer. We wish to determine whether A and B are in thermal equilibrium with each other. The thermometer (object C) is first placed in thermal contact with object A until thermal equilibrium is reached,[1] as shown in Figure 19.1a. From that moment on, the thermometer's reading remains constant, and we record this reading. The thermometer is then removed from object A and placed in thermal contact with object B, as shown in Figure 19.1b. The reading is again recorded after thermal equilibrium is reached. If the two readings are the same, then object A and object B are in thermal equilibrium with each other. If they are placed in contact with each other as in Figure 19.1c, there is no exchange of energy between them.

We can summarize these results in a statement known as the **zeroth law of thermodynamics** (the law of equilibrium):

Zeroth law of thermodynamics

> If objects A and B are separately in thermal equilibrium with a third object C, then A and B are in thermal equilibrium with each other.

This statement can easily be proved experimentally and is very important because it enables us to define temperature. We can think of **temperature** as the property that determines whether an object is in thermal equilibrium with other objects. **Two objects in thermal equilibrium with each other are at the same temperature.** Conversely, if two objects have different temperatures, then they are not in thermal equilibrium with each other.

Quick Quiz 19.1 Two objects, with different sizes, masses, and temperatures, are placed in thermal contact. Energy travels (a) from the larger object to the smaller object (b) from the object with more mass to the one with less (c) from the object at higher temperature to the object at lower temperature.

---

[1] We assume that negligible energy transfers between the thermometer and object A during the equilibrium process. Without this assumption, which is also made for the thermometer and object B, the measurement of the temperature of an object disturbs the system so that the measured temperature is different from the initial temperature of the object. In practice, whenever you measure a temperature with a thermometer, you measure the disturbed system, not the original system.

19.2 Thermometers and the Celsius Temperature Scale

Thermometers are devices that are used to measure the temperature of a system. All thermometers are based on the principle that some physical property of a system changes as the system's temperature changes. Some physical properties that change with temperature are (1) the volume of a liquid, (2) the dimensions of a solid, (3) the pressure of a gas at constant volume, (4) the volume of a gas at constant pressure, (5) the electric resistance of a conductor, and (6) the color of an object. A temperature scale can be established on the basis of any one of these physical properties.

A common thermometer in everyday use consists of a mass of liquid—usually mercury or alcohol—that expands into a glass capillary tube when heated (Fig. 19.2). In this case the physical property that changes is the volume of a liquid. Any temperature change in the range of the thermometer can be defined as being proportional to the change in length of the liquid column. The thermometer can be calibrated by placing it in thermal contact with some natural systems that remain at constant temperature. One such system is a mixture of water and ice in thermal equilibrium at atmospheric pressure. On the **Celsius temperature scale**, this mixture is defined to have a temperature of zero degrees Celsius, which is written as 0°C; this temperature is called the *ice point* of water. Another commonly used system is a mixture of water and steam in thermal equilibrium at atmospheric pressure; its temperature is 100°C, which is the *steam point* of water. Once the liquid levels in the thermometer have been established at these two points, the length of the liquid column between the two points is divided into 100 equal segments to create the Celsius scale. Thus, each segment denotes a change in temperature of one Celsius degree.

Thermometers calibrated in this way present problems when extremely accurate readings are needed. For instance, the readings given by an alcohol thermometer calibrated at the ice and steam points of water might agree with those given by a mercury thermometer only at the calibration points. Because mercury and alcohol have different thermal expansion properties, when one thermometer reads a temperature of, for example, 50°C, the other may indicate a slightly different value. The discrepancies

Figure 19.2 As a result of thermal expansion, the level of the mercury in the thermometer rises as the mercury is heated by water in the test tube.

between thermometers are especially large when the temperatures to be measured are far from the calibration points.[2]

An additional practical problem of any thermometer is the limited range of temperatures over which it can be used. A mercury thermometer, for example, cannot be used below the freezing point of mercury, which is − 39°C, and an alcohol thermometer is not useful for measuring temperatures above 85°C, the boiling point of alcohol. To surmount this problem, we need a universal thermometer whose readings are independent of the substance used in it. The gas thermometer, discussed in the next section, approaches this requirement.

19.3 The Constant-Volume Gas Thermometer and the Absolute Temperature Scale

One version of a gas thermometer is the constant-volume apparatus shown in Figure 19.3. The physical change exploited in this device is the variation of pressure of a fixed volume of gas with temperature. When the constant-volume gas thermometer was developed, it was calibrated by using the ice and steam points of water as follows. (A different calibration procedure, which we shall discuss shortly, is now used.) The flask was immersed in an ice-water bath, and mercury reservoir *B* was raised or lowered until the top of the mercury in column *A* was at the zero point on the scale. The height *h*, the difference between the mercury levels in reservoir *B* and column *A*, indicated the pressure in the flask at 0°C.

The flask was then immersed in water at the steam point, and reservoir *B* was readjusted until the top of the mercury in column *A* was again at zero on the scale; this ensured that the gas's volume was the same as it was when the flask was in the ice bath (hence, the designation "constant volume"). This adjustment of reservoir *B* gave a value for the gas pressure at 100°C. These two pressure and temperature values were then plotted, as shown in Figure 19.4. The line connecting the two points serves as a calibration curve for unknown temperatures. (Other experiments show that a linear relationship between pressure and temperature is a very good assumption.) If we wanted to measure the temperature of a substance, we would place the gas flask in thermal contact with the substance and adjust the height of reservoir *B* until the top of the mercury column in *A* is at zero on the scale. The height of the mercury column indicates the pressure of the gas; knowing the pressure, we could find the temperature of the substance using the graph in Figure 19.4.

Now let us suppose that temperatures are measured with gas thermometers containing different gases at different initial pressures. Experiments show that the thermometer readings are nearly independent of the type of gas used, as long as the gas pressure is low and the temperature is well above the point at which the gas liquefies (Fig. 19.5). The agreement among thermometers using various gases improves as the pressure is reduced.

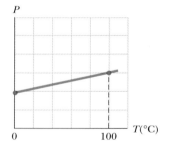

Figure 19.3 A constant-volume gas thermometer measures the pressure of the gas contained in the flask immersed in the bath. The volume of gas in the flask is kept constant by raising or lowering reservoir *B* to keep the mercury level in column *A* constant.

Figure 19.4 A typical graph of pressure versus temperature taken with a constant-volume gas thermometer. The two dots represent known reference temperatures (the ice and steam points of water).

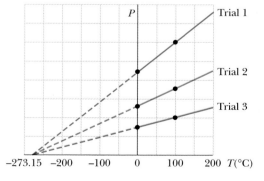

Figure 19.5 Pressure versus temperature for experimental trials in which gases have different pressures in a constant-volume gas thermometer. Note that, for all three trials, the pressure extrapolates to zero at the temperature − 273.15°C.

[2] Two thermometers that use the same liquid may also give different readings. This is due in part to difficulties in constructing uniform-bore glass capillary tubes.

If we extend the straight lines in Figure 19.5 toward negative temperatures, we find a remarkable result—**in every case, the pressure is zero when the temperature is −273.15°C!** This suggests some special role that this particular temperature must play. It is used as the basis for the **absolute temperature scale**, which sets −273.15°C as its zero point. This temperature is often referred to as **absolute zero**. The size of a degree on the absolute temperature scale is chosen to be identical to the size of a degree on the Celsius scale. Thus, the conversion between these temperatures is

$$T_C = T - 273.15 \tag{19.1}$$

where T_C is the Celsius temperature and T is the absolute temperature.

Because the ice and steam points are experimentally difficult to duplicate, an absolute temperature scale based on two new fixed points was adopted in 1954 by the International Committee on Weights and Measures. The first point is absolute zero. The second reference temperature for this new scale was chosen as the **triple point of water**, which is the single combination of temperature and pressure at which liquid water, gaseous water, and ice (solid water) coexist in equilibrium. This triple point occurs at a temperature of 0.01°C and a pressure of 4.58 mm of mercury. On the new scale, which uses the unit *kelvin*, the temperature of water at the triple point was set at 273.16 kelvins, abbreviated 273.16 K. This choice was made so that the old absolute temperature scale based on the ice and steam points would agree closely with the new scale based on the triple point. This new absolute temperature scale (also called the **Kelvin scale**) employs the SI unit of absolute temperature, the **kelvin**, which is defined to be **1/273.16 of the difference between absolute zero and the temperature of the triple point of water.**

Figure 19.6 shows the absolute temperature for various physical processes and structures. The temperature of absolute zero (0 K) cannot be achieved, although laboratory experiments incorporating the laser cooling of atoms have come very close.

What would happen to a gas if its temperature could reach 0 K (and it did not liquefy or solidify)? As Figure 19.5 indicates, the pressure it exerts on the walls of its container would be zero. In Chapter 21 we shall show that the pressure of a gas is proportional to the average kinetic energy of its molecules. Thus, according to classical physics, the kinetic energy of the gas molecules would become zero at absolute zero, and molecular motion would cease; hence, the molecules would settle out on the bottom of the container. Quantum theory modifies this prediction and shows that some residual energy, called the *zero-point energy*, would remain at this low temperature.

The Celsius, Fahrenheit, and Kelvin Temperature Scales[3]

Equation 19.1 shows that the Celsius temperature T_C is shifted from the absolute (Kelvin) temperature T by 273.15°. Because the size of a degree is the same on the two scales, a temperature difference of 5°C is equal to a temperature difference of 5 K. The two scales differ only in the choice of the zero point. Thus, the ice-point temperature on the Kelvin scale, 273.15 K, corresponds to 0.00°C, and the Kelvin-scale steam point, 373.15 K, is equivalent to 100.00°C.

A common temperature scale in everyday use in the United States is the **Fahrenheit scale**. This scale sets the temperature of the ice point at 32°F and the temperature of the steam point at 212°F. The relationship between the Celsius and Fahrenheit temperature scales is

$$T_F = \tfrac{9}{5} T_C + 32°F \tag{19.2}$$

We can use Equations 19.1 and 19.2 to find a relationship between changes in temperature on the Celsius, Kelvin, and Fahrenheit scales:

$$\Delta T_C = \Delta T = \tfrac{5}{9} \Delta T_F \tag{19.3}$$

---

[3]   Named after Anders Celsius (1701–1744), Daniel Gabriel Fahrenheit (1686–1736), and William Thomson, Lord Kelvin (1824–1907), respectively.

▲ **PITFALL PREVENTION**

19.1   A Matter of Degree

Note that notations for temperatures in the Kelvin scale do not use the degree sign. The unit for a Kelvin temperature is simply "kelvins" and not "degrees Kelvin."

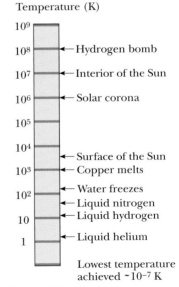

Figure 19.6 Absolute temperatures at which various physical processes occur. Note that the scale is logarithmic.

Of the three temperature scales that we have discussed, only the Kelvin scale is based on a true zero value of temperature. The Celsius and Fahrenheit scales are based on an arbitrary zero associated with one particular substance—water—on one particular planet—Earth. Thus, if you encounter an equation that calls for a temperature T or involves a ratio of temperatures, you *must* convert all temperatures to kelvins. If the equation contains a change in temperature ΔT, using Celsius temperatures will give you the correct answer, in light of Equation 19.3, but it is always *safest* to convert temperatures to the Kelvin scale.

Quick Quiz 19.2 Consider the following pairs of materials. Which pair represents two materials, one of which is twice as hot as the other? (a) boiling water at 100°C, a glass of water at 50°C (b) boiling water at 100°C, frozen methane at −50°C (c) an ice cube at −20°C, flames from a circus fire-eater at 233°C (d) No pair represents materials one of which is twice as hot as the other

---

Example 19.1 Converting Temperatures

On a day when the temperature reaches 50°F, what is the temperature in degrees Celsius and in kelvins?

Solution Substituting into Equation 19.2, we obtain

$$T_C = \tfrac{5}{9}(T_F - 32) = \tfrac{5}{9}(50 - 32)$$

$$= \boxed{10°C}$$

From Equation 19.1, we find that

$$T = T_C + 273.15 = 10°C + 273.15 = \boxed{283 \text{ K}}$$

A convenient set of weather-related temperature equivalents to keep in mind is that 0°C is (literally) freezing at 32°F, 10°C is cool at 50°F, 20°C is room temperature, 30°C is warm at 86°F, and 40°C is a hot day at 104°F.

Example 19.2 Heating a Pan of Water

A pan of water is heated from 25°C to 80°C. What is the change in its temperature on the Kelvin scale and on the Fahrenheit scale?

Solution From Equation 19.3, we see that the change in temperature on the Celsius scale equals the change on the Kelvin scale. Therefore,

$$\Delta T = \Delta T_C = 80°C - 25°C = 55°C = \boxed{55 \text{ K}}$$

From Equation 19.3, we also find that

$$\Delta T_F = \tfrac{9}{5} \Delta T_C = \tfrac{9}{5}(55°C) = \boxed{99°F}$$

19.4 Thermal Expansion of Solids and Liquids

Our discussion of the liquid thermometer makes use of one of the best-known changes in a substance: as its temperature increases, its volume increases. This phenomenon, known as **thermal expansion**, has an important role in numerous engineering applications. For example, thermal-expansion joints, such as those shown in Figure 19.7, must be included in buildings, concrete highways, railroad tracks, brick walls, and bridges to compensate for dimensional changes that occur as the temperature changes.

Thermal expansion is a consequence of the change in the *average* separation between the atoms in an object. To understand this, model the atoms as being connected by stiff springs, as discussed in Section 15.3 and shown in Figure 15.12b. At ordinary temperatures, the atoms in a solid oscillate about their equilibrium positions with an amplitude of approximately 10^{-11} m and a frequency of approximately 10^{13} Hz. The average spacing between the atoms is about 10^{-10} m. As the temperature of the solid

(a)   (b)

Figure 19.7 (a) Thermal-expansion joints are used to separate sections of roadways on bridges. Without these joints, the surfaces would buckle due to thermal expansion on very hot days or crack due to contraction on very cold days. (b) The long, vertical joint is filled with a soft material that allows the wall to expand and contract as the temperature of the bricks changes.

increases, the atoms oscillate with greater amplitudes; as a result, the average separation between them increases.[4] Consequently, the object expands.

If thermal expansion is sufficiently small relative to an object's initial dimensions, the change in any dimension is, to a good approximation, proportional to the first power of the temperature change. Suppose that an object has an initial length L_i along some direction at some temperature and that the length increases by an amount ΔL for a change in temperature ΔT. Because it is convenient to consider the fractional change in length per degree of temperature change, we define the **average coefficient of linear expansion** as

$$\alpha \equiv \frac{\Delta L/L_i}{\Delta T}$$

Experiments show that α is constant for small changes in temperature. For purposes of calculation, this equation is usually rewritten as

$$\Delta L = \alpha L_i \Delta T \tag{19.4}$$

or as

$$L_f - L_i = \alpha L_i (T_f - T_i) \tag{19.5}$$

where L_f is the final length, T_i and T_f are the initial and final temperatures, and the proportionality constant α is the average coefficient of linear expansion for a given material and has units of $(°C)^{-1}$.

It may be helpful to think of thermal expansion as an effective magnification or as a photographic enlargement of an object. For example, as a metal washer is heated (Fig. 19.8), all dimensions, including the radius of the hole, increase according to Equation 19.4. Notice that this is equivalent to saying that **a cavity in a piece of material expands in the same way as if the cavity were filled with the material.**

Table 19.1 lists the average coefficient of linear expansion for various materials. Note that for these materials α is positive, indicating an increase in length with increasing temperature. This is not always the case. Some substances—calcite ($CaCO_3$) is one example—expand along one dimension (positive α) and contract along another (negative α) as their temperatures are increased.

---

[4] More precisely, thermal expansion arises from the *asymmetrical* nature of the potential-energy curve for the atoms in a solid, as shown in Figure 15.12a. If the oscillators were truly harmonic, the average atomic separations would not change regardless of the amplitude of vibration.

Active Figure 19.8 Thermal expansion of a homogeneous metal washer. As the washer is heated, all dimensions increase. (The expansion is exaggerated in this figure.)

At the Active Figures link at http://www.pse6.com, you can compare expansions for various temperatures of the burner and materials from which the washer is made.

▲ **PITFALL PREVENTION**

19.2 Do Holes Become Larger or Smaller?

When an object's temperature is raised, every linear dimension increases in size. This includes any holes in the material, which expand in the same way as if the hole were filled with the material, as shown in Figure 19.8. Keep in mind the notion of thermal expansion as being similar to a photographic enlargement.

Table 19.1

Average Expansion Coefficients for Some Materials Near Room Temperature			
Material	**Average Linear Expansion Coefficient $(\alpha)(°C)^{-1}$**	**Material**	**Average Volume Expansion Coefficient $(\beta)(°C)^{-1}$**
Aluminum	24×10^{-6}	Alcohol, ethyl	1.12×10^{-4}
Brass and bronze	19×10^{-6}	Benzene	1.24×10^{-4}
Copper	17×10^{-6}	Acetone	1.5×10^{-4}
Glass (ordinary)	9×10^{-6}	Glycerin	4.85×10^{-4}
Glass (Pyrex)	3.2×10^{-6}	Mercury	1.82×10^{-4}
Lead	29×10^{-6}	Turpentine	9.0×10^{-4}
Steel	11×10^{-6}	Gasoline	9.6×10^{-4}
Invar (Ni–Fe alloy)	0.9×10^{-6}	Air[a] at 0°C	3.67×10^{-3}
Concrete	12×10^{-6}	Helium[a]	3.665×10^{-3}

[a] Gases do not have a specific value for the volume expansion coefficient because the amount of expansion depends on the type of process through which the gas is taken. The values given here assume that the gas undergoes an expansion at constant pressure.

Because the linear dimensions of an object change with temperature, it follows that surface area and volume change as well. The change in volume is proportional to the initial volume V_i and to the change in temperature according to the relationship

$$\Delta V = \beta V_i \Delta T \qquad (19.6)$$

where β is the **average coefficient of volume expansion**. For a solid, the average coefficient of volume expansion is three times the average linear expansion coefficient: $\beta = 3\alpha$. (This assumes that the average coefficient of linear expansion of the solid is the same in all directions—that is, the material is *isotropic*.)

To see that $\beta = 3\alpha$ for a solid, consider a solid box of dimensions ℓ, w, and h. Its volume at some temperature T_i is $V_i = \ell wh$. If the temperature changes to $T_i + \Delta T$, its volume changes to $V_i + \Delta V$, where each dimension changes according to Equation 19.4. Therefore,

$$\begin{aligned} V_i + \Delta V &= (\ell + \Delta \ell)(w + \Delta w)(h + \Delta h) \\ &= (\ell + \alpha \ell \Delta T)(w + \alpha w \Delta T)(h + \alpha h \Delta T) \\ &= \ell wh(1 + \alpha \Delta T)^3 \\ &= V_i[1 + 3\alpha \Delta T + 3(\alpha \Delta T)^2 + (\alpha \Delta T)^3] \end{aligned}$$

If we now divide both sides by V_i and isolate the term $\Delta V/V_i$, we obtain the fractional change in volume:

$$\frac{\Delta V}{V_i} = 3\alpha \Delta T + 3(\alpha \Delta T)^2 + (\alpha \Delta T)^3$$

Because $\alpha \Delta T \ll 1$ for typical values of ΔT ($< \sim 100°C$), we can neglect the terms $3(\alpha \Delta T)^2$ and $(\alpha \Delta T)^3$. Upon making this approximation, we see that

$$\frac{\Delta V}{V_i} = 3\alpha \Delta T$$

$$3\alpha = \frac{1}{V_i} \frac{\Delta V}{\Delta T}$$

Equation 19.6 shows that the right side of this expression is equal to β, and so we have $3\alpha = \beta$, the relationship we set out to prove. In a similar way, you can show that the change in area of a rectangular plate is given by $\Delta A = 2\alpha A_i \Delta T$ (see Problem 55).

As Table 19.1 indicates, each substance has its own characteristic average coefficient of expansion. For example, when the temperatures of a brass rod and a steel rod of

Steel

Brass

Room temperature

Higher temperature

(a)

On   25°C   Off   30°C

Bimetallic strip

(b)

Figure 19.9 (a) A bimetallic strip bends as the temperature changes because the two metals have different expansion coefficients. (b) A bimetallic strip used in a thermostat to break or make electrical contact.

equal length are raised by the same amount from some common initial value, the brass rod expands more than the steel rod does because brass has a greater average coefficient of expansion than steel does. A simple mechanism called a *bimetallic strip* utilizes this principle and is found in practical devices such as thermostats. It consists of two thin strips of dissimilar metals bonded together. As the temperature of the strip increases, the two metals expand by different amounts and the strip bends, as shown in Figure 19.9.

Quick Quiz 19.3 If you are asked to make a very sensitive glass thermometer, which of the following working liquids would you choose? (a) mercury (b) alcohol (c) gasoline (d) glycerin

Quick Quiz 19.4 Two spheres are made of the same metal and have the same radius, but one is hollow and the other is solid. The spheres are taken through the same temperature increase. Which sphere expands more? (a) solid sphere (b) hollow sphere (c) They expand by the same amount. (d) not enough information to say

---

Example 19.3   Expansion of a Railroad Track

A segment of steel railroad track has a length of 30.000 m when the temperature is 0.0°C.

(A) What is its length when the temperature is 40.0°C?

Solution Making use of Table 19.1 and noting that the change in temperature is 40.0°C, we find that the increase in length is

$$\Delta L = \alpha L_i \, \Delta T = [11 \times 10^{-6}(°C)^{-1}](30.000 \text{ m})(40.0°C)$$
$$= 0.013 \text{ m}$$

If the track is 30.000 m long at 0.0°C, its length at 40.0°C is

30.013 m.

(B) Suppose that the ends of the rail are rigidly clamped at 0.0°C so that expansion is prevented. What is the thermal stress set up in the rail if its temperature is raised to 40.0°C?

Solution The thermal stress will be the same as that in the situation in which we allow the rail to expand freely and then compress it with a mechanical force F back to its original length. From the definition of Young's modulus for a solid (see Eq. 12.6), we have

$$\text{Tensile stress} = \frac{F}{A} = Y\frac{\Delta L}{L_i}$$

Because Y for steel is 20×10^{10} N/m² (see Table 12.1), we have

$$\frac{F}{A} = (20 \times 10^{10} \text{ N/m}^2)\left(\frac{0.013 \text{ m}}{30.000 \text{ m}}\right) = \boxed{8.7 \times 10^7 \text{ N/m}^2}$$

What If? What if the temperature drops to −40.0°C? What is the length of the unclamped segment?

The expression for the change in length in Equation 19.4 is the same whether the temperature increases or de-creases. Thus, if there is an increase in length of 0.013 m when the temperature increases by 40°C, then there is a decrease in length of 0.013 m when the temperature decreases by 40°C. (We assume that α is constant over the entire range of temperatures.) The new length at the colder temperature is 30.000 m − 0.013 m = 29.987 m.

Example 19.4   The Thermal Electrical Short

An electronic device has been poorly designed so that two bolts attached to different parts of the device almost touch each other in its interior, as in Figure 19.10. The steel and brass bolts are at different electric potentials and if they touch, a short circuit will develop, damaging the device. (We will study electric potential in Chapter 25.) If the initial gap between the ends of the bolts is 5.0 μm at 27°C, at what temperature will the bolts touch?

Solution We can conceptualize the situation by imagining that the ends of both bolts expand into the gap between them as the temperature rises. We categorize this as a thermal expansion problem, in which the *sum* of the changes in length of the two bolts must equal the length of the initial gap between the ends. To analyze the problem, we write this condition mathematically:

$$\Delta L_{br} + \Delta L_{st} = \alpha_{br} L_{i,br} \Delta T + \alpha_{st} L_{i,st} \Delta T = 5.0 \times 10^{-6} \text{ m}$$

Solving for ΔT, we find

$$\Delta T = \frac{5.0 \times 10^{-6} \text{ m}}{\alpha_{br} L_{i,br} + \alpha_{st} L_{i,st}}$$

$$= \frac{5.0 \times 10^{-6} \text{ m}}{(19 \times 10^{-6}\text{°C}^{-1})(0.030 \text{ m}) + (11 \times 10^{-6}\text{°C}^{-1})(0.010 \text{ m})}$$

$$= 7.4\text{°C}$$

Thus, the temperature at which the bolts touch is 27°C + 7.4°C = $\boxed{34\text{°C}}$ To finalize this problem, note that this temperature is possible if the air conditioning in the building housing the device fails for a long period on a very hot summer day.

Figure 19.10 (Example 19.4) Two bolts attached to different parts of an electrical device are almost touching when the temperature is 27°C. As the temperature increases, the ends of the bolts move toward each other.

The Unusual Behavior of Water

Liquids generally increase in volume with increasing temperature and have average coefficients of volume expansion about ten times greater than those of solids. Cold water is an exception to this rule, as we can see from its density-versus-temperature curve, shown in Figure 19.11. As the temperature increases from 0°C to 4°C, water contracts and thus its density increases. Above 4°C, water expands with increasing temperature, and so its density decreases. Thus, the density of water reaches a maximum value of 1.000 g/cm³ at 4°C.

We can use this unusual thermal-expansion behavior of water to explain why a pond begins freezing at the surface rather than at the bottom. When the atmospheric temperature drops from, for example, 7°C to 6°C, the surface water also cools and consequently decreases in volume. This means that the surface water is denser than the water below it, which has not cooled and decreased in volume. As a result, the surface water sinks, and warmer water from below is forced to the surface to be cooled. When the atmospheric temperature is between 4°C and 0°C, however, the surface water expands as it cools, becoming less dense than the water below it. The mixing process

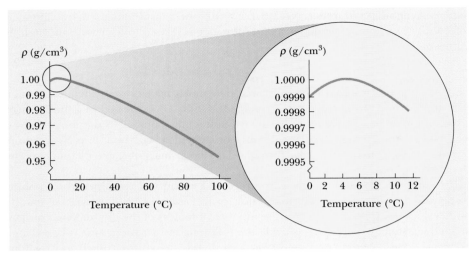

Figure 19.11 The variation in the density of water at atmospheric pressure with temperature. The inset at the right shows that the maximum density of water occurs at 4°C.

stops, and eventually the surface water freezes. As the water freezes, the ice remains on the surface because ice is less dense than water. The ice continues to build up at the surface, while water near the bottom remains at 4°C. If this were not the case, then fish and other forms of marine life would not survive.

19.5 Macroscopic Description of an Ideal Gas

The volume expansion equation $\Delta V = \beta V_i \, \Delta T$ is based on the assumption that the material has an initial volume V_i before the temperature change occurs. This is the case for solids and liquids because they have a fixed volume at a given temperature.

The case for gases is completely different. The interatomic forces within gases are very weak, and, in many cases, we can imagine these forces to be nonexistent and still make very good approximations. Note that *there is no equilibrium separation* for the atoms and, thus, no "standard" volume at a given temperature. As a result, we cannot express changes in volume ΔV in a process on a gas with Equation 19.6 because we have no defined volume V_i at the beginning of the process. For a gas, the volume is entirely determined by the container holding the gas. Thus, equations involving gases will contain the volume V as a *variable*, rather than focusing on a *change* in the volume from an initial value.

For a gas, it is useful to know how the quantities volume V, pressure P, and temperature T are related for a sample of gas of mass m. In general, the equation that interrelates these quantities, called the *equation of state*, is very complicated. However, if the gas is maintained at a very low pressure (or low density), the equation of state is quite simple and can be found experimentally. Such a low-density gas is commonly referred to as an *ideal gas*.[5]

It is convenient to express the amount of gas in a given volume in terms of the number of moles n. One **mole** of any substance is that amount of the substance that

---

[5]   To be more specific, the assumption here is that the temperature of the gas must not be too low (the gas must not condense into a liquid) or too high, and that the pressure must be low. The concept of an ideal gas implies that the gas molecules do not interact except upon collision, and that the molecular volume is negligible compared with the volume of the container. In reality, an ideal gas does not exist. However, the concept of an ideal gas is very useful because real gases at low pressures behave as ideal gases do.

contains **Avogadro's number** $N_A = 6.022 \times 10^{23}$ of constituent particles (atoms or molecules). The number of moles n of a substance is related to its mass m through the expression

$$n = \frac{m}{M} \tag{19.7}$$

where M is the molar mass of the substance. The molar mass of each chemical element is the atomic mass (from the periodic table, Appendix C) expressed in g/mol. For example, the mass of one He atom is 4.00 u (atomic mass units), so the molar mass of He is 4.00 g/mol. For a molecular substance or a chemical compound, you can add up the molar mass from its molecular formula. The molar mass of stable diatomic oxygen (O_2) is 32.0 g/mol.

Now suppose that an ideal gas is confined to a cylindrical container whose volume can be varied by means of a movable piston, as in Figure 19.12. If we assume that the cylinder does not leak, the mass (or the number of moles) of the gas remains constant. For such a system, experiments provide the following information. First, when the gas is kept at a constant temperature, its pressure is inversely proportional to its volume (Boyle's law). Second, when the pressure of the gas is kept constant, its volume is directly proportional to its temperature (the law of Charles and Gay-Lussac). These observations are summarized by the **equation of state for an ideal gas:**

> **Equation of state for an ideal gas**

$$PV = nRT \tag{19.8}$$

In this expression, known as the **ideal gas law**, R is a constant and n is the number of moles of gas in the sample. Experiments on numerous gases show that as the pressure approaches zero, the quantity PV/nT approaches the same value R for all gases. For this reason, R is called the **universal gas constant**. In SI units, in which pressure is expressed in pascals (1 Pa = 1 N/m^2) and volume in cubic meters, the product PV has units of newton·meters, or joules, and R has the value

$$R = 8.314\,\text{J/mol·K} \tag{19.9}$$

If the pressure is expressed in atmospheres and the volume in liters (1 L = 10^3 cm^3 = 10^{-3} m^3), then R has the value

$$R = 0.082\,14\ \text{L·atm/mol·K}$$

Using this value of R and Equation 19.8, we find that the volume occupied by 1 mol of any gas at atmospheric pressure and at 0°C (273 K) is 22.4 L.

The ideal gas law states that if the volume and temperature of a fixed amount of gas do not change, then the pressure also remains constant. Consider a bottle of champagne that is shaken and then spews liquid when opened, as shown in Figure 19.13.

Active Figure 19.12 An ideal gas confined to a cylinder whose volume can be varied by means of a movable piston.

At the Active Figures link at http://www.pse6.com, you can choose to keep either the temperature or the pressure constant and verify Boyle's law and the law of Charles and Gay–Lussac.

Steve Niedorf/Getty Images

Figure 19.13 A bottle of champagne is shaken and opened. Liquid spews out of the opening. A common misconception is that the pressure inside the bottle is increased due to the shaking.

A common misconception is that the pressure inside the bottle is increased when the bottle is shaken. On the contrary, because the temperature of the bottle and its contents remains constant as long as the bottle is sealed, so does the pressure, as can be shown by replacing the cork with a pressure gauge. The correct explanation is as follows. Carbon dioxide gas resides in the volume between the liquid surface and the cork. Shaking the bottle displaces some of this carbon dioxide gas into the liquid, where it forms bubbles, and these bubbles become attached to the inside of the bottle. (No new gas is generated by shaking.) When the bottle is opened, the pressure is reduced; this causes the volume of the bubbles to increase suddenly. If the bubbles are attached to the bottle (beneath the liquid surface), their rapid expansion expels liquid from the bottle. If the sides and bottom of the bottle are first tapped until no bubbles remain beneath the surface, then when the champagne is opened, the drop in pressure will not force liquid from the bottle.

The ideal gas law is often expressed in terms of the total number of molecules N. Because the total number of molecules equals the product of the number of moles n and Avogadro's number N_A, we can write Equation 19.8 as

$$PV = nRT = \frac{N}{N_A} RT$$

$$PV = Nk_B T \tag{19.10}$$

where k_B is **Boltzmann's constant**, which has the value

$$k_B = \frac{R}{N_A} = 1.38 \times 10^{-23} \text{ J/K} \tag{19.11}$$

It is common to call quantities such as P, V, and T the **thermodynamic variables** of an ideal gas. If the equation of state is known, then one of the variables can always be expressed as some function of the other two.

> ▲ **PITFALL PREVENTION**
>
> **19.3 So Many *k*'s**
>
> There are a variety of physical quantities for which the letter k is used—we have seen two previously, the force constant for a spring (Chapter 15) and the wave number for a mechanical wave (Chapter 16). Boltzmann's constant is another k, and we will see k used for thermal conductivity in Chapter 20 and for an electrical constant in Chapter 23. In order to make some sense of this confusing state of affairs, we will use a subscript for Boltzmann's constant to help us recognize it. In this book, we will see Boltzmann's constant as k_B, but keep in mind that you may see Boltzmann's constant in other resources as simply k.
>
> **Boltzmann's constant**

Quick Quiz 19.5 A common material for cushioning objects in packages is made by trapping bubbles of air between sheets of plastic. This material is more effective at keeping the contents of the package from moving around inside the package on (a) a hot day   (b) a cold day   (c) either hot or cold days.

Quick Quiz 19.6 A helium-filled rubber balloon is left in a car on a cold winter night. Compared to its size when it was in the warm car the afternoon before, the size the next morning is (a) larger   (b) smaller   (c) unchanged.

Quick Quiz 19.7 On a winter day, you turn on your furnace and the temperature of the air inside your home increases. Assuming that your home has the normal amount of leakage between inside air and outside air, the number of moles of air in your room at the higher temperature is (a) larger than before   (b) smaller than before   (c) the same as before.

Example 19.5   How Many Moles of Gas in a Container?

An ideal gas occupies a volume of 100 cm³ at 20°C and 100 Pa. Find the number of moles of gas in the container.

Solution The quantities given are volume, pressure, and temperature: $V = 100 \text{ cm}^3 = 1.00 \times 10^{-4} \text{ m}^3$, $P = 100$ Pa,

and $T = 20°C = 293$ K. Using Equation 19.8, we find that

$$n = \frac{PV}{RT} = \frac{(100 \text{ Pa})(1.00 \times 10^{-4} \text{ m}^3)}{(8.314 \text{ J/mol·K})(293 \text{ K})}$$

$$= 4.11 \times 10^{-6} \text{ mol}$$

Example 19.6   Filling a Scuba Tank

A certain scuba tank is designed to hold 66.0 ft³ of air when it is at atmospheric pressure at 22°C. When this volume of air is compressed to an absolute pressure of 3 000 lb/in.² and stored in a 10.0-L (0.350-ft³) tank, the air becomes so hot that the tank must be allowed to cool before it can be used. Before the air cools, what is its temperature? (Assume that the air behaves like an ideal gas.)

Solution If no air escapes during the compression, then the number of moles n of air remains constant; therefore, using $PV = nRT$, with n and R constant, we obtain a relationship between the initial and final values:

$$\frac{P_i V_i}{T_i} = \frac{P_f V_f}{T_f}$$

The initial pressure of the air is 14.7 lb/in.², its final pressure is 3 000 lb/in.², and the air is compressed from an initial volume of 66.0 ft³ to a final volume of 0.350 ft³. The initial temperature, converted to SI units, is 295 K. Solving for T_f, we obtain

$$T_f = \left(\frac{P_f V_f}{P_i V_i}\right) T_i = \frac{(3\ 000\ \text{lb/in.}^2)(0.350\ \text{ft}^3)}{(14.7\ \text{lb/in.}^2)(66.0\ \text{ft}^3)}\ (295\ \text{K})$$

$$= \boxed{319\ \text{K}}$$

Example 19.7   Heating a Spray Can   Interactive

A spray can containing a propellant gas at twice atmospheric pressure (202 kPa) and having a volume of 125.00 cm³ is at 22°C. It is then tossed into an open fire. When the temperature of the gas in the can reaches 195°C, what is the pressure inside the can? Assume any change in the volume of the can is negligible.

Solution We employ the same approach we used in Example 19.6, starting with the expression

$$(1) \qquad \frac{P_i V_i}{T_i} = \frac{P_f V_f}{T_f}$$

Because the initial and final volumes of the gas are assumed to be equal, this expression reduces to

$$\frac{P_i}{T_i} = \frac{P_f}{T_f}$$

Solving for P_f gives

$$(2) \qquad P_f = \left(\frac{T_f}{T_i}\right) P_i = \left(\frac{468\ \text{K}}{295\ \text{K}}\right)(202\ \text{kPa}) = \boxed{320\ \text{kPa}}$$

Obviously, the higher the temperature, the higher the pressure exerted by the trapped gas. Of course, if the pressure increases sufficiently, the can will explode. Because of this possibility, you should never dispose of spray cans in a fire.

What If? Suppose we include a volume change due to thermal expansion of the steel can as the temperature in-creases. Does this alter our answer for the final pressure significantly?

Because the thermal expansion coefficient of steel is very small, we do not expect much of an effect on our final answer. The change in the volume of the can is found using Equation 19.6 and the value for α for steel from Table 19.1:

$$\Delta V = \beta V_i\, \Delta T = 3\alpha V_i\, \Delta T$$
$$= 3(11 \times 10^{-6}\,°\text{C}^{-1})(125.00\ \text{cm}^3)(173°\text{C})$$
$$= 0.71\ \text{cm}^3$$

So the final volume of the can is 125.71 cm³. Starting from Equation (1) again, the equation for the final pressure becomes

$$P_f = \left(\frac{T_f}{T_i}\right)\left(\frac{V_i}{V_f}\right) P_i$$

This differs from Equation (2) only in the factor V_i/V_f. Let us evaluate this factor:

$$\frac{V_i}{V_f} = \frac{125.00\ \text{cm}^3}{125.71\ \text{cm}^3} = 0.994 = 99.4\%$$

Thus, the final pressure will differ by only 0.6% from the value we calculated without considering the thermal expansion of the can. Taking 99.4% of the previous final pressure, the final pressure including thermal expansion is 318 kPa.

Explore this situation at the Interactive Worked Example link at **http://www.pse6.com.**

SUMMARY

Take a practice test for this chapter by clicking on the Practice Test link at http://www.pse6.com.

Two objects are in **thermal equilibrium** with each other if they do not exchange energy when in thermal contact.

The **zeroth law of thermodynamics** states that if objects A and B are separately in thermal equilibrium with a third object C, then objects A and B are in thermal equilibrium with each other.

Temperature is the property that determines whether an object is in thermal equilibrium with other objects. **Two objects in thermal equilibrium with each other are at the same temperature.**

The SI unit of absolute temperature is the **kelvin**, which is defined to be the fraction 1/273.16 of the temperature of the triple point of water.

When the temperature of an object is changed by an amount ΔT, its length changes by an amount ΔL that is proportional to ΔT and to its initial length L_i:

$$\Delta L = \alpha L_i \Delta T \qquad (19.4)$$

where the constant α is the **average coefficient of linear expansion**. The **average coefficient of volume expansion** β for a solid is approximately equal to 3α.

An **ideal gas** is one for which PV/nT is constant. An ideal gas is described by the **equation of state,**

$$PV = nRT \qquad (19.8)$$

where n equals the number of moles of the gas, V is its volume, R is the universal gas constant (8.314 J/mol·K), and T is the absolute temperature. A real gas behaves approximately as an ideal gas if it has a low density.

QUESTIONS

1. Is it possible for two objects to be in thermal equilibrium if they are not in contact with each other? Explain.

2. A piece of copper is dropped into a beaker of water. If the water's temperature rises, what happens to the temperature of the copper? Under what conditions are the water and copper in thermal equilibrium?

3. In describing his upcoming trip to the Moon, and as portrayed in the movie *Apollo 13* (Universal, 1995), astronaut Jim Lovell said, "I'll be walking in a place where there's a 400-degree difference between sunlight and shadow." What is it that is hot in sunlight and cold in shadow? Suppose an astronaut standing on the Moon holds a thermometer in his gloved hand. Is it reading the temperature of the vacuum at the Moon's surface? Does it read any temperature? If so, what object or substance has that temperature?

4. Rubber has a negative average coefficient of linear expansion. What happens to the size of a piece of rubber as it is warmed?

5. Explain why a column of mercury in a thermometer first descends slightly and then rises when the thermometer is placed into hot water.

6. Why should the amalgam used in dental fillings have the same average coefficient of expansion as a tooth? What would occur if they were mismatched?

7. Markings to indicate length are placed on a steel tape in a room that has a temperature of 22°C. Are measurements made with the tape on a day when the temperature is 27°C too long, too short, or accurate? Defend your answer.

8. Determine the number of grams in a mole of the following gases: (a) hydrogen (b) helium (c) carbon monoxide.

9. What does the ideal gas law predict about the volume of a sample of gas at absolute zero? Why is this prediction incorrect?

10. An inflated rubber balloon filled with air is immersed in a flask of liquid nitrogen that is at 77 K. Describe what happens to the balloon, assuming that it remains flexible while being cooled.

11. Two identical cylinders at the same temperature each contain the same kind of gas and the same number of moles of gas. If the volume of cylinder A is three times greater than the volume of cylinder B, what can you say about the relative pressures in the cylinders?

12. After food is cooked in a pressure cooker, why is it very important to cool off the container with cold water before attempting to remove the lid?

13. The shore of the ocean is very rocky at a particular place. The rocks form a cave sloping upward from an underwater opening, as shown in Figure Q19.13a. (a) Inside the cave is

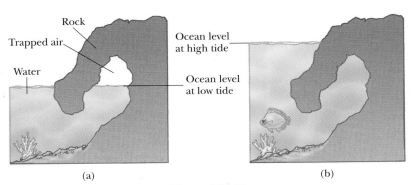

(a)   (b)

Figure Q19.13

a pocket of trapped air. As the level of the ocean rises and falls with the tides, will the level of water in the cave rise and fall? If so, will it have the same amplitude as that of the ocean? (b) **What If?** Now suppose that the cave is deeper in the water, so that it is completely submerged and filled with water at high tide, as shown in Figure Q19.13b. At low tide, will the level of the water in the cave be the same as that of the ocean?

14. In *Colonization: Second Contact* (Harry Turtledove, Ballantine Publishing Group, 1999), the Earth has been partially settled by aliens from another planet, whom humans call Lizards. Laboratory study by humans of Lizard science requires "shifting back and forth between the metric system and the one the Lizards used, which was also based on powers of ten but used different basic quantities for everything but temperature." Why might temperature be an exception?

15. The pendulum of a certain pendulum clock is made of brass. When the temperature increases, does the period of the clock increase, decrease, or remain the same? Explain.

16. An automobile radiator is filled to the brim with water while the engine is cool. What happens to the water when the engine is running and the water is heated? What do modern automobiles have in their cooling systems to prevent the loss of coolants?

17. Metal lids on glass jars can often be loosened by running hot water over them. How is this possible?

18. When the metal ring and metal sphere in Figure Q19.18 are both at room temperature, the sphere can just be passed through the ring. After the sphere is heated, it cannot be passed through the ring. Explain. **What If?** What if the ring is heated and the sphere is left at room temperature? Does the sphere pass through the ring?

Courtesy of Central Scientific Company

Figure Q19.18

PROBLEMS

1, 2, 3 = straightforward, intermediate, challenging ☐ = full solution available in the *Student Solutions Manual and Study Guide*

= coached solution with hints available at http://www.pse6.com 🖳 = computer useful in solving problem

= paired numerical and symbolic problems

Section 19.2 Thermometers and the Celsius Temperature Scale

Section 19.3 The Constant-Volume Gas Thermometer and the Absolute Temperature Scale

1. A constant-volume gas thermometer is calibrated in dry ice (that is, carbon dioxide in the solid state, which has a temperature of $-80.0°C$) and in boiling ethyl alcohol ($78.0°C$). The two pressures are 0.900 atm and 1.635 atm. (a) What Celsius value of absolute zero does the calibration yield? What is the pressure at (b) the freezing point of water and (c) the boiling point of water?

2. In a constant-volume gas thermometer, the pressure at $20.0°C$ is 0.980 atm. (a) What is the pressure at $45.0°C$? (b) What is the temperature if the pressure is 0.500 atm?

3. Liquid nitrogen has a boiling point of $-195.81°C$ at atmospheric pressure. Express this temperature (a) in degrees Fahrenheit and (b) in kelvins.

4. Convert the following to equivalent temperatures on the Celsius and Kelvin scales: (a) the normal human body temperature, $98.6°F$; (b) the air temperature on a cold day, $-5.00°F$.

5. The temperature difference between the inside and the outside of an automobile engine is $450°C$. Express this temperature difference on (a) the Fahrenheit scale and (b) the Kelvin scale.

6. On a Strange temperature scale, the freezing point of water is $-15.0°S$ and the boiling point is $+60.0°S$. Develop a *linear* conversion equation between this temperature scale and the Celsius scale.

7. The melting point of gold is $1\,064°C$, and the boiling point is $2\,660°C$. (a) Express these temperatures in kelvins. (b) Compute the difference between these temperatures in Celsius degrees and kelvins.

Section 19.4 Thermal Expansion of Solids and Liquids

Note: Table 19.1 is available for use in solving problems in this section.

8. The New River Gorge bridge in West Virginia is a steel arch bridge 518 m in length. How much does the total length of the roadway decking change between temperature extremes of $-20.0°C$ and $35.0°C$? The result indicates the

size of the expansion joints that must be built into the structure.

9. A copper telephone wire has essentially no sag between poles 35.0 m apart on a winter day when the temperature is $-20.0°C$. How much longer is the wire on a summer day when $T_C = 35.0°C$?

10. The concrete sections of a certain superhighway are designed to have a length of 25.0 m. The sections are poured and cured at 10.0°C. What minimum spacing should the engineer leave between the sections to eliminate buckling if the concrete is to reach a temperature of 50.0°C?

11. A pair of eyeglass frames is made of epoxy plastic. At room temperature (20.0°C), the frames have circular lens holes 2.20 cm in radius. To what temperature must the frames be heated if lenses 2.21 cm in radius are to be inserted in them? The average coefficient of linear expansion for epoxy is 1.30×10^{-4} (°C)$^{-1}$.

12. Each year thousands of children are badly burned by hot tap water. Figure P19.12 shows a cross-sectional view of an antiscalding faucet attachment designed to prevent such accidents. Within the device, a spring made of material with a high coefficient of thermal expansion controls a movable plunger. When the water temperature rises above a preset safe value, the expansion of the spring causes the plunger to shut off the water flow. If the initial length L of the unstressed spring is 2.40 cm and its coefficient of linear expansion is 22.0×10^{-6} (°C)$^{-1}$, determine the increase in length of the spring when the water temperature rises by 30.0°C. (You will find the increase in length to be small. For this reason actual devices have a more complicated mechanical design, to provide a greater variation in valve opening for the temperature change anticipated.)

Figure P19.12

13. The active element of a certain laser is made of a glass rod 30.0 cm long by 1.50 cm in diameter. If the temperature of the rod increases by 65.0°C, what is the increase in (a) its length, (b) its diameter, and (c) its volume? Assume that the average coefficient of linear expansion of the glass is 9.00×10^{-6} (°C)$^{-1}$.

14. Review problem. Inside the wall of a house, an L-shaped section of hot-water pipe consists of a straight horizontal piece 28.0 cm long, an elbow, and a straight vertical piece 134 cm long (Figure P19.14). A stud and a second-story

floorboard hold stationary the ends of this section of copper pipe. Find the magnitude and direction of the displacement of the pipe elbow when the water flow is turned on, raising the temperature of the pipe from 18.0°C to 46.5°C.

Figure P19.14

15. A brass ring of diameter 10.00 cm at 20.0°C is heated and slipped over an aluminum rod of diameter 10.01 cm at 20.0°C. Assuming the average coefficients of linear expansion are constant, (a) to what temperature must this combination be cooled to separate them? Is this attainable? (b) **What If?** What if the aluminum rod were 10.02 cm in diameter?

16. A square hole 8.00 cm along each side is cut in a sheet of copper. (a) Calculate the change in the area of this hole if the temperature of the sheet is increased by 50.0 K. (b) Does this change represent an increase or a decrease in the area enclosed by the hole?

17. The average coefficient of volume expansion for carbon tetrachloride is 5.81×10^{-4} (°C)$^{-1}$. If a 50.0-gal steel container is filled completely with carbon tetrachloride when the temperature is 10.0°C, how much will spill over when the temperature rises to 30.0°C?

18. At 20.0°C, an aluminum ring has an inner diameter of 5.000 0 cm and a brass rod has a diameter of 5.050 0 cm. (a) If only the ring is heated, what temperature must it reach so that it will just slip over the rod? (b) **What If?** If both are heated together, what temperature must they both reach so that the ring just slips over the rod? Would this latter process work?

19. A volumetric flask made of Pyrex is calibrated at 20.0°C. It is filled to the 100-mL mark with 35.0°C acetone. (a) What is the volume of the acetone when it cools to 20.0°C? (b) How significant is the change in volume of the flask?

20. A concrete walk is poured on a day when the temperature is 20.0°C in such a way that the ends are unable to move. (a) What is the stress in the cement on a hot day of 50.0°C? (b) Does the concrete fracture? Take Young's modulus for concrete to be 7.00×10^9 N/m² and the compressive strength to be 2.00×10^9 N/m².

21. A hollow aluminum cylinder 20.0 cm deep has an internal capacity of 2.000 L at 20.0°C. It is completely filled with turpentine and then slowly warmed to 80.0°C. (a) How much turpentine overflows? (b) If the cylinder is then cooled back to 20.0°C, how far below the cylinder's rim does the turpentine's surface recede?

22. A beaker made of ordinary glass contains a lead sphere of diameter 4.00 cm firmly attached to its bottom. At a uniform temperature of −10.0°C, the beaker is filled to the brim with 118 cm³ of mercury, which completely covers the sphere. How much mercury overflows from the beaker if the temperature is raised to 30.0°C?

23. A steel rod undergoes a stretching force of 500 N. Its cross-sectional area is 2.00 cm². Find the change in temperature that would elongate the rod by the same amount as the 500-N force does. Tables 12.1 and 19.1 are available to you.

24. The Golden Gate Bridge in San Francisco has a main span of length 1.28 km—one of the longest in the world. Imagine that a taut steel wire with this length and a cross-sectional area of 4.00×10^{-6} m² is laid on the bridge deck with its ends attached to the towers of the bridge, on a summer day when the temperature of the wire is 35.0°C. (a) When winter arrives, the towers stay the same distance apart and the bridge deck keeps the same shape as its expansion joints open. When the temperature drops to −10.0°C, what is the tension in the wire? Take Young's modulus for steel to be 20.0×10^{10} N/m². (b) Permanent deformation occurs if the stress in the steel exceeds its elastic limit of 3.00×10^8 N/m². At what temperature would this happen? (c) **What If?** How would your answers to (a) and (b) change if the Golden Gate Bridge were twice as long?

25. A certain telescope forms an image of part of a cluster of stars on a square silicon charge-coupled detector (CCD) chip 2.00 cm on each side. A star field is focused on the CCD chip when it is first turned on and its temperature is 20.0°C. The star field contains 5 342 stars scattered uniformly. To make the detector more sensitive, it is cooled to −100°C. How many star images then fit onto the chip? The average coefficient of linear expansion of silicon is 4.68×10^{-6} (°C)$^{-1}$.

Section 19.5 Macroscopic Description of an Ideal Gas

Note: Problem 8 in Chapter 1 can be assigned with this section.

26. Gas is contained in an 8.00-L vessel at a temperature of 20.0°C and a pressure of 9.00 atm. (a) Determine the number of moles of gas in the vessel. (b) How many molecules are there in the vessel?

27. An automobile tire is inflated with air originally at 10.0°C and normal atmospheric pressure. During the process, the air is compressed to 28.0% of its original volume and the temperature is increased to 40.0°C. (a) What is the tire pressure? (b) After the car is driven at high speed, the tire air temperature rises to 85.0°C and the interior volume of the tire increases by 2.00%. What is the new tire pressure (absolute) in pascals?

28. A tank having a volume of 0.100 m³ contains helium gas at 150 atm. How many balloons can the tank blow up if each filled balloon is a sphere 0.300 m in diameter at an absolute pressure of 1.20 atm?

29. An auditorium has dimensions 10.0 m × 20.0 m × 30.0 m. How many molecules of air fill the auditorium at 20.0°C and a pressure of 101 kPa?

30. Imagine a baby alien playing with a spherical balloon the size of the Earth in the outer solar system. Helium gas inside the balloon has a uniform temperature of 50.0 K due to radiation from the Sun. The uniform pressure of the helium is equal to normal atmospheric pressure on Earth. (a) Find the mass of the gas in the balloon. (b) The baby blows an additional mass of 8.00×10^{20} kg of helium into the balloon. At the same time, she wanders closer to the Sun and the pressure in the balloon doubles. Find the new temperature inside the balloon, whose volume remains constant.

31. Just 9.00 g of water is placed in a 2.00-L pressure cooker and heated to 500°C. What is the pressure inside the container?

32. One mole of oxygen gas is at a pressure of 6.00 atm and a temperature of 27.0°C. (a) If the gas is heated at constant volume until the pressure triples, what is the final temperature? (b) If the gas is heated until both the pressure and volume are doubled, what is the final temperature?

33. The mass of a hot-air balloon and its cargo (not including the air inside) is 200 kg. The air outside is at 10.0°C and 101 kPa. The volume of the balloon is 400 m³. To what temperature must the air in the balloon be heated before the balloon will lift off? (Air density at 10.0°C is 1.25 kg/m³.)

34. Your father and your little brother are confronted with the same puzzle. Your father's garden sprayer and your brother's water cannon both have tanks with a capacity of 5.00 L (Figure P19.34). Your father inserts a negligible amount of concentrated insecticide into his tank. They both pour in 4.00 L of water and seal up their tanks, so that they also contain air at atmospheric pressure. Next, each uses a hand-operated piston pump to inject more air, until the absolute pressure in the tank reaches 2.40 atm and it becomes too difficult to move the pump handle. Now each uses his device to spray out water—not air—until the stream becomes feeble, as it does when the pressure in the tank reaches 1.20 atm. Then he must pump it up again, spray again, and so on. In order to spray out all the water, each finds that he must pump up the tank three

times. This is the puzzle: most of the water sprays out as a result of the second pumping. The first and the third pumping-up processes seem just as difficult, but result in a disappointingly small amount of water coming out. Account for this phenomenon.

Figure P19.34

35. (a) Find the number of moles in one cubic meter of an ideal gas at 20.0°C and atmospheric pressure. (b) For air, Avogadro's number of molecules has mass 28.9 g. Calculate the mass of one cubic meter of air. Compare the result with the tabulated density of air.

36. The *void fraction* of a porous medium is the ratio of the void volume to the total volume of the material. The void is the hollow space within the material; it may be filled with a fluid. A cylindrical canister of diameter 2.54 cm and height 20.0 cm is filled with activated carbon having a void fraction of 0.765. Then it is flushed with an ideal gas at 25.0°C and pressure 12.5 atm. How many moles of gas are contained in the cylinder at the end of this process?

37. A cube 10.0 cm on each edge contains air (with equivalent molar mass 28.9 g/mol) at atmospheric pressure and temperature 300 K. Find (a) the mass of the gas, (b) its weight, and (c) the force it exerts on each face of the cube. (d) Comment on the physical reason why such a small sample can exert such a great force.

38. At 25.0 m below the surface of the sea ($\rho = 1\,025$ kg/m^3), where the temperature is 5.00°C, a diver exhales an air bubble having a volume of 1.00 cm^3. If the surface temperature of the sea is 20.0°C, what is the volume of the bubble just before it breaks the surface?

39. The pressure gauge on a tank registers the gauge pressure, which is the difference between the interior and exterior pressure. When the tank is full of oxygen (O_2), it contains 12.0 kg of the gas at a gauge pressure of 40.0 atm. Determine the mass of oxygen that has been withdrawn from the tank when the pressure reading is 25.0 atm. Assume that the temperature of the tank remains constant.

40. Estimate the mass of the air in your bedroom. State the quantities you take as data and the value you measure or estimate for each.

41. A popular brand of cola contains 6.50 g of carbon dioxide dissolved in 1.00 L of soft drink. If the evaporating carbon dioxide is trapped in a cylinder at 1.00 atm and 20.0°C, what volume does the gas occupy?

42. In state-of-the-art vacuum systems, pressures as low as 10^{-9} Pa are being attained. Calculate the number of molecules in a 1.00-m^3 vessel at this pressure if the temperature is 27.0°C.

43. A room of volume V contains air having equivalent molar mass M (in g/mol). If the temperature of the room is raised from T_1 to T_2, what mass of air will leave the room? Assume that the air pressure in the room is maintained at P_0.

44. A diving bell in the shape of a cylinder with a height of 2.50 m is closed at the upper end and open at the lower end. The bell is lowered from air into sea water ($\rho = 1.025$ g/cm^3). The air in the bell is initially at 20.0°C. The bell is lowered to a depth (measured to the bottom of the bell) of 45.0 fathoms or 82.3 m. At this depth the water temperature is 4.0°C, and the bell is in thermal equilibrium with the water. (a) How high does sea water rise in the bell? (b) To what minimum pressure must the air in the bell be raised to expel the water that entered?

Additional Problems

45. A student measures the length of a brass rod with a steel tape at 20.0°C. The reading is 95.00 cm. What will the tape indicate for the length of the rod when the rod and the tape are at (a) − 15.0°C and (b) 55.0°C?

46. The density of gasoline is 730 kg/m^3 at 0°C. Its average coefficient of volume expansion is 9.60×10^{-4}/°C. If 1.00 gal of gasoline occupies 0.003 80 m^3, how many extra kilograms of gasoline would you get if you bought 10.0 gal of gasoline at 0°C rather than at 20.0°C from a pump that is not temperature compensated?

47. A mercury thermometer is constructed as shown in Figure P19.47. The capillary tube has a diameter of 0.004 00 cm,

Figure P19.47 Problems 47 and 48.

and the bulb has a diameter of 0.250 cm. Neglecting the expansion of the glass, find the change in height of the mercury column that occurs with a temperature change of 30.0°C.

48. A liquid with a coefficient of volume expansion β just fills a spherical shell of volume V_i at a temperature of T_i (see Fig. P19.47). The shell is made of a material that has an average coefficient of linear expansion α. The liquid is free to expand into an open capillary of area A projecting from the top of the sphere. (a) If the temperature increases by ΔT, show that the liquid rises in the capillary by the amount Δh given by $\Delta h = (V_i/A)(\beta - 3\alpha)\Delta T$. (b) For a typical system, such as a mercury thermometer, why is it a good approximation to neglect the expansion of the shell?

49. **Review problem.** An aluminum pipe, 0.655 m long at 20.0°C and open at both ends, is used as a flute. The pipe is cooled to a low temperature but then is filled with air at 20.0°C as soon as you start to play it. After that, by how much does its fundamental frequency change as the metal rises in temperature from 5.00°C to 20.0°C?

50. A cylinder is closed by a piston connected to a spring of constant 2.00×10^3 N/m (see Fig. P19.50). With the spring relaxed, the cylinder is filled with 5.00 L of gas at a pressure of 1.00 atm and a temperature of 20.0°C. (a) If the piston has a cross-sectional area of 0.010 0 m^2 and negligible mass, how high will it rise when the temperature is raised to 250°C? (b) What is the pressure of the gas at 250°C?

250°C

20°C

h

Figure P19.50

51. A liquid has a density ρ. (a) Show that the fractional change in density for a change in temperature ΔT is $\Delta\rho/\rho = -\beta\,\Delta T$. What does the negative sign signify? (b) Fresh water has a maximum density of 1.000 0 g/cm^3 at 4.0°C. At 10.0°C, its density is 0.999 7 g/cm^3. What is β for water over this temperature interval?

52. Long-term space missions require reclamation of the oxygen in the carbon dioxide exhaled by the crew. In one method of reclamation, 1.00 mol of carbon dioxide produces 1.00 mol of oxygen and 1.00 mol of methane as a byproduct. The methane is stored in a tank under pressure and is available to control the attitude of the spacecraft by controlled venting. A single astronaut exhales 1.09 kg of carbon dioxide each day. If the methane gen-

erated in the respiration recycling of three astronauts during one week of flight is stored in an originally empty 150-L tank at -45.0°C, what is the final pressure in the tank?

53. A vertical cylinder of cross-sectional area A is fitted with a tight-fitting, frictionless piston of mass m (Fig. P19.53). (a) If n moles of an ideal gas are in the cylinder at a temperature of T, what is the height h at which the piston is in equilibrium under its own weight? (b) What is the value for h if $n = 0.200$ mol, $T = 400$ K, $A = 0.008\ 00$ m^2, and $m = 20.0$ kg?

m

Gas

h

Figure P19.53

54. A bimetallic strip is made of two ribbons of dissimilar metals bonded together. (a) First assume the strip is originally straight. As they are heated, the metal with the greater average coefficient of expansion expands more than the other, forcing the strip into an arc, with the outer radius having a greater circumference (Fig. P19.54a). Derive an expression for the angle of bending θ as a function of the initial length of the strips, their average coefficients of linear expansion, the change in temperature, and the separation of the centers of the strips ($\Delta r = r_2 - r_1$). (b) Show that the angle of bending decreases to zero when ΔT decreases to zero and also when the two average coefficients of expansion become equal. (c) **What If?** What happens if the strip is cooled? (d) Figure P19.54b shows a compact spiral bimetallic strip in a home thermostat. The equation from part (a) applies to it as well, if θ is interpreted as the angle of additional bending caused by a change in temperature. The inner end of the spiral strip is fixed, and the outer end is free to move. Assume the metals are bronze and invar, the thickness of the strip is $2\,\Delta r = 0.500$ mm, and the overall length of the spiral strip is 20.0 cm. Find the angle through which the free end of the strip turns when the temperature changes by one Celsius degree. The free end of the strip supports a capsule partly filled with mercury, visible above the strip in Figure P19.54b. When the capsule tilts, the mercury shifts from one end to the other, to make or break an electrical contact switching the furnace on or off.

Charles D. Winters

(b)

Figure P19.54

55. The rectangular plate shown in Figure P19.55 has an area A_i equal to ℓw. If the temperature increases by ΔT, each dimension increases according to the equation $\Delta L = \alpha L_i \, \Delta T$, where α is the average coefficient of linear expansion. Show that the increase in area is $\Delta A = 2\alpha A_i \, \Delta T$. What approximation does this expression assume?

Figure P19.55

56. **Review problem**. A clock with a brass pendulum has a period of 1.000 s at 20.0°C. If the temperature increases to 30.0°C, (a) by how much does the period change, and (b) how much time does the clock gain or lose in one week?

57. **Review problem**. Consider an object with any one of the shapes displayed in Table 10.2. What is the percentage increase in the moment of inertia of the object when it is heated from 0°C to 100°C if it is composed of (a) copper or (b) aluminum? Assume that the average linear expansion coefficients shown in Table 19.1 do not vary between 0°C and 100°C.

58. (a) Derive an expression for the buoyant force on a spherical balloon, submerged in water, as a function of the depth

below the surface, the volume of the balloon at the surface, the pressure at the surface, and the density of the water. (Assume water temperature does not change with depth.) (b) Does the buoyant force increase or decrease as the balloon is submerged? (c) At what depth is the buoyant force half the surface value?

59. A copper wire and a lead wire are joined together, end to end. The compound wire has an effective coefficient of linear expansion of 20.0×10^{-6} (°C)$^{-1}$. What fraction of the length of the compound wire is copper?

60. **Review problem.** Following a collision in outer space, a copper disk at 850°C is rotating about its axis with an angular speed of 25.0 rad/s. As the disk radiates infrared light, its temperature falls to 20.0°C. No external torque acts on the disk. (a) Does the angular speed change as the disk cools off? Explain why. (b) What is its angular speed at the lower temperature?

61. Two concrete spans of a 250-m-long bridge are placed end to end so that no room is allowed for expansion (Fig. P19.61a). If a temperature increase of 20.0°C occurs, what is the height y to which the spans rise when they buckle (Fig. P19.61b)?

(a)   (b)

Figure P19.61 Problems 61 and 62.

62. Two concrete spans of a bridge of length L are placed end to end so that no room is allowed for expansion (Fig. P19.61a). If a temperature increase of ΔT occurs, what is the height y to which the spans rise when they buckle (Fig. P19.61b)?

63. (a) Show that the density of an ideal gas occupying a volume V is given by $\rho = PM/RT$, where M is the molar mass. (b) Determine the density of oxygen gas at atmospheric pressure and 20.0°C.

64. (a) Use the equation of state for an ideal gas and the definition of the coefficient of volume expansion, in the form $\beta = (1/V) \, dV/dT$, to show that the coefficient of volume expansion for an ideal gas at constant pressure is given by $\beta = 1/T$, where T is the absolute temperature. (b) What value does this expression predict for β at 0°C? Compare this result with the experimental values for helium and air in Table 19.1. Note that these are much larger than the coefficients of volume expansion for most liquids and solids.

65. Starting with Equation 19.10, show that the total pressure P in a container filled with a mixture of several ideal gases is $P = P_1 + P_2 + P_3 + \cdots$, where $P_1, P_2, \ldots$, are the pressures that each gas would exert if it alone filled the container (these individual pressures are called the *partial pres-*

sures of the respective gases). This result is known as *Dalton's law of partial pressures.*

66. A sample of dry air that has a mass of 100.00 g, collected at sea level, is analyzed and found to consist of the following gases:

nitrogen (N_2) = 75.52 g

oxygen (O_2) = 23.15 g

argon (Ar) = 1.28 g

carbon dioxide (CO_2) = 0.05 g

plus trace amounts of neon, helium, methane, and other gases. (a) Calculate the partial pressure (see Problem 65) of each gas when the pressure is 1.013×10^5 Pa. (b) Determine the volume occupied by the 100-g sample at a temperature of 15.00°C and a pressure of 1.00 atm. What is the density of the air for these conditions? (c) What is the effective molar mass of the air sample?

67. Helium gas is sold in steel tanks. If the helium is used to inflate a balloon, could the balloon lift the spherical tank the helium came in? Justify your answer. Steel will rupture if subjected to tensile stress greater than its yield strength of 5×10^8 N/m². *Suggestion:* You may consider a steel shell of radius r and thickness t containing helium at high pressure and on the verge of breaking apart into two hemispheres.

68. A cylinder that has a 40.0-cm radius and is 50.0 cm deep is filled with air at 20.0°C and 1.00 atm (Fig. P19.68a). A 20.0-kg piston is now lowered into the cylinder, compressing the air trapped inside (Fig. P19.68b). Finally, a 75.0-kg man stands on the piston, further compressing the air, which remains at 20°C (Fig. P19.68c). (a) How far down (Δh) does the piston move when the man steps onto it? (b) To what temperature should the gas be heated to raise the piston and man back to h_i?

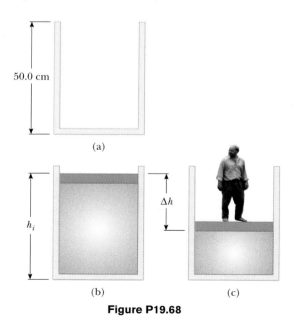

(a)

(b)   (c)

Figure P19.68

69. The relationship $L_f = L_i(1 + \alpha \Delta T)$ is an approximation that works when the average coefficient of expansion is

small. If α is large, one must integrate the relationship $dL/dT = \alpha L$ to determine the final length. (a) Assuming that the coefficient of linear expansion is constant as L varies, determine a general expression for the final length. (b) Given a rod of length 1.00 m and a temperature change of 100.0°C, determine the error caused by the approximation when $\alpha = 2.00 \times 10^{-5}$ (°C)$^{-1}$ (a typical value for a metal) and when $\alpha = 0.020\,0$ (°C)$^{-1}$ (an unrealistically large value for comparison).

70. A steel wire and a copper wire, each of diameter 2.000 mm, are joined end to end. At 40.0°C, each has an unstretched length of 2.000 m; they are connected between two fixed supports 4.000 m apart on a tabletop, so that the steel wire extends from $x = -2.000$ m to $x = 0$, the copper wire extends from $x = 0$ to $x = 2.000$ m, and the tension is negligible. The temperature is then lowered to 20.0°C. At this lower temperature, find the tension in the wire and the x coordinate of the junction between the wires. (Refer to Tables 12.1 and 19.1.)

71. **Review problem.** A steel guitar string with a diameter of 1.00 mm is stretched between supports 80.0 cm apart. The temperature is 0.0°C. (a) Find the mass per unit length of this string. (Use the value 7.86×10^3 kg/m³ for the density.) (b) The fundamental frequency of transverse oscillations of the string is 200 Hz. What is the tension in the string? (c) If the temperature is raised to 30.0°C, find the resulting values of the tension and the fundamental frequency. Assume that both the Young's modulus (Table 12.1) and the average coefficient of expansion (Table 19.1) have constant values between 0.0°C and 30.0°C.

72. In a chemical processing plant, a reaction chamber of fixed volume V_0 is connected to a reservoir chamber of fixed volume $4V_0$ by a passage containing a thermally insulating porous plug. The plug permits the chambers to be at different temperatures. The plug allows gas to pass from either chamber to the other, ensuring that the pressure is the same in both. At one point in the processing, both chambers contain gas at a pressure of 1.00 atm and a temperature of 27.0°C. Intake and exhaust valves to the pair of chambers are closed. The reservoir is maintained at 27.0°C while the reaction chamber is heated to 400°C. What is the pressure in both chambers after this is done?

73. 🖥 A 1.00-km steel railroad rail is fastened securely at both ends when the temperature is 20.0°C. As the temperature increases, the rail begins to buckle. If its shape is an arc of a vertical circle, find the height h of the center of the rail when the temperature is 25.0°C. You will need to solve a transcendental equation.

74. **Review problem.** A perfectly plane house roof makes an angle θ with the horizontal. When its temperature changes, between T_c before dawn each day to T_h in the middle of each afternoon, the roof expands and contracts uniformly with a coefficient of thermal expansion α_1. Resting on the roof is a flat rectangular metal plate with expansion coefficient α_2, greater than α_1. The length of the plate is L, measured up the slope of the roof. The component of the plate's weight perpendicular to the roof is supported by a normal force uniformly distributed over the area of the plate. The coefficient of kinetic friction between the plate and the roof is μ_k. The plate is always at the same tempera-

ture as the roof, so we assume its temperature is continuously changing. Because of the difference in expansion coefficients, each bit of the plate is moving relative to the roof below it, except for points along a certain horizontal line running across the plate. We call this the stationary line. If the temperature is rising, parts of the plate below the stationary line are moving down relative to the roof and feel a force of kinetic friction acting up the roof. Elements of area above the stationary line are sliding up the roof and on them kinetic friction acts downward parallel to the roof. The stationary line occupies no area, so we assume no force of static friction acts on the plate while the temperature is changing. The plate as a whole is very nearly in equilibrium, so the net friction force on it must be equal to the component of its weight acting down the incline. (a) Prove that the stationary line is at a distance of

$$\frac{L}{2}\left(1 - \frac{\tan\theta}{\mu_k}\right)$$

below the top edge of the plate. (b) Analyze the forces that act on the plate when the temperature is falling, and prove that the stationary line is at that same distance above the bottom edge of the plate. (c) Show that the plate steps down the roof like an inchworm, moving each day by the distance

$$\frac{L(\alpha_2 - \alpha_1)(T_h - T_c)\tan\theta}{\mu_k}$$

(d) Evaluate the distance an aluminum plate moves each day if its length is 1.20 m, if the temperature cycles between 4.00°C and 36.0°C, and if the roof has slope 18.5°,

coefficient of linear expansion 1.50×10^{-5} (°C)$^{-1}$, and coefficient of friction 0.420 with the plate. (e) **What If?** What if the expansion coefficient of the plate is less than that of the roof? Will the plate creep up the roof?

Answers to Quick Quizzes

19.1 (c). The direction of the transfer of energy depends only on temperature and not on the size of the object or on which object has more mass.

19.2 (c). The phrase "twice as hot" refers to a ratio of temperatures. When the given temperatures are converted to kelvins, only those in part (c) are in the correct ratio.

19.3 (c). Gasoline has the largest average coefficient of volume expansion.

19.4 (c). A cavity in a material expands in the same way as if it were filled with material.

19.5 (a). On a cold day, the trapped air in the bubbles is reduced in pressure, according to the ideal gas law. Thus, the volume of the bubbles may be smaller than on a hot day, and the package contents can shift more.

19.6 (b). Because of the decreased temperature of the helium, the pressure in the balloon is reduced. The atmospheric pressure around the balloon then compresses it to a smaller size until the pressure in the balloon reaches the atmospheric pressure.

19.7 (b). Because of the increased temperature, the air expands. Consequently, some of the air leaks to the outside, leaving less air in the house.

Chapter 20

Heat and the First Law of Thermodynamics

▲ *In this photograph of Bow Lake in Banff National Park, Alberta, we see evidence of water in all three phases. In the lake is liquid water, and solid water in the form of snow appears on the ground. The clouds in the sky consist of liquid water droplets that have condensed from the gaseous water vapor in the air. Changes of a substance from one phase to another are a result of energy transfer. (Jacob Taposchaner/Getty Images)*

Until about 1850, the fields of thermodynamics and mechanics were considered to be two distinct branches of science, and the law of conservation of energy seemed to describe only certain kinds of mechanical systems. However, mid-nineteenth-century experiments performed by the Englishman James Joule and others showed that there was a strong connection between the transfer of energy by heat in thermal processes and the transfer of energy by work in mechanical processes. Today we know that internal energy, which we formally define in this chapter, can be transformed to mechanical energy. Once the concept of energy was generalized from mechanics to include internal energy, the law of conservation of energy emerged as a universal law of nature.

This chapter focuses on the concept of internal energy, the processes by which energy is transferred, the first law of thermodynamics, and some of the important applications of the first law. The first law of thermodynamics is a statement of conservation of energy. It describes systems in which the only energy change is that of internal energy and the transfers of energy are by heat and work. Furthermore, the first law makes no distinction between the results of heat and the results of work. According to the first law, a system's internal energy can be changed by an energy transfer to or from the system either by heat or by work. A major difference in our discussion of work in this chapter from that in the chapters on mechanics is that we will consider work done on *deformable* systems.

20.1 Heat and Internal Energy

At the outset, it is important that we make a major distinction between internal energy and heat. **Internal energy is all the energy of a system that is associated with its microscopic components—atoms and molecules—when viewed from a reference frame at rest with respect to the center of mass of the system.** The last part of this sentence ensures that any bulk kinetic energy of the system due to its motion through space is not included in internal energy. Internal energy includes kinetic energy of random translational, rotational, and vibrational motion of molecules, potential energy within molecules, and potential energy between molecules. It is useful to relate internal energy to the temperature of an object, but this relationship is limited—we show in Section 20.3 that internal energy changes can also occur in the absence of temperature changes.

Heat is defined as the transfer of energy across the boundary of a system due to a temperature difference between the system and its surroundings. When you *heat* a substance, you are transferring energy into it by placing it in contact with surroundings that have a higher temperature. This is the case, for example, when you place a pan of cold water on a stove burner—the burner is at a higher temperature than the water, and so the water gains energy. We shall also use the term *heat* to represent the amount of energy transferred by this method.

Scientists used to think of heat as a fluid called *caloric,* which they believed was transferred between objects; thus, they defined heat in terms of the temperature

20.2 Heat, Temperature, and Internal Energy Are Different

As you read the newspaper or listen to the radio, be alert for incorrectly used phrases including the word *heat,* and think about the proper word to be used in place of *heat.* Incorrect examples include "As the truck braked to a stop, a large amount of heat was generated by friction" and "The heat of a hot summer day . . ."

James Prescott Joule
British physicist (1818–1889)

Joule received some formal education in mathematics, philosophy, and chemistry from John Dalton but was in large part self-educated. Joule's research led to the establishment of the principle of conservation of energy. His study of the quantitative relationship among electrical, mechanical, and chemical effects of heat culminated in his announcement in 1843 of the amount of work required to produce a unit of energy, called the mechanical equivalent of heat. *(By kind permission of the President and Council of the Royal Society)*

changes produced in an object during heating. Today we recognize the distinct difference between internal energy and heat. Nevertheless, we refer to quantities using names that do not quite correctly define the quantities but which have become entrenched in physics tradition based on these early ideas. Examples of such quantities are *heat capacity* and *latent heat* (Sections 20.2 and 20.3).

As an analogy to the distinction between heat and internal energy, consider the distinction between work and mechanical energy discussed in Chapter 7. The work done on a system is a measure of the amount of energy transferred to the system from its surroundings, whereas the mechanical energy of the system (kinetic plus potential) is a consequence of the motion and configuration of the system. Thus, when a person does work on a system, energy is transferred from the person to the system. It makes no sense to talk about the work *of* a system—one can refer only to the work done *on* or *by* a system when some process has occurred in which energy has been transferred to or from the system. Likewise, it makes no sense to talk about the heat *of* a system—one can refer to *heat* only when energy has been transferred as a result of a temperature difference. Both heat and work are ways of changing the energy of a system.

It is also important to recognize that the internal energy of a system can be changed even when no energy is transferred by heat. For example, when a gas in an insulated container is compressed by a piston, the temperature of the gas and its internal energy increase, but no transfer of energy by heat from the surroundings to the gas has occurred. If the gas then expands rapidly, it cools and its internal energy decreases, but no transfer of energy by heat from it to the surroundings has taken place. The temperature changes in the gas are due not to a difference in temperature between the gas and its surroundings but rather to the compression and the expansion. In each case, energy is transferred to or from the gas by *work.* The changes in internal energy in these examples are evidenced by corresponding changes in the temperature of the gas.

Units of Heat

As we have mentioned, early studies of heat focused on the resultant increase in temperature of a substance, which was often water. The early notions of heat based on caloric suggested that the flow of this fluid from one substance to another caused changes in temperature. From the name of this mythical fluid, we have an energy unit related to thermal processes, the **calorie (cal),** which is defined as **the amount of energy transfer necessary to raise the temperature of 1 g of water from 14.5°C to 15.5°C.**[1] (Note that the "Calorie," written with a capital "C" and used in describing the energy content of foods, is actually a kilocalorie.) The unit of energy in the U.S. customary system is the **British thermal unit (Btu),** which is defined as **the amount of energy transfer required to raise the temperature of 1 lb of water from 63°F to 64°F.**

Scientists are increasingly using the SI unit of energy, the *joule,* when describing thermal processes. In this textbook, heat, work, and internal energy are usually measured in joules. (Note that both heat and work are measured in energy units. Do not confuse these two means of energy *transfer* with energy itself, which is also measured in joules.)

The Mechanical Equivalent of Heat

In Chapters 7 and 8, we found that whenever friction is present in a mechanical system, some mechanical energy is lost—in other words, mechanical energy is not conserved in the presence of nonconservative forces. Various experiments show that this lost mechanical energy does not simply disappear but is transformed into internal

---

[1] Originally, the calorie was defined as the "heat" necessary to raise the temperature of 1 g of water by 1°C. However, careful measurements showed that the amount of energy required to produce a 1°C change depends somewhat on the initial temperature; hence, a more precise definition evolved.

energy. We can perform such an experiment at home by simply hammering a nail into a scrap piece of wood. What happens to all the kinetic energy of the hammer once we have finished? Some of it is now in the nail as internal energy, as demonstrated by the fact that the nail is measurably warmer. Although this connection between mechanical and internal energy was first suggested by Benjamin Thompson, it was Joule who established the equivalence of these two forms of energy.

A schematic diagram of Joule's most famous experiment is shown in Figure 20.1. The system of interest is the water in a thermally insulated container. Work is done on the water by a rotating paddle wheel, which is driven by heavy blocks falling at a constant speed. The temperature of the stirred water increases due to the friction between it and the paddles. If the energy lost in the bearings and through the walls is neglected, then the loss in potential energy associated with the blocks equals the work done by the paddle wheel on the water. If the two blocks fall through a distance h, the loss in potential energy is $2mgh$, where m is the mass of one block; this energy causes the temperature of the water to increase. By varying the conditions of the experiment, Joule found that the loss in mechanical energy $2mgh$ is proportional to the increase in water temperature ΔT. The proportionality constant was found to be approximately 4.18 J/g · °C. Hence, 4.18 J of mechanical energy raises the temperature of 1 g of water by 1°C. More precise measurements taken later demonstrated the proportionality to be 4.186 J/g · °C when the temperature of the water was raised from 14.5°C to 15.5°C. We adopt this "15-degree calorie" value:

$$1 \text{ cal} \equiv 4.186 \text{ J} \qquad (20.1)$$

This equality is known, for purely historical reasons, as the **mechanical equivalent of heat.**

Figure 20.1 Joule's experiment for determining the mechanical equivalent of heat. The falling blocks rotate the paddles, causing the temperature of the water to increase.

Example 20.1 Losing Weight the Hard Way

A student eats a dinner rated at 2 000 Calories. He wishes to do an equivalent amount of work in the gymnasium by lifting a 50.0-kg barbell. How many times must he raise the barbell to expend this much energy? Assume that he raises the barbell 2.00 m each time he lifts it and that he regains no energy when he lowers the barbell.

Solution Because 1 Calorie = 1.00×10^3 cal, the total amount of work required to be done on the barbell–Earth system is 2.00×10^6 cal. Converting this value to joules, we have

$$W = (2.00 \times 10^6 \text{ cal})(4.186 \text{ J/cal}) = 8.37 \times 10^6 \text{ J}$$

The work done in lifting the barbell a distance h is equal to mgh, and the work done in lifting it n times is $nmgh$. We equate this to the total work required:

$$W = nmgh = 8.37 \times 10^6 \text{ J}$$

$$n = \frac{W}{mgh} = \frac{8.37 \times 10^6 \text{ J}}{(50.0 \text{ kg})(9.80 \text{ m/s}^2)(2.00 \text{ m})}$$

$$= 8.54 \times 10^3 \text{ times}$$

If the student is in good shape and lifts the barbell once every 5 s, it will take him about 12 h to perform this feat. Clearly, it is much easier for this student to lose weight by dieting.

In reality, the human body is not 100% efficient. Thus, not all of the energy transformed within the body from the dinner transfers out of the body by work done on the barbell. Some of this energy is used to pump blood and perform other functions within the body. Thus, the 2 000 Calories can be worked off in less time than 12 h when these other energy requirements are included.

20.2 Specific Heat and Calorimetry

When energy is added to a system and there is no change in the kinetic or potential energy of the system, the temperature of the system usually rises. (An exception to this statement is the case in which a system undergoes a change of state—also called a *phase transition*—as discussed in the next section.) If the system consists of a sample of a substance, we find that the quantity of energy required to raise the temperature of a given mass of the substance by some amount varies from one substance to another. For example, the quantity of energy required to raise the temperature of 1 kg of water by 1°C is 4 186 J, but the quantity of energy required to raise the temperature of 1 kg of copper by 1°C is only 387 J. In the discussion that follows, we shall use heat as our example of energy transfer, but keep in mind that we could change the temperature of our system by means of any method of energy transfer.

The **heat capacity** C of a particular sample of a substance is defined as the amount of energy needed to raise the temperature of that sample by 1°C. From this definition, we see that if energy Q produces a change ΔT in the temperature of a sample, then

$$Q = C\Delta T \qquad (20.2)$$

The **specific heat** c of a substance is the heat capacity per unit mass. Thus, if energy Q transfers to a sample of a substance with mass m and the temperature of the sample changes by ΔT, then the specific heat of the substance is

Specific heat

$$c \equiv \frac{Q}{m\,\Delta T} \qquad (20.3)$$

Specific heat is essentially a measure of how thermally insensitive a substance is to the addition of energy. The greater a material's specific heat, the more energy must be added to a given mass of the material to cause a particular temperature change. Table 20.1 lists representative specific heats.

From this definition, we can relate the energy Q transferred between a sample of mass m of a material and its surroundings to a temperature change ΔT as

$$Q = mc\,\Delta T \qquad (20.4)$$

 PITFALL PREVENTION

20.3 An Unfortunate Choice of Terminology

The name *specific heat* is an unfortunate holdover from the days when thermodynamics and mechanics developed separately. A better name would be *specific energy transfer,* but the existing term is too entrenched to be replaced.

Table 20.1

Specific Heats of Some Substances at 25°C and Atmospheric Pressure		
	Specific heat c	
Substance	**J/kg·°C**	**cal/g·°C**
Elemental solids		
Aluminum	900	0.215
Beryllium	1 830	0.436
Cadmium	230	0.055
Copper	387	0.092 4
Germanium	322	0.077
Gold	129	0.030 8
Iron	448	0.107
Lead	128	0.030 5
Silicon	703	0.168
Silver	234	0.056
Other solids		
Brass	380	0.092
Glass	837	0.200
Ice (-5°C)	2 090	0.50
Marble	860	0.21
Wood	1 700	0.41
Liquids		
Alcohol (ethyl)	2 400	0.58
Mercury	140	0.033
Water (15°C)	4 186	1.00
Gas		
Steam (100°C)	2 010	0.48

For example, the energy required to raise the temperature of 0.500 kg of water by 3.00°C is $(0.500 \text{ kg})(4\,186 \text{ J/kg} \cdot °\text{C})(3.00°\text{C}) = 6.28 \times 10^3$ J. Note that when the temperature increases, Q and ΔT are taken to be positive, and energy transfers into the system. When the temperature decreases, Q and ΔT are negative, and energy transfers out of the system.

Specific heat varies with temperature. However, if temperature intervals are not too great, the temperature variation can be ignored and c can be treated as a constant.[2] For example, the specific heat of water varies by only about 1% from 0°C to 100°C at atmospheric pressure. Unless stated otherwise, we shall neglect such variations.

Measured values of specific heats are found to depend on the conditions of the experiment. In general, measurements made in a constant-pressure process are different from those made in a constant-volume process. For solids and liquids, the difference between the two values is usually no greater than a few percent and is often neglected. Most of the values given in Table 20.1 were measured at atmospheric pressure and room temperature. The specific heats for gases measured at constant pressure are quite different from values measured at constant volume (see Chapter 21).

It is interesting to note from Table 20.1 that water has the highest specific heat of common materials. This high specific heat is responsible, in part, for the moderate temperatures found near large bodies of water. As the temperature of a body of water decreases during the winter, energy is transferred from the cooling water to the air by heat, increasing the internal energy of the air. Because of the high specific heat of water, a relatively large amount of energy is transferred to the air for even modest temperature changes of the water. The air carries this internal energy landward when prevailing winds are favorable. For example, the prevailing winds on the West Coast of the United States are toward the land (eastward). Hence, the energy liberated by the Pacific Ocean as it cools keeps coastal areas much warmer than they would otherwise be. This explains why the western coastal states generally have more favorable winter weather than the eastern coastal states, where the prevailing winds do not tend to carry the energy toward land.

Conservation of Energy: Calorimetry

One technique for measuring specific heat involves heating a sample to some known temperature T_x, placing it in a vessel containing water of known mass and temperature $T_w < T_x$, and measuring the temperature of the water after equilibrium has been reached. This technique is called **calorimetry**, and devices in which this energy transfer occurs are called **calorimeters**. If the system of the sample and the water is isolated, the law of the conservation of energy requires that the amount of energy that leaves the sample (of unknown specific heat) equal the amount of energy that enters the water.[3]

PITFALL PREVENTION

20.4 Energy Can Be Transferred by Any Method

We will use Q to represent the amount of energy transferred, but keep in mind that the energy transfer in Equation 20.4 could be by *any* of the methods introduced in Chapter 7; it does not have to be heat. For example, repeatedly bending a coat hanger wire raises the temperature at the bending point by *work*.

---

[2] The definition given by Equation 20.3 assumes that the specific heat does not vary with temperature over the interval $\Delta T = T_f - T_i$. In general, if c varies with temperature over the interval, then the correct expression for Q is $Q = m\int_{T_i}^{T_f} c\,dT$.

[3] For precise measurements, the water container should be included in our calculations because it also exchanges energy with the sample. Doing so would require a knowledge of its mass and composition, however. If the mass of the water is much greater than that of the container, we can neglect the effects of the container.

PITFALL PREVENTION

20.5 Remember the Negative Sign

It is *critical* to include the negative sign in Equation 20.5. The negative sign in the equation is necessary for consistency with our sign convention for energy transfer. The energy transfer Q_{hot} has a negative value because energy is leaving the hot substance. The negative sign in the equation assures that the right-hand side is a positive number, consistent with the left-hand side, which is positive because energy is entering the cold water.

Conservation of energy allows us to write the mathematical representation of this energy statement as

$$Q_{\text{cold}} = -Q_{\text{hot}} \qquad (20.5)$$

The negative sign in the equation is necessary to maintain consistency with our sign convention for heat.

Suppose m_x is the mass of a sample of some substance whose specific heat we wish to determine. Let us call its specific heat c_x and its initial temperature T_x. Likewise, let m_w, c_w, and T_w represent corresponding values for the water. If T_f is the final equilibrium temperature after everything is mixed, then from Equation 20.4, we find that the energy transfer for the water is $m_w c_w (T_f - T_w)$, which is positive because $T_f > T_w$, and that the energy transfer for the sample of unknown specific heat is $m_x c_x (T_f - T_x)$, which is negative. Substituting these expressions into Equation 20.5 gives

$$m_w c_w (T_f - T_w) = -m_x c_x (T_f - T_x)$$

Solving for c_x gives

$$c_x = \frac{m_w c_w (T_f - T_w)}{m_x (T_x - T_f)}$$

Example 20.2 Cooling a Hot Ingot

A 0.050 0-kg ingot of metal is heated to 200.0°C and then dropped into a beaker containing 0.400 kg of water initially at 20.0°C. If the final equilibrium temperature of the mixed system is 22.4°C, find the specific heat of the metal.

Solution According to Equation 20.5, we can write

$$m_w c_w (T_f - T_w) = -m_x c_x (T_f - T_x)$$

$$(0.400 \text{ kg})(4\ 186 \text{ J/kg} \cdot °\text{C})(22.4°\text{C} - 20.0°\text{C})$$

$$= -(0.050\ 0 \text{ kg})(c_x)(22.4°\text{C} - 200.0°\text{C})$$

From this we find that

$$c_x = \boxed{453 \text{ J/kg} \cdot °\text{C}}$$

The ingot is most likely iron, as we can see by comparing this result with the data given in Table 20.1. Note that the temperature of the ingot is initially above the steam point. Thus, some of the water may vaporize when we drop the in-got into the water. We assume that we have a sealed system and that this steam cannot escape. Because the final equilibrium temperature is lower than the steam point, any steam that does result recondenses back into water.

What If? Suppose you are performing an experiment in the laboratory that uses this technique to determine the specific heat of a sample and you wish to decrease the overall uncertainty in your final result for c_x. Of the data given in the text of this example, changing which value would be most effective in decreasing the uncertainty?

Answer The largest experimental uncertainty is associated with the small temperature difference of 2.4°C for $T_f - T_w$. For example, an uncertainty of 0.1°C in each of these two temperature readings leads to an 8% uncertainty in their difference. In order for this temperature difference to be larger experimentally, the most effective change is to *decrease the amount of water*.

Example 20.3 Fun Time for a Cowboy

A cowboy fires a silver bullet with a muzzle speed of 200 m/s into the pine wall of a saloon. Assume that all the internal energy generated by the impact remains with the bullet. What is the temperature change of the bullet?

Solution The kinetic energy of the bullet is

$$K = \tfrac{1}{2} m v^2$$

Because nothing in the environment is hotter than the bullet, the bullet gains no energy by heat. Its temperature increases because the kinetic energy is transformed to extra internal energy when the bullet is stopped by the wall. The temperature change is the same as that which would take place if energy $Q = K$ were transferred by heat from a stove to the bullet. If we imagine this latter process taking place, we can calculate ΔT from Equation 20.4. Using 234 J/kg · °C as the specific heat of silver (see Table 20.1), we obtain

$$(1) \qquad \Delta T = \frac{Q}{mc} = \frac{K}{mc} = \frac{\frac{1}{2} m (200 \text{ m/s})^2}{m (234 \text{ J/kg} \cdot °\text{C})} = \boxed{85.5°\text{C}}$$

Note that the result does not depend on the mass of the bullet.

What If? Suppose that the cowboy runs out of silver bullets and fires a lead bullet at the same speed into the wall. Will the temperature change of the bullet be larger or smaller?

Answer Consulting Table 20.1, we find that the specific heat of lead is 128 J/kg · °C, which is smaller than that for silver. Thus, a given amount of energy input will raise lead to a higher temperature than silver and the final temperature

of the lead bullet will be larger. In Equation (1), we substitute the new value for the specific heat:

$$\Delta T = \frac{Q}{mc} = \frac{K}{mc} = \frac{\frac{1}{2}\cancel{m}(200 \text{ m/s})^2}{\cancel{m}(128 \text{ J/kg} \cdot °\text{C})} = 156°\text{C}$$

Note that there is no requirement that the silver and lead bullets have the same mass to determine this temperature. The only requirement is that they have the same speed.

20.3 Latent Heat

A substance often undergoes a change in temperature when energy is transferred between it and its surroundings. There are situations, however, in which the transfer of energy does not result in a change in temperature. This is the case whenever the physical characteristics of the substance change from one form to another; such a change is commonly referred to as a **phase change**. Two common phase changes are from solid to liquid (melting) and from liquid to gas (boiling); another is a change in the crystalline structure of a solid. All such phase changes involve a change in internal energy but no change in temperature. The increase in internal energy in boiling, for example, is represented by the breaking of bonds between molecules in the liquid state; this bond breaking allows the molecules to move farther apart in the gaseous state, with a corresponding increase in intermolecular potential energy.

As you might expect, different substances respond differently to the addition or removal of energy as they change phase because their internal molecular arrangements vary. Also, the amount of energy transferred during a phase change depends on the amount of substance involved. (It takes less energy to melt an ice cube than it does to thaw a frozen lake.) If a quantity Q of energy transfer is required to change the phase of a mass m of a substance, the ratio $L \equiv Q/m$ characterizes an important thermal property of that substance. Because this added or removed energy does not result in a temperature change, the quantity L is called the **latent heat** (literally, the "hidden" heat) of the substance. The value of L for a substance depends on the nature of the phase change, as well as on the properties of the substance.

From the definition of latent heat, and again choosing heat as our energy transfer mechanism, we find that the energy required to change the phase of a given mass m of a pure substance is

$$Q = \pm mL \qquad \qquad (20.6)$$

Latent heat of fusion L_f is the term used when the phase change is from solid to liquid (*to fuse* means "to combine by melting"), and **latent heat of vaporization** L_v is the term used when the phase change is from liquid to gas (the liquid "vaporizes").[4] The latent heats of various substances vary considerably, as data in Table 20.2 show. The positive sign in Equation 20.6 is used when energy enters a system, causing melting or vaporization. The negative sign corresponds to energy leaving a system, such that the system freezes or condenses.

To understand the role of latent heat in phase changes, consider the energy required to convert a 1.00-g cube of ice at −30.0°C to steam at 120.0°C. Figure 20.2 indicates the experimental results obtained when energy is gradually added to the ice. Let us examine each portion of the red curve.

 PITFALL PREVENTION

20.6 Signs Are Critical

Sign errors occur very often when students apply calorimetry equations, so we will make this point once again. For phase changes, use the correct explicit sign in Equation 20.6, depending on whether you are adding or removing energy from the substance. In Equation 20.4, there is no explicit sign to consider, but be sure that your ΔT is *always* the final temperature minus the initial temperature. In addition, make sure that you *always* include the negative sign on the right-hand side of Equation 20.5.

Latent heat

---

[4] When a gas cools, it eventually *condenses*—that is, it returns to the liquid phase. The energy given up per unit mass is called the *latent heat of condensation* and is numerically equal to the latent heat of vaporization. Likewise, when a liquid cools, it eventually solidifies, and the *latent heat of solidification* is numerically equal to the latent heat of fusion.

Table 20.2

		Latent Heat of Fusion (J/kg)		Latent Heat of Vaporization (J/kg)
Substance	Melting Point (°C)		Boiling Point (°C)	
Helium	− 269.65	5.23×10^3	− 268.93	2.09×10^4
Nitrogen	− 209.97	2.55×10^4	− 195.81	2.01×10^5
Oxygen	− 218.79	1.38×10^4	− 182.97	2.13×10^5
Ethyl alcohol	− 114	1.04×10^5	78	8.54×10^5
Water	0.00	3.33×10^5	100.00	2.26×10^6
Sulfur	119	3.81×10^4	444.60	3.26×10^5
Lead	327.3	2.45×10^4	1 750	8.70×10^5
Aluminum	660	3.97×10^5	2 450	1.14×10^7
Silver	960.80	8.82×10^4	2 193	2.33×10^6
Gold	1 063.00	6.44×10^4	2 660	1.58×10^6
Copper	1 083	1.34×10^5	1 187	5.06×10^6

Part A. On this portion of the curve, the temperature of the ice changes from − 30.0°C to 0.0°C. Because the specific heat of ice is 2 090 J/kg · °C, we can calculate the amount of energy added by using Equation 20.4:

$$Q = m_i c_i \, \Delta T = (1.00 \times 10^{-3} \text{ kg})(2\,090 \text{ J/kg} \cdot °\text{C})(30.0°\text{C}) = 62.7 \text{ J}$$

Part B. When the temperature of the ice reaches 0.0°C, the ice–water mixture remains at this temperature—even though energy is being added—until all the ice melts. The energy required to melt 1.00 g of ice at 0.0°C is, from Equation 20.6,

$$Q = m_i L_f = (1.00 \times 10^{-3} \text{ kg})(3.33 \times 10^5 \text{ J/kg}) = 333 \text{ J}$$

Thus, we have moved to the 396 J (= 62.7 J + 333 J) mark on the energy axis in Figure 20.2.

Part C. Between 0.0°C and 100.0°C, nothing surprising happens. No phase change occurs, and so all energy added to the water is used to increase its temperature. The amount of energy necessary to increase the temperature from 0.0°C to 100.0°C is

$$Q = m_w c_w \, \Delta T = (1.00 \times 10^{-3} \text{ kg})(4.19 \times 10^3 \text{ J/kg} \cdot °\text{C})(100.0°\text{C}) = 419 \text{ J}$$

Figure 20.2 A plot of temperature versus energy added when 1.00 g of ice initially at − 30.0°C is converted to steam at 120.0°C.

Part D. At 100.0°C, another phase change occurs as the water changes from water at 100.0°C to steam at 100.0°C. Similar to the ice–water mixture in part B, the water–steam mixture remains at 100.0°C—even though energy is being added—until all of the liquid has been converted to steam. The energy required to convert 1.00 g of water to steam at 100.0°C is

$$Q = m_w L_v = (1.00 \times 10^{-3} \text{ kg})(2.26 \times 10^6 \text{ J/kg}) = 2.26 \times 10^3 \text{ J}$$

Part E. On this portion of the curve, as in parts A and C, no phase change occurs; thus, all energy added is used to increase the temperature of the steam. The energy that must be added to raise the temperature of the steam from 100.0°C to 120.0°C is

$$Q = m_s c_s \, \Delta T = (1.00 \times 10^{-3} \text{ kg})(2.01 \times 10^3 \text{ J/kg} \cdot °\text{C})(20.0°\text{C}) = 40.2 \text{ J}$$

The total amount of energy that must be added to change 1 g of ice at $-30.0°$C to steam at 120.0°C is the sum of the results from all five parts of the curve, which is 3.11×10^3 J. Conversely, to cool 1 g of steam at 120.0°C to ice at $-30.0°$C, we must remove 3.11×10^3 J of energy.

Note in Figure 20.2 the relatively large amount of energy that is transferred into the water to vaporize it to steam. Imagine reversing this process—there is a large amount of energy transferred out of steam to condense it into water. This is why a burn to your skin from steam at 100°C is much more damaging than exposure of your skin to water at 100°C. A very large amount of energy enters your skin from the steam and the steam remains at 100°C for a long time while it condenses. Conversely, when your skin makes contact with water at 100°C, the water immediately begins to drop in temperature as energy transfers from the water to your skin.

We can describe phase changes in terms of a rearrangement of molecules when energy is added to or removed from a substance. (For elemental substances in which the atoms do not combine to form molecules, the following discussion should be interpreted in terms of atoms. We use the general term *molecules* to refer to both chemical compounds and elemental substances.) Consider first the liquid-to-gas phase change. The molecules in a liquid are close together, and the forces between them are stronger than those between the more widely separated molecules of a gas. Therefore, work must be done on the liquid against these attractive molecular forces if the molecules are to separate. The latent heat of vaporization is the amount of energy per unit mass that must be added to the liquid to accomplish this separation.

Similarly, for a solid, we imagine that the addition of energy causes the amplitude of vibration of the molecules about their equilibrium positions to become greater as the temperature increases. At the melting point of the solid, the amplitude is great enough to break the bonds between molecules and to allow molecules to move to new positions. The molecules in the liquid also are bound to each other, but less strongly than those in the solid phase. The latent heat of fusion is equal to the energy required per unit mass to transform the bonds among all molecules from the solid-type bond to the liquid-type bond.

As you can see from Table 20.2, the latent heat of vaporization for a given substance is usually somewhat higher than the latent heat of fusion. This is not surprising if we consider that the average distance between molecules in the gas phase is much greater than that in either the liquid or the solid phase. In the solid-to-liquid phase change, we transform solid-type bonds between molecules into liquid-type bonds between molecules, which are only slightly less strong. In the liquid-to-gas phase change, however, we break liquid-type bonds and create a situation in which the molecules of the gas essentially are not bonded to each other. Therefore, it is not surprising that more energy is required to vaporize a given mass of substance than is required to melt it.

Quick Quiz 20.3 Suppose the same process of adding energy to the ice cube is performed as discussed above, but we graph the internal energy of the system as a function of energy input. What would this graph look like?

PITFALL PREVENTION

20.7 Celsius vs. Kelvin

In equations in which T appears—for example, the ideal gas law—the Kelvin temperature *must* be used. In equations involving ΔT, such as calorimetry equations, it is *possible* to use Celsius temperatures, because a change in temperature is the same on both scales. It is *safest*, however, to *consistently* use Kelvin temperatures in all equations involving T or ΔT.

Quick Quiz 20.4 Calculate the slopes for the A, C, and E portions of Figure 20.2. Rank the slopes from least to greatest and explain what this ordering means.

PROBLEM-SOLVING HINTS

Calorimetry Problems

If you have difficulty in solving calorimetry problems, be sure to consider the following points:

- Units of measure must be consistent. For instance, if you are using specific heats measured in J/kg · °C, be sure that masses are in kilograms and temperatures are in Celsius degrees.

- Transfers of energy are given by the equation $Q = mc\,\Delta T$ only for those processes in which no phase changes occur. Use the equations $Q = \pm mL_f$ and $Q = \pm mL_v$ only when phase changes *are* taking place; be sure to select the proper sign for these equations depending on the direction of energy transfer.

- Often, errors in sign are made when the equation $Q_{cold} = -Q_{hot}$ is used. Make sure that you use the negative sign in the equation, and remember that ΔT is always the final temperature minus the initial temperature.

Example 20.4 Cooling the Steam

What mass of steam initially at 130°C is needed to warm 200 g of water in a 100-g glass container from 20.0°C to 50.0°C?

Solution The steam loses energy in three stages. In the first stage, the steam is cooled to 100°C. The energy transfer in the process is

$$Q_1 = m_s c_s\,\Delta T = m_s(2.01 \times 10^3\,\text{J/kg}\cdot°\text{C})(-30.0°\text{C})$$
$$= -m_s(6.03 \times 10^4\,\text{J/kg})$$

where m_s is the unknown mass of the steam.

In the second stage, the steam is converted to water. To find the energy transfer during this phase change, we use $Q = -mL_v$, where the negative sign indicates that energy is leaving the steam:

$$Q_2 = -m_s(2.26 \times 10^6\,\text{J/kg})$$

In the third stage, the temperature of the water created from the steam is reduced to 50.0°C. This change requires an energy transfer of

$$Q_3 = m_s c_w \Delta T = m_s(4.19 \times 10^3\,\text{J/kg}\cdot°\text{C})(-50.0°\text{C})$$
$$= -m_s(2.09 \times 10^5\,\text{J/kg})$$

Adding the energy transfers in these three stages, we obtain

$$Q_{hot} = Q_1 + Q_2 + Q_3$$
$$= -m_s[6.03 \times 10^4\,\text{J/kg} + 2.26 \times 10^6\,\text{J/kg}$$
$$+ 2.09 \times 10^5\,\text{J/kg}]$$
$$= -m_s(2.53 \times 10^6\,\text{J/kg})$$

Now, we turn our attention to the temperature increase of the water and the glass. Using Equation 20.4, we find that

$$Q_{cold} = (0.200\,\text{kg})(4.19 \times 10^3\,\text{J/kg}\cdot°\text{C})(30.0°\text{C})$$
$$+ (0.100\,\text{kg})(837\,\text{J/kg}\cdot°\text{C})(30.0°\text{C})$$
$$= 2.77 \times 10^4\,\text{J}$$

Using Equation 20.5, we can solve for the unknown mass:

$$Q_{cold} = -Q_{hot}$$
$$2.77 \times 10^4\,\text{J} = -[-m_s(2.53 \times 10^6\,\text{J/kg})]$$
$$m_s = 1.09 \times 10^{-2}\,\text{kg} = \boxed{10.9\,\text{g}}$$

What If? What if the final state of the system is water at 100°C? Would we need more or less steam? How would the analysis above change?

Answer More steam would be needed to raise the temperature of the water and glass to 100°C instead of 50.0°C. There would be two major changes in the analysis. First, we would not have a term Q_3 for the steam because the water that condenses from the steam does not cool below 100°C. Second, in Q_{cold}, the temperature change would be 80.0°C instead of 30.0°C. Thus, Q_{hot} becomes

$$Q_{hot} = Q_1 + Q_2$$
$$= -m_s(6.03 \times 10^4\,\text{J/kg} + 2.26 \times 10^6\,\text{J/kg})$$
$$= -m_s(2.32 \times 10^6\,\text{J/kg})$$

and Q_{cold} becomes

$$Q_{cold} = (0.200\,\text{kg})(4.19 \times 10^3\,\text{J/kg}\cdot°\text{C})(80.0°\text{C})$$
$$+ (0.100\,\text{kg})(837\,\text{J/kg}\cdot°\text{C})(80.0°\text{C})$$
$$= 7.37 \times 10^4\,\text{J}$$

leading to $m_s = 3.18 \times 10^{-2}\,\text{kg} = 31.8\,\text{g}$.

Example 20.5 **Boiling Liquid Helium**

Liquid helium has a very low boiling point, 4.2 K, and a very low latent heat of vaporization, 2.09×10^4 J/kg. If energy is transferred to a container of boiling liquid helium from an immersed electric heater at a rate of 10.0 W, how long does it take to boil away 1.00 kg of the liquid?

Solution Because $L_v = 2.09 \times 10^4$ J/kg, we must supply 2.09×10^4 J of energy to boil away 1.00 kg. Because

10.0 W = 10.0 J/s, 10.0 J of energy is transferred to the helium each second. From $\mathcal{P} = \Delta E/\Delta t$, the time interval required to transfer 2.09×10^4 J of energy is

$$\Delta t = \frac{\Delta E}{\mathcal{P}} = \frac{2.09 \times 10^4 \text{ J}}{10.0 \text{ J/s}} = 2.09 \times 10^3 \text{ s} \approx \boxed{35 \text{ min}}$$

20.4 Work and Heat in Thermodynamic Processes

In the macroscopic approach to thermodynamics, we describe the *state* of a system using such variables as pressure, volume, temperature, and internal energy. As a result, these quantities belong to a category called **state variables**. For any given configuration of the system, we can identify values of the state variables. It is important to note that a *macroscopic state* of an isolated system can be specified only if the system is in thermal equilibrium internally. In the case of a gas in a container, internal thermal equilibrium requires that every part of the gas be at the same pressure and temperature.

A second category of variables in situations involving energy is **transfer variables**. These variables are zero *unless* a process occurs in which energy is transferred across the boundary of the system. Because a transfer of energy across the boundary represents a change in the system, transfer variables are not associated with a given state of the system, but with a *change* in the state of the system. In the previous sections, we discussed heat as a transfer variable. For a given set of conditions of a system, there is no defined value for the heat. We can only assign a value of the heat if energy crosses the boundary by heat, resulting in a change in the system. State variables are characteristic of a system in thermal equilibrium. Transfer variables are characteristic of a process in which energy is transferred between a system and its environment.

In this section, we study another important transfer variable for thermodynamic systems—work. Work performed on particles was studied extensively in Chapter 7, and here we investigate the work done on a deformable system—a gas. Consider a gas contained in a cylinder fitted with a movable piston (Fig. 20.3). At equilibrium, the gas oc-

(a)   (b)

Figure 20.3 Work is done on a gas contained in a cylinder at a pressure *P* as the piston is pushed downward so that the gas is compressed.

cupies a volume V and exerts a uniform pressure P on the cylinder's walls and on the piston. If the piston has a cross-sectional area A, the force exerted by the gas on the piston is $F = PA$. Now let us assume that we push the piston inward and compress the gas **quasi-statically**, that is, slowly enough to allow the system to remain essentially in thermal equilibrium at all times. As the piston is pushed downward by an external force $\mathbf{F} = -F\hat{\mathbf{j}}$ through a displacement of $d\mathbf{r} = dy\hat{\mathbf{j}}$ (Fig. 20.3b), the work done on the gas is, according to our definition of work in Chapter 7,

$$dW = \mathbf{F} \cdot d\mathbf{r} = -F\hat{\mathbf{j}} \cdot dy\hat{\mathbf{j}} = -F\,dy = -PA\,dy$$

where we have set the magnitude F of the external force equal to PA because the piston is always in equilibrium between the external force and the force from the gas. For this discussion, we assume the mass of the piston is negligible. Because $A\,dy$ is the change in volume of the gas dV, we can express the work done on the gas as

$$dW = -P\,dV \qquad (20.7)$$

If the gas is compressed, dV is negative and the work done on the gas is positive. If the gas expands, dV is positive and the work done on the gas is negative. If the volume remains constant, the work done on the gas is zero. The total work done on the gas as its volume changes from V_i to V_f is given by the integral of Equation 20.7:

Work done on a gas

$$W = -\int_{V_i}^{V_f} P\,dV \qquad (20.8)$$

To evaluate this integral, one must know how the pressure varies with volume during the process.

In general, the pressure is not constant during a process followed by a gas, but depends on the volume and temperature. If the pressure and volume are known at each step of the process, the state of the gas at each step can be plotted on a graph called a ***PV diagram***, as in Figure 20.4. This type of diagram allows us to visualize a process through which a gas is progressing. The curve on a PV diagram is called the *path* taken between the initial and final states.

Note that the integral in Equation 20.8 is equal to the area under a curve on a PV diagram. Thus, we can identify an important use for PV diagrams:

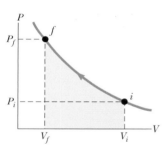

Active Figure 20.4 A gas is compressed quasi-statically (slowly) from state i to state f. The work done on the gas equals the negative of the area under the PV curve.

 At the Active Figures link at http://www.pse6.com, you can compress the piston in Figure 20.3 and see the result on the PV diagram in this figure.

> The work done on a gas in a quasi-static process that takes the gas from an initial state to a final state is the negative of the area under the curve on a PV diagram, evaluated between the initial and final states.

As Figure 20.4 suggests, for our process of compressing a gas in the cylinder, the work done depends on the particular path taken between the initial and final states. To illustrate this important point, consider several different paths connecting i and f (Fig. 20.5). In the process depicted in Figure 20.5a, the volume of the gas is first reduced from V_i to V_f at constant pressure P_i and the pressure of the gas then increases from P_i to P_f by heating at constant volume V_f. The work done on the gas along this path is $-P_i(V_f - V_i)$. In Figure 20.5b, the pressure of the gas is increased from P_i to P_f at constant volume V_i and then the volume of the gas is reduced from V_i to V_f at constant pressure P_f. The work done on the gas is $-P_f(V_f - V_i)$, which is greater than that for the process described in Figure 20.5a. It is greater because the piston is moved through the same displacement by a larger force than for the situation in Figure 20.5a. Finally, for the process described in Figure 20.5c, where both P and V change continuously, the work done on the gas has some value intermediate between the values obtained in the first two processes. To evaluate the work in this case, the function $P(V)$ must be known, so that we can evaluate the integral in Equation 20.8.

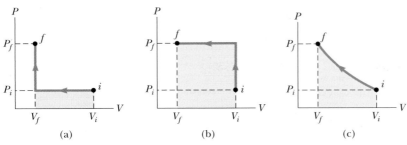

Active Figure 20.5 The work done on a gas as it is taken from an initial state to a final state depends on the path between these states.

At the Active Figures link at http://www.pse6.com, *you can choose one of the three paths and see the movement of the piston in Figure 20.3 and of a point on the PV diagram in this figure.*

The energy transfer Q into or out of a system by heat also depends on the process. Consider the situations depicted in Figure 20.6. In each case, the gas has the same initial volume, temperature, and pressure, and is assumed to be ideal. In Figure 20.6a, the gas is thermally insulated from its surroundings except at the bottom of the gas-filled region, where it is in thermal contact with an energy reservoir. An *energy reservoir* is a source of energy that is considered to be so great that a finite transfer of energy to or from the reservoir does not change its temperature. The piston is held at its initial position by an external agent—a hand, for instance. When the force holding the piston is reduced slightly, the piston rises very slowly to its final position. Because the piston is moving upward, the gas is doing work on the piston. During this expansion to the final volume V_f, just enough energy is transferred by heat from the reservoir to the gas to maintain a constant temperature T_i.

Now consider the completely thermally insulated system shown in Figure 20.6b. When the membrane is broken, the gas expands rapidly into the vacuum until it occupies a volume V_f and is at a pressure P_f. In this case, the gas does no work because it does not apply a force—no force is required to expand into a vacuum. Furthermore, no energy is transferred by heat through the insulating wall.

The initial and final states of the ideal gas in Figure 20.6a are identical to the initial and final states in Figure 20.6b, but the paths are different. In the first case, the gas does work on the piston, and energy is transferred slowly to the gas by heat. In the second case, no energy is transferred by heat, and the value of the work done is zero. Therefore, we conclude that **energy transfer by heat, like work done, depends on the initial, final, and intermediate states of the system.** In other words, because heat and work depend on the path, neither quantity is determined solely by the end points of a thermodynamic process.

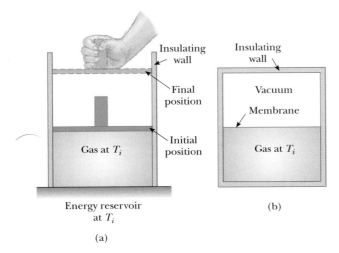

Figure 20.6 (a) A gas at temperature T_i expands slowly while absorbing energy from a reservoir in order to maintain a constant temperature. (b) A gas expands rapidly into an evacuated region after a membrane is broken.

20.5 The First Law of Thermodynamics

When we introduced the law of conservation of energy in Chapter 7, we stated that the change in the energy of a system is equal to the sum of all transfers of energy across the boundary of the system. The first law of thermodynamics is a special case of the law of conservation of energy that encompasses changes in internal energy and energy transfer by heat and work. It is a law that can be applied to many processes and provides a connection between the microscopic and macroscopic worlds.

We have discussed two ways in which energy can be transferred between a system and its surroundings. One is work done on the system, which requires that there be a macroscopic displacement of the point of application of a force. The other is heat, which occurs on a molecular level whenever a temperature difference exists across the boundary of the system. Both mechanisms result in a change in the internal energy of the system and therefore usually result in measurable changes in the macroscopic variables of the system, such as the pressure, temperature, and volume of a gas.

To better understand these ideas on a quantitative basis, suppose that a system undergoes a change from an initial state to a final state. During this change, energy transfer by heat Q to the system occurs, and work W is done on the system. As an example, suppose that the system is a gas in which the pressure and volume change from P_i and V_i to P_f and V_f. If the quantity $Q + W$ is measured for various paths connecting the initial and final equilibrium states, we find that it is the same for all paths connecting the two states. We conclude that the quantity $Q + W$ is determined completely by the initial and final states of the system, and we call this quantity the **change in the internal energy** of the system. Although Q and W both depend on the path, **the quantity $Q + W$ is independent of the path.** If we use the symbol E_{int} to represent the internal energy, then the *change* in internal energy ΔE_{int} can be expressed as[5]

First law of thermodynamics

$$\Delta E_{int} = Q + W \qquad (20.9)$$

where all quantities must have the same units of measure for energy. Equation 20.9 is known as the **first law of thermodynamics.** One of the important consequences of the first law of thermodynamics is that there exists a quantity known as internal energy whose value is determined by the state of the system. The internal energy is therefore a state variable like pressure, volume, and temperature.

When a system undergoes an infinitesimal change in state in which a small amount of energy dQ is transferred by heat and a small amount of work dW is done, the internal energy changes by a small amount dE_{int}. Thus, for infinitesimal processes we can express the first law as[6]

$$dE_{int} = dQ + dW$$

The first law of thermodynamics is an energy conservation equation specifying that the only type of energy that changes in the system is the internal energy E_{int}. Let us investigate some special cases in which this condition exists.

First, consider an *isolated system*—that is, one that does not interact with its surroundings. In this case, no energy transfer by heat takes place and the work done on

 PITFALL PREVENTION

20.8 Dual Sign Conventions

Some physics and engineering textbooks present the first law as $\Delta E_{int} = Q - W$, with a minus sign between the heat and work. The reason for this is that work is defined in these treatments as the work done *by* the gas rather than *on* the gas, as in our treatment. The equivalent equation to Equation 20.8 in these treatments defines work as $W = \int_{V_i}^{V_f} P\,dV$. Thus, if positive work is done by the gas, energy is leaving the system, leading to the negative sign in the first law.

In your studies in other chemistry or engineering courses, or in your reading of other physics textbooks, be sure to note which sign convention is being used for the first law.

---

[5] It is an unfortunate accident of history that the traditional symbol for internal energy is U, which is also the traditional symbol for potential energy, as introduced in Chapter 8. To avoid confusion between potential energy and internal energy, we use the symbol E_{int} for internal energy in this book. If you take an advanced course in thermodynamics, however, be prepared to see U used as the symbol for internal energy.

[6] Note that dQ and dW are not true differential quantities because Q and W are not state variables; however, dE_{int} is. Because dQ and dW are *inexact differentials*, they are often represented by the symbols $đQ$ and $đW$. For further details on this point, see an advanced text on thermodynamics, such as R. P. Bauman, *Modern Thermodynamics and Statistical Mechanics*, New York, Macmillan Publishing Co., 1992.

the system is zero; hence, the internal energy remains constant. That is, because $Q = W = 0$, it follows that $\Delta E_{int} = 0$, and thus $E_{int,\,i} = E_{int,\,f}$. We conclude that **the internal energy E_{int} of an isolated system remains constant.**

Next, consider the case of a system (one not isolated from its surroundings) that is taken through a **cyclic process**—that is, a process that starts and ends at the same state. In this case, the change in the internal energy must again be zero, because E_{int} is a state variable, and therefore the energy Q added to the system must equal the negative of the work W done on the system during the cycle. That is, in a cyclic process,

$$\Delta E_{int} = 0 \quad \text{and} \quad Q = -W \quad \text{(cyclic process)}$$

On a PV diagram, a cyclic process appears as a closed curve. (The processes described in Figure 20.5 are represented by open curves because the initial and final states differ.) It can be shown that **in a cyclic process, the net work done on the system per cycle equals the area enclosed by the path representing the process on a PV diagram.**

20.6 Some Applications of the First Law of Thermodynamics

The first law of thermodynamics that we discussed in the preceding section relates the changes in internal energy of a system to transfers of energy by work or heat. In this section, we consider applications of the first law to processes through which a gas is taken. As a model, we consider the sample of gas contained in the piston–cylinder apparatus in Figure 20.7. This figure shows work being done on the gas and energy transferring in by heat, so the internal energy of the gas is rising. In the following discussion of various processes, refer back to this figure and mentally alter the directions of the transfer of energy so as to reflect what is happening in the process.

Before we apply the first law of thermodynamics to specific systems, it is useful to first define some idealized thermodynamic processes. An **adiabatic process** is one during which no energy enters or leaves the system by heat—that is, $Q = 0$. An adiabatic process can be achieved either by thermally insulating the walls of the system, such as the cylinder in Figure 20.7, or by performing the process rapidly, so that there is negligible time for energy to transfer by heat. Applying the first law of thermodynamics to an adiabatic process, we see that

$$\Delta E_{int} = W \quad \text{(adiabatic process)} \tag{20.10}$$

From this result, we see that if a gas is compressed adiabatically such that W is positive, then ΔE_{int} is positive and the temperature of the gas increases. Conversely, the temperature of a gas decreases when the gas expands adiabatically.

Adiabatic processes are very important in engineering practice. Some common examples are the expansion of hot gases in an internal combustion engine, the liquefaction of gases in a cooling system, and the compression stroke in a diesel engine.

The process described in Figure 20.6b, called an **adiabatic free expansion,** is unique. The process is adiabatic because it takes place in an insulated container. Because the gas expands into a vacuum, it does not apply a force on a piston as was depicted in Figure 20.6a, so no work is done on or by the gas. Thus, in this adiabatic process, both $Q = 0$ and $W = 0$. As a result, $\Delta E_{int} = 0$ for this process, as we can see from the first law. That is, **the initial and final internal energies of a gas are equal in an adiabatic free expansion.** As we shall see in the next chapter, the internal energy of an ideal gas depends only on its temperature. Thus, we expect no change in temperature during an adiabatic free expansion. This prediction is in accord with the results of experiments performed at low pressures. (Experiments performed at high pressures for real gases show a slight change in temperature after the expansion. This change is due to intermolecular interactions, which represent a deviation from the model of an ideal gas.)

Active Figure 20.7 The first law of thermodynamics equates the change in internal energy E_{int} in a system to the net energy transfer to the system by heat Q and work W. In the situation shown here, the internal energy of the gas increases.

At the Active Figures link at http://www.pse6.com, you can choose one of the four processes for the gas discussed in this section and see the movement of the piston and of a point on a PV diagram.

A process that occurs at constant pressure is called an **isobaric process.** In Figure 20.7, an isobaric process could be established by allowing the piston to move freely so that it is always in equilibrium between the net force from the gas pushing upward and the weight of the piston plus the force due to atmospheric pressure pushing downward. In Figure 20.5, the first process in part (a) and the second process in part (b) are isobaric.

In such a process, the values of the heat and the work are both usually nonzero. The work done on the gas in an isobaric process is simply

Isobaric process

$$W = -P(V_f - V_i) \qquad \text{(isobaric process)} \qquad (20.11)$$

where P is the constant pressure.

A process that takes place at constant volume is called an **isovolumetric process.** In Figure 20.7, clamping the piston at a fixed position would ensure an isovolumetric process. In Figure 20.5, the second process in part (a) and the first process in part (b) are isovolumetric.

In such a process, the value of the work done is zero because the volume does not change. Hence, from the first law we see that in an isovolumetric process, because $W = 0$,

Isovolumetric process

$$\Delta E_{int} = Q \qquad \text{(isovolumetric process)} \qquad (20.12)$$

This expression specifies that **if energy is added by heat to a system kept at constant volume, then all of the transferred energy remains in the system as an increase in its internal energy.** For example, when a can of spray paint is thrown into a fire, energy enters the system (the gas in the can) by heat through the metal walls of the can. Consequently, the temperature, and thus the pressure, in the can increases until the can possibly explodes.

Isothermal process

A process that occurs at constant temperature is called an **isothermal process**. In Figure 20.7, this process can be established by immersing the cylinder in Figure 20.7 in an ice-water bath or by putting the cylinder in contact with some other constant-temperature reservoir. A plot of P versus V at constant temperature for an ideal gas yields a hyperbolic curve called an *isotherm*. The internal energy of an ideal gas is a function of temperature only. Hence, in an isothermal process involving an ideal gas, $\Delta E_{int} = 0$. For an isothermal process, then, we conclude from the first law that the energy transfer Q must be equal to the negative of the work done on the gas—that is, $Q = -W$. Any energy that enters the system by heat is transferred out of the system by work; as a result, no change in the internal energy of the system occurs in an isothermal process.

 PITFALL PREVENTION

20.10 $Q \neq 0$ in an Isothermal Process

Do not fall into the common trap of thinking that there must be no transfer of energy by heat if the temperature does not change, as is the case in an isothermal process. Because the cause of temperature change can be either heat *or* work, the temperature can remain constant even if energy enters the gas by heat. This can only happen if the energy entering the gas by heat leaves by work.

> ## Quick Quiz 20.5 In the last three columns of the following table, fill in the boxes with −, +, or 0. For each situation, the system to be considered is identified.
>
> | Situation | System | Q | W | ΔE_{int} |
> |---|---|---|---|---|
> | (a) Rapidly pumping up a bicycle tire | Air in the pump | | | |
> | (b) Pan of room-temperature water sitting on a hot stove | Water in the pan | | | |
> | (c) Air quickly leaking out of a balloon | Air originally in the balloon | | | |

Isothermal Expansion of an Ideal Gas

Suppose that an ideal gas is allowed to expand quasi-statically at constant temperature. This process is described by the PV diagram shown in Figure 20.8. The curve is a hyperbola (see Appendix B, Eq. B.23), and the ideal gas law with T constant indicates that the equation of this curve is $PV = $ constant.

Let us calculate the work done on the gas in the expansion from state i to state f. The work done on the gas is given by Equation 20.8. Because the gas is ideal and the process is quasi-static, we can use the expression $PV = nRT$ for each point on the path. Therefore, we have

$$W = -\int_{V_i}^{V_f} P\,dV = -\int_{V_i}^{V_f} \frac{nRT}{V}\,dV$$

Because T is constant in this case, it can be removed from the integral along with n and R:

$$W = -nRT\int_{V_i}^{V_f} \frac{dV}{V} = -nRT\ln V\Big|_{V_i}^{V_f}$$

To evaluate the integral, we used $\int(dx/x) = \ln x$. Evaluating this at the initial and final volumes, we have

$$W = nRT\ln\left(\frac{V_i}{V_f}\right) \qquad (20.13)$$

Numerically, this work W equals the negative of the shaded area under the PV curve shown in Figure 20.8. Because the gas expands, $V_f > V_i$ and the value for the work done on the gas is negative, as we expect. If the gas is compressed, then $V_f < V_i$ and the work done on the gas is positive.

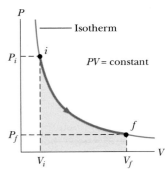

Figure 20.8 The PV diagram for an isothermal expansion of an ideal gas from an initial state to a final state. The curve is a hyperbola.

Quick Quiz 20.6 Characterize the paths in Figure 20.9 as isobaric, isovolumetric, isothermal, or adiabatic. Note that $Q = 0$ for path **B**.

Figure 20.9 (Quick Quiz 20.6) Identify the nature of paths A, B, C, and D.

Example 20.6 An Isothermal Expansion

A 1.0-mol sample of an ideal gas is kept at 0.0°C during an expansion from 3.0 L to 10.0 L.

(A) How much work is done on the gas during the expansion?

Solution Substituting the values into Equation 20.13, we have

$$W = nRT\ln\left(\frac{V_i}{V_f}\right)$$

$$= (1.0\text{ mol})(8.31\text{ J/mol}\cdot\text{K})(273\text{ K})\ln\left(\frac{3.0\text{ L}}{10.0\text{ L}}\right)$$

$$= -2.7\times10^3\text{ J}$$

(B) How much energy transfer by heat occurs with the surroundings in this process?

Solution From the first law, we find that

$$\Delta E_{int} = Q + W$$

$$0 = Q + W$$

$$Q = -W = \boxed{2.7 \times 10^3 \, J}$$

(C) If the gas is returned to the original volume by means of an isobaric process, how much work is done on the gas?

Solution The work done in an isobaric process is given by Equation 20.11. In this case, the initial volume is 10.0 L and the final volume is 3.0 L, the reverse of the situation in part (A). We are not given the pressure, so we need to incorpo-

rate the ideal gas law:

$$W = -P(V_f - V_i) = -\frac{nRT_i}{V_i}(V_f - V_i)$$

$$= -\frac{(1.0 \text{ mol})(8.31 \text{ J/mol}\cdot\text{K})(273 \text{ K})}{10.0 \times 10^{-3} \text{ m}^3}$$

$$\times (3.0 \times 10^{-3} \text{ m}^3 - 10.0 \times 10^{-3} \text{ m}^3)$$

$$= \boxed{1.6 \times 10^3 \, J}$$

Notice that we use the initial temperature and volume to determine the value of the constant pressure because we do not know the final temperature. The work done on the gas is positive because the gas is being compressed.

Example 20.7 Boiling Water

Suppose 1.00 g of water vaporizes isobarically at atmospheric pressure $(1.013 \times 10^5 \text{ Pa})$. Its volume in the liquid state is $V_i = V_{liquid} = 1.00 \text{ cm}^3$, and its volume in the vapor state is $V_f = V_{vapor} = 1\,671 \text{ cm}^3$. Find the work done in the expansion and the change in internal energy of the system. Ignore any mixing of the steam and the surrounding air—imagine that the steam simply pushes the surrounding air out of the way.

Solution Because the expansion takes place at constant pressure, the work done on the system (the vaporizing water) as it pushes away the surrounding air is, from Equation 20.11,

$$W = -P(V_f - V_i)$$

$$= -(1.013 \times 10^5 \text{ Pa})(1\,671 \times 10^{-6} \text{ m}^3 - 1.00 \times 10^{-6} \text{ m}^3)$$

$$= \boxed{-169 \, J}$$

To determine the change in internal energy, we must know the energy transfer Q needed to vaporize the water. Using Equation 20.6 and the latent heat of vaporization for water, we have

$$Q = mL_v = (1.00 \times 10^{-3} \text{ kg})(2.26 \times 10^6 \text{ J/kg}) = 2\,260 \, J$$

Hence, from the first law, the change in internal energy is

$$\Delta E_{int} = Q + W = 2\,260 \, J + (-169 \, J) = \boxed{2.09 \, kJ}$$

The positive value for ΔE_{int} indicates that the internal energy of the system increases. We see that most of the energy $(2\,090 \text{ J}/2\,260 \text{ J} = 93\%)$ transferred to the liquid goes into increasing the internal energy of the system. The remaining 7% of the energy transferred leaves the system by work done by the steam on the surrounding atmosphere.

Example 20.8 Heating a Solid

A 1.0-kg bar of copper is heated at atmospheric pressure. If its temperature increases from 20°C to 50°C,

(A) what is the work done on the copper bar by the surrounding atmosphere?

Solution Because the process is isobaric, we can find the work done on the copper bar using Equation 20.11, $W = -P(V_f - V_i)$. We can calculate the change in volume of the copper bar using Equation 19.6. Using the average linear expansion coefficient for copper given in Table 19.1, and remembering that $\beta = 3\alpha$, we obtain

$$\Delta V = \beta V_i \, \Delta T$$

$$= [5.1 \times 10^{-5} \, (°C)^{-1}](50°C - 20°C) V_i = 1.5 \times 10^{-3} V_i$$

The volume V_i is equal to m/ρ, and Table 14.1 indicates that the density of copper is $8.92 \times 10^3 \text{ kg/m}^3$. Hence,

$$\Delta V = (1.5 \times 10^{-3})\left(\frac{1.0 \text{ kg}}{8.92 \times 10^3 \text{ kg/m}^3}\right)$$

$$= 1.7 \times 10^{-7} \text{ m}^3$$

The work done on the copper bar is

$$W = -P \, \Delta V = -(1.013 \times 10^5 \text{ N/m}^2)(1.7 \times 10^{-7} \text{ m}^3)$$

$$= \boxed{-1.7 \times 10^{-2} \, J}$$

Because this work is negative, work is done *by* the copper bar on the atmosphere.

(B) What quantity of energy is transferred to the copper bar by heat?

Solution Taking the specific heat of copper from Table 20.1 and using Equation 20.4, we find that the energy transferred by heat is

$$Q = mc \, \Delta T = (1.0 \text{ kg})(387 \text{ J/kg}\cdot°C)(30°C)$$

$$= \boxed{1.2 \times 10^4 \, J}$$

(C) What is the increase in internal energy of the copper bar?

Solution From the first law of thermodynamics, we have

$$\Delta E_{\text{int}} = Q + W = 1.2 \times 10^4 \,\text{J} + (-1.7 \times 10^{-2} \,\text{J})$$

$$= 1.2 \times 10^4 \,\text{J}$$

Note that almost all of the energy transferred into the system by heat goes into increasing the internal energy of the copper bar. The fraction of energy used to do work on the surrounding atmosphere is only about 10^{-6}! Hence, when the thermal expansion of a solid or a liquid is analyzed, the small amount of work done on or by the system is usually ignored.

20.7 Energy Transfer Mechanisms

In Chapter 7, we introduced a global approach to energy analysis of physical processes through Equation 7.17, $\Delta E_{\text{system}} = \Sigma T$, where T represents energy transfer. Earlier in this chapter, we discussed two of the terms on the right-hand side of this equation, work and heat. In this section, we explore more details about heat as a means of energy transfer and consider two other energy transfer methods that are often related to temperature changes—convection (a form of matter transfer) and electromagnetic radiation.

Thermal Conduction

The process of energy transfer by heat can also be called **conduction** or **thermal conduction.** In this process, the transfer can be represented on an atomic scale as an exchange of kinetic energy between microscopic particles—molecules, atoms, and free electrons—in which less-energetic particles gain energy in collisions with more energetic particles. For example, if you hold one end of a long metal bar and insert the other end into a flame, you will find that the temperature of the metal in your hand soon increases. The energy reaches your hand by means of conduction. We can understand the process of conduction by examining what is happening to the microscopic particles in the metal. Initially, before the rod is inserted into the flame, the microscopic particles are vibrating about their equilibrium positions. As the flame heats the rod, the particles near the flame begin to vibrate with greater and greater amplitudes. These particles, in turn, collide with their neighbors and transfer some of their energy in the collisions. Slowly, the amplitudes of vibration of metal atoms and electrons farther and farther from the flame increase until, eventually, those in the metal near your hand are affected. This increased vibration is detected by an increase in the temperature of the metal and of your potentially burned hand.

The rate of thermal conduction depends on the properties of the substance being heated. For example, it is possible to hold a piece of asbestos in a flame indefinitely. This implies that very little energy is conducted through the asbestos. In general, metals are good thermal conductors, and materials such as asbestos, cork, paper, and fiberglass are poor conductors. Gases also are poor conductors because the separation distance between the particles is so great. Metals are good thermal conductors because they contain large numbers of electrons that are relatively free to move through the metal and so can transport energy over large distances. Thus, in a good conductor, such as copper, conduction takes place by means of both the vibration of atoms and the motion of free electrons.

Conduction occurs only if there is a difference in temperature between two parts of the conducting medium. Consider a slab of material of thickness Δx and cross-sectional area A. One face of the slab is at a temperature T_c, and the other face is at a temperature $T_h > T_c$ (Fig. 20.10). Experimentally, it is found that the energy Q transfers in a time interval Δt from the hotter face to the colder one. The rate $\mathcal{P} = Q/\Delta t$ at which this energy transfer occurs is found to be proportional to the cross-sectional area and the temperature difference $\Delta T = T_h - T_c$, and inversely proportional to the thickness:

A pan of boiling water sits on a stove burner. Energy enters the water through the bottom of the pan by thermal conduction.

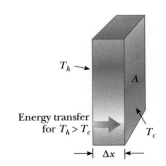

Figure 20.10 Energy transfer through a conducting slab with a cross-sectional area A and a thickness Δx. The opposite faces are at different temperatures T_c and T_h.

$$\mathcal{P} = \frac{Q}{\Delta t} \propto A \frac{\Delta T}{\Delta x}$$

Note that $\mathcal{P}$ has units of watts when Q is in joules and Δt is in seconds. This is not surprising because $\mathcal{P}$ is *power*—the rate of energy transfer by heat. For a slab of infinitesimal thickness dx and temperature difference dT, we can write the **law of thermal conduction** as

Law of thermal conduction

$$\mathcal{P} = kA \left| \frac{dT}{dx} \right| \qquad (20.14)$$

where the proportionality constant k is the **thermal conductivity** of the material and $|dT/dx|$ is the **temperature gradient** (the rate at which temperature varies with position).

Suppose that a long, uniform rod of length L is thermally insulated so that energy cannot escape by heat from its surface except at the ends, as shown in Figure 20.11. One end is in thermal contact with an energy reservoir at temperature T_c, and the other end is in thermal contact with a reservoir at temperature $T_h > T_c$. When a steady state has been reached, the temperature at each point along the rod is constant in time. In this case if we assume that k is not a function of temperature, the temperature gradient is the same everywhere along the rod and is

$$\left| \frac{dT}{dx} \right| = \frac{T_h - T_c}{L}$$

Figure 20.11 Conduction of energy through a uniform, insulated rod of length L. The opposite ends are in thermal contact with energy reservoirs at different temperatures.

Table 20.3

Thermal Conductivities	
Substance	**Thermal Conductivity (W/m · °C)**
Metals (at 25°C)	
Aluminum	238
Copper	397
Gold	314
Iron	79.5
Lead	34.7
Silver	427
Nonmetals (approximate values)	
Asbestos	0.08
Concrete	0.8
Diamond	2 300
Glass	0.8
Ice	2
Rubber	0.2
Water	0.6
Wood	0.08
Gases (at 20°C)	
Air	0.023 4
Helium	0.138
Hydrogen	0.172
Nitrogen	0.023 4
Oxygen	0.023 8

Thus the rate of energy transfer by conduction through the rod is

$$\mathcal{P} = kA\left(\frac{T_h - T_c}{L}\right) \tag{20.15}$$

Substances that are good thermal conductors have large thermal conductivity values, whereas good thermal insulators have low thermal conductivity values. Table 20.3 lists thermal conductivities for various substances. Note that metals are generally better thermal conductors than nonmetals.

For a compound slab containing several materials of thicknesses L_1, L_2, ... and thermal conductivities k_1, k_2, ..., the rate of energy transfer through the slab at steady state is

$$\mathcal{P} = \frac{A(T_h - T_c)}{\sum_i (L_i/k_i)} \tag{20.16}$$

where T_c and T_h are the temperatures of the outer surfaces (which are held constant) and the summation is over all slabs. Example 20.9 shows how this equation results from a consideration of two thicknesses of materials.

Example 20.9 Energy Transfer Through Two Slabs

Two slabs of thickness L_1 and L_2 and thermal conductivities k_1 and k_2 are in thermal contact with each other, as shown in Figure 20.12. The temperatures of their outer surfaces are T_c and T_h, respectively, and $T_h > T_c$. Determine the temperature at the interface and the rate of energy transfer by conduction through the slabs in the steady-state condition.

Solution To conceptualize this problem, notice the phrase "in the steady-state condition." We interpret this to mean that energy transfers through the compound slab at the same rate at all points. Otherwise, energy would be building up or disappearing at some point. Furthermore, the temperature will vary with position in the two slabs, most likely at different rates in each part of the compound slab. Thus, there will be some fixed temperature T at the interface

when the system is in steady state. We categorize this as a thermal conduction problem and impose the condition that the power is the same in both slabs of material. To analyze the problem, we use Equation 20.15 to express the rate at which energy is transferred through slab 1:

$$(1) \qquad \mathcal{P}_1 = k_1 A\left(\frac{T - T_c}{L_1}\right)$$

The rate at which energy is transferred through slab 2 is

$$(2) \qquad \mathcal{P}_2 = k_2 A\left(\frac{T_h - T}{L_2}\right)$$

When a steady state is reached, these two rates must be equal; hence,

$$k_1 A\left(\frac{T - T_c}{L_1}\right) = k_2 A\left(\frac{T_h - T}{L_2}\right)$$

Solving for T gives

$$(3) \qquad T = \frac{k_1 L_2 T_c + k_2 L_1 T_h}{k_1 L_2 + k_2 L_1}$$

Substituting Equation (3) into either Equation (1) or Equation (2), we obtain

$$(4) \qquad \mathcal{P} = \frac{A(T_h - T_c)}{(L_1/k_1) + (L_2/k_2)}$$

To finalize this problem, note that extension of this procedure to several slabs of materials leads to Equation 20.16.

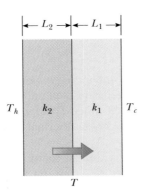

Figure 20.12 (Example 20.9) Energy transfer by conduction through two slabs in thermal contact with each other. At steady state, the rate of energy transfer through slab 1 equals the rate of energy transfer through slab 2.

What If? Suppose you are building an insulated container with two layers of insulation and the rate of energy transfer determined by Equation (4) turns out to be too high. You have enough room to increase the thickness of one of the two layers by 20%. How would you decide which layer to choose?

Answer To decrease the power as much as possible, you must increase the denominator in Equation (4) as much as possible. Whichever thickness you choose to increase, L_1 or L_2, you will increase the corresponding term L/k in the denominator by 20%. In order for this percentage change to represent the largest absolute change, you want to take 20% of the larger term. Thus, you should increase the thickness of the layer that has the larger value of L/k.

Quick Quiz 20.7 Will an ice cube wrapped in a wool blanket remain frozen for (a) a shorter length of time (b) the same length of time (c) a longer length of time than an identical ice cube exposed to air at room temperature?

Quick Quiz 20.8 You have two rods of the same length and diameter but they are formed from different materials. The rods will be used to connect two regions of different temperature such that energy will transfer through the rods by heat. They can be connected in series, as in Figure 20.13a, or in parallel, as in Figure 20.13b. In which case is the rate of energy transfer by heat larger? (a) when the rods are in series (b) when the rods are in parallel (c) The rate is the same in both cases.

(a)   (b)

Figure 20.13 (Quick Quiz 20.8) In which case is the rate of energy transfer larger?

Energy is conducted from the inside to the exterior more rapidly on the part of the roof where the snow has melted. The dormer appears to have been added and insulated. The main roof does not appear to be well insulated.

Courtesy of Dr. Albert A. Bartlett, University of Colorado, Boulder

Home Insulation

In engineering practice, the term L/k for a particular substance is referred to as the **R value** of the material. Thus, Equation 20.16 reduces to

$$\mathcal{P} = \frac{A(T_h - T_c)}{\sum_i R_i} \qquad (20.17)$$

where $R_i = L_i/k_i$. The R values for a few common building materials are given in Table 20.4. In the United States, the insulating properties of materials used in buildings are usually expressed in U.S. customary units, not SI units. Thus, in Table 20.4, measurements of R values are given as a combination of British thermal units, feet, hours, and degrees Fahrenheit.

At any vertical surface open to the air, a very thin stagnant layer of air adheres to the surface. One must consider this layer when determining the R value for a wall. The thickness of this stagnant layer on an outside wall depends on the speed of the wind. Energy loss from a house on a windy day is greater than the loss on a day when the air is calm. A representative R value for this stagnant layer of air is given in Table 20.4.

Table 20.4

Material	*R* value (ft² · °F · h/Btu)
R Values for Some Common Building Materials	
Hardwood siding (1 in. thick)	0.91
Wood shingles (lapped)	0.87
Brick (4 in. thick)	4.00
Concrete block (filled cores)	1.93
Fiberglass insulation (3.5 in. thick)	10.90
Fiberglass insulation (6 in. thick)	18.80
Fiberglass board (1 in. thick)	4.35
Cellulose fiber (1 in. thick)	3.70
Flat glass (0.125 in. thick)	0.89
Insulating glass (0.25-in. space)	1.54
Air space (3.5 in. thick)	1.01
Stagnant air layer	0.17
Drywall (0.5 in. thick)	0.45
Sheathing (0.5 in. thick)	1.32

Example 20.10   The *R* Value of a Typical Wall   `Interactive`

Calculate the total *R* value for a wall constructed as shown in Figure 20.14a. Starting outside the house (toward the front in the figure) and moving inward, the wall consists of 4 in. of brick, 0.5 in. of sheathing, an air space 3.5 in. thick, and 0.5 in. of drywall. Do not forget the stagnant air layers inside and outside the house.

Solution Referring to Table 20.4, we find that

R_1 (outside stagnant air layer) $= 0.17$ ft² · °F · h/Btu

R_2 (brick) $= 4.00$ ft² · °F · h/Btu

R_3 (sheathing) $= 1.32$ ft² · °F · h/Btu

R_4 (air space) $= 1.01$ ft² · °F · h/Btu

R_5 (drywall) $= 0.45$ ft² · °F · h/Btu

R_6 (inside stagnant air layer) $= 0.17$ ft² · °F · h/Btu

R_{total} $= 7.12$ ft² · °F · h/Btu

What If? You are not happy with this total *R* value for the wall. You cannot change the overall structure, but you can fill the air space as in Figure 20.14b. What material should you

choose to fill the air space in order to *maximize* the total *R* value?

Answer Looking at Table 20.4, we see that 3.5 in. of fiberglass insulation is over ten times as effective at insulating the wall as 3.5 in. of air. Thus, we could fill the air space with fiberglass insulation. The result is that we add 10.90 ft² · °F · h/Btu of *R* value and we lose 1.01 ft² · °F · h/Btu due to the air space we have replaced, for a total change of 10.90 ft² · °F · h/Btu − 1.01 ft² · °F · h/Btu = 9.89 ft² · °F · h/Btu. The new total *R* value is 7.12 ft² · °F · h/Btu + 9.89 ft² · °F · h/Btu = 17.01 ft² · °F · h/Btu.

Figure 20.14 (Example 20.10) An exterior house wall containing (a) an air space and (b) insulation.

 Study the *R* values of various types of common building materials at the Interactive Worked Example link at http://www.pse6.com.

Convection

At one time or another, you probably have warmed your hands by holding them over an open flame. In this situation, the air directly above the flame is heated and expands. As a result, the density of this air decreases and the air rises. This hot air warms your

Figure 20.15 Convection currents are set up in a room warmed by a radiator.

hands as it flows by. **Energy transferred by the movement of a warm substance is said to have been transferred by convection.** When the movement results from differences in density, as with air around a fire, it is referred to as *natural convection*. Air flow at a beach is an example of natural convection, as is the mixing that occurs as surface water in a lake cools and sinks (see Section 19.4). When the heated substance is forced to move by a fan or pump, as in some hot-air and hot-water heating systems, the process is called *forced convection*.

If it were not for convection currents, it would be very difficult to boil water. As water is heated in a teakettle, the lower layers are warmed first. This water expands and rises to the top because its density is lowered. At the same time, the denser, cool water at the surface sinks to the bottom of the kettle and is heated.

The same process occurs when a room is heated by a radiator. The hot radiator warms the air in the lower regions of the room. The warm air expands and rises to the ceiling because of its lower density. The denser, cooler air from above sinks, and the continuous air current pattern shown in Figure 20.15 is established.

Radiation

The third means of energy transfer that we shall discuss is **radiation**. All objects radiate energy continuously in the form of electromagnetic waves (see Chapter 34) produced by thermal vibrations of the molecules. You are likely familiar with electromagnetic radiation in the form of the orange glow from an electric stove burner, an electric space heater, or the coils of a toaster.

The rate at which an object radiates energy is proportional to the fourth power of its absolute temperature. This is known as **Stefan's law** and is expressed in equation form as

Stefan's law

$$\mathcal{P} = \sigma A e T^4 \tag{20.18}$$

where $\mathcal{P}$ is the power in watts radiated from the surface of the object, σ is a constant equal to $5.669\ 6 \times 10^{-8}\ \mathrm{W/m^2 \cdot K^4}$, A is the surface area of the object in square meters, e is the **emissivity**, and T is the surface temperature in kelvins. The value of e can vary between zero and unity, depending on the properties of the surface of the object. The emissivity is equal to the **absorptivity**, which is the fraction of the incoming radiation that the surface absorbs.

Approximately 1 340 J of electromagnetic radiation from the Sun passes perpendicularly through each 1 m² at the top of the Earth's atmosphere every second. This radiation is primarily visible and infrared light accompanied by a significant amount of ultraviolet radiation. We shall study these types of radiation in detail in Chapter 34. Some of this energy is reflected back into space, and some is absorbed by the atmosphere. However, enough energy arrives at the surface of the Earth each day to supply all our energy needs on this planet hundreds of times over—if only it could be captured and used efficiently. The growth in the number of solar energy–powered houses built in this country reflects the increasing efforts being made to use this abundant energy. Radiant energy from the Sun affects our day-to-day existence in a number of ways. For example, it influences the Earth's average temperature, ocean currents, agriculture, and rain patterns.

What happens to the atmospheric temperature at night is another example of the effects of energy transfer by radiation. If there is a cloud cover above the Earth, the water vapor in the clouds absorbs part of the infrared radiation emitted by the Earth and re-emits it back to the surface. Consequently, temperature levels at the surface remain moderate. In the absence of this cloud cover, there is less in the way to prevent this radiation from escaping into space; thus the temperature decreases more on a clear night than on a cloudy one.

As an object radiates energy at a rate given by Equation 20.18, it also absorbs electromagnetic radiation. If the latter process did not occur, an object would eventually radiate all its energy, and its temperature would reach absolute zero. The energy an object absorbs comes from its surroundings, which consist of other objects that radiate energy. If an object is at a temperature T and its surroundings are at an average temperature T_0, then the net rate of energy gained or lost by the object as a result of radiation is

$$\mathcal{P}_{\text{net}} = \sigma A e(T^4 - T_0^{\ 4}) \tag{20.19}$$

When an object is in equilibrium with its surroundings, it radiates and absorbs energy at the same rate, and its temperature remains constant. When an object is hotter than its surroundings, it radiates more energy than it absorbs, and its temperature decreases.

An **ideal absorber** is defined as an object that absorbs all the energy incident on it, and for such an object, $e = 1$. An object for which $e = 1$ is often referred to as a **black body.** We shall investigate experimental and theoretical approaches to radiation from a black body in Chapter 40. An ideal absorber is also an ideal radiator of energy. In contrast, an object for which $e = 0$ absorbs none of the energy incident on it. Such an object reflects all the incident energy, and thus is an **ideal reflector.**

The Dewar Flask

The *Dewar flask*[7] is a container designed to minimize energy losses by conduction, convection, and radiation. Such a container is used to store either cold or hot liquids for long periods of time. (A Thermos bottle is a common household equivalent of a Dewar flask.) The standard construction (Fig. 20.16) consists of a double-walled Pyrex glass vessel with silvered walls. The space between the walls is evacuated to minimize energy transfer by conduction and convection. The silvered surfaces minimize energy transfer by radiation because silver is a very good reflector and has very low emissivity. A further reduction in energy loss is obtained by reducing the size of the neck. Dewar flasks are commonly used to store liquid nitrogen (boiling point: 77 K) and liquid oxygen (boiling point: 90 K).

To confine liquid helium (boiling point: 4.2 K), which has a very low heat of vaporization, it is often necessary to use a double Dewar system, in which the Dewar flask containing the liquid is surrounded by a second Dewar flask. The space between the two flasks is filled with liquid nitrogen.

Newer designs of storage containers use "super insulation" that consists of many layers of reflecting material separated by fiberglass. All of this is in a vacuum, and no liquid nitrogen is needed with this design.

Figure 20.16 A cross-sectional view of a Dewar flask, which is used to store hot or cold substances.

Example 20.11   Who Turned Down the Thermostat?

A student is trying to decide what to wear. The surroundings (his bedroom) are at 20.0°C. If the skin temperature of the unclothed student is 35°C, what is the net energy loss from his body in 10.0 min by radiation? Assume that the emissivity of skin is 0.900 and that the surface area of the student is 1.50 m².

Solution Using Equation 20.19, we find that the net rate of energy loss from the skin is

$$\mathcal{P}_{\text{net}} = \sigma A e(T^4 - T_0^{\ 4})$$

$$= (5.67 \times 10^{-8}\,\text{W/m}^2 \cdot \text{K}^4)(1.50\,\text{m}^2)$$
$$\times (0.900)\,[(308\,\text{K})^4 - (293\,\text{K})^4] = 125\,\text{W}$$

At this rate, the total energy lost by the skin in 10 min is

$$Q = \mathcal{P}_{\text{net}}\,\Delta t = (125\,\text{W})(600\,\text{s}) = \boxed{7.5 \times 10^4\,\text{J}}$$

Note that the energy radiated by the student is roughly equivalent to that produced by two 60-W light bulbs!

---

[7] Invented by Sir James Dewar (1842–1923).

SUMMARY

Take a practice test for this chapter by clicking on the Practice Test link at http://www.pse6.com.

Internal energy is all of a system's energy that is associated with the system's microscopic components. Internal energy includes kinetic energy of random translation, rotation, and vibration of molecules, potential energy within molecules, and potential energy between molecules.

Heat is the transfer of energy across the boundary of a system resulting from a temperature difference between the system and its surroundings. We use the symbol Q for the amount of energy transferred by this process.

The **calorie** is the amount of energy necessary to raise the temperature of 1 g of water from 14.5°C to 15.5°C. The **mechanical equivalent of heat** is 1 cal = 4.186 J.

The **heat capacity** C of any sample is the amount of energy needed to raise the temperature of the sample by 1°C. The energy Q required to change the temperature of a mass m of a substance by an amount ΔT is

$$Q = mc \, \Delta T \tag{20.4}$$

where c is the **specific heat** of the substance.

The energy required to change the phase of a pure substance of mass m is

$$Q = \pm mL \tag{20.6}$$

where L is the **latent heat** of the substance and depends on the nature of the phase change and the properties of the substance. The positive sign is used if energy is entering the system, and the negative sign is used if energy is leaving.

The **work done** on a gas as its volume changes from some initial value V_i to some final value V_f is

$$W = - \int_{V_i}^{V_f} P \, dV \tag{20.8}$$

where P is the pressure, which may vary during the process. In order to evaluate W, the process must be fully specified—that is, P and V must be known during each step. In other words, the work done depends on the path taken between the initial and final states.

The **first law of thermodynamics** states that when a system undergoes a change from one state to another, the change in its internal energy is

$$\Delta E_{int} = Q + W \tag{20.9}$$

where Q is the energy transferred into the system by heat and W is the work done on the system. Although Q and W both depend on the path taken from the initial state to the final state, the quantity ΔE_{int} is path-independent.

In a **cyclic process** (one that originates and terminates at the same state), $\Delta E_{int} = 0$ and, therefore, $Q = -W$. That is, the energy transferred into the system by heat equals the negative of the work done on the system during the process.

In an **adiabatic process**, no energy is transferred by heat between the system and its surroundings ($Q = 0$). In this case, the first law gives $\Delta E_{int} = W$. That is, the internal energy changes as a consequence of work being done on the system. In the **adiabatic free expansion** of a gas $Q = 0$ and $W = 0$, and so $\Delta E_{int} = 0$. That is, the internal energy of the gas does not change in such a process.

An **isobaric process** is one that occurs at constant pressure. The work done on a gas in such a process is $W = -P(V_f - V_i)$.

An **isovolumetric process** is one that occurs at constant volume. No work is done in such a process, so $\Delta E_{int} = Q$.

An **isothermal process** is one that occurs at constant temperature. The work done on an ideal gas during an isothermal process is

$$W = nRT \ln \left(\frac{V_i}{V_f} \right) \tag{20.13}$$

Energy may be transferred by work, which we addressed in Chapter 7, and by conduction, convection, or radiation. **Conduction** can be viewed as an exchange of kinetic energy between colliding molecules or electrons. The rate of energy transfer by conduction through a slab of area A is

$$\mathcal{P} = kA\left|\frac{dT}{dx}\right| \tag{20.14}$$

where k is the **thermal conductivity** of the material from which the slab is made and $|dT/dx|$ is the **temperature gradient.** This equation can be used in many situations in which the rate of transfer of energy through materials is important.

In **convection,** a warm substance transfers energy from one location to another.

All objects emit **radiation** in the form of electromagnetic waves at the rate

$$\mathcal{P} = \sigma AeT^4 \tag{20.18}$$

An object that is hotter than its surroundings radiates more energy than it absorbs, whereas an object that is cooler than its surroundings absorbs more energy than it radiates.

QUESTIONS

1. Clearly distinguish among temperature, heat, and internal energy.

2. Ethyl alcohol has about half the specific heat of water. If equal-mass samples of alcohol and water in separate beakers are supplied with the same amount of energy, compare the temperature increases of the two liquids.

3. A small metal crucible is taken from a 200°C oven and immersed in a tub full of water at room temperature (this process is often referred to as *quenching*). What is the approximate final equilibrium temperature?

4. What is a major problem that arises in measuring specific heats if a sample with a temperature above 100°C is placed in water?

5. In a daring lecture demonstration, an instructor dips his wetted fingers into molten lead (327°C) and withdraws them quickly, without getting burned. How is this possible? (This is a dangerous experiment, which you should *NOT* attempt.)

6. What is wrong with the following statement? "Given any two objects, the one with the higher temperature contains more heat."

7. Why is a person able to remove a piece of dry aluminum foil from a hot oven with bare fingers, while a burn results if there is moisture on the foil?

8. The air temperature above coastal areas is profoundly influenced by the large specific heat of water. One reason is that the energy released when 1 m³ of water cools by 1°C will raise the temperature of a much larger volume of air by 1°C. Find this volume of air. The specific heat of air is approximately 1 kJ/kg · °C. Take the density of air to be 1.3 kg/m³.

9. Concrete has a higher specific heat than soil. Use this fact to explain (partially) why cities have a higher average nighttime temperature than the surrounding countryside. If a city is hotter than the surrounding countryside, would you expect breezes to blow from city to country or from country to city? Explain.

10. Using the first law of thermodynamics, explain why the *total* energy of an isolated system is always constant.

11. When a sealed Thermos bottle full of hot coffee is shaken, what are the changes, if any, in (a) the temperature of the coffee (b) the internal energy of the coffee?

12. Is it possible to convert internal energy to mechanical energy? Explain with examples.

13. The U.S. penny was formerly made mostly of copper and is now made of copper-coated zinc. Can a calorimetric experiment be devised to test for the metal content in a collection of pennies? If so, describe the procedure you would use.

14. Figure Q20.14 shows a pattern formed by snow on the roof of a barn. What causes the alternating pattern of snow-covered and exposed roof?

Courtesy of Dr. Albert A. Bartlett, University of Colorado, Boulder, CO

Figure Q20.14 Alternating patterns on a snow-covered roof.

15. A tile floor in a bathroom may feel uncomfortably cold to your bare feet, but a carpeted floor in an adjoining room at the same temperature will feel warm. Why?

16. Why can potatoes be baked more quickly when a metal skewer has been inserted through them?

17. A piece of paper is wrapped around a rod made half of wood and half of copper. When held over a flame, the paper in contact with the wood burns but the half in contact with the metal does not. Explain.

18. Why do heavy draperies over the windows help keep a home cool in the summer, as well as warm in the winter?

19. If you wish to cook a piece of meat thoroughly on an open fire, why should you not use a high flame? (Note that carbon is a good thermal insulator.)

20. In an experimental house, Styrofoam beads were pumped into the air space between the panes of glass in double windows at night in the winter, and pumped out to holding bins during the day. How would this assist in conserving energy in the house?

21. Pioneers stored fruits and vegetables in underground cellars. Discuss the advantages of this choice for a storage site.

22. The pioneers referred to in the last question found that a large tub of water placed in a storage cellar would prevent their food from freezing on really cold nights. Explain why this is so.

23. When camping in a canyon on a still night, one notices that as soon as the sun strikes the surrounding peaks, a breeze begins to stir. What causes the breeze?

24. Updrafts of air are familiar to all pilots and are used to keep nonmotorized gliders aloft. What causes these currents?

25. If water is a poor thermal conductor, why can its temperature be raised quickly when it is placed over a flame?

26. Why is it more comfortable to hold a cup of hot tea by the handle rather than by wrapping your hands around the cup itself?

27. If you hold water in a paper cup over a flame, you can bring the water to a boil without burning the cup. How is this possible?

28. You need to pick up a very hot cooking pot in your kitchen. You have a pair of hot pads. Should you soak them in cold water or keep them dry, to be able to pick up the pot most comfortably?

29. Suppose you pour hot coffee for your guests, and one of them wants to drink it with cream, several minutes later, and then as warm as possible. In order to have the warmest coffee, should the person add the cream just after the coffee is poured or just before drinking? Explain.

30. Two identical cups both at room temperature are filled with the same amount of hot coffee. One cup contains a metal spoon, while the other does not. If you wait for several minutes, which of the two will have the warmer coffee? Which energy transfer process explains your answer?

31. A warning sign often seen on highways just before a bridge is "Caution—Bridge surface freezes before road surface." Which of the three energy transfer processes discussed in Section 20.7 is most important in causing a bridge surface to freeze before a road surface on very cold days?

32. A professional physics teacher drops one marshmallow into a flask of liquid nitrogen, waits for the most energetic boiling to stop, fishes it out with tongs, shakes it off, pops it into his mouth, chews it up, and swallows it. Clouds of ice crystals issue from his mouth as he crunches noisily and comments on the sweet taste. How can he do this without injury? *Caution:* Liquid nitrogen can be a dangerous substance and you should *not* try this yourself. The teacher might be badly injured if he did not shake it off, if he touched the tongs to a tooth, or if he did not start with a mouthful of saliva.

33. In 1801 Humphry Davy rubbed together pieces of ice inside an ice-house. He took care that nothing in their environment was at a higher temperature than the rubbed pieces. He observed the production of drops of liquid water. Make a table listing this and other experiments or processes, to illustrate each of the following. (a) A system can absorb energy by heat, increase in internal energy, and increase in temperature. (b) A system can absorb energy by heat and increase in internal energy, without an increase in temperature. (c) A system can absorb energy by heat without increasing in temperature or in internal energy. (d) A system can increase in internal energy and in temperature, without absorbing energy by heat. (e) A system can increase in internal energy without absorbing energy by heat or increasing in temperature. (f) **What If?** If a system's temperature increases, is it necessarily true that its internal energy increases?

34. Consider the opening photograph for Part 3 on page 578. Discuss the roles of conduction, convection, and radiation in the operation of the cooling fins on the support posts of the Alaskan oil pipeline.

PROBLEMS

1, 2, 3 = straightforward, intermediate, challenging   ☐ = full solution available in the *Student Solutions Manual and Study Guide*

🌐 = coached solution with hints available at http://www.pse6.com   💻 = computer useful in solving problem

▨ = paired numerical and symbolic problems

Section 20.1   Heat and Internal Energy

1. On his honeymoon James Joule traveled from England to Switzerland. He attempted to verify his idea of the interconvertibility of mechanical energy and internal energy by measuring the increase in temperature of water that fell in a waterfall. If water at the top of an alpine waterfall has a temperature of 10.0°C and then falls 50.0 m (as at Niagara Falls), what maximum temperature at the bottom of the falls could Joule expect? He did not succeed in measuring the temperature change, partly because evaporation cooled the falling water, and also because his thermometer was not sufficiently sensitive.

2. Consider Joule's apparatus described in Figure 20.1. The mass of each of the two blocks is 1.50 kg, and the insulated tank is filled with 200 g of water. What is the increase in the temperature of the water after the blocks fall through a distance of 3.00 m?

Section 20.2   Specific Heat and Calorimetry

3. The temperature of a silver bar rises by 10.0°C when it absorbs 1.23 kJ of energy by heat. The mass of the bar is 525 g. Determine the specific heat of silver.

4. A 50.0-g sample of copper is at 25.0°C. If 1 200 J of energy is added to it by heat, what is the final temperature of the copper?

5. Systematic use of solar energy can yield a large saving in the cost of winter space heating for a typical house in the north central United States. If the house has good insulation, you may model it as losing energy by heat steadily at the rate 6 000 W on a day in April when the average exterior temperature is 4°C, and when the conventional heating system is not used at all. The passive solar energy collector can consist simply of very large windows in a room facing south. Sunlight shining in during the daytime is absorbed by the floor, interior walls, and objects in the room, raising their temperature to 38°C. As the sun goes down, insulating draperies or shutters are closed over the windows. During the period between 5:00 P.M. and 7:00 A.M. the temperature of the house will drop, and a sufficiently large "thermal mass" is required to keep it from dropping too far. The thermal mass can be a large quantity of stone (with specific heat 850 J/kg·°C) in the floor and the interior walls exposed to sunlight. What mass of stone is required if the temperature is not to drop below 18°C overnight?

6. The *Nova* laser at Lawrence Livermore National Laboratory in California is used in studies of initiating controlled nuclear fusion (Section 45.4). It can deliver a power of 1.60×10^{13} W over a time interval of 2.50 ns. Compare its energy output in one such time interval to the energy required to make a pot of tea by warming 0.800 kg of water from 20.0°C to 100°C.

7. A 1.50-kg iron horseshoe initially at 600°C is dropped into a bucket containing 20.0 kg of water at 25.0°C. What is the final temperature? (Ignore the heat capacity of the container, and assume that a negligible amount of water boils away.)

8. An aluminum cup of mass 200 g contains 800 g of water in thermal equilibrium at 80.0°C. The combination of cup and water is cooled uniformly so that the temperature decreases by 1.50°C per minute. At what rate is energy being removed by heat? Express your answer in watts.

9. An aluminum calorimeter with a mass of 100 g contains 250 g of water. The calorimeter and water are in thermal equilibrium at 10.0°C. Two metallic blocks are placed into the water. One is a 50.0-g piece of copper at 80.0°C. The other block has a mass of 70.0 g and is originally at a temperature of 100°C. The entire system stabilizes at a final temperature of 20.0°C. (a) Determine the specific heat of the unknown sample. (b) Guess the material of the unknown, using the data in Table 20.1.

10. A 3.00-g copper penny at 25.0°C drops 50.0 m to the ground. (a) Assuming that 60.0% of the change in potential energy of the penny–Earth system goes into increasing the internal energy of the penny, determine its final temperature. (b) **What If?** Does the result depend on the mass of the penny? Explain.

11. A combination of 0.250 kg of water at 20.0°C, 0.400 kg of aluminum at 26.0°C, and 0.100 kg of copper at 100°C is mixed in an insulated container and allowed to come to thermal equilibrium. Ignore any energy transfer to or from the container and determine the final temperature of the mixture.

12. If water with a mass m_h at temperature T_h is poured into an aluminum cup of mass m_{Al} containing mass m_c of water at T_c, where $T_h > T_c$, what is the equilibrium temperature of the system?

13. A water heater is operated by solar power. If the solar collector has an area of 6.00 m² and the intensity delivered by sunlight is 550 W/m², how long does it take to increase the temperature of 1.00 m³ of water from 20.0°C to 60.0°C?

14. Two thermally insulated vessels are connected by a narrow tube fitted with a valve that is initially closed. One vessel, of volume 16.8 L, contains oxygen at a temperature of 300 K and a pressure of 1.75 atm. The other vessel, of volume 22.4 L, contains oxygen at a temperature of 450 K and a pressure of 2.25 atm. When the valve is opened, the gases in the two vessels mix, and the temperature and pressure become uniform throughout. (a) What is the final temperature? (b) What is the final pressure?

Section 20.3   Latent Heat

15. How much energy is required to change a 40.0-g ice cube from ice at −10.0°C to steam at 110°C?

16. A 50.0-g copper calorimeter contains 250 g of water at 20.0°C. How much steam must be condensed into the water if the final temperature of the system is to reach 50.0°C?

17. A 3.00-g lead bullet at 30.0°C is fired at a speed of 240 m/s into a large block of ice at 0°C, in which it becomes embedded. What quantity of ice melts?

18. Steam at 100°C is added to ice at 0°C. (a) Find the amount of ice melted and the final temperature when the mass of steam is 10.0 g and the mass of ice is 50.0 g. (b) **What If?** Repeat when the mass of steam is 1.00 g and the mass of ice is 50.0 g.

19. A 1.00-kg block of copper at 20.0°C is dropped into a large vessel of liquid nitrogen at 77.3 K. How many kilograms of nitrogen boil away by the time the copper reaches 77.3 K? (The specific heat of copper is 0.092 0 cal/g·°C. The latent heat of vaporization of nitrogen is 48.0 cal/g.)

20. Assume that a hailstone at 0°C falls through air at a uniform temperature of 0°C and lands on a sidewalk also at this temperature. From what initial height must the hailstone fall in order to entirely melt on impact?

21. In an insulated vessel, 250 g of ice at 0°C is added to 600 g of water at 18.0°C. (a) What is the final temperature of the system? (b) How much ice remains when the system reaches equilibrium?

22. Review problem. Two speeding lead bullets, each of mass 5.00 g, and at temperature 20.0°C, collide head-on at speeds of 500 m/s each. Assuming a perfectly inelastic collision and no loss of energy by heat to the atmosphere, describe the final state of the two-bullet system.

Section 20.4 Work and Heat in Thermodynamic Processes

23. A sample of ideal gas is expanded to twice its original volume of $1.00 \ m^3$ in a quasi-static process for which $P = \alpha V^2$, with $\alpha = 5.00 \ atm/m^6$, as shown in Figure P20.23. How much work is done on the expanding gas?

Figure P20.23

24. (a) Determine the work done on a fluid that expands from i to f as indicated in Figure P20.24. (b) **What If?** How much work is performed on the fluid if it is compressed from f to i along the same path?

Figure P20.24

25. 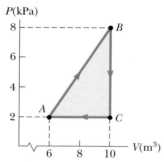 An ideal gas is enclosed in a cylinder with a movable piston on top of it. The piston has a mass of 8 000 g and an area of $5.00 \ cm^2$ and is free to slide up and down, keeping the pressure of the gas constant. How much work is done on the gas as the temperature of 0.200 mol of the gas is raised from 20.0°C to 300°C?

26. An ideal gas is enclosed in a cylinder that has a movable piston on top. The piston has a mass m and an area A and is free to slide up and down, keeping the pressure of the gas constant. How much work is done on the gas as the temperature of n mol of the gas is raised from T_1 to T_2?

27. One mole of an ideal gas is heated slowly so that it goes from the PV state (P_i, V_i) to $(3P_i, 3V_i)$ in such a way that the pressure is directly proportional to the volume. (a) How much work is done on the gas in the process? (b) How is the temperature of the gas related to its volume during this process?

Section 20.5 The First Law of Thermodynamics

28. A gas is compressed at a constant pressure of 0.800 atm from 9.00 L to 2.00 L. In the process, 400 J of energy leaves the gas by heat. (a) What is the work done on the gas? (b) What is the change in its internal energy?

29. A thermodynamic system undergoes a process in which its internal energy decreases by 500 J. At the same time, 220 J of work is done on the system. Find the energy transferred to or from it by heat.

30. A gas is taken through the cyclic process described in Figure P20.30. (a) Find the net energy transferred to the system by heat during one complete cycle. (b) **What If?** If the cycle is reversed—that is, the process follows the path $ACBA$—what is the net energy input per cycle by heat?

Figure P20.30 Problems 30 and 31.

31. Consider the cyclic process depicted in Figure P20.30. If Q is negative for the process BC and ΔE_{int} is negative for the process CA, what are the signs of Q, W, and ΔE_{int} that are associated with each process?

32. A sample of an ideal gas goes through the process shown in Figure P20.32. From A to B, the process is adiabatic; from B to C, it is isobaric with 100 kJ of energy entering the system by heat. From C to D, the process is isothermal; from D to A, it is isobaric with 150 kJ of energy leaving the system by heat. Determine the difference in internal energy $E_{int, B} - E_{int, A}$.

Figure P20.32

33. A sample of an ideal gas is in a vertical cylinder fitted with a piston. As 5.79 kJ of energy is transferred to the gas by heat to raise its temperature, the weight on the piston is adjusted so that the state of the gas changes from point A to point B along the semicircle shown in Figure P20.33. Find the change in internal energy of the gas.

Figure P20.33

Section 20.6   Some Applications of the First Law of Thermodynamics

34. One mole of an ideal gas does 3 000 J of work on its surroundings as it expands isothermally to a final pressure of 1.00 atm and volume of 25.0 L. Determine (a) the initial volume and (b) the temperature of the gas.

35. An ideal gas initially at 300 K undergoes an isobaric expansion at 2.50 kPa. If the volume increases from 1.00 m^3 to 3.00 m^3 and 12.5 kJ is transferred to the gas by heat, what are (a) the change in its internal energy and (b) its final temperature?

36. A 1.00-kg block of aluminum is heated at atmospheric pressure so that its temperature increases from 22.0°C to 40.0°C. Find (a) the work done on the aluminum, (b) the energy added to it by heat, and (c) the change in its internal energy.

37. How much work is done on the steam when 1.00 mol of water at 100°C boils and becomes 1.00 mol of steam at 100°C at 1.00 atm pressure? Assuming the steam to behave as an ideal gas, determine the change in internal energy of the material as it vaporizes.

38. An ideal gas initially at P_i, V_i, and T_i is taken through a cycle as in Figure P20.38. (a) Find the net work done on the gas per cycle. (b) What is the net energy added by heat to the system per cycle? (c) Obtain a numerical value for the net work done per cycle for 1.00 mol of gas initially at 0°C.

Figure P20.38

39. A 2.00-mol sample of helium gas initially at 300 K and 0.400 atm is compressed isothermally to 1.20 atm. Noting that the helium behaves as an ideal gas, find (a) the final volume of the gas, (b) the work done on the gas, and (c) the energy transferred by heat.

40. In Figure P20.40, the change in internal energy of a gas that is taken from A to C is +800 J. The work done on the gas along path ABC is −500 J. (a) How much energy must be added to the system by heat as it goes from A through B to C? (b) If the pressure at point A is five times that of point C, what is the work done on the system in going from C to D? (c) What is the energy exchanged with the surroundings by heat as the cycle goes from C to A along the green path? (d) If the change in internal energy in going from point D to point A is +500 J, how much energy must be added to the system by heat as it goes from point C to point D?

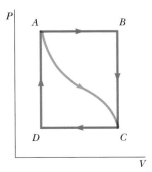

Figure P20.40

Section 20.7   Energy-Transfer Mechanisms

41. A box with a total surface area of 1.20 m^2 and a wall thickness of 4.00 cm is made of an insulating material. A 10.0-W electric heater inside the box maintains the inside temperature at 15.0°C above the outside temperature. Find the thermal conductivity k of the insulating material.

42. A glass window pane has an area of 3.00 m^2 and a thickness of 0.600 cm. If the temperature difference between its faces is 25.0°C, what is the rate of energy transfer by conduction through the window?

43. A bar of gold is in thermal contact with a bar of silver of the same length and area (Fig. P20.43). One end of the compound bar is maintained at 80.0°C while the opposite end is at 30.0°C. When the energy transfer reaches steady state, what is the temperature at the junction?

Figure P20.43

44. A thermal window with an area of 6.00 m^2 is constructed of two layers of glass, each 4.00 mm thick, and separated from each other by an air space of 5.00 mm. If the inside

surface is at 20.0°C and the outside is at −30.0°C, what is the rate of energy transfer by conduction through the window?

45. A power transistor is a solid-state electronic device. Assume that energy entering the device at the rate of 1.50 W by electrical transmission causes the internal energy of the device to increase. The surface area of the transistor is so small that it tends to overheat. To prevent overheating, the transistor is attached to a larger metal heat sink with fins. The temperature of the heat sink remains constant at 35.0°C under steady-state conditions. The transistor is electrically insulated from the heat sink by a rectangular sheet of mica measuring 8.25 mm by 6.25 mm, and 0.085 2 mm thick. The thermal conductivity of mica is equal to 0.075 3 W/m · °C. What is the operating temperature of the transistor?

46. Calculate the R value of (a) a window made of a single pane of flat glass $\frac{1}{8}$ in. thick, and (b) a thermal window made of two single panes each $\frac{1}{8}$ in. thick and separated by a $\frac{1}{4}$-in. air space. (c) By what factor is the transfer of energy by heat through the window reduced by using the thermal window instead of the single pane window?

47. The surface of the Sun has a temperature of about 5 800 K. The radius of the Sun is 6.96×10^8 m. Calculate the total energy radiated by the Sun each second. Assume that the emissivity of the Sun is 0.965.

48. A large hot pizza floats in outer space. What is the order of magnitude of (a) its rate of energy loss? (b) its rate of temperature change? List the quantities you estimate and the value you estimate for each.

49. The tungsten filament of a certain 100-W light bulb radiates 2.00 W of light. (The other 98 W is carried away by convection and conduction.) The filament has a surface area of 0.250 mm^2 and an emissivity of 0.950. Find the filament's temperature. (The melting point of tungsten is 3 683 K.)

50. At high noon, the Sun delivers 1 000 W to each square meter of a blacktop road. If the hot asphalt loses energy only by radiation, what is its equilibrium temperature?

51. The intensity of solar radiation reaching the top of the Earth's atmosphere is 1 340 W/m^2. The temperature of the Earth is affected by the so-called greenhouse effect of the atmosphere. That effect makes our planet's emissivity for visible light higher than its emissivity for infrared light. For comparison, consider a spherical object with no atmosphere, at the same distance from the Sun as the Earth. Assume that its emissivity is the same for all kinds of electromagnetic waves and that its temperature is uniform over its surface. Identify the projected area over which it absorbs sunlight and the surface area over which it radiates. Compute its equilibrium temperature. Chilly, isn't it? Your calculation applies to (a) the average temperature of the Moon, (b) astronauts in mortal danger aboard the crippled *Apollo 13* spacecraft, and (c) global catastrophe on the Earth if widespread fires should cause a layer of soot to accumulate throughout the upper atmosphere, so that most of the radiation from the Sun were absorbed there rather than at the surface below the atmosphere.

Additional Problems

52. Liquid nitrogen with a mass of 100 g at 77.3 K is stirred into a beaker containing 200 g of 5.00°C water. If the nitrogen leaves the solution as soon as it turns to gas, how much water freezes? (The latent heat of vaporization of nitrogen is 48.0 cal/g, and the latent heat of fusion of water is 79.6 cal/g.)

53. A 75.0-kg cross-country skier moves across the snow (Fig. P20.53). The coefficient of friction between the skis and the snow is 0.200. Assume that all the snow beneath his skis is at 0°C and that all the internal energy generated by friction is added to the snow, which sticks to his skis until it melts. How far would he have to ski to melt 1.00 kg of snow?

Figure P20.53

54. On a cold winter day you buy roasted chestnuts from a street vendor. Into the pocket of your down parka you put the change he gives you—coins constituting 9.00 g of copper at −12.0°C. Your pocket already contains 14.0 g of silver coins at 30.0°C. A short time later the temperature of the copper coins is 4.00°C and is increasing at a rate of 0.500°C/s. At this time, (a) what is the temperature of the silver coins, and (b) at what rate is it changing?

55. An aluminum rod 0.500 m in length and with a cross-sectional area of 2.50 cm^2 is inserted into a thermally insulated vessel containing liquid helium at 4.20 K. The rod is initially at 300 K. (a) If half of the rod is inserted into the helium, how many liters of helium boil off by the time the inserted half cools to 4.20 K? (Assume the upper half does not yet cool.) (b) If the upper end of the rod is maintained at 300 K, what is the approximate boil-off rate of liquid helium after the lower half has reached 4.20 K? (Aluminum has thermal conductivity of 31.0 J/s · cm · K at 4.2 K; ignore its temperature variation. Aluminum has a specific heat of 0.210 cal/g · °C and density of 2.70 g/cm^3. The density of liquid helium is 0.125 g/cm^3.)

56. A copper ring (with mass of 25.0 g, coefficient of linear expansion of 1.70×10^{-5} (°C)$^{-1}$, and specific heat of 9.24×10^{-2} cal/g · °C) has a diameter of 5.00 cm at its temperature of 15.0°C. A spherical aluminum shell (with mass 10.9 g, coefficient of linear expansion 2.40×10^{-5} (°C)$^{-1}$, and specific heat 0.215 cal/g·°C) has a diameter of 5.01 cm at a temperature higher than 15.0°C. The sphere is placed on top of the horizontal ring, and the two are allowed to come to thermal equilibrium without any exchange of energy with the surroundings. As soon as the sphere and ring reach thermal equilibrium, the sphere barely falls through the ring. Find (a) the equilibrium temperature, and (b) the initial temperature of the sphere.

57. A *flow calorimeter* is an apparatus used to measure the specific heat of a liquid. The technique of flow calorimetry involves measuring the temperature difference between the input and output points of a flowing stream of the liquid while energy is added by heat at a known rate. A liquid of density ρ flows through the calorimeter with volume flow rate R. At steady state, a temperature difference ΔT is established between the input and output points when energy is supplied at the rate $\mathcal{P}$. What is the specific heat of the liquid?

58. One mole of an ideal gas is contained in a cylinder with a movable piston. The initial pressure, volume, and temperature are P_i, V_i, and T_i, respectively. Find the work done on the gas for the following processes and show each process on a *PV* diagram: (a) An isobaric compression in which the final volume is half the initial volume. (b) An isothermal compression in which the final pressure is four times the initial pressure. (c) An isovolumetric process in which the final pressure is three times the initial pressure.

59. One mole of an ideal gas, initially at 300 K, is cooled at constant volume so that the final pressure is one fourth of the initial pressure. Then the gas expands at constant pressure until it reaches the initial temperature. Determine the work done on the gas.

60. **Review problem.** Continue the analysis of Problem 60 in Chapter 19. Following a collision between a large spacecraft and an asteroid, a copper disk of radius 28.0 m and thickness 1.20 m, at a temperature of 850°C, is floating in space, rotating about its axis with an angular speed of 25.0 rad/s. As the disk radiates infrared light, its temperature falls to 20.0°C. No external torque acts on the disk. (a) Find the change in kinetic energy of the disk. (b) Find the change in internal energy of the disk. (b) Find the amount of energy it radiates.

61. **Review problem.** A 670-kg meteorite happens to be composed of aluminum. When it is far from the Earth, its temperature is $-15°C$ and it moves with a speed of 14.0 km/s relative to the Earth. As it crashes into the planet, assume that the resulting additional internal energy is shared equally between the meteor and the planet, and that all of the material of the meteor rises momentarily to the same final temperature. Find this temperature. Assume that the specific heat of liquid and of gaseous aluminum is 1170 J/kg·°C.

62. An iron plate is held against an iron wheel so that a kinetic friction force of 50.0 N acts between the two pieces of metal. The relative speed at which the two surfaces slide over each other is 40.0 m/s. (a) Calculate the rate at which mechanical energy is converted to internal energy. (b) The plate and the wheel each have a mass of 5.00 kg, and each receives 50.0% of the internal energy. If the system is run as described for 10.0 s and each object is then allowed to reach a uniform internal temperature, what is the resultant temperature increase?

63. A solar cooker consists of a curved reflecting surface that concentrates sunlight onto the object to be warmed (Fig. P20.63). The solar power per unit area reaching the Earth's surface at the location is 600 W/m². The cooker faces the Sun and has a diameter of 0.600 m. Assume that 40.0% of the incident energy is transferred to 0.500 L of water in an open container, initially at 20.0°C. How long does it take to completely boil away the water? (Ignore the heat capacity of the container.)

Figure P20.63

64. Water in an electric teakettle is boiling. The power absorbed by the water is 1.00 kW. Assuming that the pressure of vapor in the kettle equals atmospheric pressure, determine the speed of effusion of vapor from the kettle's spout, if the spout has a cross-sectional area of 2.00 cm².

65. A cooking vessel on a slow burner contains 10.0 kg of water and an unknown mass of ice in equilibrium at 0°C at time $t = 0$. The temperature of the mixture is measured at various times, and the result is plotted in Figure P20.65. During the first 50.0 min, the mixture remains at 0°C. From 50.0 min to 60.0 min, the temperature increases to 2.00°C. Ignoring the heat capacity of the vessel, determine the initial mass of ice.

Figure P20.65

66. (a) In air at 0°C, a 1.60-kg copper block at 0°C is set sliding at 2.50 m/s over a sheet of ice at 0°C. Friction brings the block to rest. Find the mass of the ice that melts. To describe the process of slowing down, identify the energy input Q, the work input W, the change in internal energy ΔE_{int}, and the change in mechanical energy ΔK for the block and also for the ice. (b) A 1.60-kg block of ice at 0°C is set sliding at 2.50 m/s over a sheet of copper at 0°C. Friction brings the block to rest. Find the mass of the ice that melts. Identify Q, W, ΔE_{int}, and ΔK for the block and for the metal sheet during the process. (c) A thin 1.60-kg slab of copper at 20°C is set sliding at 2.50 m/s over an identical stationary slab at the same temperature. Friction quickly stops the motion. If no energy is lost to the environment by heat, find the change in temperature of both objects. Identify Q, W, ΔE_{int}, and ΔK for each object during the process.

67. The average thermal conductivity of the walls (including the windows) and roof of the house depicted in Figure P20.67 is 0.480 W/m·°C, and their average thickness is 21.0 cm. The house is heated with natural gas having a heat of combustion (that is, the energy provided per cubic meter of gas burned) of 9 300 kcal/m³. How many cubic meters of gas must be burned each day to maintain an inside temperature of 25.0°C if the outside temperature is 0.0°C? Disregard radiation and the energy lost by heat through the ground.

37°

5.00 m

8.00 m   10.0 m

Figure P20.67

68. A pond of water at 0°C is covered with a layer of ice 4.00 cm thick. If the air temperature stays constant at −10.0°C, how long does it take for the ice thickness to increase to 8.00 cm? *Suggestion:* Utilize Equation 20.15 in the form

$$\frac{dQ}{dt} = kA\frac{\Delta T}{x}$$

and note that the incremental energy dQ extracted from the water through the thickness x of ice is the amount required to freeze a thickness dx of ice. That is, $dQ = L\rho A\ dx$, where ρ is the density of the ice, A is the area, and L is the latent heat of fusion.

69. An ideal gas is carried through a thermodynamic cycle consisting of two isobaric and two isothermal processes as shown in Figure P20.69. Show that the net work done on the gas in the entire cycle is given by

$$W_{net} = -P_1(V_2 - V_1)\ln\frac{P_2}{P_1}$$

Figure P20.69

70. The inside of a hollow cylinder is maintained at a temperature T_a while the outside is at a lower temperature, T_b (Fig. P20.70). The wall of the cylinder has a thermal conductivity k. Ignoring end effects, show that the rate of energy conduction from the inner to the outer surface in the radial direction is

$$\frac{dQ}{dt} = 2\pi Lk\left[\frac{T_a - T_b}{\ln(b/a)}\right]$$

(*Suggestions:* The temperature gradient is dT/dr. Note that a radial energy current passes through a concentric cylinder of area $2\pi rL$.)

Figure P20.70

71. The passenger section of a jet airliner is in the shape of a cylindrical tube with a length of 35.0 m and an inner radius of 2.50 m. Its walls are lined with an insulating material 6.00 cm in thickness and having a thermal conductivity of 4.00×10^{-5} cal/s·cm·°C. A heater must maintain the interior temperature at 25.0°C while the outside temperature is −35.0°C. What power must be supplied to the heater? (Use the result of Problem 70.)

72. A student obtains the following data in a calorimetry experiment designed to measure the specific heat of aluminum:

Initial temperature of water and calorimeter:	70°C
Mass of water:	0.400 kg
Mass of calorimeter:	0.040 kg
Specific heat of calorimeter:	0.63 kJ/kg·°C
Initial temperature of aluminum:	27°C
Mass of aluminum:	0.200 kg
Final temperature of mixture:	66.3°C

Use these data to determine the specific heat of aluminum. Your result should be within 15% of the value listed in Table 20.1.

73. During periods of high activity, the Sun has more sunspots than usual. Sunspots are cooler than the rest of the luminous layer of the Sun's atmosphere (the photosphere). Paradoxically, the total power output of the active Sun is not lower than average but is the same or slightly higher than average. Work out the details of the following crude model of this phenomenon. Consider a patch of the photosphere with an area of 5.10×10^{14} m^2. Its emissivity is 0.965. (a) Find the power it radiates if its temperature is uniformly 5 800 K, corresponding to the quiet Sun. (b) To represent a sunspot, assume that 10.0% of the area is at 4 800 K and the other 90.0% is at 5 890 K. That is, a section with the surface area of the Earth is 1 000 K cooler than before and a section nine times as large is 90 K warmer. Find the average temperature of the patch. (c) Find the power output of the patch. Compare it with the answer to part (a). (The next sunspot maximum is expected around the year 2012.)

Answers to Quick Quizzes

20.1 Water, glass, iron. Because water has the highest specific heat (4 186 J/kg · °C), it has the smallest change in temperature. Glass is next (837 J/kg · °C), and iron is last (448 J/kg · °C).

20.2 Iron, glass, water. For a given temperature increase, the energy transfer by heat is proportional to the specific heat.

20.3 The figure below shows a graphical representation of the internal energy of the ice in parts A to E as a

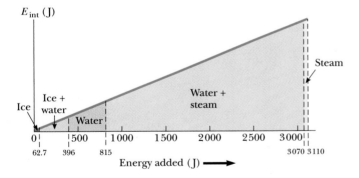

function of energy added. Notice that this graph looks quite different from Figure 20.2—it doesn't have the flat portions during the phase changes. Regardless of how the temperature is varying in Figure 20.2, the internal energy of the system simply increases linearly with energy input.

20.4 C, A, E. The slope is the ratio of the temperature change to the amount of energy input. Thus, the slope is proportional to the reciprocal of the specific heat. Water, which has the highest specific heat, has the smallest slope.

20.5

Situation	System	Q	W	ΔE_{int}
(a) Rapidly pumping up a bicycle tire	Air in the pump	0	+	+
(b) Pan of room-temperature water sitting on a hot stove	Water in the pan	+	0	+
(c) Air quickly leaking out of a balloon	Air originally in the balloon	0	−	−

(a) Because the pumping is rapid, no energy enters or leaves the system by heat. Because $W > 0$ when work is done *on* the system, it is positive here. Thus, we see that $\Delta E_{int} = Q + W$ must be positive. The air in the pump is warmer. (b) There is no work done either on or by the system, but energy transfers into the water by heat from the hot burner, making both Q and ΔE_{int} positive. (c) Again no energy transfers into or out of the system by heat, but the air molecules escaping from the balloon do work on the surrounding air molecules as they push them out of the way. Thus W is negative and ΔE_{int} is negative. The decrease in internal energy is evidenced by the fact that the escaping air becomes cooler.

20.6 A is isovolumetric, B is adiabatic, C is isothermal, and D is isobaric.

20.7 (c). The blanket acts as a thermal insulator, slowing the transfer of energy by heat from the air into the cube.

20.8 (b). In parallel, the rods present a larger area through which energy can transfer and a smaller length.

Chapter 21

The Kinetic Theory of Gases

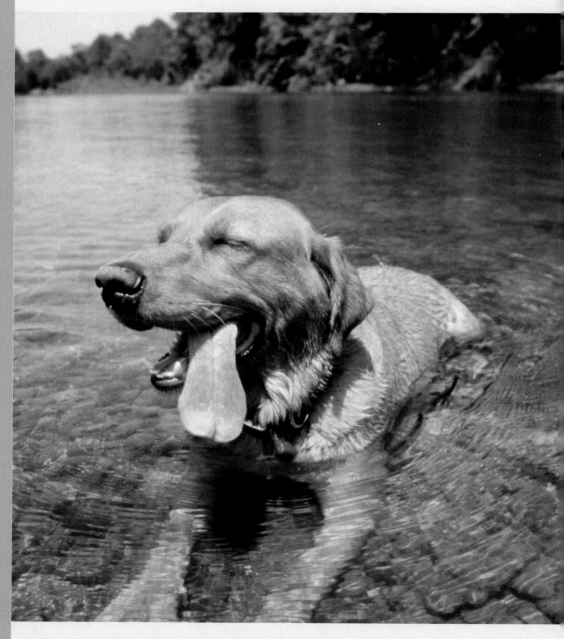

▲ *Dogs do not have sweat glands like humans. In hot weather, dogs pant to promote evaporation from the tongue. In this chapter, we show that evaporation is a cooling process based on the removal of molecules with high kinetic energy from a liquid. (Frank Oberle/Getty Images)*

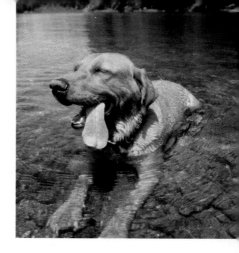

In Chapter 19 we discussed the properties of an ideal gas, using such macroscopic variables as pressure, volume, and temperature. We shall now show that such large-scale properties can be related to a description on a microscopic scale, where matter is treated as a collection of molecules. Newton's laws of motion applied in a statistical manner to a collection of particles provide a reasonable description of thermodynamic processes. To keep the mathematics relatively simple, we shall consider primarily the behavior of gases, because in gases the interactions between molecules are much weaker than they are in liquids or solids. In our model of gas behavior, called **kinetic theory**, gas molecules move about in a random fashion, colliding with the walls of their container and with each other. Kinetic theory provides us with a physical basis for our understanding of the concept of temperature.

21.1 Molecular Model of an Ideal Gas

We begin this chapter by developing a microscopic model of an ideal gas. The model shows that the pressure that a gas exerts on the walls of its container is a consequence of the collisions of the gas molecules with the walls and is consistent with the macroscopic description of Chapter 19. In developing this model, we make the following assumptions:

1. **The number of molecules in the gas is large, and the average separation between them is large compared with their dimensions.** This means that the molecules occupy a negligible volume in the container. This is consistent with the ideal gas model, in which we imagine the molecules to be point-like.

2. **The molecules obey Newton's laws of motion, but as a whole they move randomly.** By "randomly" we mean that any molecule can move in any direction with any speed. At any given moment, a certain percentage of molecules move at high speeds, and a certain percentage move at low speeds.

3. **The molecules interact only by short-range forces during elastic collisions.** This is consistent with the ideal gas model, in which the molecules exert no long-range forces on each other.

4. **The molecules make elastic collisions with the walls.**

5. **The gas under consideration is a pure substance; that is, all molecules are identical.**

Although we often picture an ideal gas as consisting of single atoms, we can assume that the behavior of molecular gases approximates that of ideal gases rather well at low pressures. Molecular rotations or vibrations have no effect, on the average, on the motions that we consider here.

For our first application of kinetic theory, let us derive an expression for the pressure of N molecules of an ideal gas in a container of volume V in terms of microscopic quantities. The container is a cube with edges of length d (Fig. 21.1). We shall first

▶ Assumptions of the molecular model of an ideal gas

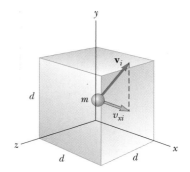

Figure 21.1 A cubical box with sides of length d containing an ideal gas. The molecule shown moves with velocity $\mathbf{v}_i$.

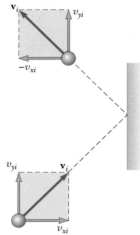

Active Figure 21.2 A molecule makes an elastic collision with the wall of the container. Its x component of momentum is reversed, while its y component remains unchanged. In this construction, we assume that the molecule moves in the xy plane.

At the Active Figures link at http://www.pse6.com, you can observe molecules within a container making collisions with the walls of the container and with each other.

focus our attention on one of these molecules of mass m, and assume that it is moving so that its component of velocity in the x direction is v_{xi} as in Figure 21.2. (The subscript i here refers to the ith molecule, not to an initial value. We will combine the effects of all of the molecules shortly.) As the molecule collides elastically with any wall (assumption 4), its velocity component perpendicular to the wall is reversed because the mass of the wall is far greater than the mass of the molecule. Because the momentum component p_{xi} of the molecule is mv_{xi} before the collision and $-mv_{xi}$ after the collision, the change in the x component of the momentum of the molecule is

$$\Delta p_{xi} = -mv_{xi} - (mv_{xi}) = -2mv_{xi}$$

Because the molecules obey Newton's laws (assumption 2), we can apply the impulse-momentum theorem (Eq. 9.8) to the molecule to give us

$$\bar{F}_{i,\text{on molecule}} \Delta t_{\text{collision}} = \Delta p_{xi} = -2mv_{xi}$$

where $\bar{F}_{i,\text{on molecule}}$ is the x component of the average force that the wall exerts on the molecule during the collision and $\Delta t_{\text{collision}}$ is the duration of the collision. In order for the molecule to make another collision with the same wall after this first collision, it must travel a distance of $2d$ in the x direction (across the container and back). Therefore, the time interval between two collisions with the same wall is

$$\Delta t = \frac{2d}{v_{xi}}$$

The force that causes the change in momentum of the molecule in the collision with the wall occurs only during the collision. However, we can average the force over the time interval for the molecule to move across the cube and back. Sometime during this time interval, the collision occurs, so that the change in momentum for this time interval is the same as that for the short duration of the collision. Thus, we can rewrite the impulse-momentum theorem as

$$\bar{F}_i \Delta t = -2mv_{xi}$$

where $\bar{F}_i$ is the average force component over the time for the molecule to move across the cube and back. Because exactly one collision occurs for each such time interval, this is also the long-term average force on the molecule, over long time intervals containing any number of multiples of Δt.

This equation and the preceding one enable us to express the x component of the long-term average force exerted by the wall on the molecule as

$$\bar{F}_i = \frac{-2mv_{xi}}{\Delta t} = \frac{-2mv_{xi}^2}{2d} = \frac{-mv_{xi}^2}{d}$$

Now, by Newton's third law, the average x component of the force exerted by the molecule on the wall is equal in magnitude and opposite in direction:

$$\bar{F}_{i,\text{on wall}} = -\bar{F}_i = -\left(\frac{-mv_{xi}^2}{d}\right) = \frac{mv_{xi}^2}{d}$$

The total average force $\bar{F}$ exerted by the gas on the wall is found by adding the average forces exerted by the individual molecules. We add terms such as that above for all molecules:

$$\bar{F} = \sum_{i=1}^{N} \frac{mv_{xi}^2}{d} = \frac{m}{d} \sum_{i=1}^{N} v_{xi}^2$$

where we have factored out the length of the box and the mass m, because assumption 5 tells us that all of the molecules are the same. We now impose assumption 1, that the number of molecules is large. For a small number of molecules, the actual force on the wall would vary with time. It would be nonzero during the short interval of a collision of a molecule with the wall and zero when no molecule happens to be hitting the wall. For a very large number of molecules, however, such as Avogadro's number, these variations in force are smoothed out, so that the average force given above is the same over *any* time interval. Thus, the *constant* force F on the wall due to the molecular collisions is

$$F = \frac{m}{d} \sum_{i=1}^{N} v_{xi}^2$$

To proceed further, let us consider how to express the average value of the square of the x component of the velocity for N molecules. The traditional average of a set of values is the sum of the values over the number of values:

$$\overline{v_x^2} = \frac{\sum_{i=1}^{N} v_{xi}^2}{N}$$

The numerator of this expression is contained in the right-hand side of the preceding equation. Thus, combining the two expressions, the total force on the wall can be written

$$F = \frac{m}{d} N\overline{v_x^2} \tag{21.1}$$

Now let us focus again on one molecule with velocity components v_{xi}, v_{yi}, and v_{zi}. The Pythagorean theorem relates the square of the speed of the molecule to the squares of the velocity components:

$$v_i^2 = v_{xi}^2 + v_{yi}^2 + v_{zi}^2$$

Hence, the average value of v^2 for all the molecules in the container is related to the average values of v_x^2, v_y^2, and v_z^2 according to the expression

$$\overline{v^2} = \overline{v_x^2} + \overline{v_y^2} + \overline{v_z^2}$$

Because the motion is completely random (assumption 2), the average values $\overline{v_x^2}$, $\overline{v_y^2}$, and $\overline{v_z^2}$ are equal to each other. Using this fact and the preceding equation, we find that

$$\overline{v^2} = 3\overline{v_x^2}$$

Thus, from Equation 21.1, the total force exerted on the wall is

$$F = \frac{N}{3}\left(\frac{m\overline{v^2}}{d}\right)$$

Using this expression, we can find the total pressure exerted on the wall:

$$P = \frac{F}{A} = \frac{F}{d^2} = \frac{1}{3}\left(\frac{N}{d^3}\, m\overline{v^2}\right) = \frac{1}{3}\left(\frac{N}{V}\right)m\overline{v^2}$$

$$P = \frac{2}{3}\left(\frac{N}{V}\right)\left(\frac{1}{2}m\overline{v^2}\right) \tag{21.2}$$

Relationship between pressure and molecular kinetic energy

This result indicates that **the pressure of a gas is proportional to the number of molecules per unit volume and to the average translational kinetic energy of the molecules, $\frac{1}{2}m\overline{v^2}$.** In analyzing this simplified model of an ideal gas, we obtain an important result that relates the macroscopic quantity of pressure to a microscopic quantity—the average value of the square of the molecular speed. Thus, we have established a key link between the molecular world and the large-scale world.

You should note that Equation 21.2 verifies some features of pressure with which you are probably familiar. One way to increase the pressure inside a container is to increase the number of molecules per unit volume N/V in the container. This is what you do when you add air to a tire. The pressure in the tire can also be increased by increasing the average translational kinetic energy of the air molecules in the tire.

This can be accomplished by increasing the temperature of that air, as we shall soon show mathematically. This is why the pressure inside a tire increases as the tire warms up during long trips. The continuous flexing of the tire as it moves along the road surface results in work done as parts of the tire distort, causing an increase in internal energy of the rubber. The increased temperature of the rubber results in the transfer of energy by heat into the air inside the tire. This transfer increases the air's temperature, and this increase in temperature in turn produces an increase in pressure.

Molecular Interpretation of Temperature

We can gain some insight into the meaning of temperature by first writing Equation 21.2 in the form

$$PV = \tfrac{2}{3}N(\tfrac{1}{2}m\overline{v^2})$$

Let us now compare this with the equation of state for an ideal gas (Eq. 19.10):

$$PV = Nk_BT$$

Recall that the equation of state is based on experimental facts concerning the macroscopic behavior of gases. Equating the right sides of these expressions, we find that

$$T = \frac{2}{3k_B}(\tfrac{1}{2}m\overline{v^2}) \qquad (21.3)$$

Temperature is proportional to average kinetic energy

This result tells us that temperature is a direct measure of average molecular kinetic energy. By rearranging Equation 21.3, we can relate the translational molecular kinetic energy to the temperature:

Average kinetic energy per molecule

$$\tfrac{1}{2}m\overline{v^2} = \tfrac{3}{2}k_BT \qquad (21.4)$$

That is, the average translational kinetic energy per molecule is $\tfrac{3}{2}k_BT$. Because $\overline{v_x^2} = \tfrac{1}{3}\overline{v^2}$, it follows that

$$\tfrac{1}{2}m\overline{v_x^2} = \tfrac{1}{2}k_BT \qquad (21.5)$$

In a similar manner, it follows that the motions in the y and z directions give us

$$\tfrac{1}{2}m\overline{v_y^2} = \tfrac{1}{2}k_BT \qquad \text{and} \qquad \tfrac{1}{2}m\overline{v_z^2} = \tfrac{1}{2}k_BT$$

Thus, each translational degree of freedom contributes an equal amount of energy, $\tfrac{1}{2}k_BT$, to the gas. (In general, a "degree of freedom" refers to an independent means by which a molecule can possess energy.) A generalization of this result, known as the **theorem of equipartition of energy**, states that

Theorem of equipartition of energy

each degree of freedom contributes $\tfrac{1}{2}k_BT$ to the energy of a system, where possible degrees of freedom in addition to those associated with translation arise from rotation and vibration of molecules.

The total translational kinetic energy of N molecules of gas is simply N times the average energy per molecule, which is given by Equation 21.4:

Total translational kinetic energy of N molecules

$$K_{\text{tot trans}} = N(\tfrac{1}{2}m\overline{v^2}) = \tfrac{3}{2}Nk_BT = \tfrac{3}{2}nRT \qquad (21.6)$$

where we have used $k_B = R/N_A$ for Boltzmann's constant and $n = N/N_A$ for the number of moles of gas. If we consider a gas in which molecules possess only translational kinetic energy, Equation 21.6 represents the internal energy of the gas. This result implies that **the internal energy of an ideal gas depends only on the temperature.** We will follow up on this point in Section 21.2.

The square root of $\overline{v^2}$ is called the *root-mean-square* (rms) *speed* of the molecules. From Equation 21.4 we find that the rms speed is

$$v_{\text{rms}} = \sqrt{\overline{v^2}} = \sqrt{\frac{3k_{\text{B}}T}{m}} = \sqrt{\frac{3RT}{M}} \qquad (21.7)$$

◄ Root-mean-square speed

where M is the molar mass in kilograms per mole and is equal to mN_{A}. This expression shows that, at a given temperature, lighter molecules move faster, on the average, than do heavier molecules. For example, at a given temperature, hydrogen molecules, whose molar mass is 2.02×10^{-3} kg/mol, have an average speed approximately four times that of oxygen molecules, whose molar mass is 32.0×10^{-3} kg/mol. Table 21.1 lists the rms speeds for various molecules at 20°C.

Table 21.1

Some rms Speeds		
Gas	**Molar mass (g/mol)**	v_{rms} **at 20°C (m/s)**
H_2	2.02	1 902
He	4.00	1 352
H_2O	18.0	637
Ne	20.2	602
N_2 or CO	28.0	511
NO	30.0	494
O_2	32.0	478
CO_2	44.0	408
SO_2	64.1	338

 PITFALL PREVENTION

21.1 The Square Root of the Square?

Notice that taking the square root of $\overline{v^2}$ does not "undo" the square because we have taken an average *between* squaring and taking the square root. While the square root of $(\overline{v})^2$ is $\overline{v}$ because the squaring is done after the averaging, the square root of $\overline{v^2}$ is *not* $\overline{v}$, but rather v_{rms}.

Example 21.1   A Tank of Helium

A tank used for filling helium balloons has a volume of 0.300 m³ and contains 2.00 mol of helium gas at 20.0°C. Assume that the helium behaves like an ideal gas.

(A) What is the total translational kinetic energy of the gas molecules?

Solution Using Equation 21.6 with $n = 2.00$ mol and $T = 293$ K, we find that

$$K_{\text{tot trans}} = \tfrac{3}{2}nRT = \tfrac{3}{2}(2.00 \text{ mol})(8.31 \text{ J/mol·K})(293 \text{ K})$$

$$= \boxed{7.30 \times 10^3 \text{ J}}$$

(B) What is the average kinetic energy per molecule?

Solution Using Equation 21.4, we find that the average kinetic energy per molecule is

$$\tfrac{1}{2}m\overline{v^2} = \tfrac{3}{2}k_{\text{B}}T = \tfrac{3}{2}(1.38 \times 10^{-23} \text{ J/K})(293 \text{ K})$$

$$= \boxed{6.07 \times 10^{-21} \text{ J}}$$

What If? What if the temperature is raised from 20.0°C to 40.0°C? Because 40.0 is twice as large as 20.0, is the total translational energy of the molecules of the gas twice as large at the higher temperature?

Answer The expression for the total translational energy depends on the temperature, and the value for the temperature must be expressed in kelvins, not in degrees Celsius. Thus, the ratio of 40.0 to 20.0 is *not* the appropriate ratio. Converting the Celsius temperatures to kelvins, 20.0°C is 293 K and 40.0°C is 313 K. Thus, the total translational energy increases by a factor of 313 K/293 K = 1.07.

Quick Quiz 21.1 Two containers hold an ideal gas at the same temperature and pressure. Both containers hold the same type of gas but container B has twice the volume of container A. The average translational kinetic energy per molecule in container B is (a) twice that for container A (b) the same as that for container A (c) half that for container A (d) impossible to determine.

Quick Quiz 21.2 Consider again the situation in Quick Quiz 21.1. The internal energy of the gas in container B is (a) twice that for container A (b) the same as that for container A (c) half that for container A (d) impossible to determine.

Quick Quiz 21.3 Consider again the situation in Quick Quiz 21.1. The rms speed of the gas molecules in container B is (a) twice that for container A (b) the same as that for container A (c) half that for container A (d) impossible to determine.

21.2 Molar Specific Heat of an Ideal Gas

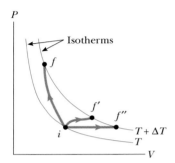

Figure 21.3 An ideal gas is taken from one isotherm at temperature T to another at temperature $T + \Delta T$ along three different paths.

Consider an ideal gas undergoing several processes such that the change in temperature is $\Delta T = T_f - T_i$ for all processes. The temperature change can be achieved by taking a variety of paths from one isotherm to another, as shown in Figure 21.3. Because ΔT is the same for each path, the change in internal energy ΔE_{int} is the same for all paths. However, we know from the first law, $Q = \Delta E_{\text{int}} - W$, that the heat Q is different for each path because W (the negative of the area under the curves) is different for each path. Thus, the heat associated with a given change in temperature does *not* have a unique value.

We can address this difficulty by defining specific heats for two processes that frequently occur: changes at constant volume and changes at constant pressure. Because the number of moles is a convenient measure of the amount of gas, we define the **molar specific heats** associated with these processes with the following equations:

$$Q = nC_V \Delta T \qquad \text{(constant volume)} \qquad (21.8)$$

$$Q = nC_P \Delta T \qquad \text{(constant pressure)} \qquad (21.9)$$

where C_V is the **molar specific heat at constant volume** and C_P is the **molar specific heat at constant pressure.** When we add energy to a gas by heat at constant pressure, not only does the internal energy of the gas increase, but work is done on the gas because of the change in volume. Therefore, the heat $Q_{\text{constant } P}$ must account for both the increase in internal energy and the transfer of energy out of the system by work. For this reason, $Q_{\text{constant } P}$ is greater than $Q_{\text{constant } V}$ for given values of n and ΔT. Thus, C_P is greater than C_V.

In the previous section, we found that the temperature of a gas is a measure of the average translational kinetic energy of the gas molecules. This kinetic energy is associated with the motion of the center of mass of each molecule. It does not include the energy associated with the internal motion of the molecule—namely, vibrations and rotations about the center of mass. This should not be surprising because the simple kinetic theory model assumes a structureless molecule.

In view of this, let us first consider the simplest case of an ideal monatomic gas, that is, a gas containing one atom per molecule, such as helium, neon, or argon. When energy is added to a monatomic gas in a container of fixed volume, all of the added energy goes into increasing the translational kinetic energy of the atoms. There is no other way to store the energy in a monatomic gas. Therefore, from Equation 21.6, we see that the internal energy E_{int} of N molecules (or n mol) of an ideal monatomic gas is

Internal energy of an ideal monatomic gas

$$E_{\text{int}} = K_{\text{tot trans}} = \tfrac{3}{2}Nk_B T = \tfrac{3}{2}nRT \qquad (21.10)$$

Note that for a monatomic ideal gas, E_{int} is a function of T only, and the functional relationship is given by Equation 21.10. In general, the internal energy of an ideal gas is a function of T only, and the exact relationship depends on the type of gas.

If energy is transferred by heat to a system at *constant volume*, then no work is done on the system. That is, $W = - \int P\,dV = 0$ for a constant-volume process. Hence, from the first law of thermodynamics, we see that

$$Q = \Delta E_{\text{int}} \tag{21.11}$$

In other words, all of the energy transferred by heat goes into increasing the internal energy of the system. A constant-volume process from i to f for an ideal gas is described in Figure 21.4, where ΔT is the temperature difference between the two isotherms. Substituting the expression for Q given by Equation 21.8 into Equation 21.11, we obtain

$$\Delta E_{\text{int}} = nC_V\,\Delta T \tag{21.12}$$

If the molar specific heat is constant, we can express the internal energy of a gas as

$$E_{\text{int}} = nC_V T$$

This equation applies to all ideal gases—to gases having more than one atom per molecule as well as to monatomic ideal gases. In the limit of infinitesimal changes, we can use Equation 21.12 to express the molar specific heat at constant volume as

$$C_V = \frac{1}{n}\frac{dE_{\text{int}}}{dT} \tag{21.13}$$

Let us now apply the results of this discussion to the monatomic gas that we have been studying. Substituting the internal energy from Equation 21.10 into Equation 21.13, we find that

$$C_V = \tfrac{3}{2}R \tag{21.14}$$

This expression predicts a value of $C_V = \tfrac{3}{2}R = 12.5 \text{ J/mol}\cdot\text{K}$ for *all* monatomic gases. This prediction is in excellent agreement with measured values of molar specific heats for such gases as helium, neon, argon, and xenon over a wide range of temperatures (Table 21.2). Small variations in Table 21.2 from the predicted values are due to the

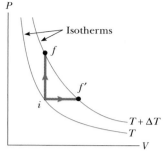

Active Figure 21.4 Energy is transferred by heat to an ideal gas in two ways. For the constant-volume path $i \rightarrow f$, all the energy goes into increasing the internal energy of the gas because no work is done. Along the constant-pressure path $i \rightarrow f'$, part of the energy transferred in by heat is transferred out by work.

 At the Active Figures link at http://www.pse6.com, *you can choose initial and final temperatures for one mole of an ideal gas undergoing constant-volume and constant pressure processes and measure Q, W, ΔE_{int}, C_V, and C_P.*

Table 21.2

Molar Specific Heats of Various Gases				
	Molar Specific Heat (J/mol · K)[a]			
Gas	**C_P**	**C_V**	**$C_P - C_V$**	**$\gamma = C_P/C_V$**
Monatomic Gases				
He	20.8	12.5	8.33	1.67
Ar	20.8	12.5	8.33	1.67
Ne	20.8	12.7	8.12	1.64
Kr	20.8	12.3	8.49	1.69
Diatomic Gases				
H_2	28.8	20.4	8.33	1.41
N_2	29.1	20.8	8.33	1.40
O_2	29.4	21.1	8.33	1.40
CO	29.3	21.0	8.33	1.40
Cl_2	34.7	25.7	8.96	1.35
Polyatomic Gases				
CO_2	37.0	28.5	8.50	1.30
SO_2	40.4	31.4	9.00	1.29
H_2O	35.4	27.0	8.37	1.30
CH_4	35.5	27.1	8.41	1.31

[a] All values except that for water were obtained at 300 K.

fact that real gases are not ideal gases. In real gases, weak intermolecular interactions occur, which are not addressed in our ideal gas model.

Now suppose that the gas is taken along the constant-pressure path $i \rightarrow f'$ shown in Figure 21.4. Along this path, the temperature again increases by ΔT. The energy that must be transferred by heat to the gas in this process is $Q = nC_P \Delta T$. Because the volume changes in this process, the work done on the gas is $W = -P \Delta V$ where P is the constant pressure at which the process occurs. Applying the first law of thermodynamics to this process, we have

$$\Delta E_{\text{int}} = Q + W = nC_P \Delta T + (-P \Delta V) \tag{21.15}$$

In this case, the energy added to the gas by heat is channeled as follows: Part of it leaves the system by work (that is, the gas moves a piston through a displacement), and the remainder appears as an increase in the internal energy of the gas. But the change in internal energy for the process $i \rightarrow f'$ is equal to that for the process $i \rightarrow f$ because E_{int} depends only on temperature for an ideal gas and because ΔT is the same for both processes. In addition, because $PV = nRT$, we note that for a constant-pressure process, $P \Delta V = nR \Delta T$. Substituting this value for $P \Delta V$ into Equation 21.15 with $\Delta E_{\text{int}} = nC_V \Delta T$ (Eq. 21.12) gives

$$nC_V \Delta T = nC_P \Delta T - nR \Delta T$$

$$C_P - C_V = R \tag{21.16}$$

This expression applies to *any* ideal gas. It predicts that the molar specific heat of an ideal gas at constant pressure is greater than the molar specific heat at constant volume by an amount R, the universal gas constant (which has the value 8.31 J/mol·K). This expression is applicable to real gases, as the data in Table 21.2 show.

Because $C_V = \frac{3}{2}R$ for a monatomic ideal gas, Equation 21.16 predicts a value $C_P = \frac{5}{2}R = 20.8$ J/mol·K for the molar specific heat of a monatomic gas at constant pressure. The ratio of these molar specific heats is a dimensionless quantity γ (Greek gamma):

Ratio of molar specific heats for a monatomic ideal gas

$$\gamma = \frac{C_P}{C_V} = \frac{5R/2}{3R/2} = \frac{5}{3} = 1.67 \tag{21.17}$$

Theoretical values of C_V, C_P and γ are in excellent agreement with experimental values obtained for monatomic gases, but they are in serious disagreement with the values for the more complex gases (see Table 21.2). This is not surprising because the value $C_V = \frac{3}{2}R$ was derived for a monatomic ideal gas and we expect some additional contribution to the molar specific heat from the internal structure of the more complex molecules. In Section 21.4, we describe the effect of molecular structure on the molar specific heat of a gas. The internal energy—and, hence, the molar specific heat—of a complex gas must include contributions from the rotational and the vibrational motions of the molecule.

In the case of solids and liquids heated at constant pressure, very little work is done because the thermal expansion is small. Consequently, C_P and C_V are approximately equal for solids and liquids.

Quick Quiz 21.4 How does the internal energy of an ideal gas change as it follows path $i \rightarrow f$ in Figure 21.4? (a) E_{int} increases. (b) E_{int} decreases. (c) E_{int} stays the same. (d) There is not enough information to determine how E_{int} changes.

Quick Quiz 21.5 How does the internal energy of an ideal gas change as it follows path $f \rightarrow f'$ along the isotherm labeled $T + \Delta T$ in Figure 21.4? (a) E_{int} increases. (b) E_{int} decreases. (c) E_{int} stays the same. (d) There is not enough information to determine how E_{int} changes.

| **Example 21.2** Heating a Cylinder of Helium |

A cylinder contains 3.00 mol of helium gas at a temperature of 300 K.

(A) If the gas is heated at constant volume, how much energy must be transferred by heat to the gas for its temperature to increase to 500 K?

Solution For the constant-volume process, we have

$$Q_1 = nC_V \, \Delta T$$

Because $C_V = 12.5 \, \text{J/mol} \cdot \text{K}$ for helium and $\Delta T = 200 \, \text{K}$, we obtain

$$Q_1 = (3.00 \, \text{mol})(12.5 \, \text{J/mol} \cdot \text{K})(200 \, \text{K})$$

$$= \boxed{7.50 \times 10^3 \, \text{J}}$$

(B) How much energy must be transferred by heat to the gas at constant pressure to raise the temperature to 500 K?

Solution Making use of Table 21.2, we obtain

$$Q_2 = nC_P \, \Delta T$$

$$= (3.00 \, \text{mol})(20.8 \, \text{J/mol} \cdot \text{K})(200 \, \text{K})$$

$$= \boxed{12.5 \times 10^3 \, \text{J}}$$

Note that this is larger than Q_1, due to the transfer of energy out of the gas by work in the constant pressure process.

21.3 Adiabatic Processes for an Ideal Gas

As we noted in Section 20.6, an **adiabatic process** is one in which no energy is transferred by heat between a system and its surroundings. For example, if a gas is compressed (or expanded) very rapidly, very little energy is transferred out of (or into) the system by heat, and so the process is nearly adiabatic. Such processes occur in the cycle of a gasoline engine, which we discuss in detail in the next chapter. Another example of an adiabatic process is the very slow expansion of a gas that is thermally insulated from its surroundings.

Suppose that an ideal gas undergoes an adiabatic expansion. At any time during the process, we assume that the gas is in an equilibrium state, so that the equation of state $PV = nRT$ is valid. As we show below, the pressure and volume of an ideal gas at any time during an adiabatic process are related by the expression

$$PV^\gamma = \text{constant} \qquad (21.18)$$

◄ Relationship between *P* and *V* for an adiabatic process involving an ideal gas

where $\gamma = C_P/C_V$ is assumed to be constant during the process. Thus, we see that all three variables in the ideal gas law—*P*, *V*, and *T*—change during an adiabatic process.

Proof That PV^γ = Constant for an Adiabatic Process

When a gas is compressed adiabatically in a thermally insulated cylinder, no energy is transferred by heat between the gas and its surroundings; thus, $Q = 0$. Let us imagine an infinitesimal change in volume dV and an accompanying infinitesimal change in temperature dT. The work done on the gas is $-P \, dV$. Because the internal energy of an ideal gas depends only on temperature, the change in the internal energy in an adiabatic process is the same as that for an isovolumetric process between the same temperatures, $dE_{\text{int}} = nC_V \, dT$ (Eq. 21.12). Hence, the first law of thermodynamics, $\Delta E_{\text{int}} = Q + W$, with $Q = 0$ becomes

$$dE_{\text{int}} = nC_V \, dT = -P \, dV$$

Taking the total differential of the equation of state of an ideal gas, $PV = nRT$, we see that

$$P \, dV + V \, dP = nR \, dT$$

Eliminating dT from these two equations, we find that

$$P \, dV + V \, dP = -\frac{R}{C_V} P \, dV$$

Figure 21.5 The *PV* diagram for an adiabatic compression. Note that $T_f > T_i$ in this process, so the temperature of the gas increases.

Substituting $R = C_P - C_V$ and dividing by PV, we obtain

$$\frac{dV}{V} + \frac{dP}{P} = -\left(\frac{C_P - C_V}{C_V}\right)\frac{dV}{V} = (1 - \gamma)\frac{dV}{V}$$

$$\frac{dP}{P} + \gamma\frac{dV}{V} = 0$$

Integrating this expression, we have

$$\ln P + \gamma \ln V = \text{constant}$$

which is equivalent to Equation 21.18:

$$PV^\gamma = \text{constant}$$

The *PV* diagram for an adiabatic compression is shown in Figure 21.5. Because $\gamma > 1$, the *PV* curve is steeper than it would be for an isothermal compression. By the definition of an adiabatic process, no energy is transferred by heat into or out of the system. Hence, from the first law, we see that ΔE_{int} is positive (work is done on the gas, so its internal energy increases) and so ΔT also is positive. Thus, the temperature of the gas increases ($T_f > T_i$) during an adiabatic compression. Conversely, the temperature decreases if the gas expands adiabatically.[1] Applying Equation 21.18 to the initial and final states, we see that

$$P_i V_i{}^\gamma = P_f V_f{}^\gamma \qquad (21.19)$$

Relationship between *T* and *V* for an adiabatic process involving an ideal gas

Using the ideal gas law, we can express Equation 21.19 as

$$T_i V_i{}^{\gamma-1} = T_f V_f{}^{\gamma-1} \qquad (21.20)$$

Example 21.3   A Diesel Engine Cylinder

Air at 20.0°C in the cylinder of a diesel engine is compressed from an initial pressure of 1.00 atm and volume of 800.0 cm³ to a volume of 60.0 cm³. Assume that air behaves as an ideal gas with $\gamma = 1.40$ and that the compression is adiabatic. Find the final pressure and temperature of the air.

Solution Conceptualize by imagining what happens if we compress a gas into a smaller volume. Our discussion above and Figure 21.5 tell us that the pressure and temperature both increase. We categorize this as a problem involving an adiabatic compression. To analyze the problem, we use Equation 21.19 to find the final pressure:

$$P_f = P_i\left(\frac{V_i}{V_f}\right)^\gamma = (1.00\text{ atm})\left(\frac{800.0\text{ cm}^3}{60.0\text{ cm}^3}\right)^{1.40}$$

$$= \boxed{37.6\text{ atm}}$$

Because $PV = nRT$ is valid throughout an ideal gas process and because no gas escapes from the cylinder,

$$\frac{P_i V_i}{T_i} = \frac{P_f V_f}{T_f}$$

$$T_f = \frac{P_f V_f}{P_i V_i}T_i = \frac{(37.6\text{ atm})(60.0\text{ cm}^3)}{(1.00\text{ atm})(800.0\text{ cm}^3)}(293\text{ K})$$

$$= 826\text{ K} = \boxed{553°C}$$

To finalize the problem, note that the temperature of the gas has increased by a factor of 2.82. The high compression in a diesel engine raises the temperature of the fuel enough to cause its combustion without the use of spark plugs.

21.4   The Equipartition of Energy

We have found that predictions based on our model for molar specific heat agree quite well with the behavior of monatomic gases but not with the behavior of complex gases (see Table 21.2). The value predicted by the model for the quantity $C_P - C_V = R$, however, is the same for all gases. This is not surprising because this difference is the result of the work done on the gas, which is independent of its molecular structure.

---

[1]   In the adiabatic free expansion discussed in Section 20.6, the temperature remains constant. This is a special process in which no work is done because the gas expands into a vacuum. In general, the temperature decreases in an adiabatic expansion in which work is done.

To clarify the variations in C_V and C_P in gases more complex than monatomic gases, let us explore further the origin of molar specific heat. So far, we have assumed that the sole contribution to the internal energy of a gas is the translational kinetic energy of the molecules. However, the internal energy of a gas includes contributions from the translational, vibrational, and rotational motion of the molecules. The rotational and vibrational motions of molecules can be activated by collisions and therefore are "coupled" to the translational motion of the molecules. The branch of physics known as *statistical mechanics* has shown that, for a large number of particles obeying the laws of Newtonian mechanics, the available energy is, on the average, shared equally by each independent degree of freedom. Recall from Section 21.1 that the equipartition theorem states that, at equilibrium, each degree of freedom contributes $\frac{1}{2}k_BT$ of energy per molecule.

Let us consider a diatomic gas whose molecules have the shape of a dumbbell (Fig. 21.6). In this model, the center of mass of the molecule can translate in the x, y, and z directions (Fig. 21.6a). In addition, the molecule can rotate about three mutually perpendicular axes (Fig. 21.6b). We can neglect the rotation about the y axis because the molecule's moment of inertia I_y and its rotational energy $\frac{1}{2}I_y\omega^2$ about this axis are negligible compared with those associated with the x and z axes. (If the two atoms are taken to be point masses, then I_y is identically zero.) Thus, there are five degrees of freedom for translation and rotation: three associated with the translational motion and two associated with the rotational motion. Because each degree of freedom contributes, on the average, $\frac{1}{2}k_BT$ of energy per molecule, the internal energy for a system of N molecules, ignoring vibration for now, is

$$E_{int} = 3N(\tfrac{1}{2}k_BT) + 2N(\tfrac{1}{2}k_BT) = \tfrac{5}{2}Nk_BT = \tfrac{5}{2}nRT$$

We can use this result and Equation 21.13 to find the molar specific heat at constant volume:

$$C_V = \frac{1}{n}\frac{dE_{int}}{dT} = \frac{1}{n}\frac{d}{dT}(\tfrac{5}{2}nRT) = \tfrac{5}{2}R \qquad (21.21)$$

From Equations 21.16 and 21.17, we find that

$$C_P = C_V + R = \tfrac{7}{2}R$$

$$\gamma = \frac{C_P}{C_V} = \frac{\tfrac{7}{2}R}{\tfrac{5}{2}R} = \frac{7}{5} = 1.40$$

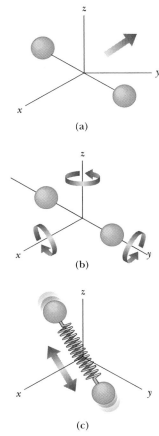

Figure 21.6 Possible motions of a diatomic molecule: (a) translational motion of the center of mass, (b) rotational motion about the various axes, and (c) vibrational motion along the molecular axis.

These results agree quite well with most of the data for diatomic molecules given in Table 21.2. This is rather surprising because we have not yet accounted for the possible vibrations of the molecule.

In the model for vibration, the two atoms are joined by an imaginary spring (see Fig. 21.6c). The vibrational motion adds two more degrees of freedom, which correspond to the kinetic energy and the potential energy associated with vibrations along the length of the molecule. Hence, classical physics and the equipartition theorem in a model that includes all three types of motion predict a total internal energy of

$$E_{int} = 3N(\tfrac{1}{2}k_BT) + 2N(\tfrac{1}{2}k_BT) + 2N(\tfrac{1}{2}k_BT) = \tfrac{7}{2}Nk_BT = \tfrac{7}{2}nRT$$

and a molar specific heat at constant volume of

$$C_V = \frac{1}{n}\frac{dE_{int}}{dT} = \frac{1}{n}\frac{d}{dT}(\tfrac{7}{2}nRT) = \tfrac{7}{2}R \qquad (21.22)$$

This value is inconsistent with experimental data for molecules such as H_2 and N_2 (see Table 21.2) and suggests a breakdown of our model based on classical physics.

It might seem that our model is a failure for predicting molar specific heats for diatomic gases. We can claim some success for our model, however, if measurements of molar specific heat are made over a wide temperature range, rather than at the

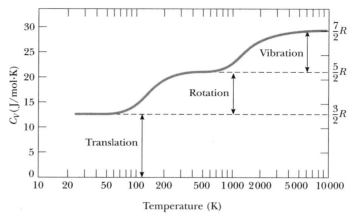

Figure 21.7 The molar specific heat of hydrogen as a function of temperature. The horizontal scale is logarithmic. Note that hydrogen liquefies at 20 K.

single temperature that gives us the values in Table 21.2. Figure 21.7 shows the molar specific heat of hydrogen as a function of temperature. There are three plateaus in the curve. The remarkable feature of these plateaus is that they are at the values of the molar specific heat predicted by Equations 21.14, 21.21, and 21.22! For low temperatures, the diatomic hydrogen gas behaves like a monatomic gas. As the temperature rises to room temperature, its molar specific heat rises to a value for a diatomic gas, consistent with the inclusion of rotation but not vibration. For high temperatures, the molar specific heat is consistent with a model including all types of motion.

Before addressing the reason for this mysterious behavior, let us make a brief remark about polyatomic gases. For molecules with more than two atoms, the vibrations are more complex than for diatomic molecules and the number of degrees of freedom is even larger. This results in an even higher predicted molar specific heat, which is in qualitative agreement with experiment. For the polyatomic gases shown in Table 21.2 we see that the molar specific heats are higher than those for diatomic gases. The more degrees of freedom available to a molecule, the more "ways" there are to store energy, resulting in a higher molar specific heat.

A Hint of Energy Quantization

Our model for molar specific heats has been based so far on purely classical notions. It predicts a value of the specific heat for a diatomic gas that, according to Figure 21.7, only agrees with experimental measurements made at high temperatures. In order to explain why this value is only true at high temperatures and why the plateaus exist in Figure 21.7, we must go beyond classical physics and introduce some quantum physics into the model. In Chapter 18, we discussed quantization of frequency for vibrating strings and air columns. This is a natural result whenever waves are subject to boundary conditions.

Quantum physics (Chapters 40 to 43) shows that atoms and molecules can be described by the physics of waves under boundary conditions. Consequently, these waves have quantized frequencies. Furthermore, in quantum physics, the energy of a system is proportional to the frequency of the wave representing the system. Hence, **the energies of atoms and molecules are quantized.**

For a molecule, quantum physics tells us that the rotational and vibrational energies are quantized. Figure 21.8 shows an **energy-level diagram** for the rotational and vibrational quantum states of a diatomic molecule. The lowest allowed state is called the **ground state**. Notice that vibrational states are separated by larger energy gaps than are rotational states.

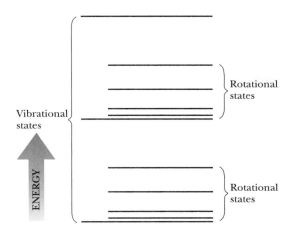

Figure 21.8 An energy-level diagram for vibrational and rotational states of a diatomic molecule. Note that the rotational states lie closer together in energy than the vibrational states.

At low temperatures, the energy that a molecule gains in collisions with its neighbors is generally not large enough to raise it to the first excited state of either rotation or vibration. Thus, even though rotation and vibration are classically allowed, they do not occur at low temperatures. All molecules are in the ground state for rotation and vibration. Thus, the only contribution to the molecules' average energy is from translation, and the specific heat is that predicted by Equation 21.14.

As the temperature is raised, the average energy of the molecules increases. In some collisions, a molecule may have enough energy transferred to it from another molecule to excite the first rotational state. As the temperature is raised further, more molecules can be excited to this state. The result is that rotation begins to contribute to the internal energy and the molar specific heat rises. At about room temperature in Figure 21.7, the second plateau has been reached and rotation contributes fully to the molar specific heat. The molar specific heat is now equal to the value predicted by Equation 21.21.

There is no contribution at room temperature from vibration, because the molecules are still in the ground vibrational state. The temperature must be raised even further to excite the first vibrational state. This happens in Figure 21.7 between 1 000 K and 10 000 K. At 10 000 K on the right side of the figure, vibration is contributing fully to the internal energy and the molar specific heat has the value predicted by Equation 21.22.

The predictions of this model are supportive of the theorem of equipartition of energy. In addition, the inclusion in the model of energy quantization from quantum physics allows a full understanding of Figure 21.7.

Quick Quiz 21.6 The molar specific heat of a diatomic gas is measured at constant volume and found to be 29.1 J/mol · K. The types of energy that are contributing to the molar specific heat are (a) translation only (b) translation and rotation only (c) translation and vibration only (d) translation, rotation, and vibration.

Quick Quiz 21.7 The molar specific heat of a gas is measured at constant volume and found to be $11R/2$. The gas is most likely to be (a) monatomic (b) diatomic   (c) polyatomic.

The Molar Specific Heat of Solids

The molar specific heats of solids also demonstrate a marked temperature dependence. Solids have molar specific heats that generally decrease in a nonlinear manner with decreasing temperature and approach zero as the temperature approaches

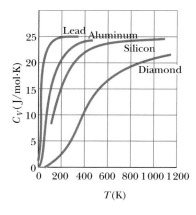

Figure 21.9 Molar specific heat of four solids. As T approaches zero, the molar specific heat also approaches zero.

absolute zero. At high temperatures (usually above 300 K), the molar specific heats approach the value of $3R \approx 25 \text{ J/mol} \cdot \text{K}$, a result known as the *DuLong–Petit law*. The typical data shown in Figure 21.9 demonstrate the temperature dependence of the molar specific heats for several solids.

We can explain the molar specific heat of a solid at high temperatures using the equipartition theorem. For small displacements of an atom from its equilibrium position, each atom executes simple harmonic motion in the x, y, and z directions. The energy associated with vibrational motion in the x direction is

$$E = \tfrac{1}{2}mv_x^2 + \tfrac{1}{2}kx^2$$

The expressions for vibrational motions in the y and z directions are analogous. Therefore, each atom of the solid has six degrees of freedom. According to the equipartition theorem, this corresponds to an average vibrational energy of $6(\tfrac{1}{2}k_B T) = 3k_B T$ per atom. Therefore, the internal energy of a solid consisting of N atoms is

Total internal energy of a solid

$$E_{\text{int}} = 3Nk_B T = 3nRT \qquad (21.23)$$

From this result, we find that the molar specific heat of a solid at constant volume is

Molar specific heat of a solid at constant volume

$$C_V = \frac{1}{n}\frac{dE_{\text{int}}}{dT} = 3R \qquad (21.24)$$

This result is in agreement with the empirical DuLong–Petit law. The discrepancies between this model and the experimental data at low temperatures are again due to the inadequacy of classical physics in describing the world at the atomic level.

21.5   The Boltzmann Distribution Law

Thus far we have considered only average values of the energies of molecules in a gas and have not addressed the distribution of energies among molecules. In reality, the motion of the molecules is extremely chaotic. Any individual molecule is colliding with others at an enormous rate—typically, a billion times per second. Each collision results in a change in the speed and direction of motion of each of the participant molecules. Equation 21.7 shows that rms molecular speeds increase with increasing temperature. What is the relative number of molecules that possess some characteristic, such as energy within a certain range?

We shall address this question by considering the **number density** $n_V(E)$. This quantity, called a *distribution function*, is defined so that $n_V(E)\, dE$ is the number of molecules per unit volume with energy between E and $E + dE$. (Note that the ratio of the number of molecules that have the desired characteristic to the total number of molecules is the probability that a particular molecule has that characteristic.) In general, the number density is found from statistical mechanics to be

$$n_V(E) = n_0 e^{-E/k_B T}$$ (21.25)   **Boltzmann distribution law**

where n_0 is defined such that $n_0\, dE$ is the number of molecules per unit volume having energy between $E = 0$ and $E = dE$. This equation, known as the **Boltzmann distribution law,** is important in describing the statistical mechanics of a large number of molecules. It states that **the probability of finding the molecules in a particular energy state varies exponentially as the negative of the energy divided by $k_B T$.** All the molecules would fall into the lowest energy level if the thermal agitation at a temperature T did not excite the molecules to higher energy levels.

Example 21.4   Thermal Excitation of Atomic Energy Levels   **Interactive**

As we discussed in Section 21.4, atoms can occupy only certain discrete energy levels. Consider a gas at a temperature of 2 500 K whose atoms can occupy only two energy levels separated by 1.50 eV, where 1 eV (electron volt) is an energy unit equal to 1.60×10^{-19} J (Fig. 21.10). Determine the ratio of the number of atoms in the higher energy level to the number in the lower energy level.

Solution Equation 21.25 gives the relative number of atoms in a given energy level. In this case, the atom has two possible energies, E_1 and E_2, where E_1 is the lower energy level. Hence, the ratio of the number of atoms in the higher energy level to the number in the lower energy level is

$$(1) \qquad \frac{n_V(E_2)}{n_V(E_1)} = \frac{n_0 e^{-E_2/k_B T}}{n_0 e^{-E_1/k_B T}} = e^{-(E_2-E_1)/k_B T}$$

In this problem, $E_2 - E_1 = 1.50$ eV, and the denominator of the exponent is

$$k_B T = (1.38 \times 10^{-23}\,\text{J/K})(2\,500\,\text{K})\left(\frac{1\,\text{eV}}{1.60 \times 10^{-19}\,\text{J}}\right)$$

$$= 0.216\,\text{eV}$$

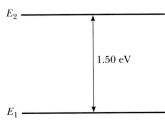

Figure 21.10 (Example 21.4) Energy-level diagram for a gas whose atoms can occupy two energy states.

Therefore, the required ratio is

$$\frac{n_V(E_2)}{n_V(E_1)} = e^{-1.50\,\text{eV}/0.216\,\text{eV}} = e^{-6.94}$$

$$= \boxed{9.64 \times 10^{-4}}$$

This result indicates that at $T = 2\,500$ K, only a small fraction of the atoms are in the higher energy level. In fact, for every atom in the higher energy level, there are about 1 000 atoms in the lower level. The number of atoms in the higher level increases at even higher temperatures, but the distribution law specifies that at equilibrium there are always more atoms in the lower level than in the higher level.

What If? What if the energy levels in Figure 21.10 were closer together in energy? Would this increase or decrease the fraction of the atoms in the upper energy level?

Answer If the excited level is lower in energy than that in Figure 21.10, it would be easier for thermal agitation to excite atoms to this level, and the fraction of atoms in this energy level would be larger. Let us see this mathematically by expressing Equation (1) as

$$r_2 = e^{-(E_2-E_1)/k_B T}$$

where r_2 is the ratio of atoms having energy E_2 to those with energy E_1. Differentiating with respect to E_2, we find

$$\frac{dr_2}{dE_2} = \frac{d}{dE_2}\left(e^{-(E_2-E_1)/k_B T}\right) = -\frac{1}{k_B T} e^{-(E_2-E_1)/k_B T} < 0$$

Because the derivative has a negative value, we see that as E_2 decreases, r_2 increases.

 At the Interactive Worked Example link at **http://www.pse6.com,** *you can investigate the effects of changing the temperature and the energy difference between the states.*

21.6   Distribution of Molecular Speeds

In 1860 James Clerk Maxwell (1831–1879) derived an expression that describes the distribution of molecular speeds in a very definite manner. His work and subsequent developments by other scientists were highly controversial because direct detection of molecules could not be achieved experimentally at that time. However, about 60 years later, experiments were devised that confirmed Maxwell's predictions.

Active Figure 21.11 The speed distribution of gas molecules at some temperature. The number of molecules having speeds in the range v to $v + dv$ is equal to the area of the shaded rectangle, $N_v dv$. The function N_v approaches zero as v approaches infinity.

At the Active Figures link at http://www.pse6.com, you can move the blue triangle and measure the number of molecules with speeds within a small range.

Let us consider a container of gas whose molecules have some distribution of speeds. Suppose we want to determine how many gas molecules have a speed in the range from, for example, 400 to 410 m/s. Intuitively, we expect that the speed distribution depends on temperature. Furthermore, we expect that the distribution peaks in the vicinity of v_{rms}. That is, few molecules are expected to have speeds much less than or much greater than v_{rms} because these extreme speeds result only from an unlikely chain of collisions.

The observed speed distribution of gas molecules in thermal equilibrium is shown in Figure 21.11. The quantity N_v, called the **Maxwell–Boltzmann speed distribution function,** is defined as follows. If N is the total number of molecules, then the number of molecules with speeds between v and $v + dv$ is $dN = N_v \, dv$. This number is also equal to the area of the shaded rectangle in Figure 21.11. Furthermore, the fraction of molecules with speeds between v and $v + dv$ is $(N_v \, dv)/N$. This fraction is also equal to the probability that a molecule has a speed in the range v to $v + dv$.

The fundamental expression that describes the distribution of speeds of N gas molecules is

$$N_v = 4\pi N \left(\frac{m}{2\pi k_B T} \right)^{3/2} v^2 e^{-mv^2/2k_B T} \tag{21.26}$$

where m is the mass of a gas molecule, k_B is Boltzmann's constant, and T is the absolute temperature.[2] Observe the appearance of the Boltzmann factor $e^{-E/k_B T}$ with $E = \frac{1}{2}mv^2$.

As indicated in Figure 21.11, the average speed is somewhat lower than the rms speed. The *most probable speed* v_{mp} is the speed at which the distribution curve reaches a peak. Using Equation 21.26, one finds that

$$v_{rms} = \sqrt{\overline{v^2}} = \sqrt{\frac{3k_B T}{m}} = 1.73 \sqrt{\frac{k_B T}{m}} \tag{21.27}$$

$$\overline{v} = \sqrt{\frac{8k_B T}{\pi m}} = 1.60 \sqrt{\frac{k_B T}{m}} \tag{21.28}$$

$$v_{mp} = \sqrt{\frac{2k_B T}{m}} = 1.41 \sqrt{\frac{k_B T}{m}} \tag{21.29}$$

Equation 21.27 has previously appeared as Equation 21.7. The details of the derivations of these equations from Equation 21.26 are left for the student (see Problems 39 and 65). From these equations, we see that

$$v_{rms} > \overline{v} > v_{mp}$$

Figure 21.12 represents speed distribution curves for nitrogen, N_2. The curves were obtained by using Equation 21.26 to evaluate the distribution function at various speeds and at two temperatures. Note that the peak in the curve shifts to the right as T increases, indicating that the average speed increases with increasing temperature, as expected. The asymmetric shape of the curves is due to the fact that the lowest speed possible is zero while the upper classical limit of the speed is infinity. (In Chapter 39, we will show that the actual upper limit is the speed of light.)

Equation 21.26 shows that the distribution of molecular speeds in a gas depends both on mass and on temperature. At a given temperature, the fraction of molecules with speeds exceeding a fixed value increases as the mass decreases. This explains why lighter molecules, such as H_2 and He, escape more readily from the Earth's atmosphere than do heavier molecules, such as N_2 and O_2. (See the discussion of escape speed in Chapter 13. Gas molecules escape even more readily from the Moon's surface than from the Earth's because the escape speed on the Moon is lower than that on the Earth.)

The speed distribution curves for molecules in a liquid are similar to those shown in Figure 21.12. We can understand the phenomenon of evaporation of a liquid from this distribution in speeds, using the fact that some molecules in the liquid are more

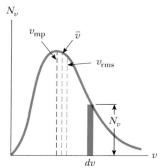

Ludwig Boltzmann
Austrian physicist (1844–1906)

Boltzmann made many important contributions to the development of the kinetic theory of gases, electromagnetism, and thermodynamics. His pioneering work in the field of kinetic theory led to the branch of physics known as statistical mechanics. *(Courtesy of AIP Niels Bohr Library, Lande Collection)*

[2]   For the derivation of this expression, see an advanced textbook on thermodynamics, such as that by R. P. Bauman, *Modern Thermodynamics with Statistical Mechanics*, New York, Macmillan Publishing Co., 1992.

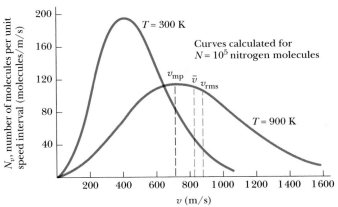

At the Active Figures link at http://www.pse6.com, *you can set the desired temperature and see the effect on the distribution curve.*

Active Figure 21.12 The speed distribution function for 10^5 nitrogen molecules at 300 K and 900 K. The total area under either curve is equal to the total number of molecules, which in this case equals 10^5. Note that $v_{rms} > \bar{v} > v_{mp}$.

energetic than others. Some of the faster-moving molecules in the liquid penetrate the surface and leave the liquid even at temperatures well below the boiling point. The molecules that escape the liquid by evaporation are those that have sufficient energy to overcome the attractive forces of the molecules in the liquid phase. Consequently, the molecules left behind in the liquid phase have a lower average kinetic energy; as a result, the temperature of the liquid decreases. Hence, evaporation is a cooling process. For example, an alcohol-soaked cloth often is placed on a feverish head to cool and comfort a patient.

Quick Quiz 21.8 Consider the qualitative shapes of the two curves in Figure 21.12, without regard for the numerical values or labels in the graph. Suppose you have two containers of gas *at the same temperature*. Container A has 10^5 nitrogen molecules and container B has 10^5 hydrogen molecules. The correct qualitative matching between the containers and the two curves in Figure 21.12 is (a) container A corresponds to the blue curve and container B to the brown curve (b) container B corresponds to the blue curve and container A to the brown curve (c) both containers correspond to the same curve.

Example 21.5   A System of Nine Particles

Nine particles have speeds of 5.00, 8.00, 12.0, 12.0, 12.0, 14.0, 14.0, 17.0, and 20.0 m/s.

(A) Find the particles' average speed.

Solution The average speed of the particles is the sum of the speeds divided by the total number of particles:

$$\bar{v} = \frac{(5.00 + 8.00 + 12.0 + 12.0 + 12.0 + 14.0 + 14.0 + 17.0 + 20.0) \text{ m/s}}{9}$$

$$= \boxed{12.7 \text{ m/s}}$$

(B) What is the rms speed of the particles?

Solution The average value of the square of the speed is

$$\overline{v^2} = \frac{\begin{array}{c}(5.00^2 + 8.00^2 + 12.0^2 + 12.0^2 + 12.0^2 \\ + 14.0^2 + 14.0^2 + 17.0^2 + 20.0^2) \text{ m}^2/\text{s}^2\end{array}}{9}$$

$$= 178 \text{ m}^2/\text{s}^2$$

Hence, the rms speed of the particles is

$$v_{rms} = \sqrt{\overline{v^2}} = \sqrt{178 \text{ m}^2/\text{s}^2} = \boxed{13.3 \text{ m/s}}$$

(C) What is the most probable speed of the particles?

Solution Three of the particles have a speed of 12.0 m/s, two have a speed of 14.0 m/s, and the remaining have different speeds. Hence, we see that the most probable speed v_{mp} is

$$\boxed{12.0 \text{ m/s.}}$$

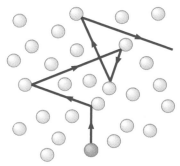

Figure 21.13 A molecule moving through a gas collides with other molecules in a random fashion. This behavior is sometimes referred to as a *random-walk process*. The mean free path increases as the number of molecules per unit volume decreases. Note that the motion is not limited to the plane of the paper.

21.7 Mean Free Path

Most of us are familiar with the fact that the strong odor associated with a gas such as ammonia may take a fraction of a minute to diffuse throughout a room. However, because average molecular speeds are typically several hundred meters per second at room temperature, we might expect a diffusion time of much less than one second. The reason for this difference is that molecules collide with one other because they are not geometrical points. Therefore, they do not travel from one side of a room to the other in a straight line. Between collisions, the molecules move with constant speed along straight lines. The average distance between collisions is called the **mean free path**. The path of an individual molecule is random and resembles that shown in Figure 21.13. As we would expect from this description, the mean free path is related to the diameter of the molecules and the density of the gas.

We now describe how to estimate the mean free path for a gas molecule. For this calculation, we assume that the molecules are spheres of diameter d. We see from Figure 21.14a that no two molecules collide unless their paths, assumed perpendicular to the page in Figure 21.14a are less than a distance d apart as the molecules approach each other. An equivalent way to describe the collisions is to imagine that one of the molecules has a diameter $2d$ and that the rest are geometrical points (Fig. 21.14b). Let us choose the large molecule to be one moving with the average speed $\bar{v}$. In a time interval Δt, this molecule travels a distance $\bar{v} \Delta t$. In this time interval, the molecule sweeps out a cylinder having a cross-sectional area πd^2 and a length $\bar{v} \Delta t$ (Fig. 21.15). Hence, the volume of the cylinder is $\pi d^2 \bar{v} \Delta t$. If n_V is the number of molecules per unit volume, then the number of point-size molecules in the cylinder is $(\pi d^2 \bar{v} \Delta t) n_V$. The molecule of equivalent diameter $2d$ collides with every molecule in this cylinder in the time interval Δt. Hence, the number of collisions in the time interval Δt is equal to the number of molecules in the cylinder, $(\pi d^2 \bar{v} \Delta t) n_V$.

The mean free path ℓ equals the average distance $\bar{v} \Delta t$ traveled in a time interval Δt divided by the number of collisions that occur in that time interval:

$$\ell = \frac{\bar{v} \Delta t}{(\pi d^2 \bar{v} \Delta t) n_V} = \frac{1}{\pi d^2 n_V}$$

Because the number of collisions in a time interval Δt is $(\pi d^2 \bar{v} \Delta t) n_V$, the number of collisions per unit time interval, or **collision frequency f,** is

$$f = \pi d^2 \bar{v} n_V$$

The inverse of the collision frequency is the average time interval between collisions, known as the **mean free time.**

Our analysis has assumed that molecules in the cylinder are stationary. When the motion of these molecules is included in the calculation, the correct results are

$$\ell = \frac{1}{\sqrt{2}\, \pi d^2 n_V} \qquad (21.30)$$

$$f = \sqrt{2}\, \pi d^2 \bar{v} n_V = \frac{\bar{v}}{\ell} \qquad (21.31)$$

Mean free path

Collision frequency

Figure 21.15 In a time interval Δt, a molecule of effective diameter $2d$ and moving to the right sweeps out a cylinder of length $\bar{v}\Delta t$ where $\bar{v}$ is its average speed. In this time interval, it collides with every point molecule within this cylinder.

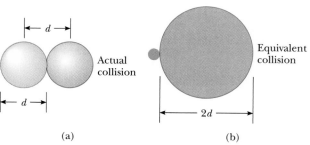

(a)   (b)

Figure 21.14 (a) Two spherical molecules, each of diameter d and moving along paths perpendicular to the page, collide if their paths are within a distance d of each other. (b) The collision between the two molecules is equivalent to a point molecule colliding with a molecule having an effective diameter of $2d$.

Example 21.6 Bouncing Around in the Air

Approximate the air around you as a collection of nitrogen molecules, each having a diameter of 2.00×10^{-10} m.

(A) How far does a typical molecule move before it collides with another molecule?

Solution Assuming that the gas is ideal, we can use the equation $PV = Nk_BT$ to obtain the number of molecules per unit volume under typical room conditions:

$$n_V = \frac{N}{V} = \frac{P}{k_B T} = \frac{1.01 \times 10^5 \text{ N/m}^2}{(1.38 \times 10^{-23} \text{ J/K})(293 \text{ K})}$$

$$= 2.50 \times 10^{25} \text{ molecules/m}^3$$

Hence, the mean free path is

$$\ell = \frac{1}{\sqrt{2}\pi d^2 n_V}$$

$$= \frac{1}{\sqrt{2}\pi(2.00 \times 10^{-10} \text{ m})^2(2.50 \times 10^{25} \text{ molecules/m}^3)}$$

$$= \boxed{2.25 \times 10^{-7} \text{ m}}$$

This value is about 10^3 times greater than the molecular diameter.

(B) On average, how frequently does one molecule collide with another?

Solution Because the rms speed of a nitrogen molecule at $20.0°C$ is 511 m/s (see Table 21.1), we know from Equations 21.27 and 21.28 that $\overline{v} = (1.60/1.73)(511 \text{ m/s}) = 473$ m/s. Therefore, the collision frequency is

$$f = \frac{\overline{v}}{\ell} = \frac{473 \text{ m/s}}{2.25 \times 10^{-7} \text{ m}} = \boxed{2.10 \times 10^9/\text{s}}$$

The molecule collides with other molecules at the average rate of about two billion times each second!

The mean free path ℓ is *not* the same as the average separation between particles. In fact, the average separation d between particles is approximately $n_V^{-1/3}$. In this example, the average molecular separation is

$$d = \frac{1}{n_V^{1/3}} = \frac{1}{(2.5 \times 10^{25})^{1/3}} = 3.4 \times 10^{-9} \text{ m}$$

SUMMARY

The pressure of N molecules of an ideal gas contained in a volume V is

$$P = \frac{2}{3}\frac{N}{V}\left(\frac{1}{2}m\overline{v^2}\right) \tag{21.2}$$

The average translational kinetic energy per molecule of a gas, $\frac{1}{2}m\overline{v^2}$, is related to the temperature T of the gas through the expression

$$\frac{1}{2}m\overline{v^2} = \frac{3}{2}k_B T \tag{21.4}$$

where k_B is Boltzmann's constant. Each translational degree of freedom (x, y, or z) has $\frac{1}{2}k_B T$ of energy associated with it.

The **theorem of equipartition of energy** states that the energy of a system in thermal equilibrium is equally divided among all degrees of freedom.

The internal energy of N molecules (or n mol) of an ideal monatomic gas is

$$E_{int} = \frac{3}{2}Nk_B T = \frac{3}{2}nRT \tag{21.10}$$

The change in internal energy for n mol of any ideal gas that undergoes a change in temperature ΔT is

$$\Delta E_{int} = nC_V \Delta T \tag{21.12}$$

where C_V is the molar specific heat at constant volume.

The molar specific heat of an ideal monatomic gas at constant volume is $C_V = \frac{3}{2}R$; the molar specific heat at constant pressure is $C_P = \frac{5}{2}R$. The ratio of specific heats is given by $\gamma = C_P/C_V = \frac{5}{3}$.

If an ideal gas undergoes an adiabatic expansion or compression, the first law of thermodynamics, together with the equation of state, shows that

$$PV^\gamma = \text{constant} \tag{21.18}$$

Take a practice test for this chapter by clicking on the Practice Test link at http://www.pse6.com.

The **Boltzmann distribution law** describes the distribution of particles among available energy states. The relative number of particles having energy between E and $E + dE$ is $n_V(E) \, dE$, where

$$n_V(E) = n_0 e^{-E/k_B T} \tag{21.25}$$

The **Maxwell–Boltzmann speed distribution function** describes the distribution of speeds of molecules in a gas:

$$N_v = 4\pi N \left(\frac{m}{2\pi k_B T} \right)^{3/2} v^2 e^{-mv^2/2k_B T} \tag{21.26}$$

This expression enables us to calculate the **root-mean-square speed,** the **average speed,** and **the most probable speed:**

$$v_{\text{rms}} = \sqrt{\overline{v^2}} = \sqrt{\frac{3k_B T}{m}} = 1.73 \sqrt{\frac{k_B T}{m}} \tag{21.27}$$

$$\overline{v} = \sqrt{\frac{8k_B T}{\pi m}} = 1.60 \sqrt{\frac{k_B T}{m}} \tag{21.28}$$

$$v_{\text{mp}} = \sqrt{\frac{2k_B T}{m}} = 1.41 \sqrt{\frac{k_B T}{m}} \tag{21.29}$$

QUESTIONS

1. Dalton's law of partial pressures states that the total pressure of a mixture of gases is equal to the sum of the partial pressures of gases making up the mixture. Give a convincing argument for this law based on the kinetic theory of gases.

2. One container is filled with helium gas and another with argon gas. If both containers are at the same temperature, which molecules have the higher rms speed? Explain.

3. A gas consists of a mixture of He and N_2 molecules. Do the lighter He molecules travel faster than the N_2 molecules? Explain.

4. Although the average speed of gas molecules in thermal equilibrium at some temperature is greater than zero, the average velocity is zero. Explain why this statement must be true.

5. When alcohol is rubbed on your body, it lowers your skin temperature. Explain this effect.

6. A liquid partially fills a container. Explain why the temperature of the liquid decreases if the container is then partially evacuated. (Using this technique, it is possible to freeze water at temperatures above 0°C.)

7. A vessel containing a fixed volume of gas is cooled. Does the mean free path of the molecules increase, decrease, or remain constant in the cooling process? What about the collision frequency?

8. A gas is compressed at a constant temperature. What happens to the mean free path of the molecules in this process?

9. If a helium-filled balloon initially at room temperature is placed in a freezer, will its volume increase, decrease, or remain the same?

10. Which is denser, dry air or air saturated with water vapor? Explain.

11. What happens to a helium-filled balloon released into the air? Will it expand or contract? Will it stop rising at some height?

12. Why does a diatomic gas have a greater energy content per mole than a monatomic gas at the same temperature?

13. An ideal gas is contained in a vessel at 300 K. If the temperature is increased to 900 K, by what factor does each one of the following change? (a) The average kinetic energy of the molecules. (b) The rms molecular speed. (c) The average momentum change of one molecule in a collision with a wall. (d) The rate of collisions of molecules with walls. (e) The pressure of the gas.

14. A vessel is filled with gas at some equilibrium pressure and temperature. Can all gas molecules in the vessel have the same speed?

15. In our model of the kinetic theory of gases, molecules were viewed as hard spheres colliding elastically with the walls of the container. Is this model realistic?

16. In view of the fact that hot air rises, why does it generally become cooler as you climb a mountain? (Note that air has low thermal conductivity.)

17. Inspecting the magnitudes of C_V and C_P for the diatomic and polyatomic gases in Table 21.2, we find that the values increase with increasing molecular mass. Give a qualitative explanation of this observation.

PROBLEMS

1, 2, 3 = straightforward, intermediate, challenging ☐ = full solution available in the *Student Solutions Manual and Study Guide*

🌐 = coached solution with hints available at http://www.pse6.com 💻 = computer useful in solving problem

▨ = paired numerical and symbolic problems

Section 21.1   Molecular Model of an Ideal Gas

1. In a 30.0-s interval, 500 hailstones strike a glass window of area 0.600 m^2 at an angle of 45.0° to the window surface. Each hailstone has a mass of 5.00 g and moves with a speed of 8.00 m/s. Assuming the collisions are elastic, find the average force and pressure on the window.

2. In a period of 1.00 s, 5.00×10^{23} nitrogen molecules strike a wall with an area of 8.00 cm^2. If the molecules move with a speed of 300 m/s and strike the wall head-on in elastic collisions, what is the pressure exerted on the wall? (The mass of one N_2 molecule is 4.68×10^{-26} kg.)

3. A sealed cubical container 20.0 cm on a side contains three times Avogadro's number of molecules at a temperature of 20.0°C. Find the force exerted by the gas on one of the walls of the container.

4. A 2.00-mol sample of oxygen gas is confined to a 5.00-L vessel at a pressure of 8.00 atm. Find the average translational kinetic energy of an oxygen molecule under these conditions.

5. A spherical balloon of volume 4 000 cm^3 contains helium at an (inside) pressure of 1.20×10^5 Pa. How many moles of helium are in the balloon if the average kinetic energy of the helium atoms is 3.60×10^{-22} J?

6. Use the definition of Avogadro's number to find the mass of a helium atom.

7. (a) How many atoms of helium gas fill a balloon having a diameter of 30.0 cm at 20.0°C and 1.00 atm? (b) What is the average kinetic energy of the helium atoms? (c) What is the root-mean-square speed of the helium atoms?

8. Given that the rms speed of a helium atom at a certain temperature is 1 350 m/s, find by proportion the rms speed of an oxygen (O_2) molecule at this temperature. The molar mass of O_2 is 32.0 g/mol, and the molar mass of He is 4.00 g/mol.

9. 🌐 A cylinder contains a mixture of helium and argon gas in equilibrium at 150°C. (a) What is the average kinetic energy for each type of gas molecule? (b) What is the root-mean-square speed of each type of molecule?

10. A 5.00-L vessel contains nitrogen gas at 27.0°C and a pressure of 3.00 atm. Find (a) the total translational kinetic energy of the gas molecules and (b) the average kinetic energy per molecule.

11. (a) Show that 1 Pa = 1 J/m^3. (b) Show that the density in space of the translational kinetic energy of an ideal gas is $3P/2$.

Section 21.2   Molar Specific Heat of an Ideal Gas

Note: You may use data in Table 21.2 about particular gases. Here we define a "monatomic ideal gas" to have molar specific heats $C_V = 3R/2$ and $C_P = 5R/2$, and a "diatomic ideal gas" to have $C_V = 5R/2$ and $C_P = 7R/2$.

12. Calculate the change in internal energy of 3.00 mol of helium gas when its temperature is increased by 2.00 K.

13. 🌐 A 1.00-mol sample of hydrogen gas is heated at constant pressure from 300 K to 420 K. Calculate (a) the energy transferred to the gas by heat, (b) the increase in its internal energy, and (c) the work done on the gas.

14. A 1.00-mol sample of air (a diatomic ideal gas) at 300 K, confined in a cylinder under a heavy piston, occupies a volume of 5.00 L. Determine the final volume of the gas after 4.40 kJ of energy is transferred to the air by heat.

15. In a constant-volume process, 209 J of energy is transferred by heat to 1.00 mol of an ideal monatomic gas initially at 300 K. Find (a) the increase in internal energy of the gas, (b) the work done on it, and (c) its final temperature.

16. A house has well-insulated walls. It contains a volume of 100 m^3 of air at 300 K. (a) Calculate the energy required to increase the temperature of this diatomic ideal gas by 1.00°C. (b) **What If?** If this energy could be used to lift an object of mass m through a height of 2.00 m, what is the value of m?

17. An incandescent lightbulb contains a volume V of argon at pressure P_i. The bulb is switched on and constant power $\mathcal{P}$ is transferred to the argon for a time interval Δt. (a) Show that the pressure P_f in the bulb at the end of this process is $P_f = P_i[1 + (\mathcal{P}\Delta t R)/(P_i V C_V)]$. (b) Find the pressure in a spherical light bulb 10.0 cm in diameter 4.00 s after it is switched on, given that it has initial pressure 1.00 atm and that 3.60 W of power is transferred to the gas.

18. A vertical cylinder with a heavy piston contains air at a temperature of 300 K. The initial pressure is 200 kPa, and the initial volume is 0.350 m^3. Take the molar mass of air as 28.9 g/mol and assume that $C_V = 5R/2$. (a) Find the specific heat of air at constant volume in units of J/kg·°C. (b) Calculate the mass of the air in the cylinder. (c) Suppose the piston is held fixed. Find the energy input required to raise the temperature of the air to 700 K. (d) **What If?** Assume again the conditions of the initial state and that the heavy piston is free to move. Find the energy input required to raise the temperature to 700 K.

19. A 1-L Thermos bottle is full of tea at 90°C. You pour out one cup and immediately screw the stopper back on. Make an order-of-magnitude estimate of the change in temperature of the tea remaining in the flask that results from the admission of air at room temperature. State the quantities you take as data and the values you measure or estimate for them.

20. A 1.00-mol sample of a diatomic ideal gas has pressure P and volume V. When the gas is heated, its pressure triples and its volume doubles. This heating process includes two steps, the first at constant pressure and the second at constant volume. Determine the amount of energy transferred to the gas by heat.

21. A 1.00-mol sample of an ideal monatomic gas is at an initial temperature of 300 K. The gas undergoes an isovolumetric process acquiring 500 J of energy by heat. It then undergoes an isobaric process losing this same amount of energy by heat. Determine (a) the new temperature of the gas and (b) the work done on the gas.

22. A vertical cylinder with a movable piston contains 1.00 mol of a diatomic ideal gas. The volume of the gas is V_i, and its temperature is T_i. Then the cylinder is set on a stove and additional weights are piled onto the piston as it moves up, in such a way that the pressure is proportional to the volume and the final volume is $2V_i$. (a) What is the final temperature? (b) How much energy is transferred to the gas by heat?

23. A container has a mixture of two gases: n_1 mol of gas 1 having molar specific heat C_1 and n_2 mol of gas 2 of molar specific heat C_2. (a) Find the molar specific heat of the mixture. (b) **What If?** What is the molar specific heat if the mixture has m gases in the amounts $n_1, n_2, n_3, \ldots, n_m$, with molar specific heats $C_1, C_2, C_3, \ldots, C_m$, respectively?

Section 21.3 Adiabatic Processes for an Ideal Gas

24. During the compression stroke of a certain gasoline engine, the pressure increases from 1.00 atm to 20.0 atm. If the process is adiabatic and the fuel–air mixture behaves as a diatomic ideal gas, (a) by what factor does the volume change and (b) by what factor does the temperature change? (c) Assuming that the compression starts with 0.016 0 mol of gas at 27.0°C, find the values of Q, W, and ΔE_{int} that characterize the process.

25. A 2.00-mol sample of a diatomic ideal gas expands slowly and adiabatically from a pressure of 5.00 atm and a volume of 12.0 L to a final volume of 30.0 L. (a) What is the final pressure of the gas? (b) What are the initial and final temperatures? (c) Find Q, W, and ΔE_{int}.

26. Air (a diatomic ideal gas) at 27.0°C and atmospheric pressure is drawn into a bicycle pump that has a cylinder with an inner diameter of 2.50 cm and length 50.0 cm. The down stroke adiabatically compresses the air, which reaches a gauge pressure of 800 kPa before entering the tire (Fig. P21.26). Determine (a) the volume of the compressed air and (b) the temperature of the compressed air. (c) **What If?** The pump is made of steel and has an inner wall that is 2.00 mm thick. Assume that 4.00 cm of the cylinder's length is allowed to come to thermal equilibrium with the air. What will be the increase in wall temperature?

27. Air in a thundercloud expands as it rises. If its initial temperature is 300 K and no energy is lost by thermal conduction on expansion, what is its temperature when the initial volume has doubled?

28. The largest bottle ever made by blowing glass has a volume of about 0.720 m³. Imagine that this bottle is filled with air that behaves as an ideal diatomic gas. The bottle is held with its opening at the bottom and rapidly submerged into the ocean. No air escapes or mixes with the water. No energy is exchanged with the ocean by heat. (a) If the final volume of the air is 0.240 m³, by what factor does the internal energy of the air increase? (b) If the bottle is submerged so that the air temperature doubles, how much volume is occupied by air?

Figure P21.26

29. A 4.00-L sample of a diatomic ideal gas with specific heat ratio 1.40, confined to a cylinder, is carried through a closed cycle. The gas is initially at 1.00 atm and at 300 K. First, its pressure is tripled under constant volume. Then, it expands adiabatically to its original pressure. Finally, the gas is compressed isobarically to its original volume. (a) Draw a PV diagram of this cycle. (b) Determine the volume of the gas at the end of the adiabatic expansion. (c) Find the temperature of the gas at the start of the adiabatic expansion. (d) Find the temperature at the end of the cycle. (e) What was the net work done on the gas for this cycle?

30. A diatomic ideal gas ($\gamma = 1.40$) confined to a cylinder is put through a closed cycle. Initially the gas is at P_i, V_i, and T_i. First, its pressure is tripled under constant volume. It then expands adiabatically to its original pressure and finally is compressed isobarically to its original volume. (a) Draw a PV diagram of this cycle. (b) Determine the volume at the end of the adiabatic expansion. Find (c) the temperature of the gas at the start of the adiabatic expansion and (d) the temperature at the end of the cycle. (e) What was the net work done on the gas for this cycle?

31. How much work is required to compress 5.00 mol of air at 20.0°C and 1.00 atm to one tenth of the original volume (a) by an isothermal process? (b) by an adiabatic process? (c) What is the final pressure in each of these two cases?

32. During the power stroke in a four-stroke automobile engine, the piston is forced down as the mixture of combustion products and air undergoes an adiabatic expansion (Fig. P21.32). Assume that (1) the engine is running at 2 500 cycles/min, (2) the gauge pressure right before the expansion is 20.0 atm, (3) the volumes of the mixture right before and after the expansion are 50.0 and 400 cm^3, respectively, (4) the time involved in the expansion is one-fourth that of the total cycle, and (5) the mixture behaves like an ideal gas with specific heat ratio 1.40. Find the average power generated during the expansion.

Before

After

Figure P21.32

Section 21.4 The Equipartition of Energy

33. Consider 2.00 mol of an ideal diatomic gas. (a) Find the total heat capacity of the gas at constant volume and at constant pressure assuming the molecules rotate but do not vibrate. (b) **What If?** Repeat, assuming the molecules both rotate and vibrate.

34. A certain molecule has f degrees of freedom. Show that an ideal gas consisting of such molecules has the following properties: (1) its total internal energy is $fnRT/2$; (2) its molar specific heat at constant volume is $fR/2$; (3) its molar specific heat at constant pressure is $(f+2)R/2$; (4) its specific heat ratio is $\gamma = C_P/C_V = (f+2)/f$.

35. In a crude model (Fig. P21.35) of a rotating diatomic molecule of chlorine (Cl$_2$), the two Cl atoms are 2.00×10^{-10} m apart and rotate about their center of mass with angular speed $\omega = 2.00 \times 10^{12}$ rad/s. What is the rotational kinetic energy of one molecule of Cl$_2$, which has a molar mass of 70.0 g/mol?

Figure P21.35

Section 21.5 The Boltzmann Distribution Law
Section 21.6 Distribution of Molecular Speeds

36. One cubic meter of atomic hydrogen at 0°C and atmospheric pressure contains approximately 2.70×10^{25} atoms. The first excited state of the hydrogen atom has an energy of 10.2 eV above the lowest energy level, called the ground state. Use the Boltzmann factor to find the number of atoms in the first excited state at 0°C and at 10 000°C.

37. Fifteen identical particles have various speeds: one has a speed of 2.00 m/s; two have speeds of 3.00 m/s; three have speeds of 5.00 m/s; four have speeds of 7.00 m/s; three have speeds of 9.00 m/s; and two have speeds of 12.0 m/s. Find (a) the average speed, (b) the rms speed, and (c) the most probable speed of these particles.

38. Two gases in a mixture diffuse through a filter at rates proportional to the gases' rms speeds. (a) Find the ratio of speeds for the two isotopes of chlorine, ^{35}Cl and ^{37}Cl, as they diffuse through the air. (b) Which isotope moves faster?

39. From the Maxwell–Boltzmann speed distribution, show that the most probable speed of a gas molecule is given by Equation 21.29. Note that the most probable speed corresponds to the point at which the slope of the speed distribution curve dN_v/dv is zero.

40. Helium gas is in thermal equilibrium with liquid helium at 4.20 K. Even though it is on the point of condensation, model the gas as ideal and determine the most probable speed of a helium atom (mass $= 6.64 \times 10^{-27}$ kg) in it.

41. **Review problem.** At what temperature would the average speed of helium atoms equal (a) the escape speed from Earth, 1.12×10^4 m/s and (b) the escape speed from the Moon, 2.37×10^3 m/s? (See Chapter 13 for a discussion of escape speed, and note that the mass of a helium atom is 6.64×10^{-27} kg.)

42. A gas is at 0°C. If we wish to double the rms speed of its molecules, to what temperature must the gas be brought?

43. Assume that the Earth's atmosphere has a uniform temperature of 20°C and uniform composition, with an effective molar mass of 28.9 g/mol. (a) Show that the number density of molecules depends on height according to

$$n_V(y) = n_0 e^{-mgy/k_B T}$$

where n_0 is the number density at sea level, where $y = 0$. This result is called the *law of atmospheres*. (b) Commercial jetliners typically cruise at an altitude of 11.0 km. Find the ratio of the atmospheric density there to the density at sea level.

44. *If you can't walk to outer space, can you at least walk halfway?* Using the law of atmospheres from Problem 43, we find that the average height of a molecule in the Earth's atmosphere is given by

$$\bar{y} = \frac{\displaystyle\int_0^\infty y n_V(y)\, dy}{\displaystyle\int_0^\infty n_V(y)\, dy} = \frac{\displaystyle\int_0^\infty y e^{-mgy/k_B T}\, dy}{\displaystyle\int_0^\infty e^{-mgy/k_B T}\, dy}$$

(a) Prove that this average height is equal to $k_B T/mg$. (b) Evaluate the average height, assuming the temperature is 10°C and the molecular mass is 28.9 u.

Section 21.7 Mean Free Path

45. In an ultra-high-vacuum system, the pressure is measured to be 1.00×10^{-10} torr (where 1 torr = 133 Pa). Assuming the molecular diameter is 3.00×10^{-10} m, the average molecular speed is 500 m/s, and the temperature is 300 K, find (a) the number of molecules in a volume of 1.00 m^3, (b) the mean free path of the molecules, and (c) the collision frequency.

46. In deep space the number density of particles can be one particle per cubic meter. Using the average temperature of 3.00 K and assuming the particle is H_2 with a diameter of 0.200 nm, (a) determine the mean free path of the particle and the average time between collisions. (b) **What If?** Repeat part (a) assuming a density of one particle per cubic centimeter.

47. Show that the mean free path for the molecules of an ideal gas is

$$\ell = \frac{k_B T}{\sqrt{2}\pi d^2 P}$$

where d is the molecular diameter.

48. In a tank full of oxygen, how many molecular diameters d (on average) does an oxygen molecule travel (at 1.00 atm and 20.0°C) before colliding with another O_2 molecule? (The diameter of the O_2 molecule is approximately 3.60×10^{-10} m.)

49. Argon gas at atmospheric pressure and 20.0°C is confined in a 1.00-m^3 vessel. The effective hard-sphere diameter of the argon atom is 3.10×10^{-10} m. (a) Determine the mean free path ℓ. (b) Find the pressure when $\ell = 1.00$ m. (c) Find the pressure when $\ell = 3.10 \times 10^{-10}$ m.

Additional Problems

50. The dimensions of a room are 4.20 m × 3.00 m × 2.50 m. (a) Find the number of molecules of air in the room at atmospheric pressure and 20.0°C. (b) Find the mass of this air, assuming that the air consists of diatomic molecules with molar mass 28.9 g/mol. (c) Find the average kinetic energy of one molecule. (d) Find the root-mean-square molecular speed. (e) On the assumption that the molar specific heat is a constant independent of temperature, we have $E_{int} = 5nRT/2$. Find the internal energy in the air. (f) **What If?** Find the internal energy of the air in the room at 25.0°C.

51. The function $E_{int} = 3.50nRT$ describes the internal energy of a certain ideal gas. A sample comprising 2.00 mol of the gas always starts at pressure 100 kPa and temperature 300 K. For each one of the following processes, determine the final pressure, volume, and temperature; the change in internal energy of the gas; the energy added to the gas by heat; and the work done on the gas. (a) The gas is heated at constant pressure to 400 K. (b) The gas is heated at constant volume to 400 K. (c) The gas is compressed at constant temperature to 120 kPa. (d) The gas is compressed adiabatically to 120 kPa.

52. Twenty particles, each of mass m and confined to a volume V, have various speeds: two have speed v; three have speed $2v$; five have speed $3v$; four have speed $4v$; three have speed

5v; two have speed 6v; one has speed 7v. Find (a) the average speed, (b) the rms speed, (c) the most probable speed, (d) the pressure the particles exert on the walls of the vessel, and (e) the average kinetic energy per particle.

53. A cylinder containing n mol of an ideal gas undergoes an adiabatic process. (a) Starting with the expression $W = -\int P\,dV$ and using the condition PV^γ = constant, show that the work done on the gas is

$$W = \left(\frac{1}{\gamma - 1}\right)(P_f V_f - P_i V_i)$$

(b) Starting with the first law of thermodynamics in differential form, prove that the work done on the gas is also equal to $nC_V(T_f - T_i)$. Show that this result is consistent with the equation in part (a).

54. As a 1.00-mol sample of a monatomic ideal gas expands adiabatically, the work done on it is −2 500 J. The initial temperature and pressure of the gas are 500 K and 3.60 atm. Calculate (a) the final temperature and (b) the final pressure. You may use the result of Problem 53.

55. A cylinder is closed at both ends and has insulating walls. It is divided into two compartments by a perfectly insulating partition that is perpendicular to the axis of the cylinder. Each compartment contains 1.00 mol of oxygen, which behaves as an ideal gas with $\gamma = 7/5$. Initially the two compartments have equal volumes, and their temperatures are 550 K and 250 K. The partition is then allowed to move slowly until the pressures on its two sides are equal. Find the final temperatures in the two compartments. You may use the result of Problem 53.

56. An air rifle shoots a lead pellet by allowing high-pressure air to expand, propelling the pellet down the rifle barrel. Because this process happens very quickly, no appreciable thermal conduction occurs, and the expansion is essentially adiabatic. Suppose that the rifle starts by admitting to the barrel 12.0 cm^3 of compressed air, which behaves as an ideal gas with $\gamma = 1.40$. The air expands behind a 1.10-g pellet and pushes on it as a piston with cross-sectional area 0.030 0 cm^2, as the pellet moves 50.0 cm along the gun barrel. The pellet emerges with muzzle speed 120 m/s. Use the result of problem 53 to find the initial pressure required.

57. **Review problem.** Oxygen at pressures much greater than 1 atm is toxic to lung cells. Assume that a deep-sea diver breathes a mixture of oxygen (O_2) and helium (He). By weight, what ratio of helium to oxygen must be used if the diver is at an ocean depth of 50.0 m?

58. A vessel contains 1.00×10^4 oxygen molecules at 500 K. (a) Make an accurate graph of the Maxwell–Boltzmann speed distribution function versus speed with points at speed intervals of 100 m/s. (b) Determine the most probable speed from this graph. (c) Calculate the average and rms speeds for the molecules and label these points on your graph. (d) From the graph, estimate the fraction of molecules with speeds in the range 300 m/s to 600 m/s.

59. The compressibility κ of a substance is defined as the fractional change in volume of that substance for a given change in pressure:

$$\kappa = -\frac{1}{V}\frac{dV}{dP}$$

(a) Explain why the negative sign in this expression ensures that κ is always positive. (b) Show that if an ideal gas is compressed isothermally, its compressibility is given by $\kappa_1 = 1/P$. (c) **What If?** Show that if an ideal gas is compressed adiabatically, its compressibility is given by $\kappa_2 = 1/\gamma P$. (d) Determine values for κ_1 and κ_2 for a monatomic ideal gas at a pressure of 2.00 atm.

60. **Review problem.** (a) Show that the speed of sound in an ideal gas is

$$v = \sqrt{\frac{\gamma R T}{M}}$$

where M is the molar mass. Use the general expression for the speed of sound in a fluid from Section 17.1, the definition of the bulk modulus from Section 12.4, and the result of Problem 59 in this chapter. As a sound wave passes through a gas, the compressions are either so rapid or so far apart that thermal conduction is prevented by a negligible time interval or by effective thickness of insulation. The compressions and rarefactions are adiabatic. (b) Compute the theoretical speed of sound in air at 20°C and compare it with the value in Table 17.1. Take $M = 28.9$ g/mol. (c) Show that the speed of sound in an ideal gas is

$$v = \sqrt{\frac{\gamma k_B T}{m}}$$

where m is the mass of one molecule. Compare it with the most probable, average, and rms molecular speeds.

61. Model air as a diatomic ideal gas with $M = 28.9$ g/mol. A cylinder with a piston contains 1.20 kg of air at 25.0°C and 200 kPa. Energy is transferred by heat into the system as it is allowed to expand, with the pressure rising to 400 kPa. Throughout the expansion, the relationship between pressure and volume is given by

$$P = C V^{1/2}$$

where C is a constant. (a) Find the initial volume. (b) Find the final volume. (c) Find the final temperature. (d) Find the work done on the air. (e) Find the energy transferred by heat.

62. *Smokin'!* A pitcher throws a 0.142-kg baseball at 47.2 m/s (Fig. P21.62). As it travels 19.4 m, the ball slows to a speed of 42.5 m/s because of air resistance. Find the change in temperature of the air through which it passes. To find the greatest possible temperature change, you may make the following assumptions: Air has a molar specific heat of $C_P = 7R/2$ and an equivalent molar mass of 28.9 g/mol. The process is so rapid that the cover of the baseball acts as thermal insulation, and the temperature of the ball itself does not change. A change in temperature happens initially only for the air in a cylinder 19.4 m in length and 3.70 cm in radius. This air is initially at 20.0°C.

63. ☐ 🖥 For a Maxwellian gas, use a computer or programmable calculator to find the numerical value of the ratio $N_v(v)/N_v(v_{mp})$ for the following values of v: $v = (v_{mp}/50)$, $(v_{mp}/10)$, $(v_{mp}/2)$, v_{mp}, $2v_{mp}$, $10v_{mp}$, and $50v_{mp}$. Give your results to three significant figures.

64. Consider the particles in a gas centrifuge, a device used to separate particles of different mass by whirling them in a circular path of radius r at angular speed ω. The force

Figure P21.62 John Lackey, the first rookie to win a World Series game 7 in 93 years, pitches for the Anaheim Angels during the final game of the 2002 World Series.

acting toward the center of the circular path on a given particle is $m\omega^2 r$. (a) Discuss how a gas centrifuge can be used to separate particles of different mass. (b) Show that the density of the particles as a function of r is

$$n(r) = n_0 e^{mr^2\omega^2/2k_B T}$$

65. Verify Equations 21.27 and 21.28 for the rms and average speed of the molecules of a gas at a temperature T. Note that the average value of v^n is

$$\overline{v^n} = \frac{1}{N} \int_0^\infty v^n N_v \, dv$$

Use the table of definite integrals in Appendix B (Table B.6).

66. On the PV diagram for an ideal gas, one isothermal curve and one adiabatic curve pass through each point. Prove that the slope of the adiabat is steeper than the slope of the isotherm by the factor γ.

67. A sample of monatomic ideal gas occupies 5.00 L at atmospheric pressure and 300 K (point A in Figure P21.67). It is heated at constant volume to 3.00 atm (point B). Then it is allowed to expand isothermally to 1.00 atm (point C) and at last compressed isobarically to its original state. (a) Find the number of moles in the sample. (b) Find the temperature at points B and C and the volume at point C. (c) Assuming that the molar specific heat does not depend on temperature, so that $E_{int} = 3nRT/2$, find the internal energy at points A, B, and C. (d) Tabulate P, V, T, and E_{int} for the states at points A, B, and C. (e) Now consider the processes $A \rightarrow B$, $B \rightarrow C$, and $C \rightarrow A$. Describe just how to carry out each process experimentally. (f) Find Q, W, and ΔE_{int} for each of the processes. (g) For the whole cycle $A \rightarrow B \rightarrow C \rightarrow A$ find Q, W, and ΔE_{int}.

68. This problem can help you to think about the size of molecules. In the city of Beijing a restaurant keeps a pot of chicken broth simmering continuously. Every morning it is topped up to contain 10.0 L of water, along with a fresh chicken, vegetables, and spices. The soup is thoroughly stirred. The molar mass of water is 18.0 g/mol. (a) Find

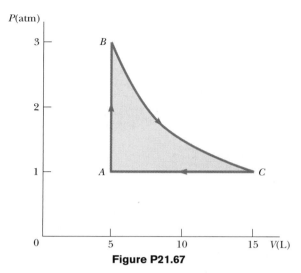

Figure P21.67

the number of molecules of water in the pot. (b) During a certain month, 90.0% of the broth was served each day to people who then emigrated immediately. Of the water molecules in the pot on the first day of the month, when was the last one likely to have been ladled out of the pot? (c) The broth has been simmering for centuries, through wars, earthquakes, and stove repairs. Suppose the water that was in the pot long ago has thoroughly mixed into the Earth's hydrosphere, of mass 1.32×10^{21} kg. How many of the water molecules originally in the pot are likely to be present in it again today?

69. Review problem. (a) If it has enough kinetic energy, a molecule at the surface of the Earth can "escape the Earth's gravitation," in the sense that it can continue to move away from the Earth forever, as discussed in Section 13.7. Using the principle of conservation of energy, show that the minimum kinetic energy needed for "escape" is mgR_E, where m is the mass of the molecule, g is the free-fall acceleration at the surface, and R_E is the radius of the Earth. (b) Calculate

the temperature for which the minimum escape kinetic energy is ten times the average kinetic energy of an oxygen molecule.

70. Using multiple laser beams, physicists have been able to cool and trap sodium atoms in a small region. In one experiment the temperature of the atoms was reduced to 0.240 mK. (a) Determine the rms speed of the sodium atoms at this temperature. The atoms can be trapped for about 1.00 s. The trap has a linear dimension of roughly 1.00 cm. (b) Approximately how long would it take an atom to wander out of the trap region if there were no trapping action?

Answers to Quick Quizzes

21.1 (b). The average translational kinetic energy per molecule is a function only of temperature.

21.2 (a). Because there are twice as many molecules and the temperature of both containers is the same, the total energy in B is twice that in A.

21.3 (b). Because both containers hold the same type of gas, the rms speed is a function only of temperature.

21.4 (a). According to Equation 21.10, E_{int} is a function of temperature only. Because the temperature increases, the internal energy increases.

21.5 (c). Along an isotherm, T is constant by definition. Therefore, the internal energy of the gas does not change.

21.6 (d). The value of 29.1 J/mol·K is $7R/2$. According to Figure 21.7, this suggests that all three types of motion are occurring.

21.7 (c). The highest possible value of C_V for a diatomic gas is $7R/2$, so the gas must be polyatomic.

21.8 (a). Because the hydrogen atoms are lighter than the nitrogen molecules, they move with a higher average speed and the distribution curve is stretched out more along the horizontal axis. See Equation 21.26 for a mathematical statement of the dependence of N_v on m.

B.C. **By John Hart**

By permission of John Hart and Creators Syndicate, Inc.

Heat Engines, Entropy, and the Second Law of Thermodynamics

▲ This cutaway image of an automobile engine shows two pistons that have work done on them by an explosive mixture of air and fuel, ultimately leading to the motion of the automobile. This apparatus can be modeled as a heat engine, which we study in this chapter. (Courtesy of Ford Motor Company)

The first law of thermodynamics, which we studied in Chapter 20, is a statement of conservation of energy. This law states that a change in internal energy in a system can occur as a result of energy transfer by heat or by work, or by both. As was stated in Chapter 20, the law makes no distinction between the results of heat and the results of work—either heat or work can cause a change in internal energy. However, there is an important distinction between heat and work that is not evident from the first law. One manifestation of this distinction is that it is impossible to design a device that, operating in a cyclic fashion, takes in energy by heat and expels an *equal* amount of energy by work. A cyclic device that takes in energy by heat and expels a *fraction* of this energy by work is possible and is called a *heat engine*.

Although the first law of thermodynamics is very important, it makes no distinction between processes that occur spontaneously and those that do not. However, only certain types of energy-conversion and energy-transfer processes actually take place in nature. The *second law of thermodynamics*, the major topic in this chapter, establishes which processes do and which do not occur. The following are examples of processes that do not violate the principle of conservation of energy if they proceed in either direction, but are observed to proceed in only one direction, governed by the second law:

- When two objects at different temperatures are placed in thermal contact with each other, the net transfer of energy by heat is always from the warmer object to the cooler object, never from the cooler to the warmer.

- A rubber ball dropped to the ground bounces several times and eventually comes to rest, but a ball lying on the ground never gathers internal energy from the ground and begins bouncing on its own.

- An oscillating pendulum eventually comes to rest because of collisions with air molecules and friction at the point of suspension. The mechanical energy of the system is converted to internal energy in the air, the pendulum, and the suspension; the reverse conversion of energy never occurs.

All these processes are *irreversible*—that is, they are processes that occur naturally in one direction only. No irreversible process has ever been observed to run backward—if it were to do so, it would violate the second law of thermodynamics.[1]

From an engineering standpoint, perhaps the most important implication of the second law is the limited efficiency of heat engines. The second law states that a machine that operates in a cycle, taking in energy by heat and expelling an equal amount of energy by work, cannot be constructed.

Lord Kelvin

British physicist and mathematician (1824–1907)

Born William Thomson in Belfast, Kelvin was the first to propose the use of an absolute scale of temperature. The Kelvin temperature scale is named in his honor. Kelvin's work in thermodynamics led to the idea that energy cannot pass spontaneously from a colder object to a hotter object. *(J. L. Charmet/SPL/Photo Researchers, Inc.)*

---

[1] Although we have never *observed* a process occurring in the time-reversed sense, it is *possible* for it to occur. As we shall see later in the chapter, however, the probability of such a process occurring is infinitesimally small. From this viewpoint, we say that processes occur with a vastly greater probability in one direction than in the opposite direction.

22.1 Heat Engines and the Second Law of Thermodynamics

A **heat engine** is a device that takes in energy by heat[2] and, operating in a cyclic process, expels a fraction of that energy by means of work. For instance, in a typical process by which a power plant produces electricity, coal or some other fuel is burned, and the high-temperature gases produced are used to convert liquid water to steam. This steam is directed at the blades of a turbine, setting it into rotation. The mechanical energy associated with this rotation is used to drive an electric generator. Another device that can be modeled as a heat engine—the internal combustion engine in an automobile—uses energy from a burning fuel to perform work on pistons that results in the motion of the automobile.

A heat engine carries some working substance through a cyclic process during which (1) the working substance absorbs energy by heat from a high-temperature energy reservoir, (2) work is done by the engine, and (3) energy is expelled by heat to a lower-temperature reservoir. As an example, consider the operation of a steam engine (Fig. 22.1), which uses water as the working substance. The water in a boiler absorbs energy from burning fuel and evaporates to steam, which then does work by expanding against a piston. After the steam cools and condenses, the liquid water produced returns to the boiler and the cycle repeats.

It is useful to represent a heat engine schematically as in Figure 22.2. The engine absorbs a quantity of energy $|Q_h|$ from the hot reservoir. For this discussion of heat engines, we will use absolute values to make all energy transfers positive and will indicate the direction of transfer with an explicit positive or negative sign. The engine does work W_{eng} (so that *negative* work $W = -W_{eng}$ is done *on* the engine), and then gives up a quantity of energy $|Q_c|$ to the cold reservoir. Because the working substance goes

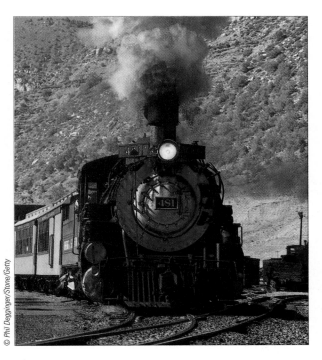

© Phil Degginger/Stone/Getty

Figure 22.1 This steam-driven locomotive runs from Durango to Silverton, Colorado. It obtains its energy by burning wood or coal. The generated energy vaporizes water into steam, which powers the locomotive. (This locomotive must take on water from tanks located along the route to replace steam lost through the funnel.) Modern locomotives use diesel fuel instead of wood or coal. Whether old-fashioned or modern, such locomotives can be modeled as heat engines, which extract energy from a burning fuel and convert a fraction of it to mechanical energy.

Active Figure 22.2 Schematic representation of a heat engine. The engine does work W_{eng}. The arrow at the top represents energy $Q_h > 0$ entering the engine. At the bottom, $Q_c < 0$ represents energy leaving the engine.

At the Active Figures link at http://www.pse6.com, you can select the efficiency of the engine and observe the transfer of energy.

---

[2] We will use heat as our model for energy transfer into a heat engine. Other methods of energy transfer are also possible in the model of a heat engine, however. For example, the Earth's atmosphere can be modeled as a heat engine, in which the input energy transfer is by means of electromagnetic radiation from the Sun. The output of the atmospheric heat engine causes the wind structure in the atmosphere.

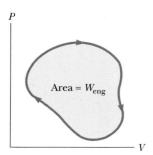

Figure 22.3 *PV* diagram for an arbitrary cyclic process taking place in an engine. The value of the net work done by the engine in one cycle equals the area enclosed by the curve.

Thermal efficiency of a heat engine

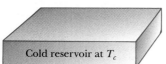

The impossible engine

Figure 22.4 Schematic diagram of a heat engine that takes in energy from a hot reservoir and does an equivalent amount of work. It is impossible to construct such a perfect engine.

through a cycle, its initial and final internal energies are equal, and so $\Delta E_{\text{int}} = 0$. Hence, from the first law of thermodynamics, $\Delta E_{\text{int}} = Q + W = Q - W_{\text{eng}}$, and with no change in internal energy, **the net work W_{eng} done by a heat engine is equal to the net energy Q_{net} transferred to it.** As we can see from Figure 22.2, $Q_{\text{net}} = |Q_h| - |Q_c|$; therefore,

$$W_{\text{eng}} = |Q_h| - |Q_c| \qquad (22.1)$$

If the working substance is a gas, **the net work done in a cyclic process is the area enclosed by the curve representing the process on a *PV* diagram.** This is shown for an arbitrary cyclic process in Figure 22.3.

The **thermal efficiency** e of a heat engine is defined as the ratio of the net work done by the engine during one cycle to the energy input at the higher temperature during the cycle:

$$e = \frac{W_{\text{eng}}}{|Q_h|} = \frac{|Q_h| - |Q_c|}{|Q_h|} = 1 - \frac{|Q_c|}{|Q_h|} \qquad (22.2)$$

We can think of the efficiency as the ratio of what you gain (work) to what you give (energy transfer at the higher temperature). In practice, all heat engines expel only a fraction of the input energy Q_h by mechanical work and consequently their efficiency is always less than 100%. For example, a good automobile engine has an efficiency of about 20%, and diesel engines have efficiencies ranging from 35% to 40%.

Equation 22.2 shows that a heat engine has 100% efficiency ($e = 1$) only if $|Q_c| = 0$—that is, if no energy is expelled to the cold reservoir. In other words, a heat engine with perfect efficiency would have to expel all of the input energy by work. On the basis of the fact that efficiencies of real engines are well below 100%, the **Kelvin–Planck form of the second law of thermodynamics** states the following:

> It is impossible to construct a heat engine that, operating in a cycle, produces no effect other than the input of energy by heat from a reservoir and the performance of an equal amount of work.

This statement of the second law means that, during the operation of a heat engine, W_{eng} can never be equal to $|Q_h|$, or, alternatively, that some energy $|Q_c|$ must be rejected to the environment. Figure 22.4 is a schematic diagram of the impossible "perfect" heat engine.

Quick Quiz 22.1 The energy input to an engine is 3.00 times greater than the work it performs. What is its thermal efficiency? (a) 3.00 (b) 1.00 (c) 0.333 (d) impossible to determine

Quick Quiz 22.2 For the engine of Quick Quiz 22.1, what fraction of the energy input is expelled to the cold reservoir? (a) 0.333 (b) 0.667 (c) 1.00 (d) impossible to determine

---

Example 22.1   The Efficiency of an Engine

An engine transfers 2.00×10^3 J of energy from a hot reservoir during a cycle and transfers 1.50×10^3 J as exhaust to a cold reservoir.

(A) Find the efficiency of the engine.

Solution The efficiency of the engine is given by Equation 22.2 as

$$e = 1 - \frac{|Q_c|}{|Q_h|} = 1 - \frac{1.50 \times 10^3 \, \text{J}}{2.00 \times 10^3 \, \text{J}} = \boxed{0.250, \text{ or } 25.0\%}$$

(B) How much work does this engine do in one cycle?

Solution The work done is the difference between the input and output energies:

$$W_{eng} = |Q_h| - |Q_c| = 2.00 \times 10^3 \, J - 1.50 \times 10^3 \, J$$

$$= \boxed{5.0 \times 10^2 \, J}$$

What If? Suppose you were asked for the power output of this engine? Do you have sufficient information to answer this question?

Answer No, you do not have enough information. The power of an engine is the *rate* at which work is done by the engine. You know how much work is done per cycle but you have no information about the time interval associated with one cycle. However, if you were told that the engine operates at 2 000 rpm (revolutions per minute), you could relate this rate to the period of rotation T of the mechanism of the engine. If we assume that there is one thermodynamic cycle per revolution, then the power is

$$\mathcal{P} = \frac{W_{eng}}{T} = \frac{5.0 \times 10^2 \, J}{\left(\frac{1}{2\,000} \, min\right)} \left(\frac{1 \, min}{60 \, s}\right) = 1.7 \times 10^4 \, W$$

22.2 Heat Pumps and Refrigerators

In a heat engine, the direction of energy transfer is from the hot reservoir to the cold reservoir, which is the natural direction. The role of the heat engine is to process the energy from the hot reservoir so as to do useful work. What if we wanted to transfer energy from the cold reservoir to the hot reservoir? Because this is not the natural direction of energy transfer, we must put some energy into a device in order to accomplish this. Devices that perform this task are called **heat pumps** or **refrigerators.** For example, we cool homes in summer using heat pumps called *air conditioners.* The air conditioner transfers energy from the cool room in the home to the warm air outside.

In a refrigerator or heat pump, the engine takes in energy $|Q_c|$ from a cold reservoir and expels energy $|Q_h|$ to a hot reservoir (Fig. 22.5). This can be accomplished only if work is done *on* the engine. From the first law, we know that the energy given up to the hot reservoir must equal the sum of the work done and the energy taken in from the cold reservoir. Therefore, the refrigerator or heat pump transfers energy from a colder body (for example, the contents of a kitchen refrigerator or the winter air outside a building) to a hotter body (the air in the kitchen or a room in the building). In practice, it is desirable to carry out this process with a minimum of work. If it could be accomplished without doing any work, then the refrigerator or heat pump would be "perfect" (Fig. 22.6). Again, the existence of such a device would be in violation of the second law of thermodynamics, which in the form of the **Clausius statement**[3] states:

▲ PITFALL PREVENTION

22.1 The First and Second Laws

Notice the distinction between the first and second laws of thermodynamics. If a gas undergoes a *one-time isothermal process* $\Delta E_{int} = Q + W = 0$. Therefore, the first law allows *all* energy input by heat to be expelled by work. In a heat engine, however, in which a substance undergoes a cyclic process, only a *portion* of the energy input by heat can be expelled by work according to the second law.

Active Figure 22.5 Schematic diagram of a heat pump, which takes in energy $Q_c > 0$ from a cold reservoir and expels energy $Q_h < 0$ to a hot reservoir. Work W is done *on* the heat pump. A refrigerator works the same way.

🚀 **At the Active Figures link at http://www.pse6.com, you can select the COP of the heat pump and observe the transfer of energy.**

---

[3] First expressed by Rudolf Clausius (1822–1888).

Impossible heat pump

Figure 22.6 Schematic diagram of an impossible heat pump or refrigerator—that is, one that takes in energy from a cold reservoir and expels an equivalent amount of energy to a hot reservoir without the input of energy by work.

> It is impossible to construct a cyclical machine whose sole effect is to transfer energy continuously by heat from one object to another object at a higher temperature without the input of energy by work.

In simpler terms, **energy does not transfer spontaneously by heat from a cold object to a hot object.** This direction of energy transfer requires an input of energy to a heat pump, which is often supplied by means of electricity.

The Clausius and Kelvin–Planck statements of the second law of thermodynamics appear, at first sight, to be unrelated, but in fact they are equivalent in all respects. Although we do not prove so here, if either statement is false, then so is the other.[4]

Heat pumps have long been used for cooling homes and buildings, and they are now becoming increasingly popular for heating them as well. The heat pump contains two sets of metal coils that can exchange energy by heat with the surroundings: one set on the outside of the building, in contact with the air or buried in the ground, and the other set in the interior of the building. In the heating mode, a circulating fluid flowing through the coils absorbs energy from the outside and releases it to the interior of the building from the interior coils. The fluid is cold and at low pressure when it is in the external coils, where it absorbs energy by heat from either the air or the ground. The resulting warm fluid is then compressed and enters the interior coils as a hot, high-pressure fluid, where it releases its stored energy to the interior air.

An air conditioner is simply a heat pump with its exterior and interior coils interchanged, so that it operates in the cooling mode. Energy is absorbed into the circulating fluid in the interior coils; then, after the fluid is compressed, energy leaves the fluid through the external coils. The air conditioner must have a way to release energy to the outside. Otherwise, the work done on the air conditioner would represent energy added to the air inside the house, and the temperature would increase. In the same manner, a refrigerator cannot cool the kitchen if the refrigerator door is left open. The amount of energy leaving the external coils (Fig. 22.7) behind or underneath the refrigerator is greater than the amount of energy removed from the food. The difference between the energy out and the energy in is the work done by the electricity supplied to the refrigerator.

The effectiveness of a heat pump is described in terms of a number called the **coefficient of performance** (COP). In the heating mode, the COP is defined as the ratio of the energy transferred to the hot reservoir to the work required to transfer that energy:

$$\text{COP (heating mode)} \equiv \frac{\text{energy transferred at high temperature}}{\text{work done on heat pump}} = \frac{|Q_h|}{W} \quad (22.3)$$

Note that the COP is similar to the thermal efficiency for a heat engine in that it is a ratio of what you gain (energy delivered to the interior of the building) to what you give (work input). Because $|Q_h|$ is generally greater than W, typical values for the COP are greater than unity. It is desirable for the COP to be as high as possible, just as it is desirable for the thermal efficiency of an engine to be as high as possible.

If the outside temperature is 25°F (-4°C) or higher, a typical value of the COP for a heat pump is about 4. That is, the amount of energy transferred to the building is about four times greater than the work done by the motor in the heat pump. However, as the outside temperature decreases, it becomes more difficult for the heat pump to extract sufficient energy from the air, and so the COP decreases. In fact, the COP can fall below unity for temperatures below about 15°F (-9°C). Thus, the use of heat pumps that extract energy from the air, while satisfactory in moderate climates, is not appropriate in areas where winter temperatures are very low. It is possible to use heat pumps in colder

Charles D. Winters

Figure 22.7 The coils on the back of a refrigerator transfer energy by heat to the air. The second law of thermodynamics states that this amount of energy must be greater than the amount of energy removed from the contents of the refrigerator, due to the input of energy by work.

---

[4] See, for example, R. P. Bauman, *Modern Thermodynamics and Statistical Mechanics*, New York, Macmillan Publishing Co., 1992.

areas by burying the external coils deep in the ground. In this case, the energy is extracted from the ground, which tends to be warmer than the air in the winter.

For a heat pump operating in the cooling mode, "what you gain" is energy removed from the cold reservoir. The most effective refrigerator or air conditioner is one that removes the greatest amount of energy from the cold reservoir in exchange for the least amount of work. Thus, for these devices we define the COP in terms of $|Q_c|$:

$$\text{COP (cooling mode)} = \frac{|Q_c|}{W} \qquad (22.4)$$

A good refrigerator should have a high COP, typically 5 or 6.

Quick Quiz 22.3 The energy entering an electric heater by electrical transmission can be converted to internal energy with an efficiency of 100%. By what factor does the cost of heating your home change when you replace your electric heating system with an electric heat pump that has a COP of 4.00? Assume that the motor running the heat pump is 100% efficient. (a) 4.00   (b) 2.00   (c) 0.500   (d) 0.250

Example 22.2   Freezing Water

A certain refrigerator has a COP of 5.00. When the refrigerator is running, its power input is 500 W. A sample of water of mass 500 g and temperature 20.0°C is placed in the freezer compartment. How long does it take to freeze the water to ice at 0°C? Assume that all other parts of the refrigerator stay at the same temperature and there is no leakage of energy from the exterior, so that the operation of the refrigerator results only in energy being extracted from the water.

Solution Conceptualize this problem by realizing that energy leaves the water, reducing its temperature and then freezing it into ice. The time interval required for this entire process is related to the rate at which energy is withdrawn from the water, which, in turn is related to the power input of the refrigerator. We categorize this problem as one in which we will need to combine our understanding of temperature changes and phase changes from Chapter 20 with our understanding of heat pumps from the current chapter. To analyze the problem, we first find the amount of energy that we must extract from 500 g of water at 20°C to turn it into ice at 0°C. Using Equations 20.4 and 20.6,

$$|Q_c| = |mc\,\Delta T + mL_f| = m|c\,\Delta T + L_f|$$
$$= (0.500\,\text{kg})[(4\,186\,\text{J/kg} \cdot {}^\circ\text{C})(20.0{}^\circ\text{C}) + 3.33 \times 10^5\,\text{J/kg}]$$
$$= 2.08 \times 10^5\,\text{J}$$

Now we use Equation 22.4 to find out how much energy we need to provide to the refrigerator to extract this much energy from the water:

$$\text{COP} = \frac{|Q_c|}{W} \longrightarrow W = \frac{|Q_c|}{\text{COP}} = \frac{2.08 \times 10^5\,\text{J}}{5.00}$$
$$W = 4.17 \times 10^4\,\text{J}$$

Using the power rating of the refrigerator, we find out the time interval required for the freezing process to occur:

$$\mathcal{P} = \frac{W}{\Delta t} \longrightarrow \Delta t = \frac{W}{\mathcal{P}} = \frac{4.17 \times 10^4\,\text{J}}{500\,\text{W}} = \boxed{83.3\,\text{s}}$$

To finalize this problem, note that this time interval is very different from that of our everyday experience; this suggests the difficulties with our assumptions. Only a small part of the energy extracted from the refrigerator interior in a given time interval will come from the water. Energy must also be extracted from the container in which the water is placed, and energy that continuously leaks into the interior from the exterior must be continuously extracted. In reality, the time interval for the water to freeze is much longer than 83.3 s.

22.3   Reversible and Irreversible Processes

In the next section we discuss a theoretical heat engine that is the most efficient possible. To understand its nature, we must first examine the meaning of reversible and irreversible processes. In a **reversible** process, the system undergoing the process can be

Figure 22.8 Adiabatic free expansion of a gas.

⚠️ **PITFALL PREVENTION**

22.2 All Real Processes Are Irreversible

The reversible process is an idealization—all real processes on Earth are irreversible.

Figure 22.9 A gas in thermal contact with an energy reservoir is compressed slowly as individual grains of sand drop onto the piston. The compression is isothermal and reversible.

returned to its initial conditions along the same path on a *PV* diagram, and every point along this path is an equilibrium state. A process that does not satisfy these requirements is **irreversible.**

All natural processes are known to be irreversible. From the endless number of examples that could be selected, let us examine the adiabatic free expansion of a gas, which was already discussed in Section 20.6, and show that it cannot be reversible. Consider a gas in a thermally insulated container, as shown in Figure 22.8. A membrane separates the gas from a vacuum. When the membrane is punctured, the gas expands freely into the vacuum. As a result of the puncture, the system has changed because it occupies a greater volume after the expansion. Because the gas does not exert a force through a displacement, it does no work on the surroundings as it expands. In addition, no energy is transferred to or from the gas by heat because the container is insulated from its surroundings. Thus, in this adiabatic process, the system has changed but the surroundings have not.

For this process to be reversible, we need to be able to return the gas to its original volume and temperature without changing the surroundings. Imagine that we try to reverse the process by compressing the gas to its original volume. To do so, we fit the container with a piston and use an engine to force the piston inward. During this process, the surroundings change because work is being done by an outside agent on the system. In addition, the system changes because the compression increases the temperature of the gas. We can lower the temperature of the gas by allowing it to come into contact with an external energy reservoir. Although this step returns the gas to its original conditions, the surroundings are again affected because energy is being added to the surroundings from the gas. If this energy could somehow be used to drive the engine that compressed the gas, then the net energy transfer to the surroundings would be zero. In this way, the system and its surroundings could be returned to their initial conditions, and we could identify the process as reversible. However, the Kelvin–Planck statement of the second law specifies that the energy removed from the gas to return the temperature to its original value cannot be completely converted to mechanical energy in the form of the work done by the engine in compressing the gas. Thus, we must conclude that the process is irreversible.

We could also argue that the adiabatic free expansion is irreversible by relying on the portion of the definition of a reversible process that refers to equilibrium states. For example, during the expansion, significant variations in pressure occur throughout the gas. Thus, there is no well-defined value of the pressure for the entire system at any time between the initial and final states. In fact, the process cannot even be represented as a path on a *PV* diagram. The *PV* diagram for an adiabatic free expansion would show the initial and final conditions as points, but these points would not be connected by a path. Thus, because the intermediate conditions between the initial and final states are not equilibrium states, the process is irreversible.

Although all real processes are irreversible, some are almost reversible. If a real process occurs very slowly such that the system is always very nearly in an equilibrium state, then the process can be approximated as being reversible. Suppose that a gas is compressed isothermally in a piston–cylinder arrangement in which the gas is in thermal contact with an energy reservoir, and we continuously transfer just enough energy from the gas to the reservoir during the process to keep the temperature constant. For example, imagine that the gas is compressed very slowly by dropping grains of sand onto a frictionless piston, as shown in Figure 22.9. As each grain lands on the piston and compresses the gas a bit, the system deviates from an equilibrium state, but is so close to one that it achieves a new equilibrium state in a relatively short time interval. Each grain added represents a change to a new equilibrium state but the differences between states are so small that we can approximate the entire process as occurring through continuous equilibrium states. We can reverse the process by slowly removing grains from the piston.

A general characteristic of a reversible process is that no dissipative effects (such as turbulence or friction) that convert mechanical energy to internal energy can be present. Such effects can be impossible to eliminate completely. Hence, it is not surprising that real processes in nature are irreversible.

22.4 The Carnot Engine

In 1824 a French engineer named Sadi Carnot described a theoretical engine, now called a **Carnot engine,** which is of great importance from both practical and theoretical viewpoints. He showed that a heat engine operating in an ideal, reversible cycle—called a **Carnot cycle**—between two energy reservoirs is the most efficient engine possible. Such an ideal engine establishes an upper limit on the efficiencies of all other engines. That is, the net work done by a working substance taken through the Carnot cycle is the greatest amount of work possible for a given amount of energy supplied to the substance at the higher temperature. **Carnot's theorem** can be stated as follows:

> No real heat engine operating between two energy reservoirs can be more efficient than a Carnot engine operating between the same two reservoirs.

To argue the validity of this theorem, imagine two heat engines operating between the *same* energy reservoirs. One is a Carnot engine with efficiency e_C, and the other is an engine with efficiency e, where we assume that $e > e_C$. The more efficient engine is used to drive the Carnot engine as a Carnot refrigerator. The output by work of the more efficient engine is matched to the input by work of the Carnot refrigerator. For the *combination* of the engine and refrigerator, no exchange by work with the surroundings occurs. Because we have assumed that the engine is more efficient than the refrigerator, the net result of the combination is a transfer of energy from the cold to the hot reservoir without work being done on the combination. According to the Clausius statement of the second law, this is impossible. Hence, the assumption that $e > e_C$ must be false. **All real engines are less efficient than the Carnot engine because they do not operate through a reversible cycle.** The efficiency of a real engine is further reduced by such practical difficulties as friction and energy losses by conduction.

To describe the Carnot cycle taking place between temperatures T_c and T_h, we assume that the working substance is an ideal gas contained in a cylinder fitted with a movable piston at one end. The cylinder's walls and the piston are thermally nonconducting. Four stages of the Carnot cycle are shown in Figure 22.10, and the PV diagram for the cycle is shown in Figure 22.11. The Carnot cycle consists of two adiabatic processes and two isothermal processes, all reversible:

1. Process $A \rightarrow B$ (Fig. 22.10a) is an isothermal expansion at temperature T_h. The gas is placed in thermal contact with an energy reservoir at temperature T_h. During the expansion, the gas absorbs energy $|Q_h|$ from the reservoir through the base of the cylinder and does work W_{AB} in raising the piston.

2. In process $B \rightarrow C$ (Fig. 22.10b), the base of the cylinder is replaced by a thermally nonconducting wall, and the gas expands adiabatically—that is, no energy enters or leaves the system by heat. During the expansion, the temperature of the gas decreases from T_h to T_c and the gas does work W_{BC} in raising the piston.

3. In process $C \rightarrow D$ (Fig. 22.10c), the gas is placed in thermal contact with an energy reservoir at temperature T_c and is compressed isothermally at temperature T_c. During this time, the gas expels energy $|Q_c|$ to the reservoir, and the work done by the piston on the gas is W_{CD}.

4. In the final process $D \rightarrow A$ (Fig. 22.10d), the base of the cylinder is replaced by a nonconducting wall, and the gas is compressed adiabatically. The temperature of the gas increases to T_h, and the work done by the piston on the gas is W_{DA}.

Sadi Carnot
French engineer (1796–1832)

Carnot was the first to show the quantitative relationship between work and heat. In 1824 he published his only work—*Reflections on the Motive Power of Heat*—which reviewed the industrial, political, and economic importance of the steam engine. In it, he defined work as "weight lifted through a height."
(J.-L. Charmet/Science Photo Library/Photo Researchers, Inc.)

▲ **PITFALL PREVENTION**

22.3 Don't Shop for a Carnot Engine

The Carnot engine is an idealization—do not expect a Carnot engine to be developed for commercial use. We explore the Carnot engine only for theoretical considerations.

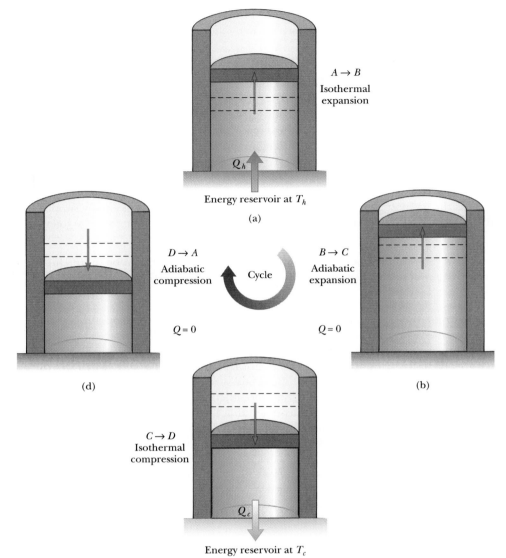

Active Figure 22.10 The Carnot cycle. (a) In process $A \rightarrow B$, the gas expands isothermally while in contact with a reservoir at T_h. (b) In process $B \rightarrow C$, the gas expands adiabatically ($Q = 0$). (c) In process $C \rightarrow D$, the gas is compressed isothermally while in contact with a reservoir at $T_c < T_h$. (d) In process $D \rightarrow A$, the gas is compressed adiabatically. The arrows on the piston indicate the direction of its motion during each process.

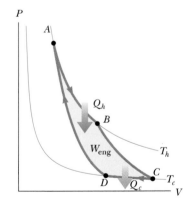

Active Figure 22.11 *PV* diagram for the Carnot cycle. The net work done W_{eng} equals the net energy transferred into the Carnot engine in one cycle, $|Q_h| - |Q_c|$. Note that $\Delta E_{int} = 0$ for the cycle.

The net work done in this reversible, cyclic process is equal to the area enclosed by the path *ABCDA* in Figure 22.11. As we demonstrated in Section 22.1, because the change in internal energy is zero, the net work W_{eng} done by the gas in one cycle equals the net energy transferred into the system, $|Q_h| - |Q_c|$. The thermal efficiency of the engine is given by Equation 22.2:

$$e = \frac{W_{eng}}{|Q_h|} = \frac{|Q_h| - |Q_c|}{|Q_h|} = 1 - \frac{|Q_c|}{|Q_h|}$$

In Example 22.3, we show that for a Carnot cycle

$$\frac{|Q_c|}{|Q_h|} = \frac{T_c}{T_h} \qquad (22.5)$$

Hence, the thermal efficiency of a Carnot engine is

$$e_C = 1 - \frac{T_c}{T_h}$$

(22.6)   **Efficiency of a Carnot engine**

This result indicates that **all Carnot engines operating between the same two temperatures have the same efficiency.**[5]

Equation 22.6 can be applied to any working substance operating in a Carnot cycle between two energy reservoirs. According to this equation, the efficiency is zero if $T_c = T_h$, as one would expect. The efficiency increases as T_c is lowered and as T_h is raised. However, the efficiency can be unity (100%) only if $T_c = 0$ K. Such reservoirs are not available; thus, the maximum efficiency is always less than 100%. In most practical cases, T_c is near room temperature, which is about 300 K. Therefore, one usually strives to increase the efficiency by raising T_h. Theoretically, a Carnot-cycle heat engine run in reverse constitutes the most effective heat pump possible, and it determines the maximum COP for a given combination of hot and cold reservoir temperatures. Using Equations 22.1 and 22.3, we see that the maximum COP for a heat pump in its heating mode is

$$COP_C \text{ (heating mode)} = \frac{|Q_h|}{W}$$

$$= \frac{|Q_h|}{|Q_h| - |Q_c|} = \frac{1}{1 - \dfrac{|Q_c|}{|Q_h|}} = \frac{1}{1 - \dfrac{T_c}{T_h}} = \frac{T_h}{T_h - T_c}$$

The Carnot COP for a heat pump in the cooling mode is

$$COP_C \text{ (cooling mode)} = \frac{T_c}{T_h - T_c}$$

As the difference between the temperatures of the two reservoirs approaches zero in this expression, the theoretical COP approaches infinity. In practice, the low temperature of the cooling coils and the high temperature at the compressor limit the COP to values below 10.

Quick Quiz 22.4 Three engines operate between reservoirs separated in temperature by 300 K. The reservoir temperatures are as follows: Engine A: $T_h = 1\,000$ K, $T_c = 700$ K; Engine B: $T_h = 800$ K, $T_c = 500$ K; Engine C: $T_h = 600$ K, $T_c = 300$ K. Rank the engines in order of theoretically possible efficiency, from highest to lowest.

---

[5] In order for the processes in the Carnot cycle to be reversible, they must be carried out infinitesimally slowly. Thus, although the Carnot engine is the most efficient engine possible, it has zero power output, because it takes an infinite time interval to complete one cycle! For a real engine, the short time interval for each cycle results in the working substance reaching a high temperature lower than that of the hot reservoir and a low temperature higher than that of the cold reservoir. An engine undergoing a Carnot cycle between this narrower temperature range was analyzed by Curzon and Ahlborn (*Am. J. Phys.*, **43**(1), 22, 1975), who found that the efficiency at maximum power output depends only on the reservoir temperatures T_c and T_h, and is given by $e_{C\text{-}A} = 1 - (T_c/T_h)^{1/2}$. The Curzon–Ahlborn efficiency $e_{C\text{-}A}$ provides a closer approximation to the efficiencies of real engines than does the Carnot efficiency.

Example 22.3 Efficiency of the Carnot Engine

Show that the efficiency of a heat engine operating in a Carnot cycle using an ideal gas is given by Equation 22.6.

Solution During the isothermal expansion (process $A \rightarrow B$ in Fig. 22.10), the temperature of the gas does not change. Thus, its internal energy remains constant. The work done on a gas during an isothermal process is given by Equation 20.13. According to the first law,

$$|Q_h| = |-W_{AB}| = nRT_h \ln \frac{V_B}{V_A}$$

In a similar manner, the energy transferred to the cold reservoir during the isothermal compression $C \rightarrow D$ is

$$|Q_c| = |-W_{CD}| = nRT_c \ln \frac{V_C}{V_D}$$

Dividing the second expression by the first, we find that

$$(1) \qquad \frac{|Q_c|}{|Q_h|} = \frac{T_c}{T_h} \frac{\ln(V_C/V_D)}{\ln(V_B/V_A)}$$

We now show that the ratio of the logarithmic quantities is unity by establishing a relationship between the ratio of volumes. For any quasi-static, adiabatic process, the temperature and volume are related by Equation 21.20:

$$T_i V_i^{\gamma-1} = T_f V_f^{\gamma-1}$$

Applying this result to the adiabatic processes $B \rightarrow C$ and $D \rightarrow A$, we obtain

$$T_h V_B^{\gamma-1} = T_c V_C^{\gamma-1}$$
$$T_h V_A^{\gamma-1} = T_c V_D^{\gamma-1}$$

Dividing the first equation by the second, we obtain

$$(V_B/V_A)^{\gamma-1} = (V_C/V_D)^{\gamma-1}$$

$$(2) \qquad \frac{V_B}{V_A} = \frac{V_C}{V_D}$$

Substituting Equation (2) into Equation (1), we find that the logarithmic terms cancel, and we obtain the relationship

$$\frac{|Q_c|}{|Q_h|} = \frac{T_c}{T_h}$$

Using this result and Equation 22.2, we see that the thermal efficiency of the Carnot engine is

$$e_C = 1 - \frac{|Q_c|}{|Q_h|} = 1 - \frac{T_c}{T_h}$$

which is Equation 22.6, the one we set out to prove.

Example 22.4 The Steam Engine

A steam engine has a boiler that operates at 500 K. The energy from the burning fuel changes water to steam, and this steam then drives a piston. The cold reservoir's temperature is that of the outside air, approximately 300 K. What is the maximum thermal efficiency of this steam engine?

Solution Using Equation 22.6, we find that the maximum thermal efficiency for any engine operating between these temperatures is

$$e_C = 1 - \frac{T_c}{T_h} = 1 - \frac{300 \text{ K}}{500 \text{ K}} = \boxed{0.400} \quad \text{or} \quad \boxed{40.0\%}$$

You should note that this is the highest *theoretical* efficiency of the engine. In practice, the efficiency is considerably lower.

What If? Suppose we wished to increase the theoretical efficiency of this engine and we could do so by increasing T_h by ΔT or by decreasing T_c by the same ΔT. Which would be more effective?

Answer A given ΔT would have a larger fractional effect on a smaller temperature, so we would expect a larger change in efficiency if we alter T_c by ΔT. Let us test this numerically. Increasing T_h by 50 K, corresponding to $T_h = 550$ K, would give a maximum efficiency of

$$e_C = 1 - \frac{T_c}{T_h} = 1 - \frac{300 \text{ K}}{550 \text{ K}} = 0.455$$

Decreasing T_c by 50 K, corresponding to $T_c = 250$ K, would give a maximum efficiency of

$$e_C = 1 - \frac{T_c}{T_h} = 1 - \frac{250 \text{ K}}{500 \text{ K}} = 0.500$$

While changing T_c is *mathematically* more effective, often changing T_h is *practically* more feasible.

Example 22.5 The Carnot Efficiency

The highest theoretical efficiency of a certain engine is 30.0%. If this engine uses the atmosphere, which has a temperature of 300 K, as its cold reservoir, what is the temperature of its hot reservoir?

Solution We use the Carnot efficiency to find T_h:

$$e_C = 1 - \frac{T_c}{T_h}$$

$$T_h = \frac{T_c}{1 - e_C} = \frac{300 \text{ K}}{1 - 0.300} = \boxed{429 \text{ K}}$$

22.5 Gasoline and Diesel Engines

In a gasoline engine, six processes occur in each cycle; five of these are illustrated in Figure 22.12. In this discussion, we consider the interior of the cylinder above the piston to be the system that is taken through repeated cycles in the operation of the engine. For a given cycle, the piston moves up and down twice. This represents a four-stroke cycle consisting of two upstrokes and two downstrokes. The processes in the cycle can be approximated by the **Otto cycle**, shown in the PV diagram in Figure 22.13. In the following discussion, refer to Figure 22.12 for the pictorial representation of the strokes and to Figure 22.13 for the significance on the PV diagram of the letter designations below:

1. During the *intake stroke* $O \rightarrow A$ (Fig. 22.12a), the piston moves downward, and a gaseous mixture of air and fuel is drawn into the cylinder at atmospheric pressure. In this process, the volume increases from V_2 to V_1. This is the energy input part of the cycle—energy enters the system (the interior of the cylinder) as potential energy stored in the fuel.

2. During the *compression stroke* $A \rightarrow B$ (Fig. 22.12b), the piston moves upward, the air–fuel mixture is compressed adiabatically from volume V_1 to volume V_2, and the temperature increases from T_A to T_B. The work done on the gas is positive, and its value is equal to the negative of the area under the curve AB in Figure 22.13.

3. In process $B \rightarrow C$, combustion occurs when the spark plug fires (Fig. 22.12c). This is not one of the strokes of the cycle because it occurs in a very short period of time while the piston is at its highest position. The combustion represents a rapid transformation from potential energy stored in chemical bonds in the fuel to internal energy associated with molecular motion, which is related to temperature. During this time, the pressure and temperature in the cylinder increase rapidly, with the temperature rising from T_B to T_C. The volume, however, remains approximately constant because of the short time interval. As a result, approximately no work is done on or by the gas. We can model this process in the PV diagram (Fig. 22.13) as

Active Figure 22.13 PV diagram for the Otto cycle, which approximately represents the processes occurring in an internal combustion engine.

At the Active Figures link at http://www.pse6.com, you can observe the Otto cycle on the PV diagram while you observe the motion of the piston and crankshaft in Figure 22.12.

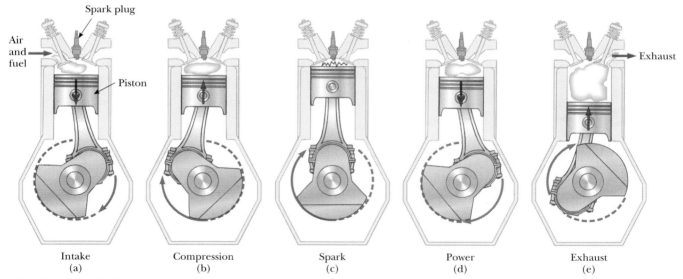

	Intake	Compression	Spark	Power	Exhaust
	(a)	(b)	(c)	(d)	(e)

Active Figure 22.12 The four-stroke cycle of a conventional gasoline engine. The arrows on the piston indicate the direction of its motion during each process. (a) In the intake stroke, air and fuel enter the cylinder. (b) The intake valve is then closed, and the air–fuel mixture is compressed by the piston. (c) The mixture is ignited by the spark plug, with the result that the temperature of the mixture increases at essentially constant volume. (d) In the power stroke, the gas expands against the piston. (e) Finally, the residual gases are expelled, and the cycle repeats.

At the Active Figures link at http://www.pse6.com, you can observe the motion of the piston and crankshaft while you also observe the cycle on the PV diagram of Figure 22.13.

that process in which the energy $|Q_h|$ enters the system. (However, in reality this process is a *conversion* of energy already in the cylinder from process $O \rightarrow A$.)

4. In the *power stroke* $C \rightarrow D$ (Fig. 22.12d), the gas expands adiabatically from V_2 to V_1. This expansion causes the temperature to drop from T_C to T_D. Work is done by the gas in pushing the piston downward, and the value of this work is equal to the area under the curve CD.

5. In the process $D \rightarrow A$ (not shown in Fig. 22.12), an exhaust valve is opened as the piston reaches the bottom of its travel, and the pressure suddenly drops for a short time interval. During this interval, the piston is almost stationary and the volume is approximately constant. Energy is expelled from the interior of the cylinder and continues to be expelled during the next process.

6. In the final process, the *exhaust stroke* $A \rightarrow O$ (Fig. 22.12e), the piston moves upward while the exhaust valve remains open. Residual gases are exhausted at atmospheric pressure, and the volume decreases from V_1 to V_2. The cycle then repeats.

If the air–fuel mixture is assumed to be an ideal gas, then the efficiency of the Otto cycle is

$$e = 1 - \frac{1}{(V_1/V_2)^{\gamma-1}} \qquad \text{(Otto cycle)} \qquad (22.7)$$

where γ is the ratio of the molar specific heats C_P/C_V for the fuel–air mixture and V_1/V_2 is the **compression ratio.** Equation 22.7, which we derive in Example 22.6, shows that the efficiency increases as the compression ratio increases. For a typical compression ratio of 8 and with $\gamma = 1.4$, we predict a theoretical efficiency of 56% for an engine operating in the idealized Otto cycle. This value is much greater than that achieved in real engines (15% to 20%) because of such effects as friction, energy transfer by conduction through the cylinder walls, and incomplete combustion of the air–fuel mixture.

Diesel engines operate on a cycle similar to the Otto cycle but do not employ a spark plug. The compression ratio for a diesel engine is much greater than that for a gasoline engine. Air in the cylinder is compressed to a very small volume, and, as a consequence, the cylinder temperature at the end of the compression stroke is very high. At this point, fuel is injected into the cylinder. The temperature is high enough for the fuel–air mixture to ignite without the assistance of a spark plug. Diesel engines are more efficient than gasoline engines because of their greater compression ratios and resulting higher combustion temperatures.

Example 22.6 Efficiency of the Otto Cycle

Show that the thermal efficiency of an engine operating in an idealized Otto cycle (see Figs. 22.12 and 22.13) is given by Equation 22.7. Treat the working substance as an ideal gas.

Solution First, let us calculate the work done on the gas during each cycle. No work is done during processes $B \rightarrow C$ and $D \rightarrow A$. The work done on the gas during the adiabatic compression $A \rightarrow B$ is positive, and the work done on the gas during the adiabatic expansion $C \rightarrow D$ is negative. The value of the net work done equals the area of the shaded region bounded by the closed curve in Figure 22.13. Because the change in internal energy for one cycle is zero, we see from the first law that the net work done during one cycle equals the net energy transfer to the system:

$$W_{\text{eng}} = |Q_h| - |Q_c|$$

Because processes $B \rightarrow C$ and $D \rightarrow A$ take place at constant volume, and because the gas is ideal, we find from the definition of molar specific heat (Eq. 21.8) that

$$|Q_h| = nC_V(T_C - T_B) \qquad \text{and} \qquad |Q_c| = nC_V(T_D - T_A)$$

Using these expressions together with Equation 22.2, we obtain for the thermal efficiency

$$(1) \qquad e = \frac{W_{\text{eng}}}{|Q_h|} = 1 - \frac{|Q_c|}{|Q_h|} = 1 - \frac{T_D - T_A}{T_C - T_B}$$

We can simplify this expression by noting that processes $A \rightarrow B$ and $C \rightarrow D$ are adiabatic and hence obey Equation 21.20. For the two adiabatic processes, then,

$$A \rightarrow B: \qquad T_A V_A^{\gamma-1} = T_B V_B^{\gamma-1}$$

$$C \rightarrow D: \qquad T_C V_C^{\gamma-1} = T_D V_D^{\gamma-1}$$

Using these equations and relying on the fact that $V_A = V_D = V_1$ and $V_B = V_C = V_2$, we find that

$$T_A V_1{}^{\gamma-1} = T_B V_2{}^{\gamma-1}$$

$$(2) \qquad T_A = T_B \left(\frac{V_2}{V_1} \right)^{\gamma-1}$$

$$T_D V_1{}^{\gamma-1} = T_C V_2{}^{\gamma-1}$$

$$(3) \qquad T_D = T_C \left(\frac{V_2}{V_1} \right)^{\gamma-1}$$

Subtracting Equation (2) from Equation (3) and rearranging, we find that

$$(4) \qquad \frac{T_D - T_A}{T_C - T_B} = \left(\frac{V_2}{V_1} \right)^{\gamma-1}$$

Substituting Equation (4) into Equation (1), we obtain for the thermal efficiency

$$(5) \qquad e = 1 - \frac{1}{(V_1/V_2)^{\gamma-1}}$$

which is Equation 22.7.

We can also express this efficiency in terms of temperatures by noting from Equations (2) and (3) that

$$\left(\frac{V_2}{V_1} \right)^{\gamma-1} = \frac{T_A}{T_B} = \frac{T_D}{T_C}$$

Therefore, Equation (5) becomes

$$(6) \qquad e = 1 - \frac{T_A}{T_B} = 1 - \frac{T_D}{T_C}$$

During the Otto cycle, the lowest temperature is T_A and the highest temperature is T_C. Therefore, the efficiency of a Carnot engine operating between reservoirs at these two temperatures, which is given by the expression $e_C = 1 - (T_A/T_C)$, is *greater* than the efficiency of the Otto cycle given by Equation (6), as expected.

Application   Models of Gasoline and Diesel Engines

We can use the thermodynamic principles discussed in this and earlier chapters to model the performance of gasoline and diesel engines. In both types of engine, a gas is first compressed in the cylinders of the engine and then the fuel–air mixture is ignited. Work is done on the gas during compression, but significantly more work is done on the piston by the mixture as the products of combustion expand in the cylinder. The power of the engine is transferred from the piston to the crankshaft by the connecting rod.

Two important quantities of either engine are the **displacement volume,** which is the volume displaced by the piston as it moves from the bottom to the top of the cylinder, and the compression ratio r, which is the ratio of the maximum and minimum volumes of the cylinder, as discussed earlier. Most gasoline and diesel engines operate with a four-stroke cycle (intake, compression, power, exhaust), in which the net work of the intake and exhaust strokes can be considered negligible. Therefore, power is developed only once for every two revolutions of the crankshaft (see Fig. 22.12).

In a diesel engine, only air (and no fuel) is present in the cylinder at the beginning of the compression. In the idealized diesel cycle of Figure 22.14, air in the cylinder undergoes an adiabatic compression from A to B. Starting at B, fuel is injected into the cylinder. The high temperature of the mixture causes combustion of the fuel–air mixture. Fuel continues to be injected in such a way that during the time interval while the fuel is being injected, the fuel–air mixture undergoes a constant-pressure expansion to an intermediate volume V_C ($B \rightarrow C$). At C, the fuel injection is cut off and the power stroke is an adiabatic expansion back to $V_D = V_A$ ($C \rightarrow D$). The exhaust valve is opened, and a constant-volume output of energy occurs ($D \rightarrow A$) as the cylinder empties.

To simplify our calculations, we assume that the mixture in the cylinder is air modeled as an ideal gas. We use specific heats c instead of molar specific heats

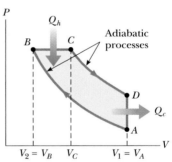

Figure 22.14 *PV* diagram for an ideal diesel engine.

C and assume constant values for air at 300 K. We express the specific heats and the universal gas constant in terms of unit masses rather than moles. Thus, $c_V = 0.718$ kJ/kg·K, $c_P = 1.005$ kJ/kg·K, $\gamma = c_P/c_V = 1.40$, and $R = c_P - c_V = 0.287$ kJ/kg·K $= 0.287$ kPa·m^3/kg·K.

A 3.00-L Gasoline Engine

Let us calculate the power delivered by a six-cylinder gasoline engine that has a displacement volume of 3.00 L operating at 4 000 rpm and having a compression ratio of $r = 9.50$. The air–fuel mixture enters a cylinder at atmospheric pressure and an ambient temperature of 27°C. During combustion, the mixture reaches a temperature of 1 350°C.

First, let us calculate the work done in an individual cylinder. Using the initial pressure $P_A = 100$ kPa, and the initial temperature $T_A = 300$ K, we calculate the initial volume and the mass of the air–fuel mixture. We know that the ratio of the initial and final volumes is the compression ratio,

$$\frac{V_A}{V_B} = r = 9.50$$

We also know that the difference in volumes is the displacement volume. The 3.00-L rating of the engine is the total displacement volume for all six cylinders. Thus, for one cylinder,

$$V_A - V_B = \frac{3.00 \text{ L}}{6} = 0.500 \times 10^{-3} \text{ m}^3$$

Solving these two equations simultaneously, we find the initial and final volumes:

$$V_A = 0.559 \times 10^{-3} \text{ m}^3 \qquad V_B = 0.588 \times 10^{-4} \text{ m}^3$$

Using the ideal gas law (in the form $PV = mRT$, because we are using the universal gas constant in terms of mass rather than moles), we can find the mass of the air–fuel mixture:

$$m = \frac{P_A V_A}{R T_A} = \frac{(100 \text{ kPa})(0.559 \times 10^{-3} \text{ m}^3)}{(0.287 \text{ kPa} \cdot \text{m}^3/\text{kg} \cdot \text{K})(300 \text{ K})}$$

$$= 6.49 \times 10^{-4} \text{ kg}$$

Process $A \rightarrow B$ (see Fig. 22.13) is an adiabatic compression, and this means that $PV^\gamma = $ constant; hence,

$$P_B V_B^\gamma = P_A V_A^\gamma$$

$$P_B = P_A \left(\frac{V_A}{V_B} \right)^\gamma = P_A(r)^\gamma = (100 \text{ kPa})(9.50)^{1.40}$$

$$= 2.34 \times 10^3 \text{ kPa}$$

Using the ideal gas law, we find that the temperature after the compression is

$$T_B = \frac{P_B V_B}{m R} = \frac{(2.34 \times 10^3 \text{ kPa})(0.588 \times 10^{-4} \text{ m}^3)}{(6.49 \times 10^{-4} \text{ kg})(0.287 \text{ kPa} \cdot \text{m}^3/\text{kg} \cdot \text{K})}$$

$$= 739 \text{ K}$$

In process $B \rightarrow C$, the combustion that transforms the potential energy in chemical bonds into internal energy of molecular motion occurs at constant volume; thus, $V_C = V_B$. Combustion causes the temperature to increase to $T_C = 1\,350°\text{C} = 1\,623 \text{ K}$. Using this value and the ideal gas law, we can calculate P_C:

$$P_C = \frac{m R T_C}{V_C}$$

$$= \frac{(6.49 \times 10^{-4} \text{ kg})(0.287 \text{ kPa} \cdot \text{m}^3/\text{kg} \cdot \text{K})(1\,623 \text{ K})}{(0.588 \times 10^{-4} \text{ m}^3)}$$

$$= 5.14 \times 10^3 \text{ kPa}$$

Process $C \rightarrow D$ is an adiabatic expansion; the pressure after the expansion is

$$P_D = P_C \left(\frac{V_C}{V_D} \right)^\gamma = P_C \left(\frac{V_B}{V_A} \right)^\gamma = P_C \left(\frac{1}{r} \right)^\gamma$$

$$= (5.14 \times 10^3 \text{ kPa}) \left(\frac{1}{9.50} \right)^{1.40} = 220 \text{ kPa}$$

Using the ideal gas law again, we find the final temperature:

$$T_D = \frac{P_D V_D}{m R} = \frac{(220 \text{ kPa})(0.559 \times 10^{-3} \text{ m}^3)}{(6.49 \times 10^{-4} \text{ kg})(0.287 \text{ kPa} \cdot \text{m}^3/\text{kg} \cdot \text{K})}$$

$$= 660 \text{ K}$$

Now that we have the temperatures at the beginning and end of each process of the cycle, we can calculate the net energy transfer and net work done in each cylinder every two cycles:

$$|Q_h| = |Q_{\text{in}}| = m c_V (T_C - T_B)$$

$$= (6.49 \times 10^{-4} \text{ kg})(0.718 \text{ kJ/kg} \cdot \text{K})(1\,623 - 739 \text{ K})$$

$$= 0.412 \text{ kJ}$$

$$|Q_c| = |Q_{\text{out}}| = m c_V (T_D - T_A)$$

$$= (6.49 \times 10^{-4} \text{ kg})(0.718 \text{ kJ/kg} \cdot \text{K})(660 \text{ K} - 300 \text{ K})$$

$$= 0.168 \text{ kJ}$$

$$W_{\text{net}} = |Q_{\text{in}}| - |Q_{\text{out}}| = 0.244 \text{ kJ}$$

From Equation 22.2, the efficiency is $e = W_{\text{net}}/|Q_{\text{in}}| = 59\%$. (We can also use Equation 22.7 to calculate the efficiency directly from the compression ratio.)

Recalling that power is delivered every other revolution of the crankshaft, we find that the net power for the six-cylinder engine operating at 4 000 rpm is

$$\mathscr{P}_{\text{net}} = 6(\tfrac{1}{2} \text{ rev})[(4\,000 \text{ rev/min})(1 \text{ min}/60 \text{ s})](0.244 \text{ kJ})$$

$$= 48.8 \text{ kW} = 65 \text{ hp}$$

A 2.00-L Diesel Engine

Let us calculate the power delivered by a four-cylinder diesel engine that has a displacement volume of 2.00 L and is operating at 3 000 rpm. The compression ratio is $r = V_A/V_B = 22.0$, and the **cutoff ratio,** which is the ratio of the volume change during the constant-pressure process $B \rightarrow C$ in Figure 22.14, is $r_c = V_C/V_B = 2.00$. The air enters each cylinder at the beginning of the compression cycle at atmospheric pressure and at an ambient temperature of 27°C.

Our model of the diesel engine is similar to our model of the gasoline engine except that now the fuel is injected at point B and the mixture self-ignites near the end of the compression cycle $A \rightarrow B$, when the temperature reaches the ignition temperature. We assume that the energy input occurs in the constant-pressure process $B \rightarrow C$, and that the expansion process continues from C to D with no further energy transfer by heat.

Let us calculate the work done in an individual cylinder that has an initial volume of $V_A = (2.00 \times 10^{-3} \text{ m}^3)/4 = 0.500 \times 10^{-3} \text{ m}^3$. Because the compression ratio is quite high, we approximate the maximum cylinder volume to be the displacement volume. Using the initial pressure $P_A = 100 \text{ kPa}$ and initial temperature $T_A = 300 \text{ K}$, we can calculate the mass of the air in the cylinder using the ideal gas law:

$$m = \frac{P_A V_A}{R T_A} = \frac{(100 \text{ kPa})(0.500 \times 10^{-3} \text{ m}^3)}{(0.287 \text{ kPa} \cdot \text{m}^3/\text{kg} \cdot \text{K})(300 \text{ K})}$$

$$= 5.81 \times 10^{-4} \text{ kg}$$

Process $A \rightarrow B$ is an adiabatic compression, so $PV^\gamma = $ constant; thus,

$$P_B V_B^\gamma = P_A V_A^\gamma$$

$$P_B = P_A \left(\frac{V_A}{V_B} \right)^\gamma = (100 \text{ kPa})(22.0)^{1.40} = 7.58 \times 10^3 \text{ kPa}$$

Using the ideal gas law, we find that the temperature of the air after the compression is

$$T_B = \frac{P_B V_B}{m R} = \frac{(7.58 \times 10^3 \text{ kPa})(0.500 \times 10^{-3} \text{ m}^3)(1/22.0)}{(5.81 \times 10^{-4} \text{ kg})(0.287 \text{ kPa} \cdot \text{m}^3/\text{kg} \cdot \text{K})}$$

$$= 1.03 \times 10^3 \text{ K}$$

Process $B \rightarrow C$ is a constant-pressure expansion; thus, $P_C = P_B$. We know from the cutoff ratio of 2.00 that the volume doubles in this process. According to the ideal gas law, a doubling of volume in an isobaric process results in a doubling of the temperature, so

$$T_C = 2T_B = 2.06 \times 10^3 \text{ K}$$

Process $C \rightarrow D$ is an adiabatic expansion; therefore,

$$P_D = P_C \left(\frac{V_C}{V_D}\right)^{\gamma} = P_C \left(\frac{V_C}{V_B}\frac{V_B}{V_D}\right)^{\gamma} = P_C \left(r_c \frac{1}{r}\right)^{\gamma}$$

$$= (7.57 \times 10^3 \text{ kPa}) \left(\frac{2.00}{22.0}\right)^{1.40}$$

$$= 264 \text{ kPa}$$

We find the temperature at D from the ideal gas law:

$$T_D = \frac{P_D V_D}{mR} = \frac{(264 \text{ kPa})(0.500 \times 10^{-3} \text{ m}^3)}{(5.81 \times 10^{-4} \text{ kg})(0.287 \text{ kPa} \cdot \text{m}^3/\text{kg} \cdot \text{K})}$$

$$= 792 \text{ K}$$

Now that we have the temperatures at the beginning and the end of each process, we can calculate the net energy transfer by heat and the net work done in each cylinder every two cycles:

$$|Q_h| = |Q_{\text{in}}| = mc_P(T_C - T_B) = 0.601 \text{ kJ}$$

$$|Q_c| = |Q_{\text{out}}| = mc_V(T_D - T_A) = 0.205 \text{ kJ}$$

$$W_{\text{net}} = |Q_{\text{in}}| - |Q_{\text{out}}| = 0.396 \text{ kJ}$$

The efficiency is $e = W_{\text{net}}/|Q_{\text{in}}| = 66\%$.

The net power for the four-cylinder engine operating at 3 000 rpm is

$$\mathcal{P}_{\text{net}} = 4(\tfrac{1}{2} \text{ rev}) \left[(3\,000 \text{ rev/min})(1 \text{ min}/60 \text{ s})\right](0.396 \text{ kJ})$$

$$= 39.6 \text{ kW} = 53 \text{ hp}$$

Modern engine design goes beyond this very simple thermodynamic treatment, which uses idealized cycles.

22.6 Entropy

The zeroth law of thermodynamics involves the concept of temperature, and the first law involves the concept of internal energy. Temperature and internal energy are both state variables—that is, they can be used to describe the thermodynamic state of a system. Another state variable—this one related to the second law of thermodynamics—is **entropy** S. In this section we define entropy on a macroscopic scale as it was first expressed by Clausius in 1865.

Entropy was originally formulated as a useful concept in thermodynamics; however, its importance grew as the field of statistical mechanics developed because the analytical techniques of statistical mechanics provide an alternative means of interpreting entropy and a more global significance to the concept. In statistical mechanics, the behavior of a substance is described in terms of the statistical behavior of its atoms and molecules. One of the main results of this treatment is that **isolated systems tend toward disorder and that entropy is a measure of this disorder.** For example, consider the molecules of a gas in the air in your room. If half of the gas molecules had velocity vectors of equal magnitude directed toward the left and the other half had velocity vectors of the same magnitude directed toward the right, the situation would be very ordered. However, such a situation is extremely unlikely. If you could actually view the molecules, you would see that they move haphazardly in all directions, bumping into one another, changing speed upon collision, some going fast and others going slowly. This situation is highly disordered.

The cause of the tendency of an isolated system toward disorder is easily explained. To do so, we distinguish between *microstates* and *macrostates* of a system. A **microstate** is a particular configuration of the individual constituents of the system. For example, the description of the ordered velocity vectors of the air molecules in your room refers to a particular microstate, and the more likely haphazard motion is another microstate—one that represents disorder. A **macrostate** is a description of the conditions of the system from a macroscopic point of view and makes use of macroscopic variables such as pressure, density, and temperature for gases.

For any given macrostate of the system, a number of microstates are possible. For example, the macrostate of a four on a pair of dice can be formed from the possible microstates 1-3, 2-2, and 3-1. It is assumed that all microstates are equally probable. However, when all possible macrostates are examined, it is found that macrostates

 PITFALL PREVENTION

22.4 Entropy Is Abstract

Entropy is one of the most abstract notions in physics, so follow the discussion in this and the subsequent sections very carefully. Do not confuse energy with entropy—even though the names sound similar, they are very different concepts.

(a)

(b)

Figure 22.15 (a) A royal flush is a highly ordered poker hand with low probability of occurring. (b) A disordered and worthless poker hand. The probability of this *particular* hand occurring is the same as that of the royal flush. There are so many worthless hands, however, that the probability of being dealt a worthless hand is much higher than that of a royal flush.

associated with disorder have far more possible microstates than those associated with order. For example, there is only one microstate associated with the macrostate of a royal flush in a poker hand of five spades, laid out in order from ten to ace (Fig. 22.15a). This is a highly ordered hand. However, there are many microstates (the set of five individual cards in a poker hand) associated with a worthless hand in poker (Fig. 22.15b).

The probability of being dealt the royal flush in spades is exactly the same as the probability of being dealt any *particular* worthless hand. Because there are so many worthless hands, however, the probability of a macrostate of a worthless hand is far larger than the probability of a macrostate of a royal flush in spades.

> **Quick Quiz 22.5** Suppose that you select four cards at random from a standard deck of playing cards and end up with a macrostate of four deuces. How many microstates are associated with this macrostate?
>
> **Quick Quiz 22.6** Suppose you pick up two cards at random from a standard deck of playing cards and end up with a macrostate of two aces. How many microstates are associated with this macrostate?

We can also imagine ordered macrostates and disordered macrostates in physical processes, not just in games of dice and poker. The probability of a system moving in time from an ordered macrostate to a disordered macrostate is far greater than the probability of the reverse, because there are more microstates in a disordered macrostate.

If we consider a system and its surroundings to include the entire Universe, then the Universe is always moving toward a macrostate corresponding to greater disorder. Because entropy is a measure of disorder, an alternative way of stating this is **the entropy of the Universe increases in all real processes.** This is yet another statement of the second law of thermodynamics that can be shown to be equivalent to the Kelvin–Planck and Clausius statements.

The original formulation of entropy in thermodynamics involves the transfer of energy by heat during a reversible process. Consider any infinitesimal process in which a system changes from one equilibrium state to another. If dQ_r is the amount of energy transferred by heat when the system follows a reversible path between the states, then the change in entropy dS is equal to this amount of energy for the reversible process divided by the absolute temperature of the system:

$$dS = \frac{dQ_r}{T} \tag{22.8}$$

We have assumed that the temperature is constant because the process is infinitesimal. Because we have claimed that entropy is a state variable, **the change in entropy during a process depends only on the end points and therefore is independent of the actual path followed. Consequently, the entropy change for an irreversible process can be determined by calculating the entropy change for a reversible process that connects the same initial and final states.**

The subscript r on the quantity dQ_r is a reminder that the transferred energy is to be measured along a reversible path, even though the system may actually have followed some irreversible path. When energy is absorbed by the system, dQ_r is positive and the entropy of the system increases. When energy is expelled by the system, dQ_r is negative and the entropy of the system decreases. Note that Equation 22.8 defines not entropy but rather the *change* in entropy. Hence, the meaningful quantity in describing a process is the *change* in entropy.

To calculate the change in entropy for a *finite* process, we must recognize that T is generally not constant. If dQ_r is the energy transferred by heat when the system follows an arbitrary reversible process between the same initial and final states as the irreversible process, then

$$\Delta S = \int_i^f dS = \int_i^f \frac{dQ_r}{T} \qquad (22.9)$$

Change in entropy for a finite process

As with an infinitesimal process, the change in entropy ΔS of a system going from one state to another has the same value for *all* paths connecting the two states. That is, the finite change in entropy ΔS of a system depends only on the properties of the initial and final equilibrium states. Thus, we are free to choose a particular reversible path over which to evaluate the entropy in place of the actual path, as long as the initial and final states are the same for both paths. This point is explored further in Section 22.7.

> **Quick Quiz 22.7** Which of the following is true for the entropy change of a system that undergoes a reversible, adiabatic process? (a) $\Delta S < 0$ (b) $\Delta S = 0$ (c) $\Delta S > 0$

> **Quick Quiz 22.8** An ideal gas is taken from an initial temperature T_i to a higher final temperature T_f along two different reversible paths: Path A is at constant pressure; Path B is at constant volume. The relation between the entropy changes of the gas for these paths is (a) $\Delta S_A > \Delta S_B$ (b) $\Delta S_A = \Delta S_B$ (c) $\Delta S_A < \Delta S_B$.

Let us consider the changes in entropy that occur in a Carnot heat engine that operates between the temperatures T_c and T_h. In one cycle, the engine takes in energy Q_h from the hot reservoir and expels energy Q_c to the cold reservoir. These energy transfers occur only during the isothermal portions of the Carnot cycle; thus, the constant temperature can be brought out in front of the integral sign in Equation 22.9. The integral then simply has the value of the total amount of energy transferred by heat. Thus, the total change in entropy for one cycle is

$$\Delta S = \frac{|Q_h|}{T_h} - \frac{|Q_c|}{T_c}$$

where the negative sign represents the fact that $|Q_c|$ is positive, but this term must represent energy leaving the engine. In Example 22.3 we showed that, for a Carnot engine,

$$\frac{|Q_c|}{|Q_h|} = \frac{T_c}{T_h}$$

Using this result in the previous expression for ΔS, we find that the total change in entropy for a Carnot engine operating in a cycle is *zero*:

$$\Delta S = 0$$

Now consider a system taken through an arbitrary (non-Carnot) reversible cycle. Because entropy is a state variable—and hence depends only on the properties of a given equilibrium state—we conclude that $\Delta S = 0$ for *any* reversible cycle. In general, we can write this condition in the mathematical form

$$\oint \frac{dQ_r}{T} = 0 \qquad (22.10)$$

where the symbol $\oint$ indicates that the integration is over a closed path.

Quasi-Static, Reversible Process for an Ideal Gas

Suppose that an ideal gas undergoes a quasi-static, reversible process from an initial state having temperature T_i and volume V_i to a final state described by T_f and V_f. Let us calculate the change in entropy of the gas for this process.

Writing the first law of thermodynamics in differential form and rearranging the terms, we have $dQ_r = dE_{int} - dW$, where $dW = -P\,dV$. For an ideal gas, recall that $dE_{int} = nC_V\,dT$ (Eq. 21.12), and from the ideal gas law, we have $P = nRT/V$. Therefore, we can express the energy transferred by heat in the process as

$$dQ_r = dE_{int} + P\,dV = nC_V\,dT + nRT\,\frac{dV}{V}$$

We cannot integrate this expression as it stands because the last term contains two variables, T and V. However, if we divide all terms by T, each of the terms on the right-hand side depends on only one variable:

$$\frac{dQ_r}{T} = nC_V\,\frac{dT}{T} + nR\,\frac{dV}{V} \qquad (22.11)$$

Assuming that C_V is constant over the process, and integrating Equation 22.11 from the initial state to the final state, we obtain

$$\Delta S = \int_i^f \frac{dQ_r}{T} = nC_V \ln \frac{T_f}{T_i} + nR \ln \frac{V_f}{V_i} \qquad (22.12)$$

This expression demonstrates mathematically what we argued earlier—ΔS depends only on the initial and final states and is independent of the path between the states. We can claim this because we have not specified the path taken between the initial and final states. We have only required that the path be reversible. Also, note in Equation 22.12 that ΔS can be positive or negative, depending on the values of the initial and final volumes and temperatures. Finally, for a cyclic process ($T_i = T_f$ and $V_i = V_f$), we see from Equation 22.12 that $\Delta S = 0$. This is further evidence that entropy is a state variable.

Example 22.7   Change in Entropy–Melting

A solid that has a latent heat of fusion L_f melts at a temperature T_m.

(A) Calculate the change in entropy of this substance when a mass m of the substance melts.

Solution Let us assume that the melting occurs so slowly that it can be considered a reversible process. In this case the temperature can be regarded as constant and equal to T_m. Making use of Equations 22.9 and that for the latent heat of fusion $Q = mL_f$ (Eq. 20.6, choosing the positive sign because energy is entering the ice), we find that

$$\Delta S = \int \frac{dQ_r}{T} = \frac{1}{T_m} \int dQ = \frac{Q}{T_m} = \boxed{\frac{mL_f}{T_m}}$$

Note that we are able to remove T_m from the integral because the process is modeled as isothermal. Note also that ΔS is positive.

(B) Estimate the value of the change in entropy of an ice cube when it melts.

Solution Let us assume an ice tray makes cubes that are about 3 cm on a side. The volume per cube is then (very roughly) 30 cm³. This much liquid water has a mass of 30 g. From Table 20.2 we find that the latent heat of fusion of ice is 3.33×10^5 J/kg. Substituting these values into our answer for part (A), we find that

$$\Delta S = \frac{mL_f}{T_m} = \frac{(0.03 \text{ kg})(3.33 \times 10^5 \text{ J/kg})}{273 \text{ K}} = \boxed{4 \times 10^1 \text{ J/K}}$$

We retain only one significant figure, in keeping with the nature of our estimations.

What If? Suppose you did not have Equation 22.9 available so that you could not calculate an entropy change. How could you argue from the statistical description of entropy that the changes in entropy for parts (A) and (B) should be positive?

Answer When a solid melts, its entropy increases because the molecules are much more disordered in the liquid state than they are in the solid state. The positive value for ΔS also means that the substance in its liquid state does not spontaneously transfer energy from itself to the surroundings and freeze because to do so would involve a spontaneous increase in order and a decrease in entropy.

22.7   Entropy Changes in Irreversible Processes

By definition, a calculation of the change in entropy for a system requires information about a reversible path connecting the initial and final equilibrium states. To calculate changes in entropy for real (irreversible) processes, we must remember that entropy (like internal energy) depends only on the *state* of the system. That is, entropy is a state variable. Hence, the change in entropy when a system moves between any two equilibrium states depends only on the initial and final states.

We can calculate the entropy change in some irreversible process between two equilibrium states by devising a reversible process (or series of reversible processes) between the same two states and computing $\Delta S = \int dQ_r/T$ for the reversible process. In irreversible processes, it is critically important that we distinguish between Q, the actual energy transfer in the process, and Q_r, the energy that would have been transferred by heat along a reversible path. Only Q_r is the correct value to be used in calculating the entropy change.

As we show in the following examples, the change in entropy for a system and its surroundings is always positive for an irreversible process. In general, the total entropy—and therefore the disorder—always increases in an irreversible process. Keeping these considerations in mind, we can state the second law of thermodynamics as follows:

> The total entropy of an isolated system that undergoes a change cannot decrease.

Furthermore, **if the process is irreversible, then the total entropy of an isolated system always increases. In a reversible process, the total entropy of an isolated system remains constant.**

When dealing with a system that is not isolated from its surroundings, remember that the increase in entropy described in the second law is that of the system *and* its surroundings. When a system and its surroundings interact in an irreversible process, the increase in entropy of one is greater than the decrease in entropy of the other. Hence, we conclude that **the change in entropy of the Universe must be greater than zero for an irreversible process and equal to zero for a reversible process.** Ultimately, the entropy of the Universe should reach a maximum value. At this value, the Universe will be in a state of uniform temperature and density. All physical, chemical, and biological processes will cease because a state of perfect disorder implies that no energy is available for doing work. This gloomy state of affairs is sometimes referred to as the heat death of the Universe.

> **Quick Quiz 22.9** True or false: The entropy change in an adiabatic process must be zero because $Q = 0$.

Entropy Change in Thermal Conduction

Let us now consider a system consisting of a hot reservoir and a cold reservoir that are in thermal contact with each other and isolated from the rest of the Universe. A process occurs during which energy Q is transferred by heat from the hot reservoir at temperature T_h to the cold reservoir at temperature T_c. The process as described is irreversible, and so we must find an equivalent reversible process. Let us assume that the objects are connected by a poor thermal conductor whose temperature spans the range from T_c to T_h. This conductor transfers energy slowly, and its state does not change during the process. Under this assumption, the energy transfer to or from each object is reversible, and we may set $Q = Q_r$.

Because the cold reservoir absorbs energy Q, its entropy increases by Q/T_c. At the same time, the hot reservoir loses energy Q, and so its entropy change is $-Q/T_h$. Because $T_h > T_c$, the increase in entropy of the cold reservoir is greater than the decrease in entropy of the hot reservoir. Therefore, the change in entropy of the system (and of the Universe) is greater than zero:

$$\Delta S_U = \frac{Q}{T_c} + \frac{-Q}{T_h} > 0$$

Example 22.8 Which Way Does the Energy Go?

A large, cold object is at 273 K, and a second large, hot object is at 373 K. Show that it is impossible for a small amount of energy—for example, 8.00 J—to be transferred spontaneously by heat from the cold object to the hot one without a decrease in the entropy of the Universe and therefore a violation of the second law.

Solution We assume that, during the energy transfer, the two objects do not undergo a temperature change. This is not a necessary assumption; we make it only to avoid complicating the situation by having to use integral calculus in our calculations. The entropy change of the hot object is

$$\Delta S_h = \frac{Q_r}{T_h} = \frac{8.00\,\text{J}}{373\,\text{K}} = 0.021\,4\,\text{J/K}$$

The cold object loses energy, and its entropy change is

$$\Delta S_c = \frac{Q_r}{T_c} = \frac{-8.00\,\text{J}}{273\,\text{K}} = -0.029\,3\,\text{J/K}$$

We consider the two objects to be isolated from the rest of the Universe. Thus, the entropy change of the Universe is just that of our two-object system, which is

$$\Delta S_U = \Delta S_c + \Delta S_h = -0.007\,9\,\text{J/K}$$

This decrease in entropy of the Universe is in violation of the second law. That is, **the spontaneous transfer of energy by heat from a cold to a hot object cannot occur.**

Suppose energy were to continue to transfer spontaneously from a cold object to a hot object, in violation of the second law. We can describe this impossible energy transfer in terms of disorder. Before the transfer, a certain degree of order is associated with the different temperatures of the objects. The hot object's molecules have a higher average energy than the cold object's molecules. If energy spontaneously transfers from the cold object to the hot object, then, over a period of time, the cold object will become colder and the hot object will become hotter. The difference in average molecular energy will become even greater; this would represent an increase in order for the system and a violation of the second law.

In comparison, the process that does occur naturally is the transfer of energy from the hot object to the cold object. In this process, the difference in average molecular energy decreases; this represents a more random distribution of energy and an increase in disorder.

Entropy Change in a Free Expansion

Let us again consider the adiabatic free expansion of a gas occupying an initial volume V_i (Fig. 22.16). In this situation, a membrane separating the gas from an evacuated region is broken, and the gas expands (irreversibly) to a volume V_f. What are the changes in entropy of the gas and of the Universe during this process?

The process is neither reversible nor quasi-static. The work done by the gas against the vacuum is zero, and because the walls are insulating, no energy is transferred by heat during the expansion. That is, $W = 0$ and $Q = 0$. Using the first law, we see that the change in internal energy is zero. Because the gas is ideal, E_{int} depends on temperature only, and we conclude that $\Delta T = 0$ or $T_i = T_f$.

To apply Equation 22.9, we cannot use $Q = 0$, the value for the irreversible process, but must instead find Q_r; that is, we must find an equivalent reversible path that shares the same initial and final states. A simple choice is an isothermal, reversible expansion in which the gas pushes slowly against a piston while energy enters the gas by heat from a reservoir to hold the temperature constant. Because T is constant in this process, Equation 22.9 gives

$$\Delta S = \int_i^f \frac{dQ_r}{T} = \frac{1}{T} \int_i^f dQ_r$$

For an isothermal process, the first law of thermodynamics specifies that $\int_i^f dQ_r$ is equal to the negative of the work done on the gas during the expansion from V_i to V_f, which is given by Equation 20.13. Using this result, we find that the entropy change for the gas is

$$\Delta S = nR \ln \frac{V_f}{V_i} \tag{22.13}$$

Figure 22.16 Adiabatic free expansion of a gas. When the membrane separating the gas from the evacuated region is ruptured, the gas expands freely and irreversibly. As a result, it occupies a greater final volume. The container is thermally insulated from its surroundings; thus, $Q = 0$.

Insulating wall

Vacuum

Membrane

Gas at T_i

Because $V_f > V_i$, we conclude that ΔS is positive. This positive result indicates that both the entropy and the disorder of the gas *increase* as a result of the irreversible, adiabatic expansion.

It is easy to see that the gas is more disordered after the expansion. Instead of being concentrated in a relatively small space, the molecules are scattered over a larger region.

Because the free expansion takes place in an insulated container, no energy is transferred by heat from the surroundings. (Remember that the isothermal, reversible expansion is only a *replacement* process that we use to calculate the entropy change for the gas; it is not the *actual* process.) Thus, the free expansion has no effect on the surroundings, and the entropy change of the surroundings is zero. Thus, the entropy change for the Universe is positive; this is consistent with the second law.

Entropy Change in Calorimetric Processes

A substance of mass m_1, specific heat c_1, and initial temperature T_c is placed in thermal contact with a second substance of mass m_2, specific heat c_2, and initial temperature $T_h > T_c$. The two substances are contained in a calorimeter so that no energy is lost to the surroundings. The system of the two substances is allowed to reach thermal equilibrium. What is the total entropy change for the system?

First, let us calculate the final equilibrium temperature T_f. Using the techniques of Section 20.2—namely, Equation 20.5, $Q_{cold} = -Q_{hot}$, and Equation 20.4, $Q = mc\,\Delta T$, we obtain

$$m_1 c_1 \Delta T_c = -m_2 c_2 \Delta T_h$$

$$m_1 c_1 (T_f - T_c) = -m_2 c_2 (T_f - T_h)$$

Solving for T_f, we have

$$T_f = \frac{m_1 c_1 T_c + m_2 c_2 T_h}{m_1 c_1 + m_2 c_2} \tag{22.14}$$

The process is irreversible because the system goes through a series of nonequilibrium states. During such a transformation, the temperature of the system at any time is not well defined because different parts of the system have different temperatures. However, we can imagine that the hot substance at the initial temperature T_h is slowly cooled to the temperature T_f as it comes into contact with a series of reservoirs differing infinitesimally in temperature, the first reservoir being at T_h and the last being at T_f. Such a series of very small changes in temperature would approximate a reversible process. We imagine doing the same thing for the cold substance. Applying Equation 22.9 and noting that $dQ = mc\,dT$ for an infinitesimal change, we have

$$\Delta S = \int_1 \frac{dQ_{cold}}{T} + \int_2 \frac{dQ_{hot}}{T} = m_1 c_1 \int_{T_c}^{T_f} \frac{dT}{T} + m_2 c_2 \int_{T_h}^{T_f} \frac{dT}{T}$$

where we have assumed that the specific heats remain constant. Integrating, we find that

$$\Delta S = m_1 c_1 \ln \frac{T_f}{T_c} + m_2 c_2 \ln \frac{T_f}{T_h} \tag{22.15}$$

where T_f is given by Equation 22.14. If Equation 22.14 is substituted into Equation 22.15, we can show that one of the terms in Equation 22.15 is always positive and the other is always negative. (You may want to verify this for yourself.) The positive term is always greater than the negative term, and this results in a positive value for ΔS. Thus, we conclude that the entropy of the Universe increases in this irreversible process.

Finally, you should note that Equation 22.15 is valid only when no mixing of different substances occurs, because a further entropy increase is associated with the increase in disorder during the mixing. If the substances are liquids or gases and mixing occurs, the result applies only if the two fluids are identical, as in the following example.

Example 22.9   **Calculating ΔS for a Calorimetric Process**

Suppose that 1.00 kg of water at 0.00°C is mixed with an equal mass of water at 100°C. After equilibrium is reached, the mixture has a uniform temperature of 50.0°C. What is the change in entropy of the system?

Solution   We can calculate the change in entropy from Equation 22.15 using the given values $m_1 = m_2 = 1.00$ kg, $c_1 = c_2 = 4\,186$ J/kg·K, $T_1 = 273$ K, $T_2 = 373$ K, and $T_f = 323$ K:

$$\Delta S = m_1 c_1 \ln \frac{T_f}{T_1} + m_2 c_2 \ln \frac{T_f}{T_2}$$

$$\Delta S = (1.00\ \text{kg})(4\,186\ \text{J/kg·K}) \ln\left(\frac{323\ \text{K}}{273\ \text{K}}\right)$$

$$+ (1.00\ \text{kg})(4\,186\ \text{J/kg·K}) \ln\left(\frac{323\ \text{K}}{373\ \text{K}}\right)$$

$$= \boxed{704\ \text{J/K} - 602\ \text{J/K} = 102\ \text{J/K}}$$

That is, as a result of this irreversible process, the increase in entropy of the cold water is greater than the decrease in entropy of the warm water. Consequently, the increase in entropy of the system is 102 J/K.

22.8 Entropy on a Microscopic Scale[6]

As we have seen, we can approach entropy by relying on macroscopic concepts. We can also treat entropy from a microscopic viewpoint through statistical analysis of molecular motions. We now use a microscopic model to investigate once again the free expansion of an ideal gas, which was discussed from a macroscopic point of view in the preceding section.

In the kinetic theory of gases, gas molecules are represented as particles moving randomly. Let us suppose that the gas is initially confined to a volume V_i, as shown in Figure 22.17a. When the partition separating V_i from a larger container is removed, the molecules eventually are distributed throughout the greater volume V_f (Fig. 22.17b). For a given uniform distribution of gas in the volume, there are a large number of equivalent microstates, and we can relate the entropy of the gas to the number of microstates corresponding to a given macrostate.

We count the number of microstates by considering the variety of molecular locations involved in the free expansion. The instant after the partition is removed (and before the molecules have had a chance to rush into the other half of the container), all the molecules are in the initial volume. We assume that each molecule occupies some microscopic volume V_m. The total number of possible locations of a single molecule in a macroscopic initial volume V_i is the ratio $w_i = V_i/V_m$, which is a huge number. We use w_i here to represent the number of *ways* that the molecule can be placed in the volume, or the number of microstates, which is equivalent to the number of available locations. We assume that the probabilities of a molecule occupying any of these locations are equal.

As more molecules are added to the system, the number of possible ways that the molecules can be positioned in the volume multiplies. For example, if we consider two molecules, for every possible placement of the first, all possible placements of the second are available. Thus, there are w_1 ways of locating the first molecule, and for each of these, there are w_2 ways of locating the second molecule. The total number of ways of locating the two molecules is $w_1 w_2$.

Neglecting the very small probability of having two molecules occupy the same location, each molecule may go into any of the V_i/V_m locations, and so the number of ways of locating N molecules in the volume becomes $W_i = w_i^N = (V_i/V_m)^N$. ($W_i$ is not to be confused with work.) Similarly, when the volume is increased to V_f, the number of ways of locating N molecules increases to $W_f = w_f^N = (V_f/V_m)^N$. The ratio of the number of ways of placing the molecules in the volume for the initial and final configurations is

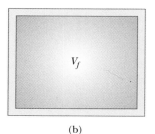

(a)

(b)

Figure 22.17 In a free expansion, the gas is allowed to expand into a region that was previously evacuated.

---

[6]   This section was adapted from A. Hudson and R. Nelson, *University Physics*, Philadelphia, Saunders College Publishing, 1990.

$$\frac{W_f}{W_i} = \frac{(V_f/V_m)^N}{(V_i/V_m)^N} = \left(\frac{V_f}{V_i}\right)^N$$

If we now take the natural logarithm of this equation and multiply by Boltzmann's constant, we find that

$$k_B \ln\left(\frac{W_f}{W_i}\right) = nN_A k_B \ln\left(\frac{V_f}{V_i}\right)$$

where we have used the equality $N = nN_A$. We know from Equation 19.11 that $N_A k_B$ is the universal gas constant R; thus, we can write this equation as

$$k_B \ln W_f - k_B \ln W_i = nR \ln\left(\frac{V_f}{V_i}\right) \tag{22.16}$$

From Equation 22.13 we know that when n mol of a gas undergoes a free expansion from V_i to V_f, the change in entropy is

$$S_f - S_i = nR \ln\left(\frac{V_f}{V_i}\right) \tag{22.17}$$

Note that the right-hand sides of Equations 22.16 and 22.17 are identical. Thus, from the left-hand sides, we make the following important connection between entropy and the number of microstates for a given macrostate:

$$S \equiv k_B \ln W \tag{22.18}$$

Entropy (microscopic definition)

The more microstates there are that correspond to a given macrostate, the greater is the entropy of that macrostate. As we have discussed previously, there are many more microstates associated with disordered macrostates than with ordered macrostates. Thus, Equation 22.18 indicates mathematically that **entropy is a measure of disorder.** Although in our discussion we used the specific example of the free expansion of an ideal gas, a more rigorous development of the statistical interpretation of entropy would lead us to the same conclusion.

We have stated that individual microstates are equally probable. However, because there are far more microstates associated with a disordered macrostate than with an ordered microstate, a disordered macrostate is much more probable than an ordered one.

Figure 22.18 shows a real-world example of this concept. There are two possible macrostates for the carnival game—winning a goldfish and winning a black fish. Because only one jar in the array of jars contains a black fish, only one possible microstate corresponds to the macrostate of winning a black fish. A large number of microstates are described by the coin's falling into a jar containing a goldfish. Thus, for the macrostate of winning a goldfish, there are many equivalent microstates. As a result, the probability of winning a goldfish is much greater than the probability of winning a black fish. If there are 24 goldfish and 1 black fish, the probability of winning the black fish is 1 in 25. This assumes that all microstates have the same probability, a situation

Figure 22.18 By tossing a coin into a jar, the carnival-goer can win the fish in the jar. It is more likely that the coin will land in a jar containing a goldfish than in the one containing the black fish.

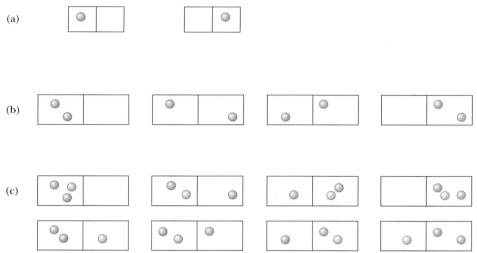

At the Active Figures link at http://www.pse6.com, *you can choose the number of molecules to put in the container and measure the probability of all of them being in the left hand side.*

Active Figure 22.19 (a) One molecule in a two-sided container has a 1-in-2 chance of being on the left side. (b) Two molecules have a 1-in-4 chance of being on the left side at the same time. (c) Three molecules have a 1-in-8 chance of being on the left side at the same time.

that may not be quite true for the situation shown in Figure 22.18. For example, if you are an accurate coin tosser and you are aiming for the edge of the array of jars, then the probability of the coin's landing in a jar near the edge is likely to be greater than the probability of its landing in a jar near the center.

Let us consider a similar type of probability problem for 100 molecules in a container. At any given moment, the probability of one molecule being in the left part of the container shown in Figure 22.19a as a result of random motion is $\frac{1}{2}$. If there are two molecules, as shown in Figure 22.19b, the probability of both being in the left part is $\left(\frac{1}{2}\right)^2$ or 1 in 4. If there are three molecules (Fig. 22.19c), the probability of all of them being in the left portion at the same moment is $\left(\frac{1}{2}\right)^3$, or 1 in 8. For 100 independently moving molecules, the probability that the 50 fastest ones will be found in the left part at any moment is $\left(\frac{1}{2}\right)^{50}$. Likewise, the probability that the remaining 50 slower molecules will be found in the right part at any moment is $\left(\frac{1}{2}\right)^{50}$. Therefore, the probability of finding this fast-slow separation as a result of random motion is the product $\left(\frac{1}{2}\right)^{50} \left(\frac{1}{2}\right)^{50} = \left(\frac{1}{2}\right)^{100}$, which corresponds to about 1 in 10^{30}. When this calculation is extrapolated from 100 molecules to the number in 1 mol of gas (6.02×10^{23}), the ordered arrangement is found to be *extremely* improbable!

Conceptual Example 22.10   Let's Play Marbles!   **Interactive**

Suppose you have a bag of 100 marbles. Fifty of the marbles are red, and 50 are green. You are allowed to draw four marbles from the bag according to the following rules. Draw one marble, record its color, and return it to the bag. Shake the bag and then draw another marble. Continue this process until you have drawn and returned four marbles. What are the possible macrostates for this set of events? What is the most likely macrostate? What is the least likely macrostate?

Solution Because each marble is returned to the bag before the next one is drawn, and the bag is shaken, the probability

of drawing a red marble is always the same as the probability of drawing a green one. All the possible microstates and macrostates are shown in Table 22.1. As this table indicates, there is only one way to draw a macrostate of four red marbles, and so there is only one microstate for that macrostate. However, there are four possible microstates that correspond to the macrostate of one green marble and three red marbles; six microstates that correspond to two green marbles and two red marbles; four microstates that correspond to three green marbles and one red marble; and one microstate that corresponds to four green marbles. The most likely, and most disordered, macrostate—two red marbles and two green marbles—corresponds to the largest number of microstates. The least likely, most ordered macrostates—four red marbles or four green marbles—correspond to the smallest number of microstates.

Table 22.1

Possible Results of Drawing Four Marbles from a Bag		
Macrostate	Possible Microstates	Total Number of Microstates
All R	RRRR	1
1G, 3R	RRRG, RRGR, RGRR, GRRR	4
2G, 2R	RRGG, RGRG, GRRG, RGGR, GRGR, GGRR	6
3G, 1R	GGGR, GGRG, GRGG, RGGG	4
All G	GGGG	1

 Explore the generation of microstates and macrostates at the Interactive Worked Example link at **http://www.pse6.com.**

Example 22.11   Adiabatic Free Expansion–One Last Time

Let us verify that the macroscopic and microscopic approaches to the calculation of entropy lead to the same conclusion for the adiabatic free expansion of an ideal gas. Suppose that an ideal gas expands to four times its initial volume. As we have seen for this process, the initial and final temperatures are the same.

(A) Using a macroscopic approach, calculate the entropy change for the gas.

(B) Using statistical considerations, calculate the change in entropy for the gas and show that it agrees with the answer you obtained in part (A).

Solution

(A) Using Equation 22.13, we have

$$\Delta S = nR \ln\left(\frac{V_f}{V_i}\right) = nR \ln\left(\frac{4V_i}{V_i}\right) = \boxed{nR \ln 4}$$

(B) The number of microstates available to a single molecule in the initial volume V_i is $w_i = V_i / V_m$. For N molecules,

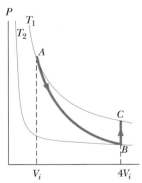

Figure 22.20 (Example 22.11) A gas expands to four times its initial volume and back to the initial temperature by means of a two-step process.

the number of available microstates is

$$W_i = w_i^N = \left(\frac{V_i}{V_m}\right)^N$$

The number of microstates for all N molecules in the final volume $V_f = 4V_i$ is

$$W_f = \left(\frac{V_f}{V_m}\right)^N = \left(\frac{4V_i}{V_m}\right)^N$$

Thus, the ratio of the number of final microstates to initial microstates is

$$\frac{W_f}{W_i} = 4^N$$

Using Equation 22.18, we obtain

$$\Delta S = k_B \ln W_f - k_B \ln W_i = k_B \ln\left(\frac{W_f}{W_i}\right)$$

$$= k_B \ln(4^N) = N k_B \ln 4 = \boxed{nR \ln 4}$$

The answer is the same as that for part (A), which dealt with macroscopic parameters.

What If? In part (A) we used Equation 22.13, which was based on a reversible isothermal process connecting the initial and final states. What if we were to choose a different reversible process? Would we arrive at the same result?

Answer We *must* arrive at the same result because entropy is a state variable. For example, consider the two-step process in Figure 22.20—a reversible adiabatic expansion from V_i to $4V_i$, $(A \rightarrow B)$ during which the temperature drops from T_1 to T_2, and a reversible isovolumetric process $(B \rightarrow C)$ that takes the gas back to the initial temperature T_1.

During the reversible adiabatic process, $\Delta S = 0$ because $Q_r = 0$. During the reversible isovolumetric process $(B \rightarrow C)$, we have from Equation 22.9,

$$\Delta S = \int_i^f \frac{dQ_r}{T} = \int_B^C \frac{nC_V\,dT}{T} = nC_V \ln\left(\frac{T_1}{T_2}\right)$$

Now, we can find the relationship of temperature T_2 to T_1 from Equation 21.20 for the adiabatic process:

$$\frac{T_1}{T_2} = \left(\frac{4V_i}{V_i}\right)^{\gamma-1} = (4)^{\gamma-1}$$

Thus,

$$\Delta S = nC_V \ln (4)^{\gamma-1} = nC_V(\gamma - 1)\ln 4$$

$$= nC_V\left(\frac{C_P}{C_V} - 1\right)\ln 4 = n(C_P - C_V)\ln 4 = nR\ln 4$$

and we do indeed obtain the exact same result for the entropy change.

SUMMARY

Take a practice test for this chapter by clicking on the Practice Test link at http://www.pse6.com.

A **heat engine** is a device that takes in energy by heat and, operating in a cyclic process, expels a fraction of that energy by means of work. The net work done by a heat engine in carrying a working substance through a cyclic process ($\Delta E_{\text{int}} = 0$) is

$$W_{\text{eng}} = |Q_h| - |Q_c| \tag{22.1}$$

where $|Q_h|$ is the energy taken in from a hot reservoir and $|Q_c|$ is the energy expelled to a cold reservoir.

The **thermal efficiency** e of a heat engine is

$$e = \frac{W_{\text{eng}}}{|Q_h|} = 1 - \frac{|Q_c|}{|Q_h|} \tag{22.2}$$

The **second law of thermodynamics** can be stated in the following two ways:

- It is impossible to construct a heat engine that, operating in a cycle, produces no effect other than the input of energy by heat from a reservoir and the performance of an equal amount of work (the Kelvin–Planck statement).

- It is impossible to construct a cyclical machine whose sole effect is to transfer energy continuously by heat from one object to another object at a higher temperature without the input of energy by work (the Clausius statement).

In a **reversible** process, the system can be returned to its initial conditions along the same path on a PV diagram, and every point along this path is an equilibrium state. A process that does not satisfy these requirements is **irreversible. Carnot's theorem** states that no real heat engine operating (irreversibly) between the temperatures T_c and T_h can be more efficient than an engine operating reversibly in a Carnot cycle between the same two temperatures.

The **thermal efficiency** of a heat engine operating in the Carnot cycle is

$$e_C = 1 - \frac{T_c}{T_h} \tag{22.6}$$

The second law of thermodynamics states that when real (irreversible) processes occur, the degree of disorder in the system plus the surroundings increases. When a process occurs in an isolated system, the state of the system becomes more disordered. The measure of disorder in a system is called **entropy** S. Thus, another way in which the second law can be stated is

- The entropy of the Universe increases in all real processes.

The **change in entropy** dS of a system during a process between two infinitesimally separated equilibrium states is

$$dS = \frac{dQ_r}{T} \tag{22.8}$$

where dQ_r is the energy transfer by heat for a reversible process that connects the initial and final states. The change in entropy of a system during an arbitrary process between an initial state and a final state is

$$\Delta S = \int_i^f \frac{dQ_r}{T} \qquad (22.9)$$

The value of ΔS for the system is the same for all paths connecting the initial and final states. The change in entropy for a system undergoing any reversible, cyclic process is zero, and when such a process occurs, the entropy of the Universe remains constant.

From a microscopic viewpoint, the entropy of a given macrostate is defined as

$$S \equiv k_B \ln W \qquad (22.18)$$

where k_B is Boltzmann's constant and W is the number of microstates of the system corresponding to the macrostate.

QUESTIONS

1. What are some factors that affect the efficiency of automobile engines?

2. In practical heat engines, which are we better able to control: the temperature of the hot reservoir or the temperature of the cold reservoir? Explain.

3. A steam-driven turbine is one major component of an electric power plant. Why is it advantageous to have the temperature of the steam as high as possible?

4. Is it possible to construct a heat engine that creates no thermal pollution? What does this tell us about environmental considerations for an industrialized society?

5. Does the second law of thermodynamics contradict or correct the first law? Argue for your answer.

6. "The first law of thermodynamics says you can't really win, and the second law says you can't even break even." Explain how this statement applies to a particular device or process; alternatively, argue against the statement.

7. In solar ponds constructed in Israel, the Sun's energy is concentrated near the bottom of a salty pond. With the proper layering of salt in the water, convection is prevented, and temperatures of 100°C may be reached. Can you estimate the maximum efficiency with which useful energy can be extracted from the pond?

8. Can a heat pump have a coefficient of performance less than unity? Explain.

9. Give various examples of irreversible processes that occur in nature. Give an example of a process in nature that is nearly reversible.

10. A heat pump is to be installed in a region where the average outdoor temperature in the winter months is − 20°C. In view of this, why would it be advisable to place the outdoor compressor unit deep in the ground? Why are heat pumps not commonly used for heating in cold climates?

11. The device shown in Figure Q22.11, called a thermoelectric converter, uses a series of semiconductor cells to convert internal energy to electric potential energy, which we will study in Chapter 25. In the photograph at the left,

Courtesy of PASCO Scientific Company

Figure Q22.11

both legs of the device are at the same temperature, and no electric potential energy is produced. However, when one leg is at a higher temperature than the other, as in the photograph on the right, electric potential energy is produced as the device extracts energy from the hot reservoir and drives a small electric motor. (a) Why does the temperature differential produce electric potential energy in this demonstration? (b) In what sense does this intriguing experiment demonstrate the second law of thermodynamics?

12. Discuss three common examples of natural processes that involve an increase in entropy. Be sure to account for all parts of each system under consideration.

13. Discuss the change in entropy of a gas that expands (a) at constant temperature and (b) adiabatically.

14. A thermodynamic process occurs in which the entropy of a system changes by -8.0 J/K. According to the second law of thermodynamics, what can you conclude about the entropy change of the environment?

15. If a supersaturated sugar solution is allowed to evaporate slowly, sugar crystals form in the container. Hence, sugar molecules go from a disordered form (in solution) to a highly ordered crystalline form. Does this process violate the second law of thermodynamics? Explain.

16. How could you increase the entropy of 1 mol of a metal that is at room temperature? How could you decrease its entropy?

17. Suppose your roommate is "Mr. Clean" and tidies up your messy room after a big party. Because your roommate is creating more order, does this represent a violation of the second law of thermodynamics?

18. Discuss the entropy changes that occur when you (a) bake a loaf of bread and (b) consume the bread.

19. "Energy is the mistress of the Universe and entropy is her shadow." Writing for an audience of general readers, argue for this statement with examples. Alternatively, argue for the view that entropy is like a decisive hands-on executive instantly determining what will happen, while energy is like a wretched back-office bookkeeper telling us how little we can afford.

20. A classmate tells you that it is just as likely for all the air molecules in the room you are both in to be concentrated in one corner (with the rest of the room being a vacuum) as it is for the air molecules to be distributed uniformly about the room in their current state. Is this a true statement? Why doesn't the situation he describes actually happen?

21. If you shake a jar full of jellybeans of different sizes, the larger beans tend to appear near the top, and the smaller ones tend to fall to the bottom. Why? Does this process violate the second law of thermodynamics?

PROBLEMS

1, 2, 3 = straightforward, intermediate, challenging ☐ = full solution available in the *Student Solutions Manual and Study Guide*

🕸 = coached solution with hints available at http://www.pse6.com 💻 = computer useful in solving problem

� = paired numerical and symbolic problems

Section 22.1 Heat Engines and the Second Law of Thermodynamics

1. A heat engine takes in 360 J of energy from a hot reservoir and performs 25.0 J of work in each cycle. Find (a) the efficiency of the engine and (b) the energy expelled to the cold reservoir in each cycle.

2. A heat engine performs 200 J of work in each cycle and has an efficiency of 30.0%. For each cycle, how much energy is (a) taken in and (b) expelled by heat?

3. A particular heat engine has a useful power output of 5.00 kW and an efficiency of 25.0%. The engine expels 8 000 J of exhaust energy in each cycle. Find (a) the energy taken in during each cycle and (b) the time interval for each cycle.

4. Heat engine X takes in four times more energy by heat from the hot reservoir than heat engine Y. Engine X delivers two times more work, and it rejects seven times more energy by heat to the cold reservoir than heat engine Y. Find the efficiency of (a) heat engine X and (b) heat engine Y.

5. A multicylinder gasoline engine in an airplane, operating at 2 500 rev/min, takes in energy 7.89×10^3 J and exhausts 4.58×10^3 J for each revolution of the crankshaft.

(a) How many liters of fuel does it consume in 1.00 h of operation if the heat of combustion is 4.03×10^7 J/L? (b) What is the mechanical power output of the engine? Ignore friction and express the answer in horsepower. (c) What is the torque exerted by the crankshaft on the load? (d) What power must the exhaust and cooling system transfer out of the engine?

6. Suppose a heat engine is connected to two energy reservoirs, one a pool of molten aluminum (660°C) and the other a block of solid mercury (-38.9°C). The engine runs by freezing 1.00 g of aluminum and melting 15.0 g of mercury during each cycle. The heat of fusion of aluminum is 3.97×10^5 J/kg; the heat of fusion of mercury is 1.18×10^4 J/kg. What is the efficiency of this engine?

Section 22.2 Heat Pumps and Refrigerators

7. A refrigerator has a coefficient of performance equal to 5.00. The refrigerator takes in 120 J of energy from a cold reservoir in each cycle. Find (a) the work required in each cycle and (b) the energy expelled to the hot reservoir.

8. A refrigerator has a coefficient of performance of 3.00. The ice tray compartment is at -20.0°C, and the room

temperature is 22.0°C. The refrigerator can convert 30.0 g of water at 22.0°C to 30.0 g of ice at -20.0°C each minute. What input power is required? Give your answer in watts.

9. In 1993 the federal government instituted a requirement that all room air conditioners sold in the United States must have an energy efficiency ratio (EER) of 10 or higher. The EER is defined as the ratio of the cooling capacity of the air conditioner, measured in Btu/h, to its electrical power requirement in watts. (a) Convert the EER of 10.0 to dimensionless form, using the conversion 1 Btu = 1 055 J. (b) What is the appropriate name for this dimensionless quantity? (c) In the 1970s it was common to find room air conditioners with EERs of 5 or lower. Compare the operating costs for 10 000-Btu/h air conditioners with EERs of 5.00 and 10.0. Assume that each air conditioner operates for 1 500 h during the summer in a city where electricity costs 10.0¢ per kWh.

Section 22.3 Reversible and Irreversible Processes
Section 22.4 The Carnot Engine

10. A Carnot engine has a power output of 150 kW. The engine operates between two reservoirs at 20.0°C and 500°C. (a) How much energy does it take in per hour? (b) How much energy is lost per hour in its exhaust?

11. One of the most efficient heat engines ever built is a steam turbine in the Ohio valley, operating between 430°C and 1 870°C on energy from West Virginia coal to produce electricity for the Midwest. (a) What is its maximum theoretical efficiency? (b) The actual efficiency of the engine is 42.0%. How much useful power does the engine deliver if it takes in 1.40×10^5 J of energy each second from its hot reservoir?

12. A heat engine operating between 200°C and 80.0°C achieves 20.0% of the maximum possible efficiency. What energy input will enable the engine to perform 10.0 kJ of work?

13. An ideal gas is taken through a Carnot cycle. The isothermal expansion occurs at 250°C, and the isothermal compression takes place at 50.0°C. The gas takes in 1 200 J of energy from the hot reservoir during the isothermal expansion. Find (a) the energy expelled to the cold reservoir in each cycle and (b) the net work done by the gas in each cycle.

14. The exhaust temperature of a Carnot heat engine is 300°C. What is the intake temperature if the efficiency of the engine is 30.0%?

15. A Carnot heat engine uses a steam boiler at 100°C as the high-temperature reservoir. The low-temperature reservoir is the outside environment at 20.0°C. Energy is exhausted to the low-temperature reservoir at the rate of 15.4 W. (a) Determine the useful power output of the heat engine. (b) How much steam will it cause to condense in the high-temperature reservoir in 1.00 h?

16. A power plant operates at a 32.0% efficiency during the summer when the sea water used for cooling is at 20.0°C. The plant uses 350°C steam to drive turbines. If the plant's efficiency changes in the same proportion as the ideal effi-

ciency, what would be the plant's efficiency in the winter, when the sea water is 10.0°C?

17. Argon enters a turbine at a rate of 80.0 kg/min, a temperature of 800°C and a pressure of 1.50 MPa. It expands adiabatically as it pushes on the turbine blades and exits at pressure 300 kPa. (a) Calculate its temperature at exit. (b) Calculate the (maximum) power output of the turning turbine. (c) The turbine is one component of a model closed-cycle gas turbine engine. Calculate the maximum efficiency of the engine.

18. An electric power plant that would make use of the temperature gradient in the ocean has been proposed. The system is to operate between 20.0°C (surface water temperature) and 5.00°C (water temperature at a depth of about 1 km). (a) What is the maximum efficiency of such a system? (b) If the useful power output of the plant is 75.0 MW, how much energy is taken in from the warm reservoir per hour? (c) In view of your answer to part (a), do you think such a system is worthwhile? Note that the "fuel" is free.

19. Here is a clever idea. Suppose you build a two-engine device such that the exhaust energy output from one heat engine is the input energy for a second heat engine. We say that the two engines are running *in series*. Let e_1 and e_2 represent the efficiencies of the two engines. (a) The overall efficiency of the two-engine device is defined as the total work output divided by the energy put into the first engine by heat. Show that the overall efficiency is given by

$$e = e_1 + e_2 - e_1e_2$$

(b) **What If?** Assume the two engines are Carnot engines. Engine 1 operates between temperatures T_h and T_i. The gas in engine 2 varies in temperature between T_i and T_c. In terms of the temperatures, what is the efficiency of the combination engine? (c) What value of the intermediate temperature T_i will result in equal work being done by each of the two engines in series? (d) What value of T_i will result in each of the two engines in series having the same efficiency?

20. A 20.0%-efficient real engine is used to speed up a train from rest to 5.00 m/s. It is known that an ideal (Carnot) engine using the same cold and hot reservoirs would accelerate the same train from rest to a speed of 6.50 m/s using the same amount of fuel. The engines use air at 300 K as a cold reservoir. Find the temperature of the steam serving as the hot reservoir.

21. A firebox is at 750 K, and the ambient temperature is 300 K. The efficiency of a Carnot engine doing 150 J of work as it transports energy between these constant-temperature baths is 60.0%. The Carnot engine must take in energy 150 J/0.600 = 250 J from the hot reservoir and must put out 100 J of energy by heat into the environment. To follow Carnot's reasoning, suppose that some other heat engine S could have efficiency 70.0%. (a) Find the energy input and wasted energy output of engine S as it does 150 J of work. (b) Let engine S operate as in part (a) and run the Carnot engine in reverse. Find the total energy the firebox puts out as both engines operate together, and the total energy trans-

ferred to the environment. Show that the Clausius statement of the second law of thermodynamics is violated. (c) Find the energy input and work output of engine S as it puts out exhaust energy of 100 J. (d) Let engine S operate as in (c) and contribute 150 J of its work output to running the Carnot engine in reverse. Find the total energy the firebox puts out as both engines operate together, the total work output, and the total energy transferred to the environment. Show that the Kelvin–Planck statement of the second law is violated. Thus our assumption about the efficiency of engine S must be false. (e) Let the engines operate together through one cycle as in part (d). Find the change in entropy of the Universe. Show that the entropy statement of the second law is violated.

22. At point *A* in a Carnot cycle, 2.34 mol of a monatomic ideal gas has a pressure of 1 400 kPa, a volume of 10.0 L, and a temperature of 720 K. It expands isothermally to point *B*, and then expands adiabatically to point *C* where its volume is 24.0 L. An isothermal compression brings it to point *D*, where its volume is 15.0 L. An adiabatic process returns the gas to point *A*. (a) Determine all the unknown pressures, volumes and temperatures as you fill in the following table:

	P	*V*	*T*
A	1 400 kPa	10.0 L	720 K
B			
C		24.0 L	
D		15.0 L	

(b) Find the energy added by heat, the work done by the engine, and the change in internal energy for each of the steps $A \rightarrow B$, $B \rightarrow C$, $C \rightarrow D$, and $D \rightarrow A$. (c) Calculate the efficiency W_{net}/Q_h. Show that it is equal to $1 - T_C/T_A$, the Carnot efficiency.

23. What is the coefficient of performance of a refrigerator that operates with Carnot efficiency between temperatures $-3.00°C$ and $+27.0°C$?

24. What is the maximum possible coefficient of performance of a heat pump that brings energy from outdoors at $-3.00°C$ into a 22.0°C house? Note that the work done to run the heat pump is also available to warm up the house.

25. An ideal refrigerator or ideal heat pump is equivalent to a Carnot engine running in reverse. That is, energy Q_c is taken in from a cold reservoir and energy Q_h is rejected to a hot reservoir. (a) Show that the work that must be supplied to run the refrigerator or heat pump is

$$W = \frac{T_h - T_c}{T_c} Q_c$$

(b) Show that the coefficient of performance of the ideal refrigerator is

$$COP = \frac{T_c}{T_h - T_c}$$

26. A heat pump, shown in Figure P22.26, is essentially an air conditioner installed backward. It extracts energy from colder air outside and deposits it in a warmer room. Suppose that the ratio of the actual energy entering the room to the work done by the device's motor is 10.0% of the theoretical maximum ratio. Determine the energy entering the room per joule of work done by the motor, given that the inside temperature is 20.0°C and the outside temperature is $-5.00°C$.

Figure P22.26

27. How much work does an ideal Carnot refrigerator require to remove 1.00 J of energy from helium at 4.00 K and reject this energy to a room-temperature (293-K) environment?

28. A refrigerator maintains a temperature of 0°C in the cold compartment with a room temperature of 25.0°C. It removes energy from the cold compartment at the rate of 8 000 kJ/h. (a) What minimum power is required to operate the refrigerator? (b) The refrigerator exhausts energy into the room at what rate?

29. If a 35.0%-efficient Carnot heat engine (Fig. 22.2) is run in reverse so as to form a refrigerator (Fig. 22.5), what would be this refrigerator's coefficient of performance?

30. Two Carnot engines have the same efficiency. One engine runs in reverse as a heat pump, and the other runs in reverse as a refrigerator. The coefficient of performance of the heat pump is 1.50 times the coefficient of performance of the refrigerator. Find (a) the coefficient of performance of the refrigerator, (b) the coefficient of performance of the heat pump, and (c) the efficiency of each heat engine.

Section 22.5 Gasoline and Diesel Engines

31. In a cylinder of an automobile engine, just after combustion, the gas is confined to a volume of 50.0 cm^3 and has an initial pressure of 3.00×10^6 Pa. The piston moves outward to a final volume of 300 cm^3, and the gas expands without energy loss by heat. (a) If $\gamma = 1.40$ for the gas, what is the final pressure? (b) How much work is done by the gas in expanding?

32. A gasoline engine has a compression ratio of 6.00 and uses a gas for which $\gamma = 1.40$. (a) What is the efficiency

of the engine if it operates in an idealized Otto cycle? (b) **What If?** If the actual efficiency is 15.0%, what fraction of the fuel is wasted as a result of friction and energy losses by heat that could by avoided in a reversible engine? (Assume complete combustion of the air–fuel mixture.)

33. A 1.60-L gasoline engine with a compression ratio of 6.20 has a useful power output of 102 hp. Assuming the engine operates in an idealized Otto cycle, find the energy taken in and the energy exhausted each second. Assume the fuel-air mixture behaves like an ideal gas with $\gamma = 1.40$.

34. The compression ratio of an Otto cycle, as shown in Figure 22.13, is $V_A/V_B = 8.00$. At the beginning A of the compression process, 500 cm^3 of gas is at 100 kPa and 20.0°C. At the beginning of the adiabatic expansion the temperature is $T_C = 750$°C. Model the working fluid as an ideal gas with $E_{int} = nC_VT = 2.50nRT$ and $\gamma = 1.40$. (a) Fill in the table below to follow the states of the gas:

	T (K)	P (kPa)	V (cm^3)	E_{int}
A	293	100	500	
B				
C	1 023			
D				
A				

(b) Fill in the table below to follow the processes:

	Q (input)	W (output)	ΔE_{int}
$A \rightarrow B$			
$B \rightarrow C$			
$C \rightarrow D$			
$D \rightarrow A$			
$ABCDA$			

(c) Identify the energy input Q_h, the energy exhaust Q_c, and the net output work W_{eng}. (d) Calculate the thermal efficiency. (e) Find the number of crankshaft revolutions per minute required for a one-cylinder engine to have an output power of 1.00 kW = 1.34 hp. Note that the thermodynamic cycle involves four piston strokes.

Section 22.6 Entropy

35. An ice tray contains 500 g of liquid water at 0°C. Calculate the change in entropy of the water as it freezes slowly and completely at 0°C.

36. At a pressure of 1 atm, liquid helium boils at 4.20 K. The latent heat of vaporization is 20.5 kJ/kg. Determine the entropy change (per kilogram) of the helium resulting from vaporization.

37. Calculate the change in entropy of 250 g of water heated slowly from 20.0°C to 80.0°C. (*Suggestion:* Note that $dQ = mc\,dT$.)

38. In making raspberry jelly, 900 g of raspberry juice is combined with 930 g of sugar. The mixture starts at room temperature, 23.0°C, and is slowly heated on a stove until it reaches 220°F. It is then poured into heated jars and allowed to cool. Assume that the juice has the same specific heat as water. The specific heat of sucrose is 0.299 cal/g · °C. Consider the heating process. (a) Which of the following terms describe(s) this process: adiabatic, isobaric, isothermal, isovolumetric, cyclic, reversible, isentropic? (b) How much energy does the mixture absorb? (c) What is the minimum change in entropy of the jelly while it is heated?

39. What change in entropy occurs when a 27.9-g ice cube at -12°C is transformed into steam at 115°C?

Section 22.7 Entropy Changes in Irreversible Processes

40. The temperature at the surface of the Sun is approximately 5 700 K, and the temperature at the surface of the Earth is approximately 290 K. What entropy change occurs when 1 000 J of energy is transferred by radiation from the Sun to the Earth?

41. A 1 500-kg car is moving at 20.0 m/s. The driver brakes to a stop. The brakes cool off to the temperature of the surrounding air, which is nearly constant at 20.0°C. What is the total entropy change?

42. A 1.00-kg iron horseshoe is taken from a forge at 900°C and dropped into 4.00 kg of water at 10.0°C. Assuming that no energy is lost by heat to the surroundings, determine the total entropy change of the horseshoe-plus-water system.

43. How fast are you personally making the entropy of the Universe increase right now? Compute an order-of-magnitude estimate, stating what quantities you take as data and the values you measure or estimate for them.

44. A rigid tank of small mass contains 40.0 g of argon, initially at 200°C and 100 kPa. The tank is placed into a reservoir at 0°C and allowed to cool to thermal equilibrium. (a) Calculate the volume of the tank. (b) Calculate the change in internal energy of the argon. (c) Calculate the energy transferred by heat. (d) Calculate the change in entropy of the argon. (e) Calculate the change in entropy of the constant-temperature bath.

45. A 1.00-mol sample of H$_2$ gas is contained in the left-hand side of the container shown in Figure P22.45, which has equal volumes left and right. The right-hand side is evacuated. When the valve is opened, the gas streams into the right-hand side. What is the final entropy change of the gas? Does the temperature of the gas change?

Figure P22.45

46. A 2.00-L container has a center partition that divides it into two equal parts, as shown in Figure P22.46. The left side contains H_2 gas, and the right side contains O_2 gas. Both gases are at room temperature and at atmospheric pressure. The partition is removed, and the gases are allowed to mix. What is the entropy increase of the system?

| 0.044 mol H_2 | 0.044 mol O_2 |

Figure P22.46

47. A 1.00-mol sample of an ideal monatomic gas, initially at a pressure of 1.00 atm and a volume of 0.025 0 m^3, is heated to a final state with a pressure of 2.00 atm and a volume of 0.040 0 m^3. Determine the change in entropy of the gas in this process.

48. A 1.00-mol sample of a diatomic ideal gas, initially having pressure P and volume V, expands so as to have pressure $2P$ and volume $2V$. Determine the entropy change of the gas in the process.

Section 22.8   Entropy on a Microscopic Scale

49. If you toss two dice, what is the total number of ways in which you can obtain (a) a 12 and (b) a 7?

50. Prepare a table like Table 22.1 for the following occurrence. You toss four coins into the air simultaneously and then record the results of your tosses in terms of the numbers of heads and tails that result. For example, HHTH and HTHH are two possible ways in which three heads and one tail can be achieved. (a) On the basis of your table, what is the most probable result of a toss? In terms of entropy, (b) what is the most ordered state and (c) what is the most disordered state?

51. Repeat the procedure used to construct Table 22.1 (a) for the case in which you draw three marbles from your bag rather than four and (b) for the case in which you draw five rather than four.

Additional Problems

52. Every second at Niagara Falls (Fig. P22.52), some 5 000 m^3 of water falls a distance of 50.0 m. What is the increase in entropy per second due to the falling water? Assume that the mass of the surroundings is so great that its temperature and that of the water stay nearly constant at 20.0°C. Suppose that a negligible amount of water evaporates.

Figure P22.52 Niagara Falls, a popular tourist attraction.

53. A house loses energy through the exterior walls and roof at a rate of 5 000 J/s = 5.00 kW when the interior temperature is 22.0°C and the outside temperature is −5.00°C. Calculate the electric power required to maintain the interior temperature at 22.0°C for the following two cases. (a) The electric power is used in electric resistance heaters (which convert all of the energy transferred in by electrical transmission into internal energy). (b) **What If?** The electric power is used to drive an electric motor that operates the compressor of a heat pump, which has a coefficient of performance equal to 60.0% of the Carnot-cycle value.

54. How much work is required, using an ideal Carnot refrigerator, to change 0.500 kg of tap water at 10.0°C into ice at −20.0°C? Assume the temperature of the freezer compartment is held at −20.0°C and the refrigerator exhausts energy into a room at 20.0°C.

55. A heat engine operates between two reservoirs at $T_2 = 600$ K and $T_1 = 350$ K. It takes in 1 000 J of energy from the higher-temperature reservoir and performs 250 J of work. Find (a) the entropy change of the Universe ΔS_U for this process and (b) the work W that could have been done by an ideal Carnot engine operating between these two reservoirs. (c) Show that the difference between the amounts of work done in parts (a) and (b) is $T_1 \Delta S_U$.

56. Two identically constructed objects, surrounded by thermal insulation, are used as energy reservoirs for a Carnot engine. The finite reservoirs both have mass m and specific heat c. They start out at temperatures T_h and T_c, where $T_h > T_c$. (a) Show that the engine will stop working when the final temperature of each object is $(T_h T_c)^{1/2}$. (b) Show that the total work done by the

Carnot engine is

$$W_{eng} = mc(T_h^{1/2} - T_c^{1/2})^2$$

57. 🌐 In 1816 Robert Stirling, a Scottish clergyman, patented the *Stirling engine*, which has found a wide variety of applications ever since. Fuel is burned externally to warm one of the engine's two cylinders. A fixed quantity of inert gas moves cyclically between the cylinders, expanding in the hot one and contracting in the cold one. Figure P22.57 represents a model for its thermodynamic cycle. Consider n mol of an ideal monatomic gas being taken once through the cycle, consisting of two isothermal processes at temperatures $3T_i$ and T_i and two constant-volume processes. Determine, in terms of n, R, and T_i, (a) the net energy transferred by heat to the gas and (b) the efficiency of the engine. A Stirling engine is easier to manufacture than an internal combustion engine or a turbine. It can run on burning garbage. It can run on the energy of sunlight and produce no material exhaust.

Figure P22.57

58. An electric power plant has an overall efficiency of 15.0%. The plant is to deliver 150 MW of power to a city, and its turbines use coal as the fuel. The burning coal produces steam that drives the turbines. This steam is then condensed to water at 25.0°C by passing it through cooling coils in contact with river water. (a) How many metric tons of coal does the plant consume each day (1 metric ton = 10^3 kg)? (b) What is the total cost of the fuel per year if the delivered price is $8.00/metric ton? (c) If the river water is delivered at 20.0°C, at what minimum rate must it flow over the cooling coils in order that its temperature not exceed 25.0°C? (*Note:* The heat of combustion of coal is 33.0 kJ/g.)

59. A power plant, having a Carnot efficiency, produces 1 000 MW of electrical power from turbines that take in steam at 500 K and reject water at 300 K into a flowing river. The water downstream is 6.00 K warmer due to the output of the power plant. Determine the flow rate of the river.

60. A power plant, having a Carnot efficiency, produces electric power $\mathcal{P}$ from turbines that take in energy from steam at temperature T_h and discharge energy at temperature T_c through a heat exchanger into a flowing river. The water downstream is warmer by ΔT due to the output of the power plant. Determine the flow rate of the river.

61. An athlete whose mass is 70.0 kg drinks 16 oz (453.6 g) of refrigerated water. The water is at a temperature of 35.0°F. (a) Ignoring the temperature change of the body that results from the water intake (so that the body is regarded as a reservoir always at 98.6°F), find the entropy increase of the entire system. (b) **What If?** Assume that the entire body is cooled by the drink and that the average specific heat of a person is equal to the specific heat of liquid water. Ignoring any other energy transfers by heat and any metabolic energy release, find the athlete's temperature after she drinks the cold water, given an initial body temperature of 98.6°F. Under these assumptions, what is the entropy increase of the entire system? Compare this result with the one you obtained in part (a).

62. A 1.00-mol sample of an ideal monatomic gas is taken through the cycle shown in Figure P22.62. The process $A \rightarrow B$ is a reversible isothermal expansion. Calculate (a) the net work done by the gas, (b) the energy added to the gas by heat, (c) the energy exhausted from the gas by heat, and (d) the efficiency of the cycle.

Figure P22.62

63. A biology laboratory is maintained at a constant temperature of 7.00°C by an air conditioner, which is vented to the air outside. On a typical hot summer day the outside temperature is 27.0°C and the air conditioning unit emits energy to the outside at a rate of 10.0 kW. Model the unit as having a coefficient of performance equal to 40.0% of the coefficient of performance of an ideal Carnot device. (a) At what rate does the air conditioner remove energy from the laboratory? (b) Calculate the power required for the work input. (c) Find the change in entropy produced by the air conditioner in 1.00 h. (d) **What If?** The outside temperature increases to 32.0°C. Find the fractional change in the coefficient of performance of the air conditioner.

64. A 1.00-mol sample of an ideal gas expands isothermally, doubling in volume. (a) Show that the work it does in ex-

panding is $W = RT \ln 2$. (b) Because the internal energy E_{int} of an ideal gas depends solely on its temperature, the change in internal energy is zero during the expansion. It follows from the first law that the energy input to the gas by heat during the expansion is equal to the energy output by work. Why does this conversion *not* violate the second law?

65. A 1.00-mol sample of a monatomic ideal gas is taken through the cycle shown in Figure P22.65. At point A, the pressure, volume, and temperature are P_i, V_i, and T_i, respectively. In terms of R and T_i, find (a) the total energy entering the system by heat per cycle, (b) the total energy leaving the system by heat per cycle, (c) the efficiency of an engine operating in this cycle, and (d) the efficiency of an engine operating in a Carnot cycle between the same temperature extremes.

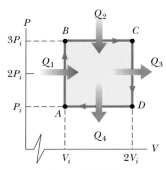

Figure P22.65

66. A sample consisting of n mol of an ideal gas undergoes a reversible isobaric expansion from volume V_i to volume $3V_i$. Find the change in entropy of the gas by calculating $\int_i^f dQ / T$ where $dQ = nC_P\, dT$.

67. A system consisting of n mol of an ideal gas undergoes two reversible processes. It starts with pressure P_i and volume V_i, expands isothermally, and then contracts adiabatically to reach a final state with pressure P_i and volume $3V_i$. (a) Find its change in entropy in the isothermal process. The entropy does not change in the adiabatic process. (b) **What If?** Explain why the answer to part (a) must be the same as the answer to Problem 66.

68. Suppose you are working in a patent office, and an inventor comes to you with the claim that her heat engine, which employs water as a working substance, has a thermodynamic efficiency of 0.61. She explains that it operates between energy reservoirs at 4°C and 0°C. It is a very complicated device, with many pistons, gears, and pulleys, and the cycle involves freezing and melting. Does her claim that $e = 0.61$ warrant serious consideration? Explain.

69. An idealized diesel engine operates in a cycle known as the *air-standard diesel cycle*, shown in Figure 22.14. Fuel is sprayed into the cylinder at the point of maximum compression, B. Combustion occurs during the expansion $B \rightarrow C$, which is modeled as an isobaric process. Show that the efficiency of an engine operating in this idealized diesel cycle is

$$e = 1 - \frac{1}{\gamma}\left(\frac{T_D - T_A}{T_C - T_B}\right)$$

70. A 1.00-mol sample of an ideal gas ($\gamma = 1.40$) is carried through the Carnot cycle described in Figure 22.11. At point A, the pressure is 25.0 atm and the temperature is 600 K. At point C, the pressure is 1.00 atm and the temperature is 400 K. (a) Determine the pressures and volumes at points A, B, C, and D. (b) Calculate the net work done per cycle. (c) Determine the efficiency of an engine operating in this cycle.

71. Suppose 1.00 kg of water at 10.0°C is mixed with 1.00 kg of water at 30.0°C at constant pressure. When the mixture has reached equilibrium, (a) what is the final temperature? (b) Take $c_P = 4.19$ kJ/kg·K for water and show that the entropy of the system increases by

$$\Delta S = 4.19 \ln\left[\left(\frac{293}{283}\right)\left(\frac{293}{303}\right)\right] \text{kJ/K}$$

(c) Verify numerically that $\Delta S > 0$. (d) Is the mixing an irreversible process?

Answers to Quick Quizzes

22.1 (c). Equation 22.2 gives this result directly.

22.2 (b). The work represents one third of the input energy. The remainder, two thirds, must be expelled to the cold reservoir.

22.3 (d). The COP of 4.00 for the heat pump means that you are receiving four times as much energy as the energy entering by electrical transmission. With four times as much energy per unit of energy from electricity, you need only one fourth as much electricity.

22.4 C, B, A. Although all three engines operate over a 300-K temperature difference, the efficiency depends on the ratio of temperatures, not the difference.

22.5 One microstate—all four deuces.

22.6 Six microstates—club–diamond, club–heart, club–spade, diamond–heart, diamond–spade, heart–spade. The macrostate of two aces is more probable than that of four deuces in Quick Quiz 22.5 because there are six times as many microstates for this particular macrostate compared to the macrostate of four deuces. Thus, in a hand of poker, two of a kind is less valuable than four of a kind.

22.7 (b). Because the process is reversible and adiabatic, $Q_r = 0$; therefore, $\Delta S = 0$.

22.8 (a). From the first law of thermodynamics, for these two reversible processes, $Q_r = \Delta E_{int} - W$. During the constant-volume process, $W = 0$, while the work W is nonzero and negative during the constant-pressure expansion. Thus, Q_r is larger for the constant-pressure process, leading to a larger value for the change in entropy. In terms of entropy as disorder, during the constant-pressure process, the gas must expand. The increase in volume results in more ways of locating the molecules of the gas in a container, resulting in a larger increase in entropy.

22.9 False. The determining factor for the entropy change is Q_r, not Q. If the adiabatic process is not reversible, the entropy change is not necessarily zero because a reversible path between the same initial and final states may involve energy transfer by heat.

Electricity and Magnetism

We now study the branch of physics concerned with electric and magnetic phenomena. The laws of electricity and magnetism have a central role in the operation of such devices as radios, televisions, electric motors, computers, high-energy accelerators, and other electronic devices. More fundamentally, the interatomic and intermolecular forces responsible for the formation of solids and liquids are electric in origin. Furthermore, such forces as the pushes and pulls between objects and the elastic force in a spring arise from electric forces at the atomic level.

Evidence in Chinese documents suggests that magnetism was observed as early as 2000 B.C. The ancient Greeks observed electric and magnetic phenomena possibly as early as 700 B.C. They found that a piece of amber, when rubbed, becomes electrified and attracts pieces of straw or feathers. The Greeks knew about magnetic forces from observations that the naturally occurring stone *magnetite* (Fe_3O_4) is attracted to iron. (The word *electric* comes from *elecktron*, the Greek word for "amber." The word *magnetic* comes from *Magnesia*, the name of the district of Greece where magnetite was first found.) In 1600, the Englishman William Gilbert discovered that electrification is not limited to amber but rather is a general phenomenon. In the years following this discovery, scientists electrified a variety of objects. Experiments by Charles Coulomb in 1785 confirmed the inverse-square law for electric forces.

It was not until the early part of the nineteenth century that scientists established that electricity and magnetism are related phenomena. In 1819, Hans Oersted discovered that a compass needle is deflected when placed near a circuit carrying an electric current. In 1831, Michael Faraday and, almost simultaneously, Joseph Henry showed that when a wire is moved near a magnet (or, equivalently, when a magnet is moved near a wire), an electric current is established in the wire. In 1873, James Clerk Maxwell used these observations and other experimental facts as a basis for formulating the laws of electromagnetism as we know them today. (*Electromagnetism* is a name given to the combined study of electricity and magnetism.) Shortly thereafter (around 1888), Heinrich Hertz verified Maxwell's predictions by producing electromagnetic waves in the laboratory. This achievement led to such practical developments as radio and television.

Maxwell's contributions to the field of electromagnetism were especially significant because the laws he formulated are basic to *all* forms of electromagnetic phenomena. His work is as important as Newton's work on the laws of motion and the theory of gravitation.

◀ *Lightning is a dramatic example of electrical phenomena occurring in nature. While we are most familiar with lightning originating from thunderclouds, it can occur in other situations, such as in a volcanic eruption (here, the Sakurajima volcano, Japan). (M. Zhilin/ M. Newman/Photo Researchers, Inc.)*

Chapter 23

Electric Fields

▲ *Mother and daughter are both enjoying the effects of electrically charging their bodies. Each individual hair on their heads becomes charged and exerts a repulsive force on the other hairs, resulting in the "stand-up" hairdos that you see here. (Courtesy of Resonance Research Corporation)*

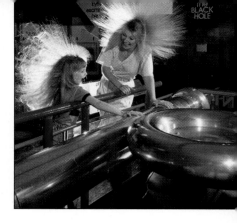

The electromagnetic force between charged particles is one of the fundamental forces of nature. We begin this chapter by describing some of the basic properties of one manifestation of the electromagnetic force, the electric force. We then discuss Coulomb's law, which is the fundamental law governing the electric force between any two charged particles. Next, we introduce the concept of an electric field associated with a charge distribution and describe its effect on other charged particles. We then show how to use Coulomb's law to calculate the electric field for a given charge distribution. We conclude the chapter with a discussion of the motion of a charged particle in a uniform electric field.

23.1 Properties of Electric Charges

A number of simple experiments demonstrate the existence of electric forces and charges. For example, after running a comb through your hair on a dry day, you will find that the comb attracts bits of paper. The attractive force is often strong enough to suspend the paper. The same effect occurs when certain materials are rubbed together, such as glass rubbed with silk or rubber with fur.

Another simple experiment is to rub an inflated balloon with wool. The balloon then adheres to a wall, often for hours. When materials behave in this way, they are said to be *electrified*, or to have become **electrically charged.** You can easily electrify your body by vigorously rubbing your shoes on a wool rug. Evidence of the electric charge on your body can be detected by lightly touching (and startling) a friend. Under the right conditions, you will see a spark when you touch, and both of you will feel a slight tingle. (Experiments such as these work best on a dry day because an excessive amount of moisture in the air can cause any charge you build up to "leak" from your body to the Earth.)

In a series of simple experiments, it was found that there are two kinds of electric charges, which were given the names **positive** and **negative** by Benjamin Franklin (1706–1790). We identify negative charge as that type possessed by electrons and positive charge as that possessed by protons. To verify that there are two types of charge, suppose a hard rubber rod that has been rubbed with fur is suspended by a sewing thread, as shown in Figure 23.1. When a glass rod that has been rubbed with silk is brought near the rubber rod, the two attract each other (Fig. 23.1a). On the other hand, if two charged rubber rods (or two charged glass rods) are brought near each other, as shown in Figure 23.1b, the two repel each other. This observation shows that the rubber and glass have two different types of charge on them. On the basis of these observations, we conclude that **charges of the same sign repel one another and charges with opposite signs attract one another.**

Using the convention suggested by Franklin, the electric charge on the glass rod is called positive and that on the rubber rod is called negative. Therefore, any charged object attracted to a charged rubber rod (or repelled by a charged glass rod) must

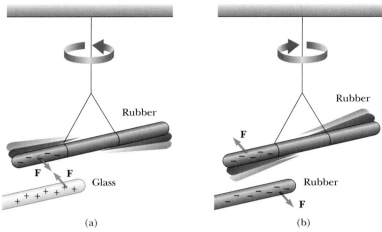

Figure 23.1 (a) A negatively charged rubber rod suspended by a thread is attracted to a positively charged glass rod. (b) A negatively charged rubber rod is repelled by another negatively charged rubber rod.

Electric charge is conserved

Figure 23.2 When a glass rod is rubbed with silk, electrons are transferred from the glass to the silk. Because of conservation of charge, each electron adds negative charge to the silk, and an equal positive charge is left behind on the rod. Also, because the charges are transferred in discrete bundles, the charges on the two objects are $\pm e$, or $\pm 2e$, or $\pm 3e$, and so on.

have a positive charge, and any charged object repelled by a charged rubber rod (or attracted to a charged glass rod) must have a negative charge.

Attractive electric forces are responsible for the behavior of a wide variety of commercial products. For example, the plastic in many contact lenses, *etafilcon*, is made up of molecules that electrically attract the protein molecules in human tears. These protein molecules are absorbed and held by the plastic so that the lens ends up being primarily composed of the wearer's tears. Because of this, the lens does not behave as a foreign object to the wearer's eye, and it can be worn comfortably. Many cosmetics also take advantage of electric forces by incorporating materials that are electrically attracted to skin or hair, causing the pigments or other chemicals to stay put once they are applied.

Another important aspect of electricity that arises from experimental observations is that **electric charge is always conserved** in an isolated system. That is, when one object is rubbed against another, charge is not created in the process. The electrified state is due to a *transfer* of charge from one object to the other. One object gains some amount of negative charge while the other gains an equal amount of positive charge. For example, when a glass rod is rubbed with silk, as in Figure 23.2, the silk obtains a negative charge that is equal in magnitude to the positive charge on the glass rod. We now know from our understanding of atomic structure that electrons are transferred from the glass to the silk in the rubbing process. Similarly, when rubber is rubbed with fur, electrons are transferred from the fur to the rubber, giving the rubber a net negative charge and the fur a net positive charge. This process is consistent with the fact that neutral, uncharged matter contains as many positive charges (protons within atomic nuclei) as negative charges (electrons).

In 1909, Robert Millikan (1868–1953) discovered that electric charge always occurs as some integral multiple of a fundamental amount of charge e (see Section 25.7). In modern terms, the electric charge q is said to be **quantized,** where q is the standard symbol used for charge as a variable. That is, electric charge exists as discrete "packets," and we can write $q = Ne$, where N is some integer. Other experiments in the same period showed that the electron has a charge $-e$ and the proton has a charge of equal magnitude but opposite sign $+e$. Some particles, such as the neutron, have no charge.

From our discussion thus far, we conclude that electric charge has the following important properties:

- There are two kinds of charges in nature; charges of opposite sign attract one another and charges of the same sign repel one another.
- Total charge in an isolated system is conserved.
- Charge is quantized.

Quick Quiz 23.1 If you rub an inflated balloon against your hair, the two materials attract each other, as shown in Figure 23.3. Is the amount of charge present in the system of the balloon and your hair after rubbing (a) less than, (b) the same as, or (c) more than the amount of charge present before rubbing?

Quick Quiz 23.2 Three objects are brought close to each other, two at a time. When objects A and B are brought together, they repel. When objects B and C are brought together, they also repel. Which of the following are true? (a) Objects A and C possess charges of the same sign. (b) Objects A and C possess charges of opposite sign. (c) All three of the objects possess charges of the same sign. (d) One of the objects is neutral. (e) We would need to perform additional experiments to determine the signs of the charges.

Charles D. Winters

Figure 23.3 (Quick Quiz 23.1) Rubbing a balloon against your hair on a dry day causes the balloon and your hair to become charged.

23.2 Charging Objects By Induction

It is convenient to classify materials in terms of the ability of electrons to move through the material:

Electrical **conductors** are materials in which some of the electrons are free electrons[1] that are not bound to atoms and can move relatively freely through the material; electrical **insulators** are materials in which all electrons are bound to atoms and cannot move freely through the material.

Materials such as glass, rubber, and wood fall into the category of electrical insulators. When such materials are charged by rubbing, only the area rubbed becomes charged, and the charged particles are unable to move to other regions of the material.

In contrast, materials such as copper, aluminum, and silver are good electrical conductors. When such materials are charged in some small region, the charge readily distributes itself over the entire surface of the material. If you hold a copper rod in your hand and rub it with wool or fur, it will not attract a small piece of paper. This might suggest that a metal cannot be charged. However, if you attach a wooden handle to the rod and then hold it by that handle as you rub the rod, the rod will remain charged and attract the piece of paper. The explanation for this is as follows: without the insulating wood, the electric charges produced by rubbing readily move from the copper through your body, which is also a conductor, and into the Earth. The insulating wooden handle prevents the flow of charge into your hand.

Semiconductors are a third class of materials, and their electrical properties are somewhere between those of insulators and those of conductors. Silicon and

---

[1]   A metal atom contains one or more outer electrons, which are weakly bound to the nucleus. When many atoms combine to form a metal, the so-called *free electrons* are these outer electrons, which are not bound to any one atom. These electrons move about the metal in a manner similar to that of gas molecules moving in a container.

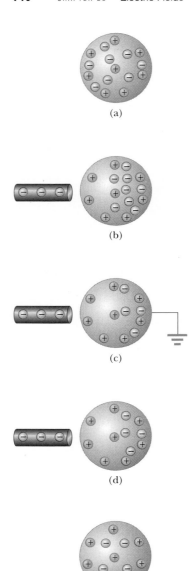

Figure 23.4 Charging a metallic object by *induction* (that is, the two objects never touch each other). (a) A neutral metallic sphere, with equal numbers of positive and negative charges. (b) The electrons on the neutral sphere are redistributed when a charged rubber rod is placed near the sphere. (c) When the sphere is grounded, some of its electrons leave through the ground wire. (d) When the ground connection is removed, the sphere has excess positive charge that is nonuniformly distributed. (e) When the rod is removed, the remaining electrons redistribute uniformly and there is a net uniform distribution of positive charge on the sphere.

germanium are well-known examples of semiconductors commonly used in the fabrication of a variety of electronic chips used in computers, cellular telephones, and stereo systems. The electrical properties of semiconductors can be changed over many orders of magnitude by the addition of controlled amounts of certain atoms to the materials.

To understand how to charge a conductor by a process known as **induction,** consider a neutral (uncharged) conducting sphere insulated from the ground, as shown in Figure 23.4a. There are an equal number of electrons and protons in the sphere if the charge on the sphere is exactly zero. When a negatively charged rubber rod is brought near the sphere, electrons in the region nearest the rod experience a repulsive force and migrate to the opposite side of the sphere. This leaves the side of the sphere near the rod with an effective positive charge because of the diminished number of electrons, as in Figure 23.4b. (The left side of the sphere in Figure 23.4b is positively charged *as if* positive charges moved into this region, but remember that it is only electrons that are free to move.) This occurs even if the rod never actually touches the sphere. If the same experiment is performed with a conducting wire connected from the sphere to the Earth (Fig. 23.4c), some of the electrons in the conductor are so strongly repelled by the presence of the negative charge in the rod that they move out of the sphere through the wire and into the Earth. The symbol ⏚ at the end of the wire in Figure 23.4c indicates that the wire is connected to **ground,** which means a reservoir, such as the Earth, that can accept or provide electrons freely with negligible effect on its electrical characteristics. If the wire to ground is then removed (Fig. 23.4d), the conducting sphere contains an excess of *induced* positive charge because it has fewer electrons than it needs to cancel out the positive charge of the protons. When the rubber rod is removed from the vicinity of the sphere (Fig. 23.4e), this induced positive charge remains on the ungrounded sphere. Note that the rubber rod loses none of its negative charge during this process.

Charging an object by induction requires no contact with the object inducing the charge. This is in contrast to charging an object by rubbing (that is, by *conduction*), which does require contact between the two objects.

A process similar to induction in conductors takes place in insulators. In most neutral molecules, the center of positive charge coincides with the center of negative charge. However, in the presence of a charged object, these centers inside each molecule in an insulator may shift slightly, resulting in more positive charge on one side of the molecule than on the other. This realignment of charge within individual molecules produces a layer of charge on the surface of the insulator, as shown in Figure 23.5a. Knowing about induction in insulators, you should be able to explain why a comb that has been rubbed through hair attracts bits of electrically neutral paper and why a balloon that has been rubbed against your clothing is able to stick to an electrically neutral wall.

Figure 23.5 (a) The charged object on the left induces a charge distribution on the surface of an insulator due to realignment of charges in the molecules. (b) A charged comb attracts bits of paper because charges in molecules in the paper are realigned.

23.3 Coulomb's Law

Charles Coulomb (1736–1806) measured the magnitudes of the electric forces between charged objects using the torsion balance, which he invented (Fig. 23.6). Coulomb confirmed that the electric force between two small charged spheres is proportional to the inverse square of their separation distance r—that is, $F_e \propto 1/r^2$. The operating principle of the torsion balance is the same as that of the apparatus used by Cavendish to measure the gravitational constant (see Section 13.2), with the electrically neutral spheres replaced by charged ones. The electric force between charged spheres A and B in Figure 23.6 causes the spheres to either attract or repel each other, and the resulting motion causes the suspended fiber to twist. Because the restoring torque of the twisted fiber is proportional to the angle through which the fiber rotates, a measurement of this angle provides a quantitative measure of the electric force of attraction or repulsion. Once the spheres are charged by rubbing, the electric force between them is very large compared with the gravitational attraction, and so the gravitational force can be neglected.

From Coulomb's experiments, we can generalize the following properties of the **electric force** between two stationary charged particles. The electric force

- is inversely proportional to the square of the separation r between the particles and directed along the line joining them;
- is proportional to the product of the charges q_1 and q_2 on the two particles;
- is attractive if the charges are of opposite sign and repulsive if the charges have the same sign;
- is a conservative force.

We will use the term **point charge** to mean a particle of zero size that carries an electric charge. The electrical behavior of electrons and protons is very well described by modeling them as point charges. From experimental observations on the electric force, we can express **Coulomb's law** as an equation giving the magnitude of the electric force (sometimes called the *Coulomb force*) between two point charges:

$$F_e = k_e \frac{|q_1||q_2|}{r^2} \qquad (23.1)$$

◄ Coulomb's law

where k_e is a constant called the **Coulomb constant.** In his experiments, Coulomb was able to show that the value of the exponent of r was 2 to within an uncertainty of a few percent. Modern experiments have shown that the exponent is 2 to within an uncertainty of a few parts in 10^{16}.

The value of the Coulomb constant depends on the choice of units. The SI unit of charge is the **coulomb** (C). The Coulomb constant k_e in SI units has the value

$$k_e = 8.987\,5 \times 10^9 \ \text{N} \cdot \text{m}^2/\text{C}^2 \qquad (23.2)$$

◄ Coulomb constant

This constant is also written in the form

$$k_e = \frac{1}{4\pi\epsilon_0} \qquad (23.3)$$

Figure 23.6 Coulomb's torsion balance, used to establish the inverse-square law for the electric force between two charges.

Table 23.1

Charge and Mass of the Electron, Proton, and Neutron		
Particle	Charge (C)	Mass (kg)
Electron (e)	$-1.602\,191\,7 \times 10^{-19}$	$9.109\,5 \times 10^{-31}$
Proton (p)	$+1.602\,191\,7 \times 10^{-19}$	$1.672\,61 \times 10^{-27}$
Neutron (n)	0	$1.674\,92 \times 10^{-27}$

Charles Coulomb

French physicist (1736–1806)

Coulomb's major contributions to science were in the areas of electrostatics and magnetism. During his lifetime, he also investigated the strengths of materials and determined the forces that affect objects on beams, thereby contributing to the field of structural mechanics. In the field of ergonomics, his research provided a fundamental understanding of the ways in which people and animals can best do work. *(Photo courtesy of AIP Niels Bohr Library/E. Scott Barr Collection)*

where the constant ϵ_0 (lowercase Greek epsilon) is known as the **permittivity of free space** and has the value

$$\epsilon_0 = 8.854\,2 \times 10^{-12}\ \mathrm{C}^2/\mathrm{N}\cdot\mathrm{m}^2 \qquad (23.4)$$

The smallest unit of charge e known in nature[2] is the charge on an electron $(-e)$ or a proton $(+e)$ and has a magnitude

$$e = 1.602\,19 \times 10^{-19}\ \mathrm{C} \qquad (23.5)$$

Therefore, 1 C of charge is approximately equal to the charge of 6.24×10^{18} electrons or protons. This number is very small when compared with the number of free electrons in 1 cm^3 of copper, which is on the order of 10^{23}. Still, 1 C is a substantial amount of charge. In typical experiments in which a rubber or glass rod is charged by friction, a net charge on the order of 10^{-6} C is obtained. In other words, only a very small fraction of the total available charge is transferred between the rod and the rubbing material.

The charges and masses of the electron, proton, and neutron are given in Table 23.1.

Quick Quiz 23.4 Object A has a charge of $+2\ \mu\mathrm{C}$, and object B has a charge of $+6\ \mu\mathrm{C}$. Which statement is true about the electric forces on the objects? (a) $F_{AB} = -3F_{BA}$ (b) $F_{AB} = -F_{BA}$ (c) $3F_{AB} = -F_{BA}$ (d) $F_{AB} = 3F_{BA}$ (e) $F_{AB} = F_{BA}$ (f) $3F_{AB} = F_{BA}$

Example 23.1   The Hydrogen Atom

The electron and proton of a hydrogen atom are separated (on the average) by a distance of approximately 5.3×10^{-11} m. Find the magnitudes of the electric force and the gravitational force between the two particles.

Solution From Coulomb's law, we find that the magnitude of the electric force is

$$F_e = k_e \frac{|e||-e|}{r^2} = (8.99 \times 10^9\ \mathrm{N}\cdot\mathrm{m}^2/\mathrm{C}^2)\frac{(1.60 \times 10^{-19}\ \mathrm{C})^2}{(5.3 \times 10^{-11}\ \mathrm{m})^2}$$

$$= \boxed{8.2 \times 10^{-8}\ \mathrm{N}}$$

Using Newton's law of universal gravitation and Table 23.1 for the particle masses, we find that the magnitude of the

gravitational force is

$$F_g = G \frac{m_e m_p}{r^2}$$

$$= (6.67 \times 10^{-11}\ \mathrm{N}\cdot\mathrm{m}^2/\mathrm{kg}^2)$$

$$\times \frac{(9.11 \times 10^{-31}\ \mathrm{kg})(1.67 \times 10^{-27}\ \mathrm{kg})}{(5.3 \times 10^{-11}\ \mathrm{m})^2}$$

$$= \boxed{3.6 \times 10^{-47}\ \mathrm{N}}$$

The ratio $F_e/F_g \approx 2 \times 10^{39}$. Thus, the gravitational force between charged atomic particles is negligible when compared with the electric force. Note the similarity of form of Newton's law of universal gravitation and Coulomb's law of electric forces. Other than magnitude, what is a fundamental difference between the two forces?

---

[2] No unit of charge smaller than e has been detected on a free particle; however, current theories propose the existence of particles called *quarks* having charges $-e/3$ and $2e/3$. Although there is considerable experimental evidence for such particles inside nuclear matter, *free* quarks have never been detected. We discuss other properties of quarks in Chapter 46 of the extended version of this text.

When dealing with Coulomb's law, you must remember that force is a vector quantity and must be treated accordingly. The law expressed in vector form for the electric force exerted by a charge q_1 on a second charge q_2, written $\mathbf{F}_{12}$, is

$$\mathbf{F}_{12} = k_e \frac{q_1 q_2}{r^2} \hat{\mathbf{r}} \qquad (23.6)$$

Vector form of Coulomb's law

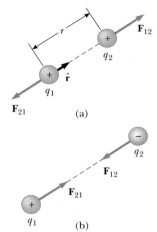

where $\hat{\mathbf{r}}$ is a unit vector directed from q_1 toward q_2, as shown in Figure 23.7a. Because the electric force obeys Newton's third law, the electric force exerted by q_2 on q_1 is equal in magnitude to the force exerted by q_1 on q_2 and in the opposite direction; that is, $\mathbf{F}_{21} = -\mathbf{F}_{12}$. Finally, from Equation 23.6, we see that if q_1 and q_2 have the same sign, as in Figure 23.7a, the product $q_1 q_2$ is positive. If q_1 and q_2 are of opposite sign, as shown in Figure 23.7b, the product $q_1 q_2$ is negative. These signs describe the *relative* direction of the force but not the *absolute* direction. A negative product indicates an attractive force, so that the charges each experience a force toward the other—thus, the force on one charge is in a direction *relative* to the other. A positive product indicates a repulsive force such that each charge experiences a force away from the other. The *absolute* direction of the force in space is not determined solely by the sign of $q_1 q_2$—whether the force on an individual charge is in the positive or negative direction on a coordinate axis depends on the location of the other charge. For example, if an x axis lies along the two charges in Figure 23.7a, the product $q_1 q_2$ is positive, but $\mathbf{F}_{12}$ points in the $+x$ direction and $\mathbf{F}_{21}$ points in the $-x$ direction.

Active Figure 23.7 Two point charges separated by a distance r exert a force on each other that is given by Coulomb's law. The force $\mathbf{F}_{21}$ exerted by q_2 on q_1 is equal in magnitude and opposite in direction to the force $\mathbf{F}_{12}$ exerted by q_1 on q_2. (a) When the charges are of the same sign, the force is repulsive. (b) When the charges are of opposite signs, the force is attractive.

Quick Quiz 23.5 Object A has a charge of $+2 \ \mu C$, and object B has a charge of $+6 \ \mu C$. Which statement is true about the electric forces on the objects? (a) $\mathbf{F}_{AB} = -3\mathbf{F}_{BA}$ (b) $\mathbf{F}_{AB} = -\mathbf{F}_{BA}$ (c) $3\mathbf{F}_{AB} = -\mathbf{F}_{BA}$ (d) $\mathbf{F}_{AB} = 3\mathbf{F}_{BA}$ (e) $\mathbf{F}_{AB} = \mathbf{F}_{BA}$ (f) $3\mathbf{F}_{AB} = \mathbf{F}_{BA}$

At the Active Figures link at http://www.pse6.com, you can move the charges to any position in two-dimensional space and observe the electric forces on them.

When more than two charges are present, the force between any pair of them is given by Equation 23.6. Therefore, the resultant force on any one of them equals the vector sum of the forces exerted by the various individual charges. For example, if four charges are present, then the resultant force exerted by particles 2, 3, and 4 on particle 1 is

$$\mathbf{F}_1 = \mathbf{F}_{21} + \mathbf{F}_{31} + \mathbf{F}_{41}$$

Example 23.2   Find the Resultant Force

Consider three point charges located at the corners of a right triangle as shown in Figure 23.8, where $q_1 = q_3 = 5.0 \ \mu C$, $q_2 = -2.0 \ \mu C$, and $a = 0.10$ m. Find the resultant force exerted on q_3.

Solution First, note the direction of the individual forces exerted by q_1 and q_2 on q_3. The force $\mathbf{F}_{23}$ exerted by q_2 on q_3 is attractive because q_2 and q_3 have opposite signs. The force $\mathbf{F}_{13}$ exerted by q_1 on q_3 is repulsive because both charges are positive.

The magnitude of $\mathbf{F}_{23}$ is

$$F_{23} = k_e \frac{|q_2||q_3|}{a^2}$$

$$= (8.99 \times 10^9 \ \text{N} \cdot \text{m}^2/\text{C}^2) \frac{(2.0 \times 10^{-6} \ \text{C})(5.0 \times 10^{-6} \ \text{C})}{(0.10 \ \text{m})^2}$$

$$= 9.0 \ \text{N}$$

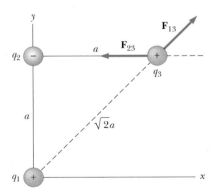

Figure 23.8 (Example 23.2) The force exerted by q_1 on q_3 is $\mathbf{F}_{13}$. The force exerted by q_2 on q_3 is $\mathbf{F}_{23}$. The resultant force $\mathbf{F}_3$ exerted on q_3 is the vector sum $\mathbf{F}_{13} + \mathbf{F}_{23}$.

In the coordinate system shown in Figure 23.8, the attractive force $\mathbf{F}_{23}$ is to the left (in the negative x direction).

The magnitude of the force $\mathbf{F}_{13}$ exerted by q_1 on q_3 is

$$F_{13} = k_e \frac{|q_1||q_3|}{(\sqrt{2}a)^2}$$

$$= (8.99 \times 10^9 \text{ N·m}^2/\text{C}^2) \frac{(5.0 \times 10^{-6} \text{ C})(5.0 \times 10^{-6} \text{ C})}{2(0.10 \text{ m})^2}$$

$$= 11 \text{ N}$$

The repulsive force $\mathbf{F}_{13}$ makes an angle of $45°$ with the x axis. Therefore, the x and y components of $\mathbf{F}_{13}$ are equal, with magnitude given by $F_{13} \cos 45° = 7.9$ N.

Combining $\mathbf{F}_{13}$ with $\mathbf{F}_{23}$ by the rules of vector addition, we arrive at the x and y components of the resultant force acting on q_3:

$$F_{3x} = F_{13x} + F_{23x} = 7.9 \text{ N} + (-9.0 \text{ N}) = -1.1 \text{ N}$$

$$F_{3y} = F_{13y} + F_{23y} = 7.9 \text{ N} + 0 = 7.9 \text{ N}$$

We can also express the resultant force acting on q_3 in unit-vector form as

$$\mathbf{F}_3 = \boxed{(-1.1\hat{\mathbf{i}} + 7.9\hat{\mathbf{j}}) \text{ N}}$$

What If? What if the signs of all three charges were changed to the opposite signs? How would this affect the result for F₃?

Answer The charge q_3 would still be attracted toward q_2 and repelled from q_1 with forces of the same magnitude. Thus, the final result for $\mathbf{F}_3$ would be exactly the same.

Example 23.3   Where Is the Resultant Force Zero?   `Interactive`

Three point charges lie along the x axis as shown in Figure 23.9. The positive charge $q_1 = 15.0 \ \mu\text{C}$ is at $x = 2.00$ m, the positive charge $q_2 = 6.00 \ \mu\text{C}$ is at the origin, and the resultant force acting on q_3 is zero. What is the x coordinate of q_3?

Solution Because q_3 is negative and q_1 and q_2 are positive, the forces $\mathbf{F}_{13}$ and $\mathbf{F}_{23}$ are both attractive, as indicated in Figure 23.9. From Coulomb's law, $\mathbf{F}_{13}$ and $\mathbf{F}_{23}$ have magnitudes

$$F_{13} = k_e \frac{|q_1||q_3|}{(2.00 - x)^2} \qquad F_{23} = k_e \frac{|q_2||q_3|}{x^2}$$

For the resultant force on q_3 to be zero, $\mathbf{F}_{23}$ must be equal in magnitude and opposite in direction to $\mathbf{F}_{13}$. Setting the magnitudes of the two forces equal, we have

$$k_e \frac{|q_2||q_3|}{x^2} = k_e \frac{|q_1||q_3|}{(2.00 - x)^2}$$

Noting that k_e and $|q_3|$ are common to both sides and so can be dropped, we solve for x and find that

$$(2.00 - x)^2 |q_2| = x^2 |q_1|$$

$$(4.00 - 4.00x + x^2)(6.00 \times 10^{-6} \text{ C}) = x^2(15.0 \times 10^{-6} \text{ C})$$

This can be reduced to the following quadratic equation:

$$3.00x^2 + 8.00x - 8.00 = 0$$

Solving this quadratic equation for x, we find that the positive root is $x = \boxed{0.775 \text{ m}}$. There is also a second root, $x = -3.44$ m. This is another location at which the magnitudes of the forces on q_3 are equal, but both forces are in the same direction at this location.

What If? Suppose charge q_3 is constrained to move only along the x axis. From its initial position at $x = 0.775$ m, it is pulled a very small distance along the x axis. When released, will it return to equilibrium or be pulled further from equilibrium? That is, is the equilibrium stable or unstable?

Answer If the charge is moved to the right, $\mathbf{F}_{13}$ becomes larger and $\mathbf{F}_{23}$ becomes smaller. This results in a net force to the right, in the same direction as the displacement. Thus, the equilibrium is *unstable*.

Note that if the charge is constrained to stay at a *fixed* x coordinate but allowed to move up and down in Figure 23.9, the equilibrium is stable. In this case, if the charge is pulled upward (or downward) and released, it will move back toward the equilibrium position and undergo oscillation.

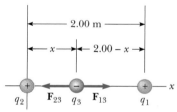

Figure 23.9 (Example 23.3) Three point charges are placed along the x axis. If the resultant force acting on q_3 is zero, then the force $\mathbf{F}_{13}$ exerted by q_1 on q_3 must be equal in magnitude and opposite in direction to the force $\mathbf{F}_{23}$ exerted by q_2 on q_3.

 At the Interactive Worked Example link at http://www.pse6.com, *you can predict where on the x axis the electric force is zero for random values of q_1 and q_2.*

Example 23.4   Find the Charge on the Spheres

Two identical small charged spheres, each having a mass of 3.0×10^{-2} kg, hang in equilibrium as shown in Figure 23.10a. The length of each string is 0.15 m, and the angle θ is $5.0°$. Find the magnitude of the charge on each sphere.

Solution Figure 23.10a helps us conceptualize this problem—the two spheres exert repulsive forces on each other. If they are held close to each other and released, they will move outward from the center and settle into the configuration in Figure 23.10a after the damped oscillations

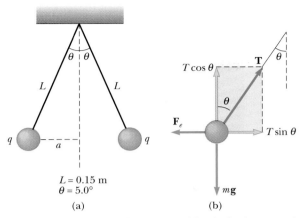

Figure 23.10 (Example 23.4) (a) Two identical spheres, each carrying the same charge q, suspended in equilibrium. (b) The free-body diagram for the sphere on the left.

due to air resistance have vanished. The key phrase "in equilibrium" helps us categorize this as an equilibrium problem, which we approach as we did equilibrium problems in Chapter 5 with the added feature that one of the forces on a sphere is an electric force. We analyze this problem by drawing the free-body diagram for the left-hand sphere in Figure 23.10b. The sphere is in equilibrium under the application of the forces **T** from the string, the electric force $\mathbf{F}_e$ from the other sphere, and the gravitational force $m\mathbf{g}$.

Because the sphere is in equilibrium, the forces in the horizontal and vertical directions must separately add up to zero:

$$(1) \qquad \sum F_x = T \sin \theta - F_e = 0$$

$$(2) \qquad \sum F_y = T \cos \theta - mg = 0$$

From Equation (2), we see that $T = mg/\cos \theta$; thus, T can be eliminated from Equation (1) if we make this substitution. This gives a value for the magnitude of the electric force F_e:

$$F_e = mg \tan \theta = (3.0 \times 10^{-2} \text{ kg})(9.80 \text{ m/s}^2) \tan(5.0°)$$

$$= 2.6 \times 10^{-2} \text{ N}$$

Considering the geometry of the right triangle in Figure 23.10a, we see that $\sin \theta = a/L$. Therefore,

$$a = L \sin \theta = (0.15 \text{ m}) \sin(5.0°) = 0.013 \text{ m}$$

The separation of the spheres is $2a = 0.026$ m.

From Coulomb's law (Eq. 23.1), the magnitude of the electric force is

$$F_e = k_e \frac{|q|^2}{r^2}$$

where $r = 2a = 0.026$ m and $|q|$ is the magnitude of the charge on each sphere. (Note that the term $|q|^2$ arises here because the charge is the same on both spheres.) This equation can be solved for $|q|^2$ to give

$$|q|^2 = \frac{F_e r^2}{k_e} = \frac{(2.6 \times 10^{-2} \text{ N})(0.026 \text{ m})^2}{8.99 \times 10^9 \text{ N} \cdot \text{m}^2/\text{C}^2} = 1.96 \times 10^{-15} \text{ C}^2$$

$$|q| = \boxed{4.4 \times 10^{-8} \text{ C}}$$

To finalize the problem, note that we found only the magnitude of the charge $|q|$ on the spheres. There is no way we could find the sign of the charge from the information given. In fact, the sign of the charge is not important. The situation will be exactly the same whether both spheres are positively charged or negatively charged.

What If? Suppose your roommate proposes solving this problem without the assumption that the charges are of equal magnitude. She claims that the symmetry of the problem is destroyed if the charges are not equal, so that the strings would make two different angles with the vertical, and the problem would be much more complicated. How would you respond?

Answer You should argue that the symmetry is not destroyed and the angles remain the same. Newton's third law requires that the electric forces on the two charges be the same, regardless of the equality or nonequality of the charges. The solution to the example remains the same through the calculation of $|q|^2$. In this situation, the value of 1.96×10^{-15} C^2 corresponds to the product $q_1 q_2$, where q_1 and q_2 are the values of the charges on the two spheres. The symmetry of the problem would be destroyed if the *masses* of the spheres were not the same. In this case, the strings would make different angles with the vertical and the problem would be more complicated.

23.4 The Electric Field

Two field forces have been introduced into our discussions so far—the gravitational force in Chapter 13 and the electric force here. As pointed out earlier, field forces can act through space, producing an effect even when no physical contact occurs between interacting objects. The gravitational field **g** at a point in space was defined in Section 13.5 to be equal to the gravitational force $\mathbf{F}_g$ acting on a test particle of mass m divided by that mass: $\mathbf{g} \equiv \mathbf{F}_g/m$. The concept of a field was developed by Michael Faraday (1791–1867) in the context of electric forces and is of such practical value that we shall devote much attention to it in the next several chapters. In this approach, an **electric field** is said to exist in the region of space around a charged object—the **source charge.** When another charged object—the **test charge**—enters this electric field, an

©Johnny Autery

This dramatic photograph captures a lightning bolt striking a tree near some rural homes. Lightning is associated with very strong electric fields in the atmosphere.

Figure 23.11 A small positive test charge q_0 placed near an object carrying a much larger positive charge Q experiences an electric field $\mathbf{E}$ directed as shown.

Definition of electric field

electric force acts on it. As an example, consider Figure 23.11, which shows a small positive test charge q_0 placed near a second object carrying a much greater positive charge Q. We define the electric field due to the source charge at the location of the test charge to be the electric force on the test charge *per unit charge*, or to be more specific

the electric field vector $\mathbf{E}$ at a point in space is defined as the electric force $\mathbf{F}_e$ acting on a positive test charge q_0 placed at that point divided by the test charge:

$$\mathbf{E} \equiv \frac{\mathbf{F}_e}{q_0} \qquad (23.7)$$

Note that $\mathbf{E}$ is the field produced by some charge or charge distribution *separate from* the test charge—it is not the field produced by the test charge itself. Also, note that the existence of an electric field is a property of its source—the presence of the test charge is not necessary for the field to exist. The test charge serves as a *detector* of the electric field.

Equation 23.7 can be rearranged as

$$\mathbf{F}_e = q\mathbf{E} \qquad (23.8)$$

where we have used the general symbol q for a charge. This equation gives us the force on a charged particle placed in an electric field. If q is positive, the force is in the same

▲ **PITFALL PREVENTION**

23.1 Particles Only

Equation 23.8 is only valid for a charged *particle*—an object of zero size. For a charged object of finite size in an electric field, the field may vary in magnitude and direction over the size of the object, so the corresponding force equation may be more complicated.

Table 23.2

Typical Electric Field Values	
Source	**E (N/C)**
Fluorescent lighting tube	10
Atmosphere (fair weather)	100
Balloon rubbed on hair	1 000
Atmosphere (under thundercloud)	10 000
Photocopier	100 000
Spark in air	>3 000 000
Near electron in hydrogen atom	5×10^{11}

Figure 23.12 (a) For a small enough test charge q_0, the charge distribution on the sphere is undisturbed. (b) When the test charge q_0' is greater, the charge distribution on the sphere is disturbed as the result of the proximity of q_0'.

direction as the field. If q is negative, the force and the field are in opposite directions. Notice the similarity between Equation 23.8 and the corresponding equation for a particle with mass placed in a gravitational field, $\mathbf{F}_g = m\mathbf{g}$ (Eq. 5.6).

The vector $\mathbf{E}$ has the SI units of newtons per coulomb (N/C). The direction of $\mathbf{E}$, as shown in Figure 23.11, is the direction of the force a positive test charge experiences when placed in the field. We say that **an electric field exists at a point if a test charge at that point experiences an electric force.** Once the magnitude and direction of the electric field are known at some point, the electric force exerted on *any* charged particle placed at that point can be calculated from Equation 23.8. The electric field magnitudes for various field sources are given in Table 23.2.

When using Equation 23.7, we must assume that the test charge q_0 is small enough that it does not disturb the charge distribution responsible for the electric field. If a vanishingly small test charge q_0 is placed near a uniformly charged metallic sphere, as in Figure 23.12a, the charge on the metallic sphere, which produces the electric field, remains uniformly distributed. If the test charge is great enough ($q_0' \gg q_0$), as in Figure 23.12b, the charge on the metallic sphere is redistributed and the ratio of the force to the test charge is different: ($F_e'/q_0' \neq F_e/q_0$). That is, because of this redistribution of charge on the metallic sphere, the electric field it sets up is different from the field it sets up in the presence of the much smaller test charge q_0.

To determine the direction of an electric field, consider a point charge q as a source charge. This charge creates an electric field at all points in space surrounding it. A test charge q_0 is placed at point P, a distance r from the source charge, as in Figure 23.13a. We imagine using the test charge to determine the direction of the electric force and therefore that of the electric field. However, the electric field does not depend on the existence of the test charge—it is established solely by the source charge. According to Coulomb's law, the force exerted by q on the test charge is

$$\mathbf{F}_e = k_e \frac{qq_0}{r^2} \hat{\mathbf{r}}$$

where $\hat{\mathbf{r}}$ is a unit vector directed from q toward q_0. This force in Figure 23.13a is directed away from the source charge q. Because the electric field at P, the position of the test charge, is defined by $\mathbf{E} = \mathbf{F}_e/q_0$, we find that at P, the electric field created by q is

$$\mathbf{E} = k_e \frac{q}{r^2} \hat{\mathbf{r}} \tag{23.9}$$

If the source charge q is positive, Figure 23.13b shows the situation with the test charge removed—the source charge sets up an electric field at point P, directed away from q. If q is negative, as in Figure 23.13c, the force on the test charge is toward the source charge, so the electric field at P is directed toward the source charge, as in Figure 23.13d.

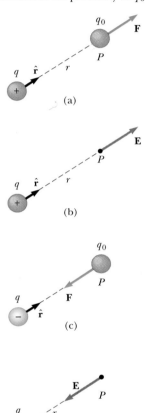

Active Figure 23.13 A test charge q_0 at point P is a distance r from a point charge q. (a) If q is positive, then the force on the test charge is directed away from q. (b) For the positive source charge, the electric field at P points radially outward from q. (c) If q is negative, then the force on the test charge is directed toward q. (d) For the negative source charge, the electric field at P points radially inward toward q.

At the Active Figures link at http://www.pse6.com, you can move point P to any position in two-dimensional space and observe the electric field due to q.

To calculate the electric field at a point P due to a group of point charges, we first calculate the electric field vectors at P individually using Equation 23.9 and then add them vectorially. In other words,

> at any point P, the total electric field due to a group of source charges equals the vector sum of the electric fields of all the charges.

This superposition principle applied to fields follows directly from the superposition property of electric forces, which, in turn, follows from the fact that we know that forces add as vectors from Chapter 5. Thus, the electric field at point P due to a group of source charges can be expressed as the vector sum

Electric field due to a finite number of point charges

$$\mathbf{E} = k_e \sum_i \frac{q_i}{r_i^2} \hat{\mathbf{r}}_i \qquad (23.10)$$

where r_i is the distance from the ith source charge q_i to the point P and $\hat{\mathbf{r}}_i$ is a unit vector directed from q_i toward P.

Quick Quiz 23.6 A test charge of $+3\ \mu C$ is at a point P where an external electric field is directed to the right and has a magnitude of 4×10^6 N/C. If the test charge is replaced with another test charge of $-3\ \mu C$, the external electric field at P (a) is unaffected (b) reverses direction (c) changes in a way that cannot be determined

---

Example 23.5 Electric Field Due to Two Charges

A charge $q_1 = 7.0\ \mu C$ is located at the origin, and a second charge $q_2 = -5.0\ \mu C$ is located on the x axis, 0.30 m from the origin (Fig. 23.14). Find the electric field at the point P, which has coordinates $(0, 0.40)$ m.

Figure 23.14 (Example 23.5) The total electric field $\mathbf{E}$ at P equals the vector sum $\mathbf{E}_1 + \mathbf{E}_2$, where $\mathbf{E}_1$ is the field due to the positive charge q_1 and $\mathbf{E}_2$ is the field due to the negative charge q_2.

Solution First, let us find the magnitude of the electric field at P due to each charge. The fields $\mathbf{E}_1$ due to the 7.0-μC charge and $\mathbf{E}_2$ due to the -5.0-μC charge are shown in Figure 23.14. Their magnitudes are

$$E_1 = k_e \frac{|q_1|}{r_1^2} = (8.99 \times 10^9 \text{ N} \cdot \text{m}^2/\text{C}^2) \frac{(7.0 \times 10^{-6} \text{ C})}{(0.40 \text{ m})^2}$$
$$= 3.9 \times 10^5 \text{ N/C}$$

$$E_2 = k_e \frac{|q_2|}{r_2^2} = (8.99 \times 10^9 \text{ N} \cdot \text{m}^2/\text{C}^2) \frac{(5.0 \times 10^{-6} \text{ C})}{(0.50 \text{ m})^2}$$
$$= 1.8 \times 10^5 \text{ N/C}$$

The vector $\mathbf{E}_1$ has only a y component. The vector $\mathbf{E}_2$ has an x component given by $E_2 \cos\theta = \frac{3}{5}E_2$ and a negative y component given by $-E_2 \sin\theta = -\frac{4}{5}E_2$. Hence, we can express the vectors as

$$\mathbf{E}_1 = 3.9 \times 10^5 \hat{\mathbf{j}} \text{ N/C}$$

$$\mathbf{E}_2 = (1.1 \times 10^5 \hat{\mathbf{i}} - 1.4 \times 10^5 \hat{\mathbf{j}}) \text{ N/C}$$

The resultant field $\mathbf{E}$ at P is the superposition of $\mathbf{E}_1$ and $\mathbf{E}_2$:

$$\mathbf{E} = \mathbf{E}_1 + \mathbf{E}_2 = (1.1 \times 10^5 \hat{\mathbf{i}} + 2.5 \times 10^5 \hat{\mathbf{j}}) \text{ N/C}$$

From this result, we find that $\mathbf{E}$ makes an angle ϕ of 66° with the positive x axis and has a magnitude of 2.7×10^5 N/C.

Example 23.6   Electric Field of a Dipole

An **electric dipole** is defined as a positive charge q and a negative charge $-q$ separated by a distance $2a$. For the dipole shown in Figure 23.15, find the electric field $\mathbf{E}$ at P due to the dipole, where P is a distance $y \gg a$ from the origin.

Solution At P, the fields $\mathbf{E}_1$ and $\mathbf{E}_2$ due to the two charges are equal in magnitude because P is equidistant from the charges. The total field is $\mathbf{E} = \mathbf{E}_1 + \mathbf{E}_2$, where

$$E_1 = E_2 = k_e \frac{q}{r^2} = k_e \frac{q}{y^2 + a^2}$$

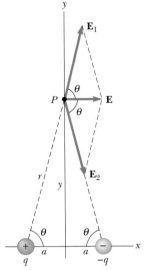

Figure 23.15 (Example 23.6) The total electric field $\mathbf{E}$ at P due to two charges of equal magnitude and opposite sign (an electric dipole) equals the vector sum $\mathbf{E}_1 + \mathbf{E}_2$. The field $\mathbf{E}_1$ is due to the positive charge q, and $\mathbf{E}_2$ is the field due to the negative charge $-q$.

The y components of $\mathbf{E}_1$ and $\mathbf{E}_2$ cancel each other, and the x components are both in the positive x direction and have the same magnitude. Therefore, $\mathbf{E}$ is parallel to the x axis and has a magnitude equal to $2E_1 \cos \theta$. From Figure 23.15 we see that $\cos \theta = a/r = a/(y^2 + a^2)^{1/2}$. Therefore,

$$E = 2E_1 \cos \theta = 2k_e \frac{q}{(y^2 + a^2)} \frac{a}{(y^2 + a^2)^{1/2}}$$

$$= k_e \frac{2qa}{(y^2 + a^2)^{3/2}}$$

Because $y \gg a$, we can neglect a^2 compared to y^2 and write

$$E \approx k_e \frac{2qa}{y^3}$$

Thus, we see that, at distances far from a dipole but along the perpendicular bisector of the line joining the two charges, the magnitude of the electric field created by the dipole varies as $1/r^3$, whereas the more slowly varying field of a point charge varies as $1/r^2$ (see Eq. 23.9). This is because at distant points, the fields of the two charges of equal magnitude and opposite sign almost cancel each other. The $1/r^3$ variation in E for the dipole also is obtained for a distant point along the x axis (see Problem 22) and for any general distant point.

The electric dipole is a good model of many molecules, such as hydrochloric acid (HCl). Neutral atoms and molecules behave as dipoles when placed in an external electric field. Furthermore, many molecules, such as HCl, are permanent dipoles. The effect of such dipoles on the behavior of materials subjected to electric fields is discussed in Chapter 26.

23.5   Electric Field of a Continuous Charge Distribution

Very often the distances between charges in a group of charges are much smaller than the distance from the group to some point of interest (for example, a point where the electric field is to be calculated). In such situations, the system of charges can be modeled as continuous. That is, the system of closely spaced charges is equivalent to a total charge that is continuously distributed along some line, over some surface, or throughout some volume.

To evaluate the electric field created by a continuous charge distribution, we use the following procedure: first, we divide the charge distribution into small elements, each of which contains a small charge Δq, as shown in Figure 23.16. Next, we use Equation 23.9 to calculate the electric field due to one of these elements at a point P. Finally, we evaluate the total electric field at P due to the charge distribution by summing the contributions of all the charge elements (that is, by applying the superposition principle).

The electric field at P due to one charge element carrying charge Δq is

$$\Delta \mathbf{E} = k_e \frac{\Delta q}{r^2} \hat{\mathbf{r}}$$

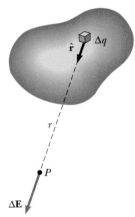

Figure 23.16 The electric field at P due to a continuous charge distribution is the vector sum of the fields $\Delta \mathbf{E}$ due to all the elements Δq of the charge distribution.

where r is the distance from the charge element to point P and $\hat{\mathbf{r}}$ is a unit vector directed from the element toward P. The total electric field at P due to all elements in the charge distribution is approximately

$$\mathbf{E} \approx k_e \sum_i \frac{\Delta q_i}{r_i^2} \, \hat{\mathbf{r}}_i$$

where the index i refers to the ith element in the distribution. Because the charge distribution is modeled as continuous, the total field at P in the limit $\Delta q_i \rightarrow 0$ is

Electric field due to a continuous charge distribution

$$\mathbf{E} = k_e \lim_{\Delta q_i \rightarrow 0} \sum_i \frac{\Delta q_i}{r_i^2} \, \hat{\mathbf{r}}_i = k_e \int \frac{dq}{r^2} \, \hat{\mathbf{r}} \qquad (23.11)$$

where the integration is over the entire charge distribution. This is a vector operation and must be treated appropriately.

We illustrate this type of calculation with several examples, in which we assume the charge is uniformly distributed on a line, on a surface, or throughout a volume. When performing such calculations, it is convenient to use the concept of a *charge density* along with the following notations:

- If a charge Q is uniformly distributed throughout a volume V, the **volume charge density** ρ is defined by

Volume charge density

$$\rho \equiv \frac{Q}{V}$$

where ρ has units of coulombs per cubic meter (C/m³).

- If a charge Q is uniformly distributed on a surface of area A, the **surface charge density** σ (lowercase Greek sigma) is defined by

Surface charge density

$$\sigma \equiv \frac{Q}{A}$$

where σ has units of coulombs per square meter (C/m²).

- If a charge Q is uniformly distributed along a line of length ℓ, the **linear charge density** λ is defined by

Linear charge density

$$\lambda \equiv \frac{Q}{\ell}$$

where λ has units of coulombs per meter (C/m).

- If the charge is nonuniformly distributed over a volume, surface, or line, the amounts of charge dq in a small volume, surface, or length element are

$$dq = \rho \, dV \qquad dq = \sigma \, dA \qquad dq = \lambda \, d\ell$$

PROBLEM-SOLVING HINTS

Finding the Electric Field

- **Units:** in calculations using the Coulomb constant k_e ($= 1/4\pi\epsilon_0$), charges must be expressed in coulombs and distances in meters.
- **Calculating the electric field of point charges:** to find the total electric field at a given point, first calculate the electric field at the point due to each individual charge. The resultant field at the point is the vector sum of the fields due to the individual charges.
- **Continuous charge distributions:** when you are confronted with problems that involve a continuous distribution of charge, the vector sums for evaluating the

total electric field at some point must be replaced by vector integrals. Divide the charge distribution into infinitesimal pieces, and calculate the vector sum by integrating over the entire charge distribution. Examples 23.7 through 23.9 demonstrate this technique.

- **Symmetry:** with both distributions of point charges and continuous charge distributions, take advantage of any symmetry in the system to simplify your calculations.

Example 23.7   The Electric Field Due to a Charged Rod

A rod of length ℓ has a uniform positive charge per unit length λ and a total charge Q. Calculate the electric field at a point P that is located along the long axis of the rod and a distance a from one end (Fig. 23.17).

Solution Let us assume that the rod is lying along the x axis, that dx is the length of one small segment, and that dq is the charge on that segment. Because the rod has a charge per unit length λ, the charge dq on the small segment is $dq = \lambda\, dx$.

The field $d\mathbf{E}$ at P due to this segment is in the negative x direction (because the source of the field carries a positive charge), and its magnitude is

$$dE = k_e \frac{dq}{x^2} = k_e \frac{\lambda\, dx}{x^2}$$

Because every other element also produces a field in the negative x direction, the problem of summing their contributions

Figure 23.17 (Example 23.7) The electric field at P due to a uniformly charged rod lying along the x axis. The magnitude of the field at P due to the segment of charge dq is $k_e\, dq/x^2$. The total field at P is the vector sum over all segments of the rod.

is particularly simple in this case. The total field at P due to all segments of the rod, which are at different distances from P, is given by Equation 23.11, which in this case becomes[3]

$$E = \int_a^{\ell + a} k_e \lambda \frac{dx}{x^2}$$

where the limits on the integral extend from one end of the rod ($x = a$) to the other ($x = \ell + a$). The constants k_e and λ can be removed from the integral to yield

$$E = k_e \lambda \int_a^{\ell + a} \frac{dx}{x^2} = k_e \lambda \left[-\frac{1}{x} \right]_a^{\ell + a}$$

$$= k_e \lambda \left(\frac{1}{a} - \frac{1}{\ell + a} \right) = \frac{k_e Q}{a(\ell + a)}$$

where we have used the fact that the total charge $Q = \lambda\ell$.

What If? Suppose we move to a point P very far away from the rod. What is the nature of the electric field at such a point?

Answer If P is far from the rod ($a \gg \ell$), then ℓ in the denominator of the final expression for E can be neglected, and $E \approx k_e Q/a^2$. This is just the form you would expect for a point charge. Therefore, at large values of a/ℓ, the charge distribution appears to be a point charge of magnitude Q—we are so far away from the rod that we cannot distinguish that it has a size. The use of the limiting technique ($a/\ell \rightarrow \infty$) often is a good method for checking a mathematical expression.

Example 23.8   The Electric Field of a Uniform Ring of Charge

A ring of radius a carries a uniformly distributed positive total charge Q. Calculate the electric field due to the ring at a point P lying a distance x from its center along the central axis perpendicular to the plane of the ring (Fig. 23.18a).

Solution The magnitude of the electric field at P due to the segment of charge dq is

$$dE = k_e \frac{dq}{r^2}$$

This field has an x component $dE_x = dE \cos\theta$ along the x axis and a component $dE_\perp$ perpendicular to the x axis. As we see in Figure 23.18b, however, the resultant field at P must lie along the x axis because the perpendicular com-

---

[3]   It is important that you understand how to carry out integrations such as this. First, express the charge element dq in terms of the other variables in the integral. (In this example, there is one variable, x, and so we made the change $dq = \lambda\, dx$.) The integral must be over scalar quantities; therefore, you must express the electric field in terms of components, if necessary. (In this example the field has only an x component, so we do not bother with this detail.) Then, reduce your expression to an integral over a single variable (or to multiple integrals, each over a single variable). In examples that have spherical or cylindrical symmetry, the single variable will be a radial coordinate.

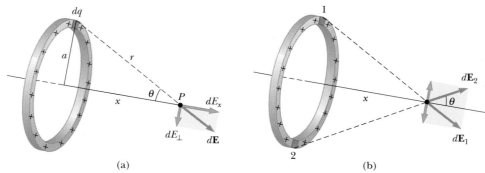

Figure 23.18 (Example 23.8) A uniformly charged ring of radius a. (a) The field at P on the x axis due to an element of charge dq. (b) The total electric field at P is along the x axis. The perpendicular component of the field at P due to segment 1 is canceled by the perpendicular component due to segment 2.

ponents of all the various charge segments sum to zero. That is, the perpendicular component of the field created by any charge element is canceled by the perpendicular component created by an element on the opposite side of the ring. Because $r = (x^2 + a^2)^{1/2}$ and $\cos \theta = x/r$, we find that

$$dE_x = dE \cos \theta = \left(k_e \frac{dq}{r^2} \right) \frac{x}{r} = \frac{k_e x}{(x^2 + a^2)^{3/2}} \, dq$$

All segments of the ring make the same contribution to the field at P because they are all equidistant from this point. Thus, we can integrate to obtain the total field at P:

$$E_x = \int \frac{k_e x}{(x^2 + a^2)^{3/2}} \, dq = \frac{k_e x}{(x^2 + a^2)^{3/2}} \int dq$$

$$= \frac{k_e x}{(x^2 + a^2)^{3/2}} \, Q$$

This result shows that the field is zero at $x = 0$. Does this finding surprise you?

What If? Suppose a negative charge is placed at the center of the ring in Figure 23.18 and displaced slightly by a distance $x \ll a$ along the x axis. When released, what type of motion does it exhibit?

Answer In the expression for the field due to a ring of charge, we let $x \ll a$, which results in

$$E_x = \frac{k_e Q}{a^3} \, x$$

Thus, from Equation 23.8, the force on a charge $-q$ placed near the center of the ring is

$$F_x = -\frac{k_e q Q}{a^3} \, x$$

Because this force has the form of Hooke's law (Eq. 15.1), the motion will be *simple harmonic*!

Example 23.9 The Electric Field of a Uniformly Charged Disk

A disk of radius R has a uniform surface charge density σ. Calculate the electric field at a point P that lies along the central perpendicular axis of the disk and a distance x from the center of the disk (Fig. 23.19).

Solution If we consider the disk as a set of concentric rings, we can use our result from Example 23.8—which gives the field created by a ring of radius a—and sum the contributions of all rings making up the disk. By symmetry, the field at an axial point must be along the central axis.

The ring of radius r and width dr shown in Figure 23.19 has a surface area equal to $2\pi r \, dr$. The charge dq on this ring is equal to the area of the ring multiplied by the surface charge density: $dq = 2\pi\sigma r \, dr$. Using this result in the equation given for E_x in Example 23.8 (with a replaced by r), we have for the field due to the ring

$$dE_x = \frac{k_e x}{(x^2 + r^2)^{3/2}} (2\pi\sigma r \, dr)$$

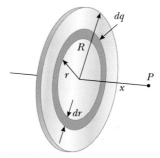

Figure 23.19 (Example 23.9) A uniformly charged disk of radius R. The electric field at an axial point P is directed along the central axis, perpendicular to the plane of the disk.

To obtain the total field at P, we integrate this expression over the limits $r = 0$ to $r = R$, noting that x is a constant.

This gives

$$E_x = k_e x \pi \sigma \int_0^R \frac{2r\,dr}{(x^2 + r^2)^{3/2}}$$

$$= k_e x \pi \sigma \int_0^R (x^2 + r^2)^{-3/2}\,d(r^2)$$

$$= k_e x \pi \sigma \left[\frac{(x^2 + r^2)^{-1/2}}{-1/2} \right]_0^R$$

$$= 2\pi k_e \sigma \left(1 - \frac{x}{(x^2 + R^2)^{1/2}} \right)$$

This result is valid for all values of $x > 0$. We can calculate the field close to the disk along the axis by assuming that $R \gg x$; thus, the expression in parentheses reduces to unity to give us the near-field approximation:

$$E_x = 2\pi k_e \sigma = \frac{\sigma}{2\epsilon_0}$$

where ϵ_0 is the permittivity of free space. In the next chapter we shall obtain the same result for the field created by a uniformly charged infinite sheet.

23.6 Electric Field Lines

We have defined the electric field mathematically through Equation 23.7. We now explore a means of representing the electric field pictorially. A convenient way of visualizing electric field patterns is to draw curved lines that are parallel to the electric field vector at any point in space. These lines, called *electric field lines* and first introduced by Faraday, are related to the electric field in a region of space in the following manner:

- The electric field vector **E** is tangent to the electric field line at each point. The line has a direction, indicated by an arrowhead, that is the same as that of the electric field vector.

- The number of lines per unit area through a surface perpendicular to the lines is proportional to the magnitude of the electric field in that region. Thus, the field lines are close together where the electric field is strong and far apart where the field is weak.

These properties are illustrated in Figure 23.20. The density of lines through surface A is greater than the density of lines through surface B. Therefore, the magnitude of the electric field is larger on surface A than on surface B. Furthermore, the fact that the lines at different locations point in different directions indicates that the field is nonuniform.

Is this relationship between strength of the electric field and the density of field lines consistent with Equation 23.9, the expression we obtained for E using Coulomb's law? To answer this question, consider an imaginary spherical surface of radius r concentric with a point charge. From symmetry, we see that the magnitude of the electric field is the same everywhere on the surface of the sphere. The number of lines N that emerge from the charge is equal to the number that penetrate the spherical surface. Hence, the number of lines per unit area on the sphere is $N/4\pi r^2$ (where the surface area of the sphere is $4\pi r^2$). Because E is proportional to the number of lines per unit area, we see that E varies as $1/r^2$; this finding is consistent with Equation 23.9.

Representative electric field lines for the field due to a single positive point charge are shown in Figure 23.21a. This two-dimensional drawing shows only the field lines that lie in the plane containing the point charge. The lines are actually directed radially outward from the charge in all directions; thus, instead of the flat "wheel" of lines shown, you should picture an entire spherical distribution of lines. Because a positive test charge placed in this field would be repelled by the positive source charge, the lines are directed radially away from the source charge. The electric field lines representing the field due to a single negative point charge are directed toward the charge (Fig. 23.21b). In either case, the lines are along the radial direction and extend all the way to infinity. Note that the lines become closer together as they approach the charge; this indicates that the strength of the field increases as we move toward the source charge.

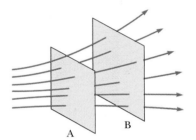

Figure 23.20 Electric field lines penetrating two surfaces. The magnitude of the field is greater on surface A than on surface B.

▲ **PITFALL PREVENTION**

23.2 Electric Field Lines are not Paths of Particles!

Electric field lines represent the field at various locations. Except in very special cases, they *do not* represent the path of a charged particle moving in an electric field.

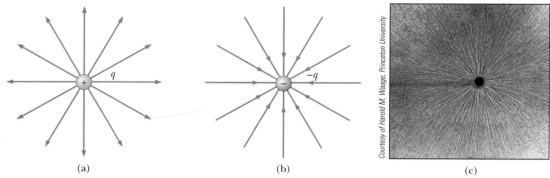

Figure 23.21 The electric field lines for a point charge. (a) For a positive point charge, the lines are directed radially outward. (b) For a negative point charge, the lines are directed radially inward. Note that the figures show only those field lines that lie in the plane of the page. (c) The dark areas are small pieces of thread suspended in oil, which align with the electric field produced by a small charged conductor at the center.

The rules for drawing electric field lines are as follows:

- The lines must begin on a positive charge and terminate on a negative charge. In the case of an excess of one type of charge, some lines will begin or end infinitely far away.
- The number of lines drawn leaving a positive charge or approaching a negative charge is proportional to the magnitude of the charge.
- No two field lines can cross.

▲ **PITFALL PREVENTION**

23.3 Electric Field Lines are not Real

Electric field lines are not material objects. They are used only as a pictorial representation to provide a qualitative description of the electric field. Only a finite number of lines from each charge can be drawn, which makes it appear as if the field were quantized and exists only in certain parts of space. The field, in fact, is continuous—existing at every point. You should avoid obtaining the wrong impression from a two-dimensional drawing of field lines used to describe a three-dimensional situation.

We choose the number of field lines starting from any positively charged object to be Cq and the number of lines ending on any negatively charged object to be $C|q|$, where C is an arbitrary proportionality constant. Once C is chosen, the number of lines is fixed. For example, if object 1 has charge Q_1 and object 2 has charge Q_2, then the ratio of number of lines is $N_2/N_1 = Q_2/Q_1$. The electric field lines for two point charges of equal magnitude but opposite signs (an electric dipole) are shown in Figure 23.22. Because the charges are of equal magnitude, the number of lines that begin at the positive charge must equal the number that terminate at the negative charge. At points very near the charges, the lines are nearly radial. The high density of lines between the charges indicates a region of strong electric field.

Figure 23.22 (a) The electric field lines for two point charges of equal magnitude and opposite sign (an electric dipole). The number of lines leaving the positive charge equals the number terminating at the negative charge. (b) The dark lines are small pieces of thread suspended in oil, which align with the electric field of a dipole.

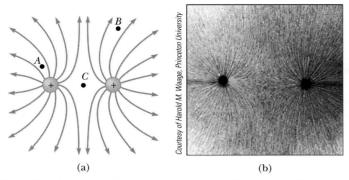

Courtesy of Harold M. Waage, Princeton University

(a)   (b)

Figure 23.23 (a) The electric field lines for two positive point charges. (The locations *A*, *B*, and *C* are discussed in Quick Quiz 23.7.) (b) Pieces of thread suspended in oil, which align with the electric field created by two equal-magnitude positive charges.

Figure 23.23 shows the electric field lines in the vicinity of two equal positive point charges. Again, the lines are nearly radial at points close to either charge, and the same number of lines emerge from each charge because the charges are equal in magnitude. At great distances from the charges, the field is approximately equal to that of a single point charge of magnitude $2q$.

Finally, in Figure 23.24 we sketch the electric field lines associated with a positive charge $+2q$ and a negative charge $-q$. In this case, the number of lines leaving $+2q$ is twice the number terminating at $-q$. Hence, only half of the lines that leave the positive charge reach the negative charge. The remaining half terminate on a negative charge we assume to be at infinity. At distances that are much greater than the charge separation, the electric field lines are equivalent to those of a single charge $+q$.

> **Quick Quiz 23.7** Rank the magnitudes of the electric field at points *A*, *B*, and *C* shown in Figure 23.23a (greatest magnitude first).

> **Quick Quiz 23.8** Which of the following statements about electric field lines associated with electric charges is false? (a) Electric field lines can be either straight or curved. (b) Electric field lines can form closed loops. (c) Electric field lines begin on positive charges and end on negative charges. (d) Electric field lines can never intersect with one another.

23.7 Motion of Charged Particles in a Uniform Electric Field

When a particle of charge q and mass m is placed in an electric field **E**, the electric force exerted on the charge is $q\mathbf{E}$ according to Equation 23.8. If this is the only force exerted on the particle, it must be the net force and causes the particle to accelerate according to Newton's second law. Thus,

$$\mathbf{F}_e = q\mathbf{E} = m\mathbf{a}$$

The acceleration of the particle is therefore

$$\mathbf{a} = \frac{q\mathbf{E}}{m} \tag{23.12}$$

If **E** is uniform (that is, constant in magnitude and direction), then the acceleration is constant. If the particle has a positive charge, its acceleration is in the direction of the

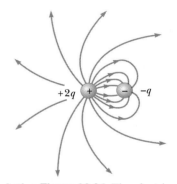

Active Figure 23.24 The electric field lines for a point charge $+2q$ and a second point charge $-q$. Note that two lines leave $+2q$ for every one that terminates on $-q$.

At the Active Figures link at http://www.pse6.com, *you can choose the values and signs for the two charges and observe the electric field lines for the configuration that you have chosen.*

electric field. If the particle has a negative charge, its acceleration is in the direction opposite the electric field.

Example 23.10 An Accelerating Positive Charge

A positive point charge q of mass m is released from rest in a uniform electric field $\mathbf{E}$ directed along the x axis, as shown in Figure 23.25. Describe its motion.

Solution The acceleration is constant and is given by $q\mathbf{E}/m$. The motion is simple linear motion along the x axis. Therefore, we can apply the equations of kinematics in one dimension (see Chapter 2):

$$x_f = x_i + v_i t + \tfrac{1}{2}at^2$$

$$v_f = v_i + at$$

$$v_f{}^2 = v_i{}^2 + 2a(x_f - x_i)$$

Choosing the initial position of the charge as $x_i = 0$ and assigning $v_i = 0$ because the particle starts from rest, the position of the particle as a function of time is

$$x_f = \tfrac{1}{2}at^2 = \frac{qE}{2m}t^2$$

The speed of the particle is given by

$$v_f = at = \frac{qE}{m}t$$

The third kinematic equation gives us

$$v_f{}^2 = 2ax_f = \left(\frac{2qE}{m}\right)x_f$$

from which we can find the kinetic energy of the charge after it has moved a distance $\Delta x = x_f - x_i$:

$$K = \tfrac{1}{2}mv_f{}^2 = \tfrac{1}{2}m\left(\frac{2qE}{m}\right)\Delta x = qE\Delta x$$

We can also obtain this result from the work–kinetic energy theorem because the work done by the electric force is $F_e\Delta x = qE\Delta x$ and $W = \Delta K$.

Figure 23.25 (Example 23.10) A positive point charge q in a uniform electric field $\mathbf{E}$ undergoes constant acceleration in the direction of the field.

The electric field in the region between two oppositely charged flat metallic plates is approximately uniform (Fig. 23.26). Suppose an electron of charge $-e$ is projected horizontally into this field from the origin with an initial velocity $v_i\hat{\mathbf{i}}$ at time $t = 0$. Because the electric field $\mathbf{E}$ in Figure 23.26 is in the positive y direction, the acceleration of the electron is in the negative y direction. That is,

$$\mathbf{a} = -\frac{eE}{m_e}\hat{\mathbf{j}} \tag{23.13}$$

Because the acceleration is constant, we can apply the equations of kinematics in two dimensions (see Chapter 4) with $v_{xi} = v_i$ and $v_{yi} = 0$. After the electron has been in the

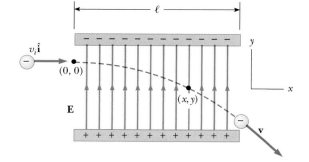

At the Active Figures link at http://www.pse6.com, you can choose the strength of the electric field and the mass and charge of the projected particle.

Active Figure 23.26 An electron is projected horizontally into a uniform electric field produced by two charged plates. The electron undergoes a downward acceleration (opposite $\mathbf{E}$), and its motion is parabolic while it is between the plates.

electric field for a time interval, the components of its velocity at time t are

$$v_x = v_i = \text{constant} \qquad (23.14)$$

$$v_y = a_y t = -\frac{eE}{m_e} t \qquad (23.15)$$

Its position coordinates at time t are

$$x_f = v_i t \qquad (23.16)$$

$$y_f = \tfrac{1}{2} a_y t^2 = -\frac{1}{2} \frac{eE}{m_e} t^2 \qquad (23.17)$$

Substituting the value $t = x_f/v_i$ from Equation 23.16 into Equation 23.17, we see that y_f is proportional to x_f^2. Hence, the trajectory is a parabola. This should not be a surprise—consider the analogous situation of throwing a ball horizontally in a uniform gravitational field (Chapter 4). After the electron leaves the field, the electric force vanishes and the electron continues to move in a straight line in the direction of $\mathbf{v}$ in Figure 23.26 with a speed $v > v_i$.

Note that we have neglected the gravitational force acting on the electron. This is a good approximation when we are dealing with atomic particles. For an electric field of 10^4 N/C, the ratio of the magnitude of the electric force eE to the magnitude of the gravitational force mg is on the order of 10^{14} for an electron and on the order of 10^{11} for a proton.

 PITFALL PREVENTION

23.4   Just Another Force

Electric forces and fields may seem abstract to you. However, once $\mathbf{F}_e$ is evaluated, it causes a particle to move according to our well-established understanding of forces and motion from Chapters 5 and 6. Keeping this link with the past in mind will help you solve problems in this chapter.

Example 23.11   An Accelerated Electron   `Interactive`

An electron enters the region of a uniform electric field as shown in Figure 23.26, with $v_i = 3.00 \times 10^6$ m/s and $E = 200$ N/C. The horizontal length of the plates is $\ell = 0.100$ m.

(A) Find the acceleration of the electron while it is in the electric field.

Solution The charge on the electron has an absolute value of 1.60×10^{-19} C, and $m_e = 9.11 \times 10^{-31}$ kg. Therefore, Equation 23.13 gives

$$\mathbf{a} = -\frac{eE}{m_e} \hat{\mathbf{j}} = -\frac{(1.60 \times 10^{-19}\ \text{C})(200\ \text{N/C})}{9.11 \times 10^{-31}\ \text{kg}} \hat{\mathbf{j}}$$

$$= -3.51 \times 10^{13} \hat{\mathbf{j}}\ \text{m/s}^2$$

(B) If the electron enters the field at time $t = 0$, find the time at which it leaves the field.

Solution The horizontal distance across the field is $\ell = 0.100$ m. Using Equation 23.16 with $x_f = \ell$, we find that the time at which the electron exits the electric field is

$$t = \frac{\ell}{v_i} = \frac{0.100\ \text{m}}{3.00 \times 10^6\ \text{m/s}} = \boxed{3.33 \times 10^{-8}\ \text{s}}$$

(C) If the vertical position of the electron as it enters the field is $y_i = 0$, what is its vertical position when it leaves the field?

Solution Using Equation 23.17 and the results from parts (A) and (B), we find that

$$y_f = \tfrac{1}{2} a_y t^2 = -\tfrac{1}{2}(3.51 \times 10^{13}\ \text{m/s}^2)(3.33 \times 10^{-8}\ \text{s})^2$$

$$= -0.019\,5\ \text{m} = \boxed{-1.95\ \text{cm}}$$

If the electron enters just below the negative plate in Figure 23.26 and the separation between the plates is less than the value we have just calculated, the electron will strike the positive plate.

 At the Interactive Worked Example link at **http://www.pse6.com,** *you can predict the required initial velocity for the exiting electron to just miss the right edge of the lower plate, for random values of the electric field.*

The Cathode Ray Tube

The example we just worked describes a portion of a cathode ray tube (CRT). This tube, illustrated in Figure 23.27, is commonly used to obtain a visual display of electronic information in oscilloscopes, radar systems, television receivers, and computer monitors. The CRT is a vacuum tube in which a beam of electrons is accelerated and deflected under the influence of electric or magnetic fields. The electron beam is

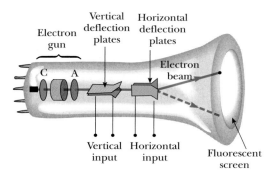

Figure 23.27 Schematic diagram of a cathode ray tube. Electrons leaving the cathode C are accelerated to the anode A. In addition to accelerating electrons, the electron gun is also used to focus the beam of electrons, and the plates deflect the beam.

produced by an assembly called an *electron gun* located in the neck of the tube. These electrons, if left undisturbed, travel in a straight-line path until they strike the front of the CRT, the "screen," which is coated with a material that emits visible light when bombarded with electrons.

In an oscilloscope, the electrons are deflected in various directions by two sets of plates placed at right angles to each other in the neck of the tube. (A television CRT steers the beam with a magnetic field, as discussed in Chapter 29.) An external electric circuit is used to control the amount of charge present on the plates. The placing of positive charge on one horizontal plate and negative charge on the other creates an electric field between the plates and allows the beam to be steered from side to side. The vertical deflection plates act in the same way, except that changing the charge on them deflects the beam vertically.

SUMMARY

Take a practice test for this chapter by clicking on the Practice Test link at http://www.pse6.com.

Electric charges have the following important properties:

- Charges of opposite sign attract one another and charges of the same sign repel one another.

- Total charge in an isolated system is conserved.

- Charge is quantized.

Conductors are materials in which electrons move freely. **Insulators** are materials in which electrons do not move freely.

Coulomb's law states that the electric force exerted by a charge q_1 on a second charge q_2 is

$$\mathbf{F}_{12} = k_e \frac{q_1 q_2}{r^2} \, \hat{\mathbf{r}} \tag{23.6}$$

where r is the distance between the two charges and $\hat{\mathbf{r}}$ is a unit vector directed from q_1 toward q_2. The constant k_e, which is called the Coulomb constant, has the value $k_e = 8.99 \times 10^9 \, \text{N} \cdot \text{m}^2/\text{C}^2$.

The smallest unit of free charge e known to exist in nature is the charge on an electron $(-e)$ or proton $(+e)$, where $e = 1.602\ 19 \times 10^{-19}$ C.

The electric field $\mathbf{E}$ at some point in space is defined as the electric force $\mathbf{F}_e$ that acts on a small positive test charge placed at that point divided by the magnitude q_0 of the test charge:

$$\mathbf{E} \equiv \frac{\mathbf{F}_e}{q_0} \tag{23.7}$$

Thus, the electric force on a charge q placed in an electric field $\mathbf{E}$ is given by

$$\mathbf{F}_e = q\mathbf{E} \tag{23.8}$$

At a distance r from a point charge q, the electric field due to the charge is given by

$$\mathbf{E} = k_e \frac{q}{r^2} \hat{\mathbf{r}} \qquad (23.9)$$

where $\hat{\mathbf{r}}$ is a unit vector directed from the charge toward the point in question. The electric field is directed radially outward from a positive charge and radially inward toward a negative charge.

The electric field due to a group of point charges can be obtained by using the superposition principle. That is, the total electric field at some point equals the vector sum of the electric fields of all the charges:

$$\mathbf{E} = k_e \sum_i \frac{q_i}{r_i^2} \hat{\mathbf{r}}_i \qquad (23.10)$$

The electric field at some point due to a continuous charge distribution is

$$\mathbf{E} = k_e \int \frac{dq}{r^2} \hat{\mathbf{r}} \qquad (23.11)$$

where dq is the charge on one element of the charge distribution and r is the distance from the element to the point in question.

Electric field lines describe an electric field in any region of space. The number of lines per unit area through a surface perpendicular to the lines is proportional to the magnitude of $\mathbf{E}$ in that region.

A charged particle of mass m and charge q moving in an electric field $\mathbf{E}$ has an acceleration

$$\mathbf{a} = \frac{q\mathbf{E}}{m} \qquad (23.12)$$

QUESTIONS

1. Explain what is meant by the term "a neutral atom." Explain what "a negatively charged atom" means.

2. A charged comb often attracts small bits of dry paper that then fly away when they touch the comb. Explain.

3. Sparks are often seen or heard on a dry day when fabrics are removed from a clothes dryer in dim light. Explain.

4. Hospital personnel must wear special conducting shoes while working around oxygen in an operating room. Why? Contrast with what might happen if people wore rubber-soled shoes.

5. Explain from an atomic viewpoint why charge is usually transferred by electrons.

6. A light, uncharged metallic sphere suspended from a thread is attracted to a charged rubber rod. After it touches the rod, the sphere is repelled by the rod. Explain.

7. A foreign student who grew up in a tropical country but is studying in the United States may have had no experience with static electricity sparks or shocks until he or she first experiences an American winter. Explain.

8. Explain the similarities and differences between Newton's law of universal gravitation and Coulomb's law.

9. A balloon is negatively charged by rubbing and then clings to a wall. Does this mean that the wall is positively charged? Why does the balloon eventually fall?

10. A light strip of aluminum foil is draped over a horizontal wooden pencil. When a rod carrying a positive charge is brought close to the foil, the two parts of the foil stand apart. Why? What kind of charge is on the foil?

11. When defining the electric field, why is it necessary to specify that the magnitude of the test charge be very small?

12. How could you experimentally distinguish an electric field from a gravitational field?

13. A large metallic sphere insulated from ground is charged with an electrostatic generator while a student standing on an insulating stool holds the sphere. Why is it safe to do this? Why would it not be safe for another person to touch the sphere after it had been charged?

14. Is it possible for an electric field to exist in empty space? Explain. Consider point A in Figure 23.23(a). Does charge exist at this point? Does a force exist at this point? Does a field exist at this point?

15. When is it valid to approximate a charge distribution by a point charge?

16. Explain why electric field lines never cross. *Suggestion*: Begin by explaining why the electric field at a particular point must have only one direction.

17. Figures 23.14 and 23.15 show three electric field vectors at the same point. With a little extrapolation, Figure

23.21 would show many electric field lines at the same point. Is it really true that "no two field lines can cross"? Are the diagrams drawn correctly? Explain your answers.

18. A free electron and a free proton are released in identical electric fields. Compare the electric forces on the two particles. Compare their accelerations.

19. Explain what happens to the magnitude of the electric field created by a point charge as r approaches zero.

20. An object with negative charge is placed in a region of space where the electric field is directed vertically upward. What is the direction of the electric force exerted on this charge?

21. A charge $4q$ is at a distance r from a charge $-q$. Compare the number of electric field lines leaving the charge $4q$ with the number entering the charge $-q$. Where do the extra lines beginning on $4q$ end?

22. Consider two equal point charges separated by some distance d. At what point (other than ∞) would a third test charge experience no net force?

23. Explain the differences between linear, surface, and volume charge densities, and give examples of when each would be used.

24. If the electron in Figure 23.26 is projected into the electric field with an arbitrary velocity $\mathbf{v}_i$ (at an arbitrary angle to $\mathbf{E}$), will its trajectory still be parabolic? Explain.

25. Would life be different if the electron were positively charged and the proton were negatively charged? Does the choice of signs have any bearing on physical and chemical interactions? Explain.

26. Why should a ground wire be connected to the metal support rod for a television antenna?

27. Suppose someone proposes the idea that people are bound to the Earth by electric forces rather than by gravity. How could you prove this idea is wrong?

28. Consider two electric dipoles in empty space. Each dipole has zero net charge. Does an electric force exist between the dipoles—that is, can two objects with zero net charge exert electric forces on each other? If so, is the force one of attraction or of repulsion?

PROBLEMS

1, 2, 3 = straightforward, intermediate, challenging ☐ = full solution available in the *Student Solutions Manual and Study Guide*

= coached solution with hints available at http://www.pse6.com 🖥 = computer useful in solving problem

= paired numerical and symbolic problems

Section 23.1 Properties of Electric Charges

1. (a) Find to three significant digits the charge and the mass of an ionized hydrogen atom, represented as H^+. *Suggestion:* Begin by looking up the mass of a neutral atom on the periodic table of the elements. (b) Find the charge and the mass of Na^+, a singly ionized sodium atom. (c) Find the charge and the average mass of a chloride ion Cl^- that joins with the Na^+ to make one molecule of table salt. (d) Find the charge and the mass of $Ca^{++} = Ca^{2+}$, a doubly ionized calcium atom. (e) You can model the center of an ammonia molecule as an N^{3-} ion. Find its charge and mass. (f) The plasma in a hot star contains quadruply ionized nitrogen atoms, N^{4+}. Find their charge and mass. (g) Find the charge and the mass of the nucleus of a nitrogen atom. (h) Find the charge and the mass of the molecular ion H_2O^-.

2. (a) Calculate the number of electrons in a small, electrically neutral silver pin that has a mass of 10.0 g. Silver has 47 electrons per atom, and its molar mass is 107.87 g/mol. (b) Electrons are added to the pin until the net negative charge is 1.00 mC. How many electrons are added for every 10^9 electrons already present?

Section 23.2 Charging Objects by Induction
Section 23.3 Coulomb's Law

3. The Nobel laureate Richard Feynman once said that if two persons stood at arm's length from each other and each person had 1% more electrons than protons, the force of repulsion between them would be enough to lift a "weight" equal to that of the entire Earth. Carry out an order-of-magnitude calculation to substantiate this assertion.

4. Two protons in an atomic nucleus are typically separated by a distance of 2×10^{-15} m. The electric repulsion force between the protons is huge, but the attractive nuclear force is even stronger and keeps the nucleus from bursting apart. What is the magnitude of the electric force between two protons separated by 2.00×10^{-15} m?

5. (a) Two protons in a molecule are separated by 3.80×10^{-10} m. Find the electric force exerted by one proton on the other. (b) How does the magnitude of this force compare to the magnitude of the gravitational force between the two protons? (c) **What If?** What must be the charge-to-mass ratio of a particle if the magnitude of the gravitational force between two of these particles equals the magnitude of electric force between them?

6. Two small silver spheres, each with a mass of 10.0 g, are separated by 1.00 m. Calculate the fraction of the electrons in one sphere that must be transferred to the other in order to produce an attractive force of 1.00×10^4 N (about 1 ton) between the spheres. (The number of electrons per atom of silver is 47, and the number of atoms per gram is Avogadro's number divided by the molar mass of silver, 107.87 g/mol.)

7. Three point charges are located at the corners of an equilateral triangle as shown in Figure P23.7. Calculate the resultant electric force on the 7.00-μC charge.

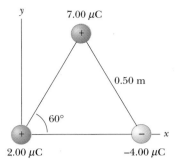

Figure P23.7 Problems 7 and 18.

8. Suppose that 1.00 g of hydrogen is separated into electrons and protons. Suppose also that the protons are placed at the Earth's north pole and the electrons are placed at the south pole. What is the resulting compressional force on the Earth?

9. Two identical conducting small spheres are placed with their centers 0.300 m apart. One is given a charge of 12.0 nC and the other a charge of −18.0 nC. (a) Find the electric force exerted by one sphere on the other. (b) **What If?** The spheres are connected by a conducting wire. Find the electric force between the two after they have come to equilibrium.

10. Two small beads having positive charges $3q$ and q are fixed at the opposite ends of a horizontal, insulating rod, extending from the origin to the point $x = d$. As shown in Figure P23.10, a third small charged bead is free to slide on the rod. At what position is the third bead in equilibrium? Can it be in stable equilibrium?

Figure P23.10

11. **Review problem.** In the Bohr theory of the hydrogen atom, an electron moves in a circular orbit about a proton, where the radius of the orbit is 0.529×10^{-10} m. (a) Find the electric force between the two. (b) If this force causes the centripetal acceleration of the electron, what is the speed of the electron?

12. **Review problem.** Two identical particles, each having charge $+ q$, are fixed in space and separated by a distance d. A third point charge $− Q$ is free to move and lies initially at rest on the perpendicular bisector of the two fixed charges a distance x from the midpoint between the two fixed charges (Fig. P23.12). (a) Show that if x is small compared with d, the motion of $− Q$ will be simple harmonic along the perpendicular bisector. Determine the period of that motion. (b) How fast will the charge $− Q$ be moving when it is at the midpoint between the two fixed charges, if initially it is released at a distance $a \ll d$ from the midpoint?

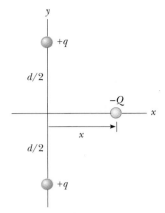

Figure P23.12

Section 23.4   The Electric Field

13. What are the magnitude and direction of the electric field that will balance the weight of (a) an electron and (b) a proton? (Use the data in Table 23.1.)

14. An object having a net charge of 24.0 μC is placed in a uniform electric field of 610 N/C directed vertically. What is the mass of this object if it "floats" in the field?

15. In Figure P23.15, determine the point (other than infinity) at which the electric field is zero.

Figure P23.15

16. An airplane is flying through a thundercloud at a height of 2 000 m. (This is a very dangerous thing to do because of updrafts, turbulence, and the possibility of electric discharge.) If a charge concentration of + 40.0 C is above the plane at a height of 3 000 m within the cloud and a charge concentration of − 40.0 C is at height 1 000 m, what is the electric field at the aircraft?

17. Two point charges are located on the x axis. The first is a charge $+ Q$ at $x = − a$. The second is an unknown charge located at $x = + 3a$. The net electric field these charges produce at the origin has a magnitude of $2k_e Q/a^2$. What are the two possible values of the unknown charge?

18. Three charges are at the corners of an equilateral triangle as shown in Figure P23.7. (a) Calculate the electric field at the position of the 2.00-μC charge due to the 7.00-μC and − 4.00-μC charges. (b) Use your answer to part (a) to determine the force on the 2.00-μC charge.

19. Three point charges are arranged as shown in Figure P23.19. (a) Find the vector electric field that the 6.00-nC and − 3.00-nC charges together create at the origin. (b) Find the vector force on the 5.00-nC charge.

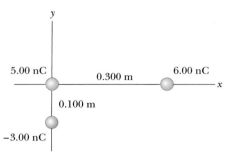

Figure P23.19

20. Two 2.00-μC point charges are located on the x axis. One is at $x = 1.00$ m, and the other is at $x = -1.00$ m. (a) Determine the electric field on the y axis at $y = 0.500$ m. (b) Calculate the electric force on a -3.00-μC charge placed on the y axis at $y = 0.500$ m.

21. Four point charges are at the corners of a square of side a as shown in Figure P23.21. (a) Determine the magnitude and direction of the electric field at the location of charge q. (b) What is the resultant force on q?

Figure P23.21

22. Consider the electric dipole shown in Figure P23.22. Show that the electric field at a *distant* point on the $+x$ axis is $E_x \approx 4k_e qa/x^3$.

Figure P23.22

23. Consider n equal positive point charges each of magnitude Q/n placed symmetrically around a circle of radius R. (a) Calculate the magnitude of the electric field at a point a distance x on the line passing through the center of the circle and perpendicular to the plane of the circle. (b) Explain why this result is identical to that of the calculation done in Example 23.8.

24. Consider an infinite number of identical charges (each of charge q) placed along the x axis at distances a, $2a$, $3a$, $4a$, . . . , from the origin. What is the electric field at the origin due to this distribution? *Suggestion:* Use the fact that

$$1 + \frac{1}{2^2} + \frac{1}{3^2} + \frac{1}{4^2} + \cdots = \frac{\pi^2}{6}$$

Section 23.5 Electric Field of a Continuous Charge Distribution

25. A rod 14.0 cm long is uniformly charged and has a total charge of -22.0 μC. Determine the magnitude and direction of the electric field along the axis of the rod at a point 36.0 cm from its center.

26. A continuous line of charge lies along the x axis, extending from $x = +x_0$ to positive infinity. The line carries charge with a uniform linear charge density λ_0. What are the magnitude and direction of the electric field at the origin?

27. A uniformly charged ring of radius 10.0 cm has a total charge of 75.0 μC. Find the electric field on the axis of the ring at (a) 1.00 cm, (b) 5.00 cm, (c) 30.0 cm, and (d) 100 cm from the center of the ring.

28. A line of charge starts at $x = +x_0$ and extends to positive infinity. The linear charge density is $\lambda = \lambda_0 x_0/x$. Determine the electric field at the origin.

29. Show that the maximum magnitude E_{max} of the electric field along the axis of a uniformly charged ring occurs at $x = a/\sqrt{2}$ (see Fig. 23.18) and has the value $Q/(6\sqrt{3}\pi\epsilon_0 a^2)$.

30. A uniformly charged disk of radius 35.0 cm carries charge with a density of 7.90×10^{-3} C/m^2. Calculate the electric field on the axis of the disk at (a) 5.00 cm, (b) 10.0 cm, (c) 50.0 cm, and (d) 200 cm from the center of the disk.

31. Example 23.9 derives the exact expression for the electric field at a point on the axis of a uniformly charged disk. Consider a disk, of radius $R = 3.00$ cm, having a uniformly distributed charge of $+5.20$ μC. (a) Using the result of Example 23.9, compute the electric field at a point on the axis and 3.00 mm from the center. **What If?** Compare this answer with the field computed from the near-field approximation $E = \sigma/2\epsilon_0$. (b) Using the result of Example 23.9, compute the electric field at a point on the axis and 30.0 cm from the center of the disk. **What If?** Compare this with the electric field obtained by treating the disk as a $+5.20$-μC point charge at a distance of 30.0 cm.

32. The electric field along the axis of a uniformly charged disk of radius R and total charge Q was calculated in Example 23.9. Show that the electric field at distances x that are large compared with R approaches that of a point charge $Q = \sigma\pi R^2$. (*Suggestion:* First show that $x/(x^2 + R^2)^{1/2} = (1 + R^2/x^2)^{-1/2}$ and use the binomial expansion $(1 + \delta)^n \approx 1 + n\delta$ when $\delta \ll 1$.)

33. A uniformly charged insulating rod of length 14.0 cm is bent into the shape of a semicircle as shown in Figure P23.33. The rod has a total charge of -7.50 μC. Find the magnitude and direction of the electric field at O, the center of the semicircle.

Figure P23.33

34. (a) Consider a uniformly charged thin-walled right circular cylindrical shell having total charge Q, radius R, and height h. Determine the electric field at a point a distance d from the right side of the cylinder as shown in Figure P23.34. (*Suggestion:* Use the result of Example 23.8 and treat the cylinder as a collection of ring charges.) (b) **What If?** Consider now a solid cylinder with the same dimensions and carrying the same charge, uniformly distributed through its volume. Use the result of Example 23.9 to find the field it creates at the same point.

Figure P23.34

35. A thin rod of length ℓ and uniform charge per unit length λ lies along the x axis, as shown in Figure P23.35. (a) Show that the electric field at P, a distance y from the rod along its perpendicular bisector, has no x component and is given by $E = 2k_e\lambda \sin \theta_0/y$. (b) **What If?** Using your result to part (a), show that the field of a rod of infinite length is $E = 2k_e\lambda/y$. (*Suggestion:* First calculate the field at P due to an element of length dx, which has a charge $\lambda\,dx$. Then change variables from x to θ, using the relationships $x = y \tan \theta$ and $dx = y \sec^2 \theta\,d\theta$, and integrate over θ.)

Figure P23.35

36. Three solid plastic cylinders all have radius 2.50 cm and length 6.00 cm. One (a) carries charge with uniform density 15.0 nC/m² everywhere on its surface. Another (b) carries charge with the same uniform density on its curved lateral surface only. The third (c) carries charge with uniform density 500 nC/m³ throughout the plastic. Find the charge of each cylinder.

37. Eight solid plastic cubes, each 3.00 cm on each edge, are glued together to form each one of the objects (i, ii, iii, and iv) shown in Figure P23.37. (a) Assuming each object carries charge with uniform density 400 nC/m³ throughout its volume, find the charge of each object. (b) Assuming each object carries charge with uniform density 15.0 nC/m² everywhere on its exposed surface, find the charge on each object. (c) Assuming charge is placed only on the edges where perpendicular surfaces meet, with uniform density 80.0 pC/m, find the charge of each object.

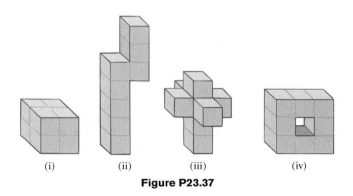

(i)   (ii)   (iii)   (iv)

Figure P23.37

Section 23.6   Electric Field Lines

38. A positively charged disk has a uniform charge per unit area as described in Example 23.9. Sketch the electric field lines in a plane perpendicular to the plane of the disk passing through its center.

39. A negatively charged rod of finite length carries charge with a uniform charge per unit length. Sketch the electric field lines in a plane containing the rod.

40. Figure P23.40 shows the electric field lines for two point charges separated by a small distance. (a) Determine the ratio q_1/q_2. (b) What are the signs of q_1 and q_2?

Figure P23.40

41. 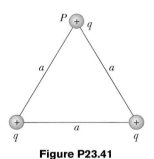 Three equal positive charges q are at the corners of an equilateral triangle of side a as shown in Figure P23.41. (a) Assume that the three charges together create an electric field. Sketch the field lines in the plane of the charges. Find the location of a point (other than ∞) where the electric field is zero. (b) What are the magnitude and direction of the electric field at P due to the two charges at the base?

Figure P23.41

Section 23.7 Motion of Charged Particles in a Uniform Electric Field

42. An electron and a proton are each placed at rest in an electric field of 520 N/C. Calculate the speed of each particle 48.0 ns after being released.

43. A proton accelerates from rest in a uniform electric field of 640 N/C. At some later time, its speed is 1.20×10^6 m/s (nonrelativistic, because v is much less than the speed of light). (a) Find the acceleration of the proton. (b) How long does it take the proton to reach this speed? (c) How far has it moved in this time? (d) What is its kinetic energy at this time?

44. A proton is projected in the positive x direction into a region of a uniform electric field $\mathbf{E} = -6.00 \times 10^5 \hat{\mathbf{i}}$ N/C at $t = 0$. The proton travels 7.00 cm before coming to rest. Determine (a) the acceleration of the proton, (b) its initial speed, and (c) the time at which the proton comes to rest.

45. 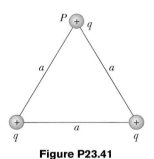 The electrons in a particle beam each have a kinetic energy K. What are the magnitude and direction of the electric field that will stop these electrons in a distance d?

46. A positively charged bead having a mass of 1.00 g falls from rest in a vacuum from a height of 5.00 m in a uniform vertical electric field with a magnitude of 1.00×10^4 N/C. The bead hits the ground at a speed of 21.0 m/s. Determine (a) the direction of the electric field (up or down), and (b) the charge on the bead.

47. A proton moves at 4.50×10^5 m/s in the horizontal direction. It enters a uniform vertical electric field with a magnitude of 9.60×10^3 N/C. Ignoring any gravitational effects, find (a) the time interval required for the proton to travel 5.00 cm horizontally, (b) its vertical displacement during the time interval in which it travels 5.00 cm horizontally, and (c) the horizontal and vertical components of its velocity after it has traveled 5.00 cm horizontally.

48. Two horizontal metal plates, each 100 mm square, are aligned 10.0 mm apart, with one above the other. They are given equal-magnitude charges of opposite sign so that a uniform downward electric field of 2 000 N/C exists in the region between them. A particle of mass 2.00×10^{-16} kg

and with a positive charge of 1.00×10^{-6} C leaves the center of the bottom negative plate with an initial speed of 1.00×10^5 m/s at an angle of 37.0° above the horizontal. Describe the trajectory of the particle. Which plate does it strike? Where does it strike, relative to its starting point?

49. Protons are projected with an initial speed $v_i = 9.55 \times 10^3$ m/s into a region where a uniform electric field $\mathbf{E} = -720\hat{\mathbf{j}}$ N/C is present, as shown in Figure P23.49. The protons are to hit a target that lies at a horizontal distance of 1.27 mm from the point where the protons cross the plane and enter the electric field in Figure P23.49. Find (a) the two projection angles θ that will result in a hit and (b) the total time of flight (the time interval during which the proton is above the plane in Figure P23.49) for each trajectory.

Figure P23.49

Additional Problems

50. Two known charges, $-12.0 \ \mu$C and $45.0 \ \mu$C, and an unknown charge are located on the x axis. The charge $-12.0 \ \mu$C is at the origin, and the charge $45.0 \ \mu$C is at $x = 15.0$ cm. The unknown charge is to be placed so that each charge is in equilibrium under the action of the electric forces exerted by the other two charges. Is this situation possible? Is it possible in more than one way? Find the required location, magnitude, and sign of the unknown charge.

51. A uniform electric field of magnitude 640 N/C exists between two parallel plates that are 4.00 cm apart. A proton is released from the positive plate at the same instant that an electron is released from the negative plate. (a) Determine the distance from the positive plate at which the two pass each other. (Ignore the electrical attraction between the proton and electron.) (b) **What If?** Repeat part (a) for a sodium ion (Na^+) and a chloride ion (Cl^-).

52. Three point charges are aligned along the x axis as shown in Figure P23.52. Find the electric field at (a) the position (2.00, 0) and (b) the position (0, 2.00).

Figure P23.52

53. A researcher studying the properties of ions in the upper atmosphere wishes to construct an apparatus with the following characteristics: Using an electric field, a beam of ions, each having charge q, mass m, and initial velocity $v\hat{\mathbf{i}}$, is turned through an angle of 90° as each ion undergoes displacement $R\hat{\mathbf{i}} + R\hat{\mathbf{j}}$. The ions enter a chamber as shown in Figure P23.53, and leave through the exit port with the same speed they had when they entered the chamber. The electric field acting on the ions is to have constant magnitude. (a) Suppose the electric field is produced by two concentric cylindrical electrodes not shown in the diagram, and hence is radial. What magnitude should the field have? **What If?** (b) If the field is produced by two flat plates and is uniform in direction, what value should the field have in this case?

Figure P23.53

54. A small, 2.00-g plastic ball is suspended by a 20.0-cm-long string in a uniform electric field as shown in Figure P23.54. If the ball is in equilibrium when the string makes a 15.0° angle with the vertical, what is the net charge on the ball?

Figure P23.54

55. ![icon] A charged cork ball of mass 1.00 g is suspended on a light string in the presence of a uniform electric field as shown in Figure P23.55. When $\mathbf{E} = (3.00\hat{\mathbf{i}} + 5.00\hat{\mathbf{j}}) \times 10^5$ N/C, the ball is in equilibrium at $\theta = 37.0°$. Find (a) the charge on the ball and (b) the tension in the string.

56. A charged cork ball of mass m is suspended on a light string in the presence of a uniform electric field as shown in Figure P23.55. When $\mathbf{E} = (A\hat{\mathbf{i}} + B\hat{\mathbf{j}})$ N/C, where A and B are positive numbers, the ball is in equilibrium at the angle θ. Find (a) the charge on the ball and (b) the tension in the string.

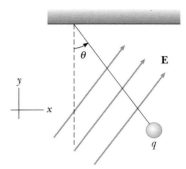

Figure P23.55 Problems 55 and 56.

57. Four identical point charges ($q = +10.0\ \mu$C) are located on the corners of a rectangle as shown in Figure P23.57. The dimensions of the rectangle are $L = 60.0$ cm and $W = 15.0$ cm. Calculate the magnitude and direction of the resultant electric force exerted on the charge at the lower left corner by the other three charges.

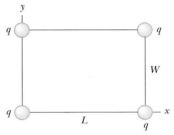

Figure P23.57

58. Inez is putting up decorations for her sister's quinceañera (fifteenth birthday party). She ties three light silk ribbons together to the top of a gateway and hangs a rubber balloon from each ribbon (Fig. P23.58). To include the

Figure P23.58

effects of the gravitational and buoyant forces on it, each balloon can be modeled as a particle of mass 2.00 g, with its center 50.0 cm from the point of support. To show off the colors of the balloons, Inez rubs the whole surface of each balloon with her woolen scarf, to make them hang separately with gaps between them. The centers of the hanging balloons form a horizontal equilateral triangle with sides 30.0 cm long. What is the common charge each balloon carries?

59. **Review problem.** Two identical metallic blocks resting on a frictionless horizontal surface are connected by a light metallic spring having a spring constant k as shown in Figure P23.59a and an unstretched length L_i. A total charge Q is slowly placed on the system, causing the spring to stretch to an equilibrium length L, as shown in Figure P23.59b. Determine the value of Q, assuming that all the charge resides on the blocks and modeling the blocks as point charges.

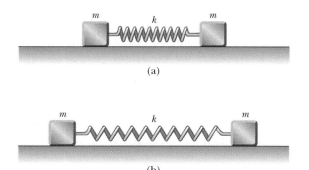

(a)

(b)

Figure P23.59

60. Consider a regular polygon with 29 sides. The distance from the center to each vertex is a. Identical charges q are placed at 28 vertices of the polygon. A single charge Q is placed at the center of the polygon. What is the magnitude and direction of the force experienced by the charge Q? (*Suggestion:* You may use the result of Problem 63 in Chapter 3.)

61. Identical thin rods of length $2a$ carry equal charges $+Q$ uniformly distributed along their lengths. The rods lie along the x axis with their centers separated by a distance $b > 2a$ (Fig. P23.61). Show that the magnitude of the force exerted by the left rod on the right one is given by

$$F = \left(\frac{k_e Q^2}{4a^2}\right) \ln\left(\frac{b^2}{b^2 - 4a^2}\right)$$

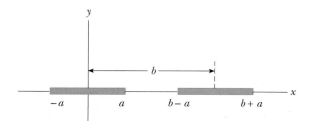

Figure P23.61

62. Two small spheres, each of mass 2.00 g, are suspended by light strings 10.0 cm in length (Fig. P23.62). A uniform electric field is applied in the x direction. The spheres have charges equal to -5.00×10^{-8} C and $+5.00 \times 10^{-8}$ C. Determine the electric field that enables the spheres to be in equilibrium at an angle $\theta = 10.0°$.

Figure P23.62

63. A line of positive charge is formed into a semicircle of radius $R = 60.0$ cm as shown in Figure P23.63. The charge per unit length along the semicircle is described by the expression $\lambda = \lambda_0 \cos \theta$. The total charge on the semicircle is 12.0 μC. Calculate the total force on a charge of 3.00 μC placed at the center of curvature.

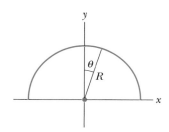

Figure P23.63

64. Three charges of equal magnitude q are fixed in position at the vertices of an equilateral triangle (Fig. P23.64). A fourth charge Q is free to move along the positive x axis

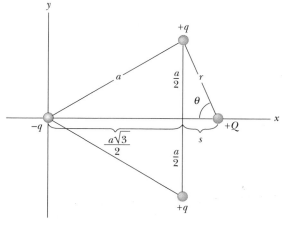

Figure P23.64

under the influence of the forces exerted by the three fixed charges. Find a value for *s* for which *Q* is in equilibrium. You will need to solve a transcendental equation.

65. Two small spheres of mass *m* are suspended from strings of length ℓ that are connected at a common point. One sphere has charge *Q*; the other has charge 2*Q*. The strings make angles θ_1 and θ_2 with the vertical. (a) How are θ_1 and θ_2 related? (b) Assume θ_1 and θ_2 are small. Show that the distance *r* between the spheres is given by

$$r \approx \left(\frac{4k_e Q^2 \ell}{mg} \right)^{1/3}$$

66. Review problem. Four identical particles, each having charge $+ q$, are fixed at the corners of a square of side *L*. A fifth point charge $- Q$ lies a distance *z* along the line perpendicular to the plane of the square and passing through the center of the square (Fig. P23.66). (a) Show that the force exerted by the other four charges on $- Q$ is

$$\mathbf{F} = - \frac{4k_e q Qz}{[z^2 + (L^2/2)]^{3/2}} \, \hat{\mathbf{k}}$$

Note that this force is directed toward the center of the square whether *z* is positive ($- Q$ above the square) or negative ($- Q$ below the square). (b) If *z* is small compared with *L*, the above expression reduces to $\mathbf{F} \approx - (\text{constant}) z \hat{\mathbf{k}}$. Why does this imply that the motion of the charge $- Q$ is simple harmonic, and what is the period of this motion if the mass of $- Q$ is *m*?

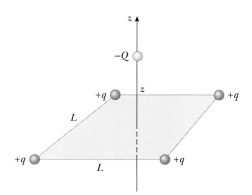

Figure P23.66

67. Review problem. A 1.00-g cork ball with charge 2.00 μC is suspended vertically on a 0.500-m-long light string in the presence of a uniform, downward-directed electric field of magnitude $E = 1.00 \times 10^5$ N/C. If the ball is displaced slightly from the vertical, it oscillates like a simple pendulum. (a) Determine the period of this oscillation. (b) Should gravity be included in the calculation for part (a)? Explain.

68. Two identical beads each have a mass *m* and charge *q*. When placed in a hemispherical bowl of radius *R* with frictionless, nonconducting walls, the beads move, and at equilibrium they are a distance *R* apart (Fig. P23.68). Determine the charge on each bead.

Figure P23.68

69. Eight point charges, each of magnitude *q*, are located on the corners of a cube of edge *s*, as shown in Figure P23.69. (a) Determine the *x*, *y*, and *z* components of the resultant force exerted by the other charges on the charge located at point *A*. (b) What are the magnitude and direction of this resultant force?

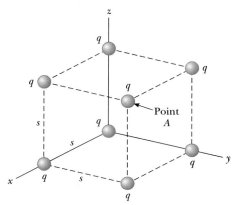

Figure P23.69 Problems 69 and 70.

70. Consider the charge distribution shown in Figure P23.69. (a) Show that the magnitude of the electric field at the center of any face of the cube has a value of $2.18k_e q/s^2$. (b) What is the direction of the electric field at the center of the top face of the cube?

71. Review problem. A negatively charged particle $- q$ is placed at the center of a uniformly charged ring, where the ring has a total positive charge *Q* as shown in Example 23.8. The particle, confined to move along the *x* axis, is displaced a small distance *x* along the axis (where $x \ll a$) and released. Show that the particle oscillates in simple harmonic motion with a frequency given by

$$f = \frac{1}{2\pi} \left(\frac{k_e q Q}{ma^3} \right)^{1/2}$$

72. A line of charge with uniform density 35.0 nC/m lies along the line $y = - 15.0$ cm, between the points with coordinates $x = 0$ and $x = 40.0$ cm. Find the electric field it creates at the origin.

73. Review problem. An electric dipole in a uniform electric field is displaced slightly from its equilibrium position, as shown in Figure P23.73, where θ is small. The separation of the charges is 2*a*, and the moment of inertia of the dipole is *I*. Assuming the dipole is released from this position, show that its angular orientation exhibits simple harmonic motion with a frequency

$$f = \frac{1}{2\pi} \sqrt{\frac{2qaE}{I}}$$

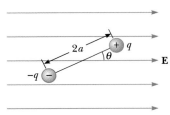

Figure P23.73

Answers to Quick Quizzes

23.1 (b). The amount of charge present in the isolated system after rubbing is the same as that before because charge is conserved; it is just distributed differently.

23.2 (a), (c), and (e). The experiment shows that A and B have charges of the same sign, as do objects B and C. Thus, all three objects have charges of the same sign. We cannot determine from this information, however, whether the charges are positive or negative.

23.3 (e). In the first experiment, objects A and B may have charges with opposite signs, or one of the objects may be neutral. The second experiment shows that B and C have charges with the same signs, so that B must be charged. But we still do not know if A is charged or neutral.

23.4 (e). From Newton's third law, the electric force exerted by object B on object A is equal in magnitude to the force exerted by object A on object B.

23.5 (b). From Newton's third law, the electric force exerted by object B on object A is equal in magnitude to the force exerted by object A on object B and in the opposite direction.

23.6 (a). There is no effect on the electric field if we assume that the source charge producing the field is not disturbed by our actions. Remember that the electric field is created by source charge(s) (unseen in this case), not the test charge(s).

23.7 *A, B, C.* The field is greatest at point *A* because this is where the field lines are closest together. The absence of lines near point *C* indicates that the electric field there is zero.

23.8 (b). Electric field lines begin and end on charges and cannot close on themselves to form loops.

Chapter 24

Gauss's Law

▲ In a table-top plasma ball, the colorful lines emanating from the sphere give evidence of strong electric fields. Using Gauss's law, we show in this chapter that the electric field surrounding a charged sphere is identical to that of a point charge. (Getty Images)

In the preceding chapter we showed how to calculate the electric field generated by a given charge distribution. In this chapter, we describe *Gauss's law* and an alternative procedure for calculating electric fields. The law is based on the fact that the fundamental electrostatic force between point charges exhibits an inverse-square behavior. Although a consequence of Coulomb's law, Gauss's law is more convenient for calculating the electric fields of highly symmetric charge distributions and makes possible useful qualitative reasoning when dealing with complicated problems.

24.1 Electric Flux

The concept of electric field lines was described qualitatively in Chapter 23. We now treat electric field lines in a more quantitative way.

Consider an electric field that is uniform in both magnitude and direction, as shown in Figure 24.1. The field lines penetrate a rectangular surface of area A, whose plane is oriented perpendicular to the field. Recall from Section 23.6 that the number of lines per unit area (in other words, the *line density*) is proportional to the magnitude of the electric field. Therefore, the total number of lines penetrating the surface is proportional to the product EA. This product of the magnitude of the electric field E and surface area A perpendicular to the field is called the **electric flux** Φ_E (uppercase Greek phi):

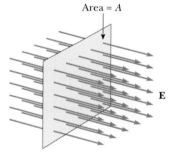

Figure 24.1 Field lines representing a uniform electric field penetrating a plane of area A perpendicular to the field. The electric flux Φ_E through this area is equal to EA.

$$\Phi_E = EA \qquad (24.1)$$

From the SI units of E and A, we see that Φ_E has units of newton-meters squared per coulomb ($\text{N} \cdot \text{m}^2/\text{C}$.) **Electric flux is proportional to the number of electric field lines penetrating some surface.**

Example 24.1 Electric Flux Through a Sphere

What is the electric flux through a sphere that has a radius of 1.00 m and carries a charge of $+1.00 \ \mu\text{C}$ at its center?

Solution The magnitude of the electric field 1.00 m from this charge is found using Equation 23.9:

$$E = k_e \frac{q}{r^2} = (8.99 \times 10^9 \ \text{N} \cdot \text{m}^2/\text{C}^2) \frac{1.00 \times 10^{-6} \ \text{C}}{(1.00 \ \text{m})^2}$$

$$= 8.99 \times 10^3 \ \text{N/C}$$

The field points radially outward and is therefore everywhere perpendicular to the surface of the sphere. The flux through the sphere (whose surface area $A = 4\pi r^2 = 12.6 \ \text{m}^2$) is thus

$$\Phi_E = EA = (8.99 \times 10^3 \ \text{N/C})(12.6 \ \text{m}^2)$$

$$= 1.13 \times 10^5 \ \text{N} \cdot \text{m}^2/\text{C}$$

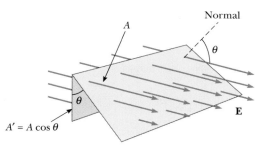

Figure 24.2 Field lines representing a uniform electric field penetrating an area A that is at an angle θ to the field. Because the number of lines that go through the area A' is the same as the number that go through A, the flux through A' is equal to the flux through A and is given by $\Phi_E = EA \cos \theta$.

If the surface under consideration is not perpendicular to the field, the flux through it must be less than that given by Equation 24.1. We can understand this by considering Figure 24.2, where the normal to the surface of area A is at an angle θ to the uniform electric field. Note that the number of lines that cross this area A is equal to the number that cross the area A', which is a projection of area A onto a plane oriented perpendicular to the field. From Figure 24.2 we see that the two areas are related by $A' = A \cos \theta$. Because the flux through A equals the flux through A', we conclude that the flux through A is

$$\Phi_E = EA' = EA \cos \theta \qquad (24.2)$$

From this result, we see that the flux through a surface of fixed area A has a maximum value EA when the surface is perpendicular to the field (when the normal to the surface is parallel to the field, that is, $\theta = 0°$ in Figure 24.2); the flux is zero when the surface is parallel to the field (when the normal to the surface is perpendicular to the field, that is, $\theta = 90°$).

We assumed a uniform electric field in the preceding discussion. In more general situations, the electric field may vary over a surface. Therefore, our definition of flux given by Equation 24.2 has meaning only over a small element of area. Consider a general surface divided up into a large number of small elements, each of area ΔA. The variation in the electric field over one element can be neglected if the element is sufficiently small. It is convenient to define a vector $\Delta \mathbf{A}_i$ whose magnitude represents the area of the ith element of the surface and whose direction is defined to be *perpendicular* to the surface element, as shown in Figure 24.3. The electric field $\mathbf{E}_i$ at the location of this element makes an angle θ_i with the vector $\Delta \mathbf{A}_i$. The electric flux $\Delta \Phi_E$ through this element is

$$\Delta \Phi_E = E_i \Delta A_i \cos \theta_i = \mathbf{E}_i \cdot \Delta \mathbf{A}_i$$

where we have used the definition of the scalar product (or dot product; see Chapter 7) of two vectors ($\mathbf{A} \cdot \mathbf{B} = AB \cos \theta$). By summing the contributions of all elements, we obtain the total flux through the surface. If we let the area of each element approach zero, then the number of elements approaches infinity and the sum is replaced by an integral. Therefore, the general definition of electric flux is[1]

$$\Phi_E = \lim_{\Delta A_i \to 0} \sum \mathbf{E}_i \cdot \Delta \mathbf{A}_i = \int_{\text{surface}} \mathbf{E} \cdot d\mathbf{A} \qquad (24.3)$$

Equation 24.3 is a *surface integral*, which means it must be evaluated over the surface in question. In general, the value of Φ_E depends both on the field pattern and on the surface.

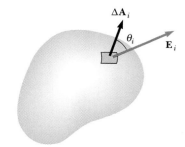

Figure 24.3 A small element of surface area ΔA_i. The electric field makes an angle θ_i with the vector $\Delta \mathbf{A}_i$, defined as being normal to the surface element, and the flux through the element is equal to $E_i \Delta A_i \cos \theta_i$.

Definition of electric flux

---

[1]   Drawings with field lines have their inaccuracies because a limited number of field lines are typically drawn in a diagram. Consequently, a small area element drawn on a diagram (depending on its location) may happen to have too few field lines penetrating it to represent the flux accurately. We stress that the basic definition of electric flux is Equation 24.3. The use of lines is only an aid for visualizing the concept.

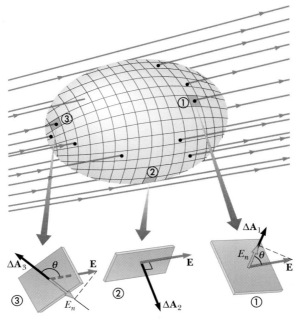

Active Figure 24.4 A closed surface in an electric field. The area vectors Δ**A**$_i$ are, by convention, normal to the surface and point outward. The flux through an area element can be positive (element ①), zero (element ②), or negative (element ③).

Karl Friedrich Gauss

German mathematician and astronomer (1777–1855)

Gauss received a doctoral degree in mathematics from the University of Helmstedt in 1799. In addition to his work in electromagnetism, he made contributions to mathematics and science in number theory, statistics, non-Euclidean geometry, and cometary orbital mechanics. He was a founder of the German Magnetic Union, which studies the Earth's magnetic field on a continual basis.

We are often interested in evaluating the flux through a *closed surface,* which is defined as one that divides space into an inside and an outside region, so that one cannot move from one region to the other without crossing the surface. The surface of a sphere, for example, is a closed surface.

Consider the closed surface in Figure 24.4. The vectors Δ**A**$_i$ point in different directions for the various surface elements, but at each point they are normal to the surface and, by convention, always point outward. At the element labeled ①, the field lines are crossing the surface from the inside to the outside and $\theta < 90°$; hence, the flux $\Delta\Phi_E = \mathbf{E} \cdot \Delta\mathbf{A}_1$ through this element is positive. For element ②, the field lines graze the surface (perpendicular to the vector Δ**A**$_2$); thus, $\theta = 90°$ and the flux is zero. For elements such as ③, where the field lines are crossing the surface from outside to inside, $180° > \theta > 90°$ and the flux is negative because $\cos\theta$ is negative. The *net* flux through the surface is proportional to the net number of lines leaving the surface, where the net number means *the number leaving the surface minus the number entering the surface.* If more lines are leaving than entering, the net flux is positive. If more lines are entering than leaving, the net flux is negative. Using the symbol $\oint$ to represent an integral over a closed surface, we can write the net flux Φ_E through a closed surface as

$$\Phi_E = \oint \mathbf{E} \cdot d\mathbf{A} = \oint E_n \, dA \qquad (24.4)$$

where E_n represents the component of the electric field normal to the surface. If the field is normal to the surface at each point and constant in magnitude, the calculation is straightforward, as it was in Example 24.1. Example 24.2 also illustrates this point.

Quick Quiz 24.1 Suppose the radius of the sphere in Example 24.1 is changed to 0.500 m. What happens to the flux through the sphere and the magnitude of the electric field at the surface of the sphere? (a) The flux and field both increase. (b) The flux and field both decrease. (c) The flux increases and the field decreases. (d) The flux decreases and the field increases. (e) The flux remains the same and the field increases. (f) The flux decreases and the field remains the same.

Quick Quiz 24.2 In a charge-free region of space, a closed container is placed in an electric field. A requirement for the total electric flux through the surface of the container to be zero is that (a) the field must be uniform, (b) the container must be symmetric, (c) the container must be oriented in a certain way, or (d) the requirement does not exist—the total electric flux is zero no matter what.

Example 24.2 Flux Through a Cube

Consider a uniform electric field **E** oriented in the x direction. Find the net electric flux through the surface of a cube of edge length ℓ, oriented as shown in Figure 24.5.

Solution The net flux is the sum of the fluxes through all faces of the cube. First, note that the flux through four of

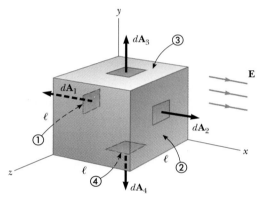

Figure 24.5 (Example 24.2) A closed surface in the shape of a cube in a uniform electric field oriented parallel to the x axis. Side ④ is the bottom of the cube, and side ① is opposite side ②.

the faces (③, ④, and the unnumbered ones) is zero because **E** is perpendicular to $d\mathbf{A}$ on these faces.

The net flux through faces ① and ② is

$$\Phi_E = \int_1 \mathbf{E} \cdot d\mathbf{A} + \int_2 \mathbf{E} \cdot d\mathbf{A}$$

For face ①, **E** is constant and directed inward but $d\mathbf{A}_1$ is directed outward ($\theta = 180°$); thus, the flux through this face is

$$\int_1 \mathbf{E} \cdot d\mathbf{A} = \int_1 E(\cos 180°)\, dA = -E \int_1 dA = -EA = -E\ell^2$$

because the area of each face is $A = \ell^2$.

For face ②, **E** is constant and outward and in the same direction as $d\mathbf{A}_2$ ($\theta = 0°$); hence, the flux through this face is

$$\int_2 \mathbf{E} \cdot d\mathbf{A} = \int_2 E(\cos 0°)\, dA = E \int_2 dA = +EA = E\ell^2$$

Therefore, the net flux over all six faces is

$$\Phi_E = -E\ell^2 + E\ell^2 + 0 + 0 + 0 + 0 = \boxed{0}$$

24.2 Gauss's Law

In this section we describe a general relationship between the net electric flux through a closed surface (often called a *gaussian surface*) and the charge enclosed by the surface. This relationship, known as *Gauss's law*, is of fundamental importance in the study of electric fields.

Let us again consider a positive point charge q located at the center of a sphere of radius r, as shown in Figure 24.6. From Equation 23.9 we know that the magnitude of the electric field everywhere on the surface of the sphere is $E = k_e q/r^2$. As noted in Example 24.1, the field lines are directed radially outward and hence are perpendicular to the surface at every point on the surface. That is, at each surface point, **E** is parallel to the vector $\Delta\mathbf{A}_i$ representing a local element of area ΔA_i surrounding the surface point. Therefore,

$$\mathbf{E} \cdot \Delta\mathbf{A}_i = E\,\Delta A_i$$

and from Equation 24.4 we find that the net flux through the gaussian surface is

$$\Phi_E = \oint \mathbf{E} \cdot d\mathbf{A} = \oint E\, dA = E \oint dA$$

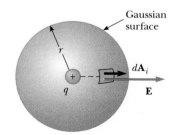

Figure 24.6 A spherical gaussian surface of radius r surrounding a point charge q. When the charge is at the center of the sphere, the electric field is everywhere normal to the surface and constant in magnitude.

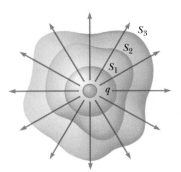

Figure 24.7 Closed surfaces of various shapes surrounding a charge q. The net electric flux is the same through all surfaces.

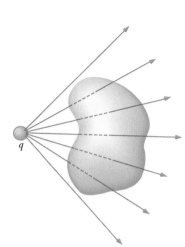

Figure 24.8 A point charge located *outside* a closed surface. The number of lines entering the surface equals the number leaving the surface.

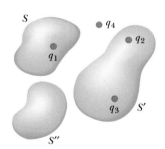

Active Figure 24.9 The net electric flux through any closed surface depends only on the charge *inside* that surface. The net flux through surface S is q_1/ϵ_0, the net flux through surface S' is $(q_2 + q_3)/\epsilon_0$, and the net flux through surface S'' is zero. Charge q_4 does not contribute to the flux through any surface because it is outside all surfaces.

At the Active Figures link at http://www.pse6.com, you can change the size and shape of a closed surface and see the effect on the electric flux of surrounding combinations of charge with that surface.

where we have moved E outside of the integral because, by symmetry, E is constant over the surface and given by $E = k_e q/r^2$. Furthermore, because the surface is spherical, $\oint dA = A = 4\pi r^2$. Hence, the net flux through the gaussian surface is

$$\Phi_E = \frac{k_e q}{r^2}(4\pi r^2) = 4\pi k_e q$$

Recalling from Section 23.3 that $k_e = 1/4\pi\epsilon_0$, we can write this equation in the form

$$\Phi_E = \frac{q}{\epsilon_0} \tag{24.5}$$

We can verify that this expression for the net flux gives the same result as Example 24.1: $\Phi_E = (1.00 \times 10^{-6} \text{ C})/(8.85 \times 10^{-12} \text{ C}^2/\text{N}\cdot\text{m}^2) = 1.13 \times 10^5 \text{ N}\cdot\text{m}^2/\text{C}$.

Note from Equation 24.5 that the net flux through the spherical surface is proportional to the charge inside. The flux is independent of the radius r because the area of the spherical surface is proportional to r^2, whereas the electric field is proportional to $1/r^2$. Thus, in the product of area and electric field, the dependence on r cancels.

Now consider several closed surfaces surrounding a charge q, as shown in Figure 24.7. Surface S_1 is spherical, but surfaces S_2 and S_3 are not. From Equation 24.5, the flux that passes through S_1 has the value q/ϵ_0. As we discussed in the preceding section, flux is proportional to the number of electric field lines passing through a surface. The construction shown in Figure 24.7 shows that the number of lines through S_1 is equal to the number of lines through the nonspherical surfaces S_2 and S_3. Therefore, we conclude that **the net flux through *any* closed surface surrounding a point charge q is given by q/ϵ_0 and is independent of the shape of that surface.**

Now consider a point charge located *outside* a closed surface of arbitrary shape, as shown in Figure 24.8. As you can see from this construction, any electric field line that enters the surface leaves the surface at another point. The number of electric field lines entering the surface equals the number leaving the surface. Therefore, we conclude that **the net electric flux through a closed surface that surrounds no charge is zero.** If we apply this result to Example 24.2, we can easily see that the net flux through the cube is zero because there is no charge inside the cube.

Let us extend these arguments to two generalized cases: (1) that of many point charges and (2) that of a continuous distribution of charge. We once again use the superposition principle, which states that **the electric field due to many charges is the vector sum of the electric fields produced by the individual charges.** Therefore, we can express the flux through any closed surface as

$$\oint \mathbf{E} \cdot d\mathbf{A} = \oint (\mathbf{E}_1 + \mathbf{E}_2 + \cdots) \cdot d\mathbf{A}$$

where $\mathbf{E}$ is the total electric field at any point on the surface produced by the vector addition of the electric fields at that point due to the individual charges. Consider the system of charges shown in Figure 24.9. The surface S surrounds only one charge, q_1; hence, the net flux through S is q_1/ϵ_0. The flux through S due to charges q_2, q_3, and q_4 outside it is zero because each electric field line that enters S at one point leaves it at another. The surface S' surrounds charges q_2 and q_3; hence, the net flux through it is $(q_2 + q_3)/\epsilon_0$. Finally, the net flux through surface S'' is zero because there is no charge inside this surface. That is, *all* the electric field lines that enter S'' at one point leave at another. Notice that charge q_4 does not contribute to the net flux through any of the surfaces because it is outside all of the surfaces.

Gauss's law, which is a generalization of what we have just described, states that the net flux through *any* closed surface is

$$\Phi_E = \oint \mathbf{E} \cdot d\mathbf{A} = \frac{q_{in}}{\epsilon_0} \qquad (24.6)$$

Gauss's law

where q_{in} represents the net charge inside the surface and $\mathbf{E}$ represents the electric field at any point on the surface.

A formal proof of Gauss's law is presented in Section 24.5. When using Equation 24.6, you should note that although the charge q_{in} is the net charge inside the gaussian surface, $\mathbf{E}$ represents the *total electric field*, which includes contributions from charges both inside and outside the surface.

In principle, Gauss's law can be solved for $\mathbf{E}$ to determine the electric field due to a system of charges or a continuous distribution of charge. In practice, however, this type of solution is applicable only in a limited number of highly symmetric situations. In the next section we use Gauss's law to evaluate the electric field for charge distributions that have spherical, cylindrical, or planar symmetry. If one chooses the gaussian surface surrounding the charge distribution carefully, the integral in Equation 24.6 can be simplified.

Quick Quiz 24.3 If the net flux through a gaussian surface is *zero*, the following four statements *could be true*. Which of the statements *must be true*? (a) There are no charges inside the surface. (b) The net charge inside the surface is zero. (c) The electric field is zero everywhere on the surface. (d) The number of electric field lines entering the surface equals the number leaving the surface.

Quick Quiz 24.4 Consider the charge distribution shown in Figure 24.9. The charges contributing to the total electric *flux* through surface S' are (a) q_1 only (b) q_4 only (c) q_2 and q_3 (d) all four charges (e) none of the charges.

Quick Quiz 24.5 Again consider the charge distribution shown in Figure 24.9. The charges contributing to the total electric *field* at a chosen point on the surface S' are (a) q_1 only (b) q_4 only (c) q_2 and q_3 (d) all four charges (e) none of the charges.

 PITFALL PREVENTION

24.1 Zero Flux is not Zero Field

We see two situations in which there is zero flux through a closed surface—either there are no charged particles enclosed by the surface or there are charged particles enclosed, but the net charge inside the surface is zero. For either situation, it is *incorrect* to conclude that the electric field on the surface is zero. Gauss's law states that the electric *flux* is proportional to the enclosed charge, not the electric *field*.

Conceptual Example 24.3   Flux Due to a Point Charge

A spherical gaussian surface surrounds a point charge q. Describe what happens to the total flux through the surface if

(A) the charge is tripled,

(B) the radius of the sphere is doubled,

(C) the surface is changed to a cube, and

(D) the charge is moved to another location inside the surface.

Solution

(A) The flux through the surface is tripled because flux is proportional to the amount of charge inside the surface.

(B) The flux does not change because all electric field lines from the charge pass through the sphere, regardless of its radius.

(C) The flux does not change when the shape of the gaussian surface changes because all electric field lines from the charge pass through the surface, regardless of its shape.

(D) The flux does not change when the charge is moved to another location inside that surface because Gauss's law refers to the total charge enclosed, regardless of where the charge is located inside the surface.

24.3   Application of Gauss's Law to Various Charge Distributions

As mentioned earlier, Gauss's law is useful in determining electric fields when the charge distribution is characterized by a high degree of symmetry. The following examples demonstrate ways of choosing the gaussian surface over which the surface integral given by Equation 24.6 can be simplified and the electric field determined. In choosing the surface, we should always take advantage of the symmetry of the charge distribution so that we can remove E from the integral and solve for it. The goal in this type of calculation is to determine a surface that satisfies one or more of the following conditions:

1. The value of the electric field can be argued by symmetry to be constant over the surface.

2. The dot product in Equation 24.6 can be expressed as a simple algebraic product $E\,dA$ because $\mathbf{E}$ and $d\mathbf{A}$ are parallel.

3. The dot product in Equation 24.6 is zero because $\mathbf{E}$ and $d\mathbf{A}$ are perpendicular.

4. The field can be argued to be zero over the surface.

All four of these conditions are used in examples throughout the remainder of this chapter.

▲ PITFALL PREVENTION

24.2   Gaussian Surfaces are not Real

A gaussian surface is an imaginary surface that you choose to satisfy the conditions listed here. It does not have to coincide with a physical surface in the situation.

Example 24.4   The Electric Field Due to a Point Charge

Starting with Gauss's law, calculate the electric field due to an isolated point charge q.

Solution A single charge represents the simplest possible charge distribution, and we use this familiar case to show how to solve for the electric field with Gauss's law. Figure 24.10 and our discussion of the electric field due to a point charge in Chapter 23 help us to conceptualize the physical situation. Because the space around the single charge has spherical symmetry, we categorize this problem as one in which there is enough symmetry to apply Gauss's law. To analyze any Gauss's law problem, we consider the details of the electric field and choose a gaussian surface that satisfies some or all of the conditions that we have listed above. We choose a spherical gaussian surface of radius r centered on the point charge, as shown in Figure 24.10. The electric field due to a positive point charge is directed radially outward by

symmetry and is therefore normal to the surface at every point. Thus, as in condition (2), $\mathbf{E}$ is parallel to $d\mathbf{A}$ at each point. Therefore, $\mathbf{E} \cdot d\mathbf{A} = E\,dA$ and Gauss's law gives

$$\Phi_E = \oint \mathbf{E} \cdot d\mathbf{A} = \oint E\,dA = \frac{q}{\epsilon_0}$$

By symmetry, E is constant everywhere on the surface, which satisfies condition (1), so it can be removed from the integral. Therefore,

$$\oint E\,dA = E \oint dA = E(4\pi r^2) = \frac{q}{\epsilon_0}$$

where we have used the fact that the surface area of a sphere is $4\pi r^2$. Now, we solve for the electric field:

$$E = \frac{q}{4\pi\epsilon_0 r^2} = k_e \frac{q}{r^2}$$

To finalize this problem, note that this is the familiar electric field due to a point charge that we developed from Coulomb's law in Chapter 23.

What If? What if the charge in Figure 24.10 were not at the center of the spherical gaussian surface?

Answer In this case, while Gauss's law would still be valid, the situation would not possess enough symmetry to evaluate the electric field. Because the charge is not at the center, the magnitude of $\mathbf{E}$ would vary over the surface of the sphere and the vector $\mathbf{E}$ would not be everywhere perpendicular to the surface.

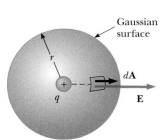

Figure 24.10 (Example 24.4) The point charge q is at the center of the spherical gaussian surface, and $\mathbf{E}$ is parallel to $d\mathbf{A}$ at every point on the surface.

Example 24.5 A Spherically Symmetric Charge Distribution

An insulating solid sphere of radius a has a uniform volume charge density ρ and carries a total positive charge Q (Fig. 24.11).

(A) Calculate the magnitude of the electric field at a point outside the sphere.

Solution Because the charge distribution is spherically symmetric, we again select a spherical gaussian surface of radius r, concentric with the sphere, as shown in Figure 24.11a. For this choice, conditions (1) and (2) are satisfied, as they were for the point charge in Example 24.4. Following the line of reasoning given in Example 24.4, we find that

$$(1) \qquad E = k_e \frac{Q}{r^2} \qquad \text{(for } r > a\text{)}$$

Note that this result is identical to the one we obtained for a point charge. Therefore, we conclude that, **for a uniformly charged sphere, the field in the region external to the sphere is** *equivalent* **to that of a point charge located at the center of the sphere.**

(B) Find the magnitude of the electric field at a point inside the sphere.

Solution In this case we select a spherical gaussian surface having radius $r < a$, concentric with the insulating sphere (Fig. 24.11b). Let us denote the volume of this smaller sphere by V'. To apply Gauss's law in this situation, it is important to recognize that the charge q_{in} within the gaussian surface of volume V' is less than Q. To calculate q_{in}, we use the fact that $q_{in} = \rho V'$:

$$q_{in} = \rho V' = \rho (\tfrac{4}{3}\pi r^3)$$

By symmetry, the magnitude of the electric field is constant everywhere on the spherical gaussian surface and is normal to the surface at each point—both conditions

(1) and (2) are satisfied. Therefore, Gauss's law in the region $r < a$ gives

$$\oint E\, dA = E \oint dA = E\,(4\pi r^2) = \frac{q_{in}}{\epsilon_0}$$

Solving for E gives

$$E = \frac{q_{in}}{4\pi\epsilon_0 r^2} = \frac{\rho(\tfrac{4}{3}\pi r^3)}{4\pi\epsilon_0 r^2} = \frac{\rho}{3\epsilon_0}\, r$$

Because $\rho = Q/\tfrac{4}{3}\pi a^3$ by definition and because $k_e = 1/4\pi\epsilon_0$, this expression for E can be written as

$$(2) \qquad E = \frac{Qr}{4\pi\epsilon_0 a^3} = k_e \frac{Q}{a^3}\, r \qquad \text{(for } r < a\text{)}$$

Note that this result for E differs from the one we obtained in part (A). It shows that $E \to 0$ as $r \to 0$. Therefore, the result eliminates the problem that would exist at $r = 0$ if E varied as $1/r^2$ inside the sphere as it does outside the sphere. That is, if $E \propto 1/r^2$ for $r < a$, the field would be infinite at $r = 0$, which is physically impossible.

What If? Suppose we approach the radial position $r = a$ from inside the sphere and from outside. Do we measure the same value of the electric field from both directions?

Answer From Equation (1), we see that the field approaches a value from the outside given by

$$E = \lim_{r \to a}\left(k_e \frac{Q}{r^2} \right) = k_e \frac{Q}{a^2}$$

From the inside, Equation (2) gives us

$$E = \lim_{r \to a}\left(k_e \frac{Q}{a^3}\, r \right) = k_e \frac{Q}{a^3}\, a = k_e \frac{Q}{a^2}$$

Thus, the value of the field is the same as we approach the surface from both directions. A plot of E versus r is shown in Figure 24.12. Note that the magnitude of the field is continuous, but the derivative of the field magnitude is not.

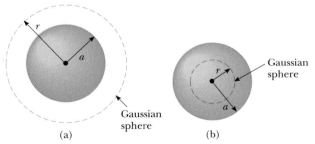

(a) (b)

Figure 24.11 (Example 24.5) A uniformly charged insulating sphere of radius a and total charge Q. (a) For points outside the sphere, a large spherical gaussian surface is drawn concentric with the sphere. In diagrams such as this, the dotted line represents the intersection of the gaussian surface with the plane of the page. (b) For points inside the sphere, a spherical gaussian surface smaller than the sphere is drawn.

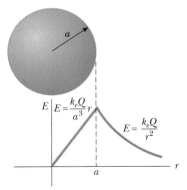

Figure 24.12 (Example 24.5) A plot of E versus r for a uniformly charged insulating sphere. The electric field inside the sphere ($r < a$) varies linearly with r. The field outside the sphere ($r > a$) is the same as that of a point charge Q located at $r = 0$.

At the Interactive Worked Example link at **http://www.pse6.com,** *you can investigate the electric field inside and outside the sphere.*

Example 24.6 The Electric Field Due to a Thin Spherical Shell

A thin spherical shell of radius a has a total charge Q distributed uniformly over its surface (Fig. 24.13a). Find the electric field at points

(A) outside and

(B) inside the shell.

Solution

(A) The calculation for the field outside the shell is identical to that for the solid sphere shown in Example 24.5a. If we construct a spherical gaussian surface of radius $r > a$ concentric with the shell (Fig. 24.13b), the charge inside this surface is Q. Therefore, the field at a point outside the shell is equivalent to that due to a point charge Q located at the center:

$$E = k_e \frac{Q}{r^2} \qquad \text{(for } r > a\text{)}$$

(B) The electric field inside the spherical shell is zero. This follows from Gauss's law applied to a spherical surface of radius $r < a$ concentric with the shell (Fig. 24.13c). Because of the spherical symmetry of the charge distribution and because the net charge inside the surface is zero—satisfaction of conditions (1) and (2) again—application of Gauss's law shows that $E = 0$ in the region $r < a$. We obtain the same results using Equation 23.11 and integrating over the charge distribution. This calculation is rather complicated. Gauss's law allows us to determine these results in a much simpler way.

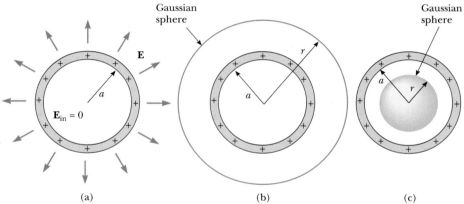

Figure 24.13 (Example 24.6) (a) The electric field inside a uniformly charged spherical shell is zero. The field outside is the same as that due to a point charge Q located at the center of the shell. (b) Gaussian surface for $r > a$. (c) Gaussian surface for $r < a$.

Example 24.7 A Cylindrically Symmetric Charge Distribution

Find the electric field a distance r from a line of positive charge of infinite length and constant charge per unit length λ (Fig. 24.14a).

Solution The symmetry of the charge distribution requires that $\mathbf{E}$ be perpendicular to the line charge and directed outward, as shown in Figure 24.14a and b. To reflect the symmetry of the charge distribution, we select a cylindrical gaussian surface of radius r and length ℓ that is coaxial with the line charge. For the curved part of this surface, $\mathbf{E}$ is constant in magnitude and perpendicular to the surface at each point—satisfaction of conditions (1) and (2). Furthermore, the flux through the ends of the gaussian cylinder is zero because $\mathbf{E}$ is parallel to these surfaces—the first application we have seen of condition (3).

We take the surface integral in Gauss's law over the entire gaussian surface. Because of the zero value of $\mathbf{E} \cdot d\mathbf{A}$ for the ends of the cylinder, however, we can restrict our attention to only the curved surface of the cylinder.

The total charge inside our gaussian surface is $\lambda \ell$. Applying Gauss's law and conditions (1) and (2), we find that for the curved surface

$$\Phi_E = \oint \mathbf{E} \cdot d\mathbf{A} = E \oint dA = EA = \frac{q_{in}}{\epsilon_0} = \frac{\lambda \ell}{\epsilon_0}$$

The area of the curved surface is $A = 2\pi r \ell$; therefore,

$$E(2\pi r \ell) = \frac{\lambda \ell}{\epsilon_0}$$

$$E = \frac{\lambda}{2\pi\epsilon_0 r} = 2k_e \frac{\lambda}{r} \qquad (24.7)$$

Thus, we see that the electric field due to a cylindrically symmetric charge distribution varies as $1/r$, whereas the field external to a spherically symmetric charge distribution varies as $1/r^2$. Equation 24.7 was also derived by integration of the field of a point charge. (See Problem 35 in Chapter 23.)

(a)

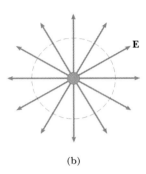

(b)

What If? What if the line segment in this example were not infinitely long?

Answer If the line charge in this example were of finite length, the result for E would not be that given by Equation 24.7. A finite line charge does not possess sufficient symmetry for us to make use of Gauss's law. This is because the magnitude of the electric field is no longer constant over the surface of the gaussian cylinder—the field near the ends of the line would be different from that far from the ends. Thus, condition (1) would not be satisfied in this situation. Furthermore, **E** is not perpendicular to the cylindrical surface at all points—the field vectors near the ends would have a component parallel to the line. Thus, condition (2) would not be satisfied. For points close to a finite line charge and far from the ends, Equation 24.7 gives a good approximation of the value of the field.

It is left for you to show (see Problem 29) that the electric field inside a uniformly charged rod of finite radius and infinite length is proportional to r.

Figure 24.14 (Example 24.7) (a) An infinite line of charge surrounded by a cylindrical gaussian surface concentric with the line. (b) An end view shows that the electric field at the cylindrical surface is constant in magnitude and perpendicular to the surface.

Example 24.8   A Plane of Charge

Find the electric field due to an infinite plane of positive charge with uniform surface charge density σ.

Solution By symmetry, **E** must be perpendicular to the plane and must have the same magnitude at all points equidistant from the plane. The fact that the direction of **E** is away from positive charges indicates that the direction of **E** on one side of the plane must be opposite its direction on the other side, as shown in Figure 24.15. A gaussian surface that reflects the symmetry is a small cylinder whose axis is perpendicular to the plane and whose ends

each have an area A and are equidistant from the plane. Because **E** is parallel to the curved surface—and, therefore, perpendicular to $d\mathbf{A}$ everywhere on the surface—condition (3) is satisfied and there is no contribution to the surface integral from this surface. For the flat ends of the cylinder, conditions (1) and (2) are satisfied. The flux through each end of the cylinder is EA; hence, the total flux through the entire gaussian surface is just that through the ends, $\Phi_E = 2EA$.

Noting that the total charge inside the surface is $q_{in} = \sigma A$, we use Gauss's law and find that the total flux through the gaussian surface is

$$\Phi_E = 2EA = \frac{q_{in}}{\epsilon_0} = \frac{\sigma A}{\epsilon_0}$$

leading to

$$E = \frac{\sigma}{2\epsilon_0} \qquad (24.8)$$

Because the distance from each flat end of the cylinder to the plane does not appear in Equation 24.8, we conclude that $E = \sigma/2\epsilon_0$ at *any* distance from the plane. That is, the field is uniform everywhere.

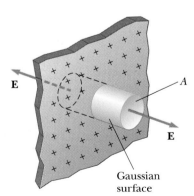

Figure 24.15 (Example 24.8) A cylindrical gaussian surface penetrating an infinite plane of charge. The flux is EA through each end of the gaussian surface and zero through its curved surface.

What If? Suppose we place two infinite planes of charge parallel to each other, one positively charged and the other negatively charged. Both planes have the same surface charge density. What does the electric field look like now?

Answer In this situation, the electric fields due to the two planes add in the region between the planes, resulting in a uniform field of magnitude σ/ϵ_0, and cancel elsewhere to give a field of zero. This is a practical way to achieve uniform electric fields, such as those needed in the CRT tube discussed in Section 23.7.

Conceptual Example 24.9 Don't Use Gauss's Law Here!

Explain why Gauss's law cannot be used to calculate the electric field near an electric dipole, a charged disk, or a triangle with a point charge at each corner.

Solution The charge distributions of all these configurations do not have sufficient symmetry to make the use of Gauss's law practical. We cannot find a closed surface surrounding any of these distributions that satisfies one or more of conditions (1) through (4) listed at the beginning of this section.

24.4 Conductors in Electrostatic Equilibrium

As we learned in Section 23.2, a good electrical conductor contains charges (electrons) that are not bound to any atom and therefore are free to move about within the material. When there is no net motion of charge within a conductor, the conductor is in **electrostatic equilibrium.** A conductor in electrostatic equilibrium has the following properties:

Properties of a conductor in electrostatic equilibrium

1. The electric field is zero everywhere inside the conductor.

2. If an isolated conductor carries a charge, the charge resides on its surface.

3. The electric field just outside a charged conductor is perpendicular to the surface of the conductor and has a magnitude σ/ϵ_0, where σ is the surface charge density at that point.

4. On an irregularly shaped conductor, the surface charge density is greatest at locations where the radius of curvature of the surface is smallest.

We verify the first three properties in the discussion that follows. The fourth property is presented here so that we have a complete list of properties for conductors in electrostatic equilibrium, but cannot be verified until Chapter 25.

We can understand the first property by considering a conducting slab placed in an external field **E** (Fig. 24.16). The electric field inside the conductor *must* be zero under the assumption that we have electrostatic equilibrium. If the field were not zero, free electrons in the conductor would experience an electric force (**F** = q**E**) and would accelerate due to this force. This motion of electrons, however, would mean that the conductor is not in electrostatic equilibrium. Thus, the existence of electrostatic equilibrium is consistent only with a zero field in the conductor.

Figure 24.16 A conducting slab in an external electric field **E**. The charges induced on the two surfaces of the slab produce an electric field that opposes the external field, giving a resultant field of zero inside the slab.

Let us investigate how this zero field is accomplished. Before the external field is applied, free electrons are uniformly distributed throughout the conductor. When the external field is applied, the free electrons accelerate to the left in Figure 24.16, causing a plane of negative charge to be present on the left surface. The movement of electrons to the left results in a plane of positive charge on the right surface. These planes of charge create an additional electric field inside the conductor that opposes the external field. As the electrons move, the surface charge densities on the left and right surfaces increase until the magnitude of the internal field equals that of the external field, resulting in a net field of zero inside the conductor. The time it takes a good conductor to reach equilibrium is on the order of 10^{-16} s, which for most purposes can be considered instantaneous.

We can use Gauss's law to verify the second property of a conductor in electrostatic equilibrium. Figure 24.17 shows an arbitrarily shaped conductor. A gaussian surface is drawn inside the conductor and can be as close to the conductor's surface as we wish. As we have just shown, the electric field everywhere inside the conductor is zero when it is in electrostatic equilibrium. Therefore, the electric field must be zero at every point on the gaussian surface, in accordance with condition (4) in Section 24.3. Thus, the net flux through this gaussian surface is zero. From this result and Gauss's law, we conclude that the net charge inside the gaussian surface is zero. Because there can be no net charge inside the gaussian surface (which is arbitrarily close to the conductor's surface), **any net charge on the conductor must reside on its surface.** Gauss's law does not indicate how this excess charge is distributed on the conductor's surface, only that it resides exclusively on the surface.

We can also use Gauss's law to verify the third property. First, note that if the field vector **E** had a component parallel to the conductor's surface, free electrons would experience an electric force and move along the surface; in such a case, the conductor would not be in equilibrium. Thus, the field vector must be perpendicular to the surface. To determine the magnitude of the electric field, we draw a gaussian surface in the shape of a small cylinder whose end faces are parallel to the surface of the conductor (Fig. 24.18). Part of the cylinder is just outside the conductor, and part is inside. The field is perpendicular to the conductor's surface from the condition of electrostatic equilibrium. Thus, we satisfy condition (3) in Section 24.3 for the curved part of the cylindrical gaussian surface—there is no flux through this part of the gaussian surface because **E** is parallel to the surface. There is no flux through the flat face of the cylinder inside the conductor because here **E** = 0; this satisfies condition (4). Hence, the net flux through the gaussian surface is that through only the flat face outside the conductor, where the field is perpendicular to the gaussian surface. Using conditions (1) and (2) for this face, the flux is *EA*, where *E* is the electric field just outside the conductor and *A* is the area of the cylinder's face. Applying Gauss's law to this surface, we obtain

$$\Phi_E = \oint E \, dA = EA = \frac{q_{in}}{\epsilon_0} = \frac{\sigma A}{\epsilon_0}$$

where we have used the fact that $q_{in} = \sigma A$. Solving for *E* gives for the electric field just outside a charged conductor

$$E = \frac{\sigma}{\epsilon_0} \qquad (24.9)$$

Figure 24.19 shows electric field lines made visible by pieces of thread floating in oil. Notice that the field lines are perpendicular to both the cylindrical conducting surface and the straight conducting surface.

Figure 24.17 A conductor of arbitrary shape. The broken line represents a gaussian surface that can be as close to the surface of the conductor as we wish.

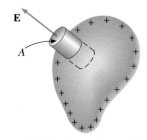

Figure 24.18 A gaussian surface in the shape of a small cylinder is used to calculate the electric field just outside a charged conductor. The flux through the gaussian surface is *EA*. Remember that **E** is zero inside the conductor.

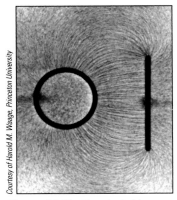

Courtesy of Harold M. Waage, Princeton University

Figure 24.19 Electric field pattern surrounding a charged conducting plate placed near an oppositely charged conducting cylinder. Small pieces of thread suspended in oil align with the electric field lines. Note that (1) the field lines are perpendicular to both conductors and (2) there are no lines inside the cylinder (*E* = 0).

Quick Quiz 24.6 Your little brother likes to rub his feet on the carpet and then touch you to give you a shock. While you are trying to escape the shock treatment, you discover a hollow metal cylinder in your basement, large enough to climb inside. In which of the following cases will you *not* be shocked? (a) You climb inside the cylinder, making contact with the inner surface, and your charged brother touches the outer metal surface. (b) Your charged brother is inside touching the inner metal surface and you are outside, touching the outer metal surface. (c) Both of you are outside the cylinder, touching its outer metal surface but not touching each other directly.

Example 24.10 **A Sphere Inside a Spherical Shell** **Interactive**

A solid conducting sphere of radius a carries a net positive charge $2Q$. A conducting spherical shell of inner radius b and outer radius c is concentric with the solid sphere and carries a net charge $-Q$. Using Gauss's law, find the electric field in the regions labeled ①, ②, ③, and ④ in Figure 24.20 and the charge distribution on the shell when the entire system is in electrostatic equilibrium.

Solution First note that the charge distributions on both the sphere and the shell are characterized by spherical symmetry around their common center. To determine the electric field at various distances r from this center, we construct a spherical gaussian surface for each of the four regions of interest. Such a surface for region ② is shown in Figure 24.20.

To find E inside the solid sphere (region ①), consider a gaussian surface of radius $r < a$. Because there can be no charge inside a conductor in electrostatic equilibrium, we see that $q_{in} = 0$; thus, on the basis of Gauss's law and symmetry, $E_1 = 0$ for $r < a$.

In region ②—between the surface of the solid sphere and the inner surface of the shell—we construct a spherical gaussian surface of radius r where $a < r < b$ and note that the charge inside this surface is $+2Q$ (the charge on the solid sphere). Because of the spherical symmetry, the electric field lines must be directed radially outward and be

constant in magnitude on the gaussian surface. Following Example 24.4 and using Gauss's law, we find that

$$E_2 A = E_2(4\pi r^2) = \frac{q_{in}}{\epsilon_0} = \frac{2Q}{\epsilon_0}$$

$$E_2 = \frac{2Q}{4\pi\epsilon_0 r^2} = \boxed{\frac{2k_e Q}{r^2}} \qquad \text{(for } a < r < b)$$

In region ④, where $r > c$, the spherical gaussian surface we construct surrounds a total charge of $q_{in} = 2Q + (-Q) = Q$. Therefore, application of Gauss's law to this surface gives

$$E_4 = \boxed{\frac{k_e Q}{r^2}} \qquad \text{(for } r > c)$$

In region ③, the electric field must be zero because the spherical shell is also a conductor in equilibrium. Figure 24.21 shows a graphical representation of the variation of electric field with r.

If we construct a gaussian surface of radius r where $b < r < c$, we see that q_{in} must be zero because $E_3 = 0$. From this argument, we conclude that the charge on the inner surface of the spherical shell must be $-2Q$ to cancel the charge $+2Q$ on the solid sphere. Because the net charge on the shell is $-Q$, we conclude that its outer surface must carry a charge $+Q$.

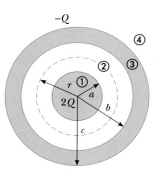

Figure 24.20 (Example 24.10) A solid conducting sphere of radius a and carrying a charge $2Q$ surrounded by a conducting spherical shell carrying a charge $-Q$.

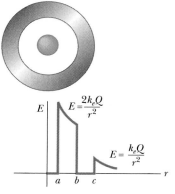

Figure 24.21 (Example 24.10) A plot of E versus r for the two-conductor system shown in Figure 24.20.

 Explore the electric field of the system in Figure 24.20 at the Interactive Worked Example link at **http://www.pse6.com.**

24.5 Formal Derivation of Gauss's Law

One way of deriving Gauss's law involves *solid angles*. Consider a spherical surface of radius r containing an area element ΔA. The solid angle $\Delta \Omega$ (Ω: uppercase Greek omega) subtended at the center of the sphere by this element is defined to be

$$\Delta\Omega \equiv \frac{\Delta A}{r^2}$$

From this equation, we see that $\Delta\Omega$ has no dimensions because ΔA and r^2 both have dimensions L^2. The dimensionless unit of a solid angle is the **steradian.** (You may want to compare this equation to Equation 10.1b, the definition of the radian.) Because the

Figure 24.22 A closed surface of arbitrary shape surrounds a point charge q. The net electric flux through the surface is independent of the shape of the surface.

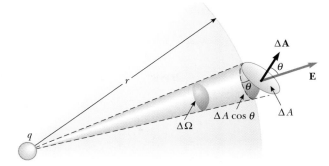

Figure 24.23 The area element ΔA subtends a solid angle $\Delta\Omega = (\Delta A \cos\theta)/r^2$ at the charge q.

surface area of a sphere is $4\pi r^2$, the total solid angle subtended by the sphere is

$$\Omega = \frac{4\pi r^2}{r^2} = 4\pi \text{ steradians}$$

Now consider a point charge q surrounded by a closed surface of arbitrary shape (Fig. 24.22). The total electric flux through this surface can be obtained by evaluating $\mathbf{E}\cdot\Delta\mathbf{A}$ for each small area element ΔA and summing over all elements. The flux through each element is

$$\Delta\Phi_E = \mathbf{E}\cdot\Delta\mathbf{A} = (E\cos\theta)\Delta A = k_e q \frac{\Delta A \cos\theta}{r^2}$$

where r is the distance from the charge to the area element, θ is the angle between the electric field $\mathbf{E}$ and $\Delta\mathbf{A}$ for the element, and $E = k_e q/r^2$ for a point charge. In Figure 24.23, we see that the projection of the area element perpendicular to the radius vector is $\Delta A \cos\theta$. Thus, the quantity $(\Delta A \cos\theta)/r^2$ is equal to the solid angle $\Delta\Omega$ that the surface element ΔA subtends at the charge q. We also see that $\Delta\Omega$ is equal to the solid angle subtended by the area element of a spherical surface of radius r. Because the total solid angle at a point is 4π steradians, the total flux through the closed surface is

$$\Phi_E = k_e q \oint \frac{dA \cos\theta}{r^2} = k_e q \oint d\Omega = 4\pi k_e q = \frac{q}{\epsilon_0}$$

Thus we have derived Gauss's law, Equation 24.6. Note that this result is independent of the shape of the closed surface and independent of the position of the charge within the surface.

SUMMARY

Electric flux is proportional to the number of electric field lines that penetrate a surface. If the electric field is uniform and makes an angle θ with the normal to a surface of area A, the electric flux through the surface is

$$\Phi_E = EA \cos\theta \qquad (24.2)$$

In general, the electric flux through a surface is

$$\Phi_E = \int_{\text{surface}} \mathbf{E}\cdot d\mathbf{A} \qquad (24.3)$$

Take a practice test for this chapter by clicking on the Practice Test link at http://www.pse6.com.

Table 24.1

Typical Electric Field Calculations Using Gauss's Law		
Charge Distribution	**Electric Field**	**Location**
Insulating sphere of radius R, uniform charge density, and total charge Q	$k_e \dfrac{Q}{r^2}$ $k_e \dfrac{Q}{R^2} r$	$r > R$ $r < R$
Thin spherical shell of radius R and total charge Q	$k_e \dfrac{Q}{r^2}$ 0	$r > R$ $r < R$
Line charge of infinite length and charge per unit length λ	$2 k_e \dfrac{\lambda}{r}$	Outside the line
Infinite charged plane having surface charge density σ	$\dfrac{\sigma}{2\epsilon_0}$	Everywhere outside the plane
Conductor having surface charge density σ	$\dfrac{\sigma}{\epsilon_0}$ 0	Just outside the conductor Inside the conductor

You should be able to apply Equations 24.2 and 24.3 in a variety of situations, particularly those in which symmetry simplifies the calculation.

Gauss's law says that the net electric flux Φ_E through any closed gaussian surface is equal to the *net* charge q_{in} inside the surface divided by ϵ_0:

$$\Phi_E = \oint \mathbf{E} \cdot d\mathbf{A} = \frac{q_{in}}{\epsilon_0} \qquad (24.6)$$

Using Gauss's law, you can calculate the electric field due to various symmetric charge distributions. Table 24.1 lists some typical results.

A conductor in **electrostatic equilibrium** has the following properties:

1. The electric field is zero everywhere inside the conductor.
2. Any net charge on the conductor resides entirely on its surface.
3. The electric field just outside the conductor is perpendicular to its surface and has a magnitude σ/ϵ_0, where σ is the surface charge density at that point.
4. On an irregularly shaped conductor, the surface charge density is greatest where the radius of curvature of the surface is the smallest.

QUESTIONS

1. The Sun is lower in the sky during the winter months than it is in the summer. How does this change the flux of sunlight hitting a given area on the surface of the Earth? How does this affect the weather?

2. If the electric field in a region of space is zero, can you conclude that no electric charges are in that region? Explain.

3. If more electric field lines leave a gaussian surface than enter it, what can you conclude about the net charge enclosed by that surface?

4. A uniform electric field exists in a region of space in which there are no charges. What can you conclude about the net electric flux through a gaussian surface placed in this region of space?

5. If the total charge inside a closed surface is known but the distribution of the charge is unspecified, can you use Gauss's law to find the electric field? Explain.

6. Explain why the electric flux through a closed surface with a given enclosed charge is independent of the size or shape of the surface.

7. Consider the electric field due to a nonconducting infinite plane having a uniform charge density. Explain why the electric field does not depend on the distance from the plane, in terms of the spacing of the electric field lines.

8. Use Gauss's law to explain why electric field lines must begin or end on electric charges. (*Suggestion:* Change the size of the gaussian surface.)

9. On the basis of the repulsive nature of the force between like charges and the freedom of motion of charge within a conductor, explain why excess charge on an isolated conductor must reside on its surface.

10. A person is placed in a large hollow metallic sphere that is insulated from ground. If a large charge is placed on the sphere, will the person be harmed upon touching the inside of the sphere? Explain what will happen if the person also has an initial charge whose sign is opposite that of the charge on the sphere.

11. Two solid spheres, both of radius R, carry identical total charges, Q. One sphere is a good conductor while the other is an insulator. If the charge on the insulating sphere is uniformly distributed throughout its interior volume, how do the electric fields outside these two spheres compare? Are the fields identical inside the two spheres?

12. A common demonstration involves charging a rubber balloon, which is an insulator, by rubbing it on your hair, and touching the balloon to a ceiling or wall, which is also an insulator. The electrical attraction between the charged balloon and the neutral wall results in the balloon sticking to the wall. Imagine now that we have two infinitely large flat sheets of insulating material. One is charged and the other is neutral. If these are brought into contact, will an attractive force exist between them, as there was for the balloon and the wall?

13. You may have heard that one of the safer places to be during a lightning storm is inside a car. Why would this be the case?

PROBLEMS

1, 2, 3 = straightforward, intermediate, challenging   ☐ = full solution available in the *Student Solutions Manual and Study Guide*

🌐 = coached solution with hints available at http://www.pse6.com   💻 = computer useful in solving problem

▨ = paired numerical and symbolic problems

Section 24.1 Electric Flux

1. An electric field with a magnitude of 3.50 kN/C is applied along the x axis. Calculate the electric flux through a rectangular plane 0.350 m wide and 0.700 m long assuming that (a) the plane is parallel to the yz plane; (b) the plane is parallel to the xy plane; (c) the plane contains the y axis, and its normal makes an angle of 40.0° with the x axis.

2. A vertical electric field of magnitude 2.00×10^4 N/C exists above the Earth's surface on a day when a thunderstorm is brewing. A car with a rectangular size of 6.00 m by 3.00 m is traveling along a roadway sloping downward at 10.0°. Determine the electric flux through the bottom of the car.

3. A 40.0-cm-diameter loop is rotated in a uniform electric field until the position of maximum electric flux is found. The flux in this position is measured to be 5.20×10^5 N·m²/C. What is the magnitude of the electric field?

4. Consider a closed triangular box resting within a horizontal electric field of magnitude $E = 7.80 \times 10^4$ N/C as shown in Figure P24.4. Calculate the electric flux through (a) the vertical rectangular surface, (b) the slanted surface, and (c) the entire surface of the box.

Figure P24.4

5. A uniform electric field $a\hat{\mathbf{i}} + b\hat{\mathbf{j}}$ intersects a surface of area A. What is the flux through this area if the surface lies (a) in the yz plane? (b) in the xz plane? (c) in the xy plane?

6. A point charge q is located at the center of a uniform ring having linear charge density λ and radius a, as shown in Figure P24.6. Determine the total electric flux through a sphere centered at the point charge and having radius R, where $R < a$.

Figure P24.6

7. A pyramid with horizontal square base, 6.00 m on each side, and a height of 4.00 m is placed in a vertical electric field of 52.0 N/C. Calculate the total electric flux through the pyramid's four slanted surfaces.

8. A cone with base radius R and height h is located on a horizontal table. A horizontal uniform field E penetrates the cone, as shown in Figure P24.8. Determine the electric flux that enters the left-hand side of the cone.

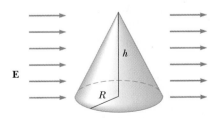

Figure P24.8

Section 24.2 Gauss's Law

9. The following charges are located inside a submarine: 5.00 μC, −9.00 μC, 27.0 μC, and −84.0 μC. (a) Calculate the net electric flux through the hull of the submarine. (b) Is the number of electric field lines leaving the submarine greater than, equal to, or less than the number entering it?

10. The electric field everywhere on the surface of a thin spherical shell of radius 0.750 m is measured to be 890 N/C and points radially toward the center of the sphere. (a) What is the net charge within the sphere's surface? (b) What can you conclude about the nature and distribution of the charge inside the spherical shell?

11. Four closed surfaces, S_1 through S_4, together with the charges $-2Q$, Q, and $-Q$ are sketched in Figure P24.11. (The colored lines are the intersections of the surfaces with the page.) Find the electric flux through each surface.

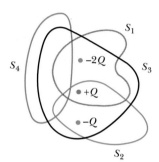

Figure P24.11

12. (a) A point charge q is located a distance d from an infinite plane. Determine the electric flux through the plane due to the point charge. (b) **What If?** A point charge q is located a *very small* distance from the center of a *very large* square on the line perpendicular to the square and going through its center. Determine the approximate electric flux through the square due to the point charge. (c) Explain why the answers to parts (a) and (b) are identical.

13. Calculate the total electric flux through the paraboloidal surface due to a uniform electric field of magnitude E_0 in the direction shown in Figure P24.13.

Figure P24.13

14. A point charge of 12.0 μC is placed at the center of a spherical shell of radius 22.0 cm. What is the total electric flux through (a) the surface of the shell and (b) any hemispherical surface of the shell? (c) Do the results depend on the radius? Explain.

15. A point charge Q is located just above the center of the flat face of a hemisphere of radius R as shown in Figure P24.15. What is the electric flux (a) through the curved surface and (b) through the flat face?

Figure P24.15

16. In the air over a particular region at an altitude of 500 m above the ground the electric field is 120 N/C directed downward. At 600 m above the ground the electric field is 100 N/C downward. What is the average volume charge density in the layer of air between these two elevations? Is it positive or negative?

17. A point charge $Q = 5.00$ μC is located at the center of a cube of edge $L = 0.100$ m. In addition, six other identical point charges having $q = -1.00$ μC are positioned symmetrically around Q as shown in Figure P24.17. Determine the electric flux through one face of the cube.

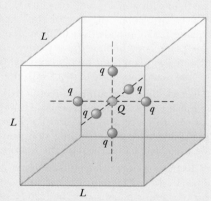

Figure P24.17 Problems 17 and 18.

18. A positive point charge Q is located at the center of a cube of edge L. In addition, six other identical negative point charges q are positioned symmetrically around Q as shown in Figure P24.17. Determine the electric flux through one face of the cube.

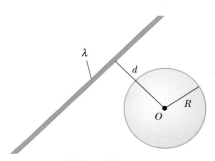

Figure P24.19

19. An infinitely long line charge having a uniform charge per unit length λ lies a distance d from point O as shown in Figure P24.19. Determine the total electric flux through the surface of a sphere of radius R centered at O resulting from this line charge. Consider both cases, where $R < d$ and $R > d$.

20. An uncharged nonconducting hollow sphere of radius 10.0 cm surrounds a 10.0-μC charge located at the origin of a cartesian coordinate system. A drill with a radius of 1.00 mm is aligned along the z axis, and a hole is drilled in the sphere. Calculate the electric flux through the hole.

21. A charge of 170 μC is at the center of a cube of edge 80.0 cm. (a) Find the total flux through each face of the cube. (b) Find the flux through the whole surface of the cube. (c) **What If?** Would your answers to parts (a) or (b) change if the charge were not at the center? Explain.

22. The line ag in Figure P24.22 is a diagonal of a cube. A point charge q is located on the extension of line ag, very close to vertex a of the cube. Determine the electric flux through each of the sides of the cube which meet at the point a.

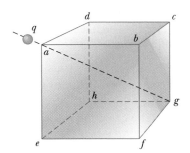

Figure P24.22

Section 24.3   Application of Gauss's Law to Various Charge Distributions

23. Determine the magnitude of the electric field at the surface of a lead-208 nucleus, which contains 82 protons and 126 neutrons. Assume the lead nucleus has a volume 208 times that of one proton, and consider a proton to be a sphere of radius 1.20×10^{-15} m.

24. A solid sphere of radius 40.0 cm has a total positive charge of 26.0 μC uniformly distributed throughout its volume. Calculate the magnitude of the electric field at (a) 0 cm, (b) 10.0 cm, (c) 40.0 cm, and (d) 60.0 cm from the center of the sphere.

25. A 10.0-g piece of Styrofoam carries a net charge of -0.700 μC and floats above the center of a large horizontal sheet of plastic that has a uniform charge density on its surface. What is the charge per unit area on the plastic sheet?

26. A cylindrical shell of radius 7.00 cm and length 240 cm has its charge uniformly distributed on its curved surface. The magnitude of the electric field at a point 19.0 cm radially outward from its axis (measured from the midpoint of the shell) is 36.0 kN/C. Find (a) the net charge on the shell and (b) the electric field at a point 4.00 cm from the axis, measured radially outward from the midpoint of the shell.

27. A particle with a charge of -60.0 nC is placed at the center of a nonconducting spherical shell of inner radius 20.0 cm and outer radius 25.0 cm. The spherical shell carries charge with a uniform density of -1.33 μC/m^3. A proton moves in a circular orbit just outside the spherical shell. Calculate the speed of the proton.

28. A nonconducting wall carries a uniform charge density of 8.60 μC/cm^2. What is the electric field 7.00 cm in front of the wall? Does your result change as the distance from the wall is varied?

29. Consider a long cylindrical charge distribution of radius R with a uniform charge density ρ. Find the electric field at distance r from the axis where $r < R$.

30. A solid plastic sphere of radius 10.0 cm has charge with uniform density throughout its volume. The electric field 5.00 cm from the center is 86.0 kN/C radially inward. Find the magnitude of the electric field 15.0 cm from the center.

31. Consider a thin spherical shell of radius 14.0 cm with a total charge of 32.0 μC distributed uniformly on its surface. Find the electric field (a) 10.0 cm and (b) 20.0 cm from the center of the charge distribution.

32. In nuclear fission, a nucleus of uranium-238, which contains 92 protons, can divide into two smaller spheres, each having 46 protons and a radius of 5.90×10^{-15} m. What is the magnitude of the repulsive electric force pushing the two spheres apart?

33. Fill two rubber balloons with air. Suspend both of them from the same point and let them hang down on strings of equal length. Rub each with wool or on your hair, so that they hang apart with a noticeable separation from each other. Make order-of-magnitude estimates of (a) the force on each, (b) the charge on each, (c) the field each creates at the center of the other, and (d) the total flux of electric field created by each balloon. In your solution state the quantities you take as data and the values you measure or estimate for them.

34. An insulating solid sphere of radius a has a uniform volume charge density and carries a total positive charge Q. A spherical gaussian surface of radius r, which shares a common center with the insulating sphere, is inflated starting from $r = 0$. (a) Find an expression for the electric flux passing through the surface of the gaussian sphere as a function of r for $r < a$. (b) Find an expression for the electric flux for $r > a$. (c) Plot the flux versus r.

35. A uniformly charged, straight filament 7.00 m in length has a total positive charge of 2.00 μC. An uncharged cardboard cylinder 2.00 cm in length and 10.0 cm in radius surrounds the filament at its center, with the filament as the axis of the cylinder. Using reasonable approximations, find (a) the electric field at the surface of the cylinder and (b) the total electric flux through the cylinder.

36. An insulating sphere is 8.00 cm in diameter and carries a 5.70-μC charge uniformly distributed throughout its interior volume. Calculate the charge enclosed by a concentric spherical surface with radius (a) $r = 2.00$ cm and (b) $r = 6.00$ cm.

37. A large flat horizontal sheet of charge has a charge per unit area of 9.00 μC/m^2. Find the electric field just above the middle of the sheet.

38. The charge per unit length on a long, straight filament is $-90.0\ \mu C/m$. Find the electric field (a) 10.0 cm, (b) 20.0 cm, and (c) 100 cm from the filament, where distances are measured perpendicular to the length of the filament.

Section 24.4 Conductors in Electrostatic Equilibrium

39. A long, straight metal rod has a radius of 5.00 cm and a charge per unit length of 30.0 nC/m. Find the electric field (a) 3.00 cm, (b) 10.0 cm, and (c) 100 cm from the axis of the rod, where distances are measured perpendicular to the rod.

40. On a clear, sunny day, a vertical electric field of about 130 N/C points down over flat ground. What is the surface charge density on the ground for these conditions?

41. A very large, thin, flat plate of aluminum of area A has a total charge Q uniformly distributed over its surfaces. Assuming the same charge is spread uniformly over the *upper* surface of an otherwise identical glass plate, compare the electric fields just above the center of the upper surface of each plate.

42. A solid copper sphere of radius 15.0 cm carries a charge of 40.0 nC. Find the electric field (a) 12.0 cm, (b) 17.0 cm, and (c) 75.0 cm from the center of the sphere. (d) **What If?** How would your answers change if the sphere were hollow?

43. A square plate of copper with 50.0-cm sides has no net charge and is placed in a region of uniform electric field of 80.0 kN/C directed perpendicularly to the plate. Find (a) the charge density of each face of the plate and (b) the total charge on each face.

44. A solid conducting sphere of radius 2.00 cm has a charge of 8.00 μC. A conducting spherical shell of inner radius 4.00 cm and outer radius 5.00 cm is concentric with the solid sphere and has a total charge of $-4.00\ \mu C$. Find the electric field at (a) $r = 1.00$ cm, (b) $r = 3.00$ cm, (c) $r = 4.50$ cm, and (d) $r = 7.00$ cm from the center of this charge configuration.

45. Two identical conducting spheres each having a radius of 0.500 cm are connected by a light 2.00-m-long conducting wire. A charge of 60.0 μC is placed on one of the conductors. Assume that the surface distribution of charge on each sphere is uniform. Determine the tension in the wire.

46. The electric field on the surface of an irregularly shaped conductor varies from 56.0 kN/C to 28.0 kN/C. Calculate the local surface charge density at the point on the surface where the radius of curvature of the surface is (a) greatest and (b) smallest.

47. A long, straight wire is surrounded by a hollow metal cylinder whose axis coincides with that of the wire. The wire has a charge per unit length of λ, and the cylinder has a net charge per unit length of 2λ. From this information, use Gauss's law to find (a) the charge per unit length on the inner and outer surfaces of the cylinder and (b) the electric field outside the cylinder, a distance r from the axis.

48. A conducting spherical shell of radius 15.0 cm carries a net charge of $-6.40\ \mu C$ uniformly distributed on its surface. Find the electric field at points (a) just outside the shell and (b) inside the shell.

49. A thin square conducting plate 50.0 cm on a side lies in the xy plane. A total charge of 4.00×10^{-8} C is placed on the plate. Find (a) the charge density on the plate, (b) the electric field just above the plate, and (c) the electric field just below the plate. You may assume that the charge density is uniform.

50. A conducting spherical shell of inner radius a and outer radius b carries a net charge Q. A point charge q is placed at the center of this shell. Determine the surface charge density on (a) the inner surface of the shell and (b) the outer surface of the shell.

51. A hollow conducting sphere is surrounded by a larger concentric spherical conducting shell. The inner sphere has charge $-Q$, and the outer shell has net charge $+3Q$. The charges are in electrostatic equilibrium. Using Gauss's law, find the charges and the electric fields everywhere.

52. A positive point charge is at a distance $R/2$ from the center of an uncharged thin conducting spherical shell of radius R. Sketch the electric field lines set up by this arrangement both inside and outside the shell.

Section 24.5 Formal Derivation of Gauss's Law

53. A sphere of radius R surrounds a point charge Q, located at its center. (a) Show that the electric flux through a circular cap of half-angle θ (Fig. P24.53) is

$$\Phi_E = \frac{Q}{2\epsilon_0}\ (1 - \cos\theta)$$

What is the flux for (b) $\theta = 90°$ and (c) $\theta = 180°$?

Figure P24.53

Additional Problems

54. A nonuniform electric field is given by the expression $\mathbf{E} = ay\hat{\mathbf{i}} + bz\hat{\mathbf{j}} + cx\hat{\mathbf{k}}$, where a, b, and c are constants. Determine the electric flux through a rectangular surface in the xy plane, extending from $x = 0$ to $x = w$ and from $y = 0$ to $y = h$.

55. A solid insulating sphere of radius a carries a net positive charge $3Q$, uniformly distributed throughout its volume. Concentric with this sphere is a conducting spherical shell with inner radius b and outer radius c, and having a net charge $-Q$, as shown in Figure P24.55. (a) Construct a spherical gaussian surface of radius $r > c$ and find the net charge enclosed by this surface. (b) What is the direction of the electric field at $r > c$? (c) Find the electric field at $r > c$. (d) Find the electric field in the region with radius r where $c > r > b$. (e) Construct a spherical gaussian surface of radius r, where $c > r > b$, and find the net charge enclosed by this surface. (f) Construct a spherical gaussian surface of radius r, where $b > r > a$, and find the net charge enclosed by this surface. (g) Find the electric field in the region $b > r > a$. (h) Construct a spherical gaussian surface of radius $r < a$, and find an expression for the net charge enclosed by this surface, as a function of r. Note that the charge inside this surface is less than $3Q$. (i) Find the electric field in the region $r < a$. (j) Determine the charge on the inner surface of the conducting shell. (k) Determine the charge on the outer surface of the conducting shell. (l) Make a plot of the magnitude of the electric field versus r.

Figure P24.55

56. Consider two identical conducting spheres whose surfaces are separated by a small distance. One sphere is given a large net positive charge while the other is given a small net positive charge. It is found that the force between them is attractive even though both spheres have net charges of the same sign. Explain how this is possible.

57. A solid, insulating sphere of radius a has a uniform charge density ρ and a total charge Q. Concentric with this sphere is an uncharged, conducting hollow sphere whose inner and outer radii are b and c, as shown in Figure P24.57. (a) Find the magnitude of the electric field in the regions $r < a$, $a < r < b$, $b < r < c$, and $r > c$. (b) Determine the induced charge per unit area on the inner and outer surfaces of the hollow sphere.

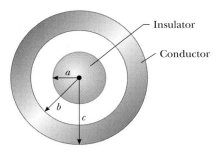

Figure P24.57 Problems 57 and 58.

58. For the configuration shown in Figure P24.57, suppose that $a = 5.00$ cm, $b = 20.0$ cm, and $c = 25.0$ cm. Furthermore, suppose that the electric field at a point 10.0 cm from the center is measured to be 3.60×10^3 N/C radially inward while the electric field at a point 50.0 cm from the center is 2.00×10^2 N/C radially outward. From this information, find (a) the charge on the insulating sphere, (b) the net charge on the hollow conducting sphere, and (c) the charges on the inner and outer surfaces of the hollow conducting sphere.

59. A particle of mass m and charge q moves at high speed along the x axis. It is initially near $x = -\infty$, and it ends up near $x = +\infty$. A second charge Q is fixed at the point $x = 0$, $y = -d$. As the moving charge passes the stationary charge, its x component of velocity does not change appreciably, but it acquires a small velocity in the y direction. Determine the angle through which the moving charge is deflected. *Suggestion:* The integral you encounter in determining v_y can be evaluated by applying Gauss's law to a long cylinder of radius d, centered on the stationary charge.

60. Review problem. An early (incorrect) model of the hydrogen atom, suggested by J. J. Thomson, proposed that a positive cloud of charge $+e$ was uniformly distributed throughout the volume of a sphere of radius R, with the electron an equal-magnitude negative point charge $-e$ at the center. (a) Using Gauss's law, show that the electron would be in equilibrium at the center and, if displaced from the center a distance $r < R$, would experience a restoring force of the form $F = -Kr$, where K is a constant. (b) Show that $K = k_e e^2 / R^3$. (c) Find an expression for the frequency f of simple harmonic oscillations that an electron of mass m_e would undergo if displaced a small distance ($< R$) from the center and released. (d) Calculate a numerical value for R that would result in a frequency of 2.47×10^{15} Hz, the frequency of the light radiated in the most intense line in the hydrogen spectrum.

61. An infinitely long cylindrical insulating shell of inner radius a and outer radius b has a uniform volume charge density ρ. A line of uniform linear charge density λ is placed along the axis of the shell. Determine the electric field everywhere.

62. Two infinite, nonconducting sheets of charge are parallel to each other, as shown in Figure P24.62. The sheet on the left has a uniform surface charge density σ, and the one

Figure P24.62

on the right has a uniform charge density $-\sigma$. Calculate the electric field at points (a) to the left of, (b) in between, and (c) to the right of the two sheets.

63. **What If?** Repeat the calculations for Problem 62 when both sheets have *positive* uniform surface charge densities of value σ.

64. A sphere of radius $2a$ is made of a nonconducting material that has a uniform volume charge density ρ. (Assume that the material does not affect the electric field.) A spherical cavity of radius a is now removed from the sphere, as shown in Figure P24.64. Show that the electric field within the cavity is uniform and is given by $E_x = 0$ and $E_y = \rho a/3\epsilon_0$. (*Suggestion:* The field within the cavity is the superposition of the field due to the original uncut sphere, plus the field due to a sphere the size of the cavity with a uniform negative charge density $-\rho$.)

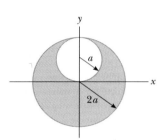

Figure P24.64

65. A uniformly charged spherical shell with surface charge density σ contains a circular hole in its surface. The radius of the hole is small compared with the radius of the sphere. What is the electric field at the center of the hole? (*Suggestion:* This problem, like Problem 64, can be solved by using the idea of superposition.)

66. A closed surface with dimensions $a = b = 0.400$ m and $c = 0.600$ m is located as in Figure P24.66. The left edge of the closed surface is located at position $x = a$. The electric field throughout the region is nonuniform and given by $\mathbf{E} = (3.0 + 2.0x^2)\hat{\mathbf{i}}$ N/C, where x is in meters. Calculate the net electric flux leaving the closed surface. What net charge is enclosed by the surface?

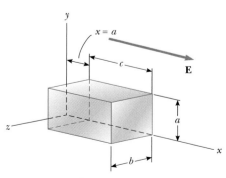

Figure P24.66

67. A solid insulating sphere of radius R has a nonuniform charge density that varies with r according to the expression $\rho = Ar^2$, where A is a constant and $r < R$ is measured from the center of the sphere. (a) Show that the magnitude of the electric field outside ($r > R$) the sphere is $E = AR^5/5\epsilon_0 r^2$. (b) Show that the magnitude of the electric field inside ($r < R$) the sphere is $E = Ar^3/5\epsilon_0$. (*Suggestion:* The total charge Q on the sphere is equal to the integral of $\rho \, dV$, where r extends from 0 to R; also, the charge q within a radius $r < R$ is less than Q. To evaluate the integrals, note that the volume element dV for a spherical shell of radius r and thickness dr is equal to $4\pi r^2 dr$.)

68. A point charge Q is located on the axis of a disk of radius R at a distance b from the plane of the disk (Fig. P24.68). Show that if one fourth of the electric flux from the charge passes through the disk, then $R = \sqrt{3}b$.

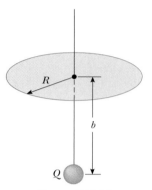

Figure P24.68

69. A spherically symmetric charge distribution has a charge density given by $\rho = a/r$, where a is constant. Find the electric field as a function of r. (*Suggestion:* The charge within a sphere of radius R is equal to the integral of $\rho \, dV$, where r extends from 0 to R. To evaluate the integral, note that the volume element dV for a spherical shell of radius r and thickness dr is equal to $4\pi r^2 dr$.)

70. An infinitely long insulating cylinder of radius R has a volume charge density that varies with the radius as

$$\rho = \rho_0 \left(a - \frac{r}{b} \right)$$

where ρ_0, a, and b are positive constants and r is the distance from the axis of the cylinder. Use Gauss's law to determine the magnitude of the electric field at radial distances (a) $r < R$ and (b) $r > R$.

71. **Review problem.** A slab of insulating material (infinite in two of its three dimensions) has a uniform positive charge density ρ. An edge view of the slab is shown in Figure P24.71. (a) Show that the magnitude of the electric field a distance x from its center and inside the slab is $E = \rho x/\epsilon_0$. (b) **What If?** Suppose an electron of charge $-e$ and mass m_e can move freely within the slab. It is released from rest at a distance x from the center. Show that the electron exhibits simple harmonic motion with a frequency

$$f = \frac{1}{2\pi} \sqrt{\frac{\rho e}{m_e \epsilon_0}}$$

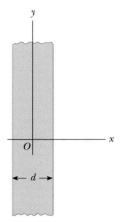

Figure P24.71 Problems 71 and 72.

72. A slab of insulating material has a nonuniform positive charge density $\rho = Cx^2$, where x is measured from the center of the slab as shown in Figure P24.71, and C is a constant. The slab is infinite in the y and z directions. Derive expressions for the electric field in (a) the exterior regions and (b) the interior region of the slab $(-d/2 < x < d/2)$.

73. (a) Using the mathematical similarity between Coulomb's law and Newton's law of universal gravitation, show that Gauss's law for gravitation can be written as

$$\oint \mathbf{g} \cdot d\mathbf{A} = -4\pi G m_{in}$$

where m_{in} is the net mass inside the gaussian surface and $\mathbf{g} = \mathbf{F}_g/m$ represents the gravitational field at any point on the gaussian surface. (b) Determine the gravitational field at a distance r from the center of the Earth where $r < R_E$, assuming that the Earth's mass density is uniform.

Answers to Quick Quizzes

24.1 (e). The same number of field lines pass through a sphere of any size. Because points on the surface of the sphere are closer to the charge, the field is stronger.

24.2 (d). All field lines that enter the container also leave the container so that the total flux is zero, regardless of the nature of the field or the container.

24.3 (b) and (d). Statement (a) is not necessarily true because an equal number of positive and negative charges could be present inside the surface. Statement (c) is not necessarily true, as can be seen from Figure 24.8: a nonzero electric field exists everywhere on the surface, but the charge is not enclosed within the surface; thus, the net flux is zero.

24.4 (c). The charges q_1 and q_4 are outside the surface and contribute zero net flux through S'.

24.5 (d). We don't need the surfaces to realize that any given point in space will experience an electric field due to all local source charges.

24.6 (a). Charges added to the metal cylinder by your brother will reside on the outer surface of the conducting cylinder. If you are on the inside, these charges cannot transfer to you from the inner surface. For this same reason, you are safe in a metal automobile during a lightning storm.

Chapter 25

Electric Potential

▲ Processes occurring during thunderstorms cause large differences in electric potential between a thundercloud and the ground. The result of this potential difference is an electrical discharge that we call lightning, such as this display over Tucson, Arizona. (© Keith Kent/ Photo Researchers, Inc.)

The concept of potential energy was introduced in Chapter 8 in connection with such conservative forces as the gravitational force and the elastic force exerted by a spring. By using the law of conservation of energy, we were able to avoid working directly with forces when solving various problems in mechanics. The concept of potential energy is also of great value in the study of electricity. Because the electrostatic force is conservative, electrostatic phenomena can be conveniently described in terms of an electric potential energy. This idea enables us to define a scalar quantity known as *electric potential*. Because the electric potential at any point in an electric field is a scalar quantity, we can use it to describe electrostatic phenomena more simply than if we were to rely only on the electric field and electric forces. The concept of electric potential is of great practical value in the operation of electric circuits and devices we will study in later chapters.

25.1 Potential Difference and Electric Potential

When a test charge q_0 is placed in an electric field $\mathbf{E}$ created by some source charge distribution, the electric force acting on the test charge is $q_0\mathbf{E}$. The force $q_0\mathbf{E}$ is conservative because the force between charges described by Coulomb's law is conservative. When the test charge is moved in the field by some external agent, the work done by the field on the charge is equal to the negative of the work done by the external agent causing the displacement. This is analogous to the situation of lifting an object with mass in a gravitational field—the work done by the external agent is mgh and the work done by the gravitational force is $-mgh$.

When analyzing electric and magnetic fields, it is common practice to use the notation $d\mathbf{s}$ to represent an infinitesimal displacement vector that is oriented tangent to a path through space. This path may be straight or curved, and an integral performed along this path is called either a *path integral* or a *line integral* (the two terms are synonymous).

For an infinitesimal displacement $d\mathbf{s}$ of a charge, the work done by the electric field on the charge is $\mathbf{F} \cdot d\mathbf{s} = q_0\mathbf{E} \cdot d\mathbf{s}$. As this amount of work is done by the field, the potential energy of the charge–field system is changed by an amount $dU = -q_0\mathbf{E} \cdot d\mathbf{s}$. For a finite displacement of the charge from point A to point B, the change in potential energy of the system $\Delta U = U_B - U_A$ is

$$\Delta U = -q_0 \int_A^B \mathbf{E} \cdot d\mathbf{s} \qquad (25.1)$$

Change in electric potential energy of a system

The integration is performed along the path that q_0 follows as it moves from A to B. Because the force $q_0\mathbf{E}$ is conservative, **this line integral does not depend on the path taken from A to B.**

For a given position of the test charge in the field, the charge–field system has a potential energy U relative to the configuration of the system that is defined as $U = 0$. Dividing the potential energy by the test charge gives a physical quantity that depends only on the source charge distribution. The potential energy per unit charge U/q_0 is independent of the value of q_0 and has a value at every point in an electric field. This quantity U/q_0 is called the **electric potential** (or simply the **potential**) V. Thus, the electric potential at any point in an electric field is

$$V = \frac{U}{q_0} \tag{25.2}$$

The fact that potential energy is a scalar quantity means that electric potential also is a scalar quantity.

As described by Equation 25.1, if the test charge is moved between two positions A and B in an electric field, the charge–field system experiences a change in potential energy. The **potential difference** $\Delta V = V_B - V_A$ between two points A and B in an electric field is defined as the change in potential energy of the system when a test charge is moved between the points divided by the test charge q_0:

$$\Delta V \equiv \frac{\Delta U}{q_0} = -\int_A^B \mathbf{E} \cdot d\mathbf{s} \tag{25.3}$$

Just as with potential energy, only *differences* in electric potential are meaningful. To avoid having to work with potential differences, however, we often take the value of the electric potential to be zero at some convenient point in an electric field.

Potential difference should not be confused with difference in potential energy. The potential difference between A and B depends only on the source charge distribution (consider points A and B *without* the presence of the test charge), while the difference in potential energy exists only if a test charge is moved between the points. **Electric potential is a scalar characteristic of an electric field, independent of any charges that may be placed in the field.**

If an external agent moves a test charge from A to B without changing the kinetic energy of the test charge, the agent performs work which changes the potential energy of the system: $W = \Delta U$. The test charge q_0 is used as a mental device to define the electric potential. Imagine an arbitrary charge q located in an electric field. From Equation 25.3, the work done by an external agent in moving a charge q through an electric field at constant velocity is

$$W = q\,\Delta V \tag{25.4}$$

Because electric potential is a measure of potential energy per unit charge, the SI unit of both electric potential and potential difference is joules per coulomb, which is defined as a **volt** (V):

$$1\,\text{V} \equiv 1\,\frac{\text{J}}{\text{C}}$$

That is, 1 J of work must be done to move a 1-C charge through a potential difference of 1 V.

Equation 25.3 shows that potential difference also has units of electric field times distance. From this, it follows that the SI unit of electric field (N/C) can also be expressed in volts per meter:

$$1\,\frac{\text{N}}{\text{C}} = 1\,\frac{\text{V}}{\text{m}}$$

Therefore, **we can interpret the electric field as a measure of the rate of change with position of the electric potential.**

A unit of energy commonly used in atomic and nuclear physics is the **electron volt** (eV), which is defined as **the energy a charge–field system gains or loses when a charge of magnitude e (that is, an electron or a proton) is moved through a potential difference of 1 V.** Because $1\,\text{V} = 1\,\text{J/C}$ and because the fundamental charge is 1.60×10^{-19} C, the electron volt is related to the joule as follows:

$$1\,\text{eV} = 1.60 \times 10^{-19}\,\text{C} \cdot \text{V} = 1.60 \times 10^{-19}\,\text{J} \tag{25.5}$$

For instance, an electron in the beam of a typical television picture tube may have a speed of 3.0×10^7 m/s. This corresponds to a kinetic energy of 4.1×10^{-16} J, which is equivalent to 2.6×10^3 eV. Such an electron has to be accelerated from rest through a potential difference of 2.6 kV to reach this speed.

Quick Quiz 25.1 In Figure 25.1, two points A and B are located within a region in which there is an electric field. The potential difference $\Delta V = V_B - V_A$ is (a) positive (b) negative (c) zero.

Quick Quiz 25.2 In Figure 25.1, a negative charge is placed at A and then moved to B. The change in potential energy of the charge–field system for this process is (a) positive (b) negative (c) zero.

25.2 Potential Differences in a Uniform Electric Field

Equations 25.1 and 25.3 hold in all electric fields, whether uniform or varying, but they can be simplified for a uniform field. First, consider a uniform electric field directed along the negative y axis, as shown in Figure 25.2a. Let us calculate the potential difference between two points A and B separated by a distance $|\mathbf{s}| = d$, where $\mathbf{s}$ is parallel to the field lines. Equation 25.3 gives

$$V_B - V_A = \Delta V = -\int_A^B \mathbf{E} \cdot d\mathbf{s} = -\int_A^B (E \cos 0°)\, ds = -\int_A^B E\, ds$$

Because E is constant, we can remove it from the integral sign; this gives

$$\Delta V = -E \int_A^B ds = -Ed \tag{25.6}$$

The negative sign indicates that the electric potential at point B is lower than at point A; that is, $V_B < V_A$. **Electric field lines always point in the direction of decreasing electric potential,** as shown in Figure 25.2a.

PITFALL PREVENTION

25.3 The Electron Volt

The electron volt is a unit of *energy*, NOT of potential. The energy of any system may be expressed in eV, but this unit is most convenient for describing the emission and absorption of visible light from atoms. Energies of nuclear processes are often expressed in MeV.

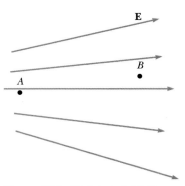

Figure 25.1 (Quick Quiz 25.1) Two points in an electric field.

Potential difference between two points in a uniform electric field

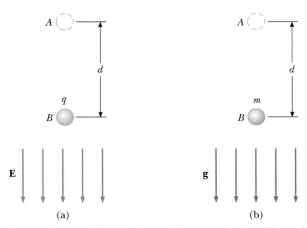

(a)   (b)

Figure 25.2 (a) When the electric field $\mathbf{E}$ is directed downward, point B is at a lower electric potential than point A. When a positive test charge moves from point A to point B, the charge–field system loses electric potential energy. (b) When an object of mass m moves downward in the direction of the gravitational field $\mathbf{g}$, the object–field system loses gravitational potential energy.

Now suppose that a test charge q_0 moves from A to B. We can calculate the change in the potential energy of the charge–field system from Equations 25.3 and 25.6:

$$\Delta U = q_0 \, \Delta V = -q_0 E d \qquad (25.7)$$

From this result, we see that if q_0 is positive, then ΔU is negative. We conclude that **a system consisting of a positive charge and an electric field loses electric potential energy when the charge moves in the direction of the field.** This means that an electric field does work on a positive charge when the charge moves in the direction of the electric field. (This is analogous to the work done by the gravitational field on a falling object, as shown in Figure 25.2b.) If a positive test charge is released from rest in this electric field, it experiences an electric force $q_0\mathbf{E}$ in the direction of $\mathbf{E}$ (downward in Fig. 25.2a). Therefore, it accelerates downward, gaining kinetic energy. **As the charged particle gains kinetic energy, the charge–field system loses an equal amount of potential energy.** This should not be surprising—it is simply conservation of energy in an isolated system as introduced in Chapter 8.

If q_0 is negative, then ΔU in Equation 25.7 is positive and the situation is reversed: **A system consisting of a negative charge and an electric field gains electric potential energy when the charge moves in the direction of the field.** If a negative charge is released from rest in an electric field, it accelerates in a direction opposite the direction of the field. In order for the negative charge to move in the direction of the field, an external agent must apply a force and do positive work on the charge.

Now consider the more general case of a charged particle that moves between A and B in a uniform electric field such that the vector $\mathbf{s}$ is not parallel to the field lines, as shown in Figure 25.3. In this case, Equation 25.3 gives

$$\Delta V = -\int_A^B \mathbf{E} \cdot d\mathbf{s} = -\mathbf{E} \cdot \int_A^B d\mathbf{s} = -\mathbf{E} \cdot \mathbf{s} \qquad (25.8)$$

where again we are able to remove $\mathbf{E}$ from the integral because it is constant. The change in potential energy of the charge–field system is

$$\Delta U = q_0 \, \Delta V = -q_0 \mathbf{E} \cdot \mathbf{s} \qquad (25.9)$$

Finally, we conclude from Equation 25.8 that all points in a plane perpendicular to a uniform electric field are at the same electric potential. We can see this in Figure 25.3, where the potential difference $V_B - V_A$ is equal to the potential difference $V_C - V_A$. (Prove this to yourself by working out the dot product $\mathbf{E} \cdot \mathbf{s}$ for $\mathbf{s}_{A \to B}$, where the angle θ between $\mathbf{E}$ and $\mathbf{s}$ is arbitrary as shown in Figure 25.3, and the dot product for $\mathbf{s}_{A \to C}$, where $\theta = 0$.) Therefore, $V_B = V_C$. **The name equipotential surface is given to any surface consisting of a continuous distribution of points having the same electric potential.**

The equipotential surfaces of a uniform electric field consist of a family of parallel planes that are all perpendicular to the field. Equipotential surfaces for fields with other symmetries are described in later sections.

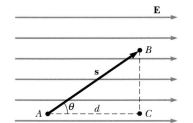

Figure 25.3 A uniform electric field directed along the positive x axis. Point B is at a lower electric potential than point A. Points B and C are at the *same* electric potential.

Change in potential energy when a charged particle is moved in a uniform electric field

Figure 25.4 (Quick Quiz 25.3) Four equipotential surfaces.

Quick Quiz 25.3 The labeled points in Figure 25.4 are on a series of equipotential surfaces associated with an electric field. Rank (from greatest to least) the work done by the electric field on a positively charged particle that moves from A to B; from B to C; from C to D; from D to E.

Quick Quiz 25.4 For the equipotential surfaces in Figure 25.4, what is the *approximate* direction of the electric field? (a) Out of the page (b) Into the page (c) Toward the right edge of the page (d) Toward the left edge of the page (e) Toward the top of the page (f) Toward the bottom of the page.

Example 25.1   The Electric Field Between Two Parallel Plates of Opposite Charge

A battery produces a specified potential difference ΔV between conductors attached to the battery terminals. A 12-V battery is connected between two parallel plates, as shown in Figure 25.5. The separation between the plates is $d = 0.30$ cm, and we assume the electric field between the plates to be

Figure 25.5 (Example 25.1) A 12-V battery connected to two parallel plates. The electric field between the plates has a magnitude given by the potential difference ΔV divided by the plate separation d.

uniform. (This assumption is reasonable if the plate separation is small relative to the plate dimensions and if we do not consider locations near the plate edges.) Find the magnitude of the electric field between the plates.

Solution The electric field is directed from the positive plate (A) to the negative one (B), and the positive plate is at a higher electric potential than the negative plate is. The potential difference between the plates must equal the potential difference between the battery terminals. We can understand this by noting that all points on a conductor in equilibrium are at the same electric potential[1]; no potential difference exists between a terminal and any portion of the plate to which it is connected. Therefore, the magnitude of the electric field between the plates is, from Equation 25.6,

$$E = \frac{|V_B - V_A|}{d} = \frac{12 \text{ V}}{0.30 \times 10^{-2} \text{ m}} = \boxed{4.0 \times 10^3 \text{ V/m}}$$

The configuration of plates in Figure 25.5 is called a *parallel-plate capacitor*, and is examined in greater detail in Chapter 26.

Example 25.2   Motion of a Proton in a Uniform Electric Field

A proton is released from rest in a uniform electric field that has a magnitude of 8.0×10^4 V/m (Fig. 25.6). The proton undergoes a displacement of 0.50 m in the direction of **E**.

(A) Find the change in electric potential between points A and B.

Solution Because the positively charged proton moves in the direction of the field, we expect it to move to a position of lower electric potential. From Equation 25.6, we have

$$\Delta V = -Ed = -(8.0 \times 10^4 \text{ V/m})(0.50 \text{ m}) = \boxed{-4.0 \times 10^4 \text{ V}}$$

(B) Find the change in potential energy of the proton–field system for this displacement.

Solution Using Equation 25.3,

$$\Delta U = q_0 \, \Delta V = e \, \Delta V$$
$$= (1.6 \times 10^{-19} \text{ C})(-4.0 \times 10^4 \text{ V})$$
$$= \boxed{-6.4 \times 10^{-15} \text{ J}}$$

The negative sign means the potential energy of the system decreases as the proton moves in the direction of the electric field. As the proton accelerates in the direction of the field, it gains kinetic energy and at the same time the system loses electric potential energy.

(C) Find the speed of the proton after completing the 0.50 m displacement in the electric field.

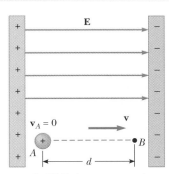

Figure 25.6 (Example 25.2) A proton accelerates from A to B in the direction of the electric field.

Solution The charge–field system is isolated, so the mechanical energy of the system is conserved:

$$\Delta K + \Delta U = 0$$
$$(\tfrac{1}{2}mv^2 - 0) + e \, \Delta V = 0$$
$$v = \sqrt{\frac{-(2e \, \Delta V)}{m}}$$
$$= \sqrt{\frac{-2(1.6 \times 10^{-19} \text{ C})(-4.0 \times 10^4 \text{ V})}{1.67 \times 10^{-27} \text{ kg}}}$$
$$= \boxed{2.8 \times 10^6 \text{ m/s}}$$

What If? **What if the situation is exactly the same as that shown in Figure 25.6, but no proton is present? Could both parts (A) and (B) of this example still be answered?**

---

[1] The electric field vanishes within a conductor in electrostatic equilibrium; thus, the path integral between any two points in the conductor must be zero. A more complete discussion of this point is given in Section 25.6.

Answer Part (A) of the example would remain exactly the same because the potential difference between points A and B is established by the source charges in the parallel plates. The potential difference does not depend on the presence of the proton, which plays the role of a test charge. Part (B) of the example would be meaningless if the proton is not present. A change in potential energy is related to a change in the charge–field system. In the absence of the proton, the system of the electric field alone does not change.

 At the Interactive Worked Example link at http://www.pse6.com, you can predict and observe the speed of the proton as it arrives at the negative plate for random values of the electric field.

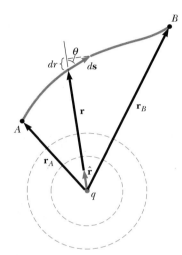

Figure 25.7 The potential difference between points A and B due to a point charge q depends *only* on the initial and final radial coordinates r_A and r_B. The two dashed circles represent intersections of spherical equipotential surfaces with the page.

 PITFALL PREVENTION

25.4 Similar Equation Warning

Do not confuse Equation 25.11 for the electric potential of a point charge with Equation 23.9 for the electric field of a point charge. Potential is proportional to $1/r$, while the field is proportional to $1/r^2$. The effect of a charge on the space surrounding it can be described in two ways. The charge sets up a vector electric field $\mathbf{E}$, which is related to the force experienced by a test charge placed in the field. It also sets up a scalar potential V, which is related to the potential energy of the two-charge system when a test charge is placed in the field.

25.3 Electric Potential and Potential Energy Due to Point Charges

In Section 23.4 we discussed the fact that an isolated positive point charge q produces an electric field that is directed radially outward from the charge. To find the electric potential at a point located a distance r from the charge, we begin with the general expression for potential difference:

$$V_B - V_A = -\int_A^B \mathbf{E} \cdot d\mathbf{s}$$

where A and B are the two arbitrary points shown in Figure 25.7. At any point in space, the electric field due to the point charge is $\mathbf{E} = k_e q \hat{\mathbf{r}}/r^2$ (Eq. 23.9), where $\hat{\mathbf{r}}$ is a unit vector directed from the charge toward the point. The quantity $\mathbf{E} \cdot d\mathbf{s}$ can be expressed as

$$\mathbf{E} \cdot d\mathbf{s} = k_e \frac{q}{r^2} \hat{\mathbf{r}} \cdot d\mathbf{s}$$

Because the magnitude of $\hat{\mathbf{r}}$ is 1, the dot product $\hat{\mathbf{r}} \cdot d\mathbf{s} = ds \cos \theta$, where θ is the angle between $\hat{\mathbf{r}}$ and $d\mathbf{s}$. Furthermore, $ds \cos \theta$ is the projection of $d\mathbf{s}$ onto $\mathbf{r}$; thus, $ds \cos \theta = dr$. That is, any displacement $d\mathbf{s}$ along the path from point A to point B produces a change dr in the magnitude of $\mathbf{r}$, the position vector of the point relative to the charge creating the field. Making these substitutions, we find that $\mathbf{E} \cdot d\mathbf{s} = (k_e q/r^2) dr$; hence, the expression for the potential difference becomes

$$V_B - V_A = -k_e q \int_{r_A}^{r_B} \frac{dr}{r^2} = k_e q \frac{1}{r} \Big]_{r_A}^{r_B}$$

$$V_B - V_A = k_e q \left[\frac{1}{r_B} - \frac{1}{r_A} \right] \qquad (25.10)$$

This equation shows us that the integral of $\mathbf{E} \cdot d\mathbf{s}$ is *independent* of the path between points A and B. Multiplying by a charge q_0 that moves between points A and B, we see that the integral of $q_0 \mathbf{E} \cdot d\mathbf{s}$ is also independent of path. This latter integral is the work done by the electric force, which tells us that the electric force is conservative (see Section 8.3). We define a field that is related to a conservative force as a **conservative field**. Thus, Equation 25.10 tells us that the electric field of a fixed point charge is conservative. Furthermore, Equation 25.10 expresses the important result that the potential difference between any two points A and B in a field created by a point charge depends only on the radial coordinates r_A and r_B. It is customary to choose the reference of electric potential for a point charge to be $V = 0$ at $r_A = \infty$. With this reference choice, the electric potential created by a point charge at any distance r from the charge is

$$V = k_e \frac{q}{r} \qquad (25.11)$$

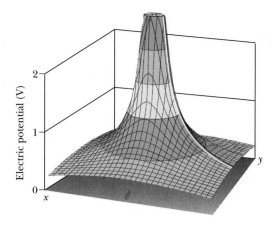

Figure 25.8 The electric potential in the plane around a single positive charge is plotted on the vertical axis. (The electric potential function for a negative charge would look like a hole instead of a hill.) The red line shows the $1/r$ nature of the electric potential, as given by Equation 25.11.

Figure 25.8 shows a plot of the electric potential on the vertical axis for a positive charge located in the xy plane. Consider the following analogy to gravitational potential: imagine trying to roll a marble toward the top of a hill shaped like the surface in Figure 25.8. Pushing the marble up the hill is analogous to pushing one positively charged object toward another positively charged object. Similarly, the electric potential graph of the region surrounding a negative charge is analogous to a "hole" with respect to any approaching positively charged objects. A charged object must be infinitely distant from another charge before the surface in Figure 25.8 is "flat" and has an electric potential of zero.

We obtain the electric potential resulting from two or more point charges by applying the superposition principle. That is, the total electric potential at some point P due to several point charges is the sum of the potentials due to the individual charges. For a group of point charges, we can write the total electric potential at P in the form

$$V = k_e \sum_i \frac{q_i}{r_i} \qquad (25.12)$$

Electric potential due to several point charges

where the potential is again taken to be zero at infinity and r_i is the distance from the point P to the charge q_i. Note that the sum in Equation 25.12 is an algebraic sum of scalars rather than a vector sum (which we use to calculate the electric field of a group of charges). Thus, it is often much easier to evaluate V than to evaluate $\mathbf{E}$. The electric potential around a dipole is illustrated in Figure 25.9. Notice the steep slope of the potential between the charges, representing a region of strong electric field.

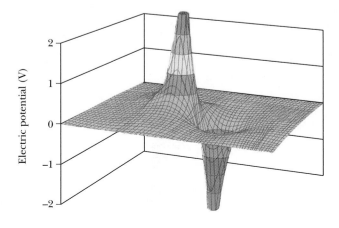

Figure 25.9 The electric potential in the plane containing a dipole.

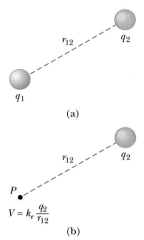

(a)

(b)

$$V = k_e \frac{q_2}{r_{12}}$$

Active Figure 25.10 (a) If two point charges are separated by a distance r_{12}, the potential energy of the pair of charges is given by $k_e q_1 q_2 / r_{12}$. (b) If charge q_1 is removed, a potential $k_e q_2 / r_{12}$ exists at point P due to charge q_2.

 At the Active Figures link at http://www.pse6.com, you can move charge q_1 or point P and see the result on the electric potential energy of the system for part (a) and the electric potential due to charge q_2 for part (b).

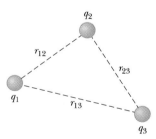

Figure 25.11 Three point charges are fixed at the positions shown. The potential energy of this system of charges is given by Equation 25.14.

⚠ **PITFALL PREVENTION**

25.5 Which Work?

There is a difference between work done *by one member of a system on another member* and work done *on a system by an external agent*. In the present discussion, we are considering the group of charges to be the system and an external agent is doing work on the system to move the charges from an infinite separation to a small separation.

We now consider the potential energy of a system of two charged particles. If V_2 is the electric potential at a point P due to charge q_2, then the work an external agent must do to bring a second charge q_1 from infinity to P without acceleration is $q_1 V_2$. This work represents a transfer of energy into the system and the energy appears in the system as potential energy U when the particles are separated by a distance r_{12} (Fig. 25.10a). Therefore, we can express the potential energy of the system as[2]

$$U = k_e \frac{q_1 q_2}{r_{12}} \tag{25.13}$$

Note that if the charges are of the same sign, U is positive. This is consistent with the fact that positive work must be done by an external agent on the system to bring the two charges near one another (because charges of the same sign repel). If the charges are of opposite sign, U is negative; this means that negative work is done by an external agent against the attractive force between the charges of opposite sign as they are brought near each other—a force must be applied opposite to the displacement to prevent q_1 from accelerating toward q_2.

In Figure 25.10b, we have removed the charge q_1. At the position that this charge previously occupied, point P, we can use Equations 25.2 and 25.13 to define a potential due to charge q_2 as $V = U/q_1 = k_e q_2 / r_{12}$. This expression is consistent with Equation 25.11.

If the system consists of more than two charged particles, we can obtain the total potential energy by calculating U for every pair of charges and summing the terms algebraically. As an example, the total potential energy of the system of three charges shown in Figure 25.11 is

$$U = k_e \left(\frac{q_1 q_2}{r_{12}} + \frac{q_1 q_3}{r_{13}} + \frac{q_2 q_3}{r_{23}} \right) \tag{25.14}$$

Physically, we can interpret this as follows: imagine that q_1 is fixed at the position shown in Figure 25.11 but that q_2 and q_3 are at infinity. The work an external agent must do to bring q_2 from infinity to its position near q_1 is $k_e q_1 q_2 / r_{12}$, which is the first term in Equation 25.14. The last two terms represent the work required to bring q_3 from infinity to its position near q_1 and q_2. (The result is independent of the order in which the charges are transported.)

Quick Quiz 25.5 A spherical balloon contains a positively charged object at its center. As the balloon is inflated to a greater volume while the charged object remains at the center, does the electric potential at the surface of the balloon (a) increase, (b) decrease, or (c) remain the same? Does the electric flux through the surface of the balloon (d) increase, (e) decrease, or (f) remain the same?

Quick Quiz 25.6 In Figure 25.10a, take q_1 to be a negative source charge and q_2 to be the test charge. If q_2 is initially positive and is changed to a charge of the same magnitude but negative, the potential at the position of q_2 due to q_1 (a) increases (b) decreases (c) remains the same.

Quick Quiz 25.7 Consider the situation in Quick Quiz 25.6 again. When q_2 is changed from positive to negative, the potential energy of the two-charge system (a) increases (b) decreases (c) remains the same.

---

[2] The expression for the electric potential energy of a system made up of two point charges, Equation 25.13, is of the *same* form as the equation for the gravitational potential energy of a system made up of two point masses, $-Gm_1 m_2 / r$ (see Chapter 13). The similarity is not surprising in view of the fact that both expressions are derived from an inverse-square force law.

Example 25.3   The Electric Potential Due to Two Point Charges

A charge $q_1 = 2.00 \ \mu C$ is located at the origin, and a charge $q_2 = -6.00 \ \mu C$ is located at $(0, 3.00)$ m, as shown in Figure 25.12a.

(A) Find the total electric potential due to these charges at the point P, whose coordinates are $(4.00, 0)$ m.

Solution For two charges, the sum in Equation 25.12 gives

$$V_P = k_e \left(\frac{q_1}{r_1} + \frac{q_2}{r_2} \right)$$

$$V_P = (8.99 \times 10^9 \ \text{N} \cdot \text{m}^2/\text{C}^2)$$

$$\times \left(\frac{2.00 \times 10^{-6} \ \text{C}}{4.00 \ \text{m}} - \frac{6.00 \times 10^{-6} \ \text{C}}{5.00 \ \text{m}} \right)$$

$$= -6.29 \times 10^3 \ \text{V}$$

(B) Find the change in potential energy of the system of two charges plus a charge $q_3 = 3.00 \ \mu C$ as the latter charge moves from infinity to point P (Fig. 25.12b).

Solution When the charge q_3 is at infinity, let us define $U_i = 0$ for the system, and when the charge is at P, $U_f = q_3 V_P$; therefore,

$$\Delta U = q_3 V_P - 0 = (3.00 \times 10^{-6} \ \text{C})(-6.29 \times 10^3 \ \text{V})$$

$$= -1.89 \times 10^{-2} \ \text{J}$$

Therefore, because the potential energy of the system has decreased, positive work would have to be done by an external agent to remove the charge from point P back to infinity.

What If? You are working through this example with a classmate and she says, "Wait a minute! In part (B), we ignored the potential energy associated with the pair of charges q_1 and q_2!" How would you respond?

Answer Given the statement of the problem, it is not necessary to include this potential energy, because part (B) asks for the *change* in potential energy of the system as q_3 is brought in from infinity. Because the configuration of charges q_1 and q_2 does not change in the process, there is no ΔU associated with these charges. However, if part (B) had asked to find the change in potential energy when *all three* charges start out infinitely far apart and are then brought to the positions in Figure 25.12b, we would need to calculate the change as follows, using Equation 25.14:

$$U = k_e \left(\frac{q_1 q_2}{r_{12}} + \frac{q_1 q_3}{r_{13}} + \frac{q_2 q_3}{r_{23}} \right)$$

$$= (8.99 \times 10^9 \ \text{N} \cdot \text{m}^2/\text{C}^2)$$

$$\times \left(\frac{(2.00 \times 10^{-6} \ \text{C})(-6.00 \times 10^{-6} \ \text{C})}{3.00 \ \text{m}} \right.$$

$$+ \frac{(2.00 \times 10^{-6} \ \text{C})(3.00 \times 10^{-6} \ \text{C})}{4.00 \ \text{m}}$$

$$\left. + \frac{(3.00 \times 10^{-6} \ \text{C})(-6.00 \times 10^{-6} \ \text{C})}{5.00 \ \text{m}} \right)$$

$$= -5.48 \times 10^{-2} \ \text{J}$$

(a)

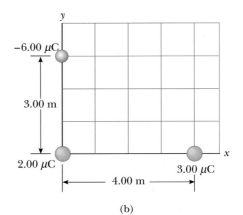

(b)

Figure 25.12 (Example 25.3) (a) The electric potential at P due to the two charges q_1 and q_2 is the algebraic sum of the potentials due to the individual charges. (b) A third charge $q_3 = 3.00 \ \mu C$ is brought from infinity to a position near the other charges.

Explore the value of the electric potential at point P and the electric potential energy of the system in Figure 25.12b at the Interactive Worked Example link at **http://www.pse6.com.**

25.4   Obtaining the Value of the Electric Field from the Electric Potential

The electric field **E** and the electric potential V are related as shown in Equation 25.3. We now show how to calculate the value of the electric field if the electric potential is known in a certain region.

From Equation 25.3 we can express the potential difference dV between two points a distance ds apart as

$$dV = -\mathbf{E} \cdot d\mathbf{s} \tag{25.15}$$

If the electric field has only one component E_x, then $\mathbf{E} \cdot d\mathbf{s} = E_x\, dx$. Therefore, Equation 25.15 becomes $dV = -E_x\, dx$, or

$$E_x = -\frac{dV}{dx} \tag{25.16}$$

That is, the x component of the electric field is equal to the negative of the derivative of the electric potential with respect to x. Similar statements can be made about the y and z components. Equation 25.16 is the mathematical statement of the fact that the electric field is a measure of the rate of change with position of the electric potential, as mentioned in Section 25.1.

Experimentally, electric potential and position can be measured easily with a voltmeter (see Section 28.5) and a meter stick. Consequently, an electric field can be determined by measuring the electric potential at several positions in the field and making a graph of the results. According to Equation 25.16, the slope of a graph of V versus x at a given point provides the magnitude of the electric field at that point.

When a test charge undergoes a displacement $d\mathbf{s}$ along an equipotential surface, then $dV = 0$ because the potential is constant along an equipotential surface. From Equation 25.15, we see that $dV = -\mathbf{E} \cdot d\mathbf{s} = 0$; thus, **E** must be perpendicular to the displacement along the equipotential surface. This shows that the **equipotential surfaces must always be perpendicular to the electric field lines passing through them.**

As mentioned at the end of Section 25.2, the equipotential surfaces for a uniform electric field consist of a family of planes perpendicular to the field lines. Figure 25.13a shows some representative equipotential surfaces for this situation.

(a)

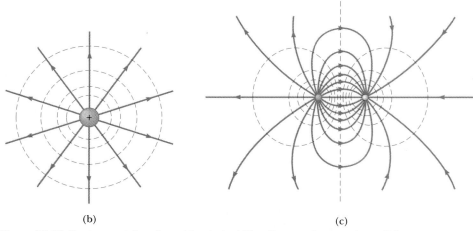

(b)   (c)

Figure 25.13 Equipotential surfaces (the dashed blue lines are intersections of these surfaces with the page) and electric field lines (red-brown lines) for (a) a uniform electric field produced by an infinite sheet of charge, (b) a point charge, and (c) an electric dipole. In all cases, the equipotential surfaces are *perpendicular* to the electric field lines at every point.

If the charge distribution creating an electric field has spherical symmetry such that the volume charge density depends only on the radial distance r, then the electric field is radial. In this case, $\mathbf{E} \cdot d\mathbf{s} = E_r \, dr$, and we can express dV in the form $dV = -E_r \, dr$. Therefore,

$$E_r = -\frac{dV}{dr} \tag{25.17}$$

For example, the electric potential of a point charge is $V = k_e q/r$. Because V is a function of r only, the potential function has spherical symmetry. Applying Equation 25.17, we find that the electric field due to the point charge is $E_r = k_e q/r^2$, a familiar result. Note that the potential changes only in the radial direction, not in any direction perpendicular to r. Thus, V (like E_r) is a function only of r. Again, this is consistent with the idea that **equipotential surfaces are perpendicular to field lines.** In this case the equipotential surfaces are a family of spheres concentric with the spherically symmetric charge distribution (Fig. 25.13b).

The equipotential surfaces for an electric dipole are sketched in Figure 25.13c.

In general, the electric potential is a function of all three spatial coordinates. If $V(r)$ is given in terms of the Cartesian coordinates, the electric field components E_x, E_y, and E_z can readily be found from $V(x, y, z)$ as the partial derivatives[3]

$$E_x = -\frac{\partial V}{\partial x} \qquad E_y = -\frac{\partial V}{\partial y} \qquad E_z = -\frac{\partial V}{\partial z} \tag{25.18}$$

Finding the electric field from the potential

For example, if $V = 3x^2 y + y^2 + yz$, then

$$\frac{\partial V}{\partial x} = \frac{\partial}{\partial x}(3x^2 y + y^2 + yz) = \frac{\partial}{\partial x}(3x^2 y) = 3y\frac{d}{dx}(x^2) = 6xy$$

Quick Quiz 25.8 In a certain region of space, the electric potential is zero everywhere along the x axis. From this we can conclude that the x component of the electric field in this region is (a) zero (b) in the $+x$ direction (c) in the $-x$ direction.

Quick Quiz 25.9 In a certain region of space, the electric field is zero. From this we can conclude that the electric potential in this region is (a) zero (b) constant (c) positive (d) negative.

Example 25.4   The Electric Potential Due to a Dipole

An electric dipole consists of two charges of equal magnitude and opposite sign separated by a distance $2a$, as shown in Figure 25.14. The dipole is along the x axis and is centered at the origin.

(A) Calculate the electric potential at point P.

Solution For point P in Figure 25.14,

$$V = k_e \sum \frac{q_i}{r_i} = k_e\left(\frac{q}{x-a} - \frac{q}{x+a}\right) = \frac{2k_e qa}{x^2 - a^2}$$

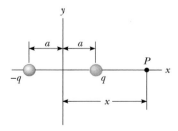

Figure 25.14 (Example 25.4) An electric dipole located on the x axis.

---

[3]   In vector notation, $\mathbf{E}$ is often written in Cartesian coordinate systems as

$$\mathbf{E} = -\nabla V = -\left(\hat{\mathbf{i}}\,\frac{\partial}{\partial x} + \hat{\mathbf{j}}\,\frac{\partial}{\partial y} + \hat{\mathbf{k}}\,\frac{\partial}{\partial z}\right)V$$

where ∇ is called the *gradient operator.*

(B) Calculate V and E_x at a point far from the dipole.

Solution If point P is far from the dipole, such that $x \gg a$, then a^2 can be neglected in the term $x^2 - a^2$ and V becomes

$$V \approx \frac{2k_e qa}{x^2} \qquad (x \gg a)$$

Using Equation 25.16 and this result, we can calculate the magnitude of the electric field at a point far from the dipole:

$$E_x = -\frac{dV}{dx} = \frac{4k_e qa}{x^3} \qquad (x \gg a)$$

(C) Calculate V and E_x if point P is located anywhere between the two charges.

Solution Using Equation 25.12,

$$V = k_e \sum \frac{q_i}{r_i} = k_e \left(\frac{q}{a - x} - \frac{q}{a + x} \right) = \frac{2k_e qx}{a^2 - x^2}$$

and using Equation 25.16,

$$E_x = -\frac{dV}{dx} = -\frac{d}{dx}\left(\frac{2k_e qx}{a^2 - x^2} \right) = -2k_e q \left(\frac{a^2 + x^2}{(a^2 - x^2)^2} \right)$$

We can check these results by considering the situation at the center of the dipole, where $x = 0$, $V = 0$, and $E_x = -2k_e q/a^2$.

What If? What if point P in Figure 25.14 happens to be located to the left of the negative charge? Would the answer to part **(A)** be the same?

Answer The potential should be negative because a point to the left of the dipole is closer to the negative charge than to the positive charge. If we redo the calculation in part (A) with P on the left side of $-q$, we have

$$V = k_e \sum \frac{q_i}{r_i} = k_e \left(\frac{q}{x + a} - \frac{q}{x - a} \right) = -\frac{2k_e qa}{x^2 - a^2}$$

Thus, the potential has the same value but is negative for points on the left of the dipole.

25.5 Electric Potential Due to Continuous Charge Distributions

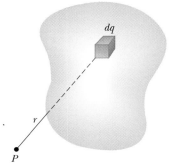

Figure 25.15 The electric potential at the point P due to a continuous charge distribution can be calculated by dividing the charge distribution into elements of charge dq and summing the electric potential contributions over all elements.

Electric potential due to a continuous charge distribution

We can calculate the electric potential due to a continuous charge distribution in two ways. If the charge distribution is known, we can start with Equation 25.11 for the electric potential of a point charge. We then consider the potential due to a small charge element dq, treating this element as a point charge (Fig. 25.15). The electric potential dV at some point P due to the charge element dq is

$$dV = k_e \frac{dq}{r} \qquad (25.19)$$

where r is the distance from the charge element to point P. To obtain the total potential at point P, we integrate Equation 25.19 to include contributions from all elements of the charge distribution. Because each element is, in general, a different distance from point P and because k_e is constant, we can express V as

$$V = k_e \int \frac{dq}{r} \qquad (25.20)$$

In effect, we have replaced the sum in Equation 25.12 with an integral. Note that this expression for V uses a particular reference: the electric potential is taken to be zero when point P is infinitely far from the charge distribution.

If the electric field is already known from other considerations, such as Gauss's law, we can calculate the electric potential due to a continuous charge distribution using Equation 25.3. If the charge distribution has sufficient symmetry, we first evaluate $\mathbf{E}$ at any point using Gauss's law and then substitute the value obtained into Equation 25.3 to determine the potential difference ΔV between any two points. We then choose the electric potential V to be zero at some convenient point.

PROBLEM-SOLVING HINTS

Calculating Electric Potential

- Remember that electric potential is a scalar quantity, so vector components do not exist. Therefore, when using the superposition principle to evaluate the electric potential at a point due to a system of point charges, simply take the algebraic sum of the potentials due to the various charges. However, you must keep track of signs. The potential is positive for positive charges and negative for negative charges.

- Just as with gravitational potential energy in mechanics, only *changes* in electric potential are significant; hence, the point where you choose the potential to be zero is arbitrary. When dealing with point charges or a charge distribution of finite size, we usually define $V = 0$ to be at a point infinitely far from the charges.

- You can evaluate the electric potential at some point P due to a continuous distribution of charge by dividing the charge distribution into infinitesimal elements of charge dq located at a distance r from P. Then, treat one charge element as a point charge, such that the potential at P due to the element is $dV = k_e dq/r$. Obtain the total potential at P by integrating dV over the entire charge distribution. In performing the integration for most problems, you must express dq and r in terms of a single variable. To simplify the integration, consider the geometry involved in the problem carefully. Study Examples 25.5 through 25.7 below for guidance.

- Another method that you can use to obtain the electric potential due to a finite continuous charge distribution is to start with the definition of potential difference given by Equation 25.3. If you know or can easily obtain **E** (from Gauss's law), then you can evaluate the line integral of $\mathbf{E} \cdot d\mathbf{s}$. This method is demonstrated in Example 25.8.

Example 25.5   Electric Potential Due to a Uniformly Charged Ring

(A) Find an expression for the electric potential at a point P located on the perpendicular central axis of a uniformly charged ring of radius a and total charge Q.

Solution Figure 25.16, in which the ring is oriented so that its plane is perpendicular to the x axis and its center is at the origin, helps us to conceptualize this problem. Because the ring consists of a continuous distribution of charge rather

than a set of discrete charges, we categorize this problem as one in which we need to use the integration technique represented by Equation 25.20. To analyze the problem, we take point P to be at a distance x from the center of the ring, as shown in Figure 25.16. The charge element dq is at a distance $\sqrt{x^2 + a^2}$ from point P. Hence, we can express V as

$$V = k_e \int \frac{dq}{r} = k_e \int \frac{dq}{\sqrt{x^2 + a^2}}$$

Because each element dq is at the same distance from point P, we can bring $\sqrt{x^2 + a^2}$ in front of the integral sign, and V reduces to

$$V = \frac{k_e}{\sqrt{x^2 + a^2}} \int dq = \frac{k_e Q}{\sqrt{x^2 + a^2}} \qquad (25.21)$$

The only variable in this expression for V is x. This is not surprising because our calculation is valid only for points along the x axis, where y and z are both zero.

(B) Find an expression for the magnitude of the electric field at point P.

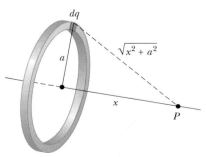

Figure 25.16 (Example 25.5) A uniformly charged ring of radius a lies in a plane perpendicular to the x axis. All elements dq of the ring are the same distance from a point P lying on the x axis.

Solution From symmetry, we see that along the x axis **E** can have only an x component. Therefore, we can use Equation 25.16:

$$E_x = -\frac{dV}{dx} = -k_e Q \frac{d}{dx} (x^2 + a^2)^{-1/2}$$

$$= -k_e Q(-\tfrac{1}{2})(x^2 + a^2)^{-3/2}(2x)$$

$$E_x = \frac{k_e Q x}{(x^2 + a^2)^{3/2}} \qquad (25.22)$$

To finalize this problem, we see that this result for the electric field agrees with that obtained by direct integration (see Example 23.8). Note that $E_x = 0$ at $x = 0$ (the center of the ring). Could you have guessed this?

Example 25.6   Electric Potential Due to a Uniformly Charged Disk

A uniformly charged disk has radius a and surface charge density σ. Find

(A) the electric potential and

(B) the magnitude of the electric field along the perpendicular central axis of the disk.

Solution (A) Again, we choose the point P to be at a distance x from the center of the disk and take the plane of the disk to be perpendicular to the x axis. We can simplify the problem by dividing the disk into a series of charged rings of infinitesimal width dr. The electric potential due to each ring is given by Equation 25.21. Consider one such ring of radius r and width dr, as indicated in Figure 25.17. The surface area of the ring is $dA = 2\pi r\,dr$. From the definition of surface charge density (see Section 23.5), we know that the charge on the ring is $dq = \sigma\,dA = \sigma 2\pi r\,dr$. Hence, the potential at the point P due to this ring is

$$dV = \frac{k_e\,dq}{\sqrt{r^2 + x^2}} = \frac{k_e \sigma 2\pi r\,dr}{\sqrt{r^2 + x^2}}$$

where x is a constant and r is a variable. To find the *total* electric potential at P, we sum over all rings making up the disk. That is, we integrate dV from $r = 0$ to $r = a$:

$$V = \pi k_e \sigma \int_0^a \frac{2r\,dr}{\sqrt{r^2 + x^2}} = \pi k_e \sigma \int_0^a (r^2 + x^2)^{-1/2}\,2r\,dr$$

This integral is of the common form $\int u^n\,du$ and has the value $u^{n+1}/(n+1)$, where $n = -\tfrac{1}{2}$ and $u = r^2 + x^2$. This gives

$$V = 2\pi k_e \sigma \left[(x^2 + a^2)^{1/2} - x\right] \qquad (25.23)$$

(B) As in Example 25.5, we can find the electric field at any axial point using Equation 25.16:

$$E_x = -\frac{dV}{dx} = 2\pi k_e \sigma \left(1 - \frac{x}{\sqrt{x^2 + a^2}}\right) \qquad (25.24)$$

The calculation of V and **E** for an arbitrary point off the axis is more difficult to perform, and we do not treat this situation in this text.

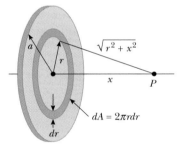

Figure 25.17 (Example 25.6) A uniformly charged disk of radius a lies in a plane perpendicular to the x axis. The calculation of the electric potential at any point P on the x axis is simplified by dividing the disk into many rings of radius r and width dr, with area $2\pi r\,dr$.

Example 25.7   Electric Potential Due to a Finite Line of Charge

A rod of length ℓ located along the x axis has a total charge Q and a uniform linear charge density $\lambda = Q/\ell$. Find the electric potential at a point P located on the y axis a distance a from the origin (Fig. 25.18).

Solution The length element dx has a charge $dq = \lambda\,dx$. Because this element is a distance $r = \sqrt{x^2 + a^2}$ from point P, we can express the potential at point P due to this element as

$$dV = k_e \frac{dq}{r} = k_e \frac{\lambda\,dx}{\sqrt{x^2 + a^2}}$$

To obtain the total potential at P, we integrate this expression over the limits $x = 0$ to $x = \ell$. Noting that k_e and λ are

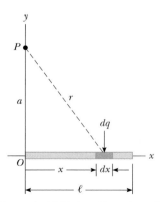

Figure 25.18 (Example 25.7) A uniform line charge of length ℓ located along the x axis. To calculate the electric potential at P, the line charge is divided into segments each of length dx and each carrying a charge $dq = \lambda\,dx$.

constants, we find that

$$V = k_e \lambda \int_0^\ell \frac{dx}{\sqrt{x^2 + a^2}} = k_e \frac{Q}{\ell} \int_0^\ell \frac{dx}{\sqrt{x^2 + a^2}}$$

This integral has the following value (see Appendix B):

$$\int \frac{dx}{\sqrt{x^2 + a^2}} = \ln \left(x + \sqrt{x^2 + a^2} \right)$$

Evaluating V, we find

$$V = \frac{k_e Q}{\ell} \ln \left(\frac{\ell + \sqrt{\ell^2 + a^2}}{a} \right) \qquad (25.25)$$

What If? What if we were asked to find the electric field at point P? Would this be a simple calculation?

Answer Calculating the electric field by means of Equation 23.11 would be a little messy. There is no symmetry to appeal to, and the integration over the line of charge would represent a vector addition of electric fields at point P. Using Equation 25.18, we could find E_y by replacing a with y in Equation 25.25 and performing the differentiation with respect to y. Because the charged rod in Figure 25.18 lies entirely to the right of $x = 0$, the electric field at point P would have an x component to the left if the rod is charged positively. We cannot use Equation 25.18 to find the x component of the field, however, because we evaluated the potential due to the rod at a specific value of x ($x = 0$) rather than a general value of x. We would need to find the potential as a function of both x and y to be able to find the x and y components of the electric field using Equation 25.25.

Example 25.8   Electric Potential Due to a Uniformly Charged Sphere

An insulating solid sphere of radius R has a uniform positive volume charge density and total charge Q.

(A) Find the electric potential at a point outside the sphere, that is, for $r > R$. Take the potential to be zero at $r = \infty$.

Solution In Example 24.5, we found that the magnitude of the electric field outside a uniformly charged sphere of radius R is

$$E_r = k_e \frac{Q}{r^2} \qquad (\text{for } r > R)$$

where the field is directed radially outward when Q is positive. This is the same as the field due to a point charge, which we studied in Section 23.4. In this case, to obtain the electric potential at an exterior point, such as B in Figure 25.19, we use Equation 25.10, choosing point A as $r = \infty$:

$$V_B - V_A = k_e Q \left[\frac{1}{r_B} - \frac{1}{r_A} \right]$$

$$V_B - 0 = k_e Q \left[\frac{1}{r_B} - 0 \right]$$

$$V_B = k_e \frac{Q}{r} \qquad (\text{for } r > R)$$

Because the potential must be continuous at $r = R$, we can use this expression to obtain the potential at the surface of the sphere. That is, the potential at a point such as C shown in Figure 25.19 is

$$V_C = k_e \frac{Q}{R} \qquad (\text{for } r = R)$$

(B) Find the potential at a point inside the sphere, that is, for $r < R$.

Solution In Example 24.5 we found that the electric field inside an insulating uniformly charged sphere is

$$E_r = \frac{k_e Q}{R^3} r \qquad (\text{for } r < R)$$

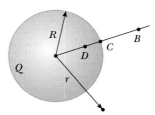

Figure 25.19 (Example 25.8) A uniformly charged insulating sphere of radius R and total charge Q. The electric potentials at points B and C are equivalent to those produced by a point charge Q located at the center of the sphere, but this is not true for point D.

We can use this result and Equation 25.3 to evaluate the potential difference $V_D - V_C$ at some interior point D:

$$V_D - V_C = -\int_R^r E_r \, dr = -\frac{k_e Q}{R^3} \int_R^r r \, dr = \frac{k_e Q}{2R^3} (R^2 - r^2)$$

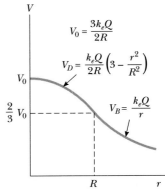

Figure 25.20 (Example 25.8) A plot of electric potential V versus distance r from the center of a uniformly charged insulating sphere of radius R. The curve for V_D inside the sphere is parabolic and joins smoothly with the curve for V_B outside the sphere, which is a hyperbola. The potential has a maximum value V_0 at the center of the sphere. We could make this graph three dimensional (similar to Figures 25.8 and 25.9) by revolving it around the vertical axis.

Substituting $V_C = k_eQ/R$ into this expression and solving for V_D, we obtain

$$V_D = \frac{k_eQ}{2R}\left(3 - \frac{r^2}{R^2}\right) \qquad \text{(for } r < R) \quad \text{(25.26)}$$

At $r = R$, this expression gives a result that agrees with that for the potential at the surface, that is, V_C. A plot of V versus r for this charge distribution is given in Figure 25.20.

25.6 Electric Potential Due to a Charged Conductor

In Section 24.4 we found that when a solid conductor in equilibrium carries a net charge, the charge resides on the outer surface of the conductor. Furthermore, we showed that the electric field just outside the conductor is perpendicular to the surface and that the field inside is zero.

We now show that **every point on the surface of a charged conductor in equilibrium is at the same electric potential.** Consider two points A and B on the surface of a charged conductor, as shown in Figure 25.21. Along a surface path connecting these points, $\mathbf{E}$ is always perpendicular to the displacement $d\mathbf{s}$; therefore $\mathbf{E} \cdot d\mathbf{s} = 0$. Using this result and Equation 25.3, we conclude that the potential difference between A and B is necessarily zero:

$$V_B - V_A = -\int_A^B \mathbf{E} \cdot d\mathbf{s} = 0$$

This result applies to any two points on the surface. Therefore, V is constant everywhere on the surface of a charged conductor in equilibrium. That is,

the surface of any charged conductor in electrostatic equilibrium is an equipotential surface. Furthermore, because the electric field is zero inside the conductor, we conclude that the electric potential is constant everywhere inside the conductor and equal to its value at the surface.

Because this is true, no work is required to move a test charge from the interior of a charged conductor to its surface.

Consider a solid metal conducting sphere of radius R and total positive charge Q, as shown in Figure 25.22a. The electric field outside the sphere is k_eQ/r^2 and points radially outward. From Example 25.8, we know that the electric potential at the interior and surface of the sphere must be k_eQ/R relative to infinity. The potential outside the sphere is k_eQ/r. Figure 25.22b is a plot of the electric potential as a function of r, and Figure 25.22c shows how the electric field varies with r.

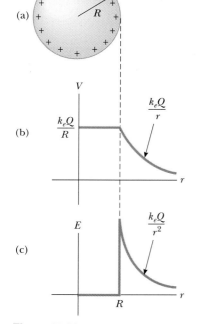

Figure 25.22 (a) The excess charge on a conducting sphere of radius R is uniformly distributed on its surface. (b) Electric potential versus distance r from the center of the charged conducting sphere. (c) Electric field magnitude versus distance r from the center of the charged conducting sphere.

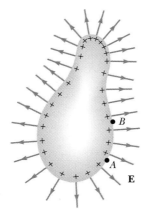

Figure 25.21 An arbitrarily shaped conductor carrying a positive charge. When the conductor is in electrostatic equilibrium, all of the charge resides at the surface, $\mathbf{E} = 0$ inside the conductor, and the direction of $\mathbf{E}$ just outside the conductor is perpendicular to the surface. The electric potential is constant inside the conductor and is equal to the potential at the surface. Note from the spacing of the positive signs that the surface charge density is nonuniform.

When a net charge is placed on a spherical conductor, the surface charge density is uniform, as indicated in Figure 25.22a. However, if the conductor is nonspherical, as in Figure 25.21, the surface charge density is high where the radius of curvature is small (as noted in Section 24.4), and it is low where the radius of curvature is large. Because the electric field just outside the conductor is proportional to the surface charge density, we see that **the electric field is large near convex points having small radii of curvature and reaches very high values at sharp points.** This is demonstrated in Figure 25.23, in which small pieces of thread suspended in oil show the electric field lines. Notice that the density of field lines is highest at the sharp tip of the left-hand conductor and at the highly curved ends of the right-hand conductor. In Example 25.9, the relationship between electric field and radius of curvature is explored mathematically.

Figure 25.24 shows the electric field lines around two spherical conductors: one carrying a net charge Q, and a larger one carrying zero net charge. In this case, the surface charge density is not uniform on either conductor. The sphere having zero net charge has negative charges induced on its side that faces the charged sphere and positive charges induced on its side opposite the charged sphere. The broken blue curves in the figure represent the cross sections of the equipotential surfaces for this charge configuration. As usual, the field lines are perpendicular to the conducting surfaces at all points, and the equipotential surfaces are perpendicular to the field lines everywhere.

> **Quick Quiz 25.10** Consider starting at the center of the left-hand sphere (sphere 1, of radius a) in Figure 25.24 and moving to the far right of the diagram, passing through the center of the right-hand sphere (sphere 2, of radius c) along the way. The centers of the spheres are a distance b apart. Draw a graph of the electric potential as a function of position relative to the center of the left-hand sphere.

 PITFALL PREVENTION

25.6 Potential May Not Be Zero

The electric potential inside the conductor is not necessarily zero in Figure 25.22, even though the electric field is zero. From Equation 25.15, we see that a zero value of the field results in no *change* in the potential from one point to another inside the conductor. Thus, the potential everywhere inside the conductor, including the surface, has the same value, which may or may not be zero, depending on where the zero of potential is defined.

Figure 25.23 Electric field pattern of a charged conducting plate placed near an oppositely charged pointed conductor. Small pieces of thread suspended in oil align with the electric field lines. The field surrounding the pointed conductor is most intense near the pointed end and at other places where the radius of curvature is small.

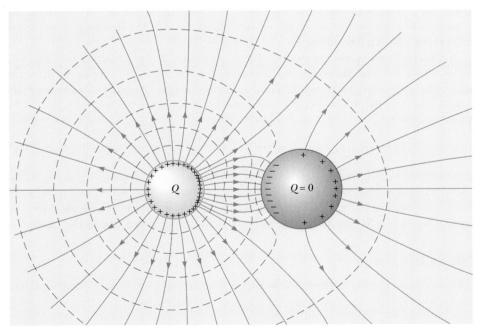

Figure 25.24 The electric field lines (in red-brown) around two spherical conductors. The smaller sphere has a net charge Q, and the larger one has zero net charge. The broken blue curves are intersections of equipotential surfaces with the page.

Example 25.9   Two Connected Charged Spheres

Two spherical conductors of radii r_1 and r_2 are separated by a distance much greater than the radius of either sphere. The spheres are connected by a conducting wire, as shown in Figure 25.25. The charges on the spheres in equilibrium are q_1 and q_2, respectively, and they are uniformly charged. Find the ratio of the magnitudes of the electric fields at the surfaces of the spheres.

Figure 25.25 (Example 25.9) Two charged spherical conductors connected by a conducting wire. The spheres are at the *same* electric potential V.

Solution Because the spheres are connected by a conducting wire, they must both be at the same electric potential:

$$V = k_e \frac{q_1}{r_1} = k_e \frac{q_2}{r_2}$$

Therefore, the ratio of charges is

$$(1) \qquad \frac{q_1}{q_2} = \frac{r_1}{r_2}$$

Because the spheres are very far apart and their surfaces uniformly charged, we can express the magnitude of the electric fields at their surfaces as

$$E_1 = k_e \frac{q_1}{r_1^2} \qquad \text{and} \qquad E_2 = k_e \frac{q_2}{r_2^2}$$

Taking the ratio of these two fields and making use of Equation (1), we find that

$$(2) \qquad \frac{E_1}{E_2} = \frac{r_2}{r_1}$$

Hence, the field is more intense in the vicinity of the smaller sphere even though the electric potentials of both spheres are the same.

A Cavity Within a Conductor

Now suppose a conductor of arbitrary shape contains a cavity as shown in Figure 25.26. Let us assume that no charges are inside the cavity. **In this case, the electric field inside the cavity must be zero** regardless of the charge distribution on the outside surface of the conductor. Furthermore, the field in the cavity is zero even if an electric field exists outside the conductor.

To prove this point, we use the fact that every point on the conductor is at the same electric potential, and therefore any two points A and B on the surface of the cavity must be at the same potential. Now imagine that a field $\mathbf{E}$ exists in the cavity and evaluate the potential difference $V_B - V_A$ defined by Equation 25.3:

$$V_B - V_A = -\int_A^B \mathbf{E} \cdot d\mathbf{s}$$

Figure 25.26 A conductor in electrostatic equilibrium containing a cavity. The electric field in the cavity is zero, regardless of the charge on the conductor.

Because $V_B - V_A = 0$, the integral of $\mathbf{E} \cdot d\mathbf{s}$ must be zero for all paths between any two points A and B on the conductor. The only way that this can be true for *all* paths is if $\mathbf{E}$ is zero *everywhere* in the cavity. Thus, we conclude that **a cavity surrounded by conducting walls is a field-free region as long as no charges are inside the cavity.**

Corona Discharge

A phenomenon known as **corona discharge** is often observed near a conductor such as a high-voltage power line. When the electric field in the vicinity of the conductor is sufficiently strong, electrons resulting from random ionizations of air molecules near the conductor accelerate away from their parent molecules. These rapidly moving electrons can ionize additional molecules near the conductor, creating more free electrons. The observed glow (or corona discharge) results from the recombination of these free electrons with the ionized air molecules. If a conductor has an irregular shape, the electric field can be very high near sharp points or edges of the conductor; consequently, the ionization process and corona discharge are most likely to occur around such points.

Corona discharge is used in the electrical transmission industry to locate broken or faulty components. For example, a broken insulator on a transmission tower has sharp edges where corona discharge is likely to occur. Similarly, corona discharge will occur at the sharp end of a broken conductor strand. Observation of these discharges is difficult because the visible radiation emitted is weak and most of the radiation is in the ultraviolet. (We will discuss ultraviolet radiation and other portions of the electromagnetic spectrum in Section 34.6.) Even use of traditional ultraviolet cameras is of little help because the radiation from the corona discharge is overwhelmed by ultraviolet radiation from the Sun. Newly developed dual-spectrum devices combine a narrowband ultraviolet camera with a visible light camera to show a daylight view of the corona discharge in the actual location on the transmission tower or cable. The ultraviolet part of the camera is designed to operate in a wavelength range in which radiation from the Sun is very weak.

25.7 The Millikan Oil-Drop Experiment

During the period from 1909 to 1913, Robert Millikan performed a brilliant set of experiments in which he measured e, the magnitude of the elementary charge on an electron, and demonstrated the quantized nature of this charge. His apparatus, diagrammed in Figure 25.27, contains two parallel metallic plates. Oil droplets from an atomizer are allowed to pass through a small hole in the upper plate. Millikan used x-rays to ionize the air in the chamber, so that freed electrons would adhere to the oil drops, giving them a negative charge. A horizontally directed light beam is used to illuminate the oil droplets, which are viewed through a telescope whose long axis is perpendicular to the light beam. When the droplets are viewed in this manner, they appear as shining stars against a dark background, and the rate at which individual drops fall can be determined.

Let us assume that a single drop having a mass m and carrying a charge q is being viewed and that its charge is negative. If no electric field is present between the plates,

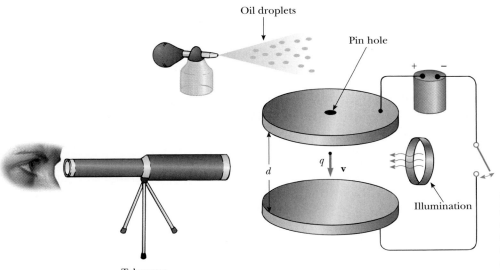

Active Figure 25.27 Schematic drawing of the Millikan oil-drop apparatus.

At the Active Figures link at http://www.pse6.com, you can do a simplified version of the experiment for yourself. You will be able to take data on a number of oil drops and determine the elementary charge from your data.

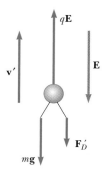

(a) Field off

(b) Field on

Figure 25.28 The forces acting on a negatively charged oil droplet in the Millikan experiment.

the two forces acting on the charge are the gravitational force $m\mathbf{g}$ acting downward[4] and a viscous drag force $\mathbf{F}_D$ acting upward as indicated in Figure 25.28a. The drag force is proportional to the drop's speed. When the drop reaches its terminal speed v, the two forces balance each other ($mg = F_D$).

Now suppose that a battery connected to the plates sets up an electric field between the plates such that the upper plate is at the higher electric potential. In this case, a third force $q\mathbf{E}$ acts on the charged drop. Because q is negative and $\mathbf{E}$ is directed downward, this electric force is directed upward, as shown in Figure 25.28b. If this force is sufficiently great, the drop moves upward and the drag force $\mathbf{F}_D'$ acts downward. When the upward electric force $q\mathbf{E}$ balances the sum of the gravitational force and the downward drag force $\mathbf{F}_D'$, the drop reaches a new terminal speed v' in the upward direction.

With the field turned on, a drop moves slowly upward, typically at rates of hundredths of a centimeter per second. The rate of fall in the absence of a field is comparable. Hence, one can follow a single droplet for hours, alternately rising and falling, by simply turning the electric field on and off.

After recording measurements on thousands of droplets, Millikan and his co-workers found that all droplets, to within about 1% precision, had a charge equal to some integer multiple of the elementary charge e:

$$q = ne \qquad n = 0, -1, -2, -3, \ldots$$

where $e = 1.60 \times 10^{-19}$ C. Millikan's experiment yields conclusive evidence that charge is quantized. For this work, he was awarded the Nobel Prize in Physics in 1923.

25.8 Applications of Electrostatics

The practical application of electrostatics is represented by such devices as lightning rods and electrostatic precipitators and by such processes as xerography and the painting of automobiles. Scientific devices based on the principles of electrostatics include electrostatic generators, the field-ion microscope, and ion-drive rocket engines.

The Van de Graaff Generator

Experimental results show that when a charged conductor is placed in contact with the inside of a hollow conductor, all of the charge on the charged conductor is transferred to the hollow conductor. In principle, the charge on the hollow conductor and its electric potential can be increased without limit by repetition of the process.

In 1929 Robert J. Van de Graaff (1901–1967) used this principle to design and build an electrostatic generator. This type of generator is used extensively in nuclear physics research. A schematic representation of the generator is given in Figure 25.29. Charge is delivered continuously to a high-potential electrode by means of a moving belt of insulating material. The high-voltage electrode is a hollow metal dome mounted on an insulating column. The belt is charged at point A by means of a corona discharge between comb-like metallic needles and a grounded grid. The needles are maintained at a positive electric potential of typically 10^4 V. The positive charge on the moving belt is transferred to the dome by a second comb of needles at point B. Because the electric field inside the dome is negligible, the positive charge on the belt is easily transferred to the conductor regardless of its potential. In practice, it is possible to increase the electric potential of the dome until electrical discharge occurs through the air. Because the "breakdown" electric field in air is about 3×10^6 V/m, a

---

[4]   There is also a buoyant force on the oil drop due to the surrounding air. This force can be incorporated as a correction in the gravitational force $m\mathbf{g}$ on the drop, so we will not consider it in our analysis.

sphere 1 m in radius can be raised to a maximum potential of 3×10^6 V. The potential can be increased further by increasing the radius of the dome and by placing the entire system in a container filled with high-pressure gas.

Van de Graaff generators can produce potential differences as large as 20 million volts. Protons accelerated through such large potential differences receive enough energy to initiate nuclear reactions between themselves and various target nuclei. Smaller generators are often seen in science classrooms and museums. If a person insulated from the ground touches the sphere of a Van de Graaff generator, his or her body can be brought to a high electric potential. The hair acquires a net positive charge, and each strand is repelled by all the others, as in the opening photograph of Chapter 23.

The Electrostatic Precipitator

One important application of electrical discharge in gases is the *electrostatic precipitator*. This device removes particulate matter from combustion gases, thereby reducing air pollution. Precipitators are especially useful in coal-burning power plants and in industrial operations that generate large quantities of smoke. Current systems are able to eliminate more than 99% of the ash from smoke.

Figure 25.30a shows a schematic diagram of an electrostatic precipitator. A high potential difference (typically 40 to 100 kV) is maintained between a wire running down the center of a duct and the walls of the duct, which are grounded. The wire is maintained at a negative electric potential with respect to the walls, so the electric field is directed toward the wire. The values of the field near the wire become high enough to cause a corona discharge around the wire; the air near the wire contains positive ions, electrons, and such negative ions as O_2^-. The air to be cleaned enters the duct and moves near the wire. As the electrons and negative ions created by the discharge are accelerated toward the outer wall by the electric field, the dirt particles in the air become charged by collisions and ion capture. Because most of the charged dirt particles are negative, they too are drawn to the duct walls by the electric field. When the duct is periodically shaken, the particles break loose and are collected at the bottom.

In addition to reducing the level of particulate matter in the atmosphere (compare Figs. 25.30b and c), the electrostatic precipitator recovers valuable materials in the form of metal oxides.

Figure 25.29 Schematic diagram of a Van de Graaff generator. Charge is transferred to the metal dome at the top by means of a moving belt. The charge is deposited on the belt at point *A* and transferred to the hollow conductor at point *B*.

(a)   (b)   (c)

Figure 25.30 (a) Schematic diagram of an electrostatic precipitator. The high negative electric potential maintained on the central coiled wire creates a corona discharge in the vicinity of the wire. Compare the air pollution when the electrostatic precipitator is (b) operating and (c) turned off.

Figure 25.31 The xerographic process: (a) The photoconductive surface of the drum is positively charged. (b) Through the use of a light source and lens, an image is formed on the surface in the form of positive charges. (c) The surface containing the image is covered with a negatively charged powder, which adheres only to the image area. (d) A piece of paper is placed over the surface and given a positive charge. This transfers the image to the paper as the negatively charged powder particles migrate to the paper. The paper is then heat-treated to "fix" the powder. (e) A laser printer operates similarly except the image is produced by turning a laser beam on and off as it sweeps across the selenium-coated drum.

Xerography and Laser Printers

The basic idea of xerography[5] was developed by Chester Carlson, who was granted a patent for the xerographic process in 1940. The unique feature of this process is the use of a photoconductive material to form an image. (A *photoconductor* is a material that is a poor electrical conductor in the dark but becomes a good electrical conductor when exposed to light.)

The xerographic process is illustrated in Figure 25.31a to d. First, the surface of a plate or drum that has been coated with a thin film of photoconductive material (usually selenium or some compound of selenium) is given a positive electrostatic charge in the dark. An image of the page to be copied is then focused by a lens onto the charged surface. The photoconducting surface becomes conducting only in areas where light strikes it. In these areas, the light produces charge carriers in the photoconductor that move the positive charge off the drum. However, positive charges remain on those areas of the photoconductor not exposed to light, leaving a latent image of the object in the form of a positive surface charge distribution.

Next, a negatively charged powder called a *toner* is dusted onto the photoconducting surface. The charged powder adheres only to those areas of the surface that contain the positively charged image. At this point, the image becomes visible. The toner (and hence the image) is then transferred to the surface of a sheet of positively charged paper.

Finally, the toner is "fixed" to the surface of the paper as the toner melts while passing through high-temperature rollers. This results in a permanent copy of the original.

A laser printer (Fig. 25.31e) operates by the same principle, with the exception that a computer-directed laser beam is used to illuminate the photoconductor instead of a lens.

---

[5] The prefix *xero-* is from the Greek word meaning "dry." Note that liquid ink is not used in xerography.

SUMMARY

Take a practice test for this chapter by clicking on the Practice Test link at http://www.pse6.com.

When a positive test charge q_0 is moved between points A and B in an electric field $\mathbf{E}$, the **change in the potential energy of the charge–field system** is

$$\Delta U = -q_0 \int_A^B \mathbf{E} \cdot d\mathbf{s} \qquad (25.1)$$

The **electric potential** $V = U/q_0$ is a scalar quantity and has the units of J/C, where $1\,\text{J/C} \equiv 1\,\text{V}$.

The **potential difference** ΔV between points A and B in an electric field $\mathbf{E}$ is defined as

$$\Delta V \equiv \frac{\Delta U}{q_0} = -\int_A^B \mathbf{E} \cdot d\mathbf{s} \qquad (25.3)$$

The potential difference between two points A and B in a uniform electric field $\mathbf{E}$, where $\mathbf{s}$ is a vector that points from A to B and is parallel to $\mathbf{E}$ is

$$\Delta V = -Ed \qquad (25.6)$$

where $d = |\mathbf{s}|$.

An **equipotential surface** is one on which all points are at the same electric potential. Equipotential surfaces are perpendicular to electric field lines.

If we define $V = 0$ at $r_A = \infty$, the electric potential due to a point charge at any distance r from the charge is

$$V = k_e \frac{q}{r} \qquad (25.11)$$

We can obtain the electric potential associated with a group of point charges by summing the potentials due to the individual charges.

The **potential energy associated with a pair of point charges** separated by a distance r_{12} is

$$U = k_e \frac{q_1 q_2}{r_{12}} \qquad (25.13)$$

This energy represents the work done by an external agent when the charges are brought from an infinite separation to the separation r_{12}. We obtain the potential energy of a distribution of point charges by summing terms like Equation 25.13 over all pairs of particles.

If we know the electric potential as a function of coordinates x, y, z, we can obtain the components of the electric field by taking the negative derivative of the electric potential with respect to the coordinates. For example, the x component of the electric field is

$$E_x = -\frac{dV}{dx} \qquad (25.16)$$

The **electric potential due to a continuous charge distribution** is

$$V = k_e \int \frac{dq}{r} \qquad (25.20)$$

Every point on the surface of a charged conductor in electrostatic equilibrium is at the same electric potential. The potential is constant everywhere inside the conductor and equal to its value at the surface.

Table 25.1 lists electric potentials due to several charge distributions.

Table 25.1

Electric Potential Due to Various Charge Distributions		
Charge Distribution	**Electric Potential**	**Location**
Uniformly charged ring of radius a	$V = k_e \dfrac{Q}{\sqrt{x^2 + a^2}}$	Along perpendicular central axis of ring, distance x from ring center
Uniformly charged disk of radius a	$V = 2\pi k_e \sigma [(x^2 + a^2)^{1/2} - x]$	Along perpendicular central axis of disk, distance x from disk center
Uniformly charged, *insulating* solid sphere of radius R and total charge Q	$V = k_e \dfrac{Q}{r}$ $\qquad r \geq R$ $\\$ $V = \dfrac{k_e Q}{2R}\left(3 - \dfrac{r^2}{R^2}\right) \qquad r < R$	
Isolated *conducting* sphere of radius R and total charge Q	$V = k_e \dfrac{Q}{r}$ $\qquad r > R$ $\\$ $V = k_e \dfrac{Q}{R}$ $\qquad r \leq R$	

QUESTIONS

1. Distinguish between electric potential and electric potential energy.

2. A negative charge moves in the direction of a uniform electric field. Does the potential energy of the charge–field system increase or decrease? Does the charge move to a position of higher or lower potential?

3. Give a physical explanation of the fact that the potential energy of a pair of charges with the same sign is positive whereas the potential energy of a pair of charges with opposite signs is negative.

4. A uniform electric field is parallel to the x axis. In what direction can a charge be displaced in this field without any external work being done on the charge?

5. Explain why equipotential surfaces are always perpendicular to electric field lines.

6. Describe the equipotential surfaces for (a) an infinite line of charge and (b) a uniformly charged sphere.

7. Explain why, under static conditions, all points in a conductor must be at the same electric potential.

8. The electric field inside a hollow, uniformly charged sphere is zero. Does this imply that the potential is zero inside the sphere? Explain.

9. The potential of a point charge is defined to be zero at an infinite distance. Why can we not define the potential of an infinite line of charge to be zero at $r = \infty$?

10. Two charged conducting spheres of different radii are connected by a conducting wire as shown in Figure 25.25. Which sphere has the greater charge density?

11. What determines the maximum potential to which the dome of a Van de Graaff generator can be raised?

12. Explain the origin of the glow sometimes observed around the cables of a high-voltage power line.

13. Why is it important to avoid sharp edges or points on conductors used in high-voltage equipment?

14. How would you shield an electronic circuit or laboratory from stray electric fields? Why does this work?

15. Two concentric spherical conducting shells of radii $a = 0.400$ m and $b = 0.500$ m are connected by a thin wire as shown in Figure Q25.15. If a total charge $Q = 10.0\ \mu$C is placed on the system, how much charge settles on each sphere?

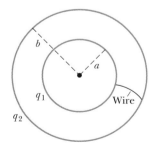

Figure Q25.15

16. Study Figure 23.4 and the accompanying text discussion of charging by induction. You may also compare to Figure

25.24. When the grounding wire is touched to the rightmost point on the sphere in Figure 23.4c, electrons are drained away from the sphere to leave the sphere positively charged. Suppose instead that the grounding wire is touched to the leftmost point on the sphere. Will electrons still drain away, moving closer to the negatively charged rod as they do so? What kind of charge, if any, will remain on the sphere?

PROBLEMS

1, 2, 3 = straightforward, intermediate, challenging ☐ = full solution available in the *Student Solutions Manual and Study Guide*

🪐 = coached solution with hints available at http://www.pse6.com 💻 = computer useful in solving problem

☐ = paired numerical and symbolic problems

Section 25.1 Potential Difference and Electric Potential

1. How much work is done (by a battery, generator, or some other source of potential difference) in moving Avogadro's number of electrons from an initial point where the electric potential is 9.00 V to a point where the potential is − 5.00 V? (The potential in each case is measured relative to a common reference point.)

2. An ion accelerated through a potential difference of 115 V experiences an increase in kinetic energy of 7.37×10^{-17} J. Calculate the charge on the ion.

3. (a) Calculate the speed of a proton that is accelerated from rest through a potential difference of 120 V. (b) Calculate the speed of an electron that is accelerated through the same potential difference.

4. What potential difference is needed to stop an electron having an initial speed of 4.20×10^5 m/s?

Section 25.2 Potential Differences in a Uniform Electric Field

5. A uniform electric field of magnitude 250 V/m is directed in the positive x direction. A $+12.0$-μC charge moves from the origin to the point $(x, y) = (20.0$ cm, 50.0 cm$)$. (a) What is the change in the potential energy of the charge–field system? (b) Through what potential difference does the charge move?

6. The difference in potential between the accelerating plates in the electron gun of a TV picture tube is about 25 000 V. If the distance between these plates is 1.50 cm, what is the magnitude of the uniform electric field in this region?

7. 🪐 An electron moving parallel to the x axis has an initial speed of 3.70×10^6 m/s at the origin. Its speed is reduced to 1.40×10^5 m/s at the point $x = 2.00$ cm. Calculate the potential difference between the origin and that point. Which point is at the higher potential?

8. Suppose an electron is released from rest in a uniform electric field whose magnitude is 5.90×10^3 V/m. (a) Through what potential difference will it have passed after moving 1.00 cm? (b) How fast will the electron be moving after it has traveled 1.00 cm?

9. A uniform electric field of magnitude 325 V/m is directed in the negative y direction in Figure P25.9. The coordinates of point A are $(- 0.200, - 0.300)$ m, and those of point B are $(0.400, 0.500)$ m. Calculate the potential difference $V_B - V_A$, using the blue path.

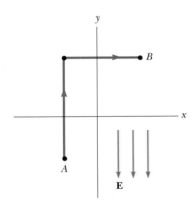

Figure P25.9

10. Starting with the definition of work, prove that at every point on an equipotential surface the surface must be perpendicular to the electric field there.

11. **Review problem.** A block having mass m and charge $+ Q$ is connected to a spring having constant k. The block lies on a frictionless horizontal track, and the system is immersed in a uniform electric field of magnitude E, directed as shown in Figure P25.11. If the block is released from rest when the spring is unstretched (at $x = 0$), (a) by what maximum amount does the spring expand? (b) What is the equilibrium position of the block? (c) Show that the block's motion is simple harmonic, and determine its period. (d) **What If?** Repeat part (a) if the coefficient of kinetic friction between block and surface is μ_k.

Figure P25.11

12. On planet Tehar, the free-fall acceleration is the same as that on Earth but there is also a strong downward electric field that is uniform close to the planet's surface. A 2.00-kg ball having a charge of 5.00 μC is thrown upward at a speed of 20.1 m/s, and it hits the ground after an interval of 4.10 s. What is the potential difference between the starting point and the top point of the trajectory?

13. An insulating rod having linear charge density $\lambda = 40.0$ μC/m and linear mass density $\mu = 0.100$ kg/m is released from rest in a uniform electric field $E = 100$ V/m directed perpendicular to the rod (Fig. P25.13). (a) Determine the speed of the rod after it has traveled 2.00 m. (b) **What If?** How does your answer to part (a) change if the electric field is not perpendicular to the rod? Explain.

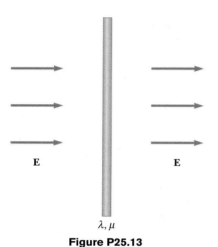

Figure P25.13

14. A particle having charge $q = +2.00$ μC and mass $m = 0.010$ 0 kg is connected to a string that is $L = 1.50$ m long and is tied to the pivot point P in Figure P25.14. The particle, string and pivot point all lie on a frictionless horizontal table. The particle is released from rest when the string makes an angle $\theta = 60.0°$ with a uniform electric field of magnitude $E = 300$ V/m. Determine the speed of the particle when the string is parallel to the electric field (point a in Fig. P25.14).

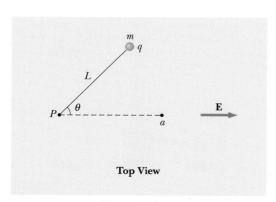

Top View

Figure P25.14

Section 25.3 Electric Potential and Potential Energy Due to Point Charges

Note: Unless stated otherwise, assume the reference level of potential is $V = 0$ at $r = \infty$.

15. (a) Find the potential at a distance of 1.00 cm from a proton. (b) What is the potential difference between two points that are 1.00 cm and 2.00 cm from a proton? (c) **What If?** Repeat parts (a) and (b) for an electron.

16. Given two 2.00-μC charges, as shown in Figure P25.16, and a positive test charge $q = 1.28 \times 10^{-18}$ C at the origin, (a) what is the net force exerted by the two 2.00-μC charges on the test charge q? (b) What is the electric field at the origin due to the two 2.00-μC charges? (c) What is the electric potential at the origin due to the two 2.00-μC charges?

Figure P25.16

17. At a certain distance from a point charge, the magnitude of the electric field is 500 V/m and the electric potential is -3.00 kV. (a) What is the distance to the charge? (b) What is the magnitude of the charge?

18. A charge $+q$ is at the origin. A charge $-2q$ is at $x = 2.00$ m on the x axis. For what finite value(s) of x is (a) the electric field zero? (b) the electric potential zero?

19. The three charges in Figure P25.19 are at the vertices of an isosceles triangle. Calculate the electric potential at the midpoint of the base, taking $q = 7.00$ μC.

Figure P25.19

20. Two point charges, $Q_1 = +5.00$ nC and $Q_2 = -3.00$ nC, are separated by 35.0 cm. (a) What is the potential energy of the pair? What is the significance of the algebraic sign of your answer? (b) What is the electric potential at a point midway between the charges?

21. *Compare this problem with Problem 57 in Chapter 23.* Four identical point charges ($q = +10.0\ \mu$C) are located on the corners of a rectangle as shown in Figure P23.57. The dimensions of the rectangle are $L = 60.0$ cm and $W = 15.0$ cm. Calculate the change in electric potential energy of the system as the charge at the lower left corner in Figure P23.57 is brought to this position from infinitely far away. Assume that the other three charges in Figure P23.57 remain fixed in position.

22. *Compare this problem with Problem 20 in Chapter 23.* Two point charges each of magnitude $2.00\ \mu$C are located on the x axis. One is at $x = 1.00$ m, and the other is at $x = -1.00$ m. (a) Determine the electric potential on the y axis at $y = 0.500$ m. (b) Calculate the change in electric potential energy of the system as a third charge of $-3.00\ \mu$C is brought from infinitely far away to a position on the y axis at $y = 0.500$ m.

23. Show that the amount of work required to assemble four identical point charges of magnitude Q at the corners of a square of side s is $5.41 k_e Q^2/s$.

24. *Compare this problem with Problem 23 in Chapter 23.* Five equal negative point charges $-q$ are placed symmetrically around a circle of radius R. Calculate the electric potential at the center of the circle.

25. *Compare this problem with Problem 41 in Chapter 23.* Three equal positive charges q are at the corners of an equilateral triangle of side a as shown in Figure P23.41. (a) At what point, if any, in the plane of the charges is the electric potential zero? (b) What is the electric potential at the point P due to the two charges at the base of the triangle?

26. **Review problem.** Two insulating spheres have radii 0.300 cm and 0.500 cm, masses 0.100 kg and 0.700 kg, and uniformly distributed charges of $-2.00\ \mu$C and $3.00\ \mu$C. They are released from rest when their centers are separated by 1.00 m. (a) How fast will each be moving when they collide? (*Suggestion:* consider conservation of energy and of linear momentum.) (b) **What If?** If the spheres were conductors, would the speeds be greater or less than those calculated in part (a)? Explain.

27. **Review problem.** Two insulating spheres have radii r_1 and r_2, masses m_1 and m_2, and uniformly distributed charges $-q_1$ and q_2. They are released from rest when their centers are separated by a distance d. (a) How fast is each moving when they collide? (*Suggestion:* consider conservation of energy and conservation of linear momentum.) (b) **What If?** If the spheres were conductors, would their speeds be greater or less than those calculated in part (a)? Explain.

28. Two particles, with charges of 20.0 nC and -20.0 nC, are placed at the points with coordinates (0, 4.00 cm) and (0, -4.00 cm), as shown in Figure P25.28. A particle with charge 10.0 nC is located at the origin. (a) Find the electric potential energy of the configuration of the

three fixed charges. (b) A fourth particle, with a mass of 2.00×10^{-13} kg and a charge of 40.0 nC, is released from rest at the point (3.00 cm, 0). Find its speed after it has moved freely to a very large distance away.

Figure P25.28

29. **Review problem.** A light unstressed spring has length d. Two identical particles, each with charge q, are connected to the opposite ends of the spring. The particles are held stationary a distance d apart and then released at the same time. The system then oscillates on a horizontal frictionless table. The spring has a bit of internal kinetic friction, so the oscillation is damped. The particles eventually stop vibrating when the distance between them is $3d$. Find the increase in internal energy that appears in the spring during the oscillations. Assume that the system of the spring and two charges is isolated.

30. Two point charges of equal magnitude are located along the y axis equal distances above and below the x axis, as shown in Figure P25.30. (a) Plot a graph of the potential at points along the x axis over the interval $-3a < x < 3a$. You should plot the potential in units of $k_e Q/a$. (b) Let the charge located at $-a$ be negative and plot the potential along the y axis over the interval $-4a < y < 4a$.

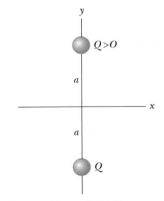

Figure P25.30

31. A small spherical object carries a charge of 8.00 nC. At what distance from the center of the object is the potential equal to 100 V? 50.0 V? 25.0 V? Is the spacing of the equipotentials proportional to the change in potential?

32. In 1911 Ernest Rutherford and his assistants Geiger and Marsden conducted an experiment in which they scattered alpha particles from thin sheets of gold. An alpha particle, having charge $+2e$ and mass 6.64×10^{-27} kg, is a product of certain radioactive decays. The results of the experiment led Rutherford to the idea that most of the mass of an atom is in a very small nucleus, with electrons in orbit around it—his planetary model of the atom. Assume an alpha particle, initially very far from a gold nucleus, is fired with a velocity of 2.00×10^7 m/s directly toward the nucleus (charge $+79e$). How close does the alpha particle get to the nucleus before turning around? Assume the gold nucleus remains stationary.

33. An electron starts from rest 3.00 cm from the center of a uniformly charged insulating sphere of radius 2.00 cm and total charge 1.00 nC. What is the speed of the electron when it reaches the surface of the sphere?

34. Calculate the energy required to assemble the array of charges shown in Figure P25.34, where $a = 0.200$ m, $b = 0.400$ m, and $q = 6.00\ \mu$C.

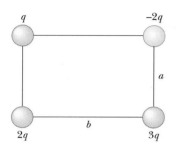

Figure P25.34

35. Four identical particles each have charge q and mass m. They are released from rest at the vertices of a square of side L. How fast is each charge moving when their distance from the center of the square doubles?

36. How much work is required to assemble eight identical point charges, each of magnitude q, at the corners of a cube of side s?

Section 25.4 Obtaining the Value of the Electric Field from the Electric Potential

37. The potential in a region between $x = 0$ and $x = 6.00$ m is $V = a + bx$, where $a = 10.0$ V and $b = -7.00$ V/m. Determine (a) the potential at $x = 0$, 3.00 m, and 6.00 m, and (b) the magnitude and direction of the electric field at $x = 0$, 3.00 m, and 6.00 m.

38. The electric potential inside a charged spherical conductor of radius R is given by $V = k_e Q/R$, and the potential outside is given by $V = k_e Q/r$. Using $E_r = -dV/dr$, derive the electric field (a) inside and (b) outside this charge distribution.

39. Over a certain region of space, the electric potential is $V = 5x - 3x^2 y + 2yz^2$. Find the expressions for the x, y, and z components of the electric field over this region. What is the magnitude of the field at the point P that has coordinates $(1, 0, -2)$ m?

40. Figure P25.40 shows several equipotential lines each labeled by its potential in volts. The distance between the lines of the square grid represents 1.00 cm. (a) Is the magnitude of the field larger at A or at B? Why? (b) What is **E** at B? (c) Represent what the field looks like by drawing at least eight field lines.

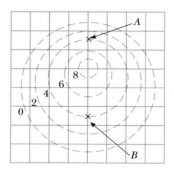

Figure P25.40

41. It is shown in Example 25.7 that the potential at a point P a distance a above one end of a uniformly charged rod of length ℓ lying along the x axis is

$$V = \frac{k_e Q}{\ell} \ln \left(\frac{\ell + \sqrt{\ell^2 + a^2}}{a} \right)$$

Use this result to derive an expression for the y component of the electric field at P. (*Suggestion:* Replace a with y.)

Section 25.5 Electric Potential Due to Continuous Charge Distributions

42. Consider a ring of radius R with the total charge Q spread uniformly over its perimeter. What is the potential difference between the point at the center of the ring and a point on its axis a distance $2R$ from the center?

43. A rod of length L (Fig. P25.43) lies along the x axis with its left end at the origin. It has a nonuniform charge density $\lambda = \alpha x$, where α is a positive constant. (a) What are the units of α? (b) Calculate the electric potential at A.

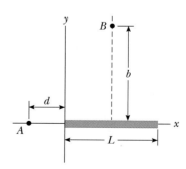

Figure P25.43 Problems 43 and 44.

44. For the arrangement described in the previous problem, calculate the electric potential at point B, which lies on the perpendicular bisector of the rod a distance b above the x axis.

45. *Compare this problem with Problem 33 in Chapter 23.* A uniformly charged insulating rod of length 14.0 cm is bent into the shape of a semicircle as shown in Figure P23.33. The rod has a total charge of $-7.50\ \mu C$. Find the electric potential at O, the center of the semicircle.

46. Calculate the electric potential at point P on the axis of the annulus shown in Figure P25.46, which has a uniform charge density σ.

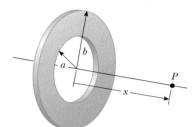

Figure P25.46

47. A wire having a uniform linear charge density λ is bent into the shape shown in Figure P25.47. Find the electric potential at point O.

Figure P25.47

Section 25.6   Electric Potential Due to a Charged Conductor

48. How many electrons should be removed from an initially uncharged spherical conductor of radius 0.300 m to produce a potential of 7.50 kV at the surface?

49. A spherical conductor has a radius of 14.0 cm and charge of 26.0 μC. Calculate the electric field and the electric potential (a) $r = 10.0$ cm, (b) $r = 20.0$ cm, and (c) $r = 14.0$ cm from the center.

50. Electric charge can accumulate on an airplane in flight. You may have observed needle-shaped metal extensions on the wing tips and tail of an airplane. Their purpose is to allow charge to leak off before much of it accumulates. The electric field around the needle is much larger than the field around the body of the airplane, and can become large enough to produce dielectric breakdown of the air, discharging the airplane. To model this process, assume that two charged spherical conductors are connected by a long conducting wire, and a charge of 1.20 μC is placed on the combination. One sphere, representing the body of the airplane, has a radius of 6.00 cm, and the other, representing the tip of the needle, has a radius of 2.00 cm. (a) What is the electric potential of each sphere? (b) What is the electric field at the surface of each sphere?

Section 25.8   Applications of Electrostatics

51. Lightning can be studied with a Van de Graaff generator, essentially consisting of a spherical dome on which charge is continuously deposited by a moving belt. Charge can be added until the electric field at the surface of the dome becomes equal to the dielectric strength of air. Any more charge leaks off in sparks, as shown in Figure P25.51. Assume the dome has a diameter of 30.0 cm and is surrounded by dry air with dielectric strength 3.00×10^6 V/m. (a) What is the maximum potential of the dome? (b) What is the maximum charge on the dome?

Figure P25.51 Problems 51 and 52.

52. The spherical dome of a Van de Graaff generator can be raised to a maximum potential of 600 kV; then additional charge leaks off in sparks, by producing dielectric breakdown of the surrounding dry air, as shown in Figure P25.51. Determine (a) the charge on the dome and (b) the radius of the dome.

Additional Problems

53. The liquid-drop model of the atomic nucleus suggests that high-energy oscillations of certain nuclei can split the nucleus into two unequal fragments plus a few neutrons. The fission products acquire kinetic energy from their mutual Coulomb repulsion. Calculate the electric potential energy (in electron volts) of two spherical fragments from a uranium nucleus having the following charges and radii: $38e$ and 5.50×10^{-15} m; $54e$ and 6.20×10^{-15} m. Assume that the charge is distributed uniformly throughout the volume of each spherical fragment and that just before separating they are at rest with their surfaces in contact. The electrons surrounding the nucleus can be ignored.

54. On a dry winter day you scuff your leather-soled shoes across a carpet and get a shock when you extend the tip of one finger toward a metal doorknob. In a dark room you see a spark perhaps 5 mm long. Make order-of-magnitude estimates of (a) your electric potential and (b) the charge on your body before you touch the doorknob. Explain your reasoning.

55. The Bohr model of the hydrogen atom states that the single electron can exist only in certain allowed orbits around the proton. The radius of each Bohr orbit is $r = n^2(0.052\ 9\ \text{nm})$ where $n = 1, 2, 3, \ldots$. Calculate the electric potential energy of a hydrogen atom when the electron (a) is in the first allowed orbit, with $n = 1$, (b) is in the second allowed orbit, $n = 2$, and (c) has escaped from the atom, with $r = \infty$. Express your answers in electron volts.

56. An electron is released from rest on the axis of a uniform positively charged ring, 0.100 m from the ring's center. If the linear charge density of the ring is $+0.100\ \mu\text{C/m}$ and the radius of the ring is 0.200 m, how fast will the electron be moving when it reaches the center of the ring?

57. As shown in Figure P25.57, two large parallel vertical conducting plates separated by distance d are charged so that their potentials are $+V_0$ and $-V_0$. A small conducting ball of mass m and radius R (where $R \ll d$) is hung midway between the plates. The thread of length L supporting the ball is a conducting wire connected to ground, so the potential of the ball is fixed at $V = 0$. The ball hangs straight down in stable equilibrium when V_0 is sufficiently small. Show that the equilibrium of the ball is unstable if V_0 exceeds the critical value $k_e d^2 mg / (4RL)$. (*Suggestion:* consider the forces on the ball when it is displaced a distance $x \ll L$.)

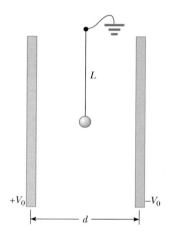

Figure P25.57

58. *Compare this problem with Problem 34 in Chapter 23.* (a) A uniformly charged cylindrical shell has total charge Q, radius R, and height h. Determine the electric potential at a point a distance d from the right end of the cylinder, as shown in Figure P25.58. (*Suggestion:* use the result of Example 25.5 by treating the cylinder as a collection of ring charges.) (b) **What If?** Use the result of Example 25.6 to solve the same problem for a solid cylinder.

Figure P25.58

59. Calculate the work that must be done to charge a spherical shell of radius R to a total charge Q.

60. Two parallel plates having charges of equal magnitude but opposite sign are separated by 12.0 cm. Each plate has a surface charge density of $36.0\ \text{nC/m}^2$. A proton is released from rest at the positive plate. Determine (a) the potential difference between the plates, (b) the kinetic energy of the proton when it reaches the negative plate, (c) the speed of the proton just before it strikes the negative plate, (d) the acceleration of the proton, and (e) the force on the proton. (f) From the force, find the magnitude of the electric field and show that it is equal to the electric field found from the charge densities on the plates.

61. A Geiger tube is a radiation detector that essentially consists of a closed, hollow metal cylinder (the cathode) of inner radius r_a and a coaxial cylindrical wire (the anode) of radius r_b (Fig. P25.61). The charge per unit length on the anode is λ, while the charge per unit length on the cathode is $-\lambda$. A gas fills the space between the electrodes. When a high-energy elementary particle passes through this space, it can ionize an atom of the gas. The strong electric field makes the resulting ion and electron accelerate in opposite directions. They strike other molecules of the gas to ionize them, producing an avalanche of electrical discharge. The pulse of electric current between the wire and the cylinder is counted by an external circuit. (a) Show that the magnitude of the potential difference between the wire and the cylinder is

$$\Delta V = 2k_e \lambda\ \ln\left(\frac{r_a}{r_b}\right)$$

(b) Show that the magnitude of the electric field in the space between cathode and anode is given by

$$E = \frac{\Delta V}{\ln(r_a/r_b)}\left(\frac{1}{r}\right)$$

where r is the distance from the axis of the anode to the point where the field is to be calculated.

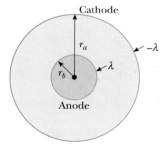

Figure P25.61 Problems 61 and 62.

62. The results of Problem 61 apply also to an electrostatic precipitator (Figures 25.30 and P25.61). An applied voltage $\Delta V = V_a - V_b = 50.0\ \text{kV}$ is to produce an electric field of magnitude 5.50 MV/m at the surface of the central wire. Assume the outer cylindrical wall has uniform radius $r_a = 0.850\ \text{m}$. (a) What should be the radius r_b of the central wire? You will need to solve a transcendental equation. (b) What is the magnitude of the electric field at the outer wall?

63. From Gauss's law, the electric field set up by a uniform line of charge is

$$\mathbf{E} = \left(\frac{\lambda}{2\pi\epsilon_0 r}\right)\hat{\mathbf{r}}$$

where $\hat{\mathbf{r}}$ is a unit vector pointing radially away from the line and λ is the linear charge density along the line. Derive an expression for the potential difference between $r = r_1$ and $r = r_2$.

64. Four balls, each with mass m, are connected by four nonconducting strings to form a square with side a, as shown in Figure P25.64. The assembly is placed on a horizontal nonconducting frictionless surface. Balls 1 and 2 each have charge q, and balls 3 and 4 are uncharged. Find the maximum speed of balls 3 and 4 after the string connecting balls 1 and 2 is cut.

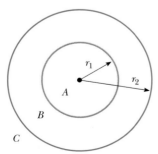

Figure P25.64

65. A point charge q is located at $x = -R$, and a point charge $-2q$ is located at the origin. Prove that the equipotential surface that has zero potential is a sphere centered at $(-4R/3, 0, 0)$ and having a radius $r = 2R/3$.

66. Consider two thin, conducting, spherical shells as shown in Figure P25.66. The inner shell has a radius $r_1 = 15.0$ cm and a charge of 10.0 nC. The outer shell has a radius $r_2 = 30.0$ cm and a charge of -15.0 nC. Find (a) the electric field $\mathbf{E}$ and (b) the electric potential V in regions A, B, and C, with $V = 0$ at $r = \infty$.

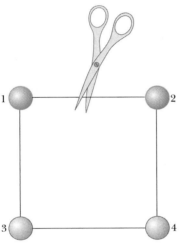

Figure P25.66

67. The x axis is the symmetry axis of a stationary uniformly charged ring of radius R and charge Q (Fig. P25.67).

A point charge Q of mass M is located initially at the center of the ring. When it is displaced slightly, the point charge accelerates along the x axis to infinity. Show that the ultimate speed of the point charge is

$$v = \left(\frac{2k_eQ^2}{MR}\right)^{1/2}$$

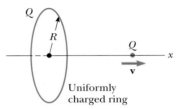

Figure P25.67

68. The thin, uniformly charged rod shown in Figure P25.68 has a linear charge density λ. Find an expression for the electric potential at P.

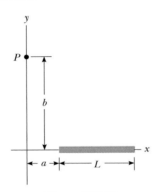

Figure P25.68

69. An electric dipole is located along the y axis as shown in Figure P25.69. The magnitude of its electric dipole moment is defined as $p = 2qa$. (a) At a point P, which is far from the dipole ($r \gg a$), show that the electric potential is

$$V = \frac{k_e p \cos\theta}{r^2}$$

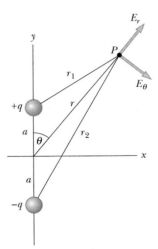

Figure P25.69

(b) Calculate the radial component E_r and the perpendicular component E_θ of the associated electric field. Note that $E_\theta = -(1/r)(\partial V/\partial \theta)$. Do these results seem reasonable for $\theta = 90°$ and $0°$? for $r = 0$? (c) For the dipole arrangement shown, express V in terms of Cartesian coordinates using $r = (x^2 + y^2)^{1/2}$ and

$$\cos \theta = \frac{y}{(x^2 + y^2)^{1/2}}$$

Using these results and again taking $r \gg a$, calculate the field components E_x and E_y.

70. When an uncharged conducting sphere of radius a is placed at the origin of an xyz coordinate system that lies in an initially uniform electric field $\mathbf{E} = E_0 \hat{\mathbf{k}}$, the resulting electric potential is $V(x, y, z) = V_0$ for points inside the sphere and

$$V(x, y, z) = V_0 - E_0 z + \frac{E_0 a^3 z}{(x^2 + y^2 + z^2)^{3/2}}$$

for points outside the sphere, where V_0 is the (constant) electric potential on the conductor. Use this equation to determine the x, y, and z components of the resulting electric field.

71. A disk of radius R (Fig. P25.71) has a nonuniform surface charge density $\sigma = Cr$, where C is a constant and r is measured from the center of the disk. Find (by direct integration) the potential at P.

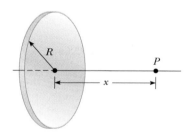

Figure P25.71

72. A solid sphere of radius R has a uniform charge density ρ and total charge Q. Derive an expression for its total electric potential energy. (*Suggestion:* imagine that the sphere is constructed by adding successive layers of concentric shells of charge $dq = (4\pi r^2 \, dr)\rho$ and use $dU = V \, dq$.)

73. ▣ Charge is uniformly distributed with a density of $100.0 \ \mu C/m^3$ throughout the volume of a cube 10.00 cm on each edge. (a) Find the electric potential at a distance of 5.000 cm from the center of one face of the cube, measured along a perpendicular to the face. Determine the potential to four significant digits. Use a numerical method that divides the cube into a sufficient number of smaller cubes, treated as point charges. Symmetry considerations will reduce the number of actual calculations. (b) **What If?** If the charge on the cube is redistributed into a uniform sphere of charge with the same center, by how much does the potential change?

Answers to Quick Quizzes

25.1 (b). When moving straight from A to B, $\mathbf{E}$ and $d\mathbf{s}$ both point toward the right. Thus, the dot product $\mathbf{E} \cdot d\mathbf{s}$ in Equation 25.3 is positive and ΔV is negative.

25.2 (a). From Equation 25.3, $\Delta U = q_0 \Delta V$, so if a negative test charge is moved through a negative potential difference, the potential energy is positive. Work must be done to move the charge in the direction opposite to the electric force on it.

25.3 $B \to C$, $C \to D$, $A \to B$, $D \to E$. Moving from B to C decreases the electric potential by 2 V, so the electric field performs 2 J of work on each coulomb of positive charge that moves. Moving from C to D decreases the electric potential by 1 V, so 1 J of work is done by the field. It takes no work to move the charge from A to B because the electric potential does not change. Moving from D to E increases the electric potential by 1 V, and thus the field does -1 J of work per unit of positive charge that moves.

25.4 (f). The electric field points in the direction of decreasing electric potential.

25.5 (b) and (f). The electric potential is inversely proportion to the radius (see Eq. 25.11). Because the same number of field lines passes through a closed surface of any shape or size, the electric flux through the surface remains constant.

25.6 (c). The potential is established only by the source charge and is independent of the test charge.

25.7 (a). The potential energy of the two-charge system is initially negative, due to the products of charges of opposite sign in Equation 25.13. When the sign of q_2 is changed, both charges are negative, and the potential energy of the system is positive.

25.8 (a). If the potential is constant (zero in this case), its derivative along this direction is zero.

25.9 (b). If the electric field is zero, there is no change in the electric potential and it must be constant. This constant value *could be* zero but does not *have to be* zero.

25.10 The graph would look like the sketch below. Notice the flat plateaus at each conductor, representing the constant electric potential inside a conductor.

Capacitance and Dielectrics

▲ All of these devices are capacitors, which store electric charge and energy. A capacitor is one type of circuit element that we can combine with others to make electric circuits. *(Paul Silverman/Fundamental Photographs)*

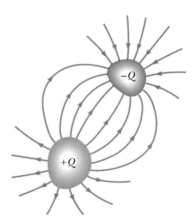

Figure 26.1 A capacitor consists of two conductors. When the capacitor is charged, the conductors carry charges of equal magnitude and opposite sign.

 PITFALL PREVENTION

26.1 Capacitance Is a Capacity

To understand capacitance, think of similar notions that use a similar word. The *capacity* of a milk carton is the volume of milk that it can store. The *heat capacity* of an object is the amount of energy an object can store per unit of temperature difference. The *capacitance* of a capacitor is the amount of charge the capacitor can store per unit of potential difference.

Definition of capacitance

In this chapter, we will introduce the first of three simple *circuit elements* that can be connected with wires to form an electric circuit. Electric circuits are the basis for the vast majority of the devices that we use in current society. We shall discuss *capacitors*—devices that store electric charge. This discussion will be followed by the study of *resistors* in Chapter 27 and *inductors* in Chapter 32. In later chapters, we will study more sophisticated circuit elements such as *diodes* and *transistors*.

Capacitors are commonly used in a variety of electric circuits. For instance, they are used to tune the frequency of radio receivers, as filters in power supplies, to eliminate sparking in automobile ignition systems, and as energy-storing devices in electronic flash units.

A capacitor consists of two conductors separated by an insulator. The capacitance of a given capacitor depends on its geometry and on the material—called a *dielectric*—that separates the conductors.

26.1 Definition of Capacitance

Consider two conductors carrying charges of equal magnitude and opposite sign, as shown in Figure 26.1. Such a combination of two conductors is called a **capacitor.** The conductors are called *plates.* A potential difference ΔV exists between the conductors due to the presence of the charges.

What determines how much charge is on the plates of a capacitor for a given voltage? Experiments show that the quantity of charge Q on a capacitor[1] is linearly proportional to the potential difference between the conductors; that is, $Q \propto \Delta V$. The proportionality constant depends on the shape and separation of the conductors.[2] We can write this relationship as $Q = C \Delta V$ if we define capacitance as follows:

The **capacitance** C of a capacitor is defined as the ratio of the magnitude of the charge on either conductor to the magnitude of the potential difference between the conductors:

$$C \equiv \frac{Q}{\Delta V} \qquad (26.1)$$

[1] Although the total charge on the capacitor is zero (because there is as much excess positive charge on one conductor as there is excess negative charge on the other), it is common practice to refer to the magnitude of the charge on either conductor as "the charge on the capacitor."

[2] The proportionality between ΔV and Q can be proved from Coulomb's law or by experiment.

Note that by definition *capacitance is always a positive quantity.* Furthermore, the charge Q and the potential difference ΔV are always expressed in Equation 26.1 as positive quantities. Because the potential difference increases linearly with the stored charge, the ratio $Q/\Delta V$ is constant for a given capacitor. Therefore, capacitance is a measure of a capacitor's ability to store charge. Because positive and negative charges are separated in the system of two conductors in a capacitor, there is electric potential energy stored in the system.

From Equation 26.1, we see that capacitance has SI units of coulombs per volt. The SI unit of capacitance is the **farad** (F), which was named in honor of Michael Faraday:

$$1\text{ F} = 1\text{ C/V}$$

The farad is a very large unit of capacitance. In practice, typical devices have capacitances ranging from microfarads (10^{-6} F) to picofarads (10^{-12} F). We shall use the symbol μF to represent microfarads. To avoid the use of Greek letters, in practice, physical capacitors often are labeled "mF" for microfarads and "mmF" for micromicrofarads or, equivalently, "pF" for picofarads.

Let us consider a capacitor formed from a pair of parallel plates, as shown in Figure 26.2. Each plate is connected to one terminal of a battery, which acts as a source of potential difference. If the capacitor is initially uncharged, the battery establishes an electric field in the connecting wires when the connections are made. Let us focus on the plate connected to the negative terminal of the battery. The electric field applies a force on electrons in the wire just outside this plate; this force causes the electrons to move onto the plate. This movement continues until the plate, the wire, and the terminal are all at the same electric potential. Once this equilibrium point is attained, a potential difference no longer exists between the terminal and the plate, and as a result no electric field is present in the wire, and the movement of electrons stops. The plate now carries a negative charge. A similar process occurs at the other capacitor plate, with electrons moving from the plate to the wire, leaving the plate positively charged. In this final configuration, the potential difference across the capacitor plates is the same as that between the terminals of the battery.

Suppose that we have a capacitor rated at 4 pF. This rating means that the capacitor can store 4 pC of charge for each volt of potential difference between the two conductors. If a 9-V battery is connected across this capacitor, one of the conductors ends up with a net charge of -36 pC and the other ends up with a net charge of $+36$ pC.

Quick Quiz 26.1 A capacitor stores charge Q at a potential difference ΔV. If the voltage applied by a battery to the capacitor is doubled to $2\,\Delta V$, (a) the capacitance falls to half its initial value and the charge remains the same (b) the capacitance and the charge both fall to half their initial values (c) the capacitance and the charge both double (d) the capacitance remains the same and the charge doubles.

26.2 Calculating Capacitance

We can derive an expression for the capacitance of a pair of oppositely charged conductors in the following manner: assume a charge of magnitude Q, and calculate the potential difference using the techniques described in the preceding chapter. We then use the expression $C = Q/\Delta V$ to evaluate the capacitance. As we might expect, we can perform this calculation relatively easily if the geometry of the capacitor is simple.

 PITFALL PREVENTION

26.2 Potential Difference is Δ*V*, not *V*

We use the symbol ΔV for the potential difference across a circuit element or a device because this is consistent with our definition of potential difference and with the meaning of the delta sign. It is a common, but confusing, practice to use the symbol V without the delta sign for a potential difference. Keep this in mind if you consult other texts.

Figure 26.2 A parallel-plate capacitor consists of two parallel conducting plates, each of area A, separated by a distance d. When the capacitor is charged by connecting the plates to the terminals of a battery, the plates carry equal amounts of charge. One plate carries positive charge, and the other carries negative charge.

▲ **PITFALL PREVENTION**

26.3 Too Many C's

Do not confuse italic C for capacitance with non-italic C for the unit coulomb.

While the most common situation is that of two conductors, a single conductor also has a capacitance. For example, imagine a spherical charged conductor. The electric field lines around this conductor are exactly the same as if there were a conducting shell of infinite radius, concentric with the sphere and carrying a charge of the same magnitude but opposite sign. Thus, we can identify the imaginary shell as the second conductor of a two-conductor capacitor. We now calculate the capacitance for this situation. The electric potential of the sphere of radius R is simply $k_e Q/R$, and setting $V = 0$ for the infinitely large shell, we have

Capacitance of an isolated charged sphere

$$C = \frac{Q}{\Delta V} = \frac{Q}{k_e Q/R} = \frac{R}{k_e} = 4\pi\epsilon_0 R \qquad (26.2)$$

This expression shows that the capacitance of an isolated charged sphere is proportional to its radius and is independent of both the charge on the sphere and the potential difference.

The capacitance of a pair of conductors depends on the geometry of the conductors. Let us illustrate this with three familiar geometries, namely, parallel plates, concentric cylinders, and concentric spheres. In these examples, we assume that the charged conductors are separated by a vacuum. The effect of a dielectric material placed between the conductors is treated in Section 26.5.

Parallel-Plate Capacitors

Two parallel metallic plates of equal area A are separated by a distance d, as shown in Figure 26.2. One plate carries a charge Q, and the other carries a charge $-Q$. Let us consider how the geometry of these conductors influences the capacity of the combination to store charge. Recall that charges of the same sign repel one another. As a capacitor is being charged by a battery, electrons flow into the negative plate and out of the positive plate. If the capacitor plates are large, the accumulated charges are able to distribute themselves over a substantial area, and the amount of charge that can be stored on a plate for a given potential difference increases as the plate area is increased. Thus, we expect the capacitance to be proportional to the plate area A.

Now let us consider the region that separates the plates. If the battery has a constant potential difference between its terminals, then the electric field between the plates must increase as d is decreased. Let us imagine that we move the plates closer together and consider the situation before any charges have had a chance to move in response to this change. Because no charges have moved, the electric field between the plates has the same value but extends over a shorter distance. Thus, the magnitude of the potential difference between the plates $\Delta V = Ed$ (Eq. 25.6) is now smaller. The difference between this new capacitor voltage and the terminal voltage of the battery now exists as a potential difference across the wires connecting the battery to the capacitor. This potential difference results in the electric field in the wires that drives more charge onto the plates, increasing the potential difference between the plates. When the potential difference between the plates again matches that of the battery, the potential difference across the wires falls back to zero, and the flow of charge stops. Thus, moving the plates closer together causes the charge on the capacitor to increase. If d is increased, the charge decreases. As a result, we expect the capacitance of the pair of plates to be inversely proportional to d.

We can verify these physical arguments with the following derivation. The surface charge density on either plate is $\sigma = Q/A$. If the plates are very close together (in comparison with their length and width), we can assume that the electric field is uniform between the plates and is zero elsewhere. According to the **What If?** feature in Example 24.8, the value of the electric field between the plates is

$$E = \frac{\sigma}{\epsilon_0} = \frac{Q}{\epsilon_0 A}$$

Because the field between the plates is uniform, the magnitude of the potential difference between the plates equals Ed (see Eq. 25.6); therefore,

$$\Delta V = Ed = \frac{Qd}{\epsilon_0 A}$$

Substituting this result into Equation 26.1, we find that the capacitance is

$$C = \frac{Q}{\Delta V} = \frac{Q}{Qd/\epsilon_0 A}$$

$$C = \frac{\epsilon_0 A}{d}$$ (26.3)   **Capacitance of parallel plates**

That is, **the capacitance of a parallel-plate capacitor is proportional to the area of its plates and inversely proportional to the plate separation,** just as we expected from our conceptual argument.

A careful inspection of the electric field lines for a parallel-plate capacitor reveals that the field is uniform in the central region between the plates, as shown in Figure 26.3a. However, the field is nonuniform at the edges of the plates. Figure 26.3b is a photograph of the electric field pattern of a parallel-plate capacitor. Note the nonuniform nature of the electric field at the ends of the plates. Such end effects can be neglected if the plate separation is small compared with the length of the plates.

Figure 26.4 shows a battery connected to a single parallel-plate capacitor with a switch in the circuit. Let us identify the circuit as a system. When the switch is closed, the battery establishes an electric field in the wires and charges flow between the wires and the capacitor. As this occurs, there is a transformation of energy within the system. Before the switch is closed, energy is stored as chemical energy in the battery. This energy is transformed during the chemical reaction that occurs within the battery when it is operating in an electric circuit. When the switch is closed, some of the chemical energy in the battery is converted to electric potential energy related to the separation of positive and negative charges on the plates. As a result, we can describe a capacitor as a device that stores energy as well as charge. We will explore this energy storage in more detail in Section 26.4.

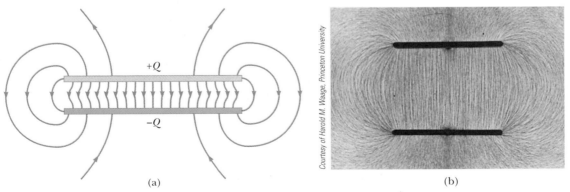

(a)   (b)

Figure 26.3 (a) The electric field between the plates of a parallel-plate capacitor is uniform near the center but nonuniform near the edges. (b) Electric field pattern of two oppositely charged conducting parallel plates. Small pieces of thread on an oil surface align with the electric field.

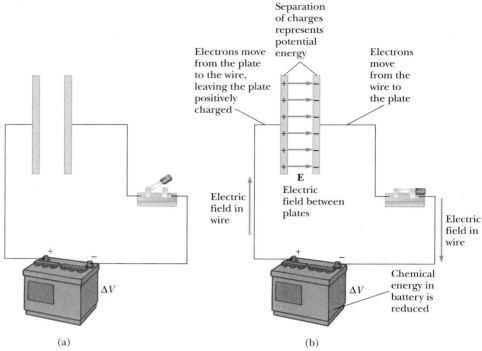

Active Figure 26.4 (a) A circuit consisting of a capacitor, a battery, and a switch. (b) When the switch is closed, the battery establishes an electric field in the wire that causes electrons to move from the left plate into the wire and into the right plate from the wire. As a result, a separation of charge exists on the plates, which represents an increase in electric potential energy of the system of the circuit. This energy in the system has been transformed from chemical energy in the battery.

Quick Quiz 26.2 Many computer keyboard buttons are constructed of capacitors, as shown in Figure 26.5. When a key is pushed down, the soft insulator between the movable plate and the fixed plate is compressed. When the key is pressed, the capacitance (a) increases, (b) decreases, or (c) changes in a way that we cannot determine because the complicated electric circuit connected to the keyboard button may cause a change in ΔV.

Figure 26.5 (Quick Quiz 26.2) One type of computer keyboard button.

Example 26.1 Parallel-Plate Capacitor

A parallel-plate capacitor with air between the plates has an area $A = 2.00 \times 10^{-4}$ m^2 and a plate separation $d = 1.00$ mm. Find its capacitance.

Solution From Equation 26.3, we find that

$$C = \frac{\epsilon_0 A}{d} = \frac{(8.85 \times 10^{-12} \text{ C}^2/\text{N} \cdot \text{m}^2)(2.00 \times 10^{-4} \text{ m}^2)}{1.00 \times 10^{-3} \text{ m}}$$

$$= 1.77 \times 10^{-12} \text{ F} = \boxed{1.77 \text{ pF}}$$

Cylindrical and Spherical Capacitors

From the definition of capacitance, we can, in principle, find the capacitance of any geometric arrangement of conductors. The following examples demonstrate the use of this definition to calculate the capacitance of the other familiar geometries that we mentioned: cylinders and spheres.

Example 26.2 The Cylindrical Capacitor

A solid cylindrical conductor of radius a and charge Q is coaxial with a cylindrical shell of negligible thickness, radius $b > a$, and charge $-Q$ (Fig. 26.6a). Find the capacitance of this cylindrical capacitor if its length is ℓ.

Solution It is difficult to apply physical arguments to this configuration, although we can reasonably expect the capacitance to be proportional to the cylinder length ℓ for the same reason that parallel-plate capacitance is proportional to plate area: stored charges have more room in which to be distributed. If we assume that ℓ is much greater than a and b, we can neglect end effects. In this case, the electric field is perpendicular to the long axis of the cylinders and is confined to the region between them (Fig. 26.6b). We must first calculate the potential difference between the two cylinders, which is given in general by

$$V_b - V_a = -\int_a^b \mathbf{E} \cdot d\mathbf{s}$$

where $\mathbf{E}$ is the electric field in the region between the cylinders. In Chapter 24, we showed using Gauss's law that the magnitude of the electric field of a cylindrical charge distribution having linear charge density λ is $E = 2k_e\lambda/r$ (Eq. 24.7). The same result applies here because, according to Gauss's law, the charge on the outer cylinder does

not contribute to the electric field inside it. Using this result and noting from Figure 26.6b that $\mathbf{E}$ is along r, we find that

$$V_b - V_a = -\int_a^b E_r\, dr = -2k_e\lambda \int_a^b \frac{dr}{r} = -2k_e\lambda \ln\left(\frac{b}{a}\right)$$

Substituting this result into Equation 26.1 and using the fact that $\lambda = Q/\ell$, we obtain

$$C = \frac{Q}{\Delta V} = \frac{Q}{(2k_e Q/\ell)\ln(b/a)} = \frac{\ell}{2k_e \ln(b/a)} \quad (26.4)$$

where ΔV is the magnitude of the potential difference between the cylinders, given by $\Delta V = |V_a - V_b| = 2k_e\lambda\ln(b/a)$, a positive quantity. As predicted, the capacitance is proportional to the length of the cylinders. As we might expect, the capacitance also depends on the radii of the two cylindrical conductors. From Equation 26.4, we see that the capacitance per unit length of a combination of concentric cylindrical conductors is

$$\frac{C}{\ell} = \frac{1}{2k_e \ln(b/a)} \quad (26.5)$$

An example of this type of geometric arrangement is a *coaxial cable*, which consists of two concentric cylindrical conductors separated by an insulator. You are likely to have a coaxial cable attached to your television set or VCR if you are a subscriber to cable television. The cable carries electrical signals in the inner and outer conductors. Such a geometry is especially useful for shielding the signals from any possible external influences.

What If? Suppose $b = 2.00a$ for the cylindrical capacitor. We would like to increase the capacitance, and we can do so by choosing to increase ℓ by 10% or by increasing a by 10%. Which choice is more effective at increasing the capacitance?

Answer According to Equation 26.4, C is proportional to ℓ, so increasing ℓ by 10% results in a 10% increase in C. For the result of the change in a, let us first evaluate C for $b = 2.00a$:

$$C = \frac{\ell}{2k_e \ln(b/a)} = \frac{\ell}{2k_e \ln(2.00)} = \frac{\ell}{2k_e(0.693)}$$

$$= 0.721\,\frac{\ell}{k_e}$$

(a)   (b)

Figure 26.6 (Example 26.2) (a) A cylindrical capacitor consists of a solid cylindrical conductor of radius a and length ℓ surrounded by a coaxial cylindrical shell of radius b. (b) End view. The electric field lines are radial. The dashed line represents the end of the cylindrical gaussian surface of radius r and length ℓ.

Now, for a 10% increase in a, the new value is $a' = 1.10a$, so

$$C' = \frac{\ell}{2k_e\ln(b/a')} = \frac{\ell}{2k_e\ln(2.00a/1.10a)}$$

$$= \frac{\ell}{2k_e\ln(2.00/1.10)} = \frac{\ell}{2k_e(0.598)} = 0.836\,\frac{\ell}{k_e}$$

The ratio of the new and old capacitances is

$$\frac{C'}{C} = \frac{0.836\,\ell/k_e}{0.721\,\ell/k_e} = 1.16$$

corresponding to a 16% increase in capacitance. Thus, it is more effective to increase a than to increase ℓ.

Note two more extensions of this problem. First, the advantage goes to increasing a only for a range of relationships between a and b. It is a valuable exercise to show that if $b > 2.85a$, increasing ℓ by 10% is more effective than increasing a (Problem 77). Second, if we increase b, we *reduce* the capacitance, so we would need to decrease b to increase the capacitance. Increasing a and decreasing b both have the effect of bringing the plates closer together, which increases the capacitance.

Example 26.3 The Spherical Capacitor

A spherical capacitor consists of a spherical conducting shell of radius b and charge $-Q$ concentric with a smaller conducting sphere of radius a and charge Q (Fig. 26.7). Find the capacitance of this device.

Solution As we showed in Chapter 24, the field outside a spherically symmetric charge distribution is radial and given by the expression k_eQ/r^2. In this case, this result applies to the field *between* the spheres ($a < r < b$). From Gauss's law we see that only the inner sphere contributes to this field. Thus, the potential difference between the spheres is

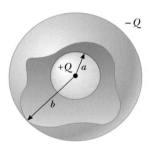

Figure 26.7 (Example 26.3) A spherical capacitor consists of an inner sphere of radius a surrounded by a concentric spherical shell of radius b. The electric field between the spheres is directed radially outward when the inner sphere is positively charged.

$$V_b - V_a = -\int_a^b E_r\,dr = -k_eQ\int_a^b \frac{dr}{r^2} = k_eQ\left[\frac{1}{r}\right]_a^b$$

$$= k_eQ\left(\frac{1}{b} - \frac{1}{a}\right)$$

The magnitude of the potential difference is

$$\Delta V = |V_b - V_a| = k_eQ\,\frac{(b-a)}{ab}$$

Substituting this value for ΔV into Equation 26.1, we obtain

$$C = \frac{Q}{\Delta V} = \frac{ab}{k_e(b-a)} \qquad (26.6)$$

What If? What if the radius b of the outer sphere approaches infinity? What does the capacitance become?

Answer In Equation 26.6, we let $b \to \infty$:

$$C = \lim_{b\to\infty}\frac{ab}{k_e(b-a)} = \frac{ab}{k_e(b)} = \frac{a}{k_e} = 4\pi\epsilon_0 a$$

Note that this is the same expression as Equation 26.2, the capacitance of an isolated spherical conductor.

26.3 Combinations of Capacitors

Two or more capacitors often are combined in electric circuits. We can calculate the equivalent capacitance of certain combinations using methods described in this section. Throughout this section, we assume that the capacitors to be combined are initially uncharged.

In studying electric circuits, we use a simplified pictorial representation called a **circuit diagram.** Such a diagram uses **circuit symbols** to represent various circuit elements. The circuit symbols are connected by straight lines that represent the wires between the circuit elements. The circuit symbols for capacitors and batteries, as well as the color codes used for them in this text, are given in Figure 26.8. The symbol for the capacitor reflects the geometry of the most common model for a capacitor—a pair of parallel plates. The positive terminal of the battery is at the higher potential and is represented in the circuit symbol by the longer line.

Capacitor symbol

Battery symbol

Switch symbol

Figure 26.8 Circuit symbols for capacitors, batteries, and switches. Note that capacitors are in blue and batteries and switches are in red.

Parallel Combination

Two capacitors connected as shown in Figure 26.9a are known as a *parallel combination* of capacitors. Figure 26.9b shows a circuit diagram for this combination of capacitors. The left plates of the capacitors are connected by a conducting wire to the positive terminal of the battery and are therefore both at the same electric potential as the positive terminal. Likewise, the right plates are connected to the negative terminal and are therefore both at the same potential as the negative terminal. Thus, **the individual potential differences across capacitors connected in parallel are the same and are equal to the potential difference applied across the combination.**

In a circuit such as that shown in Figure 26.9, the voltage applied across the combination is the terminal voltage of the battery. Situations can occur in which the parallel combination is in a circuit with other circuit elements; in such situations, we must determine the potential difference across the combination by analyzing the entire circuit.

When the capacitors are first connected in the circuit shown in Figure 26.9, electrons are transferred between the wires and the plates; this transfer leaves the left plates positively charged and the right plates negatively charged. The flow of charge ceases when the voltage across the capacitors is equal to that across the battery terminals. The capacitors reach their maximum charge when the flow of charge ceases. Let us call the maximum charges on the two capacitors Q_1 and Q_2. The *total charge* Q stored by the two capacitors is

$$Q = Q_1 + Q_2 \tag{26.7}$$

That is, **the total charge on capacitors connected in parallel is the sum of the charges on the individual capacitors.** Because the voltages across the capacitors are the same, the charges that they carry are

$$Q_1 = C_1 \, \Delta V \qquad Q_2 = C_2 \, \Delta V$$

Suppose that we wish to replace these two capacitors by one *equivalent capacitor* having a capacitance C_{eq}, as in Figure 26.9c. The effect this equivalent capacitor has on the circuit must be exactly the same as the effect of the combination of the two

$$\Delta V_1 = \Delta V_2 = \Delta V$$
$$C_{eq} = C_1 + C_2$$

(a)   (b)   (c)

Active Figure 26.9 (a) A parallel combination of two capacitors in an electric circuit in which the potential difference across the battery terminals is ΔV. (b) The circuit diagram for the parallel combination. (c) The equivalent capacitance is $C_{eq} = C_1 + C_2$.

At the Active Figures link at http://www.pse6.com, you can adjust the battery voltage and the individual capacitances to see the resulting charges and voltages on the capacitors. You can combine up to four capacitors in parallel.

individual capacitors. That is, the equivalent capacitor must store Q units of charge when connected to the battery. We can see from Figure 26.9c that the voltage across the equivalent capacitor also is ΔV because the equivalent capacitor is connected directly across the battery terminals. Thus, for the equivalent capacitor,

$$Q = C_{eq} \, \Delta V$$

Substituting these three relationships for charge into Equation 26.7, we have

$$C_{eq} \, \Delta V = C_1 \, \Delta V + C_2 \, \Delta V$$

$$C_{eq} = C_1 + C_2 \qquad \text{(parallel combination)}$$

If we extend this treatment to three or more capacitors connected in parallel, we find the equivalent capacitance to be

Capacitors in parallel

$$C_{eq} = C_1 + C_2 + C_3 + \cdots \qquad \text{(parallel combination)} \qquad (26.8)$$

Thus, **the equivalent capacitance of a parallel combination of capacitors is the algebraic sum of the individual capacitances and is greater than any of the individual capacitances.** This makes sense because we are essentially combining the areas of all the capacitor plates when we connect them with conducting wire, and capacitance of parallel plates is proportional to area (Eq. 26.3).

Series Combination

Two capacitors connected as shown in Figure 26.10a and the equivalent circuit diagram in Figure 26.10b are known as a *series combination* of capacitors. The left plate of capacitor 1 and the right plate of capacitor 2 are connected to the terminals of a battery. The other two plates are connected to each other and to nothing else; hence, they form an isolated conductor that is initially uncharged and must continue to have zero net charge. To analyze this combination, let us begin by considering the uncharged capacitors and follow what happens just after a battery is connected to the circuit. When the battery is connected, electrons are transferred out of the left plate of C_1 and into the right plate of C_2. As this negative charge accumulates on the right plate of C_2, an equivalent amount of negative charge is forced off the left plate of C_2, and this left plate therefore has an excess positive charge. The negative charge leaving

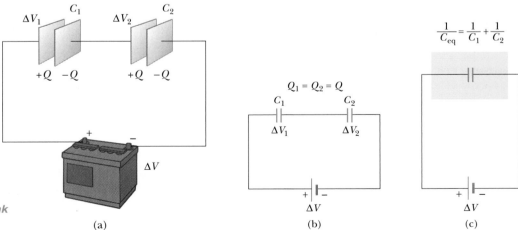

Active Figure 26.10 (a) A series combination of two capacitors. The charges on the two capacitors are the same. (b) The circuit diagram for the series combination. (c) The equivalent capacitance can be calculated from the relationship

$$\frac{1}{C_{eq}} = \frac{1}{C_1} + \frac{1}{C_2}.$$

the left plate of C_2 causes negative charges to accumulate on the right plate of C_1. As a result, all the right plates end up with a charge $-Q$, and all the left plates end up with a charge $+Q$. Thus, **the charges on capacitors connected in series are the same.**

From Figure 26.10a, we see that the voltage ΔV across the battery terminals is split between the two capacitors:

$$\Delta V = \Delta V_1 + \Delta V_2 \qquad (26.9)$$

where ΔV_1 and ΔV_2 are the potential differences across capacitors C_1 and C_2, respectively. In general, **the total potential difference across any number of capacitors connected in series is the sum of the potential differences across the individual capacitors.**

Suppose that the equivalent single capacitor in Figure 26.10c has the same effect on the circuit as the series combination when it is connected to the battery. After it is fully charged, the equivalent capacitor must have a charge of $-Q$ on its right plate and a charge of $+Q$ on its left plate. Applying the definition of capacitance to the circuit in Figure 26.10c, we have

$$\Delta V = \frac{Q}{C_{eq}}$$

Because we can apply the expression $Q = C \Delta V$ to each capacitor shown in Figure 26.10b, the potential differences across them are

$$\Delta V_1 = \frac{Q}{C_1} \qquad \Delta V_2 = \frac{Q}{C_2}$$

Substituting these expressions into Equation 26.9, we have

$$\frac{Q}{C_{eq}} = \frac{Q}{C_1} + \frac{Q}{C_2}$$

Canceling Q, we arrive at the relationship

$$\frac{1}{C_{eq}} = \frac{1}{C_1} + \frac{1}{C_2} \qquad \text{(series combination)}$$

When this analysis is applied to three or more capacitors connected in series, the relationship for the equivalent capacitance is

$$\frac{1}{C_{eq}} = \frac{1}{C_1} + \frac{1}{C_2} + \frac{1}{C_3} + \cdots \qquad \text{(series combination)} \qquad (26.10)$$

Capacitors in series

This shows that **the inverse of the equivalent capacitance is the algebraic sum of the inverses of the individual capacitances and the equivalent capacitance of a series combination is always less than any individual capacitance in the combination.**

Quick Quiz 26.3 Two capacitors are identical. They can be connected in series or in parallel. If you want the *smallest* equivalent capacitance for the combination, do you connect them in (a) series, in (b) parallel, or (c) do the combinations have the same capacitance?

Quick Quiz 26.4 Consider the two capacitors in Quick Quiz 26.3 again. Each capacitor is charged to a voltage of 10 V. If you want the largest combined potential difference across the combination, do you connect them in (a) series, in (b) parallel, or (c) do the combinations have the same potential difference?

PROBLEM-SOLVING HINTS

Capacitors

- Be careful with units. When you calculate capacitance in farads, make sure that distances are expressed in meters. When checking consistency of units, remember that the unit for electric fields can be either N/C or V/m.

- When two or more capacitors are connected in parallel, the potential difference across each is the same. The charge on each capacitor is proportional to its capacitance; hence, the capacitances can be added directly to give the equivalent capacitance of the parallel combination. The equivalent capacitance is always larger than the individual capacitances.

- When two or more capacitors are connected in series, they carry the same charge, and the sum of the potential differences equals the total potential difference applied to the combination. The sum of the reciprocals of the capacitances equals the reciprocal of the equivalent capacitance, which is always less than the capacitance of the smallest individual capacitor.

Example 26.4 Equivalent Capacitance

Interactive

Find the equivalent capacitance between a and b for the combination of capacitors shown in Figure 26.11a. All capacitances are in microfarads.

Solution Using Equations 26.8 and 26.10, we reduce the combination step by step as indicated in the figure. The 1.0-μF and 3.0-μF capacitors are in parallel and combine according to the expression $C_{eq} = C_1 + C_2 = 4.0 \ \mu F$. The 2.0-$\mu$F and 6.0-$\mu$F capacitors also are in parallel and have an equivalent capacitance of 8.0 μF. Thus, the upper branch in Figure 26.11b consists of two 4.0-μF capacitors in series, which combine as follows:

$$\frac{1}{C_{eq}} = \frac{1}{C_1} + \frac{1}{C_2} = \frac{1}{4.0 \ \mu F} + \frac{1}{4.0 \ \mu F} = \frac{1}{2.0 \ \mu F}$$

$$C_{eq} = 2.0 \ \mu F$$

The lower branch in Figure 26.11b consists of two 8.0-μF capacitors in series, which combine to yield an equivalent capacitance of 4.0 μF. Finally, the 2.0-μF and 4.0-μF capacitors in Figure 26.11c are in parallel and thus have an

equivalent capacitance of 6.0 μF.

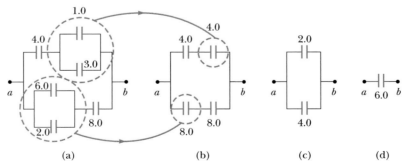

Figure 26.11 (Example 26.4) To find the equivalent capacitance of the capacitors in part (a), we reduce the various combinations in steps as indicated in parts (b), (c), and (d), using the series and parallel rules described in the text.

 Practice reducing a combination of capacitors to a single equivalent capacitance at the Interactive Worked Example link at **http://www.pse6.com.**

26.4   Energy Stored in a Charged Capacitor

Almost everyone who works with electronic equipment has at some time verified that a capacitor can store energy. If the plates of a charged capacitor are connected by a conductor, such as a wire, charge moves between each plate and its connecting wire until the capacitor is uncharged. The discharge can often be observed as a visible spark. If you should accidentally touch the opposite plates of a charged capacitor, your fingers act as a pathway for discharge, and the result is an electric shock. The degree of shock you receive depends on the capacitance and on the voltage applied to the capacitor. Such a shock could be fatal if high voltages are present, such as in the power supply of a television set. Because the charges can be stored in a capacitor even when the set is turned off, unplugging the television does not make it safe to open the case and touch the components inside.

To calculate the energy stored in the capacitor, we shall assume a charging process that is different from the actual process described in Section 26.1 but which gives the same final result. We can make this assumption because the energy in the final configuration does not depend on the actual charge-transfer process. We imagine that the charge is transferred mechanically through the space between the plates. We reach in and grab a small amount of positive charge on the plate connected to the negative terminal and apply a force that causes this positive charge to move over to the plate connected to the positive terminal. Thus, we do work on the charge as we transfer it from one plate to the other. At first, no work is required to transfer a small amount of charge dq from one plate to the other.[3] However, once this charge has been transferred, a small potential difference exists between the plates. Therefore, work must be done to move additional charge through this potential difference. As more and more charge is transferred from one plate to the other, the potential difference increases in proportion, and more work is required.

Suppose that q is the charge on the capacitor at some instant during the charging process. At the same instant, the potential difference across the capacitor is $\Delta V = q/C$. From Section 25.2, we know that the work necessary to transfer an increment of charge dq from the plate carrying charge $-q$ to the plate carrying charge q (which is at the higher electric potential) is

$$dW = \Delta V \, dq = \frac{q}{C} \, dq$$

This is illustrated in Figure 26.12. The total work required to charge the capacitor from $q = 0$ to some final charge $q = Q$ is

$$W = \int_0^Q \frac{q}{C} \, dq = \frac{1}{C} \int_0^Q q \, dq = \frac{Q^2}{2C}$$

The work done in charging the capacitor appears as electric potential energy U stored in the capacitor. Using Equation 26.1, we can express the potential energy stored in a charged capacitor in the following forms:

$$U = \frac{Q^2}{2C} = \tfrac{1}{2} Q \, \Delta V = \tfrac{1}{2} C (\Delta V)^2 \qquad (26.11)$$

Energy stored in a charged capacitor

This result applies to any capacitor, regardless of its geometry. We see that for a given capacitance, the stored energy increases as the charge increases and as the potential difference increases. In practice, there is a limit to the maximum energy (or charge) that can be stored because, at a sufficiently great value of ΔV, discharge ultimately occurs between the plates. For this reason, capacitors are usually labeled with a maximum operating voltage.

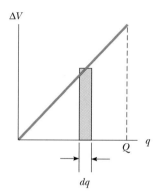

Figure 26.12 A plot of potential difference versus charge for a capacitor is a straight line having a slope $1/C$. The work required to move charge dq through the potential difference ΔV existing at the time across the capacitor plates is given approximately by the area of the shaded rectangle. The total work required to charge the capacitor to a final charge Q is the triangular area under the straight line, $W = \tfrac{1}{2} Q \Delta V$. (Don't forget that 1 V = J/C; hence, the unit for the triangular area is the joule.)

---

[3]   We shall use lowercase q for the time-varying charge on the capacitor while it is charging, to distinguish it from uppercase Q, which is the total charge on the capacitor after it is completely charged.

We can consider the energy stored in a capacitor as being stored in the electric field created between the plates as the capacitor is charged. This description is reasonable because the electric field is proportional to the charge on the capacitor. For a parallel-plate capacitor, the potential difference is related to the electric field through the relationship $\Delta V = Ed$. Furthermore, its capacitance is $C = \epsilon_0 A/d$ (Eq. 26.3). Substituting these expressions into Equation 26.11, we obtain

$$U = \frac{1}{2} \frac{\epsilon_0 A}{d} (E^2 d^2) = \frac{1}{2}(\epsilon_0 Ad)E^2 \qquad (26.12)$$

Because the volume occupied by the electric field is Ad, the *energy per unit volume* $u_E = U/Ad$, known as the *energy density*, is

$$u_E = \frac{1}{2}\epsilon_0 E^2 \qquad (26.13)$$

Energy density in an electric field

Although Equation 26.13 was derived for a parallel-plate capacitor, the expression is generally valid, regardless of the source of the electric field. That is, **the energy density in any electric field is proportional to the square of the magnitude of the electric field at a given point.**

▲ **PITFALL PREVENTION**

26.4 Not a New Kind of Energy

The energy given by Equation 26.13 is not a new kind of energy. It is familiar electric potential energy associated with a system of separated source charges. Equation 26.13 provides a new *interpretation*, or a new way of *modeling* the energy, as energy associated with the electric field, regardless of the source of the field.

Quick Quiz 26.5 You have three capacitors and a battery. In which of the following combinations of the three capacitors will the maximum possible energy be stored when the combination is attached to the battery? (a) series (b) parallel (c) Both combinations will store the same amount of energy.

Quick Quiz 26.6 You charge a parallel-plate capacitor, remove it from the battery, and prevent the wires connected to the plates from touching each other. When you pull the plates apart to a larger separation, do the following quantities increase, decrease, or stay the same? (a) C; (b) Q; (c) E between the plates; (d) ΔV; (e) energy stored in the capacitor.

Quick Quiz 26.7 Repeat Quick Quiz 26.6, but this time answer the questions for the situation in which the battery remains connected to the capacitor while you pull the plates apart.

Example 26.5 Rewiring Two Charged Capacitors   `Interactive`

Two capacitors C_1 and C_2 (where $C_1 > C_2$) are charged to the same initial potential difference ΔV_i. The charged capacitors are removed from the battery, and their plates are connected with opposite polarity as in Figure 26.13a. The switches S_1 and S_2 are then closed, as in Figure 26.13b.

(A) Find the final potential difference ΔV_f between a and b after the switches are closed.

Solution Figure 26.13 helps us conceptualize the initial and final configurations of the system. In Figure 26.13b, it might appear as if the capacitors are connected in parallel, but there is no battery in this circuit that is applying a voltage across the combination. Thus, we *cannot* categorize this as a problem in which capacitors are connected in parallel. We *can* categorize this as a problem involving an isolated system for electric charge—the left-hand plates of the capac-

Figure 26.13 (Example 26.5) (a) Two capacitors are charged to the same initial potential difference and connected together with plates of opposite sign to be in contact when the switches are closed. (b) When the switches are closed, the charges redistribute.

itors form an isolated system because they are not connected to the right-hand plates by conductors. To analyze the problem, note that the charges on the left-hand plates before the switches are closed are

$$Q_{1i} = C_1 \Delta V_i \qquad \text{and} \qquad Q_{2i} = -C_2 \Delta V_i$$

The negative sign for Q_{2i} is necessary because the charge on the left plate of capacitor C_2 is negative. The total charge Q in the system is

$$(1) \qquad Q = Q_{1i} + Q_{2i} = (C_1 - C_2)\Delta V_i$$

After the switches are closed, the total charge Q in the system remains the same but the charges on the individual capacitors change to new values Q_{1f} and Q_{2f}. Because the system is isolated,

$$(2) \qquad Q = Q_{1f} + Q_{2f}$$

The charges redistribute until the potential difference is the same across both capacitors, ΔV_f. To satisfy this requirement, the charges on the capacitors after the switches are closed are

$$Q_{1f} = C_1 \Delta V_f \qquad \text{and} \qquad Q_{2f} = C_2 \Delta V_f$$

Dividing the first equation by the second, we have

$$(3) \qquad Q_{1f} = \frac{C_1}{C_2} Q_{2f}$$

Combining Equations (2) and (3), we obtain

$$Q = Q_{1f} + Q_{2f} = \frac{C_1}{C_2} Q_{2f} + Q_{2f} = Q_{2f} \left(1 + \frac{C_1}{C_2} \right)$$

$$(4) \qquad Q_{2f} = Q \left(\frac{C_2}{C_1 + C_2} \right)$$

Using Equations (3) and (4) to find Q_{1f} in terms of Q, we have

$$(5) \qquad Q_{1f} = \frac{C_1}{C_2} Q_{2f} = \frac{C_1}{C_2} Q \left(\frac{C_2}{C_1 + C_2} \right)$$

$$= Q \left(\frac{C_1}{C_1 + C_2} \right)$$

Finally, using Equation 26.1 to find the voltage across each capacitor, we find that

$$(6) \qquad \Delta V_{1f} = \frac{Q_{1f}}{C_1} = \frac{Q[C_1/(C_1 + C_2)]}{C_1} = \frac{Q}{C_1 + C_2}$$

$$(7) \qquad \Delta V_{2f} = \frac{Q_{2f}}{C_2} = \frac{Q[C_2/(C_1 + C_2)]}{C_2} = \frac{Q}{C_1 + C_2}$$

As noted earlier, $\Delta V_{1f} = \Delta V_{2f} = \Delta V_f$.

To express ΔV_f in terms of the given quantities C_1, C_2, and ΔV_i, we substitute the value of Q from Equation (1) into either Equation (6) or (7) to obtain

$$\Delta V_f = \boxed{\left(\frac{C_1 - C_2}{C_1 + C_2} \right) \Delta V_i}$$

(B) Find the total energy stored in the capacitors before and after the switches are closed and the ratio of the final energy to the initial energy.

Solution Before the switches are closed, the total energy stored in the capacitors is

$$U_i = \tfrac{1}{2} C_1 (\Delta V_i)^2 + \tfrac{1}{2} C_2 (\Delta V_i)^2 = \boxed{\tfrac{1}{2}(C_1 + C_2)(\Delta V_i)^2}$$

After the switches are closed, the total energy stored in the capacitors is

$$U_f = \tfrac{1}{2} C_1 (\Delta V_f)^2 + \tfrac{1}{2} C_2 (\Delta V_f)^2 = \tfrac{1}{2}(C_1 + C_2)(\Delta V_f)^2$$

Using the results of part (A), we can express this as

$$U_f = \boxed{\frac{1}{2} \frac{(C_1 - C_2)^2 (\Delta V_i)^2}{(C_1 + C_2)}}$$

Therefore, the ratio of the final energy stored to the initial energy stored is

$$(8) \qquad \frac{U_f}{U_i} = \frac{\tfrac{1}{2}(C_1 - C_2)^2 (\Delta V_i)^2/(C_1 + C_2)}{\tfrac{1}{2}(C_1 + C_2)(\Delta V_i)^2}$$

$$= \boxed{\left(\frac{C_1 - C_2}{C_1 + C_2} \right)^2}$$

To finalize this problem, note that this ratio is *less* than unity, indicating that the final energy is *less* than the initial energy. At first, you might think that the law of energy conservation has been violated, but this is not the case. The "missing" energy is transferred out of the system of the capacitors by the mechanism of electromagnetic waves, as we shall see in Chapter 34.

What If? What if the two capacitors have the same capacitance? What would we expect to happen when the switches are closed?

Answer The equal-magnitude charges on the two capacitors should simply cancel each other and the capacitors will be uncharged afterward.

Let us test our results to see if this is the case mathematically. In Equation (1), because the charges are of equal magnitude and opposite sign, we see that $Q = 0$. Thus, Equations (4) and (5) show us that $Q_{1f} = Q_{2f} = 0$, consistent with our prediction. Furthermore, Equations (6) and (7) show us that $\Delta V_{1f} = \Delta V_{2f} = 0$, which is consistent with uncharged capacitors. Finally, if $C_1 = C_2$, Equation (8) shows us that $U_f = 0$, which is also consistent with uncharged capacitors.

 At the Interactive Worked Example link at **http://www.pse6.com**, *explore this situation for various initial values of the voltage and the capacitances.*

Figure 26.14 In a hospital or at an emergency scene, you might see a patient being revived with a defibrillator. The defibrillator's paddles are applied to the patient's chest, and an electric shock is sent through the chest cavity. The aim of this technique is to restore the heart's normal rhythm pattern.

One device in which capacitors have an important role is the *defibrillator* (Fig. 26.14). Up to 360 J is stored in the electric field of a large capacitor in a defibrillator when it is fully charged. The defibrillator can deliver all this energy to a patient in about 2 ms. (This is roughly equivalent to 3 000 times the power delivered to a 60-W lightbulb!) Under the proper conditions, the defibrillator can be used to stop cardiac fibrillation (random contractions) in heart attack victims. When fibrillation occurs, the heart produces a rapid, irregular pattern of beats. A fast discharge of energy through the heart can return the organ to its normal beat pattern. Emergency medical teams use portable defibrillators that contain batteries capable of charging a capacitor to a high voltage. (The circuitry actually permits the capacitor to be charged to a much higher voltage than that of the battery.) The stored energy is released through the heart by conducting electrodes, called paddles, that are placed on both sides of the victim's chest. The paramedics must wait between applications of the energy due to the time necessary for the capacitors to become fully charged. In this case and others (e.g., camera flash units and lasers used for fusion experiments), capacitors serve as energy reservoirs which can be slowly charged and then discharged quickly to provide large amounts of energy in a short pulse.

A camera's flash unit also uses a capacitor, although the total amount of energy stored is much less than that stored in a defibrillator. After the flash unit's capacitor is charged, tripping the camera's shutter causes the stored energy to be sent through a special lightbulb that briefly illuminates the subject being photographed.

26.5 Capacitors with Dielectrics

A **dielectric** is a nonconducting material, such as rubber, glass, or waxed paper. When a dielectric is inserted between the plates of a capacitor, the capacitance increases. If the dielectric completely fills the space between the plates, the capacitance increases by a dimensionless factor κ, which is called the **dielectric constant** of the material. The dielectric constant varies from one material to another. In this section, we analyze this change in capacitance in terms of electrical parameters such as electric charge, electric field, and potential difference; in Section 26.7, we shall discuss the microscopic origin of these changes.

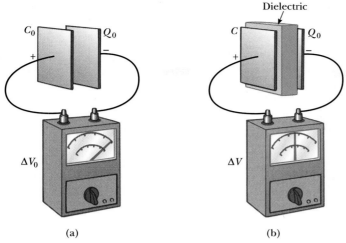

(a)   (b)

Figure 26.15 A charged capacitor (a) before and (b) after insertion of a dielectric between the plates. The charge on the plates remains unchanged, but the potential difference decreases from ΔV_0 to $\Delta V = \Delta V_0 / \kappa$. Thus, the capacitance increases from C_0 to κC_0.

We can perform the following experiment to illustrate the effect of a dielectric in a capacitor. Consider a parallel-plate capacitor that without a dielectric has a charge Q_0 and a capacitance C_0. The potential difference across the capacitor is $\Delta V_0 = Q_0/C_0$. Figure 26.15a illustrates this situation. The potential difference is measured by a *voltmeter*, which we shall study in greater detail in Chapter 28. Note that no battery is shown in the figure; also, we must assume that no charge can flow through an ideal voltmeter. Hence, there is no path by which charge can flow and alter the charge on the capacitor. If a dielectric is now inserted between the plates, as in Figure 26.15b, the voltmeter indicates that the voltage between the plates decreases to a value ΔV. The voltages with and without the dielectric are related by the factor κ as follows:

$$\Delta V = \frac{\Delta V_0}{\kappa}$$

Because $\Delta V < \Delta V_0$, we see that $\kappa > 1$.

Because the charge Q_0 on the capacitor does not change, we conclude that the capacitance must change to the value

$$C = \frac{Q_0}{\Delta V} = \frac{Q_0}{\Delta V_0/\kappa} = \kappa \frac{Q_0}{\Delta V_0}$$

$$\boxed{C = \kappa C_0} \tag{26.14}$$

That is, the capacitance *increases* by the factor κ when the dielectric completely fills the region between the plates.[4] For a parallel-plate capacitor, where $C_0 = \epsilon_0 A/d$ (Eq. 26.3), we can express the capacitance when the capacitor is filled with a dielectric as

$$C = \kappa \frac{\epsilon_0 A}{d} \tag{26.15}$$

From Equations 26.3 and 26.15, it would appear that we could make the capacitance very large by decreasing d, the distance between the plates. In practice, the lowest value of d is limited by the electric discharge that could occur through the dielectric medium separating the plates. For any given separation d, the maximum voltage that can be applied to a capacitor without causing a discharge depends on the

 PITFALL PREVENTION

26.5 Is the Capacitor Connected to a Battery?

In problems in which you are modifying a capacitor (by insertion of a dielectric, for example), you must note whether modifications to the capacitor are being made while the capacitor is connected to a battery or after it is disconnected. If the capacitor remains connected to the battery, the voltage across the capacitor necessarily remains the same. If you disconnect the capacitor from the battery before making any modifications to the capacitor, the capacitor is an isolated system and its charge remains the same.

Capacitance of a capacitor filled with a material of dielectric constant κ

---

[4]   If the dielectric is introduced while the potential difference is held constant by a battery, the charge increases to a value $Q = \kappa Q_0$. The additional charge comes from the wires attached to the capacitor, and the capacitance again increases by the factor κ.

Figure 26.16 Dielectric breakdown in air. Sparks are produced when the high voltage between the wires causes the electric field to exceed the dielectric strength of air.

© Loren Winters/Visuals Unlimited

dielectric strength (maximum electric field) of the dielectric. If the magnitude of the electric field in the dielectric exceeds the dielectric strength, then the insulating properties break down and the dielectric begins to conduct. Figure 26.16 shows the effect of exceeding the dielectric strength of air. Sparks appear between the two wires, due to ionization of atoms and recombination with electrons in the air, similar to the process that produced corona discharge in Section 25.6.

Physical capacitors have a specification called by a variety of names, including *working voltage, breakdown voltage,* and *rated voltage.* This parameter represents the largest voltage that can be applied to the capacitor without exceeding the dielectric strength of the dielectric material in the capacitor. Consequently, when selecting a capacitor for a given application, you must consider the capacitance of the device along with the expected voltage across the capacitor in the circuit, making sure that the expected voltage will be smaller than the rated voltage of the capacitor. You can see the rated voltage on several of the capacitors in the opening photograph for this chapter.

Insulating materials have values of κ greater than unity and dielectric strengths greater than that of air, as Table 26.1 indicates. Thus, we see that a dielectric provides the following advantages:

• Increase in capacitance

• Increase in maximum operating voltage

• Possible mechanical support between the plates, which allows the plates to be close together without touching, thereby decreasing d and increasing C.

Types of Capacitors

Commercial capacitors are often made from metallic foil interlaced with thin sheets of either paraffin-impregnated paper or Mylar as the dielectric material. These alternate layers of metallic foil and dielectric are rolled into a cylinder to form a small package (Fig. 26.17a). High-voltage capacitors commonly consist of a number of interwoven

Table 26.1

Approximate Dielectric Constants and Dielectric Strengths of Various Materials at Room Temperature		
Material	**Dielectric Constant κ**	**Dielectric Strength[a] (10^6 V/m)**
Air (dry)	1.000 59	3
Bakelite	4.9	24
Fused quartz	3.78	8
Mylar	3.2	7
Neoprene rubber	6.7	12
Nylon	3.4	14
Paper	3.7	16
Paraffin-impregnated paper	3.5	11
Polystyrene	2.56	24
Polyvinyl chloride	3.4	40
Porcelain	6	12
Pyrex glass	5.6	14
Silicone oil	2.5	15
Strontium titanate	233	8
Teflon	2.1	60
Vacuum	1.000 00	—
Water	80	—

[a] The dielectric strength equals the maximum electric field that can exist in a dielectric without electrical breakdown. Note that these values depend strongly on the presence of impurities and flaws in the materials.

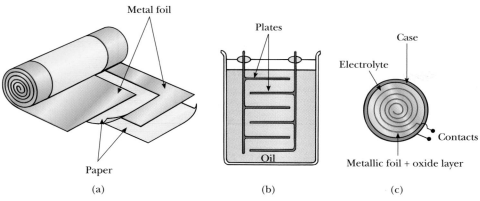

Figure 26.17 Three commercial capacitor designs. (a) A tubular capacitor, whose plates are separated by paper and then rolled into a cylinder. (b) A high-voltage capacitor consisting of many parallel plates separated by insulating oil. (c) An electrolytic capacitor.

metallic plates immersed in silicone oil (Fig. 26.17b). Small capacitors are often constructed from ceramic materials.

Often, an *electrolytic capacitor* is used to store large amounts of charge at relatively low voltages. This device, shown in Figure 26.17c, consists of a metallic foil in contact with an *electrolyte*—a solution that conducts electricity by virtue of the motion of ions contained in the solution. When a voltage is applied between the foil and the electrolyte, a thin layer of metal oxide (an insulator) is formed on the foil, and this layer serves as the dielectric. Very large values of capacitance can be obtained in an electrolytic capacitor because the dielectric layer is very thin, and thus the plate separation is very small.

Electrolytic capacitors are not reversible as are many other capacitors—they have a polarity, which is indicated by positive and negative signs marked on the device. When electrolytic capacitors are used in circuits, the polarity must be aligned properly. If the polarity of the applied voltage is opposite that which is intended, the oxide layer is removed and the capacitor conducts electricity instead of storing charge.

Variable capacitors (typically 10 to 500 pF) usually consist of two interwoven sets of metallic plates, one fixed and the other movable, and contain air as the dielectric (Fig. 26.18). These types of capacitors are often used in radio tuning circuits.

Figure 26.18 A variable capacitor. When one set of metal plates is rotated so as to lie between a fixed set of plates, the capacitance of the device changes.

Quick Quiz 26.8 If you have ever tried to hang a picture or a mirror, you know it can be difficult to locate a wooden stud in which to anchor your nail or screw. A carpenter's stud-finder is basically a capacitor with its plates arranged side by side instead of facing one another, as shown in Figure 26.19. When the device is moved over a stud, does the capacitance increase or decrease?

Figure 26.19 (Quick Quiz 26.8) A stud-finder. (a) The materials between the plates of the capacitor are the wallboard and air. (b) When the capacitor moves across a stud in the wall, the materials between the plates are the wallboard and the wood. The change in the dielectric constant causes a signal light to illuminate.

Quick Quiz 26.9 A fully charged parallel-plate capacitor remains connected to a battery while you slide a dielectric between the plates. Do the following quantities increase, decrease, or stay the same? (a) C; (b) Q; (c) E between the plates; (d) ΔV.

Example 26.6 A Paper-Filled Capacitor

A parallel-plate capacitor has plates of dimensions 2.0 cm by 3.0 cm separated by a 1.0-mm thickness of paper.

(A) Find its capacitance.

Solution Because $\kappa = 3.7$ for paper (see Table 26.1), we have

$$C = \kappa \, \frac{\epsilon_0 A}{d}$$

$$= 3.7 \left(\frac{(8.85 \times 10^{-12} \, \text{C}^2/\text{N} \cdot \text{m}^2)(6.0 \times 10^{-4} \, \text{m}^2)}{1.0 \times 10^{-3} \text{m}} \right)$$

$$= 20 \times 10^{-12} \, \text{F} = \boxed{20 \text{ pF}}$$

(B) What is the maximum charge that can be placed on the capacitor?

Solution From Table 26.1 we see that the dielectric strength of paper is 16×10^6 V/m. Because the thickness of the paper is 1.0 mm, the maximum voltage that can be applied before breakdown is

$$\Delta V_{\text{max}} = E_{\text{max}} d = (16 \times 10^6 \, \text{V/m})(1.0 \times 10^{-3} \, \text{m})$$

$$= 16 \times 10^3 \, \text{V}$$

Hence, the maximum charge is

$$Q_{\text{max}} = C \, \Delta V_{\text{max}} = (20 \times 10^{-12} \, \text{F})(16 \times 10^3 \, \text{V})$$

$$= \boxed{0.32 \ \mu\text{C}}$$

Example 26.7 Energy Stored Before and After

A parallel-plate capacitor is charged with a battery to a charge Q_0, as shown in Figure 26.20a. The battery is then removed, and a slab of material that has a dielectric constant κ is inserted between the plates, as shown in Figure 26.20b. Find the energy stored in the capacitor before and after the dielectric is inserted.

Figure 26.20 (Example 26.7) (a) A battery charges up a parallel-plate capacitor. (b) The battery is removed and a slab of dielectric material is inserted between the plates.

Solution From Equation 26.11, we see that the energy stored in the absence of the dielectric is

$$U_0 = \boxed{\frac{Q_0{}^2}{2C_0}}$$

After the battery is removed and the dielectric inserted, the *charge on the capacitor remains the same*. Hence, the energy stored in the presence of the dielectric is

$$U = \frac{Q_0{}^2}{2C}$$

But the capacitance in the presence of the dielectric is $C = \kappa C_0$, so U becomes

$$U = \frac{Q_0{}^2}{2 \kappa C_0} = \boxed{\frac{U_0}{\kappa}}$$

Because $\kappa > 1$, the final energy is less than the initial energy. We can account for the "missing" energy by noting that the dielectric, when inserted, is pulled into the device (see Section 26.7). An external agent must do negative work to keep the dielectric from accelerating. This work is simply the difference $U - U_0$. (Alternatively, the positive work done by the system on the external agent is $U_0 - U$.)

26.6   Electric Dipole in an Electric Field

We have discussed the effect on the capacitance of placing a dielectric between the plates of a capacitor. In Section 26.7, we shall describe the microscopic origin of this effect. Before we can do so, however, we need to expand upon the discussion of the electric dipole that we began in Section 23.4 (see Example 23.6). The electric dipole consists of two charges of equal magnitude and opposite sign separated by a distance $2a$, as shown in Figure 26.21. The **electric dipole moment** of this configuration is defined as the vector **p** directed from $-q$ toward $+q$ along the line joining the charges and having magnitude $2aq$:

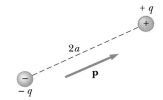

Figure 26.21 An electric dipole consists of two charges of equal magnitude and opposite sign separated by a distance of $2a$. The electric dipole moment **p** is directed from $-q$ toward $+q$.

$$p \equiv 2aq \tag{26.16}$$

Now suppose that an electric dipole is placed in a uniform electric field **E**, as shown in Figure 26.22. We identify **E** as the field *external* to the dipole, distinguishing it from the field *due to* the dipole, which we discussed in Section 23.4. The field **E** is established by some other charge distribution, and we place the dipole into this field. Let us imagine that the dipole moment makes an angle θ with the field.

The electric forces acting on the two charges are equal in magnitude ($F = qE$) and opposite in direction as shown in Figure 26.22. Thus, the net force on the dipole is zero. However, the two forces produce a net torque on the dipole; as a result, the dipole rotates in the direction that brings the dipole moment vector into greater alignment with the field. The torque due to the force on the positive charge about an axis through O in Figure 26.22 has magnitude $Fa \sin \theta$, where $a \sin \theta$ is the moment arm of F about O. This force tends to produce a clockwise rotation. The torque about O on the negative charge is also of magnitude $Fa \sin \theta$; here again, the force tends to produce a clockwise rotation. Thus, the magnitude of the net torque about O is

$$\tau = 2Fa \sin \theta$$

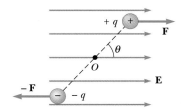

Figure 26.22 An electric dipole in a uniform external electric field. The dipole moment **p** is at an angle θ to the field, causing the dipole to experience a torque.

Because $F = qE$ and $p = 2aq$, we can express τ as

$$\tau = 2aqE \sin \theta = pE \sin \theta \tag{26.17}$$

It is convenient to express the torque in vector form as the cross product of the vectors **p** and **E**:

$$\boldsymbol{\tau} = \mathbf{p} \times \mathbf{E} \tag{26.18}$$

Torque on an electric dipole in an external electric field

We can determine the potential energy of the system—an electric dipole in an external electric field—as a function of the orientation of the dipole with respect to the field. To do this, we recognize that work must be done by an external agent to rotate the dipole through an angle so as to cause the dipole moment vector to become less aligned with the field. The work done is then stored as potential energy in the system. The work dW required to rotate the dipole through an angle $d\theta$ is $dW = \tau \, d\theta$ (Eq. 10.22). Because $\tau = pE \sin \theta$ and because the work results in an increase in the potential energy U, we find that for a rotation from θ_i to θ_f the change in potential energy of the system is

$$U_f - U_i = \int_{\theta_i}^{\theta_f} \tau \, d\theta = \int_{\theta_i}^{\theta_f} pE \sin \theta \, d\theta = pE \int_{\theta_i}^{\theta_f} \sin \theta \, d\theta$$

$$= pE[-\cos \theta]_{\theta_i}^{\theta_f} = pE(\cos \theta_i - \cos \theta_f)$$

The term that contains $\cos \theta_i$ is a constant that depends on the initial orientation of the dipole. It is convenient for us to choose a reference angle of $\theta_i = 90°$, so that $\cos \theta_i = \cos 90° = 0$. Furthermore, let us choose $U_i = 0$ at $\theta_i = 90°$ as our reference of potential energy. Hence, we can express a general value of $U = U_f$ as

$$U = -pE \cos \theta \tag{26.19}$$

We can write this expression for the potential energy of a dipole in an electric field as the dot product of the vectors **p** and **E**:

Potential energy of the system of an electric dipole in an external electric field

$$U = -\mathbf{p} \cdot \mathbf{E} \qquad (26.20)$$

To develop a conceptual understanding of Equation 26.19, compare this expression with the expression for the potential energy of the system of an object in the gravitational field of the Earth, $U = mgh$ (see Chapter 8). The gravitational expression includes a parameter associated with the object we place in the field—its mass m. Likewise, Equation 26.19 includes a parameter of the object in the electric field—its dipole moment p. The gravitational expression includes the magnitude of the gravitational field g. Similarly, Equation 26.19 includes the magnitude of the electric field E. So far, these two contributions to the potential energy expressions appear analogous. However, the final contribution is somewhat different in the two cases. In the gravitational expression, the potential energy depends on how high we lift the object, measured by h. In Equation 26.19, the potential energy depends on the angle θ through which we rotate the dipole. In both cases, we are making a change in the configuration of the system. In the gravitational case, the change involves moving an object in a *translational* sense, whereas in the electrical case, the change involves moving an object in a *rotational* sense. In both cases, however, once the change is made, the system tends to return to the original configuration when the object is released: the object of mass m falls back to the ground, and the dipole begins to rotate back toward the configuration in which it is aligned with the field. Thus, apart from the type of motion, the expressions for potential energy in these two cases are similar.

Molecules are said to be *polarized* when a separation exists between the average position of the negative charges and the average position of the positive charges in the molecule. In some molecules, such as water, this condition is always present—such molecules are called **polar molecules.** Molecules that do not possess a permanent polarization are called **nonpolar molecules.**

We can understand the permanent polarization of water by inspecting the geometry of the water molecule. In the water molecule, the oxygen atom is bonded to the hydrogen atoms such that an angle of 105° is formed between the two bonds (Fig. 26.23). The center of the negative charge distribution is near the oxygen atom, and the center of the positive charge distribution lies at a point midway along the line joining the hydrogen atoms (the point labeled × in Fig. 26.23). We can model the water molecule and other polar molecules as dipoles because the average positions of the positive and negative charges act as point charges. As a result, we can apply our discussion of dipoles to the behavior of polar molecules.

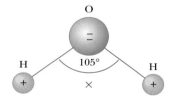

Figure 26.23 The water molecule, H_2O, has a permanent polarization resulting from its nonlinear geometry. The center of the positive charge distribution is at the point ×.

Microwave ovens take advantage of the polar nature of the water molecule. When in operation, microwave ovens generate a rapidly changing electric field that causes the polar molecules to swing back and forth, absorbing energy from the field in the process. Because the jostling molecules collide with each other, the energy they absorb from the field is converted to internal energy, which corresponds to an increase in temperature of the food.

Another household scenario in which the dipole structure of water is exploited is washing with soap and water. Grease and oil are made up of nonpolar molecules, which are generally not attracted to water. Plain water is not very useful for removing this type of grime. Soap contains long molecules called *surfactants*. In a long molecule, the polarity characteristics of one end of the molecule can be different from those at the other end. In a surfactant molecule, one end acts like a nonpolar molecule and the other acts like a polar molecule. The nonpolar end can attach to a grease or oil molecule, and the polar end can attach to a water molecule. Thus, the soap serves as a chain, linking the dirt and water molecules together. When the water is rinsed away, the grease and oil go with it.

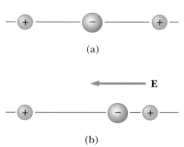

Figure 26.24 (a) A linear symmetric molecule has no permanent polarization. (b) An external electric field induces a polarization in the molecule.

A symmetric molecule (Fig. 26.24a) has no permanent polarization, but polarization can be induced by placing the molecule in an electric field. A field directed to the left, as shown in Figure 26.24b, would cause the center of the positive charge distribution to shift to the left from its initial position and the center of the negative charge distribution to shift to the right. This *induced polarization* is the effect that predominates in most materials used as dielectrics in capacitors.

Example 26.8 **The H₂O Molecule**

The water (H₂O) molecule has an electric dipole moment of 6.3×10^{-30} C·m. A sample contains 10^{21} water molecules, with the dipole moments all oriented in the direction of an electric field of magnitude 2.5×10^5 N/C. How much work is required to rotate the dipoles from this orientation ($\theta = 0°$) to one in which all the moments are perpendicular to the field ($\theta = 90°$)?

Solution The work required to rotate one molecule 90° is equal to the difference in potential energy between the 90° orientation and the 0° orientation. Using Equation 26.19,

we obtain

$$W = U_{90°} - U_{0°} = (-pE\cos 90°) - (-pE\cos 0°)$$
$$= pE = (6.3 \times 10^{-30}\ \text{C·m})(2.5 \times 10^5\ \text{N/C})$$
$$= 1.6 \times 10^{-24}\ \text{J}$$

Because there are 10^{21} molecules in the sample, the *total* work required is

$$W_{\text{total}} = (10^{21})(1.6 \times 10^{-24}\ \text{J}) = \boxed{1.6 \times 10^{-3}\ \text{J}}$$

26.7 An Atomic Description of Dielectrics

In Section 26.5 we found that the potential difference ΔV_0 between the plates of a capacitor is reduced to $\Delta V_0/\kappa$ when a dielectric is introduced. The potential difference is reduced because the magnitude of the electric field decreases between the plates. In particular, if $\mathbf{E}_0$ is the electric field without the dielectric, the field in the presence of a dielectric is

$$\mathbf{E} = \frac{\mathbf{E}_0}{\kappa} \qquad (26.21)$$

Let us first consider a dielectric made up of polar molecules placed in the electric field between the plates of a capacitor. The dipoles (that is, the polar molecules making up the dielectric) are randomly oriented in the absence of an electric field, as shown in Figure 26.25a. When an external field $\mathbf{E}_0$ due to charges on the capacitor plates is applied, a torque is exerted on the dipoles, causing them to partially align with the field, as shown in Figure 26.25b. We can now describe the dielectric as being polarized. The degree of alignment of the molecules with the electric field depends on temperature and on the magnitude of the field. In general, the alignment increases with decreasing temperature and with increasing electric field.

If the molecules of the dielectric are nonpolar, then the electric field due to the plates produces some charge separation and an *induced dipole moment*. These induced dipole moments tend to align with the external field, and the dielectric is polarized. Thus, we can polarize a dielectric with an external field regardless of whether the molecules are polar or nonpolar.

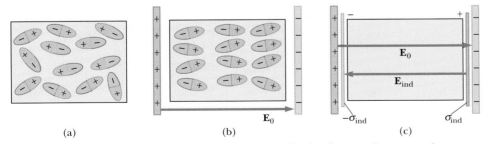

(a)   (b)   (c)

Figure 26.25 (a) Polar molecules are randomly oriented in the absence of an external electric field. (b) When an external electric field is applied, the molecules partially align with the field. (c) The charged edges of the dielectric can be modeled as an additional pair of parallel plates establishing an electric field $\mathbf{E}_{\text{ind}}$ in the direction opposite to that of $\mathbf{E}_0$.

With these ideas in mind, consider a slab of dielectric material placed between the plates of a capacitor so that it is in a uniform electric field $\mathbf{E}_0$, as shown in Figure 26.25b. The electric field due to the plates is directed to the right and polarizes the dielectric. The net effect on the dielectric is the formation of an *induced* positive surface charge density σ_{ind} on the right face and an equal-magnitude negative surface charge density $-\sigma_{ind}$ on the left face, as shown in Figure 26.25c. Because we can model these surface charge distributions as being due to parallel plates, the induced surface charges on the dielectric give rise to an induced electric field $\mathbf{E}_{ind}$ in the direction opposite the external field $\mathbf{E}_0$. Therefore, the net electric field $\mathbf{E}$ in the dielectric has a magnitude

$$E = E_0 - E_{ind} \tag{26.22}$$

In the parallel-plate capacitor shown in Figure 26.26, the external field E_0 is related to the charge density σ on the plates through the relationship $E_0 = \sigma/\epsilon_0$. The induced electric field in the dielectric is related to the induced charge density σ_{ind} through the relationship $E_{ind} = \sigma_{ind}/\epsilon_0$. Because $E = E_0/\kappa = \sigma/\kappa\epsilon_0$, substitution into Equation 26.22 gives

$$\frac{\sigma}{\kappa\epsilon_0} = \frac{\sigma}{\epsilon_0} - \frac{\sigma_{ind}}{\epsilon_0}$$

$$\sigma_{ind} = \left(\frac{\kappa - 1}{\kappa}\right)\sigma \tag{26.23}$$

Figure 26.26 Induced charge on a dielectric placed between the plates of a charged capacitor. Note that the induced charge density on the dielectric is *less* than the charge density on the plates.

Because $\kappa > 1$, this expression shows that the charge density σ_{ind} induced on the dielectric is less than the charge density σ on the plates. For instance, if $\kappa = 3$ we see that the induced charge density is two-thirds the charge density on the plates. If no dielectric is present, then $\kappa = 1$ and $\sigma_{ind} = 0$ as expected. However, if the dielectric is replaced by an electrical conductor, for which $E = 0$, then Equation 26.22 indicates that $E_0 = E_{ind}$; this corresponds to $\sigma_{ind} = \sigma$. That is, the surface charge induced on the conductor is equal in magnitude but opposite in sign to that on the plates, resulting in a net electric field of zero in the conductor (see Fig. 24.16).

We can use the existence of the induced surface charge distributions on the dielectric to explain the result of Example 26.7. As we saw there, the energy of a capacitor not connected to a battery is lowered when a dielectric is inserted between the plates; this means that negative work is done on the dielectric by the external agent inserting the dielectric into the capacitor. This, in turn, implies that a force must be acting on the dielectric that draws it into the capacitor. This force originates from the nonuniform nature of the electric field of the capacitor near its edges, as indicated in Figure 26.27.

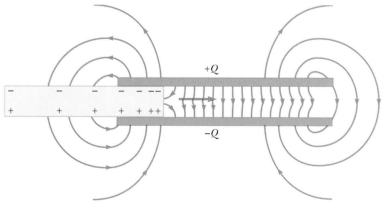

Figure 26.27 The nonuniform electric field near the edges of a parallel-plate capacitor causes a dielectric to be pulled into the capacitor. Note that the field acts on the induced surface charges on the dielectric, which are nonuniformly distributed.

The horizontal component of this *fringe field* acts on the induced charges on the surface of the dielectric, producing a net horizontal force directed into the space between the capacitor plates.

Example 26.9 Effect of a Metallic Slab

A parallel-plate capacitor has a plate separation d and plate area A. An uncharged metallic slab of thickness a is inserted midway between the plates.

(A) Find the capacitance of the device.

Solution We can solve this problem by noting that any charge that appears on one plate of the capacitor must induce a charge of equal magnitude and opposite sign on the near side of the slab, as shown in Figure 26.28a. Consequently, the net charge on the slab remains zero, and the electric field inside the slab is zero. Hence, the capacitor is equivalent to two capacitors in series, each having a plate separation $(d - a)/2$, as shown in Figure 26.28b.

Using Eq. 26.3 and the rule for adding two capacitors in series (Eq. 26.10), we obtain

$$\frac{1}{C} = \frac{1}{C_1} + \frac{1}{C_2} = \frac{1}{\left[\dfrac{\epsilon_0 A}{(d-a)/2}\right]} + \frac{1}{\left[\dfrac{\epsilon_0 A}{(d-a)/2}\right]}$$

$$C = \frac{\epsilon_0 A}{d - a}$$

Note that C approaches infinity as a approaches d. Why?

(B) Show that the capacitance of the original capacitor is unaffected by the insertion of the metallic slab if the slab is infinitesimally thin.

Solution In the result for part (A), we let $a \to 0$:

$$C = \lim_{a \to 0} \frac{\epsilon_0 A}{d - a} = \frac{\epsilon_0 A}{d}$$

which is the original capacitance.

What If? What if the metallic slab in part (A) is not midway between the plates? How does this affect the capacitance?

Answer Let us imagine that the slab in Figure 26.27a is moved upward so that the distance between the upper edge of the slab and the upper plate is b. Then, the distance between the lower edge of the slab and the lower plate is $d - b - a$. As in part (A), we find the total capacitance of the series combination:

$$\frac{1}{C} = \frac{1}{C_1} + \frac{1}{C_2} = \frac{1}{(\epsilon_0 A/b)} + \frac{1}{\epsilon_0 A/(d - b - a)}$$

$$= \frac{b}{\epsilon_0 A} + \frac{d - b - a}{\epsilon_0 A} = \frac{d - a}{\epsilon_0 A}$$

$$C = \frac{\epsilon_0 A}{d - a}$$

This is the same result as in part (A). It is independent of the value of b, so it does not matter where the slab is located. In Figure 26.28b, when the central structure is moved up or down, the decrease in plate separation of one capacitor is compensated by the increase in plate separation for the other.

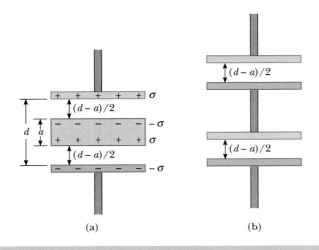

(a)   (b)

Figure 26.28 (Example 26.9) (a) A parallel-plate capacitor of plate separation d partially filled with a metallic slab of thickness a. (b) The equivalent circuit of the device in part (a) consists of two capacitors in series, each having a plate separation $(d - a)/2$.

Example 26.10 A Partially Filled Capacitor

A parallel-plate capacitor with a plate separation d has a capacitance C_0 in the absence of a dielectric. What is the capacitance when a slab of dielectric material of dielectric constant κ and thickness $\frac{1}{3}d$ is inserted between the plates (Fig. 26.29a)?

Solution In Example 26.9, we found that we could insert a metallic slab between the plates of a capacitor and consider the combination as two capacitors in series. The resulting capacitance was independent of the location of the slab. Furthermore, if the thickness of the slab approaches zero,

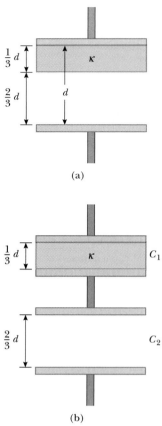

(a)

(b)

Figure 26.29 (Example 26.10) (a) A parallel-plate capacitor of plate separation d partially filled with a dielectric of thickness $d/3$. (b) The equivalent circuit of the capacitor consists of two capacitors connected in series.

then the capacitance of the system approaches the capacitance when the slab is absent. From this, we conclude that we can insert an infinitesimally thin metallic slab anywhere between the plates of a capacitor without affecting the capacitance. Thus, let us imagine sliding an infinitesimally thin metallic slab along the bottom face of the dielectric shown in Figure 26.29a. We can then consider this system to be the series combination of the two capacitors shown in Figure 26.29b: one having a plate separation $d/3$ and filled with a dielectric, and the other having a plate separation $2d/3$ and air between its plates.

From Equations 26.15 and 26.3, the two capacitances are

$$C_1 = \frac{\kappa \epsilon_0 A}{d/3} \qquad \text{and} \qquad C_2 = \frac{\epsilon_0 A}{2d/3}$$

Using Equation 26.10 for two capacitors combined in series, we have

$$\frac{1}{C} = \frac{1}{C_1} + \frac{1}{C_2} = \frac{d/3}{\kappa \epsilon_0 A} + \frac{2d/3}{\epsilon_0 A}$$

$$\frac{1}{C} = \frac{d}{3\,\epsilon_0 A}\left(\frac{1}{\kappa} + 2\right) = \frac{d}{3\,\epsilon_0 A}\left(\frac{1 + 2\kappa}{\kappa}\right)$$

$$C = \left(\frac{3\kappa}{2\kappa + 1}\right)\frac{\epsilon_0 A}{d}$$

Because the capacitance without the dielectric is $C_0 = \epsilon_0 A/d$, we see that

$$C = \left(\frac{3\kappa}{2\kappa + 1}\right) C_0$$

SUMMARY

Take a practice test for this chapter by clicking on the Practice Test link at http://www.pse6.com.

A **capacitor** consists of two conductors carrying charges of equal magnitude and opposite sign. The **capacitance** C of any capacitor is the ratio of the charge Q on either conductor to the potential difference ΔV between them:

$$C \equiv \frac{Q}{\Delta V} \qquad (26.1)$$

The capacitance depends only on the geometry of the conductors and not on an external source of charge or potential difference.

The SI unit of capacitance is coulombs per volt, or the **farad** (F), and 1 F = 1 C/V.

Capacitance expressions for various geometries are summarized in Table 26.2.

If two or more capacitors are connected in parallel, then the potential difference is the same across all of them. The equivalent capacitance of a parallel combination of capacitors is

$$C_{\text{eq}} = C_1 + C_2 + C_3 + \cdots \qquad (26.8)$$

If two or more capacitors are connected in series, the charge is the same on all of them, and the equivalent capacitance of the series combination is given by

Table 26.2

Capacitance and Geometry		
Geometry	**Capacitance**	**Equation**
Isolated sphere of radius R (second spherical conductor assumed to have infinite radius)	$C = 4\pi\epsilon_0 R$	26.2
Parallel-plate capacitor of plate area A and plate separation d	$C = \epsilon_0 \dfrac{A}{d}$	26.3
Cylindrical capacitor of length ℓ and inner and outer radii a and b, respectively	$C = \dfrac{\ell}{2k_e \ln(b/a)}$	26.4
Spherical capacitor with inner and outer radii a and b, respectively	$C = \dfrac{ab}{k_e(b-a)}$	26.6

$$\frac{1}{C_{eq}} = \frac{1}{C_1} + \frac{1}{C_2} + \frac{1}{C_3} + \cdots \qquad (26.10)$$

These two equations enable you to simplify many electric circuits by replacing multiple capacitors with a single equivalent capacitance.

Energy is stored in a capacitor because the charging process is equivalent to the transfer of charges from one conductor at a lower electric potential to another conductor at a higher potential. The energy stored in a capacitor with charge Q is

$$U = \frac{Q^2}{2C} = \tfrac{1}{2}Q\,\Delta V = \tfrac{1}{2}C(\Delta V)^2 \qquad (26.11)$$

When a dielectric material is inserted between the plates of a capacitor, the capacitance increases by a dimensionless factor κ, called the **dielectric constant:**

$$C = \kappa C_0 \qquad (26.14)$$

where C_0 is the capacitance in the absence of the dielectric. The increase in capacitance is due to a decrease in the magnitude of the electric field in the presence of the dielectric. The decrease in the magnitude of **E** arises from an internal electric field produced by aligned dipoles in the dielectric.

The **electric dipole moment p** of an electric dipole has a magnitude

$$p \equiv 2aq \qquad (26.16)$$

The direction of the electric dipole moment vector is from the negative charge toward the positive charge.

The torque acting on an electric dipole in a uniform electric field **E** is

$$\boldsymbol{\tau} = \mathbf{p} \times \mathbf{E} \qquad (26.18)$$

The potential energy of the system of an electric dipole in a uniform external electric field **E** is

$$U = -\mathbf{p} \cdot \mathbf{E} \qquad (26.20)$$

QUESTIONS

1. The plates of a capacitor are connected to a battery. What happens to the charge on the plates if the connecting wires are removed from the battery? What happens to the charge if the wires are removed from the battery and connected to each other?

2. A farad is a very large unit of capacitance. Calculate the length of one side of a square, air-filled capacitor that has a capacitance of 1 F and a plate separation of 1 m.

3. A pair of capacitors are connected in parallel while an identical pair are connected in series. Which pair would be

more dangerous to handle after being connected to the same battery? Explain.

4. If you are given three different capacitors C_1, C_2, C_3, how many different combinations of capacitance can you produce?

5. What advantage might there be in using two identical capacitors in parallel connected in series with another identical parallel pair, rather than using a single capacitor?

6. Is it always possible to reduce a combination of capacitors to one equivalent capacitor with the rules we have developed? Explain.

7. The sum of the charges on both plates of a capacitor is zero. What does a capacitor store?

8. Because the charges on the plates of a parallel-plate capacitor are opposite in sign, they attract each other. Hence, it would take positive work to increase the plate separation. What type of energy in the system changes due to the external work done in this process?

9. Why is it dangerous to touch the terminals of a high-voltage capacitor even after the applied potential difference has been turned off? What can be done to make the capacitor safe to handle after the voltage source has been removed?

10. Explain why the work needed to move a charge Q through a potential difference ΔV is $W = Q\,\Delta V$ whereas the energy stored in a charged capacitor is $U = \frac{1}{2}Q\,\Delta V$. Where does the $\frac{1}{2}$ factor come from?

11. If the potential difference across a capacitor is doubled, by what factor does the energy stored change?

12. It is possible to obtain large potential differences by first charging a group of capacitors connected in parallel and then activating a switch arrangement that in effect dis-connects the capacitors from the charging source and from each other and reconnects them in a series arrangement. The group of charged capacitors is then discharged in series. What is the maximum potential difference that can be obtained in this manner by using ten capacitors each of 500 μF and a charging source of 800 V?

13. Assume you want to increase the maximum operating voltage of a parallel-plate capacitor. Describe how you can do this for a fixed plate separation.

14. An air-filled capacitor is charged, then disconnected from the power supply, and finally connected to a voltmeter. Explain how and why the potential difference changes when a dielectric is inserted between the plates of the capacitor.

15. Using the polar molecule description of a dielectric, explain how a dielectric affects the electric field inside a capacitor.

16. Explain why a dielectric increases the maximum operating voltage of a capacitor although the physical size of the capacitor does not change.

17. What is the difference between dielectric strength and the dielectric constant?

18. Explain why a water molecule is permanently polarized. What type of molecule has no permanent polarization?

19. If a dielectric-filled capacitor is heated, how will its capacitance change? (Ignore thermal expansion and assume that the dipole orientations are temperature-dependent.)

20. If you were asked to design a capacitor where small size and large capacitance were required, what factors would be important in your design?

PROBLEMS

1, 2, 3 = straightforward, intermediate, challenging ☐ = full solution available in the *Student Solutions Manual and Study Guide*

= coached solution with hints available at http://www.pse6.com ▭ = computer useful in solving problem

= paired numerical and symbolic problems

Section 26.1 Definition of Capacitance

1. (a) How much charge is on each plate of a 4.00-μF capacitor when it is connected to a 12.0-V battery? (b) If this same capacitor is connected to a 1.50-V battery, what charge is stored?

2. Two conductors having net charges of $+10.0\ \mu$C and $-10.0\ \mu$C have a potential difference of 10.0 V between them. (a) Determine the capacitance of the system. (b) What is the potential difference between the two conductors if the charges on each are increased to $+100\ \mu$C and $-100\ \mu$C?

Section 26.2 Calculating Capacitance

3. An isolated charged conducting sphere of radius 12.0 cm creates an electric field of 4.90×10^4 N/C at a distance 21.0 cm from its center. (a) What is its surface charge density? (b) What is its capacitance?

4. (a) If a drop of liquid has capacitance 1.00 pF, what is its radius? (b) If another drop has radius 2.00 mm, what is its capacitance? (c) What is the charge on the smaller drop if its potential is 100 V?

5. Two conducting spheres with diameters of 0.400 m and 1.00 m are separated by a distance that is large compared with the diameters. The spheres are connected by a thin wire and are charged to 7.00 μC. (a) How is this total charge shared between the spheres? (Ignore any charge on the wire.) (b) What is the potential of the system of spheres when the reference potential is taken to be $V = 0$ at $r = \infty$?

6. Regarding the Earth and a cloud layer 800 m above the Earth as the "plates" of a capacitor, calculate the capacitance. Assume the cloud layer has an area of 1.00 km^2 and that the air between the cloud and the ground is pure and dry. Assume charge builds up on the cloud and on the ground until a uniform electric field of 3.00×10^6 N/C throughout the space between them makes the air break down and conduct electricity as a lightning bolt. What is the maximum charge the cloud can hold?

7. 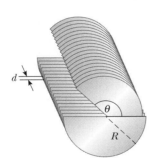 An air-filled capacitor consists of two parallel plates, each with an area of 7.60 cm^2, separated by a distance of 1.80 mm. A 20.0-V potential difference is applied to these plates. Calculate (a) the electric field between the plates, (b) the surface charge density, (c) the capacitance, and (d) the charge on each plate.

8. A 1-megabit computer memory chip contains many 60.0-f F capacitors. Each capacitor has a plate area of 21.0×10^{-12} m^2. Determine the plate separation of such a capacitor (assume a parallel-plate configuration). The order of magnitude of the diameter of an atom is 10^{-10} m $= 0.1$ nm. Express the plate separation in nanometers.

9. When a potential difference of 150 V is applied to the plates of a parallel-plate capacitor, the plates carry a surface charge density of 30.0 nC/cm^2. What is the spacing between the plates?

10. A variable air capacitor used in a radio tuning circuit is made of N semicircular plates each of radius R and positioned a distance d from its neighbors, to which it is electrically connected. As shown in Figure P26.10, a second identical set of plates is enmeshed with its plates halfway between those of the first set. The second set can rotate as a unit. Determine the capacitance as a function of the angle of rotation θ, where $\theta = 0$ corresponds to the maximum capacitance.

Figure P26.10

11. 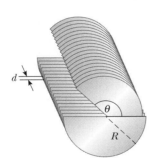 A 50.0-m length of coaxial cable has an inner conductor that has a diameter of 2.58 mm and carries a charge of 8.10 μC. The surrounding conductor has an inner diameter of 7.27 mm and a charge of -8.10 μC. (a) What is the capacitance of this cable? (b) What is the potential difference between the two conductors? Assume the region between the conductors is air.

12. A 20.0-μF spherical capacitor is composed of two concentric metal spheres, one having a radius twice as large as the

other. The region between the spheres is a vacuum. Determine the volume of this region.

13. An air-filled spherical capacitor is constructed with inner and outer shell radii of 7.00 and 14.0 cm, respectively. (a) Calculate the capacitance of the device. (b) What potential difference between the spheres results in a charge of 4.00 μC on the capacitor?

14. A small object of mass m carries a charge q and is suspended by a thread between the vertical plates of a parallel-plate capacitor. The plate separation is d. If the thread makes an angle θ with the vertical, what is the potential difference between the plates?

15. Find the capacitance of the Earth. (*Suggestion:* The outer conductor of the "spherical capacitor" may be considered as a conducting sphere at infinity where V approaches zero.)

Section 26.3 Combinations of Capacitors

16. Two capacitors, $C_1 = 5.00$ μF and $C_2 = 12.0$ μF, are connected in parallel, and the resulting combination is connected to a 9.00-V battery. (a) What is the equivalent capacitance of the combination? What are (b) the potential difference across each capacitor and (c) the charge stored on each capacitor?

17. **What If?** The two capacitors of Problem 16 are now connected in series and to a 9.00-V battery. Find (a) the equivalent capacitance of the combination, (b) the potential difference across each capacitor, and (c) the charge on each capacitor.

18. Evaluate the equivalent capacitance of the configuration shown in Figure P26.18. All the capacitors are identical, and each has capacitance C.

$$C$$
$$C \quad C$$
$$C \quad C \quad C$$

Figure P26.18

19. Two capacitors when connected in parallel give an equivalent capacitance of 9.00 pF and give an equivalent capacitance of 2.00 pF when connected in series. What is the capacitance of each capacitor?

20. Two capacitors when connected in parallel give an equivalent capacitance of C_p and an equivalent capacitance of C_s when connected in series. What is the capacitance of each capacitor?

21. 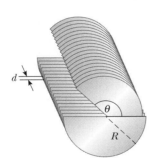 Four capacitors are connected as shown in Figure P26.21. (a) Find the equivalent capacitance between points a and b. (b) Calculate the charge on each capacitor if $\Delta V_{ab} = 15.0$ V.

Figure P26.21

22. Three capacitors are connected to a battery as shown in Figure P26.22. Their capacitances are $C_1 = 3C$, $C_2 = C$, and $C_3 = 5C$. (a) What is the equivalent capacitance of this set of capacitors? (b) State the ranking of the capacitors according to the charge they store, from largest to smallest. (c) Rank the capacitors according to the potential differences across them, from largest to smallest. (d) **What If?** If C_3 is increased, what happens to the charge stored by each of the capacitors?

Figure P26.22

23. Consider the circuit shown in Figure P26.23, where $C_1 = 6.00$ μF, $C_2 = 3.00$ μF, and $\Delta V = 20.0$ V. Capacitor C_1 is first charged by the closing of switch S_1. Switch S_1 is then opened, and the charged capacitor is connected to the uncharged capacitor by the closing of S_2. Calculate the initial charge acquired by C_1 and the final charge on each capacitor.

Figure P26.23

24. According to its design specification, the timer circuit delaying the closing of an elevator door is to have a capacitance of 32.0 μF between two points A and B. (a) When one circuit is being constructed, the inexpensive but durable capacitor installed between these two points is found to have capacitance 34.8 μF. To meet the specification, one additional capacitor can be placed between the two points. Should it be in series or in parallel with the 34.8-μF capacitor? What should be its capacitance? (b) **What If?** The next circuit comes down the assembly

line with capacitance 29.8 μF between A and B. What additional capacitor should be installed in series or in parallel in that circuit, to meet the specification?

25. A group of identical capacitors is connected first in series and then in parallel. The combined capacitance in parallel is 100 times larger than for the series connection. How many capacitors are in the group?

26. Consider three capacitors C_1, C_2, C_3, and a battery. If C_1 is connected to the battery, the charge on C_1 is 30.8 μC. Now C_1 is disconnected, discharged, and connected in series with C_2. When the series combination of C_2 and C_1 is connected across the battery, the charge on C_1 is 23.1 μC. The circuit is disconnected and the capacitors discharged. Capacitor C_3, capacitor C_1, and the battery are connected in series, resulting in a charge on C_1 of 25.2 μC. If, after being disconnected and discharged, C_1, C_2, and C_3 are connected in series with one another and with the battery, what is the charge on C_1?

27. Find the equivalent capacitance between points a and b for the group of capacitors connected as shown in Figure P26.27. Take $C_1 = 5.00$ μF, $C_2 = 10.0$ μF, and $C_3 = 2.00$ μF.

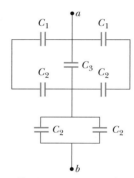

Figure P26.27 Problems 27 and 28.

28. For the network described in the previous problem, if the potential difference between points a and b is 60.0 V, what charge is stored on C_3?

29. Find the equivalent capacitance between points a and b in the combination of capacitors shown in Figure P26.29.

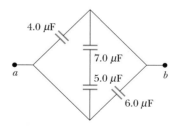

Figure P26.29

30. Some physical systems possessing capacitance continuously distributed over space can be modeled as an infinite array of discrete circuit elements. Examples are a microwave waveguide and the axon of a nerve cell. To practice analysis of an infinite array, determine the equivalent capacitance C between terminals X and Y of the infinite set of capacitors represented in Figure P26.30. Each capacitor has capacitance C_0. *Suggestion:* Imagine that the ladder is cut at the line AB, and note that the equivalent capacitance of the infinite section to the right of AB is also C.

Figure P26.30

Section 26.4 Energy Stored in a Charged Capacitor

31. (a) A 3.00-μF capacitor is connected to a 12.0-V battery. How much energy is stored in the capacitor? (b) If the capacitor had been connected to a 6.00-V battery, how much energy would have been stored?

32. The immediate cause of many deaths is ventricular fibrillation, uncoordinated quivering of the heart as opposed to proper beating. An electric shock to the chest can cause momentary paralysis of the heart muscle, after which the heart will sometimes start organized beating again. A *defibrillator* (Fig. 26.14) is a device that applies a strong electric shock to the chest over a time interval of a few milliseconds. The device contains a capacitor of several microfarads, charged to several thousand volts. Electrodes called paddles, about 8 cm across and coated with conducting paste, are held against the chest on both sides of the heart. Their handles are insulated to prevent injury to the operator, who calls, "Clear!" and pushes a button on one paddle to discharge the capacitor through the patient's chest. Assume that an energy of 300 J is to be delivered from a 30.0-μF capacitor. To what potential difference must it be charged?

33. Two capacitors, $C_1 = 25.0$ μF and $C_2 = 5.00$ μF, are connected in parallel and charged with a 100-V power supply. (a) Draw a circuit diagram and calculate the total energy stored in the two capacitors. (b) **What If?** What potential difference would be required across the same two capacitors connected in series in order that the combination stores the same amount of energy as in (a)? Draw a circuit diagram of this circuit.

34. A parallel-plate capacitor is charged and then disconnected from a battery. By what fraction does the stored energy change (increase or decrease) when the plate separation is doubled?

35. As a person moves about in a dry environment, electric charge accumulates on his body. Once it is at high voltage, either positive or negative, the body can discharge via sometimes noticeable sparks and shocks. Consider a human body well separated from ground, with the typical capacitance 150 pF. (a) What charge on the body will produce a potential of 10.0 kV? (b) Sensitive electronic devices can be destroyed by electrostatic discharge from a person. A particular device can be destroyed by a discharge releasing an energy of 250 μJ. To what voltage on the body does this correspond?

36. A uniform electric field $E = 3\,000$ V/m exists within a certain region. What volume of space contains an energy equal to 1.00×10^{-7} J? Express your answer in cubic meters and in liters.

37. A parallel-plate capacitor has a charge Q and plates of area A. What force acts on one plate to attract it toward the other plate? Because the electric field between the plates is $E = Q/A\epsilon_0$, you might think that the force is $F = QE = Q^2/A\epsilon_0$. This is wrong, because the field E includes contributions from both plates, and the field created by the positive plate cannot exert any force on the positive plate. Show that the force exerted on each plate is actually $F = Q^2/2\epsilon_0 A$. (*Suggestion:* Let $C = \epsilon_0 A/x$ for an arbitrary plate separation x; then require that the work done in separating the two charged plates be $W = \int F\,dx$.) The force exerted by one charged plate on another is sometimes used in a machine shop to hold a workpiece stationary.

38. The circuit in Figure P26.38 consists of two identical parallel metal plates connected by identical metal springs to a 100-V battery. With the switch open, the plates are uncharged, are separated by a distance $d = 8.00$ mm, and have a capacitance $C = 2.00$ μF. When the switch is closed, the distance between the plates decreases by a factor of 0.500. (a) How much charge collects on each plate and (b) what is the spring constant for each spring? (*Suggestion:* Use the result of Problem 37.)

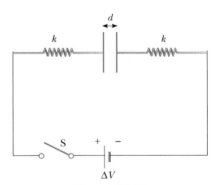

Figure P26.38

39. **Review problem.** A certain storm cloud has a potential of 1.00×10^8 V relative to a tree. If, during a lightning storm, 50.0 C of charge is transferred through this potential difference and 1.00% of the energy is absorbed by the tree, how much sap in the tree can be boiled away? Model the sap as water initially at 30.0°C. Water has a specific heat of 4 186 J/kg°C, a boiling point of 100°C, and a latent heat of vaporization of 2.26×10^6 J/kg.

40. Two identical parallel-plate capacitors, each with capacitance C, are charged to potential difference ΔV and connected in parallel. Then the plate separation in one of the capacitors is doubled. (a) Find the total energy of the system of two capacitors *before* the plate separation is doubled. (b) Find the potential difference across each capacitor *after* the plate separation is doubled. (c) Find the

total energy of the system *after* the plate separation is doubled. (d) Reconcile the difference in the answers to parts (a) and (c) with the law of conservation of energy.

41. Show that the energy associated with a conducting sphere of radius R and charge Q surrounded by a vacuum is $U = k_e Q^2 / 2R$.

42. Consider two conducting spheres with radii R_1 and R_2. They are separated by a distance much greater than either radius. A total charge Q is shared between the spheres, subject to the condition that the electric potential energy of the system has the smallest possible value. The total charge Q is equal to $q_1 + q_2$, where q_1 represents the charge on the first sphere and q_2 the charge on the second. Because the spheres are very far apart, you can assume that the charge of each is uniformly distributed over its surface. You may use the result of Problem 41. (a) Determine the values of q_1 and q_2 in terms of Q, R_1, and R_2. (b) Show that the potential difference between the spheres is zero. (We saw in Chapter 25 that two conductors joined by a conducting wire will be at the same potential in a static situation. This problem illustrates the general principle that static charge on a conductor will distribute itself so that the electric potential energy of the system is a minimum.)

Section 26.5 Capacitors with Dielectrics

43. Determine (a) the capacitance and (b) the maximum potential difference that can be applied to a Teflon-filled parallel-plate capacitor having a plate area of 1.75 cm^2 and plate separation of 0.040 0 mm.

44. (a) How much charge can be placed on a capacitor with air between the plates before it breaks down, if the area of each of the plates is 5.00 cm^2? (b) **What If?** Find the maximum charge if polystyrene is used between the plates instead of air.

45. A commercial capacitor is to be constructed as shown in Figure 26.17a. This particular capacitor is made from two strips of aluminum separated by a strip of paraffin-coated paper. Each strip of foil and paper is 7.00 cm wide. The foil is 0.004 00 mm thick, and the paper is 0.025 0 mm thick and has a dielectric constant of 3.70. What length should the strips have, if a capacitance of 9.50×10^{-8} F is desired before the capacitor is rolled up? (Adding a second strip of paper and rolling the capacitor effectively doubles its capacitance, by allowing charge storage on both sides of each strip of foil.)

46. The supermarket sells rolls of aluminum foil, of plastic wrap, and of waxed paper. Describe a capacitor made from supermarket materials. Compute order-of-magnitude estimates for its capacitance and its breakdown voltage.

47. A parallel-plate capacitor in air has a plate separation of 1.50 cm and a plate area of 25.0 cm^2. The plates are charged to a potential difference of 250 V and disconnected from the source. The capacitor is then immersed in distilled water. Determine (a) the charge on the plates before and after immersion, (b) the capacitance and potential difference after immersion, and (c) the change in energy of the capacitor. Assume the liquid is an insulator.

48. A wafer of titanium dioxide ($\kappa = 173$) of area 1.00 cm^2 has a thickness of 0.100 mm. Aluminum is evaporated on the parallel faces to form a parallel-plate capacitor. (a) Calculate the capacitance. (b) When the capacitor is charged with a 12.0-V battery, what is the magnitude of charge delivered to each plate? (c) For the situation in part (b), what are the free and induced surface charge densities? (d) What is the magnitude of the electric field?

49. Each capacitor in the combination shown in Figure P26.49 has a breakdown voltage of 15.0 V. What is the breakdown voltage of the combination?

Figure P26.49

Section 26.6 Electric Dipole in an Electric Field

50. A small rigid object carries positive and negative 3.50-nC charges. It is oriented so that the positive charge has coordinates $(-1.20$ mm, 1.10 mm) and the negative charge is at the point (1.40 mm, -1.30 mm). (a) Find the electric dipole moment of the object. The object is placed in an electric field $\mathbf{E} = (7\,800\hat{\mathbf{i}} - 4\,900\hat{\mathbf{j}})$ N/C. (b) Find the torque acting on the object. (c) Find the potential energy of the object–field system when the object is in this orientation. (d) If the orientation of the object can change, find the difference between the maximum and minimum potential energies of the system.

51. A small object with electric dipole moment $\mathbf{p}$ is placed in a nonuniform electric field $\mathbf{E} = E(x)\hat{\mathbf{i}}$. That is, the field is in the x direction and its magnitude depends on the coordinate x. Let θ represent the angle between the dipole moment and the x direction. (a) Prove that the dipole feels a net force

$$F = p\left(\frac{dE}{dx}\right)\cos\theta$$

in the direction toward which the field increases. (b) Consider a spherical balloon centered at the origin, with radius 15.0 cm and carrying charge 2.00 μC. Evaluate dE/dx at the point (16 cm, 0, 0). Assume a water droplet at this point has an induced dipole moment of $6.30\hat{\mathbf{i}}$ nC·m. Find the force on it.

Section 26.7 An Atomic Description of Dielectrics

52. A detector of radiation called a Geiger tube consists of a closed, hollow, conducting cylinder with a fine wire along its axis. Suppose that the internal diameter of the cylinder is 2.50 cm and that the wire along the axis has a diameter of 0.200 mm. The dielectric strength of the gas between the central wire and the cylinder is 1.20×10^6 V/m. Calculate the maximum potential difference that can be applied between the wire and the cylinder before breakdown occurs in the gas.

53. The general form of Gauss's law describes how a charge creates an electric field in a material, as well as in vacuum. It is

$$\oint \mathbf{E} \cdot d\mathbf{A} = \frac{q}{\epsilon}$$

where $\epsilon = \kappa \epsilon_0$ is the permittivity of the material. (a) A sheet with charge Q uniformly distributed over its area A is surrounded by a dielectric. Show that the sheet creates a uniform electric field at nearby points, with magnitude $E = Q/2A\epsilon$. (b) Two large sheets of area A, carrying opposite charges of equal magnitude Q, are a small distance d apart. Show that they create uniform electric field in the space between them, with magnitude $E = Q/A\epsilon$. (c) Assume that the negative plate is at zero potential. Show that the positive plate is at potential $Qd/A\epsilon$. (d) Show that the capacitance of the pair of plates is $A\epsilon/d = \kappa A\epsilon_0/d$.

Additional Problems

54. For the system of capacitors shown in Figure P26.54, find (a) the equivalent capacitance of the system, (b) the potential across each capacitor, (c) the charge on each capacitor, and (d) the total energy stored by the group.

Figure P26.54

55. Four parallel metal plates P_1, P_2, P_3, and P_4, each of area 7.50 cm^2, are separated successively by a distance $d = 1.19$ mm, as shown in Figure P26.55. P_1 is connected to the negative terminal of a battery, and P_2 to the positive terminal. The battery maintains a potential difference of 12.0 V. (a) If P_3 is connected to the negative terminal, what is the

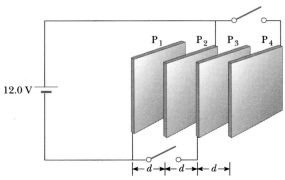

Figure P26.55

capacitance of the three-plate system $P_1P_2P_3$? (b) What is the charge on P_2? (c) If P_4 is now connected to the positive terminal of the battery, what is the capacitance of the four-plate system $P_1P_2P_3P_4$? (d) What is the charge on P_4?

56. One conductor of an overhead electric transmission line is a long aluminum wire 2.40 cm in radius. Suppose that at a particular moment it carries charge per length 1.40 μC/m and is at potential 345 kV. Find the potential 12.0 m below the wire. Ignore the other conductors of the transmission line and assume the electric field is everywhere purely radial.

57. Two large parallel metal plates are oriented horizontally and separated by a distance $3d$. A grounded conducting wire joins them, and initially each plate carries no charge. Now a third identical plate carrying charge Q is inserted between the two plates, parallel to them and located a distance d from the upper plate, as in Figure P26.57. (a) What induced charge appears on each of the two original plates? (b) What potential difference appears between the middle plate and each of the other plates? Each plate has area A.

Figure P26.57

58. A 2.00-nF parallel-plate capacitor is charged to an initial potential difference $\Delta V_i = 100$ V and then isolated. The dielectric material between the plates is mica, with a dielectric constant of 5.00. (a) How much work is required to withdraw the mica sheet? (b) What is the potential difference of the capacitor after the mica is withdrawn?

59. A parallel-plate capacitor is constructed using a dielectric material whose dielectric constant is 3.00 and whose dielectric strength is 2.00×10^8 V/m. The desired capacitance is 0.250 μF, and the capacitor must withstand a maximum potential difference of 4 000 V. Find the minimum area of the capacitor plates.

60. A 10.0-μF capacitor has plates with vacuum between them. Each plate carries a charge of magnitude 1 000 μC. A particle with charge -3.00 μC and mass 2.00×10^{-16} kg is fired from the positive plate toward the negative plate with an initial speed of 2.00×10^6 m/s. Does it reach the negative plate? If so, find its impact speed. If not, what fraction of the way across the capacitor does it travel?

61. A parallel-plate capacitor is constructed by filling the space between two square plates with blocks of three dielectric materials, as in Figure P26.61. You may assume that $\ell \gg d$. (a) Find an expression for the capacitance of the device in terms of the plate area A and d, κ_1, κ_2, and κ_3. (b) Calculate the capacitance using the values $A = 1.00$ cm^2, $d = 2.00$ mm, $\kappa_1 = 4.90$, $\kappa_2 = 5.60$, and $\kappa_3 = 2.10$.

Figure P26.61

62. A 10.0-μF capacitor is charged to 15.0 V. It is next connected in series with an uncharged 5.00-μF capacitor. The series combination is finally connected across a 50.0-V battery, as diagrammed in Figure P26.62. Find the new potential differences across the 5-μF and 10-μF capacitors.

Figure P26.62

63. (a) Two spheres have radii a and b and their centers are a distance d apart. Show that the capacitance of this system is

$$C = \frac{4\pi\epsilon_0}{\dfrac{1}{a} + \dfrac{1}{b} - \dfrac{2}{d}}$$

provided that d is large compared with a and b. (*Suggestion:* Because the spheres are far apart, assume that the potential of each equals the sum of the potentials due to each sphere, and when calculating those potentials assume that $V = k_eQ/r$ applies.) (b) Show that as d approaches infinity the above result reduces to that of two spherical capacitors in series.

64. A capacitor is constructed from two square plates of sides ℓ and separation d. A material of dielectric constant κ is inserted a distance x into the capacitor, as shown in Figure P26.64. Assume that d is much smaller than x. (a) Find the equivalent capacitance of the device. (b) Calculate the energy stored in the capacitor, letting ΔV repre-

sent the potential difference. (c) Find the direction and magnitude of the force exerted on the dielectric, assuming a constant potential difference ΔV. Ignore friction. (d) Obtain a numerical value for the force assuming that $\ell = 5.00$ cm, $\Delta V = 2\,000$ V, $d = 2.00$ mm, and the dielectric is glass ($\kappa = 4.50$). (*Suggestion:* The system can be considered as two capacitors connected in parallel.)

65. A capacitor is constructed from two square plates of sides ℓ and separation d, as suggested in Figure P26.64. You may assume that d is much less than ℓ. The plates carry charges $+Q_0$ and $-Q_0$. A block of metal has a width ℓ, a length ℓ, and a thickness slightly less than d. It is inserted a distance x into the capacitor. The charges on the plates are not disturbed as the block slides in. In a static situation, a metal prevents an electric field from penetrating inside it. The metal can be thought of as a perfect dielectric, with $\kappa \to \infty$. (a) Calculate the stored energy as a function of x. (b) Find the direction and magnitude of the force that acts on the metallic block. (c) The area of the advancing front face of the block is essentially equal to ℓd. Considering the force on the block as acting on this face, find the stress (force per area) on it. (d) For comparison, express the energy density in the electric field between the capacitor plates in terms of Q_0, ℓ, d, and ϵ_0.

66. When considering the energy supply for an automobile, the energy per unit mass of the energy source is an important parameter. Using the following data, compare the energy per unit mass (J/kg) for gasoline, lead–acid batteries, and capacitors. (The ampere A will be introduced in the next chapter as the SI unit of electric current. 1 A = 1 C/s.)
Gasoline: 126 000 Btu/gal; density = 670 kg/m^3.
Lead–acid battery: 12.0 V; 100 A · h; mass = 16.0 kg.
Capacitor: potential difference at full charge = 12.0 V; capacitance = 0.100 F; mass = 0.100 kg.

67. An isolated capacitor of unknown capacitance has been charged to a potential difference of 100 V. When the charged capacitor is then connected in parallel to an uncharged 10.0-μF capacitor, the potential difference across the combination is 30.0 V. Calculate the unknown capacitance.

68. To repair a power supply for a stereo amplifier, an electronics technician needs a 100-μF capacitor capable of withstanding a potential difference of 90 V between the plates. The only available supply is a box of five 100-μF capacitors, each having a maximum voltage capability of 50 V. Can the technician substitute a combination of these capacitors that has the proper electrical characteristics? If so, what will be the maximum voltage across any of the capacitors used? (*Suggestion:* The technician may not have to use all the capacitors in the box.)

69. A parallel-plate capacitor of plate separation d is charged to a potential difference ΔV_0. A dielectric slab of thickness d and dielectric constant κ is introduced between the plates while the battery remains connected to the plates. (a) Show that the ratio of energy stored after the dielectric is introduced to the energy stored in the empty capacitor is $U/U_0 = \kappa$. Give a physical explanation for this increase in stored energy. (b) What happens to the charge on the capacitor? (Note that this situation is not the same as in

Figure P26.64 Problems 64 and 65.

Example 26.7, in which the battery was removed from the circuit before the dielectric was introduced.)

70. A vertical parallel-plate capacitor is half filled with a dielectric for which the dielectric constant is 2.00 (Fig. P26.70a). When this capacitor is positioned horizontally, what fraction of it should be filled with the same dielectric (Fig. P26.70b) in order for the two capacitors to have equal capacitance?

(a)　　　　　　　　(b)

Figure P26.70

71. Capacitors $C_1 = 6.00 \ \mu F$ and $C_2 = 2.00 \ \mu F$ are charged as a parallel combination across a 250-V battery. The capacitors are disconnected from the battery and from each other. They are then connected positive plate to negative plate and negative plate to positive plate. Calculate the resulting charge on each capacitor.

72. Calculate the equivalent capacitance between the points a and b in Figure P26.72. Note that this is not a simple series or parallel combination. (*Suggestion:* Assume a potential difference ΔV between points a and b. Write expressions for ΔV_{ab} in terms of the charges and capacitances for the various possible pathways from a to b, and require conservation of charge for those capacitor plates that are connected to each other.)

Figure P26.72

73. The inner conductor of a coaxial cable has a radius of 0.800 mm, and the outer conductor's inside radius is 3.00 mm. The space between the conductors is filled with polyethylene, which has a dielectric constant of 2.30 and a dielectric strength of 18.0×10^6 V/m. What is the maximum potential difference that this cable can withstand?

74. You are optimizing coaxial cable design for a major manufacturer. Show that for a given outer conductor radius b, maximum potential difference capability is attained when the radius of the inner conductor is $a = b/e$ where e is the base of natural logarithms.

75. Determine the equivalent capacitance of the combination shown in Figure P26.75. (*Suggestion:* Consider the symmetry involved.)

Figure P26.75

76. Consider two long, parallel, and oppositely charged wires of radius d with their centers separated by a distance D. Assuming the charge is distributed uniformly on the surface of each wire, show that the capacitance per unit length of this pair of wires is

$$\frac{C}{\ell} = \frac{\pi \epsilon_0}{\ln[(D - d)/d]}$$

77. Example 26.2 explored a cylindrical capacitor of length ℓ and radii a and b of the two conductors. In the **What If?** section, it was claimed that increasing ℓ by 10% is more effective in terms of increasing the capacitance than increasing a by 10% if $b > 2.85a$. Verify this claim mathematically.

Answers to Quick Quizzes

26.1 (d). The capacitance is a property of the physical system and does not vary with applied voltage. According to Equation 26.1, if the voltage is doubled, the charge is doubled.

26.2 (a). When the key is pressed, the plate separation is decreased and the capacitance increases. Capacitance depends only on how a capacitor is constructed and not on the external circuit.

26.3 (a). When connecting capacitors in series, the inverses of the capacitances add, resulting in a smaller overall equivalent capacitance.

26.4 (a). When capacitors are connected in series, the voltages add, for a total of 20 V in this case. If they are combined in parallel, the voltage across the combination is still 10 V.

26.5 (b). For a given voltage, the energy stored in a capacitor is proportional to C: $U = C(\Delta V)^2/2$. Thus, you want to maximize the equivalent capacitance. You do this by connecting the three capacitors in parallel, so that the capacitances add.

26.6 (a) C decreases (Eq. 26.3). (b) Q stays the same because there is no place for the charge to flow. (c) E remains constant (see Eq. 24.8 and the paragraph following it). (d) ΔV increases because $\Delta V = Q/C$, Q is constant (part b), and C decreases (part a). (e) The energy stored in the capacitor is proportional to both Q and ΔV (Eq.

26.11) and thus increases. The additional energy comes from the work you do in pulling the two plates apart.

26.7 (a) C decreases (Eq. 26.3). (b) Q decreases. The battery supplies a constant potential difference ΔV; thus, charge must flow out of the capacitor if $C = Q/\Delta V$ is to decrease. (c) E decreases because the charge density on the plates decreases. (d) ΔV remains constant because of the presence of the battery. (e) The energy stored in the capacitor decreases (Eq. 26.11).

26.8 Increase. The dielectric constant of wood (and of all other insulating materials, for that matter) is greater than 1; therefore, the capacitance increases (Eq. 26.14). This increase is sensed by the stud-finder's special circuitry, which causes an indicator on the device to light up.

26.9 (a) C increases (Eq. 26.14). (b) Q increases. Because the battery maintains a constant ΔV, Q must increase if C increases. (c) E between the plates remains constant because $\Delta V = Ed$ and neither ΔV nor d changes. The electric field due to the charges on the plates increases because more charge has flowed onto the plates. The induced surface charges on the dielectric create a field that opposes the increase in the field caused by the greater number of charges on the plates (see Section 26.7). (d) The battery maintains a constant ΔV.

Chapter 27

Current and Resistance

▲ These power lines transfer energy from the power company to homes and businesses. The energy is transferred at a very high voltage, possibly hundreds of thousands of volts in some cases. Despite the fact that this makes power lines very dangerous, the high voltage results in less loss of power due to resistance in the wires. (Telegraph Colour Library/FPG)

Thus far our treatment of electrical phenomena has been confined to the study of charges in equilibrium situations, or *electrostatics*. We now consider situations involving electric charges that are *not* in equilibrium. We use the term *electric current,* or simply *current,* to describe the rate of flow of charge through some region of space. Most practical applications of electricity deal with electric currents. For example, the battery in a flashlight produces a current in the filament of the bulb when the switch is turned on. A variety of home appliances operate on alternating current. In these common situations, current exists in a conductor, such as a copper wire. It also is possible for currents to exist outside a conductor. For instance, a beam of electrons in a television picture tube constitutes a current.

This chapter begins with the definition of current. A microscopic description of current is given, and some of the factors that contribute to the opposition to the flow of charge in conductors are discussed. A classical model is used to describe electrical conduction in metals, and some of the limitations of this model are cited. We also define electrical resistance and introduce a new circuit element, the resistor. We conclude by discussing the rate at which energy is transferred to a device in an electric circuit.

27.1 Electric Current

In this section, we study the flow of electric charges through a piece of material. The amount of flow depends on the material through which the charges are passing and the potential difference across the material. Whenever there is a net flow of charge through some region, an electric **current** is said to exist.

It is instructive to draw an analogy between water flow and current. In many localities it is common practice to install low-flow showerheads in homes as a water-conservation measure. We quantify the flow of water from these and similar devices by specifying the amount of water that emerges during a given time interval, which is often measured in liters per minute. On a grander scale, we can characterize a river current by describing the rate at which the water flows past a particular location. For example, the flow over the brink at Niagara Falls is maintained at rates between $1\,400$ m³/s and $2\,800$ m³/s.

There is also an analogy between thermal conduction and current. In Section 20.7, we discussed the flow of energy by heat through a sample of material. The rate of energy flow is determined by the material as well as the temperature difference across the material, as described by Equation 20.14.

To define current more precisely, suppose that charges are moving perpendicular to a surface of area A, as shown in Figure 27.1. (This area could be the cross-sectional area of a wire, for example.) **The current is the rate at which charge flows through this surface.** If ΔQ is the amount of charge that passes through this area in a time interval Δt, the **average current** I_{av} is equal to the charge that passes through A per unit time:

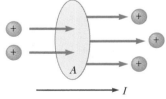

Figure 27.1 Charges in motion through an area A. The time rate at which charge flows through the area is defined as the current I. The direction of the current is the direction in which positive charges flow when free to do so.

$$I_{av} = \frac{\Delta Q}{\Delta t} \qquad (27.1)$$

If the rate at which charge flows varies in time, then the current varies in time; we define the **instantaneous current** I as the differential limit of average current:

$$I \equiv \frac{dQ}{dt} \qquad (27.2)$$

The SI unit of current is the **ampere** (A):

$$1 \text{ A} = \frac{1 \text{ C}}{1 \text{ s}} \qquad (27.3)$$

That is, 1 A of current is equivalent to 1 C of charge passing through the surface area in 1 s.

The charges passing through the surface in Figure 27.1 can be positive or negative, or both. **It is conventional to assign to the current the same direction as the flow of positive charge.** In electrical conductors, such as copper or aluminum, the current is due to the motion of negatively charged electrons. Therefore, when we speak of current in an ordinary conductor, **the direction of the current is opposite the direction of flow of electrons.** However, if we are considering a beam of positively charged protons in an accelerator, the current is in the direction of motion of the protons. In some cases—such as those involving gases and electrolytes, for instance—the current is the result of the flow of both positive and negative charges.

If the ends of a conducting wire are connected to form a loop, all points on the loop are at the same electric potential, and hence the electric field is zero within and at the surface of the conductor. Because the electric field is zero, there is no net transport of charge through the wire, and therefore there is no current. However, if the ends of the conducting wire are connected to a battery, all points on the loop are not at the same potential. The battery sets up a potential difference between the ends of the loop, creating an electric field within the wire. The electric field exerts forces on the conduction electrons in the wire, causing them to move in the wire, thus creating a current.

It is common to refer to a moving charge (positive or negative) as a mobile **charge carrier.** For example, the mobile charge carriers in a metal are electrons.

Microscopic Model of Current

We can relate current to the motion of the charge carriers by describing a microscopic model of conduction in a metal. Consider the current in a conductor of cross-sectional area A (Fig. 27.2). The volume of a section of the conductor of length Δx (the gray region shown in Fig. 27.2) is $A \Delta x$. If n represents the number of mobile charge carriers per unit volume (in other words, the charge carrier density), the number of carriers in the gray section is $nA \Delta x$. Therefore, the total charge ΔQ in this section is

$$\Delta Q = \text{number of carriers in section} \times \text{charge per carrier} = (nA \Delta x)q$$

where q is the charge on each carrier. If the carriers move with a speed v_d, the displacement they experience in the x direction in a time interval Δt is $\Delta x = v_d \Delta t$. Let us choose Δt to be the time interval required for the charges in the cylinder to move through a displacement whose magnitude is equal to the length of the cylinder. This time interval is also that required for all of the charges in the cylinder to pass through the circular area at one end. With this choice, we can write ΔQ in the form

$$\Delta Q = (nAv_d \Delta t)q$$

If we divide both sides of this equation by Δt, we see that the average current in the conductor is

$$I_{av} = \frac{\Delta Q}{\Delta t} = nqv_d A \qquad (27.4)$$

▲ **PITFALL PREVENTION**

27.1 "Current Flow" Is Redundant

The phrase *current flow* is commonly used, although it is strictly incorrect, because current *is* a flow (of charge). This is similar to the phrase *heat transfer*, which is also redundant because heat *is* a transfer (of energy). We will avoid this phrase and speak of *flow of charge* or *charge flow.*

Figure 27.2 A section of a uniform conductor of cross-sectional area A. The mobile charge carriers move with a speed v_d, and the displacement they experience in the x direction in a time interval Δt is $\Delta x = v_d \Delta t$. If we choose Δt to be the time interval during which the charges are displaced, on the average, by the length of the cylinder, the number of carriers in the section of length Δx is $nAv_d \Delta t$, where n is the number of carriers per unit volume.

Figure 27.3 A schematic representation of the zigzag motion of an electron in a conductor. The changes in direction are the result of collisions between the electron and atoms in the conductor. Note that the net motion of the electron is opposite the direction of the electric field. Because of the acceleration of the charge carriers due to the electric force, the paths are actually parabolic. However, the drift speed is much smaller than the average speed, so the parabolic shape is not visible on this scale.

The speed of the charge carriers v_d is an average speed called the **drift speed.** To understand the meaning of drift speed, consider a conductor in which the charge carriers are free electrons. If the conductor is isolated—that is, the potential difference across it is zero—then these electrons undergo random motion that is analogous to the motion of gas molecules. As we discussed earlier, when a potential difference is applied across the conductor (for example, by means of a battery), an electric field is set up in the conductor; this field exerts an electric force on the electrons, producing a current. However, the electrons do not move in straight lines along the conductor. Instead, they collide repeatedly with the metal atoms, and their resultant motion is complicated and zigzag (Fig. 27.3). Despite the collisions, the electrons move slowly along the conductor (in a direction opposite that of **E**) at the drift velocity $\mathbf{v}_d$.

We can think of the atom–electron collisions in a conductor as an effective internal friction (or drag force) similar to that experienced by the molecules of a liquid flowing through a pipe stuffed with steel wool. The energy transferred from the electrons to the metal atoms during collisions causes an increase in the vibrational energy of the atoms and a corresponding increase in the temperature of the conductor.

Quick Quiz 27.1 Consider positive and negative charges moving horizontally through the four regions shown in Figure 27.4. Rank the current in these four regions, from lowest to highest.

(a)   (b)   (c)   (d)

Figure 27.4 (Quick Quiz 27.1) Charges move through four regions.

Quick Quiz 27.2 Electric charge is conserved. As a consequence, when current arrives at a junction of wires, the charges can take either of two paths out of the junction and the numerical sum of the currents in the two paths equals the current that entered the junction. Thus, current is (a) a vector (b) a scalar (c) neither a vector nor a scalar.

Example 27.1   Drift Speed in a Copper Wire

The 12-gauge copper wire in a typical residential building has a cross-sectional area of 3.31×10^{-6} m^2. If it carries a current of 10.0 A, what is the drift speed of the electrons? Assume that each copper atom contributes one free electron to the current. The density of copper is 8.95 g/cm^3.

Solution From the periodic table of the elements in Appendix C, we find that the molar mass of copper is 63.5 g/mol. Recall that 1 mol of any substance contains Avogadro's number of atoms (6.02×10^{23}). Knowing the density of copper, we can calculate the volume occupied by 63.5 g ($= 1$ mol) of copper:

$$V = \frac{m}{\rho} = \frac{63.5 \text{ g}}{8.95 \text{ g/cm}^3} = 7.09 \text{ cm}^3$$

Because each copper atom contributes one free electron to the current, we have

$$n = \frac{6.02 \times 10^{23} \text{ electrons}}{7.09 \text{ cm}^3} \left(\frac{1.00 \times 10^6 \text{ cm}^3}{1 \text{ m}^3} \right)$$

$$= 8.49 \times 10^{28} \text{ electrons/m}^3$$

From Equation 27.4, we find that the drift speed is

$$v_d = \frac{I}{nqA}$$

where q is the absolute value of the charge on each electron. Thus,

$$v_d = \frac{I}{nqA}$$

$$= \frac{10.0 \text{ C/s}}{(8.49 \times 10^{28} \text{ m}^{-3})(1.60 \times 10^{-19} \text{ C})(3.31 \times 10^{-6} \text{ m}^2)}$$

$$= \boxed{2.22 \times 10^{-4} \text{ m/s}}$$

Example 27.1 shows that typical drift speeds are very low. For instance, electrons traveling with a speed of 2.22×10^{-4} m/s would take about 75 min to travel 1 m! In view of this, you might wonder why a light turns on almost instantaneously when a switch is thrown. In a conductor, changes in the electric field that drives the free electrons travel through the conductor with a speed close to that of light. Thus, when you flip on a light switch, electrons already in the filament of the lightbulb experience electric forces and begin moving after a time interval on the order of nanoseconds.

27.2 Resistance

In Chapter 24 we found that the electric field inside a conductor is zero. However, this statement is true *only* if the conductor is in static equilibrium. The purpose of this section is to describe what happens when the charges in the conductor are not in equilibrium, in which case there is an electric field in the conductor.

Consider a conductor of cross-sectional area A carrying a current I. The **current density** J in the conductor is defined as the current per unit area. Because the current $I = nqv_d A$, the current density is

$$J \equiv \frac{I}{A} = nqv_d \qquad (27.5)$$

where J has SI units of A/m². This expression is valid only if the current density is uniform and only if the surface of cross-sectional area A is perpendicular to the direction of the current. In general, current density is a vector quantity:

$$\mathbf{J} = nq\mathbf{v}_d \qquad (27.6)$$

From this equation, we see that current density is in the direction of charge motion for positive charge carriers and opposite the direction of motion for negative charge carriers.

A current density J and an electric field E are established in a conductor whenever a potential difference is maintained across the conductor. In some materials, the current density is proportional to the electric field:

$$\mathbf{J} = \sigma \mathbf{E} \qquad (27.7)$$

where the constant of proportionality σ is called the **conductivity** of the conductor.[1] Materials that obey Equation 27.7 are said to follow **Ohm's law,** named after Georg Simon Ohm (1789–1854). More specifically, Ohm's law states that

> for many materials (including most metals), the ratio of the current density to the electric field is a constant σ that is independent of the electric field producing the current.

Materials that obey Ohm's law and hence demonstrate this simple relationship between **E** and **J** are said to be *ohmic*. Experimentally, however, it is found that not all materials have this property. Materials and devices that do not obey Ohm's law are said to be *nonohmic*. Ohm's law is not a fundamental law of nature but rather an empirical relationship valid only for certain materials.

We can obtain an equation useful in practical applications by considering a segment of straight wire of uniform cross-sectional area A and length ℓ, as shown in

---

[1]   Do not confuse conductivity σ with surface charge density, for which the same symbol is used.

Current density

Georg Simon Ohm
German physicist (1789–1854)

Ohm, a high school teacher and later a professor at the University of Munich, formulated the concept of resistance and discovered the proportionalities expressed in Equations 27.7 and 27.8. (© Bettmann/Corbis)

PITFALL PREVENTION

27.3 We've Seen Something Like Equation 27.8 Before

In Chapter 5, we introduced Newton's second law, $\Sigma F = ma$, for a net force on an object of mass m. This can be written as

$$m = \frac{\Sigma F}{a}$$

In that chapter, we defined mass as *resistance to a change in motion in response to an external force.* Mass as resistance to changes in motion is analogous to electrical resistance to charge flow, and Equation 27.8 is analogous to the form of Newton's second law shown here.

 PITFALL PREVENTION

27.4 Equation 27.8 Is Not Ohm's Law

Many individuals call Equation 27.8 Ohm's law, but this is incorrect. This equation is simply the definition of resistance, and provides an important relationship between voltage, current, and resistance. Ohm's law is related to a linear relationship between **J** and **E** (Eq. 27.7) or, equivalently, between I and ΔV, which, from Equation 27.8, indicates that the resistance is constant, independent of the applied voltage.

Resistivity is the inverse of conductivity

Resistance of a uniform material along the length ℓ

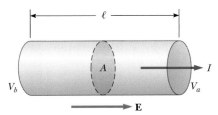

Figure 27.5 A uniform conductor of length ℓ and cross-sectional area A. A potential difference $\Delta V = V_b - V_a$ maintained across the conductor sets up an electric field **E**, and this field produces a current I that is proportional to the potential difference.

Figure 27.5. A potential difference $\Delta V = V_b - V_a$ is maintained across the wire, creating in the wire an electric field and a current. If the field is assumed to be uniform, the potential difference is related to the field through the relationship[2]

$$\Delta V = E\ell$$

Therefore, we can express the magnitude of the current density in the wire as

$$J = \sigma E = \sigma \frac{\Delta V}{\ell}$$

Because $J = I/A$, we can write the potential difference as

$$\Delta V = \frac{\ell}{\sigma} J = \left(\frac{\ell}{\sigma A}\right) I = RI$$

The quantity $R = \ell/\sigma A$ is called the **resistance** of the conductor. We can define the resistance as the ratio of the potential difference across a conductor to the current in the conductor:

$$R \equiv \frac{\Delta V}{I} \tag{27.8}$$

We will use this equation over and over again when studying electric circuits. From this result we see that resistance has SI units of volts per ampere. One volt per ampere is defined to be one **ohm** (Ω):

$$1 \ \Omega \equiv \frac{1 \ \text{V}}{1 \ \text{A}} \tag{27.9}$$

This expression shows that if a potential difference of 1 V across a conductor causes a current of 1 A, the resistance of the conductor is 1 Ω. For example, if an electrical appliance connected to a 120-V source of potential difference carries a current of 6 A, its resistance is 20 Ω.

The inverse of conductivity is **resistivity**[3] ρ:

$$\rho = \frac{1}{\sigma} \tag{27.10}$$

where ρ has the units ohm-meters ($\Omega \cdot \text{m}$). Because $R = \ell/\sigma A$, we can express the resistance of a uniform block of material along the length ℓ as

$$R = \rho \frac{\ell}{A} \tag{27.11}$$

---

[2]   This result follows from the definition of potential difference:

$$V_b - V_a = -\int_a^b \mathbf{E} \cdot d\mathbf{s} = E \int_0^\ell dx = E\ell$$

[3]   Do not confuse resistivity ρ with mass density or charge density, for which the same symbol is used.

Table 27.1

Resistivities and Temperature Coefficients of Resistivity for Various Materials		
Material	**Resistivity[a] ($\Omega \cdot$ m)**	**Temperature Coefficient[b] α[(°C)$^{-1}$]**
Silver	1.59×10^{-8}	3.8×10^{-3}
Copper	1.7×10^{-8}	3.9×10^{-3}
Gold	2.44×10^{-8}	3.4×10^{-3}
Aluminum	2.82×10^{-8}	3.9×10^{-3}
Tungsten	5.6×10^{-8}	4.5×10^{-3}
Iron	10×10^{-8}	5.0×10^{-3}
Platinum	11×10^{-8}	3.92×10^{-3}
Lead	22×10^{-8}	3.9×10^{-3}
Nichrome[c]	1.50×10^{-6}	0.4×10^{-3}
Carbon	3.5×10^{-5}	-0.5×10^{-3}
Germanium	0.46	-48×10^{-3}
Silicon	640	-75×10^{-3}
Glass	10^{10} to 10^{14}	
Hard rubber	$\sim 10^{13}$	
Sulfur	10^{15}	
Quartz (fused)	75×10^{16}	

[a] All values at 20°C.

[b] See Section 27.4.

[c] A nickel–chromium alloy commonly used in heating elements.

PITFALL PREVENTION

27.5 Resistance and Resistivity

Resistivity is property of a *substance*, while resistance is a property of an *object*. We have seen similar pairs of variables before. For example, density is a property of a substance, while mass is a property of an object. Equation 27.11 relates resistance to resistivity, and we have seen a previous equation (Equation 1.1) which relates mass to density.

Every ohmic material has a characteristic resistivity that depends on the properties of the material and on temperature. Additionally, as you can see from Equation 27.11, the resistance of a sample depends on geometry as well as on resistivity. Table 27.1 gives the resistivities of a variety of materials at 20°C. Note the enormous range, from very low values for good conductors such as copper and silver, to very high values for good insulators such as glass and rubber. An ideal conductor would have zero resistivity, and an ideal insulator would have infinite resistivity.

Equation 27.11 shows that the resistance of a given cylindrical conductor such as a wire is proportional to its length and inversely proportional to its cross-sectional area. If the length of a wire is doubled, then its resistance doubles. If its cross-sectional area is doubled, then its resistance decreases by one half. The situation is analogous to the flow of a liquid through a pipe. As the pipe's length is increased, the resistance to flow increases. As the pipe's cross-sectional area is increased, more liquid crosses a given cross section of the pipe per unit time interval. Thus, more liquid flows for the same pressure differential applied to the pipe, and the resistance to flow decreases.

Henry Leap and Jim Lehman

An assortment of resistors used in electrical circuits.

Figure 27.6 The colored bands on a resistor represent a code for determining resistance. The first two colors give the first two digits in the resistance value. The third color represents the power of ten for the multiplier of the resistance value. The last color is the tolerance of the resistance value. As an example, the four colors on the circled resistors are red $(= 2)$, black $(= 0)$, orange $(= 10^3)$, and gold $(= 5\%)$, and so the resistance value is $20 \times 10^3 \ \Omega = 20 \ k\Omega$ with a tolerance value of $5\% = 1 \ k\Omega$. (The values for the colors are from Table 27.2.)

Table 27.2

Color Coding for Resistors			
Color	**Number**	**Multiplier**	**Tolerance**
Black	0	1	
Brown	1	10^1	
Red	2	10^2	
Orange	3	10^3	
Yellow	4	10^4	
Green	5	10^5	
Blue	6	10^6	
Violet	7	10^7	
Gray	8	10^8	
White	9	10^9	
Gold		10^{-1}	5%
Silver		10^{-2}	10%
Colorless			20%

Most electric circuits use circuit elements called **resistors** to control the current level in the various parts of the circuit. Two common types of resistors are the *composition resistor*, which contains carbon, and the *wire-wound resistor*, which consists of a coil of wire. Values of resistors in ohms are normally indicated by color-coding, as shown in Figure 27.6 and Table 27.2.

Ohmic materials and devices have a linear current–potential difference relationship over a broad range of applied potential differences (Fig. 27.7a). The slope of the *I*-versus-ΔV curve in the linear region yields a value for $1/R$. Nonohmic materials have a nonlinear current–potential difference relationship. One common semiconducting device that has nonlinear *I*-versus-ΔV characteristics is the *junction diode* (Fig. 27.7b). The resistance of this device is low for currents in one direction (positive ΔV) and high for currents in the reverse direction (negative ΔV). In fact, most modern electronic devices, such as transistors, have nonlinear current–potential difference relationships; their proper operation depends on the particular way in which they violate Ohm's law.

Quick Quiz 27.3 Suppose that a current-carrying ohmic metal wire has a cross-sectional area that gradually becomes smaller from one end of the wire to the other. The current must have the same value in each section of the wire so that charge does not accumulate at any one point. How do the drift velocity and the resistance per

(a)

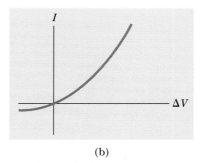

(b)

Figure 27.7 (a) The current–potential difference curve for an ohmic material. The curve is linear, and the slope is equal to the inverse of the resistance of the conductor. (b) A nonlinear current–potential difference curve for a junction diode. This device does not obey Ohm's law.

unit length vary along the wire as the area becomes smaller? (a) The drift velocity and resistance both increase. (b) The drift velocity and resistance both decrease. (c) The drift velocity increases and the resistance decreases. (d) The drift velocity decreases and the resistance increases.

Quick Quiz 27.4 A cylindrical wire has a radius r and length ℓ. If both r and ℓ are doubled, the resistance of the wire (a) increases (b) decreases (c) remains the same.

Quick Quiz 27.5 In Figure 27.7b, as the applied voltage increases, the resistance of the diode (a) increases (b) decreases (c) remains the same.

Example 27.2   The Resistance of a Conductor

Calculate the resistance of an aluminum cylinder that has a length of 10.0 cm and a cross-sectional area of 2.00×10^{-4} m². Repeat the calculation for a cylinder of the same dimensions and made of glass having a resistivity of 3.0×10^{10} Ω·m.

Solution From Equation 27.11 and Table 27.1, we can calculate the resistance of the aluminum cylinder as follows:

$$R = \rho\frac{\ell}{A} = (2.82 \times 10^{-8}\ \Omega\cdot m)\left(\frac{0.100\ m}{2.00 \times 10^{-4}\ m^2}\right)$$

$$= \boxed{1.41 \times 10^{-5}\ \Omega}$$

Similarly, for glass we find that

$$R = \rho\frac{\ell}{A} = (3.0 \times 10^{10}\ \Omega\cdot m)\left(\frac{0.100\ m}{2.00 \times 10^{-4}\ m^2}\right)$$

$$= \boxed{1.5 \times 10^{13}\ \Omega}$$

As you might guess from the large difference in resistivities, the resistances of identically shaped cylinders of aluminum and glass differ widely. The resistance of the glass cylinder is 18 orders of magnitude greater than that of the aluminum cylinder.

Example 27.3   The Resistance of Nichrome Wire   `Interactive`

(A) Calculate the resistance per unit length of a 22-gauge Nichrome wire, which has a radius of 0.321 mm.

Solution The cross-sectional area of this wire is

$$A = \pi r^2 = \pi(0.321 \times 10^{-3}\ m)^2 = 3.24 \times 10^{-7}\ m^2$$

The resistivity of Nichrome is 1.5×10^{-6} Ω·m (see Table 27.1). Thus, we can use Equation 27.11 to find the resistance per unit length:

$$\frac{R}{\ell} = \frac{\rho}{A} = \frac{1.5 \times 10^{-6}\ \Omega\cdot m}{3.24 \times 10^{-7}\ m^2} = \boxed{4.6\ \Omega/m}$$

(B) If a potential difference of 10 V is maintained across a 1.0-m length of the Nichrome wire, what is the current in the wire?

Solution Because a 1.0-m length of this wire has a resistance of 4.6 Ω, Equation 27.8 gives

$$I = \frac{\Delta V}{R} = \frac{10\ V}{4.6\ \Omega} = \boxed{2.2\ A}$$

Note from Table 27.1 that the resistivity of Nichrome wire is about 100 times that of copper. A copper wire of the same radius would have a resistance per unit length of only 0.052 Ω/m. A 1.0-m length of copper wire of the same radius would carry the same current (2.2 A) with an applied potential difference of only 0.11 V.

Because of its high resistivity and its resistance to oxidation, Nichrome is often used for heating elements in toasters, irons, and electric heaters.

 Explore the resistance of different materials at the Interactive Worked Example link at **http://www.pse6.com.**

Example 27.4   The Radial Resistance of a Coaxial Cable

Coaxial cables are used extensively for cable television and other electronic applications. A coaxial cable consists of two concentric cylindrical conductors. The region between the conductors is completely filled with silicon, as shown in Figure 27.8a, and current leakage through the silicon, in the *radial* direction, is unwanted. (The cable is designed to conduct current along its length—this is *not* the current we are considering here.) The radius of the inner conductor is $a = 0.500$ cm, the radius of the outer one is $b = 1.75$ cm, and the length is $L = 15.0$ cm.

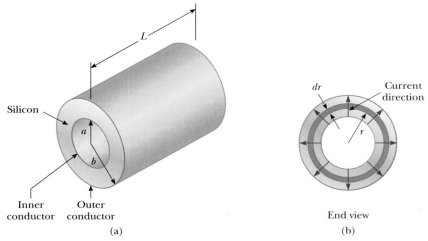

Figure 27.8 (Example 27.4) A coaxial cable. (a) Silicon fills the gap between the two conductors. (b) End view, showing current leakage.

Calculate the resistance of the silicon between the two conductors.

Solution Conceptualize by imagining two currents, as suggested in the text of the problem. The desired current is along the cable, carried within the conductors. The undesired current corresponds to charge leakage through the silicon and its direction is radial. Because we know the resistivity and the geometry of the silicon, we categorize this as a problem in which we find the resistance of the silicon from these parameters, using Equation 27.11. Because the area through which the charges pass depends on the radial position, we must use integral calculus to determine the answer.

To analyze the problem, we divide the silicon into concentric elements of infinitesimal thickness dr (Fig. 27.8b). We start by using the differential form of Equation 27.11, replacing ℓ with r for the distance variable: $dR = \rho\, dr/A$, where dR is the resistance of an element of silicon of thickness dr and surface area A. In this example, we take as our representative concentric element a hollow silicon cylinder of radius r, thickness dr, and length L, as in Figure 27.8. Any charge that passes from the inner conductor to the outer one must pass radially through this concentric element, and the area through which this charge passes is $A = 2\pi r L$. (This is the curved surface area—circumference multiplied by length—of our hollow silicon cylinder of thickness dr.) Hence, we can write the resistance of our hollow cylinder of silicon as

$$dR = \frac{\rho}{2\pi r L}\, dr$$

Because we wish to know the total resistance across the entire thickness of the silicon, we must integrate this expression from $r = a$ to $r = b$:

$$(1)\qquad R = \int_a^b dR = \frac{\rho}{2\pi L}\int_a^b \frac{dr}{r} = \frac{\rho}{2\pi L}\ln\left(\frac{b}{a}\right)$$

Substituting in the values given, and using $\rho = 640\ \Omega\cdot\text{m}$ for silicon, we obtain

$$R = \frac{640\ \Omega\cdot\text{m}}{2\pi(0.150\ \text{m})}\ln\left(\frac{1.75\ \text{cm}}{0.500\ \text{cm}}\right) = \boxed{851\ \Omega}$$

To finalize this problem, let us compare this resistance to that of the inner conductor of the cable along the 15.0-cm length. Assuming that the conductor is made of copper, we have

$$R = \rho\,\frac{\ell}{A} = (1.7 \times 10^{-8}\ \Omega\cdot\text{m})\left(\frac{0.150\ \text{m}}{\pi(5.00 \times 10^{-3}\ \text{m})^2}\right)$$
$$= 3.2 \times 10^{-5}\ \Omega$$

This resistance is much smaller than the radial resistance. As a consequence, almost all of the current corresponds to charge moving along the length of the cable, with a very small fraction leaking in the radial direction.

What If? Suppose the coaxial cable is enlarged to twice the overall diameter with two possibilities: (1) the ratio b/a is held fixed, or (2) the difference $b - a$ is held fixed. For which possibility does the leakage current between the inner and outer conductors increase when the voltage is applied between the two conductors?

Answer In order for the current to increase, the resistance must decrease. For possibility (1), in which b/a is held fixed, Equation (1) tells us that the resistance is unaffected. For possibility (2), we do not have an equation involving the difference $b - a$ to inspect. Looking at Figure 27.8b, however, we see that increasing b and a while holding the voltage constant results in charge flowing through the same thickness of silicon but through a larger overall area perpendicular to the flow. This larger area will result in lower resistance and a higher current.

27.3   A Model for Electrical Conduction

In this section we describe a classical model of electrical conduction in metals that was first proposed by Paul Drude (1863–1906) in 1900. This model leads to Ohm's law and shows that resistivity can be related to the motion of electrons in metals. Although the Drude model described here does have limitations, it nevertheless introduces concepts that are still applied in more elaborate treatments.

Consider a conductor as a regular array of atoms plus a collection of free electrons, which are sometimes called *conduction* electrons. The conduction electrons, although bound to their respective atoms when the atoms are not part of a solid, gain mobility when the free atoms condense into a solid. In the absence of an electric field, the conduction electrons move in random directions through the conductor with average speeds on the order of 10^6 m/s. The situation is similar to the motion of gas molecules confined in a vessel. In fact, some scientists refer to conduction electrons in a metal as an *electron gas*. There is no current in the conductor in the absence of an electric field because the drift velocity of the free electrons is zero. That is, on the average, just as many electrons move in one direction as in the opposite direction, and so there is no net flow of charge.

This situation changes when an electric field is applied. Now, in addition to undergoing the random motion just described, the free electrons drift slowly in a direction opposite that of the electric field, with an average drift speed v_d that is much smaller (typically 10^{-4} m/s) than their average speed between collisions (typically 10^6 m/s).

Figure 27.9 provides a crude description of the motion of free electrons in a conductor. In the absence of an electric field, there is no net displacement after many collisions (Fig. 27.9a). An electric field **E** modifies the random motion and causes the electrons to drift in a direction opposite that of **E** (Fig. 27.9b).

In our model, we assume that the motion of an electron after a collision is independent of its motion before the collision. We also assume that the excess energy acquired by the electrons in the electric field is lost to the atoms of the conductor when the electrons and atoms collide. The energy given up to the atoms increases their vibrational energy, and this causes the temperature of the conductor to increase. The temperature increase of a conductor due to resistance is utilized in electric toasters and other familiar appliances.

We are now in a position to derive an expression for the drift velocity. When a free electron of mass m_e and charge $q(=-e)$ is subjected to an electric field **E**, it experiences a force $\mathbf{F} = q\mathbf{E}$. Because this force is related to the acceleration of the electron through Newton's second law, $\mathbf{F} = m_e\mathbf{a}$, we conclude that the acceleration of the electron is

$$\mathbf{a} = \frac{q\mathbf{E}}{m_e} \tag{27.12}$$

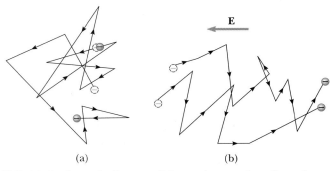

(a)   (b)

Active Figure 27.9 (a) A schematic diagram of the random motion of two charge carriers in a conductor in the absence of an electric field. The drift velocity is zero. (b) The motion of the charge carriers in a conductor in the presence of an electric field. Note that the random motion is modified by the field, and the charge carriers have a drift velocity.

At the Active Figures link at **http://www.pse6.com,** *you can adjust the electric field to see the resulting effect on the motion of an electron.*

This acceleration, which occurs for only a short time interval between collisions, enables the electron to acquire a small drift velocity. If $\mathbf{v}_i$ is the electron's initial velocity the instant after a collision (which occurs at a time that we define as $t = 0$), then the velocity of the electron at time t (at which the next collision occurs) is

$$\mathbf{v}_f = \mathbf{v}_i + \mathbf{a}t = \mathbf{v}_i + \frac{q\mathbf{E}}{m_e} t \qquad (27.13)$$

We now take the average value of $\mathbf{v}_f$ over all possible collision times t and all possible values of $\mathbf{v}_i$. If we assume that the initial velocities are randomly distributed over all possible values, we see that the average value of $\mathbf{v}_i$ is zero. The term $(q\mathbf{E}/m_e)t$ is the velocity change of the electron due to the electric field during one trip between atoms. The average value of the second term of Equation 27.13 is $(q\mathbf{E}/m_e)\tau$, where τ is the *average time interval between successive collisions*. Because the average value of $\mathbf{v}_f$ is equal to the drift velocity, we have

Drift velocity in terms of microscopic quantities

$$\overline{\mathbf{v}}_f = \mathbf{v}_d = \frac{q\mathbf{E}}{m_e} \tau \qquad (27.14)$$

We can relate this expression for drift velocity to the current in the conductor. Substituting Equation 27.14 into Equation 27.6, we find that the magnitude of the current density is

Current density in terms of microscopic quantities

$$J = nqv_d = \frac{nq^2E}{m_e} \tau \qquad (27.15)$$

where n is the number of charge carriers per unit volume. Comparing this expression with Ohm's law, $J = \sigma E$, we obtain the following relationships for conductivity and resistivity of a conductor:

Conductivity in terms of microscopic quantities

$$\sigma = \frac{nq^2\tau}{m_e} \qquad (27.16)$$

Resistivity in terms of microscopic quantities

$$\rho = \frac{1}{\sigma} = \frac{m_e}{nq^2\tau} \qquad (27.17)$$

According to this classical model, conductivity and resistivity do not depend on the strength of the electric field. This feature is characteristic of a conductor obeying Ohm's law.

The average time interval τ between collisions is related to the average distance between collisions ℓ (that is, the *mean free path*; see Section 21.7) and the average speed $\overline{v}$ through the expression

$$\tau = \frac{\ell}{\overline{v}} \qquad (27.18)$$

Example 27.5 Electron Collisions in a Wire

(A) Using the data and results from Example 27.1 and the classical model of electron conduction, estimate the average time interval between collisions for electrons in household copper wiring.

Solution From Equation 27.17, we see that

$$\tau = \frac{m_e}{nq^2\rho}$$

where $\rho = 1.7 \times 10^{-8}\ \Omega \cdot$m for copper and the carrier density is $n = 8.49 \times 10^{28}$ electrons/m³ for the wire described

in Example 27.1. Substitution of these values into the expression above gives

$$\tau = \frac{9.11 \times 10^{-31}\ \text{kg}}{(8.49 \times 10^{28}\ \text{m}^{-3})(1.6 \times 10^{-19}\ \text{C})^2 (1.7 \times 10^{-8}\ \Omega \cdot \text{m})}$$

$$= 2.5 \times 10^{-14}\ \text{s}$$

(B) Assuming that the average speed for free electrons in copper is 1.6×10^6 m/s and using the result from part (A), calculate the mean free path for electrons in copper.

Solution From Equation 27.18,

$$\ell = \bar{v}\tau = (1.6 \times 10^6 \text{ m/s})(2.5 \times 10^{-14} \text{ s})$$

$$= \boxed{4.0 \times 10^{-8} \text{ m}}$$

which is equivalent to 40 nm (compared with atomic spacings of about 0.2 nm). Thus, although the time interval between collisions is very short, an electron in the wire travels about 200 atomic spacings between collisions.

27.4 Resistance and Temperature

Over a limited temperature range, the resistivity of a conductor varies approximately linearly with temperature according to the expression

$$\rho = \rho_0[1 + \alpha(T - T_0)] \qquad (27.19)$$

Variation of ρ with temperature

where ρ is the resistivity at some temperature T (in degrees Celsius), ρ_0 is the resistivity at some reference temperature T_0 (usually taken to be 20°C), and α is the **temperature coefficient of resistivity.** From Equation 27.19, we see that the temperature coefficient of resistivity can be expressed as

$$\alpha = \frac{1}{\rho_0}\frac{\Delta\rho}{\Delta T} \qquad (27.20)$$

Temperature coefficient of resistivity

where $\Delta\rho = \rho - \rho_0$ is the change in resistivity in the temperature interval $\Delta T = T - T_0$.

The temperature coefficients of resistivity for various materials are given in Table 27.1. Note that the unit for α is degrees Celsius^{-1} [(°C)$^{-1}$]. Because resistance is proportional to resistivity (Eq. 27.11), we can write the variation of resistance as

$$R = R_0[1 + \alpha(T - T_0)] \qquad (27.21)$$

Use of this property enables us to make precise temperature measurements, as shown in Example 27.6.

Quick Quiz 27.6 When does a lightbulb carry more current: (a) just after it is turned on and the glow of the metal filament is increasing, or (b) after it has been on for a few milliseconds and the glow is steady?

Example 27.6   A Platinum Resistance Thermometer

A resistance thermometer, which measures temperature by measuring the change in resistance of a conductor, is made from platinum and has a resistance of 50.0 Ω at 20.0°C. When immersed in a vessel containing melting indium, its resistance increases to 76.8 Ω. Calculate the melting point of the indium.

Solution Solving Equation 27.21 for ΔT and using the α value for platinum given in Table 27.1, we obtain

$$\Delta T = \frac{R - R_0}{\alpha R_0} = \frac{76.8 \text{ }\Omega - 50.0 \text{ }\Omega}{[3.92 \times 10^{-3}(°C)^{-1}](50.0 \text{ }\Omega)}$$

$$= 137°C$$

Because $T_0 = 20.0°C$, we find that T, the temperature of the melting indium sample, is $\boxed{157°C}$.

For metals like copper, resistivity is nearly proportional to temperature, as shown in Figure 27.10. However, a nonlinear region always exists at very low temperatures, and the resistivity usually reaches some finite value as the temperature approaches absolute zero. This residual resistivity near absolute zero is caused primarily by the

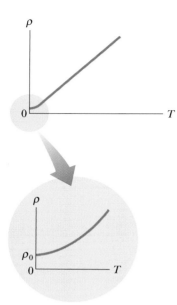

Figure 27.10 Resistivity versus temperature for a metal such as copper. The curve is linear over a wide range of temperatures, and ρ increases with increasing temperature. As T approaches absolute zero (inset), the resistivity approaches a finite value ρ_0.

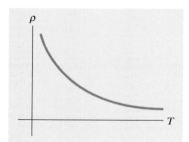

Figure 27.11 Resistivity versus temperature for a pure semiconductor, such as silicon or germanium.

collision of electrons with impurities and imperfections in the metal. In contrast, high-temperature resistivity (the linear region) is predominantly characterized by collisions between electrons and metal atoms.

Notice that three of the α values in Table 27.1 are negative; this indicates that the resistivity of these materials decreases with increasing temperature (Fig. 27.11), which is indicative of a class of materials called *semiconductors*. This behavior is due to an increase in the density of charge carriers at higher temperatures.

Because the charge carriers in a semiconductor are often associated with impurity atoms, the resistivity of these materials is very sensitive to the type and concentration of such impurities. We shall return to the study of semiconductors in Chapter 43.

27.5 Superconductors

There is a class of metals and compounds whose resistance decreases to zero when they are below a certain temperature T_c, known as the **critical temperature.** These materials are known as **superconductors.** The resistance–temperature graph for a superconductor follows that of a normal metal at temperatures above T_c (Fig. 27.12). When the temperature is at or below T_c, the resistivity drops suddenly to zero. This phenomenon was discovered in 1911 by the Dutch physicist Heike Kamerlingh-Onnes (1853–1926) as he worked with mercury, which is a superconductor below 4.2 K. Recent measurements have shown that the resistivities of superconductors below their T_c values are less than 4×10^{-25} $\Omega \cdot$m—around 10^{17} times smaller than the resistivity of copper and in practice considered to be zero.

Today thousands of superconductors are known, and as Table 27.3 illustrates, the critical temperatures of recently discovered superconductors are substantially higher than initially thought possible. Two kinds of superconductors are recognized. The more recently identified ones are essentially ceramics with high critical temperatures, whereas superconducting materials such as those observed by Kamerlingh-Onnes are metals. If a room-temperature superconductor is ever identified, its impact on technology could be tremendous.

The value of T_c is sensitive to chemical composition, pressure, and molecular structure. It is interesting to note that copper, silver, and gold, which are excellent conductors, do not exhibit superconductivity.

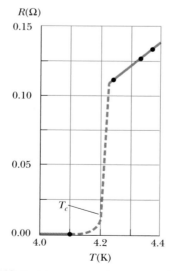

Figure 27.12 Resistance versus temperature for a sample of mercury (Hg). The graph follows that of a normal metal above the critical temperature T_c. The resistance drops to zero at T_c, which is 4.2 K for mercury.

Table 27.3

Critical Temperatures for Various Superconductors	
Material	T_c(**K**)
$HgBa_2Ca_2Cu_3O_8$	134
Tl–Ba–Ca–Cu–O	125
Bi–Sr–Ca–Cu–O	105
$YBa_2Cu_3O_7$	92
Nb_3Ge	23.2
Nb_3Sn	18.05
Nb	9.46
Pb	7.18
Hg	4.15
Sn	3.72
Al	1.19
Zn	0.88

One of the truly remarkable features of superconductors is that once a current is set up in them, it persists *without any applied potential difference* (because $R = 0$). Steady currents have been observed to persist in superconducting loops for several years with no apparent decay!

An important and useful application of superconductivity is in the development of superconducting magnets, in which the magnitudes of the magnetic field are about ten times greater than those produced by the best normal electromagnets. Such superconducting magnets are being considered as a means of storing energy. Superconducting magnets are currently used in medical magnetic resonance imaging (MRI) units, which produce high-quality images of internal organs without the need for excessive exposure of patients to x-rays or other harmful radiation.

For further information on superconductivity, see Section 43.8.

27.6 Electrical Power

If a battery is used to establish an electric current in a conductor, there is a continuous transformation of chemical energy in the battery to kinetic energy of the electrons to internal energy in the conductor, resulting in an increase in the temperature of the conductor.

In typical electric circuits, energy is transferred from a source such as a battery, to some device, such as a lightbulb or a radio receiver. Let us determine an expression that will allow us to calculate the rate of this energy transfer. First, consider the simple circuit in Figure 27.13, where we imagine energy is being delivered to a resistor. (Resistors are designated by the circuit symbol ———⋀⋀⋀———.) Because the connecting wires also have resistance, some energy is delivered to the wires and some energy to the resistor. Unless noted otherwise, we shall assume that the resistance of the wires is so small compared to the resistance of the circuit element that we ignore the energy delivered to the wires.

Imagine following a positive quantity of charge Q that is moving clockwise around the circuit in Figure 27.13 from point a through the battery and resistor back to point a. We identify the entire circuit as our system. As the charge moves from a to b through the battery, the electric potential energy of the system *increases* by an amount $Q \Delta V$ while the chemical potential energy in the battery *decreases* by the same amount. (Recall from Eq. 25.9 that $\Delta U = q \Delta V$.) However, as the charge moves from c to d through the resistor, the system *loses* this electric potential energy during collisions of electrons with atoms in the resistor. In this process, the energy is transformed to internal energy corresponding to increased vibrational motion of the atoms in the resistor. Because we have neglected the resistance of the interconnecting wires, no energy transformation occurs for paths bc and da. When the charge returns to point a, the net result is that some of the chemical energy in the battery has been delivered to the resistor and resides in the resistor as internal energy associated with molecular vibration.

The resistor is normally in contact with air, so its increased temperature will result in a transfer of energy by heat into the air. In addition, the resistor emits thermal

A small permanent magnet levitated above a disk of the superconductor YBa$_2$Cu$_3$O$_7$, which is at 77 K.

▲ PITFALL PREVENTION

27.6 Misconceptions About Current

There are several common misconceptions associated with current in a circuit like that in Figure 27.13. One is that current comes out of one terminal of the battery and is then "used up" as it passes through the resistor, leaving current in only one part of the circuit. The truth is that the current is the same *everywhere* in the circuit. A related misconception has the current coming out of the resistor being smaller than that going in, because some of the current is "used up." Another misconception has current coming out of both terminals of the battery, in opposite directions, and then "clashing" in the resistor, delivering the energy in this manner. This is not the case—the charges flow in the same rotational sense at *all* points in the circuit.

At the Active Figures link at http://www.pse6.com, you can adjust the battery voltage and the resistance to see the resulting current in the circuit and power delivered to the resistor.

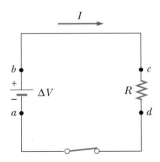

Active Figure 27.13 A circuit consisting of a resistor of resistance R and a battery having a potential difference ΔV across its terminals. Positive charge flows in the clockwise direction.

PITFALL PREVENTION

27.7 Charges Do Not Move All the Way Around a Circuit in a Short Time

Due to the very small magnitude of the drift velocity, it might take *hours* for a single electron to make one complete trip around the circuit. In terms of understanding the energy transfer in a circuit, however, it is useful to *imagine* a charge moving all the way around the circuit.

Power delivered to a device

Power delivered to a resistor

 PITFALL PREVENTION

27.8 Energy Is Not "Dissipated"

In some books, you may see Equation 27.23 described as the power "dissipated in" a resistor, suggesting that energy disappears. Instead we say energy is "delivered to" a resistor. The notion of *dissipation* arises because a warm resistor will expel energy by radiation and heat, so that energy delivered by the battery leaves the circuit. (It does not disappear!)

radiation, representing another means of escape for the energy. After some time interval has passed, the resistor reaches a constant temperature, at which time the input of energy from the battery is balanced by the output of energy by heat and radiation. Some electrical devices include *heat sinks*[4] connected to parts of the circuit to prevent these parts from reaching dangerously high temperatures. These are pieces of metal with many fins. The high thermal conductivity of the metal provides a rapid transfer of energy by heat away from the hot component, while the large number of fins provides a large surface area in contact with the air, so that energy can transfer by radiation and into the air by heat at a high rate.

Let us consider now the rate at which the system loses electric potential energy as the charge Q passes through the resistor:

$$\frac{dU}{dt} = \frac{d}{dt}(Q\,\Delta V) = \frac{dQ}{dt}\,\Delta V = I\,\Delta V$$

where I is the current in the circuit. The system regains this potential energy when the charge passes through the battery, at the expense of chemical energy in the battery. The rate at which the system loses potential energy as the charge passes through the resistor is equal to the rate at which the system gains internal energy in the resistor. Thus, the power $\mathcal{P}$, representing the rate at which energy is delivered to the resistor, is

$$\mathcal{P} = I\,\Delta V \qquad (27.22)$$

We derived this result by considering a battery delivering energy to a resistor. However, Equation 27.22 can be used to calculate the power delivered by a voltage source to *any* device carrying a current I and having a potential difference ΔV between its terminals.

Using Equation 27.22 and the fact that $\Delta V = IR$ for a resistor, we can express the power delivered to the resistor in the alternative forms

$$\mathcal{P} = I^2 R = \frac{(\Delta V)^2}{R} \qquad (27.23)$$

When I is expressed in amperes, ΔV in volts, and R in ohms, the SI unit of power is the watt, as it was in Chapter 7 in our discussion of mechanical power. The process by which power is lost as internal energy in a conductor of resistance R is often called *joule heating*[5]; this transformation is also often referred to as an I^2R loss.

When transporting energy by electricity through power lines, such as those shown in the opening photograph for this chapter, we cannot make the simplifying assumption that the lines have zero resistance. Real power lines do indeed have resistance, and power is delivered to the resistance of these wires. Utility companies seek to minimize the power transformed to internal energy in the lines and maximize the energy delivered to the consumer. Because $\mathcal{P} = I\,\Delta V$, the same amount of power can be transported either at high currents and low potential differences or at low currents and high potential differences. Utility companies choose to transport energy at low currents and high potential differences primarily for economic reasons. Copper wire is very expensive, and so it is cheaper to use high-resistance wire (that is, wire having a small cross-sectional area; see Eq. 27.11). Thus, in the expression for the power delivered to a resistor, $\mathcal{P} = I^2R$, the resistance of the wire is fixed at a relatively high value for economic considerations. The I^2R loss can be reduced by keeping the current I as low as possible, which means transferring the energy at a high voltage. In some instances, power is transported at potential differences as great as 765 kV. Once the electricity reaches your city, the potential difference is usually reduced to 4 kV by a device called a *transformer*. Another

---

[4]   This is another misuse of the word *heat* that is ingrained in our common language.

[5]   It is commonly called *joule heating* even though the process of heat does not occur. This is another example of incorrect usage of the word *heat* that has become entrenched in our language.

transformer drops the potential difference to 240 V before the electricity finally reaches your home. Of course, each time the potential difference decreases, the current increases by the same factor, and the power remains the same. We shall discuss transformers in greater detail in Chapter 33.

Demands on our dwindling energy supplies have made it necessary for us to be aware of the energy requirements of our electrical devices. Every electrical appliance carries a label that contains the information you need to calculate the appliance's power requirements. In many cases, the power consumption in watts is stated directly, as it is on a lightbulb. In other cases, the amount of current used by the device and the potential difference at which it operates are given. This information and Equation 27.22 are sufficient for calculating the power require-ment of any electrical device.

Quick Quiz 27.7 The same potential difference is applied to the two lightbulbs shown in Figure 27.14. Which one of the following statements is true? (a) The 30-W bulb carries the greater current and has the higher resistance. (b) The 30-W bulb carries the greater current, but the 60-W bulb has the higher resistance. (c) The 30-W bulb has the higher resistance, but the 60-W bulb carries the greater current. (d) The 60-W bulb carries the greater current and has the higher resistance.

Figure 27.14 (Quick Quiz 27.7) These lightbulbs operate at their rated power only when they are connected to a 120-V source.

Quick Quiz 27.8 For the two lightbulbs shown in Figure 27.15, rank the current values at points *a* through *f*, from greatest to least.

Figure 27.15 (Quick Quiz 27.8) Two lightbulbs connected across the same potential difference.

Example 27.7 Power in an Electric Heater

An electric heater is constructed by applying a potential difference of 120 V to a Nichrome wire that has a total resistance of 8.00 Ω. Find the current carried by the wire and the power rating of the heater.

Solution Because $\Delta V = IR$, we have

$$I = \frac{\Delta V}{R} = \frac{120 \text{ V}}{8.00 \text{ }\Omega} = \boxed{15.0 \text{ A}}$$

We can find the power rating using the expression $\mathcal{P} = I^2 R$:

$$\mathcal{P} = I^2 R = (15.0 \text{ A})^2 (8.00 \text{ }\Omega) = 1.80 \times 10^3 \text{ W}$$

$$\mathcal{P} = \boxed{1.80 \text{ kW}}$$

What If? What if the heater were accidentally connected to a 240-V supply? (This is difficult to do because the shape and orientation of the metal contacts in 240-V plugs are different from those in 120-V plugs.) How would this affect the current carried by the heater and the power rating of the heater?

Answer If we doubled the applied potential difference, Equation 27.8 tells us that the current would double. According to Equation 27.23, $\mathcal{P} = (\Delta V)^2/R$, the power would be four times larger.

Example 27.8   Linking Electricity and Thermodynamics

(A) What is the required resistance of an immersion heater that will increase the temperature of 1.50 kg of water from 10.0°C to 50.0°C in 10.0 min while operating at 110 V?

(B) Estimate the cost of heating the water.

Solution This example allows us to link our new understanding of power in electricity with our experience with specific heat in thermodynamics (Chapter 20). An immersion heater is a resistor that is inserted into a container of water. As energy is delivered to the immersion heater, raising its temperature, energy leaves the surface of the resistor by heat, going into the water. When the immersion heater reaches a constant temperature, the rate of energy delivered to the resistance by electrical transmission is equal to the rate of energy delivered by heat to the water.

(A) To simplify the analysis, we ignore the initial period during which the temperature of the resistor increases, and also ignore any variation of resistance with temperature. Thus, we imagine a constant rate of energy transfer for the entire 10.0 min. Setting the rate of energy delivered to the resistor equal to the rate of energy entering the water by heat, we have

$$\mathcal{P} = \frac{(\Delta V)^2}{R} = \frac{Q}{\Delta t}$$

where Q represents an amount of energy transfer by heat into the water and we have used Equation 27.23 to express

the electrical power. The amount of energy transfer by heat necessary to raise the temperature of the water is given by Equation 20.4, $Q = mc\,\Delta T$. Thus,

$$\frac{(\Delta V)^2}{R} = \frac{mc\,\Delta T}{\Delta t} \quad \longrightarrow \quad R = \frac{(\Delta V)^2\,\Delta t}{mc\,\Delta T}$$

Substituting the values given in the statement of the problem, we have

$$R = \frac{(110 \text{ V})^2(600 \text{ s})}{(1.50 \text{ kg})(4186 \text{ J/kg}\cdot°\text{C})(50.0°\text{C} - 10.0°\text{C})}$$

$$= \boxed{28.9 \ \Omega}$$

(B) Because the energy transferred equals power multiplied by time interval, the amount of energy transferred is

$$\mathcal{P}\,\Delta t = \frac{(\Delta V)^2}{R}\,\Delta t = \frac{(110 \text{ V})^2}{28.9 \ \Omega}(10.0 \text{ min})\left(\frac{1 \text{ h}}{60.0 \text{ min}}\right)$$

$$= 69.8 \text{ Wh} = 0.069 \text{ 8 kWh}$$

If the energy is purchased at an estimated price of 10.0¢ per kilowatt-hour, the cost is

$$\text{Cost} = (0.069 \text{ 8 kWh})(\$0.100/\text{kWh}) = \$0.006 \text{ 98}$$

$$\approx \boxed{0.7 \ ¢}$$

 At the Interactive Worked Example link at **http://www.pse6.com,** *you can explore the heating of the water.*

Example 27.9   Current in an Electron Beam

In a certain particle accelerator, electrons emerge with an energy of 40.0 MeV (1 MeV = 1.60×10^{-13} J). The electrons emerge not in a steady stream but rather in pulses at the rate of 250 pulses/s. This corresponds to a time interval between pulses of 4.00 ms (Fig. 27.16). Each pulse has a duration of 200 ns, and the electrons in the pulse constitute a current of 250 mA. The current is zero between pulses.

(A) How many electrons are delivered by the accelerator per pulse?

Solution We use Equation 27.2 in the form $dQ = I\,dt$ and integrate to find the charge per pulse. While the pulse is on, the current is constant; thus,

$$Q_{\text{pulse}} = I\int dt = I\,\Delta t = (250 \times 10^{-3} \text{ A})(200 \times 10^{-9} \text{ s})$$

$$= 5.00 \times 10^{-8} \text{ C}$$

Dividing this quantity of charge per pulse by the electronic charge gives the number of electrons per pulse:

Figure 27.16 (Example 27.9) Current versus time for a pulsed beam of electrons.

Electrons per pulse = $\dfrac{5.00 \times 10^{-8}\,\text{C/pulse}}{1.60 \times 10^{-19}\,\text{C/electron}}$

$= \boxed{3.13 \times 10^{11}\ \text{electrons/pulse}}$

(B) What is the average current per pulse delivered by the accelerator?

Solution Average current is given by Equation 27.1, $I_{av} = \Delta Q/\Delta t$. Because the time interval between pulses is 4.00 ms, and because we know the charge per pulse from part (A), we obtain

$$I_{av} = \frac{Q_{pulse}}{\Delta t} = \frac{5.00 \times 10^{-8}\,\text{C}}{4.00 \times 10^{-3}\,\text{s}} = \boxed{12.5\ \mu\text{A}}$$

This represents only 0.005% of the peak current, which is 250 mA.

(C) What is the peak power delivered by the electron beam?

Solution By definition, power is energy delivered per unit time interval. Thus, the peak power is equal to the energy delivered by a pulse divided by the pulse duration:

(1) $\qquad \mathcal{P}_{peak} = \dfrac{\text{pulse energy}}{\text{pulse duration}}$

$= \dfrac{(3.13 \times 10^{11}\ \text{electrons/pulse})(40.0\ \text{MeV/electron})}{2.00 \times 10^{-7}\,\text{s/pulse}}$

$\times \left(\dfrac{1.60 \times 10^{-13}\,\text{J}}{1\ \text{MeV}} \right)$

$= 1.00 \times 10^{7}\ \text{W} = \boxed{10.0\ \text{MW}}$

We could also compute this power directly. We assume that each electron has zero energy before being accelerated. Thus, by definition, each electron must go through a potential difference of 40.0 MV to acquire a final energy of 40.0 MeV. Hence, we have

(2) $\qquad \mathcal{P}_{peak} = I_{peak}\,\Delta V$

$= (250 \times 10^{-3}\,\text{A})(40.0 \times 10^{6}\,\text{V})$

$= \boxed{10.0\ \text{MW}}$

What If? What if the requested quantity in part (C) were the *average* power rather than the *peak* power?

Answer Instead of Equation (1), we would use the time interval between pulses rather than the duration of a pulse:

$\mathcal{P}_{av} = \dfrac{\text{pulse energy}}{\text{time interval between pulses}}$

$= \dfrac{(3.13 \times 10^{11}\ \text{electrons/pulse})(40.0\ \text{MeV/electron})}{4.00 \times 10^{-3}\,\text{s/pulse}}$

$\times \left(\dfrac{1.60 \times 10^{-13}\,\text{J}}{1\ \text{MeV}} \right)$

$= 500\ \text{W}$

Instead of Equation (2), we would use the average current found in part (B):

$\mathcal{P}_{av} = I_{av}\,\Delta V = (12.5 \times 10^{-6}\,\text{A})(40.0 \times 10^{6}\,\text{V})$

$= 500\ \text{W}$

Notice that these two calculations agree with each other and that the average power is much lower than the peak power.

SUMMARY

The **electric current** I in a conductor is defined as

$$I \equiv \frac{dQ}{dt} \qquad (27.2)$$

where dQ is the charge that passes through a cross section of the conductor in a time interval dt. The SI unit of current is the **ampere** (A), where 1 A = 1 C/s.

The average current in a conductor is related to the motion of the charge carriers through the relationship

$$I_{av} = nqv_d A \qquad (27.4)$$

where n is the density of charge carriers, q is the charge on each carrier, v_d is the drift speed, and A is the cross-sectional area of the conductor.

The magnitude of the **current density** J in a conductor is the current per unit area:

$$J \equiv \frac{I}{A} = nqv_d \qquad (27.5)$$

Take a practice test for this chapter by clicking on the Practice Test link at http://www.pse6.com.

The current density in an ohmic conductor is proportional to the electric field according to the expression

$$\mathbf{J} = \sigma\mathbf{E} \tag{27.7}$$

The proportionality constant σ is called the **conductivity** of the material of which the conductor is made. The inverse of σ is known as **resistivity** ρ (that is, $\rho = 1/\sigma$). Equation 27.7 is known as **Ohm's law,** and a material is said to obey this law if the ratio of its current density $\mathbf{J}$ to its applied electric field $\mathbf{E}$ is a constant that is independent of the applied field.

The **resistance** R of a conductor is defined as

$$R \equiv \frac{\Delta V}{I} \tag{27.8}$$

where ΔV is the potential difference across it, and I is the current it carries.

The SI unit of resistance is volts per ampere, which is defined to be 1 **ohm** (Ω); that is, $1\ \Omega = 1\ \text{V/A}$. If the resistance is independent of the applied potential difference, the conductor obeys Ohm's law.

For a uniform block of material of cross sectional area A and length ℓ, the resistance over the length ℓ is

$$R = \rho\frac{\ell}{A} \tag{27.11}$$

where ρ is the resistivity of the material.

In a classical model of electrical conduction in metals, the electrons are treated as molecules of a gas. In the absence of an electric field, the average velocity of the electrons is zero. When an electric field is applied, the electrons move (on the average) with a **drift velocity** $\mathbf{v}_d$ that is opposite the electric field and given by the expression

$$\mathbf{v}_d = \frac{q\mathbf{E}}{m_e}\tau \tag{27.14}$$

where τ is the average time interval between electron–atom collisions, m_e is the mass of the electron, and q is its charge. According to this model, the resistivity of the metal is

$$\rho = \frac{m_e}{nq^2\tau} \tag{27.17}$$

where n is the number of free electrons per unit volume.

The resistivity of a conductor varies approximately linearly with temperature according to the expression

$$\rho = \rho_0[1 + \alpha(T - T_0)] \tag{27.19}$$

where α is the **temperature coefficient of resistivity** and ρ_0 is the resistivity at some reference temperature T_0.

If a potential difference ΔV is maintained across a circuit element, the **power,** or rate at which energy is supplied to the element, is

$$\mathcal{P} = I\,\Delta V \tag{27.22}$$

Because the potential difference across a resistor is given by $\Delta V = IR$, we can express the power delivered to a resistor in the form

$$\mathcal{P} = I^2R = \frac{(\Delta V)^2}{R} \tag{27.23}$$

The energy delivered to a resistor by electrical transmission appears in the form of internal energy in the resistor.

QUESTIONS

1. In an analogy between electric current and automobile traffic flow, what would correspond to charge? What would correspond to current?

2. Newspaper articles often contain a statement such as "10 000 volts of electricity surged through the victim's body." What is wrong with this statement?

3. What factors affect the resistance of a conductor?

4. What is the difference between resistance and resistivity?

5. Two wires A and B of circular cross section are made of the same metal and have equal lengths, but the resistance of wire A is three times greater than that of wire B. What is the ratio of their cross-sectional areas? How do their radii compare?

6. Do all conductors obey Ohm's law? Give examples to justify your answer.

7. We have seen that an electric field must exist inside a conductor that carries a current. How is it possible in view of the fact that in electrostatics we concluded that the electric field must be zero inside a conductor?

8. A very large potential difference is not necessarily required to produce long sparks in air. With a device called *Jacob's ladder*, a potential difference of about 10 kV produces an electric arc a few millimeters long between the bottom ends of two curved rods that project upward from the power supply. (The device is seen in classic mad-scientist horror movies and in Figure Q27.8.) The arc rises, climbing the rods and getting longer and longer. It disappears when it reaches the top; then a new spark immediately forms at the bottom and the process repeats. Explain these phenomena. Why does the arc rise? Why does a new arc appear only after the previous one is gone?

9. When the voltage across a certain conductor is doubled, the current is observed to increase by a factor of three. What can you conclude about the conductor?

10. In the water analogy of an electric circuit, what corresponds to the power supply, resistor, charge, and potential difference?

11. Use the atomic theory of matter to explain why the resistance of a material should increase as its temperature increases.

12. Why might a "good" electrical conductor also be a "good" thermal conductor?

13. How does the resistance for copper and for silicon change with temperature? Why are the behaviors of these two materials different?

14. Explain how a current can persist in a superconductor without any applied voltage.

15. What single experimental requirement makes superconducting devices expensive to operate? In principle, can this limitation be overcome?

16. What would happen to the drift velocity of the electrons in a wire and to the current in the wire if the electrons could move freely without resistance through the wire?

17. If charges flow very slowly through a metal, why does it not require several hours for a light to come on when you throw a switch?

18. In a conductor, changes in the electric field that drives the electrons through the conductor propagate with a speed close to the speed of light, although the drift velocity of the electrons is very small. Explain how these statements can both be true. Does one particular electron move from one end of the conductor to the other?

19. Two conductors of the same length and radius are connected across the same potential difference. One conductor has twice the resistance of the other. To which conductor is more power delivered?

20. Two lightbulbs both operate from 120 V. One has a power of 25 W and the other 100 W. Which bulb has higher resistance? Which bulb carries more current?

21. Car batteries are often rated in ampere-hours. Does this designate the amount of current, power, energy, or charge that can be drawn from the battery?

22. If you were to design an electric heater using Nichrome wire as the heating element, what parameters of the wire could you vary to meet a specific power output, such as 1 000 W?

Figure Q27.8

PROBLEMS

1, 2, 3 = straightforward, intermediate, challenging ☐ = full solution available in the *Student Solutions Manual and Study Guide*

🪐 = coached solution with hints available at http://www.pse6.com 💻 = computer useful in solving problem

▨ = paired numerical and symbolic problems

Section 27.1 Electric Current

1. In a particular cathode ray tube, the measured beam current is 30.0 μA. How many electrons strike the tube screen every 40.0 s?

2. A teapot with a surface area of 700 cm² is to be silver plated. It is attached to the negative electrode of an electrolytic cell containing silver nitrate ($Ag^+ NO_3^-$). If the cell is powered by a 12.0-V battery and has a resistance of 1.80 Ω, how long does it take for a 0.133-mm layer of silver to build up on the teapot? (The density of silver is 10.5×10^3 kg/m³.)

3. 🪐 Suppose that the current through a conductor decreases exponentially with time according to the equation $I(t) = I_0 e^{-t/\tau}$ where I_0 is the initial current (at $t = 0$), and τ is a constant having dimensions of time. Consider a fixed observation point within the conductor. (a) How much charge passes this point between $t = 0$ and $t = \tau$? (b) How much charge passes this point between $t = 0$ and $t = 10\tau$? (c) **What If?** How much charge passes this point between $t = 0$ and $t = \infty$?

4. In the Bohr model of the hydrogen atom, an electron in the lowest energy state follows a circular path 5.29×10^{-11} m from the proton. (a) Show that the speed of the electron is 2.19×10^6 m/s. (b) What is the effective current associated with this orbiting electron?

5. A small sphere that carries a charge q is whirled in a circle at the end of an insulating string. The angular frequency of rotation is ω. What average current does this rotating charge represent?

6. The quantity of charge q (in coulombs) that has passed through a surface of area 2.00 cm² varies with time according to the equation $q = 4t^3 + 5t + 6$, where t is in seconds. (a) What is the instantaneous current through the surface at $t = 1.00$ s? (b) What is the value of the current density?

7. An electric current is given by the expression $I(t) = 100 \sin(120\pi t)$, where I is in amperes and t is in seconds. What is the total charge carried by the current from $t = 0$ to $t = (1/240)$ s?

8. Figure P27.8 represents a section of a circular conductor of nonuniform diameter carrying a current of 5.00 A. The

Figure P27.8

radius of cross section A_1 is 0.400 cm. (a) What is the magnitude of the current density across A_1? (b) If the current density across A_2 is one-fourth the value across A_1, what is the radius of the conductor at A_2?

9. The electron beam emerging from a certain high-energy electron accelerator has a circular cross section of radius 1.00 mm. (a) The beam current is 8.00 μA. Find the current density in the beam, assuming that it is uniform throughout. (b) The speed of the electrons is so close to the speed of light that their speed can be taken as $c = 3.00 \times 10^8$ m/s with negligible error. Find the electron density in the beam. (c) How long does it take for Avogadro's number of electrons to emerge from the accelerator?

10. A Van de Graaff generator produces a beam of 2.00-MeV deuterons, which are heavy hydrogen nuclei containing a proton and a neutron. (a) If the beam current is 10.0 μA, how far apart are the deuterons? (b) Is the electric force of repulsion among them a significant factor in beam stability? Explain.

11. An aluminum wire having a cross-sectional area of 4.00×10^{-6} m² carries a current of 5.00 A. Find the drift speed of the electrons in the wire. The density of aluminum is 2.70 g/cm³. Assume that one conduction electron is supplied by each atom.

Section 27.2 Resistance

12. Calculate the current density in a gold wire at 20°C, if an electric field of 0.740 V/m exists in the wire.

13. A lightbulb has a resistance of 240 Ω when operating with a potential difference of 120 V across it. What is the current in the lightbulb?

14. A resistor is constructed of a carbon rod that has a uniform cross-sectional area of 5.00 mm². When a potential difference of 15.0 V is applied across the ends of the rod, the rod carries a current of 4.00×10^{-3} A. Find (a) the resistance of the rod and (b) the rod's length.

15. 🪐 A 0.900-V potential difference is maintained across a 1.50-m length of tungsten wire that has a cross-sectional area of 0.600 mm². What is the current in the wire?

16. A conductor of uniform radius 1.20 cm carries a current of 3.00 A produced by an electric field of 120 V/m. What is the resistivity of the material?

17. Suppose that you wish to fabricate a uniform wire out of 1.00 g of copper. If the wire is to have a resistance of $R = 0.500$ Ω, and if all of the copper is to be used, what will be (a) the length and (b) the diameter of this wire?

18. Gold is the most ductile of all metals. For example, one gram of gold can be drawn into a wire 2.40 km long. What is the resistance of such a wire at 20°C? You can find the necessary reference information in this textbook.

19. (a) Make an order-of-magnitude estimate of the resistance between the ends of a rubber band. (b) Make an order-of-magnitude estimate of the resistance between the 'heads' and 'tails' sides of a penny. In each case state what quantities you take as data and the values you measure or estimate for them. (c) WARNING! Do not try this at home! What is the order of magnitude of the current that each would carry if it were connected across a 120-V power supply?

20. A solid cube of silver (density = 10.5 g/cm^3) has a mass of 90.0 g. (a) What is the resistance between opposite faces of the cube? (b) Assume each silver atom contributes one conduction electron. Find the average drift speed of electrons when a potential difference of 1.00×10^{-5} V is applied to opposite faces. The atomic number of silver is 47, and its molar mass is 107.87 g/mol.

21. A metal wire of resistance R is cut into three equal pieces that are then connected side by side to form a new wire the length of which is equal to one-third the original length. What is the resistance of this new wire?

22. Aluminum and copper wires of equal length are found to have the same resistance. What is the ratio of their radii?

23. A current density of 6.00×10^{-13} A/m^2 exists in the atmosphere at a location where the electric field is 100 V/m. Calculate the electrical conductivity of the Earth's atmosphere in this region.

24. The rod in Figure P27.24 is made of two materials. The figure is not drawn to scale. Each conductor has a square cross section 3.00 mm on a side. The first material has a resistivity of 4.00×10^{-3} $\Omega \cdot$m and is 25.0 cm long, while the second material has a resistivity of 6.00×10^{-3} $\Omega \cdot$m and is 40.0 cm long. What is the resistance between the ends of the rod?

— 25.0 cm —   — 40.0 cm —

Figure P27.24

Section 27.3   A Model for Electrical Conduction

25. If the magnitude of the drift velocity of free electrons in a copper wire is 7.84×10^{-4} m/s, what is the electric field in the conductor?

26. If the current carried by a conductor is doubled, what happens to the (a) charge carrier density? (b) current density? (c) electron drift velocity? (d) average time interval between collisions?

27. Use data from Example 27.1 to calculate the collision mean free path of electrons in copper. Assume the average thermal speed of conduction electrons is 8.60×10^5 m/s.

Section 27.4   Resistance and Temperature

28. While taking photographs in Death Valley on a day when the temperature is 58.0°C, Bill Hiker finds that a certain voltage applied to a copper wire produces a current of 1.000 A. Bill then travels to Antarctica and applies the same voltage to the same wire. What current does he register there if the temperature is -88.0°C? Assume that no change occurs in the wire's shape and size.

29. A certain lightbulb has a tungsten filament with a resistance of 19.0 Ω when cold and 140 Ω when hot. Assume that the resistivity of tungsten varies linearly with temperature even over the large temperature range involved here, and find the temperature of the hot filament. Assume the initial temperature is 20.0°C.

30. A carbon wire and a Nichrome wire are connected in series, so that the same current exists in both wires. If the combination has a resistance of 10.0 kΩ at 0°C, what is the resistance of each wire at 0°C so that the resistance of the combination does not change with temperature? The total or equivalent resistance of resistors in series is the sum of their individual resistances.

31. An aluminum wire with a diameter of 0.100 mm has a uniform electric field of 0.200 V/m imposed along its entire length. The temperature of the wire is 50.0°C. Assume one free electron per atom. (a) Use the information in Table 27.1 and determine the resistivity. (b) What is the current density in the wire? (c) What is the total current in the wire? (d) What is the drift speed of the conduction electrons? (e) What potential difference must exist between the ends of a 2.00-m length of the wire to produce the stated electric field?

32. **Review problem.** An aluminum rod has a resistance of 1.234 Ω at 20.0°C. Calculate the resistance of the rod at 120°C by accounting for the changes in both the resistivity and the dimensions of the rod.

33. What is the fractional change in the resistance of an iron filament when its temperature changes from 25.0°C to 50.0°C?

34. The resistance of a platinum wire is to be calibrated for low-temperature measurements. A platinum wire with resistance 1.00 Ω at 20.0°C is immersed in liquid nitrogen at 77 K (-196°C). If the temperature response of the platinum wire is linear, what is the expected resistance of the platinum wire at -196°C? ($\alpha_{\text{platinum}} = 3.92 \times 10^{-3}/$°C)

35. The temperature of a sample of tungsten is raised while a sample of copper is maintained at 20.0°C. At what temperature will the resistivity of the tungsten be four times that of the copper?

Section 27.6   Electrical Power

36. A toaster is rated at 600 W when connected to a 120-V source. What current does the toaster carry, and what is its resistance?

37. A Van de Graaff generator (see Figure 25.29) is operating so that the potential difference between the high-voltage electrode B and the charging needles at A is 15.0 kV. Calculate the power required to drive the belt against electrical forces at an instant when the effective current delivered to the high-voltage electrode is 500 μA.

38. In a hydroelectric installation, a turbine delivers 1 500 hp to a generator, which in turn transfers 80.0% of the mechanical energy out by electrical transmission. Under

these conditions, what current does the generator deliver at a terminal potential difference of 2 000 V?

39. What is the required resistance of an immersion heater that increases the temperature of 1.50 kg of water from 10.0°C to 50.0°C in 10.0 min while operating at 110 V?

40. One rechargeable battery of mass 15.0 g delivers to a CD player an average current of 18.0 mA at 1.60 V for 2.40 h before the battery needs to be recharged. The recharger maintains a potential difference of 2.30 V across the battery and delivers a charging current of 13.5 mA for 4.20 h. (a) What is the efficiency of the battery as an energy storage device? (b) How much internal energy is produced in the battery during one charge–discharge cycle? (b) If the battery is surrounded by ideal thermal insulation and has an overall effective specific heat of 975 J/kg°C, by how much will its temperature increase during the cycle?

41. Suppose that a voltage surge produces 140 V for a moment. By what percentage does the power output of a 120-V, 100-W lightbulb increase? Assume that its resistance does not change.

42. A 500-W heating coil designed to operate from 110 V is made of Nichrome wire 0.500 mm in diameter. (a) Assuming that the resistivity of the Nichrome remains constant at its 20.0°C value, find the length of wire used. (b) **What If?** Now consider the variation of resistivity with temperature. What power will the coil of part (a) actually deliver when it is heated to 1 200°C?

43. A coil of Nichrome wire is 25.0 m long. The wire has a diameter of 0.400 mm and is at 20.0°C. If it carries a current of 0.500 A, what are (a) the magnitude of the electric field in the wire, and (b) the power delivered to it? (c) **What If?** If the temperature is increased to 340°C and the voltage across the wire remains constant, what is the power delivered?

44. Batteries are rated in terms of ampere-hours (A·h). For example, a battery that can produce a current of 2.00 A for 3.00 h is rated at 6.00 A·h. (a) What is the total energy, in kilowatt-hours, stored in a 12.0-V battery rated at 55.0 A·h? (b) At $0.060 0 per kilowatt-hour, what is the value of the electricity produced by this battery?

45. A 10.0-V battery is connected to a 120-Ω resistor. Ignoring the internal resistance of the battery, calculate the power delivered to the resistor.

46. Residential building codes typically require the use of 12-gauge copper wire (diameter 0.205 3 cm) for wiring receptacles. Such circuits carry currents as large as 20 A. A wire of smaller diameter (with a higher gauge number) could carry this much current, but the wire could rise to a high temperature and cause a fire. (a) Calculate the rate at which internal energy is produced in 1.00 m of 12-gauge copper wire carrying a current of 20.0 A. (b) **What If?** Repeat the calculation for an aluminum wire. Would a 12-gauge aluminum wire be as safe as a copper wire?

47. An 11.0-W energy-efficient fluorescent lamp is designed to produce the same illumination as a conventional 40.0-W incandescent lightbulb. How much money does the user of the energy-efficient lamp save during 100 hours of use? Assume a cost of $0.080 0/kWh for energy from the power company.

48. We estimate that 270 million plug-in electric clocks are in the United States, approximately one clock for each person. The clocks convert energy at the average rate 2.50 W. To supply this energy, how many metric tons of coal are burned per hour in coal-fired electric generating plants that are, on average, 25.0% efficient? The heat of combustion for coal is 33.0 MJ/kg.

49. Compute the cost per day of operating a lamp that draws a current of 1.70 A from a 110-V line. Assume the cost of energy from the power company is $0.060 0/kWh.

50. **Review problem.** The heating element of a coffee maker operates at 120 V and carries a current of 2.00 A. Assuming that the water absorbs all of the energy delivered to the resistor, calculate how long it takes to raise the temperature of 0.500 kg of water from room temperature (23.0°C) to the boiling point.

51. A certain toaster has a heating element made of Nichrome wire. When the toaster is first connected to a 120-V source (and the wire is at a temperature of 20.0°C), the initial current is 1.80 A. However, the current begins to decrease as the heating element warms up. When the toaster reaches its final operating temperature, the current drops to 1.53 A. (a) Find the power delivered to the toaster when it is at its operating temperature. (b) What is the final temperature of the heating element?

52. The cost of electricity varies widely through the United States; $0.120/kWh is one typical value. At this unit price, calculate the cost of (a) leaving a 40.0-W porch light on for two weeks while you are on vacation, (b) making a piece of dark toast in 3.00 min with a 970-W toaster, and (c) drying a load of clothes in 40.0 min in a 5 200-W dryer.

53. Make an order-of-magnitude estimate of the cost of one person's routine use of a hair dryer for 1 yr. If you do not use a blow dryer yourself, observe or interview someone who does. State the quantities you estimate and their values.

Additional Problems

54. One lightbulb is marked '25 W 120 V,' and another '100 W 120 V'; this means that each bulb has its respective power delivered to it when plugged into a constant 120-V potential difference. (a) Find the resistance of each bulb. (b) How long does it take for 1.00 C to pass through the dim bulb? Is the charge different in any way upon its exit from the bulb versus its entry? (c) How long does it take for 1.00 J to pass through the dim bulb? By what mechanisms does this energy enter and exit the bulb? (d) Find how much it costs to run the dim bulb continuously for 30.0 days if the electric company sells its product at $0.070 0 per kWh. What product *does* the electric company sell? What is its price for one SI unit of this quantity?

55. A charge Q is placed on a capacitor of capacitance C. The capacitor is connected into the circuit shown in Figure P27.55, with an open switch, a resistor, and an initially uncharged capacitor of capacitance $3C$. The switch is then closed and the circuit comes to equilibrium. In terms of Q and C, find (a) the final potential difference between the plates of each capacitor, (b) the charge on each capacitor, and (c) the final energy stored in each capacitor. (d) Find the internal energy appearing in the resistor.

Figure P27.55

56. A high-voltage transmission line with a diameter of 2.00 cm and a length of 200 km carries a steady current of 1 000 A. If the conductor is copper wire with a free charge density of 8.49×10^{28} electrons/m^3, how long does it take one electron to travel the full length of the line?

57. A more general definition of the temperature coefficient of resistivity is

$$\alpha = \frac{1}{\rho}\frac{d\rho}{dT}$$

where ρ is the resistivity at temperature T. (a) Assuming that α is constant, show that

$$\rho = \rho_0 e^{\alpha(T-T_0)}$$

where ρ_0 is the resistivity at temperature T_0. (b) Using the series expansion $e^x \approx 1 + x$ for $x \ll 1$, show that the resistivity is given approximately by the expression $\rho = \rho_0[1 + \alpha(T-T_0)]$ for $\alpha(T-T_0) \ll 1$.

58. A high-voltage transmission line carries 1 000 A starting at 700 kV for a distance of 100 mi. If the resistance in the wire is 0.500 Ω/mi, what is the power loss due to resistive losses?

59. An experiment is conducted to measure the electrical resistivity of Nichrome in the form of wires with different lengths and cross-sectional areas. For one set of measurements, a student uses 30-gauge wire, which has a cross-sectional area of 7.30×10^{-8} m^2. The student measures the potential difference across the wire and the current in the wire with a voltmeter and an ammeter, respectively. For each of the measurements given in the table taken on wires of three different lengths, calculate the resistance of the wires and the corresponding values of the resistivity. What is the average value of the resistivity, and how does this value compare with the value given in Table 27.1?

L (m)	ΔV (V)	I (A)	R (Ω)	ρ ($\Omega \cdot$ m)
0.540	5.22	0.500		
1.028	5.82	0.276		
1.543	5.94	0.187		

60. An electric utility company supplies a customer's house from the main power lines (120 V) with two copper wires, each of which is 50.0 m long and has a resistance of 0.108 Ω per 300 m. (a) Find the voltage at the customer's house for a load current of 110 A. For this load current, find (b) the power the customer is receiving and (c) the electric power lost in the copper wires.

61. A straight cylindrical wire lying along the x axis has a length of 0.500 m and a diameter of 0.200 mm. It is made of a material that obeys Ohm's law with a resistivity of $\rho = 4.00 \times 10^{-8}$ $\Omega \cdot$ m. Assume that a potential of 4.00 V is maintained at $x = 0$, and that $V = 0$ at $x = 0.500$ m. Find (a) the electric field $\mathbf{E}$ in the wire, (b) the resistance of the wire, (c) the electric current in the wire, and (d) the current density $\mathbf{J}$ in the wire. Express vectors in vector notation. (e) Show that $\mathbf{E} = \rho\mathbf{J}$.

62. A straight cylindrical wire lying along the x axis has a length L and a diameter d. It is made of a material that obeys Ohm's law with a resistivity ρ. Assume that potential V is maintained at $x = 0$, and that the potential is zero at $x = L$. In terms of L, d, V, ρ, and physical constants, derive expressions for (a) the electric field in the wire, (b) the resistance of the wire, (c) the electric current in the wire, and (d) the current density in the wire. Express vectors in vector notation. (e) Prove that $\mathbf{E} = \rho\mathbf{J}$.

63. The potential difference across the filament of a lamp is maintained at a constant level while equilibrium temperature is being reached. It is observed that the steady-state current in the lamp is only one tenth of the current drawn by the lamp when it is first turned on. If the temperature coefficient of resistivity for the lamp at 20.0°C is 0.004 50 (°C)$^{-1}$, and if the resistance increases linearly with increasing temperature, what is the final operating temperature of the filament?

64. The current in a resistor decreases by 3.00 A when the voltage applied across the resistor decreases from 12.0 V to 6.00 V. Find the resistance of the resistor.

65. An electric car is designed to run off a bank of 12.0-V batteries with total energy storage of 2.00×10^7 J. (a) If the electric motor draws 8.00 kW, what is the current delivered to the motor? (b) If the electric motor draws 8.00 kW as the car moves at a steady speed of 20.0 m/s, how far will the car travel before it is "out of juice"?

66. **Review problem.** When a straight wire is heated, its resistance is given by $R = R_0[1 + \alpha(T-T_0)]$ according to Equation 27.21, where α is the temperature coefficient of resistivity. (a) Show that a more precise result, one that includes the fact that the length and area of the wire change when heated, is

$$R = \frac{R_0[1 + \alpha(T-T_0)][1 + \alpha'(T-T_0)]}{[1 + 2\alpha'(T-T_0)]}$$

where α' is the coefficient of linear expansion (see Chapter 19). (b) Compare these two results for a 2.00-m-long copper wire of radius 0.100 mm, first at 20.0°C and then heated to 100.0°C.

67. The temperature coefficients of resistivity in Table 27.1 were determined at a temperature of 20°C. What would they be at 0°C? Note that the temperature coefficient of resistivity at 20°C satisfies $\rho = \rho_0[1 + \alpha(T-T_0)]$, where ρ_0 is the resistivity of the material at $T_0 = 20$°C. The temperature

coefficient of resistivity α' at 0°C must satisfy the expression $\rho = \rho_0'[1 + \alpha'T]$, where ρ_0' is the resistivity of the material at 0°C.

68. An oceanographer is studying how the ion concentration in sea water depends on depth. She does this by lowering into the water a pair of concentric metallic cylinders (Fig. P27.68) at the end of a cable and taking data to determine the resistance between these electrodes as a function of depth. The water between the two cylinders forms a cylindrical shell of inner radius r_a, outer radius r_b, and length L much larger than r_b. The scientist applies a potential difference ΔV between the inner and outer surfaces, producing an outward radial current I. Let ρ represent the resistivity of the water. (a) Find the resistance of the water between the cylinders in terms of L, ρ, r_a, and r_b. (b) Express the resistivity of the water in terms of the measured quantities L, r_a, r_b, ΔV, and I.

Figure P27.68

69. In a certain stereo system, each speaker has a resistance of 4.00 Ω. The system is rated at 60.0 W in each channel, and each speaker circuit includes a fuse rated 4.00 A. Is this system adequately protected against overload? Explain your reasoning.

70. A close analogy exists between the flow of energy by heat because of a temperature difference (see Section 20.7) and the flow of electric charge because of a potential difference. The energy dQ and the electric charge dq can both be transported by free electrons in the conducting material. Consequently, a good electrical conductor is usually a good thermal conductor as well. Consider a thin conducting slab of thickness dx, area A, and electrical conductivity σ, with a potential difference dV between opposite faces. Show that the current $I = dq/dt$ is given by the equation on the left below:

Charge Conduction	Thermal Conduction (Eq. 20.14)
$\dfrac{dq}{dt} = \sigma A \left\| \dfrac{dV}{dx} \right\|$	$\dfrac{dQ}{dt} = kA \left\| \dfrac{dT}{dx} \right\|$

In the analogous thermal conduction equation on the right, the rate of energy flow dQ/dt (in SI units of joules per second) is due to a temperature gradient dT/dx, in a material of thermal conductivity k. State analogous rules relating the direction of the electric current to the change in potential, and relating the direction of energy flow to the change in temperature.

71. Material with uniform resistivity ρ is formed into a wedge as shown in Figure P27.71. Show that the resistance between face A and face B of this wedge is

$$R = \rho \frac{L}{w(y_2 - y_1)} \ln\left(\frac{y_2}{y_1}\right)$$

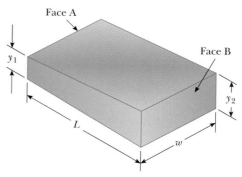

Figure P27.71

72. A material of resistivity ρ is formed into the shape of a truncated cone of altitude h as shown in Figure P27.72. The bottom end has radius b, and the top end has radius a. Assume that the current is distributed uniformly over any circular cross section of the cone, so that the current density does not depend on radial position. (The current density does vary with position along the axis of the cone.) Show that the resistance between the two ends is described by the expression

$$R = \frac{\rho}{\pi}\left(\frac{h}{ab}\right)$$

Figure P27.72

73. The dielectric material between the plates of a parallel-plate capacitor always has some nonzero conductivity σ. Let A represent the area of each plate and d the distance between them. Let κ represent the dielectric constant of the material. (a) Show that the resistance R and the capacitance C of the capacitor are related by

$$RC = \frac{\kappa \epsilon_0}{\sigma}$$

(b) Find the resistance between the plates of a 14.0-nF capacitor with a fused quartz dielectric.

74. ⬛ The current–voltage characteristic curve for a semiconductor diode as a function of temperature T is given by the equation

$$I = I_0(e^{e\,\Delta V/k_\mathrm{B}T} - 1)$$

Example 28.1   Terminal Voltage of a Battery   Interactive

A battery has an emf of 12.0 V and an internal resistance of 0.05 Ω. Its terminals are connected to a load resistance of 3.00 Ω.

(A) Find the current in the circuit and the terminal voltage of the battery.

Solution Equation 28.3 gives us the current:

$$I = \frac{\varepsilon}{R + r} = \frac{12.0 \text{ V}}{3.05 \text{ }\Omega} = \boxed{3.93 \text{ A}}$$

and from Equation 28.1, we find the terminal voltage:

$$\Delta V = \varepsilon - Ir = 12.0 \text{ V} - (3.93 \text{ A})(0.05 \text{ }\Omega) = \boxed{11.8 \text{ V}}$$

To check this result, we can calculate the voltage across the load resistance R:

$$\Delta V = IR = (3.93 \text{ A})(3.00 \text{ }\Omega) = 11.8 \text{ V}$$

(B) Calculate the power delivered to the load resistor, the power delivered to the internal resistance of the battery, and the power delivered by the battery.

Solution The power delivered to the load resistor is

$$\mathcal{P}_R = I^2R = (3.93 \text{ A})^2 (3.00 \text{ }\Omega) = \boxed{46.3 \text{ W}}$$

The power delivered to the internal resistance is

$$\mathcal{P}_r = I^2r = (3.93 \text{ A})^2 (0.05 \text{ }\Omega) = \boxed{0.772 \text{ W}}$$

Hence, the power delivered by the battery is the sum of these quantities, or 47.1 W. You should check this result, using the expression $\mathcal{P} = I\varepsilon$.

What If? As a battery ages, its internal resistance increases. Suppose the internal resistance of this battery rises to 2.00 Ω toward the end of its useful life. How does this alter the ability of the battery to deliver energy?

Answer Let us connect the same 3.00-Ω load resistor to the battery. The current in the battery now is

$$I = \frac{\varepsilon}{R + r} = \frac{12.0 \text{ V}}{(3.00 \text{ }\Omega + 2.00 \text{ }\Omega)} = 2.40 \text{ A}$$

and the terminal voltage is

$$\Delta V = \varepsilon - Ir = 12.0 \text{ V} - (2.40 \text{ A})(2.00 \text{ }\Omega) = 7.2 \text{ V}$$

Notice that the terminal voltage is only 60% of the emf. The powers delivered to the load resistor and internal resistance are

$$\mathcal{P}_R = I^2R = (2.40 \text{ A})^2 (3.00 \text{ }\Omega) = \boxed{17.3 \text{ W}}$$

$$\mathcal{P}_r = I^2r = (2.40 \text{ A})^2 (2.00 \text{ }\Omega) = 11.5 \text{ W}$$

Notice that 40% of the power from the battery is delivered to the internal resistance. In part (B), this percentage is 1.6%. Consequently, even though the emf remains fixed, the increasing internal resistance significantly reduces the ability of the battery to deliver energy.

 *At the Interactive Worked Example link at **http://www.pse6.com**, you can vary the load resistance and internal resistance, observing the power delivered to each.*

Example 28.2   Matching the Load

Show that the maximum power delivered to the load resistance R in Figure 28.2a occurs when the load resistance matches the internal resistance—that is, when $R = r$.

Solution The power delivered to the load resistance is equal to I^2R, where I is given by Equation 28.3:

$$\mathcal{P} = I^2R = \frac{\varepsilon^2 R}{(R + r)^2}$$

When $\mathcal{P}$ is plotted versus R as in Figure 28.3, we find that $\mathcal{P}$ reaches a maximum value of $\varepsilon^2/4r$ at $R = r$. When R is large, there is very little current, so that the power I^2R delivered to the load resistor is small. When R is small, the current is large and there is significant loss of power I^2r as energy is delivered to the internal resistance. When $R = r$, these effects balance to give a maximum transfer of power.

We can also prove that the power maximizes at $R = r$ by differentiating $\mathcal{P}$ with respect to R, setting the result equal

to zero, and solving for R. The details are left as a problem for you to solve (Problem 57).

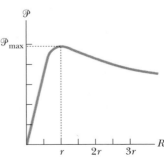

Figure 28.3 (Example 28.2) Graph of the power $\mathcal{P}$ delivered by a battery to a load resistor of resistance R as a function of R. The power delivered to the resistor is a maximum when the load resistance equals the internal resistance of the battery.

28.2 Resistors in Series and Parallel

Suppose that you and your friends are at a crowded basketball game in a sports arena and decide to leave early. You have two choices: (1) your group can exit through a single door and push your way down a long hallway containing several concession stands, each surrounded by a large crowd of people waiting to buy food or souvenirs; or (2) each member of your group can exit through a separate door in the main hall of the arena, where each will have to push his or her way through a single group of people standing by the door. In which scenario will less time be required for your group to leave the arena?

It should be clear that your group will be able to leave faster through the separate doors than down the hallway where each of you has to push through several groups of people. We could describe the groups of people in the hallway as being in *series*, because each of you must push your way through all of the groups. The groups of people around the doors in the arena can be described as being in *parallel*. Each member of your group must push through only one group of people, and each member pushes through a *different* group of people. This simple analogy will help us understand the behavior of currents in electric circuits containing more than one resistor.

When two or more resistors are connected together as are the lightbulbs in Figure 28.4a, they are said to be in *series*. Figure 28.4b is the circuit diagram for the lightbulbs, which are shown as resistors, and the battery. In a series connection, if an amount of charge Q exits resistor R_1, charge Q must also enter the second resistor R_2. (This is analogous to all members of your group pushing through each crowd in the single hallway of the sports arena.) Otherwise, charge will accumulate on the wire between the resistors. Thus, the same amount of charge passes through both resistors in a given time interval. Hence,

> for a series combination of two resistors, the currents are the same in both resistors because the amount of charge that passes through R_1 must also pass through R_2 in the same time interval.

The potential difference applied across the series combination of resistors will divide between the resistors. In Figure 28.4b, because the voltage drop[2] from a to b equals IR_1 and the voltage drop from b to c equals IR_2, the voltage drop from a to c is

$$\Delta V = IR_1 + IR_2 = I(R_1 + R_2)$$

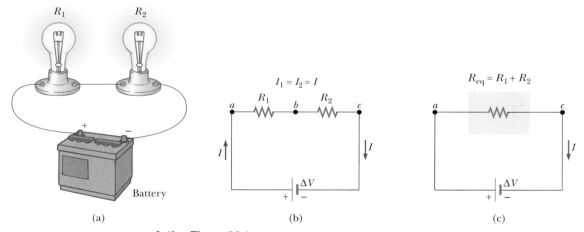

(a)   (b)   (c)

Active Figure 28.4 (a) A series connection of two lightbulbs with resistances R_1 and R_2. (b) Circuit diagram for the two-resistor circuit. The current in R_1 is the same as that in R_2. (c) The resistors replaced with a single resistor having an equivalent resistance $R_{eq} = R_1 + R_2$.

At the Active Figures link at http://www.pse6.com, you can adjust the battery voltage and resistances R_1 and R_2 to see the effect on the currents and voltages in the individual resistors.

[2]   The term *voltage drop* is synonymous with a decrease in electric potential across a resistor and is used often by individuals working with electric circuits.

The potential difference across the battery is also applied to the **equivalent resistance** R_{eq} in Figure 28.4c:

$$\Delta V = IR_{eq}$$

where we have indicated that the equivalent resistance has the same effect on the circuit because it results in the same current in the battery as the combination of resistors. Combining these equations, we see that we can replace the two resistors in series with a single equivalent resistance whose value is the *sum* of the individual resistances:

$$\Delta V = IR_{eq} = I(R_1 + R_2) \quad \longrightarrow \quad R_{eq} = R_1 + R_2 \qquad (28.5)$$

The resistance R_{eq} is equivalent to the series combination $R_1 + R_2$ in the sense that the circuit current is unchanged when R_{eq} replaces $R_1 + R_2$.

The equivalent resistance of three or more resistors connected in series is

$$R_{eq} = R_1 + R_2 + R_3 + \cdots \qquad (28.6)$$

The equivalent resistance of several resistors in series

This relationship indicates that **the equivalent resistance of a series connection of resistors is the numerical sum of the individual resistances and is always greater than any individual resistance.**

Looking back at Equation 28.3, the denominator is the simple algebraic sum of the external and internal resistances. This is consistent with the fact that internal and external resistances are in series in Figure 28.2a.

Note that if the filament of one lightbulb in Figure 28.4 were to fail, the circuit would no longer be complete (resulting in an open-circuit condition) and the second bulb would also go out. This is a general feature of a series circuit—if one device in the series creates an open circuit, all devices are inoperative.

▲ **PITFALL PREVENTION**

28.2 Lightbulbs Don't Burn

We will describe the end of the life of a lightbulb by saying that *the filament fails*, rather than by saying that the lightbulb "burns out." The word *burn* suggests a combustion process, which is not what occurs in a lightbulb.

Quick Quiz 28.2 In Figure 28.4, imagine positive charges pass first through R_1 and then through R_2. Compared to the current in R_1, the current in R_2 is (a) smaller, (b) larger, or (c) the same.

Quick Quiz 28.3 If a piece of wire is used to connect points b and c in Figure 28.4b, does the brightness of bulb R_1 (a) increase, (b) decrease, or (c) remain the same?

Quick Quiz 28.4 With the switch in the circuit of Figure 28.5 closed (left), there is no current in R_2, because the current has an alternate zero-resistance path through the switch. There is current in R_1 and this current is measured with the ammeter (a device for measuring current) at the right side of the circuit. If the switch is opened (Fig. 28.5, right), there is current in R_2. What happens to the reading on the ammeter when the switch is opened? (a) the reading goes up; (b) the reading goes down; (c) the reading does not change.

Switch closed   Switch open

Figure 28.5 (Quick Quiz 28.4) What happens when the switch is opened?

At the Active Figures link at http://www.pse6.com, *you can adjust the battery voltage and resistances R_1 and R_2 to see the effect on the currents and voltages in the individual resistors.*

Active Figure 28.6 (a) A parallel connection of two lightbulbs with resistances R_1 and R_2. (b) Circuit diagram for the two-resistor circuit. The potential difference across R_1 is the same as that across R_2. (c) The resistors replaced with a single resistor having an equivalent resistance given by Equation 28.7.

▲ **PITFALL PREVENTION**

28.3 Local and Global Changes

A local change in one part of a circuit may result in a global change throughout the circuit. For example, if a single resistance is changed in a circuit containing several resistors and batteries, the currents in all resistors and batteries, the terminal voltages of all batteries, and the voltages across all resistors may change as a result.

▲ **PITFALL PREVENTION**

28.4 Current Does Not Take the Path of Least Resistance

You may have heard a phrase like "current takes the path of least resistance" in reference to a parallel combination of current paths, such that there are two or more paths for the current to take. The phrase is incorrect. The current takes *all* paths. Those paths with lower resistance will have large currents, but even very high-resistance paths will carry *some* of the current.

Now consider two resistors connected in *parallel*, as shown in Figure 28.6. When charges reach point a in Figure 28.6b, called a *junction*, they split into two parts, with some going through R_1 and the rest going through R_2. A **junction** is any point in a circuit where a current can split (just as your group might split up and leave the sports arena through several doors, as described earlier.) This split results in less current in each individual resistor than the current leaving the battery. Because electric charge is conserved, the current I that enters point a must equal the total current leaving that point:

$$I = I_1 + I_2$$

where I_1 is the current in R_1 and I_2 is the current in R_2.

As can be seen from Figure 28.6, both resistors are connected directly across the terminals of the battery. Therefore,

> when resistors are connected in parallel, the potential differences across the resistors is the same.

Because the potential differences across the resistors are the same, the expression $\Delta V = IR$ gives

$$I = I_1 + I_2 = \frac{\Delta V}{R_1} + \frac{\Delta V}{R_2} = \Delta V\left(\frac{1}{R_1} + \frac{1}{R_2}\right) = \frac{\Delta V}{R_{eq}}$$

where R_{eq} is an equivalent single resistance which will have the same effect on the circuit as the two resistors in parallel; that is, it will draw the same current from the battery (Fig. 28.6c). From this result, we see that the equivalent resistance of two resistors in parallel is given by

$$\frac{1}{R_{eq}} = \frac{1}{R_1} + \frac{1}{R_2} \qquad (28.7)$$

or

$$R_{eq} = \frac{1}{\dfrac{1}{R_1} + \dfrac{1}{R_2}} = \frac{R_1 R_2}{R_1 + R_2}$$

An extension of this analysis to three or more resistors in parallel gives

$$\frac{1}{R_{eq}} = \frac{1}{R_1} + \frac{1}{R_2} + \frac{1}{R_3} + \cdots$$ (28.8)

The equivalent resistance of several resistors in parallel

We can see from this expression that **the inverse of the equivalent resistance of two or more resistors connected in parallel is equal to the sum of the inverses of the individual resistances. Furthermore, the equivalent resistance is always less than the smallest resistance in the group.**

Household circuits are always wired such that the appliances are connected in parallel. Each device operates independently of the others so that if one is switched off, the others remain on. In addition, in this type of connection, all of the devices operate on the same voltage.

Quick Quiz 28.5 In Figure 28.4, imagine that we add a third resistor in series with the first two. Does the current in the battery (a) increase, (b) decrease, or (c) remain the same? Does the terminal voltage of the battery (d) increase, (e) decrease, or (f) remain the same?

Quick Quiz 28.6 In Figure 28.6, imagine that we add a third resistor in parallel with the first two. Does the current in the battery (a) increase, (b) decrease, or (c) remain the same? Does the terminal voltage of the battery (d) increase, (e) decrease, or (f) remain the same?

Quick Quiz 28.7 With the switch in the circuit of Figure 28.7 open (left), there is no current in R_2. There is current in R_1 and this current is measured with the ammeter at the right side of the circuit. If the switch is closed (Fig. 28.7, right), there is current in R_2. What happens to the reading on the ammeter when the switch is closed? (a) the reading goes up; (b) the reading goes down; (c) the reading does not change.

Switch open   Switch closed
Figure 28.7 (Quick Quiz 28.7) What happens when the switch is closed?

Conceptual Example 28.3   Landscape Lights

A homeowner wishes to install 12-volt landscape lighting in his back yard. To save money, he purchases inexpensive 18-gauge cable, which has a relatively high resistance per unit length. This cable consists of two side-by-side wires separated by insulation, like the cord on an appliance. He runs a 200-foot length of this cable from the power supply to the farthest point at which he plans to position a light fixture. He attaches light fixtures across the two wires on the cable at 10-foot intervals, so the light fixtures are in parallel. Because of the cable's resistance, the brightness of the bulbs in the light fixtures is not as desired. Which problem does the homeowner have? (a) All of the bulbs glow equally less brightly than they would if lower-resistance cable had been used. (b) The brightness of the bulbs decreases as you move farther from the power supply.

Solution A circuit diagram for the system appears in Figure 28.8. The horizontal resistors (such as R_A and R_B) represent the resistance of the wires in the cable between the light fixtures while the vertical resistors (such as R_C) represent the resistance of the light fixtures themselves. Part of the terminal voltage of the power supply is dropped across resistors R_A and R_B. Thus, the voltage across light fixture R_C is less than the terminal voltage. There is a further voltage drop across resistors R_D and R_E. Consequently, the voltage across light fixture R_F is smaller than that across R_C. This continues on down the line of light fixtures, so the correct choice is (b). Each successive light fixture has a smaller voltage across it and glows less brightly than the one before.

Figure 28.8 (Conceptual Example 28.3) The circuit diagram for a set of landscape light fixtures connected in parallel across the two wires of a two-wire cable. The horizontal resistors represent resistance in the wires of the cable. The vertical resistors represent the light fixtures.

Example 28.4 Find the Equivalent Resistance

Four resistors are connected as shown in Figure 28.9a.

(A) Find the equivalent resistance between points a and c.

Solution The combination of resistors can be reduced in steps, as shown in Figure 28.9. The 8.0-Ω and 4.0-Ω resistors are in series; thus, the equivalent resistance between a and b is 12.0 Ω (see Eq. 28.5). The 6.0-Ω and 3.0-Ω resistors are in parallel, so from Equation 28.7 we find that the equivalent resistance from b to c is 2.0 Ω. Hence, the equivalent resistance from a to c is 14.0 Ω.

(B) What is the current in each resistor if a potential difference of 42 V is maintained between a and c?

Solution The currents in the 8.0-Ω and 4.0-Ω resistors are the same because they are in series. In addition, this is the same as the current that would exist in the 14.0-Ω equivalent resistor subject to the 42-V potential difference. Therefore, using Equation 27.8 ($R = \Delta V/I$) and the result from part (A), we obtain

$$I = \frac{\Delta V_{ac}}{R_{eq}} = \frac{42\ \text{V}}{14.0\ \Omega} = 3.0\ \text{A}$$

This is the current in the 8.0-Ω and 4.0-Ω resistors. When this 3.0-A current enters the junction at b, however, it splits, with part passing through the 6.0-Ω resistor (I_1) and part through the 3.0-Ω resistor (I_2). Because the potential difference is ΔV_{bc} across each of these parallel resistors, we see that $(6.0\ \Omega)I_1 = (3.0\ \Omega)I_2$, or $I_2 = 2I_1$. Using this result and the fact that $I_1 + I_2 = 3.0$ A, we find that $I_1 = 1.0$ A and

$I_2 = 2.0$ A. We could have guessed this at the start by noting that the current in the 3.0-Ω resistor has to be twice that in the 6.0-Ω resistor, in view of their relative resistances and the fact that the same voltage is applied to each of them.

As a final check of our results, note that $\Delta V_{bc} = (6.0\ \Omega)I_1 = (3.0\ \Omega)I_2 = 6.0$ V and $\Delta V_{ab} = (12.0\ \Omega)I = 36$ V; therefore, $\Delta V_{ac} = \Delta V_{ab} + \Delta V_{bc} = 42$ V, as it must.

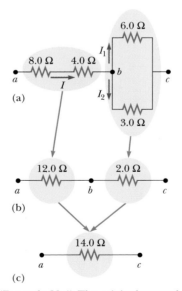

Figure 28.9 (Example 28.4) The original network of resistors is reduced to a single equivalent resistance.

Example 28.5 Finding R_{eq} by Symmetry Arguments

Consider five resistors connected as shown in Figure 28.10a. Find the equivalent resistance between points a and b.

Solution If we inspect this system of resistors, we realize that we cannot reduce it by using our rules for series and parallel

connections. We can, however, assume a current entering junction a and then apply symmetry arguments. Because of the symmetry in the circuit (all 1-Ω resistors in the outside loop), the currents in branches ac and ad must be equal; hence, the electric potentials at points c and d must be equal.

This means that $\Delta V_{cd} = 0$ and there is no current between points c and d. As a result, points c and d may be connected together without affecting the circuit, as in Figure 28.10b. Thus, the 5-Ω resistor may be removed from the circuit and

the remaining circuit then reduced as in Figures 28.10c and d. From this reduction we see that the equivalent resistance of the combination is 1 Ω. Note that the result is 1 Ω regardless of the value of the resistor connected between c and d.

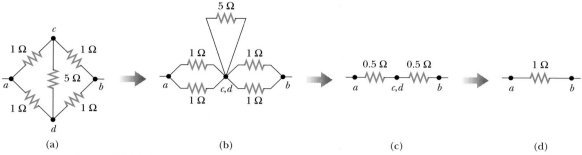

(a)   (b)   (c)   (d)

Figure 28.10 (Example 28.5) Because of the symmetry in this circuit, the 5-Ω resistor does not contribute to the resistance between points a and b and therefore can be disregarded when we calculate the equivalent resistance.

Example 28.6   Three Resistors in Parallel

Interactive

Three resistors are connected in parallel as shown in Figure 28.11a. A potential difference of 18.0 V is maintained between points a and b.

(A) Find the current in each resistor.

Solution The resistors are in parallel, and so the potential difference across each must be 18.0 V. Applying the relationship $\Delta V = IR$ to each resistor gives

$$I_1 = \frac{\Delta V}{R_1} = \frac{18.0 \text{ V}}{3.00 \ \Omega} = \boxed{6.00 \text{ A}}$$

$$I_2 = \frac{\Delta V}{R_2} = \frac{18.0 \text{ V}}{6.00 \ \Omega} = \boxed{3.00 \text{ A}}$$

$$I_3 = \frac{\Delta V}{R_3} = \frac{18.0 \text{ V}}{9.00 \ \Omega} = \boxed{2.00 \text{ A}}$$

(B) Calculate the power delivered to each resistor and the total power delivered to the combination of resistors.

Solution We apply the relationship $\mathcal{P} = I^2R$ to each resistor and obtain

3.00-Ω: $\mathcal{P}_1 = I_1{}^2R_1 = (6.00 \text{ A})^2(3.00 \ \Omega) = \boxed{108 \text{ W}}$

6.00-Ω: $\mathcal{P}_2 = I_2{}^2R_2 = (3.00 \text{ A})^2(6.00 \ \Omega) = \boxed{54.0 \text{ W}}$

9.00-Ω: $\mathcal{P}_3 = I_3{}^2R_3 = (2.00 \text{ A})^2(9.00 \ \Omega) = \boxed{36.0 \text{ W}}$

This shows that the smallest resistor receives the most power. Summing the three quantities gives a total power of 198 W.

(a)

(b)

Figure 28.11 (Example 28.6) (a) Three resistors connected in parallel. The voltage across each resistor is 18.0 V. (b) Another circuit with three resistors and a battery. Is this equivalent to the circuit in part (a) of the figure?

(C) Calculate the equivalent resistance of the circuit.

Solution We can use Equation 28.8 to find R_{eq}:

$$\frac{1}{R_{eq}} = \frac{1}{3.00\ \Omega} + \frac{1}{6.00\ \Omega} + \frac{1}{9.00\ \Omega}$$

$$R_{eq} = \frac{18.0\ \Omega}{11.0} = \boxed{1.64\ \Omega}$$

What If? **What if the circuit is as shown in Figure 28.11b instead of as in Figure 28.11a? How does this affect the calculation?**

Answer There is no effect on the calculation. The physical placement of the battery is not important. In Figure 28.11b, the battery still applies a potential difference of 18.0 V between points *a* and *b*, so the two circuits in Figure 28.11 are electrically identical.

 At the Interactive Worked Example link at **http://www.pse6.com,** *you can explore different configurations of the battery and resistors.*

Conceptual Example 28.7   Operation of a Three-Way Lightbulb

Figure 28.12 illustrates how a three-way lightbulb is constructed to provide three levels of light intensity. The socket of the lamp is equipped with a three-way switch for selecting different light intensities. The bulb contains two filaments. When the lamp is connected to a 120-V source, one filament receives 100 W of power, and the other receives 75 W. Explain how the two filaments are used to provide three different light intensities.

Solution The three light intensities are made possible by applying the 120 V to one filament alone, to the other filament alone, or to the two filaments in parallel. When switch S_1 is closed and switch S_2 is opened, current exists only in the 75-W filament. When switch S_1 is open and switch S_2 is closed, current exists only in the 100-W filament. When both switches are closed, current exists in both filaments, and the total power is 175 W.

    If the filaments were connected in series and one of them were to break, no charges could pass through the bulb, and the bulb would give no illumination, regardless of the switch position. However, with the filaments connected

in parallel, if one of them (for example, the 75-W filament) breaks, the bulb will still operate in two of the switch positions as current exists in the other (100-W) filament.

Figure 28.12 (Conceptual Example 28.7) A three-way lightbulb.

Application   Strings of Lights

Strings of lights are used for many ornamental purposes, such as decorating Christmas trees.[3] Over the years, both parallel and series connections have been used for strings of lights powered by 120 V. Series-wired bulbs are safer than parallel-wired bulbs for indoor Christmas-tree use because series-wired bulbs operate with less energy per bulb and at a lower temperature. However, if the filament of a single bulb fails (or if the bulb is removed from its socket), all the lights on the string go out. The popularity of series-wired light strings diminished because troubleshooting a failed bulb was a tedious, time-consuming chore that involved trial-and-error substitution of a good bulb in each socket along the string until the defective bulb was found.

    In a parallel-wired string, each bulb operates at 120 V. By design, the bulbs are brighter and hotter than those on

a series-wired string. As a result, these bulbs are inherently more dangerous (more likely to start a fire, for instance), but if one bulb in a parallel-wired string fails or is removed, the rest of the bulbs continue to glow. (A 25-bulb string of 4-W bulbs results in a power of 100 W; the total power becomes substantial when several strings are used.)

    A new design was developed for so-called "miniature" lights wired in series, to prevent the failure of one bulb from causing the entire string to go out. This design creates a connection (called a jumper) across the filament after it fails. When the filament breaks in one of these miniature lightbulbs, the break in the filament represents the largest resistance in the series, much larger than that of the intact filaments. As a result, most of the applied 120 V appears across the bulb with the broken filament. Inside the

---

[3]   These and other household devices, such as the three-way lightbulb in Conceptual Example 28.7 and the kitchen appliances discussed in Section 28.6, actually operate on alternating current (AC), to be introduced in Chapter 33.

lightbulb, a small jumper loop covered by an insulating material is wrapped around the filament leads. When the filament fails and 120 V appears across the bulb, an arc burns the insulation on the jumper and connects the filament leads. This connection now completes the circuit through the bulb even though its filament is no longer active (Fig. 28.13).

Suppose that all the bulbs in a 50-bulb miniature-light string are operating. A 2.40-V potential drop occurs across each bulb because the bulbs are in series. A typical power input to this style of bulb is 0.340 W. The filament resistance of each bulb at the operating temperature is $(2.40 \text{ V})^2/(0.340 \text{ W}) = 16.9 \ \Omega$. The current in each bulb is $2.40 \text{ V}/16.9 \ \Omega = 0.142$ A. When a bulb fails, the resistance across its terminals is reduced to zero because of the alternate

jumper connection mentioned in the preceding paragraph. All the other bulbs not only stay on but glow more brightly because the total resistance of the string is reduced and consequently the current in each bulb increases.

Let us assume that the resistance of a bulb remains at 16.9 Ω even though its temperature rises as a result of the increased current. If one bulb fails, the potential difference across each of the remaining bulbs increases to $120 \text{ V}/49 = 2.45$ V, the current increases from 0.142 A to 0.145 A, and the power increases to 0.355 W. As more bulbs fail, the current keeps rising, the filament of each bulb operates at a higher temperature, and the lifetime of the bulb is reduced. For this reason, you should check for failed (nonglowing) bulbs in such a series-wired string and replace them as soon as possible, in order to maximize the lifetimes of all the bulbs.

Figure 28.13 (a) Schematic diagram of a modern "miniature" holiday lightbulb, with a jumper connection to provide a current path if the filament breaks. When the filament is intact, charges flow in the filament. (b) A holiday lightbulb with a broken filament. In this case, charges flow in the jumper connection. (c) A Christmas-tree lightbulb.

28.3 Kirchhoff's Rules

As we saw in the preceding section, simple circuits can be analyzed using the expression $\Delta V = IR$ and the rules for series and parallel combinations of resistors. Very often, however, it is not possible to reduce a circuit to a single loop. The procedure for analyzing more complex circuits is greatly simplified if we use two principles called **Kirchhoff's rules:**

1. **Junction rule.** The sum of the currents entering any junction in a circuit must equal the sum of the currents leaving that junction:

$$\sum I_{\text{in}} = \sum I_{\text{out}} \tag{28.9}$$

2. **Loop rule.** The sum of the potential differences across all elements around any closed circuit loop must be zero:

$$\sum_{\substack{\text{closed} \\ \text{loop}}} \Delta V = 0 \tag{28.10}$$

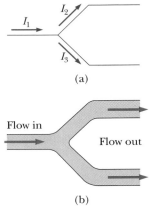

Figure 28.14 (a) Kirchhoff's junction rule. Conservation of charge requires that all charges entering a junction must leave that junction. Therefore, $I_1 = I_2 + I_3$. (b) A mechanical analog of the junction rule: the amount of water flowing out of the branches on the right must equal the amount flowing into the single branch on the left.

Figure 28.15 Rules for determining the potential differences across a resistor and a battery. (The battery is assumed to have no internal resistance.) Each circuit element is traversed from left to right.

Kirchhoff's first rule is a statement of conservation of electric charge. All charges that enter a given point in a circuit must leave that point because charge cannot build up at a point. If we apply this rule to the junction shown in Figure 28.14a, we obtain

$$I_1 = I_2 + I_3$$

Figure 28.14b represents a mechanical analog of this situation, in which water flows through a branched pipe having no leaks. Because water does not build up anywhere in the pipe, the flow rate into the pipe equals the total flow rate out of the two branches on the right.

Kirchhoff's second rule follows from the law of conservation of energy. Let us imagine moving a charge around a closed loop of a circuit. When the charge returns to the starting point, the charge–circuit system must have the same total energy as it had before the charge was moved. The sum of the increases in energy as the charge passes through some circuit elements must equal the sum of the decreases in energy as it passes through other elements. The potential energy decreases whenever the charge moves through a potential drop $-IR$ across a resistor or whenever it moves in the reverse direction through a source of emf. The potential energy increases whenever the charge passes through a battery from the negative terminal to the positive terminal.

When applying Kirchhoff's second rule in practice, we imagine *traveling* around the loop and consider changes in *electric potential*, rather than the changes in *potential energy* described in the preceding paragraph. You should note the following sign conventions when using the second rule:

- Because charges move from the high-potential end of a resistor toward the low-potential end, if a resistor is traversed in the direction of the current, the potential difference ΔV across the resistor is $-IR$ (Fig. 28.15a).

- If a resistor is traversed in the direction *opposite* the current, the potential difference ΔV across the resistor is $+IR$ (Fig. 28.15b).

- If a source of emf (assumed to have zero internal resistance) is traversed in the direction of the emf (from $-$ to $+$), the potential difference ΔV is $+\mathcal{E}$ (Fig. 28.15c). The emf of the battery increases the electric potential as we move through it in this direction.

- If a source of emf (assumed to have zero internal resistance) is traversed in the direction opposite the emf (from $+$ to $-$), the potential difference ΔV is $-\mathcal{E}$ (Fig. 28.15d). In this case the emf of the battery reduces the electric potential as we move through it.

Limitations exist on the numbers of times you can usefully apply Kirchhoff's rules in analyzing a circuit. You can use the junction rule as often as you need, so long as each time you write an equation you include in it a current that has not been used in a preceding junction-rule equation. In general, the number of times you can use the junction rule is one fewer than the number of junction points in the circuit. You can apply the loop rule as often as needed, as long as a new circuit element (resistor or battery) or a new current appears in each new equation. In general, **in order to solve a particular circuit problem, the number of independent equations you need to obtain from the two rules equals the number of unknown currents.**

Complex networks containing many loops and junctions generate great numbers of independent linear equations and a correspondingly great number of unknowns. Such situations can be handled formally through the use of matrix algebra. Computer software can also be used to solve for the unknowns.

The following examples illustrate how to use Kirchhoff's rules. In all cases, it is assumed that the circuits have reached steady-state conditions—that is, the currents in the various branches are constant. **Any capacitor acts as an open branch in a circuit;** that is, the current in the branch containing the capacitor is zero under steady-state conditions.

PROBLEM-SOLVING HINTS

Kirchhoff's Rules

- Draw a circuit diagram, and label all the known and unknown quantities. You must assign a *direction* to the current in each branch of the circuit. Although the assignment of current directions is arbitrary, you must adhere rigorously to the assigned directions when applying Kirchhoff's rules.

- Apply the junction rule to any junctions in the circuit that provide new relationships among the various currents.

- Apply the loop rule to as many loops in the circuit as are needed to solve for the unknowns. To apply this rule, you must correctly identify the potential difference as you imagine crossing each element while traversing the closed loop (either clockwise or counterclockwise). Watch out for errors in sign!

- Solve the equations simultaneously for the unknown quantities. Do not be alarmed if a current turns out to be negative; *its magnitude will be correct and the direction is opposite to that which you assigned.*

Gustav Kirchhoff
German Physicist (1824–1887)

Kirchhoff, a professor at Heidelberg, and Robert Bunsen invented the spectroscope and founded the science of spectroscopy, which we shall study in Chapter 42. They discovered the elements cesium and rubidium and invented astronomical spectroscopy. *(AIP ESVA/W.F. Meggers Collection)*

Quick Quiz 28.8 In using Kirchhoff's rules, you generally assign a separate unknown current to (a) each resistor in the circuit (b) each loop in the circuit (c) each branch in the circuit (d) each battery in the circuit.

Example 28.8 A Single-Loop Circuit

A single-loop circuit contains two resistors and two batteries, as shown in Figure 28.16. (Neglect the internal resistances of the batteries.)

(A) Find the current in the circuit.

Solution We do not need Kirchhoff's rules to analyze this simple circuit, but let us use them anyway just to see how they are applied. There are no junctions in this single-loop circuit; thus, the current is the same in all elements. Let us assume that the current is clockwise, as shown in Figure 28.16. Traversing the circuit in the clockwise direction, starting at a, we see that $a \to b$ represents a potential difference of $+ \mathcal{E}_1$, $b \to c$ represents a potential difference of $- IR_1$, $c \to d$ represents a potential difference of $- \mathcal{E}_2$, and

$d \to a$ represents a potential difference of $- IR_2$. Applying Kirchhoff's loop rule gives

$$\sum \Delta V = 0$$
$$\mathcal{E}_1 - IR_1 - \mathcal{E}_2 - IR_2 = 0$$

Solving for I and using the values given in Figure 28.16, we obtain

$$(1) \quad I = \frac{\mathcal{E}_1 - \mathcal{E}_2}{R_1 + R_2} = \frac{6.0\,\text{V} - 12\,\text{V}}{8.0\,\Omega + 10\,\Omega} = \boxed{-0.33\,\text{A}}$$

The negative sign for I indicates that the direction of the current is opposite the assumed direction. Notice that the emfs in the numerator subtract because the batteries have opposite polarities in Figure 28.16. In the denominator, the resistances add because the two resistors are in series.

(B) What power is delivered to each resistor? What power is delivered by the 12-V battery?

Solution Using Equation 27.23,

$$\mathcal{P}_1 = I^2 R_1 = (0.33\,\text{A})^2 (8.0\,\Omega) = \boxed{0.87\,\text{W}}$$

$$\mathcal{P}_2 = I^2 R_2 = (0.33\,\text{A})^2 (10\,\Omega) = \boxed{1.1\,\text{W}}$$

Hence, the total power delivered to the resistors is $\mathcal{P}_1 + \mathcal{P}_2 = 2.0\,\text{W}$.

The 12-V battery delivers power $I\mathcal{E}_2 = 4.0\,\text{W}$. Half of this power is delivered to the two resistors, as we just calculated. The other half is delivered to the 6-V battery, which is being

Figure 28.16 (Example 28.8) A series circuit containing two batteries and two resistors, where the polarities of the batteries are in opposition.

charged by the 12-V battery. If we had included the internal resistances of the batteries in our analysis, some of the power would appear as internal energy in the batteries; as a result, we would have found that less power was being delivered to the 6-V battery.

What If? What if the polarity of the 12.0-V battery were reversed? How would this affect the circuit?

Answer While we could repeat the Kirchhoff's rules calculation, let us examine Equation (1) and modify it accordingly. Because the polarities of the two batteries are now in the same direction, the signs of $\mathcal{E}_1$ and $\mathcal{E}_2$ are the same and Equation (1) becomes

$$I = \frac{\mathcal{E}_1 + \mathcal{E}_2}{R_1 + R_2} = \frac{6.0 \text{ V} + 12 \text{ V}}{8.0 \text{ } \Omega + 10 \text{ } \Omega} = 1.0 \text{ A}$$

The new powers delivered to the resistors are

$$\mathcal{P}_1 = I^2 R_1 = (1.0 \text{ A})^2 (8.0 \text{ } \Omega) = 8.0 \text{ W}$$

$$\mathcal{P}_2 = I^2 R_2 = (1.0 \text{ A})^2 (10 \text{ } \Omega) = 10 \text{ W}$$

This totals 18 W, nine times as much as in the original circuit, in which the batteries were opposing each other.

Example 28.9   Applying Kirchhoff's Rules   Interactive

Find the currents I_1, I_2, and I_3 in the circuit shown in Figure 28.17.

Solution Conceptualize by noting that we cannot simplify the circuit by the rules of adding resistances in series and in parallel. (If the 10.0-V battery were taken away, we could reduce the remaining circuit with series and parallel combinations.) Thus, we categorize this problem as one in which we must use Kirchhoff's rules. To analyze the circuit, we arbitrarily choose the directions of the currents as labeled in Figure 28.17. Applying Kirchhoff's junction rule to junction c gives

$$(1) \qquad I_1 + I_2 = I_3$$

We now have one equation with three unknowns—I_1, I_2, and I_3. There are three loops in the circuit—*abcda*, *befcb*, and *aefda*. We therefore need only two loop equations to determine the unknown currents. (The third loop equation would give no new information.) Applying Kirchhoff's loop rule to loops *abcda* and *befcb* and traversing these loops clockwise, we obtain the expressions

$$(2) \qquad abcda \quad 10.0 \text{ V} - (6.0 \text{ } \Omega) I_1 - (2.0 \text{ } \Omega) I_3 = 0$$

$$(3) \quad befcb \quad -14.0 \text{ V} + (6.0 \text{ } \Omega) I_1 - 10.0 \text{ V} - (4.0 \text{ } \Omega) I_2 = 0$$

Note that in loop *befcb* we obtain a positive value when traversing the 6.0-Ω resistor because our direction of travel is opposite the assumed direction of I_1. Expressions (1), (2), and (3) represent three independent equations with three unknowns. Substituting Equation (1) into Equation (2) gives

$$10.0 \text{ V} - (6.0 \text{ } \Omega) I_1 - (2.0 \text{ } \Omega)(I_1 + I_2) = 0$$

$$(4) \qquad 10.0 \text{ V} = (8.0 \text{ } \Omega) I_1 + (2.0 \text{ } \Omega) I_2$$

Dividing each term in Equation (3) by 2 and rearranging gives

$$(5) \qquad -12.0 \text{ V} = -(3.0 \text{ } \Omega) I_1 + (2.0 \text{ } \Omega) I_2$$

Subtracting Equation (5) from Equation (4) eliminates I_2, giving

$$22.0 \text{ V} = (11.0 \text{ } \Omega) I_1$$

$$I_1 = \boxed{2.0 \text{ A}}$$

Using this value of I_1 in Equation (5) gives a value for I_2:

$$(2.0 \text{ } \Omega) I_2 = (3.0 \text{ } \Omega) I_1 - 12.0 \text{ V}$$

$$= (3.0 \text{ } \Omega)(2.0 \text{ A}) - 12.0 \text{ V} = -6.0 \text{ V}$$

$$I_2 = \boxed{-3.0 \text{ A}}$$

Finally,

$$I_3 = I_1 + I_2 = \boxed{-1.0 \text{ A}}$$

To finalize the problem, note that I_2 and I_3 are both negative. This indicates only that the currents are opposite the direction we chose for them. However, the numerical values are correct. What would have happened had we left the current directions as labeled in Figure 28.17 but traversed the loops in the opposite direction?

Figure 28.17 (Example 28.9) A circuit containing different branches.

Practice applying Kirchhoff's rules at the Interactive Worked Example link at **http://www.pse6.com**.

Example 28.10   A Multiloop Circuit

(A) Under steady-state conditions, find the unknown currents I_1, I_2, and I_3 in the multiloop circuit shown in Figure 28.18.

Solution First note that because the capacitor represents an open circuit, there is no current between g and b along path $ghab$ under steady-state conditions. Therefore, when the charges associated with I_1 reach point g, they all go toward point b through the 8.00-V battery; hence, $I_{gb} = I_1$. Labeling the currents as shown in Figure 28.18 and applying Equation 28.9 to junction c, we obtain

$$(1) \qquad I_1 + I_2 = I_3$$

Equation 28.10 applied to loops $defcd$ and $cfgbc$, traversed clockwise, gives

$$(2) \qquad defcd \quad 4.00\ \text{V} - (3.00\ \Omega)I_2 - (5.00\ \Omega)I_3 = 0$$

$$(3) \qquad cfgbc \quad (3.00\ \Omega)I_2 - (5.00\ \Omega)I_1 + 8.00\ \text{V} = 0$$

Figure 28.18 (Example 28.10) A multiloop circuit. Kirchhoff's loop rule can be applied to *any* closed loop, including the one containing the capacitor.

From Equation (1) we see that $I_1 = I_3 - I_2$, which, when substituted into Equation (3), gives

$$(4) \qquad (8.00\ \Omega)I_2 - (5.00\ \Omega)I_3 + 8.00\ \text{V} = 0$$

Subtracting Equation (4) from Equation (2), we eliminate I_3 and find that

$$I_2 = -\frac{4.00\ \text{V}}{11.0\ \Omega} = \boxed{-0.364\ \text{A}}$$

Because our value for I_2 is negative, we conclude that the direction of I_2 is from c to f in the 3.00-Ω resistor. Despite this interpretation of the direction, however, we must continue to use this negative value for I_2 in subsequent calculations because our equations were established with our original choice of direction.

Using $I_2 = -0.364$ A in Equations (3) and (1) gives

$$I_1 = \boxed{1.38\ \text{A}} \qquad I_3 = \boxed{1.02\ \text{A}}$$

(B) What is the charge on the capacitor?

Solution We can apply Kirchhoff's loop rule to loop $bghab$ (or any other loop that contains the capacitor) to find the potential difference ΔV_{cap} across the capacitor. We use this potential difference in the loop equation without reference to a sign convention because the charge on the capacitor depends only on the magnitude of the potential difference. Moving clockwise around this loop, we obtain

$$-8.00\ \text{V} + \Delta V_{\text{cap}} - 3.00\ \text{V} = 0$$

$$\Delta V_{\text{cap}} = 11.0\ \text{V}$$

Because $Q = C\,\Delta V_{\text{cap}}$ (see Eq. 26.1), the charge on the capacitor is

$$Q = (6.00\ \mu\text{F})(11.0\ \text{V}) = \boxed{66.0\ \mu\text{C}}$$

Why is the left side of the capacitor positively charged?

28.4   *RC* Circuits

So far we have analyzed direct current circuits in which the current is constant. In DC circuits containing capacitors, the current is always in the same direction but may vary in time. A circuit containing a series combination of a resistor and a capacitor is called an **RC circuit.**

Charging a Capacitor

Figure 28.19 shows a simple series *RC* circuit. Let us assume that the capacitor in this circuit is initially uncharged. There is no current while switch S is open (Fig. 28.19b). If the switch is closed at $t = 0$, however, charge begins to flow, setting up a current in the circuit, and the capacitor begins to charge.[4] Note that during charging, charges do

---

[4]   In previous discussions of capacitors, we assumed a steady-state situation, in which no current was present in any branch of the circuit containing a capacitor. Now we are considering the case *before* the steady-state condition is realized; in this situation, charges are moving and a current exists in the wires connected to the capacitor.

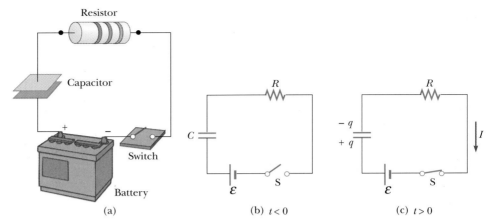

Active Figure 28.19 (a) A capacitor in series with a resistor, switch, and battery. (b) Circuit diagram representing this system at time $t < 0$, before the switch is closed. (c) Circuit diagram at time $t > 0$, after the switch has been closed.

not jump across the capacitor plates because the gap between the plates represents an open circuit. Instead, charge is transferred between each plate and its connecting wires due to the electric field established in the wires by the battery, until the capacitor is fully charged. As the plates are being charged, the potential difference across the capacitor increases. The value of the maximum charge on the plates depends on the voltage of the battery. Once the maximum charge is reached, the current in the circuit is zero because the potential difference across the capacitor matches that supplied by the battery.

To analyze this circuit quantitatively, let us apply Kirchhoff's loop rule to the circuit after the switch is closed. Traversing the loop in Fig. 28.19c clockwise gives

$$\mathcal{E} - \frac{q}{C} - IR = 0 \qquad (28.11)$$

where q/C is the potential difference across the capacitor and IR is the potential difference across the resistor. We have used the sign conventions discussed earlier for the signs on $\mathcal{E}$ and IR. For the capacitor, notice that we are traveling in the direction from the positive plate to the negative plate; this represents a decrease in potential. Thus, we use a negative sign for this potential difference in Equation 28.11. Note that q and I are *instantaneous* values that depend on time (as opposed to steady-state values) as the capacitor is being charged.

We can use Equation 28.11 to find the initial current in the circuit and the maximum charge on the capacitor. At the instant the switch is closed ($t = 0$), the charge on the capacitor is zero, and from Equation 28.11 we find that the initial current I_0 in the circuit is a maximum and is equal to

$$I_0 = \frac{\mathcal{E}}{R} \qquad \text{(current at } t = 0\text{)} \qquad (28.12)$$

At this time, the potential difference from the battery terminals appears entirely across the resistor. Later, when the capacitor is charged to its maximum value Q, charges cease to flow, the current in the circuit is zero, and the potential difference from the battery terminals appears entirely across the capacitor. Substituting $I = 0$ into Equation 28.11 gives the charge on the capacitor at this time:

$$Q = C\mathcal{E} \qquad \text{(maximum charge)} \qquad (28.13)$$

To determine analytical expressions for the time dependence of the charge and current, we must solve Equation 28.11—a single equation containing two variables, q and I. The current in all parts of the series circuit must be the same. Thus, the current in the resistance R must be the same as the current between the capacitor plates and the wires. This current is equal to the time rate of change of the charge on the capacitor plates. Thus, we substitute $I = dq/dt$ into Equation 28.11 and rearrange the equation:

$$\frac{dq}{dt} = \frac{\mathcal{E}}{R} - \frac{q}{RC}$$

To find an expression for q, we solve this separable differential equation. We first combine the terms on the right-hand side:

$$\frac{dq}{dt} = \frac{C\mathcal{E}}{RC} - \frac{q}{RC} = -\frac{q - C\mathcal{E}}{RC}$$

Now we multiply by dt and divide by $q - C\mathcal{E}$ to obtain

$$\frac{dq}{q - C\mathcal{E}} = -\frac{1}{RC} \, dt$$

Integrating this expression, using the fact that $q = 0$ at $t = 0$, we obtain

$$\int_0^q \frac{dq}{(q - C\mathcal{E})} = -\frac{1}{RC} \int_0^t dt$$

$$\ln\left(\frac{q - C\mathcal{E}}{-C\mathcal{E}}\right) = -\frac{t}{RC}$$

From the definition of the natural logarithm, we can write this expression as

$$q(t) = C\mathcal{E}(1 - e^{-t/RC}) = Q(1 - e^{-t/RC}) \qquad (28.14)$$

where e is the base of the natural logarithm and we have made the substitution from Equation 28.13.

Charge as a function of time for a capacitor being charged

We can find an expression for the charging current by differentiating Equation 28.14 with respect to time. Using $I = dq/dt$, we find that

$$I(t) = \frac{\mathcal{E}}{R} \, e^{-t/RC} \qquad (28.15)$$

Current as a function of time for a capacitor being charged

Plots of capacitor charge and circuit current versus time are shown in Figure 28.20. Note that the charge is zero at $t = 0$ and approaches the maximum value $C\mathcal{E}$ as $t \to \infty$. The current has its maximum value $I_0 = \mathcal{E}/R$ at $t = 0$ and decays exponentially to zero as $t \to \infty$. The quantity RC, which appears in the exponents of Equations 28.14 and 28.15, is called the **time constant** τ of the circuit. It represents the time interval during which the current decreases to $1/e$ of its initial value; that is, in a time interval τ, $I = e^{-1}I_0 = 0.368I_0$. In a time interval 2τ, $I = e^{-2}I_0 = 0.135I_0$, and so forth. Likewise, in a time interval τ, the charge increases from zero to $C\mathcal{E}[1 - e^{-1}] = 0.632C\mathcal{E}$.

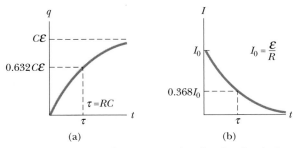

(a)   (b)

Figure 28.20 (a) Plot of capacitor charge versus time for the circuit shown in Figure 28.19. After a time interval equal to one time constant τ has passed, the charge is 63.2% of the maximum value $C\mathcal{E}$. The charge approaches its maximum value as t approaches infinity. (b) Plot of current versus time for the circuit shown in Figure 28.19. The current has its maximum value $I_0 = \mathcal{E}/R$ at $t = 0$ and decays to zero exponentially as t approaches infinity. After a time interval equal to one time constant τ has passed, the current is 36.8% of its initial value.

Active Figure 28.21 (a) A charged capacitor connected to a resistor and a switch, which is open for $t < 0$. (b) After the switch is closed at $t = 0$, a current that decreases in magnitude with time is set up in the direction shown, and the charge on the capacitor decreases exponentially with time.

At the Active Figures link at http://www.pse6.com, you can adjust the values of R and C to see the effect on the discharging of the capacitor.

Charge as a function of time for a discharging capacitor

Current as a function of time for a discharging capacitor

The following dimensional analysis shows that τ has the units of time:

$$[\tau] = [RC] = \left[\frac{\Delta V}{I} \times \frac{Q}{\Delta V} \right] = \left[\frac{Q}{Q / \Delta t} \right] = [\Delta t] = \mathrm{T}$$

Because $\tau = RC$ has units of time, the combination τ / RC is dimensionless, as it must be in order to be an exponent of e in Equations 28.14 and 28.15.

The energy output of the battery as the capacitor is fully charged is $Q\mathcal{E} = C\mathcal{E}^2$. After the capacitor is fully charged, the energy stored in the capacitor is $\frac{1}{2}Q\mathcal{E} = \frac{1}{2}C\mathcal{E}^2$, which is just half the energy output of the battery. It is left as a problem (Problem 64) to show that the remaining half of the energy supplied by the battery appears as internal energy in the resistor.

Discharging a Capacitor

Now consider the circuit shown in Figure 28.21, which consists of a capacitor carrying an initial charge Q, a resistor, and a switch. When the switch is open, a potential difference Q/C exists across the capacitor and there is zero potential difference across the resistor because $I = 0$. If the switch is closed at $t = 0$, the capacitor begins to discharge through the resistor. At some time t during the discharge, the current in the circuit is I and the charge on the capacitor is q (Fig. 28.21b). The circuit in Figure 28.21 is the same as the circuit in Figure 28.19 except for the absence of the battery. Thus, we eliminate the emf $\mathcal{E}$ from Equation 28.11 to obtain the appropriate loop equation for the circuit in Figure 28.21:

$$-\frac{q}{C} - IR = 0 \tag{28.16}$$

When we substitute $I = dq/dt$ into this expression, it becomes

$$-R\frac{dq}{dt} = \frac{q}{C}$$

$$\frac{dq}{q} = -\frac{1}{RC} \, dt$$

Integrating this expression, using the fact that $q = Q$ at $t = 0$ gives

$$\int_Q^q \frac{dq}{q} = -\frac{1}{RC} \int_0^t dt$$

$$\ln\left(\frac{q}{Q}\right) = -\frac{t}{RC}$$

$$q(t) = Qe^{-t/RC} \tag{28.17}$$

Differentiating this expression with respect to time gives the instantaneous current as a function of time:

$$I(t) = \frac{dq}{dt} = \frac{d}{dt}(Qe^{-t/RC}) = -\frac{Q}{RC}e^{-t/RC} \tag{28.18}$$

where $Q/RC = I_0$ is the initial current. The negative sign indicates that as the capacitor discharges, the current direction is opposite its direction when the capacitor was being charged. (Compare the current directions in Figs. 28.19c and 28.21b.) We see that both the charge on the capacitor and the current decay exponentially at a rate characterized by the time constant $\tau = RC$.

Quick Quiz 28.9 Consider the circuit in Figure 28.19 and assume that the battery has no internal resistance. Just after the switch is closed, the potential difference across which of the following is equal to the emf of the battery? (a) C (b) R (c) neither C nor R. After a very long time, the potential difference across which of the following is equal to the emf of the battery? (d) C (e) R (f) neither C nor R.

Quick Quiz 28.10 Consider the circuit in Figure 28.22 and assume that the battery has no internal resistance. Just after the switch is closed, the current in the battery is (a) zero (b) $\mathcal{E}/2R$ (c) $2\mathcal{E}/R$ (d) $\mathcal{E}/R$ (e) impossible to determine. After a very long time, the current in the battery is (f) zero (g) $\mathcal{E}/2R$ (h) $2\mathcal{E}/R$ (i) $\mathcal{E}/R$ (j) impossible to determine.

Figure 28.22 (Quick Quiz 28.10) How does the current vary after the switch is closed?

Conceptual Example 28.11   Intermittent Windshield Wipers

Many automobiles are equipped with windshield wipers that can operate intermittently during a light rainfall. How does the operation of such wipers depend on the charging and discharging of a capacitor?

Solution The wipers are part of an RC circuit whose time constant can be varied by selecting different values of R

through a multiposition switch. As it increases with time, the voltage across the capacitor reaches a point at which it triggers the wipers and discharges, ready to begin another charging cycle. The time interval between the individual sweeps of the wipers is determined by the value of the time constant.

Example 28.12   Charging a Capacitor in an RC Circuit

Interactive

An uncharged capacitor and a resistor are connected in series to a battery, as shown in Figure 28.23. If $\mathcal{E} = 12.0$ V, $C = 5.00$ μF, and $R = 8.00 \times 10^5$ Ω, find the time constant of the circuit, the maximum charge on the capacitor, the maximum current in the circuit, and the charge and current as functions of time.

Solution The time constant of the circuit is $\tau = RC = (8.00 \times 10^5 \ \Omega)(5.00 \times 10^{-6} \ \text{F}) = 4.00$ s. The maximum charge on the capacitor is $Q = C\mathcal{E} = (5.00 \ \mu\text{F})(12.0 \ \text{V}) = 60.0 \ \mu$C. The maximum current in the circuit is $I_0 = \mathcal{E}/R = (12.0 \ \text{V})/(8.00 \times 10^5 \ \Omega) = 15.0 \ \mu$A. Using these values and Equations 28.14 and 28.15, we find that

$$q(t) = (60.0 \ \mu\text{C})(1 - e^{-t/4.00 \ \text{s}})$$

$$I(t) = (15.0 \ \mu\text{A})e^{-t/4.00 \ \text{s}}$$

Graphs of these functions are provided in Figure 28.24.

Figure 28.23 (Example 28.12) The switch in this series RC circuit, open for times $t < 0$, is closed at $t = 0$.

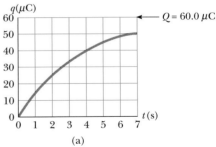

Figure 28.24 (Example 28.12) Plots of (a) charge versus time and (b) current versus time for the RC circuit shown in Figure 28.23, with $\mathcal{E} = 12.0$ V, $R = 8.00 \times 10^5$ Ω, and $C = 5.00$ μF.

 At the Interactive Worked Example link at http://www.pse6.com, you can vary R, C, and $\mathcal{E}$ and observe the charge and current as functions of time while charging or discharging the capacitor.

Example 28.13 Discharging a Capacitor in an *RC* Circuit

Consider a capacitor of capacitance C that is being discharged through a resistor of resistance R, as shown in Figure 28.21.

(A) After how many time constants is the charge on the capacitor one-fourth its initial value?

Solution The charge on the capacitor varies with time according to Equation 28.17, $q(t) = Qe^{-t/RC}$. To find the time interval during which q drops to one-fourth its initial value, we substitute $q(t) = Q/4$ into this expression and solve for t:

$$\frac{Q}{4} = Qe^{-t/RC}$$

$$\tfrac{1}{4} = e^{-t/RC}$$

Taking logarithms of both sides, we find

$$-\ln 4 = -\frac{t}{RC}$$

$$t = RC\,(\ln 4) = 1.39RC = \boxed{1.39\tau}$$

(B) The energy stored in the capacitor decreases with time as the capacitor discharges. After how many time constants is this stored energy one-fourth its initial value?

Solution Using Equations 26.11 ($U = Q^2/2C$) and 28.17, we can express the energy stored in the capacitor at any time t as

$$U = \frac{q^2}{2C} = \frac{Q^2}{2C}\,e^{-2t/RC} = U_0 e^{-2t/RC}$$

where $U_0 = Q^2/2C$ is the initial energy stored in the capacitor. As in part (A), we now set $U = U_0/4$ and solve for t:

$$\frac{U_0}{4} = U_0 e^{-2t/RC}$$

$$\tfrac{1}{4} = e^{-2t/RC}$$

Again, taking logarithms of both sides and solving for t gives

$$t = \tfrac{1}{2}\,RC\ln 4 = 0.693RC = \boxed{0.693\tau}$$

What If? What if we wanted to describe the circuit in terms of the time interval required for the charge to fall to one-half its original value, rather than by the time constant τ? This would give a parameter for the circuit called its *half-life $t_{1/2}$*. How is the half-life related to the time constant?

Answer After one half-life, the charge has fallen from Q to $Q/2$. Thus, from Equation 28.17,

$$\frac{Q}{2} = Qe^{-t_{1/2}/RC}$$

$$\tfrac{1}{2} = e^{-t_{1/2}/RC}$$

leading to

$$t_{1/2} = 0.693\tau$$

The concept of half-life will be important to us when we study nuclear decay in Chapter 44. The radioactive decay of an unstable sample behaves in a mathematically similar manner to a discharging capacitor in an *RC* circuit.

Example 28.14 Energy Delivered to a Resistor

A 5.00-μF capacitor is charged to a potential difference of 800 V and then discharged through a 25.0-kΩ resistor. How much energy is delivered to the resistor in the time interval required to fully discharge the capacitor?

Solution We shall solve this problem in two ways. The first way is to note that the initial energy in the circuit equals the energy stored in the capacitor, $C\mathcal{E}^2/2$ (see Eq. 26.11). Once the capacitor is fully discharged, the energy stored in it is zero. Because energy in an isolated system is conserved, the initial energy stored in the capacitor is transformed into internal energy in the resistor. Using the given values of C and $\mathcal{E}$, we find

$$\text{Energy} = \tfrac{1}{2}C\mathcal{E}^2 = \tfrac{1}{2}(5.00 \times 10^{-6}\,\text{F})(800\,\text{V})^2 = \boxed{1.60\,\text{J}}$$

The second way, which is more difficult but perhaps more instructive, is to note that as the capacitor discharges through the resistor, the rate at which energy is delivered to the resistor is given by I^2R, where I is the instantaneous current given by Equation 28.18. Because power is defined as the rate at which energy is transferred, we conclude that

the energy delivered to the resistor must equal the time integral of $I^2R\,dt$:

$$\text{Energy} = \int_0^\infty I^2 R\,dt = \int_0^\infty (-I_0 e^{-t/RC})^2\,R\,dt$$

To evaluate this integral, we note that the initial current I_0 is equal to $\mathcal{E}/R$ and that all parameters except t are constant. Thus, we find

$$(1) \qquad \text{Energy} = \frac{\mathcal{E}^2}{R}\int_0^\infty e^{-2t/RC}\,dt$$

This integral has a value of $RC/2$ (see Problem 35); hence, we find

$$\text{Energy} = \tfrac{1}{2}C\mathcal{E}^2$$

which agrees with the result we obtained using the simpler approach, as it must. Note that we can use this second approach to find the total energy delivered to the resistor at *any* time after the switch is closed by simply replacing the upper limit in the integral with that specific value of t.

28.5 Electrical Meters

The Galvanometer

The **galvanometer** is the main component in analog meters for measuring current and voltage. (Many analog meters are still in use although digital meters, which operate on a different principle, are currently in wide use.) Figure 28.25 illustrates the essential features of a common type called the *D'Arsonval galvanometer*. It consists of a coil of wire mounted so that it is free to rotate on a pivot in a magnetic field provided by a permanent magnet. The basic operation of the galvanometer uses the fact that a torque acts on a current loop in the presence of a magnetic field (Chapter 29). The torque experienced by the coil is proportional to the current in it: the larger the current, the greater the torque and the more the coil rotates before the spring tightens enough to stop the rotation. Hence, the deflection of a needle attached to the coil is proportional to the current. Once the instrument is properly calibrated, it can be used in conjunction with other circuit elements to measure either currents or potential differences.

The Ammeter

A device that measures current is called an **ammeter.** The charges constituting the current to be measured must pass directly through the ammeter, so the ammeter must be connected in series with other elements in the circuit, as shown in Figure 28.26. When using an ammeter to measure direct currents, you must connect it so that charges enter the instrument at the positive terminal and exit at the negative terminal.

Ideally, an ammeter should have zero resistance so that the current being measured is not altered. In the circuit shown in Figure 28.26, this condition requires that the resistance of the ammeter be much less than $R_1 + R_2$. Because any ammeter always has some internal resistance, the presence of the ammeter in the circuit slightly reduces the current from the value it would have in the meter's absence.

A typical off-the-shelf galvanometer is often not suitable for use as an ammeter, primarily because it has a resistance of about 60 Ω. An ammeter resistance this great considerably alters the current in a circuit. You can understand this by considering the following example. The current in a simple series circuit containing a 3-V battery and a 3-Ω resistor is 1 A. If you insert a 60-Ω galvanometer in this circuit to measure the current, the total resistance becomes 63 Ω and the current is reduced to 0.048 A!

A second factor that limits the use of a galvanometer as an ammeter is the fact that a typical galvanometer gives a full-scale deflection for currents on the order of 1 mA or less. Consequently, such a galvanometer cannot be used directly to measure currents greater than this value. However, it can be converted to a useful ammeter by placing a shunt resistor R_p in parallel with the galvanometer, as shown in Figure 28.27. The value of R_p must be much less than the galvanometer resistance so that most of the current to be measured is directed to the shunt resistor.

The Voltmeter

A device that measures potential difference is called a **voltmeter.** The potential difference between any two points in a circuit can be measured by attaching the terminals of the voltmeter between these points without breaking the circuit, as shown in Figure 28.28. The potential difference across resistor R_2 is measured by connecting the voltmeter in parallel with R_2. Again, it is necessary to observe the polarity of the instrument. The positive terminal of the voltmeter must be connected to the end of the resistor that is at the higher potential, and the negative terminal to the end of the resistor at the lower potential.

Figure 28.25 The principal components of a D'Arsonval galvanometer. When the coil situated in a magnetic field carries a current, the magnetic torque causes the coil to twist. The angle through which the coil rotates is proportional to the current in the coil because of the counteracting torque of the spring.

Figure 28.26 Current can be measured with an ammeter connected in series with the elements in which the measurement of a current is desired. An ideal ammeter has zero resistance.

Active Figure 28.27 A galvanometer is represented here by its internal resistance of 60 Ω. When a galvanometer is to be used as an ammeter, a shunt resistor R_p is connected in parallel with the galvanometer.

At the Active Figures link at http://www.pse6.com, you can predict the value of R_p needed to cause full-scale deflection in the circuit of Figure 28.26, and test your result.

Figure 28.28 The potential difference across a resistor can be measured with a voltmeter connected in parallel with the resistor. An ideal voltmeter has infinite resistance.

Active Figure 28.29 When the galvanometer is used as a voltmeter, a resistor R_s is connected in series with the galvanometer.

At the Active Figures link at http://www.pse6.com, *you can predict the value of R_s needed to cause full-scale deflection in the circuit of Figure 28.28, and test your result.*

An ideal voltmeter has infinite resistance so that no current exists in it. In Figure 28.28, this condition requires that the voltmeter have a resistance much greater than R_2. In practice, if this condition is not met, corrections should be made for the known resistance of the voltmeter.

A galvanometer can also be used as a voltmeter by adding an external resistor R_s in series with it, as shown in Figure 28.29. In this case, the external resistor must have a value much greater than the resistance of the galvanometer to ensure that the galvanometer does not significantly alter the voltage being measured.

28.6 Household Wiring and Electrical Safety

Household circuits represent a practical application of some of the ideas presented in this chapter. In our world of electrical appliances, it is useful to understand the power requirements and limitations of conventional electrical systems and the safety measures that prevent accidents.

In a conventional installation, the utility company distributes electric power to individual homes by means of a pair of wires, with each home connected in parallel to these wires. One wire is called the *live wire*,[5] as illustrated in Figure 28.30, and the other is called the *neutral wire*. The neutral wire is grounded; that is, its electric potential is taken to be zero. The potential difference between the live and neutral wires is about 120 V. This voltage alternates in time, and the potential of the live wire oscillates relative to ground. Much of what we have learned so far for the constant-emf situation (direct current) can also be applied to the alternating current that power companies supply to businesses and households. (Alternating voltage and current are discussed in Chapter 33.)

A meter is connected in series with the live wire entering the house to record the household's energy consumption. After the meter, the wire splits so that there are several separate circuits in parallel distributed throughout the house. Each circuit contains a circuit breaker (or, in older installations, a fuse). The wire and circuit breaker for each circuit are carefully selected to meet the current demands for that circuit. If a circuit is to carry currents as large as 30 A, a heavy wire and an appropriate circuit breaker must be selected to handle this current. A circuit used to power only lamps and small appliances often requires only 20 A. Each circuit has its own circuit breaker to provide protection for that part of the entire electrical system of the house.

Figure 28.30 Wiring diagram for a household circuit. The resistances represent appliances or other electrical devices that operate with an applied voltage of 120 V.

---

[5] *Live wire* is a common expression for a conductor whose electric potential is above or below ground potential.

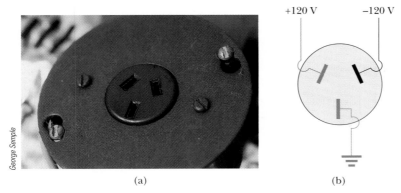

Figure 28.31 (a) An outlet for connection to a 240-V supply. (b) The connections for each of the openings in a 240-V outlet.

As an example, consider a circuit in which a toaster oven, a microwave oven, and a coffee maker are connected (corresponding to R_1, R_2, and R_3 in Fig. 28.30). We can calculate the current in each appliance by using the expression $\mathcal{P} = I\,\Delta V$. The toaster oven, rated at 1 000 W, draws a current of 1 000 W/120 V = 8.33 A. The microwave oven, rated at 1 300 W, draws 10.8 A, and the coffee maker, rated at 800 W, draws 6.67 A. If the three appliances are operated simultaneously, they draw a total current of 25.8 A. Therefore, the circuit should be wired to handle at least this much current. If the rating of the circuit breaker protecting the circuit is too small—say, 20 A—the breaker will be tripped when the third appliance is turned on, preventing all three appliances from operating. To avoid this situation, the toaster oven and coffee maker can be operated on one 20-A circuit and the microwave oven on a separate 20-A circuit.

Many heavy-duty appliances, such as electric ranges and clothes dryers, require 240 V for their operation. The power company supplies this voltage by providing a third wire that is 120 V below ground potential (Fig. 28.31). The potential difference between this live wire and the other live wire (which is 120 V above ground potential) is 240 V. An appliance that operates from a 240-V line requires half as much current compared to operating it at 120 V; therefore, smaller wires can be used in the higher-voltage circuit without overheating.

Electrical Safety

When the live wire of an electrical outlet is connected directly to ground, the circuit is completed and a short-circuit condition exists. A *short circuit* occurs when almost zero resistance exists between two points at different potentials; this results in a very large current. When this happens accidentally, a properly operating circuit breaker opens the circuit and no damage is done. However, a person in contact with ground can be electrocuted by touching the live wire of a frayed cord or other exposed conductor. An exceptionally effective (and dangerous!) ground contact is made when the person either touches a water pipe (normally at ground potential) or stands on the ground with wet feet. The latter situation represents effective ground contact because normal, nondistilled water is a conductor due to the large number of ions associated with impurities. This situation should be avoided at all cost.

Electric shock can result in fatal burns, or it can cause the muscles of vital organs, such as the heart, to malfunction. The degree of damage to the body depends on the magnitude of the current, the length of time it acts, the part of the body touched by the live wire, and the part of the body in which the current exists. Currents of 5 mA or less cause a sensation of shock but ordinarily do little or no damage. If the current is larger than about 10 mA, the muscles contract and the person may be unable to release the live wire. If a current of about 100 mA passes through the body for only a few seconds, the result can be fatal. Such a large current paralyzes the respiratory

(a)

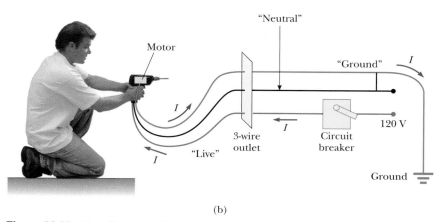

(b)

Figure 28.32 (a) A diagram of the circuit for an electric drill with only two connecting wires. The normal current path is from the live wire through the motor connections and back to ground through the neutral wire. In the situation shown, the live wire has come into contact with the drill case. As a result, the person holding the drill acts as a current path to ground and receives an electric shock. (b) This shock can be avoided by connecting the drill case to ground through a third ground wire. In this situation, the drill case remains at ground potential and no current exists in the person.

muscles and prevents breathing. In some cases, currents of about 1 A can produce serious (and sometimes fatal) burns. In practice, no contact with live wires is regarded as safe whenever the voltage is greater than 24 V.

Many 120-V outlets are designed to accept a three-pronged power cord. (This feature is required in all new electrical installations.) One of these prongs is the live wire at a nominal potential of 120 V. The second is the neutral wire, nominally at 0 V, and carries current to ground. The third, round prong is a safety ground wire that normally carries no current but is both grounded and connected directly to the casing of the appliance (see Figure 28.32). If the live wire is accidentally shorted to the casing (which can occur if the wire insulation wears off), most of the current takes the low-resistance path through the appliance to ground. In contrast, if the casing of the appliance is not properly grounded and a short occurs, anyone in contact with the appliance experiences an electric shock because the body provides a low-resistance path to ground.

Special power outlets called *ground-fault interrupters* (GFIs) are now used in kitchens, bathrooms, basements, exterior outlets, and other hazardous areas of new homes. These devices are designed to protect persons from electric shock by sensing small currents (≈ 5 mA) leaking to ground. (The principle of their operation is described in Chapter 31.) When an excessive leakage current is detected, the current is shut off in less than 1 ms.

SUMMARY

The **emf** of a battery is equal to the voltage across its terminals when the current is zero. That is, the emf is equivalent to the **open-circuit voltage** of the battery.

The **equivalent resistance** of a set of resistors connected in **series** is

$$R_{eq} = R_1 + R_2 + R_3 + \cdots \qquad (28.6)$$

The **equivalent resistance** of a set of resistors connected in **parallel** is found from the relationship

$$\frac{1}{R_{eq}} = \frac{1}{R_1} + \frac{1}{R_2} + \frac{1}{R_3} + \cdots \qquad (28.8)$$

If it is possible to combine resistors into series or parallel equivalents, the preceding two equations make it easy to determine how the resistors influence the rest of the circuit.

Circuits involving more than one loop are conveniently analyzed with the use of **Kirchhoff's rules:**

1. **Junction rule.** The sum of the currents entering any junction in an electric circuit must equal the sum of the currents leaving that junction:

$$\sum I_{in} = \sum I_{out} \qquad (28.9)$$

2. **Loop rule.** The sum of the potential differences across all elements around any circuit loop must be zero:

$$\sum_{\substack{closed \\ loop}} \Delta V = 0 \qquad (28.10)$$

The first rule is a statement of conservation of charge; the second is equivalent to a statement of conservation of energy.

When a resistor is traversed in the direction of the current, the potential difference ΔV across the resistor is $-IR$. When a resistor is traversed in the direction opposite the current, $\Delta V = +IR$. When a source of emf is traversed in the direction of the emf (negative terminal to positive terminal), the potential difference is $+\mathcal{E}$. When a source of emf is traversed opposite the emf (positive to negative), the potential difference is $-\mathcal{E}$. The use of these rules together with Equations 28.9 and 28.10 allows you to analyze electric circuits.

If a capacitor is charged with a battery through a resistor of resistance R, the charge on the capacitor and the current in the circuit vary in time according to the expressions

$$q(t) = Q(1 - e^{-t/RC}) \qquad (28.14)$$

$$I(t) = \frac{\mathcal{E}}{R} e^{-t/RC} \qquad (28.15)$$

where $Q = C\mathcal{E}$ is the maximum charge on the capacitor. The product RC is called the **time constant** τ of the circuit. If a charged capacitor is discharged through a resistor of resistance R, the charge and current decrease exponentially in time according to the expressions

$$q(t) = Qe^{-t/RC} \qquad (28.17)$$

$$I(t) = -I_0 e^{-t/RC} \qquad (28.18)$$

where Q is the initial charge on the capacitor and $I_0 = Q/RC$ is the initial current in the circuit.

Take a practice test for this chapter by clicking on the Practice Test link at http://www.pse6.com.

QUESTIONS

1. Explain the difference between load resistance in a circuit and internal resistance in a battery.

2. Under what condition does the potential difference across the terminals of a battery equal its emf? Can the terminal voltage ever exceed the emf? Explain.

3. Is the direction of current through a battery always from the negative terminal to the positive terminal? Explain.

4. How would you connect resistors so that the equivalent resistance is larger than the greatest individual resistance? Give an example involving three resistors.

5. How would you connect resistors so that the equivalent resistance is smaller than the least individual resistance? Give an example involving three resistors.

6. Given three lightbulbs and a battery, sketch as many different electric circuits as you can.

7. When resistors are connected in series, which of the following would be the same for each resistor: potential difference, current, power?

8. When resistors are connected in parallel, which of the following would be the same for each resistor: potential difference, current, power?

9. What advantage might there be in using two identical resistors in parallel connected in series with another identical parallel pair, rather than just using a single resistor?

10. An incandescent lamp connected to a 120-V source with a short extension cord provides more illumination than the same lamp connected to the same source with a very long extension cord. Explain.

11. Why is it possible for a bird to sit on a high-voltage wire without being electrocuted?

12. When can the potential difference across a resistor be positive?

13. Referring to Figure Q28.13, describe what happens to the lightbulb after the switch is closed. Assume that the capacitor has a large capacitance and is initially uncharged, and assume that the light illuminates when connected directly across the battery terminals.

14. What is the internal resistance of an ideal ammeter? Of an ideal voltmeter? Do real meters ever attain these ideals?

15. A "short circuit" is a path of very low resistance in a circuit in parallel with some other part of the circuit. Discuss the effect of the short circuit on the portion of the circuit it parallels. Use a lamp with a frayed cord as an example.

16. If electric power is transmitted over long distances, the resistance of the wires becomes significant. Why? Which method of transmission would result in less energy wasted—high current and low voltage or low current and high voltage? Explain your answer.

17. Are the two headlights of a car wired in series or in parallel? How can you tell?

18. Embodied in Kirchhoff's rules are two conservation laws. What are they?

19. Figure Q28.19 shows a series combination of three lightbulbs, all rated at 120 V with power ratings of 60 W, 75 W, and 200 W. Why is the 60-W lamp the brightest and the 200-W lamp the dimmest? Which bulb has the greatest resistance? How would their intensities differ if they were connected in parallel?

Figure Q28.19

20. A student claims that the second lightbulb in series is less bright than the first, because the first bulb uses up some of the current. How would you respond to this statement?

21. Is a circuit breaker wired in series or in parallel with the device it is protecting?

22. So that your grandmother can listen to *A Prairie Home Companion*, you take her bedside radio to the hospital where she is staying. You are required to have a maintenance worker test it for electrical safety. Finding that it develops 120 V on one of its knobs, he does not let you take it up to your grandmother's room. She complains that she has had the radio for many years and nobody has ever gotten a shock from it. You end up having to buy a new plastic radio. Is this fair? Will the old radio be safe back in her bedroom?

Figure Q28.13

23. Suppose you fall from a building and on the way down grab a high-voltage wire. If the wire supports you as you hang from it, will you be electrocuted? If the wire then breaks, should you continue to hold onto an end of the wire as you fall?

24. What advantage does 120-V operation offer over 240 V? What disadvantages?

25. When electricians work with potentially live wires, they often use the backs of their hands or fingers to move wires. Why do you suppose they use this technique?

26. What procedure would you use to try to save a person who is "frozen" to a live high-voltage wire without endangering your own life?

27. If it is the current through the body that determines how serious a shock will be, why do we see warnings of *high voltage* rather than *high current* near electrical equipment?

28. Suppose you are flying a kite when it strikes a high-voltage wire. What factors determine how great a shock you receive?

29. A series circuit consists of three identical lamps connected to a battery as shown in Figure Q28.29. When the switch S is closed, what happens (a) to the intensities of lamps A and B; (b) to the intensity of lamp C; (c) to the current in the circuit; and (d) to the voltage across the three lamps? (e) Does the power delivered to the circuit increase, decrease, or remain the same?

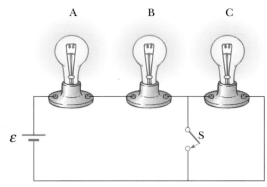

Figure Q28.29

30. If your car's headlights are on when you start the ignition, why do they dim while the car is starting?

31. A ski resort consists of a few chair lifts and several inter-connected downhill runs on the side of a mountain, with a lodge at the bottom. The lifts are analogous to batteries, and the runs are analogous to resistors. Describe how two runs can be in series. Describe how three runs can be in parallel. Sketch a junction of one lift and two runs. State Kirchhoff's junction rule for ski resorts. One of the skiers happens to be carrying a skydiver's altimeter. She never takes the same set of lifts and runs twice, but keeps passing you at the fixed location where you are working. State Kirchhoff's loop rule for ski resorts.

PROBLEMS

1, 2, 3 = straightforward, intermediate, challenging ☐ = full solution available in the *Student Solutions Manual and Study Guide*

= coached solution with hints available at http://www.pse.com 🖥 = computer useful in solving problem

= paired numerical and symbolic problems

Section 28.1 Electromotive Force

1. A battery has an emf of 15.0 V. The terminal voltage of the battery is 11.6 V when it is delivering 20.0 W of power to an external load resistor R. (a) What is the value of R? (b) What is the internal resistance of the battery?

2. (a) What is the current in a 5.60-Ω resistor connected to a battery that has a 0.200-Ω internal resistance if the terminal voltage of the battery is 10.0 V? (b) What is the emf of the battery?

3. Two 1.50-V batteries—with their positive terminals in the same direction—are inserted in series into the barrel of a flashlight. One battery has an internal resistance of 0.255 Ω, the other an internal resistance of 0.153 Ω. When the switch is closed, a current of 600 mA occurs in the lamp. (a) What is the lamp's resistance? (b) What fraction of the chemical energy transformed appears as internal energy in the batteries?

4. An automobile battery has an emf of 12.6 V and an internal resistance of 0.080 0 Ω. The headlights together present equivalent resistance 5.00 Ω (assumed constant). What is the potential difference across the headlight bulbs (a) when they are the only load on the battery and (b) when the starter motor is operated, taking an additional 35.0 A from the battery?

Section 28.2 Resistors in Series and Parallel

5. The current in a loop circuit that has a resistance of R_1 is 2.00 A. The current is reduced to 1.60 A when an additional resistor $R_2 = 3.00\ \Omega$ is added in series with R_1. What is the value of R_1?

6. (a) Find the equivalent resistance between points a and b in Figure P28.6. (b) A potential difference of 34.0 V is applied between points a and b. Calculate the current in each resistor.

Figure P28.6

7. A lightbulb marked "75 W [at] 120 V" is screwed into a socket at one end of a long extension cord, in which each of the two conductors has resistance 0.800 Ω. The other end of the extension cord is plugged into a 120-V outlet. Draw a circuit diagram and find the actual power delivered to the bulb in this circuit.

8. Four copper wires of equal length are connected in series. Their cross-sectional areas are 1.00 cm², 2.00 cm², 3.00 cm², and 5.00 cm². A potential difference of 120 V is applied across the combination. Determine the voltage across the 2.00-cm² wire.

9. Consider the circuit shown in Figure P28.9. Find (a) the current in the 20.0-Ω resistor and (b) the potential difference between points a and b.

Figure P28.9

10. For the purpose of measuring the electric resistance of shoes through the body of the wearer to a metal ground plate, the American National Standards Institute (ANSI) specifies the circuit shown in Figure P28.10. The potential difference ΔV across the 1.00-MΩ resistor is measured with a high-resistance voltmeter. (a) Show that the resistance of the footwear is given by

$$R_{shoes} = 1.00 \text{ M}\Omega \left(\frac{50.0 \text{ V} - \Delta V}{\Delta V} \right)$$

(b) In a medical test, a current through the human body should not exceed 150 μA. Can the current delivered by the ANSI-specified circuit exceed 150 μA? To decide, consider a person standing barefoot on the ground plate.

Figure P28.10

11. Three 100-Ω resistors are connected as shown in Figure P28.11. The maximum power that can safely be delivered to any one resistor is 25.0 W. (a) What is the maximum voltage that can be applied to the terminals a and b? For the voltage determined in part (a), what is the power delivered to each resistor? What is the total power delivered?

Figure P28.11

12. Using only three resistors—2.00 Ω, 3.00 Ω, and 4.00 Ω— find 17 resistance values that may be obtained by various combinations of one or more resistors. Tabulate the combinations in order of increasing resistance.

13. The current in a circuit is tripled by connecting a 500-Ω resistor in parallel with the resistance of the circuit. Determine the resistance of the circuit in the absence of the 500-Ω resistor.

14. A 6.00-V battery supplies current to the circuit shown in Figure P28.14. When the double-throw switch S is open, as shown in the figure, the current in the battery is 1.00 mA. When the switch is closed in position 1, the current in the battery is 1.20 mA. When the switch is closed in position 2, the current in the battery is 2.00 mA. Find the resistances R_1, R_2, and R_3.

Figure P28.14

15. Calculate the power delivered to each resistor in the circuit shown in Figure P28.15.

Figure P28.15

16. Two resistors connected in series have an equivalent resistance of 690 Ω. When they are connected in parallel, their equivalent resistance is 150 Ω. Find the resistance of each resistor.

17. An electric teakettle has a multiposition switch and two heating coils. When only one of the coils is switched on, the well-insulated kettle brings a full pot of water to a boil over the time interval Δt. When only the other coil is switched on, it requires a time interval of $2\Delta t$ to boil the same amount of water. Find the time interval required to boil the same amount of water if both coils are switched on (a) in a parallel connection and (b) in a series connection.

18. In Figures 28.4 and 28.6, let $R_1 = 11.0$ Ω, $R_2 = 22.0$ Ω, and let the battery have a terminal voltage of 33.0 V. (a) In the parallel circuit shown in Figure 28.6, to which resistor is more power delivered? (b) Verify that the sum of the power (I^2R) delivered to each resistor equals the power supplied by the battery ($\mathcal{P} = I\Delta V$). (c) In the series circuit, which resistor uses more power? (d) Verify that the sum of the power (I^2R) used by each resistor equals the power supplied by the battery ($\mathcal{P} = I\Delta V$). (e) Which circuit configuration uses more power?

19. Four resistors are connected to a battery as shown in Figure P28.19. The current in the battery is I, the battery emf is $\mathcal{E}$, and the resistor values are $R_1 = R$, $R_2 = 2R$, $R_3 = 4R$, $R_4 = 3R$. (a) Rank the resistors according to the potential difference across them, from largest to smallest. Note any cases of equal potential differences. (b) Determine the potential difference across each resistor in terms of $\mathcal{E}$. (c) Rank the resistors according to the current in them, from largest to smallest. Note any cases of equal currents. (d) Determine the current in each resistor in terms of I. (e) **What If?** If R_3 is increased, what happens to the current in each of the resistors? (f) In the limit that $R_3 \to \infty$, what are the new values of the current in each resistor in terms of I, the original current in the battery?

Figure P28.19

Section 28.3   Kirchhoff's Rules

Note: The currents are not necessarily in the direction shown for some circuits.

20. The ammeter shown in Figure P28.20 reads 2.00 A. Find I_1, I_2, and $\mathcal{E}$.

Figure P28.20

21. Determine the current in each branch of the circuit shown in Figure P28.21.

Figure P28.21 Problems 21, 22, and 23.

22. In Figure P28.21, show how to add just enough ammeters to measure every different current. Show how to add just enough voltmeters to measure the potential difference across each resistor and across each battery.

23. The circuit considered in Problem 21 and shown in Figure P28.21 is connected for 2.00 min. (a) Find the energy delivered by each battery. (b) Find the energy delivered to each resistor. (c) Identify the net energy transformation that occurs in the operation of the circuit and the total amount of energy transformed.

24. Using Kirchhoff's rules, (a) find the current in each resistor in Figure P28.24. (b) Find the potential difference between points c and f. Which point is at the higher potential?

Figure P28.24

25. Taking $R = 1.00 \text{ k}\Omega$ and $\mathcal{E} = 250$ V in Figure P28.25, determine the direction and magnitude of the current in the horizontal wire between a and e.

Figure P28.25

26. In the circuit of Figure P28.26, determine the current in each resistor and the voltage across the 200-Ω resistor.

Figure P28.26

27. A dead battery is charged by connecting it to the live battery of another car with jumper cables (Fig. P28.27). Determine the current in the starter and in the dead battery.

Figure P28.27

28. For the network shown in Figure P28.28, show that the resistance $R_{ab} = (27/17) \; \Omega$.

Figure P28.28

29. For the circuit shown in Figure P28.29, calculate (a) the current in the 2.00-Ω resistor and (b) the potential difference between points a and b.

Figure P28.29

30. Calculate the power delivered to each resistor shown in Figure P28.30.

Figure P28.30

Section 28.4 *RC* Circuits

31. Consider a series *RC* circuit (see Fig. 28.19) for which $R = 1.00 \text{ M}\Omega$, $C = 5.00 \; \mu\text{F}$, and $\mathcal{E} = 30.0$ V. Find (a) the time constant of the circuit and (b) the maximum charge on the capacitor after the switch is closed. (c) Find the current in the resistor 10.0 s after the switch is closed.

32. A 2.00-nF capacitor with an initial charge of 5.10 μC is discharged through a 1.30-kΩ resistor. (a) Calculate the current in the resistor 9.00 μs after the resistor is connected across the terminals of the capacitor. (b) What charge remains on the capacitor after 8.00 μs? (c) What is the maximum current in the resistor?

33. A fully charged capacitor stores energy U_0. How much energy remains when its charge has decreased to half its original value?

34. A capacitor in an *RC* circuit is charged to 60.0% of its maximum value in 0.900 s. What is the time constant of the circuit?

35. Show that the integral in Equation (1) of Example 28.14 has the value $RC/2$.

36. In the circuit of Figure P28.36, the switch S has been open for a long time. It is then suddenly closed. Determine the time constant (a) before the switch is closed and (b) after the switch is closed. (c) Let the switch be closed at $t = 0$. Determine the current in the switch as a function of time.

Figure P28.36

37. The circuit in Figure P28.37 has been connected for a long time. (a) What is the voltage across the capacitor? (b) If the battery is disconnected, how long does it take the capacitor to discharge to one tenth of its initial voltage?

1.00 Ω   8.00 Ω
1.00 μF
10.0 V
4.00 Ω   2.00 Ω

Figure P28.37

38. In places such as a hospital operating room and a factory for electronic circuit boards, electric sparks must be avoided. A person standing on a grounded floor and touching nothing else can typically have a body capacitance of 150 pF, in parallel with a foot capacitance of 80.0 pF produced by the dielectric soles of his or her shoes. The person acquires static electric charge from interactions with furniture, clothing, equipment, packaging materials, and essentially everything else. The static charge is conducted to ground through the equivalent resistance of the two shoe soles in parallel with each other. A pair of rubber-soled street shoes can present an equivalent resistance of 5 000 MΩ. A pair of shoes with special static-dissipative soles can have an equivalent resistance of 1.00 MΩ. Consider the person's body and shoes as forming an *RC* circuit with the ground. (a) How long does it take the rubber-soled shoes to reduce a 3 000-V static charge to 100 V? (b) How long does it take the static-dissipative shoes to do the same thing?

39. A 4.00-MΩ resistor and a 3.00-μF capacitor are connected in series with a 12.0-V power supply. (a) What is the time constant for the circuit? (b) Express the current in the circuit and the charge on the capacitor as functions of time.

40. Dielectric materials used in the manufacture of capacitors are characterized by conductivities that are small but not zero. Therefore, a charged capacitor slowly loses its charge by "leaking" across the dielectric. If a capacitor having capacitance *C* leaks charge such that the potential difference has decreased to half its initial ($t = 0$) value at a time *t*, what is the equivalent resistance of the dielectric?

Section 28.5 Electrical Meters

41. Assume that a galvanometer has an internal resistance of 60.0 Ω and requires a current of 0.500 mA to produce full-scale deflection. What resistance must be connected in parallel with the galvanometer if the combination is to serve as an ammeter that has a full-scale deflection for a current of 0.100 A?

42. A typical galvanometer, which requires a current of 1.50 mA for full-scale deflection and has a resistance of 75.0 Ω, may be used to measure currents of much greater values. To enable an operator to measure large currents without damage to the galvanometer, a relatively small shunt resistor is wired in parallel with the galvanometer, as suggested in Figure 28.27. Most of the current then goes through the shunt resistor. Calculate the value of the shunt resistor that allows the galvanometer to be used to measure a current of 1.00 A at full-scale deflection. (*Suggestion:* use Kirchhoff's rules.)

43. The same galvanometer described in the previous problem may be used to measure voltages. In this case a large resistor is wired in series with the galvanometer, as suggested in Figure 28.29. The effect is to limit the current in the galvanometer when large voltages are applied. Most of the potential drop occurs across the resistor placed in series. Calculate the value of the resistor that allows the galvanometer to measure an applied voltage of 25.0 V at full-scale deflection.

44. *Meter loading.* Work this problem to five-digit precision. Refer to Figure P28.44. (a) When a 180.00-Ω resistor is connected across a battery of emf 6.000 0 V and internal resistance 20.000 Ω, what is the current in the resistor? What is the potential difference across it? (b) Suppose now an ammeter of resistance 0.500 00 Ω and a voltmeter of resistance 20 000 Ω are added to the circuit as shown in Figure P28.44b. Find the reading of each. (c) **What If?** Now one terminal of one wire is moved, as shown in Figure P28.44c. Find the new meter readings.

45. Design a multirange ammeter capable of full-scale deflection for 25.0 mA, 50.0 mA, and 100 mA. Assume the meter movement is a galvanometer that has a resistance of 25.0 Ω and gives a full-scale deflection for 1.00 mA.

46. Design a multirange voltmeter capable of full-scale deflection for 20.0 V, 50.0 V, and 100 V. Assume the meter movement is a galvanometer that has a resistance of 60.0 Ω and gives a full-scale deflection for a current of 1.00 mA.

6.000 0 V
20.000 Ω

180.00 Ω

(a)   (b)   (c)

Figure P28.44

47. A particular galvanometer serves as a 2.00-V full-scale voltmeter when a 2 500-Ω resistor is connected in series with it. It serves as a 0.500-A full-scale ammeter when a 0.220-Ω resistor is connected in parallel with it. Determine the internal resistance of the galvanometer and the current required to produce full-scale deflection.

Section 28.6   Household Wiring and Electrical Safety

48. An 8.00-ft extension cord has two 18-gauge copper wires, each having a diameter of 1.024 mm. At what rate is energy delivered to the resistance in the cord when it is carrying a current of (a) 1.00 A and (b) 10.0 A?

49. An electric heater is rated at 1 500 W, a toaster at 750 W, and an electric grill at 1 000 W. The three appliances are connected to a common 120-V household circuit. (a) How much current does each draw? (b) Is a circuit with a 25.0-A circuit breaker sufficient in this situation? Explain your answer.

50. Aluminum wiring has sometimes been used instead of copper for economy. According to the National Electrical Code, the maximum allowable current for 12-gauge copper wire with rubber insulation is 20 A. What should be the maximum allowable current in a 12-gauge aluminum wire if the power per unit length delivered to the resistance in the aluminum wire is the same as that delivered in the copper wire?

51. Turn on your desk lamp. Pick up the cord, with your thumb and index finger spanning the width of the cord. (a) Compute an order-of-magnitude estimate for the current in your hand. You may assume that at a typical instant the conductor inside the lamp cord next to your thumb is at potential $\sim 10^2$ V and that the conductor next to your index finger is at ground potential (0 V). The resistance of your hand depends strongly on the thickness and the moisture content of the outer layers of your skin. Assume that the resistance of your hand between fingertip and thumb tip is $\sim 10^4$ Ω. You may model the cord as having rubber insulation. State the other quantities you measure or estimate and their values. Explain your reasoning. (b) Suppose that your body is isolated from any other charges or currents. In order-of-magnitude terms describe the potential of your thumb where it contacts the cord, and the potential of your finger where it touches the cord.

Additional Problems

52. Four 1.50-V AA batteries in series are used to power a transistor radio. If the batteries can move a charge of 240 C, how long will they last if the radio has a resistance of 200 Ω?

53. A battery has an emf of 9.20 V and an internal resistance of 1.20 Ω. (a) What resistance across the battery will extract from it a power of 12.8 W? (b) a power of 21.2 W?

54. Calculate the potential difference between points a and b in Figure P28.54 and identify which point is at the higher potential.

Figure P28.54

55. Assume you have a battery of emf $\mathcal{E}$ and three identical lightbulbs, each having constant resistance R. What is the total power delivered by the battery if the bulbs are connected (a) in series? (b) in parallel? (c) For which connection will the bulbs shine the brightest?

56. A group of students on spring break manages to reach a deserted island in their wrecked sailboat. They splash ashore with fuel, a European gasoline-powered 240-V generator, a box of North American 100-W 120-V lightbulbs, a 500-W 120-V hot pot, lamp sockets, and some insulated wire. While waiting to be rescued, they decide to use the generator to operate some lightbulbs. (a) Draw a diagram of a circuit they can use, containing the minimum number of lightbulbs with 120 V across each bulb, and no higher voltage. Find the current in the generator and its power output. (b) One student catches a fish and wants to cook it in the hot pot. Draw a diagram of a circuit containing the hot pot and the minimum number of lightbulbs with 120 V across each device, and not more. Find the current in the generator and its power output.

57. A battery has an emf $\mathcal{E}$ and internal resistance r. A variable load resistor R is connected across the terminals of the battery. (a) Determine the value of R such that the potential difference across the terminals is a maximum. (b) Determine the value of R so that the current in the circuit is a maximum. (c) Determine the value of R so that the power delivered to the load resistor is a maximum. Choosing the load resistance for maximum power transfer is a case of what is called *impedance matching* in general. Impedance matching is important in shifting gears on a bicycle, in connecting a loudspeaker to an audio amplifier, in connecting a battery charger to a bank of solar photoelectric cells, and in many other applications.

58. A 10.0-μF capacitor is charged by a 10.0-V battery through a resistance R. The capacitor reaches a potential difference of 4.00 V in a time 3.00 s after charging begins. Find R.

59. When two unknown resistors are connected in series with a battery, the battery delivers 225 W and carries a total current of 5.00 A. For the same total current, 50.0 W is delivered when the resistors are connected in parallel. Determine the values of the two resistors.

60. When two unknown resistors are connected in series with a battery, the battery delivers total power $\mathcal{P}_s$ and carries a total current of I. For the same total current, a total power

$\mathcal{P}_p$ is delivered when the resistors are connected in parallel. Determine the values of the two resistors.

61. A power supply has an open-circuit voltage of 40.0 V and an internal resistance of 2.00 Ω. It is used to charge two storage batteries connected in series, each having an emf of 6.00 V and internal resistance of 0.300 Ω. If the charging current is to be 4.00 A, (a) what additional resistance should be added in series? (b) At what rate does the internal energy increase in the supply, in the batteries, and in the added series resistance? (c) At what rate does the chemical energy increase in the batteries?

62. Two resistors R_1 and R_2 are in parallel with each other. Together they carry total current I. (a) Determine the current in each resistor. (b) Prove that this division of the total current I between the two resistors results in less power delivered to the combination than any other division. It is a general principle that *current in a direct current circuit distributes itself so that the total power delivered to the circuit is a minimum.*

63. The value of a resistor R is to be determined using the ammeter–voltmeter setup shown in Figure P28.63. The ammeter has a resistance of 0.500 Ω, and the voltmeter has a resistance of 20 000 Ω. Within what range of actual values of R will the measured values be correct to within 5.00% if the measurement is made using the circuit shown in (a) Figure P28.63a and (b) Figure P28.63b?

(a)

(b)

Figure P28.63

64. A battery is used to charge a capacitor through a resistor, as shown in Figure 28.19. Show that half the energy supplied by the battery appears as internal energy in the resistor and that half is stored in the capacitor.

65. The values of the components in a simple series RC circuit containing a switch (Fig. 28.19) are $C = 1.00\ \mu$F, $R = 2.00 \times 10^6$ Ω, and $\mathcal{E} = 10.0$ V. At the instant 10.0 s after the switch is closed, calculate (a) the charge on the capacitor, (b) the current in the resistor, (c) the rate at which energy is being stored in the capacitor, and (d) the rate at which energy is being delivered by the battery.

66. The switch in Figure P28.66a closes when $\Delta V_c > 2\,\Delta V/3$ and opens when $\Delta V_c < \Delta V/3$. The voltmeter reads a voltage as plotted in Figure P28.66b. What is the period T of the waveform in terms of R_1, R_2, and C?

(a)

(b)

Figure P28.66

67. Three 60.0-W, 120-V lightbulbs are connected across a 120-V power source, as shown in Figure P28.67. Find (a) the total power delivered to the three bulbs and (b) the voltage across each. Assume that the resistance of each bulb is constant (even though in reality the resistance might increase markedly with current).

Figure P28.67

68. Switch S has been closed for a long time, and the electric circuit shown in Figure P28.68 carries a constant current. Take $C_1 = 3.00\ \mu$F, $C_2 = 6.00\ \mu$F, $R_1 = 4.00$ kΩ, and $R_2 = 7.00$ kΩ. The power delivered to R_2 is 2.40 W. (a) Find the charge on C_1. (b) Now the switch is opened. After many milliseconds, by how much has the charge on C_2 changed?

Figure P28.68

69. Four resistors are connected in parallel across a 9.20-V battery. They carry currents of 150 mA, 45.0 mA, 14.00 mA, and 4.00 mA. (a) If the resistor with the largest resistance is replaced with one having twice the resistance, what is the ratio of the new current in the battery

to the original current? (b) **What If?** If instead the resistor with the smallest resistance is replaced with one having twice the resistance, what is the ratio of the new total current to the original current? (c) On a February night, energy leaves a house by several heat leaks, including the following: 1 500 W by conduction through the ceiling; 450 W by infiltration (air flow) around the windows; 140 W by conduction through the basement wall above the foundation sill; and 40.0 W by conduction through the plywood door to the attic. To produce the biggest saving in heating bills, which one of these energy transfers should be reduced first?

70. Figure P28.70 shows a circuit model for the transmission of an electrical signal, such as cable TV, to a large number of subscribers. Each subscriber connects a load resistance R_L between the transmission line and the ground. The ground is assumed to be at zero potential and able to carry any current between any ground connections with negligible resistance. The resistance of the transmission line itself between the connection points of different subscribers is modeled as the constant resistance R_T. Show that the equivalent resistance across the signal source is

$$R_{eq} = \tfrac{1}{2}\left[(4R_T R_L + R_T^2)^{1/2} + R_T\right]$$

Suggestion: Because the number of subscribers is large, the equivalent resistance would not change noticeably if the first subscriber cancelled his service. Consequently, the equivalent resistance of the section of the circuit to the right of the first load resistor is nearly equal to R_{eq}.

Figure P28.70

71. In Figure P28.71, suppose the switch has been closed for a time sufficiently long for the capacitor to become fully charged. Find (a) the steady-state current in each resistor and (b) the charge Q on the capacitor. (c) The switch is now opened at $t = 0$. Write an equation for the current I_{R_2} through R_2 as a function of time and (d) find the time interval required for the charge on the capacitor to fall to one-fifth its initial value.

Figure P28.71

72. A regular tetrahedron is a pyramid with a triangular base. Six 10.0-Ω resistors are placed along its six edges, with junctions at its four vertices. A 12.0-V battery is connected to any two of the vertices. Find (a) the equivalent resistance of the tetrahedron between these vertices and (b) the current in the battery.

73. The circuit shown in Figure P28.73 is set up in the laboratory to measure an unknown capacitance C with the use of a voltmeter of resistance $R = 10.0$ MΩ and a battery whose emf is 6.19 V. The data given in the table are the measured voltages across the capacitor as a function of time, where $t = 0$ represents the instant at which the switch is opened. (a) Construct a graph of $\ln(\mathcal{E}/\Delta V)$ versus t, and perform a linear least-squares fit to the data. (b) From the slope of your graph, obtain a value for the time constant of the circuit and a value for the capacitance.

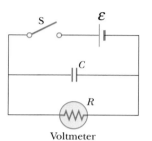

Figure P28.73

ΔV (V)	t (s)	$\ln(\mathcal{E}/\Delta V)$
6.19	0	
5.55	4.87	
4.93	11.1	
4.34	19.4	
3.72	30.8	
3.09	46.6	
2.47	67.3	
1.83	102.2	

74. The student engineer of a campus radio station wishes to verify the effectiveness of the lightning rod on the antenna mast (Fig. P28.74). The unknown resistance R_x is between points C and E. Point E is a true ground but is inaccessible for direct measurement since this stratum is several meters below the Earth's surface. Two identical rods are driven into the ground at A and B, introducing an unknown resistance R_y. The procedure is as follows. Measure resistance R_1 between points A and B, then connect A and B with a heavy conducting wire and measure resistance R_2 between

Figure P28.74

points A and C. (a) Derive an equation for R_x in terms of the observable resistances, R_1 and R_2. (b) A satisfactory ground resistance would be $R_x < 2.00\ \Omega$. Is the grounding of the station adequate if measurements give $R_1 = 13.0\ \Omega$ and $R_2 = 6.00\ \Omega$?

75. The circuit in Figure P28.75 contains two resistors, $R_1 = 2.00\ k\Omega$ and $R_2 = 3.00\ k\Omega$, and two capacitors, $C_1 = 2.00\ \mu F$ and $C_2 = 3.00\ \mu F$, connected to a battery with emf $\mathcal{E} = 120\ V$. No charge is on either capacitor before switch S is closed. Determine the charges q_1 and q_2 on capacitors C_1 and C_2, respectively, after the switch is closed. (*Suggestion:* First reconstruct the circuit so that it becomes a simple RC circuit containing a single resistor and single capacitor in series, connected to the battery, and then determine the total charge q stored in the equivalent circuit.)

Figure P28.75

76. This problem[6] illustrates how a digital voltmeter affects the voltage across a capacitor in an RC circuit. A digital voltmeter of internal resistance r is used to measure the voltage across a capacitor after the switch in Figure P28.76 is closed. Because the meter has finite resistance, part of the current supplied by the battery passes through the meter. (a) Apply Kirchhoff's rules to this circuit, and use the fact that $i_C = dq/dt$ to show that this leads to the differential equation

$$R_{eq}\frac{dq}{dt} + \frac{q}{C} = \frac{r}{r+R}\,\mathcal{E}$$

where $R_{eq} = rR/(r + R)$. (b) Show that the solution to this differential equation is

$$q = \frac{r}{r+R}\,C\mathcal{E}\,(1 - e^{-t/R_{eq}C})$$

Voltmeter

Figure P28.76

[6] After Joseph Priest, "Meter Resistance: Don't Forget It!" *The Physics Teacher*, January 2003, p. 40.

and that the voltage across the capacitor as a function of time is

$$V_C = \frac{r}{r+R}\,\mathcal{E}(1 - e^{-t/R_{eq}C})$$

(c) **What If?** If the capacitor is fully charged, and the switch is then opened, how does the voltage across the capacitor behave in this case?

Answers to Quick Quizzes

28.1 (a). Power is delivered to the internal resistance of a battery, so decreasing the internal resistance will decrease this "lost" power and increase the percentage of the power delivered to the device.

28.2 (c). In a series circuit, the current is the same in all resistors in series. Current is not "used up" as charges pass through a resistor.

28.3 (a). Connecting b to c "shorts out" bulb R_2 and changes the total resistance of the circuit from $R_1 + R_2$ to just R_1. Because the resistance of the circuit has decreased (and the emf supplied by the battery does not change), the current in the circuit increases.

28.4 (b). When the switch is opened, resistors R_1 and R_2 are in series, so that the total circuit resistance is larger than when the switch was closed. As a result, the current decreases.

28.5 (b), (d). Adding another series resistor increases the total resistance of the circuit and thus reduces the current in the circuit. The potential difference across the battery terminals increases because the reduced current results in a smaller voltage decrease across the internal resistance.

28.6 (a), (e). If the second resistor were connected in parallel, the total resistance of the circuit would decrease, and the current in the battery would increase. The potential difference across the terminals would decrease because the increased current results in a greater voltage drop across the internal resistance.

28.7 (a). When the switch is closed, resistors R_1 and R_2 are in parallel, so that the total circuit resistance is smaller than when the switch was open. As a result, the current increases.

28.8 (c). A current is assigned to a given branch of a circuit. There may be multiple resistors and batteries in a given branch.

28.9 (b), (d). Just after the switch is closed, there is no charge on the capacitor, so there is no voltage across it. Charges begin to flow in the circuit to charge up the capacitor, so that all of the voltage $\Delta V = IR$ appears across the resistor. After a long time, the capacitor is fully charged and the current drops to zero. Thus, the battery voltage is now entirely across the capacitor.

28.10 (c), (i). Just after the switch is closed, there is no charge on the capacitor. Current exists in both branches of the circuit as the capacitor begins to charge, so the right half of the circuit is equivalent to two resistances R in parallel for an equivalent resistance of $\frac{1}{2}R$. After a long time, the capacitor is fully charged and the current in the right-hand branch drops to zero. Now, current exists only in a resistance R across the battery.

Chapter 29

Magnetic Fields

▲ Magnetic fingerprinting allows fingerprints to be seen on surfaces that otherwise would not allow prints to be lifted. The powder spread on the surface is coated with an organic material that adheres to the greasy residue in a fingerprint. A magnetic "brush" removes the excess powder and makes the fingerprint visible. (James King-Holmes/Photo Researchers, Inc.)

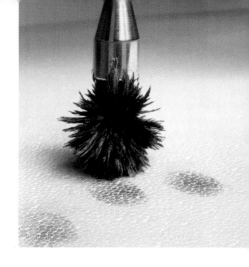

Many historians of science believe that the compass, which uses a magnetic needle, was used in China as early as the 13th century B.C., its invention being of Arabic or Indian origin. The early Greeks knew about magnetism as early as 800 B.C. They discovered that the stone magnetite (Fe_3O_4) attracts pieces of iron. Legend ascribes the name *magnetite* to the shepherd Magnes, the nails of whose shoes and the tip of whose staff stuck fast to chunks of magnetite while he pastured his flocks.

In 1269 a Frenchman named Pierre de Maricourt found that the directions of a needle near a spherical natural magnet formed lines that encircled the sphere and passed through two points diametrically opposite each other, which he called the *poles* of the magnet. Subsequent experiments showed that every magnet, regardless of its shape, has two poles, called *north* (N) and *south* (S) poles, that exert forces on other magnetic poles similar to the way that electric charges exert forces on one another. That is, like poles (N–N or S–S) repel each other, and opposite poles (N–S) attract each other.

The poles received their names because of the way a magnet, such as that in a compass, behaves in the presence of the Earth's magnetic field. If a bar magnet is suspended from its midpoint and can swing freely in a horizontal plane, it will rotate until its north pole points to the Earth's geographic North Pole and its south pole points to the Earth's geographic South Pole.[1]

In 1600 William Gilbert (1540–1603) extended de Maricourt's experiments to a variety of materials. Using the fact that a compass needle orients in preferred directions, he suggested that the Earth itself is a large permanent magnet. In 1750 experimenters used a torsion balance to show that magnetic poles exert attractive or repulsive forces on each other and that these forces vary as the inverse square of the distance between interacting poles. Although the force between two magnetic poles is otherwise similar to the force between two electric charges, electric charges can be isolated (witness the electron and proton) whereas **a single magnetic pole has never been isolated.** That is, **magnetic poles are always found in pairs.** All attempts thus far to detect an isolated magnetic pole have been unsuccessful. No matter how many times a permanent magnet is cut in two, each piece always has a north and a south pole.[2]

The relationship between magnetism and electricity was discovered in 1819 when, during a lecture demonstration, the Danish scientist Hans Christian Oersted found that an electric current in a wire deflected a nearby compass needle.[3] In the 1820s,

---

[1] Note that the Earth's geographic North Pole is magnetically a south pole, whereas its geographic South Pole is magnetically a north pole. Because *opposite* magnetic poles attract each other, the pole on a magnet that is attracted to the Earth's geographic North Pole is the magnet's *north* pole and the pole attracted to the Earth's geographic South Pole is the magnet's *south* pole.

[2] There is some theoretical basis for speculating that magnetic *monopoles*—isolated north or south poles—may exist in nature, and attempts to detect them are an active experimental field of investigation.

[3] The same discovery was reported in 1802 by an Italian jurist, Gian Dominico Romognosi, but was overlooked, probably because it was published in an obscure journal.

Hans Christian Oersted

Danish Physicist and Chemist (1777–1851)

Oersted is best known for observing that a compass needle deflects when placed near a wire carrying a current. This important discovery was the first evidence of the connection between electric and magnetic phenomena. Oersted was also the first to prepare pure aluminum. *(North Wind Picture Archives)* further connections between electricity and magnetism were demonstrated independently by Faraday and Joseph Henry (1797–1878). They showed that an electric current can be produced in a circuit either by moving a magnet near the circuit or by changing the current in a nearby circuit. These observations demonstrate that a changing magnetic field creates an electric field. Years later, theoretical work by Maxwell showed that the reverse is also true: a changing electric field creates a magnetic field.

This chapter examines the forces that act on moving charges and on current-carrying wires in the presence of a magnetic field. The source of the magnetic field is described in Chapter 30.

29.1 Magnetic Fields and Forces

In our study of electricity, we described the interactions between charged objects in terms of electric fields. Recall that an electric field surrounds any electric charge. In addition to containing an electric field, the region of space surrounding any *moving* electric charge also contains a magnetic field. A magnetic field also surrounds a magnetic substance making up a permanent magnet.

Historically, the symbol **B** has been used to represent a magnetic field, and this is the notation we use in this text. The direction of the magnetic field **B** at any location is the direction in which a compass needle points at that location. As with the electric field, we can represent the magnetic field by means of drawings with *magnetic field lines.*

Figure 29.1 shows how the magnetic field lines of a bar magnet can be traced with the aid of a compass. Note that the magnetic field lines outside the magnet point away from north poles and toward south poles. One can display magnetic field patterns of a bar magnet using small iron filings, as shown in Figure 29.2.

We can define a magnetic field **B** at some point in space in terms of the magnetic force **F**$_B$ that the field exerts on a charged particle moving with a velocity **v**, which we call the test object. For the time being, let us assume that no electric or gravitational fields are present at the location of the test object. Experiments on various charged particles moving in a magnetic field give the following results:

Properties of the magnetic force on a charge moving in a magnetic field B

- The magnitude F_B of the magnetic force exerted on the particle is proportional to the charge q and to the speed v of the particle.

- The magnitude and direction of **F**$_B$ depend on the velocity of the particle and on the magnitude and direction of the magnetic field **B**.

- When a charged particle moves parallel to the magnetic field vector, the magnetic force acting on the particle is zero.

At the Active Figures link at http://www.pse6.com, you can move the compass around and trace the magnetic field lines for yourself.

Active Figure 29.1 Compass needles can be used to trace the magnetic field lines in the region outside a bar magnet.

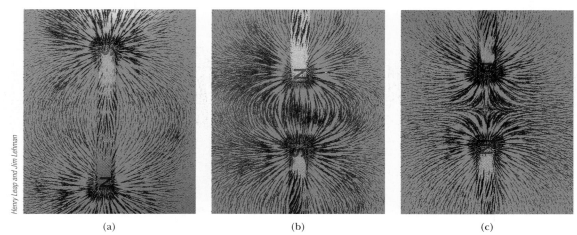

Figure 29.2 (a) Magnetic field pattern surrounding a bar magnet as displayed with iron filings. (b) Magnetic field pattern between *opposite* poles (N–S) of two bar magnets. (c) Magnetic field pattern between *like* poles (N–N) of two bar magnets.

- When the particle's velocity vector makes any angle $\theta \neq 0$ with the magnetic field, the magnetic force acts in a direction perpendicular to both **v** and **B**; that is, $\mathbf{F}_B$ is perpendicular to the plane formed by **v** and **B** (Fig. 29.3a).

- The magnetic force exerted on a positive charge is in the direction opposite the direction of the magnetic force exerted on a negative charge moving in the same direction (Fig. 29.3b).

- The magnitude of the magnetic force exerted on the moving particle is proportional to sin θ, where θ is the angle the particle's velocity vector makes with the direction of **B**.

We can summarize these observations by writing the magnetic force in the form

$$\mathbf{F}_B = q\mathbf{v} \times \mathbf{B} \qquad (29.1)$$

Vector expression for the magnetic force on a charged particle moving in a magnetic field

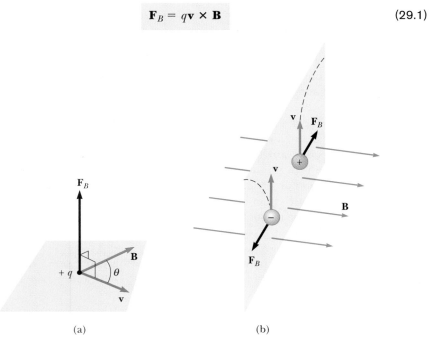

Figure 29.3 The direction of the magnetic force $\mathbf{F}_B$ acting on a charged particle moving with a velocity **v** in the presence of a magnetic field **B**. (a) The magnetic force is perpendicular to both **v** and **B**. (b) Oppositely directed magnetic forces $\mathbf{F}_B$ are exerted on two oppositely charged particles moving at the same velocity in a magnetic field. The dashed lines show the paths of the particles, which we will investigate in Section 29.4.

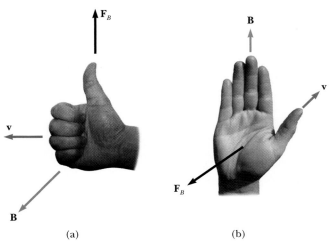

(a)   (b)

Figure 29.4 Two right-hand rules for determining the direction of the magnetic force $\mathbf{F}_B = q\mathbf{v} \times \mathbf{B}$ acting on a particle with charge q moving with a velocity $\mathbf{v}$ in a magnetic field $\mathbf{B}$. (a) In this rule, the fingers point in the direction of $\mathbf{v}$, with $\mathbf{B}$ coming out of your palm, so that you can curl your fingers in the direction of $\mathbf{B}$. The direction of $\mathbf{v} \times \mathbf{B}$, and the force on a positive charge, is the direction in which the thumb points. (b) In this rule, the vector $\mathbf{v}$ is in the direction of your thumb and $\mathbf{B}$ in the direction of your fingers. The force $\mathbf{F}_B$ on a positive charge is in the direction of your palm, as if you are pushing the particle with your hand.

which by definition of the cross product (see Section 11.1) is perpendicular to both $\mathbf{v}$ and $\mathbf{B}$. We can regard this equation as an operational definition of the magnetic field at some point in space. That is, the magnetic field is defined in terms of the force acting on a moving charged particle.

Figure 29.4 reviews two right-hand rules for determining the direction of the cross product $\mathbf{v} \times \mathbf{B}$ and determining the direction of $\mathbf{F}_B$. The rule in Figure 29.4a depends on our right-hand rule for the cross product in Figure 11.2. Point the four fingers of your right hand along the direction of $\mathbf{v}$ with the palm facing $\mathbf{B}$ and curl them toward $\mathbf{B}$. The extended thumb, which is at a right angle to the fingers, points in the direction of $\mathbf{v} \times \mathbf{B}$. Because $\mathbf{F}_B = q\mathbf{v} \times \mathbf{B}$, $\mathbf{F}_B$ is in the direction of your thumb if q is positive and opposite the direction of your thumb if q is negative. (If you need more help understanding the cross product, you should review pages 337 to 339, including Fig. 11.2.)

An alternative rule is shown in Figure 29.4b. Here the thumb points in the direction of $\mathbf{v}$ and the extended fingers in the direction of $\mathbf{B}$. Now, the force $\mathbf{F}_B$ on a positive charge extends outward from your palm. The advantage of this rule is that the force on the charge is in the direction that you would push on something with your hand—outward from your palm. The force on a negative charge is in the opposite direction. Feel free to use either of these two right-hand rules.

The magnitude of the magnetic force on a charged particle is

Magnitude of the magnetic force on a charged particle moving in a magnetic field

$$F_B = |q|vB \sin \theta \qquad (29.2)$$

where θ is the smaller angle between $\mathbf{v}$ and $\mathbf{B}$. From this expression, we see that F_B is zero when $\mathbf{v}$ is parallel or antiparallel to $\mathbf{B}$ ($\theta = 0$ or $180°$) and maximum when $\mathbf{v}$ is perpendicular to $\mathbf{B}$ ($\theta = 90°$).

There are several important differences between electric and magnetic forces:

• The electric force acts along the direction of the electric field, whereas the magnetic force acts perpendicular to the magnetic field.

• The electric force acts on a charged particle regardless of whether the particle is moving, whereas the magnetic force acts on a charged particle only when the particle is in motion.

- The electric force does work in displacing a charged particle, whereas the magnetic force associated with a steady magnetic field does no work when a particle is displaced because the force is perpendicular to the displacement.

From the last statement and on the basis of the work–kinetic energy theorem, we conclude that the kinetic energy of a charged particle moving through a magnetic field cannot be altered by the magnetic field alone. In other words, when a charged particle moves with a velocity **v** through a magnetic field, the field can alter the direction of the velocity vector but cannot change the speed or kinetic energy of the particle.

From Equation 29.2, we see that the SI unit of magnetic field is the newton per coulomb-meter per second, which is called the **tesla** (T):

$$1 \text{ T} = 1 \frac{\text{N}}{\text{C} \cdot \text{m/s}}$$

The tesla

Because a coulomb per second is defined to be an ampere, we see that

$$1 \text{ T} = 1 \frac{\text{N}}{\text{A} \cdot \text{m}}$$

A non-SI magnetic-field unit in common use, called the *gauss* (G), is related to the tesla through the conversion $1 \text{ T} = 10^4$ G. Table 29.1 shows some typical values of magnetic fields.

Quick Quiz 29.1 The north-pole end of a bar magnet is held near a positively charged piece of plastic. Is the plastic (a) attracted, (b) repelled, or (c) unaffected by the magnet?

Quick Quiz 29.2 A charged particle moves with velocity **v** in a magnetic field **B**. The magnetic force on the particle is a maximum when **v** is (a) parallel to **B**, (b) perpendicular to **B**, (c) zero.

Quick Quiz 29.3 An electron moves in the plane of this paper toward the top of the page. A magnetic field is also in the plane of the page and directed toward the right. The direction of the magnetic force on the electron is (a) toward the top of the page, (b) toward the bottom of the page, (c) toward the left edge of the page, (d) toward the right edge of the page, (e) upward out of the page, (f) downward into the page.

Table 29.1

Some Approximate Magnetic Field Magnitudes	
Source of Field	**Field Magnitude (T)**
Strong superconducting laboratory magnet	30
Strong conventional laboratory magnet	2
Medical MRI unit	1.5
Bar magnet	10^{-2}
Surface of the Sun	10^{-2}
Surface of the Earth	0.5×10^{-4}
Inside human brain (due to nerve impulses)	10^{-13}

Example 29.1 **An Electron Moving in a Magnetic Field**

An electron in a television picture tube moves toward the front of the tube with a speed of 8.0×10^6 m/s along the x axis (Fig. 29.5). Surrounding the neck of the tube are coils of wire that create a magnetic field of magnitude 0.025 T, directed at an angle of 60° to the x axis and lying in the xy plane.

(A) Calculate the magnetic force on the electron using Equation 29.2.

Solution Using Equation 29.2, we find the magnitude of the magnetic force:

$$F_B = |q|vB \sin \theta$$
$$= (1.6 \times 10^{-19} \text{ C})(8.0 \times 10^6 \text{ m/s})(0.025 \text{ T})(\sin 60°)$$
$$= \boxed{2.8 \times 10^{-14} \text{ N}}$$

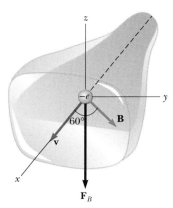

Figure 29.5 (Example 29.1) The magnetic force $\mathbf{F}_B$ acting on the electron is in the negative z direction when $\mathbf{v}$ and $\mathbf{B}$ lie in the xy plane.

Because $\mathbf{v} \times \mathbf{B}$ is in the positive z direction (from the right-hand rule) and the charge is negative, $\mathbf{F}_B$ is in the negative z direction.

(B) Find a vector expression for the magnetic force on the electron using Equation 29.1.

Solution We begin by writing a vector expression for the velocity of the electron:

$$\mathbf{v} = (8.0 \times 10^6 \,\hat{\mathbf{i}}) \text{ m/s}$$

and one for the magnetic field:

$$\mathbf{B} = (0.025 \cos 60° \,\hat{\mathbf{i}} + 0.025 \sin 60° \,\hat{\mathbf{j}}) \text{T}$$
$$= (0.013 \,\hat{\mathbf{i}} + 0.022 \,\hat{\mathbf{j}}) \text{T}$$

The force on the electron, using Equation 29.1, is

$$\mathbf{F}_B = q\mathbf{v} \times \mathbf{B}$$
$$= (-e)[(8.0 \times 10^6 \,\hat{\mathbf{i}}) \text{ m/s}] \times [(0.013 \,\hat{\mathbf{i}} + 0.022 \,\hat{\mathbf{j}})\text{T}]$$
$$= (-e)[(8.0 \times 10^6 \,\hat{\mathbf{i}}) \text{ m/s}] \times [(0.013 \,\hat{\mathbf{i}})\text{T}]$$
$$\quad + (-e)[(8.0 \times 10^6 \,\hat{\mathbf{i}}) \text{ m/s}] \times [(0.022 \,\hat{\mathbf{j}})\text{T}]$$
$$= (-e)(8.0 \times 10^6 \text{ m/s})(0.013 \text{ T})(\hat{\mathbf{i}} \times \hat{\mathbf{i}})$$
$$\quad + (-e)(8.0 \times 10^6 \text{ m/s})(0.022 \text{ T})(\hat{\mathbf{i}} \times \hat{\mathbf{j}})$$
$$= (-1.6 \times 10^{-19} \text{ C})(8.0 \times 10^6 \text{ m/s})(0.022 \text{ T}) \,\hat{\mathbf{k}}$$

where we have used Equations 11.7a and 11.7b to evaluate $\hat{\mathbf{i}} \times \hat{\mathbf{i}}$ and $\hat{\mathbf{i}} \times \hat{\mathbf{j}}$. Carrying out the multiplication, we find,

$$\mathbf{F}_B = \boxed{(-2.8 \times 10^{-14} \text{ N})\hat{\mathbf{k}}}$$

This expression agrees with the result in part (A). The magnitude is the same as we found there, and the force vector is in the negative z direction.

29.2 Magnetic Force Acting on a Current-Carrying Conductor

If a magnetic force is exerted on a single charged particle when the particle moves through a magnetic field, it should not surprise you that a current-carrying wire also experiences a force when placed in a magnetic field. This follows from the fact that the current is a collection of many charged particles in motion; hence, the resultant force exerted by the field on the wire is the vector sum of the individual forces exerted on all the charged particles making up the current. The force exerted on the particles is transmitted to the wire when the particles collide with the atoms making up the wire.

Before we continue our discussion, some explanation of the notation used in this book is in order. To indicate the direction of **B** in illustrations, we sometimes present perspective views, such as those in Figure 29.5. If **B** lies in the plane of the page or is present in a perspective drawing, we use blue vectors or blue field lines with arrowheads. In non-perspective illustrations, we depict a magnetic field perpendicular to and directed out of the page with a series of blue dots, which represent the tips of arrows coming toward you (see Fig. 29.6a). In this case, we label the field $\mathbf{B}_{out}$. If $\mathbf{B}$ is directed perpendicularly into the page, we use blue crosses, which represent the feathered tails of arrows fired away from you, as in Figure 29.6b. In this case, we label the field $\mathbf{B}_{in}$, where the subscript "in" indicates "into the page." The same notation with crosses and dots is also used for other quantities that might be perpendicular to the page, such as forces and current directions.

One can demonstrate the magnetic force acting on a current-carrying conductor by hanging a wire between the poles of a magnet, as shown in Figure 29.7a. For ease in visualization, part of the horseshoe magnet in part (a) is removed to show the end face of the south pole in parts (b), (c), and (d) of Figure 29.7. The magnetic field is directed into the page and covers the region within the shaded squares. When the current in the wire is zero, the wire remains vertical, as shown in Figure 29.7b. However, when the wire carries a current directed upward, as shown in Figure 29.7c, the wire deflects to the left. If we reverse the current, as shown in Figure 29.7d, the wire deflects to the right.

Let us quantify this discussion by considering a straight segment of wire of length L and cross-sectional area A, carrying a current I in a uniform magnetic field $\mathbf{B}$, as shown in Figure 29.8. The magnetic force exerted on a charge q moving with a drift velocity $\mathbf{v}_d$ is $q\mathbf{v}_d \times \mathbf{B}$. To find the total force acting on the wire, we multiply the force $q\mathbf{v}_d \times \mathbf{B}$ exerted on one charge by the number of charges in the segment. Because the volume of the segment is AL, the number of charges in the segment is nAL, where n is the number of charges per unit volume. Hence, the total magnetic force on the wire of length L is

$$\mathbf{F}_B = (q\mathbf{v}_d \times \mathbf{B})\, nAL$$

We can write this expression in a more convenient form by noting that, from Equation 27.4, the current in the wire is $I = nqv_d A$. Therefore,

$$\mathbf{F}_B = I\mathbf{L} \times \mathbf{B} \qquad (29.3)$$

where $\mathbf{L}$ is a vector that points in the direction of the current I and has a magnitude equal to the length L of the segment. Note that this expression applies only to a straight segment of wire in a uniform magnetic field.

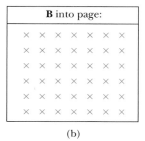

Figure 29.6 (a) Magnetic field lines coming out of the paper are indicated by dots, representing the tips of arrows coming outward. (b) Magnetic field lines going into the paper are indicated by crosses, representing the feathers of arrows going inward.

Force on a segment of current-carrying wire in a uniform magnetic field

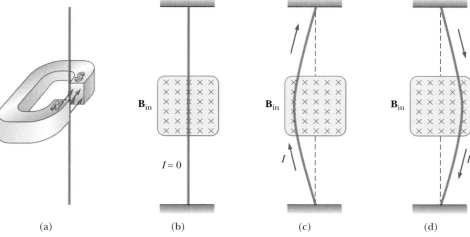

Figure 29.7 (a) A wire suspended vertically between the poles of a magnet. (b) The setup shown in part (a) as seen looking at the south pole of the magnet, so that the magnetic field (blue crosses) is directed into the page. When there is no current in the wire, it remains vertical. (c) When the current is upward, the wire deflects to the left. (d) When the current is downward, the wire deflects to the right.

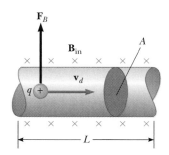

Figure 29.8 A segment of a current-carrying wire in a magnetic field $\mathbf{B}$. The magnetic force exerted on each charge making up the current is $q\mathbf{v}_d \times \mathbf{B}$ and the net force on the segment of length L is $I\mathbf{L} \times \mathbf{B}$.

Figure 29.9 A wire segment of arbitrary shape carrying a current I in a magnetic field **B** experiences a magnetic force. The magnetic force on any segment $d\mathbf{s}$ is $I\,d\mathbf{s} \times \mathbf{B}$ and is directed out of the page. You should use the right-hand rule to confirm this force direction.

Now consider an arbitrarily shaped wire segment of uniform cross section in a magnetic field, as shown in Figure 29.9. It follows from Equation 29.3 that the magnetic force exerted on a small segment of vector length $d\mathbf{s}$ in the presence of a field **B** is

$$d\mathbf{F}_B = I\,d\mathbf{s} \times \mathbf{B} \qquad (29.4)$$

where $d\mathbf{F}_B$ is directed out of the page for the directions of **B** and $d\mathbf{s}$ in Figure 29.9. We can consider Equation 29.4 as an alternative definition of **B**. That is, we can define the magnetic field **B** in terms of a measurable force exerted on a current element, where the force is a maximum when **B** is perpendicular to the element and zero when **B** is parallel to the element.

To calculate the total force $\mathbf{F}_B$ acting on the wire shown in Figure 29.9, we integrate Equation 29.4 over the length of the wire:

$$\mathbf{F}_B = I \int_a^b d\mathbf{s} \times \mathbf{B} \qquad (29.5)$$

where a and b represent the end points of the wire. When this integration is carried out, the magnitude of the magnetic field and the direction the field makes with the vector $d\mathbf{s}$ may differ at different points.

We now treat two interesting special cases involving Equation 29.5. In both cases, the magnetic field is assumed to be uniform in magnitude and direction.

Case 1. A curved wire carries a current I and is located in a uniform magnetic field **B**, as shown in Figure 29.10a. Because the field is uniform, we can take **B** outside the integral in Equation 29.5, and we obtain

$$\mathbf{F}_B = I \left(\int_a^b d\mathbf{s} \right) \times \mathbf{B} \qquad (29.6)$$

But the quantity $\int_a^b d\mathbf{s}$ represents the *vector sum* of all the length elements from a to b. From the law of vector addition, the sum equals the vector $\mathbf{L}'$, directed from a to b. Therefore, Equation 29.6 reduces to

$$\mathbf{F}_B = I\mathbf{L}' \times \mathbf{B} \qquad (29.7)$$

From this we conclude that **the magnetic force on a curved current-carrying wire in a uniform magnetic field is equal to that on a straight wire connecting the end points and carrying the same current.**

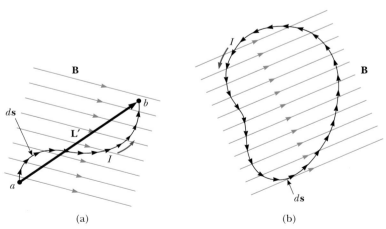

(a)   (b)

Figure 29.10 (a) A curved wire carrying a current I in a uniform magnetic field. The total magnetic force acting on the wire is equivalent to the force on a straight wire of length L' running between the ends of the curved wire. (b) A current-carrying loop of arbitrary shape in a uniform magnetic field. The net magnetic force on the loop is zero.

Case 2. An arbitrarily shaped closed loop carrying a current I is placed in a uniform magnetic field, as shown in Figure 29.10b. We can again express the magnetic force acting on the loop in the form of Equation 29.6, but this time we must take the vector sum of the length elements $d\mathbf{s}$ over the entire loop:

$$\mathbf{F}_B = I \left(\oint d\mathbf{s} \right) \times \mathbf{B}$$

Because the set of length elements forms a closed polygon, the vector sum must be zero. This follows from the procedure for adding vectors by the graphical method. Because $\oint d\mathbf{s} = 0$, we conclude that $\mathbf{F}_B = 0$; that is, **the net magnetic force acting on any closed current loop in a uniform magnetic field is zero.**

Quick Quiz 29.4 The four wires shown in Figure 29.11 all carry the same current from point A to point B through the same magnetic field. In all four parts of the figure, the points A and B are 10 cm apart. Rank the wires according to the magnitude of the magnetic force exerted on them, from greatest to least.

Figure 29.11 (Quick Quiz 29.4) Which wire experiences the greatest magnetic force?

Quick Quiz 29.5 A wire carries current in the plane of this paper toward the top of the page. The wire experiences a magnetic force toward the right edge of the page. The direction of the magnetic field causing this force is (a) in the plane of the page and toward the left edge, (b) in the plane of the page and toward the bottom edge, (c) upward out of the page, (d) downward into the page.

Example 29.2   Force on a Semicircular Conductor

A wire bent into a semicircle of radius R forms a closed circuit and carries a current I. The wire lies in the xy plane, and a uniform magnetic field is directed along the positive y axis, as shown in Figure 29.12. Find the magnitude and direction of the magnetic force acting

on the straight portion of the wire and on the curved portion.

Solution The magnetic force $\mathbf{F}_1$ acting on the straight portion has a magnitude $F_1 = ILB = 2IRB$ because $L = 2R$ and

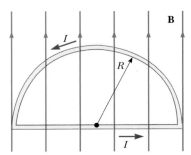

Figure 29.12 (Example 29.2) The net magnetic force acting on a closed current loop in a uniform magnetic field is zero. In the setup shown here, the magnetic force on the straight portion of the loop is $2IRB$ and directed out of the page, and the magnetic force on the curved portion is $2IRB$ directed into the page.

the wire is oriented perpendicular to **B**. The direction of $\mathbf{F}_1$ is out of the page based on the right-hand rule for the cross product **L × B**.

To find the magnetic force $\mathbf{F}_2$ acting on the curved part, we use the results of Case 1. The magnetic force on the curved portion is the same as that on a straight wire of length $2R$ carrying current I to the left. Thus, $F_2 = ILB = 2IRB$. The direction of $\mathbf{F}_2$ is into the page based on the right-hand rule for the cross product **L × B.**

Because the wire lies in the xy plane, the two forces on the loop can be expressed as

$$\mathbf{F}_1 = 2IRB\hat{\mathbf{k}}$$

$$\mathbf{F}_2 = -2IRB\hat{\mathbf{k}}$$

The net magnetic force on the loop is

$$\sum \mathbf{F} = \mathbf{F}_1 + \mathbf{F}_2 = 2IRB\hat{\mathbf{k}} - 2IRB\hat{\mathbf{k}} = 0$$

Note that this is consistent with Case 2, because the wire forms a closed loop in a uniform magnetic field.

29.3 Torque on a Current Loop in a Uniform Magnetic Field

In the preceding section, we showed how a magnetic force is exerted on a current-carrying conductor placed in a magnetic field. With this as a starting point, we now show that a torque is exerted on a current loop placed in a magnetic field. The results of this analysis will be of great value when we discuss motors in Chapter 31.

Consider a rectangular loop carrying a current I in the presence of a uniform magnetic field directed parallel to the plane of the loop, as shown in Figure 29.13a. No magnetic forces act on sides ① and ③ because these wires are parallel to the field; hence, **L × B** = 0 for these sides. However, magnetic forces do act on sides ② and ④ because these sides are oriented perpendicular to the field. The magnitude of these forces is, from Equation 29.3,

$$F_2 = F_4 = IaB$$

The direction of $\mathbf{F}_2$, the magnetic force exerted on wire ②, is out of the page in the view shown in Figure 29.13a, and that of $\mathbf{F}_4$, the magnetic force exerted on wire ④, is into the page in the same view. If we view the loop from side ③ and sight along sides ② and ④, we see the view shown in Figure 29.13b, and the two magnetic forces $\mathbf{F}_2$ and $\mathbf{F}_4$ are directed as shown. Note that the two forces point in opposite directions but are *not* directed along the same line of action. If the loop is pivoted so that it can rotate about point O, these two forces produce about O a torque that rotates the loop clockwise. The magnitude of this torque τ_{max} is

$$\tau_{\text{max}} = F_2 \frac{b}{2} + F_4 \frac{b}{2} = (IaB)\frac{b}{2} + (IaB)\frac{b}{2} = IabB$$

where the moment arm about O is $b/2$ for each force. Because the area enclosed by the loop is $A = ab$, we can express the maximum torque as

$$\tau_{\text{max}} = IAB \tag{29.8}$$

This maximum-torque result is valid only when the magnetic field is parallel to the plane of the loop. The sense of the rotation is clockwise when viewed from side ③, as indicated in Figure 29.13b. If the current direction were reversed, the force directions would also reverse, and the rotational tendency would be counterclockwise.

(a)

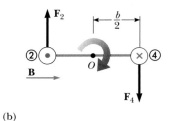

(b)

Figure 29.13 (a) Overhead view of a rectangular current loop in a uniform magnetic field. No magnetic forces are acting on sides ① and ③ because these sides are parallel to **B**. Forces are acting on sides ② and ④, however. (b) Edge view of the loop sighting down sides ② and ④ shows that the magnetic forces $\mathbf{F}_2$ and $\mathbf{F}_4$ exerted on these sides create a torque that tends to twist the loop clockwise. The purple dot in the left circle represents current in wire ② coming toward you; the purple cross in the right circle represents current in wire ④ moving away from you.

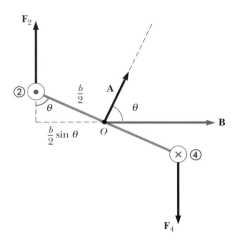

Active Figure 29.14 An end view of the loop in Figure 29.13b rotated through an angle with respect to the magnetic field. If **B** is at an angle θ with respect to vector **A**, which is perpendicular to the plane of the loop, the torque is $IAB \sin \theta$ where the magnitude of **A** is A, the area of the loop.

At the Active Figures link at http://www.pse6.com, you can choose the current in the loop, the magnetic field, and the initial orientation of the loop and observe the subsequent motion.

Now suppose that the uniform magnetic field makes an angle $\theta < 90°$ with a line perpendicular to the plane of the loop, as in Figure 29.14. For convenience, we assume that **B** is perpendicular to sides ② and ④. In this case, the magnetic forces **F**$_1$ and **F**$_3$ exerted on sides ① and ③ cancel each other and produce no torque because they pass through a common origin. However, the magnetic forces **F**$_2$ and **F**$_4$ acting on sides ② and ④ produce a torque about *any point*. Referring to the end view shown in Figure 29.14, we note that the moment arm of **F**$_2$ about the point O is equal to $(b/2) \sin \theta$. Likewise, the moment arm of **F**$_4$ about O is also $(b/2) \sin \theta$. Because $F_2 = F_4 = IaB$, the magnitude of the net torque about O is

$$\tau = F_2 \frac{b}{2} \sin \theta + F_4 \frac{b}{2} \sin \theta$$

$$= IaB \left(\frac{b}{2} \sin \theta \right) + IaB \left(\frac{b}{2} \sin \theta \right) = IabB \sin \theta$$

$$= IAB \sin \theta$$

where $A = ab$ is the area of the loop. This result shows that the torque has its maximum value IAB when the field is perpendicular to the normal to the plane of the loop ($\theta = 90°$), as we saw when discussing Figure 29.13, and is zero when the field is parallel to the normal to the plane of the loop ($\theta = 0$).

A convenient expression for the torque exerted on a loop placed in a uniform magnetic field **B** is

$$\tau = I\mathbf{A} \times \mathbf{B} \tag{29.9}$$

where **A**, the vector shown in Figure 29.14, is perpendicular to the plane of the loop and has a magnitude equal to the area of the loop. We determine the direction of **A** using the right-hand rule described in Figure 29.15. When you curl the fingers of your right hand in the direction of the current in the loop, your thumb points in the direction of **A**. As we see in Figure 29.14, the loop tends to rotate in the direction of decreasing values of θ (that is, such that the area vector **A** rotates toward the direction of the magnetic field).

The product $I\mathbf{A}$ is defined to be the **magnetic dipole moment** $\boldsymbol{\mu}$ (often simply called the "magnetic moment") of the loop:

$$\boldsymbol{\mu} = I\mathbf{A} \tag{29.10}$$

The SI unit of magnetic dipole moment is ampere-meter^2 (A $\cdot$ m^2). Using this definition, we can express the torque exerted on a current-carrying loop in a magnetic field **B** as

$$\tau = \boldsymbol{\mu} \times \mathbf{B} \tag{29.11}$$

Note that this result is analogous to Equation 26.18, $\tau = \mathbf{p} \times \mathbf{E}$, for the torque exerted on an electric dipole in the presence of an electric field **E**, where **p** is the electric dipole moment.

Figure 29.15 Right-hand rule for determining the direction of the vector **A**. The direction of the magnetic moment $\boldsymbol{\mu}$ is the same as the direction of **A**.

Torque on a current loop in a magnetic field

Magnetic dipole moment of a current loop

Torque on a magnetic moment in a magnetic field

Although we obtained the torque for a particular orientation of **B** with respect to the loop, the equation $\boldsymbol{\tau} = \boldsymbol{\mu} \times \mathbf{B}$ is valid for any orientation. Furthermore, although we derived the torque expression for a rectangular loop, the result is valid for a loop of any shape.

If a coil consists of N turns of wire, each carrying the same current and enclosing the same area, the total magnetic dipole moment of the coil is N times the magnetic dipole moment for one turn. The torque on an N-turn coil is N times that on a one-turn coil. Thus, we write $\boldsymbol{\tau} = N\boldsymbol{\mu}_{\text{loop}} \times \mathbf{B} = \boldsymbol{\mu}_{\text{coil}} \times \mathbf{B}$.

In Section 26.6, we found that the potential energy of a system of an electric dipole in an electric field is given by $U = -\mathbf{p} \cdot \mathbf{E}$. This energy depends on the orientation of the dipole in the electric field. Likewise, the potential energy of a system of a magnetic dipole in a magnetic field depends on the orientation of the dipole in the magnetic field and is given by

Potential energy of a system of a magnetic moment in a magnetic field

$$U = -\boldsymbol{\mu} \cdot \mathbf{B} \tag{29.12}$$

From this expression, we see that the system has its lowest energy $U_{\text{min}} = -\mu B$ when $\boldsymbol{\mu}$ points in the same direction as **B**. The system has its highest energy $U_{\text{max}} = +\mu B$ when $\boldsymbol{\mu}$ points in the direction opposite **B**.

Quick Quiz 29.6 Rank the magnitudes of the torques acting on the rectangular loops shown edge-on in Figure 29.16, from highest to lowest. All loops are identical and carry the same current.

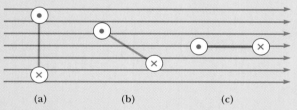

(a) (b) (c)

Figure 29.16 (Quick Quiz 29.6) Which current loop (seen edge-on) experiences the greatest torque? (Quick Quiz 29.7) Which current loop (seen edge-on) experiences the greatest net force?

Quick Quiz 29.7 Rank the magnitudes of the net forces acting on the rectangular loops shown in Figure 29.16, from highest to lowest. All loops are identical and carry the same current.

Example 29.3 The Magnetic Dipole Moment of a Coil

A rectangular coil of dimensions 5.40 cm $\times$ 8.50 cm consists of 25 turns of wire and carries a current of 15.0 mA. A 0.350-T magnetic field is applied parallel to the plane of the loop.

(A) Calculate the magnitude of its magnetic dipole moment.

Solution Because the coil has 25 turns, we modify Equation 29.10 to obtain

$$\mu_{\text{coil}} = NIA = (25)(15.0 \times 10^{-3}\,\text{A})(0.054\,0\,\text{m})(0.085\,0\,\text{m})$$

$$= \boxed{1.72 \times 10^{-3}\,\text{A} \cdot \text{m}^2}$$

(B) What is the magnitude of the torque acting on the loop?

Solution Because **B** is perpendicular to $\boldsymbol{\mu}_{\text{coil}}$, Equation 29.11 gives

$$\tau = \mu_{\text{coil}}B = (1.72 \times 10^{-3}\,\text{A} \cdot \text{m}^2)(0.350\,\text{T})$$

$$= \boxed{6.02 \times 10^{-4}\,\text{N} \cdot \text{m}}$$

Example 29.4   Satellite Attitude Control

Many satellites use coils called *torquers* to adjust their orientation. These devices interact with the Earth's magnetic field to create a torque on the spacecraft in the *x*, *y*, or *z* direction. The major advantage of this type of attitude-control system is that it uses solar-generated electricity and so does not consume any thruster fuel.

If a typical device has a magnetic dipole moment of $250\ \text{A} \cdot \text{m}^2$, what is the maximum torque applied to a satellite when its torquer is turned on at an altitude where the magnitude of the Earth's magnetic field is $3.0 \times 10^{-5}\ \text{T}$?

Solution We once again apply Equation 29.11, recognizing that the maximum torque is obtained when the magnetic dipole moment of the torquer is perpendicular to the Earth's magnetic field:

$$\tau_{\text{max}} = \mu B = (250\ \text{A} \cdot \text{m}^2)(3.0 \times 10^{-5}\ \text{T})$$

$$= \boxed{7.5 \times 10^{-3}\ \text{N} \cdot \text{m}}$$

Example 29.5   The D'Arsonval Galvanometer

An end view of a D'Arsonval galvanometer (see Section 28.5) is shown in Figure 29.17. When the turns of wire making up the coil carry a current, the magnetic field created by the magnet exerts on the coil a torque that turns it (along with its attached pointer) against the spring. Show that the angle of deflection of the pointer is directly proportional to the current in the coil.

Solution We can use Equation 29.11 to find the torque τ_m that the magnetic field exerts on the coil. If we assume that the magnetic field through the coil is perpendicular to the normal to the plane of the coil, Equation 29.11 becomes

$$\tau_m = \mu B$$

(This is a reasonable assumption because the circular cross section of the magnet ensures radial magnetic field lines.) This magnetic torque is opposed by the torque due to the spring, which is given by the rotational version of Hooke's law, $\tau_s = -\kappa\phi$, where κ is the torsional spring constant and ϕ is the angle through which the spring turns. Because the coil does not have an angular acceleration when the pointer is at rest, the sum of these torques must be zero:

$$(1) \qquad \tau_m + \tau_s = \mu B - \kappa\phi = 0$$

Equation 29.10 allows us to relate the magnetic moment of the *N* turns of wire to the current through them:

$$\mu = NIA$$

We can substitute this expression for μ in Equation (1) to obtain

$$(NIA)B - \kappa\phi = 0$$

$$\phi = \frac{NAB}{\kappa} I$$

Thus, the angle of deflection of the pointer is directly proportional to the current in the loop. The factor NAB/κ tells us that deflection also depends on the design of the meter.

Figure 29.17 (Example 29.5) Structure of a moving-coil galvanometer.

29.4   Motion of a Charged Particle in a Uniform Magnetic Field

In Section 29.1 we found that the magnetic force acting on a charged particle moving in a magnetic field is perpendicular to the velocity of the particle and that consequently the work done by the magnetic force on the particle is zero. Now consider the special case of a positively charged particle moving in a uniform magnetic field with the initial velocity vector of the particle perpendicular to the field. Let us assume that the direction of the magnetic field is into the page, as in Figure 29.18. As the particle changes the direction of its velocity in response to the magnetic force, the magnetic force remains perpendicular to the velocity. As we found in Section 6.1, if the force is always perpendicular to the velocity, the path of

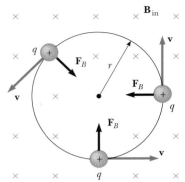

Active Figure 29.18 When the velocity of a charged particle is perpendicular to a uniform magnetic field, the particle moves in a circular path in a plane perpendicular to **B**. The magnetic force $\mathbf{F}_B$ acting on the charge is always directed toward the center of the circle.

At the Active Figures link at http://www.pse6.com, you can adjust the mass, speed, and charge of the particle and the magnitude of the magnetic field to observe the resulting circular motion.

the particle is a circle! Figure 29.18 shows the particle moving in a circle in a plane perpendicular to the magnetic field.

The particle moves in a circle because the magnetic force $\mathbf{F}_B$ is perpendicular to **v** and **B** and has a constant magnitude qvB. As Figure 29.18 illustrates, the rotation is counterclockwise for a positive charge. If q were negative, the rotation would be clockwise. We can use Equation 6.1 to equate this magnetic force to the product of the particle mass and the centripetal acceleration:

$$\sum F = ma_c$$

$$F_B = qvB = \frac{mv^2}{r}$$

$$r = \frac{mv}{qB} \tag{29.13}$$

That is, the radius of the path is proportional to the linear momentum mv of the particle and inversely proportional to the magnitude of the charge on the particle and to the magnitude of the magnetic field. The angular speed of the particle (from Eq. 10.10) is

$$\omega = \frac{v}{r} = \frac{qB}{m} \tag{29.14}$$

The period of the motion (the time interval the particle requires to complete one revolution) is equal to the circumference of the circle divided by the linear speed of the particle:

$$T = \frac{2\pi r}{v} = \frac{2\pi}{\omega} = \frac{2\pi m}{qB} \tag{29.15}$$

These results show that the angular speed of the particle and the period of the circular motion do not depend on the linear speed of the particle or on the radius of the orbit. The angular speed ω is often referred to as the **cyclotron frequency** because charged particles circulate at this angular frequency in the type of accelerator called a *cyclotron*, which is discussed in Section 29.5.

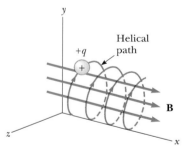

Active Figure 29.19 A charged particle having a velocity vector that has a component parallel to a uniform magnetic field moves in a helical path.

At the Active Figures link at http://www.pse6.com, you can adjust the x component of the velocity of the particle and observe the resulting helical motion.

If a charged particle moves in a uniform magnetic field with its velocity at some arbitrary angle with respect to **B**, its path is a helix. For example, if the field is directed in the x direction, as shown in Figure 29.19, there is no component of force in the x direction. As a result, $a_x = 0$, and the x component of velocity remains constant. However, the magnetic force $q\mathbf{v} \times \mathbf{B}$ causes the components v_y and v_z to change in time, and the resulting motion is a helix whose axis is parallel to the magnetic field. The projection of the path onto the yz plane (viewed along the x axis) is a circle. (The projections of the path onto the xy and xz planes are sinusoids!) Equations 29.13 to 29.15 still apply provided that v is replaced by $v_\perp = \sqrt{v_y^2 + v_z^2}$.

> **Quick Quiz 29.8** A charged particle is moving perpendicular to a magnetic field in a circle with a radius r. An identical particle enters the field, with **v** perpendicular to **B**, but with a higher speed v than the first particle. Compared to the radius of the circle for the first particle, the radius of the circle for the second particle is (a) smaller (b) larger (c) equal in size.

> **Quick Quiz 29.9** A charged particle is moving perpendicular to a magnetic field in a circle with a radius r. The magnitude of the magnetic field is increased. Compared to the initial radius of the circular path, the radius of the new path is (a) smaller (b) larger (c) equal in size.

Example 29.6 A Proton Moving Perpendicular to a Uniform Magnetic Field

A proton is moving in a circular orbit of radius 14 cm in a uniform 0.35-T magnetic field perpendicular to the velocity of the proton. Find the linear speed of the proton.

Solution From Equation 29.13, we have

$$v = \frac{qBr}{m_p} = \frac{(1.60 \times 10^{-19}\,\text{C})\,(0.35\,\text{T})\,(0.14\,\text{m})}{1.67 \times 10^{-27}\,\text{kg}}$$

$$= \boxed{4.7 \times 10^6\,\text{m/s}}$$

What If? What if an electron, rather than a proton, moves in a direction perpendicular to the same magnetic field with this same linear speed? Will the radius of its orbit be different?

Answer An electron has a much smaller mass than a proton, so the magnetic force should be able to change its velocity much easier than for the proton. Thus, we should expect the radius to be smaller. Looking at Equation 29.13, we see that r is proportional to m with q, B, and v the same for the electron as for the proton. Consequently, the radius will be smaller by the same factor as the ratio of masses m_e/m_p.

Example 29.7 Bending an Electron Beam   `Interactive`

In an experiment designed to measure the magnitude of a uniform magnetic field, electrons are accelerated from rest through a potential difference of 350 V. The electrons travel along a curved path because of the magnetic force exerted on them, and the radius of the path is measured to be 7.5 cm. (Fig. 29.20 shows such a curved beam of electrons.) If the magnetic field is perpendicular to the beam,

(A) what is the magnitude of the field?

Solution Conceptualize the circular motion of the electrons with the help of Figures 29.18 and 29.20. We categorize this problem as one involving both uniform circular motion and a magnetic force. Looking at Equation 29.13, we see that we need the speed v of the electron if we are to find the magnetic field magnitude, and v is not given. Consequently, we must find the speed of the electron based on the potential difference through which it is accelerated. Therefore, we also categorize this as a problem in conservation of mechanical energy for an isolated system. To begin analyzing the problem, we find the electron speed. For the isolated electron–electric field system, the loss of potential energy as the electron moves through the 350-V potential difference appears as an increase in the kinetic energy of the electron. Because $K_i = 0$ and $K_f = \frac{1}{2}m_e v^2$, we have

$$\Delta K + \Delta U = 0 \longrightarrow \tfrac{1}{2}m_e v^2 + (-e)\,\Delta V = 0$$

$$v = \sqrt{\frac{2e\,\Delta V}{m_e}} = \sqrt{\frac{2(1.60 \times 10^{-19}\,\text{C})\,(350\,\text{V})}{9.11 \times 10^{-31}\,\text{kg}}}$$

$$= 1.11 \times 10^7\,\text{m/s}$$

Now, using Equation 29.13, we find

$$B = \frac{m_e v}{er} = \frac{(9.11 \times 10^{-31}\,\text{kg})\,(1.11 \times 10^7\,\text{m/s})}{(1.60 \times 10^{-19}\,\text{C})\,(0.075\,\text{m})}$$

$$= \boxed{8.4 \times 10^{-4}\,\text{T}}$$

(B) What is the angular speed of the electrons?

Solution Using Equation 29.14, we find that

Henry Leap and Jim Lehman

Figure 29.20 (Example 29.7) The bending of an electron beam in a magnetic field.

$$\omega = \frac{v}{r} = \frac{1.11 \times 10^7\,\text{m/s}}{0.075\,\text{m}} = \boxed{1.5 \times 10^8\,\text{rad/s}}$$

To finalize this problem, note that the angular speed can be represented as $\omega = (1.5 \times 10^8\,\text{rad/s})(1\,\text{rev}/2\pi\,\text{rad}) = 2.4 \times 10^7\,\text{rev/s}$. The electrons travel around the circle 24 million times per second! This is consistent with the very high speed that we found in part (A).

What If? What if a sudden voltage surge causes the accelerating voltage to increase to 400 V? How does this affect the angular speed of the electrons, assuming that the magnetic field remains constant?

Answer The increase in accelerating voltage ΔV will cause the electrons to enter the magnetic field with a higher speed v. This will cause them to travel in a circle with a larger radius r. The angular speed is the ratio of v to r. Both v and r increase by the same factor, so that the effects cancel and the angular speed remains the same. Equation 29.14 is an expression for the cyclotron frequency, which is the same as the angular speed of the electrons. The cyclotron frequency depends only on the charge q, the magnetic field B, and the mass m_e, none of which have changed. Thus, the voltage surge has no effect on the angular speed. (However, in reality, the voltage surge may also increase the magnetic field if the magnetic field is powered by the same source as the accelerating voltage. In this case, the angular speed will increase according to Equation 29.14.)

 At the Interactive Worked Example link at http://www.pse6.com, you can investigate the relationship between the radius of the circular path of the electrons and the magnetic field.

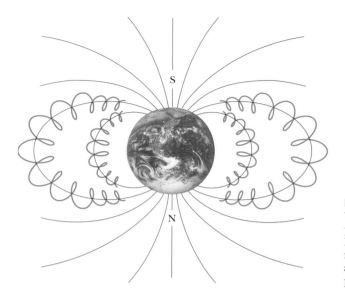

Figure 29.22 The Van Allen belts are made up of charged particles trapped by the Earth's nonuniform magnetic field. The magnetic field lines are in blue and the particle paths in red.

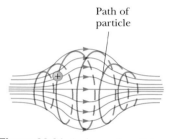

Path of particle

Figure 29.21 A charged particle moving in a nonuniform magnetic field (a magnetic bottle) spirals about the field and oscillates between the end points. The magnetic force exerted on the particle near either end of the bottle has a component that causes the particle to spiral back toward the center.

When charged particles move in a nonuniform magnetic field, the motion is complex. For example, in a magnetic field that is strong at the ends and weak in the middle, such as that shown in Figure 29.21, the particles can oscillate back and forth between two positions. A charged particle starting at one end spirals along the field lines until it reaches the other end, where it reverses its path and spirals back. This configuration is known as a *magnetic bottle* because charged particles can be trapped within it. The magnetic bottle has been used to confine a *plasma*, a gas consisting of ions and electrons. Such a plasma-confinement scheme could fulfill a crucial role in the control of nuclear fusion, a process that could supply us with an almost endless source of energy. Unfortunately, the magnetic bottle has its problems. If a large number of particles are trapped, collisions between them cause the particles to eventually leak from the system.

The Van Allen radiation belts consist of charged particles (mostly electrons and protons) surrounding the Earth in doughnut-shaped regions (Fig. 29.22). The particles, trapped by the Earth's nonuniform magnetic field, spiral around the field lines from pole to pole, covering the distance in just a few seconds. These particles originate mainly from the Sun, but some come from stars and other heavenly objects. For this reason, the particles are called *cosmic rays*. Most cosmic rays are deflected by the Earth's magnetic field and never reach the atmosphere. However, some of the particles become trapped; it is these particles that make up the Van Allen belts. When the particles are located over the poles, they sometimes collide with atoms in the atmosphere, causing the atoms to emit visible light. Such collisions are the origin of the beautiful Aurora Borealis, or Northern Lights, in the northern hemisphere and the Aurora Australis in the southern hemisphere. Auroras are usually confined to the polar regions because the Van Allen belts are nearest the Earth's surface there. Occasionally, though, solar activity causes larger numbers of charged particles to enter the belts and significantly distort the normal magnetic field lines associated with the Earth. In these situations an aurora can sometimes be seen at lower latitudes.

29.5 Applications Involving Charged Particles Moving in a Magnetic Field

A charge moving with a velocity **v** in the presence of both an electric field **E** and a magnetic field **B** experiences both an electric force $q\mathbf{E}$ and a magnetic force $q\mathbf{v} \times \mathbf{B}$. The total force (called the Lorentz force) acting on the charge is

Lorentz force

$$\mathbf{F} = q\mathbf{E} + q\mathbf{v} \times \mathbf{B} \qquad (29.16)$$

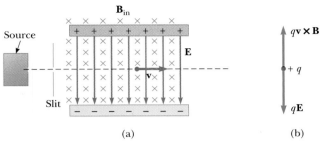

Active Figure 29.23 (a) A velocity selector. When a positively charged particle is moving with velocity **v** in the presence of a magnetic field directed into the page and an electric field directed downward, it experiences a downward electric force $q\mathbf{E}$ and an upward magnetic force $q\mathbf{v} \times \mathbf{B}$. (b) When these forces balance, the particle moves in a horizontal line through the fields.

At the Active Figures link at http://www.pse6.com, you can adjust the electric and magnetic fields to try to achieve straight line motion for the charge.

Velocity Selector

In many experiments involving moving charged particles, it is important that the particles all move with essentially the same velocity. This can be achieved by applying a combination of an electric field and a magnetic field oriented as shown in Figure 29.23. A uniform electric field is directed vertically downward (in the plane of the page in Fig. 29.23a), and a uniform magnetic field is applied in the direction perpendicular to the electric field (into the page in Fig. 29.23a). If q is positive and the velocity **v** is to the right, the magnetic force $q\mathbf{v} \times \mathbf{B}$ is upward and the electric force $q\mathbf{E}$ is downward. When the magnitudes of the two fields are chosen so that $qE = qvB$, the particle moves in a straight horizontal line through the region of the fields. From the expression $qE = qvB$, we find that

$$v = \frac{E}{B} \tag{29.17}$$

Only those particles having speed v pass undeflected through the mutually perpendicular electric and magnetic fields. The magnetic force exerted on particles moving at speeds greater than this is stronger than the electric force, and the particles are deflected upward. Those moving at speeds less than this are deflected downward.

The Mass Spectrometer

A **mass spectrometer** separates ions according to their mass-to-charge ratio. In one version of this device, known as the *Bainbridge mass spectrometer,* a beam of ions first passes through a velocity selector and then enters a second uniform magnetic field $\mathbf{B}_0$ that has the same direction as the magnetic field in the selector (Fig. 29.24). Upon entering the second magnetic field, the ions move in a semicircle of radius r before striking a detector array at P. If the ions are positively charged, the beam deflects upward, as Figure 29.24 shows. If the ions are negatively charged, the beam deflects

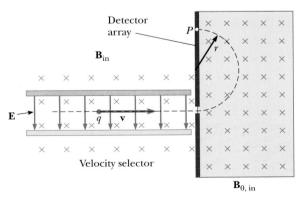

Active Figure 29.24 A mass spectrometer. Positively charged particles are sent first through a velocity selector and then into a region where the magnetic field $\mathbf{B}_0$ causes the particles to move in a semicircular path and strike a detector array at P.

At the Active Figures link at http://www.pse6.com, you can predict where particles will strike the detector array.

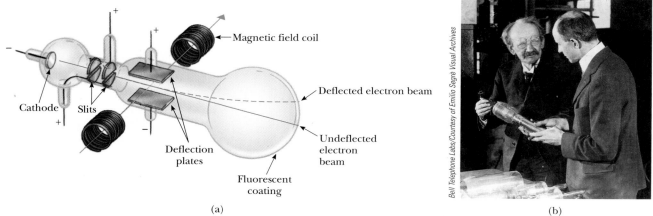

(a)

(b)

Figure 29.25 (a) Thomson's apparatus for measuring e/m_e. Electrons are accelerated from the cathode, pass through two slits, and are deflected by both an electric field and a magnetic field (directed perpendicular to the electric field). The beam of electrons then strikes a fluorescent screen. (b) J. J. Thomson *(left)* in the Cavendish Laboratory, University of Cambridge. The man on the right, Frank Baldwin Jewett, is a distant relative of John W. Jewett, Jr., co-author of this text.

downward. From Equation 29.13, we can express the ratio m/q as

$$\frac{m}{q} = \frac{rB_0}{v}$$

Using Equation 29.17, we find that

$$\frac{m}{q} = \frac{rB_0 B}{E} \qquad (29.18)$$

Therefore, we can determine m/q by measuring the radius of curvature and knowing the field magnitudes B, B_0, and E. In practice, one usually measures the masses of various isotopes of a given ion, with the ions all carrying the same charge q. In this way, the mass ratios can be determined even if q is unknown.

A variation of this technique was used by J. J. Thomson (1856–1940) in 1897 to measure the ratio e/m_e for electrons. Figure 29.25a shows the basic apparatus he used. Electrons are accelerated from the cathode and pass through two slits. They then drift into a region of perpendicular electric and magnetic fields. The magnitudes of the two fields are first adjusted to produce an undeflected beam. When the magnetic field is turned off, the electric field produces a measurable beam deflection that is recorded on the fluorescent screen. From the size of the deflection and the measured values of E and B, the charge-to-mass ratio can be determined. The results of this crucial experiment represent the discovery of the electron as a fundamental particle of nature.

Quick Quiz 29.10 Three types of particles enter a mass spectrometer like the one shown in Figure 29.24. Figure 29.26 shows where the particles strike the detector array. Rank the particles that arrive at a, b, and c by speed and m/q ratio.

Figure 29.26 (Quick Quiz 29.10) Which particles have the highest speed and which have the highest ratio of m/q?

The Cyclotron

A **cyclotron** is a device that can accelerate charged particles to very high speeds. The energetic particles produced are used to bombard atomic nuclei and thereby produce nuclear reactions of interest to researchers. A number of hospitals use cyclotron facilities to produce radioactive substances for diagnosis and treatment.

Both electric and magnetic forces have a key role in the operation of a cyclotron. A schematic drawing of a cyclotron is shown in Figure 29.27a. The charges move inside two semicircular containers D₁ and D₂, referred to as *dees*, because of their shape like the letter D. A high-frequency alternating potential difference is applied to the dees, and a uniform magnetic field is directed perpendicular to them. A positive ion released at P near the center of the magnet in one dee moves in a semicircular path (indicated by the dashed red line in the drawing) and arrives back at the gap in a time interval $T/2$, where T is the time interval needed to make one complete trip around the two dees, given by Equation 29.15. The frequency of the applied potential difference is adjusted so that the polarity of the dees is reversed in the same time interval during which the ion travels around one dee. If the applied potential difference is adjusted such that D_2 is at a lower electric potential than D_1 by an amount ΔV, the ion accelerates across the gap to D_2 and its kinetic energy increases by an amount $q\,\Delta V$. It then moves around D_2 in a semicircular path of greater radius (because its speed has increased). After a time interval $T/2$, it again arrives at the gap between the dees. By this time, the polarity across the dees has again been reversed, and the ion is given another "kick" across the gap. The motion continues so that for each half-circle trip around one dee, the ion gains additional kinetic energy equal to $q\,\Delta V$. When the radius of its path is nearly that of the dees, the energetic ion leaves the system through the exit slit. Note that the operation of the cyclotron is based on the fact that T is independent of the speed of the ion and of the radius of the circular path (Eq. 29.15).

We can obtain an expression for the kinetic energy of the ion when it exits the cyclotron in terms of the radius R of the dees. From Equation 29.13 we know that $v = qBR/m$. Hence, the kinetic energy is

$$K = \tfrac{1}{2}mv^2 = \frac{q^2 B^2 R^2}{2m} \tag{29.19}$$

When the energy of the ions in a cyclotron exceeds about 20 MeV, relativistic effects come into play. (Such effects are discussed in Chapter 39.) We observe that T increases and that the moving ions do not remain in phase with the applied potential

▲ PITFALL PREVENTION

29.1 The Cyclotron Is Not State-of-the-Art Technology

The cyclotron is important historically because it was the first particle accelerator to achieve very high particle speeds. Cyclotrons are still in use in medical applications, but most accelerators currently in research use are not cyclotrons. Research accelerators work on a different principle and are generally called *synchrotrons*.

(a)

Courtesy of Lawrence Berkeley Laboratory/University of California

(b)

Figure 29.27 (a) A cyclotron consists of an ion source at P, two dees D₁ and D₂ across which an alternating potential difference is applied, and a uniform magnetic field. (The south pole of the magnet is not shown.) The red dashed curved lines represent the path of the particles. (b) The first cyclotron, invented by E. O. Lawrence and M. S. Livingston in 1934.

difference. Some accelerators overcome this problem by modifying the period of the applied potential difference so that it remains in phase with the moving ions.

29.6 The Hall Effect

When a current-carrying conductor is placed in a magnetic field, a potential difference is generated in a direction perpendicular to both the current and the magnetic field. This phenomenon, first observed by Edwin Hall (1855–1938) in 1879, is known as the *Hall effect*. It arises from the deflection of charge carriers to one side of the conductor as a result of the magnetic force they experience. The Hall effect gives information regarding the sign of the charge carriers and their density; it can also be used to measure the magnitude of magnetic fields.

The arrangement for observing the Hall effect consists of a flat conductor carrying a current I in the x direction, as shown in Figure 29.28. A uniform magnetic field $\mathbf{B}$ is applied in the y direction. If the charge carriers are electrons moving in the negative x direction with a drift velocity $\mathbf{v}_d$, they experience an upward magnetic force $\mathbf{F}_B = q\mathbf{v}_d \times \mathbf{B}$, are deflected upward, and accumulate at the upper edge of the flat conductor, leaving an excess of positive charge at the lower edge (Fig. 29.29a). This accumulation of charge at the edges establishes an electric field in the conductor and increases until the electric force on carriers remaining in the bulk of the conductor balances the magnetic force acting on the carriers. When this equilibrium condition is reached, the electrons are no longer deflected upward. A sensitive voltmeter or potentiometer connected across the sample, as shown in Figure 29.29, can measure the potential difference—known as the **Hall voltage** ΔV_{H}—generated across the conductor.

If the charge carriers are positive and hence move in the positive x direction (for rightward current), as shown in Figures 29.28 and 29.29b, they also experience an upward magnetic force $q\mathbf{v}_d \times \mathbf{B}$. This produces a buildup of positive charge on the upper edge and leaves an excess of negative charge on the lower edge. Hence, the sign of the Hall voltage generated in the sample is opposite the sign of the Hall voltage resulting from the deflection of electrons. The sign of the charge carriers can therefore be determined from a measurement of the polarity of the Hall voltage.

In deriving an expression for the Hall voltage, we first note that the magnetic force exerted on the carriers has magnitude $qv_d B$. In equilibrium, this force is balanced by the electric force qE_{H}, where E_{H} is the magnitude of the electric field due to the charge separation (sometimes referred to as the *Hall field*). Therefore,

$$qv_d B = qE_{\mathrm{H}}$$
$$E_{\mathrm{H}} = v_d B$$

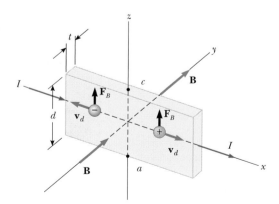

Figure 29.28 To observe the Hall effect, a magnetic field is applied to a current-carrying conductor. When I is in the x direction and $\mathbf{B}$ in the y direction, both positive and negative charge carriers are deflected upward in the magnetic field. The Hall voltage is measured between points a and c.

(a)   (b)

Figure 29.29 (a) When the charge carriers in a Hall-effect apparatus are negative, the upper edge of the conductor becomes negatively charged, and c is at a lower electric potential than a. (b) When the charge carriers are positive, the upper edge becomes positively charged, and c is at a higher potential than a. In either case, the charge carriers are no longer deflected when the edges become sufficiently charged that there is a balance on the charge carriers between the electrostatic force qE_H and the magnetic deflection force qvB.

If d is the width of the conductor, the Hall voltage is

$$\Delta V_H = E_H d = v_d B d \tag{29.20}$$

Thus, the measured Hall voltage gives a value for the drift speed of the charge carriers if d and B are known.

We can obtain the charge carrier density n by measuring the current in the sample. From Equation 27.4, we can express the drift speed as

$$v_d = \frac{I}{nqA} \tag{29.21}$$

where A is the cross-sectional area of the conductor. Substituting Equation 29.21 into Equation 29.20, we obtain

$$\Delta V_H = \frac{IBd}{nqA} \tag{29.22}$$

Because $A = td$, where t is the thickness of the conductor, we can also express Equation 29.22 as

$$\Delta V_H = \frac{IB}{nqt} = \frac{R_H IB}{t} \tag{29.23}$$ **The Hall voltage**

where $R_H = 1/nq$ is the **Hall coefficient.** This relationship shows that a properly calibrated conductor can be used to measure the magnitude of an unknown magnetic field.

Because all quantities in Equation 29.23 other than nq can be measured, a value for the Hall coefficient is readily obtainable. The sign and magnitude of R_H give the sign of the charge carriers and their number density. In most metals, the charge carriers are electrons, and the charge-carrier density determined from Hall-effect measurements is in good agreement with calculated values for such metals as lithium (Li), sodium (Na), copper (Cu), and silver (Ag), whose atoms each give up one electron to act as a current carrier. In this case, n is approximately equal to the number of conducting electrons per unit volume. However, this classical model is not valid for metals such as iron (Fe), bismuth (Bi), and cadmium (Cd) or for semiconductors. These discrepancies can be explained only by using a model based on the quantum nature of solids.

An interesting medical application related to the Hall effect is the electromagnetic blood flowmeter, first developed in the 1950s and continually improved since then. Imagine that we replace the conductor in Figure 29.29 with an artery carrying blood. The blood contains charged ions that experience electric and magnetic forces like the charge carriers in the conductor. The speed of flow of these ions can be related to the volume rate of flow of blood. Solving Equation 29.20 for the speed v_d of the ions in the blood, we obtain

$$v_d = \frac{\Delta V_H}{Bd}$$

Thus, by measuring the voltage across the artery, the diameter of the artery, and the applied magnetic field, the speed of the blood can be calculated.

Example 29.8   The Hall Effect for Copper

A rectangular copper strip 1.5 cm wide and 0.10 cm thick carries a current of 5.0 A. Find the Hall voltage for a 1.2-T magnetic field applied in a direction perpendicular to the strip.

Solution If we assume that one electron per atom is available for conduction, we can take the charge carrier density to be 8.49×10^{28} electrons/m^3 (see Example 27.1). Substituting this value and the given data into Equation 29.23 gives

$$\Delta V_H = \frac{IB}{nqt}$$

$$= \frac{(5.0 \text{ A})(1.2 \text{ T})}{(8.49 \times 10^{28} \text{ m}^{-3})(1.6 \times 10^{-19} \text{ C})(0.001\,0 \text{ m})}$$

$$\Delta V_H = \boxed{0.44 \ \mu\text{V}}$$

Such an extremely small Hall voltage is expected in good conductors. (Note that the width of the conductor is not needed in this calculation.)

What If? What if the strip has the same dimensions but is made of a semiconductor? Will the Hall voltage be smaller or larger?

Answer In semiconductors, n is much smaller than it is in metals that contribute one electron per atom to the current; hence, the Hall voltage is usually larger because it varies as the inverse of n. Currents on the order of 0.1 mA are generally used for such materials. Consider a piece of silicon that has the same dimensions as the copper strip in this example and whose value for n is 1.0×10^{20} electrons/m^3. Taking $B = 1.2$ T and $I = 0.10$ mA, we find that $\Delta V_H = 7.5$ mV. A potential difference of this magnitude is readily measured.

SUMMARY

Take a practice test for this chapter by clicking on the Practice Test link at http://www.pse6.com.

The magnetic force that acts on a charge q moving with a velocity $\mathbf{v}$ in a magnetic field $\mathbf{B}$ is

$$\mathbf{F}_B = q\mathbf{v} \times \mathbf{B} \qquad (29.1)$$

The direction of this magnetic force is perpendicular both to the velocity of the particle and to the magnetic field. The magnitude of this force is

$$F_B = |q|vB \sin \theta \qquad (29.2)$$

where θ is the smaller angle between $\mathbf{v}$ and $\mathbf{B}$. The SI unit of $\mathbf{B}$ is the **tesla** (T), where $1 \text{ T} = 1 \text{ N/A} \cdot \text{m}$.

When a charged particle moves in a magnetic field, the work done by the magnetic force on the particle is zero because the displacement is always perpendicular to the direction of the force. The magnetic field can alter the direction of the particle's velocity vector, but it cannot change its speed.

If a straight conductor of length L carries a current I, the force exerted on that conductor when it is placed in a uniform magnetic field $\mathbf{B}$ is

$$\mathbf{F}_B = I\mathbf{L} \times \mathbf{B} \qquad (29.3)$$

where the direction of $\mathbf{L}$ is in the direction of the current and $|\mathbf{L}| = L$.

If an arbitrarily shaped wire carrying a current I is placed in a magnetic field, the magnetic force exerted on a very small segment $d\mathbf{s}$ is

$$d\mathbf{F}_B = I \, d\mathbf{s} \times \mathbf{B} \qquad (29.4)$$

To determine the total magnetic force on the wire, one must integrate Equation 29.4, keeping in mind that both $\mathbf{B}$ and $d\mathbf{s}$ may vary at each point. Integration gives for the force exerted on a current-carrying conductor of arbitrary shape in a uniform magnetic field

$$\mathbf{F}_B = I\mathbf{L}' \times \mathbf{B} \tag{29.7}$$

where $\mathbf{L}'$ is a vector directed from one end of the conductor to the opposite end. Because integration of Equation 29.4 for a closed loop yields a zero result, the net magnetic force on any closed loop carrying a current in a uniform magnetic field is zero.

The **magnetic dipole moment** $\boldsymbol{\mu}$ of a loop carrying a current I is

$$\boldsymbol{\mu} = I\mathbf{A} \tag{29.10}$$

where the area vector $\mathbf{A}$ is perpendicular to the plane of the loop and $|\mathbf{A}|$ is equal to the area of the loop. The SI unit of $\boldsymbol{\mu}$ is $A \cdot m^2$.

The torque $\boldsymbol{\tau}$ on a current loop placed in a uniform magnetic field $\mathbf{B}$ is

$$\boldsymbol{\tau} = \boldsymbol{\mu} \times \mathbf{B} \tag{29.11}$$

The potential energy of the system of a magnetic dipole in a magnetic field is

$$U = -\boldsymbol{\mu} \cdot \mathbf{B} \tag{29.12}$$

If a charged particle moves in a uniform magnetic field so that its initial velocity is perpendicular to the field, the particle moves in a circle, the plane of which is perpendicular to the magnetic field. The radius of the circular path is

$$r = \frac{mv}{qB} \tag{29.13}$$

where m is the mass of the particle and q is its charge. The angular speed of the charged particle is

$$\omega = \frac{qB}{m} \tag{29.14}$$

QUESTIONS

1. At a given instant, a proton moves in the positive x direction through a magnetic field in the negative z direction. What is the direction of the magnetic force? Does the proton continue to move in the positive x direction? Explain.

2. Two charged particles are projected into a magnetic field perpendicular to their velocities. If the charges are deflected in opposite directions, what can you say about them?

3. If a charged particle moves in a straight line through some region of space, can you say that the magnetic field in that region is zero?

4. Suppose an electron is chasing a proton up this page when they suddenly enter a magnetic field perpendicular to the page. What happens to the particles?

5. How can the motion of a moving charged particle be used to distinguish between a magnetic field and an electric field? Give a specific example to justify your argument.

6. List several similarities and differences between electric and magnetic forces.

7. Justify the following statement: "It is impossible for a constant (in other words, a time-independent) magnetic field to alter the speed of a charged particle."

8. In view of your answer to Question 7, what is the role of a magnetic field in a cyclotron?

9. The electron beam in Figure Q29.9 is projected to the right. The beam deflects downward in the presence of a magnetic field produced by a pair of current-carrying coils. (a) What is the direction of the magnetic field? (b) What would happen to the beam if the magnetic field were reversed in direction?

Courtesy of Central Scientific Company

Figure Q29.9

10. A current-carrying conductor experiences no magnetic force when placed in a certain manner in a uniform magnetic field. Explain.

11. Is it possible to orient a current loop in a uniform magnetic field such that the loop does not tend to rotate? Explain.

12. Explain why it is not possible to determine the charge and the mass of a charged particle separately by measuring accelerations produced by electric and magnetic forces on the particle.

13. How can a current loop be used to determine the presence of a magnetic field in a given region of space?

14. Charged particles from outer space, called cosmic rays, strike the Earth more frequently near the poles than near the equator. Why?

15. What is the net force on a compass needle in a uniform magnetic field?

16. What type of magnetic field is required to exert a resultant force on a magnetic dipole? What is the direction of the resultant force?

17. A proton moving horizontally enters a uniform magnetic field perpendicular to the proton's velocity, as shown in Figure Q29.17. Describe the subsequent motion of the proton. How would an electron behave under the same circumstances?

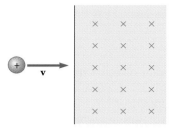

Figure Q29.17

18. In the cyclotron, why do particles having different speeds take the same amount of time to complete a one-half circle trip around one dee?

19. The *bubble chamber* is a device used for observing tracks of particles that pass through the chamber, which is immersed in a magnetic field. If some of the tracks are spirals and others are straight lines, what can you say about the particles?

20. Can a constant magnetic field set into motion an electron initially at rest? Explain your answer.

21. You are designing a magnetic probe that uses the Hall effect to measure magnetic fields. Assume that you are restricted to using a given material and that you have already made the probe as thin as possible. What, if anything, can be done to increase the Hall voltage produced for a given magnitude of magnetic field?

PROBLEMS

1, 2, 3 = straightforward, intermediate, challenging ☐ = full solution available in the *Student Solutions Manual and Study Guide*

🌐 = coached solution with hints available at http://www.pse6.com 💻 = computer useful in solving problem

▨ = paired numerical and symbolic problems

Section 29.1 Magnetic Fields and Forces

1. 🌐 Determine the initial direction of the deflection of charged particles as they enter the magnetic fields as shown in Figure P29.1.

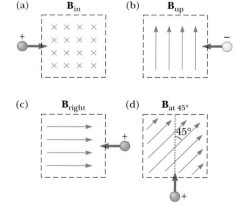

Figure P29.1

2. Consider an electron near the Earth's equator. In which direction does it tend to deflect if its velocity is directed (a) downward, (b) northward, (c) westward, or (d) southeastward?

3. An electron moving along the positive x axis perpendicular to a magnetic field experiences a magnetic deflection in the negative y direction. What is the direction of the magnetic field?

4. A proton travels with a speed of 3.00×10^6 m/s at an angle of 37.0° with the direction of a magnetic field of 0.300 T in the $+y$ direction. What are (a) the magnitude of the magnetic force on the proton and (b) its acceleration?

5. A proton moves perpendicular to a uniform magnetic field **B** at 1.00×10^7 m/s and experiences an acceleration of 2.00×10^{13} m/s² in the $+x$ direction when its velocity is in the $+z$ direction. Determine the magnitude and direction of the field.

6. An electron is accelerated through 2 400 V from rest and then enters a uniform 1.70-T magnetic field. What are (a) the maximum and (b) the minimum values of the magnetic force this charge can experience?

7. A proton moving at 4.00×10^6 m/s through a magnetic field of 1.70 T experiences a magnetic force of magnitude 8.20×10^{-13} N. What is the angle between the proton's velocity and the field?

8. At the equator, near the surface of the Earth, the magnetic field is approximately 50.0 μT northward, and the electric field is about 100 N/C downward in fair weather. Find the gravitational, electric, and magnetic forces on an electron in this environment, assuming the electron has an instantaneous velocity of 6.00×10^6 m/s directed to the east.

9. A proton moves with a velocity of $\mathbf{v} = (2\hat{\mathbf{i}} - 4\hat{\mathbf{j}} + \hat{\mathbf{k}})$ m/s in a region in which the magnetic field is $\mathbf{B} = (\hat{\mathbf{i}} + 2\hat{\mathbf{j}} - 3\hat{\mathbf{k}})$ T. What is the magnitude of the magnetic force this charge experiences?

10. An electron has a velocity of 1.20×10^4 m/s (in the positive x direction), and an acceleration of 2.00×10^{12} m/s^2 (in the positive z direction) in a uniform electric and magnetic field. If the electric field has a magnitude of 20.0 N/C (in the positive z direction), what can you determine about the magnetic field in the region? What can you not determine?

Section 29.2 Magnetic Force Acting on a Current-Carrying Conductor

11. A wire having a mass per unit length of 0.500 g/cm carries a 2.00-A current horizontally to the south. What are the direction and magnitude of the minimum magnetic field needed to lift this wire vertically upward?

12. A wire carries a steady current of 2.40 A. A straight section of the wire is 0.750 m long and lies along the x axis within a uniform magnetic field, $\mathbf{B} = 1.60\hat{\mathbf{k}}$ T. If the current is in the $+x$ direction, what is the magnetic force on the section of wire?

13. A wire 2.80 m in length carries a current of 5.00 A in a region where a uniform magnetic field has a magnitude of 0.390 T. Calculate the magnitude of the magnetic force on the wire assuming the angle between the magnetic field and the current is (a) 60.0°, (b) 90.0°, (c) 120°.

14. A conductor suspended by two flexible wires as shown in Figure P29.14 has a mass per unit length of 0.040 0 kg/m. What current must exist in the conductor in order for the tension in the supporting wires to be zero when the magnetic field is 3.60 T into the page? What is the required direction for the current?

Figure P29.14

15. **Review Problem.** A rod of mass 0.720 kg and radius 6.00 cm rests on two parallel rails (Fig. P29.15) that are $d = 12.0$ cm apart and $L = 45.0$ cm long. The rod carries a current of $I = 48.0$ A (in the direction shown) and rolls along the rails without slipping. A uniform magnetic field

of magnitude 0.240 T is directed perpendicular to the rod and the rails. If it starts from rest, what is the speed of the rod as it leaves the rails?

Figure P29.15 Problems 15 and 16.

16. **Review Problem.** A rod of mass m and radius R rests on two parallel rails (Fig. P29.15) that are a distance d apart and have a length L. The rod carries a current I (in the direction shown) and rolls along the rails without slipping. A uniform magnetic field B is directed perpendicular to the rod and the rails. If it starts from rest, what is the speed of the rod as it leaves the rails?

17. *A nonuniform magnetic field exerts a net force on a magnetic dipole.* A strong magnet is placed under a horizontal conducting ring of radius r that carries current I as shown in Figure P29.17. If the magnetic field $\mathbf{B}$ makes an angle θ with the vertical at the ring's location, what are the magnitude and direction of the resultant force on the ring?

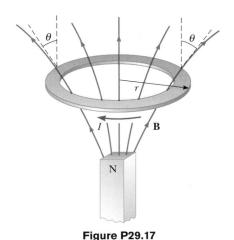

Figure P29.17

18. In Figure P29.18, the cube is 40.0 cm on each edge. Four straight segments of wire—ab, bc, cd, and da—form a closed loop that carries a current $I = 5.00$ A, in the direction shown. A uniform magnetic field of magnitude $B = 0.020\,0$ T is in the positive y direction. Determine the magnitude and direction of the magnetic force on each segment.

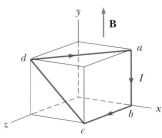

Figure P29.18

19. Assume that in Atlanta, Georgia, the Earth's magnetic field is 52.0 μT northward at 60.0° below the horizontal. A tube in a neon sign carries current 35.0 mA, between two diagonally opposite corners of a shop window, which lies in a north–south vertical plane. The current enters the tube at the bottom south corner of the window. It exits at the opposite corner, which is 1.40 m farther north and 0.850 m higher up. Between these two points, the glowing tube spells out DONUTS. Use the theorem proved as Case 1 in the text to determine the total vector magnetic force on the tube.

Section 29.3 Torque on a Current Loop in a Uniform Magnetic Field

20. A current of 17.0 mA is maintained in a single circular loop of 2.00 m circumference. A magnetic field of 0.800 T is directed parallel to the plane of the loop. (a) Calculate the magnetic moment of the loop. (b) What is the magnitude of the torque exerted by the magnetic field on the loop?

21. A small bar magnet is suspended in a uniform 0.250-T magnetic field. The maximum torque experienced by the bar magnet is 4.60×10^{-3} N·m. Calculate the magnetic moment of the bar magnet.

22. A long piece of wire with a mass of 0.100 kg and a total length of 4.00 m is used to make a square coil with a side of 0.100 m. The coil is hinged along a horizontal side, carries a 3.40-A current, and is placed in a vertical magnetic field with a magnitude of 0.010 0 T. (a) Determine the angle that the plane of the coil makes with the vertical when the coil is in equilibrium. (b) Find the torque acting on the coil due to the magnetic force at equilibrium.

23. A rectangular coil consists of $N = 100$ closely wrapped turns and has dimensions $a = 0.400$ m and $b = 0.300$ m. The coil is hinged along the y axis, and its plane makes an angle $\theta = 30.0°$ with the x axis (Fig. P29.23). What is the magnitude of the torque exerted on the coil by a uniform magnetic field $B = 0.800$ T directed along the x axis when the current is $I = 1.20$ A in the direction shown? What is the expected direction of rotation of the coil?

24. A 40.0-cm length of wire carries a current of 20.0 A. It is bent into a loop and placed with its normal perpendicular to a magnetic field with a magnitude of 0.520 T. What is the torque on the loop if it is bent into (a) an equilateral triangle? **What If?** What is the torque if the loop is (b) a square or (c) a circle? (d) Which torque is greatest?

25. A current loop with magnetic dipole moment $\boldsymbol{\mu}$ is placed in a uniform magnetic field **B**, with its moment making angle θ with the field. With the arbitrary choice of $U = 0$ for $\theta = 90°$, prove that the potential energy of the dipole–field system is $U = -\boldsymbol{\mu} \cdot \mathbf{B}$. You may imitate the discussion in Chapter 26 of the potential energy of an electric dipole in an electric field.

26. The needle of a magnetic compass has magnetic moment 9.70 mA·m². At its location, the Earth's magnetic field is 55.0 μT north at 48.0° below the horizontal. (a) Identify the orientations of the compass needle that represent minimum potential energy and maximum potential energy of the needle–field system. (b) How much work must be done on the needle to move it from the former to the latter orientation?

27. A wire is formed into a circle having a diameter of 10.0 cm and placed in a uniform magnetic field of 3.00 mT. The wire carries a current of 5.00 A. Find (a) the maximum torque on the wire and (b) the range of potential energies of the wire–field system for different orientations of the circle.

28. The rotor in a certain electric motor is a flat rectangular coil with 80 turns of wire and dimensions 2.50 cm by 4.00 cm. The rotor rotates in a uniform magnetic field of 0.800 T. When the plane of the rotor is perpendicular to the direction of the magnetic field, it carries a current of 10.0 mA. In this orientation, the magnetic moment of the rotor is directed opposite the magnetic field. The rotor then turns through one-half revolution. This process is repeated to cause the rotor to turn steadily at 3 600 rev/min. (a) Find the maximum torque acting on the rotor. (b) Find the peak power output of the motor. (c) Determine the amount of work performed by the magnetic field on the rotor in every full revolution. (d) What is the average power of the motor?

Section 29.4 Motion of a Charged Particle in a Uniform Magnetic Field

29. The magnetic field of the Earth at a certain location is directed vertically downward and has a magnitude of 50.0 μT. A proton is moving horizontally toward the west in this field with a speed of 6.20×10^6 m/s. (a) What are the direction and magnitude of the magnetic force the field exerts on this charge? (b) What is the radius of the circular arc followed by this proton?

30. A singly charged positive ion has a mass of 3.20×10^{-26} kg. After being accelerated from rest through a potential difference of 833 V, the ion enters a magnetic field of 0.920 T along a direction perpendicular to the direction

Figure P29.23

of the field. Calculate the radius of the path of the ion in the field.

31. **Review Problem.** One electron collides elastically with a second electron initially at rest. After the collision, the radii of their trajectories are 1.00 cm and 2.40 cm. The trajectories are perpendicular to a uniform magnetic field of magnitude 0.044 0 T. Determine the energy (in keV) of the incident electron.

32. A proton moving in a circular path perpendicular to a constant magnetic field takes 1.00 μs to complete one revolution. Determine the magnitude of the magnetic field.

33. A proton (charge $+e$, mass m_p), a deuteron (charge $+e$, mass $2m_p$), and an alpha particle (charge $+2e$, mass $4m_p$) are accelerated through a common potential difference ΔV. Each of the particles enters a uniform magnetic field $\mathbf{B}$, with its velocity in a direction perpendicular to $\mathbf{B}$. The proton moves in a circular path of radius r_p. Determine the radii of the circular orbits for the deuteron, r_d, and the alpha particle, r_α, in terms of r_p.

34. **Review Problem.** An electron moves in a circular path perpendicular to a constant magnetic field of magnitude 1.00 mT. The angular momentum of the electron about the center of the circle is 4.00×10^{-25} J·s. Determine (a) the radius of the circular path and (b) the speed of the electron.

35. Calculate the cyclotron frequency of a proton in a magnetic field of magnitude 5.20 T.

36. A singly charged ion of mass m is accelerated from rest by a potential difference ΔV. It is then deflected by a uniform magnetic field (perpendicular to the ion's velocity) into a semicircle of radius R. Now a doubly charged ion of mass m' is accelerated through the same potential difference and deflected by the same magnetic field into a semicircle of radius $R' = 2R$. What is the ratio of the masses of the ions?

37. A cosmic-ray proton in interstellar space has an energy of 10.0 MeV and executes a circular orbit having a radius equal to that of Mercury's orbit around the Sun (5.80 $\times$ 10^{10} m). What is the magnetic field in that region of space?

38. Figure 29.21 shows a charged particle traveling in a nonuniform magnetic field forming a magnetic bottle. (a) Explain why the positively charged particle in the figure must be moving clockwise. The particle travels along a helix whose radius decreases and whose pitch decreases as the particle moves into a stronger magnetic field. If the particle is moving to the right along the x axis, its velocity in this direction will be reduced to zero and it will be reflected from the right-hand side of the bottle, acting as a "magnetic mirror." The particle ends up bouncing back and forth between the ends of the bottle. (b) Explain qualitatively why the axial velocity is reduced to zero as the particle moves into the region of strong magnetic field at the end of the bottle. (c) Explain why the tangential velocity increases as the particle approaches the end of the bottle. (d) Explain why the orbiting particle has a magnetic dipole moment. (e) Sketch the magnetic moment and use the result of Problem 17 to explain again

how the nonuniform magnetic field exerts a force on the orbiting particle along the x axis.

39. A singly charged positive ion moving at 4.60×10^5 m/s leaves a circular track of radius 7.94 mm along a direction perpendicular to the 1.80-T magnetic field of a bubble chamber. Compute the mass (in atomic mass units) of this ion, and, from that value, identify it.

Section 29.5 Applications Involving Charged Particles Moving in a Magnetic Field

40. A velocity selector consists of electric and magnetic fields described by the expressions $\mathbf{E} = E\hat{\mathbf{k}}$ and $\mathbf{B} = B\hat{\mathbf{j}}$, with $B = 15.0$ mT. Find the value of E such that a 750-eV electron moving along the positive x axis is undeflected.

41. Singly charged uranium-238 ions are accelerated through a potential difference of 2.00 kV and enter a uniform magnetic field of 1.20 T directed perpendicular to their velocities. (a) Determine the radius of their circular path. (b) Repeat for uranium-235 ions. **What If?** How does the ratio of these path radii depend on the accelerating voltage and on the magnitude of the magnetic field?

42. Consider the mass spectrometer shown schematically in Figure 29.24. The magnitude of the electric field between the plates of the velocity selector is 2 500 V/m, and the magnetic field in both the velocity selector and the deflection chamber has a magnitude of 0.035 0 T. Calculate the radius of the path for a singly charged ion having a mass $m = 2.18 \times 10^{-26}$ kg.

43. A cyclotron designed to accelerate protons has a magnetic field of magnitude 0.450 T over a region of radius 1.20 m. What are (a) the cyclotron frequency and (b) the maximum speed acquired by the protons?

44. What is the required radius of a cyclotron designed to accelerate protons to energies of 34.0 MeV using a magnetic field of 5.20 T?

45. A cyclotron designed to accelerate protons has an outer radius of 0.350 m. The protons are emitted nearly at rest from a source at the center and are accelerated through 600 V each time they cross the gap between the dees. The dees are between the poles of an electromagnet where the field is 0.800 T. (a) Find the cyclotron frequency. (b) Find the speed at which protons exit the cyclotron and (c) their maximum kinetic energy. (d) How many revolutions does a proton make in the cyclotron? (e) For what time interval does one proton accelerate?

46. At the Fermilab accelerator in Batavia, Illinois, protons having momentum 4.80×10^{-16} kg·m/s are held in a circular orbit of radius 1.00 km by an upward magnetic field. What is the magnitude of this field?

47. The picture tube in a television uses magnetic deflection coils rather than electric deflection plates. Suppose an electron beam is accelerated through a 50.0-kV potential difference and then through a region of uniform magnetic field 1.00 cm wide. The screen is located 10.0 cm from the center of the coils and is 50.0 cm wide. When the field is turned off, the electron beam hits the center of the screen. What field magnitude is necessary to deflect the beam to the side of the screen? Ignore relativistic corrections.

Section 29.6 The Hall Effect

48. A flat ribbon of silver having a thickness $t = 0.200$ mm is used in a Hall-effect measurement of a uniform magnetic field perpendicular to the ribbon, as shown in Figure P29.48. The Hall coefficient for silver is $R_H = 0.840 \times 10^{-10}$ m³/C. (a) What is the density of charge carriers in silver? (b) If a current $I = 20.0$ A produces a Hall voltage $\Delta V_H = 15.0$ μV, what is the magnitude of the applied magnetic field?

Figure P29.48

49. A flat copper ribbon 0.330 mm thick carries a steady current of 50.0 A and is located in a uniform 1.30-T magnetic field directed perpendicular to the plane of the ribbon. If a Hall voltage of 9.60 μV is measured across the ribbon, what is the charge density of the free electrons? What effective number of free electrons per atom does this result indicate?

50. A Hall-effect probe operates with a 120-mA current. When the probe is placed in a uniform magnetic field of magnitude 0.080 0 T, it produces a Hall voltage of 0.700 μV. (a) When it is measuring an unknown magnetic field, the Hall voltage is 0.330 μV. What is the magnitude of the unknown field? (b) The thickness of the probe in the direction of **B** is 2.00 mm. Find the density of the charge carriers, each of which has charge of magnitude e.

51. In an experiment that is designed to measure the Earth's magnetic field using the Hall effect, a copper bar 0.500 cm thick is positioned along an east–west direction. If a current of 8.00 A in the conductor results in a Hall voltage of 5.10×10^{-12} V, what is the magnitude of the Earth's magnetic field? (Assume that $n = 8.49 \times 10^{28}$ electrons/m³ and that the plane of the bar is rotated to be perpendicular to the direction of **B**.)

Additional Problems

52. Assume that the region to the right of a certain vertical plane contains a vertical magnetic field of magnitude 1.00 mT, and the field is zero in the region to the left of the plane. An electron, originally traveling perpendicular to the boundary plane, passes into the region of the field. (a) Determine the time interval required for the electron to leave the "field-filled" region, noting that its path is a semicircle. (b) Find the kinetic energy of the electron if the maximum depth of penetration into the field is 2.00 cm.

53. Sodium melts at 99°C. Liquid sodium, an excellent thermal conductor, is used in some nuclear reactors to cool the reactor core. The liquid sodium is moved through pipes by pumps that exploit the force on a moving charge in a magnetic field. The principle is as follows. Assume the liquid metal to be in an electrically insulating pipe having a rectangular cross section of width w and height h. A uniform magnetic field perpendicular to the pipe affects a section of length L (Fig. P29.53). An electric current directed perpendicular to the pipe and to the magnetic field produces a current density J in the liquid sodium. (a) Explain why this arrangement produces on the liquid a force that is directed along the length of the pipe. (b) Show that the section of liquid in the magnetic field experiences a pressure increase JLB.

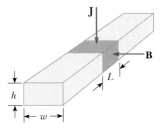

Figure P29.53

54. A 0.200-kg metal rod carrying a current of 10.0 A glides on two horizontal rails 0.500 m apart. What vertical magnetic field is required to keep the rod moving at a constant speed if the coefficient of kinetic friction between the rod and rails is 0.100?

55. Protons having a kinetic energy of 5.00 MeV are moving in the positive x direction and enter a magnetic field $\mathbf{B} = 0.050\ 0\hat{\mathbf{k}}$ T directed out of the plane of the page and extending from $x = 0$ to $x = 1.00$ m, as shown in Figure P29.55. (a) Calculate the y component of the protons' momentum as they leave the magnetic field. (b) Find the angle α between the initial velocity vector of the proton beam and the velocity vector after the beam emerges from the field. Ignore relativistic effects and note that 1 eV $= 1.60 \times 10^{-19}$ J.

Figure P29.55

56. (a) A proton moving in the $+x$ direction with velocity $\mathbf{v} = v_i\hat{\mathbf{i}}$ experiences a magnetic force $\mathbf{F} = F_i\hat{\mathbf{j}}$ in the $+y$ direction. Explain what you can and cannot infer about **B** from this information. (b) **What If?** In terms of F_i, what would be the force on a proton in the same field moving with velocity $\mathbf{v} = -v_i\hat{\mathbf{i}}$? (c) What would be the force on an electron in the same field moving with velocity $\mathbf{v} = v_i\hat{\mathbf{i}}$?

57. A positive charge $q = 3.20 \times 10^{-19}$ C moves with a velocity $\mathbf{v} = (2\hat{\mathbf{i}} + 3\hat{\mathbf{j}} - \hat{\mathbf{k}})$ m/s through a region where both a uniform magnetic field and a uniform electric field exist. (a) Calculate the total force on the moving charge (in unit-vector notation), taking $\mathbf{B} = (2\hat{\mathbf{i}} + 4\hat{\mathbf{j}} + \hat{\mathbf{k}})$ T and $\mathbf{E} = (4\hat{\mathbf{i}} - \hat{\mathbf{j}} - 2\hat{\mathbf{k}})$ V/m. (b) What angle does the force vector make with the positive x axis?

58. Review Problem. A wire having a linear mass density of 1.00 g/cm is placed on a horizontal surface that has a coefficient of kinetic friction of 0.200. The wire carries a current of 1.50 A toward the east and slides horizontally to the north. What are the magnitude and direction of the smallest magnetic field that enables the wire to move in this fashion?

59. Electrons in a beam are accelerated from rest through a potential difference ΔV. The beam enters an experimental chamber through a small hole. As shown in Figure P29.59, the electron velocity vectors lie within a narrow cone of half angle ϕ oriented along the beam axis. We wish to use a uniform magnetic field directed parallel to the axis to focus the beam, so that all of the electrons can pass through a small exit port on the opposite side of the chamber after they travel the length d of the chamber. What is the required magnitude of the magnetic field? *Hint:* Because every electron passes through the same potential difference and the angle ϕ is small, they all require the same time interval to travel the axial distance d.

Figure P29.59

60. Review Problem. A proton is at rest at the plane vertical boundary of a region containing a uniform vertical magnetic field B. An alpha particle moving horizontally makes a head-on elastic collision with the proton. Immediately after the collision, both particles enter the magnetic field, moving perpendicular to the direction of the field. The radius of the proton's trajectory is R. Find the radius of the alpha particle's trajectory. The mass of the alpha particle is four times that of the proton, and its charge is twice that of the proton.

61. The circuit in Figure P29.61 consists of wires at the top and bottom and identical metal springs in the left and right sides. The upper portion of the circuit is fixed. The wire at the bottom has a mass of 10.0 g and is 5.00 cm

Figure P29.61

long. The springs stretch 0.500 cm under the weight of the wire and the circuit has a total resistance of 12.0 Ω. When a magnetic field is turned on, directed out of the page, the springs stretch an additional 0.300 cm. What is the magnitude of the magnetic field?

62. A hand-held electric mixer contains an electric motor. Model the motor as a single flat compact circular coil carrying electric current in a region where a magnetic field is produced by an external permanent magnet. You need consider only one instant in the operation of the motor. (We will consider motors again in Chapter 31.) The coil moves because the magnetic field exerts torque on the coil, as described in Section 29.3. Make order-of-magnitude estimates of the magnetic field, the torque on the coil, the current in it, its area, and the number of turns in the coil, so that they are related according to Equation 29.11. Note that the input power to the motor is electric, given by $\mathcal{P} = I \, \Delta V$, and the useful output power is mechanical, $\mathcal{P} = \tau \omega$.

63. A nonconducting sphere has mass 80.0 g and radius 20.0 cm. A flat compact coil of wire with 5 turns is wrapped tightly around it, with each turn concentric with the sphere. As shown in Figure P29.63, the sphere is placed on an inclined plane that slopes downward to the left, making an angle θ with the horizontal, so that the coil is parallel to the inclined plane. A uniform magnetic field of 0.350 T vertically upward exists in the region of the sphere. What current in the coil will enable the sphere to rest in equilibrium on the inclined plane? Show that the result does not depend on the value of θ.

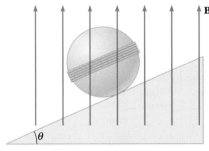

Figure P29.63

64. A metal rod having a mass per unit length λ carries a current I. The rod hangs from two vertical wires in a uniform vertical magnetic field as shown in Figure P29.64. The wires make an angle θ with the vertical when in equilibrium. Determine the magnitude of the magnetic field.

Figure P29.64

65. A cyclotron is sometimes used for carbon dating, as described in Chapter 44. Carbon-14 and carbon-12 ions are obtained from a sample of the material to be dated, and accelerated in the cyclotron. If the cyclotron has a magnetic field of magnitude 2.40 T, what is the difference in cyclotron frequencies for the two ions?

66. A uniform magnetic field of magnitude 0.150 T is directed along the positive x axis. A positron moving at 5.00×10^6 m/s enters the field along a direction that makes an angle of 85.0° with the x axis (Fig. P29.66). The motion of the particle is expected to be a helix, as described in Section 29.4. Calculate (a) the pitch p and (b) the radius r of the trajectory.

Figure P29.66

67. Consider an electron orbiting a proton and maintained in a fixed circular path of radius $R = 5.29 \times 10^{-11}$ m by the Coulomb force. Treating the orbiting charge as a current loop, calculate the resulting torque when the system is in a magnetic field of 0.400 T directed perpendicular to the magnetic moment of the electron.

68. A singly charged ion completes five revolutions in a uniform magnetic field of magnitude 5.00×10^{-2} T in 1.50 ms. Calculate the mass of the ion in kilograms.

69. A proton moving in the plane of the page has a kinetic energy of 6.00 MeV. A magnetic field of magnitude $B = 1.00$ T is directed into the page. The proton enters the magnetic field with its velocity vector at an angle $\theta = 45.0°$ to the linear boundary of the field as shown in Figure P29.69. (a) Find x, the distance from the point of entry to where the proton will leave the field. (b) Determine θ', the angle between the boundary and the proton's velocity vector as it leaves the field.

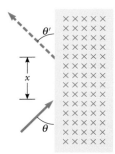

Figure P29.69

70. Table P29.70 shows measurements of a Hall voltage and corresponding magnetic field for a probe used to measure magnetic fields. (a) Plot these data, and deduce a relationship between the two variables. (b) If the measurements were taken with a current of 0.200 A and the sample is made from a material having a charge-carrier density of 1.00×10^{26}/m^3, what is the thickness of the sample?

Table P29.70

ΔV_H (μV)	B (T)
0	0.00
11	0.10
19	0.20
28	0.30
42	0.40
50	0.50
61	0.60
68	0.70
79	0.80
90	0.90
102	1.00

71. A heart surgeon monitors the flow rate of blood through an artery using an electromagnetic flowmeter (Fig. P29.71). Electrodes A and B make contact with the outer surface of the blood vessel, which has interior diameter 3.00 mm. (a) For a magnetic field magnitude of 0.040 0 T, an emf of 160 μV appears between the electrodes. Calculate the speed of the blood. (b) Verify that electrode A is positive, as shown. Does the sign of the emf depend on whether the mobile ions in the blood are predominantly positively or negatively charged? Explain.

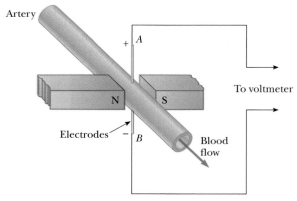

Figure P29.71

72. As shown in Figure P29.72, a particle of mass m having positive charge q is initially traveling with velocity $v\hat{\mathbf{j}}$. At the origin of coordinates it enters a region between $y = 0$ and $y = h$ containing a uniform magnetic field $B\hat{\mathbf{k}}$ directed perpendicularly out of the page. (a) What is the critical value of v such that the particle just reaches $y = h$?

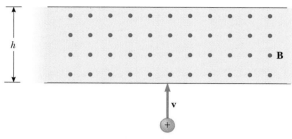

Figure P29.72

Describe the path of the particle under this condition, and predict its final velocity. (b) Specify the path the particle takes and its final velocity, if v is less than the critical value. (c) **What If?** Specify the path the particle takes and its final velocity if v is greater than the critical value.

Answers to Quick Quizzes

29.1 (c). The magnetic force exerted by a magnetic field on a charge is proportional to the charge's velocity relative to the field. If the charge is stationary, as in this situation, there is no magnetic force.

29.2 (b). The maximum value of $\sin \theta$ occurs for $\theta = 90°$.

29.3 (e). The right-hand rule gives the direction. Be sure to account for the negative charge on the electron.

29.4 (a), (b) = (c), (d). The magnitude of the force depends on the value of $\sin \theta$. The maximum force occurs when the wire is perpendicular to the field (a), and there is zero force when the wire is parallel (d). Choices (b) and (c) represent the same force because Case 1 tells us that a straight wire between A and B will have the same force on it as the curved wire.

29.5 (c). Use the right-hand rule to determine the direction of the magnetic field.

29.6 (c), (b), (a). Because all loops enclose the same area and carry the same current, the magnitude of μ is the same for all. For (c), μ points upward and is perpendicular to the magnetic field and $\tau = \mu B$, the maximum torque possible. For the loop in (a), μ points along the direction of **B** and the torque is zero. For (b), the torque is intermediate between zero and the maximum value.

29.7 (a) = (b) = (c). Because the magnetic field is uniform, there is zero net force on all three loops.

29.8 (b). The magnetic force on the particle increases in proportion to v, but the centripetal acceleration increases according to the square of v. The result is a larger radius, as we can see from Equation 29.13.

29.9 (a). The magnetic force on the particle increases in proportion to B. The result is a smaller radius, as we can see from Equation 29.13.

29.10 Speed: (a) = (b) = (c). m/q ratio, from greatest to least: (c), (b), (a). The velocity selector ensures that all three types of particles have the same speed. We cannot determine individual masses or charges, but we can rank the particles by m/q ratio. Equation 29.18 indicates that those particles traveling through the circle of greatest radius have the greatest m/q ratio.

Chapter 30

Sources of the Magnetic Field

▲ *A proposed method for launching future payloads into space is the use of rail guns, in which projectiles are accelerated by means of magnetic forces. This photo shows the firing of a projectile at a speed of over 3 km/s from an experimental rail gun at Sandia National Research Laboratories, Albuquerque, New Mexico. (Defense Threat Reduction Agency [DTRA])*

In the preceding chapter, we discussed the magnetic force exerted on a charged particle moving in a magnetic field. To complete the description of the magnetic interaction, this chapter explores the origin of the magnetic field—moving charges. We begin by showing how to use the law of Biot and Savart to calculate the magnetic field produced at some point in space by a small current element. Using this formalism and the principle of superposition, we then calculate the total magnetic field due to various current distributions. Next, we show how to determine the force between two current-carrying conductors, which leads to the definition of the ampere. We also introduce Ampère's law, which is useful in calculating the magnetic field of a highly symmetric configuration carrying a steady current.

This chapter is also concerned with the complex processes that occur in magnetic materials. All magnetic effects in matter can be explained on the basis of atomic magnetic moments, which arise both from the orbital motion of electrons and from an intrinsic property of electrons known as spin.

30.1 The Biot–Savart Law

Shortly after Oersted's discovery in 1819 that a compass needle is deflected by a current-carrying conductor, Jean-Baptiste Biot (1774–1862) and Félix Savart (1791–1841) performed quantitative experiments on the force exerted by an electric current on a nearby magnet. From their experimental results, Biot and Savart arrived at a mathematical expression that gives the magnetic field at some point in space in terms of the current that produces the field. That expression is based on the following experimental observations for the magnetic field $d\mathbf{B}$ at a point P associated with a length element $d\mathbf{s}$ of a wire carrying a steady current I (Fig. 30.1):

- The vector $d\mathbf{B}$ is perpendicular both to $d\mathbf{s}$ (which points in the direction of the current) and to the unit vector $\hat{\mathbf{r}}$ directed from $d\mathbf{s}$ toward P.
- The magnitude of $d\mathbf{B}$ is inversely proportional to r^2, where r is the distance from $d\mathbf{s}$ to P.
- The magnitude of $d\mathbf{B}$ is proportional to the current and to the magnitude ds of the length element $d\mathbf{s}$.
- The magnitude of $d\mathbf{B}$ is proportional to $\sin\theta$, where θ is the angle between the vectors $d\mathbf{s}$ and $\hat{\mathbf{r}}$.

These observations are summarized in the mathematical expression known today as the **Biot–Savart law:**

$$d\mathbf{B} = \frac{\mu_0}{4\pi}\frac{I\,d\mathbf{s}\times\hat{\mathbf{r}}}{r^2} \tag{30.1}$$

Figure 30.1 The magnetic field $d\mathbf{B}$ at a point due to the current I through a length element $d\mathbf{s}$ is given by the Biot–Savart law. The direction of the field is out of the page at P and into the page at P'.

▲ **PITFALL PREVENTION**

30.1 The Biot–Savart Law

The magnetic field described by the Biot–Savart law is the field *due to* a given current-carrying conductor. Do not confuse this field with any *external* field that may be applied to the conductor from some other source.

Biot–Savart law

where μ_0 is a constant called the **permeability of free space:**

Permeability of free space

$$\mu_0 = 4\pi \times 10^{-7}\ \text{T·m/A} \qquad (30.2)$$

Note that the field $d\mathbf{B}$ in Equation 30.1 is the field created by the current in only a small length element $d\mathbf{s}$ of the conductor. To find the *total* magnetic field $\mathbf{B}$ created at some point by a current of finite size, we must sum up contributions from all current elements $I\,d\mathbf{s}$ that make up the current. That is, we must evaluate $\mathbf{B}$ by integrating Equation 30.1:

$$\mathbf{B} = \frac{\mu_0 I}{4\pi} \int \frac{d\mathbf{s} \times \hat{\mathbf{r}}}{r^2} \qquad (30.3)$$

where the integral is taken over the entire current distribution. This expression must be handled with special care because the integrand is a cross product and therefore a vector quantity. We shall see one case of such an integration in Example 30.1.

Although we developed the Biot–Savart law for a current-carrying wire, it is also valid for a current consisting of charges flowing through space, such as the electron beam in a television set. In that case, $d\mathbf{s}$ represents the length of a small segment of space in which the charges flow.

Interesting similarities exist between Equation 30.1 for the magnetic field due to a current element and Equation 23.9 for the electric field due to a point charge. The magnitude of the magnetic field varies as the inverse square of the distance from the source, as does the electric field due to a point charge. However, the directions of the two fields are quite different. The electric field created by a point charge is radial, but the magnetic field created by a current element is perpendicular to both the length element $d\mathbf{s}$ and the unit vector $\hat{\mathbf{r}}$, as described by the cross product in Equation 30.1. Hence, if the conductor lies in the plane of the page, as shown in Figure 30.1, $d\mathbf{B}$ points out of the page at P and into the page at P'.

Another difference between electric and magnetic fields is related to the source of the field. An electric field is established by an isolated electric charge. The Biot–Savart law gives the magnetic field of an isolated current element at some point, but such an isolated current element cannot exist the way an isolated electric charge can. A current element *must* be part of an extended current distribution because we must have a complete circuit in order for charges to flow. Thus, the Biot–Savart law (Eq. 30.1) is only the first step in a calculation of a magnetic field; it must be followed by an integration over the current distribution, as in Equation 30.3.

Quick Quiz 30.1 Consider the current in the length of wire shown in Figure 30.2. Rank the points *A*, *B*, and *C*, in terms of magnitude of the magnetic field due to the current in the length element shown, from greatest to least.

Figure 30.2 (Quick Quiz 30.1) Where is the magnetic field the greatest?

Example 30.1   Magnetic Field Surrounding a Thin, Straight Conductor

Consider a thin, straight wire carrying a constant current I and placed along the x axis as shown in Figure 30.3. Determine the magnitude and direction of the magnetic field at point P due to this current.

Solution From the Biot–Savart law, we expect that the magnitude of the field is proportional to the current in the wire and decreases as the distance a from the wire to point P increases. We start by considering a length element ds located a distance r from P. The direction of the magnetic field at point P due to the current in this element is out of the page because $ds \times \hat{\mathbf{r}}$ is out of the page. In fact, because *all* of the current elements $I \, ds$ lie in the plane of the page, they all produce a magnetic field directed out of the page at point P. Thus, we have the direction of the magnetic field at point P, and we need only find the magnitude. Taking the origin at O and letting point P be along the positive y axis, with $\hat{\mathbf{k}}$ being a unit vector pointing out of the page, we see that

$$ds \times \hat{\mathbf{r}} = |ds \times \hat{\mathbf{r}}| \, \hat{\mathbf{k}} = (dx \sin \theta) \hat{\mathbf{k}}$$

where $|ds \times \hat{\mathbf{r}}|$ represents the magnitude of $ds \times \hat{\mathbf{r}}$. Because $\hat{\mathbf{r}}$ is a unit vector, the magnitude of the cross product is simply the magnitude of ds, which is the length dx. Substitution into Equation 30.1 gives

$$d\mathbf{B} = (dB)\hat{\mathbf{k}} = \frac{\mu_0 I}{4\pi} \frac{dx \sin \theta}{r^2} \hat{\mathbf{k}}$$

Because all current elements produce a magnetic field in the $\hat{\mathbf{k}}$ direction, let us restrict our attention to the magnitude of the field due to one current element, which is

$$(1) \qquad dB = \frac{\mu_0 I}{4\pi} \frac{dx \sin \theta}{r^2}$$

To integrate this expression, we must relate the variables θ, x, and r. One approach is to express x and r in terms of θ. From the geometry in Figure 30.3a, we have

$$(2) \qquad r = \frac{a}{\sin \theta} = a \csc \theta$$

Because $\tan \theta = a/(-x)$ from the right triangle in Figure 30.3a (the negative sign is necessary because ds is located at a negative value of x), we have

$$x = -a \cot \theta$$

Taking the derivative of this expression gives

$$(3) \qquad dx = a \csc^2 \theta \, d\theta$$

Substitution of Equations (2) and (3) into Equation (1) gives

$$(4) \qquad dB = \frac{\mu_0 I}{4\pi} \frac{a \csc^2 \theta \sin \theta \, d\theta}{a^2 \csc^2 \theta} = \frac{\mu_0 I}{4\pi a} \sin \theta \, d\theta$$

an expression in which the only variable is θ. We now obtain the magnitude of the magnetic field at point P by integrating Equation (4) over all elements, where the subtending angles range from θ_1 to θ_2 as defined in Figure 30.3b:

$$B = \frac{\mu_0 I}{4\pi a} \int_{\theta_1}^{\theta_2} \sin \theta \, d\theta = \frac{\mu_0 I}{4\pi a} (\cos \theta_1 - \cos \theta_2) \qquad (30.4)$$

We can use this result to find the magnetic field of *any* straight current-carrying wire if we know the geometry and hence the angles θ_1 and θ_2. Consider the special case of an infinitely long, straight wire. If we let the wire in Figure 30.3b become infinitely long, we see that $\theta_1 = 0$ and $\theta_2 = \pi$ for length elements ranging between positions $x = -\infty$ and $x = +\infty$. Because $(\cos \theta_1 - \cos \theta_2) = (\cos 0 - \cos \pi) = 2$, Equation 30.4 becomes

$$B = \frac{\mu_0 I}{2\pi a} \qquad (30.5)$$

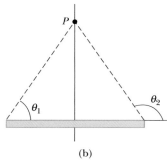

Figure 30.3 (Example 30.1) (a) A thin, straight wire carrying a current I. The magnetic field at point P due to the current in each element ds of the wire is out of the page, so the net field at point P is also out of the page. (b) The angles θ_1 and θ_2 used for determining the net field. When the wire is infinitely long, $\theta_1 = 0$ and $\theta_2 = 180°$.

Equations 30.4 and 30.5 both show that the magnitude of the magnetic field is proportional to the current and decreases with increasing distance from the wire, as we expected. Notice that Equation 30.5 has the same mathematical form as the expression for the magnitude of the electric field due to a long charged wire (see Eq. 24.7).

At the Interactive Worked Example link at **http://www.pse6.com,** *you can explore the field for different lengths of wire.*

Figure 30.4 The right-hand rule for determining the direction of the magnetic field surrounding a long, straight wire carrying a current. Note that the magnetic field lines form circles around the wire.

The result of Example 30.1 is important because a current in the form of a long, straight wire occurs often. Figure 30.4 is a perspective view of the magnetic field surrounding a long, straight current-carrying wire. Because of the symmetry of the wire, the magnetic field lines are circles concentric with the wire and lie in planes perpendicular to the wire. The magnitude of **B** is constant on any circle of radius a and is given by Equation 30.5. A convenient rule for determining the direction of **B** is to grasp the wire with the right hand, positioning the thumb along the direction of the current. The four fingers wrap in the direction of the magnetic field.

Another observation we can make in Figure 30.4 is that the magnetic field line shown has no beginning and no end. It forms a closed loop. This is a major difference between magnetic field lines and electric field lines, which begin on positive charges and end on negative charges. We will explore this feature of magnetic field lines further in Section 30.6.

Example 30.2   Magnetic Field Due to a Curved Wire Segment

Calculate the magnetic field at point O for the current-carrying wire segment shown in Figure 30.5. The wire consists of two straight portions and a circular arc of radius R, which subtends an angle θ. The arrowheads on the wire indicate the direction of the current.

Solution The magnetic field at O due to the current in the straight segments AA' and CC' is zero because $d\mathbf{s}$ is parallel to $\hat{\mathbf{r}}$ along these paths; this means that $d\mathbf{s} \times \hat{\mathbf{r}} = 0$. Each length element $d\mathbf{s}$ along path AC is at the same distance R from O, and the current in each contributes a field element $d\mathbf{B}$ directed into the page at O. Furthermore, at every point on AC, $d\mathbf{s}$ is perpendicular to $\hat{\mathbf{r}}$; hence, $|d\mathbf{s} \times \hat{\mathbf{r}}| = ds$. Using this information and Equation 30.1, we can find the magnitude of the field at O due to the current in an element of length ds:

$$dB = \frac{\mu_0 I}{4\pi} \frac{ds}{R^2}$$

Because I and R are constants in this situation, we can easily integrate this expression over the curved path AC:

$$B = \frac{\mu_0 I}{4\pi R^2} \int ds = \frac{\mu_0 I}{4\pi R^2} s = \frac{\mu_0 I}{4\pi R} \theta \qquad (30.6)$$

where we have used the fact that $s = R\theta$ with θ measured in radians. The direction of **B** is into the page at O because $d\mathbf{s} \times \hat{\mathbf{r}}$ is into the page for every length element.

What If? What if you were asked to find the magnetic field at the center of a circular wire loop of radius R that carries a current I? Can we answer this question at this point in our understanding of the source of magnetic fields?

Answer Yes, we can. We argued that the straight wires in Figure 30.5 do not contribute to the magnetic field. The only contribution is from the curved segment. If we imagine increasing the angle θ, the curved segment will become a full circle when $\theta = 2\pi$. Thus, we can find the magnetic field at the center of a wire loop by letting $\theta = 2\pi$ in Equation 30.6:

$$B = \frac{\mu_0 I}{4\pi R} 2\pi = \frac{\mu_0 I}{2R}$$

We will confirm this result as a limiting case of a more general result in Example 30.3.

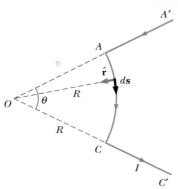

Figure 30.5 (Example 30.2) The magnetic field at O due to the current in the curved segment AC is into the page. The contribution to the field at O due to the current in the two straight segments is zero.

Example 30.3   Magnetic Field on the Axis of a Circular Current Loop   Interactive

Consider a circular wire loop of radius R located in the yz plane and carrying a steady current I, as in Figure 30.6. Calculate the magnetic field at an axial point P a distance x from the center of the loop.

Solution In this situation, every length element $d\mathbf{s}$ is perpendicular to the vector $\hat{\mathbf{r}}$ at the location of the element. Thus, for any element, $|d\mathbf{s} \times \hat{\mathbf{r}}| = (ds)(1)\sin 90° = ds$. Furthermore, all length elements around the loop are at the

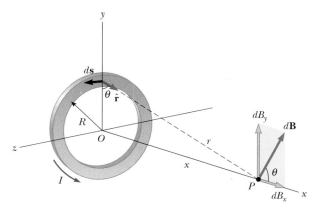

Figure 30.6 (Example 30.3) Geometry for calculating the magnetic field at a point P lying on the axis of a current loop. By symmetry, the total field **B** is along this axis.

same distance r from P, where $r^2 = x^2 + R^2$. Hence, the magnitude of $d\mathbf{B}$ due to the current in any length element $d\mathbf{s}$ is

$$dB = \frac{\mu_0 I}{4\pi} \frac{|d\mathbf{s} \times \hat{\mathbf{r}}|}{r^2} = \frac{\mu_0 I}{4\pi} \frac{ds}{(x^2 + R^2)}$$

The direction of $d\mathbf{B}$ is perpendicular to the plane formed by $\hat{\mathbf{r}}$ and $d\mathbf{s}$, as shown in Figure 30.6. We can resolve this vector into a component dB_x along the x axis and a component dB_y perpendicular to the x axis. When the components dB_y are summed over all elements around the loop, the resultant component is zero. That is, by symmetry the current in any element on one side of the loop sets up a perpendicular component of $d\mathbf{B}$ that cancels the perpendicular component set up by the current through the element diametrically opposite it. Therefore, *the resultant field at P must be along the x axis* and we can find it by integrating the components $dB_x = dB \cos \theta$. That is,

$\mathbf{B} = B_x \hat{\mathbf{i}}$ where

$$B_x = \oint dB \cos \theta = \frac{\mu_0 I}{4\pi} \oint \frac{ds \cos \theta}{x^2 + R^2}$$

and we must take the integral over the entire loop. Because θ, x, and R are constants for all elements of the loop and because $\cos \theta = R/(x^2 + R^2)^{1/2}$, we obtain

$$B_x = \frac{\mu_0 I R}{4\pi (x^2 + R^2)^{3/2}} \oint ds = \frac{\mu_0 I R^2}{2(x^2 + R^2)^{3/2}} \quad (30.7)$$

where we have used the fact that $\oint ds = 2\pi R$ (the circumference of the loop).

To find the magnetic field at the center of the loop, we set $x = 0$ in Equation 30.7. At this special point, therefore,

$$B = \frac{\mu_0 I}{2R} \quad \text{(at } x = 0) \quad (30.8)$$

which is consistent with the result of the **What If?** feature in Example 30.2.

The pattern of magnetic field lines for a circular current loop is shown in Figure 30.7a. For clarity, the lines are drawn for only one plane—one that contains the axis of the loop. Note that the field-line pattern is axially symmetric and looks like the pattern around a bar magnet, shown in Figure 30.7c.

What If? What if we consider points on the x axis very far from the loop? How does the magnetic field behave at these distant points?

Answer In this case, in which $x \gg R$, we can neglect the term R^2 in the denominator of Equation 30.7 and obtain

$$B \approx \frac{\mu_0 I R^2}{2x^3} \quad \text{(for } x \gg R) \quad (30.9)$$

Because the magnitude of the magnetic moment μ of the loop is defined as the product of current and loop area (see

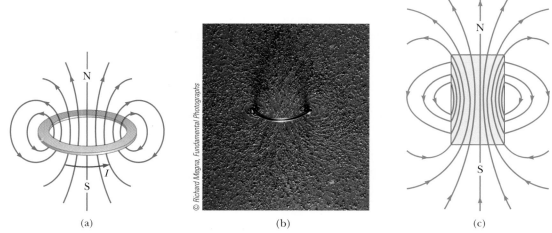

(a)   (b)   (c)

© Richard Megna, Fundamental Photographs

Figure 30.7 (Example 30.3) (a) Magnetic field lines surrounding a current loop. (b) Magnetic field lines surrounding a current loop, displayed with iron filings. (c) Magnetic field lines surrounding a bar magnet. Note the similarity between this line pattern and that of a current loop.

Eq. 29.10), $\mu = I(\pi R^2)$ for our circular loop. We can express Equation 30.9 as

$$B \approx \frac{\mu_0}{2\pi} \frac{\mu}{x^3}$$ (30.10)

This result is similar in form to the expression for the electric field due to an electric dipole, $E = k_e(2qa/y^3)$ (see Example 23.6), where $2qa = p$ is the electric dipole moment as defined in Equation 26.16.

 At the Interactive Worked Example link at **http://www.pse6.com,** *you can explore the field for different loop radii.*

30.2 The Magnetic Force Between Two Parallel Conductors

In Chapter 29 we described the magnetic force that acts on a current-carrying conductor placed in an external magnetic field. Because a current in a conductor sets up its own magnetic field, it is easy to understand that two current-carrying conductors exert magnetic forces on each other. Such forces can be used as the basis for defining the ampere and the coulomb.

Consider two long, straight, parallel wires separated by a distance a and carrying currents I_1 and I_2 in the same direction, as in Figure 30.8. We can determine the force exerted on one wire due to the magnetic field set up by the other wire. Wire 2, which carries a current I_2 and is identified arbitrarily as the source wire, creates a magnetic field $\mathbf{B}_2$ at the location of wire 1, the test wire. The direction of $\mathbf{B}_2$ is perpendicular to wire 1, as shown in Figure 30.8. According to Equation 29.3, the magnetic force on a length ℓ of wire 1 is $\mathbf{F}_1 = I_1\boldsymbol{\ell} \times \mathbf{B}_2$. Because $\boldsymbol{\ell}$ is perpendicular to $\mathbf{B}_2$ in this situation, the magnitude of $\mathbf{F}_1$ is $F_1 = I_1\ell B_2$. Because the magnitude of $\mathbf{B}_2$ is given by Equation 30.5, we see that

$$F_1 = I_1\ell B_2 = I_1\ell\left(\frac{\mu_0 I_2}{2\pi a}\right) = \frac{\mu_0 I_1 I_2}{2\pi a}\,\ell$$ (30.11)

Active Figure 30.8 Two parallel wires that each carry a steady current exert a magnetic force on each other. The field $\mathbf{B}_2$ due to the current in wire 2 exerts a magnetic force of magnitude $F_1 = I_1\ell B_2$ on wire 1. The force is attractive if the currents are parallel (as shown) and repulsive if the currents are antiparallel.

At the Active Figures link at **http://www.pse6.com,** *you can adjust the currents in the wires and the distance between them to see the effect on the force.*

The direction of $\mathbf{F}_1$ is toward wire 2 because $\boldsymbol{\ell} \times \mathbf{B}_2$ is in that direction. If the field set up at wire 2 by wire 1 is calculated, the force $\mathbf{F}_2$ acting on wire 2 is found to be equal in magnitude and opposite in direction to $\mathbf{F}_1$. This is what we expect because Newton's third law must be obeyed.[1] When the currents are in opposite directions (that is, when one of the currents is reversed in Fig. 30.8), the forces are reversed and the wires repel each other. Hence, **parallel conductors carrying currents in the same direction attract each other, and parallel conductors carrying currents in opposite directions repel each other.**

Because the magnitudes of the forces are the same on both wires, we denote the magnitude of the magnetic force between the wires as simply F_B. We can rewrite this magnitude in terms of the force per unit length:

$$\frac{F_B}{\ell} = \frac{\mu_0 I_1 I_2}{2\pi a}$$ (30.12)

The force between two parallel wires is used to define the **ampere** as follows:

Definition of the ampere

When the magnitude of the force per unit length between two long parallel wires that carry identical currents and are separated by 1 m is 2×10^{-7} N/m, the current in each wire is defined to be 1 A.

---

[1] Although the total force exerted on wire 1 is equal in magnitude and opposite in direction to the total force exerted on wire 2, Newton's third law does not apply when one considers two small elements of the wires that are not exactly opposite each other. This apparent violation of Newton's third law and of the law of conservation of momentum is described in more advanced treatments on electricity and magnetism.

The value 2×10^{-7} N/m is obtained from Equation 30.12 with $I_1 = I_2 = 1$ A and $a = 1$ m. Because this definition is based on a force, a mechanical measurement can be used to standardize the ampere. For instance, the National Institute of Standards and Technology uses an instrument called a *current balance* for primary current measurements. The results are then used to standardize other, more conventional instruments, such as ammeters.

The SI unit of charge, the **coulomb,** is defined in terms of the ampere:

When a conductor carries a steady current of 1 A, the quantity of charge that flows through a cross section of the conductor in 1 s is 1 C.

In deriving Equations 30.11 and 30.12, we assumed that both wires are long compared with their separation distance. In fact, only one wire needs to be long. The equations accurately describe the forces exerted on each other by a long wire and a straight parallel wire of limited length ℓ.

Quick Quiz 30.2 For $I_1 = 2$ A and $I_2 = 6$ A in Figure 30.8, which is true: (a) $F_1 = 3F_2$, (b) $F_1 = F_2/3$, (c) $F_1 = F_2$?

Quick Quiz 30.3 A loose spiral spring carrying no current is hung from the ceiling. When a switch is thrown so that a current exists in the spring, do the coils move (a) closer together, (b) farther apart, or (c) do they not move at all?

30.3 Ampère's Law

Oersted's 1819 discovery about deflected compass needles demonstrates that a current-carrying conductor produces a magnetic field. Figure 30.9a shows how this effect can be demonstrated in the classroom. Several compass needles are placed in a horizontal plane near a long vertical wire. When no current is

Andre-Marie Ampère
French Physicist (1775–1836)

Ampère is credited with the discovery of electromagnetism—the relationship between electric currents and magnetic fields. Ampère's genius, particularly in mathematics, became evident by the time he was 12 years old; his personal life, however, was filled with tragedy. His father, a wealthy city official, was guillotined during the French Revolution, and his wife died young, in 1803. Ampère died at the age of 61 of pneumonia. His judgment of his life is clear from the epitaph he chose for his gravestone: *Tandem Felix* (Happy at Last). *(Leonard de Selva/CORBIS)*

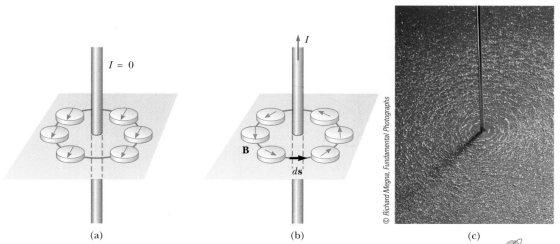

© Richard Megna, Fundamental Photographs

(a)   (b)   (c)

Active Figure 30.9 (a) When no current is present in the wire, all compass needles point in the same direction (toward the Earth's north pole). (b) When the wire carries a strong current, the compass needles deflect in a direction tangent to the circle, which is the direction of the magnetic field created by the current. (c) Circular magnetic field lines surrounding a current-carrying conductor, displayed with iron filings.

At the Active Figures link at http://www.pse6.com, you can change the value of the current to see the effect on the compasses.

present in the wire, all the needles point in the same direction (that of the Earth's magnetic field), as expected. When the wire carries a strong, steady current, the needles all deflect in a direction tangent to the circle, as in Figure 30.9b. These observations demonstrate that the direction of the magnetic field produced by the current in the wire is consistent with the right-hand rule described in Figure 30.4. When the current is reversed, the needles in Figure 30.9b also reverse.

Because the compass needles point in the direction of **B**, we conclude that the lines of **B** form circles around the wire, as discussed in the preceding section. By symmetry, the magnitude of **B** is the same everywhere on a circular path centered on the wire and lying in a plane perpendicular to the wire. By varying the current and distance a from the wire, we find that B is proportional to the current and inversely proportional to the distance from the wire, as Equation 30.5 describes.

Now let us evaluate the product $\mathbf{B} \cdot d\mathbf{s}$ for a small length element $d\mathbf{s}$ on the circular path defined by the compass needles, and sum the products for all elements over the closed circular path.[2] Along this path, the vectors $d\mathbf{s}$ and **B** are parallel at each point (see Fig. 30.9b), so $\mathbf{B} \cdot d\mathbf{s} = B\,ds$. Furthermore, the magnitude of **B** is constant on this circle and is given by Equation 30.5. Therefore, the sum of the products $B\,ds$ over the closed path, which is equivalent to the line integral of $\mathbf{B} \cdot d\mathbf{s}$, is

$$\oint \mathbf{B} \cdot d\mathbf{s} = B \oint ds = \frac{\mu_0 I}{2\pi r}(2\pi r) = \mu_0 I$$

where $\oint ds = 2\pi r$ is the circumference of the circular path. Although this result was calculated for the special case of a circular path surrounding a wire, it holds for a closed path of *any* shape (an *amperian loop*) surrounding a current that exists in an unbroken circuit. The general case, known as **Ampère's law,** can be stated as follows:

> The line integral of $\mathbf{B} \cdot d\mathbf{s}$ around any closed path equals $\mu_0 I$, where I is the total steady current passing through any surface bounded by the closed path.
>
> $$\oint \mathbf{B} \cdot d\mathbf{s} = \mu_0 I \qquad (30.13)$$

Ampère's law describes the creation of magnetic fields by all continuous current configurations, but at our mathematical level it is useful only for calculating the magnetic field of current configurations having a high degree of symmetry. Its use is similar to that of Gauss's law in calculating electric fields for highly symmetric charge distributions.

> **Quick Quiz 30.4** Rank the magnitudes of $\oint \mathbf{B} \cdot d\mathbf{s}$ for the closed paths in Figure 30.10, from least to greatest.

PITFALL PREVENTION

30.2 Avoiding Problems with Signs

When using Ampère's law, apply the following right-hand rule. Point your thumb in the direction of the current through the amperian loop. Your curled fingers then point in the direction that you should integrate around the loop in order to avoid having to define the current as negative.

Ampère's law

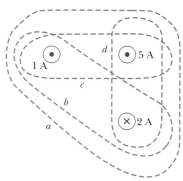

Figure 30.10 (Quick Quiz 30.4) Four closed paths around three current-carrying wires.

---

[2] You may wonder why we would choose to do this. The origin of Ampère's law is in nineteenth century science, in which a "magnetic charge" (the supposed analog to an isolated electric charge) was imagined to be moved around a circular field line. The work done on the charge was related to $\mathbf{B} \cdot d\mathbf{s}$, just as the work done moving an electric charge in an electric field is related to $\mathbf{E} \cdot d\mathbf{s}$. Thus, Ampère's law, a valid and useful principle, arose from an erroneous and abandoned work calculation!

Quick Quiz 30.5 Rank the magnitudes of $\oint \mathbf{B} \cdot d\mathbf{s}$ for the closed paths in Figure 30.11, from least to greatest.

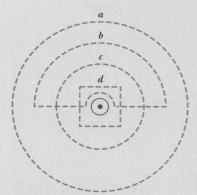

Figure 30.11 (Quick Quiz 30.5) Several closed paths near a single current-carrying wire.

Example 30.4   The Magnetic Field Created by a Long Current-Carrying Wire

A long, straight wire of radius R carries a steady current I that is uniformly distributed through the cross section of the wire (Fig. 30.12). Calculate the magnetic field a distance r from the center of the wire in the regions $r \geq R$ and $r < R$.

Solution Figure 30.12 helps us to conceptualize the wire and the current. Because the wire has a high degree of symmetry, we categorize this as an Ampère's law problem. For the $r \geq R$ case, we should arrive at the same result we obtained in Example 30.1, in which we applied the Biot–Savart law to the same situation. To analyze the problem, let us choose for our path of integration circle 1 in Figure 30.12. From symmetry, $\mathbf{B}$ must be constant in magnitude and parallel to $d\mathbf{s}$ at every point on this circle. Because the total current passing through the plane of the

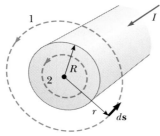

Figure 30.12 (Example 30.4) A long, straight wire of radius R carrying a steady current I uniformly distributed across the cross section of the wire. The magnetic field at any point can be calculated from Ampère's law using a circular path of radius r, concentric with the wire.

circle is I, Ampère's law gives

$$\oint \mathbf{B} \cdot d\mathbf{s} = B \oint ds = B(2\pi r) = \mu_0 I$$

$$B = \frac{\mu_0 I}{2\pi r} \qquad \text{(for } r \geq R) \qquad (30.14)$$

which is identical in form to Equation 30.5. Note how much easier it is to use Ampère's law than to use the Biot–Savart law. This is often the case in highly symmetric situations.

Now consider the interior of the wire, where $r < R$. Here the current I' passing through the plane of circle 2 is less than the total current I. Because the current is uniform over the cross section of the wire, the fraction of the current enclosed by circle 2 must equal the ratio of the area πr^2 enclosed by circle 2 to the cross-sectional area πR^2 of the wire:[3]

$$\frac{I'}{I} = \frac{\pi r^2}{\pi R^2}$$

$$I' = \frac{r^2}{R^2} I$$

Following the same procedure as for circle 1, we apply Ampère's law to circle 2:

$$\oint \mathbf{B} \cdot d\mathbf{s} = B(2\pi r) = \mu_0 I' = \mu_0 \left(\frac{r^2}{R^2} I \right)$$

---

[3]   Another way to look at this problem is to realize that the current enclosed by circle 2 must equal the product of the current density $J = I/\pi R^2$ and the area πr^2 of this circle.

$$B = \left(\frac{\mu_0 I}{2\pi R^2} \right) r \qquad (\text{for } r < R) \qquad (30.15)$$

To finalize this problem, note that this result is similar in form to the expression for the electric field inside a uniformly charged sphere (see Example 24.5). The magnitude of the magnetic field versus r for this configuration is plotted in Figure 30.13. Note that inside the wire, $B \rightarrow 0$ as $r \rightarrow 0$. Furthermore, we see that Equations 30.14 and 30.15 give the same value of the magnetic field at $r = R$, demonstrating that the magnetic field is continuous at the surface of the wire.

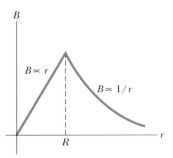

Figure 30.13 (Example 30.4) Magnitude of the magnetic field versus r for the wire shown in Figure 30.12. The field is proportional to r inside the wire and varies as $1/r$ outside the wire.

Example 30.5   The Magnetic Field Created by a Toroid

A device called a *toroid* (Fig. 30.14) is often used to create an almost uniform magnetic field in some enclosed area. The device consists of a conducting wire wrapped around a ring (a *torus*) made of a nonconducting material. For a toroid having N closely spaced turns of wire, calculate the magnetic field in the region occupied by the torus, a distance r from the center.

Solution To calculate this field, we must evaluate $\oint \mathbf{B} \cdot d\mathbf{s}$ over the circular amperian loop of radius r in the plane of Figure 30.14. By symmetry, we see that the magnitude of the field is constant on this circle and tangent to it, so $\mathbf{B} \cdot d\mathbf{s} = B \, ds$. Furthermore, the wire passes through the loop N times,

so that the total current through the loop is NI. Therefore, the right side of Equation 30.13 is $\mu_0 NI$ in this case.

Ampère's law applied to the circle gives

$$\oint \mathbf{B} \cdot d\mathbf{s} = B \oint ds = B(2\pi r) = \mu_0 NI$$

$$B = \frac{\mu_0 NI}{2\pi r} \qquad (30.16)$$

This result shows that B varies as $1/r$ and hence is *nonuniform* in the region occupied by the torus. However, if r is very large compared with the cross-sectional radius a of the torus, then the field is approximately uniform inside the torus.

For an ideal toroid, in which the turns are closely spaced, the external magnetic field is close to zero. It is not exactly zero, however. In Figure 30.14, imagine the radius r of the amperian loop to be either smaller than b or larger than c. In either case, the loop encloses zero net current, so $\oint \mathbf{B} \cdot d\mathbf{s} = 0$. We might be tempted to claim that this proves that $\mathbf{B} = 0$, but it does not. Consider the amperian loop on the right side of the toroid in Figure 30.14. The plane of this loop is perpendicular to the page, and the toroid passes through the loop. As charges enter the toroid as indicated by the current directions in Figure 30.14, they work their way counterclockwise around the toroid. Thus, a current passes through the perpendicular amperian loop! This current is small, but it is not zero. As a result, the toroid acts as a current loop and produces a weak external field of the form shown in Figure 30.7. The reason that $\oint \mathbf{B} \cdot d\mathbf{s} = 0$ for the amperian loops of radius $r < b$ and $r > c$ in the plane of the page is that the field lines are perpendicular to $d\mathbf{s}$, *not* because $\mathbf{B} = 0$.

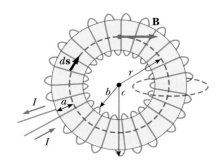

Figure 30.14 (Example 30.5) A toroid consisting of many turns of wire. If the turns are closely spaced, the magnetic field in the interior of the torus (the gold-shaded region) is tangent to the dashed circle and varies as $1/r$. The dimension a is the cross-sectional radius of the torus. The field outside the toroid is very small and can be described by using the amperian loop at the right side, perpendicular to the page.

Example 30.6   Magnetic Field Created by an Infinite Current Sheet

So far we have imagined currents carried by wires of small cross section. Let us now consider an example in which a current exists in an extended object. A thin, infinitely large sheet lying in the yz plane carries a current of linear current density $\mathbf{J}_s$. The current is in the y direction, and J_s represents the current per unit length measured along the z axis. Find the magnetic field near the sheet.

Solution This situation is similar to those involving Gauss's law (see Example 24.8). You may recall that the electric field

due to an infinite sheet of charge does not depend on distance from the sheet. Thus, we might expect a similar result here for the magnetic field.

To evaluate the line integral in Ampère's law, we construct a rectangular path through the sheet, as in Figure 30.15. The rectangle has dimensions ℓ and w, with the sides of length ℓ parallel to the sheet surface. The net current in the plane of the rectangle is $J_s \ell$. We apply Ampère's law over the rectangle and note that the two sides of length w do not contribute to the line integral because the component of $\mathbf{B}$

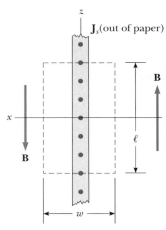

Figure 30.15 (Example 30.6) End view of an infinite current sheet lying in the yz plane, where the current is in the y direction (out of the page). This view shows the direction of **B** on both sides of the sheet.

along the direction of these paths is zero. By symmetry, the magnetic field is constant over the sides of length ℓ because every point on the infinitely large sheet is equivalent, and

hence the field should not vary from point to point. The only choices of field direction that are reasonable in this situation are either perpendicular or parallel to the sheet. However, a perpendicular field would pass *through* the current, which is inconsistent with the Biot–Savart law. Assuming a field that is constant in magnitude and parallel to the plane of the sheet, we obtain

$$\oint \mathbf{B} \cdot d\mathbf{s} = \mu_0 I = \mu_0 J_s \ell$$

$$2B\ell = \mu_0 J_s \ell$$

$$B = \mu_0 \frac{J_s}{2}$$

This result shows that the magnetic field is independent of distance from the current sheet, as we suspected. The expression for the magnitude of the magnetic field is similar in form to that for the magnitude of the electric field due to an infinite sheet of charge (Example 24.8):

$$E = \frac{\sigma}{2\epsilon_0}$$

Example 30.7   The Magnetic Force on a Current Segment

Wire 1 in Figure 30.16 is oriented along the y axis and carries a steady current I_1. A rectangular loop located to the right of the wire and in the xy plane carries a current I_2. Find the magnetic force exerted by wire 1 on the top wire of length b in the loop, labeled "Wire 2" in the figure.

Solution You may be tempted to use Equation 30.12 to obtain the force exerted on a small segment of length dx of wire 2. However, this equation applies only to two *parallel* wires and cannot be used here. The correct approach is to consider the force exerted by wire 1 on a small segment $d\mathbf{s}$ of wire 2 by using Equation 29.4. This force is given by $d\mathbf{F}_B = I\, d\mathbf{s} \times \mathbf{B}$, where $I = I_2$ and **B** is the magnetic field created by the current in wire 1 at the position of $d\mathbf{s}$. From Ampère's law, the field at a distance x

from wire 1 (see Eq. 30.14) is

$$\mathbf{B} = \frac{\mu_0 I_1}{2\pi x} (-\hat{\mathbf{k}})$$

where the unit vector $-\hat{\mathbf{k}}$ is used to indicate that the field due to the current in wire 1 at the position of $d\mathbf{s}$ points into the page. Because wire 2 is along the x axis, $d\mathbf{s} = dx\,\hat{\mathbf{i}}$, and we find that

$$d\mathbf{F}_B = \frac{\mu_0 I_1 I_2}{2\pi x} [\hat{\mathbf{i}} \times (-\hat{\mathbf{k}})]\, dx = \frac{\mu_0 I_1 I_2}{2\pi} \frac{dx}{x} \hat{\mathbf{j}}$$

Integrating over the limits $x = a$ to $x = a + b$ gives

$$\mathbf{F}_B = \frac{\mu_0 I_1 I_2}{2\pi} \ln x \Big]_a^{a+b} \hat{\mathbf{j}}$$

$$(1) \qquad \mathbf{F}_B = \frac{\mu_0 I_1 I_2}{2\pi} \ln\left(1 + \frac{b}{a}\right) \hat{\mathbf{j}}$$

The force on wire 2 points in the positive y direction, as indicated by the unit vector $\hat{\mathbf{j}}$ and as shown in Figure 30.16.

What If? What if the wire loop is moved to the left in Figure 30.16 until $a = 0$? What happens to the magnitude of the force on the wire?

Answer The force should become stronger because the loop is moving into a region of stronger magnetic field. Equation (1) shows that the force not only becomes stronger but the magnitude of the force becomes *infinite* as $a \rightarrow 0$! Thus, as the loop is moved to the left in Figure 30.16, the loop should be torn apart by the infinite upward force on the top side and the corresponding downward force on the bottom side! Furthermore, the force on the left side is

Figure 30.16 (Example 30.7) A wire on one side of a rectangular loop lying near a current-carrying wire experiences a force.

toward the left and should also become infinite. This is larger than the force toward the right on the right side because this side is still far from the wire, so the loop should be pulled into the wire with infinite force!

Does this really happen? In reality, it is impossible for $a \rightarrow 0$ because both wire 1 and wire 2 have finite sizes, so that the separation of the centers of the two wires is at least the sum of their radii.

A similar situation occurs when we re-examine the magnetic field due to a long straight wire, given by Equation 30.5. If we could move our observation point infinitesimally close to the wire, the magnetic field would become infinite! But in reality, the wire has a radius, and as soon as we enter the wire, the magnetic field starts to fall off as described by Equation 30.15 — approaching zero as we approach the center of the wire.

30.4   The Magnetic Field of a Solenoid

A **solenoid** is a long wire wound in the form of a helix. With this configuration, a reasonably uniform magnetic field can be produced in the space surrounded by the turns of wire—which we shall call the *interior* of the solenoid—when the solenoid carries a current. When the turns are closely spaced, each can be approximated as a circular loop, and the net magnetic field is the vector sum of the fields resulting from all the turns.

Figure 30.17 shows the magnetic field lines surrounding a loosely wound solenoid. Note that the field lines in the interior are nearly parallel to one another, are uniformly distributed, and are close together, indicating that the field in this space is strong and almost uniform.

If the turns are closely spaced and the solenoid is of finite length, the magnetic field lines are as shown in Figure 30.18a. This field line distribution is similar to that surrounding a bar magnet (see Fig. 30.18b). Hence, one end of the solenoid behaves like the north pole of a magnet, and the opposite end behaves like the south pole. As the length of the solenoid increases, the interior field becomes more uniform and the exterior field becomes weaker. An *ideal solenoid* is approached when the turns are closely spaced and the length is much greater than the radius of the turns. Figure 30.19 shows a longitudinal cross section of part of such a solenoid carrying a current *I*. In this case, the external field is close to zero, and the interior field is uniform over a great volume.

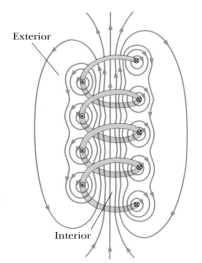

Figure 30.17 The magnetic field lines for a loosely wound solenoid.

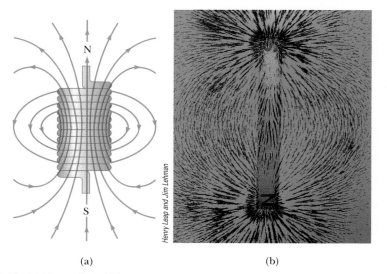

(a)   (b)

Henry Leap and Jim Lehman

Figure 30.18 (a) Magnetic field lines for a tightly wound solenoid of finite length, carrying a steady current. The field in the interior space is strong and nearly uniform. Note that the field lines resemble those of a bar magnet, meaning that the solenoid effectively has north and south poles. (b) The magnetic field pattern of a bar magnet, displayed with small iron filings on a sheet of paper.

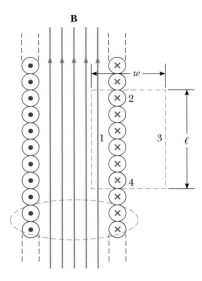

Figure 30.19 Cross-sectional view of an ideal solenoid, where the interior magnetic field is uniform and the exterior field is close to zero. Ampère's law applied to the circular path near the bottom whose plane is perpendicular to the page can be used to show that there is a weak field outside the solenoid. Ampère's law applied to the rectangular dashed path in the plane of the page can be used to calculate the magnitude of the interior field.

If we consider the amperian loop perpendicular to the page in Figure 30.19, surrounding the ideal solenoid, we see that it encloses a small current as the charges in the wire move coil by coil along the length of the solenoid. Thus, there is a nonzero magnetic field outside the solenoid. It is a weak field, with circular field lines, like those due to a line of current as in Figure 30.4. For an ideal solenoid, this is the only field external to the solenoid. We can eliminate this field in Figure 30.19 by adding a second layer of turns of wire outside the first layer, with the current carried along the axis of the solenoid in the opposite direction compared to the first layer. Then the net current along the axis is zero.

We can use Ampère's law to obtain a quantitative expression for the interior magnetic field in an ideal solenoid. Because the solenoid is ideal, **B** in the interior space is uniform and parallel to the axis, and the magnetic field lines in the exterior space form circles around the solenoid. The planes of these circles are perpendicular to the page. Consider the rectangular path of length ℓ and width w shown in Figure 30.19. We can apply Ampère's law to this path by evaluating the integral of $\mathbf{B} \cdot d\mathbf{s}$ over each side of the rectangle. The contribution along side 3 is zero because the magnetic field lines are perpendicular to the path in this region. The contributions from sides 2 and 4 are both zero, again because **B** is perpendicular to $d\mathbf{s}$ along these paths, both inside and outside the solenoid. Side 1 gives a contribution to the integral because along this path **B** is uniform and parallel to $d\mathbf{s}$. The integral over the closed rectangular path is therefore

$$\oint \mathbf{B} \cdot d\mathbf{s} = \int_{\text{path 1}} \mathbf{B} \cdot d\mathbf{s} = B \int_{\text{path 1}} ds = B\ell$$

The right side of Ampère's law involves the total current I through the area bounded by the path of integration. In this case, the total current through the rectangular path equals the current through each turn multiplied by the number of turns. If N is the number of turns in the length ℓ, the total current through the rectangle is NI. Therefore, Ampère's law applied to this path gives

$$\oint \mathbf{B} \cdot d\mathbf{s} = B\ell = \mu_0 NI$$

$$B = \mu_0 \frac{N}{\ell} I = \mu_0 nI \qquad (30.17) \qquad \textbf{Magnetic field inside a solenoid}$$

where $n = N/\ell$ is the number of turns per unit length.

We also could obtain this result by reconsidering the magnetic field of a toroid (see Example 30.5). If the radius r of the torus in Figure 30.14 containing N turns is much greater than the toroid's cross-sectional radius a, a short section of the toroid approximates a solenoid for which $n = N/2\pi r$. In this limit, Equation 30.16 agrees with Equation 30.17.

Equation 30.17 is valid only for points near the center (that is, far from the ends) of a very long solenoid. As you might expect, the field near each end is smaller than the value given by Equation 30.17. At the very end of a long solenoid, the magnitude of the field is half the magnitude at the center (see Problem 32).

> **Quick Quiz 30.6** Consider a solenoid that is very long compared to the radius. Of the following choices, the most effective way to increase the magnetic field in the interior of the solenoid is to (a) double its length, keeping the number of turns per unit length constant, (b) reduce its radius by half, keeping the number of turns per unit length constant, (c) overwrapping the entire solenoid with an additional layer of current-carrying wire.

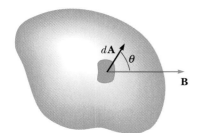

Figure 30.20 The magnetic flux through an area element dA is $\mathbf{B} \cdot d\mathbf{A} = B \, dA \cos \theta$, where $d\mathbf{A}$ is a vector perpendicular to the surface.

Definition of magnetic flux

30.5 Magnetic Flux

The flux associated with a magnetic field is defined in a manner similar to that used to define electric flux (see Eq. 24.3). Consider an element of area dA on an arbitrarily shaped surface, as shown in Figure 30.20. If the magnetic field at this element is $\mathbf{B}$, the magnetic flux through the element is $\mathbf{B} \cdot d\mathbf{A}$, where $d\mathbf{A}$ is a vector that is perpendicular to the surface and has a magnitude equal to the area dA. Therefore, the total magnetic flux Φ_B through the surface is

$$\Phi_B = \int \mathbf{B} \cdot d\mathbf{A} \tag{30.18}$$

Consider the special case of a plane of area A in a uniform field $\mathbf{B}$ that makes an angle θ with $d\mathbf{A}$. The magnetic flux through the plane in this case is

$$\Phi_B = BA \cos \theta \tag{30.19}$$

If the magnetic field is parallel to the plane, as in Figure 30.21a, then $\theta = 90°$ and the flux through the plane is zero. If the field is perpendicular to the plane, as in Figure 30.21b, then $\theta = 0$ and the flux through the plane is BA (the maximum value).

The unit of magnetic flux is $T \cdot m^2$, which is defined as a *weber* (Wb); 1 Wb = 1 $T \cdot m^2$.

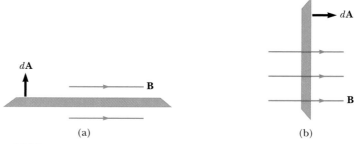

(a)   (b)

At the Active Figures link at http://www.pse6.com, *you can rotate the plane and change the value of the field to see the effect on the flux.*

Active Figure 30.21 Magnetic flux through a plane lying in a magnetic field. (a) The flux through the plane is zero when the magnetic field is parallel to the plane surface. (b) The flux through the plane is a maximum when the magnetic field is perpendicular to the plane.

Example 30.8   Magnetic Flux Through a Rectangular Loop

A rectangular loop of width a and length b is located near a long wire carrying a current I (Fig. 30.22). The distance between the wire and the closest side of the loop is c. The wire is parallel to the long side of the loop. Find the total magnetic flux through the loop due to the current in the wire.

Solution From Equation 30.14, we know that the magnitude of the magnetic field created by the wire at a distance r from the wire is

$$B = \frac{\mu_0 I}{2\pi r}$$

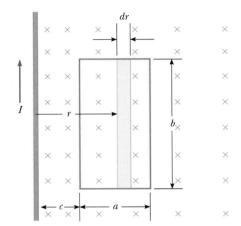

Figure 30.22 (Example 30.8) The magnetic field due to the wire carrying a current I is not uniform over the rectangular loop.

The factor $1/r$ indicates that the field varies over the loop, and Figure 30.22 shows that the field is directed into the page at the location of the loop. Because **B** is parallel to $d\mathbf{A}$ at any point within the loop, the magnetic flux through an area element dA is

$$\Phi_B = \int B \, dA = \int \frac{\mu_0 I}{2\pi r} \, dA$$

To integrate, we first express the area element (the tan region in Fig. 30.22) as $dA = b \, dr$. Because r is now the only variable in the integral, we have

$$\Phi_B = \frac{\mu_0 I b}{2\pi} \int_c^{a+c} \frac{dr}{r} = \frac{\mu_0 I b}{2\pi} \ln r \Big]_c^{a+c}$$

$$(1) \qquad = \frac{\mu_0 I b}{2\pi} \ln\left(\frac{a+c}{c}\right) = \frac{\mu_0 I b}{2\pi} \ln\left(1 + \frac{a}{c}\right)$$

What If? Suppose we move the loop in Figure 30.22 very far away from the wire. What happens to the magnetic flux?

Answer The flux should become smaller as the loop moves into weaker and weaker fields.

As the loop moves far away, the value of c is much larger than that of a, so that $a/c \rightarrow 0$. Thus, the natural logarithm in Equation (1) approaches the limit

$$\ln\left(1 + \frac{a}{c}\right) \quad \longrightarrow \quad \ln(1 + 0) = \ln(1) = 0$$

and we find that $\Phi_B \rightarrow 0$ as we expected.

 At the Interactive Worked Example link at **http://www.pse6.com,** *you can investigate the flux as the loop parameters change.*

30.6   Gauss's Law in Magnetism

In Chapter 24 we found that the electric flux through a closed surface surrounding a net charge is proportional to that charge (Gauss's law). In other words, the number of electric field lines leaving the surface depends only on the net charge within it. This property is based on the fact that electric field lines originate and terminate on electric charges.

The situation is quite different for magnetic fields, which are continuous and form closed loops. In other words, magnetic field lines do not begin or end at any point—as illustrated in Figures 30.4 and 30.23. Figure 30.23 shows the magnetic field lines of a bar magnet. Note that for any closed surface, such as the one outlined by the dashed line in Figure 30.23, the number of lines entering the surface equals the number leaving the surface; thus, the net magnetic flux is zero. In contrast, for a closed surface surrounding one charge of an electric dipole (Fig. 30.24), the net electric flux is not zero.

Gauss's law in magnetism states that

the net magnetic flux through any closed surface is always zero:

$$\oint \mathbf{B} \cdot d\mathbf{A} = 0 \qquad\qquad (30.20)$$ Gauss's law in magnetism

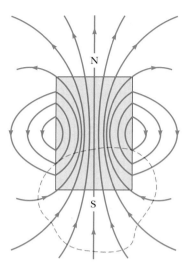

Figure 30.23 The magnetic field lines of a bar magnet form closed loops. Note that the net magnetic flux through a closed surface surrounding one of the poles (or any other closed surface) is zero. (The dashed line represents the intersection of the surface with the page.)

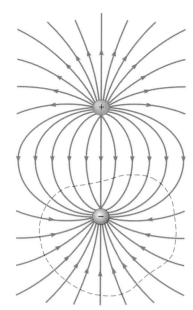

Figure 30.24 The electric field lines surrounding an electric dipole begin on the positive charge and terminate on the negative charge. The electric flux through a closed surface surrounding one of the charges is not zero.

This statement is based on the experimental fact, mentioned in the opening of Chapter 29, that isolated magnetic poles (monopoles) have never been detected and perhaps do not exist. Nonetheless, scientists continue the search because certain theories that are otherwise successful in explaining fundamental physical behavior suggest the possible existence of monopoles.

30.7 Displacement Current and the General Form of Ampère's Law

We have seen that charges in motion produce magnetic fields. When a current-carrying conductor has high symmetry, we can use Ampère's law to calculate the magnetic field it creates. In Equation 30.13, $\oint \mathbf{B} \cdot d\mathbf{s} = \mu_0 I$, the line integral is over any closed path through which the conduction current passes, where the conduction current is defined by the expression $I = dq/dt$. (In this section we use the term *conduction current* to refer to the current carried by the wire, to distinguish it from a new type of current that we shall introduce shortly.) We now show that **Ampère's law in this form is valid only if any electric fields present are constant in time.** Maxwell recognized this limitation and modified Ampère's law to include time-varying electric fields.

We can understand the problem by considering a capacitor that is being charged as illustrated in Figure 30.25. When a conduction current is present, the charge on the positive plate changes but *no conduction current exists in the gap between the plates.* Now consider the two surfaces S_1 and S_2 in Figure 30.25, bounded by the same path P. Ampère's law states that $\oint \mathbf{B} \cdot d\mathbf{s}$ around this path must equal $\mu_0 I$, where I is the total current through *any* surface bounded by the path P.

When the path P is considered as bounding S_1, $\oint \mathbf{B} \cdot d\mathbf{s} = \mu_0 I$ because the conduction current passes through S_1. When the path is considered as bounding S_2, however, $\oint \mathbf{B} \cdot d\mathbf{s} = 0$ because no conduction current passes through S_2. Thus, we have a contradictory situation that arises from the discontinuity of the current! Maxwell solved this problem by postulating an additional term on the right side

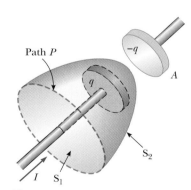

Figure 30.25 Two surfaces S_1 and S_2 near the plate of a capacitor are bounded by the same path P. The conduction current in the wire passes only through S_1. This leads to a contradiction in Ampère's law that is resolved only if one postulates a displacement current through S_2.

of Equation 30.13, which includes a factor called the **displacement current** I_d, defined as[4]

$$I_d \equiv \epsilon_0 \frac{d\Phi_E}{dt}$$ (30.21)   **Displacement current**

where ϵ_0 is the permittivity of free space (see Section 23.3) and $\Phi_E = \int \mathbf{E} \cdot d\mathbf{A}$ is the electric flux (see Eq. 24.3).

As the capacitor is being charged (or discharged), the changing electric field between the plates may be considered equivalent to a current that acts as a continuation of the conduction current in the wire. When the expression for the displacement current given by Equation 30.21 is added to the conduction current on the right side of Ampère's law, the difficulty represented in Figure 30.25 is resolved. No matter which surface bounded by the path P is chosen, either a conduction current or a displacement current passes through it. With this new term I_d, we can express the general form of Ampère's law (sometimes called the **Ampère–Maxwell law**) as[5]

$$\oint \mathbf{B} \cdot d\mathbf{s} = \mu_0(I + I_d) = \mu_0 I + \mu_0 \epsilon_0 \frac{d\Phi_E}{dt}$$ (30.22)   **Ampère–Maxwell law**

We can understand the meaning of this expression by referring to Figure 30.26. The electric flux through surface S_2 is $\Phi_E = \int \mathbf{E} \cdot d\mathbf{A} = EA$, where A is the area of the capacitor plates and E is the magnitude of the uniform electric field between the plates. If q is the charge on the plates at any instant, then $E = q/(\epsilon_0 A)$. (See Section 26.2.) Therefore, the electric flux through S_2 is simply

$$\Phi_E = EA = \frac{q}{\epsilon_0}$$

Hence, the displacement current through S_2 is

$$I_d = \epsilon_0 \frac{d\Phi_E}{dt} = \frac{dq}{dt}$$ (30.23)

That is, the displacement current I_d through S_2 is precisely equal to the conduction current I through S_1!

Figure 30.26 Because it exists only in the wires attached to the capacitor plates, the conduction current $I = dq/dt$ passes through S_1 but not through S_2. Only the displacement current $I_d = \epsilon_0 d\Phi_E/dt$ passes through S_2. The two currents must be equal for continuity.

---

[4] *Displacement* in this context does not have the meaning it does in Chapter 2. Despite the inaccurate implications, the word is historically entrenched in the language of physics, so we continue to use it.

[5] Strictly speaking, this expression is valid only in a vacuum. If a magnetic material is present, one must change μ_0 and ϵ_0 on the right-hand side of Equation 30.22 to the permeability μ_m (see Section 30.8) and permittivity ϵ characteristic of the material. Alternatively, one may include a magnetizing current I_m on the right hand side of Equation 30.22 to make Ampère's law fully general. On a microscopic scale, I_m is as real as I.

By considering surface S_2, we can identify the displacement current as the source of the magnetic field on the surface boundary. The displacement current has its physical origin in the time-varying electric field. The central point of this formalism is that

> magnetic fields are produced both by conduction currents and by time-varying electric fields.

This result was a remarkable example of theoretical work by Maxwell, and it contributed to major advances in the understanding of electromagnetism.

Quick Quiz 30.7 In an *RC* circuit, the capacitor begins to discharge. During the discharge, in the region of space between the plates of the capacitor, there is (a) conduction current but no displacement current, (b) displacement current but no conduction current, (c) both conduction and displacement current, (d) no current of any type.

Quick Quiz 30.8 The capacitor in an *RC* circuit begins to discharge. During the discharge, in the region of space between the plates of the capacitor, there is (a) an electric field but no magnetic field, (b) a magnetic field but no electric field, (c) both electric and magnetic fields, (d) no fields of any type.

Example 30.9 Displacement Current in a Capacitor

A sinusoidally varying voltage is applied across an 8.00-μF capacitor. The frequency of the voltage is 3.00 kHz, and the voltage amplitude is 30.0 V. Find the displacement current in the capacitor.

Solution The angular frequency of the source, from Equation 15.12, is given by $\omega = 2\pi f = 2\pi(3.00 \times 10^3 \text{ Hz}) = 1.88 \times 10^4 \text{ s}^{-1}$. Hence, the voltage across the capacitor in terms of t is

$$\Delta V = \Delta V_{max} \sin \omega t = (30.0 \text{ V}) \sin(1.88 \times 10^4 \ t)$$

We can use Equation 30.23 and the fact that the charge on the capacitor is $q = C \Delta V$ to find the displacement current:

$$I_d = \frac{dq}{dt} = \frac{d}{dt} (C\Delta V) = C \frac{d}{dt} (\Delta V)$$

$$= (8.00 \times 10^{-6} \text{ F}) \frac{d}{dt} [(30.0 \text{ V}) \sin(1.88 \times 10^4 \ t)]$$

$$= (4.52 \text{ A}) \cos(1.88 \times 10^4 \ t)$$

The displacement current varies sinusoidally with time and has a maximum value of 4.52 A.

30.8 Magnetism in Matter

The magnetic field produced by a current in a coil of wire gives us a hint as to what causes certain materials to exhibit strong magnetic properties. Earlier we found that a coil like the one shown in Figure 30.18 has a north pole and a south pole. In general, *any* current loop has a magnetic field and thus has a magnetic dipole moment, including the atomic-level current loops described in some models of the atom.

The Magnetic Moments of Atoms

We begin our discussion with a classical model of the atom in which electrons move in circular orbits around the much more massive nucleus. In this model, an orbiting electron constitutes a tiny current loop (because it is a moving charge),

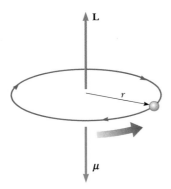

Figure 30.27 An electron moving in the direction of the gray arrow in a circular orbit of radius r has an angular momentum **L** in one direction and a magnetic moment $\boldsymbol{\mu}$ in the opposite direction. Because the electron carries a negative charge, the direction of the current due to its motion about the nucleus is opposite the direction of that motion.

and the magnetic moment of the electron is associated with this orbital motion. Although this model has many deficiencies, some of its predictions are in good agreement with the correct theory, which is expressed in terms of quantum physics.

In our classical model, we assume that an electron moves with constant speed v in a circular orbit of radius r about the nucleus, as in Figure 30.27. Because the electron travels a distance of $2\pi r$ (the circumference of the circle) in a time interval T, its orbital speed is $v = 2\pi r/T$. The current I associated with this orbiting electron is its charge e divided by T. Using $T = 2\pi/\omega$ and $\omega = v/r$, we have

$$I = \frac{e}{T} = \frac{e\omega}{2\pi} = \frac{ev}{2\pi r}$$

The magnitude of the magnetic moment associated with this current loop is $\mu = IA$, where $A = \pi r^2$ is the area enclosed by the orbit. Therefore,

$$\mu = IA = \left(\frac{ev}{2\pi r}\right)\pi r^2 = \tfrac{1}{2}evr \tag{30.24}$$

Because the magnitude of the orbital angular momentum of the electron is $L = m_e vr$ (Eq. 11.12 with $\phi = 90°$), the magnetic moment can be written as

$$\mu = \left(\frac{e}{2m_e}\right)L \tag{30.25}$$

Orbital magnetic moment

This result demonstrates that **the magnetic moment of the electron is proportional to its orbital angular momentum.** Because the electron is negatively charged, the vectors $\boldsymbol{\mu}$ and **L** point in *opposite* directions. Both vectors are perpendicular to the plane of the orbit, as indicated in Figure 30.27.

A fundamental outcome of quantum physics is that orbital angular momentum is quantized and is equal to multiples of $\hbar = h/2\pi = 1.05 \times 10^{-34}\,\text{J} \cdot \text{s}$, where h is Planck's constant (introduced in Section 11.6). The smallest nonzero value of the electron's magnetic moment resulting from its orbital motion is

$$\mu = \sqrt{2}\,\frac{e}{2m_e}\,\hbar \tag{30.26}$$

We shall see in Chapter 42 how expressions such as Equation 30.26 arise.

Because all substances contain electrons, you may wonder why most substances are not magnetic. The main reason is that in most substances, the magnetic moment of one electron in an atom is canceled by that of another electron orbiting in the opposite direction. The net result is that, for most materials, **the magnetic effect produced by the orbital motion of the electrons is either zero or very small.**

In addition to its orbital magnetic moment, an electron (as well as protons, neutrons, and other particles) has an intrinsic property called **spin** that also contributes to its magnetic moment. Classically, the electron might be viewed as

 PITFALL PREVENTION

30.3 The Electron Does Not Spin

Do not be misled; the electron is *not* physically spinning. It has an intrinsic angular momentum *as if it were spinning*, but the notion of rotation for a point particle is meaningless. Rotation applies only to a *rigid object*, with an extent in space, as in Chapter 10. Spin angular momentum is actually a relativistic effect.

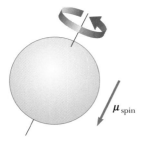

Figure 30.28 Classical model of a spinning electron. We can adopt this model to remind ourselves that electrons have an intrinsic angular momentum. The model should not be pushed too far, however—it gives an incorrect magnitude for the magnetic moment, incorrect quantum numbers, and too many degrees of freedom.

Table 30.1

Magnetic Moments of Some Atoms and Ions	
Atom or Ion	**Magnetic Moment (10^{-24} J/T)**
H	9.27
He	0
Ne	0
Ce^{3+}	19.8
Yb^{3+}	37.1

Magnetization vector M

spinning about its axis as shown in Figure 30.28, but you should be very careful with the classical interpretation. The magnitude of the angular momentum **S** associated with spin is on the same order of magnitude as the magnitude of the angular momentum **L** due to the orbital motion. The magnitude of the spin angular momentum of an electron predicted by quantum theory is

$$S = \frac{\sqrt{3}}{2} \hbar$$

The magnetic moment characteristically associated with the spin of an electron has the value

$$\mu_{\text{spin}} = \frac{e\hbar}{2m_e} \tag{30.27}$$

This combination of constants is called the **Bohr magneton μ_B**:

$$\mu_B = \frac{e\hbar}{2m_e} = 9.27 \times 10^{-24} \text{ J/T} \tag{30.28}$$

Thus, atomic magnetic moments can be expressed as multiples of the Bohr magneton. (Note that 1 J/T = 1 A·m².)

In atoms containing many electrons, the electrons usually pair up with their spins opposite each other; thus, the spin magnetic moments cancel. However, atoms containing an odd number of electrons must have at least one unpaired electron and therefore some spin magnetic moment. The total magnetic moment of an atom is the vector sum of the orbital and spin magnetic moments, and a few examples are given in Table 30.1. Note that helium and neon have zero moments because their individual spin and orbital moments cancel.

The nucleus of an atom also has a magnetic moment associated with its constituent protons and neutrons. However, the magnetic moment of a proton or neutron is much smaller than that of an electron and can usually be neglected. We can understand this by inspecting Equation 30.28 and replacing the mass of the electron with the mass of a proton or a neutron. Because the masses of the proton and neutron are much greater than that of the electron, their magnetic moments are on the order of 10^3 times smaller than that of the electron.

Magnetization Vector and Magnetic Field Strength

The magnetic state of a substance is described by a quantity called the **magnetization vector M. The magnitude of this vector is defined as the magnetic moment per unit volume of the substance.** As you might expect, the total magnetic field **B** at a point within a substance depends on both the applied (external) field $\mathbf{B}_0$ and the magnetization of the substance.

Consider a region in which a magnetic field $\mathbf{B}_0$ is produced by a current-carrying conductor. If we now fill that region with a magnetic substance, the total magnetic field **B** in the region is $\mathbf{B} = \mathbf{B}_0 + \mathbf{B}_m$, where $\mathbf{B}_m$ is the field produced by the magnetic substance.

Let us determine the relationship between $\mathbf{B}_m$ and **M**. Imagine that the field $\mathbf{B}_m$ is created by a solenoid rather than by the magnetic material. Then, $B_m = \mu_0 nI$, where I is the current in the imaginary solenoid and n is the number of turns per unit length. Let us manipulate this expression as follows:

$$B_m = \mu_0 nI = \mu_0 \frac{N}{\ell} I = \mu_0 \frac{NIA}{\ell A}$$

where N is the number of turns in length ℓ, and we have multiplied the numerator and denominator by A, the cross sectional area of the solenoid in the last step. We recognize the numerator NIA as the total magnetic moment of all the loops in length ℓ and the denominator ℓA as the volume of the solenoid associated with this length:

$$B_m = \mu_0 \frac{\mu}{V}$$

The ratio of total magnetic moment to volume is what we have defined as magnetization in the case where the field is due to a material rather than a solenoid. Thus, we can express the contribution $\mathbf{B}_m$ to the total field in terms of the magnetization vector of the substance as $\mathbf{B}_m = \mu_0 \mathbf{M}$. When a substance is placed in a magnetic field, the total magnetic field in the region is expressed as

$$\mathbf{B} = \mathbf{B}_0 + \mu_0 \mathbf{M} \qquad (30.29)$$

When analyzing magnetic fields that arise from magnetization, it is convenient to introduce a field quantity called the **magnetic field strength H** within the substance. The magnetic field strength is related to the magnetic field due to the conduction currents in wires. To emphasize the distinction between the field strength $\mathbf{H}$ and the field $\mathbf{B}$, the latter is often called the *magnetic flux density* or the *magnetic induction*. The magnetic field strength is the magnetic moment per unit volume due to currents; thus, it is similar to the vector $\mathbf{M}$ and has the same units.

Recognizing the similarity between $\mathbf{M}$ and $\mathbf{H}$, we can define $\mathbf{H}$ as $\mathbf{H} \equiv \mathbf{B}_0/\mu_0$. Thus, Equation 30.29 can be written

$$\mathbf{B} = \mu_0(\mathbf{H} + \mathbf{M}) \qquad (30.30)$$

The quantities $\mathbf{H}$ and $\mathbf{M}$ have the same units. Because $\mathbf{M}$ is magnetic moment per unit volume, its SI units are (ampere)(meter)^2/(meter)^3, or amperes per meter.

To better understand these expressions, consider the torus region of a toroid that carries a current I. If this region is a vacuum, $\mathbf{M} = 0$ (because no magnetic material is present), the total magnetic field is that arising from the current alone, and $\mathbf{B} = \mathbf{B}_0 = \mu_0 \mathbf{H}$. Because $B_0 = \mu_0 nI$ in the torus region, where n is the number of turns per unit length of the toroid, $H = B_0/\mu_0 = \mu_0 nI/\mu_0$ or

$$H = nI \qquad (30.31)$$

In this case, the magnetic field in the torus region is due only to the current in the windings of the toroid.

If the torus is now made of some substance and the current I is kept constant, $\mathbf{H}$ in the torus region remains unchanged (because it depends on the current only) and has magnitude nI. The total field $\mathbf{B}$, however, is different from that when the torus region was a vacuum. From Equation 30.30, we see that part of $\mathbf{B}$ arises from the term $\mu_0 \mathbf{H}$ associated with the current in the toroid, and part arises from the term $\mu_0 \mathbf{M}$ due to the magnetization of the substance of which the torus is made.

Classification of Magnetic Substances

Substances can be classified as belonging to one of three categories, depending on their magnetic properties. **Paramagnetic** and **ferromagnetic** materials are those made of atoms that have permanent magnetic moments. **Diamagnetic** materials are those made of atoms that do not have permanent magnetic moments.

For paramagnetic and diamagnetic substances, the magnetization vector $\mathbf{M}$ is proportional to the magnetic field strength $\mathbf{H}$. For these substances placed in an external magnetic field, we can write

$$\mathbf{M} = \chi \mathbf{H} \qquad (30.32)$$

where χ (Greek letter chi) is a dimensionless factor called the **magnetic susceptibility.** It can be considered a measure of how *susceptible* a material is to being magnetized. For paramagnetic substances, χ is positive and $\mathbf{M}$ is in the same direction as $\mathbf{H}$. For

Magnetic field strength H

Magnetic susceptibility χ

Table 30.2

Magnetic Susceptibilities of Some Paramagnetic and Diamagnetic Substances at 300 K			
Paramagnetic Substance	χ	**Diamagnetic Substance**	χ
Aluminum	2.3×10^{-5}	Bismuth	-1.66×10^{-5}
Calcium	1.9×10^{-5}	Copper	-9.8×10^{-6}
Chromium	2.7×10^{-4}	Diamond	-2.2×10^{-5}
Lithium	2.1×10^{-5}	Gold	-3.6×10^{-5}
Magnesium	1.2×10^{-5}	Lead	-1.7×10^{-5}
Niobium	2.6×10^{-4}	Mercury	-2.9×10^{-5}
Oxygen	2.1×10^{-6}	Nitrogen	-5.0×10^{-9}
Platinum	2.9×10^{-4}	Silver	-2.6×10^{-5}
Tungsten	6.8×10^{-5}	Silicon	-4.2×10^{-6}

diamagnetic substances, χ is negative and **M** is opposite **H**. The susceptibilities of some substances are given in Table 30.2.

Substituting Equation 30.32 for **M** into Equation 30.30 gives

$$\mathbf{B} = \mu_0(\mathbf{H} + \mathbf{M}) = \mu_0(\mathbf{H} + \chi\mathbf{H}) = \mu_0(1 + \chi)\mathbf{H}$$

or

$$\mathbf{B} = \mu_m\mathbf{H} \tag{30.33}$$

Magnetic permeability

where the constant μ_m is called the **magnetic permeability** of the substance and is related to the susceptibility by

$$\mu_m = \mu_0(1 + \chi) \tag{30.34}$$

Substances may be classified in terms of how their magnetic permeability μ_m compares with μ_0 (the permeability of free space), as follows:

$$\text{Paramagnetic} \qquad \mu_m > \mu_0$$

$$\text{Diamagnetic} \qquad \mu_m < \mu_0$$

Because χ is very small for paramagnetic and diamagnetic substances (see Table 30.2), μ_m is nearly equal to μ_0 for these substances. For ferromagnetic substances, however, μ_m is typically several thousand times greater than μ_0 (meaning that χ is very large for ferromagnetic substances).

Although Equation 30.33 provides a simple relationship between **B** and **H**, we must interpret it with care when dealing with ferromagnetic substances. We find that **M** is not a linear function of **H** for ferromagnetic substances. This is because the value of μ_m is not only a characteristic of the ferromagnetic substance but also depends on the previous state of the substance and on the process it underwent as it moved from its previous state to its present one. We shall investigate this more deeply after the following example.

Example 30.10   An Iron-Filled Toroid

A toroid wound with 60.0 turns/m of wire carries a current of 5.00 A. The torus is iron, which has a magnetic permeability of $\mu_m = 5\,000\mu_0$ under the given conditions. Find H and B inside the iron.

Solution Using Equations 30.31 and 30.33, we obtain

$$H = nI = (60.0 \text{ turns/m})(5.00 \text{ A}) = \boxed{300 \text{ A} \cdot \text{turns/m}}$$

$$B = \mu_m H = 5\,000\mu_0 H$$
$$= 5\,000(4\pi \times 10^{-7} \text{ T} \cdot \text{m/A})(300 \text{ A} \cdot \text{turns/m})$$
$$= \boxed{1.88 \text{ T}}$$

This value of B is 5 000 times the value in the absence of iron!

Ferromagnetism

A small number of crystalline substances exhibit strong magnetic effects called **ferromagnetism.** Some examples of ferromagnetic substances are iron, cobalt, nickel, gadolinium, and dysprosium. These substances contain permanent atomic magnetic moments that tend to align parallel to each other even in a weak external magnetic field. Once the moments are aligned, the substance remains magnetized after the external field is removed. This permanent alignment is due to a strong coupling between neighboring moments, a coupling that can be understood only in quantum-mechanical terms.

All ferromagnetic materials are made up of microscopic regions called **domains,** regions within which all magnetic moments are aligned. These domains have volumes of about 10^{-12} to 10^{-8} m^3 and contain 10^{17} to 10^{21} atoms. The boundaries between the various domains having different orientations are called **domain walls.** In an unmagnetized sample, the magnetic moments in the domains are randomly oriented so that the net magnetic moment is zero, as in Figure 30.29a. When the sample is placed in an external magnetic field $\mathbf{B}_0$, the size of those domains with magnetic moments aligned with the field grows, which results in a magnetized sample, as in Figure 30.29b. As the external field becomes very strong, as in Figure 30.29c, the domains in which the magnetic moments are not aligned with the field become very small. When the external field is removed, the sample may retain a net magnetization in the direction of the original field. At ordinary temperatures, thermal agitation is not sufficient to disrupt this preferred orientation of magnetic moments.

A typical experimental arrangement that is used to measure the magnetic properties of a ferromagnetic material consists of a torus made of the material wound with N turns of wire, as shown in Figure 30.30, where the windings are represented in black and are referred to as the *primary coil.* This apparatus is sometimes referred to as a **Rowland ring.** A *secondary coil* (the red wires in Fig. 30.30) connected to a galvanometer is used to measure the total magnetic flux through the torus. The magnetic field $\mathbf{B}$ in the torus is measured by increasing the current in the toroid from zero to I. As the current changes, the magnetic flux through the secondary coil changes by an amount BA, where A is the cross-sectional area of the toroid. As shown in Chapter 31, because of this changing flux, an emf that is proportional to the rate of change in magnetic flux is induced in the secondary coil. If the galvanometer is properly calibrated, a value for $\mathbf{B}$ corresponding to any value of the current in the primary coil can be obtained. The magnetic field $\mathbf{B}$ is measured first in the absence of the torus and then with the torus in place. The magnetic properties of the torus material are then obtained from a comparison of the two measurements.

Now consider a torus made of unmagnetized iron. If the current in the primary coil is increased from zero to some value I, the magnitude of the magnetic field strength H increases linearly with I according to the expression $H = nI$. Furthermore,

(a)

(b) $\longrightarrow \mathbf{B}_0$

(c) $\longrightarrow \mathbf{B}_0$

Figure 30.29 (a) Random orientation of atomic magnetic dipoles in the domains of an unmagnetized substance. (b) When an external field $\mathbf{B}_0$ is applied, the domains with components of magnetic moment in the same direction as $\mathbf{B}_0$ grow larger, giving the sample a net magnetization. (c) As the field is made even stronger, the domains with magnetic moment vectors not aligned with the external field become very small.

Figure 30.30 A toroidal winding arrangement used to measure the magnetic properties of a material. The torus is made of the material under study, and the circuit containing the galvanometer measures the magnetic flux.

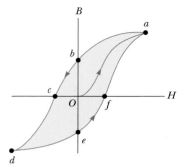

Figure 30.31 Magnetization curve for a ferromagnetic material.

the magnitude of the total field B also increases with increasing current, as shown by the curve from point O to point a in Figure 30.31. At point O, the domains in the iron are randomly oriented, corresponding to $B_m = 0$. As the increasing current in the primary coil causes the external field $\mathbf{B}_0$ to increase, the aligned domains grow in size until nearly all magnetic moments are aligned at point a. At this point the iron core is approaching *saturation*, which is the condition in which all magnetic moments in the iron are aligned.

Next, suppose that the current is reduced to zero, and the external field is consequently eliminated. The B-versus-H curve, called a **magnetization curve,** now follows the path ab in Figure 30.31. Note that at point b, $\mathbf{B}$ is not zero even though the external field $\mathbf{B}_0$ is zero. The reason is that the iron is now magnetized due to the alignment of a large number of its magnetic moments (that is, $\mathbf{B} = \mathbf{B}_m$). At this point, the iron is said to have a *remanent* magnetization.

If the current in the primary coil is reversed so that the direction of the external magnetic field is reversed, the magnetic moments reorient until the sample is again unmagnetized at point c, where $B = 0$. An increase in the reverse current causes the iron to be magnetized in the opposite direction, approaching saturation at point d in Figure 30.31. A similar sequence of events occurs as the current is reduced to zero and then increased in the original (positive) direction. In this case the magnetization curve follows the path *def*. If the current is increased sufficiently, the magnetization curve returns to point a, where the sample again has its maximum magnetization.

The effect just described, called **magnetic hysteresis,** shows that the magnetization of a ferromagnetic substance depends on the history of the substance as well as on the magnitude of the applied field. (The word *hysteresis* means "lagging behind.") It is often said that a ferromagnetic substance has a "memory" because it remains magnetized after the external field is removed. The closed loop in Figure 30.31 is referred to as a hysteresis loop. Its shape and size depend on the properties of the ferromagnetic substance and on the strength of the maximum applied field. The hysteresis loop for "hard" ferromagnetic materials is characteristically wide like the one shown in Figure 30.32a, corresponding to a large remanent magnetization. Such materials cannot be easily demagnetized by an external field. "Soft" ferromagnetic materials, such as iron, have a very narrow hysteresis loop and a small remanent magnetization (Fig. 30.32b.) Such materials are easily magnetized and demagnetized. An ideal soft ferromagnet would exhibit no hysteresis and hence would have no remanent magnetization. A ferromagnetic substance can be demagnetized by carrying it through successive hysteresis loops, due to a decreasing applied magnetic field, as shown in Figure 30.33.

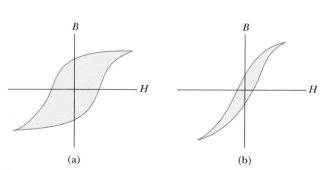

Figure 30.32 Hysteresis loops for (a) a hard ferromagnetic material and (b) a soft ferromagnetic material.

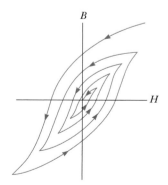

Figure 30.33 Demagnetizing a ferromagnetic material by carrying it through successive hysteresis loops.

The magnetization curve is useful for another reason: **the area enclosed by the magnetization curve represents the energy input required to take the material through the hysteresis cycle.** The energy acquired by the material in the magnetization process originates from the source of the external field—that is, the emf in the circuit of the toroidal coil. When the magnetization cycle is repeated, dissipative processes within the material due to realignment of the magnetic moments result in an increase in internal energy, made evident by an increase in the temperature of the substance. For this reason, devices subjected to alternating fields (such as AC adapters for cell phones, power tools, and so on) use cores made of soft ferromagnetic substances, which have narrow hysteresis loops and correspondingly little energy loss per cycle.

Magnetic computer disks store information by alternating the direction of **B** for portions of a thin layer of ferromagnetic material. Floppy disks have the layer on a circular sheet of plastic. Hard disks have several rigid platters with magnetic coatings on each side. Audio tapes and videotapes work the same way as floppy disks except that the ferromagnetic material is on a very long strip of plastic. Tiny coils of wire in a recording head are placed close to the magnetic material (which is moving rapidly past the head). Varying the current in the coils creates a magnetic field that magnetizes the recording material. To retrieve the information, the magnetized material is moved past a playback coil. The changing magnetism of the material induces a current in the coil, as shown in Chapter 31. This current is then amplified by audio or video equipment, or it is processed by computer circuitry.

When the temperature of a ferromagnetic substance reaches or exceeds a critical temperature called the **Curie temperature,** the substance loses its residual magnetization and becomes paramagnetic (Fig. 30.34). Below the Curie temperature, the magnetic moments are aligned and the substance is ferromagnetic. Above the Curie temperature, the thermal agitation is great enough to cause a random orientation of the moments, and the substance becomes paramagnetic. Curie temperatures for several ferromagnetic substances are given in Table 30.3.

Paramagnetism

Paramagnetic substances have a small but positive magnetic susceptibility $(0 < \chi \ll 1)$ resulting from the presence of atoms (or ions) that have permanent magnetic moments. These moments interact only weakly with each other and are randomly oriented in the absence of an external magnetic field. When a paramagnetic substance is placed in an external magnetic field, its atomic moments tend to line up with the field. However, this alignment process must compete with thermal motion, which tends to randomize the magnetic moment orientations.

Pierre Curie (1859–1906) and others since him have found experimentally that, under a wide range of conditions, the magnetization of a paramagnetic substance is proportional to the applied magnetic field and inversely proportional to the absolute temperature:

$$M = C \frac{B_0}{T} \tag{30.35}$$

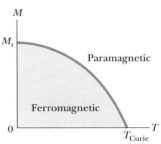

Figure 30.34 Magnetization versus absolute temperature for a ferromagnetic substance. The magnetic moments are aligned below the Curie temperature T_{Curie}, where the substance is ferromagnetic. The substance becomes paramagnetic (magnetic moments unaligned) above T_{Curie}.

Table 30.3

Curie Temperatures for Several Ferromagnetic Substance	
Substance	$T_{Curie}(\mathbf{K})$
Iron	1 043
Cobalt	1 394
Nickel	631
Gadolinium	317
Fe_2O_3	893

This relationship is known as **Curie's law** after its discoverer, and the constant C is called **Curie's constant.** The law shows that when $B_0 = 0$, the magnetization is zero, corresponding to a random orientation of magnetic moments. As the ratio of magnetic field to temperature becomes great, the magnetization approaches its saturation value, corresponding to a complete alignment of its moments, and Equation 30.35 is no longer valid.

Diamagnetism

When an external magnetic field is applied to a diamagnetic substance, a weak magnetic moment is induced in the direction opposite the applied field. This causes diamagnetic substances to be weakly repelled by a magnet. Although diamagnetism is present in all matter, its effects are much smaller than those of paramagnetism or ferromagnetism, and are evident only when those other effects do not exist.

We can attain some understanding of diamagnetism by considering a classical model of two atomic electrons orbiting the nucleus in opposite directions but with the same speed. The electrons remain in their circular orbits because of the attractive electrostatic force exerted by the positively charged nucleus. Because the magnetic moments of the two electrons are equal in magnitude and opposite in direction, they cancel each other, and the magnetic moment of the atom is zero. When an external magnetic field is applied, the electrons experience an additional magnetic force $q\mathbf{v} \times \mathbf{B}$. This added magnetic force combines with the electrostatic force to increase the orbital speed of the electron whose magnetic moment is antiparallel to the field and to decrease the speed of the electron whose magnetic moment is parallel to the field. As a result, the two magnetic moments of the electrons no longer cancel, and the substance acquires a net magnetic moment that is opposite the applied field.

As you recall from Chapter 27, a superconductor is a substance in which the electrical resistance is zero below some critical temperature. Certain types of superconductors also exhibit perfect diamagnetism in the superconducting state. As a result, an applied magnetic field is expelled by the superconductor so that the field is zero in its interior. This phenomenon is known as the **Meissner effect.** If a permanent magnet is brought near a superconductor, the two objects repel each other. This is illustrated in Figure 30.35, which shows a small permanent magnet levitated above a superconductor maintained at 77 K.

Figure 30.35 An illustration of the Meissner effect, shown by this magnet suspended above a cooled ceramic superconductor disk, has become our most visual image of high-temperature superconductivity. Superconductivity is the loss of all resistance to electrical current, and is a key to more efficient energy use. In the Meissner effect, the magnet induces superconducting current in the disk, which is cooled to $-321°F$ (77 K). The currents create a magnetic force that repels and levitates the disk.

(*Left*) Paramagnetism: liquid oxygen, a paramagnetic material, is attracted to the poles of a magnet. (*Right*) Diamagnetism: a frog is levitated in a 16-T magnetic field at the Nijmegen High Field Magnet Laboratory, Netherlands. The levitation force is exerted on the diamagnetic water molecules in the frog's body. The frog suffered no ill effects from the levitation experience.

Example 30.11   Saturation Magnetization

Estimate the saturation magnetization in a long cylinder of iron, assuming one unpaired electron spin per atom.

Solution The saturation magnetization is obtained when all the magnetic moments in the sample are aligned. If the sample contains n atoms per unit volume, then the saturation magnetization M_s has the value

$$M_s = n\mu$$

where μ is the magnetic moment per atom. Because the molar mass of iron is 55 g/mol and its density is 7.9 g/cm³,

the value of n for iron is 8.6×10^{28} atoms/m³. Assuming that each atom contributes one Bohr magneton (due to one unpaired spin) to the magnetic moment, we obtain

$$M_s = (8.6 \times 10^{28} \text{ atoms/m}^3)(9.27 \times 10^{-24}\text{ A·m}^2/\text{atom})$$

$$= \boxed{8.0 \times 10^5 \text{ A/m}}$$

This is about half the experimentally determined saturation magnetization for iron, which indicates that actually two unpaired electron spins are present per atom.

30.9   The Magnetic Field of the Earth

When we speak of a compass magnet having a north pole and a south pole, we should say more properly that it has a "north-seeking" pole and a "south-seeking" pole. By this we mean that one pole of the magnet seeks, or points to, the north geographic pole of the Earth. Because the north pole of a magnet is attracted toward the north geographic pole of the Earth, we conclude that **the Earth's south magnetic pole is located near the north geographic pole, and the Earth's north magnetic pole is located near the south geographic pole.** In fact, the configuration of the Earth's magnetic field, pictured in Figure 30.36, is very much like the one that would be achieved by burying a gigantic bar magnet deep in the interior of the Earth.

If a compass needle is suspended in bearings that allow it to rotate in the vertical plane as well as in the horizontal plane, the needle is horizontal with respect to the Earth's surface only near the equator. As the compass is moved northward, the needle rotates so that it points more and more toward the surface of the Earth. Finally, at a point near Hudson Bay in Canada, the north pole of the needle points directly downward. This site, first found in 1832, is considered to be the location of the south magnetic pole of the Earth. It is approximately 1 300 mi from the Earth's geographic North Pole, and its exact position varies slowly with time. Similarly, the north magnetic pole of the Earth is about 1 200 mi away from the Earth's geographic South Pole.

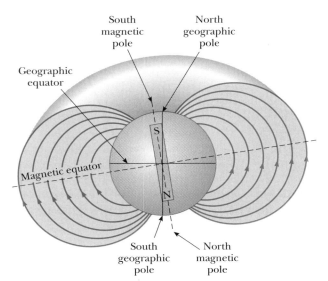

Figure 30.36 The Earth's magnetic field lines. Note that a south magnetic pole is near the north geographic pole, and a north magnetic pole is near the south geographic pole.

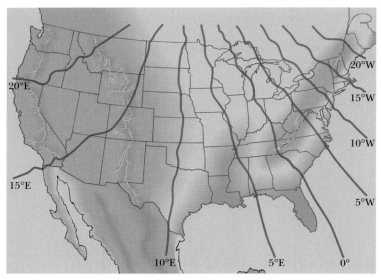

Figure 30.37 A map of the continental United States showing several lines of constant magnetic declination.

Because of this distance between the north geographic and south magnetic poles, it is only approximately correct to say that a compass needle points north. The difference between true north, defined as the geographic North Pole, and north indicated by a compass varies from point to point on the Earth, and the difference is referred to as *magnetic declination*. For example, along a line through Florida and the Great Lakes, a compass indicates true north, whereas in the state of Washington, it aligns 25° east of true north. Figure 30.37 shows some representative values of the magnetic declination for the continental United States.

Quick Quiz 30.10 If we wanted to cancel the Earth's magnetic field by running an enormous current loop around the equator, which way would the current have to to be directed: (a) east to west or (b) west to east?

Although the magnetic field pattern of the Earth is similar to the one that would be set up by a bar magnet deep within the Earth, it is easy to understand why the source of the Earth's magnetic field cannot be large masses of permanently magnetized material. The Earth does have large deposits of iron ore deep beneath its surface, but the high temperatures in the Earth's core prevent the iron from retaining any permanent magnetization. Scientists consider it more likely that the true source of the Earth's magnetic field is convection currents in the Earth's core. Charged ions or electrons circulating in the liquid interior could produce a magnetic field just as a current loop does. There is also strong evidence that the magnitude of a planet's magnetic field is related to the planet's rate of rotation. For example, Jupiter rotates faster than the Earth, and space probes indicate that Jupiter's magnetic field is stronger than ours. Venus, on the other hand, rotates more slowly than the Earth, and its magnetic field is found to be weaker. Investigation into the cause of the Earth's magnetism is ongoing.

There is an interesting sidelight concerning the Earth's magnetic field. It has been found that the direction of the field has been reversed several times during the last million years. Evidence for this is provided by basalt, a type of rock that contains iron and that forms from material spewed forth by volcanic activity on the ocean floor. As the lava cools, it solidifies and retains a picture of the Earth's magnetic field direction. The rocks are dated by other means to provide a timeline for these periodic reversals of the magnetic field.

SUMMARY

Take a practice test for this chapter by clicking on the Practice Test link at http://www.pse6.com.

The **Biot–Savart law** says that the magnetic field $d\mathbf{B}$ at a point P due to a length element $d\mathbf{s}$ that carries a steady current I is

$$d\mathbf{B} = \frac{\mu_0}{4\pi} \frac{I \, d\mathbf{s} \times \hat{\mathbf{r}}}{r^2} \qquad (30.1)$$

where μ_0 is the **permeability of free space,** r is the distance from the element to the point P, and $\hat{\mathbf{r}}$ is a unit vector pointing from $d\mathbf{s}$ toward point P. We find the total field at P by integrating this expression over the entire current distribution.

The magnetic force per unit length between two parallel wires separated by a distance a and carrying currents I_1 and I_2 has a magnitude

$$\frac{F_B}{\ell} = \frac{\mu_0 I_1 I_2}{2\pi a} \qquad (30.12)$$

The force is attractive if the currents are in the same direction and repulsive if they are in opposite directions.

Ampère's law says that the line integral of $\mathbf{B} \cdot d\mathbf{s}$ around any closed path equals $\mu_0 I$, where I is the total steady current through any surface bounded by the closed path:

$$\oint \mathbf{B} \cdot d\mathbf{s} = \mu_0 I \qquad (30.13)$$

Using Ampère's law, one finds that the magnitude of the magnetic field at a distance r from a long, straight wire carrying an electric current I is

$$B = \frac{\mu_0 I}{2\pi r} \qquad (30.14)$$

The field lines are circles concentric with the wire.

The magnitudes of the fields inside a toroid and solenoid are

$$B = \frac{\mu_0 N I}{2\pi r} \qquad \text{(toroid)} \qquad (30.16)$$

$$B = \mu_0 \frac{N}{\ell} I = \mu_0 n I \qquad \text{(solenoid)} \qquad (30.17)$$

where N is the total number of turns.

The **magnetic flux** Φ_B through a surface is defined by the surface integral

$$\Phi_B = \int \mathbf{B} \cdot d\mathbf{A} \qquad (30.18)$$

Gauss's law of magnetism states that the net magnetic flux through any closed surface is zero.

The general form of Ampère's law, which is also called the **Ampère–Maxwell law,** is

$$\oint \mathbf{B} \cdot d\mathbf{s} = \mu_0 I + \mu_0 \epsilon_0 \frac{d\Phi_E}{dt} \qquad (30.22)$$

This law describes the fact that magnetic fields are produced both by conduction currents and by changing electric fields.

When a substance is placed in an external magnetic field $\mathbf{B}_0$, the total magnetic field $\mathbf{B}$ is a combination of the external field and a magnetic field due to magnetic moments of atoms and electrons within the substance:

$$\mathbf{B} = \mathbf{B}_0 + \mu_0 \mathbf{M} \qquad (30.29)$$

where **M** is the **magnetization vector.** The magnetization vector is the magnetic moment per unit volume in the substance.

The effect of external currents on the magnetic field in a substance is described by the **magnetic field strength H** = $\mathbf{B}_0/\mu_0$. The magnetization vector is related to the magnetic field strength as follows:

$$\mathbf{M} = \chi\mathbf{H} \qquad (30.32)$$

where χ is the **magnetic susceptibility.**

Substances can be classified into one of three categories that describe their magnetic behavior. **Diamagnetic** substances are those in which the magnetization is weak and opposite the field $\mathbf{B}_0$, so that the susceptibility is negative. **Paramagnetic** substances are those in which the magnetization is weak and in the same direction as the field $\mathbf{B}_0$, so that the susceptibility is positive. In **ferromagnetic** substances, interactions between atoms cause magnetic moments to align and create a strong magnetization that remains after the external field is removed.

QUESTIONS

1. Is the magnetic field created by a current loop uniform? Explain.

2. A current in a conductor produces a magnetic field that can be calculated using the Biot–Savart law. Because current is defined as the rate of flow of charge, what can you conclude about the magnetic field produced by stationary charges? What about that produced by moving charges?

3. Explain why two parallel wires carrying currents in opposite directions repel each other.

4. Parallel current-carrying wires exert magnetic forces on each other. What about perpendicular wires? Imagine two such wires oriented perpendicular to each other, and almost touching. Does a magnetic force exist between the wires?

5. Is Ampère's law valid for all closed paths surrounding a conductor? Why is it not useful for calculating **B** for all such paths?

6. Compare Ampère's law with the Biot–Savart law. Which is more generally useful for calculating **B** for a current-carrying conductor?

7. Is the magnetic field inside a toroid uniform? Explain.

8. Describe the similarities between Ampère's law in magnetism and Gauss's law in electrostatics.

9. A hollow copper tube carries a current along its length. Why is **B** = 0 inside the tube? Is **B** nonzero outside the tube?

10. Describe the change in the magnetic field in the space enclosed by a solenoid carrying a steady current I if (a) the length of the solenoid is doubled but the number of turns remains the same and (b) the number of turns is doubled but the length remains the same.

11. A flat conducting loop is located in a uniform magnetic field directed along the x axis. For what orientation of the loop is the flux through it a maximum? A minimum?

12. What new concept did Maxwell's generalized form of Ampère's law include?

13. Many loops of wire are wrapped around a nail and the ends of the wire are connected to a battery. Identify the source of **M**, of **H**, and of **B**.

14. A magnet attracts a piece of iron. The iron can then attract another piece of iron. On the basis of domain alignment, explain what happens in each piece of iron.

15. Why does hitting a magnet with a hammer cause the magnetism to be reduced?

16. A Hindu ruler once suggested that he be entombed in a magnetic coffin with the polarity arranged so that he would be forever suspended between heaven and Earth. Is such magnetic levitation possible? Discuss.

17. Why is **M** = 0 in a vacuum? What is the relationship between **B** and **H** in a vacuum?

18. Explain why some atoms have permanent magnetic dipole moments and others do not.

19. What factors contribute to the total magnetic dipole moment of an atom?

20. Why is the susceptibility of a diamagnetic substance negative?

21. Why can the effect of diamagnetism be neglected in a paramagnetic substance?

22. Explain the significance of the Curie temperature for a ferromagnetic substance.

23. Discuss the difference among ferromagnetic, paramagnetic, and diamagnetic substances.

24. A current in a solenoid having air in the interior creates a magnetic field $\mathbf{B} = \mu_0\mathbf{H}$. Describe qualitatively what happens to the magnitude of **B** as (a) aluminum, (b) copper, and (c) iron are placed in the interior.

25. What is the difference between hard and soft ferromagnetic materials?

26. Should the surface of a computer disk be made from a hard or a soft ferromagnetic substance?

27. Explain why it is desirable to use hard ferromagnetic materials to make permanent magnets.

28. Would you expect the tape from a tape recorder to be attracted to a magnet? (Try it, but not with a recording you wish to save.)

29. Given only a strong magnet and a screwdriver, how would you first magnetize and then demagnetize the screwdriver?

30. Which way would a compass point if you were at the north magnetic pole of the Earth?

31. Figure Q30.31 shows two permanent magnets, each having a hole through its center. Note that the upper magnet is levitated above the lower one. (a) How does this occur? (b) What purpose does the pencil serve? (c) What can you say about the poles of the magnets from this observation? (d) If the upper magnet were inverted, what do you suppose would happen?

Courtesy of Central Scientific Company

Figure Q30.31

PROBLEMS

1, 2, 3 = straightforward, intermediate, challenging ☐ = full solution available in the *Student Solutions Manual and Study Guide*

🌀 = coached solution with hints available at http://www.pse6.com 💻 = computer useful in solving problem

▢ = paired numerical and symbolic problems

Section 30.1 The Biot–Savart Law

1. In Niels Bohr's 1913 model of the hydrogen atom, an electron circles the proton at a distance of 5.29×10^{-11} m with a speed of 2.19×10^6 m/s. Compute the magnitude of the magnetic field that this motion produces at the location of the proton.

2. A lightning bolt may carry a current of 1.00×10^4 A for a short period of time. What is the resulting magnetic field 100 m from the bolt? Suppose that the bolt extends far above and below the point of observation.

3. (a) A conductor in the shape of a square loop of edge length $\ell = 0.400$ m carries a current $I = 10.0$ A as in Fig. P30.3. Calculate the magnitude and direction of the magnetic field at the center of the square. (b) **What If?** If this conductor is formed into a single circular turn and carries the same current, what is the value of the magnetic field at the center?

4. Calculate the magnitude of the magnetic field at a point 100 cm from a long, thin conductor carrying a current of 1.00 A.

5. 🌀 Determine the magnetic field at a point P located a distance x from the corner of an infinitely long wire bent at a right angle, as shown in Figure P30.5. The wire carries a steady current I.

Figure P30.5

6. A conductor consists of a circular loop of radius R and two straight, long sections, as shown in Figure P30.6. The wire

Figure P30.6

Figure P30.3

lies in the plane of the paper and carries a current I. Find an expression for the vector magnetic field at the center of the loop.

7. The segment of wire in Figure P30.7 carries a current of $I = 5.00$ A, where the radius of the circular arc is $R = 3.00$ cm. Determine the magnitude and direction of the magnetic field at the origin.

Figure P30.7

8. 🖥 Consider a flat circular current loop of radius R carrying current I. Choose the x axis to be along the axis of the loop, with the origin at the center of the loop. Plot a graph of the ratio of the magnitude of the magnetic field at coordinate x to that at the origin, for $x = 0$ to $x = 5R$. It may be useful to use a programmable calculator or a computer to solve this problem.

9. Two very long, straight, parallel wires carry currents that are directed perpendicular to the page, as in Figure P30.9. Wire 1 carries a current I_1 into the page (in the $-z$ direction) and passes through the x axis at $x = +a$. Wire 2 passes through the x axis at $x = -2a$ and carries an unknown current I_2. The total magnetic field at the origin due to the current-carrying wires has the magnitude $2\mu_0 I_1/(2\pi a)$. The current I_2 can have either of two possible values. (a) Find the value of I_2 with the smaller magnitude, stating it in terms of I_1 and giving its direction. (b) Find the other possible value of I_2.

Figure P30.9

10. A very long straight wire carries current I. In the middle of the wire a right-angle bend is made. The bend forms an arc of a circle of radius r, as shown in Figure P30.10. Determine the magnetic field at the center of the arc.

Figure P30.10

11. One very long wire carries current 30.0 A to the left along the x axis. A second very long wire carries current 50.0 A to the right along the line ($y = 0.280$ m, $z = 0$). (a) Where in the plane of the two wires is the total magnetic field equal to zero? (b) A particle with a charge of -2.00 μC is moving with a velocity of $150\hat{\mathbf{i}}$ Mm/s along the line ($y = 0.100$ m, $z = 0$). Calculate the vector magnetic force acting on the particle. (c) **What If?** A uniform electric field is applied to allow this particle to pass through this region undeflected. Calculate the required vector electric field.

12. Consider the current-carrying loop shown in Figure P30.12, formed of radial lines and segments of circles whose centers are at point P. Find the magnitude and direction of **B** at P.

Figure P30.12

13. A wire carrying a current I is bent into the shape of an equilateral triangle of side L. (a) Find the magnitude of the magnetic field at the center of the triangle. (b) At a point halfway between the center and any vertex, is the field stronger or weaker than at the center?

14. Determine the magnetic field (in terms of I, a, and d) at the origin due to the current loop in Figure P30.14.

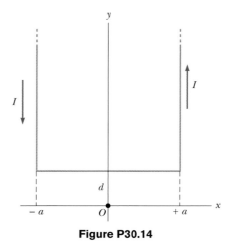

Figure P30.14

15. Two long, parallel conductors carry currents $I_1 = 3.00$ A and $I_2 = 3.00$ A, both directed into the page in Figure P30.15. Determine the magnitude and direction of the resultant magnetic field at P.

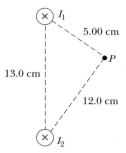

Figure P30.15

Section 30.2 The Magnetic Force Between Two Parallel Conductors

16. Two long, parallel conductors, separated by 10.0 cm, carry currents in the same direction. The first wire carries current $I_1 = 5.00$ A and the second carries $I_2 = 8.00$ A. (a) What is the magnitude of the magnetic field created by I_1 at the location of I_2? (b) What is the force per unit length exerted by I_1 on I_2? (c) What is the magnitude of the magnetic field created by I_2 at the location of I_1? (d) What is the force per length exerted by I_2 on I_1?

17. In Figure P30.17, the current in the long, straight wire is $I_1 = 5.00$ A and the wire lies in the plane of the rectangular loop, which carries the current $I_2 = 10.0$ A. The dimensions are $c = 0.100$ m, $a = 0.150$ m, and $\ell = 0.450$ m. Find the magnitude and direction of the net force exerted on the loop by the magnetic field created by the wire.

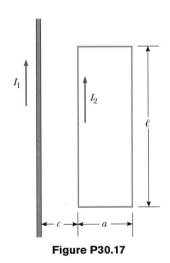

Figure P30.17

18. Two long, parallel wires are attracted to each other by a force per unit length of 320 μN/m when they are separated by a vertical distance of 0.500 m. The current in the upper wire is 20.0 A to the right. Determine the location of the line in the plane of the two wires along which the total magnetic field is zero.

19. Three long wires (wire 1, wire 2, and wire 3) hang vertically. The distance between wire 1 and wire 2 is 20.0 cm. On the left, wire 1 carries an upward current of 1.50 A. To the right, wire 2 carries a downward current of 4.00 A.

Wire 3 is located such that when it carries a certain current, each wire experiences no net force. Find (a) the position of wire 3, and (b) the magnitude and direction of the current in wire 3.

20. The unit of magnetic flux is named for Wilhelm Weber. The practical-size unit of magnetic field is named for Johann Karl Friedrich Gauss. Both were scientists at Göttingen, Germany. Along with their individual accomplishments, together they built a telegraph in 1833. It consisted of a battery and switch, at one end of a transmission line 3 km long, operating an electromagnet at the other end. (André Ampère suggested electrical signaling in 1821; Samuel Morse built a telegraph line between Baltimore and Washington in 1844.) Suppose that Weber and Gauss's transmission line was as diagrammed in Figure P30.20. Two long, parallel wires, each having a mass per unit length of 40.0 g/m, are supported in a horizontal plane by strings 6.00 cm long. When both wires carry the same current I, the wires repel each other so that the angle θ between the supporting strings is 16.0°. (a) Are the currents in the same direction or in opposite directions? (b) Find the magnitude of the current.

Figure P30.20

Section 30.3 Ampère's Law

21. Four long, parallel conductors carry equal currents of $I = 5.00$ A. Figure P30.21 is an end view of the conductors. The current direction is into the page at points A and B (indicated by the crosses) and out of the page at C and D (indicated by the dots). Calculate the magnitude and direction of the magnetic field at point P, located at the center of the square of edge length 0.200 m.

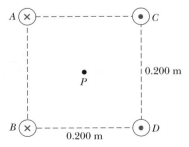

Figure P30.21

22. A long straight wire lies on a horizontal table and carries a current of 1.20 μA. In a vacuum, a proton moves parallel to the wire (opposite the current) with a constant speed of 2.30×10^4 m/s at a distance d above the wire. Determine the value of d. You may ignore the magnetic field due to the Earth.

23. Figure P30.23 is a cross-sectional view of a coaxial cable. The center conductor is surrounded by a rubber layer, which is surrounded by an outer conductor, which is surrounded by another rubber layer. In a particular application, the current in the inner conductor is 1.00 A out of the page and the current in the outer conductor is 3.00 A into the page. Determine the magnitude and direction of the magnetic field at points *a* and *b*.

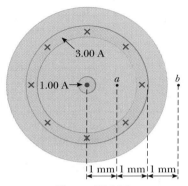

Figure P30.23

24. The magnetic field 40.0 cm away from a long straight wire carrying current 2.00 A is 1.00 μT. (a) At what distance is it 0.100 μT? (b) **What If?** At one instant, the two conductors in a long household extension cord carry equal 2.00-A currents in opposite directions. The two wires are 3.00 mm apart. Find the magnetic field 40.0 cm away from the middle of the straight cord, in the plane of the two wires. (c) At what distance is it one tenth as large? (d) The center wire in a coaxial cable carries current 2.00 A in one direction and the sheath around it carries current 2.00 A in the opposite direction. What magnetic field does the cable create at points outside?

25. A packed bundle of 100 long, straight, insulated wires forms a cylinder of radius $R = 0.500$ cm. (a) If each wire carries 2.00 A, what are the magnitude and direction of the magnetic force per unit length acting on a wire located 0.200 cm from the center of the bundle? (b) **What If?** Would a wire on the outer edge of the bundle experience a force greater or smaller than the value calculated in part (a)?

26. The magnetic coils of a tokamak fusion reactor are in the shape of a toroid having an inner radius of 0.700 m and an outer radius of 1.30 m. The toroid has 900 turns of large-diameter wire, each of which carries a current of 14.0 kA. Find the magnitude of the magnetic field inside the toroid along (a) the inner radius and (b) the outer radius.

27. Consider a column of electric current passing through plasma (ionized gas). Filaments of current within the column are magnetically attracted to one another. They can crowd together to yield a very great current density and a very strong magnetic field in a small region. Sometimes the current can be cut off momentarily by this *pinch effect*. (In a metallic wire a pinch effect is not important, because the current-carrying electrons repel one another with electric forces.) The pinch effect can be demonstrated by making an empty aluminum can carry a large current parallel to its axis. Let *R* represent the radius

of the can and *I* the upward current, uniformly distributed over its curved wall. Determine the magnetic field (a) just inside the wall and (b) just outside. (c) Determine the pressure on the wall.

28. Niobium metal becomes a superconductor when cooled below 9 K. Its superconductivity is destroyed when the surface magnetic field exceeds 0.100 T. Determine the maximum current a 2.00-mm-diameter niobium wire can carry and remain superconducting, in the absence of any external magnetic field.

29. A long cylindrical conductor of radius *R* carries a current *I* as shown in Figure P30.29. The current density *J*, however, is not uniform over the cross section of the conductor but is a function of the radius according to $J = br$, where *b* is a constant. Find an expression for the magnetic field *B* (a) at a distance $r_1 < R$ and (b) at a distance $r_2 > R$, measured from the axis.

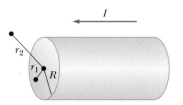

Figure P30.29

30. In Figure P30.30, both currents in the infinitely long wires are in the negative *x* direction. (a) Sketch the magnetic field pattern in the *yz* plane. (b) At what distance *d* along the *z* axis is the magnetic field a maximum?

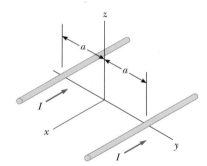

Figure P30.30

Section 30.4   The Magnetic Field of a Solenoid

31. What current is required in the windings of a long solenoid that has 1 000 turns uniformly distributed over a length of 0.400 m, to produce at the center of the solenoid a magnetic field of magnitude 1.00×10^{-4} T?

32. Consider a solenoid of length ℓ and radius *R*, containing *N* closely spaced turns and carrying a steady current *I*. (a) In terms of these parameters, find the magnetic field at a point along the axis as a function of distance *a* from the end of the solenoid. (b) Show that as ℓ becomes very long, *B* approaches $\mu_0 NI/2\ell$ at each end of the solenoid.

33. A single-turn square loop of wire, 2.00 cm on each edge, carries a clockwise current of 0.200 A. The loop is inside a solenoid, with the plane of the loop perpendicular to the magnetic field of the solenoid. The solenoid has 30 turns/cm and carries a clockwise current of 15.0 A. Find the force on each side of the loop and the torque acting on the loop.

Section 30.5   Magnetic Flux

34. Consider the hemispherical closed surface in Figure P30.34. The hemisphere is in a uniform magnetic field that makes an angle θ with the vertical. Calculate the magnetic flux through (a) the flat surface S_1 and (b) the hemispherical surface S_2.

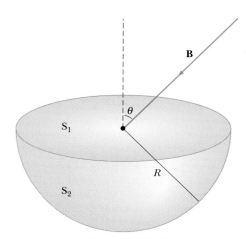

Figure P30.34

35. A cube of edge length $\ell = 2.50$ cm is positioned as shown in Figure P30.35. A uniform magnetic field given by $\mathbf{B} = (5\hat{\mathbf{i}} + 4\hat{\mathbf{j}} + 3\hat{\mathbf{k}})$T exists throughout the region. (a) Calculate the flux through the shaded face. (b) What is the total flux through the six faces?

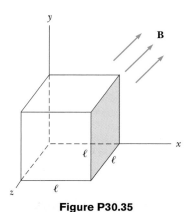

Figure P30.35

36. A solenoid 2.50 cm in diameter and 30.0 cm long has 300 turns and carries 12.0 A. (a) Calculate the flux through the surface of a disk of radius 5.00 cm that is positioned perpendicular to and centered on the axis of the solenoid, as shown in Figure P30.36a. (b) Figure P30.36b shows an enlarged end view of the same solenoid. Calculate the flux through the blue area, which is defined by an annulus that has an inner radius of 0.400 cm and outer radius of 0.800 cm.

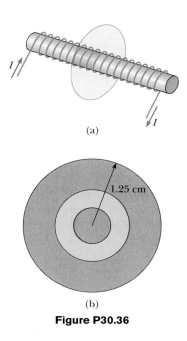

(a)

(b)

Figure P30.36

Section 30.7   Displacement Current and the General Form of Ampère's Law

37. A 0.100-A current is charging a capacitor that has square plates 5.00 cm on each side. The plate separation is 4.00 mm. Find (a) the time rate of change of electric flux between the plates and (b) the displacement current between the plates.

38. A 0.200-A current is charging a capacitor that has circular plates 10.0 cm in radius. If the plate separation is 4.00 mm, (a) what is the time rate of increase of electric field between the plates? (b) What is the magnetic field between the plates 5.00 cm from the center?

Section 30.8   Magnetism in Matter

39. In Bohr's 1913 model of the hydrogen atom, the electron is in a circular orbit of radius 5.29×10^{-11} m and its speed is 2.19×10^6 m/s. (a) What is the magnitude of the magnetic moment due to the electron's motion? (b) If the electron moves in a horizontal circle, counterclockwise as seen from above, what is the direction of this magnetic moment vector?

40. A magnetic field of 1.30 T is to be set up in an iron-core toroid. The toroid has a mean radius of 10.0 cm, and magnetic permeability of 5 000 μ_0. What current is required if the winding has 470 turns of wire? The thickness of the iron ring is small compared to 10 cm, so the field in the material is nearly uniform.

41. A toroid with a mean radius of 20.0 cm and 630 turns (see Fig. 30.30) is filled with powdered steel whose magnetic susceptibility χ is 100. The current in the windings is 3.00 A. Find B (assumed uniform) inside the toroid.

42. A particular paramagnetic substance achieves 10.0% of its saturation magnetization when placed in a magnetic field of 5.00 T at a temperature of 4.00 K. The density of magnetic atoms in the sample is 8.00×10^{27} atoms/m^3, and the magnetic moment per atom is 5.00 Bohr magnetons. Calculate the Curie constant for this substance.

43. Calculate the magnetic field strength H of a magnetized substance in which the magnetization is 0.880×10^6 A/m and the magnetic field has magnitude 4.40 T.

44. At *saturation*, when nearly all of the atoms have their magnetic moments aligned, the magnetic field in a sample of iron can be 2.00 T. If each electron contributes a magnetic moment of 9.27×10^{-24} A·m^2 (one Bohr magneton), how many electrons per atom contribute to the saturated field of iron? Iron contains approximately 8.50×10^{28} atoms/m^3.

45. (a) Show that Curie's law can be stated in the following way: The magnetic susceptibility of a paramagnetic substance is inversely proportional to the absolute temperature, according to $\chi = C\mu_0/T$, where C is Curie's constant. (b) Evaluate Curie's constant for chromium.

Section 30.9 The Magnetic Field of the Earth

46. A circular coil of 5 turns and a diameter of 30.0 cm is oriented in a vertical plane with its axis perpendicular to the horizontal component of the Earth's magnetic field. A horizontal compass placed at the center of the coil is made to deflect 45.0° from magnetic north by a current of 0.600 A in the coil. (a) What is the horizontal component of the Earth's magnetic field? (b) The current in the coil is switched off. A "dip needle" is a magnetic compass mounted so that it can rotate in a vertical north–south plane. At this location a dip needle makes an angle of 13.0° from the vertical. What is the total magnitude of the Earth's magnetic field at this location?

47. The magnetic moment of the Earth is approximately 8.00×10^{22} A·m^2. (a) If this were caused by the complete magnetization of a huge iron deposit, how many unpaired electrons would this correspond to? (b) At two unpaired electrons per iron atom, how many kilograms of iron would this correspond to? (Iron has a density of 7 900 kg/m^3, and approximately 8.50×10^{28} iron atoms/m^3.)

Additional Problems

48. The magnitude of the Earth's magnetic field at either pole is approximately 7.00×10^{-5} T. Suppose that the field fades away, before its next reversal. Scouts, sailors, and conservative politicians around the world join together in a program to replace the field. One plan is to use a current loop around the equator, without relying on magnetization of any materials inside the Earth. Determine the current that would generate such a field if this plan were carried out. (Take the radius of the Earth as $R_E = 6.37 \times 10^6$ m.)

49. A very long, thin strip of metal of width w carries a current I along its length as shown in Figure P30.49. Find the magnetic field at the point P in the diagram. The point P is in the plane of the strip at distance b away from it.

Figure P30.49

50. Suppose you install a compass on the center of the dashboard of a car. Compute an order-of-magnitude estimate for the magnetic field at this location produced by the current when you switch on the headlights. How does it compare with the Earth's magnetic field? You may suppose the dashboard is made mostly of plastic.

51. For a research project, a student needs a solenoid that produces an interior magnetic field of 0.030 0 T. She decides to use a current of 1.00 A and a wire 0.500 mm in diameter. She winds the solenoid in layers on an insulating form 1.00 cm in diameter and 10.0 cm long. Determine the number of layers of wire needed and the total length of the wire.

52. A thin copper bar of length $\ell = 10.0$ cm is supported horizontally by two (nonmagnetic) contacts. The bar carries current $I_1 = 100$ A in the $-x$ direction, as shown in Figure P30.52. At a distance $h = 0.500$ cm below one end of the bar, a long straight wire carries a current $I_2 = 200$ A in the z direction. Determine the magnetic force exerted on the bar.

Figure P30.52

53. A nonconducting ring of radius 10.0 cm is uniformly charged with a total positive charge 10.0 μC. The ring rotates at a constant angular speed 20.0 rad/s about an

axis through its center, perpendicular to the plane of the ring. What is the magnitude of the magnetic field on the axis of the ring 5.00 cm from its center?

54. A nonconducting ring of radius R is uniformly charged with a total positive charge q. The ring rotates at a constant angular speed ω about an axis through its center, perpendicular to the plane of the ring. What is the magnitude of the magnetic field on the axis of the ring a distance $R/2$ from its center?

55. Two circular coils of radius R, each with N turns, are perpendicular to a common axis. The coil centers are a distance R apart. Each coil carries a steady current I in the same direction, as shown in Figure P30.55. (a) Show that the magnetic field on the axis at a distance x from the center of one coil is

$$B = \frac{N\mu_0 I R^2}{2}\left[\frac{1}{(R^2 + x^2)^{3/2}} + \frac{1}{(2R^2 + x^2 - 2Rx)^{3/2}}\right]$$

(b) Show that dB/dx and d^2B/dx^2 are both zero at the point midway between the coils. This means the magnetic field in the region midway between the coils is uniform. Coils in this configuration are called *Helmholtz coils.*

Figure P30.55 Problems 55 and 56.

56. Two identical, flat, circular coils of wire each have 100 turns and a radius of 0.500 m. The coils are arranged as a set of Helmholtz coils (see Fig. P30.55), parallel and with separation 0.500 m. Each coil carries a current of 10.0 A. Determine the magnitude of the magnetic field at a point on the common axis of the coils and halfway between them.

57. We have seen that a long solenoid produces a uniform magnetic field directed along the axis of a cylindrical region. However, to produce a uniform magnetic field directed parallel to a *diameter* of a cylindrical region, one can use the *saddle coils* illustrated in Figure P30.57. The loops are wrapped over a somewhat flattened tube. Assume the straight sections of wire are very long. The end view of the tube shows how the windings are applied. The overall current distribution is the superposition of two overlapping circular cylinders of uniformly distributed current, one toward you and one away from you. The current density J is the same for each cylinder. The position of the axis of one cylinder is described by a position vector **a** relative to the other cylinder. Prove that the magnetic field inside the hollow tube is $\mu_0 Ja/2$ downward. *Suggestion:* The use of vector methods simplifies the calculation.

Figure P30.57 (a) General view of one turn of each saddle coil. (b) End view of the coils carrying current into the paper on the left and out of the paper on the right.

58. A very large parallel-plate capacitor carries charge with uniform charge per unit area $+\sigma$ on the upper plate and $-\sigma$ on the lower plate. The plates are horizontal and both move horizontally with speed v to the right. (a) What is the magnetic field between the plates? (b) What is the magnetic field close to the plates but outside of the capacitor? (c) What is the magnitude and direction of the magnetic force per unit area on the upper plate? (d) At what extrapolated speed v will the magnetic force on a plate balance the electric force on the plate? Calculate this speed numerically.

59. Two circular loops are parallel, coaxial, and almost in contact, 1.00 mm apart (Fig. P30.59). Each loop is 10.0 cm in radius. The top loop carries a clockwise current of 140 A. The bottom loop carries a counterclockwise current of 140 A. (a) Calculate the magnetic force exerted by the bottom loop on the top loop. (b) The upper loop has a mass of 0.021 0 kg. Calculate its acceleration, assuming that the only forces acting on it are the force in part (a) and the gravitational force. *Suggestion:* Think about how one loop looks to a bug perched on the other loop.

Figure P30.59

60. What objects experience a force in an electric field? Chapter 23 gives the answer: any electric charge, stationary or moving, other than the charge that created the field. What creates an electric field? Any electric charge, stationary or moving, as you studied in Chapter 23. What objects experience a force in a magnetic field? An electric current or a moving electric charge, other than the current or charge that created the field, as discussed in Chapter 29. What creates a magnetic field? An electric current, as you studied in Section 30.1, or a moving electric charge, as shown in this problem. (a) To display how a moving charge creates a magnetic field, consider a charge q moving with velocity **v**. Define the vector $\mathbf{r} = r\hat{\mathbf{r}}$

to lead from the charge to some location. Show that the magnetic field at that location is

$$\mathbf{B} = \frac{\mu_0}{4\pi} \frac{q\mathbf{v} \times \hat{\mathbf{r}}}{r^2}$$

(b) Find the magnitude of the magnetic field 1.00 mm to the side of a proton moving at 2.00×10^7 m/s. (c) Find the magnetic force on a second proton at this point, moving with the same speed in the opposite direction. (d) Find the electric force on the second proton.

61. *Rail guns* have been suggested for launching projectiles into space without chemical rockets, and for ground-to-air antimissile weapons of war. A tabletop model rail gun (Fig. P30.61) consists of two long parallel horizontal rails 3.50 cm apart, bridged by a bar *BD* of mass 3.00 g. The bar is originally at rest at the midpoint of the rails and is free to slide without friction. When the switch is closed, electric current is quickly established in the circuit *ABCDEA*. The rails and bar have low electric resistance, and the current is limited to a constant 24.0 A by the power supply. (a) Find the magnitude of the magnetic field 1.75 cm from a single very long straight wire carrying current 24.0 A. (b) Find the magnitude and direction of the magnetic field at point *C* in the diagram, the midpoint of the bar, immediately after the switch is closed. *Suggestion:* Consider what conclusions you can draw from the Biot–Savart law. (c) At other points along the bar *BD*, the field is in the same direction as at point *C*, but larger in magnitude. Assume that the average effective magnetic field along *BD* is five times larger than the field at *C*. With this assumption, find the magnitude and direction of the force on the bar. (d) Find the acceleration of the bar when it is in motion. (e) Does the bar move with constant acceleration? (f) Find the velocity of the bar after it has traveled 130 cm to the end of the rails.

Figure P30.61

62. 🖳 Fifty turns of insulated wire 0.100 cm in diameter are tightly wound to form a flat spiral. The spiral fills a disk surrounding a circle of radius 5.00 cm and extending to a radius 10.00 cm at the outer edge. Assume the wire carries current *I* at the center of its cross section. Approximate each turn of wire as a circle. Then a loop of current exists at radius 5.05 cm, another at 5.15 cm, and so on. Numerically calculate the magnetic field at the center of the coil.

63. Two long, parallel conductors carry currents in the same direction as shown in Figure P30.63. Conductor A carries a current of 150 A and is held firmly in position. Conductor B carries a current I_B and is allowed to slide freely up and down (parallel to A) between a set of nonconducting guides. If the mass per unit length of conductor B

is 0.100 g/cm, what value of current I_B will result in equilibrium when the distance between the two conductors is 2.50 cm?

Figure P30.63

64. Charge is sprayed onto a large nonconducting belt above the left-hand roller in Figure P30.64. The belt carries the charge with a uniform surface charge density σ as it moves with a speed v between the rollers as shown. The charge is removed by a wiper at the right-hand roller. Consider a point just above the surface of the moving belt. (a) Find an expression for the magnitude of the magnetic field **B** at this point. (b) If the belt is positively charged, what is the direction of **B**? (Note that the belt may be considered as an infinite sheet.)

Figure P30.64

65. An infinitely long straight wire carrying a current I_1 is partially surrounded by a loop as shown in Figure P30.65.

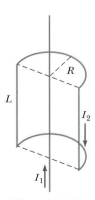

Figure P30.65

The loop has a length L, radius R, and carries a current I_2. The axis of the loop coincides with the wire. Calculate the force exerted on the loop.

66. Measurements of the magnetic field of a large tornado were made at the Geophysical Observatory in Tulsa, Oklahoma, in 1962. The tornado's field was measured to be $B = 1.50 \times 10^{-8}$ T pointing north when the tornado was 9.00 km east of the observatory. What current was carried up or down the funnel of the tornado, modeled as a long straight wire?

67. A wire is formed into the shape of a square of edge length L (Fig. P30.67). Show that when the current in the loop is I, the magnetic field at point P, a distance x from the center of the square along its axis is

$$B = \frac{\mu_0 I L^2}{2\pi(x^2 + L^2/4)\sqrt{x^2 + L^2/2}}$$

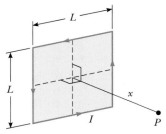

Figure P30.67

68. The force on a magnetic dipole $\boldsymbol{\mu}$ aligned with a nonuniform magnetic field in the x direction is given by $F_x = |\boldsymbol{\mu}| dB/dx$. Suppose that two flat loops of wire each have radius R and carry current I. (a) The loops are arranged coaxially and separated by a variable distance x, large compared to R. Show that the magnetic force between them varies as $1/x^4$. (b) Evaluate the magnitude of this force if $I = 10.0$ A, $R = 0.500$ cm, and $x = 5.00$ cm.

69. A wire carrying a current I is bent into the shape of an exponential spiral, $r = e^\theta$, from $\theta = 0$ to $\theta = 2\pi$ as suggested in Figure P30.69. To complete a loop, the ends of the spiral are connected by a straight wire along the x axis. Find the magnitude and direction of **B** at the origin. *Suggestions:* Use the Biot–Savart law. The angle β between a radial line and its tangent line at any point on the curve $r = f(\theta)$ is related to the function in the following way:

$$\tan \beta = \frac{r}{dr/d\theta}$$

Thus in this case $r = e^\theta$, $\tan \beta = 1$ and $\beta = \pi/4$. Therefore, the angle between $d\mathbf{s}$ and $\hat{\mathbf{r}}$ is $\pi - \beta = 3\pi/4$. Also

$$ds = \frac{dr}{\sin(\pi/4)} = \sqrt{2}\, dr$$

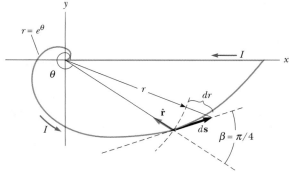

Figure P30.69

70. Table P30.70 contains data taken for a ferromagnetic material. (a) Construct a magnetization curve from the data. Remember that $\mathbf{B} = \mathbf{B}_0 + \mu_0\mathbf{M}$. (b) Determine the ratio B/B_0 for each pair of values of B and B_0, and construct a graph of B/B_0 versus B_0. (The fraction B/B_0 is called the relative permeability, and it is a measure of the induced magnetic field.)

Table P30.70

B (T)	B_0 (T)
0.2	4.8×10^{-5}
0.4	7.0×10^{-5}
0.6	8.8×10^{-5}
0.8	1.2×10^{-4}
1.0	1.8×10^{-4}
1.2	3.1×10^{-4}
1.4	8.7×10^{-4}
1.6	3.4×10^{-3}
1.8	1.2×10^{-1}

71. A sphere of radius R has a uniform volume charge density ρ. Determine the magnetic field at the center of the sphere when it rotates as a rigid object with angular speed ω about an axis through its center (Fig. P30.71).

Figure P30.71 Problems 71 and 72.

72. A sphere of radius R has a uniform volume charge density ρ. Determine the magnetic dipole moment of the sphere when it rotates as a rigid body with angular speed ω about an axis through its center (Fig. P30.71).

73. A long cylindrical conductor of radius a has two cylindrical cavities of diameter a through its entire length, as shown in Figure P30.73. A current I is directed out of the page and is uniform through a cross section of the conductor. Find the magnitude and direction of the magnetic field in terms of μ_0, I, r, and a at (a) point P_1 and (b) point P_2.

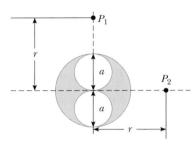

Figure P30.73

Answers to Quick Quizzes

30.1 B, C, A. Point B is closest to the current element. Point C is farther away and the field is further reduced by the $\sin\theta$ factor in the cross product $d\mathbf{s} \times \hat{\mathbf{r}}$. The field at A is zero because $\theta = 0$.

30.2 (c). $F_1 = F_2$ as required by Newton's third law. Another way to arrive at this answer is to realize that Equation 30.11 gives the same result whether the multiplication of currents is $(2\,\text{A})(6\,\text{A})$ or $(6\,\text{A})(2\,\text{A})$.

30.3 (a). The coils act like wires carrying parallel currents in the same direction and hence attract one another.

30.4 b, d, a, c. Equation 30.13 indicates that the value of the line integral depends only on the net current through each closed path. Path b encloses 1 A, path d encloses 3 A, path a encloses 4 A, and path c encloses 6 A.

30.5 b, then $a = c = d$. Paths a, c, and d all give the same nonzero value $\mu_0 I$ because the size and shape of the paths do not matter. Path b does not enclose the current, and hence its line integral is zero.

30.6 (c). The magnetic field in a very long solenoid is independent of its length or radius. Overwrapping with an additional layer of wire increases the number of turns per unit length.

30.7 (b). There can be no conduction current because there is no conductor between the plates. There is a time-varying electric field because of the decreasing charge on the plates, and the time-varying electric flux represents a displacement current.

30.8 (c). There is a time-varying electric field because of the decreasing charge on the plates. This time-varying electric field produces a magnetic field.

30.9 (a). The loop that looks like Figure 30.32a is better because the remanent magnetization at the point corresponding to point b in Figure 30.31 is greater.

30.10 (b). The lines of the Earth's magnetic field enter the planet in Hudson Bay and emerge from Antarctica; thus, the field lines resulting from the current would have to go in the opposite direction. Compare Figure 30.7a with Figure 30.36.

Calvin and Hobbes

by Bill Watterson

Chapter 31

Faraday's Law

CHAPTER OUTLINE

▲ *In a commercial electric power plant, large generators produce energy that is transferred out of the plant by electrical transmission. These generators use magnetic induction to generate a potential difference when coils of wire in the generator are rotated in a magnetic field. The source of energy to rotate the coils might be falling water, burning fossil fuels, or a nuclear reaction. (Michael Melford/Getty Images)*

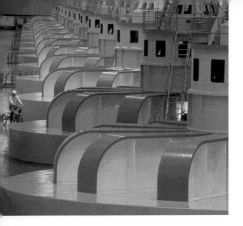

The focus of our studies in electricity and magnetism so far has been the electric fields produced by stationary charges and the magnetic fields produced by moving charges. This chapter explores the effects produced by magnetic fields that vary in time.

Experiments conducted by Michael Faraday in England in 1831 and independently by Joseph Henry in the United States that same year showed that an emf can be induced in a circuit by a changing magnetic field. The results of these experiments led to a very basic and important law of electromagnetism known as *Faraday's law of induction*. An emf (and therefore a current as well) can be induced in various processes that involve a change in a magnetic flux.

With the treatment of Faraday's law, we complete our introduction to the fundamental laws of electromagnetism. These laws can be summarized in a set of four equations called *Maxwell's equations*. Together with the *Lorentz force law*, they represent a complete theory for describing the interaction of charged objects.

Michael Faraday
British Physicist and Chemist (1791–1867)

Faraday is often regarded as the greatest experimental scientist of the 1800s. His many contributions to the study of electricity include the invention of the electric motor, electric generator, and transformer, as well as the discovery of electromagnetic induction and the laws of electrolysis. Greatly influenced by religion, he refused to work on the development of poison gas for the British military. *(By kind permission of the President and Council of the Royal Society)*

31.1 Faraday's Law of Induction

To see how an emf can be induced by a changing magnetic field, consider a loop of wire connected to a sensitive ammeter, as illustrated in Figure 31.1. When a magnet is moved toward the loop, the galvanometer needle deflects in one direction, arbitrarily shown to the right in Figure 31.1a. When the magnet is brought to rest and held stationary relative to the loop (Fig. 31.1b), no deflection is observed. When the magnet is moved away from the loop, the needle deflects in the opposite direction, as shown in Figure 31.1c. Finally, if the magnet is held stationary and the loop is moved either toward or away from it, the needle deflects. From these observations, we conclude that the loop detects that the magnet is moving relative to it and we relate this detection to a change in magnetic field. Thus, it seems that a relationship exists between current and changing magnetic field.

These results are quite remarkable in view of the fact that **a current is set up even though no batteries are present in the circuit!** We call such a current an *induced current* and say that it is produced by an *induced emf*.

Now let us describe an experiment conducted by Faraday and illustrated in Figure 31.2. A primary coil is connected to a switch and a battery. The coil is wrapped around an iron ring, and a current in the coil produces a magnetic field when the switch is closed. A secondary coil also is wrapped around the ring and is connected to a sensitive ammeter. No battery is present in the secondary circuit, and the secondary coil is not electrically connected to the primary coil. Any current detected in the secondary circuit must be induced by some external agent.

Initially, you might guess that no current is ever detected in the secondary circuit. However, something quite amazing happens when the switch in the primary circuit is either opened or thrown closed. At the instant the switch is closed, the galvanometer needle deflects in one direction and then returns to zero. At the instant the switch is opened, the needle deflects in the opposite direction and again returns to zero.

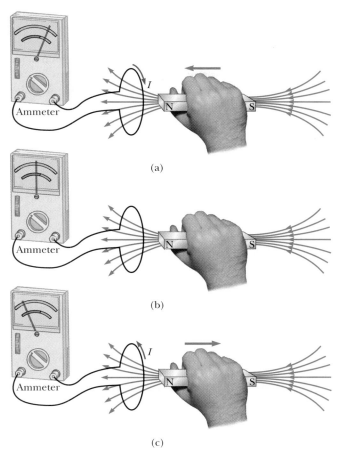

(a)

(b)

(c)

Active Figure 31.1 (a) When a magnet is moved toward a loop of wire connected to a sensitive ammeter, the ammeter deflects as shown, indicating that a current is induced in the loop. (b) When the magnet is held stationary, there is no induced current in the loop, even when the magnet is inside the loop. (c) When the magnet is moved away from the loop, the ammeter deflects in the opposite direction, indicating that the induced current is opposite that shown in part (a). Changing the direction of the magnet's motion changes the direction of the current induced by that motion.

At the Active Figures link at http://www.pse6.com, *you can move the magnet and observe the current in the ammeter.*

Finally, the galvanometer reads zero when there is either a steady current or no current in the primary circuit. The key to understanding what happens in this experiment is to note first that when the switch is closed, the current in the primary circuit produces a magnetic field that penetrates the secondary circuit. Furthermore, when

Active Figure 31.2 Faraday's experiment. When the switch in the primary circuit is closed, the ammeter in the secondary circuit deflects momentarily. The emf induced in the secondary circuit is caused by the changing magnetic field through the secondary coil.

At the Active Figures link at http://www.pse6.com, *you can open and close the switch and observe the current in the ammeter.*

the switch is closed, the magnetic field produced by the current in the primary circuit changes from zero to some value over some finite time, and this changing field induces a current in the secondary circuit.

As a result of these observations, Faraday concluded that **an electric current can be induced in a circuit (the secondary circuit in our setup) by a changing magnetic field.** The induced current exists for only a short time while the magnetic field through the secondary coil is changing. Once the magnetic field reaches a steady value, the current in the secondary coil disappears. In effect, the secondary circuit behaves as though a source of emf were connected to it for a short time. It is customary to say that **an induced emf is produced in the secondary circuit by the changing magnetic field.**

The experiments shown in Figures 31.1 and 31.2 have one thing in common: in each case, an emf is induced in the circuit when the magnetic flux through the circuit changes with time. In general,

> The emf induced in a circuit is directly proportional to the time rate of change of the magnetic flux through the circuit.

This statement, known as **Faraday's law of induction,** can be written

Faraday's law

$$\varepsilon = -\frac{d\Phi_B}{dt} \tag{31.1}$$

where $\Phi_B = \int \mathbf{B} \cdot d\mathbf{A}$ is the magnetic flux through the circuit. (See Section 30.5.)

If the circuit is a coil consisting of N loops all of the same area and if Φ_B is the magnetic flux through one loop, an emf is induced in every loop. The loops are in series, so their emfs add; thus, the total induced emf in the coil is given by the expression

$$\varepsilon = -N\frac{d\Phi_B}{dt} \tag{31.2}$$

The negative sign in Equations 31.1 and 31.2 is of important physical significance, as discussed in Section 31.3.

Suppose that a loop enclosing an area A lies in a uniform magnetic field $\mathbf{B}$, as in Figure 31.3. The magnetic flux through the loop is equal to $BA\cos\theta$; hence, the induced emf can be expressed as

$$\varepsilon = -\frac{d}{dt}(BA\cos\theta) \tag{31.3}$$

From this expression, we see that an emf can be induced in the circuit in several ways:

- The magnitude of $\mathbf{B}$ can change with time.
- The area enclosed by the loop can change with time.
- The angle θ between $\mathbf{B}$ and the normal to the loop can change with time.
- Any combination of the above can occur.

▲ **PITFALL PREVENTION**

31.1 Induced emf Requires a Change

The *existence* of a magnetic flux through an area is not sufficient to create an induced emf. There must be a *change* in the magnetic flux in order for an emf to be induced.

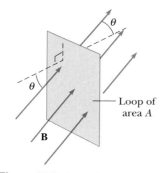

Figure 31.3 A conducting loop that encloses an area A in the presence of a uniform magnetic field $\mathbf{B}$. The angle between $\mathbf{B}$ and the normal to the loop is θ.

Quick Quiz 31.1 A circular loop of wire is held in a uniform magnetic field, with the plane of the loop perpendicular to the field lines. Which of the following will *not* cause a current to be induced in the loop? (a) crushing the loop; (b) rotating the loop about an axis perpendicular to the field lines; (c) keeping the orientation of the loop fixed and moving it along the field lines; (d) pulling the loop out of the field.

Quick Quiz 31.2 Figure 31.4 shows a graphical representation of the field magnitude versus time for a magnetic field that passes through a fixed loop and is oriented perpendicular to the plane of the loop. The magnitude of the magnetic field at any time is uniform over the area of the loop. Rank the magnitudes of the emf generated in the loop at the five instants indicated, from largest to smallest.

Figure 31.4 (Quick Quiz 31.2) The time behavior of a magnetic field through a loop.

Quick Quiz 31.3 Suppose you would like to steal power for your home from the electric company by placing a loop of wire near a transmission cable, so as to induce an emf in the loop (an illegal procedure). Should you (a) place your loop so that the transmission cable passes through your loop, or (b) simply place your loop near the transmission cable?

Some Applications of Faraday's Law

The ground fault interrupter (GFI) is an interesting safety device that protects users of electrical appliances against electric shock. Its operation makes use of Faraday's law. In the GFI shown in Figure 31.5, wire 1 leads from the wall outlet to the appliance to be protected, and wire 2 leads from the appliance back to the wall outlet. An iron ring surrounds the two wires, and a sensing coil is wrapped around part of the ring. Because the currents in the wires are in opposite directions, the net magnetic flux through the sensing coil due to the currents is zero. However, if the return current in wire 2 changes, the net magnetic flux through the sensing coil is no longer zero. (This can happen, for example, if the appliance becomes wet, enabling current to leak to ground.) Because household current is alternating (meaning that its direction keeps reversing), the magnetic flux through the sensing coil changes with time, inducing an emf in the coil. This induced emf is used to trigger a circuit breaker, which stops the current before it is able to reach a harmful level.

Another interesting application of Faraday's law is the production of sound in an electric guitar (Fig. 31.6). The coil in this case, called the *pickup coil*, is placed near the vibrating guitar string, which is made of a metal that can be magnetized. A permanent

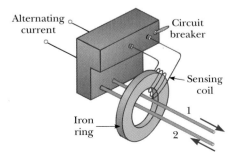

Figure 31.5 Essential components of a ground fault interrupter.

Figure 31.6 (a) In an electric guitar, a vibrating magnetized string induces an emf in a pickup coil. (b) The pickups (the circles beneath the metallic strings) of this electric guitar detect the vibrations of the strings and send this information through an amplifier and into speakers. (A switch on the guitar allows the musician to select which set of six pickups is used.)

magnet inside the coil magnetizes the portion of the string nearest the coil. When the string vibrates at some frequency, its magnetized segment produces a changing magnetic flux through the coil. The changing flux induces an emf in the coil that is fed to an amplifier. The output of the amplifier is sent to the loudspeakers, which produce the sound waves we hear.

Example 31.1   One Way to Induce an emf in a Coil

A coil consists of 200 turns of wire. Each turn is a square of side 18 cm, and a uniform magnetic field directed perpendicular to the plane of the coil is turned on. If the field changes linearly from 0 to 0.50 T in 0.80 s, what is the magnitude of the induced emf in the coil while the field is changing?

Solution The area of one turn of the coil is $(0.18 \text{ m})^2 = 0.032\,4 \text{ m}^2$. The magnetic flux through the coil at $t = 0$ is zero because $B = 0$ at that time. At $t = 0.80$ s, the magnetic flux through one turn is $\Phi_B = BA = (0.50 \text{ T})(0.032\,4 \text{ m}^2) = 0.016\,2 \text{ T} \cdot \text{m}^2$. Therefore, the magnitude of the induced emf is, from Equation 31.2,

$$|\mathcal{E}| = N\,\frac{\Delta\Phi_B}{\Delta t} = 200\,\frac{(0.016\,2 \text{ T}\cdot\text{m}^2 - 0)}{0.80 \text{ s}}$$

$$= 4.1 \text{ T}\cdot\text{m}^2/\text{s} = \boxed{4.1 \text{ V}}$$

You should be able to show that $1 \text{ T}\cdot\text{m}^2/\text{s} = 1$ V.

What If? **What if you were asked to find the magnitude of the induced current in the coil while the field is changing? Can you answer this question?**

Answer If the ends of the coil are not connected to a circuit, the answer to this question is easy—the current is zero! (Charges will move within the wire of the coil, but they cannot move into or out of the ends of the coil.) In order for a steady current to exist, the ends of the coil must be connected to an external circuit. Let us assume that the coil is connected to a circuit and that the total resistance of the coil and the circuit is 2.0 Ω. Then, the current in the coil is

$$I = \frac{\mathcal{E}}{R} = \frac{4.1 \text{ V}}{2.0 \text{ } \Omega} = 2.0 \text{ A}$$

Example 31.2   An Exponentially Decaying B Field

A loop of wire enclosing an area A is placed in a region where the magnetic field is perpendicular to the plane of the loop. The magnitude of **B** varies in time according to the expression $B = B_{max}e^{-at}$, where a is some constant. That is, at $t = 0$ the field is B_{max}, and for $t > 0$, the field decreases

exponentially (Fig. 31.7). Find the induced emf in the loop as a function of time.

Solution Because **B** is perpendicular to the plane of the loop, the magnetic flux through the loop at time $t > 0$ is

$$\Phi_B = BA \cos 0 = AB_{\max}e^{-at}$$

Because $AB_{\max}$ and a are constants, the induced emf calculated from Equation 31.1 is

$$\mathcal{E} = -\frac{d\Phi_B}{dt} = -AB_{\max}\frac{d}{dt}e^{-at}$$

$$= aAB_{\max}e^{-at}$$

This expression indicates that the induced emf decays exponentially in time. Note that the maximum emf occurs at $t = 0$, where $\mathcal{E}_{\max} = aAB_{\max}$. The plot of $\mathcal{E}$ versus t is similar to the B-versus-t curve shown in Figure 31.7.

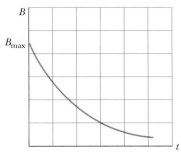

Figure 31.7 (Example 31.2) Exponential decrease in the magnitude of the magnetic field with time. The induced emf and induced current vary with time in the same way.

Conceptual Example 31.3   Which Bulb Is Shorted Out?

Two bulbs are connected to opposite sides of a circular loop of wire, as shown in Figure 31.8a. A changing magnetic field (confined to the smaller circular area shown in the figure) induces an emf in the loop that causes the two bulbs to light. When the switch is closed, the resistance-free wires connected to the switch short out bulb 2 and it goes out. What happens if the wires containing the closed switch remain connected at points a and b, but the switch and the wires are lifted up and moved to the other side of the field, as in Figure 3.18b? The wire is still connected to bulb 2 as it was before, so does it continue to stay dark?

Solution When the wire is moved to the other side, even though the connections have not changed, bulb 1 goes out and bulb 2 glows. The bulb that is shorted depends on which side of the changing field the switch is positioned! In Figure 31.8a, because the branch containing bulb 2 is infinitely more resistant than the branch containing the resistance-free switch, we can imagine removing the branch with the bulb without altering the circuit. Then we have a simple loop containing only bulb 1, which glows.

When the wire is moved, as in Figure 31.8b, there are two possible paths for current below points a and b. We can imagine removing the branch with bulb 1, leaving only a single loop with bulb 2.

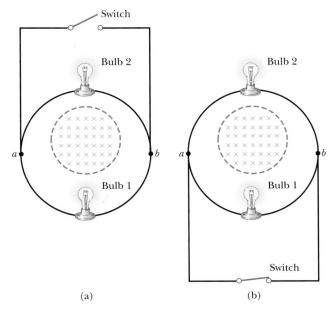

Figure 31.8 (Conceptual Example 31.3) (a) When the wire with the switch is located as shown, bulb 2 goes out when the switch is closed. (b) What happens when the switch and the wires are moved to the other side of the magnetic field?

31.2   Motional emf

In Examples 31.1 and 31.2, we considered cases in which an emf is induced in a stationary circuit placed in a magnetic field when the field changes with time. In this section we describe what is called **motional emf,** which is the emf induced in a conductor moving through a constant magnetic field.

The straight conductor of length ℓ shown in Figure 31.9 is moving through a uniform magnetic field directed into the page. For simplicity, we assume that the conductor is moving in a direction perpendicular to the field with constant velocity under the influence of some external agent. The electrons in the conductor experience a force $\mathbf{F}_B = q\mathbf{v} \times \mathbf{B}$ that is directed along the length ℓ, perpendicular to both $\mathbf{v}$ and $\mathbf{B}$ (Eq. 29.1). Under the influence of this force, the electrons move to the lower end of the conductor and accumulate there, leaving a net positive charge at the upper end. As a result of this charge separation, an electric field $\mathbf{E}$ is produced

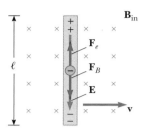

Figure 31.9 A straight electrical conductor of length ℓ moving with a velocity **v** through a uniform magnetic field **B** directed perpendicular to **v**. Due to the magnetic force on electrons, the ends of the conductor become oppositely charged. This establishes an electric field in the conductor. In steady state, the electric and magnetic forces on an electron in the wire are balanced.

inside the conductor. The charges accumulate at both ends until the downward magnetic force qvB on charges remaining in the conductor is balanced by the upward electric force qE. At this point, electrons move only with random thermal motion. The condition for equilibrium requires that

$$qE = qvB \qquad \text{or} \qquad E = vB$$

The electric field produced in the conductor is related to the potential difference across the ends of the conductor according to the relationship $\Delta V = E\ell$ (Eq. 25.6). Thus, for the equilibrium condition,

$$\Delta V = E\ell = B\ell v \qquad\qquad (31.4)$$

where the upper end of the conductor in Figure 31.9 is at a higher electric potential than the lower end. Thus, **a potential difference is maintained between the ends of the conductor as long as the conductor continues to move through the uniform magnetic field.** If the direction of the motion is reversed, the polarity of the potential difference is also reversed.

A more interesting situation occurs when the moving conductor is part of a closed conducting path. This situation is particularly useful for illustrating how a changing magnetic flux causes an induced current in a closed circuit. Consider a circuit consisting of a conducting bar of length ℓ sliding along two fixed parallel conducting rails, as shown in Figure 31.10a.

Active Figure 31.10 (a) A conducting bar sliding with a velocity **v** along two conducting rails under the action of an applied force $\mathbf{F}_{app}$. The magnetic force $\mathbf{F}_B$ opposes the motion, and a counterclockwise current I is induced in the loop. (b) The equivalent circuit diagram for the setup shown in part (a).

At the Active Figures link at http://www.pse6.com, you can adjust the applied force, the magnetic field, and the resistance to see the effects on the motion of the bar.

For simplicity, we assume that the bar has zero resistance and that the stationary part of the circuit has a resistance R. A uniform and constant magnetic field **B** is applied perpendicular to the plane of the circuit. As the bar is pulled to the right with a velocity **v** under the influence of an applied force $\mathbf{F}_{app}$, free charges in the bar experience a magnetic force directed along the length of the bar. This force sets up an induced current because the charges are free to move in the closed conducting path. In this case, the rate of change of magnetic flux through the loop and the corresponding induced motional emf across the moving bar are proportional to the change in area of the loop. If the bar is pulled to the right with a constant velocity, the work done by the applied force appears as internal energy in the resistor R. (See Section 27.6.)

Because the area enclosed by the circuit at any instant is ℓx, where x is the position of the bar, the magnetic flux through that area is

$$\Phi_B = B\ell x$$

Using Faraday's law, and noting that x changes with time at a rate $dx/dt = v$, we find that the induced motional emf is

$$\varepsilon = -\frac{d\Phi_B}{dt} = -\frac{d}{dt}(B\ell x) = -B\ell \frac{dx}{dt}$$

$$\varepsilon = -B\ell v \qquad\qquad (31.5)$$

Motional emf

Because the resistance of the circuit is R, the magnitude of the induced current is

$$I = \frac{|\varepsilon|}{R} = \frac{B\ell v}{R} \qquad\qquad (31.6)$$

The equivalent circuit diagram for this example is shown in Figure 31.10b.

Let us examine the system using energy considerations. Because no battery is in the circuit, we might wonder about the origin of the induced current and the energy delivered to the resistor. We can understand the source of this current and energy by noting that the applied force does work on the conducting bar, thereby moving charges through a magnetic field. Their movement through the field causes the charges to move along the bar with some average drift velocity, and hence a current is established. The change in energy in the system during some time interval must be equal to the transfer of energy into the system by work, consistent with the general principle of conservation of energy described by Equation 7.17.

Let us verify this mathematically. As the bar moves through the uniform magnetic field **B**, it experiences a magnetic force $\mathbf{F}_B$ of magnitude $I\ell B$ (see Section 29.2). The direction of this force is opposite the motion of the bar, to the left in Figure 31.10a. Because the bar moves with constant velocity, the applied force must be equal in magnitude and opposite in direction to the magnetic force, or to the right in Figure 31.10a. (If $\mathbf{F}_B$ acted in the direction of motion, it would cause the bar to accelerate, violating the principle of conservation of energy.) Using Equation 31.6 and the fact that $F_{app} = I\ell B$, we find that the power delivered by the applied force is

$$\mathcal{P} = F_{app}v = (I\ell B)v = \frac{B^2 \ell^2 v^2}{R} = \frac{\mathcal{E}^2}{R} \qquad (31.7)$$

From Equation 27.23, we see that this power input is equal to the rate at which energy is delivered to the resistor, so that Equation 7.17 is confirmed in this situation.

Quick Quiz 31.4 As an airplane flies from Los Angeles to Seattle, it passes through the Earth's magnetic field. As a result, a motional emf is developed between the wingtips. Which wingtip is positively charged? (a) the left wing (b) the right wing.

Quick Quiz 31.5 In Figure 31.10, a given applied force of magnitude F_{app} results in a constant speed v and a power input $\mathcal{P}$. Imagine that the force is increased so that the constant speed of the bar is doubled to $2v$. Under these conditions, the new force and the new power input are (a) $2F$ and $2\mathcal{P}$ (b) $4F$ and $2\mathcal{P}$ (c) $2F$ and $4\mathcal{P}$ (d) $4F$ and $4\mathcal{P}$.

Quick Quiz 31.6 You wish to move a rectangular loop of wire into a region of uniform magnetic field at a given speed so as to induce an emf in the loop. The plane of the loop remains perpendicular to the magnetic field lines. In which orientation should you hold the loop while you move it into the region of magnetic field in order to generate the largest emf? (a) with the long dimension of the loop parallel to the velocity vector (b) with the short dimension of the loop parallel to the velocity vector (c) either way—the emf is the same regardless of orientation.

Example 31.4 Motional emf Induced in a Rotating Bar Interactive

A conducting bar of length ℓ rotates with a constant angular speed ω about a pivot at one end. A uniform magnetic field **B** is directed perpendicular to the plane of rotation, as shown in Figure 31.11. Find the motional emf induced between the ends of the bar.

Solution Consider a segment of the bar of length dr having a velocity **v**. According to Equation 31.5, the magnitude of

the emf induced in this segment is

$$d\mathcal{E} = Bv\,dr$$

Because every segment of the bar is moving perpendicular to **B**, an emf $d\mathcal{E}$ of the same form is generated across each segment. Summing the emfs induced across all segments, which are in series, gives the total emf between the ends

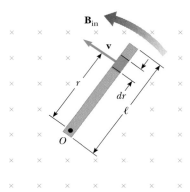

Figure 31.11 (Example 31.4) A conducting bar rotating around a pivot at one end in a uniform magnetic field that is perpendicular to the plane of rotation. A motional emf is induced across the ends of the bar.

of the bar:

$$\mathcal{E} = \int Bv \, dr$$

To integrate this expression, note that the linear speed v of an element is related to the angular speed ω through the relationship $v = r\omega$ (Eq. 10.10). Therefore, because B and ω are constants, we find that

$$\mathcal{E} = B \int v \, dr = B\omega \int_0^{\ell} r \, dr = \boxed{\tfrac{1}{2} B\omega\ell^2}$$

What If? Suppose, after reading through this example, you come up with a brilliant idea. A Ferris wheel has radial metallic spokes between the hub and the circular rim. These spokes move in the magnetic field of the Earth, so each spoke acts like the bar in Figure 31.11. You plan to use the emf generated by the rotation of the Ferris wheel to power the lightbulbs on the wheel! Will this idea work?

Answer The fact that this is not done in practice suggests that others may have thought of this idea and rejected it. Let us estimate the emf that is generated in this situation. We know the magnitude of the magnetic field of the Earth from Table 29.1, $B = 0.5 \times 10^{-4}$ T. A typical spoke on a Ferris wheel might have a length on the order of 10 m. Suppose the period of rotation is on the order of 10 s. This gives an angular speed of

$$\omega = \frac{2\pi}{T} = \frac{2\pi}{10 \text{ s}} = 0.63 \text{ s}^{-1} \sim 1 \text{ s}^{-1}$$

Assuming that the magnetic field lines of the Earth are horizontal at the location of the Ferris wheel and perpendicular to the spokes, the emf generated is

$$\mathcal{E} = \tfrac{1}{2} B\omega\ell^2 = \tfrac{1}{2} (0.5 \times 10^{-4} \text{ T}) (1 \text{ s}^{-1}) (10 \text{ m})^2$$

$$= 2.5 \times 10^{-3} \text{ V} \sim 1 \text{ mV}$$

This is a tiny emf, far smaller than that required to operate lightbulbs.

An additional difficulty is related to energy. Assuming you could find lightbulbs that operate using a potential difference on the order of millivolts, a spoke must be part of a circuit in order to provide a voltage to the bulbs. Consequently, the spoke must carry a current. Because this current-carrying spoke is in a magnetic field, a magnetic force is exerted on the spoke and the direction of the force is opposite to its direction of motion. As a result, the motor of the Ferris wheel must supply more energy to perform work against this magnetic drag force. The motor must ultimately provide the energy that is operating the lightbulbs and you have not gained anything for free!

 At the Interactive Worked Example link at **http://www.pse6.com,** you can explore the induced emf for different angular speeds and field magnitudes.

Example 31.5   Magnetic Force Acting on a Sliding Bar   `Interactive`

The conducting bar illustrated in Figure 31.12 moves on two frictionless parallel rails in the presence of a uniform magnetic field directed into the page. The bar has mass m and its length is ℓ. The bar is given an initial velocity $\mathbf{v}_i$ to the right and is released at $t = 0$.

(A) Using Newton's laws, find the velocity of the bar as a function of time.

(B) Show that the same result is reached by using an energy approach.

Solution (A) Conceptualize this situation as follows. As the bar slides to the right in Figure 31.12, a counterclockwise current is established in the circuit consisting of the bar, the rails, and the resistor. The upward current in the bar results in a magnetic force to the left on the bar as shown in the figure. As a result, the bar will slow down, so our mathematical solution should demonstrate this. The text of part (A) already categorizes this as a problem in using Newton's laws. To analyze the problem, we determine from Equation 29.3 that the magnetic force is $F_B = - I\ell B$, where the negative sign indicates that the retarding force is to the left. Because this is the *only* horizontal force acting on the bar, Newton's second law applied to motion in the

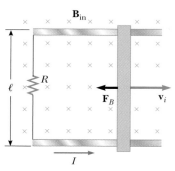

Figure 31.12 (Example 31.5) A conducting bar of length ℓ on two fixed conducting rails is given an initial velocity $\mathbf{v}_i$ to the right.

horizontal direction gives

$$F_x = ma = m\frac{dv}{dt} = -I\ell B$$

From Equation 31.6, we know that $I = B\ell v/R$, and so we can write this expression as

$$m\frac{dv}{dt} = -\frac{B^2\ell^2}{R}v$$

$$\frac{dv}{v} = -\left(\frac{B^2\ell^2}{mR}\right)dt$$

Integrating this equation using the initial condition that $v = v_i$ at $t = 0$, we find that

$$\int_{v_i}^{v}\frac{dv}{v} = \frac{-B^2\ell^2}{mR}\int_{0}^{t}dt$$

$$\ln\left(\frac{v}{v_i}\right) = -\left(\frac{B^2\ell^2}{mR}\right)t = -\frac{t}{\tau}$$

where the constant $\tau = mR/B^2\ell^2$. From this result, we see that the velocity can be expressed in the exponential form

$$(1) \qquad v = v_i e^{-t/\tau}$$

To finalize the problem, note that this expression for v indicates that the velocity of the bar decreases with time under the action of the magnetic retarding force, as we expect from our conceptualization of the problem.

(B) The text of part (B) immediately categorizes this as a problem in energy conservation. Consider the sliding bar as one system possessing kinetic energy, which decreases because energy is transferring *out* of the system by electrical transmission through the rails. The resistor is another system possessing internal energy, which rises because energy is transferring *into* this system. Because energy is not leaving the combination of two systems, the rate of energy transfer out of the bar equals the rate of energy transfer into the resistor. Thus,

$$\mathcal{P}_{\text{resistor}} = -\mathcal{P}_{\text{bar}}$$

where the negative sign is necessary because energy is leaving the bar and $\mathcal{P}_{\text{bar}}$ is a negative number. Substituting for the electrical power delivered to the resistor and the

time rate of change of kinetic energy for the bar, we have

$$I^2R = -\frac{d}{dt}(\tfrac{1}{2}mv^2)$$

Using Equation 31.6 for the current and carrying out the derivative, we find

$$\frac{B^2\ell^2v^2}{R} = -mv\frac{dv}{dt}$$

Rearranging terms gives

$$\frac{dv}{v} = -\left(\frac{B^2\ell^2}{mR}\right)dt$$

To finalize this part of the problem, note that this is the same expression that we obtained in part (A).

What If? Suppose you wished to increase the distance through which the bar moves between the time when it is initially projected and the time when it essentially comes to rest. You can do this by changing one of three variables: v_i, R, or B, by a factor of 2 or $\frac{1}{2}$. Which variable should you change in order to maximize the distance, and would you double it or halve it?

Answer Increasing v_i would make the bar move farther. Increasing R would decrease the current and, therefore, the magnetic force, making the bar move farther. Decreasing B would decrease the magnetic force and make the bar move farther. But which is most effective?

We use Equation (1) to find the distance that the bar moves by integration:

$$v = \frac{dx}{dt} = v_i e^{-t/\tau}$$

$$x = \int_{0}^{\infty}v_i e^{-t/\tau}\,dt = -v_i\tau e^{-t/\tau}\Big|_{0}^{\infty}$$

$$= -v_i\tau(0-1) = v_i\tau = v_i\left(\frac{mR}{B^2\ell^2}\right)$$

From this expression, we see that doubling v_i or R will double the distance. But changing B by a factor of $\frac{1}{2}$ causes the distance to be four times as great!

 At the *Interactive Worked Example* link at **http://www.pse6.com,** *you can study the motion of the bar after it is released.*

31.3 Lenz's Law

Faraday's law (Eq. 31.1) indicates that the induced emf and the change in flux have opposite algebraic signs. This has a very real physical interpretation that has come to be known as **Lenz's law**[1]:

> The induced current in a loop is in the direction that creates a magnetic field that opposes the change in magnetic flux through the area enclosed by the loop.

Lenz's law

[1]   Developed by the German physicist Heinrich Lenz (1804–1865).

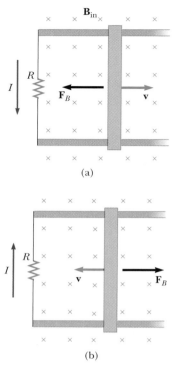

(a)

(b)

Figure 31.13 (a) As the conducting bar slides on the two fixed conducting rails, the magnetic flux due to the external magnetic field into the page through the area enclosed by the loop increases in time. By Lenz's law, the induced current must be counterclockwise so as to produce a counteracting magnetic field directed out of the page. (b) When the bar moves to the left, the induced current must be clockwise. Why?

That is, the induced current tends to keep the original magnetic flux through the circuit from changing. We shall show that this law is a consequence of the law of conservation of energy.

To understand Lenz's law, let us return to the example of a bar moving to the right on two parallel rails in the presence of a uniform magnetic field (the *external* magnetic field, Fig. 31.13a.) As the bar moves to the right, the magnetic flux through the area enclosed by the circuit increases with time because the area increases. Lenz's law states that the induced current must be directed so that the magnetic field it produces opposes the change in the external magnetic flux. Because the magnetic flux due to an external field directed into the page is increasing, the induced current, if it is to oppose this change, must produce a field directed out of the page. Hence, the induced current must be directed counterclockwise when the bar moves to the right. (Use the right-hand rule to verify this direction.) If the bar is moving to the left, as in Figure 31.13b, the external magnetic flux through the area enclosed by the loop decreases with time. Because the field is directed into the page, the direction of the induced current must be clockwise if it is to produce a field that also is directed into the page. In either case, the induced current tends to maintain the original flux through the area enclosed by the current loop.

Let us examine this situation using energy considerations. Suppose that the bar is given a slight push to the right. In the preceding analysis, we found that this motion sets up a counterclockwise current in the loop. What happens if we assume that the

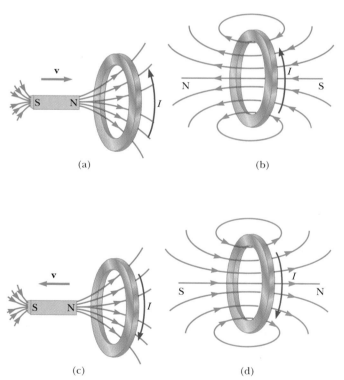

(a)   (b)

(c)   (d)

Figure 31.14 (a) When the magnet is moved toward the stationary conducting loop, a current is induced in the direction shown. The magnetic field lines shown are those due to the bar magnet. (b) This induced current produces its own magnetic field directed to the left that counteracts the increasing external flux. The magnetic field lines shown are those due to the induced current in the ring. (c) When the magnet is moved away from the stationary conducting loop, a current is induced in the direction shown. The magnetic field lines shown are those due to the bar magnet. (d) This induced current produces a magnetic field directed to the right and so counteracts the decreasing external flux. The magnetic field lines shown are those due to the induced current in the ring.

current is clockwise, such that the direction of the magnetic force exerted on the bar is to the right? This force would accelerate the rod and increase its velocity. This, in turn, would cause the area enclosed by the loop to increase more rapidly; this would result in an increase in the induced current, which would cause an increase in the force, which would produce an increase in the current, and so on. In effect, the system would acquire energy with no input of energy. This is clearly inconsistent with all experience and violates the law of conservation of energy. Thus, we are forced to conclude that the current must be counterclockwise.

Let us consider another situation, one in which a bar magnet moves toward a stationary metal loop, as in Figure 31.14a. As the magnet moves to the right toward the loop, the external magnetic flux through the loop increases with time. To counteract this increase in flux due to a field toward the right, the induced current produces its own magnetic field to the left, as illustrated in Figure 31.14b; hence, the induced current is in the direction shown. Knowing that like magnetic poles repel each other, we conclude that the left face of the current loop acts like a north pole and that the right face acts like a south pole.

If the magnet moves to the left, as in Figure 31.14c, its flux through the area enclosed by the loop decreases in time. Now the induced current in the loop is in the direction shown in Figure 31.14d because this current direction produces a magnetic field in the same direction as the external field. In this case, the left face of the loop is a south pole and the right face is a north pole.

Quick Quiz 31.7 Figure 31.15 shows a magnet being moved in the vicinity of a solenoid connected to a sensitive ammeter. The south pole of the magnet is the pole nearest the solenoid, and the ammeter indicates a clockwise (viewed from above) current in the solenoid. Is the person (a) inserting the magnet or (b) pulling it out?

Quick Quiz 31.8 Figure 31.16 shows a circular loop of wire being dropped toward a wire carrying a current to the left. The direction of the induced current in the loop of wire is (a) clockwise (b) counterclockwise (c) zero (d) impossible to determine.

Figure 31.15 (Quick Quiz 31.7)

Figure 31.16 (Quick Quiz 31.8)

Conceptual Example 31.6 Application of Lenz's Law

A metal ring is placed near a solenoid, as shown in Figure 31.17a. Find the direction of the induced current in the ring

(A) at the instant the switch in the circuit containing the solenoid is thrown closed,

(B) after the switch has been closed for several seconds, and

(C) at the instant the switch is thrown open.

Solution (A) At the instant the switch is thrown closed, the situation changes from one in which no magnetic flux exists in the ring to one in which flux exists and the magnetic field is to the left as shown in Figure 31.17b. To counteract this change in the flux, the current induced in the ring must set up a magnetic field directed from left to right in Figure 31.17b. This requires a current directed as shown.

(B) After the switch has been closed for several seconds, no change in the magnetic flux through the loop occurs; hence, the induced current in the ring is zero.

(C) Opening the switch changes the situation from one in which magnetic flux exists in the ring to one in which there is no magnetic flux. The direction of the induced current is as shown in Figure 31.17c because current in this direction

produces a magnetic field that is directed right to left and so counteracts the decrease in the flux produced by the solenoid.

Figure 31.17 (Example 31.6) A current is induced in a metal ring near a solenoid when the switch is opened or thrown closed.

Conceptual Example 31.7 A Loop Moving Through a Magnetic Field

A rectangular metallic loop of dimensions ℓ and w and resistance R moves with constant speed v to the right, as in Figure 31.18a. The loop passes through a uniform magnetic field **B** directed into the page and extending a distance $3w$ along the x axis. Defining x as the position of the right side of the loop along the x axis, plot as functions of x

(A) the magnetic flux through the area enclosed by the loop,

(B) the induced motional emf, and

(C) the external applied force necessary to counter the magnetic force and keep v constant.

Solution (A) Figure 31.18b shows the flux through the area enclosed by the loop as a function of x. Before the loop enters the field, the flux is zero. As the loop enters the field, the flux increases linearly with position until the left edge of

the loop is just inside the field. Finally, the flux through the loop decreases linearly to zero as the loop leaves the field.

(B) Before the loop enters the field, no motional emf is induced in it because no field is present (Fig. 31.18c). As the right side of the loop enters the field, the magnetic flux directed into the page increases. Hence, according to Lenz's law, the induced current is counterclockwise because it must produce its own magnetic field directed out of the page. The motional emf $-B\ell v$ (from Eq. 31.5) arises from the magnetic force experienced by charges in the right side of the loop. When the loop is entirely in the field, the change in magnetic flux is zero, and hence the motional emf vanishes. This happens because, once the left side of the loop enters the field, the motional emf induced in it cancels the motional emf present in the right side of the loop. As the right side of the loop leaves the field, the flux begins to decrease, a clockwise

Figure 31.18 (Conceptual Example 31.7) (a) A conducting rectangular loop of width w and length ℓ moving with a velocity **v** through a uniform magnetic field extending a distance $3w$. (b) Magnetic flux through the area enclosed by the loop as a function of loop position. (c) Induced emf as a function of loop position. (d) Applied force required for constant velocity as a function of loop position.

(a)

(b)

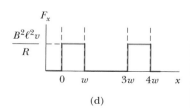

(c)

(d)

current is induced, and the induced emf is $B\ell v$. As soon as the left side leaves the field, the emf decreases to zero.

(C) The external force that must be applied to the loop to maintain this motion is plotted in Figure 31.18d. Before the loop enters the field, no magnetic force acts on it; hence, the applied force must be zero if v is constant. When the right side of the loop enters the field, the applied force necessary to maintain constant speed must be equal in magnitude and opposite in direction to the magnetic force exerted on that side. When the loop is entirely in the field, the flux through the loop is not changing with time. Hence, the net emf induced in the loop is zero, and the current also is zero. Therefore, no external force is needed to maintain the motion. Finally, as the right side leaves the field, the applied force must be equal in magnitude and opposite in direction to the magnetic force acting on the left side of the loop.

From this analysis, we conclude that power is supplied only when the loop is either entering or leaving the field. Furthermore, this example shows that the motional emf induced in the loop can be zero even when there is motion through the field! A motional emf is induced *only* when the magnetic flux through the loop *changes in time*.

31.4 Induced emf and Electric Fields

We have seen that a changing magnetic flux induces an emf and a current in a conducting loop. In our study of electricity, we related a current to an electric field that applies electric forces on charged particles. In the same way, we can relate an induced current in a conducting loop to an electric field by claiming that **an electric field is created in the conductor as a result of the changing magnetic flux.**

We also noted in our study of electricity that the existence of an electric field is independent of the presence of any test charges. This suggests that even in the absence of a conducting loop, a changing magnetic field would still generate an electric field in empty space.

This induced electric field is *nonconservative*, unlike the electrostatic field produced by stationary charges. We can illustrate this point by considering a conducting loop of radius r situated in a uniform magnetic field that is perpendicular to the plane of the loop, as in Figure 31.19. If the magnetic field changes with time, then, according to Faraday's law (Eq. 31.1), an emf $\mathcal{E} = -d\Phi_B/dt$ is induced in the loop. The induction of a current in the loop implies the presence of an induced electric field **E**, which must be tangent to the loop because this is the direction in which the charges in the wire move in response to the electric force. The work done by the electric field in moving a test charge q once around the loop is equal to $q\mathcal{E}$. Because the electric force acting on the charge is $q\mathbf{E}$, the work done by the electric field in moving the charge once around the loop is $qE(2\pi r)$, where $2\pi r$ is the circumference of the loop. These two expressions for the work done must be equal; therefore, we see that

$$q\mathcal{E} = qE(2\pi r)$$

$$E = \frac{\mathcal{E}}{2\pi r}$$

Using this result, along with Equation 31.1 and the fact that $\Phi_B = BA = \pi r^2 B$ for a circular loop, we find that the induced electric field can be expressed as

$$E = -\frac{1}{2\pi r}\frac{d\Phi_B}{dt} = -\frac{r}{2}\frac{dB}{dt} \tag{31.8}$$

If the time variation of the magnetic field is specified, we can easily calculate the induced electric field from Equation 31.8.

The emf for any closed path can be expressed as the line integral of $\mathbf{E} \cdot d\mathbf{s}$ over that path: $\mathcal{E} = \oint \mathbf{E} \cdot d\mathbf{s}$. In more general cases, E may not be constant, and the path may not be a circle. Hence, Faraday's law of induction, $\mathcal{E} = -d\Phi_B/dt$, can be written in the general form

$$\oint \mathbf{E} \cdot d\mathbf{s} = -\frac{d\Phi_B}{dt} \tag{31.9}$$

The induced electric field E in Equation 31.9 is a nonconservative field that is generated by a changing magnetic field. The field E that satisfies Equation 31.9

Figure 31.19 A conducting loop of radius r in a uniform magnetic field perpendicular to the plane of the loop. If **B** changes in time, an electric field is induced in a direction tangent to the circumference of the loop.

▲ **PITFALL PREVENTION**

31.2 Induced Electric Fields

The changing magnetic field does *not* need to be in existence at the location of the induced electric field. In Figure 31.19, even a loop outside the region of magnetic field will experience an induced electric field. For another example, consider Figure 31.8. The light bulbs glow (if the switch is open) even though the wires are outside the region of the magnetic field.

Faraday's law in general form

cannot possibly be an electrostatic field because if the field were electrostatic, and hence conservative, the line integral of $\mathbf{E} \cdot d\mathbf{s}$ over a closed loop would be zero (Section 25.1); this would be in contradiction to Equation 31.9.

Quick Quiz 31.9 In a region of space, the magnetic field increases at a constant rate. This changing magnetic field induces an electric field that (a) increases in time (b) is conservative (c) is in the direction of the magnetic field (d) has a constant magnitude.

Example 31.8 Electric Field Induced by a Changing Magnetic Field in a Solenoid

A long solenoid of radius R has n turns of wire per unit length and carries a time-varying current that varies sinusoidally as $I = I_{max} \cos \omega t$, where I_{max} is the maximum current and ω is the angular frequency of the alternating current source (Fig. 31.20).

(A) Determine the magnitude of the induced electric field outside the solenoid at a distance $r > R$ from its long central axis.

Solution First let us consider an external point and take the path for our line integral to be a circle of radius r centered on the solenoid, as illustrated in Figure 31.20. By symmetry we see that the magnitude of $\mathbf{E}$ is constant on this path and that $\mathbf{E}$ is tangent to it. The magnetic flux through the area enclosed by this path is $BA = B\pi R^2$; hence, Equation 31.9 gives

$$\oint \mathbf{E} \cdot d\mathbf{s} = -\frac{d}{dt}(B\pi R^2) = -\pi R^2 \frac{dB}{dt}$$

$$(1) \qquad \oint \mathbf{E} \cdot d\mathbf{s} = E(2\pi r) = -\pi R^2 \frac{dB}{dt}$$

The magnetic field inside a long solenoid is given by Equation 30.17, $B = \mu_0 nI$. When we substitute the expression $I = I_{max} \cos \omega t$ into this equation for B and then substitute the result into Equation (1), we find that

$$E(2\pi r) = -\pi R^2 \mu_0 nI_{max} \frac{d}{dt}(\cos \omega t)$$

$$= \pi R^2 \mu_0 nI_{max} \omega \sin \omega t$$

$$(2) \qquad E = \frac{\mu_0 nI_{max}\omega R^2}{2r} \sin \omega t \qquad (\text{for } r > R)$$

Hence, the amplitude of the electric field outside the solenoid falls off as $1/r$ and varies sinusoidally with time.

(B) What is the magnitude of the induced electric field inside the solenoid, a distance r from its axis?

Solution For an interior point ($r < R$), the flux through an integration loop is given by $B\pi r^2$. Using the same procedure as in part (A), we find that

$$E(2\pi r) = -\pi r^2 \frac{dB}{dt} = \pi r^2 \mu_0 nI_{max} \omega \sin \omega t$$

$$(3) \qquad E = \frac{\mu_0 nI_{max}\omega}{2} r \sin \omega t \qquad (\text{for } r < R)$$

This shows that the amplitude of the electric field induced inside the solenoid by the changing magnetic flux through the solenoid increases linearly with r and varies sinusoidally with time.

Figure 31.20 (Example 31.8) A long solenoid carrying a time-varying current given by $I = I_{max} \cos \omega t$. An electric field is induced both inside and outside the solenoid.

31.5 Generators and Motors

Electric generators take in energy by work and transfer it out by electrical transmission. To understand how they operate, let us consider the **alternating current** (AC) **generator.** In its simplest form, it consists of a loop of wire rotated by some external means in a magnetic field (Fig. 31.21a).

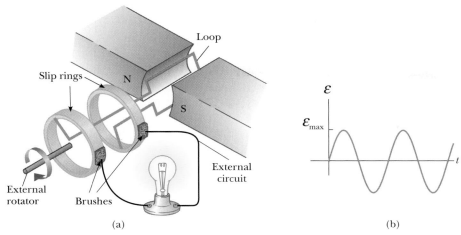

Active Figure 31.21 (a) Schematic diagram of an AC generator. An emf is induced in a loop that rotates in a magnetic field. (b) The alternating emf induced in the loop plotted as a function of time.

At the Active Figures link at http://www.pse6.com, *you can adjust the speed of rotation and the strength of the field to see the effects on the emf generated.*

In commercial power plants, the energy required to rotate the loop can be derived from a variety of sources. For example, in a hydroelectric plant, falling water directed against the blades of a turbine produces the rotary motion; in a coal-fired plant, the energy released by burning coal is used to convert water to steam, and this steam is directed against the turbine blades. As a loop rotates in a magnetic field, the magnetic flux through the area enclosed by the loop changes with time; this induces an emf and a current in the loop according to Faraday's law. The ends of the loop are connected to slip rings that rotate with the loop. Connections from these slip rings, which act as output terminals of the generator, to the external circuit are made by stationary brushes in contact with the slip rings.

Suppose that, instead of a single turn, the loop has N turns (a more practical situation), all of the same area A, and rotates in a magnetic field with a constant angular speed ω. If θ is the angle between the magnetic field and the normal to the plane of the loop, as in Figure 31.22, then the magnetic flux through the loop at any time t is

$$\Phi_B = BA \cos \theta = BA \cos \omega t$$

where we have used the relationship $\theta = \omega t$ between angular position and angular speed (see Eq. 10.3). (We have set the clock so that $t = 0$ when $\theta = 0$.) Hence, the induced emf in the coil is

$$\mathcal{E} = -N \frac{d\Phi_B}{dt} = -NAB \frac{d}{dt}(\cos \omega t) = NAB\omega \sin \omega t \qquad (31.10)$$

This result shows that the emf varies sinusoidally with time, as plotted in Figure 31.21b. From Equation 31.10 we see that the maximum emf has the value

$$\mathcal{E}_{max} = NAB\omega \qquad (31.11)$$

which occurs when $\omega t = 90°$ or $270°$. In other words, $\mathcal{E} = \mathcal{E}_{max}$ when the magnetic field is in the plane of the coil and the time rate of change of flux is a maximum. Furthermore, the emf is zero when $\omega t = 0$ or $180°$, that is, when **B** is perpendicular to the plane of the coil and the time rate of change of flux is zero.

The frequency for commercial generators in the United States and Canada is 60 Hz, whereas in some European countries it is 50 Hz. (Recall that $\omega = 2\pi f$, where f is the frequency in hertz.)

Figure 31.22 A loop enclosing an area A and containing N turns, rotating with constant angular speed ω in a magnetic field. The emf induced in the loop varies sinusoidally in time.

Quick Quiz 31.10 In an AC generator, a coil with N turns of wire spins in a magnetic field. Of the following choices, which will *not* cause an increase in the emf generated in the coil? (a) replacing the coil wire with one of lower resistance (b) spinning the coil faster (c) increasing the magnetic field (d) increasing the number of turns of wire on the coil.

Example 31.9   emf Induced in a Generator

An AC generator consists of 8 turns of wire, each of area $A = 0.090\,0$ m^2, and the total resistance of the wire is 12.0 Ω. The loop rotates in a 0.500-T magnetic field at a constant frequency of 60.0 Hz.

(A) Find the maximum induced emf.

Solution First, note that $\omega = 2\pi f = 2\pi(60.0\text{ Hz}) = 377\text{ s}^{-1}$. Thus, Equation 31.11 gives

$$\mathcal{E}_{max} = NAB\omega = 8(0.090\,0\text{ m}^2)(0.500\text{ T})(377\text{ s}^{-1})$$

$$= \boxed{136\text{ V}}$$

(B) What is the maximum induced current when the output terminals are connected to a low-resistance conductor?

Solution From Equation 27.8 and the results to part (A), we have

$$I_{max} = \frac{\mathcal{E}_{max}}{R} = \frac{136\text{ V}}{12.0\ \Omega} = \boxed{11.3\text{ A}}$$

The **direct current** (DC) **generator** is illustrated in Figure 31.23a. Such generators are used, for instance, in older cars to charge the storage batteries. The components are essentially the same as those of the AC generator except that the contacts to the rotating loop are made using a split ring called a *commutator*.

In this configuration, the output voltage always has the same polarity and pulsates with time, as shown in Figure 31.23b. We can understand the reason for this by noting that the contacts to the split ring reverse their roles every half cycle. At the same time, the polarity of the induced emf reverses; hence, the polarity of the split ring (which is the same as the polarity of the output voltage) remains the same.

A pulsating DC current is not suitable for most applications. To obtain a more steady DC current, commercial DC generators use many coils and commutators distributed so that the sinusoidal pulses from the various coils are out of phase. When these pulses are superimposed, the DC output is almost free of fluctuations.

Motors are devices into which energy is transferred by electrical transmission while energy is transferred out by work. Essentially, a motor is a generator operating

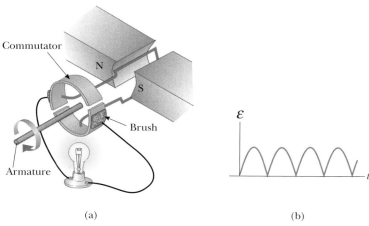

At the Active Figures link at http://www.pse6.com, you can adjust the speed of rotation and the strength of the field to see the effects on the emf generated.

Active Figure 31.23 (a) Schematic diagram of a DC generator. (b) The magnitude of the emf varies in time but the polarity never changes.

in reverse. Instead of generating a current by rotating a coil, a current is supplied to the coil by a battery, and the torque acting on the current-carrying coil causes it to rotate.

Useful mechanical work can be done by attaching the rotating coil to some external device. However, as the coil rotates in a magnetic field, the changing magnetic flux induces an emf in the coil; this induced emf always acts to reduce the current in the coil. If this were not the case, Lenz's law would be violated. The back emf increases in magnitude as the rotational speed of the coil increases. (The phrase *back emf* is used to indicate an emf that tends to reduce the supplied current.) Because the voltage available to supply current equals the difference between the supply voltage and the back emf, the current in the rotating coil is limited by the back emf.

When a motor is turned on, there is initially no back emf; thus, the current is very large because it is limited only by the resistance of the coil. As the coil begins to rotate, the induced back emf opposes the applied voltage, and the current in the coil is reduced. If the mechanical load increases, the motor slows down; this causes the back emf to decrease. This reduction in the back emf increases the current in the coil and therefore also increases the power needed from the external voltage source. For this reason, the power requirements for starting a motor and for running it are greater for heavy loads than for light ones. If the motor is allowed to run under no mechanical load, the back emf reduces the current to a value just large enough to overcome energy losses due to internal energy and friction. If a very heavy load jams the motor so that it cannot rotate, the lack of a back emf can lead to dangerously high current in the motor's wire. This is a dangerous situation, and is explored in the **What If?** section of Example 31.10.

A current application of motors in automobiles is seen in the development of *hybrid drive systems*. In these automobiles, a gasoline engine and an electric motor are combined to increase the fuel economy of the vehicle and reduce its emissions. Figure 31.24 shows the engine compartment of the Toyota Prius, which is one of a small number of hybrids available in the United States. In this automobile, power to the wheels can come from either the gasoline engine or the electric motor. In normal driving, the electric motor accelerates the vehicle from rest until it is moving at a speed of about 15 mi/h (24 km/h). During this acceleration period, the engine is not running, so that gasoline is not used and there is no emission. When a hybrid vehicle brakes, the motor acts as a generator and returns some of the kinetic energy of the vehicle back to the battery as stored energy. In a normal vehicle, this kinetic energy is simply lost as it is transformed to internal energy in the brakes and roadway.

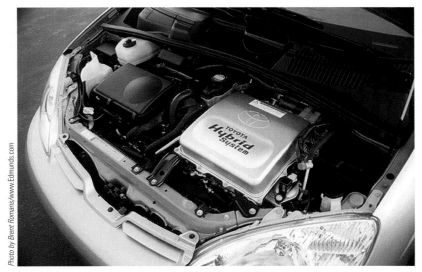

Figure 31.24 The engine compartment of the Toyota Prius, a hybrid vehicle.

Example 31.10   The Induced Current in a Motor

Assume that a motor in which the coil has a total resistance of 10 Ω is supplied by a voltage of 120 V. When the motor is running at its maximum speed, the back emf is 70 V. Find the current in the coil

(A) when the motor is turned on and

(B) when it has reached maximum speed.

Solution (A) When the motor is first turned on, the back emf is zero (because the coil is motionless). Thus, the current in the coil is a maximum and equal to

$$I = \frac{\mathcal{E}}{R} = \frac{120\text{ V}}{10\text{ }\Omega} = \boxed{12\text{ A}}$$

(B) At the maximum speed, the back emf has its maximum value. Thus, the effective supply voltage is that of the external source minus the back emf. Hence, the current is reduced to

$$I = \frac{\mathcal{E} - \mathcal{E}_{\text{back}}}{R} = \frac{120\text{ V} - 70\text{ V}}{10\text{ }\Omega} = \frac{50\text{ V}}{10\text{ }\Omega} = \boxed{5.0\text{ A}}$$

What If? Suppose that this motor is in a circular saw. You are operating the saw and the blade becomes jammed in a piece of wood so that the motor cannot turn. By what percentage does the power input to the motor increase when it is jammed?

Answer You may have everyday experiences with motors becoming warm when they are prevented from turning. This is due to the increased power input to the motor. The higher rate of energy transfer results in an increase in the internal energy of the coil, an undesirable effect. When the motor is jammed, the current is that given in part (A). Let us set up the ratio of power input to the motor when jammed to that when it is not jammed:

$$\frac{\mathcal{P}_{\text{jammed}}}{\mathcal{P}_{\text{not jammed}}} = \frac{I_{(A)}^2 R}{I_{(B)}^2 R} = \frac{I_{(A)}^2}{I_{(B)}^2}$$

where the subscripts (A) and (B) refer to the currents in parts (A) and (B) of the example. Substituting these values,

$$\frac{\mathcal{P}_{\text{jammed}}}{\mathcal{P}_{\text{not jammed}}} = \frac{(12\text{ A})^2}{(5.0\text{ A})^2} = 5.76$$

This represents a 476% increase in the input power! Such a high power input can cause the coil to become so hot that it is damaged.

31.6 Eddy Currents

Figure 31.25 Formation of eddy currents in a conducting plate moving through a magnetic field. As the plate enters or leaves the field, the changing magnetic flux induces an emf, which causes eddy currents in the plate.

As we have seen, an emf and a current are induced in a circuit by a changing magnetic flux. In the same manner, circulating currents called **eddy currents** are induced in bulk pieces of metal moving through a magnetic field. This can easily be demonstrated by allowing a flat copper or aluminum plate attached at the end of a rigid bar to swing back and forth through a magnetic field (Fig. 31.25). As the plate enters the field, the changing magnetic flux induces an emf in the plate, which in turn causes the free electrons in the plate to move, producing the swirling eddy currents. According to Lenz's law, the direction of the eddy currents is such that they create magnetic fields that oppose the change that causes the currents. For this reason, the eddy currents must produce effective magnetic poles on the plate, which are repelled by the poles of the magnet; this gives rise to a repulsive force that opposes the motion of the plate. (If the opposite were true, the plate would accelerate and its energy would increase after each swing, in violation of the law of conservation of energy.)

As indicated in Figure 31.26a, with **B** directed into the page, the induced eddy current is counterclockwise as the swinging plate enters the field at position 1. This is because the flux due to the external magnetic field into the page through the plate is increasing, and hence by Lenz's law the induced current must provide its own magnetic field out of the page. The opposite is true as the plate leaves the field at position 2, where the current is clockwise. Because the induced eddy current always produces a magnetic retarding force $\mathbf{F}_B$ when the plate enters or leaves the field, the swinging plate eventually comes to rest.

If slots are cut in the plate, as shown in Figure 31.26b, the eddy currents and the corresponding retarding force are greatly reduced. We can understand this

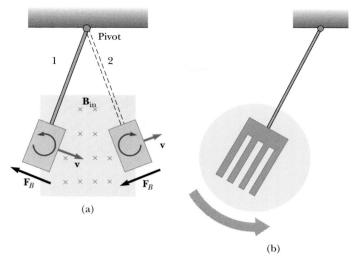

Active Figure 31.26 (a) As the conducting plate enters the field (position 1), the eddy currents are counterclockwise. As the plate leaves the field (position 2), the currents are clockwise. In either case, the force on the plate is opposite the velocity, and eventually the plate comes to rest. (b) When slots are cut in the conducting plate, the eddy currents are reduced and the plate swings more freely through the magnetic field.

Choose to let a solid or a slotted plate swing through the magnetic field and observe the effect at the Active Figures link at http://www.pse6.com.

by realizing that the cuts in the plate prevent the formation of any large current loops.

The braking systems on many subway and rapid-transit cars make use of electromagnetic induction and eddy currents. An electromagnet attached to the train is positioned near the steel rails. (An electromagnet is essentially a solenoid with an iron core.) The braking action occurs when a large current is passed through the electromagnet. The relative motion of the magnet and rails induces eddy currents in the rails, and the direction of these currents produces a drag force on the moving train. Because the eddy currents decrease steadily in magnitude as the train slows down, the braking effect is quite smooth. As a safety measure, some power tools use eddy currents to stop rapidly spinning blades once the device is turned off.

Eddy currents are often undesirable because they represent a transformation of mechanical energy to internal energy. To reduce this energy loss, conducting parts are often laminated—that is, they are built up in thin layers separated by a nonconducting material such as lacquer or a metal oxide. This layered structure increases the resistance of eddy current paths and effectively confines the currents to individual layers. Such a laminated structure is used in transformer cores (see Section 33.8) and motors to minimize eddy currents and thereby increase the efficiency of these devices.

Quick Quiz 31.11 In equal-arm balances from the early twentieth century (Fig. 31.27), it is sometimes observed that an aluminum sheet hangs from one of the arms and passes between the poles of a magnet. This causes the oscillations of the equal-arm balance to decay rapidly. In the absence of such magnetic braking, the oscillation might continue for a very long time, so that the experimenter would have to wait to take a reading. The oscillations decay because (a) the aluminum sheet is attracted to the magnet; (b) currents in the aluminum sheet set up a magnetic field that opposes the oscillations; (c) aluminum is paramagnetic.

John W. Jewett, Jr.

Figure 31.27 (Quick Quiz 31.11) In an old-fashioned equal-arm balance, an aluminum sheet hangs between the poles of a magnet.

31.7 Maxwell's Equations

We conclude this chapter by presenting four equations that are regarded as the basis of all electrical and magnetic phenomena. These equations, developed by James Clerk Maxwell, are as fundamental to electromagnetic phenomena as Newton's laws are to mechanical phenomena. In fact, the theory that Maxwell developed was more far-reaching than even he imagined because it turned out to be in agreement with the special theory of relativity, as Einstein showed in 1905.

Maxwell's equations represent the laws of electricity and magnetism that we have already discussed, but they have additional important consequences. In Chapter 34 we shall show that these equations predict the existence of electromagnetic waves (traveling patterns of electric and magnetic fields), which travel with a speed $c = 1/\sqrt{\mu_0 \epsilon_0} = 3.00 \times 10^8$ m/s, the speed of light. Furthermore, the theory shows that such waves are radiated by accelerating charges.

For simplicity, we present **Maxwell's equations** as applied to free space, that is, in the absence of any dielectric or magnetic material. The four equations are

Gauss's law

$$\oint_S \mathbf{E} \cdot d\mathbf{A} = \frac{q}{\epsilon_0}$$

(31.12)

$$\oint_S \mathbf{B} \cdot d\mathbf{A} = 0 \qquad (31.13)$$

$$\oint \mathbf{E} \cdot d\mathbf{s} = -\frac{d\Phi_B}{dt} \qquad (31.14)$$

$$\oint \mathbf{B} \cdot d\mathbf{s} = \mu_0 I + \epsilon_0 \mu_0 \frac{d\Phi_E}{dt} \qquad (31.15)$$

Equation 31.12 is Gauss's law: **the total electric flux through any closed surface equals the net charge inside that surface divided by ϵ_0.** This law relates an electric field to the charge distribution that creates it.

Equation 31.13, which can be considered Gauss's law in magnetism, states that **the net magnetic flux through a closed surface is zero.** That is, the number of magnetic field lines that enter a closed volume must equal the number that leave that volume. This implies that magnetic field lines cannot begin or end at any point. If they did, it would mean that isolated magnetic monopoles existed at those points. The fact that isolated magnetic monopoles have not been observed in nature can be taken as a confirmation of Equation 31.13.

Equation 31.14 is Faraday's law of induction, which describes the creation of an electric field by a changing magnetic flux. This law states that **the emf, which is the line integral of the electric field around any closed path, equals the rate of change of magnetic flux through any surface area bounded by that path.** One consequence of Faraday's law is the current induced in a conducting loop placed in a time-varying magnetic field.

Equation 31.15, usually called the Ampère–Maxwell law, is the generalized form of Ampère's law, and describes the creation of a magnetic field by an electric field and electric currents: **the line integral of the magnetic field around any closed path is the sum of μ_0 times the net current through that path and $\epsilon_0\mu_0$ times the rate of change of electric flux through any surface bounded by that path.**

Once the electric and magnetic fields are known at some point in space, the force acting on a particle of charge q can be calculated from the expression

$$\mathbf{F} = q\mathbf{E} + q\mathbf{v} \times \mathbf{B} \qquad (31.16)$$

This relationship is called the **Lorentz force law.** (We saw this relationship earlier as Equation 29.16.) Maxwell's equations, together with this force law, completely describe all classical electromagnetic interactions.

It is interesting to note the symmetry of Maxwell's equations. Equations 31.12 and 31.13 are symmetric, apart from the absence of the term for magnetic monopoles in Equation 31.13. Furthermore, Equations 31.14 and 31.15 are symmetric in that the line integrals of **E** and **B** around a closed path are related to the rate of change of magnetic flux and electric flux, respectively. Maxwell's equations are of fundamental importance not only to electromagnetism but to all of science. Heinrich Hertz once wrote, "One cannot escape the feeling that these mathematical formulas have an independent existence and an intelligence of their own, that they are wiser than we are, wiser even than their discoverers, that we get more out of them than we put into them."

SUMMARY

Faraday's law of induction states that the emf induced in a circuit is directly proportional to the time rate of change of magnetic flux through the circuit:

$$\varepsilon = -\frac{d\Phi_B}{dt} \qquad (31.1)$$

where $\Phi_B = \oint \mathbf{B} \cdot d\mathbf{A}$ is the magnetic flux.

Take a practice test for this chapter by clicking on the Practice Test link at http://www.pse6.com.

When a conducting bar of length ℓ moves at a velocity **v** through a magnetic field **B**, where **B** is perpendicular to the bar and to **v**, the **motional emf** induced in the bar is

$$\mathcal{E} = -B\ell v \tag{31.5}$$

Lenz's law states that the induced current and induced emf in a conductor are in such a direction as to set up a magnetic field that opposes the change that produced them.

A general form of **Faraday's law of induction** is

$$\mathcal{E} = \oint \mathbf{E} \cdot d\mathbf{s} = -\frac{d\Phi_B}{dt} \tag{31.9}$$

where **E** is the nonconservative electric field that is produced by the changing magnetic flux.

When used with the Lorentz force law, $\mathbf{F} = q\mathbf{E} + q\mathbf{v} \times \mathbf{B}$, **Maxwell's equations** describe all electromagnetic phenomena:

$$\oint_S \mathbf{E} \cdot d\mathbf{A} = \frac{q}{\epsilon_0} \tag{31.12}$$

$$\oint_S \mathbf{B} \cdot d\mathbf{A} = 0 \tag{31.13}$$

$$\oint \mathbf{E} \cdot d\mathbf{s} = -\frac{d\Phi_B}{dt} \tag{31.14}$$

$$\oint \mathbf{B} \cdot d\mathbf{s} = \mu_0 I + \epsilon_0 \mu_0 \frac{d\Phi_E}{dt} \tag{31.15}$$

QUESTIONS

1. What is the difference between magnetic flux and magnetic field?

2. A loop of wire is placed in a uniform magnetic field. For what orientation of the loop is the magnetic flux a maximum? For what orientation is the flux zero?

3. As the bar in Figure Q31.3 moves to the right, an electric field is set up directed downward in the bar. Explain why the electric field would be upward if the bar were moving to the left.

4. As the bar in Figure Q31.3 moves perpendicular to the field, is an external force required to keep it moving with constant speed?

5. The bar in Figure Q31.5 moves on rails to the right with a velocity **v**, and the uniform, constant magnetic field is directed out of the page. Why is the induced current clockwise? If the bar were moving to the left, what would be the direction of the induced current?

Figure Q31.3 Questions 3 and 4.

Figure Q31.5 Questions 5 and 6.

6. Explain why an applied force is necessary to keep the bar in Figure Q31.5 moving with a constant speed.

7. Wearing a metal bracelet in a region of strong magnetic field could be hazardous. Explain.

8. When a small magnet is moved toward a solenoid, an emf is induced in the coil. However, if the magnet is moved around inside a toroid, no measurable emf is induced. Explain.

9. How is energy produced in dams that is then transferred out by electrical transmission? (That is, how is the energy of motion of the water converted to energy that is transmitted by AC electricity?)

10. Will dropping a magnet down a long copper tube produce a current in the walls of the tube? Explain.

11. A piece of aluminum is dropped vertically downward between the poles of an electromagnet. Does the magnetic field affect the velocity of the aluminum?

12. What happens when the rotational speed of a generator coil is increased?

13. When the switch in Figure Q31.13a is closed, a current is set up in the coil and the metal ring springs upward (Fig. Q31.13b). Explain this behavior.

— Iron core
— Metal ring

Courtesy of Central Scientific Company

(a)   (b)

Figure Q31.13 Questions 13 and 14.

14. Assume that the battery in Figure Q31.13a is replaced by an AC source and the switch is held closed. If held down, the metal ring on top of the solenoid becomes hot. Why?

15. A bar magnet is held above a loop of wire in a horizontal plane, as shown in Figure Q31.15. The south end of the magnet is toward the loop of wire. The magnet is dropped toward the loop. Find the direction of the current through the resistor (a) while the magnet is falling toward the loop and (b) after the magnet has passed through the loop and moves away from it.

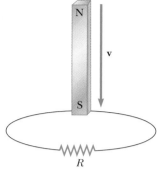

Figure Q31.15

16. Find the direction of the current in the resistor in Figure Q31.16 (a) at the instant the switch is closed, (b) after the switch has been closed for several minutes, and (c) at the instant the switch is opened.

Figure Q31.16

17. Quick Quiz 31.4 describes the emf induced between the wingtips of an airplane by its motion in the Earth's magnetic field. Can this emf be used to power a light in the passenger compartment? Explain your answer.

18. Do Maxwell's equations allow for the existence of magnetic monopoles? Explain.

19. *Induction welding* has many important industrial applications. One example is the manufacture of airtight tubes, represented in Figure Q31.19. A sheet of metal is rolled into a cylinder and forced between compression rollers to bring its edges into contact. The tube then enters a coil carrying a time-varying current. The seam is welded when induced currents around the tube raise its temperature. Typically, a sinusoidal current with a frequency of 10 kHz is used. (a) What causes a current in the tube? (b) Why is a high frequency like 10 kHz chosen, rather than the 120 Hz commonly used for power transmission? (c) Why do the induced currents raise the temperature mainly of the seam, rather than all of the metal of the tube? (d) Why is it necessary to bring the edges of the sheet together with the compression rollers before the seam can be welded?

Figure Q31.19

PROBLEMS

1, 2, 3 = straightforward, intermediate, challenging ☐ = full solution available in the *Student Solutions Manual and Study Guide*

🪐 = coached solution with hints available at http://www.pse.com   💻 = computer useful in solving problem

▨ = paired numerical and symbolic problems

Section 31.1 Faraday's Law of Induction
Section 31.3 Lenz's Law

1. A 50-turn rectangular coil of dimensions 5.00 cm × 10.0 cm is allowed to fall from a position where $B = 0$ to a new position where $B = 0.500$ T and the magnetic field is directed perpendicular to the plane of the coil. Calculate the magnitude of the average emf that is induced in the coil if the displacement occurs in 0.250 s.

2. A flat loop of wire consisting of a single turn of cross-sectional area 8.00 cm² is perpendicular to a magnetic field that increases uniformly in magnitude from 0.500 T to 2.50 T in 1.00 s. What is the resulting induced current if the loop has a resistance of 2.00 Ω?

3. A 25-turn circular coil of wire has diameter 1.00 m. It is placed with its axis along the direction of the Earth's magnetic field of 50.0 μT, and then in 0.200 s it is flipped 180°. An average emf of what magnitude is generated in the coil?

4. A rectangular loop of area A is placed in a region where the magnetic field is perpendicular to the plane of the loop. The magnitude of the field is allowed to vary in time according to $B = B_{max}e^{-t/\tau}$, where B_{max} and τ are constants. The field has the constant value B_{max} for $t < 0$. (a) Use Faraday's law to show that the emf induced in the loop is given by

$$\mathcal{E} = \frac{AB_{max}}{\tau}\,e^{-t/\tau}$$

(b) Obtain a numerical value for $\mathcal{E}$ at $t = 4.00$ s when $A = 0.160$ m², $B_{max} = 0.350$ T, and $\tau = 2.00$ s. (c) For the values of A, B_{max}, and τ given in (b), what is the maximum value of $\mathcal{E}$?

5. 🪐 A strong electromagnet produces a uniform magnetic field of 1.60 T over a cross-sectional area of 0.200 m². We place a coil having 200 turns and a total resistance of 20.0 Ω around the electromagnet. We then smoothly reduce the current in the electromagnet until it reaches zero in 20.0 ms. What is the current induced in the coil?

6. A magnetic field of 0.200 T exists within a solenoid of 500 turns and a diameter of 10.0 cm. How rapidly (that is, within what period of time) must the field be reduced to zero, if the average induced emf within the coil during this time interval is to be 10.0 kV?

7. 🪐 An aluminum ring of radius 5.00 cm and resistance 3.00×10^{-4} Ω is placed on top of a long air-core solenoid with 1 000 turns per meter and radius 3.00 cm, as shown in Figure P31.7. Over the area of the end of the solenoid, assume that the axial component of the field

produced by the solenoid is half as strong as at the center of the solenoid. Assume the solenoid produces negligible field outside its cross-sectional area. The current in the solenoid is increasing at a rate of 270 A/s. (a) What is the induced current in the ring? At the center of the ring, what are (b) the magnitude and (c) the direction of the magnetic field produced by the induced current in the ring?

Figure P31.7 Problems 7 and 8.

8. An aluminum ring of radius r_1 and resistance R is placed around the top of a long air-core solenoid with n turns per meter and smaller radius r_2 as shown in Figure P31.7. Assume that the axial component of the field produced by the solenoid over the area of the end of the solenoid is half as strong as at the center of the solenoid. Assume that the solenoid produces negligible field outside its cross-sectional area. The current in the solenoid is increasing at a rate of $\Delta I/\Delta t$. (a) What is the induced current in the ring? (b) At the center of the ring, what is the magnetic field produced by the induced current in the ring? (c) What is the direction of this field?

9. (a) A loop of wire in the shape of a rectangle of width w and length L and a long, straight wire carrying a current I lie on a tabletop as shown in Figure P31.9. (a) Determine the magnetic flux through the loop due to the current I. (b) Suppose the current is changing with time according to $I = a + bt$, where a and b are constants. Determine the emf that is induced in the loop if $b = 10.0$ A/s, $h = 1.00$ cm, $w = 10.0$ cm, and $L = 100$ cm. What is the direction of the induced current in the rectangle?

Figure P31.9 Problems 9 and 71.

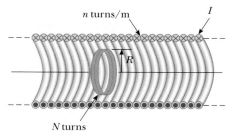

Figure P31.13

10. A coil of 15 turns and radius 10.0 cm surrounds a long solenoid of radius 2.00 cm and 1.00×10^3 turns/meter (Fig. P31.10). The current in the solenoid changes as $I = (5.00 \text{ A}) \sin(120t)$. Find the induced emf in the 15-turn coil as a function of time.

15-turn coil

Figure P31.10

11. Find the current through section PQ of length $a = 65.0$ cm in Figure P31.11. The circuit is located in a magnetic field whose magnitude varies with time according to the expression $B = (1.00 \times 10^{-3} \text{ T/s})t$. Assume the resistance per length of the wire is 0.100 Ω/m.

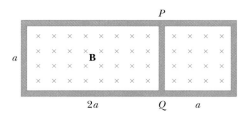

Figure P31.11

12. A 30-turn circular coil of radius 4.00 cm and resistance 1.00 Ω is placed in a magnetic field directed perpendicular to the plane of the coil. The magnitude of the magnetic field varies in time according to the expression $B = 0.010\,0t + 0.040\,0t^2$, where t is in seconds and B is in tesla. Calculate the induced emf in the coil at $t = 5.00$ s.

13. A long solenoid has $n = 400$ turns per meter and carries a current given by $I = (30.0 \text{ A})(1 - e^{-1.60t})$. Inside the solenoid and coaxial with it is a coil that has a radius of 6.00 cm and consists of a total of $N = 250$ turns of fine wire (Fig. P31.13). What emf is induced in the coil by the changing current?

14. An instrument based on induced emf has been used to measure projectile speeds up to 6 km/s. A small magnet is imbedded in the projectile, as shown in Figure P31.14. The projectile passes through two coils separated by a distance d. As the projectile passes through each coil a pulse of emf is induced in the coil. The time interval between pulses can be measured accurately with an oscilloscope, and thus the speed can be determined. (a) Sketch a graph of ΔV versus t for the arrangement shown. Consider a current that flows counterclockwise as viewed from the starting point of the projectile as positive. On your graph, indicate which pulse is from coil 1 and which is from coil 2. (b) If the pulse separation is 2.40 ms and $d = 1.50$ m, what is the projectile speed?

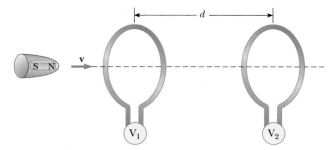

Figure P31.14

15. A coil formed by wrapping 50 turns of wire in the shape of a square is positioned in a magnetic field so that the normal to the plane of the coil makes an angle of 30.0° with the direction of the field. When the magnetic field is increased uniformly from 200 μT to 600 μT in 0.400 s, an emf of magnitude 80.0 mV is induced in the coil. What is the total length of the wire?

16. When a wire carries an AC current with a known frequency, you can use a *Rogowski coil* to determine the amplitude I_{max} of the current without disconnecting the wire to shunt the

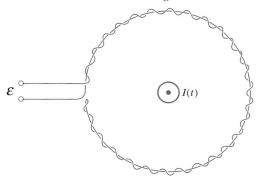

Figure P31.16

current in a meter. The Rogowski coil, shown in Figure P31.16, simply clips around the wire. It consists of a toroidal conductor wrapped around a circular return cord. The toroid has n turns per unit length and a cross-sectional area A. The current to be measured is given by $I(t) = I_{max} \sin \omega t$. (a) Show that the amplitude of the emf induced in the Rogowski coil is $\mathcal{E}_{max} = \mu_0 nA\omega I_{max}$. (b) Explain why the wire carrying the unknown current need not be at the center of the Rogowski coil, and why the coil will not respond to nearby currents that it does not enclose.

17. A toroid having a rectangular cross section ($a = 2.00$ cm by $b = 3.00$ cm) and inner radius $R = 4.00$ cm consists of 500 turns of wire that carries a sinusoidal current $I = I_{max} \sin \omega t$, with $I_{max} = 50.0$ A and a frequency $f = \omega/2\pi = 60.0$ Hz. A coil that consists of 20 turns of wire links with the toroid, as in Figure P31.17. Determine the emf induced in the coil as a function of time.

$N = 500$

a R $N' = 20$

b

Figure P31.17

18. A piece of insulated wire is shaped into a figure 8, as in Figure P31.18. The radius of the upper circle is 5.00 cm and that of the lower circle is 9.00 cm. The wire has a uniform resistance per unit length of 3.00 Ω/m. A uniform magnetic field is applied perpendicular to the plane of the two circles, in the direction shown. The magnetic field is increasing at a constant rate of 2.00 T/s. Find the magnitude and direction of the induced current in the wire.

Figure P31.18

Section 31.2   Motional emf
Section 31.3   Lenz's Law

Problem 71 in Chapter 29 can be assigned with this section.

19. An automobile has a vertical radio antenna 1.20 m long. The automobile travels at 65.0 km/h on a horizontal road

where the Earth's magnetic field is 50.0 μT directed toward the north and downward at an angle of 65.0° below the horizontal. (a) Specify the direction that the automobile should move in order to generate the maximum motional emf in the antenna, with the top of the antenna positive relative to the bottom. (b) Calculate the magnitude of this induced emf.

20. Consider the arrangement shown in Figure P31.20. Assume that $R = 6.00$ Ω, $\ell = 1.20$ m, and a uniform 2.50-T magnetic field is directed into the page. At what speed should the bar be moved to produce a current of 0.500 A in the resistor?

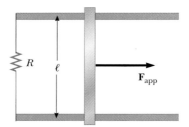

R ℓ $\mathbf{F}_{app}$

Figure P31.20 Problems 20, 21, and 22.

21. Figure P31.20 shows a top view of a bar that can slide without friction. The resistor is 6.00 Ω and a 2.50-T magnetic field is directed perpendicularly downward, into the paper. Let $\ell = 1.20$ m. (a) Calculate the applied force required to move the bar to the right at a constant speed of 2.00 m/s. (b) At what rate is energy delivered to the resistor?

22. A conducting rod of length ℓ moves on two horizontal, frictionless rails, as shown in Figure P31.20. If a constant force of 1.00 N moves the bar at 2.00 m/s through a magnetic field $\mathbf{B}$ that is directed into the page, (a) what is the current through the 8.00-Ω resistor R? (b) What is the rate at which energy is delivered to the resistor? (c) What is the mechanical power delivered by the force $\mathbf{F}_{app}$?

23. Very large magnetic fields can be produced using a procedure called *flux compression*. A metallic cylindrical tube of radius R is placed coaxially in a long solenoid of somewhat larger radius. The space between the tube and the solenoid is filled with a highly explosive material. When the explosive is set off, it collapses the tube to a cylinder of radius $r < R$. If the collapse happens very rapidly, induced current in the tube maintains the magnetic flux nearly constant inside the tube. If the initial magnetic field in the solenoid is 2.50 T, and $R/r = 12.0$, what maximum value of magnetic field can be achieved?

24. The *homopolar generator*, also called the *Faraday disk*, is a low-voltage, high-current electric generator. It consists of a rotating conducting disk with one stationary brush (a sliding electrical contact) at its axle and another at a point on its circumference, as shown in Figure P31.24. A magnetic field is applied perpendicular to the plane of the disk. Assume the field is 0.900 T, the angular speed is 3 200 rev/min, and the radius of the disk is 0.400 m. Find the emf generated between the brushes. When superconducting coils are used to produce a large magnetic field, a homopolar generator can have a power output of several

megawatts. Such a generator is useful, for example, in purifying metals by electrolysis. If a voltage is applied to the output terminals of the generator, it runs in reverse as a *homopolar motor* capable of providing great torque, useful in ship propulsion.

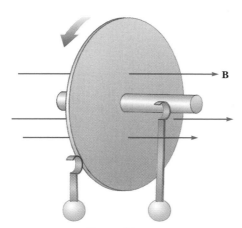

Figure P31.24

25. **Review problem.** A flexible metallic wire with linear density 3.00×10^{-3} kg/m is stretched between two fixed clamps 64.0 cm apart and held under tension 267 N. A magnet is placed near the wire as shown in Figure P31.25. Assume that the magnet produces a uniform field of 4.50 mT over a 2.00-cm length at the center of the wire, and a negligible field elsewhere. The wire is set vibrating at its fundamental (lowest) frequency. The section of the wire in the magnetic field moves with a uniform amplitude of 1.50 cm. Find (a) the frequency and (b) the amplitude of the electromotive force induced between the ends of the wire.

Figure P31.25

26. The square loop in Figure P31.26 is made of wires with total series resistance 10.0 Ω. It is placed in a uniform 0.100-T magnetic field directed perpendicularly into the plane of the paper. The loop, which is hinged at each corner, is pulled as shown until the separation between points A and B is 3.00 m. If this process takes 0.100 s, what is the average current generated in the loop? What is the direction of the current?

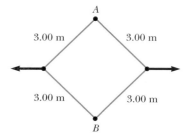

Figure P31.26

27. A helicopter (Figure P31.27) has blades of length 3.00 m, extending out from a central hub and rotating at 2.00 rev/s. If the vertical component of the Earth's magnetic field is 50.0 μT, what is the emf induced between the blade tip and the center hub?

Figure P31.27

28. Use Lenz's law to answer the following questions concerning the direction of induced currents. (a) What is the direction of the induced current in resistor R in Figure P31.28a when the bar magnet is moved to the left? (b) What is the direction of the current induced in the resistor R immediately after the switch S in Figure P31.28b

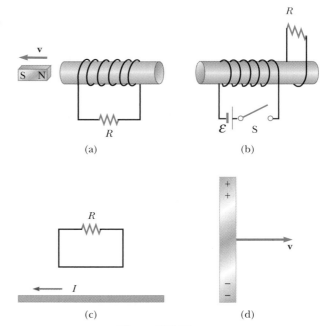

Figure P31.28

is closed? (c) What is the direction of the induced current in R when the current I in Figure P31.28c decreases rapidly to zero? (d) A copper bar is moved to the right while its axis is maintained in a direction perpendicular to a magnetic field, as shown in Figure P31.28d. If the top of the bar becomes positive relative to the bottom, what is the direction of the magnetic field?

29. A rectangular coil with resistance R has N turns, each of length ℓ and width w as shown in Figure P31.29. The coil moves into a uniform magnetic field **B** with constant velocity **v**. What are the magnitude and direction of the total magnetic force on the coil (a) as it enters the magnetic field, (b) as it moves within the field, and (c) as it leaves the field?

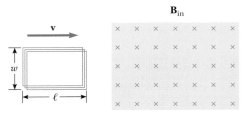

Figure P31.29

30. In Figure P31.30, the bar magnet is moved toward the loop. Is $V_a - V_b$ positive, negative, or zero? Explain.

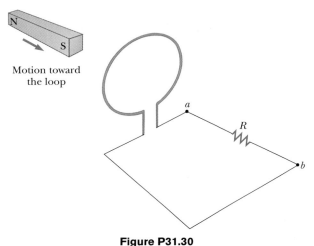

Figure P31.30

31. Two parallel rails with negligible resistance are 10.0 cm apart and are connected by a 5.00-Ω resistor. The circuit also contains two metal rods having resistances of 10.0 Ω and 15.0 Ω sliding along the rails (Fig. P31.31). The rods are pulled away from the resistor at constant speeds of 4.00 m/s and 2.00 m/s, respectively. A uniform magnetic field of magnitude 0.010 0 T is applied perpendicular to the plane of the rails. Determine the current in the 5.00-Ω resistor.

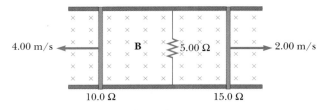

Figure P31.31

Section 31.4 Induced emf and Electric Fields

32. For the situation shown in Figure P31.32, the magnetic field changes with time according to the expression $B = (2.00t^3 - 4.00t^2 + 0.800)\,\text{T}$, and $r_2 = 2R = 5.00$ cm. (a) Calculate the magnitude and direction of the force exerted on an electron located at point P_2 when $t = 2.00$ s. (b) At what time is this force equal to zero?

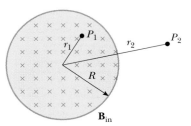

Figure P31.32 Problems 32 and 33.

33. A magnetic field directed into the page changes with time according to $B = (0.030\,0t^2 + 1.40)\,\text{T}$, where t is in seconds. The field has a circular cross section of radius $R = 2.50$ cm (Fig. P31.32). What are the magnitude and direction of the electric field at point P_1 when $t = 3.00$ s and $r_1 = 0.020\,0$ m?

34. A long solenoid with 1 000 turns per meter and radius 2.00 cm carries an oscillating current given by $I = (5.00\,\text{A})\,\sin(100\pi t)$. What is the electric field induced at a radius $r = 1.00$ cm from the axis of the solenoid? What is the direction of this electric field when the current is increasing counterclockwise in the coil?

Section 31.5 Generators and Motors

Problems 28 and 62 in Chapter 29 can be assigned with this section.

35. A coil of area 0.100 m² is rotating at 60.0 rev/s with the axis of rotation perpendicular to a 0.200-T magnetic field. (a) If the coil has 1 000 turns, what is the maximum emf generated in it? (b) What is the orientation of the coil with respect to the magnetic field when the maximum induced voltage occurs?

36. In a 250-turn automobile alternator, the magnetic flux in each turn is $\Phi_B = (2.50 \times 10^{-4}\,\text{Wb})\cos(\omega t)$, where ω is the angular speed of the alternator. The alternator is geared to rotate three times for each engine revolution. When the engine is running at an angular speed of 1 000 rev/min, determine (a) the induced emf in the alternator as a function of time and (b) the maximum emf in the alternator.

37. A long solenoid, with its axis along the x axis, consists of 200 turns per meter of wire that carries a steady current of 15.0 A. A coil is formed by wrapping 30 turns of thin wire around a circular frame that has a radius of 8.00 cm. The coil is placed inside the solenoid and mounted on an axis that is a diameter of the coil and coincides with the y axis. The coil is then rotated with an angular speed of 4.00π rad/s. (The plane of the coil is in the yz plane

at $t = 0$.) Determine the emf generated in the coil as a function of time.

38. A bar magnet is spun at constant angular speed ω around an axis as shown in Figure P31.38. A stationary flat rectangular conducting loop surrounds the magnet, and at $t = 0$, the magnet is oriented as shown. Make a qualitative graph of the induced current in the loop as a function of time, plotting counterclockwise currents as positive and clockwise currents as negative.

Figure P31.38

39. A motor in normal operation carries a direct current of 0.850 A when connected to a 120-V power supply. The resistance of the motor windings is 11.8 Ω. While in normal operation, (a) what is the back emf generated by the motor? (b) At what rate is internal energy produced in the windings? (c) **What If?** Suppose that a malfunction stops the motor shaft from rotating. At what rate will internal energy be produced in the windings in this case? (Most motors have a thermal switch that will turn off the motor to prevent overheating when this occurs.)

40. A semicircular conductor of radius $R = 0.250$ m is rotated about the axis AC at a constant rate of 120 rev/min (Fig. P31.40). A uniform magnetic field in all of the lower half of the figure is directed out of the plane of rotation and has a magnitude of 1.30 T. (a) Calculate the maximum value of the emf induced in the conductor. (b) What is the value of the average induced emf for each complete rotation? (c) **What If?** How would the answers to (a) and (b) change if **B** were allowed to extend a distance R above the axis of rotation? Sketch the emf versus time (d) when the field is as drawn in Figure P31.40 and (e) when the field is extended as described in (c).

Figure P31.40

41. The rotating loop in an AC generator is a square 10.0 cm on a side. It is rotated at 60.0 Hz in a uniform field of 0.800 T. Calculate (a) the flux through the loop as a function of time, (b) the emf induced in the loop, (c) the

current induced in the loop for a loop resistance of 1.00 Ω, (d) the power delivered to the loop, and (e) the torque that must be exerted to rotate the loop.

Section 31.6   Eddy Currents

42. Figure P31.42 represents an electromagnetic brake that uses eddy currents. An electromagnet hangs from a railroad car near one rail. To stop the car, a large current is sent through the coils of the electromagnet. The moving electromagnet induces eddy currents in the rails, whose fields oppose the change in the field of the electromagnet. The magnetic fields of the eddy currents exert force on the current in the electromagnet, thereby slowing the car. The direction of the car's motion and the direction of the current in the electromagnet are shown correctly in the picture. Determine which of the eddy currents shown on the rails is correct. Explain your answer.

Figure P31.42

43. A conducting rectangular loop of mass M, resistance R, and dimensions w by ℓ falls from rest into a magnetic field **B** as shown in Figure P31.43. During the time interval

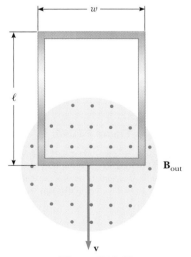

Figure P31.43

before the top edge of the loop reaches the field, the loop approaches a terminal speed v_T. (a) Show that

$$v_T = \frac{MgR}{B^2 w^2}$$

(b) Why is v_T proportional to R? (c) Why is it inversely proportional to B^2?

Section 31.7 Maxwell's Equations

44. An electron moves through a uniform electric field $\mathbf{E} = (2.50\hat{\mathbf{i}} + 5.00\hat{\mathbf{j}})$ V/m and a uniform magnetic field $\mathbf{B} = (0.400\hat{\mathbf{k}})$ T. Determine the acceleration of the electron when it has a velocity $\mathbf{v} = 10.0\hat{\mathbf{i}}$ m/s.

45. A proton moves through a uniform electric field given by $\mathbf{E} = 50.0\hat{\mathbf{j}}$ V/m and a uniform magnetic field $\mathbf{B} = (0.200\hat{\mathbf{i}} + 0.300\hat{\mathbf{j}} + 0.400\hat{\mathbf{k}})$ T. Determine the acceleration of the proton when it has a velocity $\mathbf{v} = 200\hat{\mathbf{i}}$ m/s.

Additional Problems

46. A steel guitar string vibrates (Figure 31.6). The component of magnetic field perpendicular to the area of a pickup coil nearby is given by

$$B = 50.0 \text{ mT} + (3.20 \text{ mT}) \sin(2\pi 523 \, t/\text{s})$$

The circular pickup coil has 30 turns and radius 2.70 mm. Find the emf induced in the coil as a function of time.

47. Figure P31.47 is a graph of the induced emf versus time for a coil of N turns rotating with angular speed ω in a uniform magnetic field directed perpendicular to the axis of rotation of the coil. **What If?** Copy this sketch (on a larger scale), and on the same set of axes show the graph of emf versus t (a) if the number of turns in the coil is doubled; (b) if instead the angular speed is doubled; and (c) if the angular speed is doubled while the number of turns in the coil is halved.

Figure P31.47

48. A technician wearing a brass bracelet enclosing area 0.005 00 m^2 places her hand in a solenoid whose magnetic field is 5.00 T directed perpendicular to the plane of the bracelet. The electrical resistance around the circumference of the bracelet is 0.020 0 Ω. An unexpected power failure causes the field to drop to 1.50 T in a time of 20.0 ms. Find (a) the current induced in the bracelet and (b) the power delivered to the bracelet. *Note:* As this

problem implies, you should not wear any metal objects when working in regions of strong magnetic fields.

49. Two infinitely long solenoids (seen in cross section) pass through a circuit as shown in Figure P31.49. The magnitude of **B** inside each is the same and is increasing at the rate of 100 T/s. What is the current in each resistor?

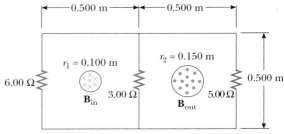

Figure P31.49

50. A conducting rod of length $\ell = 35.0$ cm is free to slide on two parallel conducting bars as shown in Figure P31.50. Two resistors $R_1 = 2.00 \, \Omega$ and $R_2 = 5.00 \, \Omega$ are connected across the ends of the bars to form a loop. A constant magnetic field $B = 2.50$ T is directed perpendicularly into the page. An external agent pulls the rod to the left with a constant speed of $v = 8.00$ m/s. Find (a) the currents in both resistors, (b) the total power delivered to the resistance of the circuit, and (c) the magnitude of the applied force that is needed to move the rod with this constant velocity.

Figure P31.50

51. Suppose you wrap wire onto the core from a roll of cellophane tape to make a coil. Describe how you can use a bar magnet to produce an induced voltage in the coil. What is the order of magnitude of the emf you generate? State the quantities you take as data and their values.

52. A bar of mass m, length d, and resistance R slides without friction in a horizontal plane, moving on parallel rails as shown in Figure P31.52. A battery that maintains a constant emf $\mathcal{E}$ is connected between the rails, and a constant magnetic field **B** is directed perpendicularly to the plane of the page. Assuming the bar starts from rest, show that at time t it moves with a speed

$$v = \frac{\mathcal{E}}{Bd}\left(1 - e^{-B^2 d^2 t / mR}\right)$$

Figure P31.52

53. **Review problem.** A particle with a mass of 2.00×10^{-16} kg and a charge of 30.0 nC starts from rest, is accelerated by a strong electric field, and is fired from a small source inside a region of uniform constant magnetic field 0.600 T. The velocity of the particle is perpendicular to the field. The circular orbit of the particle encloses a magnetic flux of 15.0 μWb. (a) Calculate the speed of the particle. (b) Calculate the potential difference through which the particle accelerated inside the source.

54. An *induction furnace* uses electromagnetic induction to produce eddy currents in a conductor, thereby raising the conductor's temperature. Commercial units operate at frequencies ranging from 60 Hz to about 1 MHz and deliver powers from a few watts to several megawatts. Induction heating can be used for welding in a vacuum enclosure, to avoid oxidation and contamination of the metal. At high frequencies, induced currents occur only near the surface of the conductor—this is the "skin effect." By creating an induced current for a short time at an appropriately high frequency, one can heat a sample down to a controlled depth. For example, the surface of a farm tiller can be tempered to make it hard and brittle for effective cutting while keeping the interior metal soft and ductile to resist breakage.

   To explore induction heating, consider a flat conducting disk of radius R, thickness b, and resistivity ρ. A sinusoidal magnetic field $B_{\text{max}} \cos \omega t$ is applied perpendicular to the disk. Assume that the frequency is so low that the skin effect is not important. Assume the eddy currents occur in circles concentric with the disk. (a) Calculate the average power delivered to the disk. (b) **What If?** By what factor does the power change when the amplitude of the field doubles? (c) When the frequency doubles? (d) When the radius of the disk doubles?

55. The plane of a square loop of wire with edge length $a = 0.200$ m is perpendicular to the Earth's magnetic field at a point where $B = 15.0$ μT, as shown in Figure P31.55. The total resistance of the loop and the wires connecting it to a sensitive ammeter is 0.500 Ω. If the loop is suddenly collapsed by horizontal forces as shown, what total charge passes through the ammeter?

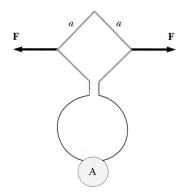

Figure P31.55

56. Magnetic field values are often determined by using a device known as a *search coil*. This technique depends on the measurement of the total charge passing through a coil in a time interval during which the magnetic flux linking the windings changes either because of the motion of the coil or because of a change in the value of B. (a) Show that as the flux through the coil changes from Φ_1 to Φ_2, the charge transferred through the coil will be given by $Q = N(\Phi_2 - \Phi_1)/R$, where R is the resistance of the coil and a sensitive ammeter connected across it and N is the number of turns. (b) As a specific example, calculate B when a 100-turn coil of resistance 200 Ω and cross-sectional area 40.0 cm^2 produces the following results. A total charge of 5.00×10^{-4} C passes through the coil when it is rotated in a uniform field from a position where the plane of the coil is perpendicular to the field to a position where the coil's plane is parallel to the field.

57. In Figure P31.57, the rolling axle, 1.50 m long, is pushed along horizontal rails at a constant speed $v = 3.00$ m/s. A resistor $R = 0.400$ Ω is connected to the rails at points a and b, directly opposite each other. (The wheels make good electrical contact with the rails, and so the axle, rails, and R form a closed-loop circuit. The only significant resistance in the circuit is R.) A uniform magnetic field $B = 0.0800$ T is vertically downward. (a) Find the induced current I in the resistor. (b) What horizontal force F is required to keep the axle rolling at constant speed? (c) Which end of the resistor, a or b, is at the higher electric potential? (d) **What If?** After the axle rolls past the resistor, does the current in R reverse direction? Explain your answer.

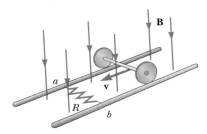

Figure P31.57

58. A conducting rod moves with a constant velocity **v** in a direction perpendicular to a long, straight wire carrying a current I as shown in Figure P31.58. Show that the magnitude of the emf generated between the ends of the rod is

$$|\mathcal{E}| = \frac{\mu_0 v I \ell}{2 \pi r}$$

In this case, note that the emf decreases with increasing r, as you might expect.

Figure P31.58

59. A circular loop of wire of radius r is in a uniform magnetic field, with the plane of the loop perpendicular to the direction of the field (Fig. P31.59). The magnetic field varies with time according to $B(t) = a + bt$, where a and b are constants. (a) Calculate the magnetic flux through the loop at $t = 0$. (b) Calculate the emf induced in the loop. (c) If the resistance of the loop is R, what is the induced current? (d) At what rate is energy being delivered to the resistance of the loop?

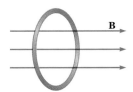

Figure P31.59

60. In Figure P31.60, a uniform magnetic field decreases at a constant rate $dB/dt = -K$, where K is a positive constant. A circular loop of wire of radius a containing a resistance R and a capacitance C is placed with its plane normal to the field. (a) Find the charge Q on the capacitor when it is fully charged. (b) Which plate is at the higher potential? (c) Discuss the force that causes the separation of charges.

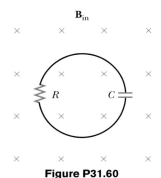

Figure P31.60

61. A rectangular coil of 60 turns, dimensions 0.100 m by 0.200 m and total resistance 10.0 Ω, rotates with angular speed 30.0 rad/s about the y axis in a region where a 1.00-T magnetic field is directed along the x axis. The rotation is initiated so that the plane of the coil is perpendicular to the direction of **B** at $t = 0$. Calculate (a) the maximum induced emf in the coil, (b) the maximum rate of change of magnetic flux through the coil, (c) the induced emf at $t = 0.050\ 0$ s, and (d) the torque exerted by the magnetic field on the coil at the instant when the emf is a maximum.

62. A small circular washer of radius 0.500 cm is held directly below a long, straight wire carrying a current of 10.0 A. The washer is located 0.500 m above the top of a table (Fig. P31.62). (a) If the washer is dropped from rest, what is the magnitude of the average induced emf in the washer from the time it is released to the moment it hits the table-top? Assume that the magnetic field is nearly constant over the area of the washer, and equal to the magnetic field at the center of the washer. (b) What is the direction of the induced current in the washer?

Figure P31.62

63. A conducting rod of length ℓ moves with velocity **v** parallel to a long wire carrying a steady current I. The axis of the rod is maintained perpendicular to the wire with the near end a distance r away, as shown in Figure P31.63. Show that the magnitude of the emf induced in the rod is

$$|\mathcal{E}| = \frac{\mu_0 Iv}{2\pi} \ln\left(1 + \frac{\ell}{r}\right)$$

Figure P31.63

64. A rectangular loop of dimensions ℓ and w moves with a constant velocity **v** away from a long wire that carries a current I in the plane of the loop (Fig. P31.64). The total resistance of the loop is R. Derive an expression that gives the current in the loop at the instant the near side is a distance r from the wire.

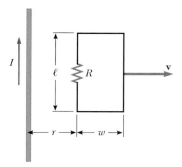

Figure P31.64

65. The magnetic flux through a metal ring varies with time t according to $\Phi_B = 3(at^3 - bt^2)$ T·m^2, with $a = 2.00$ s^{-3} and $b = 6.00$ s^{-2}. The resistance of the ring is 3.00 Ω. Determine the maximum current induced in the ring during the interval from $t = 0$ to $t = 2.00$ s.

66. **Review problem.** The bar of mass m in Figure P31.66 is pulled horizontally across parallel rails by a massless string that passes over an ideal pulley and is attached to a

suspended object of mass M. The uniform magnetic field has a magnitude B, and the distance between the rails is ℓ. The rails are connected at one end by a load resistor R. Derive an expression that gives the horizontal speed of the bar as a function of time, assuming that the suspended object is released with the bar at rest at $t = 0$. Assume no friction between rails and bar.

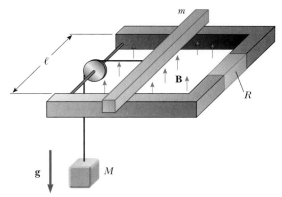

Figure P31.66

67. A solenoid wound with 2 000 turns/m is supplied with current that varies in time according to $I = (4\text{A})\sin(120\pi t)$, where t is in seconds. A small coaxial circular coil of 40 turns and radius $r = 5.00$ cm is located inside the solenoid near its center. (a) Derive an expression that describes the manner in which the emf in the small coil varies in time. (b) At what average rate is energy delivered to the small coil if the windings have a total resistance of 8.00 Ω?

68. Figure P31.68 shows a stationary conductor whose shape is similar to the letter e. The radius of its circular portion is $a = 50.0$ cm. It is placed in a constant magnetic field of 0.500 T directed out of the page. A straight conducting rod, 50.0 cm long, is pivoted about point O and rotates with a constant angular speed of 2.00 rad/s. (a) Determine the induced emf in the loop POQ. Note that the area of the loop is $\theta a^2/2$. (b) If all of the conducting material has a resistance per length of 5.00 Ω/m, what is the induced current in the loop POQ at the instant 0.250 s after point P passes point Q?

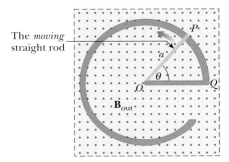

The *moving* straight rod

Figure P31.68

69. A *betatron* accelerates electrons to energies in the MeV range by means of electromagnetic induction. Electrons in a vacuum chamber are held in a circular orbit by a magnetic field perpendicular to the orbital plane. The magnetic field is gradually increased to induce an electric field around the orbit. (a) Show that the electric field is in the correct direction to make the electrons speed up. (b) Assume that the radius of the orbit remains constant. Show that the average magnetic field over the area enclosed by the orbit must be twice as large as the magnetic field at the circumference of the circle.

70. A wire 30.0 cm long is held parallel to and 80.0 cm above a long wire carrying 200 A and resting on the floor (Fig. P31.70). The 30.0-cm wire is released and falls, remaining parallel with the current-carrying wire as it falls. Assume that the falling wire accelerates at 9.80 m/s^2 and derive an equation for the emf induced in it. Express your result as a function of the time t after the wire is dropped. What is the induced emf 0.300 s after the wire is released?

Figure P31.70

71. A long, straight wire carries a current that is given by $I = I_{max} \sin(\omega t + \phi)$ and lies in the plane of a rectangular coil of N turns of wire, as shown in Figure P31.9. The quantities I_{max}, ω, and ϕ are all constants. Determine the emf induced in the coil by the magnetic field created by the current in the straight wire. Assume $I_{max} = 50.0$ A, $\omega = 200\pi$ s^{-1}, $N = 100$, $h = w = 5.00$ cm, and $L = 20.0$ cm.

72. A dime is suspended from a thread and hung between the poles of a strong horseshoe magnet as shown in Figure P31.72. The dime rotates at constant angular speed ω

Figure P31.72

about a vertical axis. Letting θ represent the angle between the direction of **B** and the normal to the face of the dime, sketch a graph of the torque due to induced currents as a function of θ for $0 < \theta < 2\pi$.

Answers to Quick Quizzes

31.1 (c). In all cases except this one, there is a change in the magnetic flux through the loop.

31.2 c, $d = e$, b, a. The magnitude of the emf is proportional to the rate of change of the magnetic flux. For the situation described, the rate of change of magnetic flux is proportional to the rate of change of the magnetic field. This rate of change is the slope of the graph in Figure 31.4. The magnitude of the slope is largest at c. Points d and e are on a straight line, so the slope is the same at each point. Point b represents a point of relatively small slope, while a is at a point of zero slope because the curve is horizontal at that point.

31.3 (b). The magnetic field lines around the transmission cable will be circular, centered on the cable. If you place your loop around the cable, there are no field lines passing through the loop, so no emf is induced. The loop must be placed next to the cable, with the plane of the loop parallel to the cable to maximize the flux through its area.

31.4 (a). The Earth's magnetic field has a downward component in the northern hemisphere. As the plane flies north, the right-hand rule illustrated in Figure 29.4 indicates that positive charge experiences a force directed toward the west. Thus, the left wingtip becomes positively charged and the right wingtip negatively charged.

31.5 (c). The force on the wire is of magnitude $F_{app} = F_B = I\ell B$, with I given by Equation 31.6. Thus, the force is proportional to the speed and the force doubles. Because $\mathcal{P} = F_{app}v$, the doubling of the force *and* the speed results in the power being four times as large.

31.6 (b). According to Equation 31.5, because B and v are fixed, the emf depends only on the length of the wire moving in the magnetic field. Thus, you want the long dimension moving through the magnetic field lines so that it is perpendicular to the velocity vector. In this case, the short dimension is parallel to the velocity vector.

31.7 (a). Because the current induced in the solenoid is clockwise when viewed from above, the magnetic field lines produced by this current point downward in Figure 31.15. Thus, the upper end of the solenoid acts as a south pole. For this situation to be consistent with Lenz's law, the south pole of the bar magnet must be approaching the solenoid.

31.8 (b). At the position of the loop, the magnetic field lines due to the wire point into the page. The loop is entering a region of stronger magnetic field as it drops toward the wire, so the flux is increasing. The induced current must set up a magnetic field that opposes this increase. To do this, it creates a magnetic field directed out of the page. By the right-hand rule for current loops, this requires a counterclockwise current in the loop.

31.9 (d). The constant rate of change of B will result in a constant rate of change of the magnetic flux. According to Equation 31.9, if $d\Phi_B/dt$ is constant, **E** is constant in magnitude.

31.10 (a). While reducing the resistance may increase the current that the generator provides to a load, it does not alter the emf. Equation 31.11 shows that the emf depends on ω, B, and N, so all other choices increase the emf.

31.11 (b). When the aluminum sheet moves between the poles of the magnet, eddy currents are established in the aluminum. According to Lenz's law, these currents are in a direction so as to oppose the original change, which is the movement of the aluminum sheet in the magnetic field. The same principle is used in common laboratory triple-beam balances. See if you can find the magnet and the aluminum sheet the next time you use a triple-beam balance.

Chapter 32

Inductance

▲ *An airport metal detector contains a large coil of wire around the frame. This coil has a property called inductance. When a passenger carries metal through the detector, the inductance of the coil changes, and the change in inductance signals an alarm to sound. (Jack Hollingsworth/Getty Images)*

In Chapter 31, we saw that an emf and a current are induced in a circuit when the magnetic flux through the area enclosed by the circuit changes with time. This phenomenon of electromagnetic induction has some practical consequences. In this chapter, we first describe an effect known as *self-induction,* in which a time-varying current in a circuit produces an induced emf opposing the emf that initially set up the time-varying current. Self-induction is the basis of the *inductor,* an electrical circuit element. We discuss the energy stored in the magnetic field of an inductor and the energy density associated with the magnetic field.

Next, we study how an emf is induced in a circuit as a result of a changing magnetic flux produced by a second circuit; this is the basic principle of *mutual induction.* Finally, we examine the characteristics of circuits that contain inductors, resistors, and capacitors in various combinations.

32.1 Self-Inductance

In this chapter, we need to distinguish carefully between emfs and currents that are caused by batteries or other sources and those that are induced by changing magnetic fields. When we use a term without an adjective (such as *emf* and *current*) we are describing the parameters associated with a physical source. We use the adjective *induced* to describe those emfs and currents caused by a changing magnetic field.

Consider a circuit consisting of a switch, a resistor, and a source of emf, as shown in Figure 32.1. When the switch is thrown to its closed position, the current does not immediately jump from zero to its maximum value $\mathcal{E}/R$. Faraday's law of electromagnetic induction (Eq. 31.1) can be used to describe this effect as follows: as the current increases with time, the magnetic flux through the circuit loop due to this current also increases with time. This increasing flux creates an induced emf in the circuit. The direction of the induced emf is such that it would cause an induced current in the loop (if the loop did not already carry a current), which would establish a magnetic field opposing the change in the original magnetic field. Thus, the direction of the induced emf is opposite the direction of the emf of the battery; this results in a gradual rather than instantaneous increase in the current to its final equilibrium value. Because of the direction of the induced emf, it is also called a *back emf,* similar to that in a motor, as discussed in Chapter 31. This effect is called **self-induction** because the changing flux through the circuit and the resultant induced emf arise from the circuit itself. The emf $\mathcal{E}_L$ set up in this case is called a **self-induced emf.**

As a second example of self-induction, consider Figure 32.2, which shows a coil wound on a cylindrical core. Assume that the current in the coil either increases or decreases with time. When the current is in the direction shown, a magnetic field directed from right to left is set up inside the coil, as seen in Figure 32.2a. As the current changes with time, the magnetic flux through the coil also changes and induces an emf in the coil. From Lenz's law, the polarity of this induced emf must be such that it opposes the change in the magnetic field from the current. If the current

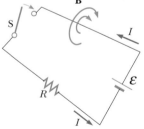

Figure 32.1 After the switch is closed, the current produces a magnetic flux through the area enclosed by the loop. As the current increases toward its equilibrium value, this magnetic flux changes in time and induces an emf in the loop.

Figure 32.2 (a) A current in the coil produces a magnetic field directed to the left. (b) If the current increases, the increasing magnetic flux creates an induced emf in the coil having the polarity shown by the dashed battery. (c) The polarity of the induced emf reverses if the current decreases.

Joseph Henry
American Physicist (1797–1878)

Henry became the first director of the Smithsonian Institution and first president of the Academy of Natural Science. He improved the design of the electromagnet and constructed one of the first motors. He also discovered the phenomenon of self-induction but failed to publish his findings. The unit of inductance, the henry, is named in his honor. *(North Wind Picture Archives)*

is increasing, the polarity of the induced emf is as pictured in Figure 32.2b, and if the current is decreasing, the polarity of the induced emf is as shown in Figure 32.2c.

To obtain a quantitative description of self-induction, we recall from Faraday's law that the induced emf is equal to the negative of the time rate of change of the magnetic flux. The magnetic flux is proportional to the magnetic field due to the current, which in turn is proportional to the current in the circuit. Therefore, **a self-induced emf is always proportional to the time rate of change of the current.** For any coil, we find that

$$\mathcal{E}_L = -L\frac{dI}{dt} \tag{32.1}$$

where L is a proportionality constant—called the **inductance** of the coil—that depends on the geometry of the coil and other physical characteristics. Combining this expression with Faraday's law, $\mathcal{E}_L = -N\,d\Phi_B/dt$, we see that the inductance of a closely spaced coil of N turns (a toroid or an ideal solenoid) carrying a current I and containing N turns is

$$L = \frac{N\Phi_B}{I} \tag{32.2}$$

◀ **Inductance of an *N*-turn coil**

where it is assumed that the same magnetic flux passes through each turn.

From Equation 32.1, we can also write the inductance as the ratio

$$L = -\frac{\mathcal{E}_L}{dI/dt} \tag{32.3}$$

◀ **Inductance**

Recall that resistance is a measure of the opposition to current ($R = \Delta V/I$); in comparison, inductance is a measure of the opposition to a *change* in current.

The SI unit of inductance is the **henry** (H), which, as we can see from Equation 32.3, is 1 volt-second per ampere:

$$1\ \text{H} = 1\ \frac{\text{V}\cdot\text{s}}{\text{A}}$$

As shown in Examples 32.1 and 32.2, the inductance of a coil depends on its geometry. This is analogous to the capacitance of a capacitor depending on the geometry of its plates, as we found in Chapter 26. Inductance calculations can be quite difficult to perform for complicated geometries; however, the examples below involve simple situations for which inductances are easily evaluated.

Quick Quiz 32.1 A coil with zero resistance has its ends labeled *a* and *b*. The potential at *a* is higher than at *b*. Which of the following could be consistent with this situation? (a) The current is constant and is directed from *a* to *b*; (b) The current is constant and is directed from *b* to *a*; (c) The current is increasing and is directed from *a* to *b*; (d) The current is decreasing and is directed from *a* to *b*; (e) The current is increasing and is directed from *b* to *a*; (f) The current is decreasing and is directed from *b* to *a*.

Example 32.1   Inductance of a Solenoid

Find the inductance of a uniformly wound solenoid having N turns and length ℓ. Assume that ℓ is much longer than the radius of the windings and that the core of the solenoid is air.

Solution We can assume that the interior magnetic field due to the current is uniform and given by Equation 30.17:

$$B = \mu_0 nI = \mu_0 \frac{N}{\ell} I$$

where $n = N/\ell$ is the number of turns per unit length. The magnetic flux through each turn is

$$\Phi_B = BA = \mu_0 \frac{NA}{\ell} I$$

where A is the cross-sectional area of the solenoid. Using this expression and Equation 32.2, we find that

$$L = \frac{N\Phi_B}{I} = \frac{\mu_0 N^2 A}{\ell} \qquad (32.4)$$

This result shows that L depends on geometry and is proportional to the square of the number of turns. Because $N = n\ell$, we can also express the result in the form

$$L = \mu_0 \frac{(n\ell)^2}{\ell} A = \mu_0 n^2 A\ell = \mu_0 n^2 V \qquad (32.5)$$

where $V = A\ell$ is the interior volume of the solenoid.

What If? What would happen to the inductance if you inserted a ferromagnetic material inside the solenoid?

Answer The inductance would increase. For a given current, the magnetic flux in the solenoid is much greater because of the increase in the magnetic field originating from the magnetization of the ferromagnetic material. For example, if the material has a magnetic permeability of $500\mu_0$, the inductance increases by a factor of 500.

Example 32.2   Calculating Inductance and emf

(A) Calculate the inductance of an air-core solenoid containing 300 turns if the length of the solenoid is 25.0 cm and its cross-sectional area is 4.00 cm^2.

Solution Using Equation 32.4, we obtain

$$L = \frac{\mu_0 N^2 A}{\ell}$$

$$= \frac{(4\pi \times 10^{-7} \text{ T} \cdot \text{m/A})(300)^2(4.00 \times 10^{-4} \text{ m}^2)}{25.0 \times 10^{-2} \text{ m}}$$

$$= 1.81 \times 10^{-4} \text{ T} \cdot \text{m}^2/\text{A} = \boxed{0.181 \text{ mH}}$$

(B) Calculate the self-induced emf in the solenoid if the current it carries is decreasing at the rate of 50.0 A/s.

Solution Using Equation 32.1 and given that $dI/dt = -50.0$ A/s, we obtain

$$\varepsilon_L = -L \frac{dI}{dt} = -(1.81 \times 10^{-4} \text{ H})(-50.0 \text{ A/s})$$

$$= \boxed{9.05 \text{ mV}}$$

32.2   *RL* Circuits

If a circuit contains a coil, such as a solenoid, the self-inductance of the coil prevents the current in the circuit from increasing or decreasing instantaneously. A circuit element that has a large self-inductance is called an **inductor** and has the circuit symbol ⎓⎓000⎓⎓. We always assume that the self-inductance of the remainder of a circuit is negligible compared with that of the inductor. Keep in mind, however, that even a circuit without a coil has some self-inductance that can affect the behavior of the circuit.

Because the inductance of the inductor results in a back emf, **an inductor in a circuit opposes changes in the current in that circuit.** The inductor attempts to keep the current the same as it was before the change occurred. If the battery voltage in the circuit is increased so that the current rises, the inductor opposes this change, and the rise is not instantaneous. If the battery voltage is decreased, the presence of the inductor results in a slow drop in the current rather than an immediate drop. Thus, the inductor causes the circuit to be "sluggish" as it reacts to changes in the voltage.

Consider the circuit shown in Figure 32.3, which contains a battery of negligible internal resistance. This is an **RL circuit** because the elements connected to the battery are a resistor and an inductor. Suppose that the switch S is open for $t < 0$ and then closed at $t = 0$. The current in the circuit begins to increase, and a back emf (Eq. 32.1) that opposes the increasing current is induced in the inductor. Because the current is increasing, dI/dt in Equation 32.1 is positive; thus, $\mathcal{E}_L$ is negative. This negative value reflects the decrease in electric potential that occurs in going from a to b across the inductor, as indicated by the positive and negative signs in Figure 32.3.

With this in mind, we can apply Kirchhoff's loop rule to this circuit, traversing the circuit in the clockwise direction:

$$\mathcal{E} - IR - L \frac{dI}{dt} = 0 \tag{32.6}$$

where IR is the voltage drop across the resistor. (We developed Kirchhoff's rules for circuits with steady currents, but they can also be applied to a circuit in which the current is changing if we imagine them to represent the circuit at one *instant* of time.) We must now look for a solution to this differential equation, which is similar to that for the RC circuit. (See Section 28.4.)

A mathematical solution of Equation 32.6 represents the current in the circuit as a function of time. To find this solution, we change variables for convenience, letting $x = (\mathcal{E}/R) - I$, so that $dx = -dI$. With these substitutions, we can write Equation 32.6 as

$$x + \frac{L}{R} \frac{dx}{dt} = 0$$

$$\frac{dx}{x} = -\frac{R}{L} dt$$

Integrating this last expression, we have

$$\int_{x_0}^{x} \frac{dx}{x} = -\frac{R}{L} \int_{0}^{t} dt$$

$$\ln \frac{x}{x_0} = -\frac{R}{L} t$$

where x_0 is the value of x at time $t = 0$. Taking the antilogarithm of this result, we obtain

$$x = x_0 e^{-Rt/L}$$

Because $I = 0$ at $t = 0$, we note from the definition of x that $x_0 = \mathcal{E}/R$. Hence, this last expression is equivalent to

$$\frac{\mathcal{E}}{R} - I = \frac{\mathcal{E}}{R} e^{-Rt/L}$$

$$I = \frac{\mathcal{E}}{R} (1 - e^{-Rt/L})$$

This expression shows how the inductor effects the current. The current does not increase instantly to its final equilibrium value when the switch is closed but instead increases according to an exponential function. If we remove the inductance in the circuit, which we can do by letting L approach zero, the exponential term becomes zero and we see that there is no time dependence of the current in this case—the current increases instantaneously to its final equilibrium value in the absence of the inductance.

We can also write this expression as

$$I = \frac{\mathcal{E}}{R} (1 - e^{-t/\tau}) \tag{32.7}$$

Active Figure 32.3 A series RL circuit. As the current increases toward its maximum value, an emf that opposes the increasing current is induced in the inductor.

At the Active Figures link at http://www.pse6.com, you can adjust the values of R and L to see the effect on the current. A graphical display as in Figure 32.4 is available.

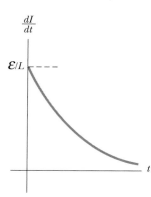

Active Figure 32.4 Plot of the current versus time for the *RL* circuit shown in Figure 32.3. The switch is open for $t < 0$ and then closed at $t = 0$, and the current increases toward its maximum value $\mathcal{E}/R$. The time constant τ is the time interval required for I to reach 63.2% of its maximum value.

Figure 32.5 Plot of dI/dt versus time for the *RL* circuit shown in Figure 32.3. The time rate of change of current is a maximum at $t = 0$, which is the instant at which the switch is closed. The rate decreases exponentially with time as I increases toward its maximum value.

At the Active Figures link at http://www.pse6.com, *you can observe this graph develop after the switch in Figure 32.3 is closed.*

where the constant τ is the **time constant** of the *RL* circuit:

Time constant of an *RL* circuit

$$\tau = \frac{L}{R} \tag{32.8}$$

Physically, τ is the time interval required for the current in the circuit to reach $(1 - e^{-1}) = 0.632 = 63.2\%$ of its final value $\mathcal{E}/R$. The time constant is a useful parameter for comparing the time responses of various circuits.

Figure 32.4 shows a graph of the current versus time in the *RL* circuit. Note that the equilibrium value of the current, which occurs as t approaches infinity, is $\mathcal{E}/R$. We can see this by setting dI/dt equal to zero in Equation 32.6 and solving for the current I. (At equilibrium, the change in the current is zero.) Thus, we see that the current initially increases very rapidly and then gradually approaches the equilibrium value $\mathcal{E}/R$ as t approaches infinity.

Let us also investigate the time rate of change of the current. Taking the first time derivative of Equation 32.7, we have

$$\frac{dI}{dt} = \frac{\mathcal{E}}{L} e^{-t/\tau} \tag{32.9}$$

From this result, we see that the time rate of change of the current is a maximum (equal to $\mathcal{E}/L$) at $t = 0$ and falls off exponentially to zero as t approaches infinity (Fig. 32.5).

Now let us consider the *RL* circuit shown in Figure 32.6. The curved lines on the switch S represent a switch that is connected either to a or b at all times. (If the switch is connected to neither a nor b, the current in the circuit suddenly stops.) Suppose that the switch has been set at position a long enough to allow the current to reach its equilibrium value $\mathcal{E}/R$. In this situation, the circuit is described by the outer loop in Figure 32.6. If the switch is thrown from a to b, the circuit is now described by just the right hand loop in Figure 32.6. Thus, we have a circuit with no battery ($\mathcal{E} = 0$). Applying Kirchhoff's loop rule to the right-hand loop at the instant the switch is thrown from a to b, we obtain

$$IR + L\frac{dI}{dt} = 0$$

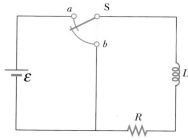

Active Figure 32.6 An *RL* circuit. When the switch S is in position a, the battery is in the circuit. When the switch is thrown to position b, the battery is no longer part of the circuit. The switch is designed so that it is never open, which would cause the current to stop.

At the Active Figures link at http://www.pse6.com, *you can adjust the values of R and L to see the effect on the current. A graphical display as in Figure 32.7 is available.*

It is left as a problem (Problem 16) to show that the solution of this differential equation is

$$I = \frac{\mathcal{E}}{R} e^{-t/\tau} = I_0 e^{-t/\tau} \tag{32.10}$$

where $\mathcal{E}$ is the emf of the battery and $I_0 = \mathcal{E}/R$ is the current at the instant at which the switch is thrown to *b*.

If the circuit did not contain an inductor, the current would immediately decrease to zero when the battery is removed. When the inductor is present, it opposes the decrease in the current and causes the current to decrease exponentially. A graph of the current in the circuit versus time (Fig. 32.7) shows that the current is continuously decreasing with time. Note that the slope *dI/dt* is always negative and has its maximum value at *t* = 0. The negative slope signifies that $\mathcal{E}_L = -L \, (dI/dt)$ is now positive.

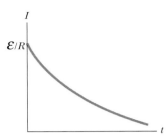

Active Figure 32.7 Current versus time for the right-hand loop of the circuit shown in Figure 32.6. For *t* < 0, the switch S is at position *a*. At *t* = 0, the switch is thrown to position *b*, and the current has its maximum value $\mathcal{E}/R$.

At the Active Figures link at http://www.pse6.com, you can observe this graph develop after the switch in Figure 32.6 is thrown to position b.

Quick Quiz 32.2 The circuit in Figure 32.8 consists of a resistor, an inductor, and an ideal battery with no internal resistance. At the instant just after the switch is closed, across which circuit element is the voltage equal to the emf of the battery? (a) the resistor (b) the inductor (c) both the inductor and resistor. After a very long time, across which circuit element is the voltage equal to the emf of the battery? (d) the resistor (e) the inductor (f) both the inductor and resistor.

Figure 32.8 (Quick Quiz 32.2)

Quick Quiz 32.3 The circuit in Figure 32.9 includes a power source that provides a sinusoidal voltage. Thus, the magnetic field in the inductor is constantly changing. The inductor is a simple air-core solenoid. The switch in the circuit is closed and the lightbulb glows steadily. An iron rod is inserted into the interior of the solenoid, which increases the magnitude of the magnetic field in the solenoid. As this happens, the brightness of the lightbulb (a) increases, (b) decreases, (c) is unaffected.

Figure 32.9 (Quick Quiz 32.3)

Quick Quiz 32.4 Two circuits like the one shown in Figure 32.6 are identical except for the value of *L*. In circuit A the inductance of the inductor is L_A, and in circuit B it is L_B. Switch S is thrown to position *a* at *t* = 0. At *t* = 10 s, the switch is thrown to position *b*. The resulting currents for the two circuits are as graphed in Figure 32.10.

If we assume that the time constant of each circuit is much less than 10 s, which of the following is true? (a) $L_A > L_B$; (b) $L_A < L_B$; (c) not enough information to tell.

Figure 32.10 (Quick Quiz 32.4)

Example 32.3 Time Constant of an *RL* Circuit

Interactive

(A) Find the time constant of the circuit shown in Figure 32.11a.

Solution The time constant is given by Equation 32.8:

$$\tau = \frac{L}{R} = \frac{30.0 \times 10^{-3}\ \text{H}}{6.00\ \Omega} = \boxed{5.00\ \text{ms}}$$

(B) The switch in Figure 32.11a is closed at $t = 0$. Calculate the current in the circuit at $t = 2.00$ ms.

Solution Using Equation 32.7 for the current as a function of time (with t and τ in milliseconds), we find that at $t = 2.00$ ms,

$$I = \frac{\mathcal{E}}{R}(1 - e^{-t/\tau}) = \frac{12.0\ \text{V}}{6.00\ \Omega}(1 - e^{-0.400}) = \boxed{0.659\ \text{A}}$$

30.0 mH

12.0 V   6.00 Ω

S

(a)

(b)

Figure 32.11 (Example 32.3) (a) The switch in this *RL* circuit is open for $t < 0$ and then closed at $t = 0$. (b) A graph of the current versus time for the circuit in part (a).

A plot of the current versus time for this circuit is given in Figure 32.11b.

(C) Compare the potential difference across the resistor with that across the inductor.

Solution At the instant the switch is closed, there is no current and thus no potential difference across the resistor. At this instant, the battery voltage appears entirely across the inductor in the form of a back emf of 12.0 V as the inductor tries to maintain the zero-current condition. (The left end of the inductor is at a higher electric potential than the right end.) As time passes, the emf across the inductor decreases and the current in the resistor (and hence the potential difference across it) increases, as shown in Figure 32.12. The sum of the two potential differences at all times is 12.0 V.

What If? In Figure 32.12, we see that the voltages across the resistor and inductor are equal at a time just before 4.00 ms. What if we wanted to delay the condition in which the voltages are equal to some later instant, such as $t = 10.0$ ms? Which parameter, *L* or *R*, would require the least adjustment, in terms of a percentage change, to achieve this?

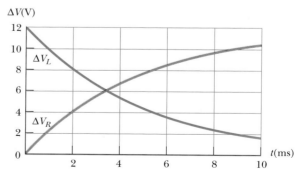

Figure 32.12 (Example 32.3) The time behavior of the potential differences across the resistor and inductor in Figure 32.11a.

Answer From Figure 32.12, we see that the voltages are equal when the voltage across the inductor has fallen to half of its original value. Thus, the time interval required for the voltages to become equal is the *half-life* $t_{1/2}$ of the decay. We introduced the half-life in the **What If?** section of Example 28.13, in describing the exponential decay in *RC* circuits, where we found that $t_{1/2} = 0.693\tau$. If we want the half-life of our *RL* circuit to be 10.0 ms, then the time constant must be

$$\tau = \frac{t_{1/2}}{0.693} = \frac{10.0 \text{ ms}}{0.693} = 14.4 \text{ ms}$$

We can achieve this time constant by holding *L* fixed and adjusting *R*; in this case,

$$\tau = \frac{L}{R} = \frac{30.0 \times 10^{-3} \text{ H}}{R} = 14.4 \text{ ms}$$

$$R = \frac{30.0 \times 10^{-3} \text{ H}}{14.4 \text{ ms}} = 2.08 \text{ }\Omega$$

This corresponds to a 65% decrease compared to the initial resistance. Now hold *R* fixed and adjust *L*:

$$\tau = \frac{L}{R} = \frac{L}{6.00 \text{ }\Omega} = 14.4 \text{ ms}$$

$$L = (6.00 \text{ }\Omega)(14.4 \text{ ms}) = 86.4 \times 10^{-3} \text{ H}$$

This represents a 188% increase in inductance! Thus, a much smaller percentage adjustment in *R* can achieve the desired effect than in *L*.

 At the Interactive Worked Example link at http://www.pse6.com, *you can explore the decay of current in an RL circuit.*

32.3 Energy in a Magnetic Field

Because the emf induced in an inductor prevents a battery from establishing an instantaneous current, the battery must provide more energy than in a circuit without the inductor. Part of the energy supplied by the battery appears as internal energy in the resistor, while the remaining energy is stored in the magnetic field of the inductor. If we multiply each term in Equation 32.6 by *I* and rearrange the expression, we have

$$I\mathcal{E} = I^2R + LI\frac{dI}{dt} \tag{32.11}$$

Recognizing $I\mathcal{E}$ as the rate at which energy is supplied by the battery and I^2R as the rate at which energy is delivered to the resistor, we see that $LI(dI/dt)$ must represent the rate at which energy is being stored in the inductor. If we let *U* denote the energy stored in the inductor at any time, then we can write the rate dU/dt at which energy is stored as

$$\frac{dU}{dt} = LI\frac{dI}{dt}$$

To find the total energy stored in the inductor, we can rewrite this expression as $dU = LI\,dI$ and integrate:

$$U = \int dU = \int_0^I LI\,dI = L\int_0^I I\,dI$$

$$U = \tfrac{1}{2}LI^2 \tag{32.12}$$

where *L* is constant and has been removed from the integral. This expression represents the energy stored in the magnetic field of the inductor when the current is *I*. Note that this equation is similar in form to Equation 26.11 for the energy stored in the electric field of a capacitor, $U = \tfrac{1}{2}C(\Delta V)^2$. In either case, we see that energy is required to establish a field.

We can also determine the energy density of a magnetic field. For simplicity, consider a solenoid whose inductance is given by Equation 32.5:

$$L = \mu_0 n^2 A\ell$$

The magnetic field of a solenoid is given by Equation 30.17:

$$B = \mu_0 nI$$

▲ PITFALL PREVENTION

32.1 Compare Energy in a Capacitor, Resistor, and Inductor

Different energy-storage mechanisms are at work in capacitors, inductors, and resistors. A capacitor stores a given amount of energy for a fixed charge on its plates; as more charge is delivered, more energy is delivered. An inductor stores a given amount of energy for constant current; as the current increases, more energy is delivered. Energy delivered to a resistor is transformed to internal energy.

Energy stored in an inductor

Substituting the expression for L and $I = B/\mu_0 n$ into Equation 32.12 gives

$$U = \tfrac{1}{2} L I^2 = \tfrac{1}{2} \mu_0 n^2 A \ell \left(\frac{B}{\mu_0 n} \right)^2 = \frac{B^2}{2\mu_0} A \ell \qquad (32.13)$$

Because $A\ell$ is the volume of the solenoid, the magnetic energy density, or the energy stored per unit volume in the magnetic field of the inductor is

Magnetic energy density

$$u_B = \frac{U}{A\ell} = \frac{B^2}{2\mu_0} \qquad (32.14)$$

Although this expression was derived for the special case of a solenoid, it is valid for any region of space in which a magnetic field exists. Note that Equation 32.14 is similar in form to Equation 26.13 for the energy per unit volume stored in an electric field, $u_E = \tfrac{1}{2}\epsilon_0 E^2$. In both cases, the energy density is proportional to the square of the field magnitude.

Quick Quiz 32.5 You are performing an experiment that requires the highest possible energy density in the interior of a very long solenoid. Which of the following increases the energy density? (More than one choice may be correct.) (a) increasing the number of turns per unit length on the solenoid (b) increasing the cross-sectional area of the solenoid (c) increasing only the length of the solenoid while keeping the number of turns per unit length fixed (d) increasing the current in the solenoid.

Example 32.4   What Happens to the Energy in the Inductor?

Consider once again the RL circuit shown in Figure 32.6. Recall that the current in the right-hand loop decays exponentially with time according to the expression $I = I_0 e^{-t/\tau}$, where $I_0 = \mathcal{E}/R$ is the initial current in the circuit and $\tau = L/R$ is the time constant. Show that all the energy initially stored in the magnetic field of the inductor appears as internal energy in the resistor as the current decays to zero.

Solution The rate dU/dt at which energy is delivered to the resistor (which is the power) is equal to $I^2 R$, where I is the instantaneous current:

$$\frac{dU}{dt} = I^2 R = (I_0 e^{-Rt/L})^2 R = I_0^2 R e^{-2Rt/L}$$

To find the total energy delivered to the resistor, we solve for dU and integrate this expression over the limits $t = 0$ to $t \rightarrow \infty$. (The upper limit is infinity because it takes an infinite amount of time for the current to reach zero.)

$$(1) \qquad U = \int_0^\infty I_0^2 R e^{-2Rt/L} \, dt = I_0^2 R \int_0^\infty e^{-2Rt/L} \, dt$$

The value of the definite integral can be shown to be $L/2R$ (see Problem 34) and so U becomes

$$U = I_0^2 R \left(\frac{L}{2R} \right) = \tfrac{1}{2} L I_0^2$$

Note that this is equal to the initial energy stored in the magnetic field of the inductor, given by Equation 32.12, as we set out to prove.

Example 32.5   The Coaxial Cable

Coaxial cables are often used to connect electrical devices, such as your stereo system, and in receiving signals in TV cable systems. Model a long coaxial cable as consisting of two thin concentric cylindrical conducting shells of radii a and b and length ℓ, as in Figure 32.13. The conducting shells carry the same current I in opposite directions. Imagine that the inner conductor carries current to a device and that the outer one acts as a return path carrying the current back to the source.

(A) Calculate the self-inductance L of this cable.

Solution Conceptualize the situation with the help of Figure 32.13. While we do not have a visible coil in this geometry,

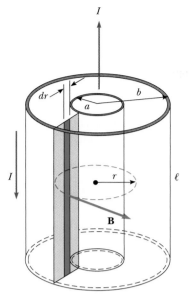

Figure 32.13 (Example 32.5) Section of a long coaxial cable. The inner and outer conductors carry equal currents in opposite directions.

measured from the common center of the shells. The magnetic field is zero outside the outer shell ($r > b$) because the net current passing through the area enclosed by a circular path surrounding the cable is zero, and hence from Ampère's law, $\oint \mathbf{B} \cdot d\mathbf{s} = 0$. The magnetic field is zero inside the inner shell because the shell is hollow and no current is present within a radius $r < a$.

The magnetic field is perpendicular to the light blue rectangle of length ℓ and width $b - a$, the cross section of interest. Because the magnetic field varies with radial position across this rectangle, we must use calculus to find the total magnetic flux. Dividing this rectangle into strips of width dr, such as the dark blue strip in Figure 32.13, we see that the area of each strip is $\ell\, dr$ and that the flux through each strip is $B\, dA = B\ell\, dr$. Hence, we find the total flux through the entire cross section by integrating:

$$\Phi_B = \int B\, dA = \int_a^b \frac{\mu_0 I}{2\pi r}\, \ell\, dr = \frac{\mu_0 I \ell}{2\pi} \int_a^b \frac{dr}{r} = \frac{\mu_0 I \ell}{2\pi} \ln\left(\frac{b}{a}\right)$$

Using this result, we find that the self-inductance of the cable is

$$L = \frac{\Phi_B}{I} = \frac{\mu_0 \ell}{2\pi} \ln\left(\frac{b}{a}\right)$$

(B) Calculate the total energy stored in the magnetic field of the cable.

Solution Using Equation 32.12 and the results to part (A) gives

$$U = \tfrac{1}{2} L I^2 = \frac{\mu_0 \ell I^2}{4\pi} \ln\left(\frac{b}{a}\right)$$

To finalize the problem, note that the inductance increases if ℓ increases, if b increases, or if a decreases. This is consistent with our conceptualization—any of these changes increases the size of the loop represented by our radial slice and through which the magnetic field passes; this increases the inductance.

imagine a thin radial slice of the coaxial cable, such as the light blue rectangle in Figure 32.13. If we consider the inner and outer conductors to be connected at the ends of the cable (above and below the figure), this slice represents one large conducting loop. The current in the loop sets up a magnetic field between the inner and outer conductors that passes through this loop. If the current changes, the magnetic field changes and the induced emf opposes the original change in the current in the conductors. We categorize this situation as one in which we can calculate an inductance, but we must return to the fundamental definition of inductance, Equation 32.2. To analyze the problem and obtain L, we must find the magnetic flux through the light blue rectangle in Figure 32.13. Ampère's law (see Section 30.3) tells us that the magnetic field in the region between the shells is due to the inner conductor and its magnitude is $B = \mu_0 I / 2\pi r$, where r is

32.4 Mutual Inductance

Very often, the magnetic flux through the area enclosed by a circuit varies with time because of time-varying currents in nearby circuits. This condition induces an emf through a process known as *mutual induction*, so called because it depends on the interaction of two circuits.

Consider the two closely wound coils of wire shown in cross-sectional view in Figure 32.14. The current I_1 in coil 1, which has N_1 turns, creates a magnetic field. Some of the magnetic field lines pass through coil 2, which has N_2 turns. The magnetic flux caused by the current in coil 1 and passing through coil 2 is represented by Φ_{12}. In analogy to Equation 32.2, we define the **mutual inductance** M_{12} of coil 2 with

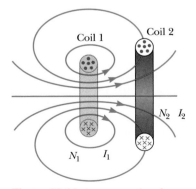

Figure 32.14 A cross-sectional view of two adjacent coils. A current in coil 1 sets up a magnetic field and some of the magnetic field lines pass through coil 2.

respect to coil 1:

Definition of mutual inductance

$$M_{12} \equiv \frac{N_2 \Phi_{12}}{I_1} \tag{32.15}$$

Mutual inductance depends on the geometry of both circuits and on their orientation with respect to each other. As the circuit separation distance increases, the mutual inductance decreases because the flux linking the circuits decreases.

If the current I_1 varies with time, we see from Faraday's law and Equation 32.15 that the emf induced by coil 1 in coil 2 is

$$\mathcal{E}_2 = -N_2 \frac{d\Phi_{12}}{dt} = -N_2 \frac{d}{dt}\left(\frac{M_{12} I_1}{N_2}\right) = -M_{12} \frac{dI_1}{dt} \tag{32.16}$$

In the preceding discussion, we assumed that the current is in coil 1. We can also imagine a current I_2 in coil 2. The preceding discussion can be repeated to show that there is a mutual inductance M_{21}. If the current I_2 varies with time, the emf induced by coil 2 in coil 1 is

$$\mathcal{E}_1 = -M_{21} \frac{dI_2}{dt} \tag{32.17}$$

In mutual induction, the emf induced in one coil is always proportional to the rate at which the current in the other coil is changing. Although the proportionality constants M_{12} and M_{21} have been treated separately, it can be shown that they are equal. Thus, with $M_{12} = M_{21} = M$, Equations 32.16 and 32.17 become

$$\mathcal{E}_2 = -M \frac{dI_1}{dt} \quad \text{and} \quad \mathcal{E}_1 = -M \frac{dI_2}{dt}$$

These two equations are similar in form to Equation 32.1 for the self-induced emf $\mathcal{E} = -L(dI/dt)$. The unit of mutual inductance is the henry.

Quick Quiz 32.6 In Figure 32.14, coil 1 is moved closer to coil 2, with the orientation of both coils remaining fixed. Because of this movement, the mutual induction of the two coils (a) increases (b) decreases (c) is unaffected.

Example 32.6 "Wireless" Battery Charger

An electric toothbrush has a base designed to hold the toothbrush handle when not in use. As shown in Figure 32.15a, the handle has a cylindrical hole that fits loosely over a matching cylinder on the base. When the handle is placed on the base, a changing current in a solenoid inside the base cylinder induces a current in a coil inside the handle. This induced current charges the battery in the handle.

We can model the base as a solenoid of length ℓ with N_B turns (Fig. 32.15b), carrying a current I, and having a cross-sectional area A. The handle coil contains N_H turns and completely surrounds the base coil. Find the mutual inductance of the system.

Solution Because the base solenoid carries a current I, the magnetic field in its interior is

$$B = \frac{\mu_0 N_B I}{\ell}$$

Because the magnetic flux Φ_{BH} through the handle's coil caused by the magnetic field of the base coil is BA, the mutual inductance is

$$M = \frac{N_H \Phi_{BH}}{I} = \frac{N_H BA}{I} = \mu_0 \frac{N_B N_H A}{\ell}$$

Wireless charging is used in a number of other "cordless" devices. One significant example is the inductive charging that is used by some electric car manufacturers that avoids direct metal-to-metal contact between the car and the charging apparatus.

(a)   (b)

Figure 32.15 (Example 32.6) (a) This electric toothbrush uses the mutual induction of solenoids as part of its battery-charging system. (b) A coil of N_H turns wrapped around the center of a solenoid of N_B turns.

32.5 Oscillations in an *LC* Circuit

When a capacitor is connected to an inductor as illustrated in Figure 32.16, the combination is an **LC circuit.** If the capacitor is initially charged and the switch is then closed, we find that both the current in the circuit and the charge on the capacitor oscillate between maximum positive and negative values. If the resistance of the circuit is zero, no energy is transformed to internal energy. In the following analysis, we neglect the resistance in the circuit. We also assume an idealized situation in which energy is not radiated away from the circuit. We shall discuss this radiation in Chapter 34, but we neglect it for now. With these idealizations—zero resistance and no radiation—the oscillations in the circuit persist indefinitely.

Assume that the capacitor has an initial charge Q_{max} (the maximum charge) and that the switch is open for $t < 0$ and then closed at $t = 0$. Let us investigate what happens from an energy viewpoint.

When the capacitor is fully charged, the energy U in the circuit is stored in the electric field of the capacitor and is equal to $Q_{max}^2/2C$ (Eq. 26.11). At this time, the current in the circuit is zero, and therefore no energy is stored in the inductor. After the switch is closed, the rate at which charges leave or enter the capacitor plates (which is also the rate at which the charge on the capacitor changes) is equal to the current in the circuit. As the capacitor begins to discharge after the switch is closed, the energy stored in its electric field decreases. The discharge of the capacitor represents a current in the circuit, and hence some energy is now stored in the magnetic field of the inductor. Thus, energy is transferred from the electric field of the capacitor to the magnetic field of the inductor. When the capacitor is fully discharged, it stores no energy. At this time, the current reaches its maximum value, and all of the energy is stored in the inductor. The current continues in the same direction, decreasing in magnitude, with the capacitor eventually becoming fully charged again but with the polarity of its plates now opposite the initial polarity. This is followed by another discharge until the circuit returns to its original state of maximum charge Q_{max} and the plate polarity shown in Figure 32.16. The energy continues to oscillate between inductor and capacitor.

Figure 32.16 A simple *LC* circuit. The capacitor has an initial charge Q_{max}, and the switch is open for $t < 0$ and then closed at $t = 0$.

The oscillations of the *LC* circuit are an electromagnetic analog to the mechanical oscillations of a block–spring system, which we studied in Chapter 15. Much of what is discussed there is applicable to *LC* oscillations. For example, we investigated the effect of driving a mechanical oscillator with an external force, which leads to the phenomenon of *resonance*. The same phenomenon is observed in the *LC* circuit. For example, a radio tuner has an *LC* circuit with a natural frequency. When the circuit is driven by the electromagnetic oscillations of a radio signal detected by the antenna, the tuner circuit responds with a large amplitude of electrical oscillation only for the station frequency that matches the natural frequency. Therefore, only the signal from one radio station is passed on to the amplifier, even though signals from all stations are driving the circuit at the same time. When you turn the knob on the radio tuner to change the station, you are changing the natural frequency of the circuit so that it will exhibit a resonance response to a different driving frequency.

A graphical description of the fields in the inductor and the capacitor in an *LC* circuit is shown in Figure 32.17. The right side of the figure shows the analogous oscillating block–spring system studied in Chapter 15. In each case, the situation is shown at intervals of one-fourth the period of oscillation *T*. The potential energy $\frac{1}{2}kx^2$ stored in a stretched spring is analogous to the electric potential energy $\frac{1}{2}C(\Delta V_{max})^2$ stored in the capacitor. The kinetic energy $\frac{1}{2}mv^2$ of the moving block is analogous to the magnetic energy $\frac{1}{2}LI^2$ stored in the inductor, which requires the presence of moving charges. In Figure 32.17a, all of the energy is stored as electric potential energy in the capacitor at $t = 0$. In Figure 32.17b, which is one fourth of a period later, all of the energy is stored as magnetic energy $\frac{1}{2}LI_{max}^2$ in the inductor, where I_{max} is the maximum current in the circuit. In Figure 32.17c, the energy in the *LC* circuit is stored completely in the capacitor, with the polarity of the plates now opposite what it was in Figure 32.17a. In parts d and e, the system returns to the initial configuration over the second half of the cycle. At times other than those shown in the figure, part of the energy is stored in the electric field of the capacitor and part is stored in the magnetic field of the inductor. In the analogous mechanical oscillation, part of the energy is potential energy in the spring and part is kinetic energy of the block.

Let us consider some arbitrary time *t* after the switch is closed, so that the capacitor has a charge $Q < Q_{max}$ and the current is $I < I_{max}$. At this time, both circuit elements store energy, but the sum of the two energies must equal the total initial energy *U* stored in the fully charged capacitor at $t = 0$:

Total energy stored in an *LC* circuit

$$U = U_C + U_L = \frac{Q^2}{2C} + \frac{1}{2}LI^2 \tag{32.18}$$

Because we have assumed the circuit resistance to be zero and we ignore electromagnetic radiation, no energy is transformed to internal energy and none is transferred out of the system of the circuit. Therefore, *the total energy of the system must remain constant in time.* This means that $dU/dt = 0$. Therefore, by differentiating Equation 32.18 with respect to time while noting that *Q* and *I* vary with time, we obtain

$$\frac{dU}{dt} = \frac{d}{dt}\left(\frac{Q^2}{2C} + \frac{1}{2}LI^2\right) = \frac{Q}{C}\frac{dQ}{dt} + LI\frac{dI}{dt} = 0 \tag{32.19}$$

We can reduce this to a differential equation in one variable by remembering that the current in the circuit is equal to the rate at which the charge on the capacitor changes: $I = dQ/dt$. From this, it follows that $dI/dt = d^2Q/dt^2$. Substitution of these relationships into Equation 32.19 gives

$$\frac{Q}{C} + L\frac{d^2Q}{dt^2} = 0$$

$$\frac{d^2Q}{dt^2} = -\frac{1}{LC}Q \tag{32.20}$$

Active Figure 32.17 Energy transfer in a resistanceless, nonradiating *LC* circuit. The capacitor has a charge Q_{max} at $t = 0$, the instant at which the switch is closed. The mechanical analog of this circuit is a block–spring system.

At the Active Figures link at http://www.pse6.com, *you can adjust the values of C and L to see the effect on the oscillating current. The block on the spring oscillates in a mechanical analog of the electrical oscillations. A graphical display as in Figure 32.18 is available, as is an energy bar graph.*

We can solve for Q by noting that this expression is of the same form as the analogous Equations 15.3 and 15.5 for a block-spring system:

$$\frac{d^2x}{dt^2} = -\frac{k}{m} x = -\omega^2 x$$

where k is the spring constant, m is the mass of the block, and $\omega = \sqrt{k/m}$. The solution of this equation has the general form (Eq. 15.6),

$$x = A \cos(\omega t + \phi)$$

where ω is the angular frequency of the simple harmonic motion, A is the amplitude of motion (the maximum value of x), and ϕ is the phase constant; the values of A and ϕ depend on the initial conditions. Because Equation 32.20 is of the same form as the differential equation of the simple harmonic oscillator, we see that it has the solution

Charge as a function of time for an ideal LC circuit

$$Q = Q_{max} \cos(\omega t + \phi) \qquad (32.21)$$

where Q_{max} is the maximum charge of the capacitor and the angular frequency ω is

Angular frequency of oscillation in an LC circuit

$$\omega = \frac{1}{\sqrt{LC}} \qquad (32.22)$$

Note that the angular frequency of the oscillations depends solely on the inductance and capacitance of the circuit. This is the *natural frequency* of oscillation of the LC circuit.

Because Q varies sinusoidally with time, the current in the circuit also varies sinusoidally. We can easily show this by differentiating Equation 32.21 with respect to time:

Current as a function of time for an ideal LC current

$$I = \frac{dQ}{dt} = -\omega Q_{max} \sin(\omega t + \phi) \qquad (32.23)$$

To determine the value of the phase angle ϕ, we examine the initial conditions, which in our situation require that at $t = 0$, $I = 0$ and $Q = Q_{max}$. Setting $I = 0$ at $t = 0$ in Equation 32.23, we have

$$0 = -\omega Q_{max} \sin \phi$$

which shows that $\phi = 0$. This value for ϕ also is consistent with Equation 32.21 and with the condition that $Q = Q_{max}$ at $t = 0$. Therefore, in our case, the expressions for Q and I are

$$Q = Q_{max} \cos \omega t \qquad (32.24)$$

$$I = -\omega Q_{max} \sin \omega t = -I_{max} \sin \omega t \qquad (32.25)$$

Graphs of Q versus t and I versus t are shown in Figure 32.18. Note that the charge on the capacitor oscillates between the extreme values Q_{max} and $-Q_{max}$, and that the current oscillates between I_{max} and $-I_{max}$. Furthermore, the current is 90° out of phase with the charge. That is, when the charge is a maximum, the current is zero, and when the charge is zero, the current has its maximum value.

Let us return to the energy discussion of the LC circuit. Substituting Equations 32.24 and 32.25 in Equation 32.18, we find that the total energy is

$$U = U_C + U_L = \frac{Q^2_{max}}{2C} \cos^2 \omega t + \tfrac{1}{2} L I^2_{max} \sin^2 \omega t \qquad (32.26)$$

This expression contains all of the features described qualitatively at the beginning of this section. It shows that the energy of the LC circuit continuously oscillates between energy stored in the electric field of the capacitor and energy stored in the magnetic

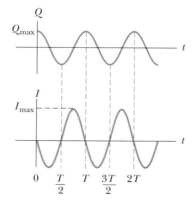

At the *Active Figures* link at http://www.pse6.com, *you can observe this graph develop for the LC circuit in Figure 32.17.*

Active Figure 32.18 Graphs of charge versus time and current versus time for a resistanceless, nonradiating LC circuit. Note that Q and I are 90° out of phase with each other.

field of the inductor. When the energy stored in the capacitor has its maximum value $Q_{max}^2/2C$, the energy stored in the inductor is zero. When the energy stored in the inductor has its maximum value $\frac{1}{2}LI_{max}^2$, the energy stored in the capacitor is zero.

Plots of the time variations of U_C and U_L are shown in Figure 32.19. The sum $U_C + U_L$ is a constant and equal to the total energy $Q_{max}^2/2C$, or $\frac{1}{2}LI_{max}^2$. Analytical verification of this is straightforward. The amplitudes of the two graphs in Figure 32.19 must be equal because the maximum energy stored in the capacitor (when $I = 0$) must equal the maximum energy stored in the inductor (when $Q = 0$). This is mathematically expressed as

$$\frac{Q_{max}^2}{2C} = \frac{LI_{max}^2}{2}$$

Using this expression in Equation 32.26 for the total energy gives

$$U = \frac{Q_{max}^2}{2C}(\cos^2 \omega t + \sin^2 \omega t) = \frac{Q_{max}^2}{2C} \quad (32.27)$$

because $\cos^2 \omega t + \sin^2 \omega t = 1$.

In our idealized situation, the oscillations in the circuit persist indefinitely; however, remember that the total energy U of the circuit remains constant only if energy transfers and transformations are neglected. In actual circuits, there is always some resistance, and hence some energy is transformed to internal energy. We mentioned at the beginning of this section that we are also ignoring radiation from the circuit. In reality, radiation is inevitable in this type of circuit, and the total energy in the circuit continuously decreases as a result of this process.

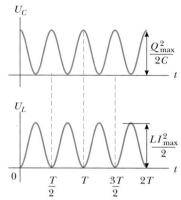

Figure 32.19 Plots of U_C versus t and U_L versus t for a resistanceless, nonradiating *LC* circuit. The sum of the two curves is a constant and equal to the total energy stored in the circuit.

Quick Quiz 32.7 At an instant of time during the oscillations of an *LC* circuit, the current is at its maximum value. At this instant, the voltage across the capacitor (a) is different from that across the inductor (b) is zero (c) has its maximum value (d) is impossible to determine

Quick Quiz 32.8 At an instant of time during the oscillations of an *LC* circuit, the current is momentarily zero. At this instant, the voltage across the capacitor (a) is different from that across the inductor (b) is zero (c) has its maximum value (d) is impossible to determine

---

Example 32.7   Oscillations in an *LC* Circuit   `Interactive`

In Figure 32.20, the capacitor is initially charged when switch S_1 is open and S_2 is closed. Switch S_2 is then opened, removing the battery from the circuit, and the capacitor remains charged. Switch S_1 is then closed, so that the capacitor is connected directly across the inductor.

(A) Find the frequency of oscillation of the circuit.

Solution Using Equation 32.22 gives for the frequency

$$f = \frac{\omega}{2\pi} = \frac{1}{2\pi\sqrt{LC}}$$

$$= \frac{1}{2\pi[(2.81 \times 10^{-3}\,\text{H})(9.00 \times 10^{-12}\,\text{F})]^{1/2}}$$

$$= 1.00 \times 10^6\,\text{Hz}$$

Figure 32.20 (Example 32.7) First the capacitor is fully charged with the switch S_1 open and S_2 closed. Then, S_2 is opened and S_1 is closed.

(B) What are the maximum values of charge on the capacitor and current in the circuit?

Solution The initial charge on the capacitor equals the maximum charge, and because $C = Q/\Delta V$, we have

$$Q_{max} = C\Delta V = (9.00 \times 10^{-12}\,\text{F})(12.0\,\text{V})$$

$$= \boxed{1.08 \times 10^{-10}\,\text{C}}$$

From Equation 32.25, we can see how the maximum current is related to the maximum charge:

$$I_{max} = \omega Q_{max} = 2\pi f Q_{max}$$

$$= (2\pi \times 10^6\,\text{s}^{-1})(1.08 \times 10^{-10}\,\text{C})$$

$$I_{max} = \boxed{6.79 \times 10^{-4}\,\text{A}}$$

(C) Determine the charge and current as functions of time.

Solution Equations 32.24 and 32.25 give the following expressions for the time variation of Q and I:

$$Q = Q_{max}\cos\omega t$$

$$= \boxed{(1.08 \times 10^{-10}\,\text{C})\cos[(2\pi \times 10^6\,\text{s}^{-1})t]}$$

$$I = -I_{max}\sin\omega t$$

$$= \boxed{(-6.79 \times 10^{-4}\,\text{A})\sin[(2\pi \times 10^6\,\text{s}^{-1})t]}$$

 At the Interactive Worked Example link at **http://www.pse6.com,** *you can study the oscillations in an LC circuit.*

Active Figure 32.21 A series *RLC* circuit. Switch S_1 is closed and the capacitor is charged. S_1 is then opened and, at $t = 0$, switch S_2 is closed.

 At the Active Figures link at **http://www.pse6.com,** *you can adjust the values of R, L, and C to see the effect on the decaying charge on the capacitor. A graphical display as in Figure 32.23a is available, as is an energy bar graph.*

32.6   The *RLC* Circuit

We now turn our attention to a more realistic circuit consisting of a resistor, an inductor, and a capacitor connected in series, as shown in Figure 32.21. Let us assume that the resistance of the resistor represents all of the resistance in the circuit. Now imagine that switch S_1 is closed and S_2 is open, so that the capacitor has an initial charge Q_{max}. Next, S_1 is opened and S_2 is closed. Once S_2 is closed and a current is established, the total energy stored in the capacitor and inductor at any time is given by Equation 32.18. However, this total energy is no longer constant, as it was in the *LC* circuit, because the resistor causes transformation to internal energy. (We continue to ignore electromagnetic radiation from the circuit in this discussion.) Because the rate of energy transformation to internal energy within a resistor is I^2R, we have

$$\frac{dU}{dt} = -I^2R$$

where the negative sign signifies that the energy U of the circuit is decreasing in time. Substituting this result into Equation 32.19 gives

$$LI\frac{dI}{dt} + \frac{Q}{C}\frac{dQ}{dt} = -I^2R \tag{32.28}$$

To convert this equation into a form that allows us to compare the electrical oscillations with their mechanical analog, we first use the fact that $I = dQ/dt$ and move all terms to the left-hand side to obtain

$$LI\frac{d^2Q}{dt^2} + I^2R + \frac{Q}{C}I = 0$$

Now we divide through by I:

$$L\frac{d^2Q}{dt^2} + IR + \frac{Q}{C} = 0$$

$$L\frac{d^2Q}{dt^2} + R\frac{dQ}{dt} + \frac{Q}{C} = 0 \tag{32.29}$$

Figure 32.22 A block–spring system moving in a viscous medium with damped harmonic motion is analogous to an *RLC* circuit.

The *RLC* circuit is analogous to the damped harmonic oscillator discussed in Section 15.6 and illustrated in Figure 32.22. The equation of motion for a damped

Table 32.1

Analogies Between Electrical and Mechanical Systems		
Electric Circuit		**One-Dimensional Mechanical System**
Charge	$Q \leftrightarrow x$	Position
Current	$I \leftrightarrow v_x$	Velocity
Potential difference	$\Delta V \leftrightarrow F_x$	Force
Resistance	$R \leftrightarrow b$	Viscous damping coefficient
Capacitance	$C \leftrightarrow 1/k$	(k = spring constant)
Inductance	$L \leftrightarrow m$	Mass
Current = time derivative of charge	$I = \dfrac{dQ}{dt} \leftrightarrow v_x = \dfrac{dx}{dt}$	Velocity = time derivative of position
Rate of change of current = second time derivative of charge	$\dfrac{dI}{dt} = \dfrac{d^2Q}{dt^2} \leftrightarrow a_x = \dfrac{dv_x}{dt} = \dfrac{d^2x}{dt^2}$	Acceleration = second time derivative of position
Energy in inductor	$U_L = \frac{1}{2}LI^2 \leftrightarrow K = \frac{1}{2}mv^2$	Kinetic energy of moving object
Energy in capacitor	$U_C = \frac{1}{2}\dfrac{Q^2}{C} \leftrightarrow U = \frac{1}{2}kx^2$	Potential energy stored in a spring
Rate of energy loss due to resistance	$I^2R \leftrightarrow bv^2$	Rate of energy loss due to friction
RLC circuit	$L\dfrac{d^2Q}{dt^2} + R\dfrac{dQ}{dt} + \dfrac{Q}{C} = 0 \leftrightarrow m\dfrac{d^2x}{dt^2} + b\dfrac{dx}{dt} + kx = 0$	Damped object on a spring

block–spring system is, from Equation 15.31,

$$m\,\frac{d^2x}{dt^2} + b\,\frac{dx}{dt} + kx = 0 \qquad (32.30)$$

Comparing Equations 32.29 and 32.30, we see that Q corresponds to the position x of the block at any instant, L to the mass m of the block, R to the damping coefficient b, and C to $1/k$, where k is the force constant of the spring. These and other relationships are listed in Table 32.1.

Because the analytical solution of Equation 32.29 is cumbersome, we give only a qualitative description of the circuit behavior. In the simplest case, when $R = 0$, Equation 32.29 reduces to that of a simple *LC* circuit, as expected, and the charge and the current oscillate sinusoidally in time. This is equivalent to removal of all damping in the mechanical oscillator.

When R is small, a situation analogous to light damping in the mechanical oscillator, the solution of Equation 32.29 is

$$Q = Q_{\max}\,e^{-Rt/2L}\cos\omega_d t \qquad (32.31)$$

where ω_d, the angular frequency at which the circuit oscillates, is given by

$$\omega_d = \left[\frac{1}{LC} - \left(\frac{R}{2L}\right)^2\right]^{1/2} \qquad (32.32)$$

That is, the value of the charge on the capacitor undergoes a damped harmonic oscillation in analogy with a block–spring system moving in a viscous medium. From Equation 32.32, we see that, when $R \ll \sqrt{4L/C}$ (so that the second term in the brackets is much smaller than the first), the frequency ω_d of the damped oscillator is close to that of the undamped oscillator, $1/\sqrt{LC}$. Because $I = dQ/dt$, it follows that the current also

At the Active Figures link at http://www.pse6.com, you can observe this graph develop for the damped RLC circuit in Figure 32.21.

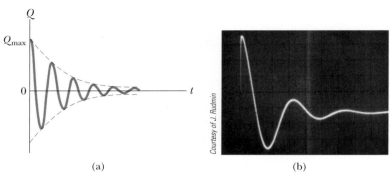

(a)   (b)

Active Figure 32.23 (a) Charge versus time for a damped *RLC* circuit. The charge decays in this way when $R > \sqrt{4L/C}$. The *Q*-versus-*t* curve represents a plot of Equation 32.31. (b) Oscilloscope pattern showing the decay in the oscillations of an *RLC* circuit.

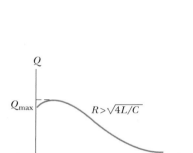

Figure 32.24 Plot of *Q* versus *t* for an overdamped *RLC* circuit, which occurs for values of $R > \sqrt{4L/C}$.

undergoes damped harmonic oscillation. A plot of the charge versus time for the damped oscillator is shown in Figure 32.23a. Note that the maximum value of *Q* decreases after each oscillation, just as the amplitude of a damped block–spring system decreases in time.

When we consider larger values of *R*, we find that the oscillations damp out more rapidly; in fact, there exists a critical resistance value $R_c = \sqrt{4L/C}$ above which no oscillations occur. A system with $R = R_c$ is said to be *critically damped*. When *R* exceeds R_c, the system is said to be *overdamped* (Fig. 32.24).

Take a practice test for this chapter by clicking on the Practice Test link at http://www.pse6.com.

SUMMARY

When the current in a coil changes with time, an emf is induced in the coil according to Faraday's law. The **self-induced emf** is

$$\mathcal{E}_L = -L\,\frac{dI}{dt} \tag{32.1}$$

where *L* is the **inductance** of the coil. Inductance is a measure of how much opposition a coil offers to a change in the current in the coil. Inductance has the SI unit of **henry** (H), where 1 H = 1 V·s/A.

The inductance of any coil is

$$L = \frac{N\Phi_B}{I} \tag{32.2}$$

where Φ_B is the magnetic flux through the coil and *N* is the total number of turns. The inductance of a device depends on its geometry. For example, the inductance of an air-core solenoid is

$$L = \frac{\mu_0 N^2 A}{\ell} \tag{32.4}$$

where *A* is the cross-sectional area, and ℓ is the length of the solenoid.

If a resistor and inductor are connected in series to a battery of emf $\mathcal{E}$, and if a switch in the circuit is open for $t < 0$ and then closed at $t = 0$, the current in the circuit varies in time according to the expression

$$I = \frac{\mathcal{E}}{R}\,(1 - e^{-t/\tau}) \tag{32.7}$$

where $\tau = L/R$ is the **time constant** of the RL circuit. That is, the current increases to an equilibrium value of $\mathcal{E}/R$ after a time interval that is long compared with τ. If the battery in the circuit is replaced by a resistanceless wire, the current decays exponentially with time according to the expression

$$I = \frac{\mathcal{E}}{R} e^{-t/\tau} \tag{32.10}$$

where $\mathcal{E}/R$ is the initial current in the circuit.

The energy stored in the magnetic field of an inductor carrying a current I is

$$U = \tfrac{1}{2} L I^2 \tag{32.12}$$

This energy is the magnetic counterpart to the energy stored in the electric field of a charged capacitor.

The energy density at a point where the magnetic field is B is

$$u_B = \frac{B^2}{2\mu_0} \tag{32.14}$$

The **mutual inductance** of a system of two coils is given by

$$M_{12} = \frac{N_2 \Phi_{12}}{I_1} = M_{21} = \frac{N_1 \Phi_{21}}{I_2} = M \tag{32.15}$$

This mutual inductance allows us to relate the induced emf in a coil to the changing source current in a nearby coil using the relationships

$$\mathcal{E}_2 = -M \frac{dI_1}{dt} \qquad \text{and} \qquad \mathcal{E}_1 = -M \frac{dI_2}{dt} \tag{32.16, 32.17}$$

In an LC circuit that has zero resistance and does not radiate electromagnetically (an idealization), the values of the charge on the capacitor and the current in the circuit vary in time according to the expressions

$$Q = Q_{\max} \cos(\omega t + \phi) \tag{32.21}$$

$$I = \frac{dQ}{dt} = -\omega Q_{\max} \sin(\omega t + \phi) \tag{32.23}$$

where $Q_{\max}$ is the maximum charge on the capacitor, ϕ is a phase constant, and ω is the angular frequency of oscillation:

$$\omega = \frac{1}{\sqrt{LC}} \tag{32.22}$$

The energy in an LC circuit continuously transfers between energy stored in the capacitor and energy stored in the inductor. The total energy of the LC circuit at any time t is

$$U = U_C + U_L = \frac{Q_{\max}^2}{2C} \cos^2 \omega t + \frac{L I_{\max}^2}{2} \sin^2 \omega t \tag{32.26}$$

At $t = 0$, all of the energy is stored in the electric field of the capacitor ($U = Q_{\max}^2/2C$). Eventually, all of this energy is transferred to the inductor ($U = L I_{\max}^2/2$). However, the total energy remains constant because energy transformations are neglected in the ideal LC circuit.

In an RLC circuit with small resistance, the charge on the capacitor varies with time according to

$$Q = Q_{\max} e^{-Rt/2L} \cos \omega_d t \tag{32.31}$$

where

$$\omega_d = \left[\frac{1}{LC} - \left(\frac{R}{2L} \right)^2 \right]^{1/2} \tag{32.32}$$

QUESTIONS

1. Why is the induced emf that appears in an inductor called a "counter" or "back" emf?

2. The current in a circuit containing a coil, resistor, and battery has reached a constant value. Does the coil have an inductance? Does the coil affect the value of the current?

3. What parameters affect the inductance of a coil? Does the inductance of a coil depend on the current in the coil?

4. How can a long piece of wire be wound on a spool so that the wire has a negligible self-inductance?

5. For the series RL circuit shown in Figure Q32.5, can the back emf ever be greater than the battery emf? Explain.

Figure Q32.5 Questions 5 and 6.

6. Suppose the switch in Figure Q32.5 has been closed for a long time and is suddenly opened. Does the current instantaneously drop to zero? Why does a spark appear at the switch contacts at the moment the switch is opened?

7. A switch controls the current in a circuit that has a large inductance. Is a spark (see Figure Q32.7) more likely to be produced at the switch when the switch is being closed or when it is being opened, or doesn't it matter? The electric arc can melt and oxidize the contact surfaces, resulting in high resistivity of the contacts and eventual destruction of the switch. Before electronic ignitions were invented, distributor contact points in automobiles had to be replaced regularly. Switches in power distribution networks and switches controlling large motors, generators, and electromagnets can suffer from arcing and can be very dangerous to operate.

8. Consider this thesis: "Joseph Henry, America's first professional physicist, caused the most recent basic change in the human view of the Universe when he discovered self-induction during a school vacation at the Albany Academy about 1830. Before that time, one could think of the Universe as composed of just one thing: matter. The energy that temporarily maintains the current after a battery is removed from a coil, on the other hand, is not energy that belongs to any chunk of matter. It is energy in the massless magnetic field surrounding the coil. With Henry's discovery, Nature forced us to admit that the Universe consists of fields as well as matter." Argue for or against the statement. What in your view comprises the Universe?

9. If the current in an inductor is doubled, by what factor does the stored energy change?

10. Discuss the similarities between the energy stored in the electric field of a charged capacitor and the energy stored in the magnetic field of a current-carrying coil.

11. What is the inductance of two inductors connected in series? Does it matter if they are solenoids or toroids?

12. The centers of two circular loops are separated by a fixed distance. For what relative orientation of the loops is their mutual inductance a maximum? a minimum? Explain.

13. Two solenoids are connected in series so that each carries the same current at any instant. Is mutual induction present? Explain.

14. In the LC circuit shown in Figure 32.16, the charge on the capacitor is sometimes zero, but at such instants the current in the circuit is not zero. How is this possible?

15. If the resistance of the wires in an LC circuit were not zero, would the oscillations persist? Explain.

16. How can you tell whether an RLC circuit is overdamped or underdamped?

17. What is the significance of critical damping in an RLC circuit?

18. Can an object exert a force on itself? When a coil induces an emf in itself, does it exert a force on itself?

Figure Q32.7

PROBLEMS

Section 32.1   Self-Inductance

1. A coil has an inductance of 3.00 mH, and the current in it changes from 0.200 A to 1.50 A in a time of 0.200 s. Find the magnitude of the average induced emf in the coil during this time.

2. A coiled telephone cord forms a spiral with 70 turns, a diameter of 1.30 cm, and an unstretched length of 60.0 cm. Determine the self-inductance of one conductor in the unstretched cord.

3. A 2.00-H inductor carries a steady current of 0.500 A. When the switch in the circuit is opened, the current is effectively zero after 10.0 ms. What is the average induced emf in the inductor during this time?

4. Calculate the magnetic flux through the area enclosed by a 300-turn, 7.20-mH coil when the current in the coil is 10.0 mA.

5. 🌀 A 10.0-mH inductor carries a current $I = I_{max} \sin \omega t$, with $I_{max} = 5.00$ A and $\omega/2\pi = 60.0$ Hz. What is the back emf as a function of time?

6. An emf of 24.0 mV is induced in a 500-turn coil at an instant when the current is 4.00 A and is changing at the rate of 10.0 A/s. What is the magnetic flux through each turn of the coil?

7. An inductor in the form of a solenoid contains 420 turns, is 16.0 cm in length, and has a cross-sectional area of 3.00 cm². What uniform rate of decrease of current through the inductor induces an emf of 175 μV?

8. The current in a 90.0-mH inductor changes with time as $I = 1.00t^2 - 6.00t$ (in SI units). Find the magnitude of the induced emf at (a) $t = 1.00$ s and (b) $t = 4.00$ s. (c) At what time is the emf zero?

9. A 40.0-mA current is carried by a uniformly wound air-core solenoid with 450 turns, a 15.0-mm diameter, and 12.0-cm length. Compute (a) the magnetic field inside the solenoid, (b) the magnetic flux through each turn, and (c) the inductance of the solenoid. (d) **What If?** If the current were different, which of these quantities would change?

10. A solenoid has 120 turns uniformly wrapped around a wooden core, which has a diameter of 10.0 mm and a length of 9.00 cm. (a) Calculate the inductance of the solenoid. (b) **What If?** The wooden core is replaced with a soft iron rod that has the same dimensions, but a magnetic permeability $\mu_m = 800\mu_0$. What is the new inductance?

11. A piece of copper wire with thin insulation, 200 m long and 1.00 mm in diameter, is wound onto a plastic tube to form a long solenoid. This coil has a circular cross section and consists of tightly wound turns in one layer. If the current in the solenoid drops linearly from 1.80 A to zero in 0.120 seconds, an emf of 80.0 mV is induced in the coil. What is the length of the solenoid, measured along its axis?

12. A toroid has a major radius R and a minor radius r, and is tightly wound with N turns of wire, as shown in Figure P32.12. If $R \gg r$, the magnetic field in the region enclosed by the wire of the torus, of cross-sectional area $A = \pi r^2$, is essentially the same as the magnetic field of a solenoid that has been bent into a large circle of radius R. Modeling the field as the uniform field of a long solenoid, show that the self-inductance of such a toroid is approximately

$$L \approx \frac{\mu_0 N^2 A}{2\pi R}$$

(An exact expression of the inductance of a toroid with a rectangular cross section is derived in Problem 64.)

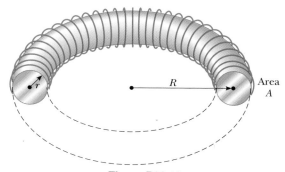

Figure P32.12

13. A self-induced emf in a solenoid of inductance L changes in time as $\mathcal{E} = \mathcal{E}_0 e^{-kt}$. Find the total charge that passes through the solenoid, assuming the charge is finite.

Section 32.2   *RL* Circuits

14. Calculate the resistance in an *RL* circuit in which $L = 2.50$ H and the current increases to 90.0% of its final value in 3.00 s.

15. A 12.0-V battery is connected into a series circuit containing a 10.0-Ω resistor and a 2.00-H inductor. How long will it take the current to reach (a) 50.0% and (b) 90.0% of its final value?

16. Show that $I = I_0 e^{-t/\tau}$ is a solution of the differential equation

$$IR + L \frac{dI}{dt} = 0$$

where $\tau = L/R$ and I_0 is the current at $t = 0$.

17. Consider the circuit in Figure P32.17, taking $\mathcal{E} = 6.00$ V, $L = 8.00$ mH, and $R = 4.00$ Ω. (a) What is the inductive time constant of the circuit? (b) Calculate the current in

Figure P32.17 Problems 17, 18, 19, and 22.

the circuit 250 μs after the switch is closed. (c) What is the value of the final steady-state current? (d) How long does it take the current to reach 80.0% of its maximum value?

18. In the circuit shown in Figure P32.17, let $L = 7.00$ H, $R = 9.00$ Ω, and $\mathcal{E} = 120$ V. What is the self-induced emf 0.200 s after the switch is closed?

19. For the RL circuit shown in Figure P32.17, let the inductance be 3.00 H, the resistance 8.00 Ω, and the battery emf 36.0 V. (a) Calculate the ratio of the potential difference across the resistor to that across the inductor when the current is 2.00 A. (b) Calculate the voltage across the inductor when the current is 4.50 A.

20. A 12.0-V battery is connected in series with a resistor and an inductor. The circuit has a time constant of 500 μs, and the maximum current is 200 mA. What is the value of the inductance?

21. An inductor that has an inductance of 15.0 H and a resistance of 30.0 Ω is connected across a 100-V battery. What is the rate of increase of the current (a) at $t = 0$ and (b) at $t = 1.50$ s?

22. When the switch in Figure P32.17 is closed, the current takes 3.00 ms to reach 98.0% of its final value. If $R = 10.0$ Ω, what is the inductance?

23. The switch in Figure P32.23 is open for $t < 0$ and then closed at time $t = 0$. Find the current in the inductor and the current in the switch as functions of time thereafter.

Figure P32.23

24. A series RL circuit with $L = 3.00$ H and a series RC circuit with $C = 3.00$ μF have equal time constants. If the two circuits contain the same resistance R, (a) what is the value of R and (b) what is the time constant?

25. A current pulse is fed to the partial circuit shown in Figure P32.25. The current begins at zero, then becomes 10.0 A between $t = 0$ and $t = 200$ μs, and then is zero once again. Determine the current in the inductor as a function of time.

Figure P32.25

26. One application of an RL circuit is the generation of time-varying high voltage from a low-voltage source, as shown in Figure P32.26. (a) What is the current in the circuit a long time after the switch has been in position a? (b) Now the switch is thrown quickly from a to b. Compute the initial voltage across each resistor and across the inductor. (c) How much time elapses before the voltage across the inductor drops to 12.0 V?

Figure P32.26

27. A 140-mH inductor and a 4.90-Ω resistor are connected with a switch to a 6.00-V battery as shown in Figure P32.27. (a) If the switch is thrown to the left (connecting the battery), how much time elapses before the current reaches 220 mA? (b) What is the current in the inductor 10.0 s after the switch is closed? (c) Now the switch is quickly thrown from a to b. How much time elapses before the current falls to 160 mA?

Figure P32.27

28. Consider two ideal inductors L_1 and L_2 that have *zero* internal resistance and are far apart, so that their magnetic fields do not influence each other. (a) Assuming these inductors

are connected in series, show that they are equivalent to a single ideal inductor having $L_{eq} = L_1 + L_2$. (b) Assuming these same two inductors are connected in parallel, show that they are equivalent to a single ideal inductor having $1/L_{eq} = 1/L_1 + 1/L_2$. (c) **What If?** Now consider two inductors L_1 and L_2 that have *nonzero* internal resistances R_1 and R_2, respectively. Assume they are still far apart so that their mutual inductance is zero. Assuming these inductors are connected in series, show that they are equivalent to a single inductor having $L_{eq} = L_1 + L_2$ and $R_{eq} = R_1 + R_2$. (d) If these same inductors are now connected in parallel, is it necessarily true that they are equivalent to a single ideal inductor having $1/L_{eq} = 1/L_1 + 1/L_2$ and $1/R_{eq} = 1/R_1 + 1/R_2$? Explain your answer.

Section 32.3 Energy in a Magnetic Field

29. Calculate the energy associated with the magnetic field of a 200-turn solenoid in which a current of 1.75 A produces a flux of 3.70×10^{-4} Wb in each turn.

30. The magnetic field inside a superconducting solenoid is 4.50 T. The solenoid has an inner diameter of 6.20 cm and a length of 26.0 cm. Determine (a) the magnetic energy density in the field and (b) the energy stored in the magnetic field within the solenoid.

31. An air-core solenoid with 68 turns is 8.00 cm long and has a diameter of 1.20 cm. How much energy is stored in its magnetic field when it carries a current of 0.770 A?

32. At $t = 0$, an emf of 500 V is applied to a coil that has an inductance of 0.800 H and a resistance of 30.0 Ω. (a) Find the energy stored in the magnetic field when the current reaches half its maximum value. (b) After the emf is connected, how long does it take the current to reach this value?

33. 🌀 On a clear day at a certain location, a 100-V/m vertical electric field exists near the Earth's surface. At the same place, the Earth's magnetic field has a magnitude of 0.500×10^{-4} T. Compute the energy densities of the two fields.

34. Complete the calculation in Example 32.4 by proving that

$$\int_0^\infty e^{-2Rt/L} \, dt = \frac{L}{2R}$$

35. An *RL* circuit in which $L = 4.00$ H and $R = 5.00$ Ω is connected to a 22.0-V battery at $t = 0$. (a) What energy is stored in the inductor when the current is 0.500 A? (b) At what rate is energy being stored in the inductor when $I = 1.00$ A? (c) What power is being delivered to the circuit by the battery when $I = 0.500$ A?

36. A 10.0-V battery, a 5.00-Ω resistor, and a 10.0-H inductor are connected in series. After the current in the circuit has reached its maximum value, calculate (a) the power being supplied by the battery, (b) the power being delivered to the resistor, (c) the power being delivered to the inductor, and (d) the energy stored in the magnetic field of the inductor.

37. A uniform electric field of magnitude 680 kV/m throughout a cylindrical volume results in a total energy of 3.40 μJ. What magnetic field over this same region stores the same total energy?

38. Assume that the magnitude of the magnetic field outside a sphere of radius R is $B = B_0(R/r)^2$, where B_0 is a constant. Determine the total energy stored in the magnetic field outside the sphere and evaluate your result for $B_0 = 5.00 \times 10^{-5}$ T and $R = 6.00 \times 10^6$ m, values appropriate for the Earth's magnetic field.

Section 32.4 Mutual Inductance

39. Two coils are close to each other. The first coil carries a time-varying current given by $I(t) = (5.00$ A$)\, e^{-0.025\,0t} \sin(377t)$. At $t = 0.800$ s, the emf measured across the second coil is -3.20 V. What is the mutual inductance of the coils?

40. Two coils, held in fixed positions, have a mutual inductance of 100 μH. What is the peak voltage in one when a sinusoidal current given by $I(t) = (10.0$ A$) \sin(1\,000t)$ is in the other coil?

41. An emf of 96.0 mV is induced in the windings of a coil when the current in a nearby coil is increasing at the rate of 1.20 A/s. What is the mutual inductance of the two coils?

42. On a printed circuit board, a relatively long straight conductor and a conducting rectangular loop lie in the same plane, as shown in Figure P31.9. Taking $h = 0.400$ mm, $w = 1.30$ mm, and $L = 2.70$ mm, find their mutual inductance.

43. Two solenoids A and B, spaced close to each other and sharing the same cylindrical axis, have 400 and 700 turns, respectively. A current of 3.50 A in coil A produces an average flux of 300 μWb through each turn of A and a flux of 90.0 μWb through each turn of B. (a) Calculate the mutual inductance of the two solenoids. (b) What is the self-inductance of A? (c) What emf is induced in B when the current in A increases at the rate of 0.500 A/s?

44. A large coil of radius R_1 and having N_1 turns is coaxial with a small coil of radius R_2 and having N_2 turns. The centers of the coils are separated by a distance x that is much larger than R_1 and R_2. What is the mutual inductance of the coils? *Suggestion:* John von Neumann proved that the same answer must result from considering the flux through the first coil of the magnetic field produced by the second coil, or from considering the flux through the second coil of the magnetic field produced by the first coil. In this problem it is easy to calculate the flux through the small coil, but it is difficult to calculate the flux through the large coil, because to do so you would have to know the magnetic field away from the axis.

45. Two inductors having self-inductances L_1 and L_2 are connected in parallel as shown in Figure P32.45a. The mutual inductance between the two inductors is M. Determine the equivalent self-inductance L_{eq} for the system (Figure P32.45b).

(a) (b)

Figure P32.45

Section 32.5   Oscillations in an *LC* Circuit

46. A 1.00-μF capacitor is charged by a 40.0-V power supply. The fully charged capacitor is then discharged through a 10.0-mH inductor. Find the maximum current in the resulting oscillations.

47. An *LC* circuit consists of a 20.0-mH inductor and a 0.500-μF capacitor. If the maximum instantaneous current is 0.100 A, what is the greatest potential difference across the capacitor?

48. In the circuit of Figure P32.48, the battery emf is 50.0 V, the resistance is 250 Ω, and the capacitance is 0.500 μF. The switch S is closed for a long time and no voltage is measured across the capacitor. After the switch is opened, the potential difference across the capacitor reaches a maximum value of 150 V. What is the value of the inductance?

Figure P32.48

49. A fixed inductance $L = 1.05\ \mu$H is used in series with a variable capacitor in the tuning section of a radiotelephone on a ship. What capacitance tunes the circuit to the signal from a transmitter broadcasting at 6.30 MHz?

50. Calculate the inductance of an *LC* circuit that oscillates at 120 Hz when the capacitance is 8.00 μF.

51. An *LC* circuit like the one in Figure 32.16 contains an 82.0-mH inductor and a 17.0-μF capacitor that initially carries a 180-μC charge. The switch is open for $t < 0$ and then closed at $t = 0$. (a) Find the frequency (in hertz) of the resulting oscillations. At $t = 1.00$ ms, find (b) the charge on the capacitor and (c) the current in the circuit.

52. The switch in Figure P32.52 is connected to point *a* for a long time. After the switch is thrown to point *b*, what are (a) the frequency of oscillation of the *LC* circuit, (b) the maximum charge that appears on the capacitor, (c) the maximum current in the inductor, and (d) the total energy the circuit possesses at $t = 3.00$ s?

Figure P32.52

53. An *LC* circuit like that in Figure 32.16 consists of a 3.30-H inductor and an 840-pF capacitor, initially carrying a 105-μC charge. The switch is open for $t < 0$ and then closed at $t = 0$. Compute the following quantities at $t = 2.00$ ms:

(a) the energy stored in the capacitor; (b) the energy stored in the inductor; (c) the total energy in the circuit.

Section 32.6   The *RLC* Circuit

54. In Figure 32.21, let $R = 7.60\ \Omega$, $L = 2.20$ mH, and $C = 1.80\ \mu$F. (a) Calculate the frequency of the damped oscillation of the circuit. (b) What is the critical resistance?

55. Consider an *LC* circuit in which $L = 500$ mH and $C = 0.100\ \mu$F. (a) What is the resonance frequency ω_0? (b) If a resistance of 1.00 kΩ is introduced into this circuit, what is the frequency of the (damped) oscillations? (c) What is the percent difference between the two frequencies?

56. Show that Equation 32.28 in the text is Kirchhoff's loop rule as applied to the circuit in Figure 32.21.

57. The energy of an *RLC* circuit decreases by 1.00% during each oscillation when $R = 2.00\ \Omega$. If this resistance is removed, the resulting *LC* circuit oscillates at a frequency of 1.00 kHz. Find the values of the inductance and the capacitance.

58. Electrical oscillations are initiated in a series circuit containing a capacitance *C*, inductance *L*, and resistance *R*. (a) If $R \ll \sqrt{4L/C}$ (weak damping), how much time elapses before the amplitude of the current oscillation falls off to 50.0% of its initial value? (b) How long does it take the energy to decrease to 50.0% of its initial value?

Additional Problems

59. **Review problem.** This problem extends the reasoning of Section 26.4, Problem 26.37, Example 30.6, and Section 32.3. (a) Consider a capacitor with vacuum between its large, closely spaced, oppositely charged parallel plates. Show that the force on one plate can be accounted for by thinking of the electric field between the plates as exerting a "negative pressure" equal to the energy density of the electric field. (b) Consider two infinite plane sheets carrying electric currents in opposite directions with equal linear current densities J_s. Calculate the force per area acting on one sheet due to the magnetic field created by the other sheet. (c) Calculate the net magnetic field between the sheets and the field outside of the volume between them. (d) Calculate the energy density in the magnetic field between the sheets. (e) Show that the force on one sheet can be accounted for by thinking of the magnetic field between the sheets as exerting a positive pressure equal to its energy density. This result for magnetic pressure applies to all current configurations, not just to sheets of current.

60. Initially, the capacitor in a series *LC* circuit is charged. A switch is closed at $t = 0$, allowing the capacitor to discharge, and at time *t* the energy stored in the capacitor is one fourth of its initial value. Determine *L*, assuming *C* is known.

61. A 1.00-mH inductor and a 1.00-μF capacitor are connected in series. The current in the circuit is described by $I = 20.0t$, where *t* is in seconds and *I* is in amperes. The capacitor initially has no charge. Determine (a) the

voltage across the inductor as a function of time, (b) the voltage across the capacitor as a function of time, and (c) the time when the energy stored in the capacitor first exceeds that in the inductor.

62. An inductor having inductance L and a capacitor having capacitance C are connected in series. The current in the circuit increases linearly in time as described by $I = Kt$, where K is a constant. The capacitor is initially uncharged. Determine (a) the voltage across the inductor as a function of time, (b) the voltage across the capacitor as a function of time, and (c) the time when the energy stored in the capacitor first exceeds that in the inductor.

63. A capacitor in a series LC circuit has an initial charge Q and is being discharged. Find, in terms of L and C, the flux through each of the N turns in the coil, when the charge on the capacitor is $Q/2$.

64. The toroid in Figure P32.64 consists of N turns and has a rectangular cross section. Its inner and outer radii are a and b, respectively. (a) Show that the inductance of the toroid is

$$ L = \frac{\mu_0 N^2 h}{2\pi} \ln \frac{b}{a} $$

(b) Using this result, compute the self-inductance of a 500-turn toroid for which $a = 10.0$ cm, $b = 12.0$ cm, and $h = 1.00$ cm. (c) **What If?** In Problem 12, an approximate expression for the inductance of a toroid with $R \gg r$ was derived. To get a feel for the accuracy of that result, use the expression in Problem 12 to compute the approximate inductance of the toroid described in part (b). Compare the result with the answer to part (b).

Figure P32.64

65. (a) A flat circular coil does not really produce a uniform magnetic field in the area it encloses, but estimate the self-inductance of a flat, compact circular coil, with radius R and N turns, by assuming that the field at its center is uniform over its area. (b) A circuit on a laboratory table consists of a 1.5-volt battery, a 270-Ω resistor, a switch, and three 30-cm-long patch cords connecting them. Suppose the circuit is arranged to be circular. Think of it as a flat coil with one turn. Compute the order of magnitude of its self-inductance and (c) of the time constant describing how fast the current increases when you close the switch.

66. A soft iron rod ($\mu_m = 800\mu_0$) is used as the core of a solenoid. The rod has a diameter of 24.0 mm and is 10.0 cm long. A 10.0-m piece of 22-gauge copper wire (diameter = 0.644 mm) is wrapped around the rod in a single uniform layer, except for a 10.0-cm length at each end, which is to

be used for connections. (a) How many turns of this wire can be wrapped around the rod? For an accurate answer you should add the diameter of the wire to the diameter of the rod in determining the circumference of each turn. Also note that the wire spirals diagonally along the surface of the rod. (b) What is the resistance of this inductor? (c) What is its inductance?

67. A wire of nonmagnetic material, with radius R, carries current uniformly distributed over its cross section. The total current carried by the wire is I. Show that the magnetic energy per unit length inside the wire is $\mu_0 I^2/16\pi$.

68. An 820-turn wire coil of resistance 24.0 Ω is placed around a 12 500-turn solenoid 7.00 cm long, as shown in Figure P32.68. Both coil and solenoid have cross-sectional areas of 1.00×10^{-4} m^2. (a) How long does it take the solenoid current to reach 63.2% of its maximum value? Determine (b) the average back emf caused by the self-inductance of the solenoid during this time interval, (c) the average rate of change in magnetic flux through the coil during this time interval, and (d) the magnitude of the average induced current in the coil.

Figure P32.68

69. At $t = 0$, the open switch in Figure P32.69 is closed. By using Kirchhoff's rules for the instantaneous currents and voltages in this two-loop circuit, show that the current in the inductor at time $t > 0$ is

$$ I(t) = \frac{\mathcal{E}}{R_1}[1 - e^{-(R'/L)t}] $$

where $R' = R_1 R_2/(R_1 + R_2)$.

Figure P32.69 Problems 69 and 70.

70. In Figure P32.69 take $\mathcal{E} = 6.00$ V, $R_1 = 5.00$ Ω, and $R_2 = 1.00$ Ω. The inductor has negligible resistance. When the switch is opened after having been closed for a long time, the current in the inductor drops to 0.250 A in 0.150 s. What is the inductance of the inductor?

71. In Figure P32.71, the switch is closed for $t < 0$, and steady-state conditions are established. The switch is opened at $t = 0$. (a) Find the initial voltage $\mathcal{E}_0$ across L just after $t = 0$. Which end of the coil is at the higher potential: a or b? (b) Make freehand graphs of the currents in R_1 and in R_2 as a function of time, treating the steady-state directions as positive. Show values before and after $t = 0$. (c) How long after $t = 0$ does the current in R_2 have the value 2.00 mA?

Figure P32.71

72. The open switch in Figure P32.72 is closed at $t = 0$. Before the switch is closed, the capacitor is uncharged, and all currents are zero. Determine the currents in L, C, and R and the potential differences across L, C, and R (a) at the instant after the switch is closed, and (b) long after it is closed.

Figure P32.72

73. To prevent damage from arcing in an electric motor, a discharge resistor is sometimes placed in parallel with the armature. If the motor is suddenly unplugged while running, this resistor limits the voltage that appears across the armature coils. Consider a 12.0-V DC motor with an armature that has a resistance of 7.50 Ω and an inductance of 450 mH. Assume the back emf in the armature coils is 10.0 V when the motor is running at normal speed. (The equivalent circuit for the armature is shown in Figure P32.73.) Calculate the maximum resistance R that limits the voltage across the armature to 80.0 V when the motor is unplugged.

Figure P32.73

74. An air-core solenoid 0.500 m in length contains 1 000 turns and has a cross-sectional area of 1.00 cm². (a) Ignoring end effects, find the self-inductance. (b) A secondary winding wrapped around the center of the solenoid has 100 turns. What is the mutual inductance? (c) The secondary winding carries a constant current of 1.00 A, and the solenoid is connected to a load of 1.00 kΩ. The constant current is suddenly stopped. How much charge flows through the load resistor?

75. The lead-in wires from a television antenna are often constructed in the form of two parallel wires (Fig. P32.75). (a) Why does this configuration of conductors have an inductance? (b) What constitutes the flux loop for this configuration? (c) Ignoring any magnetic flux inside the wires, show that the inductance of a length x of this type of lead-in is

$$L = \frac{\mu_0 x}{\pi} \ln\left(\frac{w - a}{a}\right)$$

where a is the radius of the wires and w is their center-to-center separation.

Figure P32.75

Review problems. Problems 76 through 79 apply ideas from this chapter and earlier chapters to some properties of superconductors, which were introduced in Section 27.5.

76. *The resistance of a superconductor.* In an experiment carried out by S. C. Collins between 1955 and 1958, a current was maintained in a superconducting lead ring for 2.50 yr with no observed loss. If the inductance of the ring was 3.14×10^{-8} H, and the sensitivity of the experiment was

1 part in 10^9, what was the maximum resistance of the ring? (*Suggestion:* Treat this as a decaying current in an *RL* circuit, and recall that $e^{-x} \approx 1 - x$ for small *x*.)

77. A novel method of storing energy has been proposed. A huge underground superconducting coil, 1.00 km in diameter, would be fabricated. It would carry a maximum current of 50.0 kA through each winding of a 150-turn Nb_3Sn solenoid. (a) If the inductance of this huge coil were 50.0 H, what would be the total energy stored? (b) What would be the compressive force per meter length acting between two adjacent windings 0.250 m apart?

78. *Superconducting power transmission.* The use of superconductors has been proposed for power transmission lines. A single coaxial cable (Fig. P32.78) could carry 1.00×10^3 MW (the output of a large power plant) at 200 kV, DC, over a distance of 1 000 km without loss. An inner wire of radius 2.00 cm, made from the superconductor Nb_3Sn, carries the current *I* in one direction. A surrounding superconducting cylinder, of radius 5.00 cm, would carry the return current *I*. In such a system, what is the magnetic field (a) at the surface of the inner conductor and (b) at the inner surface of the outer conductor? (c) How much energy would be stored in the space between the conductors in a 1 000-km superconducting line? (d) What is the pressure exerted on the outer conductor?

Figure P32.79

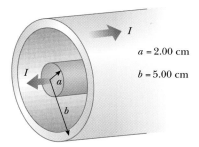

Figure P32.78

79. *The Meissner effect.* Compare this problem with Problem 65 in Chapter 26, on the force attracting a perfect dielectric into a strong electric field. A fundamental property of a Type I superconducting material is *perfect diamagnetism,* or demonstration of the *Meissner effect,* illustrated in Figure 30.35, and described as follows. The superconducting material has **B** = 0 everywhere inside it. If a sample of the material is placed into an externally produced magnetic field, or if it is cooled to become superconducting while it is in a magnetic field, electric currents appear on the surface of the sample. The currents have precisely the strength and orientation required to make the total magnetic field zero throughout the interior of the sample. The following problem will help you to understand the magnetic force that can then act on the superconducting sample.

A vertical solenoid with a length of 120 cm and a diameter of 2.50 cm consists of 1 400 turns of copper wire carrying a counterclockwise current of 2.00 A, as in Figure P32.79a. (a) Find the magnetic field in the vacuum inside the solenoid. (b) Find the energy density of the magnetic field, and note that the units J/m^3 of energy density are the same as the units N/m^2 of pressure. (c) Now a superconducting bar 2.20 cm in diameter is inserted partway into the solenoid. Its upper end is far outside the solenoid, where the magnetic field is negligible. The lower end of the bar is deep inside the solenoid. Identify the direction required for the current on the curved surface of the bar, so that the total magnetic field is zero within the bar. The field created by the supercurrents is sketched in Figure P32.79b, and the total field is sketched in Figure P32.79c. (d) The field of the solenoid exerts a force on the current in the superconductor. Identify the direction of the force on the bar. (e) Calculate the magnitude of the force by multiplying the energy density of the solenoid field times the area of the bottom end of the superconducting bar.

Answers to Quick Quizzes

32.1 (c), (f). For the constant current in (a) and (b), there is no potential difference across the resistanceless inductor. In (c), if the current increases, the emf induced in the inductor is in the opposite direction, from *b* to *a*, making *a* higher in potential than *b*. Similarly, in (f), the decreasing current induces an emf in the same direction as the current, from *b* to *a*, again making the potential higher at *a* than *b*.

32.2 (b), (d). As the switch is closed, there is no current, so there is no voltage across the resistor. After a long time, the current has reached its final value, and the inductor has no further effect on the circuit.

32.3 (b). When the iron rod is inserted into the solenoid, the inductance of the coil increases. As a result, more potential difference appears across the coil than before.

Consequently, less potential difference appears across the bulb, so the bulb is dimmer.

32.4 (b). Figure 32.10 shows that circuit B has the greater time constant because in this circuit it takes longer for the current to reach its maximum value and then longer for this current to decrease to zero after the switch is thrown to position b. Equation 32.8 indicates that, for equal resistances R_A and R_B, the condition $\tau_B > \tau_A$ means that $L_A < L_B$.

32.5 (a), (d). Because the energy density depends on the magnitude of the magnetic field, to increase the energy density, we must increase the magnetic field. For a solenoid, $B = \mu_0 nI$, where n is the number of turns per unit length. In (a), we increase n to increase the magnetic field. In (b), the change in cross-sectional area has no effect on the magnetic field. In (c), increasing the length but keeping n fixed has no effect on the magnetic field. Increasing the current in (d) increases the magnetic field in the solenoid.

32.6 (a). M_{12} increases because the magnetic flux through coil 2 increases.

32.7 (b). If the current is at its maximum value, the charge on the capacitor is zero.

32.8 (c). If the current is zero, this is the instant at which the capacitor is fully charged and the current is about to reverse direction.

Alternating Current Circuits

▲ *These large transformers are used to increase the voltage at a power plant for distribution of energy by electrical transmission to the power grid. Voltages can be changed relatively easily because power is distributed by alternating current rather than direct current. (Lester Lefkowitz/Getty Images)*

In this chapter we describe alternating current (AC) circuits. Every time we turn on a television set, a stereo, or any of a multitude of other electrical appliances in a home, we are calling on alternating currents to provide the power to operate them. We begin our study by investigating the characteristics of simple series circuits that contain resistors, inductors, and capacitors and that are driven by a sinusoidal voltage. We shall find that the maximum alternating current in each element is proportional to the maximum alternating voltage across the element. In addition, when the applied voltage is sinusoidal, the current in each element is also sinusoidal, but not necessarily in phase with the applied voltage. The primary aim of this chapter can be summarized as follows: if an AC source applies an alternating voltage to a series circuit containing resistors, inductors, and capacitors, we want to know the amplitude and time characteristics of the alternating current. We conclude the chapter with two sections concerning transformers, power transmission, and electrical filters.

33.1 AC Sources

An AC circuit consists of circuit elements and a power source that provides an alternating voltage Δv. This time-varying voltage is described by

$$\Delta v = \Delta V_{max} \sin \omega t$$

where ΔV_{max} is the maximum output voltage of the AC source, or the **voltage amplitude.** There are various possibilities for AC sources, including generators, as discussed in Section 31.5, and electrical oscillators. In a home, each electrical outlet serves as an AC source.

From Equation 15.12, the angular frequency of the AC voltage is

$$\omega = 2\pi f = \frac{2\pi}{T}$$

where f is the frequency of the source and T is the period. The source determines the frequency of the current in any circuit connected to it. Because the output voltage of an AC source varies sinusoidally with time, the voltage is positive during one half of the cycle and negative during the other half, as in Figure 33.1. Likewise, the current in any circuit driven by an AC source is an alternating current that also varies sinusoidally with time. Commercial electric-power plants in the United States use a frequency of 60 Hz, which corresponds to an angular frequency of 377 rad/s.

33.2 Resistors in an AC Circuit

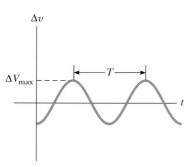

Figure 33.1 The voltage supplied by an AC source is sinusoidal with a period T.

Consider a simple AC circuit consisting of a resistor and an AC source , as shown in Figure 33.2. At any instant, the algebraic sum of the voltages around a closed loop in a circuit must be zero (Kirchhoff's loop rule). Therefore, $\Delta v + \Delta v_R = 0$, so

$$\Delta v = \Delta V_{\max} \sin \omega t$$

Active Figure 33.2 A circuit consisting of a resistor of resistance R connected to an AC source, designated by the symbol —$\sim$— .

At the Active Figures link at http://www.pse6.com, *you can adjust the resistance, the frequency, and the maximum voltage. The results can be studied with the graph and phasor diagram in Figure 33.3.*

that the magnitude of the source voltage equals the magnitude of the voltage across the resistor:

$$\Delta v = \Delta v_R = \Delta V_{\max} \sin \omega t \qquad (33.1)$$

where Δv_R is the **instantaneous voltage across the resistor.** Therefore, from Equation 27.8, $R = \Delta V/I$, the instantaneous current in the resistor is

$$i_R = \frac{\Delta v_R}{R} = \frac{\Delta V_{\max}}{R} \sin \omega t = I_{\max} \sin \omega t \qquad (33.2)$$

where $I_{\max}$ is the maximum current:

$$I_{\max} = \frac{\Delta V_{\max}}{R}$$

From Equations 33.1 and 33.2, we see that the instantaneous voltage across the resistor is

$$\Delta v_R = I_{\max} R \sin \omega t \qquad (33.3)$$

Maximum current in a resistor

Voltage across a resistor

A plot of voltage and current versus time for this circuit is shown in Figure 33.3a. At point a, the current has a maximum value in one direction, arbitrarily called the positive direction. Between points a and b, the current is decreasing in magnitude but is still in the positive direction. At b, the current is momentarily zero; it then begins to increase in the negative direction between points b and c. At c, the current has reached its maximum value in the negative direction.

The current and voltage are in step with each other because they vary identically with time. Because i_R and Δv_R both vary as $\sin \omega t$ and reach their maximum values at the same time, as shown in Figure 33.3a, they are said to be **in phase,** similar to the way that two waves can be in phase, as discussed in our study of wave motion in

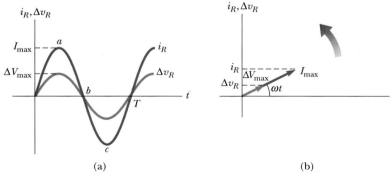

(a)   (b)

Active Figure 33.3 (a) Plots of the instantaneous current i_R and instantaneous voltage Δv_R across a resistor as functions of time. The current is in phase with the voltage, which means that the current is zero when the voltage is zero, maximum when the voltage is maximum, and minimum when the voltage is minimum. At time $t = T$, one cycle of the time-varying voltage and current has been completed. (b) Phasor diagram for the resistive circuit showing that the current is in phase with the voltage.

At the Active Figures link at http://www.pse6.com, *you can adjust the resistance, the frequency, and the maximum voltage of the circuit in Figure 33.2. The results can be studied with the graph and phasor diagram in this figure.*

Chapter 18. Thus, **for a sinusoidal applied voltage, the current in a resistor is always in phase with the voltage across the resistor.** For resistors in AC circuits, there are no new concepts to learn. Resistors behave essentially the same way in both DC and AC circuits. This will not be the case for capacitors and inductors.

To simplify our analysis of circuits containing two or more elements, we use graphical constructions called *phasor diagrams*. A **phasor** is a vector whose length is proportional to the maximum value of the variable it represents (ΔV_{max} for voltage and I_{max} for current in the present discussion) and which rotates counterclockwise at an angular speed equal to the angular frequency associated with the variable. The projection of the phasor onto the vertical axis represents the instantaneous value of the quantity it represents.

Figure 33.3b shows voltage and current phasors for the circuit of Figure 33.2 at some instant of time. The projections of the phasor arrows onto the vertical axis are determined by a sine function of the angle of the phasor with respect to the horizontal axis. For example, the projection of the current phasor in Figure 33.3b is $I_{max} \sin \omega t$. Notice that this is the same expression as Equation 33.2. Thus, we can use the projections of phasors to represent current values that vary sinusoidally in time. We can do the same with time-varying voltages. The advantage of this approach is that the phase relationships among currents and voltages can be represented as vector additions of phasors, using our vector addition techniques from Chapter 3.

In the case of the single-loop resistive circuit of Figure 33.2, the current and voltage phasors lie along the same line, as in Figure 33.3b, because i_R and Δv_R are in phase. The current and voltage in circuits containing capacitors and inductors have different phase relationships.

Quick Quiz 33.1 Consider the voltage phasor in Figure 33.4, shown at three instants of time. Choose the part of the figure that represents the instant of time at which the instantaneous value of the voltage has the largest magnitude.

Figure 33.4 (Quick Quizzes 33.1 and 33.2) A voltage phasor is shown at three instants of time.

Quick Quiz 33.2 For the voltage phasor in Figure 33.4, choose the part of the figure that represents the instant of time at which the instantaneous value of the voltage has the smallest magnitude.

For the simple resistive circuit in Figure 33.2, note that **the average value of the current over one cycle is zero.** That is, the current is maintained in the positive direction for the same amount of time and at the same magnitude as it is maintained in the negative direction. However, the direction of the current has no effect on the behavior of the resistor. We can understand this by realizing that collisions between electrons and the fixed atoms of the resistor result in an increase in the resistor's temperature. Although this temperature increase depends on the magnitude of the current, it is independent of the direction of the current.

We can make this discussion quantitative by recalling that the rate at which energy is delivered to a resistor is the power $\mathcal{P} = i^2 R$, where i is the instantaneous current in

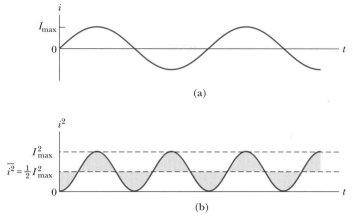

Figure 33.5 (a) Graph of the current in a resistor as a function of time. (b) Graph of the current squared in a resistor as a function of time. Notice that the gray shaded regions *under* the curve and *above* the dashed line for $I_{max}^2/2$ have the same area as the gray shaded regions *above* the curve and *below* the dashed line for $I_{max}^2/2$. Thus, the average value of i^2 is $I_{max}^2/2$.

the resistor. Because this rate is proportional to the square of the current, it makes no difference whether the current is direct or alternating—that is, whether the sign associated with the current is positive or negative. However, the temperature increase produced by an alternating current having a maximum value I_{max} is not the same as that produced by a direct current equal to I_{max}. This is because the alternating current is at this maximum value for only an instant during each cycle (Fig. 33.5a). What is of importance in an AC circuit is an average value of current, referred to as the **rms current.** As we learned in Section 21.1, the notation *rms* stands for *root-mean-square*, which in this case means the square root of the mean (average) value of the square of the current: $I_{rms} = \sqrt{\overline{i^2}}$. Because i^2 varies as $\sin^2 \omega t$ and because the average value of i^2 is $\frac{1}{2} I_{max}^2$ (see Fig. 33.5b), the rms current is[1]

$$I_{rms} = \frac{I_{max}}{\sqrt{2}} = 0.707 I_{max} \qquad (33.4)$$ **rms current**

This equation states that an alternating current whose maximum value is 2.00 A delivers to a resistor the same power as a direct current that has a value of $(0.707)(2.00 \text{ A}) = 1.41$ A. Thus, the average power delivered to a resistor that carries an alternating current is

$$\mathcal{P}_{av} = I_{rms}^2 R$$

Average power delivered to a resistor

----

[1] That the square root of the average value of i^2 is equal to $I_{max}/\sqrt{2}$ can be shown as follows. The current in the circuit varies with time according to the expression $i = I_{max} \sin \omega t$, so $i^2 = I_{max}^2 \sin^2 \omega t$. Therefore, we can find the average value of i^2 by calculating the average value of $\sin^2 \omega t$. A graph of $\cos^2 \omega t$ versus time is identical to a graph of $\sin^2 \omega t$ versus time, except that the points are shifted on the time axis. Thus, the time average of $\sin^2 \omega t$ is equal to the time average of $\cos^2 \omega t$ when taken over one or more complete cycles. That is,

$$(\sin^2 \omega t)_{av} = (\cos^2 \omega t)_{av}$$

Using this fact and the trigonometric identity $\sin^2 \theta + \cos^2 \theta = 1$, we obtain

$$(\sin^2 \omega t)_{av} + (\cos^2 \omega t)_{av} = 2(\sin^2 \omega t)_{av} = 1$$

$$(\sin^2 \omega t)_{av} = \tfrac{1}{2}$$

When we substitute this result in the expression $i^2 = I_{max}^2 \sin^2 \omega t$, we obtain $(i^2)_{av} = \overline{i^2} = I_{rms}^2 = I_{max}^2/2$, or $I_{rms} = I_{max}/\sqrt{2}$. The factor $1/\sqrt{2}$ is valid only for sinusoidally varying currents. Other waveforms, such as sawtooth variations, have different factors.

Alternating voltage is also best discussed in terms of rms voltage, and the relationship is identical to that for current:

rms voltage

$$\Delta V_{rms} = \frac{\Delta V_{max}}{\sqrt{2}} = 0.707 \, \Delta V_{max} \tag{33.5}$$

When we speak of measuring a 120-V alternating voltage from an electrical outlet, we are referring to an rms voltage of 120 V. A quick calculation using Equation 33.5 shows that such an alternating voltage has a maximum value of about 170 V. One reason we use rms values when discussing alternating currents and voltages in this chapter is that AC ammeters and voltmeters are designed to read rms values. Furthermore, with rms values, many of the equations we use have the same form as their direct current counterparts.

> **Quick Quiz 33.3** Which of the following statements might be true for a resistor connected to a sinusoidal AC source? (a) $\mathcal{P}_{av} = 0$ and $i_{av} = 0$ (b) $\mathcal{P}_{av} = 0$ and $i_{av} > 0$ (c) $\mathcal{P}_{av} > 0$ and $i_{av} = 0$ (d) $\mathcal{P}_{av} > 0$ and $i_{av} > 0$.

Example 33.1 What Is the rms Current?

The voltage output of an AC source is given by the expression $\Delta v = (200 \text{ V}) \sin \omega t$. Find the rms current in the circuit when this source is connected to a 100-Ω resistor.

Solution Comparing this expression for voltage output with the general form $\Delta v = \Delta V_{max} \sin \omega t$, we see that $\Delta V_{max} = 200$ V. Thus, the rms voltage is

$$\Delta V_{rms} = \frac{\Delta V_{max}}{\sqrt{2}} = \frac{200 \text{ V}}{\sqrt{2}} = 141 \text{ V}$$

Therefore,

$$I_{rms} = \frac{\Delta V_{rms}}{R} = \frac{141 \text{ V}}{100 \, \Omega} = \boxed{1.41 \text{ A}}$$

33.3 Inductors in an AC Circuit

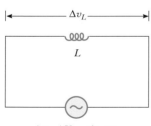

$\Delta v = \Delta V_{max} \sin \omega t$

Active Figure 33.6 A circuit consisting of an inductor of inductance L connected to an AC source.

At the Active Figures link at http://www.pse6.com, you can adjust the inductance, the frequency, and the maximum voltage. The results can be studied with the graph and phasor diagram in Figure 33.7.

Now consider an AC circuit consisting only of an inductor connected to the terminals of an AC source, as shown in Figure 33.6. If $\Delta v_L = \mathcal{E}_L = -L(di/dt)$ is the self-induced instantaneous voltage across the inductor (see Eq. 32.1), then Kirchhoff's loop rule applied to this circuit gives $\Delta v + \Delta v_L = 0$, or

$$\Delta v - L \frac{di}{dt} = 0$$

When we substitute $\Delta V_{max} \sin \omega t$ for Δv and rearrange, we obtain

$$\Delta v = L \frac{di}{dt} = \Delta V_{max} \sin \omega t \tag{33.6}$$

Solving this equation for di, we find that

$$di = \frac{\Delta V_{max}}{L} \sin \omega t \, dt$$

Integrating this expression[2] gives the instantaneous current i_L in the inductor as a function of time:

$$i_L = \frac{\Delta V_{max}}{L} \int \sin \omega t \, dt = -\frac{\Delta V_{max}}{\omega L} \cos \omega t \tag{33.7}$$

---

[2]   We neglect the constant of integration here because it depends on the initial conditions, which are not important for this situation.

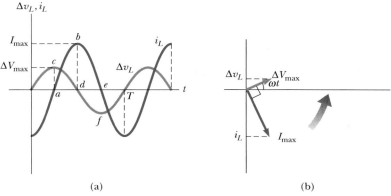

(a)   (b)

Active Figure 33.7 (a) Plots of the instantaneous current i_L and instantaneous voltage Δv_L across an inductor as functions of time. The current lags behind the voltage by 90°. (b) Phasor diagram for the inductive circuit, showing that the current lags behind the voltage by 90°.

When we use the trigonometric identity $\cos \omega t = -\sin(\omega t - \pi/2)$, we can express Equation 33.7 as

$$i_L = \frac{\Delta V_{max}}{\omega L} \sin\left(\omega t - \frac{\pi}{2}\right) \qquad (33.8)$$

Current in an inductor

Comparing this result with Equation 33.6, we see that the instantaneous current i_L in the inductor and the instantaneous voltage Δv_L across the inductor are *out* of phase by $(\pi/2)$ rad = 90°.

A plot of voltage and current versus time is provided in Figure 33.7a. In general, inductors in an AC circuit produce a current that is out of phase with the AC voltage. For example, when the current i_L in the inductor is a maximum (point *b* in Figure 33.7a), it is momentarily not changing, so the voltage across the inductor is zero (point *d*). At points like *a* and *e*, the current is zero and the rate of change of current is at a maximum. Thus, the voltage across the inductor is also at a maximum (points *c* and *f*). Note that the voltage reaches its maximum value one quarter of a period before the current reaches its maximum value. Thus, we see that

for a sinusoidal applied voltage, the current in an inductor always lags behind the voltage across the inductor by 90° (one-quarter cycle in time).

As with the relationship between current and voltage for a resistor, we can represent this relationship for an inductor with a phasor diagram as in Figure 33.7b. Notice that the phasors are at 90° to one another, representing the 90° phase difference between current and voltage.

From Equation 33.7 we see that the current in an inductive circuit reaches its maximum value when $\cos \omega t = -1$:

$$I_{max} = \frac{\Delta V_{max}}{\omega L} \qquad (33.9)$$

Maximum current in an inductor

This looks similar to the relationship between current, voltage, and resistance in a DC circuit, $I = \Delta V/R$ (Eq. 27.8). In fact, because I_{max} has units of amperes and ΔV_{max} has units of volts, ωL must have units of ohms. Therefore, ωL has the same units as resistance and is related to current and voltage in the same way as resistance. It must behave in a manner similar to resistance, in the sense that it represents opposition to the flow of charge. Notice that because ωL depends on the applied frequency ω, the inductor *reacts* differently, in terms of offering resistance to current, for different frequencies. For this reason, we define ωL as the **inductive reactance:**

Inductive reactance

$$X_L \equiv \omega L \tag{33.10}$$

and we can write Equation 33.9 as

$$I_{max} = \frac{\Delta V_{max}}{X_L} \tag{33.11}$$

The expression for the rms current in an inductor is similar to Equation 33.9, with I_{max} replaced by I_{rms} and ΔV_{max} replaced by ΔV_{rms}.

Equation 33.10 indicates that, for a given applied voltage, the inductive reactance increases as the frequency increases. This is consistent with Faraday's law—the greater the rate of change of current in the inductor, the larger is the back emf. The larger back emf translates to an increase in the reactance and a decrease in the current.

Using Equations 33.6 and 33.11, we find that the instantaneous voltage across the inductor is

Voltage across an inductor

$$\Delta v_L = -L \frac{di}{dt} = -\Delta V_{max} \sin \omega t = -I_{max} X_L \sin \omega t \tag{33.12}$$

Quick Quiz 33.4 Consider the AC circuit in Figure 33.8. The frequency of the AC source is adjusted while its voltage amplitude is held constant. The lightbulb will glow the brightest at (a) high frequencies (b) low frequencies (c) The brightness will be the same at all frequencies.

Figure 33.8 (Quick Quiz 33.4) At what frequencies will the bulb glow the brightest?

Example 33.2 A Purely Inductive AC Circuit

In a purely inductive AC circuit (see Fig. 33.6), $L = 25.0$ mH and the rms voltage is 150 V. Calculate the inductive reactance and rms current in the circuit if the frequency is 60.0 Hz.

Solution Equation 33.10 gives

$$X_L = \omega L = 2\pi f L = 2\pi(60.0 \text{ Hz})(25.0 \times 10^{-3} \text{ H})$$

$$= \boxed{9.42 \ \Omega}$$

From an rms version of Equation 33.11, the rms current is

$$I_{rms} = \frac{\Delta V_{L,rms}}{X_L} = \frac{150 \text{ V}}{9.42 \ \Omega} = \boxed{15.9 \text{ A}}$$

What If? What if the frequency increases to 6.00 kHz? What happens to the rms current in the circuit?

Answer If the frequency increases, the inductive reactance increases because the current is changing at a higher rate. The increase in inductive reactance results in a lower current. Let us calculate the new inductive reactance:

$$X_L = 2\pi(6.00 \times 10^3 \text{ Hz})(25.0 \times 10^{-3} \text{ H}) = 942 \ \Omega$$

The new current is

$$I_{rms} = \frac{150 \text{ V}}{942 \ \Omega} = 0.159 \text{ A}$$

33.4 Capacitors in an AC Circuit

Figure 33.9 shows an AC circuit consisting of a capacitor connected across the terminals of an AC source. Kirchhoff's loop rule applied to this circuit gives $\Delta v + \Delta v_C = 0$, so that the magnitude of the source voltage is equal to the magnitude of the voltage across the capacitor:

$$\Delta v = \Delta v_C = \Delta V_{max} \sin \omega t \qquad (33.13)$$

where Δv_C is the instantaneous voltage across the capacitor. We know from the definition of capacitance that $C = q/\Delta v_C$; hence, Equation 33.13 gives

$$q = C \Delta V_{max} \sin \omega t \qquad (33.14)$$

where q is the instantaneous charge on the capacitor. Because $i = dq/dt$, differentiating Equation 33.14 with respect to time gives the instantaneous current in the circuit:

$$i_C = \frac{dq}{dt} = \omega C \Delta V_{max} \cos \omega t \qquad (33.15)$$

Using the trigonometric identity

$$\cos \omega t = \sin \left(\omega t + \frac{\pi}{2} \right)$$

we can express Equation 33.15 in the alternative form

$$i_C = \omega C \Delta V_{max} \sin \left(\omega t + \frac{\pi}{2} \right) \qquad (33.16)$$

Comparing this expression with Equation 33.13, we see that the current is $\pi/2$ rad = 90° out of phase with the voltage across the capacitor. A plot of current and voltage versus time (Fig. 33.10a) shows that the current reaches its maximum value one quarter of a cycle sooner than the voltage reaches its maximum value.

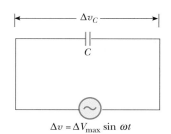

Active Figure 33.9 A circuit consisting of a capacitor of capacitance C connected to an AC source.

At the Active Figures link at http://www.pse6.com, *you can adjust the capacitance, the frequency, and the maximum voltage. The results can be studied with the graph and phasor diagram in Figure 33.10.*

Current in a capacitor

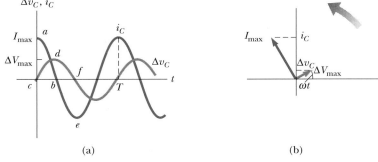

(a)   (b)

Active Figure 33.10 (a) Plots of the instantaneous current i_C and instantaneous voltage Δv_C across a capacitor as functions of time. The voltage lags behind the current by 90°. (b) Phasor diagram for the capacitive circuit, showing that the current leads the voltage by 90°.

At the Active Figures link at http://www.pse6.com, *you can adjust the capacitance, the frequency, and the maximum voltage of the circuit in Figure 33.9. The results can be studied with the graph and phasor diagram in this figure.*

Looking more closely, consider a point such as b where the current is zero. This occurs when the capacitor has just reached its maximum charge, so the voltage across the capacitor is a maximum (point d). At points such as a and e, the current is a maximum, which occurs at those instants at which the charge on the capacitor has just gone to zero and it begins to charge up with the opposite polarity. Because the charge is zero, the voltage across the capacitor is zero (points c and f). Thus, the current and voltage are out of phase.

As with inductors, we can represent the current and voltage for a capacitor on a phasor diagram. The phasor diagram in Figure 33.10b shows that

> for a sinusoidally applied voltage, the current always leads the voltage across a capacitor by 90°.

From Equation 33.15, we see that the current in the circuit reaches its maximum value when $\cos \omega t = 1$:

$$I_{max} = \omega C \Delta V_{max} = \frac{\Delta V_{max}}{(1/\omega C)} \tag{33.17}$$

As in the case with inductors, this looks like Equation 27.8, so that the denominator must play the role of resistance, with units of ohms. We give the combination $1/\omega C$ the symbol X_C, and because this function varies with frequency, we define it as the **capacitive reactance:**

Capacitive reactance

$$X_C \equiv \frac{1}{\omega C} \tag{33.18}$$

and we can write Equation 33.17 as

Maximum current in a capacitor

$$I_{max} = \frac{\Delta V_{max}}{X_C} \tag{33.19}$$

The rms current is given by an expression similar to Equation 33.19, with I_{max} replaced by I_{rms} and ΔV_{max} replaced by ΔV_{rms}.

Combining Equations 33.13 and 33.19, we can express the instantaneous voltage across the capacitor as

Voltage across a capacitor

$$\Delta v_C = \Delta V_{max} \sin \omega t = I_{max} X_C \sin \omega t \tag{33.20}$$

Equations 33.18 and 33.19 indicate that as the frequency of the voltage source increases, the capacitive reactance decreases and therefore the maximum current increases. Again, note that the frequency of the current is determined by the frequency of the voltage source driving the circuit. As the frequency approaches zero, the capacitive reactance approaches infinity, and hence the current approaches zero. This makes sense because the circuit approaches direct current conditions as ω approaches zero, and the capacitor represents an open circuit.

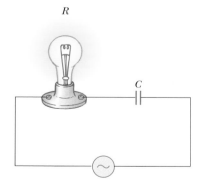

Figure 33.11 (Quick Quiz 33.5)

Quick Quiz 33.5 Consider the AC circuit in Figure 33.11. The frequency of the AC source is adjusted while its voltage amplitude is held constant. The lightbulb will glow the brightest at (a) high frequencies (b) low frequencies (c) The brightness will be same at all frequencies.

Quick Quiz 33.6 Consider the AC circuit in Figure 33.12. The frequency of the AC source is adjusted while its voltage amplitude is held constant. The lightbulb will glow the brightest at (a) high frequencies (b) low frequencies (c) The brightness will be same at all frequencies.

Figure 33.12 (Quick Quiz 33.6)

Example 33.3 A Purely Capacitive AC Circuit

An 8.00-μF capacitor is connected to the terminals of a 60.0-Hz AC source whose rms voltage is 150 V. Find the capacitive reactance and the rms current in the circuit.

Solution Using Equation 33.18 and the fact that $\omega = 2\pi f = 377 \text{ s}^{-1}$ gives

$$X_C = \frac{1}{\omega C} = \frac{1}{(377 \text{ s}^{-1})(8.00 \times 10^{-6} \text{ F})} = \boxed{332 \ \Omega}$$

Hence, from a modified Equation 33.19, the rms current is

$$I_{\text{rms}} = \frac{\Delta V_{\text{rms}}}{X_C} = \frac{150 \text{ V}}{332 \ \Omega} = \boxed{0.452 \text{ A}}$$

What If? What if the frequency is doubled? What happens to the rms current in the circuit?

Answer If the frequency increases, the capacitive reactance decreases—just the opposite as in the case of an inductor. The decrease in capacitive reactance results in an increase in the current.

Let us calculate the new capacitive reactance:

$$X_C = \frac{1}{\omega C} = \frac{1}{2(377 \text{ s}^{-1})(8.00 \times 10^{-6} \text{ F})} = 166 \ \Omega$$

The new current is

$$I_{\text{rms}} = \frac{150 \text{ V}}{166 \ \Omega} = 0.904 \text{ A}$$

33.5 The *RLC* Series Circuit

Figure 33.13a shows a circuit that contains a resistor, an inductor, and a capacitor connected in series across an alternating voltage source. As before, we assume that the applied voltage varies sinusoidally with time. It is convenient to assume that the instantaneous applied voltage is given by

$$\Delta v = \Delta V_{\text{max}} \sin \omega t$$

while the current varies as

$$i = I_{\text{max}} \sin(\omega t - \phi)$$

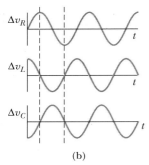

(b)

Active Figure 33.13 (a) A series circuit consisting of a resistor, an inductor, and a capacitor connected to an AC source. (b) Phase relationships for instantaneous voltages in the series *RLC* circuit.

 At the Active Figures link at http://www.pse6.com, you can adjust the resistance, the inductance, and the capacitance. The results can be studied with the graph in this figure and the phasor diagram in Figure 33.15.

where ϕ is some **phase angle** between the current and the applied voltage. Based on our discussions of phase in Sections 33.3 and 33.4, we expect that the current will generally not be in phase with the voltage in an *RLC* circuit. Our aim is to determine ϕ and I_{max}. Figure 33.13b shows the voltage versus time across each element in the circuit and their phase relationships.

First, we note that because the elements are in series, the current everywhere in the circuit must be the same at any instant. That is, **the current at all points in a series AC circuit has the same amplitude and phase.** Based on the preceding sections, we know that the voltage across each element has a different amplitude and phase. In particular, the voltage across the resistor is in phase with the current, the voltage across the inductor leads the current by 90°, and the voltage across the capacitor lags behind the current by 90°. Using these phase relationships, we can express the instantaneous voltages across the three circuit elements as

$$\Delta v_R = I_{max} R \sin \omega t = \Delta V_R \sin \omega t \tag{33.21}$$

$$\Delta v_L = I_{max} X_L \sin \left(\omega t + \frac{\pi}{2} \right) = \Delta V_L \cos \omega t \tag{33.22}$$

$$\Delta v_C = I_{max} X_C \sin \left(\omega t - \frac{\pi}{2} \right) = -\Delta V_C \cos \omega t \tag{33.23}$$

where ΔV_R, ΔV_L, and ΔV_C are the maximum voltage values across the elements:

$$\Delta V_R = I_{max} R \qquad \Delta V_L = I_{max} X_L \qquad \Delta V_C = I_{max} X_C$$

At this point, we could proceed by noting that the instantaneous voltage Δv across the three elements equals the sum

$$\Delta v = \Delta v_R + \Delta v_L + \Delta v_C$$

Although this analytical approach is correct, it is simpler to obtain the sum by examining the phasor diagram, shown in Figure 33.14. Because the current at any instant is the same in all elements, we combine the three phasor pairs shown in Figure 33.14 to obtain Figure 33.15a, in which a single phasor I_{max} is used to represent the current in each element. Because phasors are rotating vectors, we can combine the three parts of Figure 33.14 by using vector addition. To obtain the vector sum of the three voltage phasors in Figure 33.15a, we redraw the phasor diagram as in Figure 33.15b. From this diagram, we see that the vector sum of the voltage amplitudes ΔV_R, ΔV_L, and ΔV_C equals a phasor whose length is the maximum applied voltage ΔV_{max}, and which makes an angle ϕ with the current phasor I_{max}. The voltage phasors ΔV_L and ΔV_C are in opposite directions along the same line, so we can construct the difference phasor $\Delta V_L - \Delta V_C$, which is perpendicular to the phasor ΔV_R. From either one of the right triangles

(a) Resistor   (b) Inductor   (c) Capacitor

Figure 33.14 Phase relationships between the voltage and current phasors for (a) a resistor, (b) an inductor, and (c) a capacitor connected in series.

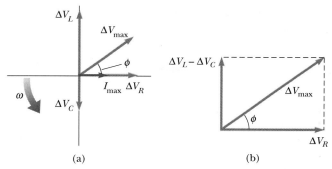

Active Figure 33.15 (a) Phasor diagram for the series *RLC* circuit shown in Figure 33.13a. The phasor ΔV_R is in phase with the current phasor I_{max}, the phasor ΔV_L leads I_{max} by 90°, and the phasor ΔV_C lags I_{max} by 90°. The total voltage ΔV_{max} makes an angle ϕ with I_{max}. (b) Simplified version of the phasor diagram shown in part (a).

At the Active Figures link at http://www.pse6.com, *you can adjust the resistance, the inductance, and the capacitance of the circuit in Figure 33.13a. The results can be studied with the graphs in Figure 33.13b and the phasor diagram in this figure.*

in Figure 33.15b, we see that

$$\Delta V_{max} = \sqrt{\Delta V_R{}^2 + (\Delta V_L - \Delta V_C)^2} = \sqrt{(I_{max} R)^2 + (I_{max} X_L - I_{max} X_C)^2}$$

$$\Delta V_{max} = I_{max} \sqrt{R^2 + (X_L - X_C)^2} \tag{33.24}$$

Therefore, we can express the maximum current as

$$I_{max} = \frac{\Delta V_{max}}{\sqrt{R^2 + (X_L - X_C)^2}}$$

Maximum current in an *RLC* circuit

Once again, this has the same mathematical form as Equation 27.8. The denominator of the fraction plays the role of resistance and is called the **impedance** *Z* of the circuit:

$$Z \equiv \sqrt{R^2 + (X_L - X_C)^2} \tag{33.25}$$

Impedance

where impedance also has units of ohms. Therefore, we can write Equation 33.24 in the form

$$\Delta V_{max} = I_{max} Z \tag{33.26}$$

We can regard Equation 33.26 as the AC equivalent of Equation 27.8. Note that the impedance and therefore the current in an AC circuit depend upon the resistance, the inductance, the capacitance, and the frequency (because the reactances are frequency-dependent).

By removing the common factor I_{max} from each phasor in Figure 33.15a, we can construct the *impedance triangle* shown in Figure 33.16. From this phasor diagram we find that the phase angle ϕ between the current and the voltage is

$$\phi = \tan^{-1}\left(\frac{X_L - X_C}{R}\right) \tag{33.27}$$

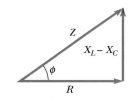

Figure 33.16 An impedance triangle for a series *RLC* circuit gives the relationship $Z = \sqrt{R^2 + (X_L - X_C)^2}$.

Phase angle

Also, from Figure 33.16, we see that $\cos \phi = R/Z$. When $X_L > X_C$ (which occurs at high frequencies), the phase angle is positive, signifying that the current lags behind the applied voltage, as in Figure 33.15a. We describe this situation by saying that the circuit is *more inductive than capacitive*. When $X_L < X_C$, the phase angle is negative, signifying that the current leads the applied voltage, and the circuit is *more capacitive than inductive*. When $X_L = X_C$, the phase angle is zero and the circuit is *purely resistive*.

Table 33.1 gives impedance values and phase angles for various series circuits containing different combinations of elements.

Table 33.1

Impedance Values and Phase Angles for Various Circuit-Element Combinations[a]		
Circuit Elements	**Impedance Z**	**Phase Angle ϕ**
R	R	$0°$
C	X_C	$-90°$
L	X_L	$+90°$
R C	$\sqrt{R^2 + X_C{}^2}$	Negative, between $-90°$ and $0°$
R L	$\sqrt{R^2 + X_L{}^2}$	Positive, between $0°$ and $90°$
R L C	$\sqrt{R^2 + (X_L - X_C)^2}$	Negative if $X_C > X_L$ Positive if $X_C < X_L$

[a] In each case, an AC voltage (not shown) is applied across the elements.

Quick Quiz 33.7 Label each part of Figure 33.17 as being $X_L > X_C$, $X_L = X_C$, or $X_L < X_C$.

Figure 33.17 (Quick Quiz 33.7) Match the phasor diagrams to the relationships between the reactances.

Example 33.4 Finding L from a Phasor Diagram

In a series RLC circuit, the applied voltage has a maximum value of 120 V and oscillates at a frequency of 60.0 Hz. The circuit contains an inductor whose inductance can be varied, a 200-Ω resistor, and a 4.00-μF capacitor. What value of L should an engineer analyzing the circuit choose such that the voltage across the capacitor lags the applied voltage by 30.0°?

Solution The phase relationships for the voltages across the elements are shown in Figure 33.18. From the figure we see that the phase angle is $\phi = -60.0°$. (The phasors representing I_{max} and ΔV_R are in the same direction.) From Equation 33.27, we find that

$$X_L = X_C + R \tan \phi$$

Substituting Equations 33.10 and 33.18 (with $\omega = 2\pi f$) into this expression gives

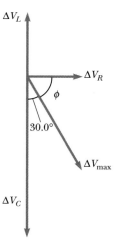

Figure 33.18 (Example 33.4) The phasor diagram for the given information.

$$2\pi f L = \frac{1}{2\pi f C} + R\tan\phi$$

$$L = \frac{1}{2\pi f}\left(\frac{1}{2\pi f C} + R\tan\phi\right)$$

Substituting the given values into the equation gives

$$L = \boxed{0.84\text{ H.}}$$

Example 33.5 Analyzing a Series *RLC* Circuit

Interactive

A series *RLC* AC circuit has $R = 425\ \Omega$, $L = 1.25$ H, $C = 3.50\ \mu\text{F}$, $\omega = 377\text{ s}^{-1}$, and $\Delta V_{max} = 150$ V.

(A) Determine the inductive reactance, the capacitive reactance, and the impedance of the circuit.

Solution The reactances are $X_L = \omega L = \boxed{471\ \Omega}$ and $X_C = 1/\omega C = \boxed{758\ \Omega}$.
The impedance is

$$Z = \sqrt{R^2 + (X_L - X_C)^2}$$
$$= \sqrt{(425\ \Omega)^2 + (471\ \Omega - 758\ \Omega)^2} = \boxed{513\ \Omega}$$

(B) Find the maximum current in the circuit.

Solution

$$I_{max} = \frac{\Delta V_{max}}{Z} = \frac{150\text{ V}}{513\ \Omega} = \boxed{0.292\text{ A}}$$

(C) Find the phase angle between the current and voltage.

Solution

$$\phi = \tan^{-1}\left(\frac{X_L - X_C}{R}\right) = \tan^{-1}\left(\frac{471\ \Omega - 758\ \Omega}{425\ \Omega}\right)$$
$$= \boxed{-34.0°}$$

Because the capacitive reactance is larger than the inductive reactance, the circuit is more capacitive than inductive. In this case, the phase angle ϕ is negative and the current leads the applied voltage.

(D) Find both the maximum voltage and the instantaneous voltage across each element.

Solution The maximum voltages are

$$\Delta V_R = I_{max}R = (0.292\text{ A})(425\ \Omega) = \boxed{124\text{ V}}$$

$$\Delta V_L = I_{max}X_L = (0.292\text{ A})(471\ \Omega) = \boxed{138\text{ V}}$$

$$\Delta V_C = I_{max}X_C = (0.292\text{ A})(758\ \Omega) = \boxed{221\text{ V}}$$

Using Equations 33.21, 33.22, and 33.23, we find that we can write the instantaneous voltages across the three elements as

$$\Delta v_R = \boxed{(124\text{ V})\sin 377t}$$

$$\Delta v_L = \boxed{(138\text{ V})\cos 377t}$$

$$\Delta v_C = \boxed{(-221\text{ V})\cos 377t}$$

What If? What if you added up the maximum voltages across the three circuit elements? Is this a physically meaningful quantity?

Answer The sum of the maximum voltages across the elements is $\Delta V_R + \Delta V_L + \Delta V_C = 484$ V. Note that this sum is much greater than the maximum voltage of the source, 150 V. The sum of the maximum voltages is a meaningless quantity because when sinusoidally varying quantities are added, *both their amplitudes and their phases* must be taken into account. We know that the maximum voltages across the various elements occur at different times. That is, the voltages must be added in a way that takes account of the different phases.

 At the Interactive Worked Example link at **http://www.pse6.com**, *you can investigate the RLC circuit for various values of the circuit elements.*

33.6 Power in an AC Circuit

Let us now take an energy approach to analyzing AC circuits, considering the transfer of energy from the AC source to the circuit. In Example 28.1 we found that the power delivered by a battery to a DC circuit is equal to the product of the current and the emf of the battery. Likewise, the instantaneous power delivered by an AC source to a circuit is the product of the source current and the applied voltage. For the *RLC* circuit shown in Figure 33.13a, we can express the instantaneous power $\mathcal{P}$ as

$$\mathcal{P} = i\,\Delta v = I_{max}\sin(\omega t - \phi)\,\Delta V_{max}\sin \omega t$$
$$= I_{max}\,\Delta V_{max}\sin \omega t \sin(\omega t - \phi) \tag{33.28}$$

This result is a complicated function of time and therefore is not very useful from a practical viewpoint. What is generally of interest is the average power over one or more cycles. Such an average can be computed by first using the trigonometric identity $\sin(\omega t - \phi) = \sin \omega t \cos \phi - \cos \omega t \sin \phi$. Substituting this into Equation 33.28 gives

$$\mathcal{P} = I_{max}\,\Delta V_{max}\sin^2 \omega t \cos \phi - I_{max}\,\Delta V_{max}\sin \omega t \cos \omega t \sin \phi \tag{33.29}$$

We now take the time average of $\mathcal{P}$ over one or more cycles, noting that I_{max}, ΔV_{max}, ϕ, and ω are all constants. The time average of the first term on the right in Equation 33.29 involves the average value of $\sin^2 \omega t$, which is $\frac{1}{2}$ (as shown in footnote 1). The time average of the second term on the right is identically zero because $\sin \omega t \cos \omega t = \frac{1}{2}\sin 2\omega t$, and the average value of $\sin 2\omega t$ is zero. Therefore, we can express the **average power** $\mathcal{P}_{av}$ as

$$\mathcal{P}_{av} = \tfrac{1}{2}I_{max}\,\Delta V_{max}\cos \phi \tag{33.30}$$

It is convenient to express the average power in terms of the rms current and rms voltage defined by Equations 33.4 and 33.5:

$$\mathcal{P}_{av} = I_{rms}\,\Delta V_{rms}\cos \phi \tag{33.31}$$

Average power delivered to an *RLC* circuit

where the quantity $\cos \phi$ is called the **power factor.** By inspecting Figure 33.15b, we see that the maximum voltage across the resistor is given by $\Delta V_R = \Delta V_{max}\cos \phi = I_{max}R$. Using Equation 33.5 and the fact that $\cos \phi = I_{max}R/\Delta V_{max}$, we find that we can express $\mathcal{P}_{av}$ as

$$\mathcal{P}_{av} = I_{rms}\,\Delta V_{rms}\cos \phi = I_{rms}\left(\frac{\Delta V_{max}}{\sqrt{2}}\right)\frac{I_{max}R}{\Delta V_{max}} = I_{rms}\frac{I_{max}R}{\sqrt{2}}$$

After making the substitution $I_{max} = \sqrt{2}\,I_{rms}$ from Equation 33.4, we have

$$\mathcal{P}_{av} = I_{rms}^2\,R \tag{33.32}$$

In words, **the average power delivered by the source is converted to internal energy in the resistor,** just as in the case of a DC circuit. When the load is purely resistive, then $\phi = 0$, $\cos \phi = 1$, and from Equation 33.31 we see that

$$\mathcal{P}_{av} = I_{rms}\,\Delta V_{rms}$$

We find that **no power losses are associated with pure capacitors and pure inductors in an AC circuit.** To see why this is true, let us first analyze the power in an AC circuit containing only a source and a capacitor. When the current begins to increase in one direction in an AC circuit, charge begins to accumulate on the capacitor, and a voltage appears across it. When this voltage reaches its maximum value, the energy stored in the capacitor is $\frac{1}{2}C(\Delta V_{max})^2$. However, this energy storage is only momentary. The capacitor is charged and discharged twice during each cycle: charge is delivered to the capacitor during two quarters of the cycle and is returned to the voltage source during the remaining two quarters. Therefore, **the average power supplied by the source is zero.** In other words, **no power losses occur in a capacitor in an AC circuit.**

Let us now consider the case of an inductor. When the current reaches its maximum value, the energy stored in the inductor is a maximum and is given by $\frac{1}{2}LI_{max}^2$. When the current begins to decrease in the circuit, this stored energy is returned to the source as the inductor attempts to maintain the current in the circuit.

Equation 33.31 shows that the power delivered by an AC source to any circuit depends on the phase—a result that has many interesting applications. For example, a factory that uses large motors in machines, generators, or transformers has a large inductive load (because of all the windings). To deliver greater power to such devices in the factory without using excessively high voltages, technicians introduce capacitance in the circuits to shift the phase.

Quick Quiz 33.8 An AC source drives an *RLC* circuit with a fixed voltage amplitude. If the driving frequency is ω_1, the circuit is more capacitive than inductive and the phase angle is $-10°$. If the driving frequency is ω_2, the circuit is more inductive than capacitive and the phase angle is $+10°$. The largest amount of power is delivered to the circuit at (a) ω_1 (b) ω_2 (c) The same amount of power is delivered at both frequencies.

Example 33.6 Average Power in an *RLC* Series Circuit

Calculate the average power delivered to the series *RLC* circuit described in Example 33.5.

Solution First, let us calculate the rms voltage and rms current, using the values of ΔV_{max} and I_{max} from Example 33.5:

$$\Delta V_{rms} = \frac{\Delta V_{max}}{\sqrt{2}} = \frac{150 \text{ V}}{\sqrt{2}} = 106 \text{ V}$$

$$I_{rms} = \frac{I_{max}}{\sqrt{2}} = \frac{0.292 \text{ A}}{\sqrt{2}} = 0.206 \text{ A}$$

Because $\phi = -34.0°$, the power factor is $\cos(-34.0°) = 0.829$; hence, the average power delivered is

$$\mathcal{P}_{av} = I_{rms} \Delta V_{rms} \cos \phi = (0.206 \text{ A})(106 \text{ V})(0.829)$$

$$= \boxed{18.1 \text{ W}}$$

We can obtain the same result using Equation 33.32.

33.7 Resonance in a Series *RLC* Circuit

A series *RLC* circuit is said to be **in resonance** when the current has its maximum value. In general, the rms current can be written

$$I_{rms} = \frac{\Delta V_{rms}}{Z} \tag{33.33}$$

where Z is the impedance. Substituting the expression for Z from Equation 33.25 into 33.33 gives

$$I_{rms} = \frac{\Delta V_{rms}}{\sqrt{R^2 + (X_L - X_C)^2}} \tag{33.34}$$

Because the impedance depends on the frequency of the source, the current in the *RLC* circuit also depends on the frequency. The frequency ω_0 at which $X_L - X_C = 0$ is called the **resonance frequency** of the circuit. To find ω_0, we use the condition $X_L = X_C$, from which we obtain $\omega_0 L = 1/\omega_0 C$, or

$$\omega_0 = \frac{1}{\sqrt{LC}} \tag{33.35}$$

Resonance frequency

This frequency also corresponds to the natural frequency of oscillation of an *LC* circuit (see Section 32.5). Therefore, the current in a series *RLC* circuit reaches its maximum value when the frequency of the applied voltage matches the natural oscillator frequency—which depends only on L and C. Furthermore, at this frequency the current is in phase with the applied voltage.

> **Quick Quiz 33.9** The impedance of a series RLC circuit at resonance is (a) larger than R (b) less than R (c) equal to R (d) impossible to determine.

A plot of rms current versus frequency for a series RLC circuit is shown in Figure 33.19a. The data assume a constant $\Delta V_{rms} = 5.0$ mV, that $L = 5.0\ \mu$H, and that $C = 2.0$ nF. The three curves correspond to three values of R. In each case, the current reaches its maximum value at the resonance frequency ω_0. Furthermore, the curves become narrower and taller as the resistance decreases.

By inspecting Equation 33.34, we must conclude that, when $R = 0$, the current becomes infinite at resonance. However, real circuits always have some resistance, which limits the value of the current to some finite value.

It is also interesting to calculate the average power as a function of frequency for a series RLC circuit. Using Equations 33.32, 33.33, and 33.25, we find that

$$\mathcal{P}_{av} = I_{rms}^2\,R = \frac{(\Delta V_{rms})^2}{Z^2}\,R = \frac{(\Delta V_{rms})^2\,R}{R^2 + (X_L - X_C)^2} \tag{33.36}$$

Because $X_L = \omega L$, $X_C = 1/\omega C$, and $\omega_0^2 = 1/LC$, we can express the term $(X_L - X_C)^2$ as

$$(X_L - X_C)^2 = \left(\omega L - \frac{1}{\omega C}\right)^2 = \frac{L^2}{\omega^2}(\omega^2 - \omega_0^2)^2$$

Using this result in Equation 33.36 gives

Average power as a function of frequency in an *RLC* circuit

$$\mathcal{P}_{av} = \frac{(\Delta V_{rms})^2\,R\omega^2}{R^2\omega^2 + L^2(\omega^2 - \omega_0^2)^2} \tag{33.37}$$

This expression shows that **at resonance, when $\omega = \omega_0$, the average power is a maximum** and has the value $(\Delta V_{rms})^2/R$. Figure 33.19b is a plot of average power

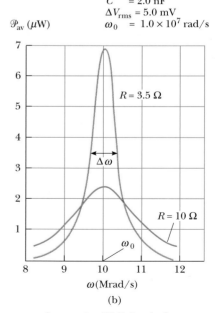

At the **Active Figures** link at http://www.pse6.com, you can adjust the resistance, the inductance, and the capacitance of the circuit in Figure 33.13a. You can then determine the current and power for a given frequency or sweep through the frequencies to generate resonance curves.

Active Figure 33.19 (a) The rms current versus frequency for a series RLC circuit, for three values of R. The current reaches its maximum value at the resonance frequency ω_0. (b) Average power delivered to the circuit versus frequency for the series RLC circuit, for two values of R.

versus frequency for two values of R in a series *RLC* circuit. As the resistance is made smaller, the curve becomes sharper in the vicinity of the resonance frequency. This curve sharpness is usually described by a dimensionless parameter known as the **quality factor,**[3] denoted by Q:

$$Q = \frac{\omega_0}{\Delta\omega}$$

Quality factor

where $\Delta\omega$ is the width of the curve measured between the two values of ω for which $\mathcal{P}_{av}$ has half its maximum value, called the *half-power points* (see Fig. 33.19b.) It is left as a problem (Problem 72) to show that the width at the half-power points has the value $\Delta\omega = R/L$, so

$$Q = \frac{\omega_0 L}{R} \qquad (33.38)$$

The curves plotted in Figure 33.20 show that a high-Q circuit responds to only a very narrow range of frequencies, whereas a low-Q circuit can detect a much broader range of frequencies. Typical values of Q in electronic circuits range from 10 to 100.

The receiving circuit of a radio is an important application of a resonant circuit. One tunes the radio to a particular station (which transmits an electromagnetic wave or signal of a specific frequency) by varying a capacitor, which changes the resonance frequency of the receiving circuit. When the resonance frequency of the circuit matches that of the incoming electromagnetic wave, the current in the receiving circuit increases. This signal caused by the incoming wave is then amplified and fed to a speaker. Because many signals are often present over a range of frequencies, it is important to design a high-Q circuit to eliminate unwanted signals. In this manner, stations whose frequencies are near but not equal to the resonance frequency give signals at the receiver that are negligibly small relative to the signal that matches the resonance frequency.

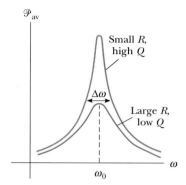

Figure 33.20 Average power versus frequency for a series *RLC* circuit. The width $\Delta\omega$ of each curve is measured between the two points where the power is half its maximum value. The power is a maximum at the resonance frequency ω_0.

Quick Quiz 33.10 An airport metal detector (see page 1003) is essentially a resonant circuit. The portal you step through is an inductor (a large loop of conducting wire) within the circuit. The frequency of the circuit is tuned to its resonance frequency when there is no metal in the inductor. Any metal on your body increases the effective inductance of the loop and changes the current in it. If you want the detector to detect a small metallic object, should the circuit have (a) a high quality factor or (b) a low quality factor?

Example 33.7 A Resonating Series *RLC* Circuit
Interactive

Consider a series *RLC* circuit for which $R = 150\ \Omega$, $L = 20.0$ mH, $\Delta V_{rms} = 20.0$ V, and $\omega = 5\ 000\ \text{s}^{-1}$. Determine the value of the capacitance for which the current is a maximum.

Solution The current has its maximum value at the resonance frequency ω_0, which should be made to match the "driving" frequency of $5\ 000\ \text{s}^{-1}$:

$$\omega_0 = 5.00 \times 10^3\ \text{s}^{-1} = \frac{1}{\sqrt{LC}}$$

$$C = \frac{1}{\omega_0^2 L} = \frac{1}{(5.00 \times 10^3\ \text{s}^{-1})^2 (20.0 \times 10^{-3}\ \text{H})}$$

$$= \boxed{2.00\ \mu\text{F}}$$

 At the Interactive Worked Example link at http://www.pse6.com, *you can explore resonance in an* **RLC** *circuit.*

---

[3] The quality factor is also defined as the ratio $2\pi E/\Delta E$ where E is the energy stored in the oscillating system and ΔE is the energy decrease per cycle of oscillation due to the resistance.

33.8 The Transformer and Power Transmission

As discussed in Section 27.6, when electric power is transmitted over great distances, it is economical to use a high voltage and a low current to minimize the I^2R loss in the transmission lines. Consequently, 350-kV lines are common, and in many areas even higher-voltage (765-kV) lines are used. At the receiving end of such lines, the consumer requires power at a low voltage (for safety and for efficiency in design). Therefore, a device is required that can change the alternating voltage and current without causing appreciable changes in the power delivered. The AC transformer is that device.

In its simplest form, the **AC transformer** consists of two coils of wire wound around a core of iron, as illustrated in Figure 33.21. (Compare this to Faraday's experiment in Figure 31.2.) The coil on the left, which is connected to the input alternating voltage source and has N_1 turns, is called the *primary winding* (or the *primary*). The coil on the right, consisting of N_2 turns and connected to a load resistor R, is called the *secondary winding* (or the *secondary*). The purpose of the iron core is to increase the magnetic flux through the coil and to provide a medium in which nearly all the magnetic field lines through one coil pass through the other coil. Eddy-current losses are reduced by using a laminated core. Iron is used as the core material because it is a soft ferromagnetic substance and hence reduces hysteresis losses. Transformation of energy to internal energy in the finite resistance of the coil wires is usually quite small. Typical transformers have power efficiencies from 90% to 99%. In the discussion that follows, we assume an *ideal transformer*, one in which the energy losses in the windings and core are zero.

First, let us consider what happens in the primary circuit. If we assume that the resistance of the primary is negligible relative to its inductive reactance, then the primary circuit is equivalent to a simple circuit consisting of an inductor connected to an AC source. Because the current is 90° out of phase with the voltage, the power factor cos ϕ is zero, and hence the average power delivered from the source to the primary circuit is zero. Faraday's law states that the voltage ΔV_1 across the primary is

Soft iron

Primary (input)   Secondary (output)

Figure 33.21 An ideal transformer consists of two coils wound on the same iron core. An alternating voltage ΔV_1 is applied to the primary coil, and the output voltage ΔV_2 is across the resistor of resistance R.

$$\Delta V_1 = -N_1 \frac{d\Phi_B}{dt} \tag{33.39}$$

where Φ_B is the magnetic flux through each turn. If we assume that all magnetic field lines remain within the iron core, the flux through each turn of the primary equals the flux through each turn of the secondary. Hence, the voltage across the secondary is

$$\Delta V_2 = -N_2 \frac{d\Phi_B}{dt} \tag{33.40}$$

Solving Equation 33.39 for $d\Phi_B/dt$ and substituting the result into Equation 33.40, we find that

$$\Delta V_2 = \frac{N_2}{N_1} \Delta V_1 \tag{33.41}$$

When $N_2 > N_1$, the output voltage ΔV_2 exceeds the input voltage ΔV_1. This setup is referred to as a *step-up transformer*. When $N_2 < N_1$, the output voltage is less than the input voltage, and we have a *step-down transformer*.

When the switch in the secondary circuit is closed, a current I_2 is induced in the secondary. If the load in the secondary circuit is a pure resistance, the induced current is in phase with the induced voltage. The power supplied to the secondary circuit must be provided by the AC source connected to the primary circuit, as shown in Figure 33.22. In an ideal transformer, where there are no losses, the power

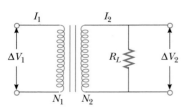

Figure 33.22 Circuit diagram for a transformer.

$I_1 \Delta V_1$ supplied by the source is equal to the power $I_2 \Delta V_2$ in the secondary circuit. That is,

$$I_1 \Delta V_1 = I_2 \Delta V_2 \qquad (33.42)$$

The value of the load resistance R_L determines the value of the secondary current because $I_2 = \Delta V_2/R_L$. Furthermore, the current in the primary is $I_1 = \Delta V_1/R_{eq}$, where

$$R_{eq} = \left(\frac{N_1}{N_2}\right)^2 R_L \qquad (33.43)$$

is the equivalent resistance of the load resistance when viewed from the primary side. From this analysis we see that a transformer may be used to match resistances between the primary circuit and the load. In this manner, maximum power transfer can be achieved between a given power source and the load resistance. For example, a transformer connected between the 1-kΩ output of an audio amplifier and an 8-Ω speaker ensures that as much of the audio signal as possible is transferred into the speaker. In stereo terminology, this is called *impedance matching*.

We can now also understand why transformers are useful for transmitting power over long distances. Because the generator voltage is stepped up, the current in the transmission line is reduced, and hence I^2R losses are reduced. In practice, the voltage is stepped up to around 230 000 V at the generating station, stepped down to around 20 000 V at a distributing station, then to 4 000 V for delivery to residential areas, and finally to 120–240 V at the customer's site.

There is a practical upper limit to the voltages that can be used in transmission lines. Excessive voltages could ionize the air surrounding the transmission lines, which could result in a conducting path to ground or to other objects in the vicinity. This, of course, would present a serious hazard to any living creatures. For this reason, a long string of insulators is used to keep high-voltage wires away from their supporting metal towers. Other insulators are used to maintain separation between wires.

Many common household electronic devices require low voltages to operate properly. A small transformer that plugs directly into the wall, like the one illustrated in Figure 33.23, can provide the proper voltage. The photograph shows the two windings wrapped around a common iron core that is found inside all these little "black boxes." This particular transformer converts the 120-V AC in the wall socket to 12.5-V AC. (Can you determine the ratio of the numbers of turns in the two coils?) Some black boxes also make use of diodes to convert the alternating current to direct current. (See Section 33.9.)

Nikola Tesla
American Physicist (1856–1943)

Tesla was born in Croatia but spent most of his professional life as an inventor in the United States. He was a key figure in the development of alternating current electricity, high-voltage transformers, and the transport of electrical power using AC transmission lines. Tesla's viewpoint was at odds with the ideas of Thomas Edison, who committed himself to the use of direct current in power transmission. Tesla's AC approach won out. *(UPI/CORBIS)*

Figure 33.23 The primary winding in this transformer is directly attached to the prongs of the plug. The secondary winding is connected to the power cord on the right, which runs to an electronic device. Many of these power-supply transformers also convert alternating current to direct current.

This transformer is smaller than the one in the opening photograph for this chapter. In addition, it is a step-down transformer. It drops the voltage from 4 000 V to 240 V for delivery to a group of residences.

Example 33.8   The Economics of AC Power

An electricity-generating station needs to deliver energy at a rate of 20 MW to a city 1.0 km away.

(A) If the resistance of the wires is 2.0 Ω and the energy costs about 10¢/kWh, estimate what it costs the utility company for the energy converted to internal energy in the wires during one day. A common voltage for commercial power generators is 22 kV, but a step-up transformer is used to boost the voltage to 230 kV before transmission.

Solution Conceptualize by noting that the resistance of the wires is in series with the resistance representing the load (homes and businesses). Thus, there will be a voltage drop in the wires, which means that some of the transmitted energy is converted to internal energy in the wires and never reaches the load. Because this is an estimate, let us categorize this as a problem in which the power factor is equal to 1. To analyze the problem, we begin by calculating I_{rms} from Equation 33.31:

$$I_{rms} = \frac{\mathcal{P}_{av}}{\Delta V_{rms}} = \frac{20 \times 10^6 \text{ W}}{230 \times 10^3 \text{ V}} = 87 \text{ A}$$

Now, we determine the rate at which energy is delivered to the resistance in the wires from Equation 33.32:

$$\mathcal{P}_{av} = I_{rms}^2 R = (87 \text{ A})^2 (2.0 \text{ Ω}) = 15 \text{ kW}$$

Over the course of a day, the energy loss due to the resistance of the wires is (15 kW)(24 h) = 360 kWh, at a cost of $36.

(B) Repeat the calculation for the situation in which the power plant delivers the energy at its original voltage of 22 kV.

Solution Again using Equation 33.31, we find

$$I_{rms} = \frac{\mathcal{P}_{av}}{\Delta V_{rms}} = \frac{20 \times 10^6 \text{ W}}{22 \times 10^3 \text{ V}} = 910 \text{ A}$$

and, from Equation 33.32,

$$\mathcal{P}_{av} = I_{rms}^2 R = (910 \text{ A})^2 (2.0 \text{ Ω}) = 1.7 \times 10^3 \text{ kW}$$

Cost per day = $(1.7 \times 10^3 \text{ kW})(24 \text{ h})(\$0.10/\text{kWh})$

$$= \quad \$4\,100$$

To finalize the example, note the tremendous savings that are possible through the use of transformers and high-voltage transmission lines. This, in combination with the efficiency of using alternating current to operate motors, led to the universal adoption of alternating current instead of direct current for commercial power grids.

33.9   Rectifiers and Filters

Portable electronic devices such as radios and compact disc (CD) players are often powered by direct current supplied by batteries. Many devices come with AC–DC converters such as that in Figure 33.23. Such a converter contains a transformer that steps the voltage down from 120 V to typically 9 V and a circuit that converts alternating current to direct current. The process of converting alternating current to direct current is called **rectification,** and the converting device is called a **rectifier.**

The most important element in a rectifier circuit is a **diode,** a circuit element that conducts current in one direction but not the other. Most diodes used in modern electronics are semiconductor devices. The circuit symbol for a diode is ──▶︎├── , where the arrow indicates the direction of the current in the diode. A diode has low resistance to current in one direction (the direction of the arrow) and high resistance to current in the opposite direction. We can understand how a diode rectifies a current by considering Figure 33.24a, which shows a diode and a resistor connected to the secondary of a transformer. The transformer reduces the voltage from 120-V AC to the lower voltage that is needed for the device having a resistance R (the load resistance). Because the diode conducts current in only one direction, the alternating current in the load resistor is reduced to the form shown by the solid curve in Figure 33.24b. The diode conducts current only when the side of the symbol containing the arrowhead has a positive potential relative to the other side. In this situation, the diode acts as a *half-wave rectifier* because current is present in the circuit during only half of each cycle.

When a capacitor is added to the circuit, as shown by the dashed lines and the capacitor symbol in Figure 33.24a, the circuit is a simple DC power supply. The time variation in the current in the load resistor (the dashed curve in Fig. 33.24b) is close to

(a)

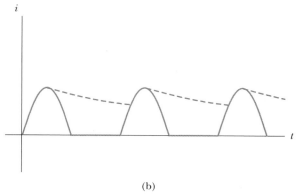

(b)

Figure 33.24 (a) A half-wave rectifier with an optional filter capacitor. (b) Current versus time in the resistor. The solid curve represents the current with no filter capacitor, and the dashed curve is the current when the circuit includes the capacitor.

being zero, as determined by the RC time constant of the circuit. As the current in the circuit begins to rise at $t = 0$ in Figure 33.24b, the capacitor charges up. When the current begins to fall, however, the capacitor discharges through the resistor, so that the current in the resistor does not fall as fast as the current from the transformer.

The RC circuit in Figure 33.24a is one example of a **filter circuit,** which is used to smooth out or eliminate a time-varying signal. For example, radios are usually powered by a 60-Hz alternating voltage. After rectification, the voltage still contains a small AC component at 60 Hz (sometimes called *ripple*), which must be filtered. By "filtered," we mean that the 60-Hz ripple must be reduced to a value much less than that of the audio signal to be amplified, because without filtering, the resulting audio signal includes an annoying hum at 60 Hz.

We can also design filters that will respond differently to different frequencies. Consider the simple series RC circuit shown in Figure 33.25a. The input voltage is across the series combination of the two elements. The output is the voltage across the resistor. A plot of the ratio of the output voltage to the input voltage as a function of the logarithm of angular frequency (see Fig. 33.25b) shows that at low frequencies ΔV_{out} is much smaller than ΔV_{in}, whereas at high frequencies the two voltages are equal. Because the circuit

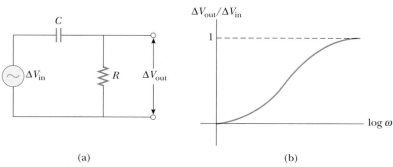

(a)   (b)

Active Figure 33.25 (a) A simple RC high-pass filter. (b) Ratio of output voltage to input voltage for an RC high-pass filter as a function of the angular frequency of the AC source.

At the Active Figures link at http://www.pse6.com, you can adjust the resistance and the capacitance of the circuit in part (a). You can then determine the output voltage for a given frequency or sweep through the frequencies to generate a curve like that in part (b).

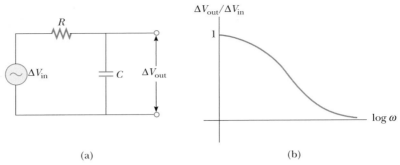

(a) (b)

Active Figure 33.26 (a) A simple *RC* low-pass filter. (b) Ratio of output voltage to input voltage for an *RC* low-pass filter as a function of the angular frequency of the AC source.

preferentially passes signals of higher frequency while blocking low-frequency signals, the circuit is called an *RC* high-pass filter. (See Problem 51 for an analysis of this filter.)

Physically, a high-pass filter works because a capacitor "blocks out" direct current and AC current at low frequencies. At low frequencies, the capacitive reactance is large and much of the applied voltage appears across the capacitor rather than across the output resistor. As the frequency increases, the capacitive reactance drops and more of the applied voltage appears across the resistor.

Now consider the circuit shown in Figure 33.26a, where we have interchanged the resistor and capacitor and the output voltage is taken across the capacitor. At low frequencies, the reactance of the capacitor and the voltage across the capacitor is high. As the frequency increases, the voltage across the capacitor drops. Thus, this is an *RC* low-pass filter. The ratio of output voltage to input voltage (see Problem 52), plotted as a function of the logarithm of ω in Figure 33.26b, shows this behavior.

You may be familiar with crossover networks, which are an important part of the speaker systems for high-fidelity audio systems. These networks use low-pass filters to direct low frequencies to a special type of speaker, the "woofer," which is designed to reproduce the low notes accurately. The high frequencies are sent to the "tweeter" speaker.

Quick Quiz 33.11 Suppose you are designing a high-fidelity system containing both large loudspeakers (woofers) and small loudspeakers (tweeters). If you wish to deliver low-frequency signals to a woofer, what device would you place in series with it? (a) an inductor (b) a capacitor (c) a resistor. If you wish to deliver high-frequency signals to a tweeter, what device would you place in series with it? (d) an inductor (e) a capacitor (f) a resistor.

SUMMARY

If an AC circuit consists of a source and a resistor, the current is in phase with the voltage. That is, the current and voltage reach their maximum values at the same time.

The **rms current** and **rms voltage** in an AC circuit in which the voltages and current vary sinusoidally are given by the expressions

$$I_{rms} = \frac{I_{max}}{\sqrt{2}} = 0.707 I_{max} \tag{33.4}$$

$$\Delta V_{rms} = \frac{\Delta V_{max}}{\sqrt{2}} = 0.707\, \Delta V_{max} \tag{33.5}$$

where I_{max} and ΔV_{max} are the maximum values.

If an AC circuit consists of a source and an inductor, the current lags behind the voltage by 90°. That is, the voltage reaches its maximum value one quarter of a period before the current reaches its maximum value.

If an AC circuit consists of a source and a capacitor, the current leads the voltage by 90°. That is, the current reaches its maximum value one quarter of a period before the voltage reaches its maximum value.

In AC circuits that contain inductors and capacitors, it is useful to define the **inductive reactance** X_L and the **capacitive reactance** X_C as

$$X_L \equiv \omega L \tag{33.10}$$

$$X_C \equiv \frac{1}{\omega C} \tag{33.18}$$

where ω is the angular frequency of the AC source. The SI unit of reactance is the ohm.

The **impedance** Z of an RLC series AC circuit is

$$Z \equiv \sqrt{R^2 + (X_L - X_C)^2} \tag{33.25}$$

This expression illustrates that we cannot simply add the resistance and reactances in a circuit. We must account for the fact that the applied voltage and current are out of phase, with the **phase angle** ϕ between the current and voltage being

$$\phi = \tan^{-1}\left(\frac{X_L - X_C}{R}\right) \tag{33.27}$$

The sign of ϕ can be positive or negative, depending on whether X_L is greater or less than X_C. The phase angle is zero when $X_L = X_C$.

The **average power** delivered by the source in an RLC circuit is

$$\mathcal{P}_{av} = I_{rms}\,\Delta V_{rms}\,\cos\phi \tag{33.31}$$

An equivalent expression for the average power is

$$\mathcal{P}_{av} = I_{rms}^2\,R \tag{33.32}$$

The average power delivered by the source results in increasing internal energy in the resistor. No power loss occurs in an ideal inductor or capacitor.

The rms current in a series RLC circuit is

$$I_{rms} = \frac{\Delta V_{rms}}{\sqrt{R^2 + (X_L - X_C)^2}} \tag{33.34}$$

A series RLC circuit is in resonance when the inductive reactance equals the capacitive reactance. When this condition is met, the current given by Equation 33.34 reaches its maximum value. The **resonance frequency** ω_0 of the circuit is

$$\omega_0 = \frac{1}{\sqrt{LC}} \tag{33.35}$$

The current in a series RLC circuit reaches its maximum value when the frequency of the source equals ω_0—that is, when the "driving" frequency matches the resonance frequency.

Transformers allow for easy changes in alternating voltage. Because energy (and therefore power) are conserved, we can write

$$I_1\,\Delta V_1 = I_2\,\Delta V_2 \tag{33.42}$$

to relate the currents and voltages in the primary and secondary windings of a transformer.

QUESTIONS

1. How can the average value of a current be zero and yet the square root of the average squared current not be zero?

2. What is the time average of the "square-wave" potential shown in Figure Q33.2? What is its rms voltage?

Figure Q33.2

3. Do AC ammeters and voltmeters read maximum, rms, or average values?

4. In the clearest terms you can, explain the statement, "The voltage across an inductor leads the current by 90°."

5. Some fluorescent lights flicker on and off 120 times every second. Explain what causes this. Why can't you see it happening?

6. Why does a capacitor act as a short circuit at high frequencies? Why does it act as an open circuit at low frequencies?

7. Explain how the mnemonic "ELI the ICE man" can be used to recall whether current leads voltage or voltage leads current in *RLC* circuits. Note that E represents emf $\mathcal{E}$.

8. Why is the sum of the maximum voltages across each of the elements in a series *RLC* circuit usually greater than the maximum applied voltage? Doesn't this violate Kirchhoff's loop rule?

9. Does the phase angle depend on frequency? What is the phase angle when the inductive reactance equals the capacitive reactance?

10. In a series *RLC* circuit, what is the possible range of values for the phase angle?

11. If the frequency is doubled in a series *RLC* circuit, what happens to the resistance, the inductive reactance, and the capacitive reactance?

12. Explain why the average power delivered to an *RLC* circuit by the source depends on the phase angle between the current and applied voltage.

13. As shown in Figure 7.5a, a person pulls a vacuum cleaner at speed v across a horizontal floor, exerting on it a force of magnitude F directed upward at an angle θ with the horizontal. At what rate is the person doing work on the cleaner? State as completely as you can the analogy between power in this situation and in an electric circuit.

14. A particular experiment requires a beam of light of very stable intensity. Why would an AC voltage be unsuitable for powering the light source?

15. Do some research to answer these questions: Who invented the metal detector? Why? Did it work?

16. What is the advantage of transmitting power at high voltages?

17. What determines the maximum voltage that can be used on a transmission line?

18. Will a transformer operate if a battery is used for the input voltage across the primary? Explain.

19. Someone argues that high-voltage power lines actually waste more energy. He points out that the rate at which internal energy is produced in a wire is given by $(\Delta V)^2/R$, where R is the resistance of the wire. Therefore, the higher the voltage, the higher the energy waste. What if anything is wrong with his argument?

20. Explain how the quality factor is related to the response characteristics of a radio receiver. Which variable most strongly influences the quality factor?

21. Why are the primary and secondary coils of a transformer wrapped on an iron core that passes through both coils?

22. With reference to Figure Q33.22, explain why the capacitor prevents a DC signal from passing between A and B, yet allows an AC signal to pass from A to B. (The circuits are said to be capacitively coupled.)

Figure Q33.22

PROBLEMS

1, 2, 3 = straightforward, intermediate, challenging   ☐ = full solution available in the *Student Solutions Manual and Study Guide*

🌀 = coached solution with hints available at http://www.pse6.com   💻 = computer useful in solving problem

▨ = paired numerical and symbolic problems

Note: Assume all AC voltages and currents are sinusoidal, unless stated otherwise.

Section 33.1   AC Sources

Section 33.2   Resistors in an AC Circuit

1. The rms output voltage of an AC source is 200 V and the operating frequency is 100 Hz. Write the equation giving the output voltage as a function of time.

2. (a) What is the resistance of a lightbulb that uses an average power of 75.0 W when connected to a 60.0-Hz power source having a maximum voltage of 170 V? (b) **What If?** What is the resistance of a 100-W bulb?

3. An AC power supply produces a maximum voltage $\Delta V_{max} = 100$ V. This power supply is connected to a 24.0-Ω resistor, and the current and resistor voltage are measured with an ideal AC ammeter and voltmeter, as shown in Figure P33.3. What does each meter read? Note that an ideal ammeter has zero resistance and that an ideal voltmeter has infinite resistance.

Figure P33.3

4. In the simple AC circuit shown in Figure 33.2, $R = 70.0$ Ω and $\Delta v = \Delta V_{max} \sin \omega t$. (a) If $\Delta v_R = 0.250 \, \Delta V_{max}$ for the first time at $t = 0.010\,0$ s, what is the angular frequency of the source? (b) What is the next value of t for which $\Delta v_R = 0.250 \, \Delta V_{max}$?

5. The current in the circuit shown in Figure 33.2 equals 60.0% of the peak current at $t = 7.00$ ms. What is the smallest frequency of the source that gives this current?

6. Figure P33.6 shows three lamps connected to a 120-V AC (rms) household supply voltage. Lamps 1 and 2 have 150-W bulbs; lamp 3 has a 100-W bulb. Find the rms current and resistance of each bulb.

Figure P33.6

7. An audio amplifier, represented by the AC source and resistor in Figure P33.7, delivers to the speaker alternating voltage at audio frequencies. If the source voltage has an amplitude of 15.0 V, $R = 8.20$ Ω, and the speaker is equivalent to a resistance of 10.4 Ω, what is the time-averaged power transferred to it?

Figure P33.7

Section 33.3   Inductors in an AC Circuit

8. An inductor is connected to a 20.0-Hz power supply that produces a 50.0-V rms voltage. What inductance is needed to keep the instantaneous current in the circuit below 80.0 mA?

9. In a purely inductive AC circuit, as shown in Figure 33.6, $\Delta V_{max} = 100$ V. (a) The maximum current is 7.50 A at 50.0 Hz. Calculate the inductance L. (b) **What If?** At what angular frequency ω is the maximum current 2.50 A?

10. An inductor has a 54.0-Ω reactance at 60.0 Hz. What is the maximum current if this inductor is connected to a 50.0-Hz source that produces a 100-V rms voltage?

11. 🌀 For the circuit shown in Figure 33.6, $\Delta V_{max} = 80.0$ V, $\omega = 65.0\pi$ rad/s, and $L = 70.0$ mH. Calculate the current in the inductor at $t = 15.5$ ms.

12. A 20.0-mH inductor is connected to a standard electrical outlet ($\Delta V_{rms} = 120$ V; $f = 60.0$ Hz). Determine the energy stored in the inductor at $t = (1/180)$ s, assuming that this energy is zero at $t = 0$.

13. **Review problem.** Determine the maximum magnetic flux through an inductor connected to a standard electrical outlet ($\Delta V_{rms} = 120$ V, $f = 60.0$ Hz).

Section 33.4   Capacitors in an AC Circuit

14. (a) For what frequencies does a 22.0-μF capacitor have a reactance below 175 Ω? (b) **What If?** Over this same frequency range, what is the reactance of a 44.0-μF capacitor?

15. What is the maximum current in a 2.20-μF capacitor when it is connected across (a) a North American electrical outlet having $\Delta V_{rms} = 120$ V, $f = 60.0$ Hz, and (b) **What If?** a European electrical outlet having $\Delta V_{rms} = 240$ V, $f = 50.0$ Hz?

16. A capacitor C is connected to a power supply that operates at a frequency f and produces an rms voltage ΔV. What is the maximum charge that appears on either of the capacitor plates?

17. What maximum current is delivered by an AC source with $\Delta V_{max} = 48.0$ V and $f = 90.0$ Hz when connected across a 3.70-μF capacitor?

18. A 1.00-mF capacitor is connected to a standard electrical outlet ($\Delta V_{rms} = 120$ V; $f = 60.0$ Hz). Determine the current in the capacitor at $t = (1/180)$ s, assuming that at $t = 0$, the energy stored in the capacitor is zero.

Section 33.5   The *RLC* Series Circuit

19. An inductor ($L = 400$ mH), a capacitor ($C = 4.43$ μF), and a resistor ($R = 500$ Ω) are connected in series. A 50.0-Hz AC source produces a peak current of 250 mA in the circuit. (a) Calculate the required peak voltage ΔV_{max}. (b) Determine the phase angle by which the current leads or lags the applied voltage.

20. At what frequency does the inductive reactance of a 57.0-μH inductor equal the capacitive reactance of a 57.0-μF capacitor?

21. A series AC circuit contains the following components: $R = 150$ Ω, $L = 250$ mH, $C = 2.00$ μF and a source with $\Delta V_{max} = 210$ V operating at 50.0 Hz. Calculate the (a) inductive reactance, (b) capacitive reactance, (c) impedance, (d) maximum current, and (e) phase angle between current and source voltage.

22. A sinusoidal voltage $\Delta v(t) = (40.0$ V$)$ sin$(100t)$ is applied to a series *RLC* circuit with $L = 160$ mH, $C = 99.0$ μF, and $R = 68.0$ Ω. (a) What is the impedance of the circuit? (b) What is the maximum current? (c) Determine the numerical values for I_{max}, ω, and ϕ in the equation $i(t) = I_{max}$ sin$(\omega t - \phi)$.

23. An *RLC* circuit consists of a 150-Ω resistor, a 21.0-μF capacitor, and a 460-mH inductor, connected in series with a 120-V, 60.0-Hz power supply. (a) What is the phase angle between the current and the applied voltage? (b) Which reaches its maximum earlier, the current or the voltage?

24. Four circuit elements—a capacitor, an inductor, a resistor, and an AC source—are connected together in various ways. First the capacitor is connected to the source, and the rms current is found to be 25.1 mA. The capacitor is disconnected and discharged, and then connected in series with the resistor and the source, making the rms current 15.7 mA. The circuit is disconnected and the capacitor discharged. The capacitor is then connected in series with the inductor and the source, making the rms current 68.2 mA. After the circuit is disconnected and the capacitor discharged, all four circuit elements are connected together in a series loop. What is the rms current in the circuit?

25. A person is working near the secondary of a transformer, as shown in Figure P33.25. The primary voltage is 120 V at 60.0 Hz. The capacitance C_s, which is the stray capacitance between the hand and the secondary winding, is 20.0 pF. Assuming the person has a body resistance to ground $R_b = 50.0$ kΩ, determine the rms voltage across the body. (*Suggestion:* Redraw the circuit with the secondary of the transformer as a simple AC source.)

5 000 V

R_b

Figure P33.25

26. An AC source with $\Delta V_{max} = 150$ V and $f = 50.0$ Hz is connected between points a and d in Figure P33.26. Calculate the maximum voltages between points (a) a and b, (b) b and c, (c) c and d, and (d) b and d.

40.0 Ω 185 mH   65.0 μF

Figure P33.26   Problems 26 and 68.

27. Draw to scale a phasor diagram showing Z, X_L, X_C, and ϕ for an AC series circuit for which $R = 300$ Ω, $C = 11.0$ μF, $L = 0.200$ H, and $f = (500/\pi)$ Hz.

28. In an *RLC* series circuit that includes a source of alternating current operating at fixed frequency and voltage, the resistance R is equal to the inductive reactance. If the plate separation of the capacitor is reduced to half of its original value, the current in the circuit doubles. Find the initial capacitive reactance in terms of R.

29. A coil of resistance 35.0 Ω and inductance 20.5 H is in series with a capacitor and a 200-V (rms), 100-Hz source. The rms current in the circuit is 4.00 A. (a) Calculate the capacitance in the circuit. (b) What is ΔV_{rms} across the coil?

Section 33.6   Power in an AC Circuit

30. The voltage source in Figure P33.30 has an output of $\Delta V_{rms} = 100$ V at an angular frequency of $\omega = 1\,000$ rad/s. Determine (a) the current in the circuit and (b) the power supplied by the source. (c) Show that the power delivered to the resistor is equal to the power supplied by the source.

50.0 mH

ΔV ⊗ 40.0 Ω

50.0 μF

Figure P33.30

31. ✍ An AC voltage of the form $\Delta v = (100 \text{ V}) \sin(1\,000t)$ is applied to a series RLC circuit. Assume the resistance is 400 Ω, the capacitance is 5.00 μF, and the inductance is 0.500 H. Find the average power delivered to the circuit.

32. A series RLC circuit has a resistance of 45.0 Ω and an impedance of 75.0 Ω. What average power is delivered to this circuit when $\Delta V_{\text{rms}} = 210$ V?

33. In a certain series RLC circuit, $I_{\text{rms}} = 9.00$ A, $\Delta V_{\text{rms}} = 180$ V, and the current leads the voltage by 37.0°. (a) What is the total resistance of the circuit? (b) Calculate the reactance of the circuit ($X_L - X_C$).

34. Suppose you manage a factory that uses many electric motors. The motors create a large inductive load to the electric power line, as well as a resistive load. The electric company builds an extra-heavy distribution line to supply you with a component of current that is 90° out of phase with the voltage, as well as with current in phase with the voltage. The electric company charges you an extra fee for "reactive volt-amps," in addition to the amount you pay for the energy you use. You can avoid the extra fee by installing a capacitor between the power line and your factory. The following problem models this solution.

In an RL circuit, a 120-V (rms), 60.0-Hz source is in series with a 25.0-mH inductor and a 20.0-Ω resistor. What are (a) the rms current and (b) the power factor? (c) What capacitor must be added in series to make the power factor 1? (d) To what value can the supply voltage be reduced, if the power supplied is to be the same as before the capacitor was installed?

35. Suppose power $\mathscr{P}$ is to be transmitted over a distance d at a voltage ΔV with only 1.00% loss. Copper wire of what diameter should be used for each of the two conductors of the transmission line? Assume the current density in the conductors is uniform.

36. A diode is a device that allows current to be carried in only one direction (the direction indicated by the arrowhead in its circuit symbol). Find in terms of ΔV and R the average power delivered to the diode circuit of Figure P33.36.

Figure P33.36

Section 33.7 Resonance in a Series *RLC* Circuit

37. An RLC circuit is used in a radio to tune into an FM station broadcasting at 99.7 MHz. The resistance in the circuit is 12.0 Ω, and the inductance is 1.40 μH. What capacitance should be used?

38. The tuning circuit of an AM radio contains an LC combination. The inductance is 0.200 mH, and the capacitor is variable, so that the circuit can resonate at any frequency between 550 kHz and 1 650 kHz. Find the range of values required for C.

39. A radar transmitter contains an LC circuit oscillating at 1.00×10^{10} Hz. (a) What capacitance will resonate with a one-turn loop of inductance 400 pH at this frequency? (b) If the capacitor has square parallel plates separated by 1.00 mm of air, what should the edge length of the plates be? (c) What is the common reactance of the loop and capacitor at resonance?

40. A series RLC circuit has components with following values: $L = 20.0$ mH, $C = 100$ nF, $R = 20.0$ Ω, and $\Delta V_{\text{max}} = 100$ V, with $\Delta v = \Delta V_{\text{max}} \sin \omega t$. Find (a) the resonant frequency, (b) the amplitude of the current at the resonant frequency, (c) the Q of the circuit, and (d) the amplitude of the voltage across the inductor at resonance.

41. A 10.0-Ω resistor, 10.0-mH inductor, and 100-μF capacitor are connected in series to a 50.0-V (rms) source having variable frequency. Find the energy that is delivered to the circuit during one period if the operating frequency is twice the resonance frequency.

42. A resistor R, inductor L, and capacitor C are connected in series to an AC source of rms voltage ΔV and variable frequency. Find the energy that is delivered to the circuit during one period if the operating frequency is twice the resonance frequency.

43. Compute the quality factor for the circuits described in Problems 22 and 23. Which circuit has the sharper resonance?

Section 33.8 The Transformer and Power Transmission

44. A step-down transformer is used for recharging the batteries of portable devices such as tape players. The turns ratio inside the transformer is 13:1 and it is used with 120-V (rms) household service. If a particular ideal transformer draws 0.350 A from the house outlet, what are (a) the voltage and (b) the current supplied to a tape player from the transformer? (c) How much power is delivered?

45. A transformer has $N_1 = 350$ turns and $N_2 = 2\,000$ turns. If the input voltage is $\Delta v(t) = (170 \text{ V}) \cos \omega t$, what rms voltage is developed across the secondary coil?

46. A step-up transformer is designed to have an output voltage of 2 200 V (rms) when the primary is connected across a 110-V (rms) source. (a) If the primary winding has 80 turns, how many turns are required on the secondary? (b) If a load resistor across the secondary draws a current of 1.50 A, what is the current in the primary, assuming ideal conditions? (c) **What If?** If the transformer actually has an efficiency of 95.0%, what is the current in the primary when the secondary current is 1.20 A?

47. In the transformer shown in Figure P33.47, the load resistor is 50.0 Ω. The turns ratio N_1:N_2 is 5:2, and the source voltage is 80.0 V (rms). If a voltmeter across the load measures 25.0 V (rms), what is the source resistance R_s?

Figure P33.47

48. The secondary voltage of an ignition transformer in a furnace is 10.0 kV. When the primary operates at an rms voltage of 120 V, the primary impedance is 24.0 Ω and the transformer is 90.0% efficient. (a) What turns ratio is required? What are (b) the current in the secondary and (c) the impedance in the secondary?

49. A transmission line that has a resistance per unit length of 4.50×10^{-4} Ω/m is to be used to transmit 5.00 MW over 400 miles $(6.44 \times 10^5$ m). The output voltage of the generator is 4.50 kV. (a) What is the line loss if a transformer is used to step up the voltage to 500 kV? (b) What fraction of the input power is lost to the line under these circumstances? (c) **What If?** What difficulties would be encountered in attempting to transmit the 5.00 MW at the generator voltage of 4.50 kV?

Section 33.9 Rectifiers and Filters

50. One particular plug-in power supply for a radio looks similar to the one shown in Figure 33.23 and is marked with the following information: Input 120 V AC 8 W Output 9 V DC 300 mA. Assume that these values are accurate to two digits. (a) Find the energy efficiency of the device when the radio is operating. (b) At what rate does the device produce wasted energy when the radio is operating? (c) Suppose that the input power to the transformer is 8.0 W when the radio is switched off and that energy costs $0.135/kWh from the electric company. Find the cost of having six such transformers around the house, plugged in for thirty-one days.

51. Consider the filter circuit shown in Figure 33.25a. (a) Show that the ratio of the output voltage to the input voltage is

$$\frac{\Delta V_{out}}{\Delta V_{in}} = \frac{R}{\sqrt{R^2 + \left(\dfrac{1}{\omega C}\right)^2}}$$

(b) What value does this ratio approach as the frequency decreases toward zero? What value does this ratio approach as the frequency increases without limit? (c) At what frequency is the ratio equal to one half?

52. Consider the filter circuit shown in Figure 33.26a. (a) Show that the ratio of the output voltage to the input

voltage is

$$\frac{\Delta V_{out}}{\Delta V_{in}} = \frac{1/\omega C}{\sqrt{R^2 + \left(\dfrac{1}{\omega C}\right)^2}}$$

(b) What value does this ratio approach as the frequency decreases toward zero? What value does this ratio approach as the frequency increases without limit? (c) At what frequency is the ratio equal to one half?

53. The RC high-pass filter shown in Figure 33.25 has a resistance $R = 0.500$ Ω. (a) What capacitance gives an output signal that has half the amplitude of a 300-Hz input signal? (b) What is the ratio $(\Delta V_{out}/\Delta V_{in})$ for a 600-Hz signal? You may use the result of Problem 51.

54. The RC low-pass filter shown in Figure 33.26 has a resistance $R = 90.0$ Ω and a capacitance $C = 8.00$ nF. Calculate the ratio $(\Delta V_{out}/\Delta V_{in})$ for an input frequency of (a) 600 Hz and (b) 600 kHz. You may use the result of Problem 52.

55. The resistor in Figure P33.55 represents the midrange speaker in a three-speaker system. Assume its resistance to be constant at 8.00 Ω. The source represents an audio amplifier producing signals of uniform amplitude $\Delta V_{in} = 10.0$ V at all audio frequencies. The inductor and capacitor are to function as a bandpass filter with $\Delta V_{out}/\Delta V_{in} = 1/2$ at 200 Hz and at 4 000 Hz. (a) Determine the required values of L and C. (b) Find the maximum value of the ratio $\Delta V_{out}/\Delta V_{in}$. (c) Find the frequency f_0 at which the ratio has its maximum value. (d) Find the phase shift between ΔV_{in} and ΔV_{out} at 200 Hz, at f_0, and at 4 000 Hz. (e) Find the average power transferred to the speaker at 200 Hz, at f_0, and at 4 000 Hz. (f) Treating the filter as a resonant circuit, find its quality factor.

Figure P33.55

Additional Problems

56. Show that the rms value for the sawtooth voltage shown in Figure P33.56 is $\Delta V_{max}/\sqrt{3}$.

Figure P33.56

57. A series RLC circuit consists of an 8.00-Ω resistor, a 5.00-μF capacitor, and a 50.0-mH inductor. A variable

frequency source applies an emf of 400 V (rms) across the combination. Determine the power delivered to the circuit when the frequency is equal to half the resonance frequency.

58. A capacitor, a coil, and two resistors of equal resistance are arranged in an AC circuit, as shown in Figure P33.58. An AC source provides an emf of 20.0 V (rms) at a frequency of 60.0 Hz. When the double-throw switch S is open, as shown in the figure, the rms current is 183 mA. When the switch is closed in position 1, the rms current is 298 mA. When the switch is closed in position 2, the rms current is 137 mA. Determine the values of R, C, and L. Is more than one set of values possible?

Figure P33.58

59. To determine the inductance of a coil used in a research project, a student first connects the coil to a 12.0-V battery and measures a current of 0.630 A. The student then connects the coil to a 24.0-V (rms), 60.0-Hz generator and measures an rms current of 0.570 A. What is the inductance?

60. **Review problem.** One insulated conductor from a household extension cord has mass per length 19.0 g/m. A section of this conductor is held under tension between two clamps. A subsection is located in a region of magnetic field of magnitude 15.3 mT perpendicular to the length of the cord. The wire carries an AC current of 9.00 A at 60.0 Hz. Determine some combination of values for the distance between the clamps and the tension in the cord so that the cord can vibrate in the lowest-frequency standing-wave vibrational state.

61. In Figure P33.61, find the rms current delivered by the 45.0-V (rms) power supply when (a) the frequency is very large and (b) the frequency is very small.

Figure P33.61

62. In the circuit shown in Figure P33.62, assume that all parameters except for C are given. (a) Find the current as a function of time. (b) Find the power delivered to the circuit. (c) Find the current as a function of time after *only* switch 1 is opened. (d) After switch 2 is *also* opened, the current and voltage are in phase. Find the capacitance C. (e) Find the impedance of the circuit when both switches are open. (f) Find the maximum energy stored in the capacitor during oscillations. (g) Find the maximum energy stored in the inductor during oscillations. (h) Now the frequency of the voltage source is doubled. Find the phase difference between the current and the voltage. (i) Find the frequency that makes the inductive reactance half the capacitive reactance.

$$\Delta v(t) = \Delta V_{max} \cos \omega t$$

Figure P33.62

63. An 80.0-Ω resistor and a 200-mH inductor are connected in *parallel* across a 100-V (rms), 60.0-Hz source. (a) What is the rms current in the resistor? (b) By what angle does the total current lead or lag behind the voltage?

64. Make an order-of-magnitude estimate of the electric current that the electric company delivers to a town (Figure P33.64) from a remote generating station. State the data you measure or estimate. If you wish, you may consider a suburban bedroom community of 20 000 people.

Eddie Hironaka/Getty Images

Figure P33.64

65. Consider a series RLC circuit having the following circuit parameters: $R = 200$ Ω, $L = 663$ mH, and $C = 26.5$ μF. The applied voltage has an amplitude of 50.0 V and a frequency of 60.0 Hz. Find the following amplitudes: (a) The current I_{max}, including its phase constant ϕ relative to the applied voltage Δv, (b) the voltage ΔV_R

across the resistor and its phase relative to the current, (c) the voltage ΔV_C across the capacitor and its phase relative to the current, and (d) the voltage ΔV_L across the inductor and its phase relative to the current.

66. A voltage $\Delta v = (100 \text{ V}) \sin \omega t$ (in SI units) is applied across a series combination of a 2.00-H inductor, a 10.0-μF capacitor, and a 10.0-Ω resistor. (a) Determine the angular frequency ω_0 at which the power delivered to the resistor is a maximum. (b) Calculate the power delivered at that frequency. (c) Determine the two angular frequencies ω_1 and ω_2 at which the power is half the maximum value. [The Q of the circuit is $\omega_0/(\omega_2 - \omega_1)$.]

67. *Impedance matching.* Example 28.2 showed that maximum power is transferred when the internal resistance of a DC source is equal to the resistance of the load. A transformer may be used to provide maximum power transfer between two AC circuits that have different impedances Z1 and Z2, where 1 and 2 are subscripts and the Z's are italic (as in the centered equation). (a) Show that the ratio of turns N_1/N_2 needed to meet this condition is

$$\frac{N_1}{N_2} = \sqrt{\frac{Z_1}{Z_2}}$$

(b) Suppose you want to use a transformer as an impedance-matching device between an audio amplifier that has an output impedance of 8.00 kΩ and a speaker that has an input impedance of 8.00 Ω. What should your N_1/N_2 ratio be?

68. A power supply with $\Delta V_{rms} = 120$ V is connected between points a and d in Figure P33.26. At what frequency will it deliver a power of 250 W?

69. Figure P33.69a shows a parallel RLC circuit, and the corresponding phasor diagram is given in Figure P33.69b. The instantaneous voltages (and rms voltages) across each of the three circuit elements are the same, and each is in phase with the current through the resistor. The currents in C and L lead or lag behind the current in the resistor, as shown in Figure P33.69b. (a) Show that the rms current delivered by the source is

$$I_{rms} = \Delta V_{rms} \left[\frac{1}{R^2} + \left(\omega C - \frac{1}{\omega L} \right)^2 \right]^{1/2}$$

(b) Show that the phase angle ϕ between ΔV_{rms} and I_{rms} is

$$\tan \phi = R \left(\frac{1}{X_C} - \frac{1}{X_L} \right)$$

70. An 80.0-Ω resistor, a 200-mH inductor, and a 0.150-μF capacitor are connected *in parallel* across a 120-V (rms) source operating at 374 rad/s. (a) What is the resonant frequency of the circuit? (b) Calculate the rms current in the resistor, inductor, and capacitor. (c) What rms current is delivered by the source? (d) Is the current leading or lagging behind the voltage? By what angle?

71. A series RLC circuit is operating at 2 000 Hz. At this frequency, $X_L = X_C = 1\,884\ \Omega$. The resistance of the circuit is 40.0 Ω. (a) Prepare a table showing the values of X_L, X_C, and Z for $f = 300, 600, 800, 1\,000, 1\,500, 2\,000, 3\,000, 4\,000, 6\,000,$ and 10 000 Hz. (b) Plot on the same set of axes X_L, X_C, and Z as a function of $\ln f$.

72. A series RLC circuit in which $R = 1.00\ \Omega$, $L = 1.00$ mH, and $C = 1.00$ nF is connected to an AC source delivering 1.00 V (rms). Make a precise graph of the power delivered to the circuit as a function of the frequency and verify that the full width of the resonance peak at half-maximum is $R/2\pi L$.

73. Suppose the high-pass filter shown in Figure 33.25 has $R = 1\,000\ \Omega$ and $C = 0.050\,0\ \mu$F. (a) At what frequency does $\Delta V_{out}/\Delta V_{in} = \frac{1}{2}$? (b) Plot $\log_{10}(\Delta V_{out}/\Delta V_{in})$ versus $\log_{10}(f)$ over the frequency range from 1 Hz to 1 MHz. (This log–log plot of gain versus frequency is known as a *Bode plot*.)

Answers to Quick Quizzes

33.1 (a). The phasor in part (a) has the largest projection onto the vertical axis.

33.2 (b). The phasor in part (b) has the smallest-magnitude projection onto the vertical axis.

33.3 (c). The average power is proportional to the rms current, which, as Figure 33.5 shows, is nonzero even though the average current is zero. Condition (a) is valid only for an open circuit, and conditions (b) and (d) can not be true because $i_{av} = 0$ if the source is sinusoidal.

33.4 (b). For low frequencies, the reactance of the inductor is small so that the current is large. Most of the voltage from the source is across the bulb, so the power delivered to it is large.

33.5 (a). For high frequencies, the reactance of the capacitor is small so that the current is large. Most of the voltage from the source is across the bulb, so the power delivered to it is large.

33.6 (b). For low frequencies, the reactance of the capacitor is large so that very little current exists in the capacitor

(a)

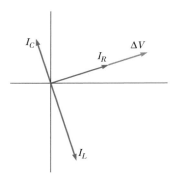

(b)

Figure P33.69

branch. The reactance of the inductor is small so that current exists in the inductor branch and the lightbulb glows. As the frequency increases, the inductive reactance increases and the capacitive reactance decreases. At high frequencies, more current exists in the capacitor branch than the inductor branch and the lightbulb glows more dimly.

33.7 (a) $X_L < X_C$. (b) $X_L = X_C$. (c) $X_L > X_C$.

33.8 (c). The cosine of $-\phi$ is the same as that of $+\phi$, so the $\cos \phi$ factor in Equation 33.31 is the same for both frequencies. The factor ΔV_{rms} is the same because the source voltage is fixed. According to Equation 33.27, changing $+\phi$ to $-\phi$ simply interchanges the values of X_L and X_C. Equation 33.25 tells us that such an interchange does not affect the impedance, so that the current I_{rms} in Equation 33.31 is the same for both frequencies.

33.9 (c). At resonance, $X_L = X_C$. According to Equation 33.25, this gives us $Z = R$.

33.10 (a). The higher the quality factor, the more sensitive the detector. As you can see from Figure 33.19, when $Q = \omega_0/\Delta\omega$ is high, a slight change in the resonance frequency (as might happen when a small piece of metal passes through the portal) causes a large change in current that can be detected easily.

33.11 (a) and (e). The current in an inductive circuit decreases with increasing frequency (see Eq. 33.9). Thus, an inductor connected in series with a woofer blocks high-frequency signals and passes low-frequency signals. The current in a capacitive circuit increases with increasing frequency (see Eq. 33.17). When a capacitor is connected in series with a tweeter, the capacitor blocks low-frequency signals and passes high-frequency signals.

Chapter 34

Electromagnetic Waves

▲ Electromagnetic waves cover a broad spectrum of wavelengths, with waves in various wavelength ranges having distinct properties. These images of the Crab Nebula show different structure for observations made with waves of various wavelengths. The images (clockwise starting from the upper left) were taken with x-rays, visible light, radio waves, and infrared waves. (upper left–NASA/CXC/SAO; upper right–Palomar Observatory; lower right–VLA/NRAO; lower left–WM Keck Observatory)

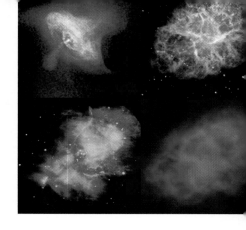

The waves described in Chapters 16, 17, and 18 are mechanical waves. By definition, the propagation of mechanical disturbances—such as sound waves, water waves, and waves on a string—requires the presence of a medium. This chapter is concerned with the properties of electromagnetic waves, which (unlike mechanical waves) can propagate through empty space.

In Section 31.7 we gave a brief description of Maxwell's equations, which form the theoretical basis of all electromagnetic phenomena. The consequences of Maxwell's equations are far-reaching and dramatic. The Ampère–Maxwell law predicts that a time-varying electric field produces a magnetic field, just as Faraday's law tells us that a time-varying magnetic field produces an electric field.

Astonishingly, Maxwell's equations also predict the existence of electromagnetic waves that propagate through space at the speed of light c. This chapter begins with a discussion of how Heinrich Hertz confirmed Maxwell's prediction when he generated and detected electromagnetic waves in 1887. That discovery has led to many practical communication systems, including radio, television, radar, and opto-electronics. On a conceptual level, Maxwell unified the subjects of light and electromagnetism by developing the idea that light is a form of electromagnetic radiation.

Next, we learn how electromagnetic waves are generated by oscillating electric charges. The waves consist of oscillating electric and magnetic fields at right angles to each other and to the direction of wave propagation. Thus, electromagnetic waves are transverse waves. The waves radiated from the oscillating charges can be detected at great distances. Furthermore, electromagnetic waves carry energy and momentum and hence can exert pressure on a surface.

The chapter concludes with a look at the wide range of frequencies covered by electromagnetic waves. For example, radio waves (frequencies of about 10^7 Hz) are electromagnetic waves produced by oscillating currents in a radio tower's transmitting antenna. Light waves are a high-frequency form of electromagnetic radiation (about 10^{14} Hz) produced by oscillating electrons in atoms.

34.1 Maxwell's Equations and Hertz's Discoveries

In his unified theory of electromagnetism, Maxwell showed that electromagnetic waves are a natural consequence of the fundamental laws expressed in the following four equations (see Section 31.7):

$$\oint \mathbf{E} \cdot d\mathbf{A} = \frac{q}{\epsilon_0} \tag{34.1}$$

$$\oint \mathbf{B} \cdot d\mathbf{A} = 0 \tag{34.2}$$

James Clerk Maxwell
Scottish Theoretical Physicist (1831–1879)

Maxwell developed the electromagnetic theory of light and the kinetic theory of gases, and explained the nature of Saturn's rings and color vision. Maxwell's successful interpretation of the electromagnetic field resulted in the field equations that bear his name. Formidable mathematical ability combined with great insight enabled him to lead the way in the study of electromagnetism and kinetic theory. He died of cancer before he was 50. (*North Wind Picture Archives*)

◀ Maxwell's equations

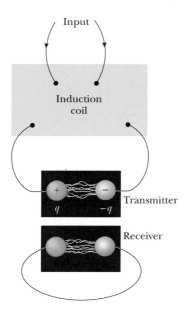

Figure 34.1 Schematic diagram of Hertz's apparatus for generating and detecting electromagnetic waves. The transmitter consists of two spherical electrodes connected to an induction coil, which provides short voltage surges to the spheres, setting up oscillations in the discharge between the electrodes. The receiver is a nearby loop of wire containing a second spark gap.

$$\oint \mathbf{E} \cdot d\mathbf{s} = -\frac{d\Phi_B}{dt} \tag{34.3}$$

$$\oint \mathbf{B} \cdot d\mathbf{s} = \mu_0 I + \mu_0 \epsilon_0 \frac{d\Phi_E}{dt} \tag{34.4}$$

In the next section we show that Equations 34.3 and 34.4 can be combined to obtain a wave equation for both the electric field and the magnetic field. In empty space, where $q = 0$ and $I = 0$, the solution to these two equations shows that the speed at which electromagnetic waves travel equals the measured speed of light. This result led Maxwell to predict that light waves are a form of electromagnetic radiation.

The experimental apparatus that Hertz used to generate and detect electromagnetic waves is shown schematically in Figure 34.1. An induction coil is connected to a transmitter made up of two spherical electrodes separated by a narrow gap. The coil provides short voltage surges to the electrodes, making one positive and the other negative. A spark is generated between the spheres when the electric field near either electrode surpasses the dielectric strength for air (3×10^6 V/m; see Table 26.1). In a strong electric field, the acceleration of free electrons provides them with enough energy to ionize any molecules they strike. This ionization provides more electrons, which can accelerate and cause further ionizations. As the air in the gap is ionized, it becomes a much better conductor, and the discharge between the electrodes exhibits an oscillatory behavior at a very high frequency. From an electric-circuit viewpoint, this is equivalent to an LC circuit in which the inductance is that of the coil and the capacitance is due to the spherical electrodes.

Because L and C are small in Hertz's apparatus, the frequency of oscillation is high, on the order of 100 MHz. (Recall from Eq. 32.22 that $\omega = 1/\sqrt{LC}$ for an LC circuit.) Electromagnetic waves are radiated at this frequency as a result of the oscillation (and hence acceleration) of free charges in the transmitter circuit. Hertz was able to detect these waves using a single loop of wire with its own spark gap (the receiver). Such a receiver loop, placed several meters from the transmitter, has its own effective inductance, capacitance, and natural frequency of oscillation. In Hertz's experiment, sparks were induced across the gap of the receiving electrodes when the frequency of the receiver was adjusted to match that of the transmitter. Thus, Hertz demonstrated that the oscillating current induced in the receiver was produced by electromagnetic waves radiated by the transmitter. His experiment is analogous to the mechanical phenomenon in which a tuning fork responds to acoustic vibrations from an identical tuning fork that is oscillating.

Heinrich Rudolf Hertz
German Physicist (1857–1894)

Hertz made his most important discovery of electromagnetic waves in 1887. After finding that the speed of an electromagnetic wave was the same as that of light, Hertz showed that electromagnetic waves, like light waves, could be reflected, refracted, and diffracted. Hertz died of blood poisoning at the age of 36. During his short life, he made many contributions to science. The hertz, equal to one complete vibration or cycle per second, is named after him. (*Hulton-Deutsch Collection/CORBIS*)

Additionally, Hertz showed in a series of experiments that the radiation generated by his spark-gap device exhibited the wave properties of interference, diffraction, reflection, refraction, and polarization, all of which are properties exhibited by light, as we shall see in Part 5 of the text. Thus, it became evident that the radio-frequency waves Hertz was generating had properties similar to those of light waves and differed only in frequency and wavelength. Perhaps his most convincing experiment was the measurement of the speed of this radiation. Waves of known frequency were reflected from a metal sheet and created a standing-wave interference pattern whose nodal points could be detected. The measured distance between the nodal points enabled determination of the wavelength λ. Using the relationship $v = \lambda f$ (Eq. 16.12), Hertz found that v was close to 3×10^8 m/s, the known speed c of visible light.

34.2 Plane Electromagnetic Waves

The properties of electromagnetic waves can be deduced from Maxwell's equations. One approach to deriving these properties is to solve the second-order differential equation obtained from Maxwell's third and fourth equations. A rigorous mathematical treatment of that sort is beyond the scope of this text. To circumvent this problem, we assume that the vectors for the electric field and magnetic field in an electromagnetic wave have a specific space–time behavior that is simple but consistent with Maxwell's equations.

To understand the prediction of electromagnetic waves more fully, let us focus our attention on an electromagnetic wave that travels in the x direction (the *direction of propagation*). In this wave, the electric field $\mathbf{E}$ is in the y direction, and the magnetic field $\mathbf{B}$ is in the z direction, as shown in Figure 34.2. Waves such as this one, in which the electric and magnetic fields are restricted to being parallel to a pair of perpendicular axes, are said to be **linearly polarized waves.** Furthermore, we assume that at any point in space, the magnitudes E and B of the fields depend upon x and t only, and not upon the y or z coordinate.

Let us also imagine that the source of the electromagnetic waves is such that a wave radiated from *any* position in the yz plane (not just from the origin as might be suggested by Figure 34.2) propagates in the x direction, and all such waves are emitted in phase. If we define a **ray** as the line along which the wave travels, then all rays for these waves are parallel. This entire collection of waves is often called a **plane wave.** A surface connecting points of equal phase on all waves, which we call a **wave front,** as introduced in Chapter 17, is a geometric plane. In comparison, a point source of radiation sends waves out radially in all directions. A surface connecting points of equal phase for this situation is a sphere, so this is called a **spherical wave.**

We can relate E and B to each other with Equations 34.3 and 34.4. In empty space, where $q = 0$ and $I = 0$, Equation 34.3 remains unchanged and Equation 34.4 becomes

$$\oint \mathbf{B} \cdot d\mathbf{s} = \epsilon_0 \mu_0 \frac{d\Phi_E}{dt} \tag{34.5}$$

Using Equations 34.3 and 34.5 and the plane-wave assumption, we obtain the following differential equations relating E and B. (We shall derive these equations formally later in this section.)

$$\frac{\partial E}{\partial x} = -\frac{\partial B}{\partial t} \tag{34.6}$$

$$\frac{\partial B}{\partial x} = -\mu_0 \epsilon_0 \frac{\partial E}{\partial t} \tag{34.7}$$

Note that the derivatives here are partial derivatives. For example, when we evaluate $\partial E/\partial x$, we assume that t is constant. Likewise, when we evaluate $\partial B/\partial t$, x is held constant. Taking the derivative of Equation 34.6 with respect to x and combining the

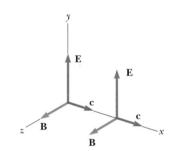

Figure 34.2 An electromagnetic wave traveling at velocity **c** in the positive x direction. The electric field is along the y direction, and the magnetic field is along the z direction. These fields depend only on x and t.

 PITFALL PREVENTION

34.1 What Is "a" Wave?

A sticky point in these types of discussions is what we mean by a *single* wave. We could define one wave as that which is emitted by a single charged particle. In practice, however, the word *wave* represents both the emission from a *single point* ("wave radiated from any position in the yz plane") and the collection of waves from *all points* on the source ("plane wave"). You should be able to use this term in both ways and to understand its meaning from the context.

result with Equation 34.7, we obtain

$$\frac{\partial^2 E}{\partial x^2} = -\frac{\partial}{\partial x}\left(\frac{\partial B}{\partial t}\right) = -\frac{\partial}{\partial t}\left(\frac{\partial B}{\partial x}\right) = -\frac{\partial}{\partial t}\left(-\mu_0\epsilon_0\frac{\partial E}{\partial t}\right)$$

$$\frac{\partial^2 E}{\partial x^2} = \mu_0\epsilon_0\frac{\partial^2 E}{\partial t^2} \tag{34.8}$$

In the same manner, taking the derivative of Equation 34.7 with respect to x and combining it with Equation 34.6, we obtain

$$\frac{\partial^2 B}{\partial x^2} = \mu_0\epsilon_0\frac{\partial^2 B}{\partial t^2} \tag{34.9}$$

Equations 34.8 and 34.9 both have the form of the general wave equation[1] with the wave speed v replaced by c, where

Speed of electromagnetic waves

$$c = \frac{1}{\sqrt{\mu_0\epsilon_0}} \tag{34.10}$$

Taking $\mu_0 = 4\pi \times 10^{-7}\,\text{T}\cdot\text{m/A}$ and $\epsilon_0 = 8.854\,19 \times 10^{-12}\,\text{C}^2/\text{N}\cdot\text{m}^2$ in Equation 34.10, we find that $c = 2.997\,92 \times 10^8$ m/s. Because this speed is precisely the same as the speed of light in empty space, we are led to believe (correctly) that light is an electromagnetic wave.

The simplest solution to Equations 34.8 and 34.9 is a sinusoidal wave, for which the field magnitudes E and B vary with x and t according to the expressions

Sinusoidal electric and magnetic fields

$$E = E_{\text{max}}\cos(kx - \omega t) \tag{34.11}$$

$$B = B_{\text{max}}\cos(kx - \omega t) \tag{34.12}$$

where E_{max} and B_{max} are the maximum values of the fields. The angular wave number is $k = 2\pi/\lambda$, where λ is the wavelength. The angular frequency is $\omega = 2\pi f$, where f is the wave frequency. The ratio ω/k equals the speed of an electromagnetic wave, c:

$$\frac{\omega}{k} = \frac{2\pi f}{2\pi/\lambda} = \lambda f = c$$

where we have used Equation 16.12, $v = c = \lambda f$, which relates the speed, frequency, and wavelength of any continuous wave. Thus, for electromagnetic waves, the wavelength and frequency of these waves are related by

$$\lambda = \frac{c}{f} = \frac{3.00 \times 10^8\,\text{m/s}}{f} \tag{34.13}$$

Figure 34.3a is a pictorial representation, at one instant, of a sinusoidal, linearly polarized plane wave moving in the positive x direction. Figure 34.3b shows how the electric and magnetic field vectors at a fixed location vary with time.

Taking partial derivatives of Equations 34.11 (with respect to x) and 34.12 (with respect to t), we find that

$$\frac{\partial E}{\partial x} = -kE_{\text{max}}\sin(kx - \omega t)$$

$$\frac{\partial B}{\partial t} = \omega B_{\text{max}}\sin(kx - \omega t)$$

---

[1] The general wave equation is of the form $(\partial^2 y/\partial x^2) = (1/v^2)(\partial^2 y/\partial t^2)$ where v is the speed of the wave and y is the wave function. The general wave equation was introduced as Equation 16.27, and it would be useful for you to review Section 16.6.

(a)   (b)

Active Figure 34.3 Representation of a sinusoidal, linearly polarized plane electromagnetic wave moving in the positive x direction with velocity **c**. (a) The wave at some instant. Note the sinusoidal variations of E and B with x. (b) A time sequence (starting at the upper left) illustrating the electric and magnetic field vectors at a fixed point in the yz plane, as seen by an observer looking in the negative x direction. The variations of E and B with t are sinusoidal.

 At the Active Figures link at http://www.pse6.com, *you can observe the wave in part (a) and the variation of the fields in part (b). In addition, you can take a "snapshot" of the wave at an instant of time and investigate the electric and magnetic fields at that instant.*

Substituting these results into Equation 34.6, we find that at any instant

$$kE_{max} = \omega B_{max}$$

$$\frac{E_{max}}{B_{max}} = \frac{\omega}{k} = c$$

Using these results together with Equations 34.11 and 34.12, we see that

$$\frac{E_{max}}{B_{max}} = \frac{E}{B} = c \qquad (34.14)$$

That is, **at every instant the ratio of the magnitude of the electric field to the magnitude of the magnetic field in an electromagnetic wave equals the speed of light.**

Finally, note that electromagnetic waves obey the superposition principle (which we discussed in Section 18.1 with respect to mechanical waves) because the differential equations involving E and B are linear equations. For example, we can add two waves with the same frequency and polarization simply by adding the magnitudes of the two electric fields algebraically.

Let us summarize the properties of electromagnetic waves as we have described them:

- The solutions of Maxwell's third and fourth equations are wave-like, with both E and B satisfying a wave equation.

- Electromagnetic waves travel through empty space at the speed of light $c = 1/\sqrt{\epsilon_0 \mu_0}$.

- The components of the electric and magnetic fields of plane electromagnetic waves are perpendicular to each other and perpendicular to the direction of wave propagation. We can summarize the latter property by saying that electromagnetic waves are transverse waves.

▲ **PITFALL PREVENTION**

34.2   E Stronger Than B?

Because the value of c is so large, some students incorrectly interpret Equation 34.14 as meaning that the electric field is much stronger than the magnetic field. Electric and magnetic fields are measured in different units, however, so they cannot be directly compared. In Section 34.3, we find that the electric and magnetic fields contribute equally to the energy of the wave.

Properties of electromagnetic waves

- The magnitudes of **E** and **B** in empty space are related by the expression $E/B = c$.
- Electromagnetic waves obey the principle of superposition.

Quick Quiz 34.1 What is the phase difference between the sinusoidal oscillations of the electric and magnetic fields in Figure 34.3? (a) 180° (b) 90° (c) 0 (d) impossible to determine.

Example 34.1   An Electromagnetic Wave   Interactive

A sinusoidal electromagnetic wave of frequency 40.0 MHz travels in free space in the x direction, as in Figure 34.4.

(A) Determine the wavelength and period of the wave.

Solution Using Equation 34.13 for light waves and given that $f = 40.0$ MHz $= 4.00 \times 10^7$ s^{-1}, we have

$$\lambda = \frac{c}{f} = \frac{3.00 \times 10^8 \text{ m/s}}{4.00 \times 10^7 \text{ s}^{-1}} = \boxed{7.50 \text{ m}}$$

The period T of the wave is the inverse of the frequency:

$$T = \frac{1}{f} = \frac{1}{4.00 \times 10^7 \text{ s}^{-1}} = \boxed{2.50 \times 10^{-8} \text{ s}}$$

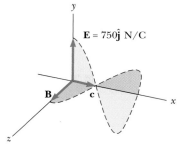

Figure 34.4 (Example 34.1) At some instant, a plane electromagnetic wave moving in the x direction has a maximum electric field of 750 N/C in the positive y direction. The corresponding magnetic field at that point has a magnitude E/c and is in the z direction.

(B) At some point and at some instant, the electric field has its maximum value of 750 N/C and is along the y axis. Calculate the magnitude and direction of the magnetic field at this position and time.

Solution From Equation 34.14 we see that

$$B_{max} = \frac{E_{max}}{c} = \frac{750 \text{ N/C}}{3.00 \times 10^8 \text{ m/s}} = \boxed{2.50 \times 10^{-6} \text{ T}}$$

Because **E** and **B** must be perpendicular to each other and perpendicular to the direction of wave propagation (x in this case), we conclude that **B** is in the z direction.

(C) Write expressions for the space–time variation of the components of the electric and magnetic fields for this wave.

Solution We can apply Equations 34.11 and 34.12 directly:

$$E = E_{max} \cos(kx - \omega t) = (750 \text{ N/C}) \cos(kx - \omega t)$$

$$B = B_{max} \cos(kx - \omega t) = (2.50 \times 10^{-6} \text{ T}) \cos(kx - \omega t)$$

where

$$\omega = 2\pi f = 2\pi(4.00 \times 10^7 \text{ s}^{-1}) = 2.51 \times 10^8 \text{ rad/s}$$

$$k = \frac{2\pi}{\lambda} = \frac{2\pi}{7.50 \text{ m}} = 0.838 \text{ rad/m}$$

 Explore electromagnetic waves of different frequencies at the Interactive Worked Example link at **http://www.pse6.com.**

Derivation of Equations 34.6 and 34.7

To derive Equation 34.6, we start with Faraday's law, Equation 34.3:

$$\oint \mathbf{E} \cdot d\mathbf{s} = -\frac{d\Phi_B}{dt}$$

Let us again assume that the electromagnetic wave is traveling in the x direction, with the electric field **E** in the positive y direction and the magnetic field **B** in the positive z direction.

Consider a rectangle of width dx and height ℓ lying in the xy plane, as shown in Figure 34.5. To apply Equation 34.3, we must first evaluate the line integral of $\mathbf{E} \cdot d\mathbf{s}$ around this rectangle. The contributions from the top and bottom of the rectangle are zero because $\mathbf{E}$ is perpendicular to $d\mathbf{s}$ for these paths. We can express the electric field on the right side of the rectangle as

$$E(x + dx, t) \approx E(x, t) + \frac{dE}{dx}\bigg]_{t\,\text{constant}} dx = E(x, t) + \frac{\partial E}{\partial x} dx$$

while the field on the left side[2] is simply $E(x, t)$. Therefore, the line integral over this rectangle is approximately

$$\oint \mathbf{E} \cdot d\mathbf{s} = [E(x + dx, t)]\ell - [E(x, t)]\ell \approx \ell \left(\frac{\partial E}{\partial x}\right) dx \qquad (34.15)$$

Because the magnetic field is in the z direction, the magnetic flux through the rectangle of area $\ell\, dx$ is approximately $\Phi_B = B\ell\, dx$. (This assumes that dx is very small compared with the wavelength of the wave.) Taking the time derivative of the magnetic flux gives

$$\frac{d\Phi_B}{dt} = \ell\, dx \frac{dB}{dt}\bigg]_{x\,\text{constant}} = \ell\, dx \frac{\partial B}{\partial t} \qquad (34.16)$$

Substituting Equations 34.15 and 34.16 into Equation 34.3 gives

$$\ell \left(\frac{\partial E}{\partial x}\right) dx = -\ell\, dx\, \frac{\partial B}{\partial t}$$

$$\frac{\partial E}{\partial x} = -\frac{\partial B}{\partial t}$$

This expression is Equation 34.6.

In a similar manner, we can verify Equation 34.7 by starting with Maxwell's fourth equation in empty space (Eq. 34.5). In this case, the line integral of $\mathbf{B} \cdot d\mathbf{s}$ is evaluated around a rectangle lying in the xz plane and having width dx and length ℓ, as in Figure 34.6. Noting that the magnitude of the magnetic field changes from $B(x, t)$ to $B(x + dx, t)$ over the width dx and that the direction in which we take the line integral is as shown in Figure 34.6, the line integral over this rectangle is found to be approximately

$$\oint \mathbf{B} \cdot d\mathbf{s} = [B(x, t)]\ell - [B(x + dx, t)]\ell \approx -\ell \left(\frac{\partial B}{\partial x}\right) dx \qquad (34.17)$$

The electric flux through the rectangle is $\Phi_E = E\ell\, dx$, which, when differentiated with respect to time, gives

$$\frac{\partial \Phi_E}{\partial t} = \ell\, dx \frac{\partial E}{\partial t} \qquad (34.18)$$

Substituting Equations 34.17 and 34.18 into Equation 34.5 gives

$$-\ell \left(\frac{\partial B}{\partial x}\right) dx = \mu_0 \epsilon_0 \ell\, dx \left(\frac{\partial E}{\partial t}\right)$$

$$\frac{\partial B}{\partial x} = -\mu_0 \epsilon_0 \frac{\partial E}{\partial t}$$

which is Equation 34.7.

<hr>

[2] Because dE/dx in this equation is expressed as the change in E with x at a given instant t, dE/dx is equivalent to the partial derivative $\partial E/\partial x$. Likewise, dB/dt means the change in B with time at a particular position x, so in Equation 34.16 we can replace dB/dt with $\partial B/\partial t$.

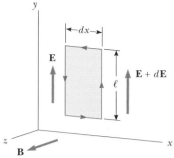

Figure 34.5 At an instant when a plane wave moving in the $+x$ direction passes through a rectangular path of width dx lying in the xy plane, the electric field in the y direction varies from $\mathbf{E}$ to $\mathbf{E} + d\mathbf{E}$. This spatial variation in $\mathbf{E}$ gives rise to a time-varying magnetic field along the z direction, according to Equation 34.6.

Figure 34.6 At an instant when a plane wave passes through a rectangular path of width dx lying in the xz plane, the magnetic field in the z direction varies from $\mathbf{B}$ to $\mathbf{B} + d\mathbf{B}$. This spatial variation in $\mathbf{B}$ gives rise to a time-varying electric field along the y direction, according to Equation 34.7.

34.3 Energy Carried by Electromagnetic Waves

Electromagnetic waves carry energy, and as they propagate through space they can transfer energy to objects placed in their path. The rate of flow of energy in an electromagnetic wave is described by a vector **S**, called the **Poynting vector,** which is defined by the expression

$$\mathbf{S} \equiv \frac{1}{\mu_0} \mathbf{E} \times \mathbf{B} \tag{34.19}$$

The magnitude of the Poynting vector represents the rate at which energy flows through a unit surface area perpendicular to the direction of wave propagation. Thus, the magnitude of the Poynting vector represents *power per unit area.* The direction of the vector is along the direction of wave propagation (Fig. 34.7). The SI units of the Poynting vector are $\text{J/s} \cdot \text{m}^2 = \text{W/m}^2$.

As an example, let us evaluate the magnitude of **S** for a plane electromagnetic wave where $|\mathbf{E} \times \mathbf{B}| = EB$. In this case,

PITFALL PREVENTION

34.3 An Instantaneous Value

The Poynting vector given by Equation 34.19 is time-dependent. Its magnitude varies in time, reaching a maximum value at the same instant as the magnitudes of **E** and **B** do. The *average* rate of energy transfer is given by Equation 34.21.

$$S = \frac{EB}{\mu_0} \tag{34.20}$$

Because $B = E/c$, we can also express this as

$$S = \frac{E^2}{\mu_0 c} = \frac{c}{\mu_0} B^2$$

These equations for S apply at any instant of time and represent the *instantaneous* rate at which energy is passing through a unit area.

What is of greater interest for a sinusoidal plane electromagnetic wave is the time average of S over one or more cycles, which is called the *wave intensity I.* (We discussed the intensity of sound waves in Chapter 17.) When this average is taken, we obtain an expression involving the time average of $\cos^2(kx - \omega t)$, which equals $\frac{1}{2}$. Hence, the average value of S (in other words, the intensity of the wave) is

Wave intensity

$$I = S_{\text{av}} = \frac{E_{\text{max}} B_{\text{max}}}{2\mu_0} = \frac{E_{\text{max}}^2}{2\mu_0 c} = \frac{c}{2\mu_0} B_{\text{max}}^2 \tag{34.21}$$

Recall that the energy per unit volume, which is the instantaneous energy density u_E associated with an electric field, is given by Equation 26.13,

PITFALL PREVENTION

34.4 Irradiance

In this discussion, intensity is defined in the same way as in Chapter 17 (as power per unit area). In the optics industry, however, power per unit area is called the *irradiance,* and radiant intensity is defined as the power in watts per solid angle (measured in steradians).

$$u_E = \tfrac{1}{2}\epsilon_0 E^2$$

and that the instantaneous energy density u_B associated with a magnetic field is given by Equation 32.14:

$$u_B = \frac{B^2}{2\mu_0}$$

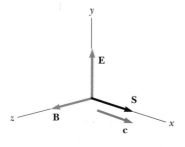

Figure 34.7 The Poynting vector **S** for a plane electromagnetic wave is along the direction of wave propagation.

Because E and B vary with time for an electromagnetic wave, the energy densities also vary with time. When we use the relationships $B = E/c$ and $c = 1/\sqrt{\epsilon_0 \mu_0}$, the expression for u_B becomes

$$u_B = \frac{(E/c)^2}{2\mu_0} = \frac{\epsilon_0 \mu_0}{2\mu_0} E^2 = \tfrac{1}{2} \epsilon_0 E^2$$

Comparing this result with the expression for u_E, we see that

$$u_B = u_E = \tfrac{1}{2} \epsilon_0 E^2 = \frac{B^2}{2\mu_0}$$

That is, **the instantaneous energy density associated with the magnetic field of an electromagnetic wave equals the instantaneous energy density associated with the electric field.** Hence, in a given volume the energy is equally shared by the two fields.

The **total instantaneous energy density** u is equal to the sum of the energy densities associated with the electric and magnetic fields:

$$u = u_E + u_B = \epsilon_0 E^2 = \frac{B^2}{\mu_0}$$

When this total instantaneous energy density is averaged over one or more cycles of an electromagnetic wave, we again obtain a factor of $\tfrac{1}{2}$. Hence, for any electromagnetic wave, the total average energy per unit volume is

$$u_{\mathrm{av}} = \epsilon_0 (E^2)_{\mathrm{av}} = \tfrac{1}{2} \epsilon_0 E_{\mathrm{max}}^2 = \frac{B_{\mathrm{max}}^2}{2\mu_0} \qquad (34.22)$$

Comparing this result with Equation 34.21 for the average value of S, we see that

$$I = S_{\mathrm{av}} = c u_{\mathrm{av}} \qquad (34.23)$$

In other words, **the intensity of an electromagnetic wave equals the average energy density multiplied by the speed of light.**

Total instantaneous energy density of an electromagnetic wave

Average energy density of an electromagnetic wave

Quick Quiz 34.2 An electromagnetic wave propagates in the $-y$ direction. The electric field at a point in space is momentarily oriented in the $+x$ direction. The magnetic field at that point is momentarily oriented in the (a) $-x$ direction (b) $+y$ direction (c) $+z$ direction (d) $-z$ direction.

Quick Quiz 34.3 Which of the following is constant for a plane electromagnetic wave? (a) magnitude of the Poynting vector (b) energy density u_E (c) energy density u_B (d) wave intensity.

Example 34.2 Fields on the Page

Estimate the maximum magnitudes of the electric and magnetic fields of the light that is incident on this page because of the visible light coming from your desk lamp. Treat the bulb as a point source of electromagnetic radiation that is 5% efficient at transforming energy coming in by electrical transmission to energy leaving by visible light.

Solution Recall from Equation 17.7 that the wave intensity I a distance r from a point source is $I = \mathscr{P}_{\mathrm{av}}/4\pi r^2$, where $\mathscr{P}_{\mathrm{av}}$ is the average power output of the source and $4\pi r^2$ is the area of a sphere of radius r centered on the source. Because

the intensity of an electromagnetic wave is also given by Equation 34.21, we have

$$I = \frac{\mathscr{P}_{\mathrm{av}}}{4\pi r^2} = \frac{E_{\mathrm{max}}^2}{2\mu_0 c}$$

We must now make some assumptions about numbers to enter in this equation. If we have a 60-W lightbulb, its output at 5% efficiency is approximately 3.0 W by visible light. (The remaining energy transfers out of the bulb by conduction and invisible radiation.) A reasonable distance from the bulb to the page might be 0.30 m. Thus, we have

$$E_{max} = \sqrt{\frac{\mu_0 c \mathcal{P}_{av}}{2\pi r^2}}$$

$$= \sqrt{\frac{(4\pi \times 10^{-7}\,\text{T·m/A})(3.00 \times 10^8\,\text{m/s})(3.0\,\text{W})}{2\pi(0.30\,\text{m})^2}}$$

$$= \boxed{45\,\text{V/m}}$$

From Equation 34.14,

$$B_{max} = \frac{E_{max}}{c} = \frac{45\,\text{V/m}}{3.00 \times 10^8\,\text{m/s}} = \boxed{1.5 \times 10^{-7}\,\text{T}}$$

This value is two orders of magnitude smaller than the Earth's magnetic field, which, unlike the magnetic field in the light wave from your desk lamp, is not oscillating.

34.4   Momentum and Radiation Pressure

Electromagnetic waves transport linear momentum as well as energy. It follows that, as this momentum is absorbed by some surface, pressure is exerted on the surface. We shall assume in this discussion that the electromagnetic wave strikes the surface at normal incidence and transports a total energy U to the surface in a time interval Δt. Maxwell showed that, if the surface absorbs all the incident energy U in this time interval (as does a black body, introduced in Section 20.7), the total momentum **p** transported to the surface has a magnitude

Momentum transported to a perfectly absorbing surface

$$p = \frac{U}{c} \qquad \text{(complete absorption)} \qquad (34.24)$$

The pressure exerted on the surface is defined as force per unit area F/A. Let us combine this with Newton's second law:

$$P = \frac{F}{A} = \frac{1}{A}\frac{dp}{dt}$$

> **PITFALL PREVENTION**
>
> **34.5   So Many p's**
>
> We have p for momentum and P for pressure, and these are both related to $\mathcal{P}$ for power! Be sure you keep these all straight.

If we now replace p, the momentum transported to the surface by radiation, from Equation 34.24, we have

$$P = \frac{1}{A}\frac{dp}{dt} = \frac{1}{A}\frac{d}{dt}\left(\frac{U}{c}\right) = \frac{1}{c}\frac{(dU/dt)}{A}$$

We recognize $(dU/dt)/A$ as the rate at which energy is arriving at the surface per unit area, which is the magnitude of the Poynting vector. Thus, the radiation pressure P exerted on the perfectly absorbing surface is

Radiation pressure exerted on a perfectly absorbing surface

$$P = \frac{S}{c} \qquad (34.25)$$

If the surface is a perfect reflector (such as a mirror) and incidence is normal, then the momentum transported to the surface in a time interval Δt is twice that given by Equation 34.24. That is, the momentum transferred to the surface by the incoming light is $p = U/c$, and that transferred by the reflected light also is $p = U/c$. Therefore,

$$p = \frac{2U}{c} \qquad \text{(complete reflection)} \qquad (34.26)$$

The momentum delivered to a surface having a reflectivity somewhere between these two extremes has a value between U/c and $2U/c$, depending on the properties of the surface. Finally, the radiation pressure exerted on a perfectly reflecting surface for normal incidence of the wave is[3]

Radiation pressure exerted on a perfectly reflecting surface

$$P = \frac{2S}{c} \qquad (34.27)$$

---

[3]   For oblique incidence on a perfectly reflecting surface, the momentum transferred is $(2U\cos\theta)/c$ and the pressure is $P = (2S\cos^2\theta)/c$ where θ is the angle between the normal to the surface and the direction of wave propagation.

Figure 34.8 An apparatus for measuring the pressure exerted by light. In practice, the system is contained in a high vacuum.

Black disk

Mirror

Light

Courtesy of NASA

Figure 34.9 Mariner 10 used its solar panels to "sail on sunlight."

Although radiation pressures are very small (about $5 \times 10^{-6} \text{ N/m}^2$ for direct sunlight), they have been measured with torsion balances such as the one shown in Figure 34.8. A mirror (a perfect reflector) and a black disk (a perfect absorber) are connected by a horizontal rod suspended from a fine fiber. Normal-incidence light striking the black disk is completely absorbed, so all of the momentum of the light is transferred to the disk. Normal-incidence light striking the mirror is totally reflected, and hence the momentum transferred to the mirror is twice as great as that transferred to the disk. The radiation pressure is determined by measuring the angle through which the horizontal connecting rod rotates. The apparatus must be placed in a high vacuum to eliminate the effects of air currents.

NASA is exploring the possibility of *solar sailing* as a low-cost means of sending spacecraft to the planets. Large sheets would experience radiation pressure from sunlight and would be used in much the way canvas sheets are used on earthbound sailboats. In 1973 NASA engineers took advantage of the momentum of the sunlight striking the solar panels of Mariner 10 (Fig. 34.9) to make small course corrections when the spacecraft's fuel supply was running low. (This procedure was carried out when the spacecraft was in the vicinity of the planet Mercury. Would it have worked as well near Pluto?)

Quick Quiz 34.4 To maximize the radiation pressure on the sails of a spacecraft using solar sailing, should the sheets be (a) very black to absorb as much sunlight as possible or (b) very shiny, to reflect as much sunlight as possible?

Quick Quiz 34.5 In an apparatus such as that in Figure 34.8, the disks are illuminated uniformly over their areas. Suppose the black disk is replaced by one with half the radius. Which of the following are different after the disk is replaced? (a) radiation pressure on the disk, (b) radiation force on the disk, (c) radiation momentum delivered to the disk in a given time interval.

Conceptual Example 34.3 Sweeping the Solar System

A great amount of dust exists in interplanetary space. Although in theory these dust particles can vary in size from molecular size to much larger, very little of the dust in our solar system is smaller than about 0.2 μm. Why?

Solution The dust particles are subject to two significant forces—the gravitational force that draws them toward the Sun and the radiation-pressure force that pushes them away from the Sun. The gravitational force is proportional to the cube of the radius of a spherical dust particle because it is proportional to the mass and therefore to the volume $4\pi r^3/3$ of the particle. The radiation pressure is proportional to the square of the radius because it depends on the planar cross section of the particle. For large particles, the gravitational force is greater than the force from radiation pressure. For particles having radii less than about 0.2 μm, the radiation-pressure force is greater than the gravitational force, and as a result these particles are swept out of the Solar System.

Example 34.4 Pressure from a Laser Pointer

Many people giving presentations use a laser pointer to direct the attention of the audience to information on a screen. If a 3.0-mW pointer creates a spot on a screen that is 2.0 mm in diameter, determine the radiation pressure on a screen that reflects 70% of the light that strikes it. The power 3.0 mW is a time-averaged value.

Solution In conceptualizing this situation, we do not expect the pressure to be very large. We categorize this as a calculation of radiation pressure using something like Equation 34.25 or Equation 34.27, but complicated by the 70% reflection. To analyze the problem, we begin by determining the magnitude of the beam's Poynting vector. We divide the time-averaged power delivered via the electromagnetic wave by the cross-sectional area of the beam:

$$S_{av} = \frac{\mathcal{P}_{av}}{A} = \frac{\mathcal{P}_{av}}{\pi r^2} = \frac{3.0 \times 10^{-3} \text{ W}}{\pi \left(\frac{2.0 \times 10^{-3} \text{ m}}{2}\right)^2} = 955 \text{ W/m}^2$$

Now we can determine the radiation pressure from the laser beam. Equation 34.27 indicates that a completely reflected beam would apply an average pressure of $P_{av} = 2S_{av}/c$. We can model the actual reflection as follows. Imagine that the surface absorbs the beam, resulting in pressure $P_{av} = S_{av}/c$. Then the surface emits the beam, resulting in additional pressure $P_{av} = S_{av}/c$. If the surface emits only a fraction f of the beam (so that f is the amount of the incident beam reflected), then the pressure due to the emitted beam is $P_{av} = f S_{av}/c$. Thus, the total pressure on the surface due to absorption and re-emission (reflection) is

$$P_{av} = \frac{S_{av}}{c} + f \frac{S_{av}}{c} = (1 + f)\frac{S_{av}}{c}$$

Notice that if $f = 1$, which represents complete reflection, this equation reduces to Equation 34.27. For a beam that is 70% reflected, the pressure is

$$P_{av} = (1 + 0.70)\frac{955 \text{ W/m}^2}{3.0 \times 10^8 \text{ m/s}} = \boxed{5.4 \times 10^{-6} \text{ N/m}^2}$$

To finalize the example, consider first the magnitude of the Poynting vector. This is about the same as the intensity of sunlight at the Earth's surface. (For this reason, it is not safe to shine the beam of a laser pointer into a person's eyes; that may be more dangerous than looking directly at the Sun.) Note also that the pressure has an extremely small value, as expected. (Recall from Section 14.2 that atmospheric pressure is approximately 10^5 N/m^2.)

What If? What if the laser pointer is moved twice as far away from the screen? Does this affect the radiation pressure on the screen?

Answer Because a laser beam is popularly represented as a beam of light with constant cross section, one might be tempted to claim that the intensity of radiation, and therefore the radiation pressure, would be independent of distance from the screen. However, a laser beam does not have a constant cross section at all distances from the source—there is a small but measurable divergence of the beam. If the laser is moved farther away from the screen, the area of illumination on the screen will increase, decreasing the intensity. In turn, this will reduce the radiation pressure.

In addition, the doubled distance from the screen will result in more loss of energy from the beam due to scattering from air molecules and dust particles as the light travels from the laser to the screen. This will further reduce the radiation pressure.

 At the Interactive Worked Example link at **http://www.pse6.com,** *you can investigate the pressure on the screen for various laser and screen parameters.*

Example 34.5 Solar Energy

As noted in the preceding example, the Sun delivers about 10^3 W/m^2 of energy to the Earth's surface via electromagnetic radiation.

(A) Calculate the total power that is incident on a roof of dimensions 8.00 m × 20.0 m.

Solution We assume that the average magnitude of the Poynting vector for solar radiation at the surface of the Earth is $S_{av} = 1\,000 \text{ W/m}^2$; this represents the power per unit area, or the light intensity. Assuming that the radiation is incident normal to the roof, we obtain

$$\mathcal{P}_{av} = S_{av} A = (1\,000 \text{ W/m}^2)(8.00 \times 20.0 \text{ m}^2)$$

$$= \boxed{1.60 \times 10^5 \text{ W}}$$

(B) Determine the radiation pressure and the radiation force exerted on the roof, assuming that the roof covering is a perfect absorber.

Solution Using Equation 34.25 with $S_{av} = 1\,000 \text{ W/m}^2$, we find that the radiation pressure is

$$P_{av} = \frac{S_{av}}{c} = \frac{1\,000 \text{ W/m}^2}{3.00 \times 10^8 \text{ m/s}} = \boxed{3.33 \times 10^{-6} \text{ N/m}^2}$$

Because pressure equals force per unit area, this corresponds to a radiation force of

$$F = P_{av} A = (3.33 \times 10^{-6} \text{ N/m}^2)(160 \text{ m}^2)$$

$$= \boxed{5.33 \times 10^{-4} \text{ N}}$$

What If? Suppose the energy striking the roof could be captured and used to operate electrical devices in the house. Could the home operate completely from this energy?

Answer The power in part (A) is large compared to the power requirements of a typical home. If this power were maintained for 24 hours per day and the energy could be absorbed and made available to electrical devices, it would provide more than enough energy for the average home. However, solar energy is not easily harnessed, and the prospects for large-scale conversion are not as bright as may appear from this calculation. For example, the efficiency of conversion from solar energy is typically 10% for photovoltaic cells, reducing the available power in part (A) by an order of magnitude. Other considerations reduce the power even further. Depending on location, the radiation will most likely not be incident normal to the roof and, even if it is (in locations near the Equator), this situation exists for only a short time near the middle of the day. No energy is available for about half of each day during the nighttime hours. Furthermore, cloudy days reduce the available energy. Finally, while energy is arriving at a large rate during the middle of the day, some of it must be stored for later use, requiring batteries or other storage devices. The result of these considerations is that complete solar operation of homes is not presently cost-effective for most homes.

34.5 Production of Electromagnetic Waves by an Antenna

Neither stationary charges nor steady currents can produce electromagnetic waves. Whenever the current in a wire changes with time, however, the wire emits electromagnetic radiation. **The fundamental mechanism responsible for this radiation is the acceleration of a charged particle. Whenever a charged particle accelerates, it must radiate energy.**

Let us consider the production of electromagnetic waves by a *half-wave antenna*. In this arrangement, two conducting rods are connected to a source of alternating voltage (such as an *LC* oscillator), as shown in Figure 34.10. The length of each rod is equal to one quarter of the wavelength of the radiation that will be emitted when the oscillator operates at frequency *f*. The oscillator forces charges to accelerate back and forth between the two rods. Figure 34.10 shows the configuration of the electric and magnetic fields at some instant when the current is upward. The electric field lines, due to the separation of charges in the upper and lower portions of the antenna, resemble those of an electric dipole. (As a result, this type of antenna is sometimes called a *dipole antenna*.) Because these charges are continuously oscillating between the two rods, the antenna can be approximated by an oscillating electric dipole. The magnetic field lines, due to the current representing the movement of charges between the ends of the antenna, form concentric circles around the antenna and are perpendicular to the electric field lines at all points. The magnetic field is zero at all points along the axis of the antenna. Furthermore, **E** and **B** are 90° out of phase in time—for example, the current is zero when the charges at the outer ends of the rods are at a maximum.

At the two points where the magnetic field is shown in Figure 34.10, the Poynting vector **S** is directed radially outward. This indicates that energy is flowing away from the antenna at this instant. At later times, the fields and the Poynting vector reverse direction as the current alternates. Because **E** and **B** are 90° out of phase at points near the dipole, the net energy flow is zero. From this, we might conclude (incorrectly) that no energy is radiated by the dipole.

However, we find that energy is indeed radiated. Because the dipole fields fall off as $1/r^3$ (as shown in Example 23.6 for the electric field of a static dipole), they are negligible at great distances from the antenna. At these great distances, something else causes a type of radiation different from that close to the antenna. The source of this radiation is the continuous induction of an electric field by the time-varying magnetic field and the induction of a magnetic field by the time-varying electric field, predicted by Equations 34.3 and 34.4. The electric and magnetic fields produced in this manner are in phase with each other and vary as $1/r$. The result is an outward flow of energy at all times.

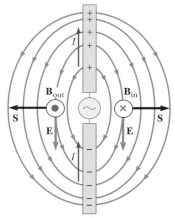

Figure 34.10 A half-wave antenna consists of two metal rods connected to an alternating voltage source. This diagram shows **E** and **B** at an arbitrary instant when the current is upward. Note that the electric field lines resemble those of a dipole (shown in Fig. 23.22).

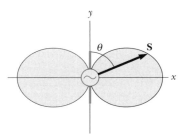

Figure 34.11 Angular dependence of the intensity of radiation produced by an oscillating electric dipole. The distance from the origin to a point on the edge of the gray shape is proportional to the intensity of radiation.

The angular dependence of the radiation intensity produced by a dipole antenna is shown in Figure 34.11. Note that the intensity and the power radiated are a maximum in a plane that is perpendicular to the antenna and passing through its midpoint. Furthermore, the power radiated is zero along the antenna's axis. A mathematical solution to Maxwell's equations for the dipole antenna shows that the intensity of the radiation varies as $(\sin^2 \theta)/r^2$, where θ is measured from the axis of the antenna.

Electromagnetic waves can also induce currents in a receiving antenna. The response of a dipole receiving antenna at a given position is a maximum when the antenna axis is parallel to the electric field at that point and zero when the axis is perpendicular to the electric field.

Quick Quiz 34.6 If the antenna in Figure 34.10 represents the source of a distant radio station, rank the following points in terms of the intensity of the radiation, from greatest to least: (a) a distance d to the right of the antenna (b) a distance $2d$ to the left of the antenna (c) a distance $2d$ in front of the antenna (out of the page) (d) a distance d above the antenna (toward the top of the page).

Quick Quiz 34.7 If the antenna in Figure 34.10 represents the source of a distant radio station, what would be the best orientation for your portable radio antenna located to the right of the figure—(a) up–down along the page, (b) left–right along the page, or (c) perpendicular to the page?

34.6 The Spectrum of Electromagnetic Waves

The various types of electromagnetic waves are listed in Figure 34.12, which shows the **electromagnetic spectrum.** Note the wide ranges of frequencies and wavelengths. No sharp dividing point exists between one type of wave and the next. Remember that **all forms of the various types of radiation are produced by the same phenomenon—accelerating charges.** The names given to the types of waves are simply for convenience in describing the region of the spectrum in which they lie.

Radio waves, whose wavelengths range from more than 10^4 m to about 0.1 m, are the result of charges accelerating through conducting wires. They are generated by such electronic devices as *LC* oscillators and are used in radio and television communication systems.

Microwaves have wavelengths ranging from approximately 0.3 m to 10^{-4} m and are also generated by electronic devices. Because of their short wavelengths, they are well suited for radar systems and for studying the atomic and molecular properties of matter. Microwave ovens are an interesting domestic application of these waves. It has been suggested that solar energy could be harnessed by beaming microwaves to the Earth from a solar collector in space.

Infrared waves have wavelengths ranging from approximately 10^{-3} m to the longest wavelength of visible light, 7×10^{-7} m. These waves, produced by molecules and room-temperature objects, are readily absorbed by most materials. The infrared (IR) energy absorbed by a substance appears as internal energy because the energy agitates the atoms of the object, increasing their vibrational or translational motion, which results in a temperature increase. Infrared radiation has practical and scientific applications in many areas, including physical therapy, IR photography, and vibrational spectroscopy.

Visible light, the most familiar form of electromagnetic waves, is the part of the electromagnetic spectrum that the human eye can detect. Light is produced by the rearrangement of electrons in atoms and molecules. The various wavelengths of visible

 PITFALL PREVENTION

34.6 "Heat Rays"

Infrared rays are often called "heat rays." This is a misnomer. While infrared radiation is used to raise or maintain temperature, as in the case of keeping food warm with "heat lamps" at a fast-food restaurant, all wavelengths of electromagnetic radiation carry energy that can cause the temperature of a system to increase. As an example, consider a potato baking in your microwave oven.

Wearing sunglasses that do not block ultraviolet (UV) light is worse for your eyes than wearing no sunglasses. The lenses of any sunglasses absorb some visible light, thus causing the wearer's pupils to dilate. If the glasses do not also block UV light, then more damage may be done to the lens of the eye because of the dilated pupils. If you wear no sunglasses at all, your pupils are contracted, you squint, and much less UV light enters your eyes. High-quality sunglasses block nearly all the eye-damaging UV light.

Figure 34.12 The electromagnetic spectrum. Note the overlap between adjacent wave types. The expanded view to the right shows details of the visible spectrum.

light, which correspond to different colors, range from red ($\lambda \approx 7 \times 10^{-7}$ m) to violet ($\lambda \approx 4 \times 10^{-7}$ m). The sensitivity of the human eye is a function of wavelength, being a maximum at a wavelength of about 5.5×10^{-7} m. With this in mind, why do you suppose tennis balls often have a yellow-green color?

Ultraviolet waves cover wavelengths ranging from approximately 4×10^{-7} m to 6×10^{-10} m. The Sun is an important source of ultraviolet (UV) light, which is the main cause of sunburn. Sunscreen lotions are transparent to visible light but absorb most UV light. The higher a sunscreen's solar protection factor (SPF), the greater the percentage of UV light absorbed. Ultraviolet rays have also been implicated in the formation of cataracts, a clouding of the lens inside the eye.

Most of the UV light from the Sun is absorbed by ozone (O_3) molecules in the Earth's upper atmosphere, in a layer called the stratosphere. This ozone shield converts lethal high-energy UV radiation to infrared radiation, which in turn warms the stratosphere. Recently, a great deal of controversy has arisen concerning the possible depletion of the protective ozone layer as a result of the chemicals emitted from aerosol spray cans and used as refrigerants.

X-rays have wavelengths in the range from approximately 10^{-8} m to 10^{-12} m. The most common source of x-rays is the stopping of high-energy electrons upon bombarding a metal target. X-rays are used as a diagnostic tool in medicine and as a treatment for certain forms of cancer. Because x-rays damage or destroy living tissues and organisms, care must be taken to avoid unnecessary exposure or overexposure. X-rays are also used in the study of crystal structure because x-ray wavelengths are comparable to the atomic separation distances in solids (about 0.1 nm).

Gamma rays are electromagnetic waves emitted by radioactive nuclei (such as ^{60}Co and ^{137}Cs) and during certain nuclear reactions. High-energy gamma rays are a component of cosmic rays that enter the Earth's atmosphere from space. They have wavelengths ranging from approximately 10^{-10} m to less than 10^{-14} m. They are highly penetrating and produce serious damage when absorbed by living tissues. Consequently, those working near such dangerous radiation must be protected with heavily absorbing materials, such as thick layers of lead.

Quick Quiz 34.8 In many kitchens, a microwave oven is used to cook food. The frequency of the microwaves is on the order of 10^{10} Hz. The wavelengths of these microwaves are on the order of (a) kilometers (b) meters (c) centimeters (d) micrometers.

Quick Quiz 34.9 A radio wave of frequency on the order of 10^5 Hz is used to carry a sound wave with a frequency on the order of 10^3 Hz. The wavelength of this radio wave is on the order of (a) kilometers (b) meters (c) centimeters (d) micrometers.

Example 34.6   A Half-Wave Antenna

A half-wave antenna works on the principle that the optimum length of the antenna is half the wavelength of the radiation being received. What is the optimum length of a car antenna when it receives a signal of frequency 94.7 MHz?

Solution Equation 34.13 tells us that the wavelength of the signal is

$$\lambda = \frac{3.00 \times 10^8 \text{ m/s}}{9.47 \times 10^7 \text{ Hz}} = \boxed{3.16 \text{ m}}$$

Thus, to operate most efficiently, the antenna should have a length of $(3.16 \text{ m})/2 = 1.58$ m. For practical reasons, car antennas are usually one-quarter wavelength in size.

SUMMARY

Take a practice test for this chapter by clicking on the Practice Test link at http://www.pse6.com.

Electromagnetic waves, which are predicted by Maxwell's equations, have the following properties:

- The electric field and the magnetic field each satisfy a wave equation. These two wave equations, which can be obtained from Maxwell's third and fourth equations, are

$$\frac{\partial^2 E}{\partial x^2} = \mu_0 \epsilon_0 \frac{\partial^2 E}{\partial t^2} \tag{34.8}$$

$$\frac{\partial^2 B}{\partial x^2} = \mu_0 \epsilon_0 \frac{\partial^2 B}{\partial t^2} \tag{34.9}$$

- The waves travel through a vacuum with the speed of light c, where

$$c = \frac{1}{\sqrt{\mu_0 \epsilon_0}} = 3.00 \times 10^8 \text{ m/s} \tag{34.10}$$

- The electric and magnetic fields are perpendicular to each other and perpendicular to the direction of wave propagation. (Hence, electromagnetic waves are transverse waves.)

- The instantaneous magnitudes of **E** and **B** in an electromagnetic wave are related by the expression

$$\frac{E}{B} = c \tag{34.14}$$

- The waves carry energy. The rate of flow of energy crossing a unit area is described by the Poynting vector **S**, where

$$\mathbf{S} \equiv \frac{1}{\mu_0} \mathbf{E} \times \mathbf{B} \qquad (34.19)$$

- Electromagnetic waves carry momentum and hence exert pressure on surfaces. If an electromagnetic wave whose Poynting vector is **S** is completely absorbed by a surface upon which it is normally incident, the radiation pressure on that surface is

$$P = \frac{S}{c} \qquad \text{(complete absorption)} \qquad (34.25)$$

If the surface totally reflects a normally incident wave, the pressure is doubled.

The electric and magnetic fields of a sinusoidal plane electromagnetic wave propagating in the positive x direction can be written

$$E = E_{max} \cos(kx - \omega t) \qquad (34.11)$$

$$B = B_{max} \cos(kx - \omega t) \qquad (34.12)$$

where ω is the angular frequency of the wave and k is the angular wave number. These equations represent special solutions to the wave equations for E and B. The wavelength and frequency of electromagnetic waves are related by

$$\lambda = \frac{c}{f} = \frac{3.00 \times 10^8 \text{ m/s}}{f} \qquad (34.13)$$

The average value of the Poynting vector for a plane electromagnetic wave has a magnitude

$$S_{av} = \frac{E_{max} B_{max}}{2\mu_0} = \frac{E_{max}^2}{2\mu_0 c} = \frac{c}{2\mu_0} B_{max}^2 \qquad (34.21)$$

The intensity of a sinusoidal plane electromagnetic wave equals the average value of the Poynting vector taken over one or more cycles.

The electromagnetic spectrum includes waves covering a broad range of wavelengths, from long radio waves at more than 10^4 m to gamma rays at less than 10^{-14} m.

QUESTIONS

1. Radio stations often advertise "instant news." If they mean that you can hear the news the instant they speak it, is their claim true? About how long would it take for a message to travel across this country by radio waves, assuming that the waves could be detected at this range?

2. Light from the Sun takes approximately 8.3 min to reach the Earth. During this time interval the Earth has continued to rotate on its axis. How far is the actual direction of the Sun from its image in the sky?

3. When light (or other electromagnetic radiation) travels across a given region, what is it that oscillates? What is it that is transported?

4. Do all current-carrying conductors emit electromagnetic waves? Explain.

5. What is the fundamental source of electromagnetic radiation?

6. If a high-frequency current is passed through a solenoid containing a metallic core, the core becomes warm due to induction. Explain why the material rises in temperature in this situation.

7. Does a wire connected to the terminals of a battery emit electromagnetic waves? Explain.

8. If you charge a comb by running it through your hair and then hold the comb next to a bar magnet, do the electric and magnetic fields produced constitute an electromagnetic wave?

9. List as many similarities and differences between sound waves and light waves as you can.

10. Describe the physical significance of the Poynting vector.

11. For a given incident energy of an electromagnetic wave, why is the radiation pressure on a perfectly reflecting surface twice as great as that on a perfect absorbing surface?

12. Before the advent of cable television and satellite dishes, city dwellers often used "rabbit ears" atop their sets (Fig. Q34.12). Certain orientations of the receiving antenna on a television set give better reception than others. Furthermore, the best orientation varies from station to station. Explain.

Figure Q34.12 Questions 12, 14, 15, and 16, and Problem 49. The V-shaped pair of long rods is the VHF antenna and the loop is the UHF antenna.

13. Often when you touch the indoor antenna on a radio or television receiver, the reception instantly improves. Why?

14. Explain how the (dipole) VHF antenna of a television set works. (See Fig. Q34.12.)

15. Explain how the UHF (loop) antenna of a television set works. (See Fig. Q34.12.)

16. Explain why the voltage induced in a UHF (loop) antenna depends on the frequency of the signal, while the voltage in a VHF (dipole) antenna does not. (See Fig. Q34.12.)

17. Electrical engineers often speak of the radiation resistance of an antenna. What do you suppose they mean by this phrase?

18. What does a radio wave do to the charges in the receiving antenna to provide a signal for your car radio?

19. An empty plastic or glass dish being removed from a microwave oven is cool to the touch. How can this be possible? (Assume that your electric bill has been paid.)

20. Why should an infrared photograph of a person look different from a photograph taken with visible light?

21. Suppose that a creature from another planet had eyes that were sensitive to infrared radiation. Describe what the alien would see if it looked around the room you are now in. In particular, what would be bright and what would be dim?

22. A welder must wear protective glasses and clothing to prevent eye damage and sunburn. What does this imply about the nature of the light produced by the welding?

23. A home microwave oven uses electromagnetic waves with a wavelength of about 12.2 cm. Some 2.4-GHz cordless telephones suffer noisy interference when a microwave oven is used nearby. Locate the waves used by both devices on the electromagnetic spectrum. Do you expect them to interfere with each other?

PROBLEMS

1, **2**, **3** = straightforward, intermediate, challenging   ☐ = full solution available in the *Student Solutions Manual and Study Guide*

〰 = coached solution with hints available at http://www.pse6.com   💻 = computer useful in solving problem

▨ = paired numerical and symbolic problems

Section 34.1   Maxwell's Equations and Hertz's Discoveries

Note: Assume that the medium is vacuum unless specified otherwise.

1. A very long, thin rod carries electric charge with the linear density 35.0 nC/m. It lies along the x axis and moves in the x direction at a speed of 15.0 Mm/s. (a) Find the electric field the rod creates at the point (0, 20.0 cm, 0). (b) Find the magnetic field it creates at the same point.

(c) Find the force exerted on an electron at this point, moving with a velocity of $(240\hat{\mathbf{i}})$ Mm/s.

Section 34.2   Plane Electromagnetic Waves

2. (a) The distance to the North Star, Polaris, is approximately 6.44×10^{18} m. If Polaris were to burn out today, in what year would we see it disappear? (b) How long does it take for sunlight to reach the Earth? (c) How long does it take for a microwave radar signal to travel from the Earth to the Moon and back? (d) How long does it take for a radio wave to travel once around the Earth in a great circle, close to the planet's surface? (e) How long

does it take for light to reach you from a lightning stroke 10.0 km away?

3. The speed of an electromagnetic wave traveling in a transparent nonmagnetic substance is $v = 1/\sqrt{\kappa\mu_0\epsilon_0}$, where κ is the dielectric constant of the substance. Determine the speed of light in water, which has a dielectric constant at optical frequencies of 1.78.

4. An electromagnetic wave in vacuum has an electric field amplitude of 220 V/m. Calculate the amplitude of the corresponding magnetic field.

5. 🖊️ Figure 34.3 shows a plane electromagnetic sinusoidal wave propagating in the x direction. Suppose that the wavelength is 50.0 m, and the electric field vibrates in the xy plane with an amplitude of 22.0 V/m. Calculate (a) the frequency of the wave and (b) the magnitude and direction of **B** when the electric field has its maximum value in the negative y direction. (c) Write an expression for **B** with the correct unit vector, with numerical values for B_{max}, k, and ω, and with its magnitude in the form

$$B = B_{max} \cos(kx - \omega t)$$

6. Write down expressions for the electric and magnetic fields of a sinusoidal plane electromagnetic wave having a frequency of 3.00 GHz and traveling in the positive x direction. The amplitude of the electric field is 300 V/m.

7. In SI units, the electric field in an electromagnetic wave is described by

$$E_y = 100 \sin(1.00 \times 10^7 x - \omega t)$$

Find (a) the amplitude of the corresponding magnetic field oscillations, (b) the wavelength λ, and (c) the frequency f.

8. Verify by substitution that the following equations are solutions to Equations 34.8 and 34.9, respectively:

$$E = E_{max} \cos(kx - \omega t)$$
$$B = B_{max} \cos(kx - \omega t)$$

9. Review problem. A standing-wave interference pattern is set up by radio waves between two metal sheets 2.00 m apart. This is the shortest distance between the plates that will produce a standing-wave pattern. What is the fundamental frequency?

10. A microwave oven is powered by an electron tube called a magnetron, which generates electromagnetic waves of frequency 2.45 GHz. The microwaves enter the oven and are reflected by the walls. The standing-wave pattern produced in the oven can cook food unevenly, with hot spots in the food at antinodes and cool spots at nodes, so a turntable is often used to rotate the food and distribute the energy. If a microwave oven intended for use with a turntable is instead used with a cooking dish in a fixed position, the antinodes can appear as burn marks on foods such as carrot strips or cheese. The separation distance between the burns is measured to be 6 cm $\pm$ 5%. From these data, calculate the speed of the microwaves.

Section 34.3 Energy Carried by Electromagnetic Waves

11. How much electromagnetic energy per cubic meter is contained in sunlight, if the intensity of sunlight at the Earth's surface under a fairly clear sky is 1 000 W/m²?

12. An AM radio station broadcasts isotropically (equally in all directions) with an average power of 4.00 kW. A dipole receiving antenna 65.0 cm long is at a location 4.00 miles from the transmitter. Compute the amplitude of the emf that is induced by this signal between the ends of the receiving antenna.

13. What is the average magnitude of the Poynting vector 5.00 miles from a radio transmitter broadcasting isotropically with an average power of 250 kW?

14. A monochromatic light source emits 100 W of electromagnetic power uniformly in all directions. (a) Calculate the average electric-field energy density 1.00 m from the source. (b) Calculate the average magnetic-field energy density at the same distance from the source. (c) Find the wave intensity at this location.

15. 🌐 A community plans to build a facility to convert solar radiation to electrical power. They require 1.00 MW of power, and the system to be installed has an efficiency of 30.0% (that is, 30.0% of the solar energy incident on the surface is converted to useful energy that can power the community). What must be the effective area of a perfectly absorbing surface used in such an installation, assuming sunlight has a constant intensity of 1 000 W/m²?

16. Assuming that the antenna of a 10.0-kW radio station radiates spherical electromagnetic waves, compute the maximum value of the magnetic field 5.00 km from the antenna, and compare this value with the surface magnetic field of the Earth.

17. 🖊️ The filament of an incandescent lamp has a 150-Ω resistance and carries a direct current of 1.00 A. The filament is 8.00 cm long and 0.900 mm in radius. (a) Calculate the Poynting vector at the surface of the filament, associated with the static electric field producing the current and the current's static magnetic field. (b) Find the magnitude of the static electric and magnetic fields at the surface of the filament.

18. One of the weapons being considered for the "Star Wars" antimissile system is a laser that could destroy ballistic missiles. When a high-power laser is used in the Earth's atmosphere, the electric field can ionize the air, turning it into a conducting plasma that reflects the laser light. In dry air at 0°C and 1 atm, electric breakdown occurs for fields with amplitudes above about 3.00 MV/m. (a) What laser beam intensity will produce such a field? (b) At this maximum intensity, what power can be delivered in a cylindrical beam of diameter 5.00 mm?

19. In a region of free space the electric field at an instant of time is $\mathbf{E} = (80.0\hat{\mathbf{i}} + 32.0\hat{\mathbf{j}} - 64.0\hat{\mathbf{k}})$ N/C and the magnetic field is $\mathbf{B} = (0.200\hat{\mathbf{i}} + 0.080\,0\hat{\mathbf{j}} + 0.290\hat{\mathbf{k}})$ μT. (a) Show that the two fields are perpendicular to each other. (b) Determine the Poynting vector for these fields.

20. Let us model the electromagnetic wave in a microwave oven as a plane traveling wave moving to the left, with an intensity of 25.0 kW/m². An oven contains two cubical containers of small mass, each full of water. One has an edge length of 6.00 cm and the other, 12.0 cm. Energy falls perpendicularly on one face of each container. The water in the smaller container absorbs 70.0% of the energy that falls on it. The water in the larger container absorbs 91.0%. (That is, the fraction 0.3 of the incoming microwave energy passes through a 6-cm thickness of water, and the fraction (0.3)(0.3) = 0.09 passes through a 12-cm thickness.) Find the temperature change of the water in each container over a time interval of 480 s. Assume that a negligible amount of energy leaves either container by heat.

21. A lightbulb filament has a resistance of 110 Ω. The bulb is plugged into a standard 120-V (rms) outlet, and emits 1.00% of the electric power delivered to it by electromagnetic radiation of frequency f. Assuming that the bulb is covered with a filter that absorbs all other frequencies, find the amplitude of the magnetic field 1.00 m from the bulb.

22. A certain microwave oven contains a magnetron that has an output of 700 W of microwave power for an electrical input power of 1.40 kW. The microwaves are entirely transferred from the magnetron into the oven chamber through a waveguide, which is a metal tube of rectangular cross section with width 6.83 cm and height 3.81 cm. (a) What is the efficiency of the magnetron? (b) Assuming that the food is absorbing all the microwaves produced by the magnetron and that no energy is reflected back into the waveguide, find the direction and magnitude of the Poynting vector, averaged over time, in the waveguide near the entrance to the oven chamber. (c) What is the maximum electric field at this point?

23. High-power lasers in factories are used to cut through cloth and metal (Fig. P34.23). One such laser has a beam diameter of 1.00 mm and generates an electric field

Figure P34.23 A laser cutting device mounted on a robot arm is being used to cut through a metallic plate.

having an amplitude of 0.700 MV/m at the target. Find (a) the amplitude of the magnetic field produced, (b) the intensity of the laser, and (c) the power delivered by the laser.

24. A 10.0-mW laser has a beam diameter of 1.60 mm. (a) What is the intensity of the light, assuming it is uniform across the circular beam? (b) What is the average energy density of the beam?

25. At one location on the Earth, the rms value of the magnetic field caused by solar radiation is 1.80 μT. From this value calculate (a) the rms electric field due to solar radiation, (b) the average energy density of the solar component of electromagnetic radiation at this location, and (c) the average magnitude of the Poynting vector for the Sun's radiation. (d) Compare the value found in part (c) to the value of the solar intensity given in Example 34.5.

Section 34.4 Momentum and Radiation Pressure

26. A 100-mW laser beam is reflected back upon itself by a mirror. Calculate the force on the mirror.

27. A radio wave transmits 25.0 W/m² of power per unit area. A flat surface of area A is perpendicular to the direction of propagation of the wave. Calculate the radiation pressure on it, assuming the surface is a perfect absorber.

28. A possible means of space flight is to place a perfectly reflecting aluminized sheet into orbit around the Earth and then use the light from the Sun to push this "solar sail." Suppose a sail of area 6.00×10^5 m² and mass 6 000 kg is placed in orbit facing the Sun. (a) What force is exerted on the sail? (b) What is the sail's acceleration? (c) How long does it take the sail to reach the Moon, 3.84×10^8 m away? Ignore all gravitational effects, assume that the acceleration calculated in part (b) remains constant, and assume a solar intensity of 1 340 W/m².

29. A 15.0-mW helium–neon laser ($\lambda = 632.8$ nm) emits a beam of circular cross section with a diameter of 2.00 mm. (a) Find the maximum electric field in the beam. (b) What total energy is contained in a 1.00-m length of the beam? (c) Find the momentum carried by a 1.00-m length of the beam.

30. Given that the intensity of solar radiation incident on the upper atmosphere of the Earth is 1 340 W/m², determine (a) the intensity of solar radiation incident on Mars, (b) the total power incident on Mars, and (c) the radiation force that acts on the planet if it absorbs nearly all of the light. (d) Compare this force to the gravitational attraction between Mars and the Sun. (See Table 13.2.)

31. A plane electromagnetic wave has an intensity of 750 W/m². A flat, rectangular surface of dimensions 50.0 cm × 100 cm is placed perpendicular to the direction of the wave. The surface absorbs half of the energy and reflects half. Calculate (a) the total energy absorbed by the surface in 1.00 min and (b) the momentum absorbed in this time.

32. A uniform circular disk of mass 24.0 g and radius 40.0 cm hangs vertically from a fixed, frictionless, horizontal hinge at a point on its circumference. A horizontal beam of electromagnetic radiation with intensity 10.0 MW/m^2 is incident on the disk in a direction perpendicular to its surface. The disk is perfectly absorbing, and the resulting radiation pressure makes the disk rotate. Find the angle through which the disk rotates as it reaches its new equilibrium position. (Assume that the radiation is *always* perpendicular to the surface of the disk.)

Section 34.5 Production of Electromagnetic Waves by an Antenna

33. Figure 34.10 shows a Hertz antenna (also known as a half-wave antenna, because its length is $\lambda/2$). The antenna is located far enough from the ground that reflections do not significantly affect its radiation pattern. Most AM radio stations, however, use a Marconi antenna, which consists of the top half of a Hertz antenna. The lower end of this (quarter-wave) antenna is connected to Earth ground, and the ground itself serves as the missing lower half. What are the heights of the Marconi antennas for radio stations broadcasting at (a) 560 kHz and (b) 1 600 kHz?

34. Two hand-held radio transceivers with dipole antennas are separated by a large fixed distance. If the transmitting antenna is vertical, what fraction of the maximum received power will appear in the receiving antenna when it is inclined from the vertical by (a) 15.0°? (b) 45.0°? (c) 90.0°?

35. Two radio-transmitting antennas are separated by half the broadcast wavelength and are driven in phase with each other. In which directions are (a) the strongest and (b) the weakest signals radiated?

36. **Review problem.** Accelerating charges radiate electromagnetic waves. Calculate the wavelength of radiation produced by a proton moving in a circle of radius R perpendicular to a magnetic field of magnitude B.

Figure P34.37 Representation of the plane electromagnetic wave radiated by an infinite current sheet lying in the *yz* plane. The vector **B** is in the *z* direction, the vector **E** is in the *y* direction, and the direction of wave motion is along *x*. Both vectors have magnitudes proportional to $\cos(kx - \omega t)$.

37. A very large flat sheet carries a uniformly distributed electric current with current per unit width J_s. Example 30.6 demonstrated that the current creates a magnetic field on both sides of the sheet, parallel to the sheet and perpendicular to the current, with magnitude $B = \frac{1}{2}\mu_0 J_s$. If the current oscillates in time according to

$$\mathbf{J}_s = J_{max}(\cos \omega t)\hat{\mathbf{j}} = J_{max}[\cos(-\omega t)]\hat{\mathbf{j}},$$

the sheet radiates an electromagnetic wave as shown in Figure P34.37. The magnetic field of the wave is described by the wave function $\mathbf{B} = \frac{1}{2}\mu_0 J_{max}[\cos(kx - \omega t)]\hat{\mathbf{k}}$. (a) Find the wave function for the electric field in the wave. (b) Find the Poynting vector as a function of *x* and *t*. (c) Find the intensity of the wave. (d) **What If?** If the sheet is to emit radiation in each direction (normal to the plane of the sheet) with intensity 570 W/m^2, what maximum value of sinusoidal current density is required?

Section 34.6 The Spectrum of Electromagnetic Waves

38. Classify waves with frequencies of 2 Hz, 2 kHz, 2 MHz, 2 GHz, 2 THz, 2 PHz, 2 EHz, 2 ZHz, and 2 YHz on the electromagnetic spectrum. Classify waves with wavelengths of 2 km, 2 m, 2 mm, 2 μm, 2 nm, 2 pm, 2 fm, and 2 am.

39. The human eye is most sensitive to light having a wavelength of 5.50×10^{-7} m, which is in the green-yellow region of the visible electromagnetic spectrum. What is the frequency of this light?

40. Compute an order-of-magnitude estimate for the frequency of an electromagnetic wave with wavelength equal to (a) your height; (b) the thickness of this sheet of paper. How is each wave classified on the electromagnetic spectrum?

41. What are the wavelengths of electromagnetic waves in free space that have frequencies of (a) 5.00×10^{19} Hz and (b) 4.00×10^9 Hz?

42. Suppose you are located 180 m from a radio transmitter. (a) How many wavelengths are you from the transmitter if the station calls itself 1 150 AM? (The AM band frequencies are in kilohertz.) (b) **What If?** What if this station is 98.1 FM? (The FM band frequencies are in megahertz.)

43. A radar pulse returns to the receiver after a total travel time of 4.00×10^{-4} s. How far away is the object that reflected the wave?

44. *This just in!* An important news announcement is transmitted by radio waves to people sitting next to their radios 100 km from the station, and by sound waves to people sitting across the newsroom, 3.00 m from the newscaster. Who receives the news first? Explain. Take the speed of sound in air to be 343 m/s.

45. The United States Navy has long proposed the construction of extremely low-frequency (ELF) communication systems. Such waves could penetrate the oceans to reach distant submarines. Calculate the length of a quarter-wavelength antenna for a transmitter generating ELF waves of frequency 75.0 Hz. How practical is this?

46. What are the wavelength ranges in (a) the AM radio band (540–1 600 kHz), and (b) the FM radio band (88.0–108 MHz)?

Additional Problems

47. Assume that the intensity of solar radiation incident on the cloudtops of the Earth is 1 340 W/m². (a) Calculate the total power radiated by the Sun, taking the average Earth–Sun separation to be 1.496×10^{11} m. (b) Determine the maximum values of the electric and magnetic fields in the sunlight at the Earth's location.

48. The intensity of solar radiation at the top of the Earth's atmosphere is 1 340 W/m². Assuming that 60% of the incoming solar energy reaches the Earth's surface and assuming that you absorb 50% of the incident energy, make an order-of-magnitude estimate of the amount of solar energy you absorb in a 60-min sunbath.

49. **Review problem.** In the absence of cable input or a satellite dish, a television set can use a dipole-receiving antenna for VHF channels and a loop antenna for UHF channels (Fig. Q34.12). The UHF antenna produces an emf from the changing magnetic flux through the loop. The TV station broadcasts a signal with a frequency f, and the signal has an electric-field amplitude E_{max} and a magnetic-field amplitude B_{max} at the location of the receiving antenna. (a) Using Faraday's law, derive an expression for the amplitude of the emf that appears in a single-turn circular loop antenna with a radius r, which is small compared with the wavelength of the wave. (b) If the electric field in the signal points vertically, what orientation of the loop gives the best reception?

50. Consider a small, spherical particle of radius r located in space a distance R from the Sun. (a) Show that the ratio F_{rad}/F_{grav} is proportional to $1/r$, where F_{rad} is the force exerted by solar radiation and F_{grav} is the force of gravitational attraction. (b) The result of part (a) means that, for a sufficiently small value of r, the force exerted on the particle by solar radiation exceeds the force of gravitational attraction. Calculate the value of r for which the particle is in equilibrium under the two forces. (Assume that the particle has a perfectly absorbing surface and a mass density of 1.50 g/cm³. Let the particle be located 3.75×10^{11} m from the Sun, and use 214 W/m² as the value of the solar intensity at that point.)

51. A dish antenna having a diameter of 20.0 m receives (at normal incidence) a radio signal from a distant source, as shown in Figure P34.51. The radio signal is a continuous sinusoidal wave with amplitude $E_{max} = 0.200\ \mu$V/m. Assume the antenna absorbs all the radiation that falls on the dish. (a) What is the amplitude of the magnetic field in this wave? (b) What is the intensity of the radiation received by this antenna? (c) What is the power received by the antenna? (d) What force is exerted by the radio waves on the antenna?

Figure P34.51

52. One goal of the Russian space program is to illuminate dark northern cities with sunlight reflected to Earth from a 200-m diameter mirrored surface in orbit. Several smaller prototypes have already been constructed and put into orbit. (a) Assume that sunlight with intensity 1 340 W/m² falls on the mirror nearly perpendicularly and that the atmosphere of the Earth allows 74.6% of the energy of sunlight to pass through it in clear weather. What is the power received by a city when the space mirror is reflecting light to it? (b) The plan is for the reflected sunlight to cover a circle of diameter 8.00 km. What is the intensity of light (the average magnitude of the Poynting vector) received by the city? (c) This intensity is what percentage of the vertical component of sunlight at Saint Petersburg in January, when the sun reaches an angle of 7.00° above the horizon at noon?

53. In 1965, Arno Penzias and Robert Wilson discovered the cosmic microwave radiation left over from the Big Bang expansion of the Universe. Suppose the energy density of this background radiation is 4.00×10^{-14} J/m³. Determine the corresponding electric field amplitude.

54. A hand-held cellular telephone operates in the 860- to 900-MHz band and has a power output of 0.600 W from an antenna 10.0 cm long (Fig. P34.54). (a) Find the average magnitude of the Poynting vector 4.00 cm from the antenna, at the location of a typical person's

Figure P34.54

Amos Morgan/Getty Images

head. Assume that the antenna emits energy with cylindrical wave fronts. (The actual radiation from antennas follows a more complicated pattern.) (b) The ANSI/IEEE C95.1-1991 maximum exposure standard is 0.57 mW/cm² for persons living near cellular telephone base stations, who would be continuously exposed to the radiation. Compare the answer to part (a) with this standard.

55. A linearly polarized microwave of wavelength 1.50 cm is directed along the positive x axis. The electric field vector has a maximum value of 175 V/m and vibrates in the xy plane. (a) Assume that the magnetic field component of the wave can be written in the form $B = B_{max} \sin(kx - \omega t)$ and give values for B_{max}, k, and ω. Also, determine in which plane the magnetic field vector vibrates. (b) Calculate the average value of the Poynting vector for this wave. (c) What radiation pressure would this wave exert if it were directed at normal incidence onto a perfectly reflecting sheet? (d) What acceleration would be imparted to a 500-g sheet (perfectly reflecting and at normal incidence) with dimensions of 1.00 m × 0.750 m?

56. The Earth reflects approximately 38.0% of the incident sunlight from its clouds and surface. (a) Given that the intensity of solar radiation is 1 340 W/m², what is the radiation pressure on the Earth, in pascals, at the location where the Sun is straight overhead? (b) Compare this to normal atmospheric pressure at the Earth's surface, which is 101 kPa.

57. An astronaut, stranded in space 10.0 m from his spacecraft and at rest relative to it, has a mass (including equipment) of 110 kg. Because he has a 100-W light source that forms a directed beam, he considers using the beam as a photon rocket to propel himself continuously toward the spacecraft. (a) Calculate how long it takes him to reach the spacecraft by this method. (b) **What If?** Suppose, instead, that he decides to throw the light source away in a direction opposite the spacecraft. If the mass of the light source is 3.00 kg and, after being thrown, it moves at 12.0 m/s relative to the recoiling astronaut, how long does it take for the astronaut to reach the spacecraft?

58. **Review problem.** A 1.00-m-diameter mirror focuses the Sun's rays onto an absorbing plate 2.00 cm in radius, which holds a can containing 1.00 L of water at 20.0°C. (a) If the solar intensity is 1.00 kW/m², what is the intensity on the absorbing plate? (b) What are the maximum magnitudes of the fields **E** and **B**? (c) If 40.0% of the energy is absorbed, how long does it take to bring the water to its boiling point?

59. Lasers have been used to suspend spherical glass beads in the Earth's gravitational field. (a) A black bead has a mass of 1.00 μg and a density of 0.200 g/cm³. Determine the radiation intensity needed to support the bead. (b) If the beam has a radius of 0.200 cm, what is the power required for this laser?

60. Lasers have been used to suspend spherical glass beads in the Earth's gravitational field. (a) A black bead has a mass

m and a density ρ. Determine the radiation intensity needed to support the bead. (b) If the beam has a radius r, what is the power required for this laser?

61. A microwave source produces pulses of 20.0-GHz radiation, with each pulse lasting 1.00 ns. A parabolic reflector with a face area of radius 6.00 cm is used to focus the microwaves into a parallel beam of radiation, as shown in Figure P34.61. The average power during each pulse is 25.0 kW. (a) What is the wavelength of these microwaves? (b) What is the total energy contained in each pulse? (c) Compute the average energy density inside each pulse. (d) Determine the amplitude of the electric and magnetic fields in these microwaves. (e) Assuming this pulsed beam strikes an absorbing surface, compute the force exerted on the surface during the 1.00-ns duration of each pulse.

Figure P34.61

62. The electromagnetic power radiated by a nonrelativistic moving point charge q having an acceleration a is

$$\mathcal{P} = \frac{q^2 a^2}{6\pi\epsilon_0 c^3}$$

where ϵ_0 is the permittivity of free space and c is the speed of light in vacuum. (a) Show that the right side of this equation has units of watts. (b) An electron is placed in a constant electric field of magnitude 100 N/C. Determine the acceleration of the electron and the electromagnetic power radiated by this electron. (c) **What If?** If a proton is placed in a cyclotron with a radius of 0.500 m and a magnetic field of magnitude 0.350 T, what electromagnetic power does this proton radiate?

63. A thin tungsten filament of length 1.00 m radiates 60.0 W of power in the form of electromagnetic waves. A perfectly absorbing surface in the form of a hollow cylinder of radius 5.00 cm and length 1.00 m is placed concentrically with the filament. Calculate the radiation pressure acting on the cylinder. (Assume that the radiation is emitted in the radial direction, and ignore end effects.)

64. The torsion balance shown in Figure 34.8 is used in an experiment to measure radiation pressure. The suspension fiber exerts an elastic restoring torque. Its torque constant is 1.00×10^{-11} N·m/degree, and the length of the horizontal rod is 6.00 cm. The beam from a 3.00-mW helium–neon laser is incident on the black disk, and the mirror disk is completely shielded.

Calculate the angle between the equilibrium positions of the horizontal bar when the beam is switched from "off" to "on."

65. A "laser cannon" of a spacecraft has a beam of cross-sectional area A. The maximum electric field in the beam is E. The beam is aimed at an asteroid that is initially moving in the direction of the spacecraft. What is the acceleration of the asteroid relative to the spacecraft if the laser beam strikes the asteroid perpendicular to its surface, and the surface is nonreflecting? The mass of the asteroid is m. Ignore the acceleration of the spacecraft.

66. A plane electromagnetic wave varies sinusoidally at 90.0 MHz as it travels along the $+x$ direction. The peak value of the electric field is 2.00 mV/m, and it is directed along the $\pm y$ direction. (a) Find the wavelength, the period, and the maximum value of the magnetic field. (b) Write expressions in SI units for the space and time variations of the electric field and of the magnetic field. Include numerical values and include subscripts to indicate coordinate directions. (c) Find the average power per unit area that this wave carries through space. (d) Find the average energy density in the radiation (in joules per cubic meter). (e) What radiation pressure would this wave exert upon a perfectly reflecting surface at normal incidence?

> *Note:* Section 20.7 introduced electromagnetic radiation as a mode of energy transfer. The following three problems use ideas introduced both there and in the current chapter.

67. Eliza is a black cat with four black kittens: Penelope, Rosalita, Sasha, and Timothy. Eliza's mass is 5.50 kg, and each kitten has mass 0.800 kg. One cool night all five sleep snuggled together on a mat, with their bodies forming one hemisphere. (a) Assuming that the purring heap has uniform density 990 kg/m^3, find the radius of the hemisphere. (b) Find the area of its curved surface. (c) Assume the surface temperature is uniformly 31.0°C and the emissivity is 0.970. Find the intensity of radiation emitted by the cats at their curved surface, and (d) the radiated power from this surface. (e) You may think of the emitted electromagnetic wave as having a single predominant frequency (of 31.2 THz). Find the amplitude of the electric field just outside the surface of the cozy pile, and (f) the amplitude of the magnetic field. (g) Are the sleeping cats charged? Are they current-carrying? Are they magnetic? Are they a radiation source? Do they glow in the dark? Give an explanation for your answers so that they do not seem contradictory. (h) **What If?** The next night the kittens all sleep alone, curling up into separate hemispheres like their mother. Find the total radiated power of the family. (For simplicity, we ignore throughout the cats' absorption of radiation from the environment.)

68. **Review problem.** (a) An elderly couple has a solar water heater installed on the roof of their house (Fig. P34.68). The heater consists of a flat closed box with

© Bill Banaszewski/Visuals Unlimited

Figure P34.68

extraordinarily good thermal insulation. Its interior is painted black, and its front face is made of insulating glass. Assume that its emissivity for visible light is 0.900 and its emissivity for infrared light is 0.700. Assume that light from the noon Sun is incident perpendicular to the glass with an intensity of 1 000 W/m^2, and that no water enters or leaves the box. Find the steady-state temperature of the interior of the box. (b) **What If?** The couple builds an identical box with no water tubes. It lies flat on the ground in front of the house. They use it as a cold frame, where they plant seeds in early spring. Assuming the same noon Sun is at an elevation angle of 50.0°, find the steady-state temperature of the interior of this box when its ventilation slots are tightly closed.

69. **Review problem.** The study of Creation suggests a Creator with an inordinate fondness for beetles and for small red stars. A small red star radiates electromagnetic waves with power 6.00×10^{23} W, which is only 0.159% of the luminosity of the Sun. Consider a spherical planet in a circular orbit around this star. Assume the emissivity of the planet is equal for infrared and for visible light. Assume the planet has a uniform surface temperature. Identify the projected area over which the planet absorbs starlight and the radiating area of the planet. If beetles thrive at a temperature of 310 K, what should be the radius of the planet's orbit?

Answers to Quick Quizzes

34.1 (c). Figure 34.3b shows that the **B** and **E** vectors reach their maximum and minimum values at the same time.

34.2 (c). The **B** field must be in the $+z$ direction in order that the Poynting vector be directed along the $-y$ direction.

34.3 (d). The first three choices are instantaneous values and vary in time. The wave intensity is an average over a full cycle.

34.4 (b). To maximize the pressure on the sails, they should be perfectly reflective, so that the pressure is given by Equation 34.27.

34.5 (b), (c). The radiation pressure (a) does not change because pressure is force per unit area. In (b), the smaller disk absorbs less radiation, resulting in a smaller force. For the same reason, the momentum in (c) is reduced.

34.6 (a), (b) = (c), (d). The closest point along the *x* axis in Figure 34.11 (choice a) will represent the highest intensity. Choices (b) and (c) correspond to points equidistant in different directions. Choice (d) is along the axis of the antenna and the intensity is zero.

34.7 (a). The best orientation is parallel to the transmitting antenna because that is the orientation of the electric field. The electric field moves electrons in the receiving antenna, thus inducing a current that is detected and amplified.

34.8 (c). Either Equation 34.13 or Figure 34.12 can be used to find the order of magnitude of the wavelengths.

34.9 (a). Either Equation 34.13 or Figure 34.12 can be used to find the order of magnitude of the wavelength.

Light and Optics

Light is basic to almost all life on the Earth. Plants convert the energy transferred by sunlight to chemical energy through photosynthesis. In addition, light is the principal means by which we are able to transmit and receive information to and from objects around us and throughout the Universe.

The nature and properties of light have been a subject of great interest and speculation since ancient times. The Greeks believed that light consisted of tiny particles *(corpuscles)* that were emitted by a light source and that these particles stimulated the perception of vision upon striking the observer's eye. Newton used this particle theory to explain the reflection and refraction (bending) of light. In 1678, one of Newton's contemporaries, the Dutch scientist Christian Huygens, was able to explain many other properties of light by proposing that light is a wave. In 1801, Thomas Young showed that light beams can interfere with one another, giving strong support to the wave theory. In 1865, Maxwell developed a brilliant theory that electromagnetic waves travel with the speed of light (see Chapter 34). By this time, the wave theory of light seemed to be firmly established.

However, at the beginning of the twentieth century, Max Planck returned to the particle theory of light to explain the radiation emitted by hot objects. Einstein then used the particle theory to explain how electrons are emitted by a metal exposed to light. Today, scientists view light as having a dual nature—that is, light exhibits characteristics of a wave in some situations and characteristics of a particle in other situations.

We shall discuss the particle nature of light in Part 6 of this text, which addresses modern physics. In Chapters 35 through 38, we concentrate on those aspects of light that are best understood through the wave model. First, we discuss the reflection of light at the boundary between two media and the refraction that occurs as light travels from one medium into another. Then, we use these ideas to study reflection and refraction as light forms images due to mirrors and lenses. Next, we describe how the lenses and mirrors used in such instruments as telescopes and microscopes help us view objects not clearly visible to the naked eye. Finally, we study the phenomena of diffraction, polarization, and interference as they apply to light. ■

◄ *The Grand Tetons in western Wyoming are reflected in a smooth lake at sunset. The optical principles that we study in this part of the book will explain the nature of the reflected image of the mountains and why the sky appears red. (David Muench/CORBIS)*

Chapter 35

The Nature of Light and the Laws of Geometric Optics

▲ This photograph of a rainbow shows a distinct secondary rainbow with the colors reversed. The appearance of the rainbow depends on three optical phenomena discussed in this chapter—reflection, refraction, and dispersion. (Mark D. Phillips/Photo Researchers, Inc.)

In this first chapter on optics, we begin by introducing two historical models for light and discussing early methods for measuring the speed of light. Next we study the fundamental phenomena of geometric optics—reflection of light from a surface and refraction as the light crosses the boundary between two media. We will also study the dispersion of light as it refracts into materials, resulting in visual displays such as the rainbow. Finally, we investigate the phenomenon of total internal reflection, which is the basis for the operation of optical fibers and the burgeoning technology of fiber optics.

35.1 The Nature of Light

Before the beginning of the nineteenth century, light was considered to be a stream of particles that either was emitted by the object being viewed or emanated from the eyes of the viewer. Newton, the chief architect of the particle theory of light, held that particles were emitted from a light source and that these particles stimulated the sense of sight upon entering the eye. Using this idea, he was able to explain reflection and refraction.

Most scientists accepted Newton's particle theory. During his lifetime, however, another theory was proposed—one that argued that light might be some sort of wave motion. In 1678, the Dutch physicist and astronomer Christian Huygens showed that a wave theory of light could also explain reflection and refraction.

In 1801, Thomas Young (1773–1829) provided the first clear demonstration of the wave nature of light. Young showed that, under appropriate conditions, light rays interfere with each other. Such behavior could not be explained at that time by a particle theory because there was no conceivable way in which two or more particles could come together and cancel one another. Additional developments during the nineteenth century led to the general acceptance of the wave theory of light, the most important resulting from the work of Maxwell, who in 1873 asserted that light was a form of high-frequency electromagnetic wave. As discussed in Chapter 34, Hertz provided experimental confirmation of Maxwell's theory in 1887 by producing and detecting electromagnetic waves.

Although the wave model and the classical theory of electricity and magnetism were able to explain most known properties of light, they could not explain some subsequent experiments. The most striking of these is the photoelectric effect, also discovered by Hertz: when light strikes a metal surface, electrons are sometimes ejected from the surface. As one example of the difficulties that arose, experiments showed that the kinetic energy of an ejected electron is independent of the light intensity. This finding contradicted the wave theory, which held that a more intense beam of light should add more energy to the electron. An explanation of the photoelectric effect was proposed by Einstein in 1905 in a theory that used the concept of quantization developed by Max Planck (1858–1947) in 1900. The quantization model assumes that the energy of a light wave is present in particles called *photons;* hence, the energy is said to be quantized. According to Einstein's theory, the energy of a photon is proportional to the frequency of the electromagnetic wave:

Energy of a photon

$$E = hf \tag{35.1}$$

where the constant of proportionality $h = 6.63 \times 10^{-34}$ J·s is Planck's constant (see Section 11.6). We will study this theory in Chapter 40.

In view of these developments, light must be regarded as having a dual nature: **Light exhibits the characteristics of a wave in some situations and the characteristics of a particle in other situations.** Light is light, to be sure. However, the question "Is light a wave or a particle?" is inappropriate. Sometimes light acts like a wave, and at other times it acts like a particle. In the next few chapters, we investigate the wave nature of light.

35.2 Measurements of the Speed of Light

Light travels at such a high speed ($c = 3.00 \times 10^8$ m/s) that early attempts to measure its speed were unsuccessful. Galileo attempted to measure the speed of light by positioning two observers in towers separated by approximately 10 km. Each observer carried a shuttered lantern. One observer would open his lantern first, and then the other would open his lantern at the moment he saw the light from the first lantern. Galileo reasoned that, knowing the transit time of the light beams from one lantern to the other, he could obtain the speed. His results were inconclusive. Today, we realize (as Galileo concluded) that it is impossible to measure the speed of light in this manner because the transit time is so much less than the reaction time of the observers.

Roemer's Method

In 1675, the Danish astronomer Ole Roemer (1644–1710) made the first successful estimate of the speed of light. Roemer's technique involved astronomical observations of one of the moons of Jupiter, Io, which has a period of revolution around Jupiter of approximately 42.5 h. The period of revolution of Jupiter around the Sun is about 12 yr; thus, as the Earth moves through 90° around the Sun, Jupiter revolves through only $(1/12)90° = 7.5°$ (Fig. 35.1).

An observer using the orbital motion of Io as a clock would expect the orbit to have a constant period. However, Roemer, after collecting data for more than a year, observed a systematic variation in Io's period. He found that the periods were longer than average when the Earth was receding from Jupiter and shorter than average when the Earth was approaching Jupiter. If Io had a constant period, Roemer should have seen it become eclipsed by Jupiter at a particular instant and should have been able to predict the time of the next eclipse. However, when he checked the time of the second eclipse as the Earth receded from Jupiter, he found that the eclipse was late. If the interval between his observations was three months, then the delay was approximately 600 s. Roemer attributed this variation in period to the fact that the distance between the Earth and Jupiter changed from one observation to the next. In three months (one quarter of the period of revolution of the Earth around the Sun), the light from Jupiter must travel an additional distance equal to the radius of the Earth's orbit.

Using Roemer's data, Huygens estimated the lower limit for the speed of light to be approximately 2.3×10^8 m/s. This experiment is important historically because it demonstrated that light does have a finite speed and gave an estimate of this speed.

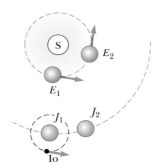

Figure 35.1 Roemer's method for measuring the speed of light. In the time interval during which the Earth travels 90° around the Sun (three months), Jupiter travels only about 7.5° (drawing not to scale).

Fizeau's Method

The first successful method for measuring the speed of light by means of purely terrestrial techniques was developed in 1849 by French physicist Armand H. L. Fizeau (1819–1896). Figure 35.2 represents a simplified diagram of Fizeau's apparatus. The basic procedure is to measure the total time interval during which light travels from some point to a distant mirror and back. If d is the distance between the light source (considered to be at the location of the wheel) and the mirror and if the time interval for one round trip is Δt, then the speed of light is $c = 2d/\Delta t$.

To measure the transit time, Fizeau used a rotating toothed wheel, which converts a continuous beam of light into a series of light pulses. The rotation of such a wheel controls what an observer at the light source sees. For example, if the pulse traveling toward the mirror and passing the opening at point A in Figure 35.2 should return to the wheel at the instant tooth B had rotated into position to cover the return path, the pulse would not reach the observer. At a greater rate of rotation, the opening at point C could move into position to allow the reflected pulse to reach the observer. Knowing the distance d, the number of teeth in the wheel, and the angular speed of the wheel, Fizeau arrived at a value of 3.1×10^8 m/s. Similar measurements made by subsequent investigators yielded more precise values for c, which led to the currently accepted value of 2.9979×10^8 m/s.

Figure 35.2 Fizeau's method for measuring the speed of light using a rotating toothed wheel. The light source is considered to be at the location of the wheel; thus, the distance d is known.

Example 35.1   Measuring the Speed of Light with Fizeau's Wheel

Assume that Fizeau's wheel has 360 teeth and is rotating at 27.5 rev/s when a pulse of light passing through opening A in Figure 35.2 is blocked by tooth B on its return. If the distance to the mirror is 7 500 m, what is the speed of light?

Solution The wheel has 360 teeth, and so it must have 360 openings. Therefore, because the light passes through opening A but is blocked by the tooth immediately adjacent to A, the wheel must rotate through an angular displacement of $(1/720)$ rev in the time interval during which the light pulse

makes its round trip. From the definition of angular speed, that time interval is

$$\Delta t = \frac{\Delta \theta}{\omega} = \frac{(1/720)\ \text{rev}}{27.5\ \text{rev/s}} = 5.05 \times 10^{-5}\ \text{s}$$

Hence, the speed of light calculated from this data is

$$c = \frac{2d}{\Delta t} = \frac{2(7\,500\ \text{m})}{5.05 \times 10^{-5}\ \text{s}} = \boxed{2.97 \times 10^8\ \text{m/s}}$$

35.3   The Ray Approximation in Geometric Optics

The field of **geometric optics** involves the study of the propagation of light, with the assumption that light travels in a fixed direction in a straight line as it passes through a uniform medium and changes its direction when it meets the surface of a different medium or if the optical properties of the medium are nonuniform in either space or time. As we study geometric optics here and in Chapter 36, we use what is called the **ray approximation.** To understand this approximation, first note that the rays of a given wave are straight lines perpendicular to the wave fronts as illustrated in Figure 35.3 for a plane wave. In the ray approximation, we assume that a wave moving through a medium travels in a straight line in the direction of its rays.

If the wave meets a barrier in which there is a circular opening whose diameter is much larger than the wavelength, as in Figure 35.4a, the wave emerging from the opening continues to move in a straight line (apart from some small edge effects); hence, the ray approximation is valid. If the diameter of the opening is on the order of the wavelength, as in Figure 35.4b, the waves spread out from the opening in all directions. This effect is called *diffraction* and will be studied in Chapter 37. Finally, if the opening is much smaller than the wavelength, the opening can be approximated as a point source of waves (Fig. 35.4c). Similar effects are seen when waves encounter an opaque object of dimension d. In this case, when $\lambda \ll d$, the object casts a sharp shadow.

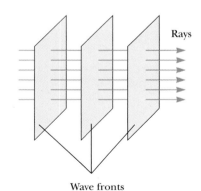

Figure 35.3 A plane wave propagating to the right. Note that the rays, which always point in the direction of the wave propagation, are straight lines perpendicular to the wave fronts.

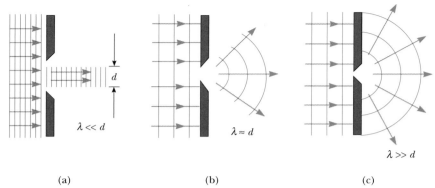

(a)   (b)   (c)

Active Figure 35.4 A plane wave of wavelength λ is incident on a barrier in which there is an opening of diameter d. (a) When $\lambda \ll d$, the rays continue in a straight-line path, and the ray approximation remains valid. (b) When $\lambda \approx d$, the rays spread out after passing through the opening. (c) When $\lambda \gg d$, the opening behaves as a point source emitting spherical waves.

At the Active Figures link at http://www.pse6.com, you can adjust the size of the opening and observe the effect on the waves passing through.

The ray approximation and the assumption that $\lambda \ll d$ are used in this chapter and in Chapter 36, both of which deal with geometric optics. This approximation is very good for the study of mirrors, lenses, prisms, and associated optical instruments, such as telescopes, cameras, and eyeglasses.

35.4 Reflection

When a light ray traveling in one medium encounters a boundary with another medium, part of the incident light is reflected. Figure 35.5a shows several rays of a beam of light incident on a smooth, mirror-like, reflecting surface. The reflected rays are parallel to each other, as indicated in the figure. The direction of a reflected ray is in the plane perpendicular to the reflecting surface that contains the

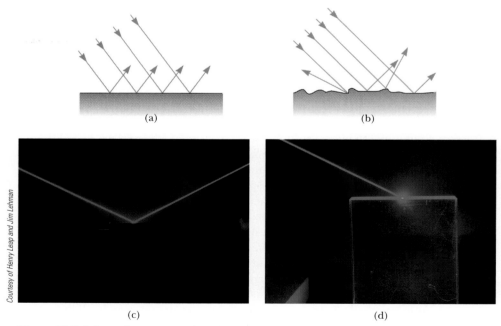

Courtesy of Henry Leap and Jim Lehman

Figure 35.5 Schematic representation of (a) specular reflection, where the reflected rays are all parallel to each other, and (b) diffuse reflection, where the reflected rays travel in random directions. (c) and (d) Photographs of specular and diffuse reflection using laser light.

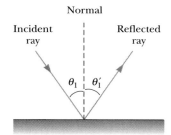

Normal

Incident
ray

Reflected
ray

θ_1 θ_1'

Active Figure 35.6 According to the law of reflection, $\theta_1' = \theta_1$. The incident ray, the reflected ray, and the normal all lie in the same plane.

At the Active Figures link at http://www.pse6.com, vary the incident angle and see the effect on the reflected ray.

incident ray. Reflection of light from such a smooth surface is called **specular reflection.** If the reflecting surface is rough, as shown in Figure 35.5b, the surface reflects the rays not as a parallel set but in various directions. Reflection from any rough surface is known as **diffuse reflection.** A surface behaves as a smooth surface as long as the surface variations are much smaller than the wavelength of the incident light.

The difference between these two kinds of reflection explains why it is more difficult to see while driving on a rainy night. If the road is wet, the smooth surface of the water specularly reflects most of your headlight beams away from your car (and perhaps into the eyes of oncoming drivers). When the road is dry, its rough surface diffusely reflects part of your headlight beam back toward you, allowing you to see the highway more clearly. In this book, we concern ourselves only with specular reflection and use the term *reflection* to mean specular reflection.

Consider a light ray traveling in air and incident at an angle on a flat, smooth surface, as shown in Figure 35.6. The incident and reflected rays make angles θ_1 and θ_1', respectively, where the angles are measured between the normal and the rays. (The normal is a line drawn perpendicular to the surface at the point where the incident ray strikes the surface.) Experiments and theory show that **the angle of reflection equals the angle of incidence:**

$$\theta_1' = \theta_1 \qquad (35.2)$$

This relationship is called the **law of reflection.**

▲ **PITFALL PREVENTION**

35.1 Subscript Notation

We use the subscript 1 to refer to parameters for the light in the initial medium. When light travels from one medium to another, we use the subscript 2 for the parameters associated with the light in the new medium. In the current discussion, the light stays in the same medium, so we only have to use the subscript 1.

Law of reflection

> **Quick Quiz 35.1** In the movies, you sometimes see an actor looking in a mirror and you can see his face in the mirror. During the filming of this scene, what does the actor see in the mirror? (a) his face (b) your face (c) the director's face (d) the movie camera (e) impossible to determine

---

Example 35.2 The Double-Reflected Light Ray **Interactive**

Two mirrors make an angle of 120° with each other, as illustrated in Figure 35.7a. A ray is incident on mirror M_1 at an angle of 65° to the normal. Find the direction of the ray after it is reflected from mirror M_2.

Solution Figure 35.7a helps conceptualize this situation. The incoming ray reflects from the first mirror, and the reflected ray is directed toward the second mirror. Thus, there is a second reflection from this latter mirror. Because the interactions with both mirrors are simple reflections, we categorize this problem as one that will require the law of reflection and some geometry. To analyze the problem, note that from the

law of reflection, we know that the first reflected ray makes an angle of 65° with the normal. Thus, this ray makes an angle of $90° - 65° = 25°$ with the horizontal.

From the triangle made by the first reflected ray and the two mirrors, we see that the first reflected ray makes an angle of 35° with M_2 (because the sum of the interior angles of any triangle is 180°). Therefore, this ray makes an angle of 55° with the normal to M_2. From the law of reflection, the second reflected ray makes an angle of 55° with the normal to M_2.

To finalize the problem, let us explore variations in the angle between the mirrors as follows.

(a)

(b)

Figure 35.7 (Example 35.2) (a) Mirrors M_1 and M_2 make an angle of 120° with each other. (b) The geometry for an arbitrary mirror angle.

What If? If the incoming and outgoing rays in Figure 35.7a are extended behind the mirror, they cross at an angle of 60°, so that the overall change in direction of the light ray is 120°. This is the same as the angle between the mirrors. What if the angle between the mirrors is changed? Is the overall change in the direction of the light ray always equal to the angle between the mirrors?

Answer Making a general statement based on one data point is always a dangerous practice! Let us investigate the change in direction for a general situation. Figure 35.7b shows the mirrors at an arbitrary angle θ and the incoming light ray striking the mirror at an arbitrary angle ϕ with respect to the normal to the mirror surface. In accordance with the law of reflection and the sum of the interior angles of a triangle, the angle γ is $180° - (90° - \phi) - \theta = 90° + \phi - \theta$. Considering the triangle highlighted in blue

in Figure 35.7b, we see that

$$\alpha + 2\gamma + 2(90° - \phi) = 180°$$

$$\alpha = 2(\phi - \gamma)$$

The change in direction of the light ray is angle β, which is $180° - \alpha$:

$$\beta = 180° - \alpha = 180° - 2(\phi - \gamma)$$
$$= 180° - 2[\phi - (90° + \phi - \theta)]$$
$$= 360° - 2\theta$$

Notice that β is not equal to θ. For $\theta = 120°$, we obtain $\beta = 120°$, which happens to be the same as the mirror angle. But this is true only for this special angle between the mirrors. For example, if $\theta = 90°$, we obtain $\beta = 180°$. In this case, the light is reflected straight back to its origin.

 Investigate this reflection situation for various mirror angles at the Interactive Worked Example link at **http://www.pse6.com.**

As discussed in the **What If?** section of the preceding example, if the angle between two mirrors is 90°, the reflected beam returns to the source parallel to its original path. This phenomenon, called *retroreflection*, has many practical applications. If a third mirror is placed perpendicular to the first two, so that the three form the corner of a cube, retroreflection works in three dimensions. In 1969, a panel of many small reflectors was placed on the Moon by the *Apollo 11* astronauts (Fig. 35.8a). A laser beam from the Earth is reflected directly back on itself and its transit time is measured. This information is used to determine the distance to the Moon with an uncertainty of 15 cm. (Imagine how difficult it would be to align a regular flat mirror so that the reflected laser beam would hit a particular location on the Earth!) A more everyday application is found in automobile taillights. Part of the plastic making up the taillight is formed into many tiny cube corners (Fig. 35.8b) so that headlight beams from cars approaching from the rear are reflected back to the drivers. Instead of cube corners, small spherical bumps are sometimes used (Fig. 35.8c). Tiny clear spheres are used in a coating material found on many road signs. Due to retroreflection from these spheres, the stop sign in Figure 35.8d appears much brighter than it would if it were simply a flat, shiny surface reflecting most of the light hitting it away from the highway.

Another practical application of the law of reflection is the digital projection of movies, television shows, and computer presentations. A digital projector makes use

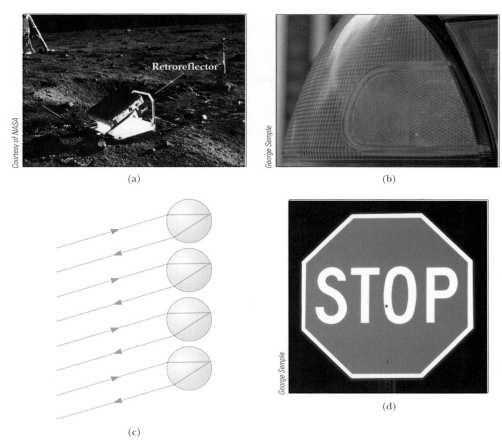

(a)

(b)

(c)

(d)

Figure 35.8 Applications of retroreflection. (a) This panel on the Moon reflects a laser beam directly back to its source on the Earth. (b) An automobile taillight has small retroreflectors that ensure that headlight beams are reflected back toward the car that sent them. (c) A light ray hitting a transparent sphere at the proper position is retroreflected. (d) This stop sign appears to glow in headlight beams because its surface is covered with a layer of many tiny retroreflecting spheres. What would you see if the sign had a mirror-like surface?

(a)

(b)

Figure 35.9 (a) An array of mirrors on the surface of a digital micromirror device. Each mirror has an area of about 16 μm^2. To provide a sense of scale, the leg of an ant appears in the photograph. (b) A close-up view of two single micromirrors. The mirror on the left is "on" and the one on the right is "off."

of an optical semiconductor chip called a *digital micromirror device*. This device contains an array of over one million tiny mirrors (Fig. 35.9a) that can be individually tilted by means of signals to an address electrode underneath the edge of the mirror. Each mirror corresponds to a pixel in the projected image. When the pixel corresponding to a given mirror is to be bright, the mirror is in the "on" position—oriented so as to reflect light from a source illuminating the array to the screen (Fig. 35.9b). When the pixel for this mirror is to be dark, the mirror is "off"—tilted so that the light is reflected away from the screen. The brightness of the pixel is determined by the total time interval during which the mirror is in the "on" position during the display of one image.

Digital movie projectors use three micromirror devices, one for each of the primary colors red, blue, and green, so that movies can be displayed with up to 35 trillion colors. Because information is stored as binary data, a digital movie does not degrade with time as does film. Furthermore, because the movie is entirely in the form of computer software, it can be delivered to theaters by means of satellites, optical discs, or optical fiber networks.

Several movies have been projected digitally to audiences and polls show that 85 percent of the viewers describe the image quality as "excellent." The first all-digital movie, from cinematography to post-production to projection, was *Star Wars Episode II: Attack of the Clones* in 2002.

35.5 Refraction

When a ray of light traveling through a transparent medium encounters a boundary leading into another transparent medium, as shown in Figure 35.10, part of the energy is reflected and part enters the second medium. The ray that enters the second medium is bent at the boundary and is said to be **refracted.** The incident ray, the reflected ray, and the refracted ray all lie in the same plane. The **angle of refraction,** θ_2 in Figure 35.10a, depends on the properties of the two media and on the angle of incidence through the relationship

$$\frac{\sin \theta_2}{\sin \theta_1} = \frac{v_2}{v_1} = \text{constant} \tag{35.3}$$

where v_1 is the speed of light in the first medium and v_2 is the speed of light in the second medium.

The path of a light ray through a refracting surface is reversible. For example, the ray shown in Figure 35.10a travels from point A to point B. If the ray originated at B, it would travel to the left along line BA to reach point A, and the reflected part would point downward and to the left in the glass.

Quick Quiz 35.2 If beam ① is the incoming beam in Figure 35.10b, which of the other four red lines are reflected beams and which are refracted beams?

From Equation 35.3, we can infer that when light moves from a material in which its speed is high to a material in which its speed is lower, as shown in Figure 35.11a, the angle of refraction θ_2 is less than the angle of incidence θ_1, and the ray is bent *toward* the normal. If the ray moves from a material in which light moves slowly to a material in which it moves more rapidly, as illustrated in Figure 35.11b, θ_2 is greater than θ_1, and the ray is bent *away* from the normal.

The behavior of light as it passes from air into another substance and then re-emerges into air is often a source of confusion to students. When light travels in air,

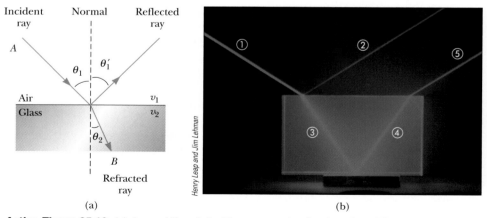

(a)   (b)

At the Active Figures link at http://www.pse6.com, vary the incident angle and see the effect on the reflected and refracted rays.

Active Figure 35.10 (a) A ray obliquely incident on an air–glass interface. The refracted ray is bent toward the normal because $v_2 < v_1$. All rays and the normal lie in the same plane. (b) Light incident on the Lucite block bends both when it enters the block and when it leaves the block.

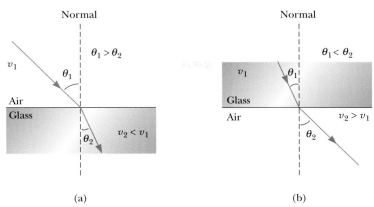

(a)   (b)

Active Figure 35.11 (a) When the light beam moves from air into glass, the light slows down on entering the glass and its path is bent toward the normal. (b) When the beam moves from glass into air, the light speeds up on entering the air and its path is bent away from the normal.

At the Active Figures link at http://www.pse6.com, light passes through three layers of material. You can vary the incident angle and see the effect on the refracted rays for a variety of values of the index of refraction (page 1104) of the three materials.

its speed is 3.00×10^8 m/s, but this speed is reduced to approximately 2×10^8 m/s when the light enters a block of glass. When the light re-emerges into air, its speed instantaneously increases to its original value of 3.00×10^8 m/s. This is far different from what happens, for example, when a bullet is fired through a block of wood. In this case, the speed of the bullet is reduced as it moves through the wood because some of its original energy is used to tear apart the wood fibers. When the bullet enters the air once again, it emerges at the speed it had just before leaving the block of wood.

To see why light behaves as it does, consider Figure 35.12, which represents a beam of light entering a piece of glass from the left. Once inside the glass, the light may encounter an electron bound to an atom, indicated as point A. Let us assume that light is absorbed by the atom; this causes the electron to oscillate (a detail represented by the double-headed vertical arrows). The oscillating electron then acts as an antenna and radiates the beam of light toward an atom at B, where the light is again absorbed. The details of these absorptions and radiations are best explained in terms of quantum mechanics (Chapter 42). For now, it is sufficient to think of light passing from one atom to another through the glass. Although light travels from one glass atom to another at 3.00×10^8 m/s, the absorption and radiation that take place cause the *average* light speed through the material to fall to about 2×10^8 m/s. Once the light emerges into the air, absorption and radiation cease and the speed of the light returns to the original value.

Figure 35.12 Light passing from one atom to another in a medium. The dots are electrons, and the vertical arrows represent their oscillations.

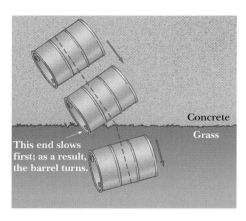

Figure 35.13 Overhead view of a barrel rolling from concrete onto grass.

A mechanical analog of refraction is shown in Figure 35.13. When the left end of the rolling barrel reaches the grass, it slows down, while the right end remains on the concrete and moves at its original speed. This difference in speeds causes the barrel to pivot, and this changes the direction of travel.

Index of Refraction

In general, the speed of light in any material is *less* than its speed in vacuum. In fact, *light travels at its maximum speed in vacuum*. It is convenient to define the **index of refraction** n of a medium to be the ratio

$$n \equiv \frac{\text{speed of light in vacuum}}{\text{speed of light in a medium}} = \frac{c}{v} \qquad (35.4)$$

From this definition, we see that the index of refraction is a dimensionless number greater than unity because v is always less than c. Furthermore, n is equal to unity for vacuum. The indices of refraction for various substances are listed in Table 35.1.

As light travels from one medium to another, its frequency does not change but its wavelength does. To see why this is so, consider Figure 35.14. Waves pass an observer at point A in medium 1 with a certain frequency and are

Index of refraction

Figure 35.14 As a wave moves from medium 1 to medium 2, its wavelength changes but its frequency remains constant.

Table 35.1

Indices of Refraction[a]			
Substance	Index of Refraction	Substance	Index of Refraction
Solids at 20°C		*Liquids at 20°C*	
Cubic zirconia	2.20	Benzene	1.501
Diamond (C)	2.419	Carbon disulfide	1.628
Fluorite (CaF_2)	1.434	Carbon tetrachloride	1.461
Fused quartz (SiO_2)	1.458	Ethyl alcohol	1.361
Gallium phosphide	3.50	Glycerin	1.473
Glass, crown	1.52	Water	1.333
Glass, flint	1.66		
Ice (H_2O)	1.309	*Gases at 0°C, 1 atm*	
Polystyrene	1.49	Air	1.000 293
Sodium chloride (NaCl)	1.544	Carbon dioxide	1.000 45

[a] All values are for light having a wavelength of 589 nm in vacuum.

incident on the boundary between medium 1 and medium 2. The frequency with which the waves pass an observer at point B in medium 2 must equal the frequency at which they pass point A. If this were not the case, then energy would be piling up at the boundary. Because there is no mechanism for this to occur, the frequency must be a constant as a light ray passes from one medium into another. Therefore, because the relationship $v = f\lambda$ (Eq. 16.12) must be valid in both media and because $f_1 = f_2 = f$, we see that

$$v_1 = f\lambda_1 \qquad \text{and} \qquad v_2 = f\lambda_2 \tag{35.5}$$

Because $v_1 \neq v_2$, it follows that $\lambda_1 \neq \lambda_2$.

We can obtain a relationship between index of refraction and wavelength by dividing the first Equation 35.5 by the second and then using Equation 35.4:

$$\frac{\lambda_1}{\lambda_2} = \frac{v_1}{v_2} = \frac{c/n_1}{c/n_2} = \frac{n_2}{n_1} \tag{35.6}$$

This gives

$$\lambda_1 n_1 = \lambda_2 n_2$$

If medium 1 is vacuum, or for all practical purposes air, then $n_1 = 1$. Hence, it follows from Equation 35.6 that the index of refraction of any medium can be expressed as the ratio

$$n = \frac{\lambda}{\lambda_n} \tag{35.7}$$

where λ is the wavelength of light in vacuum and λ_n is the wavelength of light in the medium whose index of refraction is n. From Equation 35.7, we see that because $n > 1$, $\lambda_n < \lambda$.

We are now in a position to express Equation 35.3 in an alternative form. If we replace the v_2/v_1 term in Equation 35.3 with n_1/n_2 from Equation 35.6, we obtain

$$n_1 \sin \theta_1 = n_2 \sin \theta_2 \tag{35.8}$$

The experimental discovery of this relationship is usually credited to Willebrord Snell (1591–1627) and is therefore known as **Snell's law of refraction.** We shall examine this equation further in Sections 35.6 and 35.9.

 PITFALL PREVENTION

35.3 An Inverse Relationship

The index of refraction is *inversely* proportional to the wave speed. As the wave speed v decreases, the index of refraction n increases. Thus, the higher the index of refraction of a material, the more it *slows down* light from its speed in vacuum. The more the light slows down, the more θ_2 differs from θ_1 in Equation 35.8.

Snell's law of refraction

Quick Quiz 35.3 Light passes from a material with index of refraction 1.3 into one with index of refraction 1.2. Compared to the incident ray, the refracted ray (a) bends toward the normal (b) is undeflected (c) bends away from the normal.

Quick Quiz 35.4 As light from the Sun enters the atmosphere, it refracts due to the small difference between the speeds of light in air and in vacuum. The *optical* length of the day is defined as the time interval between the instant when the top of the Sun is just visibly observed above the horizon to the instant at which the top of the Sun just disappears below the horizon. The *geometric* length of the day is defined as the time interval between the instant when a geometric straight line drawn from the observer to the top of the Sun just clears the horizon to the instant at which this line just dips below the horizon. Which is longer, (a) the optical length of a day, or (b) the geometric length of a day?

Example 35.3   An Index of Refraction Measurement

A beam of light of wavelength 550 nm traveling in air is incident on a slab of transparent material. The incident beam makes an angle of 40.0° with the normal, and the refracted beam makes an angle of 26.0° with the normal. Find the index of refraction of the material.

Solution Using Snell's law of refraction (Eq. 35.8) with these data, and taking $n_1 = 1.00$ for air, we have

$$n_1 \sin \theta_1 = n_2 \sin \theta_2$$

$$n_2 = \frac{n_1 \sin \theta_1}{\sin \theta_2} = (1.00) \frac{\sin 40.0°}{\sin 26.0°}$$

$$= \frac{0.643}{0.438} = \boxed{1.47}$$

From Table 35.1, we see that the material could be fused quartz.

Example 35.4   Angle of Refraction for Glass

A light ray of wavelength 589 nm traveling through air is incident on a smooth, flat slab of crown glass at an angle of 30.0° to the normal, as sketched in Figure 35.15. Find the angle of refraction.

Solution We rearrange Snell's law of refraction to obtain

$$\sin \theta_2 = \frac{n_1}{n_2} \sin \theta_1$$

From Table 35.1, we find that $n_1 = 1.00$ for air and $n_2 = 1.52$ for crown glass. Therefore,

$$\sin \theta_2 = \left(\frac{1.00}{1.52}\right) \sin 30.0° = 0.329$$

$$\theta_2 = \sin^{-1}(0.329) = \boxed{19.2°}$$

Because this is less than the incident angle of 30°, the refracted ray is bent toward the normal, as expected. Its

change in direction is called the *angle of deviation* and is given by $\delta = |\theta_1 - \theta_2| = 30.0° - 19.2° = 10.8°$.

Figure 35.15 (Example 35.4) Refraction of light by glass.

Example 35.5   Laser Light in a Compact Disc

A laser in a compact disc player generates light that has a wavelength of 780 nm in air.

(A) Find the speed of this light once it enters the plastic of a compact disc ($n = 1.55$).

Solution We expect to find a value less than 3.00×10^8 m/s because $n > 1$. We can obtain the speed of light in the plastic by using Equation 35.4:

$$v = \frac{c}{n} = \frac{3.00 \times 10^8 \text{ m/s}}{1.55}$$

$$v = \boxed{1.94 \times 10^8 \text{ m/s}}$$

(B) What is the wavelength of this light in the plastic?

Solution We use Equation 35.7 to calculate the wavelength in plastic, noting that we are given the wavelength in air to be $\lambda = 780$ nm:

$$\lambda_n = \frac{\lambda}{n} = \frac{780 \text{ nm}}{1.55} = \boxed{503 \text{ nm}}$$

Example 35.6   Light Passing Through a Slab   `Interactive`

A light beam passes from medium 1 to medium 2, with the latter medium being a thick slab of material whose index of refraction is n_2 (Fig. 35.16a). Show that the emerging beam is parallel to the incident beam.

Solution First, let us apply Snell's law of refraction to the upper surface:

$$(1) \qquad \sin \theta_2 = \frac{n_1}{n_2} \sin \theta_1$$

Applying this law to the lower surface gives

$$(2) \qquad \sin \theta_3 = \frac{n_2}{n_1} \sin \theta_2$$

Substituting Equation (1) into Equation (2) gives

$$\sin \theta_3 = \frac{n_2}{n_1} \left(\frac{n_1}{n_2} \sin \theta_1\right) = \sin \theta_1$$

Therefore, $\theta_3 = \theta_1$, and the slab does not alter the direction of the beam. It does, however, offset the beam parallel to itself by the distance d shown in Figure 35.16a.

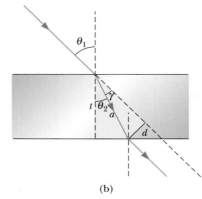

(a)   (b)

Figure 35.16 (Example 35.6) (a) When light passes through a flat slab of material, the emerging beam is parallel to the incident beam, and therefore $\theta_1 = \theta_3$. The dashed line drawn parallel to the ray coming out the bottom of the slab represents the path the light would take if the slab were not there. (b) A magnification of the area of the light path inside the slab.

What If? **What if the thickness t of the slab is doubled? Does the offset distance d also double?**

Answer Consider the magnification of the area of the light path within the slab in Figure 35.16b. The distance a is the hypotenuse of two right triangles. From the gold triangle, we see

$$a = \frac{t}{\cos \theta_2}$$

and from the blue triangle,

$$d = a \sin \gamma = a \sin(\theta_1 - \theta_2)$$

Combining these equations, we have

$$d = \frac{t}{\cos \theta_2} \sin(\theta_1 - \theta_2)$$

For a given incident angle θ_1, the refracted angle θ_2 is determined solely by the index of refraction, so the offset distance d is proportional to t. If the thickness doubles, so does the offset distance.

 Explore refraction through slabs of various thicknesses at the Interactive Worked Example link at **http://www.pse6.com.**

35.6 Huygens's Principle

In this section, we develop the laws of reflection and refraction by using a geometric method proposed by Huygens in 1678. **Huygens's principle** is a geometric construction for using knowledge of an earlier wave front to determine the position of a new wave front at some instant. In Huygens's construction,

all points on a given wave front are taken as point sources for the production of spherical secondary waves, called wavelets, which propagate outward through a medium with speeds characteristic of waves in that medium. After some time interval has passed, the new position of the wave front is the surface tangent to the wavelets.

First, consider a plane wave moving through free space, as shown in Figure 35.17a. At $t = 0$, the wave front is indicated by the plane labeled AA'. In Huygens's construction, each point on this wave front is considered a point source. For clarity, only three points on AA' are shown. With these points as sources for the wavelets, we draw circles, each of radius $c \Delta t$, where c is the speed of light in vacuum and Δt is some time interval during which the wave propagates. The surface drawn tangent to these wavelets is the plane BB', which is the wave front at a later time, and is parallel to AA'. In a similar manner, Figure 35.17b shows Huygens's construction for a spherical wave.

 PITFALL PREVENTION

35.4 Of What Use Is Huygens's Principle?

At this point, the importance of Huygens's principle may not be evident. Predicting the position of a future wave front may not seem to be very critical. However, we will use Huygens's principle in later chapters to explain additional wave phenomena for light.

Christian Huygens

Dutch Physicist and Astronomer (1629–1695)

Huygens is best known for his contributions to the fields of optics and dynamics. To Huygens, light was a type of vibratory motion, spreading out and producing the sensation of light when impinging on the eye. On the basis of this theory, he deduced the laws of reflection and refraction and explained the phenomenon of double refraction. (*Courtesy of Rijksmuseum voor de Geschiedenis der Natuurwetenschappen and Niels Bohr Library.*)

Figure 35.17 Huygens's construction for (a) a plane wave propagating to the right and (b) a spherical wave propagating to the right.

Huygens's Principle Applied to Reflection and Refraction

The laws of reflection and refraction were stated earlier in this chapter without proof. We now derive these laws, using Huygens's principle.

For the law of reflection, refer to Figure 35.18a. The line AB represents a wave front of the incident light just as ray 1 strikes the surface. At this instant, the wave at A sends out a Huygens wavelet (the circular arc centered on A) toward D. At the same time, the wave at B emits a Huygens wavelet (the circular arc centered on B) toward C. Figure 35.18a shows these wavelets after a time interval Δt, after which ray 2 strikes the surface. Because both rays 1 and 2 move with the same speed, we must have $AD = BC = c\,\Delta t$.

The remainder of our analysis depends on geometry, as summarized in Figure 35.18b, in which we isolate the triangles ABC and ADC. Note that these two triangles are congruent because they have the same hypotenuse AC and because $AD = BC$. From Figure 35.18b, we have

$$\cos \gamma = \frac{BC}{AC} \quad \text{and} \quad \cos \gamma' = \frac{AD}{AC}$$

where, comparing Figures 35.18a and 35.18b, we see that $\gamma = 90° - \theta_1$ and $\gamma' = 90° - \theta_1'$. Because $AD = BC$, we have

$$\cos \gamma = \cos \gamma'$$

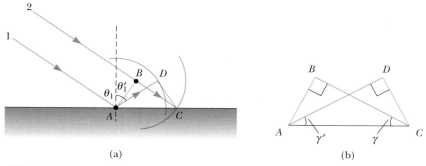

Figure 35.18 (a) Huygens's construction for proving the law of reflection. At the instant that ray 1 strikes the surface, it sends out a Huygens wavelet from A and ray 2 sends out a Huygens wavelet from B. We choose a radius of the wavelet to be $c\,\Delta t$, where Δt is the time interval for ray 2 to travel from B to C. (b) Triangle ADC is congruent to triangle ABC.

Therefore,

$$\gamma = \gamma'$$

$$90° - \theta_1 = 90° - \theta_1'$$

and

$$\theta_1 = \theta_1'$$

which is the law of reflection.

Now let us use Huygens's principle and Figure 35.19 to derive Snell's law of refraction. We focus our attention on the instant ray 1 strikes the surface and the subsequent time interval until ray 2 strikes the surface. During this time interval, the wave at A sends out a Huygens wavelet (the arc centered on A) toward D. In the same time interval, the wave at B sends out a Huygens wavelet (the arc centered on B) toward C. Because these two wavelets travel through different media, the radii of the wavelets are different. The radius of the wavelet from A is $AD = v_2 \, \Delta t$, where v_2 is the wave speed in the second medium. The radius of the wavelet from B is $BC = v_1 \, \Delta t$, where v_1 is the wave speed in the original medium.

From triangles ABC and ADC, we find that

$$\sin \theta_1 = \frac{BC}{AC} = \frac{v_1 \, \Delta t}{AC} \qquad \text{and} \qquad \sin \theta_2 = \frac{AD}{AC} = \frac{v_2 \, \Delta t}{AC}$$

If we divide the first equation by the second, we obtain

$$\frac{\sin \theta_1}{\sin \theta_2} = \frac{v_1}{v_2}$$

But from Equation 35.4 we know that $v_1 = c/n_1$ and $v_2 = c/n_2$. Therefore,

$$\frac{\sin \theta_1}{\sin \theta_2} = \frac{c/n_1}{c/n_2} = \frac{n_2}{n_1}$$

$$n_1 \sin \theta_1 = n_2 \sin \theta_2$$

which is Snell's law of refraction.

35.7 Dispersion and Prisms

An important property of the index of refraction n is that, for a given material, the index varies with the wavelength of the light passing through the material, as Figure 35.20 shows. This behavior is called **dispersion.** Because n is a function of wavelength, Snell's law of refraction indicates that light of different wavelengths is bent at different angles when incident on a refracting material.

As we see from Figure 35.20, the index of refraction generally decreases with increasing wavelength. This means that violet light bends more than red light does when passing into a refracting material. To understand the effects that dispersion can have on light, consider what happens when light strikes a prism, as shown in Figure 35.21. A ray of single-wavelength light incident on the prism from the left emerges refracted from its original direction of travel by an angle δ, called the **angle of deviation.**

Now suppose that a beam of *white light* (a combination of all visible wavelengths) is incident on a prism, as illustrated in Figure 35.22. The rays that emerge spread out in a series of colors known as the **visible spectrum.** These colors, in order of decreasing wavelength, are red, orange, yellow, green, blue, and violet. Clearly, the angle of deviation δ depends on wavelength. Violet light deviates the most, red the least, and the remaining colors in the visible spectrum fall between these extremes. Newton showed that each color has a particular angle of deviation and that the colors can be recombined to form the original white light.

The dispersion of light into a spectrum is demonstrated most vividly in nature by the formation of a rainbow, which is often seen by an observer positioned between the Sun

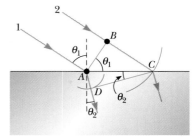

Figure 35.19 Huygens's construction for proving Snell's law of refraction. At the instant that ray 1 strikes the surface, it sends out a Huygens wavelet from A and ray 2 sends out a Huygens wavelet from B. The two wavelets have different radii because they travel in different media.

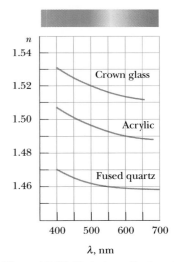

Figure 35.20 Variation of index of refraction with vacuum wavelength for three materials.

Figure 35.21 A prism refracts a single-wavelength light ray through an angle δ.

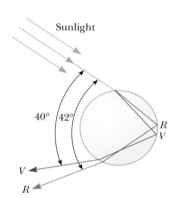

Active Figure 35.23 Path of sunlight through a spherical raindrop. Light following this path contributes to the visible rainbow.

 At the Active Figures link at http://www.pse6.com, you can vary the point at which the sunlight enters the raindrop to verify that the angles shown are the maximum angles.

▲ **PITFALL PREVENTION**

35.5 A Rainbow of Many Light Rays

Pictorial representations such as Figure 35.23 are subject to misinterpretation. The figure shows one ray of light entering the raindrop and undergoing reflection and refraction, exiting the raindrop in a range of 40° to 42° from the entering ray. This might be interpreted incorrectly as meaning that *all* light entering the raindrop exits in this small range of angles. In reality, light exits the raindrop over a much larger range of angles, from 0° to 42°. A careful analysis of the reflection and refraction from the spherical raindrop shows that the range of 40° to 42° is where the *highest-intensity light* exits the raindrop.

Figure 35.22 White light enters a glass prism at the upper left. A reflected beam of light comes out of the prism just below the incoming beam. The beam moving toward the lower right shows distinct colors. Different colors are refracted at different angles because the index of refraction of the glass depends on wavelength. Violet light deviates the most; red light deviates the least.

and a rain shower. To understand how a rainbow is formed, consider Figure 35.23. A ray of sunlight (which is white light) passing overhead strikes a drop of water in the atmosphere and is refracted and reflected as follows: It is first refracted at the front surface of the drop, with the violet light deviating the most and the red light the least. At the back surface of the drop, the light is reflected and returns to the front surface, where it again undergoes refraction as it moves from water into air. The rays leave the drop such that the angle between the incident white light and the most intense returning violet ray is 40° and the angle between the white light and the most intense returning red ray is 42°. This small angular difference between the returning rays causes us to see a colored bow.

Now suppose that an observer is viewing a rainbow, as shown in Figure 35.24. If a raindrop high in the sky is being observed, the most intense red light returning from the drop can reach the observer because it is deviated the most, but the most intense violet light passes over the observer because it is deviated the least. Hence, the observer sees this drop as being red. Similarly, a drop lower in the sky would direct the most intense violet light toward the observer and appears to be violet. (The most intense red light from this drop would pass below the eye of the observer and not be

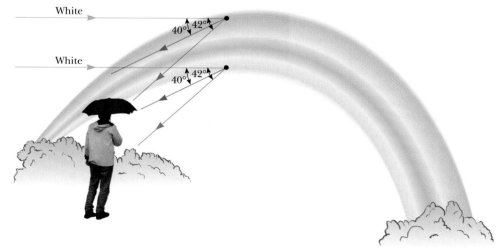

Figure 35.24 The formation of a rainbow seen by an observer standing with the Sun behind his back.

seen.) The most intense light from other colors of the spectrum would reach the observer from raindrops lying between these two extreme positions.

The opening photograph for this chapter shows a *double rainbow.* The secondary rainbow is fainter than the primary rainbow and the colors are reversed. The secondary rainbow arises from light that makes two reflections from the interior surface before exiting the raindrop. In the laboratory, rainbows have been observed in which the light makes over 30 reflections before exiting the water drop. Because each reflection involves some loss of light due to refraction out of the water drop, the intensity of these higher-order rainbows is small compared to the intensity of the primary rainbow.

Quick Quiz 35.5 Lenses in a camera use refraction to form an image on a film. Ideally, you want all the colors in the light from the object being photographed to be refracted by the same amount. Of the materials shown in Figure 35.20, which would you choose for a camera lens? (a) crown glass (b) acrylic (c) fused quartz (d) impossible to determine

---

Example 35.7 Measuring n Using a Prism

Although we do not prove it here, the minimum angle of deviation δ_{min} for a prism occurs when the angle of incidence θ_1 is such that the refracted ray inside the prism makes the same angle with the normal to the two prism faces,[1] as shown in Figure 35.25. Obtain an expression for the index of refraction of the prism material.

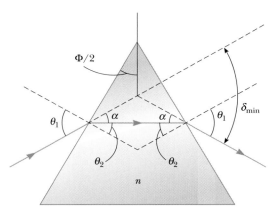

Figure 35.25 (Example 35.7) A light ray passing through a prism at the minimum angle of deviation δ_{min}.

Solution Using the geometry shown in Figure 35.25, we find that $\theta_2 = \Phi/2$, where Φ is the apex angle and

$$\theta_1 = \theta_2 + \alpha = \frac{\Phi}{2} + \frac{\delta_{min}}{2} = \frac{\Phi + \delta_{min}}{2}$$

From Snell's law of refraction, with $n_1 = 1$ because medium 1 is air, we have

$$\sin \theta_1 = n \sin \theta_2$$

$$\sin\left(\frac{\Phi + \delta_{min}}{2}\right) = n \sin(\Phi/2)$$

$$n = \frac{\sin\left(\dfrac{\Phi + \delta_{min}}{2}\right)}{\sin(\Phi/2)} \qquad (35.9)$$

Hence, knowing the apex angle Φ of the prism and measuring δ_{min}, we can calculate the index of refraction of the prism material. Furthermore, we can use a hollow prism to determine the values of n for various liquids filling the prism.

35.8 Total Internal Reflection

An interesting effect called **total internal reflection** can occur when light is directed from a medium having a given index of refraction toward one having a lower index of refraction. Consider a light beam traveling in medium 1 and meeting the boundary between medium 1 and medium 2, where n_1 is greater than n_2 (Fig. 35.26a). Various possible directions of the beam are indicated by rays 1 through 5. The refracted rays are bent away from the normal because n_1 is greater than n_2. At some particular angle of incidence θ_c, called the **critical angle,** the refracted light ray moves parallel to the boundary so that $\theta_2 = 90°$ (Fig. 35.26b).

---

[1] The details of this proof are available in texts on optics.

(a)

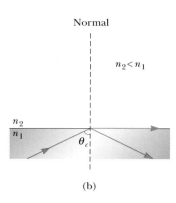

(b)

Active Figure 35.26 (a) Rays travel from a medium of index of refraction n_1 into a medium of index of refraction n_2, where $n_2 < n_1$. As the angle of incidence θ_1 increases, the angle of refraction θ_2 increases until θ_2 is 90° (ray 4). For even larger angles of incidence, total internal reflection occurs (ray 5). (b) The angle of incidence producing an angle of refraction equal to 90° is the critical angle θ_c. At this angle of incidence, all of the energy of the incident light is reflected.

At the Active Figures link at http://www.pse6.com, *you can vary the incident angle and see the effect on the refracted ray and the distribution of incident energy between the reflected and refracted rays.*

For angles of incidence greater than θ_c, the beam is entirely reflected at the boundary, as shown by ray 5 in Figure 35.26a. This ray is reflected at the boundary as it strikes the surface. This ray and all those like it obey the law of reflection; that is, for these rays, the angle of incidence equals the angle of reflection.

We can use Snell's law of refraction to find the critical angle. When $\theta_1 = \theta_c$, $\theta_2 = 90°$ and Equation 35.8 gives

$$n_1 \sin \theta_c = n_2 \sin 90° = n_2$$

Critical angle for total internal reflection

$$\sin \theta_c = \frac{n_2}{n_1} \qquad \text{(for } n_1 > n_2\text{)} \tag{35.10}$$

This equation can be used only when n_1 is greater than n_2. That is, **total internal reflection occurs only when light is directed from a medium of a given index of refraction toward a medium of lower index of refraction.** If n_1 were less than n_2, Equation 35.10 would give $\sin \theta_c > 1$; this is a meaningless result because the sine of an angle can never be greater than unity.

The critical angle for total internal reflection is small when n_1 is considerably greater than n_2. For example, the critical angle for a diamond in air is 24°. Any ray inside the diamond that approaches the surface at an angle greater than this is completely reflected back into the crystal. This property, combined with proper faceting, causes diamonds to sparkle. The angles of the facets are cut so that light is "caught" inside the crystal through multiple internal reflections. These multiple reflections give the light a long path through the medium, and substantial dispersion of colors occurs. By the time the light exits through the top surface of the crystal, the rays associated with different colors have been fairly widely separated from one another.

Cubic zirconia also has a high index of refraction and can be made to sparkle very much like a genuine diamond. If a suspect jewel is immersed in corn syrup, the difference in *n* for the cubic zirconia and that for the syrup is small, and the critical angle is therefore great. This means that more rays escape sooner, and as a result the sparkle completely disappears. A real diamond does not lose all of its sparkle when placed in corn syrup.

Quick Quiz 35.6 In Figure 35.27, five light rays enter a glass prism from the left. How many of these rays undergo total internal reflection at the slanted surface of the prism? (a) 1 (b) 2 (c) 3 (d) 4 (e) 5.

Quick Quiz 35.7 Suppose that the prism in Figure 35.27 can be rotated in the plane of the paper. In order for *all five* rays to experience total internal reflection from the slanted surface, should the prism be rotated (a) clockwise or (b) counterclockwise?

Quick Quiz 35.8 A beam of white light is incident on a crown glass–air interface as shown in Figure 35.26a. The incoming beam is rotated clockwise, so that the incident angle θ increases. Because of dispersion in the glass, some colors of light experience total internal reflection (ray 4 in Figure 35.26a) before other colors, so that the beam refracting out of the glass is no longer white. The last color to refract out of the upper surface is (a) violet (b) green (c) red (d) impossible to determine.

Figure 35.27 (Quick Quiz 35.6 and 35.7) Five nonparallel light rays enter a glass prism from the left.

Example 35.8   A View from the Fish's Eye

Find the critical angle for an air–water boundary. (The index of refraction of water is 1.33.)

Solution We can use Figure 35.26 to solve this problem, with the air above the water having index of refraction n_2 and the water having index of refraction n_1. Applying Equation 35.10, we find that

$$\sin \theta_c = \frac{n_2}{n_1} = \frac{1}{1.33} = 0.752$$

$$\theta_c = \boxed{48.8°}$$

What If? What if a fish in a still pond looks upward toward the water's surface at different angles relative to the surface, as in Figure 35.28? What does it see?

Answer Because the path of a light ray is reversible, light traveling from medium 2 into medium 1 in Figure 35.26a follows the paths shown, but in the *opposite* direction. A fish looking upward toward the water surface, as in Figure 35.28, can see out of the water if it looks toward the surface at an angle less than the critical angle. Thus, for example, when the fish's line of vision makes an angle of 40° with the normal to the surface, light from above the water reaches the fish's eye. At 48.8°, the critical angle for water, the light has to skim along the water's surface before being refracted to the fish's eye; at this angle, the fish can in principle see the whole shore of the pond. At angles greater than the critical angle, the light reaching the fish comes by means of internal reflection at the surface. Thus, at 60°, the fish sees a reflection of the bottom of the pond.

Figure 35.28 (Example 35.8) **What If?** A fish looks upward toward the water surface.

(*Left*) Strands of glass optical fibers are used to carry voice, video, and data signals in telecommunication networks. (*Right*) A bundle of optical fibers is illuminated by a laser.

Figure 35.29 Light travels in a curved transparent rod by multiple internal reflections.

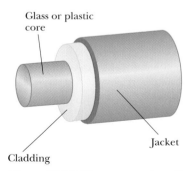

Figure 35.30 The construction of an optical fiber. Light travels in the core, which is surrounded by a cladding and a protective jacket.

Optical Fibers

Another interesting application of total internal reflection is the use of glass or transparent plastic rods to "pipe" light from one place to another. As indicated in Figure 35.29, light is confined to traveling within a rod, even around curves, as the result of successive total internal reflections. Such a light pipe is flexible if thin fibers are used rather than thick rods. A flexible light pipe is called an **optical fiber.** If a bundle of parallel fibers is used to construct an optical transmission line, images can be transferred from one point to another. This technique is used in a sizable industry known as *fiber optics.*

A practical optical fiber consists of a transparent core surrounded by a *cladding*, a material that has a lower index of refraction than the core. The combination may be surrounded by a plastic *jacket* to prevent mechanical damage. Figure 35.30 shows a cutaway view of this construction. Because the index of refraction of the cladding is less than that of the core, light traveling in the core experiences total internal reflection if it arrives at the interface between the core and the cladding at an angle of incidence that exceeds the critical angle. In this case, light "bounces" along the core of the optical fiber, losing very little of its intensity as it travels.

Any loss in intensity in an optical fiber is due essentially to reflections from the two ends and absorption by the fiber material. Optical fiber devices are particularly useful for viewing an object at an inaccessible location. For example, physicians often use such devices to examine internal organs of the body or to perform surgery without making large incisions. Optical fiber cables are replacing copper wiring and coaxial cables for telecommunications because the fibers can carry a much greater volume of telephone calls or other forms of communication than electrical wires can.

35.9 Fermat's Principle

Pierre de Fermat (1601–1665) developed a general principle that can be used to determine the path that light follows as it travels from one point to another. **Fermat's principle** states that **when a light ray travels between any two points, its path is**

the one that requires the smallest time interval. An obvious consequence of this principle is that the paths of light rays traveling in a homogeneous medium are straight lines because a straight line is the shortest distance between two points.

Let us illustrate how Fermat's principle can be used to derive Snell's law of refraction. Suppose that a light ray is to travel from point P in medium 1 to point Q in medium 2 (Fig. 35.31), where P and Q are at perpendicular distances a and b, respectively, from the interface. The speed of light is c/n_1 in medium 1 and c/n_2 in medium 2. Using the geometry of Figure 35.31, and assuming that light leaves P at $t = 0$, we see that the time at which the ray arrives at Q is

$$t = \frac{r_1}{v_1} + \frac{r_2}{v_2} = \frac{\sqrt{a^2 + x^2}}{c/n_1} + \frac{\sqrt{b^2 + (d - x)^2}}{c/n_2} \tag{35.11}$$

Figure 35.31 Geometry for deriving Snell's law of refraction using Fermat's principle.

To obtain the value of x for which t has its minimum value, we take the derivative of t with respect to x and set the derivative equal to zero:

$$\frac{dt}{dx} = \frac{n_1}{c} \frac{d}{dx} \sqrt{a^2 + x^2} + \frac{n_2}{c} \frac{d}{dx} \sqrt{b^2 + (d - x)^2}$$

$$= \frac{n_1}{c} \left(\tfrac{1}{2}\right) \frac{2x}{(a^2 + x^2)^{1/2}} + \frac{n_2}{c} \left(\tfrac{1}{2}\right) \frac{2(d - x)(-1)}{[b^2 + (d - x)^2]^{1/2}}$$

$$= \frac{n_1 x}{c(a^2 + x^2)^{1/2}} - \frac{n_2(d - x)}{c[b^2 + (d - x)^2]^{1/2}} = 0$$

or

$$\frac{n_1 x}{(a^2 + x^2)^{1/2}} = \frac{n_2(d - x)}{[b^2 + (d - x)^2]^{1/2}} \tag{35.12}$$

From Figure 35.31,

$$\sin \theta_1 = \frac{x}{(a^2 + x^2)^{1/2}} \qquad \sin \theta_2 = \frac{d - x}{[b^2 + (d - x)^2]^{1/2}}$$

Substituting these expressions into Equation 35.12, we find that

$$n_1 \sin \theta_1 = n_2 \sin \theta_2$$

which is Snell's law of refraction.

This situation is equivalent to the problem of deciding where a lifeguard who can run faster than he can swim should enter the water to help a swimmer in distress. If he enters the water too directly (in other words, at a very small value of θ_1 in Figure 35.31), the distance x is smaller than the value of x that gives the minimum value of the time interval needed for the guard to move from the starting point on the sand to the swimmer. As a result, he spends too little time running and too much time swimming. The guard's optimum location for entering the water so that he can reach the swimmer in the shortest time is at that interface point that gives the value of x that satisfies Equation 35.12.

It is a simple matter to use a similar procedure to derive the law of reflection (see Problem 65).

SUMMARY

In geometric optics, we use the **ray approximation,** in which a wave travels through a uniform medium in straight lines in the direction of the rays.

The **law of reflection** states that for a light ray traveling in air and incident on a smooth surface, the angle of reflection θ_1' equals the angle of incidence θ_1:

$$\theta_1' = \theta_1 \tag{35.2}$$

Take a practice test for this chapter by clicking on the Practice Test link at http://www.pse6.com.

Light crossing a boundary as it travels from medium 1 to medium 2 is **refracted,** or bent. The angle of refraction θ_2 is defined by the relationship

$$\frac{\sin \theta_2}{\sin \theta_1} = \frac{v_2}{v_1} = \text{constant} \tag{35.3}$$

The **index of refraction** n of a medium is defined by the ratio

$$n \equiv \frac{c}{v} \tag{35.4}$$

where c is the speed of light in a vacuum and v is the speed of light in the medium. In general, n varies with wavelength and is given by

$$n = \frac{\lambda}{\lambda_n} \tag{35.7}$$

where λ is the vacuum wavelength and λ_n is the wavelength in the medium. As light travels from one medium to another, its frequency remains the same.

Snell's law of refraction states that

$$n_1 \sin \theta_1 = n_2 \sin \theta_2 \tag{35.8}$$

where n_1 and n_2 are the indices of refraction in the two media. The incident ray, the reflected ray, the refracted ray, and the normal to the surface all lie in the same plane.

Total internal reflection occurs when light travels from a medium of high index of refraction to one of lower index of refraction. The **critical angle** θ_c for which total internal reflection occurs at an interface is given by

$$\sin \theta_c = \frac{n_2}{n_1} \qquad (\text{for } n_1 > n_2) \tag{35.10}$$

QUESTIONS

1. Light of wavelength λ is incident on a slit of width d. Under what conditions is the ray approximation valid? Under what circumstances does the slit produce enough diffraction to make the ray approximation invalid?

2. Why do astronomers looking at distant galaxies talk about looking backward in time?

3. A solar eclipse occurs when the Moon passes between the Earth and the Sun. Use a diagram to show why some areas of the Earth see a total eclipse, other areas see a partial eclipse, and most areas see no eclipse.

4. The display windows of some department stores are slanted slightly inward at the bottom. This is to decrease the glare from streetlights or the Sun, which would make it difficult for shoppers to see the display inside. Sketch a light ray reflecting from such a window to show how this technique works.

5. You take a child for walks around the neighborhood. She loves to listen to echoes from houses when she shouts or when you clap loudly. A house with a large flat front wall can produce an echo if you stand straight in front of it and reasonably far away. Draw a bird's-eye view of the situation to explain the production of the echo. Shade in the area where you can stand to hear the echo. **What If?** The child helps you to discover that a house with an L-shaped floor plan can produce echoes if you are standing in a wider range of locations. You can be standing at any reasonably distant location from which you can see the inside corner. Explain the echo in this case and draw another diagram for comparison. **What If?** What if the two wings of the house are not perpendicular? Will you and the child, standing close together, hear echoes? **What If?** What if a rectangular house and its garage have a breezeway between them, so that their perpendicular walls do not meet in an inside corner? Will this structure produce strong echoes for people in a wide range of locations? Explain your answers with diagrams.

6. The F-117A stealth fighter (Figure Q35.6) is specifically designed to be a *non*-retroreflector of radar. What aspects of its design help accomplish this? *Suggestion:* Answer the previous question as preparation for this one. Note that the bottom of the plane is flat and that all of the flat exterior panels meet at odd angles.

Figure Q35.6

7. Sound waves have much in common with light waves, including the properties of reflection and refraction. Give examples of these phenomena for sound waves.

8. Does a light ray traveling from one medium into another always bend toward the normal, as shown in Figure 35.10a? Explain.

9. As light travels from one medium to another, does the wavelength of the light change? Does the frequency change? Does the speed change? Explain.

10. A laser beam passing through a nonhomogeneous sugar solution follows a curved path. Explain.

11. A laser beam with vacuum wavelength 632.8 nm is incident from air onto a block of Lucite as shown in Figure 35.10b. The line of sight of the photograph is perpendicular to the plane in which the light moves. Find the speed, frequency, and wavelength of the light in the Lucite.

12. Suppose blue light were used instead of red light in the experiment shown in Figure 35.10b. Would the refracted beam be bent at a larger or smaller angle?

13. The level of water in a clear, colorless glass is easily observed with the naked eye. The level of liquid helium in a clear glass vessel is extremely difficult to see with the naked eye. Explain.

14. In Example 35.6 we saw that light entering a slab with parallel sides will emerge offset, but still parallel to the incoming beam. Our assumption was that the index of refraction of the material did not vary with wavelength. If the slab were made of crown glass (see Fig. 35.20), what would the outgoing beam look like?

15. Explain why a diamond sparkles more than a glass crystal of the same shape and size.

16. Explain why an oar partially in the water appears bent.

17. Total internal reflection is applied in the periscope of a submarine to let the user "see around corners." In this device, two prisms are arranged as shown in Figure Q35.17, so that an incident beam of light follows the path shown. Parallel tilted silvered mirrors could be used, but glass prisms with no silvered surfaces give higher light throughput. Propose a reason for the higher efficiency.

Figure Q35.17

18. Under certain circumstances, sound can be heard over extremely great distances. This frequently happens over a body of water, where the air near the water surface is cooler than the air higher up. Explain how the refraction of sound waves in such a situation could increase the distance over which the sound can be heard.

19. When two colors of light (X and Y) are sent through a glass prism, X is bent more than Y. Which color travels more slowly in the prism?

20. Retroreflection by transparent spheres, mentioned in Section 35.4 in the text, can be observed with dewdrops. To do so, look at the shadow of your head where it falls on dewy grass. Compare your observations to the reactions of two other people: The Renaissance artist Benvenuto Cellini described the phenomenon and his reaction in his *Autobiography*, at the end of Part One. The American philosopher Henry David Thoreau did the same in *Walden*, "Baker Farm," paragraph two. Try to find a person you know who has seen the halo—what did they think?

21. Why does the arc of a rainbow appear with red on top and violet on the bottom?

22. How is it possible that a complete circle of a rainbow can sometimes be seen from an airplane? With a stepladder, a lawn sprinkler, and a sunny day, how can you show the complete circle to children?

23. Is it possible to have total internal reflection for light incident from air on water? Explain.

24. Under what conditions is a mirage formed? On a hot day, what are we seeing when we observe "water on the road"?

PROBLEMS

1, 2, 3 = straightforward, intermediate, challenging ☐ = full solution available in the *Student Solutions Manual and Study Guide*

🌐 = coached solution with hints available at http://www.pse6.com   💻 = computer useful in solving problem

▨ = paired numerical and symbolic problems

Section 35.1   The Nature of Light
Section 35.2   Measurements of the Speed of Light

1. The *Apollo 11* astronauts set up a panel of efficient corner-cube retroreflectors on the Moon's surface. The speed of light can be found by measuring the time interval required for a laser beam to travel from Earth, reflect from the panel, and return to Earth. If this interval is measured to be 2.51 s, what is the measured speed of light? Take the center-to-center distance from Earth to Moon to be 3.84×10^8 m, and do not ignore the sizes of the Earth and Moon.

2. As a result of his observations, Roemer concluded that eclipses of Io by Jupiter were delayed by 22 min during a 6 month period as the Earth moved from the point in its orbit where it is closest to Jupiter to the diametrically opposite point where it is farthest from Jupiter. Using 1.50×10^8 km as the average radius of the Earth's orbit around the Sun, calculate the speed of light from these data.

3. In an experiment to measure the speed of light using the apparatus of Fizeau (see Fig. 35.2), the distance between light source and mirror was 11.45 km and the wheel had 720 notches. The experimentally determined value of c was 2.998×10^8 m/s. Calculate the minimum angular speed of the wheel for this experiment.

4. Figure P35.4 shows an apparatus used to measure the speed distribution of gas molecules. It consists of two slotted rotating disks separated by a distance d, with the slots displaced by the angle θ. Suppose the speed of light is measured by sending a light beam from the left through this apparatus. (a) Show that a light beam will be seen in the detector (that is, will make it through both slots) only if its speed is given by $c = \omega d/\theta$, where ω is the angular

speed of the disks and θ is measured in radians. (b) What is the measured speed of light if the distance between the two slotted rotating disks is 2.50 m, the slot in the second disk is displaced 1/60 of one degree from the slot in the first disk, and the disks are rotating at 5 555 rev/s?

Section 35.3   The Ray Approximation in Geometric Optics
Section 35.4   Reflection
Section 35.5   Refraction

Note: You may look up indices of refraction in Table 35.1.

5. A dance hall is built without pillars and with a horizontal ceiling 7.20 m above the floor. A mirror is fastened flat against one section of the ceiling. Following an earthquake, the mirror is in place and unbroken. An engineer makes a quick check of whether the ceiling is sagging by directing a vertical beam of laser light up at the mirror and observing its reflection on the floor. (a) Show that if the mirror has rotated to make an angle ϕ with the horizontal, the normal to the mirror makes an angle ϕ with the vertical. (b) Show that the reflected laser light makes an angle 2ϕ with the vertical. (c) If the reflected laser light makes a spot on the floor 1.40 cm away from the point vertically below the laser, find the angle ϕ.

6. The two mirrors illustrated in Figure P35.6 meet at a right angle. The beam of light in the vertical plane P strikes mirror 1 as shown. (a) Determine the distance the

Figure P35.4

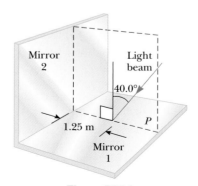

Figure P35.6

reflected light beam travels before striking mirror 2. (b) In what direction does the light beam travel after being reflected from mirror 2?

7. Two flat rectangular mirrors, both perpendicular to a horizontal sheet of paper, are set edge to edge with their reflecting surfaces perpendicular to each other. (a) A light ray in the plane of the paper strikes one of the mirrors at an arbitrary angle of incidence θ_1. Prove that the final direction of the ray, after reflection from both mirrors, is opposite to its initial direction. In a clothing store, such a pair of mirrors shows you an image of yourself as others see you, with no apparent right–left reversal. (b) **What If?** Now assume that the paper is replaced with a third flat mirror, touching edges with the other two and perpendicular to both. The set of three mirrors is called a *corner-cube reflector*. A ray of light is incident from any direction within the octant of space bounded by the reflecting surfaces. Argue that the ray will reflect once from each mirror and that its final direction will be opposite to its original direction. The *Apollo 11* astronauts placed a panel of corner cube retroreflectors on the Moon. Analysis of timing data taken with it reveals that the radius of the Moon's orbit is increasing at the rate of 3.8 cm/yr as it loses kinetic energy because of tidal friction.

8. How many times will the incident beam shown in Figure P35.8 be reflected by each of the parallel mirrors?

Mirror   Mirror

1.00 m

Incident beam

5.00°

1.00 m

Figure P35.8

9. The distance of a lightbulb from a large plane mirror is twice the distance of a person from the plane mirror. Light from the bulb reaches the person by two paths. It travels to the mirror at an angle of incidence θ, and reflects from the mirror to the person. It also travels directly to the person without reflecting off the mirror. The total distance traveled by the light in the first case is twice the distance traveled by the light in the second case. Find the value of the angle θ.

10. A narrow beam of sodium yellow light, with wavelength 589 nm in vacuum, is incident from air onto a smooth water surface at an angle of incidence of 35.0°. Determine the angle of refraction and the wavelength of the light in water.

11. *Compare this problem with the preceding problem.* A plane sound wave in air at 20°C, with wavelength 589 mm, is incident on a smooth surface of water at 25°C, at an angle of incidence of 3.50°. Determine the angle of refraction for the sound wave and the wavelength of the sound in water.

12. The wavelength of red helium–neon laser light in air is 632.8 nm. (a) What is its frequency? (b) What is its wavelength in glass that has an index of refraction of 1.50? (c) What is its speed in the glass?

13. An underwater scuba diver sees the Sun at an apparent angle of 45.0° above the horizon. What is the actual elevation angle of the Sun above the horizon?

14. A ray of light is incident on a flat surface of a block of crown glass that is surrounded by water. The angle of refraction is 19.6°. Find the angle of reflection.

15. A laser beam is incident at an angle of 30.0° from the vertical onto a solution of corn syrup in water. If the beam is refracted to 19.24° from the vertical, (a) what is the index of refraction of the syrup solution? Suppose the light is red, with vacuum wavelength 632.8 nm. Find its (b) wavelength, (c) frequency, and (d) speed in the solution.

16. Find the speed of light in (a) flint glass, (b) water, and (c) cubic zirconia.

17. A light ray initially in water enters a transparent substance at an angle of incidence of 37.0°, and the transmitted ray is refracted at an angle of 25.0°. Calculate the speed of light in the transparent substance.

18. An opaque cylindrical tank with an open top has a diameter of 3.00 m and is completely filled with water. When the afternoon Sun reaches an angle of 28.0° above the horizon, sunlight ceases to illuminate any part of the bottom of the tank. How deep is the tank?

19. A ray of light strikes a flat block of glass ($n = 1.50$) of thickness 2.00 cm at an angle of 30.0° with the normal. Trace the light beam through the glass, and find the angles of incidence and refraction at each surface.

20. Unpolarized light in vacuum is incident onto a sheet of glass with index of refraction n. The reflected and refracted rays are perpendicular to each other. Find the angle of incidence. This angle is called *Brewster's angle* or the *polarizing angle*. In this situation the reflected light is linearly polarized, with its electric field restricted to be perpendicular to the plane containing the rays and the normal.

21. When the light illustrated in Figure P35.21 passes through the glass block, it is shifted laterally by the distance d. Taking $n = 1.50$, find the value of d.

30.0°

2.00 cm

d

Figure P35.21 Problems 21 and 22.

22. Find the time interval required for the light to pass through the glass block described in the previous problem.

23. The light beam shown in Figure P35.23 makes an angle of 20.0° with the normal line NN' in the linseed oil. Determine the angles θ and θ'. (The index of refraction of linseed oil is 1.48.)

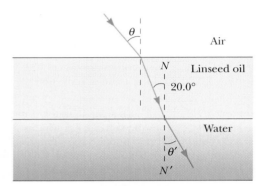

Figure P35.23

24. Three sheets of plastic have unknown indices of refraction. Sheet 1 is placed on top of sheet 2, and a laser beam is directed onto the sheets from above so that it strikes the interface at an angle of 26.5° with the normal. The refracted beam in sheet 2 makes an angle of 31.7° with the normal. The experiment is repeated with sheet 3 on top of sheet 2, and, with the same angle of incidence, the refracted beam makes an angle of 36.7° with the normal. If the experiment is repeated again with sheet 1 on top of sheet 3, what is the expected angle of refraction in sheet 3? Assume the same angle of incidence.

25. When you look through a window, by how much time is the light you see delayed by having to go through glass instead of air? Make an order-of-magnitude estimate on the basis of data you specify. By how many wavelengths is it delayed?

26. Light passes from air into flint glass. (a) What angle of incidence must the light have if the component of its velocity perpendicular to the interface is to remain constant? (b) **What If?** Can the component of velocity parallel to the interface remain constant during refraction?

27. The reflecting surfaces of two intersecting flat mirrors are at an angle θ ($0° < \theta < 90°$), as shown in Figure P35.27. For a light ray that strikes the horizontal mirror, show that the emerging ray will intersect the incident ray at an angle $\beta = 180° - 2\theta$.

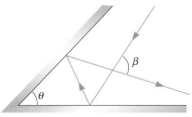

Figure P35.27

Section 35.6 Huygens's Principle

28. The speed of a water wave is described by $v = \sqrt{gd}$, where d is the water depth, assumed to be small compared to the wavelength. Because their speed changes, water waves refract when moving into a region of different depth. Sketch a map of an ocean beach on the eastern side of a landmass. Show contour lines of constant depth under water, assuming reasonably uniform slope. (a) Suppose that waves approach the coast from a storm far away to the north-northeast. Demonstrate that the waves will move nearly perpendicular to the shoreline when they reach the beach. (b) Sketch a map of a coastline with alternating bays and headlands, as suggested in Figure P35.28. Again make a reasonable guess about the shape of contour lines of constant depth. Suppose that waves approach the coast, carrying energy with uniform density along originally straight wavefronts. Show that the energy reaching the coast is concentrated at the headlands and has lower intensity in the bays.

Figure P35.28

Section 35.7 Dispersion and Prisms

Note: The apex angle of a prism is the angle between the surface at which light enters the prism and the second surface the light encounters.

29. A narrow white light beam is incident on a block of fused quartz at an angle of 30.0°. Find the angular width of the light beam inside the quartz.

30. Light of wavelength 700 nm is incident on the face of a fused quartz prism at an angle of 75.0° (with respect to the normal to the surface). The apex angle of the prism is 60.0°. Use the value of n from Figure 35.20 and calculate the angle (a) of refraction at this first surface, (b) of incidence at the second surface, (c) of refraction at the second surface, and (d) between the incident and emerging rays.

31. A prism that has an apex angle of 50.0° is made of cubic zirconia, with $n = 2.20$. What is its angle of minimum deviation?

32. A triangular glass prism with apex angle 60.0° has an index of refraction of 1.50. (a) Show that if its angle of incidence on the first surface is $\theta_1 = 48.6°$, light will pass symmetrically through the prism, as shown in

Figure 35.25. (b) Find the angle of deviation δ_{min} for $\theta_1 = 48.6°$. (c) **What If?** Find the angle of deviation if the angle of incidence on the first surface is 45.6°. (d) Find the angle of deviation if $\theta_1 = 51.6°$.

33. A triangular glass prism with apex angle $\Phi = 60.0°$ has an index of refraction $n = 1.50$ (Fig. P35.33). What is the smallest angle of incidence θ_1 for which a light ray can emerge from the other side?

Figure P35.33 Problems 33 and 34.

34. A triangular glass prism with apex angle Φ has index of refraction n. (See Fig. P35.33.) What is the smallest angle of incidence θ_1 for which a light ray can emerge from the other side?

35. [www] The index of refraction for violet light in silica flint glass is 1.66, and that for red light is 1.62. What is the angular dispersion of visible light passing through a prism of apex angle 60.0° if the angle of incidence is 50.0°? (See Fig. P35.35.)

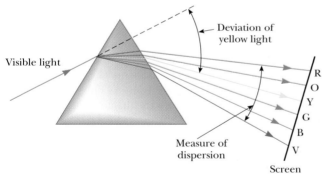

Figure P35.35

Section 35.8 Total Internal Reflection

36. For 589-nm light, calculate the critical angle for the following materials surrounded by air: (a) diamond, (b) flint glass, and (c) ice.

37. Repeat Problem 36 when the materials are surrounded by water.

38. Determine the maximum angle θ for which the light rays incident on the end of the pipe in Figure P35.38 are subject to total internal reflection along the walls of the pipe. Assume that the pipe has an index of refraction of 1.36 and the outside medium is air.

Figure P35.38

39. Consider a common mirage formed by super-heated air just above a roadway. A truck driver whose eyes are 2.00 m above the road, where $n = 1.000\ 3$, looks forward. She perceives the illusion of a patch of water ahead on the road, where her line of sight makes an angle of 1.20° below the horizontal. Find the index of refraction of the air just above the road surface. (*Suggestion:* Treat this as a problem in total internal reflection.)

40. An optical fiber has index of refraction n and diameter d. It is surrounded by air. Light is sent into the fiber along its axis, as shown in Figure P35.40. (a) Find the smallest outside radius R permitted for a bend in the fiber if no light is to escape. (b) **What If?** Does the result for part (a) predict reasonable behavior as d approaches zero? As n increases? As n approaches 1? (c) Evaluate R assuming the fiber diameter is 100 μm and its index of refraction is 1.40.

Figure P35.40

41. A large Lucite cube ($n = 1.59$) has a small air bubble (a defect in the casting process) below one surface. When a penny (diameter 1.90 cm) is placed directly over the bubble on the outside of the cube, the bubble cannot be seen by looking down into the cube at any angle. However, when a dime (diameter 1.75 cm) is placed directly over it, the bubble can be seen by looking down into the cube. What is the range of the possible depths of the air bubble beneath the surface?

42. A room contains air in which the speed of sound is 343 m/s. The walls of the room are made of concrete, in which the speed of sound is 1 850 m/s. (a) Find the critical angle for total internal reflection of sound at the concrete–air boundary. (b) In which medium must the sound be traveling in order to undergo total internal

reflection? (c) "A bare concrete wall is a highly efficient mirror for sound." Give evidence for or against this statement.

43. In about 1965, engineers at the Toro Company invented a gasoline gauge for small engines, diagrammed in Figure P35.43. The gauge has no moving parts. It consists of a flat slab of transparent plastic fitting vertically into a slot in the cap on the gas tank. None of the plastic has a reflective coating. The plastic projects from the horizontal top down nearly to the bottom of the opaque tank. Its lower edge is cut with facets making angles of 45° with the horizontal. A lawnmower operator looks down from above and sees a boundary between bright and dark on the gauge. The location of the boundary, across the width of the plastic, indicates the quantity of gasoline in the tank. Explain how the gauge works. Explain the design requirements, if any, for the index of refraction of the plastic.

Figure P35.43

Section 35.9 Fermat's Principle

44. 💻 The shoreline of a lake runs east and west. A swimmer gets into trouble 20.0 m out from shore and 26.0 m to the east of a lifeguard, whose station is 16.0 m in from the shoreline. The lifeguard takes negligible time to accelerate. He can run at 7.00 m/s and swim at 1.40 m/s. To reach the swimmer as quickly as possible, in what direction should the lifeguard start running? You will need to solve a transcendental equation numerically.

Additional Problems

45. Figure P35.45 shows a desk ornament globe containing a photograph. The flat photograph is in air, inside a vertical slot located behind a water-filled compartment having the shape of one half of a cylinder. Suppose you are looking at the center of the photograph and then rotate the globe about a vertical axis. You find that the center of the photograph disappears when you rotate the globe beyond a certain maximum angle (Fig. P35.45b). Account for this phenomenon and calculate the maximum angle. Briefly describe what you would see when you turn the globe beyond this angle.

(a)   (b)

Figure P35.45

46. A light ray enters the atmosphere of a planet where it descends vertically to the surface a distance h below. The index of refraction where the light enters the atmosphere is 1.000, and it increases linearly to the surface where it has the value n. (a) How long does it take the ray to traverse this path? (b) Compare this to the time interval required in the absence of an atmosphere.

47. A narrow beam of light is incident from air onto the surface of glass with index of refraction 1.56. Find the angle of incidence for which the corresponding angle of refraction is half the angle of incidence. (*Suggestion:* You might want to use the trigonometric identity $\sin 2\theta = 2 \sin \theta \cos \theta$.)

48. 🖥 (a) Consider a horizontal interface between air above and glass of index 1.55 below. Draw a light ray incident from the air at angle of incidence 30.0°. Determine the angles of the reflected and refracted rays and show them on the diagram. (b) **What If?** Suppose instead that the light ray is incident from the glass at angle of incidence 30.0°. Determine the angles of the reflected and refracted rays and show all three rays on a new diagram. (c) For rays incident from the air onto the air–glass surface, determine and tabulate the angles of reflection and refraction for all the angles of incidence at 10.0° intervals from 0° to 90.0°. (d) Do the same for light rays coming up to the interface through the glass.

49. 🌐 A small underwater pool light is 1.00 m below the surface. The light emerging from the water forms a circle on the water surface. What is the diameter of this circle?

50. One technique for measuring the angle of a prism is shown in Figure P35.50. A parallel beam of light is directed on the angle so that parts of the beam reflect from opposite sides. Show that the angular separation of the two reflected beams is given by $B = 2A$.

Figure P35.50

51. The walls of a prison cell are perpendicular to the four cardinal compass directions. On the first day of spring, light from the rising Sun enters a rectangular window in

the eastern wall. The light traverses 2.37 m horizontally to shine perpendicularly on the wall opposite the window. A young prisoner observes the patch of light moving across this western wall and for the first time forms his own understanding of the rotation of the Earth. (a) With what speed does the illuminated rectangle move? (b) The prisoner holds a small square mirror flat against the wall at one corner of the rectangle of light. The mirror reflects light back to a spot on the eastern wall close beside the window. How fast does the smaller square of light move across that wall? (c) Seen from a latitude of 40.0° north, the rising Sun moves through the sky along a line making a 50.0° angle with the southeastern horizon. In what direction does the rectangular patch of light on the western wall of the prisoner's cell move? (d) In what direction does the smaller square of light on the eastern wall move?

52. Figure P35.52 shows a top view of a square enclosure. The inner surfaces are plane mirrors. A ray of light enters a small hole in the center of one mirror. (a) At what angle θ must the ray enter in order to exit through the hole after being reflected once by each of the other three mirrors? (b) **What If?** Are there other values of θ for which the ray can exit after multiple reflections? If so, make a sketch of one of the ray's paths.

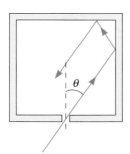

Figure P35.52

53. A hiker stands on an isolated mountain peak near sunset and observes a rainbow caused by water droplets in the air 8.00 km away. The valley is 2.00 km below the mountain peak and entirely flat. What fraction of the complete circular arc of the rainbow is visible to the hiker? (See Fig. 35.24.)

54. A 4.00-m-long pole stands vertically in a lake having a depth of 2.00 m. The Sun is 40.0° above the horizontal. Determine the length of the pole's shadow on the bottom of the lake. Take the index of refraction for water to be 1.33.

55. A laser beam strikes one end of a slab of material, as shown in Figure P35.55. The index of refraction of the slab is 1.48. Determine the number of internal reflections of the beam before it emerges from the opposite end of the slab.

Figure P35.55

56. When light is incident normally on the interface between two transparent optical media, the intensity of the reflected light is given by the expression

$$S_1' = \left(\frac{n_2 - n_1}{n_2 + n_1} \right)^2 S_1$$

In this equation S_1 represents the average magnitude of the Poynting vector in the incident light (the incident intensity), S_1' is the reflected intensity, and n_1 and n_2 are the refractive indices of the two media. (a) What fraction of the incident intensity is reflected for 589-nm light normally incident on an interface between air and crown glass? (b) **What If?** Does it matter in part (a) whether the light is in the air or in the glass as it strikes the interface?

57. Refer to Problem 56 for its description of the reflected intensity of light normally incident on an interface between two transparent media. (a) When light is normally incident on an interface between vacuum and a transparent medium of index n, show that the intensity S_2 of the transmitted light is given by $S_2/S_1 = 4n/(n + 1)^2$. (b) Light travels perpendicularly through a diamond slab, surrounded by air, with parallel surfaces of entry and exit. Apply the transmission fraction in part (a) to find the approximate overall transmission through the slab of diamond, as a percentage. Ignore light reflected back and forth within the slab.

58. **What If?** This problem builds upon the results of Problems 56 and 57. Light travels perpendicularly through a diamond slab, surrounded by air, with parallel surfaces of entry and exit. The intensity of the transmitted light is what fraction of the incident intensity? Include the effects of light reflected back and forth inside the slab.

59. The light beam in Figure P35.59 strikes surface 2 at the critical angle. Determine the angle of incidence θ_1.

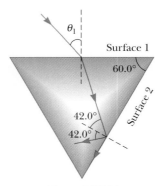

Figure P35.59

60. Builders use a leveling instrument with the beam from a fixed helium–neon laser reflecting in a horizontal plane from a small flat mirror mounted on an accurately vertical rotating shaft. The light is sufficiently bright and the rotation rate is sufficiently high that the reflected light appears as a horizontal line wherever it falls on a wall. (a) Assume the mirror is at the center of a circular grain elevator of radius R. The mirror spins with constant angular velocity ω_m. Find the speed of the spot of laser light on the wall. (b) **What If?** Assume the spinning mirror is at a perpendicular distance d from point O on a flat vertical wall. When the spot of laser light on the wall is at distance x from point O, what is its speed?

61. 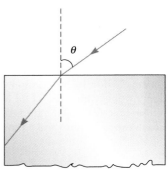 A light ray of wavelength 589 nm is incident at an angle θ on the top surface of a block of polystyrene, as shown in Figure P35.61. (a) Find the maximum value of θ for which the refracted ray undergoes total internal reflection at the left vertical face of the block. **What If?** Repeat the calculation for the case in which the polystyrene block is immersed in (b) water and (c) carbon disulfide.

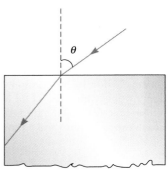

Figure P35.61

62. Refer to Quick Quiz 35.4. By how much does the duration of an optical day exceed that of a geometric day? Model the Earth's atmosphere as uniform, with index of refraction 1.000 293, a sharply defined upper surface, and depth 8 614 m. Assume that the observer is at the Earth's equator, so that the apparent path of the rising and setting Sun is perpendicular to the horizon.

63. A shallow glass dish is 4.00 cm wide at the bottom, as shown in Figure P35.63. When an observer's eye is placed as shown, the observer sees the edge of the bottom of the empty dish. When this dish is filled with water, the observer sees the center of the bottom of the dish. Find the height of the dish.

← 4.00 cm →

Figure P35.63

64. A ray of light passes from air into water. For its deviation angle $\delta = |\theta_1 - \theta_2|$ to be 10.0°, what must be its angle of incidence?

65. Derive the law of reflection (Eq. 35.2) from Fermat's principle. (See the procedure outlined in Section 35.9 for the derivation of the law of refraction from Fermat's principle.)

66. A material having an index of refraction n is surrounded by a vacuum and is in the shape of a quarter circle of radius R (Fig. P35.66). A light ray parallel to the base of the material is incident from the left at a distance L above the base and emerges from the material at the angle θ. Determine an expression for θ.

Figure P35.66

67. A transparent cylinder of radius $R = 2.00$ m has a mirrored surface on its right half, as shown in Figure P35.67. A light ray traveling in air is incident on the left side of the cylinder. The incident light ray and exiting light ray are parallel and $d = 2.00$ m. Determine the index of refraction of the material.

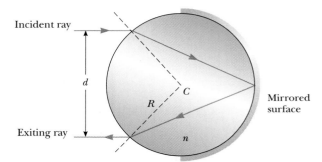

Figure P35.67

68. Suppose that a luminous sphere of radius R_1 (such as the Sun) is surrounded by a uniform atmosphere of radius R_2 and index of refraction n. When the sphere is viewed from a location far away in vacuum, what is its apparent radius? You will need to distinguish between the two cases (a) $R_2 > nR_1$ and (b) $R_2 < nR_1$.

69. A. H. Pfund's method for measuring the index of refraction of glass is illustrated in Figure P35.69. One face of a slab of thickness t is painted white, and a small hole scraped clear at point P serves as a source of diverging rays when the slab is

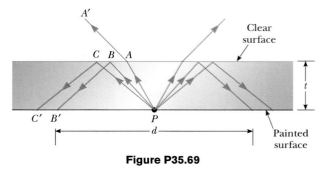

Figure P35.69

illuminated from below. Ray *PBB'* strikes the clear surface at the critical angle and is totally reflected, as are rays such as *PCC'*. Rays such as *PAA'* emerge from the clear surface. On the painted surface there appears a dark circle of diameter *d*, surrounded by an illuminated region, or halo. (a) Derive an equation for *n* in terms of the measured quantities *d* and *t*. (b) What is the diameter of the dark circle if $n = 1.52$ for a slab 0.600 cm thick? (c) If white light is used, the critical angle depends on color caused by dispersion. Is the inner edge of the white halo tinged with red light or violet light? Explain.

70. A light ray traveling in air is incident on one face of a right-angle prism of index of refraction $n = 1.50$ as shown in Figure P35.70, and the ray follows the path shown in the figure. Assuming $\theta = 60.0°$ and the base of the prism is mirrored, determine the angle ϕ made by the outgoing ray with the normal to the right face of the prism.

Incoming ray   Outgoing ray

Mirror base

Figure P35.70

71. A light ray enters a rectangular block of plastic at an angle $\theta_1 = 45.0°$ and emerges at an angle $\theta_2 = 76.0°$, as shown in Figure P35.71. (a) Determine the index of refraction of the plastic. (b) If the light ray enters the plastic at a point $L = 50.0$ cm from the bottom edge, how long does it take the light ray to travel through the plastic?

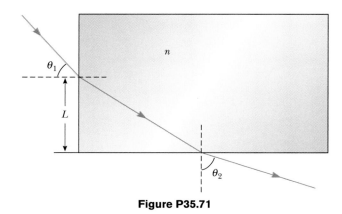

Figure P35.71

72. Students allow a narrow beam of laser light to strike a water surface. They arrange to measure the angle of refraction for selected angles of incidence and record the data shown in the accompanying table. Use the data to verify Snell's law of refraction by plotting the sine of the angle of incidence versus the sine of the angle of refraction. Use the resulting plot to deduce the index of refraction of water.

Angle of Incidence (degrees)	Angle of Refraction (degrees)
10.0	7.5
20.0	15.1
30.0	22.3
40.0	28.7
50.0	35.2
60.0	40.3
70.0	45.3
80.0	47.7

Answers to Quick Quizzes

35.1 (d). The light rays from the actor's face must reflect from the mirror and into the camera. If these light rays are reversed, light from the camera reflects from the mirror into the eyes of the actor.

35.2 Beams ② and ④ are reflected; beams ③ and ⑤ are refracted.

35.3 (c). Because the light is entering a material in which the index of refraction is lower, the speed of light is higher and the light bends away from the normal.

35.4 (a). Due to the refraction of light by air, light rays from the Sun deviate slightly downward toward the surface of the Earth as the light enters the atmosphere. Thus, in the morning, light rays from the upper edge of the Sun arrive at your eyes before the geometric line from your eyes to the top of the Sun clears the horizon. In the evening, light rays from the top of the Sun continue to arrive at your eyes even after the geometric line from your eyes to the top of the Sun dips below the horizon.

35.5 (c). An ideal camera lens would have an index of refraction that does not vary with wavelength so that all colors would be bent through the same angle by the lens. Of the three choices, fused quartz has the least variation in *n* across the visible spectrum.

35.6 (b). The two bright rays exiting the bottom of the prism on the right in Figure 35.27 result from total internal reflection at the right face of the prism. Notice that there is no refracted light exiting the slanted side for these rays. The light from the other three rays is divided into reflected and refracted parts.

35.7 (b). Counterclockwise rotation of the prism will cause the rays to strike the slanted side of the prism at a larger angle. When all five rays strike at an angle larger than the critical angle, they will all undergo total internal reflection.

35.8 (c). When the outgoing beam approaches the direction parallel to the straight side, the incident angle is approaching the critical angle for total internal reflection. The index of refraction for light at the violet end of the visible spectrum is larger than that at the red end. Thus, as the outgoing beam approaches the straight side, the violet light experiences total internal reflection first, followed by the other colors. The red light is the last to experience total internal reflection.

Chapter 36

Image Formation

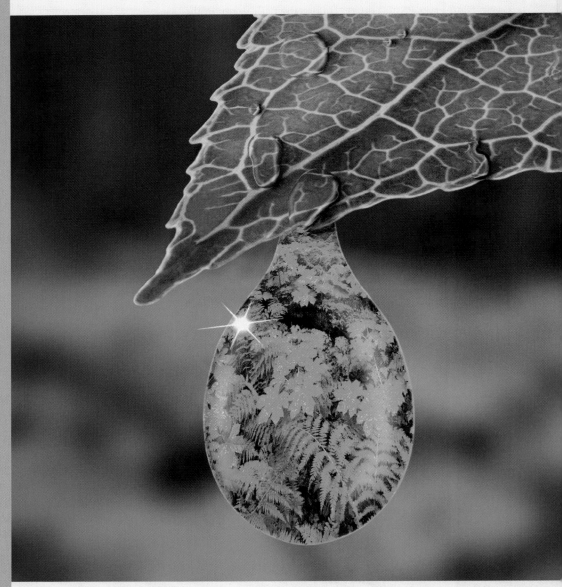

▲ The light rays coming from the leaves in the background of this scene did not form a focused image on the film of the camera that took this photograph. Consequently, the background appears very blurry. Light rays passing though the raindrop, however, have been altered so as to form a focused image of the background leaves on the film. In this chapter, we investigate the formation of images as light rays reflect from mirrors and refract through lenses. (Don Hammond/CORBIS)

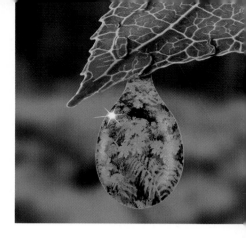

This chapter is concerned with the images that result when light rays encounter flat and curved surfaces. We find that images can be formed either by reflection or by refraction and that we can design mirrors and lenses to form images with desired characteristics. We continue to use the ray approximation and to assume that light travels in straight lines. Both of these steps lead to valid predictions in the field called *geometric optics*. In subsequent chapters, we shall concern ourselves with interference and diffraction effects—the objects of study in the field of *wave optics*.

36.1 Images Formed by Flat Mirrors

We begin by considering the simplest possible mirror, the flat mirror. Consider a point source of light placed at O in Figure 36.1, a distance p in front of a flat mirror. The distance p is called the **object distance.** Light rays leave the source and are reflected from the mirror. Upon reflection, the rays continue to diverge (spread apart). The dashed lines in Figure 36.1 are extensions of the diverging rays back to a point of intersection at I. The diverging rays appear to the viewer to come from the point I behind the mirror. Point I is called the **image** of the object at O. Regardless of the system under study, we always locate images by extending diverging rays back to a point at which they intersect. **Images are located either at a point from which rays of light *actually* diverge or at a point from which they *appear* to diverge.** Because the rays in Figure 36.1 appear to originate at I, which is a distance q behind the mirror, this is the location of the image. The distance q is called the **image distance.**

Images are classified as **real** or **virtual. A real image is formed when light rays pass through and diverge from the image point; a virtual image is formed when the light rays do not pass through the image point but only appear to diverge from that point.** The image formed by the mirror in Figure 36.1 is virtual. The image of an object seen in a flat mirror is *always* virtual. Real images can be displayed on a

Figure 36.1 An image formed by reflection from a flat mirror. The image point I is located behind the mirror a perpendicular distance q from the mirror (the image distance). The image distance has the same magnitude as the object distance p.

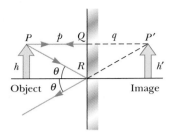

Active Figure 36.2 A geometric construction that is used to locate the image of an object placed in front of a flat mirror. Because the triangles PQR and $P'QR$ are congruent, $|p| = |q|$ and $h = h'$.

At the Active Figures link at http://www.pse6.com, you can move the object and see the effect on the image.

Lateral magnification

⚠ **PITFALL PREVENTION**

36.1 Magnification Does Not Necessarily Imply Enlargement

For optical elements other than flat mirrors, the magnification defined in Equation 36.1 can result in a number with magnitude larger *or* smaller than 1. Thus, despite the cultural usage of the word *magnification* to mean *enlargement,* the image could be smaller than the object.

screen (as at a movie), but virtual images cannot be displayed on a screen. We shall see an example of a real image in Section 36.2.

We can use the simple geometry in Figure 36.2 to examine the properties of the images of extended objects formed by flat mirrors. Even though there are an infinite number of choices of direction in which light rays could leave each point on the object, we need to choose only two rays to determine where an image is formed. One of those rays starts at P, follows a horizontal path to the mirror, and reflects back on itself. The second ray follows the oblique path PR and reflects as shown, according to the law of reflection. An observer in front of the mirror would trace the two reflected rays back to the point at which they appear to have originated, which is point P' behind the mirror. A continuation of this process for points other than P on the object would result in a virtual image (represented by a yellow arrow) behind the mirror. Because triangles PQR and $P'QR$ are congruent, $PQ = P'Q$. We conclude that **the image formed by an object placed in front of a flat mirror is as far behind the mirror as the object is in front of the mirror.**

Geometry also reveals that the object height h equals the image height h'. Let us define **lateral magnification** M of an image as follows:

$$M \equiv \frac{\text{Image height}}{\text{Object height}} = \frac{h'}{h} \tag{36.1}$$

This is a general definition of the lateral magnification for an image from any type of mirror. (This equation is also valid for images formed by lenses, which we study in Section 36.4.) For a flat mirror, $M = 1$ for any image because $h' = h$.

Finally, note that a flat mirror produces an image that has an *apparent* left–right reversal. You can see this reversal by standing in front of a mirror and raising your right hand, as shown in Figure 36.3. The image you see raises its left hand. Likewise, your hair appears to be parted on the side opposite your real part, and a mole on your right cheek appears to be on your left cheek.

This reversal is not *actually* a left–right reversal. Imagine, for example, lying on your left side on the floor, with your body parallel to the mirror surface. Now your head is on the left and your feet are on the right. If you shake your feet, the image does not shake its head! If you raise your right hand, however, the image again raises its left hand. Thus, the mirror again appears to produce a left–right reversal but in the up–down direction!

The reversal is actually a *front–back reversal,* caused by the light rays going forward toward the mirror and then reflecting back from it. An interesting exercise is to stand in front of a mirror while holding an overhead transparency in front of you so that you can read the writing on the transparency. You will also be able to read the writing on the image of the transparency. You may have had a similar experience if you have attached a transparent decal with words on it to the rear window of your car. If the

Figure 36.3 The image in the mirror of a person's right hand is reversed front to back. This makes the right hand appear to be a left hand. Notice that the thumb is on the left side of both real hands and on the left side of the image. That the thumb is not on the right side of the image indicates that there is no left-to-right reversal.

decal can be read from outside the car, you can also read it when looking into your rearview mirror from inside the car.

We conclude that the image that is formed by a flat mirror has the following properties.

- The image is as far behind the mirror as the object is in front.
- The image is unmagnified, virtual, and upright. (By upright we mean that, if the object arrow points upward as in Figure 36.2, so does the image arrow.)
- The image has front–back reversal.

Quick Quiz 36.1 In the overhead view of Figure 36.4, the image of the stone seen by observer 1 is at *C*. At which of the five points *A*, *B*, *C*, *D*, or *E* does observer 2 see the image?

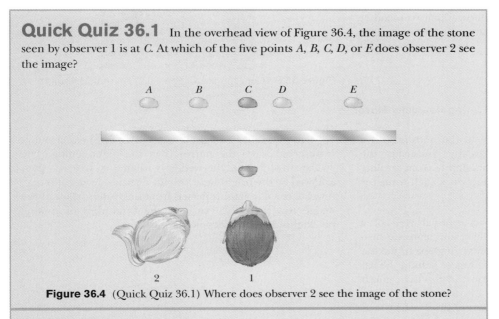

Figure 36.4 (Quick Quiz 36.1) Where does observer 2 see the image of the stone?

Quick Quiz 36.2 You are standing about 2 m away from a mirror. The mirror has water spots on its surface. True or false: It is possible for you to see the water spots and your image both in focus at the same time.

Conceptual Example 36.1   Multiple Images Formed by Two Mirrors

Two flat mirrors are perpendicular to each other, as in Figure 36.5, and an object is placed at point *O*. In this situation, multiple images are formed. Locate the positions of these images.

Solution The image of the object is at I_1 in mirror 1 and at I_2 in mirror 2. In addition, a third image is formed at I_3. This third image is the image of I_1 in mirror 2 or, equivalently, the image of I_2 in mirror 1. That is, the image at I_1 (or I_2) serves as the object for I_3. Note that to form this image at I_3, the rays reflect twice after leaving the object at *O*.

Figure 36.5 (Conceptual Example 36.1) When an object is placed in front of two mutually perpendicular mirrors as shown, three images are formed.

Conceptual Example 36.2 The Levitated Professor

The professor in the box shown in Figure 36.6 appears to be balancing himself on a few fingers, with his feet off the floor. He can maintain this position for a long time, and he appears to defy gravity. How was this illusion created?

Solution This is one of many magicians' optical illusions that make use of a mirror. The box in which the professor stands is a cubical frame that contains a flat vertical mirror positioned in a diagonal plane of the frame. The professor straddles the mirror so that one foot, which you see, is in front of the mirror, and the other foot, which you cannot see, is behind the mirror. When he raises the foot in front of the mirror, the reflection of that foot also rises, so he appears to float in air.

Courtesy of Henry Leap and Jim Lehman

Figure 36.6 (Conceptual Example 36.2) An optical illusion.

Conceptual Example 36.3 The Tilting Rearview Mirror

Most rearview mirrors in cars have a day setting and a night setting. The night setting greatly diminishes the intensity of the image in order that lights from trailing vehicles do not blind the driver. How does such a mirror work?

Solution Figure 36.7 shows a cross-sectional view of a rearview mirror for each setting. The unit consists of a reflective coating on the back of a wedge of glass. In the day setting (Fig. 36.7a), the light from an object behind the car strikes the glass wedge at point 1. Most of the light enters the wedge, refracting as it crosses the front surface, and reflects from the back surface to return to the front surface, where it is refracted again as it re-enters the air as ray *B* (for *bright*). In addition, a small portion of the light is reflected at the front surface of the glass, as indicated by ray *D* (for *dim*).

This dim reflected light is responsible for the image that is observed when the mirror is in the night setting (Fig. 36.7b). In this case, the wedge is rotated so that the path followed by the bright light (ray *B*) does not lead to the eye. Instead, the dim light reflected from the front surface of the wedge travels to the eye, and the brightness of trailing headlights does not become a hazard.

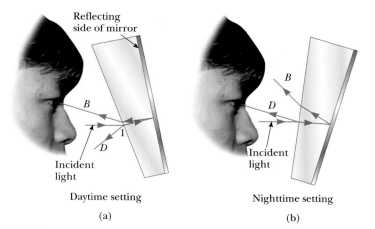

Figure 36.7 (Conceptual Example 36.3) Cross-sectional views of a rearview mirror. (a) With the day setting, the silvered back surface of the mirror reflects a bright ray *B* into the driver's eyes. (b) With the night setting, the glass of the unsilvered front surface of the mirror reflects a dim ray *D* into the driver's eyes.

Ken Kay/Fundamental Photographs

Figure 36.8 Red, blue, and green light rays are reflected by a curved mirror. Note that the three colored beams meet at a point.

36.2 Images Formed by Spherical Mirrors

Concave Mirrors

A **spherical mirror,** as its name implies, has the shape of a section of a sphere. This type of mirror focuses incoming parallel rays to a point, as demonstrated by the colored light rays in Figure 36.8. Figure 36.9a shows a cross section of a spherical mirror, with its surface represented by the solid, curved black line. (The blue band represents the structural support for the mirrored surface, such as a curved piece of glass on which the silvered surface is deposited.) Such a mirror, in which light is reflected from the inner, concave surface, is called a **concave mirror.** The mirror has a radius of curvature R, and its center of curvature is point C. Point V is the center of the spherical section, and a line through C and V is called the **principal axis** of the mirror.

Now consider a point source of light placed at point O in Figure 36.9b, where O is any point on the principal axis to the left of C. Two diverging rays that originate at O are shown. After reflecting from the mirror, these rays converge and cross at the image point I. They then continue to diverge from I as if an object were there. As a result, at point I we have a real image of the light source at O.

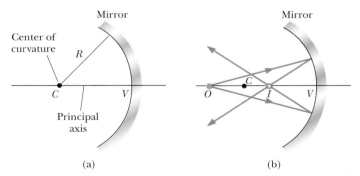

Figure 36.9 (a) A concave mirror of radius R. The center of curvature C is located on the principal axis. (b) A point object placed at O in front of a concave spherical mirror of radius R, where O is any point on the principal axis farther than R from the mirror surface, forms a real image at I. If the rays diverge from O at small angles, they all reflect through the same image point.

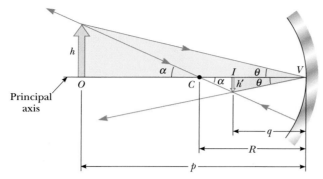

Figure 36.11 The image formed by a spherical concave mirror when the object O lies outside the center of curvature C. This geometric construction is used to derive Equation 36.4.

Figure 36.10 Rays diverging from the object at large angles from the principal axis reflect from a spherical concave mirror to intersect the principal axis at different points, resulting in a blurred image. This condition is called *spherical aberration.*

We shall consider in this section only rays that diverge from the object and make a small angle with the principal axis. Such rays are called **paraxial rays.** All paraxial rays reflect through the image point, as shown in Figure 36.9b. Rays that are far from the principal axis, such as those shown in Figure 36.10, converge to other points on the principal axis, producing a blurred image. This effect, which is called **spherical aberration,** is present to some extent for any spherical mirror and is discussed in Section 36.5.

We can use Figure 36.11 to calculate the image distance q from a knowledge of the object distance p and radius of curvature R. By convention, these distances are measured from point V. Figure 36.11 shows two rays leaving the tip of the object. One of these rays passes through the center of curvature C of the mirror, hitting the mirror perpendicular to the mirror surface and reflecting back on itself. The second ray strikes the mirror at its center (point V) and reflects as shown, obeying the law of reflection. The image of the tip of the arrow is located at the point where these two rays intersect. From the gold right triangle in Figure 36.11, we see that $\tan\theta = h/p$, and from the blue right triangle we see that $\tan\theta = -h'/q$. The negative sign is introduced because the image is inverted, so h' is taken to be negative. Thus, from Equation 36.1 and these results, we find that the magnification of the image is

$$M = \frac{h'}{h} = -\frac{q}{p} \tag{36.2}$$

We also note from the two triangles in Figure 36.11 that have α as one angle that

$$\tan\alpha = \frac{h}{p - R} \quad \text{and} \quad \tan\alpha = -\frac{h'}{R - q}$$

from which we find that

$$\frac{h'}{h} = -\frac{R - q}{p - R} \tag{36.3}$$

If we compare Equations 36.2 and 36.3, we see that

$$\frac{R - q}{p - R} = \frac{q}{p}$$

Simple algebra reduces this to

$$\frac{1}{p} + \frac{1}{q} = \frac{2}{R} \tag{36.4}$$

Mirror equation in terms of radius of curvature

This expression is called the **mirror equation.**

If the object is very far from the mirror—that is, if p is so much greater than R that p can be said to approach infinity—then $1/p \approx 0$, and we see from Equation 36.4 that $q \approx R/2$. That is, when the object is very far from the mirror, the image point is halfway between the center of curvature and the center point on the mirror, as shown in Figure 36.12a. The incoming rays from the object are essentially parallel

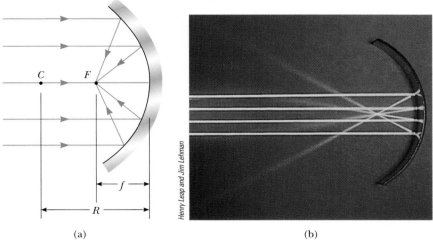

Henry Leap and Jim Lehman

(a)   (b)

Figure 36.12 (a) Light rays from a distant object ($p \rightarrow \infty$) reflect from a concave mirror through the focal point *F*. In this case, the image distance $q \approx R/2 = f$, where *f* is the focal length of the mirror. (b) Reflection of parallel rays from a concave mirror.

in this figure because the source is assumed to be very far from the mirror. We call the image point in this special case the **focal point** *F* and the image distance the **focal length** *f*, where

$$f = \frac{R}{2} \qquad (36.5)$$

In Figure 36.8, the colored beams are traveling parallel to the principal axis and the mirror reflects all three beams to the focal point. Notice that the point at which the three beams intersect and the colors add is white.

Focal length is a parameter particular to a given mirror and therefore can be used to compare one mirror with another. The mirror equation can be expressed in terms of the focal length:

$$\frac{1}{p} + \frac{1}{q} = \frac{1}{f} \qquad (36.6)$$

Notice that the focal length of a mirror depends only on the curvature of the mirror and not on the material from which the mirror is made. This is because the formation of the image results from rays reflected from the surface of the material. The situation is different for lenses; in that case the light actually passes through the material and the focal length depends on the type of material from which the lens is made.

Courtesy of Thomson Consumer Electronics

A satellite-dish antenna is a concave reflector for television signals from a satellite in orbit around the Earth. The signals are carried by microwaves that, because the satellite is so far away, are parallel when they arrive at the dish. These waves reflect from the dish and are focused on the receiver at the focal point of the dish.

▲ **PITFALL PREVENTION**

36.2   The *Focal* Point Is Not the *Focus* Point

The focal point *is usually not* the point at which the light rays focus to form an image. The focal point is determined solely by the curvature of the mirror—it does not depend on the location of the object at all. In general, an image forms at a point different from the focal point of a mirror (or a lens). The *only* exception is when the object is located infinitely far away from the mirror.

Focal length

Mirror equation in terms of focal length

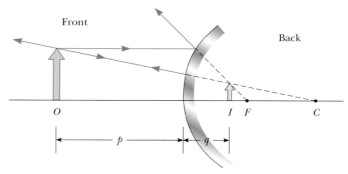

Figure 36.13 Formation of an image by a spherical convex mirror. The image formed by the real object is virtual and upright.

Convex Mirrors

Figure 36.13 shows the formation of an image by a **convex mirror**—that is, one silvered so that light is reflected from the outer, convex surface. This is sometimes called a **diverging mirror** because the rays from any point on an object diverge after reflection as though they were coming from some point behind the mirror. The image in Figure 36.13 is virtual because the reflected rays only appear to originate at the image point, as indicated by the dashed lines. Furthermore, the image is always upright and smaller than the object. This type of mirror is often used in stores to foil shoplifters. A single mirror can be used to survey a large field of view because it forms a smaller image of the interior of the store.

We do not derive any equations for convex spherical mirrors because we can use Equations 36.2, 36.4, and 36.6 for either concave or convex mirrors if we adhere to the following procedure. Let us refer to the region in which light rays move toward the mirror as the *front side* of the mirror, and the other side as the *back side*. For example, in Figures 36.11 and 36.13, the side to the left of the mirrors is the front side, and the side to the right of the mirrors is the back side. Figure 36.14 states the sign conventions for object and image distances, and Table 36.1 summarizes the sign conventions for all quantities.

Figure 36.14 Signs of p and q for convex and concave mirrors.

Ray Diagrams for Mirrors

The positions and sizes of images formed by mirrors can be conveniently determined with *ray diagrams*. These graphical constructions reveal the nature of the image and can be used to check results calculated from the mirror and magnification equations. To draw a ray diagram, we need to know the position of the object and the locations of the mirror's focal point and center of curvature. We then draw three principal rays to locate the image, as shown by the examples in Figure 36.15.

 PITFALL PREVENTION

36.3 Watch Your Signs

Success in working mirror problems (as well as problems involving refracting surfaces and thin lenses) is largely determined by proper sign choices when substituting into the equations. The best way to become adept at this is to work a multitude of problems on your own. Watching your instructor or reading the example problems is no substitute for practice.

Table 36.1

Sign Conventions for Mirrors		
Quantity	**Positive When**	**Negative When**
Object location (p)	Object is in front of mirror (real object)	Object is in back of mirror (virtual object)
Image location (q)	Image is in front of mirror (real image)	Image is in back of mirror (virtual image)
Image height (h')	Image is upright	Image is inverted
Focal length (f) and radius (R)	Mirror is concave	Mirror is convex
Magnification (M)	Image is upright	Image is inverted

(a)

(b)

Photos courtesy David Rogers

(c)

Active Figure 36.15 Ray diagrams for spherical mirrors, along with corresponding photographs of the images of candles. (a) When the object is located so that the center of curvature lies between the object and a concave mirror surface, the image is real, inverted, and reduced in size. (b) When the object is located between the focal point and a concave mirror surface, the image is virtual, upright, and enlarged. (c) When the object is in front of a convex mirror, the image is virtual, upright, and reduced in size.

At the Active Figures link at http://www.pse6.com, you can move the objects and change the focal length of the mirrors to see the effect on the images.

▲ **PITFALL PREVENTION**

36.4 We Are *Choosing* a Small Number of Rays

A *huge* number of light rays leave each point on an object (and pass through each point on an image). In a principal-ray diagram, which displays the characteristics of the image, we choose only a few rays that follow simply stated rules. Locating the image by calculation complements the diagram.

These rays all start from the same object point and are drawn as follows. We may choose any point on the object; here, we choose the top of the object for simplicity. For concave mirrors (see Figs. 36.15a and 36.15b), we draw the following three principal rays:

- Ray 1 is drawn from the top of the object parallel to the principal axis and is reflected through the focal point *F*.
- Ray 2 is drawn from the top of the object through the focal point and is reflected parallel to the principal axis.
- Ray 3 is drawn from the top of the object through the center of curvature *C* and is reflected back on itself.

The intersection of any two of these rays locates the image. The third ray serves as a check of the construction. The image point obtained in this fashion must always agree with the value of *q* calculated from the mirror equation. With concave mirrors, note what happens as the object is moved closer to the mirror. The real, inverted image in Figure 36.15a moves to the left as the object approaches the focal point. When the object is at the focal point, the image is infinitely far to the left. However, when the object lies between the focal point and the mirror surface, as shown in Figure 36.15b, the image is virtual, upright, and enlarged. This latter situation applies when you use a shaving mirror or a makeup mirror, both of which are concave. Your face is closer to the mirror than the focal point, and you see an upright, enlarged image of your face.

For convex mirrors (see Fig. 36.15c), we draw the following three principal rays:

- Ray 1 is drawn from the top of the object parallel to the principal axis and is reflected *away from* the focal point *F*.
- Ray 2 is drawn from the top of the object toward the focal point on the back side of the mirror and is reflected parallel to the principal axis.
- Ray 3 is drawn from the top of the object toward the center of curvature *C* on the back side of the mirror and is reflected back on itself.

In a convex mirror, the image of an object is always virtual, upright, and reduced in size as shown in Figure 36.15c. In this case, as the object distance decreases, the virtual image increases in size and moves away from the focal point toward the mirror as the object approaches the mirror. You should construct other diagrams to verify how image position varies with object position.

Figure 36.16 (Quick Quiz 36.4) What type of mirror is this?

Quick Quiz 36.3 You wish to reflect sunlight from a mirror onto some paper under a pile of wood in order to start a fire. Which would be the best choice for the type of mirror? (a) flat (b) concave (c) convex.

Quick Quiz 36.4 Consider the image in the mirror in Figure 36.16. Based on the appearance of this image, you would conclude that (a) the mirror is concave and the image is real. (b) the mirror is concave and the image is virtual. (c) the mirror is convex and the image is real. (d) the mirror is convex and the image is virtual.

Example 36.4 The Image formed by a Concave Mirror

Assume that a certain spherical mirror has a focal length of + 10.0 cm. Locate and describe the image for object distances of

(A) 25.0 cm,

(B) 10.0 cm, and

(C) 5.00 cm.

Solution Because the focal length is positive, we know that this is a concave mirror (see Table 36.1).

(A) This situation is analogous to that in Figure 36.15a; hence, we expect the image to be real. We find the image distance by using Equation 36.6:

$$\frac{1}{p} + \frac{1}{q} = \frac{1}{f}$$

$$\frac{1}{25.0 \text{ cm}} + \frac{1}{q} = \frac{1}{10.0 \text{ cm}}$$

$$q = \boxed{16.7 \text{ cm}}$$

The magnification of the image is given by Equation 36.2:

$$M = -\frac{q}{p} = -\frac{16.7 \text{ cm}}{25.0 \text{ cm}} = -0.668$$

The fact that the absolute value of M is less than unity tells us that the image is smaller than the object, and the negative sign for M tells us that the image is inverted. Because q is positive, the image is located on the front side of the mirror and is real.

(B) When the object distance is 10.0 cm, the object is located at the focal point. Now we find that

$$\frac{1}{10.0 \text{ cm}} + \frac{1}{q} = \frac{1}{10.0 \text{ cm}}$$

$$q = \boxed{\infty}$$

which means that rays originating from an object positioned at the focal point of a mirror are reflected so that the image is formed at an infinite distance from the mirror; that is, the rays travel parallel to one another after reflection. This is the situation in a flashlight, where the bulb filament is placed at the focal point of a reflector, producing a parallel beam of light.

(C) When the object is at $p = 5.00$ cm, it lies halfway between the focal point and the mirror surface, as shown in Figure 36.15b. Thus, we expect a magnified, virtual, upright

image. In this case, the mirror equation gives

$$\frac{1}{5.00 \text{ cm}} + \frac{1}{q} = \frac{1}{10.0 \text{ cm}}$$

$$q = \boxed{-10.0 \text{ cm}}$$

The image is virtual because it is located behind the mirror, as expected. The magnification of the image is

$$M = -\frac{q}{p} = -\left(\frac{-10.0 \text{ cm}}{5.00 \text{ cm}}\right) = +2.00$$

The image is twice as large as the object, and the positive sign for M indicates that the image is upright (see Fig. 36.15b).

What If? Suppose you set up the candle and mirror apparatus illustrated in Figure 36.15a and described in part **(A)** of the example. While adjusting the apparatus, you accidentally strike the candle with your elbow so that it begins to slide toward the mirror at velocity v_p. How fast does the image of the candle move?

Answer We solve the mirror equation, Equation 36.6, for q:

$$q = \frac{fp}{p - f}$$

Differentiating this equation with respect to time gives us the velocity of the image $v_q = dq/dt$:

$$v_q = \frac{dq}{dt} = \frac{d}{dt}\left(\frac{fp}{p-f}\right) = -\frac{f^2}{(p-f)^2}\frac{dp}{dt} = -\frac{f^2 v_p}{(p-f)^2}$$

For the object position of 25.0 cm in part (A), the velocity of the image is

$$v_q = -\frac{f^2 v_p}{(p-f)^2} = -\frac{(10.0 \text{ cm})^2 v_p}{(25.0 \text{ cm} - 10.0 \text{ cm})^2} = -0.444\, v_p$$

Thus, the speed of the image is less than that of the object in this case.

We can see two interesting behaviors of this function for v_q. First, note that the velocity is negative regardless of the value of p or f. Thus, if the object moves toward the mirror, the image moves toward the left in Figure 36.15 without regard for the side of the focal point at which the object is located or whether the mirror is concave or convex. Second, in the limit of $p \rightarrow 0$, the velocity v_q approaches $-v_p$. As the object moves very close to the mirror, the mirror looks like a plane mirror, the image is as far behind the mirror as the object is in front, and both the object and the image move with the same speed.

 Investigate the image formed for various object positions and mirror focal lengths at the Interactive Worked Example link at **http://www.pse6.com.**

Example 36.5 The Image from a Convex Mirror

An anti-shoplifting mirror, as shown in Figure 36.17, shows an image of a woman who is located 3.0 m from the mirror. The focal length of the mirror is − 0.25 m. Find

(A) the position of her image and

(B) the magnification of the image.

Solution (A) This situation is depicted in Figure 36.15c. We should expect to find an upright, reduced, virtual image. To find the image position, we use Equation 36.6:

$$\frac{1}{p} + \frac{1}{q} = \frac{1}{f} = \frac{1}{-0.25 \text{ m}}$$

$$\frac{1}{q} = \frac{1}{-0.25 \text{ m}} - \frac{1}{3.0 \text{ m}}$$

$$q = \boxed{-0.23 \text{ m}}$$

The negative value of q indicates that her image is virtual, or behind the mirror, as shown in Figure 36.15c.

(B) The magnification of the image is

$$M = -\frac{q}{p} = -\left(\frac{-0.23 \text{ m}}{3.0 \text{ m}}\right) = \boxed{+0.077}$$

The image is much smaller than the woman, and it is upright because M is positive.

Figure 36.17 (Example 36.5) Convex mirrors, often used for security in department stores, provide wide-angle viewing.

 Investigate the image formed for various object positions and mirror focal lengths at the Interactive Worked Example link at http://www.pse6.com.

36.3   Images Formed by Refraction

In this section we describe how images are formed when light rays are refracted at the boundary between two transparent materials. Consider two transparent media having indices of refraction n_1 and n_2, where the boundary between the two media is a spherical surface of radius R (Fig. 36.18). We assume that the object at O is in the medium for which the index of refraction is n_1. Let us consider the paraxial rays leaving O. As we shall see, all such rays are refracted at the spherical surface and focus at a single point I, the image point.

Figure 36.19 shows a single ray leaving point O and refracting to point I. Snell's law of refraction applied to this ray gives

$$n_1 \sin \theta_1 = n_2 \sin \theta_2$$

Because θ_1 and θ_2 are assumed to be small, we can use the small-angle approximation $\sin \theta \approx \theta$ (with angles in radians) and say that

$$n_1 \theta_1 = n_2 \theta_2$$

Now we use the fact that an exterior angle of any triangle equals the sum of the two opposite interior angles. Applying this rule to triangles OPC and PIC in Figure 36.19 gives

$$\theta_1 = \alpha + \beta$$

$$\beta = \theta_2 + \gamma$$

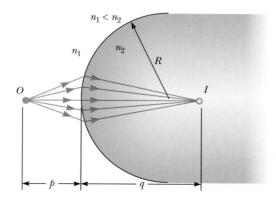

Figure 36.18 An image formed by refraction at a spherical surface. Rays making small angles with the principal axis diverge from a point object at O and are refracted through the image point I.

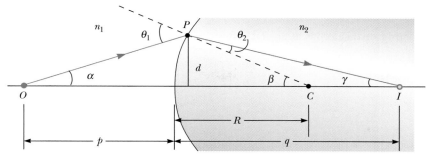

Figure 36.19 Geometry used to derive Equation 36.8, assuming that $n_1 < n_2$.

If we combine all three expressions and eliminate θ_1 and θ_2, we find that

$$n_1\alpha + n_2\gamma = (n_2 - n_1)\beta \tag{36.7}$$

From Figure 36.19, we see three right triangles that have a common vertical leg of length d. For paraxial rays (unlike the relatively large-angle ray shown in Fig. 36.19), the horizontal legs of these triangles are approximately p for the triangle containing angle α, R for the triangle containing angle β, and q for the triangle containing angle γ. In the small-angle approximation, $\tan\theta \approx \theta$, so we can write the approximate relationships from these triangles as follows:

$$\tan\alpha \approx \alpha \approx \frac{d}{p} \qquad \tan\beta \approx \beta \approx \frac{d}{R} \qquad \tan\gamma \approx \gamma \approx \frac{d}{q}$$

We substitute these expressions into Equation 36.7 and divide through by d to give

$$\frac{n_1}{p} + \frac{n_2}{q} = \frac{n_2 - n_1}{R} \tag{36.8}$$

Relation between object and image distance for a refracting surface

For a fixed object distance p, the image distance q is independent of the angle that the ray makes with the axis. This result tells us that all paraxial rays focus at the same point I.

As with mirrors, we must use a sign convention if we are to apply this equation to a variety of cases. We define the side of the surface in which light rays originate as the front side. The other side is called the back side. Real images are formed by refraction in back of the surface, in contrast with mirrors, where real images are formed in front of the reflecting surface. Because of the difference in location of real images, the refraction sign conventions for q and R are opposite the reflection sign conventions. For example, q and R are both positive in Figure 36.19. The sign conventions for spherical refracting surfaces are summarized in Table 36.2.

We derived Equation 36.8 from an assumption that $n_1 < n_2$ in Figure 36.19. This assumption is not necessary, however. Equation 36.8 is valid regardless of which index of refraction is greater.

Table 36.2

Sign Conventions for Refracting Surfaces		
Quantity	**Positive When**	**Negative When**
Object location (p)	Object is in front of surface (real object)	Object is in back of surface (virtual object)
Image location (q)	Image is in back of surface (real image)	Image is in front of surface (virtual image)
Image height (h')	Image is upright	Image is inverted
Radius (R)	Center of curvature is in back of surface	Center of curvature is in front of surface

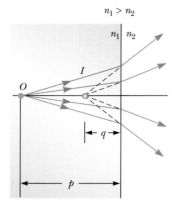

$n_1 > n_2$

Active Figure 36.20 The image formed by a flat refracting surface is virtual and on the same side of the surface as the object. All rays are assumed to be paraxial.

At the Active Figures link at http://www.pse6.com, you can move the object to see the effect on the location of the image.

Flat Refracting Surfaces

If a refracting surface is flat, then R is infinite and Equation 36.8 reduces to

$$\frac{n_1}{p} = -\frac{n_2}{q}$$

$$q = -\frac{n_2}{n_1}p \qquad (36.9)$$

From this expression we see that the sign of q is opposite that of p. Thus, according to Table 36.2, **the image formed by a flat refracting surface is on the same side of the surface as the object.** This is illustrated in Figure 36.20 for the situation in which the object is in the medium of index n_1 and n_1 is greater than n_2. In this case, a virtual image is formed between the object and the surface. If n_1 is less than n_2, the rays in the back side diverge from each other at lesser angles than those in Figure 36.20. As a result, the virtual image is formed to the left of the object.

Quick Quiz 36.5 In Figure 36.18, what happens to the image point I as the object point O is moved to the right from very far away to very close to the refracting surface? (a) It is always to the right of the surface. (b) It is always to the left of the surface. (c) It starts off to the left and at some position of O, I moves to the right of the surface. (d) It starts off to the right and at some position of O, I moves to the left of the surface.

Quick Quiz 36.6 In Figure 36.20, what happens to the image point I as the object point O moves toward the right-hand surface of the material of index of refraction n_1? (a) It always remains between O and the surface, arriving at the surface just as O does. (b) It moves toward the surface more slowly than O so that eventually O passes I. (c) It approaches the surface and then moves to the right of the surface.

Conceptual Example 36.6   Let's Go Scuba Diving!

It is well known that objects viewed under water with the naked eye appear blurred and out of focus. However, a scuba diver using a mask has a clear view of underwater objects. Explain how this works, using the facts that the indices of refraction of the cornea, water, and air are 1.376, 1.333, and 1.00029, respectively.

Solution Because the cornea and water have almost identical indices of refraction, very little refraction occurs

when a person under water views objects with the naked eye. In this case, light rays from an object focus behind the retina, resulting in a blurred image. When a mask is used, the air space between the eye and the mask surface provides the normal amount of refraction at the eye–air interface, and the light from the object focuses on the retina.

Example 36.7   Gaze into the Crystal Ball

A set of coins is embedded in a spherical plastic paperweight having a radius of 3.0 cm. The index of refraction of the plastic is $n_1 = 1.50$. One coin is located 2.0 cm from the edge of the sphere (Fig. 36.21). Find the position of the image of the coin.

Solution Because $n_1 > n_2$, where $n_2 = 1.00$ is the index of refraction for air, the rays originating from the coin are refracted away from the normal at the surface and diverge outward. Hence, the image is formed inside the paperweight and is *virtual*. Applying Equation 36.8 and noting

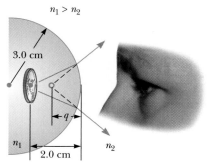

Figure 36.21 (Example 36.7) Light rays from a coin embedded in a plastic sphere form a virtual image between the surface of the object and the sphere surface. Because the object is inside the sphere, the front of the refracting surface is the *interior* of the sphere.

from Table 36.2 that R is negative, we obtain

$$\frac{n_1}{p} + \frac{n_2}{q} = \frac{n_2 - n_1}{R}$$

$$\frac{1.50}{2.0 \text{ cm}} + \frac{1}{q} = \frac{1.00 - 1.50}{-3.0 \text{ cm}}$$

$$q = \boxed{-1.7 \text{ cm}}$$

The negative sign for q indicates that the image is in front of the surface—in other words, in the same medium as the object, as shown in Figure 36.21. Being in the same medium as the object, the image must be virtual. (See Table 36.2.) The coin appears to be closer to the paperweight surface than it actually is.

Example 36.8   The One That Got Away

A small fish is swimming at a depth d below the surface of a pond (Fig. 36.22). What is the apparent depth of the fish, as viewed from directly overhead?

Solution Because the refracting surface is flat, R is infinite. Hence, we can use Equation 36.9 to determine the location of the image with $p = d$. Using the indices of refraction given in Figure 36.22, we obtain

$$q = -\frac{n_2}{n_1} p = -\frac{1.00}{1.33} d = \boxed{-0.752d}$$

Because q is negative, the image is virtual, as indicated by the dashed lines in Figure 36.22. The apparent depth is approximately three-fourths the actual depth.

What If? What if you look more carefully at the fish and measure its apparent *height*, from its upper fin to its lower fin? Is the apparent height h' of the fish different from the actual height h?

Answer Because all points on the fish appear to be fractionally closer to the observer, we would predict that the height would be smaller. If we let the distance d in Figure 36.22 be measured to the top fin and the distance to the bottom fin be $d + h$, then the images of the top and bottom of the fish are located at

$$q_{\text{top}} = -0.752d$$

$$q_{\text{bottom}} = -0.752(d + h)$$

Figure 36.22 (Example 36.8) The apparent depth q of the fish is less than the true depth d. All rays are assumed to be paraxial.

The apparent height h' of the fish is

$$h' = q_{\text{top}} - q_{\text{bottom}} = -0.752d - [-0.752(d + h)]$$

$$= 0.752h$$

and the fish appears to be approximately three-fourths its actual height.

36.4   Thin Lenses

Lenses are commonly used to form images by refraction in optical instruments, such as cameras, telescopes, and microscopes. We can use what we just learned about images formed by refracting surfaces to help us locate the image formed by a lens. We recognize that light passing through a lens experiences refraction at two surfaces. The development we shall follow is based on the notion that **the image**

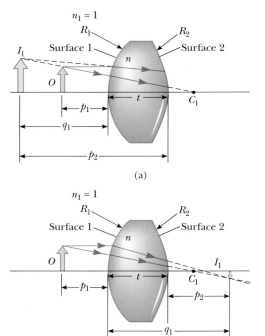

Figure 36.23 To locate the image formed by a lens, we use the virtual image at I_1 formed by surface 1 as the object for the image formed by surface 2. The point C_1 is the center of curvature of surface 1. (a) The image due to surface 1 is virtual so that I_1 is to the left of the surface. (b) The image due to surface 1 is real so that I_1 is to the right of the surface.

formed by one refracting surface serves as the object for the second surface. We shall analyze a thick lens first and then let the thickness of the lens be approximately zero.

Consider a lens having an index of refraction n and two spherical surfaces with radii of curvature R_1 and R_2, as in Figure 36.23. (Note that R_1 is the radius of curvature of the lens surface that the light from the object reaches first and that R_2 is the radius of curvature of the other surface of the lens.) An object is placed at point O at a distance p_1 in front of surface 1.

Let us begin with the image formed by surface 1. Using Equation 36.8 and assuming that $n_1 = 1$ because the lens is surrounded by air, we find that the image I_1 formed by surface 1 satisfies the equation

$$\frac{1}{p_1} + \frac{n}{q_1} = \frac{n-1}{R_1} \tag{36.10}$$

where q_1 is the position of the image due to surface 1. If the image due to surface 1 is virtual (Fig. 36.23a), q_1 is negative, and it is positive if the image is real (Fig. 36.23b).

Now we apply Equation 36.8 to surface 2, taking $n_1 = n$ and $n_2 = 1$. (We make this switch in index because the light rays approaching surface 2 are *in the material of the lens,* and this material has index n.) Taking p_2 as the object distance for surface 2 and q_2 as the image distance gives

$$\frac{n}{p_2} + \frac{1}{q_2} = \frac{1-n}{R_2} \tag{36.11}$$

We now introduce mathematically the fact that the image formed by the first surface acts as the object for the second surface. We do this by noting from Figure 36.23 that p_2, measured from surface 2, is related to q_1 as follows:

Virtual image from surface 1 (Fig. 36.23a): $p_2 = -q_1 + t$ (q_1 is negative)

Real image from surface 1 (Fig. 36.23b): $p_2 = -q_1 + t$ (q_1 is positive)

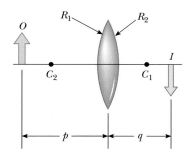

Figure 36.24 Simplified geometry for a thin lens.

where t is the thickness of the lens. For a *thin* lens (one whose thickness is small compared to the radii of curvature), we can neglect t. In this approximation, we see that $p_2 = -q_1$ for either type of image from surface 1. (If the image from surface 1 is real, the image acts as a virtual object, so p_2 is negative.) Hence, Equation 36.11 becomes

$$-\frac{n}{q_1} + \frac{1}{q_2} = \frac{1-n}{R_2}$$ (36.12)

Adding Equations 36.10 and 36.12, we find that

$$\frac{1}{p_1} + \frac{1}{q_2} = (n-1)\left(\frac{1}{R_1} - \frac{1}{R_2}\right)$$ (36.13)

For a thin lens, we can omit the subscripts on p_1 and q_2 in Equation 36.13 and call the object distance p and the image distance q, as in Figure 36.24. Hence, we can write Equation 36.13 in the form

$$\frac{1}{p} + \frac{1}{q} = (n-1)\left(\frac{1}{R_1} - \frac{1}{R_2}\right)$$ (36.14)

This expression relates the image distance q of the image formed by a thin lens to the object distance p and to the lens properties (index of refraction and radii of curvature). It is valid only for paraxial rays and only when the lens thickness is much less than R_1 and R_2.

The **focal length** f of a thin lens is the image distance that corresponds to an infinite object distance, just as with mirrors. Letting p approach ∞ and q approach f in Equation 36.14, we see that the inverse of the focal length for a thin lens is

$$\frac{1}{f} = (n-1)\left(\frac{1}{R_1} - \frac{1}{R_2}\right)$$ (36.15)   **Lens makers' equation**

This relationship is called the **lens makers' equation** because it can be used to determine the values of R_1 and R_2 that are needed for a given index of refraction and a desired focal length f. Conversely, if the index of refraction and the radii of curvature of a lens are given, this equation enables a calculation of the focal length. If the lens is immersed in something other than air, this same equation can be used, with n interpreted as the *ratio* of the index of refraction of the lens material to that of the surrounding fluid.

Using Equation 36.15, we can write Equation 36.14 in a form identical to Equation 36.6 for mirrors:

$$\frac{1}{p} + \frac{1}{q} = \frac{1}{f}$$ (36.16)   **Thin lens equation**

(a)

(b)

Figure 36.25 (*Left*) Effects of a converging lens (top) and a diverging lens (bottom) on parallel rays. (*Right*) Parallel light rays pass through (a) a converging lens and (b) a diverging lens. The focal length is the same for light rays passing through a given lens in either direction. Both focal points F_1 and F_2 are the same distance from the lens.

This equation, called the **thin lens equation,** can be used to relate the image distance and object distance for a thin lens.

Because light can travel in either direction through a lens, each lens has two focal points, one for light rays passing through in one direction and one for rays passing through in the other direction. This is illustrated in Figure 36.25 for a biconvex lens (two convex surfaces, resulting in a converging lens) and a biconcave lens (two concave surfaces, resulting in a diverging lens).

Figure 36.26 is useful for obtaining the signs of p and q, and Table 36.3 gives the sign conventions for thin lenses. Note that these sign conventions are the same as those for refracting surfaces (see Table 36.2). Applying these rules to a biconvex lens, we see that when $p > f$, the quantities p, q, and R_1 are positive, and R_2 is negative. Therefore, p, q, and f are all positive when a converging lens forms a real image of an object. For a biconcave lens, p and R_2 are positive and q and R_1 are negative, with the result that f is negative.

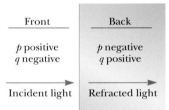

Front	Back
p positive	*p* negative
q negative	*q* positive
Incident light →	Refracted light →

Figure 36.26 A diagram for obtaining the signs of p and q for a thin lens. (This diagram also applies to a refracting surface.)

▲ **PITFALL PREVENTION**

36.5 A Lens Has Two Focal Points but Only One Focal Length

A lens has a focal point on each side, front and back. However, there is only one focal length—each of the two focal points is located the same distance from the lens (Fig. 36.25). This can be seen mathematically by interchanging R_1 and R_2 in Equation 36.15 (and changing the signs of the radii because back and front have been interchanged). As a result, the lens forms an image of an object at the same point if it is turned around. In practice this might not happen, because real lenses are not infinitesimally thin.

Table 36.3

Sign Conventions for Thin Lenses		
Quantity	**Positive When**	**Negative When**
Object location (p)	Object is in front of lens (real object)	Object is in back of lens (virtual object)
Image location (q)	Image is in back of lens (real image)	Image is in front of lens (virtual image)
Image height (h')	Image is upright	Image is inverted
R_1 and R_2	Center of curvature is in back of lens	Center of curvature is in front of lens
Focal length (f)	Converging lens	Diverging lens

Henry Leap and Jim Lehman

Various lens shapes are shown in Figure 36.27. Note that a converging lens is thicker at the center than at the edge, whereas a diverging lens is thinner at the center than at the edge.

Magnification of Images

Consider a thin lens through which light rays from an object pass. As with mirrors (Eq. 36.2), we could analyze a geometric construction to show that the lateral magnification of the image is

$$M = \frac{h'}{h} = -\frac{q}{p}$$

From this expression, it follows that when M is positive, the image is upright and on the same side of the lens as the object. When M is negative, the image is inverted and on the side of the lens opposite the object.

Ray Diagrams for Thin Lenses

Ray diagrams are convenient for locating the images formed by thin lenses or systems of lenses. They also help clarify our sign conventions. Figure 36.28 shows such diagrams for three single-lens situations.

To locate the image of a *converging* lens (Fig. 36.28a and b), the following three rays are drawn from the top of the object:

- Ray 1 is drawn parallel to the principal axis. After being refracted by the lens, this ray passes through the focal point on the back side of the lens.
- Ray 2 is drawn through the center of the lens and continues in a straight line.
- Ray 3 is drawn through the focal point on the front side of the lens (or as if coming from the focal point if $p < f$) and emerges from the lens parallel to the principal axis.

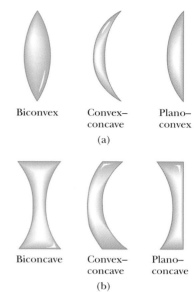

Biconvex   **Convex–concave**   **Plano–convex**

(a)

Biconcave   **Convex–concave**   **Plano–concave**

(b)

Figure 36.27 Various lens shapes. (a) Converging lenses have a positive focal length and are thickest at the middle. (b) Diverging lenses have a negative focal length and are thickest at the edges.

(a)

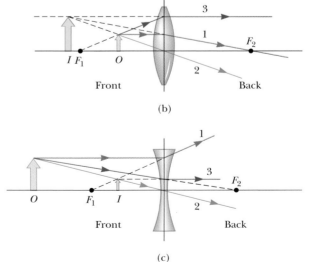

(b)

(c)

Active Figure 36.28 Ray diagrams for locating the image formed by a thin lens. (a) When the object is in front of and outside the focal point of a converging lens, the image is real, inverted, and on the back side of the lens. (b) When the object is between the focal point and a converging lens, the image is virtual, upright, larger than the object, and on the front side of the lens. (c) When an object is anywhere in front of a diverging lens, the image is virtual, upright, smaller than the object, and on the front side of the lens.

At the Active Figures link at http://www.pse6.com, *you can move the objects and change the focal length of the lenses to see the effect on the images.*

To locate the image of a *diverging* lens (Fig. 36.28c), the following three rays are drawn from the top of the object:

- Ray 1 is drawn parallel to the principal axis. After being refracted by the lens, this ray emerges directed away from the focal point on the front side of the lens.
- Ray 2 is drawn through the center of the lens and continues in a straight line.
- Ray 3 is drawn in the direction toward the focal point on the back side of the lens and emerges from the lens parallel to the principal axis.

For the converging lens in Figure 36.28a, where the object is to the left of the focal point ($p > f$), the image is real and inverted. When the object is between the focal point and the lens ($p < f$), as in Figure 36.28b, the image is virtual and upright. For a diverging lens (see Fig. 36.28c), the image is always virtual and upright, regardless of where the object is placed. These geometric constructions are reasonably accurate only if the distance between the rays and the principal axis is much less than the radii of the lens surfaces.

Note that refraction occurs only at the surfaces of the lens. A certain lens design takes advantage of this fact to produce the *Fresnel lens*, a powerful lens without great thickness. Because only the surface curvature is important in the refracting qualities of the lens, material in the middle of a Fresnel lens is removed, as shown in the cross sections of lenses in Figure 36.29. Because the edges of the curved segments cause some distortion, Fresnel lenses are usually used only in situations in which image quality is less important than reduction of weight. A classroom overhead projector often uses a Fresnel lens; the circular edges between segments of the lens can be seen by looking closely at the light projected onto a screen.

Figure 36.29 The Fresnel lens on the left has the same focal length as the thick lens on the right but is made of much less glass.

Quick Quiz 36.7 What is the focal length of a pane of window glass? (a) zero (b) infinity (c) the thickness of the glass (d) impossible to determine

Quick Quiz 36.8 Diving masks often have a lens built into the glass for divers who do not have perfect vision. This allows the individual to dive without the necessity for glasses, because the lenses in the faceplate perform the necessary refraction to provide clear vision. The proper design allows the diver to see clearly with the mask on *both* under water and in the open air. Normal eyeglasses have lenses that are curved on both the front and back surfaces. The lenses in a diving mask should be curved (a) only on the front surface (b) only on the back surface (c) on both the front and back surfaces.

Example 36.9 Images Formed by a Converging Lens   `Interactive`

A converging lens of focal length 10.0 cm forms images of objects placed

(A) 30.0 cm,

(B) 10.0 cm, and

(C) 5.00 cm from the lens.

In each case, construct a ray diagram, find the image distance and describe the image.

Solution

(A) First we construct a ray diagram as shown in Figure 36.30a. The diagram shows that we should expect a real, inverted, smaller image to be formed on the back side of the lens. The thin lens equation, Equation 36.16, can be used to find the image distance:

$$\frac{1}{p} + \frac{1}{q} = \frac{1}{f}$$

Figure 36.30 (Example 36.9) An image is formed by a converging lens. (a) The object is farther from the lens than the focal point. (b) The object is closer to the lens than the focal point.

$$\frac{1}{30.0 \text{ cm}} + \frac{1}{q} = \frac{1}{10.0 \text{ cm}}$$

$$q = \boxed{+15.0 \text{ cm}}$$

The positive sign for the image distance tells us that the image is indeed real and on the back side of the lens. The magnification of the image is

$$M = -\frac{q}{p} = -\frac{15.0 \text{ cm}}{30.0 \text{ cm}} = \boxed{-0.500}$$

Thus, the image is reduced in height by one half, and the negative sign for M tells us that the image is inverted.

(B) No calculation is necessary for this case because we know that, when the object is placed at the focal point, the image is formed at infinity. This is readily verified by substituting $p = 10.0$ cm into the thin lens equation.

(C) We now move inside the focal point. The ray diagram in Figure 36.30b shows that in this case the lens acts as a magnifying glass; that is, the image is magnified, upright, on the same side of the lens as the object, and virtual. Because the object distance is 5.00 cm, the thin lens equation gives

$$\frac{1}{5.00 \text{ cm}} + \frac{1}{q} = \frac{1}{10.0 \text{ cm}}$$

$$q = \boxed{-10.0 \text{ cm}}$$

and the magnification of the image is

$$M = -\frac{q}{p} = -\left(\frac{-10.0 \text{ cm}}{5.00 \text{ cm}}\right) = \boxed{+2.00}$$

The negative image distance tells us that the image is virtual and formed on the side of the lens from which the light is incident, the front side. The image is enlarged, and the positive sign for M tells us that the image is upright.

What If? What if the object moves right up to the lens surface, so that $p \rightarrow 0$? Where is the image?

Answer In this case, because $p \ll R$, where R is either of the radii of the surfaces of the lens, the curvature of the lens can be ignored and it should appear to have the same effect as a plane piece of material. This would suggest that the image is just on the front side of the lens, at $q = 0$. We can verify this mathematically by rearranging the thin lens equation:

$$\frac{1}{q} = \frac{1}{f} - \frac{1}{p}$$

If we let $p \rightarrow 0$, the second term on the right becomes very large compared to the first and we can neglect $1/f$. The equation becomes

$$\frac{1}{q} = -\frac{1}{p}$$

$$q = -p = 0$$

Thus, q is on the front side of the lens (because it has the opposite sign as p), and just at the lens surface.

 Investigate the image formed for various object positions and lens focal lengths at the Interactive Worked Example link at **http://www.pse6.com.**

Example 36.10 **The Case of a Diverging Lens** Interactive

Repeat Example 36.9 for a *diverging* lens of focal length 10.0 cm.

Solution

(A) We begin by constructing a ray diagram as in Figure 36.31a taking the object distance to be 30.0 cm. The diagram shows that we should expect an image that is virtual, smaller than the object, and upright. Let us now apply the thin lens equation with $p = 30.0$ cm:

$$\frac{1}{p} + \frac{1}{q} = \frac{1}{f}$$

$$\frac{1}{30.0 \text{ cm}} + \frac{1}{q} = \frac{1}{-10.0 \text{ cm}}$$

$$q = \boxed{-7.50 \text{ cm}}$$

The magnification of the image is

$$M = -\frac{q}{p} = -\left(\frac{-7.50 \text{ cm}}{30.0 \text{ cm}}\right) = \boxed{+0.250}$$

This result confirms that the image is virtual, smaller than the object, and upright.

(B) When the object is at the focal point, the ray diagram appears as in Figure 36.31b. In the thin lens equation, using $p = 10.0$ cm, we have

$$\frac{1}{10.0 \text{ cm}} + \frac{1}{q} = \frac{1}{-10.0 \text{ cm}}$$

$$q = \boxed{-5.00 \text{ cm}}$$

The magnification of the image is

$$M = -\frac{q}{p} = -\left(\frac{-5.00 \text{ cm}}{10.0 \text{ cm}}\right) = \boxed{+0.500}$$

Notice the difference between this situation and that for a converging lens. For a diverging lens, an object at the focal point does not produce an image infinitely far away.

(C) When the object is inside the focal point, at $p = 5.00$ cm, the ray diagram in Figure 36.31c shows that we expect a virtual image that is smaller than the object and upright. In

Figure 36.31 (Example 36.10) An image is formed by a diverging lens. (a) The object is farther from the lens than the focal point. (b) The object is at the focal point. (c) The object is closer to the lens than the focal point.

this case, the thin lens equation gives

$$\frac{1}{5.00 \text{ cm}} + \frac{1}{q} = \frac{1}{-10.0 \text{ cm}}$$

$$q = \boxed{-3.33 \text{ cm}}$$

and the magnification of the image is

$$M = -\left(\frac{-3.33 \text{ cm}}{5.00 \text{ cm}}\right) = \boxed{+0.667}$$

This confirms that the image is virtual, smaller than the object, and upright.

 Investigate the image formed for various object positions and lens focal lengths at the Interactive Worked Example link at **http://www.pse6.com.**

Example 36.11   A Lens Under Water

A converging glass lens ($n = 1.52$) has a focal length of 40.0 cm in air. Find its focal length when it is immersed in water, which has an index of refraction of 1.33.

Solution We can use the lens makers' equation (Eq. 36.15) in both cases, noting that R_1 and R_2 remain the same in air and water:

$$\frac{1}{f_{\text{air}}} = (n - 1)\left(\frac{1}{R_1} - \frac{1}{R_2}\right)$$

$$\frac{1}{f_{\text{water}}} = (n' - 1)\left(\frac{1}{R_1} - \frac{1}{R_2}\right)$$

where n' is the ratio of the index of refraction of glass to that of water: $n' = 1.52/1.33 = 1.14$. Dividing the first equation by the second gives

$$\frac{f_{\text{water}}}{f_{\text{air}}} = \frac{n - 1}{n' - 1} = \frac{1.52 - 1}{1.14 - 1} = 3.71$$

Because $f_{\text{air}} = 40.0$ cm, we find that

$$f_{\text{water}} = 3.71 f_{\text{air}} = 3.71(40.0 \text{ cm}) = \boxed{148 \text{ cm}}$$

The focal length of any lens is increased by a factor $(n - 1)/(n' - 1)$ when the lens is immersed in a fluid, where n' is the ratio of the index of refraction n of the lens material to that of the fluid.

Combination of Thin Lenses

If two thin lenses are used to form an image, the system can be treated in the following manner. First, the image formed by the first lens is located as if the second lens were not present. Then a ray diagram is drawn for the second lens, with the image formed by the first lens now serving as the object for the second lens. The second image formed is the final image of the system. If the image formed by the first lens lies on the back side of the second lens, then that image is treated as a **virtual object** for the second lens (that is, in the thin lens equation, p is negative). The same procedure can be extended to a system of three or more lenses. Because the magnification due to the second lens is performed on the magnified image due to the first lens, the overall magnification of the image due to the combination of lenses is the product of the individual magnifications.

Let us consider the special case of a system of two lenses of focal lengths f_1 and f_2 in contact with each other. If $p_1 = p$ is the object distance for the combination, application of the thin lens equation (Eq. 36.16) to the first lens gives

$$\frac{1}{p} + \frac{1}{q_1} = \frac{1}{f_1}$$

Light from a distant object is brought into focus by two converging lenses.

where q_1 is the image distance for the first lens. Treating this image as the object for the second lens, we see that the object distance for the second lens must be $p_2 = -q_1$. (The distances are the same because the lenses are in contact and assumed to be infinitesimally thin. The object distance is negative because the object is virtual.) Therefore, for the second lens,

$$\frac{1}{p_2} + \frac{1}{q_2} = \frac{1}{f_2}$$

$$-\frac{1}{q_1} + \frac{1}{q} = \frac{1}{f_2}$$

where $q = q_2$ is the final image distance from the second lens, which is the image distance for the combination. Adding the equations for the two lenses eliminates q_1 and gives

$$\frac{1}{p} + \frac{1}{q} = \frac{1}{f_1} + \frac{1}{f_2}$$

If we consider replacing the combination with a single lens that will form an image at the same location, we see that its focal length is related to the individual focal lengths by

Focal length for a combination of two thin lenses in contact

$$\frac{1}{f} = \frac{1}{f_1} + \frac{1}{f_2} \qquad (36.17)$$

Therefore, **two thin lenses in contact with each other are equivalent to a single thin lens having a focal length given by Equation 36.17.**

Example 36.12 Where Is the Final Image? **Interactive**

Two thin converging lenses of focal lengths $f_1 = 10.0$ cm and $f_2 = 20.0$ cm are separated by 20.0 cm, as illustrated in Figure 36.32a. An object is placed 30.0 cm to the left of lens 1. Find the position and the magnification of the final image.

Solution Conceptualize by imagining light rays passing through the first lens and forming a real image (because $p > f$) in the absence of the second lens. Figure 36.32b shows these light rays forming the inverted image I_1. Once the light rays converge to the image point, they do not stop. They continue through the image point and interact with the second lens. The rays leaving the image point behave in the same way as the rays leaving an object. Thus, the image of the first lens serves as the object of the second lens. We categorize this problem as one in which we apply the thin lens equation, but in stepwise fashion to the two lenses.

To analyze the problem, we first draw a ray diagram (Figure 36.32b) showing where the image from the first lens falls and how it acts as the object for the second lens. The location of the image formed by lens 1 is found from the thin lens equation:

$$\frac{1}{p_1} + \frac{1}{q_1} = \frac{1}{f}$$

$$\frac{1}{30.0\text{ cm}} + \frac{1}{q_1} = \frac{1}{10.0\text{ cm}}$$

$$q_1 = \boxed{+15.0\text{ cm}}$$

The magnification of this image is

$$M_1 = -\frac{q_1}{p_1} = -\frac{15.0\text{ cm}}{30.0\text{ cm}} = \boxed{-0.500}$$

The image formed by this lens acts as the object for the second lens. Thus, the object distance for the second lens is $20.0\text{ cm} - 15.0\text{ cm} = 5.00\text{ cm}$. We again apply the thin lens equation to find the location of the final image:

$$\frac{1}{5.00\text{ cm}} + \frac{1}{q_2} = \frac{1}{20.0\text{ cm}}$$

$$q_2 = \boxed{-6.67\text{ cm}}$$

The magnification of the second image is

$$M_2 = -\frac{q_2}{p_2} = -\frac{(-6.67\text{ cm})}{5.00\text{ cm}} = \boxed{+1.33}$$

Thus, the overall magnification of the system is

$$M = M_1 M_2 = (-0.500)(1.33) = \boxed{-0.667}$$

To finalize the problem, note that the negative sign on the overall magnification indicates that the final image is inverted with respect to the initial object. The fact that the absolute value of the magnification is less than one tells us that the final image is smaller than the object. The fact that q_2 is negative tells us that the final image is on the front, or left, side of lens 2. All of these conclusions are consistent with the ray diagram in Figure 36.32b.

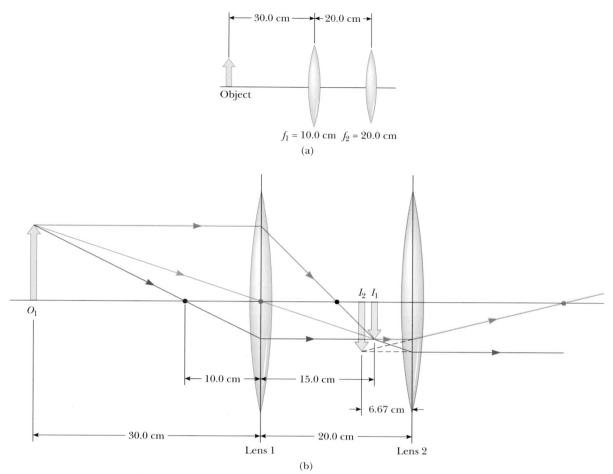

Figure 36.32 (Example 36.12) (a) A combination of two converging lenses. (b) The ray diagram showing the location of the final image due to the combination of lenses. The black dots are the focal points of lens 1 while the red dots are the focal points of lens 2.

What If? Suppose we want to create an upright image with this system of two lenses. How must the second lens be moved in order to achieve this?

Answer Because the object is farther from the first lens than the focal length of that lens, we know that the first image is inverted. Consequently, we need the second lens to invert the image once again so that the final image is upright. An inverted image is only formed by a converging lens if the object is outside the focal point. Thus, the image due to the first lens must be to the left of the focal point of the second lens in Figure 36.32b. To make this happen, we must move the second lens at least as far away from the first lens as the sum $q_1 + f_2 = 15.0 \text{ cm} + 20.0 \text{ cm} = 35.0 \text{ cm}$.

 Investigate the image formed by a combination of two lenses at the Interactive Worked Example link at **http://www.pse6.com.**

Conceptual Example 36.13   Watch Your p's and q's!

Use a spreadsheet or a similar tool to create two graphs of image distance as a function of object distance—one for a lens for which the focal length is 10 cm and one for a lens for which the focal length is − 10 cm.

Solution The graphs are shown in Figure 36.33. In each graph, a gap occurs where $p = f$, which we shall discuss. Note the similarity in the shapes—a result of the fact that image and object distances for both lenses are related according to the same equation—the thin lens equation.

The curve in the upper right portion of the $f = +10$ cm graph corresponds to an object located on the *front* side of a lens, which we have drawn as the left side of the lens in our previous diagrams. When the object is at positive infinity, a real image forms at the focal point on the back side (the positive side) of the lens, $q = f$. (The incoming rays are parallel in this case.) As the object moves closer to the lens, the image

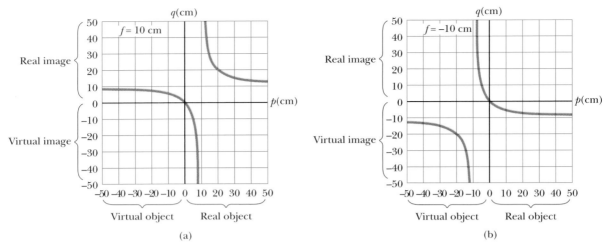

Figure 36.33 (Conceptual Example 36.13) (a) Image position as a function of object position for a lens having a focal length of + 10 cm. (b) Image position as a function of object position for a lens having a focal length of − 10 cm.

moves farther from the lens, corresponding to the upward path of the curve. This continues until the object is located at the focal point on the near side of the lens. At this point, the rays leaving the lens are parallel, making the image infinitely far away. This is described in the graph by the asymptotic approach of the curve to the line $p = f = 10$ cm.

As the object moves inside the focal point, the image becomes virtual and located near $q = -\infty$. We are now following the curve in the lower left portion of Figure 36.33a. As the object moves closer to the lens, the virtual image also moves closer to the lens. As $p \rightarrow 0$, the image distance q also approaches 0. Now imagine that we bring the object to the back side of the lens, where $p < 0$. The object is now a virtual object, so it must have been formed by some other lens. For all locations of the virtual object, the image

distance is positive and less than the focal length. The final image is real, and its position approaches the focal point as p becomes more and more negative.

The $f = -10$ cm graph shows that a distant real object forms an image at the focal point on the front side of the lens. As the object approaches the lens, the image remains virtual and moves closer to the lens. But as we continue toward the left end of the p axis, the object becomes virtual. As the position of this virtual object approaches the focal point, the image recedes toward infinity. As we pass the focal point, the image shifts from a location at positive infinity to one at negative infinity. Finally, as the virtual object continues moving away from the lens, the image is virtual, starts moving in from negative infinity, and approaches the focal point.

36.5   Lens Aberrations

Our analysis of mirrors and lenses assumes that rays make small angles with the principal axis and that the lenses are thin. In this simple model, all rays leaving a point source focus at a single point, producing a sharp image. Clearly, this is not always true. When the approximations used in this analysis do not hold, imperfect images are formed.

A precise analysis of image formation requires tracing each ray, using Snell's law at each refracting surface and the law of reflection at each reflecting surface. This procedure shows that the rays from a point object do not focus at a single point, with the result that the image is blurred. The departures of actual images from the ideal predicted by our simplified model are called **aberrations.**

Spherical Aberrations

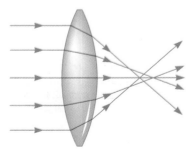

Figure 36.34 Spherical aberration caused by a converging lens. Does a diverging lens cause spherical aberration?

Spherical aberrations occur because the focal points of rays far from the principal axis of a spherical lens (or mirror) are different from the focal points of rays of the same wavelength passing near the axis. Figure 36.34 illustrates spherical aberration for parallel rays passing through a converging lens. Rays passing through points near the center of the lens are imaged farther from the lens than rays passing through points near the edges. Figure 36.10 earlier in the chapter showed a similar situation for a spherical mirror.

Many cameras have an adjustable aperture to control light intensity and reduce spherical aberration. (An aperture is an opening that controls the amount of light passing through the lens.) Sharper images are produced as the aperture size is reduced because with a small aperture only the central portion of the lens is exposed to the light; as a result, a greater percentage of the rays are paraxial. At the same time, however, less light passes through the lens. To compensate for this lower light intensity, a longer exposure time is used.

In the case of mirrors, spherical aberration can be minimized through the use of a parabolic reflecting surface rather than a spherical surface. Parabolic surfaces are not used often, however, because those with high-quality optics are very expensive to make. Parallel light rays incident on a parabolic surface focus at a common point, regardless of their distance from the principal axis. Parabolic reflecting surfaces are used in many astronomical telescopes to enhance image quality.

Chromatic Aberrations

The fact that different wavelengths of light refracted by a lens focus at different points gives rise to chromatic aberrations. In Chapter 35 we described how the index of refraction of a material varies with wavelength. For instance, when white light passes through a lens, violet rays are refracted more than red rays (Fig. 36.35). From this we see that the focal length of a lens is greater for red light than for violet light. Other wavelengths (not shown in Fig. 36.35) have focal points intermediate between those of red and violet.

Chromatic aberration for a diverging lens also results in a shorter focal length for violet light than for red light, but on the front side of the lens. Chromatic aberration can be greatly reduced by combining a converging lens made of one type of glass and a diverging lens made of another type of glass.

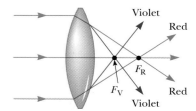

Figure 36.35 Chromatic aberration caused by a converging lens. Rays of different wavelengths focus at different points.

> **Quick Quiz 36.9** A curved mirrored surface can have (a) spherical aberration but not chromatic aberration (b) chromatic aberration but not spherical aberration (c) both spherical aberration and chromatic aberration.

36.6 The Camera

The photographic **camera** is a simple optical instrument whose essential features are shown in Figure 36.36. It consists of a light-tight chamber, a converging lens that produces a real image, and a film behind the lens to receive the image. One focuses the camera by varying the distance between lens and film. This is accomplished with an adjustable bellows in antique cameras and with some other mechanical arrangement in contemporary models. For proper focusing—which is necessary for the formation of sharp images—the lens-to-film distance depends on the object distance as well as on the focal length of the lens.

The shutter, positioned behind the lens, is a mechanical device that is opened for selected time intervals, called *exposure times*. One can photograph moving objects by using short exposure times or photograph dark scenes (with low light levels) by using long exposure times. If this adjustment were not available, it would be impossible to take stop-action photographs. For example, a rapidly moving vehicle could move enough in the time interval during which the shutter is open to produce a blurred image. Another major cause of blurred images is the movement of the camera while the shutter is open. To prevent such movement, either short exposure times or a tripod should be used, even for stationary objects. Typical shutter speeds (that is, exposure times) are $(1/30)$ s, $(1/60)$ s, $(1/125)$ s, and $(1/250)$ s. For handheld cameras,

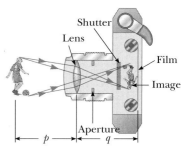

Figure 36.36 Cross-sectional view of a simple camera. Note that in reality, $p \gg q$.

the use of slower speeds can result in blurred images (due to movement), but the use of faster speeds reduces the gathered light intensity. In practice, stationary objects are normally shot with an intermediate shutter speed of $(1/60)$s.

More expensive cameras have an aperture of adjustable diameter to further control the intensity of the light reaching the film. As noted earlier, when an aperture of small diameter is used, only light from the central portion of the lens reaches the film; in this way spherical aberration is reduced.

The intensity I of the light reaching the film is proportional to the area of the lens. Because this area is proportional to the square of the diameter D, we conclude that I is also proportional to D^2. Light intensity is a measure of the rate at which energy is received by the film per unit area of the image. Because the area of the image is proportional to q^2 and $q \approx f$ (when $p \gg f$, so p can be approximated as infinite), we conclude that the intensity is also proportional to $1/f^2$, and thus $I \propto D^2/f^2$. The brightness of the image formed on the film depends on the light intensity, so we see that the image brightness depends on both the focal length and the diameter of the lens.

The ratio f/D is called the **_f_-number** of a lens:

$$f\text{-number} \equiv \frac{f}{D} \qquad (36.18)$$

Hence, the intensity of light incident on the film varies according to the following proportionality:

$$I \propto \frac{1}{(f/D)^2} \propto \frac{1}{(f\text{-number})^2} \qquad (36.19)$$

The f-number is often given as a description of the lens "speed." The lower the f-number, the wider the aperture and the higher the rate at which energy from the light exposes the film—thus, a lens with a low f-number is a "fast" lens. The conventional notation for an f-number is "$f/$" followed by the actual number. For example, "$f/4$" means an f-number of 4—it _does not_ mean to divide f by 4! Extremely fast lenses, which have f-numbers as low as approximately $f/1.2$, are expensive because it is very difficult to keep aberrations acceptably small with light rays passing through a large area of the lens. Camera lens systems (that is, combinations of lenses with adjustable apertures) are often marked with multiple f-numbers, usually $f/2.8$, $f/4$, $f/5.6$, $f/8$, $f/11$, and $f/16$. Any one of these settings can be selected by adjusting the aperture, which changes the value of D. Increasing the setting from one f-number to the next higher value (for example, from $f/2.8$ to $f/4$) decreases the area of the aperture by a factor of two. The lowest f-number setting on a camera lens corresponds to a wide-open aperture and the use of the maximum possible lens area.

Simple cameras usually have a fixed focal length and a fixed aperture size, with an f-number of about $f/11$. This high value for the f-number allows for a large **depth of field,** meaning that objects at a wide range of distances from the lens form reasonably sharp images on the film. In other words, the camera does not have to be focused.

Digital cameras are similar to the cameras we have described here except that the light does not form an image on photographic film. The image in a digital camera is formed on a _charge-coupled device_ (CCD), which digitizes the image, turning it into binary code, as we discussed for sound in Section 17.5. (The CCD is described in Section 40.2.) The digital information is then stored on a memory chip for playback on the screen of the camera, or it can be downloaded to a computer and sent to a friend or relative through the Internet.

> **Quick Quiz 36.10** A camera can be modeled as a simple converging lens that focuses an image on the film, acting as the screen. A camera is initially focused on a distant object. To focus the image of an object close to the camera, the lens must be (a) moved away from the film. (b) left where it is. (c) moved toward the film.

Example 36.14   Finding the Correct Exposure Time

The lens of a certain 35-mm camera (where 35 mm is the width of the film strip) has a focal length of 55 mm and a speed (an f-number) of $f/1.8$. The correct exposure time for this speed under certain conditions is known to be $(1/500)$ s.

(A) Determine the diameter of the lens.

Solution From Equation 36.18, we find that

$$D = \frac{f}{f\text{-number}} = \frac{55 \text{ mm}}{1.8} = \boxed{31 \text{ mm}}$$

(B) Calculate the correct exposure time if the f-number is changed to $f/4$ under the same lighting conditions.

Solution The total light energy hitting the film is proportional to the product of the intensity and the exposure time. If I is the light intensity reaching the film, then in a time interval Δt the energy per unit area received by the film is proportional to $I\Delta t$. Comparing the two situations, we require that $I_1\Delta t_1 = I_2\Delta t_2$, where Δt_1 is the correct exposure time for $f/1.8$ and Δt_2 is the correct exposure time for $f/4$. Using this result together with Equation 36.19, we find that

$$\frac{\Delta t_1}{(f_1\text{-number})^2} = \frac{\Delta t_2}{(f_2\text{-number})^2}$$

$$\Delta t_2 = \left(\frac{f_2\text{-number}}{f_1\text{-number}}\right)^2 \Delta t_1 = \left(\frac{4}{1.8}\right)^2 \left(\frac{1}{500} \text{ s}\right) \approx \boxed{\frac{1}{100} \text{ s}}$$

As the aperture size is reduced, exposure time must increase.

36.7   The Eye

Like a camera, a normal eye focuses light and produces a sharp image. However, the mechanisms by which the eye controls the amount of light admitted and adjusts to produce correctly focused images are far more complex, intricate, and effective than those in even the most sophisticated camera. In all respects, the eye is a physiological wonder.

Figure 36.37 shows the basic parts of the human eye. Light entering the eye passes through a transparent structure called the *cornea* (Fig. 36.38), behind which are a clear liquid (the *aqueous humor*), a variable aperture (the *pupil*, which is an opening in the *iris*), and the *crystalline lens*. Most of the refraction occurs at the outer surface of the eye, where the cornea is covered with a film of tears. Relatively little refraction occurs in the crystalline lens because the aqueous humor in contact with the lens has an average index of refraction close to that of the lens. The iris, which is the colored portion

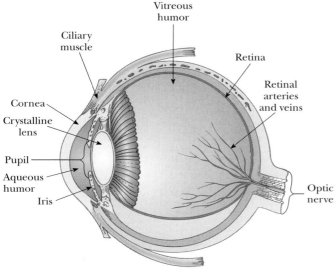

Figure 36.37 Important parts of the eye.

Figure 36.38 Close-up photograph of the cornea of the human eye.

of the eye, is a muscular diaphragm that controls pupil size. The iris regulates the amount of light entering the eye by dilating the pupil in low-light conditions and contracting the pupil in high-light conditions. The *f*-number range of the eye is from about $f/2.8$ to $f/16$.

The cornea–lens system focuses light onto the back surface of the eye, the *retina,* which consists of millions of sensitive receptors called *rods* and *cones.* When stimulated by light, these receptors send impulses via the optic nerve to the brain, where an image is perceived. By this process, a distinct image of an object is observed when the image falls on the retina.

The eye focuses on an object by varying the shape of the pliable crystalline lens through an amazing process called **accommodation.** An important component of accommodation is the *ciliary muscle,* which is situated in a circle around the rim of the lens. Thin filaments, called *zonules,* run from this muscle to the edge of the lens. When the eye is focused on a distant object, the ciliary muscle is relaxed, tightening the zonules that attach the muscle to the edge of the lens. The force of the zonules causes the lens to flatten, increasing its focal length. For an object distance of infinity, the focal length of the eye is equal to the fixed distance between lens and retina, about 1.7 cm. The eye focuses on nearby objects by tensing the ciliary muscle, which relaxes the zonules. This action allows the lens to bulge a bit, and its focal length decreases, resulting in the image being focused on the retina. All these lens adjustments take place so swiftly that we are not even aware of the change.

Accommodation is limited in that objects very close to the eye produce blurred images. The **near point** is the closest distance for which the lens can accommodate to focus light on the retina. This distance usually increases with age and has an average value of 25 cm. Typically, at age 10 the near point of the eye is about 18 cm. It increases to about 25 cm at age 20, to 50 cm at age 40, and to 500 cm or greater at age 60. The **far point** of the eye represents the greatest distance for which the lens of the relaxed eye can focus light on the retina. A person with normal vision can see very distant objects and thus has a far point that can be approximated as infinity.

Recall that the light leaving the mirror in Figure 36.8 becomes white where it comes together but then diverges into separate colors again. Because nothing but air exists at the point where the rays cross (and hence nothing exists to cause the colors to separate again), seeing white light as a result of a combination of colors must be a visual illusion. In fact, this is the case. Only three types of color-sensitive cells are present in the retina; they are called red, green, and blue cones because of the peaks of the color ranges to which they respond (Fig. 36.39). If the red and green cones are stimulated simultaneously (as would be the case if yellow light were shining on them), the brain interprets what we see as yellow. If all three types of cones are stimulated by the separate colors red, blue, and green, as in Figure 36.8, we see white. If all three types of cones are stimulated by light that contains *all* colors, such as sunlight, we again see white light.

Color televisions take advantage of this visual illusion by having only red, green, and blue dots on the screen. With specific combinations of brightness in these three

Figure 36.39 Approximate color sensitivity of the three types of cones in the retina.

primary colors, our eyes can be made to see any color in the rainbow. Thus, the yellow lemon you see in a television commercial is not really yellow, it is red and green! The paper on which this page is printed is made of tiny, matted, translucent fibers that scatter light in all directions; the resultant mixture of colors appears white to the eye. Snow, clouds, and white hair are not really white. In fact, there is no such thing as a white pigment. The appearance of these things is a consequence of the scattering of light containing all colors, which we interpret as white.

Conditions of the Eye

When the eye suffers a mismatch between the focusing range of the lens–cornea system and the length of the eye, with the result that light rays from a near object reach the retina before they converge to form an image, as shown in Figure 36.40a, the condition is known as **farsightedness** (or *hyperopia*). A farsighted person can usually see faraway objects clearly but not nearby objects. Although the near point of a normal eye is approximately 25 cm, the near point of a farsighted person is much farther away. The refracting power in the cornea and lens is insufficient to focus the light from all but distant objects satisfactorily. The condition can be corrected by placing a converging lens in front of the eye, as shown in Figure 36.40b. The lens refracts the incoming rays more toward the principal axis before entering the eye, allowing them to converge and focus on the retina.

A person with **nearsightedness** (or *myopia*), another mismatch condition, can focus on nearby objects but not on faraway objects. The far point of the nearsighted eye is not infinity and may be less than 1 m. The maximum focal length of the nearsighted eye is insufficient to produce a sharp image on the retina, and rays from a distant object converge to a focus in front of the retina. They then continue past that point, diverging before they finally reach the retina and causing blurred vision

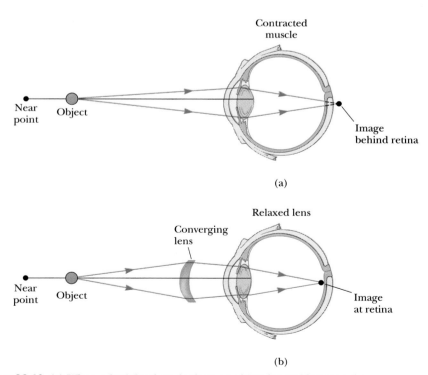

(a)

(b)

Figure 36.40 (a) When a farsighted eye looks at an object located between the near point and the eye, the image point is behind the retina, resulting in blurred vision. The eye muscle contracts to try to bring the object into focus. (b) Farsightedness is corrected with a converging lens.

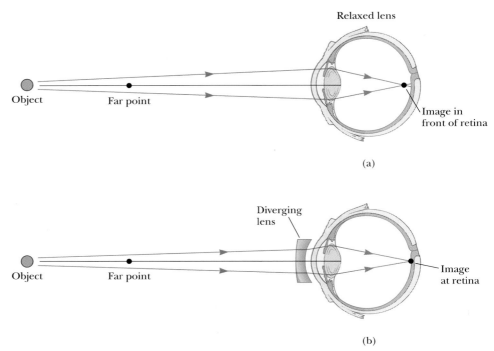

Figure 36.41 (a) When a nearsighted eye looks at an object that lies beyond the eye's far point, the image is formed in front of the retina, resulting in blurred vision. (b) Nearsightedness is corrected with a diverging lens.

(Fig. 36.41a). Nearsightedness can be corrected with a diverging lens, as shown in Figure 36.41b. The lens refracts the rays away from the principal axis before they enter the eye, allowing them to focus on the retina.

Beginning in middle age, most people lose some of their accommodation ability as the ciliary muscle weakens and the lens hardens. Unlike farsightedness, which is a mismatch between focusing power and eye length, **presbyopia** (literally, "old-age vision") is due to a reduction in accommodation ability. The cornea and lens do not have sufficient focusing power to bring nearby objects into focus on the retina. The symptoms are the same as those of farsightedness, and the condition can be corrected with converging lenses.

In the eye defect known as **astigmatism,** light from a point source produces a line image on the retina. This condition arises when either the cornea or the lens or both are not perfectly symmetric. Astigmatism can be corrected with lenses that have different curvatures in two mutually perpendicular directions.

Optometrists and ophthalmologists usually prescribe lenses[1] measured in **diopters:** the **power** P of a lens in diopters equals the inverse of the focal length in meters: $P = 1/f$. For example, a converging lens of focal length $+ 20$ cm has a power of $+ 5.0$ diopters, and a diverging lens of focal length $- 40$ cm has a power of $- 2.5$ diopters.

Quick Quiz 36.11 Two campers wish to start a fire during the day. One camper is nearsighted and one is farsighted. Whose glasses should be used to focus the Sun's rays onto some paper to start the fire? (a) either camper (b) the nearsighted camper (c) the farsighted camper.

[1] The word *lens* comes from *lentil*, the name of an Italian legume. (You may have eaten lentil soup.) Early eyeglasses were called "glass lentils" because the biconvex shape of their lenses resembled the shape of a lentil. The first lenses for farsightedness and presbyopia appeared around 1280; concave eyeglasses for correcting nearsightedness did not appear for more than 100 years after that.

Example 36.15   A Case of Nearsightedness

A particular nearsighted person is unable to see objects clearly when they are beyond 2.5 m away (the far point of this particular eye). What should the focal length be in a lens prescribed to correct this problem?

Solution The purpose of the lens in this instance is to "move" an object from infinity to a distance where it can be seen clearly. This is accomplished by having the lens produce an image at the far point. From the thin lens equation, we have

$$\frac{1}{p} + \frac{1}{q} = \frac{1}{\infty} + \frac{1}{-2.5 \text{ m}} = \frac{1}{f}$$

$$f = \boxed{-2.5 \text{ m}}$$

We use a negative sign for the image distance because the image is virtual and in front of the eye. As you should have suspected, the lens must be a diverging lens (one with a negative focal length) to correct nearsightedness.

36.8 The Simple Magnifier

The simple magnifier consists of a single converging lens. As the name implies, this device increases the apparent size of an object.

Suppose an object is viewed at some distance p from the eye, as illustrated in Figure 36.42. The size of the image formed at the retina depends on the angle θ subtended by the object at the eye. As the object moves closer to the eye, θ increases and a larger image is observed. However, an average normal eye cannot focus on an object closer than about 25 cm, the near point (Fig. 36.43a). Therefore, θ is maximum at the near point.

To further increase the apparent angular size of an object, a converging lens can be placed in front of the eye as in Figure 36.43b, with the object located at point O, just inside the focal point of the lens. At this location, the lens forms a virtual, upright, enlarged image. We define **angular magnification** m as the ratio of the angle subtended by an object with a lens in use (angle θ in Fig. 36.43b) to the angle subtended by the object placed at the near point with no lens in use (angle θ_0 in Fig. 36.43a):

$$m \equiv \frac{\theta}{\theta_0} \tag{36.20}$$

The angular magnification is a maximum when the image is at the near point of the eye—that is, when $q = -25$ cm. The object distance corresponding to this image

Figure 36.42 The size of the image formed on the retina depends on the angle θ subtended at the eye.

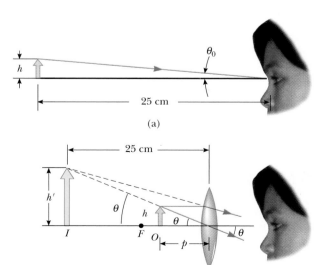

Figure 36.43 (a) An object placed at the near point of the eye ($p = 25$ cm) subtends an angle $\theta_0 \approx h/25$ at the eye. (b) An object placed near the focal point of a converging lens produces a magnified image that subtends an angle $\theta \approx h'/25$ at the eye.

distance can be calculated from the thin lens equation:

$$\frac{1}{p} + \frac{1}{-25 \text{ cm}} = \frac{1}{f}$$

$$p = \frac{25f}{25 + f}$$

where f is the focal length of the magnifier in centimeters. If we make the small-angle approximations

$$\tan \theta_0 \approx \theta_0 \approx \frac{h}{25} \quad \text{and} \quad \tan \theta \approx \theta \approx \frac{h}{p} \tag{36.21}$$

Equation 36.20 becomes

$$m_{\text{max}} = \frac{\theta}{\theta_0} = \frac{h/p}{h/25} = \frac{25}{p} = \frac{25}{25f/(25+f)}$$

$$m_{\text{max}} = 1 + \frac{25 \text{ cm}}{f} \tag{36.22}$$

Although the eye can focus on an image formed anywhere between the near point and infinity, it is most relaxed when the image is at infinity. For the image formed by the magnifying lens to appear at infinity, the object has to be at the focal point of the lens. In this case, Equations 36.21 become

$$\theta_0 \approx \frac{h}{25} \quad \text{and} \quad \theta \approx \frac{h}{f}$$

and the magnification is

$$m_{\text{min}} = \frac{\theta}{\theta_0} = \frac{25 \text{ cm}}{f} \tag{36.23}$$

A simple magnifier, also called a magnifying glass, is used to view an enlarged image of a portion of a map.

With a single lens, it is possible to obtain angular magnifications up to about 4 without serious aberrations. Magnifications up to about 20 can be achieved by using one or two additional lenses to correct for aberrations.

Example 36.16 Maximum Magnification of a Lens

What is the maximum magnification that is possible with a lens having a focal length of 10 cm, and what is the magnification of this lens when the eye is relaxed?

Solution The maximum magnification occurs when the image is located at the near point of the eye. Under these circumstances, Equation 36.22 gives

$$m_{\text{max}} = 1 + \frac{25 \text{ cm}}{f} = 1 + \frac{25 \text{ cm}}{10 \text{ cm}} = \boxed{3.5}$$

When the eye is relaxed, the image is at infinity. In this case, we use Equation 36.23:

$$m_{\text{min}} = \frac{25 \text{ cm}}{f} = \frac{25 \text{ cm}}{10 \text{ cm}} = \boxed{2.5}$$

36.9 The Compound Microscope

A simple magnifier provides only limited assistance in inspecting minute details of an object. Greater magnification can be achieved by combining two lenses in a device called a **compound microscope,** a schematic diagram of which is shown in Figure 36.44a. It consists of one lens, the *objective,* that has a very short focal length $f_o < 1$ cm and a second lens, the *eyepiece,* that has a focal length f_e of a few

Objective   Eyepiece

(a)

(b)

Active Figure 36.44 (a) Diagram of a compound microscope, which consists of an objective lens and an eyepiece lens. (b) A compound microscope. The three-objective turret allows the user to choose from several powers of magnification. Combinations of eyepieces with different focal lengths and different objectives can produce a wide range of magnifications.

At the Active Figures link at http://www.pse6.com, you can adjust the focal lengths of the objective and eyepiece lenses to see the effect on the final image.

centimeters. The two lenses are separated by a distance L that is much greater than either f_o or f_e. The object, which is placed just outside the focal point of the objective, forms a real, inverted image at I_1, and this image is located at or close to the focal point of the eyepiece. The eyepiece, which serves as a simple magnifier, produces at I_2 a virtual, enlarged image of I_1. The lateral magnification M_1 of the first image is $-q_1/p_1$. Note from Figure 36.44a that q_1 is approximately equal to L and that the object is very close to the focal point of the objective: $p_1 \approx f_o$. Thus, the lateral magnification by the objective is

$$M_o \approx -\frac{L}{f_o}$$

The angular magnification by the eyepiece for an object (corresponding to the image at I_1) placed at the focal point of the eyepiece is, from Equation 36.23,

$$m_e = \frac{25 \text{ cm}}{f_e}$$

The overall magnification of the image formed by a compound microscope is defined as the product of the lateral and angular magnifications:

$$M = M_o m_e = -\frac{L}{f_o}\left(\frac{25 \text{ cm}}{f_e}\right) \tag{36.24}$$

The negative sign indicates that the image is inverted.

The microscope has extended human vision to the point where we can view previously unknown details of incredibly small objects. The capabilities of this instrument have steadily increased with improved techniques for precision grinding of lenses. A question often asked about microscopes is: "If one were extremely patient and careful, would it be possible to construct a microscope that would enable the human eye to see an atom?" The answer is no, as long as light is used to illuminate the object. The reason is that, for an object under an optical microscope (one that uses visible light) to be seen, the object must be at least as large as a wavelength of light. Because the diameter of any atom is many times smaller than the wavelengths of visible light, the mysteries of the atom must be probed using other types of "microscopes."

The ability to use other types of waves to "see" objects also depends on wavelength. We can illustrate this with water waves in a bathtub. Suppose you vibrate your hand in the water until waves having a wavelength of about 15 cm are moving along the surface. If you hold a small object, such as a toothpick, so that it lies in the path of the waves, it does not appreciably disturb the waves; they continue along their path "oblivious" to it. Now suppose you hold a larger object, such as a toy sailboat, in the path of the 15-cm waves. In this case, the waves are considerably disturbed by the object. Because the toothpick is smaller than the wavelength of the waves, the waves do not "see" it. (The intensity of the scattered waves is low.) Because it is about the same size as the wavelength of the waves, however, the boat creates a disturbance. In other words, the object acts as the source of scattered waves that appear to come from it.

Light waves behave in this same general way. The ability of an optical microscope to view an object depends on the size of the object relative to the wavelength of the light used to observe it. Hence, we can never observe atoms with an optical microscope[2] because their dimensions are small (< 0.1 nm) relative to the wavelength of the light (< 500 nm).

36.10 The Telescope

Two fundamentally different types of **telescopes** exist; both are designed to aid in viewing distant objects, such as the planets in our Solar System. The **refracting telescope** uses a combination of lenses to form an image, and the **reflecting telescope** uses a curved mirror and a lens.

The lens combination shown in Figure 36.45a is that of a refracting telescope. Like the compound microscope, this telescope has an objective and an eyepiece. The two lenses are arranged so that the objective forms a real, inverted image of a distant object very near the focal point of the eyepiece. Because the object is essentially at infinity, this point at which I_1 forms is the focal point of the objective. The eyepiece then forms, at I_2, an enlarged, inverted image of the image at I_1. In order to provide the largest possible magnification, the image distance for the eyepiece is infinite. This means that the light rays exit the eyepiece lens parallel to the principal axis, and the image of the objective lens must form at the focal point of the eyepiece. Hence, the two lenses are separated by a distance $f_o + f_e$, which corresponds to the length of the telescope tube.

---

[2] Single-molecule near-field optic studies are routinely performed with visible light having wavelengths of about 500 nm. The technique uses very small apertures to produce images having resolutions as small as 10 nm.

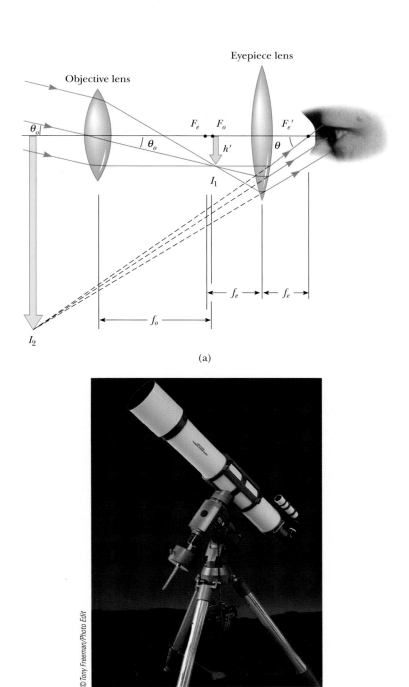

Active Figure 36.45 (a) Lens arrangement in a refracting telescope, with the object at infinity. (b) A refracting telescope.

At the Active Figures link at http://www.pse6.com, you can adjust the focal lengths of the objective and eyepiece lenses to see the effect on the final image.

The angular magnification of the telescope is given by θ/θ_o, where θ_o is the angle subtended by the object at the objective and θ is the angle subtended by the final image at the viewer's eye. Consider Figure 36.45a, in which the object is a very great distance to the left of the figure. The angle θ_o (to the *left* of the objective) subtended by the object at the objective is the same as the angle (to the *right* of the objective) subtended by the first image at the objective. Thus,

$$\tan \theta_o \approx \theta_o \approx -\frac{h'}{f_o}$$

where the negative sign indicates that the image is inverted.

The angle θ subtended by the final image at the eye is the same as the angle that a ray coming from the tip of I_1 and traveling parallel to the principal axis makes with the principal axis after it passes through the lens. Thus,

$$\tan \theta \approx \theta \approx \frac{h'}{f_e}$$

We have not used a negative sign in this equation because the final image is not inverted; the object creating this final image I_2 is I_1, and both it and I_2 point in the same direction. Hence, the angular magnification of the telescope can be expressed as

$$m = \frac{\theta}{\theta_o} = \frac{h'/f_e}{-h'/f_o} = -\frac{f_o}{f_e} \tag{36.25}$$

and we see that the angular magnification of a telescope equals the ratio of the objective focal length to the eyepiece focal length. The negative sign indicates that the image is inverted.

When we look through a telescope at such relatively nearby objects as the Moon and the planets, magnification is important. However, individual stars in our galaxy are so far away that they always appear as small points of light no matter how great the magnification. A large research telescope that is used to study very distant objects must have a great diameter to gather as much light as possible. It is difficult and expensive to manufacture large lenses for refracting telescopes. Another difficulty with large lenses is that their weight leads to sagging, which is an additional source of aberration. These problems can be partially overcome by replacing the objective with a concave mirror, which results in a reflecting telescope. Because light is reflected from the mirror and does not pass through a lens, the mirror can have rigid supports on the back side. Such supports eliminate the problem of sagging.

Figure 36.46a shows the design for a typical reflecting telescope. Incoming light rays pass down the barrel of the telescope and are reflected by a parabolic mirror at the base. These rays converge toward point A in the figure, where an image would be formed. However, before this image is formed, a small, flat mirror M reflects the light toward an opening in the side of the tube that passes into an eyepiece. This particular design is said to have a Newtonian focus because Newton developed it. Figure 36.46b

Figure 36.46 (a) A Newtonian-focus reflecting telescope. (b) A reflecting telescope. This type of telescope is shorter than that in Figure 36.45b.

shows such a telescope. Note that in the reflecting telescope the light never passes through glass (except through the small eyepiece). As a result, problems associated with chromatic aberration are virtually eliminated. The reflecting telescope can be made even shorter by orienting the flat mirror so that it reflects the light back toward the objective mirror and the light enters an eyepiece in a hole in the middle of the mirror.

The largest reflecting telescopes in the world are at the Keck Observatory on Mauna Kea, Hawaii. The site includes two telescopes with diameters of 10 m, each containing 36 hexagonally shaped, computer-controlled mirrors that work together to form a large reflecting surface. In contrast, the largest refracting telescope in the world, at the Yerkes Observatory in Williams Bay, Wisconsin, has a diameter of only 1 m.

SUMMARY

Take a practice test for this chapter by clicking on the Practice Test link at http://www.pse6.com.

The **lateral magnification** M of the image due to a mirror or lens is defined as the ratio of the image height h' to the object height h and is equal to the negative of the ratio of the image distance q to the object distance p:

$$M = \frac{h'}{h} = -\frac{q}{p} \qquad (36.1, 36.2)$$

In the paraxial ray approximation, the object distance p and image distance q for a spherical mirror of radius R are related by the **mirror equation:**

$$\frac{1}{p} + \frac{1}{q} = \frac{2}{R} = \frac{1}{f} \qquad (36.4, 36.6)$$

where $f = R/2$ is the **focal length** of the mirror.

An image can be formed by refraction from a spherical surface of radius R. The object and image distances for refraction from such a surface are related by

$$\frac{n_1}{p} + \frac{n_2}{q} = \frac{n_2 - n_1}{R} \qquad (36.8)$$

where the light is incident in the medium for which the index of refraction is n_1 and is refracted in the medium for which the index of refraction is n_2.

The inverse of the **focal length** f of a thin lens surrounded by air is given by the **lens makers' equation:**

$$\frac{1}{f} = (n - 1)\left(\frac{1}{R_1} - \frac{1}{R_2}\right) \qquad (36.15)$$

Converging lenses have positive focal lengths, and **diverging lenses** have negative focal lengths.

For a thin lens, and in the paraxial ray approximation, the object and image distances are related by the **thin lens equation:**

$$\frac{1}{p} + \frac{1}{q} = \frac{1}{f} \qquad (36.16)$$

The ratio of the focal length of a camera lens to the diameter of the lens is called the ***f*-number** of the lens:

$$f\text{-number} \equiv \frac{f}{D} \qquad (36.18)$$

The intensity of light incident on the film in the camera varies according to:

$$I \propto \frac{1}{(f/D)^2} \propto \frac{1}{(f\text{-number})^2} \qquad (36.19)$$

The maximum magnification of a single lens of focal length f used as a simple magnifier is

$$m_{max} = 1 + \frac{25 \text{ cm}}{f} \qquad (36.22)$$

The overall magnification of the image formed by a compound microscope is:

$$M = -\frac{L}{f_o}\left(\frac{25 \text{ cm}}{f_e}\right) \qquad (36.24)$$

where f_o and f_e are the focal lengths of the objective and eyepiece lenses, respectively, and L is the distance between the lenses.

The angular magnification of a refracting telescope can be expressed as

$$m = -\frac{f_o}{f_e} \qquad (36.25)$$

where f_o and f_e are the focal lengths of the objective and eyepiece lenses, respectively. The angular magnification of a reflecting telescope is given by the same expression where f_o is the focal length of the objective mirror.

QUESTIONS

1. What is wrong with the caption of the cartoon shown in Figure Q36.1?

Figure Q36.1 "Most mirrors reverse left and right. This one reverses top and bottom."

2. Consider a concave spherical mirror with a real object. Is the image always inverted? Is the image always real? Give conditions for your answers.

3. Repeat the preceding question for a convex spherical mirror.

4. Do the equations $1/p + 1/q = 1/f$ or $M = -q/p$ apply to the image formed by a flat mirror? Explain your answer.

5. Why does a clear stream, such as a creek, always appear to be shallower than it actually is? By how much is its depth apparently reduced?

6. Consider the image formed by a thin converging lens. Under what conditions is the image (a) inverted, (b) upright, (c) real, (d) virtual, (e) larger than the object, and (f) smaller than the object?

7. Repeat Question 6 for a thin diverging lens.

8. Use the lens makers' equation to verify the sign of the focal length of each of the lenses in Figure 36.27.

9. If a solid cylinder of glass or clear plastic is placed above the words LEAD OXIDE and viewed from above as shown

Figure Q36.9

in Figure Q36.9, the LEAD appears inverted but the OXIDE does not. Explain.

10. In Figure 36.28a, assume that the blue object arrow is replaced by one that is much taller than the lens. How many rays from the object will strike the lens? How many principal rays can be drawn in a ray diagram?

11. A zip-lock plastic sandwich bag filled with water can act as a crude converging lens in air. If the bag is filled with air and placed under water, is the effective lens converging or diverging?

12. Explain why a mirror cannot give rise to chromatic aberration.

13. Why do some automobile mirrors have printed on them the statement "Objects in mirror are closer than they appear"? (See Fig. P36.19.)

14. Can a converging lens be made to diverge light if it is placed into a liquid? **What If?** How about a converging mirror?

15. Explain why a fish in a spherical goldfish bowl appears larger than it really is.

16. Why do some emergency vehicles have the symbol ƎƆᴎA⅃UBMA written on the front?

17. A lens forms an image of an object on a screen. What happens to the image if you cover the top half of the lens with paper?

18. Lenses used in eyeglasses, whether converging or diverging, are always designed so that the middle of the lens curves away from the eye, like the center lenses of Figure 36.27a and b. Why?

19. Which glasses in Figure Q36.19 correct nearsightedness and which correct farsightedness?

Figure Q36.19

20. A child tries on either his hyperopic grandfather's or his myopic brother's glasses and complains that "everything looks blurry." Why do the eyes of a person wearing glasses not look blurry? (See Figure Q36.19.)

21. Consider a spherical concave mirror, with the object located to the left of the mirror beyond the focal point. Using ray diagrams, show that the image moves to the left as the object approaches the focal point.

22. In a Jules Verne novel, a piece of ice is shaped to form a magnifying lens to focus sunlight to start a fire. Is this possible?

23. The *f*-number of a camera is the focal length of the lens divided by its aperture (or diameter). How can the *f*-number of the lens be changed? How does changing this number affect the required exposure time?

24. A solar furnace can be constructed by using a concave mirror to reflect and focus sunlight into a furnace enclosure. What factors in the design of the reflecting mirror would guarantee very high temperatures?

25. One method for determining the position of an image, either real or virtual, is by means of *parallax*. If a finger or other object is placed at the position of the image, as shown in Figure Q36.25, and the finger and image are viewed simultaneously (the image is viewed through the lens if it is virtual), the finger and image have the same parallax; that is, if they are viewed from different positions, the image will appear to move along with the finger. Use this method to locate the image formed by a lens. Explain why the method works.

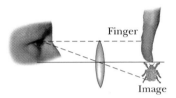

Finger

Image

Figure Q36.25

26. Figure Q36.26 shows a lithograph by M. C. Escher titled *Hand with Reflection Sphere (Self-Portrait in Spherical Mirror)*. Escher had this to say about the work: "The picture

Figure Q36.26

shows a spherical mirror, resting on a left hand. But as a print is the reverse of the original drawing on stone, it was my right hand that you see depicted. (Being left-handed, I needed my left hand to make the drawing.) Such a globe reflection collects almost one's whole surroundings in one disk-shaped image. The whole room, four walls, the floor, and the ceiling, everything, albeit distorted, is compressed into that one small circle. Your own head, or more exactly the point between your eyes, is the absolute center. No matter how you turn or twist yourself, you can't get out of that central point. You are immovably the focus, the unshakable core, of your world." Comment on the accuracy of Escher's description.

27. You can make a corner reflector by placing three flat mirrors in the corner of a room where the ceiling meets the walls. Show that no matter where you are in the room, you can see yourself reflected in the mirrors—upside down.

28. A converging lens of short focal length can take light diverging from a small source and refract it into a beam of parallel rays. A Fresnel lens, as shown in Figure 36.29, is used for this purpose in a lighthouse. A concave mirror can take light diverging from a small source and reflect it into a beam of parallel rays. Is it possible to make a Fresnel mirror? Is this an original idea, or has it already been done? *Suggestion:* Look at the walls and ceiling of an auditorium.

PROBLEMS

1, 2, 3 = straightforward, intermediate, challenging   ☐ = full solution available in the *Student Solutions Manual and Study Guide*

🌐 = coached solution with hints available at http://www.pse6.com   💻 = computer useful in solving problem

▭ = paired numerical and symbolic problems

Section 36.1   Images Formed by Flat Mirrors

1. Does your bathroom mirror show you older or younger than you actually are? Compute an order-of-magnitude estimate for the age difference, based on data that you specify.

2. In a church choir loft, two parallel walls are 5.30 m apart. The singers stand against the north wall. The organist faces the south wall, sitting 0.800 m away from it. To enable her to see the choir, a flat mirror 0.600 m wide is mounted on the south wall, straight in front of her. What width of the north wall can she see? *Suggestion:* Draw a top-view diagram to justify your answer.

3. Determine the minimum height of a vertical flat mirror in which a person 5′10″ in height can see his or her full image. (A ray diagram would be helpful.)

4. Two flat mirrors have their reflecting surfaces facing each other, with the edge of one mirror in contact with an edge of the other, so that the angle between the mirrors is α. When an object is placed between the mirrors, a number of images are formed. In general, if the angle α is such that $n\alpha = 360°$, where n is an integer, the number of images formed is $n - 1$. Graphically, find all the image positions for the case $n = 6$ when a point object is between the mirrors (but not on the angle bisector).

5. A person walks into a room with two flat mirrors on opposite walls, which produce multiple images. When the person is located 5.00 ft from the mirror on the left wall and 10.0 ft from the mirror on the right wall, find the distance from the person to the first three images seen in the mirror on the left.

6. A periscope (Figure P36.6) is useful for viewing objects that cannot be seen directly. It finds use in submarines and in watching golf matches or parades from behind a crowd of people. Suppose that the object is a distance p_1 from the upper mirror and that the two flat mirrors are separated by a distance h. (a) What is the distance of the final image from the lower mirror? (b) Is the final image real or virtual? (c) Is it upright or inverted? (d) What is its magnification? (e) Does it appear to be left–right reversed?

Figure P36.6

Section 36.2   Images Formed by Spherical Mirrors

7. A concave spherical mirror has a radius of curvature of 20.0 cm. Find the location of the image for object distances of (a) 40.0 cm, (b) 20.0 cm, and (c) 10.0 cm. For each case, state whether the image is real or virtual and upright or inverted. Find the magnification in each case.

8. At an intersection of hospital hallways, a convex mirror is mounted high on a wall to help people avoid collisions. The mirror has a radius of curvature of 0.550 m. Locate and describe the image of a patient 10.0 m from the mirror. Determine the magnification.

9. A spherical convex mirror has a radius of curvature with a magnitude of 40.0 cm. Determine the position of the virtual image and the magnification for object distances of (a) 30.0 cm and (b) 60.0 cm. (c) Are the images upright or inverted?

10. A large church has a niche in one wall. On the floor plan it appears as a semicircular indentation of radius 2.50 m. A worshiper stands on the center line of the niche, 2.00 m out from its deepest point, and whispers a prayer. Where is the sound concentrated after reflection from the back wall of the niche?

11. A concave mirror has a radius of curvature of 60.0 cm. Calculate the image position and magnification of an object placed in front of the mirror at distances of (a) 90.0 cm and (b) 20.0 cm. (c) Draw ray diagrams to obtain the image characteristics in each case.

12. A concave mirror has a focal length of 40.0 cm. Determine the object position for which the resulting image is upright and four times the size of the object.

13. A certain Christmas tree ornament is a silver sphere having a diameter of 8.50 cm. Determine an object location for which the size of the reflected image is three-fourths the size of the object. Use a principal-ray diagram to arrive at a description of the image.

14. (a) A concave mirror forms an inverted image four times larger than the object. Find the focal length of the mirror, assuming the distance between object and image is 0.600 m. (b) A convex mirror forms a virtual image half the size of the object. Assuming the distance between image and object is 20.0 cm, determine the radius of curvature of the mirror.

15. To fit a contact lens to a patient's eye, a *keratometer* can be used to measure the curvature of the front surface of the eye, the cornea. This instrument places an illuminated object of known size at a known distance p from the cornea. The cornea reflects some light from the object, forming an image of the object. The magnification M of the image is measured by using a small viewing telescope that allows comparison of the image formed by the cornea with a second calibrated image projected into the field of view by a prism arrangement. Determine the radius of curvature of the cornea for the case $p = 30.0$ cm and $M = 0.013\ 0$.

16. An object 10.0 cm tall is placed at the zero mark of a meter stick. A spherical mirror located at some point on the meter stick creates an image of the object that is upright, 4.00 cm tall, and located at the 42.0-cm mark of the meter stick. (a) Is the mirror convex or concave? (b) Where is the mirror? (c) What is the mirror's focal length?

17. A spherical mirror is to be used to form, on a screen located 5.00 m from the object, an image five times the size of the object. (a) Describe the type of mirror required. (b) Where should the mirror be positioned relative to the object?

18. A dedicated sports car enthusiast polishes the inside and outside surfaces of a hubcap that is a section of a sphere. When she looks into one side of the hubcap, she sees an image of her face 30.0 cm in back of the hubcap. She then flips the hubcap over and sees another image of her face 10.0 cm in back of the hubcap. (a) How far is her face from the hubcap? (b) What is the radius of curvature of the hubcap?

19. You unconsciously estimate the distance to an object from the angle it subtends in your field of view. This angle θ in radians is related to the linear height of the object h and to the distance d by $\theta = h/d$. Assume that you are driving a car and another car, 1.50 m high, is 24.0 m behind you. (a) Suppose your car has a flat passenger-side rearview mirror, 1.55 m from your eyes. How far from your eyes is the image of the car following you? (b) What angle does the image subtend in your field of view? (c) **What If?** Suppose instead that your car has a convex rearview mirror with a radius of curvature of magnitude 2.00 m (Fig. P36.19). How far from your eyes is the image of the car behind you? (d) What angle does the image subtend at your eyes? (e) Based on its angular size, how far away does the following car appear to be?

Figure P36.19

20. **Review Problem.** A ball is dropped at $t = 0$ from rest 3.00 m directly above the vertex of a concave mirror that has a radius of curvature of 1.00 m and lies in a horizontal plane. (a) Describe the motion of the ball's image in the mirror. (b) At what time do the ball and its image coincide?

Section 36.3   Images Formed by Refraction

21. A cubical block of ice 50.0 cm on a side is placed on a level floor over a speck of dust. Find the location of the image of the speck as viewed from above. The index of refraction of ice is 1.309.

22. A flint glass plate ($n = 1.66$) rests on the bottom of an aquarium tank. The plate is 8.00 cm thick (vertical dimension) and is covered with a layer of water ($n = 1.33$) 12.0 cm deep. Calculate the apparent thickness of the plate as viewed from straight above the water.

23. A glass sphere ($n = 1.50$) with a radius of 15.0 cm has a tiny air bubble 5.00 cm above its center. The sphere is viewed looking down along the extended radius containing the bubble. What is the apparent depth of the bubble below the surface of the sphere?

24. A simple model of the human eye ignores its lens entirely. Most of what the eye does to light happens at the outer surface of the transparent cornea. Assume that this surface has a radius of curvature of 6.00 mm, and assume that the eyeball contains just one fluid with a refractive index of 1.40. Prove that a very distant object will be imaged on the retina, 21.0 mm behind the cornea. Describe the image.

25. One end of a long glass rod ($n = 1.50$) is formed into a convex surface with a radius of curvature of 6.00 cm. An object is located in air along the axis of the rod. Find the image positions corresponding to object distances of (a) 20.0 cm, (b) 10.0 cm, and (c) 3.00 cm from the end of the rod.

26. A transparent sphere of unknown composition is observed to form an image of the Sun on the surface of the sphere opposite the Sun. What is the refractive index of the sphere material?

27. A goldfish is swimming at 2.00 cm/s toward the front wall of a rectangular aquarium. What is the apparent speed of the fish measured by an observer looking in from outside the front wall of the tank? The index of refraction of water is 1.33.

Section 36.4   Thin Lenses

28. A contact lens is made of plastic with an index of refraction of 1.50. The lens has an outer radius of curvature of $+ 2.00$ cm and an inner radius of curvature of $+ 2.50$ cm. What is the focal length of the lens?

29. The left face of a biconvex lens has a radius of curvature of magnitude 12.0 cm, and the right face has a radius of curvature of magnitude 18.0 cm. The index of refraction of the glass is 1.44. (a) Calculate the focal length of the lens. (b) **What If?** Calculate the focal length the lens has after it is turned around to interchange the radii of curvature of the two faces.

30. A converging lens has a focal length of 20.0 cm. Locate the image for object distances of (a) 40.0 cm, (b) 20.0 cm, and (c) 10.0 cm. For each case, state whether the image is real or virtual and upright or inverted. Find the magnification in each case.

31. A thin lens has a focal length of 25.0 cm. Locate and describe the image when the object is placed (a) 26.0 cm and (b) 24.0 cm in front of the lens.

32. An object located 32.0 cm in front of a lens forms an image on a screen 8.00 cm behind the lens. (a) Find the focal length of the lens. (b) Determine the magnification. (c) Is the lens converging or diverging?

33. The nickel's image in Figure P36.33 has twice the diameter of the nickel and is 2.84 cm from the lens. Determine the focal length of the lens.

Figure P36.33

34. A person looks at a gem with a jeweler's loupe—a converging lens that has a focal length of 12.5 cm. The loupe forms a virtual image 30.0 cm from the lens. (a) Determine the magnification. Is the image upright or inverted? (b) Construct a ray diagram for this arrangement.

35. Suppose an object has thickness dp so that it extends from object distance p to $p + dp$. Prove that the thickness dq of its image is given by $(- q^2/p^2)\,dp$, so that the longitudinal magnification $dq/dp = - M^2$, where M is the lateral magnification.

36. The projection lens in a certain slide projector is a single thin lens. A slide 24.0 mm high is to be projected so that its image fills a screen 1.80 m high. The slide-to-screen distance is 3.00 m. (a) Determine the focal length of the projection lens. (b) How far from the slide should the lens of the projector be placed in order to form the image on the screen?

37. An object is located 20.0 cm to the left of a diverging lens having a focal length $f = - 32.0$ cm. Determine (a) the location and (b) the magnification of the image. (c) Construct a ray diagram for this arrangement.

38. An antelope is at a distance of 20.0 m from a converging lens of focal length 30.0 cm. The lens forms an image of the animal. If the antelope runs away from the lens at a speed of 5.00 m/s, how fast does the image move? Does the image move toward or away from the lens?

39. In some types of optical spectroscopy, such as photoluminescence and Raman spectroscopy, a laser beam exits from a pupil and is focused on a sample to stimulate electromagnetic radiation from the sample. The focusing lens usually has an antireflective coating preventing any light loss. Assume a 100-mW laser is located 4.80 m from the lens, which has a focal length of 7.00 cm. (a) How far from the lens should the sample be located so that an image of the laser exit pupil is formed on the surface of the sample? (b) If the diameter of the laser exit pupil is 5.00 mm, what is the diameter of the light spot on the sample? (c) What is the light intensity at the spot?

40. Figure P36.40 shows a thin glass ($n = 1.50$) converging lens for which the radii of curvature are $R_1 = 15.0$ cm and $R_2 = -12.0$ cm. To the left of the lens is a cube having a face area of 100 cm². The base of the cube is on the axis of the lens, and the right face is 20.0 cm to the left of the lens. (a) Determine the focal length of the lens. (b) Draw the image of the square face formed by the lens. What type of geometric figure is this? (c) Determine the area of the image.

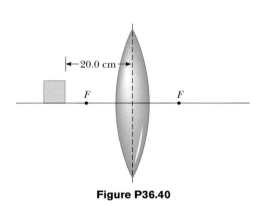

Figure P36.40

41. An object is at a distance d to the left of a flat screen. A converging lens with focal length $f < d/4$ is placed between object and screen. (a) Show that two lens positions exist that form an image on the screen, and determine how far these positions are from the object. (b) How do the two images differ from each other?

42. Figure 36.36 diagrams a cross section of a camera. It has a single lens of focal length 65.0 mm, which is to form an image on the film at the back of the camera. Suppose the position of the lens has been adjusted to focus the image of a distant object. How far and in what direction must the lens be moved to form a sharp image of an object that is 2.00 m away?

43. The South American capybara is the largest rodent on Earth; its body can be 1.20 m long. The smallest rodent is the pygmy mouse found in Texas, with an average body length of 3.60 cm. Assume that a pygmy mouse is observed by looking through a lens placed 20.0 cm from the mouse. The whole image of the mouse is the size of a capybara. Then the lens is moved a certain distance along its axis, and the image of the mouse is the same size as before! How far was the lens moved?

Section 36.5   Lens Aberrations

44. The magnitudes of the radii of curvature are 32.5 cm and 42.5 cm for the two faces of a biconcave lens. The glass has index of refraction 1.53 for violet light and 1.51 for red light. For a very distant object, locate and describe (a) the image formed by violet light, and (b) the image formed by red light.

45. Two rays traveling parallel to the principal axis strike a large plano-convex lens having a refractive index of 1.60 (Fig. P36.45). If the convex face is spherical, a ray near the edge does not pass through the focal point (spherical aberration occurs). Assume this face has a radius of curvature of 20.0 cm and the two rays are at distances $h_1 = 0.500$ cm and $h_2 = 12.0$ cm from the principal axis. Find the difference Δx in the positions where each crosses the principal axis.

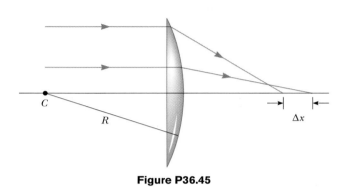

Figure P36.45

Section 36.6   The Camera

46. A camera is being used with a correct exposure at $f/4$ and a shutter speed of $(1/16)$ s. In order to photograph a rapidly moving subject, the shutter speed is changed to $(1/128)$ s. Find the new f-number setting needed to maintain satisfactory exposure.

Section 36.7   The Eye

47. A nearsighted person cannot see objects clearly beyond 25.0 cm (her far point). If she has no astigmatism and contact lenses are prescribed for her, what power and type of lens are required to correct her vision?

48. The accommodation limits for Nearsighted Nick's eyes are 18.0 cm and 80.0 cm. When he wears his glasses, he can see faraway objects clearly. At what minimum distance is he able to see objects clearly?

49. A person sees clearly when he wears eyeglasses that have a power of -4.00 diopters and sit 2.00 cm in front of his eyes. If the person wants to switch to contact lenses, which are placed directly on the eyes, what lens power should be prescribed?

Section 36.8   The Simple Magnifier
Section 36.9   The Compound Microscope
Section 36.10   The Telescope

50. A lens that has a focal length of 5.00 cm is used as a magnifying glass. (a) To obtain maximum magnification, where should the object be placed? (b) What is the magnification?

51. The distance between eyepiece and objective lens in a certain compound microscope is 23.0 cm. The focal length of the eyepiece is 2.50 cm, and that of the objective is 0.400 cm. What is the overall magnification of the microscope?

52. The desired overall magnification of a compound microscope is 140×. The objective alone produces a lateral magnification of 12.0×. Determine the required focal length of the eyepiece.

53. The Yerkes refracting telescope has a 1.00-m diameter objective lens of focal length 20.0 m. Assume it is used with an eyepiece of focal length 2.50 cm. (a) Determine the magnification of the planet Mars as seen through this telescope. (b) Are the Martian polar caps right side up or upside down?

54. Astronomers often take photographs with the objective lens or mirror of a telescope alone, without an eyepiece. (a) Show that the image size h' for this telescope is given by $h' = fh/(f - p)$ where h is the object size, f is the objective focal length, and p is the object distance. (b) **What If?** Simplify the expression in part (a) for the case in which the object distance is much greater than objective focal length. (c) The "wingspan" of the International Space Station is 108.6 m, the overall width of its solar panel configuration. Find the width of the image formed by a telescope objective of focal length 4.00 m when the station is orbiting at an altitude of 407 km.

55. Galileo devised a simple terrestrial telescope that produces an upright image. It consists of a converging objective lens and a diverging eyepiece at opposite ends of the telescope tube. For distant objects, the tube length is equal to the objective focal length minus the absolute value of the eyepiece focal length. (a) Does the user of the telescope see a real or virtual image? (b) Where is the final image? (c) If a telescope is to be constructed with a tube of length 10.0 cm and a magnification of 3.00, what are the focal lengths of the objective and eyepiece?

56. A certain telescope has an objective mirror with an aperture diameter of 200 mm and a focal length of 2 000 mm. It captures the image of a nebula on photographic film at its prime focus with an exposure time of 1.50 min. To produce the same light energy per unit area on the film, what is the required exposure time to photograph the same nebula with a smaller telescope, which has an objective with a diameter of 60.0 mm and a focal length of 900 mm?

Additional Problems

57. The distance between an object and its upright image is 20.0 cm. If the magnification is 0.500, what is the focal length of the lens that is being used to form the image?

58. The distance between an object and its upright image is d. If the magnification is M, what is the focal length of the lens that is being used to form the image?

59. Your friend needs glasses with diverging lenses of focal length − 65.0 cm for both eyes. You tell him he looks good

when he doesn't squint, but he is worried about how thick the lenses will be. Assuming the radius of curvature of the first surface is $R_1 = 50.0$ cm and the high-index plastic has a refractive index of 1.66, (a) find the required radius of curvature of the second surface. (b) Assume the lens is ground from a disk 4.00 cm in diameter and 0.100 cm thick at the center. Find the thickness of the plastic at the edge of the lens, measured parallel to the axis. *Suggestion:* Draw a large cross-sectional diagram.

60. A cylindrical rod of glass with index of refraction 1.50 is immersed in water with index 1.33. The diameter of the rod is 9.00 cm. The outer part of each end of the rod has been ground away to form each end into a hemisphere of radius 4.50 cm. The central portion of the rod with straight sides is 75.0 cm long. An object is situated in the water, on the axis of the rod, at a distance of 100 cm from the vertex of the nearer hemisphere. (a) Find the location of the final image formed by refraction at both surfaces. (b) Is the final image real or virtual? Upright or inverted? Enlarged or diminished?

61. A *zoom lens* system is a combination of lenses that produces a variable magnification while maintaining fixed object and image positions. The magnification is varied by moving one or more lenses along the axis. While multiple lenses are used in practice to obtain high-quality images, the effect of zooming in on an object can be demonstrated with a simple two-lens system. An object, two converging lenses, and a screen are mounted on an optical bench. The first lens, which is to the right of the object, has a focal length of 5.00 cm, and the second lens, which is to the right of the first lens, has a focal length of 10.0 cm. The screen is to the right of the second lens. Initially, an object is situated at a distance of 7.50 cm to the left of the first lens, and the image formed on the screen has a magnification of + 1.00. (a) Find the distance between the object and the screen. (b) Both lenses are now moved along their common axis, while the object and the screen maintain fixed positions, until the image formed on the screen has a magnification of + 3.00. Find the displacement of each lens from its initial position in (a). Can the lenses be displaced in more than one way?

62. The object in Figure P36.62 is midway between the lens and the mirror. The mirror's radius of curvature is 20.0 cm, and the lens has a focal length of − 16.7 cm. Considering only the light that leaves the object and travels first toward the mirror, locate the final image formed by this system. Is this image real or virtual? Is it upright or inverted? What is the overall magnification?

Figure P36.62

63. An object placed 10.0 cm from a concave spherical mirror produces a real image 8.00 cm from the mirror. If the object is moved to a new position 20.0 cm from the mirror, what is the position of the image? Is the latter image real or virtual?

64. In many applications it is necessary to expand or to decrease the diameter of a beam of parallel rays of light. This change can be made by using a converging lens and a diverging lens in combination. Suppose you have a converging lens of focal length 21.0 cm and a diverging lens of focal length − 12.0 cm. How can you arrange these lenses to increase the diameter of a beam of parallel rays? By what factor will the diameter increase?

65. A parallel beam of light enters a glass hemisphere perpendicular to the flat face, as shown in Figure P36.65. The magnitude of the radius is 6.00 cm, and the index of refraction is 1.560. Determine the point at which the beam is focused. (Assume paraxial rays.)

Figure P36.65

66. Review problem. A spherical lightbulb of diameter 3.20 cm radiates light equally in all directions, with power 4.50 W. (a) Find the light intensity at the surface of the bulb. (b) Find the light intensity 7.20 m away from the center of the bulb. (c) At this 7.20-m distance a lens is set up with its axis pointing toward the bulb. The lens has a circular face with a diameter 15.0 cm and has a focal length of 35.0 cm. Find the diameter of the image of the bulb. (d) Find the light intensity at the image.

67. An object is placed 12.0 cm to the left of a diverging lens of focal length − 6.00 cm. A converging lens of focal length 12.0 cm is placed a distance d to the right of the diverging lens. Find the distance d so that the final image is at infinity. Draw a ray diagram for this case.

68. An observer to the right of the mirror–lens combination shown in Figure P36.68 sees two real images that are the same size and in the same location. One image is upright

and the other is inverted. Both images are 1.50 times larger than the object. The lens has a focal length of 10.0 cm. The lens and mirror are separated by 40.0 cm. Determine the focal length of the mirror. Do not assume that the figure is drawn to scale.

69. The disk of the Sun subtends an angle of 0.533° at the Earth. What are the position and diameter of the solar image formed by a concave spherical mirror with a radius of curvature of 3.00 m?

70. Assume the intensity of sunlight is 1.00 kW/m^2 at a particular location. A highly reflecting concave mirror is to be pointed toward the Sun to produce a power of at least 350 W at the image. (a) Find the required radius R_a of the circular face area of the mirror. (b) Now suppose the light intensity is to be at least 120 kW/m^2 at the image. Find the required relationship between R_a and the radius of curvature R of the mirror. The disk of the Sun subtends an angle of 0.533° at the Earth.

71. In a darkened room, a burning candle is placed 1.50 m from a white wall. A lens is placed between candle and wall at a location that causes a larger, inverted image to form on the wall. When the lens is moved 90.0 cm toward the wall, another image of the candle is formed. Find (a) the two object distances that produce the specified images and (b) the focal length of the lens. (c) Characterize the second image.

72. Figure P36.72 shows a thin converging lens for which the radii of curvature are $R_1 = 9.00$ cm and $R_2 = -11.0$ cm. The lens is in front of a concave spherical mirror with the radius of curvature $R = 8.00$ cm. (a) Assume its focal points F_1 and F_2 are 5.00 cm from the center of the lens. Determine its index of refraction. (b) The lens and mirror are 20.0 cm apart, and an object is placed 8.00 cm to the left of the lens. Determine the position of the final image and its magnification as seen by the eye in the figure. (c) Is the final image inverted or upright? Explain.

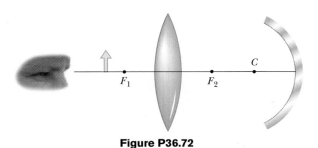

Figure P36.72

73. A compound microscope has an objective of focal length 0.300 cm and an eyepiece of focal length 2.50 cm. If an object is 3.40 mm from the objective, what is the magnification? (*Suggestion:* Use the lens equation for the objective.)

74. Two converging lenses having focal lengths of 10.0 cm and 20.0 cm are located 50.0 cm apart, as shown in

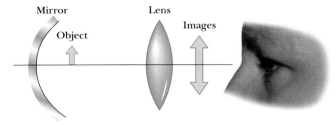

Figure P36.68

Figure P36.74. The final image is to be located between the lenses at the position indicated. (a) How far to the left of the first lens should the object be? (b) What is the overall magnification? (c) Is the final image upright or inverted?

f_1 (10.0 cm)   Object   Final image f_2 (20.0 cm)

p 31.0 cm   50.0 cm

Figure P36.74

75. A cataract-impaired lens in an eye may be surgically removed and replaced by a manufactured lens. The focal length required for the new lens is determined by the lens-to-retina distance, which is measured by a sonar-like device, and by the requirement that the implant provide for correct distant vision. (a) Assuming the distance from lens to retina is 22.4 mm, calculate the power of the implanted lens in diopters. (b) Because no accommodation occurs and the implant allows for correct distant vision, a corrective lens for close work or reading must be used. Assume a reading distance of 33.0 cm and calculate the power of the lens in the reading glasses.

76. A floating strawberry illusion is achieved with two parabolic mirrors, each having a focal length 7.50 cm, facing each other so that their centers are 7.50 cm apart (Fig.

P36.76). If a strawberry is placed on the lower mirror, an image of the strawberry is formed at the small opening at the center of the top mirror. Show that the final image is formed at that location and describe its characteristics. (*Note:* A very startling effect is to shine a flashlight beam on this image. Even at a glancing angle, the incoming light beam is seemingly reflected from the image! Do you understand why?)

77. An object 2.00 cm high is placed 40.0 cm to the left of a converging lens having a focal length of 30.0 cm. A diverging lens with a focal length of −20.0 cm is placed 110 cm to the right of the converging lens. (a) Determine the position and magnification of the final image. (b) Is the image upright or inverted? (c) **What If?** Repeat parts (a) and (b) for the case where the second lens is a converging lens having a focal length of +20.0 cm.

78. Two lenses made of kinds of glass having different refractive indices n_1 and n_2 are cemented together to form what is called an *optical doublet*. Optical doublets are often used to correct chromatic aberrations in optical devices. The first lens of a doublet has one flat side and one concave side of radius of curvature R. The second lens has two convex sides of radius of curvature R. Show that the doublet can be modeled as a single thin lens with a focal length described by

$$\frac{1}{f} = \frac{2n_2 - n_1 - 1}{R}$$

Answers to Quick Quizzes

36.1 At C. A ray traced from the stone to the mirror and then to observer 2 looks like this:

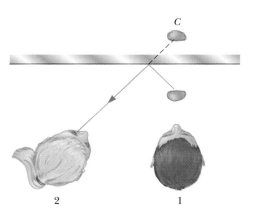

C

2   1

36.2 False. The water spots are 2 m away from you and your image is 4 m away. You cannot focus your eyes on both at the same time.

36.3 (b). A concave mirror will focus the light from a large area of the mirror onto a small area of the paper, resulting in a very high power input to the paper.

Small hole

Strawberry

Figure P36.76

36.4 (b). A convex mirror always forms an image with a magnification less than one, so the mirror must be concave. In a concave mirror, only virtual images are upright. This particular photograph is of the Hubble Space Telescope primary mirror.

36.5 (d). When O is far away, the rays refract into the material of index n_2 and converge to form a real image as in Figure 36.18. For certain combinations of R and n_2 as O moves very close to the refracting surface, the incident angle of the rays increases so much that rays are no longer refracted back toward the principal axis. This results in a virtual image as shown below:

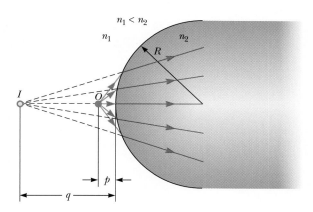

36.6 (a). No matter where O is, the rays refract into the air away from the normal and form a virtual image between O and the surface.

36.7 (b). Because the flat surfaces of the plane have infinite radii of curvature, Equation 36.15 indicates that the focal length is also infinite. Parallel rays striking the plane focus at infinity, which means that they remain parallel after passing through the glass.

36.8 (b). If there is a curve on the front surface, the refraction will differ at that surface when the mask is worn in air and water. In order for there to be no difference in refraction (for normal incidence), the front of the mask should be flat.

36.9 (a). Because the light reflecting from a mirror does not enter the material of the mirror, there is no opportunity for the dispersion of the material to cause chromatic aberration.

36.10 (a). If the object is brought closer to the lens, the image moves farther away from the lens, behind the plane of the film. In order to bring the image back up to the film, the lens is moved toward the object and away from the film.

36.11 (c). The Sun's rays must converge onto the paper. A farsighted person wears converging lenses.

Chapter 37

Interference of Light Waves

▲ The colors in many of a hummingbird's feathers are not due to pigment. The iridescence that makes the brilliant colors that often appear on the throat and belly is due to an interference effect caused by structures in the feathers. The colors will vary with the viewing angle. (RO-MA/Index Stock Imagery)

In the preceding chapter, we used light rays to examine what happens when light passes through a lens or reflects from a mirror. This discussion completed our study of *geometric optics*. Here in Chapter 37 and in the next chapter, we are concerned with *wave optics* or *physical optics*, the study of interference, diffraction, and polarization of light. These phenomena cannot be adequately explained with the ray optics used in Chapters 35 and 36. We now learn how treating light as waves rather than as rays leads to a satisfying description of such phenomena.

37.1 Conditions for Interference

In Chapter 18, we found that the superposition of two mechanical waves can be constructive or destructive. In constructive interference, the amplitude of the resultant wave at a given position or time is greater than that of either individual wave, whereas in destructive interference, the resultant amplitude is less than that of either individual wave. Light waves also interfere with each other. Fundamentally, all interference associated with light waves arises when the electromagnetic fields that constitute the individual waves combine.

As with the hummingbird feathers shown in the opening photograph, the bright colors of peacock feathers are also due to interference. In both types of birds, structures in the feathers split and recombine visible light so that interference occurs for certain colors.

If two lightbulbs are placed side by side, no interference effects are observed because the light waves from one bulb are emitted independently of those from the other bulb. The emissions from the two lightbulbs do not maintain a constant phase relationship with each other over time. Light waves from an ordinary source such as a lightbulb undergo random phase changes in time intervals less than a nanosecond. Therefore, the conditions for constructive interference, destructive interference, or some intermediate state are maintained only for such short time intervals. Because the eye cannot follow such rapid changes, no interference effects are observed. Such light sources are said to be **incoherent.**

In order to observe interference in light waves, the following conditions must be met:

- The sources must be **coherent**—that is, **they must maintain a constant phase with respect to each other.**

- The sources should be **monochromatic**—that is, of a single wavelength.

As an example, single-frequency sound waves emitted by two side-by-side loudspeakers driven by a single amplifier can interfere with each other because the two speakers are coherent—that is, they respond to the amplifier in the same way at the same time.

Conditions for interference

37.2 Young's Double-Slit Experiment

A common method for producing two coherent light sources is to use a monochromatic source to illuminate a barrier containing two small openings (usually in the shape of slits). The light emerging from the two slits is coherent because a single

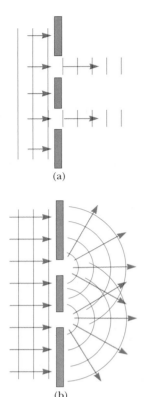

(a)

(b)

Figure 37.1 (a) If light waves did not spread out after passing through the slits, no interference would occur. (b) The light waves from the two slits overlap as they spread out, filling what we expect to be shadowed regions with light and producing interference fringes on a screen placed to the right of the slits.

At a beach in Tel Aviv, Israel, plane water waves pass through two openings in a breakwall. Notice the diffraction effect—the waves exit the openings with circular wave fronts, as in Figure 37.1b. Notice also how the beach has been shaped by the circular wave fronts.

▲ **PITFALL PREVENTION**

37.1 Interference Patterns Are Not Standing Waves

The interference pattern in Figure 37.2b shows bright and dark regions that appear similar to the antinodes and nodes of a standing-wave pattern on a string (Section 18.3). While both patterns depend on the principle of superposition, here are two major differences: (1) waves on a string propagate in only one dimension while the light-wave interference pattern exists in three dimensions; (2) the standing-wave pattern represents no net energy flow, while there is a net energy flow from the slits to the screen in an interference pattern.

source produces the original light beam and the two slits serve only to separate the original beam into two parts (which, after all, is what is done to the sound signal from the side-by-side loudspeakers at the end of the preceding section). Any random change in the light emitted by the source occurs in both beams at the same time, and as a result interference effects can be observed when the light from the two slits arrives at a viewing screen.

If the light traveled only in its original direction after passing through the slits, as shown in Figure 37.1a, the waves would not overlap and no interference pattern would be seen. Instead, as we have discussed in our treatment of Huygens's principle (Section 35.6), the waves spread out from the slits as shown in Figure 37.1b. In other words, the light deviates from a straight-line path and enters the region that would otherwise be shadowed. As noted in Section 35.3, this divergence of light from its initial line of travel is called **diffraction.**

Interference in light waves from two sources was first demonstrated by Thomas Young in 1801. A schematic diagram of the apparatus that Young used is shown in Figure 37.2a. Plane light waves arrive at a barrier that contains two parallel slits S_1 and S_2. These two slits serve as a pair of coherent light sources because waves emerging from them originate from the same wave front and therefore maintain a constant phase relationship. The light from S_1 and S_2 produces on a viewing screen a visible pattern of bright and dark parallel bands called **fringes** (Fig. 37.2b). When the light from S_1 and that from S_2 both arrive at a point on the screen such that constructive interference occurs at that location, a bright fringe appears. When the light from the two slits combines destructively at any location on the screen, a dark fringe results. Figure 37.3 is a photograph of an interference pattern produced by two coherent vibrating sources in a water tank.

Figure 37.4 shows some of the ways in which two waves can combine at the screen. In Figure 37.4a, the two waves, which leave the two slits in phase, strike the screen at the central point P. Because both waves travel the same distance, they arrive at P in phase. As a result, constructive interference occurs at this location, and a bright fringe is observed. In Figure 37.4b, the two waves also start in phase, but in this case the upper wave has to travel one wavelength farther than the lower

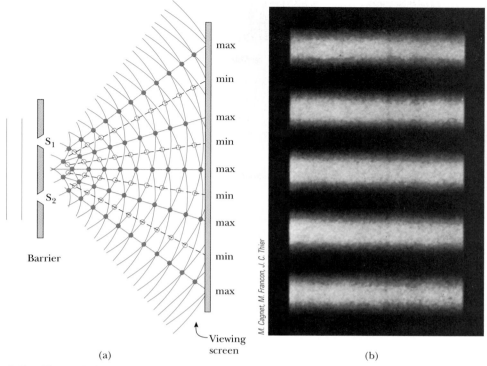

Active Figure 37.2 (a) Schematic diagram of Young's double-slit experiment. Slits S_1 and S_2 behave as coherent sources of light waves that produce an interference pattern on the viewing screen (drawing not to scale). (b) An enlargement of the center of a fringe pattern formed on the viewing screen.

Figure 37.3 An interference pattern involving water waves is produced by two vibrating sources at the water's surface. The pattern is analogous to that observed in Young's double-slit experiment. Note the regions of constructive (A) and destructive (B) interference.

At the Active Figures link at http://www.pse6.com, *you can adjust the slit separation and the wavelength of the light to see the effect on the interference pattern.*

wave to reach point Q. Because the upper wave falls behind the lower one by exactly one wavelength, they still arrive in phase at Q, and so a second bright fringe appears at this location. At point R in Figure 37.4c, however, between points P and Q, the upper wave has fallen half a wavelength behind the lower wave. This means that a trough of the lower wave overlaps a crest of the upper wave; this gives rise to destructive interference at point R. For this reason, a dark fringe is observed at this location.

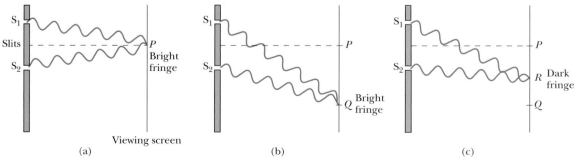

Figure 37.4 (a) Constructive interference occurs at point P when the waves combine. (b) Constructive interference also occurs at point Q. (c) Destructive interference occurs at R when the two waves combine because the upper wave falls half a wavelength behind the lower wave. (All figures not to scale.)

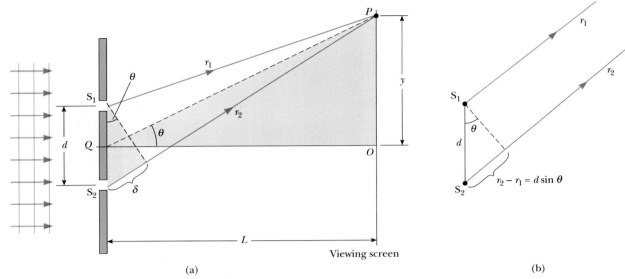

Figure 37.5 (a) Geometric construction for describing Young's double-slit experiment (not to scale). (b) When we assume that r_1 is parallel to r_2, the path difference between the two rays is $r_2 - r_1 = d \sin \theta$. For this approximation to be valid, it is essential that $L \gg d$.

We can describe Young's experiment quantitatively with the help of Figure 37.5. The viewing screen is located a perpendicular distance L from the barrier containing two slits, S_1 and S_2. These slits are separated by a distance d, and the source is monochromatic. To reach any arbitrary point P in the upper half of the screen, a wave from the lower slit must travel farther than a wave from the upper slit by a distance $d \sin \theta$. This distance is called the **path difference** δ (lowercase Greek delta). If we assume that r_1 and r_2 are parallel, which is approximately true if L is much greater than d, then δ is given by

Path difference

$$\delta = r_2 - r_1 = d \sin \theta \qquad (37.1)$$

The value of δ determines whether the two waves are in phase when they arrive at point P. If δ is either zero or some integer multiple of the wavelength, then the two waves are in phase at point P and constructive interference results. Therefore, the condition for bright fringes, or **constructive interference,** at point P is

Conditions for constructive interference

$$\delta = d \sin \theta_{\text{bright}} = m\lambda \qquad (m = 0, \pm 1, \pm 2, \ldots) \qquad (37.2)$$

The number m is called the **order number.** For constructive interference, the order number is the same as the number of wavelengths that represents the path difference between the waves from the two slits. The central bright fringe at $\theta = 0$ is called the *zeroth-order maximum.* The first maximum on either side, where $m = \pm 1$, is called the *first-order maximum,* and so forth.

When δ is an odd multiple of $\lambda/2$, the two waves arriving at point P are 180° out of phase and give rise to destructive interference. Therefore, the condition for dark fringes, or **destructive interference,** at point P is

Conditions for destructive interference

$$d \sin \theta_{\text{dark}} = (m + \tfrac{1}{2})\lambda \qquad (m = 0, \pm 1, \pm 2, \ldots) \qquad (37.3)$$

It is useful to obtain expressions for the positions along the screen of the bright and dark fringes measured vertically from O to P. In addition to our assumption that $L \gg d$, we assume $d \gg \lambda$. These can be valid assumptions because in practice L is often on the order of 1 m, d a fraction of a millimeter, and λ a fraction of a micrometer for visible light. Under these conditions, θ is small; thus, we can use the small angle approximation $\sin \theta \approx \tan \theta$. Then, from triangle OPQ in Figure 37.5a,

we see that

$$y = L \tan\theta \approx L \sin\theta \qquad (37.4)$$

Solving Equation 37.2 for $\sin\theta$ and substituting the result into Equation 37.4, we see that the positions of the bright fringes measured from O are given by the expression

$$y_{\text{bright}} = \frac{\lambda L}{d} m \qquad (m = 0, \pm 1, \pm 2, \ldots) \qquad (37.5)$$

Using Equations 37.3 and 37.4, we find that the dark fringes are located at

$$y_{\text{dark}} = \frac{\lambda L}{d} \left(m + \tfrac{1}{2}\right) \qquad (m = 0, \pm 1, \pm 2, \ldots) \qquad (37.6)$$

As we demonstrate in Example 37.1, Young's double-slit experiment provides a method for measuring the wavelength of light. In fact, Young used this technique to do just that. Additionally, his experiment gave the wave model of light a great deal of credibility. It was inconceivable that particles of light coming through the slits could cancel each other in a way that would explain the dark fringes.

 PITFALL PREVENTION

37.2 It May Not Be True That θ Is Small

The approximation $\sin\theta \approx \tan\theta$ is true to three-digit precision only for angles less than about 4°. If you are considering fringes that are far removed from the central fringe, $\tan\theta = y/L$ is still true, but the small-angle approximation may not be valid. In this case, Equations 37.5 and 37.6 cannot be used. These problems can be solved, but the geometry is not as simple.

 PITFALL PREVENTION

37.3 It May Not Be True That $L \gg d$

Equations 37.2, 37.3, 37.5, and 37.6 were developed under the assumption that $L \gg d$. This is a very common situation, but you are likely to encounter some situations in which this assumption is not valid. In those cases, the geometric construction will be more complicated, but the general approach outlined here will be similar.

Quick Quiz 37.1 If you were to blow smoke into the space between the barrier and the viewing screen of Figure 37.5a, the smoke would show (a) no evidence of interference between the barrier and the screen (b) evidence of interference everywhere between the barrier and the screen.

Quick Quiz 37.2 In a two-slit interference pattern projected on a screen, the fringes are equally spaced on the screen (a) everywhere (b) only for large angles (c) only for small angles.

Quick Quiz 37.3 Which of the following will cause the fringes in a two-slit interference pattern to move farther apart? (a) decreasing the wavelength of the light (b) decreasing the screen distance L (c) decreasing the slit spacing d (d) immersing the entire apparatus in water.

Example 37.1   Measuring the Wavelength of a Light Source   `Interactive`

A viewing screen is separated from a double-slit source by 1.2 m. The distance between the two slits is 0.030 mm. The second-order bright fringe ($m = 2$) is 4.5 cm from the center line.

(A) Determine the wavelength of the light.

Solution We can use Equation 37.5, with $m = 2$, $y_{\text{bright}} = 4.5 \times 10^{-2}$ m, $L = 1.2$ m, and $d = 3.0 \times 10^{-5}$ m:

$$\lambda = \frac{y_{\text{bright}} d}{mL} = \frac{(4.5 \times 10^{-2} \text{ m})(3.0 \times 10^{-5} \text{ m})}{2(1.2 \text{ m})}$$

$$= 5.6 \times 10^{-7} \text{ m} = \boxed{560 \text{ nm}}$$

which is in the green range of visible light.

(B) Calculate the distance between adjacent bright fringes.

Solution From Equation 37.5 and the results of part (A), we obtain

$$y_{m+1} - y_m = \frac{\lambda L}{d}(m + 1) - \frac{\lambda L}{d} m$$

$$= \frac{\lambda L}{d} = \frac{(5.6 \times 10^{-7} \text{ m})(1.2 \text{ m})}{3.0 \times 10^{-5} \text{ m}}$$

$$= 2.2 \times 10^{-2} \text{ m} = \boxed{2.2 \text{ cm}}$$

 Investigate the double-slit interference pattern at the Interactive Worked Example link at **http://www.pse6.com.**

Example 37.2   Separating Double-Slit Fringes of Two Wavelengths

A light source emits visible light of two wavelengths: $\lambda = 430$ nm and $\lambda' = 510$ nm. The source is used in a double-slit interference experiment in which $L = 1.50$ m and $d = 0.025\,0$ mm. Find the separation distance between the third-order bright fringes.

Solution Using Equation 37.5, with $m = 3$, we find that the fringe positions corresponding to these two wavelengths are

$$y_{bright} = \frac{\lambda L}{d}\,m = 3\,\frac{\lambda L}{d} = 3\,\frac{(430 \times 10^{-9}\text{ m})(1.50\text{ m})}{0.025\,0 \times 10^{-3}\text{ m}}$$

$$= 7.74 \times 10^{-2}\text{ m}$$

$$y'_{bright} = \frac{\lambda' L}{d}\,m = 3\,\frac{\lambda' L}{d} = 3\,\frac{(510 \times 10^{-9}\text{ m})(1.50\text{ m})}{0.025\,0 \times 10^{-3}\text{ m}}$$

$$= 9.18 \times 10^{-2}\text{ m}$$

Hence, the separation distance between the two fringes is

$$\Delta y = 9.18 \times 10^{-2}\text{ m} - 7.74 \times 10^{-2}\text{ m}$$

$$= 1.40 \times 10^{-2}\text{ m} = \boxed{1.40\text{ cm}}$$

What If? What if we examine the entire interference pattern due to the two wavelengths and look for overlapping fringes? Are there any locations on the screen where the bright fringes from the two wavelengths overlap exactly?

Answer We could find such a location by setting the location of any bright fringe due to λ equal to one due to λ',

using Equation 37.5:

$$\frac{\lambda L}{d}\,m = \frac{\lambda' L}{d}\,m'$$

$$\frac{\lambda}{\lambda'} = \frac{m'}{m}$$

Substituting the wavelengths, we have

$$\frac{m'}{m} = \frac{\lambda}{\lambda'} = \frac{430\text{ nm}}{510\text{ nm}} = \frac{43}{51}$$

This might suggest that the 51st bright fringe of the 430-nm light would overlap with the 43rd bright fringe of the 510-nm light. However, if we use Equation 37.5 to find the value of y for these fringes, we find

$$y = 51\,\frac{(430 \times 10^{-9}\text{ m})(1.5\text{ m})}{0.025 \times 10^{-3}\text{ m}} = 1.32\text{ m} = y'$$

This value of y is comparable to L, so that the small-angle approximation used in Equation 37.4 is *not* valid. This suggests that we should not expect Equation 37.5 to give us the correct result. If you use the exact relationship $y = L \tan\theta$, you can show that the bright fringes do indeed overlap when the same condition, $m'/m = \lambda/\lambda'$, is met (see Problem 44). Thus, the 51st fringe of the 430-nm light does overlap with the 43rd fringe of the 510-nm light, but not at the location of 1.32 m. You are asked to find the correct location as part of Problem 44.

37.3   Intensity Distribution of the Double-Slit Interference Pattern

Note that the edges of the bright fringes in Figure 37.2b are not sharp—there is a gradual change from bright to dark. So far we have discussed the locations of only the centers of the bright and dark fringes on a distant screen. Let us now direct our attention to the intensity of the light at other points between the positions of maximum constructive and destructive interference. In other words, we now calculate the distribution of light intensity associated with the double-slit interference pattern.

Again, suppose that the two slits represent coherent sources of sinusoidal waves such that the two waves from the slits have the same angular frequency ω and a constant phase difference ϕ. The total magnitude of the electric field at point P on the screen in Figure 37.6 is the superposition of the two waves. Assuming that the two waves have the same amplitude E_0, we can write the magnitude of the electric field at point P due to each wave separately as

$$E_1 = E_0 \sin \omega t \qquad \text{and} \qquad E_2 = E_0 \sin(\omega t + \phi) \tag{37.7}$$

Although the waves are in phase at the slits, *their phase difference ϕ at P depends on the path difference* $\delta = r_2 - r_1 = d \sin\theta$. A path difference of λ (for constructive interference) corresponds to a phase difference of 2π rad. A path difference of δ is the same fraction of λ as the phase difference ϕ is of 2π. We can describe this mathematically

Figure 37.6 Construction for analyzing the double-slit interference pattern. A bright fringe, or intensity maximum, is observed at O.

with the ratio

$$\frac{\delta}{\lambda} = \frac{\phi}{2\pi}$$

which gives us

$$\phi = \frac{2\pi}{\lambda}\,\delta = \frac{2\pi}{\lambda}\,d\sin\theta \qquad (37.8)$$ **Phase difference**

This equation tells us precisely how the phase difference ϕ depends on the angle θ in Figure 37.5.

Using the superposition principle and Equation 37.7, we can obtain the magnitude of the resultant electric field at point P:

$$E_P = E_1 + E_2 = E_0[\sin\omega t + \sin(\omega t + \phi)] \qquad (37.9)$$

To simplify this expression, we use the trigonometric identity

$$\sin A + \sin B = 2\sin\left(\frac{A+B}{2}\right)\cos\left(\frac{A-B}{2}\right)$$

Taking $A = \omega t + \phi$ and $B = \omega t$, we can write Equation 37.9 in the form

$$E_P = 2E_0\cos\left(\frac{\phi}{2}\right)\sin\left(\omega t + \frac{\phi}{2}\right) \qquad (37.10)$$

This result indicates that the electric field at point P has the same frequency ω as the light at the slits, but that the amplitude of the field is multiplied by the factor $2\cos(\phi/2)$. To check the consistency of this result, note that if $\phi = 0, 2\pi, 4\pi, \ldots$, then the magnitude of the electric field at point P is $2E_0$, corresponding to the condition for maximum constructive interference. These values of ϕ are consistent with Equation 37.2 for constructive interference. Likewise, if $\phi = \pi, 3\pi, 5\pi, \ldots$, then the magnitude of the electric field at point P is zero; this is consistent with Equation 37.3 for total destructive interference.

Finally, to obtain an expression for the light intensity at point P, recall from Section 34.3 that *the intensity of a wave is proportional to the square of the resultant electric field magnitude at that point* (Eq. 34.21). Using Equation 37.10, we can therefore express the light intensity at point P as

$$I \propto E_P^2 = 4E_0^2\cos^2\left(\frac{\phi}{2}\right)\sin^2\left(\omega t + \frac{\phi}{2}\right)$$

Most light-detecting instruments measure time-averaged light intensity, and the time-averaged value of $\sin^2(\omega t + \phi/2)$ over one cycle is $\frac{1}{2}$. (See Figure 33.5.) Therefore, we can write the average light intensity at point P as

$$I = I_{\max}\cos^2\left(\frac{\phi}{2}\right) \qquad (37.11)$$

where $I_{\max}$ is the maximum intensity on the screen and the expression represents the time average. Substituting the value for ϕ given by Equation 37.8 into this expression, we find that

$$I = I_{\max}\cos^2\left(\frac{\pi d\sin\theta}{\lambda}\right) \qquad (37.12)$$

Alternatively, because $\sin\theta \approx y/L$ for small values of θ in Figure 37.5, we can write Equation 37.12 in the form

$$I \approx I_{\max}\cos^2\left(\frac{\pi d}{\lambda L}\,y\right) \qquad (37.13)$$

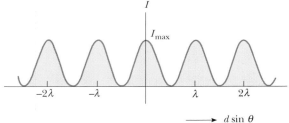

Figure 37.7 Light intensity versus $d \sin \theta$ for a double-slit interference pattern when the screen is far from the two slits ($L \gg d$).

Constructive interference, which produces light intensity maxima, occurs when the quantity $\pi\, dy/\lambda L$ is an integral multiple of π, corresponding to $y = (\lambda L/d)\, m$. This is consistent with Equation 37.5.

A plot of light intensity versus $d \sin \theta$ is given in Figure 37.7. The interference pattern consists of equally spaced fringes of equal intensity. Remember, however, that this result is valid only if the slit-to-screen distance L is much greater than the slit separation, and only for small values of θ.

Quick Quiz 37.4 At dark areas in an interference pattern, the light waves have canceled. Thus, there is zero intensity at these regions and, therefore, no energy is arriving. Consequently, when light waves interfere and form an interference pattern, (a) energy conservation is violated because energy disappears in the dark areas (b) energy transferred by the light is transformed to another type of energy in the dark areas (c) the total energy leaving the slits is distributed among light and dark areas and energy is conserved.

37.4   Phasor Addition of Waves

In the preceding section, we combined two waves algebraically to obtain the resultant wave amplitude at some point on a screen. Unfortunately, this analytical procedure becomes cumbersome when we must add several wave amplitudes. Because we shall eventually be interested in combining a large number of waves, we now describe a graphical procedure for this purpose.

Let us again consider a sinusoidal wave whose electric field component is given by

$$E_1 = E_0 \sin \omega t$$

where E_0 is the wave amplitude and ω is the angular frequency. We used phasors in Chapter 33 to analyze AC circuits, and again we find the use of phasors to be valuable

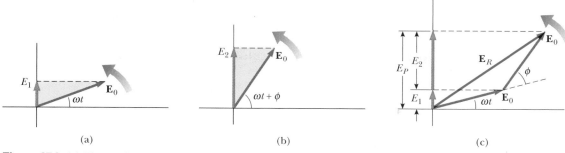

Figure 37.8 (a) Phasor diagram for the wave disturbance $E_1 = E_0 \sin \omega t$. The phasor is a vector of length E_0 rotating counterclockwise. (b) Phasor diagram for the wave $E_2 = E_0 \sin(\omega t + \phi)$. (c) The phasor $\mathbf{E}_R$ represents the combination of the waves in part (a) and (b).

in discussing wave interference. The sinusoidal wave we are discussing can be represented graphically by a phasor of magnitude E_0 rotating about the origin counterclockwise with an angular frequency ω, as in Figure 37.8a. Note that the phasor makes an angle ωt with the horizontal axis. The projection of the phasor on the vertical axis represents E_1, the magnitude of the wave disturbance at some time t. Hence, as the phasor rotates in a circle about the origin, the projection E_1 oscillates along the vertical axis.

Now consider a second sinusoidal wave whose electric field component is given by

$$E_2 = E_0 \sin(\omega t + \phi)$$

This wave has the same amplitude and frequency as E_1, but its phase is ϕ with respect to E_1. The phasor representing E_2 is shown in Figure 37.8b. We can obtain the resultant wave, which is the sum of E_1 and E_2, graphically by redrawing the phasors as shown in Figure 37.8c, in which the tail of the second phasor is placed at the tip of the first. As with vector addition, the resultant phasor $\mathbf{E}_R$ runs from the tail of the first phasor to the tip of the second. Furthermore, $\mathbf{E}_R$ rotates along with the two individual phasors at the same angular frequency ω. The projection of $\mathbf{E}_R$ along the vertical axis equals the sum of the projections of the two other phasors: $E_P = E_1 + E_2$.

It is convenient to construct the phasors at $t = 0$ as in Figure 37.9. From the geometry of one of the right triangles, we see that

$$\cos \alpha = \frac{E_R/2}{E_0}$$

which gives

$$E_R = 2E_0 \cos \alpha$$

Because the sum of the two opposite interior angles equals the exterior angle ϕ, we see that $\alpha = \phi/2$; thus,

$$E_R = 2E_0 \cos\left(\frac{\phi}{2}\right)$$

Hence, the projection of the phasor $\mathbf{E}_R$ along the vertical axis at any time t is

$$E_P = E_R \sin\left(\omega t + \frac{\phi}{2}\right) = 2E_0 \cos(\phi/2) \sin\left(\omega t + \frac{\phi}{2}\right)$$

This is consistent with the result obtained algebraically, Equation 37.10. The resultant phasor has an amplitude $2E_0 \cos(\phi/2)$ and makes an angle $\phi/2$ with the first phasor.

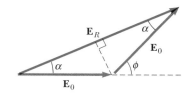

Figure 37.9 A reconstruction of the resultant phasor $\mathbf{E}_R$. From the geometry, note that $\alpha = \phi/2$.

Furthermore, the average light intensity at point P, which varies as $E_P{}^2$, is proportional to $\cos^2(\phi/2)$, as described in Equation 37.11.

We can now describe how to obtain the resultant of several waves that have the same frequency:

- Represent the waves by phasors, as shown in Figure 37.10, remembering to maintain the proper phase relationship between one phasor and the next.

- The resultant phasor $\mathbf{E}_R$ is the vector sum of the individual phasors. At each instant, the projection of $\mathbf{E}_R$ along the vertical axis represents the time variation of the resultant wave. The phase angle α of the resultant wave is the angle between $\mathbf{E}_R$ and the first phasor. From Figure 37.10, drawn for four phasors, we see that the resultant wave is given by the expression $E_P = E_R \sin(\omega t + \alpha)$.

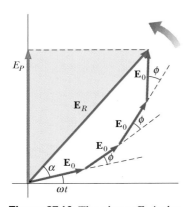

Figure 37.10 The phasor $\mathbf{E}_R$ is the resultant of four phasors of equal amplitude E_0. The phase of $\mathbf{E}_R$ with respect to the first phasor is α. The projection E_P on the vertical axis represents the combination of the four phasors.

Phasor Diagrams for Two Coherent Sources

As an example of the phasor method, consider the interference pattern produced by two coherent sources. Figure 37.11 represents the phasor diagrams for various values of the phase difference ϕ and the corresponding values of the path difference δ, which are obtained from Equation 37.8. The light intensity at a point is a maximum when $\mathbf{E}_R$ is a maximum; this occurs at $\phi = 0, 2\pi, 4\pi, \ldots$. The light intensity at some point is zero when $\mathbf{E}_R$ is zero; this occurs at $\phi = \pi, 3\pi, 5\pi, \ldots$. These results are in complete agreement with the analytical procedure described in the preceding section.

Three-Slit Interference Pattern

Using phasor diagrams, let us analyze the interference pattern caused by three equally spaced slits. We can express the electric field components at a point P on the screen caused by waves from the individual slits as

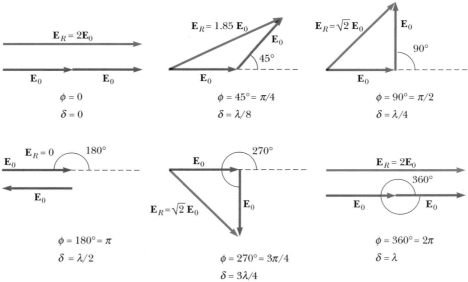

Choose any phase angle at the Active Figures link at http://www.pse6.com and see the resultant phasor.

Active Figure 37.11 Phasor diagrams for a double-slit interference pattern. The resultant phasor $\mathbf{E}_R$ is a maximum when $\phi = 0, 2\pi, 4\pi, \ldots$ and is zero when $\phi = \pi, 3\pi, 5\pi, \ldots$.

$$E_1 = E_0 \sin \omega t$$

$$E_2 = E_0 \sin(\omega t + \phi)$$

$$E_3 = E_0 \sin(\omega t + 2\phi)$$

where ϕ is the phase difference between waves from adjacent slits. We can obtain the resultant magnitude of the electric field at point P from the phasor diagram in Figure 37.12.

The phasor diagrams for various values of ϕ are shown in Figure 37.13. Note that the resultant magnitude of the electric field at P has a maximum value of $3E_0$, a condition that occurs when $\phi = 0, \pm 2\pi, \pm 4\pi, \ldots$. These points are called *primary maxima*. Such primary maxima occur whenever the three phasors are aligned as shown in Figure 37.13a. We also find secondary maxima of amplitude E_0 occurring between the primary maxima at points where $\phi = \pm \pi, \pm 3\pi, \ldots$. For these points, the wave from one slit exactly cancels that from another slit (Fig. 37.13d). This means that only light from the third slit contributes to the resultant, which consequently has a total amplitude of E_0. Total destructive interference occurs whenever the three phasors form a closed triangle, as shown in Figure 37.13c. These points where $E_R = 0$ correspond to $\phi = \pm 2\pi/3, \pm 4\pi/3, \ldots$. You should be able to construct other phasor diagrams for values of ϕ greater than π.

Figure 37.14 shows multiple-slit interference patterns for a number of configurations. For three slits, note that the primary maxima are nine times more intense than the secondary maxima as measured by the height of the curve. This is because the intensity varies as E_R^2. For N slits, the intensity of the primary maxima is N^2 times greater than that due to a single slit. As the number of slits increases, the primary maxima increase in intensity and become narrower, while the secondary maxima decrease in intensity relative to the primary maxima. Figure 37.14 also shows that as the number of slits increases, the number of secondary maxima also increases. In fact, the number of secondary maxima is always $N - 2$ where N is the number of slits. In Section 38.4 (next chapter), we shall investigate the pattern for a very large number of slits in a device called a *diffraction grating*.

Figure 37.12 Phasor diagram for three equally spaced slits.

Quick Quiz 37.5 Using Figure 37.14 as a model, sketch the interference pattern from six slits.

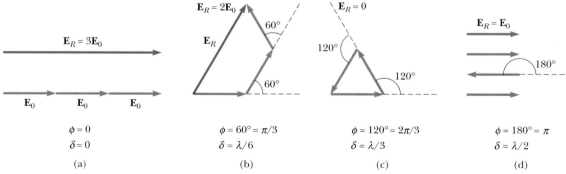

Active Figure 37.13 Phasor diagrams for three equally spaced slits at various values of ϕ. Note from (a) that there are primary maxima of amplitude $3E_0$ and from (d) that there are secondary maxima of amplitude E_0.

Choose any phase angle at the Active Figures link at **http://www.pse6.com** *and see the resultant phasor.*

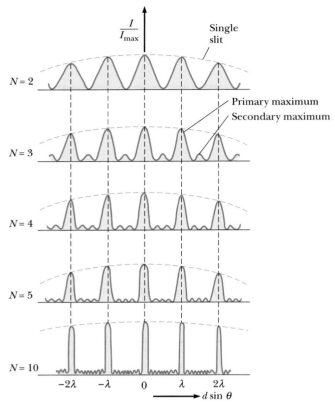

Figure 37.14 Multiple-slit interference patterns. As N, the number of slits, is increased, the primary maxima (the tallest peaks in each graph) become narrower but remain fixed in position and the number of secondary maxima increases. For any value of N, the decrease in intensity in maxima to the left and right of the central maximum, indicated by the blue dashed arcs, is due to *diffraction patterns* from the individual slits, which are discussed in Chapter 38.

37.5   Change of Phase Due to Reflection

Young's method for producing two coherent light sources involves illuminating a pair of slits with a single source. Another simple, yet ingenious, arrangement for producing an interference pattern with a single light source is known as *Lloyd's mirror*[1] (Fig. 37.15). A point light source is placed at point S close to a mirror, and a viewing screen is positioned some distance away and perpendicular to the mirror. Light waves can reach point P on the screen either directly from S to P or by the path involving reflection from the mirror. The reflected ray can be treated as a ray originating from a virtual source at point S′. As a result, we can think of this arrangement as a double-slit source with the distance between points S and S′ comparable to length d in Figure 37.5. Hence, at observation points far from the source ($L \gg d$) we expect waves from points S and S′ to form an interference pattern just like the one we see from two real coherent sources. An interference pattern is indeed observed. However, the positions of the dark and bright fringes are reversed relative to the pattern created by two real coherent sources (Young's experiment). This can only occur if the coherent sources at points S and S′ differ in phase by 180°.

To illustrate this further, consider point $P′$, the point where the mirror intersects the screen. This point is equidistant from points S and S′. If path difference alone were responsible for the phase difference, we would see a bright fringe at point $P′$ (because the path difference is zero for this point), corresponding to the central bright fringe of

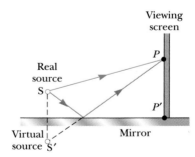

Figure 37.15 Lloyd's mirror. An interference pattern is produced at point P on the screen as a result of the combination of the direct ray (blue) and the reflected ray (brown). The reflected ray undergoes a phase change of 180°.

[1]   Developed in 1834 by Humphrey Lloyd (1800–1881), Professor of Natural and Experimental Philosophy, Trinity College, Dublin.

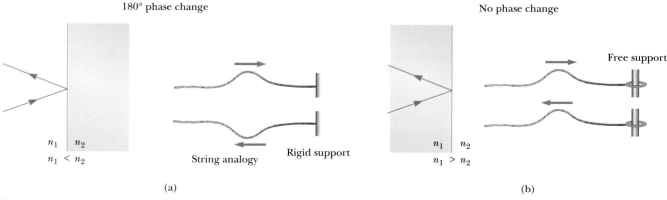

180° phase change   No phase change

Free support

n_1 n_2

$n_1 < n_2$ String analogy   Rigid support n_1 n_2

$n_1 > n_2$

(a)   (b)

Figure 37.16 (a) For $n_1 < n_2$, a light ray traveling in medium 1 when reflected from the surface of medium 2 undergoes a 180° phase change. The same thing happens with a reflected pulse traveling along a string fixed at one end. (b) For $n_1 > n_2$, a light ray traveling in medium 1 undergoes no phase change when reflected from the surface of medium 2. The same is true of a reflected wave pulse on a string whose supported end is free to move.

the two-slit interference pattern. Instead, we observe a dark fringe at point P'. From this, we conclude that a 180° phase change must be produced by reflection from the mirror. In general, **an electromagnetic wave undergoes a phase change of 180° upon reflection from a medium that has a higher index of refraction than the one in which the wave is traveling.**

It is useful to draw an analogy between reflected light waves and the reflections of a transverse wave pulse on a stretched string (Section 16.4). The reflected pulse on a string undergoes a phase change of 180° when reflected from the boundary of a denser medium, but no phase change occurs when the pulse is reflected from the boundary of a less dense medium. Similarly, an electromagnetic wave undergoes a 180° phase change when reflected from a boundary leading to an optically denser medium (defined as a medium with a higher index of refraction), but no phase change occurs when the wave is reflected from a boundary leading to a less dense medium. These rules, summarized in Figure 37.16, can be deduced from Maxwell's equations, but the treatment is beyond the scope of this text.

37.6 Interference in Thin Films

Interference effects are commonly observed in thin films, such as thin layers of oil on water or the thin surface of a soap bubble. The varied colors observed when white light is incident on such films result from the interference of waves reflected from the two surfaces of the film.

Consider a film of uniform thickness t and index of refraction n, as shown in Figure 37.17. Let us assume that the light rays traveling in air are nearly normal to the two surfaces of the film. To determine whether the reflected rays interfere constructively or destructively, we first note the following facts:

- A wave traveling from a medium of index of refraction n_1 toward a medium of index of refraction n_2 undergoes a 180° phase change upon reflection when $n_2 > n_1$ and undergoes no phase change if $n_2 < n_1$.

- The wavelength of light λ_n in a medium whose index of refraction is n (see Section 35.5) is

$$\lambda_n = \frac{\lambda}{n} \tag{37.14}$$

where λ is the wavelength of the light in free space.

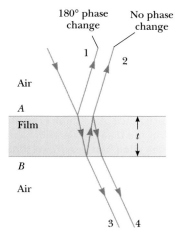

Figure 37.17 Interference in light reflected from a thin film is due to a combination of rays 1 and 2 reflected from the upper and lower surfaces of the film. Rays 3 and 4 lead to interference effects for light transmitted through the film.

Let us apply these rules to the film of Figure 37.17, where $n_{film} > n_{air}$. Reflected ray 1, which is reflected from the upper surface (A), undergoes a phase change of 180° with respect to the incident wave. Reflected ray 2, which is reflected from the lower film surface (B), undergoes no phase change because it is reflected from a medium (air) that has a lower index of refraction. Therefore, ray 1 is 180° out of phase with ray 2, which is equivalent to a path difference of $\lambda_n/2$. However, we must also consider that ray 2 travels an extra distance $2t$ before the waves recombine in the air above surface A. (Remember that we are considering light rays that are close to normal to the surface. If the rays are not close to normal, the path difference is larger than $2t$.) If $2t = \lambda_n/2$, then rays 1 and 2 recombine in phase, and the result is constructive interference. In general, the condition for *constructive* interference in thin films is[2]

$$2t = (m + \tfrac{1}{2})\lambda_n \qquad (m = 0, 1, 2, \ldots) \qquad (37.15)$$

Conditions for constructive interference in thin films

This condition takes into account two factors: (1) the difference in path length for the two rays (the term $m\lambda_n$) and (2) the 180° phase change upon reflection (the term $\lambda_n/2$). Because $\lambda_n = \lambda/n$, we can write Equation 37.15 as

$$2nt = (m + \tfrac{1}{2})\lambda \qquad (m = 0, 1, 2, \ldots) \qquad (37.16)$$

If the extra distance $2t$ traveled by ray 2 corresponds to a multiple of λ_n, then the two waves combine out of phase, and the result is destructive interference. The general equation for *destructive* interference in thin films is

Conditions for destructive interference in thin films

$$2nt = m\lambda \qquad (m = 0, 1, 2, \ldots) \qquad (37.17)$$

The foregoing conditions for constructive and destructive interference are valid when the medium above the top surface of the film is the same as the medium below the bottom surface or, if there are different media above and below the film, the index of refraction of both is less than n. If the film is placed between two different media, one with $n < n_{film}$ and the other with $n > n_{film}$, then the conditions for constructive and destructive interference are reversed. In this case, either there is a phase change of 180° for both ray 1 reflecting from surface A and ray 2 reflecting from surface B, or there is no phase change for either ray; hence, the net change in relative phase due to the reflections is zero.

Rays 3 and 4 in Figure 37.17 lead to interference effects in the light transmitted through the thin film. The analysis of these effects is similar to that of the reflected light. You are asked to explore the transmitted light in Problems 31, 36, and 37.

PITFALL PREVENTION

37.4 Be Careful with Thin Films

Be sure to include *both* effects—path length and phase change—when analyzing an interference pattern resulting from a thin film. The possible phase change is a new feature that we did not need to consider for double-slit interference. Also think carefully about the material on either side of the film. You may have situations in which there is a 180° phase change at *both* surfaces or at *neither* surface, if there are different materials on either side of the film.

> **Quick Quiz 37.6** In a laboratory accident, you spill two liquids onto water, neither of which mixes with the water. They both form thin films on the water surface. When the films become very thin as they spread, you observe that one film becomes bright and the other dark in reflected light. The film that is dark (a) has an index of refraction higher than that of water (b) has an index of refraction lower than that of water (c) has an index of refraction equal to that of water (d) has an index of refraction lower than that of the bright film.

> **Quick Quiz 37.7** One microscope slide is placed on top of another with their left edges in contact and a human hair under the right edge of the upper slide. As a result, a wedge of air exists between the slides. An interference pattern results when monochromatic light is incident on the wedge. At the left edges of the slides, there is (a) a dark fringe (b) a bright fringe (c) impossible to determine.

[2] The full interference effect in a thin film requires an analysis of an infinite number of reflections back and forth between the top and bottom surfaces of the film. We focus here only on a single reflection from the bottom of the film, which provides the largest contribution to the interference effect.

(*Left*) Interference in soap bubbles. The colors are due to interference between light rays reflected from the front and back surfaces of the thin film of soap making up the bubble. The color depends on the thickness of the film, ranging from black where the film is thinnest to magenta where it is thickest. (*Right*) A thin film of oil floating on water displays interference, as shown by the pattern of colors when white light is incident on the film. Variations in film thickness produce the interesting color pattern. The razor blade gives you an idea of the size of the colored bands.

Newton's Rings

Another method for observing interference in light waves is to place a plano-convex lens on top of a flat glass surface, as shown in Figure 37.18a. With this arrangement, the air film between the glass surfaces varies in thickness from zero at the point of contact to some value t at point P. If the radius of curvature R of the lens is much greater than the distance r, and if the system is viewed from above, a pattern of light and dark rings is observed, as shown in Figure 37.18b. These circular fringes, discovered by Newton, are called **Newton's rings.**

The interference effect is due to the combination of ray 1, reflected from the flat plate, with ray 2, reflected from the curved surface of the lens. Ray 1 undergoes a phase change of 180° upon reflection (because it is reflected from a medium of higher index of refraction), whereas ray 2 undergoes no phase change (because it is reflected from a medium of lower refractive index). Hence, the conditions for constructive and destructive interference are given by Equations 37.16 and 37.17, respectively, with $n = 1$ because the film is air.

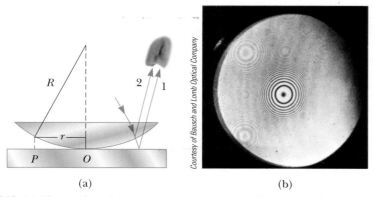

(a)   (b)

Figure 37.18 (a) The combination of rays reflected from the flat plate and the curved lens surface gives rise to an interference pattern known as Newton's rings. (b) Photograph of Newton's rings.

Figure 37.19 This asymmetrical interference pattern indicates imperfections in the lens of a Newton's-rings apparatus.

The contact point at O is dark, as seen in Figure 37.18b, because there is no path difference and the total phase change is due only to the 180° phase change upon reflection.

Using the geometry shown in Figure 37.18a, we can obtain expressions for the radii of the bright and dark bands in terms of the radius of curvature R and wavelength λ. For example, the dark rings have radii given by the expression $r \approx \sqrt{m\lambda R/n}$. The details are left as a problem for you to solve (see Problem 62). We can obtain the wavelength of the light causing the interference pattern by measuring the radii of the rings, provided R is known. Conversely, we can use a known wavelength to obtain R.

One important use of Newton's rings is in the testing of optical lenses. A circular pattern like that pictured in Figure 37.18b is obtained only when the lens is ground to a perfectly symmetric curvature. Variations from such symmetry might produce a pattern like that shown in Figure 37.19. These variations indicate how the lens must be reground and repolished to remove imperfections.

PROBLEM-SOLVING HINTS

Thin-Film Interference

You should keep the following ideas in mind when you work thin-film interference problems:

- Identify the thin film causing the interference.

- The type of interference that occurs is determined by the phase relationship between the portion of the wave reflected at the upper surface of the film and the portion reflected at the lower surface.

- Phase differences between the two portions of the wave have two causes: (1) differences in the distances traveled by the two portions and (2) phase changes that may occur upon reflection.

- When the distance traveled and phase changes upon reflection are both taken into account, the interference is constructive if the equivalent path difference between the two waves is an integral multiple of λ, and it is destructive if the path difference is $\lambda/2$, $3\lambda/2$, $5\lambda/2$, and so forth.

Example 37.3   Interference in a Soap Film

Calculate the minimum thickness of a soap-bubble film that results in constructive interference in the reflected light if the film is illuminated with light whose wavelength in free space is $\lambda = 600$ nm.

Solution The minimum film thickness for constructive interference in the reflected light corresponds to $m = 0$ in Equation 37.16. This gives $2nt = \lambda/2$, or

$$t = \frac{\lambda}{4n} = \frac{600 \text{ nm}}{4(1.33)} = \boxed{113 \text{ nm}}$$

What If? What if the film is twice as thick? Does this situation produce constructive interference?

Answer Using Equation 37.16, we can solve for the thicknesses at which constructive interference will occur:

$$t = (m + \tfrac{1}{2}) \frac{\lambda}{2n} = (2m + 1) \frac{\lambda}{4n} \qquad (m = 0, 1, 2, \ldots)$$

The allowed values of m show that constructive interference will occur for *odd* multiples of the thickness corresponding to $m = 0$, $t = 113$ nm. Thus, constructive interference will *not* occur for a film that is twice as thick.

Example 37.4   Nonreflective Coatings for Solar Cells

Solar cells—devices that generate electricity when exposed to sunlight—are often coated with a transparent, thin film of silicon monoxide (SiO, $n = 1.45$) to minimize reflective losses from the surface. Suppose that a silicon solar cell ($n = 3.5$) is coated with a thin film of silicon monoxide for this purpose (Fig. 37.20). Determine the minimum film thickness that

(b)

Figure 37.20 (Example 37.4) (a) Reflective losses from a silicon solar cell are minimized by coating the surface of the cell with a thin film of silicon monoxide. (b) The reflected light from a coated camera lens often has a reddish-violet appearance.

produces the least reflection at a wavelength of 550 nm, near the center of the visible spectrum.

Solution Figure 37.20a helps us conceptualize the path of the rays in the SiO film that result in interference in the reflected light. Based on the geometry of the SiO layer, we categorize this as a thin-film interference problem. To analyze the problem, note that the reflected light is a minimum when rays 1 and 2 in Figure 37.20a meet the condition of destructive interference. In this situation, *both* rays undergo a 180° phase change upon reflection—ray 1 from the upper SiO surface and ray 2 from the lower SiO surface. The net change in phase due to reflection is therefore zero, and the condition for a reflection minimum requires a path difference of $\lambda_n/2$, where λ_n is the wavelength of the light in SiO. Hence $2t = \lambda/2n$, where λ is the wavelength in air and n is the index of refraction of SiO. The required thickness is

$$t = \frac{\lambda}{4n} = \frac{550 \text{ nm}}{4(1.45)} = \boxed{94.8 \text{ nm}}$$

To finalize the problem, we can investigate the losses in typical solar cells. A typical uncoated solar cell has reflective losses as high as 30%; a SiO coating can reduce this value to about 10%. This significant decrease in reflective losses increases the cell's efficiency because less reflection means that more sunlight enters the silicon to create charge carriers in the cell. No coating can ever be made perfectly nonreflecting because the required thickness is wavelength-dependent and the incident light covers a wide range of wavelengths.

Glass lenses used in cameras and other optical instruments are usually coated with a transparent thin film to reduce or eliminate unwanted reflection and enhance the transmission of light through the lenses. The camera lens in Figure 37.20b has several coatings (of different thicknesses) to minimize reflection of light waves having wavelengths near the center of the visible spectrum. As a result, the little light that is reflected by the lens has a greater proportion of the far ends of the spectrum and often appears reddish-violet.

 Investigate the interference for various film properties at the Interactive Worked Example link at http://www.pse6.com.

Example 37.5   Interference in a Wedge-Shaped Film

A thin, wedge-shaped film of index of refraction n is illuminated with monochromatic light of wavelength λ, as illustrated in Figure 37.21a. Describe the interference pattern observed for this case.

Solution The interference pattern, because it is created by a thin film of variable thickness surrounded by air, is a series of alternating bright and dark parallel fringes. A dark fringe corresponding to destructive interference appears at point *O*, the apex, because here the upper reflected ray undergoes a 180° phase change while the lower one undergoes no phase change.

According to Equation 37.17, other dark minima appear when $2nt = m\lambda$; thus, $t_1 = \lambda/2n$, $t_2 = \lambda/n$, $t_3 = 3\lambda/2n$, and so on. Similarly, the bright maxima appear at locations where t satisfies Equation 37.16, $2nt = (m + \frac{1}{2})\lambda$, corresponding to thicknesses of $\lambda/4n$, $3\lambda/4n$, $5\lambda/4n$, and so on.

If white light is used, bands of different colors are observed at different points, corresponding to the different wavelengths of light (see Fig. 37.21b). This is why we see different colors in soap bubbles and other films of varying thickness.

Figure 37.21 (Example 37.5) (a) Interference bands in reflected light can be observed by illuminating a wedge-shaped film with monochromatic light. The darker areas correspond to regions where rays tend to cancel each other because of interference effects. (b) Interference in a vertical film of variable thickness. The top of the film appears darkest where the film is thinnest.

37.7 The Michelson Interferometer

The **interferometer,** invented by the American physicist A. A. Michelson (1852–1931), splits a light beam into two parts and then recombines the parts to form an interference pattern. The device can be used to measure wavelengths or other lengths with great precision because a large and precisely measurable displacement of one of the mirrors is related to an exactly countable number of wavelengths of light.

A schematic diagram of the interferometer is shown in Figure 37.22. A ray of light from a monochromatic source is split into two rays by mirror M_0, which is inclined at 45° to the incident light beam. Mirror M_0, called a *beam splitter*, transmits half the light incident on it and reflects the rest. One ray is reflected from M_0 vertically upward toward mirror M_1, and the second ray is transmitted horizontally through M_0 toward mirror M_2. Hence, the two rays travel separate paths L_1 and L_2. After reflecting from M_1 and M_2, the two rays eventually recombine at M_0 to produce an interference pattern, which can be viewed through a telescope.

The interference condition for the two rays is determined by their path length differences. When the two mirrors are exactly perpendicular to each other, the interference pattern is a target pattern of bright and dark circular fringes, similar to Newton's rings. As M_1 is moved, the fringe pattern collapses or expands, depending on the direction in which M_1 is moved. For example, if a dark circle appears at the center of the target pattern (corresponding to destructive interference) and M_1 is then moved a distance $\lambda/4$ toward M_0, the path difference changes by $\lambda/2$. What was a dark circle at the center now becomes a bright circle. As M_1 is moved an additional distance $\lambda/4$ toward M_0, the bright circle becomes a dark circle again. Thus, the fringe pattern shifts by one-half fringe each time M_1 is moved a distance $\lambda/4$. The wavelength of light is then measured by counting the number of fringe shifts for a given displacement of M_1. If the wavelength is accurately known, mirror displacements can be measured to within a fraction of the wavelength.

We will see an important historical use of the Michelson interferometer in our discussion of relativity in Chapter 39. Modern uses include the following two applications.

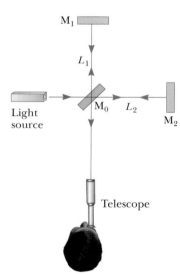

Active Figure 37.22 Diagram of the Michelson interferometer. A single ray of light is split into two rays by mirror M_0, which is called a beam splitter. The path difference between the two rays is varied with the adjustable mirror M_1. As M_1 is moved, an interference pattern changes in the field of view.

At the Active Figures link at http://www.pse6.com, *move the mirror to see the effect on the interference pattern and use the interferometer to measure the wavelength of light.*

Fourier Transform Infrared Spectroscopy (FTIR)

Spectroscopy is the study of the wavelength distribution of radiation from a sample that can be used to identify the characteristics of atoms or molecules in the sample. Infrared spectroscopy is particularly important to organic chemists in analyzing organic molecules. Traditional spectroscopy involves the use of an optical element, such as a prism (Section 35.7) or a diffraction grating (Section 38.4), which spreads out various wavelengths in a complex optical signal from the sample into different angles. In this way, the various wavelengths of radiation and their intensities in the signal can be determined. These types of devices are limited in their resolution and effectiveness because they must be scanned through the various angular deviations of the radiation.

The technique of *Fourier Transform Infrared Spectroscopy* (FTIR) is used to create a higher-resolution spectrum in a time interval of one second that may have required 30 minutes with a standard spectrometer. In this technique, the radiation from a sample enters a Michelson interferometer. The movable mirror is swept through the zero-path-difference condition and the intensity of radiation at the viewing position is recorded. The result is a complex set of data relating light intensity as a function of mirror position, called an *interferogram*. Because there is a relationship between mirror position and light intensity for a given wavelength, the interferogram contains information about all wavelengths in the signal.

In Section 18.8, we discussed Fourier analysis of a waveform. The waveform is a function that contains information about all of the individual frequency components that make up the waveform.[3] Equation 18.16 shows how the waveform is generated from the individual frequency components. Similarily, the interferogram can be analyzed by computer, in a process called a *Fourier transform*, to provide all of the wavelength components. This is the same information generated by traditional spectroscopy, but the resolution of FTIR is much higher.

Laser Interferometer Gravitational-Wave Observatory (LIGO)

Einstein's general theory of relativity (Section 39.10) predicts the existence of *gravitational waves*. These waves propagate from the site of any gravitational disturbance, which could be periodic and predictable, such as the rotation of a double star around a center of mass, or unpredictable, such as the supernova explosion of a massive star.

In Einstein's theory, gravitation is equivalent to a distortion of space. Thus, a gravitational disturbance causes an additional distortion that propagates through space in a manner similar to mechanical or electromagnetic waves. When gravitational waves from a disturbance pass by the Earth, they create a distortion of the local space. The LIGO apparatus is designed to detect this distortion. The apparatus employs a Michelson interferometer that uses laser beams with an effective path length of several kilometers. At the end of an arm of the interferometer, a mirror is mounted on a massive pendulum. When a gravitational wave passes by, the pendulum and the attached mirror move, and the interference pattern due to the laser beams from the two arms changes.

Two sites have been developed in the United States for interferometers in order to allow coincidence studies of gravitational waves. These sites are located in Richland, Washington, and Livingston, Louisiana. Figure 37.23 shows the Washington site. The two arms of the Michelson interferometer are evident in the photograph. Test runs are being performed as of the printing of this book. Cooperation with other gravitational wave detectors, such as VIRGO in Cascina, Italy, will allow detailed studies of gravitational waves.

---

[3] In acoustics, it is common to talk about the components of a complex signal in terms of frequency. In optics, it is more common to identify the components by wavelength.

LIGO Hanford Observatory

Figure 37.23 The Laser Interferometer Gravitational-Wave Observatory (LIGO) near Richland, Washington. Note the two perpendicular arms of the Michelson interferometer.

SUMMARY

Take a practice test for this chapter by clicking on the Practice Test link at http://www.pse6.com.

Interference in light waves occurs whenever two or more waves overlap at a given point. An interference pattern is observed if (1) the sources are coherent and (2) the sources have identical wavelengths.

In Young's double-slit experiment, two slits S_1 and S_2 separated by a distance d are illuminated by a single-wavelength light source. An interference pattern consisting of bright and dark fringes is observed on a viewing screen. The condition for bright fringes (**constructive interference**) is

$$\delta = d \sin\theta_{\text{bright}} = m\lambda \qquad (m = 0, \pm 1, \pm 2, \ldots) \tag{37.2}$$

The condition for dark fringes (**destructive interference**) is

$$d \sin\theta_{\text{dark}} = (m + \tfrac{1}{2})\lambda \qquad (m = 0, \pm 1, \pm 2, \ldots) \tag{37.3}$$

The number m is called the **order number** of the fringe.

The **intensity** at a point in the double-slit interference pattern is

$$I = I_{\text{max}} \cos^2\left(\frac{\pi d \sin\theta}{\lambda}\right) \tag{37.12}$$

where I_{max} is the maximum intensity on the screen and the expression represents the time average.

A wave traveling from a medium of index of refraction n_1 toward a medium of index of refraction n_2 undergoes a 180° phase change upon reflection when $n_2 > n_1$ and undergoes no phase change when $n_2 < n_1$.

The condition for constructive interference in a film of thickness t and index of refraction n surrounded by air is

$$2nt = (m + \tfrac{1}{2})\lambda \qquad (m = 0, 1, 2, \ldots) \tag{37.16}$$

where λ is the wavelength of the light in free space.

Similarly, the condition for destructive interference in a thin film surrounded by air is

$$2nt = m\lambda \qquad (m = 0, 1, 2, \ldots) \tag{37.17}$$

QUESTIONS

1. What is the necessary condition on the path length difference between two waves that interfere (a) constructively and (b) destructively?

2. Explain why two flashlights held close together do not produce an interference pattern on a distant screen.

3. If Young's double-slit experiment were performed under water, how would the observed interference pattern be affected?

4. In Young's double-slit experiment, why do we use monochromatic light? If white light is used, how would the pattern change?

5. A simple way to observe an interference pattern is to look at a distant light source through a stretched handkerchief or an opened umbrella. Explain how this works.

6. A certain oil film on water appears brightest at the outer regions, where it is thinnest. From this information, what can you say about the index of refraction of oil relative to that of water?

7. As a soap bubble evaporates, it appears black just before it breaks. Explain this phenomenon in terms of the phase changes that occur on reflection from the two surfaces of the soap film.

8. If we are to observe interference in a thin film, why must the film not be very thick (with thickness only on the order of a few wavelengths)?

9. A lens with outer radius of curvature R and index of refraction n rests on a flat glass plate. The combination is illuminated with white light from above and observed from above. Is there a dark spot or a light spot at the center of the lens? What does it mean if the observed rings are noncircular?

10. Why is the lens on a good-quality camera coated with a thin film?

11. Why is it so much easier to perform interference experiments with a laser than with an ordinary light source?

12. Suppose that reflected white light is used to observe a thin transparent coating on glass as the coating material is gradually deposited by evaporation in a vacuum. Describe color changes that might occur during the process of building up the thickness of the coating.

13. In our discussion of thin-film interference, we looked at light *reflecting* from a thin film. **What If?** Consider one light ray, the direct ray, which transmits through the film without reflecting. Consider a second ray, the reflected ray, that transmits through the first surface, reflects from the second, reflects again from the first, and then transmits out into the air, parallel to the direct ray. For normal incidence, how thick must the film be, in terms of the wavelength of light, for the outgoing rays to interfere destructively? Is it the same thickness as for reflected destructive interference?

14. Suppose you are watching television by connection to an antenna rather than a cable system. If an airplane flies near your location, you may notice wavering ghost images in the television picture. What might cause this?

PROBLEMS

1, 2, 3 = straightforward, intermediate, challenging   ☐ = full solution available in the *Student Solutions Manual and Study Guide*

🌐 = coached solution with hints available at http://www.pse6.com   💻 = computer useful in solving problem

▨ = paired numerical and symbolic problems

Section 37.1   Conditions for Interference
Section 37.2   Young's Double-Slit Experiment

Note: Problems 8, 9, 10, and 12 in Chapter 18 can be assigned with these sections.

1. A laser beam ($\lambda = 632.8$ nm) is incident on two slits 0.200 mm apart. How far apart are the bright interference fringes on a screen 5.00 m away from the double slits?

2. A Young's interference experiment is performed with monochromatic light. The separation between the slits is 0.500 mm, and the interference pattern on a screen 3.30 m away shows the first side maximum 3.40 mm from the center of the pattern. What is the wavelength?

3. 🌐 Two radio antennas separated by 300 m as shown in Figure P37.3 simultaneously broadcast identical signals at

Figure P37.3

the same wavelength. A radio in a car traveling due north receives the signals. (a) If the car is at the position of the second maximum, what is the wavelength of the signals? (b) How much farther must the car travel to encounter the next minimum in reception? (*Note:* Do not use the small-angle approximation in this problem.)

4. In a location where the speed of sound is 354 m/s, a 2 000-Hz sound wave impinges on two slits 30.0 cm apart. (a) At what angle is the first maximum located? (b) **What If?** If the sound wave is replaced by 3.00-cm microwaves, what slit separation gives the same angle for the first maximum? (c) **What If?** If the slit separation is 1.00 μm, what frequency of light gives the same first maximum angle?

5. [✎] Young's double-slit experiment is performed with 589-nm light and a distance of 2.00 m between the slits and the screen. The tenth interference minimum is observed 7.26 mm from the central maximum. Determine the spacing of the slits.

6. The two speakers of a boom box are 35.0 cm apart. A single oscillator makes the speakers vibrate in phase at a frequency of 2.00 kHz. At what angles, measured from the perpendicular bisector of the line joining the speakers, would a distant observer hear maximum sound intensity? Minimum sound intensity? (Take the speed of sound as 340 m/s.)

7. [✎] Two narrow, parallel slits separated by 0.250 mm are illuminated by green light ($\lambda = 546.1$ nm). The interference pattern is observed on a screen 1.20 m away from the plane of the slits. Calculate the distance (a) from the central maximum to the first bright region on either side of the central maximum and (b) between the first and second dark bands.

8. Light with wavelength 442 nm passes through a double-slit system that has a slit separation $d = 0.400$ mm. Determine how far away a screen must be placed in order that a dark fringe appear directly opposite both slits, with just one bright fringe between them.

9. A riverside warehouse has two open doors as shown in Figure P37.9. Its walls are lined with sound-absorbing material. A boat on the river sounds its horn. To person A the sound is loud and clear. To person B the sound is barely audible. The principal wavelength of the sound waves is 3.00 m. Assuming person B is at the position of the first minimum, determine the distance between the doors, center to center.

Figure P37.9

10. Two slits are separated by 0.320 mm. A beam of 500-nm light strikes the slits, producing an interference pattern. Determine the number of maxima observed in the angular range $-30.0° < \theta < 30.0°$.

11. Young's double-slit experiment underlies the *Instrument Landing System* used to guide aircraft to safe landings when the visibility is poor. Although real systems are more complicated than the example described here, they operate on the same principles. A pilot is trying to align her plane with a runway, as suggested in Figure P37.11a. Two radio antennas A_1 and A_2 are positioned adjacent to the runway, separated by 40.0 m. The antennas broadcast unmodulated coherent radio waves at 30.0 MHz. (a) Find the wavelength of the waves. The pilot "locks onto" the strong signal radiated along an interference maximum, and steers the plane to keep the received signal strong. If she has found the central maximum, the plane will have just the right heading to land when it reaches the runway. (b) **What If?** Suppose instead that the plane is flying along the first side maximum (Fig. P37.11b). How far to the side of the runway centerline will the plane be when it is 2.00 km from the antennas? (c) It is possible to tell the pilot she is on the wrong maximum by sending out two signals from each antenna and equipping the aircraft with a two-channel receiver. The ratio of the two frequencies must not be the ratio of small integers (such as 3/4). Explain how this two-frequency system would work, and why it would not necessarily work if the frequencies were related by an integer ratio.

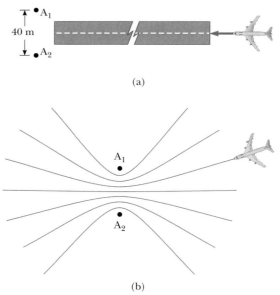

Figure P37.11

12. A student holds a laser that emits light of wavelength 633 nm. The beam passes though a pair of slits separated by 0.300 mm, in a glass plate attached to the front of the laser. The beam then falls perpendicularly on a screen, creating an interference pattern on it. The student begins to walk directly toward the screen at 3.00 m/s. The central maximum on the screen is stationary. Find the speed of the first-order maxima on the screen.

13. In Figure 37.5 let $L = 1.20$ m and $d = 0.120$ mm and assume that the slit system is illuminated with monochromatic 500-nm light. Calculate the phase difference between the two wave fronts arriving at P when (a) $\theta = 0.500°$ and (b) $y = 5.00$ mm. (c) What is the value of θ for which the phase difference is 0.333 rad? (d) What is the value of θ for which the path difference is $\lambda/4$?

14. Coherent light rays of wavelength λ strike a pair of slits separated by distance d at an angle θ_1 as shown in Figure P37.14. Assume an interference maximum is formed at an angle θ_2 a great distance from the slits. Show that $d(\sin\theta_2 - \sin\theta_1) = m\lambda$, where m is an integer.

Figure P37.14

15. In a double-slit arrangement of Figure 37.5, $d = 0.150$ mm, $L = 140$ cm, $\lambda = 643$ nm, and $y = 1.80$ cm. (a) What is the path difference δ for the rays from the two slits arriving at P? (b) Express this path difference in terms of λ. (c) Does P correspond to a maximum, a minimum, or an intermediate condition?

Section 37.3   Intensity Distribution of the Double-Slit Interference Pattern

16. The intensity on the screen at a certain point in a double-slit interference pattern is 64.0% of the maximum value. (a) What minimum phase difference (in radians) between sources produces this result? (b) Express this phase difference as a path difference for 486.1-nm light.

17. In Figure 37.5, let $L = 120$ cm and $d = 0.250$ cm. The slits are illuminated with coherent 600-nm light. Calculate the distance y above the central maximum for which the average intensity on the screen is 75.0% of the maximum.

18. Two slits are separated by 0.180 mm. An interference pattern is formed on a screen 80.0 cm away by 656.3-nm light. Calculate the fraction of the maximum intensity 0.600 cm above the central maximum.

19. Two narrow parallel slits separated by 0.850 mm are illuminated by 600-nm light, and the viewing screen is 2.80 m away from the slits. (a) What is the phase difference between the two interfering waves on a screen at a point 2.50 mm from the central bright fringe? (b) What is the ratio of the intensity at this point to the intensity at the center of a bright fringe?

20. Monochromatic coherent light of amplitude E_0 and angular frequency ω passes through three parallel slits each separated by a distance d from its neighbor. (a) Show that the time-averaged intensity as a function of the angle θ is

$$I(\theta) = I_{max}\left[1 + 2\cos\left(\frac{2\pi d \sin\theta}{\lambda}\right)\right]^2$$

(b) Determine the ratio of the intensities of the primary and secondary maxima.

Section 37.4   Phasor Addition of Waves

Note: Problems 4, 5, and 6 in Chapter 18 can be assigned with this section.

21. Marie Cornu, a physicist at the Polytechnic Institute in Paris, invented phasors in about 1880. This problem helps you to see their utility. Find the amplitude and phase constant of the sum of two waves represented by the expressions

$$E_1 = (12.0 \text{ kN/C}) \sin(15x - 4.5t)$$

and $$E_2 = (12.0 \text{ kN/C}) \sin(15x - 4.5t + 70°)$$

(a) by using a trigonometric identity (as from Appendix B), and (b) by representing the waves by phasors. (c) Find the amplitude and phase constant of the sum of the three waves represented by

$$E_1 = (12.0 \text{ kN/C}) \sin(15x - 4.5t + 70°),$$

$$E_2 = (15.5 \text{ kN/C}) \sin(15x - 4.5t - 80°),$$

and $$E_3 = (17.0 \text{ kN/C}) \sin(15x - 4.5t + 160°)$$

22. The electric fields from three coherent sources are described by $E_1 = E_0 \sin \omega t$, $E_2 = E_0 \sin(\omega t + \phi)$, and $E_3 = E_0 \sin(\omega t + 2\phi)$. Let the resultant field be represented by $E_P = E_R \sin(\omega t + \alpha)$. Use phasors to find E_R and α when (a) $\phi = 20.0°$, (b) $\phi = 60.0°$, and (c) $\phi = 120°$. (d) Repeat when $\phi = (3\pi/2)$ rad.

23. Determine the resultant of the two waves given by $E_1 = 6.0 \sin(100\pi t)$ and $E_2 = 8.0 \sin(100\pi t + \pi/2)$.

24. Suppose the slit openings in a Young's double-slit experiment have different sizes so that the electric fields and intensities from each slit are different. With $E_1 = E_{01} \sin(\omega t)$ and $E_2 = E_{02} \sin(\omega t + \phi)$, show that the resultant electric field is $E = E_0 \sin(\omega t + \theta)$, where

$$E_0 = \sqrt{E_{01}^2 + E_{02}^2 + 2E_{01}E_{02}\cos\phi}$$

and

$$\sin\theta = \frac{E_{02} \sin\phi}{E_0}$$

25. Use phasors to find the resultant (magnitude and phase angle) of two fields represented by $E_1 = 12 \sin \omega t$ and $E_2 = 18 \sin(\omega t + 60°)$. (Note that in this case the amplitudes of the two fields are unequal.)

26. Two coherent waves are described by

$$E_1 = E_0 \sin\left(\frac{2\pi x_1}{\lambda} - 2\pi ft + \frac{\pi}{6}\right)$$

$$E_2 = E_0 \sin\left(\frac{2\pi x_2}{\lambda} - 2\pi ft + \frac{\pi}{8}\right)$$

Determine the relationship between x_1 and x_2 that produces constructive interference when the two waves are superposed.

27. When illuminated, four equally spaced parallel slits act as multiple coherent sources, each differing in phase from the adjacent one by an angle ϕ. Use a phasor diagram to determine the smallest value of ϕ for which the resultant of the four waves (assumed to be of equal amplitude) is zero.

28. Sketch a phasor diagram to illustrate the resultant of $E_1 = E_{01} \sin \omega t$ and $E_2 = E_{02} \sin(\omega t + \phi)$, where $E_{02} = 1.50E_{01}$ and $\pi/6 \leq \phi \leq \pi/3$. Use the sketch and the law of cosines to show that, for two coherent waves, the resultant intensity can be written in the form $I_R = I_1 + I_2 + 2\sqrt{I_1 I_2} \cos \phi$.

29. Consider N coherent sources described as follows: $E_1 = E_0 \sin(\omega t + \phi)$, $E_2 = E_0 \sin(\omega t + 2\phi)$, $E_3 = E_0 \sin(\omega t + 3\phi)$, . . . , $E_N = E_0 \sin(\omega t + N\phi)$. Find the minimum value of ϕ for which $E_R = E_1 + E_2 + E_3 + \cdots + E_N$ is zero.

Section 37.5   Change of Phase Due to Reflection
Section 37.6   Interference in Thin Films

30. A soap bubble ($n = 1.33$) is floating in air. If the thickness of the bubble wall is 115 nm, what is the wavelength of the light that is most strongly reflected?

31. An oil film ($n = 1.45$) floating on water is illuminated by white light at normal incidence. The film is 280 nm thick. Find (a) the color of the light in the visible spectrum most strongly reflected and (b) the color of the light in the spectrum most strongly transmitted. Explain your reasoning.

32. A thin film of oil ($n = 1.25$) is located on a smooth wet pavement. When viewed perpendicular to the pavement, the film reflects most strongly red light at 640 nm and reflects no blue light at 512 nm. How thick is the oil film?

33. A possible means for making an airplane invisible to radar is to coat the plane with an antireflective polymer. If radar waves have a wavelength of 3.00 cm and the index of refraction of the polymer is $n = 1.50$, how thick would you make the coating?

34. A material having an index of refraction of 1.30 is used as an antireflective coating on a piece of glass ($n = 1.50$). What should be the minimum thickness of this film in order to minimize reflection of 500-nm light?

35. A film of MgF$_2$ ($n = 1.38$) having thickness 1.00×10^{-5} cm is used to coat a camera lens. Are any wavelengths in the visible spectrum intensified in the reflected light?

36. Astronomers observe the chromosphere of the Sun with a filter that passes the red hydrogen spectral line of wavelength 656.3 nm, called the H$_\alpha$ line. The filter consists of a transparent dielectric of thickness d held between two partially aluminized glass plates. The filter is held at a constant temperature. (a) Find the minimum value of d that produces maximum transmission of perpendicular H$_\alpha$ light, if the dielectric has an index of refraction of 1.378. (b) **What If?** If the temperature of the filter increases above the normal value, what happens to the transmitted wavelength? (Its index of refraction does not change significantly.) (c) The dielectric will also pass what near-visible wavelength? One of the glass plates is colored red to absorb this light.

37. A beam of 580-nm light passes through two closely spaced glass plates, as shown in Figure P37.37. For what minimum nonzero value of the plate separation d is the transmitted light bright?

Figure P37.37

38. When a liquid is introduced into the air space between the lens and the plate in a Newton's-rings apparatus, the diameter of the tenth ring changes from 1.50 to 1.31 cm. Find the index of refraction of the liquid.

39. An air wedge is formed between two glass plates separated at one edge by a very fine wire, as shown in Figure P37.39. When the wedge is illuminated from above by 600-nm light and viewed from above, 30 dark fringes are observed. Calculate the radius of the wire.

Figure P37.39 Problems 39 and 40.

40. Two glass plates 10.0 cm long are in contact at one end and separated at the other end by a thread 0.050 0 mm in diameter (Fig. P37.39). Light containing the two wavelengths 400 nm and 600 nm is incident perpendicularly and viewed by reflection. At what distance from the contact point is the next dark fringe?

Section 37.7   The Michelson Interferometer

41. Mirror M$_1$ in Figure 37.22 is displaced a distance ΔL. During this displacement, 250 fringe reversals (formation of successive dark or bright bands) are counted. The light being used has a wavelength of 632.8 nm. Calculate the displacement ΔL.

42. Monochromatic light is beamed into a Michelson interferometer. The movable mirror is displaced 0.382 mm, causing the interferometer pattern to reproduce itself 1 700 times. Determine the wavelength of the light. What color is it?

43. One leg of a Michelson interferometer contains an evacuated cylinder of length L, having glass plates on each end.

A gas is slowly leaked into the cylinder until a pressure of 1 atm is reached. If N bright fringes pass on the screen when light of wavelength λ is used, what is the index of refraction of the gas?

Additional Problems

44. In the **What If?** section of Example 37.2, it was claimed that overlapping fringes in a two-slit interference pattern for two different wavelengths obey the following relationship even for large values of the angle θ:

$$\frac{\lambda}{\lambda'} = \frac{m'}{m}$$

(a) Prove this assertion. (b) Using the data in Example 37.2, find the value of y on the screen at which the fringes from the two wavelengths first coincide.

45. One radio transmitter A operating at 60.0 MHz is 10.0 m from another similar transmitter B that is 180° out of phase with A. How far must an observer move from A toward B along the line connecting A and B to reach the nearest point where the two beams are in phase?

46. **Review problem.** This problem extends the result of Problem 12 in Chapter 18. Figure P37.46 shows two adjacent vibrating balls dipping into a tank of water. At distant points they produce an interference pattern of water waves, as shown in Figure 37.3. Let λ represent the wavelength of the ripples. Show that the two sources produce a standing wave along the line segment, of length d, between them. In terms of λ and d, find the number of nodes and the number of antinodes in the standing wave. Find the number of zones of constructive and of destructive interference in the interference pattern far away from the sources. Each line of destructive interference springs from a node in the standing wave and each line of constructive interference springs from an antinode.

Figure P37.46

47. Raise your hand and hold it flat. Think of the space between your index finger and your middle finger as one slit, and think of the space between middle finger and ring finger as a second slit. (a) Consider the interference resulting from sending coherent visible light perpendicularly through this pair of openings. Compute an order-of-magnitude estimate for the angle between adjacent zones of constructive interference. (b) To make the angles in the interference pattern easy to measure with a plastic protractor, you should use an electromagnetic wave with frequency of what order of magnitude? How is this wave classified on the electromagnetic spectrum?

48. In a Young's double-slit experiment using light of wavelength λ, a thin piece of Plexiglas having index of refraction n covers one of the slits. If the center point on the screen is a dark spot instead of a bright spot, what is the minimum thickness of the Plexiglas?

49. **Review problem.** A flat piece of glass is held stationary and horizontal above the flat top end of a 10.0-cm-long vertical metal rod that has its lower end rigidly fixed. The thin film of air between the rod and glass is observed to be bright by reflected light when it is illuminated by light of wavelength 500 nm. As the temperature is slowly increased by 25.0°C, the film changes from bright to dark and back to bright 200 times. What is the coefficient of linear expansion of the metal?

50. A certain crude oil has an index of refraction of 1.25. A ship dumps 1.00 m^3 of this oil into the ocean, and the oil spreads into a thin uniform slick. If the film produces a first-order maximum of light of wavelength 500 nm normally incident on it, how much surface area of the ocean does the oil slick cover? Assume that the index of refraction of the ocean water is 1.34.

51. Astronomers observe a 60.0-MHz radio source both directly and by reflection from the sea. If the receiving dish is 20.0 m above sea level, what is the angle of the radio source above the horizon at first maximum?

52. Interference effects are produced at point P on a screen as a result of direct rays from a 500-nm source and reflected rays from the mirror, as shown in Figure P37.52. Assume the source is 100 m to the left of the screen and 1.00 cm above the mirror. Find the distance y to the first dark band above the mirror.

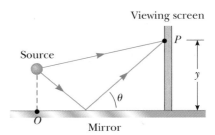

Figure P37.52

53. The waves from a radio station can reach a home receiver by two paths. One is a straight-line path from transmitter to home, a distance of 30.0 km. The second path is by reflection from the ionosphere (a layer of ionized air molecules high in the atmosphere). Assume this reflection takes place at a point midway between receiver and transmitter and that the wavelength broadcast by the radio station is 350 m. Find the minimum height of the ionospheric layer that could produce destructive interference between the direct and reflected beams. (Assume that no phase change occurs on reflection.)

54. Many cells are transparent and colorless. Structures of great interest in biology and medicine can be practically invisible to ordinary microscopy. An *interference microscope* reveals a difference in index of refraction as a shift in interference fringes, to indicate the size and shape of cell structures. The idea is exemplified in the following problem: An air wedge is formed between two glass plates in contact along one edge and slightly separated at the opposite edge. When the plates are illuminated with monochromatic light from above, the reflected light has 85 dark fringes. Calculate the number of dark fringes that appear if water ($n = 1.33$) replaces the air between the plates.

55. Measurements are made of the intensity distribution in a Young's interference pattern (see Fig. 37.7). At a particular value of y, it is found that $I/I_{\text{max}} = 0.810$ when 600-nm light is used. What wavelength of light should be used to reduce the relative intensity at the same location to 64.0% of the maximum intensity?

56. Our discussion of the techniques for determining constructive and destructive interference by reflection from a thin film in air has been confined to rays striking the film at nearly normal incidence. **What If?** Assume that a ray is incident at an angle of 30.0° (relative to the normal) on a film with index of refraction 1.38. Calculate the minimum thickness for constructive interference of sodium light with a wavelength of 590 nm.

57. The condition for constructive interference by reflection from a thin film in air as developed in Section 37.6 assumes nearly normal incidence. **What If?** Show that if the light is incident on the film at a nonzero angle ϕ_1 (relative to the normal), then the condition for constructive interference is $2nt \cos \theta_2 = (m + \frac{1}{2})\lambda$, where θ_2 is the angle of refraction.

58. (a) Both sides of a uniform film that has index of refraction n and thickness d are in contact with air. For normal incidence of light, an intensity minimum is observed in the reflected light at λ_2 and an intensity maximum is observed at λ_1, where $\lambda_1 > \lambda_2$. Assuming that no intensity minima are observed between λ_1 and λ_2, show that the integer m in Equations 37.16 and 37.17 is given by $m = \lambda_1/2(\lambda_1 - \lambda_2)$. (b) Determine the thickness of the film, assuming $n = 1.40$, $\lambda_1 = 500$ nm, and $\lambda_2 = 370$ nm.

59. Figure P37.59 shows a radio-wave transmitter and a receiver separated by a distance d and both a distance h above the ground. The receiver can receive signals both directly from the transmitter and indirectly from signals that reflect from the ground. Assume that the ground is level between the transmitter and receiver and that a 180° phase shift occurs upon reflection. Determine the longest wavelengths that interfere (a) constructively and (b) destructively.

60. A piece of transparent material having an index of refraction n is cut into the shape of a wedge as shown in Figure P37.60. The angle of the wedge is small. Monochromatic light of wavelength λ is normally incident from above, and viewed from above. Let h represent the height of the wedge and ℓ its width. Show that bright fringes occur at the positions $x = \lambda\ell(m + \frac{1}{2})/2hn$ and dark fringes occur at the positions $x = \lambda\ell m/2hn$, where $m = 0, 1, 2, \ldots$ and x is measured as shown.

Figure P37.60

61. Consider the double-slit arrangement shown in Figure P37.61, where the slit separation is d and the slit to screen distance is L. A sheet of transparent plastic having an index of refraction n and thickness t is placed over the upper slit. As a result, the central maximum of the interference pattern moves upward a distance y'. Find y'.

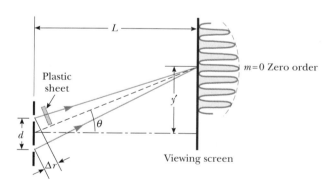

Figure P37.61

62. A plano-convex lens has index of refraction n. The curved side of the lens has radius of curvature R and rests on a flat glass surface of the same index of refraction, with a film of index n_{film} between them, as shown in Fig. 37.18a. The lens is illuminated from above by light of wavelength λ. Show that the dark Newton's rings have radii given approximately by

$$r \approx \sqrt{\frac{m\lambda R}{n_{\text{film}}}}$$

where m is an integer and r is much less than R.

Transmitter   Receiver

Figure P37.59

63. In a Newton's-rings experiment, a plano-convex glass ($n = 1.52$) lens having diameter 10.0 cm is placed on a flat plate as shown in Figure 37.18a. When 650-nm light is incident normally, 55 bright rings are observed with the last one right on the edge of the lens. (a) What is the radius of curvature of the convex surface of the lens? (b) What is the focal length of the lens?

64. A plano-concave lens having index of refraction 1.50 is placed on a flat glass plate, as shown in Figure P37.64. Its curved surface, with radius of curvature 8.00 m, is on the bottom. The lens is illuminated from above with yellow sodium light of wavelength 589 nm, and a series of concentric bright and dark rings is observed by reflection. The interference pattern has a dark spot at the center, surrounded by 50 dark rings, of which the largest is at the outer edge of the lens. (a) What is the thickness of the air layer at the center of the interference pattern? (b) Calculate the radius of the outermost dark ring. (c) Find the focal length of the lens.

Figure P37.64

65. A plano-convex lens having a radius of curvature of $r = 4.00$ m is placed on a concave glass surface whose radius of curvature is $R = 12.0$ m, as shown in Figure P37.65. Determine the radius of the 100th bright ring, assuming 500-nm light is incident normal to the flat surface of the lens.

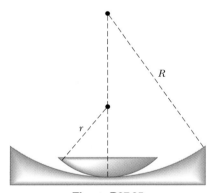

Figure P37.65

66. Use phasor addition to find the resultant amplitude and phase constant when the following three harmonic functions are combined: $E_1 = \sin(\omega t + \pi/6)$, $E_2 = 3.0 \sin(\omega t + 7\pi/2)$, and $E_3 = 6.0 \sin(\omega t + 4\pi/3)$.

67. A soap film ($n = 1.33$) is contained within a rectangular wire frame. The frame is held vertically so that the film drains downward and forms a wedge with flat faces. The thickness of the film at the top is essentially zero. The film is viewed in reflected white light with near-normal incidence, and the first violet ($\lambda = 420$ nm) interference band is observed 3.00 cm from the top edge of the film.

(a) Locate the first red ($\lambda = 680$ nm) interference band. (b) Determine the film thickness at the positions of the violet and red bands. (c) What is the wedge angle of the film?

68. Compact disc (CD) and digital video disc (DVD) players use interference to generate a strong signal from a tiny bump. The depth of a pit is chosen to be one quarter of the wavelength of the laser light used to read the disc. Then light reflected from the pit and light reflected from the adjoining flat differ in path length traveled by one-half wavelength, to interfere destructively at the detector. As the disc rotates, the light intensity drops significantly every time light is reflected from near a pit edge. The space between the leading and trailing edges of a pit determines the time between the fluctuations. The series of time intervals is decoded into a series of zeros and ones that carries the stored information. Assume that infrared light with a wavelength of 780 nm in vacuum is used in a CD player. The disc is coated with plastic having an index of refraction of 1.50. What should be the depth of each pit? A DVD player uses light of a shorter wavelength, and the pit dimensions are correspondingly smaller. This is one factor resulting in greater storage capacity on a DVD compared to a CD.

69. Interference fringes are produced using Lloyd's mirror and a 606-nm source as shown in Figure 37.15. Fringes 1.20 mm apart are formed on a screen 2.00 m from the real source S. Find the vertical distance h of the source above the reflecting surface.

70. Monochromatic light of wavelength 620 nm passes through a very narrow slit S and then strikes a screen in which are two parallel slits, S_1 and S_2, as in Figure P37.70. Slit S_1 is directly in line with S and at a distance of $L = 1.20$ m away from S, whereas S_2 is displaced a distance d to one side. The light is detected at point P on a second screen, equidistant from S_1 and S_2. When either one of the slits S_1 and S_2 is open, equal light intensities are measured at point P. When both are open, the intensity is three times larger. Find the minimum possible value for the slit separation d.

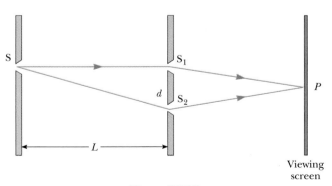

Figure P37.70

71. Slit 1 of a double slit is wider than slit 2, so that the light from 1 has an amplitude 3.00 times that of the light from 2. Show that for this situation, Equation 37.11 is replaced by the equation $I = (4I_{max}/9)(1 + 3 \cos^2 \phi/2)$.

Answers to Quick Quizzes

37.1 (b). The geometrical construction shown in Figure 37.5 is important for developing the mathematical description of interference. It is subject to misinterpretation, however, as it might suggest that the interference can only occur at the position of the screen. A better diagram for this situation is Figure 37.2, which shows *paths* of destructive and constructive interference all the way from the slits to the screen. These paths would be made visible by the smoke.

37.2 (c). Equation 37.5, which shows positions y proportional to order number m, is only valid for small angles.

37.3 (c). Equation 37.5 shows that decreasing λ or L will bring the fringes closer together. Immersing the apparatus in water decreases the wavelength so that the fringes move closer together.

37.4 (c). Conservation of energy cannot be violated. While there is no energy arriving at the location of a dark fringe, there is more energy arriving at the location of a bright fringe than there would be without the double slit.

37.5 The graph is shown in the next column. The width of the primary maxima is slightly narrower than the $N = 5$ primary width but wider than the $N = 10$ primary width. Because $N = 6$, the secondary maxima are $\frac{1}{36}$ as intense as the primary maxima.

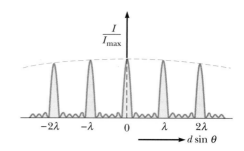

37.6 (a). One of the materials has a higher index of refraction than water, the other lower. For the material with a higher index of refraction, there is a 180° phase shift for the light reflected from the upper surface, but no such phase change from the lower surface, because the index of refraction for water on the other side is lower than that of the film. Thus, the two reflections are out of phase and interfere destructively.

37.7 (a). At the left edge, the air wedge has zero thickness and the only contribution to the interference is the 180° phase shift as the light reflects from the upper surface of the glass slide.

Diffraction Patterns and Polarization

▲ *The Hubble Space Telescope does its viewing above the atmosphere and does not suffer from the atmospheric blurring, caused by air turbulence, that plagues ground-based telescopes. Despite this advantage, it does have limitations due to diffraction effects. In this chapter we show how the wave nature of light limits the ability of any optical system to distinguish between closely spaced objects. (©Denis Scott/CORBIS)*

When plane light waves pass through a small aperture in an opaque barrier, the aperture acts as if it were a point source of light, with waves entering the shadow region behind the barrier. This phenomenon, known as diffraction, can be described only with a wave model for light, as discussed in Section 35.3. In this chapter, we investigate the features of the *diffraction pattern* that occurs when the light from the aperture is allowed to fall upon a screen.

In Chapter 34, we learned that electromagnetic waves are transverse. That is, the electric and magnetic field vectors associated with electromagnetic waves are perpendicular to the direction of wave propagation. In this chapter, we show that under certain conditions these transverse waves with electric field vectors in all possible transverse directions can be *polarized* in various ways. This means that only certain directions of the electric field vectors are present in the polarized wave.

38.1 Introduction to Diffraction Patterns

In Section 35.3 we discussed the fact that light of wavelength comparable to or larger than the width of a slit spreads out in all forward directions upon passing through the slit. We call this phenomenon *diffraction*. This behavior indicates that light, once it has passed through a narrow slit, spreads beyond the narrow path defined by the slit into regions that would be in shadow if light traveled in straight lines. Other waves, such as sound waves and water waves, also have this property of spreading when passing through apertures or by sharp edges.

Figure 38.1 The diffraction pattern that appears on a screen when light passes through a narrow vertical slit. The pattern consists of a broad central fringe and a series of less intense and narrower side fringes.

We might expect that the light passing through a small opening would simply result in a broad region of light on a screen, due to the spreading of the light as it passes through the opening. We find something more interesting, however. A **diffraction pattern** consisting of light and dark areas is observed, somewhat similar to the interference patterns discussed earlier. For example, when a narrow slit is placed between a distant light source (or a laser beam) and a screen, the light produces a diffraction pattern like that in Figure 38.1. The pattern consists of a broad, intense central band (called the **central maximum**), flanked by a series of narrower, less intense additional bands (called **side maxima** or **secondary maxima**) and a series of intervening dark bands (or **minima**). Figure 38.2 shows a diffraction pattern associated with light passing by the edge of an object. Again we see bright and dark fringes, which is reminiscent of an interference pattern.

Figure 38.3 shows a diffraction pattern associated with the shadow of a penny. A bright spot occurs at the center, and circular fringes extend outward from the shadow's edge. We can explain the central bright spot only by using the wave theory of light, which predicts constructive interference at this point. From the viewpoint of geometric optics (in which light is viewed as rays traveling in straight lines), we expect the center of the shadow to be dark because that part of the viewing screen is completely shielded by the penny.

It is interesting to point out an historical incident that occurred shortly before the central bright spot was first observed. One of the supporters of geometric optics,

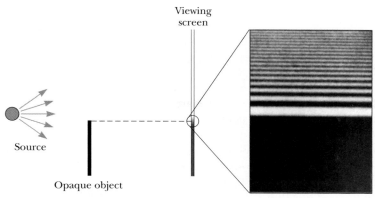

Figure 38.2 Light from a small source passes by the edge of an opaque object and continues on to a screen. A diffraction pattern consisting of bright and dark fringes appears on the screen in the region above the edge of the object.

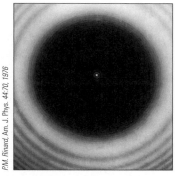

Figure 38.3 Diffraction pattern created by the illumination of a penny, with the penny positioned midway between screen and light source. Note the bright spot at the center.

Simeon Poisson, argued that if Augustin Fresnel's wave theory of light were valid, then a central bright spot should be observed in the shadow of a circular object illuminated by a point source of light. To Poisson's astonishment, the spot was observed by Dominique Arago shortly thereafter. Thus, Poisson's prediction reinforced the wave theory rather than disproving it.

38.2 Diffraction Patterns from Narrow Slits

Let us consider a common situation, that of light passing through a narrow opening modeled as a slit, and projected onto a screen. To simplify our analysis, we assume that the observing screen is far from the slit, so that the rays reaching the screen are approximately parallel. This can also be achieved experimentally by using a converging lens to focus the parallel rays on a nearby screen. In this model, the pattern on the screen is called a **Fraunhofer diffraction pattern.**[1]

Figure 38.4a shows light entering a single slit from the left and diffracting as it propagates toward a screen. Figure 38.4b is a photograph of a single-slit Fraunhofer

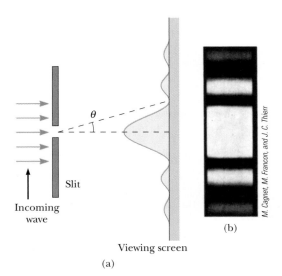

Active Figure 38.4 (a) Fraunhofer diffraction pattern of a single slit. The pattern consists of a central bright fringe flanked by much weaker maxima alternating with dark fringes. (Drawing not to scale.) (b) Photograph of a single-slit Fraunhofer diffraction pattern.

▲ PITFALL PREVENTION

38.1 Diffraction vs. Diffraction Pattern

Diffraction refers to the general behavior of waves spreading out as they pass through a slit. We used diffraction in explaining the existence of an interference pattern in Chapter 37. A *diffraction pattern* is actually a misnomer but is deeply entrenched in the language of physics. The diffraction pattern seen on a screen when a single slit is illuminated is really another interference pattern. The interference is between parts of the incident light illuminating different regions of the slit.

At the Active Figures link at http://www.pse6.com, you can adjust the slit width and the wavelength of the light to see the effect on the diffraction pattern.

[1] If the screen is brought close to the slit (and no lens is used), the pattern is a *Fresnel* diffraction pattern. The Fresnel pattern is more difficult to analyze, so we shall restrict our discussion to Fraunhofer diffraction.

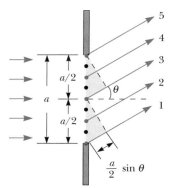

Figure 38.5 Paths of light rays that encounter a narrow slit of width a and diffract toward a screen in the direction described by angle θ. Each portion of the slit acts as a point source of light waves. The path difference between rays 1 and 3, rays 2 and 4, or rays 3 and 5 is $(a/2) \sin \theta$. (Drawing not to scale.)

PITFALL PREVENTION

38.2 Similar Equation Warning!

Equation 38.1 has exactly the same form as Equation 37.2, with d, the slit separation, used in Equation 37.2 and a, the slit width, in Equation 38.1. However, Equation 37.2 describes the *bright* regions in a two-slit interference pattern while Equation 38.1 describes the *dark* regions in a single-slit diffraction pattern. Furthermore, $m = 0$ does not represent a dark fringe in the diffraction pattern.

Condition for destructive interference for a single slit

diffraction pattern. A bright fringe is observed along the axis at $\theta = 0$, with alternating dark and bright fringes on each side of the central bright fringe.

Until now, we have assumed that slits are point sources of light. In this section, we abandon that assumption and see how the finite width of slits is the basis for understanding Fraunhofer diffraction. We can deduce some important features of this phenomenon by examining waves coming from various portions of the slit, as shown in Figure 38.5. According to Huygens's principle, **each portion of the slit acts as a source of light waves.** Hence, light from one portion of the slit can interfere with light from another portion, and the resultant light intensity on a viewing screen depends on the direction θ. Based on this analysis, we recognize that a diffraction pattern is actually an interference pattern, in which the different sources of light are different portions of the single slit!

To analyze the diffraction pattern, it is convenient to divide the slit into two halves, as shown in Figure 38.5. Keeping in mind that all the waves are in phase as they leave the slit, consider rays 1 and 3. As these two rays travel toward a viewing screen far to the right of the figure, ray 1 travels farther than ray 3 by an amount equal to the path difference $(a/2)\sin\theta$, where a is the width of the slit. Similarly, the path difference between rays 2 and 4 is also $(a/2) \sin \theta$, as is that between rays 3 and 5. If this path difference is exactly half a wavelength (corresponding to a phase difference of 180°), then the two waves cancel each other and destructive interference results. If this is true for two such rays, then it is true for any two rays that originate at points separated by half the slit width because the phase difference between two such points is 180°. Therefore, waves from the upper half of the slit interfere destructively with waves from the lower half when

$$\frac{a}{2} \sin \theta = \pm \frac{\lambda}{2}$$

or when

$$\sin \theta = \pm \frac{\lambda}{a}$$

If we divide the slit into four equal parts and use similar reasoning, we find that the viewing screen is also dark when

$$\sin \theta = \pm \frac{2\lambda}{a}$$

Likewise, we can divide the slit into six equal parts and show that darkness occurs on the screen when

$$\sin \theta = \pm \frac{3\lambda}{a}$$

Therefore, the general condition for destructive interference is

$$\sin \theta_{\text{dark}} = m \frac{\lambda}{a} \qquad m = \pm 1, \pm 2, \pm 3, \ldots \qquad (38.1)$$

This equation gives the values of θ_{dark} for which the diffraction pattern has zero light intensity—that is, when a dark fringe is formed. However, it tells us nothing about the variation in light intensity along the screen. The general features of the intensity distribution are shown in Figure 38.6. A broad central bright fringe is observed; this fringe is flanked by much weaker bright fringes alternating with dark fringes. The various dark fringes occur at the values of θ_{dark} that satisfy Equation 38.1. Each bright-fringe peak lies approximately halfway between its bordering dark-fringe minima. Note that the central bright maximum is twice as wide as the secondary maxima.

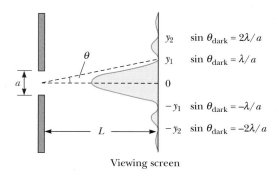

Figure 38.6 Intensity distribution for a Fraunhofer diffraction pattern from a single slit of width a. The positions of two minima on each side of the central maximum are labeled. (Drawing not to scale.)

Quick Quiz 38.1 Suppose the slit width in Figure 38.6 is made half as wide. The central bright fringe (a) becomes wider (b) remains the same (c) becomes narrower.

Quick Quiz 38.2 If a classroom door is open slightly, you can hear sounds coming from the hallway. Yet you cannot see what is happening in the hallway. Why is there this difference? (a) Light waves do not diffract through the single slit of the open doorway. (b) Sound waves can pass through the walls, but light waves cannot. (c) The open door is a small slit for sound waves, but a large slit for light waves. (d) The open door is a large slit for sound waves, but a small slit for light waves.

Example 38.1   Where Are the Dark Fringes?   `Interactive`

Light of wavelength 580 nm is incident on a slit having a width of 0.300 mm. The viewing screen is 2.00 m from the slit. Find the positions of the first dark fringes and the width of the central bright fringe.

Solution The problem statement cues us to conceptualize a single-slit diffraction pattern similar to that in Figure 38.6. We categorize this as a straightforward application of our discussion of single-slit diffraction patterns. To analyze the problem, note that the two dark fringes that flank the central bright fringe correspond to $m = \pm 1$ in Equation 38.1. Hence, we find that

$$\sin \theta_{\text{dark}} = \pm \frac{\lambda}{a} = \pm \frac{5.80 \times 10^{-7}\ \text{m}}{0.300 \times 10^{-3}\ \text{m}} = \pm 1.933 \times 10^{-3}$$

From the triangle in Figure 38.6, note that $\tan \theta_{\text{dark}} = y_1/L$. Because θ_{dark} is very small, we can use the approximation $\sin \theta_{\text{dark}} \approx \tan \theta_{\text{dark}}$; thus, $\sin \theta_{\text{dark}} \approx y_1/L$. Therefore, the positions of the first minima measured from the central axis are given by

$$y_1 \approx L \sin \theta_{\text{dark}} = (2.00\ \text{m})(\pm 1.933 \times 10^{-3})$$

$$= \pm 3.87 \times 10^{-3}\ \text{m}$$

The positive and negative signs correspond to the dark fringes on either side of the central bright fringe. Hence, the width of the central bright fringe is equal to $2|y_1| = 7.74 \times 10^{-3}\ \text{m} = $ 7.74 mm. To finalize this problem,

note that this value is much greater than the width of the slit. We finalize further by exploring what happens if we change the slit width.

What If? What if the slit width is increased by an order of magnitude to 3.00 mm? What happens to the diffraction pattern?

Answer Based on Equation 38.1, we expect that the angles at which the dark bands appear will decrease as a increases. Thus, the diffraction pattern narrows. For $a = 3.00$ mm, the sines of the angles θ_{dark} for the $m = \pm 1$ dark fringes are

$$\sin \theta_{\text{dark}} = \pm \frac{\lambda}{a} = \pm \frac{5.80 \times 10^{-7}\ \text{m}}{3.00 \times 10^{-3}\ \text{m}} = \pm 1.933 \times 10^{-4}$$

The positions of the first minima measured from the central axis are given by

$$y_1 \approx L \sin \theta_{\text{dark}} = (2.00\ \text{m})(\pm 1.933 \times 10^{-4})$$

$$= \pm 3.87 \times 10^{-4}\ \text{m}$$

and the width of the central bright fringe is equal to $2|y_1| = 7.74 \times 10^{-4}\ \text{m} = 0.774$ mm. Notice that this is *smaller* than the width of the slit.

In general, for large values of a, the various maxima and minima are so closely spaced that only a large central bright area resembling the geometric image of the slit is observed. This is very important in the performance of optical instruments such as telescopes.

Investigate the single-slit diffraction pattern at the Interactive Worked Example link at **http://www.pse6.com**.

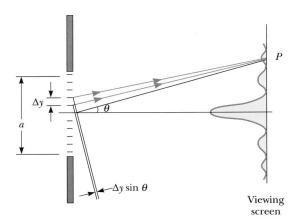

Figure 38.7 Fraunhofer diffraction pattern for a single slit. The light intensity at a distant screen is the resultant of all the incremental electric field magnitudes from zones of width Δy.

Intensity of Single-Slit Diffraction Patterns

We can use phasors to determine the light intensity distribution for a single-slit diffraction pattern. Imagine a slit divided into a large number of small zones, each of width Δy as shown in Figure 38.7. Each zone acts as a source of coherent radiation, and each contributes an incremental electric field of magnitude ΔE at some point on the screen. We obtain the total electric field magnitude E at a point on the screen by summing the contributions from all the zones. The light intensity at this point is proportional to the square of the magnitude of the electric field (Section 37.3).

The incremental electric field magnitudes between adjacent zones are out of phase with one another by an amount $\Delta \beta$, where the phase difference $\Delta \beta$ is related to the path difference $\Delta y \sin \theta$ between adjacent zones by an expression given by an argument similar to that leading to Equation 37.8:

$$\Delta \beta = \frac{2\pi}{\lambda} \Delta y \sin \theta \tag{38.2}$$

To find the magnitude of the total electric field on the screen at any angle θ, we sum the incremental magnitudes ΔE due to each zone. For small values of θ, we can assume that all the ΔE values are the same. It is convenient to use phasor diagrams for various angles, as in Figure 38.8. When $\theta = 0$, all phasors are aligned as in Figure 38.8a because all the waves from the various zones are in phase. In this case, the total electric field at the center of the screen is $E_0 = N \Delta E$, where N is the number of zones. The resultant magnitude E_R at some small angle θ is shown in Figure 38.8b, where each phasor differs in phase from an adjacent one by an amount $\Delta \beta$. In this case, E_R is the

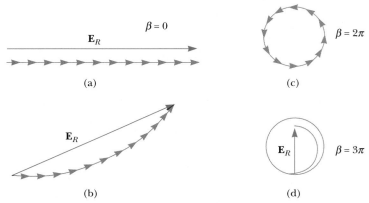

Figure 38.8 Phasor diagrams for obtaining the various maxima and minima of a single-slit diffraction pattern.

vector sum of the incremental magnitudes and hence is given by the length of the chord. Therefore, $E_R < E_0$. The total phase difference β between waves from the top and bottom portions of the slit is

$$\beta = N\,\Delta\beta = \frac{2\pi}{\lambda}\, N\,\Delta y \sin\theta = \frac{2\pi}{\lambda}\, a \sin\theta \qquad (38.3)$$

where $a = N\,\Delta y$ is the width of the slit.

As θ increases, the chain of phasors eventually forms the closed path shown in Figure 38.8c. At this point, the vector sum is zero, and so $E_R = 0$, corresponding to the first minimum on the screen. Noting that $\beta = N\,\Delta\beta = 2\pi$ in this situation, we see from Equation 38.3 that

$$2\pi = \frac{2\pi}{\lambda}\, a \sin\theta_{\text{dark}}$$

$$\sin\theta_{\text{dark}} = \frac{\lambda}{a}$$

That is, the first minimum in the diffraction pattern occurs where $\sin\theta_{\text{dark}} = \lambda/a$; this is in agreement with Equation 38.1.

At larger values of θ, the spiral chain of phasors tightens. For example, Figure 38.8d represents the situation corresponding to the second maximum, which occurs when $\beta = 360° + 180° = 540°$ (3π rad). The second minimum (two complete circles, not shown) corresponds to $\beta = 720°$ (4π rad), which satisfies the condition $\sin\theta_{\text{dark}} = 2\lambda/a$.

We can obtain the total electric-field magnitude E_R and light intensity I at any point on the screen in Figure 38.7 by considering the limiting case in which Δy becomes infinitesimal (dy) and N approaches ∞. In this limit, the phasor chains in Figure 38.8 become the curve of Figure 38.9. The arc length of the curve is E_0 because it is the sum of the magnitudes of the phasors (which is the total electric field magnitude at the center of the screen). From this figure, we see that at some angle θ, the resultant electric field magnitude E_R on the screen is equal to the chord length. From the triangle containing the angle $\beta/2$, we see that

$$\sin\frac{\beta}{2} = \frac{E_R/2}{R}$$

where R is the radius of curvature. But the arc length E_0 is equal to the product $R\beta$, where β is measured in radians. Combining this information with the previous expression gives

$$E_R = 2R \sin\frac{\beta}{2} = 2\left(\frac{E_0}{\beta}\right) \sin\frac{\beta}{2} = E_0 \left[\frac{\sin(\beta/2)}{\beta/2}\right]$$

Because the resultant light intensity I at a point on the screen is proportional to the square of the magnitude E_R, we find that

$$I = I_{\text{max}} \left[\frac{\sin(\beta/2)}{\beta/2}\right]^2 \qquad (38.4)$$

where I_{max} is the intensity at $\theta = 0$ (the central maximum). Substituting the expression for β (Eq. 38.3) into Equation 38.4, we have

$$I = I_{\text{max}} \left[\frac{\sin(\pi a \sin\theta/\lambda)}{\pi a \sin\theta/\lambda}\right]^2 \qquad (38.5)$$

From this result, we see that *minima* occur when

$$\frac{\pi a \sin\theta_{\text{dark}}}{\lambda} = m\pi$$

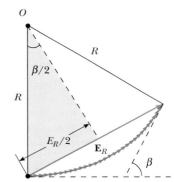

Figure 38.9 Phasor diagram for a large number of coherent sources. All the ends of the phasors lie on the circular arc of radius R. The resultant electric field magnitude E_R equals the length of the chord.

Intensity of a single-slit Fraunhofer diffraction pattern

(b)

Figure 38.10 (a) A plot of light intensity I versus $\beta/2$ for the single-slit Fraunhofer diffraction pattern. (b) Photograph of a single-slit Fraunhofer diffraction pattern.

M. Cagnet, M. Francon, and J. C. Thierr

or

Condition for intensity minima for a single slit

$$\sin \theta_{\text{dark}} = m \frac{\lambda}{a} \qquad m = \pm 1, \pm 2, \pm 3, \ldots$$

in agreement with Equation 38.1.

Figure 38.10a represents a plot of Equation 38.4, and Figure 38.10b is a photograph of a single-slit Fraunhofer diffraction pattern. Note that most of the light intensity is concentrated in the central bright fringe.

Example 38.2 Relative Intensities of the Maxima

Find the ratio of the intensities of the secondary maxima to the intensity of the central maximum for the single-slit Fraunhofer diffraction pattern.

Solution To a good approximation, the secondary maxima lie midway between the zero points. From Figure 38.10a, we see that this corresponds to $\beta/2$ values of $3\pi/2$, $5\pi/2$, $7\pi/2$, ... Substituting these values into Equation 38.4 gives for the first two ratios

$$\frac{I_1}{I_{\text{max}}} = \left[\frac{\sin(3\pi/2)}{(3\pi/2)} \right]^2 = \frac{1}{9\pi^2/4} = \boxed{0.045}$$

$$\frac{I_2}{I_{\text{max}}} = \left[\frac{\sin(5\pi/2)}{5\pi/2} \right]^2 = \frac{1}{25\pi^2/4} = \boxed{0.016}$$

That is, the first secondary maxima (the ones adjacent to the central maximum) have an intensity of 4.5% that of the central maximum, and the next secondary maxima have an intensity of 1.6% that of the central maximum.

Intensity of Two-Slit Diffraction Patterns

When more than one slit is present, we must consider not only diffraction patterns due to the individual slits but also the interference patterns due to the waves coming from different slits. Notice the curved dashed lines in Figure 37.14, which indicate a decrease in intensity of the interference maxima as θ increases. This decrease is due to a diffraction pattern. To determine the effects of both two-slit interference and a single-slit diffraction pattern from each slit, we combine Equations 37.12 and 38.5:

$$I = I_{\text{max}} \cos^2 \left(\frac{\pi d \sin \theta}{\lambda} \right) \left[\frac{\sin(\pi a \sin \theta/\lambda)}{\pi a \sin \theta/\lambda} \right]^2 \qquad (38.6)$$

Although this expression looks complicated, it merely represents the single-slit diffraction pattern (the factor in square brackets) acting as an "envelope" for a two-slit

Active Figure 38.11 The combined effects of two-slit and single-slit interference. This is the pattern produced when 650-nm light waves pass through two 3.0-μm slits that are 18 μm apart. Notice how the diffraction pattern acts as an "envelope" and controls the intensity of the regularly spaced interference maxima.

Courtesy of Central Scientific Company

At the Active Figures link at **http://www.pse6.com,** *you can adjust the slit width, slit separation, and the wavelength of the light to see the effect on the interference pattern.*

interference pattern (the cosine-squared factor), as shown in Figure 38.11. The broken blue curve in Figure 38.11 represents the factor in square brackets in Equation 38.6. The cosine-squared factor by itself would give a series of peaks all with the same height as the highest peak of the red-brown curve in Figure 38.11. Because of the effect of the square-bracket factor, however, these peaks vary in height as shown.

Equation 37.2 indicates the conditions for interference maxima as $d \sin\theta = m\lambda$, where d is the distance between the two slits. Equation 38.1 specifies that the first diffraction minimum occurs when $a \sin\theta = \lambda$, where a is the slit width. Dividing Equation 37.2 by Equation 38.1 (with $m = 1$) allows us to determine which interference maximum coincides with the first diffraction minimum:

$$\frac{d \sin\theta}{a \sin\theta} = \frac{m\lambda}{\lambda}$$

$$\frac{d}{a} = m \tag{38.7}$$

In Figure 38.11, $d/a = 18\ \mu\text{m}/3.0\ \mu\text{m} = 6$. Therefore, the sixth interference maximum (if we count the central maximum as $m = 0$) is aligned with the first diffraction minimum and cannot be seen.

Quick Quiz 38.3 Using Figure 38.11 as a starting point, make a sketch of the combined diffraction and interference pattern for 650-nm light waves striking two 3.0-μm slits located 9.0 μm apart.

Quick Quiz 38.4 Consider the central peak in the diffraction envelope in Figure 38.11. Suppose the wavelength of the light is changed to 450 nm. What happens to this central peak? (a) The width of the peak decreases and the number of interference fringes it encloses decreases. (b) The width of the peak decreases and the number of interference fringes it encloses increases. (c) The width of the peak decreases and the number of interference fringes it encloses remains the same. (d) The width of the peak increases and the number of interference fringes it encloses decreases. (e) The width of the peak increases and the number of interference fringes it encloses increases. (f) The width of the peak increases and the number of interference fringes it encloses remains the same.

38.3 Resolution of Single-Slit and Circular Apertures

The ability of optical systems to distinguish between closely spaced objects is limited because of the wave nature of light. To understand this difficulty, consider Figure 38.12, which shows two light sources far from a narrow slit of width a. The sources can be two noncoherent point sources S_1 and S_2—for example, they could be two distant stars. If no interference occurred between light passing through different parts of the slit, two distinct bright spots (or images) would be observed on the viewing screen. However, because of such interference, each source is imaged as a bright central region flanked by weaker bright and dark fringes—a diffraction pattern. What is observed on the screen is the sum of two diffraction patterns: one from S_1, and the other from S_2.

If the two sources are far enough apart to keep their central maxima from overlapping as in Figure 38.12a, their images can be distinguished and are said to be *resolved*. If the sources are close together, however, as in Figure 38.12b, the two central maxima overlap, and the images are not resolved. To determine whether two images are resolved, the following condition is often used:

When the central maximum of one image falls on the first minimum of another image, the images are said to be just resolved. This limiting condition of resolution is known as **Rayleigh's criterion.**

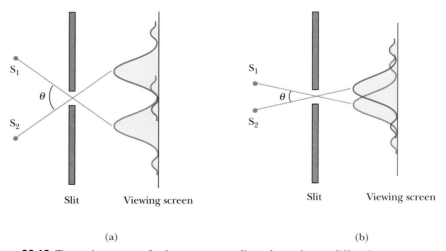

| (a) | (b) |

Figure 38.12 Two point sources far from a narrow slit each produce a diffraction pattern. (a) The angle subtended by the sources at the slit is large enough for the diffraction patterns to be distinguishable. (b) The angle subtended by the sources is so small that their diffraction patterns overlap, and the images are not well resolved. (Note that the angles are greatly exaggerated. The drawing is not to scale.)

From Rayleigh's criterion, we can determine the minimum angular separation $\theta_{\min}$ subtended by the sources at the slit in Figure 38.12 for which the images are just resolved. Equation 38.1 indicates that the first minimum in a single-slit diffraction pattern occurs at the angle for which

$$\sin\theta = \frac{\lambda}{a}$$

where a is the width of the slit. According to Rayleigh's criterion, this expression gives the smallest angular separation for which the two images are resolved. Because $\lambda \ll a$ in most situations, $\sin\theta$ is small, and we can use the approximation $\sin\theta \approx \theta$. Therefore, the limiting angle of resolution for a slit of width a is

$$\theta_{\min} = \frac{\lambda}{a} \tag{38.8}$$

where $\theta_{\min}$ is expressed in radians. Hence, the angle subtended by the two sources at the slit must be greater than λ/a if the images are to be resolved.

Many optical systems use circular apertures rather than slits. The diffraction pattern of a circular aperture, as shown in the lower half of Figure 38.13, consists of a central circular bright disk surrounded by progressively fainter bright and dark rings. Figure 38.13 shows diffraction patterns for three situations in which light from two point sources passes through a circular aperture. When the sources are far apart, their images are well resolved (Fig. 38.13a). When the angular separation of the sources satisfies Rayleigh's criterion, the images are just resolved (Fig. 38.13b). Finally, when the sources are close together, the images are said to be unresolved (Fig. 38.13c).

M. Cagnet, M. Francon, and J. C. Thierr

(a)   (b)   (c)

Figure 38.13 Individual diffraction patterns of two point sources (solid curves) and the resultant patterns (dashed curves) for various angular separations of the sources. In each case, the dashed curve is the sum of the two solid curves. (a) The sources are far apart, and the patterns are well resolved. (b) The sources are closer together such that the angular separation just satisfies Rayleigh's criterion, and the patterns are just resolved. (c) The sources are so close together that the patterns are not resolved.

Analysis shows that the limiting angle of resolution of the circular aperture is

$$\theta_{min} = 1.22 \frac{\lambda}{D} \tag{38.9}$$

where D is the diameter of the aperture. Note that this expression is similar to Equation 38.8 except for the factor 1.22, which arises from a mathematical analysis of diffraction from the circular aperture.

Quick Quiz 38.5 Cat's eyes have pupils that can be modeled as vertical slits. At night, would cats be more successful in resolving (a) headlights on a distant car, or (b) vertically-separated lights on the mast of a distant boat?

Quick Quiz 38.6 Suppose you are observing a binary star with a telescope and are having difficulty resolving the two stars. You decide to use a colored filter to maximize the resolution. (A filter of a given color transmits only that color of light.) What color filter should you choose? (a) blue (b) green (c) yellow (d) red.

Example 38.3 Limiting Resolution of a Microscope

Light of wavelength 589 nm is used to view an object under a microscope. If the aperture of the objective has a diameter of 0.900 cm,

(A) what is the limiting angle of resolution?

Solution Using Equation 38.9, we find that the limiting angle of resolution is

$$\theta_{min} = 1.22 \left(\frac{589 \times 10^{-9} \text{ m}}{0.900 \times 10^{-2} \text{ m}} \right) = \boxed{7.98 \times 10^{-5} \text{ rad}}$$

This means that any two points on the object subtending an angle smaller than this at the objective cannot be distinguished in the image.

(B) If it were possible to use visible light of any wavelength, what would be the maximum limit of resolution for this microscope?

Solution To obtain the smallest limiting angle, we have to use the shortest wavelength available in the visible spectrum. Violet light (400 nm) gives a limiting angle of resolution of

$$\theta_{min} = 1.22 \left(\frac{400 \times 10^{-9} \text{ m}}{0.900 \times 10^{-2} \text{ m}} \right) = \boxed{5.42 \times 10^{-5} \text{ rad}}$$

What If? Suppose that water ($n = 1.33$) fills the space between the object and the objective. What effect does this have on resolving power when 589-nm light is used?

Answer Because light travels more slowly in water, we know that the wavelength of the light in water is smaller than that in vacuum. Based on Equation 38.9, we expect the limiting angle of resolution to be smaller. To find the new value of the limiting angle of resolution, we first calculate the wavelength of the 589-nm light in water using Equation 35.7:

$$\lambda_{water} = \frac{\lambda_{air}}{n_{water}} = \frac{589 \text{ nm}}{1.33} = 443 \text{ nm}$$

The limiting angle of resolution at this wavelength is

$$\theta_{min} = 1.22 \left(\frac{443 \times 10^{-9} \text{ m}}{0.900 \times 10^{-2} \text{ m}} \right) = \boxed{6.00 \times 10^{-5} \text{ rad}}$$

which is indeed smaller than that calculated in part (A).

Example 38.4 Resolution of the Eye

Estimate the limiting angle of resolution for the human eye, assuming its resolution is limited only by diffraction.

Solution Let us choose a wavelength of 500 nm, near the center of the visible spectrum. Although pupil diameter varies from person to person, we estimate a daytime diameter of 2 mm. We use Equation 38.9, taking $\lambda = 500$ nm

and $D = 2$ mm:

$$\theta_{min} = 1.22 \frac{\lambda}{D} = 1.22 \left(\frac{5.00 \times 10^{-7} \text{ m}}{2 \times 10^{-3} \text{ m}} \right)$$

$$\approx 3 \times 10^{-4} \text{ rad} \approx \boxed{1 \text{ min of arc}}$$

We can use this result to determine the minimum separation distance d between two point sources that the eye can distinguish if they are a distance L from the observer (Fig. 38.14). Because θ_{min} is small, we see that

$$\sin \theta_{min} \approx \theta_{min} \approx \frac{d}{L}$$

$$d = L\theta_{min}$$

For example, if the point sources are 25 cm from the eye (the near point), then

$$d = (25 \text{ cm})(3 \times 10^{-4} \text{ rad}) = 8 \times 10^{-3} \text{ cm}$$

This is approximately equal to the thickness of a human hair.

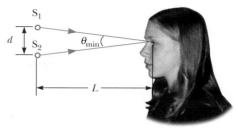

Figure 38.14 (Example 38.4) Two point sources separated by a distance d as observed by the eye.

Example 38.5   Resolution of a Telescope

The Keck telescope at Mauna Kea, Hawaii, has an effective diameter of 10 m. What is its limiting angle of resolution for 600-nm light?

Solution Because $D = 10$ m and $\lambda = 6.00 \times 10^{-7}$ m, Equation 38.9 gives

$$\theta_{min} = 1.22 \frac{\lambda}{D} = 1.22 \left(\frac{6.00 \times 10^{-7} \text{ m}}{10 \text{ m}} \right)$$

$$= 7.3 \times 10^{-8} \text{ rad} \approx \boxed{0.015 \text{ s of arc}}$$

Any two stars that subtend an angle greater than or equal to this value are resolved (if atmospheric conditions are ideal).

The Keck telescope can never reach its diffraction limit because the limiting angle of resolution is always set by atmospheric blurring at optical wavelengths. This seeing limit is usually about 1 s of arc and is never smaller than about 0.1 s of arc. (This is one of the reasons for the superiority of photographs from the Hubble Space Telescope, which views celestial objects from an orbital position above the atmosphere.)

What If? What if we consider radio telescopes? These are much larger in diameter than optical telescopes, but do they have angular resolutions that are better than optical telescopes? For example, the radio telescope at Arecibo, Puerto Rico, has a diameter of 305 m and is designed to detect radio waves of 0.75-m wavelength. How does its resolution compare to that of the Keck telescope?

Answer The increase in diameter might suggest that radio telescopes would have better resolution, but Equation 38.9 shows that θ_{min} depends on *both* diameter and wavelength. Calculating the minimum angle of resolution for the radio telescope, we find

$$\theta_{min} = 1.22 \frac{\lambda}{D} = 1.22 \left(\frac{0.75 \text{ m}}{305 \text{ m}} \right)$$

$$= 3.0 \times 10^{-3} \text{ rad} \approx 10 \text{ min of arc}$$

Notice that this limiting angle of resolution is measured in *minutes* of arc rather than the *seconds* of arc for the optical telescope. Thus, the change in wavelength more than compensates for the increase in diameter, and the limiting angle of resolution for the Arecibo radio telescope is more than 40 000 times larger (that is, *worse*) than the Keck minimum.

As an example of the effects of atmospheric blurring mentioned in Example 38.5, consider telescopic images of Pluto and its moon Charon. Figure 38.15a shows the image taken in 1978 that represents the discovery of Charon. In this photograph taken from an Earth-based telescope, atmospheric turbulence causes the image of Charon to appear only as a bump on the edge of Pluto. In comparison, Figure 38.15b shows a photograph taken with the Hubble Space Telescope. Without the problems of atmospheric turbulence, Pluto and its moon are clearly resolved.

38.4 The Diffraction Grating

The **diffraction grating,** a useful device for analyzing light sources, consists of a large number of equally spaced parallel slits. A *transmission grating* can be made by cutting parallel grooves on a glass plate with a precision ruling machine. The spaces between the grooves are transparent to the light and hence act as separate slits. A *reflection grating* can

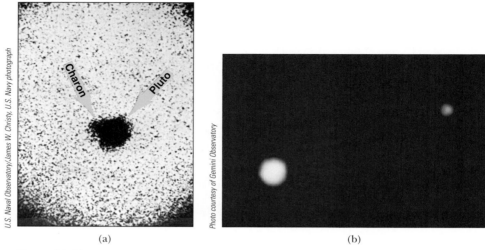

Figure 38.15 (a) The photograph on which Charon, the moon of Pluto, was discovered in 1978. From an Earth-based telescope, atmospheric blurring results in Charon appearing only as a subtle bump on the edge of Pluto. (b) A Hubble Space Telescope photo of Pluto and Charon, clearly resolving the two objects.

be made by cutting parallel grooves on the surface of a reflective material. The reflection of light from the spaces between the grooves is specular, and the reflection from the grooves cut into the material is diffuse. Thus, the spaces between the grooves act as parallel sources of reflected light, like the slits in a transmission grating. Current technology can produce gratings that have very small slit spacings. For example, a typical grating ruled with 5 000 grooves/cm has a slit spacing $d = (1/5\,000)$ cm $= 2.00 \times 10^{-4}$ cm.

A section of a diffraction grating is illustrated in Figure 38.16. A plane wave is incident from the left, normal to the plane of the grating. The pattern observed on the

▲ **PITFALL PREVENTION**

38.3 A Diffraction Grating Is an Interference Grating

As with *diffraction pattern*, *diffraction grating* is a misnomer, but is deeply entrenched in the language of physics. The diffraction grating depends on diffraction in the same way as the double slit—spreading the light so that light from different slits can interfere. It would be more correct to call it an *interference grating*, but *diffraction grating* is the name in use.

Incoming plane wave of light

P First-order maximum ($m = 1$)

Central or zeroth-order maximum ($m = 0$)

First-order maximum ($m = -1$)
P

Diffraction grating

θ

d θ

$\delta = d \sin \theta$

Figure 38.16 Side view of a diffraction grating. The slit separation is d, and the path difference between adjacent slits is $d \sin \theta$.

screen (far to the right of Figure 38.16) is the result of the combined effects of interference and diffraction. Each slit produces diffraction, and the diffracted beams interfere with one another to produce the final pattern.

The waves from all slits are in phase as they leave the slits. However, for some arbitrary direction θ measured from the horizontal, the waves must travel different path lengths before reaching the screen. From Figure 38.16, note that the path difference δ between rays from any two adjacent slits is equal to $d \sin \theta$. If this path difference equals one wavelength or some integral multiple of a wavelength, then waves from all slits are in phase at the screen and a bright fringe is observed. Therefore, the condition for *maxima* in the interference pattern at the angle θ_{bright} is

$$d \sin \theta_{bright} = m\lambda \qquad m = 0, \pm 1, \pm 2, \pm 3, \ldots \qquad (38.10)$$

Condition for interference maxima for a grating

We can use this expression to calculate the wavelength if we know the grating spacing d and the angle θ_{bright}. If the incident radiation contains several wavelengths, the mth-order maximum for each wavelength occurs at a specific angle. All wavelengths are seen at $\theta = 0$, corresponding to $m = 0$, the zeroth-order maximum. The first-order maximum ($m = 1$) is observed at an angle that satisfies the relationship $\sin \theta_{bright} = \lambda / d$; the second-order maximum ($m = 2$) is observed at a larger angle θ_{bright}, and so on.

The intensity distribution for a diffraction grating obtained with the use of a monochromatic source is shown in Figure 38.17. Note the sharpness of the principal maxima and the broadness of the dark areas. This is in contrast to the broad bright fringes characteristic of the two-slit interference pattern (see Fig. 37.7). You should also review Figure 37.14, which shows that the width of the intensity maxima decreases as the number of slits increases. Because the principal maxima are so sharp, they are much brighter than two-slit interference maxima.

A schematic drawing of a simple apparatus used to measure angles in a diffraction pattern is shown in Figure 38.18. This apparatus is a *diffraction grating spectrometer*. The light to be analyzed passes through a slit, and a collimated beam of light is incident on the grating. The diffracted light leaves the grating at angles that satisfy Equation 38.10, and a telescope is used to view the image of the slit. The wavelength can be determined by measuring the precise angles at which the images of the slit appear for the various orders.

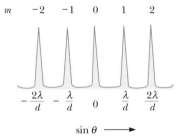

Active Figure 38.17 Intensity versus $\sin \theta$ for a diffraction grating. The zeroth-, first-, and second-order maxima are shown.

At the Active Figures link at http://www.pse6.com, *you can choose the number of slits to be illuminated to see the effect on the interference pattern.*

Active Figure 38.18 Diagram of a diffraction grating spectrometer. The collimated beam incident on the grating is spread into its various wavelength components with constructive interference for a particular wavelength occurring at the angles θ_{bright} that satisfy the equation $d \sin \theta_{bright} = m\lambda$, where $m = 0, 1, 2, \ldots$.

Use the spectrometer at the Active Figures link at http://www.pse6.com *to observe constructive interference for various wavelengths.*

Figure 38.19 A small portion of a grating light valve. The alternating reflective ribbons at different levels act as a diffraction grating, offering very-high-speed control of the direction of light toward a digital display device.

Silicon Light Machines

The spectrometer is a useful tool in *atomic spectroscopy*, in which the light from an atom is analyzed to find the wavelength components. These wavelength components can be used to identify the atom. We will investigate atomic spectra in Chapter 42 of the extended version of this text.

Another application of diffraction gratings is in the recently developed *grating light valve* (GLV), which may compete in the near future in video projection with the digital micromirror devices (DMDs) discussed in Section 35.4. The grating light valve consists of a silicon microchip fitted with an array of parallel silicon nitride ribbons coated with a thin layer of aluminum (Fig. 38.19). Each ribbon is about 20 μm long and about 5 μm wide and is separated from the silicon substrate by an air gap on the order of 100 nm. With no voltage applied, all ribbons are at the same level. In this situation, the array of ribbons acts as a flat surface, specularly reflecting incident light.

When a voltage is applied between a ribbon and the electrode on the silicon substrate, an electric force pulls the ribbon downward, closer to the substrate. Alternate ribbons can be pulled down, while those in between remain in the higher configuration. As a result, the array of ribbons acts as a diffraction grating, such that the constructive interference for a particular wavelength of light can be directed toward a screen or other optical display system. By using three such devices, one each for red, blue, and green light, full-color display is possible.

The GLV tends to be simpler to fabricate and higher in resolution than comparable DMD devices. On the other hand, DMD devices have already made an entry into the market. It will be interesting to watch this technology competition in future years.

Quick Quiz 38.7 If laser light is reflected from a phonograph record or a compact disc, a diffraction pattern appears. This is due to the fact that both devices contain parallel tracks of information that act as a reflection diffraction grating. Which device, (a) record or (b) compact disc, results in diffraction maxima that are farther apart in angle?

Quick Quiz 38.8 Ultraviolet light of wavelength 350 nm is incident on a diffraction grating with slit spacing d and forms an interference pattern on a screen a distance L away. The angular positions θ_{bright} of the interference maxima are large. The locations of the bright fringes are marked on the screen. Now red light of wavelength 700 nm is used with a diffraction grating to form another diffraction pattern on the screen. The bright fringes of this pattern will be located at the marks on the screen if

(a) the screen is moved to a distance $2L$ from the grating (b) the screen is moved to a distance $L/2$ from the grating (c) the grating is replaced with one of slit spacing $2d$ (d) the grating is replaced with one of slit spacing $d/2$ (e) nothing is changed.

Conceptual Example 38.6   A Compact Disc Is a Diffraction Grating

Light reflected from the surface of a compact disc is multicolored, as shown in Figure 38.20. The colors and their intensities depend on the orientation of the disc relative to the eye and relative to the light source. Explain how this works.

Solution The surface of a compact disc has a spiral grooved track (with adjacent grooves having a separation on the order of 1 μm). Thus, the surface acts as a reflection grating. The light reflecting from the regions between these closely spaced grooves interferes constructively only in certain directions that depend on the wavelength and on the direction of the incident light. Any section of the disc serves as a diffraction grating for white light, sending different colors in different directions. The different colors you see when viewing one section change as the light source, the disc, or you move to change the angles of incidence or diffraction.

Figure 38.20 (Conceptual Example 38.6) A compact disc observed under white light. The colors observed in the reflected light and their intensities depend on the orientation of the disc relative to the eye and relative to the light source.

Example 38.7   The Orders of a Diffraction Grating

Interactive

Monochromatic light from a helium–neon laser ($\lambda = 632.8$ nm) is incident normally on a diffraction grating containing 6 000 grooves per centimeter. Find the angles at which the first- and second-order maxima are observed.

Solution First, we must calculate the slit separation, which is equal to the inverse of the number of grooves per centimeter:

$$d = \frac{1}{6\ 000}\ \text{cm} = 1.667 \times 10^{-4}\ \text{cm} = 1\ 667\ \text{nm}$$

For the first-order maximum ($m = 1$), we obtain

$$\sin \theta_1 = \frac{\lambda}{d} = \frac{632.8\ \text{nm}}{1\ 667\ \text{nm}} = 0.379\ 6$$

$$\theta_1 = \boxed{22.31°}$$

For the second-order maximum ($m = 2$), we find

$$\sin \theta_2 = \frac{2\lambda}{d} = \frac{2(632.8\ \text{nm})}{1\ 667\ \text{nm}} = 0.759\ 2$$

$$\theta_2 = \boxed{49.39°}$$

What If? What if we look for the third-order maximum? Do we find it?

Answer For $m = 3$, we find $\sin \theta_3 = 1.139$. Because $\sin\theta$ cannot exceed unity, this does not represent a realistic solution. Hence, only zeroth-, first-, and second-order maxima are observed for this situation.

 Investigate the interference pattern from a diffraction grating at the Interactive Worked Example link at **http://www.pse6.com.**

Resolving Power of the Diffraction Grating

The diffraction grating is useful for measuring wavelengths accurately. Like the prism, the diffraction grating can be used to separate white light into its wavelength components. Of the two devices, a grating with very small slit separation is more precise if one wants to distinguish two closely spaced wavelengths.

For two nearly equal wavelengths λ_1 and λ_2 between which a diffraction grating can just barely distinguish, the **resolving power** R of the grating is defined as

Resolving power

$$R \equiv \frac{\lambda}{\lambda_2 - \lambda_1} = \frac{\lambda}{\Delta\lambda} \qquad (38.11)$$

where $\lambda = (\lambda_1 + \lambda_2)/2$ and $\Delta\lambda = \lambda_2 - \lambda_1$. Thus, a grating that has a high resolving power can distinguish small differences in wavelength. If N slits of the grating are illuminated, it can be shown that the resolving power in the mth-order diffraction is

Resolving power of a grating

$$R = Nm \qquad (38.12)$$

Thus, resolving power increases with increasing order number and with increasing number of illuminated slits.

Note that $R = 0$ for $m = 0$; this signifies that all wavelengths are indistinguishable for the zeroth-order maximum. However, consider the second-order diffraction pattern ($m = 2$) of a grating that has 5 000 rulings illuminated by the light source. The resolving power of such a grating in second order is $R = 5\,000 \times 2 = 10\,000$. Therefore, for a mean wavelength of, for example, 600 nm, the minimum wavelength separation between two spectral lines that can be just resolved is $\Delta\lambda = \lambda/R = 6.00 \times 10^{-2}$ nm. For the third-order principal maximum, $R = 15\,000$ and $\Delta\lambda = 4.00 \times 10^{-2}$ nm, and so on.

Example 38.8   Resolving Sodium Spectral Lines

When a gaseous element is raised to a very high temperature, the atoms emit radiation having discrete wavelengths. The set of wavelengths for a given element is called its *atomic spectrum* (Chapter 42). Two strong components in the atomic spectrum of sodium have wavelengths of 589.00 nm and 589.59 nm.

(A) What resolving power must a grating have if these wavelengths are to be distinguished?

Solution   Using Equation 38.11,

$$R = \frac{\lambda}{\Delta\lambda} = \frac{589.30 \text{ nm}}{589.59 \text{ nm} - 589.00 \text{ nm}} = \frac{589.30}{0.59} = \boxed{999}$$

(B) To resolve these lines in the second-order spectrum, how many slits of the grating must be illuminated?

Solution   From Equation 38.12 and the result to part (A), we find that

$$N = \frac{R}{m} = \frac{999}{2} = \boxed{500 \text{ slits}}$$

Application   Holography

One interesting application of diffraction gratings is **holography,** the production of three-dimensional images of objects. The physics of holography was developed by Dennis Gabor in 1948, and resulted in the Nobel Prize in physics for Gabor in 1971. The requirement of coherent light for holography, however, delayed the realization of holographic images from Gabor's work until the development of lasers in the 1960s. Figure 38.21 shows a hologram and the three-dimensional character of its image.

Figure 38.22 shows how a hologram is made. Light from the laser is split into two parts by a half-silvered mirror at B. One part of the beam reflects off the object to be photographed and strikes an ordinary photographic film. The other half of the beam is diverged by lens L_2, reflects from mirrors M_1 and M_2, and finally strikes the film. The two beams overlap to form an extremely complicated interference pattern on the film. Such an interference pattern can be produced only if the phase relationship of the two waves is constant throughout the exposure of the film. This condition is met by illuminating the scene with light coming through a pinhole or with coherent laser radiation. The hologram records not only the intensity of the light scattered from the object (as in a conventional photograph), but also the phase difference between the reference beam and the beam scattered from the object. Because of this phase difference, an interference pattern is formed that produces an image in which all three-dimensional information available from the perspective of any point on the hologram is preserved.

In a normal photographic image, a lens is used to focus the image so that each point on the object corresponds to a single point on the film. Notice that there is no lens used in Figure 38.22 to focus the light onto the film. Thus, light from each point on the object reaches *all* points on the film. As a result, each region of the photographic film on which the hologram is recorded contains information about all illuminated points on the object. This leads to a remarkable

Figure 38.21 In this hologram, a circuit board is shown from two different views. Notice the difference in the appearance of the measuring tape and the view through the magnifying lens.

result—if a small section of the hologram is cut from the film, the complete image can be formed from the small piece! (The quality of the image is reduced, but the entire image is present.)

A hologram is best viewed by allowing coherent light to pass through the developed film as one looks back along the direction from which the beam comes. The interference pattern on the film acts as a diffraction grating. Figure 38.23 shows two rays of light striking the film and passing through. For each ray, the $m = 0$ and $m = \pm 1$ rays in the diffraction pattern are shown emerging from the right side of the film. The $m = +1$ rays converge to

form a real image of the scene, which is not the image that is normally viewed. By extending the light rays corresponding to $m = -1$ back behind the film, we see that there is a virtual image located there, with light coming from it in exactly the same way that light came from the actual object when the film was exposed. This is the image that we see by looking through the holographic film.

Holograms are finding a number of applications. You may have a hologram on your credit card. This is a special type of hologram called a *rainbow hologram*, designed to be viewed in reflected white light.

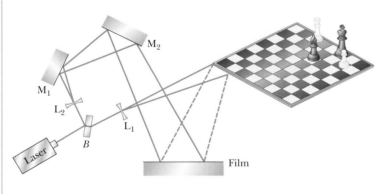

Figure 38.22 Experimental arrangement for producing a hologram.

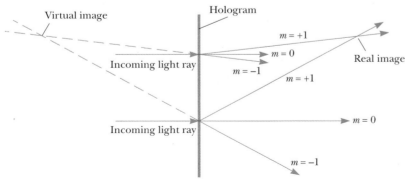

Figure 38.23 Two light rays strike a hologram at normal incidence. For each ray, outgoing rays corresponding to $m = 0$ and $m = \pm 1$ are shown. If the $m = -1$ rays are extended backward, a virtual image of the object photographed in the hologram exists on the front side of the hologram.

38.5 Diffraction of X-Rays by Crystals

In principle, the wavelength of any electromagnetic wave can be determined if a grating of the proper spacing (on the order of λ) is available. X-rays, discovered by Wilhelm Roentgen (1845–1923) in 1895, are electromagnetic waves of very short wavelength (on the order of 0.1 nm). It would be impossible to construct a grating having such a small spacing by the cutting process described at the beginning of Section 38.4. However, the atomic spacing in a solid is known to be about 0.1 nm. In 1913, Max von Laue (1879–1960) suggested that the regular array of atoms in a crystal could act as a three-dimensional diffraction grating for x-rays. Subsequent experiments confirmed this prediction. The diffraction patterns from crystals are complex because of the three-dimensional nature of crystal structure. Nevertheless, x-ray diffraction has proved to be an invaluable technique for elucidating these structures and for understanding the structure of matter.

Figure 38.24 is one experimental arrangement for observing x-ray diffraction from a crystal. A collimated beam of monochromatic x-rays is incident on a crystal. The diffracted beams are very intense in certain directions, corresponding to constructive interference from waves reflected from layers of atoms in the crystal. The diffracted beams, which can be detected by a photographic film, form an array of spots known as a *Laue pattern*, as in Figure 38.25a. One can deduce the crystalline structure by analyzing the positions and intensities of the various spots in the pattern. Fig. 38.25b shows a Laue pattern from a crystalline enzyme, using a wide range of wavelengths so that a swirling pattern results.

The arrangement of atoms in a crystal of sodium chloride (NaCl) is shown in Figure 38.26. Each unit cell (the geometric solid that repeats throughout the crystal) is a cube having an edge length a. A careful examination of the NaCl structure shows that the ions lie in discrete planes (the shaded areas in Fig. 38.26). Now suppose that an incident x-ray beam makes an angle θ with one of the planes, as in Figure 38.27. The beam can be reflected from both the upper plane and the lower one. However,

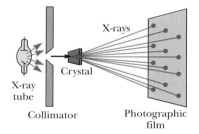

Figure 38.24 Schematic diagram of the technique used to observe the diffraction of x-rays by a crystal. The array of spots formed on the film is called a Laue pattern.

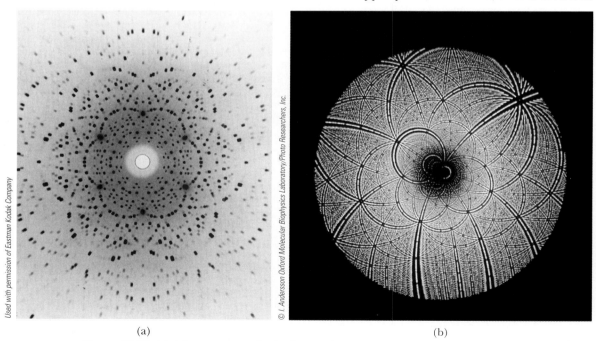

(a)   (b)

Figure 38.25 (a) A Laue pattern of a single crystal of the mineral beryl (beryllium aluminum silicate). Each dot represents a point of constructive interference. (b) A Laue pattern of the enzyme Rubisco, produced with a wide-band x-ray spectrum. This enzyme is present in plants and takes part in the process of photosynthesis. The Laue pattern is used to determine the crystal structure of Rubisco.

Figure 38.26 Crystalline structure of sodium chloride (NaCl). The blue spheres represent Cl⁻ ions, and the red spheres represent Na⁺ ions. The length of the cube edge is $a = 0.562\ 737$ nm.

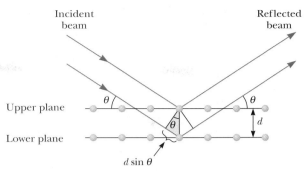

Figure 38.27 A two-dimensional description of the reflection of an x-ray beam from two parallel crystalline planes separated by a distance d. The beam reflected from the lower plane travels farther than the one reflected from the upper plane by a distance $2d \sin \theta$.

the beam reflected from the lower plane travels farther than the beam reflected from the upper plane. The effective path difference is $2d \sin \theta$. The two beams reinforce each other (constructive interference) when this path difference equals some integer multiple of λ. The same is true for reflection from the entire family of parallel planes. Hence, the condition for *constructive* interference (maxima in the reflected beam) is

$$2d \sin \theta = m\lambda \qquad m = 1, 2, 3, \ldots \qquad (38.13)$$

This condition is known as **Bragg's law,** after W. L. Bragg (1890–1971), who first derived the relationship. If the wavelength and diffraction angle are measured, Equation 38.13 can be used to calculate the spacing between atomic planes.

38.6 Polarization of Light Waves

In Chapter 34 we described the transverse nature of light and all other electromagnetic waves. Polarization, discussed in this section, is firm evidence of this transverse nature.

An ordinary beam of light consists of a large number of waves emitted by the atoms of the light source. Each atom produces a wave having some particular orientation of the electric field vector **E**, corresponding to the direction of atomic vibration. The *direction of polarization* of each individual wave is defined to be the direction in which the electric field is vibrating. In Figure 38.28, this direction happens to lie along the y axis. However, an individual electromagnetic wave could have its **E** vector in the yz plane, making any possible angle with the y axis. Because all directions of vibration from a wave source are possible, the resultant electromagnetic wave is a superposition of waves vibrating in many different directions. The result is an **unpolarized** light beam, represented in Figure 38.29a. The direction of wave propagation in this figure is perpendicular to the page. The arrows show a few possible directions of the electric field vectors for the individual waves making up the resultant beam. At any given point and at some instant of time, all these individual electric field vectors add to give one resultant electric field vector.

As noted in Section 34.2, a wave is said to be **linearly polarized** if the resultant electric field **E** vibrates in the same direction *at all times* at a particular point, as shown in Figure 38.29b. (Sometimes, such a wave is described as *plane-polarized*, or simply *polarized*.) The plane formed by **E** and the direction of propagation is called the *plane*

Bragg's law

⚠ **PITFALL PREVENTION**

38.4 Different Angles

Notice in Figure 38.27 that the angle θ is measured from the reflecting surface, rather than from the normal, as in the case of the law of reflection in Chapter 35. With slits and diffraction gratings, we also measured the angle θ from the normal to the array of slits. Because of historical tradition, the angle is measured differently in Bragg diffraction, so interpret Equation 38.13 with care.

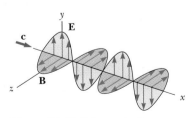

Figure 38.28 Schematic diagram of an electromagnetic wave propagating at velocity **c** in the x direction. The electric field vibrates in the xy plane, and the magnetic field vibrates in the xz plane.

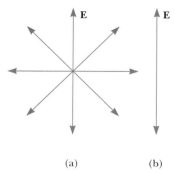

(a)   (b)

Figure 38.29 (a) A representation of an unpolarized light beam viewed along the direction of propagation (perpendicular to the page). The transverse electric field can vibrate in any direction in the plane of the page with equal probability. (b) A linearly polarized light beam with the electric field vibrating in the vertical direction.

of polarization of the wave. If the wave in Figure 38.28 represents the resultant of all individual waves, the plane of polarization is the *xy* plane.

It is possible to obtain a linearly polarized beam from an unpolarized beam by removing all waves from the beam except those whose electric field vectors oscillate in a single plane. We now discuss four processes for producing polarized light from unpolarized light.

Polarization by Selective Absorption

The most common technique for producing polarized light is to use a material that transmits waves whose electric fields vibrate in a plane parallel to a certain direction and that absorbs waves whose electric fields vibrate in all other directions.

In 1938, E. H. Land (1909–1991) discovered a material, which he called *polaroid*, that polarizes light through selective absorption by oriented molecules. This material is fabricated in thin sheets of long-chain hydrocarbons. The sheets are stretched during manufacture so that the long-chain molecules align. After a sheet is dipped into a solution containing iodine, the molecules become good electrical conductors. However, conduction takes place primarily along the hydrocarbon chains because electrons can move easily only along the chains. As a result, the molecules readily absorb light whose electric field vector is parallel to their length and allow light through whose electric field vector is perpendicular to their length.

It is common to refer to the direction perpendicular to the molecular chains as the *transmission axis*. In an ideal polarizer, all light with **E** parallel to the transmission axis is transmitted, and all light with **E** perpendicular to the transmission axis is absorbed.

Figure 38.30 represents an unpolarized light beam incident on a first polarizing sheet, called the *polarizer*. Because the transmission axis is oriented vertically in the figure, the light transmitted through this sheet is polarized vertically. A second polarizing sheet, called the *analyzer*, intercepts the beam. In Figure 38.30, the analyzer transmission axis is set at an angle θ to the polarizer axis. We call the electric field vector of the first transmitted beam $\mathbf{E}_0$. The component of $\mathbf{E}_0$ perpendicular to the analyzer axis is completely absorbed. The component of $\mathbf{E}_0$ parallel to the analyzer axis, which is allowed through by the analyzer, is $E_0 \cos\theta$. Because the intensity of the transmitted beam varies as the square of its magnitude, we conclude that the intensity of the (polarized) beam transmitted through the analyzer varies as

Malus's law

$$I = I_{max} \cos^2\theta \qquad (38.14)$$

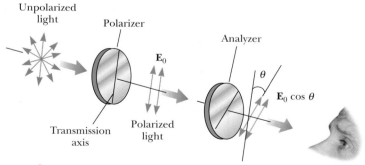

Rotate the analyzer at the Active Figures link at http://www.pse6.com, to see the effect on the transmitted light.

Active Figure 38.30 Two polarizing sheets whose transmission axes make an angle θ with each other. Only a fraction of the polarized light incident on the analyzer is transmitted through it.

(a)   (b)   (c)

Figure 38.31 The intensity of light transmitted through two polarizers depends on the relative orientation of their transmission axes. (a) The transmitted light has maximum intensity when the transmission axes are aligned with each other. (b) The transmitted light has lesser intensity when the transmission axes are at an angle of 45° with each other. (c) The transmitted light intensity is a minimum when the transmission axes are perpendicular to each other.

where I_{max} is the intensity of the polarized beam incident on the analyzer. This expression, known as **Malus's law,**[2] applies to any two polarizing materials whose transmission axes are at an angle θ to each other. From this expression, we see that the intensity of the transmitted beam is maximum when the transmission axes are parallel ($\theta = 0$ or 180°) and that it is zero (complete absorption by the analyzer) when the transmission axes are perpendicular to each other. This variation in transmitted intensity through a pair of polarizing sheets is illustrated in Figure 38.31.

Polarization by Reflection

When an unpolarized light beam is reflected from a surface, the reflected light may be completely polarized, partially polarized, or unpolarized, depending on the angle of incidence. If the angle of incidence is 0°, the reflected beam is unpolarized. For other angles of incidence, the reflected light is polarized to some extent, and for one particular angle of incidence, the reflected light is completely polarized. Let us now investigate reflection at that special angle.

Suppose that an unpolarized light beam is incident on a surface, as in Figure 38.32a. Each individual electric field vector can be resolved into two components: one parallel to the surface (and perpendicular to the page in Fig. 38.32, represented by the dots), and the other (represented by the brown arrows) perpendicular both to the first component and to the direction of propagation. Thus, the polarization of the entire beam can be described by two electric field components in these directions. It is found that the parallel component reflects more strongly than the perpendicular component, and this results in a partially polarized reflected beam. Furthermore, the refracted beam is also partially polarized.

Now suppose that the angle of incidence θ_1 is varied until the angle between the reflected and refracted beams is 90°, as in Figure 38.32b. At this particular angle of incidence, the reflected beam is completely polarized (with its electric field vector parallel to the surface), and the refracted beam is still only partially polarized. The angle of incidence at which this polarization occurs is called the **polarizing angle** θ_p.

---

[2] Named after its discoverer, E. L. Malus (1775–1812). Malus discovered that reflected light was polarized by viewing it through a calcite ($CaCO_3$) crystal.

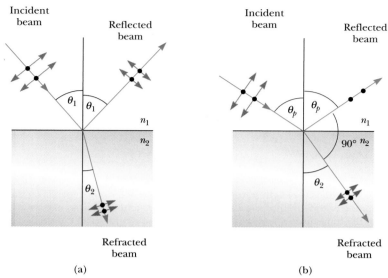

Figure 38.32 (a) When unpolarized light is incident on a reflecting surface, the reflected and refracted beams are partially polarized. (b) The reflected beam is completely polarized when the angle of incidence equals the polarizing angle θ_p, which satisfies the equation $n = \tan\theta_p$. At this incident angle, the reflected and refracted rays are perpendicular to each other.

We can obtain an expression relating the polarizing angle to the index of refraction of the reflecting substance by using Figure 38.32b. From this figure, we see that $\theta_p + 90° + \theta_2 = 180°$; thus $\theta_2 = 90° - \theta_p$. Using Snell's law of refraction (Eq. 35.8) and taking $n_1 = 1.00$ for air and $n_2 = n$, we have

$$n = \frac{\sin\theta_1}{\sin\theta_2} = \frac{\sin\theta_p}{\sin\theta_2}$$

Because $\sin\theta_2 = \sin(90° - \theta_p) = \cos\theta_p$, we can write this expression for n as $n = \sin\theta_p/\cos\theta_p$, which means that

Brewster's law

$$n = \tan\theta_p \tag{38.15}$$

This expression is called **Brewster's law,** and the polarizing angle θ_p is sometimes called **Brewster's angle,** after its discoverer, David Brewster (1781–1868). Because n varies with wavelength for a given substance, Brewster's angle is also a function of wavelength.

We can understand polarization by reflection by imagining that the electric field in the incident light sets electrons at the surface of the material in Figure 38.32b into oscillation. The component directions of oscillation are (1) parallel to the arrows shown on the refracted beam of light and (2) perpendicular to the page. The oscillating electrons act as antennas radiating light with a polarization parallel to the direction of oscillation. For the oscillations in direction (1), there is no radiation in the perpendicular direction, which is along the reflected ray (see the $\theta = 90°$ direction in Figure 34.11). For oscillations in direction (2), the electrons radiate light with a polarization perpendicular to the page (the $\theta = 0$ direction in Figure 34.11). Thus, the light reflected from the surface at this angle is completely polarized parallel to the surface.

Polarization by reflection is a common phenomenon. Sunlight reflected from water, glass, and snow is partially polarized. If the surface is horizontal, the electric field vector of the reflected light has a strong horizontal component. Sunglasses made of polarizing material reduce the glare of reflected light. The transmission axes of the lenses are oriented vertically so that they absorb the strong horizontal component of the reflected light. If you rotate sunglasses through 90 degrees, they are not as effective at blocking the glare from shiny horizontal surfaces.

Polarization by Double Refraction

Solids can be classified on the basis of internal structure. Those in which the atoms are arranged in a specific order are called *crystalline;* the NaCl structure of Figure 38.26 is just one example of a crystalline solid. Those solids in which the atoms are distributed randomly are called *amorphous*. When light travels through an amorphous material, such as glass, it travels with a speed that is the same in all directions. That is, glass has a single index of refraction. In certain crystalline materials, however, such as calcite and quartz, the speed of light is not the same in all directions. Such materials are characterized by two indices of refraction. Hence, they are often referred to as **double-refracting** or **birefringent** materials.

Upon entering a calcite crystal, unpolarized light splits into two plane-polarized rays that travel with different velocities, corresponding to two angles of refraction, as shown in Figure 38.33. The two rays are polarized in two mutually perpendicular directions, as indicated by the dots and arrows. One ray, called the **ordinary (O) ray,** is characterized by an index of refraction n_O that is the same in all directions. This means that if one could place a point source of light inside the crystal, as in Figure 38.34, the ordinary waves would spread out from the source as spheres.

The second plane-polarized ray, called the **extraordinary (E) ray,** travels with different speeds in different directions and hence is characterized by an index of refraction n_E that varies with the direction of propagation. Consider again the point source within a birefringent material, as in Figure 38.34. The source sends out an extraordinary wave having wave fronts that are elliptical in cross section. Note from Figure 38.34 that there is one direction, called the **optic axis,** along which the ordinary and extraordinary rays have the same speed, corresponding to the direction for which $n_O = n_E$. The difference in speed for the two rays is a maximum in the direction perpendicular to the optic axis. For example, in calcite, $n_O = 1.658$ at a wavelength of 589.3 nm, and n_E varies from 1.658 along the optic axis to 1.486 perpendicular to the optic axis. Values for n_O and n_E for various double-refracting crystals are given in Table 38.1.

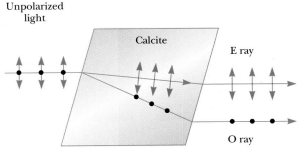

Figure 38.33 Unpolarized light incident on a calcite crystal splits into an ordinary (O) ray and an extraordinary (E) ray. These two rays are polarized in mutually perpendicular directions. (Drawing not to scale.)

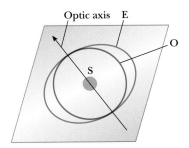

Figure 38.34 A point source S inside a double-refracting crystal produces a spherical wave front corresponding to the ordinary ray and an elliptical wave front corresponding to the extraordinary ray. The two waves propagate with the same velocity along the optic axis.

Table 38.1

Indices of Refraction for Some Double-Refracting Crystals at a Wavelength of 589.3 nm			
Crystal	n_O	n_E	n_O/n_E
Calcite ($CaCO_3$)	1.658	1.486	1.116
Quartz (SiO_2)	1.544	1.553	0.994
Sodium nitrate ($NaNO_3$)	1.587	1.336	1.188
Sodium sulfite ($NaSO_3$)	1.565	1.515	1.033
Zinc chloride ($ZnCl_2$)	1.687	1.713	0.985
Zinc sulfide (ZnS)	2.356	2.378	0.991

Figure 38.35 A calcite crystal produces a double image because it is a birefringent (double-refracting) material.

If we place a piece of calcite on a sheet of paper and then look through the crystal at any writing on the paper, we see two images, as shown in Figure 38.35. As can be seen from Figure 38.33, these two images correspond to one formed by the ordinary ray and one formed by the extraordinary ray. If the two images are viewed through a sheet of rotating polarizing glass, they alternately appear and disappear because the ordinary and extraordinary rays are plane-polarized along mutually perpendicular directions.

Some materials, such as glass and plastic, become birefringent when stressed. Suppose that an unstressed piece of plastic is placed between a polarizer and an analyzer so that light passes from polarizer to plastic to analyzer. When the plastic is unstressed and the analyzer axis is perpendicular to the polarizer axis, none of the polarized light passes through the analyzer. In other words, the unstressed plastic has no effect on the light passing through it. If the plastic is stressed, however, regions of greatest stress become birefringent and the polarization of the light passing through the plastic changes. Hence, a series of bright and dark bands is observed in the transmitted light, with the bright bands corresponding to regions of greatest stress.

Engineers often use this technique, called *optical stress analysis*, in designing structures ranging from bridges to small tools. They build a plastic model and analyze it under different load conditions to determine regions of potential weakness and failure under stress. Some examples of plastic models under stress are shown in Figure 38.36.

Polarization by Scattering

When light is incident on any material, the electrons in the material can absorb and reradiate part of the light. Such absorption and reradiation of light by electrons in the gas molecules that make up air is what causes sunlight reaching an observer on the Earth to be partially polarized. You can observe this effect—called **scattering**—by looking

(a)   (b)

Figure 38.36 (a) Strain distribution in a plastic model of a hip replacement used in a medical research laboratory. The pattern is produced when the plastic model is viewed between a polarizer and analyzer oriented perpendicular to each other. (b) A plastic model of an arch structure under load conditions observed between perpendicular polarizers. Such patterns are useful in the optimal design of architectural components.

directly up at the sky through a pair of sunglasses whose lenses are made of polarizing material. Less light passes through at certain orientations of the lenses than at others.

Figure 38.37 illustrates how sunlight becomes polarized when it is scattered. The phenomenon is similar to that creating completely polarized light upon reflection from a surface at Brewster's angle. An unpolarized beam of sunlight traveling in the horizontal direction (parallel to the ground) strikes a molecule of one of the gases that make up air, setting the electrons of the molecule into vibration. These vibrating charges act like the vibrating charges in an antenna. The horizontal component of the electric field vector in the incident wave results in a horizontal component of the vibration of the charges, and the vertical component of the vector results in a vertical component of vibration. If the observer in Figure 38.37 is looking straight up (perpendicular to the original direction of propagation of the light), the vertical oscillations of the charges send no radiation toward the observer. Thus, the observer sees light that is completely polarized in the horizontal direction, as indicated by the brown arrows. If the observer looks in other directions, the light is partially polarized in the horizontal direction.

Some phenomena involving the scattering of light in the atmosphere can be understood as follows. When light of various wavelengths λ is incident on gas molecules of diameter d, where $d \ll \lambda$, the relative intensity of the scattered light varies as $1/\lambda^4$. The condition $d \ll \lambda$ is satisfied for scattering from oxygen (O_2) and nitrogen (N_2) molecules in the atmosphere, whose diameters are about 0.2 nm. Hence, short wavelengths (blue light) are scattered more efficiently than long wavelengths (red light). Therefore, when sunlight is scattered by gas molecules in the air, the short-wavelength radiation (blue) is scattered more intensely than the long-wavelength radiation (red).

When you look up into the sky in a direction that is not toward the Sun, you see the scattered light, which is predominantly blue; hence, you see a blue sky. If you look toward the west at sunset (or toward the east at sunrise), you are looking in a direction toward the Sun and are seeing light that has passed through a large distance of air. Most of the blue light has been scattered by the air between you and the Sun. The light that survives this trip through the air to you has had much of its blue component scattered and is thus heavily weighted toward the red end of the spectrum; as a result, you see the red and orange colors of sunset.

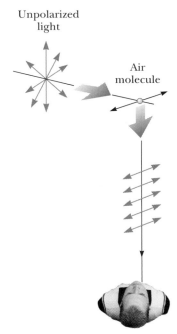

Figure 38.37 The scattering of unpolarized sunlight by air molecules. The scattered light traveling perpendicular to the incident light is plane-polarized because the vertical vibrations of the charges in the air molecule send no light in this direction.

Gary Friedman/Los Angeles Times

On the right side of this photograph is a view from the side of the freeway (cars and a truck are visible at the left) of a rocket launch from Vandenburg Air Force Base, California. The trail left by the rocket shows the effects of scattering of light by air molecules. The lower portion of the trail appears red, due to the scattering of wavelengths at the violet end of the spectrum as the light from the Sun travels through a large portion of the atmosphere to light up the trail. The upper portion of the trail is illuminated by light that has traveled through much less atmosphere and appears white.

Optical Activity

Many important applications of polarized light involve materials that display **optical activity.** A material is said to be optically active if it rotates the plane of polarization of any light transmitted through the material. The angle through which the light is rotated by a specific material depends on the length of the path through the material and on concentration if the material is in solution. One optically active material is a solution of the common sugar dextrose. A standard method for determining the concentration of sugar solutions is to measure the rotation produced by a fixed length of the solution.

Molecular asymmetry determines whether a material is optically active. For example, some proteins are optically active because of their spiral shape.

The liquid crystal displays found in most calculators have their optical activity changed by the application of electric potential across different parts of the display. Try using a pair of polarizing sunglasses to investigate the polarization used in the display of your calculator.

Quick Quiz 38.9 A polarizer for microwaves can be made as a grid of parallel metal wires about a centimeter apart. Is the electric field vector for microwaves transmitted through this polarizer (a) parallel or (b) perpendicular to the metal wires?

Quick Quiz 38.10 You are walking down a long hallway that has many light fixtures in the ceiling and a very shiny, newly waxed floor. In the floor, you see reflections of every light fixture. Now you put on sunglasses that are polarized. Some of the reflections of the light fixtures can no longer be seen (Try this!) The reflections that disappear are those (a) nearest to you (b) farthest from you (c) at an intermediate distance from you.

SUMMARY

Take a practice test for this chapter by clicking on the Practice Test link at http://www.pse6.com.

Diffraction is the deviation of light from a straight-line path when the light passes through an aperture or around an obstacle. Diffraction is due to the wave nature of light.

The **Fraunhofer diffraction pattern** produced by a single slit of width a on a distant screen consists of a central bright fringe and alternating bright and dark fringes of much lower intensities. The angles θ_{dark} at which the diffraction pattern has zero intensity, corresponding to destructive interference, are given by

$$\sin \theta_{dark} = m \frac{\lambda}{a} \qquad m = \pm 1, \pm 2, \pm 3, \ldots \qquad (38.1)$$

The intensity I of a single-slit diffraction pattern as a function of angle θ is given by the expression

$$I = I_{max} \left[\frac{\sin(\beta/2)}{\beta/2} \right]^2 \qquad (38.4)$$

where $\beta = (2\pi a \sin \theta)/\lambda$ and I_{max} is the intensity at $\theta = 0$.

Rayleigh's criterion, which is a limiting condition of resolution, states that two images formed by an aperture are just distinguishable if the central maximum of the diffraction pattern for one image falls on the first minimum of the diffraction pattern for the other image. The limiting angle of resolution for a slit of width a is $\theta_{min} = \lambda/a$, and the limiting angle of resolution for a circular aperture of diameter D is $\theta_{min} = 1.22\lambda/D$.

A **diffraction grating** consists of a large number of equally spaced, identical slits. The condition for intensity maxima in the interference pattern of a diffraction grating for normal incidence is

$$d \sin\theta_{\text{bright}} = m\lambda \qquad m = 0, \pm 1, \pm 2, \pm 3, \ldots \qquad (38.10)$$

where d is the spacing between adjacent slits and m is the order number of the diffraction pattern. The resolving power of a diffraction grating in the mth order of the diffraction pattern is

$$R = Nm \qquad (38.12)$$

where N is the number of lines in the grating that are illuminated.

When polarized light of intensity I_{max} is emitted by a polarizer and then incident on an analyzer, the light transmitted through the analyzer has an intensity equal to $I_{\text{max}} \cos^2\theta$, where θ is the angle between the polarizer and analyzer transmission axes.

In general, reflected light is partially polarized. However, reflected light is completely polarized when the angle of incidence is such that the angle between the reflected and refracted beams is 90°. This angle of incidence, called the **polarizing angle** θ_p, satisfies **Brewster's law:**

$$n = \tan\theta_p \qquad (38.15)$$

where n is the index of refraction of the reflecting medium.

QUESTIONS

1. Why can you hear around corners, but not see around corners?

2. Holding your hand at arm's length, you can readily block sunlight from reaching your eyes. Why can you not block sound from reaching your ears this way?

3. Observe the shadow of your book when it is held a few inches above a table with a small lamp several feet above the book. Why is the shadow somewhat fuzzy at the edges?

4. Knowing that radio waves travel at the speed of light and that a typical AM radio frequency is 1 000 kHz while an FM radio frequency might be 100 MHz, estimate the wavelengths of typical AM and FM radio signals. Use this information to explain why AM radio stations can fade out when you drive your car through a short tunnel or underpass, when FM radio stations do not.

5. Describe the change in width of the central maximum of the single-slit diffraction pattern as the width of the slit is made narrower.

6. John William Strutt, Lord Rayleigh (1842–1919), is known as the last person to understand all of physics and all of mathematics. He invented an improved foghorn. To warn ships of a coastline, a foghorn should radiate sound in a wide horizontal sheet over the ocean's surface. It should not waste energy by broadcasting sound upward. It should not emit sound downward, because the water in front of the foghorn would reflect that sound upward. Rayleigh's foghorn trumpet is shown in Figure Q38.6. Is it installed in the correct orientation? Decide whether the long dimension of the rectangular opening should be horizontal or vertical, and argue for your decision.

Figure Q38.6

7. Featured in the motion picture *M*A*S*H* (20th Century Fox, Aspen Productions, 1970) is a loudspeaker mounted on an exterior wall of an Army barracks. It has an approximately rectangular aperture. Its design can be thought of as based on Lord Rayleigh's foghorn trumpet, described in Question 6. Borrow or rent a copy of the movie, sketch the orientation of the loudspeaker, decide whether it is installed in the correct orientation, and argue for your decision.

8. Assuming that the headlights of a car are point sources, estimate the maximum distance from an observer to the car at which the headlights are distinguishable from each other.

9. A laser beam is incident at a shallow angle on a machinist's ruler that has a finely calibrated scale. The engraved rulings on the scale give rise to a diffraction pattern on a screen. Discuss how you can use this technique to obtain a measure of the wavelength of the laser light.

10. When you receive a chest x-ray at a hospital, the rays pass through a series of parallel ribs in your chest. Do the ribs act as a diffraction grating for x-rays?

11. Certain sunglasses use a polarizing material to reduce the intensity of light reflected from shiny surfaces. What orientation of polarization should the material have to be most effective?

12. During the "day" on the Moon (when the Sun is visible), you see a black sky and the stars can be clearly seen. During the day on the Earth, you see a blue sky with no stars. Account for this difference.

13. You can make the path of a light beam visible by placing dust in the air (perhaps by clapping two blackboard erasers in the path of the light beam). Explain why you can see the beam under these circumstances. In general, when is light visible?

14. Is light from the sky polarized? Why is it that clouds seen through Polaroid glasses stand out in bold contrast to the sky?

15. If a coin is glued to a glass sheet and this arrangement is held in front of a laser beam, the projected shadow has diffraction rings around its edge and a bright spot in the center. How is this possible?

16. How could the index of refraction of a flat piece of dark obsidian glass be determined?

17. A laser produces a beam a few millimeters wide, with uniform intensity across its width. A hair is stretched vertically across the front of the laser to cross the beam. How is the diffraction pattern it produces on a distant screen related to that of a vertical slit equal in width to the hair? How could you determine the width of the hair from measurements of its diffraction pattern?

18. A radio station serves listeners in a city to the northeast of its broadcast site. It broadcasts from three adjacent towers on a mountain ridge, along a line running east and west. Show that by introducing time delays among the signals the individual towers radiate, the station can maximize net intensity in the direction toward the city (and in the opposite direction) and minimize the signal transmitted in other directions. The towers together are said to form a *phased array*.

PROBLEMS

Section 38.2 Diffraction Patterns from Narrow Slits

1. Helium–neon laser light ($\lambda = 632.8$ nm) is sent through a 0.300-mm-wide single slit. What is the width of the central maximum on a screen 1.00 m from the slit?

2. A beam of green light is diffracted by a slit of width 0.550 mm. The diffraction pattern forms on a wall 2.06 m beyond the slit. The distance between the positions of zero intensity on both sides of the central bright fringe is 4.10 mm. Calculate the wavelength of the laser light.

3. 🪐 A screen is placed 50.0 cm from a single slit, which is illuminated with 690-nm light. If the distance between the first and third minima in the diffraction pattern is 3.00 mm, what is the width of the slit?

4. Coherent microwaves of wavelength 5.00 cm enter a long, narrow window in a building otherwise essentially opaque to the microwaves. If the window is 36.0 cm wide, what is the distance from the central maximum to the first-order minimum along a wall 6.50 m from the window?

5. Sound with a frequency 650 Hz from a distant source passes through a doorway 1.10 m wide in a sound-absorbing wall. Find the number and approximate directions of the diffraction-maximum beams radiated into the space beyond.

6. Light of wavelength 587.5 nm illuminates a single slit 0.750 mm in width. (a) At what distance from the slit should a screen be located if the first minimum in the diffraction pattern is to be 0.850 mm from the center of the principal maximum? (b) What is the width of the central maximum?

7. A beam of laser light of wavelength 632.8 nm has a circular cross section 2.00 mm in diameter. A rectangular aperture is to be placed in the center of the beam so that, when the light falls perpendicularly on a wall 4.50 m away, the central maximum fills a rectangle 110 mm wide and 6.00 mm high. The dimensions are measured between the minima bracketing the central maximum. Find the required width and height of the aperture.

8. **What If?** Assume the light in Figure 38.5 strikes the single slit at an angle β from the perpendicular direction. Show that Equation 38.1, the condition for destructive interference, must be modified to read

$$\sin \theta_{\text{dark}} = m \left(\frac{\lambda}{a} \right) - \sin \beta$$

9. A diffraction pattern is formed on a screen 120 cm away from a 0.400-mm-wide slit. Monochromatic 546.1-nm light is used. Calculate the fractional intensity I/I_{max} at a point on the screen 4.10 mm from the center of the principal maximum.

10. Coherent light of wavelength 501.5 nm is sent through two parallel slits in a large flat wall. Each slit is 0.700 μm wide. Their centers are 2.80 μm apart. The light then falls on a semicylindrical screen, with its axis at the midline between the slits. (a) Predict the direction of each interference maximum on the screen, as an angle away from the bisector of the line joining the slits. (b) Describe the pattern of light on the screen, specifying the number of bright fringes and the location of each. (c) Find the intensity of light on the screen at the center of each bright fringe, expressed as a fraction of the light intensity I_{max} at the center of the pattern.

Section 38.3 Resolution of Single-Slit and Circular Apertures

11. The pupil of a cat's eye narrows to a vertical slit of width 0.500 mm in daylight. What is the angular resolution for horizontally separated mice? Assume that the average wavelength of the light is 500 nm.

12. Find the radius a star image forms on the retina of the eye if the aperture diameter (the pupil) at night is 0.700 cm and the length of the eye is 3.00 cm. Assume the representative wavelength of starlight in the eye is 500 nm.

13. A helium–neon laser emits light that has a wavelength of 632.8 nm. The circular aperture through which the beam emerges has a diameter of 0.500 cm. Estimate the diameter of the beam 10.0 km from the laser.

14. You are vacationing in a Wonderland populated by friendly elves and a cannibalistic Cyclops that devours physics students. The elves and the Cyclops look precisely alike (everyone wears loose jeans and sweatshirts) except that each elf has two eyes, about 10.0 cm apart, and the Cyclops—you guessed it—has only one eye of about the same size as an elf's. The elves and the Cyclops are constantly at war with each other, so they rarely sleep and all have red eyes, predominantly reflecting light with a wavelength of 660 nm. From what maximum distance can you distinguish between a friendly elf and the predatory Cyclops? The air in Wonderland is always clear. Dilated with fear, your pupils have a diameter of 7.00 mm.

15. Narrow, parallel, glowing gas-filled tubes in a variety of colors form block letters to spell out the name of a night club. Adjacent tubes are all 2.80 cm apart. The tubes forming one letter are filled with neon and radiate predominantly red light with a wavelength of 640 nm. For another letter, the tubes emit predominantly violet light at 440 nm. The pupil of a dark-adapted viewer's eye is 5.20 mm in diameter. If she is in a certain range of distances away, the viewer can resolve the separate tubes of one color but not the other. Which color is easier to resolve? The viewer's distance must be in what range for her to resolve the tubes of only one color?

16. On the night of April 18, 1775, a signal was sent from the steeple of Old North Church in Boston to Paul Revere, who was 1.80 mi away: "One if by land, two if by sea." At what minimum separation did the sexton have to set the lanterns for Revere to receive the correct message about the approaching British? Assume that the patriot's pupils had a diameter of 4.00 mm at night and that the lantern light had a predominant wavelength of 580 nm.

17. The Impressionist painter Georges Seurat created paintings with an enormous number of dots of pure pigment, each of which was approximately 2.00 mm in diameter. The idea was to have colors such as red and green next to each other to form a scintillating canvas (Fig. P38.17). Outside what distance would one be unable to discern individual dots on the canvas? (Assume that $\lambda = 500$ nm and that the pupil diameter is 4.00 mm.)

Figure P38.17 *Sunday Afternoon on the Island of La Grande Jatte*, by Georges Seurat.

18. A binary star system in the constellation Orion has an angular interstellar separation of 1.00×10^{-5} rad. If $\lambda = 500$ nm, what is the smallest diameter the telescope can have to just resolve the two stars?

19. A spy satellite can consist essentially of a large-diameter concave mirror forming an image on a digital-camera detector and sending the picture to a ground receiver by radio waves. In effect, it is an astronomical telescope in orbit, looking down instead of up. Can a spy satellite read a license plate? Can it read the date on a dime? Argue for your answers by making an order-of-magnitude calculation, specifying the data you estimate.

20. A circular radar antenna on a Coast Guard ship has a diameter of 2.10 m and radiates at a frequency of 15.0 GHz. Two small boats are located 9.00 km away from the ship. How close together could the boats be and still be detected as two objects?

21. Grote Reber was a pioneer in radio astronomy. He constructed a radio telescope with a 10.0-m-diameter receiving dish. What was the telescope's angular resolution for 2.00-m radio waves?

22. When Mars is nearest the Earth, the distance separating the two planets is 88.6×10^6 km. Mars is viewed through a telescope whose mirror has a diameter of 30.0 cm. (a) If the wavelength of the light is 590 nm, what is the angular resolution of the telescope? (b) What is the smallest distance that can be resolved between two points on Mars?

Section 38.4 The Diffraction Grating

> *Note:* In the following problems, assume that the light is incident normally on the gratings.

23. White light is spread out into its spectral components by a diffraction grating. If the grating has 2 000 grooves per centimeter, at what angle does red light of wavelength 640 nm appear in first order?

24. Light from an argon laser strikes a diffraction grating that has 5 310 grooves per centimeter. The central and first-order principal maxima are separated by 0.488 m on a wall 1.72 m from the grating. Determine the wavelength of the laser light.

25. The hydrogen spectrum has a red line at 656 nm and a blue line at 434 nm. What are the angular separations between these two spectral lines obtained with a diffraction grating that has 4 500 grooves/cm?

26. A helium–neon laser ($\lambda = 632.8$ nm) is used to calibrate a diffraction grating. If the first-order maximum occurs at 20.5°, what is the spacing between adjacent grooves in the grating?

27. Three discrete spectral lines occur at angles of 10.09°, 13.71°, and 14.77° in the first-order spectrum of a grating spectrometer. (a) If the grating has 3 660 slits/cm, what are the wavelengths of the light? (b) At what angles are these lines found in the second-order spectrum?

28. Show that, whenever white light is passed through a diffraction grating of any spacing size, the violet end of the continuous visible spectrum in third order always overlaps with red light at the other end of the second-order spectrum.

29. A diffraction grating of width 4.00 cm has been ruled with 3 000 grooves/cm. (a) What is the resolving power of this grating in the first three orders? (b) If two monochromatic waves incident on this grating have a mean wavelength of 400 nm, what is their wavelength separation if they are just resolved in the third order?

30. The laser in a CD player must precisely follow the spiral track, along which the distance between one loop of the spiral and the next is only about 1.25 μm. A feedback mechanism lets the player know if the laser drifts off the track, so that the player can steer it back again. Figure P38.30 shows how a diffraction grating is used to provide information to keep the beam on track. The laser light passes through a diffraction grating just before it reaches the disk. The strong central maximum of the diffraction pattern is used to read the information in the track of pits. The two first-order side maxima are used for steering. The grating is designed so that the first-order maxima fall on the flat surfaces on both sides of the information track. Both side beams are reflected into their own detectors. As long as both beams are reflecting from smooth nonpitted surfaces, they are detected with constant high intensity. If the main beam wanders off the track, however, one of the side beams will begin to strike pits on the information track and the reflected light will diminish. This change is used with an electronic circuit to guide the beam back to

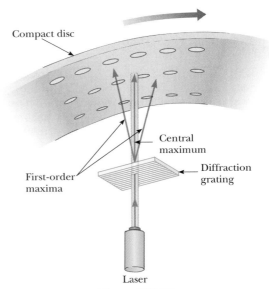

Figure P38.30

the desired location. Assume that the laser light has a wavelength of 780 nm and that the diffraction grating is positioned 6.90 μm from the disk. Assume that the first-order beams are to fall on the disk 0.400 μm on either side of the information track. What should be the number of grooves per millimeter in the grating?

31. A source emits 531.62-nm and 531.81-nm light. (a) What minimum number of grooves is required for a grating that resolves the two wavelengths in the first-order spectrum? (b) Determine the slit spacing for a grating 1.32 cm wide that has the required minimum number of grooves.

32. A diffraction grating has 4 200 rulings/cm. On a screen 2.00 m from the grating, it is found that for a particular order m, the maxima corresponding to two closely spaced wavelengths of sodium (589.0 nm and 589.6 nm) are separated by 1.59 mm. Determine the value of m.

33. A grating with 250 grooves/mm is used with an incandescent light source. Assume the visible spectrum to range in wavelength from 400 to 700 nm. In how many orders can one see (a) the entire visible spectrum and (b) the short-wavelength region?

34. A wide beam of laser light with a wavelength of 632.8 nm is directed through several narrow parallel slits, separated by 1.20 mm, and falls on a sheet of photographic film 1.40 m away. The exposure time is chosen so that the film stays unexposed everywhere except at the central region of each bright fringe. (a) Find the distance between these interference maxima. The film is printed as a transparency—it is opaque everywhere except at the exposed lines. Next, the same beam of laser light is directed through the transparency and allowed to fall on a screen 1.40 m beyond. (b) Argue that several narrow parallel bright regions, separated by 1.20 mm, will appear on the screen, as real images of the original slits. If at last the screen is removed, light will diverge from the images of the original slits with the same reconstructed wave fronts as the original slits produced. (*Suggestion:* You may find it useful to draw diagrams similar to Figure 38.16. A train of thought like this, at a soccer game, led Dennis Gabor to the invention of holography.)

Section 38.5   Diffraction of X-Rays by Crystals

35. Potassium iodide (KI) has the same crystalline structure as NaCl, with atomic planes separated by 0.353 nm. A monochromatic x-ray beam shows a first-order diffraction maximum when the grazing angle is 7.60°. Calculate the x-ray wavelength.

36. A wavelength of 0.129 nm characterizes K_α x-rays from zinc. When a beam of these x-rays is incident on the surface of a crystal whose structure is similar to that of NaCl, a first-order maximum is observed at 8.15°. Calculate the interplanar spacing based on this information.

37. If the interplanar spacing of NaCl is 0.281 nm, what is the predicted angle at which 0.140-nm x-rays are diffracted in a first-order maximum?

38. The first-order diffraction maximum is observed at 12.6° for a crystal in which the interplanar spacing is 0.240 nm. How many other orders can be observed?

39. In water of uniform depth, a wide pier is supported on pilings in several parallel rows 2.80 m apart. Ocean waves of uniform wavelength roll in, moving in a direction that makes an angle of 80.0° with the rows of posts. Find the three longest wavelengths of waves that will be strongly reflected by the pilings.

Section 38.6   Polarization of Light Waves

> Problem 34 in Chapter 34 can be assigned with this section.

40. Unpolarized light passes through two polaroid sheets. The axis of the first is vertical, and that of the second is at 30.0° to the vertical. What fraction of the incident light is transmitted?

41. Plane-polarized light is incident on a single polarizing disk with the direction of $\mathbf{E}_0$ parallel to the direction of the transmission axis. Through what angle should the disk be rotated so that the intensity in the transmitted beam is reduced by a factor of (a) 3.00, (b) 5.00, (c) 10.0?

42. Three polarizing disks whose planes are parallel are centered on a common axis. The direction of the transmission axis in each case is shown in Figure P38.42 relative to the

common vertical direction. A plane-polarized beam of light with $\mathbf{E}_0$ parallel to the vertical reference direction is incident from the left on the first disk with intensity $I_i = 10.0$ units (arbitrary). Calculate the transmitted intensity I_f when (a) $\theta_1 = 20.0°$, $\theta_2 = 40.0°$, and $\theta_3 = 60.0°$; (b) $\theta_1 = 0°$, $\theta_2 = 30.0°$, and $\theta_3 = 60.0°$.

43. The angle of incidence of a light beam onto a reflecting surface is continuously variable. The reflected ray is found to be completely polarized when the angle of incidence is 48.0°. What is the index of refraction of the reflecting material?

44. Review Problem. (a) A transparent plate with index of refraction n_2 is immersed in a medium with index n_1. Light traveling in the surrounding medium strikes the top surface of the plate at Brewster's angle. Show that if and only if the surfaces of the plate are parallel, the refracted light will strike the bottom surface of the plate at Brewster's angle for that interface. (b) **What If?** Instead of a plate, consider a prism of index of refraction n_2 separating media of different refractive indices n_1 and n_3. Is there one particular apex angle between the surfaces of the prism for which light can fall on both of its surfaces at Brewster's angle as it passes through the prism? If so, determine it.

45. The critical angle for total internal reflection for sapphire surrounded by air is 34.4°. Calculate the polarizing angle for sapphire.

46. For a particular transparent medium surrounded by air, show that the critical angle for total internal reflection and the polarizing angle are related by $\cot \theta_p = \sin \theta_c$.

47. How far above the horizon is the Moon when its image reflected in calm water is completely polarized? ($n_{\text{water}} = 1.33$)

Additional Problems

48. In Figure P38.42, suppose that the transmission axes of the left and right polarizing disks are perpendicular to each other. Also, let the center disk be rotated on the common axis with an angular speed ω. Show that if unpolarized light is incident on the left disk with an intensity I_{max}, the intensity of the beam emerging from the right disk is

$$I = \tfrac{1}{16} I_{\text{max}} (1 - \cos 4\omega t)$$

This means that the intensity of the emerging beam is modulated at a rate that is four times the rate of rotation of the center disk. [*Suggestion:* Use the trigonometric identities $\cos^2 \theta = (1 + \cos 2\theta)/2$ and $\sin^2 \theta = (1 - \cos 2\theta)/2$, and recall that $\theta = \omega t$.]

49. You want to rotate the plane of polarization of a polarized light beam by 45.0° with a maximum intensity reduction of 10.0%. (a) How many sheets of perfect polarizers do you need to achieve your goal? (b) What is the angle between adjacent polarizers?

50. Figure P38.50 shows a megaphone in use. Construct a theoretical description of how a megaphone works. You may assume that the sound of your voice radiates just through

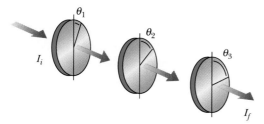

Figure P38.42 Problems 42 and 48.

Figure P38.50

the opening of your mouth. Most of the information in speech is carried not in a signal at the fundamental frequency, but in noises and in harmonics, with frequencies of a few thousand hertz. Does your theory allow any prediction that is simple to test?

51. Light from a helium–neon laser ($\lambda = 632.8$ nm) is incident on a single slit. What is the maximum width of the slit for which no diffraction minima are observed?

52. What are the approximate dimensions of the smallest object on Earth that astronauts can resolve by eye when they are orbiting 250 km above the Earth? Assume that $\lambda = 500$ nm and that a pupil diameter is 5.00 mm.

53. **Review problem.** A beam of 541-nm light is incident on a diffraction grating that has 400 grooves/mm. (a) Determine the angle of the second-order ray. (b) **What If?** If the entire apparatus is immersed in water, what is the new second-order angle of diffraction? (c) Show that the two diffracted rays of parts (a) and (b) are related through the law of refraction.

54. The *Very Large Array* (VLA) is a set of 27 radio telescope dishes in Caton and Socorro Counties, New Mexico (Fig. P38.54). The antennas can be moved apart on railroad tracks, and their combined signals give the resolving power of a synthetic aperture 36.0 km in diameter. (a) If the detectors are tuned to a frequency of 1.40 GHz, what is the angular resolution of the VLA? (b) Clouds of hydrogen radiate at this frequency. What must be the separation distance of two clouds 26 000 lightyears away at the center of the galaxy, if they are to be resolved? (c) **What If?** As the telescope looks up, a circling hawk looks down. Find the

angular resolution of the hawk's eye. Assume that the hawk is most sensitive to green light having a wavelength of 500 nm and that it has a pupil of diameter 12.0 mm. (d) A mouse is on the ground 30.0 m below. By what distance must the mouse's whiskers be separated if the hawk can resolve them?

55. A 750-nm light beam hits the flat surface of a certain liquid, and the beam is split into a reflected ray and a refracted ray. If the reflected ray is completely polarized at 36.0°, what is the wavelength of the refracted ray?

56. Iridescent peacock feathers are shown in Figure P38.56a. The surface of one microscopic barbule is composed of transparent keratin that supports rods of dark brown melanin in a regular lattice, represented in Figure P38.56b. (Your fingernails are made of keratin, and melanin is the dark pigment giving color to human skin.) In a portion of the feather that can appear turquoise, assume that the melanin rods are uniformly separated by 0.25 μm, with air between them. (a) Explain how this structure can appear blue-green when it contains no blue or green pigment.

(a)

(b)

Figure P38.56 (a) Iridescent peacock feathers. (b) Microscopic section of a feather showing dark melanin rods in a pale keratin matrix.

Figure P38.54 Some of the radio telescope dishes in the Very Large Array.

(b) Explain how it can also appear violet if light falls on it in a different direction. (c) Explain how it can present different colors to your two eyes at the same time—a characteristic of iridescence. (d) A compact disc can appear to be any color of the rainbow. Explain why this portion of the feather cannot appear yellow or red. (e) What could be different about the array of melanin rods in a portion of the feather that does appear to be red?

57. Light of wavelength 500 nm is incident normally on a diffraction grating. If the third-order maximum of the diffraction pattern is observed at 32.0°, (a) what is the number of rulings per centimeter for the grating? (b) Determine the total number of primary maxima that can be observed in this situation.

58. Light strikes a water surface at the polarizing angle. The part of the beam refracted into the water strikes a submerged glass slab (index of refraction, 1.50), as shown in Figure P38.58. The light reflected from the upper surface of the slab is completely polarized. Find the angle between the water surface and the glass slab.

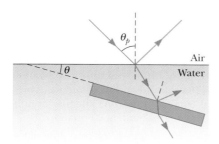

Figure P38.58

59. A beam of bright red light of wavelength 654 nm passes through a diffraction grating. Enclosing the space beyond the grating is a large screen forming one half of a cylinder centered on the grating, with its axis parallel to the slits in the grating. Fifteen bright spots appear on the screen. Find the maximum and minimum possible values for the slit separation in the diffraction grating.

60. A *pinhole camera* has a small circular aperture of diameter D. Light from distant objects passes through the aperture into an otherwise dark box, falling on a screen located a distance L away. If D is too large, the display on the screen will be fuzzy, because a bright point in the field of view will send light onto a circle of diameter slightly larger than D. On the other hand, if D is too small, diffraction will blur the display on the screen. The screen shows a reasonably sharp image if the diameter of the central disk of the diffraction pattern, specified by Equation 38.9, is equal to D at the screen. (a) Show that for monochromatic light with plane wave fronts and $L \gg D$, the condition for a sharp view is fulfilled if $D^2 = 2.44\lambda L$. (b) Find the optimum pinhole diameter for 500-nm light projected onto a screen 15.0 cm away.

61. An American standard television picture is composed of about 485 horizontal lines of varying light intensity. Assume that your ability to resolve the lines is limited only by the Rayleigh criterion and that the pupils of your eyes are 5.00 mm in diameter. Calculate the ratio of minimum viewing distance to the vertical dimension of the picture such that you will not be able to resolve the lines. Assume

that the average wavelength of the light coming from the screen is 550 nm.

62. (a) Light traveling in a medium of index of refraction n_1 is incident at an angle θ on the surface of a medium of index n_2. The angle between reflected and refracted rays is β. Show that

$$\tan \theta = \frac{n_2 \sin \beta}{n_1 - n_2 \cos \beta}$$

(*Suggestion:* Use the identity $\sin(A + B) = \sin A \cos B + \cos A \sin B$.) (b) **What If?** Show that this expression for $\tan \theta$ reduces to Brewster's law when $\beta = 90°$, $n_1 = 1$, and $n_2 = n$.

63. Suppose that the single slit in Figure 38.6 is 6.00 cm wide and in front of a microwave source operating at 7.50 GHz. (a) Calculate the angle subtended by the first minimum in the diffraction pattern. (b) What is the relative intensity I/I_{max} at $\theta = 15.0°$? (c) Assume that two such sources, separated laterally by 20.0 cm, are behind the slit. What must the maximum distance between the plane of the sources and the slit be if the diffraction patterns are to be resolved? (In this case, the approximation $\sin \theta \approx \tan \theta$ is not valid because of the relatively small value of a/λ.)

64. Two polarizing sheets are placed together with their transmission axes crossed so that no light is transmitted. A third sheet is inserted between them with its transmission axis at an angle of 45.0° with respect to each of the other axes. Find the fraction of incident unpolarized light intensity transmitted by the three-sheet combination. (Assume each polarizing sheet is ideal.)

65. Two wavelengths λ and $\lambda + \Delta\lambda$ (with $\Delta\lambda \ll \lambda$) are incident on a diffraction grating. Show that the angular separation between the spectral lines in the mth-order spectrum is

$$\Delta\theta = \frac{\Delta\lambda}{\sqrt{(d/m)^2 - \lambda^2}}$$

where d is the slit spacing and m is the order number.

66. Two closely spaced wavelengths of light are incident on a diffraction grating. (a) Starting with Equation 38.10, show that the angular dispersion of the grating is given by

$$\frac{d\theta}{d\lambda} = \frac{m}{d \cos \theta}$$

(b) A square grating 2.00 cm on each side containing 8 000 equally spaced slits is used to analyze the spectrum of mercury. Two closely spaced lines emitted by this element have wavelengths of 579.065 nm and 576.959 nm. What is the angular separation of these two wavelengths in the second-order spectrum?

67. The scale of a map is a number of kilometers per centimeter, specifying the distance on the ground that any distance on the map represents. The scale of a spectrum is its *dispersion*, a number of nanometers per centimeter, which specifies the change in wavelength that a distance across the spectrum represents. One must know the dispersion in order to compare one spectrum with another and to make a measurement of (for example) a Doppler shift. Let y represent the position relative to the center of a diffraction pattern projected onto a flat screen at distance L by a diffraction grating with slit spacing d. The dispersion is $d\lambda/dy$.

(a) Prove that the dispersion is given by

$$\frac{d\lambda}{dy} = \frac{L^2 d}{m(L^2 + y^2)^{3/2}}$$

(b) Calculate the dispersion in first order for light with a mean wavelength of 550 nm, analyzed with a grating having 8 000 rulings/cm, and projected onto a screen 2.40 m away.

68. Derive Equation 38.12 for the resolving power of a grating, $R = Nm$, where N is the number of slits illuminated and m is the order in the diffraction pattern. Remember that Rayleigh's criterion (Section 38.3) states that two wavelengths will be resolved when the principal maximum for one falls on the first minimum for the other.

69. Figure P38.69a is a three-dimensional sketch of a birefringent crystal. The dotted lines illustrate how a thin parallel-faced slab of material could be cut from the larger specimen with the optic axis of the crystal parallel to the faces of the plate. A section cut from the crystal in this manner is known as a *retardation plate*. When a beam of light is incident on the plate perpendicular to the direction of the optic axis, as shown in Figure P38.69b, the O ray and the E ray travel along a single straight line, but with different speeds. (a) Let the thickness of the plate be d and show that the phase difference between the O ray and the E ray is

$$\theta = \frac{2\pi d}{\lambda} |n_O - n_E|$$

where λ is the wavelength in air. (b) In a particular case the incident light has a wavelength of 550 nm. Find the minimum value of d for a quartz plate for which $\theta = \pi/2$. Such a plate is called a *quarter-wave plate*. (Use values of n_O and n_E from Table 38.1.)

Optic axis

(a)

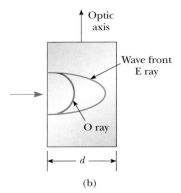

Optic axis

Wave front E ray

O ray

$\leftarrow d \rightarrow$

(b)

Figure P38.69

70. How much diffraction spreading does a light beam undergo? One quantitative answer is the *full width at half*

maximum of the central maximum of the single-slit Fraunhofer diffraction pattern. You can evaluate this angle of spreading in this problem and in the next. (a) In Equation 38.4, define $\beta/2 = \phi$ and show that, at the point where $I = 0.5I_{max}$, we must have $\sin \phi = \phi/\sqrt{2}$. (b) Let $y_1 = \sin \phi$ and $y_2 = \phi/\sqrt{2}$. Plot y_1 and y_2 on the same set of axes over a range from $\phi = 1$ rad to $\phi = \pi/2$ rad. Determine ϕ from the point of intersection of the two curves. (c) Then show that, if the fraction λ/a is not large, the angular full width at half maximum of the central diffraction maximum is $\Delta\theta = 0.886 \lambda/a$.

71. Another method to solve the equation $\phi = \sqrt{2} \sin \phi$ in Problem 70 is to guess a first value of ϕ, use a computer or calculator to see how nearly it fits, and continue to update your estimate until the equation balances. How many steps (iterations) does this take?

72. In the diffraction pattern of a single slit, described by the equation

$$I_\theta = I_{max} \frac{\sin^2(\beta/2)}{(\beta/2)^2}$$

with $\beta = (2\pi a \sin \theta)/\lambda$, the central maximum is at $\beta = 0$ and the side maxima are *approximately* at $\beta/2 = (m + \frac{1}{2})\pi$ for $m = 1, 2, 3, \ldots$. Determine more precisely (a) the location of the first side maximum, where $m = 1$, and (b) the location of the second side maximum. Observe in Figure 38.10a that the graph of intensity versus $\beta/2$ has a horizontal tangent at maxima and also at minima. You will need to solve a transcendental equation.

73. Light of wavelength 632.8 nm illuminates a single slit, and a diffraction pattern is formed on a screen 1.00 m from the slit. Using the data in the table below, plot relative intensity versus position. Choose an appropriate value for the slit width a and on the same graph used for the experimental data, plot the theoretical expression for the relative intensity

$$\frac{I_\theta}{I_{max}} = \frac{\sin^2(\beta/2)}{(\beta/2)^2}$$

What value of a gives the best fit of theory and experiment?

Relative Intensity	Position Relative to Central Maximum (mm)
1.00	0
0.95	0.8
0.80	1.6
0.60	2.4
0.39	3.2
0.21	4.0
0.079	4.8
0.014	5.6
0.003	6.5
0.015	7.3
0.036	8.1
0.047	8.9
0.043	9.7
0.029	10.5
0.013	11.3
0.002	12.1
0.000 3	12.9
0.005	13.7

0.012	14.5
0.016	15.3
0.015	16.1
0.010	16.9
0.004 4	17.7
0.000 6	18.5
0.000 3	19.3
0.003	20.2

Answers to Quick Quizzes

38.1 (a). Equation 38.1 shows that a decrease in a results in an increase in the angles at which the dark fringes appear.

38.2 (c). The space between the slightly open door and the doorframe acts as a single slit. Sound waves have wavelengths that are larger than the opening and so are diffracted and spread throughout the room you are in. Because light wavelengths are much smaller than the slit width, they experience negligible diffraction. As a result, you must have a direct line of sight to detect the light waves.

38.3 The situation is like that depicted in Figure 38.11 except that now the slits are only half as far apart. The diffraction pattern is the same, but the interference pattern is stretched out because d is smaller. Because $d/a = 3$, the $m = 3$ interference maximum coincides with the first diffraction minimum. Your sketch should look like the figure below.

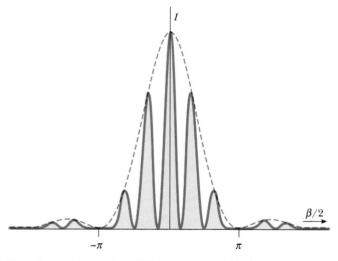

38.4 (c). In Equation 38.7, the ratio d/a is independent of wavelength, so the number of interference fringes in the central diffraction pattern peak remains the same. Equation 38.1 tells us that a decrease in wavelength causes a decrease in the width of the central peak.

38.5 (b). The effective slit width in the vertical direction of the cat's eye is larger than that in the horizontal direction. Thus, the eye has more resolving power for lights separated in the vertical direction and would be more effective at resolving the mast lights on the boat.

38.6 (a). We would like to reduce the minimum angular separation for two objects below the angle subtended by the two stars in the binary system. We can do that by reducing the wavelength of the light—this in essence makes the aperture larger, relative to the light wavelength, increasing the resolving power. Thus, we should choose a blue filter.

38.7 (b). The tracks of information on a compact disc are much closer together than on a phonograph record. As a result, the diffraction maxima from the compact disc will be farther apart than those from the record.

38.8 (c). With the doubled wavelength, the pattern will be wider. Choices (a) and (d) make the pattern even wider. From Equation 38.10, we see that choice (b) causes $\sin \theta$ to be twice as large. Because we cannot use the small angle approximation, however, a doubling of $\sin \theta$ is not the same as a doubling of θ, which would translate to a doubling of the position of a maximum along the screen. If we only consider small-angle maxima, choice (b) would work, but it does not work in the large-angle case.

38.9 (b). Electric field vectors parallel to the metal wires cause electrons in the metal to oscillate parallel to the wires. Thus, the energy from the waves with these electric field vectors is transferred to the metal by accelerating these electrons and is eventually transformed to internal energy through the resistance of the metal. Waves with electric-field vectors perpendicular to the metal wires pass through because they are not able to accelerate electrons in the wires.

38.10 (c). At some intermediate distance, the light rays from the fixtures will strike the floor at Brewster's angle and reflect to your eyes. Because this light is polarized horizontally, it will not pass through your polarized sunglasses. Tilting your head to the side will cause the reflections to reappear.

Modern Physics

A t the end of the nineteenth century, many scientists believed that they had learned most of what there was to know about physics. Newton's laws of motion and his theory of universal gravitation, Maxwell's theoretical work in unifying electricity and magnetism, the laws of thermodynamics and kinetic theory, and the principles of optics were highly successful in explaining a variety of phenomena.

As the nineteenth century turned to the twentieth, however, a major revolution shook the world of physics. In 1900 Planck provided the basic ideas that led to the formulation of the quantum theory, and in 1905 Einstein formulated his brilliant special theory of relativity. The excitement of the times is captured in Einstein's own words: "It was a marvelous time to be alive." Both ideas were to have a profound effect on our understanding of nature. Within a few decades, these two theories inspired new developments and theories in the fields of atomic physics, nuclear physics, and condensed-matter physics.

In Chapter 39 we introduce the special theory of relativity. The theory provides us with a new and deeper view of physical laws. Although the concepts underlying this theory often violate our common sense, the theory correctly predicts the results of experiments involving speeds near the speed of light. In the extended version of this textbook, *Physics for Scientists and Engineers with Modern Physics,* we cover the basic concepts of quantum mechanics and their application to atomic and molecular physics, and we introduce solid-state physics, nuclear physics, particle physics, and cosmology.

You should keep in mind that, although the physics that was developed during the twentieth century has led to a multitude of important technological achievements, the story is still incomplete. Discoveries will continue to evolve during our lifetimes, and many of these discoveries will deepen or refine our understanding of nature and the world around us. It is still a "marvelous time to be alive." ■

◀ *A portion of the accelerator tunnel at Fermilab, near Chicago, Illinois. The tunnel is circular and 1.9 km in diameter. Using electric and magnetic fields, protons and antiprotons are accelerated to speeds close to that of light and then allowed to collide head-on, in order to investigate the production of new particles. (Fermilab Photo)*

Chapter 39

Relativity

▲ Standing on the shoulders of a giant. *David Serway, son of one of the authors, watches over his children, Nathan and Kaitlyn, as they frolic in the arms of Albert Einstein at the Einstein memorial in Washington, D.C. It is well known that Einstein, the principal architect of relativity, was very fond of children. (Emily Serway)*

Our everyday experiences and observations have to do with objects that move at speeds much less than the speed of light. Newtonian mechanics was formulated by observing and describing the motion of such objects, and this formalism is very successful in describing a wide range of phenomena that occur at low speeds. However, it fails to describe properly the motion of objects whose speeds approach that of light.

Experimentally, the predictions of Newtonian theory can be tested at high speeds by accelerating electrons or other charged particles through a large electric potential difference. For example, it is possible to accelerate an electron to a speed of $0.99c$ (where c is the speed of light) by using a potential difference of several million volts. According to Newtonian mechanics, if the potential difference is increased by a factor of 4, the electron's kinetic energy is four times greater and its speed should double to $1.98c$. However, experiments show that the speed of the electron—as well as the speed of any other object in the Universe—always remains less than the speed of light, regardless of the size of the accelerating voltage. Because it places no upper limit on speed, Newtonian mechanics is contrary to modern experimental results and is clearly a limited theory.

In 1905, at the age of only 26, Einstein published his special theory of relativity. Regarding the theory, Einstein wrote:

> The relativity theory arose from necessity, from serious and deep contradictions in the old theory from which there seemed no escape. The strength of the new theory lies in the consistency and simplicity with which it solves all these difficulties . . . .[1]

Although Einstein made many other important contributions to science, the special theory of relativity alone represents one of the greatest intellectual achievements of all time. With this theory, experimental observations can be correctly predicted over the range of speeds from $v = 0$ to speeds approaching the speed of light. At low speeds, Einstein's theory reduces to Newtonian mechanics as a limiting situation. It is important to recognize that Einstein was working on electromagnetism when he developed the special theory of relativity. He was convinced that Maxwell's equations were correct, and in order to reconcile them with one of his postulates, he was forced into the revolutionary notion of assuming that space and time are not absolute.

This chapter gives an introduction to the special theory of relativity, with emphasis on some of its consequences. The special theory covers phenomena such as the slowing down of moving clocks and the contraction of moving lengths. We also discuss the relativistic forms of momentum and energy.

In addition to its well-known and essential role in theoretical physics, the special theory of relativity has practical applications, including the design of nuclear power plants and modern global positioning system (GPS) units. These devices do not work if designed in accordance with nonrelativistic principles.

[1] A. Einstein and L. Infeld, *The Evolution of Physics*, New York, Simon and Schuster, 1961.

39.1 The Principle of Galilean Relativity

To describe a physical event, we must establish a frame of reference. You should recall from Chapter 5 that an inertial frame of reference is one in which an object is observed to have no acceleration when no forces act on it. Furthermore, any system moving with constant velocity with respect to an inertial frame must also be in an inertial frame.

There is no absolute inertial reference frame. This means that the results of an experiment performed in a vehicle moving with uniform velocity will be identical to the results of the same experiment performed in a stationary vehicle. The formal statement of this result is called the **principle of Galilean relativity:**

Principle of Galilean relativity

> The laws of mechanics must be the same in all inertial frames of reference.

Let us consider an observation that illustrates the equivalence of the laws of mechanics in different inertial frames. A pickup truck moves with a constant velocity, as shown in Figure 39.1a. If a passenger in the truck throws a ball straight up, and if air effects are neglected, the passenger observes that the ball moves in a vertical path. The motion of the ball appears to be precisely the same as if the ball were thrown by a person at rest on the Earth. The law of universal gravitation and the equations of motion under constant acceleration are obeyed whether the truck is at rest or in uniform motion.

Both observers agree on the laws of physics—they each throw a ball straight up and it rises and falls back into their hand. What about the *path* of the ball thrown by the observer in the truck? Do the observers agree on the path? The observer on the ground sees the path of the ball as a parabola, as illustrated in Figure 39.1b, while, as mentioned earlier, the observer in the truck sees the ball move in a vertical path. Furthermore, according to the observer on the ground, the ball has a horizontal component of velocity equal to the velocity of the truck. Although the two observers disagree on certain aspects of the situation, they agree on the validity of Newton's laws and on such classical principles as conservation of energy and conservation of linear momentum. This agreement implies that no mechanical experiment can detect any difference between the two inertial frames. The only thing that can be detected is the relative motion of one frame with respect to the other.

> **Quick Quiz 39.1** Which observer in Figure 39.1 sees the ball's *correct* path? (a) the observer in the truck (b) the observer on the ground (c) both observers.

(a)   (b)

Figure 39.1 (a) The observer in the truck sees the ball move in a vertical path when thrown upward. (b) The Earth observer sees the path of the ball as a parabola.

Suppose that some physical phenomenon, which we call an *event,* occurs and is observed by an observer at rest in an inertial reference frame. The event's location and time of occurrence can be specified by the four coordinates (x, y, z, t). We would like to be able to transform these coordinates from those of an observer in one inertial frame to those of another observer in a frame moving with uniform relative velocity compared to the first frame. When we say an observer is "in a frame," we mean that the observer is at rest with respect to the origin of that frame.

Consider two inertial frames S and S′ (Fig. 39.2). The frame S′ moves with a constant velocity **v** along the common x and x' axes, where **v** is measured relative to S. We assume that the origins of S and S′ coincide at $t = 0$ and that an event occurs at point P in space at some instant of time. An observer in S describes the event with space–time coordinates (x, y, z, t), whereas an observer in S′ uses the coordinates (x', y', z', t') to describe the same event. As we see from the geometry in Figure 39.2, the relationships among these various coordinates can be written

$$x' = x - vt \qquad y' = y \qquad z' = z \qquad t' = t \qquad (39.1)$$

These equations are the **Galilean space–time transformation equations.** Note that time is assumed to be the same in both inertial frames. That is, within the framework of classical mechanics, all clocks run at the same rate, regardless of their velocity, so that the time at which an event occurs for an observer in S is the same as the time for the same event in S′. Consequently, the time interval between two successive events should be the same for both observers. Although this assumption may seem obvious, it turns out to be incorrect in situations where v is comparable to the speed of light.

Now suppose that a particle moves through a displacement of magnitude dx along the x axis in a time interval dt as measured by an observer in S. It follows from Equations 39.1 that the corresponding displacement dx' measured by an observer in S′ is $dx' = dx - v\,dt$, where frame S′ is moving with speed v in the x direction relative to frame S. Because $dt = dt'$, we find that

$$\frac{dx'}{dt'} = \frac{dx}{dt} - v$$

or

$$u'_x = u_x - v \qquad (39.2)$$

where u_x and u'_x are the x components of the velocity of the particle measured by observers in S and S′, respectively. (We use the symbol **u** for particle velocity rather than **v**, which is used for the relative velocity of two reference frames.) This is the **Galilean velocity transformation equation.** It is consistent with our intuitive notion of time and space as well as with our discussions in Section 4.6. As we shall soon see, however, it leads to serious contradictions when applied to electromagnetic waves.

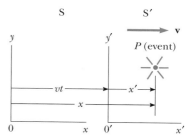

Figure 39.2 An event occurs at a point P. The event is seen by two observers in inertial frames S and S′, where S′ moves with a velocity **v** relative to S.

Galilean transformation equations

▲ **PITFALL PREVENTION**

39.1   The Relationship Between the S and S′ Frames

Many of the mathematical representations in this chapter are true *only* for the specified relationship between the S and S′ frames. The x and x' axes coincide, except that their origins are different. The y and y' axes (and the z and z' axes), are parallel, but do not coincide due to the displacement of the origin of S′ with respect to that of S. We choose the time $t = 0$ to be the instant at which the origins of the two coordinate systems coincide. If the S′ frame is moving in the positive x direction relative to S, v is positive; otherwise it is negative.

Quick Quiz 39.2 A baseball pitcher with a 90-mi/h fastball throws a ball while standing on a railroad flatcar moving at 110 mi/h. The ball is thrown in the same direction as that of the velocity of the train. Applying the Galilean velocity transformation equation, the speed of the ball relative to the Earth is (a) 90 mi/h (b) 110 mi/h (c) 20 mi/h (d) 200 mi/h (e) impossible to determine.

The Speed of Light

It is quite natural to ask whether the principle of Galilean relativity also applies to electricity, magnetism, and optics. Experiments indicate that the answer is no. Recall from Chapter 34 that Maxwell showed that the speed of light in free space is $c = 3.00 \times 10^8$ m/s. Physicists of the late 1800s thought that light waves moved through a medium called the *ether* and that the speed of light was c only in a special, absolute frame

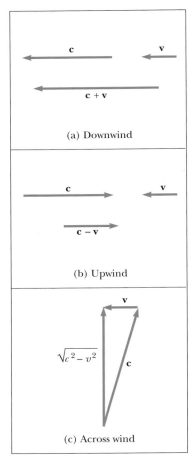

Figure 39.3 If the velocity of the ether wind relative to the Earth is **v** and the velocity of light relative to the ether is **c**, then the speed of light relative to the Earth is (a) $c + v$ in the downwind direction, (b) $c - v$ in the upwind direction, and (c) $(c^2 - v^2)^{1/2}$ in the direction perpendicular to the wind.

at rest with respect to the ether. The Galilean velocity transformation equation was expected to hold for observations of light made by an observer in any frame moving at speed v relative to the absolute ether frame. That is, if light travels along the x axis and an observer moves with velocity **v** along the x axis, the observer will measure the light to have speed $c \pm v$, depending on the directions of travel of the observer and the light.

Because the existence of a preferred, absolute ether frame would show that light was similar to other classical waves and that Newtonian ideas of an absolute frame were true, considerable importance was attached to establishing the existence of the ether frame. Prior to the late 1800s, experiments involving light traveling in media moving at the highest laboratory speeds attainable at that time were not capable of detecting differences as small as that between c and $c \pm v$. Starting in about 1880, scientists decided to use the Earth as the moving frame in an attempt to improve their chances of detecting these small changes in the speed of light.

As observers fixed on the Earth, we can take the view that we are stationary and that the absolute ether frame containing the medium for light propagation moves past us with speed v. Determining the speed of light under these circumstances is just like determining the speed of an aircraft traveling in a moving air current, or wind; consequently, we speak of an "ether wind" blowing through our apparatus fixed to the Earth.

A direct method for detecting an ether wind would use an apparatus fixed to the Earth to measure the ether wind's influence on the speed of light. If v is the speed of the ether relative to the Earth, then light should have its maximum speed $c + v$ when propagating downwind, as in Figure 39.3a. Likewise, the speed of light should have its minimum value $c - v$ when the light is propagating upwind, as in Figure 39.3b, and an intermediate value $(c^2 - v^2)^{1/2}$ in the direction perpendicular to the ether wind, as in Figure 39.3c. If the Sun is assumed to be at rest in the ether, then the velocity of the ether wind would be equal to the orbital velocity of the Earth around the Sun, which has a magnitude of approximately 3×10^4 m/s. Because $c = 3 \times 10^8$ m/s, it is necessary to detect a change in speed of about 1 part in 10^4 for measurements in the upwind or downwind directions. However, while such a change is experimentally measurable, all attempts to detect such changes and establish the existence of the ether wind (and hence the absolute frame) proved futile! We explore the classic experimental search for the ether in Section 39.2.

The principle of Galilean relativity refers only to the laws of mechanics. If it is assumed that the laws of electricity and magnetism are the same in all inertial frames, a paradox concerning the speed of light immediately arises. We can understand this by recognizing that Maxwell's equations seem to imply that the speed of light always has the fixed value 3.00×10^8 m/s in all inertial frames, a result in direct contradiction to what is expected based on the Galilean velocity transformation equation. According to Galilean relativity, the speed of light should *not* be the same in all inertial frames.

To resolve this contradiction in theories, we must conclude that either (1) the laws of electricity and magnetism are not the same in all inertial frames or (2) the Galilean velocity transformation equation is incorrect. If we assume the first alternative, then a preferred reference frame in which the speed of light has the value c must exist and the measured speed must be greater or less than this value in any other reference frame, in accordance with the Galilean velocity transformation equation. If we assume the second alternative, then we are forced to abandon the notions of absolute time and absolute length that form the basis of the Galilean space–time transformation equations.

39.2 The Michelson–Morley Experiment

The most famous experiment designed to detect small changes in the speed of light was first performed in 1881 by Albert A. Michelson (see Section 37.7) and later repeated under various conditions by Michelson and Edward W. Morley (1838–1923). We state at the outset that the outcome of the experiment contradicted the ether hypothesis.

The experiment was designed to determine the velocity of the Earth relative to that of the hypothetical ether. The experimental tool used was the Michelson interferometer, which was discussed in Section 37.7 and is shown again in Figure 39.4. Arm 2 is aligned along the direction of the Earth's motion through space. The Earth moving through the ether at speed v is equivalent to the ether flowing past the Earth in the opposite direction with speed v. This ether wind blowing in the direction opposite the direction of Earth's motion should cause the speed of light measured in the Earth frame to be $c - v$ as the light approaches mirror M_2 and $c + v$ after reflection, where c is the speed of light in the ether frame.

The two light beams reflect from M_1 and M_2 and recombine, and an interference pattern is formed, as discussed in Section 37.7. The interference pattern is observed while the interferometer is rotated through an angle of 90°. This rotation interchanges the speed of the ether wind between the arms of the interferometer. The rotation should cause the fringe pattern to shift slightly but measurably. Measurements failed, however, to show any change in the interference pattern! The Michelson–Morley experiment was repeated at different times of the year when the ether wind was expected to change direction and magnitude, but the results were always the same: **no fringe shift of the magnitude required was ever observed.**[2]

The negative results of the Michelson–Morley experiment not only contradicted the ether hypothesis but also showed that it was impossible to measure the absolute velocity of the Earth with respect to the ether frame. However, Einstein offered a postulate for his special theory of relativity that places quite a different interpretation on these null results. In later years, when more was known about the nature of light, the idea of an ether that permeates all of space was abandoned. **Light is now understood to be an electromagnetic wave, which requires no medium for its propagation.** As a result, the idea of an ether in which these waves travel became unnecessary.

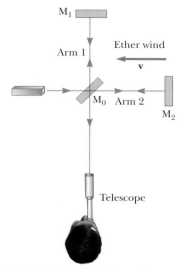

Active Figure 39.4 According to the ether wind theory, the speed of light should be $c - v$ as the beam approaches mirror M_2 and $c + v$ after reflection.

At the Active Figures link at http://www.pse6.com, you can adjust the speed of the ether wind to see the effect on the light beams if there were an ether.

Details of the Michelson–Morley Experiment

To understand the outcome of the Michelson–Morley experiment, let us assume that the two arms of the interferometer in Figure 39.4 are of equal length L. We shall analyze the situation as if there were an ether wind, because that is what Michelson and Morley expected to find. As noted above, the speed of the light beam along arm 2 should be $c - v$ as the beam approaches M_2 and $c + v$ after the beam is reflected. Thus, the time interval for travel to the right is $L/(c - v)$, and the time interval for travel to the left is $L/(c + v)$. The total time interval for the round trip along arm 2 is

$$\Delta t_{\text{arm 2}} = \frac{L}{c + v} + \frac{L}{c - v} = \frac{2Lc}{c^2 - v^2} = \frac{2L}{c}\left(1 - \frac{v^2}{c^2}\right)^{-1}$$

Now consider the light beam traveling along arm 1, perpendicular to the ether wind. Because the speed of the beam relative to the Earth is $(c^2 - v^2)^{1/2}$ in this case (see Fig. 39.3), the time interval for travel for each half of the trip is $L/(c^2 - v^2)^{1/2}$, and the total time interval for the round trip is

$$\Delta t_{\text{arm 1}} = \frac{2L}{(c^2 - v^2)^{1/2}} = \frac{2L}{c}\left(1 - \frac{v^2}{c^2}\right)^{-1/2}$$

Thus, the time difference Δt between the horizontal round trip (arm 2) and the vertical round trip (arm 1) is

$$\Delta t = \Delta t_{\text{arm 2}} - \Delta t_{\text{arm 1}} = \frac{2L}{c}\left[\left(1 - \frac{v^2}{c^2}\right)^{-1} - \left(1 - \frac{v^2}{c^2}\right)^{-1/2}\right]$$

---

[2] From an Earth observer's point of view, changes in the Earth's speed and direction of motion in the course of a year are viewed as ether wind shifts. Even if the speed of the Earth with respect to the ether were zero at some time, six months later the speed of the Earth would be 60 km/s with respect to the ether, and as a result a fringe shift should be noticed. No shift has ever been observed, however.

Because $v^2/c^2 \ll 1$, we can simplify this expression by using the following binomial expansion after dropping all terms higher than second order:

$$(1 - x)^n \approx 1 - nx \qquad (\text{for } x \ll 1)$$

In our case, $x = v^2/c^2$, and we find that

$$\Delta t = \Delta t_{\text{arm 2}} - \Delta t_{\text{arm 1}} \approx \frac{Lv^2}{c^3} \tag{39.3}$$

This time difference between the two instants at which the reflected beams arrive at the viewing telescope gives rise to a phase difference between the beams, producing an interference pattern when they combine at the position of the telescope. A shift in the interference pattern should be detected when the interferometer is rotated through 90° in a horizontal plane, so that the two beams exchange roles. This rotation results in a time difference twice that given by Equation 39.3. Thus, the path difference that corresponds to this time difference is

$$\Delta d = c\,(2\,\Delta t) = \frac{2Lv^2}{c^2}$$

Because a change in path length of one wavelength corresponds to a shift of one fringe, the corresponding fringe shift is equal to this path difference divided by the wavelength of the light:

$$\text{Shift} = \frac{2Lv^2}{\lambda c^2} \tag{39.4}$$

In the experiments by Michelson and Morley, each light beam was reflected by mirrors many times to give an effective path length L of approximately 11 m. Using this value and taking v to be equal to 3.0×10^4 m/s, the speed of the Earth around the Sun, we obtain a path difference of

$$\Delta d = \frac{2(11\text{ m})(3.0 \times 10^4\text{ m/s})^2}{(3.0 \times 10^8\text{ m/s})^2} = 2.2 \times 10^{-7}\text{ m}$$

This extra travel distance should produce a noticeable shift in the fringe pattern. Specifically, using 500-nm light, we expect a fringe shift for rotation through 90° of

$$\text{Shift} = \frac{\Delta d}{\lambda} = \frac{2.2 \times 10^{-7}\text{ m}}{5.0 \times 10^{-7}\text{ m}} \approx 0.44$$

The instrument used by Michelson and Morley could detect shifts as small as 0.01 fringe. However, **it detected no shift whatsoever in the fringe pattern.** Since then, the experiment has been repeated many times by different scientists under a wide variety of conditions, and no fringe shift has ever been detected. Thus, it was concluded that the motion of the Earth with respect to the postulated ether cannot be detected.

Many efforts were made to explain the null results of the Michelson–Morley experiment and to save the ether frame concept and the Galilean velocity transformation equation for light. All proposals resulting from these efforts have been shown to be wrong. No experiment in the history of physics received such valiant efforts to explain the absence of an expected result as did the Michelson–Morley experiment. The stage was set for Einstein, who solved the problem in 1905 with his special theory of relativity.

39.3 Einstein's Principle of Relativity

In the previous section we noted the impossibility of measuring the speed of the ether with respect to the Earth and the failure of the Galilean velocity transformation equation in the case of light. Einstein proposed a theory that boldly removed these difficulties and at the same time completely altered our notion of space and time.[3] He based his special theory of relativity on two postulates:

1. **The principle of relativity:** The laws of physics must be the same in all inertial reference frames.

2. **The constancy of the speed of light:** The speed of light in vacuum has the same value, $c = 3.00 \times 10^8$ m/s, in all inertial frames, regardless of the velocity of the observer or the velocity of the source emitting the light.

The first postulate asserts that *all* the laws of physics—those dealing with mechanics, electricity and magnetism, optics, thermodynamics, and so on—are the same in all reference frames moving with constant velocity relative to one another. This postulate is a sweeping generalization of the principle of Galilean relativity, which refers only to the laws of mechanics. From an experimental point of view, Einstein's principle of relativity means that any kind of experiment (measuring the speed of light, for example) performed in a laboratory at rest must give the same result when performed in a laboratory moving at a constant velocity with respect to the first one. Hence, no preferred inertial reference frame exists, and it is impossible to detect absolute motion.

Note that postulate 2 is required by postulate 1: if the speed of light were not the same in all inertial frames, measurements of different speeds would make it possible to distinguish between inertial frames; as a result, a preferred, absolute frame could be identified, in contradiction to postulate 1.

Although the Michelson–Morley experiment was performed before Einstein published his work on relativity, it is not clear whether or not Einstein was aware of the details of the experiment. Nonetheless, the null result of the experiment can be readily understood within the framework of Einstein's theory. According to his principle of relativity, the premises of the Michelson–Morley experiment were incorrect. In the process of trying to explain the expected results, we stated that when light traveled against the ether wind its speed was $c - v$, in accordance with the Galilean velocity transformation equation. However, if the state of motion of the observer or of the source has no influence on the value found for the speed of light, one always measures the value to be c. Likewise, the light makes the return trip after reflection from the mirror at speed c, not at speed $c + v$. Thus, the motion of the Earth does not influence the fringe pattern observed in the Michelson–Morley experiment, and a null result should be expected.

If we accept Einstein's theory of relativity, we must conclude that relative motion is unimportant when measuring the speed of light. At the same time, we shall see that we must alter our common-sense notion of space and time and be prepared for some surprising consequences. It may help as you read the pages ahead to keep in mind that our common-sense ideas are based on a lifetime of everyday experiences and not on observations of objects moving at hundreds of thousands of kilometers per second. Thus, these results will seem strange, but that is only because we have no experience with them.

39.4 Consequences of the Special Theory of Relativity

Before we discuss the consequences of Einstein's special theory of relativity, we must first understand how an observer located in an inertial reference frame describes an event. As mentioned earlier, an event is an occurrence describable by three space

Albert Einstein
German-American Physicist
(1879–1955)

Einstein, one of the greatest physicists of all times, was born in Ulm, Germany. In 1905, at the age of 26, he published four scientific papers that revolutionized physics. Two of these papers were concerned with what is now considered his most important contribution: the special theory of relativity.

In 1916, Einstein published his work on the general theory of relativity. The most dramatic prediction of this theory is the degree to which light is deflected by a gravitational field. Measurements made by astronomers on bright stars in the vicinity of the eclipsed Sun in 1919 confirmed Einstein's prediction, and as a result Einstein became a world celebrity.

Einstein was deeply disturbed by the development of quantum mechanics in the 1920s despite his own role as a scientific revolutionary. In particular, he could never accept the probabilistic view of events in nature that is a central feature of quantum theory. The last few decades of his life were devoted to an unsuccessful search for a unified theory that would combine gravitation and electromagnetism. *(AIP Niels Bohr Library)*

---

[3] A. Einstein, "On the Electrodynamics of Moving Bodies," *Ann. Physik* 17:891, 1905. For an English translation of this article and other publications by Einstein, see the book by H. Lorentz, A. Einstein, H. Minkowski, and H. Weyl, *The Principle of Relativity*, Dover, 1958.

coordinates and one time coordinate. Observers in different inertial frames will describe the same event with coordinates that have different values.

As we examine some of the consequences of relativity in the remainder of this section, we restrict our discussion to the concepts of simultaneity, time intervals, and lengths, all three of which are quite different in relativistic mechanics from what they are in Newtonian mechanics. For example, in relativistic mechanics the distance between two points and the time interval between two events depend on the frame of reference in which they are measured. That is, **in relativistic mechanics there is no such thing as an absolute length or absolute time interval.** Furthermore, **events at different locations that are observed to occur simultaneously in one frame are not necessarily observed to be simultaneous in another frame moving uniformly with respect to the first.**

Simultaneity and the Relativity of Time

A basic premise of Newtonian mechanics is that a universal time scale exists that is the same for all observers. In fact, Newton wrote that "Absolute, true, and mathematical time, of itself, and from its own nature, flows equably without relation to anything external." Thus, Newton and his followers simply took simultaneity for granted. In his special theory of relativity, Einstein abandoned this assumption.

Einstein devised the following thought experiment to illustrate this point. A boxcar moves with uniform velocity, and two lightning bolts strike its ends, as illustrated in Figure 39.5a, leaving marks on the boxcar and on the ground. The marks on the boxcar are labeled A' and B', and those on the ground are labeled A and B. An observer O' moving with the boxcar is midway between A' and B', and a ground observer O is midway between A and B. The events recorded by the observers are the striking of the boxcar by the two lightning bolts.

The light signals emitted from A and B at the instant at which the two bolts strike reach observer O at the same time, as indicated in Figure 39.5b. This observer realizes that the signals have traveled at the same speed over equal distances, and so rightly concludes that the events at A and B occurred simultaneously. Now consider the same events as viewed by observer O'. By the time the signals have reached observer O, observer O' has moved as indicated in Figure 39.5b. Thus, the signal from B' has already swept past O', but the signal from A' has not yet reached O'. In other words, O' sees the signal from B' before seeing the signal from A'. According to Einstein, *the two observers must find that light travels at the same speed.* Therefore, observer O' concludes that the lightning strikes the front of the boxcar before it strikes the back.

This thought experiment clearly demonstrates that the two events that appear to be simultaneous to observer O do not appear to be simultaneous to observer O'.

▲ PITFALL PREVENTION

39.2 Who's Right?

You might wonder which observer in Fig. 39.5 is correct concerning the two lightning strikes. *Both are correct,* because the principle of relativity states that *there is no preferred inertial frame of reference.* Although the two observers reach different conclusions, both are correct in their own reference frame because the concept of simultaneity is not absolute. This, in fact, is the central point of relativity—any uniformly moving frame of reference can be used to describe events and do physics.

Figure 39.5 (a) Two lightning bolts strike the ends of a moving boxcar. (b) The events appear to be simultaneous to the stationary observer O, standing midway between A and B. The events do not appear to be simultaneous to observer O', who claims that the front of the car is struck before the rear. Note that in (b) the leftward-traveling light signal has already passed O' but the rightward-traveling signal has not yet reached O'.

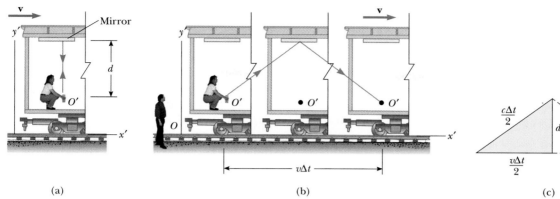

Active Figure 39.6 (a) A mirror is fixed to a moving vehicle, and a light pulse is sent out by observer O' at rest in the vehicle. (b) Relative to a stationary observer O standing alongside the vehicle, the mirror and O' move with a speed v. Note that what observer O measures for the distance the pulse travels is greater than $2d$. (c) The right triangle for calculating the relationship between Δt and Δt_p.

At the Active Figures link at http://www.pse6.com, **you can observe the bouncing of the light pulse for various speeds of the train.**

In other words,

> two events that are simultaneous in one reference frame are in general not simultaneous in a second frame moving relative to the first. That is, simultaneity is not an absolute concept but rather one that depends on the state of motion of the observer.

Einstein's thought experiment demonstrates that two observers can disagree on the simultaneity of two events. **This disagreement, however, depends on the transit time of light to the observers and, therefore, does *not* demonstrate the deeper meaning of relativity.** In relativistic analyses of high-speed situations, relativity shows that simultaneity is relative even when the transit time is subtracted out. In fact, all of the relativistic effects that we will discuss from here on will assume that we are ignoring differences caused by the transit time of light to the observers.

Time Dilation

We can illustrate the fact that observers in different inertial frames can measure different time intervals between a pair of events by considering a vehicle moving to the right with a speed v, such as the boxcar shown in Figure 39.6a. A mirror is fixed to the ceiling of the vehicle, and observer O' at rest in the frame attached to the vehicle holds a flashlight a distance d below the mirror. At some instant, the flashlight emits a pulse of light directed toward the mirror (event 1), and at some later time after reflecting from the mirror, the pulse arrives back at the flashlight (event 2). Observer O' carries a clock and uses it to measure the time interval Δt_p between these two events. (The subscript p stands for *proper*, as we shall see in a moment.) Because the light pulse has a speed c, the time interval required for the pulse to travel from O' to the mirror and back is

$$\Delta t_p = \frac{\text{distance traveled}}{\text{speed}} = \frac{2d}{c} \qquad (39.5)$$

Now consider the same pair of events as viewed by observer O in a second frame, as shown in Figure 39.6b. According to this observer, the mirror and flashlight are moving to the right with a speed v, and as a result the sequence of events appears entirely different. By the time the light from the flashlight reaches the mirror, the mirror has moved to the right a distance $v\,\Delta t/2$, where Δt is the time interval required for the light to travel from O' to the mirror and back to O' as measured by O. In other words, O concludes that, because of the motion of the vehicle, if the light is to hit the mirror, it must leave the flashlight at an angle with respect to the vertical direction. Comparing Figure 39.6a and b, we see that the light must travel farther in (b) than in (a). (Note that neither observer "knows" that he or she is moving. Each is at rest in his or her own inertial frame.)

According to the second postulate of the special theory of relativity, both observers must measure c for the speed of light. Because the light travels farther according to O, it follows that the time interval Δt measured by O is longer than the time interval Δt_p measured by O'. To obtain a relationship between these two time intervals, it is convenient to use the right triangle shown in Figure 39.6c. The Pythagorean theorem gives

$$\left(\frac{c\,\Delta t}{2}\right)^2 = \left(\frac{v\,\Delta t}{2}\right)^2 + d^2$$

Solving for Δt gives

$$\Delta t = \frac{2d}{\sqrt{c^2 - v^2}} = \frac{2d}{c\sqrt{1 - \dfrac{v^2}{c^2}}} \tag{39.6}$$

Because $\Delta t_p = 2d/c$, we can express this result as

$$\Delta t = \frac{\Delta t_p}{\sqrt{1 - \dfrac{v^2}{c^2}}} = \gamma\,\Delta t_p \tag{39.7}$$

where

$$\gamma = \frac{1}{\sqrt{1 - \dfrac{v^2}{c^2}}} \tag{39.8}$$

Because γ is always greater than unity, this result says that **the time interval Δt measured by an observer moving with respect to a clock is longer than the time interval Δt_p measured by an observer at rest with respect to the clock.** This effect is known as **time dilation.**

We can see that time dilation is not observed in our everyday lives by considering the factor γ. This factor deviates significantly from a value of 1 only for very high speeds, as shown in Figure 39.7 and Table 39.1. For example, for a speed of $0.1c$, the value of γ is 1.005. Thus, there is a time dilation of only 0.5% at one-tenth the speed of light. Speeds that we encounter on an everyday basis are far slower than this, so we do not see time dilation in normal situations.

The time interval Δt_p in Equations 39.5 and 39.7 is called the **proper time interval.** (In German, Einstein used the term *Eigenzeit*, which means "own-time.") In

Time dilation

Table 39.1

Approximate Values for γ at Various Speeds	
v/c	γ
0.001 0	1.000 000 5
0.010	1.000 05
0.10	1.005
0.20	1.021
0.30	1.048
0.40	1.091
0.50	1.155
0.60	1.250
0.70	1.400
0.80	1.667
0.90	2.294
0.92	2.552
0.94	2.931
0.96	3.571
0.98	5.025
0.99	7.089
0.995	10.01
0.999	22.37

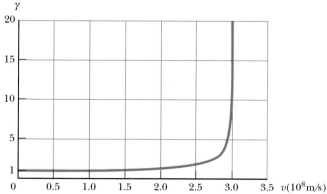

Figure 39.7 Graph of γ versus v. As the speed approaches that of light, γ increases rapidly.

general, **the proper time interval is the time interval between two events measured by an observer who sees the events occur at the same point in space.**

If a clock is moving with respect to you, the time interval between ticks of the moving clock is observed to be longer than the time interval between ticks of an identical clock in your reference frame. Thus, it is often said that a moving clock is measured to run more slowly than a clock in your reference frame by a factor γ. This is true for mechanical clocks as well as for the light clock just described. We can generalize this result by stating that all physical processes, including chemical and biological ones, are measured to slow down when those processes occur in a frame moving with respect to the observer. For example, the heartbeat of an astronaut moving through space would keep time with a clock inside the spacecraft. Both the astronaut's clock and heartbeat would be measured to slow down according to an observer on Earth comparing time intervals with his own clock (although the astronaut would have no sensation of life slowing down in the spacecraft).

Quick Quiz 39.3 Suppose the observer O' on the train in Figure 39.6 aims her flashlight at the far wall of the boxcar and turns it on and off, sending a pulse of light toward the far wall. Both O' and O measure the time interval between when the pulse leaves the flashlight and it hits the far wall. Which observer measures the proper time interval between these two events? (a) O' (b) O (c) both observers (d) neither observer.

Quick Quiz 39.4 A crew watches a movie that is two hours long in a spacecraft that is moving at high speed through space. Will an Earthbound observer, who is watching the movie through a powerful telescope, measure the duration of the movie to be (a) longer than, (b) shorter than, or (c) equal to two hours?

Strange as it may seem, time dilation is a verifiable phenomenon. An experiment reported by Hafele and Keating provided direct evidence of time dilation.[4] Time intervals measured with four cesium atomic clocks in jet flight were compared with time intervals measured by Earth-based reference atomic clocks. In order to compare these results with theory, many factors had to be considered, including periods of speeding up and slowing down relative to the Earth, variations in direction of travel, and the fact that the gravitational field experienced by the flying clocks was weaker than that experienced by the Earth-based clock. The results were in good agreement with the predictions of the special theory of relativity and can be explained in terms of the relative motion between the Earth and the jet aircraft. In their paper, Hafele and Keating stated that "Relative to the atomic time scale of the U.S. Naval Observatory, the flying clocks lost 59 ± 10 ns during the eastward trip and gained 273 ± 7 ns during the westward trip. . . . These results provide an unambiguous empirical resolution of the famous clock paradox with macroscopic clocks."

Another interesting example of time dilation involves the observation of *muons,* unstable elementary particles that have a charge equal to that of the electron and a mass 207 times that of the electron. (We will study the muon and other particles in Chapter 46.) Muons can be produced by the collision of cosmic radiation with atoms high in the atmosphere. Slow-moving muons in the laboratory have a lifetime which is measured to be the proper time interval $\Delta t_p = 2.2 \ \mu s$. If we assume that the speed of atmospheric muons is close to the speed of light, we find that these particles can travel a distance of approximately $(3.0 \times 10^8 \ \text{m/s})(2.2 \times 10^{-6} \ \text{s}) \approx 6.6 \times 10^2 \ \text{m}$ before they decay (Fig. 39.8a). Hence, they are unlikely to reach the surface of the Earth from

---

[4] J. C. Hafele and R. E. Keating, "Around the World Atomic Clocks: Relativistic Time Gains Observed," *Science,* 177:168, 1972.

▲ **PITFALL PREVENTION**

39.3 The Proper Time Interval

It is *very* important in relativistic calculations to correctly identify the observer who measures the proper time interval. The proper time interval between two events is always the time interval measured by an observer for whom the two events take place at the same position.

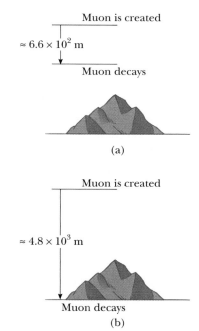

Figure 39.8 (a) Without relativistic considerations, muons created in the atmosphere and traveling downward with a speed of $0.99c$ travel only about 6.6×10^2 m before decaying with an average lifetime of $2.2 \ \mu s$. Thus, very few muons reach the surface of the Earth. (b) With relativistic considerations, the muon's lifetime is dilated according to an observer on Earth. As a result, according to this observer, the muon can travel about 4.8×10^3 m before decaying. This results in many of them arriving at the surface.

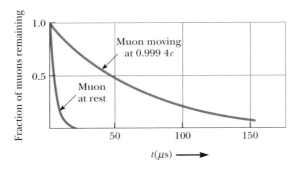

Figure 39.9 Decay curves for muons at rest and for muons traveling at a speed of 0.999 4c.

high in the atmosphere where they are produced. However, experiments show that a large number of muons *do* reach the surface. The phenomenon of time dilation explains this effect. As measured by an observer on Earth, the muons have a dilated lifetime equal to $\gamma \, \Delta t_p$. For example, for $v = 0.99c$, $\gamma \approx 7.1$ and $\gamma \, \Delta t_p \approx 16 \, \mu s$. Hence, the average distance traveled by the muons in this time as measured by an observer on Earth is approximately $(0.99)(3.0 \times 10^8 \text{ m/s})(16 \times 10^{-6} \text{ s}) \approx 4.8 \times 10^3$ m, as indicated in Figure 39.8b.

In 1976, at the laboratory of the European Council for Nuclear Research (CERN) in Geneva, muons injected into a large storage ring reached speeds of approximately 0.999 4c. Electrons produced by the decaying muons were detected by counters around the ring, enabling scientists to measure the decay rate and hence the muon lifetime. The lifetime of the moving muons was measured to be approximately 30 times as long as that of the stationary muon (Fig. 39.9), in agreement with the prediction of relativity to within two parts in a thousand.

Example 39.1   What Is the Period of the Pendulum?

The period of a pendulum is measured to be 3.00 s in the reference frame of the pendulum. What is the period when measured by an observer moving at a speed of 0.950c relative to the pendulum?

Solution To conceptualize this problem, let us change frames of reference. Instead of the observer moving at 0.950c, we can take the equivalent point of view that the observer is at rest and the pendulum is moving at 0.950c past the stationary observer. Hence, the pendulum is an example of a clock moving at high speed with respect to an observer and we can categorize this problem as one involving time dilation.

To analyze the problem, note that the proper time interval, measured in the rest frame of the pendulum, is $\Delta t_p = 3.00$ s. Because a clock moving with respect to an observer is measured to run more slowly than a stationary clock by a factor γ, Equation 39.7 gives

$$\Delta t = \gamma \, \Delta t_p = \frac{1}{\sqrt{1 - \dfrac{(0.950c)^2}{c^2}}} \, \Delta t_p = \frac{1}{\sqrt{1 - 0.902}} \, \Delta t_p$$

$$= (3.20)(3.00 \text{ s}) = \boxed{9.60 \text{ s}}$$

To finalize this problem, we see that indeed a moving pendulum is measured to take longer to complete a period

than a pendulum at rest does. The period increases by a factor of $\gamma = 3.20$. We see that this is consistent with Table 39.1, where this value lies between those for γ for $v/c = 0.94$ and $v/c = 0.96$.

What If? What if we increase the speed of the observer by 5.00%? Does the dilated time interval increase by 5.00%?

Answer Based on the highly nonlinear behavior of γ as a function of v in Figure 39.7, we would guess that the increase in Δt would be different from 5.00%. Increasing v by 5.00% gives us

$$v_{\text{new}} = (1.050 \, 0)(0.950c) = 0.997 \, 5c$$

(Because γ varies so rapidly with v when v is this large, we will keep one additional significant figure until the final answer.) If we perform the time dilation calculation again, we find that

$$\Delta t = \gamma \, \Delta t_p = \frac{1}{\sqrt{1 - \dfrac{(0.997 \, 5c)^2}{c^2}}} \, \Delta t_p = \frac{1}{\sqrt{1 - 0.995 \, 0}} \, \Delta t_p$$

$$= (14.15)(3.00 \text{ s}) = 42.5 \text{ s}$$

Thus, the 5.00% increase in speed has caused over a 300% increase in the dilated time!

Example 39.2 How Long Was Your Trip?

Suppose you are driving your car on a business trip and are traveling at 30 m/s. Your boss, who is waiting at your destination, expects the trip to take 5.0 h. When you arrive late, your excuse is that your car clock registered the passage of 5.0 h but that you were driving fast and so your clock ran more slowly than your boss's clock. If your car clock actually did indicate a 5.0-h trip, how much time passed on your boss's clock, which was at rest on the Earth?

Solution We begin by calculating γ from Equation 39.8:

$$\gamma = \frac{1}{\sqrt{1 - \dfrac{v^2}{c^2}}} = \frac{1}{\sqrt{1 - \dfrac{(3 \times 10^1 \text{ m/s})^2}{(3 \times 10^8 \text{ m/s})^2}}}$$

$$= \frac{1}{\sqrt{1 - 10^{-14}}}$$

If you try to determine this value on your calculator, you will probably obtain $\gamma = 1$. However, if we perform a binomial expansion, we can more precisely determine the value as

$$\gamma = (1 - 10^{-14})^{-1/2} \approx 1 + \tfrac{1}{2}(10^{-14}) = 1 + 5.0 \times 10^{-15}$$

This result indicates that at typical automobile speeds, γ is not much different from 1.

Applying Equation 39.7, we find Δt, the time interval measured by your boss, to be

$$\Delta t = \gamma \, \Delta t_p = (1 + 5.0 \times 10^{-15})(5.0 \text{ h})$$

$$= 5.0 \text{ h} + 2.5 \times 10^{-14} \text{ h} = \boxed{5.0 \text{ h} + 0.09 \text{ ns}}$$

Your boss's clock would be only 0.09 ns ahead of your car clock. You might want to think of another excuse!

The Twin Paradox

An intriguing consequence of time dilation is the so-called *twin paradox* (Fig. 39.10). Consider an experiment involving a set of twins named Speedo and Goslo. When they are 20 yr old, Speedo, the more adventuresome of the two, sets out on an epic journey to Planet X, located 20 ly from the Earth. (Note that 1 lightyear (ly) is the distance light travels through free space in 1 year.) Furthermore, Speedo's spacecraft is capable of reaching a speed of $0.95c$ relative to the inertial frame of his twin brother back home. After reaching Planet X, Speedo becomes homesick and immediately returns to the Earth at the same speed $0.95c$. Upon his return, Speedo is shocked to discover that Goslo has aged 42 yr and is now 62 yr old. Speedo, on the other hand, has aged only 13 yr.

At this point, it is fair to raise the following question—which twin is the traveler and which is really younger as a result of this experiment? From Goslo's frame of reference, he was at rest while his brother traveled at a high speed away from him and then came back. According to Speedo, however, he himself remained stationary while Goslo and the Earth raced away from him and then headed back. This leads to an apparent

Speedo   Goslo   Speedo   Goslo
        (a)   (b)

Figure 39.10 (a) As one twin leaves his brother on the Earth, both are the same age. (b) When Speedo returns from his journey to Planet X, he is younger than his twin Goslo.

contradiction due to the apparent symmetry of the observations. Which twin has developed signs of excess aging?

The situation in our current problem is actually not symmetrical. To resolve this apparent paradox, recall that the special theory of relativity describes observations made in inertial frames of reference moving relative to each other. Speedo, the space traveler, must experience a series of accelerations during his journey because he must fire his rocket engines to slow down and start moving back toward Earth. As a result, his speed is not always uniform, and consequently he is not in an inertial frame. Therefore, there is no paradox—only Goslo, who is always in a single inertial frame, can make correct predictions based on special relativity. During each passing year noted by Goslo, slightly less than 4 months elapses for Speedo.

Only Goslo, who is in a single inertial frame, can apply the simple time-dilation formula to Speedo's trip. Thus, Goslo finds that instead of aging 42 yr, Speedo ages only $(1 - v^2/c^2)^{1/2}(42\ \text{yr}) = 13\ \text{yr}$. Thus, according to Goslo, Speedo spends 6.5 yr traveling to Planet X and 6.5 yr returning, for a total travel time of 13 yr, in agreement with our earlier statement.

Quick Quiz 39.5 Suppose astronauts are paid according to the amount of time they spend traveling in space. After a long voyage traveling at a speed approaching c, would a crew rather be paid according to (a) an Earth-based clock, (b) their spacecraft's clock, or (c) either clock?

Length Contraction

The measured distance between two points also depends on the frame of reference. **The proper length L_p of an object is the length measured by someone at rest relative to the object.** The length of an object measured by someone in a reference frame that is moving with respect to the object is always less than the proper length. This effect is known as **length contraction.**

Consider a spacecraft traveling with a speed v from one star to another. There are two observers: one on the Earth and the other in the spacecraft. The observer at rest on the Earth (and also assumed to be at rest with respect to the two stars) measures the distance between the stars to be the proper length L_p. According to this observer, the time interval required for the spacecraft to complete the voyage is $\Delta t = L_p/v$. The passages of the two stars by the spacecraft occur at the same position for the space traveler. Thus, the space traveler measures the proper time interval Δt_p. Because of time dilation, the proper time interval is related to the Earth-measured time interval by $\Delta t_p = \Delta t/\gamma$. Because the space traveler reaches the second star in the time Δt_p, he or she concludes that the distance L between the stars is

$$L = v\,\Delta t_p = v\,\frac{\Delta t}{\gamma}$$

Because the proper length is $L_p = v\,\Delta t$, we see that

Length contraction

$$L = \frac{L_p}{\gamma} = L_p\sqrt{1 - \frac{v^2}{c^2}} \tag{39.9}$$

where $\sqrt{1 - v^2/c^2}$ is a factor less than unity. **If an object has a proper length L_p when it is measured by an observer at rest with respect to the object, then when it moves with speed v in a direction parallel to its length, its length L is measured to be shorter according to $L = L_p\sqrt{1 - v^2/c^2} = L_p/\gamma$.**

▲ **PITFALL PREVENTION**

39.4 The Proper Length

As with the proper time interval, it is *very* important in relativistic calculations to correctly identify the observer who measures the proper length. The proper length between two points in space is always the length measured by an observer at rest with respect to the points. Often the proper time interval and the proper length are *not* measured by the same observer.

For example, suppose that a meter stick moves past a stationary Earth observer with speed v, as in Figure 39.11. The length of the stick as measured by an observer in a frame attached to the stick is the proper length L_p shown in Figure 39.11a. The length of the stick L measured by the Earth observer is shorter than L_p by the factor $(1 - v^2/c^2)^{1/2}$. Note that **length contraction takes place only along the direction of motion.**

The proper length and the proper time interval are defined differently. The proper length is measured by an observer for whom the end points of the length remain fixed in space. The proper time interval is measured by someone for whom the two events take place at the same position in space. As an example of this point, let us return to the decaying muons moving at speeds close to the speed of light. An observer in the muon's reference frame would measure the proper lifetime, while an Earth-based observer would measure the proper length (the distance from creation to decay in Figure 39.8). In the muon's reference frame, there is no time dilation but the distance of travel to the surface is observed to be shorter when measured in this frame. Likewise, in the Earth observer's reference frame, there is time dilation, but the distance of travel is measured to be the proper length. Thus, when calculations on the muon are performed in both frames, the outcome of the experiment in one frame is the same as the outcome in the other frame—more muons reach the surface than would be predicted without relativistic effects.

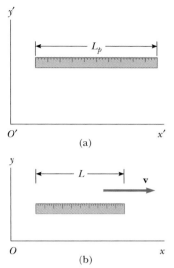

Active Figure 39.11 (a) A meter stick measured by an observer in a frame attached to the stick (that is, both have the same velocity) has its proper length L_p. (b) The stick measured by an observer in a frame in which the stick has a velocity **v** relative to the frame is measured to be shorter than its proper length L_p by a factor $(1 - v^2/c^2)^{1/2}$.

At the Active Figures link at http://www.pse6.com, you can view the meter stick from the points of view of two observers to compare the measured length of the stick.

Quick Quiz 39.6 You are packing for a trip to another star. During the journey, you will be traveling at $0.99c$. You are trying to decide whether you should buy smaller sizes of your clothing, because you will be thinner on your trip, due to length contraction. Also, you are considering saving money by reserving a smaller cabin to sleep in, because you will be shorter when you lie down. Should you (a) buy smaller sizes of clothing, (b) reserve a smaller cabin, (c) do neither of these, or (d) do both of these?

Quick Quiz 39.7 You are observing a spacecraft moving away from you. You measure it to be shorter than when it was at rest on the ground next to you. You also see a clock through the spacecraft window, and you observe that the passage of time on the clock is measured to be slower than that of the watch on your wrist. Compared to when the spacecraft was on the ground, what do you measure if the spacecraft turns around and comes *toward* you at the same speed? (a) The spacecraft is measured to be longer and the clock runs faster. (b) The spacecraft is measured to be longer and the clock runs slower. (c) The spacecraft is measured to be shorter and the clock runs faster. (d) The spacecraft is measured to be shorter and the clock runs slower.

Space–Time Graphs

It is sometimes helpful to make a *space–time graph*, in which ct is the ordinate and position x is the abscissa. The twin paradox is displayed in such a graph in Figure 39.12

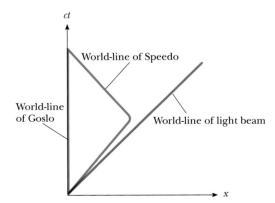

Figure 39.12 The twin paradox on a space–time graph. The twin who stays on the Earth has a world-line along the ct axis. The path of the traveling twin through space–time is represented by a world-line that changes direction.

from the point of view of Goslo. A path through space–time is called a **world-line.** At the origin, the world-lines of Speedo and Goslo coincide because the twins are in the same location at the same time. After Speedo leaves on his trip, his world-line diverges from that of his brother. Goslo's world-line is vertical because he remains fixed in location. At their reunion, the two world-lines again come together. Note that it would be impossible for Speedo to have a world-line that crossed the path of a light beam that left the Earth when he did. To do so would require him to have a speed greater than c (not possible, as shown in Sections 39.6 and 39.7).

World-lines for light beams are diagonal lines on space–time graphs, typically drawn at 45° to the right or left of vertical (assuming that the x and ct axes have the same scales), depending on whether the light beam is traveling in the direction of increasing or decreasing x. These two world-lines mean that all possible future events for Goslo and Speedo lie within two 45° lines extending from the origin. Either twin's presence at an event outside this "light cone" would require that twin to move at a speed greater than c, which we have said is not possible. Also, the only past events that Goslo and Speedo could have experienced occurred within two similar 45° world-lines that approach the origin from below the x axis.

Example 39.3   The Contraction of a Spacecraft

A spacecraft is measured to be 120.0 m long and 20.0 m in diameter while at rest relative to an observer. If this spacecraft now flies by the observer with a speed of $0.99c$, what length and diameter does the observer measure?

Solution From Equation 39.9, the length measured by the

observer is

$$L = L_p \sqrt{1 - \frac{v^2}{c^2}} = (120.0 \text{ m}) \sqrt{1 - \frac{(0.99c)^2}{c^2}} = \boxed{17 \text{ m}}$$

The diameter measured by the observer is still 20.0 m because the diameter is a dimension perpendicular to the motion and length contraction occurs only along the direction of motion.

Example 39.4   The Pole-in-the-Barn Paradox   `Interactive`

The twin paradox, discussed earlier, is a classic "paradox" in relativity. Another classic "paradox" is this: Suppose a runner moving at $0.75c$ carries a horizontal pole 15 m long toward a barn that is 10 m long. The barn has front and rear doors. An observer on the ground can instantly and simultaneously open and close the two doors by remote control. When the runner and the pole are inside the barn, the ground observer closes and then opens both doors so that the runner and pole are momentarily captured inside the barn and then proceed to exit the barn from the back door. Do both the runner and the ground observer agree that the runner makes it safely through the barn?

Solution From our everyday experience, we would be surprised to see a 15-m pole fit inside a 10-m barn. But the pole is in motion with respect to the ground observer, who measures the pole to be contracted to a length L_{pole}, where

$$L_{pole} = L_p \sqrt{1 - \frac{v^2}{c^2}} = (15 \text{ m}) \sqrt{1 - (0.75)^2} = 9.9 \text{ m}$$

Thus, the ground observer measures the pole to be slightly shorter than the barn and there is no problem with momentarily capturing the pole inside it. The "paradox" arises when we consider the runner's point of view. The runner

sees the barn contracted to

$$L_{barn} = L_p \sqrt{1 - \frac{v^2}{c^2}} = (10 \text{ m}) \sqrt{1 - (0.75)^2} = \boxed{6.6 \text{ m}}$$

Because the pole is in the rest frame of the runner, the runner measures it to have its proper length of 15 m. How can a 15-m pole fit inside a 6.6-m barn? While this is the classic question that is often asked, this is not the question we have asked, because it is not the important question. We asked *if the runner can make it safely through the barn.*

The resolution of the "paradox" lies in the relativity of simultaneity. The closing of the two doors is measured to be simultaneous by the ground observer. Because the doors are at different positions, however, they do not close simultaneously as measured by the runner. The rear door closes and then opens first, allowing the leading edge of the pole to exit. The front door of the barn does not close until the trailing edge of the pole passes by.

We can analyze this using a space-time graph. Figure 39.13a is a space–time graph from the ground observer's point of view. We choose $x = 0$ as the position of the front door of the barn and $t = 0$ as the instant at which the leading end of the pole is located at the front door of the barn. The world-lines for the two ends of the barn are separated by 10 m and are vertical because the barn is not moving relative to this observer. For the pole, we follow two tilted world-lines, one

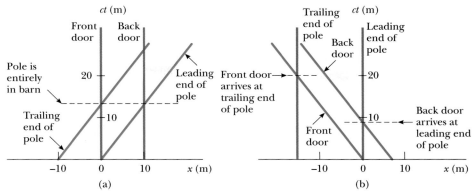

Figure 39.13 (Example 39.4) Space–time graphs for the pole-in-the-barn paradox.
(a) From the ground observer's point of view, the world-lines for the front and back
doors of the barn are vertical lines. The world-lines for the ends of the pole are
tilted and are 9.9 m apart horizontally. The front door of the barn is at $x = 0$, and
the leading end of the pole enters the front door at $t = 0$. The entire pole is inside
the barn at the time indicated by the dashed line. (b) From the runner's point of
view, the world-lines for the ends of the pole are vertical. The barn is moving in the
negative direction, so the world-lines for the front and back doors are tilted to the
left. The leading end of the pole exits the back door before the trailing end arrives
at the front door.

for each end of the moving pole. These world-lines are 9.9 m
apart horizontally, which is the contracted length seen by the
ground observer. As seen in Figure 39.13a, at one instant, the
pole is entirely within the barn.

Figure 39.13b shows the space–time graph according to
the runner. Here, the world-lines for the pole are separated by
15 m and are vertical because the pole is at rest in the runner's
frame of reference. The barn is hurtling *toward* the runner, so
the world-lines for the front and rear doors of the barn are
tilted in the opposite direction compared to Figure 39.13a.
The world-lines for the barn are separated by 6.6 m, the
contracted length as seen by the runner. Notice that the front
of the pole leaves the rear door of the barn long before the
back of the pole enters the barn. Thus, the opening of the
rear door occurs before the closing of the front door.

From the ground observer's point of view, the time at
which the trailing end of the pole enters the barn is found
from

$$\Delta t = t - 0 = t = \frac{\Delta x}{v} = \frac{9.9 \text{ m}}{0.75c} = \frac{13.2 \text{ m}}{c}$$

Thus, the pole should be completely inside the barn at a
time corresponding to $ct = 13.2$ m. This is consistent with
the point on the ct axis in Figure 39.13a where the pole is
inside the barn.

From the runner's point of view, the time at which the
leading end of the pole leaves the barn is found from

$$\Delta t = t - 0 = t = \frac{\Delta x}{v} = \frac{6.6 \text{ m}}{0.75c} = \frac{8.8 \text{ m}}{c}$$

leading to $ct = 8.8$ m. This is consistent with the point on
the ct axis in Figure 39.13b where the back door of the barn
arrives at the leading end of the pole. Finally, the time at
which the trailing end of the pole enters the front door of
the barn is found from

$$\Delta t = t - 0 = t = \frac{\Delta x}{v} = \frac{15 \text{ m}}{0.75c} = \frac{20 \text{ m}}{c}$$

This gives $ct = 20$ m, which agrees with the instant shown in
Figure 39.13b.

 Investigate the pole-in-the-barn paradox at the Interactive Worked Example link at **http://www.pse6.com.**

Example 39.5   A Voyage to Sirius

An astronaut takes a trip to Sirius, which is located a
distance of 8 lightyears from the Earth. The astronaut
measures the time of the one-way journey to be 6 yr. If the
spaceship moves at a constant speed of 0.8c, how can the 8-ly
distance be reconciled with the 6-yr trip time measured by
the astronaut?

Solution The distance of 8 ly represents the proper length
from the Earth to Sirius measured by an observer seeing
both objects nearly at rest. The astronaut sees Sirius
approaching her at 0.8c but also sees the distance

contracted to

$$\frac{8 \text{ ly}}{\gamma} = (8 \text{ ly}) \sqrt{1 - \frac{v^2}{c^2}} = (8 \text{ ly}) \sqrt{1 - \frac{(0.8c)^2}{c^2}} = 5 \text{ ly}$$

Thus, the travel time measured on her clock is

$$\Delta t = \frac{d}{v} = \frac{5 \text{ ly}}{0.8c} = 6 \text{ yr}$$

Note that we have used the value for the speed of light as
$c = 1$ ly/yr.

What If? What if this trip is observed with a very powerful telescope by a technician in Mission Control on Earth? At what time will this technician *see* that the astronaut has arrived at Sirius?

Answer The time interval that the technician will measure for the astronaut to arrive is

$$\Delta t = \frac{d}{v} = \frac{8 \text{ ly}}{0.8c} = 10 \text{ yr}$$

In order for the technician to *see* the arrival, the light from the scene of the arrival must travel back to Earth and enter

the telescope. This will require a time interval of

$$\Delta t = \frac{d}{v} = \frac{8 \text{ ly}}{c} = 8 \text{ yr}$$

Thus, the technician sees the arrival after 10 yr + 8 yr = 18 yr. Notice that if the astronaut immediately turns around and comes back home, she arrives, according to the technician, 20 years after leaving, only 2 years after he *saw* her arrive! In addition, she would have aged by only 12 years.

The Relativistic Doppler Effect

Another important consequence of time dilation is the shift in frequency found for light emitted by atoms in motion as opposed to light emitted by atoms at rest. This phenomenon, known as the Doppler effect, was introduced in Chapter 17 as it pertains to sound waves. In the case of sound, the motion of the source with respect to the medium of propagation can be distinguished from the motion of the observer with respect to the medium. Light waves must be analyzed differently, however, because they require no medium of propagation, and no method exists for distinguishing the motion of a light source from the motion of the observer.

If a light source and an observer approach each other with a relative speed v, the frequency f_{obs} measured by the observer is

$$f_{obs} = \frac{\sqrt{1 + v/c}}{\sqrt{1 - v/c}} f_{source} \tag{39.10}$$

where f_{source} is the frequency of the source measured in its rest frame. Note that this relativistic Doppler shift equation, unlike the Doppler shift equation for sound, depends only on the relative speed v of the source and observer and holds for relative speeds as great as c. As you might expect, the equation predicts that $f_{obs} > f_{source}$ when the source and observer approach each other. We obtain the expression for the case in which the source and observer recede from each other by substituting negative values for v in Equation 39.10.

The most spectacular and dramatic use of the relativistic Doppler effect is the measurement of shifts in the frequency of light emitted by a moving astronomical object such as a galaxy. Light emitted by atoms and normally found in the extreme violet region of the spectrum is shifted toward the red end of the spectrum for atoms in other galaxies—indicating that these galaxies are *receding* from us. The American astronomer Edwin Hubble (1889–1953) performed extensive measurements of this *red shift* to confirm that most galaxies are moving away from us, indicating that the Universe is expanding.

Figure 39.14 Events occur at points P and Q and are observed by an observer at rest in the S frame and another in the S′ frame, which is moving to the right with a speed v.

39.5   The Lorentz Transformation Equations

Suppose an event that occurs at some point P is reported by two observers, one at rest in a frame S and another in a frame S′ that is moving to the right with speed v as in Figure 39.14. The observer in S reports the event with space–time coordinates (x, y, z, t), while the observer in S′ reports the same event using the coordinates (x', y', z', t'). If two events occur at P and Q, Equation 39.1 predicts that $\Delta x = \Delta x'$, that is, the distance between the two points in space at which the events occur does not depend on motion of the observer. Because this is contradictory to the notion of length contraction, the Galilean transformation is not valid when v approaches the speed of light. In this section, we state the correct transformation equations that apply for all speeds in the range $0 \leq v < c$.

The equations that are valid for all speeds and enable us to transform coordinates from S to S′ are the **Lorentz transformation equations:**

$$x' = \gamma(x - vt) \qquad y' = y \qquad z' = z \qquad t' = \gamma\left(t - \frac{v}{c^2}x\right) \qquad \text{(39.11)}$$

Lorentz transformation for S → S′

These transformation equations were developed by Hendrik A. Lorentz (1853–1928) in 1890 in connection with electromagnetism. However, it was Einstein who recognized their physical significance and took the bold step of interpreting them within the framework of the special theory of relativity.

Note the difference between the Galilean and Lorentz time equations. In the Galilean case, $t = t'$, but in the Lorentz case the value for t' assigned to an event by an observer O' in the S′ frame in Figure 39.14 depends both on the time t and on the coordinate x as measured by an observer O in the S frame. This is consistent with the notion that an event is characterized by four space–time coordinates (x, y, z, t). In other words, in relativity, space and time are *not* separate concepts but rather are closely interwoven with each other.

If we wish to transform coordinates in the S′ frame to coordinates in the S frame, we simply replace v by $-v$ and interchange the primed and unprimed coordinates in Equations 39.11:

$$x = \gamma(x' + vt') \qquad y = y' \qquad z = z' \qquad t = \gamma\left(t' + \frac{v}{c^2}x'\right) \qquad \text{(39.12)}$$

Inverse Lorentz transformation for S′ → S

When $v \ll c$, the Lorentz transformation equations should reduce to the Galilean equations. To verify this, note that as v approaches zero, $v/c \ll 1$; thus, $\gamma \to 1$, and Equations 39.11 reduce to the Galilean space–time transformation equations:

$$x' = x - vt \qquad y' = y \qquad z' = z \qquad t' = t$$

In many situations, we would like to know the difference in coordinates between two events or the time interval between two events as seen by observers O and O'. We can accomplish this by writing the Lorentz equations in a form suitable for describing pairs of events. From Equations 39.11 and 39.12, we can express the differences between the four variables x, x', t, and t' in the form

$$\left.\begin{array}{l} \Delta x' = \gamma(\Delta x - v\,\Delta t) \\[2mm] \Delta t' = \gamma\left(\Delta t - \dfrac{v}{c^2}\,\Delta x\right) \end{array}\right\} S \to S' \qquad \text{(39.13)}$$

$$\left.\begin{array}{l} \Delta x = \gamma(\Delta x' + v\,\Delta t') \\[2mm] \Delta t = \gamma\left(\Delta t' + \dfrac{v}{c^2}\,\Delta x'\right) \end{array}\right\} S' \to S \qquad \text{(39.14)}$$

where $\Delta x' = x_2' - x_1'$ and $\Delta t' = t_2' - t_1'$ are the differences measured by observer O' and $\Delta x = x_2 - x_1$ and $\Delta t = t_2 - t_1$ are the differences measured by observer O. (We have not included the expressions for relating the y and z coordinates because they are unaffected by motion along the x direction.[5])

---

[5] Although relative motion of the two frames along the x axis does not change the y and z coordinates of an object, it does change the y and z velocity components of an object moving in either frame, as noted in Section 39.6.

Example 39.6   Simultaneity and Time Dilation Revisited

Use the Lorentz transformation equations in difference form to show that

(A) simultaneity is not an absolute concept and that

(B) a moving clock is measured to run more slowly than a clock that is at rest with respect to an observer.

Solution (A) Suppose that two events are simultaneous and separated in space such that $\Delta t' = 0$ and $\Delta x' \neq 0$ according to an observer O' who is moving with speed v relative to O. From the expression for Δt given in Equation 39.14, we see that in this case the time interval Δt measured by observer O is $\Delta t = \gamma v \, \Delta x'/c^2$. That is, the

time interval for the same two events as measured by O is nonzero, and so the events do not appear to be simultaneous to O.

(B) Suppose that observer O' carries a clock that he uses to measure a time interval $\Delta t'$. He finds that two events occur at the same place in his reference frame ($\Delta x' = 0$) but at different times ($\Delta t' \neq 0$). Observer O' is moving with speed v relative to O, who measures the time interval between the events to be Δt. In this situation, the expression for Δt given in Equation 39.14 becomes $\Delta t = \gamma \, \Delta t'$. This is the equation for time dilation found earlier (Eq. 39.7), where $\Delta t' = \Delta t_p$ is the proper time measured by the clock carried by observer O'. Thus, O measures the moving clock to run slow.

39.6   The Lorentz Velocity Transformation Equations

Suppose two observers in relative motion with respect to each other are both observing the motion of an object. Previously, we defined an event as occurring at an instant of time. Now, we wish to interpret the "event" as the motion of the object. We know that the Galilean velocity transformation (Eq. 39.2) is valid for low speeds. How do the observers' measurements of the velocity of the object relate to each other if the speed of the object is close to that of light? Once again S′ is our frame moving at a speed v relative to S. Suppose that an object has a velocity component u'_x measured in the S′ frame, where

$$u'_x = \frac{dx'}{dt'} \tag{39.15}$$

Using Equation 39.11, we have

$$dx' = \gamma(dx - v\,dt)$$

$$dt' = \gamma\left(dt - \frac{v}{c^2}\,dx\right)$$

Substituting these values into Equation 39.15 gives

$$u'_x = \frac{dx'}{dt'} = \frac{dx - v\,dt}{dt - \dfrac{v}{c^2}\,dx} = \frac{\dfrac{dx}{dt} - v}{1 - \dfrac{v}{c^2}\dfrac{dx}{dt}}$$

But dx/dt is just the velocity component u_x of the object measured by an observer in S, and so this expression becomes

Lorentz velocity transformation for S → S′

$$u'_x = \frac{u_x - v}{1 - \dfrac{u_x v}{c^2}} \tag{39.16}$$

If the object has velocity components along the y and z axes, the components as measured by an observer in S′ are

$$u'_y = \frac{u_y}{\gamma\left(1 - \dfrac{u_x v}{c^2}\right)} \qquad \text{and} \qquad u'_z = \frac{u_z}{\gamma\left(1 - \dfrac{u_x v}{c^2}\right)} \tag{39.17}$$

Note that u'_y and u'_z do not contain the parameter v in the numerator because the relative velocity is along the x axis.

When v is much smaller than c (the nonrelativistic case), the denominator of Equation 39.16 approaches unity, and so $u'_x \approx u_x - v$, which is the Galilean velocity transformation equation. In another extreme, when $u_x = c$, Equation 39.16 becomes

$$u'_x = \frac{c - v}{1 - \frac{cv}{c^2}} = \frac{c\left(1 - \frac{v}{c}\right)}{1 - \frac{v}{c}} = c$$

From this result, we see that a speed measured as c by an observer in S is also measured as c by an observer in S'—independent of the relative motion of S and S'. Note that this conclusion is consistent with Einstein's second postulate—that the speed of light must be c relative to all inertial reference frames. Furthermore, we find that the speed of an object can never be measured as larger than c. That is, the speed of light is the ultimate speed. We return to this point later.

To obtain u_x in terms of u'_x, we replace v by $-v$ in Equation 39.16 and interchange the roles of u_x and u'_x:

$$u_x = \frac{u'_x + v}{1 + \frac{u'_x v}{c^2}} \tag{39.18}$$

▲ PITFALL PREVENTION

39.5 What Can the Observers Agree On?

We have seen several measurements that the two observers O and O' do *not* agree on: (1) the time interval between events that take place in the same position in one of the frames, (2) the distance between two points that remain fixed in one of their frames, (3) the velocity components of a moving particle, and (4) whether two events occurring at different locations in both frames are simultaneous or not. Note that the two observers *can* agree on (1) their relative speed of motion v with respect to each other, (2) the speed c of any ray of light, and (3) the simultaneity of two events which take place at the same position *and* time in some frame.

Quick Quiz 39.8 You are driving on a freeway at a relativistic speed. Straight ahead of you, a technician standing on the ground turns on a searchlight and a beam of light moves exactly vertically upward, as seen by the technician. As you observe the beam of light, you measure the magnitude of the vertical component of its velocity as (a) equal to c (b) greater than c (c) less than c.

Quick Quiz 39.9 Consider the situation in Quick Quiz 39.8 again. If the technician aims the searchlight directly at you instead of upward, you measure the magnitude of the horizontal component of its velocity as (a) equal to c (b) greater than c (c) less than c.

Example 39.7 Relative Velocity of Two Spacecraft

Two spacecraft A and B are moving in opposite directions, as shown in Figure 39.15. An observer on the Earth measures the speed of craft A to be $0.750c$ and the speed of craft B to be $0.850c$. Find the velocity of craft B as observed by the crew on craft A.

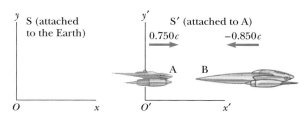

Figure 39.15 (Example 39.7) Two spacecraft A and B move in opposite directions. The speed of B relative to A is *less* than c and is obtained from the relativistic velocity transformation equation.

Solution To conceptualize this problem, we carefully identify the observers and the event. The two observers are on the Earth and on spacecraft A. The event is the motion of spacecraft B. Because the problem asks to find an observed velocity, we categorize this problem as one requiring the Lorentz velocity transformation. To analyze the problem, we note that the Earth observer makes two measurements, one of each spacecraft. We identify this observer as being at rest in the S frame. Because the velocity of spacecraft B is what we wish to measure, we identify the speed u_x as $-0.850c$. The velocity of spacecraft A is also the velocity of the observer at rest in the S' frame, which is attached to the spacecraft, relative to the observer at rest in S. Thus, $v = 0.750c$. Now we can obtain the velocity u'_x of craft B relative to craft A by using Equation 39.16:

$$u'_x = \frac{u_x - v}{1 - \dfrac{u_x v}{c^2}} = \frac{-0.850c - 0.750c}{1 - \dfrac{(-0.850c)(0.750c)}{c^2}}$$

$$= -0.977c$$

To finalize this problem, note that the negative sign indicates that craft B is moving in the negative x direction as observed by the crew on craft A. Is this consistent with your expectation from Figure 39.15? Note that the speed is less than c. That is, an object whose speed is less than c in one frame of reference must have a speed less than c in any other frame. (If the Galilean velocity transformation equation were used in this example, we would find that $u'_x = u_x - v = -0.850c - 0.750c = -1.60c$, which is impossible. The Galilean transformation equation does not work in relativistic situations.)

What If? What if the two spacecraft pass each other? Now what is their relative speed?

Answer The calculation using Equation 39.16 involves only the velocities of the two spacecraft and does not depend on their locations. After they pass each other, they have the same velocities, so the velocity of craft B as observed by the crew on craft A is the same, $-0.977c$. The only difference after they pass is that B is receding from A whereas it was approaching A before it passed.

Example 39.8   The Speeding Motorcycle

Imagine a motorcycle moving with a speed $0.80c$ past a stationary observer, as shown in Figure 39.16. If the rider

tosses a ball in the forward direction with a speed of $0.70c$ relative to himself, what is the speed of the ball relative to the stationary observer?

Solution The speed of the motorcycle relative to the stationary observer is $v = 0.80c$. The speed of the ball in the frame of reference of the motorcyclist is $u'_x = 0.70c$. Therefore, the speed u_x of the ball relative to the stationary observer is

$$u_x = \frac{u'_x + v}{1 + \dfrac{u'_x v}{c^2}} = \frac{0.70c + 0.80c}{1 + \dfrac{(0.70c)(0.80c)}{c^2}} = \boxed{0.96c}$$

Figure 39.16 (Example 39.8) A motorcyclist moves past a stationary observer with a speed of $0.80c$ and throws a ball in the direction of motion with a speed of $0.70c$ relative to himself.

Example 39.9   Relativistic Leaders of the Pack   Interactive

Two motorcycle pack leaders named David and Emily are racing at relativistic speeds along perpendicular paths, as shown in Figure 39.17. How fast does Emily recede as seen by David over his right shoulder?

Solution Figure 39.17 represents the situation as seen by a police officer at rest in frame S, who observes the following:

David: $u_x = 0.75c$ $u_y = 0$

Emily: $u_x = 0$ $u_y = -0.90c$

To calculate Emily's speed of recession as seen by David, we take S′ to move along with David and then calculate u'_x and

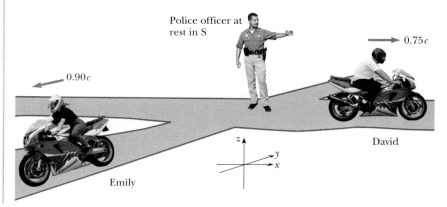

Figure 39.17 (Example 39.9) David moves to the east with a speed $0.75c$ relative to the police officer, and Emily travels south at a speed $0.90c$ relative to the officer.

u'_y for Emily using Equations 39.16 and 39.17:

$$u'_x = \frac{u_x - v}{1 - \dfrac{u_x v}{c^2}} = \frac{0 - 0.75c}{1 - \dfrac{(0)(0.75c)}{c^2}} = -0.75c$$

$$u'_y = \frac{u_y}{\gamma\left(1 - \dfrac{u_x v}{c^2}\right)} = \frac{\sqrt{1 - \dfrac{(0.75c)^2}{c^2}}\,(-0.90c)}{\left(1 - \dfrac{(0)(0.75c)}{c^2}\right)}$$

$$= -0.60c$$

Thus, the speed of Emily as observed by David is

$$u' = \sqrt{(u'_x)^2 + (u'_y)^2} = \sqrt{(-0.75c)^2 + (-0.60c)^2}$$

$$= \boxed{0.96c}$$

Note that this speed is less than c, as required by the special theory of relativity.

 Investigate this situation with various speeds of David and Emily at the Interactive Worked Example link at **http://www.pse6.com.**

39.7 Relativistic Linear Momentum and the Relativistic Form of Newton's Laws

We have seen that in order to describe properly the motion of particles within the framework of the special theory of relativity, we must replace the Galilean transformation equations by the Lorentz transformation equations. Because the laws of physics must remain unchanged under the Lorentz transformation, we must generalize Newton's laws and the definitions of linear momentum and energy to conform to the Lorentz transformation equations and the principle of relativity. These generalized definitions should reduce to the classical (nonrelativistic) definitions for $v \ll c$.

First, recall that the law of conservation of linear momentum states that when two particles (or objects that can be modeled as particles) collide, the total momentum of the isolated system of the two particles remains constant. Suppose that we observe this collision in a reference frame S and confirm that the momentum of the system is conserved. Now imagine that the momenta of the particles are measured by an observer in a second reference frame S′ moving with velocity **v** relative to the first frame. Using the Lorentz velocity transformation equation and the classical definition of linear momentum, $\mathbf{p} = m\mathbf{u}$ (where **u** is the velocity of a particle), we find that linear momentum is *not* measured to be conserved by the observer in S′. However, because the laws of physics are the same in all inertial frames, linear momentum of the system must be conserved in all frames. We have a contradiction. In view of this contradiction and assuming that the Lorentz velocity transformation equation is correct, we must modify the definition of linear momentum to satisfy the following conditions:

- The linear momentum of an isolated system must be conserved in all collisions.
- The relativistic value calculated for the linear momentum **p** of a particle must approach the classical value $m\mathbf{u}$ as **u** approaches zero.

For any particle, the correct relativistic equation for linear momentum that satisfies these conditions is

$$\mathbf{p} \equiv \frac{m\mathbf{u}}{\sqrt{1 - \dfrac{u^2}{c^2}}} = \gamma m\mathbf{u} \qquad (39.19)$$

where **u** is the velocity of the particle and m is the mass of the particle. When u is much less than c, $\gamma = (1 - u^2/c^2)^{-1/2}$ approaches unity and **p** approaches $m\mathbf{u}$. Therefore,

 PITFALL PREVENTION

39.6 Watch Out for "Relativistic Mass"

Some older treatments of relativity maintained the conservation of momentum principle at high speeds by using a model in which the mass of a particle increases with speed. You might still encounter this notion of "relativistic mass" in your outside reading, especially in older books. Be aware that this notion is no longer widely accepted and mass is considered as *invariant*, independent of speed. The mass of an object in all frames is considered to be the mass as measured by an observer at rest with respect to the object.

Definition of relativistic linear momentum

SPEED LIMIT 3×10⁸ m/s

The speed of light is the speed limit of the Universe. It is the maximum possible speed for energy transfer and for information transfer. Any object with mass must move at a lower speed.

the relativistic equation for **p** does indeed reduce to the classical expression when u is much smaller than c.

The relativistic force **F** acting on a particle whose linear momentum is **p** is defined as

$$\mathbf{F} \equiv \frac{d\mathbf{p}}{dt} \tag{39.20}$$

where **p** is given by Equation 39.19. This expression, which is the relativistic form of Newton's second law, is reasonable because it preserves classical mechanics in the limit of low velocities and is consistent with conservation of linear momentum for an isolated system (**F** = 0) both relativistically and classically.

It is left as an end-of-chapter problem (Problem 69) to show that under relativistic conditions, the acceleration **a** of a particle decreases under the action of a constant force, in which case $a \propto (1 - u^2/c^2)^{3/2}$. From this proportionality, we see that as the particle's speed approaches c, the acceleration caused by any finite force approaches zero. Hence, it is impossible to accelerate a particle from rest to a speed $u \geq c$. This argument shows that the speed of light is the ultimate speed, as noted at the end of the preceding section.

Example 39.10   Linear Momentum of an Electron

An electron, which has a mass of 9.11×10^{-31} kg, moves with a speed of $0.750c$. Find its relativistic momentum and compare this value with the momentum calculated from the classical expression.

Solution   Using Equation 39.19 with $u = 0.750c$, we have

$$p = \frac{m_e u}{\sqrt{1 - \dfrac{u^2}{c^2}}}$$

$$p = \frac{(9.11 \times 10^{-31}\ \text{kg})(0.750)(3.00 \times 10^8\ \text{m/s})}{\sqrt{1 - \dfrac{(0.750c)^2}{c^2}}}$$

$$p = \quad 3.10 \times 10^{-22}\ \text{kg·m/s}$$

The classical expression (used incorrectly here) gives

$$p_{\text{classical}} = m_e u = 2.05 \times 10^{-22}\ \text{kg·m/s}$$

Hence, the correct relativistic result is 50% greater than the classical result!

39.8   Relativistic Energy

We have seen that the definition of linear momentum requires generalization to make it compatible with Einstein's postulates. This implies that most likely the definition of kinetic energy must also be modified.

To derive the relativistic form of the work–kinetic energy theorem, let us imagine a particle moving in one dimension along the x axis. A force in the x direction causes the momentum of the particle to change according to Equation 39.20. The work done by the force F on the particle is

$$W = \int_{x_1}^{x_2} F\, dx = \int_{x_1}^{x_2} \frac{dp}{dt}\, dx \tag{39.21}$$

In order to perform this integration and find the work done on the particle and the relativistic kinetic energy as a function of u, we first evaluate dp/dt:

$$\frac{dp}{dt} = \frac{d}{dt} \frac{mu}{\sqrt{1 - \dfrac{u^2}{c^2}}} = \frac{m(du/dt)}{\left(1 - \dfrac{u^2}{c^2}\right)^{3/2}}$$

Substituting this expression for dp/dt and $dx = u\,dt$ into Equation 39.21 gives

$$W = \int_0^t \frac{m(du/dt)\,u\,dt}{\left(1 - \dfrac{u^2}{c^2}\right)^{3/2}} = m \int_0^u \frac{u}{\left(1 - \dfrac{u^2}{c^2}\right)^{3/2}}\,du$$

where we use the limits 0 and u in the integral because the integration variable has been changed from t to u. We assume that the particle is accelerated from rest to some final speed u. Evaluating the integral, we find that

$$W = \frac{mc^2}{\sqrt{1 - \dfrac{u^2}{c^2}}} - mc^2 \tag{39.22}$$

Recall from Chapter 7 that the work done by a force acting on a system consisting of a single particle equals the change in kinetic energy of the particle. Because we assumed that the initial speed of the particle is zero, we know that its initial kinetic energy is zero. We therefore conclude that the work W in Equation 39.22 is equivalent to the relativistic kinetic energy K:

$$K = \frac{mc^2}{\sqrt{1 - \dfrac{u^2}{c^2}}} - mc^2 = \gamma mc^2 - mc^2 = (\gamma - 1)mc^2 \tag{39.23}$$

Relativistic kinetic energy

This equation is routinely confirmed by experiments using high-energy particle accelerators.

At low speeds, where $u/c \ll 1$, Equation 39.23 should reduce to the classical expression $K = \frac{1}{2}mu^2$. We can check this by using the binomial expansion $(1 - \beta^2)^{-1/2} \approx 1 + \frac{1}{2}\beta^2 + \cdots$ for $\beta \ll 1$, where the higher-order powers of β are neglected in the expansion. (In treatments of relativity, β is a common symbol used to represent u/c or v/c.) In our case, $\beta = u/c$, so that

$$\gamma = \frac{1}{\sqrt{1 - \dfrac{u^2}{c^2}}} = \left(1 - \frac{u^2}{c^2}\right)^{-1/2} \approx 1 + \frac{1}{2}\frac{u^2}{c^2}$$

Substituting this into Equation 39.23 gives

$$K \approx \left[\left(1 + \frac{1}{2}\frac{u^2}{c^2}\right) - 1\right]mc^2 = \frac{1}{2}mu^2 \qquad \text{(for } u/c \ll 1)$$

which is the classical expression for kinetic energy. A graph comparing the relativistic and nonrelativistic expressions is given in Figure 39.18. In the relativistic case, the particle speed never exceeds c, regardless of the kinetic energy. The two curves are in good agreement when $u \ll c$.

Figure 39.18 A graph comparing relativistic and nonrelativistic kinetic energy of a moving particle. The energies are plotted as a function of particle speed u. In the relativistic case, u is always less than c.

The constant term mc^2 in Equation 39.23, which is independent of the speed of the particle, is called the **rest energy** E_R of the particle:

Rest energy

$$E_R = mc^2 \qquad (39.24)$$

The term γmc^2, which does depend on the particle speed, is therefore the sum of the kinetic and rest energies. We define γmc^2 to be the **total energy** E:

$$\text{Total energy} = \text{kinetic energy} + \text{rest energy}$$

$$E = K + mc^2 \qquad (39.25)$$

or

Total energy of a relativistic particle

$$E = \frac{mc^2}{\sqrt{1 - \dfrac{u^2}{c^2}}} = \gamma mc^2 \qquad (39.26)$$

The relationship $E = K + mc^2$ shows that **mass is a form of energy,** where c^2 in the rest energy term is just a constant conversion factor. This expression also shows that a small mass corresponds to an enormous amount of energy, a concept fundamental to nuclear and elementary-particle physics.

In many situations, the linear momentum or energy of a particle is measured rather than its speed. It is therefore useful to have an expression relating the total energy E to the relativistic linear momentum p. This is accomplished by using the expressions $E = \gamma mc^2$ and $p = \gamma mu$. By squaring these equations and subtracting, we can eliminate u (Problem 43). The result, after some algebra, is[6]

Energy–momentum relationship for a relativistic particle

$$E^2 = p^2c^2 + (mc^2)^2 \qquad (39.27)$$

When the particle is at rest, $p = 0$ and so $E = E_R = mc^2$.

In Section 35.1, we introduced the concept of a particle of light, called a **photon.** For particles that have zero mass, such as photons, we set $m = 0$ in Equation 39.27 and find that

$$E = pc \qquad (39.28)$$

This equation is an exact expression relating total energy and linear momentum for photons, which always travel at the speed of light (in vacuum).

Finally, note that because the mass m of a particle is independent of its motion, m must have the same value in all reference frames. For this reason, m is often called the **invariant mass.** On the other hand, because the total energy and linear momentum of a particle both depend on velocity, these quantities depend on the reference frame in which they are measured.

When we are dealing with subatomic particles, it is convenient to express their energy in electron volts (Section 25.1) because the particles are usually given this energy by acceleration through a potential difference. The conversion factor, as you recall from Equation 25.5, is

$$1 \text{ eV} = 1.60 \times 10^{-19} \text{ J}$$

For example, the mass of an electron is 9.11×10^{-31} kg. Hence, the rest energy of the electron is

$$m_e c^2 = (9.11 \times 10^{-31} \text{ kg})(3.00 \times 10^8 \text{ m/s})^2 = 8.20 \times 10^{-14} \text{ J}$$
$$= (8.20 \times 10^{-14} \text{ J})(1 \text{ eV}/1.60 \times 10^{-19} \text{ J}) = 0.511 \text{ MeV}$$

---

[6] One way to remember this relationship is to draw a right triangle having a hypotenuse of length E and legs of lengths pc and mc^2.

Quick Quiz 39.10 The following *pairs* of energies represent the rest energy and total energy of three different particles: particle 1: *E*, 2*E*; particle 2: *E*, 3*E*; particle 3: 2*E*, 4*E*. Rank the particles, from greatest to least, according to their (a) mass; (b) kinetic energy; (c) speed.

Example 39.11   The Energy of a Speedy Electron

An electron in a television picture tube typically moves with a speed $u = 0.250c$. Find its total energy and kinetic energy in electron volts.

Solution Using the fact that the rest energy of the electron is 0.511 MeV together with Equation 39.26, we have

$$E = \frac{m_e c^2}{\sqrt{1 - \dfrac{u^2}{c^2}}} = \frac{0.511 \text{ MeV}}{\sqrt{1 - \dfrac{(0.250c)^2}{c^2}}}$$

$$= 1.03(0.511 \text{ MeV}) = \boxed{0.528 \text{ MeV}}$$

This is 3% greater than the rest energy.

We obtain the kinetic energy by subtracting the rest energy from the total energy:

$$K = E - m_e c^2 = 0.528 \text{ MeV} - 0.511 \text{ MeV}$$

$$= \boxed{0.017 \text{ MeV}}$$

Example 39.12   The Energy of a Speedy Proton

(A) Find the rest energy of a proton in electron volts.

Solution Using Equation 39.24,

$$E_R = m_p c^2 = (1.67 \times 10^{-27} \text{ kg})(3.00 \times 10^8 \text{ m/s})^2$$

$$= (1.50 \times 10^{-10} \text{ J}) \left(\frac{1.00 \text{ eV}}{1.60 \times 10^{-19} \text{ J}} \right)$$

$$= \boxed{938 \text{ MeV}}$$

(B) If the total energy of a proton is three times its rest energy, what is the speed of the proton?

Solution Equation 39.26 gives

$$E = 3m_p c^2 = \frac{m_p c^2}{\sqrt{1 - \dfrac{u^2}{c^2}}}$$

$$3 = \frac{1}{\sqrt{1 - \dfrac{u^2}{c^2}}}$$

Solving for *u* gives

$$\left(1 - \frac{u^2}{c^2} \right) = \frac{1}{9}$$

$$\frac{u^2}{c^2} = \frac{8}{9}$$

$$u = \frac{\sqrt{8}}{3} c = 0.943c = \boxed{2.83 \times 10^8 \text{ m/s}}$$

(C) Determine the kinetic energy of the proton in electron volts.

Solution From Equation 39.25,

$$K = E - m_p c^2 = 3m_p c^2 - m_p c^2 = 2m_p c^2$$

Because $m_p c^2 = 938$ MeV, we see that $K = \boxed{1\,880 \text{ MeV}}$.

(D) What is the proton's momentum?

Solution We can use Equation 39.27 to calculate the momentum with $E = 3m_p c^2$:

$$E^2 = p^2 c^2 + (m_p c^2)^2 = (3m_p c^2)^2$$

$$p^2 c^2 = 9(m_p c^2)^2 - (m_p c^2)^2 = 8(m_p c^2)^2$$

$$p = \sqrt{8} \, \frac{m_p c^2}{c} = \sqrt{8} \, \frac{(938 \text{ MeV})}{c} = \boxed{2\,650 \text{ MeV}/c}$$

The unit of momentum is written MeV/*c*, which is a common unit in particle physics.

What If? In classical physics, if the momentum of a particle doubles, the kinetic energy increases by a factor of 4. What happens to the kinetic energy of the speedy proton in this example if its momentum doubles?

Answer Based on what we have seen so far in relativity, it is likely that you would predict that its kinetic energy does not increase by a factor of 4. If the momentum doubles, the new momentum is

$$p_{\text{new}} = 2 \left(\sqrt{8} \, \frac{m_p c^2}{c} \right) = 4\sqrt{2} \, \frac{m_p c^2}{c}$$

Using Equation 39.27, we find the square of the new total energy:

$$E^2_{\text{new}} = p^2_{\text{new}} c^2 + (m_p c^2)^2$$

$$E_{new}^2 = \left(4\sqrt{2}\,\frac{m_p c^2}{c}\right)^2 c^2 + (m_p c^2)^2 = 33(m_p c^2)^2$$

$$E_{new} = \sqrt{33}\,(m_p c^2) = 5.7\,m_p c^2$$

Now, using Equation 39.25, we find the new kinetic energy:

$$K_{new} = E_{new} - m_p c^2 = 5.7\,m_p c^2 - m_p c^2 = 4.7\,m_p c^2$$

Notice that this is only 2.35 times as large as the kinetic energy we found in part (C), not four times as large. In general, the factor by which the kinetic energy increases if the momentum doubles will depend on the initial momentum, but will approach 4 as the momentum approaches zero. In this latter situation, classical physics correctly describes the situation.

39.9 Mass and Energy

Equation 39.26, $E = \gamma m c^2$, which represents the total energy of a particle, suggests that even when a particle is at rest ($\gamma = 1$) it still possesses enormous energy through its mass. The clearest experimental proof of the equivalence of mass and energy occurs in nuclear and elementary particle interactions in which the conversion of mass into kinetic energy takes place. Because of this, in relativistic situations, we cannot use the principle of conservation of energy as it was outlined in Chapters 7 and 8. We must include rest energy as another form of energy storage.

This concept is important in atomic and nuclear processes, in which the change in mass is a relatively large fraction of the initial mass. For example, in a conventional nuclear reactor, the uranium nucleus undergoes *fission*, a reaction that results in several lighter fragments having considerable kinetic energy. In the case of ^{235}U, which is used as fuel in nuclear power plants, the fragments are two lighter nuclei and a few neutrons. The total mass of the fragments is less than that of the ^{235}U by an amount Δm. The corresponding energy $\Delta m c^2$ associated with this mass difference is exactly equal to the total kinetic energy of the fragments. The kinetic energy is absorbed as the fragments move through water, raising the internal energy of the water. This internal energy is used to produce steam for the generation of electrical power.

Next, consider a basic *fusion* reaction in which two deuterium atoms combine to form one helium atom. The decrease in mass that results from the creation of one helium atom from two deuterium atoms is $\Delta m = 4.25 \times 10^{-29}$ kg. Hence, the corresponding energy that results from one fusion reaction is $\Delta m c^2 = 3.83 \times 10^{-12}$ J = 23.9 MeV. To appreciate the magnitude of this result, if only 1 g of deuterium is converted to helium, the energy released is on the order of 10^{12} J! At the year 2003 cost of electrical energy, this would be worth about \$30 000. We shall present more details of these nuclear processes in Chapter 45 of the extended version of this textbook.

Example 39.13 Mass Change in a Radioactive Decay

The ^{216}Po nucleus is unstable and exhibits radioactivity (Chapter 44). It decays to ^{212}Pb by emitting an alpha particle, which is a helium nucleus, ^{4}He. Find

(A) the mass change in this decay and

(B) the energy that this represents.

Solution Using values in Table A.3, we see that the initial and final masses are

$$m_i = m(^{216}\text{Po}) = 216.001\,905\text{ u}$$

$$m_f = m(^{212}\text{Pb}) + m(^4\text{He}) = 211.991\,888\text{ u} + 4.002\,603\text{ u}$$

$$= 215.994\,491\text{ u}$$

Thus, the mass change is

$$\Delta m = 216.001\,905\text{ u} - 215.994\,491\text{ u} = 0.007\,414\text{ u}$$

$$= \boxed{1.23 \times 10^{-29}\text{ kg}}$$

(B) The energy associated with this mass change is

$$E = \Delta m c^2 = (1.23 \times 10^{-29}\text{ kg})(3.00 \times 10^8\text{ m/s})^2$$

$$= 1.11 \times 10^{-12}\text{ J} = \boxed{6.92\text{ MeV}}$$

This energy appears as the kinetic energy of the alpha particle and the ^{212}Pb nucleus after the decay.

39.10 The General Theory of Relativity

Up to this point, we have sidestepped a curious puzzle. Mass has two seemingly different properties: a *gravitational attraction* for other masses and an *inertial* property that represents a resistance to acceleration. To designate these two attributes, we use the subscripts g and i and write

$$\text{Gravitational property} \qquad F_g = m_g g$$

$$\text{Inertial property} \qquad \sum F = m_i a$$

The value for the gravitational constant G was chosen to make the magnitudes of m_g and m_i numerically equal. Regardless of how G is chosen, however, the strict proportionality of m_g and m_i has been established experimentally to an extremely high degree: a few parts in 10^{12}. Thus, it appears that gravitational mass and inertial mass may indeed be exactly proportional.

But why? They seem to involve two entirely different concepts: a force of mutual gravitational attraction between two masses, and the resistance of a single mass to being accelerated. This question, which puzzled Newton and many other physicists over the years, was answered by Einstein in 1916 when he published his theory of gravitation, known as the *general theory of relativity*. Because it is a mathematically complex theory, we offer merely a hint of its elegance and insight.

In Einstein's view, the dual behavior of mass was evidence for a very intimate and basic connection between the two behaviors. He pointed out that no mechanical experiment (such as dropping an object) could distinguish between the two situations illustrated in Figures 39.19a and 39.19b. In Figure 39.19a, a person is standing in an elevator on the surface of a planet, and feels pressed into the floor, due to the gravitational force. In Figure 39.19b, the person is in an elevator in empty space accelerating upward with $a = g$. The person feels pressed into the floor with the same force as in Figure 39.19a. In each case, an object released by the observer undergoes a downward acceleration of magnitude g relative to the floor. In Figure 39.19a, the person is in an inertial frame in a gravitational field. In Figure 39.19b, the person is in a noninertial frame accelerating in gravity-free space. Einstein's claim is that these two situations are completely equivalent.

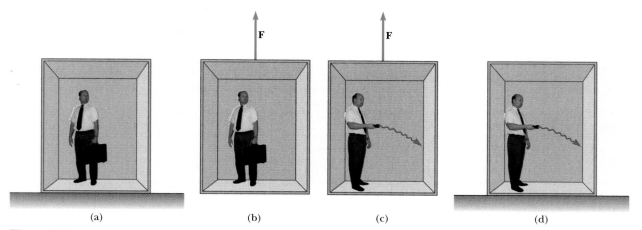

Figure 39.19 (a) The observer is at rest in a uniform gravitational field **g**, directed downward. (b) The observer is in a region where gravity is negligible, but the frame is accelerated by an external force **F** that produces an acceleration **g** directed upward. According to Einstein, the frames of reference in parts (a) and (b) are equivalent in every way. No local experiment can distinguish any difference between the two frames. (c) In the accelerating frame, a ray of light would appear to bend downward due to the acceleration of the elevator. (d) If parts (a) and (b) are truly equivalent, as Einstein proposed, then part (c) suggests that a ray of light would bend downward in a gravitational field.

Einstein carried this idea further and proposed that *no* experiment, mechanical or otherwise, could distinguish between the two cases. This extension to include all phenomena (not just mechanical ones) has interesting consequences. For example, suppose that a light pulse is sent horizontally across an elevator that is accelerating upward in empty space, as in Figure 39.19c. From the point of view of an observer in an inertial frame outside of the elevator, the light travels in a straight line while the floor of the elevator accelerates upward. According to the observer on the elevator, however, the trajectory of the light pulse bends downward as the floor of the elevator (and the observer) accelerates upward. Therefore, based on the equality of parts (a) and (b) of the figure for all phenomena, Einstein proposed that **a beam of light should also be bent downward by a gravitational field,** as in Figure 39.19d. Experiments have verified the effect, although the bending is small. A laser aimed at the horizon falls less than 1 cm after traveling 6 000 km. (No such bending is predicted in Newton's theory of gravitation.)

The two postulates of Einstein's **general theory of relativity** are

Postulates of the general theory of relativity

- All the laws of nature have the same form for observers in any frame of reference, whether accelerated or not.

- The results of experiments based on the laws of physics, as measured by an observer at rest in an inertial reference frame in a uniform gravitational field, are equivalent to those measured by an observer at rest in a uniformly accelerated reference frame in gravity-free space.

One interesting effect predicted by the general theory is that time is altered by gravity. A clock in the presence of gravity runs slower than one located where gravity is negligible. Consequently, the frequencies of radiation emitted by atoms in the presence of a strong gravitational field are *red-shifted* to lower frequencies when compared with the same emissions in the presence of a weak field. This gravitational red shift has been detected in spectral lines emitted by atoms in massive stars. It has also been verified on the Earth by comparing the frequencies of gamma rays emitted from nuclei separated vertically by about 20 m.

The second postulate suggests that a gravitational field may be "transformed away" at any point if we choose an appropriate accelerated frame of reference—a freely falling one. Einstein developed an ingenious method of describing the acceleration necessary to make the gravitational field "disappear." He specified a concept, the *curvature of space–time,* that describes the gravitational effect at every point. In fact, the curvature of space–time completely replaces Newton's gravitational theory. According to Einstein, there is no such thing as a gravitational force. Rather, the presence of a mass causes a curvature of space–time in the vicinity of the mass, and this curvature dictates the space–time path that all freely moving objects must follow. In 1979, John Wheeler summarized Einstein's general theory of relativity in a single sentence: "Space tells matter how to move and matter tells space how to curve."

As an example of the effects of curved space–time, imagine two travelers moving on parallel paths a few meters apart on the surface of the Earth and maintaining an exact northward heading along two longitude lines. As they observe each other near the equator, they will claim that their paths are exactly parallel. As they approach the North Pole, however, they notice that they are moving closer together, and they will actually meet at the North Pole. Thus, they will claim that they moved along parallel paths, but moved toward each other, *as if there were an attractive force between them.* They will make this conclusion based on their everyday experience of moving on flat surfaces. From our mental representation, however, we realize that they are walking on a curved surface, and it is the geometry of the curved surface that causes them to converge, rather than an attractive force. In a similar way, general relativity replaces the notion of forces with the movement of objects through curved space–time.

One prediction of the general theory of relativity is that a light ray passing near the Sun should be deflected in the curved space–time created by the Sun's mass. This prediction was confirmed when astronomers detected the bending of starlight near the

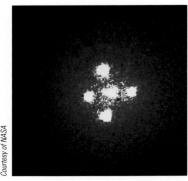

Einstein's cross. The four bright spots are images of the same galaxy that have been bent around a massive object located between the galaxy and the Earth. The massive object acts like a lens, causing the rays of light that were diverging from the distant galaxy to converge on the Earth. (If the intervening massive object had a uniform mass distribution, we would see a bright ring instead of four spots.)

Figure 39.20 Deflection of starlight passing near the Sun. Because of this effect, the Sun or some other remote object can act as a *gravitational lens.* In his general theory of relativity, Einstein calculated that starlight just grazing the Sun's surface should be deflected by an angle of 1.75 s of arc.

Sun during a total solar eclipse that occurred shortly after World War I (Fig. 39.20). When this discovery was announced, Einstein became an international celebrity.

If the concentration of mass becomes very great, as is believed to occur when a large star exhausts its nuclear fuel and collapses to a very small volume, a **black hole** may form. Here, the curvature of space–time is so extreme that, within a certain distance from the center of the black hole, all matter and light become trapped, as discussed in Section 13.7.

SUMMARY

The two basic postulates of the special theory of relativity are

- The laws of physics must be the same in all inertial reference frames.
- The speed of light in vacuum has the same value, $c = 3.00 \times 10^8$ m/s, in all inertial frames, regardless of the velocity of the observer or the velocity of the source emitting the light.

Three consequences of the special theory of relativity are

- Events that are measured to be simultaneous for one observer are not necessarily measured to be simultaneous for another observer who is in motion relative to the first.
- Clocks in motion relative to an observer are measured to run slower by a factor $\gamma = (1 - v^2/c^2)^{-1/2}$. This phenomenon is known as **time dilation.**
- The length of objects in motion are measured to be contracted in the direction of motion by a factor $1/\gamma = (1 - v^2/c^2)^{1/2}$. This phenomenon is known as **length contraction.**

To satisfy the postulates of special relativity, the Galilean transformation equations must be replaced by the **Lorentz transformation equations:**

$$x' = \gamma(x - vt) \qquad y' = y \qquad z' = z \qquad t' = \gamma\left(t - \frac{v}{c^2}x\right) \qquad (39.11)$$

where $\gamma = (1 - v^2/c^2)^{-1/2}$ and the S′ frame moves in the x direction relative to the S frame.

The relativistic form of the **velocity transformation equation** is

$$u'_x = \frac{u_x - v}{1 - \dfrac{u_x v}{c^2}} \qquad (39.16)$$

where u_x is the speed of an object as measured in the S frame and u'_x is its speed measured in the S′ frame.

Take a practice test for this chapter by clicking on the Practice Test link at http://www.pse6.com.

The relativistic expression for the **linear momentum** of a particle moving with a velocity **u** is

$$\mathbf{p} \equiv \frac{m\mathbf{u}}{\sqrt{1 - \dfrac{u^2}{c^2}}} = \gamma m\mathbf{u} \tag{39.19}$$

The relativistic expression for the **kinetic energy** of a particle is

$$K = \frac{mc^2}{\sqrt{1 - \dfrac{u^2}{c^2}}} - mc^2 = (\gamma - 1)mc^2 \tag{39.23}$$

The constant term mc^2 in Equation 39.23 is called the **rest energy** E_R of the particle:

$$E_R = mc^2 \tag{39.24}$$

The total energy E of a particle is given by

$$E = \frac{mc^2}{\sqrt{1 - \dfrac{u^2}{c^2}}} = \gamma mc^2 \tag{39.26}$$

The relativistic linear momentum of a particle is related to its total energy through the equation

$$E^2 = p^2c^2 + (mc^2)^2 \tag{39.27}$$

QUESTIONS

1. What two speed measurements do two observers in relative motion always agree on?

2. A spacecraft with the shape of a sphere moves past an observer on Earth with a speed 0.5c. What shape does the observer measure for the spacecraft as it moves past?

3. The speed of light in water is 230 Mm/s. Suppose an electron is moving through water at 250 Mm/s. Does this violate the principle of relativity?

4. Two identical clocks are synchronized. One is then put in orbit directed eastward around the Earth while the other remains on the Earth. Which clock runs slower? When the moving clock returns to the Earth, are the two still synchronized?

5. Explain why it is necessary, when defining the length of a rod, to specify that the positions of the ends of the rod are to be measured simultaneously.

6. A train is approaching you at very high speed as you stand next to the tracks. Just as an observer on the train passes you, you both begin to play the same Beethoven symphony on portable compact disc players. (a) According to you, whose CD player finishes the symphony first? (b) **What If?** According to the observer on the train, whose CD player finishes the symphony first? (c) Whose CD player really finishes the symphony first?

7. List some ways our day-to-day lives would change if the speed of light were only 50 m/s.

8. Does saying that a moving clock runs slower than a stationary one imply that something is physically unusual about the moving clock?

9. How is acceleration indicated on a space–time graph?

10. A particle is moving at a speed less than $c/2$. If the speed of the particle is doubled, what happens to its momentum?

11. Give a physical argument that shows that it is impossible to accelerate an object of mass m to the speed of light, even with a continuous force acting on it.

12. The upper limit of the speed of an electron is the speed of light c. Does that mean that the momentum of the electron has an upper limit?

13. Because mass is a measure of energy, can we conclude that the mass of a compressed spring is greater than the mass of the same spring when it is not compressed?

14. It is said that Einstein, in his teenage years, asked the question, "What would I see in a mirror if I carried it in my hands and ran at the speed of light?" How would you answer this question?

15. Some distant astronomical objects, called quasars, are receding from us at half the speed of light (or greater). What is the speed of the light we receive from these quasars?

16. Photons of light have zero mass. How is it possible that they have momentum?

17. "Newtonian mechanics correctly describes objects moving at ordinary speeds and relativistic mechanics correctly describes objects moving very fast." "Relativistic mechanics must make a smooth transition as it reduces to Newtonian mechanics in a case where the speed of an object becomes small compared to the speed of light." Argue for or against each of these two statements.

(a)

(b)

Figure Q39.18

18. Two cards have straight edges. Suppose that the top edge of one card crosses the bottom edge of another card at a small angle, as in Figure Q39.18a. A person slides the cards together at a moderately high speed. In what direction does the intersection point of the edges move? Show that it can move at a speed greater than the speed of light.

A small flashlight is suspended in a horizontal plane and set into rapid rotation. Show that the spot of light it produces on a distant screen can move across the screen at a speed greater than the speed of light. (If you use a laser pointer, as in Figure Q39.18b, make sure the direct laser light cannot enter a person's eyes.) Argue that these experiments do not invalidate the principle that no material, no energy, and no information can move faster than light moves in a vacuum.

19. Describe how the results of Example 39.7 would change if, instead of fast space vehicles, two ordinary cars were approaching each other at highway speeds.

20. Two objects are identical except that one is hotter than the other. Compare how they respond to identical forces.

21. With regard to reference frames, how does general relativity differ from special relativity?

22. Two identical clocks are in the same house, one upstairs in a bedroom, and the other downstairs in the kitchen. Which clock runs more slowly? Explain.

23. *A thought experiment.* Imagine ants living on a merry-go-round turning at relativistic speed, which is their two-dimensional world. From measurements on small circles they are thoroughly familiar with the number π. When they measure the circumference of their world, and divide it by the diameter, they expect to calculate the number $\pi = 3.141\ 59.\ .\ .\ .$ We see the merry-go-round turning at relativistic speed. From our point of view, the ants' measuring rods on the circumference are experiencing length contraction in the tangential direction; hence the ants will need some extra rods to fill that entire distance. The rods measuring the diameter, however, do not contract, because their motion is perpendicular to their lengths. As a result, the computed ratio does not agree with the number π. If you were an ant, you would say that the rest of the universe is spinning in circles, and your disk is stationary. What possible explanation can you then give for the discrepancy, in light of the general theory of relativity?

PROBLEMS

1, 2, 3 = straightforward, intermediate, challenging   ☐ = full solution available in the *Student Solutions Manual and Study Guide*

🌐 = coached solution with hints available at http://www.pse6.com   🖥 = computer useful in solving problem

▭ = paired numerical and symbolic problems

Section 39.1   The Principle of Galilean Relativity

1. A 2 000-kg car moving at 20.0 m/s collides and locks together with a 1 500-kg car at rest at a stop sign. Show that momentum is conserved in a reference frame moving at 10.0 m/s in the direction of the moving car.

2. A ball is thrown at 20.0 m/s inside a boxcar moving along the tracks at 40.0 m/s. What is the speed of the ball relative to the ground if the ball is thrown (a) forward (b) backward (c) out the side door?

3. In a laboratory frame of reference, an observer notes that Newton's second law is valid. Show that it is also valid for an observer moving at a constant speed, small compared with the speed of light, relative to the laboratory frame.

4. Show that Newton's second law is *not* valid in a reference frame moving past the laboratory frame of Problem 3 with a constant acceleration.

Section 39.2   The Michelson–Morley Experiment
Section 39.3   Einstein's Principle of Relativity
Section 39.4   Consequences of the Special Theory of Relativity

> Problem 43 in Chapter 4 can be assigned with this section.

5. How fast must a meter stick be moving if its length is measured to shrink to 0.500 m?

6. At what speed does a clock move if it is measured to run at a rate that is half the rate of a clock at rest with respect to an observer?

7. An astronaut is traveling in a space vehicle that has a speed of $0.500c$ relative to the Earth. The astronaut measures her pulse rate at 75.0 beats per minute. Signals generated by the astronaut's pulse are radioed to Earth when the vehicle is moving in a direction perpendicular to the line that connects the vehicle with an observer on the Earth. (a) What pulse rate does the Earth observer measure? (b) **What If?** What would be the pulse rate if the speed of the space vehicle were increased to $0.990c$?

8. An astronomer on Earth observes a meteoroid in the southern sky approaching the Earth at a speed of $0.800c$. At the time of its discovery the meteoroid is 20.0 ly from the Earth. Calculate (a) the time interval required for the meteoroid to reach the Earth as measured by the Earth-bound astronomer, (b) this time interval as measured by a tourist on the meteoroid, and (c) the distance to the Earth as measured by the tourist.

9. An atomic clock moves at 1 000 km/h for 1.00 h as measured by an identical clock on the Earth. How many nanoseconds slow will the moving clock be compared with the Earth clock, at the end of the 1.00-h interval?

10. A muon formed high in the Earth's atmosphere travels at speed $v = 0.990c$ for a distance of 4.60 km before it decays into an electron, a neutrino, and an antineutrino $(\mu^- \rightarrow e^- + \nu + \bar{\nu})$. (a) How long does the muon live, as measured in its reference frame? (b) How far does the Earth travel, as measured in the frame of the muon?

11. A spacecraft with a proper length of 300 m takes 0.750 μs to pass an Earth observer. Determine the speed of the spacecraft as measured by the Earth observer.

12. (a) An object of proper length L_p takes a time interval Δt to pass an Earth observer. Determine the speed of the object as measured by the Earth observer. (b) A column of tanks, 300 m long, takes 75.0 s to pass a child waiting at a street corner on her way to school. Determine the speed of the armored vehicles. (c) Show that the answer to part (a) includes the answer to Problem 11 as a special case, and includes the answer to part (b) as another special case.

13. **Review problem.** In 1963 Mercury astronaut Gordon Cooper orbited the Earth 22 times. The press stated that for each orbit he aged 2 millionths of a second less than he would have if he had remained on the Earth. (a) Assuming that he was 160 km above the Earth in a circular orbit, determine the time difference between someone on the Earth and the orbiting astronaut for the 22 orbits. You will need to use the approximation $\sqrt{1-x} \approx 1 - x/2$, for small x. (b) Did the press report accurate information? Explain.

14. For what value of v does $\gamma = 1.010\,0$? Observe that for speeds lower than this value, time dilation and length contraction are effects amounting to less than 1%.

15. A friend passes by you in a spacecraft traveling at a high speed. He tells you that his craft is 20.0 m long and that the identically constructed craft you are sitting in is 19.0 m long. According to your observations, (a) how long is your spacecraft, (b) how long is your friend's craft, and (c) what is the speed of your friend's craft?

16. The identical twins Speedo and Goslo join a migration from the Earth to Planet X. It is 20.0 ly away in a reference frame in which both planets are at rest. The twins, of the same age, depart at the same time on different spacecraft. Speedo's craft travels steadily at $0.950c$, and Goslo's at $0.750c$. Calculate the age difference between the twins after Goslo's spacecraft lands on Planet X. Which twin is the older?

17. An interstellar space probe is launched from the Earth. After a brief period of acceleration it moves with a constant velocity, with a magnitude of 70.0% of the speed of light. Its nuclear-powered batteries supply the energy to keep its data transmitter active continuously. The batteries have a lifetime of 15.0 yr as measured in a rest frame. (a) How long do the batteries on the space probe last as measured by Mission Control on the Earth? (b) How far is the probe from the Earth when its batteries fail, as measured by Mission Control? (c) How far is the probe from the Earth when its batteries fail, as measured by its built-in trip odometer? (d) For what total time interval after launch are data received from the probe by Mission Control? Note that radio waves travel at the speed of light and fill the space between the probe and the Earth at the time of battery failure.

18. **Review problem.** An alien civilization occupies a brown dwarf, nearly stationary relative to the Sun, several lightyears away. The extraterrestrials have come to love original broadcasts of *I Love Lucy*, on our television channel 2, at carrier frequency 57.0 MHz. Their line of sight to us is in the plane of the Earth's orbit. Find the difference between the highest and lowest frequencies they receive due to the Earth's orbital motion around the Sun.

19. Police radar detects the speed of a car (Fig. P39.19) as follows. Microwaves of a precisely known frequency are broadcast toward the car. The moving car reflects the microwaves with a Doppler shift. The reflected waves are received and combined with an attenuated version of the transmitted wave. Beats occur between the two microwave signals. The beat frequency is measured. (a) For an electromagnetic wave reflected back to its source from a mirror approaching at speed v, show that the reflected

wave has frequency

$$f = f_{source} \frac{c + v}{c - v}$$

where f_{source} is the source frequency. (b) When v is much less than c, the beat frequency is much smaller than the transmitted frequency. In this case use the approximation $f + f_{source} \approx 2 f_{source}$ and show that the beat frequency can be written as $f_{beat} = 2v/\lambda$. (c) What beat frequency is measured for a car speed of 30.0 m/s if the microwaves have frequency 10.0 GHz? (d) If the beat frequency measurement is accurate to ± 5 Hz, how accurate is the velocity measurement?

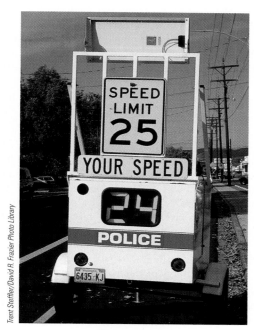

Figure P39.19

20. *The red shift.* A light source recedes from an observer with a speed v_{source} that is small compared with c. (a) Show that the fractional shift in the measured wavelength is given by the approximate expression

$$\frac{\Delta\lambda}{\lambda} \approx \frac{v_{source}}{c}$$

This phenomenon is known as the red shift, because the visible light is shifted toward the red. (b) Spectroscopic measurements of light at $\lambda = 397$ nm coming from a galaxy in Ursa Major reveal a red shift of 20.0 nm. What is the recessional speed of the galaxy?

21. A physicist drives through a stop light. When he is pulled over, he tells the police officer that the Doppler shift made the red light of wavelength 650 nm appear green to him, with a wavelength of 520 nm. The police officer writes out a traffic citation for speeding. How fast was the physicist traveling, according to his own testimony?

Section 39.5 The Lorentz Transformation Equations

22. Suzanne observes two light pulses to be emitted from the same location, but separated in time by 3.00 μs. Mark sees the emission of the same two pulses separated in time by 9.00 μs. (a) How fast is Mark moving relative to Suzanne? (b) According to Mark, what is the separation in space of the two pulses?

23. A moving rod is observed to have a length of 2.00 m and to be oriented at an angle of 30.0° with respect to the direction of motion, as shown in Figure P39.23. The rod has a speed of 0.995c. (a) What is the proper length of the rod? (b) What is the orientation angle in the proper frame?

Figure P39.23

24. An observer in reference frame S sees two events as simultaneous. Event A occurs at the point (50.0 m, 0, 0) at the instant 9:00:00 Universal time, 15 January 2004. Event B occurs at the point (150 m, 0, 0) at the same moment. A second observer, moving past with a velocity of $0.800c\hat{\mathbf{i}}$, also observes the two events. In her reference frame S', which event occurred first and what time interval elapsed between the events?

25. A red light flashes at position $x_R = 3.00$ m and time $t_R = 1.00 \times 10^{-9}$ s, and a blue light flashes at $x_B = 5.00$ m and $t_B = 9.00 \times 10^{-9}$ s, all measured in the S reference frame. Reference frame S' has its origin at the same point as S at $t = t' = 0$; frame S' moves uniformly to the right. Both flashes are observed to occur at the same place in S'. (a) Find the relative speed between S and S'. (b) Find the location of the two flashes in frame S'. (c) At what time does the red flash occur in the S' frame?

Section 39.6 The Lorentz Velocity Transformation Equations

26. A Klingon spacecraft moves away from the Earth at a speed of 0.800c (Fig. P39.26). The starship *Enterprise* pursues at a speed of 0.900c relative to the Earth. Observers on the Earth see the *Enterprise* overtaking the Klingon craft at a relative speed of 0.100c. With what speed is the *Enterprise* overtaking the Klingon craft as seen by the crew of the *Enterprise*?

Figure P39.26

27. [≈] Two jets of material from the center of a radio galaxy are ejected in opposite directions. Both jets move at 0.750c

relative to the galaxy. Determine the speed of one jet relative to the other.

28. A spacecraft is launched from the surface of the Earth with a velocity of $0.600c$ at an angle of $50.0°$ above the horizontal positive x axis. Another spacecraft is moving past, with a velocity of $0.700c$ in the negative x direction. Determine the magnitude and direction of the velocity of the first spacecraft as measured by the pilot of the second spacecraft.

Section 39.7 Relativistic Linear Momentum and the Relativistic Form of Newton's Laws

29. Calculate the momentum of an electron moving with a speed of (a) $0.010\ 0c$, (b) $0.500c$, and (c) $0.900c$.

30. The nonrelativistic expression for the momentum of a particle, $p = mu$, agrees with experiment if $u \ll c$. For what speed does the use of this equation give an error in the momentum of (a) 1.00% and (b) 10.0%?

31. A golf ball travels with a speed of 90.0 m/s. By what fraction does its relativistic momentum magnitude p differ from its classical value mu? That is, find the ratio $(p - mu)/mu$.

32. Show that the speed of an object having momentum of magnitude p and mass m is

$$u = \frac{c}{\sqrt{1 + (mc/p)^2}}$$

33. ⚛ An unstable particle at rest breaks into two fragments of unequal mass. The mass of the first fragment is 2.50×10^{-28} kg, and that of the other is 1.67×10^{-27} kg. If the lighter fragment has a speed of $0.893c$ after the breakup, what is the speed of the heavier fragment?

Section 39.8 Relativistic Energy

34. Determine the energy required to accelerate an electron from (a) $0.500c$ to $0.900c$ and (b) $0.900c$ to $0.990c$.

35. A proton in a high-energy accelerator moves with a speed of $c/2$. Use the work–kinetic energy theorem to find the work required to increase its speed to (a) $0.750c$ and (b) $0.995c$.

36. Show that, for any object moving at less than one-tenth the speed of light, the relativistic kinetic energy agrees with the result of the classical equation $K = \frac{1}{2}mu^2$ to within less than 1%. Thus for most purposes, the classical equation is good enough to describe these objects, whose motion we call *nonrelativistic*.

37. Find the momentum of a proton in MeV/c units assuming its total energy is twice its rest energy.

38. Find the kinetic energy of a 78.0-kg spacecraft launched out of the solar system with speed 106 km/s by using (a) the classical equation $K = \frac{1}{2}mu^2$. (b) **What If?** Calculate its kinetic energy using the relativistic equation.

39. ⚛ A proton moves at $0.950c$. Calculate its (a) rest energy, (b) total energy, and (c) kinetic energy.

40. A cube of steel has a volume of $1.00\ \text{cm}^3$ and a mass of 8.00 g when at rest on the Earth. If this cube is now given a speed $u = 0.900c$, what is its density as measured by a

stationary observer? Note that relativistic density is defined as E_R/c^2V.

41. An unstable particle with a mass of 3.34×10^{-27} kg is initially at rest. The particle decays into two fragments that fly off along the x axis with velocity components $0.987c$ and $-0.868c$. Find the masses of the fragments. (*Suggestion:* Conserve both energy and momentum.)

42. An object having mass 900 kg and traveling at speed $0.850c$ collides with a stationary object having mass 1 400 kg. The two objects stick together. Find (a) the speed and (b) the mass of the composite object.

43. Show that the energy–momentum relationship $E^2 = p^2c^2 + (mc^2)^2$ follows from the expressions $E = \gamma mc^2$ and $p = \gamma mu$.

44. In a typical color television picture tube, the electrons are accelerated through a potential difference of 25 000 V. (a) What speed do the electrons have when they strike the screen? (b) What is their kinetic energy in joules?

45. Consider electrons accelerated to an energy of 20.0 GeV in the 3.00-km-long Stanford Linear Accelerator. (a) What is the γ factor for the electrons? (b) What is their speed? (c) How long does the accelerator appear to them?

46. Compact high-power lasers can produce a 2.00-J light pulse of duration 100 fs, focused to a spot 1 μm in diameter. (See Mourou and Umstader, "Extreme Light," *Scientific American*, May 2002, page 81.) The electric field in the light accelerates electrons in the target material to near the speed of light. (a) What is the average power of the laser during the pulse? (b) How many electrons can be accelerated to $0.999\ 9c$ if 0.010 0% of the pulse energy is converted into energy of electron motion?

47. A pion at rest ($m_\pi = 273m_e$) decays to a muon ($m_\mu = 207m_e$) and an antineutrino ($m_{\bar{\nu}} \approx 0$). The reaction is written $\pi^- \rightarrow \mu^- + \bar{\nu}$. Find the kinetic energy of the muon and the energy of the antineutrino in electron volts. (*Suggestion:* Conserve both energy and momentum.)

48. According to observer A, two objects of equal mass and moving along the x axis collide head on and stick to each other. Before the collision, this observer measures that object 1 moves to the right with a speed of $3c/4$, while object 2 moves to the left with the same speed. According to observer B, however, object 1 is initially at rest. (a) Determine the speed of object 2 as seen by observer B. (b) Compare the total initial energy of the system in the two frames of reference.

Section 39.9 Mass and Energy

49. Make an order-of-magnitude estimate of the ratio of mass increase to the original mass of a flag, as you run it up a flagpole. In your solution explain what quantities you take as data and the values you estimate or measure for them.

50. When 1.00 g of hydrogen combines with 8.00 g of oxygen, 9.00 g of water is formed. During this chemical reaction, 2.86×10^5 J of energy is released. How much mass do the constituents of this reaction lose? Is the loss of mass likely to be detectable?

51. In a nuclear power plant the fuel rods last 3 yr before they are replaced. If a plant with rated thermal power 1.00 GW

operates at 80.0% capacity for 3.00 yr, what is the loss of mass of the fuel?

52. **Review problem.** The total volume of water in the oceans is approximately 1.40×10^9 km^3. The density of sea water is 1 030 kg/m^3, and the specific heat of the water is $4\,186$ J/(kg·°C). Find the increase in mass of the oceans produced by an increase in temperature of 10.0°C.

53. The power output of the Sun is 3.77×10^{26} W. How much mass is converted to energy in the Sun each second?

54. A gamma ray (a high-energy photon) can produce an electron (e$^-$) and a positron (e$^+$) when it enters the electric field of a heavy nucleus: $\gamma \rightarrow$ e$^+$ + e$^-$. What minimum gamma-ray energy is required to accomplish this task? (*Note:* The masses of the electron and the positron are equal.)

Section 39.10 The General Theory of Relativity

55. An Earth satellite used in the global positioning system moves in a circular orbit with period 11 h 58 min. (a) Determine the radius of its orbit. (b) Determine its speed. (c) The satellite contains an oscillator producing the principal nonmilitary GPS signal. Its frequency is 1 575.42 MHz in the reference frame of the satellite. When it is received on the Earth's surface, what is the fractional change in this frequency due to time dilation, as described by special relativity? (d) The gravitational blue shift of the frequency according to general relativity is a separate effect. The magnitude of that fractional change is given by

$$\frac{\Delta f}{f} = \frac{\Delta U_g}{mc^2}$$

where ΔU_g is the change in gravitational potential energy of an object–Earth system when the object of mass m is moved between the two points at which the signal is observed. Calculate this fractional change in frequency. (e) What is the overall fractional change in frequency? Superposed on both of these relativistic effects is a Doppler shift that is generally much larger. It can be a red shift or a blue shift, depending on the motion of a particular satellite relative to a GPS receiver (Fig. P39.55).

Figure P39.55 This global positioning system (GPS) receiver incorporates relativistically corrected time calculations in its analysis of signals it receives from orbiting satellites. This allows the unit to determine its position on the Earth's surface to within a few meters. If these corrections were not made, the location error would be about 1 km.

Additional Problems

56. An astronaut wishes to visit the Andromeda galaxy, making a one-way trip that will take 30.0 yr in the spacecraft's frame of reference. Assume that the galaxy is 2.00×10^6 ly away and that the astronaut's speed is constant. (a) How fast must he travel relative to the Earth? (b) What will be the kinetic energy of his 1 000-metric-ton spacecraft? (c) What is the cost of this energy if it is purchased at a typical consumer price for electric energy: $0.130/kWh?

57. The cosmic rays of highest energy are protons that have kinetic energy on the order of 10^{13} MeV. (a) How long would it take a proton of this energy to travel across the Milky Way galaxy, having a diameter $\sim 10^5$ ly, as measured in the proton's frame? (b) From the point of view of the proton, how many kilometers across is the galaxy?

58. An electron has a speed of $0.750c$. (a) Find the speed of a proton that has the same kinetic energy as the electron. (b) **What If?** Find the speed of a proton that has the same momentum as the electron.

59. Ted and Mary are playing a game of catch in frame S′, which is moving at $0.600c$ with respect to frame S, while Jim, at rest in frame S, watches the action (Fig. P39.59). Ted throws the ball to Mary at $0.800c$ (according to Ted) and their separation (measured in S′) is 1.80×10^{12} m. (a) According to Mary, how fast is the ball moving? (b) According to Mary, how long does it take the ball to reach her? (c) According to Jim, how far apart are Ted and Mary, and how fast is the ball moving? (d) According to Jim, how long does it take the ball to reach Mary?

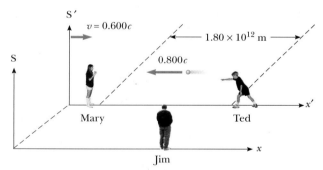

Figure P39.59

60. A rechargeable AA battery with a mass of 25.0 g can supply a power of 1.20 W for 50.0 min. (a) What is the difference in mass between a charged and an uncharged battery? (b) What fraction of the total mass is this mass difference?

61. The net nuclear fusion reaction inside the Sun can be written as $4\,^1$H $\rightarrow\,^4$He + ΔE. The rest energy of each hydrogen atom is 938.78 MeV and the rest energy of the helium-4 atom is 3 728.4 MeV. Calculate the percentage of the starting mass that is transformed to other forms of energy.

62. An object disintegrates into two fragments. One of the fragments has mass 1.00 MeV/c^2 and momentum 1.75 MeV/c in the positive x direction. The other fragment has mass 1.50 MeV/c^2 and momentum 2.00 MeV/c in the positive y direction. Find (a) the mass and (b) the speed of the original object.

63. An alien spaceship traveling at $0.600c$ toward the Earth launches a landing craft with an advance guard of purchasing agents and physics teachers. The lander travels in the same direction with a speed of $0.800c$ relative to the mother ship. As observed on the Earth, the spaceship is 0.200 ly from the Earth when the lander is launched. (a) What speed do the Earth observers measure for the approaching lander? (b) What is the distance to the Earth at the time of lander launch, as observed by the aliens? (c) How long does it take the lander to reach the Earth as observed by the aliens on the mother ship? (d) If the lander has a mass of 4.00×10^5 kg, what is its kinetic energy as observed in the Earth reference frame?

64. A physics professor on the Earth gives an exam to her students, who are in a spacecraft traveling at speed v relative to the Earth. The moment the craft passes the professor, she signals the start of the exam. She wishes her students to have a time interval T_0 (spacecraft time) to complete the exam. Show that she should wait a time interval (Earth time) of

$$T = T_0 \sqrt{\frac{1 - v/c}{1 + v/c}}$$

before sending a light signal telling them to stop. (*Suggestion:* Remember that it takes some time for the second light signal to travel from the professor to the students.)

65. Spacecraft I, containing students taking a physics exam, approaches the Earth with a speed of $0.600c$ (relative to the Earth), while spacecraft II, containing professors proctoring the exam, moves at $0.280c$ (relative to the Earth) directly toward the students. If the professors stop the exam after 50.0 min have passed on their clock, how long does the exam last as measured by (a) the students (b) an observer on the Earth?

66. Energy reaches the upper atmosphere of the Earth from the Sun at the rate of 1.79×10^{17} W. If all of this energy were absorbed by the Earth and not re-emitted, how much would the mass of the Earth increase in 1.00 yr?

67. A supertrain (proper length 100 m) travels at a speed of $0.950c$ as it passes through a tunnel (proper length 50.0 m). As seen by a trackside observer, is the train ever completely within the tunnel? If so, with how much space to spare?

68. Imagine that the entire Sun collapses to a sphere of radius R_g such that the work required to remove a small mass m from the surface would be equal to its rest energy mc^2. This radius is called the *gravitational radius* for the Sun. Find R_g. (It is believed that the ultimate fate of very massive stars is to collapse beyond their gravitational radii into black holes.)

69. A particle with electric charge q moves along a straight line in a uniform electric field $\mathbf{E}$ with a speed of u. The electric force exerted on the charge is $q\mathbf{E}$. The motion and the electric field are both in the x direction. (a) Show that the acceleration of the particle in the x direction is given by

$$a = \frac{du}{dt} = \frac{qE}{m} \left(1 - \frac{u^2}{c^2} \right)^{3/2}$$

(b) Discuss the significance of the dependence of the acceleration on the speed. (c) **What If?** If the particle starts from rest at $x = 0$ at $t = 0$, how would you proceed to find the speed of the particle and its position at time t?

70. An observer in a coasting spacecraft moves toward a mirror at speed v relative to the reference frame labeled by S in Figure P39.70. The mirror is stationary with respect to S. A light pulse emitted by the spacecraft travels toward the mirror and is reflected back to the craft. The front of the craft is a distance d from the mirror (as measured by observers in S) at the moment the light pulse leaves the craft. What is the total travel time of the pulse as measured by observers in (a) the S frame and (b) the front of the spacecraft?

Figure P39.70

71. The creation and study of new elementary particles is an important part of contemporary physics. Especially interesting is the discovery of a very massive particle. To create a particle of mass M requires an energy Mc^2. With enough energy, an exotic particle can be created by allowing a fast moving particle of ordinary matter, such as a proton, to collide with a similar target particle. Let us consider a perfectly inelastic collision between two protons: an incident proton with mass m_p, kinetic energy K, and momentum magnitude p joins with an originally stationary target proton to form a single product particle of mass M. You might think that the creation of a new product particle, nine times more massive than in a previous experiment, would require just nine times more energy for the incident proton. Unfortunately not all of the kinetic energy of the incoming proton is available to create the product particle, since conservation of momentum requires that after the collision the system as a whole still must have some kinetic energy. Only a fraction of the energy of the incident particle is thus available to create a new particle. You will determine how the energy available for particle creation depends on the energy of the moving proton. Show that the energy available to create a product particle is given by

$$Mc^2 = 2m_p c^2 \sqrt{1 + \frac{K}{2m_p c^2}}$$

From this result, when the kinetic energy K of the incident proton is large compared to its rest energy $m_p c^2$, we see that M approaches $(2m_p K)^{1/2}/c$. Thus if the energy of the incoming proton is increased by a factor of nine, the mass you can create increases only by a factor of three. This disappointing result is the main reason that most modern accelerators, such as those at CERN (in Europe), at Fermilab (near Chicago), at SLAC (at Stanford), and at DESY (in Germany), use *colliding beams*. Here the total momentum of a pair of interacting particles can be zero. The center of mass can be at rest after the collision, so in principle all of the initial kinetic energy can be used for particle creation, according to

$$Mc^2 = 2mc^2 + K = 2mc^2 \left(1 + \frac{K}{2mc^2}\right)$$

where K is the total kinetic energy of two identical colliding particles. Here if $K \gg mc^2$, we have M directly proportional to K, as we would desire. These machines are difficult to build and to operate, but they open new vistas in physics.

72. A particle of mass m moving along the x axis with a velocity component $+u$ collides head-on and sticks to a particle of mass $m/3$ moving along the x axis with the velocity component $-u$. What is the mass M of the resulting particle?

73. A rod of length L_0 moving with a speed v along the horizontal direction makes an angle θ_0 with respect to the x' axis. (a) Show that the length of the rod as measured by a stationary observer is $L = L_0[1 - (v^2/c^2)\cos^2\theta_0]^{1/2}$. (b) Show that the angle that the rod makes with the x axis is given by $\tan\theta = \gamma\tan\theta_0$. These results show that the rod is both contracted and rotated. (Take the lower end of the rod to be at the origin of the primed coordinate system.)

74. Suppose our Sun is about to explode. In an effort to escape, we depart in a spacecraft at $v = 0.800c$ and head toward the star Tau Ceti, 12.0 ly away. When we reach the midpoint of our journey from the Earth, we see our Sun explode and, unfortunately, at the same instant we see Tau Ceti explode as well. (a) In the spacecraft's frame of reference, should we conclude that the two explosions occurred simultaneously? If not, which occurred first? (b) **What If?** In a frame of reference in which the Sun and Tau Ceti are at rest, did they explode simultaneously? If not, which exploded first?

75. A ^{57}Fe nucleus at rest emits a 14.0-keV photon. Use conservation of energy and momentum to deduce the kinetic energy of the recoiling nucleus in electron volts. (Use $Mc^2 = 8.60 \times 10^{-9}$ J for the final state of the ^{57}Fe nucleus.)

76. ⌨ Prepare a graph of the relativistic kinetic energy and the classical kinetic energy, both as a function of speed, for an object with a mass of your choice. At what speed does the classical kinetic energy underestimate the experimental value by 1%? by 5%? by 50%?

Answers to Quick Quizzes

39.1 (c). While the observers' measurements differ, both are correct.

39.2 (d). The Galilean velocity transformation gives us $u_x = u'_x + v = 110$ mi/h $+ 90$ mi/h $= 200$ mi/h.

39.3 (d). The two events (the pulse leaving the flashlight and the pulse hitting the far wall) take place at different locations for both observers, so neither measures the proper time interval.

39.4 (a). The two events are the beginning and the end of the movie, both of which take place at rest with respect to the spacecraft crew. Thus, the crew measures the proper time interval of 2 h. Any observer in motion with respect to the spacecraft, which includes the observer on Earth, will measure a longer time interval due to time dilation.

39.5 (a). If their on-duty time is based on clocks that remain on the Earth, they will have larger paychecks. A shorter time interval will have passed for the astronauts in their frame of reference than for their employer back on the Earth.

39.6 (c). Both your body and your sleeping cabin are at rest in your reference frame; thus, they will have their proper length according to you. There will be no change in measured lengths of objects, including yourself, within your spacecraft.

39.7 (d). Time dilation and length contraction depend only on the relative speed of one observer relative to another, not on whether the observers are receding or approaching each other.

39.8 (c). Because of your motion toward the source of the light, the light beam has a horizontal component of velocity as measured by you. The magnitude of the vector sum of the horizontal and vertical component vectors must be equal to c, so the magnitude of the vertical component must be smaller than c.

39.9 (a). In this case, there is only a horizontal component of the velocity of the light, and you must measure a speed of c.

39.10 (a) $m_3 > m_2 = m_1$; the rest energy of particle 3 is $2E$, while it is E for particles 1 and 2. (b) $K_3 = K_2 > K_1$; the kinetic energy is the difference between the total energy and the rest energy. The kinetic energy is $4E - 2E = 2E$ for particle 3, $3E - E = 2E$ for particle 2, and $2E - E = E$ for particle 1. (c) $u_2 > u_3 = u_1$; from Equation 39.26, $E = \gamma E_R$. Solving this for the square of the particle speed u, we find $u^2 = c^2(1 - (E_R/E)^2)$. Thus, the particle with the smallest ratio of rest energy to total energy will have the largest speed. Particles 1 and 3 have the same ratio as each other, and the ratio of particle 2 is smaller.

Table A.1

| Conversion Factors |

Length

	m	cm	km	in.	ft	mi
1 meter	1	10^2	10^{-3}	39.37	3.281	6.214×10^{-4}
1 centimeter	10^{-2}	1	10^{-5}	0.393 7	3.281×10^{-2}	6.214×10^{-6}
1 kilometer	10^3	10^5	1	3.937×10^4	3.281×10^3	0.621 4
1 inch	2.540×10^{-2}	2.540	2.540×10^{-5}	1	8.333×10^{-2}	1.578×10^{-5}
1 foot	0.304 8	30.48	3.048×10^{-4}	12	1	1.894×10^{-4}
1 mile	1 609	1.609×10^5	1.609	6.336×10^4	5 280	1

Mass

	kg	g	slug	u
1 kilogram	1	10^3	6.852×10^{-2}	6.024×10^{26}
1 gram	10^{-3}	1	6.852×10^{-5}	6.024×10^{23}
1 slug	14.59	1.459×10^4	1	8.789×10^{27}
1 atomic mass unit	1.660×10^{-27}	1.660×10^{-24}	1.137×10^{-28}	1

Note: 1 metric ton = 1 000 kg.

Time

	s	min	h	day	yr
1 second	1	1.667×10^{-2}	2.778×10^{-4}	1.157×10^{-5}	3.169×10^{-8}
1 minute	60	1	1.667×10^{-2}	6.994×10^{-4}	1.901×10^{-6}
1 hour	3 600	60	1	4.167×10^{-2}	1.141×10^{-4}
1 day	8.640×10^4	1 440	24	1	2.738×10^{-5}
1 year	3.156×10^7	5.259×10^5	8.766×10^3	365.2	1

Speed

	m/s	cm/s	ft/s	mi/h
1 meter per second	1	10^2	3.281	2.237
1 centimeter per second	10^{-2}	1	3.281×10^{-2}	2.237×10^{-2}
1 foot per second	0.304 8	30.48	1	0.681 8
1 mile per hour	0.447 0	44.70	1.467	1

Note: 1 mi/min = 60 mi/h = 88 ft/s.

Force

	N	lb
1 newton	1	0.224 8
1 pound	4.448	1

continued

Table A.1

Conversion Factors *continued*			

Work, Energy, Heat

	J	ft · lb	eV
1 joule	1	0.737 6	6.242×10^{18}
1 foot-pound	1.356	1	8.464×10^{18}
1 electron volt	1.602×10^{-19}	1.182×10^{-19}	1
1 calorie	4.186	3.087	2.613×10^{19}
1 British thermal unit	1.055×10^3	7.779×10^2	6.585×10^{21}
1 kilowatt hour	3.600×10^6	2.655×10^6	2.247×10^{25}

	cal	Btu	kWh
1 joule	0.238 9	9.481×10^{-4}	2.778×10^{-7}
1 foot-pound	0.323 9	1.285×10^{-3}	3.766×10^{-7}
1 electron volt	3.827×10^{-20}	1.519×10^{-22}	4.450×10^{-26}
1 calorie	1	3.968×10^{-3}	1.163×10^{-6}
1 British thermal unit	2.520×10^2	1	2.930×10^{-4}
1 kilowatt hour	8.601×10^5	3.413×10^2	1

Pressure

	Pa	atm	
1 pascal	1	9.869×10^{-6}	
1 atmosphere	1.013×10^5	1	
1 centimeter mercury[a]	1.333×10^3	1.316×10^{-2}	
1 pound per square inch	6.895×10^3	6.805×10^{-2}	
1 pound per square foot	47.88	4.725×10^{-4}	

	cm Hg	lb/in.^{2}	lb/ft^2
1 pascal	7.501×10^{-4}	1.450×10^{-4}	2.089×10^{-2}
1 atmosphere	76	14.70	2.116×10^3
1 centimeter mercury[a]	1	0.194 3	27.85
1 pound per square inch	5.171	1	144
1 pound per square foot	3.591×10^{-2}	6.944×10^{-3}	1

[a] At 0°C and at a location where the free-fall acceleration has its "standard" value, 9.806 65 m/s^2.

Table A.2

Symbols, Dimensions, and Units of Physical Quantities				

Quantity	Common Symbol	Unit[a]	Dimensions[b]	Unit in Terms of Base SI Units
Acceleration	**a**	m/s^2	L/T^2	m/s^2
Amount of substance	n	MOLE		mol
Angle	θ, ϕ	radian (rad)	1	
Angular acceleration	α	rad/s^2	T^{-2}	s^{-2}
Angular frequency	ω	rad/s	T^{-1}	s^{-1}
Angular momentum	**L**	kg · m^2/s	ML^2/T	kg · m^2/s
Angular velocity	$\boldsymbol{\omega}$	rad/s	T^{-1}	s^{-1}
Area	A	m^2	L^2	m^2
Atomic number	Z			

continued

Table A.2

Symbols, Dimensions, and Units of Physical Quantities *continued*

Quantity	Common Symbol	Unit[a]	Dimensions[b]	Unit in Terms of Base SI Units
Capacitance	C	farad (F)	Q^2T^2/ML^2	$A^2 \cdot s^4/kg \cdot m^2$
Charge	q, Q, e	coulomb (C)	Q	$A \cdot s$
Charge density				
Line	λ	C/m	Q/L	$A \cdot s/m$
Surface	σ	C/m^2	Q/L^2	$A \cdot s/m^2$
Volume	ρ	C/m^3	Q/L^3	$A \cdot s/m^3$
Conductivity	σ	$1/\Omega \cdot m$	Q^2T/ML^3	$A^2 \cdot s^3/kg \cdot m^3$
Current	I	AMPERE	Q/T	A
Current density	$\mathbf{J}$	A/m^2	Q/T^2	A/m^2
Density	ρ	kg/m^3	M/L^3	kg/m^3
Dielectric constant	κ			
Length	ℓ, L	METER	L	m
Position	$x, y, z, \mathbf{r}$			
Displacement	$\Delta x, \Delta\mathbf{r}$			
Distance	d, h			
Electric dipole moment	$\mathbf{p}$	C·m	QL	$A \cdot s \cdot m$
Electric field	$\mathbf{E}$	V/m	ML/QT^2	$kg \cdot m/A \cdot s^3$
Electric flux	Φ_E	V·m	ML^3/QT^2	$kg \cdot m^3/A \cdot s^3$
Electromotive force	$\mathcal{E}$	volt (V)	ML^2/QT^2	$kg \cdot m^2/A \cdot s^3$
Energy	E, U, K	joule (J)	ML^2/T^2	$kg \cdot m^2/s^2$
Entropy	S	J/K	$ML^2/T^2 \cdot K$	$kg \cdot m^2/s^2 \cdot K$
Force	$\mathbf{F}$	newton (N)	ML/T^2	$kg \cdot m/s^2$
Frequency	f	hertz (Hz)	T^{-1}	s^{-1}
Heat	Q	joule (J)	ML^2/T^2	$kg \cdot m^2/s^2$
Inductance	L	henry (H)	ML^2/Q^2	$kg \cdot m^2/A^2 \cdot s^2$
Magnetic dipole moment	$\boldsymbol{\mu}$	N·m/T	QL^2/T	$A \cdot m^2$
Magnetic field	$\mathbf{B}$	tesla (T) (= Wb/m^2)	M/QT	$kg/A \cdot s^2$
Magnetic flux	Φ_B	weber (Wb)	ML^2/QT	$kg \cdot m^2/A \cdot s^2$
Mass	m, M	KILOGRAM	M	kg
Molar specific heat	C	J/mol·K		$kg \cdot m^2/s^2 \cdot mol \cdot K$
Moment of inertia	I	kg·m^2	ML^2	$kg \cdot m^2$
Momentum	$\mathbf{p}$	kg·m/s	ML/T	$kg \cdot m/s$
Period	T	s	T	s
Permeability of free space	μ_0	N/A^2 (= H/m)	ML/Q^2T	$kg \cdot m/A^2 \cdot s^2$
Permittivity of free space	ϵ_0	C^2/N·m^2 (= F/m)	Q^2T^2/ML^3	$A^2 \cdot s^4/kg \cdot m^3$
Potential	V	volt (V) (= J/C)	ML^2/QT^2	$kg \cdot m^2/A \cdot s^3$
Power	$\mathcal{P}$	watt (W) (= J/s)	ML^2/T^3	$kg \cdot m^2/s^3$
Pressure	P	pascal (Pa) (= N/m^2)	M/LT^2	$kg/m \cdot s^2$
Resistance	R	ohm (Ω) (= V/A)	ML^2/Q^2T	$kg \cdot m^2/A^2 \cdot s^3$
Specific heat	c	J/kg·K	$L^2/T^2 \cdot K$	$m^2/s^2 \cdot K$
Speed	v	m/s	L/T	m/s
Temperature	T	KELVIN	K	K
Time	t	SECOND	T	s
Torque	τ	N·m	ML^2/T^2	$kg \cdot m^2/s^2$
Velocity	$\mathbf{v}$	m/s	L/T	m/s
Volume	V	m^3	L^3	m^3
Wavelength	λ	m	L	m
Work	W	joule (J) (= N·m)	ML^2/T^2	$kg \cdot m^2/s^2$

[a] The base SI units are given in uppercase letters.

[b] The symbols M, L, T, and Q denote mass, length, time, and charge, respectively.

Table A.3

Table of Atomic Masses[a]						
Atomic Number Z	Element	Symbol	Chemical Atomic Mass (u)	Mass Number (*Indicates Radioactive) A	Atomic Mass (u)	Percent Abundance	Half-Life (If Radioactive) $T_{1/2}$
0	(Neutron)	n		1*	1.008 665		10.4 min
1	Hydrogen	H	1.007 94	1	1.007 825	99.988 5	
	Deuterium	D		2	2.014 102	0.011 5	
	Tritium	T		3*	3.016 049		12.33 yr
2	Helium	He	4.002 602	3	3.016 029	0.000 137	
				4	4.002 603	99.999 863	
				6*	6.018 888		0.81 s
3	Lithium	Li	6.941	6	6.015 122	7.5	
				7	7.016 004	92.5	
				8*	8.022 487		0.84 s
4	Beryllium	Be	9.012 182	7*	7.016 929		53.3 days
				9	9.012 182	100	
				10*	10.013 534		1.5×10^6 yr
5	Boron	B	10.811	10	10.012 937	19.9	
				11	11.009 306	80.1	
				12*	12.014 352		0.020 2 s
6	Carbon	C	12.010 7	10*	10.016 853		19.3 s
				11*	11.011 434		20.4 min
				12	12.000 000	98.93	
				13	13.003 355	1.07	
				14*	14.003 242		5 730 yr
				15*	15.010 599		2.45 s
7	Nitrogen	N	14.006 7	12*	12.018 613		0.011 0 s
				13*	13.005 739		9.96 min
				14	14.003 074	99.632	
				15	15.000 109	0.368	
				16*	16.006 101		7.13 s
				17*	17.008 450		4.17 s
8	Oxygen	O	15.999 4	14*	14.008 595		70.6 s
				15*	15.003 065		122 s
				16	15.994 915	99.757	
				17	16.999 132	0.038	
				18	17.999 160	0.205	
				19*	19.003 579		26.9 s
9	Fluorine	F	18.998 403 2	17*	17.002 095		64.5 s
				18*	18.000 938		109.8 min
				19	18.998 403	100	
				20*	19.999 981		11.0 s
				21*	20.999 949		4.2 s
10	Neon	Ne	20.179 7	18*	18.005 697		1.67 s
				19*	19.001 880		17.2 s
				20	19.992 440	90.48	
				21	20.993 847	0.27	
				22	21.991 385	9.25	
				23*	22.994 467		37.2 s
11	Sodium	Na	22.989 77	21*	20.997 655		22.5 s
				22*	21.994 437		2.61 yr

continued

Table A.3

Table of Atomic Masses[a] *continued*

Atomic Number Z	Element	Symbol	Chemical Atomic Mass (u)	Mass Number (*Indicates Radioactive) A	Atomic Mass (u)	Percent Abundance	Half-Life (If Radioactive) $T_{1/2}$
(11)	Sodium			23	22.989 770	100	
				24*	23.990 963		14.96 h
12	Magnesium	Mg	24.305 0	23*	22.994 125		11.3 s
				24	23.985 042	78.99	
				25	24.985 837	10.00	
				26	25.982 593	11.01	
				27*	26.984 341		9.46 min
13	Aluminum	Al	26.981 538	26*	25.986 892		7.4×10^5 yr
				27	26.981 539	100	
				28*	27.981 910		2.24 min
14	Silicon	Si	28.085 5	28	27.976 926	92.229 7	
				29	28.976 495	4.683 2	
				30	29.973 770	3.087 2	
				31*	30.975 363		2.62 h
				32*	31.974 148		172 yr
15	Phosphorus	P	30.973 761	30*	29.978 314		2.50 min
				31	30.973 762	100	
				32*	31.973 907		14.26 days
				33*	32.971 725		25.3 days
16	Sulfur	S	32.066	32	31.972 071	94.93	
				33	32.971 458	0.76	
				34	33.967 869	4.29	
				35*	34.969 032		87.5 days
				36	35.967 081	0.02	
17	Chlorine	Cl	35.452 7	35	34.968 853	75.78	
				36*	35.968 307		3.0×10^5 yr
				37	36.965 903	24.22	
18	Argon	Ar	39.948	36	35.967 546	0.336 5	
				37*	36.966 776		35.04 days
				38	37.962 732	0.063 2	
				39*	38.964 313		269 yr
				40	39.962 383	99.600 3	
				42*	41.963 046		33 yr
19	Potassium	K	39.098 3	39	38.963 707	93.258 1	
				40*	39.963 999	0.011 7	1.28×10^9 yr
				41	40.961 826	6.730 2	
20	Calcium	Ca	40.078	40	39.962 591	96.941	
				41*	40.962 278		1.0×10^5 yr
				42	41.958 618	0.647	
				43	42.958 767	0.135	
				44	43.955 481	2.086	
				46	45.953 693	0.004	
				48	47.952 534	0.187	
21	Scandium	Sc	44.955 910	41*	40.969 251		0.596 s
				45	44.955 910	100	
22	Titanium	Ti	47.867	44*	43.959 690		49 yr
				46	45.952 630	8.25	

continued

Table A.3

Table of Atomic Masses[a] *continued*						
Atomic Number Z	Element	Symbol	Chemical Atomic Mass (u)	Mass Number (*Indicates Radioactive) A	Atomic Mass (u)	Percent Abundance	Half-Life (If Radioactive) $T_{1/2}$
(22)	Titanium			47	46.951 764	7.44	
				48	47.947 947	73.72	
				49	48.947 871	5.41	
				50	49.944 792	5.18	
23	Vanadium	V	50.941 5	48*	47.952 254		15.97 days
				50*	49.947 163	0.250	1.5×10^{17} yr
				51	50.943 964	99.750	
24	Chromium	Cr	51.996 1	48*	47.954 036		21.6 h
				50	49.946 050	4.345	
				52	51.940 512	83.789	
				53	52.940 654	9.501	
				54	53.938 885	2.365	
25	Manganese	Mn	54.938 049	54*	53.940 363		312.1 days
				55	54.938 050	100	
26	Iron	Fe	55.845	54	53.939 615	5.845	
				55*	54.938 298		2.7 yr
				56	55.934 942	91.754	
				57	56.935 399	2.119	
				58	57.933 280	0.282	
				60*	59.934 077		1.5×10^{6} yr
27	Cobalt	Co	58.933 200	59	58.933 200	100	
				60*	59.933 822		5.27 yr
28	Nickel	Ni	58.693 4	58	57.935 348	68.076 9	
				59*	58.934 351		7.5×10^{4} yr
				60	59.930 790	26.223 1	
				61	60.931 060	1.139 9	
				62	61.928 349	3.634 5	
				63*	62.929 673		100 yr
				64	63.927 970	0.925 6	
29	Copper	Cu	63.546	63	62.929 601	69.17	
				65	64.927 794	30.83	
30	Zinc	Zn	65.39	64	63.929 147	48.63	
				66	65.926 037	27.90	
				67	66.927 131	4.10	
				68	67.924 848	18.75	
				70	69.925 325	0.62	
31	Gallium	Ga	69.723	69	68.925 581	60.108	
				71	70.924 705	39.892	
32	Germanium	Ge	72.61	70	69.924 250	20.84	
				72	71.922 076	27.54	
				73	72.923 459	7.73	
				74	73.921 178	36.28	
				76	75.921 403	7.61	
33	Arsenic	As	74.921 60	75	74.921 596	100	
34	Selenium	Se	78.96	74	73.922 477	0.89	
				76	75.919 214	9.37	
				77	76.919 915	7.63	

continued

Table A.3

Table of Atomic Masses[a] *continued*						
Atomic Number Z	Element	Symbol	Chemical Atomic Mass (u)	Mass Number (*Indicates Radioactive) A	Atomic Mass (u)	Percent Abundance	Half-Life (If Radioactive) $T_{1/2}$
(34)	Selenium			78	77.917 310	23.77	
				79*	78.918 500		$\leq 6.5 \times 10^4$ yr
				80	79.916 522	49.61	
				82*	81.916 700	8.73	1.4×10^{20} yr
35	Bromine	Br	79.904	79	78.918 338	50.69	
				81	80.916 291	49.31	
36	Krypton	Kr	83.80	78	77.920 386	0.35	
				80	79.916 378	2.28	
				81*	80.916 592		2.1×10^5 yr
				82	81.913 485	11.58	
				83	82.914 136	11.49	
				84	83.911 507	57.00	
				85*	84.912 527		10.76 yr
				86	85.910 610	17.30	
37	Rubidium	Rb	85.467 8	85	84.911 789	72.17	
				87*	86.909 184	27.83	4.75×10^{10} yr
38	Strontium	Sr	87.62	84	83.913 425	0.56	
				86	85.909 262	9.86	
				87	86.908 880	7.00	
				88	87.905 614	82.58	
				90*	89.907 738		29.1 yr
39	Yttrium	Y	88.905 85	89	88.905 848	100	
40	Zirconium	Zr	91.224	90	89.904 704	51.45	
				91	90.905 645	11.22	
				92	91.905 040	17.15	
				93*	92.906 476		1.5×10^6 yr
				94	93.906 316	17.38	
				96	95.908 276	2.80	
41	Niobium	Nb	92.906 38	91*	90.906 990		6.8×10^2 yr
				92*	91.907 193		3.5×10^7 yr
				93	92.906 378	100	
				94*	93.907 284		2×10^4 yr
42	Molybdenum	Mo	95.94	92	91.906 810	14.84	
				93*	92.906 812		3.5×10^3 yr
				94	93.905 088	9.25	
				95	94.905 842	15.92	
				96	95.904 679	16.68	
				97	96.906 021	9.55	
				98	97.905 408	24.13	
				100	99.907 477	9.63	
43	Technetium	Tc		97*	96.906 365		2.6×10^6 yr
				98*	97.907 216		4.2×10^6 yr
				99*	98.906 255		2.1×10^5 yr
44	Ruthenium	Ru	101.07	96	95.907 598	5.54	
				98	97.905 287	1.87	
				99	98.905 939	12.76	
				100	99.904 220	12.60	

continued

Table A.3

Table of Atomic Massesᵃ *continued*						
Atomic Number Z	**Element**	**Symbol**	**Chemical Atomic Mass (u)**	**Mass Number (*Indicates Radioactive) A**	**Atomic Mass (u)**	**Percent Abundance**	**Half-Life (If Radioactive) $T_{1/2}$**
(44)	Ruthenium			101	100.905 582	17.06	
				102	101.904 350	31.55	
				104	103.905 430	18.62	
45	Rhodium	Rh	102.905 50	103	102.905 504	100	
46	Palladium	Pd	106.42	102	101.905 608	1.02	
				104	103.904 035	11.14	
				105	104.905 084	22.33	
				106	105.903 483	27.33	
				107*	106.905 128		6.5×10^6 yr
				108	107.903 894	26.46	
				110	109.905 152	11.72	
47	Silver	Ag	107.868 2	107	106.905 093	51.839	
				109	108.904 756	48.161	
48	Cadmium	Cd	112.411	106	105.906 458	1.25	
				108	107.904 183	0.89	
				109*	108.904 986		462 days
				110	109.903 006	12.49	
				111	110.904 182	12.80	
				112	111.902 757	24.13	
				113*	112.904 401	12.22	9.3×10^{15} yr
				114	113.903 358	28.73	
				116	115.904 755	7.49	
49	Indium	In	114.818	113	112.904 061	4.29	
				115*	114.903 878	95.71	4.4×10^{14} yr
50	Tin	Sn	118.710	112	111.904 821	0.97	
				114	113.902 782	0.66	
				115	114.903 346	0.34	
				116	115.901 744	14.54	
				117	116.902 954	7.68	
				118	117.901 606	24.22	
				119	118.903 309	8.59	
				120	119.902 197	32.58	
				121*	120.904 237		55 yr
				122	121.903 440	4.63	
				124	123.905 275	5.79	
51	Antimony	Sb	121.760	121	120.903 818	57.21	
				123	122.904 216	42.79	
				125*	124.905 248		2.7 yr
52	Tellurium	Te	127.60	120	119.904 020	0.09	
				122	121.903 047	2.55	
				123*	122.904 273	0.89	1.3×10^{13} yr
				124	123.902 820	4.74	
				125	124.904 425	7.07	
				126	125.903 306	18.84	
				128*	127.904 461	31.74	$>8 \times 10^{24}$ yr
				130*	129.906 223	34.08	$\leq 1.25 \times 10^{21}$ yr

continued

Table A.3

Table of Atomic Masses[a] *continued*						
Atomic Number Z	Element	Symbol	Chemical Atomic Mass (u)	Mass Number (*Indicates Radioactive) A	Atomic Mass (u)	Percent Abundance	Half-Life (If Radioactive) $T_{1/2}$
53	Iodine	I	126.904 47	127	126.904 468	100	
				129*	128.904 988		1.6×10^7 yr
54	Xenon	Xe	131.29	124	123.905 896	0.09	
				126	125.904 269	0.09	
				128	127.903 530	1.92	
				129	128.904 780	26.44	
				130	129.903 508	4.08	
				131	130.905 082	21.18	
				132	131.904 145	26.89	
				134	133.905 394	10.44	
				136*	135.907 220	8.87	$\geq 2.36 \times 10^{21}$ yr
55	Cesium	Cs	132.905 45	133	132.905 447	100	
				134*	133.906 713		2.1 yr
				135*	134.905 972		2×10^6 yr
				137*	136.907 074		30 yr
56	Barium	Ba	137.327	130	129.906 310	0.106	
				132	131.905 056	0.101	
				133*	132.906 002		10.5 yr
				134	133.904 503	2.417	
				135	134.905 683	6.592	
				136	135.904 570	7.854	
				137	136.905 821	11.232	
				138	137.905 241	71.698	
57	Lanthanum	La	138.905 5	137*	136.906 466		6×10^4 yr
				138*	137.907 107	0.090	1.05×10^{11} yr
				139	138.906 349	99.910	
58	Cerium	Ce	140.116	136	135.907 144	0.185	
				138	137.905 986	0.251	
				140	139.905 434	88.450	
				142*	141.909 240	11.114	$>5 \times 10^{16}$ yr
59	Praseodymium	Pr	140.907 65	141	140.907 648	100	
60	Neodymium	Nd	144.24	142	141.907 719	27.2	
				143	142.909 810	12.2	
				144*	143.910 083	23.8	2.3×10^{15} yr
				145	144.912 569	8.3	
				146	145.913 112	17.2	
				148	147.916 888	5.7	
				150*	149.920 887	5.6	$>1 \times 10^{18}$ yr
61	Promethium	Pm		143*	142.910 928		265 days
				145*	144.912 744		17.7 yr
				146*	145.914 692		5.5 yr
				147*	146.915 134		2.623 yr
62	Samarium	Sm	150.36	144	143.911 995	3.07	
				146*	145.913 037		1.0×10^8 yr
				147*	146.914 893	14.99	1.06×10^{11} yr
				148*	147.914 818	11.24	7×10^{15} yr

continued

Table A.3

Table of Atomic Masses[a] *continued*

Atomic Number Z	Element	Symbol	Chemical Atomic Mass (u)	Mass Number (*Indicates Radioactive) A	Atomic Mass (u)	Percent Abundance	Half-Life (If Radioactive) $T_{1/2}$
(62)	Samarium			149*	148.917 180	13.82	$> 2 \times 10^{15}$ yr
				150	149.917 272	7.38	
				151*	150.919 928		90 yr
				152	151.919 728	26.75	
				154	153.922 205	22.75	
63	Europium	Eu	151.964	151	150.919 846	47.81	
				152*	151.921 740		13.5 yr
				153	152.921 226	52.19	
				154*	153.922 975		8.59 yr
				155*	154.922 889		4.7 yr
64	Gadolinium	Gd	157.25	148*	147.918 110		75 yr
				150*	149.918 656		1.8×10^{6} yr
				152*	151.919 788	0.20	1.1×10^{14} yr
				154	153.920 862	2.18	
				155	154.922 619	14.80	
				156	155.922 120	20.47	
				157	156.923 957	15.65	
				158	157.924 100	24.84	
				160	159.927 051	21.86	
65	Terbium	Tb	158.925 34	159	158.925 343	100	
66	Dysprosium	Dy	162.50	156	155.924 278	0.06	
				158	157.924 405	0.10	
				160	159.925 194	2.34	
				161	160.926 930	18.91	
				162	161.926 795	25.51	
				163	162.928 728	24.90	
				164	163.929 171	28.18	
67	Holmium	Ho	164.930 32	165	164.930 320	100	
				166*	165.932 281		1.2×10^{3} yr
68	Erbium	Er	167.6	162	161.928 775	0.14	
				164	163.929 197	1.61	
				166	165.930 290	33.61	
				167	166.932 045	22.93	
				168	167.932 368	26.78	
				170	169.935 460	14.93	
69	Thulium	Tm	168.934 21	169	168.934 211	100	
				171*	170.936 426		1.92 yr
70	Ytterbium	Yb	173.04	168	167.933 894	0.13	
				170	169.934 759	3.04	
				171	170.936 322	14.28	
				172	171.936 378	21.83	
				173	172.938 207	16.13	
				174	173.938 858	31.83	
				176	175.942 568	12.76	
71	Lutecium	Lu	174.967	173*	172.938 927		1.37 yr
				175	174.940 768	97.41	
				176*	175.942 682	2.59	3.78×10^{10} yr

continued

Table A.3

			Table of Atomic Masses[a] *continued*			

Atomic Number Z	Element	Symbol	Chemical Atomic Mass (u)	Mass Number (*Indicates Radioactive) A	Atomic Mass (u)	Percent Abundance	Half-Life (If Radioactive) $T_{1/2}$
72	Hafnium	Hf	178.49	174*	173.940 040	0.16	2.0×10^{15} yr
				176	175.941 402	5.26	
				177	176.943 220	18.60	
				178	177.943 698	27.28	
				179	178.945 815	13.62	
				180	179.946 549	35.08	
73	Tantalum	Ta	180.947 9	180*	179.947 466	0.012	8.152 h
				181	180.947 996	99.988	
74	Tungsten (Wolfram)	W	183.84	180	179.946 706	0.12	
				182	181.948 206	26.50	
				183	182.950 224	14.31	
				184*	183.950 933	30.64	$>3 \times 10^{17}$ yr
				186	185.954 362	28.43	
75	Rhenium	Re	186.207	185	184.952 956	37.40	
				187*	186.955 751	62.60	4.4×10^{10} yr
76	Osmium	Os	190.23	184	183.952 491	0.02	
				186*	185.953 838	1.59	2.0×10^{15} yr
				187	186.955 748	1.96	
				188	187.955 836	13.24	
				189	188.958 145	16.15	
				190	189.958 445	26.26	
				192	191.961 479	40.78	
				194*	193.965 179		6.0 yr
77	Iridium	Ir	192.217	191	190.960 591	37.3	
				193	192.962 924	62.7	
78	Platinum	Pt	195.078	190*	189.959 930	0.014	6.5×10^{11} yr
				192	191.961 035	0.782	
				194	193.962 664	32.967	
				195	194.964 774	33.832	
				196	195.964 935	25.242	
				198	197.967 876	7.163	
79	Gold	Au	196.966 55	197	196.966 552	100	
80	Mercury	Hg	200.59	196	195.965 815	0.15	
				198	197.966 752	9.97	
				199	198.968 262	16.87	
				200	199.968 309	23.10	
				201	200.970 285	13.18	
				202	201.970 626	29.86	
				204	203.973 476	6.87	
81	Thallium	Tl	204.383 3	203	202.972 329	29.524	
				204*	203.973 849		3.78 yr
				205	204.974 412	70.476	
		(Ra E″)		206*	205.976 095		4.2 min
		(Ac C″)		207*	206.977 408		4.77 min
		(Th C″)		208*	207.982 005		3.053 min
		(Ra C″)		210*	209.990 066		1.30 min

continued

Table A.3

| Table of Atomic Masses[a] *continued* | | | | | | |

Atomic Number Z	Element	Symbol	Chemical Atomic Mass (u)	Mass Number (*Indicates Radioactive) A	Atomic Mass (u)	Percent Abundance	Half-Life (If Radioactive) $T_{1/2}$
82	Lead	Pb	207.2	202*	201.972 144		5×10^4 yr
				204*	203.973 029	1.4	$\geq 1.4 \times 10^{17}$ yr
				205*	204.974 467		1.5×10^7 yr
				206	205.974 449	24.1	
				207	206.975 881	22.1	
				208	207.976 636	52.4	
		(Ra D)		210*	209.984 173		22.3 yr
		(Ac B)		211*	210.988 732		36.1 min
		(Th B)		212*	211.991 888		10.64 h
		(Ra B)		214*	213.999 798		26.8 min
83	Bismuth	Bi	208.980 38	207*	206.978 455		32.2 yr
				208*	207.979 727		3.7×10^5 yr
				209	208.980 383	100	
		(Ra E)		210*	209.984 105		5.01 days
		(Th C)		211*	210.987 258		2.14 min
				212*	211.991 272		60.6 min
		(Ra C)		214*	213.998 699		19.9 min
				215*	215.001 832		7.4 min
84	Polonium	Po		209*	208.982 416		102 yr
		(Ra F)		210*	209.982 857		138.38 days
		(Ac C′)		211*	210.986 637		0.52 s
		(Th C′)		212*	211.988 852		0.30 μs
		(Ra C′)		214*	213.995 186		164 μs
		(Ac A)		215*	214.999 415		0.001 8 s
		(Th A)		216*	216.001 905		0.145 s
		(Ra A)		218*	218.008 966		3.10 min
85	Astatine	At		215*	214.998 641		≈ 100 μs
				218*	218.008 682		1.6 s
				219*	219.011 297		0.9 min
86	Radon	Rn					
		(An)		219*	219.009 475		3.96 s
		(Tn)		220*	220.011 384		55.6 s
		(Rn)		222*	222.017 570		3.823 days
87	Francium	Fr					
		(Ac K)		223*	223.019 731		22 min
88	Radium	Ra					
		(Ac X)		223*	223.018 497		11.43 days
		(Th X)		224*	224.020 202		3.66 days
		(Ra)		226*	226.025 403		1 600 yr
		(Ms Th$_1$)		228*	228.031 064		5.75 yr
89	Actinium	Ac		227*	227.027 747		21.77 yr
		(Ms Th$_2$)		228*	228.031 015		6.15 h
90	Thorium	Th	232.038 1				
		(Rd Ac)		227*	227.027 699		18.72 days
		(Rd Th)		228*	228.028 731		1.913 yr
				229*	229.031 755		7 300 yr
		(Io)		230*	230.033 127		75.000 yr

continued

Table A.3

Table of Atomic Masses[a] *continued*						
Atomic Number Z	**Element**	**Symbol**	**Chemical Atomic Mass (u)**	**Mass Number (*Indicates Radioactive) A**	**Atomic Mass (u)**	**Percent Abundance**	**Half-Life (If Radioactive) $T_{1/2}$**
(90)	Thorium	(UY)		231*	231.036 297		25.52 h
		(Th)		232*	232.038 050	100	1.40×10^{10} yr
		(UX$_1$)		234*	234.043 596		24.1 days
91	Protactinium	Pa	231.035 88	231*	231.035 879		32.760 yr
		(Uz)		234*	234.043 302		6.7 h
92	Uranium	U	238.028 9	232*	232.037 146		69 yr
				233*	233.039 628		1.59×10^5 yr
				234*	234.040 946	0.005 5	2.45×10^5 yr
		(Ac U)		235*	235.043 923	0.720 0	7.04×10^8 yr
				236*	236.045 562		2.34×10^7 yr
		(UI)		238*	238.050 783	99.274 5	4.47×10^9 yr
93	Neptunium	Np		235*	235.044 056		396 days
				236*	236.046 560		1.15×10^5 yr
				237*	237.048 167		2.14×10^6 yr
94	Plutonium	Pu		236*	236.046 048		2.87 yr
				238*	238.049 553		87.7 yr
				239*	239.052 156		2.412×10^4 yr
				240*	240.053 808		6 560 yr
				241*	241.056 845		14.4 yr
				242*	242.058 737		3.73×10^6 yr
				244*	244.064 198		8.1×10^7 yr

[a] Chemical atomic masses are from T. B. Coplen, "Atomic Weights of the Elements 1999," a technical report to the International Union of Pure and Applied Chemistry, and published in *Pure and Applied Chemistry*, 73(4), 667–683, 2001. Atomic masses of the isotopes are from G. Audi and A. H. Wapstra, "The 1995 Update to the Atomic Mass Evaluation," *Nuclear Physics*, A595, vol. 4, 409–480, December 25, 1995. Percent abundance values are from K. J. R. Rosman and P. D. P. Taylor, "Isotopic Compositions of the Elements 1999", a technical report to the International Union of Pure and Applied Chemistry, and published in *Pure and Applied Chemistry*, 70(1), 217–236, 1998.

Appendix B • Mathematics Review

These appendices in mathematics are intended as a brief review of operations and methods. Early in this course, you should be totally familiar with basic algebraic techniques, analytic geometry, and trigonometry. The appendices on differential and integral calculus are more detailed and are intended for those students who have difficulty applying calculus concepts to physical situations.

B.1 Scientific Notation

Many quantities that scientists deal with often have very large or very small values. For example, the speed of light is about 300 000 000 m/s, and the ink required to make the dot over an i in this textbook has a mass of about 0.000 000 001 kg. Obviously, it is very cumbersome to read, write, and keep track of numbers such as these. We avoid this problem by using a method dealing with powers of the number 10:

$$10^0 = 1$$
$$10^1 = 10$$
$$10^2 = 10 \times 10 = 100$$
$$10^3 = 10 \times 10 \times 10 = 1000$$
$$10^4 = 10 \times 10 \times 10 \times 10 = 10\ 000$$
$$10^5 = 10 \times 10 \times 10 \times 10 \times 10 = 100\ 000$$

and so on. The number of zeros corresponds to the power to which 10 is raised, called the **exponent** of 10. For example, the speed of light, 300 000 000 m/s, can be expressed as 3×10^8 m/s.

In this method, some representative numbers smaller than unity are

$$10^{-1} = \frac{1}{10} = 0.1$$

$$10^{-2} = \frac{1}{10 \times 10} = 0.01$$

$$10^{-3} = \frac{1}{10 \times 10 \times 10} = 0.001$$

$$10^{-4} = \frac{1}{10 \times 10 \times 10 \times 10} = 0.000\ 1$$

$$10^{-5} = \frac{1}{10 \times 10 \times 10 \times 10 \times 10} = 0.000\ 01$$

In these cases, the number of places the decimal point is to the left of the digit 1 equals the value of the (negative) exponent. Numbers expressed as some power of 10 multiplied by another number between 1 and 10 are said to be in **scientific notation.** For example, the scientific notation for 5 943 000 000 is 5.943×10^9 and that for 0.000 083 2 is 8.32×10^{-5}.

When numbers expressed in scientific notation are being multiplied, the following general rule is very useful:

$$10^n \times 10^m = 10^{n+m} \tag{B.1}$$

where n and m can be *any* numbers (not necessarily integers). For example, $10^2 \times 10^5 = 10^7$. The rule also applies if one of the exponents is negative: $10^3 \times 10^{-8} = 10^{-5}$.

When dividing numbers expressed in scientific notation, note that

$$\frac{10^n}{10^m} = 10^n \times 10^{-m} = 10^{n-m} \qquad \text{(B.2)}$$

Exercises

With help from the above rules, verify the answers to the following:

1. $86\ 400 = 8.64 \times 10^4$
2. $9\ 816\ 762.5 = 9.816\ 762\ 5 \times 10^6$
3. $0.000\ 000\ 039\ 8 = 3.98 \times 10^{-8}$
4. $(4 \times 10^8)(9 \times 10^9) = 3.6 \times 10^{18}$
5. $(3 \times 10^7)(6 \times 10^{-12}) = 1.8 \times 10^{-4}$
6. $\dfrac{75 \times 10^{-11}}{5 \times 10^{-3}} = 1.5 \times 10^{-7}$
7. $\dfrac{(3 \times 10^6)(8 \times 10^{-2})}{(2 \times 10^{17})(6 \times 10^5)} = 2 \times 10^{-18}$

B.2 Algebra

Some Basic Rules

When algebraic operations are performed, the laws of arithmetic apply. Symbols such as x, y, and z are usually used to represent quantities that are not specified, what are called the **unknowns.**

First, consider the equation

$$8x = 32$$

If we wish to solve for x, we can divide (or multiply) each side of the equation by the same factor without destroying the equality. In this case, if we divide both sides by 8, we have

$$\frac{8x}{8} = \frac{32}{8}$$

$$x = 4$$

Next consider the equation

$$x + 2 = 8$$

In this type of expression, we can add or subtract the same quantity from each side. If we subtract 2 from each side, we obtain

$$x + 2 - 2 = 8 - 2$$

$$x = 6$$

In general, if $x + a = b$, then $x = b - a$.

Now consider the equation

$$\frac{x}{5} = 9$$

If we multiply each side by 5, we are left with x on the left by itself and 45 on the right:

$$\left(\frac{x}{5}\right)(5) = 9 \times 5$$

$$x = 45$$

In all cases, *whatever operation is performed on the left side of the equality must also be performed on the right side.*

The following rules for multiplying, dividing, adding, and subtracting fractions should be recalled, where a, b, and c are three numbers:

	Rule	Example
Multiplying	$\left(\dfrac{a}{b}\right)\left(\dfrac{c}{d}\right) = \dfrac{ac}{bd}$	$\left(\dfrac{2}{3}\right)\left(\dfrac{4}{5}\right) = \dfrac{8}{15}$
Dividing	$\dfrac{(a/b)}{(c/d)} = \dfrac{ad}{bc}$	$\dfrac{2/3}{4/5} = \dfrac{(2)(5)}{(4)(3)} = \dfrac{10}{12}$
Adding	$\dfrac{a}{b} \pm \dfrac{c}{d} = \dfrac{ad \pm bc}{bd}$	$\dfrac{2}{3} - \dfrac{4}{5} = \dfrac{(2)(5) - (4)(3)}{(3)(5)} = -\dfrac{2}{15}$

Exercises

In the following exercises, solve for x:

Answers

1. $a = \dfrac{1}{1+x}$ $x = \dfrac{1-a}{a}$

2. $3x - 5 = 13$ $x = 6$

3. $ax - 5 = bx + 2$ $x = \dfrac{7}{a-b}$

4. $\dfrac{5}{2x+6} = \dfrac{3}{4x+8}$ $x = -\dfrac{11}{7}$

Powers

When powers of a given quantity x are multiplied, the following rule applies:

$$x^n x^m = x^{n+m} \tag{B.3}$$

For example, $x^2 x^4 = x^{2+4} = x^6$.

When dividing the powers of a given quantity, the rule is

$$\frac{x^n}{x^m} = x^{n-m} \tag{B.4}$$

For example, $x^8/x^2 = x^{8-2} = x^6$.

A power that is a fraction, such as $\frac{1}{3}$, corresponds to a root as follows:

$$x^{1/n} = \sqrt[n]{x} \tag{B.5}$$

For example, $4^{1/3} = \sqrt[3]{4} = 1.5874$. (A scientific calculator is useful for such calculations.)

Finally, any quantity x^n raised to the mth power is

$$(x^n)^m = x^{nm} \tag{B.6}$$

Table B.1 summarizes the rules of exponents.

Table B.1

Rules of Exponents
$x^0 = 1$
$x^1 = x$
$x^n x^m = x^{n+m}$
$x^n/x^m = x^{n-m}$
$x^{1/n} = \sqrt[n]{x}$
$(x^n)^m = x^{nm}$

Exercises

Verify the following:

1. $3^2 \times 3^3 = 243$

2. $x^5 x^{-8} = x^{-3}$

3. $x^{10}/x^{-5} = x^{15}$

4. $5^{1/3} = 1.709\ 975$ (Use your calculator.)

5. $60^{1/4} = 2.783\ 158$ (Use your calculator.)

6. $(x^4)^3 = x^{12}$

Factoring

Some useful formulas for factoring an equation are

$$ax + ay + az = a(x + y + z) \qquad \text{common factor}$$

$$a^2 + 2ab + b^2 = (a + b)^2 \qquad \text{perfect square}$$

$$a^2 - b^2 = (a + b)(a - b) \qquad \text{differences of squares}$$

Quadratic Equations

The general form of a quadratic equation is

$$ax^2 + bx + c = 0 \tag{B.7}$$

where x is the unknown quantity and a, b, and c are numerical factors referred to as **coefficients** of the equation. This equation has two roots, given by

$$x = \frac{-b \pm \sqrt{b^2 - 4ac}}{2a} \tag{B.8}$$

If $b^2 \geq 4ac$, the roots are real.

Example 1

The equation $x^2 + 5x + 4 = 0$ has the following roots corresponding to the two signs of the square-root term:

$$x = \frac{-5 \pm \sqrt{5^2 - (4)(1)(4)}}{2(1)} = \frac{-5 \pm \sqrt{9}}{2} = \frac{-5 \pm 3}{2}$$

$$x_+ = \frac{-5 + 3}{2} = -1 \qquad x_- = \frac{-5 - 3}{2} = -4$$

where x_+ refers to the root corresponding to the positive sign and x_- refers to the root corresponding to the negative sign.

Exercises

Solve the following quadratic equations:

Answers

1. $x^2 + 2x - 3 = 0$ $x_+ = 1$ $x_- = -3$

2. $2x^2 - 5x + 2 = 0$ $x_+ = 2$ $x_- = \frac{1}{2}$

3. $2x^2 - 4x - 9 = 0$ $x_+ = 1 + \sqrt{22}/2$ $x_- = 1 - \sqrt{22}/2$

Linear Equations

A linear equation has the general form

$$y = mx + b \tag{B.9}$$

where m and b are constants. This equation is referred to as being linear because the graph of y versus x is a straight line, as shown in Figure B.1. The constant b, called the **y-intercept,** represents the value of y at which the straight line intersects the y axis. The constant m is equal to the **slope** of the straight line. If any two points on the straight line are specified by the coordinates (x_1, y_1) and (x_2, y_2), as in Figure B.1, then

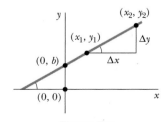

Figure B.1

the slope of the straight line can be expressed as

$$\text{Slope} = \frac{y_2 - y_1}{x_2 - x_1} = \frac{\Delta y}{\Delta x} \tag{B.10}$$

Note that m and b can have either positive or negative values. If $m > 0$, the straight line has a *positive* slope, as in Figure B.1. If $m < 0$, the straight line has a *negative* slope. In Figure B.1, both m and b are positive. Three other possible situations are shown in Figure B.2.

Figure B.2

Exercises

1. Draw graphs of the following straight lines:
   (a) $y = 5x + 3$ (b) $y = -2x + 4$ (c) $y = -3x - 6$

2. Find the slopes of the straight lines described in Exercise 1.

Answers   (a) 5   (b) -2 (c) -3

3. Find the slopes of the straight lines that pass through the following sets of points:
   (a) $(0, -4)$ and $(4, 2)$ (b) $(0, 0)$ and $(2, -5)$ (c) $(-5, 2)$ and $(4, -2)$

Answers   (a) $3/2$ (b) $-5/2$ (c) $-4/9$

Solving Simultaneous Linear Equations

Consider the equation $3x + 5y = 15$, which has two unknowns, x and y. Such an equation does not have a unique solution. For example, note that $(x = 0, y = 3)$, $(x = 5, y = 0)$, and $(x = 2, y = 9/5)$ are all solutions to this equation.

If a problem has two unknowns, a unique solution is possible only if we have *two* equations. In general, if a problem has n unknowns, its solution requires n equations. In order to solve two simultaneous equations involving two unknowns, x and y, we solve one of the equations for x in terms of y and substitute this expression into the other equation.

Example 2

Solve the following two simultaneous equations:

(1) $5x + y = -8$

(2) $2x - 2y = 4$

Solution From Equation (2), $x = y + 2$. Substitution of this into Equation (1) gives

$$5(y + 2) + y = -8$$

$$6y = -18$$

$$y = \boxed{-3}$$

$$x = y + 2 = \boxed{-1}$$

Alternate Solution Multiply each term in Equation (1) by the factor 2 and add the result to Equation (2):

$$10x + 2y = -16$$

$$\underline{2x - 2y = 4}$$

$$12x = -12$$

$$x = \boxed{-1}$$

$$y = x - 2 = \boxed{-3}$$

Two linear equations containing two unknowns can also be solved by a graphical method. If the straight lines corresponding to the two equations are plotted in a conventional coordinate system, the intersection of the two lines represents the solution. For example, consider the two equations

$$x - y = 2$$

$$x - 2y = -1$$

These are plotted in Figure B.3. The intersection of the two lines has the coordinates $x = 5$, $y = 3$. This represents the solution to the equations. You should check this solution by the analytical technique discussed above.

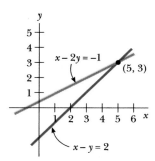

Figure B.3

Exercises

Solve the following pairs of simultaneous equations involving two unknowns:

Answers

1. $x + y = 8$ $x = 5, y = 3$
   $x - y = 2$

2. $98 - T = 10a$ $T = 65, a = 3.3$
   $T - 49 = 5a$

3. $6x + 2y = 6$ $x = 2, y = -3$
   $8x - 4y = 28$

Logarithms

Suppose that a quantity x is expressed as a power of some quantity a:

$$x = a^y \tag{B.11}$$

The number a is called the **base** number. The **logarithm** of x with respect to the base a is equal to the exponent to which the base must be raised in order to satisfy the expression $x = a^y$:

$$y = \log_a x \tag{B.12}$$

Conversely, the **antilogarithm** of y is the number x:

$$x = \text{antilog}_a y \tag{B.13}$$

In practice, the two bases most often used are base 10, called the *common* logarithm base, and base $e = 2.718\,282$, called Euler's constant or the *natural* logarithm base. When common logarithms are used,

$$y = \log_{10} x \qquad (\text{or } x = 10^y) \tag{B.14}$$

When natural logarithms are used,

$$y = \ln x \qquad (\text{or } x = e^y) \tag{B.15}$$

For example, $\log_{10} 52 = 1.716$, so that $\text{antilog}_{10} 1.716 = 10^{1.716} = 52$. Likewise, $\ln 52 = 3.951$, so antiln $3.951 = e^{3.951} = 52$.

In general, note that you can convert between base 10 and base e with the equality

$$\ln x = (2.302\,585) \log_{10} x \tag{B.16}$$

Finally, some useful properties of logarithms are

$$\left.\begin{array}{r}
\log(ab) = \log a + \log b \\
\log(a/b) = \log a - \log b \\
\log(a^n) = n \log a \\
\ln e = 1 \\
\ln e^a = a \\
\ln\left(\dfrac{1}{a}\right) = -\ln a
\end{array}\right\} \text{any base}$$

B.3 Geometry

The **distance** d between two points having coordinates (x_1, y_1) and (x_2, y_2) is

$$d = \sqrt{(x_2 - x_1)^2 + (y_2 - y_1)^2} \tag{B.17}$$

Radian measure: The arc length s of a circular arc (Fig. B.4) is proportional to the radius r for a fixed value of θ (in radians):

$$s = r\theta$$
$$\theta = \frac{s}{r} \tag{B.18}$$

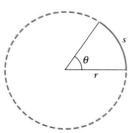

Figure B.4

Table B.2 gives the areas and volumes for several geometric shapes used throughout this text:

Table B.2

Shape	Area or Volume	Shape	Area or Volume
Rectangle	Area $= \ell w$	Sphere	Surface area $= 4\pi r^2$ Volume $= \frac{4\pi r^3}{3}$
Circle	Area $= \pi r^2$ (Circumference $= 2\pi r$)	Cylinder	Lateral surface area $= 2\pi r\ell$ Volume $= \pi r^2\ell$
Triangle	Area $= \frac{1}{2}bh$	Rectangular box	Surface area $= 2(\ell h + \ell w + hw)$ Volume $= \ell wh$

The equation of a **straight line** (Fig. B.5) is

$$y = mx + b \tag{B.19}$$

where b is the y intercept and m is the slope of the line.

The equation of a **circle** of radius R centered at the origin is

$$x^2 + y^2 = R^2 \tag{B.20}$$

The equation of an **ellipse** having the origin at its center (Fig. B.6) is

$$\frac{x^2}{a^2} + \frac{y^2}{b^2} = 1 \tag{B.21}$$

Figure B.5

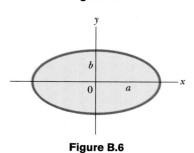

Figure B.6

where a is the length of the semimajor axis (the longer one) and b is the length of the semiminor axis (the shorter one).

The equation of a **parabola** the vertex of which is at $y = b$ (Fig. B.7) is

$$y = ax^2 + b \qquad \text{(B.22)}$$

The equation of a **rectangular hyperbola** (Fig. B.8) is

$$xy = \text{constant} \qquad \text{(B.23)}$$

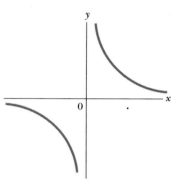

Figure B.7

B.4 Trigonometry

That portion of mathematics based on the special properties of the right triangle is called trigonometry. By definition, a right triangle is one containing a 90° angle. Consider the right triangle shown in Figure B.9, where side a is opposite the angle θ, side b is adjacent to the angle θ, and side c is the hypotenuse of the triangle. The three basic trigonometric functions defined by such a triangle are the sine (sin), cosine (cos), and tangent (tan) functions. In terms of the angle θ, these functions are defined by

$$\sin \theta \equiv \frac{\text{side opposite } \theta}{\text{hypotenuse}} = \frac{a}{c} \qquad \text{(B.24)}$$

$$\cos \theta \equiv \frac{\text{side adjacent to } \theta}{\text{hypotenuse}} = \frac{b}{c} \qquad \text{(B.25)}$$

$$\tan \theta \equiv \frac{\text{side opposite } \theta}{\text{side adjacent to } \theta} = \frac{a}{b} \qquad \text{(B.26)}$$

Figure B.8

The Pythagorean theorem provides the following relationship among the sides of a right triangle:

$$c^2 = a^2 + b^2 \qquad \text{(B.27)}$$

From the above definitions and the Pythagorean theorem, it follows that

$$\sin^2 \theta + \cos^2 \theta = 1$$

$$\tan \theta = \frac{\sin \theta}{\cos \theta}$$

The cosecant, secant, and cotangent functions are defined by

$$\csc \theta \equiv \frac{1}{\sin \theta} \qquad \sec \theta \equiv \frac{1}{\cos \theta} \qquad \cot \theta \equiv \frac{1}{\tan \theta}$$

The relationships below follow directly from the right triangle shown in Figure B.9:

$$\sin \theta = \cos(90° - \theta)$$

$$\cos \theta = \sin(90° - \theta)$$

$$\cot \theta = \tan(90° - \theta)$$

Some properties of trigonometric functions are

$$\sin(-\theta) = -\sin \theta$$

$$\cos(-\theta) = \cos \theta$$

$$\tan(-\theta) = -\tan \theta$$

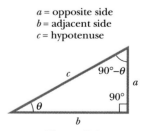

a = opposite side
b = adjacent side
c = hypotenuse

Figure B.9

The following relationships apply to *any* triangle, as shown in Figure B.10:

$$\alpha + \beta + \gamma = 180°$$

Law of cosines
$$a^2 = b^2 + c^2 - 2bc \cos \alpha$$
$$b^2 = a^2 + c^2 - 2ac \cos \beta$$
$$c^2 = a^2 + b^2 - 2ab \cos \gamma$$

Law of sines
$$\frac{a}{\sin \alpha} = \frac{b}{\sin \beta} = \frac{c}{\sin \gamma}$$

Figure B.10

Table B.3 lists a number of useful trigonometric identities.

Table B.3

Some Trigonometric Identities	
$\sin^2 \theta + \cos^2 \theta = 1$	$\csc^2 \theta = 1 + \cot^2 \theta$
$\sec^2 \theta = 1 + \tan^2 \theta$	$\sin^2 \dfrac{\theta}{2} = \tfrac{1}{2}(1 - \cos \theta)$
$\sin 2\theta = 2 \sin \theta \cos \theta$	$\cos^2 \dfrac{\theta}{2} = \tfrac{1}{2}(1 + \cos \theta)$
$\cos 2\theta = \cos^2 \theta - \sin^2 \theta$	$1 - \cos \theta = 2 \sin^2 \dfrac{\theta}{2}$
$\tan 2\theta = \dfrac{2 \tan \theta}{1 - \tan^2 \theta}$	$\tan \dfrac{\theta}{2} = \sqrt{\dfrac{1 - \cos \theta}{1 + \cos \theta}}$

$$\sin(A \pm B) = \sin A \cos B \pm \cos A \sin B$$
$$\cos(A \pm B) = \cos A \cos B \mp \sin A \sin B$$
$$\sin A \pm \sin B = 2 \sin[\tfrac{1}{2}(A \pm B)]\cos[\tfrac{1}{2}(A \mp B)]$$
$$\cos A + \cos B = 2 \cos[\tfrac{1}{2}(A + B)]\cos[\tfrac{1}{2}(A - B)]$$
$$\cos A - \cos B = 2 \sin[\tfrac{1}{2}(A + B)]\sin[\tfrac{1}{2}(B - A)]$$

Example 3

Consider the right triangle in Figure B.11, in which $a = 2$, $b = 5$, and c is unknown. From the Pythagorean theorem, we have

$$c^2 = a^2 + b^2 = 2^2 + 5^2 = 4 + 25 = 29$$

$$c = \sqrt{29} = \boxed{5.39}$$

To find the angle θ, note that

$$\tan \theta = \frac{a}{b} = \frac{2}{5} = 0.400$$

From a table of functions or from a calculator, we have

$$\theta = \tan^{-1}(0.400) = \boxed{21.8°}$$

where $\tan^{-1}(0.400)$ is the notation for "angle whose tangent is 0.400," sometimes written as arctan (0.400).

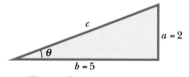

Figure B.11 (Example 3).

Exercises

1. In Figure B.12, identify (a) the side opposite θ (b) the side adjacent to ϕ. Then find (c) $\cos \theta$ (d) $\sin \phi$ (e) $\tan \phi$.

   Answers   (a) 3   (b) 3   (c) $\tfrac{4}{5}$ (d) $\tfrac{4}{5}$ (e) $\tfrac{4}{3}$

2. In a certain right triangle, the two sides that are perpendicular to each other are 5 m and 7 m long. What is the length of the third side?

   Answer   8.60 m

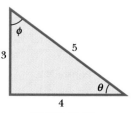

Figure B.12

3. A right triangle has a hypotenuse of length 3 m, and one of its angles is 30°. What is the length of (a) the side opposite the 30° angle (b) the side adjacent to the 30° angle?

Answers   (a) 1.5 m   (b) 2.60 m

B.5   Series Expansions

$$(a + b)^n = a^n + \frac{n}{1!}a^{n-1}b + \frac{n(n-1)}{2!}a^{n-2}b^2 + \cdots$$

$$(1 + x)^n = 1 + nx + \frac{n(n-1)}{2!}x^2 + \cdots$$

$$e^x = 1 + x + \frac{x^2}{2!} + \frac{x^3}{3!} + \cdots$$

$$\ln(1 \pm x) = \pm x - \tfrac{1}{2}x^2 \pm \tfrac{1}{3}x^3 - \cdots$$

$$\sin x = x - \frac{x^3}{3!} + \frac{x^5}{5!} - \cdots$$

$$\cos x = 1 - \frac{x^2}{2!} + \frac{x^4}{4!} - \cdots$$

$$\tan x = x + \frac{x^3}{3} + \frac{2x^5}{15} + \cdots \qquad |x| < \pi/2$$

$\left. \right\}$ x in radians

For $x \ll 1$, the following approximations can be used[1]:

$$(1 + x)^n \approx 1 + nx \qquad \sin x \approx x$$

$$e^x \approx 1 + x \qquad \cos x \approx 1$$

$$\ln(1 \pm x) \approx \pm x \qquad \tan x \approx x$$

B.6   Differential Calculus

In various branches of science, it is sometimes necessary to use the basic tools of calculus, invented by Newton, to describe physical phenomena. The use of calculus is fundamental in the treatment of various problems in Newtonian mechanics, electricity, and magnetism. In this section, we simply state some basic properties and "rules of thumb" that should be a useful review to the student.

First, a **function** must be specified that relates one variable to another (such as a coordinate as a function of time). Suppose one of the variables is called y (the dependent variable), the other x (the independent variable). We might have a function relationship such as

$$y(x) = ax^3 + bx^2 + cx + d$$

If a, b, c, and d are specified constants, then y can be calculated for any value of x. We usually deal with continuous functions, that is, those for which y varies "smoothly" with x.

The **derivative** of y with respect to x is defined as the limit, as Δx approaches zero, of the slopes of chords drawn between two points on the y versus x curve. Mathematically, we write this definition as

$$\frac{dy}{dx} = \lim_{\Delta x \to 0} \frac{\Delta y}{\Delta x} = \lim_{\Delta x \to 0} \frac{y(x + \Delta x) - y(x)}{\Delta x} \tag{B.28}$$

where Δy and Δx are defined as $\Delta x = x_2 - x_1$ and $\Delta y = y_2 - y_1$ (Fig. B.13). It is important to note that dy/dx *does not* mean dy divided by dx, but is simply a notation of the limiting process of the derivative as defined by Equation B.28.

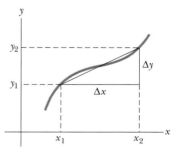

Figure B.13

---

[1]   The approximations for the functions $\sin x$, $\cos x$, and $\tan x$ are for $x \le 0.1$ rad.

A useful expression to remember when $y(x) = ax^n$, where a is a *constant* and n is *any* positive or negative number (integer or fraction), is

$$\frac{dy}{dx} = nax^{n-1} \tag{B.29}$$

If $y(x)$ is a polynomial or algebraic function of x, we apply Equation B.29 to *each* term in the polynomial and take $d[\text{constant}]/dx = 0$. In Examples 4 through 7, we evaluate the derivatives of several functions.

Special Properties of the Derivative

A. Derivative of the product of two functions   If a function $f(x)$ is given by the product of two functions, say, $g(x)$ and $h(x)$, then the derivative of $f(x)$ is defined as

$$\frac{d}{dx} f(x) = \frac{d}{dx} [g(x)h(x)] = g\frac{dh}{dx} + h\frac{dg}{dx} \tag{B.30}$$

B. Derivative of the sum of two functions   If a function $f(x)$ is equal to the sum of two functions, then the derivative of the sum is equal to the sum of the derivatives:

$$\frac{d}{dx} f(x) = \frac{d}{dx} [g(x) + h(x)] = \frac{dg}{dx} + \frac{dh}{dx} \tag{B.31}$$

C. Chain rule of differential calculus   If $y = f(x)$ and $x = g(z)$, then dy/dz can be written as the product of two derivatives:

$$\frac{dy}{dz} = \frac{dy}{dx}\frac{dx}{dz} \tag{B.32}$$

D. The second derivative   The second derivative of y with respect to x is defined as the derivative of the function dy/dx (the derivative of the derivative). It is usually written

$$\frac{d^2y}{dx^2} = \frac{d}{dx}\left(\frac{dy}{dx}\right) \tag{B.33}$$

Example 4

Suppose $y(x)$ (that is, y as a function of x) is given by

$$y(x) = ax^3 + bx + c$$

where a and b are constants. Then it follows that

$$y(x + \Delta x) = a(x + \Delta x)^3 + b(x + \Delta x) + c$$
$$y(x + \Delta x) = a(x^3 + 3x^2\Delta x + 3x\Delta x^2 + \Delta x^3) + b(x + \Delta x) + c$$

so

$$\Delta y = y(x + \Delta x) - y(x) = a(3x^2\Delta x + 3x\Delta x^2 + \Delta x^3) + b\Delta x$$

Substituting this into Equation B.28 gives

$$\frac{dy}{dx} = \lim_{\Delta x \to 0}\frac{\Delta y}{\Delta x} = \lim_{\Delta x \to 0}[3ax^2 + 3x\Delta x + \Delta x^2] + b$$

$$\frac{dy}{dx} = \boxed{3ax^2 + b}$$

Example 5

Find the derivative of

$$y(x) = 8x^5 + 4x^3 + 2x + 7$$

Solution Applying Equation B.29 to each term independently, and remembering that d/dx (constant) $= 0$, we have

$$\frac{dy}{dx} = 8(5)x^4 + 4(3)x^2 + 2(1)x^0 + 0$$

$$\frac{dy}{dx} = \boxed{40x^4 + 12x^2 + 2}$$

Example 6

Find the derivative of $y(x) = x^3/(x+1)^2$ with respect to x.

Solution We can rewrite this function as $y(x) = x^3(x+1)^{-2}$ and apply Equation B.30:

$$\frac{dy}{dx} = (x+1)^{-2}\frac{d}{dx}(x^3) + x^3\frac{d}{dx}(x+1)^{-2}$$

$$= (x+1)^{-2}3x^2 + x^3(-2)(x+1)^{-3}$$

$$\frac{dy}{dx} = \frac{3x^2}{(x+1)^2} - \frac{2x^3}{(x+1)^3}$$

Example 7

A useful formula that follows from Equation B.30 is the derivative of the quotient of two functions. Show that

$$\frac{d}{dx}\left[\frac{g(x)}{h(x)}\right] = \frac{h\dfrac{dg}{dx} - g\dfrac{dh}{dx}}{h^2}$$

Solution We can write the quotient as gh^{-1} and then apply Equations B.29 and B.30:

$$\frac{d}{dx}\left(\frac{g}{h}\right) = \frac{d}{dx}(gh^{-1}) = g\frac{d}{dx}(h^{-1}) + h^{-1}\frac{d}{dx}(g)$$

$$= -gh^{-2}\frac{dh}{dx} + h^{-1}\frac{dg}{dx}$$

$$= \frac{h\dfrac{dg}{dx} - g\dfrac{dh}{dx}}{h^2}$$

Some of the more commonly used derivatives of functions are listed in Table B.4.

B.7 Integral Calculus

We think of integration as the inverse of differentiation. As an example, consider the expression

$$f(x) = \frac{dy}{dx} = 3ax^2 + b \tag{B.34}$$

which was the result of differentiating the function

$$y(x) = ax^3 + bx + c$$

in Example 4. We can write Equation B.34 as $dy = f(x)\,dx = (3ax^2 + b)\,dx$ and obtain $y(x)$ by "summing" over all values of x. Mathematically, we write this inverse operation

$$y(x) = \int f(x)\,dx$$

For the function $f(x)$ given by Equation B.34, we have

$$y(x) = \int (3ax^2 + b)\,dx = ax^3 + bx + c$$

where c is a constant of the integration. This type of integral is called an *indefinite integral* because its value depends on the choice of c.

A general **indefinite integral** $I(x)$ is defined as

$$I(x) = \int f(x)\,dx \tag{B.35}$$

where $f(x)$ is called the *integrand* and $f(x) = dI(x)/dx$.

For a *general continuous* function $f(x)$, the integral can be described as the area under the curve bounded by $f(x)$ and the x axis, between two specified values of x, say, x_1 and x_2, as in Figure B.14.

The area of the blue element is approximately $f(x_i)\,\Delta x_i$. If we sum all these area elements from x_1 and x_2 and take the limit of this sum as $\Delta x_i \rightarrow 0$, we obtain the *true*

Table B.4

Derivative for Several Functions
$\dfrac{d}{dx}(a) = 0$
$\dfrac{d}{dx}(ax^n) = nax^{n-1}$
$\dfrac{d}{dx}(e^{ax}) = ae^{ax}$
$\dfrac{d}{dx}(\sin ax) = a\cos ax$
$\dfrac{d}{dx}(\cos ax) = -a\sin ax$
$\dfrac{d}{dx}(\tan ax) = a\sec^2 ax$
$\dfrac{d}{dx}(\cot ax) = -a\csc^2 dx$
$\dfrac{d}{dx}(\sec x) = \tan x\sec x$
$\dfrac{d}{dx}(\csc x) = -\cot x\csc x$
$\dfrac{d}{dx}(\ln ax) = \dfrac{1}{x}$

Note: The symbols a and n represent constants.

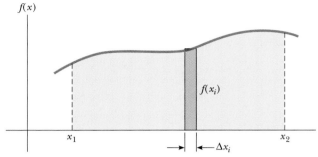

Figure B.14

area under the curve bounded by $f(x)$ and x, between the limits x_1 and x_2:

$$\text{Area} = \lim_{\Delta x \to 0} \sum_i f(x_i)\,\Delta x_i = \int_{x_1}^{x_2} f(x)\,dx \tag{B.36}$$

Integrals of the type defined by Equation B.36 are called **definite integrals.**

One common integral that arises in practical situations has the form

$$\int x^n\,dx = \frac{x^{n+1}}{n+1} + c \qquad (n \neq -1) \tag{B.37}$$

This result is obvious, being that differentiation of the right-hand side with respect to x gives $f(x) = x^n$ directly. If the limits of the integration are known, this integral becomes a *definite integral* and is written

$$\int_{x_1}^{x_2} x^n\,dx = \frac{x^{n+1}}{n+1}\bigg|_{x_1}^{x_2} = \frac{x_2^{n+1} - x_1^{n+1}}{n+1} \qquad (n \neq -1) \tag{B.38}$$

Examples

1. $\displaystyle\int_0^a x^2\,dx = \frac{x^3}{3}\bigg]_0^a = \frac{a^3}{3}$

2. $\displaystyle\int_0^b x^{3/2}\,dx = \frac{x^{5/2}}{5/2}\bigg]_0^b = \tfrac{2}{5}b^{5/2}$

3. $\displaystyle\int_3^5 x\,dx = \frac{x^2}{2}\bigg]_3^5 = \frac{5^2 - 3^2}{2} = 8$

Partial Integration

Sometimes it is useful to apply the method of *partial integration* (also called "integrating by parts") to evaluate certain integrals. The method uses the property that

$$\int u\,dv = uv - \int v\,du \tag{B.39}$$

where u and v are *carefully* chosen so as to reduce a complex integral to a simpler one. In many cases, several reductions have to be made. Consider the function

$$I(x) = \int x^2 e^x dx$$

This can be evaluated by integrating by parts twice. First, if we choose $u = x^2$, $v = e^x$, we obtain

$$\int x^2 e^x\,dx = \int x^2\,d(e^x) = x^2 e^x - 2\int e^x x\,dx + c_1$$

Now, in the second term, choose $u = x$, $v = e^x$, which gives

$$\int x^2 e^x \, dx = x^2 e^x - 2x e^x + 2\int e^x \, dx + c_1$$

or

$$\int x^2 e^x \, dx = x^2 e^x - 2x e^x + 2e^x + c_2$$

The Perfect Differential

Another useful method to remember is the use of the *perfect differential*, in which we look for a change of variable such that the differential of the function is the differential of the independent variable appearing in the integrand. For example, consider the integral

$$I(x) = \int \cos^2 x \, \sin x \, dx$$

This becomes easy to evaluate if we rewrite the differential as $d(\cos x) = -\sin x \, dx$. The integral then becomes

$$\int \cos^2 x \, \sin x \, dx = -\int \cos^2 x \, d(\cos x)$$

If we now change variables, letting $y = \cos x$, we obtain

$$\int \cos^2 x \, \sin x \, dx = -\int y^2 dy = -\frac{y^3}{3} + c = -\frac{\cos^3 x}{3} + c$$

Table B.5 lists some useful indefinite integrals. Table B.6 gives Gauss's probability integral and other definite integrals. A more complete list can be found in various handbooks, such as *The Handbook of Chemistry and Physics*, CRC Press.

Table B.5

Some Indefinite Integrals (An arbitrary constant should be added to each of these integrals.)	
$\int x^n \, dx = \dfrac{x^{n+1}}{n+1}$ (provided $n \neq -1$)	$\int \dfrac{dx}{\sqrt{a^2 - x^2}} = \sin^{-1}\dfrac{x}{a} = -\cos^{-1}\dfrac{x}{a}$ $(a^2 - x^2 > 0)$
$\int \dfrac{dx}{x} = \int x^{-1} dx = \ln x$	$\int \dfrac{dx}{\sqrt{x^2 \pm a^2}} = \ln(x + \sqrt{x^2 \pm a^2})$
$\int \dfrac{dx}{a + bx} = \dfrac{1}{b}\ln(a + bx)$	$\int \dfrac{x \, dx}{\sqrt{a^2 - x^2}} = -\sqrt{a^2 - x^2}$
$\int \dfrac{x dx}{a + bx} = \dfrac{x}{b} - \dfrac{a}{b^2}\ln(a + bx)$	$\int \dfrac{x \, dx}{\sqrt{x^2 \pm a^2}} = \sqrt{x^2 \pm a^2}$
$\int \dfrac{dx}{x(x + a)} = -\dfrac{1}{a}\ln\dfrac{x + a}{x}$	$\int \sqrt{a^2 - x^2} \, dx = \frac{1}{2}\left(x\sqrt{a^2 - x^2} + a^2\sin^{-1}\dfrac{x}{a}\right)$
$\int \dfrac{dx}{(a + bx)^2} = -\dfrac{1}{b(a + bx)}$	$\int x\sqrt{a^2 - x^2} \, dx = -\frac{1}{3}(a^2 - x^2)^{3/2}$
$\int \dfrac{dx}{a^2 + x^2} = \dfrac{1}{a}\tan^{-1}\dfrac{x}{a}$	$\int \sqrt{x^2 \pm a^2} \, dx = \frac{1}{2}[x\sqrt{x^2 \pm a^2} \pm a^2\ln(x + \sqrt{x^2 \pm a^2})]$
$\int \dfrac{dx}{a^2 - x^2} = \dfrac{1}{2a}\ln\dfrac{a + x}{a - x}$ $(a^2 - x^2 > 0)$	$\int x(\sqrt{x^2 \pm a^2}) \, dx = \frac{1}{3}(x^2 \pm a^2)^{3/2}$
$\int \dfrac{dx}{x^2 - a^2} = \dfrac{1}{2a}\ln\dfrac{x - a}{x + a}$ $(x^2 - a^2 > 0)$	$\int e^{ax} \, dx = \dfrac{1}{a}e^{ax}$
$\int \dfrac{x \, dx}{a^2 \pm x^2} = \pm\frac{1}{2}\ln(a^2 \pm x^2)$	$\int \ln ax \, dx = (x \ln ax) - x$

continued

Table B.5

Some Indefinite Integrals (An arbitrary constant should be added to each of these integrals.) *continued*

$$\int x e^{ax}\, dx = \frac{e^{ax}}{a^2}(ax - 1)$$

$$\int \frac{dx}{a + be^{cx}} = \frac{x}{a} - \frac{1}{ac}\ln(a + be^{cx})$$

$$\int \sin ax\, dx = -\frac{1}{a}\cos ax$$

$$\int \cos ax\, dx = \frac{1}{a}\sin ax$$

$$\int \tan ax\, dx = -\frac{1}{a}\ln(\cos ax) = \frac{1}{a}\ln(\sec ax)$$

$$\int \cot ax\, dx = \frac{1}{a}\ln(\sin ax)$$

$$\int \sec ax\, dx = \frac{1}{a}\ln(\sec ax + \tan ax) = \frac{1}{a}\ln\left[\tan\left(\frac{ax}{2} + \frac{\pi}{4}\right)\right]$$

$$\int \csc ax\, dx = \frac{1}{a}\ln(\csc ax - \cot ax) = \frac{1}{a}\ln\left(\tan\frac{ax}{2}\right)$$

$$\int \sin^2 ax\, dx = \frac{x}{2} - \frac{\sin 2ax}{4a}$$

$$\int \cos^2 ax\, dx = \frac{x}{2} + \frac{\sin 2ax}{4a}$$

$$\int \frac{dx}{\sin^2 ax} = -\frac{1}{a}\cot ax$$

$$\int \frac{dx}{\cos^2 ax} = \frac{1}{a}\tan ax$$

$$\int \tan^2 ax\, dx = \frac{1}{a}(\tan ax) - x$$

$$\int \cot^2 ax\, dx = -\frac{1}{a}(\cot ax) - x$$

$$\int \sin^{-1} ax\, dx = x(\sin^{-1} ax) + \frac{\sqrt{1 - a^2 x^2}}{a}$$

$$\int \cos^{-1} ax\, dx = x(\cos^{-1} ax) - \frac{\sqrt{1 - a^2 x^2}}{a}$$

$$\int \frac{dx}{(x^2 + a^2)^{3/2}} = \frac{x}{a^2\sqrt{x^2 + a^2}}$$

$$\int \frac{x\, dx}{(x^2 + a^2)^{3/2}} = -\frac{1}{\sqrt{x^2 + a^2}}$$

Table B.6

Gauss's Probability Integral and Other Definite Integrals

$$\int_0^\infty x^n e^{-ax}\, dx = \frac{n!}{a^{n+1}}$$

$$I_0 = \int_0^\infty e^{-ax^2}\, dx = \frac{1}{2}\sqrt{\frac{\pi}{a}} \quad \text{(Gauss's probability integral)}$$

$$I_1 = \int_0^\infty x e^{-ax^2}\, dx = \frac{1}{2a}$$

$$I_2 = \int_0^\infty x^2 e^{-ax^2}\, dx = -\frac{dI_0}{da} = \frac{1}{4}\sqrt{\frac{\pi}{a^3}}$$

$$I_3 = \int_0^\infty x^3 e^{-ax^2}\, dx = -\frac{dI_1}{da} = \frac{1}{2a^2}$$

$$I_4 = \int_0^\infty x^4 e^{-ax^2}\, dx = \frac{d^2 I_0}{da^2} = \frac{3}{8}\sqrt{\frac{\pi}{a^5}}$$

$$I_5 = \int_0^\infty x^5 e^{-ax^2}\, dx = \frac{d^2 I_1}{da^2} = \frac{1}{a^3}$$

$$\vdots$$

$$I_{2n} = (-1)^n \frac{d^n}{da^n} I_0$$

$$I_{2n+1} = (-1)^n \frac{d^n}{da^n} I_1$$

B.8 Propagation of Uncertainty

In laboratory experiments, a common activity is to take measurements that act as raw data. These measurements are of several types—length, time interval, temperature, voltage, etc.—and are taken by a variety of instruments. Regardless of the measurement and the quality of the instrumentation, **there is always uncertainty associated with a physical measurement.** This uncertainty is a combination of that associated with the instrument and that related to the system being measured. An example of the former is the inability to exactly determine the position of a length measurement between the lines on a meter stick. An example of uncertainty related to the system being measured is the variation of temperature within a sample of water so that a single temperature for the sample is difficult to determine.

Uncertainties can be expressed in two ways. **Absolute uncertainty** refers to an uncertainty expressed in the same units as the measurement. Thus, a length might be expressed as (5.5 ± 0.1) cm, as was the length of the computer disk label in Section 1.7. The uncertainty of ± 0.1 cm by itself is not descriptive enough for some purposes, however. This is a large uncertainty if the measurement is 1.0 cm, but it is a small uncertainty if the measurement is 100 m. To give a more descriptive account of the uncertainty, **fractional uncertainty** or **percent uncertainty** is used. In this type of description, the uncertainty is divided by the actual measurement. Thus, the length of the computer disk label could be expressed as

$$\ell = 5.5 \text{ cm} \pm \frac{0.1 \text{ cm}}{5.5 \text{ cm}} = 5.5 \text{ cm} \pm 0.018 \qquad \text{(fractional uncertainty)}$$

or as

$$\ell = 5.5 \text{ cm} \pm 1.8\% \qquad \text{(percent uncertainty)}$$

When combining measurements in a calculation, the uncertainty in the final result is larger than the uncertainty in the individual measurements. This is called **propagation of uncertainty** and is one of the challenges of experimental physics. As a calculation becomes more complicated, there is increased propagation of uncertainty and the uncertainty in the value of the final result can grow to be quite large.

There are simple rules that can provide a reasonable estimate of the uncertainty in a calculated result:

Multiplication and division: When measurements with uncertainties are multiplied or divided, add the *percent uncertainties* to obtain the percent uncertainty in the result.

Example: The Area of a Rectangular Plate

$$A = \ell w = (5.5 \text{ cm} \pm 1.8\%) \times (6.4 \text{ cm} \pm 1.6\%) = 35 \text{ cm}^2 \pm 3.4\%$$
$$= (35 \pm 1) \text{ cm}^2$$

Addition and subtraction: When measurements with uncertainties are added or subtracted, add the *absolute uncertainties* to obtain the absolute uncertainty in the result.

Example: A Change in Temperature

$$\Delta T = T_2 - T_1 = (99.2 \pm 1.5)^\circ\text{C} - (27.6 \pm 1.5)^\circ\text{C} = (71.6 \pm 3.0)^\circ\text{C}$$
$$= 71.6^\circ\text{C} \pm 4.2\%$$

Powers: If a measurement is taken to a power, the percent uncertainty is multiplied by that power to obtain the percent uncertainty in the result.

Example: The Volume of a Sphere

$$V = \tfrac{4}{3}\pi r^3 = \tfrac{4}{3}\pi (6.20 \text{ cm} \pm 2.0\%)^3 = 998 \text{ cm}^3 \pm 6.0\%$$
$$= (998 \pm 60) \text{ cm}^3$$

Notice that uncertainties in a calculation always add. As a result, an experiment involving a subtraction should be avoided if possible. This is especially true if the measurements being subtracted are close together. The result of such a calculation is a small difference in the measurements and uncertainties that add together. It is possible that the uncertainty in the result could be larger than the result itself!

Appendix C • Periodic Table of the Elements

		Transition elements
Group I	Group II	

Legend (key box):

Symbol — **Ca** 20 — Atomic number
Atomic mass † — 40.078
$4s^2$ — Electron configuration

Group I	Group II								
H 1 1.007 9 $1s$									
Li 3 6.941 $2s^1$	**Be** 4 9.0122 $2s^2$								
Na 11 22.990 $3s^1$	**Mg** 12 24.305 $3s^2$								
K 19 39.098 $4s^1$	**Ca** 20 40.078 $4s^2$	**Sc** 21 44.956 $3d^14s^2$	**Ti** 22 47.867 $3d^24s^2$	**V** 23 50.942 $3d^34s^2$	**Cr** 24 51.996 $3d^54s^1$	**Mn** 25 54.938 $3d^54s^2$	**Fe** 26 55.845 $3d^64s^2$	**Co** 27 58.933 $3d^74s^2$	
Rb 37 85.468 $5s^1$	**Sr** 38 87.62 $5s^2$	**Y** 39 88.906 $4d^15s^2$	**Zr** 40 91.224 $4d^25s^2$	**Nb** 41 92.906 $4d^45s^1$	**Mo** 42 95.94 $4d^55s^1$	**Tc** 43 (98) $4d^55s^2$	**Ru** 44 101.07 $4d^75s^1$	**Rh** 45 102.91 $4d^85s^1$	
Cs 55 132.91 $6s^1$	**Ba** 56 137.33 $6s^2$	57-71*	**Hf** 72 178.49 $5d^26s^2$	**Ta** 73 180.95 $5d^36s^2$	**W** 74 183.84 $5d^46s^2$	**Re** 75 186.21 $5d^56s^2$	**Os** 76 190.23 $5d^66s^2$	**Ir** 77 192.2 $5d^76s^2$	
Fr 87 (223) $7s^1$	**Ra** 88 (226) $7s^2$	89-103**	**Rf** 104 (261) $6d^27s^2$	**Db** 105 (262) $6d^37s^2$	**Sg** 106 (266)	**Bh** 107 (264)	**Hs** 108 (269)	**Mt** 109 (268)	

*Lanthanide series

La 57 138.91 $5d^16s^2$	**Ce** 58 140.12 $5d^14f^16s^2$	**Pr** 59 140.91 $4f^36s^2$	**Nd** 60 144.24 $4f^46s^2$	**Pm** 61 (145) $4f^56s^2$	**Sm** 62 150.36 $4f^66s^2$
Ac 89 (227) $6d^17s^2$	**Th** 90 232.04 $6d^27s^2$	**Pa** 91 231.04 $5f^26d^17s^2$	**U** 92 238.03 $5f^36d^17s^2$	**Np** 93 (237) $5f^46d^17s^2$	**Pu** 94 (244) $5f^66d^07s^2$

**Actinide series

□ Atomic mass values given are averaged over isotopes in the percentages in which they exist in nature.
† For an unstable element, mass number of the most stable known isotope is given in parentheses.
†† Elements 110, 111, 112, and 114 have not yet been named.
††† For a description of the atomic data, visit *physics.nist.gov/atomic*

			Group III	Group IV	Group V	Group VI	Group VII	Group 0
							H 1 1.007 9 $1s^1$	**He** 2 4.002 6 $1s^2$
			B 5 10.811 $2p^1$	**C** 6 12.011 $2p^2$	**N** 7 14.007 $2p^3$	**O** 8 15.999 $2p^4$	**F** 9 18.998 $2p^5$	**Ne** 10 20.180 $2p^6$
			Al 13 26.982 $3p^1$	**Si** 14 28.086 $3p^2$	**P** 15 30.974 $3p^3$	**S** 16 32.066 $3p^4$	**Cl** 17 35.453 $3p^5$	**Ar** 18 39.948 $3p^6$
Ni 28 58.693 $3d^84s^2$	**Cu** 29 63.546 $3d^{10}4s^1$	**Zn** 30 65.39 $3d^{10}4s^2$	**Ga** 31 69.723 $4p^1$	**Ge** 32 72.61 $4p^2$	**As** 33 74.922 $4p^3$	**Se** 34 78.96 $4p^4$	**Br** 35 79.904 $4p^5$	**Kr** 36 83.80 $4p^6$
Pd 46 106.42 $4d^{10}$	**Ag** 47 107.87 $4d^{10}5s^1$	**Cd** 48 112.41 $4d^{10}5s^2$	**In** 49 114.82 $5p^1$	**Sn** 50 118.71 $5p^2$	**Sb** 51 121.76 $5p^3$	**Te** 52 127.60 $5p^4$	**I** 53 126.90 $5p^5$	**Xe** 54 131.29 $5p^6$
Pt 78 195.08 $5d^96s^1$	**Au** 79 196.97 $5d^{10}6s^1$	**Hg** 80 200.59 $5d^{10}6s^2$	**Tl** 81 204.38 $6p^1$	**Pb** 82 207.2 $6p^2$	**Bi** 83 208.98 $6p^3$	**Po** 84 (209) $6p^4$	**At** 85 (210) $6p^5$	**Rn** 86 (222) $6p^6$
110†† (271)	111†† (272)	112†† (285)		114†† (289)				

Eu 63 151.96 $4f^76s^2$	**Gd** 64 157.25 $5d^14f^76s^2$	**Tb** 65 158.93 $5d^14f^86s^2$	**Dy** 66 162.50 $4f^{10}6s^2$	**Ho** 67 164.93 $4f^{11}6s^2$	**Er** 68 167.26 $4f^{12}6s^2$	**Tm** 69 168.93 $4f^{13}6s^2$	**Yb** 70 173.04 $4f^{14}6s^2$	**Lu** 71 174.97 $5d^14f^{14}6s^2$
Am 95 (243) $5f^76d^07s^2$	**Cm** 96 (247) $5f^76d^17s^2$	**Bk** 97 (247) $5f^86d^17s^2$	**Cf** 98 (251) $5f^{10}6d^07s^2$	**Es** 99 (252) $5f^{11}6d^07s^2$	**Fm** 100 (257) $5f^{12}6d^07s^2$	**Md** 101 (258) $5f^{13}6d^07s^2$	**No** 102 (259) $6d^07s^2$	**Lr** 103 (262) $6d^17s^2$

Appendix D • SI Units

Table D.1

SI Units		
	SI Base Unit	
Base Quantity	**Name**	**Symbol**
Length	Meter	m
Mass	Kilogram	kg
Time	Second	s
Electric current	Ampere	A
Temperature	Kelvin	K
Amount of substance	Mole	mol
Luminous intensity	Candela	cd

Table D.2

Some Derived SI Units				
Quantity	**Name**	**Symbol**	**Expression in Terms of Base Units**	**Expression in Terms of Other SI Units**
Plane angle	radian	rad	m/m	
Frequency	hertz	Hz	s^{-1}	
Force	newton	N	$kg \cdot m/s^2$	J/m
Pressure	pascal	Pa	$kg/m \cdot s^2$	N/m^2
Energy; work	joule	J	$kg \cdot m^2/s^2$	$N \cdot m$
Power	watt	W	$kg \cdot m^2/s^3$	J/s
Electric charge	coulomb	C	$A \cdot s$	
Electric potential	volt	V	$kg \cdot m^2/A \cdot s^3$	W/A
Capacitance	farad	F	$A^2 \cdot s^4/kg \cdot m^2$	C/V
Electric resistance	ohm	Ω	$kg \cdot m^2/A^2 \cdot s^3$	V/A
Magnetic flux	weber	Wb	$kg \cdot m^2/A \cdot s^2$	$V \cdot s$
Magnetic field	tesla	T	$kg/A \cdot s^2$	
Inductance	henry	H	$kg \cdot m^2/A^2 \cdot s^2$	$T \cdot m^2/A$

Appendix E • Nobel Prizes

All Nobel Prizes in physics are listed (and marked with a P), as well as relevant Nobel Prizes in Chemistry (C). The key dates for some of the scientific work are supplied; they often antedate the prize considerably.

1901 (P) *Wilhelm Roentgen* for discovering x-rays (1895).

1902 (P) *Hendrik A. Lorentz* for predicting the Zeeman effect and *Pieter Zeeman* for discovering the Zeeman effect, the splitting of spectral lines in magnetic fields.

1903 (P) *Antoine-Henri Becquerel* for discovering radioactivity (1896) and *Pierre* and *Marie Curie* for studying radioactivity.

1904 (P) *Lord Rayleigh* for studying the density of gases and discovering argon.
(C) *William Ramsay* for discovering the inert gas elements helium, neon, xenon, and krypton, and placing them in the periodic table.

1905 (P) *Philipp Lenard* for studying cathode rays, electrons (1898–1899).

1906 (P) *J. J. Thomson* for studying electrical discharge through gases and discovering the electron (1897).

1907 (P) *Albert A. Michelson* for inventing optical instruments and measuring the speed of light (1880s).

1908 (P) *Gabriel Lippmann* for making the first color photographic plate, using interference methods (1891).
(C) *Ernest Rutherford* for discovering that atoms can be broken apart by alpha rays and for studying radioactivity.

1909 (P) *Guglielmo Marconi* and *Carl Ferdinand Braun* for developing wireless telegraphy.

1910 (P) *Johannes D. van der Waals* for studying the equation of state for gases and liquids (1881).

1911 (P) *Wilhelm Wien* for discovering Wien's law giving the peak of a blackbody spectrum (1893).
(C) *Marie Curie* for discovering radium and polonium (1898) and isolating radium.

1912 (P) *Nils Dalén* for inventing automatic gas regulators for lighthouses.

1913 (P) *Heike Kamerlingh Onnes* for the discovery of superconductivity and liquefying helium (1908).

1914 (P) *Max T. F. von Laue* for studying x-rays from their diffraction by crystals, showing that x-rays are electromagnetic waves (1912).
(C) *Theodore W. Richards* for determining the atomic weights of sixty elements, indicating the existence of isotopes.

1915 (P) *William Henry Bragg* and *William Lawrence Bragg*, his son, for studying the diffraction of x-rays in crystals.

1917 (P) *Charles Barkla* for studying atoms by x-ray scattering (1906).

1918 (P) *Max Planck* for discovering energy quanta (1900).

1919 (P) *Johannes Stark*, for discovering the Stark effect, the splitting of spectral lines in electric fields (1913).

1920 (P) *Charles-Édouard Guillaume* for discovering invar, a nickel–steel alloy with low coefficient of expansion.
(C) *Walther Nernst* for studying heat changes in chemical reactions and formulating the third law of thermodynamics (1918).

1921 (P) *Albert Einstein* for explaining the photoelectric effect and for his services to theoretical physics (1905).
(C) *Frederick Soddy* for studying the chemistry of radioactive substances and discovering isotopes (1912).

1922   (P) *Niels Bohr* for his model of the atom and its radiation (1913).
(C) *Francis W. Aston* for using the mass spectrograph to study atomic weights, thus discovering 212 of the 287 naturally occurring isotopes.

1923   (P) *Robert A. Millikan* for measuring the charge on an electron (1911) and for studying the photoelectric effect experimentally (1914).

1924   (P) *Karl M. G. Siegbahn* for his work in x-ray spectroscopy.

1925   (P) *James Franck* and *Gustav Hertz* for discovering the Franck–Hertz effect in electron–atom collisions.

1926   (P) *Jean-Baptiste Perrin* for studying Brownian motion to validate the discontinuous structure of matter and measure the size of atoms.

1927   (P) *Arthur Holly Compton* for discovering the Compton effect on x-rays, their change in wavelength when they collide with matter (1922), and *Charles T. R. Wilson* for inventing the cloud chamber, used to study charged particles (1906).

1928   (P) *Owen W. Richardson* for studying the thermionic effect and electrons emitted by hot metals (1911).

1929   (P) *Louis Victor de Broglie* for discovering the wave nature of electrons (1923).

1930   (P) *Chandrasekhara Venkata Raman* for studying Raman scattering, the scattering of light by atoms and molecules with a change in wavelength (1928).

1932   (P) *Werner Heisenberg* for creating quantum mechanics (1925).

1933   (P) *Erwin Schrödinger* and *Paul A. M. Dirac* for developing wave mechanics (1925) and relativistic quantum mechanics (1927).
(C) *Harold Urey* for discovering heavy hydrogen, deuterium (1931).

1935   (P) *James Chadwick* for discovering the neutron (1932).
(C) *Irène* and *Frédéric Joliot-Curie* for synthesizing new radioactive elements.

1936   (P) *Carl D. Anderson* for discovering the positron in particular and antimatter in general (1932) and *Victor F. Hess* for discovering cosmic rays.
(C) *Peter J. W. Debye* for studying dipole moments and diffraction of x-rays and electrons in gases.

1937   (P) *Clinton Davisson* and *George Thomson* for discovering the diffraction of electrons by crystals, confirming de Broglie's hypothesis (1927).

1938   (P) *Enrico Fermi* for producing the transuranic radioactive elements by neutron irradiation (1934–1937).

1939   (P) *Ernest O. Lawrence* for inventing the cyclotron.

1943   (P) *Otto Stern* for developing molecular-beam studies (1923) and using them to discover the magnetic moment of the proton (1933).

1944   (P) *Isidor I. Rabi* for discovering nuclear magnetic resonance in atomic and molecular beams.
(C) *Otto Hahn* for discovering nuclear fission (1938).

1945   (P) *Wolfgang Pauli* for discovering the exclusion principle (1924).

1946   (P) *Percy W. Bridgman* for studying physics at high pressures.

1947   (P) *Edward V. Appleton* for studying the ionosphere.

1948   (P) *Patrick M. S. Blackett* for studying nuclear physics with cloud-chamber photographs of cosmic-ray interactions.

1949   (P) *Hideki Yukawa* for predicting the existence of mesons (1935).

1950   (P) *Cecil F. Powell* for developing the method of studying cosmic rays with photographic emulsions and discovering new mesons.

1951   (P) *John D. Cockcroft* and *Ernest T. S. Walton* for transmuting nuclei in an accelerator (1932).
(C) *Edwin M. McMillan* for producing neptunium (1940) and *Glenn T. Seaborg* for producing plutonium (1941) and further transuranic elements.

1952   (P) *Felix Bloch* and *Edward Mills Purcell* for discovering nuclear magnetic resonance in liquids and gases (1946).

1953   (P) *Frits Zernike* for inventing the phase-contrast microscope, which uses interference to provide high contrast.

1954   (P) *Max Born* for interpreting the wave function as a probability (1926) and other quantum-mechanical discoveries and *Walther Bothe* for developing the coincidence method to study subatomic particles (1930–1931), producing, in particular, the particle interpreted by Chadwick as the neutron.

1955 (P) *Willis E. Lamb, Jr.,* for discovering the Lamb shift in the hydrogen spectrum (1947) and *Polykarp Kusch* for determining the magnetic moment of the electron (1947).

1956 (P) *John Bardeen, Walter H. Brattain,* and *William Shockley* for inventing the transistor (1956).

1957 (P) *T.-D. Lee* and *C.-N. Yang* for predicting that parity is not conserved in beta decay (1956).

1958 (P) *Pavel A. Čerenkov* for discovering Čerenkov radiation (1935) and *Ilya M. Frank* and *Igor Tamm* for interpreting it (1937).

1959 (P) *Emilio G. Segrè* and *Owen Chamberlain* for discovering the antiproton (1955).

1960 (P) *Donald A. Glaser* for inventing the bubble chamber to study elementary particles (1952).

(C) *Willard Libby* for developing radiocarbon dating (1947).

1961 (P) *Robert Hofstadter* for discovering internal structure in protons and neutrons and *Rudolf L. Mössbauer* for discovering the Mössbauer effect of recoilless gamma-ray emission (1957).

1962 (P) *Lev Davidovich Landau* for studying liquid helium and other condensed matter theoretically.

1963 (P) *Eugene P. Wigner* for applying symmetry principles to elementary-particle theory and *Maria Goeppert Mayer* and *J. Hans D. Jensen* for studying the shell model of nuclei (1947).

1964 (P) *Charles H. Townes, Nikolai G. Basov,* and *Alexandr M. Prokhorov* for developing masers (1951–1952) and lasers.

1965 (P) *Sin-itiro Tomonaga, Julian S. Schwinger,* and *Richard P. Feynman* for developing quantum electrodynamics (1948).

1966 (P) *Alfred Kastler* for his optical methods of studying atomic energy levels.

1967 (P) *Hans Albrecht Bethe* for discovering the routes of energy production in stars (1939).

1968 (P) *Luis W. Alvarez* for discovering resonance states of elementary particles.

1969 (P) *Murray Gell-Mann* for classifying elementary particles (1963).

1970 (P) *Hannes Alfvén* for developing magnetohydrodynamic theory and *Louis Eugène Félix Néel* for discovering antiferromagnetism and ferrimagnetism (1930s).

1971 (P) *Dennis Gabor* for developing holography (1947).

(C) *Gerhard Herzberg* for studying the structure of molecules spectroscopically.

1972 (P) *John Bardeen, Leon N. Cooper,* and *John Robert Schrieffer* for explaining superconductivity (1957).

1973 (P) *Leo Esaki* for discovering tunneling in semiconductors, *Ivar Giaever* for discovering tunneling in superconductors, and *Brian D. Josephson* for predicting the Josephson effect, which involves tunneling of paired electrons (1958–1962).

1974 (P) *Anthony Hewish* for discovering pulsars and *Martin Ryle* for developing radio interferometry.

1975 (P) *Aage N. Bohr, Ben R. Mottelson,* and *James Rainwater* for discovering why some nuclei take asymmetric shapes.

1976 (P) *Burton Richter* and *Samuel C. C. Ting* for discovering the J/psi particle, the first charmed particle (1974).

1977 (P) *John H. Van Vleck, Nevill F. Mott,* and *Philip W. Anderson* for studying solids quantum-mechanically.

(C) *Ilya Prigogine* for extending thermodynamics to show how life could arise in the face of the second law.

1978 (P) *Arno A. Penzias* and *Robert W. Wilson* for discovering the cosmic background radiation (1965) and *Pyotr Kapitsa* for his studies of liquid helium.

1979 (P) *Sheldon L. Glashow, Abdus Salam,* and *Steven Weinberg* for developing the theory that unified the weak and electromagnetic forces (1958–1971).

1980 (P) *Val Fitch* and *James W. Cronin* for discovering CP (charge-parity) violation (1964), which possibly explains the cosmological dominance of matter over antimatter.

1981 (P) *Nicolaas Bloembergen* and *Arthur L. Schawlow* for developing laser spectroscopy and *Kai M. Siegbahn* for developing high-resolution electron spectroscopy (1958).

1982 (P) *Kenneth G. Wilson* for developing a method of constructing theories of phase transitions to analyze critical phenomena.

1983 (P) *William A. Fowler* for theoretical studies of astrophysical nucleosynthesis and *Subramanyan Chandrasekhar* for studying physical processes of importance to stellar structure and evolution, including the prediction of white dwarf stars (1930).

1984 (P) *Carlo Rubbia* for discovering the W and Z particles, verifying the electroweak unification, and *Simon van der Meer*, for developing the method of stochastic cooling of the CERN beam that allowed the discovery (1982–1983).

1985 (P) *Klaus von Klitzing* for the quantized Hall effect, relating to conductivity in the presence of a magnetic field (1980).

1986 (P) *Ernst Ruska* for inventing the electron microscope (1931), and *Gerd Binnig* and *Heinrich Rohrer* for inventing the scanning-tunneling electron microscope (1981).

1987 (P) *J. Georg Bednorz* and *Karl Alex Müller* for the discovery of high-temperature superconductivity (1986).

1988 (P) *Leon M. Lederman, Melvin Schwartz,* and *Jack Steinberger* for a collaborative experiment that led to the development of a new tool for studying the weak nuclear force, which affects the radioactive decay of atoms.

1989 (P) *Norman Ramsay* for various techniques in atomic physics; and *Hans Dehmelt* and *Wolfgang Paul* for the development of techniques for trapping single-charge particles.

1990 (P) *Jerome Friedman, Henry Kendall* and *Richard Taylor* for experiments important to the development of the quark model.

1991 (P) *Pierre-Gilles de Gennes* for discovering that methods developed for studying order phenomena in simple systems can be generalized to more complex forms of matter, in particular to liquid crystals and polymers.

1992 (P) *George Charpak* for developing detectors that trace the paths of evanescent subatomic particles produced in particle accelerators.

1993 (P) *Russell Hulse* and *Joseph Taylor* for discovering evidence of gravitational waves.

1994 (P) *Bertram N. Brockhouse* and *Clifford G. Shull* for pioneering work in neutron scattering.

1995 (P) *Martin L. Perl* and *Frederick Reines* for discovering the tau particle and the neutrino, respectively.

1996 (P) *David M. Lee, Douglas C. Osheroff,* and *Robert C. Richardson* for developing a superfluid using helium-3.

1997 (P) *Steven Chu, Claude Cohen-Tannoudji,* and *William D. Phillips* for developing methods to cool and trap atoms with laser light.

1998 (P) *Robert B. Laughlin, Horst L. Störmer,* and *Daniel C. Tsui* for discovering a new form of quantum fluid with fractionally charged excitations.

1999 (P) *Gerardus 'T Hooft* and *Martinus J. G. Veltman* for studies in the quantum structure of electroweak interactions in physics.

2000 (P) *Zhores I. Alferov* and *Herbert Kroemer* for developing semiconductor heterostructures used in high-speed electronics and optoelectronics and *Jack St. Clair Kilby* for participating in the invention of the integrated circuit.

2001 (P) *Eric A. Cornell, Wolfgang Ketterle,* and *Carl E. Wieman* for the achievement of Bose–Einstein condensation in dilute gases of alkali atoms.

2002 (P) *Raymond Davis Jr.* and *Masatoshi Koshiba* for the detection of cosmic neutrinos and *Riccardo Giacconi* for contributions to astrophysics that led to the discovery of cosmic x-ray sources.

Answers to Odd-Numbered Problems

CHAPTER 1

1. 0.141 nm

3. 2.15×10^4 kg/m^3

5. $4\pi\rho(r_2^3 - r_1^3)/3$

7. (a) 4.00 u $= 6.64 \times 10^{-24}$ g   (b) 55.9 u $= 9.28 \times 10^{-23}$ g
(c) 207 u $= 3.44 \times 10^{-22}$ g

9. 8.72×10^{11} atom/s

11. (a) 72.6 kg   (b) 7.82×10^{26} atoms

13. No.

15. (b) only

17. The units of G are m^3/kg $\cdot$ s^2

19. 9.19 nm/s

21. 1.39×10^3 m^2

23. (a) 0.071 4 gal/s   (b) 2.70×10^{-4} m^3/s   (c) 1.03 h

25. 11.4×10^3 kg/m^3

27. 667 lb/s

29. (a) 190 yr   (b) 2.32×10^4 times

31. 151 μm

33. 1.00×10^{10} lb

35. (a) 2.07 mm   (b) 8.62×10^{13} times as large

37. 5.0 m

39. 2.86 cm

41. $\sim 10^6$ balls

43. $\sim 10^7$

45. $\sim 10^2$ kg; $\sim 10^3$ kg

47. $\sim 10^2$ tuners

49. (a) (346 ± 13) m^2 (b) (66.0 ± 1.3) m

51. $(1.61 \pm 0.17) \times 10^3$ kg/m^3

53. 31 556 926.0 s

55. 5.2 m^3, 3%

57. 2.57×10^{-10} m

59. $0.579\,t\,\text{ft}^3/\text{s} + 1.19 \times 10^{-9}\,t^2\,\text{ft}^3/\text{s}^2$

61. 3.41 m

63. 0.449%

65. (a) 0.529 cm/s   (b) 11.5 cm/s

67. 1×10^{10} gal/yr

69. $\sim 10^{11}$ stars

71. (a) 3.16×10^7 s/yr   (b) 6.05×10^{10} yr

CHAPTER 2

1. (a) 2.30 m/s   (b) 16.1 m/s   (c) 11.5 m/s

3. (a) 5 m/s   (b) 1.2 m/s   (c) -2.5 m/s   (d) -3.3 m/s   (e) 0

5. (a) 3.75 m/s   (b) 0

7. (a) -2.4 m/s   (b) -3.8 m/s   (c) 4.0 s

9. (a) 5.0 m/s   (b) -2.5 m/s   (c) 0   (d) 5.0 m/s

11. 1.34×10^4 m/s^2

13. (a) 52.4 ft/s, 55.0 ft/s, 55.5 ft/s, 57.4 ft/s   (b) 0.598 ft/s^2

15. (a) 2.00 m   (b) -3.00 m/s   (c) -2.00 m/s^2

17. (a) 1.3 m/s^2 (b) 2.0 m/s^2 at 3 s
(c) at $t = 6$ s and for $t > 10$ s   (d) -1.5 m/s^2 at 8 s

19. 2.74×10^5 m/s^2, which is 2.79×10^4 g

21. -16.0 cm/s^2

23. (a) 4.53 s   (b) 14.1 m/s

25. (a) 2.56 m   (b) -3.00 m/s

27. (a) 20.0 s   (b) no

29. 3.10 m/s

31. (a) -202 m/s^2 (b) 198 m

33. (a) 4.98×10^{-9} s   (b) 1.20×10^{15} m/s^2

35. (a) v_c/t_m (c) $v_c t_0/2$ (d) $v_c t_0$ (e) yes, no

37. (a) 3.00 m/s   (b) 6.00 s   (c) -0.300 m/s^2
(d) 2.05 m/s

39. 31 s

41. \$99.3/h

43. (a) 10.0 m/s up   (b) 4.68 m/s down

45. (a) 2.17 s   (b) -21.2 m/s   (c) 2.23 s

47. (a) 29.4 m/s   (b) 44.1 m

49. (a) 7.82 m   (b) 0.782 s

51. 7.96 s

53. (a) $a_x(t) = a_{xi} + Jt$, $v_x(t) = v_{xi} + a_{xi}t + (1/2)Jt^2$,
$x(t) = x_i + v_{xi}t + (1/2)a_{xi}t^2 + (1/6)Jt^3$

55. (a) $a = -(10.0 \times 10^7 \text{ m/s}^3)t + 3.00 \times 10^5 \text{ m/s}^2$;
$x = -(1.67 \times 10^7 \text{ m/s}^3)t^3 + (1.50 \times 10^5 \text{ m/s}^2)t^2$
(b) 3.00×10^{-3} s (c) 450 m/s (d) 0.900 m

59. (a) Acela steadily cruises out of the city center at 45 mi/h. In
less than a minute it smoothly speeds up to 150 mi/h; then
its speed is nudged up to 170 mi/h. Next it smoothly slows to
a very low speed, which it maintains as it rolls into a railroad
yard. When it stops, it immediately begins backing up and
smoothly speeds up to 50 mi/h in reverse, all in less than
seven minutes after it started.   (b) 2.2 mi/h/s $= 0.98$ m/s^2
(c) 6.7 mi

61. 48.0 mm

63. (a) 15.0 s   (b) 30.0 m/s   (c) 225 m

65. (a) 5.43 m/s^2 and 3.83 m/s^2 (b) 10.9 m/s and 11.5 m/s
(c) Maggie by 2.62 m

67. $\sim 10^3$ m/s^2

69. (a) 3.00 s   (b) -15.3 m/s   (c) 31.4 m/s down and 34.8 m/s down

71. (c) v_{boy}^2/h, 0   (d) v_{boy}, 0

73. (a) 5.46 s   (b) 73.0 m
(c) $v_{Stan} = 22.6$ m/s, $v_{Kathy} = 26.7$ m/s

75. $0.577v$

CHAPTER 3

1. $(-2.75, -4.76)$ m

3. (a) 2.24 m   (b) 2.24 m at 26.6°

5. $y = 1.15$; $r = 2.31$

7. 70.0 m

9. 310 km at 57° S of W

11. (a) 10.0 m   (b) 15.7 m   (c) 0

13. (a) $\sim 10^5$ m   vertically   upward   (b) $\sim 10^3$ m   vertically upward

15. (a) 5.2 m at 60°   (b) 3.0 m at 330°   (c) 3.0 m at 150°
(d) 5.2 m at 300°

17. approximately 420 ft at $-3°$

19. 47.2 units at 122°

21. (a) $(-11.1\hat{i} + 6.40\hat{j})$ m   (b) $(1.65\hat{i} + 2.86\hat{j})$ cm
(c) $(-18.0\hat{i} - 12.6\hat{j})$ in.

23. (a) 5.00 blocks at 53.1° N of E   (b) 13.0 blocks

25. 358 m at 2.00° S of E

27. (a)

(b) $\mathbf{C} = 5.00\hat{i} + 4.00\hat{j}$ or 6.40 at 38.7°; $\mathbf{D} = -1.00\hat{i} + 8.00\hat{j}$ or 8.06 at 97.2°

29. 196 cm at 345°

31. (a) $2.00\hat{i} - 6.00\hat{j}$ (b) $4.00\hat{i} + 2.00\hat{j}$ (c) 6.32
(d) 4.47   (e) 288°; 26.6°

33. 9.48 m at 166°

35. (a) 185 N at 77.8° from the $+x$ axis
(b) $(-39.3\hat{i} - 181\hat{j})$ N

37. $\mathbf{A} + \mathbf{B} = (2.60\hat{i} + 4.50\hat{j})$ m

39. $|\mathbf{B}| = 7.81$, $\theta_x = 59.2°$, $\theta_y = 39.8°$, $\theta_z = 67.4°$

41. (a) $8.00\hat{i} + 12.0\hat{j} - 4.00\hat{k}$ (b) $2.00\hat{i} + 3.00\hat{j} - 1.00\hat{k}$
(c) $-24.0\hat{i} - 36.0\hat{j} + 12.0\hat{k}$

43. (a) 5.92 m is the magnitude of $(5.00\hat{i} - 1.00\hat{j} - 3.00\hat{k})$ m
(b) 19.0 m is the magnitude of $(4.00\hat{i} - 11.0\hat{j} - 15.0\hat{k})$ m

45. 157 km

47. (a) $-3.00\hat{i} + 2.00\hat{j}$ (b) 3.61 at 146°
(c) $3.00\hat{i} - 6.00\hat{j}$

49. (a) $49.5\hat{i} + 27.1\hat{j}$ (b) 56.4 units at 28.7°

51. 1.15°

53. (a) 2.00, 1.00, 3.00   (b) 3.74   (c) $\theta_x = 57.7°$,
$\theta_y = 74.5°$, $\theta_z = 36.7°$

55. 2.29 km

57. (a) 11.2 m   (b) 12.9 m at 36.4°

59. 240 m at 237°

61. 390 mi/h at 7.37° north of east

63. (a) zero   (b) zero

65. 106°

CHAPTER 4

1. (a) 4.87 km at 209° from east   (b) 23.3 m/s   (c) 13.5 m/s at 209°

3. 2.50 m/s

5. (a) $(2.00\hat{i} + 3.00\hat{j})$ m/s²
(b) $(3.00t + t^2)\hat{i}$ m $+ (1.50t^2 - 2.00t)\hat{j}$ m

7. (a) $(0.800\hat{i} - 0.300\hat{j})$ m/s²   (b) 339°
(c) $(360\hat{i} - 72.7\hat{j})$ m, $-15.2°$

9. (a) $x = 0.010\ 0$ m, $y = 2.41 \times 10^{-4}$ m
(b) $\mathbf{v} = (1.84 \times 10^7\hat{i} + 8.78 \times 10^5\hat{j})$ m/s
(c) $v = 1.85 \times 10^7$ m/s
(d) $\theta = 2.73°$

11. (a) $3.34\hat{i}$ m/s   (b) $-50.9°$

13. (a) 20.0°   (b) 3.05 s

15. 53.1°

17. (a) 22.6 m   (b) 52.3 m   (c) 1.18 s

19. (a) The ball clears by 0.889 m while
(b) descending

21. (a) 18.1 m/s   (b) 1.13 m   (c) 2.79 m

23. 9.91 m/s

25. (a) 30.3 m/s   (b) 2.09 s

27. 377 m/s²

29. 10.5 m/s, 219 m/s² inward

31. (a) 6.00 rev/s   (b) 1.52 km/s²
(c) 1.28 km/s²

33. 1.48 m/s² inward and 29.9° backward

35. (a) 13.0 m/s²   (b) 5.70 m/s   (c) 7.50 m/s²

37. $\theta = \tan^{-1}(1/4\pi) = 4.55°$

39. (a) 57.7 km/h at 60.0° west of vertical
(b) 28.9 km/h downward

41. 2.02×10^3 s; 21.0% longer

43. $t_{Alan} = \dfrac{2L/c}{1 - v^2/c^2}$, $t_{Beth} = \dfrac{2L/c}{\sqrt{1 - v^2/c^2}}$. Beth returns first.

45. 15.3 m

47. (a) 101 m/s   (b) 32 700 ft   (c) 20.6 s
(d) 180 m/s

49. 54.4 m/s²

51. (a) 41.7 m/s   (b) 3.81 s   (c) $(34.1\hat{i} - 13.4\hat{j})$ m/s;
36.7 m/s

53. (a) 25.0 m/s²; 9.80 m/s²

(b)

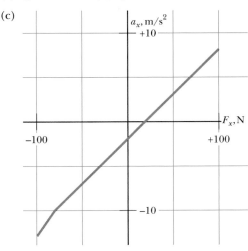

(c) 26.8 m/s² inward at 21.4° below the horizontal

55. (a) 26.6°   (b) 0.949

57. (a) 0.600 m   (b) 0.402 m   (c) 1.87 m/s² toward center   (d) 9.80 m/s² down

59. (a) 6.80 km   (b) 3.00 km vertically above the impact point   (c) 66.2°

61. (a) 46.5 m/s   (b) −77.6°   (c) 6.34 s

63. (a) 20.0 m/s, 5.00 s   (b) $(16.0\hat{\mathbf{i}} − 27.1\hat{\mathbf{j}})$ m/s   (c) 6.53 s   (d) $24.5\hat{\mathbf{i}}$ m

65. (a) 22.9 m/s   (b) 360 m from the base of the cliff   (c) $\mathbf{v} = (114\hat{\mathbf{i}} − 44.3\hat{\mathbf{j}})$ m/s

67. (a) 43.2 m   (b) $(9.66\hat{\mathbf{i}} − 25.5\hat{\mathbf{j}})$ m/s

69. (a) 4.00 km/h   (b) 4.00 km/h

71. Safe distances are less than 270 m or greater than 3.48×10^3 m from the western shore.

CHAPTER 5

1. (a) 1/3   (b) 0.750 m/s²

3. $(6.00\hat{\mathbf{i}} + 15.0\hat{\mathbf{j}})$ N; 16.2 N

5. (a) $(2.50\hat{\mathbf{i}} + 5.00\hat{\mathbf{j}})$ N   (b) 5.59 N

7. (a) 3.64×10^{-18} N   (b) 8.93×10^{-30} N is 408 billion times smaller

9. 2.38 kN

11. (a) 5.00 m/s² at 36.9°   (b) 6.08 m/s² at 25.3°

13. (a) $\sim 10^{-22}$ m/s²   (b) $\sim 10^{-23}$ m

15. (a) 15.0 lb up   (b) 5.00 lb up   (c) 0

17. 613 N

21. (a) 49.0 N   (b) 98.0 N   (c) 24.5 N

23. 8.66 N east

25. 3.73 m

27. A is in compression 3.83 kN and B is in tension 3.37 kN

29. 950 N

31. (a) $F_x > 19.6$ N   (b) $F_x \leq − 78.4$ N

(c)

33. (a) 706 N   (b) 814 N   (c) 706 N   (d) 648 N

35. (a) 0.404   (b) 45.8 lb

37. (a) 256 m   (b) 42.7 m

39. (a) 1.10 s   (b) 0.875 s

41. (a) 1.78 m/s²   (b) 0.368   (c) 9.37 N   (d) 2.67 m/s

43. 37.8 N

45. (a)

(b) 27.2 N, 1.29 m/s²

47. $\mu_k = (3/5)\tan\theta$

49. (a) 8.05 N   (b) 53.2 N   (c) 42.0 N

51. (a)

(b) 0.408 m/s^2 (c) 83.3 N

53. (a) $F_A = mg(\sin\theta - \mu_s \cos\theta)$
(b) $F_B = mg(\sin\theta - \mu_s \cos\theta)/(\cos\theta + \mu_s \sin\theta)$
(c) A's job is easier
(d) B's job is easier

55. (a) $Mg/2$, $Mg/2$, $Mg/2$, $3Mg/2$, Mg (b) $Mg/2$

57. (b)

θ	0	15°	30°	45°	60°
P(N)	40.0	46.4	60.1	94.3	260

59. (a) 19.3°   (b) 4.21 N

61. $(M + m_1 + m_2)(m_2 g/m_1)$

63. (a) $m_2 g\left[\dfrac{m_1 M}{m_1 M + m_2(m_1 + M)}\right]$

(b) $\dfrac{m_2 g(M + m_1)}{m_1 M + m_2(m_1 + M)}$

(c) $\dfrac{m_1 m_2 g}{m_1 M + m_2(m_1 + M)}$

(d) $\dfrac{M m_2 g}{m_1 M + m_2(m_1 + M)}$

65. (c) 3.56 N

67. (a) $T = f/(2\sin\theta)$ (b) 410 N

69. (a) 30.7°   (b) 0.843 N

71. 0.060 0 m

73. (a) $T_1 = \dfrac{2mg}{\sin\theta_1}$

$T_2 = \dfrac{mg}{\sin\theta_2} = \dfrac{mg}{\sin[\tan^{-1}(\frac{1}{2}\tan\theta_1)]}$

$T_3 = 2mg/\tan\theta_1$

(b) $\theta_2 = \tan^{-1}\left(\dfrac{\tan\theta_1}{2}\right)$

CHAPTER 6

1. Any speed up to 8.08 m/s

3. (a) 8.32×10^{-8} N toward the nucleus
(b) 9.13×10^{22} m/s^2 inward

5. (a) static friction   (b) 0.085 0

7. $v \le 14.3$ m/s

9. (a) 68.6 N toward the center of the circle and 784 N up
(b) 0.857 m/s^2

11. (a) 108 N   (b) 56.2 N

13. (a) 4.81 m/s   (b) 700 N up

15. No. The jungle lord needs a vine of tensile strength 1.38 kN.

17. 3.13 m/s

19. (a) 2.49×10^4 N up   (b) 12.1 m/s

21. (a) 3.60 m/s^2 (b) zero   (c) An observer in the car (a noninertial frame) claims an 18.0-N force toward the left and an 18.0-N force toward the right. An inertial observer (outside the car) claims only an 18.0-N force toward the right.

23. (a) 17.0°   (b) 5.12 N

25. (a) 491 N   (b) 50.1 kg   (c) 2.00 m/s^2

27. (a) $v = [2(a - \mu_k g)\ell]^{1/2}$;   (b) $v' = (2\mu_k g\ell/v)$, where $v = [2(a - \mu_k g)\ell]^{1/2}$

29. 93.8 N

31. 0.092 7°

33. (a) 32.7 s^{-1} (b) 9.80 m/s^2 down   (c) 4.90 m/s^2 down

35. 3.01 N up

37. (a) 1.47 N·s/m   (b) 2.04×10^{-3} s   (c) 2.94×10^{-2} N

39. (a) 0.034 7 s^{-1} (b) 2.50 m/s   (c) $a = -cv$

41. (a) $x = k^{-1}\ln(1 + kv_0 t)$ (b) $v = v_0 e^{-kx}$

43. $\sim 10^1$ N

45. (a) 13.7 m/s down

(b)

t (s)	x (m)	v (m/s)
0	0	0
0.2	0	−1.96
0.4	−0.392	−3.88
... 1.0	−3.77	−8.71
... 2.0	−14.4	−12.56
... 4.0	−41.0	−13.67

47. (a) 49.5 m/s down and 4.95 m/s down

(b)

t (s)	y (m)	v (m/s)
0	1 000	0
... 1	995	−9.7
... 2	980	−18.6
... 10	674	−47.7
... 10.1	671	−16.7
... 12	659	−4.95
... 145	0	−4.95

49. (a) 2.33×10^{-4} kg/m   (b) 57.0 m/s   (c) 44.9 m/s. The second trajectory is higher and shorter than the first. In both cases, the ball attains maximum height when it has covered 56% of its horizontal range, and attains minimum speed a little later. The impact speeds are also similar, 30 m/s and 29 m/s.

51. (a) 11.5 kN   (b) 14.1 m/s = 50.9 km/h

53. (a) 0.016 2 kg/m   (b) $\frac{1}{2}D\rho A$ (c) 0.778   (d) 1.5%
(e) For stacked coffee filters falling in air at terminal speed, the graph of resistive force as a function of squared speed demonstrates that the force is proportional to the speed squared, within the experimental uncertainty, estimated as ± 2%. This proportionality agrees with that predicted by the theoretical equation $R = \frac{1}{2}D\rho Av^2$. The value of the constant slope of the graph implies that the drag coefficient for coffee filters is $D = 0.78 \pm 2\%$.

55. $g(\cos \phi \tan \theta - \sin \phi)$

57. (b) 732 N down at the equator and 735 N down at the poles

59. (a) 967 lb   (b) − 647 lb (pilot must be strapped in)
(c) Speed and radius of path can be adjusted so that $v^2 = gR$.

61. (a) 1.58 m/s²   (b) 455 N   (c) 329 N   (d) 397 N upward and 9.15° inward

63. (a) 5.19 m/s   (b) $T = 555$ N

$T \cos 28.0°$

$T \sin 28.0°$

490 N

65. (b) 2.54 s; 23.6 rev/min

67. (a) $v_{min} = \sqrt{\dfrac{Rg(\tan \theta - \mu_s)}{1 + \mu_s \tan \theta}}$, $v_{max} = \sqrt{\dfrac{Rg(\tan \theta + \mu_s)}{1 - \mu_s \tan \theta}}$
(b) $\mu_s = \tan \theta$ (c) 8.57 m/s $\le v \le$ 16.6 m/s

69. (a) 0.013 2 m/s   (b) 1.03 m/s   (c) 6.87 m/s

71. 12.8 N

73. $\Sigma \mathbf{F} = -k m \mathbf{v}$

CHAPTER 7

1. (a) 31.9 J   (b) 0   (c) 0   (d) 31.9 J

3. − 4.70 kJ

5. 28.9

7. (a) 16.0 J   (b) 36.9°

9. (a) 11.3°   (b) 156°   (c) 82.3°

11. (a) 24.0 J   (b) − 3.00 J   (c) 21.0 J

13. (a) 7.50 J   (b) 15.0 J   (c) 7.50 J   (d) 30.0 J

15. (a) 0.938 cm   (b) 1.25 J

17. (a) 0.768 m   (b) 1.68×10^5 J

19. 12.0 J

21. (a) 0.020 4 m   (b) 720 N/m

23. kg/s²

25. (a) 33.8 J   (b) 135 J

27. 878 kN up

29. (a) 4.56 kJ   (b) 6.34 kN   (c) 422 km/s²   (d) 6.34 kN

31. (a) 650 J   (b) 588 J   (c) 0   (d) 0   (e) 62.0 J
(f) 1.76 m/s

33. (a) − 168 J   (b) 184 J   (c) 500 J   (d) 148 J
(e) 5.65 m/s

35. 2.04 m

37. 875 W

39. (a) 20.6 kJ   (b) 686 W

41. $46.2

43. (a) 423 mi/gal   (b) 776 mi/gal

45. (a) 0.013 5 gal   (b) 73.8   (c) 8.08 kW

47. 2.92 m/s

49. (a) $(2 + 24t^2 + 72t^4)$ J   (b) $12t$ m/s²; $48t$ N
(c) $(48t + 288t^3)$ W   (d) 1 250 J

51. $k_1 x_{max}^2/2 + k_2 x_{max}^3/3$

53. (a) $\sqrt{2W/m}$ (b) W/d

55. (b) 240 W

57. (a) 1.38×10^4 J   (b) 3.02×10^4 W

59. (a) $\mathcal{P} = 2Mgv_T$ (b) $\mathcal{P} = 24Mgv_T$

61. (a) 4.12 m   (b) 3.35 m

63. 1.68 m/s

65. -1.37×10^{-21} J

67. 0.799 J

69. (b) For a block of weight w pushed over a rough horizontal surface at constant velocity, $b = \mu_k$. For a load pulled vertically upward at constant velocity, $b = 1$.

CHAPTER 8

1. (a) 259 kJ, 0, − 259 kJ   (b) 0, − 259 kJ, − 259 kJ

3. 22.0 kW

5. (a) $v = (3gR)^{1/2}$ (b) 0.098 0 N down

7. (a) 1.47 m/s   (b) 1.35 m/s

9. (a) 2.29 m/s   (b) 1.98 m/s

11. 10.2 m

13. (a) 4.43 m/s   (b) 5.00 m

15. 5.49 m/s

17. (a) 18.5 km, 51.0 km   (b) 10.0 MJ

19. (a) 25.8 m   (b) 27.1 m/s²

21. (a) − 196 J   (b) − 196 J   (c) − 196 J. The force is conservative.

23. (a) 125 J   (b) 50.0 J   (c) 66.7 J   (d) Nonconservative. The results differ.

25. (a) − 9.00 J; no; the force is conservative.   (b) 3.39 m/s
(c) 9.00 J

27. 26.5 m/s

29. 6.92 m/s

31. 3.74 m/s

33. (a) -160 J   (b) 73.5 J   (c) 28.8 N   (d) 0.679

35. (a) 1.40 m/s   (b) 4.60 cm after release   (c) 1.79 m/s

37. (a) 0.381 m   (b) 0.143 m   (c) 0.371 m

39. (a) $a_x = -\mu_k gx/L$ (b) $v = (\mu_k gL)^{1/2}$

41. (a) 40.0 J   (b) -40.0 J   (c) 62.5 J

43. (A/r^2) away from the other particle

45. (a) $+$ at Ⓑ, $-$ at Ⓓ, 0 at Ⓐ, Ⓒ, and Ⓔ   (b) Ⓒ stable; Ⓐ and Ⓔ unstable

(c)

47. (b)

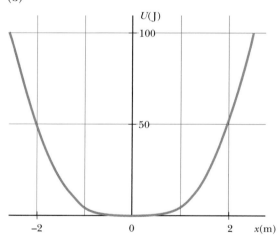

Equilibrium at $x = 0$

(c) 0.823 m/s

49. $\sim 10^3$ W peak or $\sim 10^2$ W sustainable

51. 48.2°

53. (a) 0.225 J   (b) $\Delta E_{mech} = -0.363$ J   (c) No; the normal force changes in a complicated way.

55. (a) 23.6 cm   (b) 5.90 m/s² up the incline; no.
(c) Gravitational potential energy is transformed into kinetic energy plus elastic potential energy and then entirely into elastic potential energy.

57. 0.328

59. 1.24 m/s

61. (a) 0.400 m   (b) 4.10 m/s   (c) The block stays on the track.

63. $(h/5)(4\sin^2\theta + 1)$

65. (a) 6.15 m/s   (b) 9.87 m/s

67. (a) 11.1 m/s   (b) 19.6 m/s²   upward   (c) 2.23×10^3 N upward   (d) 1.01×10^3 J   (e) 5.14 m/s   (f) 1.35 m (g) 1.39 s

69. (b) 1.44 m   (c) 0.400 m   (d) No. A very strong wind pulls the string out horizontally. The largest possible equilibrium height is equal to L.

73. (a) $2.5R$

CHAPTER 9

1. (a) $(9.00\hat{\mathbf{i}} - 12.0\hat{\mathbf{j}})$ kg·m/s   (b) 15.0 kg·m/s at 307°

3. $\sim 10^{-23}$ m/s

5. (b) $p = \sqrt{2mK}$

7. (a) 13.5 N·s   (b) 9.00 kN   (c) 18.0 kN

9. 260 N normal to the wall

11. (a) $(9.05\hat{\mathbf{i}} + 6.12\hat{\mathbf{j}})$ N·s   (b) $(377\hat{\mathbf{i}} + 255\hat{\mathbf{j}})$ N

13. 15.0 N in the direction of the initial velocity of the exiting water stream

15. 65.2 m/s

17. 301 m/s

19. (a) 2.50 m/s   (b) 37.5 kJ   (c) Each process is the time-reversal of the other. The same momentum conservation equation describes both.

21. (a) $v_{gx} = 1.15$ m/s   (b) $v_{px} = -0.346$ m/s

23. (a) 0.284   (b) 115 fJ and 45.4 fJ

25. 91.2 m/s

27. (a) 2.24 m/s to the right   (b) No. Coupling order makes no difference.

29. $v_{orange} = 3.99$ m/s, $v_{yellow} = 3.01$ m/s

31. $v_{green} = 7.07$ m/s, $v_{blue} = 5.89$ m/s

33. 2.50 m/s at $-60.0°$

35. $(3.00\hat{\mathbf{i}} - 1.20\hat{\mathbf{j}})$ m/s

37. (a) $(-9.33\hat{\mathbf{i}} - 8.33\hat{\mathbf{j}})$ Mm/s   (b) 439 fJ

39. 0.006 73 nm from the oxygen nucleus along the bisector of the angle

41. $\mathbf{r}_{CM} = (11.7\hat{\mathbf{i}} + 13.3\hat{\mathbf{j}})$ cm

43. (a) 15.9 g   (b) 0.153 m

45. (a) $(1.40\hat{\mathbf{i}} + 2.40\hat{\mathbf{j}})$ m/s   (b) $(7.00\hat{\mathbf{i}} + 12.0\hat{\mathbf{j}})$ kg·m/s

47. 0.700 m

49. (a) 39.0 MN   (b) 3.20 m/s² up

51. (a) 442 metric tons   (b) 19.2 metric tons

53. 4.41 kg

55. (a) $1.33\hat{\mathbf{i}}$ m/s   (b) $-235\hat{\mathbf{i}}$ N   (c) 0.680 s   (d) $-160\hat{\mathbf{i}}$ N·s and $+160\hat{\mathbf{i}}$ N·s   (e) 1.81 m   (f) 0.454 m   (g) -427 J (h) $+107$ J   (i) Equal friction forces act through different distances on person and cart, to do different amounts of work on them. The total work on both together, -320 J, becomes $+320$ J of extra internal energy in this perfectly inelastic collision.

57. 240 s

59. (a) 0; inelastic   (b) $(-0.250\hat{\mathbf{i}} + 0.750\hat{\mathbf{j}} - 2.00\hat{\mathbf{k}})$ m/s; perfectly inelastic   (c) either $a = -6.74$ with $\mathbf{v} = -0.419\hat{\mathbf{k}}$ m/s or $a = 2.74$ with $\mathbf{v} = -3.58\hat{\mathbf{k}}$ m/s

61. (a) $v_i = v(m + \rho V)/m$ (b) The cart slows with constant acceleration and eventually comes to rest.

63. (a) $m/M = 0.403$ (b) No changes; no difference.

65. (a) 6.29 m/s   (b) 6.16 m/s

67. (a) 100 m/s   (b) 374 J

69. (a) $(20.0\hat{\mathbf{i}} + 7.00\hat{\mathbf{j}})$ m/s   (b) $4.00\hat{\mathbf{i}}$ m/s^2 (c) $4.00\hat{\mathbf{i}}$ m/s^2
  (d) $(50.0\hat{\mathbf{i}} + 35.0\hat{\mathbf{j}})$ m   (e) 600 J   (f) 674 J   (g) 674 J

71. $(3Mgx/L)\hat{\mathbf{j}}$

73. $\dfrac{m_1(R + \ell/2)}{(m_1 + m_2)}$

CHAPTER 10

1. (a) 5.00 rad, 10.0 rad/s, 4.00 rad/s^2 (b) 53.0 rad, 22.0 rad/s, 4.00 rad/s^2

3. (a) 4.00 rad/s^2 (b) 18.0 rad

5. (a) 5.24 s   (b) 27.4 rad

7. 50.0 rev

9. (a) 7.27×10^{-5} rad/s   (b) 2.57×10^4 s = 428 min

11. $\sim 10^7$ rev

13. (a) 8.00 rad/s   (b) 8.00 m/s, $a_r = -64.0$ m/s^2, $a_t = 4.00$ m/s^2 (c) 9.00 rad

15. (a) 25.0 rad/s   (b) 39.8 rad/s^2 (c) 0.628 s

17. (a) 126 rad/s   (b) 3.77 m/s   (c) 1.26 km/s^2 (d) 20.1 m

19. (a) $\omega(2h^3/g)^{1/2}$ (b) 0.011 6 m   (c) Yes; the deflection is only 0.02% of the original height.

21. (a) 143 kg·m^2 (b) 2.57 kJ

23. $11mL^2/12$

25. 5.80 kg·m^2; the height makes no difference

29. $(23/48)MR^2\omega^2$

31. -3.55 N·m

33. 8.02×10^3 N

35. (a) 24.0 N·m   (b) 0.035 6 rad/s^2 (c) 1.07 m/s^2

37. (a) 0.309 m/s^2 (b) 7.67 N and 9.22 N

39. 21.5 N

41. 24.5 km

43. (a) 1.59 m/s   (b) 53.1 rad/s

45. (a) 11.4 N, 7.57 m/s^2, 9.53 m/s down   (b) 9.53 m/s

49. (a) $2(Rg/3)^{1/2}$ (b) $4(Rg/3)^{1/2}$ (c) $(Rg)^{1/2}$

51. (a) 500 J   (b) 250 J   (c) 750 J

53. (a) $\frac{2}{3}g\sin\theta$ for the disk, larger than $\frac{1}{2}g\sin\theta$ for the hoop   (b) $\frac{1}{3}\tan\theta$

55. 1.21×10^{-4} kg·m^2; height is unnecessary

57. $\frac{1}{3}\ell$

59. (a) 4.00 J   (b) 1.60 s   (c) yes

61. (a) $(3g/L)^{1/2}$ (b) $3g/2L$ (c) $-\frac{3}{2}g\hat{\mathbf{i}} - \frac{3}{4}g\hat{\mathbf{j}}$
  (d) $-\frac{3}{2}Mg\hat{\mathbf{i}} + \frac{1}{4}Mg\hat{\mathbf{j}}$

63. -0.322 rad/s^2

65. (b) $2gM(\sin\theta - \mu\cos\theta)(m + 2M)^{-1}$

67. (a) $\sim -10^{-22}$ s^{-2} (b) $\sim -10^{16}$ N·m   (c) $\sim 10^{13}$ m

71. (a) 118 N and 156 N   (b) 1.17 kg·m^2

73. (a) $\alpha = -0.176$ rad/s^2 (b) 1.29 rev   (c) 9.26 rev

75. (a) 61.2 J   (b) 50.8 J

79. (a) $2.70R$ (b) $F_x = -20mg/7, F_y = -mg$

81. $\sim 10^1$ m

83. (a) $(3gh/4)^{1/2}$ (b) $(3gh/4)^{1/2}$

85. (c) $(8Fd/3M)^{1/2}$

87. $\mathbf{F}_1$ to right, $\mathbf{F}_2$ no rolling, $\mathbf{F}_3$ and $\mathbf{F}_4$ to left

CHAPTER 11

1. $-7.00\hat{\mathbf{i}} + 16.0\hat{\mathbf{j}} - 10.0\hat{\mathbf{k}}$

3. (a) $-17.0\hat{\mathbf{k}}$ (b) 70.6°

5. 0.343 N·m horizontally north

7. 45.0°

9. $F_3 = F_1 + F_2$; no

11. $(17.5\hat{\mathbf{k}})$ kg·m^2/s

13. $(60.0\hat{\mathbf{k}})$ kg·m^2/s

15. $mvR[\cos(vt/R) + 1]\hat{\mathbf{k}}$

17. (a) zero   (b) $[-mv_i^3 \sin^2\theta\cos\theta/2g]\hat{\mathbf{k}}$
  (c) $[-2mv_i^3 \sin^2\theta\cos\theta/g]\hat{\mathbf{k}}$ (d) The downward gravitational force exerts a torque in the $-z$ direction.

19. $-m\ell gt\cos\theta\hat{\mathbf{k}}$

23. (a) 0.360 kg·m^2/s   (b) 0.540 kg·m^2/s

25. (a) 0.433 kg·m^2/s   (b) 1.73 kg·m^2/s

27. (a) 1.57×10^8 kg·m^2/s   (b) 6.26×10^3 s = 1.74 h

29. 7.14 rev/min

31. (a) 9.20 rad/s   (b) 9.20 rad/s

33. (a) 0.360 rad/s counterclockwise   (b) 99.9 J

35. (a) $mv\ell$ down   (b) $M/(M + m)$

37. (a) $\omega = 2mv_id/(M + 2m)R^2$ (b) No; some mechanical energy changes into internal energy

39. $\sim 10^{-13}$ rad/s

41. 5.45×10^{22} N·m

43. 7.50×10^{-11} s

45. (a) $7md^2/3$ (b) $mgd\,\hat{\mathbf{k}}$ (c) $3g/7d$ counterclockwise
  (d) $2g/7$ upward   (e) mgd (f) $\sqrt{6g/7d}$ (g) $m\sqrt{14gd^3/3}$
  (h) $\sqrt{2gd/21}$

47. 0.910 km/s

49. (a) v_ir_i/r (b) $T = (mv_i^2r_i^2)r^{-3}$ (c) $\frac{1}{2}mv_i^2(r_i^2/r^2 - 1)$
  (d) 4.50 m/s, 10.1 N, 0.450 J

51. (a) 3 750 kg·m^2/s   (b) 1.88 kJ   (c) 3 750 kg·m^2/s
  (d) 10.0 m/s   (e) 7.50 kJ   (f) 5.62 kJ

53. An increase of 0.550 s

55. $4[ga(\sqrt{2} - 1)/3]^{1/2}$

CHAPTER 12

1. 10.0 N up; 6.00 N·m counterclockwise

3. $[(m_1 + m_b)d + (m_1\ell/2)]/m_2$

5. (3.85 cm, 6.85 cm)

7. (−1.50 m, −1.50 m)

9. 177 kg

11. 8.33%

13. (a) $f_s = 268$ N, $n = 1\,300$ N   (b) 0.324

15. (a) 1.04 kN at 60.0°   (b) $(370\hat{\mathbf{i}} + 900\hat{\mathbf{j}})$ N

17. 2.94 kN on each rear wheel and 4.41 kN on each front wheel

19. (a) 29.9 N   (b) 22.2 N

21. (a) 1.73 rad/s²   (b) 1.56 rad/s
    (c) $(−4.72\hat{\mathbf{i}} + 6.62\hat{\mathbf{j}})$ kN   (d) $38.9\hat{\mathbf{j}}$ kN

23. 2.82 m

25. 88.2 N and 58.8 N

27. 4.90 mm

29. 10×10^{10} N/m²

31. 23.8 μm

33. (a) 3.14×10^4 N   (b) 6.28×10^4 N

35. 1.65×10^8 N/m²

37. 0.860 mm

39. $n_A = 5.98 \times 10^5$ N, $n_B = 4.80 \times 10^5$ N

41. 9.00 ft

43. (a)

(b) $T = 343$ N; $R_x = 171$ N to the right, $R_y = 683$ N up
(c) 5.13 m

45. (a) $T = F_g(L + d)/\sin\theta\,(2L + d)$
    (b) $R_x = F_g(L + d)\cot\theta/(2L + d)$; $R_y = F_gL/(2L + d)$

47. $\mathbf{F}_A = (−6.47 \times 10^5\hat{\mathbf{i}} + 1.27 \times 10^5\hat{\mathbf{j}})$ N,
    $\mathbf{F}_B = 6.47 \times 10^5\hat{\mathbf{i}}$ N

49. 5.08 kN; $R_x = 4.77$ kN, $R_y = 8.26$ kN

51. $T = 2.71$ kN, $R_x = 2.65$ kN

53. (a) 20.1 cm to the left of the front edge; $\mu_k = 0.571$
    (b) 0.501 m

55. (a) $M = (m/2)(2\mu_s\sin\theta − \cos\theta)(\cos\theta − \mu_s\sin\theta)^{-1}$
    (b) $R = (m + M)g(1 + \mu_s^2)^{1/2}$;
    $F = g[M^2 + \mu_s^2(m + M)^2]^{1/2}$

57. (a) 133 N   (b) $n_A = 429$ N and $n_B = 257$ N
    (c) $R_x = 133$ N and $R_y = −257$ N

59. 66.7 N

63. 1.09 m

65. (a) 4 500 N   (b) 4.50×10^6 N/m²   (c) The board will break.

67. 5.73 rad/s

69. $n_A = 11.0$ kN, $n_E = 3.67$ kN; $F_{AB} = F_{DE} = 7.35$ kN compression; $F_{AC} = F_{CE} = 6.37$ kN compression; $F_{BC} = F_{CD} = 4.24$ kN tension; $F_{BD} = 8.49$ kN compression

71. (a) $P_y = (F_g/L)(d − ah/g)$ (b) 0.306 m
    (c) $(−306\hat{\mathbf{i}} + 5.53\hat{\mathbf{j}})$ N

73. Decrease h, increase d

CHAPTER 13

1. $\sim 10^{-7}$ N toward you

3. (a) 2.50×10^{-5} N toward the 500-kg object
    (b) between the objects and 0.245 m from the 500-kg object

5. $(−100\hat{\mathbf{i}} + 59.3\hat{\mathbf{j}})$ pN

7. 7.41×10^{-10} N

9. 0.613 m/s² toward the Earth

11. (a) 3.46×10^8 m   (b) 3.34×10^{-3} m/s² toward the Earth

13. 1.26×10^{32} kg

15. 1.90×10^{27} kg

17. 35.2 AU

19. 8.92×10^7 m

21. After 393 yr, Mercury would be farther from the Sun than Pluto

23. $\mathbf{g} = (Gm/\ell^2)(\frac{1}{2} + \sqrt{2})$ toward the opposite corner

25. $\mathbf{g} = 2MGr(r^2 + a^2)^{-3/2}$ toward the center of mass

27. 4.17×10^{10} J

29. (a) 1.84×10^9 kg/m³   (b) 3.27×10^6 m/s²
    (c) $−2.08 \times 10^{13}$ J

31. (a) $−1.67 \times 10^{-14}$ J   (b) at the center

33. 1.66×10^4 m/s

37. (a) 5.30×10^3 s   (b) 7.79 km/s   (c) 6.43×10^9 J

39. 469 MJ

41. 15.6 km/s

43. (b) 1.00×10^7 m   (c) 1.00×10^4 m/s

45. (a) 0.980   (b) 127 yr   (c) $−2.13 \times 10^{17}$ J

49. (b) $2[Gm^3(1/2r − 1/R)]^{1/2}$

51. (b) 1.10×10^{32} kg

53. (a) $−7.04 \times 10^4$ J   (b) $−1.57 \times 10^5$ J   (c) 13.2 m/s

55. 7.79×10^{14} kg

57. $\omega = 0.057\,2$ rad/s or 1 rev in 110 s

59. $v_{esc} = (8\pi G\rho/3)^{1/2}R$

61. (a) $m_2(2G/d)^{1/2}(m_1 + m_2)^{-1/2}$ and
    $m_1(2G/d)^{1/2}(m_1 + m_2)^{-1/2}$;
    relative speed $(2G/d)^{1/2}(m_1 + m_2)^{1/2}$
    (b) 1.07×10^{32} J and 2.67×10^{31} J

63. (a) 8.50×10^8 J   (b) 2.71×10^9 J

65. (a) 200 Myr   (b) $\sim 10^{41}$ kg; $\sim 10^{11}$ stars

67. $(GM_E/4R_E)^{1/2}$

71.

t (s)	x (m)	y (m)	v_x (m/s)	v_y (m/s)
0	0	12 740 000	5 000	0
10	50 000	12 740 000	4 999.9	−24.6
20	99 999	12 739 754	4 999.7	−49.1
30	149 996	12 739 263	4 999.4	−73.7...

The object does not hit the Earth; its minimum radius is $1.33R_E$ as shown in the diagram below. Its period is 1.09×10^4 s. A circular orbit would require speed 5.60 km/s.

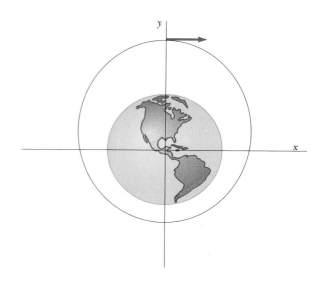

CHAPTER 14

1. 0.111 kg

3. 6.24 MPa

5. 5.27×10^{18} kg

7. 1.62 m

9. 7.74×10^{-3} m^2

11. 271 kN horizontally backward

13. $P_0 + \frac{1}{2}\rho d\sqrt{g^2 + a^2}$

15. 0.722 mm

17. 10.5 m; no; some alcohol and water evaporate

19. 98.6 kPa

21. (a) 1.57 Pa, 1.55×10^{-2} atm, 11.8 mm Hg   (b) The fluid level in the tap should rise.   (c) Blockage of flow of the cerebrospinal fluid

23. 0.258 N

25. (a) 9.80 N   (b) 6.17 N

27. (a) $1.017\ 9 \times 10^3$ N down, $1.029\ 7 \times 10^3$ N up   (b) 86.2 N

29. (a) 7.00 cm   (b) 2.80 kg

33. 1 430 m^3

35. 1 250 kg/m^3 and 500 kg/m^3

37. 1.28×10^4 m^2

39. (a) 17.7 m/s   (b) 1.73 mm

41. 31.6 m/s

43. 0.247 cm

45. (a) 1 atm + 15.0 MPa   (b) 2.95 m/s   (c) 4.34 kPa

47. 2.51×10^{-3} m^3/s

49. 103 m/s

51. (a) 4.43 m/s
    (b) The siphon can be no higher than 10.3 m.

53. 12.6 m/s

55. 1.91 m

59. 0.604 m

63. 17.3 N and 31.7 N

65. 90.04%

67. 758 Pa

69. 4.43 m/s

71. (a) 1.25 cm   (b) 13.8 m/s

73. (a) 3.307 g   (b) 3.271 g   (c) 3.48×10^{-4} N

75. (c) 1.70 m^2

CHAPTER 15

1. (a) The motion repeats precisely.   (b) 1.82 s
   (c) No, the force is not in the form of Hooke's law.

3. (a) 1.50 Hz, 0.667 s   (b) 4.00 m   (c) π rad
   (d) 2.83 m

5. (b) 18.8 cm/s, 0.333 s   (c) 178 cm /s^2, 0.500 s
   (d) 12.0 cm

7. (a) 2.40 s   (b) 0.417 Hz   (c) 2.62 rad/s

9. 40.9 N/m

11. (a) 40.0 cm/s, 160 cm/s^2 (b) 32.0 cm/s, -96.0 cm/s^2
    (c) 0.232 s

13. 0.628 m/s

15. (a) 0.542 kg   (b) 1.81 s   (c) 1.20 m/s^2

17. 2.23 m/s

19. (a) 28.0 mJ   (b) 1.02 m/s   (c) 12.2 mJ   (d) 15.8 mJ

21. (a) E increases by a factor of 4.   (b) v_{max} is doubled.
    (c) a_{max} is doubled.   (d) Period is unchanged.

23. 2.60 cm and -2.60 cm

25. (b) 0.628 s

27. (a) 35.7 m   (b) 29.1 s

29. $\sim 10^0$ s

31. Assuming simple harmonic motion, (a) 0.820 m/s
    (b) 2.57 rad/s^2 (c) 0.641 N. More precisely,
    (a) 0.817 m/s   (b) 2.54 rad/s^2 (c) 0.634 N

35. 0.944 kg·m^2

39. (a) 5.00×10^{-7} kg·m^2 (b) 3.16×10^{-4} N·m/rad

41. 1.00×10^{-3} s^{-1}

43. (a) 7.00 Hz   (b) 2.00%   (c) 10.6 s

45. (a) 1.00 s   (b) 5.09 cm

47. 318 N

49. 1.74 Hz

51. (a) $2Mg$; $Mg(1 + y/L)$
    (b) $T = (4\pi/3)(2L/g)^{1/2}$; 2.68 s

53. 6.62 cm

55. 9.19×10^{13} Hz

57. (a)

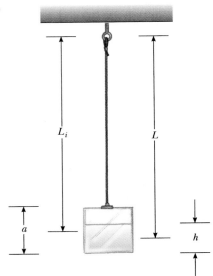

(b) $\dfrac{dT}{dt} = \dfrac{\pi(dM/dt)}{2\rho a^2 g^{1/2}[L_i + (dM/dt)\,t/2\rho a^2]^{1/2}}$

(c) $T = 2\pi g^{-1/2}\,[L_i + (dM/dt)\,t/2\rho a^2]^{1/2}$

59. $f = (2\pi L)^{-1}\,(gL + kh^2/M)^{1/2}$

61. (b) 1.23 Hz

63. (a) 3.00 s (b) 14.3 J (c) 25.5°

65. If the cyclist goes over them at one certain speed, the washboard bumps can excite a resonance vibration of the bike, so large in amplitude as to make the rider lose control. $\sim 10^1$ m

73. For $\theta_{max} = 5.00°$ there is precise agreement. For $\theta_{max} = 100°$ there are large differences, and the period is 23% greater than small-angle period.

75. (b) after 42.1 min

CHAPTER 16

1. $y = 6\,[(x - 4.5t)^2 + 3]^{-1}$

3. (a) left (b) 5.00 m/s

5. 184 km

7. 0.319 m

9. 2.00 cm, 2.98 m, 0.576 Hz, 1.72 m/s

11. (a) 3.77 m/s (b) 118 m/s^2

13. (a) 0.250 m (b) 40.0 rad/s (c) 0.300 rad/m (d) 20.9 m (e) 133 m/s (f) $+x$

15. (a) $y = (8.00\text{ cm})\sin(7.85x + 6\pi t)$ (b) $y = (8.00\text{ cm})\sin(7.85x + 6\pi t - 0.785)$

17. (a) -1.51 m/s, 0 (b) 16.0 m, 0.500 s, 32.0 m/s

19. (a) 0.500 Hz, 3.14 rad/s (b) 3.14 rad/m (c) $(0.100\text{ m})\sin(3.14x/\text{m} - 3.14t/\text{s})$ (d) $(0.100\text{ m})\sin(-3.14t/\text{s})$ (e) $(0.100\text{ m})\sin(4.71\text{ rad} - 3.14t/\text{s})$ (f) 0.314 m/s

21. 80.0 N

23. 520 m/s

25. 1.64 m/s^2

27. 13.5 N

29. 185 m/s

31. 0.329 s

33. (a) s and kg·m/s^2 (b) time interval (period) and force (tension)

37. 55.1 Hz

39. (a) 62.5 m/s (b) 7.85 m (c) 7.96 Hz (d) 21.1 W

41. $\sqrt{2}\mathscr{P}_0$

43. (a) $A = 40$ (b) $A = 7.00$, $B = 0$, $C = 3.00$. One can take the dot product of the given equation with each one of $\hat{\mathbf{i}}, \hat{\mathbf{j}},$ and $\hat{\mathbf{k}}$. (c) $A = 0$, $B = 7.00$ mm, $C = 3.00/\text{m}$, $D = 4.00/\text{s}$, $E = 2.00$. Consider the average value of both sides of the given equation to find A. Then consider the maximum value of both sides to find B. One can evaluate the partial derivative of both sides of the given equation with respect to x and separately with respect to t to obtain equations yielding C and D upon chosen substitutions for x and t. Then substitute $x = 0$ and $t = 0$ to obtain E.

47. ~ 1 min

49. (a) $(3.33\hat{\mathbf{i}})$ m/s (b) -5.48 cm (c) 0.667 m, 5.00 Hz (d) 11.0 m/s

51. 0.456 m/s

53. (a) 39.2 N (b) 0.892 m (c) 83.6 m/s

55. (a) 179 m/s (b) 17.7 kW

57. 0.084 3 rad

61. (a) $(0.707)2(L/g)^{1/2}$ (b) $L/4$

63. 3.86×10^{-4}

65. (b) 31.6 m/s

67. (a) $\dfrac{\mu\omega^3}{2k}A_0^2 e^{-2bx}$ (b) $\dfrac{\mu\omega^3}{2k}A_0^2$ (c) e^{-2bx}

69. (a) $\mu_0 + (\mu_L - \mu_0)x/L$

CHAPTER 17

1. 5.56 km

3. 7.82 m

5. (a) 826 m (b) 1.47 s

7. 5.67 mm

9. 1.50 mm to 75.0 μm

11. (a) 2.00 μm, 40.0 cm, 54.6 m/s (b) -0.433 μm (c) 1.72 mm/s

13. $\Delta P = (0.200\text{ N/m}^2)\sin(62.8\,x/\text{m} - 2.16 \times 10^4\,t/\text{s})$

15. 5.81 m

19. 66.0 dB

21. (a) 3.75 W/m^2 (b) 0.600 W/m^2

23. (a) 2.34 m and 0.390 m (b) 0.161 N/m^2 for both notes (c) 4.25×10^{-7} m and 7.09×10^{-8} m (d) The wavelengths and displacement amplitudes would be larger by a factor of 1.09. The answer to (b) is unchanged.

25. (a) 1.32×10^{-4} W/m^2 (b) 81.2 dB

27. (a) 0.691 m (b) 691 km

29. 65.6 dB

31. (a) 65.0 dB (b) 67.8 dB (c) 69.6 dB

33. (a) 30.0 m (b) 9.49×10^5 m

35. (a) 332 J (b) 46.4 dB

37. (a) 338 Hz (b) 483 Hz

39. 26.4 m/s

41. 19.3 m

43. (a) 0.364 m (b) 0.398 m (c) 941 Hz (d) 938 Hz

45. 2.82×10^8 m/s

47. (a) 56.3 s (b) 56.6 km farther along

49. 22.3° left of center

51. $f \sim$ 300 Hz, $\lambda \sim 10^0$ m, duration $\sim 10^{-1}$ s

55. 6.01 km

57. (a) 55.8 m/s (b) 2 500 Hz

59. 1 204.2 Hz

61. 1.60

63. 2.34 m

65. (a) 0.948° (b) 4.40°

67. 1.34×10^4 N

69. (b) 531 Hz

71. (a) 6.45 (b) 0

CHAPTER 18

1. (a) −1.65 cm (b) −6.02 cm (c) 1.15 cm

3. (a) $+x, -x$ (b) 0.750 s (c) 1.00 m

5. (a) 9.24 m (b) 600 Hz

7. (a) zero (b) 0.300 m

9. (a) 2 (b) 9.28 m and 1.99 m

11. (a) 156° (b) 0.058 4 cm

13. 15.7 m, 31.8 Hz, 500 m/s

15. At 0.089 1 m, 0.303 m, 0.518 m, 0.732 m, 0.947 m, 1.16 m from one speaker

17. (a) 4.24 cm (b) 6.00 cm (c) 6.00 cm
    (d) 0.500 cm, 1.50 cm, 2.50 cm

19. 0.786 Hz, 1.57 Hz, 2.36 Hz, 3.14 Hz

21. (a) 350 Hz (b) 400 kg

23. 1.27 cm

25. (a) reduced by 1/2 (b) reduced by $1/\sqrt{2}$
    (c) increased by $\sqrt{2}$

27. (a) 163 N (b) 660 Hz

29. $\dfrac{Mg}{4Lf^2 \tan\theta}$

31. (a) 3 loops (b) 16.7 Hz (c) 1 loop

33. (a) 3.66 m/s (b) 0.200 Hz

35. 9.00 kHz

37. (a) 0.357 m (b) 0.715 m

39. 57.6 Hz

41. n(206 Hz) for $n = 1$ to 9 and n(84.5 Hz) for $n = 2$ to 23

43. 50.0 Hz, 1.70 m

45. (a) 350 m/s (b) 1.14 m

47. (a) 162 Hz (b) 1.06 m

49. (a) 1.59 kHz (b) odd-numbered harmonics
    (c) 1.11 kHz

51. 5.64 beats/s

53. (a) 1.99 beats/s (b) 3.38 m/s

55. The second harmonic of E is close to the third harmonic of A, and the fourth harmonic of C$^\#$ is close to the fifth harmonic of A.

57. (a) 34.8 m/s (b) 0.977 m

59. 3.85 m/s away from the station or 3.77 m/s toward the station

61. 21.5 m

63. (a) 59.9 Hz (b) 20.0 cm

65. (a) 1/2 (b) $[n/(n+1)]^2 T$ (c) 9/16

67. $y_1 + y_2 = 11.2 \sin(2.00x - 10.0t + 63.4°)$

69. (a) 78.9 N (b) 211 Hz

CHAPTER 19

1. (a) −274°C (b) 1.27 atm (c) 1.74 atm

3. (a) −320°F (b) 77.3 K

5. (a) 810°F (b) 450 K

7. (a) 1 337 K, 2 993 K (b) 1 596°C = 1 596 K

9. 3.27 cm

11. 55.0°C

13. (a) 0.176 mm (b) 8.78 μm (c) 0.093 0 cm^3

15. (a) −179°C (attainable)
    (b) −376°C (below 0 K, unattainable)

17. 0.548 gal

19. (a) 99.8 mL
    (b) about 6% of the volume change of the acetone

21. (a) 99.4 cm^3 (b) 0.943 cm

23. 1.14°C

25. 5 336 images

27. (a) 400 kPa (b) 449 kPa

29. 1.50×10^{29} molecules

31. 1.61 MPa = 15.9 atm

33. 472 K

35. (a) 41.6 mol (b) 1.20 kg, nearly in agreement with the tabulated density

37. (a) 1.17 g (b) 11.5 mN (c) 1.01 kN
    (d) The molecules must be moving very fast.

39. 4.39 kg

41. 3.55 L

43. $m_1 - m_2 = \dfrac{P_0 V M}{R}\left(\dfrac{1}{T_1} - \dfrac{1}{T_2}\right)$

45. (a) 94.97 cm (b) 95.03 cm

47. 3.55 cm

49. It falls by 0.094 3 Hz

51. (a) Expansion makes density drop. (b) $5 \times 10^{-5}/°C$

53. (a) $h = nRT/(mg + P_0A)$ (b) 0.661 m

55. We assume that $\alpha \Delta T$ is much less than 1.

57. (a) 0.340% (b) 0.480%

59. 0.750

61. 2.74 m

63. (b) 1.33 kg/m^3

67. No. Steel is not strong enough.

69. (a) $L_f = L_i e^{\alpha \Delta T}$ (b) 2.00×10^{-4}%; 59.4%

71. (a) 6.17×10^{-3} kg/m (b) 632 N (c) 580 N; 192 Hz

73. 4.54 m

CHAPTER 20

1. $(10.0 + 0.117)°C$

3. 0.234 kJ/kg $\cdot$ °C

5. 1.78×10^4 kg

7. 29.6°C

9. (a) 0.435 cal/g $\cdot$ °C (b) beryllium

11. 23.6°C

13. 50.7 ks

15. 1.22×10^5 J

17. 0.294 g

19. 0.414 kg

21. (a) 0°C (b) 114 g

23. -1.18 MJ

25. -466 J

27. (a) $-4P_iV_i$ (b) It is proportional to the square of the volume, according to $T = (P_i/nRV_i)V^2$

29. $Q = -720$ J

31.

	Q	W	ΔE_{int}
BC	$-$	0	$-$
CA	$-$	$+$	$-$
AB	$+$	$-$	$+$

33. 3.60 kJ

35. (a) 7.50 kJ (b) 900 K

37. -3.10 kJ; 37.6 kJ

39. (a) 0.041 0 m^3 (b) $+5.48$ kJ (c) -5.48 kJ

41. 2.22×10^{-2} W/m $\cdot$ °C

43. 51.2°C

45. 67.9°C

47. 3.77×10^{26} J/s

49. 3.49×10^3 K

51. 277 K = 4°C

53. 2.27 km

55. (a) 16.8 L (b) 0.351 L/s

57. $c = \mathcal{P}/\rho R \Delta T$

59. -1.87 kJ

61. 5.87×10^4 °C

63. 5.31 h

65. 1.44 kg

67. 38.6 m^3/d

71. 9.32 kW

73. (a) 3.16×10^{22} W (b) 5.78×10^3 K, 0.327% less than 5 800 K (c) 3.17×10^{22} W, 0.408% larger

CHAPTER 21

1. 0.943 N; 1.57 Pa

3. 3.65×10^4 N

5. 3.32 mol

7. (a) 3.54×10^{23} atoms (b) 6.07×10^{-21} J (c) 1.35 km/s

9. (a) 8.76×10^{-21} J for both (b) 1.62 km/s for helium and 514 m/s for argon

13. (a) 3.46 kJ (b) 2.45 kJ (c) -1.01 kJ

15. (a) 209 J (b) zero (c) 317 K

17. 1.18 atm

19. Between 10^{-2} and 10^{-3} °C

21. (a) 316 K (b) 200 J

23. (a) $C = \dfrac{n_1 C_1 + n_2 C_2}{n_1 + n_2}$ (b) $C = \dfrac{\sum\limits_{i=1}^{m} n_i C_i}{\sum\limits_{i=1}^{m} n_i}$

25. (a) 1.39 atm (b) 366 K, 253 K (c) 0, -4.66 kJ, -4.66 kJ

27. 227 K

29. (a)

(b) 8.77 L (c) 900 K (d) 300 K (e) -336 J

31. (a) 28.0 kJ (b) 46.1 kJ (c) Isothermal process, $P_f = 10.0$ atm; adiabatic process, $P_f = 25.1$ atm

33. (a) 9.95 cal/K, 13.9 cal/K (b) 13.9 cal/K, 17.9 cal/K

35. 2.33×10^{-21} J

37. (a) 6.80 m/s (b) 7.41 m/s (c) 7.00 m/s

41. (a) 2.37×10^4 K (b) 1.06×10^3 K

43. (b) 0.278

45. (a) 3.21×10^{12} molecules   (b) 779 km
(c) 6.42×10^{-4} s^{-1}

49. (a) 9.36×10^{-8} m   (b) 9.36×10^{-8} atm   (c) 302 atm

51. (a) 100 kPa, 66.5 L, 400 K, 5.82 kJ, 7.48 kJ, -1.66 kJ

(b) 133 kPa, 49.9 L, 400 K, 5.82 kJ, 5.82 kJ, 0

(c) 120 kPa, 41.6 L, 300 K, 0, -910 J, $+910$ J

(d) 120 kPa, 43.3 L, 312 K, 722 J, 0, $+722$ J

55. 510 K and 290 K

57. 0.623

59. (a) Pressure increases as volume decreases
(d) 0.500 atm^{-1}, 0.300 atm^{-1}

61. (a) 0.514 m^3 (b) 2.06 m^3 (c) 2.38×10^3 K
(d) -480 kJ   (e) 2.28 MJ

63. 1.09×10^{-3}; 2.69×10^{-2}; 0.529; 1.00; 0.199; 1.01×10^{-41};
1.25×10^{-1082}

67. (a) 0.203 mol   (b) $T_B = T_C = 900$ K, $V_C = 15.0$ L

(c, d)	P, atm	V, L	T, K	E_{int}, kJ
A	1.00	5.00	300	0.760
B	3.00	5.00	900	2.28
C	1.00	15.0	900	2.28
A	1.00	5.00	300	0.760

(e) Lock the piston in place and put the cylinder into an oven at 900 K. Keep the gas in the oven while gradually letting the gas expand to lift a load on the piston as far as it can. Move the cylinder from the oven back to the 300-K room and let the gas cool and contract.

(f, g)	Q, kJ	W, kJ	ΔE_{int}, kJ
AB	1.52	0	1.52
BC	1.67	-1.67	0
CD	-2.53	$+1.01$	-1.52
$ABCA$	0.656	-0.656	0

69. 1.60×10^4 K

CHAPTER 22

1. (a) 6.94%   (b) 335 J

3. (a) 10.7 kJ   (b) 0.533 s

5. (a) 29.4 L/h   (b) 185 hp   (c) 527 N·m
(d) 1.91×10^5 W

7. (a) 24.0 J   (b) 144 J

9. (a) 2.93   (b) coefficient of performance for a refrigerator
(c) $300 is twice as large as $150

11. (a) 67.2%   (b) 58.8 kW

13. (a) 741 J   (b) 459 J

15. (a) 4.20 W   (b) 31.2 g

17. (a) 564 K   (b) 212 kW   (c) 47.5%

19. (b) $1 - T_c/T_h$ (c) $(T_c + T_h)/2$ (d) $(T_h T_c)^{1/2}$

21. (a) 214 J, 64.3 J
(b) -35.7 J, -35.7 J. The net effect is the transport of energy by heat from the cold to the hot reservoir without expenditure of external work.   (c) 333 J, 233 J

(d) 83.3 J, 83.3 J, 0. The net effect is converting energy, taken in by heat, entirely into energy output by work in a cyclic process.
(e) -0.111 J/K. The entropy of the Universe has decreased.

23. 9.00

27. 72.2 J

29. 1.86

31. (a) 244 kPa   (b) 192 J

33. 146 kW, 70.8 kW

35. -610 J/K

37. 195 J/K

39. 236 J/K

41. 1.02 kJ/K

43. $\sim 10^0$ W/K from metabolism; much more if you are using high-power electric appliances or an automobile, or if your taxes are paying for a war.

45. 5.76 J/K; temperature is constant if the gas is ideal

47. 18.4 J/K

49. (a) 1   (b) 6

51. (a)

Result	Number of Ways to Draw
All R	1
2 R, 1 G	3
1R, 2 G	3
All G	1

(b)

Result	Number of Ways to Draw
All R	1
4R, 1G	5
3R, 2G	10
2R, 3G	10
1R, 4G	5
All G	1

53. (a) 5.00 kW   (b) 763 W

55. (a) 0.476 J/K   (b) 417 J   (c) $W_{\text{net}} = T_1 \Delta S_U = 167$ J

57. (a) $2nRT_i \ln 2$ (b) 0.273

59. 5.97×10^4 kg/s

61. (a) 3.19 cal/K   (b) 98.19°F, 2.59 cal/K

63. (a) 8.48 kW   (b) 1.52 kW   (c) 1.09×10^4 J/K
(d) COP drops by 20.0%

65. (a) $10.5nRT_i$ (b) $8.50nRT_i$ (c) 0.190   (d) 0.833

67. (a) $nC_P \ln 3$
(b) Both ask for the change in entropy between the same two states of the same system. Entropy is a function of state. The change in entropy does not depend on path, but only on original and final states.

71. (a) 20.0°C   (c) $\Delta S = +4.88$ J/K   (d) Yes

CHAPTER 23

1. (a) $+160$ zC, 1.01 u   (b) $+160$ zC, 23.0 u
(c) -160 zC, 35.5 u   (d) $+320$ zC, 40.1 u
(e) -480 zC, 14.0 u   (f) $+640$ zC, 14.0 u
(g) $+1.12$ aC, 14.0 u   (h) -160 zC, 18.0 u

3. The force is $\sim 10^{26}$ N.

5. (a) 1.59 nN away from the other
   (b) 1.24×10^{36} times larger
   (c) 8.61×10^{-11} C/kg

7. 0.872 N at 330°

9. (a) 2.16×10^{-5} N toward the other   (b) 8.99×10^{-7} N away from the other

11. (a) 82.2 nN   (b) 2.19 Mm/s

13. (a) 55.8 pN/C down   (b) 102 nN/C up

15. 1.82 m to the left of the negative charge

17. $-9Q$ and $+27Q$

19. (a) $(-0.599\hat{\mathbf{i}} - 2.70\hat{\mathbf{j}})$ kN/C   (b) $(-3.00\hat{\mathbf{i}} - 13.5\hat{\mathbf{j}})$ μN

21. (a) $5.91 k_e q/a^2$ at 58.8°   (b) $5.91 k_e q^2/a^2$ at 58.8°

23. (a) $[k_e Qx/(R^2 + x^2)^{3/2}]\hat{\mathbf{i}}$ (b) As long as the charge is symmetrically placed, the number of charges does not matter. A continuous ring corresponds to n becoming larger without limit.

25. 1.59×10^6 N/C toward the rod

27. (a) $6.64\hat{\mathbf{i}}$ MN/C   (b) $24.1\hat{\mathbf{i}}$ MN/C   (c) $6.40\hat{\mathbf{i}}$ MN/C
    (d) $0.664\hat{\mathbf{i}}$ MN/C, taking the axis of the ring as the x axis

31. (a) 93.6 MN/C; the near-field approximation is 104 MN/C, about 11% high   (b) 0.516 MN/C; the point-charge approximation is 0.519 MN/C, about 0.6% high

33. $-21.6\hat{\mathbf{i}}$ MN/C

37. (a) 86.4 pC for each
    (b) 324 pC, 459 pC, 459 pC, 432 pC
    (c) 57.6 pC, 106 pC, 154 pC, 96.0 pC

39.

41. (a)

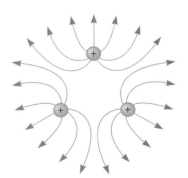

The field is zero at the center of the triangle.
(b) $1.73 k_e q\hat{\mathbf{j}}/a^2$

43. (a) 61.3 Gm/s^2 (b) 19.5 μs   (c) 11.7 m   (d) 1.20 fJ

45. K/ed in the direction of motion

47. (a) 111 ns   (b) 5.68 mm   (c) $(450\hat{\mathbf{i}} + 102\hat{\mathbf{j}})$ km/s

49. (a) 36.9°, 53.1°   (b) 167 ns, 221 ns

51. (a) 21.8 μm   (b) 2.43 cm

53. (a) mv^2/qR (b) $mv^2/2^{1/2}qR$ oriented at 135° to the x axis

55. (a) 10.9 nC   (b) 5.44 mN

57. 40.9 N at 263°

59. $Q = 2L \sqrt{\dfrac{k(L - L_i)}{k_e}}$

63. $-707\hat{\mathbf{j}}$ mN

65. (a) $\theta_1 = \theta_2$

67. (a) 0.307 s   (b) Yes. Ignoring gravity makes a difference of 2.28%.

69. (a) $\mathbf{F} = 1.90(k_e q^2/s^2)(\hat{\mathbf{i}} + \hat{\mathbf{j}} + \hat{\mathbf{k}})$ (b) $\mathbf{F} = 3.29(k_e q^2/s^2)$ in the direction away from the diagonally opposite vertex

CHAPTER 24

1. (a) 858 N·m^2/C   (b) 0   (c) 657 N·m^2/C

3. 4.14 MN/C

5. (a) aA (b) bA (c) 0

7. 1.87 kN·m^2/C

9. (a) -6.89 MN·m^2/C   (b) The number of lines entering exceeds the number leaving by 2.91 times or more.

11. $-Q/\epsilon_0$ for S_1; 0 for S_2; $-2Q/\epsilon_0$ for S_3; 0 for S_4

13. $E_0 \pi r^2$

15. (a) $+Q/2\epsilon_0$ (b) $-Q/2\epsilon_0$

17. -18.8 kN·m^2/C

19. 0 if $R \le d$; $\dfrac{2\lambda}{\epsilon_0} \sqrt{R^2 - d^2}$ if $R > d$

21. (a) 3.20 MN·m^2/C   (b) 19.2 MN·m^2/C   (c) The answer to (a) could change, but the answer to (b) would stay the same.

23. 2.33×10^{21} N/C

25. -2.48 μC/m^2

27. 5.94×10^5 m/s

29. $\mathbf{E} = \rho r/2\epsilon_0$ away from the axis

31. (a) 0   (b) 7.19 MN/C away from the center

33. (a) $\sim$1 mN   (b) $\sim$100 nC   (c) $\sim$10 kN/C
    (d) $\sim$10 kN·m^2/C

35. (a) 51.4 kN/C outward   (b) 646 N·m^2/C

37. 508 kN/C up

39. (a) 0   (b) 5 400 N/C outward   (c) 540 N/C outward

41. $\mathbf{E} = Q/2\epsilon_0 A$ vertically upward in each case if $Q > 0$

43. (a) $+708$ nC/m^2 and -708 nC/m^2 (b) $+177$ nC and -177 nC

45. 2.00 N

47. (a) $-\lambda$, $+3\lambda$ (b) $3\lambda/2\pi\epsilon_0 r$ radially outward

49. (a) 80.0 nC/m^2 on each face   (b) $9.04\hat{\mathbf{k}}$ kN/C
    (c) $-9.04\hat{\mathbf{k}}$ kN/C

51. $\mathbf{E} = 0$ inside the sphere and within the material of the shell. $\mathbf{E} = k_e Q/r^2$ radially inward between the sphere and the shell. $\mathbf{E} = 2k_e Q/r^2$ radially outward outside the shell. Charge $-Q$ resides on the outer surface of the sphere.

$+Q$ is on the inner surface of the shell. $+2Q$ is on the outer surface of the shell.

53. (b) $Q/2\epsilon_0$ (c) Q/ϵ_0

55. (a) $+2Q$ (b) radially outward   (c) $2k_eQ/r^2$ (d) 0   (e) 0
 (f) $3Q$ (g) $3k_eQ/r^2$ radially   outward   (h) $3Qr^3/a^3$
 (i) $3k_eQr/a^3$ radially   outward   (j) $-3Q$ (k) $+2Q$
 (l) See below.

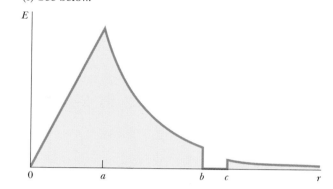

57. (a) $\rho r/3\epsilon_0$; $Q/4\pi\epsilon_0 r^2$; 0; $Q/4\pi\epsilon_0 r^2$, all radially outward
 (b) $-Q/4\pi b^2$ and $+Q/4\pi c^2$

59. $\theta = \tan^{-1}[qQ/(2\pi\epsilon_0 dmv^2)]$

61. For $r < a$, $\mathbf{E} = \lambda/2\pi\epsilon_0 r$ radially outward. For $a < r < b$,
 $\mathbf{E} = [\lambda + \rho\pi(r^2 - a^2)]/2\pi\epsilon_0 r$ radially outward. For $r > b$,
 $\mathbf{E} = [\lambda + \rho\pi(b^2 - a^2)]/2\pi\epsilon_0 r$ radially outward.

63. (a) σ/ϵ_0 away from both plates   (b) 0   (c) σ/ϵ_0 away from both plates

65. $\sigma/2\epsilon_0$ radially outward

69. $\mathbf{E} = a/2\epsilon_0$ radially outward

73. (b) $\mathbf{g} = GM_E r/R_E{}^3$ radially inward

CHAPTER 25

1. 1.35 MJ

3. (a) 152 km/s   (b) 6.49 Mm/s

5. (a) -600 μJ   (b) -50.0 V

7. 38.9 V; the origin

9. $+260$ V

11. (a) $2QE/k$ (b) QE/k (c) $2\pi\sqrt{m/k}$ (d) $2(QE - \mu_k mg)/k$

13. (a) 0.400 m/s   (b) the same

15. (a) 1.44×10^{-7} V   (b) -7.19×10^{-8} V
 (c) -1.44×10^{-7} V, $+7.19 \times 10^{-8}$ V

17. (a) 6.00 m   (b) -2.00 μC

19. -11.0 MV

21. 8.95 J

25. (a) no point at a finite distance from the charges
 (b) $2k_e q/a$

27. (a) $v_1 = \sqrt{\dfrac{2m_2 k_e q_1 q_2}{m_1(m_1 + m_2)}\left(\dfrac{1}{r_1 + r_2} - \dfrac{1}{d}\right)}$

 $v_2 = \sqrt{\dfrac{2m_1 k_e q_1 q_2}{m_2(m_1 + m_2)}\left(\dfrac{1}{r_1 + r_2} - \dfrac{1}{d}\right)}$

 (b) faster than calculated in (a)

29. $5k_e q^2/9d$

31. 0.720 m, 1.44 m, 2.88 m. No. The radii of the equipotentials are inversely proportional to the potential.

33. 7.26 Mm/s

35. $\left[\left(1 + \sqrt{\dfrac{1}{8}}\right)\dfrac{k_e q^2}{mL}\right]^{1/2}$

37. (a) 10.0 V, -11.0 V, -32.0 V
 (b) 7.00 N/C in the $+x$ direction

39. $\mathbf{E} = (-5 + 6xy)\hat{\mathbf{i}} + (3x^2 - 2z^2)\hat{\mathbf{j}} - 4yz\hat{\mathbf{k}}$; 7.07 N/C

41. $E_y = \dfrac{k_e Q}{y\sqrt{\ell^2 + y^2}}$

43. (a) C/m^2 (b) $k_e\alpha[L - d\ln(1 + L/d)]$

45. -1.51 MV

47. $k_e\lambda(\pi + 2\ln 3)$

49. (a) 0, 1.67 MV   (b) 5.84 MN/C away, 1.17 MV
 (c) 11.9 MN/C away, 1.67 MV

51. (a) 450 kV   (b) 7.51 μC

53. 253 MeV

55. (a) -27.2 eV   (b) -6.80 eV   (c) 0

59. $k_e Q^2/2R$

63. $V_2 - V_1 = (-\lambda/2\pi\epsilon_0)\ln(r_2/r_1)$

69. (b) $E_r = (2k_e p\cos\theta)/r^3$; $E_\theta = (k_e p\sin\theta)/r^3$; yes; no
 (c) $V = k_e py(x^2 + y^2)^{-3/2}$;
 $\mathbf{E} = 3k_e pxy(x^2 + y^2)^{-5/2}\hat{\mathbf{i}} + k_e p(2y^2 - x^2)(x^2 + y^2)^{-5/2}\hat{\mathbf{j}}$

71. $V = \pi k_e C\left[R\sqrt{x^2 + R^2} + x^2\ln\left(\dfrac{x}{R + \sqrt{x^2 + R^2}}\right)\right]$

73. (a) 8 876 V   (b) 112 V

CHAPTER 26

1. (a) 48.0 μC   (b) 6.00 μC

3. (a) 1.33 μC/m^2 (b) 13.3 pF

5. (a) 5.00 μC on the larger and 2.00 μC on the smaller sphere   (b) 89.9 kV

7. (a) 11.1 kV/m toward the negative plate.
 (b) 98.3 nC/m^2 (c) 3.74 pF   (d) 74.7 pC

9. 4.42 μm

11. (a) 2.68 nF   (b) 3.02 kV

13. (a) 15.6 pF   (b) 256 kV

15. 708 μF

17. (a) 3.53 μF   (b) 6.35 V and 2.65 V   (c) 31.8 μC on each

19. 6.00 pF and 3.00 pF

21. (a) 5.96 μF   (b) 89.5 μC on 20 μF, 63.2 μC on 6 μF, 26.3 μC on 15 μF and on 3 μF

23. 120 μC; 80.0 μC and 40.0 μC

25. 10

27. 6.04 μF

29. 12.9 μF

31. (a) 216 μJ   (b) 54.0 μJ

33. (a) Circuit diagram:

Stored energy = 0.150 J
(b) Potential difference = 268 V
Circuit diagram:

35. (a) 1.50 μC   (b) 1.83 kV

39. 9.79 kg

43. (a) 81.3 pF   (b) 2.40 kV

45. 1.04 m

47. (a) 369 pC   (b) 118 pF, 3.12 V   (c) -45.5 nJ

49. 22.5 V

51. (b) -8.78×10^6 N/C·m; $-5.53 \times 10^{-2}\,\hat{\mathbf{i}}$ N

55. (a) 11.2 pF   (b) 134 pC   (c) 16.7 pF   (d) 66.9 pC

57. (a) $-2Q/3$ on upper plate, $-Q/3$ on lower plate
(b) $2Qd/3\epsilon_0 A$

59. 0.188 m^2

61. (a) $C = \dfrac{\epsilon_0 A}{d}\left(\dfrac{\kappa_1}{2} + \dfrac{\kappa_2 \kappa_3}{\kappa_2 + \kappa_3}\right)$ (b) 1.76 pF

63. (b) $1/C$ approaches $\dfrac{1}{4\pi\epsilon_0 a} + \dfrac{1}{4\pi\epsilon_0 b}$

65. (a) $Q_0^2\, d(\ell - x)/(2\ell^3\epsilon_0)$ (b) $Q_0^2\, d/(2\ell^3\epsilon_0)$ to the right
(c) $Q_0^2/(2\ell^4\epsilon_0)$ (d) $Q_0^2/(2\ell^4\epsilon_0)$

67. 4.29 μF

69. (a) The additional energy comes from work done by the electric field in the wires as it forces more charge onto the already-charged plates.   (b) $Q/Q_0 = \kappa$

71. 750 μC on C_1 and 250 μC on C_2

73. 19.0 kV

75. $\frac{4}{3}C$

CHAPTER 27

1. 7.50×10^{15} electrons

3. (a) $0.632\,I_0\tau$ (b) $0.999\,95\,I_0\tau$ (c) $I_0\tau$

5. $q\omega/2\pi$

7. 0.265 C

9. (a) 2.55 A/m^2 (b) 5.31×10^{10} m^{-3} (c) 1.20×10^{10} s

11. 0.130 mm/s

13. 500 mA

15. 6.43 A

17. (a) 1.82 m   (b) 280 μm

19. (a) $\sim 10^{18}\ \Omega$ (b) $\sim 10^{-7}\ \Omega$ (c) ~ 100 aA, ~ 1 GA

21. $R/9$

23. $6.00 \times 10^{-15}/\Omega\cdot$m

25. 0.181 V/m

27. 21.2 nm

29. $1.44 \times 10^{3\circ}$C

31. (a) 31.5 n$\Omega\cdot$m   (b) 6.35 MA/m^2 (c) 49.9 mA
(d) 659 μm/s   (e) 0.400 V

33. 0.125

35. 67.6°C

37. 7.50 W

39. 28.9 Ω

41. 36.1%

43. (a) 5.97 V/m   (b) 74.6 W   (c) 66.1 W

45. 0.833 W

47. \$0.232

49. 26.9 cents/d

51. (a) 184 W   (b) 461°C

53. $\sim$\$1

55. (a) $Q/4C$ (b) $Q/4$ and $3Q/4$ (c) $Q^2/32C$ and $3Q^2/32C$ (d) $3Q^2/8C$

59. Experimental resistivity = 1.47 $\mu\Omega\cdot$m $\pm$ 4%, in agreement with 1.50 $\mu\Omega\cdot$m

61. (a) $(8.00\hat{\mathbf{i}})$ V/m   (b) 0.637 Ω (c) 6.28 A
(d) $(200\hat{\mathbf{i}})$ MA/m^2

63. 2 020°C

65. (a) 667 A   (b) 50.0 km

67. Material $\alpha' = \alpha/(1 - 20\alpha)$

Material	$\alpha' = \alpha/(1 - 20\alpha)$
Silver	$4.1 \times 10^{-3}/$°C
Copper	$4.2 \times 10^{-3}/$°C
Gold	$3.6 \times 10^{-3}/$°C
Aluminum	$4.2 \times 10^{-3}/$°C
Tungsten	$4.9 \times 10^{-3}/$°C
Iron	$5.6 \times 10^{-3}/$°C
Platinum	$4.25 \times 10^{-3}/$°C
Lead	$4.2 \times 10^{-3}/$°C
Nichrome	$0.4 \times 10^{-3}/$°C
Carbon	$-0.5 \times 10^{-3}/$°C
Germanium	$-24 \times 10^{-3}/$°C
Silicon	$-30 \times 10^{-3}/$°C

69. No. The fuses should pass no more than 3.87 A.

73. (b) 1.79 PΩ

75. (a) $\dfrac{\epsilon_0 \ell}{2d}(\ell + 2x + \kappa\ell - 2\kappa x)$

(b) $\dfrac{\epsilon_0 \ell v\, \Delta V\, (\kappa - 1)}{d}$ clockwise

CHAPTER 28

1. (a) 6.73 Ω (b) 1.97 Ω

3. (a) 4.59 Ω (b) 8.16%

5. 12.0 Ω

7. Circuit diagram:

power 73.8 W

9. (a) 227 mA   (b) 5.68 V

11. (a) 75.0 V   (b) 25.0 W, 6.25 W, and 6.25 W; 37.5 W

13. 1.00 kΩ

15. 14.2 W to 2 Ω, 28.4 W to 4 Ω, 1.33 W to 3 Ω, 4.00 W to 1 Ω

17. (a) $\Delta t_p = 2\,\Delta t/3$ (b) $\Delta t_s = 3\,\Delta t$

19. (a) $\Delta V_4 > \Delta V_3 > \Delta V_1 > \Delta V_2$
(b) $\Delta V_1 = \mathcal{E}/3$, $\Delta V_2 = 2\mathcal{E}/9$, $\Delta V_3 = 4\mathcal{E}/9$, $\Delta V_4 = 2\mathcal{E}/3$
(c) $I_1 > I_4 > I_2 = I_3$ (d) $I_1 = I$, $I_2 = I_3 = I/3$, $I_4 = 2I/3$
(e) I_4 increases while I_1, I_2, and I_3 decrease
(f) $I_1 = 3I/4$, $I_2 = I_3 = 0$, $I_4 = 3I/4$

21. 846 mA down in the 8-Ω resistor; 462 mA down in the middle branch; 1.31 A up in the right-hand branch

23. (a) -222 J and 1.88 kJ   (b) 687 J, 128 J, 25.6 J, 616 J, 205 J
(c) 1.66 kJ of chemical energy is transformed into internal energy

25. 50.0 mA from a to e

27. starter 171 A; battery 0.283 A

29. (a) 909 mA   (b) -1.82 V $= V_b - V_a$

31. (a) 5.00 s   (b) 150 μC   (c) 4.06 μA

33. $U_0/4$

37. (a) 6.00 V   (b) 8.29 μs

39. (a) 12.0 s   (b) $I(t) = (3.00\ \mu\text{A})\,e^{-t/12.0\ \text{s}}$;
$q(t) = (36.0\ \mu\text{C})\,(1 - e^{-t/12.0\ \text{s}})$

41. 0.302 Ω

43. 16.6 kΩ

45.

47. 145 Ω, 0.756 mA

49. (a) 12.5 A, 6.25 A, 8.33 A   (b) No; together they would require 27.1 A.

51. (a) $\sim 10^{-14}$ A   (b) $V_h/2 + (\sim 10^{-10}\ \text{V})$ and $V_h/2 - (\sim 10^{-10}\ \text{V})$, where V_h is the potential of the live wire, $\sim 10^2$ V

53. (a) either 3.84 Ω or 0.375 Ω (b) impossible

55. (a) $\mathcal{E}^2/3R$ (b) $3\mathcal{E}^2/R$ (c) in the parallel connection

57. (a) $R \rightarrow \infty$ (b) $R \rightarrow 0$ (c) $R = r$

59. 6.00 Ω; 3.00 Ω

61. (a) 4.40 Ω (b) 32.0 W, 9.60 W, 70.4 W   (c) 48.0 W

63. (a) $R \leq 1\,050\ \Omega$ (b) $R \geq 10.0\ \Omega$

65. (a) 9.93 μC   (b) 33.7 nA   (c) 334 nW   (d) 337 nW

67. (a) 40.0 W   (b) 80.0 V, 40.0 V, 40.0 V

69. (a) 0.991   (b) 0.648   (c) Insulation should be added to the ceiling.

71. (a) 0 in 3 kΩ and 333 μA in 12 kΩ and 15 kΩ (b) 50.0 μC
(c) $(278\ \mu\text{A})\,e^{-t/180\ \text{ms}}$ (d) 290 ms

73. (a) $\ln(\mathcal{E}/\Delta V) = (0.011\,8)t + 0.088\,2$ (b) 84.7 s, 8.47 μF

75. $q_1 = (240\ \mu\text{C})\,(1 - e^{-1\,000t/6})$; $q_2 = (360\ \mu\text{C})\,(1 - e^{-1\,000t/6})$

CHAPTER 29

1. (a) up   (b) out of the plane of the paper   (c) no deflection   (d) into the plane of the paper

3. negative z direction

5. $(-20.9\hat{\mathbf{j}})$ mT

7. 48.9° or 131°

9. 2.34 aN

11. 0.245 T east

13. (a) 4.73 N   (b) 5.46 N   (c) 4.73 N

15. 1.07 m/s

17. $2\pi rIB \sin\theta$ up

19. 2.98 μN west

21. 18.4 mA$\cdot$m^2

23. 9.98 N$\cdot$m clockwise as seen looking down from above

27. (a) 118 μN$\cdot$m   (b) $-118\ \mu\text{J} \leq U \leq 118\ \mu\text{J}$

29. (a) 49.6 aN south   (b) 1.29 km

31. 115 keV

33. $r_\alpha = r_d = \sqrt{2}\,r_p$

35. 4.98×10^8 rad/s

37. 7.88 pT

39. $m = 2.99$ u, either $^3_1\text{H}^+$ or $^3_2\text{He}^+$

41. (a) 8.28 cm   (b) 8.23 cm; ratio is independent of both ΔV and B

43. (a) 4.31×10^7 rad/s   (b) 51.7 Mm/s

45. (a) 7.66×10^7 rad/s   (b) 26.8 Mm/s   (c) 3.76 MeV
(d) 3.13×10^3 rev   (e) 257 μs

47. 70.1 mT

49. 1.28×10^{29} m^{-3}, 1.52

51. 43.3 μT

53. (a) The electric current experiences a magnetic force.

55. (a) -8.00×10^{-21} kg$\cdot$m/s   (b) 8.90°

57. (a) $(3.52\hat{\mathbf{i}} - 1.60\hat{\mathbf{j}})$ aN   (b) 24.4°

59. $(2\pi/d)(2m_e\,\Delta V/e)^{1/2}$

61. 0.588 T

63. 0.713 A counterclockwise as seen from above

65. 438 kHz

67. 3.70×10^{-24} N·m

69. (a) 0.501 m (b) 45.0°

71. (a) 1.33 m/s (b) Positive ions moving toward you in magnetic field to the right feel upward magnetic force, and migrate upward in the blood vessel. Negative ions moving toward you feel downward magnetic force and accumulate at the bottom of this section of vessel. Thus both species can participate in the generation of the emf.

CHAPTER 30

1. 12.5 T

3. (a) 28.3 μT into the paper (b) 24.7 μT into the paper

5. $\dfrac{\mu_0 I}{4\pi x}$ into the paper

7. 26.2 μT into the paper

9. (a) $2I_1$ out of the page (b) $6I_1$ into the page

11. (a) along the line $(y = -0.420$ m, $z = 0)$
  (b) $(-34.7\hat{\mathbf{j}})$ mN (c) $(17.3\hat{\mathbf{j}})$ kN/C

13. (a) $4.5\,\dfrac{\mu_0 I}{\pi L}$ (b) stronger

15. $(-13.0\hat{\mathbf{j}})\ \mu$T

17. $(-27.0\hat{\mathbf{i}})\ \mu$N

19. (a) 12.0 cm to the left of wire 1 (b) 2.40 A, downward

21. 20.0 μT toward the bottom of the page

23. 200 μT toward the top of the page; 133 μT toward the bottom of the page

25. (a) 6.34 mN/m inward (b) greater

27. (a) 0 (b) $\dfrac{\mu_0 I}{2\pi R}$ tangent to the wall in a

  counterclockwise sense (c) $\dfrac{\mu_0 I^2}{(2\pi R)^2}$ inward

29. (a) $\frac{1}{3}\mu_0 b r_1^2$ (b) $\dfrac{\mu_0 b R^3}{3r_2}$

31. 31.8 mA

33. 226 μN away from the center of the loop, 0

35. (a) 3.13 mWb (b) 0

37. (a) 11.3 GV·m/s (b) 0.100 A

39. (a) 9.27×10^{-24} A·m² (b) down

41. 0.191 T

43. 2.62 MA/m

45. (b) 6.45×10^4 K·A/T·m

47. (a) 8.63×10^{45} electrons (b) 4.01×10^{20} kg

49. $\dfrac{\mu_0 I}{2\pi w}\ln\left(1 + \dfrac{w}{b}\right)\hat{\mathbf{k}}$

51. 12 layers, 120 m

53. 143 pT away along the axis

59. (a) 2.46 N up (b) 107 m/s² up

61. (a) 274 μT (b) $(-274\hat{\mathbf{j}})\ \mu$T (c) $(1.15\hat{\mathbf{i}})$ mN
  (d) $(0.384\hat{\mathbf{i}})$ m/s² (e) acceleration is constant
  (f) $(0.999\hat{\mathbf{i}})$ m/s

63. 81.7 A

65. $\dfrac{\mu_0 I_1 I_2 L}{\pi R}$ to the right

69. $\dfrac{\mu_0 I}{4\pi}(1 - e^{-2\pi})$ out of the plane of the paper

71. $\frac{1}{3}\rho\mu_0\omega R^2$

73. (a) $\dfrac{\mu_0 I(2r^2 - a^2)}{\pi r(4r^2 - a^2)}$ to the left (b) $\dfrac{\mu_0 I(2r^2 + a^2)}{\pi r(4r^2 + a^2)}$ toward
  the top of the page

CHAPTER 31

1. 500 mV

3. 9.82 mV

5. 160 A

7. (a) 1.60 A counterclockwise (b) 20.1 μT (c) up

9. (a) $(\mu_0 IL/2\pi)\ln(1 + w/h)$ (b) $-4.80\ \mu$V; current is
  counterclockwise

11. 283 μA upward

13. $(68.2$ mV$)\,e^{-1.6t}$, tending to produce counterclockwise
  current

15. 272 m

17. $(0.422$ V$)\cos \omega t$

19. (a) eastward (b) 458 μV

21. (a) 3.00 N to the right (b) 6.00 W

23. 360 T

25. (a) 233 Hz (b) 1.98 mV

27. 2.83 mV

29. (a) $F = N^2 B^2 w^2 v/R$ to the left (b) 0 (c) $F = N^2 B^2 w^2 v/R$
  to the left

31. 145 μA

33. 1.80 mN/C upward and to the left, perpendicular to r_1

35. (a) 7.54 kV (b) The plane of the coil is parallel to **B**.

37. $(28.6$ mV$)\sin(4\pi t)$

39. (a) 110 V (b) 8.53 W (c) 1.22 kW

41. (a) $(8.00$ mWb$)\cos(377t)$ (b) $(3.02$ V$)\sin(377t)$
  (c) $(3.02$ A$)\sin(377t)$ (d) $(9.10$ W$)\sin^2(377t)$
  (e) $(24.1$ mN·m$)\sin^2(377t)$

43. (b) Larger R makes current smaller, so the loop must travel faster to maintain equality of magnetic force and weight. (c) The magnetic force is proportional to the product of field and current, while the current is itself proportional to field. If B becomes two times smaller, the speed must become four times larger to compensate.

45. $(-2.87\hat{\mathbf{j}} + 5.75\hat{\mathbf{k}})$ Gm/s²

47. (a) Doubling N doubles the amplitude. (b) Doubling ω doubles the amplitude and halves the period. (c) Doubling ω and halving N leaves the amplitude the same and cuts the period in half.

49. 62.3 mA down through 6.00 Ω, 860 mA down through 5.00 Ω, 923 mA up through 3.00 Ω

51. $\sim 10^{-4}$ V, by reversing a 20-turn coil of diameter 3 cm in 0.1 s in a field of 10^{-3} T

53. (a) 254 km/s   (b) 215 V

55. 1.20 μC

57. (a) 0.900 A   (b) 0.108 N   (c) b (d) no

59. (a) $a\pi r^2$ (b) $-b\pi r^2$ (c) $-b\pi r^2/R$ (d) $b^2\pi^2 r^4/R$

61. (a) 36.0 V   (b) 600 mWb/s   (c) 35.9 V   (d) 4.32 N·m

65. 6.00 A

67. (a) $(1.19\ \text{V})\cos(120\pi t)$ (b) 88.5 mW

71. $(-87.1\ \text{mV})\cos(200\pi t + \phi)$

CHAPTER 32

1. 19.5 mV

3. 100 V

5. $(18.8\ \text{V})\cos(377t)$

7. -0.421 A/s

9. (a) 188 μT   (b) 33.3 nT·m^2 (c) 0.375 mH   (d) B and Φ_B are proportional to current; L is independent of current

11. 0.750 m

13. $\mathcal{E}_0/k^2 L$

15. (a) 0.139 s   (b) 0.461 s

17. (a) 2.00 ms   (b) 0.176 A   (c) 1.50 A   (d) 3.22 ms

19. (a) 0.800   (b) 0

21. (a) 6.67 A/s   (b) 0.332 A/s

23. $(500\ \text{mA})(1 - e^{-10t/s})$, 1.50 A $- (0.25\ \text{A})\,e^{-10t/s}$

25. 0 for $t < 0$; $(10\ \text{A})(1 - e^{-10\,000t})$ for $0 < t < 200$ μs; $(63.9\ \text{A})\,e^{-10\,000t}$ for $t > 200$ μs

27. (a) 5.66 ms   (b) 1.22 A   (c) 58.1 ms

29. 0.064 8 J

31. 2.44 μJ

33. 44.2 nJ/m^3 for the **E**-field and 995 μJ/m^3 for the **B**-field

35. (a) 0.500 J   (b) 17.0 W   (c) 11.0 W

37. 2.27 mT

39. 1.73 mH

41. 80.0 mH

43. (a) 18.0 mH   (b) 34.3 mH   (c) -9.00 mV

45. $(L_1 L_2 - M^2)/(L_1 + L_2 - 2M)$

47. 20.0 V

49. 608 pF

51. (a) 135 Hz   (b) 119 μC   (c) -114 mA

53. (a) 6.03 J   (b) 0.529 J   (c) 6.56 J

55. (a) 4.47 krad/s   (b) 4.36 krad/s   (c) 2.53%

57. $L = 199$ mH; $C = 127$ nF

59. (b) $\mu_0 J_s^2/2$ away from the other sheet   (c) $\mu_0 J_s$ and zero   (d) $\mu_0 J_s^2/2$

61. (a) -20.0 mV   (b) $-(10.0\ \text{MV/s}^2)\,t^2$ (c) 63.2 μs

63. $(Q/2N)(3L/C)^{1/2}$

65. (a) $L \approx (\pi/2)N^2\mu_0 R$ (b) ~ 100 nH   (c) ~ 1 ns

71. (a) 72.0 V; b
      (b)

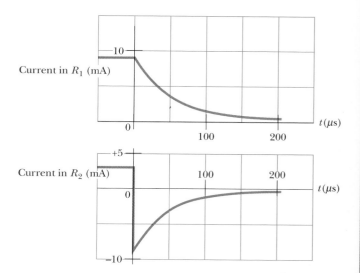

      (c) 75.2 μs

73. 300 Ω

75. (a) It creates a magnetic field.   (b) The long narrow rectangular area between the conductors encloses all of the magnetic flux.

77. (a) 62.5 GJ   (b) 2 000 N

79. (a) 2.93 mT up   (b) 3.42 Pa   (c) clockwise as seen from above   (d) up   (e) 1.30 mN

CHAPTER 33

1. $\Delta v(t) = (283\ \text{V})\sin(628t)$

3. 2.95 A, 70.7 V

5. 14.6 Hz

7. 3.38 W

9. (a) 42.4 mH   (b) 942 rad/s

11. 5.60 A

13. 0.450 Wb

15. (a) 141 mA   (b) 235 mA

17. 100 mA

19. (a) 194 V   (b) current leads by 49.9°

21. (a) 78.5 Ω   (b) 1.59 kΩ   (c) 1.52 kΩ   (d) 138 mA
      (e) $-84.3°$

23. (a) 17.4°   (b) voltage leads the current

25. 1.88 V

27.

(b)

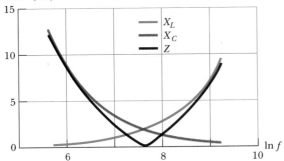

29. (a) either 123 nF or 124 nF   (b) 51.5 kV

31. 8.00 W

33. (a) 16.0 Ω   (b) -12.0 Ω

35. $\sqrt{\dfrac{800\,\rho \mathscr{P} d}{\pi (\Delta V)^2}}$

37. 1.82 pF

39. (a) 633 fF   (b) 8.46 mm   (c) 25.1 Ω

41. 242 mJ

43. 0.591 and 0.987; the circuit in Problem 23

45. 687 V

47. 87.5 Ω

49. (a) 29.0 kW   (b) 5.80×10^{-3} (c) If the generator were limited to 4 500 V, no more than 17.5 kW could be delivered to the load, never 5 000 kW.

51. (b) 0; 1   (c) $f_h = (10.88RC)^{-1}$

53. (a) 613 μF   (b) 0.756

55. (a) 580 μH and 54.6 μF   (b) 1   (c) 894 Hz   (d) ΔV_{out} leads ΔV_{in} by 60.0° at 200 Hz. ΔV_{out} and ΔV_{in} are in phase at 894 Hz. ΔV_{out} lags ΔV_{in} by 60.0° at 4 000 Hz.   (e) 1.56 W, 6.25 W, 1.56 W   (f) 0.408

57. 56.7 W

59. 99.6 mH

61. (a) 225 mA   (b) 450 mA

63. (a) 1.25 A   (b) Current lags voltage by 46.7°.

65. (a) 200 mA; voltage leads by 36.8°   (b) 40.0 V; $\phi = 0°$ (c) 20.0 V; $\phi = -90.0°$ (d) 50.0 V; $\phi = +90.0°$

67. (b) 31.6

71.

f (Hz)	X_L (Ω)	X_C (Ω)	Z (Ω)
300	283	12 600	12 300
600	565	6 280	5 720
800	754	4 710	3 960
1 000	942	3 770	2 830
1 500	1 410	2 510	1 100
2 000	1 880	1 880	40.0
3 000	2 830	1 260	1 570
4 000	3 770	942	2 830
6 000	5 650	628	5 020
10 000	9 420	377	9 040

73. (a) 1.84 kHz
    (b)

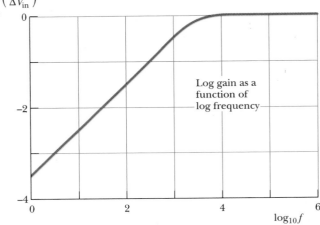

CHAPTER 34

1. (a) $(3.15\hat{\mathbf{j}})$ kN/C   (b) $(525\hat{\mathbf{k}})$ nT   (c) $(-483\hat{\mathbf{j}})$ aN

3. 2.25×10^8 m/s

5. (a) 6.00 MHz   (b) $(-73.3\hat{\mathbf{k}})$ nT
    (c) $\mathbf{B} = [(-73.3\hat{\mathbf{k}})$ nT$]\cos(0.126x - 3.77 \times 10^7 t)$

7. (a) 0.333 μT   (b) 0.628 μm   (c) 477 THz

9. 75.0 MHz

11. 3.33 μJ/m^3

13. 307 μW/m^2

15. 3.33×10^3 m^2

17. (a) 332 kW/m^2 radially inward   (b) 1.88 kV/m and 222 μT

19. (a) $\mathbf{E} \cdot \mathbf{B} = 0$ (b) $(11.5\hat{\mathbf{i}} - 28.6\hat{\mathbf{j}})$ W/m^2

21. 29.5 nT

23. (a) 2.33 mT   (b) 650 MW/m^2 (c) 510 W

25. (a) 540 V/m   (b) 2.58 μJ/m^3 (c) 773 W/m^2 (d) 77.3% of the intensity in Example 34.5

27. 83.3 nPa

29. (a) 1.90 kN/C   (b) 50.0 pJ   (c) 1.67×10^{-19} kg·m/s

31. (a) 11.3 kJ   (b) 1.13×10^{-4} kg·m/s

33. (a) 134 m   (b) 46.9 m

35. (a) away along the perpendicular bisector of the line segment joining the antennas   (b) along the extensions of the line segment joining the antennas

37. (a) $\mathbf{E} = \frac{1}{2}\mu_0 c J_{max}[\cos(kx - \omega t)]\hat{\mathbf{j}}$

   (b) $\mathbf{S} = \frac{1}{4}\mu_0 c J_{max}^2[\cos^2(kx - \omega t)]\hat{\mathbf{i}}$

   (c) $I = \dfrac{\mu_0 c J_{max}^2}{8}$ (d) 3.48 A/m

39. 545 THz

41. (a) 6.00 pm   (b) 7.50 cm

43. 60.0 km

45. 1.00 Mm = 621 mi; not very practical

47. (a) 3.77×10^{26} W   (b) 1.01 kV/m and 3.35 μT

49. (a) $2\pi^2 r^2 f B_{max} \cos\theta$, where θ is the angle between the magnetic field and the normal to the loop   (b) The loop should be in the vertical plane containing the line of sight to the transmitter.

51. (a) 6.67×10^{-16} T   (b) 5.31×10^{-17} W/m^2
   (c) 1.67×10^{-14} W   (d) 5.56×10^{-23} N

53. 95.1 mV/m

55. (a) $B_{max} = 583$ nT, $k = 419$ rad/m, $\omega = 126$ Grad/s; **B** vibrates in xz plane   (b) $\mathbf{S}_{av} = (40.6\hat{\mathbf{i}})$ W/m^2
   (c) 271 nPa   (d) $(406\hat{\mathbf{i}})$ nm/s^2

57. (a) 22.6 h   (b) 30.6 s

59. (a) 8.32×10^7 W/m^2 (b) 1.05 kW

61. (a) 1.50 cm   (b) 25.0 μJ   (c) 7.37 mJ/m^3
   (d) 40.8 kV/m, 136 μT   (e) 83.3 μN

63. 637 nPa

65. $\epsilon_0 E^2 A/2m$

67. (a) 16.1 cm   (b) 0.163 m^2 (c) 470 W/m^2 (d) 76.8 W
   (e) 595 N/C   (f) 1.98 μT   (g) The cats are nonmagnetic and carry no macroscopic charge or current. Oscillating charges within molecules make them emit infrared radiation.   (h) 119 W

69. 4.77 Gm

CHAPTER 35

1. 299.5 Mm/s

3. 114 rad/s

5. (c) 0.055 7 °

9. 23.3°

11. 15.4°; 2.56 m

13. 19.5° above the horizon

15. (a) 1.52   (b) 417 nm   (c) 474 THz   (d) 198 Mm/s

17. 158 Mm/s

19. 30.0° and 19.5° at entry; 19.5° and 30.0° at exit

21. 3.88 mm

23. 30.4° and 22.3°

25. $\sim 10^{-11}$ s; between 10^3 and 10^4 wavelengths

29. 0.171°

31. 86.8°

33. 27.9°

35. 4.61°

37. (a) 33.4°   (b) 53.4°   (c) There is no critical angle.

39. 1.000 08

41. 1.08 cm $< d <$ 1.17 cm

43. Skylight incident from above travels down the plastic. If the index of refraction of the plastic is greater than 1.41, the rays close in direction to the vertical are totally reflected from the side walls of the slab and from both facets at the bottom of the plastic, where it is not immersed in gasoline. This light returns up inside the plastic and makes it look bright. Where the plastic is immersed in gasoline, total internal reflection is frustrated and the downward-propagating light passes from the plastic out into the gasoline. Little light is reflected up, and the gauge looks dark.

45. Scattered light leaving the photograph in all forward horizontal directions in air is gathered by refraction into a fan in the water of half-angle 48.6°. At larger angles you see things on the other side of the globe, reflected by total internal reflection at the back surface of the cylinder.

47. 77.5°

49. 2.27 m

51. (a) 0.172 mm/s   (b) 0.345 mm/s   (c) northward at 50.0° below the horizontal   (d) northward at 50.0° below the horizontal

53. 62.2%

55. 82 reflections

57. (b) 68.5%

59. 27.5°

61. (a) It always happens.   (b) 30.3°   (c) It cannot happen.

63. 2.36 cm

67. 1.93

69. (a) $n = [1 + (4t/d)^2]^{1/2}$ (b) 2.10 cm   (c) violet

71. (a) 1.20   (b) 3.40 ns

CHAPTER 36

1. $\sim 10^{-9}$ s younger

3. 35.0 in.

5. 10.0 ft, 30.0 ft, 40.0 ft

7. (a) 13.3 cm, -0.333, real and inverted   (b) 20.0 cm, -1.00, real and inverted   (c) no image is formed

9. (a) -12.0 cm; 0.400   (b) -15.0 cm; 0.250   (c) upright

11. (a) $q = 45.0$ cm; $M = -0.500$ (b) $q = -60.0$ cm; $M = 3.00$ (c) Image (a) is real, inverted, and diminished. Image (b) is virtual, upright, and enlarged. The ray diagrams are like Figures 36.15a and 36.15b, respectively.

13. At 0.708 cm in front of the reflecting surface. Image is virtual, upright, and diminished.

15. 7.90 mm

17. (a) a concave mirror with radius of curvature 2.08 m
   (b) 1.25 m from the object

19. (a) 25.6 m   (b) 0.058 7 rad   (c) 2.51 m   (d) 0.023 9 rad
   (e) 62.8 m from your eyes

21. 38.2 cm below the top surface of the ice

23. 8.57 cm

25. (a) 45.0 cm   (b) −90.0 cm   (c) −6.00 cm

27. 1.50 cm/s

29. (a) 16.4 cm   (b) 16.4 cm

31. (a) 650 cm from the lens on the opposite side from the object; real, inverted, enlarged   (b) 600 cm from the lens on the same side as the object; virtual, upright, enlarged

33. 2.84 cm

37. (a) − 12.3 cm, to the left of the lens   (b) 0.615
   (c)

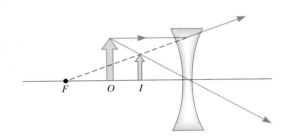

39. (a) 7.10 cm   (b) 0.074 0 mm   (c) 23.3 MW/m²

41. (a) $p = \dfrac{d}{2} \pm \sqrt{\dfrac{d^2}{4} - fd}$ (b) Both images are real and inverted. One is enlarged, the other diminished.

43. 1.24 cm

45. 21.3 cm

47. − 4.00 diopters, a diverging lens

49. − 3.70 diopters

51. − 575

53. (a) − 800   (b) image is inverted

55. (a) virtual   (b) infinity   (c) 15.0 cm, − 5.00 cm

57. − 40.0 cm

59. (a) 23.1 cm   (b) 0.147 cm

61. (a) 67.5 cm   (b) The lenses can be displaced in two ways. The first lens can be displaced 1.28 cm farther away from the object, and the second lens 17.7 cm toward the object. Alternatively, the first lens can be displaced 0.927 cm toward the object and the second lens 4.44 cm toward the object.

63. $q = 5.71$ cm; real

65. 0.107 m to the right of the vertex of the hemispherical face

67. 8.00 cm

Ray diagram:

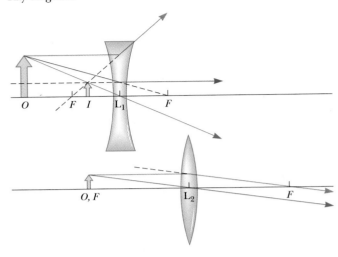

69. 1.50 m in front of the mirror; 1.40 cm (inverted)

71. (a) 30.0 cm and 120 cm   (b) 24.0 cm   (c) real, inverted, diminished with $M = - 0.250$

73. − 75.0

75. (a) 44.6 diopters   (b) 3.03 diopters

77. (a) 20.0 cm to the right of the second lens, − 6.00
   (b) inverted   (c) 6.67 cm to the right of the second lens, − 2.00, inverted

CHAPTER 37

1. 1.58 cm

3. (a) 55.7 m   (b) 124 m

5. 1.54 mm

7. (a) 2.62 mm   (b) 2.62 mm

9. 11.3 m

11. (a) 10.0 m   (b) 516 m   (c) Only the runway centerline is a maximum for the interference patterns for both frequencies. If the frequencies were related by a ratio of small integers k/ℓ, the plane could by mistake fly along the kth side maximum of one signal where it coincides with the ℓth side maximum of the other.

13. (a) 13.2 rad   (b) 6.28 rad   (c) 0.012 7 degree
   (d) 0.059 7 degree

15. (a) 1.93 μm   (b) 3.00λ   (c) maximum

17. 48.0 μm

19. (a) 7.95 rad   (b) 0.453

21. (a) and (b) 19.7 kN/C at 35.0°   (c) 9.36 kN/C at 169°

23. 10.0 sin(100πt + 0.927)

25. 26.2 sin(ωt + 36.6°)

27. π/2

29. 360°/N

31. (a) green   (b) violet

33. 0.500 cm

35. no reflection maxima in the visible spectrum

37. 290 nm

39. 4.35 μm

41. 39.6 μm

43. $1 + N\lambda/2L$

45. 1.25 m

47. (a) $\sim 10^{-3}$ degree   (b) $\sim 10^{11}$ Hz, microwave

49. 20.0×10^{-6} °C^{-1}

51. 3.58°

53. 1.62 km

55. 421 nm

59. (a) $2(4h^2 + d^2)^{1/2} - 2d$ (b) $(4h^2 + d^2)^{1/2} - d$

61. $y' = (n - 1)tL/d$

63. (a) 70.6 m   (b) 136 m

65. 1.73 cm

67. (a) 4.86 cm from the top   (b) 78.9 nm and 128 nm
(c) 2.63×10^{-6} rad

69. 0.505 mm

CHAPTER 38

1. 4.22 mm

3. 0.230 mm

5. three maxima, at 0° and near 46° on both sides

7. 51.8 μm wide and 949 μm high

9. 0.016 2

11. 1.00 mrad

13. 3.09 m

15. violet; between 186 m and 271 m

17. 13.1 m

19. Neither. It can resolve objects no closer than several centimeters apart.

21. 0.244 rad = 14.0°

23. 7.35°

25. 5.91° in first order, 13.2° in second order, 26.5° in third order

27. (a) 478.7 nm, 647.6 nm, and 696.6 nm   (b) 20.51°, 28.30°, and 30.66°

29. (a) 12 000, 24 000, 36 000   (b) 11.1 pm

31. (a) 2 800 grooves   (b) 4.72 μm

33. (a) 5 orders   (b) 10 orders in the short-wavelength region

35. 93.4 pm

37. 14.4°

39. 5.51 m, 2.76 m, 1.84 m

41. (a) 54.7°   (b) 63.4°   (c) 71.6°

43. 1.11

45. 60.5°

47. 36.9° above the horizon

49. (a) 6   (b) 7.50°

51. 632.8 nm

53. (a) 25.6°   (b) 19.0°

55. 545 nm

57. (a) 3.53×10^3 grooves/cm   (c) Eleven maxima

59. 4.58 μm $< d <$ 5.23 μm

61. 15.4

63. (a) 41.8°   (b) 0.593   (c) 0.262 m

67. (b) 3.77 nm/cm

69. (b) 15.3 μm

71. $\phi = 1.391\ 557\ 4$ after seventeen steps or fewer

73. $a = 99.5$ μm $\pm 1\%$

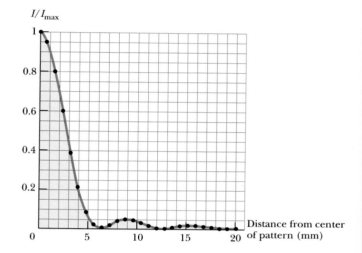

CHAPTER 39

5. $0.866c$

7. (a) 64.9/min   (b) 10.6/min

9. 1.54 ns

11. $0.800c$

13. (a) 39.2 μs   (b) accurate to one digit

15. (a) 20.0 m   (b) 19.0 m   (c) $0.312c$

17. (a) 21.0 yr   (b) 14.7 ly   (c) 10.5 ly   (d) 35.7 yr

19. (c) 2.00 kHz   (d) $\pm 0.075\ 0$ m/s ≈ 0.2 mi/h

21. $0.220c = 6.59 \times 10^7$ m/s

23. (a) 17.4 m   (b) 3.30°

25. (a) 2.50×10^8 m/s   (b) 4.97 m   (c) -1.33×10^{-8} s

27. $0.960c$

29. (a) 2.73×10^{-24} kg·m/s   (b) 1.58×10^{-22} kg·m/s
(c) 5.64×10^{-22} kg·m/s

31. 4.50×10^{-14}

33. $0.285c$

35. (a) 5.37×10^{-11} J   (b) 1.33×10^{-9} J

37. 1.63×10^3 MeV/c

39. (a) 938 MeV   (b) 3.00 GeV   (c) 2.07 GeV

41. 8.84×10^{-28} kg and 2.51×10^{-28} kg

45. (a) 3.91×10^4 (b) $u = 0.999\ 999\ 999\ 7c$ (c) 7.67 cm

47. 4.08 MeV and 29.6 MeV

49. $\sim 10^{-15}$

51. 0.842 kg

53. 4.19×10^9 kg/s

55. (a) 26.6 Mm   (b) 3.87 km/s   (c) -8.34×10^{-11}
(d) 5.29×10^{-10} (e) $+4.46 \times 10^{-10}$

57. (a) a few hundred seconds   (b) $\sim 10^8$ km

59. (a) $0.800c$ (b) 7.50 ks   (c) 1.44 Tm, $-0.385c$ (d) 4.88 ks

61. 0.712%

63. (a) $0.946c$ (b) 0.160 ly   (c) 0.114 yr   (d) 7.50×10^{22} J

65. (a) 76.0 min   (b) 52.1 min

67. yes, with 18.8 m to spare

69. (b) For u small compared to c, the relativistic expression agrees with the classical expression. As u approaches c, the acceleration approaches zero, so that the object can never reach or surpass the speed of light.
(c) Perform $\int (1 - u^2/c^2)^{-3/2} du = (qE/m) \int dt$ to obtain $u = qEct(m^2c^2 + q^2E^2t^2)^{-1/2}$ and then $\int dx = \int qEct(m^2c^2 + q^2E^2t^2)^{-1/2} dt$ to obtain $x = (c/qE)[(m^2c^2 + q^2E^2t^2)^{1/2} - mc]$

75. 1.82×10^{-3} eV

Credits

Photographs

This page constitutes an extension of the copyright page. We have made every effort to trace the ownership of all copyrighted material and to secure permission from copyright holders. In the event of any question arising as to the use of any material, we will be pleased to make the necessary corrections in future printings. Thanks are due to the following authors, publishers, and agents for permission to use the material indicated.

Chapter 1. **xxiv:** Courtesy of NASA **2:** elektraVision/Index Stock Imagery **6:** top left, Courtesy of National Institute of Standards and Technology, U.S. Dept. of Commerce; top right, Courtesy of National Institute of Standards and Technology, U.S. Dept. of Commerce **13:** Phil Boorman/Getty Images **19:** Sylvain Grandadam/Photo Researchers, Inc.

Chapter 2. **23:** George Lepp/Getty Images **26:** Ken White/Allsport/Getty Images **46:** North Wind Picture Archive **52:** bottom left, Courtesy U.S. Air Force; bottom right, Photri, Inc. **53:** George Semple **55:** Courtesy Amtrak Nec Media Relations

Chapter 3. **58:** Mark Wagner/Getty Images

Chapter 4. **77:** © Arndt/Premium Stock/PictureQuest **79:** Mark C. Burnett/Photo Researchers, Inc. **84:** The Telegraph Colour Library/Getty Images **86:** Mike Powell/Allsport/Getty Images **88:** Central Scientific Company **103:** bottom left, McKinney/Getty Images **104:** top left, Jed Jacobsohn/Allsport/Getty Images; top right, Bill Lee/Dembinsky Photo Associates; bottom left, Sam Sargent/Liaison International; bottom right, Courtesy of NASA **106:** Courtesy of NASA

Chapter 5. **111:** © Steve Raymer/CORBIS **114:** Giraudon/Art Resource **119:** Courtesy of NASA **120:** John Gillmoure/corbisstockmarket.com **122:** © John Elk III/Stock, Boston Inc./PictureQuest **139:** Roger Violet, Mill Valley, CA, University Science Books, 1992

Chapter 6. **150:** © Paul Hardy/CORBIS **151:** Mike Powell/Allsport/Getty Images **152:** © Tom Carroll/Index Stock Imagery/PictureQuest **157:** Robin Smith/Getty Images **165:** Jump Run Productions/Getty Images **166:** Charles D. Winters **174:** Frank Cezus/Getty Images **178:** Color Box/Getty Images

Chapter 7. **181:** Billy Hustace/Getty Images **183:** all, Charles D. Winters **184:** Gerard Vandystadt/Photo Researchers, Inc. **198:** all except top right, George Semple **198:** top right, Digital Vision/Getty Images **213:** Ron Chapple/Getty Images

Chapter 8. **217:** Harold E. Edgerton/Courtesy of Palm Press, Inc. **242:** Gamma **250:** Engraving from Scientific American, July 1888

Chapter 9. **251:** Mark Cooper/corbisstockmarket.com **257:** Courtesy of Saab **259:** top, Harold & Esther Edgerton Foundation 2002, courtesy of Palm Press, Inc; bottom, © Tim Wright/CORBIS **263:** No credit available **264:** Courtesy of Central Scientific Company **276:** Richard Megna/Fundamental Photographs **278:** Courtesy of NASA **279:** Courtesy of NASA **281:** Bill Stormont/corbisstockmarket.com **286:** Eye Ubiquitous/CORBIS

Chapter 10. **292:** Courtesy Tourism Malaysia **299:** George Semple **316:** Henry Leap and Jim Lehman **323:** Bruce Ayers/Getty Images **327:** John Lawrence/Getty Images **329:** Jerry Wachter/Photo Researchers, Inc.

Chapter 11. **336:** Otto Gruele/Getty Images **346:** both, © 1998 David Madison **354:** Gerard Lacz/NHPA

Chapter 12. **362:** John W. Jewett, Jr. **366:** Charles D. Winters

Chapter 13. **389:** University of Arizona/JPL/NASA **394:** Courtesy of NASA **410:** Courtesy H. Ford, et al., & NASA **417:** Courtesy of NASA

Chapter 14. **420:** Austin MacRae **422:** Earl Young/Getty Images **424:** right, David Frazier **427:** © Hulton-Deutsch Collection/CORBIS **430:** Geraldine Prentice/Getty Images **431:** Andy Sachs/Getty Images **431:** Werner Wolff/stockphoto.com **432:** George Semple **434:** Bettmann/CORBIS **435:** Courtesy of Central Scientific Company **438:** Galen Rowell/Peter Arnold, Inc. **439:** bottom left, Pamela Zilly/Getty Images **439:** top right, Henry Leap and Jim Lehman **439:** bottom right, Henry Leap and Jim Lehman **444:** George Semple **445:** Stan Osolinski/Dembinsky Photo Associates **447:** The Granger Collection **448:** Courtesy of Jeanne Maier

Chapter 15. **450:** Don Bonsey/Getty Images **452:** both, www.comstock.com **465:** top left, Courtesy of Ford Motor Company; bottom left, © Link/Visuals Unlimited **474:** both, UPI-Bettmann/CORBIS **478:** Telegraph Colour Library/Getty Images **479:** George Semple

Chapter 16. **486:** Kathy Ferguson Johnson/PhotoEdit/PictureQuest **509:** Gregg Adams/Getty Images

Chapter 17. **512:** Getty Images **518:** Courtesy Kenneth Burger Museum Archives/Kent State University **524:** Courtesy of the Educational Development Center, Newton, MA **527:** © 1973 Kim Vandiver and Harold E. Edgerton/Courtesy of Palm Press, Inc. **528:** www.comstock.com

Bohr Library. **1110:** David Parker/Science Photo Library/Photo Researchers, Inc. **1113:** Courtesy of Henry Leap and Jim Lehman **1114:** left, Dennis O'Clair/Getty Images; right, Hank Morgan/Photo Researchers, Inc. **1117:** Courtesy of U.S. Air Force, Langley Air Force Base **1120:** Roy Atkeson Photo Library **1122:** Courtesy Edwin Lo

Chapter 36. **1126:** Don Hammond/CORBIS **1128:** George Semple **1130:** Courtesy of Henry Leap/Jim Lehman **1131:** Ken Kay/Fundamental Photographs **1133:** top, Henry Leap/Jim Lehman **1133:** Courtesy of Thompson Consumer Electronics **1135:** David Rogers **1136:** NASA **1138:** © 1990 Paul Silverman/Fundamental Photographs **1144:** Henry Leap/Jim Lehman **1149:** Henry Leap/Jim Lehman **1155:** From Lennart Nilsson, in collaboration with Jan Lindberg, *Behold Man: A Photographic Journey of Discovery Inside the Body,* Boston, Little, Brown & Co. 1974. **1160:** George Semple **1161:** © Tony Freeman/PhotoEdit **1163:** © Tony Freeman/PhotoEdit **1164:** Orion® Sky View Pro™ 6 EQ Reflector, Orion Telescopes & Binoculars. Copyright © 2003 OrionTelescopes & Binoculars. All rights reserved. "Orion" is the trademark and service mark of Optronic Technologies, Inc. 800-676-1343, www.telescope.com **1166:** © 2003 by Sidney Harris. All rights reserved. **1166:** Richard Megna/Fundamental Photographs **1167:** George Semple **1167:** M.C. Escher/Cordon Art-Baarn-Holland. All Rights Reserved. **1169:** THE FAR SIDE © 1985 FARWORKS, INC. Used by permission. All rights reserved. **1174:** © Michael Levin/Opti-Gone Associates

Chapter 37. **1176:** RO-MA/Index Stock Imagery **1177:** © Kolar, Richard/Animals Animals/Earth Scenes **1178:** Courtesy of Sabina Zigman/Benjamin Cardozo High School, and by permission of *The Physics Teacher,* vol. 37, January 1999, p. 55. **1179:** Photograph from M. Cagnet, M. Francon, and J.C. Thierr, *Atlas of Optical Phenomena,* Berlin, Springer-Verlag, 1962. **1179:** Richard Megna/Fundamental Photographs **1184:** Photograph from M. Cagnet, M. Francon, and J.C. Thierr, *Atlas of Optical Phenomena,* Berlin, Springer-Verlag, 1962. **1191:** left, Dr. Jeremy Burgess/Science Photo Library/Photo Researchers, Inc. **1191:** right, Peter Aprahamian/Science Photo Library/Photo Researchers, Inc. **1191:** Courtesy Bausch and Lomb Optical Company **1192:** From Physical Science Study Committee, College Physics, Lexington, MA, Heath, 1968. **1193:** Kristen Brochmann/Fundamental Photographs **1194:** Richard Megna/Fundamental Photographs **1196:** LIGO Hanford Observatory **1201:** Courtesy of Central Scientific Supply

Chapter 38. **1205:** Denis Scott/CORBIS **1207:** right, P.M. Rinard, *Am. J. Phys.* 44:70, 1976 **1207:** From M. Cagnet, M. Francon, and J.C. Thierr, *Atlas of Optical Phenomena,* Berlin, Springer-Verlag, 1962, plate 18. **1212:** From M. Cagnet, M. Francon, and J.C. Thierr, *Atlas of Optical Phenomena,* Berlin, Springer-Verlag, 1962, plate 18. **1213:** Photograph courtesy of Central Scientific Company **1215:** From M. Cagnet, M. Francon, and J.C. Thierr, *Atlas of Optical Phenomena,* Berlin, Springer-Verlag, 1962, plate 16. **1218:** left, U.S. Naval Observatory/James W. Christy; right, Photo courtesy of Gemini Observatory **1220:** Courtesy Silicon Light Machines

1221: Kristen Brochmann/Fundamental Photographs **1223:** Photo by Ronald R. Erickson; hologram by Nicklaus Philips **1224:** bottom left, Used with permission of Eastman Kodak Company; bottom right, I. Andersson, Oxford Molecular Biophysics Laboratory/Science Photo Library/Photo Researchers, Inc. **1227:** Henry Leap/Jim Lehman **1230:** left, Henry Leap/Jim Lehman **1230:** bottom: a) Sepp Seitz 1981; b) Peter Aprahamian/Science Photo Library/Photo Researchers, Inc. **1231:** Photograph by Gary Friedman. Copyright © 2003, Los Angeles Times. Reprinted with permission. **1235:** SuperStock **1238:** top, Susan Allen Sigma/Getty Images **1238:** bottom, left: Riccardo Giovanni and Martha Haynes, Cornell University **1238:** center, right: Diane Schiumo 1988, Fundamental Photographs **1241:** © 2003 by Sidney Harris. All rights reserved.

Chapter 39. **1242:** Fermilab Photo, Courtesy of Fermi National Accelerator Laboratory **1244:** Emily Serway **1251:** AIP Niels Bohr Library **1275:** Courtesy NASA **1279:** Trent Steffler/David R. Frazier Photolibrary **1281:** Photo courtesy of Garmin Ltd. **1283:** © 2003 by Sidney Harris. All rights reserved.

Tables and Illustrations

This page constitutes an extension of the copyright page. We have made every effort to trace the ownership of all copyrighted material and to secure permission from copyright holders. In the event of any question arising as to the use of any material, we will be pleased to make the necessary corrections in future printings. Thanks are due to the following authors, publishers, and agents for permission to use the material indicated.

Chapter 2. **56:** By permission of John Hart and Creators Syndicate, Inc.

Chapter 13. **415:** By permission of John Hart and Creators Syndicate, Inc.

Chapter 14. **449:** Calvin & Hobbes © 1992 Watterson. Reprinted with permission of Universal Press Syndicate. All rights reserved.

Chapter 21. **666:** By permission of John Hart and Creators Syndicate, Inc.

Chapter 30. **966:** Calvin and Hobbes © Watterson. Reprinted with permission of Universal Press Syndicate. All rights reserved.

Chapter 36. **1166:** © 2003 Sidney Harris **1169:** The Far Side® by Gary Larson © 1985 FarWorks, Inc. All Rights Reserved. Used with permission.

Chapter 38. **1241:** © 2003 by Sidney Harris

Chapter 39. **1283:** © 2003 by Sidney Harris

Index

Locator notes: **boldface** indicates a definition; *italics* indicates a figure; *t* indicates a table

Standard Abbreviations and Symbols for Units

Symbol	Unit	Symbol	Unit
A	ampere	K	kelvin
u	atomic mass unit	kg	kilogram
atm	atmosphere	kmol	kilomole
Btu	British thermal unit	L	liter
C	coulomb	lb	pound
°C	degree Celsius	ly	lightyear
cal	calorie	m	meter
d	day	min	minute
eV	electron volt	mol	mole
°F	degree Fahrenheit	N	newton
F	farad	Pa	pascal
ft	foot	rad	radian
G	gauss	rev	revolution
g	gram	s	second
H	henry	T	tesla
h	hour	V	volt
hp	horsepower	W	watt
Hz	hertz	Wb	weber
in.	inch	yr	year
J	joule	Ω	ohm

Mathematical Symbols Used in the Text and Their Meaning

Symbol	Meaning
$=$	is equal to
$\equiv$	is defined as
$\neq$	is not equal to
$\propto$	is proportional to
$\sim$	is on the order of
$>$	is greater than
$<$	is less than
$\gg (\ll)$	is much greater (less) than
$\approx$	is approximately equal to
Δx	the change in x
$\sum_{i=1}^{N} x_i$	the sum of all quantities x_i from $i = 1$ to $i = N$
$\lvert x \rvert$	the magnitude of x (always a nonnegative quantity)
$\Delta x \rightarrow 0$	Δx approaches zero
$\dfrac{dx}{dt}$	the derivative of x with respect to t
$\dfrac{\partial x}{\partial t}$	the partial derivative of x with respect to t
$\displaystyle\int$	integral